Current Index to Statistics

APPLICATIONS, METHODS AND THEORY

Volume 21

Featuring articles appearing during 1995

Editors

KLAUS HINKELMANN
Department of Statistics
Virginia Tech
Blacksburg, Virginia 24061-0439 USA

MICHAEL J. WICHURA
Department of Statistics
University of Chicago
Chicago, Illinois 60637-4931 USA

RONALD A. THISTED
Department of Statistics
University of Chicago
Chicago, Illinois 60637-4931 USA

Editorial assistants
LINDA BREEDING
ERINN BREEDING

Founding editor
BRIAN L. JOINER (1975-79)

Sponsored by

The AMERICAN STATISTICAL ASSOCIATION and
The INSTITUTE OF MATHEMATICAL STATISTICS

CURRENT INDEX TO STATISTICS
c/o American Statistical Association
1429 Duke Street
Alexandria, VA 223140-3402 USA

Printed in the United States of America

ISBN 1-883276-34-9 (Vol. 21)
ISSN 0364-1228

CONTENTS

CURRENT INDEX TO STATISTICS

1. INTRODUCTION

This annual series aspires to provide comprehensive indexing coverage on a timely basis for the field of statistics. Statistics is viewed in a very broad sense and ranges from probability theory to such topics as how to increase the response rate of mail surveys. Due to space constraints, however, many articles from supporting disciplines that are of interest to some statisticians cannot be covered. Decisions concerning inclusion often are difficult. The criteria for inclusion involve consideration of whether or not the topic falls into the domain of another discipline rather than what proportion of "statisticians" would be interested in the topic.

The present volume contains approximately 8100 entries, 79% of which appeared during 1995. Also included are articles with publication dates from 1990 through 1994 that did not appear in previous volumes. About 5400 of the entries came from the core journals listed on the inside covers of this volume. The contributing editors and editorial collaborators selected nearly 1600 articles from approximately 400 non-core journals and other sources listed in Section 4. Another 260 entries are books. The remaining entries came from proceedings. This volume of *CIS* contains no entries taken from *Statistical Theory and Methods Abstracts (STMA)* or from edited books; the omitted entries will appear in the next volume.

Items considered for indexing in *CIS* include journal articles, books and edited books, articles from edited books and conference proceedings (provided that the proceedings are available to non-conference participants), book and software reviews, prepared discussions of articles, and correction notes. Items such as notices and advertisements, editorials, letters to the editor, abstracts, meeting announcements, theses and dissertations, and technical reports are not knowingly indexed.

Contributions by readers from non-standard sources are vital if coverage is to continue to be extensive. Copies of relevant articles or publications are welcomed for future issues of *CIS*. Please include the name of the journal, volume, year, inclusive page numbers, and useful keywords not in the title. The publisher's name and address are also needed. All such submissions are subjected to the usual inclusion criteria. Anyone wishing to contribute in a more formal way is urged to contact the editor for further information.

We wish to thank the Contributing Editors and Editorial Collaborators, the Management Committee, and the ASA and IMS offices for their support. Special thanks go to Edward E. Gbur for overhauling the journal and publisher abbreviation files. Finally, we wish to thank the Editorial Assistants, Linda and Erinn Breeding, for their painstaking work on this volume.

1.1 HOW TO USE THE AUTHOR INDEX

Articles by more than one person are listed under each of the authors. Each entry is followed by the names of any co-authors, the title of the article, and its location. Two long dashes in multiauthor entries indicate the position of the author being indexed in the author list. This convention preserves the original ordering of the author's names on the publication. Two long dashes also appear whenever an author has more than one article included in this index.

For edited volumes, book and software reviews, and translated works, parenthetical notations *(Ed, Rev, Auth, Transl)* indicate the role of each individual. In the case of multiple names associated with a role, the designation is made only once, after the last individual in that role.

Authors' names are written in inverted form, with the family name preceding any given names and suffixes such as *Jr, Sr, II, and III*. The following prefixes (in upper or lower case) are considered to be part of the family name: *de, del, den, der, des, di, la, le, st, van, vander, von, vonder.* Thus, *Wim J. van der Linden* will be found under *van der Linden, Wim J.*; *S. Das Gupta* under *Gupta, S. Das*; and *Paul W. Mielke Jr* under *Mielke, Paul W., Jr.* Hyphenated names and names with an included apostrophe are considered as one word. Thus, *D. Vere-Jones* is indexed under *Vere-Jones, D.*

In different journals, authors are sometimes listed with their given names in full or with one or more of their given names shortened to initials. Authors whose names are followed by suffixes sometimes omit the suffix. *CIS* records the author's name as it appears in the journal. As a result, listings for the same author may not appear consecutively in the index.

The entries for each person are sorted so that original works authored or edited by that person are at the top of the listing, followed by things (e.g., book reviews) which that person wrote about someone else, followed by things written about that person by someone else. Within each of these three categories the entries are subsorted alphabetically by name, with the name of the person being indexed sorting before any other name; then in reverse order by year; and finally by journal/publisher abbreviation.

1.2 HOW TO USE THE SUBJECT INDEX

Unlike the Author Index, which always gives the full listing for each work, the Subject Index shortens each listing, as necessary, to fit on one line. The shortening process goes through the following steps: first, authors' given names are shortened to initials; then the list of authors is truncated by dropping names off the end and appending *et al.*; then the role indicator of the sole remaining author is discarded; and then that author's initials are discarded. If the listing is still too long, the title is truncated at one or both ends, with the omitted words replaced by the notation "...". The full title is always given in the Author Index, which should be consulted for the complete citation.

The Subject Index has two basic types of entries — entries resulting from an important word in the title of the article, and entries resulting from a keyword or phrase. Keywords are printed in boldface on the gutter; subsequent lines containing titles of articles to which the keyword has been assigned are indented. On average, an article will appear 4 times in the Subject Index.

Entries indexed by a number or by a Greek letter are found under the English spelling. Thus, "2×2 table" is alphabetized under "two" and "β" is found under "beta". British spellings of words such as *neighbourhood* in a title have not been respelled. Standard statistical phrases such as *likelihood ratio* or *generalized linear model* are indexed only under the first word of the phrase.

Keywords supplied by journals come in a wide variety of forms, many of which are not suitable for this type of indexing system. They have been edited and used selectively. A set of guidelines used for keyword construction in *CIS* is available from the editor. In addition, an article in *The American Statistician* (1995, volume 49, pages 29-33) discusses keywords and their role in bibliographic work.

The Subject Index is sorted by important words in the title, considering in turn the word on the gutter, then the words to its right, and then the words to its left in reverse order. Conjunctions and prepositions other than *with, within, and without* are passed over when sorting, and so is punctuation. Plural nouns ending in "s" are changed to singular prior to sorting. Hyphenated words are considered as a series of separate words. As a result, phrases such as *cross-over design* and *crossover design* may not appear consecutively in the Subject Index.

In using the Subject Index it is helpful to think of words an author might have used in the title of an article on the topic of interest. Once an initial article is found, there are several ways to expand the list of references:

- Look for articles by the same author.
- Use other words in the title as search words.
- Use possible synonyms or antonyms of search words. Use the special *See* and *See also* entries, if appropriate.
- Go to the actual publication and look at the reference list.

1.3 OTHER DETAILS ABOUT USAGE

A. TRANSLATIONS

All titles and keywords are given in English. Whenever possible, translations of non-English titles provided by the journals are used. When the original article is in another language, the language is indicated in parentheses after the title.

B. FORMAT FOR THE LOCATION FIELD

The location field of each journal article supplies the following information:

- Last two digits of the year of publication
- Abbreviation of the journal name
- Volume number
- Beginning page number
- Ending page number

For journals in which each issue is renumbered beginning with page 1, the issue number is included immediately after the volume number. Whenever the ending page number or other location information is not known, a "?" is printed. A "+" following the ending page number indicates that the article is continued on some later page in that issue of the journal. If the entry is a book, the location field gives the year, publisher abbreviation, and number of pages.

C. ABBREVIATIONS

All abbreviations for journal names, edited books, and conference proceedings used to construct this printed volume are listed in Section 4. Names of journals which were searched but did not contain any appropriate articles are also listed. Also shown are the full names of journals and in most cases, the abbreviation for the publishers' name. Publisher abbreviations are listed in Section 5.

D. MISCELLANEOUS

For core journals, the years and issues of the articles covered in the current volume are given in parentheses following the title on the inside of the front and back covers. Also included on the inside back cover is the list of proceedings abbreviations which have been indexed in this volume.

Corrigenda to articles appearing in the current volume are given at the end of the location field in the Author Index. Separate "Correction to" entries are used if the original article has appeared in a previous *CIS* volume. The journal volume and page numbers of the original article are given after the title in this form of correction entry.

If an article was published with discussion, the pages on which the discussion appears are given in parentheses at the end of the location field of the article in the Author Index. Prepared discussions are also included in the Author Index under the names of the discussants. Discussion information appears at the end of the title in the Subject Index. Since some titles are shortened in the Subject Index, the Author Index should always be consulted for the complete citation.

Four new core journals have been added in this volume. Three of them, *Bernoulli (Bernoull)*, *Lifetime Data Analysis (LiDaAnal)*, and *Student (Student)* began publication in 1995, the other *Environmental and Ecological Statistics (EnvEcoSt)* in 1994.

1.4 COMPUTER READABLE FORMS OF CIS

The *CIS* records for the years 1975–1995 are also available from several sources in computer readable form. Computer searching of such a database has several advantages over the use of printed volumes. Among these are the ability to search multiple years simultaneously, the ability to search on the union or intersection of several keywords and phrases or other words such as an author's name, and the ability to produce permanent printed or computer file copies of the retrieved references for bibliographic purposes.

A computer readable version of the *CIS* records for the years 1975–1995 is available as an alternative form to the printed volume from *Current Index to Statistics* under the sponsorship of the American Statistical Association and the Institute of Mathematical Statistics. The 1995 edition of the *CIS Extended Database* contains additional material not found in any of the *CIS* printed volumes. Information on the *CIS Extended Database* can be obtained from the IMS Business Office, 3401 Investment Boulevard, Suite 7, Hayward, CA 94545 U.S.A. or from the American Statistical Association, 1429 Duke Street, Alexandria, VA 22314-3402 U.S.A.

In addition, the *CIS* records for this period are scheduled to be included in the *MathSci* database, sponsored by the American Mathematical Society. The *MathSci* database is available through several commercial online information retrieval services. Information about it can be obtained from the American Mathematical Society, 201 Charles Street, P.O. Box 6248, Providence, RI 02940, U.S.A.

AUTHOR INDEX

2. AUTHOR INDEX

Aalen, Odd O.; Bjertness, Espen; Sønju, Torleif: Analysis of dependent survival data applied to lifetimes of amalgam fillings. 95 StatMed 14 1819-1829

Aaronson, Dustin: Pinkleton, Bruce; Reagan, Joey; — —; Ramo, Eduard: Does "I'm not selling anything" increase response rates in telephone surveys?. 94 ASASrvyRsr 1242-1247

Aaronson, Jon; Keane, Michael: Isomorphism of random walks. 94 IsrelJMa 87 37-63

— —; Liggett, Thomas; Picco, Pierre: Equivalence of renewal sequences and isomorphism of random walks. 94 IsrelJMa 87 65-76

Abadir, K. M.: The joint density of two functionals of Brownian motion. 95 MaMethSt 4 449-462

Abadir, Karim M.: A new test for nonstationarity against the stable alternative. 95 EconTher 11 81-104

— —: The limiting distribution of the t ratio under a unit root. 95 EconTher 11 775-793

Abay, A.: Extremes of interarrival times of a Poisson process under conditioning. 95 ApplMath 23 73-82

Abbott, Andrew: Comment on "Measuring the agreement between sequences". 95 SocMethR 24 232-243

Abbott, Robert D.: Comment on "Statistical issues in public health". 94 ASAProEpid 125-126

Abd-Elfattah, Abd-Allah M.: Ashour, Samir K.; — —: The sampling distribution of some goodness-of-fit test statistics using Burr distribution and censored samples. 94 EgpStJ 38 147-163

Abdel-Hameed, M.: Correction to "Inspection and maintenance policies of devices subject to deterioration" (V19 p917-931). 95 AdvAppPr 27 584-584

Abdel-Hameed, Mohamed; Nakhi, Yasmin: Optimal control of dams. 95 AplStMDA 11 25-34

Abdous, Belkacem: Computationally efficient classes of higher-order kernel functions. 95 CandJSt 23 21-27

— —; Remillard, Bruno: Relating quantiles and expectiles under weighted-symmetry. 95 AnInStMa 47 371-384

Abdullah, M. E.; El-Sayed, Yousry M.: Design of crossover microcomputer program and application on drug bioequivalence data. 95 CmpBiomd 48 237-239

Abdullah, Mokhtar B.: Detection of influential observations in functional errors-in-variables models. 95 CommStA 24 1585-1595

— —: One-step GM-estimators for orthogonal regression. 95 JStCmpSm 52 301-309

Abel, P.: A stochastic target costing model (German). 95 ZAngMaMk 75 557-558S

Abel, R. J. R.; Mills, W. H.: Some new BIBDs with $k = 6$ and $\lambda = 1$. 95 JComDsgn 3 381-391

Abelson, Robert P.: Statistics as principled argument. 95 ErlbaumA:NJ 232p

Abetti, Frank: Anderson, DeAnn; — —; Savage, Phillip: Process improvement utilizing computer simulation: Case study. 95 ASQCTekCon 713-724

Abidi, H.; Pontier, J.; Borms, J.; Duquet, W.: The benefit of modelling growth curves for longitudinal data (French). 95 RStApp 43/3 55-72

Abkowitz, Janis L.: Newton, Michael A.; Guttorp, Peter; Catlin, Sandra; Assunção, Renato; — —: Stochastic modeling of early hematopoiesis. 95 JASA 90 1146-1155

Abma, Joyce C.: Guadagno, Mary Ann; — —; Chandra, Anjani; London, Kathy A.; Peterson, Linda S.; Piccinino, Linda J.; Wilson, Jacqueline B.: Women's reports of sensitive health conditions, attitudes, and behaviors. 94 ASAProGovt 82-87

Abou-El-ata, M. O.; Hariri, A. M. A.: Point estimation and confidence intervals of the M/M/2/N queue with balking and heterogeneity. 95 AJMMgSci 15 35-55

Abou-Nasr, M. A.; Sid-Ahmed, M. A.: Fast learning and efficient memory utilization with a prototype based neural classifier. 95 PattrnRc 28 581-593

Abouammoh, A. M.; Hendi, M. I.; Khalique, A.: Shock models with NBRU properties. 95 PakJStA 11 35-47

Abraham, B.: Rai, S. N.; — —; Peiris, S.: Analysis of short time series with an over-dispersion model. 95 CommStA 24 335-348

Abraham, Romain: On the connected components of the support of super Brownian motion and of its exit measure. 95 StocProc 60 227-245

Abrams, Keith: Collett, D. (Auth); — — (Rev): Review of "Modelling survival data in medical research". 95 JAppStat 22 434-434

Abramson, Bruce; Clemen, Robert: Probability forecasting. 95 IntJFor 11 1-4

— —; Finizza, Anthony: Probabilistic forecasts from probabilistic models: A case study in the oil market. 95 IntJFor 11 63-72

Absalom, Sandra T.: Yu, Ching C.; — —; Plunkert, Patricia A.: An evaluation study of forecasting models. 94 ASAProGovt 173-177

Abt, Markus; Strehler, Erwin; Rosenbusch, Bernd; Sterzik, Karl: Predicting the success of in vitro fertilization: Conventional semen analysis compared to the hamster ova penetration test. 95 StatMed 14 1807-1818

Abughalous, Mansour M.; Bansal, Naveen K.: On selecting the best natural exponential families with quadratic variance function. 95 StPrLet 25 341-349

Accardi, L.: — — (Ed); Lindblad, G. (Rev): Review of "Quantum probability and related topics (Volume 7)". 95 Metrika 42 268-269

Accomando, William P.; Trost, Donald C.; Finman, Jeffrey S.: Moving from multinational clinical trials to supra-trials. 94 ASAProBiop 232-237

Acero, Juan Garrido: Katz, Richard W.; — —: Sensitivity analysis of extreme precipitation events. 94 JClimlgy 14 985-999

Achcar, Jorge Alberto: Inferences for accelerated life tests considering a bivariate exponential distribution. 95 MaOpfStS 26 269-283

Ackerson, Lynn: Nonparametric statistics to investigate discrimination. 94 ASAProStEd 39-41

Adachi, Kohei: Statistical classification procedures for polygraph tests of guilty knowledge. 95 Behvmtrk 22 49-66

Adachi, Kohichi: Feng, Wei; — —; Kowada, Masasi: Dynamically optimal replacement policy for a shock model in a Markov random environment. 94 JOpRsJpn 37 255-270

Adams, A.: Shiffler, R.; — — (Auth); Ziegel, Eric R. (Rev): Review of "Introductory business statistics with computer applications". 95 Technmcs 37 239-240

Adams, Benjamin M.; Lin, Winnie: Forecast-based monitoring schemes using combined EWMA-Shewhart control charts. 94 ASAProQlPr 104-108

Mansfield, Edward R.; — —: Teaching the standard deviations of linear combinations using graphics. 94 ASAProStEd 80-83

Pappanastos, Edward A.; — —: A comparison of robust control charts. 94 ASAProQlPr 316-320

Superville, Claude R.; — —: Autocorrelated data: Control charts versus tracking signals. 94 ASAProQlPr 194-199

Adams, G. J.; Goodwin, G. C.: Parameter estimation for periodic ARMA models. 95 JTimSrAn 16 127-145

Adams, John; Burnam, Audrey: Federal block grants: What do statisticians have to contribute?. 95 Chance 8/4 40-42

— — (Auth); Senn, Stephen (Rev): Review of "Risk". 95 StatMed 14 1727-1728

Adams, Raymond J.: Wilson, Mark; — —: Rasch models for item bundles. 95 Psymtrka 60 181-198

Adams, Robert: Bennett, R. E.; Ward, W. C. (Ed); — — (Rev): Review of "Construction versus choice in cognitive measurement". 95 Ststcian 44 137-137

Adang, Pim; Melenberg, Bertrand: Nonnegativity constraints and intratemporal uncertainty in a multi-good life-cycle model. 95 JApEcmtx 10 1-15

Adashek, Daniel: Schwartz, Steven H.; Klein, Robert E.; — —: Veterans living in group quarters: What data from the 1990 Census show. 94 ASAProGovt 13-18

Adatia, A.: Best linear unbiased estimator of the Rayleigh scale parameter based on fairly large censored samples. 95 IEEEReli 44 302-309

Adcock, C. J.: The Bayesian approach to determination of sample sizes - Some comments on the paper by Joseph, Wolfson and du Berger. 95 Ststcian 44 155-161

Adcock, Chris: Pesaran, M. Hashem; Potter, Simon M. (Ed); — — (Rev): Review of "Nonlinear dynamics chaos and econometrics". 95 IntJFor 11 599-601

Adeleke, B. L.: Ipinyomi, R. A.; — —: Correction to "Nested row-column designs" (V38 p197-202). 95 Ststcian 44 547-547

Adell, José Carlos; Dopazo, Joaquín: Monte Carlo simulation in phylogenies: An application to test the constancy of evolutionary rates. 94 JMolEvol 38 305-309

Ades, A. E.: Wade, A. M.; — —; Salt, A. T.; Jayatunga, R.; Sonksen, P. M.: Age-related standards for ordinal data: Modelling the changes in visual acuity from 2 to 9 years of age. 95 StatMed 14 257-266

Adhikari, A.: Freedman, D.; Pisani, R.; Purves, R.; — — (Auth); Mizera, Ivan (Rev): Review of "Statistics (Second edition)". 95 JAppStat 22 321-322

Adke, S. R.: Mukherjee, S. P.; Chaudhuri, A.; Basu, S. K. (Ed); — — (Rev): Review of "Essays on probability and statistics: Festschrift in honour of Professor Anil Kumar Bhattacharyya". 95 CalcutSt 45 131-132

Adler, André: Explicit stable strong laws of large numbers. 94 CalcutSt 44 141-149

Adler, Robert J.; Samorodnitsky, Gennady: Super fractional Brownian motion, fractional super Brownian motion and related self-similar (super) processes. 95 AnlsProb 23 743-766

Mikosch, Thomas; Gadrich, Tamar; Klüppelberg, Claudia; — —: Parameter estimation for ARMA models with infinite variance innovations. 95 AnlsStat 23 305-326

Mytnik, Leonid; — —: Bisexual branching diffusions. 95 AdvAppPr 27 980-1018

Adler, Robert K.: Battles, Stephanie J.; — —; Olsen, Jay K.: Demand-side management: Utility programs offered and participants in the residential, commercial, and manufacturing sectors of the U.S. economy. 94 ASAProBuEc 20-27

Aebischer, N. J.: Investigating the effects of hunting on the survival of British pigeons and doves by analysis of ringing recoveries. 95 JAppStat 22 923-934

Aerts, M.; Veraverbeke, N.: Bootstrapping a nonparametric polytomous regression model. 95 MaMethSt 4 189-200

Aerts, Marc; Janssen, Paul: Weighted bootstrapping for U-quantiles. 95 StPrLet 22 317-323

Afsarinejad, Kasra: Repeated measurements designs with unequal period sizes. 94 JItalStS 3 161-168

Agard, David B.: The GRIP: A performance statistic for tournament golf. 94 ASAProSpts 1-6

Agarwal, K. N.: Jaggi, Seema; — —: Augmented partial diallel design for estimating combining ability of parents. 95 BiomtrcJ 37 879-887

Agarwal, R. C.; Kathuria, O. P.: The effect of measurement errors under survey conditions on least squares estimator of regression coefficients. 94 JIAgSt 46 290-297

Agee, Ernest: Gluhovsky, Alexander; — —: Reply to comment on "A definitive approach to turbulence statistical studies in planetary boundary layers". 95 JAtmsphr 52 3197-3198

Aggarwal, K. K.: — — (Auth); Khamis, Imad H. (Rev): Review of "Reliability engineering". 95 Technmcs 37 234-235

Aggarwal, M. L.; Chowdhury, S. Roy: Linear graphs for 2^{k-p} fractional factorial under split plot design. 94 JStatRsr 28 73-87

Aggarwal, Raj; Mohanty, Sunil; Song, Frank: Are survey forecasts of macroeconomic variables rational?. 95 JBusUChi 68 99-119

Aggoun, Lakhdar: Martin boundaries for discrete Markov chains. 95 StocAnAp 13 257-268

— —; Elliott, Robert J.: Finite-dimensional models for hidden Markov chains. 95 AdvAppPr 27 146-160

Elliott, Robert J.; — —; Moore, John B.: Hidden Markov models: Estimation and control. 95 Springer:Brln:NY 361p

Elliott, Robert J.; — —; Moore, John B. (Auth); Goodwin, C. G. (Rev): Review of "Hidden Markov models: Estimation and control". 95 ShrtBkRv 15/2 29-29

Agnihotri, Monica: Singh, R. Karan; — —; Verma, Sanjay: A generalized class of estimators utilizing Lindley-like mean correction in linear regression model. 94 Metron 52/3 147-160

Agras, Jean: Opsomer, Jean-Didier; — —; Carpi, Anthony; Rodrigues, Geraldo: An application of locally weighted regression to airborne mercury deposition around an incinerator site. 95 EnvrMtrc 6 205-219

Agrawal, M. C.; Sthapit, A. B.: Unbiased ratio-type variance estimation. 95 StPrLet 25 361-364

Agrawal, Rajeev: Sample mean based index policies with $O(\log n)$ regret for the multi-armed bandit problem. 95 AdvAppPr 27 1054-1078

Agrawal, Rekha: Upadhyay, Satyanshu Kumar; — —: Bayesian prediction limits for the mean of future observations from the inverse Gaussian distribution. 94 Metron 52/1 101-111

Agresti, Alan: Logit models and related quasi-symmetric log-linear models for comparing responses to similar items in a survey. 95 SocMethR 24 68-95

— —; Ghosh, Atalanta; Bini, Matilde: Raking kappa: Describing potential impact of marginal distributions on measures of agreement. 95 BiomtrcJ 37 811-820

Kim, Donguk; — —: Improved exact inference about conditional association in three-way contingency tables. 95 JASA 90 632-639

Agrò, G.: Maximum likelihood estimation for the exponential power function parameters. 95 CommStB 24 523-536

Aguirre-Torres, Victor; Rios-Curil, Alejandro: The effect and adjustment of complex surveys on chi-squared goodness-of-fit tests: Some Monte Carlo evidence. 94 ASASrvyRsr 602-607

Ahlbrandt, Robert; Iyer, Hari K.; Vecchia, Dominic F.: Optimum design of serial measurement trees. 95 JStPlInf 48 379-390

Ahlén, Anders: Öhrn, Kenth; — —; Sternad, Mikael: A probabilistic approach to multivariable robust filtering and open-loop control. 95 IEEEAuCn 40 405-418

Ahmad, Ibrahim A.: On multivariate kernel estimation for samples from weighted distributions. 95 StPrLet 22 121-129

Ahmad, Munir: Estimation of parameters of the incomplete bivariate hyper-Poisson distributions. 92 ProIrStC1 157-162

Ahmad, Naseer: Noor, Islam; — —: Exact and approximate relative efficiency of generalized ridge-type shrinkage estimator. 95 PakJStA 11 17-22

Ahmad, S. E.; Tomkins, R. J.: Estimating lognormal means using uncertain prior information. 95 PakJStA 11 67-92

Ahmadi, Susan: Ford, Ian; Norrie, John; — —: Model inconsistency, illustrated by the Cox proportional hazards model. 95 StatMed 14 735-746

Ahmed, M.: Das, M. N.; Giri, N.; — —: Designs through recoding of varietal and level codes. 95 StPrLet 23 371-380

Ahmed, Mohiy El-Din M.: Ashour, Samir K.; Jones, Peter W.; — —: Discriminant analysis for social-economic and cultural factors affecting diarrhea diseases in children under two years in Egypt. 95 EgpStJ 39 1-11

Ahmed, N. U.; Ding, X.: A semilinear McKean-Vlasov stochastic evolution equation in Hilbert space. 95 StocProc 60 65-85

— —; Yeap, T. H.: Performance analysis of slotted ALOHA protocol for optical fiber local area networks with star topology. 95 StocAnAp 13 389-410

Zhu, Qingxin; — —: Dynamic programming equations of stochastic optimal control in Banach space. 95 StocAnAp 13 369-387

Ahmed, S. E.: Improved shrinkage estimation of relative potency. 95 BiomtrcJ 37 627-638

— —: A pooling methodology for coefficient of variation. 95 SankhyaB 57 57-75

— —: Improved estimation of the coefficient of variation. 94 JAppStat 21 565-573

Ahn, Byung Chul: Choi, In; — —: Testing for cointegration in a system of equations. 95 EconTher 11 952-983

Ahn, Chul; Odom-Maryon, Tamara: Estimation of a common odds ratio under binary cluster sampling. 95 StatMed 14 1567-1576

Ahn, Chul W.; Collins, Mark: Estimation of the size of a closed population. 94 ASAProSoSt 112-115 (Disc: p122-123)

Ahn, Hongshik; Chen, James J.: Generation of over-dispersed and under-dispersed binomial variates. 95 JCmpGrSt 4 55-64

— —; Chen, James J.: Modeling developmental toxic effects through recursive partitioning. 94 ASAProBiop 130-135

— —; Kodell, Ralph L.: Estimation and testing of tumor incidence rates in experiments lacking cause-of-death data. 95 BiomtrcJ 37 745-763

Ahn, Hyungsok; Kohatsu-Higa, Arturo: The Euler scheme for anticipating stochastic differential equations. 95 Stochatx 54 247-269

Ahn, Hyungtaik: Nonparametric two-stage estimation of conditional choice probabilities in a binary choice model under uncertainty. 95 JEconmtx 67 337-378

Ahn, Kwang-Il: On the use of coefficient of variation for uncertainty analysis in fault tree analysis. 95 RelEngSS 47 229-230

Park, Chang K.; — —: A new approach for measuring uncertainty importance and distributional sensitivity in probabilistic safety assessment. 94 RelEngSS 46 253-261

Ahn, Seung C.; Schmidt, Peter: A separability result for GMM estimation, with applications to GLS prediction and conditional moment tests. 95 EcomtrRv 14 19-34

— —; Schmidt, Peter: Efficient estimation of models for dynamic panel data. 95 JEconmtx 68 5-27

Ahn, Sung K.: Reduced rank estimation of vector autoregressive models with cointegration and codependence. 94 ASAProBuEc 151-156

Cho, Sinsup; Park, Young J.; — —: Unit root tests for seasonal models with deterministic trends. 95 StPrLet 25 27-35

Ahnn, Sang; Anderson, Stewart J.: Sample size determination for comparing more than two survival distributions. 95 StatMed 14 2273-2282

Ahrens, H.: Rasch, D.; Tiku, M. L.; Sumpf, D. (Ed); — — (Rev. German): Review of "Elsevier's dictionary of biometry". 95 BiomtrcJ 37 896-896

Ahsanullah, M.: Record values, random record models and concomitants. 94 JStatRsr 28 89-109

Ahsanullah, Mohammad: Wesolowski, Jacek; — —: Conditional specifications of multivariate Pareto and Student distributions. 95 CommStA 24 1023-1031

Ahuja, Elizabeth M.: The role of VA administrative files in the National Survey of Veterans. 94 ASAProGovt 25-27

Aickin, Mikel: Lindsey, J. K. (Auth); — — (Rev): Review of "Models for repeated measurements". 95 AJEpidm 141 382-383

Aihara, Shin Ichi; Bagchi, Arunabha: Nonlinear smoothing for random fields. 95 StocProc 55 143-158

Aiken, Leona S.: Kreft, Ita G. G.; de Leeuw, Jan; — —: The effect of different forms of centering in hierarchical linear models. 95 MultBehR 30 1-21

Airault, Helene: Malliavin, Paul; — —; Kay, Leslie; Letac, Gerard: Integration and probability. 95 Springer:Brln:NY 322p

Malliavin, Paul; — —; Kay, Leslie; Letac, Gerard (Auth); Barnett, C. (Rev): Review of "Integration and probability". 95 ShrtBkRv 15/3 46-46

Ait-Hennani, L.: Bertrand-Retali, M.; — —: Uniform convergence of density estimators on spheres. 95 ApplMath 22 427-446

Aitken, C. G. G.; Shaw, J.; Talbot, M.: Contaminant detection in the visual inspection of seed samples. 95 ApplStat 44 431-440

Bross, I. D. (Auth); — — (Rev): Review of "Fifty years of folly and fraud 'in the name of science': From crossroads to the health care crisis". 95 Biomtrcs 51 1587-1587

— — (Auth); Balding, D. J. (Rev): Review of "Statistics and the evaluation of evidence for forensic scientists". 95 ShrtBkRv 15/3 41-41

Aitkin, Murray: Comment on "The statistics of linear models: Back to basics" (V4 p221-234). 95 StatComp 5 85-86

— —: Probability model choice in single samples from exponential families using Poisson log-linear modelling, and model comparison using Bayes and posterior Bayes factors. 95 StatComp 5 113-120

Akahira, M.; Takeuchi, K.: Non-regular statistical estimation. 95 Springer:Brln:NY 183p

Akahira, Masafumi: A higher order approximation to a percentage point of the non-central t-distribution. 95 CommStB 24 595-605

— —: The Bhattacharyya type bound for the asymptotic variance and the sequential discretized likelihood estimation procedure. 95 SqtlAnly 14 193-204

— —; Sato, Michikazu; Torigoe, Norio: On the new approximation to non-central t-distributions. 95 JJapnStS 25 1-18

— —; Takeuchi, Kei; Koike, Ken-ichi: Unbiased estimation in sequential binomial sampling. 92 RStAJp 39/4 1-13

Sato, Michikazu; — —: Information inequalities for the minimax risk. 95 JJapnStS 25 151-158

Akahori, Jirô: Some formulae for a new type of path-dependent option. 95 AnlsApPr 5 383-388

Akazawa, Kouhei: Nakamura, Tsuyoshi; — —: Computer program for the proportional hazards measurement error model. 94 CmpBiomd 45 203-212

Akdeniz, Fikri; Kaçranlar, Selahattin: On the almost unbiased generalized Liu estimator and unbiased estimation of the bias and MSE. 95 CommStA 24 1789-1797

Akerhielm, Karen: Adding value to the value-added educational production function specification. 94 ASAProSoSt 192-195 (Disc: p213-215)

Akers, R. Patrick: Preisler, Haiganoush K.; — —: Autoregressive-type models for the analysis of bark beetle tracks. 95 Biomtrcs 51 259-267

Aki, Sigeo; Hirano, Katuomi: Joint distributions of numbers of success-runs and failures until the first consecutive k successes. 95 AnInStMa 47 225-235

Uchida, Masayuki; — —: Sooner and later waiting time problems in a two-state Markov chain. 95 AnInStMa 47 415-433

Akin, Don: Biemer, Paul; — —: The efficiency of list-assisted random digit dialing sampling schemes for single and dual frame surveys. 94 ASASrvyRsr 1-10

Akman, H. Olcay: Gupta, Ramesh C.; — —: Bayes estimation in a mixture inverse Gaussian model. 95 AnInStMa 47 493-503

Gupta, Ramesh C.; — —: Mean residual life function for certain types of non-monotonic ageing. 95 CommStSM 11 219-225

Gupta, Ramesh C.; — —: Correction to "Mean residual life function for certain types of non-monotonic ageing" (V11 p219-225). 95 CommStSM 11 561-562

Gupta, Ramesh C.; — —: On the reliability studies of a weighted inverse Gaussian model. 95 JStPlInf 48 69-83

Akritas, Michael G.; Murphy, Susan A.; LaValley, Michael P.: The Theil-Sen estimator with doubly censored data and applications to astronomy. 95 JASA 90 170-177

Al-Hussaini, A. N.; Tang, Lingqi: Ito formulas for Skorohod and Skorohod-Stratonovich integrals in the two-parameter case. 95 StocAnAp 13 411-436

Al-Hussaini, Essam K.; Jaheen, Zeinhum F.: Bayesian prediction bounds for the Burr type XII failure model. 95 CommStA 24 1829-1842

Alabert, Aureli; Ferrante, Marco; Nualart, David: Markov field property of stochastic differential equations. 95 AnlsProb 23 1262-1288

Aladin, N. V.; Plotnikov, I. S.; Potts, W. T. W.: The Aral Sea desiccation and possible ways of rehabilitating and conserving its northern part. 95 EnvrMtrc 6 17-29

Alahmadi, Abdulhadi M.: Torsney, Ben; — —: Designing for minimally dependent observations. 95 StSinica 5 499-514

Alam, Khursheed; Williams, Calvin L.: A multivariate goodness-of-fit test for stochastically ordered distributions. 95 BiomtrcJ 37 945-956

Alamatsaz, M. H.: Generalized unimodality. 92 ProIrStC1 1-10

Alameddine, Amine: Sherali, Hanif D.; — —; Glickman, Theodore S.: Biconvex models and algorithms for risk management problems. 94 AJMMgSci 14 197-228

Alarie, Yves: Boylstein, Lee Ann; Anderson, Stewart J.; — —: Comparison of smoothing methods when applied to respiratory data collected in mice exposed to an aerosolized machining fluid. 94 ASAProBiop 136-141

Albers, Willem: A two-stage rank test using density estimation. 95 AnInStMa 47 675-691

Albert, Jim: Teaching inference about proportions using Bayes and discrete models. 95 JStEd 3/3 58kb (jse/v3n3/albert)

— —; Chib, Siddhartha: Bayesian residual analysis for binary response regression models. 95 Biomtrka 82 747-759

Albert, José Ramon G.; Baxter, Laurence A.: Applications of the EM algorithm to the analysis of life length data. 95 ApplStat 44 323-341

Albert, Paul S.; McShane, Lisa M.: A generalized estimating equations approach for spatially correlated binary data: Applications to the analysis of neuroimaging data. 95 Biomtrcs 51 627-638

Longford, Nicholas T. (Auth); — — (Rev): Review of "Random coefficient models". 95 JASA 90 801-802

Albert, Richard H.: Liu, Peng T.; — —; Springer, Janet A.: Is one sample enough?. 94 ASAProBays 18-22

Springer, Janet A.; Liu, Peng T.; — —; Chow, Lien P.: Construction of a spectrum lifetable. 94 ASAProSoSt 82-87

Alberti, Nicholas S.: Sweet, Elizabeth M.; — —: Coverage implications of using the term 'usual residence'. 94 ASASrvyRsr 319-324

Albeverio, Sergio; Wu, Jiang-Lun: Stochastic anticipative calculus for functionals of a Gaussian generalized random field. 95 Stochatx 54 317-337

Albin, J. M. P.: Upper and lower classes for L^2- and L^p-norms of Brownian motion and norms of α-stable motion. 95 StocProc 58 91-103

Alcover, R.: Romero, R.; Ferrer, A.; Capilla, C.; Zunica, L.; Balasch, S.; Serra, V.; — —: Teaching statistics to engineers: An innovative pedagogical experience. 95 JStEd 3/1 33kb (jse/v3n1/romero)

Aldous, D.; Diaconis, P.: Hammersley's interacting particle process and longest increasing subsequences. 95 ProbTher 103 199-213

Aldous, David: Darwin's log: A toy model of speciation and extinction. 95 JAppProb 32 279-295

Alegre, Antoni; Claramunt, M. Mercé: Allocation of solvency cost in group annuities: Actuarial principles and cooperative game theory. 95 InsrneME 17 19-34

Alekperov, Sh. O.; Kruglov, V. M.: Convergence of convolutions of concentration functions to degenerate, normal, and Poisson laws. 94 ThProbAp 39 197-216

Alex, M.; Steinebach, J.: Invariance principles for renewal processes and some applications. 95 ThPrMaSt 50 23-54

Alexander, Charles H.: Chand, Nanak; — —: Imputing income for an N-person consumer unit. 94 ASASrvyRsr 412-416

Alexander, K. S.: Correction to "The rate of convergence of the mean length of the longest common subsequence" (V4 p1074-1082). 95 AnlsApPr 5 327-327

Alexander, Kenneth S.: Percolation and minimal spanning forests in infinite graphs. 95 AnlsProb 23 87-104

Alexander, Lois: Reinventing the population census: Balancing privacy and efficiency. In memory of Roger A. Herriot. 94 ASAProGovt 1-2

Alexander, Melvin T.: Montgomery, Douglas C.; Runger, George; — — (Rev): Review of "Applied statistics and probability for engineers". 95 Technmcs 37 455-457

Alexandrou, V. A.: Koutras, M. V.; — —: Runs, scans and urn model distributions: A unified Markov chain approach. 95 AnInStMa 47 743-766

Alfa, Attahiru Sule: Diamond, Jeffrey E.; — —: Matrix analytical methods for M/PH/1 retrial queues. 95 CommStSM 11 447-470

Alfsen, K. H.: Use of macroeconomic models in analysis of environmental problems in Norway, and consequences for environmental statistics. 92 StatJUN 9 51-72

Algina, James; Blair, R. Clifford; Coombs, William T.: A maximum test for scale: Type I error rates and power. 95 JEdStat 20 27-39

— —; Oshima, T. C.: An improved general approximation test for the main effect in a split-plot design. 95 BJMaStPs 48 149-160

Algoet, Paul H.: Correction to "Universal schemes for prediction, gambling and portfolio selection" (V20 p901-941). 95 AnlsProb 23 474-478

Alho, Juha M.: Analysis of sample based capture-recapture experiments. 94 JOffStat 10 245-256

— —; Valtonen, Esko: Interval estimation of inverse dose-response. 95 Biomtrcs 51 491-501

Ali Mousa, M. A. M.: Empirical Bayes estimators for the Burr type XII accelerated life testing model based on type-2 censored data. 95 JStCmpSm 52 95-103

Ali, A. K. A.; Lin, C. Y.: Diagonalization of variance-covariance matrices using F-G algorithm. 95 EgpStJ 39 12-25

Ali, Mukhtar M.: Probability models on horse race outcomes. 94 ASAProSpts 7-11

Alimov, D.: Markov functionals of an ergodic Markov process. 94 ThProbAp 39 504-512

— —: Asymptotically ergodic Markov functionals of an ergodic process. 94 ThProbAp 39 537-546

Alioum, Ahmadou; Commenges, Daniel: A direct approach for correcting AIDS incidence: Variance formula and comparison with other methods. 95 StatMed 14 27-38

Allan, R. N.: Román, J.; — —: Stochastic simulation and Monte Carlo methods applied to the assessment of hydro-thermal generating system operation. 94 Top 2 1-24

Alldredge, J. R.: Ratkowsky, D. A.; Evans, M. A.; — — (Auth); Gillespie, Brenda (Rev): Review of "Crossover experiments: Design, analysis and application". 95 JAppStat 22 188-188

Ratkowsky, D. A.; Evans, M. A.; — — (Auth); Walker, Stephen (Rev): Review of "Cross-over experiments: Design, analysis, and application". 95 JRSS-A 158 200-200

Alldredge, J. Richard: Ratkowsky, David A.; Evans, Marc A.; — — (Auth); Berghold, Andrea (Rev): Review of "Cross-over experiments. Design, analysis and application". 95 CmpStQ 10 99-100

Allen, F. L.: Panter, D. M.; — —: Using best linear unbiased predictions to enhance breeding for yield in soybean: I. Choosing parents. 95 CropSci 35 397-405

Allen, Linda J. S.: Martin, Clyde F.; — —; Stamp, Mark: Urn model simulations of a sexually transmitted disease epidemic. 95 AplMaCmp 71 179-199

Allen, Nancy L.; Donoghue, John R.: Detecting differential item functioning: Current methods and continuing problems. 94 ASAProSoSt 17-26 (Disc: p27-29)

Allen, P. Geoffrey; Morzuch, Bernard J.: Comparing probability forecasts derived from theoretical distributions. 95 IntJFor 11 147-157

Allen, Rich: Forecasting crop acreages and yields in the face of and in spite of floods. 94 ASAProEnvr 187-191

Allen, Stephen H.: Kraft, Kathy M.; Johnson, Douglas H.; Samuelson, Jack M.; — —: Using known populations of pronghorn to evaluate sampling plans and estimators. 95 JWildlMg 59 129-137

Allen, Tiwanda M.; Petroni, Rita J.: Mover nonresponse adjustment research for the Survey of Income and Program Participation. 94 ASASrvyRsr 662-667 (Disc: p680-682)

Singh, Rajendra P.; Petroni, Rita J.; — —: Oversampling in panel surveys. 94 ASASrvyRsr 674-679 (Disc: p680-682)

Allenby, Greg M.; Arora, Neeraj; Ginter, James L.: Incorporating prior knowledge into the analysis of conjoint studies. 95 JMarRsr 32 152-162
— —; Lenk, Peter J.: Reassessing brand loyalty, price sensitivity, and merchandising effects on consumer brand choice. 95 JBES 13 281-289

Allende, Renato: Small, Christopher G.; McLeish, D. L. (Auth); Marshall, Guillermo; — — (Rev): Review of "Hilbert space methods in probability and statistical inference". 95 JASA 90 1491-1492

Allende, Sira; Bouza, Carlos; Romero, Isidro: Fitting a linear regression model by combining least squares and least absolute value estimation. 95 Questo 19 107-121

Allin, Paul: Bartholomew, David; Moore, Peter; Smith, Fred; — —: The measurement of unemployment in the UK. 95 JRSS-A 158 363-405 (Disc: p405-417)

Allison, David B.: When is it worth measuring a covariate in a randomized clinical trial?. 95 JCCPsych 63 339-343

Allison, L.; Wallace, C. S.: The posterior probability distribution of alignments and its application to parameter estimation of evolutionary trees and to optimization of multiple alignments. 94 JMolEvol 39 418-430

Alloway, James A., Jr: Assessment of an innovative statistics PC laboratory. 94 ASAProStEd 75-79

Almahmeed, Mohammad A.: Shaarawy, Samir M.; — —: Bayesian classification with multivariate autoregressive sources that might have different orders. 95 CommStB 24 567-581

Almekinders, Geert J.: Exchange rate policy and the (un)conditional variance of the DM/$-rate. 95 JNOknSt 214 146-153

Almhana, J.: Pham, T.; — —: The generalized Gamma distribution: Its hazard rate and stress-strength model. 95 IEEEReli 44 392-397

Almiñana, M.; Pastor, J. T.: Two new heuristics for the location set covering problem. 94 Top 2 315-328

Almond, Russell G.: Graphic belief modeling. 95 ChapmnHall:Lnd 350p
— —: Comment on "A probabilistic and statistical view of fuzzy methods". 95 Technmcs 37 267-270

Alper, Paul; Raymond, Robert L.: Imposing views, imposing shoes: A statistician as a sole model. 95 AmerStat 49 317-319

Alpérovitch, A.: Tzourio, C.; Maccario, J.; — —: Estimation of the proportion of genetic cases in late-onset diseases: An application to Alzheimer's disease. 95 StatMed 14 1353-1363

Alpuim, M. T.; Catkan, N. A.; Hüsler, J.: Extremes and clustering on nonstationary max-AR(1) sequences. 95 StocProc 56 171-184

Alsberg, Bjørn K.: Speed improvement of multivariate algorithms by the method of postponed basis matrix multiplication. Part III. The use of asymmetric and generalized eigenvalue equations. 95 Chemolab 30 223-225

Alsmeyer, Gerold: Random walks with stochastically bounded increments: Renewal theory. 95 MatNach 172 13-31

Alsup, Fred; Watson, Ricky M. (Auth); McLaughlin, Gregory C. (Rev): Review of "Practical statistical process control: A tool for quality manufacturing". 95 Technmcs 37 461-462

Alt, R.: Lamotte, J. L.; — —: Comparison of simulated annealing algorithms for image restoration. 94 MaCmpSim 37 1-15

Altekar, Maneesha; Lucas, James M.; Schuenemeyer, John H.: Shift detection in two-way arrays. 95 CommStB 24 711-731
— —; Lucas, James M.; Schuenemeyer, John H.: Shift detection in two-way arrays. 94 ASAProPEng 76-80
Babcock, Laura M.; — —: Use of statistical design to develop flame retardant polymer formulations. 95 Chemolab 29 141-146

Althoen, S. C.: Frank, H.; — — (Auth); Guiard, V. (Rev): Review of "Statistics: Concepts and Applications". 95 BiomtrcJ 37 639-639

Altland, Hank W.: Callanan, Terrance P.; — —: Providing effective training industry - An experience with robust design. 94 ASAProStEd 126-131

Altman, D. G.; Lausen, B.; Sauerbrei, W.; Schumacher, M.: Dangers of using "optimal" cutpoints in the evaluation of prognostic factors. 94 JNCncrI 86 829-835

Altman, Douglas: Statistics in the medical literature: 1. 95 StatMed 14 873-874

Altman, N. S.; Casella, G.: Nonparametric empirical Bayes growth curve analysis. 95 JASA 90 508-515

Altman, Naomi; Léger, Christian: Bandwidth selection for kernel distribution function estimation. 95 JStPlInf 46 195-214
— —; Léger, Christian: Cross-validation, the bootstrap, and related methods for tuning parameter selection. 94 ASAProStCp 41-50
— —; Léger, Christian: On the optimality of prediction based selection criteria and the convergence rates of estimators. 94 ASAProStCp 112-117

Altmann, Gabriel: Zörnig, Peter; — —: Unified representation of Zipf distributions. 95 CmpStDA 19 461-473

Altmann, Michael: Kaplan, E. H.; Brandeau, M. L. (Ed); — — (Rev): Review of "Modeling the AIDS epidemic: Planning, policy, and prediction". 95 Chance 8/4 36-39

Altpeter, Ekkehardt: Validation studies in epidemiology. 95 Student 1 35-49

Altschul, Stephen F.: A protein alignment scoring system sensitive at all evolutionary distances. 93 JMolEvol 36 290-300

Alvarez, M.; Cuevas, C. M.; Escudero, L. F.; de la Fuente, J. L.; García, C.; Prieto, F. J.: Network planning under uncertainty with an application to hydropower generation. 94 Top 2 25-58

Alves, M. I. Fraga: Estimation of the tail parameter in the domain of attraction of an extremal distribution. 95 JStPlInf 45 143-173

Alves, Wayne: Hwang, Lie-Ju; — —: Analysis of neurologic outcomes using ordinal regression. 94 ASAProBiop 352-354

Alvo, Mayer; Cabilio, Paul: Rank correlation methods for missing data. 95 CandJSt 23 345-358
— —; Cabilio, Paul: Approximate and exact distributions of rank tests for balanced incomplete block designs. 95 CommStA 24 3073-3121
— —; Cabilio, Paul: Testing ordered alternatives in the presence of incomplete data. 95 JASA 90 1015-1024

Alwan, Layth C.; Roberts, Harry V.: The problem of misplaced control limits. 95 ApplStat 44 269-278 (Disc: p289-306)

Aly, Emad-Eldin A. A.; Kochar, Subhash C.: A CUSUM test for symmetry against positive biasedness. 94 CalcutSt 44 157-163

Amanatidis, G. T.: Kambezidis, H. D.; Tulleken, R.; — —; Paliatsos, A. G.; Asimakopoulos, D. N.: Statistical evaluation of selected air pollutants in Athens, Greece. 95 EnvrMtrc 6 349-361

Amaratunga, Dhammika; Ferreira, Irene: Nonlinear model selection for finding receptor sites. 94 ASAProBiop 126-129

Amari, Shun-ichi: Information geometry of the EM and em algorithms for neural networks. 95 NuralNet 8 1379-1408

Ambagaspitiya, R. S.: A family of discrete distributions. 95 InsrneME 16 107-127

Ambartzumian, R. V.; Oganian, V. K.: Measure generation by Euler functionals. 95 AdvAppPr 27 606-626

Amemiya, Yasuo: Papadopoulos, Savas; — —: Asymptotic robustness for the structural equation analysis of several populations. 94 ASAProBuEc 65-70

Amer, R.; Carreras, F.: Cooperation indices and coalitional value. 95 Top 3 117-135

Amin, Raid W.; Reynolds, Marion R., Jr; Bakir, Saad: Nonparametric quality control charts based on the sign statistic. 95 CommStA 24 1597-1623
— —; Wolff, Hans: The behaviour of EWMA-type quality control schemes in the case of mixture alternatives. 95 SqtlAnly 14 157-177

Amir, Madjid; Knight, Frank B.: Law of the iterated logarithm and local variations at zero of the sticky Brownian motion. 95 StPrLet 23 289-295

Amirkhalkhali, Saleh; Ogwang, Tomson; Amirkhalkhali, Samad; Rao, U. L. Gouranga: On the forecasting performance of estimators for a structural equation in a large system: Some Monte Carlo results. 95 Ststcian 44 343-351

Amirkhalkhali, Samad: Amirkhalkhali, Saleh; Ogwang, Tomson; — —; Rao, U. L. Gouranga: On the forecasting performance of estimators for a structural equation in a large system: Some Monte Carlo results. 95 Ststcian 44 343-351

Amoako-Adu, Ben: Raj, Baldev; — —: Cointegration analysis of stock prices, inflation and real economic activity. 94 ASAProBuEc 190-195

Amos, Christopher I.: Robust variance-components approach for assessing genetic linkage in pedigrees. 94 AJHuGene 54 535-543

Amrhein, Peter: An example of a two-sided Wilcoxon signed rank test which is not unbiased. 95 AnInStMa 47 167-170
— —: Minimax estimation of proportions under random sample size. 95 JASA 90 1107-1111

Amsler, Christine; Lee, Junsoo: An LM test for a unit root in the presence of a structural change. 95 EconTher 11 359-368

Amy, Carolyn K.; Strong, Ann: Teaching the concepts and tools of variation using body temperature. 95 QulPr 28/3 168-168

An, Anthony B.; Breidt, F. Jay; Fuller, Wayne A.: Regression weighting methods for SIPP data. 94 ASASrvyRsr 434-439 (Disc: p440-442)
Leonard, Kelli A.; — —; Nusser, Sarah M.; Breidt, F. Jay: Approximating the variance of the survey regression estimator using poststratification. 94 ASASrvyRsr 222-227

An, Baoshe: Yang, Zhenhai; — —: Parameter estimation based on binary data (Chinese). 95 ChJApPSt 10 142-147

An, Hongzhi; Hickernell, F. J.; Zhu, Lixing: Uniformly consistent estimation in random regression model (Chinese). 95 KToChina 39 1829-1832

Ananda, Malwane M. A.: Generalized F-tests for unbalanced nested designs under heteroscedasticity. 95 AnInStMa 47 731-742

Anantharam, Venkat; Konstantopoulos, Takis: A functional central limit theorem for the jump counts of Markov processes with an application to Jackson networks. 95 AdvAppPr 27 476-509

Anantharaman, V.: Subramanian, R.; — —: Reliability analysis of a complex standby redundant system. 95 RelEngSS 48 57-70

Anastassiou, G. A.: Central limit theorem, weak law of large numbers for martingales in Banach spaces, and weak invariance principle - A quantitative study. 95 JMultiAn 52 158-180

Andelt, William F.: Hein, Eric W.; — —: Estimating coyote density from mark-resight surveys. 95 JWildlMg 59 164-169

Andersen, E. B.: Heinen, Ton (Auth); — — (Rev): Review of "Discrete latent variable models". 95 Psymtrka 60 449-450

Andersen, Erling B.: Residual analysis in the polytomous Rasch model. 95 Psymtrka 60 375-393

Andersen, H. H.; Højberre, M.; Sørensen, D.; Eriksen, P. S.: Linear and graphical models for the multivariate complex normal distribution. 95 Springer:Brln:NY 183p

Andersen, John M.: Runway landing distribution and tests for normality. 94 ASAProBays 150-152

Andersen, P. K.; Borgan, Ø.; Gill, R. D.; Keiding, N.: Statistical models based on counting processes. 95 Springer:Brln:NY 767p

Commenges, D.; — —: Score test of homogeneity for survival data. 95 LiDaAnal 1 145-156 (Disc: p157-159)

— —; Borgan, O.; Gill, R. D.; Keiding, N. (Auth); Milbrodt, Hartmut (Rev. German): Review of "Statistical models based on counting processes". 95 AllgeStA 79 221-221

— —; Borgan, Ø.; Gill, R. D.; Keiding, N. (Auth); Hjort, Nils Lid (Rev): Review of "Statistical models based on counting processes". 95 StNeerla 49 247-249

— —; Borgan, Ø.; Gill, R. D.; Keiding, N. (Auth); McKeague, Ian W. (Rev): Review of "Statistical models based on counting processes". 95 SIAMRvw 37 475-476

Andersen, Per Kragh; Rønn, Birgitte B.: A nonparametric test for comparing two samples where all observations are either left- or right-censored. 95 Biomtrcs 51 323-329

Anderson, A. J. B.: Marazzi, A. (Auth); — — (Rev): Review of "Algorithms, routines and S functions for robust statistics". 95 Biomtrcs 51 385-386

Anderson, C. W.; Turkman, K. F.: Sums and maxima of stationary sequences with heavy tailed distributions. 95 SankhyaA 57 1-10

Mole, N.; — —; Nadarajah, S.; Wright, C.: A generalized Pareto distribution model for high concentrations in short-range atmospheric dispersion. 95 EnvrMtrc 6 595-606

Haßmann, S.; Reiss, R.-D.; Thomas, M. (Auth); — — (Rev): Review of "XTREMES". 95 CmpStQ 10 91-98

Anderson, Carolyn J.; Wasserman, Stanley: Log-multiplicative models for valued social relations. 95 SocMethR 24 96-127

Anderson, Charles; Pal, Nabendu: A note on admissibility when precision is unbounded. 95 AnlsStat 23 593-597

Anderson, Christine M.; Wu, C. F. Jeff: Measuring location effects from factorial experiments with a directional response. 95 IntStRvw 63 345-363

— —; Wu, C. F. Jeff: Studying noise factors and dispersion of directional data. 94 ASAProPEng 265-268

Anderson, D. A.; Federer, W. T.: Representation and bounds for the general fractional factorial. 95 CommStA 24 363-373

Anderson, D. R.: Burnham, K. P.; — —; White, G. C.: Selection among open population capture-recapture models when capture probabilities are heterogeneous. 95 JAppStat 22 611-624

Anderson, Dan; Carlson, Richard R.; Stukel, Janice: The clinical indicators: Measurements of healthcare delivery. 94 ASAProQlPr 55-59

Anderson, David R.; Southwell, Colin: Estimates of macropod density from line transect surveys relative to analyst expertise. 95 JWildlMg 59 852-857

— —; White, Gary C.; Burnham, Kenneth P.: Some specialized risk assessment methodologies for vertebrate populations. 95 EnvEcoSt 2 91-115

Burnham, Kenneth P.; White, Gary C.; — —: Model selection strategy in the analysis of capture-recapture data. 95 Biomtrcs 51 888-898

Martin, Thomas E.; Clobert, Jean; — —: Return rates in studies of life history evolution: Are biases large?. 95 JAppStat 22 863-875

Wilson, Kenneth R.; — —: Continuous-time capture-recapture population estimation when capture probabilities vary over time. 95 EnvEcoSt 2 55-69

Anderson, DeAnn; Abetti, Frank; Savage, Phillip: Process improvement utilizing computer simulation: Case study. 95 ASQCTekCon 713-724

Anderson, Donald A.: BuHamra, Sana; — —: The spectrum of the information matrix for s^n parallel flats fractions when s is prime. 95 JStPlInf 47 333-345

Anderson, Elizabeth A.: Judgmental and statistical methods of peak electric load management. 95 IntJFor 11 295-305

Anderson, Garnet L.; Fleming, Thomas R.: Model misspecification in proportional hazards regression. 95 Biomtrka 82 527-541

Anderson, I.; Finizio, N. J.: Cyclically resolvable designs and triple whist tournaments. 93 JComDsgn 1 347-358

Anderson, Jon E.; Dayton, J. David: Instructional regression modules using XLISP-STAT. 95 JStEd 3/1 37kb (jse/v3n1/anderson)

— —; Louis, Thomas A.: Survival analysis using a scale change random effects model. 95 JASA 90 669-679

Anderson, Keaven M.: Cole, Bernard F.; Gelber, Richard D.; — —; Goldhirsch, Aron: Cost-effectiveness analysis using quality-adjusted survival. 94 ASAProBiop 9-18

Anderson, Margo; Fienberg, Stephen E.: Black, white, and shades of gray (and brown and yellow). 95 Chance 8/1 15-18

Anderson, N. H.; Titterington, D. M.: A comparison of two statistics for detecting clustering in one dimension. 95 JStCmpSm 53 103-125

Anderson, Oliver D.: Comment on "Three curious properties of the sample variance and autocovariance for stationary processes with unknown mean" (V47 p274-276). 95 AmerStat 49 118-118

— —: The precision of approximations to moments of sample moments. 95 JStCmpSm 52 163-165

— —: An interesting non-linear transformation of the theoretical autocorrelation function. 95 JStCmpSm 52 165-167

— —: Finite lengthed time-series properties versus asymptotic theory in approximating sample serial correlation structure: Part 2. 95 JStCmpSm 52 167-171

Anderson, R. L.: Zarnoch, S. J.; — —; Sheffield, R. M.: Using the β-binomial distribution to characterize forest health. 95 CanJForR 25 462-469

Anderson, Stewart J.; Jones, Richard H.: Smoothing splines for longitudinal data. 95 StatMed 14 1235-1248

Ahnn, Sang; — —: Sample size determination for comparing more than two survival distributions. 95 StatMed 14 2273-2282

Boylstein, Lee Ann; — —; Alarie, Yves: Comparison of smoothing methods when applied to respiratory data collected in mice exposed to an aerosolized machining fluid. 94 ASAProBiop 136-141

Dignam, James J.; Weissfeld, Lisa A.; — —: Methods for bounding the marginal survival distribution. 95 StatMed 14 1985-1998

Anderson, T. W.: Goodness-of-fit tests for probability distributions and spectral distributions. 95 PrEngInf 9 27-37

— —; Fang, K. T.; Olkin, I. (Ed): Multivariate analysis and its applications. 95 Arnold:Lnd:MD 471p

— — (Auth); Fernandez-Macho, F. Javier (Rev): Review of "The statistical analysis of time series". 95 JAppStat 22 548-550

— — (Auth); Walker, A. M. (Rev): Review of "The statistical analysis of time series". 95 JTimSrAn 16 357-358

Anderson, V. L.: Lorenzen, T. J.; — — (Auth); Conniffe, D. (Rev): Review of "Design of experiments: A no-name approach". 95 JRSS-A 158 195-197

Lorenzen, T. J.; — — (Auth); Holle, Rolf (Rev): Review of "Design of experiments. A no-name approach". 95 CmpStDA 19 101-102

Anderson, Virgil L.: Lorenzen, Thomas J.; — — (Auth); O'Brien, C. J. (Rev): Review of "Design of experiments: A no-name approach". 95 AstrlJSt 37 125-127

Andersson, Bertil: Waller, Lance A.; Turnbull, Bruce W.; Gustafsson, Göran; Hjalmars, Ulf; — —: Detection and assessment of clusters of disease: An application to nuclear power plant facilities and childhood leukaemia in Sweden. 95 StatMed 14 3-16

Andersson, Claes; Nordberg, Lennart: A method for variance estimation of non-linear functions of totals in surveys - Theory and software implementation. 94 JOffStat 10 395-405

Andersson, Håkan; Djehiche, Boualem: Limit theorems for multitype epidemics. 95 StocProc 56 57-75

Andersson, Steen A.; Madigan, David; Perlman, Michael D.; Triggs, Christopher M.: On the relation between conditional independence models determined by finite distributive lattices and by directed acyclic graphs. 95 JStPlInf 48 25-46

— —; Perlman, Michael D.: Unbiasedness of the likelihood ratio test for lattice conditional independence models. 95 JMultiAn 53 1-17

Andersson, Steen Arne; Perlman, Michael D.: Testing lattice conditional independence models. 95 JMultiAn 53 18-38

Andes, Nancy; Davis, Jack E.: Linking public health data using geographic information system techniques: Alaskan community characteristics and infant mortality. 95 StatMed 14 481-490 (Disc: p687-689)

Andjel, Enrique D.; Mountford, Thomas S.; Schonmann, Roberto H.: Equivalence of exponential decay rates for bootstrap percolation like cellular automata. 95 AIHenrPB 31 13-25

Andreasen, N. C.: Arndt, S.; Tyrrell, G.; Wollson, R. F.; Flaum, M.; — —: Effects of errors in a multicenter medical study: Preventing misinterpreted data. 94 JPsyctrc 28 447-459

Andres, T. H.: Saltelli, A.; — —; Homma, T.: Sensitivity analysis of model output. Performance of the iterated fractional factorial design method. 95 CmpStDA 20 387-407

Andrews, C. Webster: Cummins, David J.; — —: Iteratively reweighted partial least squares: A performance analysis by Monte Carlo simulation. 95 JChemtrc 9 489-507

Andrews, Donald W. K.: Nonparametric kernel estimation for semiparametric models. 95 EconTher 11 560-596

— —; Ploberger, Werner: Admissibility of the likelihood ratio test when a nuisance parameter is present only under the alternative. 95 AnlsStat 23 1609-1629

Andrich, David: Distinctive and incompatible properties of two common classes of IRT models for graded responses. 95 ApPsychM 19 101-119

— —: Models for measurement, precision, and the nondichotomization of graded responses. 95 Psymtrka 60 7-26 (Disc: p27-46)

— —: Reply to comment on "Models for measurement, precision, and the nondichotomization of graded responses". 95 Psymtrka 60 37-46

Clogg, Clifford C.; Shihadeh, Edward S. (Auth); — — (Rev): Review of "Statistical models for ordinal variables". 95 JASA 90 1131-1131

Wainer, Howard; Dorans, Neil J.; Flaugher, Ronald; Green, Bert F.; Mislevy, Robert; Steinberg, Lynne; Thissen, David (Auth); — — (Rev): Review of "Computerized adaptive testing: A primer". 95 Psymtrka 60 615-620

Andronico, Maurizio; Casale, Salvatore; La Corte, Aurelio: A method for the generation of correlated random processes. 95 JCmpInT 3 9-18

Anello, Charles: Statistical issues in postmarketing surveillance at FDA. 94 ASAProBiop 452-459

Angeli, A.; Cocchi, D.; Pasquini, L.; Samoggia, A.: Family structure and income: A first analysis on cross-section data (Italian). 94 Statstca 54 435-452

Angelis, L.; Moyssiadis, C.; Kageyama, S.: Methods of constructing A-efficient BTIUB designs. 93 UtilMath 44 5-15

Angermeier, Paul L.; Smogor, Roy A.: Estimating number of species and relative abundances in stream-fish communities: Effects of sampling effort and discontinuous spatial distributions. 95 JFishRBC 52 936-949

Ensign, William E.; — —; Dolloff, C. Andrew: Use of line transect methods to estimate abundance of benthic stream fishes. 95 JFishRBC 52 213-222

Angers, Jean-François: Belzile, Éric; — —: Choice of noninformative priors for the variance components of an unbalanced one-way random model. 95 CommStA 24 1325-1341

Angle, John: Frequency spikes in income distributions. 94 ASAProBuEc 265-270

Angleton, George M.; Bonham, Charles D.: Least squares regression vs. geometric mean regression for ecotoxicology studies. 95 AplMaCmp 72 21-32

Angrist, Joshua D.: Introduction to the JBES symposium on program and policy evaluation. 95 JBES 13/2 133-136

— —; Imbens, Guido W.: Two-stage least squares estimation of average causal effects in models with variable treatment intensity. 95 JASA 90 431-442

— —; Krueger, Alan B.: Split-sample instrumental variables estimates of the return to schooling. 95 JBES 13/2 225-235

Anisimov, Vladimir N.: Yakovlev, Andrej Yu.; Tsodikov, Alexandr D.; — —: A new model of aging: Specific versions and their application. 95 BiomtrcJ 37 435-448

Annand, M. Christina: Campana, Steven E.; — —; McMillan, James I.: Graphical and statistical methods for determining the consistency of age determinations. 95 TrAFish 124 131-138

Anstrom, Kevin J.; Quade, Dana: Testing whether a count variable belongs to the Katz family of distributions. 94 ASAProStCp 194-199

Antal, Peter: Enlargement of obstacles for the simple random walk. 95 AnlsProb 23 1061-1101

Anthony, Robert G.: Rosenberg, Daniel K.; Overton, W. Scott; — —: Estimation of animal abundance when capture probabilities are low and heterogeneous. 95 JWildlMg 59 252-261

Antoch, Jaromir; Hušková, Marie; Veraverbeke, Noël: Change-point problem and bootstrap. 95 JNnparSt 5 123-144

Antoch, Jaromír; Ekblom, Hakan: Recursive robust regression computational aspects and comparison. 95 CmpStDA 19 115-128

— —; Hájek, Petr; Jiroušek, Radim: Obituary: Tomáš Havránek. 95 CmpStDA 19 111-113

Anton, Vicente Nuñez: Kmietowicz, Z. W.; Yannoulis, Y. (Auth); — — (Rev): Review of "Statistical tables for economic, business and social studies (Second edition)". 95 JAppStat 22 439-440

Antón, Vicente Núñez: Späth, Helmuth (Auth); — — (Rev): Review of "Mathematical algorithms for linear regression". 95 JAppStat 22 318-319

Antoni, A.; Dhorne, T.: Information from subjects by variables tables (French). 95 RStApp 43/4 43-61

Antoniadis, A.; Oppenheim, G. (Ed): Wavelets and statistics. 95 Springer:Brln:NY 410p

Aoshima, Makoto: In heteroscedastic simultaneous inference, the heteroscedastic method II (based on Healy's two-stage sampling) is more economical than the original HM (which is based on Chatterjee's sampling). 94 AJMMgSci 14 167-196

Apian Software: — — (Auth); Lachenbruch, Peter A. (Rev): Review of "Survey Pro for Windows (Version 1.8)". 95 Chance 8/4 58-59

Appa, Gautam: Williams, H. P. (Auth); — — (Rev): Review of "Model solving in mathematical programming". 95 JRSS-A 158 202-202

Appel, Martin V.; Cole, Ronald: Spoken language recognition of the Year 2000 Census Questionnaire Feasibility Test. 94 ASASrvyRsr 1310-1315

Gillman, Daniel W.; — —: Developing metadata database at the Census Bureau. 94 ASASrvyRsr 781-784

Nicholls, William L., II; — —: New CASIC technologies at the U.S. Bureau of the Census. 94 ASASrvyRsr 757-762

Appelo, C. A. J.: van Veldhuizen, M.; — —; Griffioen, J.: The (least squares) quotient algorithm as a rapid tool for obtaining sorption isotherms from column elution curves. 95 WaterReR 31 849-857

Applebaugh, W.; Chadha, S. S.: Optimal lengths of left-turn lanes. 94 ASAProStEd 102-103

Appleton, David R.: The uniform, logistic, and lambda distributions. 95 CommStB 24 947-951

— —: What do we mean by a statistical model?. 95 StatMed 14 185-197

— —: May the best man win?. 95 Ststcian 44 529-538

Arabie, Phipps: Hubert, Lawrence; — —: Iterative projection strategies for the least squares fitting of tree structures to proximity data. 95 BJMaStPs 48 281-317

Hubert, Lawrence; — —: The approximation of two-mode proximity matrices by sums of order-constrained matrices. 95 Psymtrka 60 573-605

Mirkin, Boris; — —; Hubert, Lawrence J.: Additive two-mode clustering: The error-variance approach revisited. 95 JClass 12 243-263

Eliashberg, J.; Lilien, G. (Ed); — — (Rev): Review of "Handbook in operations research and management science: Marketing". 95 JClass 12 143-144

Aragon, Jorge: Rabinowitz, Daniel; Tsiatis, Anastasios; — —: Regression with interval-censored data. 95 Biomtrka 82 501-513

Aragón, Jorge; Quiróz, Adolfo J.: Rank regression for current status data. 95 StPrLet 24 251-256

Arai, Fumihito: Ishigami, Hideyuki; Fukuda, Toshio; Shibata, Takanori; — —: Structure optimization of fuzzy neural network by genetic algorithm. 95 FuzzySet 71 257-264

Aras, Girish; Gardiner, Joseph: Fixed diameter confidence ellipsoid in time-sequential models. 95 SqtlAnly 14 29-39

Arasu, K. T.; Ray-Chaudhuri, D. K.; Singhi, N. M.: Simple designs. 93 JCISS 18 130-135

— —; Sehgal, Surinder K.: Some new difference sets. 95 JCombThA 69 170-172

Arbekov, I. M.; Vinokurov, V. I.; Ryazanov, B. V.: On the problem of detecting random trajectories. 94 ThProbAp 39 547-557

Arbia, Giuseppe: A bird's eye view of spatial data and applications. 94 QStMaA 16 5-16

— —: Selection techniques in sampling spatial units. 94 QStMaA 16 81-91

Arbuckle, James; BMDP Statistical Software Inc; Scientific Software International (Auth); Hox, Joop J. (Rev): Review of "AMOS, EQS, and LISREL for Windows: A comparative review". 95 StrEqMod 2 79-91

Arcangeli, Rémi; Ycart, Bernard: Almost sure convergence of smoothing D^m-splines for noisy data. 93 NumeriMa 66 281-294

Arcas, Antoni: Capdevila, Carles; — —: Empirical significance test of the goodness-of-fit for some pyramidal clustering procedures. 95 Questo 19 123-130

Capdevila, Carles; — —: Indexed pyramids and pyramidal dissimilarities (Spanish). 95 Questo 19 131-151

Arce, Constantino: — — (Auth); Maydeu-Olivares, Albert (Rev): Review of "Multidimensional scaling. A multivariate technique for the analysis of proximity and preference data" (Spanish). 95 JClass 12 137-139

Archambault, S.; Moore, M.: Statistical inference for Voronoi images. 95 CandJSt 23 11-20

Archer, G. E. B.; Titterington, D. M.: The iterative image space reconstruction algorithm (ISRA) as an alternative to the EM algorithm for solving positive linear inverse problems. 95 StSinica 5 77-96

Arcones, Miguel A.: The asymptotic accuracy of the bootstrap of U-quantiles. 95 AnlsStat 23 1802-1822

— —: The law of the iterated logarithm for empirical processes under absolute regularity. 95 JTherPrb 8 433-451

— —: On the law of the iterated logarithm for Gaussian processes. 95 JTherPrb 8 877-903

— —: A Bernstein-type inequality for U-statistics and U-processes. 95 StPrLet 22 239-247

— —: On the central limit theorem for U-statistics under absolute regularity. 95 StPrLet 24 245-249

— —: Asymptotic normality of multivariate trimmed means. 95 StPrLet 25 45-53

— —: Limit theorems for nonlinear functionals of a stationary Gaussian sequence of vectors. 94 AnlsProb 22 2242-2274

— —; Giné, Evarist: On the law of the iterated logarithm for canonical U-statistics and processes. 95 StocProc 58 217-245

Ardanuy, R.: Stochastic differential games: The linear quadratic zero sum case. 95 SankhyaA 57 161-165

Arellano, Consuelo; Pantula, Sastry G.: Testing for trend stationarity versus difference stationarity. 95 JTimSrAn 16 147-164

Arellano, Manuel; Bover, Olympia: Another look at the instrumental variable estimation of error-components models. 95 JEconmtx 68 29-51

Arellano-Valle, Reinaldo B.; Bolfarine, Heleno: On some characterizations of the t-distribution. 95 StPrLet 25 79-85

Arfi, Mounir: Nonparametric drift estimation from ergodic samples. 95 JNnparSt 5 381-389

Arghami, N. R.; Billard, L.: A sequential dual of Stein's two-sample procedure. 92 ProIrStC1 11-18

Argyelan, J.; Nassar, R.: A non-linear model of an open flow adsorption process. 95 EnvrMtrc 6 195-203

Arjas, Elja; Holmberg, Jan: Marked point process framework for living probabilistic safety assessment and risk follow-up. 95 RelEngSS 49 59-73

— —; Liu, Liping: Assessing the losses caused by an industrial intervention: A hierarchical Bayesian approach. 95 ApplStat 44 357-368

Arkonac, Seyhan Z.; Higgins, Matthew L.: A Monte Carlo study of tests for non-nested models estimated by generalized method of moments. 95 CommStB 24 745-763

Arlinghaus, Sandra L.; Griffith, Daniel A. (Ed): Practical handbook of spatial statistics. 95 CRCPress:FL 336p

Armanino, C.: Forina, M.; Drava, G.; — —; Boggia, R.; Lanteri, S.; Leardi, R.; Corti, P.; Conti, P.; Giangiacomo, R.; Galliena, C.; Bigoni, R.; Quartari, I.; Serra, C.; Ferri, D.; Leoni, O.; Lazzeri, L.: Transfer of calibration function in near-infrared spectroscopy. 95 Chemolab 27 189-203

Arminger, G.; Clogg, C. C.; Sobel, M. E. (Ed); Kemp, G. C. R. (Rev): Review of "Handbook of statistical modeling for the social and behavioral sciences". 95 Biomtrcs 51 1586-1587

Arminger, Gerhard; Clogg, Clifford C.; Sobel, Michael E.: Handbook of statistical modeling for the social and behavioral sciences. 95 PlenumPubl:NY 618p

Ziegler, Andreas; Bachleitner, Reinhard; — —: Pseudo-maximum-likelihood-estimation and regression diagnostics for count data: Suicide by tourism? (German). 95 AllgeStA 79 170-195

— —; Clogg, Clifford C.; Sobel, Michael E. (Ed); Dudgeon, Paul (Rev): Review of "Handbook of statistical modeling for the social and behavioural sciences". 95 Chance 8/4 32-36

Armitage, P.: Berry, G.; — —: Mid-P confidence intervals: A brief review. 95 Ststcian 44 417-423

Armitage, Peter: David Byar's approach to the analysis of survival data with covariates. 95 ConClnTr 16 259-269

— —: Before and after Bradford Hill: Some trends in medical statistics. 95 JRSS-A 158 143-153

Armour, Len: Barnett, Neil S.; Gomm, Ian S.; — —: Location of the optimal sampling point for the quality assessment of continuous streams. 95 AstrlJSt 37 145-152

Armstrong, Ben G.: Comparing standardized mortality ratios. 95 AnEpidm 5 60-64

Armstrong, J. Scott; Fildes, Robert: Comment on "On the limitations of comparing mean square forecast errors" (V12 p617-637). 95 JForcast 14 67-71

Yokum, J. Thomas; — —: Beyond accuracy: Comparison of criteria used to select forecasting methods. 95 IntJFor 11 591-597

Armstrong, Michael J.: Joint reliability-importance of components. 95 IEEEReli 44 408-412

Armstrong, Ronald D.: Sklar, Michael G.; — —: A piecewise linear approximation procedure for L_p norm curve fitting. 95 JStCmpSm 52 323-335

Arnab, Raghunath: Optimal estimation of a finite population total under randomized response surveys. 95 MaOpfStS 27 175-180

Arnason, A. Neil; Schwarz, Carl J.: POPAN-4: Enhancements to a system for the analysis of mark-recapture data from open populations. 95 JAppStat 22 785-800

Arnaudon, M.; Paycha, S.: Factorisation of semi-martingales on principal fibre bundles and the Faddeev-Popov procedure in gauge theories. 95 Stochatx 53 81-107

Arndt, S.; Tyrrell, G.; Wollson, R. F.; Flaum, M.; Andreasen, N. C.: Effects of errors in a multicenter medical study: Preventing misinterpreted data. 94 JPsyctrc 28 447-459

Arnedt, Cheryl Mercado: Frankovic, Kathleen A.; — —; Ramnath, Bala: Interactive polling and Americans' comfort level with technology. 94 ASASrvyRsr 1316-1321

Arnholt, Alan T.; Hebert, Jaimie L.: Estimating the mean with known coefficient of variation. 95 AmerStat 49 367-369

Arnold, B. F.: Mittag, H. J. (Auth); — — (Rev): Review of "Quality control charts" (German). 95 StHefte 36 96-96

Arnold, Barry C.: Bayesian inference for conditionally specified models. 94 ASAProBays 165-168

— —; Castillo, Enrique; Sarabia, José María: General conditional specification models. 95 CommStA 24 1-11

— —; Groeneveld, Richard A.: Measuring skewness with respect to the mode. 95 AmerStat 49 34-38

Arnold, Bernhard F.: Statistical tests optimally meeting certain fuzzy requirements on the power function and on the sample size. 95 FuzzySet 75 365-372

Arnold, Greg: Waddell, Peter J.; Penny, David; Hendy, Michael D.; — —: The sampling distributions and covariance matrix of phylogenetic spectra. 94 MolBioEv 11 630-642

Arnold, J. Tim: Dickey, David A.; — —: Teaching statistics with data of historical significance: Galileo's gravity and motion experiments. 95 JStEd 3/1 20kb (jse/v3n1/datasets.dickey)

Arnold, Jesse C.; Lemon, G. H.; Norton, J. M.: Sample size requirements for rare or abundant attribute sampling. 95 CommStB 24 131-148

— —; Reynolds, Marion R.: CUSUM charts with variable sample size and variable interval size. 94 ASAProQlPr 87-92

Sethuraman, Shanthi; Reynolds, Marion R., Jr; — —: Process control strategies using variable sampling intervals. 94 ASAProQlPr 109-114

Arnold, Ludwig; Imkeller, Peter: Furstenberg-Khasminskii formulas for Lyapunov exponents via anticipative calculus. 95 Stochatx 54 127-168

— —; Scheutzow, Michael: Perfect cocycles through stochastic differential equations. 95 ProbTher 101 65-88

Arnold, Steven F.: Macchiavelli, Raúl; — —: Difficulties with the use of penalized likelihood criteria in antedependence and polynomial models. 95 CommStA 24 501-522

Arntzen, Halvard: Optimal choice of production capacity in a random market. 95 Stochatx 55 87-120

Arntzenius, Frank: A heuristic for conceptual change. 95 PhilSci 62 357-369

Arora, Neeraj: Allenby, Greg M.; — —; Ginter, James L.: Incorporating prior knowledge into the analysis of conjoint studies. 95 JMarRsr 32 152-162

Arous, G. B.; Guionnet, A.: Large deviations for Langevin spin glass dynamics. 95 ProbTher 102 455-509

Arratia, Richard; Stark, Dudley; Tavaré, Simon: Total variation asymptotics for Poisson process approximations of logarithmic combinatorial assemblies. 95 AnlsProb 23 1347-1388

Arron, Geoffrey: Camacho, Fernando; — —: Effects of the regulators paclobutrazol and flurprimidol on the growth of terminal sprouts formed on trimmed silver maple trees. 95 CandJSt 23 312-322 (Disc: p329-332)

Arslan, Olcay; Constable, Patrick D. L.; Kent, John T.: Convergence behavior of the EM algorithm for the multivariate t-distribution. 95 CommStA 24 2981-3000

Artalejo, J. R.; Falin, J. I.: Stochastic decomposition for retrial queues. 94 Top 2 329-342

Martin, M.; — —: Analysis of an M/G/1 queue with two types of impatient units. 95 AdvAppPr 27 840-861

Artaza, E.: Iznaga, N.; Núñez, G.; Solozabal, J.; Morales, A.; — —; Rubio, R.; Cardenas, E.: A personal computer-based system for parallel line analysis of bioassay data. 95 CmpBiomd 47 167-175

Artiles-Léon, Noel; Novoa-Ramírez, Clara M.; Domenech, Carolina: Improving fabric finishing through experimental design. 95 ASQCTekCon 952-962

Arumainayagam, S. D.; Soundararajan, V.: Construction and evaluation of matched quick switching systems. 95 JAppStat 22 245-251

Arumainayagam, S. Devaraj; Soundararajan, V.: Quick switching double sampling system indexed by the crossover point. 95 CommStB 24 765-773

— —; Soundararajan, V.: Construction and selection of quick switching double-sampling system: Sample size tightening. 95 JAppStat 22 105-119

Arunachalam, V.: A genetic basis behind combining ability and breeding values in mono- and di-genic systems. 95 BiomtrcJ 37 217-228

Asha, G.: Nair, N. Unnikrishnan; — —: Characteristic properties of a bivariate geometric law. 94 JIndnStA 32 111-116

Ashenfelter, Orley; Ashmore, David; Lalonde, Robert: Bordeaux wine vintage quality and the weather. 95 Chance 8/4 7-14

Ashmore, David: Ashenfelter, Orley; — —; Lalonde, Robert: Bordeaux wine vintage quality and the weather. 95 Chance 8/4 7-14

Ashour, Samir K.; Abd-Elfattah, Abd-Allah M.: The sampling distribution of some goodness-of-fit test statistics using Burr distribution and censored samples. 94 EgpStJ 38 147-163

— —; Jones, Peter W.; Ahmed, Mohiy El-Din M.: Discriminant analysis for social-economic and cultural factors affecting diarrhea diseases in children under two years in Egypt. 95 EgpStJ 39 1-11

Asimakopoulos, D. N.: Kambezidis, H. D.; Tulleken, R.; Amanatidis, G. T.; Paliatsos, A. G.; — —: Statistical evaluation of selected air pollutants in Athens, Greece. 95 EnvrMtrc 6 349-361

Asmussen, Marjorie A.; Basten, Christopher J.: Sampling theory for cytonuclear disequilibria. 94 Genetcs 138 1351-1363

Asmussen, Søren: Stationary distributions for fluid flow models with or without Brownian noise. 95 CommStSM 11 21-49

— —; Frey, Andreas; Rolski, Tomasz; Schmidt, Volker: Does Markov-modulation increase the risk?. 95 AstinB 25 49-66

— —; Nielsen, Hanne Mandrup: Ruin probabilities via local adjustment coefficients. 95 JAppProb 32 736-755

— —; Schmidt, Volker: Ladder height distributions with marks. 95 StocProc 58 105-119

Aspandiiarov, Sanjar: On the convergence of 2-dimensional Markov chains in quadrants with boundary reflection. 95 Stochatx 53 275-303

Assaf, David; Shepp, Larry; Yao, Yi-Ching: Optimal tracking of Brownian motion. 95 Stochatx 54 233-245

Assani, I.; Lesigne, E.; Rudolph, D.: Wiener-Wintner return-times ergodic theorem. 95 IsrelJMa 92 375-395

Asselain, Bernard: Tsodikov, Alexander D.; — —; Fourque, Alain; Hoang, Thu; Yakovlev, Andrej Yu.: Discrete strategies of cancer post-treatment surveillance. Estimation and optimization problems. 95 Biomtrcs 51 437-447

Assimakopoulos, V.: A successive filtering technique for identifying long-term trends. 95 JForcast 14 35-43

Assing, S.; Manthey, R.: The behavior of solutions of stochastic differential inequalities. 95 ProbTher 103 493-514

Assunção, Renato: Newton, Michael A.; Guttorp, Peter; Catlin, Sandra; — —; Abkowitz, Janis L.: Stochastic modeling of early hematopoiesis. 95 JASA 90 1146-1155

Asthana, Sheel: Chaturvedi, Anoop; Hasegawa, Hikaru; — —: Bayesian predictive analysis of the linear regression model with and Edgeworth series prior distribution. 95 CommStA 24 2469-2484

Asymont, I. M.; Fayolle, G.; Menshikov, M. V.: Random walks in a quarter plane with zero drifts: Transience and recurrence. 95 JAppProb 32 941-955

Atchley, William R.: Xu, Shizhong; — —; Fitch, Walter M.: Phylogenetic inference under the pure drift model. 94 MolBioEv 11 949-960

Athreya, K. B.; Vidyashankar, A. N.: Large deviation rates for branching processes. II. The multitype case. 95 AnlsApPr 5 566-576

— —; Weerasinghe, A. P. N.: Ergodic theorems for transient one-dimensional diffusions. 95 StocProc 58 173-185

Atkinson, A. C.: Beyond response surfaces: Recent developments in optimum experimental design. 95 Chemolab 28 35-47

— —: Comment on "Constrained optimization of experimental design". 95 MaOpfStS 26 149-150

— —; Cook, R. D.: D-optimum designs for heteroscedastic linear models. 95 JASA 90 204-212

— —; Demetrio, C. G. B.; Zocchi, S. S.: Optimum dose levels when males and females differ in response. 95 ApplStat 44 213-226

— —; Donev, A. N. (Auth); Laycock, P. J. (Rev): Review of "Optimum experimental designs". 95 Biomtrcs 51 379-379

Atkinson, Anthony: Barnett, V.; Lewis, T. (Auth); — — (Rev): Review of "Outliers in statistical data (Third edition)". 95 JRSS-A 158 630-631

Brown, P. J. (Auth); — — (Rev): Review of "Measurement, regression, and calibration". 95 JRSS-A 158 187-187

Atkinson, E. Neely: Interactive dynamic graphics for exploratory survival analysis. 95 AmerStat 49 77-84

Atlas, Robert S.: Overall, John E.; — —: Power of a test that is robust against variance heterogeneity. 95 PsychRpt 77 155-159

Atrostic, B. K.: A multiple imputation approach to microsimulation. 94 ASASrvyRsr 529-534

Attal, Stéphane: Representation of Wiener space endomorphisms preserving martingales (French). 95 AIHenrPB 31 467-484

Attanasio, Massimo; Wong, Weng Kee: Transformations for stabilizing spread in a time series model: A comparative study. 95 JStCmpSm 52 55-70

Attfield, C. L. F.: A Bartlett adjustment to the likelihood ratio test for a system of equations. 95 JEconmtx 66 207-223

Atwood, Corwin L.; Bryan, Mark F.: Two-stage sampling for bounding a mean. 95 EnvrMtrc 6 73-83

Atzeni, M. G.: Mayer, D. G.; — —; Swain, A. J.; Stuart, M. A.: Models for the spatial dispersal of insect pests. 95 EnvrMtrc 6 497-503

Aubin, Christophe; Fougére, Denis; Husson, Emmanuel; Ivaldi, Marc: Real-time pricing of electricity for residential customers: Econometric analysis of an experiment. 95 JApEcmtx 10 171-191S

Aubin, Elisete C. Q.: Cordeiro, Gauss M.; Cribari-Neto, Francisco; — —; Ferrari, Silvia L. P.: Bartlett corrections for one-parameter exponential family models. 95 JStCmpSm 53 211-231

Aubin, J. P.; Prato, G. Da: Stochastic Nagumo's viability theorem. 95 StocAnAp 13 1-11

Auclair, Paul F.: Poston, Catherine A.; Barr, David R.; — —: Solving the ranking and selection indifference-zone formulation for normal distributions using computer software. 94 ASAProStCp 290-291

Avi-Itzhak, Benjamin: Winch, Janice Kim; — —: Ordering of tandem constant-service stations to minimize in-process stock cost. 95 PrEnglnf 9 457-473

Avi-Itzhak, Hadar I.; Van Mieghem, Jan A.; Rub, Leonardo: Multiple subclass pattern recognition: A maximin correlation approach. 95 IEEEPAMI 17 418-431

Axelrod, Michael C.; Glosup, Jeffrey G.: Durbin-Watson revisited. 94 ASAProStCp 176-181

Glosup, Jeffrey G.; — —: Use of the AIC with the EM algorithm: A demonstration of a probability model selection technique. 94 ASAProPEng 120-125

Axenovich, T. I.: Ginsburg, E. Kh.; — —; Babenko, V. N.: Comparison of asymptotic tests of the Hardy-Weinberg distribution. 91 AnHuGene 55 329-338

Ayala, Guillermo; Simó, Amelia: Bivariate random closed sets and nerve fibre degeneration. 95 AdvAppPr 27 293-305

Ayers, Christopher V.: Putting the pieces together: Integrated implementation of SPC. 94 ASAProPEng 291-294

Aykroyd, R. G.: Partition models in the analysis of autoradiographic images. 95 ApplStat 44 441-454

Ayton, P.: Wright, G.; — — (Ed); Lee, Peter M. (Rev): Review of "Subjective probability". 95 JRSS-A 158 637-638

Ayuthya, Rampai Suksawasdi Na; Böhning, Danmar: Traffic accident mapping in Bangkok metropolis: A case study. 95 StatMed 14 2445-2458

Azad, Sam: Mocan, H. Naci; — —: Accuracy and rationality of state General Fund Revenue forecasts: Evidence from panel data. 95 IntJFor 11 417-427

Azaïs, J.-M.: Bailey, R. A.; — —; Monod, H.: Are neighbour methods preferable to analysis of variance for completely systematic designs? 'Silly designs are silly!'. 95 Biomtrka 82 655-659

Azari, A. S.; Müller, H. G.: [Algorithm AS 297] Orthogonal polynomial and hybrid estimators for nonparametric regression. 95 ApplStat 44 264-268

— —; Tsai, Chih-Ling: Non-parametric regression approach for model checking on the two-phase regression problem. 94 JAppStat 21 597-606

Azen, Stanley P.: Lapuerta, Pablo; — —; LaBree, Laurie: Use of neural networks in predicting the risk of coronary artery disease. 95 CmpBimdR 28 38-52

Markus, Richard A.; Frank, Jess; Groshen, Susan; — —: An alternative approach to the optimal design of an LD50 bioassay. 95 StatMed 14 841-852

Azimi-Sadjadi, Mahmood R.: Bannour, Sami; — —: Principal component extraction using recursive least squares learning. 95 IEEENrlN 6 457-469

Azoff, E. Michael: — — (Auth); Meade, Nigel (Rev): Review of "Neural network time series forecasting of financial markets". 95 IntJFor 11 601-602

Azzalini, A.: Logistic regression and other discrete data models for serially correlated observations. 94 JItalStS 3 169-179

Azzam, Abdul-Mordy H.: Ilallawa, Abdel M.; — —: A new method for generating the design matrix of a linear regression model. 95 EgpStJ 39 106-119

Baade, Ingrid A.; Pettitt, Anthony N.: Tests of loglinear and linear relative risks for Cox's model. 95 Biomtrcs 51 1502-1513

Baba, M. Y.: Rodrigues, J.; — —: Bayesian estimation of a simple regression model with measurement errors. 94 RBrzlPrS 8 107-118

Babcock, Laura M.; Altekar, Maneesha: Use of statistical design to develop flame retardant polymer formulations. 95 Chemolab 29 141-146

Babenko, V. N.: Ginsburg, E. Kh.; Axenovich, T. I.; — —: Comparison of asymptotic tests of the Hardy-Weinberg distribution. 91 AnHuGene 55 329-338

Babiker, Hamza A.: Hill, William G.; — —; Ranford-Cartwright, Lisa C.; Walliker, David: Estimation of inbreeding coefficients from genotypic data on multiple alleles, and application to estimation of clonality in malaria parasites. 95 GeneRsr 65 53-61

Babron, Marie-Claude: Dizier, Marie-Hélène; — —; Clerget-Darpoux, Françoise: Interactive effect of two candidate genes in a disease: Extension of the marker-association-segregation χ^2 method. 94 AJHuGene 55 1042-1049

Babu, Gutti Jogesh: Bootstrap for nonstandard cases. 95 JStPlInf 43 197-203

— —: Absolute continuity of the distributions of additive arithmetic functions. 95 SankhyaA 57 29-32

Serinko, Regis J.; — —: Asymptotics of k-mean clustering under non-i.i.d. sampling. 95 StPrLet 24 57-66

Baccelli, F.; Bremaud, P.: Elements of queueing theory: Palm-martingale calculus and stochastic recurrences. 94 Springer:Brln:NY 256p

Baccelli, François; Foss, Serguei: On the saturation rule for the stability of queues. 95 JAppProb 32 494-507

— —; Klein, Maurice; Zuyev, Sergei: Perturbation analysis of functionals of random measures. 95 AdvAppPr 27 306-325

Bacchetti, Peter; Segal, Mark R.: Survival trees with time-dependent covariates: Application to estimating changes in the incubation period of AIDS. 95 LiDaAnal 1 35-47

Baccin, Maria C.; Ferrante, Marco: On a stochastic delay difference equation with boundary conditions and its Markov property. 95 StocProc 60 131-146

Bachleitner, Reinhard: Ziegler, Andreas; — —; Arminger, Gerhard: Pseudo-maximum-likelihood-estimation and regression diagnostics for count data: Suicide by tourism? (German). 95 AllgeStA 79 170-195

Bachmaier, M.; Precht, M.: Robust confidence intervals for contrasts based upon a likelihood ratio test. 95 StHefte 36 215-236

Backeljau, T.: van Dongen, S.; — —: One- and two-sample tests for single-locus inbreeding coefficients using the bootstrap. 95 Heredity 74 129-135

Bacon, Donald R.: A maximum likelihood approach to correlational outlier identification. 95 MultBehR 30 125-148

— —; Sauer, Paul L.; Young, Murray: Composite reliability in structural equations modeling. 95 EdPsMeas 55 394-406

Bacro, J. N.; Brito, M.: Weak limiting behaviour of a simple tail Pareto-index estimator. 95 JStPlInf 45 7-19

Baddeley, A. J.; Cruz-Orive, L. M.: The Rao-Blackwell theorem in stereology and some counterexamples. 95 AdvAppPr 27 2-19

— —; van Lieshout, M. N. M.: Area-interaction point processes. 95 AnInStMa 47 601-619

Cabo, A. J.; — —: Line transects, covariance functions and set convergence. 95 AdvAppPr 27 585-605

Badrigal, J. L.; Richards, D. O.; Free, J. C.; Lee, T. N.; Lawson, J. S.; Hawks, V. D.: Q^2: A TQM course for undergraduates. 94 ASAProQlPr 172-180

Bae, Jongsig; Levental, Shlomo: A uniform CLT for continuous martingales. 95 JKoreaSt 24 225-231

— —; Levental, Shlomo: Uniform CLT for Markov chains and its invariance principle: A martingale approach. 95 JTherPrb 8 549-570

Bae, Whasoo: A study on a basis for the selection of a design for quadratic model fits fearing a cubic bias in multiple response case. 95 JKoreaSt 24 31-44

Baek, Hoh Yoo: Kim, Byung Hwee; Koh, Tae Wook; — —: Estimators with nondecreasing risk in a multivariate normal distribution. 95 JKoreaSt 24 257-266

Baenziger, P. Stephen: Stroup, Walter W.; — —; Mulitze, Dieter K.: Removing spatial variation from wheat yield trials: A comparison of methods. 94 CropSci 34 62-66

Baeza-Baeza, Juan José: Xie, Yu-Long; — —; Ramis-Ramos, Guillermo: Kinetic spectrophotometric resolution of binary mixtures using three-way partial least squares. 95 Chemolab 27 211-220

Baeza-Yates, Ricardo A.: Fringe analysis revisited. 95 CmpSurvy 27 109-119

Bagai, Isha; Deshpande, J. V.: Nonparametric tests for the two sample problem with two independent competing risks. 95 JNnparSt 5 21-31

— —; Rao, B. L. S. Prakasa: Kernel-type density and failure rate estimation for associated sequences. 95 AnInStMa 47 253-266

— —; Rao, B. L. S. Prakasa: Kernel type density estimates for positive valued random variables. 95 SankhyaA 57 56-67

Bagby, Richard J.: Calculating normal probabilities. 95 AmMaMnth 102 46-49

Bagchi, Arunabha: Aihara, Shin Ichi; — —: Nonlinear smoothing for random fields. 95 StocProc 55 143-158

Baglivo, Jenny: Waterloo Maple Software Inc (Auth); — — (Rev): Review of "Maple V for the Macintosh (Version 5.0.2)". 95 AmerStat 49 86-92
Wolfram Research Inc (Auth); — — (Rev): Review of "Mathematica for the Macintosh (Version 2.0.4)". 95 AmerStat 49 86-92

Bagozzi, Richard P.: Baumgartner, Hans; — —: Specification, estimation, and testing of moment structure models based on latent variates involving interactions among the exogenous constructs. 95 SocMethR 24 187-213

Bagui, S. C.; Mehra, K. L.; Rao, M. S.: A nearest neighbour classification rule for multiple observations based on a sub-sample approach. 95 SankhyaA 57 316-332

Bahrampour, Abbas: Becker, Niels G.; — —; Dietz, Klaus: Threshold parameters for epidemics in different community settings. 95 MaBisci 129 189-208

Bai, D. S.; Choi, I. S.: $\bar{X}$ and R control charts for skewed populations. 95 JQualTek 27 120-131
— —; Chun, Y. R.; Kim, J. G.: Failure-censored accelerated life test sampling plans for Weibull distribution under expected test time constraint. 95 RelEngSS 50 61-68

Bai, Do Sun; Choi, In Su; Hwang, Yong Keun: Field data analyses for products with multiple-modes of failure (Korean). 95 KorApS 8/1 89-104
— —; Kwon, Hyuck Moo: Economic design of a two-stage screening procedure with a prescribed outgoing quality. 95 Metrika 42 1-18
— —; Yun, Hyung Je; Choi, In Su: Field data analyses for repairable products (Korean). 95 KorApS 8/2 133-145

Bai, Jun; Jakeman, Anthony J.; McAleer, Michael: Empirical models for evaluating errors in fitting extremes of a probability distribution. 95 MaCmpSim 39 1-7

Bai, Jushan: Least absolute deviation estimation of a shift. 95 EconTher 11 403-436
— —: Weak convergence of the sequential empirical processes of residuals in ARMA models. 94 AnlsStat 22 2051-2061

Bai, Z. D.: Silverstein, Jack W.; — —: On the empirical distribution of eigenvalues of a class of large dimensional random matrices. 95 JMultiAn 54 175-192

Bai, Zhidong; Chen, Xiru; Krishnaiah, P. R.; Zhao, Lingcheng: Asymptotic properties of the EVLP estimation for superimposed exponential signals in noise (Chinese). 95 ChJApPSt 10 148-163
Cheng, Ph. E.; — —: Optimal strong convergence rates in nonparametric regression. 95 MaMethSt 4 405-420

Bailar, Barbara A.: Discussion of "Estimation in the presence of measurement error". 95 IntStRvw 63 142-143

Bailar, John C., III: A larger perspective. 95 AmerStat 49 10-11 (Disc: p18-23)
— —: How statisticians can help the news media do a better job. 95 Chance 8/1 24-24
— —: How the news media can help statisticians do a better job. 95 Chance 8/1 25-25+

Bailer, A. John: Piegorsch, Walter W.; — —: Statistical approaches for analyzing mutational spectra: Some recommendations for categorical data. 94 Genetcs 136 403-416

Bailey, Michael; Bowden, John; Callahan, Alexander J.: Managing ship performance of naval gunfire support using statistical process control. 94 MilOpRes 1 5-11

Bailey, R. A.; Azaïs, J.-M.; Monod, H.: Are neighbour methods preferable to analysis of variance for completely systematic designs? 'Silly designs are silly!'. 95 Biomtrka 82 655-659
— —; Monod, H.; Morgan, J. P.: Construction and optimality of affine-resolvable designs. 95 Biomtrka 82 187-200
— —; Preece, D. A.; Rowley, C. A.: Randomization for a balanced superimposition of one Youden square on another. 95 JRSS-B 57 459-469

Bailey, Steven P.: How to deal with the process capability "catch-22". 95 ASQCTekCon 1105-1116

Bailey, T. C.: Powell, K. J.; Sapatinas, T.; — —; Krzanowski, W. J.: Application of wavelets to the pre-processing of underwater sounds. 95 StatComp 5 265-273

Bailey, Trevor C.; Gatrell, Anthony C.: Interactive spatial data analysis. 95 Wiley:NY:UK 432p

Baillie, Stephen R.: Uses of ringing data for the conservation and management of bird populations: A ringing scheme perspective. 95 JAppStat 22 967-987

Bain, Lee J.; Gan, Gaoxiong: Expectations of the maximum of sums and the sum of maxima for correlated normal variables. 95 CommStB 24 775-780
Gan, Gaoxiong; — —: Distribution of order statistics for discrete parents with applications to censored sampling. 95 JStPlInf 44 37-46

Bain, Raymond P.: Rosenberger, William F.; Lachin, John M.; — —: Nonparametric test of stochastic ordering for multiple longitudinal measures. 95 JBiophSt 5 235-243

Bajorski, Piotr: On some one-sided tests in the two-sample problem. 95 CommStA 24 75-86

Bakeman, R.: — — (Auth); Goda, David (Rev): Review of "Understanding social science statistics: A spreadsheet approach". 95 Ststcian 44 539-540

Bakeman, Roger; Quera, Vicenç: Analyzing interaction: Sequential analysis with SDIS and GSEQ. 95 CambridgeUniv 166p
— —; Quera, Vicenç: Log-linear approaches to lag-sequential analysis when consecutive codes may and cannot repeat. 95 PsychBu 118 272-284
— —; Robinson, Byron F. (Auth); Rice, Janet C. (Rev): Review of "Understanding log-linear analysis with ILOG: An interactive approach". 95 EdPsMeas 55 350-355

Baker, Reg; Bradburn, Norman; Johnson, Robert: CAPI: An experimental evaluation. 94 ASASrvyRsr 851-856 (Disc: p861-862)

Baker, Regina M.: Bound, John; Jaeger, David A.; — —: Problems with instrumental variables estimation when the correlation between the instruments and the endogenous explanatory variable is weak. 95 JASA 90 443-450

Baker, Rose D.: Two permutation tests of equality of variances. 95 StatComp 5 289-296
— —; Stevens, Robert H.: A random-effects model for analysis of infectious disease final-state data. 95 Biomtrcs 51 956-968

Baker, S. G.; Lindeman, K. S.: Correction to "The paired availability design: A proposal for evaluating epidural analgesia during labor" (V13 p2269-2278). 95 StatMed 14 1841-1841

Baker, Stuart G.: Evaluating multiple diagnostic tests with partial verification. 95 Biomtrcs 51 330-337
— —: Marginal regression for repeated binary data with outcome subject to non-ignorable non-response. 95 Biomtrcs 51 1042-1052

Bakir, Saad: Amin, Raid W.; Reynolds, Marion R., Jr; — —: Nonparametric quality control charts based on the sign statistic. 95 CommStA 24 1597-1623

Baksalary, J. K.; Gnot, S.; Kageyama, S.: Best estimation of variance components with arbitrary kurtosis in two-way layouts mixed models. 95 JStPlInf 44 65-75

Baksalary, Jerzy K.; Markiewicz, Augustyn; Rao, C. Radhakrishna: Admissible linear estimation in the general Gauss-Markov model with respect to an arbitrary quadratic risk function. 95 JStPlInf 44 341-347

Balabushkin, A. N.: Approximation to the exit probability of a continuous Gaussian process over a U-shaped boundary of increasing curvature. 95 JAppProb 32 429-442

Balagtas, Cecile C.; Becker, Mark P.; Lang, Joseph B.: Marginal modelling of categorical data from crossover experiments. 95 ApplStat 44 63-77

Balakrishnan, A. V.: Introduction to random processes in engineering. 95 Wiley:NY:UK 402p

Balakrishnan, N.: — — (Ed): Recent advances in life-testing and reliability. 95 CRCPress:FL 672p
— —; Balasubramanian, K.: Order statistics from non-identical power function random variables. 95 CommStA 24 1443-1454
— —; Gupta, Shanti S.; Panchapakesan, S.: Estimation of the location and scale parameters of the extreme value distribution based on multiply type-II censored samples. 95 CommStA 24 2105-2125
— —; Gupta, Shanti S.; Panchapakesan, S.: Estimation of the mean and standard deviation of the logistic distribution based on multiply type-II censored samples. 95 MaOpfStS 27 127-142
— —; Sandhu, R. A.: A simple simulational algorithm for generating progressive Type-II censored samples. 95 AmerStat 49 229-230
— —; Sandhu, R. A.: Recurrence relations for single and product moments of order statistics from a generalized half logistic distribution with applications to inference. 95 JStCmpSm 52 385-398
Balasubramanian, K.; — —: On a class of multivariate distributions closed under concomitance of order statistics. 95 StPrLet 23 239-242
Chen, Gemai; — —: A general purpose approximate goodness-of-fit test. 95 JQualTek 27 154-161
Johnson, Norman L.; Kotz, Samuel; — —: Continuous univariate distributions. Volume 2 (Second edition). 95 Wiley:NY:UK 719p
Viveros, Román; Balasubramanian, K.; — —: Reply to comment on "Binomial and negative binomial analogues under correlated Bernoulli trials". 95 AmerStat 49 325-326
Barnett, V.; Lewis, T. (Auth); — — (Rev): Review of "Outliers in statistical data (Third edition)". 95 Biomtrcs 51 381-381
— — (Ed); Howlader, Hatema A. (Rev): Review of "Handbook of the logistic distribution". 95 JAppStat 22 189-191
Johnson, N. L.; Kotz, S.; — — (Auth); Chatfield, Chris (Rev): Review of "Continuous univariate distributions, Volume 1 (Second edition)". 95 Ststcian 44 544-545
Johnson, N. L.; Kotz, S.; — — (Auth); Kocherlakota, S. (Rev): Review of "Continuous univariate distributions, Volumes 1 and 2 (Second edition)". 95 Biomtrcs 51 1582-1583
Johnson, N. L.; Kotz, S.; — — (Auth); Omori, Yasuhiro (Rev): Review of "Continuous univariate distributions, Volume 1 (Second edition)". 95 JASA 90 1490-1491
Johnson, N. L.; Kotz, S.; — — (Auth); Tricker, A. R. (Rev): Review of "Continuous univariate distributions: Volume I (Second edition)". 95 JAppStat 22 323-323
Johnson, Norman L.; Kotz, Samuel; — — (Auth); Famoye, Felix (Rev): Review of "Continuous univariate distributions, Volume I (Second edition)". 95 Technmcs 37 466-466

Balamurali, S.: Suresh, K. K.; — —: Construction and selection of tightened-normal-tightened plans indexed by maximum allowable percentage defective. 94 JAppStat 21 589-595

Balasch, S.: Romero, R.; Ferrer, A.; Capilla, C.; Zunica, L.; — —; Serra, V.; Alcover, R.: Teaching statistics to engineers: An innovative pedagogical experience. 95 JStEd 3/1 33kb (jse/v3n1/romero)

Balasooriya, Uditha: Failure-censored reliability sampling plans for the exponential distribution. 95 JStCmpSm 52 337-349

Balasubramanian, K.; Balakrishnan, N.: On a class of multivariate distributions closed under concomitance of order statistics. 95 StPrLet 23 239-242

Balakrishnan, N.; — —: Order statistics from non-identical power function random variables. 95 CommStA 24 1443-1454

Viveros, Román; — —; Balakrishnan, N.: Reply to comment on "Binomial and negative binomial analogues under correlated Bernoulli trials". 95 AmerStat 49 325-326

Balbi, Simona: Influence and stability in non symmetrical correspondence analysis. 94 Metron 52/3 111-128

Balch, Alfred H.: Moving to higher dimensions 3 ≠ 2. 94 ASAProBiop 45-50

Choudhury, Japobrata; — —: Assessing the robustness of clinical trial results. 94 ASAProBiop 484-486

Balding, D. J.: Bruno, W. J.; Knill, E.; — —; Bruce, D. C.; Doggett, N. A.; Sawhill, W. W.; Stallings, R. L.; Whittaker, C. C.; Torney, D. C.: Efficient pooling designs for library screening. 95 Genomics 26 21-30

Aitken, C. G. G. (Auth); — — (Rev): Review of "Statistics and the evaluation of evidence for forensic scientists". 95 ShrtBkRv 15/3 41-41

Balding, David J.: Estimating products in forensic identification using DNA profiles. 95 JASA 90 839-844

— —; Donnelly, Peter: Inference in forensic identification. 95 JRSS-A 158 21-40 (Disc: p40-53)

Baldwin, Alison K.: Rasinski, Kenneth A.; — —; Willis, Gordon B.; Jobe, Jared B.: Risk and loss perceptions associated with survey reporting of sensitive behaviors. 94 ASASrvyRsr 497-502

Balk, Bert M.: Axiomatic price index theory: A survey. 95 IntStRvw 63 69-93

Ball, F. G.; Davies, S. S.: Statistical inference for a two-state Markov model of a single ion channel, incorporating time interval omission. 95 JRSS-B 57 269-287

Ball, Frank: A note on variation in birth processes. 95 MaScitst 20 50-55

— —; Clancy, Damian: The final outcome and temporal solution of a carrier-borne epidemic model. 95 JAppProb 32 304-315

— —; Clancy, Damian: The final outcome of an epidemic model with several different types of infective in a large population. 95 JAppProb 32 579-590

— —; Donnelly, Peter: Strong approximations for epidemic models. 95 StocProc 55 1-21

Ballantyne, Roy: Sparks, Ross; — —: Quality control methods in large-scale assessment procedures using 'double-marking' or 'partial double-marking'. 95 AstrlJSt 37 283-298

Ballerini, Rocco: Nonconstant levels for t-year return periods for two-dimensional Poisson processes with periodic intensity. 95 StPrLet 22 43-47

Ballou, Dale; Podgursky, Michael: Teacher quality in public and private schools. 94 ASAProSoSt 196-200 (Disc: p213-215)

Bally, Vlad; Millet, Annie; Sanz-Solé, Marta: Approximation and support theorem in Hölder norm for parabolic stochastic partial differential equations. 95 AnlsProb 23 178-222

Baltagi, Badi H.; Li, Qi: Testing AR(1) against MA(1) disturbances in an error component model. 95 JEconmtx 68 133-151

Raj, Baldev; — — (Ed); Lenz, H.-J. (Rev): Review of "Panel data analysis". 95 CmpStDA 19 481-483

Baltazar-Aban, Inmaculada; Peña, Edsel A.: Properties of hazard-based residuals and implications in model diagnostics. 95 JASA 90 185-197

Bandeen-Roche, Karen J.: Liang, Kung-Yee; Self, Steven G.; — —; Zeger, Scott L.: Some recent developments for regression analysis of multivariate failure time data. 95 LiDaAnal 1 403-415

Kateman, G.; Buydens, L. (Auth); — — (Rev): Review of "Quality control in analytic chemistry (2nd ed)". 95 Technmcs 37 115-116

Bandopadhyay, T.: Giri, N. C. (Auth); — — (Rev): Review of "Introduction to probability and statistics (Second edition)". 94 CalcutSt 44 267-268

Bandt, Christoph: Krause, Ekkehard; — —; Schulz, Anne; Schulz, Holger: Fractal exponents for the upper airways of mammalian lungs. 95 CmpStDA 20 583-590

Bandyopadhyay, Sati P.; Brown, Lawrence D.; Richardson, Gordon D.: Analysts' use of earnings forecasts in predicting stock returns: Forecast horizon effects. 95 IntJFor 11 429-445

Bandyopadhyay, Uttam; Biswas, Atanu: Some nonparametric group sequential-type tests for two population problems. 95 CalcutSt 45 73-91

— —; Chattopadhyay, Gopaldeb: Progressive censoring under inverse sampling for nonparametric two-sample problems. 95 SqtlAnly 14 1-28

Banerjee, P. K.: Fifty years of Calcutta Statistical Association. 95 CalcutSt 45 1-23

Banerjee, S.; Kageyama, S.: Methods of constructing nested partially balanced incomplete block designs. 93 UtilMath 43 3-6

Banker, R. D.; Cooper, W. W.: Validation and generalization of DEA and its uses. 94 Top 2 249-314

Bankier, Michael; Fillion, Jean-Marc; Luc, Manchi; Nadeau, Christian: Imputing numeric and qualitative variables simultaneously. 94 ASASrvyRsr 242-247

Banks, R. B.: Growth and diffusion phenomena: Mathematical frameworks and applications. 94 Springer:Brln:NY 455p

Bannour, Sami; Azimi-Sadjadi, Mahmood R.: Principal component extraction using recursive least squares learning. 95 IEEENrlN 6 457-469

Bansal, Naveen K.: Abughalous, Mansour M.; — —: On selecting the best natural exponential families with quadratic variance function. 95 StPrLet 25 341-349

Bansal, Ravi; Gallant, A. Ronald; Hussey, Robert; Tauchen, George: Nonparametric estimation of structural models for high-frequency currency market data. 95 JEconmtx 66 251-287

Bapat, R. B.: — — (Auth); Drygas, H. (Rev): Review of "Linear algebra and linear models". 95 Metrika 42 78-78

Bar-Hen, Avner; Daudin, J.-J.: Generalization of the Mahalanobis distance in the mixed case. 95 JMultiAn 53 332-342

Bar-Lev, S. K.; Parlar, M.; Perry, D.: Optimal sequential decisions for incomplete identification of group-testable items. 95 SqtlAnly 14 41-57

Barabesi, L.; Fattorini, L.: A note on bandwidth selections for kernel density estimators at endpoints. 94 Metron 52/3 43-56

Barabesi, Lucio; Fattorini, Lorenzo: Approximation of density functions by orthogonal series with grouped data. 94 JItalStS 3 181-200

Baran, Gary S.: Optimal estimation of sigma for X charts. 94 ASAProQlPr 321-326

Barbe, P.; Bertail, P.: The weighted bootstrap. 95 Springer:Brln:NY 230p

— —; Bertail, P. (Auth); Young, G. A. (Rev): Review of "The weighted bootstrap". 95 ShrtBkRv 15/3 45-45

Barbolosi, D.; Faivre, C.: Metrical properties of some random variables connected with the continued fraction expansion. 95 IndgMatN 6 257-265

Barbour, A. D.; Chen, Louis H. Y.; Choi, K. P.: Poisson approximation for unbounded functions, I: Independent summands. 95 StSinica 5 749-766

— —; Grübel, R.: The first divisible sum. 95 JTherPrb 8 39-47

— —; Holst, L.; Janson, S. (Auth); Papaioannou, T. (Rev): Review of "Poisson approximation". 95 Biomtrcs 51 382-383

— —; Holst, Lars; Janson, Svante (Auth); Zubkov, A. M. (Rev): Review of "Poisson approximation". 94 ThProbAp 39 720-721

Barbour, Andrew D.; Chryssaphinou, Ourania; Roos, Malgorzata: Compound Poisson approximation in reliability theory. 95 IEEEReli 44 398-402

Barbour, Garth: Irvine, John M.; — —: Texture analysis of imagery data for waste site characterization. 94 ASAProEnvr 122-126

Barchielli, A.; Holevo, A. S.: Constructing quantum measurement processes via classical stochastic calculus. 95 StocProc 58 293-317

Barden, Dennis: Le, Huiling; — —: Itô correction terms for the radial parts of semimartingales on manifolds. 95 ProbTher 101 133-146

Bardhan, Indrajit: Synthetic replication of American contingent claims when portfolios are constrained. 95 StocProc 57 149-165

— —: Further applications of a general rate conservation law. 95 StocProc 60 113-130

— —; Sigman, Karl: Stationary regimes for inventory processes. 95 StocProc 56 77-86

Barendregt, L. G.; van Pul, M. C.: On the estimation of the parameters for the Littlewood model in software reliability. 95 StNeerla 49 165-184

Barkan, Philip: Hinckley, C. Martin; — —: The role of variation, mistakes, and complexity in producing nonconformities. 95 JQualTek 27 242-249

Barker, Anne M.: Visions, tactics, and TQM for academic programs in quality. 94 ASAProQlPr 294-294

Barker, Peter D. R.: Turcott, Robert G.; — —; Teich, Malvin C.: Long-duration correlation in the sequence of action potentials in an insect visual interneuron. 95 JStCmpSm 52 253-271

Barker, Terry: Hargreaves, Colin (Ed); — — (Rev): Review of "Macroeconomic modelling of the long run". 95 JApEcmtx 10 211-212

Barlow, Richard E.; Tsai, Peisung: Foundational issues concerning the analysis of censored data. 95 LiDaAnal 1 27-34

— —; Clarotti, Carlo A.; Spizzichino, Fabio (Ed); Gruber, Marvin H. J. (Rev): Review of "Reliability and decision making". 95 Technmcs 37 346-347

Barmi, Hammou El: A linear-quadratic distributional identity. 95 StPrLet 24 33-37

— —; Dykstra, Richard: Testing for and against a set of linear inequality constraints in a multinomial setting. 95 CandJSt 23 131-143

— —; Kochar, Subhash C.: Likelihood ratio tests for bivariate symmetry against ordered alternatives in a square contingency table. 95 StPrLet 22 167-173

Barnard, George A.: Pivotal models and the fiducial argument. 95 IntStRvw 63 309-323

Barnard, John; Meng, Xiao-Li: Exploring cross-match estimators with multiply-imputed data sets. 94 ASASrvyRsr 894-899 (Disc: p906-907)

Barnard, Roger W.; Chanda, Kamal C.: An application of the Schur-Cohn algorithm to time series analysis. 95 JTimSrAn 16 445-449

Barndorff-Nielsen, O. E.: Quasi profile and directed likelihoods from estimating functions. 95 AnInStMa 47 461-464

— —: Stable and invariant adjusted profile likelihood and directed likelihood for curved exponential models. 95 Biomtrka 82 489-499

— —; Cox, D. R. (Auth); Corcoran, Stephen (Rev): Review of "Inference and asymptotics". 95 Ststcian 44 279-279

— —; Cox, D. R. (Auth); DiCiccio, Thomas J. (Rev): Review of "Inference and asymptotics". 95 JASA 90 1129-1130

— —; Cox, D. R. (Auth); Hingley, Peter J. (Rev): Review of "Inference and asymptotics". 95 Ststcian 44 407-408

— —; Cox, D. R. (Auth); Pfanzagl, J. (Rev): Review of "Inference and asymptotics". 95 Metrika 42 275-275

Basu, Asit P.: Sun, Kai; — —: A characterization of a bivariate geometric distribution. 95 StPrLet 23 307-311

Basu, Ayanendranath: Basu, Srabashi; — —: Comparison of several goodness-of-fit tests for the kappa statistic based on exact power and coverage probability. 95 StatMed 14 347-356

Basu, Srabashi; Raychaudhuri, Aparna; — —: Improving precision through modelling: An illustration with hierarchical kappa. 95 CommStB 24 399-408

Park, Chanseok; — —; Basu, Srabashi: Robust minimum distance inference based on combined distances. 95 CommStB 24 653-673

Basu, Parantap: Vinod, H. D.; — —: Forecasting consumption, income and real interest rates from alternative state space models. 95 IntJFor 11 217-231

Basu, S. K.: Mukherjee, S. P.; Chaudhuri, A.; — — (Ed); Adke, S. R. (Rev): Review of "Essays on probability and statistics: Festschrift in honour of Professor Anil Kumar Bhattacharyya". 95 CalcutSt 45 131-132

Mukherjee, S. P.; Chaudhuri, A.; — — (Ed); Gupta, S. Das (Rev): Review of "Essays on probability and statistics: Festschrift in honour of Professor Anil Kumar Bhattacharyya". 95 SankhyaA 57 166-166

Basu, Sanjib: Ranges of posterior probability over a distribution band. 95 JStPlInf 44 149-166

— —: Sensitivity of posterior quantities to the sampling distribution: More than one observation. 94 ASAProBays 106-111

— —: Local sensitivity, functional derivatives and nonlinear posterior quantities. 94 ASAProBays 175-180

— —; DasGupta, Anirban: Robustness of standard confidence intervals for location parameters under departure from normality. 95 AnlsStat 23 1433-1442

— —; DasGupta, Anirban: Robust Bayesian analysis with distribution bands. 95 StatDecn 13 333-349

Basu, Srabashi; Basu, Ayanendranath: Comparison of several goodness-of-fit tests for the kappa statistic based on exact power and coverage probability. 95 StatMed 14 347-356

— —; Landis, J. Richard: Model-based estimation of population attributable risk under cross-sectional sampling. 95 AJEpidm 141 1338-1343

— —; Landis, J. Richard: Model-based estimation of population attributable risk under cross-sectional sampling. 95 AJEpidm 142 1338-1343

— —; Raychaudhuri, Aparna; Basu, Ayanendranath: Improving precision through modelling: An illustration with hierarchical kappa. 95 CommStB 24 399-408

Park, Chanseok; Basu, Ayanendranath; — —: Robust minimum distance inference based on combined distances. 95 CommStB 24 653-673

Basu, Sujit K.: Mitra, Murari; — —: Change point estimation in non-monotonic aging models. 95 AnInStMa 47 483-491

Batagelj, Vladimir; Bren, Matevz: Comparing resemblance measures. 95 JClass 12 73-90

Bates, Douglas M.; Martin, Sandra J.: Design considerations for nonlinear calibration. 94 ASAProPEng 12-17

Pinheiro, José C.; — —: Approximations to the log-likelihood function in the nonlinear mixed-effects model. 95 JCmpGrSt 4 12-35

Pinheiro, José C.; — —; Lindstrom, Mary J.: Model building in nonlinear mixed effects models. 94 ASAProBiop 1-8

Bates, Nancy; Gerber, Eleanor: A typology of temporary mobility patterns and reporting of usual residence. 94 ASASrvyRsr 331-336

Bather, John: Poisson search models and the evaluation of stopping rules. 95 SqtlAnly 14 205-216

Batra, S.; König, M. F.; Cruz-Orive, L. M.: Unbiased estimation of capillary length from vertical slices. 95 JMicro 178 152-159

Batra, Satish: Thelin, Bernt; Lundstedt, Torbjörn; Lundgren, Rolf; Olsson, Arne; — —: Classification of Estradurin batches: Correlation between 31P NMR and a biological duration test for batch approval. 95 Chemolab 27 135-145

Batson, Robert G.: Gholston, Sampson; — —: Case study: Precision improvement of a digital ball comparator via parameter design. 94 ASAProPEng 302-307

Battles, Stephanie J.; Adler, Robert K.; Olsen, Jay K.: Demand-side management: Utility programs offered and participants in the residential, commercial, and manufacturing sectors of the U.S. economy. 94 ASAProBuEc 20-27

Bauchau, V.; van Noordwijk, A. J.: Comparison of survival estimates obtained from three different methods of recapture in the same population of the great tit. 95 JAppStat 22 1031-1037

Bauer, Christina; Schneider, Rolf: Extremal problems for geometric probabilities involving convex bodies. 95 AdvAppPr 27 20-34

Bauer, P.; Röhmel, J.: An adaptive method for establishing a dose-response relationship. 95 StatMed 14 1595-1607

Bauer, Peter: Hilgers, Ralf-Dieter; — —: Optimal designs for mixture amount experiments. 95 JStPlInf 48 241-246

Baumgartner, Cornelia: The polynomial model in the multivariate analysis of variance. 95 BiomtrcJ 37 299-318

Baumgartner, Hans; Bagozzi, Richard P.: Specification, estimation, and testing of moment structure models based on latent variates involving interactions among the exogenous constructs. 95 SocMethR 24 187-213

Baur, M. P.: Knapp, M.; Seuchter, S. A.; — —: Two-locus disease models with two marker loci: The power of affected-sib-pair tests. 94 AJHuGene 55 1030-1041

Knapp, M.; Seuchter, S. A.; — —: Linkage analysis in nuclear families. 1: Optimality criteria for affected sib-pair tests. 94 HumHrdty 44 37-43

Knapp, M.; Seuchter, S. A.; — —: Linkage analysis in nuclear families. 2: Relationship between affected sib-pair tests and lod score analysis. 94 HumHrdty 44 44-51

Baushev, A. N.: Weak convergence of probability measures in L_p spaces (Russian). 94 VLenUM /4 15-20

Baxley, Robert V., Jr: An application of variable sampling interval control charts. 95 JQualTek 27 275-282

Baxt, William G.: Application of artificial neural networks to clinical medicine. 95 Lancet 346 1135-1138

Baxter, J. R.; Rosenthal, Jeffrey S.: Rates of convergence for everywhere-positive Markov chains. 95 StPrLet 22 333-338

Baxter, Laurence A.: Estimation subject to block censoring. 95 IEEEReli 44 489-495

— —; Li, Linxiong: Nonparametric confidence intervals for the renewal function with censored data. 95 JNnparSt 4 317-326

Albert, José Ramon G.; — —: Applications of the EM algorithm to the analysis of life length data. 95 ApplStat 44 323-341

Puterman, Martin L. (Auth); — — (Rev): Review of "Markov decision processes: Discrete stochastic dynamic programming". 95 Technmcs 37 353-353

Baxter, M. J.: Standardization and transformation in principal component analysis, with applications to archaeometry. 95 ApplStat 44 513-527

Bayard, Steven; Jinot, Jennifer; Brown, Kenneth: Passive smoking and lung cancer: The U.S. EPS's weight-of-evidence analysis, with emphasis on the epidemiology studies. 94 ASAProEnvr 56-64 (Disc: p74-76)

— —; Jinot, Jennifer; Flatman, George: Comment on "Uncertainties in lung cancer risk estimates reported for exposure to environmental tobacco smoke". 95 EnvrMtrc 6 413-418

Bayarri, M. J.; Font, B.: A (Bayesian) note on non-random samples from finite populations. 94 ASAProBays 246-251

Bayer, M. M.: Kaandorp, J. A. (Auth); Howroyd, J. D.; — — (Rev): Review of "Fractal modelling: Growth and form in biology". 95 Biomtrcs 51 794-795

Bayne, C. K.: Ma, C. Y.; — —: Differentiation of aroclors in environmental samples using linear discrimination. 94 ASAProEnvr 142-143

Bayne, Charles K.: Haswell, Stephen John (Ed); — — (Rev): Review of "Practical guide to chemometrics". 95 Technmcs 37 230-231

Beamonte, Eduardo; Bermúdez, José D.: Comparison of Gamma survival curves (Spanish). 95 Questo 19 171-186

Bean, Judy A.: Lu, Ying; — —: On the sample size for one-sided equivalence of sensitivities based upon McNemar's test. 95 StatMed 14 1831-1839

Bean, N. G.; Gibbens, R. J.; Zachary, S.: Asymptotic analysis of single resource loss systems in heavy traffic, with applications to integrated networks. 95 AdvAppPr 27 273-292

— —; Taylor, P. G.: Maximal profit dimensioning and tariffing of loss networks. 95 PrEngInf 9 323-340

Beatty, Paul: Understanding the standardized/non-standardized interviewing controversy. 95 JOffStat 11 147-160

— —; Schechter, Susan: An examination of mode effects in cognitive laboratory research. 94 ASASrvyRsr 1275-1280

Schechter, Susan; — —; Block, Alan: Cognitive issues and methodological implications in the development and testing of a traffic safety questionnaire. 94 ASASrvyRsr 1215-1219

Beauchamp, Stephen: Burrows, William R.; Benjamin, Mario; — —; Lord, Edward R.; McCollor, Douglas; Thomson, Bruce: CART decision-tree statistical analysis and prediction of summer season maximum surface ozone for the Vancouver, Montreal, and Atlantic regions of Canada. 95 JApMetrl 34 1848-1862

Beauregard, Michael R.; Mikulak, Raymond J.; Olson, Barbara A. (Auth); Belunis, Nancy (Rev): Review of "A Practical guide to statistical quality improvement: Opening up the statistical toolbox". 95 Technmcs 37 233-234

— —; Mikulak, Raymond J.; Olson, Barbara A. (Auth); Richards, John D. (Rev): Review of "A practical guide to statistical quality improvement: Opening up the statistical toolbox". 95 JQualTek 27 177-178

Beaver, Robert J.: Mendenhall, William; — — (Auth); Halvorsen, Katherine Taylor (Rev): Review of "Introduction to probability and statistics (Ninth edition)". 95 AmerStat 49 398-398

Bechhofer, Robert E.; Santner, Thomas J.; Goldsman, David M.: Design and analysis of experiments for statistical selection, screening, and multiple comparisons. 95 Wiley:NY:UK 325p

Bechtel, Gordon G.; Ofir, Chezy; Khuri, André I.: Replicated paired comparisons at the individual level. 95 BJMaStPs 48 115-127

Bechtold, William A.: Reich, Robin M.; Czaplewski, Raymond L.; — —: Spatial cross-correlation of undisturbed, natural shortleaf pine stands in northern Georgia. 94 EnvEcoSt 1 201-217

Beck, Anatole: Baston, Vic; — —: Generalizations in the linear search problem. 95 IsrelJMa 90 301-323

Beck, James: Davies, Glenn M.; Koch, Gary G.; — —: Statistical strategies for event rate comparisons in dental studies. 94 ASAProBiop 420-424

Beck, M. B.: Jakeman, A. J.; — —; McAleer, M. J. (Ed); Keesman, Karel (Rev): Review of "Modelling change in environmental systems". 95 EnvrMtrc 6 425-427

Becka, Michael: Urfer, Wolfgang; — —: Statistical data analysis for identifying kinetic processes. 94 ASAProEnvr 14-18

Becker, Betsy Jane: Corrections to "Using results from replicated studies to estimate linear models" (V17 p341-362). 95 JEdStat 20 100-102

Becker, G.; Camarinopoulos, L.; Zioutas, G.: An inhomogeneous state graph model and application for a phased mission and tolerable downtime problem. 95 RelEngSS 49 51-57

Becker, Mark P.: Balagtas, Cecile C.; — —; Lang, Joseph B.: Marginal modelling of categorical data from crossover experiments. 95 ApplStat 44 63-77

Ten Have, Thomas R.; — —: Multivariate contingency tables and the analysis of exchangeability. 95 Biomtrcs 51 1001-1016

Becker, Niels G.; Bahrampour, Abbas; Dietz, Klaus: Threshold parameters for epidemics in different community settings. 95 MaBisci 129 189-208

— —; Dietz, Klaus: The effect of household distribution on transmission and control of highly infectious diseases. 95 MaBisci 127 207-219

Rouderfer, Vladimir; — —: Assessment of two-dose vaccination schedules: Availability for vaccination and catch-up strategies. 95 MaBisci 129 41-66

Becker, Patricia C.: The Census Bureau, its funders, and its customers. 94 ASAProSoSt 1-2

— —: Comment on "Statistical advances in the estimation and projection of demographic and social characteristics". 94 ASAProSoSt 122-123

Becker, Richard A.; Clark, Linda A.; Lambert, Diane: Graphical analysis of data on networks. 94 ASAProStGr 21-24

Bedi, Pradeep Kumar: An alternative estimator in Midzuno scheme for multiple characteristics. 95 CommStB 24 17-30

Bedrick, Edward J.: A note on the attenuation of correlation. 95 BJMaStPs 48 271-280

Bedwell, Michael: Directory enquiries?. 95 TeachgSt 17/1 10-12

Beg, Ismat; Shahzad, Naseer: Random fixed points for random multivalued operators defined on unbounded sets in Banach spaces. 95 StocAnAp 13 269-277

Beg, M. A.: Hanif, Muhammad; — —; Khawaja, Iffat: Sampling with unequal probabilities: A historical review. 92 ProIrStC1 19-37

Begg, Andy: Statistics and the mathematical processes. 95 TeachgSt 17/2 40-43

Begg, Melissa Dowd: Paik, Myunghee Cho; — —; El-Sadr, Wafaa; Gorman, Jack; Stein, Zena: Difference in clinical implications of CD4 counts among HIV-infected homosexual men and injection drug using men and women. 95 StatMed 14 1889-1900

Begleiter, Henri: Wang, Wenyu; — —; Porjesz, Bernice: Localization of multiple dipole sources of human cerebral potentials based on SCD and a model selection criterion. 94 ASAProBiop 217-222

Behboodian, J.: Fazli, K.; — —: Skorohod's theorem and convergence of types. 95 StPrLet 24 243-244

Behrens, John T.: Yu, Chong Ho; — —: Identification of misconceptions in learning statistical power with dynamic graphics as a remedial tool. 94 ASAProStEd 242-246

Beilin, L. J.: Cox, K.; Puddey, I. B.; Burke, V.; McAleer, M.; — —; Morton, A. R.: Effects of exercise and calorie restriction on ambulatory blood pressure in overweight, sedentary males: A time series analysis. 95 EnvrMtrc 6 523-528

Beirlant, J.; Einmahl, J. H. J.: Asymptotic confidence intervals for the length of the shortt under random censoring. 95 StNeerla 49 1-8

— —; Mason, D. M.: On the asymptotic normality of L_p-norms of empirical functionals. 95 MaMethSt 4 1-19

Beirlant, Jan; Broniatowski, Michel; Teugels, Jozef L.; Vynckier, Petra: The mean residual life function at great age: Applications to tail estimation. 95 JStPlInf 45 21-48

Bekaert, Geert: The time variation of expected returns and volatility in foreign-exchange markets. 95 JBES 13 397-408

Bekker, A.; Roux, J. J. J.: Bayesian multivariate normal analysis with a Wishart prior. 95 CommStA 24 2485-2497

Bel, Liliane; Robbiano, Luc: Extension of a uniqueness theorem for Zakai's equation (French). 95 Stochatx 52 227-245

Belanger, A.; Larrivee, D.: New approach for constructing Canadian working life tables, 1889-1987. 92 StatJUN 9 27-50

Belanger, Albert J.: D'Agostino, Ralph B.; — —; Markson, Elizabeth W.; Kelly-Hayes, Maggie; Wolf, Philip A.: Development of health risk appraisal functions in the presence of multiple indicators: The Framingham study nursing home institutionalization model. 95 StatMed 14 1757-1770

Belavkin, V. P.: Measurement and filtering of quantum diffusion II. 94 ThProbAp 39 363-378

Belforte, G.: Uncertainty evaluation for estimates from recursive projection algorithms. 95 IJAdCSgP 9 97-106

Belin, Thomas R.; Rubin, Donald B.: A method for calibrating false match rates in record linkage. 95 JASA 90 694-707

— —; Rubin, Donald B.: The analysis of repeated-measures data on schizophrenic reaction times using mixture models. 95 StatMed 14 747-768

Belitser, E. N.; Levit, B. Y.: On minimax filtering over ellipsoids. 95 MaMethSt 4 259-273

Belknap, J. K.: Empirical estimates of Bonferroni corrections for use in chromosome mapping studies with the BXD recombinant inbred strains. 92 BehavGen 22 677-684

Bell, P. A.: Kokic, P. N.; — —: Optimal winsorizing cutoffs for a stratified finite population estimator. 94 JOffStat 10 419-435

Bell, Robert M.: Sampling and statistical estimation in the decennial census. 94 ASASrvyRsr 71-79 (Disc: p90-95)

Klein, Stephen P.; — —: How will the NCAA's new standards affect minority student-athletes?. 95 Chance 8/3 18-21

Bellacicco, Antonio: Image processing and object classification with fuzzy logic (Italian). 94 QStMaA 16 19-26

Bellavance, F.; Tardif, S.: A nonparametric approach to the analysis of three-treatment three-period crossover designs. 95 Biomtrka 82 865-875

Bellegarda, Eveline J.; Bellegarda, Jerome R.; Nahamoo, David; Nathan, Krishna S.: A fast statistical mixture algorithm for on-line handwriting recognition. 94 IEEEPAMI 16 1227-1233

Bellegarda, Jerome R.: Bellegarda, Eveline J.; — —; Nahamoo, David; Nathan, Krishna S.: A fast statistical mixture algorithm for on-line handwriting recognition. 94 IEEEPAMI 16 1227-1233

Bellhouse, D. R.: Estimation of correlation in randomized response. 95 SurvyM 21 13-19

— —; Rao, J. N. K.: Analysis of domain means in complex surveys. 94 ASASrvyRsr 29-34

Mangat, N. S.; Singh, R.; Singh, S.; — —; Kashani, H. B.: On efficiency of using distinct respondents in a randomized response survey. 95 SurvyM 21 21-23

Belli, Robert F.: Jay, Gina M.; — —; Lepkowski, James M.: Quality of last doctor visit reports: A comparison of medical record and survey data. 94 ASASrvyRsr 362-367 (Disc: p379-381)

Jay, Gina M.; — —; Lepkowski, James M.: Use of last physician visit to characterize health care utilization. 94 ASASrvyRsr 368-372 (Disc: p379-381)

Bello, A. L.: Missing values, imputation and error rate estimators in discriminant analysis. 95 AstrlJSt 37 95-104

— —: Imputation techniques in regression analysis: Looking closely at their implementation. 95 CmpStDA 20 45-57

Bello, Abdul L.: Gradshteyn, I. S.; Ryzhik, I. M. (Auth); Jeffrey, A. (Ed); — — (Rev): Review of "Tables of integrals, series, and products (Fifth edition)". 95 Ststcian 44 287-288

Bellomo, N.; Brzezniak, Z.; de Socio, L. M. (Auth); Elworthy, K. D. (Rev): Review of "Nonlinear stochastic evolution problems in applied sciences". 95 Metrika 42 140-141

— —; Brzezniak, Z.; de Socio, L. M. (Auth); Renshaw, Eric (Rev): Review of "Nonlinear stochastic evolution problems in applied sciences". 94 BLndnMaS 26 409-410

Bellow, Michael E.: Large domain satellite based estimators of crop planted area. 94 ASASrvyRsr 194-199

Belov, N. G.: On the main directions of CIS Statistical Committee activities in organizing statistics according to market economy conditions. 93 StatJUN 10 381-388

Belsley, D. A.: — — (Ed); Smith, Murray D. (Rev): Review of "Computational techniques for econometrics and economic analysis". 95 Ststcian 44 540-540

— — (Ed); Uebe, Götz (Rev): Review of "Computational techniques for econometrics and economic analysis". 95 CmpStDA 19 359-360

Belton, Willie J., Jr: Cebula, Richard J.; — —; Chou, Ray Y.; Barth, James R.: Deposit insurance and bank failures: A preliminary analysis. 94 ASAProBuEc 293-297

Belunis, Nancy: Beauregard, Michael R.; Mikulak, Raymond J.; Olson, Barbara A. (Auth); — — (Rev): Review of "A Practical guide to statistical quality improvement: Opening up the statistical toolbox". 95 Technmcs 37 233-234

Belzil, Christian: Estimating models of labor market transitions based on alternative specifications of job search theory. 95 JStPlInf 43 243-255

Belzile, Éric; Angers, Jean François: Choice of noninformative priors for the variance components of an unbalanced one-way random model. 95 CommStA 24 1325-1341

Bence, James R.: Analysis of short time series: Correcting for autocorrelation. 95 EcolgMon 65 628-639

Bench, Jill; Clark, Colleen; Dufour, Johane; Kaushal, Ritu: Implementing computer-assisted interviewing for the Canadian Labour Force Survey. 94 ASAProGovt 93-98

Benedetti, Carlo: On some inequalities related to the range of some statistical indices. 94 Metron 52/1 189-215

Benedetti, Roberto: A geometric approach to spatial time series analysis (Italian). 94 QStMaA 16 27-43

— —; Palma, Daniela: Optimal sampling designs for dependent spatial units. 95 EnvrMtrc 6 101-114

Beneš, Viktor: On second-order formulas in anisotropic stereology. 95 AdvAppPr 27 326-343

Bengtsson, E.: Pedersen, F.; — —; Nordin, B.: An extended strategy for exploratory multivariate image analysis including noise considerations. 95 JChemtrc 9 389-409

Benhamou, Ellen: Estéve, Jacques; — —; Raymond, Luc (Ed): Statistical methods in cancer research. Volume 4: Descriptive epidemiology. 95 OxfordUniv 320p

Benichou, Jacques: A complete analysis of variability for estimates of absolute risk from a population-based case-control study on breast cancer. 95 BiomtrcJ 37 3-24

— —; Gail, Mitchell H.: Methods of inference for estimates of absolute risk derived from population-based case-control studies. 95 Biomtrcs 51 182-194

Bening, V. E.: A formula for deficiency: One sample L- and R-tests. I. 95 MaMethSt 4 167-188

— —: A formula for deficiency: One-sample L- and R-tests. II. 95 MaMethSt 4 274-293

Benjamin, Mario: Burrows, William R.; — —; Beauchamp, Stephen; Lord, Edward R.; McCollor, Douglas; Thomson, Bruce: CART decision-tree statistical analysis and prediction of summer season maximum surface ozone for the Vancouver, Montreal, and Atlantic regions of Canada. 95 JApMetrl 34 1848-1862

Benjamini, Itai; Kesten, Harry: Percolation of arbitrary words in $\{0,1\}^N$. 95 AnlsProb 23 1024-1060

— —; Pemantle, Robin; Peres, Yuval: Martin capacity for Markov chains. 95 AnlsProb 23 1332-1346

Benjamini, Yoav; Hochberg, Yosef: Controlling the false discovery rate: A practical and powerful approach to multiple testing. 95 JRSS-B 57 289-300

— —; Krieger, Abba M.: On the consistency of tests of symmetry. 95 JNnparSt 4 283-292

Bennett, F. E.; Yin, J.: Some results on $(v,\{5,w^*\})$-PBDs. 95 JComDsgn 3 455-468

Bennett, Jay: Cost-effectiveness in public education. 95 Chance 8/1 38-41

Bennett, Jay M.: MVP, LVP, and PGP: A statistical analysis of Toronto in the World Series. 94 ASAProSpts 66-71

Bennett, R. E.; Ward, W. C. (Ed); Adams, Robert (Rev): Review of "Construction versus choice in cognitive measurement". 95 Ststcian 44 137-137

Bennett, S.; Radalowicz, A.; Vella, V.; Tomkins, A.: A computer simulation of household sampling schemes for health surveys in developing countries. 94 IJEpidem 23 1282-1291

Benneyan, James C.; Kaminsky, Frank C.: Successfully applying SPC to improve health care: Barriers to improving quality and reducing liability. 95 ASQCTekCon 570-578

— —; Kaminsky, Frank C.: Another view on how to measure health care quality. 95 QulPr 28/2 120-123

Beno, M.: Komornik, J.; Mikulecky, M.; — —: Additivity test for composition of binomial effects in chromosome aberrations after radiation injury. 95 StatMed 14 1941-1947

Benois, O.; Kipnis, C.; Landim, C.: Large deviations from the hydrodynamical limit of mean zero asymmetric zero range processes. 95 StocProc 55 65-89

Benoist, D.; Tourbier, Y.; Germain-Tourbier, S.: Experimental design: Construction and analysis (French). 95 Lavoisier:Paris ?p

Bensberg, Frank: Taudes, Alfred; Reepmeyer, Jan-Armin; — — (Auth); Weinberger, Volker (Rev): Review of "UNIX: Introduction for DOS-users" (German). 95 CmpStDA 19 360-361

Benski, Claudio: Applicability of nine numerical techniques for detecting active factors in unreplicated experimental designs. 94 ASAProStCp 214-217

Bensoussan, Alain; Flandoli, Franco: Stochastic inertial manifold. 95 Stochatx 53 13-39

Bentkus, Vidmantas; Götze, Friedrich; Zitikis, Ričardas: Lower estimates of the convergence rate for U-statistics. 94 AnlsProb 22 1707-1714

Bentler, P. M.: Lee, Sik-Yum; Poon, Wai-Yin; — —: A two-stage estimation of structural equation models with continuous and polytomous variables. 95 BJMaStPs 48 339-358

Bentler, Peter M.: Chan, Wai; Yung, Yiu-Fai; — —: A note on using an unbiased weight matrix in the ADF test statistic. 95 MultBehR 30 453-459

Zhu, Li-Xing; Fang, Kai-Tai; Bhatti, M. I.; — —: Testing sphericity of a high-dimensional distribution based on bootstrap approximation. 95 PakJStA 11 49-65

Bentley, Gillian R.: Murphy, Susan A.; — —; O'Hanesian, Mary Ann: An analysis for menstrual data with time-varying covariates. 95 StatMed 14 1843-1857

Benveniste, Albert: Juditsky, Anatoli; Hjalmarsson, Håkan; — —; Delyon, Bernard; Ljung, Lennart; Sjöberg, Jonas; Zhang, Qinghua: Nonlinear black-box models in system identification: Mathematical foundations. 95 Autmatca 31 1725-1750

Sjöberg, Jonas; Zhang, Qinghua; Ljung, Lennart; — —; Delyon, Bernard; Glorennec, Pierre-Yves; Hjalmarsson, Håkan; Juditsky, Anatoli: Nonlinear black-box modeling in system identification: A unified overview. 95 Autmatca 31 1691-1724

Bera, Anil K.; Ng, Pin T.: Tests for normality using estimated score function. 95 JStCmpSm 52 273-287

— —; Ra, Sungsup: A test for the presence of conditional heteroskedasticity within ARCH-M framework. 95 EcomtrRv 14 473-485

Beran, Jan: Maximum likelihood estimation of the differencing parameter for invertible short and long memory autoregressive integrated moving average models. 95 JRSS-B 57 659-672

— —; Gasser, Theo: Testing equality of variances for paired time series. 95 JTimSrAn 16 165-176

Beran, R.; Millar, P. W.: Minimum distance estimation in random coefficient regression models. 94 AnlsStat 22 1976-1992

Beran, R. J.: Hjorth, J. S. Urban (Auth); — — (Rev): Review of "Computer intensive statistical methods". 95 SIAMRvw 37 459-460

Beran, Rudolf: Prediction in random coefficient regression. 95 JStPlInf 43 205-213

— —: Stein confidence sets and the bootstrap. 95 StSinica 5 109-127

Rieder, Helmut (Auth); — — (Rev): Review of "Robust asymptotic statistics". 95 JASA 90 1128-1129

Bérard, H.: Use of the generalized edit and imputation system for the Canadian Survey of Family Expenditures: A feasibility study. 94 ASAProSoSt 178-183

Bercovici, H.; Voiculescu, D.: Superconvergence to the central limit and failure of the Cramér theorem for free random variables. 95 ProbTher 103 215-222

Berg, Bernd A.: Locating global minima in optimization problems by a random-cost approach. 93 Nature 361 708-710

Berg, Sven: Vogt, W. P. (Auth); — — (Rev): Review of "Dictionary of statistics and methodology. A nontechnical guide for the social sciences". 94 JOffStat 10 225-226

Berge, Jos M. F. ten: Suppressing permutations or rigid planar rotations: A remedy against nonoptimal varimax rotations. 95 Psymtrka 60 437-446

Nunnally, J. C.; Bernstein, I. H. (Auth); — — (Rev): Review of "Psychometric theory". 95 Psymtrka 60 313-315

— — (Auth); Takane, Yoshio (Rev): Review of "Least squares optimization in multivariate analysis". 95 Psymtrka 60 447-448

Berger, Arthur W.; Whitt, Ward: Maximum values in queueing processes. 95 PrEngInf 9 375-409

Berger, J. O.: Wang, Y.; Boukai, B.; — —: A Bayesian testing procedure with valid conditional frequentist interpretation. 94 ASAProBays 187-195

Berger, James O.; Brown, Lawrence D.; Wolpert, Robert L.: A unified conditional frequentist and Bayesian test for fixed and sequential simple hypothesis testing. 94 AnlsStat 22 1787-1807

Berger, Martijn P. F.; van der Linden, Wim J.: The design of tests with different optimality criteria (Dutch). 95 TOnderwR 20 79-92

Berger, Roger L.: Liu, Huimei; — —: Uniformly more powerful, one-sided tests for hypotheses about linear inequalities. 95 AnlsStat 23 55-72

Berger, Roxane du: Joseph, Lawrence; Wolfson, David B.; — —: Sample size calculations for binomial proportions via highest posterior density intervals. 95 Ststcian 44 143-154

Joseph, Lawrence; Wolfson, David B.; — —: Some comments on Bayesian sample size determination. 95 Ststcian 44 167-171

Bergeret, F.; Chandon, Y.: Search for a sequence of optimal tests to control manufacturing costs (French). 95 RStApp 43/3 21-33

Berghold, Andrea: Ratkowsky, David A.; Evans, Marc A.; Alldredge, J. Richard (Auth); — — (Rev): Review of "Cross-over experiments. Design, analysis and application". 95 CmpStQ 10 99-100

Berglund, Anders: Lindgren, Fredrik; Geladi, Paul; — —; Sjöström, Michael; Wold, Svante: Interactive variable selection (IVS) for PLS. Part II: Chemical applications. 95 JChemtrc 9 331-342

Bergman, Aviv; Otto, Sarah P.; Feldman, Marcus W.: On the evolution of recombination in haploids and diploids II. Stochastic models. 95 Complxty 1 49-57

Bergman, Lars R.: Pretesting procedures at Statistics Sweden's Measurement, Evaluation and Development Laboratory. 95 JOffStat 11 309-323

— —; Kristiansson, Karl-Erik; Olofsson, Anita; Säfström, Margaretha: Decentralised CATI versus paper and pencil interviewing: Effects on the results in the Swedish labour force surveys. 94 JOffStat 10 181-195

Bergman, Richard N.: Bradley, David C.; Steil, Garry M.; — —: OOPSEG: A data smoothing program for quantitation and isolation of random measurement error. 95 CmpBiomd 46 67-77

Bergonzini, J.-Cl.: Analysis and planning of experimental block designs (French). 95 Masson:NY:Paris 372p

Bergström, Reinhold: The relationship between manufacturing production and different business survey series in Sweden 1968-1992. 95 IntJFor 11 379-393

Berk, J. M.; Bikker, J. A.: International interdependence of business cycles in the manufacturing industry: The use of leading indicators for forecasting and analysis. 95 JForcast 14 1-23

Berk, Kenneth N.; Booth, David E.: Seeing a curve in multiple regression. 95 Technmcs 37 385-398

Berk, Marc L.; Edwards, W. Sherman; Cantor, Joel C.: Validating access to care. 94 ASAProSoSt 240-244

Schur, Claudia L.; — —; Mueller, Curt D.: Developing a prescription drug expenditure model: The effect of alternative assumptions. 94 ASAProSoSt 245-250

Berkes, István: On the almost sure central limit theorem and domains of attraction. 95 ProbTher 102 1-17

Berkey, C. S.; Hoaglin, D. C.; Mosteller, F.; Colditz, G. A.: A random-effects regression model for meta-analysis. 95 StatMed 14 395-411

Berlanga, Mauro: Dawson, Deanna K.; Sauer, John R.; Wood, Paul A.; — —; Wilson, Marcia H.; Robbins, Chandler S.: Estimating bird species richness from capture and count data. 95 JAppStat 22 1063-1068

Berlin, Jesse A.: Benefits of heterogeneity in meta-analysis of data from epidemiologic studies. 95 AJEpidm 142 383-387

Draper, David; Gaver, Donald, P., Jr; Goel, Prem K.; Greenhouse, Joel B.; Hedges, Larry V.; Morris, Carl N.; Tucker, John R.; Waternaux, Christine M. (Auth); — — (Rev): Review of "Combining information: Statistical issues and opportunities for research". 95 StatMed 14 1611-1612

Berliner, L. Mark: MacEachern, Steven N.; — —: Asymptotic inference for dynamical systems observed with error. 95 JStPlInf 46 277-292

Berlinet, Alain: A note on variance reduction. 95 StPrLet 25 357-360

— —; Devroye, Luc; Györfi, László: Asymptotic normality of L_1-error in density estimation. 95 MaOpfStS 26 329-343

Bermúdez, José D.: Beamonte, Eduardo; — —: Comparison of Gamma survival curves (Spanish). 95 Questo 19 171-186

Bernadac, Evelyne: Random continued fractions and inverse Gaussian distribution on a symmetric cone. 95 JTherPrb 8 221-258

Bernard, Andrew B.; Durlauf, Steven N.: Convergence in international output. 95 JApEcmtx 10 97-108

Bernardinelli, L.; Clayton, D.; Montomoli, C.: Bayesian estimates of disease maps: How important are priors?. 95 StatMed 14 2411-2431

— —; Clayton, D.; Pascutto, C.; Montomoli, C.; Ghislandi, M.; Songini, M.: Bayesian analysis of space-time variation in disease risk. 95 StatMed 14 2433-2443

Bernardo, J. M.; Smith, A. F. M. (Auth); Lindley, D. V. (Rev): Review of "Bayesian theory". 95 JRSS-A 158 340-341

Bernardo, José M.; Smith, Adrian F. M. (Auth); Piccinato, Ludovico (Rev): Review of "Bayesian theory". 94 JItalStS 3 155-160

Bernardo, Rex: Genetic models for predicting maize single-cross performance in unbalanced yield trial data. 95 CropSci 35 141-147

— —: Prediction of maize single-cross performance using RFLPs and information from related hybrids. 94 CropSci 34 20-25

Berndt, Ernst R.; Griliches, Zvi; Rappaport, Neal J.: Econometric estimates of price indexes for personal computers in the 1990's. 95 JEconmtx 68 243-268

— —; Morrison, Catherine J.: High-tech capital formation and economic performance in U.S. manufacturing industries: An exploratory analysis. 95 JEconmtx 65 9-43

Bernhard, P.: Basar, T.; — —: H^∞-optimal control and related minimax design problems - A dynamic game approach (Second edition). 95 Birkhs:Basl:MA 411p

Basar, T.; — — (Auth); Goodwin, G. C. (Rev): Review of "H^∞-optimal control and related minimax design problems - A dynamic game approach (Second edition)". 96 ShrtBkRv 16/1 11-11

Bernstein, Hank: Mendelsohn, Susan L.; Walker, Chip; — —: Conducting global consumer research: Considerations and pitfalls. 94 ASASrvyRsr 1196-1198

Bernstein, I. H.: Nunnally, J. C.; — — (Auth); Berge, Jos M. F. ten (Rev): Review of "Psychometric theory". 95 Psymtrka 60 313-315

Bernt, K.; Theus, M. (Rev): Review of "Mercator for Windows 4.0". 95 CmpStQ 10 307-308

Berred, Mohamed: K-record values and the extreme-value index. 95 JStPlInf 45 49-63

Berridge, Damon M.: Modelling ordinal recurrent events. 95 JStPlInf 47 71-78

Berrondo-Agrell, M.; Fourastié, J.: Understanding probability (French). 94 Hachette:Fr 160p

Berry, Donald A.; Eick, Stephen G.: Adaptive assignment versus balanced randomization in clinical trials: A decision analysis. 95 StatMed 14 231-246

Lewis, Roger J.; — —: Bayesian group sequential clinical trial designs based on a quadratic loss function.. 94 ASAProBays 23-28

Rosner, Gary L.; — —: A Bayesian group sequential design for a multiple arm randomized clinical trial. 95 StatMed 14 381-394

Berry, G.: Environmental mesothelioma incidence, time since exposure to asbestos and level of exposure. 95 EnvrMtrc 6 221-228

— —; Armitage, P.: Mid-P confidence intervals: A brief review. 95 Ststcian 44 417-423

Berry, Sandra: Thran, Sara L.; Downes-Le Guin, Theodore; — —: Improving productivity in an annual survey of physicians. 94 ASASrvyRsr 577-582

Bertail, P.: Barbe, P.; — —: The weighted bootstrap. 95 Springer:Brln:NY 230p

Barbe, P.; — — (Auth); Young, G. A. (Rev): Review of "The weighted bootstrap". 95 ShrtBkRv 15/3 45-45

Berthod, Marc: Urago, Sabine; Zerubia, Josiane; — —: A Markovian model for contour grouping. 95 PattrnRc 28 683-693

Berthouex, P. M.: Johnson, Richard A.; Gan, D. Robert; — —: Goodness-of-fit using very small but related samples with application to censored data estimation of PCB contamination. 95 EnvrMtrc 6 341-348

Berthouex, Paul Mac; Brown, Linfield C.: Statistics for environmental engineers. 94 CRCPress:FL 352p

Bertin, Emile; Theodorescu, Radu: On a two-parameter family of discrete unimodal random variables. 95 SankhyaA 57 41-55

— —; Theodorescu, Radu: Preserving unimodality by mixing. 95 StPrLet 25 281-288

Bertino, Salvatore: The minimum of the expected value of the product of three random variables in the Fréchet class. 94 JItalStS 3 201-211

Bertoin, J.; Doney, R. A.: On conditioning a random walk to stay nonnegative. 94 AnlsProb 22 2152-2167

Bertoin, Jean: Lévy processes that can creep downwards never increase. 95 AIHenrPB 31 379-391

— —: Sample path behaviour in connection with generalized arcsine laws. 95 ProbTher 103 317-327

— —: On the local rate of growth of Lévy processes with no positive jumps. 95 StocProc 55 91-100

Bertolino, Francesco; Piccinato, Ludovico; Racugno, Walter: Multiple Bayes factors for testing hypotheses. 95 JASA 90 213-219

Bertolucci, Darryl: Greenblatt, Janet C.; — —: Characteristics of persons who reported driving under the influence of alcohol or other drugs. 94 ASASrvyRsr 547-552

Bertrand, Pierre: Adaptive control of partially observed linear stochastic systems. 95 Stochatx 54 21-51

Bertrand-Retali, M.; Ait-Hennani, L.: Uniform convergence of density estimators on spheres. 95 ApplMath 22 427-446

Besag, Julian: Comment on "Markov chains for exploring posterior distributions". 94 AnlsStat 22 1734-1741

— —; Green, Peter; Higdon, David; Mengersen, Kerrie: Bayesian computation and stochastic systems. 95 StatSci 10 3-41 (Disc: p41-66)

— —; Green, Peter; Higdon, David; Mengersen, Kerrie: Reply to comments on "Bayesian computation and stochastic systems". 95 StatSci 10 58-66

— —; Kooperberg, Charles: On conditional and intrinsic autoregressions. 95 Biomtrka 82 733-746

Besl, John: Kale, Balkrishna; — —; Palit, Charles; Voss, Paul: Updating age-sex-specific net migration rates with limited data. 94 ASAProSoSt 116-121 (Disc: p122-123)

Besse, P. C.; Ferraty, F.: A fixed effect curvilinear model. 95 CmpStQ 10 339-351

Best, D. J.: Rayner, J. C. W.; — —: Smooth tests for the bivariate Poisson distribution. 95 AstrlJSt 37 233-245

Rayner, J. C. W.; — —; Mathews, K. L.: Interpreting the skewness coefficient. 95 CommStA 24 593-600

Best, N. G.: Gilks, W. R.; — —; Tan, K. K. C.: Adaptive rejection Metropolis sampling within Gibbs sampling. 95 ApplStat 44 455-472

Bethea, Robert M.; Duran, Benjamin S.; Boullion, Thomas L.: Statistical methods for engineers and scientists. 95 Dekker:NY 672p

— —; Duran, Benjamin S.; Boullion, Thomas L. (Auth); Nelson, Lloyd S. (Rev): Review of "Statistical methods for engineers and scientists (Third edition)". 95 JQualTek 27 392-393

— —; Rhinehart, R. Russell (Auth); Shanmugam, Ramalingam (Rev): Review of "Applied engineering statistics". 95 JQualTek 27 173-173

Betne, Prabha; Katti, S. K.: Optimal design for Langmuir with two inputs. 94 ASAProBiop 425-429

— —; Katti, S. K.: Transforms for numerical estimation in Langmuir. 94 ASAProQlPr 286-290

Betounes, David: Trace operators, Feynman distributions, and multiparameter white noise. 95 JTherPrb 8 119-138

Betteley, G.; Mettrick, N.; Sweeney, E.; Wilson, D. (Auth); Shade, John (Rev): Review of "Using statistics in industry". 95 Ststcian 44 541-542

Bettonvil, Bert: Factor screening by sequential bifurcation. 95 CommStB 24 165-185

Beus, Carolyn M.: Davis, Alan B.; Larimore, Walter L.; Fellingham, Gilbert W.; — —: Using subsequent studies to recover estimability. 94 ASAProEpid 112-117 (Disc: p125-126)

Bewley, Ronald; Yang, Minxian: Tests for cointegration based on canonical correlation analysis. 95 JASA 90 990-996

— —; Yang, Minxian: Testing for cointegration: The effects of mis-specifying the lag length. 95 MaCmpSim 39 251-255

Beyer, Bruce K.: Introduction to motor activity data (reproductive toxicology). 94 ASAProBiop 482-483

Bez, K.: Das, G.; — —: Preliminary test estimators in double sampling with two auxiliary variables. 95 CommStA 24 1211-1226

Bhandary, Madhusudan: Detection of outliers in the linear model when the regression parameters are stochastic. 95 CommStA 24 153-173

— —: Detection of outliers in growth curve models. 95 CommStA 24 1923-1940

Bhapkar, Vasant P.: Completeness and optimality of marginal likelihood estimating equations. 95 CommStA 24 945-952

Bhat, B. V. Rajarama; Parthasarathy, K. R.: Markov dilations of nonconservative dynamical semigroups and a quantum boundary theory. 95 AIHenrPB 31 601-651

Bhattacharya, Bhaskar: Restricted tests for and against the increasing failure rate ordering on multinomial parameters. 95 StPrLet 25 309-316

— —; Dykstra, Richard: A general duality approach to I-projections. 95 JStPlInf 47 203-216

Bhattacharya, Partha P.; Georgiadis, Leonidas; Tsoucas, Pantelis: Problems of adaptive optimization in multiclass M/GI/1 queues with Bernoulli feedback. 95 MaOpRsr 20 355-380

Bhattacharya, Rabi; Lee, Chanho: On geometric ergodicity of nonlinear autoregressive models. 95 StPrLet 22 311-315

Bhattacharya, Rabi N.; Götze, Friedrich: Time-scales for Gaussian approximation and its breakdown under a hierarchy of periodic spatial heterogeneities. 95 Bernoull 1 81-123

— —; Lee, Chanho: Ergodicity of nonlinear first order autoregressive models. 95 JTherPrb 8 207-219

Bhattacharyya, Gouri K.: Sen, Ananda; — —: Inference procedures for the linear failure rate model. 95 JStPlInf 46 59-76

Bhattacharyya, P.: Ganesan, L.; — —: A statistical design of experiments approach for texture description. 95 PattrnRc 28 99-105

Bhattacharyya, S.: Vijayan, K.; Mukhopadhyay, P.; — —: On non-negative unbiased estimation of quadratic forms in finite population sampling. 95 AstrlJSt 37 169-178

Bhatti, M. I.; Barry, A. M.: On testing for equicorrelation in random coefficient models. 95 BiomtrcJ 37 383-391

Zhu, Li-Xing; Fang, Kai-Tai; — —; Bentler, Peter M.: Testing sphericity of a high-dimensional distribution based on bootstrap approximation. 95 PakJStA 11 49-65

Bhatti, M. Ishaq: Optimal testing for equicorrelated linear regression models. 95 StHefte 36 299-312

Bhaumik, Dulal K.: Majorization and D-optimality for complete block designs under correlations. 95 JRSS-B 57 139-143

— —: Optimality in the competing effect model. 95 SankhyaB 57 48-56

— —; Mathew, Thomas: Minimaxity of randomized optimal designs with respect to a general optimality criterion. 95 SankhyaB 57 122-127

Dasgupta, Ratan; — —: Upper and lower tolerance limits of atmospheric ozone level and extreme value distribution. 95 SankhyaB 57 182-199

Bhimasankaram, P.; Sengupta, D.; Ramanathan, S.: Recursive inference in a general linear model. 95 SankhyaA 57 227-255

Bhuyan, K. C.; Nair, G. A.: Structural changes in the relationships of body dimensions of four species of pulmonate slugs (Mollusca : Gastropoda : Pulmonata). 95 BiomtrcJ 37 1005-1015

Bian, G.: Robust Bayesian estimators in a one-way ANOVA model. 95 Test 4 115-135

Bibby, Bo Martin; Sørensen, Michael: Martingale estimation functions for discretely observed diffusion processes. 95 Bernoull 1 17-39

Bibby, Katherine; Davies, Neville: STEPS for learning statistics. 95 TeachgSt 17/3 107-110

Bichteler, K.; Lin, S. J.: On the stochastic Fubini theorem. 95 Stochatx 54 271-279

Bickart, Barbara: Czaja, Ronald; Blair, Johnny; — —; Eastman, Elizabeth: Respondent strategies for recall of crime victimization incidents. 94 JOffStat 10 257-276

Bickel, Peter J.: What academia needs. 95 AmerStat 49 5-6 (Disc: p18-23)

— —; Ritov, Ya'acov: An exponential inequality for U-statistics with applications to testing. 95 PrEngInf 9 39-52

Bickmore, Wendy A.: Carothers, Andrew D.; — —: Models of DNA replication timing in interphase nuclei: An exercise in inferring process from state. 95 Biomtrcs 51 750-755

Bieler, Gayle S.; Williams, Rick L.: Cluster sampling techniques in quantal response teratology and developmental toxicity studies. 95 Biomtrcs 51 764-776

Biemer, Paul: Discussion of "Estimation in the presence of measurement error". 95 IntStRvw 63 143-147

— —; Akin, Don: The efficiency of list-assisted random digit dialing sampling schemes for single and dual frame surveys. 94 ASASrvyRsr 1-10

— —; Caspar, Rachel: Continuous quality improvement for survey operations: Some general principles and applications. 94 JOffStat 10 307-326

Biemer, Paul P.: O'Reilly, James M.; Hubbard, Michael L.; Lessler, Judith T.; — —; Turner, Charles F.: Audio and video computer assisted self-interviewing: Preliminary tests of new technologies for data collection. 94 JOffStat 10 197-214

Bienias, Julia L.; Lassman, David M.; Scheleur, Scott A.; Hogan, Howard: Improving outlier detection in two establishment surveys. 94 ASASrvyRsr 1140-1143

Bier, V. M.; Yi, W.: The performance of precursor-based estimators for rare event frequencies. 95 RelEngSS 50 241-251

Bier, Vicki M.: On the concept of perfect aggregation in Bayesian estimation. 94 RelEngSS 46 271-281 (Disc: p287-289)

— —; Yi, Woojune: A Bayesian method for analyzing dependencies in precursor data. 95 IntJFor 11 25-41

Bierbrauer, Jürgen: Bounds on orthogonal arrays and resilient functions. 95 JComDsgn 3 179-183

Biernat, Janusz: Discussion of compensating faults in modelling of NMR system reliability. 95 RelEngSS 47 221-228

Biggeri, Annibale; Marchi, Marco: Case-control designs for the detection of spatial clusters of diseases. 95 EnvrMtrc 6 385-393

Cislaghi, Cesare; — —; Braga, Mario; Lagazio, Corrado; Marchi, Marco: Exploratory tools for disease mapping in geographical epidemiology. 95 StatMed 14 2363-2381

Biggerstaff, B.: Mengersen, K. L.; Tweedie, R. L.; — —: The impact of method choice on meta-analysis. 95 AstrlJSt 37 19-44

Biggins, J. D.; Yue, K. K.: A likelihood approach to scaling examination marks. 95 AstrlJSt 37 273-282

Bignell, C. M.: Dunlop, Peter J.; — —; Jackson, J. F.; Hibbert, D. Brynn: Chemometric analysis of gas chromatographic data of oils from Eucalyptus species. 95 Chemolab 30 59-67

Bigoni, R.: Forina, M.; Drava, G.; Armanino, C.; Boggia, R.; Lanteri, S.; Leardi, R.; Corti, P.; Conti, P.; Giangiacomo, R.; Galliena, C.; — —; Quartari, I.; Serra, C.; Ferri, D.; Leoni, O.; Lazzeri, L.: Transfer of calibration function in near-infrared spectroscopy. 95 Chemolab 27 189-203

Bikker, J. A.: Berk, J. M.; — —: International interdependence of business cycles in the manufacturing industry: The use of leading indicators for forecasting and analysis. 95 JForcast 14 1-23

Bilgi, Shobha: Nabar, S. P.; — —: Cumulative sum control charts for the inverse Gaussian distribution. 94 JIndnStA 32 9-14

Billard, L.; Meshkani, M. R.: Estimation of a stationary Markov chain. 95 JASA 90 307-315

Arghami, N. R.; — —: A sequential dual of Stein's two-sample procedure. 92 ProIrStCl 11-18

Dayananda, P. W. A.; — —; Chakraborty, S.: Estimation of rate parameter and its relationship with latent and infectious periods in plant disease epidemics. 95 Biomtrcs 51 284-292

Lange, N.; Ryan, L.; — —; Brillinger, D. R.; Conquest, L.; Greenhouse, J. (Ed); Herrendörfer, G. (Rev): Review of "Case studies in biometry". 95 BiomtrcJ 37 640-640

Billard, Lynne: The roads travelled. 95 Biomtrcs 51 1-11

Lange, Nick; Ryan, Louise; — —; Brillinger, David R.; Conquest, Loveday; Greenhouse, Joel (Ed); Hilton, Joan (Rev): Review of "Case studies in biometry". 95 JASA 90 800-800

Billeaud, Kathleen: Thissen, David; Pommerich, Mary; — —; Williams, Valerie S. L.: Item response theory for scores on tests including polytomous items with ordered responses. 95 ApPsychM 19 39-49

Billheimer, Dean; Guttorp, Peter: Zooplankton proportion estimates from non-uniform sample volumes. 95 EnvEcoSt 2 117-124

Billingsley, Patrick: Probability and measure (Third edition). 95 Wiley:NY:UK 593p

— — (Auth); Krämer, Walter (Rev): Review of "Probability and measure (Third edition)". 95 CmpStDA 20 703-703

Bilodeau, Martin: Minimax estimators of the mean vector in normal mixed linear models. 95 JMultiAn 52 73-82

Bilotti-Aliaga, Martha: Exploring probability and statistics through hands-on experiments and cooperative learning. 94 ASAProStEd 150-155

Biltoft, C.: Higson, H. L.; Griffiths, R. F.; Jones, C. D.; — —: Effect of atmospheric stability on concentration fluctuations and wake retention times for dispersion in the vicinity of an isolated building. 95 EnvrMtrc 6 571-581

Binder, David A.; Kovacevic, Milorad S.: Estimating some measures of income inequality from survey data: An application of the estimating equations approach. 95 SurvyM 21 137-145

— —; Michaud, Sylvie; Poirier, Claude: Model-based reweighting for nonresponse adjustment. 94 ASASrvyRsr 989-994

Cox, Brenda G.; — —; Chinnappa, B. Nanjamma; Christianson, Anders; Colledge, Michael J.; Kott, Phillip S. (Ed): Business survey methods. 95 Wiley:NY:UK 732p

Bingham, N. H.: The sample mid-range and symmetrized extremal laws. 95 StPrLet 23 281-288

Dembo, A.; Zeitouni, O. (Auth); — — (Rev): Review of "Large deviations techniques and applications". 94 BLndnMaS 26 508-510

Kelly, F. P. (Ed); — — (Rev): Review of "Probability, statistics and optimization: A tribute to Peter Whittle". 95 JRSS-A 158 353-354

Bingham, Stephen F.: Williford, William O.; Krol, William F.; — —; Collins, Joseph F.; Weiss, David G.: The multicenter clinical trials coordinating center statistician: 'More than a consultant'. 95 AmerStat 49 221-225

Bini, Matilde: Agresti, Alan; Ghosh, Atalanta; — —: Raking kappa: Describing potential impact of marginal distributions on measures of agreement. 95 BiomtrcJ 37 811-820

Binswanger, K.; Embrechts, P.: Longest runs in coin tossing. 94 InsrneME 15 139-149

Biradar, R. S.: Singh, H. P.; — —: A class of unbiased ratio estimators in two-phase sampling. 94 Statstca 54 349-359

Birch, Jeffrey B.: Ten suggestions for effectively teaching short courses to heterogeneous groups. 95 AmerStat 49 190-195

— —: Ten suggestions for effectively teaching short courses to heterogeneous groups. 94 ASAProStEd 33-38

Birgé, Lucien; Massart, Pascal: Estimation of integral functionals of a density. 95 AnlsStat 23 11-29

Birkett, Nicholas J.: Comment on "When will nondifferential misclassification of an exposure reverse the direction of a trend?" (V140 p565-571). 95 AJEpidm 142 783-784

Birman, Alexander; Gail, H. Richard; Hantler, Sidney L.; Rosberg, Zvi; Sidi, Moshe: An optimal service policy for buffer systems. 95 JACM 42 641-657

Bischoff, Wolfgang: Determinant formulas with applications to designing when the observations are correlated. 95 AnInStMa 47 385-399

— —: Comment on "Constrained optimization of experimental design". 95 MaOpfStS 26 150-153

— —: Lower bounds for the efficiency of designs with respect to the D-criterion when the observations are correlated. 95 MaOpfStS 27 27-44

— —; Fieger, Werner; Wulfert, Stefanie: Minimax- and Γ-minimax estimation of a bounded normal mean under LINEX loss. 95 StatDecn 13 287-298

Bisgaard, Soren: Vardeman, Stephen B. (Auth); — — (Rev): Review of "Statistics for engineering problem solving". 95 AmerStat 49 320-321

Bisgaard, Søren; Fuller, Howard T.: Sample size estimates for 2^{k-p} designs with binary responses. 95 JQualTek 27 344-354

Bishir, John; Roberds, James: Analysis of failure time in clonally propagated plant populations. 95 MaBisci 125 109-125

Bishop, D. Timothy: Spence, M. Anne; — —; Boehnke, Michael; Elston, Robert C.; Falk, Catherine; Hodge, Susan E.: Methodological issues in linkage analyses for psychiatric disorders: Secular trends, assortative mating, bilineal pedigrees. 93 HumHrdty 43 166-172

Bishwal, J. P. N.: Mishra, M. N.; — —: Approximate maximum likelihood estimation for diffusion processes from discrete observations. 95 Stochatx 52 1-13

Bismarck, M.; Gajek, L.: An upper bound for the power of a test. 95 CommStA 24 813-828

Bissell, A. F.: Weighted cumulative sums for text analysis using word counts. 95 JRSS-A 158 525-545

Bissell, D.: — — (Auth); Davies, Neville (Rev): Review of "Statistical methods for SPC and TQM". 95 JRSS-A 158 341-342

— — (Auth); Sheil, J. (Rev): Review of "Statistical methods for SPC and TQM". 95 Ststcian 44 279-281

Biswas, Atanu: Bandyopadhyay, Uttam; — —: Some nonparametric group sequential-type tests for two population problems. 95 CalcutSt 45 73-91

Biswas, Suddhendu: Applied stochastic processes: A biostatistical and population oriented approach. 95 Wiley:NY:UK 427p

— —; Chitkara, Sunita: On the estimation of true treatment response when patients enter staggeredly in a clinical trial. 95 BiomtrcJ 37 571-580

— —; Thomas, Mariamma: On a continuous branching process and birth-death and immigration models for testing the efficacy of some control programmes for AIDS. 95 BiomtrcJ 37 235-249

Bithell, J. F.: The choice of test for detecting raised disease risk near a point source. 95 StatMed 14 2309-2322

Bitmead, Robert R.: Blackmore, Perry A.; — —: Duality between the discrete-time Kalman filter and LQ control law. 95 IEEEAuCn 40 1442-1444

Bjertness, Espen: Aalen, Odd O.; — —; Sønju, Torleif: Analysis of dependent survival data applied to lifetimes of amalgam fillings. 95 StatMed 14 1819-1829

Björnsson, H.: Pollock, K. H.; Jones, C. M.; Brown, T. L. (Auth); — — (Rev): Review of "Angler survey methods and their applications in fisheries management". 95 Biomtrcs 51 385-385

Black, R. J.: DON Econmetrics (Auth); — — (Rev): Review of "SIMPC 3.81". 95 JApEcmtx 10 469-476

Black, Stephen; Mansouri, H.: On exact distributions of rank tests for ordered alternatives in block designs. 95 CmpStDA 20 265-274

Blackburn, McKinley L.; Neumark, David: Are OLS estimates of the return to schooling biased downward? another look. 95 RvEconSt 77 217-229

Blackburn, Robert M.; Siltanen, Janet; Jarman, Jennifer: The measurement of occupational gender segregation: Current problems and a new approach. 95 JRSS-A 158 319-331

Blackmore, Perry A.; Bitmead, Robert R.: Duality between the discrete-time Kalman filter and LQ control law. 95 IEEEAuCn 40 1442-1444

Blackwell, D.: Minimax vs. Bayes prediction. 95 PrEngInf 9 53-58

Blaga, Petru P.: Hoffman-Jørgensen, J. (Auth); — — (Rev): Review of "Probability with a view toward statistics (Volume 1)". 95 JAppStat 22 320-321

Hoffmann-Jørgensen, J. (Auth); — — (Rev): Review of "Probability with a view toward statistics (Volume 2)". 95 JAppStat 22 321-321

Blair, Clifford R.; Hochberg, Yosef: Improved Bonferroni procedures for testing overall and pairwise homogeneity hypotheses. 95 JStCmpSm 51 281-289

Blair, Johnny: Czaja, Ronald; — —; Bickart, Barbara; Eastman, Elizabeth: Respondent strategies for recall of crime victimization incidents. 94 JOffStat 10 257-276

Blair, R. Clifford: Algina, James; — —; Coombs, William T.: A maximum test for scale: Type I error rates and power. 95 JEdStat 20 27-39

Blais, Lucie: Suissa, Samy; — —: Binary regression with continuous outcomes. 95 StatMed 14 247-255

Blake, Jonathan E.; Collings, Bruce Jay: An evaluation of three methods for ranking paired contestants. 94 ASAProSpts 31-35

Blanc, Hans: van den Hout, Wilbert; — —: Development and justification of the power-series algorithm for BMAP-systems. 95 CommStSM 11 471-496

Blanchard, Fletcher A.: What should you do on the first day? Motivating and setting goals. 94 ASAProStEd 63-65

Blanchard, John L.: A construction for Steiner 3-designs. 95 JCombThA 71 60-66

Blanciforti, Laura A.: Garner, Thesia I.; — —: Household income reporting: An analysis of U.S. Consumer Expenditure Survey data. 94 JOffStat 10 69-91

Blass, Richard: Tucker, Clyde; Kojetin, Brian; — —; Tucker, Ronald: Measuring the effects of the transition on field performance. 94 ASASrvyRsr 348-353 (Disc: p360-361)

Błaszczyszyn, B.: Factorial moment expansion for stochastic systems. 95 StocProc 56 321-335

Błaszczyszyn, Bartłomiej; Frey, Andreas; Schmidt, Volker: Light-traffic approximations for Markov-modulated multi-server queues. 95 CommStSM 11 423-445

Blattberg, Robert C.: Kim, Byung-Do; — —; Rossi, Peter E.: Modeling the distribution of price sensitivity and implications for optimal retail pricing. 95 JBES 13 291-303

Blattenberger, Gail; Fowles, Richard: Road closure to mitigate avalanche danger: A case study for Little Cottonwood Canyon. 95 IntJFor 11 159-174

Blauwet, Mary Beth; Morgan, Timothy M.: Changing risk factor relationships in aging populations due to self-selection statistical bias (selective survival bias). 94 ASAProEpid 76-81

Blazek, L. W.: Using CMM data for process evaluation and improvement. 94 ASAProPEng 26-34

Bleher, Pavel M.; Lebowitz, Joel L.: Variance of number of lattice points in random narrow elliptic strip. 95 AIHenrPB 31 27-58

Blettner, Maria: Vach, Werner; — —: Logistic regression with incompletely observed categorical covariates - Investigating the sensitivity against violation of the missing at random assumption. 95 StatMed 14 1315-1329

Blien, Uwe: Müller, Walter; — —; Wirth, Heike: Identification risks of microdata: Evidence from experimental studies. 95 SocMethR 24 131-157

Blinder-Weiner, S.: Heller, I.; Isakov, A.; — —; Topilsky, M.: Bayesian classification of vasculitis: A simulation study. 95 MInfoMed 34 259-265

Blizard, David A.: Recombinant-inbred strains: General methodological considerations relevant to the study of complex characters. 92 BehavGen 22 621-634

Bloch, Daniel A.: Tubert-Bitter, Pascale; — —; Raynauld, Jean-Pierre: Comparing the bivariate effects of toxicity and efficacy of treatments. 95 StatMed 14 1129-1141 (Disc: p1143-1143)

Bloch, Sylvan C.: Spreadsheet analysis for engineers and scientists. 95 Wiley:-NY:UK 336p

Block, Alan: Schechter, Susan; Beatty, Paul; — —: Cognitive issues and methodological implications in the development and testing of a traffic safety questionnaire. 94 ASASrvyRsr 1215-1219

Blodgett, Robert: Kim, Jay Jong-Ik; Zaslavsky, Alan; — —: Between-state heterogeneity of undercount rates and surrogate variables in the 1990 U.S. Census. 95 SurvyM 21 55-62

Blodgett, Robert J.: Rank tests with weighted data. 94 ASAProStCp 200-204

Bloom, W. R.; Heyer, H.: Harmonic analysis of probability measures on hypergroups. 94 DGruyter:NY:Grm 601p

— —; Heyer, H. (Auth); Barnett, C. (Rev): Review of "Harmonic analysis of probability measures on hypergroups". 95 ShrtBkRv 15/3 49-49

Bloomfield, Peter: Epperson, David L.; Davis, Jerry M.; — —; Karl, Thomas R.; McNab, Alan L.; Gallo, Kevin P.: Estimating the urban bias of surface shelter temperatures using upper-air and satellite data. Part I: Development of models predicting surface shelter temperatures. 95 JApMetrl 34 340-357

Lund, Robert; Hurd, Harry; — —; Smith, Richard: Climatological time series with periodic correlation. 95 JClimate 8 2787-2809

Blouin, David C.: What effects should be included in the model?. 94 ASAProStEd 117-120

Blount, Douglas: Density-dependent limits for a nonlinear reaction-diffusion model. 94 AnlsProb 22 2040-2070

Blower, G.: Jurek, Z. J.; Mason, J. D. (Auth); — — (Rev): Review of "Operator-limit distributions in probability theory". 95 JRSS-A 158 352-353

Bloxom, Bruce; Pashley, Peter J.; Nicewander, W. Alan; Yan, Duanli: Linking to a large-scale assessment: An empirical evaluation. 95 JEdStat 20 1-26

Blumenthal, Saul; Dahiya, Ram C.: Estimation of sample size with grouped data. 95 JStPlInf 44 95-115

Bluvband, Zigmund; Grabov, Pavel; Ingman, Dov: New approach to process control. 95 ASQCTekCon 369-377

Blyth, Colin R.; Staudte, Robert G.: Estimating statistical hypotheses. 95 StPrLet 23 45-52

Blyth, S.: Correction to "Local divergence and association" (V81 p579-584). 95 Biomtrka 82 667-667

Blyth, Stephen: The dead of the Gulag: An experiment in statistical investigation. 95 ApplStat 44 307-321

Blythe, Stephen; Busenberg, Stavros; Castillo-Chavez, Carlos: Affinity in paired event probability. 95 MaBisci 128 265-284

BMDP Statistical Software Inc: — — (Auth); Getson, Pamela (Rev): Review of "BMDP New System for Windows (Version 1.11)". 95 Chance 8/4 53-58

Arbuckle, James; — —; Scientific Software International (Auth); Hox, Joop J. (Rev): Review of "AMOS, EQS, and LISREL for Windows: A comparative review". 95 StrEqMod 2 79-91

Boardman, Thomas J.: Reply to comment on "The statistician who changed the world: W. Edwards Deming, 1900-1993". 95 AmerStat 49 325-325

— —: W. Edwards Deming's red bead experiment fosters profound knowledge. 95 Stats /13 3-9

Boček, P.; Lachout, P.: Linear programming approach to LMS-estimation. 95 CmpStDA 19 129-134

Bock, Hans H.: Methodological comments relating to the presented classificatory and data-analytic methods. 95 CmpStQ 10 43-51

Böckenholt, Ulf: van Duijn, Marijtje A. J.; — —: Mixture models for the analysis of repeated count data. 95 ApplStat 44 473-485

Bodin, Jean-Louis: Official statistics and democracy. 94 ASAProGovt 170-172

Boedeker, Wolfgang: Drescher, Karsten; — —: Assessment of the combined effects of substances: The relationship between concentration addition and independent action. 95 Biomtrcs 51 716-730

Boehnke, M.: Lange, K.; — —: Bayesian methods and optimal experimental design for gene mapping by radiation hybrids. 92 AnHuGene 56 119-144

Boehnke, Michael: Limits of resolution of genetic linkage studies: Implications for the positional cloning of human disease genes. 94 AJHuGene 54 379-390

Spence, M. Anne; Bishop, D. Timothy; — —; Elston, Robert C.; Falk, Catherine; Hodge, Susan E.: Methodological issues in linkage analyses for psychiatric disorders: Secular trends, assortative mating, bilineal pedigrees. 93 HumHrdty 43 166-172

Boente, Graciela; Fraiman, Ricardo: Asymptotic distribution of data-driven smoothers in density and regression estimation under dependence. 95 CandJSt 23 383-397

— —; Fraiman, Ricardo: Asymptotic distribution of smoothers based on local means and local medians under dependence. 95 JMultiAn 54 77-90

Boer, Rob: van Oortmarssen, G. J.; — —; Habbema, J. D. F.: Modelling issues in cancer screening. 95 StMedRes 4 33-54

Boes, Duane C.: Chebaane, Mohamed; Salas, Jose D.; — —: Product periodic autoregressive processes for modeling intermittent monthly streamflows. 95 WaterReR 31 1513-1518

Bofinger, Eve; Bofinger, Mark: Equivalence with respect to a control: Stepwise tests. 95 JRSS-B 57 721-733

Bofinger, Mark: Bofinger, Eve; — —: Equivalence with respect to a control: Stepwise tests. 95 JRSS-B 57 721-733

Bogdan, Malgorzata: Data driven versions of Pearson's chi-square test for uniformity. 95 JStCmpSm 52 217-237

Bogen, Karen; Moore, Jeffrey C.; Marquis, Kent H.: Can we get respondents to use their personal income records. 94 ASASrvyRsr 1252-1257

Marquis, Kent H.; Moore, Jeffrey C.; — —: An experiment to reduce measurement error in the SIPP: Preliminary results. 94 ASASrvyRsr 725-729

Bogen, Kenneth T.: Methods to approximate joint uncertainty and variability in risk. 95 RiskAnal 15 411-419

Boggia, R.: Forina, M.; Drava, G.; Armanino, C.; — —; Lanteri, S.; Leardi, R.; Corti, P.; Conti, P.; Giangiacomo, R.; Galliena, C.; Bigoni, R.; Quartari, I.; Serra, C.; Ferri, D.; Leoni, O.; Lazzeri, L.: Transfer of calibration function in near-infrared spectroscopy. 95 Chemolab 27 189-203

Bohlin, Torsten; Graebe, Stefan F.: Issues in nonlinear stochastic grey box identification. 95 IJAdCSgP 9 465-490

Böhning, D.: Jørgensen, B. (Auth); — — (Rev): Review of "The theory of linear models". 95 Biomtrcs 51 791-792

Böhning, Dankmar: A review of reliable maximum likelihood algorithms for semiparametric mixture models. 95 JStPlInf 47 5-28

Böhning, Danmar: Ayuthya, Rampai Suksawasdi Na; — —: Traffic accident mapping in Bangkok metropolis: A case study. 95 StatMed 14 2445-2458

Bohoris, George A.: Parametric statistical techniques for the comparative analysis of censored reliability data: A review. 95 RelEngSS 48 149-155

Boissel, Jean-Pierre; Durieu, Isabelle; Girard, Pascal; Nony, Patrice; Chauvin, Franck; Haugh, Margaret: Dose-ranging trials: Guidelines for data collection and standardized descriptions. 95 ConClnTr 16 319-330

Girard, Pascal; Laporte-Simitsidis, Silvy; Mismetti, Patrick; Decousus, Hervé; — —: Influence of confounding factors on designs for dose-effect relationship estimates. 95 StatMed 14 987-1005 (Disc: p1007-1008)

Bojdecki, Tomasz; Gorostiza, Luis G.: Self-intersection local time for Gaussian $S'(R^d)$-processes: Existence, path continuity and examples. 95 StocProc 60 191-226

Böker, Fred; Dannenberg, Olaf: What can P-values say about the quality of test statistics for comparative simulation studies? (German). 95 AllgeStA 79 233-251

Boland, Philip J.: Discussion of the symposium on the teaching of statistics in higher education in Ireland. 95 Ststcian 44 55-56

Boldin, M. V.: On nonparametric sign tests in multiparameter autoregression. 95 MaMethSt 4 435-448

Boldrighini, C.; Soloveitchik, M.: Drift and diffusion for a mechanical system. 95 ProbTher 103 349-379

Bolfarine, Heleno; Galea Rojas, Manuel: Maximum likelihood estimation of simultaneous pairwise linear structural relationships. 95 BiomtrcJ 37 673-689

— —; Galea Rojas, Manuel: Structural comparative calibration using the EM algorithm. 95 JAppStat 22 277-292

— —; Galea-Rojas, Manuel: Comment on "Functional comparative calibration using an EM algorithm" (V48 p1263-1271). 95 Biomtrcs 51 1579-1580

Arellano-Valle, Reinaldo B.; — —: On some characterizations of the t-distribution. 95 StPrLet 25 79-85

Leite, José G.; — —: Bayesian estimation of the number of equiprobable classes in a population. 94 RBrzlPrS 8 127-133

Bolger, Fergus; Harvey, Nigel: Judging the probability that the next point in an observed time-series will be below, or above, a given value. 95 JForcast 14 597-607

Bolker, B. M.: Grenfell, B. T.; Kleczkowski, A.; Gilligan, C. A.; — —: Spatial heterogeneity, nonlinear dynamics and chaos in infectious diseases. 95 StMedRes 4 160-183

Bolland, Thomas W.: Successful customers of statistics at a distant learning site. 94 ASAProQlPr 300-304

Bollen, Kenneth A.; Guilkey, David K.; Mroz, Thomas A.: Binary outcomes and endogenous explanatory variables: Tests and solutions with an application to the demand for contraceptive use in Tunisia. 95 Dmogrphy 32 111-131

Bollerslev, Tim; Rossi, Peter E.: Obituary: Daniel B. Nelson, 1959-1995. 95 JBES 13 361-364

Bollobás, Béla; Brightwell, Graham: Random high-dimensional orders. 95 AdvAppPr 27 161-184

Bolognese, James A.; Nguyen, Ha H.: Estimating "non-detects" for inclusion in linear regression. 94 ASAProBiop 414-419

Bolognese, Ralph A.: Vogel, Richard M.; — —: Storage-reliability-resilience-yield relations for over-year water supply systems. 95 WaterReR 31 645-654

Bolstad, William M.: The multiprocess dynamic Poisson model. 95 JASA 90 227-232

Bolthausen, Erwin; Deuschel, Jean-Dominique; Tamura, Yozo: Laplace approximations for large deviations of nonreversible Markov processes. The nondegenerate case. 95 AnlsProb 23 236-267

Bolton, S.: — — (Auth); Enas, Nathan (Rev): Review of "Pharmaceutical statistics: Practical and clinical applications (Second edition)". 95 StMedRes 4 267-267

Bonarini, Andrea; Bontempi, Gianluca: A qualitative simulation approach for fuzzy dynamical models. 94 ACMMdSim 4 285-313

Bondarenko, M. V.; Vishnyakov, A. N.: On the Russian stock exchange. 94 ThProbAp 39 188-194

Bondesson, Lennart: Factorization theory for probability distributions. 95 ScanActJ 44-53

Boneh, Ruth: Cohen, Ayala; Makov, Udi; — —: Computer intensive methods for inference on parameters of complex models - A Bayesian alternative. 95 JStPlInf 47 79-94

Bonham, Carl; Cohen, Richard: Comment on "Testing the rationality of price forecasts: New evidence from panel data" (V80 p714-735). 95 AmEcRv 85 284-289

Bonham, Charles D.: Angleton, George M.; — —: Least squares regression vs. geometric mean regression for ecotoxicology studies. 95 AplMaCmp 72 21-32

Bonham, G. H.: Health and environment: Public health decisions based on evidence, psychology and politics. 95 EnvrMtrc 6 311-318

Boniface, D.: Experiment design and statistical methods for behavioural and social research. 95 ChapmnHall:Lnd 272p

Bonissone, Piero P.: Comment on "A probabilistic and statistical view of fuzzy methods". 95 Technmcs 37 262-266

Bonneu, M.: Unspecified link functions in generalized linear models (French). 95 RStApp 43/2 49-66

Bonney, George E.: Li, Zhaohai; — —; Lathrop, G. M.; Rao, D. C.: Genetic analysis combining path analysis with regressive models: The TAU model of multifactorial transmission. 94 HumHrdty 44 305-311

Bonney, George Ebow: Compound regressive models for family data. 92 HumHrdty 42 28-41

Bontempi, Gianluca: Bonarini, Andrea; — —: A qualitative simulation approach for fuzzy dynamical models. 94 ACMMdSim 4 285-313

Booker, Jane: Kingman, J. F. C. (Auth); — — (Rev): Review of "Poisson processes". 95 Technmcs 37 124-124

Booksh, Karl; Henshaw, John M.; Burgess, Lloyd W.; Kowalski, Bruce R.: A second-order standard addition method with application to calibration of a kinetics-spectroscopic sensor for quantitation of trichloroethylene. 95 JChemtrc 9 263-282

Boomsma, A.: — — (Auth); Dalgleish, Lenard I. (Rev): Review of "BOJA: A program for bootstrap and jackknife analysis". 95 StatComp 5 165-174

Boomsma, Dorret I.: Dolan, Conor V.; Molenaar, Peter C. M.; — —: Simultaneous genetic analysis of means and covariance structure: Pearson-Lawley selection rules. 94 BehavGen 24 17-24

Boos, Dennis D.; Brownie, Cavell: ANOVA and rank tests when the number of treatments is large. 95 StPrLet 23 183-191

Booth, Alison L.; Satchell, Stephen E.: The hazards of doing a PhD: An analysis of completion and withdrawal rates of British PhD students in the 1980s. 95 JRSS-A 158 297-318

Booth, David E.: Berk, Kenneth N.; — —: Seeing a curve in multiple regression. 95 Technmcs 37 385-398

Booth, James: A note on the accuracy of two saddlepoint tail probability approximations. 94 ASAProPEng 56-58

Booth, James G.; Butler, Ronald W.; Huzurbazar, S.; Wood, Andrew T. A.: Saddlepoint approximations for P-values of some tests of covariance matrices. 95 JStCmpSm 53 165-180

— —; Wood, Andrew T. A.: An example in which the Lugannani-Rice saddlepoint formula fails. 95 StPrLet 23 53-61

Boots, B.: Okabe, A.; — —; Sugihara, K. (Auth); Lorz, U. (Rev): Review of "Spatial tessellations. Concepts and applications of Voronoi diagrams". 95 MaOpfStS 26 183-183

Boreham, John: Obituary: Geoffrey Penrice, 1923-1994. 95 JRSS-A 158 180-181

Borgan, O.: Andersen, P. K.; — —; Gill, R. D.; Keiding, N. (Auth); Milbrodt, Hartmut (Rev. German): Review of "Statistical models based on counting processes". 95 AllgeStA 79 221-221

Borgan, Ø.; Goldstein, L.; Langholz, B.: Methods for the analysis of sampled cohort data in the Cox proportional hazards model. 95 AnlsStat 23 1749-1778

Andersen, P. K.; — —; Gill, R. D.; Keiding, N.: Statistical models based on counting processes. 95 Springer:Brln:NY 767p

Andersen, P. K.; — —; Gill, R. D.; Keiding, N. (Auth); Hjort, Nils Lid (Rev): Review of "Statistical models based on counting processes". 95 StNeerla 49 247-249

Andersen, P. K.; — —; Gill, R. D.; Keiding, N. (Auth); McKeague, Ian W. (Rev): Review of "Statistical models based on counting processes". 95 SIAMRvw 37 475-476

Borgan, Ørnulf: Langholz, Bryan; — —: Counter-matching: A stratified nested case-control sampling method. 95 Biomtrka 82 69-79

Borghesi, Joanna L.: Delpizzo, Vincent; — —: Exposure measurement errors, risk estimate and statistical power in case-control studies using dichotomous analysis of a continuous exposure variable. 95 IJEpidem 24 851-862

Borhani, Hossein: Use of exact and asymptotic tests in labor discrimination cases. 94 ASAProBuEc 277-280

Borkar, V. S.: Probability theory. An advanced course. 95 Springer:Brln:NY 145p

Borkowski, John J.: Finding maximum G-criterion values for central composite designs on the hypercube. 95 CommStA 24 2041-2058

— —: Minimum, maximum, and average spherical prediction variances for central composite and Box-Behnken designs. 95 CommStA 24 2581-2600

— —: Spherical prediction-variance properties of central composite and Box-Behnken designs. 95 Technmcs 37 399-410

Farnum, Nicholas R. (Auth); — — (Rev): Review of "Modern statistical quality control and improvement". 95 JQualTek 27 91-93

Tague, Nancy R. (Auth); — — (Rev): Review of "The quality toolbox". 95 JQualTek 27 391-392

Borms, J.: Abidi, H.; Pontier, J.; — —; Duquet, W.: The benefit of modelling growth curves for longitudinal data (French). 95 RStApp 43/3 55-72

Born, Patricia: Viscusi, W. Kip; — —: The general-liability reform experiments and the distribution of insurance-market outcomes. 95 JBES 13/2 183-188

Borodin, A. N.; Freidlin, M. I.: Fast oscillating random perturbations of dynamical systems with conservation laws. 95 AIHenrPB 31 485-525

Borovcnik, Manfred: Gibbons, Jean Dickinson; Chakraborti, Subhabrata (Auth); — — (Rev): Review of "Nonparametric statistical inference". 95 CmpStQ 10 100-103

Borovkov, K. A.: On crossing times for multidimensional walks with skip-free components. 95 JAppProb 32 991-1006

— —: Analysis of transient effects for branching processes. 94 ThProbAp 39 379-395

Lindvall, T. (Auth); — — (Rev): Review of "Lectures on the coupling method". 94 ThProbAp 39 721-723

Borovkov, Konstantin; Pfeifer, Dietmar: On record indices and record times. 95 JStPlInf 45 65-79

Borovskich, Yu. V.: Koroljuk, V. S.; — — (Auth); Glendinning, R. H. (Rev): Review of "Theory of U-statistics". 95 Ststcian 44 411-411

Koroljuk, V. S.; — — (Auth); Janssen, Paul (Rev): Review of "Theory of U-statistics". 95 JASA 90 804-805

Borovskikh, Yu. V.; Korolyuk, V. S.: Random permanents. 94 VSP:Neth 194p

Borowiak, Dale: Jorgensen, Bent (Auth); — — (Rev): Review of "The theory of linear models". 95 Technmcs 37 228-229

Borowick, Kent: Moore, Kris; Trower, Jonathan; — —: How birth order influences individual characteristics. 95 Chance 8/4 20-24

Moore, Kris; Trower, Jonathan; — —: An empirical study of birth order. 94 ASAProBuEc 347-350

Borst, S. C.: A globally gated polling system with a dormant server. 95 PrEngInf 9 239-254

Bosch, Ronald J.; Ye, Yinyu; Woodworth, George G.: A convergent algorithm for quantile regression with smoothing splines. 95 CmpStDA 19 613-630

Bose, A.; Kaj, I.: Diffusion approximation for an age-structured population. 95 AnlsApPr 5 140-157

Bose, Arup: Estimating the asymptotic dispersion of the L_1 median. 95 AnInStMa 47 267-271

— —; Boukai, Benzion: Bias corrected sequential estimation for the mean of NEF-PVF distributions. 95 SqtlAnly 14 307-319

— —; Mukhopadhyay, Nitis: A note on accelerated sequential estimation of the mean of NEF-PVF distributions. 95 AnInStMa 47 99-104

— —; Mukhopadhyay, Nitis: Sequential interval estimation via replicated piecewise stopping times and accelerated stopping times in a class of two-parameter exponential family of distributions. 95 SqtlAnly 14 287-305

— —; Mukhopadhyay, Nitis: Sequential estimation of the mean of an exponential distribution via replicated piecewise stopping number. 95 StatDecn 13 351-361

Bose, Sudip; Slud, Eric: Maximin efficiency-robust tests and some extensions. 95 JStPlInf 46 105-121

Boshuizen, Frans A.; Gouweleeuw, José M.: A continuous-time job search model: General renewal processes. 95 CommStSM 11 349-369

Bosq, D.; Guégan, D.: Nonparametric estimation of the chaotic function and the invariant measure of a dynamical system. 95 StPrLet 25 201-212

Bosq, Denis: Asymptotic estimation of a non-linear infinite filter. Application to the estimation of Volterra series. 95 JNnparSt 5 61-73

— —: Optimal asymptotic quadratic error of density estimators for strong mixing or chaotic data. 95 StPrLet 22 339-347

Bossaerts, Peter: The econometrics of learning in financial markets. 95 EconTher 11 151-189

Bossche, A.: Sherwin, D. J.; — — (Auth); Phillips, Michael J. (Rev): Review of "The reliability, availability and productiveness of systems". 95 Ststcian 44 412-413

Bostick, Robert M.: Grambsch, Patricia M.; Randall, Bryan L.; — —; Potter, John D.; Louis, Thomas A.: Modeling the labeling index distribution: An application of functional data analysis. 95 JASA 90 813-821

Boswijk, H. Peter: Efficient inference on cointegration parameters in structural error correction models. 95 JEconmtx 69 133-158 (Disc: p159-175)

— —: Reply to comment on "Efficient inference on cointegration parameters in structural error correction models". 95 JEconmtx 69 173-175

— —; Franses, Philip Hans: Periodic cointegration: Representation and inference. 95 RvEconSt 77 436-454

Bosworth, D. L.; Wilson, R. A.; Young, A. (Auth); Lievesley, Denise (Rev): Review of "Reviews of United Kingdom statistical sources, Volume 27, Research and development". 95 JRSS-A 158 631-632

Botman, Steven L.: Estimation and analysis effects resulting from a sample size reduction in a large national survey. 94 ASASrvyRsr 1168-1172

— —; Jack, Susan S.: Combining National Health Interview Survey datasets: Issues and approaches. 95 StatMed 14 669-677 (Disc: p679-693)

Johnson, Ayah E.; — —; Basiotis, Peter: Nonresponse in federal demographic surveys: 1981-1991. 94 ASASrvyRsr 983-988

Kalsbeek, William D.; — —; Massey, James T.; Liu, Pao-Wen: Cost-efficiency and the number of allowable call attempts in the National Health Interview Survey. 94 JOffStat 10 133-152

Bottani, Norberto: Creating a system of international education indicators: The OECD INES project. 94 ASAProGovt 153-158

Boucher, Hélène: Kim, KyungMann; — —; Tsiatis, Anastasios A.: Design and analysis of group sequential logrank tests in maximum duration versus information trials. 95 Biomtrcs 51 988-1000

Tsiatis, Anastasios A.; — —; Kim, Kyungmann: Sequential methods for parametric survival models. 95 Biomtrka 82 165-173

Boucherie, Richard J.: Product forms based on backward traffic equations. 95 JAppProb 32 508-518

Bouckaert, A.: Ritter, C.; — —; van Lierde, M.; Theunissen, I.: Modelling mortality and morbidity of newborns. 95 StatMed 14 2191-2204

Boucneau, M.: Darius, P.; — —; De Greef, P.; De Feber, E.; Froeschl, K.: Modelling metadata. 93 StatJUN 10 171-180

Bougerol, Philippe; Elie, Laure: Existence of positive harmonic functions on groups and on covering manifolds. 95 AIHenrPB 31 59-80

Boukai, B.: A useful property of order statistics and its application to UMVU estimation. 95 StHefte 36 279-283

Wang, Y.; — —; Berger, J. O.: A Bayesian testing procedure with valid conditional frequentist interpretation. 94 ASAProBays 187-195

Boukai, Benzion: Bose, Arup; — —: Bias corrected sequential estimation for the mean of NEF-PVF distributions. 95 SqtlAnly 14 307-319

Boulaevskaia, Vera: Predicting dew formation. 95 Stats /12 21-22

Boularan, J.; Ferre, L.; Vieu, P.: A nonparametric model for unbalanced longitudinal data with application to geophysical data. 95 CmpStQ 10 285-298

Boularan, Joël; Ferré, Louis; Vieu, Philippe: Location of particular points in non-parametric regression analysis. 95 AstrlJSt 37 161-168

Bouleau, Nicolas; Lepingle, Dominique (Auth); Spall, James C., Nelson, J. Barry (Rev): Review of "Numerical methods for stochastic processes". 95 Technmcs 37 236-237

Boulerice, Bernard; Ducharme, Gilles R.: A note on smooth tests of goodness of fit for location-scale families. 95 Biomtrka 82 437-438

Boullion, Thomas L.: Bethea, Robert M.; Duran, Benjamin S.; — —: Statistical methods for engineers and scientists. 95 Dekker:NY 672p

Bethea, Robert M.; Duran, Benjamin S.; — — (Auth); Nelson, Lloyd S. (Rev): Review of "Statistical methods for engineers and scientists (Third edition)". 95 JQualTek 27 392-393

Boumans, Marcel: Frisch on testing of business cycle theories. 95 JEconmtx 67 129-147

Bound, John; Jaeger, David A.; Baker, Regina M.: Problems with instrumental variables estimation when the correlation between the instruments and the endogenous explanatory variable is weak. 95 JASA 90 443-450

Bourdeau, Marc; Burtschy, Bernard: Syntactic identification of ARMA processes. 94 ASAProBuEc 169-171

Bourguignon, B.: de Aguiar, P. F.; — —; Khots, M. S.; Massart, D. L.; Phan-Than-Luu, R.: D-optimal designs. 95 Chemolab 30 199-210

Bousfield, Marie V.: Estimation of fertility rates for small areas in postcensal years. 94 ASAProSoSt 106-111 (Disc: p122-123)

Bouza, Carlos: Allende, Sira; — —; Romero, Isidro: Fitting a linear regression model by combining least squares and least absolute value estimation. 95 Questo 19 107-121

Bouza, Carlos N.: Linear rank tests derived from a superpopulation model. 95 BiomtrcJ 37 497-506

Bove, Giuseppe: Some methods for the simultaneous analysis of data matrices and the factorial invariance problem. 94 Metron 52/3 73-87

Bover, Olympia: Arellano, Manuel; — —: Another look at the instrumental variable estimation of error-components models. 95 JEconmtx 68 29-51

Bovier, Anton; Gayrard, Véronique; Picco, Pierre: Large deviation principles for the Hopfield model and the Kac-Hopfield model. 95 ProbTher 101 511-546

Bowden, David C.; Kufeld, Roland C.: Generalized mark-sight population size estimation applied to Colorado moose. 95 JWildlMg 59 840-851

Pojar, Thomas M.; — —; Gill, R. Bruce: Aerial counting experiments to estimate pronghorn density and herd structure. 95 JWildlMg 59 117-128

Bowden, John: Bailey, Michael; — —; Callahan, Alexander J.: Managing ship performance of naval gunfire support using statistical process control. 94 MilOpRes 1 5-11

Bowerman, Bruce L.; O'Connell, Richard T.: Liberal education in business statistics. 94 ASAProStEd 259-262

Bowker, Albert H.; Olkin, Ingram; Veinott, Arthur F., Jr: Gerald J. Lieberman. 95 PrEngInf 9 3-26

Bowlus, Audra J.; Kiefer, Nicholas M.; Neumann, George R.: Estimation of equilibrium wage distributions with heterogeneity. 95 JApEcmtx 10 119-131S

Bowman, Adrian: Cleveland, W. S. (Auth); — — (Rev): Review of "Visualizing data". 95 JRSS-A 158 344-345

Bowman, Adrian W.: Young, Stuart G.; — —: Non-parametric analysis of covariance. 95 Biomtrcs 51 920-931

Bowman, Dale; Chen, James J.; George, E. Olusegun: Estimating variance functions in developmental toxicity studies. 95 Biomtrcs 51 1523-1528

— —; George, E. Olusegun: A saturated model for analyzing exchangeable binary data: Applications to clinical and developmental toxicity studies. 95 JASA 90 871-879

George, E. Olusegun; — —: A full likelihood procedure for analysing exchangeable binary data. 95 Biomtrcs 51 512-523

Bowman, Daryl T.; Rawlings, John O.: Establishing a rejection procedure for crop performance data. 95 AgronmyJ 87 147-151

Bowman, K. O.; Shenton, L. R.: Estimation problems associated with the three parameter Gamma distribution. 95 CommStA 24 1355-1376

Bown, Fred: Chase, Warren; — — (Auth); Muthulakshmi, S. (Rev): Review of "General statistics". 95 JAppStat 22 322-323

Box, G.; Jenkins, G.; Reinsel, G. (Auth); Ziegel, Eric R. (Rev): Review of "Time series analysis: Forecasting and control". 95 Technmcs 37 238-239

Box, George E. P.; Luceño, Alberto: Discrete proportional-integral control with constrained adjustment. 95 Ststcian 44 479-495

Boxe, Arup; Dasgupta, Ratan: On some asymptotic properties of U statistics and one-sided estimates. 94 AnlsProb 22 1715-1724

Boyarsky, Gregory: Grady, James J.; — —: The effects of an outlier while modeling the covariance matrix for incomplete longitudinal data. 94 ASAProBiop 441-445

Boyd, Amy Hancock: A minimum cost exponential CUSUM test. 95 CommStA 24 2929-2958

Boyd, James C.: Harris, Eugene K.; — —: Statistical bases of reference values in laboratory medicine. 95 Dekker:NY 384p

Boyle, Peter: Doll, R.; Fraumeni, J. F., Jr; Muir, C. S. (Ed); — — (Rev): Review of "Trends in cancer incidence and mortality". 95 StatMed 14 2507-2508

Boylstein, Lee Ann; Anderson, Stewart J.; Alarie, Yves: Comparison of smoothing methods when applied to respiratory data collected in mice exposed to an aerosolized machining fluid. 94 ASAProBiop 136-141

Boys, R. J.: Glazebrook, K. D.; — —: A class of Bayesian models for optimal exploration. 95 JRSS-B 57 705-720

Braak, Cajo J. F. ter: Non-linear methods for multivariate statistical calibration and their use in palaeoecology: A comparison of inverse (k-nearest neighbours, partial least squares and weighted averaging partial least squares) and classical approaches. 95 Chemolab 28 165-180

Braak, Cajo J. F. Ter: De Vries, S.; — —: Prediction error in partial least squares regression: A critique on the deviation used in the Unscrambler. 95 Chemolab 30 239-245

Bradburn, Norman: Baker, Reg; — —; Johnson, Robert: CAPI: An experimental evaluation. 94 ASASrvyRsr 851-856 (Disc: p861-862)

Bradburn, Norman M.: Stefey, Duane L.; — — (Ed); Wolter, Kirk M. (Rev): Review of "Counting people in the information age". 95 Chance 8/3 44-49

Bradley, A. Allen: Brockway, Charles G.; — —: Errors in streamflow drought statistics reconstructed from tree ring data. 95 WaterReR 31 2279-2293

Bradley, David C.; Steil, Garry M.; Bergman, Richard N.: OOPSEG: A data smoothing program for quantitation and isolation of random measurement error. 95 CmpBiomd 46 67-77

Bradley, Richard C.: On a theorem of K. Schmidt. 95 StPrLet 24 9-12

Bradstreet, Thomas E.: Favorite data sets from early phases of drug research - Part 3. 94 ASAProStEd 247-252

Bradu, Dan; Hawkins, Douglas M.: An Anscombe type robust regression statistic. 95 CmpStDA 20 355-386

Brady, Mark F.: Priore, Roger L.; — —; Lawrence, David D.: Power of detecting the effect of a continuous variable in a Cox model. 94 ASAProBiop 200-205

Brady, Michael: Shapiro, Larry S.; — —: Rejecting outliers and estimating errors in an orthogonal outliers and estimating errors in an orthogonal-regression framework. 95 PTLnAMa 350 407-439

Braga, Mario: Cislaghi, Cesare; Biggeri, Annibale; — —; Lagazio, Corrado; Marchi, Marco: Exploratory tools for disease mapping in geographical epidemiology. 95 StatMed 14 2363-2381

Brainerd, Barron: A generalization of the hypergeometric distribution suggested by a model for the structure of a play. 95 MaScitst 20 33-39

Braithwaite, G. R.: Mann, P. S. (Auth); — — (Rev): Review of "Introductory statistics". 95 JRSS-A 158 354-354

Brakhage, David H.: Smith, David R.; Conroy, Michael J.; — —: Efficiency of adaptive cluster sampling for estimating density of wintering waterfowl. 95 Biomtrcs 51 777-788

Brakstad, Frode: The feasibility of latent variables applied to GC-MS data. 95 Chemolab 29 157-176

Bramwell, Andrew: Primary data. 95 TeachgSt 17/3 88-89

Branco, Joao: Pires, Ana; — —; Picado, Ana; Mendonça, Elsa: Metabolic responses in the assessment of pollution effects. 95 EnvrMtrc 6 155-163

Brand, Richard J.: Chasalow, Scott D.; — —: [Algorithm AS 299] Generation of simplex lattice points. 95 ApplStat 44 534-545

Brandeau, M. L.: Kaplan, E. H.; — — (Ed); Altmann, Michael (Rev): Review of "Modeling the AIDS epidemic: Planning, policy, and prediction". 95 Chance 8/4 36-39

Brandt, Andreas: Last, Günter; — —: Marked point processes on the real line: The dynamic approach. 95 Springer:Brln:NY 490p

Last, Günter; — — (Auth); Embrechts, P. A. L. (Rev): Review of "Marked point processes on the real line: The dynamic approach". 96 ShrtBkRv 16/1 10-10

Brännäs, Kurt: Prediction and control for a time-series count data model. 95 IntJFor 11 263-270

De Gooijer, Jan G.; — —: Invertibility of non-linear time series models. 95 CommStA 24 2701-2714

Brannigan, Lawrence H.; Grieshaber, Mark V.; Schnur, Dora M.: Use experimental design in organic synthesis. 95 ChemTech 25 29-35

Brant, Larry J.: Morrell, Christopher H.; Pearson, Jay D.; Carter, H. Ballentine; — —: Estimating unknown transition times using a piecewise nonlinear mixed-effects model in men with prostate cancer. 95 JASA 90 45-53

Brashers, Dale E.: Jackson, Sally; — — (Auth); Iacobucci, Dawn (Rev): Review of "Random factors in ANOVA". 95 JMarRsr 32 238-239

Brassesco, S.; De Masi, A.; Presutti, E.: Brownian fluctuations of the interface in the $D = 1$ Ginzburg-Landau equation with noise. 95 AIHenrPB 31 81-118

Braun, H.: — — (Ed); Ziegel, Eric R. (Rev): Review of "The collected works of John W. Tukey, Volume VIII: Multiple comparisons". 95 Technmcs 37 242-242

Braun, H. I.: — — (Ed); Martin, M. A. (Rev): Review of "The collected works of John W. Tukey: Volume VIII Multiple comparisons 1948-1983". 95 Biomtrcs 51 380-381

— — (Ed); Young, Karen (Rev): Review of "The collected works of John W. Tukey, Volume 8, Multiple comparisons: 1948-1983". 95 JRSS-A 158 632-632

Braun, W. John: An illustration of bootstrapping using video lottery terminal data. 95 JStEd 3/2 18kb (jse/v3n2/datasets.braun)

Braverman, Michael; Samorodnitsky, Gennady: Functionals of infinitely divisible stochastic processes with exponential tails. 95 StocProc 56 207-231

Bravo, José I.; De Fuentes, Íñigo; Fernández, Arturo J.: A semi-parametric estimation of a survival function from incomplete and doubly censored data. 95 CommStA 24 2735-2752

Bray, Robert M.: Iachan, Ronaldo; Dennis, Michael L.; — —: Estimating the prevalence of substance abuse among people who are homeless. 94 ASASrvyRsr 1094-1099

Bredfeldt, Hans; Foster, Suzy; Scrutchins, Sam: Extending the use of IRS's tax compliance measurement program to develop estimates for small subpopulations using double sampling. 94 ASAProGovt 128-133

Bregman, Arie; Fuss, Melvyn; Regev, Haim: The production and cost structure of Israeli industry: Evidence from individual firm data. 95 JEconmtx 65 45-81

Brehm, John: — — (Auth); Parsley, Teresa L. (Rev): Review of "The Phantom respondents: Opinion surveys and political representation". 94 JOffStat 10 351-355

Breidt, F. Jay: Markov chain designs for one-per-stratum sampling. 95 SurvyM 21 63-70

— —; Davis, Richard A.; Dunsmuir, William T. M.: Improved bootstrap prediction intervals for autoregressions. 95 JTimSrAn 16 177-200

An, Anthony B.; — —; Fuller, Wayne A.: Regression weighting methods for SIPP data. 94 ASASrvyRsr 434-439 (Disc: p440-442)

Leonard, Kelli A.; An, Anthony B.; Nusser, Sarah M.; — —: Approximating the variance of the survey regression estimator using poststratification. 94 ASASrvyRsr 222-227

McVey, Anita; — —; Fuller, Wayne A.: Two-phase estimation by imputation. 94 ASASrvyRsr 1053-1058

Breiman, Leo: Better subset regression using the nonnegative garrote. 95 Technmcs 37 373-384

Breitung, J.: Modified stationarity tests with improved power in small samples. 95 StHefte 36 77-95

Breitung, K. W.: Asymptotic approximations for probability integrals. 94 Springer:Brln:NY 146p

— — (Auth); Plachky, D. (Rev): Review of "Asymptotic approximations for probability integrals". 95 StatDecn 13 399-399

— — (Auth); Veraverbeke, N. (Rev): Review of "Asymptotic approximations for probability integrals". 95 ShrtBkRv 15/2 26-26

Brekke, Kris: Little, Thomas A.; — —: Partition of variation: A new method for σ reduction. 95 ASQCTekCon 208-216

Bremaud, P.: Baccelli, F.; — —: Elements of queueing theory: Palm-martingale calculus and stochastic recurrences. 94 Springer:Brln:NY 256p

Bremer, Ronald: Barnett, Vic; Lewis, Toby (Auth); — — (Rev): Review of "Outliers in statistical data (3rd ed)". 95 Technmcs 37 117-118

Bren, Matevz: Batagelj, Vladimir; — —: Comparing resemblance measures. 95 JClass 12 73-90

Brennan, R. L.: Kolen, M. J.; — —: Test equating: Methods and practices. 95 Springer:Brln:NY 333p

Kolen, M. J.; — — (Auth); Ramsay, J. O. (Rev): Review of "Test equating: Methods and practices". 96 ShrtBkRv 16/1 6-6

Brenner, Hermann: Alternative approaches for estimating prevalence in epidemiologic surveys with two waves of respondents. 95 AJEpidm 142 1236-1245

Brereton, Richard G.; Gurden, Stephen P.; Groves, John A.: Use of eigenvalues for determining the number of components in window factor analysis of spectroscopic and chromatographic data. 95 Chemolab 27 73-87

Elbergali, Abdallah K.; — —: Influence of predicted elution regions on the performance of methods for evolutionary factor analysis as applied to high-performance liquid chromatography. 95 Chemolab 27 55-71

— — (Ed); Nemeth, Margaret A. (Rev): Review of "Multivariate pattern recognition in chemometrics, illustrated by case studies". 95 Technmcs 37 116-116

Breslaw, Jon A.: Nonparametric estimation of welfare changes. 95 EcomtrRv 14 163-182

— —; Smith, J. Barry: A simple and efficient method for estimating the magnitude and precision of welfare changes. 95 JApEcmtx 10 313-327

— —; Smith, J. Barry: Measuring welfare changes when quantity is constrained. 95 JBES 13 95-103

Breslow, Norman E.; Lin, Xihong: Bias correction in generalised linear mixed models with a single component of dispersion. 95 Biomtrka 82 81-91

Bresnahan, Timothy F.; Trajtenberg, M.: General purpose technologies: 'Engines of growth'?. 95 JEconmtx 65 83-108

Bretagnolle, Jean; Klopotowski, Andrzej: On the existence of sequences of s by s independent exchangeable or stationary random variables (French). 95 AIHenrPB 31 325-350

Breton, Michèle; Hachem, Saeb El: A scenario aggregation algorithm for the solution of stochastic dynamic minimax problems. 95 Stochatx 53 305-322

Brette, Stéphane; Idier, Jérôme; Mohammad-Djafari, Ali: Scale invariant Markov models for linear inverse problems. 94 ASAProBays 266-270

Bretting, P.: Crossa, J.; Hernandez, C. M.; — —; Eberhart, S. A.; Taba, S.: Statistical genetic considerations for maintaining germ plasm collections. 93 ThApGene 86 673-678

Brettle, Raymond P.: Fielding, Katherine L.; — —; Gore, Sheila M.; O'Brien, Fiona; Wyld, Rona; Robertson, J. Roy; Weightman, Rosina: Heterosexual transmission of HIV analysed by generalized estimating equations. 95 StatMed 14 1365-1378

Brewer, Ken: Arm-waving mathematics: Sound, if not rigorous. 95 TeachgSt 17/3 94-96

Brick, J. Michael; Waksberg, Joseph; Kulp, Dale; Starer, Amy: Bias in list-assisted telephone samples. 95 PubOpinQ 59 218-235

— —; Waksberg, Joseph; Kulp, Dale; Starer, Amy: Bias in list-assisted telephone samples. 94 ASASrvyRsr 1293-1298

Kalton, Graham; — —: Weighting schemes for household panel surveys. 95 SurvyM 21 33-44

Kalton, Graham; — —: Weighting schemes for household panel surveys. 94 ASASrvyRsr 785-790

Brick, Michael; Waksberg, Joseph; Keeter, Scott: Evaluating the use of data on interruptions in telephone service for nontelephone households. 94 ASASrvyRsr 19-28

Brick, Mike: Rizzo, Lou; Kalton, Graham; — —: Adjusting for panel nonresponse in the Survey of Income and Program Participation. 94 ASASrvyRsr 422-427 (Disc: p440-442)

Bright, L.; Taylor, P. G.: Calculating the equilibrium distribution in level dependent quasi-birth-and-death processes. 95 CommStSM 11 497-525

Brightwell, Graham; Ott, Teunis J.; Winkler, Peter: Target shooting with programmed random variables. 95 AnlsApPr 5 834-853

Bollobás, Béla; — —: Random high-dimensional orders. 95 AdvAppPr 27 161-184

Bril, Gordon: Forecasting hurricane tracks using the Kalman filter. 95 EnvrMtrc 6 7-16

Brill, P. H.: Huang, M. L.; — —: Estimation in systems of M/G/c/c queues. 94 ASAProPEng 151-156

Brillinger, D. R.: Lange, N.; Ryan, L.; Billard, L.; — —; Conquest, L.; Greenhouse, J. (Ed); Herrendörfer, G. (Rev): Review of "Case studies in biometry". 95 BiomtrcJ 37 640-640

Brillinger, David R.: Lange, Nick; Ryan, Louise; Billard, Lynne; — —; Conquest, Loveday; Greenhouse, Joel (Ed); Hilton, Joan (Rev): Review of "Case studies in biometry". 95 JASA 90 800-800

Brimacombe, Michael: Bayesian inference and likelihood selection: Empirically conditioned likelihood. 94 ASAProBays 40-45

— —: Conditional Bayesian significance levels: Hierarchical models. 94 ASAProBays 169-174

Bring, Johan: Comment on "Stopping a clinical trial very early because of toxicity: Summarizing the evidence" (V14 p286-295). 95 ConClnTr 16 131-132

Brink, Anders; Westerlund, Tapio: The joint problem of model structure determination and parameter estimation in quantitative IR spectroscopy. 95 Chemolab 29 29-36

Bristol, David R.: P-value adjustments for subgroup analyses of binary variables. 94 ASAProBiop 120-125

Brito, M.: Bacro, J. N.; — —: Weak limiting behaviour of a simple tail Pareto-index estimator. 95 JStPlInf 45 7-19

Brizzi, M.: Barnett, V.; Turkman, K. F. (Ed); — — (Rev. Italian): Review of "Statistics for the environment". 94 Statstca 54 397-398

Bro, Rasmus: Algorithm for finding an interpretable simple neural network solution using PLS. 95 JChemtrc 9 423-430

Brockett, Patrick L.: MacMinn, Richard D.; — —: Corporate spin-offs as a value enhancing technique when faced with legal liability. 95 InsrneME 16 63-68

Brockway, Charles G.; Bradley, A. Allen: Errors in streamflow drought statistics reconstructed from tree ring data. 95 WaterReR 31 2279-2293

Brockwell, P. J.; Stramer, O.: On the approximation of continuous time threshold ARMA processes. 95 AnInStMa 47 1-20

Brockwell, Peter J.: A note on the embedding of discrete-time ARMA processes. 95 JTimSrAn 16 451-460

Donahue, Rafe M. J.; — —; Davis, Richard: On permissible correlations for locally correlated stationary processes. 95 StPrLet 22 49-53

Shitan, Mahendran; — —: An asymptotic test for separability of a spatial autoregressive model. 95 CommStA 24 2027-2040

Broderick, G.; Paris, J.; Valade, J. L.: A composite representation of pulp quality. 95 Chemolab 29 19-28

Brodsky, Marc D.: Hughes, Arthur L.; — —: Variance estimation of drug abuse episodes using the bootstrap. 94 ASASrvyRsr 212-217

Brodziak, Jon K. T.: Hayes, Daniel B.; — —; O'Gorman, Joseph B.: Efficiency and bias of estimators and sampling designs for determining length-weight relationships of fish. 95 JFishRBC 52 84-92

Broemeling, Lyle D.: Cook, Peyton; — —: Bayesian statistics using Mathematica. 95 AmerStat 49 70-76

Bromaghin, Jeffrey F.: Correction to "Sample size determination for interval estimation of multinomial probabilities" (V47 p203-206). 95 AmerStat 49 403-403

Bromberg, Judith: Wan, Jim Y.; Wang, Whedy; — —: A SAS macro for residual deviance of ordinal regression analysis. 94 CmpBiomd 45 307-310

Broniatowski, Michel: Beirlant, Jan; — —; Teugels, Jozef L.; Vynckier, Petra: The mean residual life function at great age: Applications to tail estimation. 95 JStPlInf 45 21-48

Bronson, Richard: Jacobsen, Chanoch; — —: Computer simulations and empirical testing of sociological theory. 95 SocMethR 23 479-506

Brooker, Peter I.: Sevcik, Karel; — —: Application of the gnostical theory of uncertain data to spatial estimation, identification and quantification. 95 EnvrMtrc 6 457-463

Brookmeyer, R.: Wang, M.-C.; — —; Jewell, N. P.: Reply to comment on "Statistical models for prevalent cohort data". 95 Biomtrcs 51 374-375

— —; Gail, M. H. (Auth); Ghani, Azra (Rev): Review of "AIDS epidemiology: A quantitative approach". 95 JRSS-A 158 343-343

Brookmeyer, Ron: A short method for constructing an abridged life table. 95 AJEpidm 141 991-992

; Yasui, Yutaka: Statistical analysis of passive surveillance disease registry data. 95 Biomtrcs 51 831-842

Liao, Jiangang; — —: An empirical Bayes approach to smoothing in backcalculation of HIV infection rates. 95 Biomtrcs 51 579-588

— —; Gail, Mitchell H. (Auth); DeGruttola, Victor (Rev): Review of "AIDS epidemiology: A quantitative approach". 95 JASA 90 797-798

— —; Gail, Mitchell H. (Auth); Farrington, C. P. (Rev): Review of "AIDS epidemiology: A quantitative approach". 95 AJEpidm 141 180-181

Brooks, Maria Mori; Hallstrom, Al; Peckova, Monika: A simulation study used to design the sequential monitoring plan for a clinical trial. 95 StatMed 14 2227-2237

Brooks, Robert: The robustness of point optimal testing for Rosenberg random regression coefficients. 95 EcomtrRv 14 35-53

Brooks, S. P.; Morgan, B. J. T.: Optimization using simulated annealing. 95 Ststcian 44 241-257

Brooks, Stephen: White, D. J. (Auth); — — (Rev): Review of "Markov decision processes". 95 Ststcian 44 292-292

Brooks, Stephen P.: [Algorithm AS 298] A hybrid optimization algorithm. 95 ApplStat 44 530-533

Brorsen, B. Wade: Liu, Shi-Miin; — —: Maximum likelihood estimation of a GARCH-stable model. 95 JApEcmtx 10 273-285

Brosh, Menashe: Kamgar-Parsi, Behzad; Kamgar-Parsi, Behrooz; — —: Distribution and moments of the weighted sum of uniform random variables, with applications in reducing Monte Carlo simulations. 95 JStCmpSm 52 399-414

Bross, I. D.: — — (Auth); Aitken, C. G. G. (Rev): Review of "Fifty years of folly and fraud 'in the name of science': From crossroads to the health care crisis". 95 Biomtrcs 51 1587-1587

Brossard, Hubert: Image of "Made in Switzerland" (French). 95 Student 1 51-61

Brouaye, F.: Choosing between two unknown intervals: An empirical Bayes testing approach to a classification problem. 95 CommStA 24 3163-3181

Brown, Bryan W.; Walker, Mary Beth: Stochastic specification in random production models of cost-minimizing firms. 95 JEconmtx 66 175-205

Ligeralde, Antonio V.; — —: Band covariance matrix estimation using restricted residuals: A Monte Carlo analysis. 94 IntEconR 35 751-767

van Dijk, Herman K.; Monfort, Alain; — — (Ed): Econometric inference using simulation techniques. 95 Wiley:NY:UK 256p

Brown, Byron Wm., Jr: Hornberger, John C.; — —; Halpern, Jerry: Designing a cost-effective clinical trial. 95 StatMed 14 2249-2259

Brown, D.; Rothery, P. (Auth); Kemp, A. W. (Rev): Review of "Computing examples supplement to models in biology: Mathematics, statistics and computing". 95 Biomtrcs 51 792-793

— —; Rothery, P. (Auth); Kemp, A. W. (Rev): Review of "Models in biology: Mathematics, statistics and computing". 95 Biomtrcs 51 792-793

Brown, Edward F.: Saleh, A. K. Md. E.; Hassanein, Khatab M.; — —: Optimum spacings, ABLUE and tests of hypothesis in simple linear model with extreme-value errors. 94 JStatRsr 28 53-63

Brown, Eugene F.; Johnson, William H.: Stratification designs for the housing sample of the consumer price index. 94 ASAProGovt 140-143

Brown, J. W.; Hedayat, A. S.; Parker, E. T.: A pair of orthogonal Latin squares of order 10 with four shared parallel transversals. 93 JCISS 18 113-115

Brown, Jason: Efron, B.; Tibshirani, R. J. (Auth); — — (Rev): Review of "An introduction to the bootstrap". 95 JQualTek 27 90-91

Brown, Jason J.: A bootstrap algorithm for the isotropic random sphere. 95 AdvAppPr 27 627-641

— —: A central limit theorem for the isotropic random sphere. 95 AdvAppPr 27 642-651

Brown, Kenneth: Bayard, Steven; Jinot, Jennifer; — —: Passive smoking and lung cancer: The U.S. EPS's weight-of-evidence analysis, with emphasis on the epidemiology studies. 94 ASAProEnvr 56-64 (Disc: p74-76)

Brown, L. D.; Cohen, Arthur: Complete classes for confidence set estimation. 95 StSinica 5 291-301

Brown, Lawrence D.; Casella, George; Hwang, J. T. Gene: Optimal confidence sets, bioequivalence, and the Limaçon of Pascal. 95 JASA 90 880-889

Bandyopadhyay, Sati P.; — —; Richardson, Gordon D.: Analysts' use of earnings forecasts in predicting stock returns: Forecast horizon effects. 95 IntJFor 11 429-445

Berger, James O.; — —; Wolpert, Robert L.: A unified conditional frequentist and Bayesian test for fixed and sequential simple hypothesis testing. 94 AnlsStat 22 1787-1807

Brown, Linfield C.: Berthouex, Paul Mac; — —: Statistics for environmental engineers. 94 CRCPress:FL 352p

Brown, Mark: The distribution of total variation distance, with applications to simultaneous confidence intervals. 95 CmpOpRsh 22 373-381

Brown, Michael W.: Smith, David R.; Reinecke, Kenneth J.; Conroy, Michael J.; — —; Nassar, James R.: Factors affecting visibility rate of waterfowl surveys in the Mississippi Alluvial Valley. 95 JWildlMg 59 515-527

Brown, Morton B.: Mauger, David T.; — —; Kushler, Robert H.: A comparison of methods that characterize pulses in a time series. 95 StatMed 14 311-325

Efron, B.; Tibshirani, Robert J. (Auth); — — (Rev): Review of "An introduction to the bootstrap". 95 StMedRes 4 91-92

Brown, Norman R.: Conrad, Frederick G.; — —: Strategies for estimating category frequency: Effects of abstractness and distinctiveness. 94 ASASrvyRsr 1345-1350

Brown, P. J.: Cleave, N.; — —; Payne, C. D.: Evaluation of methods for ecological inference. 95 JRSS-A 158 55-72

Gelman, A.; Carlin, J. B.; Stern, H. S.; Rubin, D. B. (Auth); — — (Rev): Review of "Bayesian data analysis". 96 ShrtBkRv 16/1 5-5

— — (Auth); Atkinson, Anthony (Rev): Review of "Measurement, regression, and calibration". 95 JRSS-A 158 187-187

— —; Fearn, Tom (Rev): Review of "Measurement, regression, and calibration". 95 StMedRes 4 265-266

Brown, Patrick O.: Dupuis, Josée; — —; Siegmund, David: Statistical methods for linkage analysis of complex traits from high-resolution maps of identity by descent. 95 Genetcs 140 843-856

Brown, Philip J.: Diagnostics and selection for linearity and interaction in calibration with pure and multicomponent spectra. 95 Chemolab 28 139-148

Brown, Steven D.: Has the chemometrics revolution ended? Some views on the past, present and future of chemometrics. 95 Chemolab 30 49-58

Brown, T. L.: Pollock, K. H.; Jones, C. M.; — — (Auth); Björnsson, H. (Rev): Review of "Angler survey methods and their applications in fisheries management". 95 Biomtrcs 51 385-385

Brown, Timothy C.; Xia, Aihua: On metrics in point process approximation. 95 Stochatx 52 247-263

— —; Xia, Aihua: On Stein-Chen factors for Poisson approximation. 95 StPrLet 23 327-332

Browne, Richard H.: On the use of a pilot sample for sample size determination. 95 StatMed 14 1933-1940

Browne, Sid; Bunge, John: Random record processes and state dependent thinning. 95 StocProc 55 131-142

Brownie, Cavell: Boos, Dennis D.; — —: ANOVA and rank tests when the number of treatments is large. 95 StPrLet 23 183-191

Etchison, Tonya; — —; Pantula, Sastry G.: A portmanteau test for spatial ARMA models. 95 Biomtrcs 51 1536-1542

Kendall, William L.; Pollock, Kenneth H.; — —: A likelihood-based approach to capture-recapture estimation of demographic parameters under the robust design. 95 Biomtrcs 51 293-308

Brownson, Ross C.: Thacker, Stephen B.; Stroup, Donna F.; Rothenberg, Richard B.; — —: Public health surveillance for chronic conditions: A scientific basis for decisions. 95 StatMed 14 629-641 (Disc: p695-696)

Broze, Laurence; Gouriéroux, Christian; Szafarz, Ariane: Solutions of multivariate rational expectations models. 95 EconTher 11 229-257

Bruce, D. C.: Bruno, W. J.; Knill, E.; Balding, D. J.; — —; Doggett, N. A.; Sawhill, W. W.; Stallings, R. L.; Whittaker, C. C.; Torney, D. C.: Efficient pooling designs for library screening. 95 Genomics 26 21-30

Bruce, Peter: — — (Auth); Reeves, Jane (Rev): Review of "Resampling stats (Version 3.14)". 95 TeachgSt 17/3 101-103

Brüederl, Josef; Diekmann, Andreas: The log-logistic rate model: Two generalizations with an application to demographic data. 95 SocMethR 24 158-186

Brugger, Richard M.: Vardeman, Stephen B. (Auth); — — (Rev): Review of "Statistics for engineering problem solving". 95 Technmcs 37 348-348

Brunelle, Rocco: Huster, William; — —; Enas, Greg: Essential efficacy data analysis for a 2×2 crossover study. 94 ASAProBiop 503-509

Brunelli, Lina; Cicchitelli, Guiseppe (Ed); Holmes, Peter (Rev): Review of "Proceedings of the First Scientific Meeting of the International Association for Statistical Education". 95 AmerStat 49 276-276

Brunelli, R.; Messelodi, S.: Robust estimation of correlation with applications to computer vision. 95 PattrnRc 28 833-841

Brunham, Robert C.: Nagelkerke, Nico J. D.; Moses, Stephen; Plummer, Francis A.; — —; Fish, David: Logistic regression in case-control studies: The effect of using independent as dependent variables. 95 StatMed 14 769-775

Brunier, Hazel: Whitehead, John; — —: Bayesian decision procedures for dose determining experiments. 95 StatMed 14 885-893 (Disc: p897-899)

Brunner, Edgar; Puri, Madan L.; Sun, Shan: Nonparametric methods for stratified two-sample designs with application to multiclinic trials. 95 JASA 90 1004-1014

Brunner, Lawrence J.: Using the Gibbs sampler to simulate from the Bayes estimate of a decreasing density. 95 CommStA 24 215-226

— —: Bayesian linear regression with error terms that have symmetric unimodal densities. 95 JNnparSt 4 335-348

Bruno, John; Coffman, Edward G., Jr; Downey, Peter: Scheduling independent tasks to minimize the makespan on identical machines. 95 PrEngInf 9 447-456

Bruno, W. J.; Knill, E.; Balding, D. J.; Bruce, D. C.; Doggett, N. A.; Sawhill, W. W.; Stallings, R. L.; Whittaker, C. C.; Torney, D. C.: Efficient pooling designs for library screening. 95 Genomics 26 21-30

Bryan, Mark F.: Atwood, Corwin L.; — —: Two-stage sampling for bounding a mean. 95 EnvrMtrc 6 73-83

Bryant, Edward C.: Marker, David A.; — —; Moriarity, Chris: National Employer Health Insurance Survey sample design. 94 ASASrvyRsr 264-269

Bryant, John; Day, Roger: Incorporating toxicity considerations into the design of two-stage phase II clinical trials. 95 Biomtrcs 51 1372-1383

Shannon, William D.; — —; Logan, Theodore F.; Day, Roger: An application of decision theory to patient screening for an autologous tumour vaccine trial. 95 StatMed 14 2099-2110

Bryc, W.: The normal distribution - Characterizations and applications. 95 Springer:Brln:NY 139p

— —: The normal distribution: Characterizations with applications. 95 Springer:Brln:NY 140p

— — (Auth); Leslie, J. R. (Rev): Review of "The normal distribution - Characterizations and applications". 96 ShrtBkRv 16/1 4-4

Bryc, Włodzimierz; Dembo, Amir: On large deviations of empirical measures for stationary Gaussian processes. 95 StocProc 58 23-34

Brydges, D. C.; Slade, G.: The diffusive phase of a model of self-interacting walks. 95 ProbTher 103 285-315

Bryukhan, Fedor F.; Diab, Roseanne D.: Wind energy resource estimation of the upper atmosphere over Southern Africa. 95 JApMetrl 34 2565-2571

Brzeskwiniewicz, Henryk: On the PBB designs for multiresponse experiments. 95 BiomtrcJ 37 191-198

— —: On the E-optimality of some two-way elimination of heterogeneity designs. 95 StPrLet 25 9-13

Wagner, Wiesław; — —: Normality testing in a mixed model of a two-way nested classification. 95 BiomtrcJ 37 889-895

Brzezniak, Z.: Bellomo, N.; — —; de Socio, L. M. (Auth); Elworthy, K. D. (Rev): Review of "Nonlinear stochastic evolution problems in applied sciences". 95 Metrika 42 140-141

Bellomo, N.; — —; de Socio, L. M. (Auth); Renshaw, Eric (Rev): Review of "Nonlinear stochastic evolution problems in applied sciences". 94 BLndnMaS 26 409-410

Brzeźniak, Zdzisław; Flandoli, Franco: Almost sure approximation of Wong-Zakai type for stochastic partial differential equations. 95 StocProc 55 329-358

Bshouty, D.: On a characterization of variance functions of natural exponential families. 95 MaMethSt 4 92-98

Buchanan, Bruce S.: Sinha, Atanu R.; — —: Assessing the stability of principal components using regression. 95 Psymtrka 60 355-369

Buchinsky, Moshe: Quantile regression, Box-Cox transformation model, and the U.S. wage structure, 1963-1987. 95 JEconmtx 65 109-154

— —: Estimating the asymptotic covariance matrix for quantile regression models. A Monte Carlo study. 95 JEconmtx 68 303-338

Buck, James R.; Pignatiello, Joseph J., Jr; Raz, Tzvi: Bayesian Poisson rate estimation using parameter maps. 95 IIETrn 27 405-412

Buckland, Stephen T.: Comment on "Inference from a deterministic population dynamics model for Bowhead whales". 95 JASA 90 417-420

Buckle, D. J.: Bayesian inference for stable distributions. 95 JASA 90 605-613

Buckley, James J.; Hayashi, Yoichi: Neural nets for fuzzy systems. 95 FuzzySet 71 265-276

Buderer, Nancy Fenn; Peterson, Edward: Confidence intervals and hypothesis testing for screening parameters in the case of multiple observations per subject. 94 ASAProStEd 175-179

Budescu, David V.: Moxey, Linda M.; Sanford, Anthony J. (Auth); — —; Wallsten, Thomas S. (Rev): Review of "Communicating quantities: A psychological perspective". 95 Chance 8/2 38-40

Budowle, Bruce: Chakraborty, Ranajit; Zhong, Yixi; Jin, Li; — —: Nondetectability of restriction fragments and independence of DNA fragment sizes within and between loci in RFLP typing of DNA. 94 AJHuGene 54 391-401

Bugaighis, Mohamed M.: Exchange of censorship types and its impact on the estimation of parameters of a Weibull regression model. 95 IEEEReli 44 496-499

Bugner, Emmanuel: Feinberg, Max; — —; Theiller, Gédéon; Houba, Victor; Kadijk, Fetze: Expression of the reference value for proficiency tests. 95 JChemtrc 9 197-209

BuHamra, Sana; Anderson, Donald A.: The spectrum of the information matrix for s^n parallel flats fractions when s is prime. 95 JStPlInf 47 333-345

Bühlmann, Peter: The blockwise bootstrap for general empirical processes of stationary sequences. 95 StocProc 58 247-265

— —: Moving-average representation of autoregressive approximations. 95 StocProc 60 331-342

Buja, Andreas: Cook, Dianne; — —; Cabrera, Javier; Hurley, Catherine: Grand tour and projection pursuit. 95 JCmpGrSt 4 155-172

Donnell, Deborah J.; — —; Stuetzle, Werner: Analysis of additive dependencies and concurvities using smallest additive principal components. 94 AnlsStat 22 1635-1673

Hastie, Trevor; — —; Tibshirani, Robert: Penalized discriminant analysis. 95 AnlsStat 23 73-102

Bulfield, Grahame: Keightley, Peter D.; — —: Detection of quantitative trait loci from frequency changes of marker alleles under selection. 93 GeneRsr 62 195-203

Bullard, Peter R.: Peters, A. K. (Auth); — — (Rev): Review of "Understanding probability and statistics - A book of problems". 95 Ststcian 44 412-412

Buncher, Ralph: Pan, U. WinKey; Clark, Scott; — —; Johnson, Dale: Implementing a database management system for an H.U.D. lead-based paint abatement study. 94 ASAProEnvr 162-166

Bunck, Christine M.; Chen, Chiu-Lan; Pollock, Kenneth H.: Robustness of survival estimates from ratio-telemetry studies with uncertain relocation of individuals. 95 JWildlMg 59 790-794

Pollock, Kenneth H.; — —; Winterstein, Scott R.; Chen, Chiu-Lan: A capture-recapture survival analysis model for radio-tagged animals. 95 JAppStat 22 661-672

Bunge, J.; Fitzpatrick, M.; Handley, J.: Comparison of three estimators of the number of species. 95 JAppStat 22 45-59

Bunge, John: Browne, Sid; — —: Random record processes and state dependent thinning. 95 StocProc 55 131-142

Büning, H.: Weichselberger, A. (Auth); — — (Rev): Review of "A new nonparametric goodness-of-fit test for location alternatives" (German). 95 Metrika 42 145-146

Büning, Herbert: — — (Auth); Rieder, Helmut (Rev): Review of "Robust and adaptive tests" (German). 95 StHefte 36 272-272

Buonaccorsi, John P.: Prediction in the presence of measurement error: General discussion and an example predicting defoliation. 95 Biomtrcs 51 1562-1569

Jovanovic, Borko D.; Hosmer, David W.; — —: Equivalence of several methods for efficient best subsets selection in generalized linear models. 95 CmpStDA 20 59-64

Stanek, Edward J., III; — —: Comment on "The mixed-model ANOVA: The truth, the computer packages, the books. Part I: Balanced data" (V47 p48-59). 95 AmerStat 49 323-324

Buratti, Marco: Improving two theorems of Bose on difference families. 95 JComDsgn 3 15-24

Burch, Thomas K.: Estimating the Goodman, Keyfitz, Pullum kinship equations: An alternative procedure. 95 MaPopStd 5 161-170

Burdick, Donald S.: An introduction to tensor products with applications to multiway data analysis. 95 Chemolab 28 229-237

Burdick, Elisabeth: Colditz, Graham A.; — —; Mosteller, Frederick: Heterogeneity in meta-analysis of data from epidemiologic studies: A commentary. 95 AJEpidm 142 371-382

Burdick, Richard K.: Gui, Rongde; Graybill, Franklin A.; — —; Ting, Naitee: Confidence intervals on ratios of linear combinations for non-disjoint sets of expected mean squares. 95 JStPlInf 48 215-227

Yu, Qing-Ling; — —: Confidence intervals on variance components in regression models with balanced $(Q-1)$-fold nested error structure. 95 CommStA 24 1151-1167

Burdzy, Krzysztof; Toby, Ellen: A Skorohod-type lemma and a decomposition of reflected Brownian motion. 95 AnlsProb 23 586-604

Bass, Richard F.; — —: Conditioned Brownian motion in planar domains. 95 ProbTher 101 479-493

Burgess, Bonnie B.: Patchen, Robert H.; — —; Cralley, Marla D.: Prospects for measuring the child radio audience in the U.S. using the seven-day personal diary method. 94 ASASrvyRsr 1220-1225

Burgess, L.; Street, D. J.: Algorithm for constructing orthogonal main effect plans. 94 UtilMath 46 33-48

Burgess, Lloyd W.: Booksh, Karl; Henshaw, John M.; — —; Kowalski, Bruce R.: A second-order standard addition method with application to calibration of a kinetics-spectroscopic sensor for quantitation of trichloroethylene. 95 JChemtrc 9 263-282

Burillo, P.: Bustince, H.; — —: Correlation of interval-valued intuitionistic fuzzy sets. 95 FuzzySet 74 237-244

Buring, Julie E.; Lee, I-Min: Confounding in epidemiologic research. 95 AJPubHlh 85 164-165

Burke, John; Mohadjer, Leyla; Green, Jim; Waksberg, Joseph; Kirsch, Irwin S.; Kolstad, Andrew: Composite estimation in national and state surveys. 94 ASASrvyRsr 873-878

Burke, Murray D.; Yuen, Kam C.: Goodness-of-fit tests for the Cox model via bootstrap method. 95 JStPlInf 47 237-256

Burke, V.: Cox, K.; Puddey, I. B.; — —; McAleer, M.; Beilin, L. J.; Morton, A. R.: Effects of exercise and calorie restriction on ambulatory blood pressure in overweight, sedentary males: A time series analysis. 95 EnvrMtrc 6 523-528

Burkhardt, Harold E.: Applications of statistical ecology in forestry - Past accomplishments and future outlook. 94 ASAProEnvr 10-13

Burkhart, Harold E.: Schabenberger, Oliver; Gregoire, Timothy G.; — —: Comment on "Multistate models for monitoring individual trees in permanent observation plots". 94 EnvEcoSt 1 192-196

Burl, Michael: Chien, Steve; Gratch, Jonathan; — —: On the efficient allocation of resources for hypothesis evaluation: A statistical approach. 95 IEEEPAMI 17 652-665

Burman, P.; Nolan, D.: A general Akaike-type criterion for model selection in robust regression. 95 Biomtrka 82 877-886

Burnam, Audrey: Adams, John; — —: Federal block grants: What do statisticians have to contribute?. 95 Chance 8/4 40-42

Burnett, R. T.: Duddek, C.; Le, N. D.; Zidek, J. V.; — —: Multivariate imputation in cross-sectional analysis of health effects associated with air pollution. 95 EnvEcoSt 2 191-212

Burnett, Richard T.; Ross, W. H.; Krewski, Daniel: Non-linear mixed regression models. 95 EnvrMtrc 6 85-99

Burnham, K. P.; Anderson, D. R.; White, G. C.: Selection among open population capture-recapture models when capture probabilities are heterogeneous. 95 JAppStat 22 611-624

Patil, G. P.; Rao, C. R. (Ed); — — (Rev): Review of "Handbook of statistics, Volume 12: Environmental statistics". 95 Biomtrcs 51 1584-1585

Burnham, Kenneth P.; White, Gary C.; Anderson, David R.: Model selection strategy in the analysis of capture-recapture data. 95 Biomtrcs 51 888-898

Anderson, David R.; White, Gary C.; — —: Some specialized risk assessment methodologies for vertebrate populations. 95 EnvEcoSt 2 91-115

Burns, Kevin C.: Smith, Dennis E.; — —: Estimating a percentile of a contaminant concentration distribution when all observations are less than a detection limit. 94 ASAProEnvr 139-141

Burns, Patrick J.: Fraley, Chris; — —: Large-scale estimation of variance and covariance components. 95 SIAMSSC 16 192-209

Burpee, Joseph E.: Policy micro analysis and simulation using APL. 94 ASAProGovt 178-182

Burr, Jeffrey A.; Massagli, Michael P.; Mutchler, Jan E.; Pienta, Amy: Labor force status and transitions among older African American and white men. 94 ASAProSoSt 129-134

Burrell, Quentin: Hoffmann-Jørgensen, J. (Auth); — — (Rev): Review of "Probability with a view toward statistics". 95 Ststcian 44 542-543

Burrows, William R.; Benjamin, Mario; Beauchamp, Stephen; Lord, Edward R.; McCollor, Douglas; Thomson, Bruce: CART decision-tree statistical analysis and prediction of summer season maximum surface ozone for the Vancouver, Montreal, and Atlantic regions of Canada. 95 JApMetrl 34 1848-1862

Burt, Jameson: Perry, Charles; — —; Iwig, William: Redrawing the 1993 Farm Cost and Returns Survey list sample to reduce overlap with three other 1993 surveys and the 1992 FCRS. 94 ASASrvyRsr 632-637

Burt, Oscar R.: Pooling experiments over time to quantify the dynamics of dryland winter wheat yield response. 95 AgronmyJ 87 156-164

Burton, Joe W.; Carver, Brett F.: Selection among S_1 families vs. selfed half-sib or full-sib families in autogamous crops. 93 CropSci 33 21-28

Burton, Kenneth M.: Desante, David F.; — —; Saracco, James F.; Walker, Brett L.: Productivity indices and survival rate estimates from MAPS, a continent-wide programme of constant-effort mist-netting in North America. 95 JAppStat 22 935-947

Burton, Robert M.; Choi, Kyeong; Meester, Ronald: Stationary straight-line representations of stationary random graphs. 95 IndgMatN 6 267-277

— —; Rösler, Uwe: An L_2 convergence theorem for random affine mappings. 95 JAppProb 32 183-192

— —; Steif, Jeffrey E.: Quite weak Bernoulli with exponential rate and percolation for random fields. 95 StocProc 58 35-55

Burtschy, Bernard: Bourdeau, Marc; — —: Syntactic identification of ARMA processes. 94 ASAProBuEc 169-171

Busenberg, Stavros: Blythe, Stephen; — —; Castillo-Chavez, Carlos: Affinity in paired event probability. 95 MaBisci 128 265-284

Bushman, Brad J.; Wang, Morgan C.: A procedure for combining sample correlation coefficients and vote counts to obtain an estimate and a confidence interval for the population correlation coefficient. 95 PsychBu 117 530-546

Business Forecast Systems Inc.: — — (Auth); Weller, Barry R. (Rev): Review of "Forecast pro for windows extended edition (XE) (Version 2)". 95 IntJFor 11 175-187

Busk, Patricia L.: Improving the quality of teaching statistics: Techniques from research on classroom learning and assessment. 94 ASAProQlPr 164-171

Buss, Arnold H.; Lawrence, Stephen R.: Mean-variance interactions in process improvement and capacity design. 95 IIETrn 27 339-347

Bustince, H.; Burillo, P.: Correlation of interval-valued intuitionistic fuzzy sets. 95 FuzzySet 74 237-244

Butani, Shail; Cahoon, Lawrence; Fay, Robert; Kostanich, Donna: Measurement of different design effects. 94 ASASrvyRsr 342-347 (Disc: p360-361)

Butani, Shail J.: Phipps, Polly A.; — —; Chun, Young I.: Research on establishment-survey questionnaire design. 95 JBES 13 337-346

Butler, Ronald W.: Booth, James G.; — —; Huzurbazar, S.; Wood, Andrew T. A.: Saddlepoint approximations for P-values of some tests of covariance matrices. 95 JStCmpSm 53 165-180

Butov, A. A.: Martingale methods for random walks in a one-dimensional random environment. 94 ThProbAp 39 558-572

Butterworth, William: Coe, Paul R.; — —: Optimal stopping in 'The Showcase Showdown'. 95 AmerStat 49 271-275

Buydens, L.: de Weijer, A. P.; — —; Kateman, G.; Heuvel, H. M.: Spectral curve fitting of infrared spectra obtained from semi-crystalline polyester yarns. 95 Chemolab 28 149-164

Kateman, G.; — — (Auth); Bandeen-Roche, Karen J. (Rev): Review of "Quality control in analytic chemistry (2nd ed)". 95 Technmcs 37 115-116

Buydens, L. M. C.: Derks, E. P. P. A.; Pastor, M. S. Sánchez; — —: Robustness analysis of radial base function and multi-layered feed-forward neural network models. 95 Chemolab 28 49-60

Buydens, Lutgarde M. C.; Schoenmakers, Peter J. (Ed); Pretsch, E. (Rev): Review of "Intelligent software for chemical analysis". 95 Chemolab 27 148-149

Buzas, J. S.: Stefanski, L. A.; — —: Instrumental variable estimation in binary regression measurement error variables. 95 JASA 90 541-550

Byar, David P.; Freedman, Laurence S.; Herzberg, Agnes M.: Identifying which sets of parameters are simultaneously estimable in an incomplete factorial design. 95 Ststcian 44 451-456

Byers, J. D.; Peel, D. A.: Forecasting industrial production using non-linear methods. 95 JForcast 14 325-336

Byrne, Barbara M.: Strategies in testing for an invariant second-order factor structure: A comparison of EQS and LISREL. 95 StrEqMod 2 53-72

— —: Structural equation modeling with EQS and EQS/Windows. Basic concepts, applications, and programming. 94 SagePubl:CA:Lnd 288p

— — (Auth); Crowley, Susan L. (Rev): Review of "Structural equation modeling with EQS and EQS/Windows". 95 EdPsMeas 55 135-140

— — (Auth); Hancock, Gregory R. (Rev): Review of "Structural equation modeling with EQS and EQS/Windows: Basic concepts, applications, and programming". 95 StrEqMod 2 73-78

Caballero, María Emilia; Fernández, Begoña; Nualart, David: Smoothness of distributions for solutions of anticipating stochastic differential equations. 95 Stochatx 52 303-322

Cabanettes, A.: Goulard, M.; Pagés, L.; — —: Marked point process: Using correlation functions to explore a spatial data set. 95 BiomtrcJ 37 837-853

Cabilio, Paul: Alvo, Mayer; — —: Rank correlation methods for missing data. 95 CandJSt 23 345-358

Alvo, Mayer; — —: Approximate and exact distributions of rank tests for balanced incomplete block designs. 95 CommStA 24 3073-3121

Alvo, Mayer; — —: Testing ordered alternatives in the presence of incomplete data. 95 JASA 90 1015-1024

Cabo, A. J.; Baddeley, A. J.: Line transects, covariance functions and set convergence. 95 AdvAppPr 27 585-605

Cabrera, Javier: Cook, Dianne; Buja, Andreas; — —; Hurley, Catherine: Grand tour and projection pursuit. 95 JCmpGrSt 4 155-172

Cacciari, Mario: Montanari, Gian Carlo; Cavallini, Andrea; Tommasini, Laura; — —; Contin, Alfredo: Comparison of random generators for Monte Carlo estimates of Weibull parameters. 95 Metron 53 55-77

Cacoullos, T.; Papathanasiou, V.: A generalization of covariance identity and related characterizations. 95 MaMethSt 4 106-113

Cadell, Diane: DiGaetano, Ralph; Ward, John; — —; Ryaboy, Lana: Sample design of the National Pregnancy and Health Survey. 94 ASASrvyRsr 1100-1104

Cadenas, J. M.; Verdegay, J. L.: PROBO: An interactive system in fuzzy linear programming. 95 FuzzySet 76 319-332

Cadigan, N. G.: Local influence in structural equation models. 95 StrEqMod 2 13-30

Cadigan, Noel G.: Myers, Ransom A.; — —: Statistical analysis of catch-at-age data with correlated errors. 95 JFishRBC 52 1265-1273

Cadima, Jorge; Jolliffe, Ian T.: Loadings and correlations in the interpretation of principal components. 95 JAppStat 22 203-214

Cadish, Marc: Macher, Christian L.; — —; Clerc, Jean-Thomas; Pretsch, Ernö: Hypermedia - A new concept for information management. 95 Chemolab 28 213-228

Cahoon, Lawrence: Butani, Shail; — —; Fay, Robert; Kostanich, Donna: Measurement of different design effects. 94 ASASrvyRsr 342-347 (Disc: p360-361)

Cai, Jianwen; Prentice, Ross L.: Estimating equations for hazard ratio parameters based on correlated failure time data. 95 Biomtrka 82 151-164

Cain, Michael: Reply to comment on "The moment-generating function of the minimum of bivariate normal random variables". 95 AmerStat 49 240-240

— —; Janssen, Christian: Real estate price prediction under asymmetric loss. 95 AnInStMa 47 401-414

Cairns, Andrew J. G.: The present value of a series of cashflows: Convergence in a random environment. 95 AstinB 25 81-94

Cairoli, R.; Dalang, Robert C.: Sequential stochastic optimization. 95 Wiley:NY:UK 350p

Calabria, R.; Pulcini, G.: Bayes prediction of number of failures in Weibull samples. 95 CommStA 24 487-499

Caldwell, Carol V.; Monsell, Brian C.; Piesto, Nancy A.; Shimberg, Michael Z.: Using seasonal adjustment models to correct for panel imbalance in the Monthly Wholesale Trade Survey. 94 ASAProBuEc 83-88

Caliński, Tadeusz: On the randomization theory of experiments in nested block designs. 94 ListyBio 31 45-77

Callahan, Alexander J.: Bailey, Michael; Bowden, John; — —: Managing ship performance of naval gunfire support using statistical process control. 94 MilOpRes 1 5-11

Callanan, Terrance P.; Altland, Hank W.: Providing effective training industry - An experience with robust design. 94 ASAProStEd 126-131

Calvin, James A.; Dykstra, Richard L.: REML estimation of covariance matrices with restricted parameter spaces. 95 JASA 90 321-329

Calvo, E.; Tijs, S. H.; Valenciano, F.; Zarzuelo, J. M.: On the axiomatization of the τ-value. 95 Top 3 35-46

Camacho, Fernando; Arron, Geoffrey: Effects of the regulators paclobutrazol and flurprimidol on the growth of terminal sprouts formed on trimmed silver maple trees. 95 CandJSt 23 312-322 (Disc: p329-332)

Camarinopoulos, L.: Becker, G.; — —; Zioutas, G.: An inhomogeneous state graph model and application for a phased mission and tolerable downtime problem. 95 RelEngSS 49 51-57

Cambanis, S.; Fakhre-Zakeri, I.: On prediction of heavy-tailed autoregressive sequences: Forward versus reversed time. 94 ThProbAp 39 217-233

Cambien, F.: Tiret, L.; Nicaud, V.; Ehnholm, C.; Havekes, L.; Menzel, H. J.; Ducimetiere, P.; — —: Inference of the strength of genotype-disease association from studies comparing offspring with and without parental history of disease. 93 AnHuGene 57 141-149

Cameron, Joseph M.; Rosenblatt, Joan R.: Obituary: Churchill Eisenhart, 1913-1994. 95 AmerStat 49 243-244

Camilli, Gregory: The relationship between Fisher's exact test and Pearson's chi-square test: A Bayesian perspective. 95 Psymtrka 60 305-312

Campana, Steven E.; Annand, M. Christina; McMillan, James I.: Graphical and statistical methods for determining the consistency of age determinations. 95 TrAFish 124 131-138

Campanelli, Pamela; Channell, Joanna; McAulay, Liz; Renouf, Antoinette; Thomas, Roger: A cognitive and linguistic exploration of the meaning of training: Implications for survey design. 94 ASASrvyRsr 1281-1286

Campbell, Alec D.: Schulman, Daniel A.; — —; Kostello, Eric C.: Data analysis using S-Plus. 95 SocMethR 23 392-413

Campbell, Bryan; Ghysels, Eric: Federal budget projections: A nonparametric assessment of bias and efficiency. 95 RvEconSt 77 17-31

Campbell, J. M.: Comment on "Statistical values, quality, and certification" (V48 p65-87). 95 AmerStat 49 115-116

Campbell, L. L.; McKellips, A. L.; Wittke, P. H.: Distributions and expectations of singular random variables. 95 JAppProb 32 1032-1040

Camper, A. K.: Hamilton, M. A.; Johnson, K. R.; — —; Stoodley, P.; Harkin, G. J.; Gillis, R. J.; Shope, P. A.: Analysis of bacterial spatial patterns at the initial stage of biofilm formation. 95 BiomtrcJ 37 393–408

Campisi, B.: Vojnovic, D.; — —; Mattei, A.; Favretto, L.: Experimental mixture design to ameliorate the sensory quality evaluation of extra virgin olive oils. 95 Chemolab 27 205–210

Campling, B. G.: Lam, Y. M.; Pym, J.; — —: Nonlinear models for assessing combinations of chemotherapeutic drugs with resistance modifiers. 94 ASAProBiop 394–398

Campodónico, Sylvia; Singpurwalla, Nozer D.: Inference and predictions from Poisson point processes incorporating expert knowledge. 95 JASA 90 220–226

Canavos, G.; Miller, D. (Auth); Ziegel, Eric R. (Rev): Review of "Introduction to modern business statistics". 95 Technmcs 37 239–240

Canavos, George C.; Miller, Don M. (Auth); Delaney, Nancy Jo (Rev): Review of "Modern business statistics". 95 AmerStat 49 397–397

Cannings, C.: Lewis, C. M.; — —: The number of loci needed for ELOD calculations. 92 AnHuGene 56 59–69

Cannon, Ann; Cressie, Noel: Temporal analogues to spatial K functions. 95 BiomtrcJ 37 351–373

Cano, Juan A.: Moreno, Elías; — —: Classes of bidimensional priors specified on a collection of sets: Bayesian robustness. 95 JStPlInf 46 325–334

Canova, Fabio; Hansen, Bruce E.: Are seasonal patterns constant over time? A test for seasonal stability. 95 JBES 13 237–252

Cans, C.; Lavergne, C.: Logistic regression versus a generalized additive model: A practical example (French). 95 RStApp 43/2 77–90

Cantet, R. J. C.: Lo, L. L.; Fernando, R. L.; — —; Grossman, M.: Theory for modelling means and covariances in a two-breed population with dominance inheritance. 95 ThApGene 90 49–62

Cantor, Alan B.: A test of the association of a time-dependent state variable to survival. 95 CmpBiomd 46 101–105

Cantor, David: Discussion of "What respondents learn from questionnaires: The survey interview and the logic of conversation". 95 IntStRvw 63 173–177

— —; Gaertner, Gregory; Keil, Linda: The validity of self-reported drug use among the incarcerated. 94 ASASrvyRsr 1088–1093

Cantor, Joel C.: Berk, Marc L.; Edwards, W. Sherman; — —: Validating access to care. 94 ASAProSoSt 240–244

Cantwell, Patrick: Lent, Janice; Miller, Stephen; — —: Composite weights for the Current Population Survey. 94 ASASrvyRsr 867–872

Cao, Guangxia: The definition of the niche by fuzzy set theory. 95 EcolgMdl 77 65–71

Cao, Jin: Lawless, Jerry; Hu, Joan; — —: Methods for the estimation of failure distributions and rates from automobile warranty data. 95 LiDaAnal 1 227–240

Cao, R.: García-Jurado, I.; González-Manteiga, W.; Prada-Sánchez, J. M.; Febrero-Bande, M.; — —: Predicting using Box-Jenkins, nonparametric, and bootstrap techniques. 95 Technmcs 37 303–310

Cao, Ricardo; Cuevas, Antonio; Fraiman, Ricardo: Minimum distance density-based estimation. 95 CmpStDA 20 611–631

Cao, Xi-Ren: Event coupling and performance sensitivity analysis of generalized semi-Markov processes. 95 AdvAppPr 27 741–769

— —: State aggregation and discrete-state Markov chains embedded in a class of point processes. 95 JAppProb 32 39–51

Capasso, Vincenzo; De Giosa, Marcello; Mininni, Rosamaria: Asymptotic properties of the maximum likelihood estimators of parameters of a spatial counting process modelling crystallization of polymers. 95 StocAnAp 13 279–294

Capdevila, Carles; Arcas, Antoni: Empirical significance test of the goodness-of-fit for some pyramidal clustering procedures. 95 Questo 19 123–130

— —; Arcas, Antoni: Indexed pyramids and pyramidal dissimilarities (Spanish). 95 Questo 19 131–151

Capéraà, Philippe; Rivest, Louis-Paul: On the variance of the trimmed mean. 95 StPrLet 22 79–85

Capilla, C.: Romero, R.; Ferrer, A.; — —; Zunica, L.; Balasch, S.; Serra, V.; Alcover, R.: Teaching statistics to engineers: An innovative pedagogical experience. 95 JStEd 3/1 33kb (jse/v3n1/romero)

Capiński, Marek; Cutland, Nigel J.: Foias and Hope statistical solutions for stochastic Navier-Stokes equations. 95 Stochatx 52 193–205

Capitaine, Mireille: Onsager-Machlup functional for some smooth norms on Wiener space. 95 ProbTher 102 189–201

Capizzi, Thomas P.; Zhang, Ji: Testing the hypothesis that matters for multiple primary endpoints. 94 ASAProBiop 460–465

Caraballo, Indhira: Deliz, José R.; — —: Injection molding optimization through DOE. 95 ASQCTekCon 963–969

Carangelo, Richard M.: Clearly illustrate multiple measurables with the web chart. 95 QulPr 28/1 136–136

Carberry, P. S.: Jones, P. N.; — —: Optimal use of data in crop model development and validation. 95 EnvrMtrc 6 491–496

Carbonez, A.; Györfi, L.; van der Meulen, E. C.: Partitioning-estimates of a regression function under random censoring. 95 StatDecn 13 21–37

Cardenas, E.: Iznaga, N.; Núñez, G.; Solozabal, J.; Morales, A.; Artaza, E.; Rubio, R.; — —: A personal computer-based system for parallel line analysis of bioassay data. 95 CmpBiomd 47 167–175

Cardenas, Valerie: Raz, Jonathan; — —; Fletcher, Daniel: Frequency domain estimation of covariate effects in multichannel brain evoked potential data. 95 Biomtrcs 51 448–460

Cardon, L. R.: Fulker, D. W.; Cherny, S. S.; — —: Multipoint interval mapping of quantitative trait loci, using sib pairs. 95 AJHuGene 56 1224–1233

Cardon, Lon R.; Fulker, David W.: The power of interval mapping of quantitative trait loci, using selected sib pairs. 94 AJHuGene 55 825–833

— —; Fulker, David W.: A model of developmental change in hierarchical phenotypes with application to specific cognitive abilities. 94 BehavGen 24 1–16

Fulker, David W.; — —: A sib-pair approach to interval mapping of quantitative trait loci. 94 AJHuGene 54 1092–1103

Cardoso, Domingos: Esteves, Jorge Sá; Craveirinha, José; — —: Computing Erlang-B function derivatives in the number of servers. 95 CommStSM 11 311–331

Carfagna, Elisabetta; Di Bacco, Mario; Luiselli, Donata: An example of the use of Bayesian analysis in human biology. 94 ASAProBays 252–255

Carlin, B. P.: Ghosh, M.; — —; Srivastava, M. S.: Probability matching priors for linear calibration. 95 Test 4 333–357

Carlin, Bradley P.; Chib, Siddhartha: Bayesian model choice via Markov chain Monte Carlo methods. 95 JRSS-B 57 473–484

Gelfand, Alan E.; — —: Comment on "Bayesian computation and stochastic systems". 95 StatSci 10 43–46

Gelfand, Alan E.; Sahu, Sujit K.; — —: Efficient parametrisations for normal linear mixed models. 95 Biomtrka 82 479–488

Carlin, J.: Gelman, A.; — —; Stern, H.; Rubin, D.: Bayesian data analysis. 95 ChapmnHall:Lnd 552p

Carlin, J. B.: Gelman, A.; — —; Stern, H. S.; Rubin, D. B. (Auth); Brown, P. J. (Rev): Review of "Bayesian data analysis". 96 ShrtBkRv 16/1 5–5

Carling, Kenneth: Starting values for the iterative maximum likelihood estimator in survival analysis. 95 JAppStat 22 531–535

— —; Jacobson, Tor: Modeling unemployment duration in a dependent competing risks framework: Identification and estimation. 95 LiDaAnal 1 111–122

Carlson, Barbara Lepidus; Cohen, Steven B.; Johnson, Ayah E.: Family unit constructs, dynamics, and analysis in the household component of the NMES. 94 ASAProSoSt 257–262

— —; Hall, John W.: Weighting sample data when multiple sample frames are used. 94 ASASrvyRsr 882–887

Cohen, Steven B.; — —: A comparison of household and medical provider reported expenditures in the 1987 NMES. 94 JOffStat 10 3–29

Carlson, James E.: Muraki, Eiji; — —: Full-information factor analysis for polytomous item responses. 95 ApPsychM 19 73–90

Carlson, Lynda T.: New directions for establishment surveys. 94 ASASrvyRsr 170–177

Carlson, Richard R.: Anderson, Dan; — —; Stukel, Janice: The clinical indicators: Measurements of healthcare delivery. 94 ASAProQlPr 55–59

Carlson, Robert G.: Wang, Jichuan; Fisher, James H.; Siegal, Harvey A.; Falck, Russel S.; — —: Influence of measurement errors on HIV risk behavior analysis: A case study examining condom use among drug users. 95 StrEqMod 2 319–334

Carmona, R. A.; Molchanov, S. A.: Stationary parabolic Anderson model and intermittency. 95 ProbTher 102 433–453

Carmona, Sara C.: An asymptotic problem for a reaction-diffusion equation with a fast diffusion component. 95 Stochatx 52 43–80

— —: Wave front propagation for a Cauchy problem with a fast component. 95 Stochatx 52 81–105

Carnal, H.; Fel'dman, G. M.: Phase retrieval for probability measures on abelian groups. 95 JTherPrb 8 717–725

Carniello, Michael L.: Mandel, John (Auth); — — (Rev): Review of "Analysis of two-way layouts". 95 Technmcs 37 461–461

Caroni, C.; Prescott, P.: On Rohlf's method for the detection of outliers in multivariate data. 95 JMultiAn 52 295–307

— —; Prescott, P.: Union-intersection testing for outliers in multivariate normal data. 95 JStCmpSm 51 185–196

Carothers, Andrew D.: Methods for maternal age-standardization of the incidence of congenital abnormalities. 95 StatMed 14 1797–1806

— —; Bickmore, Wendy A.: Models of DNA replication timing in interphase nuclei: An exercise in inferring process from state. 95 Biomtrcs 51 750–755

Carpenter, James: Kinnear, P. R.; Gray, C. D. (Auth); — — (Rev): Review of "SPSS for Windows made simple". 95 JAppStat 22 436–437

Kinnear, P. R.; Gray, C. D. (Auth); — — (Rev): Review of "SPSS/PC+ made simple". 95 JAppStat 22 436–437

Westfall, P. H.; Young, S. S. (Auth); — — (Rev): Review of "Resampling-based multiple testing". 95 JAppStat 22 434–435

Carpenter, Mark; Hebert, Jaimie L.: Estimating the minimum and maximum location parameters for two IG-exponential scale mixtures. 95 CommStA 24 1227-1233

— —; Pal, Nabendu: Estimation of the smaller and larger of two exponential location parameters with a common unknown scale parameter. 95 CalcutSt 45 61-71

Carpenter, S. R.: Stow, C. A.; — —; Cottingham, K. L.: Resource vs. ratio-dependent consumer-resource models: A Bayesian perspective. 95 Ecology 76 1986-1990

Carpi, Anthony: Opsomer, Jean-Didier; Agras, Jean; — —; Rodrigues, Geraldo: An application of locally weighted regression to airborne mercury deposition around an incinerator site. 95 EnvrMtrc 6 205-219

Carré, André: Viel, Jean-François; Pobel, Dominique; — —: Incidence of leukaemia in young people around the La Hague nuclear waste reprocessing plant: A sensitivity analysis. 95 StatMed 14 2459-2472

Carreras, F.: Amer, R.; — —: Cooperation indices and coalitional value. 95 Top 3 117-135

Carrière, J. F.: Comment on "A large sample test for one-parameter families of copulas" (V23 p1311-1317). 95 CommStA 24 2959-2962

Carriere, Jacques F.: Dependent decrement theory. 95 TrSActur 46 45-66 (Disc: p67-74)

Carriére, Jacques F.: Removing cancer when it is correlated with other causes of death. 95 BiomtrcJ 37 339-350

Carriere, K. C.: Keselman, H. J.; — —; Lix, Lisa M.: Robust and powerful nonorthogonal analyses. 95 Psymtrka 60 395-418

Carrizosa, Emilio; Plastria, Frank: The determination of a "least quantile of squares regression line" for all quantiles. 95 CmpStDA 20 467-479

Carroll, J. Douglas: Minimax length links of a dissimilarity matrix and minimum spanning trees. 95 Psymtrka 60 371-374

— —; Corter, James E.: A graph-theoretic method for organizing overlapping clusters into trees, multiple trees, or extended trees. 95 JClass 12 283-313

— —; Green, Paul E.: Psychometric methods in marketing research: Part 1, Conjoint analysis. 95 JMarRsr 32 385-391

— —; Winsberg, Suzanne: Fitting an extended INDSCAL model to three-way proximity data. 95 JClass 12 57-71

Carroll, R.: Landin, R.; Freedman, L.; — —: Correction to "Adjusting for time trends when estimating the relationship between dietary intake obtained from a food frequency questionnaire and true average intake" (V51 p169-181). 95 Biomtrcs 51 1196-1196

Carroll, R. J.; Knickerbocker, R. K.; Wang, C. Y.: Dimension reduction in a semiparametric regression model with errors in covariates. 95 AnlsStat 23 161-181

— —; Li, Ker-Chau: Binary regressors in dimension reduction models: A new look at treatment comparisons. 95 StSinica 5 667-688

— —; Wang, Suojin; Wang, C. Y.: Prospective analysis of logistic case-control studies. 95 JASA 90 157-169

Wang, C. Y.; — —: On robust logistic case-control studies with response-dependent weights. 95 JStPlInf 43 331-340

Carroll, Raymond J.; Ruppert, David; Stefanski, Leonard A.: Measurement error in nonlinear models. 95 ChapmnHall:Lnd 336p

Gutierrez, Roberto G.; — —; Wang, Naisyin; Lee, Gyung-Hee; Taylor, Brian H.: Analysis of tomato root initiation using a normal mixture distribution. 95 Biomtrcs 51 1461-1468

Kim, Mimi Y.; Pasternack, Bernard S.; — —; Koenig, Karen L.; Toniolo, Paolo G.: Estimating the reliability of an exposure variable in the presence of confounders. 95 StatMed 14 1437-1446

Landin, Richard; Freedman, Laurence S.; — —: Adjusting for time trends when estimating the relationship between dietary intake obtained from a food frequency questionnaire and true average intake. 95 Biomtrcs 51 169-181

— —; Ruppert, David; Stefanski, Leonard A. (Auth); Draper, N. R. (Rev): Review of "Measurement error in nonlinear models". 96 ShrtBkRv 16/1 5-5

Carroll, Steven S.; Day, Gerald N.; Cressie, Noel; Carroll, Thomas R.: Spatial modelling of snow water equivalent using airborne and ground-based snow data. 95 EnvrMtrc 6 127-139

— —; Monsees, Geoffrey A.: The effect of scaling input variables on the direction of steepest ascent. 94 ASAProQlPr 267-269

Carroll, Thomas R.: Carroll, Steven S.; Day, Gerald N.; Cressie, Noel; — —: Spatial modelling of snow water equivalent using airborne and ground-based snow data. 95 EnvrMtrc 6 127-139

Carter, H. Ballentine: Morrell, Christopher H.; Pearson, Jay D.; — —; Brant, Larry J.: Estimating unknown transition times using a piecewise nonlinear mixed-effects model in men with prostate cancer. 95 JASA 90 45-53

Carter, Lucette; Mougeot, Mathilde: Using computers to teach theoretical statistics. 95 TeachgSt 17/2 44-46

Carter, Nancy J.: Schwertman, Neil C.; — —: A more practical Scheffé-type multiple comparison procedure for commonly encountered numbers of comparisons. 95 JStCmpSm 53 181-196

Carter, Robert C.: Damos, Diane L.; — —: Information-processing slope scores can be reliable!. 95 HumanFac 37 424-429

Carter, W. Hans, Jr: Gennings, Chris; — —: Utilizing concentration-response data from individual components to detect statistically significant departures from additivity in chemical mixtures. 95 Biomtrcs 51 1264-1277

Cartwright, Nancy: Probabilities and experiments. 95 JEconmtx 67 47-59

Carvajal, S. S. R.: Gupta, B. C.; — —: Search designs of resolution 3.1(2) for 2^m factorials. 93 JCISS 18 247-254

Carvalho, José L.: Nerlove, Marc; Grether, David M.; — —: Analysis of economic time series. A synthesis (Revised edition). 95 Academic:NY:Lnd 468p

Carver, Brett F.: Burton, Joe W.; — —: Selection among S_1 families vs. selfed half-sib or full-sib families in autogamous crops. 93 CropSci 33 21-28

Carverhill, Andrew: A simplified exposition of the Health, Jarrow and Morton model. 95 Stochatx 53 227-240

Casady, Robert J.; Dorfman, Alan; Wang, Suojin: Confidence intervals for sub-domain means and totals. 94 ASASrvyRsr 1036-1041

Casale, Salvatore: Andronico, Maurizio; — —; La Corte, Aurelio: A method for the generation of correlated random processes. 95 JCmpInT 3 9-18

Casas, José Miguel; Guijarro, Marta: Asymptotic design unbiasedness and robustness of linear estimation under superpopulation models: A selection procedure (Spanish). 94 Questo 18 173-188

Casassas, E.: Tauler, R.; Izquierdo-Ridorsa, A.; Gargallo, R.; — —: Application of a new multivariate curve resolution procedure to the simultaneous analysis of several spectroscopic titrations of the copper(II)-polyinosinic acid system. 95 Chemolab 27 163-174

Cascini, Egidio; Cascini, Leonardo: Forecasting spring flow time series. 94 JItalStS 3 1-23

Cascini, Leonardo: Cascini, Egidio; — —: Forecasting spring flow time series. 94 JItalStS 3 1-23

Casella, G.: Altman, N. S.; — —: Nonparametric empirical Bayes growth curve analysis. 95 JASA 90 508-515

Searle, S. R.; — —; McCulloch, C. E. M. (Auth); Cavedon, G. (Rev): Review of "Variance components". 95 Metron 53 217-218

Casella, George; DiCiccio, Thomas J.; Wells, Martin T.: Comment on "Inference". 95 StatSci 10 179-185

Brown, Lawrence D.; — —; Hwang, J. T. Gene: Optimal confidence sets, bioequivalence, and the Limaçon of Pascal. 95 JASA 90 880-889

Goutis, Constantinos; — —: Frequentist post-data inference. 95 IntStRvw 63 325-344

Goutis, Constantinos; — —: Improved invariant set estimation for general-scale families. 95 JStPlInf 44 327-340

Casey, William E.: Three dimensional modeling of quality concepts. 95 ASQCTek-Con 1134-1141

Casgrain, Philippe: Legendre, Pierre; Lapointe, François-Joseph; — —: Modeling brain evolution from behavior: A permutational regression approach. 94 Evolutn 48 1487-1499

Cashman, Erin: Kojetin, Brian; Tucker, Clyde; — —: Response to a government survey as political participation: The relation of economic and political conditions to refusal rates in the Current Population Survey. 94 ASASrvyRsr 1236-1241

Casin, Ph.: Discriminant analysis for longitudinal data matrices (French). 95 RStApp 43/3 73-91

Caspar, Rachel: Biemer, Paul; — —: Continuous quality improvement for survey operations: Some general principles and applications. 94 JOffStat 10 307-326

Caspary, W.; Wichmann, K. (Auth); Trenkler, G. (Rev): Review of "Linear models. Algebraic foundation and statistical applications" (German). 95 StHefte 36 190-190

Casti, J. L.; Karlqvist, A. (Ed); Gani, J. (Rev): Review of "Cooperation and conflict in general evolutionary processes". 95 Biomtrcs 51 1191-1193

Casti, John L.: Bell curves and monkey languages. 95 Complxty 1 12-15

Castillo, A.: Hernandez, C. M.; Crossa, J.; — —: The area under the function: An index for selecting desirable genotypes. 93 ThApGene 87 409-415

Castillo, Enrique; Hadi, Ali S.: A method for estimating parameters and quantiles of distributions of continuous random variables. 95 CmpStDA 20 421-439

— —; Hadi, Ali S.: Modeling lifetime data with application to fatigue models. 95 JASA 90 1041-1054

— —; Hadi, Ali S.: Parameters and quantiles estimation for continuously distributed random variables. 94 ASAProStCp 284-289

Arnold, Barry C.; — —; Sarabia, José María: General conditional specification models. 95 CommStA 24 1-11

Castillo-Chavez, Carlos: Blythe, Stephen; Busenberg, Stavros; — —: Affinity in paired event probability. 95 MaBisci 128 265-284

Castrup, Howard: Analyzing uncertainty for risk management. 95 ASQCTekCon 120-135

Catalano, Paul J.; Ryan, Louise M.; Senchaudhuri, Pralay: Developmental toxicity modeling for risk assessment. 94 ASAProEnvr 38-45

Catchpole, E. A.: MATLAB: An environment for analyzing ring-recovery and recapture data. 95 JAppStat 22 801-816

— —; Freeman, S. N.; Morgan, B. J. T.: Modelling age variation in survival and reporting rates for recovery models. 95 JAppStat 22 597-609

Catkan, N. A.: Alpuim, M. T.; — —; Hüsler, J.: Extremes and clustering on nonstationary max-AR(1) sequences. 95 StocProc 56 171-184

Catlin, Sandra: Newton, Michael A.; Guttorp, Peter; — —; Assunção, Renato; Abkowitz, Janis L.: Stochastic modeling of early hematopoiesis. 95 JASA 90 1146-1155

Cattiaux, P.; Léonard, C.: Correction to "Minimization of the Kullback information of diffusion processes" (V30 p83-132). 95 AIHenrPB 31 705-707

Caudill, Samuel P.; Hill, Robert H.: Multiple hypothesis tests in multiple investigations. 95 StatMed 14 585-589 (Disc: p701-702)

Smith, S. Jay; — —; Steinberg, Karen K.; Thacker, Stephen B.: On combining dose-response data from epidemiological studies by meta-analysis. 95 StatMed 14 531-544 (Disc: p679-685)

Caudill, Steven B.: Dichotomous choice models and limited independent variables. 95 Ststcian 44 405-406

— —; Ford, Jon M.; Gropper, Daniel M.: Frontier estimation and firm-specific inefficiency measures in the presence of heteroscedasticity. 95 JBES 13 105-111

— —; Ford, Jon M.; Kaserman, David L.: Certificate-of-need regulation and the diffusion of innovations: A random coefficient model. 95 JApEcmtx 10 73-78

Caulcutt, R.: Achieving quality improvement: A practical guide. 95 ChapmnHall:-Lnd 248p

— — (Auth); Riedel, M. (Rev): Review of "Statistics in research and development (Second edition)". 95 MaOpfStS 26 180-180

Caulcutt, Roland: The rights and wrongs of control charts. 95 ApplStat 44 279-288 (Disc: p289-306)

Caulkins, Jonathan P.: Merz, Jon F.; — —: Propensity to abuse - Propensity to murder?. 95 Chance 8/2 14-14

Cavallini, Andrea: Montanari, Gian Carlo; — —; Tommasini, Laura; Cacciari, Mario; Contin, Alfredo: Comparison of random generators for Monte Carlo estimates of Weibull parameters. 95 Metron 53 55-77

Cavanagh, Christopher L.; Elliott, Graham; Stock, James H.: Inference in models with nearly integrated regressors. 95 EconTher 11 1131-1147

Cavedon, G.: Searle, S. R.; Casella, G.; McCulloch, C. E. M. (Auth); — — (Rev): Review of "Variance components". 95 Metron 53 217-218

Cazzola, Alberto: A profile of the female cycle length. 94 Statstca 54 453-479

Cebula, Richard J.; Belton, Willie J., Jr; Chou, Ray Y.; Barth, James R.: Deposit insurance and bank failures: A preliminary analysis. 94 ASAProBuEc 293-297

Ceccarelli, S.: Singh, M.; — —; Hamblin, J.: Estimation of heritability from varietal trials data. 93 ThApGene 86 437-441

Cecco, Kevin: Characteristics of responding and nonresponding households from the 1990 Decennial Census. 94 ASAProGovt 76-81

Ceci, C.; Mazliak, L.: Controlled trees. 94 RBrzlPrS 8 93-105

Cederlund, Ann E.: Confidentiality issues in videotaped data collection. 94 ASASrvyRsr 1248-1251

Čekanavičius, V.: On asymptotic expansions in the first uniform Kolmogorov theorem. 95 StPrLet 25 145-151

Celeux, Gilles; Govaert, Gérard: Gaussian parsimonious clustering models. 95 PattrnRc 28 781-793

Cellier, Dominique; Fourdrinier, Dominique: Shrinkage estimators under spherical symmetry for the general linear model. 95 JMultiAn 52 338-351

— —; Fourdrinier, Dominique: Shrinkage positive rule estimators for spherically symmetric distributions. 95 JMultiAn 53 194-209

Centeno, Maria de Lourdes: The effect of the retention limit on the risk reserve. 95 AstinB 25 67-74

Ceranka, B.; Krzyszkowska, J.: Reinforced block designs with two groups of treatments. 94 ListyBio 31 17-25

Cervetz, C.: — — (Auth); Ziegel, Eric R. (Rev): Review of "Developing new products with TQM". 95 Technmcs 37 241-241

Cha, Stephen S.: Jung, Sin-Ho; Wieand, Sam; — —: A statistic for comparing two correlated markers which are prognostic for time to an event. 95 StatMed 14 2217-2225

Chadha, Janice; Katz, Barry: Success in renal transplantation: A correlational study. 94 ASAProSoSt 251-256

Chadha, S. S.: Applebaugh, W.; — —: Optimal lengths of left-turn lanes. 94 ASAProStEd 102-103

Chadjiconstantinidis, S.; Moyassiadis, C.: The A-optimal saturated weighing design for $N = 17$ observations. 92 UtilMath 42 3-13

Chae, K. C.; Lee, C. S.: Probabilities and mechanics of a cuboctahedral die (Korean). 95 KorApS 8/2 179-185

Kim, T. S.; — —: Probabilities of hands in 1 person seven-card stud (Korean). 95 KorApS 8/2 163-177

Chai, Feng-Shun: Construction and optimality of nearly linear trend-free designs. 95 JStPlInf 48 113-129

Chai, John J.; Chen, Chung: Joint effects of errors of measurement and grouping on the estimated product-moment correlation. 95 CommStA 24 283-294

Chakak, Abderrahmane; Koehler, Kenneth J.: A strategy for constructing multivariate distributions. 95 CommStB 24 537-550

Chakraborti, S.: Kao, Nancy L.; — —: Small sample null distribution of two nonparametric tests for comparing treatments with a control in a two-way layout. 95 CommStB 24 149-164

Gibbons, J. D.; — — (Auth); Williams, Calvin L. (Rev): Review of "Nonparametric statistical inference (Third edition)". 95 JAppStat 22 193-194

Chakraborti, Subhabrata: Gibbons, Jean Dickinson; — — (Auth); Borovcnik, Manfred (Rev): Review of "Nonparametric statistical inference". 95 CmpStQ 10 100-103

Chakraborty, Ranajit; Zhong, Yixi: Statistical power of an exact test of Hardy-Weinberg proportions of genotypic data at a multiallelic locus. 94 HumHrdty 44 1-9

— —; Zhong, Yixi; Jin, Li; Budowle, Bruce: Nondetectability of restriction fragments and independence of DNA fragment sizes within and between loci in RFLP typing of DNA. 94 AJHuGene 54 391-401

Chakraborty, S.: Dayananda, P. W. A.; Billard, L.; — —: Estimation of rate parameter and its relationship with latent and infectious periods in plant disease epidemics. 95 Biomtrcs 51 284-292

Chakravarti, A.: Li, C. C.; Weeks, D. E.; — —: Similarity of DNA fingerprints due to chance and relatedness. 93 HumHrdty 43 45-52

Chakravarti, Aravinda: Li, Ching Chun; — —: DNA profile similarity in a subdivided population. 94 HumHrdty 44 100-109

Chakravarty, Alaka G.; Harkins, Ralph D.: Some approaches to the evaluation of laboratory data to identify potential adverse events. 94 ASAProBiop 73-76

Challenor, Peter: Fisher, N. I.; Lewis, T.; Embleton, B. J. J. (Auth); — — (Rev): Review of "Statistical analysis of spherical data". 95 Ststcian 44 286-287

Chambers, R. L.; Dorfman, A. H.: Robust sample survey inference via bootstrapping and bias correction: The case of the ratio estimator. 94 ASAProStCp 51-60

Chambers, Ray L.: Kokic, Philip N.; Moon, Lynelle; Gooday, Jane; — —: Estimating temporal farm income distributions using spatial smoothing techniques. 95 AstrlJSt 37 129-143

Chamblee, Ronald F.: Harris, Kenneth W.; — —; Winter, Francine D.: Evaluation of an automated industry and occupation coding system. 94 ASAProSoSt 174-177

Chambless, Lloyd E.: Lyles, Robert H.; — —: Effects of model misspecification in the estimation of variance components and intraclass correlation for paired data. 95 StatMed 14 1693-1706

Champ, Charles W.: Lowry, Cynthia A.; — —; Woodall, William H.: The performance of control charts for monitoring process variation. 95 CommStB 24 409-437

Kiemele, Mark J.; Schmidt, Stephen R. (Auth); — — (Rev): Review of "Basic statistics tools for continuous improvement (Third Edition)". 95 JQualTek 27 174-175

Kotz, Samuel; Johnson, Norman L. (Auth); — — (Rev): Review of "Process capability indices". 95 Technmcs 37 233-233

Champely, S.; Deleuze, Ch.: Co-inertia of two random variables in Hilbert space and B-spline approximations with a forestry application (French). 95 RStApp 43/1 91-107

Champely, Stéphane: Thioulouse, Jean; Chessel, Daniel; — —: Multivariate analysis of spatial patterns: A unified approach to local and global structures. 95 EnvEcoSt 2 1-14

Chan, Chorkin: Huo, Qiang; — —: Contextual vector quantization for speech recognition with discrete hidden Markov Model. 95 PattrnRc 28 513-517

Chan, Grace; Hall, Peter; Poskitt, D. S.: Periodogram-based estimators of fractal properties. 95 AnlsStat 23 1684-1711

Chan, Ivan Siu-Fung: Zelterman, Daniel; — —; Mielke, Paul W., Jr: Exact tests of significance in higher dimensional tables. 95 AmerStat 49 357-361

Chan, J.: Kung, H. C.; Liu, X. H.; — —: Socioeconomic status and suicide. 94 ASAProSoSt 235-239

Chan, K. S.; Ledolter, Johannes: Monte Carlo EM estimation for time series models involving counts. 95 JASA 90 242-252

Chan, Ki-Kan: Liu, Lon-Mu; — —; Montgomery, Alan L.; Muller, Mervin E.: A system-independent graphical user interface for statistical software. 95 CmpStDA 19 23-44

Chan, Kung Sik; Geyer, Charles J.: Comment on "Markov chains for exploring posterior distributions". 94 AnlsStat 22 1747-1758

Chan, L. K.; Mak, T. K.: A regression approach for discovering small variation around a target. 95 ApplStat 44 369-377

Chan, Lai K.; Li, Guo-Ying: Bivariate control charts for testing $\bar{\mathbf{X}}$ data patterns. 95 CandJSt 23 85-99

Chan, Michael: Levitan, Emanuel; — —; Herman, Gabor T.: Image-modeling Gibbs priors. 95 CVGIPGr 57 117-130

Chan, N. N.: On a downdating formula for regression. 95 JStPlInf 46 347-350

Chan, Nai N.; Li, Kim-Hung: A matrix identity and its applications to equivalent hypotheses in linear models. 95 CommStA 24 2769-2777

Chan, Ngai Hang; Terrin, Norma: Inference for unstable long-memory processes with applications to fractional unit root autoregressions. 95 AnlsStat 23 1662-1683

Chan, Onn; Naiman, Daniel Q.: Some simple U-statistic tests for uniformity on certain homogeneous spaces. 95 JMultiAn 54 210-226

Chan, Terence: Some asymptotic results related to the law of iterated logarithm for Brownian motion. 95 JTherPrb 8 643-667

Chan, Wai; Yung, Yiu-Fai; Bentler, Peter M.: A note on using an unbiased weight matrix in the ADF test statistic. 95 MultBehR 30 453-459

Dunn, G.; Everitt, B.; Pickles, A. (Auth); — — (Rev): Review of "Modelling covariances and latent variables using EQS". 95 SIAMRvw 37 462-464

Chan, Wai-sum: Outliers and financial time series modelling: A cautionary note. 95 MaCmpSim 39 425-430

— —: On large-sample tests concerning proportions. 95 TeachgSt 17/1 16-16

Chan, Wai-Sum: Understanding the effect of time series outliers on sample autocorrelations. 95 Test 4 179-186

Chan, Yiu-Man: Approximate confidence interval estimation of $P(Y < X)$. 94 ASAProPEng 146–150

Chand, Nanak; Alexander, Charles H.: Imputing income for an N-person consumer unit. 94 ASASrvyRsr 412–416

Chanda, Kamal C.: Large sample analysis of autoregressive moving-average models with errors in variables. 95 JTimSrAn 16 1–15

Barnard, Roger W.; — —: An application of the Schur-Cohn algorithm to time series analysis. 95 JTimSrAn 16 445–449

Chandon, Y.: Bergeret, F.; — —: Search for a sequence of optimal tests to control manufacturing costs (French). 95 RStApp 43/3 21–33

Chandra, Anjani: Guadagno, Mary Ann; Abma, Joyce C.; — —; London, Kathy A.; Peterson, Linda S.; Piccinino, Linda J.; Wilson, Jacqueline B.: Women's reports of sensitive health conditions, attitudes, and behaviors. 94 ASAProGovt 82–87

Chandra, S.: Efficiency of check-plot designs in unreplicated field trials. 94 ThApGene 88 618–620

Chandrasekaran, S.; English, J. R.; Disney, R. L.: Modeling and analysis of EWMA control schemes with variance-adjusted control limits. 95 IIETrn 27 282–290

Chang, Cheng-Tao: Nedelman, Jerry R.; Karara, Adel H.; — —; Gibiansky, Ekaterina; McDonald, Scott; Robinson, William T.: Inferring systemic exposure from a pharmacokinetic screen: Model-free and model-based approaches. 95 StatMed 14 955–968 (Disc: p969–970)

Chang, Ching-Hui: Pal, Nabendu; Sinha, Bikas K.; Chaudhuri, Gopal; — —: Estimation of a multivariate normal mean vector and local improvements. 95 MaOpfStS 26 1–17

Chang, Der-Shin; Lin, Guan-Chyun: Stochastic regression model with heteroscedastic disturbance. 95 AnInStMa 47 351–369

Chang, Derek K.: A game with four players. 95 StPrLet 23 111–115

Chang, Fu-Chuen: Huang, Mong-Na Lo; — —; Wong, Weng Kee: D-optimal designs for polynomial regression without an intercept. 95 StSinica 5 441–458

Chang, G.-H.: Mansouri, H.; — —: A comparative study of some rank tests for interaction. 95 CmpStDA 19 85–96

Chang, Hsiu-Fen: Huang, Mong-Na; — —: Marginally restricted constrained optimal designs. 95 SankhyaB 57 128–141

Chang, Hui S.: Hsing, Yu; — —: Household saving behavior in Taiwan: Tests of functional forms and cointegration. 94 ASAProBuEc 123–128

Chang, Hyejung: Haining, Robert P. (Auth); — — (Rev): Review of "Spatial data analysis in the social and environmental sciences". 95 Chance 8/1 42–43

Chang, I-Shou; Hsiung, Chao A.: Asymptotic consistency of the maximum likelihood estimate in positron emission tomography and applications. 94 AnlsStat 22 1871–1883

Chang, Michael: Ghosh, Dhiren; Kaufman, Steven F.; Smith, Wray; — —: Optimal periodicity of a survey: Sampling error, data deterioration, and cost. 94 ASASrvyRsr 1122–1127

Chang, Myron N.; Gould, A. Lawrence; Snapinn, Steven M.: P-values for group sequential testing. 95 Biomtrka 82 650–654

Chang, Shing I.: An integrated scheme for process monitoring and diagnosis. 95 ASQCTekCon 725–732

Chang, T. C.; Gan, F. F.: A cumulative sum control chart for monitoring process variance. 95 JQualTek 27 109–119

— —; Gan, F. F.: Correction to "A cumulative sum control chart for monitoring process variance" (V27 p109-119). 95 JQualTek 27 402–402

Chang, Ted; Ko, Daijin: M-estimates of rigid body motion on the sphere and in Euclidean space. 95 AnlsStat 23 1823–1847

Chang, Tiao J.; Teoh, Choo B.: Use of the kriging method for studying characteristics of ground water droughts. 95 WaterReB 31 1001–1007

Chang, Xiang Qian; Moore, Charles N.: The equivalence of two conditions for weight functions for martingales. 95 StPrLet 23 129–133

Chang, Xuejiang: Chen, Min; — —: The asymptotic properties of the estimation of the inverse correlation function (Chinese). 95 AMaAPknC 17 23–45

Chang, Yoosoon; Phillips, Peter C. B.: Time series regression with mixtures of integrated processes. 95 EconTher 11 1033–1094

Chang, Yuan-chin Ivan: Estimation in some binary regression models with prescribed accuracy. 95 JStPlInf 44 313–325

Chang, Yuan-Tsung: Two-stage James-Stein estimators of the mean based on prior knowledge. 95 CommStA 24 2211–2227

Changnon, Stanley A.; Gabriel, K. Ruben; Westcott, Nancy E.; Czys, Robert R.: Exploratory analysis of seeding effects on rainfall: Illinois 1989. 95 JApMetrl 34 1215–1224

Channell, Joanna: Campanelli, Pamela; — —; McAulay, Liz; Renouf, Antoinette; Thomas, Roger: A cognitive and linguistic exploration of the meaning of training: Implications for survey design. 94 ASASrvyRsr 1281–1286

Chao, Anne: Hwang, W.-D.; — —: Quantifying the effects of unequal catchabilities on Jolly-Seber estimators via sample coverage. 95 Biomtrcs 51 128–141

Chao, Li: Huang, Jianyi; — —: Assessment methods in international student classrooms. 94 ASAProQlPr 291–293

Chao, M. T.; Lo, S. H.: Some statistical mean value theorems related to the bootstrap. 95 StSinica 5 129–139

Chao, Min-Te: Chen, Kani; — —; Lo, Shaw-Hwa: On strong uniform consistency of the Lynden-Bell estimator for truncated data. 95 AnlsStat 23 440–449

Chao, Xiuli; Dai, Liyi: A monotonicity result for a single-server loss system. 95 JAppProb 32 1112–1117

— —; Pinedo, Michael: On queueing networks with signals and history-dependent routing. 95 PrEngInf 9 341–354

Chaouche, A.: Parent, E.; — —; Girard, Ph.: In support of Bayesian statistics for quality control by attribute. Part I: Simple sampling plans (French). 95 RStApp 43/4 5–18

Chapman, David W.: Optimum sample design for personal-visit establishment surveys. 94 ASASrvyRsr 704–707

Chapman, G. R.: Perturbation analysis for an orthogonal least-squares estimator. 95 CandJSt 23 101–108

Chapman, G. Robert; Chen, Gemai; Kim, Peter T.: Assessing geometric integrity through spherical regression techniques. 95 StSinica 5 173–191 (Disc: p191-220)

— —; Chen, Gemai; Kim, Peter T.: Reply to comments on "Assessing geometric integrity through spherical regression techniques". 95 StSinica 5 215–220

Chappell, David: Hopkins, Kenneth D.; — —: Quick power estimates for comparing proportions. 94 EdPsMeas 54 903–912

Chappell, Rick; Nondahl, David M.; Fowler, John F.: Modeling dose and local control in radiotherapy. 95 JASA 90 829–838

Orzack, Steven Hecht; — —: Sample-size determination for parent-offspring regression of a composite trait. 95 Biomtrcs 51 822–830

Wang, Yuedong; Wahba, Grace; — —; Gu, Chong: Simulation studies of smoothing parameter estimates and Bayesian confidence intervals in Bernoulli SS ANOVA models. 95 CommStB 24 1037–1059

Charlesworth, Brian: Evolution in age-structured populations (Second edition). 94 CambridgeUniv 319p

Charlier, E.; Melenberg, B.; van Soest, A. H. O.: A smoothed maximum score estimator for the binary choice panel data model with an application to labour force participation. 95 StNeerla 49 324–342

Charnes, John M.; Gitlow, Howard S.: Using control charts to corroborate bribery in jai alai. 95 AmerStat 49 386–389

Chasalow, Scott D.; Brand, Richard J.: [Algorithm AS 299] Generation of simplex lattice points. 95 ApplStat 44 534–545

Chase, Warren; Bown, Fred (Auth); Muthulakshmi, S. (Rev): Review of "General statistics". 95 JAppStat 22 322–323

Chatfield, C.: Hand, D. J.; Daly, F.; Lunn, A. D.; McConway, K.; Ostrowski, E. (Ed); — — (Rev): Review of "A handbook of small data sets". 95 JRSS-A 158 349–349

Chatfield, Chris: Positive or negative?. 95 IntJFor 11 501–502

— —: Model uncertainty, data mining and statistical inference. 95 JRSS-A 158 419–444 (Disc: p444-466)

Johnson, N. L.; Kotz, S.; Balakrishnan, N. (Auth); — — (Rev): Review of "Continuous univariate distributions, Volume 1 (Second edition)". 95 Ststcian 44 544–545

Chatfield, Christopher: Problem solving. A statistician's guide (Second edition). 95 ChapmnHall:Lnd 344p

Chattamvelli, R.: A note on the noncentral beta distribution function. 95 AmerStat 49 231–234

— —: On the doubly noncentral F distribution. 95 CmpStDA 20 481–489

— —; Jones, M. C.: Recurrence relations for noncentral density, distribution functions and inverse moments. 95 JStCmpSm 52 289–299

— —; Shanmugam, R.: Efficient computation of the noncentral χ^2 distribution. 95 CommStB 24 675–689

Chatterjee, K.; Mukerjee, R.: Search designs for searching several three-factor interactions in general factorials. 92 UtilMath 42 181–191

Mukerjee, R.; — —: An application of Hadamard matrices for the construction of main effect plus two plans for 2^m factorials. 94 UtilMath 45 213–218

Chatterjee, Kashinath; Mukerjee, Rahul: D-optimal saturated main effect plans for $2 \times s_2 \times s_3$ factorials. 93 JCISS 18 116–122

Chatterjee, Rabikar: Desarbo, Wayne S.; Ramaswamy, Venkatram; — —: Analyzing constant-sum multiple criterion data: A segment-level approach. 95 JMarRsr 32 222–232

Chatterjee, Samprit; Handcock, Mark S.; Simonoff, Jeffrey S.: A casebook for a first course in statistics and data analysis. 95 Wiley:NY:UK 314p

Chatterjee, Sangit; Hawkes, James: Apparent decline as a sign of improvement? or, Can less be more?. 95 TeachgSt 17/3 82–83

— —; Laudato, Matthew: Gender and performance in athletics. 94 ASAProSpts 36–41

Chattopadhyay, Asis Kumar: Cohen, A. Clifford (Auth); — — (Rev): Review of "Truncated and censored samples. Theory and applications". 95 CalcutSt 45 133–133

Chattopadhyay, Gopaldeb: Bandyopadhyay, Uttam; — —: Progressive censoring under inverse sampling for nonparametric two-sample problems. 95 SqtlAnly 14 1–28

Chattopadhyay, Saibal: Mukhopadhyay, Nitis; — —: Comparing exponential clinical trials by combining individual sequential experiments available from independent studies. 94 Metron 52/1 17–40

Chaturvedi, Ajit; Gupta, Rahul: Second-order approximations to two classes of sequential estimation procedures. 95 MaOpfStS 27 159–173

Singh, R. Karan; — —; Kumar, Somesh: Two stage and sequential procedures for estimating a linear function of k multinormal means. 94 JIndnStA 32 1–8

Chaturvedi, Anoop: A note on the Stein rule estimation in linear models with nonscalar error covariance matrix. 95 SankhyaB 57 158–165

— —; Hasegawa, Hikaru; Asthana, Sheel: Bayesian predictive analysis of the linear regression model with and Edgeworth series prior distribution. 95 CommStA 24 2469–2484

Chatwin, P. C.; Lewis, D. M.; Sullivan, Paul J.: Turbulent dispersion and the beta distribution. 95 EnvrMtrc 6 395–402

— —; Sullivan, P. J.: Some targets for turbulent diffusion research. 95 EnvrMtrc 6 685–687

Lewis, D. M.; — —: A new model PDF for contaminants dispersing in the atmosphere. 95 EnvrMtrc 6 583–593

Zimmerman, William B.; — —: Statistical fluctuations due to microscale mixing in a diffusion layer. 95 EnvrMtrc 6 665–675

Chaubey, Yogendra P.; Talukder, Enayet: A likelihood ratio type test for heteroscedasticity. 94 ASAProStCp 170–175

Chaudhary, Mohammad A.: Relative efficiency of two two-stage sample designs. 94 ASASrvyRsr 228–231

Chaudhuri, A.: Mukherjee, S. P.; — —; Basu, S. K. (Ed); Adke, S. R. (Rev): Review of "Essays on probability and statistics: Festschrift in honour of Professor Anil Kumar Bhattacharyya". 95 CalcutSt 45 131–132

Mukherjee, S. P.; — —; Basu, S. K. (Ed); Gupta, S. Das (Rev): Review of "Essays on probability and statistics: Festschrift in honour of Professor Anil Kumar Bhattacharyya". 95 SankhyaA 57 166–166

Chaudhuri, Arijit; Maiti, Tapabrata: Asymptotic design-cum-model approach for convex weighting of direct and indirect components of small domain predictors. 95 CommStA 24 2803–2813

Chaudhuri, Gopal: Pal, Nabendu; Sinha, Bikas K.; — —; Chang, Ching-Hui: Estimation of a multivariate normal mean vector and local improvements. 95 MaOpfStS 26 1–17

Chaudhuri, Probal; Dewanji, Anup: On a likelihood-based approach in nonparametric smoothing and cross-validation. 95 StPrLet 22 7–15

— —; Lo, Wen-Da; Loh, Wei-Yin; Yang, Ching-Ching: Generalized regression trees. 95 StSinica 5 641–666

— —; Mykland, Per A.: On efficient designing of nonlinear experiments. 95 StSinica 5 421–440

Chauveau, Didier: A stochastic EM algorithm for mixtures with censored data. 95 JStPlInf 46 1–25

Chauvin, Franck: Boissel, Jean-Pierre; Durieu, Isabelle; Girard, Pascal; Nony, Patrice; — —; Haugh, Margaret: Dose-ranging trials: Guidelines for data collection and standardized descriptions. 95 ConClnTr 16 319–330

Cheadle, Allen: Diehr, Paula; Martin, Donald; Koepsell, Thomas; — —; Wagner, Edward; Psaty, Bruce: Optimal survey design for community-intervention evaluations: Cohort, cross-sectional, or both?. 94 ASAProEpid 34–43

Diehr, Paula; Martin, Donald C.; Koepsell, Thomas; — —: Breaking the matches in a paired t-test for community interventions when the number of pairs is small. 95 StatMed 14 1491–1504

Koepsell, Thomas D.; Diehr, Paula H.; — —; Kristal, Alan: Symposium on community intervention trials. 95 AJEpidm 142 594–599

Chebaane, Mohamed; Salas, Jose D.; Boes, Duane C.: Product periodic autoregressive processes for modeling intermittent monthly streamflows. 95 WaterReR 31 1513–1518

Cheeseman, Peter: Comment on "A probabilistic and statistical view of fuzzy methods". 95 Technmcs 37 282–283

Chellappa, Rama: Image understanding and statistics: What each can do for the other?. 94 ASAProStGr 32–41

Chen, Anyue; Renshaw, Eric: Markov branching processes regulated by emigration and large immigration. 95 StocProc 57 339–359

Chen, Bee-Lian; Gallo, Paul P.: Boundary calculation issues in clinical trials with two-sided stopping criteria. 94 ASAProBiop 115–119

Chen, C. L.; Swallow, W. H.: Sensitivity analysis of variable-size group testing and its related continuous models. 95 BiomtrcJ 37 173–181

Chen, Cathy W. S.; Lee, Jack C.: Bayesian inference of threshold autoregressive models. 95 JTimSrAn 16 483–492

Chen, Chiu-Lan: Bunck, Christine M.; — —; Pollock, Kenneth H.: Robustness of survival estimates from ratio-telemetry studies with uncertain relocation of individuals. 95 JWildlMg 59 790–794

Pollock, Kenneth H.; Bunck, Christine M.; Winterstein, Scott R.; — —: A capture-recapture survival analysis model for radio-tagged animals. 95 JAppStat 22 661–672

Chen, Chung: Chai, John J.; — —: Joint effects of errors of measurement and grouping on the estimated product-moment correlation. 95 CommStA 24 283–294

Chen, E. J.; Gambino, J.; Laniel, N.; Lindeyer, J.: Design and estimation issues for income in the redesign of the Canadian Labour Force Survey. 94 ASASrvyRsr 919–924

Chen, Gemai: Generalized log-normal distributions with reliability application. 95 CmpStDA 19 309–319

— —: Geometric quality inspection. 94 ASAProPEng 50–55

— —; Balakrishnan, N.: A general purpose approximate goodness-of-fit test. 95 JQualTek 27 154–161

Chapman, G. Robert; — —; Kim, Peter T.: Assessing geometric integrity through spherical regression techniques. 95 StSinica 5 173–191 (Disc: p191-220)

Chapman, G. Robert; — —; Kim, Peter T.: Reply to comments on "Assessing geometric integrity through spherical regression techniques". 95 StSinica 5 215–220

Chen, Guijing; Zhao, Zhongbo; Ding, Yuanyao: Parametric and semiparametric estimates in general truncated families (Chinese). 95 JSysSMaS 14 228–241

Li, Ningning; — —: A kind of rank test about intersection point of two continuous populations (Chinese). 95 KToChina 39 1543–1546

Chen, Hanfeng: Tests following transformations. 95 AnlsStat 23 1587–1593

— —; Su, John Q.: Simultaneous confidence bands for comparing diagnostic markers. 95 CommStA 24 375–393

Kouider, Elies; — —: Concavity of Box-Cox log-likelihood function. 95 StPrLet 25 171–175

Chen, Hong: Fluid approximations and stability of multiclass queueing networks: Work-conserving disciplines. 95 AnlsApPr 5 637–665

Chen, Hsiu-Hsi: Duffy, Stephen W.; — —; Tabar, Laszlo; Day, Nicholas E.: Estimation of mean sojourn time in breast cancer screening using a Markov chain model of both entry to and exit from the preclinical detectable phase. 95 StatMed 14 1531–1543

Chen, Huifen; Schmeiser, Bruce W.: Monte Carlo estimation for guaranteed-coverage non-normal tolerance intervals. 95 JStCmpSm 51 223–238

Chen, Hung: Asymptotically efficient estimation in semiparametric generalized linear models. 95 AnlsStat 23 1102–1129

Chen, James J.; Hwang, Jing-Shiang; Tsong, Yi: Estimation of the shelf-life of drugs with mixed effects models. 95 JBiophSt 5 131–140

Ahn, Hongshik; — —: Generation of over-dispersed and under-dispersed binomial variates. 95 JCmpGrSt 4 55–64

Ahn, Hongshik; — —: Modeling developmental toxic effects through recursive partitioning. 94 ASAProBiop 130–135

Bowman, Dale; — —; George, E. Olusegun: Estimating variance functions in developmental toxicity studies. 95 Biomtrcs 51 1523–1528

Chen, Jiading; Hickernell, Fred J.: A kind of asymptotically optimal sequential test under complex supposition. 95 SciSPkng 24 673–683

Lin, Huanbin; — —: The confidence limits for the reliability in regression model. 95 ActScNat 30 23–29

Chen, Jiahua: Optimal rate of convergence for finite mixture models. 95 AnlsStat 23 221–233

Chen, Jian-Shen; Jennrich, Robert I.: Diagnostics for linearization confidence intervals in nonlinear regression. 95 JASA 90 1068–1074

— —; Jennrich, Robert I.: Transformations for improving linearization confidence intervals in nonlinear regression. 95 JASA 90 1271–1276

— —; Jennrich, Robert I.: The profile Z plot and confidence intervals in maximum likelihood analysis. 94 ASAProStCp 124–129

Chen, Jianbao: Wang, Xuewen; Zhan, Jinlong; — —: All admissible linear estimators of regression coefficients and parameters in variance component models. 95 AMaPeknE 37 653–662

Chen, K.; Heyde, C. C.: On asymptotic optimality of estimating functions. 95 JStPlInf 48 101–112

Chen, Kani; Chao, Min-Te; Lo, Shaw-Hwa: On strong uniform consistency of the Lynden-Bell estimator for truncated data. 95 AnlsStat 23 440–449

Chen, L. H. Y.; Roos, M.: Compound Poisson approximation for unbounded functions on a group, with application to large deviations. 95 ProbTher 103 515–528

Chen, Ling: A one-sided trimmed mean estimator for the mean of positively skewed distributions. 95 AstrlJSt 37 313–322

— —: A minimum cost estimator for the mean of positively skewed distributions with applications to estimation of exposure to contaminated soils. 95 EnvrMtrc 6 181–193

— —: Testing the mean of skewed distributions. 95 JASA 90 767–772

— —; Mi, Jie: Estimation of non-occurrence probability in periodic model. 95 CommStA 24 1493–1505

— —; Shapiro, Samuel S.: An alternative test for normality based on normalized spacings. 95 JStCmpSm 53 269–287

Chen, Louis H. Y.: Barbour, A. D.; — —; Choi, K. P.: Poisson approximation for unbounded functions, I: Independent summands. 95 StSinica 5 749–766

Chen, M. H.: Wang, P. C.; — —; Wei, S. C.: To analyze binary data from mixed-level orthogonal arrays. 95 CmpStDA 20 689–695

Chen, Maoxue; Liu, Jinshan: On optimality of the covariance improved estimate: A simulation study (Chinese). 95 ChJApPSt 10 1–4

Chen, Mei-Ching: Li, Jufang; — —; Schreuder, Hans T.; Gregoire, Tim G.: Forestry applications of saddle-point approximations to construct confidence intervals for population means. 95 Biomtrcs 51 61–72

Chen, Min; Chang, Xuejiang: The asymptotic properties of the estimation of the inverse correlation function (Chinese). 95 AMaAPknC 17 23–45

Chen, Ming-Hui; Nandram, Balgobin: A Gibbs sampling approach for predicting shelf-life of a military ration with sensory data. 94 ASAProBays 84–89

— —; Schmeiser, Bruce W.: Random-direction interior-point Markov chains: A family of black-box samplers. 94 ASAProBays 1–6

Yang, Ruoyong; — —: Bayesian analysis for random coefficient regression models using noninformative priors. 95 JMultiAn 55 283–311

Chen, Ping: Zhang, Dixin; — —: Randomly weighted approximation and its application to empirical process (Chinese). 95 KToChina 39 2232–2234

Chen, Pinyuen: Mukhopadhyay, Nitis; Solanky, Tumulesh K. S. (Auth); — — (Rev): Review of "Multistage selection and ranking procedures: Second-order asymptotics". 95 JASA 90 1130–1131

Chen, Qiansheng: A new method of hierarchical cluster of sample-difference sequence method (Chinese). 95 KToChina 39 97–101

Chen, R.; Hawaleshka, O.; Strong, D. R.: Analytical approach for determining the cost optimum raw material quality level. 95 JAppStat 22 375–388

Chen, R. W.; Hu, H.: A note on the distribution of occurrences of a pattern in a string. 95 SochwJMa 21 471–473

Chen, Rong: Threshold variable selection in open-loop threshold autoregressive models. 95 JTimSrAn 16 461–481

— —; Liu, Jun S.; Tsay, Ruey S.: Additivity tests for nonlinear autoregression. 95 Biomtrka 82 369–383

Chen, Sy-Mien; Hsu, Nai-Feng: The asymptotic distribution of the process capability index C_{pmk}. 95 CommStA 24 1279–1291

Chen, T. H.; Hornberger, G. M.; Jakeman, A. J.; Swank, W. T.: The performance of different loss models in the simulation of streamflow. 95 EnvrMtrc 6 479–484

Chen, Toly; Wang, Mao-Jiun J.: Applying fuzzy set approach to signal detection theory. 95 FuzzySet 72 39–49

Chen, Tsing-Chang: Wikle, Christopher K.; Sherman, Peter J.; — —: Identifying periodic components in atmospheric data using a family of minimum variance spectral estimators. 95 JClimate 8 2352–2363

Chen, Tuhao; Seneta, E.: A note on bivariate Dawson-Sankoff-type bounds. 95 StPrLet 24 99–104

Chen, Xiru: Some results in consistency of L-S estimates (Chinese). 95 KToChina 39 1164–1167

Bai, Zhidong; — —; Krishnaiah, P. R.; Zhao, Lingcheng: Asymptotic properties of the EVLP estimation for superimposed exponential signals in noise (Chinese). 95 ChJApPSt 10 148–163

Gao, Jiti; — —; Zhao, Lingcheng: Asymptotic normality of a class of estimators in partial linear models. 95 AMaPeknE 37 256–268

Shi, Peide; — —: A note on the optimal global rates of convergence of M-estimator for nonparametric regression (Chinese). 95 ChinaAMA 15 171–176

Chen, Y.; Jackson, D. A.: Robust estimation of mean and variance in fisheries. 95 TrAFish 124 401–412

— —; Paloheimo, J. E.: A robust regression analysis of recruitment in fisheries. 95 JFishRBC 52 993–1006

Chen, Yan Qiu; Nixon, Mark S.; Thomas, David W.: Statistical geometrical features for texture classification. 95 PattrnRc 28 537–552

Chen, Yu Qing: Xiang, Qing; — —: On the size of the multiplier groups of cyclic difference sets. 95 JCombThA 69 168–169

Chen, Yu-Min: Takada, Hirokazu; Ullah, Aman; — —: Estimation of the seemingly unrelated regression model when the error covariance matrix is singular. 95 JAppStat 22 517–530

Chen, Yuan: Wang, Lingling; — —: Statistical analysis of life test and accelerated life test under type 1 censoring of a lognormal distribution (Chinese). 95 AMaAPknC 17 109–117

Chen, Yueh H.; Lin, Winston T.: Testing the arbitrage pricing theory and non-stationarity of the risk premium: An error components non-linear approach. 94 ASAProBuEc 178–183

Chen, Z. Q.; Williams, R. J.; Zhao, Z.: On the existence of positive solutions for semilinear elliptic equations with Neumann boundary conditions. 95 ProbTher 101 251–276

Chen, Zhang-Guang: Xie, Yu-Long; Liang, Yi-Zeng; — —; Huang, Zi-Hong; Yu, Ru-Qin: A nonlinearity tracking analysis algorithm for treatment of non-linearity in multivariate calibration. 95 Chemolab 27 21–32

Chen, Zhiqiang: Bounds for the breakdown point of the simplicial median. 95 JMultiAn 55 1–13

— —: Robustness of the half-space median. 95 JStPlInf 46 175–181

Cheng, B.: Heathcote, C. R.; Rachev, S. T.; — —: Testing multivariate symmetry. 95 JMultiAn 54 91–112

Cheng, Bing; Kay, Jim: A regularization method for one-dimensional edge detection and edge-preserving smoothing. 95 CmpStQ 10 53–69

Cheng, Cheng: The Bernstein polynomial estimator of a smooth quantile function. 95 StPrLet 24 321–330

Cheng, Ching-Shui: Complete class results for the moment matrices of designs over permutation-invariant sets. 95 AnlsStat 23 41–54

— —: Some projection properties of orthogonal arrays. 95 AnlsStat 23 1223–1233

— —: Optimal regression designs under random block-effects models. 95 StSinica 5 485–497

— —: Neighbor-balanced orthogonal arrays. 93 JCISS 18 61–67

— —; Li, Ker-Chau: A study of the method of principal Hessian direction for analysis of data from designed experiments. 95 StSinica 5 617–639

Cheng, Chuen-Sheng: A cumulative score control scheme for detecting process shifts. 95 CommStA 24 755–774

Cheng, K. F.: Chu, C. K.; — —: Nonparametric regression estimates using misclassified binary responses. 95 Biomtrka 82 315–325

Cheng, Mary: Roe, Denise J.; — —; Elsberry, Victor A.; Nolan, Paul E.: Estimating parameters in non-linear mixed effect models: A comparison using clinical data in children. 94 ASAProBiop 313–318

Cheng, P. E.: Chu, C. K.; — —: Nonparametric regression estimation with missing data. 95 JStPlInf 48 85–99

Cheng, Peter H.; Meng, Cliff Y. K.: A new formula for tail probabilities of Dunnett's T with unequal sample sizes. 95 CommStA 24 523–532

Cheng, Ph. E.; Bai, Zhidong: Optimal strong convergence rates in nonparametric regression. 95 MaMethSt 4 405–420

Cheng, Philip E.: A note on strong convergence rates in nonparametric regression. 95 StPrLet 24 357–364

Liou, Michelle; — —: Asymptotic standard error of equipercentile equating. 95 JEdStat 20 259–286

Liou, Michelle; — —: Equipercentile equating via data-imputation techniques. 95 Psymtrka 60 119–136

Cheng, Ping: Hong, Shengyan; — —: Convergence rates for parameter estimation in semi-parametric regression models (Chinese). 95 ChJApPSt 10 62–71

Zhu, Li-Xing; — —: The asymptotic tail behaviours of projection pursuit-type Kolmogorov statistics. 95 StPrLet 25 253–263

Cheng, R. C. H.; Traylor, L.: Non-regular maximum likelihood problems. 95 JRSS-B 57 3–24 (Disc: p24-44)

Cheng, Richard; Milligan, Glenn W.: Hierarchical clustering algorithms with influence detection. 95 EdPsMeas 55 237–244

— —; Milligan, Glenn W.: Mapping influence regions in hierarchical clustering. 95 MultBehR 30 547–576

Cheng, Rong: Liu, Jun S.; — —: Blind deconvolution via sequential imputations. 95 JASA 90 567–576

Cheng, S. C.; Wei, L. J.; Ying, Z.: Analysis of transformation models with censored data. 95 Biomtrka 82 835–845

Cheng, Shidong: The asymptotic distributions for sums of order statistics (I). 95 ActScNat 30 323–338

Cheng, Yiming: Optimum plans for step-stress accelerated life testing (Chinese). 95 ChJApPSt 10 52–61

Chergui, Jamal D.: Estimation and tests of the discrete probability law based on the empirical generating function (two dimensional case). 94 RACEMdrd 88 245–286

Cherian, Varghese I.: Glencross, Michael J.; — —: Attitudes toward applied statistics of postgraduate students in education in the Lebowa region of South Africa. 95 PsychRpt 77 315–322

Chernoff, Herman; Lander, Eric: Asymptotic distribution of the likelihood ratio test that a mixture of two binomials is a single binomial. 95 JStPlInf 43 19–40

Chernov, N. I.: Limit theorems and Markov approximations for chaotic dynamical systems. 95 ProbTher 101 321–362

Cherny, S. S.; DeFries, J. C.; Fulker, D. W.: Multiple regression analysis of twin data: A model-fitting approach. 92 BehavGen 22 489–498

Fulker, D. W.; — —; Cardon, L. R.: Multipoint interval mapping of quantitative trait loci, using sib pairs. 95 AJHuGene 56 1224–1233

Rodriguez, L. A.; Fulker, D. W.; — —: A maximum-likelihood model-fitting approach to conducting a Hayman analysis of diallel tables with complete or missing data. 93 BehavGen 23 69–76

Chesher, Andrew; Smith, Richard J.: Bartlett corrections to likelihood ratio tests. 95 Biomtrka 82 433–436

Chesney, Marc; Elliott, Robert J.: Estimating the instantaneous volatility and covariance of risky assets. 95 AplStMDA 11 51–58

Chessel, D.: Torre, F.; — —: The joint structure of two totally matched tables (French). 95 RStApp 43/1 109–121

Chessel, Daniel: Thioulouse, Jean; — —; Champely, Stéphane: Multivariate analysis of spatial patterns: A unified approach to local and global structures. 95 EnvEcoSt 2 1–14

Chester, Paul: Scott, Stuart; Stamas, George; Sullivan, Thomas J.; — —: Seasonal adjustment of hybrid economic time series. 94 ASAProBuEc 163–168

Chetwynd, A. G.: Diggle, P. J.; — —; Häggkvist, R.; Morris, S. E.: Second-order analysis of space-time clustering. 95 StMedRes 4 124–136

Cheung, Yin-Wong; Lai, Kon S.: Lag order and critical values of the augmented Dickey-Fuller test. 95 JBES 13 277–280

— —; Lai, Kon S.: Estimating finite sample critical values for unit root tests using pure random walk processes: A note. 95 JTimSrAn 16 493–498

Chhikara, R. S.: Goyal, R. C.; Singh, Randhir; — —: Estimation of crop yield using post-stratification based on satellite data. 94 JIAgSt 46 210–222

Chi, George Y. H.; Hung, James; Dubey, Satya D.; Lipicky, Raymond J.: Dose response studies and special populations. 94 ASAProBiop 88–93

Hung, H. M. James; — —; O'Neill, Robert T.: Efficacy evaluation for monotherapies in two-by-two factorial trials. 95 Biomtrcs 51 1483–1493

Nuri, Walid A.; — —: Using the GEE method to construct linear models for ordinal responses. 94 ASAProBiop 82–87

Chiang, Kok Leong: Chua, Tin Chiu; — —: An alternative estimator for randomized response sampling with continuous distributions from a dichotomous population. 95 CommStA 24 2919–2928

Chiano, M. N.; Yates, J. R. W.: Bootstrapping in human genetic linkage. 94 AnHuGene 58 129–143

Chiao, Chih-Hua: Tseng, Sheng-Tsaing; Hamada, Michael; — —: Using degradation data to improve fluorescent lamp reliability. 95 JQualTek 27 363-369

Chib, Siddhartha: Marginal likelihood from the Gibbs output. 95 JASA 90 1313-1321

— —; Greenberg, Edward: Understanding the Metropolis-Hastings algorithm. 95 AmerStat 49 327-335

— —; Greenberg, Edward: Hierarchical analysis of SUR models with extensions to correlated serial errors and time-varying parameter models. 95 JEconmtx 68 339-360

Albert, Jim; — —: Bayesian residual analysis for binary response regression models. 95 Biomtrka 82 747-759

Carlin, Bradley P.; — —: Bayesian model choice via Markov chain Monte Carlo methods. 95 JRSS-B 57 473-484

Chien, Steve; Gratch, Jonathan; Burl, Michael: On the efficient allocation of resources for hypothesis evaluation: A statistical approach. 95 IEEEPAMI 17 652-665

Chien, Wei-Ting Kary; Kuo, Way: Extensions of the Kaplan-Meier estimator. 95 CommStB 24 953-964

— —; Kuo, Way: Modeling and maximizing burn-in effectiveness. 95 IEEEReli 44 19-25

Chiera, C.: Zeephongsekul, P.; — —: Optimal software release policy based on a two-person game of timing. 95 JAppProb 32 470-481

Chikuse, Yasuko; Watson, Geoffrey S.: Large sample asymptotic theory of tests for uniformity on the Grassmann manifold. 95 JMultiAn 54 18-31

Chilvers, Clair E. D.: Controversies in cancer epidemiology. 95 StatMed 14 151-159

Chinchilli, V. M.: Gunsolley, J. C.; Getchell, C.; — —: Small sample characteristics of generalized estimating equations. 95 CommStB 24 869-878

Chinchilli, Vernon M.; Esinhart, James D.: The assessment of individual bioequivalence via tolerance intervals. 94 ASAProBiop 29-32

— —; Esinhart, James D.; Miller, W. Greg: Partial likelihood analysis of within-unit variances in repeated measurement experiments. 95 Biomtrcs 51 205-216

Have, Thomas R. Ten; — —: The analysis of bioequivalence with respect to TMAX under a 2×2 crossover design. 95 JBiophSt 5 185-199

Schwartz, Pamela F.; Gennings, Chris; — —: Threshold models for combination data from reproductive and developmental experiments. 95 JASA 90 862-870

Chinnappa, B. Nanjamma: Cox, Brenda G.; Binder, David A.; — —; Christianson, Anders; Colledge, Michael J.; Kott, Phillip S. (Ed): Business survey methods. 95 Wiley:NY:UK 732p

Chintagunta, Pradeep K.; Vilcassim, Naufel J.: A two-period repeated game advertising investment model for oligopolistic markets with an application to the beer industry. 95 DecisnSc 26 531-559

Chiou, Paul: A pooling procedure for the exponential distribution. 95 CommStA 24 445-458

— —; Han, Chien-Pai: Conditional interval estimation of the exponential location parameter following rejection of a pre-test. 95 CommStA 24 1481-1492

— —; Han, Chien-Pai: Interval estimation of error variance following a preliminary test in one-way random model. 95 CommStB 24 817-824

— —; Han, Chien-Pai: Conditional interval estimation of the exponential scale parameter following rejection of a preliminary test. 94 JStatRsr 28 65-72

Chiplonkar, S. A.: Prayag, V. R.; — —; Paranjpe, S. A.: Nested ANOVA for repeated surveys with autocorrelated errors. 95 BiomtrcJ 37 565-570

Chirinko, Robert S.: Nonconvexities, labor hoarding, technology shocks, and procyclical productivity. A structural econometric analysis. 95 JEconmtx 66 61-98

Chissom, Brad S.: Song, Qiang; Leland, Robert P.; — —: A new fuzzy time-series model of fuzzy number observations. 95 FuzzySet 73 341-348

Chistyakov, G. P.: The exact-order estimates of decompositional stability of Linnik's L class distributions. 94 ThProbAp 39 315-322

Chitkara, Sunita: Biswas, Suddhendu; — —: On the estimation of true treatment response when patients enter staggeredly in a clinical trial. 95 BiomtrcJ 37 571-580

Chiu, S. N.: Limit theorems for the time of completion of Johnson-Mehl tessellations. 95 AdvAppPr 27 889-910

Cowan, Richard; — —; Holst, Lars: A limit theorem for the replication time of a DNA molecule. 95 JAppProb 32 296-303

Cho, Dongchul: West, Kenneth D.; — —: The predictive ability of several models of exchange rate volatility. 95 JEconmtx 69 367-391

Cho, Gyo-Young: Song, Jae-Kee; — —: A note on percentile residual life. 95 SankhyaA 57 333-335

Cho, JoongJae; Han, JeongHye: Statistical estimation for generalized logit model of nominal type with bootstrap method. 95 JKoreaSt 24 1-18

Cho, Junghwan; Gemperline, Paul J.: Pattern recognition analysis of near-infrared spectra by robust distance method. 95 JChemtrc 9 169-178

Cho, Nhansook: Weak convergence of stochastic integrals driven by martingale measure. 95 StocProc 59 55-79

Cho, S.: Lee, Y. H.; — —; Kim, W. C.; Park, B. U.: On estimating integrated squared spectral density derivatives. 95 JStPlInf 48 165-184

Cho, Sinsup; Park, Young J.; Ahn, Sung K.: Unit root tests for seasonal models with deterministic trends. 95 StPrLet 25 27-35

Park, Young J.; — —: LaGrange multiplier test for multiple unit roots. 94 ASAProBuEc 147-150

Park, Young Jin; — —: Relations between the Lagrange multiplier tests and t-statistics for seasonal unit roots. 95 JKoreaSt 24 77-90

Choe, M. K.: Retherford, R. D.; — — (Auth); Fearn, Tom (Rev): Review of "Statistical models for causal analysis". 95 JRSS-A 158 200-201

Choi, Byoung-Chul; Lee, Eun-joo: Three-level skip-lot sampling plan split by two stages (Korean). 95 KorApS 8/2 55-64

Choi, ByoungSeon: Kim, Chul Eung; — —: A Sanov-type proof of the joint sufficiency of the sample mean and the sample variance. 95 JKoreaSt 24 563-568

Choi, Daiwoo: Criterion of discrete unimodal mixtures (Korean). 95 KorApS 8/1 159-167

Choi, Hyun Jip: Hong, Chong Sun; — —; Lee, Chong Cheol: Uniformly minimum variance unbiased estimation for distributions with support depending on two parameters. 95 JKoreaSt 24 45-64

Choi, I. S.: Bai, D. S.; — —: $\bar{X}$ and R control charts for skewed populations. 95 JQualTek 27 120-131

Choi, In; Ahn, Byung Chul: Testing for cointegration in a system of equations. 95 EconTher 11 952-983

Choi, In Su: Bai, Do Sun; — —; Hwang, Yong Keun: Field data analyses for products with multiple-modes of failure (Korean). 95 KorApS 8/1 89-104

Bai, Do Sun; Yun, Hyung Je; — —: Field data analyses for repairable products (Korean). 95 KorApS 8/2 133-145

Choi, Jai Won; Curtin, Lester R.: Sample surveys at National Center for Health Statistics. 94 ASASrvyRsr 1162-1167

Choi, K. P.: Barbour, A. D.; Chen, Louis H. Y.; — —: Poisson approximation for unbounded functions, I: Independent summands. 95 StSinica 5 749-766

Choi, Kyeong: Burton, Robert M.; — —; Meester, Ronald: Stationary straight-line representations of stationary random graphs. 95 IndgMatN 6 267-277

Choi, S. C.; Lu, I. L.: Effect of non-random missing data mechanisms in clinical trials. 95 StatMed 14 2675-2684

Choi, Sang-Il: Silverstein, Jack W.; — —: Analysis of the limiting spectral distribution of large dimensional random matrices. 95 JMultiAn 54 295-309

Choi, Seung Hoe: Kim, Hae Kyung; — —: Asymptotic properties of nonlinear least absolute deviation estimators. 95 JKoreaSt 24 127-139

Choi, Seungmook; Wohar, Mark E.: The expectations theory of interest rates: Cointegration and factor decomposition. 95 IntJFor 11 253-262

Choi, Sung C.; Germanson, Teresa P.; Barnes, Thomas Y.: A simple measure in defining optimal strata in clinical trials. 95 ConClnTr 16 164-171

Choi, Yong-Seok: Resistant h-plot for a sample variance-covariance matrix. 95 JKoreaSt 24 407-417

— —; Shim, Hee-Jeong: Applications of the supplementary principal component analysis for the 1982-1992 Korean pro baseball data (Korean). 95 KorApS 8/1 51-60

Choi, Young Hun: Asymptotically distribution-free procedure in a two-way layout. 95 JKoreaSt 24 375-387

— —: Simulation analysis of type I error and power for F test and rank transformed F test in 2×2 factorial ANOVA (Korean). 95 KorApS 8/2 87-97

Chojnowska-Michalik, A.; Goldys, B.: Existence, uniqueness and invariant measures for stochastic semilinear equations on Hilbert spaces. 95 ProbTher 102 331-356

Chong, Edwin K. P.: Wang, I-Jeng; — —; Kulkarni, Sanjeev R.: On equivalence of some noise conditions for stochastic approximation algorithms. 95 PIEEEDecCt 3849-3854

Chopra, D. V.; Dios, R.: On the number of constraints for balanced arrays of strength four. 93 JCISS 18 203-208

Chou, Ching-Sung: On some inequalities of local time. 95 JTherPrb 8 17-22

Chou, Fee Seng: Özekici, Süleyman; Li, Jingwen; — —: Waiting time in M/G/1 queues with impolite arrival disciplines. 95 PrEngInf 9 255-267

Chou, Jine-phone: Admissibility of conjugate Bayes estimators for the mean of a negative binomial distribution. 95 StatDecn 13 301-306

Chou, Ray Y.; Ng, Victor K.; Pi, Lynn K.: Cointegration of international stock market indices. 94 ASAProBuEc 196-201

Cebula, Richard J.; Belton, Willie J., Jr; — —; Barth, James R.: Deposit insurance and bank failures: A preliminary analysis. 94 ASAProBuEc 293-297

Choudhry, G. H.: A data collection methodology for the 2001 Canadian Census. 94 ASASrvyRsr 1134-1139

Choudhry, G. Hussain; Colledge, Michael J.; Mayda, Jacqueline E.: Redesign of the Hungarian Retail Trade and Catering Survey: Methodological issues. 94 ASAProBuEc 71-76

Choudhury, Askar H.; Power, Simon; St. Louis, Robert D.: A computationally attractive method for estimating the linear regression model with autoregressive-moving average disturbances. 94 Metron 52/1 41-51

Choudhury, Gagan L.; Leung, Kin K.; Whitt, Ward: An algorithm to compute blocking probabilities in multi-rate multi-class multi-resource loss models. 95 AdvAppPr 27 1104-1143

— —; Leung, Kin K.; Whitt, Ward: Calculating normalization constants of closed queuing networks by numerically inverting their generating functions. 95 JACM 42 935-970

Choudhury, Japobrata; Balch, Alfred H.: Assessing the robustness of clinical trial results. 94 ASAProBiop 484-486

Chow, Andrew: Minkin, Salomon; — —: Assessing sensitivity in suspension to cytosine arabinoside: Statistical analysis, associations with clinical outcome and experimental design. 95 StatMed 14 1081-1095 (Disc: p1097-1097)

Chow, Lien P.: Springer, Janet A.; Liu, Peng T.; Albert, Richard H.; — —: Construction of a spectrum lifetable. 94 ASAProSoSt 82-87

Chow, S.-C.; Liu, J.-P. (Auth); Day, S. (Rev): Review of "Statistical design and analysis in pharmaceutical science: Validation, process controls, and stability". 95 Biomtrcs 51 1190-1191

Chow, Shein-Chung: Pharmaceutical validation - Assay and process. 94 ASAProBiop 285-290

— —; Liu, Jen-Pei: Statistical design and analysis in pharmaceutical science: Validation, process controls, and stability. 95 Dekker:NY 584p

Ju, Huey L.; — —: On stability designs in drug shelf-life estimation. 95 JBiophSt 5 201-214

Ki, Fanny Y. C.; Liu, Jen-Pei; Wang, Wenping; — —: The impact of outlying subjects on decision of bioequivalence. 95 JBiophSt 5 71-94

Tse, Siu-Keung; — —: On model selection for standard curve in assay development. 95 JBiophSt 5 285-296

Wang, Wenping; — —; Wei, William W. S.: On likelihood distance for outliers detection. 95 JBiophSt 5 307-322

Wang, Wenping; Hsuan, Francis; — —: Patient compliance and the fluctuation of the serum drug concentration. 94 ASAProBiop 94-99

— —; Liu, Jen-Pei (Auth); Lagakos, S. W. (Rev): Review of "Statistical design and analysis in pharmaceutical science". 95 ShrtBkRv 15/3 42-42

Wang, Song-Gui; — — (Auth); Dean, Angela (Rev): Review of "Advanced linear models: Theory and applications". 95 Technmcs 37 119-120

Wang, Song-Gui; — — (Auth); Martin, Frank B. (Rev): Review of "Advanced linear models". 95 JASA 90 802-803

Chow, William: CUSUM charts-for-variance procedure and its ARL approximation. 94 ASAProQlPr 93-97

Chowdhury, S. Roy: Aggarwal, M. L.; — —: Linear graphs for 2^{k-p} fractional factorial under split plot design. 94 JStatRsr 28 73-87

Choy, S. T. B.: Yeh, Lam; — —: Bayesian variable sampling plans for the exponential distribution with uniformly distributed random censoring. 95 JStPlInf 47 277-293

Christ, Ralf; Steinebach, Josef: Estimating the adjustment coefficient in an ARMA(p,q) risk model. 95 InsrneME 17 149-161

Christen, J. A.: Reply to comment on "Summarizing a set of radiocarbon determinations: A robust approach". 95 ApplStat 44 408-408

Christensen, A.: Wang, C. S.; Gianola, D.; Sorensen, D. A.; Jensen, J.; — —; Rutledge, J. J.: Response to selection for litter size in Danish Landrace pigs: A Bayesian analysis. 94 ThApGene 88 220-230

Christensen, Morgens: Lundbye-Christensen, Søren; — —: Modelling the growth of human placental lactogen in maternal serum during late pregnancy. 95 RBrzlPrS 9 1-24

Christensen, Ronald: Comment on "Karl Pearson and R. A. Fisher on statistical tests: A 1935 exchange from Nature" (V48 p2-11). 95 AmerStat 49 400-400

Christiansen, Cindy: Morris, Carl; — —: Fitting Weibull duration models with random effects. 95 LiDaAnal 1 347-359

Christianson, Anders: Cox, Brenda G.; Binder, David A.; Chinnappa, B. Nanjamma; — —; Colledge, Michael J.; Kott, Phillip S. (Ed): Business survey methods. 95 Wiley:NY:UK 732p

Christie, Olav H. J.: Introduction to multivariate methodology, an alternative way?. 95 Chemolab 29 177-188

Christopher, Susan R.; Kupper, Lawrence L.: On the effects of predictor misclassification in multiple linear regression analysis. 95 CommStA 24 13-37

Christy, Alfred A.: Lindblom, Torbjörn; — —; Libnau, Fred O.: Quantitative determination of stabiliser in a single base propellant by chemometric analysis of Fourier transform infrared spectra. 95 Chemolab 29 243-254

Chromy, James R.: What is sample design efficiency? Let's let the customer decide. 94 ASAProQlPr 226-231

Chryssaphinou, Ourania: Barbour, Andrew D.; — —; Roos, Malgorzata: Compound Poisson approximation in reliability theory. 95 IEEEReli 44 398-402

Chu, C. K.: The effect of the smoothness of the regression function on a bandwidth selector in nonparametric regression. 95 CommStB 24 61-77

— —: Bandwidth selection in nonparametric regression with general errors. 95 JStPlInf 44 265-275

— —; Cheng, K. F.: Nonparametric regression estimates using misclassified binary responses. 95 Biomtrka 82 315-325

— —; Cheng, P. E.: Nonparametric regression estimation with missing data. 95 JStPlInf 48 85-99

Chu, Chia-Shang J.; Hornik, Kurt; Kuan, Chung-Ming: MOSUM tests for parameter constancy. 95 Biomtrka 82 603-617

Chu, Chia-Shang James: Detecting parameter shift in GARCH models. 95 EcomtrRv 14 241-266

— —; Hornik, Kurt; Kuan, Chung-Ming: The moving-estimates test for parameter stability. 95 EconTher 11 699-720

Chu, Chih-Kang: Godtliebsen, Fred; — —: Estimation of the number of true gray levels, their values, and relative frequencies in a noisy image. 95 JASA 90 890-899

Chu, David P. T.; Sutradhar, Brajendra C.: On Cochran's and Hartley's tests for homogeneity of variances when observations are autocorrelated. 95 CommStB 24 327-347

Chua, Tin Chiu; Chiang, Kok Leong: An alternative estimator for randomized response sampling with continuous distributions from a dichotomous population. 95 CommStA 24 2919-2928

Chuang-Stein, Christy; Tong, Donald M.: Multiple comparisons procedures for comparing several treatments with a control based on binary data. 95 StatMed 14 2509-2522

Chui, Charles K.: — — (Auth); Stettner, Haro (Rev): Review of "Multivariate splines". 95 CmpStQ 10 104-106

Chukova, S.: Dimitrov, B.; — —; Khalil, Z.: Definitions, characterizations and structural properties of probability distributions similar to the exponential. 95 JStPlInf 43 271-287

Chun, Y. R.: Bai, D. S.; — —; Kim, J. G.: Failure-censored accelerated life test sampling plans for Weibull distribution under expected test time constraint. 95 RelEngSS 50 61-68

Chun, Young I.; Devens, Richard; Mikkelson, Gordon; Yule, James: Designing a customer satisfaction survey. 94 ASAProQlPr 204-209

— —; Stone, Deborah: Using an establishment survey response model, and focus groups in the redesign of an establishment survey questionnaire. 94 ASASrvyRsr 1363-1367

Phipps, Polly A.; Butani, Shail J.; — —: Research on establishment-survey questionnaire design. 95 JBES 13 337-346

Chung, K. L.; Zhao, Z.: From Brownian motion to Schrödinger equation. 95 Springer:Brln:NY 287p

— —; Zhao, Z. (Auth); Nagasawa, M. (Rev): Review of "From Brownian motion to Schrödinger equation". 95 ShrtBkRv 15/3 49-49

Chung, Kai Lai: Green, brown, and probability. 95 WrldSciPub:Sing 120p

Chung, S. H.: Pulford, G. W.; Gallant, J. C.; Kennedy, R. A.; — —: Evaluation and estimation of various Markov models with applications to membrane channel kinetics. 95 BiomtrcJ 37 39-63

Chung, Younghae: Davis, Charles S.; — —: Randomization model methods for evaluating treatment efficacy in multicenter clinical trials. 95 Biomtrcs 51 1163-1174

Chung, Younshik; Kim, Chansoo; Dey, Dipak K.: Simultaneous estimation of Poisson means under weighted entropy loss. 94 CalcutSt 44 165-174

Chuprunov, A. N.: On refinement of Banach-valued limit theorems for stable laws. 94 ThProbAp 39 657-662

Church, J. D.: Cobb, E. Benton; — —: A discrete delta method for estimating standard errors for estimators in quantal bioassay. 95 BiomtrcJ 37 691-699

Churchill, G. A.; Doerge, R. W.: Empirical threshold values for quantitative trait mapping. 94 Genetcs 138 963-971

Churchill, Gary A.: Thorne, Jeffrey L.; — —: Estimation and reliability of molecular sequence alignments. 95 Biomtrcs 51 100-113

von Haeseler, Arndt; — —: Network models for sequence evolution. 93 JMolEvol 37 77-85

Chute, C. G.; Yang, Y.: An overview of statistical methods for the classification and retrieval of patient events. 95 MInfoMed 34 104-110

Ciapponi, Tamara M.: Ferguson, Gwyn R.; — —: A response model for the international price program. 94 ASASrvyRsr 1007-1012

Ciardo, Gianfranco; Leemis, Lawrence M.; Nicol, David: On the minimum of independent geometrically distributed random variables. 95 StPrLet 23 313-326

Cicchitelli, Guiseppe: Brunelli, Lina; — — (Ed); Holmes, Peter (Rev): Review of "Proceedings of the First Scientific Meeting of the International Association for Statistical Education". 95 AmerStat 49 276-276

Çinar, A.: Raich, A. C.; — —: Multivariate statistical methods for monitoring continuous processes: Assessment of discrimination power of disturbance models and diagnosis of multiple disturbances. 95 Chemolab 30 37-48

Çinar, Ali: Nonlinear time series models for multivariable dynamic processes. 95 Chemolab 30 147-158

Cioczek-Georges, Renata; Mandelbrot, Benoit B.; Samorodnitsky, Gennady; Taqqu, Murad S.: Stable fractal sums of pulses: The cylindrical case. 95 Bernoull 1 201-216

— —; Taqqu, Murad S.: Necessary conditions for the existence of conditional moments of stable random variables. 95 StocProc 56 233-246

— —; Taqqu, Murad S.: Form of the conditional variance for stable random variables. 95 StSinica 5 351-361

Ciozek-Georges, R.; Mandelbrot, B. B.: A class of micropulses and antipersistent fractional Brownian motion. 95 StocProc 60 1-18

Cipra, Tomas: Romera, Rosario; — —: On practical implementation of robust Kalman filtering. 95 CommStB 24 461-488

Cislaghi, Cesare; Biggeri, Annibale; Braga, Mario; Lagazio, Corrado; Marchi, Marco: Exploratory tools for disease mapping in geographical epidemiology. 95 StatMed 14 2363-2381

Ciszek, William: Hopke, Philip K.; Li, Chong Le; — —; Landsberger, Sheldon: The use of bootstrapping to estimate conditional probability fields for source locations of airborne pollutants. 95 Chemolab 30 69-79

Citro, Constance F.: Time for a change: A tale of two indicators. 95 Chance 8/3 50-51

— —; Cohen, Michael L.: Comment on "Survey of Income and Program Participation". 94 ASASrvyRsr 440-442

Claessens, Stijn: Kroner, Kenneth F.; Kneafsey, Kevin P.; — —: Forecasting volatility in commodity markets. 95 JForcast 14 77-95

Clancy, Damian: Ball, Frank; — —: The final outcome and temporal solution of a carrier-borne epidemic model. 95 JAppProb 32 304-315

Ball, Frank; — —: The final outcome of an epidemic model with several different types of infective in a large population. 95 JAppProb 32 579-590

Claramunt, M. Mercé: Alegre, Antoni; — —: Allocation of solvency cost in group annuities: Actuarial principles and cooperative game theory. 95 InsrneME 17 19-34

Clark, Colleen: Bench, Jill; — —; Dufour, Johane; Kaushal, Ritu: Implementing computer-assisted interviewing for the Canadian Labour Force Survey. 94 ASAProGovt 93-98

Clark, Cynthia Z. F.: Agricultural statistics on the environment. 94 ASAProEnvr 127-132

— —: Organizational needs and management issues for establishment survey programs. 94 ASASrvyRsr 178-182

Clark, Gordon M.: Rossetti, Manuel D.; — —: Estimating expected sojourn times for a Markov renewal process. 95 CommStA 24 553-579

Clark, Jon R.: Dillman, Don A.; — —; Sinclair, Michael D.: How prenotice letters, stamped return envelopes and reminder postcards affect mailback response rates for census questionnaires. 95 SurvyM 21 159-165

Dillman, Don A.; Singer, Eleanor; — —; Treat, James B.: Effects of benefits appeals, mandatory appeals and variations in statements of confidentiality on completion rates for census questionnaires. 94 ASASrvyRsr 1333-1338

Clark, Linda A.: Becker, Richard A.; — —; Lambert, Diane: Graphical analysis of data on networks. 94 ASAProStGr 21-24

Clark, S.: Watson, S. M.; Redfern, E.; — —; Tight, M.; Davies, N.: An influence method for outlier detection applied to time series traffic data. 95 JAppStat 22 135-149

Clark, Scott: Pan, U. WinKey; — —; Buncher, Ralph; Johnson, Dale: Implementing a database management system for an H.U.D. lead-based paint abatement study. 94 ASAProEnvr 162-166

Clarke, B.: Wasserman, L.; — —: Information tradeoff. 95 Test 4 19-38

Clarke, G. P. Y.: Hohls, T.; — —: Additive and non-additive genetic correlations. I. Properties of sampling distributions. 95 BiomtrcJ 37 919-935

Hohls, T.; — —; Shanahan, P. E.; Gevers, H. O.: Additive and non-additive genetic correlations. II. Application to data obtained from a 12×12 diallel cross. 95 BiomtrcJ 37 937-943

Hohls, T.; — —; Shanahan, P. E.; Gevers, H. O.: An unbiased estimator for identifying lines useful for the improvement of elite single crosses, based on a combining ability model. 95 Heredity 75 155-163

Clarke, Liz; Mole, Nils: Modelling the evolution of moments of contaminant concentration in turbulent flows. 95 EnvrMtrc 6 607-617

Clarke, M. R. B.: Kontoghiorghes, E. J.; — —: An alternative approach for the numerical solution of seemingly unrelated regression equations models. 95 CmpStDA 19 369-377

Clarke, Michael J.: Stewart, Lesley A.; — —: Practical methodology of meta-analyses (overviews) using updated individual patient data. 95 StatMed 14 2057-2079

Clarke, Stephen R.; Norman, John M.: Home ground advantage of individual clubs in English soccer. 95 Ststcian 44 509-521

Clarkson, Douglas B.; Donnell, Deborah; Minstrell, Jim; Hunt, Earl; Madigan, David; Traynor, Charles: VITAL: An intelligent tutoring system for statistics. 94 ASAProStEd 88-93

Clarkson, Kenneth L.: Las Vegas algorithms for linear and integer programming when the dimension is small. 95 JACM 42 488-499

Clarkson, S. P.; Williams, W. H.: How much does poor reading lower math scores?. 94 ASAProStEd 203-208

Clarkson, Thomas W.: Yakir, Benjamin; Cox, Christopher; Tanner, Martin A.; — —; Shamlaye, C.: Seasonal changes in methylmercury levels in human hair. 94 ASAProBays 35-39

Clarotti, C. A.: Comment on "Probabilistic interpretation of command and control signals: Bayesian updating of the probability of nuclear attack" (V47 p27-36). 95 RelEngSS 48 157-158

Clarotti, Carlo A.: Barlow, Richard E.; — —; Spizzichino, Fabio (Ed); Gruber, Marvin H. J. (Rev): Review of "Reliability and decision making". 95 Technmcs 37 346-347

Clayton, D.: Bernardinelli, L.; — —; Montomoli, C.: Bayesian estimates of disease maps: How important are priors?. 95 StatMed 14 2411-2431

Bernardinelli, L.; — —; Pascutto, C.; Montomoli, C.; Ghislandi, M.; Songini, M.: Bayesian analysis of space-time variation in disease risk. 95 StatMed 14 2433-2443

— —; Hills, M. (Auth); Dewey, M. (Rev): Review of "Statistical models in epidemiology". 95 JRSS-A 158 343-344

— —; Hills, M. (Auth); Ziegel, Eric R. (Rev): Review of "Statistical models in epidemiology". 95 Technmcs 37 240-240

Clayton, David; Hills, Michael (Auth); Greenland, Sander (Rev): Review of "Statistical models in epidemiology". 95 AJEpidm 141 182-183

— —; Hills, Michael (Auth); Hughes, A. O. (Rev): Review of "Statistical models in epidemiology". 95 StatMed 14 104-105

— —; Hills, Michael (Auth); Selvin, Steve (Rev): Review of "Statistical models in epidemiology". 95 JASA 90 798-799

Clayton, Howard R.: A comparison of bootstrap approaches for evaluating dollar unit samples in auditing. 94 ASAProStCp 205-208

Clayton, Murray K.; Hudelson, Brian D.: Confidence intervals for autocorrelations based on cyclic samples. 95 JASA 90 753-757

Ginebra, Josep; — —: Response surface bandits. 95 JRSS-B 57 771-784

Clayton, Richard L.: Rosen, Richard J.; — —; Pivetz, Lynda L.: Converting mail reporters to touchtone data entry. 94 ASASrvyRsr 763-768

Cleave, N.; Brown, P. J.; Payne, C. D.: Evaluation of methods for ecological inference. 95 JRSS-A 158 55-72

Clegg, D.: Eckschlager, K.; Danzer, K. (Auth); — — (Rev): Review of "Information theory in analytical chemistry". 95 Chemolab 28 207-208

Clemen, Robert: Abramson, Bruce; — —: Probability forecasting. 95 IntJFor 11 1-4

Clemen, Robert T.; Murphy, Allan H.; Winkler, Robert L.: Screening probability forecasts: Contrasts between choosing and combining. 95 IntJFor 11 133-145

DeWispelare, Aaron R.; Herren, L. Tandy; — —: The use of probability elicitation in the high-level nuclear waste regulation program. 95 IntJFor 11 5-24

Clements, Michael P.; Hendry, David F.: Forecasting in cointegrated systems. 95 JApEcmtx 10 127-146

— —; Hendry, David F.: Reply to comment on "On the limitations of comparing mean square forecast errors". 95 JForcast 14 73-75

Clements, Philip J.: Lachenbruch, Peter A.; — —; He, Weizhong: On encoding values for data recorded as $X > C$. 95 BiomtrcJ 37 855-867

Cleophas, Ton J. M.: Reply to comment on "Interaction biases in two-period crossover studies: A modified analysis to test with more sensitivity" (V35 p181-191). 95 BiomtrcJ 37 507-510

Clerc, Jean-Thomas: Macher, Christian L.; Cadish, Marc; — —; Pretsch, Ernö: Hypermedia - A new concept for information management. 95 Chemolab 28 213-228

Clerget-Darpoux, Françoise: Dizier, Marie-Hélène; Babron, Marie-Claude; — —: Interactive effect of two candidate genes in a disease: Extension of the marker-association-segregation χ^2 method. 94 AJHuGene 55 1042-1049

Cléroux, Robert; Lazraq, Aziz; Lepage, Yves: Vector correlation based on ranks and a nonparametric test of no association between vectors. 95 CommStA 24 713-733

Cleveland, W. S.: — — (Auth); Bowman, Adrian (Rev): Review of "Visualizing data". 95 JRSS-A 158 344-345

Cleveland, William S.: The elements of graphing data (Revised edition). 95 ChapmnHall:Lnd 304p

— — (Auth); Nelson, Lloyd S. (Rev): Review of "The elements of graphing data (Second edition)". 95 JQualTek 27 176-177

— — (Auth); Royston, Patrick (Rev): Review of "Visualizing data". 95 StatMed 14 1395-1396

— — (Auth); Weiss, Robert E. (Rev): Review of "The elements of graphing data (Revised edition)". 95 JASA 90 1488-1489

Cliff, A. D.: Analysing geographically-related disease data. 95 StMedRes 4 93-101

— —; Haggett, P.; Smallman-Raynor, M. R.; Stroup, D. F.; Williamson, G. D.: The application of multidimensional scaling methods to epidemiological data. 95 StMedRes 4 102-123

Clifford, A. J.: Song, K. S.; Müller, H. G.; — —; Furr, H. C.; Olson, J. A.: Estimating derivatives of pharmacokinetic response curves with varying bandwidths. 95 Biomtrcs 51 12-20

Clobert, J.: Capture-recapture and evolutionary ecology: A difficult wedding?. 95 JAppStat 22 989-1008

Legendre, S.; — —: ULM, a software for conservation and evolutionary biologists. 95 JAppStat 22 817-834

Clobert, Jean: Martin, Thomas E.; — —; Anderson, David R.: Return rates in studies of life history evolution: Are biases large?. 95 JAppStat 22 863-875

Clogg, C. C.: Arminger, G.; — —; Sobel, M. E. (Ed); Kemp, G. C. R. (Rev): Review of "Handbook of statistical modeling for the social and behavioral sciences". 95 Biomtrcs 51 1586-1587

Clogg, Clifford C.: Arminger, Gerhard; — —; Sobel, Michael E.: Handbook of statistical modeling for the social and behavioral sciences. 95 PlenumPubl:NY 618p

— —; Shihadeh, Edward S. (Auth); Andrich, David (Rev): Review of "Statistical models for ordinal variables". 95 JASA 90 1131-1131

— —; Shihadeh, Edward S. (Auth); Richards, John D. (Rev): Review of "Statistical models for ordinal variables". 95 JQualTek 27 394-394

— —; Shihadeh, Edward S. (Auth); Von Tress, Mark (Rev): Review of "Statistical methods for ordinal variables". 95 Technmcs 37 118-119

Arminger, Gerhard; — —; Sobel, Michael E. (Ed); Dudgeon, Paul (Rev): Review of "Handbook of statistical modeling for the social and behavioural sciences". 95 Chance 8/4 32-36

Clyde, Merlise A.; DeSimone, Heather; Parmigiani, Giovanni: A comparison of algorithms for sampling models. 94 ASAProBays 211-216

Coad, D. S.: Sequential allocation rules for multi-armed clinical trials. 95 JStCmpSm 52 239-251

Coakley, Clint W.; Heise, Mark A.: A new version of the sign test in the presence of ties. 94 ASAProEnvr 144-149

Cobb, E. Benton; Church, J. D.: A discrete delta method for estimating standard errors for estimators in quantal bioassay. 95 BiomtrcJ 37 691-699

Cobb, George W.: Moore, David S.; — —; Garfield, Joan; Meeker, William Q.: Statistics education fin de Siécle. 95 AmerStat 49 250-260

Cocchi, D.: Angeli, A.; — —; Pasquini, L.; Samoggia, A.: Family structure and income: A first analysis on cross-section data (Italian). 94 Statstca 54 435-452

Cochran, J. K.: Torng, J. C.-C.; — —; Montgomery, D. C.; Lawrence, F. P.: Correction to "Implementing statistically constrained economic EWMA control charts" (V27 p257-264). 95 JQualTek 27 402-402

Cochran, Jeffery K.: Montgomery, Douglas C.; Torng, James C.-C.; — —; Lawrence, Frederick P.: Statistically constrained economic design of the EWMA control chart. 95 JQualTek 27 250-256

Torng, James C.-C.; — —; Montgomery, Douglas C.; Lawrence, Frederick P.: Implementing statistically constrained economic EWMA control charts. 95 JQualTek 27 257-264

Cochran, Steven J.; Defina, Robert H.: Can purchasing power parity help forecast the dollar?. 95 JForcast 14 523-532

Cochran, W. George; Lee, Jung-Soon; Potthoff, Jürgen: Stochastic Volterra equations with singular kernels. 95 StocProc 56 337-349

Cockayne, E. J.; Hartnell, B. L.; Hedetniemi, S. T.; Laskar, R.: Perfect domination in graphs. 93 JCISS 18 136-148

Cockerham, C. Clark; Weir, B. S.: Estimation of gene flow from F-statistics. 93 Evolutn 47 855-863

Cocozza-Thivent, C.; Roussignol, M.: Coupling methods in reliability (French). 95 AIHenrPB 31 119-141

Coe, Paul R.; Butterworth, William: Optimal stopping in 'The Showcase Showdown'. 95 AmerStat 49 271-275

Coelho, Carlos: Brief note on the approximate distribution of the dispersion matrix of a weighted sample from a p-variate normal distribution. 94 JIndnStA 32 77-81

Coelho, Robby; Grunow, Bryan; Myers, Casey; Rasp, John: On the odds of an extended hitting streak. 94 ASAProSpts 46-49

Coenegracht, P. M. J.: Duineveld, C. A. A.; — —: Multicriteria steepest ascent in a design space consisting of both mixture and process variables. 95 Chemolab 30 23-36

Coffey, John: A linear birth-and-death predator-prey process. 95 JAppProb 32 274-277

Coffey, W. T.; Kalmykov, Yu. P.: The Langevin equation and its applications to stochastic problems in physics, chemistry and electrical engineering. 95 WrldSciPub:Sing 500p

Coffman, E. G., Jr; Gilbert, E. N.; Greenberg, A. G.; Leighton, F. T.; Robert, Philippe; Stolyar, A. L.: Queues served by a rotating ring. 95 CommStSM 11 371-394

— —; Poonen, Bjorn; Winkler, Peter: Packing random intervals. 95 ProbTher 102 105-121

— —; Puhalskii, A. A.; Reiman, M. I.: Polling systems with zero switchover times: A heavy-traffic averaging principle. 95 AnlsApPr 5 681-719

Coffman, Edward G., Jr: Bruno, John; — —; Downey, Peter: Scheduling independent tasks to minimize the makespan on identical machines. 95 PrEngInf 9 447-456

Cohen, A. Clifford: — — (Auth); Chattopadhyay, Asis Kumar (Rev): Review of "Truncated and censored samples. Theory and applications". 95 CalcutSt 45 133-133

Cohen, Arthur; Fygenson, Mendel: Testing homogeneity of uniform scale distributions against two-sided and one-sided alternatives. 95 JASA 90 1062-1067

— —; Kemperman, J. H. B.; Sackrowitz, H. B.: Unbiased testing in exponential family regression. 94 AnlsStat 22 1931-1946

— —; Sackrowitz, H. B.: Inadmissibility of some tests for order-restricted alternatives. 95 StPrLet 24 153-156

— —; Sackrowitz, H. B.; Samuel-Cahn, E.: Cone order association. 95 JMultiAn 55 320-330

— —; Sackrowitz, Harold B.: On stochastic ordering of random vectors. 95 JAppProb 32 960-965

— —; Sackrowitz, Harold B.; Samuel-Cahn, Ester: Constructing tests for normal order-restricted inference. 95 Bernoull 1 321-333

Brown, L. D.; — —: Complete classes for confidence set estimation. 95 StSinica 5 291-301

Cohen, Ayala; Barnett, Ofra: Assessing goodness-of-fit of parametric regression models for lifetime data - Graphical methods. 95 StatMed 14 1785-1795

— —; Makov, Udi; Boneh, Ruth: Computer intensive methods for inference on parameters of complex models - A Bayesian alternative. 95 JStPlInf 47 79-94

Rom, Meir; — —: Estimation in the polytomous logistic regression model. 95 JStPlInf 43 341-353

Rosen, Ori; — —: Constructing a bootstrap confidence interval for the unknown concentration in radioimmunoassay. 95 StatMed 14 935-952 (Disc: p953-953)

Cohen, Bernard L.: Comment on "Divergent biases in ecologic and individual level studies" (V11 p1209-1223). 95 StatMed 14 327-328

Cohen, Bob: Trippe, Carole; — —: Moving math from the mainframe to the PC: A better policy tool. 94 ASAProGovt 34-38 (Disc: p51-52)

Cohen, Michael: Dandekar, Ramesh; — —; Kirkendall, Nancy; Kydes, Andy: Applicability of localized linear regression techniques to develop response surface models. 94 ASAProPEng 87-91

Cohen, Michael L.: Comment on "Microsimulating federal program choices". 94 ASAProGovt 51-52

Citro, Constance F.; — —: Comment on "Survey of Income and Program Participation". 94 ASASrvyRsr 440-442

Cohen, Michael P.: Integrated sampling of education institutions. 94 ASASrvyRsr 638-640

Cohen, Richard: Bonham, Carl; — —: Comment on "Testing the rationality of price forecasts: New evidence from panel data" (V80 p714-735). 95 AmEcRv 85 284-289

Cohen, Steven B.; Carlson, Barbara Lepidus: A comparison of household and medical provider reported expenditures in the 1987 NMES. 94 JOffStat 10 3-29

— —; Potter, D. E. B.: An estimation strategy for the combined population represented by the NMES Household and Nursing Home Surveys. 94 ASASrvyRsr 276-281

Carlson, Barbara Lepidus; — —; Johnson, Ayah E.: Family unit constructs, dynamics, and analysis in the household component of the NMES. 94 ASAProSoSt 257-262

Cohen, Uri: Weissman, Ishay; — —: The extremal index and clustering of high values for derived stationary sequences. 95 JAppProb 32 972-981

Cohn, Elchanan; Cohn, Sharon: Mathematics background and achievement in principles of economics. 94 ASAProBuEc 271-276

Cohn, Sharon: Cohn, Elchanan; — —: Mathematics background and achievement in principles of economics. 94 ASAProBuEc 271-276

Colbourn, Charles J.; Cusack, Charles A.; Kreher, Donald L.: Partial Steiner triple systems with equal-sized holes. 95 JCombThA 70 56-65

Colditz, G. A.: Berkey, C. S.; Hoaglin, D. C.; Mosteller, F.; — —: A random-effects regression model for meta-analysis. 95 StatMed 14 395-411

Colditz, Graham A.; Burdick, Elisabeth; Mosteller, Frederick: Heterogeneity in meta-analysis of data from epidemiologic studies: A commentary. 95 AJEpidm 142 371-382

Cole, Bernard F.; Gelber, Richard D.; Anderson, Keaven M.; Goldhirsch, Aron: Cost-effectiveness analysis using quality-adjusted survival. 94 ASAProBiop 9-18

— —; Gelber, Richard D.; Goldhirsch, Aron: A quality-adjusted survival meta-analysis of adjuvant chemotherapy for premenopausal breast cancer. 95 StatMed 14 1771-1784

— —; Lee, Mei-Ling T.; Whitmore, G. Alex; Zaslavsky, Alan M.: An empirical Bayes model for Markov-dependent binary sequences with randomly missing observations. 95 JASA 90 1364-1372

Gelber, Richard D.; — —; Gelber, Shari; Goldhirsch, Aron: Comparing treatments using quality-adjusted survival: The Q-TWiST method. 95 AmerStat 49 161-169

Cole, Ronald: Appel, Martin V.; — —: Spoken language recognition of the Year 2000 Census Questionnaire Feasibility Test. 94 ASASrvyRsr 1310-1315

Coleman, Daniel A.: A fixed-width interval for $1/\beta$ in simple linear regression. 95 JStPlInf 44 291-312

Coleman, Joan L.; James, Penny L.: The reliability of the locality pay estimates. 94 ASASrvyRsr 1031-1035

Coleman, R.: Isaac, R. (Auth); — — (Rev): Review of "The pleasures of probability". 95 ShrtBkRv 15/3 42-42

Neuts, Marcel F. (Auth); — — (Rev): Review of "Algorithmic probability. A collection of problems". 96 ShrtBkRv 16/1 4-4

Coleman, Rodney: Galambos, J.; Gani, J. (Ed); — — (Rev): Review of "Studies in applied probability: Papers in honour of Lajos Takács". 95 JRSS-A 158 634-635

Stroock, D. W. (Auth); — — (Rev): Review of "Probability theory, and analytic view". 95 JRSS-A 158 356-357

Coleman, S. C.: Reichardt, Charles S.; — —: The criteria for convergent and discriminant validity in a multitrait-multimethod matrix. 95 MultBehR 30 513-538

Colleau, J. J.: Verrier, E.; — —; Foulley, J. L.: Long-term effects of selection based on the animal model BLUP in a finite population. 93 ThApGene 87 446-454

Colledge, Michael J.: Choudhry, G. Hussain; — —; Mayda, Jacqueline E.: Redesign of the Hungarian Retail Trade and Catering Survey: Methodological issues. 94 ASAProBuEc 71-76

Cox, Brenda G.; Binder, David A.; Chinnappa, B. Nanjamma; Christianson, Anders; — —; Kott, Phillip S. (Ed): Business survey methods. 95 Wiley:NY:UK 732p

Collet, Pierre; Martínez, Servet; San Martín, Jaime: Asymptotic laws for one-dimensional diffusions conditioned to nonabsorption. 95 AnlsProb 23 1300-1314

Cook, R. Dennis: Graphics for studying net effects of regression predictors. 95 StSinica 5 689-708

— —: Using dimension-reduction subspaces to identify important inputs in models of physical systems. 94 ASAProPEng 18-25

Wetzel, N.; — —: Exploring regression structure with graphics. 94 ASAProStGr 42-47

— —; Weisberg, Sandford (Auth); Cook, Dianne (Rev): Review of "An introduction to regression graphics". 95 JASA 90 1126-1127

— —; Weisberg, Sanford (Auth); Goldman, Aaron (Rev): Review of "An introduction to regression graphics". 95 Technmcs 37 343-344

Cook, Richard J.: Interim analyses in 2 × 2 crossover trials. 95 Biomtrcs 51 932-945

— —: The design and analysis of randomized trials with recurrent events. 95 StatMed 14 2081-2098

— —; Farewell, Vern T.: Conditional inference for subject-specific and marginal agreement: Two families of agreement measures. 95 CandJSt 23 333-344

Cooke, Fred: Pradel, Roger; Cooch, Evan; — —: Transient animals in a resident population of snow geese: Local emigration or heterogeneity?. 95 JAppStat 22 695-710

Viallefont, Anne; Cooch, Evan G.; — —: Estimation of trade-offs with capture-recapture models: A case study on the lesser snow goose. 95 JAppStat 22 847-861

Cooley, Mary E.: Short, Thomas H.; Moriarty, Helene; — —: Readability of educational materials for patients with cancer. 95 JStEd 3/2 15kb (jse/v3n2/datasets.short)

Coombs, William T.: Algina, James; Blair, R. Clifford; — —: A maximum test for scale: Type I error rates and power. 95 JEdStat 20 27-39

Coons, Lori; LaLonde, Steven: Robust design when variability in control/noise factors can't be controlled during experimentation. 94 ASAProPEng 65-70

Cooper, David B.: Subrahmonia, Jayashree; Keren, Daniel; — —: Recognition of objects in 2D images or 3D range data based on high degree implicit polynomial curves or surfaces, algebraic invariants, and asymptotic Bayesian methods. 94 ASAProPEng 44-49

Cooper, H.; Hedges, L. (Ed); Ziegel, Eric R. (Rev): Review of "The handbook of research synthesis". 95 Technmcs 37 242-242

— —; Hedges, L. V. (Ed); Howel, D. (Rev): Review of "The handbook of research synthesis". 95 Biomtrcs 51 1585-1586

Cooper, R. I.: Fleming, C.; Nellis, J. G. (Auth); — — (Rev): Review of "Principles of applied statistics". 95 JRSS-A 158 348-349

Cooper, W. W.: Banker, R. D.; — —: Validation and generalization of DEA and its uses. 94 Top 2 249-314

Copas, J. B.: Local likelihood based on kernel censoring. 95 JRSS-B 57 221-235

Cope, Ronald T.: Zeng, Lingjia; — —: Standard error of linear equating for the counterbalanced design. 95 JEdStat 20 337-348

Copeland, Jack: Henderson, William G.; Moritz, Thomas; Goldman, Steven; — —; Sethi, Gulshan: Use of cumulative meta-analysis in the design, monitoring, and final analysis of a clinical trial: A case study. 95 ConClnTr 16 331-341

Copeland, Karen A. F.: Evans, Merran; Hastings, Nicholas; Peacock, Brian (Auth); — — (Rev): Review of "Statistical distributions". 95 JQualTek 27 94-95

Hinkelmann, Klaus; Kempthorne, Oscar (Auth); — — (Rev): Review of "Design and analysis of experiments, Volume 1: Introduction to experimental design". 95 JQualTek 27 175-176

Latzko, William J.; Saunders, David M. (Auth); — — (Rev): Review of "Four days with Dr. Deming". 95 JQualTek 27 270-270

Mann, Prem S. (Auth); — — (Rev): Review of "Introductory statistics (Second edition)". 95 JQualTek 27 390-391

Panizzon, Gaetano (Auth); — — (Rev): Review of "Bilateral plans for inspection and process control by variables". 95 JQualTek 27 88-88

Stanford, John L.; Vardeman, Stephen B. (Ed); — — (Rev): Review of "Statistical methods for physical science". 95 JQualTek 27 180-181

Copeland, Kennon R.: Comment on "Current Population Survey". 94 ASASrvyRsr 360-361

Copete, J. L.: Senar, J. C.; — —: Mediterranean house sparrows (Passer domesticus) are not used to freezing temperatures: An analysis of survival rates. 95 JAppStat 22 1069-1074

Corcoran, Stephen: Barndorff-Nielsen, O. E.; Cox, D. R. (Auth); — — (Rev): Review of "Inference and asymptotics". 95 Ststcian 44 279-279

Lindsey, J. K. (Auth); — — (Rev): Review of "Models for repeated measurements". 95 JAppStat 22 435-435

Corcuera, J. M.: Oller, J. M.; — —: Intrinsic analysis of statistical estimation. 95 AnlsStat 23 1562-1581

Cordeiro, Gauss M.: Performance of a Bartlett-type modification for the deviance. 95 JStCmpSm 51 385-403

— —; Cribari-Neto, Francisco; Aubin, Elisete C. Q.; Ferrari, Silvia L. P.: Bartlett corrections for one-parameter exponential family models. 95 JStCmpSm 53 211-231

Paula, Gilberto A.; — —: Bias correction and improved residuals for non-exponential family nonlinear models. 95 CommStA 24 1193-1210

Edwards, A. W. F. (Auth); — — (Rev): Review of "Likelihood". 95 JAppStat 22 191-192

Coria, José C.: Pankratz, Peter C.; — —: Reporting negative contamination levels on silicon wafers - Is this nonsense?. 94 ASAProPEng 71-75

Corle, Donald K.: Green, Sylvan B.; — —; Gail, Mitchell H.; Mark, Steven D.; Pee, David; Freedman, Laurence S.; Graubard, Barry I.; Lynn, William R.: Interplay between design and analysis for behavioral intervention trials with community as the unit of randomization. 95 AJEpidm 142 587-593

Corley, Thomas: Wang, Chamont; — —: The longest chute is the worse chute: Comments and additions to "Mr. Markov plays chutes and ladders". 94 ASAProStEd 104-106

Corneal, Sherry E.: Hershberger, Scott L.; — —; Molenaar, Peter C. M.: Dynamic factor analysis: An application to emotional response patterns underlying daughter/father and stepdaughter/stepfather relationships. 95 StrEqMod 2 31-52

Cornelius, P. L.: Statistical tests and retention of terms in the additive main effects and multiplicative interaction model for cultivar trials. 93 CropSci 33 1186-1193

— —; Seyedsadr, M.; Crossa, J.: Using the shifted multiplicative model to search for "separability" in crop cultivar trials. 92 ThApGene 84 161-172

— —; Van Sanford, D. A.; Seyedsadr, M. S.: Clustering cultivars into groups without rank-change interactions. 93 CropSci 33 1193-1200

Van Sanford, D. A.; Pfeiffer, T. W.; — —: Selection index based on genetic correlations among environments. 93 CropSci 33 1244-1248

Cornelius, Paul L.: Crossa, José; — —; Sayre, Ken; Ortiz-Monasterio R., Jóse Iván: A shifted multiplicative model fusion method for grouping environments without cultivar rank change. 95 CropSci 35 54-62

Cornell, John A.: Fitting models to data from mixture experiments containing other factors. 95 JQualTek 27 13-33

— —: Wagon wheel designs with residual effects from neighboring treatments. 93 JCISS 18 255-272

— —; Randles, Ron H.; Vining, G. Geoff: Challenges in teaching short courses by an academic department: The University of Florida experience. 95 AmerStat 49 195-200

Cornia, Gary C.: Nelson, Ray D.; — —: Intrayear monitoring of state tax collections and revenue forecasts. 94 ASAProBuEc 339-344

Correa, Juan C.: A new estimator of entropy. 95 CommStA 24 2439-2449

Correa-Villaseñor, Adolfo; Stewart, Walter F.; Franco-Marina, Francisco; Seacat, Hui: Bias from nondifferential misclassification in case-control studies with three exposure levels. 95 Epidmlgy 6 276-281

Corsini, G.; Gini, F.; Greco, M. V.; Verrazzani, L.: Cramér-Rao bounds and estimation of the parameters of the Gumbel distribution. 95 IEEEAES 31 1202-1204

Corter, James E.: Carroll, J. Douglas; — —: A graph-theoretic method for organizing overlapping clusters into trees, multiple trees, or extended trees. 95 JClass 12 283-313

Cortessis, Victoria: Thomas, Duncan C.; — —: A Gibbs sampling approach to linkage analysis. 92 HumHrdty 42 63-76

Corti, P.: Forina, M.; Drava, G.; Armanino, C.; Boggia, R.; Lanteri, S.; Leardi, R.; — —; Conti, P.; Giangiacomo, R.; Galliena, C.; Bigoni, R.; Quartari, I.; Serra, C.; Ferri, D.; Leoni, O.; Lazzeri, L.: Transfer of calibration function in near-infrared spectroscopy. 95 Chemolab 27 189-203

Costa, M.: Duo, Q. (Auth); — — (Rev. Italian): Review of "The formation of econometrics". 94 Statstca 54 540-541

Costa, Michele: Dynamic component detection in a multifactor model for stock returns. 94 JItalStS 3 25-36

Costantino, R. F.: Dennis, Brian; Desharnais, Robert A.; Cushing, J. M.; — —: Nonlinear demographic dynamics: Mathematical models, statistical methods, and biological experiments. 95 EcolgMon 65 261-281

Coste, Joël; Fermanian, Jacques; Venot, Alain: Methodological and statistical problems in the construction of composite measurement scales: A survey of six medical and epidemiological journals. 95 StatMed 14 331-345

— —; Walter, Eric; Venot, Alain: A new approach to selection and weighting of items in evaluative composite measurement scales. 95 StatMed 14 2565-2580

Cota, Albert A.; Evans, Charles R.; Longman, R. Stewart; Dion, Kenneth L.; Kilik, Lindy: Using and misusing factor analysis to explore group cohesion. 95 JClPsych 51 308-316

Cothern, C. R.; Ross, N. P. (Ed); Ziegel, Eric R. (Rev): Review of "Environmental statistics, assessment, and forecasting". 95 Technmcs 37 128-128

Cothern, C. Richard: — — (Ed): Handbook for environmental risk decisionmaking. 95 CRCPress:FL 416p

— —; Ross, N. Phillip (Ed): Environmental statistics, assessment, and forecasting. 94 CRCPress:FL 432p

Cottingham, K. L.: Stow, C. A.; Carpenter, S. R.; — —: Resource vs. ratio-dependent consumer-resource models: A Bayesian perspective. 95 Ecology 76 1986-1990

Coughlin, P. J.: — — (Auth); Tanguiane, A. S. (Rev): Review of "Probabilistic voting theory". 95 JClass 12 145-146

Coupal, Louis: Joseph, Lawrence; Gyorkos, Theresa W.; — —: Bayesian estimation of disease prevalence and the parameters of diagnostic tests in the absence of a gold standard. 95 AJEpidm 141 263-272

Crossa, José; Cornelius, Paul L.; Sayre, Ken; Ortiz-Monasterio R., Jóse Iván: A shifted multiplicative model fusion method for grouping environments without cultivar rank change. 95 CropSci 35 54-62

Crosson, William L.: Duchon, Claude E.; Renkevens, Thomas M.; — —: Estimation of daily area-average rainfall during the CaPE experiment in Central Florida. 95 JApMetrl 34 2704-2714

Croucher, John: Scoring patterns in rugby league. 95 TeachgSt 17/2 47-49

Croucher, John S.: Replaying the 1994 Wimbledon men's singles final. 95 NZndSt 30 2-8

Crouchley, R.: Pickles, A.; Neale, M.; Simonoff, E.; Rutter, M.; Hewitt, J.; Meyer, J.; — —; Silberg, J.; Eaves, L.: A simple method for censored age-of-onset data subject to recall bias: Mothers' reports of age of puberty in male twins. 94 BehavGen 24 457-468

Crouchley, Robert: A random-effects model for ordered categorical data. 95 JASA 90 489-498

— —; Pickles, Andrew R.: Multivariate survival models for repeated and correlated events. 95 JStPlInf 47 95-110

Pickles, Andrew; — —: A comparison of frailty models for multivariate survival data. 95 StatMed 14 1447-1461

Crouse, Robert H.; Jin, Chun; Hanumara, R. C.: Unbiased ridge estimation with prior information and ridge trace. 95 CommStA 24 2341-2354

Croux, Christophe; Rousseeuw, Peter J.: Highly robust regression methods which are also efficient. 94 ASAProStEd 1-5

Hössjer, Ola; — —: Generalizing univariate signed rank statistics for testing and estimating a multivariate location parameter. 95 JNnparSt 4 293-308

Rousseeuw, Peter J.; — —; Hössjer, Ola: Sensitivity functions and numerical analysis of the repeated median slope. 95 CmpStQ 10 71-90

Crowder, Martin: On the use of a working correlation matrix in using generalised linear models for repeated measures. 95 Biomtrka 82 407-410

— —: Analysis of fracture toughness data. 95 LiDaAnal 1 59-71

Taylor, H. M.; Karlin, S. (Auth); — — (Rev): Review of "An introduction to stochastic modelling". 95 Ststcian 44 291-292

Crowder, Stephen V.: Woodall, William H.; — —; Wade, Mark R.: Comment on "Q charts". 95 JQualTek 27 328-332

Crowley, John: LeBlanc, Michael; — —: Step-function covariate effects in the proportional-hazards model. 95 CandJSt 23 109-129

LeBlanc, Michael; — —: Semiparametric regression functionals. 95 JASA 90 95-105

Piantadosi, Steven; — —: An implicitly defined parametric model for censored survival data and covariates. 95 Biomtrcs 51 249-258

Crowley, Susan L.: Byrne, Barbara M. (Auth); — — (Rev): Review of "Structural equation modeling with EQS and EQS/Windows". 95 EdPsMeas 55 135-140

Crowther, N. A. S.: Matthews, G. B.; — —: A maximum likelihood estimation procedure when modelling in terms of constraints. 95 SoAfrStJ 29 29-50

Cruickshank, Scott: Minor, James M.; Namini, Hamid; — —: From here to there (dosing to efficacy) with neural nets. 94 ASAProBiop 334-339

Crump, Kenny S.: Calculation of benchmark doses from continuous data. 95 RiskAnal 15 79-89

Crusio, Wim E.: Bi- and multi-variate analyses of diallel crosses: A tool for the genetic dissection of neurobehavioral phenotypes. 93 BehavGen 23 59-68

Cruz-Orive, L. M.; Howard, C. V.: Estimation of individual feature surface area with the vertical spatial grid. 95 JMicro 178 146-151

Baddeley, A. J.; — —: The Rao-Blackwell theorem in stereology and some counterexamples. 95 AdvAppPr 27 2-19

Batra, S.; König, M. F.; — —: Unbiased estimation of capillary length from vertical slices. 95 JMicro 178 152-159

Cryer, J.; Miller, R. (Auth); Ziegel, Eric R. (Rev): Review of "Statistics for business: Data analysis and modelling". 95 Technmcs 37 239-240

Csáki, E.; Földes, A.: On the logarithmic average of additive functionals. 95 StPrLet 22 261-268

Csáki, Endre; Csörgő, Miklós: On additive functionals of Markov chains. 95 JTherPrb 8 905-919

— —; Csörgő, Miklós; Földes, Antónia; Révész, Pál: Global Strassen-type theorems for iterated Brownian motions. 95 StocProc 59 321-341

Csenki, A.: An integral equation approach to the interval reliability of systems modelled by finite semi-Markov processes. 95 RelEngSS 47 37-45

Tichy, M. (Auth); — — (Rev): Review of "Applied methods of structural reliability". 95 Ststcian 44 414-415

Csenki, Attila: Mission availability for repairable semi-Markov systems: Analytical results and computational implementation. 95 MaOpfStS 26 75-87

— —: The number of visits to a subset of the state space by a discrete-parameter semi-Markov process. 95 StPrLet 22 71-77

Cserháti, Tibor; Forgács, Esther: Use of canonical correlation analysis for the evaluation of chromatographic retention data. 95 Chemolab 28 305-313

Csörgő, Miklós; Horváth, Lajos: On the distance between smoothed empirical and quantile processes. 95 AnlsStat 23 113-131

— —; Lin, Zheng-Yan; Shao, Qi-Man: On moduli of continuity for local times of Gaussian processes. 95 StocProc 58 1-21

Csáki, Endre; — —: On additive functionals of Markov chains. 95 JTherPrb 8 905-919

Csáki, Endre; — —; Földes, Antónia; Révész, Pál: Global Strassen-type theorems for iterated Brownian motions. 95 StocProc 59 321-341

Kalpazidou, S. L. (Auth); — — (Rev): Review of "Cycle representations of Markov processes". 95 ShrtBkRv 15/3 47-47

Csörgő, Sándor; Mason, David M.: Intermediate sums and stochastic compactness of maxima. 95 JStPlInf 45 81-90

— —; Mielniczuk, Jan: Density estimation under long-range dependence. 95 AnlsStat 23 990-999

— —; Mielniczuk, Jan: Nonparametric regression under long-range dependent normal errors. 95 AnlsStat 23 1000-1014

— —; Mielniczuk, Jan: Distant long-range dependent sums and regression estimation. 95 StocProc 59 143-155

— —; Simons, Gordon: Precision calculation of distributions for trimmed sums. 95 AnlsApPr 5 854-873

Cuadras, C. M.; Fortiana, J.: A continuous metric scaling solution for a random variable. 95 JMultiAn 52 1-14

Cubadda, Gianluca: A note on testing for seasonal cointegration using principal components in the frequency domain. 95 JTimSrAn 16 499-508

— —: Testing for cointegration at any frequency using spectral methods. 94 JItalStS 3 37-50

Cubanski, David; Cyganski, David: Multivariate classification through adaptive Delaunay-based C^0 spline approximation. 95 IEEEPAMI 17 403-417

Cudeck, Robert: O'Dell, Lisa L.; — —: Comment on "Testing for equivalence of measurement scales: Simple structure and metric invariance reconsidered". 95 MultBehR 30 77-81

Cuesta-Albertos, J. A.; Domínguez-Menchero, J. S.; Matrán, C.: Consistency of L_p-best monotone approximations. 95 JStPlInf 47 295-318

Cuevas, Antonio; Romo, Juan: On the estimation of the influence curve. 95 CandJSt 23 1-9

Cao, Ricardo; — —; Fraiman, Ricardo: Minimum distance density-based estimation. 95 CmpStDA 20 611-631

Cuevas, C. M.: Alvarez, M.; — —; Escudero, L. F.; de la Fuente, J. L.; García, C.; Prieto, F. J.: Network planning under uncertainty with an application to hydropower generation. 94 Top 2 25-58

Cui, Haiyan; Stein, Alfred; Myers, Don E.: Extension of spatial information, Bayesian kriging and updating of prior variogram parameters. 95 EnvrMtrc 6 373-384

Cui, Hengiian; Tian, Yubin: Estimation of projection absolute median deviation and its application (Chinese). 95 JSysSMaS 14 63-72

Cui, Hengjian; Li, Guoying: Estimate of covariance matrix with additional information (Chinese). 95 KToChina 39 1354-1357

Cui, Lirong; Hawkes, Alan G.; Jalali, Assad: The increasing failure rate property of consecutive-k-out-of-n systems. 95 PrEngInf 9 217-225

Cullis, Brian R.: Gilmour, Arthur R.; Thompson, Robin; — —: Average information REML: An efficient algorithm for variance parameter estimation in linear mixed models. 95 Biomtrcs 51 1440-1450

Gogel, Beverley J.; — —; Verbyla, Arūnus P.: REML estimation of multiplicative effects in multienvironment variety trials. 95 Biomtrcs 51 744-749

Cummins, David J.; Andrews, C. Webster: Iteratively reweighted partial least squares: A performance analysis by Monte Carlo simulation. 95 JChemtrc 9 489-507

Cummins, J. David; Lamm-Tennant, Joan: Capital structure and the cost of equity capital in the property-liability insurance industry. 94 InsrneME 15 187-201

Cunningham, Patricia: Edwards, W. Sherman; — —: Scheduling initial calls in a telephone survey of businesses. 94 ASASrvyRsr 1368-1373

Cupples, L. Adrienne; Gagnon, David R.; Ramaswamy, Ratna; D'Agostino, Ralph B.: Age-adjusted survival curves with application in the Framingham study. 95 StatMed 14 1731-1744

Farrer, Lindsay A.; — —: Estimating the probability for major gene Alzheimer disease. 94 AJHuGene 54 374-383

Curbelo, Rosario: López, Carlos; González, Juan F.; — —: Principal component analysis of rain gauge data: Treating missing observations (Spanish). 94 Estd 46 55-83

Curnow, R. N.: Martínez, O.; — —: Estimating the locations and the sizes of the effects of quantitative trait loci using flanking markers. 92 ThApGene 85 480-488

Curnow, Robert N.: Martínez, Octavio; — —: Missing markers when estimating quantitative trait loci using regression mapping. 94 Heredity 73 198-206

Curren, Tom: So many problems, so little time. 95 Stats /12 7-9

Currie, Iain D.: Maximum likelihood estimation and Mathematica. 95 ApplStat 44 379-394

Curry, David J.; Divakar, Suresh; Mathur, Sharat K.; Whiteman, Charles H.: BVAR as a category management tool: An illustration and comparison with alternative techniques. 95 JForcast 14 181-199

Curtin, Lester R.: Choi, Jai Won; — —: Sample surveys at National Center for Health Statistics. 94 ASASrvyRsr 1162-1167

Curtis, D.; Gurling, H. M. D.: Using a dummy quantitative variable to deal with multiple affection categories in genetic linkage analysis. 91 AnHuGene 55 321-327

— —; Sham, P. C.: Using risk calculation to implement an extended relative pair analysis. 94 AnHuGene 58 151-162

Curtis, David; Gurling, Hugh: A procedure for combining two-point lod scores into a summary multipoint map. 93 HumHrdty 43 173-185

Curtsinger, James W.: Zelterman, Daniel; — —: Survival curves subjected to occasional insults. 95 Biomtrcs 51 1140-1146

Cusack, Charles A.: Colbourn, Charles J.; — —; Kreher, Donald L.: Partial Steiner triple systems with equal-sized holes. 95 JCombThA 70 56-65

Cushing, J. M.: Dennis, Brian; Desharnais, Robert A.; — —; Costantino, R. F.: Nonlinear demographic dynamics: Mathematical models, statistical methods, and biological experiments. 95 EcolgMon 65 261-281

Cuthbertson, Keith; Hall, Stephen G.; Taylor, Mark P. (Auth); Ericsson, Neil R.; Irons, John S. (Rev): Review of "Applied econometric techniques". 95 EcomtrRv 14 121-132

Cutland, Nigel J.; Kopp, P. Ekkehard; Willinger, Walter: From discrete to continuous stochastic calculus. 95 Stochatx 52 173-192

Capiński, Marek; — —: Foias and Hope statistical solutions for stochastic Navier-Stokes equations. 95 Stochatx 52 193-205

Cuzick, J.: Elliott, P.; — —; English, D.; Stern, R. (Ed); Dunn, Graham (Rev): Review of "Geographical and environmental epidemiology: Methods for small-area studies". 95 StMedRes 4 184-184

Cuzick, Jack: A strong law for weighted sums of i.i.d. random variables. 95 JTherPrb 8 625-641

— —; Giné, Evarist; Zinn, Joel: Laws of large numbers for quadratic forms, maxima of products and truncated sums of i.i.d. random variables. 95 AnlsProb 23 292-333

Cvitanić, J.: Soner, H. M.; Shreve, S. E.; — —: There is no nontrivial hedging portfolio for option pricing with transaction costs. 95 AnlsApPr 5 327-355

CWDE Software: — — (Auth); Royle, Gwen (Rev): Review of "World Development Database". 95 TeachgSt 17/2 76-76

Ćwik, Jan; Mielniczuk, Jan: Nonparametric rank discrimination method. 95 CmpStDA 19 59-74

Cyganski, David: Cubanski, David; — —: Multivariate classification through adaptive Delaunay-based C^0 spline approximation. 95 IEEEPAMI 17 403-417

Czachor, James: Wharton, Robert; — —: An upper bound for the proportion of process output not meeting specifications. 94 ASAProQlPr 264-266

Czaja, Ronald; Blair, Johnny; Bickart, Barbara; Eastman, Elizabeth: Respondent strategies for recall of crime victimization incidents. 94 JOffStat 10 257-276

Czajka, John L.: Income stratification in panel surveys: Issues in design and estimation. 94 ASASrvyRsr 791-796

Jacobson, Jonathan E.; — —: Using panel data to simulate the impact of policy changes on women's labor supply and welfare participation. 94 ASAProGovt 45-50 (Disc: p51-52)

Czajka, Stanisław: Statistical methods for the analysis of a series of multiple experiments carried out in incomplete block design (Polish). 95 ListyBio 32 101-129

Czaplewski, Raymond L.: Reich, Robin M.; — —; Bechtold, William A.: Spatial cross-correlation of undisturbed, natural shortleaf pine stands in northern Georgia. 94 EnvEcoSt 1 201-217

Czitrom, A. Veronica: Statistics in the semiconductor industry: Implementation strategy. 94 ASAProPEng 1-5 (Disc: p9-11)

Czys, Robert R.: Changnon, Stanley A.; Gabriel, K. Ruben; Westcott, Nancy E.; — —: Exploratory analysis of seeding effects on rainfall: Illinois 1989. 95 JApMetrl 34 1215-1224

D'Agostino, Ralph B.: Reply to comment on "A technique for summarizing longitudinal data". 95 StatMed 14 1145-1146

— —; Belanger, Albert J.; Markson, Elizabeth W.; Kelly-Hayes, Maggie; Wolf, Philip A.: Development of health risk appraisal functions in the presence of multiple indicators: The Framingham study nursing home institutionalization model. 95 StatMed 14 1757-1770

Cupples, L. Adrienne; Gagnon, David R.; Ramaswamy, Ratna; — —: Age-adjusted survival curves with application in the Framingham study. 95 StatMed 14 1731-1744

Collett, D. (Auth); — — (Rev): Review of "Modelling survival data in medical research". 95 JASA 90 1125-1126

Fisher, Lloyd D.; van Belle, Gerald (Auth); — — (Rev): Review of "Biostatistics: A methodology for the health sciences". 95 JASA 90 799-800

D'Agostino, Ralph B., Jr; Sparrow, David; Weiss, Scott; Rosner, Bernard: Longitudinal models for analysis of respiratory function. 95 StatMed 14 2205-2216

D'Aristotile, Anthony J.: The nearest neighbor random walk on subspaces of a vector space and rate of convergence. 95 JTherPrb 8 321-346

Dabrowska, Dorota M.: Nonparametric regression with censored covariates. 95 JMultiAn 54 253-283

— —: Estimation of transition probabilities and bootstrap in a semiparametric Markov renewal model. 95 JNnparSt 5 237-259

Dabrowski, André Robert; Zoglat, Abdelhak: Strong invariance principles for triangular arrays of weakly dependent random variables. 95 CandJSt 23 299-310

Daduna, H.; Szekli, R.: Dependencies in Markovian networks. 95 AdvAppPr 27 226-254

Daffer, Peter: Şchiopu-Kratina, Ioana; — —: Convergence of weighted sums and laws of large numbers in $D([0,1];E)$. 95 JMultiAn 53 279-292

Dagenais, Marcel G.: The performance of alternative estimators for the probit model with first-order serial correlation. 95 JStCmpSm 51 343-351

Dagum, Camilo: Human capital, income and wealth distribution models and their applications to the U.S.A.. 94 ASAProBuEc 253-258

Dagum, E. B.: Sutradhar, B. C.; MacNeil, I. B.; — —: A simple test for stable seasonality. 95 JStPlInf 43 157-167

Dagum, P.; Galper, A.; Horvitz, E.; Seiver, A.: Uncertain reasoning and forecasting. 95 IntJFor 11 73-87

Dahel, S.: Mean estimation for a partitioned multinormal distribution. 95 CommStA 24 601-606

Dahiya, Ram C.: Blumenthal, Saul; — —: Estimation of sample size with grouped data. 95 JStPlInf 44 95-115

Dahlberg, Steve: Liu, Ping-Yu; — —: Design and analysis of multiarm clinical trials with survival endpoints. 95 ConClnTr 16 119-130

Dahlhaus, Rainer: Efficient location and regression estimation for long range dependent regression models. 95 AnlsStat 23 1029-1047

Dahm, P. Fred; Gail, Mitchell H.; Rosenberg, Philip S.; Pee, David: Determining the value of additional surrogate exposure data for improving the estimate of an odds ratio. 95 StatMed 14 2581-2598

Dai, J. G.: On positive Harris recurrence of multiclass queueing networks: A unified approach via fluid limit models. 95 AnlsApPr 5 49-77

Dai, Liyi: Chao, Xiuli; — —: A monotonicity result for a single-server loss system. 95 JAppProb 32 1112-1117

Daigle, Gaétan: Rivest, Louis-Paul; Potvin, François; Crépeau, Hélène; — —: Statistical methods for aerial surveys using the double-count technique to correct visibility bias. 95 Biomtrcs 51 461-470

Dalabehera, M.; Sahoo, L. N.: Efficiencies of six almost unbiased ratio estimators under a particular model. 95 StHefte 36 61-67

Dalang, Robert C.: Cairoli, R.; — —: Sequential stochastic optimization. 95 Wiley:NY:UK 350p

Dale, A.; Marsh, C. (Ed); Dorling, Daniel (Rev): Review of "The 1991 Census User's Guide". 95 JRSS-A 158 188-189

Dale, Andrew I.: — — (Transl): Pierre-Simon Laplace, philosophical essay on probabilities. 95 Springer:Brln:NY 270p

Dalén, Jörgen: Quantifying errors in the Swedish consumer price index. 95 JOffStat 11 261-275

— —: Reply to comment on "Computing elementary aggregates in the Swedish Consumer Price Index". 94 JOffStat 10 107-108

— —; Ohlsson, Esbjörn: Variance estimation in the Swedish consumer price index. 95 JBES 13 347-356

Dalenius, Tore: Controlling invasion of privacy in surveys of change over time - A non-technical review. 95 JOffStat 11 107-113

Daley, D. J.: Two-moment scaling formulae for aggregating examination marks. 95 AstrlJSt 37 253-272

— —; Rolski, T.: Correction to "Light traffic approximations in many-server queues" (V24 p202-218). 95 AdvAppPr 27 584-584

Dalgleish, Lenard I.: Boomsma, A. (Auth); — — (Rev): Review of "BOJA: A program for bootstrap and jackknife analysis". 95 StatComp 5 165-174

Dall'aglio, M.: Conigliani, C.; — —; Pacifico, M. Perone; Salinetti, G.: Robust statistics: From classical to Bayesian analysis. 94 Metron 52/3 89-109

Dallal, Gerard E.: Comment on "Statistical values, quality, and certification" (V48 p65-87). 95 AmerStat 49 116-116

Daly, F.: Jones, M. C.; — —: Density probability plots. 95 CommStB 24 911-927

Hand, D.; — —; Lunn, A.; McConway, K.; Ostrowski, E. (Ed); Ziegel, Eric R. (Rev): Review of "A handbook of small data sets". 95 Technmcs 37 242-242

Hand, D. J.; — —; Lunn, A. D. (Auth); Derr, J. (Rev): Review of "A handbook of small data sets". 95 Biomtrcs 51 1586-1586

Hand, D. J.; — —; Lunn, A. D.; McConway, K.; Ostrowski, E. (Ed); Chatfield, C. (Rev): Review of "A handbook of small data sets". 95 JRSS-A 158 349-349

Hand, D. J.; — —; Lunn, A. D.; McConway, K. J.; Ostrowski, E. (Ed); Schimek, Michael G. (Rev): Review of "Handbook of small data sets". 95 CmpStQ 10 197-197

Hand, D. J.; — —; Lunn, A. D.; McConway, K. J.; Ostrowski, E. (Ed); Street, Deborah J. (Rev): Review of "A handbook of small data sets". 95 AstrlJSt 37 251-251

Hand, D. J.; — —; Lunn, A. D.; McConway, K. J.; Ostrowski, E. (Auth); Trenkler, Dietrich (Rev): Review of "A handbook of small data sets". 95 CmpStDA 19 101-101

Daly, Mark J.: Kruglyak, Leonid; — —; Lander, Eric S.: Rapid multipoint linkage analysis of recessive traits in nuclear families, including homozygosity mapping. 95 AJHuGene 56 519-527

Damaraju, C. V.: Mazumdar, Sati; Mattison, Donald R.; — —: Issues and approaches for assessing risks from reproductive toxicants: DBCP as a case study. 95 SankhyaB 57 223-236

Damien, Paul; Laud, Purushottam W.; Smith, Adrian F. M.: Approximate random variate generation from infinitely divisible distributions with applications to Bayesian inference. 95 JRSS-B 57 547-563

Damos, Diane L.; Carter, Robert C.: Information-processing slope scores can be reliable!. 95 HumanFac 37 424-429

Dancik, B. P.: Xie, C. Y.; Yeh, F. C.; — —; Strobeck, C.: Joint estimation of immigration and mating system parameters in gymnosperms using the EM algorithm. 91 ThApGene 83 137-140

Dandekar, Ramesh; Cohen, Michael; Kirkendall, Nancy; Kydes, Andy: Applicability of localized linear regression techniques to develop response surface models. 94 ASAProPEng 87-91

Danehy, Timothy J.; Lock, Robin H.: CHODR - Using statistics to predict college hockey. 95 Stats /13 10-14

Danell, Ö.: Jansson, G.; — —: Needs and benefits of empirical power transformations for production and quality traits in forest tree breeding. 93 ThApGene 87 487-497

Daniel, Wayne W.: — — (Auth); Enderlein, G. (Rev): Review of "Biostatistics - A foundations for analysis in the health sciences" (German). 95 BiomtrcJ 37 744-744

Dannenberg, Olaf: Böker, Fred; — —: What can P-values say about the quality of test statistics for comparative simulation studies? (German). 95 AllgeStA 79 233-251

Danzer, K.: Eckschlager, K.; — — (Auth); Clegg, D. (Rev): Review of "Information theory in analytical chemistry". 95 Chemolab 28 207-208

Daone, David P.; Tracy, Ronald L.; Mathieson, Kieran D.: Visualizing power and type I and II error. 94 ASAProStEd 66-69

Dargahi-Noubary, G. R.: Stochastic modeling and identification of seismic records based on established deterministic formulations. 95 JTimSrAn 16 201-219

Darius, P.; Boucneau, M.; De Greef, P.; De Feber, E.; Froeschl, K.: Modelling metadata. 93 StatJUN 10 171-180

Darling, R. W. R.: Constructing gamma-martingales with prescribed limit, using backwards SDE. 95 AnlsProb 23 1234-1261

Darroch, J. N.: McCloud, P. I.; — —: An analysis of categorical repeated measurements in the presence of an external factor. 95 Biomtrka 82 47-55

Darvasi, A.; Soller, M.: Optimum spacing of genetic markers for determining linkage between marker loci and quantitative trait loci. 94 ThApGene 89 351-357

Das, Ashish: Dey, Aloke; Shah, Kirti R.; — —: Optimal block designs with minimal and nearly minimal number of units. 95 StSinica 5 547-558

Gupta, Sudhir; — —; Kageyama, Sanpei: Single replicate orthogonal block designs for circulant partial diallel crosses. 95 CommStA 24 2601-2607

Das, G.; Bez, K.: Preliminary test estimators in double sampling with two auxiliary variables. 95 CommStA 24 1211-1226

Das, Kalyan; Sutradhar, Brajendra C.: Score tests for fixed effects and overdispersion components in nonlinear models for repeated measures. 95 JMultiAn 54 126-146

— —; Sutradhar, Brajendra C.: Correction to "Score tests for fixed effects and overdispersion components in nonlinear models for repeated measures" (V54 p126-146). 95 JMultiAn 55 349-349

Sutradhar, Brajendra C.; — —: Analysing exponential family based messy data. 95 CommStA 24 2683-2699

Sutradhar, Brajendra C.; — —: Tests for overdispersion in longitudinal data in the presence of outliers. 95 SankhyaA 57 299-315

Das, M. N.; Giri, N.; Ahmed, M.: Designs through recoding of varietal and level codes. 95 StPrLet 23 371-380

Das, Shubhabrata; Sen, Pranab Kumar: Simultaneous spike-trains and stochastic dependence. 95 SankhyaB 57 32-47

DasGupta, A.; Ghosh, J. K.; Zen, M. M.: A new general method for constructing confidence sets in arbitrary dimensions: With applications. 95 AnlsStat 23 1408-1432

DasGupta, Anirban: Basu, Sanjib; — —: Robustness of standard confidence intervals for location parameters under departure from normality. 95 AnlsStat 23 1433-1442

Basu, Sanjib; — —: Robust Bayesian analysis with distribution bands. 95 StatDecn 13 333-349

Oh, Hyun Sook; — —: Robustness of Stein's two stage procedure. 95 SqtlAnly 14 321-334

Vidakovic, Brani; — —: Lower bounds on Bayes risks for estimating a normal variance: With applications. 95 CandJSt 23 269-282

Dasgupta, Ratan; Bhaumik, Dulal K.: Upper and lower tolerance limits of atmospheric ozone level and extreme value distribution. 95 SankhyaB 57 182-199

Boxe, Arup; — —: On some asymptotic properties of U statistics and one-sided estimates. 94 AnlsProb 22 1715-1724

Dasgupta, U.: Mukherjee, S. P.; — —: Estimation of production function using quadratic programming. 94 JIAgSt 46 223-230

Daskalakis, Constantine: Fitzmaurice, Garrett M.; Laird, Nan M.; Zahner, Gwendolyn E. P.; — —: Bivariate logistic regression analysis of childhood psychopathology ratings using multiple informants. 95 AJEpidm 142 1194-1203

Dassios, Angelos: The distribution of the quantile of a Brownian motion with drift and the pricing of related path-dependent options. 95 AnlsApPr 5 389-398

Davis, M. H. A. (Auth); — — (Rev): Review of "Markov models and optimization". 95 JASA 90 1134-1135

Datamost Corporation: — — (Auth); Lachenbruch, Peter A. (Rev): Review of "StatMost". 95 Chance 8/2 56-61

Datta, G. S.; Ghosh, J. K.: Noninformative priors for maximal invariant parameter in group models. 95 Test 4 95-114

— —; Lahiri, P.: Robust hierarchical Bayes estimation of small area characteristics in the presence of covariates and outliers. 95 JMultiAn 54 310-328

Datta, Gauri Sankar; Ghosh, Jayanta Kumar: On priors providing frequentist validity for Bayesian inference. 95 Biomtrka 82 37-45

— —; Ghosh, Malay: Some remarks on noninformative priors. 95 JASA 90 1357-1363

— —; Ghosh, Malay: Hierarchical Bayes estimators of the error variance in one-way ANOVA models. 95 JStPlInf 45 399-411

Datta, Somnath: A minimax optimal estimator for continuous monotone densities. 95 JStPlInf 46 183-193

— —: Consistency of the MLE for a general sequential design problem. 95 SankhyaA 57 88-99

— —: Limit theory and bootstrap for explosive and partially explosive autoregression. 95 StocProc 57 285-304

— —: On a modified bootstrap for certain asymptotically nonnormal statistics. 95 StPrLet 24 91-98

— —; McCormick, William P.: Bootstrap inference for a first-order autoregression with positive innovations. 95 JASA 90 1289-1300

— —; McCormick, William P.: Some continuous Edgeworth expansions for Markov chains with applications to bootstrap. 95 JMultiAn 52 83-106

— —; Sriram, T. N.: A modified bootstrap for branching processes with immigration. 95 StocProc 56 275-294

Datta, Sujay: Mukhopadhyay, Nitis; — —: On fine-tuning a purely sequential procedure and associated second-order properties. 95 SankhyaA 57 100-117

Mukhopadhyay, Nitis; — —: On fine-tuned bounded risk sequential point estimation of the mean of an exponential distribution. 95 SoAfrStJ 29 9-27

Daudin, J.-J.: Bar-Hen, Avner; — —: Generalization of the Mahalanobis distance in the mixed case. 95 JMultiAn 53 332-342

Daurés, Jean-Pierre: Lehmann, Michel; — —; Mottet, Nicolas; Navratil, Henri: Comparison between exact and parametric distributions of multiple inter-raters agreement coefficient. 95 CmpBiomd 47 113-121

David, H. A.: First (?) occurrence of common terms in mathematical statistics. 95 AmerStat 49 121-133

— —: On recurrence relations for order statistics. 95 StPrLet 24 133-138

David, Herbert T.: Yoo, Seongmo; — —: Three remarks on Pitman domination. 95 JKoreaSt 24 361-373

David, Israel: A sequential assignment match process with general renewal arrival times. 95 PrEngInf 9 475-492

David, Katerina M.: Papageorgiou, H.; — —: The structure of compounded trivariate binomial distributions. 95 BiomtrcJ 37 81-95

Papageorgiou, H.; — —: On a class of bivariate compounded Poisson distributions. 95 StPrLet 23 93-104

David, Martin: Desktop documentation: Individual income tax microdata, SOL. 94 ASASrvyRsr 775-780

Davidian, M.; Giltinan, D. M.: Nonlinear models for repeated measurement data.. 95 ChapmnHall:Lnd 359p

— —; Giltinan, D. M. (Auth); Hand, D. J. (Rev): Review of "Nonlinear models for repeated measurement data.". 96 ShrtBkRv 16/1 6-7

Davidian, Marie; Giltinan, David M.: Nonlinear models for repeated measurement data. 95 ChapmnHall:Lnd 384p

Davidson, James: White, H. (Auth); — — (Rev): Review of "Estimation, inference and specification analysis". 95 JApEcmtx 10 477-479

Davidson, R.; Mackinnon, J. G. (Auth); Smith, Richard J. (Rev): Review of "Estimation and inference in econometrics". 95 JApEcmtx 10 339-341

Davidson, Roger: Official labour statistics: A historical perspective. 95 JRSS-A 158 165-173

Davidson, Russell; MacKinnon, James G. (Auth); Zinde-Walsh, Victoria (Rev): Review of "Estimation and inference in econometrics". 95 EconTher 11 631-635

Davies, Anthony; Lahiri, Kajal: A new framework for analyzing survey forecasts using three-dimensional panel data. 95 JEconmtx 68 205-227

Davies, Glenn M.; Koch, Gary G.; Beck, James: Statistical strategies for event rate comparisons in dental studies. 94 ASAProBiop 420-424

Davies, Leslie: — — (Auth); Schiferl, Elizabeth A. (Rev): Review of "Efficiency in research, development and production: The statistical design and analysis of chemical experiments". 95 Technmcs 37 457-458

Davies, N.: Watson, S. M.; Redfern, E.; Clark, S.; Tight, M.; — —: An influence method for outlier detection applied to time series traffic data. 95 JAppStat 22 135-149

Davies, Neville; Dawson, Ed; Gustafson, Helen; Pettitt, A. N.: Testing for randomness in stream ciphers using the binary derivative. 95 StatComp 5 307-310

Bibby, Katherine; — —: STEPS for learning statistics. 95 TeachgSt 17/3 107-110

Bissell, D. (Auth); — — (Rev): Review of "Statistical methods for SPC and TQM". 95 JRSS-A 158 341-342

Everitt, B. S. (Auth); — — (Rev): Review of "A handbook of statistical analysis using S-Plus". 95 JRSS-A 158 633-634

Harvey, A. C. (Auth); — — (Rev): Review of "Time series models (Second edition)". 95 JRSS-A 158 191-193

Janacek, G.; Swift, L. (Auth); — — (Rev): Review of "Time series - Forecasting, simulation, applications". 95 JRSS-A 158 193-195

Pole, A.; West, M.; Harrison, J. (Auth); — — (Rev): Review of "Applied Bayesian forecasting and time series analysis". 95 JRSS-A 158 635-636

Davies, P. L.: Data features. 95 StNeerla 49 185-245

Davies, Richard B.: dos Santos, Dirley M.; — —; Francis, Brian: Nonparametric hazard versus nonparametric frailty distribution in modelling recurrence of breast cancer. 95 JStPlInf 47 111-127

Davies, S. S.: Ball, F. G.; — —: Statistical inference for a two-state Markov model of a single ion channel, incorporating time interval omission. 95 JRSS-B 57 269-287

Davis, Alan B.; Larimore, Walter L.: Multiplicative risk model for predicting rates. 94 ASAProEpid 123-124 (Disc: p125-126)

— —; Larimore, Walter L.; Fellingham, Gilbert W.; Beus, Carolyn M.: Using subsequent studies to recover estimability. 94 ASAProEpid 112-117 (Disc: p125-126)

Davis, Burgess; McDonald, David: An elementary proof of the local central limit theorem. 95 JTherPrb 8 693-701

Davis, Charles S.; Chung, Younghae: Randomization model methods for evaluating treatment efficacy in multicenter clinical trials. 95 Biomtrcs 51 1163-1174

Davis, Darwin J.: Saniga, Erwin M.; — —; McWilliams, Thomas P.: Economic, statistical, and economic-statistical design of attribute charts. 95 JQualTek 27 56-73

Davis, J. M.: Pietrogrande, M. C.; Dondi, F.; Felinger, A.; — —: Statistical study of peak overlapping in multicomponent chromatograms: Importance of the retention pattern. 95 Chemolab 28 239-258

Davis, Jack E.: Andes, Nancy; — —: Linking public health data using geographic information system techniques: Alaskan community characteristics and infant mortality. 95 StatMed 14 481-490 (Disc: p687-689)

Davis, Jerry M.: Epperson, David L.; — —; Bloomfield, Peter; Karl, Thomas R.; McNab, Alan L.; Gallo, Kevin P.: Estimating the urban bias of surface shelter temperatures using upper-air and satellite data. Part I: Development of models predicting surface shelter temperatures. 95 JApMetrl 34 340-357

Davis, Joseph C.: Hughes-Oliver, Jacqueline M.; Lu, Jye-Chyi; — —; Gyurcsik, Ronald S.: Robust improvement of a semi-conductor fabrication process via spatial modeling of the mean and covariance surfaces. 94 ASAProPEng 235-244

Davis, M. H. A.: — — (Auth); Collins, Sean (Rev): Review of "Markov models and optimization". 95 JAppStat 22 186-187

— — (Auth); Dassios, Angelos (Rev): Review of "Markov models and optimization". 95 JASA 90 1134-1135

— — (Auth); Hoffman, Lorrie L. (Rev): Review of "Markov models and optimization". 95 Technmcs 37 236-236

— — (Auth); Lenhart, Suzanne (Rev): Review of "Markov models and optimization". 95 SIAMRvw 37 262-263

Davis, Mary C.: Wallman, Katherine; Groves, Robert M.; Parsons, Joseph; — —; Lapham, Susan J.: Increasing statistical awareness through graduate training: The joint program in survey methodology. 94 ASAProSoSt 39-44

Davis, Peter; Scott, Alastair: The effect of interviewer variance on domain comparisons. 95 SurvyM 21 99-106

Davis, Randy L.; Koch, Gary G.: Survey and conditional logistic regression methods for the analysis of a multiperiod crossover study to compare two treatments. 94 ASAProBiop 446-451

Davis, Richard: Donahue, Rafe M. J.; Brockwell, Peter J.; — —: On permissible correlations for locally correlated stationary processes. 95 StPrLet 22 49-53

Davis, Richard A.; Hsing, Tailen: Point process and partial sum convergence for weakly dependent random variables with infinite variance. 95 AnlsProb 23 879-917

— —; Huang, Dawei; Yao, Yi-Ching: Testing for a change in the parameter values and order of an autoregressive model. 95 AnlsStat 23 282-304

Breidt, F. Jay; — —; Dunsmuir, William T. M.: Improved bootstrap prediction intervals for autoregressions. 95 JTimSrAn 16 177-200

Davis, Robert; Rosegrant, Susan; Watkins, Michael: Managing the link between measurement and compensation. 95 QulPr 28/2 101-106

Davis, Ron: Lipshutz, Robert J.; Taverner, Fred; Hennessy, Kevin; Hartzell, George; — —: DNA sequence confidence estimation. 94 Genomics 19 417-424

Davis, T. M. E.; Martin, R. B.: Models of parasite clearance following treatment of falciparum malaria in humans. 95 EnvrMtrc 6 529-534

Davis, T. P.: Analysis of an experiment aimed at improving the reliability of transmission centre shafts. 95 LiDaAnal 1 275-306

Davison, A. C.; Hinkley, D. V.; Worton, B. J.: Accurate and efficient construction of bootstrap likelihoods. 95 StatComp 5 257-264

Dawdy, David R.; Gupta, Vijay K.: Multiscaling and skew separation in regional floods. 95 WaterReR 31 2761-2767

Dawes, Robyn M.: — — (Auth); Sherman, Malcolm J. (Rev): Review of "House of cards: Psychology and psychotherapy built on myth". 95 JASA 90 1492-1492

Dawid, A. P.; DeGroot, M. H.; Mortera, J.: Coherent combination of experts' opinions. 95 Test 4 263-293 (Disc: p293-313)

— —; Goutis, C.: Comment on "Inference". 95 StatSci 10 185-186

Dawkins, Brian P.: Investigating the geometry of a p-dimensional data set. 95 JASA 90 350-359

Dawson, Deanna K.; Sauer, John R.; Wood, Paul A.; Berlanga, Mauro; Wilson, Marcia H.; Robbins, Chandler S.: Estimating bird species richness from capture and count data. 95 JAppStat 22 1063-1068

Dawson, Donald A.; Fleischmann, Klaus: Super-Brownian motions in higher dimensions with absolutely continuous measure states. 95 JTherPrb 8 179-206

— —; Fleischmann, Klaus; Li, Yi; Mueller, Carl: Singularity of super-Brownian local time at a point catalyst. 95 AnlsProb 23 37-55

Dawson, Ed: Davies, Neville; — —; Gustafson, Helen; Pettitt, A. N.: Testing for randomness in stream ciphers using the binary derivative. 95 StatComp 5 307-310

Dawson, Ree: Lavori, Philip W.; — —; Shera, David: A multiple imputation strategy for clinical trials with truncation of patient data. 95 StatMed 14 1913-1925

Dawson, Robert J. MacG.: The 'unusual episode' data revisited. 95 JStEd 3/3 23kb

Day, Gerald N.: Carroll, Steven S.; — —; Cressie, Noel; Carroll, Thomas R.: Spatial modelling of snow water equivalent using airborne and ground-based snow data. 95 EnvrMtrc 6 127-139

Day, Martin V.: On the exit law from saddle points. 95 StocProc 60 287-311

Day, N. E.; Duffy, S. W.: Comment on "Surrogate markers in AIDS and cancer trials" (V13 p1423-1435). 95 StatMed 14 2619-2620

— —; Gore, S. M.; De Angelis, D.: Acquired immune deficiency syndrome predictions for England and Wales (1992-97): Sensitivity analysis, information, decision. 95 JRSS-A 158 505-524

Day, Nicholas E.: Duffy, Stephen W.; Chen, Hsiu-Hsi; Tabar, Laszlo; — —: Estimation of mean sojourn time in breast cancer screening using a Markov chain model of both entry to and exit from the preclinical detectable phase. 95 StatMed 14 1531-1543

Day, Roger: Bryant, John; — —: Incorporating toxicity considerations into the design of two-stage phase II clinical trials. 95 Biomtrcs 51 1372-1383

Shannon, William D.; Bryant, John; Logan, Theodore F.; — —: An application of decision theory to patient screening for an autologous tumour vaccine trial. 95 StatMed 14 2099-2110

Day, S.: Chow, S.-C.; Liu, J.-P. (Auth); — — (Rev): Review of "Statistical design and analysis in pharmaceutical science: Validation, process controls, and stability". 95 Biomtrcs 51 1190-1191

Day, Simon: Mandel, J. (Auth); — — (Rev): Review of "Analysis of two-way layouts". 95 Ststcian 44 545-546

Dayan, Peter: Hinton, Geoffrey E.; — —; Neal, Radford M.; Zemel, Richard S.: Using neural networks to learn intractable generative models. 94 ASAProStCp 11-20

Dayananda, P. W. A.; Billard, L.; Chakraborty, S.: Estimation of rate parameter and its relationship with latent and infectious periods in plant disease epidemics. 95 Biomtrcs 51 284-292

Daykin, C. D.; Pentikainen, T.; Pesonen, M. (Auth); Vajda, S. (Rev): Review of "Practical risk theory for actuaries". 95 Ststcian 44 282-283

— —; Pentikäinen, T.; Pesonen, E. (Auth); Hardy, M. R. (Rev): Review of "Practical risk theory for actuaries". 95 JAppStat 22 547-548

— —; Pentikäinen, T.; Pesonen, M. (Auth); Coutts, Stewart M. (Rev): Review of "Practical risk theory for actuaries". 95 JRSS-A 158 633-633

— —; Pentikäinen, T.; Pesonen, M. (Auth); Neuhaus, Walther (Rev): Review of "Practical risk theory for actuaries". 95 AstinB 25 75-76

Dayton, C. Mitchell: Huang, Chuen-Chuen Joyce; — —: Detecting patterns of bivariate mean vectors using model-selection criteria. 95 BJMaStPs 48 129-147

Dayton, J. David: Anderson, Jon E.; — —: Instructional regression modules using XLISP-STAT. 95 JStEd 3/1 37kb (jse/v3n1/anderson)

de Aguiar, P. F.; Bourguignon, B.; Khots, M. S.; Massart, D. L.; Phan-Than-Luu, R.: D-optimal designs. 95 Chemolab 30 199-210

De Angelis, D.: Day, N. E.; Gore, S. M.; — —: Acquired immune deficiency syndrome predictions for England and Wales (1992-97): Sensitivity analysis, information, decision. 95 JRSS-A 158 505-524

De Ayala, R. J.; Schafer, William D.; Sava-Bolesta, Monica: An investigation of the standard errors of expected a posteriori ability estimates. 95 BJMaStPs 48 385-405

De Boeck, Paul: Van Mechelen, Iven; — —; Rosenberg, Seymour: The conjunctive model of hierarchical classes. 95 Psymtrka 60 505-521

de Boer, I. J. M.; Hoeschele, I.: Genetic evaluation methods for populations with dominance and inbreeding. 93 ThApGene 86 245-258

de Braganca Pereira, Carlos Alberto: Mimoso Paulino, Carlos Daniel; — —: On identifiability of parametric statistical models. 94 JItalStS 3 125-151

de Ceballos, Federico: Luceño, Alberto; — —: Describing extra-binomial variation with partially correlated models. 95 CommStA 24 1637-1653

De Cesare, L.; Posa, D.: A simulation technique of a non-Gaussian spatial process. 95 CmpStDA 20 543-555

de Falguerolles, A.: Correction to "Comment on 'An extension of Yule's Q to multivariate binary data'" (V51 p1184-1185). 95 Biomtrcs 51 1589-1589

de Falguerolles, Antoine: Comment on "An extension of Yule's Q to multivariate binary data" (V50 p847-852). 95 Biomtrcs 51 1184-1185

— —; Jmel, Said; Whittaker, Joe: Correspondence analysis and association models constrained by a conditional independence graph. 95 Psymtrka 60 161-180

De Feber, E.: Darius, P.; Boucneau, M.; De Greef, P.; — —; Froeschl, K.: Modelling metadata. 93 StatJUN 10 171-180

De Fuentes, Íñigo: Bravo, José I.; — —; Fernández, Arturo J.: A semi-parametric estimation of a survival function from incomplete and doubly censored data. 95 CommStA 24 2735-2752

De Giosa, Marcello; Mininni, Rosamaria: On the Doléans function of set-indexed submartingales. 95 StPrLet 24 71-75

Capasso, Vincenzo; — —; Mininni, Rosamaria: Asymptotic properties of the maximum likelihood estimators of parameters of a spatial counting process modelling crystallization of polymers. 95 StocAnAp 13 279-294

De Gooijer, Jan G.: Cross-validation criteria for covariance structures. 95 CommStB 24 1-16

— —: Obituary: Oliver Duncan Anderson, 1940-1995. 95 IntJFor 11 195-196

— —; Brännäs, Kurt: Invertibility of non-linear time series models. 95 CommStA 24 2701-2714

De Greef, P.: Darius, P.; Boucneau, M.; — —; De Feber, E.; Froeschl, K.: Modelling metadata. 93 StatJUN 10 171-180

de Haan, L.: A unified criterion for the domain of attraction of extreme-value distributions. 94 ThProbAp 39 323-329

de Haan, Laurens; Huang, Xin: Large quantile estimation in a multivariate setting. 95 JMultiAn 53 247-263

de Ita, Celia: Ezcurra, Exequiel; — —: Mexico's policy framework for environmental reporting. 94 ASAProEnvr 19-25 (Disc: p36-37)

De Jong, Piet; Shephard, Neil: The simulation smoother for time series models. 95 Biomtrka 82 339-350

de Jong, R. M.: Laws of large numbers for dependent heterogeneous processes. 95 EconTher 11 347-358

de Jong, Sijmen: PLS shrinks. 95 JChemtrc 9 323-326

de Jongh, P. J.; de Wet, T.; Fresen, J. L.: An experimental design for evaluating sonar detection performance. 95 SoAfrStJ 29 79-103

de Koning, G. J.; Weller, J. I.: Efficiency of direct selection on quantitative trait loci for a two-trait breeding objective. 94 ThApGene 88 669-677

de Korvin, André: Shipley, Margaret F.; — —: Rough set theory fuzzy belief functions related to statistical confidence: Application and evaluation for golf course closing. 95 StocAnAp 13 487-502

de la Fuente, David; Pino, Raúl: Comparative analysis of univariate and transfer-function forecasting using neural networks and Box-Jenkins techniques (Spanish). 95 Questo 19 187-215

de la Fuente, J. L.: Alvarez, M.; Cuevas, C. M.; Escudero, L. F.; — —; García, C.; Prieto, F. J.: Network planning under uncertainty with an application to hydropower generation. 94 Top 2 25-58

De la Horra, Julian; Fernandez, Carmen: Sensitivity to prior independence via Farlie-Gumbel-Morgenstern model. 95 CommStA 24 987-996

de la Peña, V. H.; Montgomery-Smith, S. J.; Szulga, Jerzy: Contraction and decoupling inequalities for multilinear forms and U-statistics. 94 AnlsProb 22 1745-1765

de la Peña, Victor H.: Correction to "A bound on the moment generating function of a sum of dependent variables with an application to simple random sampling without replacement" (V30 p197-211). 95 AIHenrPB 31 703-704

— —; Montgomery-Smith, S. J.: Decoupling inequalities for the tail probabilities of multivariate U-statistics. 95 AnlsProb 23 806-816

de la Puente, Manuel; Wobus, Peter: An item nonresponse and log-linear analysis of the Spanish Language Forms Availability Test. 94 ASASrvyRsr 583-588

de la Rue, Thierry: Mean motion and Gaussian dynamical system (French). 95 ProbTher 102 45-56

de la Sen, M.: Luo, N.; — —: Robust stabilization of internally delayed uncertain systems via sliding mode control. 95 Top 3 1-34

de Leeuw, Jan; Kreft, Ita G. G.: Questioning multilevel models. 95 JEdStat 20 171-189 (Disc: p201-227, 239-240)

— —; Kreft, Ita G. G.: Reply to comments on "Questioning multilevel models". 95 JEdStat 20 239-240

Kreft, Ita G. G.; — —; Aiken, Leona S.: The effect of different forms of centering in hierarchical linear models. 95 MultBehR 30 1-21

Takane, Yoshio; Kiers, Henk A. L.; — —: Component analysis with different sets of constraints on different dimensions. 95 Psymtrka 60 259-280

de León, D. González: Lorieux, M.; Perrier, X.; Goffinet, B.; Lanaud, C.; — —: Maximum-likelihood models for mapping genetic markers showing segregation distortion. 2. F_2 populations. 95 ThApGene 90 81-89

de Macedo, Marcia M. A.: Nobre, Flavio F.; — —: Feasibility of contour mapping epidemiological data with missing values. 95 StatMed 14 605-613 (Disc: p697-699)

De Marchi, Neil: Kim, Jinbang; — —; Morgan, Mary S.: Empirical model particularities and belief in the natural rate hypothesis. 95 JEconmtx 67 81-102

De Masi, A.: Brassesco, S.; — —; Presutti, E.: Brownian fluctuations of the interface in the $D = 1$ Ginzburg-Landau equation with noise. 95 AIHenrPB 31 81-118

de Moraes, Aipore R.; Dunsmore, Ian R.: Predictive comparisons in ordinal models. 95 CommStA 24 2145-2164

De Mori, Renato: Kuhn, Roland; — —: The application of semantic classification trees to natural language understanding. 95 IEEEPAMI 17 449-460

de Sá, Paula Gonzaga: Ferrari, Pablo A.; Frigessi, Arnoldo; — —: Fast approximate maximum a posteriori restoration of multicolour images. 95 JRSS-B 57 485-500

de Socio, L. M.: Bellomo, N.; Brzezniak, Z.; — — (Auth); Elworthy, K. D. (Rev): Review of "Nonlinear stochastic evolution problems in applied sciences". 95 Metrika 42 140-141

Bellomo, N.; Brzezniak, Z.; — — (Auth); Renshaw, Eric (Rev): Review of "Nonlinear stochastic evolution problems in applied sciences". 94 BLndnMaS 26 409-410

de Souza, Daniel I., Jr; Lamberson, Leonard R.: Bayesian Weibull reliability estimation. 95 IIETrn 27 311-320

De Stavola, B. L.: Gehan, E. A.; Lemak, N. A. (Auth); — — (Rev): Review of "Statistics in medical research. Developments in clinical trials". 95 ShrtBkRv 15/2 22-22

de Turckheim, Elisabeth: Prum, Bernard; Rodolphe, François; — —: Finding words with unexpected frequencies in deoxyribonucleic acid sequences. 95 JRSS-B 57 205-220

De Vries, S.; Braak, Cajo J. F. Ter: Prediction error in partial least squares regression: A critique on the deviation used in the Unscrambler. 95 Chemolab 30 239-245

De Vylder, F.; Goovaerts, M. J.: A note on the solution of practical ruin problems. 94 InsrneME 15 181-186

— —; Marceau, E.: Explicit analytic ruin probabilities for bounded claims. 95 InsrneME 16 79-105

de Waal, D. J.; Groenewald, P. C. N.: Bayesian estimation of ground-water levels. 95 SoAfrStJ 29 65-78

— —; Groenewald, P. C. N.; Kemp, C. J.: Perturbation of the normal model in linear regression. 95 SoAfrStJ 29 105-130

— —; Kemp, C. J.: A Bayesian model for estimating the failure rate for different groups. 95 SoAfrStJ 29 131-154

de Weijer, A. P.; Buydens, L.; Kateman, G.; Heuvel, H. M.: Spectral curve fitting of infrared spectra obtained from semi-crystalline polyester yarns. 95 Chemolab 28 149-164

de Wet, T.: de Jongh, P. J.; — —; Fresen, J. L.: An experimental design for evaluating sonar detection performance. 95 SoAfrStJ 29 79-103

de Wolf, Virginia A.: Enhancing researcher access to economic microdata at the Bureau of Labor Statistics. 94 ASAProGovt 164-169

De, Anindya K.; Ghosh, Jayanta K.: A decision rule for dimension in the context of MANOVA. 95 SankhyaA 57 256-266

De, Moloy: On a class of admissible estimators of the common mean of two univariate normal populations. 95 CalcutSt 45 103-109

Dean, A. M.: Kao, L.-J.; Yates, P. K.; Lewis, S. M.; — —: Approximate information matrices for estimating a given set of contrasts. 95 StSinica 5 593-598

Dean, Angela: Wang, Song-Gui; Chow, Shein-Chung (Auth); — — (Rev): Review of "Advanced linear models: Theory and applications". 95 Technmcs 37 119-120

Dean, C. B.: Modified pseudo-likelihood estimator of the overdispersion parameter in Poisson mixture models. 94 JAppStat 21 523-532

— —; Eaves, D. M.; Martinez, C. J.: A comment on the use of empirical covariance matrices in the analysis of count data. 95 JStPlInf 48 197-205

Deaton, Angus; Laroque, Guy: Estimating a nonlinear rational expectations commodity price model with unobservable state variables. 95 JApEcmtx 10 9-40S

Deaver, David: Using guardbands to justify TURs less than 4:1. 95 ASQCTekCon 136-141

DeCaluwe, Craig L.: Hillis, Margaret A.; Minkus, Gail W.; — —; Myers, Michael N.; Schafer, Edward T.: An improved approach to product quality through testing. 95 ATTTJ 74/3 40-48

Décarie, Robert: Nadeau, Simon; — —; Lambert, Daniel; St-Georges, Mario: Nonlinear modeling of muskrat use of habitat. 95 JWildlMg 59 110-117

Decousus, Hervé: Girard, Pascal; Laporte-Simitsidis, Silvy; Mismetti, Patrick; — —; Boissel, Jean-Pierre: Influence of confounding factors on designs for dose-effect relationship estimates. 95 StatMed 14 987-1005 (Disc: p1007-1008)

Decreusefond, L.; Üstünel, A. S.: The Beneš equation and stochastic calculus of variations. 95 StocProc 57 273-284

Dee, Dick P.: On-line estimation of error covariance parameters for atmospheric data assimilation. 95 MWeathr 123 1128-1145

Deelstra, G.; Delbaen, F.: Long-term returns in stochastic interest rate models. 95 InsrneME 17 163-169

Defina, Robert H.: Cochran, Steven J.; — —: Can purchasing power parity help forecast the dollar?. 95 JForcast 14 523-532

DeFries, J. C.: Cherny, S. S.; — —; Fulker, D. W.: Multiple regression analysis of twin data: A model-fitting approach. 92 BehavGen 22 489-498

Degan, Donald J.: Vondracek, Bruce; — —: Among-and within-transect variability in estimates of shad abundance made with hydroacoustics. 95 NAJFishM 15 933-939

DeGroot, M. H.: Dawid, A. P.; — —; Mortera, J.: Coherent combination of experts' opinions. 95 Test 4 263-293 (Disc: p293-313)

DeGroot, Morris H.; Fienberg, Stephen E.; Kadane, Joseph B.: Statistics and the law. 95 Wiley:NY:UK 512p

DeGruttola, Victor: Tsiatis, A. A.; — —; Wulfsohn, M. S.: Modeling the relationship of survival to longitudinal data measured with error. Applications to survival and CD4 counts in patients with AIDS. 95 JASA 90 27-37

Brookmeyer, Ron; Gail, Mitchell H. (Auth); — — (Rev): Review of "AIDS epidemiology: A quantitative approach". 95 JASA 90 797-798

Dehay, Dominique: Asymptotic behavior of estimators of cyclic functional parameters for some nonstationary processes. 95 StatDecn 13 273-286

Deheuvels, Paul; Lifshits, Mikhail A.: Necessary and sufficient conditions for the Strassen law of the iterated logarithm in nonuniform topologies. 94 AnlsProb 22 1838-1856

— —; Mason, David M.: On the fractal nature of empirical increments. 95 AnlsProb 23 355-387

— —; Mason, David M.: Nonstandard local empirical processes indexed by sets. 95 JStPlInf 45 91-112

Dekker, Jacqueline M.: Schouten, Evert G.; — —; Kok, Frans J.; LeCessie, Saskia; van Houwelingen, Hans C.; Pool, Jan: Reply to comment on "Risk ratio and rate ratio estimation in case-cohort designs: Hypertension and cardiovascular mortality". 95 StatMed 14 1610-1610

del Castillo, Enrique: Comment on "Q charts". 95 JQualTek 27 316-321

Delaney, Nancy Jo: Canavos, George C.; Miller, Don M. (Auth); — — (Rev): Review of "Modern business statistics". 95 AmerStat 49 397-397

Delbaen, F.; Schachermayer, W.: Arbitrage possibilities in Bessel processes and their relations to local martingales. 95 ProbTher 102 357-366

Deelstra, G.; — —: Long-term returns in stochastic interest rate models. 95 InsrneME 17 163-169

Delbaen, Freddy; Schachermayer, Walter: The no-arbitrage property under a change of numéraire. 95 Stochatx 53 213-226

Delecroix, M.; Simioni, M.; Thomas-Agnan, C.: A shape constrained smoother: Simulation study. 95 CmpStQ 10 155-175

Deleuze, Ch.: Champely, S.; — —: Co-inertia of two random variables in Hilbert space and B-spline approximations with a forestry application (French). 95 RStApp 43/1 91-107

Delgado, Rosario; Sanz-Solé, Marta: Green formulas in anticipating stochastic calculus. 95 StocProc 57 113-148

Deliz, José R.; Caraballo, Indhira: Injection molding optimization through DOE. 95 ASQCTekCon 963-969

Dellaportas, Petros: Random variate transformations in the Gibbs sampler: Issues of efficiency and convergence. 95 StatComp 5 133-140

— —; Stephens, David A.: Bayesian analysis of errors-in-variables regression models. 95 Biomtrcs 51 1085-1095

Delpizzo, Vincent; Borghesi, Joanna L.: Exposure measurement errors, risk estimate and statistical power in case-control studies using dichotomous analysis of a continuous exposure variable. 95 IJEpidem 24 851-862

Delyon, Bernard: Juditsky, Anatoli; Hjalmarsson, Håkan; Benveniste, Albert; — —; Ljung, Lennart; Sjöberg, Jonas; Zhang, Qinghua: Nonlinear black-box models in system identification: Mathematical foundations. 95 Autmatca 31 1725-1750

Sjöberg, Jonas; Zhang, Qinghua; Ljung, Lennart; Benveniste, Albert; — —; Glorennec, Pierre-Yves; Hjalmarsson, Håkan; Juditsky, Anatoli: Nonlinear black-box modeling in system identification: A unified overview. 95 Autmatca 31 1691-1724

Dembo, A.; Zeitouni, O. (Auth); Bingham, N. H. (Rev): Review of "Large deviations techniques and applications". 94 BLndnMaS 26 508-510

Dembo, Amir; Karlin, Samuel; Zeitouni, Ofer: Critical phenomena for sequence matching with scoring. 94 AnlsProb 22 1993-2021

— —; Karlin, Samuel; Zeitouni, Ofer: Limit distribution of maximal non-aligned two-sequence segmental score. 94 AnlsProb 22 2022-2039

— —; Mayer-Wolf, Eddy; Zeitouni, Ofer: Exact behavior of Gaussian seminorms. 95 StPrLet 23 275-280

— —; Zajic, Tim: Large deviations: From empirical mean and measure to partial sums process. 95 StocProc 57 191-224

Bryc, Włodzimierz; — —: On large deviations of empirical measures for stationary Gaussian processes. 95 StocProc 58 23-34

Demetrio, C. G. B.: Atkinson, A. C.; — —; Zocchi, S. S.: Optimum dose levels when males and females differ in response. 95 ApplStat 44 213-226

DeMets, David L.: Gange, Stephen J.; Linton, Kathryn L. P.; Scott, Alastair J.; — —; Klein, Ronald: A comparison of methods for correlated ordinal measures with ophthalmic applications. 95 StatMed 14 1961-1974

Lee, Jae Won; — —: Group sequential comparison of changes: Ad-hoc versus more exact method. 95 Biomtrcs 51 21-30

Dempster, M. A. H.; Ye, J. J.: Impulse control of piecewise deterministic Markov processes. 95 AnlsApPr 5 399-423

Deng, Li: Sun, Don X.; — —; Wu, C. F. J.: Topics on hidden Markov models and their applications in speech recognition. 94 ASAProStCp 90-99

Deng, Lih-Yuan; Lin, Dennis K. J.: Criteria for supersaturated designs. 94 ASAProQlPr 124-128

Deng, Weian: Iyengar, S. Sitharama; — —: An efficient edge detection algorithm using relaxation labeling technique. 95 PattrnRc 28 519-536

Denham, M. C.: Implementing partial least squares. 95 StatComp 5 191-202

Deniau, Cl.: Viano, M. C.; — —; Oppenheim, G.: Long-range dependence and mixing for discrete time fractional processes. 95 JTimSrAn 16 323-338

Dennis, Brian; Desharnais, Robert A.; Cushing, J. M.; Costantino, R. F.: Nonlinear demographic dynamics: Mathematical models, statistical methods, and biological experiments. 95 EcolgMon 65 261-281

Dennis, Michael L.: Iachan, Ronaldo; — —; Bray, Robert M.: Estimating the prevalence of substance abuse among people who are homeless. 94 ASASrvyRsr 1094-1099

Deo, Rohit; Fuller, Wayne A.: Symmetric tests for cointegration. 94 ASAProBuEc 105-110

Deppa, Brant D.: Teaching undergraduate students modern methods of data analysis. 94 ASAProStEd 209-214

Depuydt, C. E.: Schoonjans, F.; Zalata, A.; — —; Comhaire, F. H.: MedCalc: A new computer program for medical statistics. 95 CmpBiomd 48 257-262

Derderian, Edna: Ghosh, Subir; — —: Determination of robust design against noise factors and in presence of signal factors. 95 CommStB 24 309-326

Deriyeva, O. M.: Some properties of measures corresponding to random fields of diffusion type on the plane. 95 ThPrMaSt 50 71-76

Derks, E. P. P. A.; Pastor, M. S. Sánchez; Buydens, L. M. C.: Robustness analysis of radial base function and multi-layered feed-forward neural network models. 95 Chemolab 28 49-60

Derksen, R. W.: Calibration error and the measurement of turbulence statistics. 95 EnvrMtrc 6 651-658

Derkx, Bert H. F.: Houwing-Duistermaat, Jeanine J.; — —; Rosendaal, Frits R.; van Houwelingen, Hans C.: Testing familial aggregation. 95 Biomtrcs 51 1292-1301

Derman, Cyrus; Ross, Sheldon: Reliability for a system with dependent components. 95 PrEngInf 9 59-63

— —; Ross, Sheldon: An improved estimator of σ in quality control. 95 PrEngInf 9 411-415

Dermoune, A.: Differential calculus relative to some point processes. 95 StPrLet 24 233-242

— —; Kutoyants, Yu.: Expansion of distribution of maximum likelihood estimate for misspecified diffusion type observation. 95 Stochatx 52 121-145

Dern, Philip: Hung, H. M. James; — —: On combining information from randomized clinical studies for combination therapy. 94 ASAProBiop 77-81

Derr, J.: Hand, D. J.; Daly, F.; Lunn, A. D. (Auth); — — (Rev): Review of "A handbook of small data sets". 95 Biomtrcs 51 1586-1586

Derr, Janice A.: Statistical consulting practicum: Opportunity for authentic learning. 94 ASAProStEd 140-143

Derzko, Gérard: Lecoutre, Bruno; — —; Grouin, Jean-Marie: Bayesian predictive approach for inference about proportions. 95 StatMed 14 1057-1063 (Disc: p1065-1065)

Desai, U. B.: Kapoor, S.; Mundkur, P. Y.; — —: Depth and image recovery using a MRF model. 94 IEEEPAMI 16 1117-1122

Desante, David F.: Suggestions for future directions for studies of marked migratory landbirds from the perspective of a practitioner in population management and conservation. 95 JAppStat 22 949-965

— —; Burton, Kenneth M.; Saracco, James F.; Walker, Brett L.: Productivity indices and survival rate estimates from MAPS, a continent-wide programme of constant-effort mist-netting in North America. 95 JAppStat 22 935-947

Desarbo, Wayne S.; Ramaswamy, Venkatram; Chatterjee, Rabikar: Analyzing constant-sum multiple criterion data: A segment-level approach. 95 JMarRsr 32 222-232

DeSarbo, Wayne S.: Wedel, Michel; — —: A mixture likelihood approach for generalized linear models. 95 JClass 12 21-55

Desarbo, Wayne S.: Wedel, Michel; Kamakura, Wagner A.; — —; Hofstede, Frenkel Ter: Implications for asymmetry, nonproportionality, and heterogeneity in brand switching from piece-wise exponential mixture hazard models. 95 JMarRsr 32 457-462

DeSarbo, Wayne S.: Young, Martin R.; — —: A parametric procedure for ultrametric tree estimation from conditional rank order proximity data. 95 Psymtrka 60 47-75

Desharnais, Robert A.: Dennis, Brian; — —; Cushing, J. M.; Costantino, R. F.: Nonlinear demographic dynamics: Mathematical models, statistical methods, and biological experiments. 95 EcolgMon 65 261-281

Deshmukh, S. R.: Joshi, B. H.; Naik-Nimbalkar, U. V.; — —; Dharmadhikari, A. D.: Prediction of failure - A case study. 95 JAppStat 22 485-493

Kale, Mandar; — —: Maximum likelihood estimation in birth-death process allowing catastrophes. 95 CommStA 24 1863-1873

Deshpande, J. V.; Gore, A. P.; Shanubhogue, A.: Statistical analysis of nonnormal data. 95 Wiley:NY:UK 240p

— —; Karia, S. R.: Confidence bounds for the joint survival function in the dependent competing risks setup. 95 CommStA 24 2315-2327

Bagai, Isha; — —: Nonparametric tests for the two sample problem with two independent competing risks. 95 JNnparSt 5 21-31

Deshpande, Jayant V.; Fareed, T. P. Muhammad: A note on conditionally unbiased estimation after selection. 95 StPrLet 22 17-23
— —; Sengupta, Debasis: Testing the hypothesis of proportional hazards in two populations. 95 Biomtrka 82 251-261

DeSimone, Heather: Clyde, Merlise A.; — —; Parmigiani, Giovanni: A comparison of algorithms for sampling models. 94 ASAProBays 211-216

Desmond, Anthony F.; Yang, Zhenlin: Shortest prediction intervals for the Birnbaum-Saunders distribution. 95 CommStA 24 1383-1401

Despotis, D. K.: Zopounidis, C.; — —; Stavropoulou, E.: Multiattribute evaluation of Greek banking performance. 95 AplStMDA 11 97-107

Dette, H.; Heiligers, B.; Studden, W. J.: Minimax designs in linear regression models. 95 AnlsStat 23 30-40
Kari, A. F. (Auth); — — (Rev): Review of "Probability". 95 Metrika 42 275-276

Dette, Holger: Optimal designs for identifying the degree of a polynomial regression. 95 AnlsStat 23 1248-1266
— —: Comment on "Constrained optimization of experimental design". 95 MaOpfStS 26 153-161
— —: Correction to "Bayesian D-optimal and model robust designs in linear regression models" (V25 p27-46). 95 MaOpfStS 26 285-286
— —; Gefeller, Olaf: The impact of different definitions of nearest neighbour distances for censored data on nearest neighbour kernel estimators of the hazard rate. 95 JNnparSt 4 271-282
— —; Studden, William J.: Optimal designs for polynomial regression when the degree is not known. 95 StSinica 5 459-473

Deuschel, Jean-Dominique; Zeitouni, Ofer: Limiting curves for I.I.D. records. 95 AnlsProb 23 852-878
Bolthausen, Erwin; — —; Tamura, Yozo: Laplace approximations for large deviations of nonreversible Markov processes. The nondegenerate case. 95 AnlsProb 23 236-267

Deutsch, C.; Journel, A. (Auth); Ziegel, Eric R. (Rev): Review of "GSLIB: Geostatistical software library and user's guide". 95 Technmcs 37 126-126

Devaux, M. F.: Vigneau, E.; — —; Safar, M.: Application of Procrustean methods to mid- and near-infrared spectral data. 95 JChemtrc 9 125-135

Devens, Richard: Chun, Young I.; — —; Mikkelson, Gordon; Yule, James: Designing a customer satisfaction survey. 94 ASAProQlPr 204-209

Deville, Jean-Claude; Särndal, Carl-Erik: Variance estimation for the regression imputed Horvitz-Thompson estimator. 94 JOffStat 10 381-394

Devine, Colum: McClean, Sally; — —: A nonparametric maximum likelihood estimator for incomplete renewal data. 95 Biomtrka 82 791-803

Devinney, Timothy: Cooil, Bruce; — —; McCormick, Douglas: Models for automobile insurance accidents and losses. 94 ASAProBuEc 217-222

Devlin, Bernie; Fienberg, Stephen E.; Resnick, Daniel P.; Roeder, Kathryn: Wringing the bell curve: A cautionary tale about the relationships among race, genes, and IQ. 95 Chance 8/3 27-36
Herrnstein, R. J.; Murray, C. (Auth); — —; Fienberg, Stephen E.; Resnick, Daniel P.; Roeder, Kathryn (Rev): Review of "The bell curve: Intelligence and class structure in American life". 95 JASA 90 1483-1488

Devooght, J.; Tombuyses, B.: The use of the component influence graph to reduce the size of Markovian availability problems. 94 RelEngSS 46 237-244
Tombuyses, B.; — —: Solving Markovian systems of O.D.E. for availability and reliability calculations. 95 RelEngSS 48 47-55

Devroye, Luc: Another proof of a slow convergence result of Birgé. 95 StPrLet 23 63-67
— —; Lugosi, Gábor: Lower bounds in pattern recognition and learning. 95 PattrnRc 28 1011-1018
Berlinet, Alain; — —; Györfi, László: Asymptotic normality of L_1-error in density estimation. 95 MaOpfStS 26 329-343

Dewanji, Anup: Chaudhuri, Probal; — —: On a likelihood-based approach in nonparametric smoothing and cross-validation. 95 StPrLet 22 7-15
Krewski, Daniel; Smythe, Robert T.; — —; Szyszkowicz, Mieczyslaw: Bootstrapping an empirical Bayes estimator of the distribution of historical controls in carcinogen bioassay. 95 JStPlInf 43 215-228

Dewey, M.: Clayton, D.; Hills, M. (Auth); — — (Rev): Review of "Statistical models in epidemiology". 95 JRSS-A 158 343-344

DeWispelare, Aaron R.; Herren, L. Tandy; Clemen, Robert T.: The use of probability elicitation in the high-level nuclear waste regulation program. 95 IntJFor 11 5-24

Dexter, Franklin: Statistical analysis of drug interactions in anesthesia. 95 JThBio 172 305-314

Dey, A.: Some orthogonal arrays with variable symbols. 93 JCISS 18 209-215

Dey, Aloke; Shah, Kirti R.; Das, Ashish: Optimal block designs with minimal and nearly minimal number of units. 95 StSinica 5 547-558

Dey, Dipak K.; Kuo, Lynn; Sahu, Sujit K.: A Bayesian predictive approach to determining the number of components in a mixture distribution. 95 StatComp 5 297-305
Chung, Younshik; Kim, Chansoo; — —: Simultaneous estimation of Poisson means under weighted entropy loss. 94 CalcutSt 44 165-174
Gelfand, Alan E.; Mallick, Bani K.; — —: Modeling expert opinion arising as a partial probabilistic specification. 95 JASA 90 598-604
Peng, Fengchun; — —: Bayesian analysis of outlier problems using divergence measures. 95 CandJSt 23 199-213
Ravishanker, Nalini; — —; Wu, Lilian S. Y.: Forecasting, planning and contemporaneous outlier analysis for IBM regional revenue based on shrinkage estimation. 94 ASAProBuEc 211-216
Ravishanker, Nalini; Wu, Lilian S.-Y.; — —: Shrinkage estimation in time series using a bootstrapped covariance estimate. 95 JStCmpSm 53 259-267

Deza, Michel; Laurent, Monique: Variety of hypercube embeddings of the equidistant metric and designs. 93 JCISS 18 289-315

Dezhbakhsh, Hashem; Thursby, Jerry G.: A Monte Carlo comparison of tests based on the Durbin-Watson statistic with other autocorrelation tests in dynamic models. 95 EcomtrRv 14 347-365

Dhaene, Jan; Vandebroek, Martina: Recursions for the individual model. 95 InsrneME 16 31-38

Dhar, Sunil K.: Extension of a negative multinomial model. 95 CommStA 24 39-57
— —; Jiang, Xulun: Probability bounds on the finite sum of the binary sequence of order k. 95 JAppProb 32 1014-1027

Dhariyal, I. D.; Rathore, R. K. S.; Srivastava, Tanuja; Munshi, P.; Rastogi, Richa: Design of optimal statistical filter for the discrete convolution backprojection method. 94 AJMMgSci 14 229-265

Dhariyal, Ishwari D.: Misra, Neeraj; — —: Some inadmissibility results for estimating ordered uniform scale parameters. 95 CommStA 24 675-685

Dharmadhikari, A. D.: Joshi, B. H.; Naik-Nimbalkar, U. V.; Deshmukh, S. R.; — —: Prediction of failure - A case study. 95 JAppStat 22 485-493

Dharmapala, Dhammika; McAleer, Michael: Prediction and accommodation in econometric modelling. 95 EnvrMtrc 6 551-556

Dhorne, T.: Antoni, A.; — —: Information from subjects by variables tables (French). 95 RStApp 43/4 43-61

Dhrymes, P. J.: Topics in advanced econometrics. Volume 2: Linear and nonlinear simultaneous equations. 94 Springer:Brln:NY 401p

Di Bacco, Mario; Regoli, Giuliana: An alternative axiomatic approach to expected utility theory. 95 Metron 53 21-29
— —; Regoli, Giuliana: Likelihood as an index of fit. 94 JItalStS 3 243-253
Carfagna, Elisabetta; — —; Luiselli, Donata: An example of the use of Bayesian analysis in human biology. 94 ASAProBays 252-255

Di Consiglio, Loredana: Kaur, Amarjot; — —; Patil, Ganapati P.; Taillie, Charles: Propagation functions for monitoring distributional changes. 95 EnvEcoSt 2 239-269

Di Crescenzo, A. G.; Giorno, V.; Nobile, A. G.; Ricciardi, L. M.: On a symmetry-based constructive approach to probability densities for two-dimensional diffusion processes. 95 JAppProb 32 316-336

Di Masi, G. B.; Kabanov, Yu. M.: A first order approximation for the convergence of distributions of the Cox processes with fast Markov switchings. 95 Stochatx 54 211-219
— —; Kabanov, Yu. M.; Runggaldier, W. J.: Mean-variance hedging of options on stocks with Markov volatilities. 94 ThProbAp 39 172-182
— —; Stettner, Ł.: Bayesian ergodic adaptive control of discrete time Markov processes. 95 Stochatx 54 301-316

Diab, Roseanne D.: Bryukhan, Fedor F.; — —: Wind energy resource estimation of the upper atmosphere over Southern Africa. 95 JApMetrl 34 2565-2571

Diaconis, P.: Aldous, D.; — —: Hammersley's interacting particle process and longest increasing subsequences. 95 ProbTher 103 199-213

Diamond, Jeffrey E.; Alfa, Attahiru Sule: Matrix analytical methods for M/PH/1 retrial queues. 95 CommStSM 11 447-470

Diamond, Neil T.: Some properties of a foldover design. 95 AstrlJSt 37 345-352

DiCiccio, Thomas J.; Romano, Joseph P.: On bootstrap procedures for second-order accurate confidence limits in parametric models. 95 StSinica 5 141-160
Casella, George; — —; Wells, Martin T.: Comment on "Inference". 95 StatSci 10 179-185
Barndorff-Nielsen, O. E.; Cox, D. R. (Auth); — — (Rev): Review of "Inference and asymptotics". 95 JASA 90 1129-1130

Dick, Peter: Modelling net undercoverage in the 1991 Canadian Census. 95 SurvyM 21 45-54

Dickey, David A.; Arnold, J. Tim: Teaching statistics with data of historical significance: Galileo's gravity and motion experiments. 95 JStEd 3/1 20kb (jse/v3n1/datasets.dickey)

Dickinson, Teresa: Keep track of your measures by using a measures data base. 95 QulPr 28/9 168-168

Dickson, David C. M.; Egídio dos Reis, Alfredo D.; Waters, Howard R.: Some stable algorithms in ruin theory and their applications. 95 AstinB 25 153-175

Dickson, R. A.: Mardia, K. V.; Dryden, I. L.; Hurn, M. A.; Li, Q.; Millner, P. A.; — —: Familial spinal shape. 94 JAppStat 21 623-641

Diebold, Francis X.; Mariano, Roberto S.: Comparing predictive accuracy. 95 JBES 13 253-263

Diebolt, Jean: A nonparametric test for the regression function: Asymptotic theory. 95 JStPlInf 44 1-17

Diehr, Paula; Martin, Donald; Koepsell, Thomas; Cheadle, Allen; Wagner, Edward; Psaty, Bruce: Optimal survey design for community-intervention evaluations: Cohort, cross-sectional, or both?. 94 ASAProEpid 34-43
— —; Martin, Donald C.; Koepsell, Thomas; Cheadle, Allen: Breaking the matches in a paired t-test for community interventions when the number of pairs is small. 95 StatMed 14 1491-1504

Diehr, Paula H.: Koepsell, Thomas D.; — —; Cheadle, Allen; Kristal, Alan: Symposium on community intervention trials. 95 AJEpidm 142 594-599

Diekmann, Andreas: Brüederl, Josef; — —: The log-logistic rate model: Two generalizations with an application to demographic data. 95 SocMethR 24 158-186

Dielman, Lynn; Couper, Mick P.: Data quality in a CAPI survey: Keying errors. 95 JOffStat 11 141-146

Dielman, Terry E.; Rose, Elizabeth L.: A bootstrap approach to hypothesis testing in least absolute value regression. 95 CmpStDA 20 119-130

— —; Rose, Elizabeth L.: Pre-test estimators in least absolute value regression with autocorrelated errors. 94 ASAProBuEc 95-100

Diem, Kenneth L.: Pugesek, Bruce H.; Nations, Chris; — —; Pradel, Roger: Mark-resighting analysis of a California gull population. 95 JAppStat 22 625-639

Dietrich, C. R.: A simple and efficient space domain implementation of the turning bands method. 95 WaterReR 31 147-156

Dietrich, Duane L.: Hassett, Thomas F.; — —; Szidarovszky, Ferenc: Time-varying failure rates in the availability and reliability analysis of repairable systems. 95 IEEEReli 44 155-160

Dietrich, Frank H., II: McClave, James T.; — — (Auth); Halvorsen, Katherine (Rev): Review of "Statistics (Sixth edition)". 95 AmerStat 49 322-322

Dietz, Elizabeth: Reducing nonsampling error in the Bureau of Labor Statistics Occupational Compensation Surveys. 94 ASASrvyRsr 1042-1047

Dietz, Klaus: Becker, Niels G.; — —: The effect of household distribution on transmission and control of highly infectious diseases. 95 MaBisci 127 207-219

Becker, Niels G.; Bahrampour, Abbas; — —: Threshold parameters for epidemics in different community settings. 95 MaBisci 129 189-208

Diewert, W. E.; Wales, T. J.: Flexible functional forms and tests of homogeneous separability. 95 JEconmtx 67 259-302

— —; Nakamura, A. (Ed); Krämer, Walter (Rev. German): Review of "Essays in index number theory (Volume I)". 95 AllgeStA 79 97-97

DiGaetano, Ralph; Ward, John; Cadell, Diane; Ryaboy, Lana: Sample design of the National Pregnancy and Health Survey. 94 ASASrvyRsr 1100-1104

Diggle, P.: Barry, J.; — —: Choosing the smoothing parameter in a Fourier approach to nonparametric deconvolution of a density estimate. 95 JNnparSt 4 223-232

— — (Auth); Janacek, G. (Rev): Review of "Time series: A biostatistical introduction". 95 JTimSrAn 16 355-356

Diggle, P. J.; Chetwynd, A. G.; Häggkvist, R.; Morris, S. E.: Second-order analysis of space-time clustering. 95 StMedRes 4 124-136

— —; Liang, K.-Y.; Zeger, S. L. (Auth); Goldstein, Harvey (Rev): Review of "Analysis of longitudinal data". 95 JRSS-A 158 345-346

Diggle, Peter J.: Kelsall, Julia E.; — —: Kernel estimation of relative risk. 95 Bernoull 1 3-16

Kelsall, Julia E.; — —: Non-parametric estimation of spatial variation in relative risk. 95 StatMed 14 2335-2342

— —; Liang, Kung-Yee; Zeger, Scott L. (Auth); Verbyla, A. P. (Rev): Review of "Analysis of longitudinal data". 95 AstrlJSt 37 367-368

— —; Liang, Yung-Yee; Zeger, Scott L. (Auth); Ten Have, Thomas R. (Rev): Review of "Analysis of longitudinal data". 95 JASA 90 1123-1124

Dignam, James J.; Weissfeld, Lisa A.; Anderson, Stewart J.: Methods for bounding the marginal survival distribution. 95 StatMed 14 1985-1998

Dijkstra, Theo K.: Pyrrho's lemma, or have it your way. 95 Metrika 42 119-125

Dijkstra, Wil; Taris, Toon: Measuring the agreement between sequences. 95 SocMethR 24 214-231 (Disc: p232-243)

Dikta, G.: Asymptotic normality under the Koziol-Green model. 95 CommStA 24 1537-1549

Dillbeck, Michael C.: Time series analysis of improved quality of life in Canada: Social change, collective consciousness, and the TM-Sidhi program. 95 PsychRpt 76 1171-1193

Dillman, Don A.; Clark, Jon R.; Sinclair, Michael D.: How prenotice letters, stamped return envelopes and reminder postcards affect mailback response rates for census questionnaires. 95 SurvyM 21 159-165

— —; Dolsen, Dana E.; Machlis, Gary E.: Increasing response to personally-delivered mail-back questionnaires. 95 JOffStat 11 129-139

— —; Singer, Eleanor; Clark, Jon R.; Treat, James B.: Effects of benefits appeals, mandatory appeals and variations in statements of confidentiality on completion rates for census questionnaires. 94 ASASrvyRsr 1333-1338

Sangster, Roberta L.; Rockwood, Todd H.; — —: The influence of administration mode on responses to numeric rating scales. 94 ASASrvyRsr 708-712

Dilworth, S. J.; Koldobsky, A. L.: The Fourier transform of order statistics with applications to Lorentz spaces. 95 IsrelJMa 92 411-425

Dimitrov, B.; Chukova, S.; Khalil, Z.: Definitions, characterizations and structural properties of probability distributions similar to the exponential. 95 JStPlInf 43 271-287

Dimitrov, Boyan; von Collani, Elart: Contorted uniform and Pareto distributions. 95 StPrLet 23 157-164

Ding, Lin; Velicer, Wayne F.; Harlow, Lisa L.: Effects of estimation methods, number of indicators per factor, and improper solutions on structural equation modeling fit indices. 95 StrEqMod 2 119-144

Ding, Shuliang: Some results on estimating the regression parameters in seemingly unrelated regression system (SURS) (Chinese). 95 ChJApPSt 10 243-252

Ding, X.: Ahmed, N. U.; — —: A semilinear McKean-Vlasov stochastic evolution equation in Hilbert space. 95 StocProc 60 65-85

Ding, Ye: Computing back-calculation estimates of AIDS epidemic. 95 StatMed 14 1505-1512

Ding, Yuanyao: Chen, Guijing; Zhao, Zhongbo; — —: Parametric and semiparametric estimates in general truncated families (Chinese). 95 JSysSMaS 14 228-241

Dingman, S. Lawrence; Lawlor, Stephen C.: Estimating low-flow quantiles from drainage-basin characteristics in New Hampshire and Vermont. 95 WaterReB 31 243-256

Dinwoodie, I. H.: A probability inequality for the occupation measure of a reversible Markov chain. 95 AnlsApPr 5 37-43

— —: Stationary exponential families. 95 AnlsStat 23 327-337

— —: Occupation measure for random walk on the circle. 95 JTherPrb 8 669-677

— —: A bound on the rate of convergence for the discrete Gibbs sampler. 95 PrEngInf 9 211-215

— —; Ney, Peter: Occupation measures for Markov chains. 95 JTherPrb 8 679-691

Dion, J. P.; Yanev, N. M.: Central limit theorem for martingales in BGWR branching processes with some statistical applications. 95 MaMethSt 4 344-358

Dion, Jean-Pierre; Ferland, René: Absolute continuity, singular measures and asymptotics for estimators. 95 JStPlInf 43 235-242

Dion, Kenneth L.: Cota, Albert A.; Evans, Charles R.; Longman, R. Stewart; — —; Kilik, Lindy: Using and misusing factor analysis to explore group cohesion. 95 JClPsych 51 308-316

Dios, R.: Chopra, D. V.; — —: On the number of constraints for balanced arrays of strength four. 93 JCISS 18 203-208

Dippo, Cathryn S.; Kostanich, Donna L.; Polivka, Anne E.: Effects of methodological change in the Current Population Survey. 94 ASASrvyRsr 260-263

DiSantostefano, Rachael L.; Muller, Keith E.: A comparison of power approximations for Satterthwaite's test. 95 CommStB 24 583-593

— —; Muller, Keith E.: A comparison of power approximations for Satterthwaite's test. 94 ASAProStCp 278-283

Disney, R. L.: Chandrasekaran, S.; English, J. R.; — —: Modeling and analysis of EWMA control schemes with variance-adjusted control limits. 95 IIETrn 27 282-290

Divakar, Suresh: Curry, David J.; — —; Mathur, Sharat K.; Whiteman, Charles H.: BVAR as a category management tool: An illustration and comparison with alternative techniques. 95 JForcast 14 181-199

Dixon, Mark J.; Tawn, Jonathan A.: A semi-parametric model for multivariate extreme values. 95 StatComp 5 215-225

Dizier, Marie-Hélène; Babron, Marie-Claude; Clerget-Darpoux, Françoise: Interactive effect of two candidate genes in a disease: Extension of the marker-association-segregation χ^2 method. 94 AJHuGene 55 1042-1049

Djehiche, Boualem; Kaj, Ingemar: The rate function for some measure-valued jump processes. 95 AnlsProb 23 1414-1438

Andersson, Håkan; — —: Limit theorems for multitype epidemics. 95 StocProc 56 57-75

Djojosugito, Rianto A.: A cubic smoothing spline based lack of fit test for nonlinear regression models. 95 CommStA 24 2183-2197

Doane, David P.: Mathieson, Kieran; — —; Tracy, Ronald L.: A program for visualizing comparisons between two normal distributions. 95 JStEd 3/1 23kb (jse/v3n1/mathieson)

Tracy, Ronald L.; Mathieson, Kieran; — —: Visualizing and describing the shape of distributions. 94 ASAProStEd 70-74

Dobbins, Thomas: Snavely, Duane; — —: An exploratory data analysis of ambulatory blood pressure monitoring as screening criteria. 94 ASAProBiop 51-55

Dobrić, V.; Yukich, J. E.: Asymptotics for transportation cost in high dimensions. 95 JTherPrb 8 97-118

Dobrow, Robert P.; Fill, James Allen: On the Markov chain for the move-to-root rule for binary search trees. 95 AnlsApPr 5 1-19

— —; Fill, James Allen: Rates of convergence for the move-to-root Markov chain for binary search trees. 95 AnlsApPr 5 20-36

Dobson, Andrew: Grenfell, Bryan; — — (Ed): Ecology of infectious diseases in natural populations. 95 CambridgeUniv 532p

Dobson, S. M.; Goddard, J. A.: The demand for professional league football in England and Wales, 1925-92. 95 Ststcian 44 259-277

Dodd, F. J.; Donegan, H. A.; McMaster, T. B. M.: Reassessment of consistency criteria in judgment matrices. 95 Ststcian 44 31-41

Dodge, Yadolah; Jurečková, Jana: Estimation of quantile density function based on regression quantiles. 95 StPrLet 23 73-78

Dods, Johannah L.; Giles, David E. A.: Alternative strategies for 'augmenting' the Dickey-Fuller test: Size-robustness in the face of pre-testing. 95 JStCmpSm 53 243-258

Dodson, Bryan: Optimization of reliability verification test strategies. 95 ASQCTekCon 1039-1045

— —: Weibull analysis (with software). 95 ASQualCntrl:WI 256p

— — (Auth); Padgett, W. J. (Rev): Review of "Weibull analysis". 95 JQualTek 27 395-396

— — (Auth); Ziegel, Eric R. (Rev): Review of "Weibull analysis". 95 Technmcs 37 347-348

Doerge, R. W.: Testing for linkage: Phase known/unknown. 95 JHeredty 86 61-62

Churchill, G. A.; — —: Empirical threshold values for quantitative trait mapping. 94 Genetcs 138 963-971

Doerschuk, Peter C.: Cramer-Rao bounds for discrete-time nonlinear filtering problems. 95 IEEEAuCn 40 1465-1469

Wu, Chi-hsin; — —: Cluster expansions for the deterministic computation of Bayesian estimators based on Markov random fields. 95 IEEEPAMI 17 275-293

Wu, Chi-hsin; — —: Tree approximations to Markov random fields. 95 IEEEPAMI 17 391-402

Does, R. J. M. M.: Tan, E. S.; Imbos, Tj.; — —; Theunissen, M.: An optimal, unbiased classification rule for mastery testing based on longitudinal data. 95 EdPsMeas 55 595-612

Does, Ronald J. M. M.: Roes, Kit C. B.; — —: Shewhart-type charts in nonstandard situations. 95 Technmcs 37 15-24 (Disc: p24-35; Reply: P36-40)

Doganaksoy, Necip: Determining the duration of a demonstration life-test before all units fail. 95 IEEEReli 44 26-30

Stephens, Kenneth S. (Auth); — — (Rev): Review of "Volume 4 - How to perform skip-lot and chain sampling (Second edition)". 95 JQualTek 27 394-395

Ushakov, Igor A.; Harrison, Robert A. (Ed); — — (Rev): Review of "Handbook on reliability engineering". 95 JASA 90 804-804

Doggett, N. A.: Bruno, W. J.; Knill, E.; Balding, D. J.; Bruce, D. C.; — —; Sawhill, W. W.; Stallings, R. L.; Whittaker, C. C.; Torney, D. C.: Efficient pooling designs for library screening. 95 Genomics 26 21-30

Doksum, Kjell; Samarov, Alexander: Nonparametric estimation of global functionals and a measure of the explanatory power of covariates in regression. 95 AnlsStat 23 1443-1473

Doksum, Kjell A.; Normand, Sharon-Lise T.: Gaussian models for degradation processes - Part I: Methods for the analysis of biomarker data. 95 LiDaAnal 1 131-144

Dokuchaev, N. G.: Distributions of Itô processes: Estimates for the density and for conditional expectations of integral functionals. 94 ThProbAp 39 662-670

Dolan, Conor V.; Molenaar, Peter C. M.; Boomsma, Dorret I.: Simultaneous genetic analysis of means and covariance structure: Pearson-Lawley selection rules. 94 BehavGen 24 17-24

Doll, R.; Fraumeni, J. F., Jr; Muir, C. S. (Ed); Boyle, Peter (Rev): Review of "Trends in cancer incidence and mortality". 95 StatMed 14 2507-2508

Doll, Sir Richard: Sir Austin Bradford Hill: A personal view of his contribution to epidemiology. 95 JRSS-A 158 155-163

Dolloff, C. Andrew: Ensign, William E.; Angermeier, Paul L.; — —: Use of line transect methods to estimate abundance of benthic stream fishes. 95 JFishRBC 52 213-222

Dolsen, Dana E.: Dillman, Don A.; — —; Machlis, Gary E.: Increasing response to personally-delivered mail-back questionnaires. 95 JOffStat 11 129-139

Domenech, Carolina: Artiles-Léon, Noel; Novoa-Ramírez, Clara M.; — —: Improving fabric finishing through experimental design. 95 ASQCTekCon 952-962

Dominé, Marco: Moments of the first-passage time of a Wiener process with drift between two elastic barriers. 95 JAppProb 32 1007-1013

Domínguez-Menchero, J. S.: Cuesta-Albertos, J. A.; — —; Matrán, C.: Consistency of L_p-best monotone approximations. 95 JStPlInf 47 295-318

Domsta, J.; Grabski, F.: The first exit of almost strongly recurrent semi-Markov processes. 95 ApplMath 23 285-304

DON Econmetrics: — — (Auth); Black, R. J. (Rev): Review of "SIMPC 3.81". 95 JApEcmtx 10 469-476

Donahue, Rafe M. J.; Brockwell, Peter J.; Davis, Richard: On permissible correlations for locally correlated stationary processes. 95 StPrLet 22 49-53

Donald, Stephen G.: Two-step estimation of heteroskedastic sample selection models. 95 JEconmtx 65 347-380

Dondi, F.: Pietrogrande, M. C.; — —; Felinger, A.; Davis, J. M.: Statistical study of peak overlapping in multicomponent chromatograms: Importance of the retention pattern. 95 Chemolab 28 239-258

Donegan, H. A.: Dodd, F. J.; — —; McMaster, T. B. M.: Reassessment of consistency criteria in judgment matrices. 95 Ststcian 44 31-41

Donegan, Kevin: Comment on "Wanna bet? On gambling strategies that may or may not work in a casino" (V47 p108-111). 95 AmerStat 49 402-402

Donev, A. N.: Atkinson, A. C.; — — (Auth); Laycock, P. J. (Rev): Review of "Optimum experimental designs". 95 Biomtrcs 51 379-379

Doney, R. A.: Spitzer's condition and ladder variables in random walks. 95 ProbTher 101 577-580

Bertoin, J.; — —: On conditioning a random walk to stay nonnegative. 94 AnlsProb 22 2152-2167

Dong, Jianping; Simonoff, Jeffrey S.: A geometric combination estimator for d-dimensional ordinal sparse contingency tables. 95 AnlsStat 23 1143-1159

Donnell, Deborah: Clarkson, Douglas B.; — —; Minstrell, Jim; Hunt, Earl; Madigan, David; Traynor, Charles: VITAL: An intelligent tutoring system for statistics. 94 ASAProStEd 88-93

Donnell, Deborah J.; Buja, Andreas; Stuetzle, Werner: Analysis of additive dependencies and concurvities using smallest additive principal components. 94 AnlsStat 22 1635-1673

Donnelly, Christl A.: The spatial analysis of covariates in a study of environmental epidemiology. 95 StatMed 14 2393-2409

— —; Laird, Nan M.; Ware, James H.: Prediction and creation of smooth curves for temporally correlated longitudinal data. 95 JASA 90 984-989

Donnelly, Peter: Balding, David J.; — —: Inference in forensic identification. 95 JRSS-A 158 21-40 (Disc: p40-53)

Ball, Frank; — —: Strong approximations for epidemic models. 95 StocProc 55 1-21

Donner, Allan: Eliasziw, Michael; — —: Evaluation of a proposed modification to the pairwise estimator of a parent-offspring correlation. 95 SankhyaB 57 151-157

Simpson, Judy M.; Klar, Neil; — —: Accounting for cluster randomization: A review of primary prevention trials, 1990 through 1993. 95 AJPubHlh 85 1378-1383

Donoghue, John R.: The effects of within-group covariance structure on recovery in cluster analysis I. The bivariate case. 95 MultBehR 30 227-254

— —: Univariate screening measures for cluster analysis. 95 MultBehR 30 385-427

Allen, Nancy L.; — —: Detecting differential item functioning: Current methods and continuing problems. 94 ASAProSoSt 17-26 (Disc: p27-29)

Donoho, David L.; Johnstone, Iain M.: Adapting to unknown smoothness via wavelet shrinkage. 95 JASA 90 1200-1224

— —; Johnstone, Iain M.; Kerkyacharian, Gérard; Picard, Dominique: Wavelet shrinkage: Asymptopia?. 95 JRSS-B 57 301-337 (Disc: p337-369)

Donthu, Naveen: Rust, Roland T.; — —: Capturing geographically localized misspecification error in retail store choice models. 95 JMarRsr 32 103-110

Dopazo, Joaquín: Estimating errors and confidence intervals for branch lengths in phylogenetic trees by a bootstrap approach. 94 JMolEvol 38 300-304

Adell, José Carlos; — —: Monte Carlo simulation in phylogenies: An application to test the constancy of evolutionary rates. 94 JMolEvol 38 305-309

Dorans, Neil J.: Potenza, Maria T.; — —: DIF assessment for polytomously scored items: A framework for classification and evaluation. 95 ApPsychM 19 23-37

Wainer, Howard; — —; Flaugher, Ronald; Green, Bert F.; Mislevy, Robert; Steinberg, Lynne; Thissen, David (Auth); Andrich, David (Rev): Review of "Computerized adaptive testing: A primer". 95 Psymtrka 60 615-620

Doray, Louis G.; Luong, Andrew: Quadratic distance estimators for the zeta family. 95 InsrneME 16 255-260

Dorfman, A. H.: Chambers, R. L.; — —: Robust sample survey inference via bootstrapping and bias correction: The case of the ratio estimator. 94 ASAProStCp 51-60

Dorfman, Alan: Casady, Robert J.; — —; Wang, Suojin: Confidence intervals for sub-domain means and totals. 94 ASASrvyRsr 1036-1041

Dorfman, Jeffrey H.: A numerical Bayesian test for cointegration of AR processes. 95 JEconmtx 66 289-324

— —; Havenner, Arthur: Model specification tests for balanced representation state space models. 95 CommStA 24 97-119

Dorinski, Suzanne M.; Huang, Hertz: Use of administrative data in SIPP longitudinal estimation. 94 ASASrvyRsr 656-661 (Disc: p680-682)

Dorkenoo, K. M.; Mathieu, J.-R.: Generalized linear models with a small scale parameter. 95 MaOpfStS 27 107-113

Dorling, Daniel: Dale, A.; Marsh, C. (Ed); — — (Rev): Review of "The 1991 Census User's Guide". 95 JRSS-A 158 188-189

Dorn, G. J. W.: Inductive countersupport. 95 ZeitWiss 26 187-189

Dorogovtsev, A. A.: Stochastic analysis and random maps in Hilbert space. 94 VSP:Neth 110p

Dorsey, Robert E.; Mayer, Walter J.: Genetic algorithms for estimation problems with multiple optima, nondifferentiability, and other irregular features. 95 JBES 13 53-66

dos Santos Silva, Isabell: Swerdlow, Anthony; — — (Auth); Yuen, Peter; Machin, David (Rev): Review of "Atlas of cancer incidence in England and Wales: 1968-85". 95 StatMed 14 1725-1726

dos Santos, Dirley M.; Davies, Richard B.; Francis, Brian: Nonparametric hazard versus nonparametric frailty distribution in modelling recurrence of breast cancer. 95 JStPlInf 47 111-127

Doss, Hani: Comment on "Markov chains for exploring posterior distributions". 94 AnlsStat 22 1728-1734

— —: Bayesian nonparametric estimation for incomplete data via successive substitution sampling. 94 AnlsStat 22 1763-1786

Li, Gang; — —: An approach to nonparametric regression for life history data using local linear fitting. 95 AnlsStat 23 787-823

Dosse, M. Bennani: Multidimensional scaling of a three-way table (French). 95 RStApp 43/4 63-75

Dougherty, Edward R.: Loce, Robert P.; — —: Mean-absolute-error representation and optimization of computational-morphological filters. 95 CVGIPGr 57 27-37

Douglas, J. B.: Comment on "Estimation and testing for the parameters of the bivariate Poisson populations based on a 2 by 2 sample" (V61 p361-367). 95 IntStRvw 63 117-118

Doukhan, P.; Massart, P.; Rio, E.: Invariance principles for absolutely regular empirical processes. 95 AIHenrPB 31 393-427

Dowlatabadi, Hadi: Tschang, F. Ted; — —: A Bayesian technique for refining the uncertainty in global energy model forecasts. 95 IntJFor 11 43-61

Downes-Le Guin, Theodore: Thran, Sara L.; — —; Berry, Sandra: Improving productivity in an annual survey of physicians. 94 ASASrvyRsr 577-582

Downey, Peter: Bruno, John; Coffman, Edward G., Jr; — —: Scheduling independent tasks to minimize the makespan on identical machines. 95 PrEngInf 9 447-456

Downing, Darryl J.: Cromwell, Jeff B.; Labys, Walter C.; Terraza, Michel (Auth); — — (Rev): Review of "Univariate tests for time series models". 95 Technmcs 37 346-346

Doyle, Pat; Farley, Dean: Alternative strategies for imputing premiums and predicting expenditures under health care reform. 94 ASASrvyRsr 517-522

Doyle, Suzanne R.: Overall, John E.; — —: Reply to comment on "Sample size estimation for repeated measurement designs". 95 ConClnTr 16 451-452

Dragalin, V.: An optimal sequential selection procedure. 95 CommStA 24 429-443

Drake, Christiana; McQuarrie, Allan: A note on the bias due to omitted confounders. 95 Biomtrka 82 633-638

Drake, Laurel G.: Margavio, Thomas M.; Conerly, Michael D.; Woodall, William H.; — —: Alarm rates for quality control charts. 95 StPrLet 24 219-224

Drakopoulos, John A.: Probabilities, possibilities, and fuzzy sets. 95 FuzzySet 75 1-15

— —: Sigmoidal theory. 95 FuzzySet 76 349-363

Drapeau, Richard A.: TQM in undergraduate business statistics: What are the customers' needs?. 94 ASAProQlPr 270-272

Draper, David: Inference and hierarchical modeling in the social sciences. 95 JEdStat 20 115-147 (Disc: p201-233)

— —: Reply to comments on "Inference and hierarchical modeling in the social sciences". 95 JEdStat 20 228-233

— —: Assessment and propagation of model uncertainty. 95 JRSS-B 57 45-70 (Disc: p71-97)

— —; Gaver, Donald, P., Jr; Goel, Prem K.; Greenhouse, Joel B.; Hedges, Larry V.; Morris, Carl N.; Tucker, John R.; Waternaux, Christine M. (Auth); Berlin, Jesse A. (Rev): Review of "Combining information: Statistical issues and opportunities for research". 95 StatMed 14 1611-1612

Draper, Gerald: Richardson, Sylvia; Monfort, Christine; Green, Martyn; — —; Muirhead, Colin: Spatial variation of natural radiation and childhood leukaemia incidence in Great Britain. 95 StatMed 14 2487-2501

Draper, N. R.: Carroll, Raymond J.; Ruppert, David; Stefanski, Leonard A. (Auth); — — (Rev): Review of "Measurement error in nonlinear models". 96 ShrtBkRv 16/1 5-5

Maritz, J. S. (Auth); — — (Rev): Review of "Distribution-free statistical methods (Second edition)". 95 ShrtBkRv 15/3 45-45

Rao, C. Radhakrishna; Toutenburg, Helge (Auth); — — (Rev): Review of "Linear models: Least squares and alternatives". 96 ShrtBkRv 16/1 7-8

Draper, Norman R.; Guttman, Irwin: Confidence intervals versus regions. 95 Ststcian 44 399-403

— —; Lin, Dennis K. J.: Characterizing projected designs: Repeat and mirror-image runs. 95 CommStA 24 775-795

Lin, Dennis K. J.; — —: Screening properties of certain two-level designs. 95 Metrika 42 99-118

Ying, Lisa H.; Pukelsheim, Friedrich; — —: Slope rotatability over all directions designs. 95 JAppStat 22 331-341

Ying, Lisa H.; Pukelsheim, Friedrich; — —: Slope rotatability over all directions designs for $k \geq 4$. 95 JAppStat 22 343-354

Drappo, Silvia: Rotondi, Renata, — —: A clustering method for global optimization based on the kth nearest neighbour. 95 StatComp 5 317-326

Drasgow, Fritz: Comment on "Testing for equivalence of measurement scales: Simple structure and metric invariance reconsidered". 95 MultBehR 30 83-85

Drava, G.: Forina, M.; — —; Armanino, C.; Boggia, R.; Lanteri, S.; Leardi, R.; Corti, P.; Conti, P.; Giangiacomo, R.; Galliena, C.; Bigoni, R.; Quartari, I.; Serra, C.; Ferri, D.; Leoni, O.; Lazzeri, L.: Transfer of calibration function in near-infrared spectroscopy. 95 Chemolab 27 189-203

Drava, Giuliana: Sergent, Michelle; Mathieu, Didier; Phan-Tan-Luu, Roger; — —: Correct and incorrect use of multilinear regression. 95 Chemolab 27 153-162

Drescher, Karsten; Boedeker, Wolfgang: Assessment of the combined effects of substances: The relationship between concentration addition and independent action. 95 Biomtrcs 51 716-730

Drew, David; Steyne, Dave: The world's 100 largest cities: A case study of inequality. 95 TeachgSt 17/1 17-21

Drezner, Tammy: Drezner, Zvi; — —: Comment on "A binomial test of model fit" (V1 p268-273). 95 StrEqMod 2 144-145

Drezner, Zvi: Multirelation - A correlation among more than two variables. 95 CmpStDA 19 283-292

— —; Drezner, Tammy: Comment on "A binomial test of model fit" (V1 p268-273). 95 StrEqMod 2 144-145

— —; Wesolowsky, G. O.: Multivariate screening procedures for quality cost minimization. 95 IIETrn 27 300-304

Driscoll, Charles T.; Letterman, Raymond D.: Factors regulating residual aluminium concentrations in treated waters. 95 EnvrMtrc 6 287-305 (Disc: p305-309)

Driscoll, John C.: Rotemberg, Julio J.; — —; Poterba, James M.: Money, output, and prices: Evidence from a new monetary aggregate. 95 JBES 13 67-83

Driscoll, Michael F.; Krasnicka, Barbara: An accessible proof of Craig's theorem in the general case. 95 AmerStat 49 59-62

Droge, B.: Miller, A. J. (Auth); — — (Rev): Review of "Subset selection in regression". 95 MaOpfStS 26 179-179

Ross, G. J. S. (Auth); — — (Rev): Review of "Nonlinear estimation". 95 MaOpfStS 26 181-182

Droge, Bernd; Georg, Thomas: On selecting the smoothing parameter of least squares regression estimates using the minimax regret approach. 95 StatDecn 13 1-20

Dror, Shuki; Faraggi, David; Reiser, Benjamin: Dynamic treatment allocation adjusting for prognostic factors for more than two treatments. 95 Biomtrcs 51 1338-1343

Drovandi, Simona: A structural model with latent variables used to analyze the links between modernization and fertility: The case of Brazil (Italian). 94 Statstca 54 521-538

Drozdov, Aleksey: Stability of a class of stochastic integro-differential equations. 95 StocAnAp 13 517-530

Drum, Melinda L.: Jørgensen, B. (Auth); — — (Rev): Review of "The theory of linear models". 95 StMedRes 4 266-267

Drummey, Kevin Ward: How many dinosaurs are there? Or estimating population size using the capture/recapture method. 94 ASAProStEd 215-217

Dryden, I. L.: Mardia, K. V.; — —; Hurn, M. A.; Li, Q.; Millner, P. A.; Dickson, R. A.: Familial spinal shape. 94 JAppStat 21 623-641

Drygas, H.: Bapat, R. B. (Auth); — — (Rev): Review of "Linear algebra and linear models". 95 Metrika 42 78-78

Dshalalow, Jewgeni H.: — — (Ed): Advances in queuing. Models, methods, and problems. 95 CRCPress:FL 384p

Du, Charles; Pinedo, Michael: A convexity property of a Markov-modulated queueing loss system. 95 PrEngInf 9 193-199

Du, Qing: A monotonicity result for a single-server queue subject to a Markov-modulated Poisson process. 95 JAppProb 32 1103-1111

Dua, Pami; Ray, Subhash C.: A BVAR model for the Connecticut economy. 95 JForcast 14 167-180

— —; Smyth, David J.: Forecasting US home sales using BVAR models and survey data on households' buying attitudes for homes. 95 JForcast 14 217-227

Duan, Xiaoping; Kageyama, Sanpei: Constructions of group divisible designs. 95 JJapnStS 25 121-128

Dubas, Françoise: Empirical influence functions for the Cox regression model. 95 Student 1 87-98

Dube, M.; Singh, V.: Predictive efficiency of improved estimators in restricted regression models. 95 Test 4 323-331

Dubey, P. K.: Pandey, B. N.; Malik, H. J.; — —: Bayesian shrinkage estimators for a measure of dispersion of an inverse Gaussian distribution. 95 CommStA 24 2261-2270

Dubey, S. D.: Huque, M. F.; Harrison, F. D.; Sankoh, A. J.; — —: Longitudinal data analysis for heartburn clinical trials. 94 ASAProBiop 182-187

Dubey, Satya D.: Chi, George Y. H.; Hung, James; — —; Lipicky, Raymond J.: Dose response studies and special populations. 94 ASAProBiop 88-93

Dubinko, T. Yu.: A robust estimate of the location parameter for a normal distribution with asymmetrical littering (Russian). 95 VLenUM /2 94-96

Dubreuil, Guylaine; Tremblay, Johanne: The use of generalized raking procedures to improve the quality of small domain estimation. 94 ASASrvyRsr 293-298

Ducharme, Gilles R.: Uniqueness of the least-distances estimator in regression models with multivariate response. 95 CandJSt 23 421-424

— —; Gannoun, Ali; Guertin, Marie-Claude; Jéquier, Jean-Claude: Reference values obtained by kernel-based estimation of quantile regressions. 95 Biomtrcs 51 1105-1116

Boulerice, Bernard; — —: A note on smooth tests of goodness of fit for location-scale families. 95 Biomtrka 82 437-438

Duchon, Claude E.; Renkevens, Thomas M.; Crosson, William L.: Estimation of daily area-average rainfall during the CaPE experiment in Central Florida. 95 JApMetrl 34 2704-2714

Ducimetiere, P.: Tiret, L.; Nicaud, V.; Ehnholm, C.; Havekes, L.; Menzel, H. J.; — —; Cambien, F.: Inference of the strength of genotype-disease association from studies comparing offspring with and without parental history of disease. 93 AnHuGene 57 141-149

Duddek, C.; Le, N. D.; Zidek, J. V.; Burnett, R. T.: Multivariate imputation in cross-sectional analysis of health effects associated with air pollution. 95 EnvEcoSt 2 191-212

Dudewicz, Edward J.; van der Meulen, Edward C.; SriRam, M. G.; Teoh, Nick K. W.: Entropy-based random number evaluation. 95 AJMMgSci 15 115-153

Dudgeon, Paul: Arminger, Gerhard; Clogg, Clifford C.; Sobel, Michael E. (Ed); — — (Rev): Review of "Handbook of statistical modeling for the social and behavioural sciences". 95 Chance 8/4 32-36

Duever, Thomas A.: Rey, Timothy D.; — —: Modeling nonlinear systems, bridging theory and technology. 94 ASAProQlPr 310-315

Duffer, Allen; Lessler, Judith; Weeks, Michael; Mosher, William: Effects of incentive payments on response rates and field costs in a pretest of a national CAPI survey. 94 ASASrvyRsr 1386-1392

Duffie, Darrell; Ma, Jin; Yong, Jiongmin: Black's consol rate conjecture. 95 AnlsApPr 5 356-382

Duffy, David L.: Biometrical genetic analysis of the cotwin control design. 94 BehavGen 24 341-344

Duffy, Roger E.: Pareto analysis and trend charts: A powerful duo. 95 QulPr 28/11 152-152

Duffy, S. W.: Day, N. E.; — —: Comment on "Surrogate markers in AIDS and cancer trials" (V13 p1423-1435). 95 StatMed 14 2619-2620

Duffy, Stephen W.; Chen, Hsiu-Hsi; Tabar, Laszlo; Day, Nicholas E.: Estimation of mean sojourn time in breast cancer screening using a Markov chain model of both entry to and exit from the preclinical detectable phase. 95 StatMed 14 1531-1543

Dufour, Jean-Marie; Nsiri, Saïd; Tessier, David: Parsimonious autoregressive conditions for noncausality in multivariate ARMA models. 94 ASAProBuEc 129-134

— —; Torrés, Olivier: Some exact inference procedures for stationary and nonstationary autoregressive processes. 94 ASAProBuEc 135-140

Dufour, Johane: Bench, Jill; Clark, Colleen; — —; Kaushal, Ritu: Implementing computer-assisted interviewing for the Canadian Labour Force Survey. 94 ASAProGovt 93-98

Duineveld, C. A. A.; Coenegracht, P. M. J.: Multicriteria steepest ascent in a design space consisting of both mixture and process variables. 95 Chemolab 30 23-36

Dukes, M. D. G.; Palutikof, J. P.: Estimation of extreme wind speeds with very long return periods. 95 JApMetrl 34 1950-1961

Dulaney, Wayne P.; Lengnick, Laura L.; Hart, Galen F.: Use of geostatistical techniques in the design of an agricultural field experiment. 94 ASASrvyRsr 183-187

Dümbgen, Lutz: Minimax tests for convex cones. 95 AnInStMa 47 155-165

— —: Likelihood ratio tests for principal components. 95 JMultiAn 52 245-258

— —: A simple proof and refinement of Wielandt's eigenvalue inequality. 95 StPrLet 25 113-115

DuMouchel, William: Comment on "Meta-analysis for dose-response models". 95 StatMed 14 679-685

— —; Lane, Thomas: Comment on "Guiding data analysts with visual statistical strategies". 95 JCmpGrSt 4 251-256

DuMouchel, William H.; Lane, Thomas P.: Construction of mixture-process designs via the Franklin-Bailey algorithm. 94 ASAProQlPr 7-16

Duncan, J. W.; Gross, A. C. (Auth); Eklöf, Jan A. (Rev): Review of "Statistics for the 21st century. Proposals for improving statistics for better decision making". 95 JOffStat 11 225-229

— —; Gross, A. C. (Auth); Swires-Hennessy, E. (Rev): Review of "Statistics for the 21st century: Proposals for improving statistics for better decision making". 95 Ststcian 44 138-139

Duncan, Susan C.: Duncan, Terry E.; — —: Modeling the processes of development via latent variable growth curve methodology. 95 StrEqMod 2 187-213

Duncan, Terry E.; — —; Hops, Hyman; Stoolmiller, Mike: An analysis of the relationship between parent and adolescent marijuana use via generalized estimating equation methodology. 95 MultBehR 30 317-339

Duncan, Terry E.; Duncan, Susan C.: Modeling the processes of development via latent variable growth curve methodology. 95 StrEqMod 2 187-213

— —; Duncan, Susan C.; Hops, Hyman; Stoolmiller, Mike: An analysis of the relationship between parent and adolescent marijuana use via generalized estimating equation methodology. 95 MultBehR 30 317-339

Dunlop, Peter J.; Bignell, C. M.; Jackson, J. F.; Hibbert, D. Brynn: Chemometric analysis of gas chromatographic data of oils from Eucalyptus species. 95 Chemolab 30 59-67

Dunn, Charles L.: Critical values and powers for tests of uniformity of directions under multivariate normality. 95 CommStA 24 2541-2560

Dunn, G.; Everitt, B.: Clinical biostatistics: An introduction to evidence-based medicine. 95 Arnold:Lnd:MD 154p

— —; Everitt, B. (Auth); Lagakos, S. W. (Rev): Review of "Clinical biostatistics: An introduction to evidence-based medicine". 95 ShrtBkRv 15/3 41-42

— —; Everitt, B.; Pickles, A. (Auth); Chan, Wai (Rev): Review of "Modelling covariances and latent variables using EQS". 95 SIAMRvw 37 462-464

— —; Everitt, B.; Pickles, A. (Auth); Schimek, Michael G. (Rev): Review of "Modelling covariances and latent variables using EQS". 95 CmpStQ 10 196-196

Dunn, Graham; Everitt, Brian: Clinical biostatistics. An introduction to evidence-based medicine. 95 Wiley:NY:UK 154p

Everitt, Brian S.; — —: Applied multivariate data analysis. 95 Wiley:NY:UK 304p

Pickles, Andrew; — —; Vázquez-Barquero, José Luis: Screening for stratification in two-phase ('two-stage') epidemiological surveys. 95 StMedRes 4 73-89

Pickles, Andrew; — —; Vázquez-Barquero, José Luis: Reply to comment on "Screening for stratification in two-phase ('two-stage') epidemiological surveys". 95 StMedRes 4 263-263

Elliott, P.; Cuzick, J.; English, D.; Stern, R. (Ed); — — (Rev): Review of "Geographical and environmental epidemiology: Methods for small-area studies". 95 StMedRes 4 184-184

Dunn, W. J., III: Computational methods for the estimation of solution properties of organic pollutants. 95 Chemolab 29 1-10

Dunne, Adrian: Decision and detection limits for linear homoscedastic assays. 95 StatMed 14 1949-1959

Dunnett, Charles W.; Tamhane, Ajit C.: Step-up multiple testing of parameters with unequally correlated estimates. 95 Biomtrcs 51 217-227

Dunsmore, Ian R.: de Moraes, Aipore R.; — —: Predictive comparisons in ordinal models. 95 CommStA 24 2145-2164

Geisser, S. (Auth); — — (Rev): Review of "Predictive inference: An introduction". 95 JRSS-A 158 191-191

Dunsmuir, William T. M.: Breidt, F. Jay; Davis, Richard A.; — —: Improved bootstrap prediction intervals for autoregressions. 95 JTimSrAn 16 177-200

Düntsch, Ivo; Gediga, Günther: Skills and knowledge structures. 95 BJMaStPs 48 9-27

Duo, Q.: — — (Auth); Costa, M. (Rev. Italian): Review of "The formation of econometrics". 94 Statstca 54 540-541

Dupont, F.: Alternative adjustments where there are several levels of auxiliary information. 95 SurvyM 21 125-135

Dupuis, Jerome A.: Bayesian estimation of movement and survival probabilities from capture-recapture data. 95 Biomtrka 82 761-772

Dupuis, Josée; Brown, Patrick O.; Siegmund, David: Statistical methods for linkage analysis of complex traits from high-resolution maps of identity by descent. 95 Genetcs 140 843-856

Duquet, W.: Abidi, H.; Pontier, J.; Borms, J.; — —: The benefit of modelling growth curves for longitudinal data (French). 95 RStApp 43/3 55-72

Duran, Benjamin S.: Bethea, Robert M.; — —; Boullion, Thomas L.: Statistical methods for engineers and scientists. 95 Dekker:NY 672p

Bethea, Robert M.; — —; Boullion, Thomas L. (Auth); Nelson, Lloyd S. (Rev): Review of "Statistical methods for engineers and scientists (Third edition)". 95 JQualTek 27 392-393

Durham, S.: Lee, Shiowjen; — —; Lynch, J.: On the calculation of the reliability of general load sharing systems. 95 JAppProb 32 777-792

Durham, Stephen D.: Flournoy, Nancy; — —; Rosenberger, William F.: Toxicity in sequential dose-response experiments. 95 SqtlAnly 14 217-227

Durham, Todd A.: Kalsbeek, William D.; — —: Nonresponse and its effects in a followup telephone survey of low income women. 94 ASASrvyRsr 943-948

Durieu, Isabelle: Boissel, Jean-Pierre; — —; Girard, Pascal; Nony, Patrice; Chauvin, Franck; Haugh, Margaret: Dose-ranging trials: Guidelines for data collection and standardized descriptions. 95 ConClnTr 16 319-330

Durlauf, Steven N.; Johnson, Paul A.: Multiple regimes and cross-country growth behaviour. 95 JApEcmtx 10 365-384

Bernard, Andrew B.; — —: Convergence in international output. 95 JApEcmtx 10 97-108

Durrett, R.: Cox, J. T.; — —: Hybrid zones and voter model interfaces. 95 Bernoull 1 343-370

Durrett, Rick; Schinazi, Rinaldo: Intermediate phase for the contact process on a tree. 95 AnlsProb 23 668-673

Dutka, Solomon; Frankel, Lester R.: The maintenance of ethical principles in marketing research. 94 ASASrvyRsr 143-145

Dutta, K.; Goswami, V.: A comparative study of multidimensional sampling plans. 94 JIAgSt 46 254-266

Duval, Marie-Claude: Mantel, Harold; Laniel, Normand; — —; Marion, Jocelyne: Cost modelling of alternative sample designs for rural areas in the Canadian Labour Force Survey. 94 ASASrvyRsr 925-930

Dwyer, Greg; Elkinton, Joseph S.: Host dispersal and the spatial spread of insect pathogens. 95 Ecology 76 1262-1275

Dwyer, James H.: MacKinnon, David P.; Warsi, Ghulam; — —: A simulation study of mediated effect measures. 95 MultBehR 30 41-62

Dyer, Danny; Pierce, Rebecca L.: Bayesian HPM intervals for the number of nonconforming items in a lot. 95 CmpStDA 20 159-172

Dyer, James S.: Wright, M. Keith; Stokes, Lynne; — —: Reliability and coherence of causal, diagnostic, and joint subjective probabilities. 95 DecisnSc 26 691-709

Wright, M. Keith; Stokes, Lynne; — —: Reliability and coherence of causal, diagnostic, and joint subjective probabilities. 94 DecisnSc 25 691-709

Dyke, G. V.; Lane, P. W.; Jenkyn, J. F.: Sensitivity (stability) analysis of multiple variety trials, with special reference to data expressed as proportions or percentages. 95 ExpAgric 31 75-87

Dyke, George: Obituary: Frank Yates, 1902-1994. 95 JRSS-A 158 333-338

Dykstra, Richard; Kochar, Subhash; Robertson, Tim: Likelihood ratio tests for symmetry against one-sided alternatives. 95 AnInStMa 47 719-730

— —; Kochar, Subhash; Robertson, Tim: Inference for likelihood ratio ordering in the two-sample problem. 95 JASA 90 1034-1040

— —; Kochar, Subhash; Robertson, Tim: Likelihood based inference for cause specific hazard rates under order restrictions. 95 JMultiAn 54 163-174

Barmi, Hammou El; — —: Testing for and against a set of linear inequality constraints in a multinomial setting. 95 CandJSt 23 131-143

Bhattacharya, Bhaskar; — —: A general duality approach to I-projections. 95 JStPlInf 47 203-216

Dykstra, Richard L.: Calvin, James A.; — —: REML estimation of covariance matrices with restricted parameter spaces. 95 JASA 90 321-329

Dynkin, E. B.; Kuznetsov, S. E.: Markov snakes and superprocesses. 95 ProbTher 103 433-473

Dzhvarsheishvili, I. A.: On amarts with continuous time. 94 ThProbAp 39 512-519

Dziubdziela, Wiesław: On the limit distributions of sums of mixing Bernoulli random variables. 95 StPrLet 23 179-182

Eales, John D.; Jennison, Christopher: Optimal two-sided group sequential tests. 95 SqtlAnly 14 273-286

Earl, Elizabeth: Poyer, David A.; — —: A statistical analysis of structural differences in minority household electricity demand. 94 ASAProGovt 221-226

Earman, J.: — — (Auth); Neuberg, Leland G. (Rev): Review of "Bayes or bust? - A critical examination of Bayesian confirmation theory". 95 Ststcian 44 283-284

Earman, John: — — (Auth); Worrall, John (Rev): Review of "Bayes or bust? A critical examination of Bayesian confirmation theory". 94 PhilSci 61 672-673

Easterling, David R.; Peterson, Thomas C.: A new method for detecting undocumented discontinuities in climatological time series. 95 JClimlgy 15 369-377

Eastman, Elizabeth: Czaja, Ronald; Blair, Johnny; Bickart, Barbara; — —: Respondent strategies for recall of crime victimization incidents. 94 JOffStat 10 257-276

Eastwood, Vera R.: Steinebach, Josef; — —: On extreme value asymptotics for increments of renewal processes. 95 JStPlInf 45 301-312

Eatherall, A.: Sefton, C. E. M.; Whitehead, P. G.; — —; Littlewood, I. G.; Jakeman, A. J.: Dynamic response characteristics of the Plynlimon catchments and preliminary analysis of relationships to physical descriptors. 95 EnvrMtrc 6 465-472

Eaves, B. Curtis; Rothblum, Uriel G.: A discounted-cost continuous-time flexible manufacturing and operator scheduling model solved by deconvexification over time. 95 PrEngInf 9 65-98

Eaves, D. M.: Dean, C. B.; — —; Martinez, C. J.: A comment on the use of empirical covariance matrices in the analysis of count data. 95 JStPlInf 48 197-205

Eaves, L.: Pickles, A.; Neale, M.; Simonoff, E.; Rutter, M.; Hewitt, J.; Meyer, J.; Crouchley, R.; Silberg, J.; — —: A simple method for censored age-of-onset data subject to recall bias: Mothers' reports of age of puberty in male twins. 94 BehavGen 24 457-468

Eaves, L. J.; Silberg, J. L.; Hewitt, K.; Rutter, M.; Meyer, J. M.; Neale, M. C.; Pickles, A.: Analyzing twin resemblance in multisymptom data: Genetic applications of a latent class model for symptoms of conduct disorder in juvenile boys. 93 BehavGen 23 5-20

Heath, A. C.; Kessler, R. C.; Neale, M. C.; Hewitt, J. K.; — —; Kendler, K. S.: Testing hypotheses about direction of causation using cross-sectional family data. 93 BehavGen 23 29-50

Eaves, Lindon; Meyer, Joanne: Locating human quantitative trait loci: Guidelines for the selection of sibling pairs for genotyping. 94 BehavGen 24 443-456

Eaves, Lindon J.: Effect of genetic architecture on the power of human linkage studies to resolve the contribution of quantitative trait loci. 94 Heredity 72 175-192

— —; Neale, Michael C.; Meyer, Joanne M.: A model for comparative ratings in studies of within-family differences. 91 BehavGen 21 531-536

Neale, Michael C.; — —: Estimating and controlling for the effects of volunteer bias with pairs of relatives. 93 BehavGen 23 271-278

Neale, Michael C.; — —; Kendler, Kenneth S.: The power of the classical twin study to resolve variation in threshold traits. 94 BehavGen 24 239-258

Eberhardt, L. L.: Using the Lotka-Leslie model for sea otters. 95 JWildlMg 59 222-227

Eberhardt, Mark S.: Madans, Jennifer H.; Reuben, Cynthia A.; Rothwell, Sandra T.; — —: Differences in morbidity measures and risk factor identification using multiple data sources: The case of coronary heart disease. 95 StatMed 14 643-653 (Disc: p691-692)

Eberhart, S. A.: Crossa, J.; Hernandez, C. M.; Bretting, P.; — —; Taba, S.: Statistical genetic considerations for maintaining germ plasm collections. 93 ThApGene 86 673-678

Eberlein, Ernst; Keller, Ulrich: Hyperbolic distributions in finance. 95 Bernoull 1 281-299

Ebneshahrashoob, M.; Sobel, Milton: Dirichlet analysis for inverse multinomial with both quota and quota-free cells. 95 CmpStDA 19 293-307

Ebrahimi, Nader; Pellerey, Franco: New partial ordering of survival functions based on the notion of uncertainty. 95 JAppProb 32 202-211

Ramalingam, T.; — —: Testing for independence between hitting times of bivariate PQD processes. 95 SankhyaA 57 202-220

Soofi, Ehsan S.; — —; Habibullah, Mohamed: Information distinguishability with application to analysis of failure data. 95 JASA 90 657-668

Eccles, Diana M.: Shields, Denis C.; Marlow, Angela J.; Houlston, Richard S.; — —; Morton, Newton E.: Prediction of genetic risks from segregation analyses of morbid risks. 94 HumHrdty 44 52-55

Eccleston, J. A.: Saunders, I. W.; — —; Martin, R. J.: An algorithm for the design of 2^p factorial experiments on continuous processes. 95 AstrlJSt 37 353-365

Eckberg, Douglas Lee: Estimates of early twentieth-century U.S. homicide rates: An econometric forecasting approach. 95 Dmogrphy 32 1-16

Eckes, George P.; Orr, Carolyn: A train the trainer program for design of experiments. 95 ASQCTekCon 348-355

Eckoff, Ronald D.: Assessing public health consequences of a flood disaster. 94 ASAProEnvr 192-193

Eckschlager, K.; Danzer, K. (Auth); Clegg, D. (Rev): Review of "Information theory in analytical chemistry". 95 Chemolab 28 207-208

Eddy, William F.; Mockus, Audris: An interactive image index. 94 ASAProStGr 1-6

— —; Que, Shingo: Dynamic three-dimensional display of U.S. air traffic. 95 JCmpGrSt 4 261-280

Edelman, Alan; Kostlan, Eric: How many zeros of a random polynomial are real?. 95 BAmMaSoc 32 1-37

Edgeman, Rick L.: Iman, Ronald L. (Auth); — — (Rev): Review of "A data-based approach to statistics". 95 JQualTek 27 88-89

Edgerton, Richard T.: A mathematical investigation of quality control. 95 MathTchr 88 200-202

Edgington, Eugene S.: Randomization tests (Third edition). 95 Dekker:NY 424p

Edissonov, Ivan: The new ARSTI optimization method: Adaptive random search with translating intervals. 94 AJMMgSci 14 143-166

Edmonson, R. N.: Mandel, J. (Auth); — — (Rev): Review of "Analysis of two-way layouts". 95 Biomtrcs 51 1188-1189

Edmonston, Barry; Lee, Sharon M.; Passel, Jeffrey S.: U.S. Population projections for national origin groups: Taking into account ethnicity and exogamy. 94 ASAProSoSt 100-105 (Disc: p122-123)

— —; Schultze, Charles (Ed); Wolter, Kirk M. (Rev): Review of "Modernizing the U.S. Census". 95 Chance 8/3 44-49

Edwardes, Michael D. deB.: A confidence interval for $\Pr(X < Y) - \Pr(X > Y)$ estimated from simple cluster samples. 95 Biomtrcs 51 571-578

— —: Comment on "Risk ratio and rate ratio estimation in case-cohort designs: Hypertension and cardiovascular mortality" (V12 p1733-1745). 95 StatMed 14 1609-1609

Edwards, A. W. F.: Fiducial inference and the fundamental theorem of natural selection. 95 Biomtrcs 51 799-809

— — (Auth); Cordeiro, Gauss M. (Rev): Review of "Likelihood". 95 JAppStat 22 191-192

— — (Auth); Smith, C. A. B. (Rev): Review of "Likelihood (Expanded edition)". 93 AnHuGene 57 315-321

Edwards, Brad: Counting gays and lesbians: A case for "new" survey methods. 94 ASASrvyRsr 1305-1309

Edwards, D.: Correction to "Review of 'Resampling-based multiple testing'" (V27 p272-273). 95 JQualTek 27 403-403

— —: Introduction to graphical modelling. 95 Springer:Brln:NY 274p

Edwards, Don: Westfall, Peter H.; Young, S. Stanley (Auth); — — (Rev): Review of "Resampling-based multiple testing". 95 JQualTek 27 272-273

Spurrier, John D.; — —; Thombs, Lori A. (Auth); Hartlaub, Bradley A. (Rev): Review of "Elementary statistics laboratory manual (MS-DOS Version)". 95 AmerStat 49 397-398

Edwards, J. H.: The two locus problem. 94 AnHuGene 58 145-150

Edwards, Lloyd J.: Effect of measurement error in testing separate linear models. 95 JStCmpSm 52 205-216

; Hamilton, Scott A.. Errors-in-variables and the Box-Cox transformation. 95 CmpStDA 20 131-140

Edwards, W. Sherman; Cunningham, Patricia: Scheduling initial calls in a telephone survey of businesses. 94 ASASrvyRsr 1368-1373

Berk, Marc L.; — —; Cantor, Joel C.: Validating access to care. 94 ASAProSoSt 240-244

Efird, Jimmy Thomas; Pardo, Francisco: A non-parametric two-sample survival test based on a single occupancy Fermi-Dirac model for the discrete range distribution. 94 ASAProBiop 166-170

Efromovich, Sam: Sequential nonparametric estimation with assigned risk. 95 AnlsStat 23 1376-1392

Efron, B.; Tibshirani, R. J. (Auth); Brown, Jason (Rev): Review of "An introduction to the bootstrap". 95 JQualTek 27 90-91

— —; Tibshirani, R. J. (Auth); Härdle, W. (Rev): Review of "An introduction to the bootstrap". 95 Metrika 42 267-268

— —; Tibshirani, R. J. (Auth); Marriott, Paul (Rev): Review of "An introduction to the bootstrap". 95 JRSS-A 158 347-347

— —; Tibshirani, Robert J. (Auth); Brown, Morton B. (Rev): Review of "An introduction to the bootstrap". 95 StMedRes 4 91-92

Efron, Bradley; Tibshirani, Robert J. (Auth); Grimshaw, Scott D. (Rev): Review of "An introduction to the bootstrap". 95 Technmcs 37 341-341

Egge, Eric; Foley, Sean; Haskins, Loren; Johnson, Roger: A data analysis based elementary statistics course. 94 ASAProStEd 144-149

Eggermont, P. P. B.; LaRiccia, V. N.: Maximum smoothed likelihood density estimation for inverse problems. 95 AnlsStat 23 199-220

— —; Lariccia, V. N.: Maximum smoothed likelihood density estimation. 95 JNnparSt 4 211-222

Eghball, Bahman; Power, James F.: Fractal description of temporal yield variability of 10 crops in the United States. 95 AgronmyJ 87 152-156

Egidi, V.; Sabbadini, L. L.; Zaccarin, S.: Woman roles between family and employment: Empirical evidence from individual data (Italian). 94 Statstca 54 411-433

Egídio dos Reis, Alfredo D.: Dickson, David C. M.; — —; Waters, Howard R.: Some stable algorithms in ruin theory and their applications. 95 AstinB 25 153-175

Egorov, V. A.; Pozdnyakov, V. I.: The law of the iterated logarithm for trimmed sums (Russian). 94 VLenUM /4 34-38

Ehm, W.; Mammen, E.; Müller, D. W.: Power robustification of approximately linear tests. 95 JASA 90 1025-1033

Ehnholm, C.: Tiret, L.; Nicaud, V.; — —; Havekes, L.; Menzel, H. J.; Ducimetiere, P.; Cambien, F.: Inference of the strength of genotype-disease association from studies comparing offspring with and without parental history of disease. 93 AnHuGene 57 141-149

Eichelsbacher, Peter; Löwe, Matthias: A large deviation principle for m-variate von Mises-statistics and U-statistics. 95 JTherPrb 8 807-824

Eichenauer-Herrmann, Jürgen: Pseudorandom number generation by nonlinear methods. 95 IntStRvw 63 247-255

— —; Niederreiter, Harald: Digital inversive pseudorandom numbers. 94 ACMMdSim 4 339-349

Eick, Stephen G.: Berry, Donald A.; — —: Adaptive assignment versus balanced randomization in clinical trials: A decision analysis. 95 StatMed 14 231-246

Einmahl, J. H. J.: Beirlant, J.; — —: Asymptotic confidence intervals for the length of the short$t$ under random censoring. 95 StNeerla 49 1-8

Einmahl, John H. J.: A Bahadur-Kiefer theorem beyond the largest observation. 95 JMultiAn 55 29-38

— —; Ruymgaart, Frits H.: Tail processes under heavy random censorship applications. 95 JNnparSt 5 223-235

Einmahl, U.; Goodman, V.: Clustering behavior of finite variance partial sum processes. 95 ProbTher 102 547-565

Einmahl, Uwe: On the cluster set problem for the generalized law of the iterated logarithm in Euclidean space. 95 AnlsProb 23 817-851

— —; Mason, David M.: A universal Chung-type law of the iterated logarithm. 94 AnlsProb 22 1803-1825

Eisele, Jeffrey R.; Woodroofe, Michael B.: Central limit theorems for doubly adaptive biased coin designs. 95 AnlsStat 23 234-254

Eisenberg, Bennett: Baryshnikov, Yuliy; — —; Stengle, Gilbert: A necessary and sufficient condition for the existence of the limiting probability of a tie for first place. 95 StPrLet 23 203-209

Eisenhower, Donna; Schmitt, Angela: A records-survey comparison of eligibility and health care utilization measures for Medicaid beneficiaries: Adult and child reports. 94 ASASrvyRsr 713-718

Eisenhower, Donna L.: Design-oriented focus groups and cognitive laboratories: A comparison. 94 ASASrvyRsr 1374-1379

Hardy, Linda P.; — —: Developing methods of collecting the occupation of persons with college degrees. 94 ASAProGovt 110-115

Ekblom, Hakan: Antoch, Jaromír; — —: Recursive robust regression computational aspects and comparison. 95 CmpStDA 19 115-128

Ekholm, Anders; Smith, Peter W. F.; McDonald, John W.: Marginal regression analysis of a multivariate binary response. 95 Biomtrka 82 847-854

Eklöf, Jan A.: Duncan, J. W.; Gross, A. C. (Auth); — — (Rev): Review of "Statistics for the 21st century. Proposals for improving statistics for better decision making". 95 JOffStat 11 225-229

El-Arishy, S.: Useful relationship between the log-survival function and truncated moments, with application. 95 StHefte 36 145-154

El-Din, Moustafa Mohamed Mohie; Mahmoud, Mohamed Abdel Wahab; Sultan, Khalaf Salman: On moment generating function of order statistics for doubly truncated exponential distribution. 95 Metron 53 171-183

El-Gamal, Mahmoud A.; Grether, David M.: Are people Bayesian? Uncovering behavioral strategies. 95 JASA 90 1137-1145

El-Helbawy, Abdalla T.; Hassan, Tawfik: On the Wald, Lagrangian multiplier and likelihood ratio tests when the information matrix is singular. 94 JItalStS 3 51-60

El-Sadr, Wafaa: Paik, Myunghee Cho; Begg, Melissa Dowd; — —; Gorman, Jack; Stein, Zena: Difference in clinical implications of CD4 counts among HIV-infected homosexual men and injection drug using men and women. 95 StatMed 14 1889-1900

El-Saidi, M. A.: Ghitany, M. E.; — —; Khalil, Z.: Characterization of a general class of life-testing models. 95 JAppProb 32 548-553

El-Saidi, Mohammed A.: A symmetric extended logistic model with applications to experimental toxicity data. 95 BiomtrcJ 37 205-216

El-Sayed, Khaled M. F.: Moustafa, Magdi S.; — —: Iterative solution of the $M^{[x]}/C_2/S/N$ queues. 94 AplMaCmp 66 313-323

El-Sayed, Yousry M.: Abdullah, M. E.; — —: Design of crossover microcomputer program and application on drug bioequivalence data. 95 CmpBiomd 48 237-239

El-Shaarawi, A. H.: Trend detection and estimation with environmental applications. 95 MaCmpSim 39 441-447

Elbergali, Abdallah K.; Brereton, Richard G.: Influence of predicted elution regions on the performance of methods for evolutionary factor analysis as applied to high-performance liquid chromatography. 95 Chemolab 27 55-71

Elfessi, Abdulaziz: Pal, Nabendu; — —: Improved estimation of a multivariate normal mean vector and the dispersion matrix: How one affects the other. 95 SankhyaA 57 267-286

Elian, Silvia Nagib: Rodrigues, Josemar; — —: Maximum likelihood prediction in finite populations. 95 RBrzlPrS 9 43-54

Eliashberg, J.; Lilien, G. (Ed); Arabie, Phipps (Rev): Review of "Handbook in operations research and management science: Marketing". 95 JClass 12 143-144

Eliason, Scott R.: Modeling manifest and latent dimensions of association in two-way cross-classifications. 95 SocMethR 24 30-67

Hagenaars, Jacques A. (Auth); — — (Rev): Review of "Categorical longitudinal data: Log-linear panel, trend, and cohort analysis". 95 JEdStat 20 103-104

Eliasziw, Michael; Donner, Allan: Evaluation of a proposed modification to the pairwise estimator of a parent-offspring correlation. 95 SankhyaB 57 151-157

Elie, Laure: Bougerol, Philippe; — —: Existence of positive harmonic functions on groups and on covering manifolds. 95 AlHenrPB 31 59-80

Elizundia, José M.; Hau, Guadalupe: Using SPC in two Mexican poultry processing plants. 95 ASQCTekCon 579-587

— —; Hau, Guadalupe: SCP implementation in a Mexican poultry processing plant. 94 ASAProPEng 295-301

Elkinton, Joseph S.: Dwyer, Greg; — —: Host dispersal and the spatial spread of insect pathogens. 95 Ecology 76 1262-1275

Ellenberg, Jonas H.: Some perspectives on the career of Samuel W. Greehouse: The first 75 years. 95 StatMed 14 1615-1619

Ellickson, B.: — — (Auth); Sheer, Andrew (Rev): Review of "Competitive equilibrium: Theory and applications". 95 JRSS-A 158 347-348

Elliott, Graham: Cavanagh, Christopher L.; — —; Stock, James H.: Inference in models with nearly integrated regressors. 95 EconTher 11 1131-1147

Elliott, P.; Cuzick, J.; English, D.; Stern, R. (Ed); Dunn, Graham (Rev): Review of "Geographical and environmental epidemiology: Methods for small-area studies". 95 StMedRes 4 184-184

Elliott, Paul; Martuzzi, Marco; Shaddick, Gavin: Spatial statistical methods in environmental epidemiology: A critique. 95 StMedRes 4 137-159

Elliott, Robert J.: Recursive estimation for hidden Markov models: A dependent case. 95 StocAnAp 13 437-460

— —; Aggoun, Lakhdar; Moore, John B.: Hidden Markov models: Estimation and control. 95 Springer:Brln:NY 361p

Aggoun, Lakhdar; — —: Finite-dimensional models for hidden Markov chains. 95 AdvAppPr 27 146-160

Chesney, Marc; — —: Estimating the instantaneous volatility and covariance of risky assets. 95 AplStMDA 11 51-58

— —; Aggoun, Lakhdar; Moore, John B. (Auth); Goodwin, C. G. (Rev): Review of "Hidden Markov models: Estimation and control". 95 ShrtBkRv 15/2 29-29

Ellis, Richard S.: An overview of the theory of large deviations and applications to statistical mechanics. 95 ScanActJ 97-142

Ellis, Steven P.: Dimension of the singular sets of plane-fitters. 95 AnlsStat 23 490-501

— —: A note on the smoothness of L^1-estimators for the linear model. 95 SankhyaA 57 221-226

Ellis, Yukiko: Schwede, Laurie; — —: Exploring associations between subjective and objective assessments of household membership. 94 ASASrvyRsr 325-330

Elomaa, Matti; Lochmüller, Charles H.; Kaljurand, Mihkel; Koel, Mihkel: Application of evolving factor analysis in thermochromatography. 95 Chemolab 30 129-132

Elsberry, Victor A.: Roe, Denise J.; Cheng, Mary; — —; Nolan, Paul E.: Estimating parameters in non-linear mixed effect models: A comparison using clinical data in children. 94 ASAProBiop 313-318

Elsen, J. M.: Le Roy, Pascale; — —: Simple test statistics for major gene detection: A numerical comparison. 92 ThApGene 83 635-644

Elsen, Jean-Michel: Haley, Chris S.; Knott, Sara A.; — —: Mapping quantitative trait loci in crosses between outbred lines using least squares. 94 Genetcs 136 1195-1207

Elshamy, M.: Randomly perturbed vibrations. 95 JAppProb 32 417-428

Elson, Paul J.: Use of the fast Fourier transform algorithm in the calculation of the operating characteristics of group sequential clinical trials. 95 CmpStDA 20 491-498

Elster, Clemens; Neumaier, Arnold: Screening by conference designs. 95 Biomtrka 82 589-602

Elston, D. A.; Proe, M. F.: Smoothing regression coefficients in an overspecified regression model with interrelated explanatory variables. 95 ApplStat 44 395-406

Patton, D. R. (Auth); — — (Rev): Review of "Wildlife habitat relationships in forested ecosystems". 95 Biomtrcs 51 384-385

Elston, R. C.: Fernando, R. L.; Stricker, C.; — —: An efficient algorithm to compute the posterior genotypic distribution for every member of a pedigree without loops. 93 ThApGene 87 89-93

Elston, Robert C.; George, Varghese T.; Severtson, Forrestt: The Elston-Stewart algorithm for continuous genotypes and environmental factors. 92 HumHrdty 42 16-27

Spence, M. Anne; Bishop, D. Timothy; Boehnke, Michael; — —; Falk, Catherine; Hodge, Susan E.: Methodological issues in linkage analyses for psychiatric disorders: Secular trends, assortative mating, bilineal pedigrees. 93 HumHrdty 43 166-172

Eltinge, J. L.; Yansaneh, I. S.; Paulin, G. D.: Assessment of reported differences between expenditures and low incomes in the U.S. Consumer Expenditure Survey. 94 ASASrvyRsr 96-105

Elton, R. A.: Lee, E. T. (Auth); — — (Rev): Review of "Statistical methods for survival data analysis (Second edition)". 95 Biomtrcs 51 383-383

Elwood, J. M.: MacNeill, I. B.; — —; Miller, D.; Mao, Y.: Trends in mortality from melanoma in Canada and prediction of future rates. 95 StatMed 14 821-839

Elworthy, K. D.: Prato, G. Da; — —; Zabczyk, J.: Strong Feller property for stochastic semilinear equations. 95 StocAnAp 13 35-45

Bellomo, N.; Brzezniak, Z.; de Socio, L. M. (Auth); — — (Rev): Review of "Nonlinear stochastic evolution problems in applied sciences". 95 Metrika 42 140-141

— —; Everitt, W. N.; Lee, E. B. (Ed); Zacharias, K. (Rev): Review of "Differential equations, dynamical systems, and control science". 95 BiomtrcJ 37 127-128

— —; Ikeda, N. (Ed); Glendinning, R. H. (Rev): Review of "Asymptotic problems in probability theory: Stochastic models and diffusions on fractals". 95 Ststcian 44 139-140

— —; Ikeda, N. (Ed); Wakolbinger, A. (Rev): Review of "Asymptotic problems in probability theory: Stochastic models and diffusions on fractals". 95 Metrika 42 69-70

Ikeda, N.; — — (Auth); Simakani, J. E. (Rev): Review of "Asymptotic problems in probability theory: Wiener functionals and fractionals". 95 Ststcian 44 289-289

Embleton, B. J. J.: Fisher, N. I.; Lewis, T.; — — (Auth); Challenor, Peter (Rev): Review of "Statistical analysis of spherical data". 95 Ststcian 44 286-287

Embrechts, P.; Rogers, L. C. G.; Yor, M.: A proof of Dassios' representation of the α-quantile of Brownian motion with drift. 95 AnlsApPr 5 757-767

Binswanger, K.; — —: Longest runs in coin tossing. 94 InsrneME 15 139-149

Embrechts, P. A. L.: Bass, R. F. (Auth); — — (Rev): Review of "Probabilistic techniques in analysis". 95 ShrtBkRv 15/3 43-43

Gut, A. (Auth); — — (Rev): Review of "An intermediate course in probability". 96 ShrtBkRv 16/1 4-4

Guttorp, Peter (Auth); — — (Rev): Review of "Stochastic modeling of scientific data". 96 ShrtBkRv 16/1 1-2

Last, Günter; Brandt, Andreas (Auth); — — (Rev): Review of "Marked point processes on the real line: The dynamic approach". 96 ShrtBkRv 16/1 10-10

Rogers, L. C. G.; Williams, David (Auth); — — (Rev): Review of "Diffusions, Markov processes, and martingales. Volume 1: Foundations (Second edition)". 95 ShrtBkRv 15/2 28-28

Embrechts, Paul: Risk theory of the second and third kind. 95 ScanActJ 35-43

— —; Samorodnitsky, Gennady: Sample quantiles of heavy tailed stochastic processes. 95 StocProc 59 217-233

Hoffmann-Jørgensen, J. (Auth); — — (Rev): Review of "Probability with a view toward statistics. Volumes 1 and 2". 95 JASA 90 806-806

Emerson, Scott S.: Stopping a clinical trial very early based on unplanned interim analyses: A group sequential approach. 95 Biomtrcs 51 1152-1162

Enas, Greg: Huster, William; Brunelle, Rocco; — —: Essential efficacy data analysis for a 2×2 crossover study. 94 ASAProBiop 503-509

Enas, Gregory G.; Goldstein, David J.: Defining, monitoring and combining safety information in clinical trials. 95 StatMed 14 1099-1111 (Disc: p1113-1116)

Enas, Nathan: Bolton, S. (Auth); — — (Rev): Review of "Pharmaceutical statistics: Practical and clinical applications (Second edition)". 95 StMedRes 4 267-267

Enchautegui, María E.: Impacts of immigrants on employment and wage growth: Evidence of positive effects of U.S. Immigration. 94 ASAProSoSt 30-38

Endean, Sarah: School Mathematics Project (Auth); — — (Rev): Review of "Statistics in action SMP 16-19 Mathematics". 95 TeachgSt 17/2 79-79

Enderlein, G.: Daniel, Wayne W. (Auth); — — (Rev): Review of "Biostatistics - A foundations for analysis in the health sciences" (German). 95 BiomtrcJ 37 744-744

Kreienbrock, L.; Schach, S. (Auth); — — (Rev. German): Review of "Epidemiological methods" (German). 95 BiomtrcJ 37 1016-1016

Endo, Hideki: Correspondence analysis of an artificial binary cylinder data. 95 StPrLet 25 231-240

Okamoto, Masashi; — —: Spherical trait in correspondence analysis of artificial data. 95 JJapnStS 25 183-191

Endrényi, László: Tóthfalusi, László; — —: Heteroscedastic nonlinear regression when there are possible outliers. 94 ASAProStCp 239-244

Engel, B.: Meuwissen, T. H. E.; — —; van der Werf, J. H. J.: Maximizing selection efficiency for categorical traits. 95 JAnmlSci 73 1933-1939

Engel, Joachim; Gasser, Theo: A minimax result for a class of nonparametric density estimators. 95 JNnparSt 4 327-334

Herrmann, Eva; Wand, M. P.; — —; Gasser, Theo: A bandwidth selector for bivariate kernel regression. 95 JRSS-B 57 171-180

Kneip, Alois; — —: Model estimation in nonlinear regression under shape invariance. 95 AnlsStat 23 551-570

England, Amy M.; Hubbell, Katie A.; Judkins, David R.; Ryaboy, Svetlana: Imputation of medical cost and payment data. 94 ASASrvyRsr 406-411

Engle, Robert F.; Kroner, Kenneth F.: Multivariate simultaneous generalized ARCH. 95 EconTher 11 122-150

English, D.: Elliott, P.; Cuzick, J.; — —; Stern, R. (Ed); Dunn, Graham (Rev): Review of "Geographical and environmental epidemiology: Methods for small-area studies". 95 StMedRes 4 184-184

English, J. R.: Chandrasekaran, S.; — —; Disney, R. L.: Modeling and analysis of EWMA control schemes with variance-adjusted control limits. 95 IIETrn 27 282-290

Englund, Evan J.; Heravi, Naser: Phased sampling for soil remediation. 94 EnvEcoSt 1 247-263

Englund, Jan-Eric: Using a diagnostic test in a finite population. 94 MaScitst 19 108-118

Ennadifi, G.: Asymptotic behaviour of record times for renewal paced records. 95 StatDecn 13 379-397

Ennadifi, Gratiane: Strong approximation of the number of renewal paced record times. 95 JStPlInf 45 113-132

Ensign, William E.; Angermeier, Paul L.; Dolloff, C. Andrew: Use of line transect methods to estimate abundance of benthic stream fishes. 95 JFishRBC 52 213-222

Ensor, Katherine B.: Harvey, Andrew C. (Auth); — — (Rev): Review of "Time series models". 95 Technmcs 37 464-465

Entsuah, Richard: Software for handling dropouts in clinical trials. 94 ASAProBiop 188-193

Epperson, David L.; Davis, Jerry M.; Bloomfield, Peter; Karl, Thomas R.; McNab, Alan L.; Gallo, Kevin P.: Estimating the urban bias of surface shelter temperatures using upper-air and satellite data. Part I: Development of models predicting surface shelter temperatures. 95 JApMetrl 34 340-357

Epps, T. W.: A test of fit for lattice distributions. 95 CommStA 24 1455-1479

Epstein, Joan: Wright, Douglas; Gfroerer, Joseph; — —: Ratio estimation of hard-core drug use. 94 ASASrvyRsr 541-546

Epstein, Joan F.; Gfroerer, Joseph C.: Estimating substance abuse treatment need from the NHSDA. 94 ASASrvyRsr 559-564

Epstein, Leonardo D.; Muñoz, Alvaro: A parametric model for survival and intermediate event times. 94 ASAProEpid 91-96

Epstein, Richard A.: The theory of gambling and statistical logic (Revised edition). 95 Academic:NY:Lnd 450p

Erickson, Roy V.; Fabian, Václav; Mařík, Jan: An optimum design for estimating the first derivative. 95 AnlsStat 23 1234-1247

Ericsson, Neil R.: Comment on "Efficient inference on cointegration parameters in structural error correction models". 95 JEconmtx 69 159-171

Cuthbertson, Keith; Hall, Stephen G.; Taylor, Mark P. (Auth); — —; Irons, John S. (Rev): Review of "Applied econometric techniques". 95 EcomtrRv 14 121-132

Eriksen, P. S.: Andersen, H. H.; Højbjerre, M.; Sørensen, D.; — —: Linear and graphical models for the multivariate complex normal distribution. 95 Springer:Brln:NY 183p

Erkel-Rousse, H.: Detection of multicollinearity in an ordinary linear model: Some elements of an informed use of the Belsley, Kuh and Welsch indicators (French). 95 RStApp 43/4 19-42

Ermakov, M. S.: Minimax nonparametric testing of hypotheses on the distribution density. 94 ThProbAp 39 396-416

Ermakov, S. M.; Wang, Xiao-Qun: Random invariant cubature formulas (Russian). 94 VLenUM /4 109-112

Ernst, Lawrence R.; Ikeda, Michael M.: A reduced-size transportation algorithm for maximizing the overlap between surveys. 95 SurvyM 21 147-157

Erraoui, M.; Ouknine, Y.: Limit theorems for anticipative stochastic differential equations (French). 95 Stochatx 53 1-11

Erritt, John: Obituary: Sir John Boreham, 1925-1994. 95 JRSS-A 158 182-184

Escobar, Luis A.; Meeker, William Q., Jr: Planning accelerated life tests with two or more experimental factors. 95 Technmcs 37 411-427

Meeker, William Q.; — —: Teaching about approximate confidence regions based on maximum likelihood estimation. 95 AmerStat 49 48-53

Escobar, Michael D.: Nonparametric Bayesian methods in hierarchical models. 95 JStPlInf 43 97-106

— —; West, Mike: Bayesian density estimation and inference using mixtures. 95 JASA 90 577-588

Escolano, Sylvie; Golmard, Jean-Louis; Mallet, Alain: A fully nonparametric approach in survival models with explanatory variables. 95 CommStA 24 3027-3054

Escribano, Alvaro: PcGive Professional 8. 95 JApEcmtx 10 79-86

Escudero, L. F.; Kamesam, P. V.: On solving stochastic production planning problems via scenario modelling. 95 Top 3 69-95

Alvarez, M.; Cuevas, C. M.; — —; de la Fuente, J. L.; García, C.; Prieto, F. J.: Network planning under uncertainty with an application to hydropower generation. 94 Top 2 25-58

Esinhart, James D.: Chinchilli, Vernon M.; — —: The assessment of individual bioequivalence via tolerance intervals. 94 ASAProBiop 29-32

Chinchilli, Vernon M.; — —; Miller, W. Greg: Partial likelihood analysis of within-unit variances in repeated measurement experiments. 95 Biomtrcs 51 205-216

Eskridge, K. M.: Statistical analysis of disease reaction data using nonparametric methods. 95 HortSci 30 478-481

Eslinger, Paul W.: Woodward, Wayne A.; Whitney, Paul; — —: Minimum Hellinger distance estimation of mixture proportions. 95 JStPlInf 48 303-319

Espa, Giuseppe: Point pattern identification: An alternative approach using distances between observed events (Italian). 94 QStMaA 16 117-125

— —; Terpessi, Cristina: Short and medium term rain forecasting in the Oristano basin (Italian). 94 QStMaA 16 145-160

Espasa, Antoni; Peña, Daniel: The decomposition of forecast in seasonal ARIMA models. 95 JForcast 14 565-583

Esposito, Richard; Fox, James K.; Lin, Dongyow; Tidemann, Kevin: ARIES: Visual techniques for statistical data investigation at the Bureau of Labor Statistics. 94 ASAProStGr 7-10

Esteban, M. D.; Morales, D.: Estimating a survival function with doubly censored data in a proportional hazard model. 95 AplStMDA 11 145-157

Estevao, V.; Hidiroglou, M. A.; Särndal, C. E.: Methodological principles for a generalized estimation system at Statistics Canada. 95 JOffStat 11 181-204

— —; Patak, Z.; Särndal, C.: Maximizing the use of auxiliary information for calibration and regression estimation. 94 ASASrvyRsr 42-48

Estéve, Jacques; Benhamou, Ellen; Raymond, Luc (Ed): Statistical methods in cancer research. Volume 4: Descriptive epidemiology. 95 OxfordUniv 320p

Esteves da Silva, Joaquim C. G.; Machado, Adélio A. S. C.: Evolving factor analysis of synchronous fluorescence spectra of humic substances in the presence of Cu(II). 95 Chemolab 27 115-128

Esteves, Jorge Sá; Craveirinha, José; Cardoso, Domingos: Computing Erlang-B function derivatives in the number of servers. 95 CommStSM 11 311-331

Estey, Elihu H.: Thall, Peter F.; Simon, Richard M.; — —: Bayesian sequential monitoring designs for single-arm clinical trials with multiple outcomes. 95 StatMed 14 357-379

Etchison, Tonya; Brownie, Cavell; Pantula, Sastry G.: A portmanteau test for spatial ARMA models. 95 Biomtrcs 51 1536-1542

Etheridge, Alison: — — (Ed): Stochastic partial differential equations. 95 CambridgeUniv 347p

Ethier, S. N.; Krone, Stephen M.: Comparing Fleming-Viot and Dawson-Watanabe processes. 95 StocProc 60 171-190

Ette, Ene I.; Sun, He; Ludden, Thomas M.: Design of population pharmacokinetic studies. 94 ASAProBiop 487-492

Etzioni, R.; Self, S. G.: Correction to "On the catch-up time method for analyzing cancer screening trials" (V51 p31-43). 95 Biomtrcs 51 1589-1589

Etzioni, Ruth; Self, Steven G.: On the catch-up time method for analyzing cancer screening trials. 95 Biomtrcs 51 31-43

Self, Steven G.; — —: A likelihood ratio test for cancer screening trials. 95 Biomtrcs 51 44-50

Etzioni, Ruth D.; Connor, Robert J.; Prorok, Philip C.; Self, Steven G.: Design and analysis of cancer screening trials. 95 StMedRes 4 3-17

Evanoff, Javon: Haskin, F. Eric; Radke, George E., Jr; — —: Algorithms for analyzing the probability of M-out-of-N events. 95 RelEngSS 47 153-162

Evans, B. Timothy: Smith, Robert T., Jr; — —; Fowler, John L.: The Commodity Flow Survey. 94 ASAProBuEc 89-94

Evans, Charles R.: Cota, Albert A.; — —; Longman, R. Stewart; Dion, Kenneth L.; Kilik, Lindy: Using and misusing factor analysis to explore group cohesion. 95 JClPsych 51 308-316

Evans, Geoffrey: Heath, Anthony; — —; Payne, Clive: Modelling the class-party relationship in Britain, 1964-92. 95 JRSS-A 158 563-574

Evans, Gwyn: Median and quartiles from grouped data. 95 TeachgSt 17/3 99-100

Evans, K. Franklin; Turk, Joseph; Wong, Takmeng; Stephens, Graeme L.: A Bayesian approach to microwave precipitation profile retrieval. 95 JApMetrl 34 260-279

Evans, M.; Hastings, N.; Peacock, B. (Auth); Trenkler, Götz (Rev): Review of "Statistical distributions (Second edition)". 95 CmpStDA 19 483-484

Evans, M. A.: Ratkowsky, D. A.; — —; Alldredge, J. R. (Auth); Gillespie, Brenda (Rev): Review of "Crossover experiments: Design, analysis and application". 95 JAppStat 22 188-188

Ratkowsky, D. A.; — —; Alldredge, J. R. (Auth); Walker, Stephen (Rev): Review of "Cross-over experiments: Design, analysis, and application". 95 JRSS-A 158 200-200

Evans, Marc A.: Ratkowsky, David A.; — —; Alldredge, J. Richard (Auth); Berghold, Andrea (Rev): Review of "Cross-over experiments. Design, analysis and application". 95 CmpStQ 10 99-100

Evans, Martin G.: Kanetkar, Vinay; — —; Everell, Shirley Anne; Irvine, Diane; Millman, Zeeva: The effect of scale changes on meta-analysis of multiplicative and main effects models. 95 EdPsMeas 55 206-224

Evans, Merran; Hastings, Nicholas; Peacock, Brian (Auth); Copeland, Karen A. F. (Rev): Review of "Statistical distributions". 95 JQualTek 27 94-95

Evans, R.: Mšicki, D.; — —: Integrated probabilistic data association-finite resolution. 95 Autmatca 31 559-570

Evans, Thomas D.: Zimmerman, Tamara Sue; — —; Tiller, Richard B.: State unemployment rate time series models. 94 ASASrvyRsr 1077-1082

Everell, Shirley Anne: Kanetkar, Vinay; Evans, Martin G.; — —; Irvine, Diane; Millman, Zeeva: The effect of scale changes on meta-analysis of multiplicative and main effects models. 95 EdPsMeas 55 206-224

Everitt, B.: Dunn, G.; — —: Clinical biostatistics: An introduction to evidence-based medicine. 95 Arnold:Lnd:MD 154p

Dunn, G.; — — (Auth); Lagakos, S. W. (Rev): Review of "Clinical biostatistics: An introduction to evidence-based medicine". 95 ShrtBkRv 15/3 41-42

Dunn, G.; — —; Pickles, A. (Auth); Chan, Wai (Rev): Review of "Modelling covariances and latent variables using EQS". 95 SIAMRvw 37 462-464

Dunn, G.; — —; Pickles, A. (Auth); Schimek, Michael G. (Rev): Review of "Modelling covariances and latent variables using EQS". 95 CmpStQ 10 196-196

Everitt, B. S.: The Cambridge dictionary of statistics in the medical sciences. 95 CambridgeUniv 274p

— —: The analysis of repeated measures: A practical review with examples. 95 Ststcian 44 113-135

— — (Auth); Davies, Neville (Rev): Review of "A handbook of statistical analysis using S-Plus". 95 JRSS-A 158 633-634

— — (Auth); Herrendörfer, G. (Rev): Review of "The analysis of contingency tables". 95 BiomtrcJ 37 104-104

— — (Auth); Kimber, Alan (Rev): Review of "A handbook of statistical analyses using S-Plus". 95 Ststcian 44 408-409

— — (Auth); Ruxton, Graeme D. (Rev): Review of "A handbook of statistical analyses using S-Plus". 95 JAppStat 22 545-545

Everitt, Brian: The Eurovision song contest. 95 Stats /13 19-22

Dunn, Graham; — —: Clinical biostatistics. An introduction to evidence-based medicine. 95 Wiley:NY:UK 154p

Everitt, Brian S.; Dunn, Graham: Applied multivariate data analysis. 95 Wiley:-NY:UK 304p

— — (Auth); Ko, Daijin (Rev): Review of "A handbook of statistical analyses using S-Plus". 95 Technmcs 37 232-233

— — (Auth); Venables, W. N. (Rev): Review of "A handbook of statistical analyses using S-Plus". 95 AstrlJSt 37 369-372

— — (Auth); Williams, Calvin (Rev): Review of "A handbook of statistical analyses using S-Plus". 95 JQualTek 27 89-90

Everitt, W. N.: Elworthy, K. D.; — —; Lee, E. B. (Ed); Zacharias, K. (Rev): Review of "Differential equations, dynamical systems, and control science". 95 BiomtrcJ 37 127-128

Evinger, Suzann: How shall we measure our nation's diversity?. 95 Chance 8/1 7-14

Evmiridis, Nickolaos P.: Papamichael, Emmanuel M.; — —; Thanasoulias, Nickolaos; Stefanou, Despina: Design of experiments for a precise estimation of the calibration curve of periodate in its determination by flow-injection analysis and chemiluminescence detection.. 95 Chemolab 30 227-237

Ewens, W. J.: Obituary: Motoo Kimura, 1924-1994. 95 JAppProb 32 1142-1144

Ewens, Warren J.; Spielman, Richard S.: The transmission/disequilibrium test: History, subdivision, and admixture. 95 AJHuGene 57 455-464

Excoffier, Laurent; Smouse, Peter E.: Using allele frequencies and geographic subdivision to reconstruct gene trees within a species: Molecular variance parsimony. 94 Genetcs 136 343-359

Ezcurra, Exequiel; de Ita, Celia: Mexico's policy framework for environmental reporting. 94 ASAProEnvr 19-25 (Disc: p36-37)

Ezzati, Trena M.; Hoffman, Keith; Judkins, David R.; Massey, James T.; Moore, Thomas F.: A dual frame design for sampling elderly minorities and persons with disabilities. 95 StatMed 14 571-583 (Disc: p695-696)

Ezzati-Rice, Trena M.; Khare, Meena: Modeling of response propensity in the Third National Health and Nutrition Examination Survey. 94 ASASrvyRsr 955-959

Khare, Meena; — —; Mohadjer, Leyla K.; Waksberg, Joseph: An evaluation of nonresponse bias in NHANES III (1988-91). 94 ASASrvyRsr 949-954

Fabian, Václav: Erickson, Roy V.; — —; Mařík, Jan: An optimum design for estimating the first derivative. 95 AnlsStat 23 1234-1247

Fabrizius, M. A.; Openshaw, S. J.: Methods to evaluate populations for alleles to improve an elite hybrid. 94 ThApGene 88 653-661

Fahimi, Mansour: Poststratification of pooled survey data. 94 ASASrvyRsr 879-881

Fahrmeir, L.; Tutz, G. (Auth); Glosup, Jeffrey (Rev): Review of "Multivariate statistical modelling based on generalized linear models". 95 Technmcs 37 350-351

Fairchild, Charles K.: Modeling trends in emergency room drug abuse episodes. 94 ASASrvyRsr 305-309

Faivre, C.: Barbolosi, D.; — —: Metrical properties of some random variables connected with the continued fraction expansion. 95 IndgMatN 6 257-265

Fakhre-Zakeri, I.: Cambanis, S.; — —: On prediction of heavy-tailed autoregressive sequences: Forward versus reversed time. 94 ThProbAp 39 217–233

Falck, Russel S.: Wang, Jichuan; Fisher, James H.; Siegal, Harvey A.; — —; Carlson, Robert G.: Influence of measurement errors on HIV risk behavior analysis: A case study examining condom use among drug users. 95 StrEqMod 2 319–334

Falconer, K. J.; Xiao, Y. M.: Average densities of the image and zero set of stable processes. 95 StocProc 55 271–283

Falin, G.: An error bound for a continuous time approximation of a time-sharing queue. 95 CommStSM 11 163–169

Falin, Gennadi: Heavy traffic analysis of a random walk on a lattice semi-strip. 95 CommStSM 11 395–409

Falin, J. I.: Artalejo, J. R.; — —: Stochastic decomposition for retrial queues. 94 Top 2 329–342

Falissard, B.: Displaying a correlation matrix on the unit sphere in R^3 (French). 95 RStApp 43/2 35–48

Faliva, Mario; Zoia, M. Grazia: Detecting and testing causality in linear econometric models. 94 JItalStS 3 61–76

Falk, A.: Wisdom updated. 95 PhilSci 62 389–403

Falk, Catherine: Spence, M. Anne; Bishop, D. Timothy; Boehnke, Michael; Elston, Robert C.; — —; Hodge, Susan E.: Methodological issues in linkage analyses for psychiatric disorders: Secular trends, assortative mating, bilineal pedigrees. 93 HumHrdty 43 166–172

Falk, Eric: Rotz, Wendy; Mulrow, Jeri; — —: The 1995 Taxpayer Compliance Measurement Program (TCMP) sample redesign - A case history. 94 ASASrvyRsr 699–703

Falk, Michael: LAN of extreme order statistics. 95 AnInStMa 47 693–717

— —: Some best parameter estimates for distributions with finite endpoint. 95 MaOpfStS 27 115–125

— —; Hüsler, Jurg; Reiss, Rolf-Dieter (Auth); Mikosch, Thomas (Rev): Review of "Law of small numbers: Extremes and rare events". 95 JASA 90 1133–1134

Falk, R.: — — (Auth); Shotter, M. V. (Rev): Review of "Understanding probability and statistics: A book of problems". 95 Biomtrcs 51 386–386

— — (Auth); Woodward, P. W. (Rev): Review of "Understanding probability and statistics: A book of problems". 95 Ststcian 44 284–285

Famoye, Felix: Generalized binomial regression model. 95 BiomtrcJ 37 581–594

— —; Consul, P. C.: Bivariate generalized Poisson distribution with some applications. 95 Metrika 42 127–138

Consul, P. C.; — —: On the generalized negative binomial distribution. 95 CommStA 24 459–472

Johnson, Norman L.; Kotz, Samuel; Balakrishnan, N. (Auth); — — (Rev): Review of "Continuous univariate distributions, Volume I (Second edition)". 95 Technmcs 37 466–466

Famulari, Melissa: A household-based, nonparametric test of demand theory. 95 RvEconSt 77 372–382

Fan, Jianqing; Gijbels, Irène: Adaptive order polynomial fitting: Bandwidth robustification and bias reduction. 95 JCmpGrSt 4 213–227

— —; Gijbels, Irène: Data-driven bandwidth selection in local polynomial fitting: Variable bandwidth and spatial adaptation. 95 JRSS-B 57 371–394

— —; Heckman, Nancy E.; Wand, M. P.: Local polynomial kernel regression for generalized linear models and quasi-likelihood functions. 95 JASA 90 141–150

Fan, Jincheng; Wang, Nin; Mei, Changlin: The random weighting approximation for the distribution of the partial sum of a stationary linear series (Chinese). 95 AMaAPknC 17 65–75

Fan, Rocky Y. K.; Field, Christopher A.: Approximations for marginal densities of M-estimators. 95 CandJSt 23 185–197

Fan, Tsai-Hung: Bayesian bootstrap clones for finite state Markov chains. 95 JStCmpSm 53 289–298

Fan, Yanqin: Bootstrapping a consistent nonparametric goodness-of-fit test. 95 EcomtrRv 14 367–382

— —: Average derivative estimation with errors-in-variables. 95 JNnparSt 4 395–407

— —; Gencay, Ramazan: A consistent nonparametric test of symmetry in linear regression models. 95 JASA 90 551–557

Fang, K. T.: Anderson, T. W.; — —; Olkin, I. (Ed): Multivariate analysis and its applications. 95 Arnold:Lnd:MD 471p

Fang, K.-T.; Wang, Y. (Auth); Ripley, B. D. (Rev): Review of "Number-theoretic methods in statistics". 95 JRSS-A 158 189–190

Fang, Kai-Tai: Li, Run-Ze; — —: Estimation of scale matrix of elliptically contoured matrix distributions. 95 StPrLet 24 289–297

Pan, Jian-Xin; — —: Multiple outlier detection in growth curve model with unstructured covariance matrix. 95 AnInStMa 47 137–153

Wang, Yuchung; Lin, Dennis K. J.; — —: Designing outer array points. 95 JQualTek 27 226–241

Zhu, Li-Xing; — —; Bhatti, M. I.; Bentler, Peter M.: Testing sphericity of a high-dimensional distribution based on bootstrap approximation. 95 PakJStA 11 49–65

Zhu, Li-Xing; Wong, Hoi Lam; — —: A test for multivariate normality based on sample entropy and projection pursuit. 95 JStPlInf 45 373–385

— —; Wang, Y. (Auth); Ginebra, Josep (Rev): Review of "Number-theoretic methods in statistics". 95 JASA 90 1134–1134

Fang, Kaitai; Li, Jiukun: Some new results of uniform design (Chinese). 95 KToChina 39 1921–1924

Fang, Xiangzhong: The law of iterated logarithms of maximum likelihood estimate for Weibull processes (Chinese). 95 ChJApPSt 10 314–319

Faraggi, David; Simon, Richard: Maximum likelihood neural network prediction models. 95 BiomtrcJ 37 713–725

— —; Simon, Richard: The maximum likelihood neural network as a statistical classification model. 95 JStPlInf 46 93–104

— —; Simon, Richard: A neural network model for survival data. 95 StatMed 14 73–82

Dror, Shuki; — —; Reiser, Benjamin: Dynamic treatment allocation adjusting for prognostic factors for more than two treatments. 95 Biomtrcs 51 1338–1343

Faraway, Julian J.; Sun, Jiayang: Simultaneous confidence bands for linear regression with heteroscedastic errors. 95 JASA 90 1094–1098

Färe, Rolf; Grosskopf, Shawna: Nonparametric tests of regularity, Farrell efficiency, and goodness-of-fit. 95 JEconmtx 69 415–425

Farebrother, Richard William: On hyperplanes of closest fit. 95 CmpStDA 19 53–58

Fareed, T. P. Muhammad: Deshpande, Jayant V.; — —: A note on conditionally unbiased estimation after selection. 95 StPrLet 22 17–23

Farewell, V. T.: Lancaster, H. O. (Auth); — — (Rev): Review of "Quantitative methods in biological and medical sciences. A historical essay". 95 ShrtBkRv 15/2 22–22

Farewell, Vern T.: Cook, Richard J.; — —: Conditional inference for subject-specific and marginal agreement: Two families of agreement measures. 95 CandJSt 23 333–344

Faria, A. E.; Souza, R. C.: A re-evaluation of the quasi-Bayes approach to the linear combination of forecasts. 95 JForcast 14 533–542

Farley, Dean: Doyle, Pat; — —: Alternative strategies for imputing premiums and predicting expenditures under health care reform. 94 ASASrvyRsr 517–522

Farnum, Nicholas R.: Comment on "Q charts". 95 JQualTek 27 322–323

— — (Auth); Borkowski, John J. (Rev): Review of "Modern statistical quality control and improvement". 95 JQualTek 27 91–93

Farquhar, John W.: Fortmann, Stephen P.; Flora, June A.; Winkleby, Marilyn A.; Schooler, Caroline; Taylor, C. Barr; — —: Community intervention trials: Reflections on the Stanford Five-City Project experience. 95 AJEpidm 142 576–586

Farrer, Lindsay A.; Cupples, L. Adrienne: Estimating the probability for major gene Alzheimer disease. 94 AJHuGene 54 374–383

Farrington, C. P.: Relative incidence estimation from case series for vaccine safety evaluation. 95 Biomtrcs 51 228–235

Brookmeyer, Ron; Gail, Mitchell H. (Auth); — — (Rev): Review of "AIDS epidemiology: A quantitative approach". 95 AJEpidm 141 180–181

Fattinger, Karin E.; Sheiner, Lewis B.; Verotta, Davide: A new method to explore the distribution of interindividual random effects in non-linear mixed effects models. 95 Biomtrcs 51 1236–1251

Fattorini, L.: Barabesi, L.; — —: A note on bandwidth selections for kernel density estimators at endpoints. 94 Metron 52/3 43–56

Fattorini, Lorenzo: Barabesi, Lucio; — —: Approximation of density functions by orthogonal series with grouped data. 94 JItalStS 3 181–200

Favretto, L.: Vojnovic, D.; Campisi, B.; Mattei, A.; — —: Experimental mixture design to ameliorate the sensory quality evaluation of extra virgin olive oils. 95 Chemolab 27 205–210

Fay, Robert: Butani, Shail; Cahoon, Lawrence; — —; Kostanich, Donna: Measurement of different design effects. 94 ASASrvyRsr 342–347 (Disc: p360-361)

Fay, Robert E.: Comment on "Alternative methods for the 2000 census". 94 ASASrvyRsr 90–92

— —: Analyzing imputed survey data sets with model assisted estimators. 94 ASASrvyRsr 900–905 (Disc: p906-907)

Fayolle, G.; Malyshev, V. A.; Menshikov, M. V.: Topics in the constructive theory of countable Markov chains. 95 CambridgeUniv 169p

Asymont, I. M.; — —; Menshikov, M. V.: Random walks in a quarter plane with zero drifts: Transience and recurrence. 95 JAppProb 32 941–955

— —; Malyshev, V. A.; Menshikov, M. V. (Auth); Kozek, A. S. (Rev): Review of "Topics in the constructive theory of countable Markov chains". 96 ShrtBkRv 16/1 9–10

Fazli, K.; Behboodian, J.: Skorohod's theorem and convergence of types. 95 StPrLet 24 243–244

Fearn, Dean H.; Nebenzahl, Elliott: Using power transformations when approximating quantiles. 95 CommStA 24 1073–1093

Fearn, Tom: Mertens, Bart; — —; Thompson, Michael: The efficient cross-validation of principal components applied to principal component regression. 95 StatComp 5 227–235

Brown, P. J.; — — (Rev): Review of "Measurement, regression, and calibration". 95 StMedRes 4 265–266

Retherford, R. D.; Choe, M. K. (Auth); — — (Rev): Review of "Statistical models for causal analysis". 95 JRSS-A 158 200–201

Febrero-Bande, M.: García-Jurado, I.; González-Manteiga, W.; Prada-Sánchez, J. M.; — —; Cao, R.: Predicting using Box-Jenkins, nonparametric, and bootstrap techniques. 95 Technmcs 37 303–310

Federer, W. T.; Scully, B. T.: A parsimonious statistical design and breeding procedure for evaluating and selecting desirable characteristics over environments. 93 ThApGene 86 612-620

Anderson, D. A.; — —: Representation and bounds for the general fractional factorial. 95 CommStA 24 363-373

Schwager, S. J.; Mutschler, M. A.; — —; Scully, B. T.: The effect of linkage on sample size determination for multiple trait selection. 93 ThApGene 86 964-974

— — (Auth); Rudolph, P. E. (Rev): Review of "Statistical design and analysis for intercropping experiments. Volume I: Two crops". 95 BiomtrcJ 37 204-204

Federer, Walter T.: A simple procedure for constructing experiment designs with incomplete blocks of sizes 2 and 3. 95 BiomtrcJ 37 899-907

Fedorov, V. V.; Hackl, P.; Müller, W. G.: Optimal and practicable designs for measuring plaque pH-profiles. 95 StatMed 14 2609-2617

Lindman, H. R. (Auth); — — (Rev): Review of "Analysis of variance in experimental design". 95 JQualTek 27 268-268

Shao, J.; Tu, D. (Auth); — — (Rev): Review of "The jackknife and bootstrap". 96 ShrtBkRv 16/1 7-7

Vapnik, V. N. (Auth); — — (Rev): Review of "The nature of statistical learning theory". 96 ShrtBkRv 16/1 3-3

Fedorov, Valery: Cook, Dennis; — —: Constrained optimization of experimental design. 95 MaOpfStS 26 129-148 (Disc: p149-178)

Cook, Dennis; — —: Reply to comments on "Constrained optimization of experimental design". 95 MaOpfStS 26 175-178

Fedrizzi, Michele: Pereira, Ricardo Alberto Marques; — —: Fuzzy surface reconstruction. 94 QStMaA 16 71-77

Fei, Heliang; Kong, Fanhui; Tang, Yingcai: Estimation for two-parameter Weibull distribution and extreme-value distribution under multiply type-II censoring. 95 CommStA 24 2087-2104

— —; Sinha, Bimal K.; Wu, Zhong: Estimation of parameters in two-parameter Weibull and extreme-value distributions using ranked set sample. 94 JStatRsr 28 149-161

Tang, Yincai; — —: Maximum likelihood estimation in the life testing for the two-parameter exponential distribution with replacement (Chinese). 95 ChJApPSt 10 399-404

Feigin, Paul D.; Resnick, Sidney I.; Stărică, Cătălin: Testing for independence in heavy tailed and positive innovation time series. 95 CommStSM 11 587-612

Feild, Thomas C.: Smith, Mary E.; — —: Veterans and VA system users classified by federal poverty guidelines and thresholds and the VA means test. 94 ASAProGovt 28-33

Feinberg, Eugene A.; Shwartz, Adam: Constrained Markov decision models with weighted discounted rewards. 95 MaOpRsr 20 302-320

Feinberg, Max; Bugner, Emmanuel; Theiller, Gédéon; Houba, Victor; Kadijk, Fetze: Expression of the reference value for proficiency tests. 95 JChemtrc 9 197-209

Feingold, Eleanor; Lamb, Neil E.; Sherman, Stephanie L.: Methods for genetic linkage analysis using trisomies. 95 AJHuGene 56 475-483

Feinstein, Alvan R.: Peduzzi, Peter; Concato, John; — —: Importance of events per independent variable in proportional hazards regression analysis II. Accuracy and precision of regression estimates. 95 JClEpid 48 1503-1510

Fel'dman, G. M.: Carnal, H.; — —: Phase retrieval for probability measures on abelian groups. 95 JTherPrb 8 717-725

Feldman, Henry A.: Zucker, David M.; Lakatos, Edward; Webber, Larry S.; Murray, David M.; McKinlay, Sonja M.; — —; Kelder, Steve H.; Nader, Philip R.: Statistical design of the child and adolescent trial for cardiovascular health (CATCH): Implications of cluster randomization. 95 ConClnTr 16 96-118

Feldman, Marcus W.: Bergman, Aviv; Otto, Sarah P.; — —: On the evolution of recombination in haploids and diploids II. Stochastic models. 95 Complxty 1 49-57

Felinger, A.: Pietrogrande, M. C.; Dondi, F.; — —; Davis, J. M.: Statistical study of peak overlapping in multicomponent chromatograms: Importance of the retention pattern. 95 Chemolab 28 239-258

Fellegi, I. P.: Marketing of statistics in Canada. 91 StatJUN 8 295-306

Fellegi, Ivan P.: Statistical statesmanship in official statistics. 94 ASAProGovt 7-8

— —: Comment on "Alternative methods for the 2000 census". 94 ASASrvyRsr 93-95

Fellingham, G. W.; Raghunathan, T. E.: Sensitivity of point and interval estimates to distributional assumptions in longitudinal data analysis of small samples. 95 CommStB 24 617-630

Fellingham, Gilbert W.; Wise, Morgan D.: EM for longitudinal data from multiple studies. 95 StSinica 5 97-107

Davis, Alan B.; Larimore, Walter L.; — —; Beus, Carolyn M.: Using subsequent studies to recover estimability. 94 ASAProEpid 112-117 (Disc: p125-126)

Peck, Robert R.; Madrigal, J. L.; — —: Ranking teams competing in pairwise comparisons using the dynamic linear model. 94 ASAProBays 112-116 (Disc: p134-135)

Feng, Shui: Phase transitions of some non-linear stochastic models. 95 JAppProb 32 193-201

— —: Nonlinear master equation of multitype particle systems. 95 StocProc 57 247-271

— —: Large deviations for empirical process of mean-field interacting particle system with unbounded jumps. 94 AnlsProb 22 2122-2151

Feng, Wei; Adachi, Kohichi; Kowada, Masasi: Dynamically optimal replacement policy for a shock model in a Markov random environment. 94 JOpRsJpn 37 255-270

Feng, Yu: Wei, Bocheng; — —: On calculating the moments of M-estimator in nonlinear regression (Chinese). 95 AMaAPknC 17 437-447

Fenstad, Grete: Hjort, Nils Lid; — —: Second-order asymptotics for the number of times an estimator is more than ε from its target value. 95 JStPlInf 48 261-275

Ferdous, J.; Uddin, Md. Borhan; Pandey, M.: Reliability estimation with Weibull inter failure times. 95 RelEngSS 50 285-296

Ferger, D.: Nonparametric tests for nonstandard change-point problems. 95 AnlsStat 23 1848-1861

Ferger, Dietmar: The joint distribution of the running maximum and its location of D-valued Markov processes. 95 JAppProb 32 842-845

Ferguson, D. P.: SAS use in data editing. 91 StatJUN 8 167-174

Ferguson, Gwyn R.; Ciapponi, Tamara M.: A response model for the international price program. 94 ASASrvyRsr 1007-1012

Ferguson, T. S.: A class of symmetric bivariate uniform distributions. 95 StHefte 36 31-40

Ferland, René: Dion, Jean-Pierre; — —: Absolute continuity, singular measures and asymptotics for estimators. 95 JStPlInf 43 235-242

Ferlay, J.: Parkin, D. M.; Muir, C. S.; Whelan, S. L.; Gao, Y. T.; — —; Powell, J. (Ed); Hill, Catherine (Rev): Review of "Cancer incidence in five continents. Volume VI". 95 StatMed 14 104-104

Fermanian, Jacques: Coste, Joël; — —; Venot, Alain: Methodological and statistical problems in the construction of composite measurement scales: A survey of six medical and epidemiological journals. 95 StatMed 14 331-345

Fernández, Arturo J.: Bravo, José I.; De Fuentes, Íñigo; — —: A semi-parametric estimation of a survival function from incomplete and doubly censored data. 95 CommStA 24 2735-2752

Fernández, Begoña: Caballero, María Emilia; — —; Nualart, David: Smoothness of distributions for solutions of anticipating stochastic differential equations. 95 Stochatx 52 303-322

Fernandez, Carmen: De la Horra, Julian; — —: Sensitivity to prior independence via Farlie-Gumbel-Morgenstern model. 95 CommStA 24 987-996

Fernández, Carmen; Osiewalski, Jacek; Steel, Mark F. J.: Modeling and inference with v-spherical distributions. 95 JASA 90 1331-1340

Fernández, M. C. Ortiz; Herrero Gutiérrez, A.; Pastor, M. S. Sánchez; Sarabia, L. A.; Íñiguez Crespo, M.: The UNEQ, PLS and MLF neural network methods in the modelling and prediction of the colour of young red wines from the Denomination of Origin 'Rioja'. 95 Chemolab 28 273-285

Fernandez-Macho, F. Javier: Anderson, T. W. (Auth); — — (Rev): Review of "The statistical analysis of time series". 95 JAppStat 22 548-550

Fernando, R. L.; Stricker, C.; Elston, R. C.: An efficient algorithm to compute the posterior genotypic distribution for every member of a pedigree without loops. 93 ThApGene 87 89-93

Lo, L. L.; — —; Cantet, R. J. C.; Grossman, M.: Theory for modelling means and covariances in a two-breed population with dominance inheritance. 95 ThApGene 90 49-62

Lo, L. L.; — —; Grossman, M.: Covariance between relatives in multibreed populations: Additive model. 93 ThApGene 87 423-430

Ferrall, Christopher: Interactive statistics tutorials in Stata. 95 JStEd 3/3 28kb (jse/v3n3/ferrall)

Ferrándiz, J.; Montes, F.: On statistical inference for random closed sets in compact subsets of R^d. 95 AdvAppPr 27 344-349

Ferrándiz, Juan; López, Antonio; Llopis, Agustín; Morales, María; Tejerizo, María Luisa: Spatial interaction between neighbouring counties: Cancer mortality data in Valencia (Spain). 95 Biomtrcs 51 665-678

Ferrante, M.; Nualart, D.: Markov field property for stochastic differential equations with boundary conditions. 95 Stochatx 55 55-69

Ferrante, Marco; Giummolé, Federica: Finite dimensional filters for a discrete-time nonlinear system with generalized Gaussian white noise. 95 Stochatx 53 195-211

Alabert, Aureli; — —; Nualart, David: Markov field property of stochastic differential equations. 95 AnlsProb 23 1262-1288

Baccin, Maria C.; — —: On a stochastic delay difference equation with boundary conditions and its Markov property. 95 StocProc 60 131-146

Ferrari, P. A.; Galves, A.; Liggett, T. M.: Exponential waiting time for filling a large interval in the symmetric simple exclusion process. 95 AIHenrPB 31 155-175

— —; Kesten, H.; Martinez, S.; Picco, P.: Existence of quasi-stationary distributions. A renewal dynamical approach. 95 AnlsProb 23 501-521

— —; Kipnis, C.: Second class particles in the rarefaction fan. 95 AIHenrPB 31 143-154

Ferrari, Pablo A.; Frigessi, Arnoldo; de Sá, Paula Gonzaga: Fast approximate maximum a posteriori restoration of multicolour images. 95 JRSS-B 57 485-500

Ferrari, Silvia L. P.: Cordeiro, Gauss M.; Cribari-Neto, Francisco; Aubin, Elisete C. Q.; — —: Bartlett corrections for one-parameter exponential family models. 95 JStCmpSm 53 211-231

Cribari-Neto, Francisco; — —: Second order asymptotics for score tests in generalised linear models. 95 Biomtrka 82 426-432

Cribari-Neto, Francisco; — —: An improved Lagrange multiplier test for heteroskedasticity. 95 CommStB 24 31-44

Ferraris, Victor A.: An analysis of the thoracic surgical literature. 94 ASAProQlPr 158-163

Ferraro, David L.: Paulin, Geoffrey D.; — —: Do expenditures explain income?: A study of variables for income imputation. 94 ASASrvyRsr 106-115

Ferraty, F.: Besse, P. C.; — —: A fixed effect curvilinear model. 95 CmpStQ 10 339-351

Ferre, L.: Boularan, J.; — —; Vieu, P.: A nonparametric model for unbalanced longitudinal data with application to geophysical data. 95 CmpStQ 10 285-298

Ferré, Louis: Selection of components in principal component analysis: A comparison of methods. 95 CmpStDA 19 669-682

— —: Improvement of some multidimensional estimates by reduction of dimensionality. 95 JMultiAn 54 147-162

Boularan, Joël; — —; Vieu, Philippe: Location of particular points in non-parametric regression analysis. 95 AstrlJSt 37 161-168

Ferreira, Helena: Extremes of a random number of variables from periodic sequences. 95 JStPlInf 45 133-141

Ferreira, Irene: Amaratunga, Dhammika; — —: Nonlinear model selection for finding receptor sites. 94 ASAProBiop 126-129

Ferrer, A.: Romero, R.; — —; Capilla, C.; Zunica, L.; Balasch, S.; Serra, V.; Alcover, R.: Teaching statistics to engineers: An innovative pedagogical experience. 95 JStEd 3/1 33kb (jse/v3n1/romero)

Ferretti, Nélida; Kelmansky, Diana; Yohai, Victor: A goodness-of-fit test based on ranks for ARMA models. 95 CommStA 24 295-318

Ferri, D.: Forina, M.; Drava, G.; Armanino, C.; Boggia, R.; Lanteri, S.; Leardi, R.; Corti, P.; Conti, P.; Giangiacomo, R.; Galliena, C.; Bigoni, R.; Quartari, I.; Serra, C.; — —; Leoni, O.; Lazzeri, L.: Transfer of calibration function in near-infrared spectroscopy. 95 Chemolab 27 189-203

Ferryanto, Sg.: On estimation of the Walsh-Fourier spectral density of two dimensional strictly homogeneous random fields. 95 JNnparSt 5 391-407

Ferryman, Thomas A.; Press, S. James: Comparison of image reconstruction algorithms. 94 ASAProBays 159-164

Ferson, Wayne E.; Foerster, Stephen R.: Finite sample properties of the generalized method of moments in tests of conditional asset pricing models. 94 JFinEcon 36 29-55

Fessler, Jeffrey A.: Hero, Alfred O.; — —: Convergence in norm for alternating expectation-maximization (EM) type algorithms. 95 StSinica 5 41-54

Feu, Chris du: 1st - Class!. 95 TeachgSt 17/1 13-15

— —: Tree slugs. 95 TeachgSt 17/2 52-53

— —: Birds of Britain. 95 TeachgSt 17/3 104-106

Hawksworth, D. L. (Ed); — — (Rev): Review of "Biodiversity - Measurement and estimation". 95 TeachgSt 17/3 111-111

Longman Logotron (Auth); — — (Rev): Review of "Pinpoint - Datafiles". 95 TeachgSt 17/2 75-76

Feuer, Bernice I.; Lepeniotis, Stefanos: Free radical polymerization of nonlinear optical side chain methacrylate copolymers in dimethylsulfoxide - An experimental design. 95 Chemolab 29 113-124

Feyen, Jan: Ramos, José; Mallants, Dirk; — —: State space identification of linear deterministic rainfall-runoff models. 95 WaterReR 31 1519-1531

Fiebig, Denzil G.: Bartels, Robert; — —: Optimal design in end-use metering experiments. 95 MaCmpSim 39 305-309

Fieger, W.: Tong, Y. L. (Auth); — — (Rev): Review of "The multivariate normal distribution". 95 Metrika 42 75-76

Fieger, Werner: Bischoff, Wolfgang; — —; Wulfert, Stefanie: Minimax- and Γ-minimax estimation of a bounded normal mean under LINEX loss. 95 StatDecn 13 287-298

Field, Christopher A.: Fan, Rocky Y. K.; — —: Approximations for marginal densities of *M*-estimators. 95 CandJSt 23 185-197

Fielding, A.: Hutchison, T. P.; Lai, C. D. (Auth); — — (Rev): Review of "The data sets from CBDEA". 95 Ststcian 44 288-289

Kessler, J. T.; Kelsbeck, W. D. (Auth); — — (Rev): Review of "Non-sampling errors in surveys". 95 Ststcian 44 289-290

Fielding, Antony: Hansen, M. H.; Hurwitz, W. N.; Madow, W. G. (Auth); — — (Rev): Review of "Sample survey methods and theory (Volume I, 2)". 95 JRSS-A 158 350-350

Nolan, B. (Auth); — — (Rev): Review of "Data analysis: An introduction". 95 Ststcian 44 546-547

Fielding, Katherine L.; Brettle, Raymond P.; Gore, Sheila M.; O'Brien, Fiona; Wyld, Rona; Robertson, J. Roy; Weightman, Rosina: Heterosexual transmission of HIV analysed by generalized estimating equations. 95 StatMed 14 1365-1378

Fields, Paul J.: Martino, Joseph P. (Auth); — — (Rev): Review of "Technological forecasting for decision making (Third edition)". 95 Intrfc 25/1 137-139

Fienberg, Stephen E.: Conflicts between the needs for access to statistical information and demands for confidentiality. 94 JOffStat 10 115-132

— —; Jazairi, Nuri: Stanley Warner's contributions to statistically balanced information technology. 95 SurvyM 21 7-12

— —; Martin, Margaret E.; Straf, Miron L.: The Committee on National Statistics: Fostering interactions between statisticians in academia and government. 95 IntStRvw 63 257-269

Anderson, Margo; — —: Black, white, and shades of gray (and brown and yellow). 95 Chance 8/1 15-18

DeGroot, Morris H.; — —; Kadane, Joseph B.: Statistics and the law. 95 Wiley:NY:UK 512p

Devlin, Bernie; — —; Resnick, Daniel P.; Roeder, Kathryn: Wringing the bell curve: A cautionary tale about the relationships among race, genes, and IQ. 95 Chance 8/3 27-36

Herrnstein, R. J.; Murray, C. (Auth); Devlin, Bernie; — —; Resnick, Daniel P.; Roeder, Kathryn (Rev): Review of "The bell curve: Intelligence and class structure in American life". 95 JASA 90 1483-1488

Filar, Jerzy A.; Krass, Dmitry; Ross, Keith W.: Percentile performance criteria for limiting average Markov decision processes. 95 IEEEAuCn 40 2-10

Fildes, Robert; Makridakis, Spyros: The impact of empirical accuracy studies on time series analysis and forecasting. 95 IntStRvw 63 289-308

Armstrong, J. Scott; — —: Comment on "On the limitations of comparing mean square forecast errors" (V12 p617-637). 95 JForcast 14 67-71

Fill, Heinz D.; Stedinger, Jery R.: L moment and probability plot correlation coefficient goodness-of-fit tests for the Gumbel distribution and impact of autocorrelation. 95 WaterReR 31 225-229

Fill, James Allen: Dobrow, Robert P.; — —: On the Markov chain for the move-to-root rule for binary search trees. 95 AnlsApPr 5 1-19

Dobrow, Robert P.; — —: Rates of convergence for the move-to-root Markov chain for binary search trees. 95 AnlsApPr 5 20-36

Fillion, Jean-Marc: Bankier, Michael; — —; Luc, Manchi; Nadeau, Christian: Imputing numeric and qualitative variables simultaneously. 94 ASASrvyRsr 242-247

Filloon, Thomas G.: Estimating the minimum therapeutically effective dose of a compound via regression modelling and percentile estimation. 95 StatMed 14 925-932 (Disc: p933-933)

Findley, David F.; Parzen, Emanuel: A conversation with Hirotugu Akaike. 95 StatSci 10 104-117

Janacek, Gareth; Swift, Louise (Auth); — — (Rev): Review of "Time series: Forecasting, simulation, applications". 95 SIAMRvw 37 253-257

Fine, Terrence L.: Sadrolhefazi, Amir; — —: Finite-dimensional distributions and tail behavior in stationary interval-valued probability models. 94 AnlsStat 22 1840-1870

Finizio, N. J.: Anderson, I.; — —: Cyclically resolvable designs and triple whist tournaments. 93 JComDsgn 1 347-358

Finizza, Anthony: Abramson, Bruce; — —: Probabilistic forecasts from probabilistic models: A case study in the oil market. 95 IntJFor 11 63-72

Fink, Arlene: How to design surveys. 95 SagePubl:CA:Lnd 96p

— —: How to sample in surveys. 95 SagePubl:CA:Lnd 96p

— —: How to analyze survey data. 95 SagePubl:CA:Lnd 112p

— —: How to report on surveys. 95 SagePubl:CA:Lnd 112p

— —: How to ask survey questions. 95 SagePubl:CA:Lnd 120p

— —: The survey handbook. 95 SagePubl:CA:Lnd 144p

Fink, Steven: Kasprzyk, Daniel; Gruber, Kerry; Salvucci, Sameena; Saba, Mehrdad; Zhang, Fan; — —: Some data issues in school-based surveys. 94 ASASrvyRsr 815-820

Finkelstein, Dianne M.; Schoenfeld, David A. (Ed): AIDS clinical trials: Guidelines for design and analysis. 95 Wiley:NY:UK 472p

Finkelstein, M.; Scheiberg, S.; Tucker, H. G.: Integral transforms with infinitely divisible kernels. 94 ThProbAp 39 670-676

Finkelstein, Mark: Estimating the frequency distribution of the numbers bet on the California lottery. 95 AplMaCmp 69 195-207

— —; Scheinberg, Stephen: The storage complexity of some estimators of the median. 94 AplMaCmp 60 15-24

— —; Tucker, Howard G.: The Wilcoxon rank-sum test in dichotomous regression analysis. 92 ProIrStC1 243-259

Finkle, William D.: Greenland, Sander; — —: A critical look at methods for handling missing covariates in epidemiologic regression analyses. 95 AJEpidm 141 1255-1264

Greenland, Sander; — —: A critical look at methods for handling missing covariates in epidemiologic regression analyses. 95 AJEpidm 142 1255-1264

Finman, Jeffrey S.: Accomando, William P.; Trost, Donald C.; — —: Moving from multinational clinical trials to supra-trials. 94 ASAProBiop 232-237

Finn, Jeremy D.: Olkin, Ingram; — —: Correlations redux. 95 PsychBu 118 155-164

Finn, Lynda M.; Garhart, Casey A.; Reynard, Susan E.: Getting statistical tools to the masses: Common problems in training design and execution. 95 ASQCTekCon 356-364

Finnegan, G. Paul; Heavlin, William D.: Application of Hoaglin's G and H curves to process specifications. 94 ASAProQlPr 146-150

Finney, David J.: Statistical science and effective scientific communication. 95 JAppStat 22 293-308

Firinguetti, Luis: Crivelli, Ana; — —; Montaño, Rosa; Muñóz, Margarita: Confidence intervals in ridge regression by bootstrapping the dependent variable: A simulation study. 95 CommStB 24 631-652

Firth, David: Correction to "Bias reduction of maximum likelihood estimates" (V80 p27-38). 95 Biomtrka 82 667-667

Fischbeck, P. S.: Paté-Cornell, M. E.; — —: Reply to comment on "Probabilistic interpretation of command and control signals: Bayesian updating of the probability of nuclear attack". 95 RelEngSS 48 159-160

Fischbeck, Paul S.: Paté-Cornell, M. Elisabeth; — —: Probabilistic interpretation of command and control signals Bayesian updating of the probability of nuclear attack. 95 RelEngSS 47 27-36

Fischer, Donald P.: Huggins, Vicki J.; — —: The redesign of the Survey of Income and Program Participation. 94 ASASrvyRsr 668-673 (Disc: p680-682)

Fischer, G. H.; Molenaar, I. W.: Rasch models: Foundations, recent developments, and applications. 95 Springer:Brln:NY 436p

Fischer, Gerhard H.: Some neglected problems in IRT. 95 Psymtrka 60 459-487

Fischer, J.: Hierarchical cluster analysis. 95 CmpStQ 10 21-27

Fish, David: Nagelkerke, Nico J. D.; Moses, Stephen; Plummer, Francis A.; Brunham, Robert C.; — —: Logistic regression in case-control studies: The effect of using independent as dependent variables. 95 StatMed 14 769-775

Fisher, Alan C.: Hwang, Ching C.; Wang, Antonia C.; — —: The application of Cox regression models to the analysis of a double-blind comparative trial and subsequent open label extension in HIV-infected subjects. 94 ASAProBiop 194-199

Fisher, Franklin M.: The production-theoretic measurement of input price and quantity indices. 95 JEconmtx 65 155-174

Fisher, Gordon; Sim, Ah Boon: Some finite sample theory for bootstrap regression estimates. 95 JStPlInf 43 289-300

Fisher, James H.: Wang, Jichuan; — —; Siegal, Harvey A.; Falck, Russel S.; Carlson, Robert G.: Influence of measurement errors on HIV risk behavior analysis: A case study examining condom use among drug users. 95 StrEqMod 2 319-334

Fisher, Lloyd D.; van Belle, Gerald (Auth); D'Agostino, Ralph B. (Rev): Review of "Biostatistics: A methodology for the health sciences". 95 JASA 90 799-800

— —; van Belle, Gerald (Auth); Machin, David (Rev): Review of "Biostatistics: A methodology for the health sciences". 95 StatMed 14 878-879

Fisher, N. I.; Sen, P. K.: The collected works of Wassily Hoeffding. 94 Springer:Brln:NY 658p

— — (Auth); Kumar, Kuldeep (Rev): Review of "Statistical analysis of circular data". 95 Ststcian 44 285-286

— —; Lewis, T.; Embleton, B. J. J. (Auth); Challenor, Peter (Rev): Review of "Statistical analysis of spherical data". 95 Ststcian 44 286-287

— — (Auth); Prentice, M. J. (Rev): Review of "Statistical analysis of circular data". 95 JRSS-A 158 190-190

— — (Auth); Wehrly, Thomas E. (Rev): Review of "Statistical analysis of circular data". 95 Technmcs 37 229-230

Fisher, Nicholas I.: Hall, Peter; — —; Hoffmann, Branka: On the nonparametric estimation of covariance functions. 94 AnlsStat 22 2115-2134

Lewis, Toby; — —: Estimating the angle between the mean directions of two spherical distributions. 95 AstrlJSt 37 179-191

— — (Auth); Lange, Nicholas (Rev): Review of "Statistical analysis of circular data". 95 JASA 90 801-801

Fisher, Robin: Logistic regression analysis of CPS overlap survey split panel data. 94 ASASrvyRsr 620-625

Thompson, Jenny; — —: Two-sample McNemar tests for complex survey data. 94 ASASrvyRsr 608-613

Fishman, G.: Monte Carlo: Concepts, algorithms, and applications. 95 Springer:-Brln:NY 600p

Fitch, David J.; Flores, Rafael; Matute, Jorge: Are the biases and inefficiencies of the EPI method of sampling relatively unimportant with reasonable between village variances?. 94 ASASrvyRsr 288-292

Fitch, Walter M.: Xu, Shizhong; Atchley, William R.; — —: Phylogenetic inference under the pure drift model. 94 MolBioEv 11 949-960

Fitzmaurice, Garret M.: Lipsitz, Stuart R.; — —; Sleeper, Lynn; Zhao, L. P.: Estimation methods for the joint distribution of repeated binary observations. 95 Biomtrcs 51 562-570

Fitzmaurice, Garrett: Lipsitz, Stuart R.; — —: Reply to comment on "An extension of Yule's Q to multivariate binary data". 95 Biomtrcs 51 1185-1185

Fitzmaurice, Garrett M.: A caveat concerning independence estimating equations with multivariate binary data. 95 Biomtrcs 51 309-317

— —; Laird, Nan M.: Regression models for a bivariate discrete and continuous outcome with clustering. 95 JASA 90 845-852

— —; Laird, Nan M.; Zahner, Gwendolyn E. P.; Daskalakis, Constantine: Bivariate logistic regression analysis of childhood psychopathology ratings using multiple informants. 95 AJEpidm 142 1194-1203

— —; Lipsitz, Stuart R.: A model for binary time series data with serial odds ratio patterns. 95 ApplStat 44 51-61

— —; Molenberghs, Geert; Lipsitz, Stuart R.: Regression models for longitudinal binary responses with informative drop-outs. 95 JRSS-B 57 691-704

Fitzpatrick, Ben G.: Statistical tests of fit in estimation problems for structured population modeling. 95 QApplMa 53 105-128

Fitzpatrick, M.: Bunge, J.; — —; Handley, J.: Comparison of three estimators of the number of species. 95 JAppStat 22 45-59

Fitzsimmons, P. J.; Getoor, R. K.: Occupation time distributions for Lévy bridges and excursions. 95 StocProc 58 73-89

Flak, T.; Schmid, W.: Extreme sums of strictly stationary sequences of M-dependent variables. 95 SankhyaA 57 186-201

Flanagan, R.: Ramsay, J. O.; Wang, X.; — —: A functional data analysis of the pinch force of human fingers. 95 ApplStat 44 17-30

Flandoli, Franco; Gatarek, Dariusz: Martingale and stationary solutions for stochastic Navier-Stokes equations. 95 ProbTher 102 367-391

— —; Tortorelli, Vincenzo M.: Time discretization of Ornstein-Uhlenbeck equations and stochastic Navier-Stokes equations with a generalized noise. 95 Stochatx 55 141-165

Bensoussan, Alain; — —: Stochastic inertial manifold. 95 Stochatx 53 13-39

Brzeźniak, Zdzisław; — —: Almost sure approximation of Wong-Zakai type for stochastic partial differential equations. 95 StocProc 55 329-358

Flandre, Philippe; O'Quigley, John: A two-stage procedure for survival studies with surrogate endpoints. 95 Biomtrcs 51 969-976

Flatman, George: Bayard, Steven; Jinot, Jennifer; — —: Comment on "Uncertainties in lung cancer risk estimates reported for exposure to environmental tobacco smoke". 95 EnvrMtrc 6 413-418

Flatman, George T.: Comment on "Risk assessment of lung cancer due to environmental tobacco smoke (ETS)". 94 ASAProEnvr 74-76

Flaugher, Ronald: Wainer, Howard; Dorans, Neil J.; — —; Green, Bert F.; Mislevy, Robert; Steinberg, Lynne; Thissen, David (Auth); Andrich, David (Rev): Review of "Computerized adaptive testing: A primer". 95 Psymtrka 60 615-620

Flaum, M.: Arndt, S.; Tyrrell, G.; Wollson, R. F.; — —; Andreasen, N. C.: Effects of errors in a multicenter medical study: Preventing misinterpreted data. 94 JPsyctrc 28 447-459

Fleischer, I.; Joffe, A.: Ratio ergodicity for non-homogeneous Markov chains in general state spaces. 95 JTherPrb 8 31-37

Fleischer, Karlheinz: Two tests of pseudo random number generators for independence and uniform distribution. 95 JStCmpSm 52 311-322

Fleischer, W.: Nagel, M.; Wernecke, K. D.; — — (Auth); Bartkowiak, Anna (Rev): Review of "Computer-aided data analysis" (German). 95 ListyBio 32 47-49

Fleischmann, Klaus; Le Gall, Jean-François: A new approach to the single point catalytic super-Brownian motion. 95 ProbTher 102 63-82

Dawson, Donald A.; — —: Super-Brownian motions in higher dimensions with absolutely continuous measure states. 95 JTherPrb 8 179-206

Dawson, Donald A.; — —; Li, Yi; Mueller, Carl: Singularity of super-Brownian local time at a point catalyst. 95 AnlsProb 23 37-55

Fleming, C.; Nellis, J. G. (Auth); Cooper, R. I. (Rev): Review of "Principles of applied statistics". 95 JRSS-A 158 348-349

Fleming, Philip J.; Schaeffer, Dennis; Simon, Burton: Efficient Monte-Carlo simulation of a product-form model for a cellular system with dynamic resource sharing. 95 ACMMdSim 5 3-21

Fleming, Thomas R.: Anderson, Garnet L.; — —: Model misspecification in proportional hazards regression. 95 Biomtrka 82 527-541

Grambsch, Patricia M.; Therneau, Terry M.; — —: Diagnostic plots to reveal functional form for covariates in multiplicative intensity models. 95 Biomtrcs 51 1469-1482

Fletcher, Daniel: Raz, Jonathan; Cardenas, Valerie; — —: Frequency domain estimation of covariate effects in multichannel brain evoked potential data. 95 Biomtrcs 51 448-460

Fletcher, Mike: Play your cards right. 95 TeachgSt 17/2 74-74

Flint, Paul L.; Pollock, Kenneth H.; Thomas, Dana; Sedinger, James S.: Estimating prefledging survival : Allowing for brood mixing and dependence among brood mates. 95 JWildlMg 59 448-455

Flora, June A.: Fortmann, Stephen P.; — —; Winkleby, Marilyn A.; Schooler, Caroline; Taylor, C. Barr; Farquhar, John W.: Community intervention trials: Reflections on the Stanford Five-City Project experience. 95 AJEpidm 142 576-586

Flores, Rafael: Fitch, David J.; — —; Matute, Jorge: Are the biases and inefficiencies of the EPI method of sampling relatively unimportant with reasonable between village variances?. 94 ASASrvyRsr 288-292

Florit, Carme; Nualart, David: A local criterion for smoothness of densities and application to the supremum of the Brownian sheet. 95 StPrLet 22 25-31

Flournoy, Nancy: Life and hard times of a statistician: Lynne Billard. 95 Stats /12 16-20

— —; Durham, Stephen D.; Rosenberger, William F.: Toxicity in sequential dose-response experiments. 95 SqtlAnly 14 217-227

Flury, Bernard: Li, Luning; — —: Uniqueness of principal points for univariate distributions. 95 StPrLet 25 323-327

Flury, Bernard D.; Nel, Daan G.; Pienaar, Inet: Simultaneous detection of shift in means and variances. 95 JASA 90 1474-1481

Neuenschwander, Beat E.; — —: Common canonical variates. 95 Biomtrka 82 553-560

Tarpey, Thaddeus; Li, Luning; — —: Principal points and self-consistent points of elliptical distributions. 95 AnlsStat 23 103-112

Flygare, A.-M.; Barrlund, A.: Efficient implementation of some contextual classification methods. 95 CmpStQ 10 327-338

Foerster, Stephen R.: Ferson, Wayne E.; — —: Finite sample properties of the generalized method of moments in tests of conditional asset pricing models. 94 JFinEcon 36 29-55

Földes, A.: Csáki, E.; — —: On the logarithmic average of additive functionals. 95 StPrLet 22 261-268

Földes, Antónia: Csáki, Endre; Csörgő, Miklós; — —; Révész, Pál: Global Strassen-type theorems for iterated Brownian motions. 95 StocProc 59 321-341

Foley, Marianne: White, Donald B.; — —: Statistical modeling of long term cumulative effects. 94 ASAProBiop 328-333

Foley, Sean: Egge, Eric; — —; Haskins, Loren; Johnson, Roger: A data analysis based elementary statistics course. 94 ASAProStEd 144-149

Folks, J. Leroy: Woldie, Mammo; — —: Calibration for inverse Gaussian regression. 95 CommStA 24 2609-2620

Follmann, Dean: Multivariate tests for multiple endpoints in clinical trials. 95 StatMed 14 1163-1175

— —; Wu, Margaret: An approximate generalized linear model with random effects for informative missing data. 95 Biomtrcs 51 151-168

Follmann, Dean A.: Proschan, Michael A.; — —: Multiple comparisons with control in a single experiment versus separate experiments: Why do we feel differently?. 95 AmerStat 49 144-149

Föllmer, Hans; Protter, Philip; Shiryayev, Albert N.: Quadratic covariation and an extension of Itô's formula. 95 Bernoull 1 149-169

Folsom, Ralph E.; Liu, Jun: Small area estimation for the National Household Survey of Drug Abuse. 94 ASASrvyRsr 565-570

— —; Witt, Michael B.: Testing a new attrition nonresponse adjustment method for SIPP. 94 ASASrvyRsr 428-433 (Disc: p440-442)

Fong, Daniel Y. T.; Yip, Paul S. F.: A note on information loss in analyzing a mixture model of count data. 95 CommStA 24 3197-3209

Fong, Wai Mun; Ouliaris, Sam: Spectral tests of the martingale hypothesis for exchange rates. 95 JApEcmtx 10 255-271

Font, B.: Bayarri, M. J.; — —: A (Bayesian) note on non-random samples from finite populations. 94 ASAProBays 246-251

Foolad, M. R.; Jones, R. A.: Models to estimate maternally controlled genetic variation in quantitative seed characters. 92 ThApGene 83 360-366

Forbes, Andrew B.; Santner, Thomas J.: Estimators of odds ratio regression parameters in matched case-control studies with covariate measurement error. 95 JASA 90 1075-1084

Ford, Ian; Norrie, John; Ahmadi, Susan: Model inconsistency, illustrated by the Cox proportional hazards model. 95 StatMed 14 735-746

Ford, Jon M.: Caudill, Steven B.; — —; Gropper, Daniel M.: Frontier estimation and firm-specific inefficiency measures in the presence of heteroscedasticity. 95 JBES 13 105-111

Caudill, Steven B.; — —; Kaserman, David L.: Certificate-of-need regulation and the diffusion of innovations: A random coefficient model. 95 JApEcmtx 10 73-78

Foresi, Silverio; Peracchi, Franco: The conditional distribution of excess returns: An empirical analysis. 95 JASA 90 451-466

Forgács, Esther: Cserháti, Tibor; — —: Use of canonical correlation analysis for the evaluation of chromatographic retention data. 95 Chemolab 28 305-313

Forina, M.; Drava, G.; Armanino, C.; Boggia, R.; Lanteri, S.; Leardi, R.; Corti, P.; Conti, P.; Giangiacomo, R.; Galliena, C.; Bigoni, R.; Quartari, I.; Serra, C.; Ferri, D.; Leoni, O.; Lazzeri, L.: Transfer of calibration function in near-infrared spectroscopy. 95 Chemolab 27 189-203

— —; Lanteri, S.; Sarabia, L.: Distance and class space in the UNEQ class-modeling technique. 95 JChemtrc 9 69-89

Meloun, M.; Militky, J.; — — (Auth); Parczewski, Andrzej (Rev): Review of "Chemometrics for analytical chemistry, Volume 2: PC-aided regression and related methods". 95 Chemolab 29 326-327

Forina, Michele: Meloun, Milan; Militky, Jiri; — — (Auth); LaLonde, Steven M. (Rev): Review of "Chemometrics for analytical chemistry Volume 1: PC-aided statistical data analysis". 95 Technmcs 37 114-115

Formann, Anton K.: Small-sample comparison of the exact and asymptotic upper tail probabilities of chi-squared goodness-of-fit statistics: Pearson's χ^2, likelihood ratio, and power-divergence statistic ($\lambda = 2/3$). 95 JStCmpSm 51 369-384

Fornari, F.; Mele, A.: A two-factor model of the term structure with optimal filtering behaviour. 94 Statstca 54 293-312

Forster, D.; Snow, R. W.: An assessment of the use of hand-held computers during demographic surveys in developing countries. 95 SurvyM 21 179-184

Forte, B.; Vrscay, E. R.: Solving the inverse problem for measures using iterated function systems: A new approach. 95 AdvAppPr 27 800-820

Forthofer, Ronald N.; Lee, Eun Sul: Introduction to biostatistics. A guide to design, analysis, and discovery. 95 Academic:NY:Lnd 567p

Fortiana, J.: Cuadras, C. M.; — —: A continuous metric scaling solution for a random variable. 95 JMultiAn 52 1-14

Fortin, Nicole M.: Heterogeneity biases, distributional effects, and aggregate consumption. 95 JApEcmtx 10 287-311

Fortmann, Stephen P.; Flora, June A.; Winkleby, Marilyn A.; Schooler, Caroline; Taylor, C. Barr; Farquhar, John W.: Community intervention trials: Reflections on the Stanford Five-City Project experience. 95 AJEpidm 142 576-586

Fosam, E. B.; Sapatinas, T.: Characterisations of some income distributions based on multiplicative damage models. 95 AstrlJSt 37 89-93

Jones, K. F.; Taylor, N. W.; — —: Statistical approaches for the reliability of structural designs. 95 JAppStat 22 443-458

Foss, Serguei: Baccelli, François; — —: On the saturation rule for the stability of queues. 95 JAppProb 32 494-507

Foster, Dean P.; George, Edward I.: The risk inflation criterion for multiple regression. 94 AnlsStat 22 1947-1975

Nelson, Daniel B.; — —: Filtering and forecasting with misspecified ARCH models II: Making the right forecast with the wrong model. 95 JEconmtx 67 303-335

Foster, F. Douglas; Viswanathan, S.: Can speculative trading explain the volume-volatility relation?. 95 JBES 13 379-396

Foster, Jeremy J.: — — (Auth); Hawkins, Anne (Rev): Review of "SPSS/PC+ and SPSS for Windows - A beginner's guide to data analysis (Second edition)". 95 TeachgSt 17/2 79-79

Foster, Peter: A comparative study of some bias correction techniques for kernel-based density estimators. 95 JStCmpSm 51 137-152

Foster, Robert D.: Wallace, Jean E.; — —; Kendall, John E.: Improving control of polymer viscosity using time-series experimentation and statistical process control. 94 ASAProPEng 320-325

Foster, Suzy: Bredfeldt, Hans; — —; Scrutchins, Sam: Extending the use of IRS's tax compliance measurement program to develop estimates for small subpopulations using double sampling. 94 ASAProGovt 128-133

Fotopoulos, S. B.: Kim, Hag-Min; — —: Characterization study of a stochastic process associated with times in queueing systems. 94 ASAProPEng 163-168

Fotopoulos, Stergios B.: Rustagi, Jagdish S. (Auth); — — (Rev): Review of "Optimization techniques in statistics". 95 Technmcs 37 466-467

Fougére, Denis: Aubin, Christophe; — —; Husson, Emmanuel; Ivaldi, Marc: Real-time pricing of electricity for residential customers: Econometric analysis of an experiment. 95 JApEcmtx 10 171-191S

Foulley, J. L.: Verrier, E.; Colleau, J. J.; — —: Long-term effects of selection based on the animal model BLUP in a finite population. 93 ThApGene 87 446-454

Fountain, Robert L.: A method to detect and estimate population heterogeneity. 94 ASAProEpid 82-87

Fourastié, J.: Berrondo-Agrell, M.; — —: Understanding probability (French). 94 Hachette:Fr 160p

Fourati, S.: Representation of measures for additive functionals between two times (French). 95 AIHenrPB 31 527-544

Fourdrinier, Dominique; Robert, Christian P.: Intrinsic losses for empirical Bayes estimation: A note on normal and Poisson cases. 95 StPrLet 23 35-44

— —; Wells, Martin T.: Estimation of a loss function for spherically symmetric distributions in the general linear model. 95 AnlsStat 23 571-592

— —; Wells, Martin T.: Loss estimation for spherically symmetric distributions. 95 JMultiAn 53 311-331

Cellier, Dominique; — —: Shrinkage estimators under spherical symmetry for the general linear model. 95 JMultiAn 52 338-351

Cellier, Dominique; — —: Shrinkage positive rule estimators for spherically symmetric distributions. 95 JMultiAn 53 194-209

Fourque, Alain: Tsodikov, Alexander D.; Asselain, Bernard; — —; Hoang, Thu; Yakovlev, Andrej Yu.: Discrete strategies of cancer post-treatment surveillance. Estimation and optimization problems. 95 Biomtrcs 51 437-447

Foust, Laura B.: Grimes, James A.; — —: Establishing release limits with a random slopes stability model. 94 ASAProBiop 498-502

Fowler, Floyd Jackson, Jr: Comment on "Redesign of the National Health Interview Survey questionnaire". 94 ASASrvyRsr 379-381

Fowler, John F.: Chappell, Rick; Nondahl, David M.; — —: Modeling dose and local control in radiotherapy. 95 JASA 90 829-838

Fowler, John L.: Smith, Robert T., Jr; Evans, B. Timothy; — —: The Commodity Flow Survey. 94 ASAProBuEc 89-94

Fowles, Richard: Blattenberger, Gail; — —: Road closure to mitigate avalanche danger: A case study for Little Cottonwood Canyon. 95 IntJFor 11 159-174

Fox, James K.: Esposito, Richard; — —; Lin, Dongyow; Tidemann, Kevin: ARIES: Visual techniques for statistical data investigation at the Bureau of Labor Statistics. 94 ASAProStGr 7-10

Fox, John; Friendly, Michael: Data analysis using APL2 and APL2STAT. 95 SocMethR 23 282-302

— —; Stine, Robert: Introduction to the special issue on computing environments. 95 SocMethR 23 267-281

Fox, Richard: Van de Geer, John P. (Auth); — — (Rev): Review of "Multivariate analysis of categorical data: Theory and applications. Volumes 1 and 2". 95 JMarRsr 32 111-112

Fraas, John W.: Newman, Isadore; — —; Norfolk, Timothy: Reply to comments on "A binomial test of model fit". 95 StrEqMod 2 155-162

Fraiman, Ricardo: Boente, Graciela; — —: Asymptotic distribution of data-driven smoothers in density and regression estimation under dependence. 95 CandJSt 23 383-397

Boente, Graciela; — —: Asymptotic distribution of smoothers based on local means and local medians under dependence. 95 JMultiAn 54 77-90

Cao, Ricardo; Cuevas, Antonio; — —: Minimum distance density-based estimation. 95 CmpStDA 20 611-631

Fraley, Chris; Burns, Patrick J.: Large-scale estimation of variance and covariance components. 95 SIAMSSC 16 192-209

Franchet, Y.: Metadata and official statistics. 93 StatJUN 10 97-100

Francis, B. J.: Seeber, G. U. H.; — —; Hatzinger, R.; Steckel-Berger, G.: Statistical modelling. 95 Springer:Brln:NY 327p

Francis, Brian: dos Santos, Dirley M.; Davies, Richard B.; — —: Nonparametric hazard versus nonparametric frailty distribution in modelling recurrence of breast cancer. 95 JStPlInf 47 111-127

Francis, Charles M.: Estimating survival rates from recoveries of birds ringed as young: A case study. 95 JAppStat 22 567-577

— —: How useful are recoveries of North American passerines for survival analyses?. 95 JAppStat 22 1075-1081

Franco, M.; Ruiz, J. M.: On characterization of continuous distributions with adjacent order statistics. 95 MaOpfStS 26 375-385

Franco-Marina, Francisco: Correa-Villaseñor, Adolfo; Stewart, Walter F.; — —; Seacat, Hui: Bias from nondifferential misclassification in case-control studies with three exposure levels. 95 Epidmlgy 6 276-281

Francos, Joseph M.; Meiri, A. Zvi; Porat, Boaz: A Wold-like decomposition of two-dimensional discrete homogeneous random fields. 95 AnlsApPr 5 248-260

Frangos, Chris C.; Schucany, William R.: Improved bootstrap confidence intervals in certain toxicological experiments. 95 CommStA 24 829-844

Frank, Ellen: Kamlet, Mark S.; Paul, Nancy; Greenhouse, Joel; Kupfer, David; — —; Wade, Martcia: Cost utility analysis of maintenance treatment for recurrent depression. 95 ConClnTr 16 17-40

Frank, H.; Althoen, S. C. (Auth); Guiard, V. (Rev): Review of "Statistics: Concepts and Applications". 95 BiomtrcJ 37 639-639

Frank, I. E.; Todeschini, R. (Auth); Rius, F. Xavier (Rev): Review of "The data analysis handbook". 95 Chemolab 29 327-327

Frank, Ildiko E.: Modern nonlinear regression methods. 95 Chemolab 27 1-19

Frank, Jess: Markus, Richard A.; — —; Groshen, Susan; Azen, Stanley P.: An alternative approach to the optimal design of an LD50 bioassay. 95 StatMed 14 841-852

Frank, Ove; Snijders, Tom: Estimating the size of hidden populations using snowball sampling. 94 JOffStat 10 53-67

Frankel, Lester R.: Dutka, Solomon; — —: The maintenance of ethical principles in marketing research. 94 ASASrvyRsr 143-145

Frankel, Martin R.: Kish, Leslie; — —; Verma, Vijay; Kaćiroti, Niko: Design effects for correlated $(P_i - P_j)$. 95 SurvyM 21 117-124

Franklin, LeRoy A.: Determining sample size for lower confidence limits for C_p and C_{pk}. 94 ASAProQlPr 23-27

Frankovic, Kathleen A.; Arnedt, Cheryl Mercado; Ramnath, Bala: Interactive polling and Americans' comfort level with technology. 94 ASASrvyRsr 1316-1321

Franses, Philip Hans: The effects of seasonally adjusting a periodic autoregressive process. 95 CmpStDA 19 683-704

— —: A vector of quarters representation for bivariate time series. 95 EcomtrRv 14 55-63

— —: A differencing test. 95 EcomtrRv 14 183-193

— —: On periodic autoregressions and structural breaks in seasonal time series. 95 EnvrMtrc 6 451-455

— —; Kloek, Teun: A periodic cointegration model of quarterly consumption. 95 AplStMDA 11 159-166

— —; Paap, Richard: Moving average filters and periodic integration. 95 MaCmpSim 39 245-249

Boswijk, H. Peter; — —: Periodic cointegration: Representation and inference. 95 RvEconSt 77 436-454

Fraser, Martin D.; Bartolucci, Alfred A.; Smith, William A.; Singh, Karen P.: A Bayesian decision procedure for selecting prognostic variables associated with survival for data in which censoring is prevalent. 95 BiomtrcJ 37 463-479

Fraumeni, J. F., Jr: Doll, R.; — —; Muir, C. S. (Ed); Boyle, Peter (Rev): Review of "Trends in cancer incidence and mortality". 95 StatMed 14 2507-2508

Frazier, Richard L.: Jackson, Betty J.; — —; King, Karen E.; Schwanz, Dennis J.: Improving the coverage of private elementary-secondary schools. 94 ASASrvyRsr 833-838

Freddi, Augusto: On the comparison of probability laws of a subset of one-dimensional real variates (Italian). 95 Metron 53 147-161

Free, J. C.: Badrigal, J. L.; Richards, D. O.; — —; Lee, T. N.; Lawson, J. S.; Hawks, V. D.: Q^2: A TQM course for undergraduates. 94 ASAProQlPr 172-180

Freedman, D.; Pisani, R.; Purves, R.; Adhikari, A. (Auth); Mizera, Ivan (Rev): Review of "Statistics (Second edition)". 95 JAppStat 22 321-322

Freedman, L.: Landin, R.; — —; Carroll, R.: Correction to "Adjusting for time trends when estimating the relationship between dietary intake obtained from a food frequency questionnaire and true average intake" (V51 p169-181). 95 Biomtrcs 51 1196-1196

Freedman, Laurence S.: Byar, David P.; — —; Herzberg, Agnes M.: Identifying which sets of parameters are simultaneously estimable in an incomplete factorial design. 95 Ststcian 44 451-456

Green, Sylvan B.; Corle, Donald K.; Gail, Mitchell H.; Mark, Steven D.; Pee, David; — —; Graubard, Barry I.; Lynn, William R.: Interplay between design and analysis for behavioral intervention trials with community as the unit of randomization. 95 AJEpidm 142 587-593

Landin, Richard; — —; Carroll, Raymond J.: Adjusting for time trends when estimating the relationship between dietary intake obtained from a food frequency questionnaire and true average intake. 95 Biomtrcs 51 169-181

Freeman, Katherine: Teaching critical review of the medical literature: A course in clinical trials. 94 ASAProStEd 156-161

Freeman, P. R.; Smith, A. F. M. (Ed); Queen, Catriona (Rev): Review of "Aspects of uncertainty: A tribute to Dennis Lindley". 95 JAppStat 22 541-541

Freeman, S. N.: Catchpole, E. A.; — —; Morgan, B. J. T.: Modelling age variation in survival and reporting rates for recovery models. 95 JAppStat 22 597-609

Freeman, Stephen N.: Lebreton, Jean-Dominique; Morgan, Byron J. T.; Pradel, Roger; — —: A simultaneous survival rate analysis of dead recovery and live recapture data. 95 Biomtrcs 51 1418-1428

Freeny, Anne: Mandel, John (Auth); — — (Rev): Review of "Analysis of two-way layouts". 95 JASA 90 1127-1128

Freeny, Anne E.; Landwehr, James M.: Graphical analysis for a large designed experiment. 95 Technmcs 37 1-14

Frees, Edward W.: Assessing cross-sectional correlation in panel data. 95 JEconmtx 69 393-414

— —: Semiparametric estimation of warranty costs. 95 JNnparSt 5 103-122

Freidlin, M. I.: Borodin, A. N.; — —: Fast oscillating random perturbations of dynamical systems with conservation laws. 95 AIHenrPB 31 485-525

— — (Ed); Plachky, D. (Rev): Review of "The Dynkin festschrift - Markov processes and their applications". 95 StatDecn 13 400-402

Freimer, Marshall: Mudholkar, Govind S.; Srivastava, Deo Kumar; — —: The exponentiated Weibull family: A reanalysis of the bus-motor-failure data. 95 Technmcs 37 436-445

French, Martyn A. H.: Segal, Mark R.; James, Ian R.; — —; Mallal, Simon A.: Statistical issues in the evaluation of markers of HIV progression. 95 IntStRvw 63 179-197

French, Simon: Wiper, Michael P.; — —: Combining experts' opinions using a normal-Wishart model. 95 JForcast 14 25-34

Fresen, J. L.: de Jongh, P. J.; de Wet, T.; — —: An experimental design for evaluating sonar detection performance. 95 SoAfrStJ 29 79-103

Frey, Andreas: Asmussen, Søren; — —; Rolski, Tomasz; Schmidt, Volker: Does Markov-modulation increase the risk?. 95 AstinB 25 49-66

Błaszczyszyn, Bartłomiej; — —; Schmidt, Volker: Light-traffic approximations for Markov-modulated multi-server queues. 95 CommStSM 11 423-445

Frey, Béatrice: Forecast of unemployment in Switzerland. 95 Student 1 1-15

Frey, Carolin; Frey, Michael: Analysis of assays of the joint action of three drugs. 94 ASAProBiop 33-38

Frey, James H.: Lavrakas, Paul J. (Auth); — — (Rev): Review of "Telephone survey methods: Sampling, selection and supervision (Second edition)". 94 JOffStat 10 349-351

Frey, Michael: A multivariate tolerance interpretation of assays of joint action. 94 ASAProBiop 39-44

Frey, Carolin; — —: Analysis of assays of the joint action of three drugs. 94 ASAProBiop 33-38

Kedem, Benjamin (Auth); — — (Rev): Review of "Time series analysis by higher order crossings". 95 Technmcs 37 235-235

Frey, William H.: Poverty migration for U.S. States: Immigration impacts. 94 ASAProSoSt 135-140

Frick, H.: Comparing trials with multiple outcomes: The multivariate one-sided hypothesis with unknown covariances. 95 BiomtrcJ 37 909-917

Frick, Joachim: Rendtel, Ulrich; Wagner, Gert; — —: A method for checking longitudinal weights in panel surveys - The case of the German Socio-Economic Panel (German). 95 AllgeStA 79 252-277

Fricker, Ronald D., Jr: The inherent difficulties in applying TQM to a particular service industry. 94 ASAProQlPr 181-188

Friedl, Herwig; Tilg, Norbert: Variance estimates in logistic regression using the bootstrap. 95 CommStA 24 473-486

Friedman, Daniel; Sunder, Shyam (Auth); Lamoureux, Christopher G. (Rev): Review of "Experimental methods: A primer for economists". 95 JFinance 50 1341-1345

Friedman, Jerome H.; Roosen, Charles B.: An introduction to multivariate adaptive regression splines. 95 StMedRes 4 197-217

Land, Stephanie R.; — —: Adaptive signal regression. 94 ASAProStCp 100-105

Friendly, Michael: Conceptual and visual models for categorical data. 95 AmerStat 49 153-160

Fox, John; — —: Data analysis using APL2 and APL2STAT. 95 SocMethR 23 282-302

Fries, Gerhard; Woodburn, R. Louise: The challenges of preparing sensitive data for public release. 94 ASASrvyRsr 1184-1189

Frieze, Alan; Pittel, Boris G.: Probabilistic analysis of an algorithm in the theory of markets in indivisible goods. 95 AnlsApPr 5 768-808

Frigessi, Arnoldo: Comment on "Bayesian computation and stochastic systems". 95 StatSci 10 41-43

Ferrari, Pablo A.; — —; de Sá, Paula Gonzaga: Fast approximate maximum a posteriori restoration of multicolour images. 95 JRSS-B 57 485-500

Fritz, J.; Rüdiger, B.: Time dependent critical fluctuations of a one dimensional local mean field model. 95 ProbTher 103 381-407

Froeschl, K.: Darius, P.; Boucneau, M.; De Greef, P.; De Feber, E.; — —: Modelling metadata. 93 StatJUN 10 171-180

Froeschl, K. A.: Towards an operative view of semantic metadata. 93 StatJUN 10 181-194

Froeschl, Karl Anton: A formal model evaluation approach to the analysis of treatment effects in paired sample data. 95 CmpStDA 19 493-517

From, Steven G.: A two-stage procedure for estimating the parameters of some discrete distributions. 95 JStCmpSm 53 23-39

— —: A weighted least-squares procedure for estimating the parameters of Altham's multiplicative generalization of the binomial distribution. 95 StPrLet 25 193-199

Frühwirth-Schnatter, Sylvia: Bayesian model discrimination and Bayes factors for linear Gaussian state space models. 95 JRSS-B 57 237-246

Frydman, Halina: Semiparametric estimation in a three-state duration-dependent Markov model from interval-censored observations with application to AIDS data. 95 Biomtrcs 51 502-511

— —: Nonparametric estimation of a Markov 'illness-death' process from interval-censored observations, with application to diabetes survival data. 95 Biomtrka 82 773-789

— —; Hussain, Taukir: Statistical inference for switching Brownian motions. 95 SankhyaA 57 11-28

Fu, J. C.; Koutras, M. V.: Reliability bounds for coherent structures with independent components. 95 StPrLet 22 137-148

Fu, James; Sproule, Robert: A generalization of the binomial distribution. 95 CommStA 24 2645-2658

Fu, James C.: Exact and limiting distributions of the number of successions in a random permutation. 95 AnInStMa 47 435-446

Fu, Michael C.; Hu, Jian-Qiang: On unbounded hazard rates for smoothed perturbation analysis. 95 JAppProb 32 659-667

— —; Hu, Jian-Qiang: Sensitivity analysis for Monte Carlo simulation of option pricing. 95 PrEngInf 9 417-446

Fu, Y. B.; Ritland, K.: On estimating the linkage of marker genes to viability genes controlling inbreeding depression. 94 ThApGene 88 925-932

Fu, Yun-Xin: A phylogenetic estimator of effective population size or mutation rate. 94 Genetcs 136 685-692

— —: Estimating effective population size or mutation rate using the frequencies of mutations of various classes in a sample of DNA sequences. 94 Genetcs 138 1375-1386

Gu, Xun; — —; Li, Wen-Hsiung: Maximum likelihood estimation of the heterogeneity of substitution rate among nucleotide sites. 95 MolBioEv 12 546-557

Fuentes, Montserrat: Gallego, F. Javier; — —: Geographic sampling with a distance threshold. 94 QStMaA 16 93-101

Fuji-Hara, Ryoh: Kuriki, Shinji; — —: Balanced arrays of strength two and nested (r, λ)-designs. 94 JComDsgn 2 407-414

Fujii, Yoshinori: Yanagawa, Takashi; — —: Projection-method Mantel-Haenszel estimator for K $2 \times J$ tables. 95 JASA 90 649-656

Fujii, Yoshio: Yamamoto, Sumiyasu; — —; Hyodo, Yoshifumi; Yumiba, Hiromu: Classification of orthogonal 24-run 2^5 factorial designs derivable from saturated two-symbol orthogonal arrays of strength 2, size 24 and index 6. 95 TRUMTk 31/1 39-53

Yamamoto, Sumiyasu; — —; Mitsuoka, Motohiro: Three-symbol orthogonal arrays of strength 2 and index 2 having maximal constraints: Computational study. 93 JCISS 18 329-342

Fujikoshi, Yasunori: Seo, Takashi; Kanda, Takashi; — —: The effects of nonnormality on tests for dimensionality in canonical correlation and MANOVA models. 95 JMultiAn 52 325-337

Fujisawa, Hironori: A note on the maximum likelihood estimators for multivariate normal distribution with monotone data. 95 CommStA 24 1377-1382

Fukuda, Toshio: Ishigami, Hideyuki; — —; Shibata, Takanori; Arai, Fumihito: Structure optimization of fuzzy neural network by genetic algorithm. 95 FuzzySet 71 257-264

Fukushima, M.: Ma, Z. M.; Röckner, M. (Auth); — — (Rev): Review of "Introduction to the non-symmetric Dirichlet forms". 95 Metrika 42 70-72

Fukushima, Masao: Sagara, Nobuko; — —: A hybrid method for solving the nonlinear least squares problem with linear inequality constraints. 94 JOpRsJpn 38 55-69

Füle, Erika: On ecological regression and ridge estimation. 95 CommStB 24 385-398

Fulker, D. W.; Cherny, S. S.; Cardon, L. R.: Multipoint interval mapping of quantitative trait loci, using sib pairs. 95 AJHuGene 56 1224-1233

Cherny, S. S.; DeFries, J. C.; — —: Multiple regression analysis of twin data: A model-fitting approach. 92 BehavGen 22 489-498

Rodriguez, L. A.; — —; Cherny, S. S.: A maximum-likelihood model-fitting approach to conducting a Hayman analysis of diallel tables with complete or missing data. 93 BehavGen 23 69-76

Fulker, David W.; Cardon, Lon R.: A sib-pair approach to interval mapping of quantitative trait loci. 94 AJHuGene 54 1092-1103

Cardon, Lon R.; — —: The power of interval mapping of quantitative trait loci, using selected sib pairs. 94 AJHuGene 55 825-833

Cardon, Lon R.; — —: A model of developmental change in hierarchical phenotypes with application to specific cognitive abilities. 94 BehavGen 24 1-16

Fulkerson, Bill: Michie, D.; Spiegelhalter, D. J.; Taylor, C. C. (Ed); — — (Rev): Review of "Machine learning, neural and statistical classification". 95 Technmcs 37 459-459

Fuller, Howard T.: Bisgaard, Søren; — —: Sample size estimates for 2^{k-p} designs with binary responses. 95 JQualTek 27 344-354

Fuller, J. David: Lachtermacher, Gerson; — —: Backpropagation in time-series forecasting. 95 JForcast 14 381-393

Fuller, Wayne A.: Estimation in the presence of measurement error. 95 IntStRvw 63 121-141 (Disc: p142-147)

— —: Introduction to statistical time series (Second edition). 95 Wiley:NY:UK 698p

An, Anthony B.; Breidt, F. Jay; — —: Regression weighting methods for SIPP data. 94 ASASrvyRsr 434-439 (Disc: p440-442)

Deo, Rohit; — —: Symmetric tests for cointegration. 94 ASAProBuEc 105-110

McVey, Anita; Breidt, F. Jay; — —: Two-phase estimation by imputation. 94 ASASrvyRsr 1053-1058

Park, Heon Jin; — —: Alternative estimators and unit root tests for the autoregressive process. 95 JTimSrAn 16 415-429

Fullerton, Thomas M., Jr; West, Carol Taylor: Assessing the historical accuracy of regional employment forecasts. 94 ASAProBuEc 333-338

Funaki, T.: The scaling limit for a stochastic PDE and the separation of phases. 95 ProbTher 102 221-288

— —; Surgailis, D.; Woyczynski, W. A.: Gibbs-Cox random fields and Burgers turbulence. 95 AnlsApPr 5 461-492

Fung, W. K.; Kwan, C. W.: Sensitivity analysis in factor analysis: Difference between using covariance and correlation matrices. 95 Psymtrka 60 607-614

Fung, Wing K.: A cautionary note on the use of generalized Cook-type measures. 95 CmpStDA 19 321-326

— —: Detecting influential observations for estimated probabilities in multiple discriminant analysis. 95 CmpStDA 20 557-568

— —: Graphical summaries for influence of multiple observations. 95 CommStA 24 415-427

— —: Diagnostics in linear discriminant analysis. 95 JASA 90 952-956

— —: A stepwise method for the identification of multiple outliers and influential observations. 95 SoAfrStJ 29 51-64

— —: A statistical analysis of dependence of voter shares under a two-vote system: The Hong Kong experience. 94 ASAProSoSt 76-81

Furino, Steven: Existence results for near resolvable designs. 95 JComDsgn 3 101-113

— —; Greig, Malcolm; Miao, Ying; Yin, Jianxing: Resolvable BIBDs with block size 8 and index 7. 94 JComDsgn 2 101-116

Furino, Steven C.; Mullin, Ronald C.: Block designs with large holes and α-resolvable BIBDs. 93 JComDsgn 1 101-112

Lee, Therese C. Y.; — —: A translation of J. X. Lu's "An existence theory for resolvable balanced incomplete block designs". 95 JComDsgn 3 321-340

Furnival, George M.: Gregoire, Timothy G.; Valentine, Harry T.; — —: Sampling methods to estimate foliage and other characteristics of individual trees. 95 Ecology 76 1181-1194

Furr, H. C.: Song, K. S.; Müller, H. G.; Clifford, A. J.; — —; Olson, J. A.: Estimating derivatives of pharmacokinetic response curves with varying bandwidths. 95 Biomtrcs 51 12-20

Furstenberg, Hillel; Peres, Yuval; Weiss, Benjamin: Perfect filtering and double disjointness. 95 AIHenrPB 31 453-465

Furukawa, Y.: Urahama, K.; — —: Gradient descent learning of nearest neighbor classifiers with outlier rejection. 95 PattrnRc 28 761-768

Fuss, Melvyn: Bregman, Arie; — —; Regev, Haim: The production and cost structure of Israeli industry: Evidence from individual firm data. 95 JEconmtx 65 45-81

Futrell, David: When quality is a matter of taste, use reliability indexes. 95 QulPr 28/5 81-86

Fygenson, Mendel: Cohen, Arthur; — —: Testing homogeneity of uniform scale distributions against two-sided and one-sided alternatives. 95 JASA 90 1062-1067

Gabbard, Susan M.; Nakamoto, Jorge: Evaluating translations of survey instruments in Spanish and English: The 1990 Spanish Language Census long form. 94 ASASrvyRsr 1232-1235

Gabler, Siegfried; Stenger, Horst: Improving the RHC-strategy. 95 StHefte 36 327-336

Gabriel, K. Ruben: Changnon, Stanley A.; — —; Westcott, Nancy E.; Czys, Robert R.: Exploratory analysis of seeding effects on rainfall: Illinois 1989. 95 JApMetrl 34 1215-1224

Verboon, Peter; — —: Generalized Procrustes analysis with iterative weighting to achieve resistance. 95 BJMaStPs 48 57-73

Gadrich, Tamar: Mikosch, Thomas; — —; Klüppelberg, Claudia; Adler, Robert J.: Parameter estimation for ARMA models with infinite variance innovations. 95 AnlsStat 23 305-326

Gaertner, Gregory: Cantor, David; — —; Keil, Linda: The validity of self-reported drug use among the incarcerated. 94 ASASrvyRsr 1088-1093

Gaffke, Norbert; Heiligers, Berthold: Computing optimal approximate invariant designs for cubic regression on multidimensional balls and cubes. 95 JStPlInf 47 347-376

— —; Heiligers, Berthold: Optimal and robust invariant designs for cubic multiple regression. 95 Metrika 42 29-48

— —; Heiligers, Berthold: Algorithms for optimal design with application to multiple polynomial regression. 95 Metrika 42 173-190

Gaggermeier, Karl: Ritter, Gunter; Gallegos, María Teresa; — —: Automatic context-sensitive karyotyping of human chromosomes based on elliptically symmetric statistical distributions. 95 PattrnRc 28 823-831

Gagnon, David R.: Cupples, L. Adrienne; — —; Ramaswamy, Ratna; D'Agostino, Ralph B.: Age-adjusted survival curves with application in the Framingham study. 95 StatMed 14 1731-1744

Gagnon, J. H.: Laumann, E. O.; — —; Michael, R. T.; Michaels, S. (Auth); Wasserman, Stanley (Rev): Review of "The social organization of sexuality". 95 Chance 8/2 40-42

Gagnon, Robert C.: Lee, Kwan R.; — —: Calibration and stability analysis with a simple mixed linear model when the experiment is of a split-plot type. 94 ASAProBiop 435-440

Gail, H. Richard: Birman, Alexander; — —; Hantler, Sidney L.; Rosberg, Zvi; Sidi, Moshe: An optimal service policy for buffer systems. 95 JACM 42 641-657

Gail, M. H.: Brookmeyer, R.; — — (Auth); Ghani, Azra (Rev): Review of "AIDS epidemiology: A quantitative approach". 95 JRSS-A 158 343-343

Gail, Mitchell H.: Dave Byar's contribution to epidemiology. 95 ConClnTr 16 230-248

Benichou, Jacques; — —: Methods of inference for estimates of absolute risk derived from population-based case-control studies. 95 Biomtrcs 51 182-194

Dahm, P. Fred; — —; Rosenberg, Philip S.; Pee, David: Determining the value of additional surrogate exposure data for improving the estimate of an odds ratio. 95 StatMed 14 2581-2598

Green, Sylvan B.; Corle, Donald K.; — —; Mark, Steven D.; Pee, David; Freedman, Laurence S.; Graubard, Barry I.; Lynn, William R.: Interplay between design and analysis for behavioral intervention trials with community as the unit of randomization. 95 AJEpidm 142 587-593

Brookmeyer, Ron; — — (Auth); DeGruttola, Victor (Rev): Review of "AIDS epidemiology: A quantitative approach". 95 JASA 90 797-798

Brookmeyer, Ron; — — (Auth); Farrington, C. P. (Rev): Review of "AIDS epidemiology: A quantitative approach". 95 AJEpidm 141 180-181

Gaines, Arthur G.: Solow, Andrew R.; — —: An empirical Bayes approach to monitoring water quality. 95 EnvrMtrc 6 1-5

Gaines, Leonard M.: Measuring relative electricity cost sensitivity in manufacturing establishments in New York State. 94 ASAProGovt 183-188

— —; González, María E.: Is there a place in ASA for statisticians who work for state/provincial or local governments?. 94 ASAProGovt 149-152

Gaines, Steven D.: Rice, William R.; — —: Extending nondirectional heterogeneity tests to evaluate simply ordered alternative hypotheses. 94 PNatAcSc 91 225-226

Gajek, L.: Note on unbiased estimability of the larger of two mean values. 95 ApplMath 23 239-245

— —; Kałuszka, M.: Lower bounds for the asymptotic Bayes risk in the scale model (with an application to the second-order minimax estimation). 94 AnlsStat 22 1831-1839

Bismarck, M.; — —: An upper bound for the power of a test. 95 CommStA 24 813-828

Gajek, Lesław; Kałuszka, Marek: Nonexponential applications of a global Cramér-Rao inequality. 95 MaOpfStS 26 111-122

Gajjar, A. V.: Patel, M. N.; — —: Some results on maximum likelihood estimators of parameters of exponential distribution under type I progressive censoring with changing failure rates. 95 CommStA 24 2421-2435

Gajjar, Yogesh; Mehta, Cyrus; Patel, Nitin (Auth); Smith, Peter W. F. (Rev): Review of "LOGXACT-TURBO (Version 1.1)". 95 ApplStat 44 139-141

Gakis, K. G.; Rhee, H. K.; Sivazlian, B. D.: Distributions and first moments of the busy and idle periods in controllable M/G/1 queueing models with simple and dyadic policy. 95 StocAnAp 13 47-81

Gal, Iddo: Statistical tools and statistical literacy: The case of the average. 95 TeachgSt 17/3 97-99

Galambos, J.: The development of the mathematical theory of extremes in the past half century. 94 ThProbAp 39 234-248

— —; Gani, J. (Ed); Coleman, Rodney (Rev): Review of "Studies in applied probability: Papers in honour of Lajos Takács". 95 JRSS-A 158 634-635

— —; Gani, J. (Ed); Haigh, J. (Rev): Review of "Studies in applied probability: Papers in honour of Lajos Takács". 95 Biomtrcs 51 1187-1187

Galambos, Janos: Advanced probability theory (Second edition). 95 Dekker:NY 461p

— —; Xu, Yuan: Bivariate extension of the method of polynomials for Bonferroni-type inequalities. 95 JMultiAn 52 131-139

Galatsanos, Nikolas P.: Zhu, Wenwu; — —; Katsaggelos, Aggelos K.: Regularized multichannel restoration using cross-validation. 95 CVGIPGr 57 38-54

Galbraith, John W.; Zinde-Walsh, Victoria: Transforming the error-components model for estimation with general ARMA disturbances. 95 JEconmtx 66 349-355

Gale, William A.: Cognition, computation, and formal systems: Some of Tomás Havránek's interests and disambiguating word senses. 95 CmpStDA 19 135-148

Galea Rojas, Manuel: Bolfarine, Heleno; — —: Maximum likelihood estimation of simultaneous pairwise linear structural relationships. 95 BiomtrcJ 37 673-689

Bolfarine, Heleno; — —: Structural comparative calibration using the EM algorithm. 95 JAppStat 22 277-292

Galea-Rojas, Manuel: Bolfarine, Heleno; — —: Comment on "Functional comparative calibration using an EM algorithm" (V48 p1263-1271). 95 Biomtrcs 51 1579-1580

Gallagher, Neal B.: Wise, Barry M.; Holt, Bradley R.; — —; Lee, Samuel: A comparison of neural networks, non-linear biased regression and a genetic algorithm for dynamic model identification. 95 Chemolab 30 81-89

Gallais, A.: Analysis of variance components of testcross progenies in an autotetraploid species and consequences for recurrent selection with a tester. 92 ThApGene 83 353-359

Naciri, Y.; — —: Predicting mean and variance of all possible lines and hybrids from designs with partially inbred progenies. 94 ThApGene 89 693-697

Gallant, A. Ronald: Bansal, Ravi; — —; Hussey, Robert; Tauchen, George: Nonparametric estimation of structural models for high-frequency currency market data. 95 JEconmtx 66 251-287

Gallant, J. C.: Pulford, G. W.; — —; Kennedy, R. A.; Chung, S. H.: Evaluation and estimation of various Markov models with applications to membrane channel kinetics. 95 BiomtrcJ 37 39-63

Gallego, F. Javier; Fuentes, Montserrat: Geographic sampling with a distance threshold. 94 QStMaA 16 93-101

Gallegos, María Teresa: Ritter, Gunter; — —; Gaggermeier, Karl: Automatic context-sensitive karyotyping of human chromosomes based on elliptically symmetric statistical distributions. 95 PattrnRc 28 823-831

Galliena, C.: Forina, M.; Drava, G.; Armanino, C.; Boggia, R.; Lanteri, S.; Leardi, R.; Corti, P.; Conti, P.; Giangiacomo, R.; — —; Bigoni, R.; Quartari, I.; Serra, C.; Ferri, D.; Leoni, O.; Lazzeri, L.: Transfer of calibration function in near-infrared spectroscopy. 95 Chemolab 27 189-203

Gallo, Giampiero M.: Otranto, Edoardo; — —: Regression diagnostic techniques to detect balanced space-to-time ratios in STARMA models. 94 Metron 52/3 129-145

Gallo, Kevin P.: Epperson, David L.; Davis, Jerry M.; Bloomfield, Peter; Karl, Thomas R.; McNab, Alan L.; — —: Estimating the urban bias of surface shelter temperatures using upper-air and satellite data. Part I: Development of models predicting surface shelter temperatures. 95 JApMetrl 34 340-357

Gallo, Paul P.: Chen, Bee-Lian; — —: Boundary calculation issues in clinical trials with two-sided stopping criteria. 94 ASAProBiop 115-119

Kianifard, Farid; — —: Poisson regression analysis in clinical research. 95 JBiophSt 5 115-129

Longford, Nicholas T. (Auth); — — (Rev): Review of "Random coefficient methods". 95 Technmcs 37 120-120

Galper, A.: Dagum, P.; — —; Horvitz, E.; Seiver, A.: Uncertain reasoning and forecasting. 95 IntJFor 11 73-87

Galves, A.: Ferrari, P. A.; — —; Liggett, T. M.: Exponential waiting time for filling a large interval in the symmetric simple exclusion process. 95 AIHenrPB 31 155-175

Gambino, J.: Chen, E. J.; — —; Laniel, N.; Lindeyer, J.: Design and estimation issues for income in the redesign of the Canadian Labour Force Survey. 94 ASASrvyRsr 919-924

Gámiz-Pérez, M. Luz: Pérez-Ocón, Rafael; — —: On the HNBUE property in a class of correlated cumulative shock models. 95 AdvAppPr 27 1186-1188

Pérez-Ocón, Rafael; — —: Conditions on the arrival process to obtain HNBUE survival using a shock model. 95 CommStA 24 931-944

Gan, D. Robert: Johnson, Richard A.; — —; Berthouex, P. M.: Goodness-of-fit using very small but related samples with application to censored data estimation of PCB contamination. 95 EnvrMtrc 6 341-348

Gan, F. F.: Joint monitoring of process mean and variance using exponentially weighted moving average control charts. 95 Technmcs 37 446-453

Chang, T. C.; — —: A cumulative sum control chart for monitoring process variance. 95 JQualTek 27 109-119

Chang, T. C.; — —: Correction to "A cumulative sum control chart for monitoring process variance" (V27 p109-119). 95 JQualTek 27 402-402

Gan, Gaoxiong; Bain, Lee J.: Distribution of order statistics for discrete parents with applications to censored sampling. 95 JStPlInf 44 37-46

Bain, Lee J.; — —: Expectations of the maximum of sums and the sum of maxima for correlated normal variables. 95 CommStB 24 775-780

Gan, Nianci; Pantula, Sastry G.: Testing for deterministic trends in correlated data. 94 ASAProBuEc 141-146

Gan, Thian Yew: A time series model for estimating the weekly winter maximum temperature of northwest territories. 95 JApMetrl 34 847-860

Gana, Rajaram: Ridge regression estimation of the linear probability model. 95 JAppStat 22 537-539

Gandolfi, A.: van den Berg, J.; — —: A triangle inequality for covariances of binary FKG random variables. 95 AnlsApPr 5 322-326

Ganesan, L.; Bhattacharyya, P.: A statistical design of experiments approach for texture description. 95 PattrnRc 28 99-105

Gange, Stephen J.: Generating multivariate categorical variates using the iterative proportional fitting algorithm. 95 AmerStat 49 134–138

— —; Linton, Kathryn L. P.; Scott, Alastair J.; DeMets, David L.; Klein, Ronald: A comparison of methods for correlated ordinal measures with ophthalmic applications. 95 StatMed 14 1961–1974

Gangopadhyay, Ashis K.: A note on the asymptotic behavior of conditional extremes. 95 StPrLet 25 163–170

Ganguly, S. S.; Naik-Nimbalkar, U. V.: Analysis of ordinal data in a study of endometrial cancer under a matched pairs case-control design. 95 StatMed 14 1545–1552

Gangwar, D. K.: Katyar, N. P.; Singh, Housila P.; — —: A class of almost unbiased regression-type estimators for finite population mean applying jackknife technique. 94 JIAgSt 46 183–189

Gani, J.; Yakowitz, Sid: Error bounds for deterministic approximations to Markov processes, with applications to epidemic models. 95 JAppProb 32 1063–1076

Barndorff-Nielsen, O. E.; Jensen, J. L.; Kendall, W. S. (Ed); — — (Rev): Review of "Networks and chaos - Statistical and probabilistic aspects". 95 Metrika 42 271–272

Casti, J. L.; Karlqvist, A. (Ed); — — (Rev): Review of "Cooperation and conflict in general evolutionary processes". 95 Biomtrcs 51 1191–1193

Galambos, J.; — — (Ed); Coleman, Rodney (Rev): Review of "Studies in applied probability: Papers in honour of Lajos Takács". 95 JRSS-A 158 634–635

Galambos, J.; — — (Ed); Haigh, J. (Rev): Review of "Studies in applied probability: Papers in honour of Lajos Takács". 95 Biomtrcs 51 1187–1187

Gannoun, Ali: Ducharme, Gilles R.; — —; Guertin, Marie-Claude; Jéquier, Jean-Claude: Reference values obtained by kernel-based estimation of quantile regressions. 95 Biomtrcs 51 1105–1116

Gansky, Stuart A.; Koch, Gary G.: Evaluation of alternative strategies for managing stratification factors in the analysis of randomized clinical trials. 94 ASAProBiop 405–410

Ganter, Barbara: Schwarz, Carl James; — —: Estimating the movement among staging areas of the barnacle goose (Branta leucopsis). 95 JAppStat 22 711–724

Gao, A. J.: Yin, G.; Zhu, Y. M.; — —: On characterization of a recursively defined process: Necessary and sufficient conditions, sensitivity. 95 Stochatx 52 265–282

Gao, Feng; Sacks, Jerome; Welch, William J.: Predicting ozone levels and trends. 94 ASAProEnvr 77–81

Wahba, Grace; Johnson, Donald R.; — —; Gong, Jianjian: Adaptive tuning of numerical weather prediction models: Randomized GCV in three- and four-dimensional data assimilation. 95 MWeathr 11 3358–3369

Gao, Jiti: Asymptotic theory for partly linear models. 95 CommStA 24 1985–2009

— —: Asymptotic properties of some estimators for partly linear stationary autoregressive models. 95 CommStA 24 2011–2026

— —: The laws of the iterated logarithm of some estimates in partly linear models. 95 StPrLet 25 153–162

— —; Chen, Xiru; Zhao, Lingcheng: Asymptotic normality of a class of estimators in partial linear models. 95 AMaPeknE 37 256–268

— —; Liang, Hua: Asymptotic normality of pseudo-LS estimator for partly linear autoregression models. 95 StPrLet 23 27–34

Gao, Min; Minkin, Salomon: Nonlinear mixed effect model for estimating population characteristics of drug combination effect. 94 ASAProBiop 160–165

Gao, Ping: The lifetime of conditioned diffusions associated with some degenerate elliptic operators. 95 StocProc 57 319–337

Gao, Sujuan; Smith, T. M. F.: On the nonexistence of a global nonnegative minimum bias invariant quadratic estimator of variance components. 95 StPrLet 25 117–120

Gao, Y. T.: Parkin, D. M.; Muir, C. S.; Whelan, S. L.; — —; Ferlay, J.; Powell, J. (Ed); Hill, Catherine (Rev): Review of "Cancer incidence in five continents. Volume VI". 95 StatMed 14 104–104

Gaposhkin, V. F.: On the strong law of large numbers for blockwise independent and blockwise orthogonal random variables. 94 ThProbAp 39 677–684

García Rubio, E.; Peirats, V.: Evaluation of data editing procedures: Results of a simulation approach. 91 StatJUN 8 175–190

García, C.: Alvarez, M.; Cuevas, C. M.; Escudero, L. F.; de la Fuente, J. L.; — —; Prieto, F. J.: Network planning under uncertainty with an application to hydropower generation. 94 Top 2 25–58

Garcia, Eva Ferreira: Oltman, D. Olson; Lackritz, J. R. (Auth); — — (Rev): Review of "Statistics for business and economics". 95 JAppStat 22 437–437

Garcia, Nancy Lopes: Birth and death processes as projections of higher-dimensional Poisson processes. 95 AdvAppPr 27 911–930

— —: Maximum likelihood estimation and suboptimal stopping time for a Poisson point process. 95 RBrzlPrS 9 67–82

— —: Approximations of large population epidemic models. 95 StocProc 60 147–160

Garcia-Diaz, A.; Phillips, D. T.: Principles of experimental design and analysis. 94 ChapmnHall:Lnd 384p

García-Jurado, I.; González-Manteiga, W.; Prada-Sánchez, J. M.; Febrero-Bande, M.; Cao, R.: Predicting using Box-Jenkins, nonparametric, and bootstrap techniques. 95 Technmcs 37 303–310

Gardenier, John S.: Problems, trade-offs, and solutions for CAPI surveys. 94 ASASrvyRsr 857–860 (Disc: p861-862)

Gardenier, Turkan K.: A simulation-oriented approach to disparate impact. 94 ASAProStEd 180–185

Gardiner, Joseph: Aras, Girish; — —: Fixed diameter confidence ellipsoid in time-sequential models. 95 SqtlAnly 14 29–39

Gardiner, Joseph C.: Rahbar, Mohammad H.; — —: Nonparametric estimation of regression parameters from censored data with a discrete covariate. 95 StPrLet 24 13–20

Garel, Bernard; Hallin, Marc: Local asymptotic normality of multivariate ARMA processes with a linear trend. 95 AnInStMa 47 551–579

Garfield, Joan: How students learn statistics. 95 IntStRvw 63 25–34

— —: Reflections on the past 15 years. 95 TeachgSt 17/2 77–78

Moore, David S.; Cobb, George W.; — —; Meeker, William Q.: Statistics education fin de Siécle. 95 AmerStat 49 250–260

Garfield, Joan B.: Comment on "Statistics education". 95 AmerStat 49 18–20

Garfinkel, Simson L.; Mahoney, Michael K. (Auth); Lenz, H.-J. (Rev): Review of "NeXTSTEP programming, Step one: Object-oriented applications". 95 CmpStDA 19 712–713

Gargallo, R.: Tauler, R.; Izquierdo-Ridorsa, A.; — —; Casassas, E.: Application of a new multivariate curve resolution procedure to the simultaneous analysis of several spectroscopic titrations of the copper(II)-polyinosinic acid system. 95 Chemolab 27 163–174

Gargiullo, Paul M.; Rothenberg, Richard B.; Wilson, Hoyt G.: Confidence intervals, hypothesis tests, and sample sizes for the prevented fraction in cross-sectional studies. 95 StatMed 14 51–72

Garhart, Casey A.: Finn, Lynda M.; — —; Reynard, Susan E.: Getting statistical tools to the masses: Common problems in training design and execution. 95 ASQCTekCon 356–364

Garibaldi, Ubaldo; Penco, Maria Antonietta: A predictive approach to contingency tables (Italian). 95 Metron 53 79–113

Garin, Araceli: Hildebrand, D. K.; Ott, L. (Auth); — — (Rev): Review of "Statistical thinking for managers (Third edition)". 95 JAppStat 22 438–439

Garner, Thesia I.; Blanciforti, Laura A.: Household income reporting: An analysis of U.S. Consumer Expenditure Survey data. 94 JOffStat 10 69–91

Garnier, J.: Stochastic invariant imbedding: Application to stochastic differential equations with boundary conditions. 95 ProbTher 103 249–271

Garratt, Mark: Kuhfeld, Warren F.; Tobias, Randall D.; — —: Efficient experimental design with marketing research applications. 94 JMarRsr 31 545–557

Garside, S.: Rothery, P.; Martin, A. R.; Hiby, L.; — —: Estimating foetal growth rate from cross-sectional samples of field data with application to pilot-whales. 95 ApplStat 44 163–172

Garson, Glenn I.; Moser, E. Barry: Aggregation and the Pearson chi-square statistic for homogeneous proportions and distributions in ecology. 95 Ecology 76 2258–2269

Garthwaite, Paul H.; Yu, Keming; Hope, Peter B.: Bayesian analysis of a multiple-recapture model. 95 CommStA 24 2229–2247

Garven, Frances: Reed, Margaret; — —: Minitab menus - Do they help or hinder?. 95 TeachgSt 17/1 22–25

Garvey, Paul R.: A family of joint probability models for cost and schedule uncertainties. 95 JCostAnl /1 156–200

Gåsemyr, Jørund; Natvig, Bent: Using expert opinions in Bayesian prediction of component lifetimes in a shock model. 95 MaOpRsr 20 227–242

Gaskell, George: O'Muircheartaigh, Colm; — —: The basis of norms for vague quantifiers. 94 ASASrvyRsr 1327–1332

O'Muircheartaigh, Colm; — —; Wright, Daniel B.: Weighing anchors: Verbal and numeric labels for response scales. 95 JOffStat 11 295–307

Gaskins, John N.: Kumar, V.; Leone, Robert P.; — —: Aggregate and disaggregate sector forecasting using consumer confidence measures. 95 IntJFor 11 361–377

Gasparini, Mauro: Exact multivariate Bayesian bootstrap distributions of moments. 95 AnlsStat 23 762–768

Gasser, Robert P. H.; Richards, W. Graham: An introduction to statistical thermodynamics. 95 WrldSciPub:Sing 180p

Gasser, Theo; Kneip, Alois: Searching for structure in curve samples. 95 JASA 90 1179–1188

Beran, Jan; — —: Testing equality of variances for paired time series. 95 JTimSrAn 16 165–176

Engel, Joachim; — —: A minimax result for a class of nonparametric density estimators. 95 JNnparSt 4 327–334

Herrmann, Eva; Wand, M. P.; Engel, Joachim; — —: A bandwidth selector for bivariate kernel regression. 95 JRSS-B 57 171–180

Gastaldi, Tommaso: Interval estimation from a life-test through a sequential Kolmogorov-Smirnov type procedure. 95 IEEEReli 44 470–476

Gasteiger, Johann: Zupan, Jure; Novič, Marjana; — —: Neural networks with counter-propagation learning strategy used for modelling. 95 Chemolab 27 175–187

Gastwirth, Joseph L.: Statistical issues arising in equal employment litigation. 94 ASAProBuEc 44-53 (Disc: p54-54)
— —; Greenhouse, Samuel W.: Biostatistical concepts and methods in the legal setting. 95 StatMed 14 1641-1653
Sinclair, Michael D.; — —: On designs for measuring prevalence rates and error rates from diagnostic testing procedures. 94 ASAProBiop 148-153

Gatarek, Dariusz: Flandoli, Franco; — —: Martingale and stationary solutions for stochastic Navier-Stokes equations. 95 ProbTher 102 367-391
Prato, Giuseppe Da; — —: Stochastic Burgers equation with correlated noise. 95 Stochatx 52 29-41

Gates, Charles E.: What really is experimental error in block designs?. 95 AmerStat 49 362-363

Gather, Ursula; Tomkins, Jim: On stability of intermediate order statistics. 95 JStPlInf 45 175-183

Gatrell, Anthony C.: Bailey, Trevor C.; — —: Interactive spatial data analysis. 95 Wiley:NY:UK 432p

Gatsonis, C.; Hodges, J. S.; Kass, R. E.; Singpurwalla, N. D. (Ed): Case studies in Bayesian statistics. Volume II. 95 Springer:Brln:NY 364p

Gauchi, J.-P.: Use of PLS regression for the analysis of experimental designs in formulation chemistry (French). 95 RStApp 43/1 65-89
Tenenhaus, M.; — —; Ménardo, C.: PLS regression and applications (French). 95 RStApp 43/1 7-63

Gauderman, W. James: A method for simulating familial disease data with variable age at onset and genetic and environmental effects. 95 StatComp 5 237-243

Gauglitz, Günter: Kraus, Gerolf; — —: Optical reflectometric gas sensing: Classification of hydrocarbon vapours by pattern recognition applied to RIFS sensor signals. 95 Chemolab 30 211-221

Gauss, C. F.: — — (Auth); Stewart, G. W. (Transl); Stigler, S. M. (Rev): Review of "Theory of the combination of observations least subject to error". 96 ShrtBkRv 16/1 41-41

Gauss, Karl Friedrich: — — (Auth); Stewart, G. W. (Transl): Theory of the combination of observations least subject to errors. Part one, part two, supplement. 95 SocApMathSim:PA 220p

Gaut, Brandon S.; Weir, B. S.: Detecting substitution-rate heterogeneity among regions of a nucleotide sequence. 94 MolBioEv 11 620-629
Muse, Spencer V.; — —: A likelihood approach for comparing synonymous and nonsynonymous nucleotide substitution rates, with application to the chloroplast genome. 94 MolBioEv 11 715-724

Gaver, Donald P.; Jacobs, Patricia A.: Asymptotic properties of stochastic greedy bin-packing. 95 CommStSM 11 333-348

Gaver, Donald, P., Jr: Draper, David; — —; Goel, Prem K.; Greenhouse, Joel B.; Hedges, Larry V.; Morris, Carl N.; Tucker, John R.; Waternaux, Christine M. (Auth); Berlin, Jesse A. (Rev): Review of "Combining information: Statistical issues and opportunities for research". 95 StatMed 14 1611-1612

Gavrikov, Vladimir; Stoyan, Dietrich: The use of marked point processes in ecological and environmental forest studies. 95 EnvEcoSt 2 331-344

Gavrin, Jonathan: Madigan, David; — —; Raftery, Adrian E.: Eliciting prior information to enhance the predictive performance of Bayesian graphical models. 95 CommStA 24 2271-2292

Gay, S. L.: Stereological equations for phases within particles. 95 JMicro 179 297-305

Gayrard, Véronique: Bovier, Anton; — —; Picco, Pierre: Large deviation principles for the Hopfield model and the Kac-Hopfield model. 95 ProbTher 101 511-546

Gbur, Edward E., Jr; Trumbo, Bruce E.: Key words and phrases - The key to scholarly visibility and efficiency in an information explosion. 95 AmerStat 49 29-33

Gbur, Philip M.; Kindred, Machell; Mersch, Michael L.: Master address file: Update methodology and quality improvement program. 94 ASASrvyRsr 748-753 (Disc: p754-756)

Gediga, Günther: Düntsch, Ivo; — —: Skills and knowledge structures. 95 BJMaStPs 48 9-27

Gefeller, Olaf: Dette, Holger; — —: The impact of different definitions of nearest neighbour distances for censored data on nearest neighbour kernel estimators of the hazard rate. 95 JNnparSt 4 271-282

Gehan, E. A.; Lemak, N. A.: Statistics in medical research. Developments in clinical trials. 94 PlenumPubl:NY 214p
— —; Lemak, N. A. (Auth); De Stavola, B. L. (Rev): Review of "Statistics in medical research. Developments in clinical trials". 95 ShrtBkRv 15/2 22-22

Gehan, Edmund A.; Lemak, Noreen A.: Statistics in medical research: Developments in clinical trials. 94 PlenumPubl:NY 224p

Geisser, S.: — — (Auth); Dunsmore, Ian R. (Rev): Review of "Predictive inference: An introduction". 95 JRSS-A 158 191-191

Geisser, Seymour; Johnson, Wesley: Testing independence when the form of the bivariate distribution is unspecified. 95 StatMed 14 1621-1639
— — (Auth); Lawera, Martin (Rev): Review of "Predictive inference: An introduction". 95 Technmcs 37 121-121
— — (Auth); Mouhab, Abderrahamane (Rev): Review of "Predictive inference: An introduction". 95 JASA 90 1489-1490

Geladi, Paul: Lindgren, Fredrik; — —; Berglund, Anders; Sjöström, Michael; Wold, Svante: Interactive variable selection (IVS) for PLS. Part II: Chemical applications. 95 JChemtrc 9 331-342
Rännar, Stefan; — —; Lindgren, Fredrik; Wold, Svante: A PLS kernel algorithm for data sets with many variables and few objects. Part II: Cross-validation, missing data and examples. 95 JChemtrc 9 459-470

Gelber, Richard D.; Cole, Bernard F.; Gelber, Shari; Goldhirsch, Aron: Comparing treatments using quality-adjusted survival: The Q-TWiST method. 95 AmerStat 49 161-169
Cole, Bernard F.; — —; Anderson, Keaven M.; Goldhirsch, Aron: Cost-effectiveness analysis using quality-adjusted survival. 94 ASAProBiop 9-18
Cole, Bernard F.; — —; Goldhirsch, Aron: A quality-adjusted survival meta-analysis of adjuvant chemotherapy for premenopausal breast cancer. 95 StatMed 14 1771-1784

Gelber, Shari: Gelber, Richard D.; Cole, Bernard F.; — —; Goldhirsch, Aron: Comparing treatments using quality-adjusted survival: The Q-TWiST method. 95 AmerStat 49 161-169

Gelbrich, Matthias: Kloeden, Peter E.; Platen, Eckhard (Auth); — —; Römisch, Werner (Rev): Review of "Numerical solution of stochastic differential equations". 95 SIAMRvw 37 272-275

Gelfand, Alan E.; Carlin, Bradley P.: Comment on "Bayesian computation and stochastic systems". 95 StatSci 10 43-46
— —; Mallick, Bani K.: Bayesian analysis of proportional hazards models built from monotone functions. 95 Biomtrcs 51 843-852
— —; Mallick, Bani K.; Dey, Dipak K.: Modeling expert opinion arising as a partial probabilistic specification. 95 JASA 90 598-604
— —; Mukhopadhyay, Saurabh: On nonparametric Bayesian inference for the distribution of a random sample. 95 CandJSt 23 411-420
— —; Sahu, Sujit K.; Carlin, Bradley P.: Efficient parametrisations for normal linear mixed models. 95 Biomtrka 82 479-488

Gelman, A.; Carlin, J.; Stern, H.; Rubin, D.: Bayesian data analysis. 95 ChapmnHall:Lnd 552p
— —; Carlin, J. B.; Stern, H. S.; Rubin, D. B. (Auth); Brown, P. J. (Rev): Review of "Bayesian data analysis". 96 ShrtBkRv 16/1 5-5

Gelman, Andrew: Method of moments using Monte Carlo simulation. 95 JCmpGrSt 4 36-54

Geman, Hélyette; Karoui, Nicole El; Rochet, Jean-Charles: Changes of numéraire, changes of probability measure and option pricing. 95 JAppProb 32 443-458

Geman, Stuart: Künsch, Hans; — —; Kehagias, Athanasios: Hidden Markov random fields. 95 AnlsApPr 5 577-602

Gemperline, Paul J.: Cho, Junghwan; — —: Pattern recognition analysis of near-infrared spectra by robust distance method. 95 JChemtrc 9 169-178

Gencay, Ramazan: Fan, Yanqin; — —: A consistent nonparametric test of symmetry in linear regression models. 95 JASA 90 551-557

Genest, C.; Ghoudi, K.; Rivest, L.-P.: A semiparametric estimation procedure of dependence parameters in multivariate families of distributions. 95 Biomtrka 82 543-552

Geng, Zhi: Guo, Jianhua; — —: Collapsibility of logistic regression coefficients. 95 JRSS-B 57 263-267

Gennings, Chris; Carter, W. Hans, Jr: Utilizing concentration-response data from individual components to detect statistically significant departures from additivity in chemical mixtures. 95 Biomtrcs 51 1264-1277
Schwartz, Pamela F.; — —; Chinchilli, Vernon M.: Threshold models for combination data from reproductive and developmental experiments. 95 JASA 90 862-870

Gentle, James E.: Valliant, Richard; — —: An application of mathematical programming to sample allocation. 94 ASASrvyRsr 683-688

Gentleman, Jane F.; Park, Evelyn: The probability of divorce, based on the age difference of the married couple. 94 ASAProSoSt 216-221
Ng, Edward; — —: Life expectancy in Canada: An empirical comparison of different estimation methods. 94 ASAProSoSt 88-93

Genz, Alan; Monahan, John: Random integration rules for statistical computation. 94 ASAProStCp 182-187

Geoffard, Pierre-Yves; Philipson, Tomas: The empirical content of canonical models of infectious diseases: The proportional hazard specification. 95 Biomtrka 82 101-111

Georg, Thomas: Droge, Bernd; — —: On selecting the smoothing parameter of least squares regression estimates using the minimax regret approach. 95 StatDecn 13 1-20

Georgakis, Christos: Ku, Wenfu; Storer, Robert H.; — —: Disturbance detection and isolation by dynamic principal component analysis. 95 Chemolab 30 179-196

George, E. Olusegun; Bowman, Dale: A full likelihood procedure for analysing exchangeable binary data. 95 Biomtrcs 51 512-523
Bowman, Dale; — —: A saturated model for analyzing exchangeable binary data: Applications to clinical and developmental toxicity studies. 95 JASA 90 871-879
Bowman, Dale; Chen, James J.; — —: Estimating variance functions in developmental toxicity studies. 95 Biomtrcs 51 1523-1528

George, Edward I.: Foster, Dean P.; — —: The risk inflation criterion for multiple regression. 94 AnlsStat 22 1947-1975

George, Karen J.; McKean, Joseph W.; Schucany, William R.; Sheather, Simon J.: A comparison of confidence intervals from R-estimators in regression. 95 JStCmpSm 53 13-22

George, Peter J.; Oksanen, Ernest H.; Veall, Michael R.: Analytic and bootstrap approaches to testing a market saturation hypothesis. 95 MaCmpSim 39 311-315

George, Rani: Kaplan, David; — —: A study of the power associated with testing factor mean differences under violations of factorial invariance. 95 StrEqMod 2 101-118

George, Robert; Goel, Prem K.: Hierarchical Bayesian models and Γ-minimaxity. 94 ASAProBays 96-101

George, Stephen L.: Statistical assessment of the utility of tumor markers. 94 ASAProEpid 8-13

George, Varghese: Shahane, Aditi; — —; Johnson, William D.: Effect of bivariate regression toward the mean in uncontrolled clinical trials. 95 CommStA 24 2165-2181

George, Varghese T.: Elston, Robert C.; — —; Severtson, Forrestt: The Elston-Stewart algorithm for continuous genotypes and environmental factors. 92 HumHrdty 42 16-27

Georgiadis, Leonidas: Bhattacharya, Partha P.; — —; Tsoucas, Pantelis: Problems of adaptive optimization in multiclass M/GI/1 queues with Bernoulli feedback. 95 MaOpRsr 20 355-380

Gerber, Eleanor: Bates, Nancy; — —: A typology of temporary mobility patterns and reporting of usual residence. 94 ASASrvyRsr 331-336

Gerber, Eleanor R.: Hidden assumptions: The use of vignettes in cognitive interviewing. 94 ASASrvyRsr 1269-1274

Gerber, H. U.: Kaas, R.; — —: Some alternatives for the individual model. 94 InsrneME 15 127-132

Gerber, Hans U.: A teacher's remark on exact credibility. 95 AstinB 25 189-192

— —; Shiu, Elias S. W.: Option pricing by Esscher transforms. 95 TrSActur 46 99-140 (Disc: p141-191)

— —; Shiu, Elias S. W.: From perpetual strangles to Russian options. 94 InsrneME 15 121-126

Germain-Tourbier, S.: Benoist, D.; Tourbier, Y.; — —: Experimental design: Construction and analysis (French). 95 Lavoisier:Paris ?p

Germanson, Teresa P.: Choi, Sung C.; — —; Barnes, Thomas Y.: A simple measure in defining optimal strata in clinical trials. 95 ConClnTr 16 164-171

Gerontidis, Ioannis I.: Markov population replacement processes. 95 AdvAppPr 27 711-740

Kalashnikov, V. V. (Auth); — — (Rev): Review of "Mathematical methods in queuing theory". 95 Ststcian 44 409-410

Gersch, Will; Stone, David: Multivariate autoregressive time series modeling: One scalar autoregressive model at-a-time. 95 CommStA 24 2715-2733

Gerstein, Dean: Johnson, Robert; Malloy, Kay; — —; Suter, Natalie; Nisenbaum, Susan: Increasing response rates in follow-up surveys of drug treatment patients. 94 ASASrvyRsr 1263-1268

Gertsbakh, I.: On the Fisher information in type-I censored and quantal response data. 95 StPrLet 23 297-306

Kordonsky, Kh. B.; — —: System state monitoring and lifetime scales - II. 95 RelEngSS 49 145-154

Gertsbakh, I. B.: Kordonsky, Kh. B.; — —: System state monitoring and lifetime scales - I. 95 RelEngSS 47 1-14

Geske, Mark X.; Godbole, Anant P.; Schaffner, Andrew A.; Skolnick, Allison M.; Wallstrom, Garrick L.: Compound Poisson approximations for word patterns under Markovian hypotheses. 95 JAppProb 32 877-892

Gessler, J. R.: — — (Auth); Schlittgen, R. (Rev. German): Review of "Statistical graphics" (German). 95 AllgeStA 79 223-224

Gessley, Mary K.; Kelly, E. Thomas; Ponikowski, Chester H.: Introducing births into the employment cost index survey. 94 ASASrvyRsr 797-802

Getchell, C.: Gunsolley, J. C.; — —; Chinchilli, V. M.: Small sample characteristics of generalized estimating equations. 95 CommStB 24 869-878

Getoor, R. K.: Fitzsimmons, P. J.; — —: Occupation time distributions for Lévy bridges and excursions. 95 StocProc 58 73-89

Getson, Pamela: BMDP Statistical Software Inc (Auth); — — (Rev): Review of "BMDP New System for Windows (Version 1.11)". 95 Chance 8/4 53-58

Gevers, H. O.: Hohls, T.; Clarke, G. P. Y.; Shanahan, P. E.; — —: Additive and non-additive genetic correlations. II. Application to data obtained from a 12×12 diallel cross. 95 BiomtrcJ 37 937-943

Hohls, T.; Clarke, G. P. Y.; Shanahan, P. E.; — —: An unbiased estimator for identifying lines useful for the improvement of elite single crosses, based on a combining ability model. 95 Heredity 75 155-163

Geweke, John; Keane, Michael: Recursively simulating multinomial multiperiod probit probabilities. 94 ASAProBuEc 1-6

Geyer, Alois L. J.: Wagner, Udo; — —: A maximum entropy method for inverting Laplace transforms of probability density functions. 95 Biomtrka 82 887-892

Geyer, Charles J.: Conditioning in Markov chain Monte Carlo. 95 JCmpGrSt 4 148-154

— —: Comment on "Bayesian computation and stochastic systems". 95 StatSci 10 46-48

— —: On the asymptotics of constrained M-estimation. 94 AnlsStat 22 1993-2010

— —; Thompson, Elizabeth A.: Annealing Markov chain Monte Carlo with applications to ancestral inference. 95 JASA 90 909-920

Chan, Kung Sik; — —: Comment on "Markov chains for exploring posterior distributions". 94 AnlsStat 22 1747-1758

Gfroerer, Joseph: Wright, Douglas; — —; Epstein, Joan: Ratio estimation of hard-core drug use. 94 ASASrvyRsr 541-546

Gfroerer, Joseph C.: Colliver, James D.; — —: Predictors of drug abuse treatment utilization among frequent cocaine users. 94 ASASrvyRsr 553-558

Epstein, Joan F.; — —: Estimating substance abuse treatment need from the NHSDA. 94 ASASrvyRsr 559-564

Ghani, Azra: Brookmeyer, R.; Gail, M. H. (Auth); — — (Rev): Review of "AIDS epidemiology: A quantitative approach". 95 JRSS-A 158 343-343

Gharraph, M. K.: Shoukry, E. M.; — —; Hassan, N. A.: Multi-channel bi-level queueing system with heterogeneous servers. 94 EgpStJ 38 208-221

Ghazal, G. A.; Mira, S. I.: The exact finite sample properties of the estimators of coefficients in mixed dummy-variable error components model. 95 EgpStJ 39 26-50

Ghislandi, M.: Bernardinelli, L.; Clayton, D.; Pascutto, C.; Montomoli, C.; — —; Songini, M.: Bayesian analysis of space-time variation in disease risk. 95 StatMed 14 2433-2443

Ghitany, M. E.; El-Saidi, M. A.; Khalil, Z.: Characterization of a general class of life-testing models. 95 JAppProb 32 548-553

— —; Maller, R.; Zhou, S.: Estimating the proportion of immunes in censored samples: A simulation study. 95 StatMed 14 39-49

Gholston, Sampson; Batson, Robert G.: Case study: Precision improvement of a digital ball comparator via parameter design. 94 ASAProPEng 302-307

Ghorai, J. K.: Large sample confidence interval for the quantiles in the proportional hazards model of random censorship. 95 JStCmpSm 53 91-102

Ghorbanzadeh, Dariush: A test for the detection of a change in mean in a Gaussian model (French). 95 RStApp 43/2 67-76

Ghosal, Subhashis; Samanta, Tapas: Asymptotic behaviour of Bayes estimates and posterior distributions in multiparameter nonregular cases. 95 MaMethSt 4 361-388

Ghosal, Sugata; Mehrotra, Rajiv: Range surface characterization and segmentation using neural networks. 95 PattrnRc 28 711-727

Ghosh, Atalanta: Agresti, Alan; — —; Bini, Matilde: Raking kappa: Describing potential impact of marginal distributions on measures of agreement. 95 BiomtrcJ 37 811-820

Ghosh, D. K.; Joshi, K. S.: Construction of variance balanced designs through triangular PBIB designs. 95 CalcutSt 45 111-118

— —; Karmokar, P. K.: Some methods for constructing efficiency balanced designs. 94 JIAgSt 46 305-309

Shah, S. M.; — —; Patel, A. H.: Construction of α-resolvable BIBDs using 1-resolvable and α-affine resolvable BIBDs. 93 UtilMath 43 219-224

Ghosh, Dhiren; Kaufman, Steven F.; Smith, Wray; Chang, Michael: Optimal periodicity of a survey: Sampling error, data deterioration, and cost. 94 ASASrvyRsr 1122-1127

Ghosh, J. K.; Mukerjee, Rahul: On perturbed ellipsoidal and highest posterior density regions with approximate frequentist validity. 95 JRSS-B 57 761-769

DasGupta, A.; — —; Zen, M. M.: A new general method for constructing confidence sets in arbitrary dimensions: With applications. 95 AnlsStat 23 1408-1432

Datta, G. S.; — —: Noninformative priors for maximal invariant parameter in group models. 95 Test 4 95-114

Hall, W. J.; Wijsman, R. A.; — —: Correction to "The relationship between sufficiency and invariance with applications in sequential analysis" (V36 p575-614). 95 AnlsStat 23 705-705

Ghosh, Jayanta K.; Mukerjee, Rahul: Frequentist validity of highest posterior density regions in the presence of nuisance parameters. 95 StatDecn 13 131-139

De, Anindya K.; — —: A decision rule for dimension in the context of MANOVA. 95 SankhyaA 57 256-266

Ghosh, Jayanta Kumar: Datta, Gauri Sankar; — —: On priors providing frequentist validity for Bayesian inference. 95 Biomtrka 82 37-45

Ghosh, M.; Carlin, B. P.; Srivastava, M. S.: Probability matching priors for linear calibration. 95 Test 4 333-357

Ghosh, Malay: Inconsistent maximum likelihood estimators for the Rasch model. 95 StPrLet 23 165-170

— —; Hoekstra, Robert M.: A.P.O. rules in hierarchical Bayes regression models. 95 SqtlAnly 14 99-115

Datta, Gauri Sankar; — —: Some remarks on noninformative priors. 95 JASA 90 1357-1363

Datta, Gauri Sankar; — —: Hierarchical Bayes estimators of the error variance in one-way ANOVA models. 95 JStPlInf 45 399-411

Mukhopadhyay, Saurabh; — —: On the uniform approximation of Laplace's prior by t-priors in location problems. 95 JMultiAn 54 284-294

Ghosh, Subir: Sequential construction of new main effect plans with higher revealing power for 2^7 factorials. 93 JCISS 18 217-231

— —; Derderian, Edna: Determination of robust design against noise factors and in presence of signal factors. 95 CommStB 24 309-326

Ghoshal, J.: Ray, Kumar S.; — —: Neuro-linguistic approach to pattern recognition. 95 FuzzySet 74 277-305

Ghoudi, K.: Genest, C.; — —; Rivest, L.-P.: A semiparametric estimation procedure of dependence parameters in multivariate families of distributions. 95 Biomtrka 82 543-552

Ghoudi, Kilani: Asymptotics of multivariate randomness statistics. 95 JMultiAn 55 340-348

Ghysels, Eric: Campbell, Bryan; — —: Federal budget projections: A nonparametric assessment of bias and efficiency. 95 RvEconSt 77 17-31

Giacomin, G.: Onset and structure of interfaces in a Kawasaki + Glauber interacting particle system. 95 ProbTher 103 1-24

Giangiacomo, R.: Forina, M.; Drava, G.; Armanino, C.; Boggia, R.; Lanteri, S.; Leardi, R.; Corti, P.; Conti, P.; — —; Galliena, C.; Bigoni, R.; Quartari, I.; Serra, C.; Ferri, D.; Leoni, O.; Lazzeri, L.: Transfer of calibration function in near-infrared spectroscopy. 95 Chemolab 27 189-203

Gianini, Jacqueline: Nash, John C.; Quon, Tony K.; — —: Statistical issues in spreadsheet software. 94 ASAProStEd 238-241

Giannakis, Georgios B.: Sadler, Brian M.; — —; Shamsunder, Sanyogita: Noise subspace techniques in non-Gaussian noise using cumulants. 95 IEEEAES 31 1009-1018

Gianola, D.: Wang, C. S.; — —; Sorensen, D. A.; Jensen, J.; Christensen, A.; Rutledge, J. J.: Response to selection for litter size in Danish Landrace pigs: A Bayesian analysis. 94 ThApGene 88 220-230

Gibbens, R. J.: Bean, N. G.; — —; Zachary, S.: Asymptotic analysis of single resource loss systems in heavy traffic, with applications to integrated networks. 95 AdvAppPr 27 273-292

Gibbons, J. D.; Chakraborti, S. (Auth); Williams, Calvin L. (Rev): Review of "Nonparametric statistical inference (Third edition)". 95 JAppStat 22 193-194

Gibbons, Jean Dickinson; Chakraborti, Subhabrata (Auth); Borovcnik, Manfred (Rev): Review of "Nonparametric statistical inference". 95 CmpStQ 10 100-103

— — (Auth); Nelson, Lloyd S. (Rev): Review of "Nonparametric measures of association". 95 JQualTek 27 93-93

Gibbons, Robert: — — (Auth); Ziegel, Eric R. (Rev): Review of "Statistical methods for groundwater monitoring". 95 Technmcs 37 231-232

Gibbons, Robert D.: Some statistical and conceptual issues in the detection of low-level environmental pollutants. 95 EnvEcoSt 2 125-145 (Disc: p145-167)

Gibiansky, Ekaterina: Nedelman, Jerry R.; Karara, Adel H.; Chang, Cheng-Tao; — —; McDonald, Scott; Robinson, William T.: Inferring systemic exposure from a pharmacokinetic screen: Model-free and model-based approaches. 95 StatMed 14 955-968 (Disc: p969-970)

Gibson, G. J.: Glasbey, C. A.; Horgan, G. W.; — —; Hitchcock, D.: Fish shape analysis using landmarks. 95 BiomtrcJ 37 481-495

Gijbels, Irène; Mielniczuk, Jan: Asymptotic properties of kernel estimators of the Radon-Nikodym derivative with applications to discriminant analysis. 95 StSinica 5 261-278

Fan, Jianqing; — —: Adaptive order polynomial fitting: Bandwidth robustification and bias reduction. 95 JCmpGrSt 4 213-227

Fan, Jianqing; — —: Data-driven bandwidth selection in local polynomial fitting: Variable bandwidth and spatial adaptation. 95 JRSS-B 57 371-394

Gilbert, Christopher L.: Modelling market fundamentals: A model of the aluminium market. 95 JApEcmtx 10 385-410

Gilbert, E. N.: Coffman, E. G., Jr; — —; Greenberg, A. G.; Leighton, F. T.; Robert, Philippe; Stolyar, A. L.: Queues served by a rotating ring. 95 CommStSM 11 371-394

Gilbert, Paul D.: Combining VAR estimation and state space model reduction for simple good predictions. 95 JForcast 14 229-250

Giles, David E. A.: Dods, Johannah L.; — —: Alternative strategies for 'augmenting' the Dickey-Fuller test: Size-robustness in the face of pre-testing. 95 JStCmpSm 53 243-258

Gilks, W. R.; Best, N. G.; Tan, K. K. C.: Adaptive rejection Metropolis sampling within Gibbs sampling. 95 ApplStat 44 455-472

Roberts, G. O.; Sahu, S. K.; — —: Comment on "Bayesian computation and stochastic systems". 95 StatSci 10 49-51

Gill, Amar Nath: Kumar, Narinder; — —; Mehta, Gobind P.: Selection procedures for scale parameters when the populations have a common known quantile. 94 Metron 52/3 171-196

Gill, Len: Maddala, G. S.; Rao, C. R.; Vinod, H. D. (Ed); — — (Rev): Review of "Handbook of statistics (Volume 11) Econometrics". 95 IntJFor 11 189-191

Gill, R. Bruce: Pojar, Thomas M.; Bowden, David C.; — —: Aerial counting experiments to estimate pronghorn density and herd structure. 95 JWildlMg 59 117-128

Gill, R. D.: Andersen, P. K.; Borgan, Ø.; — —; Keiding, N.: Statistical models based on counting processes. 95 Springer:Brln:NY 767p

Andersen, P. K.; Borgan, O.; — —; Keiding, N. (Auth); Milbrodt, Hartmut (Rev. German): Review of "Statistical models based on counting processes". 95 AllgeStA 79 221-221

Andersen, P. K.; Borgan, Ø.; — —; Keiding, N. (Auth); Hjort, Nils Lid (Rev): Review of "Statistical models based on counting processes". 95 StNeerla 49 247-249

Andersen, P. K.; Borgan, Ø.; — —; Keiding, N. (Auth); McKeague, Ian W. (Rev): Review of "Statistical models based on counting processes". 95 SIAMRvw 37 475-476

Gill, Richard D.; Levit, Boris Y.: Applications of the van Trees inequality: A Bayesian Cramér-Rao bound. 95 Bernoull 1 59-79

— —; van der Laan, Mark J.; Wellner, Jon A.: Inefficient estimators of the bivariate survival function for three models. 95 AIHenrPB 31 545-597

Gilles, F. H.: Tavaré, C. J.; Sobel, E. L.; — —: Misclassification of a prognostic dichotomous variable: Sample size and parameter estimate adjustment. 95 StatMed 14 1307-1314

Gillespie, Brenda: Ratkowsky, D. A.; Evans, M. A.; Alldredge, J. R. (Auth); — — (Rev): Review of "Crossover experiments: Design, analysis and application". 95 JAppStat 22 188-188

Gillespie, David F.: McGartland Rubio, Doris; — —: Problems with error in structural equation models. 95 StrEqMod 2 367-378

Gillett, Raphael: The expected power of F and t tests conditional on information from an earlier experiment. 95 BJMaStPs 48 371-384

Gilligan, C. A.: Grenfell, B. T.; Kleczkowski, A.; — —; Bolker, B. M.: Spatial heterogeneity, nonlinear dynamics and chaos in infectious diseases. 95 StMedRes 4 160-183

Gillis, R. J.: Hamilton, M. A.; Johnson, K. R.; Camper, A. K.; Stoodley, P.; Harkin, G. J.; — —; Shope, P. A.: Analysis of bacterial spatial patterns at the initial stage of biofilm formation. 95 BiomtrcJ 37 393-408

Gillman, Daniel W.; Appel, Martin V.: Developing metadata database at the Census Bureau. 94 ASASrvyRsr 781-784

Gilmour, Arthur R.; Thompson, Robin; Cullis, Brian R.: Average information REML: An efficient algorithm for variance parameter estimation in linear mixed models. 95 Biomtrcs 51 1440-1450

Gilmour, S. G.; Mead, R.: Stopping rules for sequences of factorial designs. 95 ApplStat 44 343-355

Giltinan, D. M.: Davidian, M.; — —: Nonlinear models for repeated measurement data.. 95 ChapmnHall:Lnd 359p

Davidian, M.; — — (Auth); Hand, D. J. (Rev): Review of "Nonlinear models for repeated measurement data.". 96 ShrtBkRv 16/1 6-7

Giltinan, David M.: Davidian, Marie; — —: Nonlinear models for repeated measurement data. 95 ChapmnHall:Lnd 384p

Gilula, Zvi; Haberman, Shelby J.: Prediction functions for categorical panel data. 95 AnlsStat 23 1130-1142

— —; Haberman, Shelby J.: Dispersion of categorical variables and penalty functions: Derivation, estimation, and comparability. 95 JASA 90 1447-1452

Gimbrere, C. H.: Schouten, L. J.; Straatman, H.; Kiemeney, L. A.; — —; Verbeek, A. L.: The capture-recapture method for estimation of cancer registry completeness: A useful tool?. 94 IJEpidem 23 1111-1116

Giné, Evarist: Arcones, Miguel A.; — —: On the law of the iterated logarithm for canonical U-statistics and processes. 95 StocProc 58 217-245

Cuzick, Jack; — —; Zinn, Joel: Laws of large numbers for quadratic forms, maxima of products and truncated sums of i.i.d. random variables. 95 AnlsProb 23 292-333

Ginebra, Josep; Clayton, Murray K.: Response surface bandits. 95 JRSS-B 57 771-784

Fang, Kai-Tai; Wang, Y. (Auth); — — (Rev): Review of "Number-theoretic methods in statistics". 95 JASA 90 1134-1134

Gini, F.: Corsini, G.; — —; Greco, M. V.; Verrazzani, L.: Cramér-Rao bounds and estimation of the parameters of the Gumbel distribution. 95 IEEEAES 31 1202-1204

Ginsburg, E. Kh.; Axenovich, T. I.; Babenko, V. N.: Comparison of asymptotic tests of the Hardy-Weinberg distribution. 91 AnHuGene 55 329-338

Ginter, James L.: Allenby, Greg M.; Arora, Neeraj; — —: Incorporating prior knowledge into the analysis of conjoint studies. 95 JMarRsr 32 152-162

Giorgi, G. M.; Mondani, R.: Sampling distribution of the Bonferroni inequality index from exponential population. 95 SankhyaB 57 10-18

Giorgi, Giovanni Maria; Mondani, Riccardo: The exact sampling distribution of the Bonferroni concentration index. 94 Metron 52/3 5-41

Giorno, V.: Di Crescenzo, A. G.; — —; Nobile, A. G.; Ricciardi, L. M.: On a symmetry-based constructive approach to probability densities for two-dimensional diffusion processes. 95 JAppProb 32 316-336

Giovagnoli, Alessandra; Wynn, H. P.: Multivariate dispersion orderings. 95 StPrLet 22 325-332

Giovannelli, Jean-François: Idier, Jérôme; — —; Querleux, Bernard: Bayesian time-varying AR spectral estimation for ultrasound attenuation measurement in biological tissues. 94 ASAProBays 256-261

Girard, D. A.: The fast Monte-Carlo cross-validation and C_L procedures: Comments, new results and application to image recovery problems. 95 CmpStQ 10 205-231 (Disc: p233-258)

Girard, Pascal; Laporte-Simitsidis, Silvy; Mismetti, Patrick; Decousus, Hervé; Boissel, Jean-Pierre: Influence of confounding factors on designs for dose-effect relationship estimates. 95 StatMed 14 987-1005 (Disc: p1007-1008)

Boissel, Jean-Pierre; Durieu, Isabelle; — —; Nony, Patrice; Chauvin, Franck; Haugh, Margaret: Dose-ranging trials: Guidelines for data collection and standardized descriptions. 95 ConClnTr 16 319-330

Girard, Ph.: Parent, E.; Chaouche, A.; — —: In support of Bayesian statistics for quality control by attribute. Part I: Simple sampling plans (French). 95 RStApp 43/4 5-18

Giri, N.: Das, M. N.; — —; Ahmed, M.: Designs through recoding of varietal and level codes. 95 StPrLet 23 371-380

Giri, N. C.: — — (Auth); Bandopadhyay, T. (Rev): Review of "Introduction to probability and statistics (Second edition)". 94 CalcutSt 44 267-268

Giri, Narayan: Jóhannesson, Benedikt; — —: On approximations involving the beta distribution. 95 CommStB 24 489-503

Giri, Narayan C.: Group invariance in statistical inference. 95 WrldSciPub:Sing 250p

Giribone, Pietro: Response surface methodology for non-equispaced experimental configurations applied to a discrete and stochastic simulator of a manufacturing company. 94 JAppStat 21 549-564

Girko, V. L.: Canonical spectral equation. 94 ThProbAp 39 685-691

— —; Repin, K. Yu.: The quarter-of-circumference law. 95 ThPrMaSt 50 67-70

Girko, Vyacheslav L.: Statistical analysis of observations of increasing dimension. 95 Kluwer:MA:Hllnd 286p

Girón, Javier: Moreno, Elías; — —: Prior elicitation and Bayesian robustness in a noisy Poisson process. 94 ASAProBays 217-220

Gitlow, Howard S.: Charnes, John M.; — —: Using control charts to corroborate bribery in jai alai. 95 AmerStat 49 386-389

Giudici, Paolo: Bayes factors for zero partial covariances. 95 JStPlInf 46 161-174

Giuffra, Luis: Risch, Neil; — —: Model misspecification and multipoint linkage analysis. 92 HumHrdty 42 77-92

Giummolé, Federica: Ferrante, Marco; — —: Finite dimensional filters for a discrete-time nonlinear system with generalized Gaussian white noise. 95 Stochatx 53 195-211

Givaruangsawat, Sumalee: Weerakkody, Govinda J.; — —: Estimating the correlation coefficient in the presence of correlated observations from a bivariate normal population. 95 CommStA 24 1705-1719

Givens, Geof H.: Consistency of the local kernel density estimator. 95 StPrLet 25 55-61

Raftery, Adrian E.; — —; Zeh, Judith E.: Inference from a deterministic population dynamics model for Bowhead whales. 95 JASA 90 402-416 (Disc: p417-430)

Raftery, Adrian E.; — —; Zeh, Judith E.: Reply to comments on "Inference from a deterministic population dynamics model for Bowhead whales". 95 JASA 90 427-430

Glaccum, John F.: Spot check your SPC applications. 95 QulPr 28/5 184-184

Glad, Ingrid K.; Sebastiani, Giovanni: A Bayesian approach to synthetic magnetic resonance imaging. 95 Biomtrka 82 237-250

Hjort, Nils Lid; — —: Nonparametric density estimation with a parametric start. 95 AnlsStat 23 882-904

Gladkov, Yu. M.: Volkov, V. Ya.; — —: Reconfigurable combined forecasts in a non-stationary inflationary environment. 95 JForcast 14 395-403

Glasbey, C. A.: Imputation of missing values in spatio-temporal solar radiation data. 95 EnvrMtrc 6 363-371

— —; Horgan, G. W.: Image analysis for the biological sciences. 95 Wiley:NY:UK 218p

— —; Horgan, G. W.; Gibson, G. J.; Hitchcock, D.: Fish shape analysis using landmarks. 95 BiomtrcJ 37 481-495

Glasbey, Chris: Comment on "Analysis of longitudinal data with unequally spaced observations and time-dependent correlated errors" (V50 p445-456). 95 Biomtrcs 51 375-376

Glaser, R. E.: Weibull accelerated life testing with unreported early failures. 95 IEEEReli 44 31-36

Glaser, Ronald: Huh, Myung-Hoe; Critchlow, Douglas E.; Verducci, Joseph S.; Klecolt-Glaser, Janice; — —; Malarkey, William B.: A symmetric analysis of paired rankings with application to temporal patterns of hormonal concentration. 95 Biomtrcs 51 1361-1371

Glass, J. C.; McKillop, D. G.; Hyndman, N.: Efficiency in the provision of university teaching and research: An empirical analysis of UK universities. 95 JApEcmtx 10 61-72

Glasserman, Paul; Kou, Shing-Gang: Analysis of an importance sampling estimator for tandem queues. 95 ACMMdSlm 5 22-42

— —; Kou, Shing-Gang: Limits of first passage times to rare sets in regenerative processes. 95 AnlsApPr 5 424-445

— —; Yao, David D.: Stochastic vector difference equations with stationary coefficients. 95 JAppProb 32 851-866

— —; Yao, David D. (Auth); Haas, Peter J. (Rev): Review of "Monotone structure in discrete-event systems". 95 JASA 90 808-809

— —; Yao, David D. (Auth); Raghavachari, Madabhushi (Rev): Review of "Monotone structure in discrete-event systems". 95 Technmcs 37 467-468

Glaz, Joseph: Testing procedures for detecting multiple clusters. 94 ASAProEpid 59-63

Sison, Christina P.; — —: Correction to "Simultaneous confidence intervals and sample size determination for multinomial proportions" (V90 p366-369). 95 JASA 90 1494-1494

Sison, Cristina P.; — —: Simultaneous confidence intervals and sample size determination for multinomial proportions. 95 JASA 90 366-369

Wallenstein, Sylvan; Naus, Joseph; — —: Power of the scan statistics. 94 ASAProEpid 70-75

Glazebrook, K. D.; Boys, R. J.: A class of Bayesian models for optimal exploration. 95 JRSS-B 57 705-720

— —; Greatrix, S.: On transforming an index for generalised bandit problems. 95 JAppProb 32 168-182

— —; Owen, R. W.: On the value of adaptive solutions to stochastic scheduling problems. 95 MaOpRsr 20 65-89

Glencross, Michael J.; Cherian, Varghese I.: Attitudes toward applied statistics of postgraduate students in education in the Lebowa region of South Africa. 95 PsychRpt 77 315-322

Glendinning, R. H.: Model selection for infinite variance time series. 95 CommStA 24 889-910

Elworthy, K. D.; Ikeda, N. (Ed); — — (Rev): Review of "Asymptotic problems in probability theory: Stochastic models and diffusions on fractals". 95 Ststcian 44 139-140

Koroljuk, V. S.; Borovskich, Yu. V. (Auth); — — (Rev): Review of "Theory of *U*-statistics". 95 Ststcian 44 411-411

Glickman, Lawrence T.: Yi, Qilong; — —: Computer simulation analysis of Sartwell's incubation period model for diseases with uncertain etiology - The effect of competing risk. 95 AJEpidm 142 363-368

Glickman, Mark E.; Stern, Hal S.: Inference from an autoregressive state-space model via iterative simulation. 94 ASAProSpts 72-76

Glickman, Theodore S.: Sherali, Hanif D.; Alameddine, Amine; — —: Biconvex models and algorithms for risk management problems. 94 AJMMgSci 14 197-228

Glimm, Ekkehard: Jones, R. H. (Auth); — — (Rev): Review of "Longitudinal data with serial correlation: A state-space approach". 95 BiomtrcJ 37 172-172

Glinsman, Lisha; Hayek, Lee-Ann: Statistical evaluation of renaissance art objects. 95 Stats /13 15-18

Glonek, G. F. V.; McCullagh, P.: Multivariate logistic models. 95 JRSS-B 57 533-546

Glorennec, Pierre-Yves: Sjöberg, Jonas; Zhang, Qinghua; Ljung, Lennart; Benveniste, Albert; Delyon, Bernard; — —; Hjalmarsson, Håkan; Juditsky, Anatoli: Nonlinear black-box modeling in system identification: A unified overview. 95 Autmatca 31 1691-1724

Glorfeld, Louis W.: An improvement on Horn's parallel analysis methodology for selecting the correct number of factors to retain. 95 EdPsMeas 55 377-393

Glosup, Jeffrey: Fahrmeir, L.; Tutz, G. (Auth); — — (Rev): Review of "Multivariate statistical modelling based on generalized linear models". 95 Technmcs 37 350-351

Glosup, Jeffrey G.; Axelrod, Michael C.: Use of the AIC with the EM algorithm: A demonstration of a probability model selection technique. 94 ASAProPEng 120-125

Axelrod, Michael C.; — —: Durbin-Watson revisited. 94 ASAProStCp 176-181

Glover, Fred; Kingman, Darwin; Phillips, Nancy V. (Auth); Hansen, Pierre (Rev): Review of "Network models in optimization and their applications in practice". 95 JClass 12 150-152

Gluhovsky, Alexander; Agee, Ernest: Reply to comment on "A definitive approach to turbulence statistical studies in planetary boundary layers". 95 JAtmsphr 52 3197-3198

Glynn, Peter W.; Iglehart, Donald L.: A martingale approach to regenerative simulation. 95 PrEngInf 9 123-131

— —; L'Ecuyer, Pierre: Likelihood ratio gradient estimation for stochastic recursions. 95 AdvAppPr 27 1019-1053

Gnanadesikan, R.; Kettenring, J. R.; Tsao, S. L.: Weighting and selection of variables for cluster analysis. 95 JClass 12 113-136

Gnedenko, B. V.; Ushakov, I. A.: Probabilistic reliability engineering. 95 Wiley:NY:UK 518p

— —; Ushakov, I. A. (Auth); Goodwin, G. C. (Rev): Review of "Probabilistic reliability engineering". 96 ShrtBkRv 16/1 8-8

Gnedin, Alexander V.: On a class of exchangeable sequences. 95 StPrLet 25 351-355

— —; Krengel, Ulrich: A stochastic game of optimal stopping and order selection. 95 AnlsApPr 5 310-321

Gnot, S.; Trenkler, G.; Zmyślony, R.: Nonnegative minimum biased quadratic estimation in the linear regression models. 95 JMultiAn 54 113-125

Baksalary, J. K.; — —; Kageyama, S.: Best estimation of variance components with arbitrary kurtosis in two-way layouts mixed models. 95 JStPlInf 44 65-75

Goble, William M.: — — (Auth); Tortorella, Michael (Rev): Review of "Evaluating control systems reliability: Techniques and applications". 95 Technmcs 37 344-345

Gobreial, Hany S.: Simplification of Milstar microwave and digital cost analysis model. 95 JParmc 15/1 92-117

Goda, David: Bakeman, R. (Auth); — — (Rev): Review of "Understanding social science statistics: A spreadsheet approach". 95 Ststcian 44 539-540

Godambe, V. P.: Estimation of parameters in survey sampling: Optimality. 95 CandJSt 23 227-243

— —: Comment on "Inference". 95 StatSci 10 173-174

Godbole, Anant P.: Geske, Mark X.; — —; Schaffner, Andrew A.; Skolnick, Allison M.; Wallstrom, Garrick L.: Compound Poisson approximations for word patterns under Markovian hypotheses. 95 JAppProb 32 877-892

Goddard, J. A.: Dobson, S. M.; — —: The demand for professional league football in England and Wales, 1925-92. 95 Ststcian 44 259-277

Goddard, M. E.: A mixed model for analyses of data on multiple genetic markers. 92 ThApGene 83 878-886

Godfrey, L. G.; Orme, C. D.: The sensitivity of some general checks to omitted variables in the linear model. 94 IntEconR 35 489-506

Godsey, E. B.: Panizzon, Gaetano (Auth); — — (Rev): Review of "Bilateral plans for inspection and process control by variables". 95 Technmcs 37 463-464

Godtliebsen, Fred; Chu, Chih-Kang: Estimation of the number of true gray levels, their values, and relative frequencies in a noisy image. 95 JASA 90 890-899

Goedegebuure, R. V.; Overgaag, P. H. A.: A classification of types of use of official statistics. 91 StatJUN 8 217-232

Goel, Prem K.; Padilla, Maria L. R.: Generalized Hellinger transforms as information measures. 94 ASAProBays 78-83

George, Robert; — —: Hierarchical Bayesian models and Γ-minimaxity. 94 ASAProBays 96-101

Draper, David; Gaver, Donald, P., Jr; — —; Greenhouse, Joel B.; Hedges, Larry V.; Morris, Carl N.; Tucker, John R.; Waternaux, Christine M. (Auth); Berlin, Jesse A. (Rev): Review of "Combining information: Statistical issues and opportunities for research". 95 StatMed 14 1611-1612

Klein, John P.; — — (Auth); Schlittgen, Rainer (Rev): Review of "Survival analysis: State of the art". 95 CmpStDA 20 592-593

Goetghebeur, Els; Ryan, Louise: Analysis of competing risks survival data when some failure types are missing. 95 Biomtrka 82 821-833

Goetghebeur, Els J. T.; Pocock, Stuart J.: Detection and estimation of *J*-shaped risk-response relationships. 95 JRSS-A 158 107-121

Goffinet, B.: Lorieux, M.; Perrier, X.; — —; Lanaud, C.; de León, D. González: Maximum-likelihood models for mapping genetic markers showing segregation distortion. 2. F_2 populations. 95 ThApGene 90 81-89

Mangin, B.; — —; Rebaï, A.: Constructing confidence intervals for QTL location. 94 Genetcs 138 1301-1308

Goffinet, Bruno: Rebai, Ahmed; — —; Mangin, Brigitte: Comparing power of different methods for QTL detection. 95 Biomtrcs 51 87-99

Rebaï, Ahmed; — —; Mangin, Brigitte: Approximate thresholds of interval mapping tests for QTL detection. 94 Genetcs 138 235-240

Gogel, Beverley J.; Cullis, Brian R.; Verbyla, Arūnus P.: REML estimation of multiplicative effects in multienvironment variety trials. 95 Biomtrcs 51 744-749

Gohout, W.: Mittag, H.-J. (Auth); — — (Rev. German): Review of "Quality control charts" (German). 95 AllgeStA 79 105-106

Gokhale, D. V.; Klein, Ruben: Analysis of contingency tables by marginal scores. 95 RBrzlPrS 9 25-42

— —; Klein, Ruben: Association models for contingency tables via minimum discrimination information approach and correspondence analysis. 95 RBrzlPrS 9 83-98

Goldenberg, Karen L.: Answering questions, questioning answers: Evaluating data quality in an establishment survey. 94 ASASrvyRsr 1357-1362

— —; Moore, Susan E.; Rosen, Richard J.: Commercial payroll software and the quality of employment data. 94 ASASrvyRsr 1025-1030

Goldhirsch, Aron: Cole, Bernard F.; Gelber, Richard D.; — —: A quality-adjusted survival meta-analysis of adjuvant chemotherapy for premenopausal breast cancer. 95 StatMed 14 1771-1784

Cole, Bernard F.; Gelber, Richard D.; Anderson, Keaven M.; — —: Cost-effectiveness analysis using quality-adjusted survival. 94 ASAProBiop 9-18

Gelber, Richard D.; Cole, Bernard F.; Gelber, Shari; — —: Comparing treatments using quality-adjusted survival: The Q-TWiST method. 95 AmerStat 49 161-169

Goldie, Charles M.: Comment on "The Luria-Delbrück distribution" (V30 p991-994, V31 p822-828). 95 JAppProb 32 840-841

— —; Resnick, Sidney I.: Many multivariate records. 95 StocProc 59 185-216

Port, Sidney C. (Auth); — — (Rev): Review of "Theoretical probability for applications". 95 JASA 90 806-808

Golding, Brian: Maximum-likelihood estimates of selection coefficients from DNA sequence data. 93 Evolutn 47 1420-1431

Goldman, Aaron: Cook, R. Dennis; Weisberg, Sanford (Auth); — — (Rev): Review of "An introduction to regression graphics". 95 Technmcs 37 343-344

Goldman, Nick: Statistical tests of models of DNA substitution. 93 JMolEvol 36 182-198

— —: Simple diagnostic statistical tests of models for DNA substitution. 93 JMolEvol 37 650-661

Goldman, Noreen: Social factors and health: The causation-selection issue revisited. 94 PNatAcSc 91 1251-1255

Rodríguez, Germán; — —: An assessment of estimation procedures for multilevel models with binary responses. 95 JRSS-A 158 73-89

Goldman, Steven: Henderson, William G.; Moritz, Thomas; — —; Copeland, Jack; Sethi, Gulshan: Use of cumulative meta-analysis in the design, monitoring, and final analysis of a clinical trial: A case study. 95 ConClnTr 16 331-341

Goldschmidt, H. M. J.; Lent, R. W.: From data to information: How to define the context?. 95 Chemolab 28 181-192

Goldsman, David M.: Bechhofer, Robert E.; Santner, Thomas J.; — —: Design and analysis of experiments for statistical selection, screening, and multiple comparisons. 95 Wiley:NY:UK 325p

Goldstein, Darlene R.: A combined test of linkage heterogeneity. 94 AJHuGene 55 841-848

Goldstein, David B.: Pollock, David D.; — —: A comparison of two methods for constructing evolutionary distances from a weighted contribution of transition and transversion differences. 95 MolBioEv 12 713-717

Goldstein, David J.: Enas, Gregory G.; — —: Defining, monitoring and combining safety information in clinical trials. 95 StatMed 14 1099-1111 (Disc: p1113-1116)

Goldstein, Harvey: Multilevel statistical models (Second edition). 95 Arnold:-Lnd:MD 192p

— —: Comment on "Hierarchical linear models". 95 JEdStat 20 201-204

— —; Healy, Michael J. R.: The graphical presentation of a collection of means. 95 JRSS-A 158 175-177

Diggle, P. J.; Liang, K.-Y.; Zeger, S. L. (Auth); — — (Rev): Review of "Analysis of longitudinal data". 95 JRSS-A 158 345-346

— — (Auth); Hand, D. J. (Rev): Review of "Multilevel statistical models". 95 ShrtBkRv 15/3 43-44

Goldstein, L.: Borgan, Ø.; — —; Langholz, B.: Methods for the analysis of sampled cohort data in the Cox proportional hazards model. 95 AnlsStat 23 1749-1778

Goldstein, Larry; Waterman, Michael S.: Poisson, compound Poisson and process approximations for testing statistical significance in sequence comparisons. 92 BMathBio 54 785-812

Goldstein, Michael; Wooff, David A.: Bayes linear computation: Concepts, implementation and programs. 95 StatComp 5 327-341

Goldstein, Sheldon: Antisymmetric functionals of reversible Markov processes. 95 AIHenrPB 31 177-190

Goldys, B.: Chojnowska-Michalik, A.; — —: Existence, uniqueness and invariant measures for stochastic semilinear equations on Hilbert spaces. 95 ProbTher 102 331-356

Golec, Janusz: Stochastic averaging principle for systems with pathwise uniqueness. 95 StocAnAp 13 307-322

Göller, Gernot: Mager, Harry; — —: Analysis of pseudo-profiles in organ pharmacokinetics and toxicokinetics. 95 StatMed 14 1009-1024

Gollin, Albert: Horvitz, Dan; Koshland, Daniel; Rubin, Donald; — —; Sawyer, Tom; Tanur, Judith M.: Pseudo-opinion polls: SLOP or useful data?. 95 Chance 8/2 16-25

Golmard, Jean-Louis: Escolano, Sylvie; — —; Mallet, Alain: A fully nonparametric approach in survival models with explanatory variables. 95 CommStA 24 3027-3054

Gombay, E.: Nonparametric truncated sequential change-point detection. 95 StatDecn 13 71-82

Gomel, Jessica N.: Reise, Steven P.; — —: Modeling qualitative variation within latent trait dimensions: Application of mixed-measurement to personality assessment. 95 MultBehR 30 341-358

Gomeni, Roberto: Mentré, France; — —: A two-step iterative algorithm for estimation in nonlinear mixed-effect models with an evaluation in population pharmacokinetics. 95 JBiophSt 5 141-158

Gómez-Corral, A.: Martín, M.; — —: On the M/G/1 retrial queueing system with linear control policy. 95 Top 3 285-305

Gomm, Ian S.: Barnett, Neil S.; — —; Armour, Len: Location of the optimal sampling point for the quality assessment of continuous streams. 95 AstrlJSt 37 145-152

Gong, Jianjian: Wahba, Grace; Johnson, Donald R.; Gao, Feng; — —: Adaptive tuning of numerical weather prediction models: Randomized GCV in three- and four-dimensional data assimilation. 95 MWeathr 11 3358-3369

Gong, Wei: Weideman, William E.; Manry, Michael T.; Yau, Hung-Chun; — —: Comparisons of a neural network and a nearest-neighbor classifier via the numeric handprint recognition problem. 95 IEEENrlN 6 1524-1530

Gong, Wei-Bo: Nananukul, Soracha; — —: The mean waiting time of a GI/G/1 queue in light traffic via random thinning. 95 JAppProb 32 256-266

Gong, Xiaofeng; Richman, Michael B.: On the application of cluster analysis to growing season precipitation data in North America east of the Rockies. 95 JClimate 8 897-931

Gonin, R.: Considerations in the analysis of a single arm bone marrow transplant clinical trial. 94 ASAProBiop 493-497

Gonzales, L.: McDonald, L. L.; — —; Manly, B. F. J.: Using selection functions to describe changes in environmental variables. 95 EnvEcoSt 2 225-237

Gonzalez Manteiga, W.; Vilar Fernandez, J. M.: Testing linear regression models using non-parametric regression estimators when errors are non-independent. 95 CmpStDA 20 521-541

González Manteiga, Wenceslao: Smooth linear regression using sample-determined bandwidth. 95 SankhyaA 57 79-87

González, Elizabeth: López, Carlos; — —; Goyret, Jorge: Principal component analysis of rain gauge data: Detecting outliers (Spanish). 94 Estd 46 25-54

González, Juan F.: López, Carlos; — —; Curbelo, Rosario: Principal component analysis of rain gauge data: Treating missing observations (Spanish). 94 Estd 46 55-83

González, María E.: Gaines, Leonard M.; — —: Is there a place in ASA for statisticians who work for state/provincial or local governments?. 94 ASAProGovt 149-152

Gonzalez, Maria Elena: Shettle, Carolyn F.; Guenther, Patricia M.; Kaspryzk, Daniel; — —: Investigating nonresponse in federal surveys. 94 ASASrvyRsr 972-976

González, Pilar; Moral, Paz: An analysis of the international tourism demand in Spain. 95 IntJFor 11 233-251

González-Andujar, J. L.; Perry, J. N.: Models for the herbicidal control of the seed bank of Avena sterilis: The effects of spatial and temporal heterogeneity and of dispersal. 95 JAppEcol 32 578-587

González-Manteiga, W.: García-Jurado, I.; — —; Prada-Sánchez, J. M.; Febrero-Bande, M.; Cao, R.: Predicting using Box-Jenkins, nonparametric, and bootstrap techniques. 95 Technmcs 37 303-310

Gonzalo, Jesus; Granger, Clive: Estimation of common long-memory components in cointegrated systems. 95 JBES 13 27-35

Good, I. J.: A comment concerning physical numerology. 95 JStCmpSm 52 190-191

— —: Reliability always depends on probability of course. 95 JStCmpSm 52 192-193

— —: Dual density functions. 95 JStCmpSm 52 193-194

Good, Ph.: — — (Auth); Rudolph, P. E. (Rev): Review of "Permutation tests. A practical guide to resampling methods for testing hypotheses". 95 BiomtrcJ 37 150-150

Good, Philip: — — (Auth); Niknian, Minoo (Rev): Review of "Permutation tests: A practical guide to resampling methods for testing hypotheses". 95 Technmcs 37 341-342

Goodall, Gerald: Don't get t out of proportion!. 95 TeachgSt 17/2 50-51

Gooday, Jane: Kokic, Philip N.; Moon, Lynelle; — —; Chambers, Ray L.: Estimating temporal farm income distributions using spatial smoothing techniques. 95 AstrlJSt 37 129-143

Goodfriend, Jason H.: Rubinstein, Reuven Y.; Shapiro, Alexander (Auth); — — (Rev): Review of "Discrete event systems: Sensitivity analysis and stochastic optimization by the score function method". 95 Technmcs 37 124-125

Goodman, Steven N.: Confounding and effect-modification. 95 AJEpidm 141 1111-1112

— —: Reply to comment on "Statistical significance testing in the American journal of epidemiology, 1970-1990". 95 AJEpidm 142 103-103

— —; Zahurak, Marianna L.; Piantadosi, Steven: Some practical improvements in the continual reassessment method for phase I studies. 95 StatMed 14 1149-1161

Goodman, V.: Einmahl, U.; — —: Clustering behavior of finite variance partial sum processes. 95 ProbTher 102 547-565

Goodwin, C. G.: Elliott, Robert J.; Aggoun, Lakhdar; Moore, John B. (Auth); — — (Rev): Review of "Hidden Markov models: Estimation and control". 95 ShrtBkRv 15/2 29-29

Goodwin, G. C.: Adams, G. J.; — —: Parameter estimation for periodic ARMA models. 95 JTimSrAn 16 127-145

Basar, T.; Bernhard, P. (Auth); — — (Rev): Review of "H^∞-optimal control and related minimax design problems - A dynamic game approach (Second edition)". 96 ShrtBkRv 16/1 11-11

Gnedenko, B. V.; Ushakov, I. A. (Auth); — — (Rev): Review of "Probabilistic reliability engineering". 96 ShrtBkRv 16/1 8-8

Sethi, S. P.; Zhang, Q. (Auth); — — (Rev): Review of "Hierarchical decision making in stochastic manufacturing systems". 95 ShrtBkRv 15/2 29-29

Goodwin, Graham C.: Ninness, Brett; — —: Estimation of model quality. 95 Autmatca 31 1771-1797

Goovaerts, M. J.: De Vylder, F.; — —: A note on the solution of practical ruin problems. 94 InsrneME 15 181-186

Goradia, Tushar M.; Lange, Kenneth; Miller, Perry L.; Nadkarni, Prakash M.: Fast computation of genetic likelihoods on human pedigree data. 92 HumHrdty 42 42-62

Gordienko, E.; Hernández-Lerma, O.: Average cost Markov control processes with weighted norms: Existence of canonical policies. 95 ApplMath 23 199-218

— —; Hernández-Lerma, O.: Average cost Markov control processes with weighted norms: Value iteration. 95 ApplMath 23 219-237

Gordon, A. D.: Local transformation of facial features. 95 JAppStat 22 179-184

Manton, K. G.; Woodbury, M. A.; Tolley, H. D. (Auth); — — (Rev): Review of "Statistical applications using fuzzy sets". 95 Biomtrcs 51 793-794

Gordon, Louis; Pollak, Moshe: A robust surveillance scheme for stochastically ordered alternatives. 95 AnlsStat 23 1350-1375

Gordon, Robert D.: Statistics in the semiconductor industry: Challenges in volume manufacturing. 94 ASAProPEng 6-8 (Disc: p9-11)

Gordon, Sue: A theoretical approach to understanding learners of statistics. 95 JStEd 3/3 60kb (jse/v3n3/gordon)

Gore, A. P.: Deshpande, J. V.; — —; Shanubhogue, A.: Statistical analysis of nonnormal data. 95 Wiley:NY:UK 240p

Gore, S. D.; Patil, G. P.: Identifying extremely large values using composite sample data. 94 EnvEcoSt 1 227-240 (Disc: p240-245)

Gore, S. M.: Day, N. E.; — —; De Angelis, D.: Acquired immune deficiency syndrome predictions for England and Wales (1992-97): Sensitivity analysis, information, decision. 95 JRSS-A 158 505-524

Gore, Sheila M.: Fielding, Katherine L.; Brettle, Raymond P.; — —; O'Brien, Fiona; Wyld, Rona; Robertson, J. Roy; Weightman, Rosina: Heterosexual transmission of HIV analysed by generalized estimating equations. 95 StatMed 14 1365-1378

Gorelick, Steven M.: Harvey, Charles F.; — —: Temporal moment-generating equations: Modeling transport and mass transfer in heterogeneous aquifers. 95 WaterReR 31 1895-1911

Gorman, Jack: Paik, Myunghee Cho; Begg, Melissa Dowd; El-Sadr, Wafaa; — —; Stein, Zena: Difference in clinical implications of CD4 counts among HIV-infected homosexual men and injection drug using men and women. 95 StatMed 14 1889-1900

Gorostiza, Luis G.; Hochberg, Kenneth J.; Wakolbinger, Anton: Persistence of a critical super-2 process. 95 JAppProb 32 534-540

Bojdecki, Tomasz; — —: Self-intersection local time for Gaussian $S'(R^d)$-processes: Existence, path continuity and examples. 95 StocProc 60 191-226

Gort, Steven M.; Hoogerbrugge, Ronald: A user-friendly spreadsheet program for calibration using weighted regression. 95 Chemolab 28 193-199

Goswami, A.; Sengupta, Arindam: Time-spaced polynomial martingales generated by a discrete-time martingale. 95 JTherPrb 8 417-432

Goswami, V.: Dutta, K.; — —: A comparative study of multidimensional sampling plans. 94 JIAgSt 46 254-266

Gottschau, Adam; Hogh, Birthe: Interval censored survival data and multistate compartmental models in the analysis of first appearance of Plasmodium falciparum parasites in infants. 95 StatMed 14 2727-2736

Götze, F.; Hipp, C.: Asymptotic distribution of statistics in time series. 94 AnlsStat 22 2062-2088

Götze, Friedrich: Bentkus, Vidmantas; — —; Zitikis, Ričardas: Lower estimates of the convergence rate for U-statistics. 94 AnlsProb 22 1707-1714

Bhattacharya, Rabi N.; — —: Time-scales for Gaussian approximation and its breakdown under a hierarchy of periodic spatial heterogeneities. 95 Bernoull 1 81-123

Goudie, I. B. J.: A plant-capture approach for achieving complete coverage of a population. 95 CommStA 24 1293-1305

Goulard, M.; Pagés, L.; Cabanettes, A.: Marked point process: Using correlation functions to explore a spatial data set. 95 BiomtrcJ 37 837-853

Gould, A. Lawrence: Planning and revising the sample size for a trial. 95 StatMed 14 1039-1051 (Disc: p1053-1055)

Chang, Myron N.; — —; Snapinn, Steven M.: P-values for group sequential testing. 95 Biomtrka 82 650-654

Shih, Weichung Joseph; — —: Re-evaluating design specifications of longitudinal clinical trials without unblinding when the key response is rate of change. 95 StatMed 14 2239-2248

Gould, William: The Pentium bug and statistical software. 95 Chance 8/2 9-13

Gould, William W.; Rogers, William H.: Quantile regression as an alternative to robust regression. 94 ASAProStEd 12-16

Goupy, Jacques: Kempthorne, Oscar; Hinkelmann, Klaus (Auth); — — (Rev): Review of "Design and analysis of experiments. Volume 1. Introduction to experimental design". 95 Chemolab 28 203-205

Gourieroux, Christian; Monfort, Alain (Auth); Vuong, Quang (Transl): Statistics and econometric models. Volumes 1 and 2. 95 CambridgeUniv 1062p

Gouriéroux, Christian; Monfort, Alain: Testing, encompassing, and simulating dynamic econometric models. 95 EconTher 11 195-228

Broze, Laurence; — —; Szafarz, Ariane: Solutions of multivariate rational expectations models. 95 EconTher 11 229-257

Goutis, C.: Dawid, A. P.; — —: Comment on "Inference". 95 StatSci 10 185-186

Goutis, Constantinos; Casella, George: Frequentist post-data inference. 95 IntStRvw 63 325-344

— —; Casella, George: Improved invariant set estimation for general-scale families. 95 JStPlInf 44 327-340

Johaadien, Hassan; — —: Approximate maximum likelihood estimation in the presence of extra-Poisson variation. 95 CommStA 24 1625-1636

Gouweleeuw, José M.: Boshuizen, Frans A.; — —: A continuous-time job search model: General renewal processes. 95 CommStSM 11 349-369

Govaert, Gérard: Celeux, Gilles; — —: Gaussian parsimonious clustering models. 95 PattrnRc 28 781-793

Govindaraju, K.: Kuralmani, V.; — —: Modified tables for the selection of double sampling attribute plan indexed by AQL and LQL. 95 CommStA 24 1897-1921

Govindarajulu, Z.: A brief survey of LMP rank tests for random effects in one and two-way models. 95 CommStA 24 59-73

Yu, Qiqing; — —: Admissibility and minimaxity of the UMVU estimator of $P\{X < Y\}$. 95 AnlsStat 23 598-607

Gowda, K. Chidananda; Ravi, T. V.: Divisive clustering of symbolic objects using the concepts of both similarity and dissimilarity. 95 PattrnRc 28 1277-1282

Gower, John: Comment on "The statistics of linear models: Back to basics" (V4 p221-234). 95 StatComp 5 91-92

Goyal, R. C.; Singh, Randhir; Chhikara, R. S.: Estimation of crop yield using post-stratification based on satellite data. 94 JIAgSt 46 210-222

Goyret, Jorge: López, Carlos; González, Elizabeth; — —: Principal component analysis of rain gauge data: Detecting outliers (Spanish). 94 Estd 46 25-54

Grabov, Pavel: Bluvband, Zigmund; — —; Ingman, Dov: New approach to process control. 95 ASQCTekCon 369-377

Green, P. J.; Silverman, B. W. (Auth); Lange, Nicholas (Rev): Review of "Nonparametric regression and generalized linear models: A roughness penalty approach". 95 JASA 90 1489-1489

— —; Silverman, B. W. (Auth); Qui, Chunfu (Rev): Review of "Nonparametric regression and generalized linear models: A roughness penalty approach". 95 Technmcs 37 342-343

Green, Paul E.: Carroll, J. Douglas; — —: Psychometric methods in marketing research: Part 1, Conjoint analysis. 95 JMarRsr 32 385-391

Green, Peter: Besag, Julian; — —; Higdon, David; Mengersen, Kerrie: Bayesian computation and stochastic systems. 95 StatSci 10 3-41 (Disc: p41-66)

Besag, Julian; — —; Higdon, David; Mengersen, Kerrie: Reply to comments on "Bayesian computation and stochastic systems". 95 StatSci 10 58-66

Green, Peter J.: Reversible jump Markov chain Monte Carlo computation and Bayesian model determination. 95 Biomtrka 82 711-732

Green, Sylvan B.; Corle, Donald K.; Gail, Mitchell H.; Mark, Steven D.; Pee, David; Freedman, Laurence S.; Graubard, Barry I.; Lynn, William R.: Interplay between design and analysis for behavioral intervention trials with community as the unit of randomization. 95 AJEpidm 142 587-593

Greenberg, A. G.: Coffman, E. G., Jr; Gilbert, E. N.; — —; Leighton, F. T.; Robert, Philippe; Stolyar, A. L.: Queues served by a rotating ring. 95 CommStSM 11 371-394

Greenberg, David A.: Vieland, Veronica J.; — —; Hodge, Susan E.: Adequacy of single-locus approximations for linkage analyses of oligogenic traits: Extension to multigenerational pedigree structures. 93 HumHrdty 43 329-336

Greenberg, Edward: Chib, Siddhartha; — —: Understanding the Metropolis-Hastings algorithm. 95 AmerStat 49 327-335

Chib, Siddhartha; — —: Hierarchical analysis of SUR models with extensions to correlated serial errors and time-varying parameter models. 95 JEconmtx 68 339-360

Greenblatt, Janet C.; Bertolucci, Darryl: Characteristics of persons who reported driving under the influence of alcohol or other drugs. 94 ASASrvyRsr 547-552

Greene, Jody M.: Iachan, Ronald; Ringwalt, Chris L.; — —: Substance abuse among runaway and homeless youth. 94 ASASrvyRsr 1083-1087

Greene, Laura Knapp: Greene, William H.; — —; Seaks, Terry G.: Estimating the functional form of the independent variables in probit models. 95 ApplEcon 27 193-196

Greene, William H.; Greene, Laura Knapp; Seaks, Terry G.: Estimating the functional form of the independent variables in probit models. 95 ApplEcon 27 193-196

Greenhouse, J.: Lange, N.; Ryan, L.; Billard, L.; Brillinger, D. R.; Conquest, L.; — — (Ed); Herrendörfer, G. (Rev): Review of "Case studies in biometry". 95 BiomtrcJ 37 640-640

Greenhouse, Joel: Kamlet, Mark S.; Paul, Nancy; — —; Kupfer, David; Frank, Ellen; Wade, Martcia: Cost utility analysis of maintenance treatment for recurrent depression. 95 ConClnTr 16 17-40

Lange, Nick; Ryan, Louise; Billard, Lynne; Brillinger, David R.; Conquest, Loveday; — — (Ed); Hilton, Joan (Rev): Review of "Case studies in biometry". 95 JASA 90 800-800

Greenhouse, Joel B.; Wasserman, Larry: Robust Bayesian methods for monitoring clinical trials. 95 StatMed 14 1379-1391

Draper, David; Gaver, Donald, P., Jr; Goel, Prem K.; — —; Hedges, Larry V.; Morris, Carl N.; Tucker, John R.; Waternaux, Christine M. (Auth); Berlin, Jesse A. (Rev): Review of "Combining information: Statistical issues and opportunities for research". 95 StatMed 14 1611-1612

Greenhouse, Samuel W.: Gastwirth, Joseph L.; — —: Biostatistical concepts and methods in the legal setting. 95 StatMed 14 1641-1653

Greenland, Betsy S.; Stokes, S. Lynne: Repetitive testing in the presence of inspection errors. 95 Technmcs 37 102-111

Greenland, Sander: Reply to comment on "Statistical significance testing in the American journal of epidemiology, 1970-1990". 95 AJEpidm 142 102-103

— —: Reply to comment on "A critical look at some popular meta-analytic methods". 95 AJEpidm 142 1008-1009

— —: Dose-response and trend analysis in epidemiology: Alternatives to categorical analysis. 95 Epidmlgy 6 356-365

— —: Avoiding power loss associated with categorization and ordinal scores in dose-response and trend analysis. 95 Epidmlgy 6 450-454

— —: Reply to comment on "Divergent biases in ecologic and individual level studies". 95 StatMed 14 328-330

— —; Finkle, William D.: A critical look at methods for handling missing covariates in epidemiologic regression analyses. 95 AJEpidm 141 1255-1264

— —; Finkle, William D.: A critical look at methods for handling missing covariates in epidemiologic regression analyses. 95 AJEpidm 142 1255-1264

Joffe, Marshall M.; — —: Standardized estimates from categorical regression models. 95 StatMed 14 2131-2141

Weinberg, Clarice R.; Umbach, David M.; — —: Reply to comments on "When will nondifferential misclassification of an exposure reverse the direction of a trend?". 95 AJEpidm 142 784-784

Clayton, David; Hills, Michael (Auth); — — (Rev): Review of "Statistical models in epidemiology". 95 AJEpidm 141 182-183

Greenwood, P. E.; Wefelmeyer, W.: Efficiency of empirical estimators for Markov chains. 95 AnlsStat 23 132-143

Greer, Daniel A.: Quality award evaluation. 94 ASAProQlPr 247-249

— —: Graphics for regression analysis display and diagnostics. 94 ASAProStGr 66-70

— —; Halteman, Edward J.: SPC: A new look at what to teach. 94 ASAProBays 153-156

Gregoire, Tim G.: Li, Jufang; Chen, Mei-Ching; Schreuder, Hans T.; — —: Forestry applications of saddle-point approximations to construct confidence intervals for population means. 95 Biomtrcs 51 61-72

Gregoire, Timothy G.; Monkevich, Nicholas S.: The reflection method of line intercept sampling to eliminate boundary bias. 94 EnvEcoSt 1 219-226

— —; Schabenberger, Oliver; Barrett, James P.: Linear modelling of irregularly spaced, unbalanced, longitudinal data from permanent-plot measurements. 95 CanJForR 25 137-156

— —; Valentine, Harry T.; Furnival, George M.: Sampling methods to estimate foliage and other characteristics of individual trees. 95 Ecology 76 1181-1194

Schabenberger, Oliver; — —; Burkhart, Harold E.: Comment on "Multistate models for monitoring individual trees in permanent observation plots". 94 EnvEcoSt 1 192-196

Gregory, John; Hughes, H. R.: A generalized approximation theory for quadratic forms: Application to randomized spline type Sturm-Liouville problems. 95 JTherPrb 8 703-715

— —; Hughes, H. R.: A generalized approximation theory for quadratic forms II: Application to randomized spline type focal/conjugate point problems. 95 JTherPrb 8 963-971

— —; Lin, Cantian (Auth); Prvan, Tania (Rev): Review of "Constrained optimization in the calculus of variations and optimal control theory". 95 AstrlJSt 37 250-250

Greig, Malcolm: Furino, Steven; — —; Miao, Ying; Yin, Jianxing: Resolvable BIBDs with block size 8 and index 7. 94 JComDsgn 2 101-116

Grenander, U.: — — (Auth); Keenan, Daniel (Rev): Review of "General pattern theory: A mathematical study of regular structures". 95 SIAMRvw 37 258-261

— — (Auth); Ripley, B. D. (Rev): Review of "General pattern theory: A mathematical theory of regular structures". 95 JRSS-A 158 635-635

Grenander, Ulf: A survey of the life and works of Harald Cramér. 95 ScanActJ 2-5

— —; Keenan, Daniel: Understanding variable shapes in 3-D. 95 CommStSM 11 51-78

Grenfell, B. T.; Kleczkowski, A.; Gilligan, C. A.; Bolker, B. M.: Spatial heterogeneity, nonlinear dynamics and chaos in infectious diseases. 95 StMedRes 4 160-183

Grenfell, Bryan; Dobson, Andrew (Ed): Ecology of infectious diseases in natural populations. 95 CambridgeUniv 532p

Grether, David M.: El-Gamal, Mahmoud A.; — —: Are people Bayesian? Uncovering behavioral strategies. 95 JASA 90 1137-1145

Nerlove, Marc; — —; Carvalho, José L.: Analysis of economic time series. A synthesis (Revised edition). 95 Academic:NY:Lnd 468p

Greven, Andreas: Cox, J. T.; — —; Shiga, Tokuzo: Finite and infinite systems of interacting diffusions. 95 ProbTher 103 165-197

Cranston, Michael; — —: Coupling and harmonic functions in the case of continuous time Markov processes. 95 StocProc 60 261-286

Gribble, Steve: Rowe, Geoff; — —: Income statistics from survey data: Effects of respondent rounding. 94 ASAProBuEc 77-82

Griep, M. I.; Wakeling, I. N.; Vankeerberghen, P.; Massart, D. L.: Comparison of semirobust and robust partial least squares procedures. 95 Chemolab 29 37-50

Grieshaber, Mark V.: Brannigan, Lawrence H.; — —; Schnur, Dora M.: Use experimental design in organic synthesis. 95 ChemTech 25 29-35

Grieve, Andrew P.: Extending a Bayesian analysis of the two-period crossover to accommodate missing data. 95 Biomtrka 82 277-286

Griffin, Deborah H.: Martin, Elizabeth A.; — —: The role of questionnaire design in reducing census coverage error. 94 ASASrvyRsr 736-741 (Disc: p754-756)

Griffin, Richard A.: An application of stochastic optimization to combine two files. 94 ASASrvyRsr 485-490

Griffioen, J.: van Veldhuizen, M.; Appelo, C. A. J.; — —: The (least squares) quotient algorithm as a rapid tool for obtaining sorption isotherms from column elution curves. 95 WaterReR 31 849-857

Griffith, Daniel A.; Sone, Akio: Trade-offs associated with normalizing constant computational simplifications for estimating spatial statistical models. 95 JStCmpSm 51 165-183

Arlinghaus, Sandra L.; — — (Ed): Practical handbook of spatial statistics. 95 CRCPress:FL 336p

Griffith, G. R.: Vere, D. T.; — —: Modifying quantitative forecasts of livestock production using expert judgements. 95 JForcast 14 453-464

Griffiths, R. C.; Tavaré, Simon: Unrooted genealogical tree probabilities in the infinitely-many-sites model. 95 MaBisci 127 77-98

Griffiths, R. F.: Higson, H. L.; — —; Jones, C. D.; Biltoft, C.: Effect of atmospheric stability on concentration fluctuations and wake retention times for dispersion in the vicinity of an isolated building. 95 EnvrMtrc 6 571-581

Griffiths, W. E.; Hill, R. C.; Judge, G. G. (Auth); Phillips, Robert F. (Rev): Review of "Learning and practicing econometrics". 95 IntJFor 11 331-333

Griggs, Kenneth: Remus, William; O'Connor, Marcus; — —: Does reliable information improve the accuracy of judgemental forecasts?. 95 IntJFor 11 285-293

Griliches, Zvi; Regev, Haim: Firm productivity in Israeli industry 1979-1988. 95 JEconmtx 65 175-203

Berndt, Ernst R.; — —; Rappaport, Neal J.: Econometric estimates of price indexes for personal computers in the 1990's. 95 JEconmtx 68 243-268

Grill, K.: A lim inf result for the increments of the Wiener process under the L_2-norm. 95 ProbTher 102 567-573

Grima, Pere: Prat, Albert; — —: Determination of design factor values to obtain robust products: Application to circuits with inductance and resistance (Spanish). 95 Questo 19 153-169

Grimes, James A.; Foust, Laura B.: Establishing release limits with a random slopes stability model. 94 ASAProBiop 498-502

Grimm, Laurence G.: Howell, D. C. (Auth); — — (Rev): Review of "Statistical methods for psychology (Third edition)". 95 JAppStat 22 194-195

Grimshaw, Scott D.: Efron, Bradley; Tibshirani, Robert J. (Auth); — — (Rev): Review of "An introduction to the bootstrap". 95 Technmcs 37 341-341

Grimson, Roger C.: Disease cluster tests based on the maximum cell frequency. 94 ASAProEpid 64-69

Grishechkin, S. A.: Interoutput times in an M/G/1/PS queue with instant feedback. 94 ThProbAp 39 692-696

Grize, Yves L.: A review of robust process design approaches. 95 JChemtrc 9 239-262

Grize, Yves-L.: Robust parameter design: A systematic approach. 94 ASAProPEng 217-224

Grizzle, James E.: Raghunathan, Trivellore E.; — —: A split questionnaire survey design. 95 JASA 90 55-63

Groeneboom, P.; Wellner, J. A. (Auth); Plachky, D. (Rev): Review of "Information bounds and nonparametric maximum likelihood estimation". 95 StatDecn 13 94-95

Groeneboom, Piet; Jongbloed, Geurt: Isotonic estimation and rates of convergence in Wicksell's problem. 95 AnlsStat 23 1518-1542

Groenen, Patrick J. F.; Mathar, Rudolf; Heiser, Willem J.: The majorization approach to multidimensional scaling for Minkowski distances. 95 JClass 12 3-19

Groeneveld, Richard A.: Arnold, Barry C.; — —: Measuring skewness with respect to the mode. 95 AmerStat 49 34-38

Groenewald, P. C. N.: de Waal, D. J.; — —: Bayesian estimation of ground-water levels. 95 SoAfrStJ 29 65-78

de Waal, D. J.; — —; Kemp, C. J.: Perturbation of the normal model in linear regression. 95 SoAfrStJ 29 105-130

Plessis, J. L. du; van der Merwe, A. J.; — —: Reference priors for the multivariate calibration problem. 95 SoAfrStJ 29 155-168

Grondin, Chantal; Michaud, Sylvie: Data quality of income data using computer assisted interviewing: The experience of the Canadian survey of labour and income dynamics. 94 ASASrvyRsr 839-844 (Disc: p861-862)

Grondona, Martin O.; Cressie, Noel: Residuals based estimators of the covariogram. 95 MaOpfStS 26 209-218

Gropper, Daniel M.: Caudill, Steven B.; Ford, Jon M.; — —: Frontier estimation and firm-specific inefficiency measures in the presence of heteroscedasticity. 95 JBES 13 105-111

Grosclaude, Pascal: Environmental indices for the United States. 95 Student 1 109-123

Grosek, Otokar: On a reconstruction of a Markov chain. 95 JCISS 20 85-93

Groshen, Susan: Markus, Richard A.; Frank, Jess; — —; Azen, Stanley P.: An alternative approach to the optimal design of an LD50 bioassay. 95 StatMed 14 841-852

Grosof, Miriam Schapiro; Sardy, Hyman: Are the problems the problem in introductory statistics?. 94 ASAProStEd 192-196

Gross, A. C.: Duncan, J. W.; — — (Auth); Eklöf, Jan A. (Rev): Review of "Statistics for the 21st century. Proposals for improving statistics for better decision making". 95 JOffStat 11 225-229

Duncan, J. W.; — — (Auth); Swires-Hennessy, E. (Rev): Review of "Statistics for the 21st century: Proposals for improving statistics for better decision making". 95 Ststcian 44 138-139

Gross, Alan J.: Uncertainties in lung cancer risk estimates reported for exposure to environmental tobacco smoke. 95 EnvrMtrc 6 403-412 (Disc: p413-424)

— —: Reply to comments on "Uncertainties in lung cancer risk estimates reported for exposure to environmental tobacco smoke". 95 EnvrMtrc 6 423-424

Tu, Dongsheng; — —: Accurate confidence intervals for the ratio of specific occurrence/exposure rates in risk and survival analysis. 95 BiomtrcJ 37 611-626

Gross, Alan L.; Torres-Quevedo, Rocio: Estimating correlations with missing data, a Bayesian approach. 95 Psymtrka 60 341-354

Grosskopf, Shawna: Färe, Rolf; — —: Nonparametric tests of regularity, Farrell efficiency, and goodness-of-fit. 95 JEconmtx 69 415-425

Grossman, M.: Lo, L. L.; Fernando, R. L.; — —: Covariance between relatives in multibreed populations: Additive model. 93 ThApGene 87 423-430

Lo, L. L.; Fernando, R. L.; Cantet, R. J. C.; — —: Theory for modelling means and covariances in a two-breed population with dominance inheritance. 95 ThApGene 90 49-62

Grouin, Jean-Marie: Lecoutre, Bruno; Derzko, Gérard; — —: Bayesian predictive approach for inference about proportions. 95 StatMed 14 1057-1063 (Disc: p1065-1065)

Grove, John S.; Sharp, Dan S.; Wilkins, Lynne R.: Regression with random error in X: Increasing power of the test by non-linear adjustment for measurement error. 94 ASAProEpid 143-146

Groves, John A.: Brereton, Richard G.; Gurden, Stephen P.; — —: Use of eigenvalues for determining the number of components in window factor analysis of spectroscopic and chromatographic data. 95 Chemolab 27 73-87

Groves, Robert M.; Couper, Mick P.: Theoretical motivation for post-survey nonresponse adjustment in household surveys. 95 JOffStat 11 93-106

Wallman, Katherine; — —; Parsons, Joseph; Davis, Mary C.; Lapham, Susan J.: Increasing statistical awareness through graduate training: The joint program in survey methodology. 94 ASAProSoSt 39-44

Grübel, R.: Barbour, A. D.; — —: The first divisible sum. 95 JTherPrb 8 39-47

Gruber, Jonathan; Hanratty, Maria: The labor-market effects of introducing national health insurance: Evidence from Canada. 95 JBES 13/2 163-173

Gruber, Kerry: Kaspryzk, Daniel; — —; Salvucci, Sameena; Saba, Mehrdad; Zhang, Fan; Fink, Steven: Some data issues in school-based surveys. 94 ASASrvyRsr 815-820

Gruber, Marvin H. J.: The comparison of the efficiency of Bayes and empirical Bayes estimators. 94 ASAProPEng 132-135

Barlow, Richard E.; Clarotti, Carlo A.; Spizzichino, Fabio (Ed); — — (Rev): Review of "Reliability and decision making". 95 Technmcs 37 346-347

Grudzień, Z.; Szynal, D.: Characterizations of distributions by moments of order statistics when the sample size is random. 95 ApplMath 23 305-318

Gruenewald, Paul J.: Millar, Alexander B.; — —: The development of spatial models for community evaluation. 94 ASAProSoSt 227-231

Gruet, J.-C.; Shi, Z.: Some asymptotic results for exponential functionals of Brownian motion. 95 JAppProb 32 930-940

Grung, Bjørn; Kvalheim, Olav M.: Resolution of multicomponent profiles with partial selectivity. A comparison of direct methods. 95 Chemolab 29 75-87

— —; Kvalheim, Olav M.: Detection and quantification of embedded minor analytes in three-way multicomponent profiles by evolving projections and internal rank annihilation. 95 Chemolab 29 213-221

— —; Kvalheim, Olav M.: Rank mapping of three-way multicomponent profiles. 95 Chemolab 29 223-232

Grunow, Bryan: Coelho, Robby; — —; Myers, Casey; Rasp, John: On the odds of an extended hitting streak. 94 ASAProSpts 46-49

Grzegorzewski, P.: The robustness against dependence of nonparametric tests for the two-sample location problem. 95 ApplMath 22 469-476

Gu, Chong: Smoothing spline density estimation: Conditional distribution. 95 StSinica 5 709-726

Wang, Yuedong; Wahba, Grace; Chappell, Rick; — —: Simulation studies of smoothing parameter estimates and Bayesian confidence intervals in Bernoulli SS ANOVA models. 95 CommStB 24 1037-1059

Gu, Hai-Yan: Studies on optimum condition-monitor cycle time policies with general repair result. 94 RelEngSS 46 245-252

Gu, Minggao: Convergence of increments for cumulative hazard function in a mixed censorship-truncation model with application to hazard estimators. 95 StPrLet 23 135-139

— —; Ying, Zhiliang: Group sequential methods for survival data using partial likelihood score processes with covariate adjustment. 95 StSinica 5 793-804

Gu, Minghong: Jiang, Changjian; Pan, Xuebiao; — —: The use of mixture models to detect effects of major genes on quantitative characters in a plant breeding experiment. 94 Genetcs 136 383-394

Gu, Xun; Fu, Yun-Xin; Li, Wen-Hsiung: Maximum likelihood estimation of the heterogeneity of substitution rate among nucleotide sites. 95 MolBioEv 12 546-557

Guadagno, Mary Ann; Abma, Joyce C.; Chandra, Anjani; London, Kathy A.; Peterson, Linda S.; Piccinino, Linda J.; Wilson, Jacqueline B.: Women's reports of sensitive health conditions, attitudes, and behaviors. 94 ASAProGovt 82-87

Guan, Yingnan; Zhang, Chongqi: I_λ-optimal designs of parameter estimation for the additive mixture model of 2-degree (Chinese). 95 AMaAPknC 17 161-172

Guay, M.: Curvature measures for multiresponse regression models. 95 Biomtrka 82 411-417

Guégan, D.: Bosq, D.; — —: Nonparametric estimation of the chaotic function and the invariant measure of a dynamical system. 95 StPrLet 25 201-212

Guenther, Patricia M.; Kott, Phillip S.: Problems in estimating usual dietary intakes. 94 ASAProEpid 44-48

Shettle, Carolyn F.; — —; Kaspryzk, Daniel; Gonzalez, Maria Elena: Investigating nonresponse in federal surveys. 94 ASASrvyRsr 972-976

Guerrero, V. M.; Martínez, J.: A recursive ARIMA-based procedure for disaggregating a time series variable using concurrent data. 95 Test 4 359-376

Gurling, Hugh: Curtis, David; — —: A procedure for combining two-point lod scores into a summary multipoint map. 93 HumHrdty 43 173-185

Gustafson, Helen: Davies, Neville; Dawson, Ed; — —; Pettitt, A. N.: Testing for randomness in stream ciphers using the binary derivative. 95 StatComp 5 307-310

Gustafson, Paul: A Bayesian analysis of bivariate survival data from a multicentre cancer clinical trial. 95 StatMed 14 2523-2535

Gustafsson, Bertil; Kreiss, Heinz-Otto; Oliger, Joseph: Time dependent problems and difference methods. 95 Wiley:NY:UK 642p

Gustafsson, Fredrik; Hjalmarsson, Håkan: Twenty-one ML estimators for model selection. 95 Autmatca 31 1377-1392

Gustafsson, Göran: Waller, Lance A.; Turnbull, Bruce W.; — —; Hjalmars, Ulf; Andersson, Bertil: Detection and assessment of clusters of disease: An application to nuclear power plant facilities and childhood leukaemia in Sweden. 95 StatMed 14 3-16

Gut, A.: An intermediate course in probability. 95 Springer:Brln:NY 278p

— — (Auth); Embrechts, P. A. L. (Rev): Review of "An intermediate course in probability". 96 ShrtBkRv 16/1 4-4

Gutiérrez Jáimez, R.; Román, P. Román; Ruiz, F. Torres: A note on the Volterra integral equation for the first-passage-time probability density. 95 JAppProb 32 635-648

Gutierrez, Ramon: Valderrama, Mariano J.; Jimenez, Francisco; — —; Martinez-Almecija, Alfredo: Estimation and filtering on a doubly stochastic Poisson process. 95 AplStMDA 11 13-24

Gutierrez, Roberto G.; Carroll, Raymond J.; Wang, Naisyin; Lee, Gyung-Hee; Taylor, Brian H.: Analysis of tomato root initiation using a normal mixture distribution. 95 Biomtrcs 51 1461-1468

Gutiérrez-Peña, E.; Smith, A. F. M.: Conjugate parameterizations for natural exponential families. 95 JASA 90 1347-1356

Guttman, I.: Reiser, B.; — —; Lin, Dennis K. J.; Guess, Frank M.; Usher, John S.: Bayesian inference for masked system lifetime data. 95 ApplStat 44 79-90

Guttman, Irwin; Lin, Dennis K. J.: Robust recursive estimation for correlated observations. 95 StPrLet 23 79-92

— —; Lin, Dennis K. J.; Reiser, B.; Usher, John S.: Dependent masking and system life data analysis: Bayesian inference for two-component systems. 95 LiDaAnal 1 87-100

Draper, Norman R.; — —: Confidence intervals versus regions. 95 Ststcian 44 399-403

Olkin, Ingram; — —; Philips, Robert: Estimating the number of aberrant laboratories. 95 PrEnglnf 9 133-150

Sinha, S. K.; — —: Acceptance sampling in presence of misclassification - A Bayesian approach. 94 JIndnStA 32 83-93

Guttorp, P.: Sampson, P. D.; — —; Meiring, W.: Spatio-temporal analysis of regional ozone data for operational evaluation of an air quality model. 94 ASAProEnvr 46-55

— — (Auth); Panaretos, J. (Rev): Review of "Statistical inference for branching processes". 95 Biomtrcs 51 383-384

Guttorp, Peter: Stochastic modeling of scientific data. 95 ChapmnHall:Lnd 392p

Billheimer, Dean; — —: Zooplankton proportion estimates from non-uniform sample volumes. 95 EnvEcoSt 2 117-124

Newton, Michael A.; — —; Catlin, Sandra; Assunção, Renato; Abkowitz, Janis L.: Stochastic modeling of early hematopoiesis. 95 JASA 90 1146-1155

— — (Transl): Three papers on the history of branching processes. 95 IntStRvw 63 233-245

— — (Auth); Basawa, Ishwar V. (Rev): Review of "Statistical inference for branching processes". 95 JStPlInf 47 393-394

— — (Auth); Embrechts, P. A. L. (Rev): Review of "Stochastic modeling of scientific data". 96 ShrtBkRv 16/1 1-2

Güven, B.: Asymptotic properties of maximum likelihood estimation in the mixed analysis of variance model. 95 StHefte 36 175-182

Güven, Bilgehan: Maximum likelihood estimation in simple linear regression with one fold nested error. 95 CommStA 24 121-130

Guyon, X.: Random fields on a network. Modeling, statistics, and applications. 95 Springer:Brln:NY 255p

— —: Random fields on a network: Modelling, statistics, and applications. 95 Springer:Brln:NY 272p

— — (Auth); Ludeña, C. (Transl); Ripley, B. D. (Rev): Review of "Random fields on a network. Modeling, statistics, and applications". 96 ShrtBkRv 16/1 10-10

Gyöngy, I.: On the support of the solutions of stochastic differential equations. 94 ThProbAp 39 519-523

Gyöngy, István; Nualart, David: Implicit scheme for quasi-linear parabolic partial differential equations perturbed by space-time white noise. 95 StocProc 58 57-72

— —; Nualart, David; Sanz-Sole, Marta: Approximation and support theorems in modulus spaces. 95 ProbTher 101 495-509

Györfi, L.: Carbonez, A.; — —; van der Meulen, E. C.: Partitioning-estimates of a regression function under random censoring. 95 StatDecn 13 21-37

Györfi, László: Berlinet, Alain; Devroye, Luc; — —: Asymptotic normality of L_1-error in density estimation. 95 MaOpfStS 26 329-343

Gyorkos, Theresa W.: Joseph, Lawrence; — —; Coupal, Louis: Bayesian estimation of disease prevalence and the parameters of diagnostic tests in the absence of a gold standard. 95 AJEpidm 141 263-272

Gyurcsik, Ronald S.: Hughes-Oliver, Jacqueline M.; Lu, Jye-Chyi; Davis, Joseph C.; — —: Robust improvement of a semi-conductor fabrication process via spatial modeling of the mean and covariance surfaces. 94 ASAProPEng 235-244

H-Akman, Olcay: Gupta, Ramesh C.; Gupta, Rameshwar D.; — —: Estimation of reliability under ordered restriction on the parameters. 95 CommStA 24 1799-1812

Haag, Uwe: Koch, Armin; — — (Ed): The Statistical Software Guide '94/95. 95 CmpStDA 19 237-262

Haaland, P. D.: Graham, R. C. (Auth); — — (Rev): Review of "Large data analysis for the chemical sciences: A guide to statistical techniques". 95 Chemolab 27 245-246

Haaland, Perry: Nychka, Doug; Gray, Gerry; — —; Martin, David; O'Connell, Michael: A nonparametric regression approach to syringe grading for quality improvement. 95 JASA 90 1171-1178

Haaland, Perry D.; O'Connell, Michael A.: Inference for effect-saturated fractional factorials. 95 Technmcs 37 82-93

Haas, Peter J.: Glasserman, Paul; Yao, David D. (Auth); — — (Rev): Review of "Monotone structure in discrete-event systems". 95 JASA 90 808-809

Haas, Timothy C.: Local prediction of a spatio-temporal process with an application to wet sulfate deposition. 95 JASA 90 1189-1199

— —: Nonstationary multivariate spatial prediction. 94 ASAProEnvr 194-203

Habbema, J. D. F.: van Oortmarssen, G. J.; Boer, Rob; — —: Modelling issues in cancer screening. 95 StMedRes 4 33-54

Haber, Michael; Halloran, M. Elizabeth; Longini, Ira M., Jr; Watelet, Luc: Estimation of vaccine efficacy in non-randomly mixing populations. 95 BiomtrcJ 37 25-38

Haberle, Katharine Rowse: Kaminsky, Frank C.; — —: Probability distributions for the Deming cost models. 95 JQualTek 27 355-362

Haberman, S.: Renshaw, A. E.; — —: On the graduations associated with a multiple state model for permanent health insurance. 95 InsrneME 17 1-17

Haberman, Shelby J.: Computation of maximum likelihood estimates in association models. 95 JASA 90 1438-1446

Gilula, Zvi; — —: Prediction functions for categorical panel data. 95 AnlsStat 23 1130-1142

Gilula, Zvi; — —: Dispersion of categorical variables and penalty functions: Derivation, estimation, and comparability. 95 JASA 90 1447-1452

Manton, Kenneth G.; Woodbury, Max A.; Tolley, H. Dennis (Auth); — — (Rev): Review of "Statistical applications using fuzzy sets". 95 JASA 90 1131-1133

Haberman, Steven: Pension funding with time delays and the optimal spread period. 95 AstinB 25 177-187

— —; Schmool, Marlena: Estimates of the British Jewish population 1984-88. 95 JRSS-A 158 547-562

— —; Sung, Joo-Ho: Dynamic approaches to pension funding. 94 InsrneME 15 151-162

Habibullah, Mohamed; Rogers, James; Voss, Theresa: Issues related to the bioequivalence of drug formulations. 94 ASAProStCp 222-224

Soofi, Ehsan S.; Ebrahimi, Nader; — —: Information distinguishability with application to analysis of failure data. 95 JASA 90 657-668

Hachem, Saeb El: Breton, Michèle; — —: A scenario aggregation algorithm for the solution of stochastic dynamic minimax problems. 95 Stochatx 53 305-322

Hackett, Christine A.; Weller, Joel I.: Genetic mapping of quantitative trait loci for traits with ordinal distributions. 95 Biomtrcs 51 1252-1263

Hackl, P.: Fedorov, V. V.; — —; Müller, W. G.: Optimal and practicable designs for measuring plaque pH-profiles. 95 StatMed 14 2609-2617

Haddad, John N.: On the closed form of the likelihood function of the first order moving average model. 95 Biomtrka 82 232-234

Haddad, P. R.: Mulholland, M.; Hibbert, D. B.; — —; Parslov, P.: A comparison of classification in artificial intelligence, induction versus a self-organising neural networks. 95 Chemolab 30 117-128

Mulholland, M.; Hibbert, D. B.; — —; Sammut, C.: Application of the C4.5 classifier to building an expert system for ion chromatography. 95 Chemolab 27 95-104

Hadgu, Alula: Analysis of three mouth rinses in inhibiting the development of supragingival dental plaque. 94 ASAProEpid 101-106

Hadi, Ali S.; Jones, Wendell D.; Ling, Robert F.: A unifying representation of some case-deletion influence measures in univariate and multivariate linear regression. 95 JStPlInf 46 123-135

— —; Simonoff, Jeffrey S.: Reply to comment on "Procedures for the identification of multiple outliers in linear models". 95 JASA 90 811-812

Castillo, Enrique; — —: A method for estimating parameters and quantiles of distributions of continuous random variables. 95 CmpStDA 20 421-439

Castillo, Enrique; — —: Modeling lifetime data with application to fatigue models. 95 JASA 90 1041-1054

Castillo, Enrique; — —: Parameters and quantiles estimation for continuously distributed random variables. 94 ASAProStCp 284-289

Hadjicostas, Petros: Tan, Kok Hooi; — —: Some properties of a limiting distribution in Quicksort. 95 StPrLet 25 87-94

Haesloop, Dan: Wise, Barry M.; — —: A non-linear dynamic model identification challenge problem. 95 Chemolab 30 91-96

Hagenaars, Jacques A.: — — (Auth); Eliason, Scott R. (Rev): Review of "Categorical longitudinal data: Log-linear panel, trend, and cohort analysis". 95 JEdStat 20 103-104

— — (Auth); Hershberger, Scott L. (Rev): Review of "Loglinear models with latent variables". 95 StrEqMod 2 172-174

Haggard, Lois M.; Gray, Donald Z.: Improving validity of parental reports of child immunization status in a telephone survey. 94 ASASrvyRsr 1258-1262

Romuald, Kristina S.; — —: The effect of varying the respondent selection script on respondent self-selection in RDD telephone surveys. 94 ASASrvyRsr 1299-1304

Haggett, P.: Cliff, A. D.; — —; Smallman-Raynor, M. R.; Stroup, D. F.; Williamson, G. D.: The application of multidimensional scaling methods to epidemiological data. 95 StMedRes 4 102-123

Häggkvist, R.: Diggle, P. J.; Chetwynd, A. G.; — —; Morris, S. E.: Second-order analysis of space-time clustering. 95 StMedRes 4 124-136

Häggström, Olle: On the relation between finite range potentials and subshifts of finite type. 95 ProbTher 101 469-478

— —: Random-cluster measures and uniform spanning trees. 95 StocProc 59 267-275

Hahn, Gerald J.: Deming's impact on industrial statistics: Some reflections. 95 AmerStat 49 336-341

Hahn, Jinyong: Bootstrapping quantile regression estimators. 95 EconTher 11 105-121

Hahn, Marjorie G.: Shao, Yongzhao; — —: Limit theorems for the logarithm of sample spacings. 95 StPrLet 24 121-132

Hahn, Marjorie G. W.; Klass, Michael J.: Uniform local probability approximations: Improvements on Berry-Esseen. 95 AnlsProb 23 446-463

Haider, Adnan M.: The expectation-maximization method, the extended-maximum likelihood method, and incomplete multivariate normal data. 94 ASAProQlPr 281-285

Haie-Meder, Christine: Lancar, Rémi; Kramar, Andrew; — —: Non-parametric methods for analysing recurrent complications of varying severity. 95 StatMed 14 2701-2712

Haigh, J.: Galambos, J.; Gani, J. (Ed); — — (Rev): Review of "Studies in applied probability: Papers in honour of Lajos Takács". 95 Biomtrcs 51 1187-1187

Haight, Robert G.: Comparing extinction risk and economic cost in wildlife conservation planning. 95 EcolgApp 5 767-775

Haiman, George; Kiki, Maxime; Puri, Madan L.: Extremes of Markov sequences. 95 JStPlInf 45 185-201

Hainer, Peter C.: "Strivin', gettin' over, and keepin' people out of our business": Pushing household boundaries in survey data collection. 94 ASASrvyRsr 337-341

Haines, Linda M.: A geometric approach to optimal design for one-parameter non-linear models. 95 JRSS-B 57 575-598

Mukhopadhyay, Saurabh; — —: Bayesian D-optimal designs for the exponential growth model. 95 JStPlInf 44 385-397

Haining, Robert P.: — — (Auth); Chang, Hyejung (Rev): Review of "Spatial data analysis in the social and environmental sciences". 95 Chance 8/1 42-43

Hainsworth, Timothy J.: Kent, John T.; — —: Confidence intervals for the noncentral chi-squared distribution. 95 JStPlInf 46 147-159

Hájek, P.; Sochorová, A.; Zvárová, J.: GUHA for personal computers. 95 CmpStDA 19 149-153

Hájek, Petr: Antoch, Jaromír; — —; Jiroušek, Radim: Obituary: Tomáš Havránek. 95 CmpStDA 19 111-113

Halavin, James J.; Mathiason, David J.: Designing and running undergraduate statistical laboratories. 94 ASAProStEd 132-136

Hale, Michael; Reeve, Russell: Planning a randomized concentration controlled trial with a binary response. 94 ASAProBiop 319-323

Reeve, Russell; — —: Results and efficiency of Bayesian dose adjustment in a clinical trial with binary endpoint. 94 ASAProBiop 324-327

Haley, Chris S.; Knott, Sara A.; Elsen, Jean-Michel: Mapping quantitative trait loci in crosses between outbred lines using least squares. 94 Genetcs 136 1195-1207

Hall, Bronwyn H.; Mariesse, Jacques: Exploring the relationship between R and D and productivity in French manufacturing firms. 95 JEconmtx 65 263-293

Hall, Charles B.: Liao, J. G.; — —: Fast and accurate computation of the exact confidence limits for the common odds ratio in several 2×2 tables. 95 JCmpGrSt 4 173-179

Hall, John W.: Carlson, Barbara Lepidus; — —: Weighting sample data when multiple sample frames are used. 94 ASASrvyRsr 882-887

Hall, P.: The bootstrap and Edgeworth expansion. 95 Springer:Brln:NY 352p

Hall, Peter: On the biases of error estimators in prediction problems. 95 StPrLet 24 257-262

— —; Fisher, Nicholas I.; Hoffmann, Branka: On the nonparametric estimation of covariance functions. 94 AnlsStat 22 2115-2134

— —; Horowitz, Joel L.; Jing, Bing-Yi: On blocking rules for the bootstrap with dependent data. 95 Biomtrka 82 561-574

— —; Hu, Tien Chung; Marron, J. S.: Improved variable window kernel estimates of probability densities. 95 AnlsStat 23 1-10

— —; Jing, Bing-Yi: Uniform coverage bounds for confidence intervals and Berry-Esseen theorems for Edgeworth expansion. 95 AnlsStat 23 363-375

— —; Mammen, Enno: On general resampling algorithms and their performance in distribution estimation. 94 AnlsStat 22 2011-2030

— —; Marron, J. Stephen; Titterington, D. M.: On partial local smoothing rules for curve estimation. 95 Biomtrka 82 575-587

— —; Patil, Prakash: Formulae for mean integrated squared error of nonlinear wavelet-based density estimators. 95 AnlsStat 23 905-928

— —; Patil, Prakash: On wavelet methods for estimating smooth functions. 95 Bernoull 1 41-58

— —; Wolff, Rodney C. L.: Properties of invariant distributions and Lyapunov exponents for chaotic logistic maps. 95 JRSS-B 57 439-452

— —; Wolff, Rodney C. L.: Estimators of integrals of powers of density derivatives. 95 StPrLet 24 105-110

Chan, Grace; — —; Poskitt, D. S.: Periodogram-based estimators of fractal properties. 95 AnlsStat 23 1684-1711

Hall, Stephen G.: Cuthbertson, Keith; — —; Taylor, Mark P. (Auth); Ericsson, Neil R.; Irons, John S. (Rev): Review of "Applied econometric techniques". 95 EcomtrRv 14 121-132

Hall, Timothy: Kloeden, Peter E.; Platen, Eckhard; Schurz, Henri (Auth); — — (Rev): Review of "Numerical solution of SDE through computer experiments". 95 Technmcs 37 125-125

Hall, W. J.; Wijsman, R. A.; Ghosh, J. K.: Correction to "The relationship between sufficiency and invariance with applications in sequential analysis" (V36 p575-614). 95 AnlsStat 23 705-705

Hallahan, Charlie: Data analysis using SAS. 95 SocMethR 23 373-391

Hallin, Marc; Puri, Madan L.: A multivariate Wald-Wolfowitz rank test against serial dependence. 95 CandJSt 23 55-65

Garel, Bernard; — —: Local asymptotic normality of multivariate ARMA processes with a linear trend. 95 AnInStMa 47 551-579

Hallman, William K.: Weinstein, Neil D.; Sandman, Peter M.; — —: Testing a visual display to explain small probabilities. 94 RiskAnal 14 895-896

Halloran, M. Elizabeth: Haber, Michael; — —; Longini, Ira M., Jr; Watelet, Luc: Estimation of vaccine efficacy in non-randomly mixing populations. 95 BiomtrcJ 37 25-38

Hallstrom, Al; McBride, Ruth; Moore, Rich: Toward vital status sweeps: A case history in sequential monitoring. 95 StatMed 14 1927-1931

Brooks, Maria Mori; — —; Peckova, Monika: A simulation study used to design the sequential monitoring plan for a clinical trial. 95 StatMed 14 2227-2237

Halousková, Alena: Kárný, Miroslav; — —; Nedoma, Petr: Recursive approximation by ARX model: A tool for grey box modelling. 95 IJAdCSgP 9 525-546

Halpern, Jerry: Hornberger, John C.; Brown, Byron Wm., Jr; — —: Designing a cost-effective clinical trial. 95 StatMed 14 2249-2259

Halsey, Neal A.: Moulton, Lawrence H.; — —: A mixture model with detection limits for regression analyses of antibody response to vaccine. 95 Biomtrcs 51 1570-1578

Moulton, Lawrence H.; Holt, Elizabeth A.; Job, Jayakaran S.; — —: Percentile regression analysis of correlated antibody responses. 95 StatMed 14 777-787

Halteman, Edward J.: Greer, Daniel A.; — —: SPC: A new look at what to teach. 94 ASAProBays 153-156

Halvorsen, Katherine: Iman, Ronald L. (Auth); — — (Rev): Review of "A data-based approach to statistics". 95 AmerStat 49 321-321

McClave, James T.; Dietrich, Frank H., II (Auth); — — (Rev): Review of "Statistics (Sixth edition)". 95 AmerStat 49 322-322

Halvorsen, Katherine Taylor: Mendenhall, William; Beaver, Robert J. (Auth); — — (Rev): Review of "Introduction to probability and statistics (Ninth edition)". 95 AmerStat 49 398-398

Hamada, M.; Wu, C. F.: Analysis of censored data from fractionated experiments: A Bayesian approach. 95 JASA 90 467-477

— —; Wu, C. F. J.: The treatment of related experimental factors by sliding levels. 95 JQualTek 27 45-55

Hamada, Michael: Tseng, Sheng-Tsaing; — —; Chiao, Chih-Hua: Using degradation data to improve fluorescent lamp reliability. 95 JQualTek 27 363-369

Hamada, Noboru: A survey of recent work on characterization of minihypers in $PG(t,q)$ and nonbinary linear codes meeting the Griesmer bound. 93 JCISS 18 161-191

Hamada, Toshio: An exponential two-armed bandit problem with one arm known under batch sampling. 95 JJapnStS 25 205-216

Hamadéne, S.; Lepeltier, J. P.: Backward equations, stochastic control and zero-sum stochastic differential games. 95 Stochatx 54 221-231

Hamblin, J.: Singh, M.; Ceccarelli, S.; — —: Estimation of heritability from varietal trials data. 93 ThApGene 86 437-441

Hambly, B. M.: On constant tail behaviour for the limiting random variable in a supercritical branching process. 95 JAppProb 32 267-273

Hamdy, H. I.; Son, M. S.: Accelerated sequential procedure to estimate the mean of unknown distribution. 94 ASASrvyRsr 1173-1178

Hamdy, M. I.: Distribution-free confidence intervals for $\Pr(X < Y)$ based on independent samples of X and Y. 95 CommStB 24 1005-1017

Hamedani, G. G.: On symmetric sum of sub-independent random variables. 95 CalcutSt 45 119-124

Hamel, Denis; Massé, Jean-Claude: A Monte Carlo study of the robustness of prediction in growth curve models. 95 JStCmpSm 52 33-53

Hamill, Thomas M.: Wilks, Daniel S.; — —: Potential economic value of ensemble-based surface weather forecasts. 95 MWeathr 11 3565-3575

Hamilton, David; Wu, Ka Ho: Confidence regions for parameters in the AR(1) model. 95 JTimSrAn 16 249-265

Hamilton, David C.; Lesperance, Mary L.: A comparison of methods for univariate and multivariate acceptance sampling by variables. 95 Technmcs 37 329-339

Hamilton, James D.: — — (Auth); Hansen, Bruce E. (Rev): Review of "Time series analysis". 95 EconTher 11 625-630

— — (Auth); Marquez, Jamie (Rev): Review of "Time series analysis". 95 IntJFor 11 494-495

Hamilton, M. A.; Johnson, K. R.; Camper, A. K.; Stoodley, P.; Harkin, G. J.; Gillis, R. J.; Shope, P. A.: Analysis of bacterial spatial patterns at the initial stage of biofilm formation. 95 BiomtrcJ 37 393-408

Hamilton, Scott A.: Edwards, Lloyd J.; — —: Errors-in-variables and the Box-Cox transformation. 95 CmpStDA 20 131-140

Hammack, William: Sharp inequalities for the distribution of a stochastic integral in which the integrator is a bounded submartingale. 95 AnlsProb 23 223-235

Hammerstrom, Thomas: Tsong, Yi; — —: Statistical issues in drug quality control based on dissolution testing. 94 ASAProBiop 295-300

Tsong, Yi; — —; Lin, Karl; Ong, Ting Eng: Dissolution test acceptance sampling plans. 95 JBiophSt 5 171-183

Hammitt, James K.: Can more information increase uncertainty?. 95 Chance 8/3 15-17

Hammond, Sean M.; Lienert, Gustav A.: Modified phi correlation coefficients for the multivariate analysis of ordinally scaled variables. 95 EdPsMeas 55 225-236

Hamza, K.; Klebaner, F. C.: Conditions for integrability of Markov chains. 95 JAppProb 32 541-547

Hamza, Kais: The smallest uniform upper bound on the distance between the mean and the median of the binomial and Poisson distributions. 95 StPrLet 23 21-25

Han, Chien-Pai; Hawkins, D. L.: On discriminant function using adaptive estimators of means. 94 ASAProBuEc 61-64

Chiou, Paul; — —: Conditional interval estimation of the exponential location parameter following rejection of a pre-test. 95 CommStA 24 1481-1492

Chiou, Paul; — —: Interval estimation of error variance following a preliminary test in one-way random model. 95 CommStB 24 817-824

Chiou, Paul; — —: Conditional interval estimation of the exponential scale parameter following rejection of a preliminary test. 94 JStatRsr 28 65-72

Han, Geun-Shik; Warde, William D.: Correlations in randomized response surveys. 94 ASASrvyRsr 491-496

Han, JeongHye: Cho, JoongJae; — —: Statistical estimation for generalized logit model of nominal type with bootstrap method. 95 JKoreaSt 24 1-18

Han, Kyung Soo: Kim, Woo-Chul; — —: On the performance of empirical Bayes simultaneous interval estimates for all pairwise comparisons. 95 JKoreaSt 24 161-181

Han, Sang-Tae; Huh, Myung-Hoe: Biplot of ranked data. 95 JKoreaSt 24 439-451

Hancock, Gregory R.: Klockars, Alan J.; — —; McAweeney, Mary J.: Power of unweighted and weighted versions of simultaneous and sequential multiple-comparison procedures. 95 PsychBu 118 300-307

Byrne, Barbara M. (Auth); — — (Rev): Review of "Structural equation modeling with EQS and EQS/Windows: Basic concepts, applications, and programming". 95 StrEqMod 2 73-78

Hand, D.; Daly, F.; Lunn, A.; McConway, K.; Ostrowski, E. (Ed); Ziegel, Eric R. (Rev): Review of "A handbook of small data sets". 95 Technmcs 37 242-242

Hand, D. J.: Data, metadata and information. 93 StatJUN 10 143-152

Davidian, M.; Giltinan, D. M. (Auth); — — (Rev): Review of "Nonlinear models for repeated measurement data.". 96 ShrtBkRv 16/1 6-7

Goldstein, Harvey (Auth); — — (Rev): Review of "Multilevel statistical models". 95 ShrtBkRv 15/3 43-44

Huberty, C. J. (Auth); — — (Rev): Review of "Applied discriminant analysis". 95 Biomtrcs 51 1187-1188

Krzanowski, W. (Ed); — — (Rev): Review of "Recent advances in descriptive multivariate analysis". 95 ShrtBkRv 15/3 44-44

Lindsey, J. K. (Auth); — — (Rev): Review of "Modelling frequency and count data". 96 ShrtBkRv 16/1 2-2

Lindsey, James K. (Auth); — — (Rev): Review of "Introductory statistics: A modelling approach". 96 ShrtBkRv 16/1 2-2

— —; Daly, F.; Lunn, A. D. (Auth); Derr, J. (Rev): Review of "A handbook of small data sets". 95 Biomtrcs 51 1586-1586

— —; Daly, F.; Lunn, A. D.; McConway, K.; Ostrowski, E. (Ed); Chatfield, C. (Rev): Review of "A handbook of small data sets". 95 JRSS-A 158 349-349

— —; Daly, F.; Lunn, A. D.; McConway, K. J.; Ostrowski, E. (Ed); Schimek, Michael G. (Rev): Review of "Handbook of small data sets". 95 CmpStQ 10 197-197

— —; Daly, F.; Lunn, A. D.; McConway, K. J.; Ostrowski, E. (Ed); Street, Deborah J. (Rev): Review of "A handbook of small data sets". 95 AstrlJSt 37 251-251

— —; Daly, F.; Lunn, A. D.; McConway, K. J.; Ostrowski, E. (Auth); Trenkler, Dietrich (Rev): Review of "A handbook of small data sets". 95 CmpStDA 19 101-101

— — (Ed); Spiegelhalter, David (Rev): Review of "Artificial intelligence frontiers in statistics: AI and statistics III". 95 JAppStat 22 186-186

— — (Ed); Ziegel, Eric R. (Rev): Review of "Artificial intelligence frontiers in statistics". 95 Technmcs 37 127-127

— — (Ed); Ziegel, Eric R. (Rev): Review of "AI and computer power: The impact of statistics". 95 Technmcs 37 241-242

Hand, David: An episode in the history of measurement. 95 TeachgSt 17/2 54-56

Hand, David J.: Comment on "Screening for stratification in two-phase ('two-stage') epidemiological surveys" (V4 p73-89). 95 StMedRes 4 263-263

Handcock, Mark S.: Chatterjee, Samprit; — —; Simonoff, Jeffrey S.: A casebook for a first course in statistics and data analysis. 95 Wiley:NY:UK 314p

Lunneborg, Clifford E. (Auth); — — (Rev): Review of "Modeling experimental and observational data". 95 AmerStat 49 238-238

Händel, Peter: Stoica, Petre; — —; Nehoral, Arye: Improved sequential MUSIC. 95 IEEEAES 31 1230-1239

Handelman, David: Space-time boundaries for random walks obtained from diffuse measures. 94 IsrelJMa 86 107-156

Handley, J.: Bunge, J.; Fitzpatrick, M.; — —: Comparison of three estimators of the number of species. 95 JAppStat 22 45-59

Hanfelt, John J.; Liang, Kung-yee: Approximate likelihood ratios for general estimating functions. 95 Biomtrka 82 461-477

Hanif, Muhammad; Beg, M. A.; Khawaja, Iffat: Sampling with unequal probabilities: A historical review. 92 ProIrStCl 19-37

Hanley, James A.: Comment on "Uncertainties in lung cancer risk estimates reported for exposure to environmental tobacco smoke". 95 EnvrMtrc 6 419-421

— —; MacGibbon, Brenda: The Poisson distribution and "multiple events" cell occupancy problems. 95 Biomtrcs 51 1557-1561

Hannan, E. J.; Kavalieris, L.: A note on ARMA parameterisation. 95 AstrlJSt 37 105-110

Hanratty, Maria: Gruber, Jonathan; — —: The labor-market effects of introducing national health insurance: Evidence from Canada. 95 JBES 13/2 163-173

Hansen, Bruce E.: Rethinking the univariate approach to unit root testing - Using covariates to increase power. 95 EconTher 11 1148-1171

Canova, Fabio; — —: Are seasonal patterns constant over time? A test for seasonal stability. 95 JBES 13 237-252

Hamilton, James D. (Auth); — — (Rev): Review of "Time series analysis". 95 EconTher 11 625-630

Hansen, G.: Hillmer, M. (Auth); — — (Rev. German): Review of "A causal analysis with latent variables of macroeconomic interdependencies" (German). 95 AllgeStA 79 224-227

Hansen, M. H.; Hurwitz, W. N.; Madow, W. G. (Auth); Fielding, Antony (Rev): Review of "Sample survey methods and theory (Volume I, 2)". 95 JRSS-A 158 350-350

— —; Hurwitz, W. N.; Madow, W. G. (Auth); Schwaiger, M. (Rev. German): Review of "Sample survey methods and theory (Volume I and II)". 95 AllgeStA 79 335-337

Hansen, Pierre: Glover, Fred; Kingman, Darwin; Phillips, Nancy V. (Auth); — — (Rev): Review of "Network models in optimization and their applications in practice". 95 JClass 12 150-152

Hansen, Sue Ellen: Couper, Mick P.; Sadosky, Sally A.; — —: Measuring interviewer performance using CAPI. 94 ASASrvyRsr 845-850

Hanson, Allen R.: Kumar, Rakesh; — —: Robust methods for estimating pose and a sensitivity analysis. 94 CVGIPIm 60 313-342

Hantler, Sidney L.: Birman, Alexander; Gail, H. Richard; — —; Rosberg, Zvi; Sidi, Moshe: An optimal service policy for buffer systems. 95 JACM 42 641-657

Hanumara, R. C.: Crouse, Robert H.; Jin, Chun; — —: Unbiased ridge estimation with prior information and ridge trace. 95 CommStA 24 2341-2354

Hanusz, Zofia: Relative potency of two preparations in two-way elimination of heterogeneity designs with multivariate responses. 95 Biomtrcs 51 1133-1139

Happacher, Maximilian: Exact and approximate D-optimal designs in polynomial regression. 95 Metrika 42 19-27

Hapuarachchi, K. P.; Wronski, A.: Control charts for non-response rates in the Canadian Labour Force Survey. 94 ASASrvyRsr 937-942

Haq, M. Safiul; Ng, V. Ming: A linear model with a special class of covariance structure. 94 JStatRsr 28 189-198

Harbron, Chris: Heuristic algorithms for finding inexpensive elimination schemes. 95 StatComp 5 275-287

Hardgrave, Bill C.: Wilson, Rick L.; — —: Predicting graduate student success in an MBA program: Regression versus classification. 95 EdPsMeas 55 186-195

Hardin, J. Michael: Conerly, Michael D.; — —; Waller, Lance A.: When is Satterthwaite's approximation adequate?. 94 ASAProStCp 209-213

Waller, Lance A.; Turnbull, Bruce W.; — —: Obtaining distribution functions by numerical inversion of characteristic functions with applications. 95 AmerStat 49 346-350

Hardin, James W.: North, Gerald R.; Kim, Kwang-Y.; Shen, Samuel S. P.; — —: Detection of forced climate signals. Part I: Filter theory. 95 JClimate 8 401-408

Härdle, W.; Huet, S.; Jolivet, E.: Better bootstrap confidence intervals for regression curve estimation. 95 MaOpfStS 26 287-306

— —; Klinke, S.; Turlach, B. A. (Ed): XploRe: An interactive statistical computing environment. 95 Springer:Brln:NY 387p

— —; Marron, J. S.: Fast and simple scatterplot smoothing. 95 CmpStDA 20 1-17

— —; Park, B. U.: Testing increasing dispersion. 95 CmpStDA 19 641-653

— —; Park, B. U.; Tsybakov, A. B.: Estimation of non-sharp support boundaries. 95 JMultiAn 55 205-218

— —; Steiger, W.: [Algorithm AS 296] Optimal median smoothing. 95 ApplStat 44 258-264

— —; Tsybakov, A. B.: Additive nonparametric regression on principal components. 95 JNnparSt 5 157-184

Efron, B.; Tibshirani, R. J. (Auth); — — (Rev): Review of "An introduction to the bootstrap". 95 Metrika 42 267-268

— —; Klinke, S.; Turlach, B. A. (Ed); Murdoch, D. J. (Rev): Review of "XploRe: An interactive statistical computing environment". 96 ShrtBkRv 16/1 3-3

Härdle, Wolfgang; Kirman, Alan: Nonclassical demand - A model-free examination of price-quantity relations in the Marseille fish market. 95 JEconmtx 67 227-257

Hardy, Linda P.; Eisenhower, Donna L.: Developing methods of collecting the occupation of persons with college degrees. 94 ASAProGovt 110-115

Hardy, M. R.: Daykin, C. D.; Pentikäinen, T.; Pesonen, E. (Auth); — — (Rev): Review of "Practical risk theory for actuaries". 95 JAppStat 22 547-548

Hare, Lynne; Hoerl, Roger W.; Snee, Ronald D.: Statistical thinking for business. 95 ASQCTekCon 787-788

Hare, Lynne B.; Hoerl, Roger W.; Hromi, John D.; Snee, Ronald D.: The role of statistical thinking in management. 95 QulPr 28/2 53-60

Kuehl, Robert O. (Auth); — — (Rev): Review of "Statistical principles of research design and analysis". 95 JQualTek 27 96-97

Hargreaves, Colin: — — (Ed); Barker, Terry (Rev): Review of "Macroeconomic modelling of the long run". 95 JApEcmtx 10 211-212

Hariri, A. M. A.: Abou-El-ata, M. O.; — —: Point estimation and confidence intervals of the M/M/2/N queue with balking and heterogeneity. 95 AJMMgSci 15 35-55

Harkin, G. J.: Hamilton, M. A.; Johnson, K. R.; Camper, A. K.; Stoodley, P.; — —; Gillis, R. J.; Shope, P. A.: Analysis of bacterial spatial patterns at the initial stage of biofilm formation. 95 BiomtrcJ 37 393-408

Harkins, Ralph D.: Chakravarty, Alaka G.; — —: Some approaches to the evaluation of laboratory data to identify potential adverse events. 94 ASAProBiop 73-76

Harlow, Lisa L.: Ding, Lin; Velicer, Wayne F.; — —: Effects of estimation methods, number of indicators per factor, and improper solutions on structural equation modeling fit indices. 95 StrEqMod 2 119-144

Haroun, Ahmed: Shaarawy, Samir; — —: Bayesian classification with ARMA sources. 94 EgpStJ 38 165-176

Haroun, Ahmed H.: Goodness-of-fit for multivariate linear regression models. 94 EgpStJ 38 138-146

Harrell, Frank E., Jr: Herndon, James E., II; — —: The restricted cubic spline as baseline hazard in the proportional hazards model with step function time-dependent covariables. 95 StatMed 14 2119-2129

Harrington, Peter de B.: Wabuyele, Busolo Wa; — —: Quantitative comparison of bidirectional and optimal associative memories for background prediction of spectra. 95 Chemolab 29 51-61

Harris, David C.: King, Maxwell L.; — —: The application of the Durbin-Watson test to the dynamic regression model under normal and non-normal errors. 95 EcomtrRv 14 487-510

Harris, Eugene K.; Boyd, James C.: Statistical bases of reference values in laboratory medicine. 95 Dekker:NY 384p

Harris, Kenneth W.; Chamblee, Ronald F.; Winter, Francine D.: Evaluation of an automated industry and occupation coding system. 94 ASAProSoSt 174-177

Harris, T. Robert; Klassen, Albert D.; Wilsnack, Richard W.; Wilsnack, Sharon C.: A model for the interaction of partnership dissolution and relationship quality as predicators of remission of problem drinking in women. 94 ASAProSoSt 232-234

Harris, Thomas R.: Lambert, David K.; — —: The use of conditional probability functions in range data analysis and simulation. 93 JRngeMgt 46 157-160

Harrison, F. D.: Huque, M. F.; — —; Sankoh, A. J.; Dubey, S. D.: Longitudinal data analysis for heartburn clinical trials. 94 ASAProBiop 182-187

Harrison, J.: Pole, A.; West, M.; — — (Auth); Davies, Neville (Rev): Review of "Applied Bayesian forecasting and time series analysis". 95 JRSS-A 158 635-636

Harrison, Peter G.; Pitel, Edwige: Response time distributions in tandem g-networks. 95 JAppProb 32 224-246

Harrison, Robert A.: Ushakov, Igor A.; — — (Ed); Doganaksoy, Necip (Rev): Review of "Handbook on reliability engineering". 95 JASA 90 804-804

Harrison, Robert F.: Cross, Simon S.; — —; Kennedy, R. Lee: Introduction to neural networks. 95 Lancet 346 1075-1079

Harrop, Anne; Plewis, Ian: Two decades of family change: Secondary analysis of continuous government surveys. 95 JRSS-A 158 91-106

Hart, Galen F.: Dulaney, Wayne P.; Lengnick, Laura L.; — —: Use of geostatistical techniques in the design of an agricultural field experiment. 94 ASASrvyRsr 183-187

Harter, H. L.: — — (Auth); Hušková, M. (Rev): Review of "The chronological annotated bibliography of order statistics (Volumes IV-VII)". 95 MaOpfStS 26 182-182

Hartfiel, D. J.: Indeterminate Markov systems. 95 AplMaCmp 72 51-59

Hartigan, J. A.: Kiuchi, Amy S.; — —; Holford, Theodore R.; Rubinstein, Pablo; Stevens, Cladd E.: Change points in the series of T4 counts prior to AIDS. 95 Biomtrcs 51 236-248

Hartlaub, Bradley A.: Spurrier, John D.; Edwards, Don; Thombs, Lori A. (Auth); — — (Rev): Review of "Elementary statistics laboratory manual (MS-DOS Version)". 95 AmerStat 49 397-398

Hartnell, B. L.: Cockayne, E. J.; — —; Hedetniemi, S. T.; Laskar, R.: Perfect domination in graphs. 93 JCISS 18 136-148

Hartzell, George: Lipshutz, Robert J.; Taverner, Fred; Hennessy, Kevin; — —; Davis, Ron: DNA sequence confidence estimation. 94 Genomics 19 417-424

Harvey, A. C.: — — (Auth); Davies, Neville (Rev): Review of "Time series models (Second edition)". 95 JRSS-A 158 191-193

Harvey, Andrew C.: — — (Auth); Ensor, Katherine B. (Rev): Review of "Time series models". 95 Technmcs 37 464-465

Harvey, Charles F.; Gorelick, Steven M.: Temporal moment-generating equations: Modeling transport and mass transfer in heterogeneous aquifers. 95 WaterReR 31 1895-1911

Harvey, Nigel: Bolger, Fergus; — —: Judging the probability that the next point in an observed time-series will be below, or above, a given value. 95 JForcast 14 597-607

Harvill, Jane L.; Newton, H. Joseph: Saddlepoint approximations for the difference of order statistics. 95 Biomtrka 82 226-231

— —; Newton, H. Joseph: Using symbolic math to evaluate saddlepoint approximations for the difference of order statistics. 95 CommStB 24 781-791

Hary, I.: Schwartz, H. J.; Mosler, C.; — —; Pielert, V.: Factors affecting spatial preferences in settlement site selection in migratory pastoralism. 95 EnvrMtrc 6 485-490

Hasan, M. N.; Koenker, R. W.: Robust rank tests of the unit root hypothesis based on regression rank scores. 94 ASAProBuEc 35-43

Hasegawa, Hikaru: On risk comparisons of some estimators in a linear regression model under Inagaki's loss function and nonnormal error terms. 95 CommStA 24 1655-1685

— —: On small sample properties of Zellner's estimator for the case of two SUR equations with compound normal disturbances. 95 CommStB 24 45-59

Chaturvedi, Anoop; — —; Asthana, Sheel: Bayesian predictive analysis of the linear regression model with and Edgeworth series prior distribution. 95 CommStA 24 2469-2484

Hasegawa, Masami; Kishino, Hirohisa: Accuracies of the simple methods for estimating the bootstrap probability of a maximum-likelihood tree. 94 MolBioEv 11 142-145

Hasegawa, Toshiharu: Takine, Tetsuya; — —: On the single-server queue with independent MAP/G and M/GI input streams. 95 CommStSM 11 227-234

Takine, Tetsuya; Sengupta, Bhaskar; — —: A conformance measure for queues. 95 CommStSM 11 645-670

Hashimoto, Hideo: On the optimal extraction rate of a non-renewable resource with an extraction capacity. 95 EnvrMtrc 6 547-550

Haskin, F. Eric; Radke, George E., Jr; Evanoff, Javon: Algorithms for analyzing the probability of M-out-of-N events. 95 RelEngSS 47 153-162

Haskins, Loren: Egge, Eric; Foley, Sean; — —; Johnson, Roger: A data analysis based elementary statistics course. 94 ASAProStEd 144-149

Haslett, Stephen; Telfar, Grant: Sample design for auditing error rates in claims data: A case study. 95 NZndSt 30 23-36

Hasman, A.: Muijtjens, A. M. M.; — — (Auth); Schulz, K. (Rev): Review of "Bayes" (German). 95 CmpStQ 10 309-309

Hassan, A.: Zaki, M.; Tohamy, F. El; — —: Landsat image interpretation. 95 CmpStDA 20 75-97

Hassan, N. A.: Shoukry, E. M.; Gharraph, M. K.; — —: Multi-channel bi-level queueing system with heterogeneous servers. 94 EgpStJ 38 208-221

Hassan, Tawfik: Numerical solution of likelihood equations for a first-order response surface model for mixture paired comparisons. 94 EgpStJ 38 196-207

El-Helbawy, Abdalla T.; — —: On the Wald, Lagrangian multiplier and likelihood ratio tests when the information matrix is singular. 94 JItalStS 3 51-60

Hassanein, Khatab M.: Saleh, A. K. Md. E.; — —; Brown, Edward F.: Optimum spacings, ABLUE and tests of hypothesis in simple linear model with extreme-value errors. 94 JStatRsr 28 53-63

Hasselblad, Vic; Hedges, Larry V.: Meta-analysis of screening and diagnostic tests. 95 PsychBu 117 167-178

Hassett, Nancy C.: Saavedra, Pedro J.; — —; Weir, Paula: Comparison of national and regional gasoline prices from two surveys. 94 ASASrvyRsr 1179-1183

Hassett, Thomas F.; Dietrich, Duane L.; Szidarovszky, Ferenc: Time-varying failure rates in the availability and reliability analysis of repairable systems. 95 IEEEReli 44 155-160

Hassler, Uwe: The periodogram regression: Correction and comments. 95 CommStA 24 3137-3146

— —; Wolters, Jürgen: Long memory in inflation rates: International evidence. 95 JBES 13 37-45

Haßmann, S.; Reiss, R.-D.; Thomas, M. (Auth); Anderson, C. W. (Rev): Review of "XTREMES". 95 CmpStQ 10 91-98

Hastie, Trevor; Buja, Andreas; Tibshirani, Robert: Penalized discriminant analysis. 95 AnlsStat 23 73-102

— —; Tibshirani, Robert: Generalized additive models for medical research. 95 StMedRes 4 187-196

Hastings, N.: Evans, M.; — —; Peacock, B. (Auth); Trenkler, Götz (Rev): Review of "Statistical distributions (Second edition)". 95 CmpStDA 19 483-484

Hastings, Nicholas: Evans, Merran; — —; Peacock, Brian (Auth); Copeland, Karen A. F. (Rev): Review of "Statistical distributions". 95 JQualTek 27 94-95

Haswell, Stephen John: — — (Ed); Bayne, Charles K. (Rev): Review of "Practical guide to chemometrics". 95 Technmcs 37 230-231

Hatcher, Jeff: Norris, Douglas A.; — —: The impact of interviewer characteristics on response in a national survey of violence against women. 94 ASASrvyRsr 449-454

Hatzinger, R.: Seeber, G. U. H.; Francis, B. J.; — —; Steckel-Berger, G.: Statistical modelling. 95 Springer:Brln:NY 327p

Hau, Guadalupe: Elizundia, José M.; — —: Using SPC in two Mexican poultry processing plants. 95 ASQCTekCon 579-587

Elizundia, José M.; — —: SCP implementation in a Mexican poultry processing plant. 94 ASAProPEng 295-301

Haugh, Larry D.; Pond, Andrew D.: Adjusted individual control charts for short runs. 95 ASQCTekCon 1117-1126

Haugh, Margaret: Boissel, Jean-Pierre; Durieu, Isabelle; Girard, Pascal; Nony, Patrice; Chauvin, Franck; — —: Dose-ranging trials: Guidelines for data collection and standardized descriptions. 95 ConClnTr 16 319-330

Haughton, D.: Sinai, Y. (Auth); — — (Transl): Probability theory: An introductory course. 95 Springer:Brln:NY 138p

Sinai, Y. (Auth); — — (Transl); Sedgwick, Philip (Rev): Review of "Probability theory: An introductory course". 95 Ststcian 44 142-142

Haukka, Jari K.: Correction for covariate measurement error in generalized linear models - A bootstrap approach. 95 Biomtrcs 51 1127-1132

Hauser, Michael A.; Reschenhofer, Erhard: Estimation of the fractionally differencing parameter with the R/S method. 95 CmpStDA 20 569-579

Hausken, Elvie Germino: West, Jerry; — —: Different approaches to counting early childhood program participation. 94 ASAProSoSt 141-146

Hausman, J. A.; Newey, W. K.; Powell, J. L.: Nonlinear errors in variables: Estimation of some Engel curves. 95 JEconmtx 65 205-233

Have, Thomas R. Ten; Chinchilli, Vernon M.: The analysis of bioequivalence with respect to TMAX under a 2×2 crossover design. 95 JBiophSt 5 185-199

— —; Landis, J. Richard; Weaver, Susan L.: Association models for periodontal disease progression: A comparison of methods for clustered binary data. 95 StatMed 14 413-429

Havekes, L.: Tiret, L.; Nicaud, V.; Ehnholm, C.; — —; Menzel, H. J.; Ducimetiere, P.; Cambien, F.: Inference of the strength of genotype-disease association from studies comparing offspring with and without parental history of disease. 93 AnHuGene 57 141-149

Havenner, Arthur: Dorfman, Jeffrey H.; — —: Model specification tests for balanced representation state space models. 95 CommStA 24 97-119

Haverkort, Boudewijn R.; Meeuwissen, Adrianus M. H.: Sensitivity and uncertainty analysis of Markov-reward models. 95 IEEEReli 44 147-154

Hawaleshka, O.: Chen, R.; — —; Strong, D. R.: Analytical approach for determining the cost optimum raw material quality level. 95 JAppStat 22 375-388

Hawes, Catherine: Iannacchione, Vincent; Wildfire, Judith; Lux, Linda; — —: Covering the population of unlicensed board and care homes: Problems and solutions. 94 ASASrvyRsr 270-275

Hawkes, Alan G.: Cui, Lirong; — —; Jalali, Assad: The increasing failure rate property of consecutive-k-out-of-n systems. 95 PrEngInf 9 217-225

Hawkes, James: Chatterjee, Sangit; — —: Apparent decline as a sign of improvement? or, Can less be more?. 95 TeachgSt 17/3 82-83

Hawkins, Anne: Foster, Jeremy J. (Auth); — — (Rev): Review of "SPSS/PC+ and SPSS for Windows - A beginner's guide to data analysis (Second edition)". 95 TeachgSt 17/2 79-79

Hawkins, D. L.: Han, Chien-Pai; — —: On discriminant function using adaptive estimators of means. 94 ASAProBuEc 61-64

Hawkins, Del I.: Kanji, Gopal K. (Auth); — — (Rev): Review of "100 statistical tests". 95 JMarRsr 32 112-113

Hawkins, Douglas M.: Convergence of the feasible solution algorithm for least median of squares regression. 95 CmpStDA 19 519-538

— —: Comment on "Q charts". 95 JQualTek 27 324-327

Bradu, Dan; — —: An Anscombe type robust regression statistic. 95 CmpStDA 20 355-386

Hawks, V. D.: Badrigal, J. L.; Richards, D. O.; Free, J. C.; Lee, T. N.; Lawson, J. S.; — —: Q^2: A TQM course for undergraduates. 94 ASAProQlPr 172-180

Hawksworth, D. L.: — — (Ed); Feu, Chris du (Rev): Review of "Biodiversity - Measurement and estimation". 95 TeachgSt 17/3 111-111

Hayakawa, T.: Mathai, A. M.; Provost, S. B.; — —: Bilinear forms and zonal polynomials. 95 Springer:Brln:NY 376p

Hayashi, Chikio; Hayashi, Fumi: Comparative study of national character (Japanese). 95 PrInStMa 43 27-80

Hayashi, Fumi: Hayashi, Chikio; — —: Comparative study of national character (Japanese). 95 PrInStMa 43 27-80

Hayashi, T.; Ukai, Y.: Change in genetic variance under selection in a self-fertilizing population. 94 Genetcs 136 693-704

— —; Ukai, Y.: Detection of additive and dominance effects of QTLs in interval mapping of F_2 RFLP data. 94 ThApGene 87 1021-1027

Hayashi, Yoichi: Buckley, James J.; — —: Neural nets for fuzzy systems. 95 FuzzySet 71 265-276

Haybittle, J.; Yuen, P.; Machin, D.: Multiple comparisons in disease mapping. 95 StatMed 14 2503-2505

Hayek, Lee-Ann: Glinsman, Lisha; — —: Statistical evaluation of renaissance art objects. 95 Stats /13 15-18

Hayes, Bob: — — (Auth); Ziegel, Eric R. (Rev): Review of "Measuring customer satisfaction". 95 Technmcs 37 128-129

Hayes, Daniel B.; Brodziak, Jon K. T.; O'Gorman, Joseph B.: Efficiency and bias of estimators and sampling designs for determining length-weight relationships of fish. 95 JFishRBC 52 84-92

Hays, Paul B.: Wu, Dong L.; — —; Skinner, Wilbert R.: A least squares method for spectral analysis of space-time series. 95 JAtmsphr 52 3501-3511

Hayya, Jack C.; Xu, Susan H.; Ramasesh, Ranga V.; He, Xin X.: Order crossover in inventory systems. 95 CommStSM 11 279-309

Hazelton, Martin: Improved Monte Carlo inference for models with additive error. 95 StatComp 5 343-350

Hazod, W.: On the limit behavior of products of a random number of group valued random variables. 94 ThProbAp 39 249-263

Jurek, Z. H.; Mason, D. J. (Auth); — — (Rev): Review of "Operator-limit distributions in probability theory". 95 Metrika 42 265-267

He, Kun: On estimating domain totals over a subpopulation. 95 AnInStMa 47 637-644

— —: On estimating a linear combination of strata means with random sample sizes. 95 JMultiAn 55 39-60

He, Liangcai: He, Zhongshi; — —: The convergence theorem and limit of the iterative algorithm for choosing k value in ridge regression estimation (Chinese). 95 AMaAPknC 17 59-64

He, Min: Wu, Qiguang; — —: Several notes on admissible estimators of parameters and losses under truncated distribution families (Chinese). 95 JSysSMaS 14 168-176

He, Qi-Ming: Differentiability of the matrices R and G in the matrix-analytic method. 95 CommStSM 11 123-132

— —; Jewkes, E.: Flow time in the MAP/G/1 queue with customer batching and setup times. 95 CommStSM 11 691-711

He, Weizhong: Lachenbruch, Peter A.; Clements, Philip J.; — —: On encoding values for data recorded as $X > C$. 95 BiomtrcJ 37 855-867

He, Xin X.: Hayya, Jack C.; Xu, Susan H.; Ramasesh, Ranga V.; — —: Order crossover in inventory systems. 95 CommStSM 11 279-309

He, Xuming: Shi, Peide; — —: The optimal rates of convergence on nonparametric quantile regression estimate on dependent data (Chinese). 95 KToChina 39 1537-1539

He, Zhongshi; He, Liangcai: The convergence theorem and limit of the iterative algorithm for choosing k value in ridge regression estimation (Chinese). 95 AMaAPknC 17 59-64

Heagy, W. K.; Sullivan, P. J.: Fixed-point values of contaminant cloud dilution. 95 EnvrMtrc 6 637-641

Healy, M. J. R.: Frank Yates, 1902-1994 - The work of a statistician. 95 IntStRvw 63 271-288

Wang, Chamant (Auth); — — (Rev): Review of "Sense and nonsense of statistical inference". 95 StatMed 14 103-104

Healy, Michael J. R.: Goldstein, Harvey; — —: The graphical presentation of a collection of means. 95 JRSS-A 158 175-177

Hearn, William S.: Pollock, Kenneth H.; Conroy, Michael J.; — —: Separation of hunting and natural mortality using ring-return models: An overview. 95 JAppStat 22 557-566

Heath, A. C.; Kessler, R. C.; Neale, M. C.; Hewitt, J. K.; Eaves, L. J.; Kendler, K. S.: Testing hypotheses about direction of causation using cross-sectional family data. 93 BehavGen 23 29-50

Sham, P. C.; Walters, E. E.; Neale, M. C.; — —; MacLean, C. J.; Kendler, K. S.: Logistic regression analysis of twin data: Estimation of parameters of the multifactorial liability-threshold model. 94 BehavGen 24 229-238

Heath, Anthony; Evans, Geoffrey; Payne, Clive: Modelling the class-party relationship in Britain, 1964-92. 95 JRSS-A 158 563-574

Heathcote, C. R.; Rachev, S. T.; Cheng, B.: Testing multivariate symmetry. 95 JMultiAn 54 91-112

Heavlin, William D.: Variance components and computer experiments. 94 ASAProPEng 103-108

— —; Krivokapić, Zoran: Virtual manufacturing: Design, calibration, and analysis. 94 ASAProPEng 274-279

Finnegan, G. Paul; — —: Application of Hoaglin's G and H curves to process specifications. 94 ASAProQlPr 146-150

Hebert, Jaimie L.: Arnholt, Alan T.; — —: Estimating the mean with known coefficient of variation. 95 AmerStat 49 367-369

Carpenter, Mark; — —: Estimating the minimum and maximum location parameters for two IG-exponential scale mixtures. 95 CommStA 24 1227-1233

Heckler, Charles E.: Huberty, Carl J. (Auth); — — (Rev): Review of "Applied discriminant analysis". 95 Technmcs 37 349-350

Heckman, Nancy E.: Fan, Jianqing; — —; Wand, M. P.: Local polynomial kernel regression for generalized linear models and quasi-likelihood functions. 95 JASA 90 141-150

Hedayat, A. S.; Majumdar, Dibyen: Generating desirable sampling plans by the technique of trade-off in experimental design. 95 JStPlInf 44 237-247

— —; Stufken, J.; Zhang, W. G.: Virtually balanced incomplete block designs for $v = 22$, $k = 8$, and $\lambda = 4$. 95 JComDsgn 3 195-201

— —; Stufken, J.; Zhang, W. G.: Contingently and virtually balanced incomplete block designs and their efficiencies under various optimality criteria. 95 StSinica 5 575-591

Brown, J. W.; — —; Parker, E. T.: A pair of orthogonal Latin squares of order 10 with four shared parallel transversals. 93 JCISS 18 113-115

— —; Sinha, B. K. (Auth); Rust, Keith (Rev): Review of "Design and inference in finite population sampling". 94 JOffStat 10 226-228

Hedayat, Samad: Statistical designs for drug developments. 92 ProIrStC1 39-50

Hedetniemi, S. T.: Cockayne, E. J.; Hartnell, B. L.; — —; Laskar, R.: Perfect domination in graphs. 93 JCISS 18 136-148

Hedges, L.: Cooper, H.; — — (Ed); Ziegel, Eric R. (Rev): Review of "The handbook of research synthesis". 95 Technmcs 37 242-242

Hedges, L. V.: Cooper, H.; — — (Ed); Howel, D. (Rev): Review of "The handbook of research synthesis". 95 Biomtrcs 51 1585-1586

Hedges, Larry V.: Hasselblad, Vic; — —: Meta-analysis of screening and diagnostic tests. 95 PsychBu 117 167-178

Vevea, Jack L.; — —: A general linear model for estimating effect size in the presence of publication bias. 95 Psymtrka 60 419-435

Draper, David; Gaver, Donald, P., Jr; Goel, Prem K.; Greenhouse, Joel B.; — —; Morris, Carl N.; Tucker, John R.; Waternaux, Christine M. (Auth); Berlin, Jesse A. (Rev): Review of "Combining information: Statistical issues and opportunities for research". 95 StatMed 14 1611-1612

Heerink, Nico: — — (Auth); Ioannides, Chris (Rev): Review of "Population growth, income distribution and economic development". 95 IntJFor 11 333-334

Heft, Marc W.: O'Brien, Ralph G.; — —: Discrimination indexes and models for studying sensory functioning in aging. 95 JAppStat 22 9-27

Heiberger, Richard M.: Raghavarao, Damaraju; — —: Variance balanced generalized nested row-column designs with empty diagonal cells. 93 JCISS 18 42-52

Marazzi, A. (Auth); Iglewicz, Boris; — —; Moore, Dirk; Tan, Yun (Rev): Review of "Algorithms, routines, and S functions, for robust statistics". 95 Technmcs 37 351-353

Heidelberger, Philip: Fast simulation of rare events in queueing and reliability models. 95 ACMMdSim 5 43-85

Heij, Christiaan: Roorda, Berend; — —: Global total least squares modeling of multivariable time series. 95 IEEEAuCn 40 50-63

Heiligers, B.: Dette, H.; — —; Studden, W. J.: Minimax designs in linear regression models. 95 AnlsStat 23 30-40

Heiligers, Berthold; Sinha, Bikas Kumar: Optimality aspects of Agrawal's designs: Part II. 95 StSinica 5 599-604

Gaffke, Norbert; — —: Computing optimal approximate invariant designs for cubic regression on multidimensional balls and cubes. 95 JStPlInf 47 347-376

Gaffke, Norbert; — —: Optimal and robust invariant designs for cubic multiple regression. 95 Metrika 42 29-48

Gaffke, Norbert; — —: Algorithms for optimal design with application to multiple polynomial regression. 95 Metrika 42 173-190

Hein, Eric W.; Andelt, William F.: Estimating coyote density from mark-resight surveys. 95 JWildlMg 59 164-169

Heinen, Ton: — — (Auth); Andersen, E. B. (Rev): Review of "Discrete latent variable models". 95 Psymtrka 60 449-450

Heiner, Karl W.: Heiner, Karlyn S.; — —: Tree rings in the Northern Shawangunks. 95 Chance 8/1 30-37

Heiner, Karlyn S.; Heiner, Karl W.: Tree rings in the Northern Shawangunks. 95 Chance 8/1 30-37

Heinrich, G.; Jensen, U.: Parameter estimation for a bivariate lifetime distribution in reliability with multivariate extensions. 95 Metrika 42 49-65

Heinrich, L.: Rao, C. Radhakrishna; Shanbhag, D. N. (Auth); — — (Rev): Review of "Choquet-Deny type functional equations with applications to stochastic models". 95 BiomtrcJ 37 726-726

Heinrich, Lothar; Schüle, Eik: Generation of the typical cell of a non-Poissonian Johnson-Mehl tessellation. 95 CommStSM 11 541-560

Heise, Mark A.: Coakley, Clint W.; — —: A new version of the sign test in the presence of ties. 94 ASAProEnvr 144-149

Heiser, Willem J.: Groenen, Patrick J. F.; Mathar, Rudolf; — —: The majorization approach to multidimensional scaling for Minkowski distances. 95 JClass 12 3-19

Heisey, Dennis M.; Nordheim, Erik V.: Modelling age-specific survival in nesting studies, using a general approach for doubly-censored and truncated data. 95 Biomtrcs 51 51-60

Heitjan, Daniel F.: Bayesian analysis of a protocol for sampling preserved tumor specimens. 95 JASA 90 38-44

Helland, Inge S.: Simple counterexamples against the conditionality principle. 95 AmerStat 49 351-356

— —; Næs, Tormod; Isaksson, Tomas: Related versions of the multiplicative scatter correction method for preprocessing spectroscopic data. 95 Chemolab 29 233-241

Heller, I.; Isakov, A.; Blinder-Weiner, S.; Topilsky, M.: Bayesian classification of vasculitis: A simulation study. 95 MInfoMed 34 259-265

Helm, Werner E.: Comment on "Statistical values, quality, and certification" (V48 p65-70). 95 AmerStat 49 240-241

Helms, Ronald W.: Grady, James J.; — —: Model selection techniques for the covariance matrix for incomplete longitudinal data. 95 StatMed 14 1397-1416

Smith, Fraser B.; — —: EM mixed model analysis of data from informatively censored normal distributions. 95 Biomtrcs 51 425-436

Helton, J. C.; Johnson, J. D.; McKay, M. D.; Shiver, A. W.; Sprung, J. L.: Robustness of an uncertainty and sensitivity analysis of early exposure results with the MACCS reactor accident consequence model. 95 RelEngSS 48 129-148

Hembree, G. Barry: Recursive estimation of the weighting factor for EWMA control charts from autocorrelated data. 94 ASAProQlPr 39-43

Hemmi, Isao: Parameter estimation in Cox's regression model for competing risks with missing failure types. 95 JJapnStS 25 81-96

Henderson, Robin: Problems and prediction in survival-data analysis. 95 StatMed 14 161-184

Henderson, W.; Pearce, C. E. M.; Taylor, P. G.; van Dijk, N. M.: Insensitivity in discrete-time generalized semi-Markov processes allowing multiple events and probabilistic service scheduling. 95 AnlsApPr 5 78-96

Henderson, William G.; Moritz, Thomas; Goldman, Steven; Copeland, Jack; Sethi, Gulshan: Use of cumulative meta-analysis in the design, monitoring, and final analysis of a clinical trial: A case study. 95 ConClnTr 16 331-341

Hendi, M. I.: Abouammoh, A. M.; — —; Khalique, A.: Shock models with NBRU properties. 95 PakJStA 11 35-47

Hendrickson, John A., Jr: On the enumeration of rectangular (0, 1)-matrices. 95 JStCmpSm 51 291-313

Hendry, David: Dynamic econometrics. 95 OxfordUniv 904p

Hendry, David F.: On the interactions of unit roots and exogeneity. 95 EcomtrRv 14 383-419

Clements, Michael P.; — —: Forecasting in cointegrated systems. 95 JApEcmtx 10 127-146

Clements, Michael P.; — —: Reply to comment on "On the limitations of comparing mean square forecast errors". 95 JForcast 14 73-75

Hendy, Michael D.: Waddell, Peter J.; Penny, David; — —; Arnold, Greg: The sampling distributions and covariance matrix of phylogenetic spectra. 94 MolBioEv 11 630-642

Hengartner, Nicolas W.; Stark, Philip B.: Finite-sample confidence envelopes for shape-restricted densities. 95 AnlsStat 23 525-550

Hennemuth, R. C.: Taillie, C.; Patil, G. P.; — —: Modelling and analysis of recruitment distributions. 95 EnvEcoSt 2 315-329

Hennessy, Kevin: Lipshutz, Robert J.; Taverner, Fred; — —; Hartzell, George; Davis, Ron: DNA sequence confidence estimation. 94 Genomics 19 417-424

Hennig, Christian: Efficient high-breakdown-point estimators in robust regression: Which function to choose?. 95 StatDecn 13 221-241

Henrion, M.: Morgan, M.; — — (Auth); Ziegel, Eric R. (Rev): Review of "Uncertainty". 95 Technmcs 37 237-238

Henshaw, John M.: Booksh, Karl; — —; Burgess, Lloyd W.; Kowalski, Bruce R.: A second-order standard addition method with application to calibration of a kinetics-spectroscopic sensor for quantitation of trichloroethylene. 95 JChemtrc 9 263-282

Henson, Donald Earl; Ries, Lynn A.: The relative survival rate. 95 Cancer 76 1687-1688

Henze, Norbert; Klar, Bernhard: Bootstrap based goodness of fit tests for the generalized Poisson model. 95 CommStA 24 1875-1896

Heravi, Naser: Englund, Evan J.; — —: Phased sampling for soil remediation. 94 EnvEcoSt 1 247-263

Herkenrath, U.: Premium adjustment by generalized adaptive exponential smoothing. 94 InsrneME 15 203-217

Herman, Gabor T.: Levitan, Emanuel; Chan, Michael; — —: Image-modeling Gibbs priors. 95 CVGIPGr 57 117-130

Hernández Bastida, A.: Vázquez-Polo, F. J.; — —: Behaviour of the posterior error rate with partial prior information in auditing. 95 JAppStat 22 469-476

Hernandez, C. M.; Crossa, J.; Castillo, A.: The area under the function: An index for selecting desirable genotypes. 93 ThApGene 87 409-415

Crossa, J.; — —; Bretting, P.; Eberhart, S. A.; Taba, S.: Statistical genetic considerations for maintaining germ plasm collections. 93 ThApGene 86 673-678

Hernández-Lerma, O.; Lasserre, J.-B.: Discrete-time Markov control processes: Basic optimality criteria. 95 Springer:Brln:NY 216p

Gordienko, E.; — —: Average cost Markov control processes with weighted norms: Existence of canonical policies. 95 ApplMath 23 199-218

Gordienko, E.; — —: Average cost Markov control processes with weighted norms: Value iteration. 95 ApplMath 23 219-237

Hernández-Pajares, M.: Murtagh, F.; — —: The Kohonen self-organizing map method: An assessment. 95 JClass 12 165-190

Herndon, James E., II; Harrell, Frank E., Jr: The restricted cubic spline as baseline hazard in the proportional hazards model with step function time-dependent covariables. 95 StatMed 14 2119-2129

Hero, Alfred O.; Fessler, Jeffrey A.: Convergence in norm for alternating expectation-maximization (EM) type algorithms. 95 StSinica 5 41-54

Herouin, Eric: Lamouroux, Nicolas; Souchon, Yves; — —: Predicting velocity frequency distributions in stream reaches. 95 WaterReR 31 2367-2375

Herren, L. Tandy: DeWispelare, Aaron R.; — —; Clemen, Robert T.: The use of probability elicitation in the high-level nuclear waste regulation program. 95 IntJFor 11 5-24

Herrendörfer, G.: Rudolph, P. E.; — —: Optimal experimental design and accuracy of parameter estimation for nonlinear regression models used in long-term selection. 95 BiomtrcJ 37 183-190

Everitt, B. S. (Auth); — — (Rev): Review of "The analysis of contingency tables". 95 BiomtrcJ 37 104-104

Lange, N.; Ryan, L.; Billard, L.; Brillinger, D. R.; Conquest, L.; Greenhouse, J. (Ed); — — (Rev): Review of "Case studies in biometry". 95 BiomtrcJ 37 640-640

Herrero Gutiérrez, A.: Fernández, M. C. Ortiz; — —; Pastor, M. S. Sánchez; Sarabia, L. A.; Íñiguez Crespo, M.: The UNEQ, PLS and MLF neural network methods in the modelling and prediction of the colour of young red wines from the Denomination of Origin 'Rioja'. 95 Chemolab 28 273-285

Herrero-Gutiérrez, A.: Ortiz-Fernández, M. C.; — —: Regression by least median squares, a methodological contribution to titration analysis. 95 Chemolab 27 231-243

Herrin, Gary D.: Radson, Darrell; — —: Augmenting a factorial experiment when one factor is an uncontrollable random variable: A case study. 95 Technmcs 37 70-81

Scheaffer, Richard; McClave, James T. (Auth); — — (Rev): Review of "Probability and statistics for engineers (Fourth edition)". 95 AmerStat 49 238-239

Herring, Susan; Thoni, Janice: Statistics before college. 94 ASAProStEd 218-221

Herrmann, D. J.: White, A. A.; Pickle, L. W.; — —; Croner, C. M.; Wilson, B. F.: Map design preferences associated with professional discipline. 94 ASAProStGr 54-59

Herrmann, Eva; Wand, M. P.; Engel, Joachim; Gasser, Theo: A bandwidth selector for bivariate kernel regression. 95 JRSS-B 57 171-180

Herrnstein, R. J.; Murray, C. (Auth); Devlin, Bernie; Fienberg, Stephen E.; Resnick, Daniel P.; Roeder, Kathryn (Rev): Review of "The bell curve: Intelligence and class structure in American life". 95 JASA 90 1483-1488

Hershberger, Scott L.; Corneal, Sherry E.; Molenaar, Peter C. M.: Dynamic factor analysis: An application to emotional response patterns underlying daughter/father and stepdaughter/stepfather relationships. 95 StrEqMod 2 31-52

Hagenaars, Jacques A. (Auth); — — (Rev): Review of "Loglinear models with latent variables". 95 StrEqMod 2 172-174

Wickens, Thomas D. (Auth); — — (Rev): Review of "The geometry of multivariate statistics". 95 StrEqMod 2 379-382

Hertz-Picciotto, Irva: Epidemiology and quantitative risk assessment: A bridge from science to policy. 95 AJPubHlh 85 484-491

Hertzberg, Vicki Stover; Rice, Carol H.; Linz, Douglas H.; Pinney, Susan M.: Occupational epidemiology in the era of total quality: Challenges for exposure assessment. 94 ASAProEpid 139-142

Hertzberg, Vicky Stover: Comment on "Combining National Health Interview Survey datasets: Issues and approaches". 95 StatMed 14 693-693

Hervé, Jean Christophe: Pierrat, Jean Claude; Houllier, François; — —; Salas González, Raúl: Estimating the mean of the r largest values in a finite population: Application to forest inventory (French). 95 Biomtrcs 51 679-686

Herzberg, Agnes M.: Byar, David P.; Freedman, Laurence S.; — —: Identifying which sets of parameters are simultaneously estimable in an incomplete factorial design. 95 Ststcian 44 451-456

Herzog, Thomas N.: Comment on "The statistician who changed the world: W. Edwards Deming, 1900-1993" (V48 p179-187). 95 AmerStat 49 324-325

Hess, Kenneth R.: Graphical methods for assessing violations of the proportional hazards assumption in Cox regression. 95 StatMed 14 1707-1723

Hesse, Christian H.: Deconvolving a density from contaminated dependent observations. 95 AnInStMa 47 645-663

— —: Deconvolving a density from partially contaminated observations. 95 JMultiAn 55 246-260

Hesselager, O.: Kaas, R.; — —: Ordering claim size distributions and mixed Poisson probabilities. 95 InsrneME 17 193-201

Hesselager, Ole: Modelling of discretized claim numbers in loss reserving. 95 AstinB 25 119-135

— —: Order relations for some distributions. 95 InsrneME 16 129-134

Hestbeck, Jay B.: Bias in transition-specific survival and movement probabilities estimated using capture-recapture data. 95 JAppStat 22 737-750

— —: Population study and management of Atlantic Flyway Canada geese. 95 JAppStat 22 877-890

Hesterberg, Tim: Correction to "Saddlepoint quantiles and distribution curves, with bootstrap applications" (V9 p207-211). 95 CmpStQ 10 193-193

— —: Tail-specific linear approximations for efficient bootstrap simulations. 95 JCmpGrSt 4 113-133

— —: Weighted average importance sampling and defensive mixture distributions. 95 Technmcs 37 185-194

Hesterberg, Tim C.: Short, Thomas H.; Rossman, Allan J.; — —: Bayesian analyses of the Bradley-Terry model. 94 ASAProBays 117-122 (Disc: p134-135)

Hestir, Kevin; Williams, Stanley C.: Supports of doubly stochastic measures. 95 Bernoull 1 217-243

Hettmansperger, T. P.: McKean, J. W.; Sheather, S. J.; — —: The interpretability of LMS and LTS residual plots. 94 ASAProStEd 17-26

Hettmansperger, Thomas P.: The ranked-set sample sign test. 95 JNnparSt 4 263-270

Heubach, Silvia; Watkins, Joseph: A stochastic model for the movement of a white blood cell. 95 AdvAppPr 27 443-475

Heuch, Ivar: York, Jeremy; Madigan, David; — —; Lie, Rolv Terje: Birth defects registered by double sampling: A Bayesian approach incorporating covariates and model uncertainty. 95 ApplStat 44 227-242

Heuvel, H. M.: de Weijer, A. P.; Buydens, L.; Kateman, G.; — —: Spectral curve fitting of infrared spectra obtained from semi-crystalline polyester yarns. 95 Chemolab 28 149-164

Hewett, John E.; Petroski, Gregory F.; Li, Yiqun: Tables of critical values for two group interim analyses. 95 CommStB 24 825-852

Hewitt, J.: Pickles, A.; Neale, M.; Simonoff, E.; Rutter, M.; — —; Meyer, J.; Crouchley, R.; Silberg, J.; Eaves, L.: A simple method for censored age-of-onset data subject to recall bias: Mothers' reports of age of puberty in male twins. 94 BehavGen 24 457-468

Hewitt, J. K.: Heath, A. C.; Kessler, R. C.; Neale, M. C.; — —; Eaves, L. J.; Kendler, K. S.: Testing hypotheses about direction of causation using cross-sectional family data. 93 BehavGen 23 29-50

Hewitt, K.: Eaves, L. J.; Silberg, J. L.; — —; Rutter, M.; Meyer, J. M.; Neale, M. C.; Pickles, A.: Analyzing twin resemblance in multisymptom data: Genetic applications of a latent class model for symptoms of conduct disorder in juvenile boys. 93 BehavGen 23 5-20

Heyde, C. C.: — — (Ed): Branching processes. Proceedings of the First World Congress. 95 Springer:Brln:NY 180p

Chen, K.; — —: On asymptotic optimality of estimating functions. 95 JStPlInf 48 101-112

Révész, P. (Auth); — — (Rev): Review of "Random walks of infinitely many particles". 95 ShrtBkRv 15/3 47-47

Zacks, Shelemyahu (Auth); — — (Rev): Review of "Stochastic visibility in random fields". 95 ShrtBkRv 15/3 48-48

Heyde, Chris: A conversation with Joe Gani. 95 StatSci 10 214-230

Heyer, H.: — — (Ed): Probability measures on groups and related structures XI. 95 WrldSciPub:Sing 460p

Bloom, W. R.; — —: Harmonic analysis of probability measures on hypergroups. 94 DGruyter:NY:Grm 601p

Mosler, K.; Scarsini, M. (Ed); — — (Rev): Review of "Stochastic orders and decision under risk". 95 Metrika 42 139-139

Bloom, W. R.; — — (Auth); Barnett, C. (Rev): Review of "Harmonic analysis of probability measures on hypergroups". 95 ShrtBkRv 15/3 49-49

Heyman, Daniel P.: A decomposition theorem for infinite stochastic matrices. 95 JAppProb 32 893-901

Hibbert, D. B.: Mulholland, M.; — —; Haddad, P. R.; Parslov, P.: A comparison of classification in artificial intelligence, induction versus a self-organising neural networks. 95 Chemolab 30 117-128

Mulholland, M.; — —; Haddad, P. R.; Sammut, C.: Application of the C4.5 classifier to building an expert system for ion chromatography. 95 Chemolab 27 95-104

Hibbert, D. Brynn: Dunlop, Peter J.; Bignell, C. M.; Jackson, J. F.; — —: Chemometric analysis of gas chromatographic data of oils from Eucalyptus species. 95 Chemolab 30 59-67

Hiby, L.: Rothery, P.; Martin, A. R.; — —; Garside, S.: Estimating foetal growth rate from cross-sectional samples of field data with application to pilot-whales. 95 ApplStat 44 163-172

Hickernell, F. J.: An, Hongzhi; — —; Zhu, Lixing: Uniformly consistent estimation in random regression model (Chinese). 95 KToChina 39 1829-1832

Hickernell, Fred J.: Chen, Jiading; — —: A kind of asymptotically optimal sequential test under complex supposition. 95 SciSPkng 24 673-683

Hida, Takeyuki; Kuo, Hui-Hsuing; Potthoff, Jürgen; Streit, Ludwig (Auth); Øksendal, Bernt (Rev): Review of "White noise. An infinite dimensional calculus". 95 BLndnMaS 27 189-191

Hidalgo, Javier: A nonparametric conditional moment test for structural stability. 95 EconTher 11 671-698

Højsgaard, Søren; Thiesson, Bo: BIFROST - Block recursive models induced from relevant knowledge, observations, and statistical techniques. 95 CmpStDA 19 155-175

Holden, Ken: Vector autoregression modelling and forecasting. 95 JForcast 14 159-166

Wallis, Kenneth F. (Ed); — — (Rev): Review of "Macroeconomic modelling Vols I and II". 95 IntJFor 11 334-335

Holder, Roger L.: Nevill, Alan M.; — —: Comment on "Spurious correlation and the fallacy of the ratio standard revisited" (V156 p379-392). 95 JRSS-A 158 619-621

Holevo, A. S.: Barchielli, A.; — —: Constructing quantum measurement processes via classical stochastic calculus. 95 StocProc 58 293-317

Holford, Theodore R.; Stack, Catharine: Study design for epidemiologic studies with measurement error. 95 StMedRes 4 339-358

Concato, John; Peduzzi, Peter; — —: Importance of events per independent variable in proportional hazards analysis I. Background, goals, and general strategy. 95 JClEpid 48 1495-1501

Kiuchi, Amy S.; Hartigan, J. A.; — —; Rubinstein, Pablo; Stevens, Cladd E.: Change points in the series of T4 counts prior to AIDS. 95 Biomtrcs 51 236-248

Holiday, David B.: Near optimal weights in nonparametric regression under some common restrictions. 95 StPrLet 22 33-42

Holland, Burt: Rom, Dror M.; — —: A new closed multiple testing procedure for hierarchical families of hypotheses. 95 JStPlInf 46 265-275

Holland, P. W.; Wainer, H. (Ed); Mellenbergh, Gideon J. (Rev): Review of "Differential item functioning". 95 Psymtrka 60 621-623

Holland, Paul W.: Comment on "Defining, detecting, and correcting test and item bias". 94 ASAProSoSt 27-29

Hollander, Myles; Marshall, Albert W.: A conversation with Frank Proschan. 95 StatSci 10 118-133

— —; Peña, Edsel A.: Dynamic reliability models with conditional proportional hazards. 95 LiDaAnal 1 377-401

Holle, Rolf: Lorenzen, T. J.; Anderson, V. L. (Auth); — — (Rev): Review of "Design of experiments. A no-name approach". 95 CmpStDA 19 101-102

Hollingsworth, Caroline: Beginning statistics with cuisenaire rods. 95 TeachgSt 17/2 38-39

Holly, Alberto: A random linear functional approach to efficiency bounds. 95 JEconmtx 65 235-261

Holm, Sture: Confidence sets of fixed size with predetermined confidence level. 95 CommStA 24 1521-1536

Holmberg, Jan: Arjas, Elja; — —: Marked point process framework for living probabilistic safety assessment and risk follow-up. 95 RelEngSS 49 59-73

Holmes, Peter: Brunelli, Lina; Cicchitelli, Guiseppe (Ed); — — (Rev): Review of "Proceedings of the First Scientific Meeting of the International Association for Statistical Education". 95 AmerStat 49 276-276

Holmes, Talmage M.: Roberson, Paula K.; Shema, Sarah J.; Mundfrom, Daniel J.; — —: Evaluating change and preference questions using paired Likert scale data. 94 ASAProSoSt 189-191

Holmgren, Eric B.: The P-P plot as a method for comparing treatment effects. 95 JASA 90 360-365

Holmquist, Björn: Simulation of multimodal circular distributions of wrapped Cauchy type and of von Mises type. 95 CommStB 24 607-615

Holst, J.: Sadegh, P.; — —; Madsen, H.; Melgaard, H.: Experiment design for grey box identification. 95 IJAdCSgP 9 491-507

Holst, L.: Barbour, A. D.; — —; Janson, S. (Auth); Papaioannou, T. (Rev): Review of "Poisson approximation". 95 Biomtrcs 51 382-383

Holst, Lars: Cowan, Richard; Chiu, S. N.; — —: A limit theorem for the replication time of a DNA molecule. 95 JAppProb 32 296-303

Barbour, A. D.; — —; Janson, Svante (Auth); Zubkov, A. M. (Rev): Review of "Poisson approximation". 94 ThProbAp 39 720-721

Holt, Albert; Kaufman, Steven; Scheuren, Fritz; Smith, Wray: Intersurvey consistency in school surveys. 94 ASASrvyRsr 1105-1110 (Disc: p1128-1129)

Holt, Bradley R.: Wise, Barry M.; — —; Gallagher, Neal B.; Lee, Samuel: A comparison of neural networks, non-linear biased regression and a genetic algorithm for dynamic model identification. 95 Chemolab 30 81-89

Holt, Elizabeth A.: Moulton, Lawrence H.; — —; Job, Jayakaran S.; Halsey, Neal A.: Percentile regression analysis of correlated antibody responses. 95 StatMed 14 777-787

Holton, Jack: Rezayat, Fahimeh; — —: Neural network controller and quality improvement: An application. 94 ASAProQlPr 33-38

Homble, Patrick; McCormick, William P.: Weak limit results for the extremes of a class of shot noise processes. 95 JAppProb 32 707-726

Homma, T.: Saltelli, A.; Andres, T. H.; — —: Sensitivity analysis of model output. Performance of the iterated fractional factorial design method. 95 CmpStDA 20 387-407

Hong, Changkon; Kim, Choongrak: Existence and uniqueness of the smoothest density with prescribed moments. 95 JKoreaSt 24 233-242

Hong, Chong Sun; Choi, Hyun Jip; Lee, Chong Cheol: Uniformly minimum variance unbiased estimation for distributions with support depending on two parameters. 95 JKoreaSt 24 45-64

Hong, Dug Hun: A central limit theorem with random indices for martingale difference sequences. 95 JKoreaSt 24 219-224

— —; Hwang, Seok Yoon: Chow-Robbins "fair" games problem for pairwise independent random variables with multidimensional indices. 95 BIMaASnc 23 7-11

— —; Hwang, Seok Yoon: Correlation of intuitionistic fuzzy sets in probability spaces. 95 FuzzySet 75 77-81

— —; Kim, Hye Kyung; Kim, Ju Young: A weak convergence theorem for mixingale arrays. 95 JKoreaSt 24 273-280

— —; Oh, Kwang Sik: On the weak law of large numbers for arrays. 95 StPrLet 22 55-57

— —; Oh, Kwang Sik; Park, Hee Joo: On a stopping rule for the random walks with time stationary random distribution function. 95 JKoreaSt 24 293-301

Hong, Ki-Hak: Lee, Gi-Sung; — —: Unrelated question model with quantitative attribute by stratified double sampling (Korean). 95 KorApS 8/1 27-38

Lee, Hwa-Young; — —: A study on the efficiency comparison of the black-box method with the randomized response technique (Korean). 95 KorApS 8/2 27-41

Hong, Shengyan; Cheng, Ping: Convergence rates for parameter estimation in semi-parametric regression models (Chinese). 95 ChJApPSt 10 62-71

Hoogendoorn, A. W.: Sikkel, D.; — —: Models for monthly penetrations with incomplete panel data. 95 StNeerla 49 378-391

Hoogerbrugge, Ronald: Gort, Steven M.; — —: A user-friendly spreadsheet program for calibration using weighted regression. 95 Chemolab 28 193-199

Hook, Ernest B.: Zelterman, Daniel; — —; Regal, Ronald R.: Bayesian estimates of a population size. 94 ASAProBays 29-34

Hope, Peter B.: Garthwaite, Paul H.; Yu, Keming; — —: Bayesian analysis of a multiple-recapture model. 95 CommStA 24 2229-2247

Hopke, Philip K.: How to start surfing the Internet. 95 Chemolab 30 1-9

— —; Li, Chong Le; Ciszek, William; Landsberger, Sheldon: The use of bootstrapping to estimate conditional probability fields for source locations of airborne pollutants. 95 Chemolab 30 69-79

Hopkins, Kenneth D.; Chappell, David: Quick power estimates for comparing proportions. 94 EdPsMeas 54 903-912

Hoppe, F. M.: — — (Ed); Read, C. B. (Rev): Review of "Multiple comparisons, selection, and applications in biometry: A festschrift in honor of Charles W. Dunnett". 95 Biomtrcs 51 380-380

Hoppe, Fred M.: — — (Ed); Horn, Manfred (Rev): Review of "Multiple comparisons, selection and application in biometry". 95 CmpStQ 10 103-103

— — (Ed); Littler, Ray (Rev): Review of "Multiple comparisons, selection, and applications in biometry - A festschrift in honor of Charles W. Dunnett". 95 AstrlJSt 37 251-252

Hoppensteadt, F.; Salehi, H.; Skorokhod, A. V.: Randomly perturbed Volterra integral equations and some applications. 95 Stochatx 54 89-125

Hops, Hyman: Duncan, Terry E.; Duncan, Susan C.; — —; Stoolmiller, Mike: An analysis of the relationship between parent and adolescent marijuana use via generalized estimating equation methodology. 95 MultBehR 30 317-339

Hörchner, Uwe; Kalivas, John H.: Simulated-annealing-based optimization algorithms: Fundamentals and wavelength selection applications. 95 JChemtrc 9 283-308

Hordijk, A.; Loeve, J. A.: Undiscounted Markov decision chains with partial information: An algorithm for computing a locally optimal periodic policy. 94 ZeitOpRs 40 163-181

Horgan, G. W.: Glasbey, C. A.; — —: Image analysis for the biological sciences. 95 Wiley:NY:UK 218p

Glasbey, C. A.; — —; Gibson, G. J.; Hitchcock, D.: Fish shape analysis using landmarks. 95 BiomtrcJ 37 481-495

Hörmann, Wolfgang: A rejection technique for sampling from T-concave distributions. 95 ACMMaSfw 21 182-193

Horn, Manfred: Hoppe, Fred M. (Ed); — — (Rev): Review of "Multiple comparisons, selection and application in biometry". 95 CmpStQ 10 103-103

Hornberger, G. M.: Chen, T. H.; — —; Jakeman, A. J.; Swank, W. T.: The performance of different loss models in the simulation of streamflow. 95 EnvrMtrc 6 479-484

Hornberger, John C.; Brown, Byron Wm., Jr; Halpern, Jerry: Designing a cost-effective clinical trial. 95 StatMed 14 2249-2259

Hornik, Kurt: Chu, Chia-Shang J.; — —; Kuan, Chung-Ming: MOSUM tests for parameter constancy. 95 Biomtrka 82 603-617

Chu, Chia-Shang James; — —; Kuan, Chung-Ming: The moving-estimates test for parameter stability. 95 EconTher 11 699-720

Kuan, Chung-Ming; — —: The generalized fluctuation test: A unifying view. 95 EcomtrRv 14 135-161

Horowitz, Joel L.: Hall, Peter; — —; Jing, Bing-Yi: On blocking rules for the bootstrap with dependent data. 95 Biomtrka 82 561-574

Markatou, Marianthi; — —: Robust scale estimation in the error-components model using the empirical characteristic function. 95 CandJSt 23 369-381

Markatou, Marianthi; — —; Lenth, Russell V.: Robust scale estimation based on the empirical characteristic function. 95 StPrLet 25 185-192

Horsthemke, Benedikt; Rüttermann, Markus: On the maximum entropy principle for a class of stochastic processes. 95 StocProc 56 117-132

Horstmann, Vibeke: On confidence regions induced by the Wilcoxon rank sum test. 95 StPrLet 23 333-338

Horváth, Lajos: Kac's representation from an asymptotic viewpoint. 95 JStPlInf 46 27-45

— —: Detecting changes in linear regressions. 95 MaOpfStS 26 189-208

— —; Khoshnevisan, Davar: Weight functions and pathwise local central limit theorems. 95 StocProc 59 105-123

— —; Shao, Qi-Man: Limit theorems for the union-intersection test. 95 JStPlInf 44 133-148

Csörgő, Miklós; — —: On the distance between smoothed empirical and quantile processes. 95 AnlsStat 23 113-131

Horvath, Michael T. K.; Watson, Mark W.: Testing for cointegration when some of the cointegrating vectors are prespecified. 95 EconTher 11 984-1014

Horvatić-Baldasar, K.: Matulić-Bedenić, I.; — —; Kramer, E.: Construction of new symmetric designs with parameters (66, 26, 10). 95 JComDsgn 3 405-410

Horvitz, Dan; Koshland, Daniel; Rubin, Donald; Gollin, Albert; Sawyer, Tom; Tanur, Judith M.: Pseudo-opinion polls: SLOP or useful data?. 95 Chance 8/2 16-25

Horvitz, E.: Dagum, P.; Galper, A.; — —; Seiver, A.: Uncertain reasoning and forecasting. 95 IntJFor 11 73-87

Hoshino, Naoto; Miyazaki, Haruo; Seki, Yoichi: On the level probabilities for useful partially ordered alternatives in the analysis of variance. 95 CommStA 24 2059-2071

Hoshmand, A. Reza: Experimental research design and analysis. A practical approach for agricultural and natural sciences. 94 CRCPress:FL 416p

Hosking, J. R. M.; Wallis, J. R.: A comparison of unbiased and plotting-position estimators of L moments. 95 WaterReR 31 2019-2025

Höskuldsson, Agnar: A combined theory for PCA and PLS. 95 JChemtrc 9 91-123

Hosmane, Balakrishna; Shu, Vincent; Morris, David: Finite sample properties of van Elteren test. 94 ASAProBiop 430-434

Hosmer, David W.; Lemeshow, Stanley: Confidence interval estimates of an index of quality performance based on logistic regression models. 95 StatMed 14 2161-2172

Jovanovic, Borko D.; — —; Buonaccorsi, John P.: Equivalence of several methods for efficient best subsets selection in generalized linear models. 95 CmpStDA 20 59-64

Jovanovic, Borko D.; — —; Murray, James: A method for fitting Poisson regression model with extra zeros. 94 ASAProEpid 97-100

Hossain, Anwar: Howlader, Hatem A.; — —: On Bayesian estimation and prediction from Rayleigh based on type II censored data. 95 CommStA 24 2249-2259

Hossain, Syed A.: Uddin, Nizam; — —: E-optimal nested row-column designs with unequal replications of treatments. 95 SankhyaB 57 112-121

Hössjer, Ola: Exact computation of the least trimmed squares estimate in simple linear regression. 95 CmpStDA 19 265-282

— —; Croux, Christophe: Generalizing univariate signed rank statistics for testing and estimating a multivariate location parameter. 95 JNnparSt 4 293-308

— —; Mielniczuk, Jan: Delta method for long-range dependent observations. 95 JNnparSt 5 75-82

— —; Rousseeuw, Peter J.; Ruts, Ida: The repeated median intercept estimator: Influence function and asymptotic normality. 95 JMultiAn 52 45-72

— —; Ruppert, David: Asymptotics for the transformation kernel density estimator. 95 AnlsStat 23 1198-1222

Rousseeuw, Peter J.; Croux, Christophe; — —: Sensitivity functions and numerical analysis of the repeated median slope. 95 CmpStQ 10 71-90

Høst, Gudmund; Omre, Henning; Switzer, Paul: Spatial interpolation errors for monitoring data. 95 JASA 90 853-861

Hostetter, Susan C.: Linking individuals in a capital gains panel for tax policy analysis. 94 ASASrvyRsr 461-466

Houba, Victor: Feinberg, Max; Bugner, Emmanuel; Theiller, Gédéon; — —; Kadijk, Fetze: Expression of the reference value for proficiency tests. 95 JChemtrc 9 197-209

Houdré, Christian: Reconstruction of band limited processes from irregular samples. 95 AnlsProb 23 674-696

— —: On the almost sure convergence of series of stationary and related nonstationary variables. 95 AnlsProb 23 1204-1218

— —: Some applications of covariance identities and inequalities to functions of multivariate normal variables. 95 JASA 90 965-968

— —; Kagan, Abram: Variance inequalities for functions of Gaussian variables. 95 JTherPrb 8 23-30

— —; Kedem, Benjamin: A note on autocovariance estimation in the presence of discrete spectra. 95 StPrLet 24 1-8

— —; Pérez-Abreu, Víctor: Covariance identities and inequalities for functionals on Wiener and Poisson spaces. 95 AnlsProb 23 400-419

Hougaard, Philip: Frailty models for survival data. 95 LiDaAnal 1 255-273

— —: Nonlinear regression and curved exponential families. Improvement of the approximation to the asymptotic distribution. 95 Metrika 42 191-202

Houllier, François: Pierrat, Jean Claude; — —; Hervé, Jean Christophe; Salas González, Raúl: Estimating the mean of the r largest values in a finite population: Application to forest inventory (French). 95 Biomtrcs 51 679-686

Houlston, Richard S.: Shields, Denis C.; Marlow, Angela J.; — —; Eccles, Diana M.; Morton, Newton E.: Prediction of genetic risks from segregation analyses of morbid risks. 94 HumHrdty 44 52-55

Houshmand, Ali A.; Panganamamula, Srinivasarao: An analytical approximation and a neural network model for optimal sample size in vendor selection. 95 JStCmpSm 53 65-78

Houston, Mark: Woodruff, David; — —: Growth rate reliability in longitudinal measurement. 94 EdPsMeas 54 897-902

Houwing-Duistermaat, Jeanine J.; Derkx, Bert H. F.; Rosendaal, Frits R.; van Houwelingen, Hans C.: Testing familial aggregation. 95 Biomtrcs 51 1292-1301

Hove, Helge; Liang, Yi-Zeng; Kvalheim, Olav M.: Trimmed object projections: A nonparametric robust latent-structure decomposition method. 95 Chemolab 27 33-40

Howard, C. V.: Cruz-Orive, L. M.; — —: Estimation of individual feature surface area with the vertical spatial grid. 95 JMicro 178 146-151

Howel, D.: Cooper, H.; Hedges, L. V. (Ed); — — (Rev): Review of "The handbook of research synthesis". 95 Biomtrcs 51 1585-1586

Howel, Denise: Kleinbaum, D. G. (Auth); — — (Rev): Review of "Logistic regression: A self learning text". 95 Ststcian 44 410-411

Howell, D. C.: — — (Auth); Grimm, Laurence G. (Rev): Review of "Statistical methods for psychology (Third edition)". 95 JAppStat 22 194-195

Howlader, Hatem A.; Hossain, Anwar: On Bayesian estimation and prediction from Rayleigh based on type II censored data. 95 CommStA 24 2249-2259

Howlader, Hatema A.: Balakrishnan, N. (Ed); — — (Rev): Review of "Handbook of the logistic distribution". 95 JAppStat 22 189-191

Howroyd, J. D.: Kaandorp, J. A. (Auth); — —; Bayer, M. M. (Rev): Review of "Fractal modelling: Growth and form in biology". 95 Biomtrcs 51 794-795

Hox, Joop J.: Arbuckle, James; BMDP Statistical Software Inc; Scientific Software International (Auth); — — (Rev): Review of "AMOS, EQS, and LISREL for Windows: A comparative review". 95 StrEqMod 2 79-91

Hoyert, Donna L.; Singh, Gopal K.; Rosenberg, Harry M.: Sources of data on socio-economic differential mortality in the United States. 95 JOffStat 11 233-260

Høyland, A.: Rausand, M. (Auth); Sherwin, David (Rev): Review of "System reliability theory - Models and statistical methods". 95 Ststcian 44 543-544

Hoyle, Rick H.: — — (Ed): Structural equation modeling. Concepts, issues, and applications. 95 SagePubl:CA:Lnd 336p

Hromi, John D.: Hare, Lynne B.; Hoerl, Roger W.; — —; Snee, Ronald D.: The role of statistical thinking in management. 95 QulPr 28/2 53-60

Hsiao, Cheng; Mountain, Dean C.; Illman, Kathleen Ho: A Bayesian integration of end-use metering and conditional-demand analysis. 95 JBES 13 315-326

Hsieh, F. Y.: A cautionary note on the analysis of extreme data with Cox regression. 95 AmerStat 49 226-228

Hsieh, Fushing: The empirical process approach for semiparametric two-sample models with heterogeneous treatment effect. 95 JRSS-B 57 735-748

— —: A regression approach for estimating the center of symmetry. 95 StPrLet 22 157-160

— —; Turnbull, Bruce W.: A note on the local asymptotically minimax rate for estimating a crossing point in a diagnostic marker problem. 95 StPrLet 24 181-185

Robins, James M.; — —; Newey, Whitney: Semiparametric efficient estimation of a conditional density with missing or mismeasured covariates. 95 JRSS-B 57 409-424

Hsing, Tailen: A note on the asymptotic independence of the sum and maximum of strongly mixing stationary random variables. 95 AnlsProb 23 938-947

— —: Limit theorems for stable processes with application to spectral density estimation. 95 StocProc 57 39-71

— —: On the asymptotic independence of the sum and rare values of weakly dependent stationary random variables. 95 StocProc 60 49-63

Davis, Richard A.; — —: Point process and partial sum convergence for weakly dependent random variables with infinite variance. 95 AnlsProb 23 879-917

Hsing, Yu; Chang, Hui S.: Household saving behavior in Taiwan: Tests of functional forms and cointegration. 94 ASAProBuEc 123-128

Hsiung, Chao A.: Chang, I-Shou; — —: Asymptotic consistency of the maximum likelihood estimate in positron emission tomography and applications. 94 AnlsStat 22 1871-1883

Hsu, Guang-Hui; Yuan, Xue-Ming: First passage times and their algorithms for Markov processes. 95 CommStSM 11 195-209

Hsu, Jin-Wen Yang: Sampling behaviour in estimating predictive validity in the context of selection and latent variable modelling: A Monte Carlo study. 95 BJMaStPs 48 75-97

Hsu, John S. J.: Generalized Laplacian approximations in Bayesian inference. 95 CandJSt 23 399-410

Hsu, Kuo-lin; Gupta, Hoshin Vijai; Sorooshian, Soroosh: Artificial neural network modeling of the rainfall-runoff process. 95 WaterReR 31 2517-2530

Hsu, Louis M.: Comment on "A binomial test of model fit" (V1 p268-273). 95 StrEqMod 2 152-154

Hsu, Margaretha M.; Wilkens, Henry T.: Advertising-related aspirin consumption among under 45 and over 45 consumers. 94 ASAProBiop 276-278

Hsu, Nai-Feng: Chen, Sy-Mien; — —: The asymptotic distribution of the process capability index C_{pmk}. 95 CommStA 24 1279-1291

Hsu, Y.-S.: An optimal grouping for estimating a nondecreasing function with categorical data. 95 SochwJMa 21 253-261

Hsuan, Francis: Hwang, Dar-Shong; — —: A delta procedure for the analysis of crossover and related designs in drug trials. 94 ASAProBiop 307-312

Wang, Wenping; — —; Chow, Shein-Chung: Patient compliance and the fluctuation of the serum drug concentration. 94 ASAProBiop 94-99

Hu, Albert; Sachs, Emanuel: Run by run control: A Bayesian approach to the control of semiconductor fabrication processes. 94 ASAProPEng 245-254

Hu, Baiding: A note on calculating the Gini index. 95 MaCmpSim 39 353-358

Hu, Feifang; Zidek, James V.: A bootstrap based on the estimating equations of the linear model. 95 Biomtrka 82 263-275

Hu, H.: Chen, R. W.; — —: A note on the distribution of occurrences of a pattern in a string. 95 SochwJMa 21 471-473

Hu, Inchi; Rukhin, Andrew L.: A lower bound for error probability in change-point estimation. 95 StSinica 5 319-331

Hu, Jian-Qiang: Fu, Michael C.; — —: On unbounded hazard rates for smoothed perturbation analysis. 95 JAppProb 32 659-667

Fu, Michael C.; — —: Sensitivity analysis for Monte Carlo simulation of option pricing. 95 PrEngInf 9 417-446

Hu, Joan: Lawless, Jerry; — —; Cao, Jin: Methods for the estimation of failure distributions and rates from automobile warranty data. 95 LiDaAnal 1 227-240

Hu, Qing; Yang, Jingyu; Zhang, Qian; Liu, Ke; Shen, Xiaojun: An automatic seal imprint verification approach. 95 PattrnRc 28 1251-1266

Hu, Shuhe: Sums of double arrays of LP-mixingale sequence and its application in linear models (Chinese). 95 AMaAPknC 17 248-254

— —: Strong consistency of a semiparametric regression estimate under fixed design. 95 AMaPeknE 37 393-401

Hu, Stephen N.: Construction of simple and observable Venn diagrams for any finite number of sets. 94 ASAProStGr 60-65

Hu, Taizhong; Joe, Harry: Monotonicity of positive dependence with time for stationary reversible Markov chains. 95 PrEngInf 9 227-237

Hu, Tien Chung: Hall, Peter; — —; Marron, J. S.: Improved variable window kernel estimates of probability densities. 95 AnlsStat 23 1-10

Hu, Xiaomi: A note on the symmetry of normal mean hypotheses and its implications. 95 StPrLet 25 15-20

Hu, Y.; Peng, S.: Solution of forward-backward stochastic differential equations. 95 ProbTher 103 273-283

— —; Pierre-Loti-Viaud, D.; Shi, Z.: Laws of the iterated logarithm for iterated Wiener processes. 95 JTherPrb 8 303-319

— —; Shi, Z.: Sojourns and future infima of planar Brownian motion. 95 ProbTher 103 329-348

— —; Shi, Z.: The Csörgő-Révész modulus of non-differentiability of iterated Brownian motion. 95 StocProc 58 267-279

Hu, Yijun: A note on the relationship between the rate functions for stationary dependent random sequences in τ-topology and the relative entropy. 95 StocProc 56 351-355

Hu, Yiming; Woyczynski, W. A.: Limit behavior of quadratic forms of moving averages and statistical solutions of the Burgers' equation. 95 JMultiAn 52 15-44

Hua, Fan Ai: On the l^p convergence of martingales associated with coverage (French). 95 JAppProb 32 668-678

Huang, Chuen-Chuen Joyce; Dayton, C. Mitchell: Detecting patterns of bivariate mean vectors using model-selection criteria. 95 BJMaStPs 48 129-147

Huang, Dawei: Davis, Richard A.; — —; Yao, Yi-Ching: Testing for a change in the parameter values and order of an autoregressive model. 95 AnlsStat 23 282-304

Huang, Deng-Yuan; Lee, Ren Fen: Selecting the largest capability index from several quality control processes. 95 JStPlInf 46 335-346

Huang, Hertz: Dorinski, Suzanne M.; — —: Use of administrative data in SIPP longitudinal estimation. 94 ASASrvyRsr 656-661 (Disc: p680-682)

Huang, J.; Wellner, J. A.: Asymptotic normality of the NPMLE of linear functionals for interval censored data, case 1. 95 StNeerla 49 153-163

Huang, J. S.; Lin, G. D.: Characterization of distributions using equilibrium transformations. 95 SankhyaA 57 179-185

Huang, Jianyi; Chao, Li: Assessment methods in international student classrooms. 94 ASAProQlPr 291-293

Huang, M. L.; Brill, P. H.: Estimation in systems of M/G/c/c queues. 94 ASAProPEng 151-156

Huang, Mong-Na; Chang, Hsiu-Fen: Marginally restricted constrained optimal designs. 95 SankhyaB 57 128-141

Huang, Mong-Na Lo; Chang, Fu-Chuen; Wong, Weng Kee: D-optimal designs for polynomial regression without an intercept. 95 StSinica 5 441-458

Huang, Su-Yun: Empirical Bayes testing procedures in some nonexponential families using asymmetric Linex loss function. 95 JStPlInf 46 293-309

Huang, Xiang-Ning; Li, Bai-Bing: A general structure for the unbiased estimate of the parameter's estimable function in linear regression. 95 StocAnAp 13 83-90

Huang, Xin: de Haan, Laurens; — —: Large quantile estimation in a multivariate setting. 95 JMultiAn 53 247-263

Huang, Ying; Veinott, Arthur F., Jr: Markov branching decision chains with interest-rate-dependent rewards. 95 PrEngInf 9 99-121

— —; Wang, Mei-Cheng: Estimating the occurrence rate for prevalent survival data in competing risks models. 95 JASA 90 1406-1415

Huang, Zhimin; Li, Susan: The role of cardinal utilities in multiple objective programming. 94 AJMMgSci 14 301-325

Huang, Zi-Hong: Xie, Yu-Long; Liang, Yi-Zeng; Chen, Zhang-Guang; — —; Yu, Ru-Qin: A nonlinearity tracking analysis algorithm for treatment of non-linearity in multivariate calibration. 95 Chemolab 27 21-32

Hubbard, Michael L.: O'Reilly, James M.; — —; Lessler, Judith T.; Biemer, Paul P.; Turner, Charles F.: Audio and video computer assisted self-interviewing: Preliminary tests of new technologies for data collection. 94 JOffStat 10 197-214

Hubbell, Katie A.: England, Amy M.; — —; Judkins, David R.; Ryaboy, Svetlana: Imputation of medical cost and payment data. 94 ASASrvyRsr 406-411

Huber, George P.; Van de Ven, Andrew H. (Ed): Longitudinal field research methods. Studying processes of organizational change. 95 SagePubl:CA:Lnd 400p

Huber, Peter J.: Robustness: Where are we now?. 95 Student 1 75-86

Huber, Thomas M.; Nagel, Matthias: Look first, then model: Exploring the data using the interactive graphics of PC-ISP. 95 CmpStQ 10 29-36

Hubert, Lawrence; Arabie, Phipps: Iterative projection strategies for the least squares fitting of tree structures to proximity data. 95 BJMaStPs 48 281-317

— —; Arabie, Phipps: The approximation of two-mode proximity matrices by sums of order-constrained matrices. 95 Psymtrka 60 573-605

Hubert, Lawrence J.: Mirkin, Boris; Arabie, Phipps; — —: Additive two-mode clustering: The error-variance approach revisited. 95 JClass 12 243-263

Huberty, C. J.: — — (Auth); Cox, Michael A. A. (Rev): Review of "Applied discriminant analysis". 95 JAppStat 22 433-434

— — (Auth); Hand, D. J. (Rev): Review of "Applied discriminant analysis". 95 Biomtrcs 51 1187-1188

Huberty, Carl J.: — — (Auth); Heckler, Charles E. (Rev): Review of "Applied discriminant analysis". 95 Technmcs 37 349-350

— — (Auth); Thompson, Bruce (Rev): Review of "Applied discriminant analysis". 95 EdPsMeas 55 340-350

Huda, S.; Khan, I. H.; Mukerjee, Rahul: On optimal designs with restricted circular string property. 95 CmpStDA 19 75-83

Hudelson, Brian D.: Clayton, Murray K.; — —: Confidence intervals for autocorrelations based on cyclic samples. 95 JASA 90 753-757

Hudson, Brian: Using IT to investigate world statistics. 95 TeachgSt 17/2 57-60

Hudson, W. N.; Mason, J. D.; Veeh, J. A.: Operator stable probability measures: An overview. 94 ThProbAp 39 264-276

Huebner, M.; Rozovskii, B. L.: On asymptotic properties of maximum likelihood estimators for parabolic stochastic PDE's. 95 ProbTher 103 143-163

Huet, S.: Härdle, W.; — —; Jolivet, E.: Better bootstrap confidence intervals for regression curve estimation. 95 MaOpfStS 26 287-306

Huggins, R. M.: A law of large numbers for the bifurcating autoregressive process. 95 CommStSM 11 273-278

Huggins, Richard: Yip, Paul; — —: Some aspects of inference for chain binomial models. 95 StocAnAp 13 355-367

Huggins, Vicki J.; Fischer, Donald P.: The redesign of the Survey of Income and Program Participation. 94 ASASrvyRsr 668-673 (Disc: p680-682)

Hughes, A. O.: Clayton, David; Hills, Michael (Auth); — — (Rev): Review of "Statistical models in epidemiology". 95 StatMed 14 104-105

Hughes, Arthur L.; Brodsky, Marc D.: Variance estimation of drug abuse episodes using the bootstrap. 94 ASASrvyRsr 212-217

Hughes, Barry D.: Random walks and random environments. Volume 1: Random walks. 95 OxfordUniv 654p

— — (Auth); Révész, P. (Rev): Review of "Random walks and random environments. Volume 1: Random walks". 95 ShrtBkRv 15/3 47-47

Hughes, Bruce: Ramos, Magdalena; — —; McKay, Betty: Results from a feasibility test of an alternative automated data capture methodology during the 1992 Census of Agriculture. 94 ASASrvyRsr 769-774

Hughes, G.; Madden, L. V.: Using the beta-binomial distribution to describe aggregated patterns of disease incidence. 93 Phytopgy 83 759-762

Hughes, H. R.: Gregory, John; — —: A generalized approximation theory for quadratic forms: Application to randomized spline type Sturm-Liouville problems. 95 JTherPrb 8 703-715

Gregory, John; — —: A generalized approximation theory for quadratic forms II: Application to randomized spline type focal/conjugate point problems. 95 JTherPrb 8 963-971

Hughes, Michael D.; Thompson, Simon G.; Pocock, Stuart J.: Optimal sequential screening guidelines for quantitative risk factors measured with error. 95 JASA 90 19-26

Lin, Hung-Mo; — —: Use of historical marker data for assessing treatment effects in phase I/II trials when subject selection is determined by baseline marker level. 95 Biomtrcs 51 1053-1063

Hughes-Oliver, Jacqueline M.: Curvature adjusted optimal nonlinear designs for correlated responses. 94 ASAProBuEc 111-116

— —; Lu, Jye-Chyi; Davis, Joseph C.; Gyurcsik, Ronald S.: Robust improvement of a semi-conductor fabrication process via spatial modeling of the mean and covariance surfaces. 94 ASAProPEng 235-244

Hwang, J. T. Gene: Fieller's problems and resampling techniques. 95 StSinica 5 161–171

— —; Liu, Hung-kung: Application of empirical linear prediction to quality assurance in industrial manufacturing. 94 ASAProPEng 175–180

— —; Liu, Hung-kung: Asymptotic theory for prediction through empirical linear models. 94 ASAProPEng 207–216

Brown, Lawrence D.; Casella, George; — —: Optimal confidence sets, bioequivalence, and the Limaçon of Pascal. 95 JASA 90 880–889

Hwang, Jing-Shiang: Chen, James J.; — —; Tsong, Yi: Estimation of the shelf-life of drugs with mixed effects models. 95 JBiophSt 5 131–140

Hwang, Lie-Ju; Alves, Wayne: Analysis of neurologic outcomes using ordinal regression. 94 ASAProBiop 352–354

Hwang, Ruey-Ching: Asymptotic properties of locally weighted regression. 95 JNnparSt 5 303–310

Hwang, Seok Yoon: Hong, Dug Hun; — —: Chow-Robbins "fair" games problem for pairwise independent random variables with multidimensional indices. 95 BIMaASnc 23 7–11

Hong, Dug Hun; — —: Correlation of intuitionistic fuzzy sets in probability spaces. 95 FuzzySet 75 77–81

Hwang, Sun Young: A study on a regression model with nonlinear time series errors (Korean). 95 KorApS 8/2 187–200

Hwang, W.-D.; Chao, Anne: Quantifying the effects of unequal catchabilities on Jolly-Seber estimators via sample coverage. 95 Biomtrcs 51 128–141

Hwang, Yong Keun: Bai, Do Sun; Choi, In Su; — —: Field data analyses for products with multiple-modes of failure (Korean). 95 KorApS 8/1 89–104

Hwarng, H. Brian: Proper and effective training of a pattern recognizer for cyclic data. 95 IIETrn 27 746–756

Hyánek, J.; Moliński, K.; Szwaczkowski, T.: Estimation of parameters in linear model with the presence of selection. 95 ListyBio 32 51–60

Hyatt, Raymond R., Jr; Rhodes, William: The price and purity of cocaine: The relationship to emergency room visits and death, and to drug use among arrestees. 95 StatMed 14 655–668

Hylleberg, Svend: Tests for seasonal unit roots - General to specific or specific to general?. 95 JEconmtx 69 5–25

Hyndman, N.: Glass, J. C.; McKillop, D. G.; — —: Efficiency in the provision of university teaching and research: An empirical analysis of UK universities. 95 JApEcmtx 10 61–72

Hyndman, Rob J.: Highest-density forecast regions for non-linear and non-normal time series models. 95 JForcast 14 431–441

Hyne, V.: Kearsey, M. J.; — —: QTL analysis: A simple 'marker-regression' approach. 94 ThApGene 89 698–702

Hyodo, Yoshifumi: Kuwada, Masahide; — —: Analysis of variance of balanced fractional $r^m \times s^n$ factorial designs of resolution V. 95 JJapnStS 25 129–141

Yamamoto, Sumiyasu; Fujii, Yoshio; — —; Yumiba, Hiromu: Classification of orthogonal 24-run 2^5 factorial designs derivable from saturated two-symbol orthogonal arrays of strength 2, size 24 and index 6. 95 TRUMTk 31/1 39–53

Iachan, Ronald; Ringwalt, Chris L.; Greene, Jody M.: Substance abuse among runaway and homeless youth. 94 ASASrvyRsr 1083–1087

Iachan, Ronaldo; Dennis, Michael L.; Bray, Robert M.: Estimating the prevalence of substance abuse among people who are homeless. 94 ASASrvyRsr 1094–1099

— —; Kemp, Suzanne, S.: Visitor sample surveys. 95 SurvyM 21 89–96

Iachine, Ivan A.: Yashin, Anatoli I.; Vaupel, James W.; — —: Correlated individual frailty: An advantageous approach to survival analysis of bivariate data. 95 MaPopStd 5 145–159

Iacobini, Alberto: Panizzon, G. (Auth); — — (Rev): Review of "Bilateral plans for inspection and process control by variables". 94 Metron 52/1 216–217

Iacobucci, Dawn: Jackson, Sally; Brashers, Dale E. (Auth); — — (Rev): Review of "Random factors in ANOVA". 95 JMarRsr 32 238–239

Iancu, C.: Puterman, M. L. (Auth); — — (Rev): Review of "Markov decision processes: Discrete stochastic dynamic programming". 95 JAppStat 22 547–547

Iannacchione, Vincent; Wildfire, Judith; Lux, Linda; Hawes, Catherine: Covering the population of unlicensed board and care homes: Problems and solutions. 94 ASASrvyRsr 270–275

Ibarra, J. I.: A new approach for the determination of horizontal wind direction fluctuations. 95 JApMetrl 34 1942–1949

Ibrahim, Joseph G.: Laud, Purushottam W.; — —: Predictive model selection. 95 JRSS-B 57 247–262

Ibrahim, Mohammad: Katti, S. K.; — —: Towards a better understanding of Bayesian methods. 94 ASAProBays 142–147

Ichikawa, Masanori; Konishi, Sadanori: Application of the bootstrap methods in factor analysis. 95 Psymtrka 60 77–93

Idier, Jérôme; Giovannelli, Jean-François; Querleux, Bernard: Bayesian time-varying AR spectral estimation for ultrasound attenuation measurement in biological tissues. 94 ASAProBays 256–261

Brette, Stéphane; — —; Mohammad-Djafari, Ali: Scale invariant Markov models for linear inverse problems. 94 ASAProBays 266–270

Iemma, A. F.: Palm, R.; — —: Some alternatives to classical regression in the case of collinearity (French). 95 RStApp 43/2 5–33

Igaki, Nobuko; Sumita, Ushio; Kowada, Masashi: Analysis of Markov renewal shock models. 95 JAppProb 32 821–831

Iglehart, Donald L.: Glynn, Peter W.; — —: A martingale approach to regenerative simulation. 95 PrEngInf 9 123–131

Iglewicz, Boris: Comparison of estimators for sigma in quality control charts. 94 ASAProQlPr 28–32

Marazzi, A. (Auth); — —; Heiberger, Richard M.; Moore, Dirk; Tan, Yun (Rev): Review of "Algorithms, routines, and S functions, for robust statistics". 95 Technmcs 37 351–353

Ihara, Masamori; Kano, Yutaka: Identifiability of full, marginal, and conditional factor analysis models. 95 StPrLet 23 343–350

Iiyama, Yuriko; Nishimura, Kazuo; Sibuya, Masaaki: Power of record-breaking test (Japanese). 95 JpJAppSt 24 13–26

Ikeda, Michael M.: Ernst, Lawrence R.; — —: A reduced-size transportation algorithm for maximizing the overlap between surveys. 95 SurvyM 21 147–157

Ikeda, N.; Elworthy, K. D. (Auth); Simakani, J. E. (Rev): Review of "Asymptotic problems in probability theory: Wiener functionals and fractionals". 95 Ststcian 44 289–289

Elworthy, K. D.; — — (Ed); Glendinning, R. H. (Rev): Review of "Asymptotic problems in probability theory: Stochastic models and diffusions on fractals". 95 Ststcian 44 139–140

Elworthy, K. D.; — — (Ed); Wakolbinger, A. (Rev): Review of "Asymptotic problems in probability theory: Stochastic models and diffusions on fractals". 95 Metrika 42 69–70

Ilallawa, Abdel M.; Azzam, Abdul-Mordy H.: A new method for generating the design matrix of a linear regression model. 95 EgpStJ 39 106–119

Illman, Kathleen Ho: Hsiao, Cheng; Mountain, Dean C.; — —: A Bayesian integration of end-use metering and conditional-demand analysis. 95 JBES 13 315–326

Iltis, Michael: Sharp asymptotics of large deviations in R^d. 95 JTherPrb 8 501–522

Im, Mi-Jeong; Kwon, Woo-Joo; Lee, Joo-Ho: Two-stage Latin hypercube sampling and its application (Korean). 95 KorApS 8/2 99–108

Iman, Ronald L.: New paradigms for the statistics profession. 95 JASA 90 1–6

— — (Auth); Edgeman, Rick L. (Rev): Review of "A data-based approach to statistics". 95 JQualTek 27 88–89

— — (Auth); Halvorsen, Katherine (Rev): Review of "A data-based approach to statistics". 95 AmerStat 49 321–321

Imbens, Guido; van der Klaauw, Wilbert: Evaluating the cost of conscription in The Netherlands. 95 JBES 13/2 207–215

Lancaster, Tony; — —: Optimal stock/flow panels. 95 JEconmtx 66 325–348

Imbens, Guido W.: Angrist, Joshua D.; — —: Two-stage least squares estimation of average causal effects in models with variable treatment intensity. 95 JASA 90 431–442

Imbos, Tj.: Tan, E. S.; — —; Does, R. J. M. M.; Theunissen, M.: An optimal, unbiased classification rule for mastery testing based on longitudinal data. 95 EdPsMeas 55 595–612

Imkeller, Peter; Perez-Abreu, Victor; Vives, Josep: Chaos expansions of double intersection local time of Brownian motion in R^d and renormalization. 95 StocProc 56 1–34

Arnold, Ludwig; — —: Furstenberg-Khasminskii formulas for Lyapunov exponents via anticipative calculus. 95 Stochatx 54 127–168

Imlahi, A.: Functional laws of the iterated logarithm for records. 95 JStPlInf 45 215–324

Imrey, Peter B.: Reply to comments on "Statistical values, quality, and certification". 95 AmerStat 49 117–118

— —: Reply to comment on "Statistical values, quality, and certification". 95 AmerStat 49 241–241

Inagaki, Atsushi: Entromin and entromax: Criteria for factor rotation based on entropy (Japanese). 94 JpJBeh 21/2 10–20

Infante, A. M.: The difficulties of estimation of dispersion parameters in linear models - An illustration. 95 StHefte 36 183–189

Ingersoll, Richard M.: Teacher shortages and teacher quality. 94 ASAProSoSt 201–206 (Disc: p213–215)

Ingman, Dov: Bluvband, Zigmund; Grabov, Pavel; — —: New approach to process control. 95 ASQCTekCon 369–377

Ingster, Yu. I.: Minimax testing of hypotheses on the distribution density for ellipsoids in l_p. 94 ThProbAp 39 417–436

Íñiguez Crespo, M.: Fernández, M. C. Ortiz; Herrero Gutiérrez, A.; Pastor, M. S. Sánchez; Sarabia, L. A.; — —: The UNEQ, PLS and MLF neural network methods in the modelling and prediction of the colour of young red wines from the Denomination of Origin 'Rioja'. 95 Chemolab 28 273–285

Inman, Henry F.: Reply to comment on "Karl Pearson and R. A. Fisher on statistical tests: A 1935 exchange from Nature". 95 AmerStat 49 400–401

Inoue, Hiroshi: Randomly weighted sums for exchangeable fuzzy random variables. 95 FuzzySet 69 347–354

— —; Taylor, Robert L.: A SLLN for arrays of rowwise exchangeable fuzzy random sets. 95 StocAnAp 13 461–470

Insua, David Ríos; Ruggeri, Fabrizio; Vidakovic, Brani: Some results on posterior regret Γ-minimax estimation. 95 StatDecn 13 315–331

International Statistical Institute: — — (Auth); Tanis, Elliot A. (Rev): Review of "Proceedings of the Fourth International Conference on Teaching Statistics". 95 AmerStat 49 219-220

International Working Group for Disease Monitoring and Forecasting: Capture-recapture and multiple-record systems estimation 1: History and theoretical development. 95 AJEpidm 141 1047-1058

— —: Capture-recapture and multiple-record systems estimation II: Applications in human diseases. 95 AJEpidm 141 1059-1068

Intrator, Orna; Kooperberg, Charles: Trees and splines in survival analysis. 95 StMedRes 4 237-261

Inuiguchi, Masahiro; Sakawa, Masatoshi: A possibilistic linear program is equivalent to a stochastic linear program in a special case. 95 FuzzySet 76 309-317

Ioannatos, Petros E.: Censored regression estimation under unobserved heterogeneity: A stochastic parameter approach. 95 JBES 13 327-335

Ioannides, Chris: Heerink, Nico (Auth); — — (Rev): Review of "Population growth, income distribution and economic development". 95 IntJFor 11 333-334

Ioannidis, C.; Laws, J.; Matthews, K.; Morgan, B.: Business cycle analysis and forecasting with a structural vector autoregression model for Wales. 95 JForcast 14 251-265

Ioannidis, Christos: Silver, Mick; — —: European inflation and its variability. 95 Ststcian 44 309-322

Ioannidis, Evangelos E.: On the behavior of a Capon-type spectral density estimator. 94 AnlsStat 22 2089-2114

Ioffe, Dmitry: Exact large deviation bounds up to T_c for the Ising model in two dimensions. 95 ProbTher 102 313-330

Ionin, Yury J.; Shrikhande, Mohan S.: Five-designs with three intersection numbers. 95 JCombThA 69 36-50

Ipinyomi, R. A.; Adeleke, B. L.: Correction to "Nested row-column designs" (V38 p197-202). 95 Ststcian 44 547-547

Irons, John S.: Cuthbertson, Keith; Hall, Stephen G.; Taylor, Mark P. (Auth); Ericsson, Neil R.; — — (Rev): Review of "Applied econometric techniques". 95 EcomtrRv 14 121-132

Irtel, Hans: An extension of the concept of specific objectivity. 95 Psymtrka 60 115-118

Irvine, Diane: Kanetkar, Vinay; Evans, Martin G.; Everell, Shirley Anne; — —; Millman, Zeeva: The effect of scale changes on meta-analysis of multiplicative and main effects models. 95 EdPsMeas 55 206-224

Irvine, John M.: A correction for 'end effects' in subjective rating experiments. 94 ASAProSoSt 184-188

— —; Barbour, Garth: Texture analysis of imagery data for waste site characterization. 94 ASAProEnvr 122-126

Isaac, R.: — — (Auth); Coleman, R. (Rev): Review of "The pleasures of probability". 95 ShrtBkRv 15/3 42-42

Isaac, R. E.: The pleasures of probability. 95 Springer:Brln:NY 241p

Isakov, A.: Heller, I.; — —; Blinder-Weiner, S.; Topilsky, M.: Bayesian classification of vasculitis: A simulation study. 95 MInfoMed 34 259-265

Isaksson, T.: Hildrum, K.; — —; Naes, T.; Tandberg, A. (Ed); Ziegel, Eric R. (Rev): Review of "Near infra-red spectroscopy: Bridging the gap between data analysis and NIR applications". 95 Technmcs 37 126-127

Isaksson, Tomas: Helland, Inge S.; Næs, Tormod; — —: Related versions of the multiplicative scatter correction method for preprocessing spectroscopic data. 95 Chemolab 29 233-241

Iscoe, I.; McDonald, D.: Asymptotics of exit times for Markov jump processes. II: Applications to Jackson networks. 94 AnlsProb 22 2168-2182

Isham, Valerie: Stochastic models of host-macroparasite interaction. 95 AnlsApPr 5 720-740

Ishibuchi, H.: Tanaka, H.; — —; Yoshikawa, S.: Exponential possibility regression analysis. 95 FuzzySet 69 305-318

Ishigami, Hideyuki; Fukuda, Toshio; Shibata, Takanori; Arai, Fumihito: Structure optimization of fuzzy neural network by genetic algorithm. 95 FuzzySet 71 257-264

Ishiguro, Makio: Kato, Hiroko; Naniwa, Sadao; — —: A multivariate stochastic model with non-stationary trend component. 95 AplStMDA 11 77-95

Islam, Ali S.: Paul, Sudhir R.; — —: Analysis of proportions in the presence of over-/under-dispersion. 95 Biomtrcs 51 1400-1410

Islam, M. N.: A stochastic model for surname evolution. 95 BiomtrcJ 37 119-126

Islam, Towhidul: Meade, Nigel; — —: Forecasting with growth curves: An empirical comparison. 95 IntJFor 11 199-215

Meade, Nigel; — —: Prediction intervals for growth curve forecasts. 95 JForcast 14 413-430

Ismail, Moshira A.: On the loss of information through the use of hybrid censoring. 95 EgpStJ 39 65-88

Isogai, Eiichi; Uno, Chikara: On the sequential point estimation of the mean of a gamma distribution. 95 StPrLet 22 287-293

Isogawa, Yoshiko: Slope-range and generalized least squares: Estimation of the structural relationship with replicated observations, unequal means and variances. 95 JJapnStS 25 57-70

— —: A note on a unified test for multivariate linear relationships. 95 StPrLet 24 147-151

Ito, Hiroshi: Yamazaki, Genji; — —: Optimal order for two servers in tandem. 95 AnInStMa 47 31-48

Ivaldi, Marc: Aubin, Christophe; Fougére, Denis; Husson, Emmanuel; — —: Real-time pricing of electricity for residential customers: Econometric analysis of an experiment. 95 JApEcmtx 10 171-191S

Ivanoff, B. G.; Weber, N. C.: Functional limit theorems for row and column exchangeable arrays. 95 JMultiAn 55 133-148

Ivanoff, B. Gail; Merzbach, Ely: Stopping and set-indexed local martingales. 95 StocProc 57 83-98

Ivanoff, Gail B.; Merzbach, Ely; Schiopu-Kratina, Ioana: Lattices of random sets and progressivity. 95 StPrLet 22 97-102

Ivchenko, G. I.; Mirakhmedov, Sh. A.: Large deviations and intermediate efficiency of decomposable statistics in a multinomial scheme. 95 MaMethSt 4 294-311

Ives, Anthony R.: Measuring resilience in stochastic systems. 95 EcolgMon 65 217-233

— —: Predicting the response of populations to environmental change. 95 Ecology 76 926-941

Iwig, William: Perry, Charles; Burt, Jameson; — —: Redrawing the 1993 Farm Cost and Returns Survey list sample to reduce overlap with three other 1993 surveys and the 1992 FCRS. 94 ASASrvyRsr 632-637

Iyengar, R. N.; Basak, P.: Regionalization of Indian monsoon rainfall and long-term variability signals. 94 JClimlgy 14 1095-1114

Iyengar, Raghavan J.; Yoo, Jaesin S.; Zafra, Pablo M. R.: An application of some patterned matrices to the principal component analysis of linked variables. 95 PakJStA 11 23-33

Iyengar, S. Sitharama; Deng, Weian: An efficient edge detection algorithm using relaxation labeling technique. 95 PattrnRc 28 519-536

Iyer, Hari K.: Ahlbrandt, Robert; — —; Vecchia, Dominic F.: Optimum design of serial measurement trees. 95 JStPlInf 48 379-390

Iyer, Hariharan K.: Graybill, Franklin A.; — — (Auth); Barrett, Bruce (Rev): Review of "Regression analysis: Concepts and applications". 95 Technmcs 37 229-229

Graybill, Franklin A.; — — (Auth); Swain, James J. (Rev): Review of "Regression analysis: Concepts and applications". 95 JQualTek 27 179-180

Graybill, Franklin A.; — — (Auth); Winner, Larry (Rev): Review of "Regression analysis - Concepts and applications". 95 AmerStat 49 114-114

Iznaga, N.; Núñez, G.; Solozabal, J.; Morales, A.; Artaza, E.; Rubio, R.; Cardenas, E.: A personal computer-based system for parallel line analysis of bioassay data. 95 CmpBiomd 47 167-175

Izquierdo-Ridorsa, A.: Tauler, R.; — —; Gargallo, R.; Casassas, E.: Application of a new multivariate curve resolution procedure to the simultaneous analysis of several spectroscopic titrations of the copper(II)-polyinosinic acid system. 95 Chemolab 27 163-174

Izsák, János; Papp, László: Application of the quadratic entropy indices for diversity studies of drosophilid assemblages. 95 EnvEcoSt 2 213-224

Jaccard, James; Wan, Choi K.: Measurement error in the analysis of interaction effects between continuous predictors using multiple regression: Multiple indicator and structural equation approaches. 95 PsychBu 117 348-357

Jack, Susan S.: Botman, Steven L.; — —: Combining National Health Interview Survey datasets: Issues and approaches. 95 StatMed 14 669-677 (Disc: p679-693)

Jacka, S. D.; Roberts, G. O.: Weak convergence of conditioned processes on a countable state space. 95 JAppProb 32 902-916

Jackson, Betty J.; Frazier, Richard L.; King, Karen E.; Schwanz, Dennis J.: Improving the coverage of private elementary-secondary schools. 94 ASASrvyRsr 833-838

Jackson, D. A.: Chen, Y.; — —: Robust estimation of mean and variance in fisheries. 95 TrAFish 124 401-412

Jackson, David: Tennis in Lilliput: A fable on sports and psychology. 95 Chance 8/3 37-40

Jackson, Donald A.: Reply to comment on "Stopping rules in principal components analysis: A comparison of heuristical and statistical approaches". 95 Ecology 76 644-645

Jackson, J. Edward: Basilevsky, Alexander (Auth); — — (Rev): Review of "Statistical factor analysis and related methods: Theory and applications". 95 Technmcs 37 227-228

— — (Auth); Millsap, Roger E. (Rev): Review of "A user's guide to principal components". 95 JEdStat 20 105-107

Jackson, J. F.: Dunlop, Peter J.; Bignell, C. M.; — —; Hibbert, D. Brynn: Chemometric analysis of gas chromatographic data of oils from Eucalyptus species. 95 Chemolab 30 59-67

Jackson, Sally; Brashers, Dale E. (Auth); Iacobucci, Dawn (Rev): Review of "Random factors in ANOVA". 95 JMarRsr 32 238-239

Jacob, Pierre; Oliveira, Paulo Eduardo: A general approach to nonparametric histogram estimation. 95 MaOpfStS 27 73-92

— —; Suquet, Charles: Estimating the edge of a Poisson process by orthogonal series. 95 JStPlInf 46 215-234

Jacobs, K.: Sinai, Yakov G. (Auth); — — (Rev): Review of "Probability theory". 95 Metrika 42 77-78

Jacobs, Patricia A.: Gaver, Donald P.; — —: Asymptotic properties of stochastic greedy bin-packing. 95 CommStSM 11 333-348

Jacobsen, Chanoch; Bronson, Richard: Computer simulations and empirical testing of sociological theory. 95 SocMethR 23 479-506

Jacobsen, Martin; Keiding, Niels: Coarsening at random in general sample spaces and random censoring in continuous time. 95 AnlsStat 23 774-786

Jacobson, Jonathan E.; Czajka, John L.: Using panel data to simulate the impact of policy changes on women's labor supply and welfare participation. 94 ASAProGovt 45-50 (Disc: p51-52)

Jacobson, Shawn: Evaluation of alternative rent estimators for the consumer price index. 94 ASAProGovt 134-139

Jacobson, Sheldon H.: Analyzing the M/M/1 queue in frequency domain experiments. 95 AplMaCmp 69 185-194

Jacobson, Tor: On the determination of lag order in vector autoregressions of cointegrated systems. 95 CmpStQ 10 177-192

Carling, Kenneth; — —: Modeling unemployment duration in a dependent competing risks framework: Identification and estimation. 95 LiDaAnal 1 111-122

Jacqmin, Hélène: Commenges, Daniel; — —; Letenneur, Luc; Van Duijn, Cornelia M.: Score test for familial aggregation in probands studies: Application to Alzheimer's disease. 95 Biomtrcs 51 542-551

Jacqmin-Gadda, Hélène; Commenges, Daniel: Tests of homogeneity for generalized linear models. 95 JASA 90 1237-1246

Jacquez, Geoffrey M.: The map comparison problem: Tests for the overlap of geographic boundaries. 95 StatMed 14 2343-2361

— —: Disease cluster tests for uncertain health events. 94 ASAProEnvr 110-115

Waller, Lance A.; — —: Disease models implicit in statistical tests of disease clustering. 95 Epidmlgy 6 584-590

Jacquot, S.; Royer, G.: Ergodicity of stochastic plates. 95 ProbTher 102 19-44

Jacroux, Mike; Majumdar, Dibyen; Shah, Kirti R.: Efficient block designs in the presence of trends. 95 StSinica 5 605-615

Jaeger, David A.: Bound, John; — —; Baker, Regina M.: Problems with instrumental variables estimation when the correlation between the instruments and the endogenous explanatory variable is weak. 95 JASA 90 443-450

Jagers, Peter: Branching processes as population dynamics. 95 Bernoull 1 191-200

Jaggi, Seema; Agarwal, K. N.: Augmented partial diallel design for estimating combining ability of parents. 95 BiomtrcJ 37 879-887

Jaggia, Sanjiv; Thosar, Satish: Contested tender offers: An estimate of the hazard function. 95 JBES 13 113-119

Jaheen, Zeinhum F.: Al-Hussaini, Essam K.; — —: Bayesian prediction bounds for the Burr type XII failure model. 95 CommStA 24 1829-1842

Jain, J. P.: Muralidharan, K.; — —: Gene flow technique for stage-structured populations. 95 BiomtrcJ 37 229-234

Jain, Kamlesh: Mittag, H.-J.; Rinne, H. (Auth); — — (Rev): Review of "Statistical methods of quality assurance". 95 Technmcs 37 234-234

Jain, Kanchan; Nanda, Asok K.: On multivariate weighted distributions. 95 CommStA 24 2517-2539

Jain, R. K.: Jain, Sudha; — —: Problem of estimation in the analysis of reliability data. 94 ASAProPEng 136-139

Jain, Sudha; Jain, R. K.: Problem of estimation in the analysis of reliability data. 94 ASAProPEng 136-139

Jakeman, A. J.: Chen, T. H.; Hornberger, G. M.; — —; Swank, W. T.: The performance of different loss models in the simulation of streamflow. 95 EnvrMtrc 6 479-484

Mulquiney, J. E.; Norton, J. P.; — —; Taylor, J. A.: Random walks in the Kalman filter: Implications for greenhouse gas flux deductions. 95 EnvrMtrc 6 473-478

Sefton, C. E. M.; Whitehead, P. G.; Eatherall, A.; Littlewood, I. G.; — —: Dynamic response characteristics of the Plynlimon catchments and preliminary analysis of relationships to physical descriptors. 95 EnvrMtrc 6 465-472

— —; Beck, M. B.; McAleer, M. J. (Ed); Keesman, Karel (Rev): Review of "Modelling change in environmental systems". 95 EnvrMtrc 6 425-427

Jakeman, Anthony J.: Bai, Jun; — —; McAleer, Michael: Empirical models for evaluating errors in fitting extremes of a probability distribution. 95 MaCmpSim 39 1-7

Jakeman, Eric; Phayre, Sean; Renshaw, Eric: The evolution and measurement of a population of pairs. 95 JAppProb 32 1048-1062

Jalali, Assad; Pemberton, John: Mixture models for time series. 95 JAppProb 32 123-138

Cui, Lirong; Hawkes, Alan G.; — —: The increasing failure rate property of consecutive-k-out-of-n systems. 95 PrEngInf 9 217-225

Jalava, Juho-Pertti: Taavitsainen, Veli-Matti; — —: Soft and harder multivariate modelling in developing the properties of titanium dioxide pigments. 95 Chemolab 29 307-319

James, Ian R.: Comment on "Applications of multiple imputation to the analysis of censored regression data" (V47 p1297-1309). 95 Biomtrcs 51 358-362

— —; Hogan, Stephen J.; Mallal, Simon A.: Analysis of doubly censored data and the HIV incubation period distribution in Western Australia. 95 AstrlJSt 37 299-312

Segal, Mark R.; — —; French, Martyn A. H.; Mallal, Simon A.: Statistical issues in the evaluation of markers of HIV progression. 95 IntStRvw 63 179-197

James, Penny L.: Coleman, Joan L.; — —: The reliability of the locality pay estimates. 94 ASASrvyRsr 1031-1035

Jamshidian, Mortaza: Graphical data analysis in linear regression. 92 ProIrStC1 51-122

Jan, H. W.: Wang, P. C.; — —: Designing two-level factorial experiments using orthogonal arrays when the run order is important. 95 Ststcian 44 379-388

Jana, Prasanta Kumar; Roy, Dilip: Estimation of reliability under stress-strength model in a bivariate exponential set-up. 94 CalcutSt 44 175-181

Janacek, G.: Diggle, P. (Auth); — — (Rev): Review of "Time series: A biostatistical introduction". 95 JTimSrAn 16 355-356

— —; Swift, L. (Auth); Davies, Neville (Rev): Review of "Time series - Forecasting, simulation, applications". 95 JRSS-A 158 193-195

Janacek, Gareth; Swift, Louise (Auth); Findley, David F. (Rev): Review of "Time series: Forecasting, simulation, applications". 95 SIAMRvw 37 253-257

Janardan, K. G.: Environmental risk assessment model for low dose extrapolation. 95 CommStA 24 2621-2633

— —; Vasudeva, R.; Mohan, N. R.: On some special birth processes in biology. 95 CommStA 24 1033-1038

Jang, Dae Heung: A survey of misuses of statistical graphics in Korean mass media and a study for statistical and institutional solutions to misuses (Korean). 95 KorApS 8/2 1-26

Janicki, A.: Computer simulation of a nonlinear model for electrical circuits with α-stable noise. 95 ApplMath 23 95-105

— —; Weron, A. (Auth); Lee, P. M. (Rev): Review of "Simulation and chaotic behaviour of α-stable stochastic processes". 95 JRSS-A 158 351-352

Janicki, Aleksander; Weron, Aleksander: Simulation and chaotic behavior of α-stable stochastic processes. 95 Dekker:NY 376p

— —; Weron, Aleksander: Computer simulation of attractors in stochastic models with α-stable noise. 95 MaCmpSim 39 9-19

Janis, Stuart J.: Statistical process control in chemical batch processes (things are seldom what they seem). 94 ASAProPEng 308-313

Janko, Zvonimir: On symmetric designs with parameters (176, 50, 14). 95 JCombThA 72 310-314

Janowski, Steven A.: Exact solution of the totally asymmetric simple exclusion process: Shock profiles. 94 RBrzlPrS 8 85-91

Jansen, Margo G. H.: van Duijn, Marijtje A. J.; — —: Modeling repeated count data: Some extensions of the Rasch Poisson counts model. 95 JEdStat 20 241-258

Jansen, R. A. R. G.; Oud, J. H. L.: Longitudinal LISREL model estimation from incomplete panel data using the EM algorithm and the Kalman smoother. 95 StNeerla 49 362-377

Jansen, R. C.: A general mixture model for mapping quantitative trait loci by using molecular markers. 92 ThApGene 85 252-260

Jansen, Ritsert C.: Controlling the type I and type II errors in mapping quantitative trait loci. 94 Genetcs 138 871-881

— —; Stam, Piet: High resolution of quantitative traits into multiple loci via interval mapping. 94 Genetcs 136 1447-1455

Jansen, U.: Patil, G. P.; Rao, C. R. (Ed); — — (Rev): Review of "Multivariate environmental statistics. North-Holland Series in Statistics and Probability Volume 6". 95 BiomtrcJ 37 255-256

Patil, G. P.; Rao, C. R. (Ed); — — (Rev): Review of "Environmental statistics: Handbook of statistics Volume 12". 95 BiomtrcJ 37 511-512

Janson, S.: Barbour, A. D.; Holst, L.; — — (Auth); Papaioannou, T. (Rev): Review of "Poisson approximation". 95 Biomtrcs 51 382-383

Janson, Svante: Barbour, A. D.; Holst, Lars; — — (Auth); Zubkov, A. M. (Rev): Review of "Poisson approximation". 94 ThProbAp 39 720-721

Janssen, A.; Mason, D. M. (Auth); Jurečková, Jana (Rev): Review of "Non-standard rank tests". 95 SIAMRvw 37 261-262

Janssen, Arnold: Principal component decomposition of non-parametric tests. 95 ProbTher 101 193-209

— —: Sums of independent triangular arrays and extreme order statistics. 94 AnlsProb 22 1766-1793

Janssen, Christian: Cain, Michael; — —: Real estate price prediction under asymmetric loss. 95 AnInStMa 47 401-414

Janssen, J.; Skiadas, C. H.: Dynamic modelling of life table data. 95 AplStMDA 11 35-49

Janssen, Jacques; Skiadas, Christos H.; Zopounidis, Constantin (Ed): Advances in stochastic modelling and data analysis. 95 Kluwer:MA:Hllnd 440p

Janssen, Jeannette C. M.: On even and odd latin squares. 95 JCombThA 69 173-181

Janssen, P.; Marron, J. S.; Veraverbeke, N.; Sarle, W.: Scale measures for bandwidth selection. 95 JNnparSt 5 359-380

Janssen, Paul: Aerts, Marc; — —: Weighted bootstrapping for U-quantiles. 95 StPrLet 22 317-323

Koroljuk, V. S.; Borovskich, Yu. V. (Auth); — — (Rev): Review of "Theory of U-statistics". 95 JASA 90 804-805

Jansson, G.; Danell, Ö.: Needs and benefits of empirical power transformations for production and quality traits in forest tree breeding. 93 ThApGene 87 487-497

Janž, M.; Lachout, P.: A central limit theorem for stationary random fields. 95 MaMethSt 4 463-471

Jarjoura, David; Speight, Joan D.: Statistical consulting as alchemy. 94 ASAProStEd 165-169

Jarman, Jennifer: Blackburn, Robert M.; Siltanen, Janet; — —: The measurement of occupational gender segregation: Current problems and a new approach. 95 JRSS-A 158 319–331

Jaro, Matthew A.: Probabilistic linkage of large public health data files. 95 StatMed 14 491–498 (Disc: p687-689)

Jarque, C. M.; McKenzie, C. R.: Testing for multivariate normality in simultaneous equations models. 95 MaCmpSim 39 323–328

Jarrett, Jeffrey: Review of quality control and improvement statistical software system. 95 CmpStDA 20 333–339

Jarvis, Robert L.: Sheaffer, Susan E.; — —: Bias in Canada goose population size estimates from sighting data. 95 JWildlMg 59 464–473

Jaworski, Wojciech: On the asymptotic and invariant σ-algebras of random walks on locally compact groups. 95 ProbTher 101 147–171

Jay, Gina M.; Belli, Robert F.; Lepkowski, James M.: Quality of last doctor visit reports: A comparison of medical record and survey data. 94 ASASrvyRsr 362–367 (Disc: p379-381)

— —; Belli, Robert F.; Lepkowski, James M.: Use of last physician visit to characterize health care utilization. 94 ASASrvyRsr 368–372 (Disc: p379-381)

Jayakumar, K.: Pillai, R. N.; — —: Discrete Mittag-Leffler distributions. 95 StPrLet 23 271–274

Jayasuriya, Bodhini: Testing for polynomial regression using nonparametric regression techniques. 94 ASAProBuEc 241–246

Jayatunga, R.: Wade, A. M.; Ades, A. E.; Salt, A. T.; — —; Sonksen, P. M.: Age-related standards for ordinal data: Modelling the changes in visual acuity from 2 to 9 years of age. 95 StatMed 14 257–266

Jazairi, Nuri: Fienberg, Stephen E.; — —: Stanley Warner's contributions to statistically balanced information technology. 95 SurvyM 21 7–12

Jeanpierre, M.: Hazard and probabilities of unknown genotypes. 92 AnHuGene 56 325–330

Jeffers, J. N. R.: The statistician and the computer. 95 JAppStat 22 227–234

Jeffrey, A.: Gradshteyn, I. S.; Ryzhik, I. M. (Auth); — — (Ed); Bello, Abdul L. (Rev): Review of "Tables of integrals, series, and products (Fifth edition)". 95 Ststcian 44 287–288

Jeganathan, P.: Some aspects of asymptotic theory with applications to time series models. 95 EconTher 11 818–887

Jelenkowska, T. H.: A Bayesian analysis of the multivariate mixed linear model. 95 CommStA 24 3183–3196

Jenkins, Cleo R.; Wetzel, Angela-Jo: The 1991-92 teacher follow-up survey reinterview and extensive reconciliation. 94 ASASrvyRsr 821–826

Jenkins, G.: Box, G.; — —; Reinsel, G. (Auth); Ziegel, Eric R. (Rev): Review of "Time series analysis: Forecasting and control". 95 Technmcs 37 238–239

Jenkyn, J. F.: Dyke, G. V.; Lane, P. W.; — —: Sensitivity (stability) analysis of multiple variety trials, with special reference to data expressed as proportions or percentages. 95 ExpAgric 31 75–87

Jennison, Christopher; Sheehan, Nuala: Theoretical and empirical properties of the genetic algorithm as a numerical optimizer. 95 JCmpGrSt 4 296–318

Eales, John D.; — —: Optimal two-sided group sequential tests. 95 SqtlAnly 14 273–286

Hurn, Merrilee; — —: A study of simulated annealing and a revised cascade algorithm for image reconstruction. 95 StatComp 5 175–190

Jennrich, Robert I.: Chen, Jian-Shen; — —: Diagnostics for linearization confidence intervals in nonlinear regression. 95 JASA 90 1068–1074

Chen, Jian-Shen; — —: Transformations for improving linearization confidence intervals in nonlinear regression. 95 JASA 90 1271–1276

Chen, Jian-Shen; — —: The profile Z plot and confidence intervals in maximum likelihood analysis. 94 ASAProStCp 124–129

Jensen, D. R.: Efficiency indices for second-order designs. 95 BiomtrcJ 37 65–80

— —: Efficiency comparisons of central composite designs. 95 JStCmpSm 52 177–183

— —: Power of Hotelling's T^2 in Gaussian mixtures. 95 JStCmpSm 52 184–187

— —: Local bounds on power of T^2 in scale families. 95 JStCmpSm 52 187–190

Jensen, Eva B. Vedel: Rotational versions of the Crofton formula. 95 AdvAppPr 27 87–96

Jensen, J.: Wang, C. S.; Gianola, D.; Sorensen, D. A.; — —; Christensen, A.; Rutledge, J. J.: Response to selection for litter size in Danish Landrace pigs: A Bayesian analysis. 94 ThApGene 88 220–230

Jensen, J. L.: Barndorff-Nielsen, O. E.; — —; Kendall, W. S. (Ed); Gani, J. (Rev): Review of "Networks and chaos - Statistical and probabilistic aspects". 95 Metrika 42 271–272

Jensen, J. L.: Barndorff-Nielsen, O. E.; — —; Kendall, W. S. (Auth); Mallows, Colin (Rev): Review of "Networks and chaos: Statistical and probabilistic aspects". 95 JASA 90 1133–1133

Barndorff-Nielsen, O. E.; — —; Kendall, W. S. (Ed); Tong, Howell (Rev): Review of "Networks and chaos - Statistical and probabilistic aspects". 95 JRSS-A 158 629–630

Jensen, Jens Ledet: Saddlepoint approximations. 95 ClrndnPr:UK 329p

— —: Correction to "A large deviation-type approximation for the 'Box class' of likelihood ratio criteria" (V86 p437-440). 95 JASA 90 812–812

— —: Asymptotic expansions at work. 95 ScanActJ 143–152

Jensen, Mark J.: A Monte Carlo study on two methods of calculating the MLE's covariance matrix in a seemingly unrelated nonlinear regression. 95 EcomtrRv 14 315–330

Jensen, U.: Heinrich, G.; — —: Parameter estimation for a bivariate lifetime distribution in reliability with multivariate extensions. 95 Metrika 42 49–65

Jeon, Jong Woo: Kim, Kyu Seong; — —; Park, Hong Nai: A study on the sample design of the fishery household economy survey (Korean). 95 KorApS 8/2 43–54

Jeon, Jong-Woo: Kim, Choongrak; — —: Detecting influential observations on the smoothing parameter in nonparametric regression. 95 JKoreaSt 24 495–506

Jeong, Hyean-Seok; Yum, Bong-Jin: Type-I censored life test plans in the exponential case. 95 CommStB 24 187–205

Jeong, MiSook: Kim, SungHo; — —: A bootstrap approach for factor numbers in binary data (Korean). 95 KorApS 8/2 201–216

Jeong, Yousceek: Optimal designs for experiments with a large number of factors. 95 JStPlInf 46 381–393

— —: Two-stage parameter designs for quality improvement. 94 ASAProQlPr 215–219

Jéquier, Jean-Claude: Ducharme, Gilles R.; Gannoun, Ali; Guertin, Marie-Claude; — —: Reference values obtained by kernel-based estimation of quantile regressions. 95 Biomtrcs 51 1105–1116

Jernigan, Robert W.; O'Connell, Julie A.: Influence on smoothness in penalized likelihood regression for binary data. 94 ASAProStCp 118–123

Jewell, N. P.: Wang, M.-C.; Brookmeyer, R.; — —: Reply to comment on "Statistical models for prevalent cohort data". 95 Biomtrcs 51 374–375

Jewell, Nicholas P.; Kimber, Alan C.; Lee, Mei-Ling Ting; Whitmore, G. A. (Ed): Lifetime data: Models in reliability and survival analysis. 95 Kluwer:MA:Hllnd 424p

— —; van der Laan, Mark: Generalizations of current status data with applications. 95 LiDaAnal 1 101–109

Jewell, William S.: The effects of quantized data. 95 AplStMDA 11 201–216

Jewkes, E.: He, Qi-Ming; — —: Flow time in the MAP/G/1 queue with customer batching and setup times. 95 CommStSM 11 691–711

Jha, V. D.: Khurana, A. P.; — —: Some recurrence relations between product moments of order statistics of a doubly truncated Pareto distribution. 95 SankhyaB 57 1–9

Ji, Shuixin: Wang, Y. H.; — —: Limit theorems for the number of occurrences of consecutive k successes in n Markovian trials. 95 JAppProb 32 727–735

Ji, Xiaoyan; Raghavachari, M.; Wallace, William A.: Approximate reasoning in belief networks: Statistical validity, accuracy, and determination of sample size. 94 ASAProBays 128–133 (Disc: p134-135)

Jia, Xinwei; Rao, M. Bhaskara: A new modeling approach for joint action of drugs. 94 ASAProBiop 154–159

— —; Rao, M. Bhaskara: An alternative model for joint action of two drugs. 94 ASAProQlPr 305–309

Jiang, Changjian; Pan, Xuebiao; Gu, Minghong: The use of mixture models to detect effects of major genes on quantitative characters in a plant breeding experiment. 94 Genetcs 136 383–394

— —; Zeng, Zhao-Bang: Multiple trait analysis of genetic mapping for quantitative trait loci. 95 Genetcs 140 1111–1127

Jiang, Qi; Succop, Paul A.: A study of the log-log and log-additive models for the relationship between blood lead and environmental lead. 94 ASAProEnvr 88–91

Jiang, R.; Murthy, D. N. P.: Modeling failure-data by mixture of 2 Weibull distributions: A graphical approach. 95 IEEEReli 44 477–488

— —; Murthy, D. N. P.: Reliability modeling involving two Weibull distributions. 95 RelEngSS 47 187–198

Jiang, Tiefeng; Rao, M. Bhaskara; Wang, Xiangchen: Large deviations for moving average processes. 95 StocProc 59 309–320

Li, Deli; Rao, M. Bhaskara; — —; Wang, Xiangchen: Complete convergence and almost sure convergence of weighted sums of random variables. 95 JTherPrb 8 49–76

Jiang, Xulun: Dhar, Sunil K.; — —: Probability bounds on the finite sum of the binary sequence of order k. 95 JAppProb 32 1014–1027

Jiang, Yong: Nishimura, Shoichi; — —: An M/G/1 vacation model with two service modes. 95 PrEngInf 9 355–374

Jimenez, Francisco: Valderrama, Mariano J.; — —; Gutierrez, Ramon; Martinez-Almecija, Alfredo: Estimation and filtering on a doubly stochastic Poisson process. 95 AplStMDA 11 13–24

Jin, Chun: Shrinkage estimation of a mean vector in a two-sample problem. 95 CommStA 24 2881–2888

Crouse, Robert H.; — —; Hanumara, R. C.: Unbiased ridge estimation with prior information and ridge trace. 95 CommStA 24 2341–2354

Jin, Hua: Zheng, Zhongguo; — —: Lower confidence limit in censored case. 95 JSysSMSE 7 40–44

Jin, Kun; Speed, T. P.; Thomson, Glenys: Tests of random mating for a highly polymorphic locus: Application to HLA data. 95 Biomtrcs 51 1064–1076

Jin, Li: Chakraborty, Ranajit; Zhong, Yixi; — —; Budowle, Bruce: Nondetectability of restriction fragments and independence of DNA fragment sizes within and between loci in RFLP typing of DNA. 94 AJHuGene 54 391–401

Jing, Bing-Yi: Some resampling procedures under symmetry. 95 AstrlJSt 37 337-344

— —: Two-sample empirical likelihood method. 95 StPrLet 24 315-319

Hall, Peter; — —: Uniform coverage bounds for confidence intervals and Berry-Esseen theorems for Edgeworth expansion. 95 AnlsStat 23 363-375

Hall, Peter; Horowitz, Joel L.; — —: On blocking rules for the bootstrap with dependent data. 95 Biomtrka 82 561-574

Jing, Yihua: The effect of thyroxine and prolactin on the molting of newts: A factorial study. 95 Stats /12 24-27

Jinno, Kiyokatsu: Obituary: Yoshikatsu Miyashita, 1945-1995. 95 Chemolab 30 197-198

Jinot, Jennifer: Bayard, Steven; — —; Brown, Kenneth: Passive smoking and lung cancer: The U.S. EPS's weight-of-evidence analysis, with emphasis on the epidemiology studies. 94 ASAProEnvr 56-64 (Disc: p74-76)

Bayard, Steven; — —; Flatman, George: Comment on "Uncertainties in lung cancer risk estimates reported for exposure to environmental tobacco smoke". 95 EnvrMtrc 6 413-418

Jiroušek, Radim; Přeučil, Stanislav: On the effective implementation of the iterative proportional fitting procedure. 95 CmpStDA 19 177-189

Antoch, Jaromír; Hájek, Petr; — —: Obituary: Tomáš Havránek. 95 CmpStDA 19 111-113

Jmel, S.: Least squares estimation for association models with linear constraints on the parameters (French). 95 RStApp 43/3 93-104

Jmel, Said: de Falguerolles, Antoine; — —; Whittaker, Joe: Correspondence analysis and association models constrained by a conditional independence graph. 95 Psymtrka 60 161-180

Jo, Jinnam; Hinkelmann, Klaus: Some properties of Box-Behnken designs. 93 JCISS 18 273-287

Joag-dev, Kumar; Kochar, Subhash; Proschan, Frank: A general composition theorem and its applications to certain partial orderings of distributions. 95 StPrLet 22 111-119

Job, Jayakaran S.: Moulton, Lawrence H.; Holt, Elizabeth A.; — —; Halsey, Neal A.: Percentile regression analysis of correlated antibody responses. 95 StatMed 14 777-787

Jobe, Jared B.: Rasinski, Kenneth A.; Baldwin, Alison K.; Willis, Gordon B.; — —: Risk and loss perceptions associated with survey reporting of sensitive behaviors. 94 ASASrvyRsr 497-502

Tourangeau, Roger; — —; Pratt, William F.; Rasinski, Kenneth: Design and results of the Women's Health Study. 94 ASASrvyRsr 49-58

Jocelyn, Wisner: Rahim, M. A.; — —: Sample allocation in multivariate stratified design: An alternative to convex programming. 94 ASASrvyRsr 689-692

Jöckel, K.-H.: van der Linde, A.; Witzko, K.-H.; — —: Spatial-temporal analysis of mortality using splines. 95 Biomtrcs 51 1352-1360

Joe, Harry: Approximations to multivariate normal rectangle probabilities based on conditional expectations. 95 JASA 90 957-964

Hu, Taizhong; — —: Monotonicity of positive dependence with time for stationary reversible Markov chains. 95 PrEngInf 9 227-237

Joffe, A.: Fleischer, I.; — —: Ratio ergodicity for non-homogeneous Markov chains in general state spaces. 95 JTherPrb 8 31-37

Joffe, Marshall M.; Greenland, Sander: Standardized estimates from categorical regression models. 95 StatMed 14 2131-2141

Johaadien, Hassan; Goutis, Constantinos: Approximate maximum likelihood estimation in the presence of extra-Poisson variation. 95 CommStA 24 1625-1636

Jóhannesson, Benedikt; Giri, Narayan: On approximations involving the beta distribution. 95 CommStB 24 489-503

Johansen, Søren: A statistical analysis of cointegration for I(2) variables. 95 EconTher 11 25-59

— —: Identifying restrictions of linear equations with applications to simultaneous equations and cointegration. 95 JEconmtx 69 111-132

Johansson, Per: Tests for serial correlation and overdispersion in a count data regression model. 95 JStCmpSm 53 153-164

John, J. A.; Williams, E. R.: Cyclic and computer generated designs (Second edition). 95 ChapmnHall:Lnd 272p

John, P. W. M.; Johnson, M. E.; Moore, L. M.; Ylvisaker, D.: Minimax distance designs in two-level factorial experiments. 95 JStPlInf 44 249-263

John, Peter W. M.: Optimal foldover designs. 95 CommStA 24 1821-1827

— —: Patterns in residuals for 2^n designs. 95 JAppStat 22 253-259

Johnson, Ayah E.; Botman, Steven L.; Basiotis, Peter: Nonresponse in federal demographic surveys: 1981-1991. 94 ASASrvyRsr 983-988

Carlson, Barbara Lepidus; Cohen, Steven B.; — —: Family unit constructs, dynamics, and analysis in the household component of the NMES. 94 ASAProSoSt 257-262

Johnson, Christopher H.: Synthetic estimates for small areas in PRAMS. 94 ASASrvyRsr 1065-1070

Johnson, Dale: Pan, U. WinKey; Clark, Scott; Buncher, Ralph; — —: Implementing a database management system for an H.U.D. lead-based paint abatement study. 94 ASAProEnvr 162-166

Johnson, Dallas E.: Remmenga, Marta D.; — —: A comparison of inference procedures in unbalanced split-plot designs. 95 JStCmpSm 51 353-367

Johnson, Donald R.: Wahba, Grace; — —; Gao, Feng; Gong, Jianjian: Adaptive tuning of numerical weather prediction models: Randomized GCV in three- and four-dimensional data assimilation. 95 MWeathr 11 3358-3369

Johnson, Douglas H.: Statistical sirens: The allure of nonparametrics. 95 Ecology 76 1998-2000

Kraft, Kathy M.; — —; Samuelson, Jack M.; Allen, Stephen H.: Using known populations of pronghorn to evaluate sampling plans and estimators. 95 JWildlMg 59 129-137

Johnson, G. D.: Myers, W. L.; — —; Patil, G. P.: Rapid mobilization of spatial/temporal information in the context of natural catastrophes. 94 ASAProStGr 25-31

Johnson, I. R.: Modelling: Keeping plant science and agriculture students interested. 95 EnvrMtrc 6 517-522

Johnson, J. D.: Helton, J. C.; — —; McKay, M. D.; Shiver, A. W.; Sprung, J. L.: Robustness of an uncertainty and sensitivity analysis of early exposure results with the MACCS reactor accident consequence model. 95 RelEngSS 48 129-148

Johnson, K. R.: Hamilton, M. A.; — —; Camper, A. K.; Stoodley, P.; Harkin, G. J.; Gillis, R. J.; Shope, P. A.: Analysis of bacterial spatial patterns at the initial stage of biofilm formation. 95 BiomtrcJ 37 393-408

Johnson, Keith L.: Kasperski, William J.; Schneider, Helmut; — —: Analyzing unreplicated factorials using the n smallest contrasts. 95 JQualTek 27 377-386

Johnson, M. E.: John, P. W. M.; — —; Moore, L. M.; Ylvisaker, D.: Minimax distance designs in two-level factorial experiments. 95 JStPlInf 44 249-263

Johnson, N. L.; Kotz, S.; Balakrishnan, N. (Auth); Chatfield, Chris (Rev): Review of "Continuous univariate distributions, Volume 1 (Second edition)". 95 Ststcian 44 544-545

— —; Kotz, S.; Balakrishnan, N. (Auth); Kocherlakota, S. (Rev): Review of "Continuous univariate distributions, Volumes 1 and 2 (Second edition)". 95 Biomtrcs 51 1582-1583

— —; Kotz, S.; Balakrishnan, N. (Auth); Omori, Yasuhiro (Rev): Review of "Continuous univariate distributions, Volume 1 (Second edition)". 95 JASA 90 1490-1491

— —; Kotz, S.; Balakrishnan, N. (Auth); Tricker, A. R. (Rev): Review of "Continuous univariate distributions: Volume I (Second edition)". 95 JAppStat 22 323-323

— —; Kotz, S.; Kemp, A. W. (Auth); Consul, P. C. (Rev): Review of "Univariate discrete distributions". 95 Metrika 42 72-73

Johnson, Norman L.; Kotz, Samuel: Use of moments in studies of limit distributions arising from iterated random subdivisions of an interval. 95 StPrLet 24 111-119

— —; Kotz, Samuel; Balakrishnan, N.: Continuous univariate distributions. Volume 2 (Second edition). 95 Wiley:NY:UK 719p

— —; Kotz, Samuel; Balakrishnan, N. (Auth); Famoye, Felix (Rev): Review of "Continuous univariate distributions, Volume I (Second edition)". 95 Technmcs 37 466-466

Kotz, Samuel; — — (Auth); Champ, Charles W. (Rev): Review of "Process capability indices". 95 Technmcs 37 233-233

Kotz, Samuel; — — (Auth); Lovelace, Cynthia (Rev): Review of "Process capability indices". 95 JQualTek 27 270-271

Kotz, Samuel; — — (Auth); Walker, Esteban (Rev): Review of "Process capability indices". 95 JASA 90 1127-1127

Johnson, Paul A.: Durlauf, Steven N.; — —: Multiple regimes and cross-country growth behaviour. 95 JApEcmtx 10 365-384

Johnson, Richard: Krzanowski, W. J.; Marriott, F. H. C. (Auth); — — (Rev): Review of "Multivariate analysis, Part 1: Distributions, ordination, and inference". 95 JASA 90 1490-1490

Johnson, Richard A.; Gan, D. Robert; Berthouex, P. M.: Goodness-of-fit using very small but related samples with application to censored data estimation of PCB contamination. 95 EnvrMtrc 6 341-348

— —; Sim, Songyong: Nonparametric tests for ordered c-sample contaminated distributions. 95 CommStA 24 861-879

Johnson, Robert; Malloy, Kay; Gerstein, Dean; Suter, Natalie; Nisenbaum, Susan: Increasing response rates in follow-up surveys of drug treatment patients. 94 ASASrvyRsr 1263-1268

Baker, Reg; Bradburn, Norman; — —: CAPI: An experimental evaluation. 94 ASASrvyRsr 851-856 (Disc: p861-862)

Johnson, Roger: A multiple regression project. 95 TeachgSt 17/2 64-66

Egge, Eric; Foley, Sean; Haskins, Loren; — —: A data analysis based elementary statistics course. 94 ASAProStEd 144-149

Johnson, Thomas: Hoff, Jennifer; — —; Lu, J. C.: The statisticians role in quality improvement partnerships. 94 ASAProQIPr 252-257

Johnson, Valen E.: Wilson, Alyson G.; — —: Using features to model prior structural information. 94 ASAProPEng 255-264

Johnson, Wesley: Geisser, Seymour; — —: Testing independence when the form of the bivariate distribution is unspecified. 95 StatMed 14 1621-1639

Johnson, William D.; May, Warren L.: Combining 2 × 2 tables that contain structural zeros. 95 StatMed 14 1901-1911

— —; Mercante, Donald E.: Analyzing multivariate data in crossover designs using permutation tests. 94 ASAProBiop 244-249

Crawford, Ann Marie K.; — —: Statistical considerations for multivariate repeated measures with patterned covariance. 94 ASAProBiop 61-66

May, Warren L.; — —: Testing quasi-independence in stratified triangular contingency tables. 95 BiomtrcJ 37 531-546

May, Warren L.; — —: Some applications of the analysis of multivariate normal data with missing observations. 95 JBiophSt 5 215-228

Shahane, Aditi; George, Varghese; — —: Effect of bivariate regression toward the mean in uncontrolled clinical trials. 95 CommStA 24 2165-2181

Xie, Xuemin; — —: The rank tests for a completely randomized design with multivariate longitudinal responses and a randomized complete block design with multivariate responses. 94 ASAProBiop 250-255

Johnson, William H.: Brown, Eugene F.; — —: Stratification designs for the housing sample of the consumer price index. 94 ASAProGovt 140-143

Johnston, Roger G.; Schroder, Shayla D.; Mallawaaratchy, A. Rajika: Statistical artifacts in the ratio of discrete quantities. 95 AmerStat 49 285-291

Johnstone, Iain M.: Donoho, David L.; — —: Adapting to unknown smoothness via wavelet shrinkage. 95 JASA 90 1200-1224

Donoho, David L.; — —; Kerkyacharian, Gérard; Picard, Dominique: Wavelet shrinkage: Asymptopia?. 95 JRSS-B 57 301-337 (Disc: p337-369)

Joiner, Brian L.: — — (Auth); Ziegel, Eric R. (Rev): Review of "Fourth generation management: The new business consciousness". 95 Technmcs 37 113-113

Jolayemi, Joel K.: A power function model for determining sample sizes of the operations of multivariate control charts. 95 CmpStDA 20 633-641

Jolivet, E.: Härdle, W.; Huet, S.; — —: Better bootstrap confidence intervals for regression curve estimation. 95 MaOpfStS 26 287-306

Jolliffe, Flavia: Ross, S. M. (Auth); — — (Rev): Review of "Introduction to probability models (Fifth edition)". 95 JRSS-A 158 201-202

Jolliffe, I. T.; Jones, B.; Morgan, B. J. T.: Identifying influential observations in hierarchical cluster analysis. 95 JAppStat 22 61-80

Jolliffe, Ian T.: Sample sizes and the central limit theorem: The Poisson distribution as an illustration. 95 AmerStat 49 269-269

— —: Rotation of principal components: Choice of normalization constraints. 95 JAppStat 22 29-35

Cadima, Jorge; — —: Loadings and correlations in the interpretation of principal components. 95 JAppStat 22 203-214

Joly, S.; Le Calvé, G.: Three-way distances. 95 JClass 12 191-205

Jonathan, P.: Krzanowski, W. J.; — —; McCarthy, W. V.; Thomas, M. R.: Discriminant analysis with singular covariance matrices: Methods and applications to spectroscopic data. 95 ApplStat 44 101-115

Jones, Albyn C.: What do you get when you combine a Pentium and a research grant?. 95 Chance 8/2 7-8

Jones, B.; Lewis, J. A.: The case for cross-over trials in phase III. 95 StatMed 14 1025-1038 (Disc: p1053-1055)

Jolliffe, I. T.; — —; Morgan, B. J. T.: Identifying influential observations in hierarchical cluster analysis. 95 JAppStat 22 61-80

Jones, Bruce L.: A stochastic population model for high demand CCRCs. 95 InsrneME 16 69-77

Jones, Byron: Senn, S. (Auth); — — (Rev): Review of "Cross-over trials in clinical research". 95 JAppStat 22 317-318

Jones, C. D.: Higson, H. L.; Griffiths, R. F.; — —; Biltoft, C.: Effect of atmospheric stability on concentration fluctuations and wake retention times for dispersion in the vicinity of an isolated building. 95 EnvrMtrc 6 571-581

Jones, C. M.; Robson, D. S.; Lakkis, H. D.; Kressel, J.: Properties of catch rates used in analysis of angler surveys. 95 TrAFish 124 911-928

Pollock, K. H.; — —; Brown, T. L. (Auth); Björnsson, H. (Rev): Review of "Angler survey methods and their applications in fisheries management". 95 Biomtrcs 51 385-385

Jones, Clifton T.: A dynamic analysis of interfuel substitution in U.S. industrial energy demand. 95 JBES 13 459-465

Jones, David R.: Meta-analysis: Weighing the evidence. 95 StatMed 14 137-149

Lewis, John A.; — —; Röhmel, Joachim: Biostatistical methodology in clinical trials - A European guideline. 95 StatMed 14 1655-1657

Jones, K. F.; Marrs, I.; Young, J.; Townend, C.: Investigation and improvement of a process for vacuum-formed ceramic fibre composites, using a fractional 2^n experimental design. 95 JAppStat 22 459-467

— —; Taylor, N. W.; Fosam, E. B.: Statistical approaches for the reliability of structural designs. 95 JAppStat 22 443-458

Jones, Lyle V.; Williams, Valerie S. L.: Multiple comparisons for education data. 94 ASAProSoSt 6-8

Jones, M. C.: On higher order kernels. 95 JNnparSt 5 215-221

— —: Local and variable bandwidths and local linear regression. 95 MaOpfStS 27 65-71

— —: On two recent papers of Y. Kanazawa. 95 StPrLet 24 269-271

— —; Daly, F.: Density probability plots. 95 CommStB 24 911-927

— —; Linton, O.; Nielsen, J. P.: A simple bias reduction method for density estimation. 95 Biomtrka 82 327-338

Chattamvelli, R.; — —: Recurrence relations for noncentral density, distribution functions and inverse moments. 95 JStCmpSm 52 289-299

Wand, M. P.; — — (Auth); Mackisack, Margaret S. (Rev): Review of "Kernel smoothing". 95 AstrlJSt 37 368-369

Wand, M. P.; — — (Auth); Ramsay, J. O. (Rev): Review of "Kernel smoothing". 95 ShrtBkRv 15/2 27-27

Jones, M. E.; Thomas, S. M.; Rogers, A.: Luria-Delbrück fluctuation experiments: Design and analysis. 94 Genetcs 136 1209-1216

Jones, M.-C.: Wand, M.-P.; — —: Kernel smoothing. 95 ChapmnHall:Lnd 224p

Jones, P. N.; Carberry, P. S.: Optimal use of data in crop model development and validation. 95 EnvrMtrc 6 491-496

Jones, Peter W.: Ashour, Samir K.; — —; Ahmed, Mohiy El-Din M.: Discriminant analysis for social-economic and cultural factors affecting diarrhea diseases in children under two years in Egypt. 95 EgpStJ 39 1-11

Jones, R. A.: Foolad, M. R.; — —: Models to estimate maternally controlled genetic variation in quantitative seed characters. 92 ThApGene 83 360-366

Jones, R. H.: — — (Auth); Glimm, Ekkehard (Rev): Review of "Longitudinal data with serial correlation: A state-space approach". 95 BiomtrcJ 37 172-172

— — (Auth); Laird, N. M. (Rev): Review of "Longitudinal data with serial correlation: A state-space approach". 95 Metrika 42 68-69

— — (Auth); Lange, N. (Rev): Review of "Longitudinal data with serial correlation: A state-space approach". 95 Biomtrcs 51 1193-1194

— — (Auth); Schlittgen, R. (Rev. German): Review of "Longitudinal data with serial correlation: A state-space approach". 95 AllgeStA 79 332-333

— — (Auth); Shumway, Robert H. (Rev): Review of "Longitudinal data with serial correlation: A state-space approach". 95 JStPlInf 47 395-397

Jones, R. Hughes: Preece, D. A.; Vowden, B. J.; — —; Roger, C. A.; Vowden, C. J.: Choreographing designs. 95 MaScitst 20 15-32

Jones, Richard H.: Anderson, Stewart J.; — —: Smoothing splines for longitudinal data. 95 StatMed 14 1235-1248

Marshall, Guillermo; — —: Multi-state models and diabetic retinopathy. 95 StatMed 14 1975-1983

Marshall, Guillermo; Guo, Wensheng; — —: MARKOV: A computer program for multi-state Markov models with covariables. 95 CmpBiomd 47 147-156

Jones, Wendell D.: Hadi, Ali S.; — —; Ling, Robert F.: A unifying representation of some case-deletion influence measures in univariate and multivariate linear regression. 95 JStPlInf 46 123-135

Jongbloed, Geurt: Groeneboom, Piet; — —: Isotonic estimation and rates of convergence in Wicksell's problem. 95 AnlsStat 23 1518-1542

Joo, Sang Yeol: Kim, Gwan Young; — —; Kim, Yun Kyong: A sharp result of random upper functions for Levy processes. 95 JKoreaSt 24 65-76

Jordan, B. P.: MacPhee, I. M.; — —: Optimal search for a moving target. 95 PrEngInf 9 159-182

Jordan, Michael I.; Xu, Lei: Convergence results for the EM approach to mixtures of experts architectures. 95 NuralNet 8 1409-1431

Jordan, Paul: Estimation of tolerance limits from reference data. 95 CmpStDA 19 655-668

Jordan, Scott M.; Krishnamoorthy, K.: Confidence regions for the common mean vector of several multivariate normal populations. 95 CandJSt 23 283-297

— —; Krishnamoorthy, K.: On combining independent tests in linear models. 95 StPrLet 23 117-122

Jorde, Per Erik; Ryman, Nils: Temporal allele frequency change and estimation of effective size in populations with overlapping generations. 95 Genetcs 139 1077-1090

Jorgensen, B.: — — (Auth); Snell, Joyce (Rev): Review of "The theory of linear models". 95 JRSS-A 158 195-195

Jørgensen, B.: — — (Auth); Böhning, D. (Rev): Review of "The theory of linear models". 95 Biomtrcs 51 791-792

— — (Auth); Drum, Melinda L. (Rev): Review of "The theory of linear models". 95 StMedRes 4 266-267

— — (Auth); Martin, R. J. (Rev): Review of "The theory of linear models". 95 JAppStat 22 543-544

Jorgensen, Bent: — — (Auth); Borowiak, Dale (Rev): Review of "The theory of linear models". 95 Technmcs 37 228-229

— — (Auth); Prvan, Tania (Rev): Review of "The theory of linear models". 95 AstrlJSt 37 248-248

Jørgenson, B.: — — (Auth); Neuhaus, G. (Rev): Review of "The theory of linear models". 95 Metrika 42 276-277

Joseph, Lawrence; Gyorkos, Theresa W.; Coupal, Louis: Bayesian estimation of disease prevalence and the parameters of diagnostic tests in the absence of a gold standard. 95 AJEpidm 141 263-272

— —; Wolfson, David B.; Berger, Roxane du: Sample size calculations for binomial proportions via highest posterior density intervals. 95 Ststcian 44 143-154

— —; Wolfson, David B.; Berger, Roxane du: Some comments on Bayesian sample size determination. 95 Ststcian 44 167-171

Joseph, Nathan Lael: Cointegration, error-correction models, and forecasting using realigned foreign exchange rates. 95 JForcast 14 499-522

Joseph, Stephen; Lewis, Christopher Alan: Factor analysis of the Center for Epidemiological Studies-Depression Scale. 95 PsychRpt 76 40-42

Joshi, B. H.; Naik-Nimbalkar, U. V.; Deshmukh, S. R.; Dharmadhikari, A. D.: Prediction of failure - A case study. 95 JAppStat 22 485-493

Kakihara, Yûichirô: Vector harmonizable processes: Wold and Cramér decompositions. 95 StocAnAp 13 531-541

Kakiuchi, Itsuro; Kimura, Miyoshi: Majorization methods on hyperplanes and their applications. 95 JStPlInf 47 217-235

Kala, Radosław: On perpendicularity in designed experiments. 95 ListyBio 32 93-99

Kalamkar, V. A.: Minification processes with discrete marginals. 95 JAppProb 32 692-706

Kalashnikov, V. V.: — — (Auth); Gerontidis, Ioannis I. (Rev): Review of "Mathematical methods in queuing theory". 95 Ststcian 44 409-410

— —; Zolotarev, V. M. (Ed); Weron, A. (Rev): Review of "Stability problems for stochastic models". 95 Metrika 42 272-273

Kalbfleisch, J. D.: Sun, J.; — —: Estimation of the mean function of point processes based on panel count data. 95 StSinica 5 279-289

Kaldor, J. M.: Evaluating causality in the AIDS epidemic: The role of mathematical models. 95 EnvrMtrc 6 535-540

Kale, B. K.; Sebastian, G.: Approximate estimation of non-identifiable parameters in a convolution. 95 StPrLet 25 373-378

Kale, Balkrishna; Besl, John; Palit, Charles; Voss, Paul: Updating age-sex-specific net migration rates with limited data. 94 ASAProSoSt 116-121 (Disc: p122-123)

Kale, Mandar; Deshmukh, S. R.: Maximum likelihood estimation in birth-death process allowing catastrophes. 95 CommStA 24 1863-1873

Kalirajan, K. P.; Obwona, M. B.: Estimating individual response coefficients in varying coefficients regression models. 95 JAppStat 22 477-484

— — (Auth); Senroy, Sugata (Rev): Review of "Applied econometrics". 95 CalcutSt 45 135-136

Kalivas, John H.: Hörchner, Uwe; — —: Simulated-annealing-based optimization algorithms: Fundamentals and wavelength selection applications. 95 JChemtrc 9 283-308

Kaljurand, M.: Koel, M.; — —; Lochmüller, C. H.; Moebus, M.: Factor analytical resolution of simultaneous, dynamic thermal process in thermochromatography of oil shale. 95 Chemolab 30 173-178

Kaljurand, Mihkel: Elomaa, Matti; Lochmüller, Charles H.; — —; Koel, Mihkel: Application of evolving factor analysis in thermochromatography. 95 Chemolab 30 129-132

Kallenberg, L. C. M.: Survey of linear programming for standard and nonstandard Markovian control problems. Part II: Applications. 94 ZeitOpRs 40 127-143

Kallenberg, O.: Random arrays and functionals with multivariate rotational symmetries. 95 ProbTher 103 91-141

Kallenberg, Wilbert C. M.; Koning, Alex J.: On Wieand's theorem. 95 StPrLet 25 121-132

— —; Ledwina, Teresa: Consistency and Monte Carlo simulation of a data driven version of smooth goodness-of-fit tests. 95 AnlsStat 23 1594-1608

Kallianpur, G.; Xiong, J.: Diffusion approximation of nuclear space-valued stochastic differential equations driven by Poisson random measures. 95 AnlsApPr 5 493-517

Kalmykov, Yu. P.: Coffey, W. T.; — —: The Langevin equation and its applications to stochastic problems in physics, chemistry and electrical engineering. 95 WrldSciPub:Sing 500p

Kalpazidou, S.: On the rotational dimension of stochastic matrices. 95 AnlsProb 23 966-975

Kalpazidou, S. L.: Cycle representations of Markov processes. 95 Springer:Brln:-NY 194p

— — (Auth); Csörgő, Miklós (Rev): Review of "Cycle representations of Markov processes". 95 ShrtBkRv 15/3 47-47

Kalpić, Damir: Automated coding of census data. 94 JOffStat 10 449-463

Kalsbeek, William D.; Botman, Steven L.; Massey, James T.; Liu, Pao-Wen: Cost-efficiency and the number of allowable call attempts in the National Health Interview Survey. 94 JOffStat 10 133-152

— —; Durham, Todd A.: Nonresponse and its effects in a followup telephone survey of low income women. 94 ASASrvyRsr 943-948

Kalton, Graham; Brick, J. Michael: Weighting schemes for household panel surveys. 95 SurvyM 21 33-44

— —; Brick, J. Michael: Weighting schemes for household panel surveys. 94 ASASrvyRsr 785-790

Rizzo, Lou; — —; Brick, Mike: Adjusting for panel nonresponse in the Survey of Income and Program Participation. 94 ASASrvyRsr 422-427 (Disc: p440-442)

Kalton, Nigel; Saab, Elias; Montgomery-Smith, Stephen (Ed): Interaction between functional analysis, harmonic analysis, and probability. 95 Dekker:NY 494p

Kałuszka, M.: Estimates of some probabilities in multidimensional convex records. 95 ApplMath 23 1-11

Gajek, L.; — —: Lower bounds for the asymptotic Bayes risk in the scale model (with an application to the second-order minimax estimation). 94 AnlsStat 22 1831-1839

Kałuszka, Marek: Gajek, Lesław; — —: Nonexponential applications of a global Cramér-Rao inequality. 95 MaOpfStS 26 111-122

Kamakura, Wagner A.: Wedel, Michel; — —; Desarbo, Wayne S.; Hofstede, Frenkel Ter: Implications for asymmetry, nonproportionality, and heterogeneity in brand switching from piece-wise exponential mixture hazard models. 95 JMarRsr 32 457-462

Kambezidis, H. D.; Tulleken, R.; Amanatidis, G. T.; Paliatsos, A. G.; Asimakopoulos, D. N.: Statistical evaluation of selected air pollutants in Athens, Greece. 95 EnvrMtrc 6 349-361

Kamesam, P. V.: Escudero, L. F.; — —: On solving stochastic production planning problems via scenario modelling. 95 Top 3 69-95

Kamgar-Parsi, Behrooz: Kamgar-Parsi, Behzad; — —; Brosh, Menashe: Distribution and moments of the weighted sum of uniform random variables, with applications in reducing Monte Carlo simulations. 95 JStCmpSm 52 399-414

Kamgar-Parsi, Behzad; Kamgar-Parsi, Behrooz; Brosh, Menashe: Distribution and moments of the weighted sum of uniform random variables, with applications in reducing Monte Carlo simulations. 95 JStCmpSm 52 399-414

Kamidi, R. E.: Statistical adjustment of maize grain yield for sub-optimal plot stands. 95 ExpAgric 31 299-306

Kaminsky, Frank C.; Haberle, Katharine Rowse: Probability distributions for the Deming cost models. 95 JQualTek 27 355-362

Benneyan, James C.; — —: Successfully applying SPC to improve health care: Barriers to improving quality and reducing liability. 95 ASQCTekCon 570-578

Benneyan, James C.; — —: Another view on how to measure health care quality. 95 QulPr 28/2 120-123

Kamlet, Mark S.; Paul, Nancy; Greenhouse, Joel; Kupfer, David; Frank, Ellen; Wade, Martcia: Cost utility analysis of maintenance treatment for recurrent depression. 95 ConClnTr 16 17-40

Kamps, U.: — — (Auth); Plachky, D. (Rev): Review of "A concept of generalized order statistics". 95 StatDecn 13 402-403

Kamps, Udo: Recurrence relations for moments of record values. 95 JStPlInf 45 225-234

— —: A concept of generalized order statistics. 95 JStPlInf 48 1-23

Kan, Stephen H.: Metrics and models in software quality engineering. 95 ASQualCntrl:WI 344p

Kanata, Yakichi: Maeda, Yutaka; Hirano, Hiroaki; — —: A learning rule of neural networks via simultaneous perturbation and its hardware implementation. 95 Autmatca 31 251-259

Kanatani, Kenichi: Statistical foundation for hypothesis testing of image data. 94 CVGIPlm 60 382-391

Kanda, Takashi: Seo, Takashi; — —; Fujikoshi, Yasunori: The effects of non-normality on tests for dimensionality in canonical correlation and MANOVA models. 95 JMultiAn 52 325-337

Kandel, Abraham; Martins, Alejandro; Pacheco, Roberto: Comment on "A probabilistic and statistical view of fuzzy methods". 95 Technmcs 37 276-281

Römer, Christoph; — —: Statistical tests for fuzzy data. 95 FuzzySet 72 1-26

Manton, Kenneth G.; Woodbury, Max A.; Tolley, H. Dennis (Auth); — — (Rev): Review of "Statistical applications using fuzzy sets". 95 Technmcs 37 237-237

Kane, S. P.: On confounding in factorial experiments. 94 JIAgSt 46 330-333

Kanetkar, Vinay; Evans, Martin G.; Everell, Shirley Anne; Irvine, Diane; Millman, Zeeva: The effect of scale changes on meta-analysis of multiplicative and main effects models. 95 EdPsMeas 55 206-224

Kang, Phillee: Dependent F ratios sharing a common denominator in ANOVA table. 95 JKoreaSt 24 111-125

Kaniovski, Yu.; Pflug, G.: Non-standard limit theorems for urn models and stochastic approximation procedures. 95 CommStSM 11 79-102

Kanji, Gopal K.: — — (Auth); Hawkins, Del I. (Rev): Review of "100 statistical tests". 95 JMarRsr 32 112-113

— — (Auth); Sridhara, R. (Rev): Review of "100 statistical tests". 95 StMedRes 4 269-269

Kanjilal, P. P.: Sarkar, A.; — —: On a method of identification of best subset model from full AR-model. 95 CommStA 24 1551-1567

Kano, Yutaka: Ihara, Masamori; — —: Identifiability of full, marginal, and conditional factor analysis models. 95 StPrLet 23 343-350

Kantam, R. R. L.; Rao, A. Vasudeva; Narasimham, V. L.: Linear unbiased estimation of scale parameter by absolute values of order statistics in logistic distribution. 94 JStatRsr 28 199-219

Kanter, Marek: How to measure the error of an M-estimate and report it conservatively. 95 JMultiAn 54 32-76

Kao, Chiang; Wong, J. Y.: Several extensively tested random number generators. 95 CmpOpRsh 22 1035-1039

Kao, Chihwa; Ross, Stephan L.: A CUSUM test in the linear regression model with serially correlated disturbances. 95 EcomtrRv 14 331-346

Kao, L.-J.; Yates, P. K.; Lewis, S. M.; Dean, A. M.: Approximate information matrices for estimating a given set of contrasts. 95 StSinica 5 593-598

Kao, Nancy L.; Chakraborti, S.: Small sample null distribution of two nonparametric tests for comparing treatments with a control in a two-way layout. 95 CommStB 24 149-164

Kapatou, Alexandra; Reynolds, Marion R., Jr: Multivariate nonparametric control charts using small samples. 94 ASAProQlPr 241-246

Kaplan, David: The impact of BIB spiraling-induced missing data patterns on goodness-of-fit tests in factor analysis. 95 JEdStat 20 69-82

— —; George, Rani: A study of the power associated with testing factor mean differences under violations of factorial invariance. 95 StrEqMod 2 101-118

Kaplan, E. H.; Brandeau, M. L. (Ed); Altmann, Michael (Rev): Review of "Modeling the AIDS epidemic: Planning, policy, and prediction". 95 Chance 8/4 36-39

Kaplan, Mark: Decision theory as philosophy. 95 CambridgeUniv 240p

Kaufman, Steven: Properties of the Schools and Staffing Survey's bootstrap variance estimator. 94 ASASrvyRsr 1116-1121 (Disc: p1128-1129)

Holt, Albert; — —; Scheuren, Fritz; Smith, Wray: Intersurvey consistency in school surveys. 94 ASASrvyRsr 1105-1110 (Disc: p1128-1129)

Kaufman, Steven F.: Ghosh, Dhiren; — —; Smith, Wray; Chang, Michael: Optimal periodicity of a survey: Sampling error, data deterioration, and cost. 94 ASASrvyRsr 1122-1127

Kaufmann, E.; Reiss, R.-D.: Approximation rates for multivariate exceedances. 95 JStPlInf 45 235-245

Kaufmann, Edgar: Von Mises conditions, δ-neighborhoods and rates of convergence for maxima. 95 StPrLet 25 63-70

Kaufmann, Peter: Multivariate optimization strategy for liquid chromatography. II. Exploring and finding optimal conditions in the search area of the multidimensional solvent space. 95 Chemolab 27 105-114

Kaufmann, Pirmin: Weber, Rudolf O.; — —: Automated classification scheme for wind fields. 95 JApMetrl 34 1133-1141

Kaur, A.; Patil, G. P.; Sinha, A. K.; Taillie, C.: Ranked set sampling: An annotated bibliography. 95 EnvEcoSt 2 25-54

Kaur, Amarjot; Di Consiglio, Loredana; Patil, Ganapati P.; Taillie, Charles: Propagation functions for monitoring distributional changes. 95 EnvEcoSt 2 239-269

Kaushal, Ritu: Bench, Jill; Clark, Colleen; Dufour, Johane; — —: Implementing computer-assisted interviewing for the Canadian Labour Force Survey. 94 ASAProGovt 93-98

Kavalieris, L.: Hannan, E. J.; — —: A note on ARMA parameterisation. 95 AstrlJSt 37 105-110

Kawasaki, Hidemi; Miyakawa, Masami: An ANOVA model and interaction element for three-way layout without replication. 95 JJapnStS 25 143-150

Kay, Jim: Cheng, Bing; — —: A regularization method for one-dimensional edge detection and edge-preserving smoothing. 95 CmpStQ 10 53-69

Kay, Leslie: Malliavin, Paul; Airault, Helene; — —; Letac, Gerard: Integration and probability. 95 Springer:Brln:NY 322p

Malliavin, Paul; Airault, Helene; — —; Letac, Gerard (Auth); Barnett, C. (Rev): Review of "Integration and probability". 95 ShrtBkRv 15/3 46-46

Kay, Steven M.: — — (Auth); Sengupta, Sailes K. (Rev): Review of "Fundamentals of statistical signal processing: Estimation theory". 95 Technmcs 37 465-466

Kayargadde, Vishwakumara; Martens, Jean-Bernard: Estimation of edge parameters and image blur using polynomial transforms. 94 CVGIPGr 56 442-461

Kazempour, Kazem: Impact of stratification imbalance on probability of type I error. 95 AmerStat 49 170-174

Kazi-Aoual, Frédérique; Hitier, Simon; Sabatier, Robert; Lebreton, Jean-Dominique: Refined approximations to permutation tests for multivariate inference. 95 CmpStDA 20 643-656

Ke, Huei-Yaw: Evaluation of operational reliability from ground test data. 95 RelEngSS 49 103-107

Keane, Michael: Aaronson, Jon; — —: Isomorphism of random walks. 94 IsrelJMa 87 37-63

Geweke, John; — —: Recursively simulating multinomial multiperiod probit probabilities. 94 ASAProBuEc 1-6

Keane, Michael P.; Runkle, David E.: Reply to comment on "Testing the rationality of price forecasts: New evidence from panel data". 95 AmEcRv 85 290-290

Kearns, Brian: Núñez, Megan; — —; Ronen, Amit: Perceptions of God's role in the origin of life on earth. 95 Stats /13 28-29

Kearsey, M. J.; Hyne, V.: QTL analysis: A simple 'marker-regression' approach. 94 ThApGene 89 698-702

Keating, Kim A.: Mitigating elevation-induced errors in satellite telemetry locations. 95 JWildlMg 59 801-808

Keats, J. Bert; Miskulin, John D.; Runger, George C.: Statistical process control scheme design. 95 JQualTek 27 214-225

Simpson, James R.; — —: Trade-off analysis versus constrained optimization with an economic control chart model. 95 IIETrn 27 765-772

Simpson, James R.; — —: A sensitivity study and comparison of the CUSUM versus Shewhart control charts using an economic model. 94 ASAProQlPr 98-103

Kececioglu, Dimitri; Sun, Feng-Bin: A general discrete-time dynamic programming model for the opportunistic replacement policy and its application to ball-bearing systems. 95 RelEngSS 47 175-185

Kedem, Benjamin: Houdré, Christian; — —: A note on autocovariance estimation in the presence of discrete spectra. 95 StPrLet 24 1-8

— — (Auth); Frey, Michael (Rev): Review of "Time series analysis by higher order crossings". 95 Technmcs 37 235-235

Keel, Alex; Müller, Heinz H.: Efficient portfolios in the asset liability context. 95 AstinB 25 33-48

Keeler, Carolyn M.; Steinhorst, R. Kirk: Using small groups to promote active learning in the introductory statistics course: A report from the field. 95 JStEd 3/2 28kb (jse/v3n2/keeler)

Steinhorst, R. Kirk; — —: Developing material for introductory statistics courses from a conceptual, active learning viewpoint. 95 JStEd 3/3 33kb (jse/v3n3/steinhorst)

Keen, K. J.: Estimating the correlation among siblings. 93 AnHuGene 57 297-305

Keenan, Daniel: Grenander, Ulf; — —: Understanding variable shapes in 3-D. 95 CommStSM 11 51-78

Grenander, U. (Auth); — — (Rev): Review of "General pattern theory: A mathematical study of regular structures". 95 SIAMRvw 37 258-261

Keenan, Thomas M.: A system for measuring short-term producibility. 95 ASQCTekCon 50-56

Keene, Oliver N.: The log transformation is special. 95 StatMed 14 811-819

Keener, Robert; Lerche, Hans Rudolf; Woodroofe, Michael: A nonlinear parking problem. 95 SqtlAnly 14 247-272

Keesman, K. J.: Model parameter and modelling error identification in non-linear conceptual models from bounded-error data. 95 EnvrMtrc 6 445-449

Keesman, Karel: Jakeman, A. J.; Beck, M. B.; McAleer, M. J. (Ed); — — (Rev): Review of "Modelling change in environmental systems". 95 EnvrMtrc 6 425-427

Keeter, Scott: Estimating telephone noncoverage bias with a telephone survey. 95 PubOpinQ 59 196-217

Brick, Michael; Waksberg, Joseph; — —: Evaluating the use of data on interruptions in telephone service for nontelephone households. 94 ASASrvyRsr 19-28

Kehagias, Athanasios: Künsch, Hans; Geman, Stuart; — —: Hidden Markov random fields. 95 AnlsApPr 5 577-602

Keiding, N.: Andersen, P. K.; Borgan, Ø.; Gill, R. D.; — —: Statistical models based on counting processes. 95 Springer:Brln:NY 767p

Andersen, P. K.; Borgan, O.; Gill, R. D.; — — (Auth); Milbrodt, Hartmut (Rev. German): Review of "Statistical models based on counting processes". 95 AllgeStA 79 221-221

Andersen, P. K.; Borgan, Ø.; Gill, R. D.; — — (Auth); Hjort, Nils Lid (Rev): Review of "Statistical models based on counting processes". 95 StNeerla 49 247-249

Andersen, P. K.; Borgan, Ø.; Gill, R. D.; — — (Auth); McKeague, Ian W. (Rev): Review of "Statistical models based on counting processes". 95 SIAMRvw 37 475-476

Keiding, Niels: Historical controls and modern survival analysis. 95 LiDaAnal 1 19-25

Jacobsen, Martin; — —: Coarsening at random in general sample spaces and random censoring in continuous time. 95 AnlsStat 23 774-786

Klein, John P.; — —; Kreiner, Svend: Graphical models for panel studies, illustrated on data from the Framingham Heart Study. 95 StatMed 14 1265-1290

Keightley, Peter D.; Bulfield, Grahame: Detection of quantitative trait loci from frequency changes of marker alleles under selection. 93 GeneRsr 62 195-203

Keil, Linda: Cantor, David; Gaertner, Gregory; — —: The validity of self-reported drug use among the incarcerated. 94 ASASrvyRsr 1088-1093

Keizer, L. C. P.: van Eeuwijk, F. A.; — —: On the use of diagnostic biplots in model screening for genotype by environment tables. 95 StatComp 5 141-153

Kelder, Steve H.: Zucker, David M.; Lakatos, Edward; Webber, Larry S.; Murray, David M.; McKinlay, Sonja M.; Feldman, Henry A.; — —; Nader, Philip R.: Statistical design of the child and adolescent trial for cardiovascular health (CATCH): Implications of cluster randomization. 95 ConClnTr 16 96-118

Kelle, Udo: — — (Ed): Computer-aided qualitative data analysis. Theory, methods, and practice. 95 SagePubl:CA:Lnd 224p

Keller, L. Robin; Sarin, Rakesh K.: Fair processes for societal decisions involving distributional inequalities. 95 RiskAnal 15 49-59

Keller, Paul A.; Runger, George C.: Achieving measurable improvement in service management. 95 ASQCTekCon 330-337

Keller, Ulrich: Eberlein, Ernst; — —: Hyperbolic distributions in finance. 95 Bernoull 1 281-299

Keller, Wouter J.: Changes in statistical technology. 95 JOffStat 11 115-127

Keller-McNulty, S.: Higgins, J.; — — (Auth); Huzurbazar, Aparna V. (Rev): Review of "Concepts in probability and stochastic modeling". 95 Technmcs 37 467-467

Keller-McNulty, Sallie: Error-free least squares based on multiple homomorphic images. 94 ASAProStCp 33-40

Higgins, James J.; — — (Auth); Nickerson, David M. (Rev): Review of "Concepts in probability and stochastic modeling". 95 AmerStat 49 321-321

Kellerman, David: Evaluating coverage in the public library statistics program. 94 ASAProGovt 70-75

Kelly, E. Thomas: Gessley, Mary K.; — —; Ponikowski, Chester H.: Introducing births into the employment cost index survey. 94 ASASrvyRsr 797-802

Kelly, F. P.: — — (Ed); Bingham, N. H. (Rev): Review of "Probability, statistics and optimization: A tribute to Peter Whittle". 95 JRSS-A 158 353-354

Kelly, Mary; Miller, Sylvia: The improvement of the Employment Cost Index Survey: A total quality approach within the Bureau of Labor Statistics. 94 ASAProQlPr 210-214

Kelly-Hayes, Maggie: D'Agostino, Ralph B.; Belanger, Albert J.; Markson, Elizabeth W.; — —; Wolf, Philip A.: Development of health risk appraisal functions in the presence of multiple indicators: The Framingham study nursing home institutionalization model. 95 StatMed 14 1757-1770

Kelmansky, Diana: Ferretti, Nélida; — —; Yohai, Victor: A goodness-of-fit test based on ranks for ARMA models. 95 CommStA 24 295-318

Kelsall, Julia E.; Diggle, Peter J.: Kernel estimation of relative risk. 95 Bernoull 1 3-16

— —; Diggle, Peter J.: Non-parametric estimation of spatial variation in relative risk. 95 StatMed 14 2335-2342

Kelsbeck, W. D.: Kessler, J. T.; — — (Auth); Fielding, A. (Rev): Review of "Non-sampling errors in surveys". 95 Ststcian 44 289-290

Kelt, Douglas A.: Mehlman, David W.; Shepherd, Ursula L.; — —: Comment on "Stopping rules in principal components analysis: A comparison of heuristical and statistical approaches" (V74 p2204-2214). 95 Ecology 76 640-643

Kemp, A. W.: Barnett, V.; Turkman, K. F. (Ed); — — (Rev): Review of "Statistics for the environment 2: Water related issues". 95 Biomtrcs 51 795-796

Brown, D.; Rothery, P. (Auth); — — (Rev): Review of "Computing examples supplement to models in biology: Mathematics, statistics and computing". 95 Biomtrcs 51 792-793

Brown, D.; Rothery, P. (Auth); — — (Rev): Review of "Models in biology: Mathematics, statistics and computing". 95 Biomtrcs 51 792-793

Kingman, J. F. C. (Auth); — — (Rev): Review of "Poisson processes". 94 BLndnMaS 26 612-614

Rasch, D.; Tiku, M. L.; Sumpf, D. (Ed); — — (Rev): Review of "Elsevier's dictionary of biometry". 95 Biomtrcs 51 1186-1187

Johnson, N. L.; Kotz, S.; — — (Auth); Consul, P. C. (Rev): Review of "Univariate discrete distributions". 95 Metrika 42 72-73

Kemp, C. D.: Lehtonen, Risto; Pahkinen, Erkki J. (Auth); — — (Rev): Review of "Practical methods for design and analysis of complex surveys". 95 ShrtBkRv 15/2 24-25

Stuart, A.; Ord, J. K. (Auth); — — (Rev): Review of "Kendall's advanced theory of statistics, Volume I: Distribution theory (Sixth edition)". 95 Biomtrcs 51 1583-1584

Kemp, C. J.: de Waal, D. J.; — —: A Bayesian model for estimating the failure rate for different groups. 95 SoAfrStJ 29 131-154

de Waal, D. J.; Groenewald, P. C. N.; — —: Perturbation of the normal model in linear regression. 95 SoAfrStJ 29 105-130

Kemp, G. C. R.: Arminger, G.; Clogg, C. C.; Sobel, M. E. (Ed); — — (Rev): Review of "Handbook of statistical modeling for the social and behavioral sciences". 95 Biomtrcs 51 1586-1587

Kemp, Suzanne, S.: Iachan, Ronaldo; — —: Visitor sample surveys. 95 SurvyM 21 89-96

Kemperman, J. H. B.: Cohen, Arthur; — —; Sackrowitz, H. B.: Unbiased testing in exponential family regression. 94 AnlsStat 22 1931-1946

Kempthorne, O.: Hinkelmann, K.; — — (Auth); Rasch, D. (Rev): Review of "Design and analysis of experiments, Volume I: Introduction to experimental design". 95 Biomtrcs 51 378-379

Hinkelmann, K.; — — (Auth); White, Lynda (Rev): Review of "Design and analysis of experiments. Volume I, Introduction to experimental design". 95 JRSS-A 158 351-351

Kempthorne, Oscar; Hinkelmann, Klaus (Auth); Goupy, Jacques (Rev): Review of "Design and analysis of experiments. Volume 1. Introduction to experimental design". 95 Chemolab 28 203-205

Hinkelmann, Klaus; — — (Auth); Copeland, Karen A. F. (Rev): Review of "Design and analysis of experiments, Volume 1: Introduction to experimental design". 95 JQualTek 27 175-176

Hinkelmann, Klaus; — — (Auth); Toman, Blaza (Rev): Review of "Design and analysis of experiments, Volume 1: Introduction to experimental design". 95 Technmcs 37 348-349

Kendal, Wayne S.: A probabilistic model for the variance to mean power law in ecology. 95 EcolgMdl 80 293-297

Kendall, David G.: How to look at objects in a five-dimensional shape space: Looking at geodesics. 95 AdvAppPr 27 35-43

— —: The foundation of the Bernoulli Society. 95 Bernoull 1 1-2

— —: How to look at objects in a 5-dimensional shape space. I: Looking at distributions. 94 ThProbAp 39 277-281

Kendall, John E.: Wallace, Jean E.; Foster, Robert D.; — —: Improving control of polymer viscosity using time-series experimentation and statistical process control. 94 ASAProPEng 320-325

Kendall, W. S.: Barndorff-Nielsen, O. E.; Jensen, J. L.; — — (Ed); Gani, J. (Rev): Review of "Networks and chaos - Statistical and probabilistic aspects". 95 Metrika 42 271-272

Barndorff-Nielsen, O. E.; Jensen, J. L.; — — (Auth); Mallows, Colin (Rev): Review of "Networks and chaos: Statistical and probabilistic aspects". 95 JASA 90 1133-1133

Barndorff-Nielsen, O. E.; Jensen, J. L.; — — (Ed); Tong, Howell (Rev): Review of "Networks and chaos - Statistical and probabilistic aspects". 95 JRSS-A 158 629-630

Kendall, Wilfrid S.: The radial part of a γ-martingale and a non-implosion theorem. 95 AnlsProb 23 479-500

Kendall, William L.; Nichols, James D.: On the use of secondary capture-recapture samples to estimate temporary emigration and breeding proportions. 95 JAppStat 22 751-762

— —; Pollock, Kenneth H.; Brownie, Cavell: A likelihood-based approach to capture-recapture estimation of demographic parameters under the robust design. 95 Biomtrcs 51 293-308

Nichols, James D.; — —: The use of multi-state capture-recapture models to address questions in evolutionary ecology. 95 JAppStat 22 835-846

Kendler, K. S.: Heath, A. C.; Kessler, R. C.; Neale, M. C.; Hewitt, J. K.; Eaves, L. J.; — —: Testing hypotheses about direction of causation using cross-sectional family data. 93 BehavGen 23 29-50

Sham, P. C.; Walters, E. E.; Neale, M. C.; Heath, A. C.; MacLean, C. J.; — —: Logistic regression analysis of twin data: Estimation of parameters of the multifactorial liability-threshold model. 94 BehavGen 24 229-238

Kendler, Kenneth S.: Neale, Michael C.; Eaves, Lindon J.; — —: The power of the classical twin study to resolve variation in threshold traits. 94 BehavGen 24 239-258

Kenessey, Zoltan: ISI: Towards the 21st century. 95 JOffStat 11 11-20

Kennedy, B. W.: Villanueva, B.; — —: Index versus tandem selection after repeated generations of selection. 93 ThApGene 85 706-712

Kennedy, Peter E.: Randomization tests in econometrics. 95 JBES 13 85-94

Kennedy, R. A.: Pulford, G. W.; Gallant, J. C.; — —; Chung, S. H.: Evaluation and estimation of various Markov models with applications to membrane channel kinetics. 95 BiomtrcJ 37 39-63

Kennedy, R. Lee: Cross, Simon S.; Harrison, Robert F.; — —: Introduction to neural networks. 95 Lancet 346 1075-1079

Kennedy, William J.: Wang, Morgan C.; — —: A self-validating numerical method for computation of central and non-central F probabilities and percentiles. 95 StatComp 5 155-163

Kennickell, Arthur B.; McManus, Douglas A.: Multiple imputation of the 1983 and 1989 waves of the SCF. 94 ASASrvyRsr 523-528

Kenny, David A.; Zautra, Alex: The trait-state-error model for multiwave data. 95 JCCPsych 63 52-59

Kent, John T.; Hainsworth, Timothy J.: Confidence intervals for the noncentral chi-squared distribution. 95 JStPlInf 46 147-159

Arsian, Olcay; Constable, Patrick D. L.; — —: Convergence behavior of the EM algorithm for the multivariate t-distribution. 95 CommStA 24 2981-3000

Kepner, James L.: Kinney, John J.; — —: Generalised geometric random variables. 95 TeachgSt 17/1 7-9

Keramidas, Elaine M.; Lee, Jack C.: Selection of a covariance structure for growth curves. 95 BiomtrcJ 37 783-797

Keren, Daniel: Subrahmonia, Jayashree; — —; Cooper, David B.: Recognition of objects in 2D images or 3D range data based on high degree implicit polynomial curves or surfaces, algebraic invariants, and asymptotic Bayesian methods. 94 ASAProPEng 44-49

Kerim, Abdel; Saffar, A. El: RAS and least squares methods versus other mathematical techniques in developing and updating input-output tables (Arabic). 95 EgpStJ 39 120-143

Kerkyacharian, Gérard: Donoho, David L.; Johnstone, Iain M.; — —; Picard, Dominique: Wavelet shrinkage: Asymptopia?. 95 JRSS-B 57 301-337 (Disc: p337-369)

Kersting, G.; Klebaner, F. C.: Sharp conditions for nonexplosions and explosions in Markov jump processes. 95 AnlsProb 23 268-272

Keselman, H. J.; Carriere, K. C.; Lix, Lisa M.: Robust and powerful nonorthogonal analyses. 95 Psymtrka 60 395-418

— —; Keselman, Joanne C.; Lix, Lisa M.: The analysis of repeated measurements: Univariate tests, multivariate tests, or both?. 95 BJMaStPs 48 319-338

— —; Lix, Lisa M.: Improved repeated measures stepwise multiple comparison procedures. 95 JEdStat 20 83-99

Lix, Lisa M.; — —: Approximate degrees of freedom tests: A unified perspective on testing for mean equality. 95 PsychBu 117 547-560

Keselman, Joanne C.: Keselman, H. J.; — —; Lix, Lisa M.: The analysis of repeated measurements: Univariate tests, multivariate tests, or both?. 95 BJMaStPs 48 319-338

Kesidis, George: Analog optimization with Wong's stochastic neural network. 95 IEEENrlN 6 258-260

Kessler, J. T.; Kelsbeck, W. D. (Auth); Fielding, A. (Rev): Review of "Non-sampling errors in surveys". 95 Ststcian 44 289-290

Kessler, R. C.: Heath, A. C.; — —; Neale, M. C.; Hewitt, J. K.; Eaves, L. J.; Kendler, K. S.: Testing hypotheses about direction of causation using cross-sectional family data. 93 BehavGen 23 29-50

Kesten, H.: Ferrari, P. A.; — —; Martinez, S.; Picco, P.: Existence of quasi-stationary distributions. A renewal dynamical approach. 95 AnlsProb 23 501-521

Kesten, Harry: A ratio limit theorem for (sub) Markov chains on $\{1, 2, \ldots\}$ with bounded jumps. 95 AdvAppPr 27 652-691

— —: Branching random walk with a critical branching part. 95 JTherPrb 8 921-962

— —; Schonmann, Roberto H.: On some growth models with a small parameter. 95 ProbTher 101 435-468

Benjamini, Itai; — —: Percolation of arbitrary words in $\{0, 1\}^N$. 95 AnlsProb 23 1024-1060

Ketelle, Richard H.: Wolf, Dennis A.; Tardiff, Mark F.; — —: Characterizing groundwater at Oak Ridge National Laboratory hazardous waste sites. 94 ASAProEnvr 116-121

Kettenring, J. R.: Gnanadesikan, R.; — —; Tsao, S. L.: Weighting and selection of variables for cluster analysis. 95 JClass 12 113-136

Keuzenkamp, Hugo A.; Barten, Anton P.: Rejection without falsification: On the history of testing the homogeneity condition in the theory of consumer demand. 95 JEconmtx 67 103-127

— —; Magnus, Jan R.: Editors' introduction - The significance of testing in econometrics. 95 JEconmtx 67 1-3

— —; Magnus, Jan R.: On tests and significance in econometrics. 95 JEconmtx 67 5-24

Khalfin, L. A.; Klebanov, L. B.: A solution of the computer tomography paradox and estimating the distances between the densities of measures with the same marginals. 94 AnlsProb 22 2235-2241

Khalikulov, S. I.: On approximation of homogeneous random fields by fields with bounded spectrum. 95 ThPrMaSt 50 129-132

Khalil, Z.: Dimitrov, B.; Chukova, S.; — —: Definitions, characterizations and structural properties of probability distributions similar to the exponential. 95 JStPlInf 43 271-287

Ghitany, M. E.; El-Saidi, M. A.; — —: Characterization of a general class of life-testing models. 95 JAppProb 32 548-553

Khalique, A.: Abouammoh, A. M.; Hendi, M. I.; — —: Shock models with NBRU properties. 95 PakJStA 11 35-47

Khamis, Imad H.: Aggarwal, K. K. (Auth); — — (Rev): Review of "Reliability engineering". 95 Technmcs 37 234-235

Khan, I. H.: Huda, S.; — —; Mukerjee, Rahul: On optimal designs with restricted circular string property. 95 CmpStDA 19 75-83

Khan, Rasul A.: Detecting changes in probabilities of a multi-component process. 95 SqtlAnly 14 375-388

Khan, Shahjahan: Inference about the parameters of the multilinear model with equi-correlated responses. 94 JStatRsr 28 163-176

Khan-Malek, Richard C.: Kringle, Robert O.; — —: A statistical assessment of the recommendations from a conference on analytical methods validation in bioavailability, bioequivalence and pharmacokinetic studies. 94 ASAProBiop 510-514

Khare, Meena; Ezzati-Rice, Trena M.; Mohadjer, Leyla K.; Waksberg, Joseph: An evaluation of nonresponse bias in NHANES III (1988-91). 94 ASASrvyRsr 949-954

Ezzati-Rice, Trena M.; — —: Modeling of response propensity in the Third National Health and Nutrition Examination Survey. 94 ASASrvyRsr 955-959

Khatree, Ravindra: Gupta, Rameshwar D.; — —: Some comments on maximum likelihood estimators of scale parameters or its square. 94 JIndnStA 32 123-131

Khattree, Ravindra; Naik, Dayanand N.: Statistical tests for random effects in staggered nested designs. 95 JAppStat 22 495-505

Khawaja, Iffat: Hanif, Muhammad; Beg, M. A.; — —: Sampling with unequal probabilities: A historical review. 92 ProIrStCl 19-37

Kherwa, G. Ram; Prasad, Jagdish: A note on generalized binary proper efficiency-balanced block designs. 95 CommStA 24 2797-2801

Khinkis, Leonid A.; Greco, William R.: Critical review of response-surface models of synergy. 94 ASAProBiop 389-393

Khoshnevisan, Davar: On the distribution of bubbles of the Brownian sheet. 95 AnlsProb 23 786-805

— —; Lewis, Thomas M.: The favorite point of a Poisson process. 95 StocProc 57 19-38

Bass, Richard F.; — —: Laws of the iterated logarithm for local times of the empirical process. 95 AnlsProb 23 388-399

Horváth, Lajos; — —: Weight functions and pathwise local central limit theorems. 95 StocProc 59 105-123

Khots, M. S.: de Aguiar, P. F.; Bourguignon, B.; — —; Massart, D. L.; Phan-Than-Luu, R.: D-optimal designs. 95 Chemolab 30 199-210

Khurana, A. P.; Jha, V. D.: Some recurrence relations between product moments of order statistics of a doubly truncated Pareto distribution. 95 SankhyaB 57 1-9

Khuri, A. I.: A measure to evaluate the closeness of Satterthwaite's approximation. 95 BiomtrcJ 37 547-563

Khuri, André I.: A test to detect inadequacy of Satterthwaite's approximation in balanced mixed models. 95 MaOpfStS 27 45-54

— —: Response surface methodology within the framework of GLM. 93 JCISS 18 193-202

Bechtel, Gordon G.; Ofir, Chezy; — —: Replicated paired comparisons at the individual level. 95 BJMaStPs 48 115-127

Khurshid, Anwer; Sahai, Haredeo: A bibliography on telephone survey methodology. 95 JOffStat 11 325-367

Sahai, Hardeo; — —: Statistics in epidemiology. Methods, techniques and applications. 95 CRCPress:FL 352p

Sahai, Hardeo; — —: On analysis of epidemiological data involving a 2×2 contingency table: An overview of Fisher's exact test and Yates' correction for continuity. 95 JBiophSt 5 43-70

Khusanbayev, Ya. M.; Rakhimov, G. M.: On approximation of sums of conditionally independent random variables. 95 ThPrMaSt 50 133-142

Khush, G. S.: Kumar, Ish; — —: A general method of detecting additive, dominance and epistatic variation for metrical traits. V. Triple test cross analysis. 94 Heredity 72 563-569

Ki, Fanny Y. C.; Liu, Jen-Pei; Wang, Wenping; Chow, Shein-Chung: The impact of outlying subjects on decision of bioequivalence. 95 JBiophSt 5 71-94

Kianifard, Farid; Gallo, Paul P.: Poisson regression analysis in clinical research. 95 JBiophSt 5 115-129

Kleinbaum, D. G. (Auth); — — (Rev): Review of "Logistic regression: A self-learning text". 95 Technmcs 37 116-117

Vach, W. (Auth); — — (Rev): Review of "Logistic regression with missing values in the covariates". 95 Technmcs 37 460-461

Kidd, Martin; Laubscher, Nico F.: Robust confidence intervals for scale and its application to the Rayleigh distribution. 95 SoAfrStJ 29 199-217

Kiefer, Nicholas: Kapteyn, Arie; — —; Rust, John: Introduction: The microeconometrics of dynamic decision making. 95 JApEcmtx 10 1-7S

Kiefer, Nicholas M.: Bowlus, Audra J.; — —; Neumann, George R.: Estimation of equilibrium wage distributions with heterogeneity. 95 JApEcmtx 10 119-131S

Kiemele, Mark J.; Schmidt, Stephen R. (Auth); Champ, Charles W. (Rev): Review of "Basic statistics tools for continuous improvement (Third Edition)". 95 JQualTek 27 174-175

Kiemeney, L. A.: Schouten, L. J.; Straatman, H.; — —; Gimbrere, C. H.; Verbeek, A. L.: The capture-recapture method for estimation of cancer registry completeness: A useful tool?. 94 IJEpidem 23 1111-1116

Kiers, H. A. L.: Krijnen, W. P.; — —: An efficient algorithm for weighted PCA. 95 CmpStQ 10 299-306

Kiers, Henk A. L.: Maximization of sums of quotients of quadratic forms and some generalizations. 95 Psymtrka 60 221-245

— —; Smilde, Age K.: Some theoretical results on second-order calibration methods for data with and without rank overlap. 95 JChemtrc 9 179-195

Takane, Yoshio; — —; de Leeuw, Jan: Component analysis with different sets of constraints on different dimensions. 95 Psymtrka 60 259-280

Kifer, Y.: Multidimensional random subshifts of finite type and their large deviations. 95 ProbTher 103 223-248

Kiffe, Thomas R.: Matis, James H.; Zheng, Qi; — —: Describing the spread of biological populations using stochastic compartmental models with births. 95 MaBisci 126 215-247

Kiki, Maxime: Haiman, George; — —; Puri, Madan L.: Extremes of Markov sequences. 95 JStPlInf 45 185-201

Kikuchi, Kenichi; Mayekawa, Shin-ichi: On the sampling distribution of swap-rate. 95 Behvmtrk 22 185-204

Kikuchi, Yasuki: Yanagawa, Takashi; — —: Determination of the no-observed-adverse-effect level by the AIC. 95 SankhyaB 57 285-297

Kilik, Lindy: Cota, Albert A.; Evans, Charles R.; Longman, R. Stewart; Dion, Kenneth L.; — —: Using and misusing factor analysis to explore group cohesion. 95 JClPsych 51 308-316

Kim, Byung Hwee; Koh, Tae Wook; Baek, Hoh Yoo: Estimators with nondecreasing risk in a multivariate normal distribution. 95 JKoreaSt 24 257-266

Kim, Byung Soo; Oh, Kyong Joo; Park, Cheolyong: Overdispersion in count data - A review (Korean). 95 KorApS 8/2 147-161

Lee, Sunho; Park, Cheolyong; — —: Tests for detecting overdispersion in Poisson models. 95 CommStA 24 2405-2420

Kim, Byung-Do; Blattberg, Robert C.; Rossi, Peter E.: Modeling the distribution of price sensitivity and implications for optimal retail pricing. 95 JBES 13 291-303

Kim, Chansoo: Chung, Younshik; — —; Dey, Dipak K.: Simultaneous estimation of Poisson means under weighted entropy loss. 94 CalcutSt 44 165-174

Kim, Choongrak; Jeon, Jong-Woo: Detecting influential observations on the smoothing parameter in nonparametric regression. 95 JKoreaSt 24 495-506

Hong, Changkon; — —: Existence and uniqueness of the smoothest density with prescribed moments. 95 JKoreaSt 24 233-242

Kim, Chul Eung; Choi, ByoungSeon: A Sanov-type proof of the joint sufficiency of the sample mean and the sample variance. 95 JKoreaSt 24 563-568

Kim, Dong Hee; Lim, Dong Hoon: Rank tests for parallelism of regression lines against umbrella alternatives. 95 JNnparSt 5 289-302

— —; Lim, Dong Hoon: An asymptotically distribution-free test for parallelism of regression lines against umbrella alternatives (Korean). 95 KorApS 8/1 105-117

Kim, Dong K.; Taylor, Jeremy M. G.: The restricted EM algorithm for maximum likelihood estimation under linear restrictions on the parameters. 95 JASA 90 708-716

Kim, Dongcheol: The errors in the variables problem in the cross-section of expected stock returns. 95 JFinance 50 1605-1634

Kim, Dongjae; Wolfe, Douglas A.: Distribution-free partially sequential tests for treatments versus control setting. 95 BiomtrcJ 37 259-271

Kim, Donguk; Agresti, Alan: Improved exact inference about conditional association in three-way contingency tables. 95 JASA 90 632-639

Kim, Gwan Young; Joo, Sang Yeol; Kim, Yun Kyong: A sharp result of random upper functions for Levy processes. 95 JKoreaSt 24 65-76

Kim, Hae Kyung; Choi, Seung Hoe: Asymptotic properties of nonlinear least absolute deviation estimators. 95 JKoreaSt 24 127-139

Kim, Hag-Min; Fotopoulos, S. B.: Characterization study of a stochastic process associated with times in queueing systems. 94 ASAProPEng 163-168

Kim, Hea-Jung: On a balanced quadratic classification rule. 95 CommStA 24 607-623

— —: On a balanced classification rule. 95 JKoreaSt 24 453-470

Kim, Heon Joo: Sohn, Joong Kweon; — —: Bayesian inference for the two-parameter exponential models: Type-II censored case. 95 JKoreaSt 24 313-335

Kim, Hye Kyung: Hong, Dug Hun; — —; Kim, Ju Young: A weak convergence theorem for mixingale arrays. 95 JKoreaSt 24 273-280

Kim, Hyung Jun: Kim, Jinho; — —: The advantage of playing first in Go. 95 Chance 8/3 22-26

Kim, J. G.: Bai, D. S.; Chun, Y. R.; — —: Failure-censored accelerated life test sampling plans for Weibull distribution under expected test time constraint. 95 RelEngSS 50 61-68

Kim, Jay: Oh, Sungsoo; Paletz, Dave; — —; Salyers, Eddie: Development of edit parameters for 1992 economic census enterprise reports. 94 ASASrvyRsr 1144-1149

Kim, Jay Jong-Ik; Zaslavsky, Alan; Blodgett, Robert: Between-state heterogeneity of undercount rates and surrogate variables in the 1990 U.S. Census. 95 SurvyM 21 55-62

Kim, Jeong-Ran: Kim, Joo-Han; — —: Two-sample chi-square test for randomly censored data (Korean). 95 KorApS 8/2 109-119

Kim, Ji-Hyun: Conditional bootstrap methods for censored survival data. 95 JKoreaSt 24 197-218

— —: A sampling scheme for the estimation of low proportion (Korean). 95 KorApS 8/1 1-7

Kim, Jinbang; De Marchi, Neil; Morgan, Mary S.: Empirical model particularities and belief in the natural rate hypothesis. 95 JEconmtx 67 81-102

Kim, Jinheum; Song, Moon Sup: A goodness-of-fit test for the additive risk model with a binary covariate. 95 JKoreaSt 24 537-549

Kim, Jinho; Kim, Hyung Jun: The advantage of playing first in Go. 95 Chance 8/3 22-26

Kim, John S.; Mattson, LeRoy T.: Use of field data in reliability studies of medical devices. 94 ASAProQlPr 200-203

Kim, Joo-Han; Kim, Jeong-Ran: Two-sample chi-square test for randomly censored data (Korean). 95 KorApS 8/2 109-119

Kim, Ju Young: Hong, Dug Hun; Kim, Hye Kyung; — —: A weak convergence theorem for mixingale arrays. 95 JKoreaSt 24 273-280

Kim, Keeyoung: Park, YouSung; Lee, Haeyong; — —: Flows and some extreme values in multiple server open Jackson network. 95 JKoreaSt 24 389-405

Kim, Kwang-Y.: North, Gerald R.; — —; Shen, Samuel S. P.; Hardin, James W.: Detection of forced climate signals. Part I: Filter theory. 95 JClimate 8 401-408

Kim, Kyeong Soo; Lee, Byeong Gi: GA-based optimal dimensioning of three-level traffic shaper for statistical multiplexing in ATM networks. 95 AnlsTele 50 624-631

Kim, Kyu Seong; Jeon, Jong Woo; Park, Hong Nai: A study on the sample design of the fishery household economy survey (Korean). 95 KorApS 8/2 43-54

Kim, KyungMann: SEQPWR and SEQOPR: Computer programs for design of maximum information trials based on group sequential logrank tests. 95 CmpBiomd 46 143-153

Kim, Kyungmann: A bivariate cumulative probit regression model for ordered categorical data. 95 StatMed 14 1341-1352

Kim, KyungMann; Boucher, Hélène; Tsiatis, Anastasios A.: Design and analysis of group sequential logrank tests in maximum duration versus information trials. 95 Biomtrcs 51 988-1000

Kim, Kyungmann: Tsiatis, Anastasios A.; Boucher, Hélène; — —: Sequential methods for parametric survival models. 95 Biomtrka 82 165-173

Kim, KyungMann: Williamson, John M.; — —; Lipsitz, Stuart R.: Analyzing bivariate ordinal data using a global odds ratio. 95 JASA 90 1432-1437

Kim, Mimi Y.; Pasternack, Bernard S.; Carroll, Raymond J.; Koenig, Karen L.; Toniolo, Paolo G.: Estimating the reliability of an exposure variable in the presence of confounders. 95 StatMed 14 1437-1446

Kim, Minbo; Hill, R. Carter: Shrinkage estimation in nonlinear regression. The Box-Cox transformation. 95 JEconmtx 66 1-33

Kim, Myung Geun: Local influence in multivariate regression. 95 CommStA 24 1271-1278

— —: The influence in comparing covariance matrices. 95 CommStA 24 1431-1441

Kim, Peter T.: Chapman, G. Robert; Chen, Gemai; — —: Assessing geometric integrity through spherical regression techniques. 95 StSinica 5 173-191 (Disc: p191-220)

Chapman, G. Robert; Chen, Gemai; — —: Reply to comments on "Assessing geometric integrity through spherical regression techniques". 95 StSinica 5 215-220

Kim, S. S.; Park, S. H.: Dynamic plots for displaying the roles of variables and observations in regression model. 95 CmpStDA 19 401-418

Kim, Soon-ock; Song, Moon Sup: Distribution-free tests for symmetry based on the linear signed rank statistics. 95 JNnparSt 5 311-321

Kim, Sung Lai: Asymptotic properties of the stopping times in a certain sequential procedure. 95 JKoreaSt 24 337-347

Kim, SungHo; Jeong, MiSook: A bootstrap approach for factor numbers in binary data (Korean). 95 KorApS 8/2 201-216

Kim, T. S.; Chae, K. C.: Probabilities of hands in 1 person seven-card stud (Korean). 95 KorApS 8/2 163-177

Kim, Tae Yoon; Cox, Dennis D.: Asymptotic behaviors of some measures of accuracy in nonparametric curve estimation with dependent observations. 95 JMultiAn 53 67-93

Cox, Dennis D.; — —: Moment bounds for mixing random variables useful in nonparametric function estimation. 95 StocProc 56 151-158

Kim, Tae-sung: A note on stationary linearly positive quadrant dependent sequences. 95 JKoreaSt 24 249-256

Kim, Tae-Sung; Seo, Hye-Young: A note on some negative dependence notions. 95 CommStA 24 845-858

— —; Seo, Hye-Young: An invariance principle for stationary strong mixing random fields. 95 JKoreaSt 24 281-292

Kim, Taeil; Taylor, Lowell J.: The employment effect in retail trade of California's 1988 minimum wage increase. 95 JBES 13/2 175-182

Kim, W. C.: Lee, Y. H.; Cho, S.; — —; Park, B. U.: On estimating integrated squared spectral density derivatives. 95 JStPlInf 48 165-184

Kim, Woo-Chul; Han, Kyung Soo: On the performance of empirical Bayes simultaneous interval estimates for all pairwise comparisons. 95 JKoreaSt 24 161-181

Kim, Woochul; Ko, Duk-Hyun: On the performance of iterated wild bootstrap interval estimation of the mean response. 95 JKoreaSt 24 551-562

Kim, Yun Kyong: Kim, Gwan Young; Joo, Sang Yeol; — —: A sharp result of random upper functions for Levy processes. 95 JKoreaSt 24 65-76

Kim, Yung-Hwan: Lee, Sukhoon; Park, Nae-Hyun; — —: A clustering method using the Coulomb energy network (Korean). 95 KorApS 8/1 39-50

Kimber, Alan: Everitt, B. S. (Auth); — — (Rev): Review of "A handbook of statistical analyses using S-Plus". 95 Ststcian 44 408-409

Kimber, Alan C.: Jewell, Nicholas P.; — —; Lee, Mei-Ling Ting; Whitmore, G. A. (Ed): Lifetime data: Models in reliability and survival analysis. 95 Kluwer:MA:Hllnd 424p

Kimber, Hilary: The 'Golden Egg'. 95 TeachgSt 17/2 34-37

Kimble, David L.: Estimation of monetary value of uninterrupted telephone service. 94 ASAProBuEc 202-204

Spaulding, JoAnn K.; — —: A decision theory model for stocking switch packs. 94 ASAProBuEc 205-210

Kimura, Miyoshi: Kakiuchi, Itsuro; — —: Majorization methods on hyperplanes and their applications. 95 JStPlInf 47 217-235

Kinack, Mark: Sunter, Deborah; — —: Redesigning the Canadian Labour Force Survey questionnaire: Development and testing. 94 ASASrvyRsr 908-912

Kinack, Mark D.: Rowland, John H.; — —: A comparison of two methods of automated industry coding. 94 ASASrvyRsr 1130-1133

Kinateder, Kimberly K. J.: Corner Markov processes. 95 JTherPrb 8 539-547

Kindel, Carrol Benner: Electronic data collection at the National Center for Education Statistics: Successes in the collection of library data. 94 JOffStat 10 93-102

Kindred, Machell: Gbur, Philip M.; — —; Mersch, Michael L.: Master address file: Update methodology and quality improvement program. 94 ASASrvyRsr 748-753 (Disc: p754-756)

King, Bruce: Creighton, J. H. C. (Auth); — — (Rev): Review of "A first course in probability models and statistical inference". 95 AmerStat 49 113-113

King, Karen E.: Jackson, Betty J.; Frazier, Richard L.; — —; Schwanz, Dennis J.: Improving the coverage of private elementary-secondary schools. 94 ASASrvyRsr 833-838

King, Karen Ellen; Kaufman, Steve: Estimation issues related to the student component of the SASS. 94 ASASrvyRsr 1111-1115 (Disc: p1128-1129)

King, Maxwell L.; Harris, David C.: The application of the Durbin-Watson test to the dynamic regression model under normal and non-normal errors. 95 EcomtrRv 14 487-510

Granger, Clive W. J.; — —; White, Halbert: Comments on testing economic theories and the use of model selection criteria. 95 JEconmtx 67 173-187

Kinghorn, Brian P.: van Arendonk, Johan A. M.; Tier, Bruce; — —: Use of multiple genetic markers in prediction of breeding values. 94 Genetcs 137 319 329

Kingkade, W. Ward; Vasin, Sergey A.: Welfare consequences of mortality dynamics in the former Soviet Republics: Length of working life. 94 ASAProSoSt 94-99

Kingman, Darwin: Glover, Fred; — —; Phillips, Nancy V. (Auth); Hansen, Pierre (Rev): Review of "Network models in optimization and their applications in practice". 95 JClass 12 150-152

Kingman, J. F. C.: — — (Auth); Booker, Jane (Rev): Review of "Poisson processes". 95 Technmcs 37 124-124

— — (Auth); Kemp, A. W. (Rev): Review of "Poisson processes". 94 BLndnMaS 26 612-614

Kinnear, P. R.; Gray, C. D. (Auth); Carpenter, James (Rev): Review of "SPSS for Windows made simple". 95 JAppStat 22 436-437

— —; Gray, C. D. (Auth); Carpenter, James (Rev): Review of "SPSS/PC+ made simple". 95 JAppStat 22 436-437

Kinney, John J.; Kepner, James L.: Generalised geometric random variables. 95 TeachgSt 17/1 7-9

Kintner, Hallie J.: A private sector view of census 2000. 94 ASAProSoSt 3-5

Kipnis, C.; Landim, C.; Olla, S.: Macroscopic properties of a stationary non-equilibrium distribution for a non-gradient interacting particle system. 95 AIHenrPB 31 191-221

— —; Léonard, C.: Large deviations of an asymmetric hydrodynamical system of independent particles (French). 95 AIHenrPB 31 223-248

Benois, O.; — —; Landim, C.: Large deviations from the hydrodynamical limit of mean zero asymmetric zero range processes. 95 StocProc 55 65-89

Ferrari, P. A.; — —: Second class particles in the rarefaction fan. 95 AIHenrPB 31 143-154

Kirby, S. P. J.: Theobald, C. M.; — —: Discrimination using covariance selection models in the automated allocation of human chromosomes. 95 Ststcian 44 235-240

Kirk, Roger E.: Experimental design: Procedures for the behavioral sciences (Third edition). 95 BrooksCole:CA 921p

Kirkendall, Nancy: Dandekar, Ramesh; Cohen, Michael; — —; Kydes, Andy: Applicability of localized linear regression techniques to develop response surface models. 94 ASAProPEng 87-91

Kirkpatrick, Mark; Slatkin, Montgomery: Searching for evolutionary patterns in the shape of a phylogenetic tree. 93 Evolutn 47 1171-1181

Kirman, Alan: Härdle, Wolfgang; — —: Nonclassical demand - A model-free examination of price-quantity relations in the Marseille fish market. 95 JEconmtx 67 227-257

Kirmani, S. N. U. A.; Gupta, Ramesh C.: Some results on randomly stopped minimal repair processes. 95 CommStSM 11 631-644

Kochar, Subhash C.; — —: Some results on normalized spacings from restricted families of distributions. 95 JStPlInf 46 47-57

Kirsch, Irwin S.: Burke, John; Mohadjer, Leyla; Green, Jim; Waksberg, Joseph; — —; Kolstad, Andrew: Composite estimation in national and state surveys. 94 ASASrvyRsr 873-878

Kirzhner, Valery M.: Korol, Abraham B.; Ronin, Yefim I.; — —: Interval mapping of quantitative trait loci employing correlated trait complexes. 95 Genetcs 140 1137-1147

Kiser, Terry L.: Schwertman, Neil C.; — —: The role of geometry in pairwise and mutual independence. 95 AmerStat 49 56-59

Kish, Leslie: Methods for design effects. 95 JOffStat 11 55-77

— —: Survey sampling. 95 Wiley:NY:UK 661p

— —; Frankel, Martin R.; Verma, Vijay; Kaćiroti, Niko: Design effects for correlated $(P_i - P_j)$. 95 SurvyM 21 117-124

Kishino, Hirohisa: Hasegawa, Masami; — —: Accuracies of the simple methods for estimating the bootstrap probability of a maximum-likelihood tree. 94 MolBioEv 11 142-145

Kitaev, Mikhail Yu.; Rykov, Vladimir V.: Controlled queueing systems. 95 CRCPress:FL 304p

Kitamura, Yuichi: Estimation of cointegrated systems with I(2) processes. 95 EconTher 11 1-24

— —; Phillips, Peter C. B.: Efficient IV estimation in nonstationary regression. 95 EconTher 11 1095-1130

Kitanidis, Peter K.: Quasi-linear geostatistical theory for inversing. 95 WaterReR 31 2411-2419

Kitsos, Christos P.; Müller, Christine H.: Robust linear calibration. 95 MaOpfStS 27 93-106

— —; Müller, Werner G. (Ed): MODA 4 - Advances in model-oriented data analysis. 95 Physica:Ger 297p

Kiuchi, Amy S.; Hartigan, J. A.; Holford, Theodore R.; Rubinstein, Pablo; Stevens, Cladd E.: Change points in the series of T4 counts prior to AIDS. 95 Biomtrcs 51 236-248

Kiviet, J. F.: van den Doel, I. T.; — —: Neglected dynamics in panel data models; consequences and detection in finite samples. 95 StNeerla 49 343-361

Kiviet, Jan F.: On bias, inconsistency, and efficiency of various estimators in dynamic panel data models. 95 JEconmtx 68 53-78

— —; Phillips, Garry D. A.; Schipp, Bernhard: The bias of OLS, GLS, and ZEF estimators in dynamic seemingly unrelated regression models. 95 JEconmtx 69 241-266

Kjærulff, Uffe: dHugin: A computational system for dynamic time-sliced Bayesian networks. 95 IntJFor 11 89-111

Klar, Bernhard: Henze, Norbert; — —: Bootstrap based goodness of fit tests for the generalized Poisson model. 95 CommStA 24 1875-1896

Klar, Neil: Simpson, Judy M.; — —; Donner, Allan: Accounting for cluster randomization: A review of primary prevention trials, 1990 through 1993. 95 AJPubHlh 85 1378-1383

Klar, R.: Opitz, O.; Lausen, B.; — — (Ed); Windham, M. P. (Rev): Review of "Information and classification: Concepts, methods and applications". 95 JClass 12 157-158

Klass, Michael J.; Zhang, Cun-Hui: On the almost sure minimal growth rate of partial sum maxima. 94 AnlsProb 22 1857-1878

Hahn, Marjorie G. W.; — —: Uniform local probability approximations: Improvements on Berry-Esseen. 95 AnlsProb 23 446-463

Klassen, Albert D.: Harris, T. Robert; — —; Wilsnack, Richard W.; Wilsnack, Sharon C.: A model for the interaction of partnership dissolution and relationship quality as predicators of remission of problem drinking in women. 94 ASAProSoSt 232-234

Klauer, K. C.: Krijnen, W. P. (Auth); — — (Rev): Review of "The analysis of three-way arrays by constrained PARAFAC methods". 95 JClass 12 147-149

Klebaner, F. C.: Hamza, K.; — —: Conditions for integrability of Markov chains. 95 JAppProb 32 541-547

Kersting, G.; — —: Sharp conditions for nonexplosions and explosions in Markov jump processes. 95 AnlsProb 23 268-272

Klebanov, L. B.; Rachev, S. T.: The method of moments in computer tomography. 95 MaScitst 20 1-14

Khalfin, L. A.; — —: A solution of the computer tomography paradox and estimating the distances between the densities of measures with the same marginals. 94 AnlsProb 22 2235-2241

Klecolt-Glaser, Janice: Huh, Myung-Hoe; Critchlow, Douglas E.; Verducci, Joseph S.; — —; Glaser, Ronald; Malarkey, William B.: A symmetric analysis of paired rankings with application to temporal patterns of hormonal concentration. 95 Biomtrcs 51 1361-1371

Kleczkowski, A.: Grenfell, B. T.; — —; Gilligan, C. A.; Bolker, B. M.: Spatial heterogeneity, nonlinear dynamics and chaos in infectious diseases. 95 StMedRes 4 160-183

Kleijnen, Jack P. C.; Van Groenendaal, Willem J. H.: Two-stage versus sequential sample-size determination in regression analysis of simulation experiments. 95 AJMMgSci 15 83-114

Klein, John P.; Keiding, Niels; Kreiner, Svend: Graphical models for panel studies, illustrated on data from the Framingham Heart Study. 95 StatMed 14 1265-1290

Moeschberger, M. L.; — —: Statistical methods for dependent competing risks. 95 LiDaAnal 1 195-204

Wang, S. T.; — —; Moeschberger, M. L.: Semi-parametric estimation of covariate effects using the positive stable frailty model. 95 AplStMDA 11 121-133

Zheng, Ming; — —: Estimates of marginal survival for dependent competing risks based on an assumed copula. 95 Biomtrka 82 127-138

— —; Goel, Prem K. (Auth); Schlittgen, Rainer (Rev): Review of "Survival analysis: State of the art". 95 CmpStDA 20 592-593

Klein, Maurice: Baccelli, François; — —; Zuyev, Sergei: Perturbation analysis of functionals of random measures. 95 AdvAppPr 27 306-325

Klein, Philip A.: Niemira, Michael P.; — — (Auth); Mullineux, Andy W. (Rev): Review of "Forecasting financial and economic cycles". 95 IntJFor 11 191-192

Klein, Robert E.: Schwartz, Steven H.; — —; Adashek, Daniel: Veterans living in group quarters: What data from the 1990 Census show. 94 ASAProGovt 13-18

Klein, Ronald: Gange, Stephen J.; Linton, Kathryn L. P.; Scott, Alastair J.; DeMets, David L.; — —: A comparison of methods for correlated ordinal measures with ophthalmic applications. 95 StatMed 14 1961-1974

Klein, Ruben: Gokhale, D. V.; — —: Analysis of contingency tables by marginal scores. 95 RBrzlPrS 9 25-42

Gokhale, D. V.; — —: Association models for contingency tables via minimum discrimination information approach and correspondence analysis. 95 RBrzlPrS 9 83-98

Klein, Stephen P.; Bell, Robert M.: How will the NCAA's new standards affect minority student-athletes?. 95 Chance 8/3 18-21

Kleinbaum, D. G.: Survival analysis. A self-learning text. 95 Springer:Brln:NY 250p

— — (Auth); Howel, Denise (Rev): Review of "Logistic regression: A self learning text". 95 Ststcian 44 410-411

— — (Auth); Kianifard, Farid (Rev): Review of "Logistic regression: A self-learning text". 95 Technmcs 37 116-117

Kleinbaum, David G.: Teaching short courses in biostatistics and epidemiology. 95 AmerStat 49 200-203

— —: Teaching short courses in biostatistics and epidemiology. 94 ASAProStEd 29-32

Kleinberg, Norman L.: A note on finite securities market models. 95 StocAnAp 13 543-554

Kleinman, D. L.: Shakeri, M.; Pattipati, K. R.; — —: Optimal measurement scheduling for state estimation. 95 IEEEAES 31 716-729

Kleinschmidt, Peter; Mitterreiter, Ilse: Improved chromosome classification using monotonic functions of Mahalanobis distance and the transportation method. 94 ZeitOpRs 40 305-323

Klemm, Rebecca; Stroup, Donna: Meta-analysis for public health professionals. 94 ASAProEpid 107-111 (Disc: p125-126)

Klemm, Rebecca J.: Krumhaus, Susan G.; — —: An overview of the National Survey of Veterans. 94 ASAProGovt 19-21

Klensin, John C.: When the metadata exceed the data: Data management with uncertain data. 95 StatComp 5 73-84

Klesov, O. I.: The law of large numbers for multiple sums of independent identically distributed random variables. 95 ThPrMaSt 50 77-87

Kline, Paul: Krzanowski, Wojtek J.; — —: Cross-validation for choosing the number of important components in principal component analysis. 95 MultBehR 30 149-165

Klinke, S.: Härdle, W.; — —; Turlach, B. A. (Ed): XploRe: An interactive statistical computing environment. 95 Springer:Brln:NY 387p

Härdle, W.; — —; Turlach, B. A. (Ed); Murdoch, D. J. (Rev): Review of "XploRe: An interactive statistical computing environment". 96 ShrtBkRv 16/1 3 3

Klockars, Alan J.; Hancock, Gregory R.; McAweeney, Mary J.: Power of unweighted and weighted versions of simultaneous and sequential multiple-comparison procedures. 95 PsychBu 118 300-307

Kloeden, Peter E.; Platen, Eckhard (Auth); Gelbrich, Matthias; Römisch, Werner (Rev): Review of "Numerical solution of stochastic differential equations". 95 SIAMRvw 37 272-275

— —; Platen, Eckhard; Schurz, Henri (Auth); Hall, Timothy (Rev): Review of "Numerical solution of SDE through computer experiments". 95 Technmcs 37 125-125

Kloek, Teun: Franses, Philip Hans; — —: A periodic cointegration model of quarterly consumption. 95 AplStMDA 11 159-166

Klopotowski, Andrzej: Bretagnolle, Jean; — —: On the existence of sequences of s by s independent exchangeable or stationary random variables (French). 95 AIHenrPB 31 325-350

Klotz, Jerome H.: Updating simple linear regression. 95 StSinica 5 399-403

Klüppelberg, C.; Mikosch, T.: Delay in claim settlement and ruin probability approximations. 95 ScanActJ 154-168

Klüppelberg, Claudia; Mikosch, Thomas: Explosive Poisson shot noise processes with applications to risk reserves. 95 Bernoull 1 125-147

Mikosch, Thomas; Gadrich, Tamar; — —; Adler, Robert J.: Parameter estimation for ARMA models with infinite variance innovations. 95 AnlsStat 23 305-326

Kmietowicz, Z. W.: Significance of the sample lead: Some theoretical and empirical results. 95 CommStA 24 1507-1519

— —: Exact distribution of the sample lead. 95 JAppStat 22 151-161

— —: Correction to "The significance of the lead in opinion polls" (V17 p9-30). 95 JAppStat 22 551-551

— —: Correction to "Exact distribution of the sample lead" (V22 p151-161). 95 JAppStat 22 552-552

— —: Accuracy of indices of industrial production in developing countries. 95 Ststcian 44 295-307

— —: The exact distribution of the sample lead: Further results. 95 Ststcian 44 457-467

— —; Yannoulis, Y. (Auth); Anton, Vicente Nuñez (Rev): Review of "Statistical tables for economic, business and social studies (Second edition)". 95 JAppStat 22 439-440

Knapp, M.; Seuchter, S. A.; Baur, M. P.: Two-locus disease models with two marker loci: The power of affected-sib-pair tests. 94 AJHuGene 55 1030-1041

— —; Seuchter, S. A.; Baur, M. P.: Linkage analysis in nuclear families. 1: Optimality criteria for affected sib-pair tests. 94 HumHrdty 44 37-43

— —; Seuchter, S. A.; Baur, M. P.: Linkage analysis in nuclear families. 2: Relationship between affected sib-pair tests and lod score analysis. 94 HumHrdty 44 44-51

Knapp, R. B.: Shaffer, R. E.; Small, G. W.; Combs, R. J.; — —; Kroutil, R. T.: Experimental design protocol for the pattern recognition analysis of bandpass filtered Fourier transform infrared interferograms. 95 Chemolab 29 89-108

Knaub, James R., Jr: Relative standard error for a ratio of variables at an aggregate level under model sampling. 94 ASASrvyRsr 310-312

Kneafsey, Kevin P.: Kroner, Kenneth F.; — —; Claessens, Stijn: Forecasting volatility in commodity markets. 95 JForcast 14 77-95

Kneip, Alois; Engel, Joachim: Model estimation in nonlinear regression under shape invariance. 95 AnlsStat 23 551-570

Gasser, Theo; — —: Searching for structure in curve samples. 95 JASA 90 1179-1188

Knessl, Charles; Tier, Charles: Asymptotic properties of first passage times for tandem Jackson networks I: Buildup of large queue lengths. 95 CommStSM 11 133-162

Knezevic, J.: — — (Auth); Lynch, James (Rev): Review of "Reliability, maintainability and supportability: A probabilistic approach". 95 Technmcs 37 122-122

Knickerbocker, R. K.: Carroll, R. J.; — —; Wang, C. Y.: Dimension reduction in a semiparametric regression model with errors in covariates. 95 AnlsStat 23 161-181

Knight, Frank B.: Amir, Madjid; — —: Law of the iterated logarithm and local variations at zero of the sticky Brownian motion. 95 StPrLet 23 289-295

Knight, J. L.: Hisamatsu, H.; — —; Maekawa, K.: The exact distribution of the OLS and GLS estimators in regression with an integrated regressor and correlated errors - Comparison of numerical and Monte Carlo integration. 95 MaCmpSim 39 273-277

Knight, Keith: Bivariate autoregressive processes with finite and infinite variance errors. 94 ASAProBuEc 28-34

Knill, E.: Bruno, W. J.; — —; Balding, D. J.; Bruce, D. C.; Doggett, N. A.; Sawhill, W. W.; Stallings, R. L.; Whittaker, C. C.; Torney, D. C.: Efficient pooling designs for library screening. 95 Genomics 26 21-30

Knott, Michael: Cracknell, David; — —: The measurement of price elasticities - The BT experience. 95 IntJFor 11 321-329

Knott, S. A.: Prediction of the power of detection of marker-quantitative trait locus linkages using analysis of variance. 94 ThApGene 89 318-322

Knott, Sara A.: Haley, Chris S.; — —; Elsen, Jean-Michel: Mapping quantitative trait loci in crosses between outbred lines using least squares. 94 Genetcs 136 1195-1207

Knüsel, Leo: On the accuracy of the statistical distributions in GAUSS. 95 CmpStDA 20 699-702

Ko, Daijin: Chang, Ted; — —: M-estimates of rigid body motion on the sphere and in Euclidean space. 95 AnlsStat 23 1823-1847

Everitt, Brian S. (Auth); — — (Rev): Review of "A handbook of statistical analyses using S-Plus". 95 Technmcs 37 232-233

Spector, Phil (Auth); — — (Rev): Review of "An introduction to S and S-Plus". 95 Technmcs 37 232-232

Ko, Duk-Hyun: Kim, Woochul; — —: On the performance of iterated wild bootstrap interval estimation of the mean response. 95 JKoreaSt 24 551-562

Kobus, Maria: Generalized Poisson distributions as limits of sums for arrays of dependent random vectors. 95 JMultiAn 52 199-244

Koç, Cemal: Recurring-with-carry sequences. 95 JAppProb 32 966-971

Koch, Armin; Haag, Uwe (Ed): The Statistical Software Guide '94/95. 95 CmpStDA 19 237-262

Koch, Gary G.: Davies, Glenn M.; — —; Beck, James: Statistical strategies for event rate comparisons in dental studies. 94 ASAProBiop 420-424

Davis, Randy L.; — —: Survey and conditional logistic regression methods for the analysis of a multiperiod crossover study to compare two treatments. 94 ASAProBiop 446-451

Gansky, Stuart A.; — —: Evaluation of alternative strategies for managing stratification factors in the analysis of randomized clinical trials. 94 ASAProBiop 405-410

Marques, Eliana H. de F.; — —: A model for distributions and correlation structure of multivariate polytomous data from a two-stage cluster sample. 94 ASAProBiop 346-351

Kochar, Subhash: Dykstra, Richard; — —; Robertson, Tim: Likelihood ratio tests for symmetry against one-sided alternatives. 95 AnInStMa 47 719-730

Dykstra, Richard; — —; Robertson, Tim: Inference for likelihood ratio ordering in the two-sample problem. 95 JASA 90 1034-1040

Dykstra, Richard; — —; Robertson, Tim: Likelihood based inference for cause specific hazard rates under order restrictions. 95 JMultiAn 54 163-174

Joag-dev, Kumar; — —; Proschan, Frank: A general composition theorem and its applications to certain partial orderings of distributions. 95 StPrLet 22 111-119

Kochar, Subhash C.; Kirmani, S. N. U. A.: Some results on normalized spacings from restricted families of distributions. 95 JStPlInf 46 47-57

Aly, Emad-Eldin A. A.; — —: A CUSUM test for symmetry against positive biasedness. 94 CalcutSt 44 157-163

Barmi, Hammou El; — —: Likelihood ratio tests for bivariate symmetry against ordered alternatives in a square contingency table. 95 StPrLet 22 167-173

Kocherlakota, S.: Johnson, N. L.; Kotz, S.; Balakrishnan, N. (Auth); — — (Rev): Review of "Continuous univariate distributions, Volumes 1 and 2 (Second edition)". 95 Biomtrcs 51 1582-1583

Kocherlakota, Subrahmaniam: Discrete bivariate weighted distributions under multiplicative weight function. 95 CommStA 24 533-551

Kockläuner, Gerhard: A note on beta coefficients in SPSS. 95 AllgeStA 79 115-116

Kodell, Ralph L.: Ahn, Hongshik; — —: Estimation and testing of tumor incidence rates in experiments lacking cause-of-death data. 95 BiomtrcJ 37 745-763

Koehler, Kenneth J.; Symanowski, James T.: Constructing multivariate distributions with specific marginal distributions. 95 JMultiAn 55 261-282

Chakak, Abderrahmane; — —: A strategy for constructing multivariate distributions. 95 CommStB 24 537-550

Morel, Jorge G.; — —: A one-step Gauss-Newton estimator for modelling categorical data with extraneous variation. 95 ApplStat 44 187-200

Koel, M.; Kaljurand, M.; Lochmüller, C. H.; Moebus, M.: Factor analytical resolution of simultaneous, dynamic thermal process in thermochromatography of oil shale. 95 Chemolab 30 173-178

Koel, Mihkel: Elomaa, Matti; Lochmüller, Charles H.; Kaljurand, Mihkel; — —: Application of evolving factor analysis in thermochromatography. 95 Chemolab 30 129-132

Koell, Christophe: Asymptotic optimality of the Wald sequential test. 95 SqtlAnly 14 341-360

Koenig, Karen L.: Kim, Mimi Y.; Pasternack, Bernard S.; Carroll, Raymond J.; — —; Toniolo, Paolo G.: Estimating the reliability of an exposure variable in the presence of confounders. 95 StatMed 14 1437-1446

Koenker, R. W.: Hasan, M. N.; — —: Robust rank tests of the unit root hypothesis based on regression rank scores. 94 ASAProBuEc 35-43

Koepsell, Thomas: Diehr, Paula; Martin, Donald; — —; Cheadle, Allen; Wagner, Edward; Psaty, Bruce: Optimal survey design for community-intervention evaluations: Cohort, cross-sectional, or both?. 94 ASAProEpid 34-43

Diehr, Paula; Martin, Donald C.; — —; Cheadle, Allen: Breaking the matches in a paired t-test for community interventions when the number of pairs is small. 95 StatMed 14 1491-1504

Koepsell, Thomas D.; Diehr, Paula H.; Cheadle, Allen; Kristal, Alan: Symposium on community intervention trials. 95 AJEpidm 142 594-599

Koesmarno, H. K.: Modelling and simulation of barley kernel growth. 95 EnvrMtrc 6 511-516

Kofman, D.; Korezlioglu, H.: Some EATA properties for marked point processes. 95 JAppProb 32 922-929

Kogan, Y.; Liptser, R.; Shenfild, M.: State-dependent Beneš buffer model with fast loading and output rates. 95 AnlsApPr 5 97-120

Kogut, Sarah J. H.: A comparison of several outcome measures in cost-effectiveness studies. 94 ASAProBiop 267-271

Koh, Tae Wook: Kim, Byung Hwee; — —; Baek, Hoh Yoo: Estimators with nondecreasing risk in a multivariate normal distribution. 95 JKoreaSt 24 257-266

Kohatsu-Higa, Arturo: Ahn, Hyungsok; — —: The Euler scheme for anticipating stochastic differential equations. 95 Stochatx 54 247-269

Kohn, Nina E.: Lesser, Martin L.; — —; Napolitano, Barbara A.; Pahwa, Savita: The FU-PLOT: A graphical method for visualizing the timing of follow-up in longitudinal studies. 95 AmerStat 49 139-143

Köhne, Stephan: Westfall, Peter H.; Young, S. Stanley (Auth); — —; Pigeot, Iris (Rev): Review of "Resampling-based multiple testing. Examples and methods for p-value adjustment". 95 CmpStDA 20 235-236

Koike, Ken-ichi: Akahira, Masafumi; Takeuchi, Kei; — —: Unbiased estimation in sequential binomial sampling. 92 RStAJp 39/4 1-13

Kojetin, Brian; Tucker, Clyde; Cashman, Erin: Response to a government survey as political participation: The relation of economic and political conditions to refusal rates in the Current Population Survey. 94 ASASrvyRsr 1236-1241

Tucker, Clyde; — —; Blass, Richard; Tucker, Ronald: Measuring the effects of the transition on field performance. 94 ASASrvyRsr 348-353 (Disc: p360-361)

Kojetin, Brian A.: Characteristics of nonrespondents to the Current Population Survey (CPS) and Consumer Expenditure Interview Survey (CEIS). 94 ASASrvyRsr 966-971

Kok, Frans J.: Schouten, Evert G.; Dekker, Jacqueline M.; — —; LeCessie, Saskia; van Houwelingen, Hans C.; Pool, Jan: Reply to comment on "Risk ratio and rate ratio estimation in case-cohort designs: Hypertension and cardiovascular mortality". 95 StatMed 14 1610-1610

Kokic, P. N.; Bell, P. A.: Optimal winsorizing cutoffs for a stratified finite population estimator. 94 JOffStat 10 419-435

Kokic, Philip N.; Moon, Lynelle; Gooday, Jane; Chambers, Ray L.: Estimating temporal farm income distributions using spatial smoothing techniques. 95 AstrlJSt 37 129-143

Kokoszka, Piotr S.; Taqqu, Murad S.: Fractional ARIMA with stable innovations. 95 StocProc 60 19-47

Kolassa, John E.: A comparison of size and power calculations for the Wilcoxon statistic for ordered categorical data. 95 StatMed 14 1577-1581

— —: Edgeworth approximations for rank sum test statistics. 95 StPrLet 24 169-171

Koldobsky, A. L.: Dilworth, S. J.; — —: The Fourier transform of order statistics with applications to Lorentz spaces. 95 IsrelJMa 92 411-425

Kolen, M. J.; Brennan, R. L.: Test equating: Methods and practices. 95 Springer:Brln:NY 333p

— —; Brennan, R. L. (Auth); Ramsay, J. O. (Rev): Review of "Test equating: Methods and practices". 96 ShrtBkRv 16/1 6-6

Kollo, Tõnu; von Rosen, Dietrich: Approximating by the Wishart distribution. 95 AnInStMa 47 767-783

— —; von Rosen, Dietrich: Minimal moments and cumulants of symmetric matrices: An application to the Wishart distribution. 95 JMultiAn 55 149-164

Kolonko, M.: A piecewise Markovian model for simulated annealing with stochastic cooling schedules. 95 JAppProb 32 649-658

Kolstad, Andrew: Burke, John; Mohadjer, Leyla; Green, Jim; Waksberg, Joseph; Kirsch, Irwin S.; — —: Composite estimation in national and state surveys. 94 ASASrvyRsr 873-878

Komarov, S. V.; Krasulina, T. P.: Using stochastic approximation to find the extinction probability of a Galton-Watson branching process. 94 VLenUM /3 34-39

Komazawa, Tsutomu; Tsuchiya, Takahiro: Exploratory data analysis for Hayashi's quantification method III by graphics: Its application of ordinal structure analysis to the Ninth Nationwide Survey of the Japanese National Character (Japanese). 95 PrInStMa 43 161-176

Komornik, J.; Mikulecky, M.; Beno, M.: Additivity test for composition of binomial effects in chromosome aberrations after radiation injury. 95 StatMed 14 1941-1947

Konakov, V. D.; Piterbarg, V. I.: High level excursions of Gaussian fields and the weakly optimal choice of the smoothing parameter. I. 95 MaMethSt 4 421-434

Konev, Victor; Lai, Tze Leung: Estimators with prescribed precision in stochastic regression models. 95 SqtlAnly 14 179-192

Kong, Augustine: Liu, Jun S.; Wong, Wing H.; — —: Covariance structure and convergence rate of the Gibbs sampler with various scans. 95 JRSS-B 57 157-169

Kong, Fanhui: Fei, Heliang; — —; Tang, Yingcai: Estimation for two-parameter Weibull distribution and extreme-value distribution under multiply type-II censoring. 95 CommStA 24 2087-2104

König, M. F.: Batra, S.; — —; Cruz-Orive, L. M.: Unbiased estimation of capillary length from vertical slices. 95 JMicro 178 152-159

Koning, Alex J.: Kallenberg, Wilbert C. M.; — —: On Wieand's theorem. 95 StPrLet 25 121-132

Koning, Pierre; Ridder, Geert; Van den Berg, Gerard J.: Structural and frictional unemployment in an equilibrium search model with heterogeneous agents. 95 JApEcmtx 10 133-151S

Konishi, Sadanori: Ichikawa, Masanori; — —: Application of the bootstrap methods in factor analysis. 95 Psymtrka 60 77-93

Konno, Norio: Asymptotic behavior of basic contact process with rapid stirring. 95 JTherPrb 8 833-876

Konno, Toshiharu: Makino, Toji; — —: On the use of circular tetrads in preference polygon (Japanese). 94 JpJBeh 21/2 1-9

Konno, Y.: Shiraishi, T.; — —: On construction of improved estimators in multiple-design multivariate linear models under general restriction. 95 AnInStMa 47 665-674

Konno, Yoshihiko: Estimation of a normal covariance matrix with incomplete data under Stein's loss. 95 JMultiAn 52 308-324

Kono, C.: Woodward, J. T.; — —; Madsen, L. L.; Zasadzinski, J. A.: Inherent bias in correlation averaged images. 95 JMicro 178 86-92

Konold, Clifford: Confessions of a coin flipper and would-be instructor. 95 AmerStat 49 203-209

— —: Issues in assessing conceptual understanding in probability and statistics. 95 JStEd 3/1 27kb (jse/v3n1/konold)

Konschnik, Carl A.: The impact of dropping the business area sample as a means of covering birth employers and nonemployers in the Current Business Surveys. 94 ASASrvyRsr 995-1000

Steel, Philip M.; — —: Administrative record matching for the 1992 economic censuses. 94 ASASrvyRsr 473-478

Konstantopoulos, Takis: Ballot theorems revisited. 95 StPrLet 24 331-338

— —; Zazanis, Michael: A discrete-time proof of Neveu's exchange formula. 95 JAppProb 32 917-921

Anantharam, Venkat; — —: A functional central limit theorem for the jump counts of Markov processes with an application to Jackson networks. 95 AdvAppPr 27 476-509

Kontoghiorghes, E. J.; Clarke, M. R. B.: An alternative approach for the numerical solution of seemingly unrelated regression equations models. 95 CmpStDA 19 369-377

Kontovasilis, Kimon P.; Mitrou, Nikolas M.: Markov-modulated traffic with nearly complete decomposability characteristics and associated fluid queueing models. 95 AdvAppPr 27 1144-1185

Koop, G.; Steel, M. F. J.; Osiewalski, J.: Posterior analysis of stochastic frontier models using Gibbs sampling. 95 CmpStQ 10 353-373

Koop, Gary; Osiewalski, Jacek; Steel, Mark F. J.: Bayesian long-run prediction in time series models. 95 JEconmtx 69 61-80

— —; Poirier, Dale J.: An empirical investigation of Wagner's hypothesis by using a model occurrence framework. 95 JRSS-A 158 123-141

Kooperberg, Charles; O'Sullivan, Finbarr: Persistent oscillation patterns in climatological fields. 94 ASASrvyRsr 200-205

— —; Stone, Charles J.; Truong, Young K.: Hazard regression. 95 JASA 90 78-94

— —; Stone, Charles J.; Truong, Young K.: Logspline estimation of a possibly mixed spectral distribution. 95 JTimSrAn 16 359-388

— —; Stone, Charles J.; Truong, Young K.: Rate of convergence for logspline spectral density estimation. 95 JTimSrAn 16 389-401

Besag, Julian; — —: On conditional and intrinsic autoregressions. 95 Biomtrka 82 733-746

Intrator, Orna; — —: Trees and splines in survival analysis. 95 StMedRes 4 237-261

Kopociński, B.: The largest square of successes covering the origin for Bernoulli trials on the lattice. 95 StNeerla 49 31-39

Kopp, P. Ekkehard: Cutland, Nigel J.; — —; Willinger, Walter: From discrete to continuous stochastic calculus. 95 Stochatx 52 173-192

Kordonsky, Kh. B.; Gertsbakh, I.: System state monitoring and lifetime scales - II. 95 RelEngSS 49 145-154

— —; Gertsbakh, I. B.: System state monitoring and lifetime scales - I. 95 RelEngSS 47 1-14

Koreisha, Sergio G.; Pukkila, Tarmo: A comparison between different order-determination criteria for identification of ARIMA models. 95 JBES 13 127-131

— —; Pukkila, Tarmo: The identification of seasonal autoregressive models. 95 JTimSrAn 16 267-290

Korezlioglu, H.: Kofman, D.; — —: Some EATA properties for marked point processes. 95 JAppProb 32 922-929

Korilis, Yannis A.; Lazar, Aurel A.: On the existence of equilibria in noncooperative optimal flow control. 95 JACM 42 584-613

Korn, Edward L.; Graubard, Barry I.: Examples of differing weighted and unweighted estimates from a sample survey. 95 AmerStat 49 291-295

— —; Graubard, Barry I.: Analysis of large health surveys: Accounting for the sampling design. 95 JRSS-A 158 263-295

— —; Graubard, Barry I.: Variance estimation for superpopulation parameters: Should one use with-replacement estimators?. 94 ASASrvyRsr 124-132

Korol, Abraham B.; Ronin, Yefim I.; Kirzhner, Valery M.: Interval mapping of quantitative trait loci employing correlated trait complexes. 95 Genetcs 140 1137-1147

Korolev, V. Yu.: Convergence of random sequences with independent random indices I.. 94 ThProbAp 39 282-297

Koroljuk, V. S.; Borovskich, Yu. V. (Auth); Glendinning, R. H. (Rev): Review of "Theory of U-statistics". 95 Ststcian 44 411-411

— —; Borovskich, Yu. V. (Auth); Janssen, Paul (Rev): Review of "Theory of U-statistics". 95 JASA 90 804-805

Korolyuk, V. S.: Borovskikh, Yu. V.; — —: Random permanents. 94 VSP:Neth 194p

— — (Auth); Svishchuk, A. V. (Rev): Review of "Stochastic models for systems" (Russian). 95 ThPrMaSt 50 147-147

Korolyuk, Vladimir S.; Swishchuk, Anatoly V.: Evolution of systems in random media. 95 CRCPress:FL 368p

Kőrösi, Gábor; Lovrics, László; Mátyás, László: Aggregation and the long run properties of economic time series. 95 MaCmpSim 39 279-286

Korostelev, A. P.; Simar, L.; Tsybakov, A. B.: Efficient estimation of monotone boundaries. 95 AnlsStat 23 476-489

Kortram, R. A.; van Rooij, A. C. M.; Lenstra, A. J.; Ridder, G.: Constructive identification of the mixed proportional hazards model. 95 StNeerla 49 269-281

Koschnick, Klaus-Uwe; Sharma, Bhu Dev: A generalization of the Sharma-Kaushik class of metrics. 93 JCISS 18 401-407

Koshland, Daniel: Horvitz, Dan; — —; Rubin, Donald; Gollin, Albert; Sawyer, Tom; Tanur, Judith M.: Pseudo-opinion polls: SLOP or useful data?. 95 Chance 8/2 16-25

Kosonen, Peter; Winne, Philip H.: Effects of teaching statistical laws on reasoning about everyday problems. 95 JEdPsych 87 33-46

Kostanich, Donna: Butani, Shail; Cahoon, Lawrence; Fay, Robert; — —: Measurement of different design effects. 94 ASASrvyRsr 342-347 (Disc: p360-361)

Thibaudeau, Yves; — —; Parmer, Randall: The influence of implicit methodologies in the analysis of the CCO and the CPS. 94 ASASrvyRsr 730-735

Kostanich, Donna L.: Dippo, Cathryn S.; — —; Polivka, Anne E.: Effects of methodological change in the Current Population Survey. 94 ASASrvyRsr 260-263

Kostello, Eric C.: Schulman, Daniel A.; Campbell, Alec D.; — —: Data analysis using S-Plus. 95 SocMethR 23 392-413

Kostlan, Eric: Edelman, Alan; — —: How many zeros of a random polynomial are real?. 95 BAmMaSoc 32 1-37

Kotelenez, Peter: A class of quasilinear stochastic partial differential equations of McKean-Vlasov type with mass conservation. 95 ProbTher 102 159-188

Koti, Kallappa M.: On the normal approximation for the sample p-quantile. 94 ASAProStCp 188-193

Kott, Phillip S.: Developing an estimation strategy for the 1993 AMS pesticide data program. 94 ASAProEnvr 133-138

— —: Regression analysis of repeated survey data (with available software). 94 ASASrvyRsr 116-123

— —: Reweighting and variance estimation for the characteristics of business owners survey. 94 JOffStat 10 407-418

Cox, Brenda G.; Binder, David A.; Chinnappa, B. Nanjamma; Christianson, Anders; Colledge, Michael J.; — — (Ed): Business survey methods. 95 Wiley:NY:UK 732p

Guenther, Patricia M.; — —: Problems in estimating usual dietary intakes. 94 ASAProEpid 44-48

Kötting, J.: Urfer, W.; Schwarzenbach, F. H.; — —; Müller, P.: Multistate models for monitoring individual trees in permanent observation plots. 94 EnvEcoSt 1 171-192 (Disc: p192-199)

Urfer, W.; Scwarzenbach, F. H.; — —; Müller, P.: Reply to comments on "Multistate models for monitoring individual trees in permanent observation plots". 94 EnvEcoSt 1 197-199

Kotz, S.: Johnson, N. L.; — —; Balakrishnan, N. (Auth); Chatfield, Chris (Rev): Review of "Continuous univariate distributions, Volume 1 (Second edition)". 95 Ststcian 44 544-545

Johnson, N. L.; — —; Balakrishnan, N. (Auth); Kocherlakota, S. (Rev): Review of "Continuous univariate distributions, Volumes 1 and 2 (Second edition)". 95 Biomtrcs 51 1582-1583

Johnson, N. L.; — —; Balakrishnan, N. (Auth); Omori, Yasuhiro (Rev): Review of "Continuous univariate distributions, Volume 1 (Second edition)". 95 JASA 90 1490-1491

Johnson, N. L.; — —; Balakrishnan, N. (Auth); Tricker, A. R. (Rev): Review of "Continuous univariate distributions: Volume I (Second edition)". 95 JAppStat 22 323-323

Johnson, N. L.; — —; Kemp, A. W. (Auth); Consul, P. C. (Rev): Review of "Univariate discrete distributions". 95 Metrika 42 72-73

Kotz, Samuel; Ostrovskii, Iosif: Correction to "Characteristic functions of a class of elliptical distributions" (V49 p164-178). 95 JMultiAn 52 352-354

Johnson, Norman L.; — —: Use of moments in studies of limit distributions arising from iterated random subdivisions of an interval. 95 StPrLet 24 111-119

Johnson, Norman L.; — —; Balakrishnan, N.: Continuous univariate distributions. Volume 2 (Second edition). 95 Wiley:NY:UK 719p

— —; Johnson, Norman L. (Auth); Champ, Charles W. (Rev): Review of "Process capability indices". 95 Technmcs 37 233-233

— —; Johnson, Norman L. (Auth); Lovelace, Cynthia (Rev): Review of "Process capability indices". 95 JQualTek 27 270-271

— —; Johnson, Norman L. (Auth); Walker, Esteban (Rev): Review of "Process capability indices". 95 JASA 90 1127-1127

Johnson, Norman L.; — —; Balakrishnan, N. (Auth); Famoye, Felix (Rev): Review of "Continuous univariate distributions, Volume 1 (Second edition)". 95 Technmcs 37 466-466

Kou, Shing-Gang: Glasserman, Paul; — —: Analysis of an importance sampling estimator for tandem queues. 95 ACMMdSim 5 22-42

Glasserman, Paul; — —: Limits of first passage times to rare sets in regenerative processes. 95 AnlsApPr 5 424-445

Kouider, Elies; Chen, Hanfeng: Concavity of Box-Cox log-likelihood function. 95 StPrLet 25 171-175

Koukouvinos, Christos: Optimal weighing designs and some new weighing matrices. 95 StPrLet 25 37-44

Koul, Hira L.; Saleh, A. K. Md. E.: Autoregression quantiles and related rank-scores processes. 95 AnlsStat 23 670-689

— —; Zhu, Zhiwei: Bahadur-Kiefer representations for GM-estimators in autoregression models. 95 StocProc 57 167-189

Yang, Song; — —: Minimum distance estimation of the center of symmetry with randomly censored data. 95 Metrika 42 79-97

Kouritzin, Michael A.: Strong approximation for cross-covariances of linear variables with long-range dependence. 95 StocProc 60 343-353

Kourouklis, S.: Estimation of an exponential quantile under Pitman's measure of closeness. 95 CandJSt 23 257-268

Kourouklis, Stavros: Estimating powers of the scale parameter of an exponential distribution with unknown location under Pitman's measure of closeness. 95 JStPlInf 48 185-195

Kourti, Theodora; MacGregor, John F.: Process analysis, monitoring and diagnosis, using multivariate projection methods. 95 Chemolab 28 3-21

Koutras, M. V.; Alexandrou, V. A.: Runs, scans and urn model distributions: A unified Markov chain approach. 95 AnInStMa 47 743-766

— —; Papadopoulos, G. K.; Papastavridis, S. G.: Runs on a circle. 95 JAppProb 32 396-404

Fu, J. C.; — —: Reliability bounds for coherent structures with independent components. 95 StPrLet 22 137-148

Koutsoukos, Antonis; Slud, Eric: Inaccuracy rates and Hodges-Lehmann large deviation rates for parametric inferences with nuisance parameters. 95 JStPlInf 48 47-68

Kovacevic, Milorad S.: Binder, David A.; — —: Estimating some measures of income inequality from survey data: An application of the estimating equations approach. 95 SurvyM 21 137-145

Kovačević, Milorad S.; Liu, Tzen-Ping: Statistical matching of survey datafiles: A simulation study. 94 ASASrvyRsr 479-484

— —; Yung, Wesley; Pandher, Gurupdesh S.: Resampling inference for quantile shares. 94 ASASrvyRsr 206-211

Kovar, John G.: Comment on "Imputation". 94 ASASrvyRsr 906-907

Kowada, Masashi: Igaki, Nobuko; Sumita, Ushio; — —: Analysis of Markov renewal shock models. 95 JAppProb 32 821-831

Kowada, Masasi: Feng, Wei; Adachi, Kohichi; — —: Dynamically optimal replacement policy for a shock model in a Markov random environment. 94 JOpRsJpn 37 255-270

Kowaka, M.: — — (Ed); Ziegel, Eric R. (Rev): Review of "Introduction to life prediction of industrial plant materials". 95 Technmcs 37 457-457

Kowalski, Bruce: Tauler, Romá; Smilde, Age; — —: Selectivity, local rank, three-way data analysis and ambiguity in multivariate curve resolution. 95 JChemtrc 9 31-58

Kowalski, Bruce R.: Booksh, Karl; Henshaw, John M.; Burgess, Lloyd W.; — —: A second-order standard addition method with application to calibration of a kinetics-spectroscopic sensor for quantitation of trichloroethylene. 95 JChemtrc 9 263-282

Kowalski, Jacek P.: Complete classes of tests for regularly varying distributions. 95 AnInStMa 47 321-350

— —: Complete classes of tests for some nonparametric hypotheses. 95 StatDecn 13 167-179

Kozachenko, Yu. V.; Mokliatchouk, O. M.: Pre-Gaussian random vectors and their applications. 95 ThPrMaSt 50 89-98

Kozarovitzky, Eugene: Molchanov, Ilya S.; Omey, Edward; — —: An elementary renewal theorem for random compact convex sets. 95 AdvAppPr 27 931-942

Kozek, A. S.: Fayolle, G.; Malyshev, V. A.; Menshikov, M. V. (Auth); — — (Rev): Review of "Topics in the constructive theory of countable Markov chains". 96 ShrtBkRv 16/1 9-10

Kozek, Andrzej S.: A rule of thumb (not only) for gamblers. 95 StocProc 55 169-181

Koziara, Jan: The order determination methods for a multivariate autoregressive model (Polish). 95 ListyBio 32 1-14

Kraemer, Helena Chmura: Statistical issues in assessing comorbidity. 95 StatMed 14 721-733

Krafft, Olaf: Comment on "Constrained optimization of experimental design". 95 MaOpfStS 26 162-162

— —; Schaefer, Martin: Exact Elfving-minimax designs for quadratic regression. 95 StSinica 5 475-484

Kraft, Kathy M.; Johnson, Douglas H.; Samuelson, Jack M.; Allen, Stephen H.: Using known populations of pronghorn to evaluate sampling plans and estimators. 95 JWildlMg 59 129-137

Kragten, J.: A standard scheme for calculating numerically standard deviations and confidence intervals. 95 Chemolab 28 89–97

Krajka, A.; Szynal, D.: On monotone dependence functions of the quantile type. 95 ApplMath 23 51–72

Kramar, Andrew: Lancar, Rémi; — —; Haie-Meder, Christine: Non-parametric methods for analysing recurrent complications of varying severity. 95 StatMed 14 2701–2712

Krambeck, H.-J.: Application and abuse of statistical methods in mathematical modelling in limnology. 95 EcolgMdl 78 7–15

Kramer, E.: Matulić-Bedenić, I.; Horvatić-Baldasar, K.; — —: Construction of new symmetric designs with parameters (66, 26, 10). 95 JComDsgn 3 405–410

Kramer, E. S.; Mathon, Rudolf: Proper $S(t,K,v)$'s for $t \geq 3$, $\mu \leq 16$, $|K| > 1$ and their extensions. 95 JComDsgn 3 411–425

Kramer, Mark A.: Tan, Jonathan S.; — —: Hazard function modeling using cross validation: From data collection to model selection. 95 RelEngSS 49 155–169

Krämer, Walter: Billingsley, Patrick (Auth); — — (Rev): Review of "Probability and measure (Third edition)". 95 CmpStDA 20 703–703

Diewert, W. E.; Nakamura, A. (Ed); — — (Rev. German): Review of "Essays in index number theory (Volume I)". 95 AllgeStA 79 97–97

Maddala, G. S.; Rao, C. R.; Vinod, H. D. (Ed); — — (Rev. German): Review of "Econometrics (Handbook of statistics) (Volume II)". 95 AllgeStA 79 101–102

Kramkov, D. O.; Mordecky, E.: Integral option. 94 ThProbAp 39 162–172

— —; Shiryaev, A. N.: On the rational pricing of the "Russian option" for the symmetrical binomial model of a (B,S)-market. 94 ThProbAp 39 153–162

Kabanov, Yu. M.; — —: Large financial markets: Asymptotic arbitrage and contiguity. 94 ThProbAp 39 182–187

Kabanov, Yu. M.; — —: No-arbitrage and equivalent martingale measures: An elementary proof of the Harrison-Pliska theorem. 94 ThProbAp 39 523–527

Shiryaev, A. N.; Kabanov, Yu. M.; — —; Mel'nikov, A. V.: Toward the theory of pricing of options of both European and American types. II. Continuous time.. 94 ThProbAp 39 61–102

Kramkov, O. D.: Shiryaev, A. N.; Kabanov, Yu. M.; — —; Mel'nikov, A. V.: Toward the theory of pricing of options of both European and American types. I. Discrete time.. 94 ThProbAp 39 14–60

Krasnicka, Barbara: Driscoll, Michael F.; — —: An accessible proof of Craig's theorem in the general case. 95 AmerStat 49 59–62

Krass, Dmitry: Filar, Jerzy A.; — —; Ross, Keith W.: Percentile performance criteria for limiting average Markov decision processes. 95 IEEEAuCn 40 2–10

Krasulina, T. P.: Komarov, S. V.; — —: Using stochastic approximation to find the extinction probability of a Galton-Watson branching process. 94 VLenUM /3 34–39

Kratz, M.: Hüsler, J.; — —: Rate of Poisson approximation of the number of exceedances of nonstationary normal sequences. 95 StocProc 55 301–313

Kratzke, Diem-Tran: West, Sandra A.; — —; Robertson, Kenneth W.: Variance estimators for variables that have both observed and imputed values. 94 ASASrvyRsr 394–399

Kraus, Gerolf; Gauglitz, Günter: Optical reflectometric gas sensing: Classification of hydrocarbon vapours by pattern recognition applied to RIFS sensor signals. 95 Chemolab 30 211–221

Krause, Andreas: Electronic services in statistics. 95 CmpStDA 19 595–604

Schmid, Bernhard; Polasek, Wolfgang; Weiner, Jacob; — —; Stoll, Peter: Modeling of discontinuous relationships in biology with censored regression. 94 ANatrls 143 494–507

Krause, Ekkehard; Bandt, Christoph; Schulz, Anne; Schulz, Holger: Fractal exponents for the upper airways of mammalian lungs. 95 CmpStDA 20 583–590

Krause, Ulf: Möttönen, Jyrki; Oja, Hannu; — —; Rantakallio, Paula: Application of random coefficient regression model to myopia data: A case study. 95 BiomtrcJ 37 657–672

Krawchuk, S.: Singh, M. P.; Tambay, J. L.; — —: The National Population Health Survey: Design and issues. 94 ASASrvyRsr 803–808

Krebs, Dagmar: Non-response to sensitive questions: Nationalism in Germany. 94 ASASrvyRsr 1199–1203

Krebs, W. B.: Brownian motion on the continuum tree. 95 ProbTher 101 421–433

Kreft, Ita G. G.; de Leeuw, Jan; Aiken, Leona S.: The effect of different forms of centering in hierarchical linear models. 95 MultBehR 30 1–21

de Leeuw, Jan; — —: Questioning multilevel models. 95 JEdStat 20 171–189 (Disc: p201-227, 239-240)

de Leeuw, Jan; — —: Reply to comments on "Questioning multilevel models". 95 JEdStat 20 239–240

Kreher, Donald L.: Colbourn, Charles J.; Cusack, Charles A.; — —: Partial Steiner triple systems with equal-sized holes. 95 JCombThA 70 56–65

Kreienbrock, L.; Schach, S. (Auth); Enderlein, G. (Rev. German): Review of "Epidemiological methods" (German). 95 BiomtrcJ 37 1016–1016

Kreiner, Svend: Klein, John P.; Keiding, Niels; — —: Graphical models for panel studies, illustrated on data from the Framingham Heart Study. 95 StatMed 14 1265–1290

Kreiss, Heinz-Otto: Gustafsson, Bertil; — —; Oliger, Joseph: Time dependent problems and difference methods. 95 Wiley:NY:UK 642p

Krengel, Ulrich: Gnedin, Alexander V.; — —: A stochastic game of optimal stopping and order selection. 95 AnlsApPr 5 310–321

Krenzke, Thomas: Slanta, John; — —: Applying the Lavallée and Hidiroglou method to obtain stratification boundaries for the Census Bureau's Annual Capital Expenditures Survey. 94 ASASrvyRsr 693–698

Kressel, J.: Jones, C. M.; Robson, D. S.; Lakkis, H. D.; — —: Properties of catch rates used in analysis of angler surveys. 95 TrAFish 124 911–928

Krewski, D.; Wang, Y.; Bartlett, S.; Krishnan, K.: Uncertainty, variability, and sensitivity analysis in physiological pharmacokinetic models. 95 JBiophSt 5 245–271

Krewski, Daniel; Smythe, Robert T.; Dewanji, Anup; Szyszkowicz, Mieczyslaw: Bootstrapping an empirical Bayes estimator of the distribution of historical controls in carcinogen bioassay. 95 JStPlInf 43 215–228

Burnett, Richard T.; Ross, W. H.; — —: Non-linear mixed regression models. 95 EnvrMtrc 6 85–99

Wang, Yong; — —; Bartlett, Sheryl; Krishnan, Kannan: Analysis of uncertainty and sensitivity in physiologically based pharmacokinetic models. 94 ASAProBiop 100–105

Krieger, Abba M.: Benjamini, Yoav; — —: On the consistency of tests of symmetry. 95 JNnparSt 4 283–292

Krijnen, W. P.; Kiers, H. A. L.: An efficient algorithm for weighted PCA. 95 CmpStQ 10 299–306

— — (Auth); Klauer, K. C. (Rev): Review of "The analysis of three-way arrays by constrained PARAFAC methods". 95 JClass 12 147–149

Kriman, Vladimir: Analysis of the sensitivity of the GI/GI/m/B queue to buffer size: The score function method. 95 CommStSM 11 171–194

Kringle, Robert O.; Khan-Malek, Richard C.: A statistical assessment of the recommendations from a conference on analytical methods validation in bioavailability, bioequivalence and pharmacokinetic studies. 94 ASAProBiop 510–514

Krishnaiah, P. R.: Bai, Zhidong; Chen, Xiru; — —; Zhao, Lingcheng: Asymptotic properties of the EVLP estimation for superimposed exponential signals in noise (Chinese). 95 ChJApPSt 10 148–163

Krishnamoorthy, K.; Shah, A. K.: Correction to "Testing equality of several normal means to a specified standard" (V27 p132-138). 95 JQualTek 27 402–402

— —; Shah, Arvind K.: Testing equality of several normal means to a specified standard. 95 JQualTek 27 132–138

Jordan, Scott M.; — —: Confidence regions for the common mean vector of several multivariate normal populations. 95 CandJSt 23 283–297

Jordan, Scott M.; — —: On combining independent tests in linear models. 95 StPrLet 23 117–122

Krishnamurty, Goteti Bala; Kasovia-Schmitt, Patricia; Ostroff, David J. (Auth); Hilden, Jørgen (Rev): Review of "Statistics. An interactive text for the health and life sciences". 95 StatMed 14 2621–2622

Krishnan, K.: Krewski, D.; Wang, Y.; Bartlett, S.; — —: Uncertainty, variability, and sensitivity analysis in physiological pharmacokinetic models. 95 JBiophSt 5 245–271

Krishnan, Kannan: Wang, Yong; Krewski, Daniel; Bartlett, Sheryl; — —: Analysis of uncertainty and sensitivity in physiologically based pharmacokinetic models. 94 ASAProBiop 100–105

Kristal, Alan: Koepsell, Thomas D.; Diehr, Paula H.; Cheadle, Allen; — —: Symposium on community intervention trials. 95 AJEpidm 142 594–599

Kristensen, Leif: Mann, Jakob; Lenschow, Donald H.; — —: Comment on "A definitive approach to turbulence statistical studies in planetary boundary layers" (V51 p1682-1690). 95 JAtmsphr 52 3194–3196

Kristiansson, Karl-Erik: Bergman, Lars R.; — —; Olofsson, Anita; Säfström, Margaretha: Decentralised CATI versus paper and pencil interviewing: Effects on the results in the Swedish labour force surveys. 94 JOffStat 10 181–195

Krivokapić, Zoran: Heavlin, William D.; — —: Virtual manufacturing: Design, calibration, and analysis. 94 ASAProPEng 274–279

Křivý, Ivan; Tvrdík, Josef: The controlled random search algorithm in optimizing regression models. 95 CmpStDA 20 229–234

Kroese, A. H.; van der Meulen, E. A.; Poortema, K.; Schaafsma, W.: Distributional inference. 95 StNeerla 49 63–82

Krol, William F.: Williford, William O.; — —; Bingham, Stephen F.; Collins, Joseph F.; Weiss, David G.: The multicenter clinical trials coordinating center statistician: 'More than a consultant'. 95 AmerStat 49 221–225

Krone, Stephen M.: Conditioned superprocesses and their weighted occupation times. 95 StPrLet 22 59–69

Ethier, S. N.; — —: Comparing Fleming-Viot and Dawson-Watanabe processes. 95 StocProc 60 171–190

Kroner, Kenneth F.; Kneafsey, Kevin P.; Claessens, Stijn: Forecasting volatility in commodity markets. 95 JForcast 14 77–95

Engle, Robert F.; — —: Multivariate simultaneous generalized ARCH. 95 EconTher 11 122–150

Kronmal, Richard A.: Reply to comments on "Spurious correlation and the fallacy of the ratio standard revisited". 95 JRSS-A 158 623–625

Kroutil, R. T.: Shaffer, R. E.; Small, G. W.; Combs, R. J.; Knapp, R. B.; — —: Experimental design protocol for the pattern recognition analysis of bandpass filtered Fourier transform infrared interferograms. 95 Chemolab 29 89–108

Krueger, Alan B.: Angrist, Joshua D.; — —: Split-sample instrumental variables estimates of the return to schooling. 95 JBES 13/2 225–235

Krug, T.; Martin, R. J.: Information loss on the dependence parameter due to incomplete data from a one-parameter spatial Gaussian process. 95 CommStA 24 3055-3072

Kruglov, V. M.: Alekperov, Sh. O.; — —: Convergence of convolutions of concentration functions to degenerate, normal, and Poisson laws. 94 ThProbAp 39 197-216

Kruglyak, Leonid; Daly, Mark J.; Lander, Eric S.: Rapid multipoint linkage analysis of recessive traits in nuclear families, including homozygosity mapping. 95 AJHuGene 56 519-527

— —; Lander, Eric S.: High-resolution genetic mapping of complex traits. 95 AJHuGene 56 1212-1223

— —; Lander, Eric S.: Complete multipoint sib-pair analysis of qualitative and quantitative traits. 95 AJHuGene 57 439-454

— —; Lander, Eric S.: A nonparametric approach for mapping quantitative trait loci. 95 Genetcs 139 1421-1428

Krumhaus, Susan G.; Klemm, Rebecca J.: An overview of the National Survey of Veterans. 94 ASAProGovt 19-21

Krus, David J.: Liang, Kun-Hsia; — —; Webb, James M.: K-fold crossvalidation in canonical analysis. 95 MultBehR 30 539-545

Krylov, N. V.; Kuznetsov, S. E.; Skorokhod, A. V.; Yushkevich, A. A.: E. B. Dynkin's 70th birthday. 94 ThProbAp 39 654-656

Krzanowski, W.: — — (Ed); Hand, D. J. (Rev): Review of "Recent advances in descriptive multivariate analysis". 95 ShrtBkRv 15/3 44-44

Krzanowski, W. J.: Selection of variables, and assessment of their performance, in mixed-variable discriminant analysis. 95 CmpStDA 19 419-431

— —: Orthogonal canonical variates for discrimination and classification. 95 JChemtrc 9 509-520

— —: Correction to "Ordination in the presence of group structure, for general multivariate data" (VII p195-207). 95 JClass 12 2-2

— — (Ed): Recent advances in descriptive multivariate analysis. 95 OxfordUniv 376p

— —; Jonathan, P.; McCarthy, W. V.; Thomas, M. R.: Discriminant analysis with singular covariance matrices: Methods and applications to spectroscopic data. 95 ApplStat 44 101-115

— —; Marriott, F. H. C.: Multivariate analysis. Part 2: Classification, covariance structures and repeated measurements. 95 Arnold:Lnd:MD 280p

Powell, K. J.; Sapatinas, T.; Bailey, T. C.; — —: Application of wavelets to the pre-processing of underwater sounds. 95 StatComp 5 265-273

— —; Marriott, F. H. C. (Auth); Johnson, Richard (Rev): Review of "Multivariate analysis, Part 1: Distributions, ordination, and inference". 95 JASA 90 1490-1490

— —; Marriott, F. H. C. (Auth); Shanmugam, Ramalingam (Rev): Review of "Multivariate analysis Part 1: Distributions, ordination and inference". 95 Technmcs 37 351-351

Krzanowski, Wojtek J.; Kline, Paul: Cross-validation for choosing the number of important components in principal component analysis. 95 MultBehR 30 149-165

Krzyszkowska, J.: Ceranka, B.; — —: Reinforced block designs with two groups of treatments. 94 ListyBio 31 17-25

Krzysztofowicz, Roman: Long, Dou; — —: A family of bivariate densities constructed from marginals. 95 JASA 90 739-746

Kshirsagar, Anant M.; Smith, William Boyce: Growth curves. 95 Dekker:NY 359p

Park, Pil S.; — —: Effect of some modifications in standard discriminant analysis. 95 CommStB 24 807-816

Ku, Walter: Wei, Paul; Zeidler, James; — —: Analysis of multiframe target detection using pixel statistics. 95 IEEEAES 31 238-247

Ku, Wenfu; Storer, Robert H.; Georgakis, Christos: Disturbance detection and isolation by dynamic principal component analysis. 95 Chemolab 30 179-196

Kuan, Chung-Ming; Hornik, Kurt: The generalized fluctuation test: A unifying view. 95 EcomtrRv 14 135-161

— —; Liu, Tung: Forecasting exchange rates using feedforward and recurrent neural networks. 95 JApEcmtx 10 347-364

Chu, Chia-Shang J.; Hornik, Kurt; — —: MOSUM tests for parameter constancy. 95 Biomtrka 82 603-617

Chu, Chia-Shang James; Hornik, Kurt; — —: The moving-estimates test for parameter stability. 95 EconTher 11 699-720

Nunes, Luis C.; — —; Newbold, Paul: Spurious break. 95 EconTher 11 736-749

Kubicka, Ewa; Kubicki, Grzegorz; McMorris, F. R.: An algorithm to find agreement subtrees. 95 JClass 12 91-99

Kubicki, Grzegorz: Kubicka, Ewa; — —; McMorris, F. R.: An algorithm to find agreement subtrees. 95 JClass 12 91-99

Kubokawa, Tatsuya: Estimation of variance components in mixed linear models. 95 JMultiAn 53 210-236

— —; Saleh, A. K. Md. E.: Estimation of location and scale parameters under order restrictions. 94 JStatRsr 28 41-51

Küchenhoff, H.: The identification of logistic regression models with errors in the variables. 95 StHefte 36 41-48

Kuehl, Robert O.: — — (Auth); Hare, Lynne B. (Rev): Review of "Statistical principles of research design and analysis". 95 JQualTek 27 96-97

Kuelbs, J.; Li, W. V.; Shao, Qi-man: Small ball probabilities for Gaussian processes with stationary increments under Hölder norms. 95 JTherPrb 8 361-386

Kuelbs, James; Li, Wenbo V.; Talagrand, Michel: Lim inf results for Gaussian samples and Chung's functional LIL. 94 AnlsProb 22 1879-1903

Kufeld, Roland C.: Bowden, David C.; — —: Generalized mark-sight population size estimation applied to Colorado moose. 95 JWildlMg 59 840-851

Kugler, D. L.: Larsen, D. P.; Urquhart, N. S.; — —: Regional scale trend monitoring of indicators of trophic condition of lakes. 95 WaterReB 31 117-140

Kuhfeld, Warren F.; Tobias, Randall D.; Garratt, Mark: Efficient experimental design with marketing research applications. 94 JMarRsr 31 545-557

Kühn, Andreas: Roebruck, Peter; — —: Comparison of tests and sample size formulae for proving therapeutic equivalence based on the difference of binomial probabilities. 95 StatMed 14 1583-1594

Kuhn, Roland; De Mori, Renato: The application of semantic classification trees to natural language understanding. 95 IEEEPAMI 17 449-460

Kuk, A. Y. C.: Wang, Julian (Transl); — — (Rev): Review of "Essentials of statistical methods". 95 AstrlJSt 37 127-128

Kuk, Anthony Y. C.: Asymptotically unbiased estimation in generalized linear models with random effects. 95 JRSS-B 57 395-407

— —: Modelling paired comparison data with large numbers of draws and large variability of draw percentages among players. 95 Ststcian 44 523-528

Kulasekera, K. B.: Comparison of regression curves using quasi-residuals. 95 JASA 90 1085-1093

— —: A bound on the L_1-error of a nonparametric density estimator with censored data. 95 StPrLet 23 233-238

Kulczycki, G. W.: Wassell, J. T.; — —; Moyer, E. S.: Frailty models of manufacturing effects. 95 LiDaAnal 1 161-170

Kulczycki, Gregory W.: Wassell, James T.; — —; Moyer, Ernest S.: Modelling frailty in the manufacture of respiratory protective devices. 94 ASAProEpid 133-138

Kulinskaya, Elena: Large sample results for tests of association II. Multinomial and stratified sampling. 95 CommStA 24 1121-1150

— —: Coefficients of the asymptotic distribution of the Kolmogorov-Smirnov statistic when parameters are estimated. 95 JNnparSt 5 43-60

Kulkarni, Pandurang M.; Shah, Arvind K.: Testing the equality of several binomial proportions to a prespecified standard. 95 StPrLet 25 213-219

Kulkarni, Sanjeev R.; Zeitouni, Ofer: A general classification rule for probability measures. 95 AnlsStat 23 1393-1407

Wang, I-Jeng; Chong, Edwin K. P.; — —: On equivalence of some noise conditions for stochastic approximation algorithms. 95 PIEEEDecCt 3849-3854

Kulkarni, Vidyadhar G.: Modeling and analysis of stochastic systems. 95 ChapmnHall:Lnd 624p

— — (Auth); Leslie, J. R. (Rev): Review of "Modeling and analysis of stochastic systems". 96 ShrtBkRv 16/1 4-5

Kulldorff, Martin; Nagarwalla, Neville: Spatial disease clusters: Detection and inference. 95 StatMed 14 799-810

Kulmann, Hermann: Notes on the computation of the exact distribution function of the χ^2-and related test statistics in the equiprobable case. 95 CmpStDA 19 707-710

Kulp, Dale: Brick, J. Michael; Waksberg, Joseph; — —; Starer, Amy: Bias in list-assisted telephone samples. 95 PubOpinQ 59 218-235

Brick, J. Michael; Waksberg, Joseph; — —; Starer, Amy: Bias in list-assisted telephone samples. 94 ASASrvyRsr 1293-1298

Kulp, Dale W.: Dynamics of 'list-assisted' random digit dialing (RDD) frame coverage. 94 ASASrvyRsr 11-18

Kumar, B. Krishna: Parthasarathy, P. R.; — —: Stochastic compartmental models with Prendville growth mechanisms. 95 MaBisci 125 51-60

Kumar, C. Satheesh: Moothathu, T. S.; — —: Some properties of the stuttering Poisson distribution. 95 CalcutSt 45 125-130

Kumar, Ish; Khush, G. S.: A general method of detecting additive, dominance and epistatic variation for metrical traits. V. Triple test cross analysis. 94 Heredity 72 563-569

Kumar, Kuldeep: Fisher, N. I. (Auth); — — (Rev): Review of "Statistical analysis of circular data". 95 Ststcian 44 285-286

Kumar, Narinder; Gill, Amar Nath; Mehta, Gobind P.: Selection procedures for scale parameters when the populations have a common known quantile. 94 Metron 52/3 171-196

Misra, Neeraj; — —: Selecting the best exponential population when the scale parameters are bounded. 95 MaOpfStS 27 151-158

Kumar, Rakesh; Hanson, Allen R.: Robust methods for estimating pose and a sensitivity analysis. 94 CVGIPIm 60 313-342

Kumar, Somesh; Sharma, Divakar: Estimating the common location. 95 MaOpfStS 26 231-240

Singh, R. Karan; Chaturvedi, Ajit; — —: Two stage and sequential procedures for estimating a linear function of k multinormal means. 94 JIndnStA 32 1-8

Kumar, V.; Leone, Robert P.; Gaskins, John N.: Aggregate and disaggregate sector forecasting using consumer confidence measures. 95 IntJFor 11 361-377

Kumar, V. S. Sampath; Vasantha, P.: Continuous inspection of Markov processes with a clearance interval. 95 JAppStat 22 427-432

Kumbhakar, Subal C.; Hjalmarsson, Lennart: Labour-use efficiency in Swedish social insurance offices. 95 JApEcmtx 10 33-47

Kuncheva, Ludmila I.: Mitra, Sushmita; — —: Improving classification performance using fuzzy MLP and two-level selective partitioning of the feature space. 95 FuzzySet 70 1-13

Kundra, Inderjit: Time-series models to forecast nuclear net generation of electricity. 94 ASAProBuEc 327-332

Kundu, Debasis: Consistency of the undamped exponential signal model on a restricted parameter space. 95 CommStA 24 241-251

— —; Kundu, Rameswar: Consistent estimates of super imposed exponential signals when some observations are missing. 95 JStPlInf 44 205-218

Kundu, Rameswar: Kundu, Debasis; — —: Consistent estimates of super imposed exponential signals when some observations are missing. 95 JStPlInf 44 205-218

Kung, H. C.; Liu, X. H.; Chan, J.: Socioeconomic status and suicide. 94 ASAProSoSt 235-239

Künsch, Hans; Geman, Stuart; Kehagias, Athanasios: Hidden Markov random fields. 95 AnlsApPr 5 577-602

Kunte, S.: Unnikrishnan, N. K.; — —: Optimality of an analogue of Basu's estimator under a double sampling design. 95 SankhyaB 57 103-111

Kuo, Hui-Hsuing: Hida, Takeyuki; — —; Potthoff, Jürgen; Streit, Ludwig (Auth); Øksendal, Bernt (Rev): Review of "White noise. An infinite dimensional calculus". 95 BLndnMaS 27 189-191

Kuo, Lynn; Mallick, Bani: Nonparametric Bayesian inference for accelerated failure time model. 94 ASAProBays 90-95

— —; Yang, Tae: Bayesian computation of software reliability. 95 JCmpGrSt 4 65-82

— —; Yang, Tae: A unified approach to the nonhomogeneous Poisson process in reliability modeling. 94 ASAProPEng 169-174

Dey, Dipak K.; — —; Sahu, Sujit K.: A Bayesian predictive approach to determining the number of components in a mixture distribution. 95 StatComp 5 297-305

Kuo, Way: Chien, Wei-Ting Kary; — —: Extensions of the Kaplan-Meier estimator. 95 CommStB 24 953-964

Chien, Wei-Ting Kary; — —: Modeling and maximizing burn-in effectiveness. 95 IEEEReli 44 19-25

Kupfer, David: Kamlet, Mark S.; Paul, Nancy; Greenhouse, Joel; — —; Frank, Ellen; Wade, Martcia: Cost utility analysis of maintenance treatment for recurrent depression. 95 ConClnTr 16 17-40

Kupper, Lawrence L.: Christopher, Susan R.; — —: On the effects of predictor misclassification in multiple linear regression analysis. 95 CommStA 24 13-37

Kuralmani, V.; Govindaraju, K.: Modified tables for the selection of double sampling attribute plan indexed by AQL and LQL. 95 CommStA 24 1897-1921

Kurchenko, O. O.: Some inequalities for the Orlicz norm of a linear form of independent pre-Gaussian random variables. 95 ThPrMaSt 50 99-102

Kuriki, Shinji; Fuji-Hara, Ryoh: Balanced arrays of strength two and nested (r,λ)-designs. 94 JComDsgn 2 407-414

Kurtanjek, Želimir: Structure of principal component based neural network models of dynamic systems. 95 JCmpInT 3 1-8

Kurtz, Thomas G.; Pardoux, Étienne; Protter, Philip: Stratonovich stochastic differential equations driven by general semimartingales. 95 AIHenrPB 31 351-377

Kuruppu, Janaki: Katti, S. K.; — —: Teacher evaluation forms: Their validity or lack there of. 94 ASAProStEd 94-99

Kurz, Ludwik: Trivedi, Yagnesh C.; — —: Image restoration using recursive estimators. 95 IEEESyst 25 1470-1482

Kusaka, Yasuo; Suzuki, Hisatoshi: An analysis of new product diffusion process. 94 JOpRsJpn 38 188-211

Kushler, Robert: Control charts for customer satisfaction survey data. 94 ASAProQlPr 334-340

Kushler, Robert H.: Mauger, David T.; Brown, Morton B.; — —: A comparison of methods that characterize pulses in a time series. 95 StatMed 14 311-325

Kushner, Harold J.; Yang, J.: Analysis of adaptive step-size SA algorithms for parameter tracking. 95 IEEEAuCn 40 1403-1410

— —; Yang, Jichuan: Stochastic approximation with averaging and feedback: Rapidly convergent "on-line" algorithms. 95 IEEEAuCn 40 24-34

Kusuoka, Shigeo: Limit theorem on option replication cost with transaction costs. 95 AnlsApPr 5 198-221

Kutoyants, Yu.; Vostrikova, L.: On non-consistency of estimators. 95 Stochatx 53 53-80

Dermoune, A.; — —: Expansion of distribution of maximum likelihood estimate for misspecified diffusion type observation. 95 Stochatx 52 121-145

Kuwada, Masahide: Analysis of variance of partially balanced fractional $2^{m_1+m_2}$ factorial designs of resolution IV. 93 JCISS 18 343-367

— —; Hyodo, Yoshifumi: Analysis of variance of balanced fractional $r^m \times s^n$ factorial designs of resolution V. 95 JJapnStS 25 129-141

Kuwata, Kaihei: Katayama, Ryu; — —; Kajitani, Yuji; Watanabe, Masahide: Embedding dimension estimation of chaotic time series using self-generating radial basis function network. 95 FuzzySet 71 311-327

Kuznetsov, S. E.: Dynkin, E. B.; — —: Markov snakes and superprocesses. 95 ProbTher 103 433-473

Krylov, N. V.; — —; Skorokhod, A. V.; Yushkevich, A. A.: E. B. Dynkin's 70th birthday. 94 ThProbAp 39 654-656

Kvalheim, O. M.: Massart, D. L.; Lewi, P. J. (Auth); — — (Rev): Review of "Video course: Principal components". 95 Chemolab 27 147-147

Kvalheim, Olav M.: Grung, Bjørn; — —: Resolution of multicomponent profiles with partial selectivity. A comparison of direct methods. 95 Chemolab 29 75-87

Grung, Bjørn; — —: Detection and quantification of embedded minor analytes in three-way multicomponent profiles by evolving projections and internal rank annihilation. 95 Chemolab 29 213-221

Grung, Bjørn; — —: Rank mapping of three-way multicomponent profiles. 95 Chemolab 29 223-232

Hove, Helge; Liang, Yi-Zeng; — —: Trimmed object projections: A nonparametric robust latent-structure decomposition method. 95 Chemolab 27 33-40

Kvam, Paul H.; Martz, Harry F.: Bayesian inference in a discrete shock model using confounded common cause data. 95 RelEngSS 48 19-25

Kvasnička, V.; Pospíchal, J.: Messy simulated annealing. 95 JChemtrc 9 309-322

Kwan, Andy C. C.: Nebebe, Fassil; — —: Detecting 'potential' outliers in time series: Some Monte Carlo evidence on dummy variable plot approaches. 95 JStPlInf 43 87-95

Kwan, C. W.: Fung, W. K.; — —: Sensitivity analysis in factor analysis: Difference between using covariance and correlation matrices. 95 Psymtrka 60 607-614

Kwon, Hyuck Moo: Bai, Do Sun; — —: Economic design of a two-stage screening procedure with a prescribed outgoing quality. 95 Metrika 42 1-18

Kwon, Woo-Joo: Im, Mi-Jeong; — —; Lee, Joo-Ho: Two-stage Latin hypercube sampling and its application (Korean). 95 KorApS 8/2 99-108

Kwon, Youngmee: Reflected Brownian motion in a cone: Semimartingale property. 95 ProbTher 101 211-226

Kwong, Koon-Shing: Evaluation of one-sided percentage points of the singular multivariate normal distribution. 95 JStCmpSm 51 121-135

Kyburg, Henry, Jr: Maher, Patrick (Auth); — — (Rev): Review of "Betting on theories". 95 PhilSci 62 343-345

Kydes, Andy: Dandekar, Ramesh; Cohen, Michael; Kirkendall, Nancy; — —: Applicability of localized linear regression techniques to develop response surface models. 94 ASAProPEng 87-91

Kyriakidis, E. G.: Optimal control of two competing diseases or species. 95 MaScitst 20 56-66

L'Ecuyer, Pierre: Glynn, Peter W.; — —: Likelihood ratio gradient estimation for stochastic recursions. 95 AdvAppPr 27 1019-1053

La Corte, Aurelio: Andronico, Maurizio; Casale, Salvatore; — —: A method for the generation of correlated random processes. 95 JCmpInT 3 9-18

La Du, Terence J.; Tanaka, J. S.: Incremental fit index changes for nested structural equation models. 95 MultBehR 30 289-316

La Rotonda, L.: Macchiato, M. F.; — —; Lapenna, V.; Ragosta, M.: Time modelling and spatial clustering of daily ambient temperature: An application in Southern Italy. 95 EnvrMtrc 6 31-53

La Tona, Letizia: A jackknife estimator of the scale parameter of L_p-normal distribution (Italian). 94 Metron 52/1 181-188

Laaksonen, S.: Adjustment methods for non-response and their application to Finnish income data. 92 StatJUN 9 125-138

Labouvie, Erich; Ruetsch, Charles: Testing for equivalence of measurement scales: Simple structure and metric invariance reconsidered. 95 MultBehR 30 63-76 (Disc: p77-123)

— —; Ruetsch, Charles: Reply to comments on "Testing for equivalence of measurement scales: Simple structure and metric invariance reconsidered". 95 MultBehR 30 107-111

— —; Ruetsch, Charles: Reply to comments on "Testing for equivalence of measurement scales: Simple structure and metric invariance reconsidered". 95 MultBehR 30 121-123

LaBrecque, John F.: Lemke, Jon H.; — —; Rubenstein, Linda M.: Analysis of the efficacy of rhIL-1ra for the treatment of sepsis. 94 ASAProBiop 340-345

LaBree, Laurie: Lapuerta, Pablo; Azen, Stanley P.; — —: Use of neural networks in predicting the risk of coronary artery disease. 95 CmpBimdR 28 38-52

Labropulu, F.; Sullivan, P. J.: Mean-square values of concentration in a contaminant cloud. 95 EnvrMtrc 6 619-625

— —; Sullivan, P. J.; Ye, H.: A contaminant-dilution heuristic model. 95 EnvrMtrc 6 677-684

Labys, Walter C.: Cromwell, Jeff B.; — —; Terraza, Michel (Auth); Downing, Darryl J. (Rev): Review of "Univariate tests for time series models". 95 Technmcs 37 346-346

Lacaze, B.: Stationarity of independent sequences. 95 StPrLet 23 9-11

Lachenbruch, Peter A.; Clements, Philip J.; He, Weizhong: On encoding values for data recorded as $X > C$. 95 BiomtrcJ 37 855-867

Apian Software (Auth); — — (Rev): Review of "Survey Pro for Windows (Version 1.8)". 95 Chance 8/4 58-59

Datamost Corporation (Auth); — — (Rev): Review of "StatMost". 95 Chance 8/2 56-61

GraphPad Software (Auth); — — (Rev): Review of "InStat". 95 Chance 8/3 57-60

Minitab Inc (Auth); — — (Rev): Review of "Minitab: Release 10 for Windows". 95 Chance 8/1 58-61

Lachin, John M.: Lan, K. K. Gordon; — —: Martingales without tears. 95 LiDaAnal 1 361-375

Lan, K. K. Gordon; Rosenberger, William F.; — —: Sequential monitoring of survival data with the Wilcoxon statistic. 95 Biomtrcs 51 1175-1183

Rosenberger, William F.; — —; Bain, Raymond P.: Nonparametric test of stochastic ordering for multiple longitudinal measures. 95 JBiophSt 5 235-243

Lachout, P.: Boček, P.; — —: Linear programming approach to LMS-estimation. 95 CmpStDA 19 129-134

Janž, M.; — —: A central limit theorem for stationary random fields. 95 MaMethSt 4 463-471

Lachtermacher, Gerson; Fuller, J. David: Backpropagation in time-series forecasting. 95 JForcast 14 381-393

Lackritz, J. R.: Oltman, D. Olson; — — (Auth); Garcia, Eva Ferreira (Rev): Review of "Statistics for business and economics". 95 JAppStat 22 437-437

Ladalla, Joseph N.: On fitting a modified limiting normal to the posterior distribution. 95 JStPlInf 43 229-234

Ladde, G. S.; Medhin, N. G.: Derivation of optimality conditions for a stochastic control problem. 95 StocAnAp 13 165-176

Lagakos, S. W.: Chow, Shein-Chung; Liu, Jen-Pei (Auth); — — (Rev): Review of "Statistical design and analysis in pharmaceutical science". 95 ShrtBkRv 15/3 42-42

Dunn, G.; Everitt, B. (Auth); — — (Rev): Review of "Clinical biostatistics: An introduction to evidence-based medicine". 95 ShrtBkRv 15/3 41-42

Lagazio, Corrado: Cislaghi, Cesare; Biggeri, Annibale; Braga, Mario; — —; Marchi, Marco: Exploratory tools for disease mapping in geographical epidemiology. 95 StatMed 14 2363-2381

Lagona, Francesco: Interaction levels in an autoregressive graphical model. 94 Metron 52/1 113-141

Lahiri, Kajal: Davies, Anthony; — —: A new framework for analyzing survey forecasts using three-dimensional panel data. 95 JEconmtx 68 205-227

Lahiri, P.; Rao, J. N. K.: Robust estimation of mean squared error of small area estimators. 95 JASA 90 758-766

Datta, G. S.; — —: Robust hierarchical Bayes estimation of small area characteristics in the presence of covariates and outliers. 95 JMultiAn 54 310-328

Lahiri, S. N.: On the asymptotic behaviour of the moving block bootstrap for normalized sums of heavy-tail random variables. 95 AnlsStat 23 1331-1349

Lai, C. D.: Construction of bivariate distributions by a generalised trivariate reduction technique. 95 StPrLet 25 265-270

Hutchison, T. P.; — — (Auth); Fielding, A. (Rev): Review of "The data sets from CBDEA". 95 Ststcian 44 288-289

Lai, Dejian: Comment on "A simple test for spatial pattern in regional health data" (V13 p1037-1044). 95 StatMed 14 1393-1393

Lai, Kon S.: Cheung, Yin-Wong; — —: Lag order and critical values of the augmented Dickey-Fuller test. 95 JBES 13 277-280

Cheung, Yin-Wong; — —: Estimating finite sample critical values for unit root tests using pure random walk processes: A note. 95 JTimSrAn 16 493-498

Lai, Tze Leung: Sequential changepoint detection in quality control and dynamical systems. 95 JRSS-B 57 613-644 (Disc: p645-658)

— —: Asymptotic properties of nonlinear least squares estimates in stochastic regression models. 94 AnlsStat 22 1917-1930

— —; Wang, Julia Qizhi: Asymptotic expansions for the distributions of stopped random walks and first passage times. 94 AnlsProb 22 1957-1992

— —; Ying, Zhiliang; Zheng, Zukang: Asymptotic normality of a class of adaptive statistics with applications to synthetic data methods for censored regression. 95 JMultiAn 52 259-279

Konev, Victor; — —: Estimators with prescribed precision in stochastic regression models. 95 SqtlAnly 14 179-192

Laihonen, A.: Marketing of statistical service: Some Finnish experiences. 91 StatJUN 8 307-312

Laird, N. M.: Jones, R. H. (Auth); — — (Rev): Review of "Longitudinal data with serial correlation: A state-space approach". 95 Metrika 42 68-69

Laird, Nan M.: Donnelly, Christl A.; — —; Ware, James H.: Prediction and creation of smooth curves for temporally correlated longitudinal data. 95 JASA 90 984-989

Fitzmaurice, Garrett M.; — —: Regression models for a bivariate discrete and continuous outcome with clustering. 95 JASA 90 845-852

Fitzmaurice, Garrett M.; — —; Zahner, Gwendolyn E. P.; Daskalakis, Constantine: Bivariate logistic regression analysis of childhood psychopathology ratings using multiple informants. 95 AJEpidm 142 1194-1203

Wang-Clow, Fong; Lange, Mary; — —; Ware, James H.: A simulation study of estimators for rates of change in longitudinal studies with attrition. 95 StatMed 14 283-297

Lakatos, Edward: Zucker, David M.; — —; Webber, Larry S.; Murray, David M.; McKinlay, Sonja M.; Feldman, Henry A.; Kelder, Steve H.; Nader, Philip R.: Statistical design of the child and adolescent trial for cardiovascular health (CATCH): Implications of cluster randomization. 95 ConClnTr 16 96-118

Lakkis, H. D.: Jones, C. M.; Robson, D. S.; — —; Kressel, J.: Properties of catch rates used in analysis of angler surveys. 95 TrAFish 124 911-928

Lakner, Peter: Utility maximization with partial information. 95 StocProc 56 247-273

Lall, Upmanu; Mann, Michael: The Great Salt Lake: A barometer of low-frequency climatic variability. 95 WaterReR 31 2503-2515

Rajagopalan, Balaji; — —: A kernel estimator for discrete distributions. 95 JNnparSt 4 409-426

Lalley, Steven P.: Return probabilities for random walk on a half-line. 95 JTherPrb 8 571-599

Lalonde, Robert: Ashenfelter, Orley; Ashmore, David; — —: Bordeaux wine vintage quality and the weather. 95 Chance 8/4 7-14

LaLonde, Steven; Noble, William: Approaches to robust design when the response is multivariate. 94 ASAProPEng 187-192

Coons, Lori; — —: Robust design when variability in control/noise factors can't be controlled during experimentation. 94 ASAProPEng 65-70

LaLonde, Steven M.: Meloun, Milan; Militky, Jiri; Forina, Michele (Auth); — — (Rev): Review of "Chemometrics for analytical chemistry Volume 1: PC-aided statistical data analysis". 95 Technmcs 37 114-115

Lam, James: Yeung, Iris; — —: Comparison of analog and digital simulation on some continuous time systems. 94 ASAProStEd 186-191

Lam, K.: Li, W. K.; — —: Modelling asymmetry in stock returns by a threshold autoregressive conditional heteroscedastic model. 95 Ststcian 44 333-341

Lam, Ming Long: Comment on "Algorithm AS 226: Computing non central beta probabilities" (V36 p241-244). 95 ApplStat 44 551-552

Lam, Y. M.; Pym, J.; Campling, B. G.: Nonlinear models for assessing combinations of chemotherapeutic drugs with resistance modifiers. 94 ASAProBiop 394-398

Lamb, Neil E.: Feingold, Eleanor; — —; Sherman, Stephanie L.: Methods for genetic linkage analysis using trisomies. 95 AJHuGene 56 475-483

Lamberson, Leonard R.: de Souza, Daniel I., Jr; — —: Bayesian Weibull reliability estimation. 95 IIETrn 27 311-320

Lambert, Daniel: Nadeau, Simon; Décarie, Robert; — —; St-Georges, Mario: Nonlinear modeling of muskrat use of habitat. 95 JWildlMg 59 110-117

Lambert, David K.; Harris, Thomas R.: The use of conditional probability functions in range data analysis and simulation. 93 JRngeMgt 46 157-160

Lambert, Diane; Roeder, Kathryn: Overdispersion diagnostics for generalized linear models. 95 JASA 90 1225-1236

Becker, Richard A.; Clark, Linda A.; — —: Graphical analysis of data on networks. 94 ASAProStGr 21-24

Lambert, P.: Lindsey, J. K.; — —: Dynamic generalized linear models and repeated measurements. 95 JStPlInf 47 129-139

Lamken, E. R.: The existence of doubly resolvable $(v,3,2)$-BIBDs. 95 JCombThA 72 50-76

— —: The existence of doubly near resolvable $(v,3,2)$-BIBDs. 94 JComDsgn 2 427-440

Lamm-Tennant, Joan: Cummins, J. David; — —: Capital structure and the cost of equity capital in the property-liability insurance industry. 94 InsrneME 15 187-201

Lamotte, J. L.; Alt, R.: Comparison of simulated annealing algorithms for image restoration. 94 MaCmpSim 37 1-15

Lamoureux, Christopher G.: Friedman, Daniel; Sunder, Shyam (Auth); — — (Rev): Review of "Experimental methods: A primer for economists". 95 JFinance 50 1341-1345

Lamouroux, Nicolas; Souchon, Yves; Herouin, Eric: Predicting velocity frequency distributions in stream reaches. 95 WaterReR 31 2367-2375

Lan, K. K. Gordon; Lachin, John M.: Martingales without tears. 95 LiDaAnal 1 361-375

— —; Rosenberger, William F.; Lachin, John M.: Sequential monitoring of survival data with the Wilcoxon statistic. 95 Biomtrcs 51 1175-1183

Lana, X.; Mills, G. Fernández: Minimum sample size for synoptic weather type classification. Application to winter period data recorded on the Catalan coast (North-east Spain). 94 JClimlgy 14 1051-1060

Lanaud, C.: Lorieux, M.; Perrier, X.; Goffinet, B.; — —; de León, D. González: Maximum-likelihood models for mapping genetic markers showing segregation distortion. 2. F_2 populations. 95 ThApGene 90 81-89

Lancar, Rémi; Kramar, Andrew; Haie-Meder, Christine: Non-parametric methods for analysing recurrent complications of varying severity. 95 StatMed 14 2701-2712

Lancaster, H. O.: Quantitative methods in biological and medical sciences. A historical essay. 94 Springer:Brln:NY 297p

— — (Auth); Farewell, V. T. (Rev): Review of "Quantitative methods in biological and medical sciences. A historical essay". 95 ShrtBkRv 15/2 22-22

— — (Auth); Mayo, Oliver (Rev): Review of "Quantitative methods in biological and medical science - A historical essay". 95 AstrlJSt 37 372-374

Lancaster, Tony; Imbens, Guido: Optimal stock/flow panels. 95 JEconmtx 66 325-348

Land, Stephanie R.; Friedman, Jerome H.: Adaptive signal regression. 94 ASAProStCp 100-105

Lander, Eric: Chernoff, Herman; — —: Asymptotic distribution of the likelihood ratio test that a mixture of two binomials is a single binomial. 95 JStPlInf 43 19-40

Lander, Eric S.: Kruglyak, Leonid; — —: High-resolution genetic mapping of complex traits. 95 AJHuGene 56 1212-1223

Kruglyak, Leonid; — —: Complete multipoint sib-pair analysis of qualitative and quantitative traits. 95 AJHuGene 57 439-454

Kruglyak, Leonid; — —: A nonparametric approach for mapping quantitative trait loci. 95 Genetcs 139 1421-1428

Kruglyak, Leonid; Daly, Mark J.; — —: Rapid multipoint linkage analysis of recessive traits in nuclear families, including homozygosity mapping. 95 AJHuGene 56 519-527

Landim, C.: Benois, O.; Kipnis, C.; — —: Large deviations from the hydrodynamical limit of mean zero asymmetric zero range processes. 95 StocProc 55 65-89

Kipnis, C.; — —; Olla, S.: Macroscopic properties of a stationary non-equilibrium distribution for a non-gradient interacting particle system. 95 AIHenrPB 31 191-221

Saada, E.; — —: Forward to the special issue in honor of Claude Kipnis (French). 95 AIHenrPB 31 3-12

Landin, R.; Freedman, L.; Carroll, R.: Correction to "Adjusting for time trends when estimating the relationship between dietary intake obtained from a food frequency questionnaire and true average intake" (V51 p169-181). 95 Biomtrcs 51 1196-1196

Landin, Richard; Freedman, Laurence S.; Carroll, Raymond J.: Adjusting for time trends when estimating the relationship between dietary intake obtained from a food frequency questionnaire and true average intake. 95 Biomtrcs 51 169-181

Landis, J. Richard: Basu, Srabashi; — —: Model-based estimation of population attributable risk under cross-sectional sampling. 95 AJEpidm 141 1338-1343

Basu, Srabashi; — —: Model-based estimation of population attributable risk under cross-sectional sampling. 95 AJEpidm 142 1338-1343

Have, Thomas R. Ten; — —; Weaver, Susan L.: Association models for periodontal disease progression: A comparison of methods for clustered binary data. 95 StatMed 14 413-429

Landsberger, Sheldon: Hopke, Philip K.; Li, Chong Le; Ciszek, William; — —: The use of bootstrapping to estimate conditional probability fields for source locations of airborne pollutants. 95 Chemolab 30 69-79

Landwehr, James M.: Freeny, Anne E.; — —: Graphical analysis for a large designed experiment. 95 Technmcs 37 1-14

Lane, P. W.: Dyke, G. V.; — —; Jenkyn, J. F.: Sensitivity (stability) analysis of multiple variety trials, with special reference to data expressed as proportions or percentages. 95 ExpAgric 31 75-87

Lane, Peter W.: Nelder, John A.; — —: The computer analysis of factorial experiments: In memoriam - Frank Yates. 95 AmerStat 49 382-385

Lane, Thomas: DuMouchel, William; — —: Comment on "Guiding data analysts with visual statistical strategies". 95 JCmpGrSt 4 251-256

Lane, Thomas P.: DuMouchel, William H.; — —: Construction of mixture-process designs via the Franklin-Bailey algorithm. 94 ASAProQlPr 7-16

Lang, Joseph B.: Balagtas, Cecile C.; Becker, Mark P.; — —: Marginal modelling of categorical data from crossover experiments. 95 ApplStat 44 63-77

Langaris, C.: Katsaros, A.; — —: An N-class structured priority queue with vacations. 95 CommStSM 11 235-248

Lange, K.; Boehnke, M.: Bayesian methods and optimal experimental design for gene mapping by radiation hybrids. 92 AnHuGene 56 119-144

Lange, Kenneth: A gradient algorithm locally equivalent to the EM algorithm. 95 JRSS-B 57 425-437

— —: A quasi-Newton acceleration of the EM algorithm. 95 StSinica 5 1-18

Goradia, Tushar M.; — —; Miller, Perry L.; Nadkarni, Prakash M.: Fast computation of genetic likelihoods on human pedigree data. 92 HumHrdty 42 42-62

Lange, Mary: Wang-Clow, Fong; — —; Laird, Nan M.; Ware, James H.: A simulation study of estimators for rates of change in longitudinal studies with attrition. 95 StatMed 14 283-297

Lange, N.: Jones, R. H. (Auth); — — (Rev): Review of "Longitudinal data with serial correlation: A state-space approach". 95 Biomtrcs 51 1193-1194

— —; Ryan, L.; Billard, L.; Brillinger, D. R.; Conquest, L.; Greenhouse, J. (Ed); Herrendörfer, G. (Rev): Review of "Case studies in biometry". 95 BiomtrcJ 37 640-640

Lange, Nicholas: Fisher, Nicholas I. (Auth); — — (Rev): Review of "Statistical analysis of circular data". 95 JASA 90 801-801

Green, P. J.; Silverman, B. W. (Auth); — — (Rev): Review of "Nonparametric regression and generalized linear models: A roughness penalty approach". 95 JASA 90 1489-1489

Lange, Nick; Ryan, Louise; Billard, Lynne; Brillinger, David R.; Conquest, Loveday; Greenhouse, Joel (Ed); Hilton, Joan (Rev): Review of "Case studies in biometry". 95 JASA 90 800-800

Langholz, B.: Borgan, Ø.; Goldstein, L.; — —: Methods for the analysis of sampled cohort data in the Cox proportional hazards model. 95 AnlsStat 23 1749-1778

Langholz, Bryan; Borgan, Ørnulf: Counter-matching: A stratified nested case-control sampling method. 95 Biomtrka 82 69-79

Witte, John S.; Thomas, Duncan C.; — —: Comment on "Statistical significance testing in the American journal of epidemiology, 1970-1990" (V139 p1047-1052). 95 AJEpidm 142 101-101

Langsrud, Øyvind; Næs, Tormod: On the structure of PLS in orthogonal designs. 95 JChemtrc 9 483-487

Laniel, N.: Chen, E. J.; Gambino, J.; — —; Lindeyer, J.: Design and estimation issues for income in the redesign of the Canadian Labour Force Survey. 94 ASASrvyRsr 919-924

Laniel, Normand; Mohl, Chris: Analysis of urban cluster size in the Canadian Labour Force Survey. 94 ASASrvyRsr 931-936

Mantel, Harold; — —; Duval, Marie-Claude; Marion, Jocelyne: Cost modelling of alternative sample designs for rural areas in the Canadian Labour Force Survey. 94 ASASrvyRsr 925-930

Mian, Ijaz U. H.; — —: Sample allocation for the Canadian Labour Force Survey. 94 ASASrvyRsr 913-918

Lanteri, S.: Forina, M.; — —; Sarabia, L.: Distance and class space in the UNEQ class-modeling technique. 95 JChemtrc 9 69-89

Forina, M.; Drava, G.; Armanino, C.; Boggia, R.; — —; Leardi, R.; Corti, P.; Conti, P.; Giangiacomo, R.; Galliena, C.; Bigoni, R.; Quartari, I.; Serra, C.; Ferri, D.; Leoni, O.; Lazzeri, L.: Transfer of calibration function in near-infrared spectroscopy. 95 Chemolab 27 189-203

Lao, Chang S.: Statistical considerations for survival analysis from medical device clinical studies. 95 JBiophSt 5 159-170

— —: Statistical analysis for estimating relative risk of first cancer progression by monitoring carcinoembryonic antigen (CEA) levels for stage IV breast cancer patients. 94 ASAProBiop 206-210

Lapenna, V.: Macchiato, M. F.; La Rotonda, L.; — —; Ragosta, M.: Time modelling and spatial clustering of daily ambient temperature: An application in Southern Italy. 95 EnvrMtrc 6 31-53

Lapham, Susan J.: Wallman, Katherine; Groves, Robert M.; Parsons, Joseph; Davis, Mary C.; — —: Increasing statistical awareness through graduate training: The joint program in survey methodology. 94 ASAProSoSt 39-44

Lapointe, François-Joseph: Comparison tests for dendrograms: A comparative evaluation. 95 JClass 12 265-282

Legendre, Pierre; — —; Casgrain, Philippe: Modeling brain evolution from behavior: A permutational regression approach. 94 Evolutn 48 1487-1499

Laporte-Simitsidis, Silvy: Girard, Pascal; — —; Mismetti, Patrick; Decousus, Hervé; Boissel, Jean-Pierre: Influence of confounding factors on designs for dose-effect relationship estimates. 95 StatMed 14 987-1005 (Disc: p1007-1008)

Lapuerta, Pablo; Azen, Stanley P.; LaBree, Laurie: Use of neural networks in predicting the risk of coronary artery disease. 95 CmpBimdR 28 38-52

Larholt, Kay M.; Sampson, Allan R.: Effects of heteroscedasticity upon certain analyses when regression lines are not parallel. 95 Biomtrcs 51 731-737

LaRiccia, V. N.: Eggermont, P. P. B.; — —: Maximum smoothed likelihood density estimation for inverse problems. 95 AnlsStat 23 199-220

Lariccia, V. N.: Eggermont, P. P. B.; — —: Maximum smoothed likelihood density estimation. 95 JNnparSt 4 211-222

Larimore, Walter L.: Davis, Alan B.; — —: Multiplicative risk model for predicting rates. 94 ASAProEpid 123-124 (Disc: p125-126)

Davis, Alan B.; — —; Fellingham, Gilbert W.; Beus, Carolyn M.: Using subsequent studies to recover estimability. 94 ASAProEpid 112-117 (Disc: p125-126)

Laroche, Etienne: Hypercontractivity for spin systems with infinite support (French). 95 ProbTher 101 89-132

Larocque, Denis; Tardif, Serge: Development of an optimal test based on weighted rankings. 95 CommStB 24 367-383

Laroque, Guy: Deaton, Angus; — —: Estimating a nonlinear rational expectations commodity price model with unobservable state variables. 95 JApEcmtx 10 9-40S

Larrivee, D.: Belanger, A.; — —: New approach for constructing Canadian working life tables, 1889-1987. 92 StatJUN 9 27-50

Larsen, D. P.; Urquhart, N. S.; Kugler, D. L.: Regional scale trend monitoring of indicators of trophic condition of lakes. 95 WaterReB 31 117-140

Larsen, Greg: Metcalfe, Andrew V. (Auth); — — (Rev): Review of "Statistics in engineering: A practical approach". 95 Technmcs 37 455-455

Larsen, Michael D.: Data augmentation with Bayesian iterative proportional fitting applied to a Census Bureau latent-class problem. 94 ASAProGovt 116-121

Larson, Martin G.: Assessment of cardiovascular risk factors in the elderly: The Framingham heart study. 95 StatMed 14 1745-1756

Larsson, Rolf: The asymptotic distributions of some test statistics in near-integrated AR processes. 95 EconTher 11 306-330

Laskar, R.: Cockayne, E. J.; Hartnell, B. L.; Hedetniemi, S. T.; — —: Perfect domination in graphs. 93 JCISS 18 136-148

Laskey, Kathryn B.: Comment on "Bayesian analysis for pairwise comparisons, binary sequences, and belief networks". 94 ASAProBays 134-135

Laskey, Kathryn Blackmond: Sensitivity analysis for probability assessments in Bayesian networks. 95 IEEESyst 25 901-909

Lasserre, J.-B.: Hernández-Lerma, O.; — —: Discrete-time Markov control processes: Basic optimality criteria. 95 Springer:Brln:NY 216p

Lassman, David M.: Bienias, Julia L.; — —; Scheleur, Scott A.; Hogan, Howard: Improving outlier detection in two establishment surveys. 94 ASASrvyRsr 1140-1143

Last, Günter; Brandt, Andreas: Marked point processes on the real line: The dynamic approach. 95 Springer:Brln:NY 490p

— —; Brandt, Andreas (Auth); Embrechts, P. A. L. (Rev): Review of "Marked point processes on the real line: The dynamic approach". 96 ShrtBkRv 16/1 10-10

Last, John M.: — — (Ed): A dictionary of epidemiology (Third edition). 95 OxfordUniv 208p

Lathrop, G. M.: Li, Zhaohai; Bonney, George E.; — —; Rao, D. C.: Genetic analysis combining path analysis with regressive models: The TAU model of multifactorial transmission. 94 HumHrdty 44 305-311

Lathrop, G. Mark: Weeks, Daniel E.; — —; Ott, Jurg: Multipoint mapping under genetic interference. 93 HumHrdty 43 86-97

Weeks, Daniel E.; Ott, Jurg; — —: Detection of genetic interference: Simulation studies and mouse data. 94 Genetcs 136 1217-1226

Latouche, Guy; Ramaswami, V.: Expected passage times in homogeneous quasi-birth-and-death processes. 95 CommStSM 11 103-122

Latour, R.: Tricot, J.-M.; — —: An approach towards the comparison of profiles using a generalized concentration coefficient (French). 95 RStApp 43/3 35-53

Latzko, William J.; Saunders, David M. (Auth); Copeland, Karen A. F. (Rev): Review of "Four days with Dr. Deming". 95 JQualTek 27 270-270

Laubscher, Nico F.: Kidd, Martin; — —: Robust confidence intervals for scale and its application to the Rayleigh distribution. 95 SoAfrStJ 29 199-217

Laud, Purushottam W.; Ibrahim, Joseph G.: Predictive model selection. 95 JRSS-B 57 247-262

Damien, Paul; — —; Smith, Adrian F. M.: Approximate random variate generation from infinitely divisible distributions with applications to Bayesian inference. 95 JRSS-B 57 547-563

Laudato, Matthew: Chatterjee, Sangit; — —: Gender and performance in athletics. 94 ASAProSpts 36-41

Laumann, E. O.; Gagnon, J. H.; Michael, R. T.; Michaels, S. (Auth); Wasserman, Stanley (Rev): Review of "The social organization of sexuality". 95 Chance 8/2 40-42

Laurent, Kirsty: Woodward, Mark; — —; Tunstall-Pedoe, Hugh: An analysis of risk factors for prevalent coronary heart disease by using the proportional odds model. 95 Ststcian 44 69-80

Laurent, Monique: Deza, Michel; — —: Variety of hypercube embeddings of the equidistant metric and designs. 93 JCISS 18 289-315

Lauritzen, Steffen L.: The EM algorithm for graphical association models with missing data. 95 CmpStDA 19 191-201

Lausen, B.: Altman, D. G.; — —; Sauerbrei, W.; Schumacher, M.: Dangers of using "optimal" cutpoints in the evaluation of prognostic factors. 94 JNCncrI 86 829-835

Opitz, O.; — —; Klar, R. (Ed); Windham, M. P. (Rev): Review of "Information and classification: Concepts, methods and applications". 95 JClass 12 157-158

LaValle, Steven M.; Hutchinson, Seth A.: A framework for constructing probability distributions on the space of image segmentations. 95 CVGIPIm 61 203-230

— —; Hutchinson, Seth A.: A Bayesian segmentation methodology for parametric image models. 95 IEEEPAMI 17 211-217

Lavallée, Pierre: Cross-sectional weighting of longitudinal surveys of individuals and households using the weight share method. 95 SurvyM 21 25-32

Michaud, Sylvie; — —: Comment on "Sample design and weighting research in panel surveys". 94 ASASrvyRsr 680-682

LaValley, Michael P.: Akritas, Michael G.; Murphy, Susan A.; — —: The Theil-Sen estimator with doubly censored data and applications to astronomy. 95 JASA 90 170-177

LaVange, Lisa M.: Shah, Babubbai V.; — —: Mixed models for survey data. 94 ASAProStEd 111-116

Lavergne, C.: Cans, C.; — —: Logistic regression versus a generalized additive model: A practical example (French). 95 RStApp 43/2 77-90

Lavergne, Christian: A jackknife method for estimation of variance components. 95 MaOpfStS 27 1-13

Lavine, Barry K.: Minitab Inc (Auth); — — (Rev): Review of "SCAN: Software for chemometric analysis (Release 1 for Windows)". 95 Chemolab 30 253-255

Lavine, M.: On an approximate likelihood for quantiles. 95 Biomtrka 82 220-222

— —; Mockus, A.: A nonparametric Bayes method for isotonic regression. 95 JStPlInf 46 235-248

Laviolette, Michael: Fuzzy logic: Great hope or grating hype?. 95 Chance 8/4 15-19

— —; Seaman, John W., Jr; Barrett, J. Douglas; Woodall, William H.: A probabilistic and statistical view of fuzzy methods. 95 Technmcs 37 249-261 (Disc: p262-292)

— —; Seaman, John W., Jr; Barrett, J. Douglas; Woodall, William H.: Reply to comments on "A probabilistic and statistical view of fuzzy methods". 95 Technmcs 37 287-292

Lavori, Philip W.; Dawson, Ree; Shera, David: A multiple imputation strategy for clinical trials with truncation of patient data. 95 StatMed 14 1913-1925

Lavrakas, Paul J.: Schejbal, Judith A.; — —: Coverage error and cost issues in small area telephone surveys. 94 ASASrvyRsr 1287-1292

— — (Auth); Frey, James H. (Rev): Review of "Telephone survey methods: Sampling, selection and supervision (Second edition)". 94 JOffStat 10 349-351

Law, Kenneth S.: The use of Fisher's Z in Schmidt-Hunter-type meta-analyses. 95 JEdStat 20 287-306

Lawal, H. Bayo; Upton, G. J. G.: An algorithm for fitting models to $N \times N$ contingency tables having ordered categories. 95 CommStB 24 793-805

Lawera, Martin: Geisser, Seymour (Auth); — — (Rev): Review of "Predictive inference: An introduction". 95 Technmcs 37 121-121

Lawler, G. F.: Introduction to stochastic calculus. 95 ChapmnHall:Lnd 176p

— — (Auth); Riddler-Rowe, C. J. (Rev): Review of "Introduction to stochastic calculus". 96 ShrtBkRv 16/1 8-8

Lawler, Gregory F.: Subdiffusive fluctuations for internal diffusion limited aggregation. 95 AnlsProb 23 71-86

— —: Introduction to stochastic processes. 95 ChapmnHall:Lnd 196p

Weiss, G. H. (Auth); — — (Rev): Review of "Aspects and applications of the random walk". 95 SIAMRvw 37 470-471

Lawless, J. F.: The analysis of recurrent events for multiple subjects. 95 ApplStat 44 487-498

— —; Nadeau, C.: Some simple robust methods for the analysis of recurrent events. 95 Technmcs 37 158-168

Lawless, Jerry; Hu, Joan; Cao, Jin: Methods for the estimation of failure distributions and rates from automobile warranty data. 95 LiDaAnal 1 227-240

Qin, Jing; — —: Estimating equations, empirical likelihood and constraints on parameters. 95 CandJSt 23 145-159

Lawlor, Stephen C.: Dingman, S. Lawrence; — —: Estimating low-flow quantiles from drainage-basin characteristics in New Hampshire and Vermont. 95 WaterReB 31 243-256

Lawrance, A. J.: Deletion influence and masking in regression. 95 JRSS-B 57 181-189

— —: Testing regression lack of fit without replication: A tutorial around Minitab's XLOF procedures. 94 JAppStat 21 541-548

Lawrence Erlbaum Associates Inc: — — (Auth); Wood, Phillip K. (Rev): Review of "TETRAD II: Tools for causal modeling". 95 StrEqMod 2 277-287

Lawrence, Charles E.: Liu, Jun S.; Neuwald, Andrew F.; — —: Bayesian models for multiple local sequence alignment and Gibbs sampling strategies. 95 JASA 90 1156-1170

Lawrence, David; McDavitt, Chris: Significance editing in the Australian Survey of Average Weekly Earnings. 94 JOffStat 10 437-447

Lawrence, David D.: Priore, Roger L.; Brady, Mark F.; — —: Power of detecting the effect of a continuous variable in a Cox model. 94 ASAProBiop 200-205

Lawrence, F. P.: Torng, J. C.-C.; Cochran, J. K.; Montgomery, D. C.; — —: Correction to "Implementing statistically constrained economic EWMA control charts" (V27 p257-264). 95 JQualTek 27 402-402

Lawrence, Frederick P.: Montgomery, Douglas C.; Torng, James C.-C.; Cochran, Jeffery K.; — —: Statistically constrained economic design of the EWMA control chart. 95 JQualTek 27 250-256

Torng, James C.-C.; Cochran, Jeffery K.; Montgomery, Douglas C.; — —: Implementing statistically constrained economic EWMA control charts. 95 JQualTek 27 257-264

Lawrence, Michael; O'Connor, Marcus: The anchor and adjustment heuristic in time-series forecasting. 95 JForcast 14 443-451

Lawrence, Stephen R.: Buss, Arnold H.; — —: Mean-variance interactions in process improvement and capacity design. 95 IIETrn 27 339-347

Lawrimore, Kay W.: Marcis, John G.; — —: Men's jobs and women's jobs: A pilot study. 94 ASAProBuEc 247-252

Laws, J.: Ioannidis, C.; — —; Matthews, K.; Morgan, B.: Business cycle analysis and forecasting with a structural vector autoregression model for Wales. 95 JForcast 14 251-265

Lawson, Andrew B.: MCMC methods for putative pollution source problems in environmental epidemiology. 95 StatMed 14 2473-2485

— —; Viel, Jean-François: Tests for directional space-time interaction in epidemiological data. 95 StatMed 14 2383-2391

Waller, Lance A.; — —: The power of focused tests to detect disease clustering. 95 StatMed 14 2289-2308

Lawson, J. S.: Badrigal, J. L.; Richards, D. O.; Free, J. C.; Lee, T. N.; — —; Hawks, V. D.: Q^2: A TQM course for undergraduates. 94 ASAProQlPr 172-180

Lawton, Richard: Morgan, Carolyn; — —: Repeated-measures analysis of serum lipid PCB data. 94 ASAProEnvr 92-97

Laycock, P. J.; Rowley, P. J.: A method for generating and labelling all regular fractions or blocks for q^{n-m}-designs. 95 JRSS-B 57 191-204

Atkinson, A. C.; Donev, A. N. (Auth); — — (Rev): Review of "Optimum experimental designs". 95 Biomtrcs 51 379-379

Laywine, C.: Complete sets of orthogonal frequency squares and affine resolvable designs. 93 UtilMath 43 161-170

Lazar, Aurel A.: Korilis, Yannis A.; — —: On the existence of equilibria in noncooperative optimal flow control. 95 JACM 42 584-613

Lazraq, Aziz: Cléroux, Robert; — —; Lepage, Yves: Vector correlation based on ranks and a nonparametric test of no association between vectors. 95 CommStA 24 713-733

Lazzeri, L.: Forina, M.; Drava, G.; Armanino, C.; Boggia, R.; Lanteri, S.; Leardi, R.; Corti, P.; Conti, P.; Giangiacomo, R.; Galliena, C.; Bigoni, R.; Quartari, I.; Serra, C.; Ferri, D.; Leoni, O.; — —: Transfer of calibration function in near-infrared spectroscopy. 95 Chemolab 27 189-203

Lee, Jae Won: Two-sample rank tests for acceleration in cure models. 95 StatMed 14 2111-2118

— —; DeMets, David L.: Group sequential comparison of changes: Ad-hoc versus more exact method. 95 Biomtrcs 51 21-30

— —; Sather, Harland N.: Group sequential methods for comparison of cure rates in clinical trials. 95 Biomtrcs 51 756-763

Lee, Jennifer: Niland, Joyce C.; Odom-Maryon, Tamara L.; — —; Tilley, Barbara C.: A survey of biostatistical consulting units throughout North America. 95 AmerStat 49 183-189

Lee, Jiyeon; Lee, Eui Yong: A random shock model for a linearly deteriorating system. 95 JKoreaSt 24 471-479

Lee, JongHyup; Shin, DongWan: Testing for a unit root in an ARIMA$(p,1,q)$ signal observed with measurement error. 95 JKoreaSt 24 481-493

Lee, Joo-Ho: Im, Mi-Jeong; Kwon, Woo-Joo; — —: Two-stage Latin hypercube sampling and its application (Korean). 95 KorApS 8/2 99-108

Lee, Jung-Soon: Cochran, W. George; — —; Potthoff, Jürgen: Stochastic Volterra equations with singular kernels. 95 StocProc 56 337-349

Lee, Junsoo; Schmidt, Peter: Unit root tests based on instrumental variables estimation. 94 IntEconR 35 449-462

Amsler, Christine; — —: An LM test for a unit root in the presence of a structural change. 95 EconTher 11 359-368

Lee, Kay-O: Ryu, Jea-Bok; — —; Lee, Gi-Sung: A practical plan of randomized response technique (Korean). 95 KorApS 8/1 9-26

Lee, Kwan R.; Gagnon, Robert C.: Calibration and stability analysis with a simple mixed linear model when the experiment is of a split-plot type. 94 ASAProBiop 435-440

Lee, Lung-fei: Asymptotic bias in simulated maximum likelihood estimation of discrete choice models. 95 EconTher 11 437-483

— —: Semiparametric maximum likelihood estimation of polychotomous and sequential choice models. 95 JEconmtx 65 381-428

— —: The computation of opportunity costs in polychotomous choice models with selectivity. 95 RvEconSt 77 423-435

— —; Sepanski, Jungsywan H.: Estimation of linear and nonlinear errors-in-variables models using validation data. 95 JASA 90 130-140

Lee, Lung-Fei: Sepanski, Jungsywan H.; — —: Semiparametric estimation of nonlinear errors-in-variables models with validation study. 95 JNnparSt 4 365-394

Lee, Mei-Ling T.: Cole, Bernard F.; — —; Whitmore, G. Alex; Zaslavsky, Alan M.: An empirical Bayes model for Markov-dependent binary sequences with randomly missing observations. 95 JASA 90 1364-1372

Lee, Mei-Ling Ting: Jewell, Nicholas P.; Kimber, Alan C.; — —; Whitmore, G. A. (Ed): Lifetime data: Models in reliability and survival analysis. 95 Kluwer:MA:Hllnd 424p

Lee, Min Soo: Park, Sung Hyun; — —: Slope-rotatability over all directions in third order response surface models. 95 JKoreaSt 24 519-536

Lee, Myoung-Jae: Semi-parametric estimation of simultaneous equations with limited dependent variables: A case study of female labour supply. 95 JApEcmtx 10 187-200

Lee, P. M.: Janicki, A.; Weron, A. (Auth); — — (Rev): Review of "Simulation and chaotic behaviour of α-stable stochastic processes". 95 JRSS-A 158 351-352

Lee, P. N.: Lung cancer and ETS: Is the epidemiologic evidence conclusive?. 94 ASAProEnvr 65-73 (Disc: p74-76)

Lee, Peter M.: Wright, G.; Ayton, P. (Ed); — — (Rev): Review of "Subjective probability". 95 JRSS-A 158 637-638

Lee, Ren Fen: Huang, Deng-Yuan; — —: Selecting the largest capability index from several quality control processes. 95 JStPlInf 46 335-346

Lee, Richard J.: Oehlert, Gary W.; — —; Van Orden, Drew: Statistical analysis of asbestos fibre counts. 95 EnvrMtrc 6 115-126

Lee, Robert Edwin; Spurrier, John D.: Distribution-free multiple comparisons between successive treatments. 95 JNnparSt 5 261-273

— —; Spurrier, John D.: Successive comparisons between ordered treatments. 95 JStPlInf 43 323-330

Lee, Robin: Estimating telephone demand: Validation of previous models. 94 ASAProQlPr 82-86

Lee, Robyn B.; Zeichner, Brian: Statistical analysis of pests resistance data. 94 ASAProEnvr 179-180

Lee, Ronald: Tuljapurkar, Shripad; — —: Comment on "Inference from a deterministic population dynamics model for Bowhead whales". 95 JASA 90 424-425

Lee, Samuel: Wise, Barry M.; Holt, Bradley R.; Gallagher, Neal B.; — —: A comparison of neural networks, non-linear biased regression and a genetic algorithm for dynamic model identification. 95 Chemolab 30 81-89

Lee, Sangyeol: A note on the strong mixing property for a random coefficient autoregressive process. 95 JKoreaSt 24 243-248

— —: A trimmed mean of location of an AR(∞) stationary process. 95 JStPlInf 48 131-140

Lee, Seungyeoun; Wolfe, Robert A.: A class of linear nonparametric rank tests for dependent censoring. 95 CommStA 24 687-712

Lee, Sharon M.: Edmonston, Barry; — —; Passel, Jeffrey S.: U.S. Population projections for national origin groups: Taking into account ethnicity and exogamy. 94 ASAProSoSt 100-105 (Disc: p122-123)

Lee, Shiowjen; Durham, S.; Lynch, J.: On the calculation of the reliability of general load sharing systems. 95 JAppProb 32 777-792

Lee, Sho Rong: Tan, Wai-Yuan; — —; Tang, Si Chin: Characterization of HIV infection and seroconversion by a stochastic model of the HIV epidemic. 95 MaBisci 126 81-123

Lee, Sik-Yum; Poon, Wai-Yin: Estimation of factor scores in a two-level confirmatory factor analysis model. 95 CmpStDA 20 275-284

— —; Poon, Wai-Yin; Bentler, P. M.: A two-stage estimation of structural equation models with continuous and polytomous variables. 95 BJMaStPs 48 339-358

Wang, S. J.; — —: A geometric approach of the generalized least-squares estimation in analysis of covariance structures. 95 StPrLet 24 39-47

Lee, Stephen M. S.; Young, G. Alastair: Asymptotic iterated bootstrap confidence intervals. 95 AnlsStat 23 1301-1330

Lee, Suk-hoon: Park, Hee-chang; — —: Optimal design of partially step-stress life testing for the series systems (Korean). 95 KorApS 8/2 121-132

Lee, Sukhoon; Park, Nae-Hyun; Kim, Yung-Hwan: A clustering method using the Coulomb energy network (Korean). 95 KorApS 8/1 39-50

Lee, Sunho; Park, Cheolyong; Kim, Byung Soo: Tests for detecting overdispersion in Poisson models. 95 CommStA 24 2405-2420

Lee, T. N.: Badrigal, J. L.; Richards, D. O.; Free, J. C.; — —; Lawson, J. S.; Hawks, V. D.: Q^2: A TQM course for undergraduates. 94 ASAProQlPr 172-180

Lee, Therese C. Y.; Furino, Steven C.: A translation of J. X. Lu's "An existence theory for resolvable balanced incomplete block designs". 95 JComDsgn 3 321-340

Lee, Timothy H.; Searls, Donald T.: Bias correction in risk assessment when logistic regression is used with an unevenly proportioned sample between risk and non-risk groups. 94 ASAProBuEc 304-309

Lee, Tzong-Yow: Large deviations for independent random walks on the line. 95 AnlsProb 23 1315-1331

Lee, Y. H.; Cho, S.; Kim, W. C.; Park, B. U.: On estimating integrated squared spectral density derivatives. 95 JStPlInf 48 165-184

Lee, Yong Goo: Multivariate pHd analysis (Korean). 95 KorApS 8/1 61-74

Lee, Yoong-Sin: Graphical demonstration of an optimality property of the median. 95 AmerStat 49 369-372

Leemis, Lawrence M.: Ciardo, Gianfranco; — —; Nicol, David: On the minimum of independent geometrically distributed random variables. 95 StPrLet 23 313-326

Lefebvre, Mario: Solutions to first passage time problems by the method of separation of variables (French). 95 JAppProb 32 337-348

Lefèvre, Claude; Utev, Sergey: Poisson approximation for the final state of a generalized epidemic process. 95 AnlsProb 23 1139-1162

— —; Utev, Sergey: On a conjecture for the non-existence of the expectation of randomly stopped sums. 95 JAppProb 32 1138-1141

Lefkopoulou, Myrto; Zelen, Marvin: Intermediate clinical events, surrogate markers and survival. 95 LiDaAnal 1 73-85

Legler, Julie M.; — —; Ryan, Louise M.: Efficiency and power of tests for multiple binary outcomes. 95 JASA 90 680-693

Lefkovitch, L. P.: Jui, Perry Y.; — —: Selecting the size of a diallel cross experiment. 92 ThApGene 85 21-25

Legendre, Pierre; Lapointe, François-Joseph; Casgrain, Philippe: Modeling brain evolution from behavior: A permutational regression approach. 94 Evolutn 48 1487-1499

Legendre, S.; Clobert, J.: ULM, a software for conservation and evolutionary biologists. 95 JAppStat 22 817-834

Léger, Christian: Altman, Naomi; — —: Bandwidth selection for kernel distribution function estimation. 95 JStPlInf 46 195-214

Altman, Naomi; — —: Cross-validation, the bootstrap, and related methods for tuning parameter selection. 94 ASAProStCp 41-50

Altman, Naomi; — —: On the optimality of prediction based selection criteria and the convergence rates of estimators. 94 ASAProStCp 112-117

Legler, Julie: "I always wanted to be a statistician" or The accidental statistician. 95 Stats /12 10-13

Legler, Julie M.; Lefkopoulou, Myrto; Ryan, Louise M.: Efficiency and power of tests for multiple binary outcomes. 95 JASA 90 680-693

Legreton, Jean-Dominique: The future of population dynamic studies using marked individuals: A statistician's perspective. 95 JAppStat 22 1009-1030

Lehmann, Michel; Daurés, Jean-Pierre; Mottet, Nicolas; Navratil, Henri: Comparison between exact and parametric distributions of multiple inter-raters agreement coefficient. 95 CmpBiomd 47 113-121

Lehnen, Sandra: Implementing quality in public education: A case study of human resources management. 94 ASAProQlPr 60-65

Lehoczky, John: Modernizing statistics Ph.D. programs. 95 AmerStat 49 12-17 (Disc: p18-23)

Lehtonen, R.; Pahkinen, E. J. (Auth); Plachky, D. (Rev): Review of "Practical methods for design and analysis of complex surveys". 95 StatDecn 13 399-400

Lehtonen, Risto; Pahkinen, Erkki J.: Practical methods for design and analysis of complex surveys. 95 Wiley:NY:UK 337p

— —; Pahkinen, Erkki J. (Auth); Kemp, C. D. (Rev): Review of "Practical methods for design and analysis of complex surveys". 95 ShrtBkRv 15/2 24-25

Lei, Ding-Hwa; Vardeman, Stephen B.: LRT method of constructing a two-sided acceptance region for variables acceptance sampling and its comparison with other methods. 94 ASAProPEng 115-119

Lei, Xingye; Peng, Y. B.; Wright, F. T.: Testing homogeneity of normal means with a simply ordered alternative and dependent observations. 95 CmpStDA 20 173-183

Leibowitz, Daniela; Leibowitz, Elia M.: An algorithm for a period search in a sparsely covered time series at a fixed phase. 95 JTimSrAn 16 221-236

Leibowitz, Elia M.: Leibowitz, Daniela; — —: An algorithm for a period search in a sparsely covered time series at a fixed phase. 95 JTimSrAn 16 221-236

Leighton, F. T.: Coffman, E. G., Jr; Gilbert, E. N.; Greenberg, A. G.; — —; Robert, Philippe; Stolyar, A. L.: Queues served by a rotating ring. 95 CommStSM 11 371-394

Leitch, Gordon; Tanner, J. Ernest: Professional economic forecasts: Are they worth their costs?. 95 JForcast 14 143-157

Leitch, Roger D.: Reliability analysis for engineers. An introduction. 95 OxfordUniv 248p

Leite, José G.; Bolfarine, Heleno: Bayesian estimation of the number of equiprobable classes in a population. 94 RBrzlPrS 8 127-133

Leitner, M.: Hyper-weakly harmonizable processes and operator families. 95 StocAnAp 13 471-485

Leiva, Ricardo Aníbal: Bayesian prediction of the number of change points in the piecewise linear model. 94 JItalStS 3 271-289

Leland, Robert P.: Song, Qiang; — —; Chissom, Brad S.: A new fuzzy time-series model of fuzzy number observations. 95 FuzzySet 73 341-348

Leland, Will E.: Willinger, Walter; Taqqu, Murad S.; — —; Wilson, Daniel V.: Self-similarity in high-speed packet traffic: Analysis and modeling of ethernet traffic measurements. 95 StatSci 10 67-85

Lemak, N. A.: Gehan, E. A.; — —: Statistics in medical research. Developments in clinical trials. 94 PlenumPubl:NY 214p

Gehan, E. A.; — — (Auth); De Stavola, B. L. (Rev): Review of "Statistics in medical research. Developments in clinical trials". 95 ShrtBkRv 15/2 22-22

Lemak, Noreen A.: Gehan, Edmund A.; — —: Statistics in medical research: Developments in clinical trials. 94 PlenumPubl:NY 224p

Lemdani, Mohamed; Pons, Odile: Tests for genetic linkage and homogeneity. 95 Biomtrcs 51 1033-1041

Lemeshow, Stanley: Hosmer, David W.; — —: Confidence interval estimates of an index of quality performance based on logistic regression models. 95 StatMed 14 2161-2172

Lemke, Jon H.; LaBrecque, John F.; Rubenstein, Linda M.: Analysis of the efficacy of rhIL-1ra for the treatment of sepsis. 94 ASAProBiop 340-345

Lemon, G. H.: Arnold, Jesse C.; — —; Norton, J. M.: Sample size requirements for rare or abundant attribute sampling. 95 CommStB 24 131-148

Lenarduzzi, L.: Optimal neighborhoods for local regression in the bivariate case. 95 JNnparSt 4 233-248

Lengnick, Laura L.: Dulaney, Wayne P.; — —; Hart, Galen F.: Use of geostatistical techniques in the design of an agricultural field experiment. 94 ASASrvyRsr 183-187

Lengyel, Tamás: On the probability of reaching a given head to tail ratio. 95 JCISS 20 273-278

Lenhart, Suzanne: Davis, M. H. A. (Auth); — — (Rev): Review of "Markov models and optimization". 95 SIAMRvw 37 262-263

Lenk, Peter J.: Allenby, Greg M.; — —: Reassessing brand loyalty, price sensitivity, and merchandising effects on consumer brand choice. 95 JBES 13 281-289

Lennon, Jack J.; Turner, John R. G.: Predicting the spatial distribution of climate: Temperature in Great Britain. 95 JAnmlEco 64 370-392

Lenschow, Donald H.: Mann, Jakob; — —; Kristensen, Leif: Comment on "A definitive approach to turbulence statistical studies in planetary boundary layers" (V51 p1682-1690). 95 JAtmsphr 52 3194-3196

Lenstra, A. J.: Kortram, R. A.; van Rooij, A. C. M.; — —; Ridder, G.: Constructive identification of the mixed proportional hazards model. 95 StNeerla 49 269-281

Lent, Janice; Miller, Stephen; Cantwell, Patrick: Composite weights for the Current Population Survey. 94 ASASrvyRsr 867-872

Lent, R. W.: Goldschmidt, H. M. J.; — —: From data to information: How to define the context?. 95 Chemolab 28 181-192

Lenth, R. V.; Markatou, M.; Tsimikas, J.: Robust tests based on the sample characteristic function. 95 AstrlJSt 37 45-60

Lenth, Russell V.: Markatou, Marianthi; Horowitz, Joel L.; — —: Robust scale estimation based on the empirical characteristic function. 95 StPrLet 25 185-192

Hjorth, J. S. Urban (Auth); — — (Rev): Review of "Computer intensive statistical methods: Validation, model selection, and bootstrap". 95 Technmcs 37 458-459

Lenz, H.-J.: Condra, Lloyd W. (Auth); — — (Rev): Review of "Reliability improvement with design of experiments". 95 CmpStDA 19 605-605

Garfinkel, Simson L.; Mahoney, Michael K. (Auth); — — (Rev): Review of "NeXTSTEP programming, Step one: Object-oriented applications". 95 CmpStDA 19 712-713

Michalewicz, Z. (Ed); — — (Rev): Review of "Statistical and scientific databases". 95 AllgeStA 79 103-105

Michalewicz, Zbigniew (Ed); — — (Rev): Review of "Statistical and scientific databases". 95 CmpStDA 19 479-481

Morgenthaler, S.; Ronchetti, E.; Stahel, W. A. (Ed); — — (Rev): Review of "New directions in statistical data analysis and robustness". 95 CmpStDA 19 711-712

Raj, Baldev; Baltagi, Badi H. (Ed); — — (Rev): Review of "Panel data analysis". 95 CmpStDA 19 481-483

Léon, Jens: Hühn, Manfred; — —: Nonparametric analysis of cultivar performance trials: Experimental results and comparison of different procedures based on ranks. 95 AgronmyJ 87 627-632

Léonard, C.: On large deviations for particle systems associated with spatially homogeneous Boltzmann type equations. 95 ProbTher 101 1-44

Cattiaux, P.; — —: Correction to "Minimization of the Kullback information of diffusion processes" (V30 p83-132). 95 AIHenrPB 31 705-707

Kipnis, C.; — —: Large deviations of an asymmetric hydrodynamical system of independent particles (French). 95 AIHenrPB 31 223-248

Léonard, Christian: Large deviations for long range interacting particle systems with jumps. 95 AIHenrPB 31 289-323

Leonard, Kelli A.; An, Anthony B.; Nusser, Sarah M.; Breidt, F. Jay: Approximating the variance of the survey regression estimator using poststratification. 94 ASASrvyRsr 222-227

Leone, Robert P.: Kumar, V.; — —; Gaskins, John N.: Aggregate and disaggregate sector forecasting using consumer confidence measures. 95 IntJFor 11 361-377

Leong, P. M.; Morgenthaler, S.: Plots of DNA data. 94 ASAProStGr 11-20

Leoni, O.: Forina, M.; Drava, G.; Armanino, C.; Boggia, R.; Lanteri, S.; Leardi, R.; Corti, P.; Conti, P.; Giangiacomo, R.; Galliena, C.; Bigoni, R.; Quartari, I.; Serra, C.; Ferri, D.; — —; Lazzeri, L.: Transfer of calibration function in near-infrared spectroscopy. 95 Chemolab 27 189-203

Lepage, Yves: Cléroux, Robert; Lazraq, Aziz; — —: Vector correlation based on ranks and a nonparametric test of no association between vectors. 95 CommStA 24 713-733

Tricot, Jean-Marie; — —: Weighted concordance analysis. 95 BiomtrcJ 37 727-743

Lepeltier, J. P.: Hamadéne, S.; — —: Backward equations, stochastic control and zero-sum stochastic differential games. 95 Stochatx 54 221-231

Lepeniotis, Stefanos: Feuer, Bernice I.; — —: Free radical polymerization of nonlinear optical side chain methacrylate copolymers in dimethylsulfoxide - An experimental design. 95 Chemolab 29 113-124

Lepeniotis, Stefanos S.; Vigezzi, Michael J.: Lowering manufacturing cost of material by formulating it through statistical modeling and design. 95 Chemolab 29 133-139

Lepingle, Dominique: Bouleau, Nicolas; — — (Auth); Spall, James C.; Nelson, J. Barry (Rev): Review of "Numerical methods for stochastic processes". 95 Technmcs 37 236-237

Lepkowski, James M.: Jay, Gina M.; Belli, Robert F.; — —: Quality of last doctor visit reports: A comparison of medical record and survey data. 94 ASASrvyRsr 362-367 (Disc: p379-381)

Jay, Gina M.; Belli, Robert F.; — —: Use of last physician visit to characterize health care utilization. 94 ASASrvyRsr 368-372 (Disc: p379-381)

Leppard, P.: O'Neill, T. J.; Tallis, G. M.; — —: A review of the technical features of breast cancer screening illustrated by a specific model using South Australian cancer registry data. 95 StMedRes 4 55-72

Lepskii, O. V.; Spokoiny, V. G.: Local adaptation to inhomogeneous smoothness: Resolution level. 95 MaMethSt 4 239-258

Lerche, Hans Rudolf: Keener, Robert; — —; Woodroofe, Michael: A nonlinear parking problem. 95 SqtlAnly 14 247-272

Lesaffre, E.: Svensson, E. (Auth); — — (Rev): Review of "Analysis of systematic and random differences between paired ordinal categorical data". 95 Biomtrcs 51 794-794

Lesaffre, Emmanuel: Rudolfer, Stephan M.; Watson, Peter C.; — —: Are ordinal models useful for classification? A revised analysis. 95 JStCmpSm 52 105-132

Lesch, Scott M.; Strauss, David J.; Rhoades, James D.: Spatial prediction of soil salinity using electromagnetic induction techniques 1. Statistical prediction models: A comparison of multiple linear regression and cokriging. 95 WaterReR 31 373-386

— —; Strauss, David J.; Rhoades, James D.: Spatial prediction of soil salinity using electromagnetic induction techniques. 2. An efficient spatial sampling algorithm suitable for multiple linear regression model identification and estimation. 95 WaterReR 31 387-398

Lesigne, E.: Assani, I.; — —; Rudolph, D.: Wiener-Wintner return-times ergodic theorem. 95 IsrelJMa 92 375-395

Leslie, J. R.: Bryc, W. (Auth); — — (Rev): Review of "The normal distribution - Characterizations and applications". 96 ShrtBkRv 16/1 4-4

Kulkarni, Vidyadhar G. (Auth); — — (Rev): Review of "Modeling and analysis of stochastic systems". 96 ShrtBkRv 16/1 4-5

Lesperance, Mary L.: Hamilton, David C.; — —: A comparison of methods for univariate and multivariate acceptance sampling by variables. 95 Technmcs 37 329-339

Lindsay, Bruce G.; — —: A review of semiparametric mixture models. 95 JStPlInf 47 29-39

Lesser, Martin L.; Kohn, Nina E.; Napolitano, Barbara A.; Pahwa, Savita: The FU-PLOT: A graphical method for visualizing the timing of follow-up in longitudinal studies. 95 AmerStat 49 139-143

— —; Parker, Robert A.: The biostatistician in medical research: Allocating time and effort. 95 StatMed 14 1683-1692

Lesser, Virginia M.: Ramsey, Fred L.; — —: Conditional and unconditional estimators of population size. 95 EnvEcoSt 2 181-190

Ott, Wayne R. (Auth); — — (Rev): Review of "Environmental statistics and data analysis". 95 AmerStat 49 398-399

Lessler, Judith: Duffer, Allen; — —; Weeks, Michael; Mosher, William: Effects of incentive payments on response rates and field costs in a pretest of a national CAPI survey. 94 ASASrvyRsr 1386-1392

Lessler, Judith T.; Weeks, Michael F.; O'Reilly, James M.: Results from the National Survey of Family Growth cycle V pretest. 94 ASASrvyRsr 64-70

O'Reilly, James M.; Hubbard, Michael L.; — —; Biemer, Paul P.; Turner, Charles F.: Audio and video computer assisted self-interviewing: Preliminary tests of new technologies for data collection. 94 JOffStat 10 197-214

Letac, G.; Massam, H.: Craig-Sakamoto's theorem for the Wishart distributions on symmetric cones. 95 AnInStMa 47 785-799

Letac, Gerard: Malliavin, Paul; Airault, Helene; Kay, Leslie; — —: Integration and probability. 95 Springer:Brln:NY 322p

Malliavin, Paul; Airault, Helene; Kay, Leslie; — — (Auth); Barnett, C. (Rev): Review of "Integration and probability". 95 ShrtBkRv 15/3 46-46

Letac, Gérard; Seshadri, Vanamamalai: A random continued fraction in R^{d+1} with an inverse Gaussian distribution. 95 Bernoull 1 381-393

Letchford, S.: Statistics for accountants. 95 ChapmnHall:Lnd 288p

Letenneur, Luc: Commenges, Daniel; Jacqmin, Hélène; — —; Van Duijn, Cornelia M.: Score test for familial aggregation in probands studies: Application to Alzheimer's disease. 95 Biomtrcs 51 542-551

Leti, Giuseppe: Obituary: Amato Herzel, 1923-1994. 94 Metron 52/1 5-15

Letterman, Raymond D.: Driscoll, Charles T.; — —: Factors regulating residual aluminium concentrations in treated waters. 95 EnvrMtrc 6 287-305 (Disc: p305-309)

Leung, C. Y.: Correction to "The location linear discriminant with covariates" (V23 p1027-1046). 95 CommStB 24 1061-1062

Leung, Kin K.: Choudhury, Gagan L.; — —; Whitt, Ward: An algorithm to compute blocking probabilities in multi-rate multi-class multi-resource loss models. 95 AdvAppPr 27 1104-1143

Choudhury, Gagan L.; — —; Whitt, Ward: Calculating normalization constants of closed queuing networks by numerically inverting their generating functions. 95 JACM 42 935-970

Leuridan, Christophe: Control of the H^p norm of a martingale for maxima of local times (French). 95 AnlsProb 23 1289-1299

— —: Ray-Knight theorems and the Itô measure for Brownian motion over the torus R/Z (French). 95 Stochatx 53 109-128

Levental, Shlomo: Bae, Jongsig; — —: A uniform CLT for continuous martingales. 95 JKoreaSt 24 225-231

Bae, Jongsig; — —: Uniform CLT for Markov chains and its invariance principle: A martingale approach. 95 JTherPrb 8 549-570

Levin, Bruce: Accounting for the effects of both group- and individual-level variables in community-level studies. 95 AJPubHlh 85 163-164

Levine, David M.; Ramsey, Patricia P.: Reengineering the business statistics course: Meeting the customer needs of the twenty-first century. 94 ASAProStEd 121-122

Levine, Phillip B.; Zimmerman, David J.: The benefit of additional high-school math and science classes for young men and women. 95 JBES 13/2 137-149

Levit, B. Y.: Belitser, E. N.; — —: On minimax filtering over ellipsoids. 95 MaMethSt 4 259-273

Levit, Boris Y.: Gill, Richard D.; — —: Applications of the van Trees inequality: A Bayesian Cramér-Rao bound. 95 Bernoull 1 59-79

Levitan, Emanuel; Chan, Michael; Herman, Gabor T.: Image-modeling Gibbs priors. 95 CVGIPGr 57 117-130

Levitin, Daniel J.: Russell, Gerald S.; — —: An expanded table of probability values for Rao's spacing test. 95 CommStB 24 879-888

Levy, Philip: Charles Spearman's contributions to test theory. 95 BJMaStPs 48 221-235

Lewbel, Arthur: Consistent nonparametric hypothesis tests with an application to Slutsky symmetry. 95 JEconmtx 67 379-401

Lewi, P. J.: Massart, D. L.; — — (Auth); Kvalheim, O. M. (Rev): Review of "Video course: Principal components". 95 Chemolab 27 147-147

Lewi, Paul J.: Pattern recognition, reflections from a chemometric point of view. 95 Chemolab 28 23-33

Pepperrell, Catherine (Auth); — — (Rev): Review of "Three-dimensional chemical similarity searching". 95 Chemolab 28 205-207

Lewis, C. M.; Cannings, C.: The number of loci needed for ELOD calculations. 92 AnHuGene 56 59-69

Lewis, Christopher Alan: Joseph, Stephen; — —: Factor analysis of the Center for Epidemiological Studies-Depression Scale. 95 PsychRpt 76 40-42

Lewis, D. M.; Chatwin, P. C.: A new model PDF for contaminants dispersing in the atmosphere. 95 EnvrMtrc 6 583-593

Chatwin, P. C.; — —; Sullivan, Paul J.: Turbulent dispersion and the beta distribution. 95 EnvrMtrc 6 395-402

Lewis, J. A.: Jones, B.; — —: The case for cross-over trials in phase III. 95 StatMed 14 1025-1038 (Disc: p1053-1055)

Lewis, John A.: Statistical issues in the regulation of medicines. 95 StatMed 14 127-136

— —; Jones, David R.; Röhmel, Joachim: Biostatistical methodology in clinical trials - A European guideline. 95 StatMed 14 1655-1657

Lewis, Paul J.; Wolfgang, Glenn S.; Vacca, E. Ann: 1992 Census of Agriculture coverage evaluation estimation. 94 ASASrvyRsr 1019-1024

Wolfgang, Glenn S.; — —; Vacca, E. Ann: Cognitive research for the 1997 Census of Agriculture report form. 94 ASASrvyRsr 503-508

Lewis, Roger J.; Berry, Donald A.: Bayesian group sequential clinical trial designs based on a quadratic loss function.. 94 ASAProBays 23-28

Lewis, S. M.: Kao, L.-J.; Yates, P. K.; — —; Dean, A. M.: Approximate information matrices for estimating a given set of contrasts. 95 StSinica 5 593-598

Lewis, T.: Barnett, V.; — — (Auth); Atkinson, Anthony (Rev): Review of "Outliers in statistical data (Third edition)". 95 JRSS-A 158 630-631

Barnett, V.; — — (Auth); Balakrishnan, N. (Rev): Review of "Outliers in statistical data (Third edition)". 95 Biomtrcs 51 381-381

Barnett, V.; — — (Auth); Pincus, R. (Rev): Review of "Outliers in statistical data (Third edition)". 95 BiomtrcJ 37 256-256

Fisher, N. I.; — —; Embleton, B. J. J. (Auth); Challenor, Peter (Rev): Review of "Statistical analysis of spherical data". 95 Ststcian 44 286-287

Lewis, Thomas M.: Khoshnevisan, Davar; — —: The favorite point of a Poisson process. 95 StocProc 57 19-38

Lewis, Toby; Fisher, Nicholas I.: Estimating the angle between the mean directions of two spherical distributions. 95 AstrlJSt 37 179-191

Barnett, Vic; — —: Outliers in statistical data (Third edition). 94 Wiley:NY:UK 584p

Barnett, Vic; — — (Auth); Bremer, Ronald (Rev): Review of "Outliers in statistical data (3rd ed)". 95 Technmcs 37 117-118

Lewontin, R. C.: The detection of linkage disequilibrium in molecular sequence data. 95 Genetcs 140 377-388

Li, Bai-Bing: Huang, Xiang-Ning; — —: A general structure for the unbiased estimate of the parameter's estimable function in linear regression. 95 StocAnAp 13 83-90

Li, Bing: Lindsay, Bruce G.; — —: Comment on "Inference". 95 StatSci 10 175-177

Murphy, Susan; — —: Projected partial likelihood and its application to longitudinal data. 95 Biomtrka 82 399-406

Li, C. C.; Weeks, D. E.; Chakravarti, A.: Similarity of DNA fingerprints due to chance and relatedness. 93 HumHrdty 43 45-52

Li, Ching Chun; Chakravarti, Aravinda: DNA profile similarity in a subdivided population. 94 HumHrdty 44 100-109

Li, Chong Le: Hopke, Philip K.; — —; Ciszek, William; Landsberger, Sheldon: The use of bootstrapping to estimate conditional probability fields for source locations of airborne pollutants. 95 Chemolab 30 69-79

Li, Deli; Rao, M. Bhaskara; Jiang, Tiefeng; Wang, Xiangchen: Complete convergence and almost sure convergence of weighted sums of random variables. 95 JTherPrb 8 49-76

— —; Rao, M. Bhaskara; Wang, Xiang Chen: On the strong law of large numbers and the law of the logarithm for weighted sums of independent random variables with multidimensional indices. 95 JMultiAn 52 181-198

Wang, Xaingchen; Rao, M. Bhaskara; — —: Some results on strong limit theorems for (LB)-space-valued random variables. 95 StPrLet 23 247-251

Li, E.: Hunt, D. N.; — —: Loss of efficiency in spoilt experiments. 95 JAppStat 22 309-315

Li, Gang: Nonparametric likelihood ratio estimation of probabilities for truncated data. 95 JASA 90 997-1003

— —: On nonparametric likelihood ratio estimation of survival probabilities for censored data. 95 StPrLet 25 95-104

— —; Doss, Hani: An approach to nonparametric regression for life history data using local linear fitting. 95 AnlsStat 23 787-823

Li, Guibin: Comparison of various estimates of differencing order in an ARIMA model (Chinese). 95 ChJApPSt 10 353-362

— —: Identification of fractional differencing autoregressive models. 95 CommStA 24 2635-2643

Li, Guo-Ying: Chan, Lai K.; — —: Bivariate control charts for testing $\bar{\mathbf{X}}$ data patterns. 95 CandJSt 23 85-99

Li, Guoying; Liu, Wanrong; Zha, Wenxing: Projection pursuit Neyman goodness-of-fit test and its bootstrap approximation (Chinese). 95 KToChina 39 678-681

Cui, Hengjian; — —: Estimate of covariance matrix with additional information (Chinese). 95 KToChina 39 1354-1357

Shi, Peide; — —: Asymptotic normality of L_1-norm estimators for parametric components in a partly linear model (Chinese). 95 ChinaAMA 15 478-484

Shi, Peide; — —: Asymptotics of the minimum L_1-norm estimates in a partly linear model. 95 JSysSMSE 7 67-77

Shi, Peide; — —: Asymptotic normality of the M-estimator of parametric components in partial linear models (Chinese). 95 KToChina 39 1-2

Shi, Peide; — —: A note on the convergent rates of M-estimates for a partly linear model. 95 MaOpfStS 26 27-47

Zhang, Dixin; — —: Bootstrap approximation for generalized U-processes (Chinese). 95 KToChina 39 1249-1252

Li, Haijun; Shaked, Moshe: On the first passage times for Markov processes with monotone convex transition kernels. 95 StocProc 58 205-216

Li, Hongyi: The power of modified instrumental variables tests for unit roots. 95 Biomtrka 82 660-666

Li, Huan; Zhu, Yixin: On M/G/1 queues with exhaustive service and generalized vacations. 95 AdvAppPr 27 510-531

Li, Jian-Hua; Watanabe, Hiroshi: A Bayesian study of the correlation between standard scores of twins. 94 Behvmtrk 21 121-130

Li, Jingwen; Ou, Jihong: Characterizing the idle-period distribution of GI/G/1 queues. 95 JAppProb 32 247-255

Özekici, Süleyman; — —; Chou, Fee Seng: Waiting time in M/G/1 queues with impolite arrival disciplines. 95 PrEngInf 9 255-267

Li, Jiukun: Fang, Kaitai; — —: Some new results of uniform design (Chinese). 95 KToChina 39 1921-1924

Li, Jufang; Chen, Mei-Ching; Schreuder, Hans T.; Gregoire, Tim G.: Forestry applications of saddle-point approximations to construct confidence intervals for population means. 95 Biomtrcs 51 61-72

Li, Ker-Chau: Carroll, R. J.; — —: Binary regressors in dimension reduction models: A new look at treatment comparisons. 95 StSinica 5 667-688

Cheng, Ching-Shui; — —: A study of the method of principal Hessian direction for analysis of data from designed experiments. 95 StSinica 5 617-639

Li, Kim-Hung: Chan, Nai N.; — —: A matrix identity and its applications to equivalent hypotheses in linear models. 95 CommStA 24 2769-2777

Li, Linxiong: Baxter, Laurence A.; — —: Nonparametric confidence intervals for the renewal function with censored data. 95 JNnparSt 4 317-326

Li, Luning; Flury, Bernard: Uniqueness of principal points for univariate distributions. 95 StPrLet 25 323-327

Tarpey, Thaddeus; — —; Flury, Bernard D.: Principal points and self-consistent points of elliptical distributions. 95 AnlsStat 23 103-112

Li, Lushu: Random fuzzy sets and fuzzy martingales. 95 FuzzySet 69 181-192

Li, Ningning; Chen, Guijing: A kind of rank test about intersection point of two continuous populations (Chinese). 95 KToChina 39 1543-1546

Li, Q.: Mardia, K. V.; Dryden, I. L.; Hurn, M. A.; — —; Millner, P. A.; Dickson, R. A.: Familial spinal shape. 94 JAppStat 21 623-641

Li, Qi; Stengos, Thanasis: Adaptive estimation in the panel data error component model with heteroskedasticity of unknown form. 94 IntEconR 35 981-1000

Baltagi, Badi H.; — —: Testing AR(1) against MA(1) disturbances in an error component model. 95 JEconmtx 68 133-151

Li, Run-Ze; Fang, Kai-Tai: Estimation of scale matrix of elliptically contoured matrix distributions. 95 StPrLet 24 289-297

Li, Susan: Huang, Zhimin; — —: The role of cardinal utilities in multiple objective programming. 94 AJMMgSci 14 301-325

Li, Ta-Hsin: Blind deconvolution of linear systems with multilevel nonstationary inputs. 95 AnlsStat 23 690-704

Li, Teng; Sinha, Bimal K.: Tests of ordered hypotheses for gamma scale parameters. 95 JStPlInf 43 311-321

— —; Sinha, Bimal K.: Tests of ordered hypotheses for gamma scale parameters. 95 JStPlInf 45 387-397

Li, W. K.; Lam, K.: Modelling asymmetry in stock returns by a threshold autoregressive conditional heteroscedastic model. 95 Ststcian 44 333-341

Hui, Y. V.; — —: On fractionally differenced periodic processes. 95 SankhyaB 57 19-31

Wong, H.; — —: Portmanteau test for conditional heteroscedasticity, using ranks of squared residuals. 95 JAppStat 22 121-134

Li, W. M.: Wu, W. R.; — —: A new approach for mapping quantitative trait loci using complete genetic marker linkage maps. 94 ThApGene 89 535-539

Li, W. V.: Kuelbs, J.; — —; Shao, Qi-man: Small ball probabilities for Gaussian processes with stationary increments under Hölder norms. 95 JTherPrb 8 361-386

Li, Weixin: Lu, Changyu; — —: Admissibility of linear estimator in linear models with respect to an incomplete ellipsoidal restrictions. 95 AMaPeknE 37 289-295

Li, Wen-Hsiung: Gu, Xun; Fu, Yun-Xin; — —: Maximum likelihood estimation of the heterogeneity of substitution rate among nucleotide sites. 95 MolBioEv 12 546-557

Zharkikh, Andrey; — —: Statistical properties of bootstrap estimation of phylogenetic variability from nucleotide sequences: II. Four taxa without a molecular clock. 92 JMolEvol 35 356-366

Li, Wenbo V.: Kuelbs, James; — —; Talagrand, Michel: Lim inf results for Gaussian samples and Chung's functional LIL. 94 AnlsProb 22 1879-1903

Li, Wenlian; Wu, C. F. J.: An integrated method of parameter design and tolerance design. 94 ASAProQlPr 44-48

Li, Xihe; Liu, Rongbing: Basic concepts of the fuzzied statistical decision and its superiority (Chinese). 95 AMaAPknC 17 512-521

Li, Xunjing; Tang, Shanjian: General necessary conditions for partially observed optimal stochastic controls. 95 JAppProb 32 1118-1137

Li, Yi: Dawson, Donald A.; Fleischmann, Klaus; — —; Mueller, Carl: Singularity of super-Brownian local time at a point catalyst. 95 AnlsProb 23 37-55

Li, Yiqun: Hewett, John E.; Petroski, Gregory F.; — —: Tables of critical values for two group interim analyses. 95 CommStB 24 825-852

Li, Yuan Hwang: LTDOMAIN: A look-up table for the corresponding estimated one-parameter logistic model scale score and the unbiased domain score for each number-correct score. 95 ApPsychM 19 50-50

— —: Lord_DIF: A computer program to compute Lord's DIF statistics. 95 ApPsychM 19 72-72

Li, Yuanzhang; Saxena, K. M. Lal: Optimal robust Bayes estimation. 95 JStPlInf 46 365-380

Li, Zhaohai: A multiplicative random effects model for meta-analysis with application to estimation of a mixture component. 95 Biomtrcs 51 864-873

— —; Bonney, George E.; Lathrop, G. M.; Rao, D. C.: Genetic analysis combining path analysis with regressive models: The TAU model of multifactorial transmission. 94 HumHrdty 44 305-311

Lian, Ie-Bin: Asymptotics of forced-in variables from screening processes. 95 CommStA 24 131-151

Liang, H.: A note on asymptotic normality for nonparametric multiple regression: The fixed design case. 95 SochwJMa 21 395-399

Liang, Hua: Bahadur asymptotic efficiency in a semi-parametric regression model. 95 ChinaAMB 15 129-140

— —: On the smallest possible asymptotically efficient variance in semiparametric models. 95 JSysSMSE 7 29-33

— —: Second-order asymptotic efficiency of PMLE in generalized linear models. 95 StPrLet 24 273-279

— —: On Bahadur asymptotic efficiency of the maximum likelihood estimator for a generalized semiparametric model. 95 StSinica 5 363-371

— —; Yuan, Yahua: Second order asymptotic efficiency for $Y = \theta_1 + g(T) + \epsilon$ (Chinese). 95 ChJApPSt 10 133-141

Gao, Jiti; — —: Asymptotic normality of pseudo-LS estimator for partly linear autoregression models. 95 StPrLet 23 27-34

Liang, K.-Y.: Diggle, P. J.; — —; Zeger, S. L. (Auth); Goldstein, Harvey (Rev): Review of "Analysis of longitudinal data". 95 JRSS-A 158 345-346

Liang, Kun-Hsia; Krus, David J.; Webb, James M.: K-fold crossvalidation in canonical analysis. 95 MultBehR 30 539-545

Liang, Kung-Yee; Self, Steven G.; Bandeen-Roche, Karen J.; Zeger, Scott L.: Some recent developments for regression analysis of multivariate failure time data. 95 LiDaAnal 1 403-415

— —; Zeger, Scott L.: Reply to comments on "Inference". 95 StatSci 10 196-199

Liang, Kung-yee: Hanfelt, John J.; — —: Approximate likelihood ratios for general estimating functions. 95 Biomtrka 82 461-477

Liang, Kung-Yee: Diggle, Peter J.; — —; Zeger, Scott L. (Auth); Verbyla, A. P. (Rev): Review of "Analysis of longitudinal data". 95 AstrlJSt 37 367-368

Liang, Kyung-Yee; Zeger, Scott L.: Inference based on estimating functions in the presence of nuisance parameters. 95 StatSci 10 158-173 (Disc: p173-199)

Liang, Yi-Zeng: Hove, Helge; — —; Kvalheim, Olav M.: Trimmed object projections: A nonparametric robust latent-structure decomposition method. 95 Chemolab 27 33-40

Xie, Yu-Long; — —; Chen, Zhang-Guang; Huang, Zi-Hong; Yu, Ru-Qin: A nonlinearity tracking analysis algorithm for treatment of non-linearity in multivariate calibration. 95 Chemolab 27 21-32

Liang, Yung-Yee: Diggle, Peter J.; — —; Zeger, Scott L. (Auth); Ten Have, Thomas R. (Rev): Review of "Analysis of longitudinal data". 95 JASA 90 1123-1124

Liao, J. G.; Hall, Charles B.: Fast and accurate computation of the exact confidence limits for the common odds ratio in several 2×2 tables. 95 JCmpGrSt 4 173-179

Liao, Jiangang; Brookmeyer, Ron: An empirical Bayes approach to smoothing in backcalculation of HIV infection rates. 95 Biomtrcs 51 579-588

Liao, M.; Zheng, W. A.: Radial part of Brownian motion on a Riemannian manifold. 95 AnlsProb 23 173-177

Libnau, Fred O.: Lindblom, Torbjörn; Christy, Alfred A.; — —: Quantitative determination of stabiliser in a single base propellant by chemometric analysis of Fourier transform infrared spectra. 95 Chemolab 29 243-254

Lie, Rolv Terje: York, Jeremy; Madigan, David; Heuch, Ivar; — —: Birth defects registered by double sampling: A Bayesian approach incorporating covariates and model uncertainty. 95 ApplStat 44 227-242

Liebhart, Jerzy: Bartkowiak, Anna; — —: Estimation of the spirometric residual volume (RV) by a regression built from Gower distances. 95 BiomtrcJ 37 131-149

Liebhold, Andrew M.: Williams, David W.; — —: Detection of delayed density dependence: Effects of autocorrelation in an exogenous factor. 95 Ecology 76 1005-1008

Liebscher, E.: Strong convergence of sums of φ-mixing random variables. 95 MaMethSt 4 216-229

Lienert, Gustav A.: Hammond, Sean M.; — —: Modified phi correlation coefficients for the multivariate analysis of ordinally scaled variables. 95 EdPsMeas 55 225-236

Lievesley, Denise: Bosworth, D. L.; Wilson, R. A.; Young, A. (Auth); — — (Rev): Review of "Reviews of United Kingdom statistical sources, Volume 27, Research and development". 95 JRSS-A 158 631-632

Lifshits, M. A.: Gaussian random functions. 95 Kluwer:MA:Hllnd 352p

Lifshits, Mikhail A.: Deheuvels, Paul; — —: Necessary and sufficient conditions for the Strassen law of the iterated logarithm in nonuniform topologies. 94 AnlsProb 22 1838-1856

Ligeralde, Antonio V.; Brown, Bryan W.: Band covariance matrix estimation using restricted residuals: A Monte Carlo analysis. 94 IntEconR 35 751-767

Liggett, T. M.: Ferrari, P. A.; Galves, A.; — —: Exponential waiting time for filling a large interval in the symmetric simple exclusion process. 95 AIHenrPB 31 155-175

Liggett, Thomas: Aaronson, Jon; — —; Picco, Pierre: Equivalence of renewal sequences and isomorphism of random walks. 94 IsrelJMa 87 65-76

Liggett, Thomas M.: Survival of discrete time growth models, with applications to oriented percolation. 95 AnlsApPr 5 613-636

— —: Improved upper bounds for the contact process critical value. 95 AnlsProb 23 697-723

— —; Petersen, Peter: The law of large numbers and $\sqrt{2}$. 95 AmMaMnth 102 31-35

Liggett, Walter: Functional errors-in-variables models in measurement optimization experiments. 94 ASAProPEng 193-199

Lii, Keh-Shin; Masry, Elias: On the selection of random sampling schemes for the spectral estimation of continuous time processes. 95 JTimSrAn 16 291-311

— —; Tsou, Tai-Houn: Bispectral analysis of randomly sampled data. 95 JTimSrAn 16 43-66

Lilien, G.: Eliashberg, J.; — — (Ed); Arabie, Phipps (Rev): Review of "Handbook in operations research and management science: Marketing". 95 JClass 12 143-144

Lilleleht, Vilju: Rattiste, Kalev; — —: Survival rates of breeding common gulls in Estonia. 95 JAppStat 22 1057-1062

Lim, Dong Hoon: Kim, Dong Hee; — —: Rank tests for parallelism of regression lines against umbrella alternatives. 95 JNnparSt 5 289-302

Kim, Dong Hee; — —: An asymptotically distribution-free test for parallelism of regression lines against umbrella alternatives (Korean). 95 KorApS 8/1 105-117

Lim, Jae-Hak; Park, Dong Ho: Trend change in mean residual life. 95 IEEEReli 44 291-296

Lim, K. S.: An algorithm for determining the chaos of TAR(1). 95 CommStB 24 505-522

Lin, C.; Wallis, W. D.; Zhu, L.: Generalized 4-profiles of Hadamard matrices. 93 JCISS 18 397-400

Lin, C. H.: Sukhatme, Shashikala; — —: Exact powers of the Wilcoxon test and the logrank test under random censoring. 95 JStPlInf 48 291-301

Lin, C. Thomas: Mudholkar, Govind S.; Srivastava, Deo Kumar; — —: Some p-variate adaptations of the Shapiro-Wilk test of normality. 95 CommStA 24 953-985

Lin, C. Y.: Ali, A. K. A.; — —: Diagonalization of variance-covariance matrices using F-G algorithm. 95 EgpStJ 39 12-25

Lin, Cantian; Wallis, W. D.: Barker sequences and circulant Hadamard matrices. 93 JCISS 18 19-25

Gregory, John; — — (Auth); Prvan, Tania (Rev): Review of "Constrained optimization in the calculus of variations and optimal control theory". 95 AstrlJSt 37 250-250

Lin, Chen-Sin: Single machine stochastic scheduling: Minimizing the number of tardy jobs. 94 AJMMgSci 14 289-300

Lin, D. Y.; Ying, Zhiliang: Semiparametric analysis of general additive-multiplicative hazard models for counting processes. 95 AnlsStat 23 1712-1734

— —; Ying, Zhiliang: Semiparametric inference for the accelerated life model with time-dependent covariates. 95 JStPlInf 44 47-63

Lin, Dennis K. J.: Generating Systematic supersaturated designs. 95 Technmcs 37 213-225

— —: Correction to "Another look at first-order saturated designs: The p-efficient designs" (V35 p384-292). 95 Technmcs 37 473-473

— —; Draper, Norman R.: Screening properties of certain two-level designs. 95 Metrika 42 99-118

— —; Tu, Wanzhu: Dual response surface optimization. 95 JQualTek 27 34-39

Deng, Lih-Yuan; — —: Criteria for supersaturated designs. 94 ASAProQlPr 124-128

Draper, Norman R.; — —: Characterizing projected designs: Repeat and mirror-image runs. 95 CommStA 24 775-795

Guttman, Irwin; — —: Robust recursive estimation for correlated observations. 95 StPrLet 23 79-92

Guttman, Irwin; — —; Reiser, B.; Usher, John S.: Dependent masking and system life data analysis: Bayesian inference for two-component systems. 95 LiDaAnal 1 87-100

Reiser, B.; Guttman, I.; — —; Guess, Frank M.; Usher, John S.: Bayesian inference for masked system lifetime data. 95 ApplStat 44 79-90

Wang, Yuchung; — —; Fang, Kai-Tai: Designing outer array points. 95 JQualTek 27 226-241

Lin, Dongyow: Esposito, Richard; Fox, James K.; — —; Tidemann, Kevin: ARIES: Visual techniques for statistical data investigation at the Bureau of Labor Statistics. 94 ASAProStGr 7-10

Lin, G. D.: Huang, J. S.; — —: Characterization of distributions using equilibrium transformations. 95 SankhyaA 57 179-185

Lin, Ginnie: Reiser, Mark; — —: A goodness-of-fit test for the latent class model when expected frequencies are small. 94 ASAProSoSt 49-54

Lin, Guan-Chyun: Chang, Der-Shin; — —: Stochastic regression model with heteroscedastic disturbance. 95 AnInStMa 47 351-369

Lin, Guoying: Shi, Peide; — —: Global convergence rates of B-spline M-estimators in nonparametric regression. 95 StSinica 5 303-318

Lin, Gwo Dong: On weak convergence within a Lorenz ordering family of distributions. 95 JAppProb 32 1028-1031

Lin, Hua: Yatawara, Nihal; — —: Multiple outlier detection in time series. 94 ASAProBuEc 310-315

Lin, Huanbin; Chen, Jiading: The confidence limits for the reliability in regression model. 95 ActScNat 30 23-29

Lin, Hung-Mo; Hughes, Michael D.: Use of historical marker data for assessing treatment effects in phase I/II trials when subject selection is determined by baseline marker level. 95 Biomtrcs 51 1053-1063

Lin, I-fen; Schaeffer, Nora Cate: Using survey participants to estimate the impact of nonparticipation. 95 PubOpinQ 59 236-258

Lin, Jin-Lung: Granger, Clive W. J.; — —: Causality in the long run. 95 EconTher 11 530-536

Lin, Jinn-Tyan: A simple approximation for the bivariate normal integral. 95 PrEnglnf 9 317-321

Lin, Karl: Tsong, Yi; Hammerstrom, Thomas; — —; Ong, Ting Eng: Dissolution test acceptance sampling plans. 95 JBiophSt 5 171-183

Lin, S.; Thompson, E.; Wijsman, E.: An algorithm for Monte Carlo estimation of genotype probabilities on complex pedigrees. 94 AnHuGene 58 343-357

Lin, S. J.: Stochastic analysis of fractional Brownian motions. 95 Stochatx 55 121-140

— —: On L^p stochastic representations. 95 StPrLet 24 225-227

Bichteler, K.; — —: On the stochastic Fubini theorem. 95 Stochatx 54 271-279

Lin, Shang P.; Tang, Dei-In: A hierarchical active constraints search algorithm for optimal scaling of ordered categorical responses. 95 JStCmpSm 53 197-209

Tang, Dei-In; — —: On improving some methods for multiple endpoints. 94 ASAProBiop 466-474

Lin, Shili: A scheme for constructing an irreducible markov chain for pedigree data. 95 Biomtrcs 51 318-322

— —; Thompson, Elizabeth; Wijsman, Ellen: Finding noncommunicating sets for Markov chain Monte Carlo estimations on pedigrees. 94 AJHuGene 54 695-704

Lin, Tsae-Yun Daphne: Applicability of matrix and bracket approach to stability study design. 94 ASAProBiop 142-147

Lin, Wen-Ling: Japan's financial deregulation and linkage of the Gensaki and Euroyen deposit markets. 95 JApEcmtx 10 447-467

Lin, Winnie: Adams, Benjamin M.; — —: Forecast-based monitoring schemes using combined EWMA-Shewhart control charts. 94 ASAProQlPr 104-108

Lin, Winston T.: Chen, Yueh H.; — —: Testing the arbitrage pricing theory and non-stationarity of the risk premium: An error components non-linear approach. 94 ASAProBuEc 178-183

Lin, Xihong: Breslow, Norman E.; — —: Bias correction in generalised linear mixed models with a single component of dispersion. 95 Biomtrka 82 81-91

Lin, Yachen: Asymptotic confidence bands for normal distributions. 95 CommStA 24 641-652

Lin, Z. Y.: On large increments of infinite series of Ornstein-Uhlenbeck processes. 95 StocProc 60 161-169

Lin, Zheng-Yan: Csörgő, Miklós; — —; Shao, Qi-Man: On moduli of continuity for local times of Gaussian processes. 95 StocProc 58 1-21

Linacre, Susan: Planning the methodology work program in a statistical agency. 95 JOffStat 11 41-53

Linda, Stephen B.: Rodgers, James A., Jr; — —; Nesbitt, Stephen A.: Comparing aerial estimates with ground counts of nests in wood stork colonies. 95 JWildlMg 59 656-666

Lindberg, Mark S.; Sedinger, James S.; Rexstad, Eric A.: Estimating nest site fidelity of adult female black brant with multi-state modeling and geographic information systems. 95 JAppStat 22 725-735

Lindblad, G.: Accardi, L. (Ed); — — (Rev): Review of "Quantum probability and related topics (Volume 7)". 95 Metrika 42 268-269

Lindblom, Torbjörn; Christy, Alfred A.; Libnau, Fred O.: Quantitative determination of stabiliser in a single base propellant by chemometric analysis of Fourier transform infrared spectra. 95 Chemolab 29 243-254

Lindeboom, M.: van den Berg, G. J.; — —: Durations in panel data subject to attrition: A note on estimation in the case of a stock sample. 95 StNeerla 49 282-293

Lindeman, K. S.: Baker, S. G.; — —: Correction to "The paired availability design: A proposal for evaluating epidural analgesia during labor" (V13 p2269-2278). 95 StatMed 14 1841-1841

Lindeyer, J.: Chen, E. J.; Gambino, J.; Laniel, N.; — —: Design and estimation issues for income in the redesign of the Canadian Labour Force Survey. 94 ASASrvyRsr 919-924

Lindgren, Bernard W.: — — (Auth); Katzenbeisser, W. (Rev): Review of "Statistical theory (Fourth edition)". 95 CmpStQ 10 103-104

Lindgren, Bruce R.: Le, Chap T.; — —: Construction and comparison of two receiver operating characteristic curves derived from the same samples. 95 BiomtrcJ 37 869-877

Lindgren, Fredrik; Geladi, Paul; Berglund, Anders; Sjöström, Michael; Wold, Svante: Interactive variable selection (IVS) for PLS. Part II: Chemical applications. 95 JChemtrc 9 331-342

Rännar, Stefan; Geladi, Paul; — —; Wold, Svante: A PLS kernel algorithm for data sets with many variables and few objects. Part II: Cross-validation, missing data and examples. 95 JChemtrc 9 459-470

Lindgren, Georg; Rychlik, Igor: How reliable are contour curves? Confidence sets for level contours. 95 Bernoull 1 301-319

Lindley, D. V.; Scott, W. F.: New Cambridge statistical tables (Second edition). 95 CambridgeUniv 96p

Bernardo, J. M.; Smith, A. F. M. (Auth); — — (Rev): Review of "Bayesian theory". 95 JRSS-A 158 340-341

Lindman, H. R.: — — (Auth); Fedorov, V. V. (Rev): Review of "Analysis of variance in experimental design". 95 JQualTek 27 268-268

Lindqvist, Bo: On the uncertainty of a system reliability estimate. 94 RelEngSS 46 263-269

Lindsay, Bruce G.: Mixture models: Theory, geometry and applications. 95 IMathStat:CA 163p

— —; Lesperance, Mary L.: A review of semiparametric mixture models. 95 JStPlInf 47 29-39

— —; Li, Bing: Comment on "Inference". 95 StatSci 10 175-177

Lindsey, George: Comments on modelling baseball. 94 ASAProSpts 21-24

Lindsey, J. K.: Fitting parametric counting processes by using log-linear models. 95 ApplStat 44 201-212

— —: Modelling frequency and count data. 95 ClrndnPr:UK 291p

— —: Comment on "The statistics of linear models: Back to basics" (V4 p221-234). 95 StatComp 5 87-89

— —; Lambert, P.: Dynamic generalized linear models and repeated measurements. 95 JStPlInf 47 129-139

Svensson, Elisabeth (Auth); — — (Rev): Review of "Analysis of systematic and random differences between paired ordinal categorical data". 95 StatMed 14 1726-1727

— — (Auth); Aickin, Mikel (Rev): Review of "Models for repeated measurements". 95 AJEpidm 141 382-383

— — (Auth); Corcoran, Stephen (Rev): Review of "Models for repeated measurements". 95 JAppStat 22 435-435

— — (Auth); Hand, D. J. (Rev): Review of "Modelling frequency and count data". 96 ShrtBkRv 16/1 2-2

— — (Auth); Lipsitz, Stuart R. (Rev): Review of "Models for repeated measurements". 95 JASA 90 1124-1125

Lindsey, James K.: Introductory statistics: A modelling approach. 95 OxfordUniv 214p

— — (Auth); Hand, D. J. (Rev): Review of "Introductory statistics: A modelling approach". 96 ShrtBkRv 16/1 2-2

Lindstrom, K.: A macro-editing application developed for PC-SAS. 91 StatJUN 8 155-166

Lindstrom, Mary J.: Self-modelling with random shift and scale parameters and a free-knot spline shape function. 95 StatMed 14 2009-2021

Pinheiro, José C.; Bates, Douglas M.; — —: Model building in nonlinear mixed effects models. 94 ASAProBiop 1-8

Lindstrøm, Tom; Øksendal, Bernt: Stability properties of stochastic partial differential equations. 95 StocAnAp 13 177-204

Lindvall, T.: Rao, C. R.; Shanbhag, D. N. (Auth); — — (Rev): Review of "Choquet-Deny type functional equations with applications to stochastic models". 95 ShrtBkRv 15/3 48-49

— — (Auth); Borovkov, K. A. (Rev): Review of "Lectures on the coupling method". 94 ThProbAp 39 721-723

Lindvall, Torgny: — — (Auth); O'Connell, Neil (Rev): Review of "Lectures on the coupling method". 95 AnlsProb 23 1456-1460

Ling, Chiahua: Pal, Nabendu; — —: Improved minimax estimation of powers of the variance of a multivariate normal distribution under the entropy loss function. 95 StPrLet 24 205-211

Ling, Robert F.: Hadi, Ali S.; Jones, Wendell D.; — —: A unifying representation of some case-deletion influence measures in univariate and multivariate linear regression. 95 JStPlInf 46 123-135

Link, William A.; Sauer, John R.: Estimation and confidence intervals for empirical mixing distributions. 95 Biomtrcs 51 810-821

Linssen, H. N.: van Berkum, E. E. M.; — —; Overdijk, D. A.: The reduction and ancillary classification of observations. 95 JStPlInf 45 347-372

Linton, Kathryn L. P.: Gange, Stephen J.; — —; Scott, Alastair J.; DeMets, David L.; Klein, Ronald: A comparison of methods for correlated ordinal measures with ophthalmic applications. 95 StatMed 14 1961-1974

Linton, O.: Jones, M. C.; — —; Nielsen, J. P.: A simple bias reduction method for density estimation. 95 Biomtrka 82 327-338

Linton, Oliver; Nielsen, Jens Perch: A kernel method of estimating structured nonparametric regression based on marginal integration. 95 Biomtrka 82 93-100

Linton, Oliver B.: Nielsen, Jens P.; — —: Kernel estimation in a nonparametric marker dependent hazard model. 95 AnlsStat 23 1735-1748

Linz, Douglas H.: Hertzberg, Vicki Stover; Rice, Carol H.; — —; Pinney, Susan M.: Occupational epidemiology in the era of total quality: Challenges for exposure assessment. 94 ASAProEpid 139-142

Liou, Michelle; Cheng, Philip E.: Asymptotic standard error of equipercentile equating. 95 JEdStat 20 259-286

— —; Cheng, Philip E.: Equipercentile equating via data-imputation techniques. 95 Psymtrka 60 119-136

Lipicky, Raymond J.: Chi, George Y. H.; Hung, James; Dubey, Satya D.; — —: Dose response studies and special populations. 94 ASAProBiop 88-93

Lipscomb, John R.: Multi-process control charts. 94 ASAProPEng 314-319

Lipshutz, Robert J.; Taverner, Fred; Hennessy, Kevin; Hartzell, George; Davis, Ron: DNA sequence confidence estimation. 94 Genomics 19 417-424

Lipsitz, Stuart R.; Fitzmaurice, Garret M.; Sleeper, Lynn; Zhao, L. P.: Estimation methods for the joint distribution of repeated binary observations. 95 Biomtrcs 51 562-570

— —; Fitzmaurice, Garrett: Reply to comment on "An extension of Yule's Q to multivariate binary data". 95 Biomtrcs 51 1185-1185

— —; Parzen, Michael: Sample size calculations for non-randomized studies. 95 Ststcian 44 81-90

Fitzmaurice, Garrett M.; — —: A model for binary time series data with serial odds ratio patterns. 95 ApplStat 44 51-61

Fitzmaurice, Garrett M.; Molenberghs, Geert; — —: Regression models for longitudinal binary responses with informative drop-outs. 95 JRSS-B 57 691-704

Williamson, John M.; Kim, KyungMann; — —: Analyzing bivariate ordinal data using a global odds ratio. 95 JASA 90 1432-1437

Lindsey, J. K. (Auth); — — (Rev): Review of "Models for repeated measurements". 95 JASA 90 1124-1125

Liptser, R.: Kogan, Y.; — —; Shenfild, M.: State-dependent Beneš buffer model with fast loading and output rates. 95 AnlsApPr 5 97-120

Lisagor, L.: Malmborg, E.; — —: Implementing a statistical metainformation system. 93 StatJUN 10 195-200

Liski, Erkki P.; Nummi, Tapio: Prediction and inverse estimation in repeated-measures models. 95 JStPlInf 47 141-151

Little, Jani: McNown, Robert; Rogers, Andrei; — —: Simplicity and complexity in extrapolative population forecasting models. 95 MaPopStd 5 235-257

Little, M. P.: Are two mutations sufficient to cause cancer? Some generalizations of the two-mutation model of carcinogenesis of Moolgavkar, Venzon, and Knudson, and of the multistage model of Armitage and Doll. 95 Biomtrcs 51 1278-1291

Little, R. J. A.: Correction to "Statistical analysis of masked data" (V9 p407-426). 94 JOffStat 10 469-469

Little, Roderick J. A.: Modeling the drop-out mechanism in repeated-measures studies. 95 JASA 90 1112-1121

Little, Thomas A.; Brekke, Kris: Partition of variation: A new method for σ reduction. 95 ASQCTekCon 208-216

Littler, Ray: Hoppe, Fred M. (Ed); — — (Rev): Review of "Multiple comparisons, selection, and applications in biometry - A festschrift in honor of Charles W. Dunnett". 95 AstrlJSt 37 251-252

Littlewood, I. G.: Sefton, C. E. M.; Whitehead, P. G.; Eatherall, A.; — —; Jakeman, A. J.: Dynamic response characteristics of the Plynlimon catchments and preliminary analysis of relationships to physical descriptors. 95 EnvrMtrc 6 465-472

Litvak, Eugene: Tu, Xin Ming; — —; Pagano, Marcello: On the informativeness and accuracy of pooled testing in estimating prevalence of a rare disease: Application to HIV screening. 95 Biomtrka 82 287-297

Litwin, Mark S.: How to measure survey reliability and validity. 95 SagePubl:CA:-Lnd 96p

Liu, Aiyi: Covariates selection and parameter estimation in growth curve models. 95 AMaPeknE 37 362-372

Liu, Ben-Chieh: Public library statistics: An interstate service analysis. 94 ASAProGovt 215-220

Liu, Chuanhai: Missing data imputation using the multivariate t distribution. 95 JMultiAn 53 139-158

— —; Rubin, Donald B.: ML estimation of the t distribution using EM and its extensions, ECM and ECME. 95 StSinica 5 19-39

Liu, Chui Yu; Sambol, Nancy C.: Pharmacodynamic analysis of analgesic clinical trials using empirical methods. 94 ASAProBiop 171-176

Liu, D.; Yamashita, Y.; Ogawa, H.: Pattern recognition in the presence of noise. 95 PattrnRc 28 989-995

Liu, Dongsheng: Wong, Chi Song; — —: Moments of generalized Wishart distributions. 95 JMultiAn 52 280-294

Liu, Guoxin: Liu, Wen; — —: A class of strong laws for functionals of countable nonhomogeneous Markov chains. 95 StPrLet 22 87-96

Liu, Honghu; Weiss, Robert E.; Wenger, Neil S.: Model selection in repeated measures random effects models. 94 ASAProBiop 371-376

Liu, Huimei; Berger, Roger L.: Uniformly more powerful, one-sided tests for hypotheses about linear inequalities. 95 AnlsStat 23 55-72

Liu, Hung-kung: Hwang, J. T. Gene; — —: Application of empirical linear prediction to quality assurance in industrial manufacturing. 94 ASAProPEng 175-180

Hwang, J. T. Gene; — —: Asymptotic theory for prediction through empirical linear models. 94 ASAProPEng 207-216

Liu, J. P.: Weng, C. S. Wayne; — —: Some pitfalls in sample size estimation for an anti-infective study. 94 ASAProBiop 56-60

Liu, J.-P.: Chow, S.-C.; — — (Auth); Day, S. (Rev): Review of "Statistical design and analysis in pharmaceutical science: Validation, process controls, and stability". 95 Biomtrcs 51 1190-1191

Liu, Jen-Pei: Use of the repeated cross-over designs in assessing bioequivalence. 95 StatMed 14 1067-1078 (Disc: p1079-1080)

— —; Weng, Chung-Sing: Bias of two one-sided tests procedures in assessment of bioequivalence. 95 StatMed 14 853-861

Chow, Shein-Chung; — —: Statistical design and analysis in pharmaceutical science: Validation, process controls, and stability. 95 Dekker:NY 584p

Ki, Fanny Y. C.; — —; Wang, Wenping; Chow, Shein-Chung: The impact of outlying subjects on decision of bioequivalence. 95 JBiophSt 5 71-94

Chow, Shein-Chung; — — (Auth); Lagakos, S. W. (Rev): Review of "Statistical design and analysis in pharmaceutical science". 95 ShrtBkRv 15/3 42-42

Liu, Jeng-Fu: On time-sequential test for a class of distributions. 95 StSinica 5 251-260

Liu, Jinshan: A method of undetermined coefficients in estimating two seemingly unrelated regression equations (Chinese). 95 ChJApPSt 10 78-83

Chen, Maoxue; — —: On optimality of the covariance improved estimate: A simulation study (Chinese). 95 ChJApPSt 10 1-4

Liu, Jun: Folsom, Ralph E.; — —: Small area estimation for the National Household Survey of Drug Abuse. 94 ASASrvyRsr 565-570

Liu, Jun S.; Cheng, Rong: Blind deconvolution via sequential imputations. 95 JASA 90 567-576

— —; Neuwald, Andrew F.; Lawrence, Charles E.: Bayesian models for multiple local sequence alignment and Gibbs sampling strategies. 95 JASA 90 1156-1170

— —; Wong, Wing H.; Kong, Augustine: Covariance structure and convergence rate of the Gibbs sampler with various scans. 95 JRSS-B 57 157-169

Chen, Rong; — —; Tsay, Ruey S.: Additivity tests for nonlinear autoregression. 95 Biomtrka 82 369-383

Liu, Ke: Hu, Qing; Yang, Jingyu; Zhang, Qian; — —; Shen, Xiaojun: An automatic seal imprint verification approach. 95 PattrnRc 28 1251-1266

Liu, Liping: Arjas, Elja; — —: Assessing the losses caused by an industrial intervention: A hierarchical Bayesian approach. 95 ApplStat 44 357-368

Liu, Lon-Mu; Chan, Ki-Kan; Montgomery, Alan L.; Muller, Mervin E.: A system-independent graphical user interface for statistical software. 95 CmpStDA 19 23-44

Liu, Pao-Wen: Kalsbeek, William D.; Botman, Steven L.; Massey, James T.; — —: Cost-efficiency and the number of allowable call attempts in the National Health Interview Survey. 94 JOffStat 10 133-152

Liu, Peng T.; Albert, Richard H.; Springer, Janet A.: Is one sample enough?. 94 ASAProBays 18-22

Springer, Janet A.; — —; Albert, Richard H.; Chow, Lien P.: Construction of a spectrum lifetable. 94 ASAProSoSt 82-87

Liu, Ping-Yu; Dahlberg, Steve: Design and analysis of multiarm clinical trials with survival endpoints. 95 ConClnTr 16 119-130

Liu, Regina Y.: Control charts for multivariate processes. 95 JASA 90 1380-1387

— —; Singh, Kesar: Using i.i.d. bootstrap inference for general non-i.i.d. models. 95 JStPlInf 43 67-75

Liu, Rongbing: Li, Xihe; — —: Basic concepts of the fuzzied statistical decision and its superiority (Chinese). 95 AMaAPknC 17 512-521

Liu, S.: Neudecker, H.; — —; Polasek, W.: The Hadamard product and some of its applications in statistics. 95 MaOpfStS 26 365-373

Liu, Shi-Miin; Brorsen, B. Wade: Maximum likelihood estimation of a GARCH-stable model. 95 JApEcmtx 10 273-285

Liu, Shu-Ing: Bayesian multiperiod forecasts for ARX models. 95 AnInStMa 47 211-224

— —: Comparison of forecasts for ARMA models between a random coefficient approach and a Bayesian approach. 95 CommStA 24 319-333

Liu, Shuangzhe; Neudecker, Heinz: Matrix-trace Cauchy-Schwarz inequalities and applications in canonical correlation analysis. 95 StHefte 36 287-298

— —; Neudecker, Heinz: A V optimal design for Scheffé's polynomial model. 95 StPrLet 23 253-258

Neudecker, Heinz; Polasek, Wolfgang; — —: The heteroskedastic linear regression model and the Hadamard product. A note. 95 JEconmtx 68 361-366

Liu, Tung: Kuan, Chung-Ming; — —: Forecasting exchange rates using feedforward and recurrent neural networks. 95 JApEcmtx 10 347-364

Liu, Tzen-Ping: Kovačević, Milorad S.; — —: Statistical matching of survey datafiles: A simulation study. 94 ASASrvyRsr 479-484

Liu, W.: Fixed-width simultaneous confidence intervals for all-pairwise comparisons. 95 CmpStDA 20 35-44

— —: A random-size subset approach to the selection problem. 95 JStPlInf 48 153-164

Liu, W. Z.; White, A. P.: A comparison of nearest neighbour and tree-based methods of non-parametric discriminant analysis. 95 JStCmpSm 53 41-50

Liu, Wanrong: Li, Guoying; — —; Zha, Wenxing: Projection pursuit Neyman goodness-of-fit test and its bootstrap approximation (Chinese). 95 KToChina 39 678-681

Liu, Wei: A group sequential procedure for all-pairwise comparisons of k treatments based on the range statistic. 95 Biomtrcs 51 946-955

— —: On the construction of exact lower confidence bounds on the distance between any two ranked location parameters of k populations. 95 CommStA 24 2073-2085

— —: On the existence of tests uniformly more powerful than the likelihood ratio test. 95 JStPlInf 44 19-35

Liu, Wen; Liu, Guoxin: A class of strong laws for functionals of countable nonhomogeneous Markov chains. 95 StPrLet 22 87-96

— —; Liu, Zikuan: Likelihood ratio and a class of strong laws for the sequence of integer-valued random variables. 95 StPrLet 22 249-256

— —; Wang, Zhongzhi: An extension of a theorem on gambling systems. 95 JMultiAn 55 125-132

— —; Yang, Weiguo: Some limit properties of nonhomogeneous Markov information sources. 95 JCISS 20 279-292

— —; Yang, Weiguo: A limit theorem for the entropy density of nonhomogeneous Markov information source. 95 StPrLet 22 295-301

Liu, X. H.: Kung, H. C.; — —; Chan, J.: Socioeconomic status and suicide. 94 ASAProSoSt 235-239

Liu, Xijian: Symmetric two-particle exclusion-eating processes. 95 AnlsProb 23 1439-1455

Liu, Ying-Hsing; Yang, Ming-Chung: Bounded risk conditions in simultaneous estimation of independent Poisson means. 95 JStPlInf 47 319-331

Liu, Yuanlong; Schutz, Robert W.: Overtime in the National Hockey League: Is it a valid tie-breaking procedure?. 94 ASAProSpts 55-60

Liu, Z. J.; Rao, C. Radhakrishna: Asymptotic distribution of statistics based on quadratic entropy and bootstrapping. 95 JStPlInf 43 1-18

Liu, Zikuan: Liu, Wen; — —: Likelihood ratio and a class of strong laws for the sequence of integer-valued random variables. 95 StPrLet 22 249-256

Lix, Lisa M.; Keselman, H. J.: Approximate degrees of freedom tests: A unified perspective on testing for mean equality. 95 PsychBu 117 547-560

Keselman, H. J.; — —: Improved repeated measures stepwise multiple comparison procedures. 95 JEdStat 20 83-99

Keselman, H. J.; Carriere, K. C.; — —: Robust and powerful nonorthogonal analyses. 95 Psymtrka 60 395-418

Keselman, H. J.; Keselman, Joanne C.; — —: The analysis of repeated measurements: Univariate tests, multivariate tests, or both?. 95 BJMaStPs 48 319-338

Ljung, L.: Sjöberg, J.; — —: Overtraining, regularization and searching for a minimum, with application to neural networks. 95 IJContrl 62 1391-1407

Ljung, Lennart: Guo, Lei; — —: Performance analysis of general tracking algorithms. 95 IEEEAuCn 40 1388-1402

Juditsky, Anatoli; Hjalmarsson, Håkan; Benveniste, Albert; Delyon, Bernard; — —; Sjöberg, Jonas; Zhang, Qinghua: Nonlinear black-box models in system identification: Mathematical foundations. 95 Autmatca 31 1725-1750

Sjöberg, Jonas; Zhang, Qinghua; — —; Benveniste, Albert; Delyon, Bernard; Glorennec, Pierre-Yves; Hjalmarsson, Håkan; Juditsky, Anatoli: Nonlinear black-box modeling in system identification: A unified overview. 95 Autmatca 31 1691-1724

Llopis, Agustín: Ferrándiz, Juan; López, Antonio; — —; Morales, María; Tejerizo, María Luisa: Spatial interaction between neighbouring counties: Cancer mortality data in Valencia (Spain). 95 Biomtrcs 51 665-678

Lloyd, Christopher J.: Accurate confidence intervals from recapture experiments. 95 CommStA 24 1235-1253

Lo, Andrew W.; Wang, Jiang: Implementing option pricing models when assest returns are predictable. 95 JFinance 50 87-129

Lo, L. L.; Fernando, R. L.; Cantet, R. J. C.; Grossman, M.: Theory for modelling means and covariances in a two-breed population with dominance inheritance. 95 ThApGene 90 49-62

— —; Fernando, R. L.; Grossman, M.: Covariance between relatives in multibreed populations: Additive model. 93 ThApGene 87 423-430

Lo, S. H.: Chao, M. T.; — —: Some statistical mean value theorems related to the bootstrap. 95 StSinica 5 129-139

Lo, Shaw-Hwa: Chen, Kani; Chao, Min-Te; — —: On strong uniform consistency of the Lynden-Bell estimator for truncated data. 95 AnlsStat 23 440-449

Lo, Wen-Da: Chaudhuri, Probal; — —; Loh, Wei-Yin; Yang, Ching-Ching: Generalized regression trees. 95 StSinica 5 641-666

Loce, Robert P.; Dougherty, Edward R.: Mean-absolute-error representation and optimization of computational-morphological filters. 95 CVGIPGr 57 27-37

Lochmüller, C. H.: Koel, M.; Kaljurand, M.; — —; Moebus, M.: Factor analytical resolution of simultaneous, dynamic thermal process in thermochromatography of oil shale. 95 Chemolab 30 173-178

Lochmüller, Charles H.: Elomaa, Matti; — —; Kaljurand, Mihkel; Koel, Mihkel: Application of evolving factor analysis in thermochromatography. 95 Chemolab 30 129-132

Lochner, Robert H.: Developing a balanced set of measures in health care. 95 ASQCTekCon 293-300

Lock, M. D.: Tartar, M. E.; — — (Auth); Nason, Guy (Rev): Review of "Model-free curve estimation". 95 SIAMRvw 37 270-271

Lock, Michael D.: Tarter, Michael E.; — — (Auth); Wharton, Robert M. (Rev): Review of "Model-free curve estimation". 95 Technmcs 37 460-460

Lock, Robin H.: Tricks of the trade: Getting going the first day. 94 ASAProStEd 49-52

Danehy, Timothy J.; — —: CHODR - Using statistics to predict college hockey. 95 Stats /13 10-14

Locks, Mitchell O.: Reliability, maintainability, and availability assessment (Second edition). 95 ASQualCntrl:WI 348p

Loeve, J. A.: Hordijk, A.; — —: Undiscounted Markov decision chains with partial information: An algorithm for computing a locally optimal periodic policy. 94 ZeitOpRs 40 163-181

Logan, Theodore F.: Shannon, William D.; Bryant, John; — —; Day, Roger: An application of decision theory to patient screening for an autologous tumour vaccine trial. 95 StatMed 14 2099-2110

Loh, Wei-Liem: On linear discriminant analysis with adaptive ridge classification rules. 95 JMultiAn 53 264-278

Loh, Wei-Yin: Chaudhuri, Probal; Lo, Wen-Da; — —; Yang, Ching-Ching: Generalized regression trees. 95 StSinica 5 641-666

Zheng, Xiaodong; — —: Consistent variable selection in linear models. 95 JASA 90 151-156

Zheng, Xiaodong; — —: Bootstrapping binomial confidence intervals. 95 JStPlInf 43 355-380

Lohr, Sharon L.: Hasse diagrams in statistical consulting and teaching. 95 AmerStat 49 376-381

— —: Optimal Bayesian design of experiments for the one-way random effects model. 95 Biomtrka 82 175-186

Lomnitz, C.: Fundamentals of earthquake prediction. 94 Wiley:NY:UK 326p

— — (Auth); Vere-Jones, D. (Rev): Review of "Fundamentals of earthquake prediction". 95 ShrtBkRv 15/2 22-22

London, Kathy A.: Guadagno, Mary Ann; Abma, Joyce C.; Chandra, Anjani; — —; Peterson, Linda S.; Piccinino, Linda J.; Wilson, Jacqueline B.: Women's reports of sensitive health conditions, attitudes, and behaviors. 94 ASAProGovt 82-87

Long, Dou; Krzysztofowicz, Roman: A family of bivariate densities constructed from marginals. 95 JASA 90 739-746

Long, J. Scott; Noss, Brian: Data analysis using GAUSS and Markov. 95 SocMethR 23 303-328

Long, Jeffrey C.; Williams, Robert C.; Urbanek, Margrit: An E-M algorithm and testing strategy for multiple-locus haplotypes. 95 AJHuGene 56 799-810

Longford, N.: Models for uncertainty in educational testing. 95 Springer:Brln:NY 300p

Longford, N. T.: Methods for uncertainty in educational testing. 95 Springer:-Brln:NY 285p

— —: Score adjustment for subjectively rated responses. 94 ASAProBays 123-127 (Disc: p134-135)

— — (Auth); Ramsay, J. O. (Rev): Review of "Methods for uncertainty in educational testing". 96 ShrtBkRv 16/1 5-6

— — (Auth); Rodriguez-Poo, Juan M. (Rev): Review of "Random coefficient models". 95 JAppStat 22 544-545

Longford, Nicholas T.: Comment on "Hierarchical linear models". 95 JEdStat 20 205-209

— — (Auth); Albert, Paul S. (Rev): Review of "Random coefficient models". 95 JASA 90 801-802

— — (Auth); Gallo, Paul P. (Rev): Review of "Random coefficient methods". 95 Technmcs 37 120-120

Longini, Ira M., Jr: Haber, Michael; Halloran, M. Elizabeth; — —; Watelet, Luc: Estimation of vaccine efficacy in non-randomly mixing populations. 95 BiomtrcJ 37 25-38

Longman Logotron: — — (Auth); Feu, Chris du (Rev): Review of "Pinpoint - Datafiles". 95 TeachgSt 17/2 75-76

— — (Auth); Graham, Alan (Rev): Review of "Junior Pinpoint". 95 TeachgSt 17/2 75-75

Longman, R. Stewart: Cota, Albert A.; Evans, Charles R.; — —; Dion, Kenneth L.; Kilik, Lindy: Using and misusing factor analysis to explore group cohesion. 95 JClPsych 51 308-316

Longton, Barbara; Sailer, Peter; Wilson, Janette: How sales of capital assets were affected by the Tax Reform Act of 1986. 94 ASAProGovt 201-205 (Disc: p212-214)

Looney, Stephen W.: How to use tests for univariate normality to assess multivariate normality. 95 AmerStat 49 64-70

Grady, Christopher S.; — —; Steiner, R. Prasaad: A study of biostatistics requirements in medical schools in the United States. 94 ASAProStEd 263-266

López, Antonio: Ferrándiz, Juan; — —; Llopis, Agustín; Morales, María; Tejerizo, María Luisa: Spatial interaction between neighbouring counties: Cancer mortality data in Valencia (Spain). 95 Biomtrcs 51 665-678

López, Carlos; González, Elizabeth; Goyret, Jorge: Principal component analysis of rain gauge data: Detecting outliers (Spanish). 94 Estd 46 25-54

— —; González, Juan F.; Curbelo, Rosario: Principal component analysis of rain gauge data: Treating missing observations (Spanish). 94 Estd 46 55-83

Lord, Edward R.: Burrows, William R.; Benjamin, Mario; Beauchamp, Stephen; — —; McCollor, Douglas; Thomson, Bruce: CART decision-tree statistical analysis and prediction of summer season maximum surface ozone for the Vancouver, Montreal, and Atlantic regions of Canada. 95 JApMetrl 34 1848-1862

Lorenzato, Stefan: Rocke, David M.; — —: A two-component model for measurement error in analytical chemistry. 95 Technmcs 37 176-184

Lorenzen, Gunter: A new family of goodness-of-fit statistics for discrete multivariate data. 95 StPrLet 25 301-307

Lorenzen, T. J.; Anderson, V. L. (Auth); Conniffe, D. (Rev): Review of "Design of experiments: A no-name approach". 95 JRSS-A 158 195-197

— —; Anderson, V. L. (Auth); Holle, Rolf (Rev): Review of "Design of experiments. A no-name approach". 95 CmpStDA 19 101-102

Lorenzen, Thomas J.; Anderson, Virgil L. (Auth); O'Brien, C. J. (Rev): Review of "Design of experiments: A no-name approach". 95 AstrlJSt 37 125-127

Lorieux, M.; Perrier, X.; Goffinet, B.; Lanaud, C.; de León, D. González: Maximum-likelihood models for mapping genetic markers showing segregation distortion. 2. F_2 populations. 95 ThApGene 90 81-89

Lorz, U.: Okabe, A.; Boots, B.; Sugihara, K. (Auth); — — (Rev): Review of "Spatial tessellations. Concepts and applications of Voronoi diagrams". 95 MaOpfStS 26 183-183

Łoskot, Krzysztof; Rudnicki, Ryszard: Limit theorems for stochastically perturbed dynamical systems. 95 JAppProb 32 459-469

Lothian, Jack: The relationship between profitability, size and concentration in Canadian corp.. 94 ASAProBuEc 287-292

Lotito, S.: Hühn, M.; — —; Piepho, H. P.: Relationships between genotype × environment interactions and rank orders for a set of genotypes tested in different environments. 93 ThApGene 86 943-950

Louchard, Guy: Some distributed algorithms revisited. 95 CommStSM 11 563-586

Loughin, Thomas M.: A residual bootstrap for regression parameters in proportional hazards models. 95 JStCmpSm 52 367-384

— —; Noble, William: A permutation test for effects in an unreplicated factorial design. 94 ASAProPEng 181-186

Louis, Thomas A.: Anderson, Jon E.; — —: Survival analysis using a scale change random effects model. 95 JASA 90 669-679

Grambsch, Patricia M.; Randall, Bryan L.; Bostick, Roberd M.; Potter, John D.; — —: Modeling the labeling index distribution: An application of functional data analysis. 95 JASA 90 813-821

Shih, Joanna H.; — —: Inferences on the association parameter in copula models for bivariate survival data. 95 Biomtrcs 51 1384-1399

Shih, Joanna H.; — —: Assessing gamma frailty models for clustered failure time data. 95 LiDaAnal 1 205-220

Loukas, S.: Papageorgiou, H.; — —: A bivariate discrete Charlier series distribution. 95 BiomtrcJ 37 105-117

Lovelace, Cynthia: Kotz, Samuel; Johnson, Norman L. (Auth); — — (Rev): Review of "Process capability indices". 95 JQualTek 27 270-271

Lovell, C. A. K.: Linear programming approaches to the measurement and analysis of productive efficiency. 94 Top 2 175-248

Lovie, A. D.: Who discovered Spearman's rank correlation?. 95 BJMaStPs 48 255-269

Lovie, P.; — —: The cold equations: Spearman and Wilson on factor indeterminacy. 95 BJMaStPs 48 237-253

Lovie, P.; Lovie, A. D.: The cold equations: Spearman and Wilson on factor indeterminacy. 95 BJMaStPs 48 237-253

Lovie, Pat: NCSS Statistical Software (Auth); — — (Rev): Review of "NCSS 6.0 for Windows". 95 BJMaStPs 48 407-408

Lovie, Sandy: Minitab Inc (Auth); — — (Rev): Review of "MINITAB Release 10 for Windows". 95 BJMaStPs 48 205-207

Loviscach, J.: A generalized notion of variation applied to Markov chains and Anosov maps. 95 ProbTher 103 553-570

Lovrics, László: Kőrösi, Gábor; — —; Mátyás, László: Aggregation and the long run properties of economic time series. 95 MaCmpSim 39 279-286

Low, Mark G.: Bias-variance tradeoffs in functional estimation problems. 95 AnlsStat 23 824-835

Löwe, Matthias: Eichelsbacher, Peter; — —: A large deviation principle for m-variate von Mises-statistics and U-statistics. 95 JTherPrb 8 807-824

Lowe, Norma: Katz, Alex S.; Katz, Simon; — —: Bootstrap based analysis of neural network's accuracy and prediction of clinical outcome. 94 ASAProBiop 177-181

Lowry, Cynthia A.; Champ, Charles W.; Woodall, William H.: The performance of control charts for monitoring process variation. 95 CommStB 24 409-437

— —; Montgomery, Douglas C.: A review of multivariate control charts. 95 IIETrn 27 800-810

Lozano, J.; Novič, M.; Rius, F. X.; Zupan, J.: Modelling metabolic energy by neural networks. 95 Chemolab 28 61-72

Lu, Chang-yu: Admissibility of inhomogeneous linear estimators in linear models with respect to incomplete ellipsoidal restrictions. 95 CommStA 24 1737-1742

Lu, Changyu; Li, Weixin: Admissibility of linear estimator in linear models with respect to an incomplete ellipsoidal restrictions. 95 AMaPeknE 37 289-295

Lu, Hsiu Ying: PC-SAS program for estimating Hühn's nonparametric stability statistics. 95 AgronmyJ 87 888-891

Lu, I. L.: Choi, S. C.; — —: Effect of non-random missing data mechanisms in clinical trials. 95 StatMed 14 2675-2684

Lu, J. C.: Hoff, Jennifer; Johnson, Thomas; — —: The statisticians role in quality improvement partnerships. 94 ASAProQlPr 252-257

Lu, Jye-Chy: Mesenbrink, Peter; — —: Model and transformation selection in fractionated experiments with complex aliasing. 94 ASAProQlPr 118-123

Lu, Jye-Chyi: Hughes-Oliver, Jacqueline M.; — —; Davis, Joseph C.; Gyurcsik, Ronald S.: Robust improvement of a semi-conductor fabrication process via spatial modeling of the mean and covariance surfaces. 94 ASAProPEng 235-244

Lu, Ming-Wei; Rudy, Richard J.: A practical method to prioritize problem investigations. 95 ASQCTekCon 57-65

Lu, Wuewen: Estimation of the parameters in the location scale model under randomly right-censoring (Chinese). 95 ChJApPSt 10 272-277

Lu, Ying; Bean, Judy A.: On the sample size for one-sided equivalence of sensitivities based upon McNemar's test. 95 StatMed 14 1831-1839

— —; Malani, Hina Mehta: Analysis of animal carcinogenicity experiments with multiple tumor types. 95 Biomtrcs 51 73-86

Lubinsky, David J.: Young, Forrest W.; — —: Guiding data analysts with visual statistical strategies. 95 JCmpGrSt 4 229-250 (Disc: p251-260)

Lubrano, Michel: Testing for unit roots in a Bayesian framework. 95 JEconmtx 69 81-109

Luc, Manchi: Bankier, Michael; Fillion, Jean-Marc; — —; Nadeau, Christian: Imputing numeric and qualitative variables simultaneously. 94 ASASrvyRsr 242-247

Lucas, André: Unit root tests based on M estimators. 95 EconTher 11 331-346

— —: An outlier robust unit root test with an application to the extended Nelson-Plosser data. 95 JEconmtx 66 153-173

Hoek, Henk; — —; van Dijk, Herman K.: Classical and Bayesian aspects of robust unit root inference. 95 JEconmtx 69 27-59

Lucas, James M.: Altekar, Maneesha; — —; Schuenemeyer, John H.: Shift detection in two-way arrays. 95 CommStB 24 711-731

Altekar, Maneesha; — —; Schuenemeyer, John H.: Shift detection in two-way arrays. 94 ASAProPEng 76-80

Lucchetti, R.: Models in differences with measurement errors (Italian). 93 GiornEco 52 403-415

Luceño, Alberto: A family of partially correlated Poisson models for overdispersion. 95 CmpStDA 20 511-520

— —: Choosing the EWMA parameter in engineering process control. 95 JQualTek 27 162-168

— —; de Ceballos, Federico: Describing extra-binomial variation with partially correlated models. 95 CommStA 24 1637-1653

Box, George E. P.; — —: Discrete proportional-integral control with constrained adjustment. 95 Ststcian 44 479-495

Luchian, Henri; Stamate, Daniel: Answer-perturbation techniques for the protection of statistical databases. 95 StatComp 5 203-213

Ludden, Thomas M.: Ette, Ene I.; Sun, He; — —: Design of population pharmacokinetic studies. 94 ASAProBiop 487-492

Ludeña, C.: Guyon, X. (Auth); — — (Transl); Ripley, B. D. (Rev): Review of "Random fields on a network. Modeling, statistics, and applications". 96 ShrtBkRv 16/1 10-10

Lugosi, Gábor: Improved upper bounds for probabilities of uniform deviations. 95 StPrLet 25 71-77

Devroye, Luc; — —: Lower bounds in pattern recognition and learning. 95 PattrnRc 28 1011-1018

Lui, Kung-Jong: Sample size calculation for intraclass correlation in the presence of random loss of sampled patients. 95 BiomtrcJ 37 97-103

— —: Confidence intervals for the risk ratio in cohort studies under inverse sampling. 95 BiomtrcJ 37 965-971

— —: Comment on "A Wilcoxon-type test for trend" (V4 p87-90). 95 StatMed 14 445-446

— —: Confidence limits for the population prevalence rate based on the negative binomial distribution. 95 StatMed 14 1471-1477

— —: Notes on conditional confidence limits under inverse sampling. 95 StatMed 14 2051-2056

Luinge, H. J.; van der Maas, J. H.; Visser, T.: Partial least squares regression as a multivariate tool for the interpretation of infrared spectra. 95 Chemolab 28 129-138

Luiselli, Donata: Carfagna, Elisabetta; Di Bacco, Mario; — —: An example of the use of Bayesian analysis in human biology. 94 ASAProBays 252-255

Lumley, Thomas: Efficient execution of Stone's likelihood ratio tests for disease clustering. 95 CmpStDA 20 499-510

Lumsdaine, Robin L.: Finite-sample properties of the maximum likelihood estimator in GARCH(1, 1) and IGARCH(1, 1) models: A Monte Carlo investigation. 95 JBES 13 1-10

Lund, Eiliv: Comparison of additive and multiplicative models for reproductive risk factors and post-menopausal breast cancer. 95 StatMed 14 267-274

Lund, Robert; Hurd, Harry; Bloomfield, Peter; Smith, Richard: Climatological time series with periodic correlation. 95 JClimate 8 2787-2809

Lundbye-Christensen, Søren; Christensen, Morgens: Modelling the growth of human placental lactogen in maternal serum during late pregnancy. 95 RBrzlPrS 9 1-24

Lundgren, Rolf: Thelin, Bernt; Lundstedt, Torbjörn; — —; Olsson, Arne; Batra, Satish: Classification of Estradurin batches: Correlation between 31P NMR and a biological duration test for batch approval. 95 Chemolab 27 135-145

Lundstedt, Torbjörn; Thelin, Bernt: A multivariate strategy for optimizing a two-step process. 95 Chemolab 29 255-261

Thelin, Bernt; — —; Lundgren, Rolf; Olsson, Arne; Batra, Satish: Classification of Estradurin batches: Correlation between 31P NMR and a biological duration test for batch approval. 95 Chemolab 27 135-145

Lunn, A.: Hand, D.; Daly, F.; — —; McConway, K.; Ostrowski, E. (Ed); Ziegel, Eric R. (Rev): Review of "A handbook of small data sets". 95 Technmcs 37 242-242

Lunn, A. D.: Hand, D. J.; Daly, F.; — — (Auth); Derr, J. (Rev): Review of "A handbook of small data sets". 95 Biomtrcs 51 1586-1586

Hand, D. J.; Daly, F.; — —; McConway, K.; Ostrowski, E. (Ed); Chatfield, C. (Rev): Review of "A handbook of small data sets". 95 JRSS-A 158 349-349

Hand, D. J.; Daly, F.; — —; McConway, K. J.; Ostrowski, E. (Ed); Schimek, Michael G. (Rev): Review of "Handbook of small data sets". 95 CmpStQ 10 197-197

Hand, D. J.; Daly, F.; — —; McConway, K. J.; Ostrowski, E. (Ed); Street, Deborah J. (Rev): Review of "A handbook of small data sets". 95 AstrlJSt 37 251-251

Hand, D. J.; Daly, F.; — —; McConway, K. J.; Ostrowski, E. (Auth); Trenkler, Dietrich (Rev): Review of "A handbook of small data sets". 95 CmpStDA 19 101-101

Lunn, Mary; McNeil, Don: Applying Cox regression to competing risks. 95 Biomtrcs 51 524-532

Lunneborg, Clifford E.: — — (Auth); Handcock, Mark S. (Rev): Review of "Modeling experimental and observational data". 95 AmerStat 49 238-238

Luo, N.; de la Sen, M.: Robust stabilization of internally delayed uncertain systems via sliding mode control. 95 Top 3 1-34

Luong, Andrew: Doray, Louis G.; — —: Quadratic distance estimators for the zeta family. 95 InsrneME 16 255-260

Lupu, Mihaela; Todor, Dorin: A singular value decomposition based algorithm for multicomponent exponential fitting of NMR relaxation signals. 95 Chemolab 29 11-17

Lurie, Philip M.: A review of five statistical packages for Windows. 95 AmerStat 49 99-107

Luschgy, Harald: Local asymptotic quadraticity of stochastic process models based on stopping times. 95 StocProc 57 305-317

Lütkepohl, H.: Reinsel, C. (Auth); — — (Rev): Review of "Elements of multivariate time series analysis". 95 Metrika 42 147-148

LuValle, Michael J.: Meeker, William Q.; — —: An accelerated life test model based on reliability kinetics. 95 Technmcs 37 133-146

Lux, Linda: Iannacchione, Vincent; Wildfire, Judith; — —; Hawes, Catherine: Covering the population of unlicensed board and care homes: Problems and solutions. 94 ASASrvyRsr 270-275

Luxhoj, James T.; Shyur, Huan-Jyh: Reliability curve fitting for aging helicopter components. 95 RelEngSS 48 229-234

Lyles, Robert H.; Chambless, Lloyd E.: Effects of model misspecification in the estimation of variance components and intraclass correlation for paired data. 95 StatMed 14 1693-1706

Lynch, J.: Lee, Shiowjen; Durham, S.; — —: On the calculation of the reliability of general load sharing systems. 95 JAppProb 32 777-792

Lynch, James: Knezevic, J. (Auth); — — (Rev): Review of "Reliability, maintainability and supportability: A probabilistic approach". 95 Technmcs 37 122-122

Lynch, John T.: Shevchenko, Ivan P.; — —; Mattie, Angela S.; Reed-Fourquet, Lori L.: Verification of information in a large medical database using linkages with external databases. 95 StatMed 14 511-530 (Disc: p687-689)

Lynch, Mary Jo: Public library statistics: Two systems compared. 94 ASAProGovt 53-57

Lynch, Thomas B.: Variance reduction in Ueno's method and cylinder sampling for forest volume estimation. 95 CanJForR 25 1783-1794

Lynn, Peter; Taylor, Bridget: On the bias and variance of samples of individuals: Comparison of the electoral registers and postcode address file as sampling frames. 95 Ststcian 44 173-194

Lynn, William R.: Green, Sylvan B.; Corle, Donald K.; Gail, Mitchell H.; Mark, Steven D.; Pee, David; Freedman, Laurence S.; Graubard, Barry I.; — —: Interplay between design and analysis for behavioral intervention trials with community as the unit of randomization. 95 AJEpidm 142 587-593

Lyon, John; Tsai, Chih-Ling: Improved tests for the first-order autoregressive model with heteroscedasticity. 95 JStCmpSm 52 71-83

Lyons, Russell; Pemantle, Robin; Peres, Yuval: Conceptual proofs of $L\log L$ criteria for mean behavior of branching processes. 95 AnlsProb 23 1125-1138

Ma, C. Y.; Bayne, C. K.: Differentiation of aroclors in environmental samples using linear discrimination. 94 ASAProEnvr 142-143

Ma, Chunsheng: Yue, Xinnian; — —: Multivariate l_p-norm symmetric distributions. 95 StPrLet 24 281-288

Ma, Guangqin: Cox, Christopher; — —: Asymptotic confidence bands for generalized nonlinear regression models. 95 Biomtrcs 51 142-150

Ma, Jin: Duffie, Darrell; — —; Yong, Jiongmin: Black's consol rate conjecture. 95 AnlsApPr 5 356-382

Ma, Z. M.; Röckner, M. (Auth); Fukushima, M. (Rev): Review of "Introduction to the non-symmetric Dirichlet forms". 95 Metrika 42 70-72

Maccario, J.: Tzourio, C.; — —; Alpérovitch, A.: Estimation of the proportion of genetic cases in late-onset diseases: An application to Alzheimer's disease. 95 StatMed 14 1353-1363

Macchiato, M. F.; La Rotonda, L.; Lapenna, V.; Ragosta, M.: Time modelling and spatial clustering of daily ambient temperature: An application in Southern Italy. 95 EnvrMtrc 6 31-53

Macchiavelli, Raúl; Arnold, Steven F.: Difficulties with the use of penalized likelihood criteria in antedependence and polynomial models. 95 CommStA 24 501-522

MacDonald, I. L.; Zucchini, W.: Hidden Markov and other models for discrete-valued time series. 95 ChapmnHall:Lnd 180p

MacDonald, Iain L.: Hidden Markov models and animal behaviour. 95 BiomtrcJ 37 701-712

MacEachern, S. N.; Notz, W. I.; Whittinghill, D. C.; Zhu, Y.: Robustness to the unavailability of data in the linear model, with applications. 95 JStPlInf 48 207-213

MacEachern, Steven N.; Berliner, L. Mark: Asymptotic inference for dynamical systems observed with error. 95 JStPlInf 46 277-292

MacGibbon, Brenda: Hanley, James A.; — —: The Poisson distribution and "multiple events" cell occupancy problems. 95 Biomtrcs 51 1557-1561

MacGillivray, H.: Sen, P. K.; Singer, J. M. (Auth); — — (Rev): Review of "Large sample methods in statistics". 95 Biomtrcs 51 381-382

MacGregor, John F.: Kourti, Theodora; — —: Process analysis, monitoring and diagnosis, using multivariate projection methods. 95 Chemolab 28 3-21

Nomikos, Paul; — —: Multi-way partial least squares in monitoring batch processes. 95 Chemolab 30 97-108

Nomikos, Paul; — —: Multivariate SPC charts for monitoring batch processes. 95 Technmcs 37 41-59

Machado, Adélio A. S. C.: Esteves da Silva, Joaquim C. G.; — —: Evolving factor analysis of synchronous fluorescence spectra of humic substances in the presence of Cu(II). 95 Chemolab 27 115-128

Macher, Christian L.; Cadish, Marc; Clerc, Jean-Thomas; Pretsch, Ernö: Hypermedia - A new concept for information management. 95 Chemolab 28 213-228

Machihara, Fumiaki: A G/SM/1 queue with vacations depending on service times. 95 CommStSM 11 671-690

Machin, D.: Haybittle, J.; Yuen, P.; — —: Multiple comparisons in disease mapping. 95 StatMed 14 2503-2505

Machin, David: Parmar, Mahesh K. B.; — —: Survival analysis: A practical approach. 95 Wiley:NY:UK 255p

Fisher, Lloyd D.; van Belle, Gerald (Auth); — — (Rev): Review of "Biostatistics: A methodology for the health sciences". 95 StatMed 14 878-879

Swerdlow, Anthony; dos Santos Silva, Isabell (Auth); Yuen, Peter; — — (Rev): Review of "Atlas of cancer incidence in England and Wales: 1968-85". 95 StatMed 14 1725-1726

Parmar, Mahesh K. B.; — — (Auth); Juritz, J. M. (Rev): Review of "Survival analysis: A practical approach". 96 ShrtBkRv 16/1 2-3

Mächler, Martin B.: Variational solution of penalized likelihood problems and smooth curve estimation. 95 AnlsStat 23 1496-1517

Machlis, Gary E.: Dillman, Don A.; Dolsen, Dana E.; — —: Increasing response to personally-delivered mail-back questionnaires. 95 JOffStat 11 129-139

Machnes, Yaffa: Deductible insurance and production. 95 InsrneME 17 119-123

Mack, Thomas: Which stochastic model is underlying the chain ladder method?. 94 InsrneME 15 133-138

Mackenzie, Gilbert: Regression models for repeated medical random counts. 95 Ststcian 44 365-372

Mackinnon, David P.; Scribner, Richard; Taft, Katharine A.: Development and applications of a city-level alcohol availability and alcohol problems database. 95 StatMed 14 591-604 (Disc: p697-699)

MacKinnon, David P.; Warsi, Ghulam; Dwyer, James H.: A simulation study of mediated effect measures. 95 MultBehR 30 41-62

Mackinnon, J. G.: Davidson, R.; — — (Auth); Smith, Richard J. (Rev): Review of "Estimation and inference in econometrics". 95 JApEcmtx 10 339-341

MacKinnon, James G.: Davidson, Russell; — — (Auth); Zinde-Walsh, Victoria (Rev): Review of "Estimation and inference in econometrics". 95 EconTher 11 631-635

Mackisack, Margaret S.: Wand, M. P.; Jones, M. C. (Auth); — — (Rev): Review of "Kernel smoothing". 95 AstrlJSt 37 368-369

MacLean, C. J.: Sham, P. C.; Walters, E. E.; Neale, M. C.; Heath, A. C.; — —; Kendler, K. S.: Logistic regression analysis of twin data: Estimation of parameters of the multifactorial liability-threshold model. 94 BehavGen 24 229-238

Maclure, Malcolm: Mittleman, Murray A.; — —; Robins, James M.: Control sampling strategies for case-crossover studies: An assessment of relative efficiency. 95 AJEpidm 142 91-98

MacMinn, Richard D.; Brockett, Patrick L.: Corporate spin-offs as a value enhancing technique when faced with legal liability. 95 InsrneME 16 63-68

MacNeil, I. B.: Sutradhar, B. C.; — —; Dagum, E. B.: A simple test for stable seasonality. 95 JStPlInf 43 157-167

MacNeill, I. B.; Elwood, J. M.; Miller, D.; Mao, Y.: Trends in mortality from melanoma in Canada and prediction of future rates. 95 StatMed 14 821-839

MacNeill, Ian B.: Trends in statistical ecology and environmental statistics. 94 ASAProEnvr 8-9

MacPhee, I. M.; Jordan, B. P.: Optimal search for a moving target. 95 PrEngInf 9 159-182

Madan, K. C.: A bulk input queue with a stand-by. 95 SoAfrStJ 29 1-7

— —: A queueing system with random failures and delayed repairs. 94 JIndnStA 32 39-48

Madan, V. K.; Sinha, R. K.: Digital smoothing of census data employing Fourier transforms. 95 CmpStDA 20 285-294

Madans, Jennifer H.; Reuben, Cynthia A.; Rothwell, Sandra T.; Eberhardt, Mark S.: Differences in morbidity measures and risk factor identification using multiple data sources: The case of coronary heart disease. 95 StatMed 14 643-653 (Disc: p691-692)

Maddala, G. S.; Nimalendran, M.: An unobserved component panel data model to study the effect of earnings surprises on stock prices, trading volumes, and spreads. 95 JEconmtx 68 229-242

Joutz, Frederick L.; — —; Trost, Robert P.: An integrated Bayesian vector autoregression and error correction model for forecasting electricity consumption and prices. 95 JForcast 14 287-310

— —; Rao, C. R.; Vinod, H. D. (Ed); Gill, Len (Rev): Review of "Handbook of statistics (Volume 11) Econometrics". 95 IntJFor 11 189-191

— —; Rao, C. R.; Vinod, H. D. (Ed); Krämer, Walter (Rev. German): Review of "Econometrics (Handbook of statistics) (Volume II)". 95 AllgeStA 79 101-102

Madden, L. V.: Hughes, G.; — —: Using the beta-binomial distribution to describe aggregated patterns of disease incidence. 93 Phytopgy 83 759-762

Madden, Roland A.: Weber, Rudolf O.; — —: Optimal averaging for the determination of global mean temperature: Experiments with model data. 95 JClimate 8 418-430

Madi, Mohamed T.: On the invariant estimation of a normal variance ratio. 95 JStPlInf 44 349-357

Madigan, David; Gavrin, Jonathan; Raftery, Adrian E.: Eliciting prior information to enhance the predictive performance of Bayesian graphical models. 95 CommStA 24 2271-2292

— —; York, Jeremy: Bayesian graphical models for discrete data. 95 IntStRvw 63 215-232

Andersson, Steen A.; — —; Perlman, Michael D.; Triggs, Christopher M.: On the relation between conditional independence models determined by finite distributive lattices and by directed acyclic graphs. 95 JStPlInf 48 25-46

Clarkson, Douglas B.; Donnell, Deborah; Minstrell, Jim; Hunt, Earl; — —; Traynor, Charles: VITAL: An intelligent tutoring system for statistics. 94 ASAProStEd 88-93

York, Jeremy; — —; Heuch, Ivar; Lie, Rolv Terje: Birth defects registered by double sampling: A Bayesian approach incorporating covariates and model uncertainty. 95 ApplStat 44 227-242

Madow, W. G.: Hansen, M. H.; Hurwitz, W. N.; — — (Auth); Fielding, Antony (Rev): Review of "Sample survey methods and theory (Volume I, 2)". 95 JRSS-A 158 350-350

Hansen, M. H.; Hurwitz, W. N.; — — (Auth); Schwaiger, M. (Rev. German): Review of "Sample survey methods and theory (Volume I and II)". 95 AllgeStA 79 335-337

Madras, Neal; Tanny, David: Oscillating random walk with a moving boundary. 94 IsrelJMa 88 333-365

Granovsky, Boris L.; — —: The noisy voter model. 95 StocProc 55 23-43

Madrigal, J. L.: Peck, Robert R.; — —; Fellingham, Gilbert W.: Ranking teams competing in pairwise comparisons using the dynamic linear model. 94 ASAProBays 112-116 (Disc: p134-135)

Madsen, H.: Sadegh, P.; Holst, J.; — —; Melgaard, H.: Experiment design for grey box identification. 95 IJAdCSgP 9 491-507

Madsen, L. L.: Woodward, J. T.; Kono, C.; — —; Zasadzinski, J. A.: Inherent bias in correlation averaged images. 95 JMicro 178 86-92

Madsen, Richard W.: Secondary students' concepts of probability. 95 TeachgSt 17/3 90-93

Maechler, M.: Polasek, W. (Auth); — — (Rev): Review of "EDA. Exploratory data analysis. Introduction to descriptive statistics (Second edition) (German)". 95 ShrtBkRv 15/2 23-23

Maeda, Tadahiko: A causal model of some aspects and determinants of satisfaction of the Japanese - An application of covariance structure analysis to the Survey of Japanese National Character (Japanese). 95 PrInStMa 43 141-160

Maeda, Yutaka; Hirano, Hiroaki; Kanata, Yakichi: A learning rule of neural networks via simultaneous perturbation and its hardware implementation. 95 Autmatca 31 251-259

Maeder, Marcel; Molloy, Kirsten Jane: Lifetime resolved fluorescence detection in chromatography. Global analysis of emission decay surfaces. 95 Chemolab 28 99-107

Maekawa, K.: Hisamatsu, H.; Knight, J. L.; — —: The exact distribution of the OLS and GLS estimators in regression with an integrated regressor and correlated errors - Comparison of numerical and Monte Carlo integration. 95 MaCmpSim 39 273-277

Maekawa, Koichi: Srivastava, V. K.; — —: Efficiency properties of feasible generalized least squares estimators in SURE models under non-normal disturbances. 95 JEconmtx 66 99-121

Maesono, Yoshihiko: On the normal approximations of studentized U-statistic. 95 JJapnStS 25 19-33

Magagnoli, U.: Panizzon, G. (Auth); — — (Rev. Italian): Review of "Bilateral plans for inspection and process control by variables". 94 Statstca 54 539-540

Magalhães, Marcos N.; McNickle, Donald C.; Salles, Maria Creusa B.: Departure processes from a loss system with two stations and one server. 94 RBrzlPrS 8 135-146

Magdalinos, Michael A.; Symeonides, Spyridon D.: Alternative size corrections for some GLS test statistics. The case of the AR(1) model. 95 JEconmtx 66 35-59

Magee, Elvin: Schnell, Daniel J.; — —; Sheridan, John R.: A regression method for analysing ordinal data from intervention trials. 95 StatMed 14 1177-1189

Magel, Rhonda C.: Mahrer, Joanne M.; — —: A comparison of tests for the k-sample, non-decreasing alternative. 95 StatMed 14 863-871

Mager, Harry; Göller, Gernot: Analysis of pseudo-profiles in organ pharmacokinetics and toxicokinetics. 95 StatMed 14 1009-1024

Mager, Peter P.: Diagnostics statistics in QSAR. 95 JChemtrc 9 211-221

Magiera, Ryszard: Bayes sequential estimation procedures in exponential-type processes for a polynomial cost function. 95 MaOpfStS 26 61-73

Magnac, Thierry; Robin, Jean-Marc; Visser, Michael: Analysing incomplete individual employment histories using indirect inference. 95 JApEcmtx 10 153-169S

Magnus, Jan R.: Keuzenkamp, Hugo A.; — —: Editors' introduction - The significance of testing in econometrics. 95 JEconmtx 67 1-3

Keuzenkamp, Hugo A.; — —: On tests and significance in econometrics. 95 JEconmtx 67 5-24

Mahajan, P. K.; Gupta, J. P.; Singh, R.: Determination of optimum strata boundaries for scrambled randomized response. 94 Statstca 54 375-381

Maher, Patrick: — — (Auth); Kyburg, Henry, Jr (Rev): Review of "Betting on theories". 95 PhilSci 62 343-345

Mahler, Ronald P. S.: Combining ambiguous evidence with respect to ambiguous a priori knowledge. Part II: Fuzzy logic. 95 FuzzySet 75 319-354

Mahmoud, Hosam M.; Orlandić, Ratko: Toward a formal derivation of the expected behavior of prefix b-trees. 95 PrEngInf 9 183-192

Mahmoud, Mohamed Abdel Wahab: El-Din, Moustafa Mohamed Mohie; — —; Sultan, Khalaf Salman: On moment generating function of order statistics for doubly truncated exponential distribution. 95 Metron 53 171-183

Mahoney, Michael K.: Garfinkel, Simson L.; — — (Auth); Lenz, H.-J. (Rev): Review of "NeXTSTEP programming, Step one: Object-oriented applications". 95 CmpStDA 19 712-713

Mahrer, Joanne M.; Magel, Rhonda C.: A comparison of tests for the k-sample, non-decreasing alternative. 95 StatMed 14 863-871

Maiboroda, R. E.: Some problems of statistics of distribution mixtures with variable concentrations. 94 ThProbAp 39 703-707

Maiti, Tapabrata: Chaudhuri, Arijit; — —: Asymptotic design-cum-model approach for convex weighting of direct and indirect components of small domain predictors. 95 CommStA 24 2803-2813

Majetić, Dubravko: Dynamic neural network for prediction and identification of nonlinear dynamic systems. 95 JCmpInT 3 99-106

Majumdar, A. A. K.: A best-choice problem with three stops and random termination. 95 SankhyaB 57 76-84

Majumdar, Dibyen: Hedayat, A. S.; — —: Generating desirable sampling plans by the technique of trade-off in experimental design. 95 JStPlInf 44 237-247

Jacroux, Mike; — —; Shah, Kirti R.: Efficient block designs in the presence of trends. 95 StSinica 5 605-615

Majumdar, Suman: Consistency of posterior mixtures in the Gaussian family on a Hilbert space and its applications. 95 JMultiAn 55 187-204

Mak, T. K.: Chan, L. K.; — —: A regression approach for discovering small variation around a target. 95 ApplStat 44 369-377

Mak, Tak K.: Regression models with incidental parameters. 95 JRSS-B 57 453-457

Makhno, S. Ya.: A theorem on large deviations for one class of diffusion processes. 94 ThProbAp 39 437-446

Makino, Toji; Konno, Toshiharu: On the use of circular tetrads in preference polygon (Japanese). 94 JpJBeh 21/2 1-9

Makov, Udi: Cohen, Ayala; — —; Boneh, Ruth: Computer intensive methods for inference on parameters of complex models - A Bayesian alternative. 95 JStPlInf 47 79-94

Makov, Udi E.: Loss robustness via Fisher-weighted squared-error loss function. 95 InsrneME 16 1-6

Makridakis, Spyros: Fildes, Robert; — —: The impact of empirical accuracy studies on time series analysis and forecasting. 95 IntStRvw 63 289-308

Malani, Hina Mehta: A modification of the redistribution to the right algorithm using disease markers. 95 Biomtrka 82 515-526

Lu, Ying; — —: Analysis of animal carcinogenicity experiments with multiple tumor types. 95 Biomtrcs 51 73-86

Malarkey, William B.: Huh, Myung-Hoe; Critchlow, Douglas E.; Verducci, Joseph S.; Klecolt-Glaser, Janice; Glaser, Ronald; — —: A symmetric analysis of paired rankings with application to temporal patterns of hormonal concentration. 95 Biomtrcs 51 1361-1371

Malcolm, Edith: Rawson, Nigel S. B.; — —: Validity of the recording of ischaemic heart disease and chronic obstructive pulmonary disease in the Saskatchewan health care datafiles. 95 StatMed 14 2627-2643

Malec, Donald: Selecting multiple-objective fixed-cost sample designs using an admissibility criterion. 95 JStPlInf 48 229-240

— —; Parsons, Van L.: State estimates for the National Health Interview Survey using a random digit dialing supplement. 94 ASASrvyRsr 1190-1195

Malik, Athar B.: McKay, Ruth B.; Robison, Edwin L.; — —: Touch-tone data entry for household surveys: Research findings and possible applications. 94 ASASrvyRsr 509-511

Malik, H. J.: Pandey, B. N.; — —; Dubey, P. K.: Bayesian shrinkage estimators for a measure of dispersion of an inverse Gaussian distribution. 95 CommStA 24 2261-2270

Malinovskii, V. K.: Correction to "Asymptotic expansions in the problem of sequential estimation of an autoregressive parameter" (V37 p381-383). 94 ThProbAp 39 719-719

Mallal, Simon A.: James, Ian R.; Hogan, Stephen J.; — —: Analysis of doubly censored data and the HIV incubation period distribution in Western Australia. 95 AstrlJSt 37 299-312

Segal, Mark R.; James, Ian R.; French, Martyn A. H.; — —: Statistical issues in the evaluation of markers of HIV progression. 95 IntStRvw 63 179-197

Mallants, Dirk: Ramos, José; — —; Feyen, Jan: State space identification of linear deterministic rainfall-runoff models. 95 WaterReR 31 1519-1531

Mallawaaratchy, A. Rajika: Johnston, Roger G.; Schroder, Shayla D.; — —: Statistical artifacts in the ratio of discrete quantities. 95 AmerStat 49 285-291

Maller, R.: Ghitany, M. E.; — —; Zhou, S.: Estimating the proportion of immunes in censored samples: A simulation study. 95 StatMed 14 39-49

Maller, R. A.; Zhou, Sean: Testing for the presence of immune or cured individuals in censored survival data. 95 Biomtrcs 51 1197-1205

Zhou, S.; — —: The likelihood ratio test for the presence of immunes in a censored sample. 95 MaOpfStS 27 181-201

Zhou, S.; — —; Speed, T. P.: On a shared allele test of random mating. 95 AstrlJSt 37 61-72

Mallet, Alain: Escolano, Sylvie; Golmard, Jean-Louis; — —: A fully nonparametric approach in survival models with explanatory variables. 95 CommStA 24 3027-3054

Malliavin, Paul; Airault, Helene; Kay, Leslie; Letac, Gerard: Integration and probability. 95 Springer:Brln:NY 322p

— —; Airault, Helene; Kay, Leslie; Letac, Gerard (Auth); Barnett, C. (Rev): Review of "Integration and probability". 95 ShrtBkRv 15/3 46-46

Mallick, Bani: Kuo, Lynn; — —: Nonparametric Bayesian inference for accelerated failure time model. 94 ASAProBays 90-95

Mallick, Bani K.: Gelfand, Alan E.; — —: Bayesian analysis of proportional hazards models built from monotone functions. 95 Biomtrcs 51 843-852

Gelfand, Alan E.; — —; Dey, Dipak K.: Modeling expert opinion arising as a partial probabilistic specification. 95 JASA 90 598-604

Mallows, Colin: Barndorff-Nielsen, O. E.; Jensen, J. L.; Kendall, W. S. (Auth); — — (Rev): Review of "Networks and chaos: Statistical and probabilistic aspects". 95 JASA 90 1133-1133

Mallows, Colin L.: More comments on C_p. 95 Technmcs 37 362-372

Malloy, Kay: Johnson, Robert; — —; Gerstein, Dean; Suter, Natalie; Nisenbaum, Susan: Increasing response rates in follow-up surveys of drug treatment patients. 94 ASASrvyRsr 1263-1268

Malmborg, E.; Lisagor, L.: Implementing a statistical metainformation system. 93 StatJUN 10 195-200

Malric, M.: The Lévy transform and zeros in Brownian motion (French). 95 ProbTher 101 227-236

Malý, Marek: Jurečková, Jana; — —: The asymptotics for studentized k-step M-estimators of location. 95 SqtlAnly 14 229-245

Malyshev, V. A.: Fayolle, G.; — —; Menshikov, M. V.: Topics in the constructive theory of countable Markov chains. 95 CambridgeUniv 169p

Fayolle, G., — —; Menshikov, M. V. (Auth); Kozek, A. S. (Rev): Review of "Topics in the constructive theory of countable Markov chains". 96 ShrtBkRv 16/1 9-10

Mammen, E.: On qualitative smoothness of kernel density estimates. 95 MaOpfStS 26 253-267

— —; Tsybakov, A. B.: Asymptotical minimax recovery of sets with smooth boundaries. 95 AnlsStat 23 502-524

Ehm, W.; — —; Müller, D. W.: Power robustification of approximately linear tests. 95 JASA 90 1025-1033

Mammen, Enno: Hall, Peter; — —: On general resampling algorithms and their performance in distribution estimation. 94 AnlsStat 22 2011-2030

Manatunga, Amita: Collett, D. (Auth); — — (Rev): Review of "Modelling survival data in medical research". 95 StMedRes 4 268-268

Mandal, A.: Mukherjee, S. P.; — —: Secretary selection problem with a chance constraint. 94 JIndnStA 32 29-34

Mandel, J.: — — (Auth); Day, Simon (Rev): Review of "Analysis of two-way layouts". 95 Ststcian 44 545-546

— — (Auth); Edmonson, R. N. (Rev): Review of "Analysis of two-way layouts". 95 Biomtrcs 51 1188-1189

Mandel, John: The analysis of two-way layouts. 95 ChapmnHall:Lnd 176p
— — (Auth); Carniello, Michael L. (Rev): Review of "Analysis of two-way layouts". 95 Technmcs 37 461-461
— — (Auth); Freeny, Anne (Rev): Review of "Analysis of two-way layouts". 95 JASA 90 1127-1128
— — (Auth); Nelson, Lloyd S. (Rev): Review of "Analysis of two-way layouts". 95 JQualTek 27 172-173

Mandelbaum, Avi: Kaspi, Haya; — —: Lévy bandits: Multi-armed bandits driven by Lévy processes. 95 AnlsApPr 5 541-565

Mandelbrot, B. B.: Ciozek-Georges, R.; — —: A class of micropulses and antipersistent fractional Brownian motion. 95 StocProc 60 1-18

Mandelbrot, Benoit B.: Cioczek-Georges, Renata; — —; Samorodnitsky, Gennady; Taqqu, Murad S.: Stable fractal sums of pulses: The cylindrical case. 95 Bernoull 1 201-216

Mandeli, John P.: Construction of asymmetrical orthogonal arrays having factors with a large non-prime power number of levels. 95 JStPlInf 47 377-391

Mandjes, Michel; Ridder, Ad: Finding the conjugate of Markov fluid processes. 95 PrEngInf 9 297-315

Mandy, David M.; Martins-Filho, Carlos: A unified approach to asymptotic equivalence of Aitken and feasible Aitken instrumental variables estimators. 94 IntEconR 35 957-979

Mangasarian, Olvi L.: Trippi, Robert R.; Turban, Efraim (Ed); — — (Rev): Review of "Neural networks in finance and investing". 95 Intrfc 25/1 141-142

Mangat, N. S.; Singh, R.; Singh, S.; Bellhouse, D. R.; Kashani, H. B.: On efficiency of using distinct respondents in a randomized response survey. 95 SurvyM 21 21-23
Singh, R.; Singh, S.; — —; Tracy, D. S.: An improved two stage randomized response strategy. 95 StHefte 36 265-271
Singh, S.; — —; Singh, R.: On estimation of mean/total of stigmatized quantitative variable. 94 Statstca 54 383-386

Mangat, Naurang Singh: Singh, Ravindra; Singh, Sarjinder; — —: Mail survey design for sensitive quantitative variable. 95 Metron 53 43-54

Mangin, B.; Goffinet, B.; Rebaï, A.: Constructing confidence intervals for QTL location. 94 Genetcs 138 1301-1308

Mangin, Brigitte: Rebai, Ahmed; Goffinet, Bruno; — —: Comparing power of different methods for QTL detection. 95 Biomtrcs 51 87-99
Rebaï, Ahmed; Goffinet, Bruno; — —: Approximate thresholds of interval mapping tests for QTL detection. 94 Genetcs 138 235-240

Manly, B. F. J.: McDonald, L. L.; Gonzales, L.; — —: Using selection functions to describe changes in environmental variables. 95 EnvEcoSt 2 225-237
— —; McDonald, L. L.; Thomas, D. L. (Auth); North, Philip M. (Rev): Review of "Resource selection by animals: Statistical design and analysis for field studies". 95 JAppStat 22 192-193

Manly, Bryan F. J.: Measuring selectivity from multiple choice feeding-preference experiments. 95 Biomtrcs 51 709-715
— —: Multivariate statistical methods. A primer (Second edition). 95 ChapmnHall:Lnd 232p
— —: A note on the analysis of species co-occurrences. 95 Ecology 76 1109-1115
— —: Randomization tests to compare means with unequal variation. 95 SankhyaB 57 200-222

Mann, Jakob; Lenschow, Donald H.; Kristensen, Leif: Comment on "A definitive approach to turbulence statistical studies in planetary boundary layers" (V51 p1682-1690). 95 JAtmsphr 52 3194-3196

Mann, Michael: Lall, Upmanu; — —: The Great Salt Lake: A barometer of low-frequency climatic variability. 95 WaterReR 31 2503-2515

Mann, P.: — — (Auth); Ziegel, Eric R. (Rev): Review of "Statistics for business and economics". 95 Technmcs 37 239-240

Mann, P. S.: — — (Auth); Braithwaite, G. R. (Rev): Review of "Introductory statistics". 95 JRSS-A 158 354-354

Mann, Prem S.: — — (Auth); Copeland, Karen A. F. (Rev): Review of "Introductory statistics (Second edition)". 95 JQualTek 27 390-391
— — (Auth); Vinci, E. (Rev): Review of "Introductory statistics" (Italian). 94 Metron 52/3 211-212

Mannan, M.; Nassar, R.: Size and power of test statistics for gene correlation in 2×2 contingency tables. 95 BiomtrcJ 37 409-433

Manne, Rolf: On the resolution problem in hyphenated chromatography. 95 Chemolab 27 89-94

Manry, Michael T.: Weideman, William E.; — —; Yau, Hung-Chun; Gong, Wei: Comparisons of a neural network and a nearest-neighbor classifier via the numeric handprint recognition problem. 95 IEEENrlN 6 1524-1530

Mansfield, Edward R.; Adams, Benjamin M.: Teaching the standard deviations of linear combinations using graphics. 94 ASAProStEd 80-83

Mansouri, H.; Chang, G.-H.: A comparative study of some rank tests for interaction. 95 CmpStDA 19 85-96
Black, Stephen; — —: On exact distributions of rank tests for ordered alternatives in block designs. 95 CmpStDA 20 265-274

Mantani, Akihisa: Morimune, Kimio; — —: Estimating the rank of co-integration when the order of a vector autoregression is unknown. 95 MaCmpSim 39 265-271

Mantel, Harold; Laniel, Normand; Duval, Marie-Claude; Marion, Jocelyne: Cost modelling of alternative sample designs for rural areas in the Canadian Labour Force Survey. 94 ASASrvyRsr 925-930

Manthey, R.: Assing, S.; — —: The behavior of solutions of stochastic differential inequalities. 95 ProbTher 103 493-514

Manthey, Ralf: A strong existence theorem for stochastic reaction-diffusion equations with continuous coefficients. 95 Stochatx 52 15-27

Manton, K. G.; Woodbury, M. A.; Tolley, H. D. (Auth); Gordon, A. D. (Rev): Review of "Statistical applications using fuzzy sets". 95 Biomtrcs 51 793-794

Manton, Kenneth G.; Woodbury, Max A.; Tolley, H. Dennis (Auth); Haberman, Shelby J. (Rev): Review of "Statistical applications using fuzzy sets". 95 JASA 90 1131-1133
— —; Woodbury, Max A.; Tolley, H. Dennis (Auth); Kandel, Abraham (Rev): Review of "Statistical applications using fuzzy sets". 95 Technmcs 37 237-237

Mantrala, Anand: McDonald, James B.; — —: The distribution of personal income: Revisited. 95 JApEcmtx 10 201-204

Mantrala, Murali K.: Hutchinson, J. Wesley; Raman, Kalyan; — —: Finding choice alternatives in memory: Probability models of brand name recall. 94 JMarRsr 31 441-461

Manuel, C. M.; Tejada, J.: Reciprocal cooperation in the prisoner's dilemma repeated with random horizon. 95 Top 3 97-116

Mao, Jian-Hua; Wheldon, Tom E.: A stochastic model for multistage tumorigenesis in developing and adult mice. 95 MaBisci 129 95-110

Mao, Xuerong: Adapted solutions of backward stochastic differential equations with non-Lipschitz coefficients. 95 StocProc 58 281-292

Mao, Y.: MacNeill, I. B.; Elwood, J. M.; Miller, D.; — —: Trends in mortality from melanoma in Canada and prediction of future rates. 95 StatMed 14 821-839

Mapstone, Bruce D.: Scalable decision rules for environmental impact studies: Effect size, Type I, and Type II errors. 95 EcolgApp 5 401-410

Maragos, Petros: Yang, Ping-Fai; — —: Min-max classifiers: Learnability, design and application. 95 PattrnRc 28 879-899

Maravalle, Maurizio: Simeone, Bruno: A spanning tree heuristic for regional clustering. 95 CommStA 24 625-639
— —; Simeone, Bruno: Correction to "A spanning tree heuristic for regional clustering" (V24 p625-639). 95 CommStA 24 2437-2437

Marazzi, A.: — — (Auth); Anderson, A. J. B. (Rev): Review of "Algorithms, routines and S functions for robust statistics". 95 Biomtrcs 51 385-386
— — (Auth); Iglewicz, Boris; Heiberger, Richard M.; Moore, Dirk; Tan, Yun (Rev): Review of "Algorithms, routines, and S functions, for robust statistics". 95 Technmcs 37 351-353

Marceau, E.: De Vylder, F.; — —: Explicit analytic ruin probabilities for bounded claims. 95 InsrneME 16 79-105

Marchant, John H.: Peach, William J.; Crick, Humphrey Q. P.; — —: The demography of the decline in the British willow warbler population. 95 JAppStat 22 905-922

Marchi, Marco: Biggeri, Annibale; — —: Case-control designs for the detection of spatial clusters of diseases. 95 EnvrMtrc 6 385-393
Cislaghi, Cesare; Biggeri, Annibale; Braga, Mario; Lagazio, Corrado; — —: Exploratory tools for disease mapping in geographical epidemiology. 95 StatMed 14 2363-2381

Marcis, John G.; Lawrimore, Kay W.: Men's jobs and women's jobs: A pilot study. 94 ASAProBuEc 247-252

Marcus, Michael B.; Rosen, Jay: Logarithmic averages for the local times of recurrent random walks and Levy processes. 95 StocProc 59 175-184

Marden, John I.: Analyzing and modeling rank data. 95 ChapmnHall:Lnd 330p
— —; Muyot, Maria Emerita T.: Rank tests for main and interaction effects in analysis of variance. 95 JASA 90 1388-1398

Marder, Karen: Tang, Ming-Xin; Tsai, Wei-Yann; — —; Mayeux, Richard: Linear rank tests for doubly censored data. 95 StatMed 14 2555-2563

Mardia, K. V.; Dryden, I. L.; Hurn, M. A.; Li, Q.; Millner, P. A.; Dickson, R. A.: Familial spinal shape. 94 JAppStat 21 623-641
— — (Ed); Prum, B. (Rev): Review of "The art of statistical science". 95 MaOpfStS 26 180-181

Marengo, E.: Todeschini, R.; Gramatica, P.; Provenzani, R.; — —: Weighted holistic invariant molecular descriptors. Part 2. Theory development and applications on modeling physicochemical properties of polyaromatic hydrocarbons. 95 Chemolab 27 221-229

Margavio, Thomas M.; Conerly, Michael D.; Woodall, William H.; Drake, Laurel G.: Alarm rates for quality control charts. 95 StPrLet 24 219-224

Margevich, D. E.: Windig, W.; — —; McKenna, W. P.: A novel tool for two-dimensional correlation spectroscopy. 95 Chemolab 28 109-128

Margolin, Barry H.: Sen, Pranab Kumar; — —: Inhalation toxicology: Awareness, identifiability, statistical perspectives and risk assessment. 95 SankhyaB 57 252-276

Mariano, Roberto S.; Tanizaki, Hisashi: Prediction of final data with use of preliminary and/or revised data. 95 JForcast 14 351-380
Diebold, Francis X.; — —: Comparing predictive accuracy. 95 JBES 13 253-263

Mariesse, Jacques: Hall, Bronwyn H.; — —: Exploring the relationship between R and D and productivity in French manufacturing firms. 95 JEconmtx 65 263-293

Mařík, Jan: Erickson, Roy V.; Fabian, Václav; — —: An optimum design for estimating the first derivative. 95 AnlsStat 23 1234-1247

Marinos, Aristides Th.; Tzonou, Anastasia J.; Karantzas, Menelaos E.: Experimental quantiles of epidemiological indices in case-control studies with non-differential misclassification. 95 StatMed 14 1291-1306

Marion, Jocelyne: Mantel, Harold; Laniel, Normand; Duval, Marie-Claude; — —: Cost modelling of alternative sample designs for rural areas in the Canadian Labour Force Survey. 94 ASASrvyRsr 925-930

Mariotto, Angela B.: Verdecchia, Arduino; — —: A back-calculation method to estimate the age and period HIV infection intensity, considering the susceptible population. 95 StatMed 14 1513-1530

Maris, Eric: Psychometric latent response models. 95 Psymtrka 60 523-547

Maritz, J. S.: A permutation paired test allowing for missing values. 95 AstrlJSt 37 153-159

— —: Distribution-free statistical methods (Second edition). 95 ChapmnHall:-Lnd 255p

— —: Distribution free statistical methods (Second edition). 95 ChapmnHall:-Lnd 272p

— — (Auth); Draper, N. R. (Rev): Review of "Distribution-free statistical methods (Second edition)". 95 ShrtBkRv 15/3 45-45

Mark, Steven D.: Green, Sylvan B.; Corle, Donald K.; Gail, Mitchell H.; — —; Pee, David; Freedman, Laurence S.; Graubard, Barry I.; Lynn, William R.: Interplay between design and analysis for behavioral intervention trials with community as the unit of randomization. 95 AJEpidm 142 587-593

Markatou, M.: Lenth, R. V.; — —; Tsimikas, J.: Robust tests based on the sample characteristic function. 95 AstrlJSt 37 45-60

Markatou, Marianthi; Horowitz, Joel L.: Robust scale estimation in the error-components model using the empirical characteristic function. 95 CandJSt 23 369-381

— —; Horowitz, Joel L.; Lenth, Russell V.: Robust scale estimation based on the empirical characteristic function. 95 StPrLet 25 185-192

Marker, David A.; Bryant, Edward C.; Moriarity, Chris: National Employer Health Insurance Survey sample design. 94 ASASrvyRsr 264-269

Markham, Ina S.: Combining neural networks and statistical predictions to solve the classification problem in discriminant analysis. 95 DecisnSc 26 229-242

Markiewicz, Augustyn: Baksalary, Jerzy K.; — —; Rao, C. Radhakrishna: Admissible linear estimation in the general Gauss-Markov model with respect to an arbitrary quadratic risk function. 95 JStPlInf 44 341-347

Markson, Elizabeth W.: D'Agostino, Ralph B.; Belanger, Albert J.; — —; Kelly-Hayes, Maggie; Wolf, Philip A.: Development of health risk appraisal functions in the presence of multiple indicators: The Framingham study nursing home institutionalization model. 95 StatMed 14 1757-1770

Markus, Richard A.; Frank, Jess; Groshen, Susan; Azen, Stanley P.: An alternative approach to the optimal design of an LD50 bioassay. 95 StatMed 14 841-852

Marlow, Angela J.: Shields, Denis C.; — —; Houlston, Richard S.; Eccles, Diana M.; Morton, Newton E.: Prediction of genetic risks from segregation analyses of morbid risks. 94 HumHrdty 44 52-55

Marlow, Norman A.; Tortorella, Michael: Some general characteristics of two-state reliability models for maintained systems. 95 JAppProb 32 805-820

Marmol, Francesc: Spurious regressions between $I(d)$ processes. 95 JTimSrAn 16 313-321

Marohn, F.: Neglecting observations in Gaussian sequences of statistical experiments. 95 StatDecn 13 83-92

Maron, J. S.; Tsybakov, A. B.: Visual error criteria for qualitative smoothing. 95 JASA 90 499-507

Maronna, Ricardo A.; Yohai, Victor J.: The behavior of the Stahel-Donoho robust multivariate estimator. 95 JASA 90 330-341

Marquardt, Donald W.: The missing linkage in ISO 9001: What's being done about it?. 95 ASQCTekCon 1056-1063

Marques, Eliana H. de F.; Koch, Gary G.: A model for distributions and correlation structure of multivariate polytomous data from a two-stage cluster sample. 94 ASAProBiop 346-351

Marquez, Jamie: Hamilton, James D. (Auth); — — (Rev): Review of "Time series analysis". 95 IntJFor 11 494-495

Marquis, Kent H.; Moore, Jeffrey C.; Bogen, Karen: An experiment to reduce measurement error in the SIPP: Preliminary results. 94 ASASrvyRsr 725-729

Bogen, Karen; Moore, Jeffrey C.; — —: Can we get respondents to use their personal income records. 94 ASASrvyRsr 1252-1257

Marr, Ray L.: Some simple graphs for simultaneous comparisons. 94 ASAProPEng 109-114

Marriott, F. H. C.: Krzanowski, W. J.; — —: Multivariate analysis. Part 2: Classification, covariance structures and repeated measurements. 95 Arnold:Lnd:MD 280p

Krzanowski, W. J.; — — (Auth); Johnson, Richard (Rev): Review of "Multivariate analysis, Part 1: Distributions, ordination, and inference". 95 JASA 90 1490-1490

Krzanowski, W. J.; — — (Auth); Shanmugam, Ramalingam (Rev): Review of "Multivariate analysis Part 1: Distributions, ordination and inference". 95 Technmcs 37 351-351

Marriott, Paul: Efron, B.; Tibshirani, R. J. (Auth); — — (Rev): Review of "An introduction to the bootstrap". 95 JRSS-A 158 347-347

Murray, Michael K.; Rice, John W. (Auth); — — (Rev): Review of "Differential geometry and statistics". 95 BLndnMaS 27 619-620

Marron, J. S.: Hall, Peter; Hu, Tien Chung; — —: Improved variable window kernel estimates of probability densities. 95 AnlsStat 23 1-10

Härdle, W.; — —: Fast and simple scatterplot smoothing. 95 CmpStDA 20 1-17

Janssen, P.; — —; Veraverbeke, N.; Sarle, W.: Scale measures for bandwidth selection. 95 JNnparSt 5 359-380

Marron, J. Stephen: Hall, Peter; — —; Titterington, D. M.: On partial local smoothing rules for curve estimation. 95 Biomtrka 82 575-587

Marrs, I.: Jones, K. F.; — —; Young, J.; Townend, C.: Investigation and improvement of a process for vacuum-formed ceramic fibre composites, using a fractional 2^n experimental design. 95 JAppStat 22 459-467

Marsaglia, G.: Random variables with independent integer and fractional parts. 95 StNeerla 49 133-137

Marschner, Ian C.: Computation of age-specific HIV incidence estimates using the EM algorithm. 95 JStCmpSm 53 299-312

Marsh, C.: Dale, A.; — — (Ed); Dorling, Daniel (Rev): Review of "The 1991 Census User's Guide". 95 JRSS-A 158 188-189

Marsh, Catherine: Skinner, Chris; — —; Openshaw, Stan; Wymer, Colin: Disclosure control for census microdata. 94 JOffStat 10 31-51

Marsh, Gary M.: Stone, Roslyn A.; — —; Owens, Ada D.; Quinn, Margaret M.; Smith, Thomas J.: Retrospective estimation of airborne fiber exposure: Statistical aspects. 94 ASAProEpid 127-132

Marsh, Herbert W.: $\Delta 2$ and $\chi^2 I2$ fit indices for structural equation models: A brief note of clarification. 95 StrEqMod 2 246-254

Marshall, Albert W.; Olkin, Ingram: Multivariate exponential and geometric distributions with limited memory. 95 JMultiAn 53 110-125

Hollander, Myles; — —: A conversation with Frank Proschan. 95 StatSci 10 118-133

Marshall, Guillermo; Guo, Wensheng; Jones, Richard H.: MARKOV: A computer program for multi-state Markov models with covariables. 95 CmpBiomd 47 147-156

— —; Jones, Richard H.: Multi-state models and diabetic retinopathy. 95 StatMed 14 1975-1983

Guo, Wensheng; — —: ORDMKV: A computer program fitting proportional odds model for multi-state Markov process. 95 CmpBiomd 46 257-263

Small, Christopher G.; McLeish, D. L. (Auth); — —; Allende, Renato (Rev): Review of "Hilbert space methods in probability and statistical inference". 95 JASA 90 1491-1492

Marshall, Roger J.: A program to implement a search method for identification of clinical subgroups. 95 StatMed 14 2645-2659

Martens, Jean-Bernard: Kayargadde, Vishwakumara; — —: Estimation of edge parameters and image blur using polynomial transforms. 94 CVGIPGr 56 442-461

Martikainen, Alexander: On the strong law of large numbers for sums of pairwise independent random variables. 95 StPrLet 25 21-26

Martín Andrés, A.; Tejedor, I. Herranz: Is Fisher's exact test very conservative?. 95 CmpStDA 19 579-591

— —; Tejedor, I. Herranz; Mato, A. Silva: The Wilcoxon, Spearman, Fisher, χ^2, Student and Pearson tests and 2×2 tables. 95 Ststcian 44 441-450

Silva Mato, A.; — —: Optimal unconditional tables for comparing two independent proportions. 95 BiomtrcJ 37 821-836

Martin, A. R.: Rothery, P.; — —; Hiby, L.; Garside, S.: Estimating foetal growth rate from cross-sectional samples of field data with application to pilot-whales. 95 ApplStat 44 163-172

Martin, Antoinette Ware: Osmint, Jeffrey B.; — —: Response in federally sponsored establishment surveys. 94 ASASrvyRsr 977-982

Martin, Clyde; Shubov, Victor: Natural exponential families of probability distributions and exponential-polynomial approximation. 93 AplMaCmp 59 275-297

Martin, Clyde F.; Allen, Linda J. S.; Stamp, Mark: Urn model simulations of a sexually transmitted disease epidemic. 95 AplMaCmp 71 179-199

Martin, David: Nychka, Doug; Gray, Gerry; Haaland, Perry; — —; O'Connell, Michael: A nonparametric regression approach to syringe grading for quality improvement. 95 JASA 90 1171-1178

Martin, Donald: Diehr, Paula; — —; Koepsell, Thomas; Cheadle, Allen; Wagner, Edward; Psaty, Bruce: Optimal survey design for community-intervention evaluations: Cohort, cross-sectional, or both?. 94 ASAProEpid 34-43

Martin, Donald C.: Diehr, Paula; — —; Koepsell, Thomas; Cheadle, Allen: Breaking the matches in a paired t-test for community interventions when the number of pairs is small. 95 StatMed 14 1491-1504

Martin, E. B.; Scott, E. M.: The detection of change in spatial processes. 95 EnvrMtrc 6 141-154

Martin, Elizabeth A.; Griffin, Deborah H.: The role of questionnaire design in reducing census coverage error. 94 ASASrvyRsr 736-741 (Disc: p754-756)

Martin, Frank B.: Wang, Song-Gui; Chow, Shein-Chung (Auth); — — (Rev): Review of "Advanced linear models". 95 JASA 90 802-803

Martin, M.; Artalejo, J. R.: Analysis of an M/G/1 queue with two types of impatient units. 95 AdvAppPr 27 840-861

Martín, M.; Gómez-Corral, A.: On the M/G/1 retrial queueing system with linear control policy. 95 Top 3 285-305

Martin, M. A.: Braun, H. I. (Ed); — — (Rev): Review of "The collected works of John W. Tukey: Volume VIII Multiple comparisons 1948-1983". 95 Biomtrcs 51 380-381

Martin, Margaret; Suber, Marsha: Statistical education for quality managers. 94 ASAProBays 148–149

Martin, Margaret E.: Fienberg, Stephen E.; — —; Straf, Miron L.: The Committee on National Statistics: Fostering interactions between statisticians in academia and government. 95 IntStRvw 63 257–269

Martin, R. B.: Davis, T. M. E.; — —: Models of parasite clearance following treatment of falciparum malaria in humans. 95 EnvrMtrc 6 529–534

Martin, R. J.: Some results on randomization when observations are spatially dependent and the analysis uses generalized least-squares. 95 CommStA 24 2199–2210

Krug, T.; — —: Information loss on the dependence parameter due to incomplete data from a one-parameter spatial Gaussian process. 95 CommStA 24 3055–3072

Saunders, I. W.; Eccleston, J. A.; — —: An algorithm for the design of 2^p factorial experiments on continuous processes. 95 AstrlJSt 37 353–365

Jørgensen, B. (Auth); — — (Rev): Review of "The theory of linear models". 95 JAppStat 22 543–544

Martin, S. W.: Shoukri, M. M.; — —; Mian, I. U. H.: Maximum likelihood estimation of the kappa coefficient from models of matched binary responses. 95 StatMed 14 83–99

Martin, Sandra J.: Bates, Douglas M.; — —: Design considerations for nonlinear calibration. 94 ASAProPEng 12–17

Martin, Thomas E.; Clobert, Jean; Anderson, David R.: Return rates in studies of life history evolution: Are biases large?. 95 JAppStat 22 863–875

Martin-Löf, A.: — — (Ed): Harald Cramer collected works. Volumes 1 and 2. 94 Springer:Brln:NY 2165p

Martin-Löf, Anders: Harald Cramér and insurance mathematics. 95 AplStMDA 11 271–276

— — (Ed); Uebe, Götz (Rev): Review of "H. Cramér: Collected works. Volumes 1 and 2". 95 AllgeStA 79 332–332

Martinez, C. J.: Dean, C. B.; Eaves, D. M.; — —: A comment on the use of empirical covariance matrices in the analysis of count data. 95 JStPlInf 48 197–205

Swartz, T. B.; Villegas, C.; — —: On semiparametric pivotal Bayesian inference for quantiles. 95 CommStA 24 2499–2515

Martínez, J.: Guerrero, V. M.; — —: A recursive ARIMA-based procedure for disaggregating a time series variable using concurrent data. 95 Test 4 359–376

Martínez, O.; Curnow, R. N.: Estimating the locations and the sizes of the effects of quantitative trait loci using flanking markers. 92 ThApGene 85 480–488

Martínez, Octavio; Curnow, Robert N.: Missing markers when estimating quantitative trait loci using regression mapping. 94 Heredity 73 198–206

Martinez, S.: Ferrari, P. A.; Kesten, H.; — —; Picco, P.: Existence of quasi-stationary distributions. A renewal dynamical approach. 95 AnlsProb 23 501–521

Martínez, Servet; Vares, Maria Eulália: A Markov chain associated with the minimal quasi-stationary distribution of birth-death chains. 95 JAppProb 32 25–38

Collet, Pierre; — —; San Martín, Jaime: Asymptotic laws for one-dimensional diffusions conditioned to nonabsorption. 95 AnlsProb 23 1300–1314

Martinez-Almecija, Alfredo: Valderrama, Mariano J.; Jimenez, Francisco; Gutierrez, Ramon; — —: Estimation and filtering on a doubly stochastic Poisson process. 95 AplStMDA 11 13–24

Martino, Joseph P.: — — (Auth); Fields, Paul J. (Rev): Review of "Technological forecasting for decision making (Third edition)". 95 Intrfc 25/1 137–139

Martins, Alejandro: Kandel, Abraham; — —; Pacheco, Roberto: Comment on "A probabilistic and statistical view of fuzzy methods". 95 Technmcs 37 276–281

Martins, Emí P.: Estimating the rate of phenotypic evolution from comparative data. 94 ANatrls 144 193–209

Martins-Filho, Carlos: Mandy, David M.; — —: A unified approach to asymptotic equivalence of Aitken and feasible Aitken instrumental variables estimators. 94 IntEconR 35 957–979

Martinsek, Adam T.: Estimating a slope parameter in regression with prescribed proportional accuracy. 95 StatDecn 13 363–377

Martuzzi, Marco: Elliott, Paul; — —; Shaddick, Gavin: Spatial statistical methods in environmental epidemiology: A critique. 95 StMedRes 4 137–159

Martz, H. F.: Silver, R. N.; — —: Applications of quantum entropy to statistics. 94 ASAProStCp 61–70

Martz, Harry F.; Picard, Richard R.: Uncertainty in Poisson event counts and exposure time in rate estimation. 95 RelEngSS 48 181–190

Kvam, Paul H.; — —: Bayesian inference in a discrete shock model using confounded common cause data. 95 RelEngSS 48 19–25

Marubini, E.; Valsecchi, M. G.: Analysing survival data from clinical trials and observational studies. 94 Wiley:NY:UK 414p

— —; Valsecchi, M. G. (Auth); Juritz, J. M. (Rev): Review of "Analysing survival data from clinical trials and observational studies". 95 ShrtBkRv 15/2 23–23

— —; Valsecchi, M. G. (Auth); McLachlan, G. J. (Rev): Review of "Analysing survival data from clinical trials and observational studies". 95 Biomtrcs 51 1191–1191

Marubini, Ettore; Valsecchi, Maria Grazia: Analyzing survival data from clinical trials and observational studies. 95 Wiley:NY:UK 424p

Maryak, John L.; Spall, James C.; Silberman, Geoffrey L.: Uncertainties for recursive estimators in nonlinear state-space models, with applications to epidemiology. 95 Autmatca 31 1889–1892

Marzjarani, Morteza: Dealing with uncertain knowledge in knowledge base systems. 94 ASAProStCp 136–140

Marzluff, John M.: Schueck, Linda S.; — —: Influence of weather on conclusions about effects of human activities on raptors. 95 JWildlMg 59 674–682

Masani, P. R.: The aftermath of Cramér's work on stochastic processes. 95 ScanActJ 66–96

Mase, Shigeru: Stereological estimation of particle size distributions. 95 AdvAppPr 27 350–366

— —: Consistency of the maximum pseudo-likelihood estimator of continuous state space Gibbsian processes. 95 AnlsApPr 5 603–612

Mashayekhi, M.: On symmetrization of product measures. 95 MaMethSt 4 334–343

Mason, D. J.: Jurek, Z. H.; — — (Auth); Hazod, W. (Rev): Review of "Operator-limit distributions in probability theory". 95 Metrika 42 265–267

Mason, D. M.: Beirlant, J.; — —: On the asymptotic normality of L_p-norms of empirical functionals. 95 MaMethSt 4 1–19

Janssen, A.; — — (Auth); Jurečková, Jana (Rev): Review of "Non-standard rank tests". 95 SIAMRvw 37 261–262

Mason, David M.: A universal one-sided law of the iterated logarithm. 94 AnlsProb 22 1826–1837

Csörgő, Sándor; — —: Intermediate sums and stochastic compactness of maxima. 95 JStPlInf 45 81–90

Deheuvels, Paul; — —: On the fractal nature of empirical increments. 95 AnlsProb 23 355–387

Deheuvels, Paul; — —: Nonstandard local empirical processes indexed by sets. 95 JStPlInf 45 91–112

Einmahl, Uwe; — —: A universal Chung-type law of the iterated logarithm. 94 AnlsProb 22 1803–1825

Mason, J. D.: Hudson, W. N.; — —; Veeh, J. A.: Operator stable probability measures: An overview. 94 ThProbAp 39 264–276

Jurek, Z. J.; — — (Auth); Blower, G. (Rev): Review of "Operator-limit distributions in probability theory". 95 JRSS-A 158 352–353

Mason, J. David: Jurek, Zbigniew J.; — — (Auth); Veeh, Jerry Alan (Rev): Review of "Operator-limit distributions in probability theory". 95 BAmMaSoc 32 278–280

Mason, R. L.: Tracy, N. D.; Young, J. C.; — —: A bivariate control chart with meaningful principal components. 94 ASAProBays 157–158

Tracy, N. D.; Young, J. C.; — —: Monitoring a time-dependent multivariate process. 94 ASAProQlPr 155–157

Mason, Robert L.; Tracy, Nola D.; Young, John C.: Decomposition of T^2 for multivariate control chart interpretation. 95 JQualTek 27 99–108

Tracy, Nola D.; Young, John C.; — —: A bivariate control chart for paired measurements. 95 JQualTek 27 370–376

Mason, Thomas J.: The development of the series of U.S. cancer atlases: Implications for future epidemiologic research. 95 StatMed 14 473–479 (Disc: p697–699)

Mason, William M.: Comment on "Hierarchical linear models". 95 JEdStat 20 221–227

Masry, Elias; Tjøstheim, Dag: Nonparametric estimation and identification of nonlinear ARCH time series. 95 EconTher 11 258–289

Lii, Keh-Shin; — —: On the selection of random sampling schemes for the spectral estimation of continuous time processes. 95 JTimSrAn 16 291–311

Massagli, Michael P.: Burr, Jeffrey A.; — —; Mutchler, Jan E.; Pienta, Amy: Labor force status and transitions among older African American and white men. 94 ASAProSoSt 129–134

Massam, H.: Letac, G.; — —: Craig-Sakamoto's theorem for the Wishart distributions on symmetric cones. 95 AnInStMa 47 785–799

Massam, Hélène: An exact decomposition for a class of exponential dispersion models: Application to sequential testing. 95 CandJSt 23 67–84

Massart, D. L.: de Aguiar, P. F.; Bourguignon, B.; Khots, M. S.; — —; Phan-Than-Luu, R.: D-optimal designs. 95 Chemolab 30 199–210

Griep, M. I.; Wakeling, I. N.; Vankeerberghen, P.; — —: Comparison of semirobust and robust partial least squares procedures. 95 Chemolab 29 37–50

Vankeerberghen, P.; Smeyers-Verbeke, J.; Leardi, R.; Karr, C. L.; — —: Robust regression and outlier detection for non-linear models using genetic algorithms. 95 Chemolab 28 73–87

Walczak, B.; — —: Robust principal components regression as a detection tool for outliers. 95 Chemolab 27 41–54

— —; Lewi, P. J. (Auth); Kvalheim, O. M. (Rev): Review of "Video course: Principal components". 95 Chemolab 27 147–147

Massart, P.: Doukhan, P.; — —; Rio, E.: Invariance principles for absolutely regular empirical processes. 95 AlHenrPB 31 393–427

Massart, Pascal: Birgé, Lucien; — —: Estimation of integral functionals of a density. 95 AnlsStat 23 11–29

Massé, Bruno: Invariance principle for the deviation between the probability content and the interior point proportion of a random convex hull. 95 JAppProb 32 1041–1047

— —; Viano, Marie-Claude: Explicit and exponential bounds for a test on the coefficient of an AR (1) model. 95 StPrLet 25 365–371

Massé, Jean-Claude: Hamel, Denis; — —: A Monte Carlo study of the robustness of prediction in growth curve models. 95 JStCmpSm 52 33-53

Massey, Catherine J.: Silver, N. Clayton; Wadiak, Diane L.; — —: A Microsoft FORTRAN 77 program for testing the differences among independent first-order partial correlations. 95 EdPsMeas 55 245-248

Massey, James T.: Ezzati, Trena M.; Hoffman, Keith; Judkins, David R.; — —; Moore, Thomas F.: A dual frame design for sampling elderly minorities and persons with disabilities. 95 StatMed 14 571-583 (Disc: p695-696)

Kalsbeek, William D.; Botman, Steven L.; — —; Liu, Pao-Wen: Cost-efficiency and the number of allowable call attempts in the National Health Interview Survey. 94 JOffStat 10 133-152

Masters, Timothy: Neural, novel and hybrid algorithms for time series prediction. 95 Wiley:NY:UK 544p

Masulis, Ronald W.; Ng, Victor K.: Overnight and daytime stock-return dynamics on the London stock exchange: The impacts of "big bang" and the 1987 stock-market crash. 95 JBES 13 365-378

Masuyama, Motosaburo: On the refinement of the LGC model. 93 RStAJp 40/1 38-41

Matejcik, Frank J.; Nelson, Barry L.: One-stage multiple comparisons with the best for unequal variances. 94 ASAProBuEc 55-60

Mateos, A.: Ríos-Insua, S.; Pachón, J. G.; — —: A method of multiobjective decision making using a vector value function. 94 Questo 18 153-172

Mathai, A. M.; Provost, S. B.; Hayakawa, T.: Bilinear forms and zonal polynomials. 95 Springer:Brln:NY 376p

— — (Auth); Wright, Peter (Rev): Review of "A handbook of generalized special functions for statistical and physical sciences". 95 AstrlJSt 37 249-250

Mathar, Rudolf: Groenen, Patrick J. F.; — —; Heiser, Willem J.: The majorization approach to multidimensional scaling for Minkowski distances. 95 JClass 12 3-19

Mathes, Harald: Schneeweiss, Hans; — —: Factor analysis and principal components. 95 JMultiAn 55 105-124

Matheson, A. C.: Williams, E. R.; — — (Auth); Nester, Marks R. (Rev): Review of "Experimental design and analysis for use in tree improvement". 95 AstrlJSt 37 247-248

Williams, E. R.; — — (Auth); Smith, R. I. (Rev): Review of "Experimental design and analysis for use in tree improvement". 95 Biomtrcs 51 1190-1190

Mathew, Thomas: Bhaumik, Dulal K.; — —: Minimaxity of randomized optimal designs with respect to a general optimality criterion. 95 SankhyaB 57 122-127

Zhou, Leping; — —: A note on combining independent tests for a common mean. 93 JCISS 18 79-84

Mathews, K. L.: Rayner, J. C. W.; Best, D. J.; — —: Interpreting the skewness coefficient. 95 CommStA 24 593-600

Mathiason, David J.: Halavin, James J.; — —: Designing and running undergraduate statistical laboratories. 94 ASAProStEd 132-136

Mathieson, Kieran; Doane, David P.; Tracy, Ronald L.: A program for visualizing comparisons between two normal distributions. 95 JStEd 3/1 23kb (jse/v3n1/mathieson)

Tracy, Ronald L.; — —; Doane, David P.: Visualizing and describing the shape of distributions. 94 ASAProStEd 70-74

Mathieson, Kieran D.: Daone, David P.; Tracy, Ronald L.; — —: Visualizing power and type I and II error. 94 ASAProStEd 66-69

Mathieu, Didier: Sergent, Michelle; — —; Phan-Tan-Luu, Roger; Drava, Giuliana: Correct and incorrect use of multilinear regression. 95 Chemolab 27 153-162

Mathieu, J.-R.: Dorkenoo, K. M.; — —: Generalized linear models with a small scale parameter. 95 MaOpfStS 27 107-113

Mathieu, P.: Barlow-Yor inequalities for local times of the intersection of two planar Brownian motions (French). 95 ProbTher 102 83-103

Mathieu, Pierre: Spectra, exit times and long time asymptotics in the zero-white-noise limit. 95 Stochatx 55 1-20

— —: Limit theorems for diffusions with a random potential. 95 StocProc 60 103-111

Mathiowetz, Nancy A.: Moeller, John F.; — —: Problems of screening for poverty status. 94 JOffStat 10 327-337

Mathon, Rudolf: Kramer, E. S.; — —: Proper $S(t,K,v)$'s for $t \geq 3$, $\mu \leq 16$, $|K| > 1$ and their extensions. 95 JComDsgn 3 411-425

Mathur, Sharat K.: Curry, David J.; Divakar, Suresh; — —; Whiteman, Charles H.: BVAR as a category management tool: An illustration and comparison with alternative techniques. 95 JForcast 14 181-199

Matis, James H.; Zheng, Qi; Kiffe, Thomas R.: Describing the spread of biological populations using stochastic compartmental models with births. 95 MaBisci 126 215-247

Mato, A. Silva: Martín Andrés, A.; Tejedor, I. Herranz; — —: The Wilcoxon, Spearman, Fisher, χ^2, Student and Pearson tests and 2×2 tables. 95 Ststcian 44 441-450

Matrán, C.: Cuesta-Albertos, J. A.; Domínguez-Menchero, J. S.; — —: Consistency of L_p-best monotone approximations. 95 JStPlInf 47 295-318

Mattei, A.: Vojnovic, D.; Campisi, B.; — —; Favretto, L.: Experimental mixture design to ameliorate the sensory quality evaluation of extra virgin olive oils. 95 Chemolab 27 205-210

Matthes, K.: Resnick, Sidney I. (Auth); — — (Rev): Review of "Adventures in stochastic processes" (German). 95 BiomtrcJ 37 272-272

van der Weide, J. A. M. (Auth); — — (Rev): Review of "Stochastic processes and point processes of excursions". 95 BiomtrcJ 37 764-764

Matthes, Klaus: Reiss, R.-D. (Auth); — — (Rev): Review of "A course on point processes". 95 BiomtrcJ 37 182-182

Matthews, D.: McLachlan, G. J.; McLaren, C. E.; — —: An algorithm for the likelihood ratio test of one versus two components in a normal mixture model fitted to grouped and truncated data. 95 CommStB 24 965-985

Matthews, D. E.: Rai, S. N.; — —: The analysis of incomplete data using stochastic covariates. 95 CandJSt 23 29-42

Matthews, G. B.; Crowther, N. A. S.: A maximum likelihood estimation procedure when modelling in terms of constraints. 95 SoAfrStJ 29 29-50

Matthews, Holly: Wiens, Brian; — —: Parameters for analysis of wound healing in early clinical studies in humans. 94 ASAProBiop 377-382

Matthews, J. N. S.; Morris, K. P.: An application of Bradley-Terry-type models to the measurement of pain. 95 ApplStat 44 243-255

Mehrabi, Yadollah; — —: Likelihood-based methods for bias reduction in limiting dilution assays. 95 Biomtrcs 51 1543-1549

Matthews, John N. S.: Small clinical trials: Are they all bad?. 95 StatMed 14 115-126

Matthews, K.: Ioannidis, C.; Laws, J.; — —; Morgan, B.: Business cycle analysis and forecasting with a structural vector autoregression model for Wales. 95 JForcast 14 251-265

Mattie, Angela S.: Shevchenko, Ivan P.; Lynch, John T.; — —; Reed-Fourquet, Lori L.: Verification of information in a large medical database using linkages with external databases. 95 StatMed 14 511-530 (Disc: p687-689)

Mattison, Donald R.: Mazumdar, Sati; — —; Damaraju, C. V.: Issues and approaches for assessing risks from reproductive toxicants: DBCP as a case study. 95 SankhyaB 57 223-236

Mattner, L.: Mean unbiased medians, median unbiased means, and symmetric random walks. 95 MaMethSt 4 99-105

Mattson, LeRoy T.: Kim, John S.; — —: Use of field data in reliability studies of medical devices. 94 ASAProQlPr 200-203

Matulić-Bedenić, I.; Horvatić-Baldasar, K.; Kramer, E.: Construction of new symmetric designs with parameters (66, 26, 10). 95 JComDsgn 3 405-410

Matute, Jorge: Fitch, David J.; Flores, Rafael; — —: Are the biases and inefficiencies of the EPI method of sampling relatively unimportant with reasonable between village variances?. 94 ASASrvyRsr 288-292

Mátyás, László: Kőrösi, Gábor; Lovrics, László; — —: Aggregation and the long run properties of economic time series. 95 MaCmpSim 39 279-286

Mauger, David T.; Brown, Morton B.; Kushler, Robert H.: A comparison of methods that characterize pulses in a time series. 95 StatMed 14 311-325

Mauger, Elizabeth A.; Wolfe, Robert A.; Port, Friedrich K.: Transient effects in the Cox proportional hazards regression model. 95 StatMed 14 1553-1565

Mauldin, R. Daniel; Monticino, Michael G.: Randomly generated distributions. 95 IsrelJMa 91 215-237

Mauricio, José Alberto: Exact maximum likelihood estimation of stationary vector ARMA models. 95 JASA 90 282-291

May, Warren L.; Johnson, William D.: Testing quasi-independence in stratified triangular contingency tables. 95 BiomtrcJ 37 531-546

— —; Johnson, William D.: Some applications of the analysis of multivariate normal data with missing observations. 95 JBiophSt 5 215-228

Johnson, William D.; — —: Combining 2×2 tables that contain structural zeros. 95 StatMed 14 1901-1911

Mayda, Jacqueline E.: Choudhry, G. Hussain; Colledge, Michael J.; — —: Redesign of the Hungarian Retail Trade and Catering Survey: Methodological issues. 94 ASAProBuEc 71-76

Maydeu-Olivares, Albert: Arce, Constantino (Auth); — — (Rev): Review of "Multidimensional scaling. A multivariate technique for the analysis of proximity and preference data" (Spanish). 95 JClass 12 137-139

Mayekawa, Shin-ichi: An efficient algorithm for feed-forward neural network regression analysis. 95 Behvmtrk 22 67-90

Kikuchi, Kenichi; — —: On the sampling distribution of swap-rate. 95 Behvmtrk 22 185-204

Mayer, D. G.; Atzeni, M. G.; Swain, A. J.; Stuart, M. A.: Models for the spatial dispersal of insect pests. 95 EnvrMtrc 6 497-503

Mayer, Walter J.: Dorsey, Robert E.; — —: Genetic algorithms for estimation problems with multiple optima, nondifferentiability, and other irregular features. 95 JBES 13 53-66

Mayer-Wolf, Eddy: Dembo, Amir; — —; Zeitouni, Ofer: Exact behavior of Gaussian seminorms. 95 StPrLet 23 275-280

Mayeux, Richard: Tang, Ming-Xin; Tsai, Wei-Yann; Marder, Karen; — —: Linear rank tests for doubly censored data. 95 StatMed 14 2555-2563

Mayo, Oliver: Lancaster, H. O. (Auth); — — (Rev): Review of "Quantitative methods in biological and medical science - A historical essay". 95 AstrlJSt 37 372-374

Mazliak, L.: Ceci, C.; — —: Controlled trees. 94 RBrzlPrS 8 93-105

Mazumdar, Mainak: Rajgopal, Jayant; — —: Designing component test plans for series system reliability via mathematical programming. 95 Technmcs 37 195-212

Mazumdar, Ravi R.: Guillemin, Fabrice M.; — —: Excursions of the workload process in G/GI/1 queues. 95 StocProc 58 139-154

Mazumdar, Sati; Mattison, Donald R.; Damaraju, C. V.: Issues and approaches for assessing risks from reproductive toxicants: DBCP as a case study. 95 SankhyaB 57 223-236

Mazzuchi, Thomas A.; Soyer, Refik; Soofi, Ehsan: Lifetime modeling via Bayesian information distinguishability. 94 ASAProBays 238-245

McAleer, M.: Cox, K.; Puddey, I. B.; Burke, V.; — —; Beilin, L. J.; Morton, A. R.: Effects of exercise and calorie restriction on ambulatory blood pressure in overweight, sedentary males: A time series analysis. 95 EnvrMtrc 6 523-528

McAleer, M. J.: Jakeman, A. J.; Beck, M. B.; — — (Ed); Keesman, Karel (Rev): Review of "Modelling change in environmental systems". 95 EnvrMtrc 6 425-427

McAleer, Michael: The significance of testing empirical non-nested models. 95 JEconmtx 67 149-171

Bai, Jun; Jakeman, Anthony J.; — —: Empirical models for evaluating errors in fitting extremes of a probability distribution. 95 MaCmpSim 39 1-7

Dharmapala, Dhammika; — —: Prediction and accommodation in econometric modelling. 95 EnvrMtrc 6 551-556

Smith, Jeremy; — —: Alternative procedures for converting qualitative response data to quantitative expectations: An application to Australian manufacturing. 95 JApEcmtx 10 165-185

Smith, Jeremy; — —: The performance of alternative estimators in models with generated regressors when the expectations equation has reduced explanatory power. 95 MaCmpSim 39 343-346

McAllister, Murdoch K.; Pikitch, Ellen K.; Punt, Andre E.; Hilborn, Ray: A Bayesian approach to stock assessment and harvest decisions using the sampling/importance resampling algorithm. 94 JFishRBC 51 2673-2687

McAulay, Liz: Campanelli, Pamela; Channell, Joanna; — —; Renouf, Antoinette; Thomas, Roger: A cognitive and linguistic exploration of the meaning of training: Implications for survey design. 94 ASASrvyRsr 1281-1286

McAweeney, Mary J.: Klockars, Alan J.; Hancock, Gregory R.; — —: Power of unweighted and weighted versions of simultaneous and sequential multiple-comparison procedures. 95 PsychBu 118 300-307

McBride, Ruth: Hallstrom, Al; — —; Moore, Rich: Toward vital status sweeps: A case history in sequential monitoring. 95 StatMed 14 1927-1931

McCabe, B. P. M.; Tremayne, A. R.: Testing a time series for difference stationarity. 95 AnlsStat 23 1015-1028

McCabe, G. P.: Moore, D. S.; — — (Auth); Kaplon, Howard S. (Rev): Review of "Introduction to the practice of statistics (Second edition)". 95 JAppStat 22 196-197

McCall, Brian P.: The impact of unemployment insurance benefit levels on recipiency. 95 JBES 13/2 189-198

— —: Identifying state dependence in duration models. 94 ASAProBuEc 14-16 (Disc: p17-19)

McCarthy, W. V.: Krzanowski, W. J.; Jonathan, P.; — —; Thomas, M. R.: Discriminant analysis with singular covariance matrices: Methods and applications to spectroscopic data. 95 ApplStat 44 101-115

McClave, J.: Scheaffer, R.; — — (Auth); Ziegel, Eric R. (Rev): Review of "Probability and statistics for engineers". 95 Technmcs 37 239-239

McClave, James T.; Dietrich, Frank H., II (Auth); Halvorsen, Katherine (Rev): Review of "Statistics (Sixth edition)". 95 AmerStat 49 322-322

Scheaffer, Richard; — — (Auth); Herrin, Gary D. (Rev): Review of "Probability and statistics for engineers (Fourth edition)". 95 AmerStat 49 238-239

Scheaffer, Richard L.; — — (Auth); Rigdon, Steven E. (Rev): Review of "Probability and statistics for engineers". 95 JQualTek 27 178-179

McClean, Sally: Discussion on the symposium on the teaching of statistics in higher education in Ireland. 95 Ststcian 44 56-57

— —; Devine, Colum: A nonparametric maximum likelihood estimator for incomplete renewal data. 95 Biomtrka 82 791-803

Robinson, Stephen; — —: Simulation of wind speed for the modelling of pine saw-fly mortality. 95 Ststcian 44 469-477

McCloud, P. I.; Darroch, J. N.: An analysis of categorical repeated measurements in the presence of an external factor. 95 Biomtrka 82 47-55

McCollor, Douglas: Burrows, William R.; Benjamin, Mario; Beauchamp, Stephen; Lord, Edward R.; — —; Thomson, Bruce: CART decision-tree statistical analysis and prediction of summer season maximum surface ozone for the Vancouver, Montreal, and Atlantic regions of Canada. 95 JApMetrl 34 1848-1862

McConway, K.: Hand, D.; Daly, F.; Lunn, A.; — —; Ostrowski, E. (Ed); Ziegel, Eric R. (Rev): Review of "A handbook of small data sets". 95 Technmcs 37 242-242

Hand, D. J.; Daly, F.; Lunn, A. D.; — —; Ostrowski, E. (Ed); Chatfield, C. (Rev): Review of "A handbook of small data sets". 95 JRSS-A 158 349-349

McConway, K. J.: Hand, D. J.; Daly, F.; Lunn, A. D.; — —; Ostrowski, E. (Ed); Schimek, Michael G. (Rev): Review of "Handbook of small data sets". 95 CmpStQ 10 197-197

Hand, D. J.; Daly, F.; Lunn, A. D.; — —; Ostrowski, E. (Ed); Street, Deborah J. (Rev): Review of "A handbook of small data sets". 95 AstrlJSt 37 251-251

Hand, D. J.; Daly, F.; Lunn, A. D.; — —; Ostrowski, E. (Auth); Trenkler, Dietrich (Rev): Review of "A handbook of small data sets". 95 CmpStDA 19 101-101

McCormick, Douglas: Cooil, Bruce; Devinney, Timothy; — —: Models for automobile insurance accidents and losses. 94 ASAProBuEc 217-222

McCormick, William P.: Datta, Somnath; — —: Bootstrap inference for a first-order autoregression with positive innovations. 95 JASA 90 1289-1300

Datta, Somnath; — —: Some continuous Edgeworth expansions for Markov chains with applications to bootstrap. 95 JMultiAn 52 83-106

Homble, Patrick; — —: Weak limit results for the extremes of a class of shot noise processes. 95 JAppProb 32 707-726

McCoy, E. J.: Walden, A. T.; — —; Percival, D. B.: The effective bandwidth of a multitaper spectral estimator. 95 Biomtrka 82 201-214

McCrohan, Kevin F.; Sugrue, Timothy F.: Segmentation in the informal economy. 94 ASAProBuEc 259-264

McCullagh, P.: Glonek, G. F. V.; — —: Multivariate logistic models. 95 JRSS-B 57 533-546

McCullagh, Peter: Comment on "Inference". 95 StatSci 10 177-179

Shun, Zhenming; — —: Laplace approximation of high dimensional integrals. 95 JRSS-B 57 749-760

McCulloch, C. E.; Searle, S. R.: On an identity derived from unbiasedness in linear models. 95 AmerStat 49 39-42

McCulloch, C. E. M.: Searle, S. R.; Casella, G.; — — (Auth); Cavedon, G. (Rev): Review of "Variance components". 95 Metron 53 217-218

McCulloch, Charles E.: Natarajan, Ranjini; — —: A note on the existence of the posterior distribution for a class of mixed models for binomial responses. 95 Biomtrka 82 639-643

McCullough, B. D.: Vinod, H. D.; — —: Estimating cointegration parameters: An application of the double bootstrap. 95 JStPlInf 43 147-156

McDavitt, Chris: Lawrence, David; — —: Significance editing in the Australian Survey of Average Weekly Earnings. 94 JOffStat 10 437-447

McDermott, Michael P.: Mudholkar, Govind S.; — —; Mudholkar, Anil: Robust finite-intersection tests for homogeneity of ordered variances. 95 JStPlInf 43 185-195

McDiarmid, Colin: Minimal positions in a branching random walk. 95 AnlsApPr 5 128-139

McDonald, D.: Iscoe, I.; — —: Asymptotics of exit times for Markov jump processes. II: Applications to Jackson networks. 94 AnlsProb 22 2168-2182

McDonald, David: Davis, Burgess; — —: An elementary proof of the local central limit theorem. 95 JTherPrb 8 693-701

McDonald, J. W.; Smith, P. W. F.: Exact conditional tests of quasi-independence for triangular contingency tables: Estimating attained significance levels. 95 ApplStat 44 143-151

McDonald, James B.; Mantrala, Anand: The distribution of personal income: Revisited. 95 JApEcmtx 10 201-204

— —; Xu, Yexiao J.: A generalization of the beta distribution with applications. 95 JEconmtx 66 133-152

— —; Xu, Yexiao J.: Correction to "A generalization of the beta distribution with applications" (V66 p133-152). 95 JEconmtx 69 427-428

— —; Xu, Yexiao J.: Some comparisons and applications of semi-parametric and flexible parametric estimation procedures in econometric models. 94 ASAProBuEc 229-234

McDonald, John W.: Ekholm, Anders; Smith, Peter W. F.; — —: Marginal regression analysis of a multivariate binary response. 95 Biomtrka 82 847-854

Smith, Peter W. F.; — —: Exact conditional tests for incomplete contingency tables: Estimating attained significance levels. 95 StatComp 5 253-256

McDonald, L. L.; Gonzales, L.; Manly, B. F. J.: Using selection functions to describe changes in environmental variables. 95 EnvEcoSt 2 225-237

Manly, B. F. J.; — —; Thomas, D. L. (Auth); North, Philip M. (Rev): Review of "Resource selection by animals: Statistical design and analysis for field studies". 95 JAppStat 22 192-193

McDonald, R. P.: Comment on "Testing for equivalence of measurement scales: Simple structure and metric invariance reconsidered". 95 MultBehR 30 113-115

McDonald, Roderick P.: Comment on "Testing for equivalence of measurement scales: Simple structure and metric invariance reconsidered". 95 MultBehR 30 87-88

— —; Mok, Magdalena M.-C.: Goodness-of-fit in item response models. 95 MultBehR 30 23-40

McDonald, Scott: Nedelman, Jerry R.; Karara, Adel H.; Chang, Cheng-Tao; Gibiansky, Ekaterina; — —; Robinson, William T.: Inferring systemic exposure from a pharmacokinetic screen: Model-free and model-based approaches. 95 StatMed 14 955-968 (Disc: p969-970)

McGarry, Tim; Schutz, Robert W.: Analysis of the 1986 and 1994 World Cup soccer tournament. 94 ASAProSpts 61-65

McGartland Rubio, Doris; Gillespie, David F.: Problems with error in structural equation models. 95 StrEqMod 2 367-378

McGilchrist, C. A.; Ragunathan, T. E.: Generalised components of variance models. 95 CommStA 24 349-361

— —; Yau, K. K. W.: The derivation of BLUP, ML, REML estimation methods for generalised linear mixed models. 95 CommStA 24 2963-2980

— —; Yau, K. K. W.: Estimation of survival with time dependent frailty. 94 ASAProEpid 88-90

Wu, Y.; — —: Recursive computation for repeated measures models with patterned variance matrices. 95 JStCmpSm 51 333-341

Wu, Yuan; — —: Heterogeneous variances in repeated measures. 95 BiomtrcJ 37 523-530

Meester, Ronald: Equality of critical densities in continuum percolation. 95 JAppProb 32 90-104

Burton, Robert M.; Choi, Kyeong; — —: Stationary straight-line representations of stationary random graphs. 95 IndgMatN 6 267-277

Meesters, A. G. C. A.: Weijers, E. P.; van Delden, A.; Vugts, H. F.; — —: Characteristics of convective turbulence in the surface layer investigated by principal component analysis. 95 JApMetrl 34 528-541

Meeuwissen, Adrianus M. H.: Haverkort, Boudewijn R.; — —: Sensitivity and uncertainty analysis of Markov-reward models. 95 IEEEReli 44 147-154

Meffe, Gary K.: Sheldon, Andrew L.; — —: Path analysis of collective properties and habitat relationships of fish assemblages in coastal plain streams. 95 JFishRBC 52 23-33

Megiddo, N.: — — (Ed): Essays in game theory: In honor of Michael Maschler. 94 Springer:Brln:NY 203p

Mehdi, J.: — — (Auth); Warde, William D. (Rev): Review of "Statistical methods: An introductory text". 95 AmerStat 49 113-114

Mehlman, David W.; Shepherd, Ursula L.; Kelt, Douglas A.: Comment on "Stopping rules in principal components analysis: A comparison of heuristical and statistical approaches" (V74 p2204-2214). 95 Ecology 76 640-643

Mehra, K. L.: Bagui, S. C.; — —; Rao, M. S.: A nearest neighbour classification rule for multiple observations based on a sub-sample approach. 95 SankhyaA 57 316-332

Mehrabi, Yadollah; Matthews, J. N. S.: Likelihood-based methods for bias reduction in limiting dilution assays. 95 Biomtrcs 51 1543-1549

Mehrotra, Devan V.: Robust elementwise estimation of a dispersion matrix. 95 Biomtrcs 51 1344-1351

Mehrotra, Rajiv: Ghosal, Sugata; — —: Range surface characterization and segmentation using neural networks. 95 PattrnRc 28 711-727

Mehta, Cyrus: Gajjar, Yogesh; — —; Patel, Nitin (Auth); Smith, Peter W. F. (Rev): Review of "LOGXACT-TURBO (Version 1.1)". 95 ApplStat 44 139-141

Mehta, Cyrus R.; Patel, Nitin R.: Exact logistic regression: Theory and examples. 95 StatMed 14 2143-2160

Mehta, Gobind P.: Kumar, Narinder; Gill, Amar Nath; — —: Selection procedures for scale parameters when the populations have a common known quantile. 94 Metron 52/3 171-196

Mei, Changlin: Fan, Jincheng; Wang, Nin; — —: The random weighting approximation for the distribution of the partial sum of a stationary linear series (Chinese). 95 AMaAPknC 17 65-75

Meier, Peter C.; Zünd, Richard E. (Auth); Hillyer, Martin (Rev): Review of "Statistical methods in analytic chemistry". 95 Technmcs 37 113-114

Meiri, A. Zvi: Francos, Joseph M.; — —; Porat, Boaz: A Wold-like decomposition of two-dimensional discrete homogeneous random fields. 95 AnlsApPr 5 248-260

Meiring, W.: Sampson, P. D.; Guttorp, P.; — —: Spatio-temporal analysis of regional ozone data for operational evaluation of an air quality model. 94 ASAProEnvr 46-55

Mejza, Iwona: Some split-plot designs in series of experiments. 94 ListyBio 31 3-16

Mejza, Stanisław: Obituary: Czesław Platt, 1925-1995 (Polish). 95 ListyBio 32 140-140

— —: On modelling of experiments in natural sciences. 94 ListyBio 31 79-100

Kachlicka, Danuta; — —: Cox's design in repeated row-column design with split plots and a control. 95 ListyBio 32 81-92

Pukelsheim, Friedrich (Auth); — — (Rev. Polish): Review of "Optimal design of experiments". 94 ListyBio 31 121-125

Meko, David; Graybill, Donald A.: Tree-ring reconstruction of Upper Gila River discharge. 95 WaterReB 31 605-616

Mel'nikov, A. V.: Shiryaev, A. N.; Kabanov, Yu. M.; Kramkov, D. O.; — —: Toward the theory of pricing of options of both European and American types. II. Continuous time.. 94 ThProbAp 39 61-102

Shiryaev, A. N.; Kabanov, Yu. M.; Kramkov, O. D.; — —: Toward the theory of pricing of options of both European and American types. I. Discrete time.. 94 ThProbAp 39 14-60

Mele, A.: Fornari, F.; — —: A two-factor model of the term structure with optimal filtering behaviour. 94 Statstca 54 293-312

Méléard, Sylvie: Graham, Carl; — —: Dynamic asymptotic results for a generalized star-shaped loss network. 95 AnlsApPr 5 666-680

Melenberg, B.: Charlier, E.; — —; van Soest, A. H. O.: A smoothed maximum score estimator for the binary choice panel data model with an application to labour force participation. 95 StNeerla 49 324-342

Melenberg, Bertrand: Adang, Pim; — —: Nonnegativity constraints and intratemporal uncertainty in a multi-good life-cycle model. 95 JApEcmtx 10 1-15

Melgaard, H.: Sadegh, P.; Holst, J.; Madsen, H.; — —: Experiment design for grey box identification. 95 IJAdCSgP 9 491-507

Melilli, Eugenio: A measure of dispersion based on Prohorov's distance. 95 Metron 53 115-134

Mellenbergh, Gideon J.: Conceptual notes on models for discrete polytomous item responses. 95 ApPsychM 19 91-100

Holland, P. W.; Wainer, H. (Ed); — — (Rev): Review of "Differential item functioning". 95 Psymtrka 60 621-623

Meloche, J.: Tactical aspects of search and rescue operations. 95 StPrLet 25 271-279

Melolidakis, Costis: Katehakis, Michael N.; — —: On the optimal maintenance of systems and control of arrivals in queues. 95 StocAnAp 13 137-164

Meloun, M.; Militky, J.; Forina, M. (Auth); Parczewski, Andrzej (Rev): Review of "Chemometrics for analytical chemistry, Volume 2: PC-aided regression and related methods". 95 Chemolab 29 326-327

Meloun, Milan; Militky, Jiri; Forina, Michele (Auth); LaLonde, Steven M. (Rev): Review of "Chemometrics for analytical chemistry Volume 1: PC-aided statistical data analysis". 95 Technmcs 37 114-115

Melton, Kim I.: Improving the presentation of business statistics. 94 ASAProQlPr 273-275

Ménardo, C.: Tenenhaus, M.; Gauchi, J.-P.; — —: PLS regression and applications (French). 95 RStApp 43/1 7-63

Mendelsohn, Susan L.; Walker, Chip; Bernstein, Hank: Conducting global consumer research: Considerations and pitfalls. 94 ASASrvyRsr 1196-1198

Mendenhall, William; Beaver, Robert J. (Auth); Halvorsen, Katherine Taylor (Rev): Review of "Introduction to probability and statistics (Ninth edition)". 95 AmerStat 49 398-398

Méndez-Ramírez, Ignacio: Pérez-Salvador, Blanca Rosa; — —: Estimating the population mean using supplementary dephased information. 95 CommStB 24 733-743

Mendonça, Elsa: Pires, Ana; Branco, Joao; Picado, Ana; — —: Metabolic responses in the assessment of pollution effects. 95 EnvrMtrc 6 155-163

Menéndez, M.; Morales, D.; Pardo, L.; Vajda, I.: Divergence-based estimation and testing of statistical models of classification. 95 JMultiAn 54 329-354

Menéndez, M. L.; Morales, D.; Pardo, L.; Salicrú, M.: Asymptotic behaviour and statistical applications of divergence measures in multinomial populations: A unified study. 95 StHefte 36 1-29

— —; Pardo, L.; Morales, D.; Salicrú, M.: Some statistical applications of (r,s)-directed divergences. 92 UtilMath 42 115-127

Menéndez, María Luisa: Pardo, Leandro; Salicrú, Miguel; — —; Morales, Domingo: A test for homogeneity of variances based on Shannon's entropy. 95 Metron 53 135-146

Meng, Cliff Y. K.: Cheng, Peter H.; — —: A new formula for tail probabilities of Dunnett's T with unequal sample sizes. 95 CommStA 24 523-532

Meng, Fan C.: Some further results on ranking the importance of system components. 95 RelEngSS 47 97-101

Meng, Qingguo: Meng, Qingyuan; Zhao, Bintao; — —: The essential theory of the quadratic forms of the Wishart matrix and its application (Chinese). 95 ChJApPSt 10 337-343

Meng, Qingyuan; Zhao, Bintao; Meng, Qingguo: The essential theory of the quadratic forms of the Wishart matrix and its application (Chinese). 95 ChJApPSt 10 337-343

Meng, Xiao-Li: Comment on "A characterization of the Poisson distribution and the probability of winning a game" (V48 p294-298). 95 AmerStat 49 401-401

— —; Zaslavsky, Alan M.: Single observation unbiased priors. 94 ASAProBays 46-51

Barnard, John; — —: Exploring cross-match estimators with multiply-imputed data sets. 94 ASASrvyRsr 894-899 (Disc: p906-907)

Pedlow, Steven; — —: Exploring hypothesis-testing procedures with multiply-imputed data under unequal fractions of missing information. 94 ASASrvyRsr 595-599

van Dyk, David A.; — —: Permuting CM steps within the ECM algorithm. 94 ASAProStCp 130-135

van Dyk, David A.; — —; Rubin, Donald B.: Maximum likelihood estimation via the ECM algorithm: Computing the asymptotic variance. 95 StSinica 5 55-75

Mengersen, K. L.; Tweedie, R. L.; Biggerstaff, B.: The impact of method choice on meta-analysis. 95 AstrlJSt 37 19-44

Tweedie, R. L.; — —: Meta-analytic approaches to dose-response relationships, with application in studies of lung cancer and exposure to environmental tobacco smoke. 95 StatMed 14 545-569 (Disc: p679-685)

Mengersen, Kerrie: Besag, Julian; Green, Peter; Higdon, David; — —: Bayesian computation and stochastic systems. 95 StatSci 10 3-41 (Disc: p41-66)

Besag, Julian; Green, Peter; Higdon, David; — —: Reply to comments on "Bayesian computation and stochastic systems". 95 StatSci 10 58-66

Menshikov, M. V.: Asymont, I. M.; Fayolle, G.; — —: Random walks in a quarter plane with zero drifts: Transience and recurrence. 95 JAppProb 32 941-955

Fayolle, G.; Malyshev, V. A.; — —: Topics in the constructive theory of countable Markov chains. 95 CambridgeUniv 169p

Fayolle, G.; Malyshev, V. A.; — — (Auth); Kozek, A. S. (Rev): Review of "Topics in the constructive theory of countable Markov chains". 96 ShrtBkRv 16/1 9-10

Mentré, France; Gomeni, Roberto: A two-step iterative algorithm for estimation in nonlinear mixed-effect models with an evaluation in population pharmacokinetics. 95 JBiophSt 5 141-158

Mentzer, John T.; Kahn, Kenneth B.: Forecasting technique familiarity, satisfaction, usage, and application. 95 JForcast 14 465-476

Menzefricke, Ulrich: On the performance of the Gibbs sampler for the multivariate normal distribution. 95 CommStA 24 191-213

Menzel, H. J.: Tiret, L.; Nicaud, V.; Ehnholm, C.; Havekes, L.; — —; Ducimetiere, P.; Cambien, F.: Inference of the strength of genotype-disease association from studies comparing offspring with and without parental history of disease. 93 AnHuGene 57 141-149

Mercante, Donald E.: Johnson, William D.; — —: Analyzing multivariate data in crossover designs using permutation tests. 94 ASAProBiop 244-249

Mercer, Robert L.: Nádas, Arthur; — —: Hidden Markov models and their estimation. 94 ASAProStCp 80-89

Meredith, M. P.: Stehman, S. V.; — —: Practical analysis of factorial experiments in forestry. 95 CanJForR 25 446-461

Meredith, William: Comment on "Testing for equivalence of measurement scales: Simple structure and metric invariance reconsidered". 95 MultBehR 30 89-94

— —: Comment on "Testing for equivalence of measurement scales: Simple structure and metric invariance reconsidered". 95 MultBehR 30 117-117

Mersch, Michael L.: Gbur, Philip M.; Kindred, Machell; — —: Master address file: Update methodology and quality improvement program. 94 ASASrvyRsr 748-753 (Disc: p754-756)

Mertens, Bart; Fearn, Tom; Thompson, Michael: The efficient cross-validation of principal components applied to principal component regression. 95 StatComp 5 227-235

Merz, Jon F.; Caulkins, Jonathan P.: Propensity to abuse - Propensity to murder?. 95 Chance 8/2 14-14

Merzbach, Ely: Ivanoff, B. Gail; — —: Stopping and set-indexed local martingales. 95 StocProc 57 83-98

Ivanoff, Gail B.; — —; Schiopu-Kratina, Ioana: Lattices of random sets and progressivity. 95 StPrLet 22 97-102

Mesenbrink, Peter; Lu, Jye-Chy: Model and transformation selection in fractionated experiments with complex aliasing. 94 ASAProQlPr 118-123

Meshkani, M. R.: Billard, L.; — —: Estimation of a stationary Markov chain. 95 JASA 90 307-315

Messelodi, S.: Brunelli, R.; — —: Robust estimation of correlation with applications to computer vision. 95 PattrnRc 28 833-841

Metcalfe, A. V.: Statistics in engineering. 95 ChapmnHall:Lnd 256p

Metcalfe, Andrew V.: — — (Auth); Larsen, Greg (Rev): Review of "Statistics in engineering: A practical approach". 95 Technmcs 37 455-455

— — (Auth); Prvan, Tania (Rev): Review of "Statistics in engineering: A practical approach". 95 AstrlJSt 37 128-128

Metha, Cyrus R.: Senchaudhuri, Pralay; — —; Patel, Nitin R.: Estimating exact p values by the method of control variates or Monte Carlo rescue. 95 JASA 90 640-648

Mettrick, N.: Betteley, G.; — —; Sweeney, E.; Wilson, D. (Auth); Shade, John (Rev): Review of "Using statistics in industry". 95 Ststcian 44 541-542

Meuwissen, T. H. E.; Engel, B.; van der Werf, J. H. J.: Maximizing selection efficiency for categorical traits. 95 JAnmlSci 73 1933-1939

Meyer, Bruce D.: Natural and quasi-experiments in economics. 95 JBES 13/2 151-161

Meyer, Denny: Disadvantages of moving averages. 95 NZndSt 30 9-22

Meyer, J.: Pickles, A.; Neale, M.; Simonoff, E.; Rutter, M.; Hewitt, J.; — —; Crouchley, R.; Silberg, J.; Eaves, L.: A simple method for censored age-of-onset data subject to recall bias: Mothers' reports of age of puberty in male twins. 94 BehavGen 24 457-468

Meyer, J. M.: Eaves, L. J.; Silberg, J. L.; Hewitt, K.; Rutter, M.; — —; Neale, M. C.; Pickles, A.: Analyzing twin resemblance in multisymptom data: Genetic applications of a latent class model for symptoms of conduct disorder in juvenile boys. 93 BehavGen 23 5-20

Meyer, Joanne: Eaves, Lindon; — —: Locating human quantitative trait loci: Guidelines for the selection of sibling pairs for genotyping. 94 BehavGen 24 443-456

Meyer, Joanne M.: Eaves, Lindon J.; Neale, Michael C.; — —: A model for comparative ratings in studies of within-family differences. 91 BehavGen 21 531-536

Meyer, Ruth K.; Nachtsheim, Christopher J.: The coordinate-exchange algorithm for constructing exact optimal experimental designs. 95 Technmcs 37 60-69

Meyre, Thierry; Werner, Wendelin: On the occupation times of cones by Brownian motion. 95 ProbTher 101 409-419

Mi, Jie: Bathtub failure rate and upside-down bathtub mean residual life. 95 IEEEReli 44 388-391

— —: Limiting behavior of some measures of system availability. 95 JAppProb 32 482-493

Chen, Ling; — —: Estimation of non-occurrence probability in periodic model. 95 CommStA 24 1493-1505

Miamee, Abol G.; Schröder, Bernd S. W.: On completeness of the spectral domain of harmonizable processes. 95 ProbTher 101 303-309

Mian, I. U. H.: Shoukri, M. M.; Martin, S. W.; — —: Maximum likelihood estimation of the kappa coefficient from models of matched binary responses. 95 StatMed 14 83-99

Mian, Ijaz U. H.; Laniel, Normand: Sample allocation for the Canadian Labour Force Survey. 94 ASASrvyRsr 913-918

Miao, Baiqi; Zhang, Shuguang: The exponential convergence rates of the projection residue of U-statistics and their application (Chinese). 95 ChJApPSt 10 253-264

Miao, Y.: Kageyama, S.; — —: A construction of resolvable designs. 94 UtilMath 46 49-54

Miao, Ying: Furino, Steven; Greig, Malcolm; — —; Yin, Jianxing: Resolvable BIBDs with block size 8 and index 7. 94 JComDsgn 2 101-116

Michael, R. T.: Laumann, E. O.; Gagnon, J. H.; — —; Michaels, S. (Auth); Wasserman, Stanley (Rev): Review of "The social organization of sexuality". 95 Chance 8/2 40-42

Michaels, S.: Laumann, E. O.; Gagnon, J. H.; Michael, R. T.; — — (Auth); Wasserman, Stanley (Rev): Review of "The social organization of sexuality". 95 Chance 8/2 40-42

Michalewicz, Z.: — — (Ed); Lenz, H.-J. (Rev): Review of "Statistical and scientific databases". 95 AllgeStA 79 103-105

Michalewicz, Zbigniew: — — (Ed); Lenz, H.-J. (Rev): Review of "Statistical and scientific databases". 95 CmpStDA 19 479-481

Michaud, Sylvie; Lavallée, Pierre: Comment on "Sample design and weighting research in panel surveys". 94 ASASrvyRsr 680-682

Binder, David A.; — —; Poirier, Claude: Model-based reweighting for nonresponse adjustment. 94 ASASrvyRsr 989-994

Grondin, Chantal; — —: Data quality of income data using computer assisted interviewing: The experience of the Canadian survey of labour and income dynamics. 94 ASASrvyRsr 839-844 (Disc: p861-862)

Michie, D.; Spiegelhalter, D. J.; Taylor, C. C. (Ed); Fulkerson, Bill (Rev): Review of "Machine learning, neural and statistical classification". 95 Technmcs 37 459-459

Michna, Zbigniew; Rychlik, Igor: The expected number of level crossings for certain symmetric α-stable processes. 95 CommStSM 11 1-19

Miclo, Laurent: Remarks about the ergodicity of simulated annealing algorithms on a graph (French). 95 StocProc 58 329-360

Mielke, Paul W., Jr: Zelterman, Daniel; Chan, Ivan Siu-Fung; — —: Exact tests of significance in higher dimensional tables. 95 AmerStat 49 357-361

Mielniczuk, Jan: Csörgő, Sándor; — —: Density estimation under long-range dependence. 95 AnlsStat 23 990-999

Csörgő, Sándor; — —: Nonparametric regression under long-range dependent normal errors. 95 AnlsStat 23 1000-1014

Csörgő, Sándor; — —: Distant long-range dependent sums and regression estimation. 95 StocProc 59 143-155

Ćwik, Jan; — —: Nonparametric rank discrimination method. 95 CmpStDA 19 59-74

Gijbels, Irène; — —: Asymptotic properties of kernel estimators of the Radon-Nikodym derivative with applications to discriminant analysis. 95 StSinica 5 261-278

Hössjer, Ola; — —: Delta method for long-range dependent observations. 95 JNnparSt 5 75-82

Miescke, K. J.; Shi, Dylan: Optimum Markov replacement policies for using the most reliable components. 95 JStPlInf 45 331-346

Mijnheer, Joop: On the influence of the extremal term in a (simple) U-statistic. 95 ThPrMaSt 50 103-106

Mikami, T.: Correction to "Limit theorems on the exit problems for small random perturbations of dynamical systems I" (V46 p79-106). 95 Stochatx 52 323-323

Mikkelson, Gordon: Chun, Young I.; Devens, Richard; — —; Yule, James: Designing a customer satisfaction survey. 94 ASAProQlPr 204-209

Mikl, Jaromir: Muse, Alison G.; — —; Smith, Perry F.: Evaluating the quality of anonymous record linkage using deterministic procedures with the New York State AIDS registry and a hospital discharge file. 95 StatMed 14 499-509 (Disc: p687-689)

Mikosch, T.: Correction to "Functional limit theorems for random quadratic forms" (V37 p81-98). 95 StocProc 59 353-353

Klüppelberg, C.; — —: Delay in claim settlement and ruin probability approximations. 95 ScanActJ 154-168

Mikosch, Thomas; Gadrich, Tamar; Klüppelberg, Claudia; Adler, Robert J.: Parameter estimation for ARMA models with infinite variance innovations. 95 AnlsStat 23 305-326

Klüppelberg, Claudia; — —: Explosive Poisson shot noise processes with applications to risk reserves. 95 Bernoull 1 125-147

Falk, Michael; Hüsler, Jurg; Reiss, Rolf-Dieter (Auth); — — (Rev): Review of "Law of small numbers: Extremes and rare events". 95 JASA 90 1133-1134

Samorodnitsky, Gennady; Taqqu, Murad S. (Auth); — — (Rev): Review of "Stable non-Gaussian random processes: Stochastic models with infinite variance". 95 JASA 90 805-806

Mikulak, Raymond J.: Beauregard, Michael R.; — —; Olson, Barbara A. (Auth); Belunis, Nancy (Rev): Review of "A Practical guide to statistical quality improvement: Opening up the statistical toolbox". 95 Technmcs 37 233-234

Beauregard, Michael R.; — —; Olson, Barbara A. (Auth); Richards, John D. (Rev): Review of "A practical guide to statistical quality improvement: Opening up the statistical toolbox". 95 JQualTek 27 177-178

Mikulecky, M.: Komornik, J.; — —; Beno, M.: Additivity test for composition of binomial effects in chromosome aberrations after radiation injury. 95 StatMed 14 1941-1947

Mikulski, Piotr W.: Monsour, Michael J.; — —: Moments of the limiting distribution for the boundary case in the first order autoregressive process. 94 AJMMgSci 14 327-347

Milan, Luis; Whittaker, Joe: Application of the parametric bootstrap to models that incorporate a singular value decomposition. 95 ApplStat 44 31-49

Milanese, M.: Kacewicz, B.; — —: Optimality properties in finite sample l_1 identification with bounded noise. 95 IJAdCSgP 9 87–96

Milbrodt, Hartmut: Andersen, P. K.; Borgan, O.; Gill, R. D.; Keiding, N. (Auth); — — (Rev. German): Review of "Statistical models based on counting processes". 95 AllgeStA 79 221–221

Miles, Matthew B.: Weitzman, Eben A.; — —: Computer programs for qualitative data analysis. A software sourcebook. 95 SagePubl:CA:Lnd 372p

Miles, R. E.: A heuristic proof of a long-standing conjecture of D. G. Kendall concerning the shapes of certain large random polygons. 95 AdvAppPr 27 397–417

Militky, J.: Meloun, M.; — —; Forina, M. (Auth); Parczewski, Andrzej (Rev): Review of "Chemometrics for analytical chemistry, Volume 2: PC-aided regression and related methods". 95 Chemolab 29 326–327

Militky, Jiri: Meloun, Milan; — —; Forina, Michele (Auth); LaLonde, Steven M. (Rev): Review of "Chemometrics for analytical chemistry Volume 1: PC-aided statistical data analysis". 95 Technmcs 37 114–115

Milito, Rodolfo A.: On arrival driven queueing models: Admission control, traffic policing, abandonments, and correlated arrivals. 95 AplMaCmp 70 77–94

Millar, Alexander B.; Gruenewald, Paul J.: The development of spatial models for community evaluation. 94 ASAProSoSt 227–231

Millar, David B.: A new metric for data clustering. 94 ASAProStCp 251–256

Millar, P. W.: Beran, R.; — —: Minimum distance estimation in random coefficient regression models. 94 AnlsStat 22 1976–1992

Miller, A. J.: — — (Auth); Droge, B. (Rev): Review of "Subset selection in regression". 95 MaOpfStS 26 179–179

Miller, Charles E.: The use of chemometric techniques in process analytical method development and operation. 95 Chemolab 30 11–22

Miller, Curtis: A CLT for the periodograms of a ρ^*-mixing random field. 95 StocProc 60 313–330

Miller, D.: MacNeill, I. B.; Elwood, J. M.; — —; Mao, Y.: Trends in mortality from melanoma in Canada and prediction of future rates. 95 StatMed 14 821–839

Canavos, G.; — — (Auth); Ziegel, Eric R. (Rev): Review of "Introduction to modern business statistics". 95 Technmcs 37 239–240

Miller, Don M.: Canavos, George C.; — — (Auth); Delaney, Nancy Jo (Rev): Review of "Modern business statistics". 95 AmerStat 49 397–397

Miller, John P.; Newbold, Paul: Uncertainty about the persistence of economic shocks. 95 JBES 13 435–440

— —; Newbold, Paul: A generalized variance ratio test of ARIMA$(p, 1, q)$ model specification. 95 JTimSrAn 16 403–413

Miller, Michael B.: Coefficient alpha: A basic introduction from the perspectives of classical test theory and structural equation modeling. 95 StrEqMod 2 255–273

The MathWorks Inc (Auth); — — (Rev): Review of "MATLAB Version 4.2". 95 StrEqMod 2 175–185

Miller, Michael E.: Analysing categorical responses obtained from large clusters. 95 ApplStat 44 173–186

Miller, Perry L.: Goradia, Tushar M.; Lange, Kenneth; — —; Nadkarni, Prakash M.: Fast computation of genetic likelihoods on human pedigree data. 92 HumHrdty 42 42–62

Miller, R.: Cryer, J.; — — (Auth); Ziegel, Eric R. (Rev): Review of "Statistics for business: Data analysis and modelling". 95 Technmcs 37 239–240

Miller, Stephen: Lent, Janice; — —; Cantwell, Patrick: Composite weights for the Current Population Survey. 94 ASASrvyRsr 867–872

Miller, Sylvia: Kelly, Mary; — —: The improvement of the Employment Cost Index Survey: A total quality approach within the Bureau of Labor Statistics. 94 ASAProQlPr 210–214

Miller, Thomas W.: Model selection in tree-structured regression. 94 ASAProStCp 158–163

Miller, W. Greg: Chinchilli, Vernon M.; Esinhart, James D.; — —: Partial likelihood analysis of within-unit variances in repeated measurement experiments. 95 Biomtrcs 51 205–216

Millet, Annie; Smoleński, Włodzmierz: Small perturbations of Gaussian regressors. 95 StPrLet 24 21–31

Bally, Vlad; — —; Sanz-Solé, Marta: Approximation and support theorem in Hölder norm for parabolic stochastic partial differential equations. 95 AnlsProb 23 178–222

Milligan, Glenn W.: Cheng, Richard; — —: Hierarchical clustering algorithms with influence detection. 95 EdPsMeas 55 237–244

Cheng, Richard; — —: Mapping influence regions in hierarchical clustering. 95 MultBehR 30 547–576

Milliken, George A.: Comment on "Multiple hypothesis tests in multiple investigations". 95 StatMed 14 701–702

Millman, Zeeva: Kanetkar, Vinay; Evans, Martin G.; Everell, Shirley Anne; Irvine, Diane; — —: The effect of scale changes on meta-analysis of multiplicative and main effects models. 95 EdPsMeas 55 206–224

Millner, P. A.: Mardia, K. V.; Dryden, I. L.; Hurn, M. A.; Li, Q.; — —; Dickson, R. A.: Familial spinal shape. 94 JAppStat 21 623–641

Mills, G. Fernández: Lana, X.; — —: Minimum sample size for synoptic weather type classification. Application to winter period data recorded on the Catalan coast (North-east Spain). 94 JClimlgy 14 1051–1060

Mills, Terence C.: Modelling skewness and kurtosis in the London Stock Exchange FT-SE index return distributions. 95 Ststcian 44 323–332

Coutts, J. Andrew; — —; Roberts, Jennifer: Testing cumulative prediction errors in event study methodology. 95 JForcast 14 107–115

— — (Auth); Lee, Bong-Soo (Rev): Review of "The econometric modelling of financial time series". 95 JFinance 50 387–390

Mills, W. H.: Abel, R. J. R.; — —: Some new BIBDs with $k = 6$ and $\lambda = 1$. 95 JComDsgn 3 381–391

Millsap, Roger E.: Measurement invariance, predictive invariance, and the duality paradox. 95 MultBehR 30 577–605

Jackson, J. Edward (Auth); — — (Rev): Review of "A user's guide to principal components". 95 JEdStat 20 105–107

Milstein, G. N.: Numerical integration of stochastic differential equations. 95 Kluwer:MA:Hllnd 184p

Mimoso Paulino, Carlos Daniel; de Braganca Pereira, Carlos Alberto: On identifiability of parametric statistical models. 94 JItalStS 3 125–151

— —; Pereira, Carlos Alberto De Bragança: Bayesian methods for categorical data under informative general censoring. 95 Biomtrka 82 439–446

Min, Chung-ki: Forecasting the adoptions of new consumer durable goods. 94 ASAProBays 262–265

Zellner, Arnold; — —: Gibbs sampler convergence criteria. 95 JASA 90 921–927

Minami, Mihoko: Ogawa, Naoko; — —; Sibuya, Masaaki: Characteristics of the linear smoothing spline (Japanese). 95 JpJAppSt 24 27–41

Mininni, Rosamaria: Capasso, Vincenzo; De Giosa, Marcello; — —: Asymptotic properties of the maximum likelihood estimators of parameters of a spatial counting process modelling crystallization of polymers. 95 StocAnAp 13 279–294

De Giosa, Marcello; — —: On the Doléans function of set-indexed submartingales. 95 StPrLet 24 71–75

Minitab Inc: — — (Auth); Lachenbruch, Peter A. (Rev): Review of "Minitab: Release 10 for Windows". 95 Chance 8/1 58–61

— — (Auth); Lavine, Barry K. (Rev): Review of "SCAN: Software for chemometric analysis (Release 1 for Windows)". 95 Chemolab 30 253–255

— — (Auth); Lovie, Sandy (Rev): Review of "MINITAB Release 10 for Windows". 95 BJMaStPs 48 205–207

Minkin, Salomon; Chow, Andrew: Assessing sensitivity in suspension to cytosine arabinoside: Statistical analysis, associations with clinical outcome and experimental design. 95 StatMed 14 1081–1095 (Disc: p1097-1097)

Gao, Min; — —: Nonlinear mixed effect model for estimating population characteristics of drug combination effect. 94 ASAProBiop 160–165

Minkkinen, Pentti: Estimation of variance components from the results of interlaboratory comparisons. 95 Chemolab 29 263–270

Minkus, Gail W.: Hillis, Margaret A.; — —; DeCaluwe, Craig L.; Myers, Michael N.; Schafer, Edward T.: An improved approach to product quality through testing. 95 ATTTJ 74/3 40–48

Minlos, A.: Zhizhina, E. A.; — —: A local limit theorem for nonhomogeneous random walk on a lattice. 94 ThProbAp 39 490–503

Minor, James M.; Namini, Hamid; Cruickshank, Scott: From here to there (dosing to efficacy) with neural nets. 94 ASAProBiop 334–339

Minstrell, Jim: Clarkson, Douglas B.; Donnell, Deborah; — —; Hunt, Earl; Madigan, David; Traynor, Charles: VITAL: An intelligent tutoring system for statistics. 94 ASAProStEd 88–93

Mira, S. I.: Ghazal, G. A.; — —: The exact finite sample properties of the estimators of coefficients in mixed dummy-variable error components model. 95 EgpStJ 39 26–50

Mirakhmedov, Sh. A.: Ivchenko, G. I.; — —: Large deviations and intermediate efficiency of decomposable statistics in a multinomial scheme. 95 MaMethSt 4 294–311

Miranda, Mario J.; Schnitkey, Gary D.: An empirical model of asset replacement in dairy production. 95 JApEcmtx 10 41–55S

Mirkin, Boris; Arabie, Phipps; Hubert, Lawrence J.: Additive two-mode clustering: The error-variance approach revisited. 95 JClass 12 243–263

Miroshin, R. N.: The power terms of the asymptotic series for moments of the number of zeros of the Wong process on a large time interval (Russian). 95 VLenUM /2 36–42

Mirowski, Philip: Three ways to think about testing in econometrics. 95 JEconmtx 67 25–46

— — (Ed); Stigler, Stephen M. (Rev): Review of "Edgeworth on chance, economic hazard, and statistics". 95 JASA 90 803–804

Mishkin, Frederic S.: Nonstationarity of regressors and tests on real-interest-rate behavior. 95 JBES 13 47–51

Mishra, M. N.; Bishwal, J. P. N.: Approximate maximum likelihood estimation for diffusion processes from discrete observations. 95 Stochatx 52 1–13

— —; Rao, B. L. S. Prakasa: On L_1-consistency of kernel-type density estimator for stationary Markov processes. 95 CommStA 24 581–591

Mishra, Satya N.: Selukar, Rajesh S.; Shah, Arvind K.; — —: A simple approach to simultaneous comparison of normal means. 95 JStCmpSm 52 174–177

Misiewicz, Jolanta K.: Infinite divisibility of sub-stable processes. Part I. Geometry of sub-spaces of L_α-space. 95 StocProc 56 100–116

Miskulin, John D.: Keats, J. Bert; — —; Runger, George C.: Statistical process control scheme design. 95 JQualTek 27 214–225

Mislevy, Robert: Wainer, Howard; Dorans, Neil J.; Flaugher, Ronald; Green, Bert F.; — —; Steinberg, Lynne; Thissen, David (Auth); Andrich, David (Rev): Review of "Computerized adaptive testing: A primer". 95 Psymtrka 60 615-620

Mismetti, Patrick: Girard, Pascal; Laporte-Simitsidis, Silvy; — —; Decousus, Hervé; Boissel, Jean-Pierre: Influence of confounding factors on designs for dose-effect relationship estimates. 95 StatMed 14 987-1005 (Disc: p1007-1008)

Misra, Neeraj; Dhariyal, Ishwari D.: Some inadmissibility results for estimating ordered uniform scale parameters. 95 CommStA 24 675-685

— —; Kumar, Narinder: Selecting the best exponential population when the scale parameters are bounded. 95 MaOpfStS 27 151-158

Vijayasree, G.; — —; Singh, Harshinder: Componentwise estimation of ordered parameters of k (≥ 2) exponential populations. 95 AnInStMa 47 287-307

Misra, Sheela: Singh, R. Karan; — —; Pandey, S. K.: A generalized class of estimators in linear regression models with multivariate-t distributed error. 95 StPrLet 23 171-178

Mita, Haruyoshi: Tail probability of the sequential maximum likelihood estimator for the exponential class. 95 JJapnStS 25 173-182

Mital, D. P.: Suganthan, P. N.; Teoh, E. K.; — —: Pattern recognition by graph matching using the Potts MFT neural networks. 95 PattrnRc 28 997-1009

Mitchell, A. F. S.: Robert, C. P. (Auth); — — (Rev): Review of "The Bayesian choice: A decision-theoretic motivation". 95 ShrtBkRv 15/2 27-27

Mitchell, Ann F. S.: Correction to "A note on posterior moments for a normal mean with double-exponential prior" (V56 p605-610). 95 JRSS-B 57 471-471

Mitchell, Melanie: Genetic algorithms: An overview. 95 Complxty 1 31-39

Mitchell, Sam: Toward a defensible bootstrapping. 95 PhilSci 62 241-260

Mitchell, Susan; Mooney, Geraldine; Shettle, Carolyn: The impact of using priority mail in a mixed mode survey. 94 ASAProGovt 88-92

Mitchell, Toby J.; Morris, Max D.; Ylvisaker, Donald: Two-level fractional factorials and Bayesian prediction. 95 StSinica 5 559-573

Morris, Max D.; — —: Exploratory designs for computational experiments. 95 JStPlInf 43 381-402

Mitra, Murari; Basu, Sujit K.: Change point estimation in non-monotonic aging models. 95 AnInStMa 47 483-491

Mitra, S.: Boundary condition sets limits on Brass's model life table parameters. 94 ASAProSoSt 55-59

Mitra, Sushmita; Kuncheva, Ludmila I.: Improving classification performance using fuzzy MLP and two-level selective partitioning of the feature space. 95 FuzzySet 70 1-13

Mitrou, Nikolas M.: Kontovasilis, Kimon P.; — —: Markov-modulated traffic with nearly complete decomposability characteristics and associated fluid queueing models. 95 AdvAppPr 27 1144-1185

Mitsuoka, Motohiro: Yamamoto, Sumiyasu; Fujii, Yoshio; — —: Three-symbol orthogonal arrays of strength 2 and index 2 having maximal constraints: Computational study. 93 JCISS 18 329-342

Mittag, H. J.: — — (Auth); Arnold, B. F. (Rev): Review of "Quality control charts" (German). 95 StHefte 36 96-96

Mittag, H.-J.: — — (Auth); Gohout, W. (Rev. German): Review of "Quality control charts" (German). 95 AllgeStA 79 105-106

— —; Rinne, H. (Auth); Jain, Kamlesh (Rev): Review of "Statistical methods of quality assurance". 95 Technmcs 37 234-234

— —; Rinne, H. (Auth); Nelson, Lloyd S. (Rev): Review of "Statistical methods of quality assurance". 95 JQualTek 27 95-96

Mittag, Hans-Joachim: Measurement error effect on control chart performance. 95 ASQCTekCon 66-73

Mitterreiter, Ilse: Kleinschmidt, Peter; — —: Improved chromosome classification using monotonic functions of Mahalanobis distance and the transportation method. 94 ZeitOpRs 40 305-323

Mittleman, Murray A.: Estimation of exposure prevalence in a population at risk using data from cases and an external estimate of the relative risk. 95 Epidmlgy 6 551-553

— —; Maclure, Malcolm; Robins, James M.: Control sampling strategies for case-crossover studies: An assessment of relative efficiency. 95 AJEpidm 142 91-98

Miyakawa, Masami: An interpretation of the interaction terms in Mandel's ANOVA model from Hirotsu's interaction elements. 93 RStAJp 40/1 1-9

Kawasaki, Hidemi; — —: An ANOVA model and interaction element for three-way layout without replication. 95 JJapnStS 25 143-150

Miyano, Hisao: Hierarchical clustering method using probabilistic similarity measure. 95 Behvmtrk 22 145-153

Miyazaki, Haruo: Hoshino, Naoto; — —; Seki, Yoichi: On the level probabilities for useful partially ordered alternatives in the analysis of variance. 95 CommStA 24 2059-2071

Miyazawa, Masakiyo: Note on generalizations of Mecke's formula and extensions of $H = \lambda G$. 95 JAppProb 32 105-122

— —; Schassberger, Rolf; Schmidt, Volker: On the structure of an insensitive generalized semi-Markov process with reallocation and point-process input. 95 AdvAppPr 27 203-225

Takahashi, Yoshitaka; — —: Relationship between queue length and waiting time in a stationary discrete-time queue. 95 CommStSM 11 249-271

Mizera, Ivan: Freedman, D.; Pisani, R.; Purves, R.; Adhikari, A. (Auth); — — (Rev): Review of "Statistics (Second edition)". 95 JAppStat 22 321-322

Mizon, Grayham E.: A simple message for autocorrelation correctors: Don't. 95 JEconmtx 69 267-288

Mocan, H. Naci; Azad, Sam: Accuracy and rationality of state General Fund Revenue forecasts: Evidence from panel data. 95 IntJFor 11 417-427

Mockus, A.: Lavine, M.; — —: A nonparametric Bayes method for isotonic regression. 95 JStPlInf 46 235-248

Mockus, Audris: Eddy, William F.; — —: An interactive image index. 94 ASAProStGr 1-6

Modarres, R.: Hurley, C.; — —: Low-storage quantile estimation. 95 CmpStQ 10 311-325

Moeanaddin, Rahim: Computational aspects of linear and non-linear time series analysis. 92 ProIrStCl 123-155

Moebus, M.: Koel, M.; Kaljurand, M.; Lochmüller, C. H.; — —: Factor analytical resolution of simultaneous, dynamic thermal process in thermochromatography of oil shale. 95 Chemolab 30 173-178

Moeller, John F.: Prescription drug coverage and the demand for outpatient prescription drugs by the elderly. 94 ASAProSoSt 147-152

— —; Mathiowetz, Nancy A.: Problems of screening for poverty status. 94 JOffStat 10 327-337

Moeschberger, M. L.; Klein, John P.: Statistical methods for dependent competing risks. 95 LiDaAnal 1 195-204

Wang, S. T.; Klein, John P.; — —: Semi-parametric estimation of covariate effects using the positive stable frailty model. 95 AplStMDA 11 121-133

Moffatt, Peter G.: Grouped Poisson regression models: Theory and an application to public house visit frequency. 95 CommStA 24 2779-2796

Moh, Jenlong; Shih, Frank Y.: A general purpose model for image operations based on multilayer perceptrons. 95 PattrnRc 28 1083-1090

Mohadjer, Leyla: Burke, John; — —; Green, Jim; Waksberg, Joseph; Kirsch, Irwin S.; Kolstad, Andrew: Composite estimation in national and state surveys. 94 ASASrvyRsr 873-878

Mohadjer, Leyla K.: Khare, Meena; Ezzati-Rice, Trena M.; — —; Waksberg, Joseph: An evaluation of nonresponse bias in NHANES III (1988-91). 94 ASASrvyRsr 949-954

Mohammad-Djafari, Ali: Brette, Stéphane; Idier, Jérôme; — —: Scale invariant Markov models for linear inverse problems. 94 ASAProBays 266-270

Mohan, N. R.: Janardan, K. G.; Vasudeva, R.; — —: On some special birth processes in biology. 95 CommStA 24 1033-1038

Mohanty, S.; Sahoo, J.: A note on improving the ratio method of estimation through linear transformation using certain known population parameters. 95 SankhyaB 57 93-102

Sahoo, J.; Sahoo, L. N.; — —: Empirical comparison of some regression and regression-type strategies. 95 StHefte 36 337-347

Mohanty, Sunil: Aggarwal, Raj; — —; Song, Frank: Are survey forecasts of macroeconomic variables rational?. 95 JBusUChi 68 99-119

Mohanty, Suprasana: Sahoo, Jagatananda; Sahoo, Lokanath; — —: Unequal probability sampling using a transformed auxiliary variable. 94 Metron 52/1 71-83

Mohie EL-Din, M. M.; Sultan, K. S.: Recurrence relations for expectations of functions of order statistics for doubly truncated distributions and their applications. 95 CommStA 24 997-1010

Mohl, Chris: Laniel, Normand; — —: Analysis of urban cluster size in the Canadian Labour Force Survey. 94 ASASrvyRsr 931-936

Moitra, Soumyo: Weerahandi, Samaradasa; — —: Using survey data to predict adoption and switching for services. 95 JMarRsr 32 85-96

Mok, Magdalena M.-C.: McDonald, Roderick P.; — —: Goodness-of-fit in item response models. 95 MultBehR 30 23-40

Mokliatchouk, O. M.: Kozachenko, Yu. V.; — —: Pre-Gaussian random vectors and their applications. 95 ThPrMaSt 50 89-98

Moklyachuk, M. P.: Minimax interpolation of random fields that are time homogeneous and isotropic on a sphere. 95 ThPrMaSt 50 107-115

Mokros, Jan; Russell, Susan Jo: Children's concepts of average and representativeness. 95 JRsrMaEd 26 20-39

Molchanov, Ilya S.: Statistics of the Boolean Model: From the estimation of means to the estimation of distributions. 95 AdvAppPr 27 63-86

— —; Omey, Edward; Kozarovitzky, Eugene: An elementary renewal theorem for random compact convex sets. 95 AdvAppPr 27 931-942

Molchanov, S. A.: Carmona, R. A.; — —: Stationary parabolic Anderson model and intermittency. 95 ProbTher 102 433-453

Mole, N.; Anderson, C. W.; Nadarajah, S.; Wright, C.: A generalized Pareto distribution model for high concentrations in short-range atmospheric dispersion. 95 EnvrMtrc 6 595-606

Mole, Nils: The α-β model for concentration moments in turbulent flows. 95 EnvrMtrc 6 559-569

Clarke, Liz; — —: Modelling the evolution of moments of contaminant concentration in turbulent flows. 95 EnvrMtrc 6 607-617

Molenaar, I. W.: Fischer, G. H.; — —: Rasch models: Foundations, recent developments, and applications. 95 Springer:Brln:NY 436p

Molenaar, J.: Visser, H.; — —: Trend estimation and regression analysis in climatological time series: An application of structural time series models and the Kalman filter. 95 JClimate 8 969-979

Molenaar, Peter C. M.: Dolan, Conor V.; — —; Boomsma, Dorret I.: Simultaneous genetic analysis of means and covariance structure: Pearson-Lawley selection rules. 94 BehavGen 24 17-24

Hershberger, Scott L.; Corneal, Sherry E.; — —: Dynamic factor analysis: An application to emotional response patterns underlying daughter/father and stepdaughter/stepfather relationships. 95 StrEqMod 2 31-52

Molenberghs, Geert: Fitzmaurice, Garrett M.; — —; Lipsitz, Stuart R.: Regression models for longitudinal binary responses with informative drop-outs. 95 JRSS-B 57 691-704

Molina, R.: On the hierarchical Bayesian approach to image restoration: Applications to astronomical images. 94 IEEEPAMI 16 1122-1128

Molińska, Anna; Moliński, Krzysztof: Improvement of ANOVA estimators of variance components in block designs with random treatments. 95 StPrLet 22 281-285

Moliński, K.: Hyánek, J.; — —; Szwaczkowski, T.: Estimation of parameters in linear model with the presence of selection. 95 ListyBio 32 51-60

Moliński, Krzysztof: Molińska, Anna; — —: Improvement of ANOVA estimators of variance components in block designs with random treatments. 95 StPrLet 22 281-285

Møller, Christian Max: A counting process approach to stochastic interest. 95 InsrneME 17 181-192

— —: Stochastic differential equations for ruin probabilities. 95 JAppProb 32 74-89

Møller, Jesper: Generation of Johnson-Mehl crystals and comparative analysis of models for random nucleation. 95 AdvAppPr 27 367-383

Högmander, Harri; — —: Estimating distribution maps from atlas data using methods of statistical image analysis. 95 Biomtrcs 51 393-404

Møller, Susanne: An extension of the continual reassessment methods using a preliminary up-and-down design in a dose finding study in cancer patients, in order to investigate a greater range of doses. 95 StatMed 14 911-922 (Disc: p923-923)

Mollison, Denis: — — (Ed): Epidemic models: Their structure and relation to data. 95 CambridgeUniv 441p

Molloy, Kirsten Jane: Maeder, Marcel; — —: Lifetime resolved fluorescence detection in chromatography. Global analysis of emission decay surfaces. 95 Chemolab 28 99-107

Monahan, John: Genz, Alan; — —: Random integration rules for statistical computation. 94 ASAProStCp 182-187

Monat, Pascale; Stricker, Christophe: Föllmer-Schweizer decomposition and mean-variance hedging for general claims. 95 AnlsProb 23 605-628

Moncaup, S.; Sarda, P.; Vieu, P.: Nonparametric estimators of an intensity function applied to meteorological data (French). 95 RStApp 43/4 77-88

Mondani, R.: Giorgi, G. M.; — —: Sampling distribution of the Bonferroni inequality index from exponential population. 95 SankhyaB 57 10-18

Mondani, Riccardo: Giorgi, Giovanni Maria; — —: The exact sampling distribution of the Bonferroni concentration index. 94 Metron 52/3 5-41

Monfort, Alain: Gourieroux, Christian; — — (Auth); Vuong, Quang (Transl): Statistics and econometric models. Volumes 1 and 2. 95 CambridgeUniv 1062p

Gouriéroux, Christian; — —: Testing, encompassing, and simulating dynamic econometric models. 95 EconTher 11 195-228

van Dijk, Herman K.; — —; Brown, Bryan W. (Ed): Econometric inference using simulation techniques. 95 Wiley:NY:UK 256p

Monfort, Christine: Richardson, Sylvia; — —; Green, Martyn; Draper, Gerald; Muirhead, Colin: Spatial variation of natural radiation and childhood leukaemia incidence in Great Britain. 95 StatMed 14 2487-2501

Monkevich, Nicholas S.: Gregoire, Timothy G.; — —: The reflection method of line intercept sampling to eliminate boundary bias. 94 EnvEcoSt 1 219-226

Monod, H.: Bailey, R. A.; — —; Morgan, J. P.: Construction and optimality of affine-resolvable designs. 95 Biomtrka 82 187-200

Bailey, R. A.; Azaïs, J.-M.; — —: Are neighbour methods preferable to analysis of variance for completely systematic designs? 'Silly designs are silly!'. 95 Biomtrka 82 655-659

Monrad, Ditlev; Rootzén, Holger: Small values of Gaussian processes and functional laws of the iterated logarithm. 95 ProbTher 101 173-192

Monsees, Geoffrey A.: Carroll, Steven S.; — —: The effect of scaling input variables on the direction of steepest ascent. 94 ASAProQlPr 267-269

Monsell, Brian C.: Caldwell, Carol V.; — —; Piesto, Nancy A.; Shimberg, Michael Z.: Using seasonal adjustment models to correct for panel imbalance in the Monthly Wholesale Trade Survey. 94 ASAProBuEc 83-88

Monsour, Michael J.; Mikulski, Piotr W.: Moments of the limiting distribution for the boundary case in the first order autoregressive process. 94 AJMMgSci 14 327-347

Montanari, Gian Carlo; Cavallini, Andrea; Tommasini, Laura; Cacciari, Mario; Contin, Alfredo: Comparison of random generators for Monte Carlo estimates of Weibull parameters. 95 Metron 53 55-77

Montaño, Rosa: Crivelli, Ana; Firinguetti, Luis; — —; Muñóz, Margarita: Confidence intervals in ridge regression by bootstrapping the dependent variable: A simulation study. 95 CommStB 24 631-652

Montero, J.; Pearman, A. D.; Tejada, J.: Fuzzy multicriteria decision support for budget allocation in the transport sector. 95 Top 3 47-68

Montes, F.: Ferrándiz, J.; — —: On statistical inference for random closed sets in compact subsets of R^d. 95 AdvAppPr 27 344-349

Montgomery, Alan L.: Liu, Lon-Mu; Chan, Ki-Kan; — —; Muller, Mervin E.: A system-independent graphical user interface for statistical software. 95 CmpStDA 19 23-44

Montgomery, D. C.: Torng, J. C.-C.; Cochran, J. K.; — —; Lawrence, F. P.: Correction to "Implementing statistically constrained economic EWMA control charts" (V27 p257-264). 95 JQualTek 27 402-402

— —; Runger, G. C. (Auth); Snell, Joyce (Rev): Review of "Applied statistics and probability for engineers". 95 JRSS-A 158 355-355

Montgomery, Douglas C.; Torng, James C.-C.; Cochran, Jeffery K.; Lawrence, Frederick P.: Statistically constrained economic design of the EWMA control chart. 95 JQualTek 27 250-256

Lowry, Cynthia A.; — —: A review of multivariate control charts. 95 IIETrn 27 800-810

Myers, Raymond H.; — —: Response surface methodology: Process and product optimization using designed experiments. 95 Wiley:NY:UK 720p

Prabhu, Sharad S.; — —; Runger, George C.: A design tool to evaluate average time to signal properties of adaptive $\bar{X}$ charts. 95 JQualTek 27 74-83

Torng, James C.-C.; Cochran, Jeffery K.; — —; Lawrence, Frederick P.: Implementing statistically constrained economic EWMA control charts. 95 JQualTek 27 257-264

— —; Runger, George; Alexander, Melvin T. (Rev): Review of "Applied statistics and probability for engineers". 95 Technmcs 37 455-457

Montgomery-Smith, S. J.: de la Peña, V. H.; — —; Szulga, Jerzy: Contraction and decoupling inequalities for multilinear forms and U-statistics. 94 AnlsProb 22 1745-1765

de la Peña, Victor H.; — —: Decoupling inequalities for the tail probabilities of multivariate U-statistics. 95 AnlsProb 23 806-816

Montgomery-Smith, Stephen: Kalton, Nigel; Saab, Elias; — — (Ed): Interaction between functional analysis, harmonic analysis, and probability. 95 Dekker:-NY 494p

Monti, Katherine L.: Folded empirical distribution function curves - Mountain plots. 95 AmerStat 49 342-345

Monticino, Michael G.: Mauldin, R. Daniel; — —: Randomly generated distributions. 95 IsrelJMa 91 215-237

Montjoy, Robert S.: Privacy issues and the National Voter Registration Act of 1993. 94 ASAProSoSt 170-173

Montomoli, C.: Bernardinelli, L.; Clayton, D.; — —: Bayesian estimates of disease maps: How important are priors?. 95 StatMed 14 2411-2431

Bernardinelli, L.; Clayton, D.; Pascutto, C.; — —; Ghislandi, M.; Songini, M.: Bayesian analysis of space-time variation in disease risk. 95 StatMed 14 2433-2443

Mooijaart, Ab: van der Heijden, Peter G. M.; — —: Some new log-bilinear models for the analysis of asymmetry in a square contingency table. 95 SocMethR 24 7-29

Moon, Lynelle: Kokic, Philip N.; — —; Gooday, Jane; Chambers, Ray L.: Estimating temporal farm income distributions using spatial smoothing techniques. 95 AstrlJSt 37 129-143

Moon, Myung-Sang; Gunst, Richard F.: Polynomial measurement error modeling. 95 CmpStDA 19 1-21

Moon, S. E.: Ryoo, S. B.; — —: Detection of warming effects due to industrialization: An accumulated intervention model with an application in Pohang, Korea. 95 JApMetrl 34 1466-1471

Moonesinghe, Ramal; Wright, F. T.: Testing homogeneity with an ordered alternative in a two-factor layout by combining p-values. 95 AnInStMa 47 505-523

Mooney, Geraldine: Mitchell, Susan; — —; Shettle, Carolyn: The impact of using priority mail in a mixed mode survey. 94 ASAProGovt 88-92

Moore, Charles N.: Chang, Xiang Qian; — —: The equivalence of two conditions for weight functions for martingales. 95 StPrLet 23 129-133

Moore, D. S.; McCabe, G. P. (Auth); Kaplon, Howard S. (Rev): Review of "Introduction to the practice of statistics (Second edition)". 95 JAppStat 22 196-197

Moore, David: On the job: A day in the life of a professor. 95 Stats /13 26-27

— — (Ed); Rolph, John E. (Rev): Review of "Statistics: Decisions through data". 95 Chance 8/1 43-44

Moore, David S.; Cobb, George W.; Garfield, Joan; Meeker, William Q.: Statistics education fin de Siécle. 95 AmerStat 49 250-260

Moore, Dirk: Marazzi, A. (Auth); Iglewicz, Boris; Heiberger, Richard M.; — —; Tan, Yun (Rev): Review of "Algorithms, routines, and S functions, for robust statistics". 95 Technmcs 37 351-353

Moore, Jeffrey C.: Bogen, Karen; — —; Marquis, Kent H.: Can we get respondents to use their personal income records. 94 ASASrvyRsr 1252-1257

Marquis, Kent H.; — —; Bogen, Karen: An experiment to reduce measurement error in the SIPP: Preliminary results. 94 ASASrvyRsr 725-729

Von Thurn, Dawn R.; — —: Results from a cognitive exploration of the 1993 American Housing Survey. 94 ASASrvyRsr 1210-1214

Moore, John B.: Elliott, Robert J.; Aggoun, Lakhdar; — —: Hidden Markov models: Estimation and control. 95 Springer:Brln:NY 361p

Elliott, Robert J.; Aggoun, Lakhdar; — — (Auth); Goodwin, C. G. (Rev): Review of "Hidden Markov models: Estimation and control". 95 ShrtBkRv 15/2 29-29

Moore, Kris; Trower, Jonathan; Borowick, Kent: How birth order influences individual characteristics. 95 Chance 8/4 20-24

— —; Trower, Jonathan; Borowick, Kent: An empirical study of birth order. 94 ASAProBuEc 347-350

Moore, L. M.: John, P. W. M.; Johnson, M. E.; — —; Ylvisaker, D.: Minimax distance designs in two-level factorial experiments. 95 JStPlInf 44 249-263

Moore, M.: Archambault, S.; — —: Statistical inference for Voronoi images. 95 CandJSt 23 11-20

Moore, Peter: Bartholomew, David; — —; Smith, Fred; Allin, Paul: The measurement of unemployment in the UK. 95 JRSS-A 158 363-405 (Disc: p405-417)

Moore, Peter G.: A national lottery in the UK. 94 JAppStat 21 607-622

Moore, Rich: Hallstrom, Al; McBride, Ruth; — —: Toward vital status sweeps: A case history in sequential monitoring. 95 StatMed 14 1927-1931

Moore, Susan E.: Goldenberg, Karen L.; — —; Rosen, Richard J.: Commercial payroll software and the quality of employment data. 94 ASASrvyRsr 1025-1030

Moore, Thomas F.: Ezzati, Trena M.; Hoffman, Keith; Judkins, David R.; Massey, James T.; — —: A dual frame design for sampling elderly minorities and persons with disabilities. 95 StatMed 14 571-583 (Disc: p695-696)

Moothathu, T. S.; Kumar, C. Satheesh: Some properties of the stuttering Poisson distribution. 95 CalcutSt 45 125-130

Moral, Paz: González, Pilar; — —: An analysis of the international tourism demand in Spain. 95 IntJFor 11 233-251

Morales, A.: Iznaga, N.; Núñez, G.; Solozabal, J.; — —; Artaza, E.; Rubio, R.; Cardenas, E.: A personal computer-based system for parallel line analysis of bioassay data. 95 CmpBiomd 47 167-175

Morales, D.; Pardo, L.; Vajda, I.: Asymptotic divergence of estimates of discrete distributions. 95 JStPlInf 48 347-369

Esteban, M. D.; — —: Estimating a survival function with doubly censored data in a proportional hazard model. 95 AplStMDA 11 145-157

Menéndez, M.; — —; Pardo, L.; Vajda, I.: Divergence-based estimation and testing of statistical models of classification. 95 JMultiAn 54 329-354

Menéndez, M. L.; — —; Pardo, L.; Salicrú, M.: Asymptotic behaviour and statistical applications of divergence measures in multinomial populations: A unified study. 95 StHefte 36 1-29

Menéndez, M. L.; Pardo, L.; — —; Salicrú, M.: Some statistical applications of (r,s)-directed divergences. 92 UtilMath 42 115-127

Morales, Domingo: Pardo, Leandro; Salicrú, Miguel; Menéndez, María Luisa; — —: A test for homogeneity of variances based on Shannon's entropy. 95 Metron 53 135-146

Morales, María: Ferrándiz, Juan; López, Antonio; Llopis, Agustín; — —; Tejerizo, María Luisa: Spatial interaction between neighbouring counties: Cancer mortality data in Valencia (Spain). 95 Biomtrcs 51 665-678

Morante, S.: Colosimo, A.; — —; Parisi, V.; Rossi, G. C.: An improved method for detection of words with unusual occurrence frequency in nucleotidic sequences. 93 JThBio 165 659-672

Mordecky, E.: Kramkov, D. O.; — —: Integral option. 94 ThProbAp 39 162-172

Morel, Jorge G.; Koehler, Kenneth J.: A one-step Gauss-Newton estimator for modelling categorical data with extraneous variation. 95 ApplStat 44 187-200

Moreno, Eías; Girón, Javier: Prior elicitation and Bayesian robustness in a noisy Poisson process. 94 ASAProBays 217-220

Moreno, Elías; Cano, Juan A.: Classes of bidimensional priors specified on a collection of sets: Bayesian robustness. 95 JStPlInf 46 325-334

Moreno-Gonzalez, J.: Estimates of marker-associated QTL effects in Monte Carlo backcross generations using multiple regression. 92 ThApGene 85 423-434

— —: Genetic models to estimate additive and non-additive effects of marker-associated QTL using multiple regression techniques. 92 ThApGene 85 435-444

Morgan, B.: Ioannidis, C.; Laws, J.; Matthews, K.; — —: Business cycle analysis and forecasting with a structural vector autoregression model for Wales. 95 JForcast 14 251-265

Morgan, B. J. T.: Brooks, S. P.; — —: Optimization using simulated annealing. 95 Ststcian 44 241-257

Catchpole, E. A.; Freeman, S. N.; — —: Modelling age variation in survival and reporting rates for recovery models. 95 JAppStat 22 597-609

Jolliffe, I. T.; Jones, B.; — —: Identifying influential observations in hierarchical cluster analysis. 95 JAppStat 22 61-80

Morgan, Byron: Young, Philip; — —; Sonksen, Peter; Till, Sebastian; Williams, Charles: Using a mixture model to predict the occurrence of diabetic retinopathy. 95 StatMed 14 2599-2608

Morgan, Byron J. T.; Ray, Andrew P. G.: Non-uniqueness and inversions in cluster analysis. 95 ApplStat 44 117-134

Lebreton, Jean-Dominique; — —; Pradel, Roger; Freeman, Stephen N.: A simultaneous survival rate analysis of dead recovery and live recapture data. 95 Biomtrcs 51 1418-1428

Morgan, Carolyn; Lawton, Richard: Repeated-measures analysis of serum lipid PCB data. 94 ASAProEnvr 92-97

Morgan, Charles J.: Siegel, Andrew F.; — —: Statistics and data analysis: An introduction (Second edition). 95 Wiley:NY:UK 656p

Morgan, J. P.: Bailey, R. A.; Monod, H.; — —: Construction and optimality of affine-resolvable designs. 95 Biomtrka 82 187-200

Morgan, John P.; Uddin, Nizam: Optimal, non-binary, variance balanced designs. 95 StSinica 5 535-546

Morgan, M.; Henrion, M. (Auth); Ziegel, Eric R. (Rev): Review of "Uncertainty". 95 Technmcs 37 237-238

Morgan, Mary S.: Kim, Jinbang; De Marchi, Neil; — —: Empirical model particularities and belief in the natural rate hypothesis. 95 JEconmtx 67 81-102

— — (Auth); Neuberg, Leland Gerson (Rev): Review of "The history of econometric ideas". 95 EconTher 11 371-383 (Disc: p384-397)

Morgan, Timothy M.: Blauwet, Mary Beth; — —: Changing risk factor relationships in aging populations due to self-selection statistical bias (selective survival bias). 94 ASAProEpid 76-81

Tudor, Gail E.; — —: Analyzing survival data in conjunction with time-dependent surrogate endpoints in clinical trials. 95 CommStA 24 2293-2313

Morgenthaler, S.: Leong, P. M.; — —: Plots of DNA data. 94 ASAProStGr 11-20

— —; Ronchetti, E.; Stahel, W. A. (Ed); Lenz, H.-J. (Rev): Review of "New directions in statistical data analysis and robustness". 95 CmpStDA 19 711-712

Móri, Tamás F.: Covering with blocks in the non-symmetric case. 95 JTherPrb 8 139-164

Moriarity, Chris: Marker, David A.; Bryant, Edward C.; — —: National Employer Health Insurance Survey sample design. 94 ASASrvyRsr 264-269

Moriarty, Helene: Short, Thomas H.; — —; Cooley, Mary E.: Readability of educational materials for patients with cancer. 95 JStEd 3/2 15kb (jse/v3n2/datasets.short)

Móricz, F.: Maximal probability inequalities for stochastic processes. 95 MatNach 172 239-248

Morii, Fujiki: An image thresholding method using a minimum weighted squared-distortion criterion. 95 PattrnRc 28 1063-1071

Morikawa, Toshihiko; Yoshida, Michihiro: A useful testing strategy in phase III trials: Combined test of superiority and test of equivalence. 95 JBiophSt 5 297-306

Morimune, Kimio; Mantani, Akihisa: Estimating the rank of co-integration when the order of a vector autoregression is unknown. 95 MaCmpSim 39 265-271

Moritz, Thomas: Henderson, William G.; — —; Goldman, Steven; Copeland, Jack; Sethi, Gulshan: Use of cumulative meta-analysis in the design, monitoring, and final analysis of a clinical trial: A case study. 95 ConClnTr 16 331-341

Morozan, T.: Stability and control for linear systems with jump Markov perturbations. 95 StocAnAp 13 91-110

— —: Stability radii of some stochastic differential equations. 95 Stochatx 54 281-291

Morrell, Christopher H.; Pearson, Jay D.; Carter, H. Ballentine; Brant, Larry J.: Estimating unknown transition times using a piecewise nonlinear mixed-effects model in men with prostate cancer. 95 JASA 90 45-53

— —; Peyrot, Mark F.; Varga, Nicholas: An analysis of voting coalitions in the New York colonial legislature. 94 ASAProSoSt 66-70

Morrice, Douglas J.; Song, Mark: An approximation method for a single server queue with cyclically indexed arrival and service rates. 95 AplMaCmp 71 247-269

Morris, Carl; Christiansen, Cindy: Fitting Weibull duration models with random effects. 95 LiDaAnal 1 347-359

Morris, Carl N.: Comment on "Statistics education". 95 AmerStat 49 21-23

— —: Hierarchical models for educational data: An overview. 95 JEdStat 20 190-200

Draper, David; Gaver, Donald, P., Jr; Goel, Prem K.; Greenhouse, Joel B.; Hedges, Larry V.; — —; Tucker, John R.; Waternaux, Christine M. (Auth); Berlin, Jesse A. (Rev): Review of "Combining information: Statistical issues and opportunities for research". 95 StatMed 14 1611-1612

Morris, David: Hosmane, Balakrishna; Shu, Vincent; — —: Finite sample properties of van Elteren test. 94 ASAProBiop 430-434

Morris, K. P.: Matthews, J. N. S.; — —: An application of Bradley-Terry-type models to the measurement of pain. 95 ApplStat 44 243-255

Morris, M. D.; Solomon, A. D.: Design and analysis for an inverse problem arising from an advection-dispersion process. 95 Technmcs 37 293-302

Morris, Max D.; Mitchell, Toby J.: Exploratory designs for computational experiments. 95 JStPlInf 43 381-402

Mitchell, Toby J.; — —; Ylvisaker, Donald: Two-level fractional factorials and Bayesian prediction. 95 StSinica 5 559-573

Morris, P.: Introduction to game theory. 94 Springer:Brln:NY 230p

Morris, S. E.: Diggle, P. J.; Chetwynd, A. G.; Häggkvist, R.; — —: Second-order analysis of space-time clustering. 95 StMedRes 4 124-136

Morrison, Catherine J.: Berndt, Ernst R.; — —: High-tech capital formation and economic performance in U.S. manufacturing industries: An exploratory analysis. 95 JEconmtx 65 9-43

Morrison, D. F.: Nagaraja, H. N.; Sen, P. K.; — — (Ed): Statistical theory and applications: Papers in honor of Herbert A. David. 95 Springer:Brln:NY 334p

Morrison, Kent E.: Cosine products, Fourier transforms, and random sums. 95 AmMaMnth 102 716-724

Morry, Marietta: Pfeffermann, Danny; — —; Wong, Paul: Estimation of the variances of X-11 ARIMA seasonally adjusted estimators for a multiplicative decomposition and heteroscedastic variances. 95 IntJFor 11 271-283

Morse, David T.: Spatz, Chris (Auth); — — (Rev): Review of "Basic statistics: Tales of distributions (Fifth edition)". 95 EdPsMeas 55 140-146

Mortell, Robert R.; Runger, George C.: Statistical process control of multiple stream processes. 95 JQualTek 27 1-12

Mortera, J.: Dawid, A. P.; DeGroot, M. H.; — —: Coherent combination of experts' opinions. 95 Test 4 263-293 (Disc: p293-313)

Morton, A. R.: Cox, K.; Puddey, I. B.; Burke, V.; McAleer, M.; Beilin, L. J.; — —: Effects of exercise and calorie restriction on ambulatory blood pressure in overweight, sedentary males: A time series analysis. 95 EnvrMtrc 6 523-528

Morton, N. E.; Shields, D. C.; Collins, A.: Genetic epidemiology of complex phenotypes. 91 AnHuGene 55 301-314

Morton, Newton E.: Shields, Denis C.; Marlow, Angela J.; Houlston, Richard S.; Eccles, Diana M.; — —: Prediction of genetic risks from segregation analyses of morbid risks. 94 HumHrdty 44 52-55

Morzuch, Bernard J.: Allen, P. Geoffrey; — —: Comparing probability forecasts derived from theoretical distributions. 95 IntJFor 11 147-157

Moser, E. Barry: Garson, Glenn I.; — —: Aggregation and the Pearson chi-square statistic for homogeneous proportions and distributions in ecology. 95 Ecology 76 2258-2269

Moses, Lincoln E.: Some of David Byar's work: In appreciation. 95 ConClnTr 16 216-229

Moses, O. Douglas: On the reliability of indicators of learning curve model accuracy. 95 JCostAnl /1 63-101

Moses, Stephen: Nagelkerke, Nico J. D.; — —; Plummer, Francis A.; Brunham, Robert C.; Fish, David: Logistic regression in case-control studies: The effect of using independent as dependent variables. 95 StatMed 14 769-775

Mosher, William: Duffer, Allen; Lessler, Judith; Weeks, Michael; — —: Effects of incentive payments on response rates and field costs in a pretest of a national CAPI survey. 94 ASASrvyRsr 1386-1392

Mosher, William D.; Pratt, William F.: CAPI, event histories, and incentives in the SNFG cycle 5 pretest. 94 ASASrvyRsr 59-63

Mosier, Michael C.; Graybill, Franklin A.: Approximate confidence intervals for contrasts in the balanced three factor mixed model. 95 BiomtrcJ 37 273-297

Moskowitz, Herbert: Seppala, Tomi; — —; Plante, Robert; Tang, Jen: Statistical process control via the subgroup bootstrap. 95 JQualTek 27 139-153

Tsai, Hsien-Tang; — —; Tang, Jen: A one-sided single screening procedure based on individual unit misclassification error. 95 IIETrn 27 695-706

Wardell, Don G.; — —; Plante, Robert D.: Reply to comment on "Run-length distribution of special-cause control charts for correlated processes". 95 Technmcs 37 243-245

Mosler, C.: Schwartz, H. J.; — —; Hary, I.; Pielert, V.: Factors affecting spatial preferences in settlement site selection in migratory pastoralism. 95 EnvrMtrc 6 485-490

Mosler, K.; Scarsini, M. (Ed); Heyer, H. (Rev): Review of "Stochastic orders and decision under risk". 95 Metrika 42 139-139

Mostaghimi, Mehdi: Evaluating the performance of a turning point prediction model. 94 ASAProBuEc 321-326

Mosteller, F.: Berkey, C. S.; Hoaglin, D. C.; — —; Colditz, G. A.: A random-effects regression model for meta-analysis. 95 StatMed 14 395-411

Mosteller, Frederick: Colditz, Graham A.; Burdick, Elisabeth; — —: Heterogeneity in meta-analysis of data from epidemiologic studies: A commentary. 95 AJEpidm 142 371-382

Motro, U.; Soller, M.: Sequential sampling in determining linkage between marker loci and quantitative trait loci. 93 ThApGene 85 658-664

Motsa, A. I.: Velikii, Yu. A.; — —; Sil'vestrov, D. S.: Dual processes and ergodic type theorems for Markov chains in the scheme of series. 94 ThProbAp 39 642-653

Mottet, Nicolas: Lehmann, Michel; Daurés, Jean-Pierre; — —; Navratil, Henri: Comparison between exact and parametric distributions of multiple inter-raters agreement coefficient. 95 CmpBiomd 47 113-121

Möttönen, Jyrki; Oja, Hannu: Multivariate spatial sign and rank methods. 95 JNnparSt 5 201-213

— —; Oja, Hannu; Krause, Ulf; Rantakallio, Paula: Application of random coefficient regression model to myopia data: A case study. 95 BiomtrcJ 37 657-672

Motulsky, Harvey: Intuitive biostatistics. 95 OxfordUniv 384p

Motwani, Rajeev; Raghavan, Prabhakar: Randomized algorithms. 95 CambridgeUniv 490p

Mouchart, Michel; Scheihing, Eliana: Approximate Bayesian inferences: Numerical experiences when measuring the loss of information. 94 ASAProBays 221-226

Mougeot, Mathilde: Carter, Lucette; — —: Using computers to teach theoretical statistics. 95 TeachgSt 17/2 44-46

Mouhab, Abderrahamane: Geisser, Seymour (Auth); — — (Rev): Review of "Predictive inference: An introduction". 95 JASA 90 1489-1490

Moulton, Brent R.: Interarea indexes of the cost of shelter using hedonic quality adjustment techniques. 95 JEconmtx 68 181-204

Moulton, Lawrence H.; Halsey, Neal A.: A mixture model with detection limits for regression analyses of antibody response to vaccine. 95 Biomtrcs 51 1570-1578

— —; Holt, Elizabeth A.; Job, Jayakaran S.; Halsey, Neal A.: Percentile regression analysis of correlated antibody responses. 95 StatMed 14 777-787

Mountain, Dean C.: Hsiao, Cheng; — —; Illman, Kathleen Ho: A Bayesian integration of end-use metering and conditional-demand analysis. 95 JBES 13 315-326

Mountford, T.; Prabhakar, B.: On the weak convergence of departures from an infinite series of ·/M/1 queues. 95 AnlsApPr 5 121-127

Mountford, T. S.: Critical length for semi-oriented bootstrap percolation. 95 StocProc 56 185-205

— —: Exponential convergence for attractive reversible subcritical nearest particle systems. 95 StocProc 59 235-249

Mountford, Thomas S.: Andjel, Enrique D.; — —; Schonmann, Roberto H.: Equivalence of exponential decay rates for bootstrap percolation like cellular automata. 95 AIHenrPB 31 13-25

Moustafa, Magdi S.: The conjugate gradient method for queueing networks. 95 AplMaCmp 71 289-294

— —; El-Sayed, Khaled M. F.: Iterative solution of the $M^{[x]}/C_2/S/N$ queues. 94 AplMaCmp 66 313-323

Mowers, R. P.: Wright, A. J.; — —: Multiple regression for molecular-marker, quantitative trait data from large F_2 populations. 94 ThApGene 89 305-312

Moxey, Linda M.; Sanford, Anthony J. (Auth); Budescu, David V.; Wallsten, Thomas S. (Rev): Review of "Communicating quantities: A psychological perspective". 95 Chance 8/2 38-40

Moyassiadis, C.: Chadjiconstantinidis, S.; — —: The A-optimal saturated weighing design for $N = 17$ observations. 92 UtilMath 42 3-13

Moyeed, R. A.: Spline smoother as a dynamic linear model. 95 AstrlJSt 37 193-204

Moyer, E. S.: Wassell, J. T.; Kulczycki, G. W.; — —: Frailty models of manufacturing effects. 95 LiDaAnal 1 161-170

Moyer, Ernest S.: Wassell, James T.; Kulczycki, Gregory W.; — —: Modelling frailty in the manufacture of respiratory protective devices. 94 ASAProEpid 133-138

Moyssiadis, C.: Angelis, L.; — —; Kageyama, S.: Methods of constructing A-efficient BTIUB designs. 93 UtilMath 44 5-15

Mroz, Thomas A.: Bollen, Kenneth A.; Guilkey, David K.; — —: Binary outcomes and endogenous explanatory variables: Tests and solutions with an application to the demand for contraceptive use in Tunisia. 95 Dmogrphy 32 111-131

Mšicki, D.; Evans, R.: Integrated probabilistic data association-finite resolution. 95 Autmatca 31 559-570

Mücklich, F.: Ohser, J.; — —: Stereology for some classes of polyhedrons. 95 AdvAppPr 27 384-396

Mudambi, Ram; Taylor, Larry W.: Some non-parametric tests for duration dependence: An application to UK business cycle data. 95 JAppStat 22 163-177

Mudholkar, Anil: Mudholkar, Govind S.; McDermott, Michael P.; — —: Robust finite-intersection tests for homogeneity of ordered variances. 95 JStPlInf 43 185-195

Mudholkar, Govind S.; McDermott, Michael P.; Mudholkar, Anil: Robust finite-intersection tests for homogeneity of ordered variances. 95 JStPlInf 43 185-195

— —; Srivastava, Deo Kumar; Freimer, Marshall: The exponentiated Weibull family: A reanalysis of the bus-motor-failure data. 95 Technmcs 37 436-445

— —; Srivastava, Deo Kumar; Lin, C. Thomas: Some p-variate adaptations of the Shapiro-Wilk test of normality. 95 CommStA 24 953-985

Sarkar, Ila C.; — —; Raubertas, Richard F.: An approximation to the distribution of the resultant from a Fisher distribution. 95 CommStB 24 227-241

Mueller, Carl: Dawson, Donald A.; Fleischmann, Klaus; Li, Yi; — —: Singularity of super-Brownian local time at a point catalyst. 95 AnlsProb 23 37-55

Mueller, Curt D.: Schur, Claudia L.; Berk, Marc L.; — —: Developing a prescription drug expenditure model: The effect of alternative assumptions. 94 ASAProSoSt 245-250

Mueller, R. O.: Basic principles of structural equation modeling: An introduction of LISREL and EQS. 95 Springer:Brln:NY 229p

Muijtjens, A. M. M.; Hasman, A. (Auth); Schulz, K. (Rev): Review of "Bayes" (German). 95 CmpStQ 10 309-309

Muir, C. S.: Doll, R.; Fraumeni, J. F., Jr; — — (Ed); Boyle, Peter (Rev): Review of "Trends in cancer incidence and mortality". 95 StatMed 14 2507-2508

Parkin, D. M.; — —; Whelan, S. L.; Gao, Y. T.; Ferlay, J.; Powell, J. (Ed); Hill, Catherine (Rev): Review of "Cancer incidence in five continents. Volume VI". 95 StatMed 14 104-104

Muir, W.; Nyquist, W. E.; Xu, S.: Alternative partitioning of the genotype-by-environment interaction. 92 ThApGene 84 193-200

Muir, W. M.: Xu, Shizhong; — —: Selection index updating. 92 ThApGene 83 451-458

Muirhead, Colin: Richardson, Sylvia; Monfort, Christine; Green, Martyn; Draper, Gerald; — —: Spatial variation of natural radiation and childhood leukaemia incidence in Great Britain. 95 StatMed 14 2487-2501

Mukerjee, Hari: A simple combinatorial proof of a result by Robertson and Pillers. 95 StPrLet 24 173-175

— —; Tu, Renjin: Order-restricted inferences in linear regression. 95 JASA 90 717-728

Muotka, Timo; Penttinen, Antti: Detecting small-scale spatial patterns in lotic predator-prey relationships: Statistical methods and a case study. 94 JFishRBC 51 2210-2218

Muradoğlu, Gulnur: Önkal, Dilek; — —: Effects of feedback on probabilistic forecasts of stock prices. 95 IntJFor 11 307-319

Muraki, Eiji; Carlson, James E.: Full-information factor analysis for polytomous item responses. 95 ApPsychM 19 73-90

Muralidharan, K.; Jain, J. P.: Gene flow technique for stage-structured populations. 95 Biomtrcj 37 229-234

Murdoch, D. J.: Härdle, W.; Klinke, S.; Turlach, B. A. (Ed); — — (Rev): Review of "XploRe: An interactive statistical computing environment". 96 ShrtBkRv 16/1 3-3

Murphy, Allan H.: Clemen, Robert T.; — —; Winkler, Robert L.: Screening probability forecasts: Contrasts between choosing and combining. 95 IntJFor 11 133-145

Murphy, S. A.: Asymptotic theory for the frailty model. 95 AnlsStat 23 182-198

— —: Likelihood ratio-based confidence intervals in survival analysis. 95 JASA 90 1399-1405

Murphy, Susan; Li, Bing: Projected partial likelihood and its application to longitudinal data. 95 Biomtrka 82 399-406

Murphy, Susan A.; Bentley, Gillian R.; O'Hanesian, Mary Ann: An analysis for menstrual data with time-varying covariates. 95 StatMed 14 1843-1857

Akritas, Michael G.; — —; LaValley, Michael P.: The Theil-Sen estimator with doubly censored data and applications to astronomy. 95 JASA 90 170-177

Murray, C.: Herrnstein, R. J.; — — (Auth); Devlin, Bernie; Fienberg, Stephen E.; Resnick, Daniel P.; Roeder, Kathryn (Rev): Review of "The bell curve: Intelligence and class structure in American life". 95 JASA 90 1483-1488

Murray, David M.: Design and analysis of community trials: Lessons from the Minnesota Heart Health Program. 95 AJEpidm 142 569-575

Zucker, David M.; Lakatos, Edward; Webber, Larry S.; — —; McKinlay, Sonja M.; Feldman, Henry A.; Kelder, Steve H.; Nader, Philip R.: Statistical design of the child and adolescent trial for cardiovascular health (CATCH): Implications of cluster randomization. 95 ConClnTr 16 96-118

Murray, J. D.: Mathematical biology. 93 Springer:Brln:NY 767p

Murray, James: Jovanovic, Borko D.; Hosmer, David W.; — —: A method for fitting Poisson regression model with extra zeros. 94 ASAProEpid 97-100

Murray, Michael K.; Rice, John W. (Auth); Marriott, Paul (Rev): Review of "Differential geometry and statistics". 95 BLndnMaS 27 619-620

Murray, Thomas M.: Yellman, Ted W.; — —: Comment on "The meaning of probability in probabilistic safety analysis" (V45 p261-269). 95 RelEngSS 49 201-205

Murtagh, F.; Hernández-Pajares, M.: The Kohonen self-organizing map method: An assessment. 95 JClass 12 165-190

Murtaugh, Paul A.: ROC curves with multiple marker measurements. 95 Biomtrcs 51 1514-1522

Murthy, D. N. P.: Jiang, R.; — —: Modeling failure-data by mixture of 2 Weibull distributions: A graphical approach. 95 IEEEReli 44 477-488

Jiang, R.; — —: Reliability modeling involving two Weibull distributions. 95 RelEngSS 47 187-198

Murthy, M. N. Narasimha: Raju, C.; — —: Procedures and tables for the construction and selection of chain sampling plans ChSP-4 (c_1, c_r) - Part 4. 95 JAppStat 22 261-271

Raju, C.; — —: Minimum risks chain sampling plans (ChSP-4 (c_1, c_2)) indexed by acceptable quality level and limiting quality level. 95 JAppStat 22 389-426

Muse, Alison G.; Mikl, Jaromir; Smith, Perry F.: Evaluating the quality of anonymous record linkage using deterministic procedures with the New York State AIDS registry and a hospital discharge file. 95 StatMed 14 499-509 (Disc: p687-689)

Muse, Spencer V.: Evolutionary analyses of DNA sequences subject to constraints on secondary structure. 95 Genetcs 139 1429-1439

— —; Gaut, Brandon S.: A likelihood approach for comparing synonymous and nonsynonymous nucleotide substitution rates, with application to the chloroplast genome. 94 MolBioEv 11 715-724

Musiela, Marek: General framework for pricing derivative securities. 95 StocProc 55 227-251

Mutchler, Jan E.: Burr, Jeffrey A.; Massagli, Michael P.; — —; Pienta, Amy: Labor force status and transitions among older African American and white men. 94 ASAProSoSt 129-134

Muthaloo, Subarau: Ong, S. H.; — —: A class of discrete distributions suited to fitting very long-tailed data. 95 CommStB 24 929-945

Muthén, Bengt O.; Satorra, Albert: Technical aspects of Muthén's LISCOMP approach to estimation of latent variable relations with a comprehensive measurement model. 95 Psymtrka 60 489-503

Waller, Niels G.; — —: Genetic Tobit factor analysis: Quantitative genetic modeling with censored data. 92 BehavGen 22 265-292

Muthulakshmi, S.: Chase, Warren; Bown, Fred (Auth); — — (Rev): Review of "General statistics". 95 JAppStat 22 322-323

Mutschler, M. A.: Schwager, S. J.; — —; Federer, W. T.; Scully, B. T.: The effect of linkage on sample size determination for multiple trait selection. 93 ThApGene 86 964-974

Muttlak, Hassen A.: Parameters estimation in a simple linear regression using rank set sampling. 95 Biomtrcj 37 799-810

Muyot, Maria Emerita T.: Marden, John I.; — —: Rank tests for main and interaction effects in analysis of variance. 95 JASA 90 1388-1398

Myers, Casey: Coelho, Robby; Grunow, Bryan; — —; Rasp, John: On the odds of an extended hitting streak. 94 ASAProSpts 46-49

Myers, Don E.: Cui, Haiyan; Stein, Alfred; — —: Extension of spatial information, Bayesian kriging and updating of prior variogram parameters. 95 EnvrMtrc 6 373-384

Myers, Jerome L.; Well, Arnold D.: Research design and statistical analysis. 95 ErlbaumA:NJ 728p

Myers, Michael N.: Hillis, Margaret A.; Minkus, Gail W.; DeCaluwe, Craig L.; — —; Schafer, Edward T.: An improved approach to product quality through testing. 95 ATTTJ 74/3 40-48

Myers, Ransom A.; Cadigan, Noel G.: Statistical analysis of catch-at-age data with correlated errors. 95 JFishRBC 52 1265-1273

Myers, Raymond H.; Montgomery, Douglas C.: Response surface methodology: Process and product optimization using designed experiments. 95 Wiley:-NY:UK 720p

Myers, W. L.; Johnson, G. D.; Patil, G. P.: Rapid mobilization of spatial/temporal information in the context of natural catastrophes. 94 ASAProStGr 25-31

Mykland, P. A.: Embedding and asymptotic expansions for martingales. 95 ProbTher 103 475-492

Mykland, Per; Tierney, Luke; Yu, Bin: Regeneration in Markov chain samplers. 95 JASA 90 233-241

Mykland, Per A.: Chaudhuri, Probal; — —: On efficient designing of nonlinear experiments. 95 StSinica 5 421-440

Mykland, Per Aslak: Dual likelihood. 95 AnlsStat 23 396-421

— —: Martingale expansions and second order inference. 95 AnlsStat 23 707-731

Mytnik, Leonid; Adler, Robert J.: Bisexual branching diffusions. 95 AdvAppPr 27 980-1018

Nabar, S. P.; Bilgi, Shobha: Cumulative sum control charts for the inverse Gaussian distribution. 94 JIndnStA 32 9-14

Nabar, Smita: A sequential sampling plan for the scale parameter of the inverse Gaussian distribution. 94 JIndnStA 32 117-121

Nabeya, Seiji; Perron, Pierre: Approximations to some exact distributions in the first order autoregressive model with dependent errors. 95 EcomtrRv 14 421-457

Nachtsheim, Christopher J.: Meyer, Ruth K.; — —: The coordinate-exchange algorithm for constructing exact optimal experimental designs. 95 Technmcs 37 60-69

Naciri, Y.; Gallais, A.: Predicting mean and variance of all possible lines and hybrids from designs with partially inbred progenies. 94 ThApGene 89 693-697

Nadarajah, S.: Mole, N.; Anderson, C. W.; — —; Wright, C.: A generalized Pareto distribution model for high concentrations in short-range atmospheric dispersion. 95 EnvrMtrc 6 595-606

Nádas, Arthur; Mercer, Robert L.: Hidden Markov models and their estimation. 94 ASAProStCp 80-89

Nadeau, C.: Lawless, J. F.; — —: Some simple robust methods for the analysis of recurrent events. 95 Technmcs 37 158-168

Nadeau, Christian: Bankier, Michael; Fillion, Jean-Marc; Luc, Manchi; — —: Imputing numeric and qualitative variables simultaneously. 94 ASASrvyRsr 242-247

Nadeau, Simon; Décarie, Robert; Lambert, Daniel; St-Georges, Mario: Nonlinear modeling of muskrat use of habitat. 95 JWildlMg 59 110-117

Nader, Philip R.: Zucker, David M.; Lakatos, Edward; Webber, Larry S.; Murray, David M.; McKinlay, Sonja M.; Feldman, Henry A.; Kelder, Steve H.; — —: Statistical design of the child and adolescent trial for cardiovascular health (CATCH): Implications of cluster randomization. 95 ConClnTr 16 96-118

Nadkarni, Prakash M.: Goradia, Tushar M.; Lange, Kenneth; Miller, Perry L.; — —: Fast computation of genetic likelihoods on human pedigree data. 92 HumHrdty 42 42-62

Naes, T.: Hildrum, K.; Isaksson, T.; — —; Tandberg, A. (Ed); Ziegel, Eric R. (Rev): Review of "Near infra-red spectroscopy: Bridging the gap between data analysis and NIR applications". 95 Technmcs 37 126-127

Næs, Tormod: Helland, Inge S.; — —; Isaksson, Tomas: Related versions of the multiplicative scatter correction method for preprocessing spectroscopic data. 95 Chemolab 29 233-241

Langsrud, Øyvind; — —: On the structure of PLS in orthogonal designs. 95 JChemtrc 9 483-487

Nagabhushanam, A.: Panda, S. K.; — —: Fuzzy data distortion. 95 CmpStDA 19 553-562

Nagaev, A. V.: Some properties of convex hulls generated by homogeneous Poisson point processes in an unbounded convex domain. 95 AnInStMa 47 21-29

— —: Asymptotics of the Bayes risk in discrimination between normal and uniform location-scale families. 95 MaMethSt 4 312-333

Nagar, D. K.: Gupta, A. K.; — —: On testing structure of mean vector of compound symmetric Gaussian model. 95 CommStA 24 3123-3136

Nagaraja, H. N.; Sen, P. K.; Morrison, D. F. (Ed): Statistical theory and applications: Papers in honor of Herbert A. David. 95 Springer:Brln:NY 334p

Joshi, S. N.; — —: Joint distribution of maxima of concomitants of subsets of order statistics. 95 Bernoull 1 245-255

Nagaraja, Haikady N.: Raqab, Mohammad Z.; — —: On some predictors of future order statistics. 95 Metron 53 185-204

Nagarsenker, Brahamanand; Nagarsenker, Panna: Statistical analysis appropriate for K samples of two variables. 94 ASAProStCp 218-221

Nagarsenker, Panna: Nagarsenker, Brahamanand; — —: Statistical analysis appropriate for K samples of two variables. 94 ASAProStCp 218-221

Nagarwalla, Neville: Kulldorff, Martin; — —: Spatial disease clusters: Detection and inference. 95 StatMed 14 799-810

Nagasawa, M.: Chung, K. L.; Zhao, Z. (Auth); — — (Rev): Review of "From Brownian motion to Schrödinger equation". 95 ShrtBkRv 15/3 49-49

Nagata, Yasushi: The relationship between the improvement on the point estimation and the improvement on the interval estimation for the disturbance variance in a linear regression model. 95 CommStA 24 1687-1704

— —: On normal approximation to the distribution of the estimated process capability index $\hat{C}_{pk}$. 95 JJapnStS 25 159-171

Takada, Yoshikazu; — —: Fixed-width sequential confidence interval for the mean of a gamma distribution. 95 JStPlInf 44 277-289

Nagel, M.; Wernecke, K. D.; Fleischer, W. (Auth); Bartkowiak, Anna (Rev): Review of "Computer-aided data analysis" (German). 95 ListyBio 32 47-49

Nagel, Matthias: Huber, Thomas M.; — —: Look first, then model: Exploring the data using the interactive graphics of PC-ISP. 95 CmpStQ 10 29-36

Nagel, W.: Stoyan, D.; Stoyan, H. (Auth); — — (Rev): Review of "Fractals, random shapes and point fields. Methods of geometrical statistics". 95 BiomtrcJ 37 978-978

Nagelkerke, Nico J. D.: Statistical inference in stationary populations. 95 Biomtrcs 51 347-350

— —; Moses, Stephen; Plummer, Francis A.; Brunham, Robert C.; Fish, David: Logistic regression in case-control studies: The effect of using independent as dependent variables. 95 StatMed 14 769-775

Nagle, Daniel T.; Saniie, Jafar: Performance analysis of linearly combined order statistic CFAR detectors. 95 IEEEAES 31 522-533

Nahamoo, David: Bellegarda, Eveline J.; Bellegarda, Jerome R.; — —; Nathan, Krishna S.: A fast statistical mixture algorithm for on-line handwriting recognition. 94 IEEEPAMI 16 1227-1233

Nahar, S.: Rahman, M. S.; — —: General formulae for the determination of moments and cumulants. 95 PakJStA 11 11-15

Naik, Dayanand N.: Khattree, Ravindra; — —: Statistical tests for random effects in staggered nested designs. 95 JAppStat 22 495-505

Naik-Nimbalkar, U. V.; Rajarshi, M. B.: Filtering and smoothing via estimating functions. 95 JASA 90 301-306

Ganguly, S. S.; — —: Analysis of ordinal data in a study of endometrial cancer under a matched pairs case-control design. 95 StatMed 14 1545-1552

Joshi, B. H.; — —; Deshmukh, S. R.; Dharmadhikari, A. D.: Prediction of failure - A case study. 95 JAppStat 22 485-493

Naiman, Daniel Q.: Chan, Onn; — —: Some simple U-statistic tests for uniformity on certain homogeneous spaces. 95 JMultiAn 54 210-226

Nair, G. A.: Bhuyan, K. C.; — —: Structural changes in the relationships of body dimensions of four species of pulmonate slugs (Mollusca : Gastropoda : Pulmonata). 95 BiomtrcJ 37 1005-1015

Nair, K. R. Muraleedharan: Padamadan, Veenus; — —: Characterization of a bivariate Pareto distribution. 94 JIndnStA 32 15-20

Nair, N. Unnikrishnan; Asha, G.: Characteristic properties of a bivariate geometric law. 94 JIndnStA 32 111-116

Unnithan, V. K. G.; — —: Minimum variance stratification. 95 CommStB 24 275-284

Nakajima, Hisao: Higashi, Masahiko; — —: Indirect effects in ecological interaction networks 1. The chain rule approach. 95 MaBisci 130 99-128

Nakamoto, Jorge: Gabbard, Susan M.; — —: Evaluating translations of survey instruments in Spanish and English: The 1990 Spanish Language Census long form. 94 ASASrvyRsr 1232-1235

Nakamura, A.: Diewert, W. E.; — — (Ed); Krämer, Walter (Rev. German): Review of "Essays in index number theory (Volume I)". 95 AllgeStA 79 97-97

Nakamura, Takashi: Bayesian logit cohort models with age-by-period interaction effects and an over-dispersion parameter and their application to the data from the study of the Japanese national character (Japanese). 95 PrInStMa 43 99-119

Nakamura, Tsuyoshi; Akazawa, Kouhei: Computer program for the proportional hazards measurement error model. 94 CmpBiomd 45 203-212

Nakashima, Eiji: Analysis of count data using power variance function. 95 JJapnStS 25 193-204

Nakhi, Yasmin: Abdel-Hameed, Mohamed; — —: Optimal control of dams. 95 AplStMDA 11 25-34

Nam, Eunwoo; Rosalsky, Andrew: On the almost certain rate of convergence of series of independent random variables. 95 JKoreaSt 24 91-109

Nam, Ho Soo; Song, Moon Sup; Shin, Bong Sup: A robust test for the parallelism of two regression lines (Korean). 95 KorApS 8/2 77-86

Nam, Jun-mo: Confidence limits for the ratio of two binomial proportions based on likelihood scores: Non-iterative method. 95 BiomtrcJ 37 375-379

— —: Simple test for the Hardy-Weinberg law for HLA data with no observed double blanks. 95 Biomtrcs 51 354-357

— —: Interval estimation and significance testing for cyclic trends in seasonality studies. 95 Biomtrcs 51 1411-1417

Nam, Jun-Mo: Sample size determination in stratified trials to establish the equivalence of two treatments. 95 StatMed 14 2037-2049

Namini, Hamid: Minor, James M.; — —; Cruickshank, Scott: From here to there (dosing to efficacy) with neural nets. 94 ASAProBiop 334-339

Nananukul, Soracha; Gong, Wei-Bo: The mean waiting time of a GI/G/1 queue in light traffic via random thinning. 95 JAppProb 32 256-266

Nanda, Asok K.: Jain, Kanchan; — —: On multivariate weighted distributions. 95 CommStA 24 2517-2539

Nandram, B.: Bayesian cuboid prediction intervals: An application to tensile-strength prediction. 95 JStPlInf 44 167-180

Nandram, Balgobin: Bayesian predictive inference for multivariate sample surveys. 94 JOffStat 10 167-179

Chen, Ming-Hui; — —: A Gibbs sampling approach for predicting shelf-life of a military ration with sensory data. 94 ASAProBays 84-89

Rashid, M. Mushfiqur; — —: A rank-based analysis of one-way repeated measures designs with a changing covariate. 95 JStPlInf 46 249-263

Rashid, M. Mushfiqur; — —: Rank-based predictive inference for the finite population mean of a small area. 94 ASASrvyRsr 1059-1064

Naniwa, Sadao: Kato, Hiroko; — —; Ishiguro, Makio: A multivariate stochastic model with non-stationary trend component. 95 AplStMDA 11 77-95

Napolitano, Barbara A.: Lesser, Martin L.; Kohn, Nina E.; — —; Pahwa, Savita: The FU-PLOT: A graphical method for visualizing the timing of follow-up in longitudinal studies. 95 AmerStat 49 139-143

Naranjo, Joshua D.: Witt, Lee D.; — —; McKean, Joseph W.: Influence functions for rank-based procedures in the linear model. 95 JNnparSt 5 339-358

Narasimham, V. L.: Kantam, R. R. L.; Rao, A. Vasudeva; — —: Linear unbiased estimation of scale parameter by absolute values of order statistics in logistic distribution. 94 JStatRsr 28 199-219

Victorbabu, B. Re.; — —: A new type of slope rotatable central composite design. 94 JIAgSt 46 315-317

Nascimento Silva, P. L. D.; Skinner, C. J.: Estimating distribution functions with auxiliary information using poststratification. 95 JOffStat 11 277-294

Nash, John C.: The software taxi: An easy-to-use tool for statistical presentations. 94 ASAProStCp 269-271

— —; Nash, Mary M.: Managing risks. 95 Chance 8/4 25-31

— —; Quon, Tony K.; Gianini, Jacqueline: Statistical issues in spreadsheet software. 94 ASAProStEd 238-241

The Soft Warehouse Inc (Auth); — — (Rev): Review of "DERIVE (Version 2.55, Extended memory Version 2.55XM, and Version 2.60)". 95 AmerStat 49 93-99

Nash, Mary M.: Nash, John C.; — —: Managing risks. 95 Chance 8/4 25-31

Nason, Guy: Three-dimensional projection pursuit. 95 ApplStat 44 411-430

Tartar, M. E.; Lock, M. D. (Auth); — — (Rev): Review of "Model-free curve estimation". 95 SIAMRvw 37 270-271

Nassar, James R.: Smith, David R.; Reinecke, Kenneth J.; Conroy, Michael J.; Brown, Michael W.; — —: Factors affecting visibility rate of waterfowl surveys in the Mississippi Alluvial Valley. 95 JWildlMg 59 515-527

Nassar, R.: Argyelan, J.; — —: A non-linear model of an open flow adsorption process. 95 EnvrMtrc 6 195-203

Mannan, M.; — —: Size and power of test statistics for gene correlation in 2×2 contingency tables. 95 BiomtrcJ 37 409-433

Natarajan, Jayalakshmi; Strawderman, William E.: Sequential estimation of the variance of a normal distribution. 95 SqtlAnly 14 361-374

Natarajan, Ranjini; McCulloch, Charles E.: A note on the existence of the posterior distribution for a class of mixed models for binomial responses. 95 Biomtrka 82 639-643

Nathan, Krishna S.: Bellegarda, Eveline J.; Bellegarda, Jerome R.; Nahamoo, David; — —: A fast statistical mixture algorithm for on-line handwriting recognition. 94 IEEEPAMI 16 1227-1233

Nations, Chris: Pugesek, Bruce H.; — —; Diem, Kenneth L.; Pradel, Roger: Mark-resighting analysis of a California gull population. 95 JAppStat 22 625-639

Natvig, Bent: Gåsemyr, Jørund; — —: Using expert opinions in Bayesian prediction of component lifetimes in a shock model. 95 MaOpRsr 20 227-242

Naus, Joseph: Wallenstein, Sylvan; — —; Glaz, Joseph: Power of the scan statistics. 94 ASAProEpid 70-75

Navarro, Alfredo: Schindler, Eric; — —: Census plus: An alternative coverage methodology. 94 ASASrvyRsr 248-253

Navarro, J.: Ruiz, J. M.; — —: Characterization of discrete distributions using expected values. 95 StHefte 36 237-252

Navidi, W. C.: Bootstrapping a method of phylogenetic inference. 95 JStPlInf 43 169-184

Navratil, Henri: Lehmann, Michel; Daurés, Jean-Pierre; Mottet, Nicolas; — —: Comparison between exact and parametric distributions of multiple inter-raters agreement coefficient. 95 CmpBiomd 47 113-121

Nawata, Kazumitsu: Estimation of sample-selection models by the maximum likelihood method. 95 MaCmpSim 39 299-303

Nayak, Tapan Kumar: Estimating the parameter of a selected population. 95 CalcutSt 45 93-101

NCSS Statistical Software: — — (Auth); Lovie, Pat (Rev): Review of "NCSS 6.0 for Windows". 95 BJMaStPs 48 407-408

Neal, Radford M.: Hinton, Geoffrey E.; Dayan, Peter; — —; Zemel, Richard S.: Using neural networks to learn intractable generative models. 94 ASAProStCp 11-20

Neale, M.: Pickles, A.; — —; Simonoff, E.; Rutter, M.; Hewitt, J.; Meyer, J.; Crouchley, R.; Silberg, J.; Eaves, L.: A simple method for censored age-of-onset data subject to recall bias: Mothers' reports of age of puberty in male twins. 94 BehavGen 24 457-468

Neale, M. C.: Eaves, L. J.; Silberg, J. L.; Hewitt, K.; Rutter, M.; Meyer, J. M.; — —; Pickles, A.: Analyzing twin resemblance in multisymptom data: Genetic applications of a latent class model for symptoms of conduct disorder in juvenile boys. 93 BehavGen 23 5-20

Heath, A. C.; Kessler, R. C.; — —; Hewitt, J. K.; Eaves, L. J.; Kendler, K. S.: Testing hypotheses about direction of causation using cross-sectional family data. 93 BehavGen 23 29-50

Sham, P. C.; Walters, E. E.; — —; Heath, A. C.; MacLean, C. J.; Kendler, K. S.: Logistic regression analysis of twin data: Estimation of parameters of the multifactorial liability-threshold model. 94 BehavGen 24 229-238

Neale, Michael C.; Eaves, Lindon J.: Estimating and controlling for the effects of volunteer bias with pairs of relatives. 93 BehavGen 23 271-278

— —; Eaves, Lindon J.; Kendler, Kenneth S.: The power of the classical twin study to resolve variation in threshold traits. 94 BehavGen 24 239-258

Eaves, Lindon J.; — —; Meyer, Joanne M.: A model for comparative ratings in studies of within-family differences. 91 BehavGen 21 531-536

Nebebe, Fassil; Kwan, Andy C. C.: Detecting 'potential' outliers in time series: Some Monte Carlo evidence on dummy variable plot approaches. 95 JStPlInf 43 87-95

Nebenzahl, Elliott: Fearn, Dean H.; — —: Using power transformations when approximating quantiles. 95 CommStA 24 1073-1093

Nedelman, Jerry R.; Karara, Adel H.; Chang, Cheng-Tao; Gibiansky, Ekaterina; McDonald, Scott; Robinson, William T.: Inferring systemic exposure from a pharmacokinetic screen: Model-free and model-based approaches. 95 StatMed 14 955-968 (Disc: p969-970)

Neden, Linda R.: Sloan, Jeff A.; Schwarz, Carl J.; — —: Dataset on silver-maple-tree growth regulators. 95 CandJSt 23 325-329 (Disc: p329-332)

Nedoma, Petr: Kárný, Miroslav; Halousková, Alena; — —: Recursive approximation by ARX model: A tool for grey box modelling. 95 IJAdCSgP 9 525-546

Neerchal, Nagaraj K.; Shafer, Ronald W.: Spatial analysis of benthic data from the Chesapeake Bay. 94 ASAProEnvr 204-208

Sengupta, Soma; — —: Classification of grab samples using composite sampling: An improved strategy. 94 CalcutSt 44 195-208

Nehoral, Arye: Stoica, Petre; Händel, Peter; — —: Improved sequential MUSIC. 95 IEEEAES 31 1230-1239

Nei, Masatoshi: Ota, Tatsuya; — —: Variance and covariances of the numbers of synonymous and nonsynonymous substitutions per site. 94 MolBioEv 11 613-619

Rzhetsky, Andrey; — —: Unbiased estimates of the number of nucleotide substitutions when substitution rate varies among different sites. 94 JMolEvol 38 295-299

Rzhetsky, Andrey; — —: Statistical properties of the ordinary least-squares, generalized least-squares, and minimum-evolution methods of phylogenetic inference. 92 JMolEvol 35 367-375

Sitnikova, Tatyana; Rzhetsky, Andrey; — —: Interior-branch and bootstrap tests of phylogenetic trees. 95 MolBioEv 12 319-333

Neill, W. A.: Rhoades, D. A.; — —: The effect of station drift on the estimation of shifts in meteorological records due to site changes. 95 JClimlgy 15 207-220

Nel, Daan G.: Flury, Bernard D.; — —; Pienaar, Inet: Simultaneous detection of shift in means and variances. 95 JASA 90 1474-1481

Nelder, J. A.: Reply to comments on "The statistics of linear models: Back to basics". 95 StatComp 5 109-111

Nelder, John A.; Lane, Peter W.: The computer analysis of factorial experiments: In memoriam - Frank Yates. 95 AmerStat 49 382-385

Nellis, J. G.: Fleming, C.; — — (Auth); Cooper, R. I. (Rev): Review of "Principles of applied statistics". 95 JRSS-A 158 348-349

Nelson, Barry L.: Matejcik, Frank J.; — —: One-stage multiple comparisons with the best for unequal variances. 94 ASAProBuEc 55-60

Nelson, Daniel B.; Foster, Dean P.: Filtering and forecasting with misspecified ARCH models II: Making the right forecast with the wrong model. 95 JEconmtx 67 303-335

Nelson, J. Barry: Bouleau, Nicolas; Lepingle, Dominique (Auth); Spall, James C.; — — (Rev): Review of "Numerical methods for stochastic processes". 95 Technmcs 37 236-237

Nelson, L. S.: Correction to "Tests on early failures - The precedence life test" (V25 p140-143). 95 JQualTek 27 402-402

— —: Correction to "Using nested designs: II. Confidence limits for standard deviations" (V27 p265-265). 95 JQualTek 27 403-403

Nelson, Lloyd S.: Single test that two samples came from the same normal population. 95 JQualTek 27 84-87

— —: Using nested designs: I. Estimation of standard deviations. 95 JQualTek 27 169-171

— —: Using nested designs: II. Confidence limits for standard deviations. 95 JQualTek 27 265-267

— —: The usefulness of Monte Carlo tests. 95 JQualTek 27 387-389

Bethea, Robert M.; Duran, Benjamin S.; Boullion, Thomas L. (Auth); — — (Rev): Review of "Statistical methods for engineers and scientists (Third edition)". 95 JQualTek 27 392-393

Cleveland, William S. (Auth); — — (Rev): Review of "The elements of graphing data (Second edition)". 95 JQualTek 27 176-177

Gibbons, Jean Dickinson (Auth); — — (Rev): Review of "Nonparametric measures of association". 95 JQualTek 27 93-93

Mandel, John (Auth); — — (Rev): Review of "Analysis of two-way layouts". 95 JQualTek 27 172-173

Mittag, H.-J.; Rinne, H. (Auth); — — (Rev): Review of "Statistical methods of quality assurance". 95 JQualTek 27 95-96

Roberts, George W. (Ed); — — (Rev): Review of "Quality planning, control, and improvement in research and development". 95 JQualTek 27 271-272

Nelson, R.: Probability, stochastic processes, and queueing theory: The mathematics of computer performance modelling. 95 Springer:Brln:NY 592p

Nelson, Randolph: Philips, Thomas K.; — —: The moment bound is tighter than Chernoff's bound for positive tail probabilities. 95 AmerStat 49 175-178

Nelson, Ray D.; Cornia, Gary C.: Intrayear monitoring of state tax collections and revenue forecasts. 94 ASAProBuEc 339-344

Nelson, Wayne: Confidence limits for recurrence data - Applied to cost or number of product repairs. 95 Technmcs 37 147-157

Nematollahi, N.: Parsian, A.; — —: On the admissibility of ordered Poisson parameter under the entropy loss function. 95 CommStA 24 2451-2467

Nemenyi, Peter: Comment on "Recursive schemes for calculating cumulative binomial and Poisson probabilities" (V48 p136-138). 95 AmerStat 49 402-403

Németh, Lou: Zaciewski, Robert D.; — —: The multi-vari chart: An underutilized quality tool. 95 QulPr 28/10 81-83

Nemeth, Margaret A.: Brereton, Richard G. (Ed); — — (Rev): Review of "Multivariate pattern recognition in chemometrics, illustrated by case studies". 95 Technmcs 37 116-116

Nerlove, Marc; Grether, David M.; Carvalho, José L.: Analysis of economic time series. A synthesis (Revised edition). 95 Academic:NY:Lnd 468p

Nesbitt, Stephen A.: Rodgers, James A., Jr; Linda, Stephen B.; — —: Comparing aerial estimates with ground counts of nests in wood stork colonies. 95 JWildlMg 59 656-666

Nesselroade, John R.: Comment on "Testing for equivalence of measurement scales: Simple structure and metric invariance reconsidered". 95 MultBehR 30 95-99

— —: Comment on "Testing for equivalence of measurement scales: Simple structure and metric invariance reconsidered". 95 MultBehR 30 119-120

Nester, Marks R.: Williams, E. R.; Matheson, A. C. (Auth); — — (Rev): Review of "Experimental design and analysis for use in tree improvement". 95 AstrlJSt 37 247-248

Neuberg, Leland G.: Randomization and cause and effect inference in social experimentation. 94 ASAProSoSt 153-158

Earman, J. (Auth); — — (Rev): Review of "Bayes or bust? - A critical examination of Bayesian confirmation theory". 95 Ststcian 44 283-284

Neuberg, Leland Gerson: Morgan, Mary S. (Auth); — — (Rev): Review of "The history of econometric ideas". 95 EconTher 11 371-383 (Disc: p384-397)

Neudecker, H.; Liu, S.; Polasek, W.: The Hadamard product and some of its applications in statistics. 95 MaOpfStS 26 365-373

Neudecker, Heinz; Polasek, Wolfgang; Liu, Shuangzhe: The heteroskedastic linear regression model and the Hadamard product. A note. 95 JEconmtx 68 361-366

Liu, Shuangzhe; — —: Matrix-trace Cauchy-Schwarz inequalities and applications in canonical correlation analysis. 95 StHefte 36 287-298

Liu, Shuangzhe; — —: A *V*-optimal design for Scheffé's polynomial model. 95 StPrLet 23 253-258

Neuenschwander, Beat E.; Flury, Bernard D.: Common canonical variates. 95 Biomtrka 82 553-560

Neuenschwander, Daniel: A Gaussian central limit theorem for trimmed products on simply connected step 2-nilpotent lie groups. 95 JTherPrb 8 165-174

— —; Schott, René: On the local and asymptotic behavior of Brownian motion on simply connected nilpotent Lie groups. 95 JTherPrb 8 795-806

Neufelder, Ann M.: — — (Auth); Soyer, Refik (Rev): Review of "Ensuring software reliability". 95 Technmcs 37 122-123

Neuhaus, G.: Jørgenson, B. (Auth); — — (Rev): Review of "The theory of linear models". 95 Metrika 42 276-277

Neuhaus, Walther: Balanced credibility estimation. 95 AplStMDA 11 217-230

Daykin, C. D.; Pentikäinen, T.; Pesonen, M. (Auth); — — (Rev): Review of "Practical risk theory for actuaries". 95 AstinB 25 75-76

Neumaier, Arnold: Elster, Clemens; — —: Screening by conference designs. 95 Biomtrka 82 589-602

Neuman, Rosalind: Ott, Jurg (Auth); — — (Rev): Review of "Analysis of human genetic linkage (Revised edition)". 93 BehavGen 23 299-300

Neumann, George R.: Bowlus, Audra J.; Kiefer, Nicholas M.; — —: Estimation of equilibrium wage distributions with heterogeneity. 95 JApEcmtx 10 119-131S

Neumann, M. H.; Spokoiny, V. G.: On the efficiency of wavelet estimators under arbitrary error distributions. 95 MaMethSt 4 137-166

Neumark, David; Wascher, William: Minimum wage effects on employment and school enrollment. 95 JBES 13/2 199-206

Blackburn, McKinley L.; — —: Are OLS estimates of the return to schooling biased downward? another look. 95 RvEconSt 77 217-229

Neuts, M. F.; Pearce, C. E. M.: A central limit theorem for a class of Markovian packet streams under heavy traffic conditions. 95 CommStSM 11 210-218

Neuts, Marcel F.: Algorithmic probability. A collection of problems. 95 ChapmnHall:Lnd 448p

— — (Auth); Coleman, R. (Rev): Review of "Algorithmic probability. A collection of problems". 96 ShrtBkRv 16/1 4-4

Neuwald, Andrew F.: Liu, Jun S.; — —; Lawrence, Charles E.: Bayesian models for multiple local sequence alignment and Gibbs sampling strategies. 95 JASA 90 1156-1170

Nevill, Alan M.; Holder, Roger L.: Comment on "Spurious correlation and the fallacy of the ratio standard revisited" (V156 p379-392). 95 JRSS-A 158 619-621

Nevzorov, Valery: Asymptotic distributions of records in nonstationary schemes. 95 JStPlInf 45 261-273

Newbold, Paul: Miller, John P.; — —: Uncertainty about the persistence of economic shocks. 95 JBES 13 435-440

Miller, John P.; — —: A generalized variance ratio test of ARIMA$(p,1,q)$ model specification. 95 JTimSrAn 16 403-413

Nunes, Luis C.; Kuan, Chung-Ming; — —: Spurious break. 95 EconTher 11 736-749

Newey, W. K.: Hausman, J. A.; — —; Powell, J. L.: Nonlinear errors in variables: Estimation of some Engel curves. 95 JEconmtx 65 205-233

Newey, Whitney: Robins, James M.; Hsieh, Fushing; — —: Semiparametric efficient estimation of a conditional density with missing or mismeasured covariates. 95 JRSS-B 57 409-424

Newman, C. M.; Stein, D. L.: Random walk in a strongly inhomogeneous environment and invasion percolation. 95 AIHenrPB 31 249-261

Newman, Charles M.; Piza, Marcelo S. T.: Divergence of shape fluctuations in two dimensions. 95 AnlsProb 23 977-1005

Newman, Isadore; Fraas, John W.; Norfolk, Timothy: Reply to comments on "A binomial test of model fit". 95 StrEqMod 2 155-162

Newton, H. Joseph: Harvill, Jane L.; — —: Saddlepoint approximations for the difference of order statistics. 95 Biomtrka 82 226-231

Harvill, Jane L.; — —: Using symbolic math to evaluate saddlepoint approximations for the difference of order statistics. 95 CommStB 24 781-791

Newton, J.: van den Berg, J. L. (Auth); — — (Rev): Review of "Sojourn times in feedback and processor sharing queues". 95 JRSS-A 158 186-187

Newton, Michael A.: A diffuse prior limit in semiparametric binary regression. 94 ASAProBays 181-186

— —; Guttorp, Peter; Catlin, Sandra; Assunção, Renato; Abkowitz, Janis L.: Stochastic modeling of early hematopoiesis. 95 JASA 90 1146-1155

Ney, Hermann: On the probabilistic interpretation of neural network classifiers and discriminative training criteria. 95 IEEEPAMI 17 107-119

Ney, Peter: Dinwoodie, I. H.; — —: Occupation measures for Markov chains. 95 JTherPrb 8 679-691

Neytchev, Plamen Nikolov: SWEEP operator for least-squares subject to linear constraints. 95 CmpStDA 20 599-609

Ng, Edward; Gentleman, Jane F.: Life expectancy in Canada: An empirical comparison of different estimation methods. 94 ASAProSoSt 88-93

Ng, Kai W.: Zhu, Li-Xing; — —: M-type estimators of regression function with applications. 95 StPrLet 25 133-144

Zhu, Li-Xing; — —: Asymptotics of sliced inverse regression. 95 StSinica 5 727-736

Ng, Pin T.: Finite sample properties of adaptive regression estimators. 95 EcomtrRv 14 267-297

Bera, Anil K.; — —: Tests for normality using estimated score function. 95 JStCmpSm 52 273-287

Ng, Serena: Testing for homogeneity in demand systems when the regressors are nonstationary. 95 JApEcmtx 10 147-163

— —; Perron, Pierre: Unit root tests in ARMA models with data-dependent methods for the selection of the truncation lag. 95 JASA 90 268-281

Predicta Software Inc (Auth); — — (Rev): Review of "Coint 2.0". 95 JApEcmtx 10 205-210

Ng, Tie-Hua: Conventional null hypothesis testing in active control equivalence studies. 95 ConClnTr 16 356-358

Ng, V. Ming: Haq, M. Safiul; — —: A linear model with a special class of covariance structure. 94 JStatRsr 28 189-198

Ng, Victor K.: Chou, Ray Y.; — —; Pi, Lynn K.: Cointegration of international stock market indices. 94 ASAProBuEc 196-201

Masulis, Ronald W.; — —: Overnight and daytime stock-return dynamics on the London stock exchange: The impacts of "big bang" and the 1987 stock-market crash. 95 JBES 13 365-378

Nguyen, Ha H.: Bolognese, James A.; — —: Estimating "non-detects" for inclusion in linear regression. 94 ASAProBiop 414-419

Nguyen, Trang: Sidow, Arend; — —; Speed, Terence P.: Estimating the fraction of invariable codons with a capture-recapture method. 92 JMolEvol 35 253-260

Nguyen, Truc T.: Conditional distributions and characterizations of multivariate stable distribution. 95 JMultiAn 53 181-193

Niang, N.; Saporta, G.: Average run length of a moving average control chart to detect shifts in the mean of a process (French). 95 RStApp 43/3 5-20

Nicaud, V.: Tiret, L.; — —; Ehnholm, C.; Havekes, L.; Menzel, H. J.; Ducimetiere, P.; Cambien, F.: Inference of the strength of genotype-disease association from studies comparing offspring with and without parental history of disease. 93 AnHuGene 57 141-149

Nicewander, W. Alan: Bloxom, Bruce; Pashley, Peter J.; — —; Yan, Duanli: Linking to a large-scale assessment: An empirical evaluation. 95 JEdStat 20 1-26

Nicholls, William L., II; Appel, Martin V.: New CASIC technologies at the U.S. Bureau of the Census. 94 ASASrvyRsr 757-762

Nichols, James D.; Kendall, William L.: The use of multi-state capture-recapture models to address questions in evolutionary ecology. 95 JAppStat 22 835-846

Kendall, William L.; — —: On the use of secondary capture-recapture samples to estimate temporary emigration and breeding proportions. 95 JAppStat 22 751-762

Nick, Todd G.: Improving the quality of journals while improving consulting. 94 ASAProQlPr 232-234

Nickerson, David M.: Comment on "Construction of a conservative confidence region from projections of an exact confidence region in multiple linear regression" (V48 p120-124). 95 AmerStat 49 324-324

Higgins, James J.; Keller-McNulty, Sallie (Auth); — — (Rev): Review of "Concepts in probability and stochastic modeling". 95 AmerStat 49 321-321

Nicol, David: Ciardo, Gianfranco; Leemis, Lawrence M.; — —: On the minimum of independent geometrically distributed random variables. 95 StPrLet 23 313-326

Niederliński, A.: An upper bound for the recursive least squares estimation error. 95 IEEEAuCn 40 1655-1661

Niederreiter, H.; Shiue, P. J.-S. (Ed): Monte Carlo and quasi-Monte Carlo methods in scientific computing. 95 Springer:Brln:NY 372p

Niederreiter, Harald: Eichenauer-Herrmann, Jürgen; — —: Digital inversive pseudorandom numbers. 94 ACMMdSim 4 339-349

Nielsen, Hanne Mandrup: Asmussen, Søren; — —: Ruin probabilities via local adjustment coefficients. 95 JAppProb 32 736-755

Nielsen, J. Aase; Sandmann, Klaus: Equity-linked life insurance: A model with stochastic interest rates. 95 InsrneME 16 225-253

Nielsen, J. P.: Jones, M. C.; Linton, O.; — —: A simple bias reduction method for density estimation. 95 Biomtrka 82 327-338

Nielsen, Jens P.; Linton, Oliver B.: Kernel estimation in a nonparametric marker dependent hazard model. 95 AnlsStat 23 1735-1748

Nielsen, Jens Perch: Linton, Oliver; — —: A kernel method of estimating structured nonparametric regression based on marginal integration. 95 Biomtrka 82 93-100

Niemira, Michael P.; Klein, Philip A. (Auth); Mullineux, Andy W. (Rev): Review of "Forecasting financial and economic cycles". 95 IntJFor 11 191-192

Niemiro, W.: Estimation of nuisance parameters for inference based on least absolute deviations. 95 ApplMath 22 515-529

Niemiro, Wojciech: Tail events of simulated annealing Markov chains. 95 JAppProb 32 867-876

— —; Pokarowski, Piotr: Tail events of some nonhomogeneous Markov chains. 95 AnlsApPr 5 261-293

Nienhuis, J.: Tivang, J. G.; — —; Smith, O. S.: Estimation of sampling variance of molecular marker data using the bootstrap procedure. 94 ThApGene 89 259-264

Nieto, Fabio H.: Note on the estimation of missing data in a time series using the dual autocorrelation function (Spanish). 94 Estd 46 85-103

— —; Guerrero, Victor M.: Kalman filter for singular and conditional state-space models when the system state and the observational error are correlated. 95 StPrLet 22 303-310

Niezgoda, M.: Stępniak, C.; — —: Inverting covariance matrices in unbalanced hierarchical models. 95 JStCmpSm 51 215-221

Nigam, A. K.: Saxena, Sanjay Kumar; — —; Shukla, N. D.: Variance estimation for combined ratio estimator. 95 SankhyaB 57 85-92

Niggemeyer, B.; Radtke, M.; Reich, A.: Applications of risk theory and multivariate analysis in insurance practice. 95 AplStMDA 11 231-244

Niinimaa, A.; Oja, H.: On the influence functions of certain bivariate medians. 95 JRSS-B 57 565-574

Nikabadze, A. M.: McKeague, Ian W.; — —; Sun, Yanqing: On omnibus test for independence of a survival time from a covariate. 95 AnlsStat 23 450-475

Nikiforov, I. V.: Basseville, M.; — — (Auth); Peach, Nigel (Rev): Review of "Detection of abrupt changes: Theory and applications". 95 JRSS-A 158 185-186

Nikitin, Yakov: Asymptotic efficiency of nonparametric tests. 95 CambridgeUniv 290p

Niknian, Minoo: Good, Philip (Auth); — — (Rev): Review of "Permutation tests: A practical guide to resampling methods for testing hypotheses". 95 Technmcs 37 341-342

Nikulin, M. S.; Voinov, V. G.: On the problem of the means of weighted normal populations. 95 Questo 19 93-106

Voinov, V. G.; — —: A review of the results on the Stein approach for estimators improvement. 95 Questo 19 65-91

Niland, Joyce C.; Odom-Maryon, Tamara L.; Lee, Jennifer; Tilley, Barbara C.: A survey of biostatistical consulting units throughout North America. 95 AmerStat 49 183-189

Niland, Laura J.: Ward, Joe H., Jr; — —: Empowering high school students to exploit statistical models and software for research projects. 94 ASAProStEd 44-48

Nilsson, Nils-Otto: Säll, Torbjörn; — —: The robustness of recombination frequency estimates in intercrosses with dominant markers. 94 Genetcs 137 589-596

Nimalendran, M.: Maddala, G. S.; — —: An unobserved component panel data model to study the effect of earnings surprises on stock prices, trading volumes, and spreads. 95 JEconmtx 68 229-242

Ninness, Brett; Goodwin, Graham C.: Estimation of model quality. 95 Autmatca 31 1771-1797

Nirel, Ronit; Rosenfeld, Daniel: Estimation of the effect of operational seeding on rain amounts in Israel. 95 JApMetrl 34 2220-2229

Nisenbaum, Susan: Johnson, Robert; Malloy, Kay; Gerstein, Dean; Suter, Natalie; — —: Increasing response rates in follow-up surveys of drug treatment patients. 94 ASASrvyRsr 1263-1268

Nishimura, Kazuo: Iiyama, Yuriko; — —; Sibuya, Masaaki: Power of record-breaking test (Japanese). 95 JpJAppSt 24 13-26

Nishimura, Shoichi; Jiang, Yong: An M/G/1 vacation model with two service modes. 95 PrEngInf 9 355-374

Nishisato, Shizuhiko: Graphical representation of quantified categorical data: Its inherent problems. 95 JStPlInf 43 121-132

Nishiyama, Yoichi: Local asymptotic normality of a sequential model for marked point processes and its applications. 95 AnInStMa 47 195-209

Nisihira, Sigeki: Social values do change? (Japanese). 95 PrInStMa 43 81-97

Nissel, Muriel: Social trends and social change. 95 JRSS-A 158 491-504

Niu, Xufeng: Asymptotic properties of maximum likelihood estimates in a class of space-time regression models. 95 JMultiAn 55 82-104

— —; Tiao, George C.: Modeling satellite ozone data. 95 JASA 90 969-983

Nixon, Mark S.: Chen, Yan Qiu; — —; Thomas, David W.: Statistical geometrical features for texture classification. 95 PattrnRc 28 537-552

Nobel, Andrew: A counterexample concerning uniform ergodic theorems for a class of functions. 95 StPrLet 24 165-168

Nobile, A. G.: Di Crescenzo, A. G.; Giorno, V.; — —; Ricciardi, L. M.: On a symmetry-based constructive approach to probability densities for two-dimensional diffusion processes. 95 JAppProb 32 316-336

Noble, John M.: Almost sure behaviour of evolution with random potential. 95 StocAnAp 13 323-334

Noble, Nicholas R.: Quantitative Micro Software (Auth); — — (Rev): Review of "EViews - Micro TSP for Windows". 95 IntJFor 11 497-499

Noble, William: LaLonde, Steven; — —: Approaches to robust design when the response is multivariate. 94 ASAProPEng 187-192

Loughin, Thomas M.; — —: A permutation test for effects in an unreplicated factorial design. 94 ASAProPEng 181-186

Nobre, Flavio F.; de Macedo, Marcia M. A.: Feasibility of contour mapping epidemiological data with missing values. 95 StatMed 14 605-613 (Disc: p697-699)

Nodjiram, Sagombaye: Analysis of a contingency table with negative numbers: Application of the analysis of a generalised Burt sub-table (French). 95 CAnDnnes 20 111-120

Nolan, B.: — — (Auth); Fielding, Antony (Rev): Review of "Data analysis: An introduction". 95 Ststcian 44 546-547

Nolan, D.: Burman, P.; — —: A general Akaike-type criterion for model selection in robust regression. 95 Biomtrka 82 877-886

Nolan, Frank: Apportionment to whole numbers. 95 NZndSt 30 40-47

Nolan, James R.: Integrating cases into the business statistics course. 94 ASAProStEd 123-125

Nolan, John P.; Rajput, Balram: Calculation of multidimensional stable densities. 95 CommStB 24 551-566

Nolan, Paul E.: Roe, Denise J.; Cheng, Mary; Elsberry, Victor A.; — —: Estimating parameters in non-linear mixed effect models: A comparison using clinical data in children. 94 ASAProBiop 313-318

Noll, Dominikus: Rates of convergence for best entropy estimates. 95 StatDecn 13 141-165

Nollau, V.: Inequalities for variances of some functions of random variables. 95 StHefte 36 163-174

Nomikos, Paul; MacGregor, John F.: Multi-way partial least squares in monitoring batch processes. 95 Chemolab 30 97-108

— —; MacGregor, John F.: Multivariate SPC charts for monitoring batch processes. 95 Technmcs 37 41-59

Nonaka, Yasuo: Pan, Zhijie; — —: Importance analysis for the systems with common cause failures. 95 RelEngSS 50 297-300

Nondahl, David M.: Chappell, Rick; — —; Fowler, John F.: Modeling dose and local control in radiotherapy. 95 JASA 90 829-838

Nony, Patrice: Boissel, Jean-Pierre; Durieu, Isabelle; Girard, Pascal; — —; Chauvin, Franck; Haugh, Margaret: Dose-ranging trials: Guidelines for data collection and standardized descriptions. 95 ConClnTr 16 319-330

Noor, Islam; Ahmad, Naseer: Exact and approximate relative efficiency of generalized ridge-type shrinkage estimator. 95 PakJStA 11 17-22

Noorbaloochi, S.: Shafie, K.; — —: Asymmetric unbiased estimation in location families. 95 StatDecn 13 307-314

Norberg, Ragnar: A time-continuous Markov chain interest model with applications to insurance. 95 AplStMDA 11 245-256

— —: Differential equations for moments of present values in life insurance. 95 InsrneME 17 171-180

Nordberg, Lennart: Andersson, Claes; — —: A method for variance estimation of non-linear functions of totals in surveys - Theory and software implementation. 94 JOffStat 10 395-405

Nordbotten, S.: Statistical meta-knowledge and -data. 93 StatJUN 10 101-112

Nordbrock, Earl: Design and analysis of stability studies. 94 ASAProBiop 291-294

Nordheim, Erik V.: Heisey, Dennis M.; — —: Modelling age-specific survival in nesting studies, using a general approach for doubly-censored and truncated data. 95 Biomtrcs 51 51-60

Nordin, B.: Pedersen, F.; Bengtsson, E.; — —: An extended strategy for exploratory multivariate image analysis including noise considerations. 95 JChemtrc 9 389-409

Norfolk, Timothy: Newman, Isadore; Fraas, John W.; — —: Reply to comments on "A binomial test of model fit". 95 StrEqMod 2 155-162

Nørgaard, Lars: Direct standardisation in multi wavelength fluorescence spectroscopy. 95 Chemolab 29 283-293

Norin, N. V.: Itô formula for an extended stochastic integral with nonanticipating kernel. 94 ThProbAp 39 573-592

Norman, John M.: Clarke, Stephen R.; — —: Home ground advantage of individual clubs in English soccer. 95 Ststcian 44 509-521

Normand, Sharon-Lise T.: Meta-analysis software: A comparative review. 95 AmerStat 49 298-309

Doksum, Kjell A.; — —: Gaussian models for degradation processes - Part I: Methods for the analysis of biomarker data. 95 LiDaAnal 1 131-144

Norrie, John: Ford, Ian; — —; Ahmadi, Susan: Model inconsistency, illustrated by the Cox proportional hazards model. 95 StatMed 14 735-746

Norris, Douglas A.; Hatcher, Jeff: The impact of interviewer characteristics on response in a national survey of violence against women. 94 ASASrvyRsr 449-454

Norris, James L., III: Sample design for observing classes of objects. 95 CommStA 24 3147-3162

— —; Pollock, Kenneth H.: A capture-recapture model with heterogeneity and behavioural response. 95 EnvEcoSt 2 305-313

North, Gerald R.; Kim, Kwang-Y.; Shen, Samuel S. P.; Hardin, James W.: Detection of forced climate signals. Part I: Filter theory. 95 JClimate 8 401-408

North, Philip M.: Manly, B. F. J.; McDonald, L. L.; Thomas, D. L. (Auth); — — (Rev): Review of "Resource selection by animals: Statistical design and analysis for field studies". 95 JAppStat 22 192-193

Northridge, Mary E.: Public health methods - Attributable risk as a link between causality and public health action. 95 AJPubHlh 85 1202-1204

Northrup, David A.: Underreporting of harassment in public places. 94 ASASrvyRsr 455-460

Northshield, S.: On the spectrum and Martin boundary of homogeneous spaces. 95 StPrLet 22 275-279

Norton, H. James: Teaching undergraduates and medical residents how to critique journal articles. 94 ASAProStEd 162-164

Norton, J. M.: Arnold, Jesse C.; Lemon, G. H.; — —: Sample size requirements for rare or abundant attribute sampling. 95 CommStB 24 131-148

Norton, J. P.: Mulquiney, J. E.; — —; Jakeman, A. J.; Taylor, J. A.: Random walks in the Kalman filter: Implications for greenhouse gas flux deductions. 95 EnvrMtrc 6 473-478

Nortvedt, Ragnar; Tuene, Stig: Multivariate evaluation of feed for Atlantic halibut. 95 Chemolab 29 271-282

Norvaiša, R.: The Strassen law of the iterated logarithm in Banach function spaces. 95 StPrLet 25 1-8

Norvaiša, Rimas; Samorodnitsky, Gennady: Stable processes with sample paths in Orlicz spaces. 94 AnlsProb 22 1904-1929

Noss, Brian: Long, J. Scott; — —: Data analysis using GAUSS and Markov. 95 SocMethR 23 303-328

Notz, W. I.: MacEachern, S. N.; — —; Whittinghill, D. C.; Zhu, Y.: Robustness to the unavailability of data in the linear model, with applications. 95 JStPlInf 48 207-213

O'Sullivan, Finbarr: A study of least squares and maximum likelihood for image reconstruction in positron emission tomography. 95 AnlsStat 23 1267-1300
Kooperberg, Charles; — —: Persistent oscillation patterns in climatological fields. 94 ASASrvyRsr 200-205

Oakes, D.: Rosenbaum, P. R. (Auth); — — (Rev): Review of "Observational studies". 96 ShrtBkRv 16/1 1-1

Oakes, David: Multiple time scales in survival analysis. 95 LiDaAnal 1 7-18

Oberhelman, Dennis: Kadiyala, K. Rao; — —: Estimation and inference in simultaneous equations. 94 ASAProBuEc 235-240

Obuchowski, Nancy A.; Rockette, Howard E.: Hypothesis testing of diagnostic accuracy for multiple readers and multiple tests: An ANOVA approach with dependent observations. 95 CommStB 24 285-308

Obwona, M. B.: Kalirajan, K. P.; — —: Estimating individual response coefficients in varying coefficients regression models. 95 JAppStat 22 477-484

Ocaña, Jordi; Vegas, Esteban: Variance reduction for Bernoulli response variables in simulation. 95 CmpStDA 19 631-640
Sanchez, Alex; — —; Utzet, Frederic: Sampling theory, estimation, and significance testing for Prevosti's estimate of genetic distance. 95 Biomtrcs 51 1216-1235

Oddy, V. H.: McPhee, M. J.; Ransom, K. P.; — —: Calibration of empirical models: Simulation of pasture growth and decay. 95 EnvrMtrc 6 505-510

Odell, Patricia M.; Schumacher, Phyllis S.: Attitudes and achievement in freshman math classes. 94 ASAProStEd 197-202

Oden, Neal: Adjusting Moran's I for population density. 95 StatMed 14 17-26

Odom-Maryon, Tamara: Ahn, Chul; — —: Estimation of a common odds ratio under binary cluster sampling. 95 StatMed 14 1567-1576

Odom-Maryon, Tamara L.: Niland, Joyce C.; — —; Lee, Jennifer; Tilley, Barbara C.: A survey of biostatistical consulting units throughout North America. 95 AmerStat 49 183-189

Oehler, Karen L.; Gray, Robert M.: Combining image compression and classification using vector quantization. 95 IEEEPAMI 17 461-473

Oehlert, Gary W.: The ability of wet deposition networks to detect temporal trends. 95 EnvrMtrc 6 327-339
— —; Lee, Richard J.; Van Orden, Drew: Statistical analysis of asbestos fibre counts. 95 EnvrMtrc 6 115-126

Ofir, Chezy: Bechtel, Gordon G.; — —; Khuri, André I.: Replicated paired comparisons at the individual level. 95 BJMaStPs 48 115-127

Ofori-Nyarko, S.: Gupta, A. K.; — —: Improved estimation of generalized variance and precision. 95 MaOpfStS 26 99-109

Ofori-Nyarko, Samuel: Gupta, A. K.; — —: Improved minimax estimators of normal covariance and precision matrices. 95 MaOpfStS 26 19-25

Öfversten, Jukka: Estimation in mixed models via layer triangular transformation. 95 CmpStDA 20 657-667

Oganian, V. K.: Ambartzumian, R. V.; — —: Measure generation by Euler functionals. 95 AdvAppPr 27 606-626

Ogasawara, Haruhiko: Structural model of ability distribution in the item response theory. 95 Behvmtrk 22 37-48
— —: The gamma-gamma regression for the distribution model of event times. 95 JpnPsyRs 37 70-79

Ogata, Yosihiko: Evaluation of probability forecasts of events. 95 IntJFor 11 539-541

Ogawa, H.: Liu, D.; Yamashita, Y.; — —: Pattern recognition in the presence of noise. 95 PattrnRc 28 989-995

Ogawa, Naoko; Minami, Mihoko; Sibuya, Masaaki: Characteristics of the linear smoothing spline (Japanese). 95 JpJAppSt 24 27-41

Ogwang, Tomson: Amirkhalkhali, Saleh; — —; Amirkhalkhali, Samad; Rao, U. L. Gouranga: On the forecasting performance of estimators for a structural equation in a large system: Some Monte Carlo results. 95 Ststcian 44 343-351

Oh, H. Lock: Hinkins, Susan; — —; Scheuren, Fritz: Inverse sampling design algorithms. 94 ASASrvyRsr 626-631

Oh, Hyun Sook; DasGupta, Anirban: Robustness of Stein's two stage procedure. 95 SqtlAnly 14 321-334

Oh, Kwang Sik: Hong, Dug Hun; — —: On the weak law of large numbers for arrays. 95 StPrLet 22 55-57
Hong, Dug Hun; — —; Park, Hee Joo: On a stopping rule for the random walks with time stationary random distribution function. 95 JKoreaSt 24 293-301

Oh, Kyong Joo: Kim, Byung Soo; — —; Park, Cheolyong: Overdispersion in count data - A review (Korean). 95 KorApS 8/2 147-161

Oh, Myongsik: On maximum likelihood estimation of cell probabilities in $2 \times k$ contingency tables under negative dependence restrictions with various sampling schemes. 95 CommStA 24 2127-2143

Oh, Sungsoo; Paletz, Dave; Kim, Jay; Salyers, Eddie: Development of edit parameters for 1992 economic census enterprise reports. 94 ASASrvyRsr 1144-1149

Ohlsson, Esbjörn: Dalén, Jörgen; — —: Variance estimation in the Swedish consumer price index. 95 JBES 13 347-356

Öhman, Marie-Louise: Estimating location parameters using censored rank statistics. 95 BiomtrcJ 37 767-782

Öhrn, Kenth; Ahlén, Anders; Sternad, Mikael: A probabilistic approach to multivariable robust filtering and open-loop control. 95 IEEEAuCn 40 405-418

Ohser, J.; Mücklich, F.: Stereology for some classes of polyhedrons. 95 AdvAppPr 27 384-396

Ohtani, K.: Generalized ridge regression estimators under the LINEX loss function. 95 StHefte 36 99-110

Ohtani, Kazuhiro: Srivastava, Viren K.; — —: A comparison of interval constrained least squares and mixed regression estimators. 95 CommStA 24 395-413

Oja, H.: Niinimaa, A.; — —: On the influence functions of certain bivariate medians. 95 JRSS-B 57 565-574

Oja, Hannu: Möttönen, Jyrki; — —: Multivariate spatial sign and rank methods. 95 JNnparSt 5 201-213
Möttönen, Jyrki; — —; Krause, Ulf; Rantakallio, Paula: Application of random coefficient regression model to myopia data: A case study. 95 BiomtrcJ 37 657-672

Okabe, A.; Boots, B.; Sugihara, K. (Auth); Lorz, U. (Rev): Review of "Spatial tessellations. Concepts and applications of Voronoi diagrams". 95 MaOpfStS 26 183-183

Okabe, Yasunori; Ootsuka, Takashi: Application of the theory of KM_2O-Langevin equations to the non-linear prediction problem for the one-dimensional strictly stationary time series. 95 JMaJapan 47 349-367

Okada, Mihoko; O'Brien, Myles: IntroStat: A hypertext-based design for an electronic textbook to introduce biomedical statistics. 95 CmpBiomd 46 265-276

Okamoto, Masashi: Correspondence analysis of an artificial torus data. 94 Behvmtrk 21 149-161
— —; Endo, Hideki: Spherical trait in correspondence analysis of artificial data. 95 JJapnStS 25 183-191

Okamoto, Yasuharu: Analysis of semantic space by probabilistic unfolding model (Japanese). 94 JpJBeh 21/1 66-74

Oksanen, Ernest H.: George, Peter J.; — —; Veall, Michael R.: Analytic and bootstrap approaches to testing a market saturation hypothesis. 95 MaCmpSim 39 311-315

Øksendal, B.: Stochastic differential equations: An introduction with applications (Fourth edition). 95 Springer:Brln:NY 271p

Øksendal, Bernt: Lindstrøm, Tom; — —: Stability properties of stochastic partial differential equations. 95 StocAnAp 13 177-204
Hida, Takeyuki; Kuo, Hui-Hsuing; Potthoff, Jürgen; Streit, Ludwig (Auth); — — (Rev): Review of "White noise. An infinite dimensional calculus". 95 BLndnMaS 27 189-191

Oldford, R. W.: A physical device for demonstrating confounding, blocking, and the role of randomization in uncovering a causal relationship. 95 AmerStat 49 210-216

Oliger, Joseph: Gustafsson, Bertil; Kreiss, Heinz-Otto; — —: Time dependent problems and difference methods. 95 Wiley:NY:UK 642p

Oliveira, Paulo Eduardo: Jacob, Pierre; — —: A general approach to nonparametric histogram estimation. 95 MaOpfStS 27 73-92

Olken, Frank; Rotem, Doron: Random sampling from databases: A survey. 95 StatComp 5 25-42
— —; Rotem, Doron: Sampling from spatial databases. 95 StatComp 5 43-57

Olkin, I.: Anderson, T. W.; Fang, K. T.; — — (Ed): Multivariate analysis and its applications. 95 Arnold:Lnd:MD 471p

Olkin, Ingram: Meta-analysis: Reconciling the results of independent studies. 95 StatMed 14 457-472
— —; Finn, Jeremy D.: Correlations redux. 95 PsychBu 118 155-164
— —; Guttman, Irwin; Philips, Robert: Estimating the number of aberrant laboratories. 95 PrEngInf 9 133-150
— —; Shaw, Douglas V.: Meta-analysis and its applications in horticultural science. 95 HortSci 30 1343-1348
— —; Viana, Marlos: Correlation analysis of extreme observations from a multivariate normal distribution. 95 JASA 90 1373-1379
Bowker, Albert H.; — —; Veinott, Arthur F., Jr: Gerald J. Lieberman. 95 PrEngInf 9 3-26
Marshall, Albert W.; — —: Multivariate exponential and geometric distributions with limited memory. 95 JMultiAn 53 110-125

Olla, S.: Kipnis, C.; Landim, C.; — —: Macroscopic properties of a stationary non-equilibrium distribution for a non-gradient interacting particle system. 95 AIHenrPB 31 191-221

Oller, J. M.; Corcuera, J. M.: Intrinsic analysis of statistical estimation. 95 AnlsStat 23 1562-1581

Oller, Josep M.: Villarroya, Angel; Ríos, Martín; — —: Discriminant analysis algorithm based on a distance function and on a Bayesian decision. 95 Biomtrcs 51 908-919

Olley, Steven: Pakes, Ariel; — —: A limit theorem for a smooth class of semiparametric estimators. 95 JEconmtx 65 295-332

Olofsson, Anita: Bergman, Lars R.; Kristiansson, Karl-Erik; — —; Säfström, Margaretha: Decentralised CATI versus paper and pencil interviewing: Effects on the results in the Swedish labour force surveys. 94 JOffStat 10 181-195

Olsder, Geert Jan: Basar, Tamer; — —: Dynamic noncooperative game theory (Second edition). 95 Academic:NY:Lnd 519p

Olsen, Jay K.: Battles, Stephanie J.; Adler, Robert K.; — —: Demand-side management: Utility programs offered and participants in the residential, commercial, and manufacturing sectors of the U.S. economy. 94 ASAProBuEc 20-27

Olsen, L.: — — (Auth); Taylor, S. J. (Rev): Review of "Random geometrically graph directed self-similar multifractals". 95 Metrika 42 277-278

Olsen, Lars: — — (Auth); Patzschke, Norbert (Rev): Review of "Random graph directed self-similar multifractals". 95 BiomtrcJ 37 374-374

Olshan, Andrew F.: Savitz, David A.; — —: Multiple comparisons and related issues in the interpretation of epidemiologic data. 95 AJEpidm 141 904-908

Savitz, David A.; — —: Multiple comparisons and related issues in the interpretation of epidemiologic data. 95 AJEpidm 142 904-908

Olson, Barbara A.: Beauregard, Michael R.; Mikulak, Raymond J.; — — (Auth); Belunis, Nancy (Rev): Review of "A Practical guide to statistical quality improvement: Opening up the statistical toolbox". 95 Technmcs 37 233-234

Beauregard, Michael R.; Mikulak, Raymond J.; — — (Auth); Richards, John D. (Rev): Review of "A practical guide to statistical quality improvement: Opening up the statistical toolbox". 95 JQualTek 27 177-178

Olson, J. A.: Song, K. S.; Müller, H. G.; Clifford, A. J.; Furr, H. C.; — —: Estimating derivatives of pharmacokinetic response curves with varying bandwidths. 95 Biomtrcs 51 12-20

Olson, J. M.: Testing the Hardy-Weinberg law across strata. 93 AnHuGene 57 291-295

Olson, Jane M.: Multipoint linkage analysis using sib pairs: An interval mapping approach for dichotomous outcomes. 95 AJHuGene 56 788-798

— —; Wijsman, Ellen M.: Design and sample-size considerations in the detection of linkage disequilibrium with a disease locus. 94 AJHuGene 55 574-580

Olsson, Arne: Thelin, Bernt; Lundstedt, Torbjörn; Lundgren, Rolf; — —; Batra, Satish: Classification of Estradurin batches: Correlation between 31P NMR and a biological duration test for batch approval. 95 Chemolab 27 135-145

Olszak, Michael; Ritschard, Gilbert: The behaviour of nominal and ordinal partial association measures. 95 Ststcian 44 195-212

Oltman, D. Olson; Lackritz, J. R. (Auth); Garcia, Eva Ferreira (Rev): Review of "Statistics for business and economics". 95 JAppStat 22 437-437

Oluyede, Broderick O.: Estimation and testing of multinomial parameters under order restrictions. 94 ASAProQlPr 276-280

Oman, Samuel D.: Checking the assumptions in mixed-model analysis of variance: A residual analysis approach. 95 CmpStDA 20 309-330

Omar, Rumana Z.; Stallard, Nigel; Whitehead, John: A parametric multistate model for the analysis of carcinogenicity experiments. 95 LiDaAnal 1 327-346

Omey, E.: On a subclass of regularly varying functions. 95 JStPlInf 45 275-290

Omey, Edward: Molchanov, Ilya S.; — —; Kozarovitzky, Eugene: An elementary renewal theorem for random compact convex sets. 95 AdvAppPr 27 931-942

Omori, Yasuhiro: Johnson, N. L.; Kotz, S.; Balakrishnan, N. (Auth); — — (Rev): Review of "Continuous univariate distributions, Volume 1 (Second edition)". 95 JASA 90 1490-1491

Omre, Henning: Høst, Gudmund; — —; Switzer, Paul: Spatial interpolation errors for monitoring data. 95 JASA 90 853-861

Ong, S. H.: Characterization of the negative binomial and gamma distributions. 95 BiomtrcJ 37 251-254

— —: Computation of probabilities of a generalized log-series and related distributions. 95 CommStA 24 253-271

— —: Computation of bivariate gamma and inverted beta distribution functions. 95 JStCmpSm 51 153-163

— —: Some stochastic models leading to the convolution of two binomial variables. 95 StPrLet 22 161-166

— —; Muthaloo, Subarau: A class of discrete distributions suited to fitting very long-tailed data. 95 CommStB 24 929-945

Ong, Ting Eng: Tsong, Yi; Hammerstrom, Thomas; Lin, Karl; — —: Dissolution test acceptance sampling plans. 95 JBiophSt 5 171-183

Onishi, Haruo: A user-knowledge-based variable selection method for limited information maximum likelihood using principal components. 95 CmpStDA 19 379-399

Önkal, Dilek; Muradoğlu, Gulnur: Effects of feedback on probabilistic forecasts of stock prices. 95 IntJFor 11 307-319

Oommen, B. J.: Valiveti, R. S.; — —; Zgierski, J. R.: Adaptive linear list reorganization under a generalized query system. 95 JAppProb 32 793-804

Ootsuka, Takashi: Okabe, Yasunori; — —: Application of the theory of KM_2O-Langevin equations to the non-linear prediction problem for the one-dimensional strictly stationary time series. 95 JMaJapan 47 349-367

Openshaw, S. J.: Fabrizius, M. A.; — —: Methods to evaluate populations for alleles to improve an elite hybrid. 94 ThApGene 88 653-661

Openshaw, Stan: Developing automated and smart spatial pattern exploration tools for geographical information systems applications. 95 Ststcian 44 3-16

Skinner, Chris; Marsh, Catherine; — —; Wymer, Colin: Disclosure control for census microdata. 94 JOffStat 10 31-51

Opitz, O.; Lausen, B.; Klar, R. (Ed); Windham, M. P. (Rev): Review of "Information and classification: Concepts, methods and applications". 95 JClass 12 157-158

Oppenheim, G.: Antoniadis, A.; — — (Ed): Wavelets and statistics. 95 Springer:-Brln:NY 410p

Viano, M. C.; Deniau, Cl.; — —: Long-range dependence and mixing for discrete time fractional processes. 95 JTimSrAn 16 323-338

Opsomer, Jean-Didier; Agras, Jean; Carpi, Anthony; Rodrigues, Geraldo: An application of locally weighted regression to airborne mercury deposition around an incinerator site. 95 EnvrMtrc 6 205-219

Ord, J. K.: Scazzero, Joseph A.; — —: Interval estimates for the optimum point of a quadratic logistic curve - A comparison of different estimation methods via simulation. 95 CommStB 24 207-226

Stuart, A.; — — (Auth); Kemp, C. D. (Rev): Review of "Kendall's advanced theory of statistics, Volume I: Distribution theory (Sixth edition)". 95 Biomtrcs 51 1583-1584

Ord, J. Keith: Comment on "Reflections on making statistics more effective in business schools (MSMESB)". 94 ASAProBuEc 319-320

Orkin, Michael; Kakigi, Richard: What is the worth of free casino credit?. 95 AmMaMnth 102 3-8

Orlandić, Ratko: Mahmoud, Hosam M.; — —: Toward a formal derivation of the expected behavior of prefix b-trees. 95 PrEngInf 9 183-192

Orme, C. D.: Godfrey, L. G.; — —: The sensitivity of some general checks to omitted variables in the linear model. 94 IntEconR 35 489-506

Orme, Chris: On the use of artificial regressions in certain microeconometric models. 95 EconTher 11 290-305

Orr, Carolyn: Eckes, George P.; — —: A train the trainer program for design of experiments. 95 ASQCTekCon 348-355

Orr, David B.: Fundamentals of applied statistics. A conceptual approach. 95 ChapmnHall:Lnd 250p

Ortiz-Fernández, M. C.; Herrero-Gutiérrez, A.: Regression by least median squares, a methodological contribution to titration analysis. 95 Chemolab 27 231-243

Ortiz-Monasterio R., Jóse Iván: Crossa, José; Cornelius, Paul L.; Sayre, Ken; — —: A shifted multiplicative model fusion method for grouping environments without cultivar rank change. 95 CropSci 35 54-62

Orzack, Steven Hecht; Chappell, Rick: Sample-size determination for parent-offspring regression of a composite trait. 95 Biomtrcs 51 822-830

Osada, Hirofumi; Saitoh, Toshifumi: An invariance principle for non-symmetric Markov processes and reflecting diffusions in random domains. 95 ProbTher 101 45-63

Osahan, S. S.: Tracy, D. S.; — —: A partial randomized response strategy. 95 Test 4 315-321

Osborne, M. R.; Smyth, G. K.: A modified prony algorithm for exponential function fitting. 95 SIAMSSC 16 119-138

Oshima, T. C.: Algina, James; — —: An improved general approximation test for the main effect in a split-plot design. 95 BJMaStPs 48 149-160

Osiewalski, J.: Koop, G.; Steel, M. F. J.; — —: Posterior analysis of stochastic frontier models using Gibbs sampling. 95 CmpStQ 10 353-373

Osiewalski, Jacek: Fernández, Carmen; — —; Steel, Mark F. J.: Modeling and inference with v-spherical distributions. 95 JASA 90 1331-1340

Koop, Gary; — —; Steel, Mark F. J.: Bayesian long-run prediction in time series models. 95 JEconmtx 69 61-80

Oskrochi, G. Reza: Analysis of censored correlated observations. 95 JStPlInf 47 165-180

Osman, Magued I.: Shafy, Mona Abdel; — —: The compound Gompertz distribution. 95 EgpStJ 39 89-105

Osmint, Jeffrey B.; Martin, Antoinette Ware: Response in federally sponsored establishment surveys. 94 ASASrvyRsr 977-982

Ostroff, David J.: Krishnamurty, Goteti Bala; Kasovia-Schmitt, Patricia; — — (Auth); Hilden, Jørgen (Rev): Review of "Statistics. An interactive text for the health and life sciences". 95 StatMed 14 2621-2622

Ostrovskii, E. I.; Tsykunova, S. Yu.: Asymptotic properties of distribution of maximum of Gaussian nonstationary process occurring in covariance statistic. 94 ThProbAp 39 527-534

Ostrovskii, Iosif: Kotz, Samuel; — —: Correction to "Characteristic functions of a class of elliptical distributions" (V49 p164-178). 95 JMultiAn 52 352-354

Ostrowski, E.: Hand, D.; Daly, F.; Lunn, A.; McConway, K.; — — (Ed); Ziegel, Eric R. (Rev): Review of "A handbook of small data sets". 95 Technmcs 37 242-242

Hand, D. J.; Daly, F.; Lunn, A. D.; McConway, K.; — — (Ed); Chatfield, C. (Rev): Review of "A handbook of small data sets". 95 JRSS-A 158 349-349

Hand, D. J.; Daly, F.; Lunn, A. D.; McConway, K. J.; — — (Ed); Schimek, Michael G. (Rev): Review of "Handbook of small data sets". 95 CmpStQ 10 197-197

Hand, D. J.; Daly, F.; Lunn, A. D.; McConway, K. J.; — — (Ed); Street, Deborah J. (Rev): Review of "A handbook of small data sets". 95 AstrlJSt 37 251-251

Hand, D. J.; Daly, F.; Lunn, A. D.; McConway, K. J.; — — (Auth); Trenkler, Dietrich (Rev): Review of "A handbook of small data sets". 95 CmpStDA 19 101-101

Ota, Tatsuya; Nei, Masatoshi: Variance and covariances of the numbers of synonymous and nonsynonymous substitutions per site. 94 MolBioEv 11 613-619

Otranto, Edoardo; Gallo, Giampiero M.: Regression diagnostic techniques to detect balanced space-to-time ratios in STARMA models. 94 Metron 52/3 129-145

Ott, Jurg: Detecting marker inconsistencies in human gene mapping. 93 HumHrdty 43 25-30

Leal, Suzanne M.; — —: A likelihood approach to calculating risk support intervals. 94 AJHuGene 54 913-917

Weeks, Daniel E.; — —; Lathrop, G. Mark: Detection of genetic interference: Simulation studies and mouse data. 94 Genetcs 136 1217-1226

Weeks, Daniel E.; Lathrop, G. Mark; — —: Multipoint mapping under genetic interference. 93 HumHrdty 43 86-97

— — (Auth); Neuman, Rosalind (Rev): Review of "Analysis of human genetic linkage (Revised edition)". 93 BehavGen 23 299-300

Ott, L.: Hildebrand, D. K.; — — (Auth); Garin, Araceli (Rev): Review of "Statistical thinking for managers (Third edition)". 95 JAppStat 22 438-439

Ott, Teunis J.: Brightwell, Graham; — —; Winkler, Peter: Target shooting with programmed random variables. 95 AnlsApPr 5 834-853

Ott, Wayne R.: Environmental statistics and data analysis. 95 CRCPress:FL 336p

— — (Auth); Lesser, Virginia M. (Rev): Review of "Environmental statistics and data analysis". 95 AmerStat 49 398-399

Otto, Gordon H.; Thompson, Russell G.: The search for best practices in the U.S. petroleum industry. 94 ASAProQlPr 70-75

Otto, Sarah P.: Bergman, Aviv; — —; Feldman, Marcus W.: On the evolution of recombination in haploids and diploids II. Stochastic models. 95 Complxty 1 49-57

Ou, Jihong: Li, Jingwen; — —: Characterizing the idle-period distribution of GI/G/1 queues. 95 JAppProb 32 247-255

Oud, J. H. L.: Jansen, R. A. R. G.; — —: Longitudinal LISREL model estimation from incomplete panel data using the EM algorithm and the Kalman smoother. 95 StNeerla 49 362-377

Ouknine, Y.: Erraoui, M.; — —: Limit theorems for anticipative stochastic differential equations (French). 95 Stochatx 53 1-11

Ouknine, Youssef; Rutkowski, Marek: Local times of functions of continuous semimartingales. 95 StocAnAp 13 211-231

Ould-Saïd, Elias: Lecoutre, Jean Pierre; — —: Hazard rate estimation for strong-mixing and censored processes. 95 JNnparSt 5 83-89

Lecoutre, Jean-Pierre; — —: Convergence of the conditional Kaplan-Meier estimate under strong mixing. 95 JStPlInf 44 359-369

Ouliaris, Sam: Fong, Wai Mun; — —: Spectral tests of the martingale hypothesis for exchange rates. 95 JApEcmtx 10 255-271

Overall, John E.; Atlas, Robert S.: Power of a test that is robust against variance heterogeneity. 95 PsychRpt 77 155-159

— —; Doyle, Suzanne R.: Reply to comment on "Sample size estimation for repeated measurement designs". 95 ConClnTr 16 451-452

Overbeck, Ludger; Röckner, Michael; Schmuland, Byron: An analytic approach to Fleming-Viot processes with interactive selection. 95 AnlsProb 23 1-36

Overdijk, D. A.: van Berkum, E. E. M.; Linssen, H. N.; — —: The reduction and ancillary classification of observations. 95 JStPlInf 45 347-372

Overgaag, P. H. A.: Goedegebuure, R. V.; — —: A classification of types of use of official statistics. 91 StatJUN 8 217-232

Overton, W. Scott; Stehman, S. V.: Design implications of anticipated data uses for comprehensive environmental monitoring programmes. 95 EnvEcoSt 2 287-303

— —; Stehman, Stephen V.: The Horvitz-Thompson as a unifying perspective for probability sampling: With examples from natural resource sampling. 95 AmerStat 49 261-268

Rosenberg, Daniel K.; — —; Anthony, Robert G.: Estimation of animal abundance when capture probabilities are low and heterogeneous. 95 JWildlMg 59 252-261

Owen, Art B.: Nonparametric likelihood confidence bands for a distribution function. 95 JASA 90 516-521

Owen, Guillermo: Game theory (Third edition). 95 Academic:NY:Lnd 472p

Owen, R. W.: Glazebrook, K. D.; — —: On the value of adaptive solutions to stochastic scheduling problems. 95 MaOpRsr 20 65-89

Owen, Robin E.; Packer, Laurence: Estimation of the proportion of diploid males in populations of Hymenoptera. 94 Heredity 72 219-227

Owens, Ada D.: Stone, Roslyn A.; Marsh, Gary M.; — —; Quinn, Margaret M.; Smith, Thomas J.: Retrospective estimation of airborne fiber exposure: Statistical aspects. 94 ASAProEpid 127-132

Owens, Stephen D.: A study in library structure and organization and their relationship to the census of governments. 94 ASAProGovt 64-69

Oxley, Les; Roberts, Colin J.: Modelling the supply of fishing effort. 95 EnvrMtrc 6 541-545

Oya, Kosuke: Size corrections for the Wald statistic in structural equation models. 95 MaCmpSim 39 317-322

Özekici, Süleyman; Li, Jingwen; Chou, Fee Seng: Waiting time in M/G/1 queues with impolite arrival disciplines. 95 PrEngInf 9 255-267

Ozgur, Ceyhun: Strasser, Sandra E.; — —: Undergraduate business statistics: A survey of topics and teaching methods. 95 Intrfc 25/3 95-103

Paap, Richard: Franses, Philip Hans; — —: Moving average filters and periodic integration. 95 MaCmpSim 39 245-249

Pacheco, A.: Prabhu, N. U.; — —: A storage model for data communication systems. 94 ThProbAp 39 604-627

Pacheco, Roberto: Kandel, Abraham; Martins, Alejandro; — —: Comment on "A probabilistic and statistical view of fuzzy methods". 95 Technmcs 37 276-281

Pachón, J. G.: Ríos-Insua, S.; — —; Mateos, A.: A method of multiobjective decision making using a vector value function. 94 Questo 18 153-172

Pacifico, M. Perone: Conigliani, C.; Dall'aglio, M.; — —; Salinetti, G.: Robust statistics: From classical to Bayesian analysis. 94 Metron 52/3 89-109

Packer, Laurence: Owen, Robin E.; — —: Estimation of the proportion of diploid males in populations of Hymenoptera. 94 Heredity 72 219-227

Padamadan, Veenus; Nair, K. R. Muraleedharan: Characterization of a bivariate Pareto distribution. 94 JIndnStA 32 15-20

Paddle, G. M.: Obituary: Albert Baines, 1927-1994. 95 JRSS-A 158 179-180

Padgett, W. J.: Gulati, Sneh; — —: Nonparametric function estimation from inversely sampled record-breaking data. 95 CandJSt 23 359-368

Gulati, Sneh; — —: Smooth nonparametric function estimation from inversely sampled record-breaking data. 94 ASAProPEng 140-145

Dodson, Bryan (Auth); — — (Rev): Review of "Weibull analysis". 95 JQualTek 27 395-396

Padilla, Maria L. R.: Goel, Prem K.; — —: Generalized Hellinger transforms as information measures. 94 ASAProBays 78-83

Pados, Dimitris A.; Papantoni-Kazakos, P.: New nonleast-squares neural network learning algorithms for hypothesis testing. 95 IEEENrlN 6 596-609

Pagan, Adrian: The ET interview: Gregory C. Chow. 95 EconTher 11 597-624

Pagano, Marcello: Tu, Xin Ming; Litvak, Eugene; — —: On the informativeness and accuracy of pooled testing in estimating prevalence of a rare disease: Application to HIV screening. 95 Biomtrka 82 287-297

Page, Dominique; Tremolieres, Raymond: Optimal dynamic equilibrium in insurance policies. 95 AplStMDA 11 181-198

Pagel, Mark: Seeking the evolutionary regression coefficient: An analysis of what comparative methods measure. 93 JThBio 164 191-205

Pagés, L.: Goulard, M.; — —; Cabanettes, A.: Marked point process: Using correlation functions to explore a spatial data set. 95 BiomtrcJ 37 837-853

Pahkinen, E. J.: Lehtonen, R.; — — (Auth); Plachky, D. (Rev): Review of "Practical methods for design and analysis of complex surveys". 95 StatDecn 13 399-400

Pahkinen, Erkki J.: Lehtonen, Risto; — —: Practical methods for design and analysis of complex surveys. 95 Wiley:NY:UK 337p

Lehtonen, Risto; — — (Auth); Kemp, C. D. (Rev): Review of "Practical methods for design and analysis of complex surveys". 95 ShrtBkRv 15/2 24-25

Pahwa, Savita: Lesser, Martin L.; Kohn, Nina E.; Napolitano, Barbara A.; — —: The FU-PLOT: A graphical method for visualizing the timing of follow-up in longitudinal studies. 95 AmerStat 49 139-143

Pai, Jeffrey S.; Ravishanker, Nalini: Bayesian analysis of fractionally differenced ARIMA processes using Gibbs sampling. 94 ASAProBays 69-74

Paik, Myunghee Cho: The EM algorithm for the quasi-likelihood regression model. 95 CommStA 24 1403-1430

— —; Begg, Melissa Dowd; El-Sadr, Wafaa; Gorman, Jack; Stein, Zena: Difference in clinical implications of CD4 counts among HIV-infected homosexual men and injection drug using men and women. 95 StatMed 14 1889-1900

Pak, Rojin: Robustness of minimum disparity estimators in linear regression models. 95 JKoreaSt 24 349-360

Pakes, Anthony G.: Quasi-stationary laws for Markov processes: Examples of an always proximate absorbing state. 95 AdvAppPr 27 120-145

— —: Characterization of discrete laws via mixed sums and Markov branching processes. 95 StocProc 55 285-300

Pakes, Ariel; Olley, Steven: A limit theorem for a smooth class of semiparametric estimators. 95 JEconmtx 65 295-332

Pal, Nabendu; Elfessi, Abdulaziz: Improved estimation of a multivariate normal mean vector and the dispersion matrix: How one affects the other. 95 SankhyaA 57 267-286

— —; Ling, Chiahua: Improved minimax estimation of powers of the variance of a multivariate normal distribution under the entropy loss function. 95 StPrLet 24 205-211

— —; Sinha, Bikas K.; Chaudhuri, Gopal; Chang, Ching-Hui: Estimation of a multivariate normal mean vector and local improvements. 95 MaOpfStS 26 1-17

Anderson, Charles; — —: A note on admissibility when precision is unbounded. 95 AnlsStat 23 593-597

Carpenter, Mark; — —: Estimation of the smaller and larger of two exponential location parameters with a common unknown scale parameter. 95 CalcutSt 45 61-71

Paletz, Dave: Oh, Sungsoo; — —; Kim, Jay; Salyers, Eddie: Development of edit parameters for 1992 economic census enterprise reports. 94 ASASrvyRsr 1144-1149

Paliatsos, A. G.: Kambezidis, H. D.; Tulleken, R.; Amanatidis, G. T.; — —; Asimakopoulos, D. N.: Statistical evaluation of selected air pollutants in Athens, Greece. 95 EnvrMtrc 6 349-361

Palit, Charles: Kale, Balkrishna; Besl, John; — —; Voss, Paul: Updating age-sex-specific net migration rates with limited data. 94 ASAProSoSt 116-121 (Disc: p122-123)

Palit, Charles D.: An evaluation of an estimation procedure from a DSRDD telephone sample design or how well do sample weights work. 94 ASASrvyRsr 647-652

Pallas, Michelle: Rocke, David M.; — —: Prediction of automotive emissions from gasoline composition. 94 ASAProPEng 269-273

Pallini, Andrea: Estimating probabilities from invariant permutation distributions. 94 JItalStS 3 77-91

— —: Bahadur exact slopes for a class of combinations of dependent tests. 94 Metron 52/1 53-65

Palm, Franz C.: Comment on "Bayesian model selection and prediction with empirical applications". 95 JEconmtx 69 333-335

Palm, R.; Iemma, A. F.: Some alternatives to classical regression in the case of collinearity (French). 95 RStApp 43/2 5-33

Palma, Daniela: On spatial analysis and map transformations. 94 QStMaA 16 127-143

Benedetti, Roberto; — —: Optimal sampling designs for dependent spatial units. 95 EnvrMtrc 6 101-114

Paloheimo, J. E.: Chen, Y.; — —: A robust regression analysis of recruitment in fisheries. 95 JFishRBC 52 993-1006

Palutikof, J. P.: Dukes, M. D. G.; — —: Estimation of extreme wind speeds with very long return periods. 95 JApMetrl 34 1950-1961

Pan, Guohua; Wolfe, Douglas A.: Distribution-free test based on ranks for comparing groups with umbrella orderings. 95 JNnparSt 5 409-416

Pan, Jian-Xin; Fang, Kai-Tai: Multiple outlier detection in growth curve model with unstructured covariance matrix. 95 AnInStMa 47 137-153

Pan, U. WinKey; Clark, Scott; Buncher, Ralph; Johnson, Dale: Implementing a database management system for an H.U.D. lead-based paint abatement study. 94 ASAProEnvr 162-166

Pan, Xuebiao: Jiang, Changjian; — —; Gu, Minghong: The use of mixture models to detect effects of major genes on quantitative characters in a plant breeding experiment. 94 Genetcs 136 383-394

Pan, Zhijie; Nonaka, Yasuo: Importance analysis for the systems with common cause failures. 95 RelEngSS 50 297-300

Panaretos, J.: Guttorp, P. (Auth); — — (Rev): Review of "Statistical inference for branching processes". 95 Biomtrcs 51 383-384

Panchapakesan, S.: Balakrishnan, N.; Gupta, Shanti S.; — —: Estimation of the location and scale parameters of the extreme value distribution based on multiply type-II censored samples. 95 CommStA 24 2105-2125

Balakrishnan, N.; Gupta, Shanti S.; — —: Estimation of the mean and standard deviation of the logistic distribution based on multiply type-II censored samples. 95 MaOpfStS 27 127-142

Pancheva, E.: On a problem of a Khinchin-type decomposition theorem for extreme values. 94 ThProbAp 39 329-336

Panda, S. K.; Nagabhushanam, A.: Fuzzy data distortion. 95 CmpStDA 19 553-562

Pandey, Arvind; Suchindran, C. M.: Some analytical models to estimate maternal age at birth using age-specific fertility rates. 95 SankhyaB 57 142-150

Pandey, B. N.; Malik, H. J.; Dubey, P. K.: Bayesian shrinkage estimators for a measure of dispersion of an inverse Gaussian distribution. 95 CommStA 24 2261-2270

Pandey, M.: Ferdous, J.; Uddin, Md. Borhan; — —: Reliability estimation with Weibull inter failure times. 95 RelEngSS 50 285-296

Pandey, R. N.: Ray, S. N.; — —: Role of administrative bodies in environmental studies. 95 SankhyaB 57 237-251

Pandey, S. K.: Singh, R. K.; — —; Srivastava, V. K.: Correction to "A generalized class of shrinkage estimators in linear regression when disturbances are not normal" (V23 p2029-2046). 95 CommStA 24 859-859

Singh, R. Karan; Misra, Sheela; — —: A generalized class of estimators in linear regression models with multivariate-t distributed error. 95 StPrLet 23 171-178

Pandher, Gurupdesh S.: Kovačević, Milorad S.; Yung, Wesley; — —: Resampling inference for quantile shares. 94 ASASrvyRsr 206-211

Panganamamula, Srinivasarao: Houshmand, Ali A.; — —: An analytical approximation and a neural network model for optimal sample size in vendor selection. 95 JStCmpSm 53 65-78

Panizzon, G.: — — (Auth); Iacobini, Alberto (Rev): Review of "Bilateral plans for inspection and process control by variables". 94 Metron 52/1 216-217

— — (Auth); Magagnoli, U. (Rev. Italian): Review of "Bilateral plans for inspection and process control by variables". 94 Statstca 54 539-540

Panizzon, Gaetano: — — (Auth); Copeland, Karen A. F. (Rev): Review of "Bilateral plans for inspection and process control by variables". 95 JQualTek 27 88-88

— — (Auth); Godsey, E. B. (Rev): Review of "Bilateral plans for inspection and process control by variables". 95 Technmcs 37 463-464

Panjer, Harry H.; Wang, Shaun: Computational aspects of Sundt's generalized class. 95 AstinB 25 5-18

Pankratz, Peter C.; Coria, José C.: Reporting negative contamination levels on silicon wafers - Is this nonsense?. 94 ASAProPEng 71-75

Panter, D. M.; Allen, F. L.: Using best linear unbiased predictions to enhance breeding for yield in soybean: I. Choosing parents. 95 CropSci 35 397-405

Pantula, Sastry G.: Arellano, Consuelo; — —: Testing for trend stationarity versus difference stationarity. 95 JTimSrAn 16 147-164

Etchison, Tonya; Brownie, Cavell; — —: A portmanteau test for spatial ARMA models. 95 Biomtrcs 51 1536-1542

Gan, Nianci; — —: Testing for deterministic trends in correlated data. 94 ASAProBuEc 141-146

Ramos, Rogelio Q.; — —: Estimation of nonlinear random coefficient models. 95 StPrLet 24 49-56

Papadatos, N.: Intermediate order statistics with applications to nonparametric estimation. 95 StPrLet 22 231-238

— —; Papathanasiou, V.: Distance in variation and a Fisher-type information. 95 MaMethSt 4 230-237

Papadatos, Nickos: Maximum variance of order statistics. 95 AnInStMa 47 185-193

Papadopoulos, Alex S.; Hoekstra, Robert M.: Some robustness issues in the Bayesian analysis of a Weibull failure model. 95 JStCmpSm 52 85-94

Papadopoulos, G. K.: Koutras, M. V.; — —; Papastavridis, S. G.: Runs on a circle. 95 JAppProb 32 396-404

Papadopoulos, Savas; Amemiya, Yasuo: Asymptotic robustness for the structural equation analysis of several populations. 94 ASAProBuEc 65-70

Papageorgiou, H.: Some remarks on the D compound Poisson distribution. 95 StHefte 36 371-375

— —; David, Katerina M.: The structure of compounded trivariate binomial distributions. 95 BiomtrcJ 37 81-95

— —; David, Katerina M.: On a class of bivariate compounded Poisson distributions. 95 StPrLet 23 93-104

— —; Loukas, S.: A bivariate discrete Charlier series distribution. 95 BiomtrcJ 37 105-117

Papaioannou, T.: Barbour, A. D.; Holst, L.; Janson, S. (Auth); — — (Rev): Review of "Poisson approximation". 95 Biomtrcs 51 382-383

Papamichael, Emmanuel M.; Evmiridis, Nickolaos P.; Thanasoulias, Nickolaos; Stefanou, Despina: Design of experiments for a precise estimation of the calibration curve of periodate in its determination by flow-injection analysis and chemiluminescence detection.. 95 Chemolab 30 227-237

Papantoni-Kazakos, P.: Pados, Dimitris A.; — —: New nonleast-squares neural network learning algorithms for hypothesis testing. 95 IEEENrlN 6 596-609

Papastavridis, S. G.: Koutras, M. V.; Papadopoulos, G. K.; — —: Runs on a circle. 95 JAppProb 32 396-404

Papathanasiou, V.: A characterization of the Pearson system of distributions and the associated orthogonal polynomials. 95 AnInStMa 47 171-176

Cacoullos, T.; — —: A generalization of covariance identity and related characterizations. 95 MaMethSt 4 106-113

Papadatos, N.; — —: Distance in variation and a Fisher-type information. 95 MaMethSt 4 230-237

Papp, László: Izsák, János; — —: Application of the quadratic entropy indices for diversity studies of drosophilid assemblages. 95 EnvEcoSt 2 213-224

Papp, Zoltan: Perez Wilson, Mario (Auth); — — (Rev): Review of "Multi-vari chart and analysis - A pre-experimentation technique". 95 Technmcs 37 121-121

Pappanastos, Edward A.; Adams, Benjamin M.: A comparison of robust control charts. 94 ASAProQlPr 316-320

Paranjpe, S. A.: Prayag, V. R.; Chiplonkar, S. A.; — —: Nested ANOVA for repeated surveys with autocorrelated errors. 95 BiomtrcJ 37 565-570

Parczewski, Andrzej: Meloun, M.; Militky, J.; Forina, M. (Auth); — — (Rev): Review of "Chemometrics for analytical chemistry, Volume 2: PC-aided regression and related methods". 95 Chemolab 29 326-327

Pardo, Francisco: Efird, Jimmy Thomas; — —: A non-parametric two-sample survival test based on a single occupancy Fermi-Dirac model for the discrete range distribution. 94 ASAProBiop 166-170

Pardo, L.: Menéndez, M.; Morales, D.; — —; Vajda, I.: Divergence-based estimation and testing of statistical models of classification. 95 JMultiAn 54 329-354

Menéndez, M. L.; — —; Morales, D.; Salicrú, M.: Some statistical applications of (r,s)-directed divergences. 92 UtilMath 42 115-127

Menéndez, M. L.; Morales, D.; — —; Salicrú, M.: Asymptotic behaviour and statistical applications of divergence measures in multinomial populations: A unified study. 95 StHefte 36 1-29

Morales, D.; — —; Vajda, I.: Asymptotic divergence of estimates of discrete distributions. 95 JStPlInf 48 347-369

Pardo, Leandro; Salicrú, Miguel; Menéndez, María Luisa; Morales, Domingo: A test for homogeneity of variances based on Shannon's entropy. 95 Metron 53 135-146

Pardoux, Étienne: Kurtz, Thomas G.; — —; Protter, Philip: Stratonovich stochastic differential equations driven by general semimartingales. 95 AIHenrPB 31 351-377

Parent, E.; Chaouche, A.; Girard, Ph.: In support of Bayesian statistics for quality control by attribute. Part I: Simple sampling plans (French). 95 RStApp 43/4 5-18

Parigi, G.; Schlitzer, G.: Quarterly forecasts of the Italian business cycle by means of monthly economic indicators. 95 JForcast 14 117-141

Parikh, Kirit S.: Statistics: A key technology for development - How well have we used the key?. 95 CalcutSt 45 25-59

Paris, J.: Broderick, G.; — —; Valade, J. L.: A composite representation of pulp quality. 95 Chemolab 29 19-28

Parisi, V.: Colosimo, A.; Morante, S.; — —; Rossi, G. C.: An improved method for detection of words with unusual occurrence frequency in nucleotidic sequences. 93 JThBio 165 659-672

Park, B. U.: Härdle, W.; — —: Testing increasing dispersion. 95 CmpStDA 19 641-653
Härdle, W.; — —; Tsybakov, A. B.: Estimation of non-sharp support boundaries. 95 JMultiAn 55 205-218
Lee, Y. H.; Cho, S.; Kim, W. C.; — —: On estimating integrated squared spectral density derivatives. 95 JStPlInf 48 165-184

Park, Byung-Gu: Lee, Eui Yong; Song, Mun Sup; — —: A diffusion model for a system subject to random shocks. 95 JKoreaSt 24 141-147

Park, Chang K.; Ahn, Kwang-Il: A new approach for measuring uncertainty importance and distributional sensitivity in probabilistic safety assessment. 94 RelEngSS 46 253-261

Park, Chanseok; Basu, Ayanendranath; Basu, Srabashi: Robust minimum distance inference based on combined distances. 95 CommStB 24 653-673

Park, Cheolyong: Some remarks on the chi-squared test with both margins fixed. 95 CommStA 24 653-661
Kim, Byung Soo; Oh, Kyong Joo; — —: Overdispersion in count data - A review (Korean). 95 KorApS 8/2 147-161
Lee, Sunho; — —; Kim, Byung Soo: Tests for detecting overdispersion in Poisson models. 95 CommStA 24 2405-2420

Park, D. K.; Shah, K. R.: On connectedness of row-column designs. 95 CommStA 24 87-96

Park, Dong Ho: Lim, Jae-Hak; — —: Trend change in mean residual life. 95 IEEEReli 44 291-296

Park, Evelyn: Gentleman, Jane F.; — —: The probability of divorce, based on the age difference of the married couple. 94 ASAProSoSt 216-221

Park, Hee Joo: Hong, Dug Hun; Oh, Kwang Sik; — —: On a stopping rule for the random walks with time stationary random distribution function. 95 JKoreaSt 24 293-301

Park, Hee-chang; Lee, Suk-hoon: Optimal design of partially step-stress life testing for the series systems (Korean). 95 KorApS 8/2 121-132

Park, Heon Jin; Fuller, Wayne A.: Alternative estimators and unit root tests for the autoregressive process. 95 JTimSrAn 16 415-429

Park, Hong Nai: Kim, Kyu Seong; Jeon, Jong Woo; — —: A study on the sample design of the fishery household economy survey (Korean). 95 KorApS 8/2 43-54

Park, Hyo-Il; Park, Sang-Gue: Quantile estimation of treatment effect for the two-sample problem with right censored data. 95 StPrLet 24 139-145

Park, Nae-Hyun: Lee, Sukhoon; — —; Kim, Yung-Hwan: A clustering method using the Coulomb energy network (Korean). 95 KorApS 8/1 39-50

Park, Pil S.; Kshirsagar, Anant M.: Effect of some modifications in standard discriminant analysis. 95 CommStB 24 807-816

Park, PooGyeon; Kailath, Thomas: New square-root algorithms for Kalman filtering. 95 IEEEAuCn 40 895-899
— —; Kailath, Thomas: Square-root RTS smoothing algorithms. 95 IJContrl 62 1049-1060

Park, S. H.: Kim, S. S.; — —: Dynamic plots for displaying the roles of variables and observations in regression model. 95 CmpStDA 19 401-418

Park, Sang-Gue: Park, Hyo-Il; — —: Quantile estimation of treatment effect for the two-sample problem with right censored data. 95 StPrLet 24 139-145

Park, Sung Hyun; Lee, Min Soo: Slope-rotatability over all directions in third order response surface models. 95 JKoreaSt 24 519-536

Park, Taesung; Shin, Min-Woong: A practical extension of the generalized estimating equation approach for longitudinal data. 95 CommStA 24 2561-2579

Park, Young J.; Cho, Sinsup: LaGrange multiplier test for multiple unit roots. 94 ASAProBuEc 147-150
Cho, Sinsup; — —; Ahn, Sung K.: Unit root tests for seasonal models with deterministic trends. 95 StPrLet 25 27-35

Park, Young Jin; Cho, Sinsup: Relations between the Lagrange multiplier tests and t-statistics for seasonal unit roots. 95 JKoreaSt 24 77-90

Park, YouSung: Asymptotic distributions of maximum queue lengths for M/G/1 and GI/M/1 systems. 95 JKoreaSt 24 19-29
— —; Lee, Haeyong; Kim, Keeyoung: Flows and some extreme values in multiple server open Jackson network. 95 JKoreaSt 24 389-405

Parker, E. T.: Brown, J. W.; Hedayat, A. S.; — —: A pair of orthogonal Latin squares of order 10 with four shared parallel transversals. 93 JCISS 18 113-115

Parker, Gary: A second order stochastic differential equation for the force of interest. 95 InsrneME 16 211-224

Parker, June M.: Rao, M. M. (Auth); — — (Rev): Review of "Conditional measures and applications". 95 BLndnMaS 27 612-613

Parker, R. L.: Stark, P. B.; — —: Bounded-variable least-squares: An algorithm and applications. 95 CmpStQ 10 129-141

Parker, Robert A.: Lesser, Martin L.; — —: The biostatistician in medical research: Allocating time and effort. 95 StatMed 14 1683-1692

Parkin, D. M.; Muir, C. S.; Whelan, S. L.; Gao, Y. T.; Ferlay, J.; Powell, J. (Ed); Hill, Catherine (Rev): Review of "Cancer incidence in five continents. Volume VI". 95 StatMed 14 104-104

Parlange, Marc B.: Katz, Richard W.; — —: Generalizations of chain-dependent processes: Application to hourly precipitation. 95 WaterReR 31 1331-1341

Parlar, M.: Bar-Lev, S. K.; — —; Perry, D.: Optimal sequential decisions for incomplete identification of group-testable items. 95 SqtlAnly 14 41-57

Parmar, Mahesh K. B.; Machin, David: Survival analysis: A practical approach. 95 Wiley:NY:UK 255p
— —; Machin, David (Auth); Juritz, J. M. (Rev): Review of "Survival analysis: A practical approach". 96 ShrtBkRv 16/1 2-3

Parmer, Randall: Thibaudeau, Yves; Kostanich, Donna; — —: The influence of implicit methodologies in the analysis of the CCO and the CPS. 94 ASASrvyRsr 730-735
Thompson, Jenny; — —: A comparison of nonparametric methods with parametric methods for the CPS CATI/CAPI mode effects analysis. 94 ASASrvyRsr 614-619

Parmigiani, Giovanni: Clyde, Merlise A.; DeSimone, Heather; — —: A comparison of algorithms for sampling models. 94 ASAProBays 211-216
Müller, Peter; — —: Optimal design via curve fitting of Monte Carlo experiments. 95 JASA 90 1322-1330

Parnes, Milton: Schultz, Delray J.; — —; Srinivasan, R.: Further applications of D-complete designs to group testing. 93 JCISS 18 31-41

Parsian, A.; Nematollahi, N.: On the admissibility of ordered Poisson parameter under the entropy loss function. 95 CommStA 24 2451-2467

Parsley, Teresa L.: Brehm, John (Auth); — — (Rev): Review of "The Phantom respondents: Opinion surveys and political representation". 94 JOffStat 10 351-355

Parslov, P.: Mulholland, M.; Hibbert, D. B.; Haddad, P. R.; — —: A comparison of classification in artificial intelligence, induction versus a self-organising neural networks. 95 Chemolab 30 117-128

Parsons, Joseph: Wallman, Katherine; Groves, Robert M.; — —; Davis, Mary C.; Lapham, Susan J.: Increasing statistical awareness through graduate training: The joint program in survey methodology. 94 ASAProSoSt 39-44

Parsons, P. Ellen: LeClere, Felicia B.; — —: Social correlates of response bias in the National Health Interview Survey: Lessons for the redesign. 94 ASASrvyRsr 373-378 (Disc: p379-381)

Parsons, Van L.: Malec, Donald; — —: State estimates for the National Health Interview Survey using a random digit dialing supplement. 94 ASASrvyRsr 1190-1195

Partanen, Päivi; Penttinen, Antti: Spatial mark-recapture method in the estimation of crayfish population size. 95 BiomtrcJ 37 979-994

Parthasarathy, K. R.: Bhat, B. V. Rajarama; — —: Markov dilations of nonconservative dynamical semigroups and a quantum boundary theory. 95 AIHenrPB 31 601-651

Parthasarathy, P. R.; Kumar, B. Krishna: Stochastic compartmental models with Prendville growth mechanisms. 95 MaBisci 125 51-60

Partrat, Christian: Compound model for two dependent kinds of claim. 94 InsrneME 15 219-231

Paruolo, P.; Pillati, M.: A nonlinear model for the conditional expectations of asset returns. 94 Statstca 54 329-347

Paruolo, Paolo: The role of the drift in I(2) systems. 94 JItalStS 3 93-123

Parzen, Emanuel: Findley, David F.; — —: A conversation with Hirotugu Akaike. 95 StatSci 10 104-117

Parzen, Michael: Lipsitz, Stuart R.; — —: Sample size calculations for non-randomized studies. 95 Ststcian 44 81-90

Pascual, Miguel A.; Hilborn, Ray: Conservation of harvested populations in fluctuating environments: The case of the Serengeti wildebeest. 95 JAppEcol 32 468-480

Pascutto, C.: Bernardinelli, L.; Clayton, D.; — —; Montomoli, C.; Ghislandi, M.; Songini, M.: Bayesian analysis of space-time variation in disease risk. 95 StatMed 14 2433-2443

Pashley, Peter J.: Bloxom, Bruce; — —; Nicewander, W. Alan; Yan, Duanli: Linking to a large-scale assessment: An empirical evaluation. 95 JEdStat 20 1-26

Pasquini, L.: Angeli, A.; Cocchi, D.; — —; Samoggia, A.: Family structure and income: A first analysis on cross-section data (Italian). 94 Statstca 54 435-452

Passaquindici, Maria: Hypergeometric series $\Phi(a,b,z)$: Further developments (Italian). 94 Metron 52/3 197-210

Passel, Jeffrey S.: Edmonston, Barry; Lee, Sharon M.; — —: U.S. Population projections for national origin groups: Taking into account ethnicity and exogamy. 94 ASAProSoSt 100-105 (Disc: p122-123)

Pasternack, Bernard S.: Kim, Mimi Y.; — —; Carroll, Raymond J.; Koenig, Karen L.; Toniolo, Paolo G.: Estimating the reliability of an exposure variable in the presence of confounders. 95 StatMed 14 1437-1446

Pastor, J. T.: Almiñana, M.; — —: Two new heuristics for the location set covering problem. 94 Top 2 315-328

Pastor, M. S. Sánchez: Derks, E. P. P. A.; — —; Buydens, L. M. C.: Robustness analysis of radial base function and multi-layered feed-forward neural network models. 95 Chemolab 28 49-60
Fernández, M. C. Ortiz; Herrero Gutiérrez, A.; — —; Sarabia, L. A.; Íñiguez Crespo, M.: The UNEQ, PLS and MLF neural network methods in the modelling and prediction of the colour of young red wines from the Denomination of Origin 'Rioja'. 95 Chemolab 28 273-285

Patak, Z.; Hidiroglou, M. A.: Raking ratio estimation: An application to the Canadian Monthly Retail Trade Survey. 94 ASASrvyRsr 299-304

Estevao, V.; — —; Särndal, C.: Maximizing the use of auxiliary information for calibration and regression estimation. 94 ASASrvyRsr 42-48

Patchen, Robert H.; Burgess, Bonnie B.; Cralley, Marla D.: Prospects for measuring the child radio audience in the U.S. using the seven-day personal diary method. 94 ASASrvyRsr 1220-1225

Paté-Cornell, M. E.; Fischbeck, P. S.: Reply to comment on "Probabilistic interpretation of command and control signals: Bayesian updating of the probability of nuclear attack". 95 RelEngSS 48 159-160

Paté-Cornell, M. Elisabeth; Fischbeck, Paul S.: Probabilistic interpretation of command and control signals Bayesian updating of the probability of nuclear attack. 95 RelEngSS 47 27-36

Patel, A. H.: Shah, S. M.; Ghosh, D. K.; — —: Construction of α-resolvable BIBDs using 1-resolvable and α-affine resolvable BIBDs. 93 UtilMath 43 219-224

Patel, M. N.; Gajjar, A. V.: Some results on maximum likelihood estimators of parameters of exponential distribution under type I progressive censoring with changing failure rates. 95 CommStA 24 2421-2435

Patel, Nitin: Gajjar, Yogesh; Mehta, Cyrus; — — (Auth); Smith, Peter W. F. (Rev): Review of "LOGXACT-TURBO (Version 1.1)". 95 ApplStat 44 139-141

Patel, Nitin R.: Mehta, Cyrus R.; — —: Exact logistic regression: Theory and examples. 95 StatMed 14 2143-2160

Senchaudhuri, Pralay; Metha, Cyrus R.; — —: Estimating exact p values by the method of control variates or Monte Carlo rescue. 95 JASA 90 640-648

Patil, G. P.: Statistical ecology and related ecological statistics - 25 years. 95 EnvEcoSt 2 81-89

— —: Editorial: Composite sampling. 95 EnvEcoSt 2 169-179

— —: Editorial: Ranked set sampling. 95 EnvEcoSt 2 271-285

— —: Statistical ecology and related ecological statistics - 25 years. 94 ASAProEnvr 1-7

— —; Sinha, A. K.; Taillie, C.: Finite population corrections for ranked set sampling. 95 AnInStMa 47 621-636

Gore, S. D.; — —: Identifying extremely large values using composite sample data. 94 EnvEcoSt 1 227-240 (Disc: p240-245)

Kaur, A.; — —; Sinha, A. K.; Taillie, C.: Ranked set sampling: An annotated bibliography. 95 EnvEcoSt 2 25-54

Myers, W. L.; Johnson, G. D.; — —: Rapid mobilization of spatial/temporal information in the context of natural catastrophes. 94 ASAProStGr 25-31

Taillie, C.; — —; Hennemuth, R. C.: Modelling and analysis of recruitment distributions. 95 EnvEcoSt 2 315-329

Talwalker, S.; — —; Taillie, C.: Qualitative and quantitative assessment of the risk from the exposure to fetotoxic chemical compounds. 95 EnvEcoSt 2 71-78 (Disc: p78-79)

Talwalker, S.; — —; Taillie, C.: Dose-response models for the probabilities of differential toxic effects in developmental toxicity studies. 94 JStatRsr 28 221-232

— —; Rao, C. R. (Ed); Burnham, K. P. (Rev): Review of "Handbook of statistics, Volume 12: Environmental statistics". 95 Biomtrcs 51 1584-1585

— —; Rao, C. R. (Ed); Jansen, U. (Rev): Review of "Multivariate environmental statistics. North-Holland Series in Statistics and Probability Volume 6". 95 BiomtrcJ 37 255-256

— —; Rao, C. R. (Ed); Jansen, U. (Rev): Review of "Environmental statistics: Handbook of statistics Volume 12". 95 BiomtrcJ 37 511-512

— —; Rao, C. R. (Ed); Seber, G. A. F. (Rev): Review of "Multivariate environmental statistics". 95 Biomtrcs 51 1189-1190

Patil, Ganapati P.: Kaur, Amarjot; Di Consiglio, Loredana; — —; Taillie, Charles: Propagation functions for monitoring distributional changes. 95 EnvEcoSt 2 239-269

Patil, Prakash: Hall, Peter; — —: Formulae for mean integrated squared error of nonlinear wavelet-based density estimators. 95 AnlsStat 23 905-928

Hall, Peter; — —: On wavelet methods for estimating smooth functions. 95 Bernoull 1 41-58

Paton, D.: Some effects of questionnaire design on reported rates of victimization. 94 ASASrvyRsr 443-448

Patterson, K. D.: Forecasting the final vintage of real personal disposable income: A state space approach. 95 IntJFor 11 395-405

— —: A state space approach to forecasting the final vintage of revised data with an application to the index of industrial production. 95 JForcast 14 337-350

Pattipati, K. R.: Shakeri, M.; — —; Kleinman, D. L.: Optimal measurement scheduling for state estimation. 95 IEEEAES 31 716-729

Pattison, P.: — — (Auth); Prescott, P. (Rev): Review of "Algebraic models for social networks". 95 JRSS-A 158 197-198

Patton, D. R.: — — (Auth); Elston, D. A. (Rev): Review of "Wildlife habitat relationships in forested ecosystems". 95 Biomtrcs 51 384-385

Patzschke, Norbert: Olsen, Lars (Auth); — — (Rev): Review of "Random graph directed self-similar multifractals". 95 BiomtrcJ 37 374-374

Paul, Nancy: Kamlet, Mark S.; — —; Greenhouse, Joel; Kupfer, David; Frank, Ellen; Wade, Martcia: Cost utility analysis of maintenance treatment for recurrent depression. 95 ConClnTr 16 17-40

Paul, Sudhir R.; Islam, Ali S.: Analysis of proportions in the presence of over-/under-dispersion. 95 Biomtrcs 51 1400-1410

Paula, Gilberto A.: Influence and residuals in restricted generalized linear models. 95 JStCmpSm 51 315-331

— —; Cordeiro, Gauss M.: Bias correction and improved residuals for non-exponential family nonlinear models. 95 CommStA 24 1193-1210

— —; Sen, Pranab K.: One-sided tests in generalized linear models with parallel regression lines. 95 Biomtrcs 51 1494-1501

Paulauskas, V.: Nualart, D. (Auth); — — (Rev): Review of "The Malliavin calculus and related topics". 96 ShrtBkRv 16/1 8-8

Paulin, G. D.: Eltinge, J. L.; Yansaneh, I. S.; — —: Assessment of reported differences between expenditures and low incomes in the U.S. Consumer Expenditure Survey. 94 ASASrvyRsr 96-105

Paulin, Geoffrey D.; Ferraro, David L.: Do expenditures explain income?: A study of variables for income imputation. 94 ASASrvyRsr 106-115

Paulsen, Jostein: Optimal per claim deductibility in insurance with the possibility of risky investments. 95 InsrneME 17 133-147

Pausche, Joan M.: 1994 National Census Test: Quantitative comparisons of coverage. 94 ASASrvyRsr 313-318

Pavešić, Nikola: Continuous speech recognition by a network of hidden Markov models. 95 JCmpInT 3 193-205

Paycha, S.: Arnaudon, M.; — —: Factorisation of semi-martingales on principal fibre bundles and the Faddeev-Popov procedure in gauge theories. 95 Stochatx 53 81-107

Payne, C. D.: Cleave, N.; Brown, P. J.; — —: Evaluation of methods for ecological inference. 95 JRSS-A 158 55-72

Payne, Clive; Payne, Joan; O'Muircheartaigh, Colm: Testing the tightening bond hypothesis: An application of new models for analysing trends in multiway tables. 95 JStPlInf 47 181-201

Heath, Anthony; Evans, Geoffrey; — —: Modelling the class-party relationship in Britain, 1964-92. 95 JRSS-A 158 563-574

Payne, Joan: Payne, Clive; — —; O'Muircheartaigh, Colm: Testing the tightening bond hypothesis: An application of new models for analysing trends in multiway tables. 95 JStPlInf 47 181-201

Pazdan, James L.: Graves, Thomas S.; — —: A permutation test analogue to Tarone's test for trend in survival analysis. 95 JStCmpSm 53 79-89

Pázman, Anrej: Comment on "Constrained optimization of experimental design". 95 MaOpfStS 26 167-168

Peach, Nigel: Basseville, M.; Nikiforov, I. V. (Auth); — — (Rev): Review of "Detection of abrupt changes: Theory and applications". 95 JRSS-A 158 185-186

Peach, William J.; Crick, Humphrey Q. P.; Marchant, John H.: The demography of the decline in the British willow warbler population. 95 JAppStat 22 905-922

Peacock, B.: Evans, M.; Hastings, N.; — — (Auth); Trenkler, Götz (Rev): Review of "Statistical distributions (Second edition)". 95 CmpStDA 19 483-484

Peacock, Brian: Evans, Merran; Hastings, Nicholas; — — (Auth); Copeland, Karen A. F. (Rev): Review of "Statistical distributions". 95 JQualTek 27 94-95

Pearce, C. E. M.: Henderson, W.; — —; Taylor, P. G.; van Dijk, N. M.: Insensitivity in discrete-time generalized semi-Markov processes allowing multiple events and probabilistic service scheduling. 95 AnlsApPr 5 78-96

Neuts, M. F.; — —: A central limit theorem for a class of Markovian packet streams under heavy traffic conditions. 95 CommStSM 11 210-218

Pearce, S. C.: Some design problems in crop experimentation. I. The use of blocks. 95 ExpAgric 31 191-203

— —: Some design problems in crop experimentation. II. Multiple blocking systems. 95 ExpAgric 31 279-290

— —: Some design problems in crop experimentation. III. Non-orthogonality. 95 ExpAgric 31 409-422

— —: Cubic lattices. 95 JAppStat 22 355-373

Pearl, Judea: Causal diagrams for empirical research. 95 Biomtrka 82 669-688 (Disc: p688-710)

Pearman, A. D.: Montero, J.; — —; Tejada, J.: Fuzzy multicriteria decision support for budget allocation in the transport sector. 95 Top 3 47-68

Pearson, Jay D.: Morrell, Christopher H.; — —; Carter, H. Ballentine; Brant, Larry J.: Estimating unknown transition times using a piecewise nonlinear mixed-effects model in men with prostate cancer. 95 JASA 90 45-53

Peck, Robert R.; Madrigal, J. L.; Fellingham, Gilbert W.: Ranking teams competing in pairwise comparisons using the dynamic linear model. 94 ASAProBays 112-116 (Disc: p134-135)

Peckham, Scott D.: New results for self-similar trees with applications to river networks. 95 WaterReR 31 1023-1029

Peckova, Monika: Brooks, Maria Mori; Hallstrom, Al; — —: A simulation study used to design the sequential monitoring plan for a clinical trial. 95 StatMed 14 2227-2237

Pedersen, Asger Roer: Consistency and asymptotic normality of an approximate maximum likelihood estimator for discretely observed diffusion processes. 95 Bernoull 1 257-279

Pedersen, F.; Bengtsson, E.; Nordin, B.: An extended strategy for exploratory multivariate image analysis including noise considerations. 95 JChemtrc 9 389-409

Pedlow, Steven; Meng, Xiao-Li: Exploring hypothesis-testing procedures with multiply-imputed data under unequal fractions of missing information. 94 ASASrvyRsr 595-599

Peduzzi, Peter; Concato, John; Feinstein, Alvan R.: Importance of events per independent variable in proportional hazards regression analysis II. Accuracy and precision of regression estimates. 95 JClEpid 48 1503-1510

Concato, John; — —; Holford, Theodore R.: Importance of events per independent variable in proportional hazards analysis I. Background, goals, and general strategy. 95 JClEpid 48 1495-1501

Pee, David: Dahm, P. Fred; Gail, Mitchell H.; Rosenberg, Philip S.; — —: Determining the value of additional surrogate exposure data for improving the estimate of an odds ratio. 95 StatMed 14 2581-2598

Green, Sylvan B.; Corle, Donald K.; Gail, Mitchell H.; Mark, Steven D.; — —; Freedman, Laurence S.; Graubard, Barry I.; Lynn, William R.: Interplay between design and analysis for behavioral intervention trials with community as the unit of randomization. 95 AJEpidm 142 587-593

Peel, D. A.: Byers, J. D.; — —: Forecasting industrial production using non-linear methods. 95 JForcast 14 325-336

Peeva, V.: Marketing of statistical information and services. 91 StatJUN 8 339-344

Peirats, V.: Garcia Rubio, E.; — —: Evaluation of data editing procedures: Results of a simulation approach. 91 StatJUN 8 175-190

Peiris, M. Shelton: Some aspects of forecasting with vector moving average processes. 94 CalcutSt 44 151-156

Peiris, S.: Rai, S. N.; Abraham, B.; — —: Analysis of short time series with an over-dispersion model. 95 CommStA 24 335-348

Peköz, Erol A.; Ross, Sheldon M.: A simple derivation of exact reliability formulas for linear and circular consecutive-k-of-n: F systems. 95 JAppProb 32 554-557

Peligard, Magda; Suresh, Ram: Estimation of variance of partial sums of an associated sequence of random variables. 95 StocProc 56 307-319

Peligrad, Magda; Shao, Qi-Man: Estimation of the variance of partial sums for ρ-mixing random variables. 95 JMultiAn 52 140-157

— —; Shao, Qi-Man: A note on the almost sure central limit theorem for weakly dependent random variables. 95 StPrLet 22 131-136

Pelizzon, L.; Weber, G.: The Italian term structure of interest rates and the currency devaluation of September 1992: A factor-ARCH analysis. 94 Statstca 54 313-327

Pellerey, Franco: On the preservation of some orderings of risks under convolution. 95 InsrneME 16 23-30

Ebrahimi, Nader; — —: New partial ordering of survival functions based on the notion of uncertainty. 95 JAppProb 32 202-211

Pemantle, Robin: Tree-indexed processes. 95 StatSci 10 200-213

— —; Peres, Yuval: Critical random walk in random environment on trees. 95 AnlsProb 23 105-140

— —; Peres, Yuval: Galton-Watson trees with the same mean have the same polar sets. 95 AnlsProb 23 1102-1124

Benjamini, Itai; — —; Peres, Yuval: Martin capacity for Markov chains. 95 AnlsProb 23 1332-1346

Lyons, Russell; — —; Peres, Yuval: Conceptual proofs of $L \log L$ criteria for mean behavior of branching processes. 95 AnlsProb 23 1125-1138

Pemberton, John: Jalali, Assad; — —: Mixture models for time series. 95 JAppProb 32 123-138

Peña, Daniel: Forecasting growth with time series models. 95 JForcast 14 97-105

— —; Yohai, Victor J.: The detection of influential subsets in linear regression by using an influence matrix. 95 JRSS-B 57 145-156

— —; Yohai, Victor J.: Correction to "The detection of influential subsets in linear regression by using an influence matrix" (V57 p145-156). 95 JRSS-B 57 611-611

Espasa, Antoni; — —: The decomposition of forecast in seasonal ARIMA models. 95 JForcast 14 565-583

Peña, Edsel A.; Rohatgi, Vijay K.; Székely, Gábor J.: Inference for a general type II censorship model. 95 MaOpfStS 26 241-252

Baltazar-Aban, Inmaculada; — —: Properties of hazard-based residuals and implications in model diagnostics. 95 JASA 90 185-197

Hollander, Myles; — —: Dynamic reliability models with conditional proportional hazards. 95 LiDaAnal 1 377-401

Penco, Maria Antonietta: Garibaldi, Ubaldo; — —: A predictive approach to contingency tables (Italian). 95 Metron 53 79-113

Pendleton, Grey W.; Sauer, John R.: Delineating bird populations using ring recoveries. 95 JAppStat 22 1049-1055

Penev, Spiridon: Saddlepoint approximation in the linear structural relationship model. 95 CommStB 24 349-366

Raykov, Tenko; — —: Comment on "A binomial test of model fit" (V1 p268-273). 95 StrEqMod 2 146-151

Peng, Anrong; Pieczynski, Wojciech: Adaptive mixture estimation and unsupervised local Bayesian image segmentation. 95 CVGIPGr 57 389-399

Peng, Fengchun; Dey, Dipak K.: Bayesian analysis of outlier problems using divergence measures. 95 CandJSt 23 199-213

Peng, S.: Hu, Y.; — —: Solution of forward-backward stochastic differential equations. 95 ProbTher 103 273-283

Peng, Y. B.: Lei, Xingye; — —; Wright, F. T.: Testing homogeneity of normal means with a simply ordered alternative and dependent observations. 95 CmpStDA 20 173-183

Penny, David: Waddell, Peter J.; — —; Hendy, Michael D.; Arnold, Greg: The sampling distributions and covariance matrix of phylogenetic spectra. 94 MolBioEv 11 630-642

Penny, Richard N.: Särndel, C.-E.; Swensson, B.; Wretman, J. (Auth); — — (Rev): Review of "Model assisted survey sampling". 95 Ststcian 44 140-142

Penrose, Mathew D.: Single linkage clustering and continuum percolation. 95 JMultiAn 53 94-109

— —: Generalised two-sample U-statistics and a two-species reaction-diffusion model. 95 StocProc 55 57-64

Pentikainen, T.: Daykin, C. D.; — —; Pesonen, M. (Auth); Vajda, S. (Rev): Review of "Practical risk theory for actuaries". 95 Ststcian 44 282-283

Pentikäinen, T.: Daykin, C. D.; — —; Pesonen, E. (Auth); Hardy, M. R. (Rev): Review of "Practical risk theory for actuaries". 95 JAppStat 22 547-548

Daykin, C. D.; — —; Pesonen, M. (Auth); Coutts, Stewart M. (Rev): Review of "Practical risk theory for actuaries". 95 JRSS-A 158 633-633

Daykin, C. D.; — —; Pesonen, M. (Auth); Neuhaus, Walther (Rev): Review of "Practical risk theory for actuaries". 95 AstinB 25 75-76

Penttinen, Antti: Muotka, Timo; — —: Detecting small-scale spatial patterns in lotic predator-prey relationships: Statistical methods and a case study. 94 JFishRBC 51 2210-2218

Partanen, Päivi; — —: Spatial mark-recapture method in the estimation of crayfish population size. 95 BiomtrcJ 37 979-994

Pepe, Margaret Sullivan: Reilly, Marie; — —: A mean score method for missing and auxiliary covariate data in regression models. 95 Biomtrka 82 299-314

Zhou, Haibo; — —: Auxiliary covariate data in failure time regression. 95 Biomtrka 82 139-149

Pepperrell, Catherine: — — (Auth); Lewi, Paul J. (Rev): Review of "Three-dimensional chemical similarity searching". 95 Chemolab 28 205-207

Peracchi, Franco; Welch, Finis: How representative are matched cross-sections? Evidence from the Current Population Survey. 95 JEconmtx 68 153-179

Foresi, Silverio; — —: The conditional distribution of excess returns: An empirical analysis. 95 JASA 90 451-466

Percival, D. A.; Walden, A. T. (Auth); Swift, A. Louise (Rev): Review of "Spectral analysis for physical applications: Multitaper and conventional techniques". 95 JRSS-A 158 198-199

Percival, D. B.: Walden, A. T.; McCoy, E. J.; — —: The effective bandwidth of a multitaper spectral estimator. 95 Biomtrka 82 201-214

Percival, Donald P.: On estimation of the wavelet variance. 95 Biomtrka 82 619-631

Percus, Jerome K.: Percus, Ora E.; — —: String matching for the novice. 94 AmMaMnth 101 944-947

Percus, Ora E.; Percus, Jerome K.: String matching for the novice. 94 AmMaMnth 101 944-947

— —; Whitlock, Paula A.: Theory and application of Marsaglia's monkey test for pseudorandom number generators. 95 ACMMdSim 5 87-100

Pereira, Carlos Alberto De Bragança: Mimoso Paulino, Carlos Daniel; — —: Bayesian methods for categorical data under informative general censoring. 95 Biomtrka 82 439-446

Pereira, Ricardo Alberto Marques; Fedrizzi, Michele: Fuzzy surface reconstruction. 94 QStMaA 16 71-77

Pereira-Mendoza, Lionel: Graphing in the primary school: Algorithm versus comprehension. 95 TeachgSt 17/1 2-6

— — (Ed); Watkins, Ann E. (Rev): Review of "Introducing data analysis in the schools: Who should teach it and how?". 95 AmerStat 49 62-63

Peres, Yuval: Benjamini, Itai; Pemantle, Robin; — —: Martin capacity for Markov chains. 95 AnlsProb 23 1332-1346

Furstenberg, Hillel; — —; Weiss, Benjamin: Perfect filtering and double disjointness. 95 AIHenrPB 31 453-465

Lyons, Russell; Pemantle, Robin; — —: Conceptual proofs of $L \log L$ criteria for mean behavior of branching processes. 95 AnlsProb 23 1125-1138

Pemantle, Robin; — —: Critical random walk in random environment on trees. 95 AnlsProb 23 105-140

Pemantle, Robin; — —: Galton-Watson trees with the same mean have the same polar sets. 95 AnlsProb 23 1102-1124

Perez Wilson, Mario: — — (Auth); Papp, Zoltan (Rev): Review of "Multi-vari chart and analysis - A pre-experimentation technique". 95 Technmcs 37 121-121

Perez-Abreu, Victor: Imkeller, Peter; — —; Vives, Josep: Chaos expansions of double intersection local time of Brownian motion in R^d and renormalization. 95 StocProc 56 1-34

Pérez-Abreu, Víctor: Houdré, Christian; — —: Covariance identities and inequalities for functionals on Wiener and Poisson spaces. 95 AnlsProb 23 400-419

Pérez-Ocón, Rafael; Gámiz-Pérez, M. Luz: On the HNBUE property in a class of correlated cumulative shock models. 95 AdvAppPr 27 1186-1188

— —; Gámiz-Pérez, M. Luz: Conditions on the arrival process to obtain HNBUE survival using a shock model. 95 CommStA 24 931-944

Pérez-Salvador, Blanca Rosa; Méndez-Ramirez, Ignacio: Estimating the population mean using supplementary dephased information. 95 CommStB 24 733-743

Pericchi, L. R.: Sansó, B.; — —: On near ignorance classes. 94 RBrzlPrS 8 119-126

Phillips, Garry D. A.: Kiviet, Jan F.; — —; Schipp, Bernhard: The bias of OLS, GLS, and ZEF estimators in dynamic seemingly unrelated regression models. 95 JEconmtx 69 241-266

Phillips, Joan M.: Sudman, Seymour; — —: Cognitive aspects of organizational reporting. 94 ASASrvyRsr 1351-1356

Phillips, Michael J.: Sweeting, Trevor J.; — —: An application of nonstandard asymptotics to the analysis of repairable-systems data. 95 Technmcs 37 428-435

Sherwin, D. J.; Bossche, A. (Auth); — — (Rev): Review of "The reliability, availability and productiveness of systems". 95 Ststcian 44 412-413

Phillips, Nancy V.: Glover, Fred; Kingman, Darwin; — — (Auth); Hansen, Pierre (Rev): Review of "Network models in optimization and their applications in practice". 95 JClass 12 150-152

Phillips, Peter C. B.: Trending multiple time series. 95 EconTher 11 811-817

— —: Robust nonstationary regression. 95 EconTher 11 912-951

— —: Nonstationary time series and cointegration. 95 JApEcmtx 10 87-94

— —: Bayesian model selection and prediction with empirical applications. 95 JEconmtx 69 289-331 (Disc: p333-365)

— —: Reply to comments on "Bayesian model selection and prediction with empirical applications". 95 JEconmtx 69 351-365

Chang, Yoosoon; — —: Time series regression with mixtures of integrated processes. 95 EconTher 11 1033-1094

Kitamura, Yuichi; — —: Efficient IV estimation in nonstationary regression. 95 EconTher 11 1095-1130

Phillips, Robert F.: Griffiths, W. E.; Hill, R. C.; Judge, G. G. (Auth); — — (Rev): Review of "Learning and practicing econometrics". 95 IntJFor 11 331-333

Phipps, Polly A.; Butani, Shail J.; Chun, Young I.: Research on establishment-survey questionnaire design. 95 JBES 13 337-346

Pi, Lynn K.: Chou, Ray Y.; Ng, Victor K.; — —: Cointegration of international stock market indices. 94 ASAProBuEc 196-201

Piantadosi, Steven; Crowley, John: An implicitly defined parametric model for censored survival data and covariates. 95 Biomtrcs 51 249-258

Goodman, Steven N.; Zahurak, Marianna L.; — —: Some practical improvements in the continual reassessment method for phase I studies. 95 StatMed 14 1149-1161

Wijesinha, Manel Cooray; — —: Dose-response models with covariates. 95 Biomtrcs 51 977-987

Piau, Didier: Brownian martingales and Sakai's conjecture (French). 95 AIHenrPB 31 429-452

Picado, Ana: Pires, Ana; Branco, Joao; — —; Mendonça, Elsa: Metabolic responses in the assessment of pollution effects. 95 EnvrMtrc 6 155-163

Picard, Dominique: Donoho, David L.; Johnstone, Iain M.; Kerkyacharian, Gérard; — —: Wavelet shrinkage: Asymptopia?. 95 JRSS-B 57 301-337 (Disc: p337-369)

Picard, Richard R.: Martz, Harry F.; — —: Uncertainty in Poisson event counts and exposure time in rate estimation. 95 RelEngSS 48 181-190

Picardello, Massimo A.; Woess, Wolfgang: The full Martin boundary of the bi-tree. 94 AnlsProb 22 2203-2222

Piccinato, Ludovico: Bertolino, Francesco; — —; Racugno, Walter: Multiple Bayes factors for testing hypotheses. 95 JASA 90 213-219

Bernardo, José M.; Smith, Adrian F. M. (Auth); — — (Rev): Review of "Bayesian theory". 94 JItalStS 3 155-160

Piccinino, Linda J.: Guadagno, Mary Ann; Abma, Joyce C.; Chandra, Anjani; London, Kathy A.; Peterson, Linda S.; — —; Wilson, Jacqueline B.: Women's reports of sensitive health conditions, attitudes, and behaviors. 94 ASAProGovt 82-87

Picco, P.: Ferrari, P. A.; Kesten, H.; Martinez, S.; — —: Existence of quasi-stationary distributions. A renewal dynamical approach. 95 AnlsProb 23 501-521

Picco, Pierre: Aaronson, Jon; Liggett, Thomas; — —: Equivalence of renewal sequences and isomorphism of random walks. 94 IsrelJMa 87 65-76

Bovier, Anton; Gayrard, Véronique; — —: Large deviation principles for the Hopfield model and the Kac-Hopfield model. 95 ProbTher 101 511-546

Pickle, L. W.: White, A. A.; — —; Herrmann, D. J.; Croner, C. M.; Wilson, B. F.: Map design preferences associated with professional discipline. 94 ASAProStGr 54-59

Pickle, Linda W.; White, Andrew A.: Effects of the choice of age-adjustment method on maps of death rates. 95 StatMed 14 615-627 (Disc: p697-699)

Pickles, A.; Neale, M.; Simonoff, E.; Rutter, M.; Hewitt, J.; Meyer, J.; Crouchley, R.; Silberg, J.; Eaves, L.: A simple method for censored age-of-onset data subject to recall bias: Mothers' reports of age of puberty in male twins. 94 BehavGen 24 457-468

Eaves, L. J.; Silberg, J. L.; Hewitt, K.; Rutter, M.; Meyer, J. M.; Neale, M. C.; — —: Analyzing twin resemblance in multisymptom data: Genetic applications of a latent class model for symptoms of conduct disorder in juvenile boys. 93 BehavGen 23 5-20

Dunn, G.; Everitt, B.; — — (Auth); Chan, Wai (Rev): Review of "Modelling covariances and latent variables using EQS". 95 SIAMRvw 37 462-464

Dunn, G.; Everitt, B.; — — (Auth); Schimek, Michael G. (Rev): Review of "Modelling covariances and latent variables using EQS". 95 CmpStQ 10 196-196

Pickles, Andrew; Crouchley, Robert: A comparison of frailty models for multivariate survival data. 95 StatMed 14 1447-1461

— —; Dunn, Graham; Vázquez-Barquero, José Luis: Screening for stratification in two-phase ('two-stage') epidemiological surveys. 95 StMedRes 4 73-89

— —; Dunn, Graham; Vázquez-Barquero, José Luis: Reply to comment on "Screening for stratification in two-phase ('two-stage') epidemiological surveys". 95 StMedRes 4 263-263

Pickles, Andrew R.: Crouchley, Robert; — —: Multivariate survival models for repeated and correlated events. 95 JStPlInf 47 95-110

Pieczynski, Wojciech: Peng, Anrong; — —: Adaptive mixture estimation and unsupervised local Bayesian image segmentation. 95 CVGIPGr 57 389-399

Piedmonte, Marion R.: Qu, Yinsheng; — —; Medendorp, Sharon V.: Latent variable models for clustered ordinal data. 95 Biomtrcs 51 268-275

Piegorsch, Walter W.; Bailer, A. John: Statistical approaches for analyzing mutational spectra: Some recommendations for categorical data. 94 Genetcs 136 403-416

Pielert, V.: Schwartz, H. J.; Mosler, C.; Hary, I.; — —: Factors affecting spatial preferences in settlement site selection in migratory pastoralism. 95 EnvrMtrc 6 485-490

Pienaar, Inet: Flury, Bernard D.; Nel, Daan G.; — —: Simultaneous detection of shift in means and variances. 95 JASA 90 1474-1481

Pienta, Amy: Burr, Jeffrey A.; Massagli, Michael P.; Mutchler, Jan E.; — —: Labor force status and transitions among older African American and white men. 94 ASAProSoSt 129-134

Piepho, H. P.: Robustness of statistical tests for multiplicative terms in the additive main effects and multiplicative interaction model for cultivar trials. 95 ThApGene 90 438-443

Hühn, M.; Lotito, S.; — —: Relationships between genotype × environment interactions and rank orders for a set of genotypes tested in different environments. 93 ThApGene 86 943-950

Piepho, H.-P.: Implication of correlations among some common stability statistics - A Monte Carlo simulations. 95 ThApGene 90 457-461

— —: Best linear unbiased prediction (BLUP) for regional yield trials: A comparison to additive main effects and multiplicative interaction (AMMI) analysis. 94 ThApGene 89 647-654

Piepho, Hans-Peter: Assessing cultivar adaptability by multiple comparison with the best. 95 AgronmyJ 87 1225-1227

— —: A robust test for homoscedasticity in a two-way layout. 95 BiomtrcJ 37 151-160

— —: Detecting and handling heteroscedasticity in yield trial data. 95 CommStB 24 243-274

Pierce, Donald A.; Peters, Dawn: Correction to "Higher-order asymptotics and the likelihood principle: One-parameter models" (V81 p1-10). 95 Biomtrka 82 235-235

Pierce, R. C.: Barnes, Ron; — —: A study of various control chart patterns and their statistical interpretation. 94 ASAProStEd 137-139

Pierce, Rebecca L.: Dyer, Danny; — —: Bayesian HPM intervals for the number of nonconforming items in a lot. 95 CmpStDA 20 159-172

Pierrat, Jean Claude; Houllier, François; Hervé, Jean Christophe; Salas González, Raúl: Estimating the mean of the r largest values in a finite population: Application to forest inventory (French). 95 Biomtrcs 51 679-686

Pierre Loti Viaud, Daniel: Random perturbations of recursive sequences with an application to an epidemic model. 95 JAppProb 32 559-578

Pierre-Loti-Viaud, D.: Hu, Y.; — —; Shi, Z.: Laws of the iterated logarithm for iterated Wiener processes. 95 JTherPrb 8 303-319

Pierse, R. G.; Snell, A. J.: Temporal aggregation and the power of tests for a unit root. 95 JEconmtx 65 333-345

Piesto, Nancy A.: Caldwell, Carol V.; Monsell, Brian C.; — —; Shimberg, Michael Z.: Using seasonal adjustment models to correct for panel imbalance in the Monthly Wholesale Trade Survey. 94 ASAProBuEc 83-88

Pietrogrande, M. C.; Dondi, F.; Felinger, A.; Davis, J. M.: Statistical study of peak overlapping in multicomponent chromatograms: Importance of the retention pattern. 95 Chemolab 28 239-258

Pigeot, Iris: Westfall, Peter H.; Young, S. Stanley (Auth); Köhne, Stephan; — — (Rev): Review of "Resampling-based multiple testing. Examples and methods for p-value adjustment". 95 CmpStDA 20 235-236

Pignatiello, Joseph J., Jr: Buck, James R.; — —; Raz, Tzvi: Bayesian Poisson rate estimation using parameter maps. 95 IIETrn 27 405-412

Pike, M. C.; Spicer, D. V.: Hormones, breast cell proliferation and breast cancer risk. 94 ASAProEpid 24-33

Pikitch, Ellen K.: McAllister, Murdoch K.; — —; Punt, Andre E.; Hilborn, Ray: A Bayesian approach to stock assessment and harvest decisions using the sampling/importance resampling algorithm. 94 JFishRBC 51 2673-2687

Pillai, R. N.; Jayakumar, K.: Discrete Mittag-Leffler distributions. 95 StPrLet 23 271-274

Pillati, M.: Paruolo, P.; — —: A nonlinear model for the conditional expectations of asset returns. 94 Statstca 54 329-347

Pincus, R.: Barnett, V.; Lewis, T. (Auth); — — (Rev): Review of "Outliers in statistical data (Third edition)". 95 BiomtrcJ 37 256-256

Cook, R. D.; Weisberg, S. (Auth); — — (Rev): Review of "An introduction to regression graphics". 95 BiomtrcJ 37 298-298

Pinder, Jonathan P.; Shoesmith, Gary L.: Modelling multivariate cointegrated systems: Insights from non-linear dynamics. 95 JForcast 14 311-324

Pinedo, Michael: Chao, Xiuli; — —: On queueing networks with signals and history-dependent routing. 95 PrEngInf 9 341-354

Du, Charles; — —: A convexity property of a Markov-modulated queueing loss system. 95 PrEngInf 9 193-199

Pinelis, Iosif: Optimum bounds for the distributions of martingales in Banach spaces. 94 AnlsProb 22 1679-1706

Ping, Robert A., Jr: A parsimonious estimating technique for interaction and quadratic latent variables. 95 JMarRsr 32 336-347

Pinheiro, José C.; Bates, Douglas M.: Approximations to the log-likelihood function in the nonlinear mixed-effects model. 95 JCmpGrSt 4 12-35

— —; Bates, Douglas M.; Lindstrom, Mary J.: Model building in nonlinear mixed effects models. 94 ASAProBiop 1-8

Pinkleton, Bruce; Reagan, Joey; Aaronson, Dustin; Ramo, Eduard: Does "I'm not selling anything" increase response rates in telephone surveys?. 94 ASASrvyRsr 1242-1247

Pinney, Susan M.: Hertzberg, Vicki Stover; Rice, Carol H.; Linz, Douglas H.; — —: Occupational epidemiology in the era of total quality: Challenges for exposure assessment. 94 ASAProEpid 139-142

Pino, Raúl: de la Fuente, David; — —: Comparative analysis of univariate and transfer-function forecasting using neural networks and Box-Jenkins techniques (Spanish). 95 Questo 19 187-215

Pinsker, Ilia: Grechanovsky, Eugene; — —: Conditional p-values for the F-statistic in a forward selection procedure. 95 CmpStDA 20 239-263

Pinsky, Carl: Whaley, Fredrick S.; Schoenfelder, John R.; — —: Comparing a standard diagnostic procedure to a second procedure used in combination with the standard procedure. 94 ASAProBiop 272-275

Piper, S. E.: A model of the ring-recovery reporting process for the Cape Griffon Gyps coprotheres. 95 JAppStat 22 641-659

Pires, Ana; Branco, Joao; Picado, Ana; Mendonça, Elsa: Metabolic responses in the assessment of pollution effects. 95 EnvrMtrc 6 155-163

Pisanets, S. I.: Two limit theorems for diffusion type stochastic equations. 94 ThProbAp 39 708-713

Pisani, R.: Freedman, D.; — —; Purves, R.; Adhikari, A. (Auth); Mizera, Ivan (Rev): Review of "Statistics (Second edition)". 95 JAppStat 22 321-322

Pischke, Jörn-Steffen: Measurement error and earnings dynamics: Some estimates from the PSID validation study. 95 JBES 13 305-314

Pistone, Giovanni; Sempi, Carlo: An infinite-dimensional geometric structure on the space of all the probability measures equivalent to a given one. 95 AnlsStat 23 1543-1561

Pitacco, Ermanno: Actuarial models for pricing disability benefits: Towards a unifying approach. 95 InsrneME 16 39-62

Pitard, Francis: — — (Auth); Ziegel, Eric R. (Rev): Review of "Pierre Gy's sampling theory and sampling practice (Second edition)". 95 Technmcs 37 125-126

Pitel, Edwige: Harrison, Peter G.; — —: Response time distributions in tandem g-networks. 95 JAppProb 32 224-246

Piterbarg, V. I.: Konakov, V. D.; — —: High level excursions of Gaussian fields and the weakly optimal choice of the smoothing parameter. I. 95 MaMethSt 4 421-434

Pitman, J.: — — (Auth); Williams, R. J. (Rev): Review of "Probability". 95 Metrika 42 143-144

Pitman, Jim: Exchangeable and partially exchangeable random partitions. 95 ProbTher 102 145-158

Pitt, Loren D.; Robeva, Raina; Wang, Dao Yi: An error analysis for the numerical calculation of certain random integrals: Part 1. 95 AnlsApPr 5 171-197

Pittel, Boris G.: Frieze, Alan; — —: Probabilistic analysis of an algorithm in the theory of markets in indivisible goods. 95 AnlsApPr 5 768-808

Pitts, S. M.: Perros, H. G. (Auth); — — (Rev): Review of "Queueing networks with blocking: Exact and approximate solutions". 95 ShrtBkRv 15/3 46-46

Pivetz, Lynda L.: Rosen, Richard J.; Clayton, Richard L.; — —: Converting mail reporters to touchtone data entry. 94 ASASrvyRsr 763-768

Piza, Marcelo S. T.: Newman, Charles M.; — —: Divergence of shape fluctuations in two dimensions. 95 AnlsProb 23 977-1005

Plachky, D.: Some measure theoretical characterizations of compactness of metric spaces. 95 SankhyaA 57 154 160

Breitung, K. W. (Auth); — — (Rev): Review of "Asymptotic approximations for probability integrals". 95 StatDecn 13 399-399

Freidlin, M. I. (Ed); — — (Rev): Review of "The Dynkin festschrift - Markov processes and their applications". 95 StatDecn 13 400-402

Groeneboom, P.; Wellner, J. A. (Auth); — — (Rev): Review of "Information bounds and nonparametric maximum likelihood estimation". 95 StatDecn 13 94-95

Kamps, U. (Auth); — — (Rev): Review of "A concept of generalized order statistics". 95 StatDecn 13 402-403

Lehtonen, R.; Pahkinen, E. J. (Auth); — — (Rev): Review of "Practical methods for design and analysis of complex surveys". 95 StatDecn 13 399-400

Puterman, M. L. (Auth); — — (Rev): Review of "Markov decision processes: Discrete stochastic dynamic programming". 95 StatDecn 13 93-93

Rieder, H. (Auth); — — (Rev): Review of "Robust asymptotic statistics". 95 StatDecn 13 93-94

Plackett, Robin: Obituary: Churchill Eisenhart, 1913-1994. 95 JRSS-A 158 338-338

Plaetke, Rosemarie; Schachtel, Gabriel A.: Two statistical tests for meiotic breakpoint analysis. 95 AJHuGene 56 508-518

Plante, Robert: Seppala, Tomi; Moskowitz, Herbert; — —; Tang, Jen: Statistical process control via the subgroup bootstrap. 95 JQualTek 27 139-153

Plante, Robert D.: Wardell, Don G.; Moskowitz, Herbert; — —: Reply to comment on "Run-length distribution of special-cause control charts for correlated processes". 95 Technmcs 37 243-245

Plastria, Frank: Carrizosa, Emilio; — —: The determination of a "least quantile of squares regression line" for all quantiles. 95 CmpStDA 20 467-479

Platen, Eckhard: Kloeden, Peter E.; — — (Auth); Gelbrich, Matthias; Römisch, Werner (Rev): Review of "Numerical solution of stochastic differential equations". 95 SIAMRvw 37 272-275

Kloeden, Peter E.; — —; Schurz, Henri (Auth); Hall, Timothy (Rev): Review of "Numerical solution of SDE through computer experiments". 95 Technmcs 37 125-125

Plessis, J. L. du; van der Merwe, A. J.: A Bayesian approach to multivariate and conditional calibration. 95 CmpStDA 19 539-552

— —; van der Merwe, A. J.: Bayesian calibration in meat science research-estimation of the pH in postmortem muscle. 95 JStCmpSm 53 1-11

— —; van der Merwe, A. J.; Groenewald, P. C. N.: Reference priors for the multivariate calibration problem. 95 SoAfrStJ 29 155-168

Plewis, Ian: Harrop, Anne; — —: Two decades of family change: Secondary analysis of continuous government surveys. 95 JRSS-A 158 91-106

Ploberger, Werner: Andrews, Donald W. K.; — —: Admissibility of the likelihood ratio test when a nuisance parameter is present only under the alternative. 95 AnlsStat 23 1609-1629

Plotnikov, I. S.: Aladin, N. V.; — —; Potts, W. T. W.: The Aral Sea desiccation and possible ways of rehabilitating and conserving its northern part. 95 EnvrMtrc 6 17-29

Plummer, Francis A.: Nagelkerke, Nico J. D.; Moses, Stephen; — —; Brunham, Robert C.; Fish, David: Logistic regression in case-control studies: The effect of using independent as dependent variables. 95 StatMed 14 769-775

Plunkert, Patricia A.: Yu, Ching C.; Absalom, Sandra T.; — —: An evaluation study of forecasting models. 94 ASAProGovt 173-177

Pobel, Dominique: Viel, Jean-François; — —; Carré, André: Incidence of leukaemia in young people around the La Hague nuclear waste reprocessing plant: A sensitivity analysis. 95 StatMed 14 2459-2472

Pocock, Stuart J.: Life as an academic medical statistician and how to survive it. 95 StatMed 14 209-222

Goetghebeur, Els J. T.; — —: Detection and estimation of J-shaped risk-response relationships. 95 JRSS-A 158 107-121

Hughes, Michael D.; Thompson, Simon G.; — —: Optimal sequential screening guidelines for quantitative risk factors measured with error. 95 JASA 90 19-26

Podehl, M.: User requirements for statistical metainformation. 93 StatJUN 10 113-120

Podgursky, Michael: Ballou, Dale; — —: Teacher quality in public and private schools. 94 ASAProSoSt 196-200 (Disc: p213-215)

Poirier, Claude: Binder, David A.; Michaud, Sylvie; — —: Model-based reweighting for nonresponse adjustment. 94 ASASrvyRsr 989-994

Whitridge, Patricia J.; — —: Evaluation of the Canadian Potato Area Estimation Program. 94 ASASrvyRsr 1150-1155

Poirier, Dale J.: Koop, Gary; — —: An empirical investigation of Wagner's hypothesis by using a model occurrence framework. 95 JRSS-A 158 123-141

Pojar, Thomas M.; Bowden, David C.; Gill, R. Bruce: Aerial counting experiments to estimate pronghorn density and herd structure. 95 JWildlMg 59 117-128

Pokarowski, Piotr: Niemiro, Wojciech; — —: Tail events of some nonhomogeneous Markov chains. 95 AnlsApPr 5 261-293

Polasek, W.: EDA. Exploratory data analysis. Introduction to descriptive statistics (Second edition) (German). 94 Springer:Brln:NY 345p

Neudecker, H.; Liu, S.; — —: The Hadamard product and some of its applications in statistics. 95 MaOpfStS 26 365-373

— — (Auth); Maechler, M. (Rev): Review of "EDA. Exploratory data analysis. Introduction to descriptive statistics (Second edition) (German)". 95 ShrtBkRv 15/2 23-23

Polasek, Wolfgang: Neudecker, Heinz; — —; Liu, Shuangzhe: The heteroskedastic linear regression model and the Hadamard product. A note. 95 JEconmtx 68 361-366

Schmid, Bernhard; — —; Weiner, Jacob; Krause, Andreas; Stoll, Peter: Modeling of discontinuous relationships in biology with censored regression. 94 ANatrls 143 494-507

Pole, A.; West, M.; Harrison, J. (Auth); Davies, Neville (Rev): Review of "Applied Bayesian forecasting and time series analysis". 95 JRSS-A 158 635-636

Polisicchio, M.: The global concentration measure ζ and the principle of transfers: An empirical analysis (Italian). 93 GiornEco 52 291-308

Politis, Dimitris N.; Romano, Joseph P.: Bias-corrected nonparametric spectral estimation. 95 JTimSrAn 16 67-103

— —; Romano, Joseph P.: Large sample confidence regions based on subsamples under minimal assumptions. 94 AnlsStat 22 2031-2050

Polivka, Anne E.: Dippo, Cathryn S.; Kostanich, Donna L.; — —: Effects of methodological change in the Current Population Survey. 94 ASASrvyRsr 260-263

Power, James F.: Eghball, Bahman; — —: Fractal description of temporal yield variability of 10 crops in the United States. 95 AgronmyJ 87 152-156

Power, Simon: Monte Carlo response surfaces: A comparative approach. 95 JStPlInf 43 301-310

Choudhury, Askar H.; — —; St. Louis, Robert D.: A computationally attractive method for estimating the linear regression model with autoregressive-moving average disturbances. 94 Metron 52/1 41-51

Powers, Michael R.: A theory of risk, return and solvency. 95 InsrneME 17 101-118

Poyer, David A.; Earl, Elizabeth: A statistical analysis of structural differences in minority household electricity demand. 94 ASAProGovt 221-226

Pozdnyakov, V. I.: The strong law of large numbers for trimmed sums (Russian). 94 VLenUM /2 20-25

Egorov, V. A.; — —: The law of the iterated logarithm for trimmed sums (Russian). 94 VLenUM /4 34-38

Prabhakar, B.: Mountford, T.; — —: On the weak convergence of departures from an infinite series of ·/M/1 queues. 95 AnlsApPr 5 121-127

Prabhu, N. U.; Pacheco, A.: A storage model for data communication systems. 94 ThProbAp 39 604-627

Prabhu, Sharad: Runger, George C.; Willemain, Thomas R.; — —: Average run lengths for CUSUM control charts applied to residuals. 95 CommStA 24 273-282

Prabhu, Sharad S.; Montgomery, Douglas C.; Runger, George C.: A design tool to evaluate average time to signal properties of adaptive $\bar{X}$ charts. 95 JQualTek 27 74-83

Prada-Sánchez, J. M.: García-Jurado, I.; González-Manteiga, W.; — —; Febrero-Bande, M.; Cao, R.: Predicting using Box-Jenkins, nonparametric, and bootstrap techniques. 95 Technmcs 37 303-310

Pradel, R.; Reboulet, A.-M.; Viallefont, A.: Testing hypotheses and estimating survival from capture histories with CR. 95 JAppStat 22 775-784

Pradel, Roger; Cooch, Evan; Cooke, Fred: Transient animals in a resident population of snow geese: Local emigration or heterogeneity?. 95 JAppStat 22 695-710

Lebreton, Jean-Dominique; Morgan, Byron J. T.; — —; Freeman, Stephen N.: A simultaneous survival rate analysis of dead recovery and live recapture data. 95 Biomtrcs 51 1418-1428

Pugesek, Bruce H.; Nations, Chris; Diem, Kenneth L.; — —: Mark-resighting analysis of a California gull population. 95 JAppStat 22 625-639

Prasad, Jagdish: Kherwa, G. Ram; — —: A note on generalized binary proper efficiency-balanced block designs. 95 CommStA 24 2797-2801

Prat, Albert; Grima, Pere: Determination of design factor values to obtain robust products: Application to circuits with inductance and resistance (Spanish). 95 Questo 19 153-169

Prato, G. Da: Invariant measures for white noise driven stochastic partial differential equations. 95 StocAnAp 13 295-305

— —; Elworthy, K. D.; Zabczyk, J.: Strong Feller property for stochastic semilinear equations. 95 StocAnAp 13 35-45

— —; Tudor, Constantin: Periodic and almost periodic solutions for semilinear stochastic equations. 95 StocAnAp 13 13-33

— —; Zabczyk, J.: Convergence to equilibrium for classical and quantum spin systems. 95 ProbTher 103 529-552

Aubin, J. P.; — —: Stochastic Nagumo's viability theorem. 95 StocAnAp 13 1-11

Prato, Giuseppe Da; Gatarek, Dariusz: Stochastic Burgers equation with correlated noise. 95 Stochatx 52 29-41

Pratt, William F.: Mosher, William D.; — —: CAPI, event histories, and incentives in the SNFG cycle 5 pretest. 94 ASASrvyRsr 59-63

Tourangeau, Roger; Jobe, Jared B.; — —; Rasinski, Kenneth: Design and results of the Women's Health Study. 94 ASASrvyRsr 49-58

Prayag, V. R.; Chiplonkar, S. A.; Paranjpe, S. A.: Nested ANOVA for repeated surveys with autocorrelated errors. 95 BiomtrcJ 37 565-570

Precht, M.: Bachmaier, M.; — —: Robust confidence intervals for contrasts based upon a likelihood ratio test. 95 StHefte 36 215-236

Predicta Software Inc: — — (Auth); Ng, Serena (Rev): Review of "Coint 2.0". 95 JApEcmtx 10 205-210

Preece, D. A.; Vowden, B. J.; Jones, R. Hughes; Roger, C. A.; Vowden, C. J.: Choreographing designs. 95 MaScitst 20 15-32

Bailey, R. A.; — —; Rowley, C. A.: Randomization for a balanced superimposition of one Youden square on another. 95 JRSS-B 57 459-469

Preece, Peter F. W.; Read, Kenneth L. Q.: A stage-theory model of cognitive development. 95 BJMaStPs 48 1-7

Pregibon, Daryl: Comment on "Guiding data analysts with visual statistical strategies". 95 JCmpGrSt 4 257-258

Preisler, Haiganoush K.; Akers, R. Patrick: Autoregressive-type models for the analysis of bark beetle tracks. 95 Biomtrcs 51 259-267

Prentice, M. J.: Fisher, N. I. (Auth); — — (Rev): Review of "Statistical analysis of circular data". 95 JRSS-A 158 190-190

Prentice, Ross L.: On the role, design, and analysis of disease prevention trials. 95 ConClnTr 16 249-258

— —: Design issues in cohort studies. 95 StMedRes 4 273-292

— —; Sheppard, Lianne: Aggregate data studies of disease risk factors. 95 Biomtrka 82 113-125

Cai, Jianwen; — —: Estimating equations for hazard ratio parameters based on correlated failure time data. 95 Biomtrka 82 151-164

Sheppard, Lianne; — —: On the reliability and precision of within- and between-population estimates of relative rate parameters. 95 Biomtrcs 51 853-863

Prescott, P.: Caroni, C.; — —: On Rohlf's method for the detection of outliers in multivariate data. 95 JMultiAn 52 295-307

Caroni, C.; — —: Union-intersection testing for outliers in multivariate normal data. 95 JStCmpSm 51 185-196

Pattison, P. (Auth); — — (Rev): Review of "Algebraic models for social networks". 95 JRSS-A 158 197-198

Presnell, Brett: Rosalsky, Andrew; Stoyanov, Jordan; — —: An ergodic-type theorem á la Feller for nonintegrable strictly stationary continuous time processes. 95 StocAnAp 13 555-572

Press, S. James: Sample-audit tax assessment for businesses: What's fair?. 95 JBES 13 357-359

— —; Shigemasu, K.: Posterior distribution for the number of factors. 94 ASAProBays 75-77

Ferryman, Thomas A.; — —: Comparison of image reconstruction algorithms. 94 ASAProBays 159-164

Preston, Ian: Sampling distributions of relative poverty statistics. 95 ApplStat 44 91-99

Preston, Samuel H.: Shrestha, Laura B.; — —: Consistency of census and vital registration data on older Americans: 1970-1990. 95 SurvyM 21 167-177

Presutti, E.: Brassesco, S.; De Masi, A.; — —: Brownian fluctuations of the interface in the $D = 1$ Ginzburg-Landau equation with noise. 95 AIHenrPB 31 81-118

Pretsch, E.: Buydens, Lutgarde M. C.; Schoenmakers, Peter J. (Ed); — — (Rev): Review of "Intelligent software for chemical analysis". 95 Chemolab 27 148-149

Pretsch, Ernö: Macher, Christian L.; Cadish, Marc; Clerc, Jean-Thomas; — —: Hypermedia - A new concept for information management. 95 Chemolab 28 213-228

Přeučil, Stanislav: Jiroušek, Radim; — —: On the effective implementation of the iterative proportional fitting procedure. 95 CmpStDA 19 177-189

Prieto, F. J.: Alvarez, M.; Cuevas, C. M.; Escudero, L. F.; de la Fuente, J. L.; García, C.; — —: Network planning under uncertainty with an application to hydropower generation. 94 Top 2 25-58

Prieto, Francisco J.: Juan, Jesus; — —: A subsampling method for the computation of multivariate estimators with high breakdown point. 95 JCmpGrSt 4 319-334

Priore, Roger L.; Brady, Mark F.; Lawrence, David D.: Power of detecting the effect of a continuous variable in a Cox model. 94 ASAProBiop 200-205

Proctor, C. H.: Developing a concept of population in survey sampling consistent with that in experimental statistics. 94 ASASrvyRsr 653-655

Proe, M. F.: Elston, D. A.; — —: Smoothing regression coefficients in an overspecified regression model with interrelated explanatory variables. 95 ApplStat 44 395-406

Pronzato, L.: Walter, E.; — —: Identifying parametric models for experimental data (French). 94 Masson:NY:Paris 384p

Prorok, Philip C.: Etzioni, Ruth D.; Connor, Robert J.; — —; Self, Steven G.: Design and analysis of cancer screening trials. 95 StMedRes 4 3-17

Xu, Jian-Lun; — —: Non-parametric estimation of the post-lead-time survival distribution of screen-detected cancer cases. 95 StatMed 14 2715-2725

Proschan, Frank: Joag-dev, Kumar; Kochar, Subhash; — —: A general composition theorem and its applications to certain partial orderings of distributions. 95 StPrLet 22 111-119

Proschan, Michael A.; Follmann, Dean A.: Multiple comparisons with control in a single experiment versus separate experiments: Why do we feel differently?. 95 AmerStat 49 144-149

— —; Hunsberger, Sally A.: Designed extension of studies based on conditional power. 95 Biomtrcs 51 1315-1324

Protter, Philip: Föllmer, Hans; — —; Shiryayev, Albert N.: Quadratic covariation and an extension of Itô's formula. 95 Bernoull 1 149-169

Kurtz, Thomas G.; Pardoux, Étienne; — —: Stratonovich stochastic differential equations driven by general semimartingales. 95 AIHenrPB 31 351-377

Provencher, Bill: An investigation of the harvest decision of timber firms in the south-east United States. 95 JApEcmtx 10 57-74S

Provenzani, R.: Todeschini, R.; Gramatica, P.; — —; Marengo, E.: Weighted holistic invariant molecular descriptors. Part 2. Theory development and applications on modeling physicochemical properties of polyaromatic hydrocarbons. 95 Chemolab 27 221-229

Provost, S. B.: Mathai, A. M.; — —; Hayakawa, T.: Bilinear forms and zonal polynomials. 95 Springer:Brln:NY 376p

Provost, Serge B.; Rudiuk, Edmund M.: The sampling distribution of the serial correlation coefficient. 95 AJMMgSci 15 57-81

Prum, B.: Mardia, K. V. (Ed); — — (Rev): Review of "The art of statistical science". 95 MaOpfStS 26 180-181

Prum, Bernard; Rodolphe, François; de Turckheim, Elisabeth: Finding words with unexpected frequencies in deoxyribonucleic acid sequences. 95 JRSS-B 57 205-220

Pruska, Krystyna: Bayesian estimation of a structural changes model with dependent Bernoulli variable. 95 ListyBio 32 15-21

Prvan, Tania: Gregory, John; Lin, Cantian (Auth); — — (Rev): Review of "Constrained optimization in the calculus of variations and optimal control theory". 95 AstrlJSt 37 250-250

Jorgensen, Bent (Auth); — — (Rev): Review of "The theory of linear models". 95 AstrlJSt 37 248-248

Metcalfe, Andrew V. (Auth); — — (Rev): Review of "Statistics in engineering: A practical approach". 95 AstrlJSt 37 128-128

Psaty, Bruce: Diehr, Paula; Martin, Donald; Koepsell, Thomas; Cheadle, Allen; Wagner, Edward; — —: Optimal survey design for community-intervention evaluations: Cohort, cross-sectional, or both?. 94 ASAProEpid 34-43

Puddey, I. B.: Cox, K.; — —; Burke, V.; McAleer, M.; Beilin, L. J.; Morton, A. R.: Effects of exercise and calorie restriction on ambulatory blood pressure in overweight, sedentary males: A time series analysis. 95 EnvrMtrc 6 523-528

Pugesek, Bruce H.; Nations, Chris; Diem, Kenneth L.; Pradel, Roger: Mark-resighting analysis of a California gull population. 95 JAppStat 22 625-639

— —, Tomer, Adrian: Determination of selection gradients using multiple regression versus structural equation models (SEM). 95 BiomtrcJ 37 449-462

Puhalskii, A. A.: Coffman, E. G., Jr; — —; Reiman, M. I.: Polling systems with zero switchover times: A heavy-traffic averaging principle. 95 AnlsApPr 5 681-719

Pukelsheim, F.; Wilhelm, A.: Comment on "Constrained optimization of experimental design". 95 MaOpfStS 26 168-172

— — (Auth); Titterington, D. M. (Rev): Review of "Optimal design of experiments". 95 Metrika 42 74-75

Pukelsheim, Friedrich; Sinha, Bikas Kumar: Optimal block designs revisited: An approximate theory detour. 95 StPrLet 24 375-380

Ying, Lisa H.; — —; Draper, Norman R.: Slope rotatability over all directions designs. 95 JAppStat 22 331-341

Ying, Lisa H.; — —; Draper, Norman R.: Slope rotatability over all directions designs for $k \geq 4$. 95 JAppStat 22 343-354

— — (Auth); Mejza, Stanisław (Rev. Polish): Review of "Optimal design of experiments". 94 ListyBio 31 121-125

— — (Auth); Tobias, Randall D. (Rev): Review of "Optimal design of experiments". 95 JQualTek 27 93-94

Pukkila, Tarmo: Koreisha, Sergio G.; — —: A comparison between different order-determination criteria for identification of ARIMA models. 95 JBES 13 127-131

Koreisha, Sergio G.; — —: The identification of seasonal autoregressive models. 95 JTimSrAn 16 267-290

Pulcini, G.: Calabria, R.; — —: Bayes prediction of number of failures in Weibull samples. 95 CommStA 24 487-499

Pulford, G. W.; Gallant, J. C.; Kennedy, R. A.; Chung, S. H.: Evaluation and estimation of various Markov models with applications to membrane channel kinetics. 95 BiomtrcJ 37 39-63

Punt, Andre E.: McAllister, Murdoch K.; Pikitch, Ellen K.; — —; Hilborn, Ray: A Bayesian approach to stock assessment and harvest decisions using the sampling/importance resampling algorithm. 94 JFishRBC 51 2673-2687

Puri, Madan L.: Brunner, Edgar; — —; Sun, Shan: Nonparametric methods for stratified two-sample designs with application to multiclinic trials. 95 JASA 90 1004-1014

Haiman, George; Kiki, Maxime; — —: Extremes of Markov sequences. 95 JStPlInf 45 185-201

Hallin, Marc; — —: A multivariate Wald-Wolfowitz rank test against serial dependence. 95 CandJSt 23 55-65

Taniguchi, Masanobu; — —: Higher order asymptotic theory for normalizing transformations of maximum likelihood estimators. 95 AnInStMa 47 581-600

Purkayastha, Sumitra: An almost sure representation of sample circular median. 95 JStPlInf 46 77-91

— —: On the asymptotic efficiency of the sample circular median. 95 StatDecn 13 243-252

— —; Srivastava, M. S.: Asymptotic distributions of some test criteria for the covariance matrix in elliptical distributions under local alternatives. 95 JMultiAn 55 165-186

Purohit, S. G.: Testing whether overhauls renew the system. 94 JIndnStA 32 57-68

Pursey, Stuart: Current and future approaches to editing Canadian trade import data. 94 ASAProGovt 105-109

Purves, R.: Freedman, D.; Pisani, R.; — —; Adhikari, A. (Auth); Mizera, Ivan (Rev): Review of "Statistics (Second edition)". 95 JAppStat 22 321-322

Pusz, J.: A regression characterisation of the Meixner hypergeometric distribution. 95 AstrlJSt 37 83-87

Puterman, M. L.: — — (Auth); Collins, Sean (Rev): Review of "Markov decision processes: Discrete stochastic dynamic programming". 95 JAppStat 22 545-546

— — (Auth); Iancu, C. (Rev): Review of "Markov decision processes: Discrete stochastic dynamic programming". 95 JAppStat 22 547-547

— — (Auth); Plachky, D. (Rev): Review of "Markov decision processes: Discrete stochastic dynamic programming". 95 StatDecn 13 93-93

— — (Auth); Whittle, P. (Rev): Review of "Markov decision processes". 95 JRSS-A 158 636-637

Puterman, Martin L.: — — (Auth); Baxter, Laurence A. (Rev): Review of "Markov decision processes: Discrete stochastic dynamic programming". 95 Technmcs 37 353-353

Putnam, Bluford H.; Quintana, Jose Mario: New Bayesian statistical approaches to estimating and evaluating models of exchange rates determination. 94 ASAProBays 232-237

Puzey, Anthony S.: A reappraisal of the principle underlying the conventional actuarial estimator of q_x. 95 InsrneME 17 125-132

Pyke, R.: Petrov, V. V. (Auth); — — (Rev): Review of "Limit theorems of probability theory: Sequences of independent random variables". 95 ShrtBkRv 15/3 46-46

Pym, J.: Lam, Y. M.; — —; Campling, B. G.: Nonlinear models for assessing combinations of chemotherapeutic drugs with resistance modifiers. 94 ASAProBiop 394-398

Pyrczak, Fred: Making sense of statistics. A conceptual overview. 95 PyrczakPubl:CA 128p

Qian, Jiahe: Optimally weighted means in stratified sampling. 94 ASASrvyRsr 863-866

Qian, Weimin: Spline smoothed bootstrapping for kernel estimator (Chinese). 95 AMaAPknC 17 265-277

Qiao, Hongzhu; Tsokos, Chris P.: Estimation of the three parameter Weibull probability distribution. 95 MaCmpSim 39 173-185

— —; Tsokos, Chris P.: Parameter estimation of the Weibull probability distribution. 94 MaCmpSim 37 47-55

Qin, Jing; Lawless, Jerry: Estimating equations, empirical likelihood and constraints on parameters. 95 CandJSt 23 145-159

Qin, Yongsong: Wang, Chengming; — —: Linear empirical Bayes estimation applied in multiple linear regression (Chinese). 95 ChJApPSt 10 371-378

Qing, Hong: General admissibility of linear estimates of regression coefficients in growth curve model (Chinese). 95 ChJApPSt 10 265-271

Qingji, Yang: Itô's formula for multidimensional convex functions. 95 Stochatx 52 157-171

Qiying, Wang: The strong law of U-statistics with ϕ^*-mixing samples. 95 StPrLet 23 151-155

Qu, Yinsheng; Piedmonte, Marion R.; Medendorp, Sharon V.: Latent variable models for clustered ordinal data. 95 Biomtrcs 51 268-275

Quade, Dana: Anstrom, Kevin J.; — —: Testing whether a count variable belongs to the Katz family of distributions. 94 ASAProStCp 194-199

Quan, Hui; Yu, Kai F.: Sequential testing with interval censored data. 95 JStCmpSm 51 239-247

Quantitative Micro Software: — — (Auth); Noble, Nicholas R. (Rev): Review of "EViews - Micro TSP for Windows". 95 IntJFor 11 497-499

Quarshie, Leonard B.; Shindo, Hisakazu: A control chart for attributes and its applications. 95 ASQCTekCon 378-388

Quartari, I.: Forina, M.; Drava, G.; Armanino, C.; Boggia, R.; Lanteri, S.; Leardi, R.; Corti, P.; Conti, P.; Giangiacomo, R.; Galliena, C.; Bigoni, R.; — —; Serra, C.; Ferri, D.; Leoni, O.; Lazzeri, L.: Transfer of calibration function in near-infrared spectroscopy. 95 Chemolab 27 189-203

Quastel, Jeremy: Large deviations from a hydrodynamic scaling limit for a nongradient system. 95 AnlsProb 23 724-742

Que, Shingo: Eddy, William F.; — —: Dynamic three-dimensional display of U.S. air traffic. 95 JCmpGrSt 4 261-280

Queen, Catriona: Freeman, P. R.; Smith, A. F. M. (Ed); — — (Rev): Review of "Aspects of uncertainty: A tribute to Dennis Lindley". 95 JAppStat 22 541-541

Quera, Vicenç: Bakeman, Roger; — —: Analyzing interaction: Sequential analysis with SDIS and GSEQ. 95 CambridgeUniv 166p

Bakeman, Roger; — —: Log-linear approaches to lag-sequential analysis when consecutive codes may and cannot repeat. 95 PsychBu 118 272-284

Querleux, Bernard: Idier, Jérôme; Giovannelli, Jean-François; — —: Bayesian time-varying AR spectral estimation for ultrasound attenuation measurement in biological tissues. 94 ASAProBays 256-261

Quesada-Molina, J. J.; Rodríguez-Lallena, J. A.: Bivariate copulas with quadratic sections. 95 JNnparSt 5 323-337

Quesenberry, C. P.: Correction to "On properties of Q charts for variables" (V27 p184-203). 95 JQualTek 27 402-402

— —: Correction to "SPC Q charts for start-up processes and short or long runs" (V23 p213-224). 95 JQualTek 27 402-402

Quesenberry, Charles P.: On properties of Q charts for variables. 95 JQualTek 27 184-203

— —: On properties of binomial Q charts for attributes. 95 JQualTek 27 204-213

— —: On properties of Poisson Q charts for attributes. 95 JQualTek 27 293-303 (Disc: p316-343)

— —: Geometric Q charts for high quality processes. 95 JQualTek 27 304-315 (Disc: p316-343)

— —: Reply to comments on "Q charts". 95 JQualTek 27 333-343

Qui, Chunfu: Green, P. J.; Silverman, B. W. (Auth); — — (Rev): Review of "Nonparametric regression and generalized linear models: A roughness penalty approach". 95 Technmcs 37 342-343

Quinn, Linda M.: Statistical ethics. 95 Stats /12 3-6

Quinn, Margaret M.: Stone, Roslyn A.; Marsh, Gary M.; Owens, Ada D.; — —; Smith, Thomas J.: Retrospective estimation of airborne fiber exposure: Statistical aspects. 94 ASAProEpid 127-132

Quinn, Terrance J., II: Karunamuni, Rohana J.; — —: Bayesian estimation of animal abundance for line transect sampling. 95 Biomtrcs 51 1325-1337

Quintana, Jose Mario: Putnam, Bluford H.; — —: New Bayesian statistical approaches to estimating and evaluating models of exchange rates determination. 94 ASAProBays 232-237

Quintos, Carmela E.: Sustainability of the deficit process with structural shifts. 95 JBES 13 409-417

Quiróz, Adolfo J.: Aragón, Jorge; — —: Rank regression for current status data. 95 StPrLet 24 251-256

Quon, Tony K.: Nash, John C.; — —; Gianini, Jacqueline: Statistical issues in spreadsheet software. 94 ASAProStEd 238-241

Ra, Sungsup: Bera, Anil K.; — —: A test for the presence of conditional heteroskedasticity within ARCH-M framework. 95 EcomtrRv 14 473-485

Rabinowitz, Daniel; Tsiatis, Anastasios; Aragon, Jorge: Regression with interval-censored data. 95 Biomtrka 82 501-513

Rachev, S. T.; Ruschendorf, L.: Models for option prices. 94 ThProbAp 39 120-152

— —; Rüschendorf, L.: Probability metrics and recursive algorithms. 95 AdvAppPr 27 770-799

Heathcote, C. R.; — —; Cheng, B.: Testing multivariate symmetry. 95 JMultiAn 54 91-112

Klebanov, L. B.; — —: The method of moments in computer tomography. 95 MaScitst 20 1-14

Rachev, Svetlozar T.; Samorodnitsky, Gennady: Limit laws for a stochastic process and random recursion arising in probabilistic modelling. 95 AdvAppPr 27 185-202

— —; Wu, Chufang; Yakovlev, Andrej Yu.: A bivariate limiting distribution of tumor latency time. 95 MaBisci 127 127-147

— — (Auth); Constantin, G. (Rev): Review of "Probability metrics and the stability of stochastic models". 95 BLndnMaS 27 87-89

Racine-Poon, Amy: Wakefield, Jon; — —: An application of Bayesian population pharmacokinetic/pharmacodynamic models to dose recommendation. 95 StatMed 14 971-986

Račkauskas, Alfredas: On the rate of convergence in the martingale CLT. 95 StPrLet 23 221-226

Racugno, Walter: Bertolino, Francesco; Piccinato, Ludovico; — —: Multiple Bayes factors for testing hypotheses. 95 JASA 90 213-219

Radalowicz, A.: Bennett, S.; — —; Vella, V.; Tomkins, A.: A computer simulation of household sampling schemes for health surveys in developing countries. 94 IJEpidem 23 1282-1291

Radke, George E., Jr: Haskin, F. Eric; — —; Evanoff, Javon: Algorithms for analyzing the probability of M-out-of-N events. 95 RelEngSS 47 153-162

Radson, Darrell; Herrin, Gary D.: Augmenting a factorial experiment when one factor is an uncontrollable random variable: A case study. 95 Technmcs 37 70-81

Radtke, M.: Niggemeyer, B.; — —; Reich, A.: Applications of risk theory and multivariate analysis in insurance practice. 95 AplStMDA 11 231-244

Rady, El-Houssainy A.: Partitioning error spaces in linear models (error contrasts). 95 EgpStJ 39 51-64

Raeside, Robert: Shearer, Peter (Auth); — — (Rev): Review of "Business forecasting and planning". 95 IntJFor 11 192-193

Rafanelli, Maurizio: Aggregate statistical data: Models for their representation. 95 StatComp 5 3-24

Raftery, Adrian E.; Givens, Geof H.; Zeh, Judith E.: Inference from a deterministic population dynamics model for Bowhead whales. 95 JASA 90 402-416 (Disc: p417-430)

— —; Givens, Geof H.; Zeh, Judith E.: Reply to comments on "Inference from a deterministic population dynamics model for Bowhead whales". 95 JASA 90 427-430

Kass, Robert E.; — —: Bayes factors. 95 JASA 90 773-795

Madigan, David; Gavrin, Jonathan; — —: Eliciting prior information to enhance the predictive performance of Bayesian graphical models. 95 CommStA 24 2271-2292

Raghavachari, M.: Ji, Xiaoyan; — —; Wallace, William A.: Approximate reasoning in belief networks: Statistical validity, accuracy, and determination of sample size. 94 ASAProBays 128-133 (Disc: p134-135)

Raghavachari, Madabhushi: Glasserman, Paul; Yao, David D. (Auth); — — (Rev): Review of "Monotone structure in discrete-event systems". 95 Technmcs 37 467-468

Raghavan, Prabhakar: Motwani, Rajeev; — —: Randomized algorithms. 95 CambridgeUniv 490p

Raghavarao, Damaraju; Heiberger, Richard M.: Variance balanced generalized nested row-column designs with empty diagonal cells. 93 JCISS 18 42-52

Raghunathan, T. E.: Fellingham, G. W.; — —: Sensitivity of point and interval estimates to distributional assumptions in longitudinal data analysis of small samples. 95 CommStB 24 617-630

Raghunathan, Trivellore E.; Grizzle, James E.: A split questionnaire survey design. 95 JASA 90 55-63

Ragosta, M.: Macchiato, M. F.; La Rotonda, L.; Lapenna, V.; — —: Time modelling and spatial clustering of daily ambient temperature: An application in Southern Italy. 95 EnvrMtrc 6 31-53

Ragunathan, T. E.: McGilchrist, C. A.; — —: Generalised components of variance models. 95 CommStA 24 349-361

Rahbar, Mohammad H.: Sequential fixed width confidence intervals for regression parameters from censored data with a discrete covariate. 95 SqtlAnly 14 143-156

— —; Gardiner, Joseph C.: Nonparametric estimation of regression parameters from censored data with a discrete covariate. 95 StPrLet 24 13-20

Rahim, M. A.; Jocelyn, Wisner: Sample allocation in multivariate stratified design: An alternative to convex programming. 94 ASASrvyRsr 689-692

Rahimov, I.: Random sums and branching stochastic processes. 95 Springer:-Brln:NY 195p

— — (Auth); Teugels, J. L. (Rev): Review of "Random sums and branching stochastic processes". 95 ShrtBkRv 15/3 48-48

Rahlfs, Volker W.; Zimmermann, Helmuth: Nonparametric methods in standard statistics packages: The case of the Friedman analysis. 95 CmpStDA 20 101-110

Rahman, M. S.; Nahar, S.: General formulae for the determination of moments and cumulants. 95 PakJStA 11 11-15

Rahman, Mohammad A.: Effects of observed correlation structure among tumor types on experimentwise false positive rate in animal carcinogenicity studies. 95 JBiophSt 5 273-283

Rai, Anil: Variance estimation techniques in categorical data analysis for survey data. 94 JIAgSt 46 147-155

Rai, S. N.; Abraham, B.; Peiris, S.: Analysis of short time series with an over-dispersion model. 95 CommStA 24 335-348

— —; Matthews, D. E.: The analysis of incomplete data using stochastic covariates. 95 CandJSt 23 29-42

Raich, A. C.; Çinar, A.: Multivariate statistical methods for monitoring continuous processes: Assessment of discrimination power of disturbance models and diagnosis of multiple disturbances. 95 Chemolab 30 37-48

Raj, Baldev: Institutional hypothesis of the long-run income velocity of money and parameter stability of the equilibrium relationship. 95 JApEcmtx 10 233-253

— —; Amoako-Adu, Ben: Cointegration analysis of stock prices, inflation and real economic activity. 94 ASAProBuEc 190-195

— —; Baltagi, Badi H. (Ed); Lenz, H.-J. (Rev): Review of "Panel data analysis". 95 CmpStDA 19 481-483

Rajagopalan, Balaji; Lall, Upmanu: A kernel estimator for discrete distributions. 95 JNnparSt 4 409-426

Rajan, Murli: Vakil, Farok; — —: A test of the three moment international capital asset pricing model. 94 ASAProBuEc 184-189

Rajarshi, M. B.: Naik-Nimbalkar, U. V.; — —: Filtering and smoothing via estimating functions. 95 JASA 90 301-306

Rajeev, B.; Yor, M.: Local times and almost sure convergence of semi-martingales. 95 AIHenrPB 31 653-667

Rajgopal, Jayant; Mazumdar, Mainak: Designing component test plans for series system reliability via mathematical programming. 95 Technmcs 37 195-212

Rajput, Balram: Nolan, John P.; — —: Calculation of multidimensional stable densities. 95 CommStB 24 551-566

Raju, C.; Murthy, M. N. Narasimha: Procedures and tables for the construction and selection of chain sampling plans ChSP-4 (c_1, c_r) - Part 4. 95 JAppStat 22 261-271

— —; Murthy, M. N. Narasimha: Minimum risks chain sampling plans (ChSP-4 (c_1, c_2)) indexed by acceptable quality level and limiting quality level. 95 JAppStat 22 389-426

Rakhimov, G. M.: Khusanbayev, Ya. M.; — —: On approximation of sums of conditionally independent random variables. 95 ThPrMaSt 50 133-142

Raktoe, B. L.; Pesotan, H.: Some aspects on the design and analysis of fractional factorial designs with linear models containing nuisance parameters. 93 JCISS 18 369-395

Ralescu, S. S.: Strong approximation theorems for integrated kernel quantiles. 95 MaMethSt 4 201-215

Ralescu, Stefan S.: The law of the iterated logarithm for the multivariate nearest neighbor density estimators. 95 JMultiAn 53 159-179

Ramachandran, D.; Rüschendorf, L.: A general duality theorem for marginal problems. 95 ProbTher 101 311-319

Ramachandran, K. M.: Nearly optimal controls for singularly perturbed wideband noise systems. 95 StocAnAp 13 335-353

Ramalingam, T.; Ebrahimi, Nader: Testing for independence between hitting times of bivariate PQD processes. 95 SankhyaA 57 202-220

Raman, Kalyan: Hutchinson, J. Wesley; — —; Mantrala, Murali K.: Finding choice alternatives in memory: Probability models of brand name recall. 94 JMarRsr 31 441-461

Ramanathan, S.: Bhimasankaram, P.; Sengupta, D.; — —: Recursive inference in a general linear model. 95 SankhyaA 57 227-255

Ramasesh, Ranga V.: Hayya, Jack C.; Xu, Susan H.; — —; He, Xin X.: Order crossover in inventory systems. 95 CommStSM 11 279-309

Ramaswami, V.: Latouche, Guy; — —: Expected passage times in homogeneous quasi-birth-and-death processes. 95 CommStSM 11 103-122

Ramaswamy, Ratna: Cupples, L. Adrienne; Gagnon, David R.; — —; D'Agostino, Ralph B.: Age-adjusted survival curves with application in the Framingham study. 95 StatMed 14 1731-1744

Ramaswamy, Venkatram: Desarbo, Wayne S.; — —; Chatterjee, Rabikar: Analyzing constant-sum multiple criterion data: A segment-level approach. 95 JMarRsr 32 222-232

Ramesh, N. I.: Statistical analysis on Markov-modulated Poisson processes. 95 EnvrMtrc 6 165-179

Ramis-Ramos, Guillermo: Xie, Yu-Long; Baeza-Baeza, Juan José; — —: Kinetic spectrophotometric resolution of binary mixtures using three-way partial least squares. 95 Chemolab 27 211-220

Ramnath, Bala: Frankovic, Kathleen A.; Arnedt, Cheryl Mercado; — —: Interactive polling and Americans' comfort level with technology. 94 ASASrvyRsr 1316-1321

Ramo, Eduard: Pinkleton, Bruce; Reagan, Joey; Aaronson, Dustin; — —: Does "I'm not selling anything" increase response rates in telephone surveys?. 94 ASASrvyRsr 1242-1247

Ramos, José; Mallants, Dirk; Feyen, Jan: State space identification of linear deterministic rainfall-runoff models. 95 WaterReR 31 1519-1531

Ramos, Magdalena; Hughes, Bruce; McKay, Betty: Results from a feasibility test of an alternative automated data capture methodology during the 1992 Census of Agriculture. 94 ASASrvyRsr 769-774

Ramos, Rogelio Q.; Pantula, Sastry G.: Estimation of nonlinear random coefficient models. 95 StPrLet 24 49-56

Ramsay, J. O.: A similarity-based smoothing approach to nondimensional item analysis. 95 Psymtrka 60 323-339

— —; Wang, X.; Flanagan, R.: A functional data analysis of the pinch force of human fingers. 95 ApplStat 44 17-30

Kolen, M. J.; Brennan, R. L. (Auth); — — (Rev): Review of "Test equating: Methods and practices". 96 ShrtBkRv 16/1 6-6

Longford, N. T. (Auth); — — (Rev): Review of "Methods for uncertainty in educational testing". 96 ShrtBkRv 16/1 5-6

Wand, M. P.; Jones, M. C. (Auth); — — (Rev): Review of "Kernel smoothing". 95 ShrtBkRv 15/2 27-27

Ramsey, Fred L.; Lesser, Virginia M.: Conditional and unconditional estimators of population size. 95 EnvEcoSt 2 181-190

— —; Schafer, Daniel W.: Teaching graduate service courses through case studies. 94 ASAProStEd 27-28

Ramsey, Patricia P.: Levine, David M.; — —: Reengineering the business statistics course: Meeting the customer needs of the twenty-first century. 94 ASAProStEd 121-122

Ramsey, Philip H.; — —: Correction to "Updated version of the critical values of the standardized fourth moment" (V44 p231-241). 95 JStCmpSm 53 127-127

Ramsey, Philip H.; Ramsey, Patricia P.: Correction to "Updated version of the critical values of the standardized fourth moment" (V44 p231-241). 95 JStCmpSm 53 127-127

Rananand, Nol: Markov approximations to D-BMAPs: Information theoretic bounds on queueing performance. 95 CommStSM 11 713-734

Rancourt, Eric; Särndal, Carl; Lee, Hyunshik: Estimation of the variance in the presence of nearest neighbour imputation. 94 ASASrvyRsr 888-893 (Disc: p906-907)

Lee, Hyunshik; — —; Särndal, Carl E.: Experiments with variance estimation from survey data with imputed values. 94 JOffStat 10 231-243

Rand, Daniel R.: Selecting efficient experiment designs for mixed 2-level and 3-level factors. 94 ASAProPEng 332-338

Randall, Bryan L.: Grambsch, Patricia M.; — —; Bostick, Roberd M.; Potter, John D.; Louis, Thomas A.: Modeling the labeling index distribution: An application of functional data analysis. 95 JASA 90 813-821

Randles, Ron H.: Cornell, John A.; — —; Vining, G. Geoff: Challenges in teaching short courses by an academic department: The University of Florida experience. 95 AmerStat 49 195-200

Randles, Ronald H.: Sprent, P. (Auth); — — (Rev): Review of "Applied nonparametric statistical methods (Second edition)". 95 JAppStat 22 189-189

Randolph, Paul; Şahinoğlu, Mehmet: A stopping rule for a compound Poisson random variable. 95 AplStMDA 11 135-143

Ranford-Cartwright, Lisa C.: Hill, William G.; Babiker, Hamza A.; — —; Walliker, David: Estimation of inbreeding coefficients from genotypic data on multiple alleles, and application to estimation of clonality in malaria parasites. 95 GeneRsr 65 53-61

Rännar, Stefan; Geladi, Paul; Lindgren, Fredrik; Wold, Svante: A PLS kernel algorithm for data sets with many variables and few objects. Part II: Cross-validation, missing data and examples. 95 JChemtrc 9 459-470

Sjöström, Michael; — —; Wieslander, Åke: Polypeptide sequence property relationships in Escherichia coli based on auto cross covariances. 95 Chemolab 29 295-305

Ransom, K. P.: McPhee, M. J.; — —; Oddy, V. H.: Calibration of empirical models: Simulation of pasture growth and decay. 95 EnvrMtrc 6 505-510

Rantakallio, Paula: Möttönen, Jyrki; Oja, Hannu; Krause, Ulf; — —: Application of random coefficient regression model to myopia data: A case study. 95 BiomtrcJ 37 657-672

Rao, A. Vasudeva: Kantam, R. R. L.; — —; Narasimham, V. L.: Linear unbiased estimation of scale parameter by absolute values of order statistics in logistic distribution. 94 JStatRsr 28 199-219

Rao, B. L. S. Prakasa: Consistent estimation of density-weighted average derivative by orthogonal series method. 95 StPrLet 22 205-212

— —; Sen, Arusharka: Limit distributions of conditional U-statistics. 95 JTherPrb 8 261-301

— —; Sen, Arusharka: An application of the 'martingale methods' to nonparametric regression based on censored data. 95 StatDecn 13 201-220

Bagai, Isha; — —: Kernel-type density and failure rate estimation for associated sequences. 95 AnInStMa 47 253-266

Bagai, Isha; — —: Kernel type density estimates for positive valued random variables. 95 SankhyaA 57 56-67

Mishra, M. N.; — —: On L_1-consistency of kernel-type density estimator for stationary Markov processes. 95 CommStA 24 581-591

Rao, C. R.: Comment on "Random thoughts on survival". 95 Student 1 127-128

— —; Shanbhag, D. N.: Choquet-Deny type functional equations with applications to stochastic models. 94 Wiley:NY:UK 290p

— —; Zhao, L. C.: Convergence theorems for empirical cumulative quantile regression functions. 95 MaMethSt 4 81-91

— —; Zhao, L. C.: Correction to "Convergence theorems for empirical cumulative quantile regression functions" (V4 p81-91). 95 MaMethSt 4 359-359

— — (Ed); Schimek, Michael G. (Rev): Review of "Handbook of statistics Volume 9. Computational statistics". 95 CmpStQ 10 194-195

— —; Shanbhag, D. N. (Auth); Lindvall, T. (Rev): Review of "Choquet-Deny type functional equations with applications to stochastic models". 95 ShrtBkRv 15/3 48-49

Maddala, G. S.; — —; Vinod, H. D. (Ed); Gill, Len (Rev): Review of "Handbook of statistics (Volume 11) Econometrics". 95 IntJFor 11 189-191

Maddala, G. S.; — —; Vinod, H. D. (Ed); Krämer, Walter (Rev. German): Review of "Econometrics (Handbook of statistics) (Volume II)". 95 AllgeStA 79 101-102

Patil, G. P.; — — (Ed); Burnham, K. P. (Rev): Review of "Handbook of statistics, Volume 12: Environmental statistics". 95 Biomtrcs 51 1584-1585

Patil, G. P.; — — (Ed); Jansen, U. (Rev): Review of "Multivariate environmental statistics. North-Holland Series in Statistics and Probability Volume 6". 95 BiomtrcJ 37 255-256

Patil, G. P.; — — (Ed); Jansen, U. (Rev): Review of "Environmental statistics: Handbook of statistics Volume 12". 95 BiomtrcJ 37 511-512

Patil, G. P.; — — (Ed); Seber, G. A. F. (Rev): Review of "Multivariate environmental statistics". 95 Biomtrcs 51 1189-1190

Rao, C. Radhakrishna: A review of canonical coordinates and an alternative to correspondence analysis using Hellinger distance. 95 Questo 19 23-63

— —; Mukerjee, Rahul: Comparison of Bartlett-type adjustments for the efficient score statistic. 95 JStPlInf 46 137-146

— —; Mukerjee, Rahul: On posterior credible sets based on the score statistic. 95 StSinica 5 781-791

— —; Shanbhag, D. N.: Characterizations based on regression properties: Improved versions of recent results. 95 SankhyaA 57 167-178

— —; Toutenburg, Helge: Linear models: Least squares and alternatives. 95 Springer:Brln:NY 352p

— —; Zhao, L. C.: Strassen's law of the iterated logarithm for the lorenz curves. 95 JMultiAn 54 239-252

— —; Zhao, L. C.: Recent contributions to censored regression models. 95 Metrika 42 203-213

Baksalary, Jerzy K.; Markiewicz, Augustyn; — —: Admissible linear estimation in the general Gauss-Markov model with respect to an arbitrary quadratic risk function. 95 JStPlInf 44 341-347

Liu, Z. J.; — —: Asymptotic distribution of statistics based on quadratic entropy and bootstrapping. 95 JStPlInf 43 1-18

— —; Shanbhag, D. N. (Auth); Heinrich, L. (Rev): Review of "Choquet-Deny type functional equations with applications to stochastic models". 95 BiomtrcJ 37 726-726

— —; Toutenburg, Helge (Auth); Draper, N. R. (Rev): Review of "Linear models: Least squares and alternatives". 96 ShrtBkRv 16/1 7-8

Ray-Chaudhuri, D. K.: Arasu, K. T.; — —; Singhi, N. M.: Simple designs. 93 JCISS 18 130-135

Raychaudhuri, Aparna: Basu, Srabashi; — —; Basu, Ayanendranath: Improving precision through modelling: An illustration with hierarchical kappa. 95 CommStB 24 399-408

Rayens, William: Hinkle, John; — —: Partial least squares and compositional data: Problems and alternatives. 95 Chemolab 30 159-172

Raykov, Tenko: Multivariate structural modeling of plasticity in fluid intelligence of aged adults. 95 MultBehR 30 255-287

— —; Penev, Spiridon: Comment on "A binomial test of model fit" (V1 p268-273). 95 StrEqMod 2 146-151

— —; Widaman, Keith F.: Issues in applied structural equation modeling research. 95 StrEqMod 2 289-318

Raymond, Luc: Estéve, Jacques; Benhamou, Ellen; — — (Ed): Statistical methods in cancer research. Volume 4: Descriptive epidemiology. 95 OxfordUniv 320p

Raymond, Robert L.: Alper, Paul; — —: Imposing views, imposing shoes: A statistician as a sole model. 95 AmerStat 49 317-319

Raynauld, Jean-Pierre: Tubert-Bitter, Pascale; Bloch, Daniel A.; — —: Comparing the bivariate effects of toxicity and efficacy of treatments. 95 StatMed 14 1129-1141 (Disc: p1143-1143)

Rayner, J. C. W.; Best, D. J.: Smooth tests for the bivariate Poisson distribution. 95 AstrlJSt 37 233-245

— —; Best, D. J.; Mathews, K. L.: Interpreting the skewness coefficient. 95 CommStA 24 593-600

Raz, Jonathan; Cardenas, Valerie; Fletcher, Daniel: Frequency domain estimation of covariate effects in multichannel brain evoked potential data. 95 Biomtrcs 51 448-460

Raz, Tzvi: Buck, James R.; Pignatiello, Joseph J., Jr; — —: Bayesian Poisson rate estimation using parameter maps. 95 IIETrn 27 405-412

Read, C. B.: Hoppe, F. M. (Ed); — — (Rev): Review of "Multiple comparisons, selection, and applications in biometry: A festschrift in honor of Charles W. Dunnett". 95 Biomtrcs 51 380-380

Read, Kenneth L. Q.: Preece, Peter F. W.; — —: A stage-theory model of cognitive development. 95 BJMaStPs 48 1-7

Reagan, Joey: Pinkleton, Bruce; — —; Aaronson, Dustin; Ramo, Eduard: Does "I'm not selling anything" increase response rates in telephone surveys?. 94 ASASrvyRsr 1242-1247

Reagan, Patricia B.: Stasny, Elizabeth A.; — —: Adjusting pension fund membership for "don't know" responses. 94 ASAProBuEc 298-303

Rebaï, A.: Mangin, B.; Goffinet, B.; — —: Constructing confidence intervals for QTL location. 94 Genetcs 138 1301-1308

Rebai, Ahmed; Goffinet, Bruno; Mangin, Brigitte: Comparing power of different methods for QTL detection. 95 Biomtrcs 51 87-99

Rebaï, Ahmed; Goffinet, Bruno; Mangin, Brigitte: Approximate thresholds of interval mapping tests for QTL detection. 94 Genetcs 138 235-240

Reboulet, A.-M.: Pradel, R.; — —; Viallefont, A.: Testing hypotheses and estimating survival from capture histories with CR. 95 JAppStat 22 775-784

Reboussin, David M.: Flexible monitoring of multi-arm clinical trials with repeated measurements. 94 ASAProBiop 111-114

Recchia, C.: Solow, A. R.; Smith, W. K.; — —: A conditional test of independence of Markov chains. 95 BiomtrcJ 37 973-977

Redfern, E.: Watson, S. M.; — —; Clark, S.; Tight, M.; Davies, N.: An influence method for outlier detection applied to time series traffic data. 95 JAppStat 22 135-149

Reece, Jack E.; Spagon, Patrick D.: Linked training and consulting in the semiconductor industry: The SEMATECH approach. 94 ASAProStEd 228-233

Reed, J. F.; Stark, D. B.: Robust analysis of variance: A simulation study. 95 JAppStat 22 87-104

Reed, J. F., III; Stark, D. B.: Monte Carlo generation of S_B and S_U distributions. 95 CmpBiomd 47 157-165

Reed, Margaret; Garven, Frances: Minitab menus - Do they help or hinder?. 95 TeachgSt 17/1 22-25

Reed-Fourquet, Lori L.: Shevchenko, Ivan P.; Lynch, John T.; Mattie, Angela S.; — —: Verification of information in a large medical database using linkages with external databases. 95 StatMed 14 511-530 (Disc: p687-689)

Reepmeyer, Jan-Armin: Taudes, Alfred; — —; Bensberg, Frank (Auth); Weinberger, Volker (Rev): Review of "UNIX: Introduction for DOS-users" (German). 95 CmpStDA 19 360-361

Rees, D.-G.: Essential statistics (Third edition). 94 ChapmnHall:Lnd 276p

Rees, Mike: Stirzaker, D. (Auth); — — (Rev): Review of "Elementary probability". 95 JRSS-A 158 356-356

Rees, Rolf S.: Two new direct product-type constructions for resolvable group-divisible designs. 93 JComDsgn 1 15-26

Reese, R. Allan: Relations - Who needs 'em?. 95 Chance 8/2 52-55

— —: Means what he says, says what he means. 95 Chance 8/4 49-52

van Pul, M. C. J. (Auth); — — (Rev): Review of "Statistical analysis of software reliability models". 95 JRSS-A 158 199-199

Reeve, Russell; Hale, Michael: Results and efficiency of Bayesian dose adjustment in a clinical trial with binary endpoint. 94 ASAProBiop 324-327

Hale, Michael; — —: Planning a randomized concentration controlled trial with a binary response. 94 ASAProBiop 319-323

Reeves, Jane: Bruce, Peter (Auth); — — (Rev): Review of "Resampling stats (Version 3.14)". 95 TeachgSt 17/3 101-103

Reeves, Jaxk H.: Heterogeneity in the substitution process of amino acid sites of proteins coded for by mitochondrial DNA. 92 JMolEvol 35 17-31

Refenes, Apostolos-Paul: — — (Ed); Meade, Nigel (Rev): Review of "Neural networks in the capital markets". 95 IntJFor 11 602-603

— — (Ed); Trippi, Robert R. (Rev): Review of "Neural networks in the capital markets". 95 IntJFor 11 603-604

Regal, Ronald R.: Zelterman, Daniel; Hook, Ernest B.; — —: Bayesian estimates of a population size. 94 ASAProBays 29-34

Regev, Haim: Bregman, Arie; Fuss, Melvyn; — —: The production and cost structure of Israeli industry: Evidence from individual firm data. 95 JEconmtx 65 45-81

Griliches, Zvi; — —: Firm productivity in Israeli industry 1979-1988. 95 JEconmtx 65 175-203

Regoli, Giuliana: Di Bacco, Mario; — —: An alternative axiomatic approach to expected utility theory. 95 Metron 53 21-29

Di Bacco, Mario; — —: Likelihood as an index of fit. 94 JItalStS 3 243-253

Rehberg, J.: Yakubov, S. (Auth); — — (Rev): Review of "Completeness of root functions of regular differential operators". 95 BiomtrcJ 37 118-118

Reich, A.: Niggemeyer, B.; Radtke, M.; — —: Applications of risk theory and multivariate analysis in insurance practice. 95 AplStMDA 11 231-244

Reich, Robin M.; Czaplewski, Raymond L.; Bechtold, William A.: Spatial cross-correlation of undisturbed, natural shortleaf pine stands in northern Georgia. 94 EnvEcoSt 1 201-217

Reichardt, Charles S.; Coleman, S. C.: The criteria for convergent and discriminant validity in a multitrait-multimethod matrix. 95 MultBehR 30 513-538

Reid, D. H.: Operational time and a fundamental problem of insurance in a data-rich environment. 95 AplStMDA 11 257-269

Reid, N.: The roles of conditioning in inference. 95 StatSci 10 138-157 (Disc: p173-199)

— —: Reply to comments on "Inference". 95 StatSci 10 193-196

— —: Comment on "Statistical issues arising in equal employment litigation". 94 ASAProBuEc 54-54

Schwarz, C.; — —: Comment on "Effects of growth regulators on silver maple trees: A case study". 95 CandJSt 23 329-332

Reilly, Marie; Pepe, Margaret Sullivan: A mean score method for missing and auxiliary covariate data in regression models. 95 Biomtrka 82 299-314

Reilly, Terence: A necessary and sufficient condition for identification of confirmatory factor analysis models of factor complexity one. 95 SocMethR 23 421-441

O'Brien, Robert M.; — —: Equality in constraints and metric-setting measurement models. 95 StrEqMod 2 1-12

Reiman, M. I.: Coffman, E. G., Jr; Puhalskii, A. A.; — —: Polling systems with zero switchover times: A heavy-traffic averaging principle. 95 AnlsApPr 5 681-719

Reimers, Hans-Eggert: Interval forecasting in cointegrated systems. 95 StHefte 36 349-369

Rein, Steven R.: A nonlinear state-space model of Chinook salmon population dynamics. 94 ASAProEnvr 150-155

Reinecke, Kenneth J.: Smith, David R.; — —; Conroy, Michael J.; Brown, Michael W.; Nassar, James R.: Factors affecting visibility rate of waterfowl surveys in the Mississippi Alluvial Valley. 95 JWildlMg 59 515-527

Reinert, Gesine: A weak law of large numbers for empirical measures via Stein's method. 95 AnlsProb 23 334-354

Reinsdorf, Marshall: Comment on "Computing elementary aggregates in the Swedish Consumer Price Index" (V8 p129-147). 94 JOffStat 10 103-107

Reinsel, C.: — — (Auth); Lütkepohl, H. (Rev): Review of "Elements of multivariate time series analysis". 95 Metrika 42 147-148

Reinsel, G.: Box, G.; Jenkins, G.; — — (Auth); Ziegel, Eric R. (Rev): Review of "Time series analysis: Forecasting and control". 95 Technmcs 37 238-239

Reinsel, Gregory C.: Finite sample forecast results for vector autoregressive moving average models. 95 JForcast 14 405-412

Yap, Sook Fwe; — —: Estimation and testing for unit roots in a partially nonstationary vector autoregressive moving average model. 95 JASA 90 253-267

Yap, Sook Fwe; — —: Results on estimation and testing for a unit root in the nonstationary autoregressive moving-average model. 95 JTimSrAn 16 339-353

— — (Auth); Timm, Neil H. (Rev): Review of "Elements of multivariate time series analysis". 95 Technmcs 37 118-118

Reise, Steven P.; Gomel, Jessica N.: Modeling qualitative variation within latent trait dimensions: Application of mixed-measurement to personality assessment. 95 MultBehR 30 341-358

Waller, Niels G.; — —: Genetic and environmental influences on item response pattern scalability. 92 BehavGen 22 135-152

Reiser, B.; Guttman, I.; Lin, Dennis K. J.; Guess, Frank M.; Usher, John S.: Bayesian inference for masked system lifetime data. 95 ApplStat 44 79-90

Guttman, Irwin; Lin, Dennis K. J.; — —; Usher, John S.: Dependent masking and system life data analysis: Bayesian inference for two-component systems. 95 LiDaAnal 1 87-100

Reiser, Benjamin: Dror, Shuki; Faraggi, David; — —: Dynamic treatment allocation adjusting for prognostic factors for more than two treatments. 95 Biomtrcs 51 1338-1343

Reiser, Mark; Lin, Ginnie: A goodness-of-fit test for the latent class model when expected frequencies are small. 94 ASAProSoSt 49-54

Reiss, R.-D.: Kaufmann, E.; — —: Approximation rates for multivariate exceedances. 95 JStPlInf 45 235-245

— — (Auth); Matthes, Klaus (Rev): Review of "A course on point processes". 95 BiomtrcJ 37 182-182

Haßmann, S.; — —; Thomas, M. (Auth); Anderson, C. W. (Rev): Review of "XTREMES". 95 CmpStQ 10 91-98

Reiss, Rolf-Dieter: Falk, Michael; Hüsler, Jurg; — — (Auth); Mikosch, Thomas (Rev): Review of "Law of small numbers: Extremes and rare events". 95 JASA 90 1133-1134

Rekab, Kamel: Tahir, Mohamed; — —: Asymptotically optimal empirical Bayes procedures for selecting good uniform $U(0,\theta_i)$ populations. 95 StocAnAp 13 233-247

Remillard, Bruno: Abdous, Belkacem; — —: Relating quantiles and expectiles under weighted-symmetry. 95 AnInStMa 47 371-384

Rémillard, Bruno: On Chung's law of the iterated logarithm for some stochastic integrals. 94 AnlsProb 22 1794-1802

Remmenga, Marta D.; Johnson, Dallas E.: A comparison of inference procedures in unbalanced split-plot designs. 95 JStCmpSm 51 353-367

Remus, William; O'Connor, Marcus; Griggs, Kenneth: Does reliable information improve the accuracy of judgemental forecasts?. 95 IntJFor 11 285-293

Ren, Jian-Jian: Generalized Cramér-von Mises tests of goodness of fit for doubly censored data. 95 AnInStMa 47 525-549

— —; Sen, Pranab Kumar: Hadamard differentiability on $D[0,1]^p$. 95 JMultiAn 55 14-28

Rencher, Alvin C.: Methods of multivariate analysis. 95 Wiley:NY:UK 627p

Rendtel, Ulrich; Wagner, Gert; Frick, Joachim: A method for checking longitudinal weights in panel surveys - The case of the German Socio-Economic Panel (German). 95 AllgeStA 79 252-277

Renkevens, Thomas M.: Duchon, Claude E.; — —; Crosson, William L.: Estimation of daily area-average rainfall during the CaPE experiment in Central Florida. 95 JApMetrl 34 2704-2714

Renouf, Antoinette: Campanelli, Pamela; Channell, Joanna; McAulay, Liz; — —; Thomas, Roger: A cognitive and linguistic exploration of the meaning of training: Implications for survey design. 94 ASASrvyRsr 1281-1286

Renshaw, A. E.; Haberman, S.: On the graduations associated with a multiple state model for permanent health insurance. 95 InsrneME 17 1-17

Renshaw, Eric: Chen, Anyue; — —: Markov branching processes regulated by emigration and large immigration. 95 StocProc 57 339-359

Jakeman, Eric; Phayre, Sean; — —: The evolution and measurement of a population of pairs. 95 JAppProb 32 1048-1062

Barnsley, M. F. (Auth); — — (Rev): Review of "Fractals everywhere". 95 JRSS-A 158 339-340

Bellomo, N.; Brzezniak, Z.; de Socio, L. M. (Auth); — — (Rev): Review of "Nonlinear stochastic evolution problems in applied sciences". 94 BLndnMaS 26 409-410

— — (Auth); Zeh, Judy (Rev): Review of "Modelling biological populations in space and time". 95 JASA 90 800-801

Repin, K. Yu.: Girko, V. L.; — —: The quarter-of-circumference law. 95 ThPrMaSt 50 67-70

Reschenhofer, Erhard: Use of phase for the detection of hidden periodicities. 95 BiomtrcJ 37 957-964

Hauser, Michael A.; — —: Estimation of the fractionally differencing parameter with the R/S method. 95 CmpStDA 20 569-579

Resnick, Daniel P.: Devlin, Bernie; Fienberg, Stephen E.; — —; Roeder, Kathryn: Wringing the bell curve: A cautionary tale about the relationships among race, genes, and IQ. 95 Chance 8/3 27-36

Herrnstein, R. J.; Murray, C. (Auth); Devlin, Bernie; Fienberg, Stephen E.; — —; Roeder, Kathryn (Rev): Review of "The bell curve: Intelligence and class structure in American life". 95 JASA 90 1483-1488

Resnick, Sidney; Stărică, Cătălin: Consistency of Hill's estimator for dependent data. 95 JAppProb 32 139-167

Resnick, Sidney I.: Feigin, Paul D.; — —; Stărică, Cătălin: Testing for independence in heavy tailed and positive innovation time series. 95 CommStSM 11 587-612

Goldie, Charles M.; — —: Many multivariate records. 95 StocProc 59 185-216

— — (Auth); Matthes, K. (Rev): Review of "Adventures in stochastic processes" (German). 95 BiomtrcJ 37 272-272

Retherford, R. D.; Choe, M. K. (Auth); Fearn, Tom (Rev): Review of "Statistical models for causal analysis". 95 JRSS-A 158 200-201

Reuben, Cynthia A.: Madans, Jennifer H.; — —; Rothwell, Sandra T.; Eberhardt, Mark S.: Differences in morbidity measures and risk factor identification using multiple data sources: The case of coronary heart disease. 95 StatMed 14 643-653 (Disc: p691-692)

Révész, P.: Strong theorems on the extreme values of stationary Poisson processes. 95 JStPlInf 45 291-300

— —: Random walks of infinitely many particles. 94 WrldSciPub:Sing 191p

Hughes, Barry D. (Auth); — — (Rev): Review of "Random walks and random environments. Volume 1: Random walks". 95 ShrtBkRv 15/3 47-47

— — (Auth); Heyde, C. C. (Rev): Review of "Random walks of infinitely many particles". 95 ShrtBkRv 15/3 47-47

Révész, Pál: Csáki, Endre; Csörgő, Miklós; Földes, Antónia; — —: Global Strassen-type theorems for iterated Brownian motions. 95 StocProc 59 321-341

Rexstad, Eric A.: Lindberg, Mark S.; Sedinger, James S.; — —: Estimating nest site fidelity of adult female black brant with multi-state modeling and geographic information systems. 95 JAppStat 22 725-735

Schmutz, Joel A.; Ward, David H.; Sedinger, James S.; — —: Survival estimation and the effects of dependency among animals. 95 JAppStat 22 673-681

Rey, Timothy D.; Duever, Thomas A.: Modeling nonlinear systems, bridging theory and technology. 94 ASAProQlPr 310-315

Reynard, Susan E.: Finn, Lynda M.; Garhart, Casey A.; — —: Getting statistical tools to the masses: Common problems in training design and execution. 95 ASQCTekCon 356-364

Reynolds, Marion R.: Arnold, Jesse C.; — —: CUSUM charts with variable sample size and variable interval size. 94 ASAProQlPr 87-92

Reynolds, Marion R., Jr: Evaluating properties of variable sampling interval control charts. 95 SqtlAnly 14 59-97

Amin, Raid W.; — —; Bakir, Saad: Nonparametric quality control charts based on the sign statistic. 95 CommStA 24 1597-1623

Kapatou, Alexandra; — —: Multivariate nonparametric control charts using small samples. 94 ASAProQlPr 241-246

Sethuraman, Shanthi; — —; Arnold, Jesse C.: Process control strategies using variable sampling intervals. 94 ASAProQlPr 109-114

Rezayat, Fahimeh: On the use of an SPSA-based model-free controller in quality improvement. 95 Autmatca 31 913-915

— —; Holton, Jack: Neural network controller and quality improvement: An application. 94 ASAProQlPr 33-38

Rhee, H. K.: Gakis, K. G.; — —; Sivazlian, B. D.: Distributions and first moments of the busy and idle periods in controllable M/G/1 queueing models with simple and dyadic policy. 95 StocAnAp 13 47-81

Rhee, WanSoo T.: On rates of convergence for common subsequences and first passage time. 95 AnlsApPr 5 44-48

Rhinehart, R. Russell: Bethea, Robert M.; — — (Auth); Shanmugam, Ramalingam (Rev): Review of "Applied engineering statistics". 95 JQualTek 27 173-173

Rhoades, D. A.; Neill, W. A.: The effect of station drift on the estimation of shifts in meteorological records due to site changes. 95 JClimlgy 15 207-220

Rhoades, James D.: Lesch, Scott M.; Strauss, David J.; — —: Spatial prediction of soil salinity using electromagnetic induction techniques 1. Statistical prediction models: A comparison of multiple linear regression and cokriging. 95 WaterReR 31 373-386

Lesch, Scott M.; Strauss, David J.; — —: Spatial prediction of soil salinity using electromagnetic induction techniques. 2. An efficient spatial sampling algorithm suitable for multiple linear regression model identification and estimation. 95 WaterReR 31 387-398

Rhodes, William: Hyatt, Raymond R., Jr; — —: The price and purity of cocaine: The relationship to emergency room visits and death, and to drug use among arrestees. 95 StatMed 14 655-668

Riboli, Elio: Kaaks, Rudolf; — —; van Staveren, Wija: Calibration of dietary intake measurements in prospective cohort studies. 95 AJEpidm 141 548-556

Kaaks, Rudolf; — —; van Staveren, Wija: Sample size requirements for calibration studies of dietary intake measurements in prospective cohort investigations. 95 AJEpidm 141 557-565

Kaaks, Rudolf; — —; van Staveren, Wija: Sample size requirements for calibration studies of dietary intake measurements in prospective cohort investigations. 95 AJEpidm 142 557-565

Ricciardi, L. M.: Di Crescenzo, A. G.; Giorno, V.; Nobile, A. G.; — —: On a symmetry-based constructive approach to probability densities for two-dimensional diffusion processes. 95 JAppProb 32 316-336

Rice, Carol H.: Hertzberg, Vicki Stover; — —; Linz, Douglas H.; Pinney, Susan M.: Occupational epidemiology in the era of total quality: Challenges for exposure assessment. 94 ASAProEpid 139-142

Rice, Janet C.: Bakeman, Roger; Robinson, Byron F. (Auth); — — (Rev): Review of "Understanding log-linear analysis with ILOG: An interactive approach". 95 EdPsMeas 55 350-355

Rice, John: — — (Auth); Ziegel, Eric R. (Rev): Review of "Mathematical statistics and data analysis (Second edition)". 95 Technmcs 37 127-127

Rice, John W.: Murray, Michael K.; — — (Auth); Marriott, Paul (Rev): Review of "Differential geometry and statistics". 95 BLndnMaS 27 619-620

Rice, William R.; Gaines, Steven D.: Extending nondirectional heterogeneity tests to evaluate simply ordered alternative hypotheses. 94 PNatAcSc 91 225-226

Richard, Jean-François: Comment on "Bayesian model selection and prediction with empirical applications". 95 JEconmtx 69 337-349

Richards, D. O.: Badrigal, J. L.; — —; Free, J. C.; Lee, T. N.; Lawson, J. S.; Hawks, V. D.: Q^2: A TQM course for undergraduates. 94 ASAProQlPr 172-180

Richards, Donald St. P.: Exact asymptotics for some probability distributions on compact manifolds. 95 AnlsStat 23 1582-1586
Gupta, Rameshwar D.; — —: Multivariate Liouville distributions, IV. 95 JMultiAn 54 1-17
Richards, John D.: Beauregard, Michael R.; Mikulak, Raymond J.; Olson, Barbara A. (Auth); — — (Rev): Review of "A practical guide to statistical quality improvement: Opening up the statistical toolbox". 95 JQualTek 27 177-178
Clogg, Clifford C.; Shihadeh, Edward S. (Auth); — — (Rev): Review of "Statistical models for ordinal variables". 95 JQualTek 27 394-394
Richards, Laura J.: Schnute, Jon T.; — —: The influence of error on population estimates from catch-age models. 95 JFishRBC 52 2063-2077
Richards, W. Graham: Gasser, Robert P. H.; — —: An introduction to statistical thermodynamics. 95 WrldSciPub:Sing 180p
Richardson, Alice M.; Welsh, Alan H.: Robust restricted maximum likelihood in mixed linear models. 95 Biomtrcs 51 1429-1439
Richardson, Gordon D.: Bandyopadhyay, Sati P.; Brown, Lawrence D.; — —: Analysts' use of earnings forecasts in predicting stock returns: Forecast horizon effects. 95 IntJFor 11 429-445
Richardson, Sylvia; Monfort, Christine; Green, Martyn; Draper, Gerald; Muirhead, Colin: Spatial variation of natural radiation and childhood leukaemia incidence in Great Britain. 95 StatMed 14 2487-2501
Watier, Laurence; — —: Modelling of an epidemiological time series by a threshold autoregressive model. 95 Ststcian 44 353-364
Richman, Michael B.: Gong, Xiaofeng; — —: On the application of cluster analysis to growing season precipitation data in North America east of the Rockies. 95 JClimate 8 897-931
Richter, W.-D.: Zoefel, P. (Auth); — — (Rev): Review of "Univariate analyses of variance" (German). 95 MaOpfStS 26 179-180
Rickman, Dan S.: A Bayesian analysis of the use of pooled coefficients in a structural regional economic model. 95 IntJFor 11 477-490
Ridder, Ad: Mandjes, Michel; — —: Finding the conjugate of Markov fluid processes. 95 PrEngInf 9 297-315
Ridder, G.: Kortram, R. A.; van Rooij, A. C. M.; Lenstra, A. J.; — —: Constructive identification of the mixed proportional hazards model. 95 StNeerla 49 269-281
Ridder, Geert: Koning, Pierre; — —; Van den Berg, Gerard J.: Structural and frictional unemployment in an equilibrium search model with heterogeneous agents. 95 JApEcmtx 10 133-151S
Riddler-Rowe, C. J.: Lawler, G. F. (Auth); — — (Rev): Review of "Introduction to stochastic calculus". 96 ShrtBkRv 16/1 8-8
Ride, David J.: A practical method of estimating toxic loads in the presence of concentration fluctuations. 95 EnvrMtrc 6 643-650
Ridout, M. S.: Three-stage designs for seed testing experiments. 95 ApplStat 44 153-162
Smith, D. M.; — —: Comment on "Algorithm AS 189: Maximum likelihood estimation of the parameters of the beta binomial distribution" (V32 p196-204). 95 ApplStat 44 545-547
Rieck, James R.: Parameter estimation for the Birnbaum-Saunders distribution based on symmetrically censored samples. 95 CommStA 24 1721-1736
Riedel, M.: Caulcutt, R. (Auth); — — (Rev): Review of "Statistics in research and development (Second edition)". 95 MaOpfStS 26 180-180
Rieder, H.: — — (Auth); Plachky, D. (Rev): Review of "Robust asymptotic statistics". 95 StatDecn 13 93-94
Rieder, Helmut: Büning, Herbert (Auth); — — (Rev): Review of "Robust and adaptive tests" (German). 95 StHefte 36 272-272
— — (Auth); Beran, Rudolf (Rev): Review of "Robust asymptotic statistics". 95 JASA 90 1128-1129
Rieder, Ulrich; Weishaupt, Jürgen: Customer scheduling with incomplete information. 95 PrEngInf 9 269-284
Ries, Lynn A.: Henson, Donald Earl; — —: The relative survival rate. 95 Cancer 76 1687-1688
Rifkin, Robert D.: Effects of correlated and uncorrelated measurement error on linear regression and correlation in medical method comparison studies. 95 StatMed 14 789-798
Rigdon, Edward E.: A necessary and sufficient identification rule for structural models estimated in practice. 95 MultBehR 30 359-383
Rigdon, Steven E.: An integral equation for the in-control average run length of a multivariate exponentially weighted moving average control chart. 95 JStCmpSm 52 351-365
— —: A double-integral equation for the average run length of a multivariate exponentially weighted moving average control chart. 95 StPrLet 24 365-373
Scheaffer, Richard L.; McClave, James T. (Auth); — — (Rev): Review of "Probability and statistics for engineers". 95 JQualTek 27 178-179
Riley, Janet: Setting objectives in scientific studies. 95 TeachgSt 17/2 61-63
Barnett, Vic; — —: Statistics for environmental change. 95 ExpAgric 31 117-130
Ringwalt, Chris L.: Iachan, Ronald; — —; Greene, Jody M.: Substance abuse among runaway and homeless youth. 94 ASASrvyRsr 1083-1087
Rinne, H.: Mittag, H.-J.; — — (Auth); Jain, Kamlesh (Rev): Review of "Statistical methods of quality assurance". 95 Technmcs 37 234-234
Mittag, H.-J.; — — (Auth); Nelson, Lloyd S. (Rev): Review of "Statistical methods of quality assurance". 95 JQualTek 27 95-96
Rio, E.: Doukhan, P.; Massart, P.; — —: Invariance principles for absolutely regular empirical processes. 95 AIHenrPB 31 393-427
Rio, Emmanuel: A maximal inequality and dependent Marcinkiewicz-Zygmund strong laws. 95 AnlsProb 23 918-937
— —: The functional law of the iterated logarithm for stationary strongly mixing sequences. 95 AnlsProb 23 1188-1203
Ríos, Martín: Villarroya, Angel; — —; Oller, Josep M.: Discriminant analysis algorithm based on a distance function and on a Bayesian decision. 95 Biomtrcs 51 908-919
Rios-Curil, Alejandro: Aguirre-Torres, Victor; — —: The effect and adjustment of complex surveys on chi-squared goodness-of-fit tests: Some Monte Carlo evidence. 94 ASASrvyRsr 602-607
Ríos-Insua, S.; Pachón, J. G.; Mateos, A.: A method of multiobjective decision making using a vector value function. 94 Questo 18 153-172
Ripley, B. D.; Solomon, P. J.: Comment on "Statistical models for prevalent cohort data" (V49 p1-11). 95 Biomtrcs 51 373-374
Venables, W. N.; — —: Modern applied statistics with S-Plus. 94 Springer:Brln:-NY 462p
Fang, K.-T.; Wang, Y. (Auth); — — (Rev): Review of "Number-theoretic methods in statistics". 95 JRSS-A 158 189-190
Grenander, U. (Auth); — — (Rev): Review of "General pattern theory: A mathematical theory of regular structures". 95 JRSS-A 158 635-635
Guyon, X. (Auth); Ludeña, C. (Transl); — — (Rev): Review of "Random fields on a network. Modeling, statistics, and applications". 96 ShrtBkRv 16/1 10-10
Shedler, G. S. (Auth); — — (Rev): Review of "Regenerative stochastic simulation". 94 BLndnMaS 26 410-411
Venables, W. N.; — — (Auth); Smith, J. T. (Rev): Review of "Modern applied statistics with S-Plus". 95 ShrtBkRv 15/2 24-24
Risch, Neil; Giuffra, Luis: Model misspecification and multipoint linkage analysis. 92 HumHrdty 42 77-92
Rishel, Raymond: Tracking the output of a poorly known Markov process. 95 Stochatx 52 147-156
Rising, William: Geometric Markov chains. 95 JAppProb 32 349-374
Ritland, K.: Fu, Y. B.; — —: On estimating the linkage of marker genes to viability genes controlling inbreeding depression. 94 ThApGene 88 925-932
Ritov, Ya'acov: PM algorithms for calculating minimum χ^2 estimators with partially observed sub-tables. 95 CmpStDA 20 19-33
Bickel, Peter J.; — —: An exponential inequality for U-statistics with applications to testing. 95 PrEngInf 9 39-52
Ritschard, Gilbert: Olszak, Michael; — —: The behaviour of nominal and ordinal partial association measures. 95 Ststcian 44 195-212
Ritter, C.; Bouckaert, A.; van Lierde, M.; Theunissen, I.: Modelling mortality and morbidity of newborns. 95 StatMed 14 2191-2204
Ritter, Gunter; Gallegos, María Teresa; Gaggermeier, Karl: Automatic context-sensitive karyotyping of human chromosomes based on elliptically symmetric statistical distributions. 95 PattrnRc 28 823-831
Ritter, Klaus; Wasilkowski, Grzegorz W.; Woźniakowski, Henryk: Multivariate integration and approximation for random fields satisfying Sacks-Ylvisaker conditions. 95 AnlsApPr 5 518-540
Riu, J.; Rius, F. X.: Univariate regression models with errors in both axes. 95 JChemtrc 9 343-362
Rius, F. X.: Lozano, J.; Novič, M.; — —; Zupan, J.: Modelling metabolic energy by neural networks. 95 Chemolab 28 61-72
Riu, J.; — —: Univariate regression models with errors in both axes. 95 JChemtrc 9 343-362
Rius, F. Xavier: Frank, I. E.; Todeschini, R. (Auth); — — (Rev): Review of "The data analysis handbook". 95 Chemolab 29 327-327
Rivest, L.-P.: Genest, C.; Ghoudi, K.; — —: A semiparametric estimation procedure of dependence parameters in multivariate families of distributions. 95 Biomtrka 82 543-552
Rivest, Louis-Paul: Comment on "Assessing geometric integrity through spherical regression techniques". 95 StSinica 5 204-210
— —; Hurtubise, Daniel: On Searl's winsorized mean for skewed populations. 95 SurvyM 21 107-116
— —; Potvin, François; Crépeau, Hélène; Daigle, Gaétan: Statistical methods for aerial surveys using the double-count technique to correct visibility bias. 95 Biomtrcs 51 461-470
Capéraà, Philippe; — —: On the variance of the trimmed mean. 95 StPrLet 22 79-85
Rizzi, Alfredo: In memory of Tommaso Salvemini. 95 Metron 53 3-20
— —: Distribution of the grammatical forms in the Italian language (Italian). 94 Statstca 54 275-291
— —; Vichi, Maurizio: Representation, synthesis, variability and data preprocessing of a three-way data set. 95 CmpStDA 19 203-222
Rizzo, Lou; Kalton, Graham; Brick, Mike: Adjusting for panel nonresponse in the Survey of Income and Program Participation. 94 ASASrvyRsr 422-427 (Disc: p440-442)
Robb, William H.: Resampling variance estimates for complex survey designs - A simulation study. 94 ASASrvyRsr 218-221
Robbiano, Luc: Bel, Liliane; — —: Extension of a uniqueness theorem for Zakai's equation (French). 95 Stochatx 52 227-245

Robbins, Chandler S.: Dawson, Deanna K.; Sauer, John R.; Wood, Paul A.; Berlanga, Mauro; Wilson, Marcia H.; — —: Estimating bird species richness from capture and count data. 95 JAppStat 22 1063-1068

Roberds, James: Bishir, John; — —: Analysis of failure time in clonally propagated plant populations. 95 MaBisci 125 109-125

Roberson, Paula K.; Shema, Sarah J.; Mundfrom, Daniel J.; Holmes, Talmage M.: Evaluating change and preference questions using paired Likert scale data. 94 ASAProSoSt 189-191

Robert, C. P.: The Bayesian choice: A decision-theoretic motivation. 94 Springer:Brln:NY 436p

— — (Auth); Mitchell, A. F. S. (Rev): Review of "The Bayesian choice: A decision-theoretic motivation". 95 ShrtBkRv 15/2 27-27

Robert, Christian P.: Simulation of truncated normal variables. 95 StatComp 5 121-125

— —: Comment on "Markov chains for exploring posterior distributions". 94 AnlsStat 22 1742-1747

Fourdrinier, Dominique; — —: Intrinsic losses for empirical Bayes estimation: A note on normal and Poisson cases. 95 StPrLet 23 35-44

Robert, Philippe: Coffman, E. G., Jr; Gilbert, E. N.; Greenberg, A. G.; Leighton, F. T.; — —; Stolyar, A. L.: Queues served by a rotating ring. 95 CommStSM 11 371-394

Roberts, Charles E., Jr: Sets of mutually orthogonal Latin squares with "like subsquares," II. 95 JCombThA 70 159-164

Roberts, Colin J.: Oxley, Les; — —: Modelling the supply of fishing effort. 95 EnvrMtrc 6 541-545

Roberts, G. O.; Sahu, S. K.; Gilks, W. R.: Comment on "Bayesian computation and stochastic systems". 95 StatSci 10 49-51

— —; Shortland, C. F.: The hazard rate tangent approximation for boundary hitting times. 95 AnlsApPr 5 446-460

Jacka, S. D.; — —: Weak convergence of conditioned processes on a countable state space. 95 JAppProb 32 902-916

Roberts, George W.: — — (Ed); Nelson, Lloyd S. (Rev): Review of "Quality planning, control, and improvement in research and development". 95 JQualTek 27 271-272

Roberts, Harry V.: Reflections on making statistics more effective in business schools (MSMESB). 94 ASAProBuEc 316-318 (Disc: p319-320)

Alwan, Layth C.; — —: The problem of misplaced control limits. 95 ApplStat 44 269-278 (Disc: p289-306)

Roberts, Jennifer: Coutts, J. Andrew; Mills, Terence C.; — —: Testing cumulative prediction errors in event study methodology. 95 JForcast 14 107-115

Roberts, Leigh A.: On the existence of moments of ratios of quadratic forms. 95 EconTher 11 750-774

Robertson, Chris: Modelling recall - Clustering and order effects. 95 BJMaStPs 48 29-50

Robertson, J. Roy: Fielding, Katherine L.; Brettle, Raymond P.; Gore, Sheila M.; O'Brien, Fiona; Wyld, Rona; — —; Weightman, Rosina: Heterosexual transmission of HIV analysed by generalized estimating equations. 95 StatMed 14 1365-1378

Robertson, Kenneth: Tou, Albert; — —: A study of imputation effects on estimation and variance estimation in the Occupational Employment Statistics Survey. 94 ASAProGovt 99-104

Robertson, Kenneth W.: West, Sandra A.; Kratzke, Diem-Tran; — —: Variance estimators for variables that have both observed and imputed values. 94 ASASrvyRsr 394-399

Robertson, Tim: Dykstra, Richard; Kochar, Subhash; — —: Likelihood ratio tests for symmetry against one-sided alternatives. 95 AnInStMa 47 719-730

Dykstra, Richard; Kochar, Subhash; — —: Inference for likelihood ratio ordering in the two-sample problem. 95 JASA 90 1034-1040

Dykstra, Richard; Kochar, Subhash; — —: Likelihood based inference for cause specific hazard rates under order restrictions. 95 JMultiAn 54 163-174

Robeva, Raina: Pitt, Loren D.; — —; Wang, Dao Yi: An error analysis for the numerical calculation of certain random integrals: Part 1. 95 AnlsApPr 5 171-197

Robin, Jean-Marc: Magnac, Thierry; — —; Visser, Michael: Analysing incomplete individual employment histories using indirect inference. 95 JApEcmtx 10 153-169S

Robins, James M.: An analytic method for randomized trials with informative censoring: Part I. 95 LiDaAnal 1 241-254

— —: An analytic method for randomized trials with informative censoring: Part II. 95 LiDaAnal 1 417-434

— —; Hsieh, Fushing; Newey, Whitney: Semiparametric efficient estimation of a conditional density with missing or mismeasured covariates. 95 JRSS-B 57 409-424

— —; Rotnitzky, Andrea: Semiparametric efficiency in multivariate regression models with missing data. 95 JASA 90 122-129

— —; Rotnitzky, Andrea; Zhao, Lue Ping: Analysis of semiparametric regression models for repeated outcomes in the presence of missing data. 95 JASA 90 106-121

Mittleman, Murray A.; Maclure, Malcolm; — —: Control sampling strategies for case-crossover studies: An assessment of relative efficiency. 95 AJEpidm 142 91-98

Rotnitzky, Andrea; — —: Semiparametric regression estimation in the presence of dependent censoring. 95 Biomtrka 82 805-820

Robinson, Byron F.: Bakeman, Roger; — — (Auth); Rice, Janet C. (Rev): Review of "Understanding log-linear analysis with ILOG: An interactive approach". 95 EdPsMeas 55 350-355

Robinson, Geoff: Veevers, Alan; — —: Obituary: Richard A. Stone, 1961-1994. 95 JRSS-A 158 627-628

Robinson, Michael E.; Tawn, Jonathan A.: Statistics for exceptional athletics records. 95 ApplStat 44 499-511

Robinson, P. M.: Log-periodogram regression of time series with long range dependence. 95 AnlsStat 23 1048-1072

— —: Gaussian semiparametric estimation of long range dependence. 95 AnlsStat 23 1630-1661

— —: Nearest-neighbour estimation of semiparametric regression models. 95 JNnparSt 5 33-41

— —: Obituary: E. J. Hannan, 1921-1994. 95 JRSS-A 158 181-182

— —: The approximate distribution of nonparametric regression estimates. 95 StPrLet 23 193-201

— —: An invariance property of optimal spectral bandwidths. 95 StPrLet 24 345-346

Robinson, Stephen; McClean, Sally: Simulation of wind speed for the modelling of pine saw-fly mortality. 95 Ststcian 44 469-477

Robinson, William T.: Nedelman, Jerry R.; Karara, Adel H.; Chang, Cheng-Tao; Gibiansky, Ekaterina; McDonald, Scott; — —: Inferring systemic exposure from a pharmacokinetic screen: Model-free and model-based approaches. 95 StatMed 14 955-968 (Disc: p969-970)

Robison, Edwin L.: McKay, Ruth B.; — —; Malik, Athar B.: Touch-tone data entry for household surveys: Research findings and possible applications. 94 ASASrvyRsr 509-511

Robison, Leslie L.: Sakkinen, Pamela A.; Severson, Richard K.; Ross, Julie A.; — —: Random-digit dialing for control selection in childhood cancer studies: The geographic proximity and demographics within matched sets. 95 AJPubHlh 85 555-557

Robson, D. S.: Jones, C. M.; — —; Lakkis, H. D.; Kressel, J.: Properties of catch rates used in analysis of angler surveys. 95 TrAFish 124 911-928

Rocha-Martinez, José M.; Shaked, Moshe: A discrete-time model of failure and repair. 95 AplStMDA 11 167-180

Rochet, Jean-Charles: Geman, Hélyette; Karoui, Nicole El; — —: Changes of numéraire, changes of probability measure and option pricing. 95 JAppProb 32 443-458

Rochon, James: Comment on "Sample size estimation for repeated measurement designs" (V15 p100-123, 165-172). 95 ConClnTr 16 449-450

— —: Supplementing the intent-to-treat analysis: Accounting for covariates observed postrandomization in clinical trials. 95 JASA 90 292-300

Rocke, David M.; Lorenzato, Stefan: A two-component model for measurement error in analytical chemistry. 95 Technmcs 37 176-184

— —; Pallas, Michelle: Prediction of automotive emissions from gasoline composition. 94 ASAProPEng 269-273

Rockette, Howard E.: Obuchowski, Nancy A.; — —: Hypothesis testing of diagnostic accuracy for multiple readers and multiple tests: An ANOVA approach with dependent observations. 95 CommStB 24 285-308

Röckner, M.: Ma, Z. M.; — — (Auth); Fukushima, M. (Rev): Review of "Introduction to the non-symmetric Dirichlet forms". 95 Metrika 42 70-72

Röckner, Michael: Overbeck, Ludger; — —; Schmuland, Byron: An analytic approach to Fleming-Viot processes with interactive selection. 95 AnlsProb 23 1-36

Rockwood, Todd H.: Sangster, Roberta L.; — —; Dillman, Don A.: The influence of administration mode on responses to numeric rating scales. 94 ASASrvyRsr 708-712

Rodean, Howard C.: Turbulent diffusion as a stochastic Lagrangian process. 95 EnvrMtrc 6 659-663

Rodgers, Jack; Singh, Karan: Using microsimulation modeling techniques to estimate the effects of health reform on individual businesses. 94 ASAProGovt 39-44 (Disc: p51-52)

Rodgers, James A., Jr; Linda, Stephen B.; Nesbitt, Stephen A.: Comparing aerial estimates with ground counts of nests in wood stork colonies. 95 JWildlMg 59 656-666

Rodgers, Joseph Lee; McGue, Matt: A simple algebraic demonstration of the validity of DeFries-Fulker analysis in unselected samples with multiple kinship levels. 94 BehavGen 24 259-262

Rodolphe, François: Prum, Bernard; — —; de Turckheim, Elisabeth: Finding words with unexpected frequencies in deoxyribonucleic acid sequences. 95 JRSS-B 57 205-220

Rodrigues, Eliane R.: Convergence to stationary state for a Markov move-to-front scheme. 95 JAppProb 32 768-776

— —: The performance of the move-to-front scheme under some particular forms of Markov requests. 95 JAppProb 32 1089-1102

Rodrigues, Geraldo: Opsomer, Jean-Didier; Agras, Jean; Carpi, Anthony; — —: An application of locally weighted regression to airborne mercury deposition around an incinerator site. 95 EnvrMtrc 6 205-219

Rodrigues, J.; Baba, M. Y.: Bayesian estimation of a simple regression model with measurement errors. 94 RBrzlPrS 8 107-118

Rodrigues, Josemar; Elian, Silvia Nagib: Maximum likelihood prediction in finite populations. 95 RBrzlPrS 9 43-54

Rodríguez, Germán; Goldman, Noreen: An assessment of estimation procedures for multilevel models with binary responses. 95 JRSS-A 158 73-89

Rodríguez, Juan C.; Viollaz, Aldo J.: A Cramér-von Mises type goodness-of-fit test with asymmetric weight function. 95 CommStA 24 1095-1120

Rodriguez, L. A.; Fulker, D. W.; Cherny, S. S.: A maximum-likelihood model-fitting approach to conducting a Hayman analysis of diallel tables with complete or missing data. 93 BehavGen 23 69-76

Rodriguez, Robert; Tobias, Randall; Wolfinger, Russell: Comment on "The statistics of linear models: Back to basics" (V4 p221-234). 95 StatComp 5 97-101

Rodríguez-Lallena, J. A.: Quesada-Molina, J. J.; — —: Bivariate copulas with quadratic sections. 95 JNnparSt 5 323-337

Rodriguez-Poo, Juan M.: Longford, N. T. (Auth); — — (Rev): Review of "Random coefficient models". 95 JAppStat 22 544-545

Roe, Denise J.; Cheng, Mary; Elsberry, Victor A.; Nolan, Paul E.: Estimating parameters in non-linear mixed effect models: A comparison using clinical data in children. 94 ASAProBiop 313-318

Roebruck, Peter; Kühn, Andreas: Comparison of tests and sample size formulae for proving therapeutic equivalence based on the difference of binomial probabilities. 95 StatMed 14 1583-1594

Roeder, Kathryn: Devlin, Bernie; Fienberg, Stephen E.; Resnick, Daniel P.; — —: Wringing the bell curve: A cautionary tale about the relationships among race, genes, and IQ. 95 Chance 8/3 27-36

Lambert, Diane; — —: Overdispersion diagnostics for generalized linear models. 95 JASA 90 1225-1236

Herrnstein, R. J.; Murray, C. (Auth); Devlin, Bernie; Fienberg, Stephen E.; Resnick, Daniel P.; — — (Rev): Review of "The bell curve: Intelligence and class structure in American life". 95 JASA 90 1483-1488

Roes, Kit C. B.; Does, Ronald J. M. M.: Shewhart-type charts in nonstandard situations. 95 Technmcs 37 15-24 (Disc: p24-35; Reply: P36-40)

Roesch, Francis A.: Spatial analysis for monitoring forest health. 94 ASAProEnvr 104-109

Roger, C. A.: Preece, D. A.; Vowden, B. J.; Jones, R. Hughes; — —; Vowden, C. J.: Choreographing designs. 95 MaScitst 20 15-32

Rogers, A.: Jones, M. E.; Thomas, S. M.; — —: Luria-Delbrück fluctuation experiments: Design and analysis. 94 Genetcs 136 1209-1216

Rogers, Andrei: McNown, Robert; — —; Little, Jani: Simplicity and complexity in extrapolative population forecasting models. 95 MaPopStd 5 235-257

Rogers, James: Habibullah, Mohamed; — —; Voss, Theresa: Issues related to the bioequivalence of drug formulations. 94 ASAProStCp 222-224

Rogers, L. C. G.; Shi, Z.: The value of an Asian option. 95 JAppProb 32 1077-1088

— —; Williams, David: Diffusions, Markov processes, and martingales. Volume 1: Foundations (Second edition). 94 Wiley:NY:UK 386p

Embrechts, P.; — —; Yor, M.: A proof of Dassios' representation of the α-quantile of Brownian motion with drift. 95 AnlsApPr 5 757-767

— —; Williams, David (Auth); Embrechts, P. A. L. (Rev): Review of "Diffusions, Markov processes, and martingales. Volume 1: Foundations (Second edition)". 95 ShrtBkRv 15/2 28-28

Rogers, William H.: Gould, William W.; — —: Quantile regression as an alternative to robust regression. 94 ASAProStEd 12-16

Roginsky, Allen: Average error in the central limit theorem for the cumulative processes. 95 StPrLet 24 199-204

Rogosa, David; Saner, Hilary: Longitudinal data analysis examples with random coefficient models. 95 JEdStat 20 149-170 (Disc: p201-227, 234-238)

— —; Saner, Hilary: Reply to comments on "Longitudinal data analysis examples with random coefficient models". 95 JEdStat 20 234-238

Rohatgi, Vijay K.; Szekely, Gabor; Zsigri, G.: On aging life distribution classes: Further probability properties and tests. 94 JStatRsr 28 111-122

Peña, Edsel A.; — —; Székely, Gábor J.: Inference for a general type II censorship model. 95 MaOpfStS 26 241-252

Rohlf, F.-J.: Sokal, R.-S.; — —: Biometry: The principles and practice of statistics in biological research. 95 Freeman:NY 918p

Röhmel, J.: Bauer, P.; — —: An adaptive method for establishing a dose-response relationship. 95 StatMed 14 1595-1607

Röhmel, Joachim: Re-evaluation of decisions on the recommended dose from phase III dose finding trials. 95 StatMed 14 895-895 (Disc: p897-899)

Lewis, John A.; Jones, David R.; — —: Biostatistical methodology in clinical trials - A European guideline. 95 StatMed 14 1655-1657

Rojo, Javier: Characterization of some concepts of aging. 95 IEEEReli 44 285-290

— —: On the weak convergence of certain estimators of stochastically ordered survival functions. 95 JNnparSt 4 349-363

— —: Nonparametric quantile estimation under order constraints. 95 JNnparSt 5 185-200

Rolph, John E.: Moore, David (Ed); — — (Rev): Review of "Statistics: Decisions through data". 95 Chance 8/1 43-44

Rolski, T.: Daley, D. J.; — —: Correction to "Light traffic approximations in many-server queues" (V24 p202-218). 95 AdvAppPr 27 584-584

Rolski, Tomasz: Asmussen, Søren; Frey, Andreas; — —; Schmidt, Volker: Does Markov-modulation increase the risk?. 95 AstinB 25 49-66

Rom, Dror: Hochberg, Yosef; — —: Extensions of multiple testing procedures based on Simes' test. 95 JStPlInf 48 141-152

Rom, Dror M.; Holland, Burt: A new closed multiple testing procedure for hierarchical families of hypotheses. 95 JStPlInf 46 265-275

Rom, Meir; Cohen, Ayala: Estimation in the polytomous logistic regression model. 95 JStPlInf 43 341-353

Román, J.; Allan, R. N.: Stochastic simulation and Monte Carlo methods applied to the assessment of hydro-thermal generating system operation. 94 Top 2 1-24

Román, P. Román: Gutiérrez Jáimez, R.; — —; Ruiz, F. Torres: A note on the Volterra integral equation for the first-passage-time probability density. 95 JAppProb 32 635-648

Romano, Joseph P.; Thombs, Lori A.: Inference for autocorrelations by resampling. 94 ASAProStCp 1-10

DiCiccio, Thomas J.; — —: On bootstrap procedures for second-order accurate confidence limits in parametric models. 95 StSinica 5 141-160

Politis, Dimitris N.; — —: Bias-corrected nonparametric spectral estimation. 95 JTimSrAn 16 67-103

Politis, Dimitris N.; — —: Large sample confidence regions based on subsamples under minimal assumptions. 94 AnlsStat 22 2031-2050

Romeo, Charles J.: Measurement error or endogeneity: Sorting out simultaneity. 94 ASAProBays 196-201

Römer, Christoph; Kandel, Abraham: Statistical tests for fuzzy data. 95 FuzzySet 72 1-26

Romera, Rosario; Cipra, Tomas: On practical implementation of robust Kalman filtering. 95 CommStB 24 461-488

Romero, Isidro: Allende, Sira; Bouza, Carlos; — —: Fitting a linear regression model by combining least squares and least absolute value estimation. 95 Questo 19 107-121

Romero, R.; Ferrer, A.; Capilla, C.; Zunica, L.; Balasch, S.; Serra, V.; Alcover, R.: Teaching statistics to engineers: An innovative pedagogical experience. 95 JStEd 3/1 33kb (jse/v3n1/romero)

Romeu, Jorge Luis: A statistical comparison of Cuban socioeconomic development. 94 ASAProSoSt 159-163

Römisch, Werner: Kloeden, Peter E.; Platen, Eckhard (Auth); Gelbrich, Matthias; — — (Rev): Review of "Numerical solution of stochastic differential equations". 95 SIAMRvw 37 272-275

Romo, Juan: Cuevas, Antonio; — —: On the estimation of the influence curve. 95 CandJSt 23 1-9

Romuald, Kristina S.; Haggard, Lois M.: The effect of varying the respondent selection script on respondent self-selection in RDD telephone surveys. 94 ASASrvyRsr 1299-1304

Ronchetti, E.: Morgenthaler, S.; — —; Stahel, W. A. (Ed); Lenz, H.-J. (Rev): Review of "New directions in statistical data analysis and robustness". 95 CmpStDA 19 711-712

Rondel, R. K.; Varley, S. A.; Webb, C. F. (Ed); Therasse, P. (Rev): Review of "Clinical data management". 95 StatMed 14 877-877

Ronen, Amit: Núñez, Megan; Kearns, Brian; — —: Perceptions of God's role in the origin of life on earth. 95 Stats /13 28-29

Ronin, Yefim I.: Korol, Abraham B.; — —; Kirzhner, Valery M.: Interval mapping of quantitative trait loci employing correlated trait complexes. 95 Genetcs 140 1137-1147

Rønn, Birgitte B.: Andersen, Per Kragh; — —: A nonparametric test for comparing two samples where all observations are either left- or right-censored. 95 Biomtrcs 51 323-329

Roorda, Berend; Heij, Christiaan: Global total least squares modeling of multivariable time series. 95 IEEEAuCn 40 50-63

Roos, Bero: A semigroup approach to Poisson approximation with respect to the point metric. 95 StPrLet 24 305-314

Roos, M.: Chen, L. H. Y.; — —: Compound Poisson approximation for unbounded functions on a group, with application to large deviations. 95 ProbTher 103 515-528

Roos, Malgorzata: Barbour, Andrew D.; Chryssaphinou, Ourania; — —: Compound Poisson approximation in reliability theory. 95 IEEEReli 44 398-402

Roosen, Charles B.: Friedman, Jerome H.; — —: An introduction to multivariate adaptive regression splines. 95 StMedRes 4 197-217

Rootzén, Holger: Extreme value theory for stochastic processes. 95 ScanActJ 54-65

Monrad, Ditlev; — —: Small values of Gaussian processes and functional laws of the iterated logarithm. 95 ProbTher 101 173-192

Rorstad, K.: The use of powerful workstations in statistical analysis and model building. 91 StatJUN 8 279-288

Rosage, Laura R.: Self-employed nonfilers: The accuracy of "enforcement-secured" returns. 94 ASAProGovt 206-211 (Disc: p212-214)

Rosalsky, Andrew; Stoyanov, Jordan; Presnell, Brett: An ergodic-type theorem á la Feller for nonintegrable strictly stationary continuous time processes. 95 StocAnAp 13 555-572

Nam, Eunwoo; — —: On the almost certain rate of convergence of series of independent random variables. 95 JKoreaSt 24 91-109

Rosberg, Zvi: Birman, Alexander; Gail, H. Richard; Hantler, Sidney L.; — —; Sidi, Moshe: An optimal service policy for buffer systems. 95 JACM 42 641-657

Rose, Elizabeth L.: Dielman, Terry E.; — —: A bootstrap approach to hypothesis testing in least absolute value regression. 95 CmpStDA 20 119-130

Dielman, Terry E.; — —: Pre-test estimators in least absolute value regression with autocorrelated errors. 94 ASAProBuEc 95-100

Van de Geer, John P. (Auth); — — (Rev): Review of "Multivariate analysis of categorical data: Theory and applications. Volumes 1 and 2". 95 StrEqMod 2 274-276

Rose, Kenneth A.: Smith, Eric P.; — —: Model goodness-of-fit analysis using regression and related techniques. 95 EcolgMdl 77 49-64

Rose, Kenneth H.: A performance measurement model. 95 QulPr 28/2 63-66

Rosegrant, Susan: Davis, Robert; — —; Watkins, Michael: Managing the link between measurement and compensation. 95 QulPr 28/2 101-106

Rosen, Jay: Marcus, Michael B.; — —: Logarithmic averages for the local times of recurrent random walks and Levy processes. 95 StocProc 59 175-184

Rosen, Ori; Cohen, Ayala: Constructing a bootstrap confidence interval for the unknown concentration in radioimmunoassay. 95 StatMed 14 935-952 (Disc: p953-953)

Rosen, Richard J.; Clayton, Richard L.; Pivetz, Lynda L.: Converting mail reporters to touchtone data entry. 94 ASASrvyRsr 763-768

Goldenberg, Karen L.; Moore, Susan E.; — —: Commercial payroll software and the quality of employment data. 94 ASASrvyRsr 1025-1030

Rosenbaum, P. R.: Observational studies. 95 Springer:Brln:NY 230p

— — (Auth); Oakes, D. (Rev): Review of "Observational studies". 96 ShrtBkRv 16/1 1-1

Rosenbaum, Paul R.: Quantiles in nonrandom samples and observational studies. 95 JASA 90 1424-1431

Silber, Jeffrey H.; — —; Ross, Richard N.: Comparing the contributions of groups of predictors: Which outcomes vary with hospital rather than patient characteristics?. 95 JASA 90 7-18

Rosenbaum, W. L.: Sterling, T. D.; — —; Weinkam, J. J.: Publication decisions revisited: The effect of the outcome of statistical tests on the decision to publish and vice versa. 95 AmerStat 49 108-112

Rosenbaum, Wilfred L.; Sterling, Theodor D.; Weinkam, James J.: Correcting standardized rate ratios for imprecise classification of a polychotomous exposure variable with limited data. 95 AJEpidm 141 442-445

— —; Sterling, Theodor D.; Weinkam, James J.: Correcting standardized rate ratios for imprecise classification of a polychotomous exposure variable with limited data. 95 AJEpidm 142 442-445

Rosenberg, Daniel K.; Overton, W. Scott; Anthony, Robert G.: Estimation of animal abundance when capture probabilities are low and heterogeneous. 95 JWildlMg 59 252-261

Rosenberg, Harry M.: Hoyert, Donna L.; Singh, Gopal K.; — —: Sources of data on socio-economic differential mortality in the United States. 95 JOffStat 11 233-260

Rosenberg, Philip S.: Hazard function estimation using B-splines. 95 Biomtrcs 51 874-887

Dahm, P. Fred; Gail, Mitchell H.; — —; Pee, David: Determining the value of additional surrogate exposure data for improving the estimate of an odds ratio. 95 StatMed 14 2581-2598

Rosenberg, Seymour: Van Mechelen, Iven; De Boeck, Paul; — —: The conjunctive model of hierarchical classes. 95 Psymtrka 60 505-521

Rosenberger, William F.; Lachin, John M.; Bain, Raymond P.: Nonparametric test of stochastic ordering for multiple longitudinal measures. 95 JBiophSt 5 235-243

Flournoy, Nancy; Durham, Stephen D.; — —: Toxicity in sequential dose-response experiments. 95 SqtlAnly 14 217-227

Lan, K. K. Gordon; — —; Lachin, John M.: Sequential monitoring of survival data with the Wilcoxon statistic. 95 Biomtrcs 51 1175-1183

Rosenblatt, Joan R.: Cameron, Joseph M.; — —: Obituary: Churchill Eisenhart, 1913-1994. 95 AmerStat 49 243-244

Rosenblatt, Murray: Prediction and non-Gaussian autoregressive stationary sequences. 95 AnlsApPr 5 239-247

Rosenbusch, Bernd: Abt, Markus; Strehler, Erwin; — —; Sterzik, Karl: Predicting the success of in vitro fertilization: Conventional semen analysis compared to the hamster ova penetration test. 95 StatMed 14 1807-1818

Rosendaal, Frits R.: Houwing-Duistermaat, Jeanine J.; Derkx, Bert H. F.; — —; van Houwelingen, Hans C.: Testing familial aggregation. 95 Biomtrcs 51 1292-1301

Rosenfeld, Barry; Turkheimer, Eric: Multidimensional representation of decision-making in chronic schizophrenics. 95 MultBehR 30 199-211

Rosenfeld, Daniel: Nirel, Ronit; — —: Estimation of the effect of operational seeding on rain amounts in Israel. 95 JApMetrl 34 2220-2229

Rosenthal, Jeffrey: Correction to "Minorization condition and convergence rates for Markov chain Monte Carlo" (V90 p558-566). 95 JASA 90 1136-1136

Rosenthal, Jeffrey S.: Rates of convergence for Gibbs sampling for variance component models. 95 AnlsStat 23 740-761

— —: Minorization conditions and convergence rates for Markov chain Monte Carlo. 95 JASA 90 558-566

— —: Convergence rates for Markov chains. 95 SIAMRvw 37 387-405

Baxter, J. R.; — —: Rates of convergence for everywhere-positive Markov chains. 95 StPrLet 22 333-338

Hoffman, John R.; — —: Convergence of independent particle systems. 95 StocProc 56 295-305

Rosenthal, Robert: Writing meta-analytic reviews. 95 PsychBu 118 183-192

Rosenzweig, Michael L.: Species diversity in space and time. 95 CambridgeUniv 456p

Rosinski, Jan: On the structure of stationary stable processes. 95 AnlsProb 23 1163-1187

Rosiński, Jan: Remarks on strong exponential integrability of vector-valued random series and triangular arrays. 95 AnlsProb 23 464-473

Roskam, Edw. E.: Comment on "Models for measurement, precision, and the nondichotomization of graded responses". 95 Psymtrka 60 27-35

Rösler, U.: Distributions slanted to the right. 95 StNeerla 49 83-93

Rösler, Uwe: Burton, Robert M.; — —: An L_2 convergence theorem for random affine mappings. 95 JAppProb 32 183-192

Rosner, Bernard: D'Agostino, Ralph B., Jr; Sparrow, David; Weiss, Scott; — —: Longitudinal models for analysis of respiratory function. 95 StatMed 14 2205-2216

Rosner, Gary L.; Berry, Donald A.: A Bayesian group sequential design for a multiple arm randomized clinical trial. 95 StatMed 14 381-394

Ross, G. J. S.: — — (Auth); Droge, B. (Rev): Review of "Nonlinear estimation". 95 MaOpfStS 26 181-182

Ross, Julie A.: Sakkinen, Pamela A.; Severson, Richard K.; — —; Robison, Leslie L.: Random-digit dialing for control selection in childhood cancer studies: The geographic proximity and demographics within matched sets. 95 AJPubHlh 85 555-557

Ross, Keith W.: Filar, Jerzy A.; Krass, Dmitry; — —: Percentile performance criteria for limiting average Markov decision processes. 95 IEEEAuCn 40 2-10

Ross, N. P.: Cothern, C. R.; — — (Ed); Ziegel, Eric R. (Rev): Review of "Environmental statistics, assessment, and forecasting". 95 Technmcs 37 128-128

Ross, N. Phillip: What the government needs. 95 AmerStat 49 7-9 (Disc: p18-23)

Cothern, C. Richard; — — (Ed): Environmental statistics, assessment, and forecasting. 94 CRCPress:FL 432p

Ross, Richard N.: Silber, Jeffrey H.; Rosenbaum, Paul R.; — —: Comparing the contributions of groups of predictors: Which outcomes vary with hospital rather than patient characteristics?. 95 JASA 90 7-18

Ross, S. M.: — — (Auth); Jolliffe, Flavia (Rev): Review of "Introduction to probability models (Fifth edition)". 95 JRSS-A 158 201-202

Ross, Sheldon: Derman, Cyrus; — —: Reliability for a system with dependent components. 95 PrEngInf 9 59-63

Derman, Cyrus; — —: An improved estimator of σ in quality control. 95 PrEngInf 9 411-415

Ross, Sheldon M.: Extending Blackwell's strengthening of Azuma's martingale inequality. 95 PrEngInf 9 493-496

Peköz, Erol A.; — —: A simple derivation of exact reliability formulas for linear and circular consecutive-k-of-n: F systems. 95 JAppProb 32 554-557

Ross, Stephan L.: Kao, Chihwa; — —: A CUSUM test in the linear regression model with serially correlated disturbances. 95 EcomtrRv 14 331-346

Ross, W. H.: Burnett, Richard T.; — —; Krewski, Daniel: Non-linear mixed regression models. 95 EnvrMtrc 6 85-99

Rossana, Robert J.; Seater, John J.: Temporal aggregation and economic time series. 95 JBES 13 441-451

Rossberg, H.-J.: Limit theorems involving restricted convergence. 94 ThProbAp 39 298-314

Rossberg, Hans-Joachim: Correction to "Characterization of the standard normal probability density within the class of self-adjoint (self-reciprocal) densities" (V22 p651-654). 95 MaOpfStS 26 387-387

Rossetti, Manuel D.; Clark, Gordon M.: Estimating expected sojourn times for a Markov renewal process. 95 CommStA 24 553-579

Rossi, G. C.: Colosimo, A.; Morante, S.; Parisi, V.; — —: An improved method for detection of words with unusual occurrence frequency in nucleotidic sequences. 93 JThBio 165 659-672

Rossi, Peter E.: Bollerslev, Tim; : Obituary: Daniel B. Nelson, 1959 1995. 95 JBES 13 361-364

Kim, Byung-Do; Blattberg, Robert C.; — —: Modeling the distribution of price sensitivity and implications for optimal retail pricing. 95 JBES 13 291-303

Rossman, Allan J.; Short, Thomas H.: Conditional probability and education reform: Are they compatible?. 95 JStEd 3/2 23kb (jse/v3n2/rossman)

Short, Thomas H.; — —; Hesterberg, Tim C.: Bayesian analyses of the Bradley-Terry model. 94 ASAProBays 117-122 (Disc: p134-135)

Wardrop, Robert L. (Auth); — — (Rev): Review of "Statistics: Learning in the presence of variation". 95 AmerStat 49 237-238

Rost, D.: A simulation study of the weighted k-means cluster procedure. 95 JStCmpSm 53 51-63

Rotar, Vladimir: Probability theory. 95 WrldSciPub:Sing 350p

Rotem, Doron: Olken, Frank; — —: Random sampling from databases: A survey. 95 StatComp 5 25-42

Olken, Frank; — —: Sampling from spatial databases. 95 StatComp 5 43-57

Rotemberg, Julio J.; Driscoll, John C.; Poterba, James M.: Money, output, and prices: Evidence from a new monetary aggregate. 95 JBES 13 67-83

Roters, Markus: An optimal stopping problem for random walks with non-zero drift. 95 JAppProb 32 956-959

— —: A best possible improvement of Wald's equation for stochastic processes. 95 Stochatx 54 293-300

Rothblum, Uriel G.: Eaves, B. Curtis; — —: A discounted-cost continuous-time flexible manufacturing and operator scheduling model solved by deconvexification over time. 95 PrEngInf 9 65-98

Rothenberg, Lori F.: VanBrackle, Lewis; VanBrackle, Anita S.; — —: Students' internalized representation of simple linear regression. 94 ASAProStEd 100-101

Rothenberg, Richard B.: Gargiullo, Paul M.; — —; Wilson, Hoyt G.: Confidence intervals, hypothesis tests, and sample sizes for the prevented fraction in cross-sectional studies. 95 StatMed 14 51-72

Thacker, Stephen B.; Stroup, Donna F.; — —; Brownson, Ross C.: Public health surveillance for chronic conditions: A scientific basis for decisions. 95 StatMed 14 629-641 (Disc: p695-696)

Rothery, P.; Martin, A. R.; Hiby, L.; Garside, S.: Estimating foetal growth rate from cross-sectional samples of field data with application to pilot-whales. 95 ApplStat 44 163-172

Brown, D.; — — (Auth); Kemp, A. W. (Rev): Review of "Computing examples supplement to models in biology: Mathematics, statistics and computing". 95 Biomtrcs 51 792-793

Brown, D.; — — (Auth); Kemp, A. W. (Rev): Review of "Models in biology: Mathematics, statistics and computing". 95 Biomtrcs 51 792-793

Rothwell, Geoffrey: Rust, John; — —: Optimal response to a shift in regulatory regime: The case of the US nuclear power industry. 95 JApEcmtx 10 75-118S

Rothwell, Sandra T.: Madans, Jennifer H.; Reuben, Cynthia A.; — —; Eberhardt, Mark S.: Differences in morbidity measures and risk factor identification using multiple data sources: The case of coronary heart disease. 95 StatMed 14 643-653 (Disc: p691-692)

Rotnitzky, Andrea; Robins, James M.: Semiparametric regression estimation in the presence of dependent censoring. 95 Biomtrka 82 805-820

Robins, James M.; — —: Semiparametric efficiency in multivariate regression models with missing data. 95 JASA 90 122-129

Robins, James M.; — —; Zhao, Lue Ping: Analysis of semiparametric regression models for repeated outcomes in the presence of missing data. 95 JASA 90 106-121

Rotondi, Renata; Drappo, Silvia: A clustering method for global optimization based on the kth nearest neighbour. 95 StatComp 5 317-326

Rotz, Wendy; Mulrow, Jeri; Falk, Eric: The 1995 Taxpayer Compliance Measurement Program (TCMP) sample redesign - A case history. 94 ASASrvyRsr 699-703

Rouderfer, Vladimir; Becker, Niels G.: Assessment of two-dose vaccination schedules: Availability for vaccination and catch-up strategies. 95 MaBisci 129 41-66

Rouncefield, Mary: The statistics of poverty and inequality. 95 JStEd 3/2 10kb (jse/v3n2/datasets.rouncefield)

Graham, Alan (Auth); — — (Rev): Review of "Statistics an introduction". 95 TeachgSt 17/3 111-111

Rouse, Cecilia Elena: Democratization of diversion? The effect of community colleges on educational attainment. 95 JBES 13/2 217-224

Roussas, George G.: Asymptotic normality of a smooth estimate of a random field distribution function under association. 95 StPrLet 24 77-90

Rousseeuw, Peter J.: Comment on "A probabilistic and statistical view of fuzzy methods". 95 Technmcs 37 283-286

— —; Croux, Christophe; Hössjer, Ola: Sensitivity functions and numerical analysis of the repeated median slope. 95 CmpStQ 10 71-90

Croux, Christophe; — —: Highly robust regression methods which are also efficient. 94 ASAProStEd 1-5

Hössjer, Ola; — —; Ruts, Ida: The repeated median intercept estimator: Influence function and asymptotic normality. 95 JMultiAn 52 45-72

Roussignol, M.: Cocozza-Thivent, C.; — —: Coupling methods in reliability (French). 95 AIHenrPB 31 119-141

Rousson, Valentin: Axiomatic of descriptive statistics (French). 95 Student 1 17-34

Routledge, Rick; Tsao, Min: Uniform validity of saddlepoint expansion on compact sets. 95 CandJSt 23 425-431

Roux, J. J. J.: Bekker, A.; — —: Bayesian multivariate normal analysis with a Wishart prior. 95 CommStA 24 2485-2497

Rowe, Geoff; Gribble, Steve: Income statistics from survey data: Effects of respondent rounding. 94 ASAProBuEc 77-82

Rowland, John H.; Kinack, Mark D.: A comparison of two methods of automated industry coding. 94 ASASrvyRsr 1130-1133

Rowley, C. A.: Bailey, R. A.; Preece, D. A.; — —: Randomization for a balanced superimposition of one Youden square on another. 95 JRSS-B 57 459-469

Rowley, P. J.: Laycock, P. J.; — —: A method for generating and labelling all regular fractions or blocks for q^{n-m}-designs. 95 JRSS-B 57 191-204

Roy, Dilip: Jana, Prasanta Kumar; — —: Estimation of reliability under stress-strength model in a bivariate exponential set-up. 94 CalcutSt 44 175-181

Roy, Laurent: Measuring and reporting nonresponse rates in agricultural surveys. 94 ASASrvyRsr 1156-1161

Roy, Nilanjana: Ullah, Aman; Srivastava, Virendra K.; — —: Moments of the function of non-normal random vector with applications to econometric estimators and test statistics. 95 EcomtrRv 14 459-471

Roy, Tapon: Robust non-linear regression analysis. 95 JChemtrc 9 451-457

Royall, Richard M.: Tsou, Tsung-Shan; — —: Robust likelihoods. 95 JASA 90 316-320

Royce, Daniel; Schreiner, Irwin: Improving coverage in a national survey of teachers. 94 ASASrvyRsr 827-832

Royen, Thomas: On some central and non-central multivariate chi-square distributions. 95 StSinica 5 373-397

Royer, G.: Jacquot, S.; — —: Ergodicity of stochastic plates. 95 ProbTher 102 19-44

Royle, Gwen: CWDE Software (Auth); — — (Rev): Review of "World Development Database". 95 TeachgSt 17/2 76-76

Roynette, B.: An instability phenomenon of the nonlinear stochastic differential equation (French). 95 Stochatx 53 175-194

Royston, Patrick: Comment on "Algorithm AS 181: The W-test for normality" (V31 p176-180). 95 ApplStat 44 547-551

— —: Calculation of unconditional and conditional reference intervals for foetal size and growth from longitudinal measurements. 95 StatMed 14 1417-1436

— —; Thompson, Simon G.: Comparing non-nested regression models. 95 Biomtrcs 51 114-127

Cleveland, William S. (Auth); — — (Rev): Review of "Visualizing data". 95 StatMed 14 1395-1396

Różański, R.: On a strongly consistent estimator of the squared L_2-norm of a function. 95 ApplMath 23 279-284

Rozovskii, B. L.: Huebner, M.; — —: On asymptotic properties of maximum likelihood estimators for parabolic stochastic PDE's. 95 ProbTher 103 143-163

Rub, Leonardo: Avi-Itzhak, Hadar I.; Van Mieghem, Jan A.; — —: Multiple subclass pattern recognition: A maximin correlation approach. 95 IEEEPAMI 17 418-431

Rubenstein, Linda M.: Lemke, Jon H.; LaBrecque, John F.; — —: Analysis of the efficacy of rhIL-1ra for the treatment of sepsis. 94 ASAProBiop 340-345

Ruberg, Stephen J.: Dose response studies. I. Some design considerations. 95 JBiophSt 5 1-14

— —: Dose response studies. II. Analysis and interpretation. 95 JBiophSt 5 15-42

Rubin, D.: Gelman, A.; Carlin, J.; Stern, H.; — —: Bayesian data analysis. 95 ChapmnHall:Lnd 552p

Rubin, D. B.: Gelman, A.; Carlin, J. B.; Stern, H. S.; — — (Auth); Brown, P. J. (Rev): Review of "Bayesian data analysis". 96 ShrtBkRv 16/1 5-5

Rubin, Donald: Horvitz, Dan; Koshland, Daniel; — —; Gollin, Albert; Sawyer, Tom; Tanur, Judith M.: Pseudo-opinion polls: SLOP or useful data?. 95 Chance 8/2 16-25

Rubin, Donald B.; Stern, Hal S.; Vehovar, Vasja: Handling "Don't know" survey responses: The case of the Slovenian plebiscite. 95 JASA 90 822-828

Belin, Thomas R.; — —: A method for calibrating false match rates in record linkage. 95 JASA 90 694-707

Belin, Thomas R.; — —: The analysis of repeated-measures data on schizophrenic reaction times using mixture models. 95 StatMed 14 747-768

Liu, Chuanhai; — —: ML estimation of the t distribution using EM and its extensions, ECM and ECME. 95 StSinica 5 19-39

van Dyk, David A.; Meng, Xiao-Li; — —: Maximum likelihood estimation via the ECM algorithm: Computing the asymptotic variance. 95 StSinica 5 55-75

Rubin, Herman; Song, Kai-Sheng: Exact computation of the asymptotic efficiency of maximum likelihood estimators of a discontinuous signal in a Gaussian white noise. 95 AnlsStat 23 732-739

Rubinstein, Pablo: Kiuchi, Amy S.; Hartigan, J. A.; Holford, Theodore R.; — —; Stevens, Cladd E.: Change points in the series of T4 counts prior to AIDS. 95 Biomtrcs 51 236-248

Rubinstein, Reuven Y.; Shapiro, Alexander (Auth); Goodfriend, Jason H. (Rev): Review of "Discrete event systems: Sensitivity analysis and stochastic optimization by the score function method". 95 Technmcs 37 124-125

Rubio, R.: Iznaga, N.; Núñez, G.; Solozabal, J.; Morales, A.; Artaza, E.; — —; Cardenas, E.: A personal computer-based system for parallel line analysis of bioassay data. 95 CmpBiomd 47 167-175

Rubshtein, Ben-Zion: Convolutions of random measures on compact groups. 95 JTherPrb 8 523-538

Rüdiger, B.: Fritz, J.; — —: Time dependent critical fluctuations of a one dimensional local mean field model. 95 ProbTher 103 381-407

Rudiuk, Edmund M.: Provost, Serge B.; — —: The sampling distribution of the serial correlation coefficient. 95 AJMMgSci 15 57-81

Rudnicki, Ryszard: Łoskot, Krzysztof; — —: Limit theorems for stochastically perturbed dynamical systems. 95 JAppProb 32 459-469

Rudolfer, Stephan M.; Watson, Peter C.; Lesaffre, Emmanuel: Are ordinal models useful for classification? A revised analysis. 95 JStCmpSm 52 105-132

Rudolph, D.: Assani, I.; Lesigne, E.; — —: Wiener-Wintner return-times ergodic theorem. 95 IsrelJMa 92 375-395

Rudolph, P. E.; Herrendörfer, G.: Optimal experimental design and accuracy of parameter estimation for nonlinear regression models used in long-term selection. 95 BiomtrcJ 37 183-190

Federer, W. T. (Auth); — — (Rev): Review of "Statistical design and analysis for intercropping experiments. Volume I: Two crops". 95 BiomtrcJ 37 204-204

Good, Ph. (Auth); — — (Rev): Review of "Permutation tests. A practical guide to resampling methods for testing hypotheses". 95 BiomtrcJ 37 150-150

Rudy, Richard J.: Lu, Ming-Wei; — —: A practical method to prioritize problem investigations. 95 ASQCTekCon 57-65

Rue, Havard: New loss functions in Bayesian imaging. 95 JASA 90 900-908

Rueda, Cristina; Salvador, Bonifacio: Reduction of risk using restricted estimators. 95 CommStA 24 1011-1022

Ruetsch, Charles: Labouvie, Erich; — —: Testing for equivalence of measurement scales: Simple structure and metric invariance reconsidered. 95 MultBehR 30 63-76 (Disc: p77-123)

Labouvie, Erich; — —: Reply to comments on "Testing for equivalence of measurement scales: Simple structure and metric invariance reconsidered". 95 MultBehR 30 107-111

Labouvie, Erich; — —: Reply to comments on "Testing for equivalence of measurement scales: Simple structure and metric invariance reconsidered". 95 MultBehR 30 121-123

Ruggeri, Fabrizio; Wasserman, Larry: Density based classes of priors: Infinitesimal properties and approximations. 95 JStPlInf 46 311-324

Insua, David Ríos; — —; Vidakovic, Brani: Some results on posterior regret Γ-minimax estimation. 95 StatDecn 13 315-331

Ruiz, Alberto: A nonparametric bound for the Bayes error. 95 PattrnRc 28 921-930

Ruiz, F. Torres: Gutiérrez Jáimez, R.; Román, P. Román; — —: A note on the Volterra integral equation for the first-passage-time probability density. 95 JAppProb 32 635-648

Ruiz, J. M.; Navarro, J.: Characterization of discrete distributions using expected values. 95 StHefte 36 237-252

Franco, M.; — —: On characterization of continuous distributions with adjacent order statistics. 95 MaOpfStS 26 375-385

Rukhin, A. L.; Shi, J.: Recursive procedures for multiple decisions: Finite time memory and stepwise maximum likelihood procedure. 95 StHefte 36 155-162

Rukhin, Andrew L.: Admissibility: Survey of a concept in progress. 95 IntStRvw 63 95-115

— —; Vajda, Igor: Adaptive decision making for stochastic processes. 95 JStPlInf 45 313-329

Hu, Inchi; — —: A lower bound for error probability in change-point estimation. 95 StSinica 5 319-331

Runger, G. C.: Montgomery, D. C.; — — (Auth); Snell, Joyce (Rev): Review of "Applied statistics and probability for engineers". 95 JRSS-A 158 355-355

Runger, George: Montgomery, Douglas C.; — —; Alexander, Melvin T. (Rev): Review of "Applied statistics and probability for engineers". 95 Technmcs 37 455-457

Runger, George C.; Willemain, Thomas R.: Model-based and model-free control of autocorrelated processes. 95 JQualTek 27 283-292

— —; Willemain, Thomas R.; Prabhu, Sharad: Average run lengths for CUSUM control charts applied to residuals. 95 CommStA 24 273-282

Keats, J. Bert; Miskulin, John D.; — —: Statistical process control scheme design. 95 JQualTek 27 214-225

Keller, Paul A.; — —: Achieving measurable improvement in service management. 95 ASQCTekCon 330-337

Mortell, Robert R.; — —: Statistical process control of multiple stream processes. 95 JQualTek 27 1-12

Prabhu, Sharad S.; Montgomery, Douglas C.; — —: A design tool to evaluate average time to signal properties of adaptive $\bar{X}$ charts. 95 JQualTek 27 74-83

Runggaldier, W. J.: Di Masi, G. B.; Kabanov, Yu. M.; — —: Mean-variance hedging of options on stocks with Markov volatilities. 94 ThProbAp 39 172-182

Runkle, David E.: Keane, Michael P.; — —: Reply to comment on "Testing the rationality of price forecasts: New evidence from panel data". 95 AmEcRv 85 290-290

Ruppert, D.; Sheather, S. J.; Wand, M. P.: An effective bandwidth selector for local least squares regression. 95 JASA 90 1257-1270

Ruppert, David: Carroll, Raymond J.; — —; Stefanski, Leonard A.: Measurement error in nonlinear models. 95 ChapmnHall:Lnd 336p

Hössjer, Ola; — —: Asymptotics for the transformation kernel density estimator. 95 AnlsStat 23 1198-1222

Nychka, Douglas; — —: Nonparametric transformations for both sides of a regression model. 95 JRSS-B 57 519-532

Wang, Naisyin; — —: Nonparametric estimation of the transformation in the transform-both-sides regression model. 95 JASA 90 522-534

Carroll, Raymond J.; — —; Stefanski, Leonard A. (Auth); Draper, N. R. (Rev): Review of "Measurement error in nonlinear models". 96 ShrtBkRv 16/1 5-5

Ruschendorf, L.: Rachev, S. T.; — —: Models for option prices. 94 ThProbAp 39 120-152

Rüschendorf, L.: Rachev, S. T.; — —: Probability metrics and recursive algorithms. 95 AdvAppPr 27 770-799

Ramachandran, D.; — —: A general duality theorem for marginal problems. 95 ProbTher 101 311-319

Rüschendorf, Ludger: Convergence of the iterative proportional fitting procedure. 95 AnlsStat 23 1160-1174

Rushton, Lesley: Steenland, Kyle (Ed); — — (Rev): Review of "Case studies in occupational epidemiology". 95 StatMed 14 1148-1148

Russell, Edmund L., III: Figure skating: What is a 5.8 really?. 94 ASAProSpts 25-30

Russell, Gerald S.; Levitin, Daniel J.: An expanded table of probability values for Rao's spacing test. 95 CommStB 24 879-888

Russell, Susan Jo: Mokros, Jan; — —: Children's concepts of average and representativeness. 95 JRsrMaEd 26 20-39

Russo, Francesco; Vallois, Pierre: The generalized covariation process and Ito formula. 95 StocProc 59 81-104

Rust, John; Rothwell, Geoffrey: Optimal response to a shift in regulatory regime: The case of the US nuclear power industry. 95 JApEcmtx 10 75-118S

Kapteyn, Arie; Kiefer, Nicholas; — —: Introduction: The microeconometrics of dynamic decision making. 95 JApEcmtx 10 1-7S

Rust, Keith: Hedayat, A. S.; Sinha, B. K. (Auth); — — (Rev): Review of "Design and inference in finite population sampling". 94 JOffStat 10 226-228

Rust, Keith F.: Continuous measurement alternatives to census data collection. 94 ASASrvyRsr 80-89 (Disc: p90-95)

Rust, Roland T.; Donthu, Naveen: Capturing geographically localized misspecification error in retail store choice models. 95 JMarRsr 32 103-110

Cooil, Bruce; — —: General estimators for the reliability of qualitative data. 95 Psymtrka 60 199-220

Rustagi, Jagdish S.: — — (Auth); Fotopoulos, Stergios B. (Rev): Review of "Optimization techniques in statistics". 95 Technmcs 37 466-467

Rutchik, Robert H.: Energy information administration guidelines for statistical graphs. 94 ASAProStGr 48-53

Rutkowski, Marek: Left and right linear innovations for a multivariate S α S random variable. 95 StPrLet 22 175-184

Ouknine, Youssef; — —: Local times of functions of continuous semimartingales. 95 StocAnAp 13 211-231

Rutledge, J. J.: Wang, C. S.; Gianola, D.; Sorensen, D. A.; Jensen, J.; Christensen, A.; — —: Response to selection for litter size in Danish Landrace pigs: A Bayesian analysis. 94 ThApGene 88 220-230

Ruts, Ida: Hössjer, Ola; Rousseeuw, Peter J.; — —: The repeated median intercept estimator: Influence function and asymptotic normality. 95 JMultiAn 52 45-72

Rutter, Carolyn; Williamson, G. David: Small area analysis for public health data. 94 ASAProEpid 118-122 (Disc: p125-126)

Rutter, M.: Eaves, L. J.; Silberg, J. L.; Hewitt, K.; — —; Meyer, J. M.; Neale, M. C.; Pickles, A.: Analyzing twin resemblance in multisymptom data: Genetic applications of a latent class model for symptoms of conduct disorder in juvenile boys. 93 BehavGen 23 5-20

Pickles, A.; Neale, M.; Simonoff, E.; — —; Hewitt, J.; Meyer, J.; Crouchley, R.; Silberg, J.; Eaves, L.: A simple method for censored age-of-onset data subject to recall bias: Mothers' reports of age of puberty in male twins. 94 BehavGen 24 457-468

Rüttermann, Markus: Horsthemke, Benedikt; — —: On the maximum entropy principle for a class of stochastic processes. 95 StocProc 56 117-132

Ruxton, Graeme D.: Everitt, B. S. (Auth); — — (Rev): Review of "A handbook of statistical analyses using S-Plus". 95 JAppStat 22 545-545

Ruymgaart, Frits H.: Einmahl, John H. J.; — —: Tail processes under heavy random censorship applications. 95 JNnparSt 5 223-235

Ryaboy, Lana: DiGaetano, Ralph; Ward, John; Cadell, Diane; — —: Sample design of the National Pregnancy and Health Survey. 94 ASASrvyRsr 1100-1104

Ryaboy, Svetlana: England, Amy M.; Hubbell, Katie A.; Judkins, David R.; — —: Imputation of medical cost and payment data. 94 ASASrvyRsr 406-411

Ryan, L.: Lange, N.; — —; Billard, L.; Brillinger, D. R.; Conquest, L.; Greenhouse, J. (Ed); Herrendörfer, G. (Rev): Review of "Case studies in biometry". 95 BiomtrcJ 37 640-640

Ryan, Louise: Goetghebeur, Els; — —: Analysis of competing risks survival data when some failure types are missing. 95 Biomtrka 82 821-833

Lange, Nick; — —; Billard, Lynne; Brillinger, David R.; Conquest, Loveday; Greenhouse, Joel (Ed); Hilton, Joan (Rev): Review of "Case studies in biometry". 95 JASA 90 800-800

Ryan, Louise M.: Comment on "Inference". 95 StatSci 10 189-193

Catalano, Paul J.; — —; Senchaudhuri, Pralay: Developmental toxicity modeling for risk assessment. 94 ASAProEnvr 38-45

Legler, Julie M.; Lefkopoulou, Myrto; — —: Efficiency and power of tests for multiple binary outcomes. 95 JASA 90 680-693

Ryan, Thomas P.: Comment on "Procedures for the identification of multiple outliers in linear models"(V88 p1264-72). 95 JASA 90 811-811

Ryazanov, B. V.: Arbekov, I. M.; Vinokurov, V. I.; — —: On the problem of detecting random trajectories. 94 ThProbAp 39 547-557

Rychlik, Igor: Lindgren, Georg; — —: How reliable are contour curves? Confidence sets for level contours. 95 Bernoull 1 301-319

Michna, Zbigniew; — —: The expected number of level crossings for certain symmetric α-stable processes. 95 CommStSM 11 1-19

Rychlik, T.: A class of unbiased kernel estimates of a probability density function. 95 ApplMath 22 485-497

Rychlik, Tomasz: Bounds for order statistics based on dependent variables with given nonidentical distributions. 95 StPrLet 23 351-358

Rycraft, Michael: Electrifying statistics for your students. 95 TeachgSt 17/2 67-69

Rycroft, Robert S.: Student editions of forecasting software: A survey. 95 IntJFor 11 337-351

Rydén, Tobias: Estimating the order of hidden Markov models. 95 MaOpfStS 26 345-354

— —: Consistent and asymptotically normal parameter estimates for hidden Markov models. 94 AnlsStat 22 1884-1895

Rykov, Vladimir V.: Kitaev, Mikhail Yu.; — —: Controlled queueing systems. 95 CRCPress:FL 304p

Ryman, Nils: Jorde, Per Erik; — —: Temporal allele frequency change and estimation of effective size in populations with overlapping generations. 95 Genetcs 139 1077-1090

Ryoo, S. B.; Moon, S. E.: Detection of warming effects due to industrialization: An accumulated intervention model with an application in Pohang, Korea. 95 JApMetrl 34 1466-1471

Ryu, Jea-Bok; Lee, Kay-O; Lee, Gi-Sung: A practical plan of randomized response technique (Korean). 95 KorApS 8/1 9-26

Ryu, Keunkwan: Analysis of a continuous-time proportional hazard model using discrete duration data. 95 EcomtrRv 14 299-313

Ryzhik, I. M.: Gradshteyn, I. S.; — — (Auth); Jeffrey, A. (Ed); Bello, Abdul L. (Rev): Review of "Tables of integrals, series, and products (Fifth edition)". 95 Ststcian 44 287-288

Rzhetsky, Andrey; Nei, Masatoshi: Unbiased estimates of the number of nucleotide substitutions when substitution rate varies among different sites. 94 JMolEvol 38 295-299

— —; Nei, Masatoshi: Statistical properties of the ordinary least-squares, generalized least-squares, and minimum-evolution methods of phylogenetic inference. 92 JMolEvol 35 367-375

Sitnikova, Tatyana; — —; Nei, Masatoshi: Interior-branch and bootstrap tests of phylogenetic trees. 95 MolBioEv 12 319-333

Saab, Elias: Kalton, Nigel; — —; Montgomery-Smith, Stephen (Ed): Interaction between functional analysis, harmonic analysis, and probability. 95 Dekker:-NY 494p

Saada, E.; Landim, C.: Forward to the special issue in honor of Claude Kipnis (French). 95 AIHenrPB 31 3-12

Saada, Ellen: A voter model in random environment (French). 95 AIHenrPB 31 263-271

Saari, Donald G.: A chaotic exploration of aggregation paradoxes. 95 SIAMRvw 37 37-52

Saavedra, Pedro J.; Hassett, Nancy C.; Weir, Paula: Comparison of national and regional gasoline prices from two surveys. 94 ASASrvyRsr 1179-1183

O'Colmain, Benita J.; — —; Weir, Paula: Variance estimation for EIA-877 propane prices. 94 ASASrvyRsr 232-235

Weir, Paula; — —: On-highway Diesel Fuel Price Survey. 94 ASAProGovt 144-148

Saba, Mehrdad: Kasprzyk, Daniel; Gruber, Kerry; Salvucci, Sameena; — —; Zhang, Fan; Fink, Steven: Some data issues in school-based surveys. 94 ASASrvyRsr 815-820

Sabatier, Robert: Kazi-Aoual, Frédérique; Hitier, Simon; — —; Lebreton, Jean-Dominique: Refined approximations to permutation tests for multivariate inference. 95 CmpStDA 20 643-656

Sabbadini, L. L.: Egidi, V.; — —; Zaccarin, S.: Woman roles between family and employment: Empirical evidence from individual data (Italian). 94 Statstca 54 411-433

Sabre, R.: Spectral density estimation for stationary stable random fields. 95 ApplMath 23 107-133

Sachs, Emanuel: Hu, Albert; — —: Run by run control: A Bayesian approach to the control of semiconductor fabrication processes. 94 ASAProPEng 245-254

Sackrowitz, H. B.: Cohen, Arthur; — —: Inadmissibility of some tests for order-restricted alternatives. 95 StPrLet 24 153-156

Cohen, Arthur; — —; Samuel-Cahn, E.: Cone order association. 95 JMultiAn 55 320-330

Cohen, Arthur; Kemperman, J. H. B.; — —: Unbiased testing in exponential family regression. 94 AnlsStat 22 1931-1946

Sackrowitz, Harold B.: Cohen, Arthur; — —: On stochastic ordering of random vectors. 95 JAppProb 32 960-965

Cohen, Arthur; — —; Samuel-Cahn, Ester: Constructing tests for normal order-restricted inference. 95 Bernoull 1 321-333

Sacks, Jerome: Gao, Feng; — —; Welch, William J.: Predicting ozone levels and trends. 94 ASAProEnvr 77-81

Sadegh, P.; Holst, J.; Madsen, H.; Melgaard, H.: Experiment design for grey box identification. 95 IJAdCSgP 9 491-507

Sadler, Brian M.; Giannakis, Georgios B.; Shamsunder, Sanyogita: Noise subspace techniques in non-Gaussian noise using cumulants. 95 IEEEAES 31 1009-1018

Sadosky, Sally A.: Couper, Mick P.; — —; Hansen, Sue Ellen: Measuring interviewer performance using CAPI. 94 ASASrvyRsr 845-850

Sadowsky, John S.: The probability of large queue lengths and waiting times in a heterogeneous multiserver queue II: Positive recurrence and logarithmic limits. 95 AdvAppPr 27 567-583

— —; Szpankowski, Wojciech: The probability of large queue lengths and waiting times in a heterogeneous multiserver queue I: Tight limits. 95 AdvAppPr 27 532-566

Sadrolhefazi, Amir; Fine, Terrence L.: Finite-dimensional distributions and tail behavior in stationary interval-valued probability models. 94 AnlsStat 22 1840-1870

Safar, M.: Vigneau, E.; Devaux, M. F.; — —: Application of Procrustean methods to mid- and near-infrared spectral data. 95 JChemtrc 9 125-135

Saffar, A. El: Kerim, Abdel; — —: RAS and least squares methods versus other mathematical techniques in developing and updating input-output tables (Arabic). 95 EgpStJ 39 120-143

Säfström, Margaretha: Bergman, Lars R.; Kristiansson, Karl-Erik; Olofsson, Anita; — —: Decentralised CATI versus paper and pencil interviewing: Effects on the results in the Swedish labour force surveys. 94 JOffStat 10 181-195

Sagara, Nobuko; Fukushima, Masao: A hybrid method for solving the nonlinear least squares problem with linear inequality constraints. 94 JOpRsJpn 38 55-69

Sagitov, Serik: A key limit theorem for critical branching processes. 95 StocProc 56 87-100

Sahai, Hardeo; Khurshid, Anwer: Statistics in epidemiology. Methods, techniques and applications. 95 CRCPress:FL 352p

— —; Khurshid, Anwer: On analysis of epidemiological data involving a 2×2 contingency table: An overview of Fisher's exact test and Yates' correction for continuity. 95 JBiophSt 5 43-70

— —; Vega, Juan: Training teachers to teach statistics in the secondary schools of Puerto Rico. 94 ASAProStEd 253-258

Sahai, Haredeo: Khurshid, Anwer; — —: A bibliography on telephone survey methodology. 95 JOffStat 11 325-367

Şahinoğlu, Mehmet: Randolph, Paul; — —: A stopping rule for a compound Poisson random variable. 95 AplStMDA 11 135-143

Sahoo, J.; Sahoo, L. N.: A note on the efficiencies of two almost unbiased predictive ratio estimators. 95 StHefte 36 273-278

— —; Sahoo, L. N.; Mohanty, S.: Empirical comparison of some regression and regression-type strategies. 95 StHefte 36 337-347

Mohanty, S.; — —: A note on improving the ratio method of estimation through linear transformation using certain known population parameters. 95 SankhyaB 57 93-102

Sahoo, Jagatananda; Sahoo, Lokanath; Mohanty, Suprasana: Unequal probability sampling using a transformed auxiliary variable. 94 Metron 52/1 71-83

Sahoo, L. N.: Dalabehera, M.; — —: Efficiencies of six almost unbiased ratio estimators under a particular model. 95 StHefte 36 61-67

Sahoo, J.; — —: A note on the efficiencies of two almost unbiased predictive ratio estimators. 95 StHefte 36 273-278

Sahoo, J.; — —; Mohanty, S.: Empirical comparison of some regression and regression-type strategies. 95 StHefte 36 337-347

Sahoo, Lokanath: Sahoo, Jagatananda; — —; Mohanty, Suprasana: Unequal probability sampling using a transformed auxiliary variable. 94 Metron 52/1 71-83

Sahu, S. K.: Roberts, G. O.; — —; Gilks, W. R.: Comment on "Bayesian computation and stochastic systems". 95 StatSci 10 49-51

Sahu, Sujit K.: Dey, Dipak K.; Kuo, Lynn; — —: A Bayesian predictive approach to determining the number of components in a mixture distribution. 95 StatComp 5 297-305

Gelfand, Alan E.; — —; Carlin, Bradley P.: Efficient parametrisations for normal linear mixed models. 95 Biomtrka 82 479-488

Saijets, M.: The unified file system and its metadata part. 93 StatJUN 10 201-208

Saikkonen, Pentti: Problems with the asymptotic theory of maximum likelihood estimation in integrated and cointegrated systems. 95 EconTher 11 888-911

— —: Dependent versions of a central limit theorem for the squared length of a sample mean. 95 StPrLet 22 185-194

Sailer, Peter: Longton, Barbara; — —; Wilson, Janette: How sales of capital assets were affected by the Tax Reform Act of 1986. 94 ASAProGovt 201-205 (Disc: p212-214)

Saito, Takayuki: Relationships between redundancy analysis and canonical regression analysis. 95 JJapnStS 25 111-119

Saitoh, Toshifumi: Osada, Hirofumi; — —: An invariance principle for non-symmetric Markov processes and reflecting diffusions in random domains. 95 ProbTher 101 45-63

Sakamoto, Yosiyuki: A statistical research of the Japanese national character: General trend of opinion over past forty years (Japanese). 95 PrInStMa 43 5-26

Sakata, Shinichi; White, Halbert: An alternative definition of finite-sample breakdown point with applications to regression model estimators. 95 JASA 90 1099-1106

Sakawa, Masatoshi: Inuiguchi, Masahiro; — —: A possibilistic linear program is equivalent to a stochastic linear program in a special case. 95 FuzzySet 76 309-317

Sakkinen, Pamela A.; Severson, Richard K.; Ross, Julie A.; Robison, Leslie L.: Random-digit dialing for control selection in childhood cancer studies: The geographic proximity and demographics within matched sets. 95 AJPubHlh 85 555-557

Salas González, Raúl: Pierrat, Jean Claude; Houllier, François; Hervé, Jean Christophe; — —: Estimating the mean of the r largest values in a finite population: Application to forest inventory (French). 95 Biomtrcs 51 679-686

Salas, Jose D.: Chebaane, Mohamed; — —; Boes, Duane C.: Product periodic autoregressive processes for modeling intermittent monthly streamflows. 95 WaterReR 31 1513-1518

Salau, M. O.: On the numerical implementation of the generalized least squares procedure for ARMA estimation. 95 CommStB 24 111-130

Saldanha, Richard A.: Hjorth, J. S. U. (Auth); — — (Rev): Review of "Computer intensive statistical methods: Validation, model selection and bootstrap". 95 Ststcian 44 288-288

Saleh, A. K. Md. E.: Obituary: Qazi Motahar Husain, 1897-1981. 94 JStatRsr 28 233-236

— —; Hassanein, Khatab M.; Brown, Edward F.: Optimum spacings, ABLUE and tests of hypothesis in simple linear model with extreme-value errors. 94 JStatRsr 28 53-63

Koul, Hira L.; — —: Autoregression quantiles and related rank-scores processes. 95 AnlsStat 23 670-689

Kubokawa, Tatsuya; — —: Estimation of location and scale parameters under order restrictions. 94 JStatRsr 28 41-51

Saleh, Amina A.: Approximating the characteristic function of the Student's t distribution. 94 EgpStJ 38 177-195

Salehi, H.: Hoppensteadt, F.; — —; Skorokhod, A. V.: Randomly perturbed Volterra integral equations and some applications. 95 Stochatx 54 89-125

Salemi, Michael K.: Revealed preference of the Federal Reserve: Using inverse-control theory to interpret the policy equation of a vector autoregression. 95 JBES 13 419-433

Salicrú, M.: Menéndez, M. L.; Morales, D.; Pardo, L.; — —: Asymptotic behaviour and statistical applications of divergence measures in multinomial populations: A unified study. 95 StHefte 36 1-29

Menéndez, M. L.; Pardo, L.; Morales, D.; — —: Some statistical applications of (r,s)-directed divergences. 92 UtilMath 42 115-127

Salicrú, Miguel: Pardo, Leandro; — —; Menéndez, María Luisa; Morales, Domingo: A test for homogeneity of variances based on Shannon's entropy. 95 Metron 53 135-146

Salinetti, G.: Conigliani, C.; Dall'aglio, M.; Pacifico, M. Perone; — —: Robust statistics: From classical to Bayesian analysis. 94 Metron 52/3 89-109

Säll, Torbjörn; Nilsson, Nils-Otto: The robustness of recombination frequency estimates in intercrosses with dominant markers. 94 Genetcs 137 589-596

Salles, Maria Creusa B.: Magalhães, Marcos N.; McNickle, Donald C.; — —: Departure processes from a loss system with two stations and one server. 94 RBrzlPrS 8 135-146

Salminen, P.: Kaj, I.; — —: On the ultimate value of local time of one-dimensional super-Brownian motion. 95 StocProc 59 21-42

Saloff-Coste, L.: Precise estimates on the rate at which certain diffusions tend to equilibrium. 94 MaZeit 217 641-677

Salt, A. T.: Wade, A. M.; Ades, A. E.; — —; Jayatunga, R.; Sonksen, P. M.: Age-related standards for ordinal data: Modelling the changes in visual acuity from 2 to 9 years of age. 95 StatMed 14 257-266

Saltelli, A.; Andres, T. H.; Homma, T.: Sensitivity analysis of model output. Performance of the iterated fractional factorial design method. 95 CmpStDA 20 387-407

Saltelli, Andrea; Sobol, Ilya M.: About the use of rank transformation in sensitivity analysis of model output. 95 RelEngSS 50 225-239

Salvador, Bonifacio: Rueda, Cristina; — —: Reduction of risk using restricted estimators. 95 CommStA 24 1011-1022

Salvaggio, Antonino: Non-AIDS related mortality in the human immunodeficiency virus infection. 95 ListyBio 32 23-33

Salvatierra, Stella M.: Viollaz, Aido J.; Sfer, Ana M.; — —: An approximation of the unconditional error rates of the sample linear discriminant function. 95 CommStA 24 1941-1969

Salvucci, Sameena: Kasprzyk, Daniel; Gruber, Kerry; — —; Saba, Mehrdad; Zhang, Fan; Fink, Steven: Some data issues in school-based surveys. 94 ASASrvyRsr 815-820

Salyers, Eddie: Oh, Sungsoo; Paletz, Dave; Kim, Jay; — —: Development of edit parameters for 1992 economic census enterprise reports. 94 ASASrvyRsr 1144-1149

Samanta, P. K.: Vesely, W. E.; Uryasev, S. P.; — —: Failure of emergency diesel generators: A population analysis using empirical Bayes methods. 94 RelEngSS 46 221-229

Samanta, Tapas: Ghosal, Subhashis; — —: Asymptotic behaviour of Bayes estimates and posterior distributions in multiparameter nonregular cases. 95 MaMethSt 4 361-388

Samarov, Alexander: Doksum, Kjell; — —: Nonparametric estimation of global functionals and a measure of the explanatory power of covariates in regression. 95 AnlsStat 23 1443-1473

Sambol, Nancy C.: Liu, Chui Yu; — —: Pharmacodynamic analysis of analgesic clinical trials using empirical methods. 94 ASAProBiop 171-176

Samejima, Fumiko: Acceleration model in the heterogeneous case of the general graded response model. 95 Psymtrka 60 549-572

Samiei, Hossein: Pesaran, M. Hashem; — —: Forecasting ultimate resource recovery. 95 IntJFor 11 543-555

Sammut, C.: Mulholland, M.; Hibbert, D. B.; Haddad, P. R.; — —: Application of the C4.5 classifier to building an expert system for ion chromatography. 95 Chemolab 27 95-104

Samoggia, A.: Angeli, A.; Cocchi, D.; Pasquini, L.; — —: Family structure and income: A first analysis on cross-section data (Italian). 94 Statstca 54 435-452

Samorodnitsky, Gennady: Association of infinitely divisible random vectors. 95 StocProc 55 45-55

— —; Taqqu, Murad S.: Lévy measures of infinitely divisible random vectors and Slepian inequalities. 94 AnlsProb 22 1930-1956

Adler, Robert J.; — —: Super fractional Brownian motion, fractional super Brownian motion and related self-similar (super) processes. 95 AnlsProb 23 743-766

Braverman, Michael; — —: Functionals of infinitely divisible stochastic processes with exponential tails. 95 StocProc 56 207-231

Cioczek-Georges, Renata; Mandelbrot, Benoit B.; — —; Taqqu, Murad S.: Stable fractal sums of pulses: The cylindrical case. 95 Bernoull 1 201-216

Embrechts, Paul; — —: Sample quantiles of heavy tailed stochastic processes. 95 StocProc 59 217-233

Norvaiša, Rimas; — —: Stable processes with sample paths in Orlicz spaces. 94 AnlsProb 22 1904-1929

Rachev, Svetlozar T.; — —: Limit laws for a stochastic process and random recursion arising in probabilistic modelling. 95 AdvAppPr 27 185-202

— —; Taqqu, Murad S. (Auth); Mikosch, Thomas (Rev): Review of "Stable non-Gaussian random processes: Stochastic models with infinite variance". 95 JASA 90 805-806

Sampson, Allan R.: Barnhart, Huiman X.; — —: Multiple population models for multivariate random length data - With applications in clinical trials. 95 Biomtrcs 51 195-204

Larholt, Kay M.; — —: Effects of heteroscedasticity upon certain analyses when regression lines are not parallel. 95 Biomtrcs 51 731-737

Sampson, P. D.; Guttorp, P.; Meiring, W.: Spatio-temporal analysis of regional ozone data for operational evaluation of an air quality model. 94 ASAProEnvr 46-55

Samuel-Cahn, E.: Cohen, Arthur; Sackrowitz, H. B.; — —: Cone order association. 95 JMultiAn 55 320-330

Samuel-Cahn, Ester: The best-choice secretary problem with random freeze on jobs. 95 StocProc 55 315-327

Cohen, Arthur; Sackrowitz, Harold B.; — —: Constructing tests for normal order-restricted inference. 95 Bernoull 1 321-333

Samuelson, Jack M.: Kraft, Kathy M.; Johnson, Douglas H.; — —; Allen, Stephen H.: Using known populations of pronghorn to evaluate sampling plans and estimators. 95 JWildlMg 59 129-137

San Martín, Jaime: Collet, Pierre; Martínez, Servet; — —: Asymptotic laws for one-dimensional diffusions conditioned to nonabsorption. 95 AnlsProb 23 1300-1314

Sánchez de Naranjo, M. V.: A central limit theorem for non-linear functionals of stationary Gaussian vector processes. 95 StPrLet 22 223-230

Sanchez, Alex; Ocaña, Jordi; Utzet, Frederic: Sampling theory, estimation, and significance testing for Prevosti's estimate of genetic distance. 95 Biomtrcs 51 1216-1235

Sánchez, M. S.; Sarabia, L. A.: Efficiency of multi-layered feed-forward neural networks on classification in relation to linear discriminant analysis, quadratic discriminant analysis and regularized discriminant analysis. 95 Chemolab 28 287-303

Sanchez, Marcus John: Advantages of developing, testing and accessing complex statistical tools in an ASSIST environment. 94 ASAProStCp 141-146

Sandhu, R. A.: Balakrishnan, N.; — —: A simple simulational algorithm for generating progressive Type-II censored samples. 95 AmerStat 49 229-230

Balakrishnan, N.; — —: Recurrence relations for single and product moments of order statistics from a generalized half logistic distribution with applications to inference. 95 JStCmpSm 52 385-398

Sandman, Peter M.: Weinstein, Neil D.; — —; Hallman, William K.: Testing a visual display to explain small probabilities. 94 RiskAnal 14 895-896

Sandmann, Klaus: Nielsen, J. Aase; — —: Equity-linked life insurance: A model with stochastic interest rates. 95 InsrneME 16 225-253

Sands, Barbara N.: Trosset, Michael W.; — —: On the choice of a discrepancy functional for model selection. 95 CommStA 24 2841-2863

Saner, Hilary: Rogosa, David; — —: Longitudinal data analysis examples with random coefficient models. 95 JEdStat 20 149-170 (Disc: p201-227, 234-238)

Rogosa, David; — —: Reply to comments on "Longitudinal data analysis examples with random coefficient models". 95 JEdStat 20 234-238

Sanford, Anthony J.: Moxey, Linda M.; — — (Auth); Budescu, David V.; Wallsten, Thomas S. (Rev): Review of "Communicating quantities: A psychological perspective". 95 Chance 8/2 38-40

Sangster, Roberta L.; Rockwood, Todd H.; Dillman, Don A.: The influence of administration mode on responses to numeric rating scales. 94 ASASrvyRsr 708-712

Saniga, Erwin M.; Davis, Darwin J.; McWilliams, Thomas P.: Economic, statistical, and economic-statistical design of attribute charts. 95 JQualTek 27 56-73

Saniie, Jafar: Nagle, Daniel T.; — —: Performance analysis of linearly combined order statistic CFAR detectors. 95 IEEEAES 31 522-533

Sankoff, David: Constantinescu, Mariana; — —: An efficient algorithm for supertrees. 95 JClass 12 101-112

Sankoh, A. J.; Huque, M. F.: Comment on "Procedure for comparing samples with multiple endpoints" (V40 p1079-1087). 95 Biomtrcs 51 1580-1581

Huque, M. F.; Harrison, F. D.; — —; Dubey, S. D.: Longitudinal data analysis for heartburn clinical trials. 94 ASAProBiop 182-187

Sankoh, Abdul J.: Small-sample properties of finite population ratios. 95 CommStB 24 897-909

Sansó, B.; Pericchi, L. R.: On near ignorance classes. 94 RBrzlPrS 8 119-126

Sansó, Bruno: Pericchi, Luis R.; — —: A note on bounded influence in Bayesian analysis. 95 Biomtrka 82 223-225

Sansom, John: Rainfall discrimination and spatial variation using breakpoint data. 95 JClimate 8 624-636

Santner, Thomas J.: Bechhofer, Robert E.; — —; Goldsman, David M.: Design and analysis of experiments for statistical selection, screening, and multiple comparisons. 95 Wiley:NY:UK 325p

Forbes, Andrew B.; — —: Estimators of odds ratio regression parameters in matched case-control studies with covariate measurement error. 95 JASA 90 1075-1084

Sanz-Sole, Marta: Gyöngy, István; Nualart, David; — —: Approximation and support theorems in modulus spaces. 95 ProbTher 101 495-509

Sanz-Solé, Marta: Bally, Vlad; Millet, Annie; — —: Approximation and support theorem in Hölder norm for parabolic stochastic partial differential equations. 95 AnlsProb 23 178-222

Delgado, Rosario; — —: Green formulas in anticipating stochastic calculus. 95 StocProc 57 113-148

Sapatinas, T.: Fosam, E. B.; — —: Characterisations of some income distributions based on multiplicative damage models. 95 AstrlJSt 37 89-93

Powell, K. J.; — —; Bailey, T. C.; Krzanowski, W. J.: Application of wavelets to the pre-processing of underwater sounds. 95 StatComp 5 265-273

Sapatinas, Theofanis: Identifiability of mixtures of power-series distributions and related characterizations. 95 AnInStMa 47 447-459

— —: Characterizations of probability distributions based on discrete p-monotonicity. 95 StPrLet 24 339-344

Saphire, Diane G.: Spirer, Herbert; Spirer, Louise (Auth); — — (Rev): Review of "Data analysis for monitoring human rights". 95 AmerStat 49 322-322

Sapio, Bartolomeo: SEARCH (Scenario evaluation and analysis through repeated cross impact handling): A new method for scenario analysis with an application to the Videotel service in Italy. 95 IntJFor 11 113-131

Saporta, G.: Niang, N.; — —: Average run length of a moving average control chart to detect shifts in the mean of a process (French). 95 RStApp 43/3 5-20

Sapra, Sunil K.: Comment on "The moment-generating function of the minimum of bivariate normal random variables" (V48 p124-125). 95 AmerStat 49 240-240

Sarabia, José María: The centered normal conditionals distribution. 95 CommStA 24 2889-2900

Arnold, Barry C.; Castillo, Enrique; — —: General conditional specification models. 95 CommStA 24 1-11

Sarabia, L.: Forina, M.; Lanteri, S.; — —: Distance and class space in the UNEQ class-modeling technique. 95 JChemtrc 9 69-89

Sarabia, L. A.: Fernández, M. C. Ortiz; Herrero Gutiérrez, A.; Pastor, M. S. Sánchez; — —; Íñiguez Crespo, M.: The UNEQ, PLS and MLF neural network methods in the modelling and prediction of the colour of young red wines from the Denomination of Origin 'Rioja'. 95 Chemolab 28 273-285

Sánchez, M. S.; — —: Efficiency of multi-layered feed-forward neural networks on classification in relation to linear discriminant analysis, quadratic discriminant analysis and regularized discriminant analysis. 95 Chemolab 28 287-303

Saracco, James F.: Desante, David F.; Burton, Kenneth M.; — —; Walker, Brett L.: Productivity indices and survival rate estimates from MAPS, a continent-wide programme of constant-effort mist-netting in North America. 95 JAppStat 22 935-947

Saran, Jagdish: Joint distributions based on the number of runs and some other two-sample rank order statistics for arbitrary sample sizes. 95 JStPlInf 47 257-275

Sarantis, Nicholas; Stewart, Chris: Structural, VAR and BVAR models of exchange rate determination: A comparison of their forecasting performance. 95 JForcast 14 201-215

Sarda, P.: Moncaup, S.; — —; Vieu, P.: Nonparametric estimators of an intensity function applied to meteorological data (French). 95 RStApp 43/4 77-88

Sardy, Hyman: Grosof, Miriam Schapiro; — —: Are the problems the problem in introductory statistics?. 94 ASAProStEd 192-196

Sarin, Rakesh K.: Keller, L. Robin; — —: Fair processes for societal decisions involving distributional inequalities. 95 RiskAnal 15 49-59

Sarkar, A.; Kanjilal, P. P.: On a method of identification of best subset model from full AR-model. 95 CommStA 24 1551-1567

Sarkar, Ila C.; Mudholkar, Govind S.; Raubertas, Richard F.: An approximation to the distribution of the resultant from a Fisher distribution. 95 CommStB 24 227-241

Sarkar, Sahadeb: Shin, Dong Wan; — —: Estimation of the multivariate autoregressive moving average having parameter restrictions and an application to rotational sampling. 95 JTimSrAn 16 431-444

Shin, Dong Wan; — —: Estimation for stationary AR(1) models with nonconsecutively observed samples. 95 SankhyaA 57 287-298

Sarkar, Sanat K.; Snapinn, Steve; Wang, Wenjin: On improving the min test for the analysis of combination drug trials. 95 JStCmpSm 51 197-213

Särkkä, Aila: Pseudo-likelihood approach for Gibbs point processes in connection with field observations. 95 MaOpfStS 26 89-97

Sarle, W.: Janssen, P.; Marron, J. S.; Veraverbeke, N.; — —: Scale measures for bandwidth selection. 95 JNnparSt 5 359-380

Särndal, C.: Estevao, V.; Patak, Z.; — —: Maximizing the use of auxiliary information for calibration and regression estimation. 94 ASASrvyRsr 42-48

Särndal, C. E.: Estevao, V.; Hidiroglou, M. A.; — —: Methodological principles for a generalized estimation system at Statistics Canada. 95 JOffStat 11 181-204

Särndal, C.-E.: Obituary: Stanley L. Warner, 1928-1992. 95 SurvyM 21 3-6

Särndal, Carl: Rancourt, Eric; — —; Lee, Hyunshik: Estimation of the variance in the presence of nearest neighbour imputation. 94 ASASrvyRsr 888-893 (Disc: p906-907)

Särndal, Carl E.: Lee, Hyunshik; Rancourt, Eric; — —: Experiments with variance estimation from survey data with imputed values. 94 JOffStat 10 231-243

Särndal, Carl-Erik: Deville, Jean-Claude; — —: Variance estimation for the regression imputed Horvitz-Thompson estimator. 94 JOffStat 10 381-394

Särndel, C.-E.; Swensson, B.; Wretman, J. (Auth); Penny, Richard N. (Rev): Review of "Model assisted survey sampling". 95 Ststcian 44 140-142

Sartor, Francis; Walckiers, Denise: Estimate of disease prevalence using drug consumption data. 95 AJEpidm 141 782-787

Sasieni, Peter D.: Efficiently weighted estimating equations with application to proportional excess hazards. 95 LiDaAnal 1 49-57

Sataev, E. A.: Asymptotically Markov stochastic 1. 94 ThProbAp 39 448-464

Satchell, Stephen E.: Booth, Alison L.; — —: The hazards of doing a PhD: An analysis of completion and withdrawal rates of British PhD students in the 1980s. 95 JRSS-A 158 297-318

Satchell, Steve; Timmermann, Allan: On the optimality of adaptive expectations: Muth revisited. 95 IntJFor 11 407-416

— —; Timmermann, Allan: An assessment of the economic value of non-linear foreign exchange rate forecasts. 95 JForcast 14 477-497

Sather, Harland N.: Lee, Jae Won; — —: Group sequential methods for comparison of cure rates in clinical trials. 95 Biomtrcs 51 756-763

Sato, Hiroshi; Tamashiro, Masakazu: Absolute continuity of one-sided random translations. 95 StocProc 58 187-204

Sato, K.: Multimodal convolutions of unimodal infinitely divisible distributions. 94 ThProbAp 39 336-347

Sato, Michikazu: Comment on "Unbiased estimation for sequential multinomial sampling plans" (V12 p253-259). 95 SqtlAnly 14 335-340

— —; Akahira, Masafumi: Information inequalities for the minimax risk. 95 JJapnStS 25 151-158

Akahira, Masafumi; — —; Torigoe, Norio: On the new approximation to non-central t-distributions. 95 JJapnStS 25 1-18

Sato, Mika; Sato, Yoshiharu: A general fuzzy clustering model based on aggregation operators. 95 Behvmtrk 22 115-128

Sato, Tosiya: A further look at the Cochran-Mantel-Haenszel risk difference. 95 ConClnTr 16 359-361

Sato, Yoshiharu: Sato, Mika; — —: A general fuzzy clustering model based on aggregation operators. 95 Behvmtrk 22 115-128

Suzukawa, Akio; — —: A generalization of testing independence of sets of variates. 95 JJapnStS 25 71-80

Satorra, Albert: Muthén, Bengt O.; — —: Technical aspects of Muthén's LISCOMP approach to estimation of latent variable relations with a comprehensive measurement model. 95 Psymtrka 60 489-503

Satten, Glen A.: Upper and lower bound distributions that give simultaneous confidence intervals for quantiles. 95 JASA 90 747-752

Sauer, John R.: Dawson, Deanna K.; — —; Wood, Paul A.; Berlanga, Mauro; Wilson, Marcia H.; Robbins, Chandler S.: Estimating bird species richness from capture and count data. 95 JAppStat 22 1063-1068

Link, William A.; — —: Estimation and confidence intervals for empirical mixing distributions. 95 Biomtrcs 51 810-821

Pendleton, Grey W.; — —: Delineating bird populations using ring recoveries. 95 JAppStat 22 1049-1055

Sauer, Paul L.: Bacon, Donald R.; — —; Young, Murray: Composite reliability in structural equations modeling. 95 EdPsMeas 55 394-406

Sauerbrei, W.: Altman, D. G.; Lausen, B.; — —; Schumacher, M.: Dangers of using "optimal" cutpoints in the evaluation of prognostic factors. 94 JNCncrI 86 829-835

Saunders, David M.: Latzko, William J.; — — (Auth); Copeland, Karen A. F. (Rev): Review of "Four days with Dr. Deming". 95 JQualTek 27 270-270

Saunders, I. W.; Eccleston, J. A.; Martin, R. J.: An algorithm for the design of 2^p factorial experiments on continuous processes. 95 AstrlJSt 37 353-365

Sava-Bolesta, Monica: De Ayala, R. J.; Schafer, William D.; — —: An investigation of the standard errors of expected a posteriori ability estimates. 95 BJMaStPs 48 385-405

Savage, Phillip: Anderson, DeAnn; Abetti, Frank; — —: Process improvement utilizing computer simulation: Case study. 95 ASQCTekCon 713-724

Savchuk, V. P.: Estimation of structures reliability for non-precise limit state models and vague data. 95 RelEngSS 47 47-58

Savitz, David A.; Olshan, Andrew F.: Multiple comparisons and related issues in the interpretation of epidemiologic data. 95 AJEpidm 141 904-908

— —; Olshan, Andrew F.: Multiple comparisons and related issues in the interpretation of epidemiologic data. 95 AJEpidm 142 904-908

Poole, Charles; — —: Reply to comment on "Statistical significance testing in the American journal of epidemiology, 1970-1990". 95 AJEpidm 142 102-102

Savoca, Elizabeth: Factors affecting hospital admissions and length of stay: Some preliminary results. 94 ASAProSoSt 222-226

Sawhill, W. W.: Bruno, W. J.; Knill, E.; Balding, D. J.; Bruce, D. C.; Doggett, N. A.; — —; Stallings, R. L.; Whittaker, C. C.; Torney, D. C.: Efficient pooling designs for library screening. 95 Genomics 26 21-30

Sawyer, Tom: Horvitz, Dan; Koshland, Daniel; Rubin, Donald; Gollin, Albert; — —; Tanur, Judith M.: Pseudo-opinion polls: SLOP or useful data?. 95 Chance 8/2 16-25

Saxena, K. M. Lal: Li, Yuanzhang; — —: Optimal robust Bayes estimation. 95 JStPlInf 46 365-380

Saxena, K. P.: Tiwari, Gayatri; — —: On bias and efficiency of a preliminary test estimator of variance in a normal distribution. 95 Metron 53 163-169

Saxena, M. C.: Singh, M.; Saxena, N. P.; — —: Estimation of pinnule area in chickpea leaves: An errors-in-variables approach. 95 JAppStat 22 37-43

Saxena, N. P.: Singh, M.; — —; Saxena, M. C.: Estimation of pinnule area in chickpea leaves: An errors-in-variables approach. 95 JAppStat 22 37-43

Saxena, Sanjay Kumar; Nigam, A. K.; Shukla, N. D.: Variance estimation for combined ratio estimator. 95 SankhyaB 57 85-92

Sayre, Ken: Crossa, José; Cornelius, Paul L.; — —; Ortiz-Monasterio R., Jóse Iván: A shifted multiplicative model fusion method for grouping environments without cultivar rank change. 95 CropSci 35 54-62

Scallan, A. J.: Asymmetric tolerance distributions: An application of two-piece distributions. 95 JAppStat 22 81-85

Scarsini, M.: Mosler, K.; — — (Ed); Heyer, H. (Rev): Review of "Stochastic orders and decision under risk". 95 Metrika 42 139-139

Scarsini, Marco: Bassan, Bruno; — —: Finite exchangeability and linear regression. 95 StPrLet 23 105-110

Scazzero, Joseph A.; Ord, J. K.: Interval estimates for the optimum point of a quadratic logistic curve - A comparison of different estimation methods via simulation. 95 CommStB 24 207-226

Schaafsma, W.: Kroese, A. H.; van der Meulen, E. A.; Poortema, K.; — —: Distributional inference. 95 StNeerla 49 63-82

Schäbe, H.; Tiedge, J.: Kernel estimation for characteristics of pure jump processes. 95 StHefte 36 131-144

Schabenberger, Oliver: Nonlinear mixed effects growth models for repeated measures in ecology. 94 ASAProEnvr 156-161

— —; Gregoire, Timothy G.; Burkhart, Harold E.: Comment on "Multistate models for monitoring individual trees in permanent observation plots". 94 EnvEcoSt 1 192-196

Gregoire, Timothy G.; — —; Barrett, James P.: Linear modelling of irregularly spaced, unbalanced, longitudinal data from permanent-plot measurements. 95 CanJForR 25 137-156

Schach, S.: Kreienbrock, L.; — — (Auth); Enderlein, G. (Rev. German): Review of "Epidemiological methods" (German). 95 BiomtrcJ 37 1016-1016

Schachermayer, W.: Delbaen, F.; — —: Arbitrage possibilities in Bessel processes and their relations to local martingales. 95 ProbTher 102 357-366

Schachermayer, Walter: Delbaen, Freddy; — —: The no-arbitrage property under a change of numéraire. 95 Stochatx 53 213-226

Schachtel, Gabriel A.: Plaetke, Rosemarie; — —: Two statistical tests for meiotic breakpoint analysis. 95 AJHuGene 56 508-518

Schaefer, Martin: Krafft, Olaf; — —: Exact Elfving-minimax designs for quadratic regression. 95 StSinica 5 475-484

Schaeffer, Dennis: Fleming, Philip J.; — —; Simon, Burton: Efficient Monte-Carlo simulation of a product-form model for a cellular system with dynamic resource sharing. 95 ACMMdSim 5 3-21

Schaeffer, Nora Cate: A decade of questions. 95 JOffStat 11 79-92

Lin, I-fen; — —: Using survey participants to estimate the impact of nonparticipation. 95 PubOpinQ 59 236-258

Schafer, Daniel W.: Ramsey, Fred L.; — —: Teaching graduate service courses through case studies. 94 ASAProStEd 27-28

Schafer, Edward T.: Hillis, Margaret A.; Minkus, Gail W.; DeCaluwe, Craig L.; Myers, Michael N.; — —: An improved approach to product quality through testing. 95 ATTTJ 74/3 40-48

Schäfer, H.: Correction to "Efficient confidence bounds for ROC curves" (V13 p1551-1561). 95 StatMed 14 875-875

Schafer, Joseph L.: Analysis of incomplete multivariate data by simulation. 95 ChapmnHall:Lnd 300p

Schafer, William D.: De Ayala, R. J.; — —; Sava-Bolesta, Monica: An investigation of the standard errors of expected a posteriori ability estimates. 95 BJMaStPs 48 385-405

Schaffner, Andrew A.: Geske, Mark X.; Godbole, Anant P.; — —; Skolnick, Allison M.; Wallstrom, Garrick L.: Compound Poisson approximations for word patterns under Markovian hypotheses. 95 JAppProb 32 877-892

Schaible, W. L.: Indirect estimators in U.S. federal programs. 95 Springer:Brln:NY 195p

Schaid, Daniel J.: Genetic relative risk methods using cases and their parents. 94 ASAProEpid 1-7

— —; Sommer, Steve S.: Comparison of statistics for candidate-gene association studies using cases and parents. 94 AJHuGene 55 402-409

Schall, Robert: Assessment of individual and population bioequivalence using the probability that bioavailabilities are similar. 95 Biomtrcs 51 615-626

Schassberger, R.: van den Berg, J. L. (Auth); — — (Rev): Review of "Sojourn times in feedback and processor sharing queues". 95 Metrika 42 146-147

Schassberger, Rolf: Miyazawa, Masakiyo; — —; Schmidt, Volker: On the structure of an insensitive generalized semi-Markov process with reallocation and point-process input. 95 AdvAppPr 27 203-225

Scheaffer, R.; McClave, J. (Auth); Ziegel, Eric R. (Rev): Review of "Probability and statistics for engineers". 95 Technmcs 37 239-239

Scheaffer, Richard; McClave, James T. (Auth); Herrin, Gary D. (Rev): Review of "Probability and statistics for engineers (Fourth edition)". 95 AmerStat 49 238-239

Scheaffer, Richard L.; McClave, James T. (Auth); Rigdon, Steven E. (Rev): Review of "Probability and statistics for engineers". 95 JQualTek 27 178-179

Schechter, Susan; Beatty, Paul; Block, Alan: Cognitive issues and methodological implications in the development and testing of a traffic safety questionnaire. 94 ASASrvyRsr 1215-1219

Beatty, Paul; — —: An examination of mode effects in cognitive laboratory research. 94 ASASrvyRsr 1275-1280

Scheerens, Jaap: International indicators on teachers within the framework of the OECD-INES. 94 ASAProGovt 159-163

Scheffer, Carel L.: The rank of the present excursion. 95 StocProc 55 101-118

Scheffler, Hans-Peter: Moments of measures attracted to operator semi-stable laws. 95 StPrLet 24 187-192

Scheiberg, S.: Finkelstein, M.; — —; Tucker, H. G.: Integral transforms with infinitely divisible kernels. 94 ThProbAp 39 670-676

Scheible, Paul: Ponikowski, Chester H.; — —; Wiatrowski, William: The survey of 1992 health expenditures: A quick response survey. 94 ASASrvyRsr 282-287

Scheiblechner, Hartmann: Isotonic ordinal probabilistic models (ISOP). 95 Psymtrka 60 281-304

Scheihing, Eliana: Mouchart, Michel; — —: Approximate Bayesian inferences: Numerical experiences when measuring the loss of information. 94 ASAProBays 221-226

Scheinberg, Stephen: Finkelstein, Mark; — —: The storage complexity of some estimators of the median. 94 AplMaCmp 60 15-24

Scheiner, S.; Gurevitch, J. (Ed); Ziegel, Eric R. (Rev): Review of "Design and analysis of ecological experiments". 95 Technmcs 37 128-128

Scheinerman, Edward R.; Wilf, Herbert S.: The rectilinear crossing number of a complete graph and Sylvester's "Four point problem" of geometric probability. 94 AmMaMnth 101 939-943

Schejbal, Judith A.; Lavrakas, Paul J.: Coverage error and cost issues in small area telephone surveys. 94 ASASrvyRsr 1287-1292

Scheleur, Scott A.: Bienias, Julia L.; Lassman, David M.; — —; Hogan, Howard: Improving outlier detection in two establishment surveys. 94 ASASrvyRsr 1140-1143

Schell, Michael J.: Singh, Bahadur; — —; Wright, F. T.: Approximations to the powers of the likelihood ratio tests: The loop ordering and slippage alternatives. 95 CommStB 24 91-109

Schervish, M. J.: Theory of statistics. 95 Springer:Brln:NY 702p

Scheuren, Fritz: Ten years of the Journal of Official Statistics. 95 JOffStat 11 5-9

Hinkins, Susan; Oh, H. Lock; — —: Inverse sampling design algorithms. 94 ASASrvyRsr 626-631

Holt, Albert; Kaufman, Steven; — —; Smith, Wray: Intersurvey consistency in school surveys. 94 ASASrvyRsr 1105-1110 (Disc: p1128-1129)

Scheutzow, Michael: Arnold, Ludwig; — —: Perfect cocycles through stochastic differential equations. 95 ProbTher 101 65-88

Schick, Anton: Correction to "On efficient estimation in regression models" (V21 p1486-1521). 95 AnlsStat 23 1862-1863

Schield, Milo: Random sampling versus representative samples. 94 ASAProStEd 107-110

Schiepers, J.: On the choice of equivalence scales. 92 StatJUN 9 111-124

Schiferl, Elizabeth A.: Davies, Leslie (Auth); — — (Rev): Review of "Efficiency in research, development and production: The statistical design and analysis of chemical experiments". 95 Technmcs 37 457-458

Schilling, Edward G.: Statistical standards: A global perspective. 94 JOffStat 10 339-346

Schimek, Michael G.: Dunn, G.; Everitt, B.; Pickles, A. (Auth); — — (Rev): Review of "Modelling covariances and latent variables using EQS". 95 CmpStQ 10 196-196

Hand, D. J.; Daly, F.; Lunn, A. D.; McConway, K. J.; Ostrowski, E. (Ed); — — (Rev): Review of "Handbook of small data sets". 95 CmpStQ 10 197-197

Rao, C. R. (Ed); — — (Rev): Review of "Handbook of statistics Volume 9. Computational statistics". 95 CmpStQ 10 194-195

Schinazi, Rinaldo: Durrett, Rick; — —: Intermediate phase for the contact process on a tree. 95 AnlsProb 23 668-673

Schindler, Eric; Navarro, Alfredo: Census plus: An alternative coverage methodology. 94 ASASrvyRsr 248-253

Schiopu-Kratina, Ioana: Ivanoff, Gail B.; Merzbach, Ely; — —: Lattices of random sets and progressivity. 95 StPrLet 22 97-102

Şchiopu-Kratina, Ioana; Daffer, Peter: Convergence of weighted sums and laws of large numbers in $D([0,1];E)$. 95 JMultiAn 53 279-292

Schipp, Bernhard: Kiviet, Jan F.; Phillips, Garry D. A.; — —: The bias of OLS, GLS, and ZEF estimators in dynamic seemingly unrelated regression models. 95 JEconmtx 69 241-266

Schirm, Allen L.: The relative accuracy of direct and indirect estimates of state poverty rates. 94 ASASrvyRsr 1071-1076

Schlichting, Kurt C.; Tuckel, Peter S.: A spatial analysis of contextual effects on voter participation in the 1992 presidential election. 94 ASASrvyRsr 1339-1344

Schlittgen, R.: Gessler, J. R. (Auth); — — (Rev. German): Review of "Statistical graphics" (German). 95 AllgeStA 79 223-224

Jones, R. H. (Auth); — — (Rev. German): Review of "Longitudinal data with serial correlation: A state-space approach". 95 AllgeStA 79 332-333

Westfall, P. H.; Young, S. S. (Auth); — — (Rev. German): Review of "Resampling - Base multiple testing". 95 AllgeStA 79 340-340

Schlittgen, Rainer: Klein, John P.; Goel, Prem K. (Auth); — — (Rev): Review of "Survival analysis: State of the art". 95 CmpStDA 20 592-593

Schlitzer, G.: Parigi, G.; — —: Quarterly forecasts of the Italian business cycle by means of monthly economic indicators. 95 JForcast 14 117-141

Schmeiser, Bruce W.: Chen, Huifen; — —: Monte Carlo estimation for guaranteed-coverage non-normal tolerance intervals. 95 JStCmpSm 51 223-238

Chen, Ming-Hui; — —: Random-direction interior-point Markov chains: A family of black-box samplers. 94 ASAProBays 1-6

Schmid, Bernhard; Polasek, Wolfgang; Weiner, Jacob; Krause, Andreas; Stoll, Peter: Modeling of discontinuous relationships in biology with censored regression. 94 ANatrls 143 494-507

Schmid, Friedrich: Measuring interdistributional income inequality - Inferential procedures (German). 95 AllgeStA 79 278-288

— —: Bounding shares and quantiles from grouped data when group means are known. 94 Metron 52/1 143-153

— —; Trede, Mark: A distribution free test for the two sample problem for general alternatives. 95 CmpStDA 20 409-419

Schmid, W.: On the run length of a Shewhart chart for correlated data. 95 StHefte 36 111-130

Flak, T.; — —: Extreme sums of strictly stationary sequences of M-dependent variables. 95 SankhyaA 57 186-201

Schmidli, H.: Cramér-Lundberg approximations for ruin probabilities of risk processes perturbed by diffusion. 95 InsrneME 16 135-149

— —: Reduced rank regression: With applications to quantitative structure-activity relationships. 95 Springer:Brln:NY 179p

Schmidli, Hanspeter: Barndorff-Nielsen, Ole E.; — —: Saddlepoint approximations for the probability of ruin in finite time. 95 ScanActJ 169-186

Schmidt, Karsten; Stahlecker, Peter: Reducing the maximum risk of regression estimators by polyhedral projection. 95 JStCmpSm 52 1-15

Schmidt, M.: Karr, A. F. (Auth); — — (Rev. German): Review of "Probability". 95 AllgeStA 79 333-334

Schmidt, Peter: Ahn, Seung C.; — —: A separability result for GMM estimation, with applications to GLS prediction and conditional moment tests. 95 EcomtrRv 14 19-34

Ahn, Seung C.; — —: Efficient estimation of models for dynamic panel data. 95 JEconmtx 68 5-27

Lee, Junsoo; — —: Unit root tests based on instrumental variables estimation. 94 IntEconR 35 449-462

Schmidt, R.: Optimizing the number of customers in a closed waiting queue with several customer classes (German). 95 ZAngMaMk 75 571-572S

Schmidt, Stephen R.: Kiemele, Mark J.; — — (Auth); Champ, Charles W. (Rev): Review of "Basic statistics tools for continuous improvement (Third Edition)". 95 JQualTek 27 174-175

Schmidt, Volker: Asmussen, Søren; — —: Ladder height distributions with marks. 95 StocProc 58 105-119

Asmussen, Søren; Frey, Andreas; Rolski, Tomasz; — —: Does Markov-modulation increase the risk?. 95 AstinB 25 49-66

Błaszczyszyn, Bartłomiej; Frey, Andreas; — —: Light-traffic approximations for Markov-modulated multi-server queues. 95 CommStSM 11 423-445

Miyazawa, Masakiyo; Schassberger, Rolf; — —: On the structure of an insensitive generalized semi-Markov process with reallocation and point-process input. 95 AdvAppPr 27 203-225

Schmitt, Angela: Eisenhower, Donna; — —: A records-survey comparison of eligibility and health care utilization measures for Medicaid beneficiaries: Adult and child reports. 94 ASASrvyRsr 713-718

Schmitz, N.: Sen, P. K.; Singer, J. M. (Auth); — — (Rev): Review of "Large sample methods in statistics". 95 Metrika 42 274-274

— — (Auth); Cressie, N. (Rev): Review of "Optimally sequentially planned decision procedures". 95 Metrika 42 141-143

Schmitz, Norbert: Comment on "Stein's two-stage procedure and exact consistency" (82ScanActJ p110-122). 95 ScanActJ 187-187

Schmitz, S.-F. H.; Schwager, S. J.; Pollak, E. J.: Determining the minimum sample size required to obtain sufficient progeny with a desired genotype at two quantitative trait loci. 93 ThApGene 87 136-144

Schmool, Marlena: Haberman, Steven; — —: Estimates of the British Jewish population 1984-88. 95 JRSS-A 158 547-562

Schmuland, B.: Karunamuni, R. J.; — —: A robust generalized Bayes estimator of a multivariate normal mean. 95 MaMethSt 4 472-482

Schmuland, Byron: Overbeck, Ludger; Röckner, Michael; — —: An analytic approach to Fleming-Viot processes with interactive selection. 95 AnlsProb 23 1-36

Schmutz, Joel A.; Ward, David H.; Sedinger, James S.; Rexstad, Eric A.: Survival estimation and the effects of dependency among animals. 95 JAppStat 22 673-681

Schneeweiss, Hans; Mathes, Harald: Factor analysis and principal components. 95 JMultiAn 55 105-124

Schneeweiß, Hans: — — (Auth); Trenkler, Dietrich (Rev): Review of "Ökonometrie (Fourth edition)". 95 CmpStDA 19 360-360

Schneeweiss, Winfrid G.: Mean time to finish a randomly disturbed 2-phase mission. 95 IEEEReli 44 310-314

Schneider, Helmut: Kasperski, William J.; — —; Johnson, Keith L.: Analyzing unreplicated factorials using the n smallest contrasts. 95 JQualTek 27 377-386

Schneider, Michael T.: Creating experimenters by design. 95 ASQCTekCon 365-366

Schneider, Rolf: Bauer, Christina; — —: Extremal problems for geometric probabilities involving convex bodies. 95 AdvAppPr 27 20-34

Schnell, Daniel J.; Magee, Elvin; Sheridan, John R.: A regression method for analysing ordinal data from intervention trials. 95 StatMed 14 1177-1189

Schnieper, René: On the estimation of the credibility factor: A Bayesian approach. 95 AstinB 25 137-151

Schnitkey, Gary D.: Miranda, Mario J.; — —: An empirical model of asset replacement in dairy production. 95 JApEcmtx 10 41-55S

Schnur, Dora M.: Brannigan, Lawrence H.; Grieshaber, Mark V.; — —: Use experimental design in organic synthesis. 95 ChemTech 25 29-35

Schnute, Jon T.; Richards, Laura J.: The influence of error on population estimates from catch-age models. 95 JFishRBC 52 2063-2077

Schoenfeld, David A.: Finkelstein, Dianne M.; — — (Ed): AIDS clinical trials: Guidelines for design and analysis. 95 Wiley:NY:UK 472p

Schoenfelder, John R.: Whaley, Fredrick S.; — —; Pinsky, Carl: Comparing a standard diagnostic procedure to a second procedure used in combination with the standard procedure. 94 ASAProBiop 272-275

Schoenmakers, Peter J.: Buydens, Lutgarde M. C.; — — (Ed); Pretsch, E. (Rev): Review of "Intelligent software for chemical analysis". 95 Chemolab 27 148-149

Scholz, F. W.; Tosch, T. J.: Small sample uni- and multi-variate control charts for means. 94 ASAProQlPr 17-22

Schöniger, Michael: von Haeseler, Arndt; — —: Modelling evolution of autocorrelated sequences. 95 CmpStDA 20 457-458

Schonmann, Roberto H.: Andjel, Enrique D.; Mountford, Thomas S.; — —: Equivalence of exponential decay rates for bootstrap percolation like cellular automata. 95 AIHenrPB 31 13-25

Kesten, Harry; — —: On some growth models with a small parameter. 95 ProbTher 101 435-468

School Mathematics Project: — — (Auth); Endean, Sarah (Rev): Review of "Statistics in action SMP 16-19 Mathematics". 95 TeachgSt 17/2 79-79

Schooler, Caroline: Fortmann, Stephen P.; Flora, June A.; Winkleby, Marilyn A.; — —; Taylor, C. Barr; Farquhar, John W.: Community intervention trials: Reflections on the Stanford Five-City Project experience. 95 AJEpidm 142 576-586

Schoonjans, F.; Zalata, A.; Depuydt, C. E.; Comhaire, F. H.: MedCalc: A new computer program for medical statistics. 95 CmpBiomd 48 257-262

Schork, M. Anthony: Schork, Nicholas J.; — —: The relative efficiency and power of small-pedigree studies of the heritability of a quantitative trait. 93 HumHrdty 43 1-11

Schork, Nicholas J.; Schork, M. Anthony: The relative efficiency and power of small-pedigree studies of the heritability of a quantitative trait. 93 HumHrdty 43 1-11

Schott, René: Neuenschwander, Daniel; — —: On the local and asymptotic behavior of Brownian motion on simply connected nilpotent Lie groups. 95 JTherPrb 8 795-806

Schouten, Evert G.; Dekker, Jacqueline M.; Kok, Frans J.; LeCessie, Saskia; van Houwelingen, Hans C.; Pool, Jan: Reply to comment on "Risk ratio and rate ratio estimation in case-cohort designs: Hypertension and cardiovascular mortality". 95 StatMed 14 1610-1610

Schouten, Hubert J. A.: Adaptive biased urn randomization in small strata when blinding is impossible. 95 Biomtrcs 51 1529-1535

Schouten, L. J.; Straatman, H.; Kiemeney, L. A.; Gimbrere, C. H.; Verbeek, A. L.: The capture-recapture method for estimation of cancer registry completeness: A useful tool?. 94 IJEpidem 23 1111-1116

Schreiner, Irwin: Royce, Daniel; — —: Improving coverage in a national survey of teachers. 94 ASASrvyRsr 827-832

Schreuder, Hans T.: Li, Jufang; Chen, Mei-Ching; — —; Gregoire, Tim G.: Forestry applications of saddle-point approximations to construct confidence intervals for population means. 95 Biomtrcs 51 61-72

Schröder, Bernd S. W.: Miamee, Abol G.; — —: On completeness of the spectral domain of harmonizable processes. 95 ProbTher 101 303-309

Schroder, Shayla D.: Johnston, Roger G.; — —; Mallawaaratchy, A. Rajika: Statistical artifacts in the ratio of discrete quantities. 95 AmerStat 49 285-291

Schröer, Gunar; Trenkler, Dietrich: Exact and randomization distributions of Kolmogorov-Smirnov tests two or three samples. 95 CmpStDA 20 185-202

Schucany, William R.: Adaptive bandwidth choice for kernel regression. 95 JASA 90 535-540

— —; Wang, Suojin: Bootstrap methods for nonstandard mixtures. 94 ASAProPEng 59-64

Frangos, Chris C.; — —: Improved bootstrap confidence intervals in certain toxicological experiments. 95 CommStA 24 829-844

George, Karen J.; McKean, Joseph W.; — —; Sheather, Simon J.: A comparison of confidence intervals from R-estimators in regression. 95 JStCmpSm 53 13-22

Schueck, Linda S.; Marzluff, John M.: Influence of weather on conclusions about effects of human activities on raptors. 95 JWildlMg 59 674-682

Schuenemeyer, John H.: Altekar, Maneesha; Lucas, James M.; — —: Shift detection in two-way arrays. 95 CommStB 24 711-731

Altekar, Maneesha; Lucas, James M.; — —: Shift detection in two-way arrays. 94 ASAProPEng 76-80

Schüle, Eik: Heinrich, Lothar; — —: Generation of the typical cell of a non-Poissonian Johnson-Mehl tessellation. 95 CommStSM 11 541-560

Schulman, Daniel A.; Campbell, Alec D.; Kostello, Eric C.: Data analysis using S-Plus. 95 SocMethR 23 392-413

Schultz, Delray J.; Parnes, Milton; Srinivasan, R.: Further applications of D-complete designs to group testing. 93 JCISS 18 31-41

Schultze, Charles: Edmonston, Barry; — — (Ed); Wolter, Kirk M. (Rev): Review of "Modernizing the U.S. Census". 95 Chance 8/3 44-49

Schulz, Anne: Krause, Ekkehard; Bandt, Christoph; — —; Schulz, Holger: Fractal exponents for the upper airways of mammalian lungs. 95 CmpStDA 20 583-590

Schulz, Henrik; Tetzlaff, Klaus; Vogel, Lydia: Simple encoding of infrared spectra for pattern recognition. 1: Statistical examination of the effectiveness and information content by factor analysis. 95 JChemtrc 9 143-168

Schulz, Holger: Krause, Ekkehard; Bandt, Christoph; Schulz, Anne; — —: Fractal exponents for the upper airways of mammalian lungs. 95 CmpStDA 20 583-590

Schulz, K.: Muijtjens, A. M. M.; Hasman, A. (Auth); — — (Rev): Review of "Bayes" (German). 95 CmpStQ 10 309-309

Schum, David A.: Evidential foundations of probabilistic reasoning. 94 Wiley:-NY:UK 545p

Schumacher, M.: Altman, D. G.; Lausen, B.; Sauerbrei, W.; — —: Dangers of using "optimal" cutpoints in the evaluation of prognostic factors. 94 JNCncrI 86 829-835

Schumacher, Martin: Graf, Erika; — —: An investigation on measures of explained variation in survival analysis. 95 Ststcian 44 497-507

Schumacher, Phyllis S.: Odell, Patricia M.; — —: Attitudes and achievement in freshman math classes. 94 ASAProStEd 197-202

Schur, Claudia L.; Berk, Marc L.; Mueller, Curt D.: Developing a prescription drug expenditure model: The effect of alternative assumptions. 94 ASAProSoSt 245-250

Schurz, Henri: Kloeden, Peter E.; Platen, Eckhard; — — (Auth); Hall, Timothy (Rev): Review of "Numerical solution of SDE through computer experiments". 95 Technmcs 37 125-125

Schuster, E. F.: Kaigh, W. D.; — —: SRS runs and spacings tests to assess randomness. 94 ASAProStCp 272-277

Schuster, Ernst: Preference of equivalence tests with standardized mean difference demonstrated by an application. 95 CmpStDA 19 353-356

Schutz, Robert W.: The systematic study of "statistics in sports": Do we need a framework?. 94 ASAProSpts 16-20

Liu, Yuanlong; — —: Overtime in the National Hockey League: Is it a valid tie-breaking procedure?. 94 ASAProSpts 55-60

McGarry, Tim; — —: Analysis of the 1986 and 1994 World Cup soccer tournament. 94 ASAProSpts 61-65

Schwabe, R.: Comment on "Constrained optimization of experimental design". 95 MaOpfStS 26 172-173

Schwabe, Rainer: Experimental design for additive linear models. 94 ASAProPEng 280-284

— —; Wierich, Werner: D-optimal designs of experiments with non-interacting factors. 95 JStPlInf 44 371-384

Schwager, S. J.; Mutschler, M. A.; Federer, W. T.; Scully, B. T.: The effect of linkage on sample size determination for multiple trait selection. 93 ThApGene 86 964-974

Schmitz, S.-F. H.; — —; Pollak, E. J.: Determining the minimum sample size required to obtain sufficient progeny with a desired genotype at two quantitative trait loci. 93 ThApGene 87 136-144

Schwaiger, M.: Hansen, M. H.; Hurwitz, W. N.; Madow, W. G. (Auth); — — (Rev. German): Review of "Sample survey methods and theory (Volume I and II)". 95 AllgeStA 79 335-337

Schwanz, Dennis J.: Jackson, Betty J.; Frazier, Richard L.; King, Karen E.; — —: Improving the coverage of private elementary-secondary schools. 94 ASASrvyRsr 833-838

Schwartz, H. J.; Mosler, C.; Hary, I.; Pielert, V.: Factors affecting spatial preferences in settlement site selection in migratory pastoralism. 95 EnvrMtrc 6 485-490

Schwartz, Harvey A.: Comment on "Data linkage". 95 StatMed 14 687-689

Schwartz, Pamela F.; Gennings, Chris; Chinchilli, Vernon M.: Threshold models for combination data from reproductive and developmental experiments. 95 JASA 90 862-870

Schwartz, Steven H.; Klein, Robert E.; Adashek, Daniel: Veterans living in group quarters: What data from the 1990 Census show. 94 ASAProGovt 13-18

Schwarz, C.; Reid, N.: Comment on "Effects of growth regulators on silver maple trees: A case study". 95 CandJSt 23 329-332

Schwarz, Carl J.: Reply to comment on "Comment on 'The mixed-model ANOVA: The truth, the computer packages, the books. Part I: Balanced data'". 95 AmerStat 49 324-324

Arnason, A. Neil; — —: POPAN-4: Enhancements to a system for the analysis of mark-recapture data from open populations. 95 JAppStat 22 785-800

Sloan, Jeff A.; — —; Neden, Linda R.: Dataset on silver-maple-tree growth regulators. 95 CandJSt 23 325-329 (Disc: p329-332)

Schwarz, Carl James; Ganter, Barbara: Estimating the movement among staging areas of the barnacle goose (Branta leucopsis). 95 JAppStat 22 711-724

Schwarz, Norbert: What respondents learn from questionnaires: The survey interview and the logic of conversation. 95 IntStRvw 63 153-168 (Disc: p169-177)

Schwarzenbach, F. H.: Urfer, W.; — —; Kötting, J.; Müller, P.: Multistate models for monitoring individual trees in permanent observation plots. 94 EnvEcoSt 1 171-192 (Disc: p192-199)

Schwede, Laurie; Ellis, Yukiko: Exploring associations between subjective and objective assessments of household membership. 94 ASASrvyRsr 325-330

Schweder, Tore: Comment On "Inference from a deterministic population dynamics model for Bowhead whales". 95 JASA 90 420-423

Schweitzer, D.: Zwinderman, A. H.; van Houwelingen, J. C.; — —: Prediction of the next hypercalcemia free period: Application of random effect models with selection on first event. 95 StNeerla 49 310-323

Schweitzer, Walter: Transformation for experimental design in off-line quality control (German). 95 AllgeStA 79 117-139

Schweizer, Karl: Perspectives for investigating data of a psychological survey. 95 CmpStQ 10 1-8

Schweizer, Martin: Variance-optimal hedging in discrete time. 95 MaOpRsr 20 1-64

— —: On the minimal martingale measure and the Föllmer-Schweizer decomposition. 95 StocAnAp 13 573-599

Schwertman, Neil C.; Carter, Nancy J.: A more practical Scheffé-type multiple comparison procedure for commonly encountered numbers of comparisons. 95 JStCmpSm 53 181-196

— —; Kiser, Terry L.: The role of geometry in pairwise and mutual independence. 95 AmerStat 49 56-59

Scientific Software International: — — (Auth); Hutchinson, Susan R. (Rev): Review of "LISCOMP 1.1". 95 StrEqMod 2 383-386

Arbuckle, James; BMDP Statistical Software Inc; — — (Auth); Hox, Joop J. (Rev): Review of "AMOS, EQS, and LISREL for Windows: A comparative review". 95 StrEqMod 2 79-91

Sclove, Stanley L.; Sherman, Simon A.: Cluster analysis dihedral angles of amino acid residues in proteins. 94 ASAProBiop 399-404

Scollnik, David P. M.: Bayesian analysis of two overdispersed Poisson models. 95 Biomtrcs 51 1117-1126

— —: Bayesian analysis of an intervened Poisson distribution. 95 CommStA 24 735-754

— —: Bayesian analysis of two overdispersed Poisson regression models. 95 CommStA 24 2901-2918

Scot, D.: McLachlan, G. J.; — —: Asymptotic relative efficiency of the linear discriminant function under partial nonrandom classification of the training data. 95 JStCmpSm 52 415-426

Scott, Alastair: Davis, Peter; — —: The effect of interviewer variance on domain comparisons. 95 SurvyM 21 99-106

Scott, Alastair J.: Gange, Stephen J.; Linton, Kathryn L. P.; — —; DeMets, David L.; Klein, Ronald: A comparison of methods for correlated ordinal measures with ophthalmic applications. 95 StatMed 14 1961-1974

Scott, D. W.: — — (Auth); Silverman, B. W. (Rev): Review of "Multivariate density estimation". 95 Metrika 42 66-68

Scott, Donald R.: Bounds for molecular weights of organic compounds from low resolution mass spectra. 95 Chemolab 27 129-133

Scott, E. M.: Martin, E. B.; — —: The detection of change in spatial processes. 95 EnvrMtrc 6 141-154

Collett, D. (Auth); — — (Rev): Review of "Modelling survival data in medical research". 95 Ststcian 44 137-138

Scott, Stuart; Stamas, George; Sullivan, Thomas J.; Chester, Paul: Seasonal adjustment of hybrid economic time series. 94 ASAProBuEc 163-168

Scott, W. F.: Lindley, D. V.; — —: New Cambridge statistical tables (Second edition). 95 CambridgeUniv 96p

Scribner, Richard: Mackinnon, David P.; — —; Taft, Katharine A.: Development and applications of a city-level alcohol availability and alcohol problems database. 95 StatMed 14 591-604 (Disc: p697-699)

Scrutchins, Sam: Bredfeldt, Hans; Foster, Suzy; — —: Extending the use of IRS's tax compliance measurement program to develop estimates for small subpopulations using double sampling. 94 ASAProGovt 128-133

Scully, B. T.: Federer, W. T.; — —: A parsimonious statistical design and breeding procedure for evaluating and selecting desirable characteristics over environments. 93 ThApGene 86 612-620

Schwager, S. J.; Mutschler, M. A.; Federer, W. T.; — —: The effect of linkage on sample size determination for multiple trait selection. 93 ThApGene 86 964-974

Scwarzenbach, F. H.: Urfer, W.; — —; Kötting, J.; Müller, P.: Reply to comments on "Multistate models for monitoring individual trees in permanent observation plots". 94 EnvEcoSt 1 197-199

Seacat, Hui: Correa-Villaseñor, Adolfo; Stewart, Walter F.; Franco-Marina, Francisco; — —: Bias from nondifferential misclassification in case-control studies with three exposure levels. 95 Epidmlgy 6 276-281

Seaks, Terry G.: Greene, William H.; Greene, Laura Knapp; — —: Estimating the functional form of the independent variables in probit models. 95 ApplEcon 27 193-196

Seal, K. C.: National sample surveys for development planning - An overview. 94 CalcutSt 44 239-250

Seaman, John W., Jr: Laviolette, Michael; — —; Barrett, J. Douglas; Woodall, William H.: A probabilistic and statistical view of fuzzy methods. 95 Technmcs 37 249-261 (Disc: p262-292)

Laviolette, Michael; — —; Barrett, J. Douglas; Woodall, William H.: Reply to comments on "A probabilistic and statistical view of fuzzy methods". 95 Technmcs 37 287-292

Turner, Danny W.; Young, Dean M.; — —: A Kolmogorov inequality for the sum of independent Bernoulli random variables with unequal means. 95 StPrLet 23 243-245

Searle, S. R.: McCulloch, C. E.; — —: On an identity derived from unbiasedness in linear models. 95 AmerStat 49 39-42

— —; Casella, G.; McCulloch, C. E. M. (Auth); Cavedon, G. (Rev): Review of "Variance components". 95 Metron 53 217-218

Searle, Shayle R.: An overview of variance component estimation. 95 Metrika 42 215-230

— —: Comment on "The statistics of linear models: Back to basics" (V4 p221-234). 95 StatComp 5 103-107

Searls, Donald T.: Lee, Timothy H.; — —: Bias correction in risk assessment when logistic regression is used with an unevenly proportioned sample between risk and non-risk groups. 94 ASAProBuEc 304-309

Seater, John J.: Rossana, Robert J.; — —: Temporal aggregation and economic time series. 95 JBES 13 441-451

Seaver, William L.; Snyder, Richard: Process improvement in telecommunications with multivariate methods. 94 ASAProQlPr 235-240

Sebastian, G.: Kale, B. K.; — —: Approximate estimation of non-identifiable parameters in a convolution. 95 StPrLet 25 373-378

Sebastiani, Giovanni: Glad, Ingrid K.; — —: A Bayesian approach to synthetic magnetic resonance imaging. 95 Biomtrka 82 237-250

Seber, G. A. F.: Patil, G. P.; Rao, C. R. (Ed); — — (Rev): Review of "Multivariate environmental statistics". 95 Biomtrcs 51 1189-1190

Sedgwick, Philip: Sinai, Y. (Auth); Haughton, D. (Transl); — — (Rev): Review of "Probability theory: An introductory course". 95 Ststcian 44 142-142

Sedinger, James S.: Flint, Paul L.; Pollock, Kenneth H.; Thomas, Dana; — —: Estimating prefledging survival : Allowing for brood mixing and dependence among brood mates. 95 JWildlMg 59 448-455

Lindberg, Mark S.; — —; Rexstad, Eric A.: Estimating nest site fidelity of adult female black brant with multi-state modeling and geographic information systems. 95 JAppStat 22 725-735

Schmutz, Joel A.; Ward, David H.; — —; Rexstad, Eric A.: Survival estimation and the effects of dependency among animals. 95 JAppStat 22 673-681

See, K.; Smith, Eric P.: Multiplicative effects in three-way analysis of variance. 94 ASAProStCp 225-230

Seeber, G. U. H.: Computer algebra in statistics: Some examples. 94 ASAProStCp 147-151

— —; Francis, B. J.; Hatzinger, R.; Steckel-Berger, G.: Statistical modelling. 95 Springer:Brln:NY 327p

Sefton, C. E. M.; Whitehead, P. G.; Eatherall, A.; Littlewood, I. G.; Jakeman, A. J.: Dynamic response characteristics of the Plynlimon catchments and preliminary analysis of relationships to physical descriptors. 95 EnvrMtrc 6 465-472

Segal, Mark R.: Extending the elements of tree-structured regression. 95 StMedRes 4 219-236

— —; James, Ian R.; French, Martyn A. H.; Mallal, Simon A.: Statistical issues in the evaluation of markers of HIV progression. 95 IntStRvw 63 179-197

Bacchetti, Peter; — —: Survival trees with time-dependent covariates: Application to estimating changes in the incubation period of AIDS. 95 LiDaAnal 1 35-47

Segel, Lee A.: Grappling with complexity. 95 Complxty 1 18-25

Sehgal, Surinder K.: Arasu, K. T.; — —: Some new difference sets. 95 JCombThA 69 170-172

Seiver, A.: Dagum, P.; Galper, A.; Horvitz, E.; — —: Uncertain reasoning and forecasting. 95 IntJFor 11 73-87

Seki, Yoichi: Hoshino, Naoto; Miyazaki, Haruo; — —: On the level probabilities for useful partially ordered alternatives in the analysis of variance. 95 CommStA 24 2059-2071

Sekiya, Yuri: Taneichi, Nobuhiro; — —: Power approximations of the test of homogeneity for multinomial populations. 95 JJapnStS 25 97-109

Self, S. G.: Etzioni, R.; — —: Correction to "On the catch-up time method for analyzing cancer screening trials" (V51 p31-43). 95 Biomtrcs 51 1589-1589

Self, Steven G.; Etzioni, Ruth: A likelihood ratio test for cancer screening trials. 95 Biomtrcs 51 44-50

Etzioni, Ruth; — —: On the catch-up time method for analyzing cancer screening trials. 95 Biomtrcs 51 31-43

Etzioni, Ruth D.; Connor, Robert J.; Prorok, Philip C.; — —: Design and analysis of cancer screening trials. 95 StMedRes 4 3-17

Liang, Kung-Yee; — —; Bandeen-Roche, Karen J.; Zeger, Scott L.: Some recent developments for regression analysis of multivariate failure time data. 95 LiDaAnal 1 403-415

Sellke, Thomas M.: How many iid samples does it take to see all the balls in a box?. 95 AnlsApPr 5 294-309

Selukar, Rajesh S.; Shah, Arvind K.; Mishra, Satya N.: A simple approach to simultaneous comparison of normal means. 95 JStCmpSm 52 174-177

Selvin, Steve: Statistical analysis of epidemiological data (Second edition). 95 OxfordUniv 448p

Clayton, David; Hills, Michael (Auth); — — (Rev): Review of "Statistical models in epidemiology". 95 JASA 90 798-799

— — (Auth); Smeeton, Nigel (Rev): Review of "Practical biostatistical methods". 95 AJEpidm 141 229-230

— — (Auth); Ziegel, Eric R. (Rev): Review of "Statistical analysis of epidemiologic data". 95 Technmcs 37 240-241

— — (Auth); Ziegel, Eric R. (Rev): Review of "Practical biostatistical methods". 95 Technmcs 37 241-241

Sempi, Carlo: Pistone, Giovanni; — —: An infinite-dimensional geometric structure on the space of all the probability measures equivalent to a given one. 95 AnlsStat 23 1543-1561

Sen, A.: Srivastava, M.: Regression analysis. Theory, methods, and applications. 90 Springer:Brln:NY 347p

Sen, Ananda: Comment on "A characterization of the Poisson distribution and the probability of winning a game" (V48 p294-298). 95 AmerStat 49 401-401

— —; Bhattacharyya, Gouri K.: Inference procedures for the linear failure rate model. 95 JStPlInf 46 59-76

Sen, Arusharka: Rao, B. L. S. Prakasa; — —: Limit distributions of conditional *U*-statistics. 95 JTherPrb 8 261-301

Rao, B. L. S. Prakasa; — —: An application of the 'martingale methods' to nonparametric regression based on censored data. 95 StatDecn 13 201-220

Sen, P.: Wang, J.; Yang, J. B.; — —: Safety analysis and synthesis using fuzzy sets and evidential reasoning. 95 RelEngSS 47 103-118

Sen, P. K.: Fisher, N. I.; — —: The collected works of Wassily Hoeffding. 94 Springer:Brln:NY 658p

Nagaraja, H. N.; — —; Morrison, D. F. (Ed): Statistical theory and applications: Papers in honor of Herbert A. David. 95 Springer:Brln:NY 334p

— —; Singer, J. M. (Auth); MacGillivray, H. (Rev): Review of "Large sample methods in statistics". 95 Biomtrcs 51 381-382

— —; Singer, J. M. (Auth); Schmitz, N. (Rev): Review of "Large sample methods in statistics". 95 Metrika 42 274-274

Sen, Pranab K.: Paula, Gilberto A.; — —: One-sided tests in generalized linear models with parallel regression lines. 95 Biomtrcs 51 1494-1501

Tsai, Ming-Tien; — —: A test of quasi-independence in ordinal triangular contingency tables. 95 StSinica 5 767-780

Sen, Pranab Kumar: Statistical analysis of some reliability models: Parametrics, semi-parametrics and nonparametrics. 95 JStPlInf 43 41-66

— —: Bridging the biostatistics-epidemiology gap: The Bangladesh task. 94 JStatRsr 28 21-39

— —; Margolin, Barry H.: Inhalation toxicology: Awareness, identifiability, statistical perspectives and risk assessment. 95 SankhyaB 57 252-276

Das, Shubhabrata; — —: Simultaneous spike-trains and stochastic dependence. 95 SankhyaB 57 32-47

Ren, Jian-Jian; — —: Hadamard differentiability on $D[0,1]^p$. 95 JMultiAn 55 14-28

Wada, Cicilia Yuko; — —: Restricted alternative test in a parametric model with competing risk data. 95 JStPlInf 44 193-203

Senar, J. C.; Copete, J. L.: Mediterranean house sparrows (Passer domesticus) are not used to freezing temperatures: An analysis of survival rates. 95 JAppStat 22 1069-1074

Senchaudhuri, Pralay; Metha, Cyrus R.; Patel, Nitin R.: Estimating exact p values by the method of control variates or Monte Carlo rescue. 95 JASA 90 640-648

Catalano, Paul J.; Ryan, Louise M.; — —: Developmental toxicity modeling for risk assessment. 94 ASAProEnvr 38-45

Seneta, E.: A semi-deterministic generalization of a (2×2) contingency table. 95 ThPrMaSt 50 117-121

Chen, Tuhao; — —: A note on bivariate Dawson-Sankoff-type bounds. 95 StPrLet 24 99-104

Sengupta, Arindam: Goswami, A.; — —: Time-spaced polynomial martingales generated by a discrete-time martingale. 95 JTherPrb 8 417-432

Sengupta, Bhaskar: Takine, Tetsuya; — —; Hasegawa, Toshiharu: A conformance measure for queues. 95 CommStSM 11 645-670

Takine, Tetsuya; — —; Yeung, Raymond W.: A generalization of the matrix M/G/1 paradigm for Markov chains with a tree structure. 95 CommStSM 11 411-421

Sengupta, D.: Bhimasankaram, P.; — —; Ramanathan, S.: Recursive inference in a general linear model. 95 SankhyaA 57 227-255

Sengupta, Debasis: Reliability bounds for the L-class and Laplace order. 95 JAppProb 32 832-835

— —: Optimal choice of a new observation in a linear model. 95 SankhyaA 57 137-153

Deshpande, Jayant V.; — —: Testing the hypothesis of proportional hazards in two populations. 95 Biomtrka 82 251-261

Sengupta, J. K.: — — (Auth); Shanmugam, Ramalingam (Rev): Review of "Econometrics of information and efficiency". 95 Ststcian 44 290-291

Sengupta, S.: Mukerjee, Rahul; — —: Optimal estimation of a finite multidimensional population total. 95 JStPlInf 48 339-346

Sengupta, Sailes K.: Kay, Steven M. (Auth); — — (Rev): Review of "Fundamentals of statistical signal processing: Estimation theory". 95 Technmcs 37 465-466

Sengupta, Soma; Neerchal, Nagaraj K.: Classification of grab samples using composite sampling: An improved strategy. 94 CalcutSt 44 195-208

Senn, S.: — — (Auth); Jones, Byron (Rev): Review of "Cross-over trials in clinical research". 95 JAppStat 22 317-318

Senn, Stephen: In defence of analysis of covariance: A reply to Chambless and Roeback. 95 StatMed 14 2283-2285

— —: A personal view of some controversies in allocating treatment to patients in clinical trials. 95 StatMed 14 2661-2674

Adams, John (Auth); — — (Rev): Review of "Risk". 95 StatMed 14 1727-1728

Senroy, Sugata: Kalirajan, K. P. (Auth); — — (Rev): Review of "Applied econometrics". 95 CalcutSt 45 135-136

Seo, H. S.: Nonparametric test on dimensionality of explanatory variables (Korean). 95 KorApS 8/2 65-75

Seo, Hye-Young: Kim, Tae-Sung; — —: A note on some negative dependence notions. 95 CommStA 24 845-858

Kim, Tae-Sung; — —: An invariance principle for stationary strong mixing random fields. 95 JKoreaSt 24 281-292

Seo, Takashi; Kanda, Takashi; Fujikoshi, Yasunori: The effects of nonnormality on tests for dimensionality in canonical correlation and MANOVA models. 95 JMultiAn 52 325-337

Sepanski, Jungsywan H.; Lee, Lung-Fei: Semiparametric estimation of nonlinear errors-in-variables models with validation study. 95 JNnparSt 4 365-394

Lee, Lung-fei; — —: Estimation of linear and nonlinear errors-in-variables models using validation data. 95 JASA 90 130-140

Seppala, Tomi; Moskowitz, Herbert; Plante, Robert; Tang, Jen: Statistical process control via the subgroup bootstrap. 95 JQualTek 27 139-153

Seppäläinen, Timo: Maximum entropy principles for disordered spins. 95 ProbTher 101 547-576

Sergent, Michelle; Mathieu, Didier; Phan-Tan-Luu, Roger; Drava, Giuliana: Correct and incorrect use of multilinear regression. 95 Chemolab 27 153-162

Serinko, Regis J.; Babu, Gutti Jogesh: Asymptotics of k-mean clustering under non-i.i.d. sampling. 95 StPrLet 24 57-66

Serlet, Laurent: Some dimension results for super-Brownian motion. 95 ProbTher 101 371-391

— —: On the Hausdorff measure of multiple points and collision points of super-Brownian motion. 95 Stochatx 54 169-198

Serletis, Apostolos: Random walks, breaking trend functions, and the chaotic structure of the velocity of money. 95 JBES 13 453-458

Serra, C.: Forina, M.; Drava, G.; Armanino, C.; Boggia, R.; Lanteri, S.; Leardi, R.; Corti, P.; Conti, P.; Giangiacomo, R.; Galliena, C.; Bigoni, R.; Quartari, I.; — —; Ferri, D.; Leoni, O.; Lazzeri, L.: Transfer of calibration function in near-infrared spectroscopy. 95 Chemolab 27 189-203

Serra, V.: Romero, R.; Ferrer, A.; Capilla, C.; Zunica, L.; Balasch, S.; — —; Alcover, R.: Teaching statistics to engineers: An innovative pedagogical experience. 95 JStEd 3/1 33kb (jse/v3n1/romero)

Seshadri, Vanamamalai: Letac, Gérard; — —: A random continued fraction in R^{d+1} with an inverse Gaussian distribution. 95 Bernoull 1 381-393

Sethi, Gulshan: Henderson, William G.; Moritz, Thomas; Goldman, Steven; Copeland, Jack; — —: Use of cumulative meta-analysis in the design, monitoring, and final analysis of a clinical trial: A case study. 95 ConClnTr 16 331-341

Sethi, Ishwar K.: Skrzypek, J.; Karplus, W. (Ed); — — (Rev): Review of "Neural networks in vision and pattern recognition". 95 JClass 12 140-142

Sethi, S. P.; Zhang, Q.: Hierarchical decision making in stochastic manufacturing systems. 94 Birkhs:Basl:MA 19p

— —; Zhang, Q. (Auth); Goodwin, G. C. (Rev): Review of "Hierarchical decision making in stochastic manufacturing systems". 95 ShrtBkRv 15/2 29-29

Sethuraman, Jayaram: Gurland, John; — —: How pooling failure data may reverse increasing failure rates. 95 JASA 90 1416-1423

Sethuraman, S.; Basawa, I. V.: Maximum likelihood estimation for a fractionally differenced autoregressive model on a two-dimensional lattice. 95 JStPlInf 44 219-235

Sethuraman, Shanthi; Reynolds, Marion R., Jr; Arnold, Jesse C.: Process control strategies using variable sampling intervals. 94 ASAProQlPr 109-114

Seuc, Armando H.: Cochran Q test with Turbo BASIC. 95 CmpBiomd 46 29-34

Seuchter, S. A.: Knapp, M.; — —; Baur, M. P.: Two-locus disease models with two marker loci: The power of affected-sib-pair tests. 94 AJHuGene 55 1030-1041

Knapp, M.; — —; Baur, M. P.: Linkage analysis in nuclear families. 1: Optimality criteria for affected sib-pair tests. 94 HumHrdty 44 37-43

Knapp, M.; — —; Baur, M. P.: Linkage analysis in nuclear families. 2: Relationship between affected sib-pair tests and lod score analysis. 94 HumHrdty 44 44-51

Séva, Marc: On the local limit theorem for non-uniformly ergodic Markov chains. 95 JAppProb 32 52-62

Sevcik, Karel; Brooker, Peter I.: Application of the gnostical theory of uncertain data to spatial estimation, identification and quantification. 95 EnvrMtrc 6 457-463

Severini, Thomas A.: Information and conditional inference. 95 JASA 90 1341-1346

— —: Comment on "Inference". 95 StatSci 10 187-189

Severson, Richard K.: Sakkinen, Pamela A.; — —; Ross, Julie A.; Robison, Leslie L.: Random-digit dialing for control selection in childhood cancer studies: The geographic proximity and demographics within matched sets. 95 AJPubHlh 85 555-557

Severtson, Forrestt: Elston, Robert C.; George, Varghese T.; — —: The Elston-Stewart algorithm for continuous genotypes and environmental factors. 92 HumHrdty 42 16-27

Seyedsadr, M.: Cornelius, P. L.; — —; Crossa, J.: Using the shifted multiplicative model to search for "separability" in crop cultivar trials. 92 ThApGene 84 161-172

Seyedsadr, M. S.: Cornelius, P. L.; Van Sanford, D. A.; — —: Clustering cultivars into groups without rank-change interactions. 93 CropSci 33 1193-1200

Sfer, Ana M.: Viollaz, Aido J.; — —; Salvatierra, Stella M.: An approximation of the unconditional error rates of the sample linear discriminant function. 95 CommStA 24 1941-1969

Shaarawy, Samir; Haroun, Ahmed: Bayesian classification with ARMA sources. 94 EgpStJ 38 165-176

Shaarawy, Samir M.; Almahmeed, Mohammad A.: Bayesian classification with multivariate autoregressive sources that might have different orders. 95 CommStB 24 567-581

Shaban, Abdel Magid M.: Comparison between the MINQUE and the PSD-MINQMBE in the unbalanced one way random effects model. 94 EgpStJ 38 222-230

— —: ANOVA, MINQUE, PSD-MINQMBE, CANOVA and CMINQUE in estimating variance components. 94 Statstca 54 481-489

Shaddick, Gavin: Elliott, Paul; Martuzzi, Marco; — —: Spatial statistical methods in environmental epidemiology: A critique. 95 StMedRes 4 137-159

Shade, John: Betteley, G.; Mettrick, N.; Sweeney, E.; Wilson, D. (Auth); — — (Rev): Review of "Using statistics in industry". 95 Ststcian 44 541-542

Shafer, Ronald W.: Neerchal, Nagaraj K.; — —: Spatial analysis of benthic data from the Chesapeake Bay. 94 ASAProEnvr 204-208

Shaffer, R. E.; Small, G. W.; Combs, R. J.; Knapp, R. B.; Kroutil, R. T.: Experimental design protocol for the pattern recognition analysis of bandpass filtered Fourier transform infrared interferograms. 95 Chemolab 29 89-108

Shafi, Bahman: A comparison of two SAS procedures in decomposing the structure of two-way tables. 94 ASAProStCp 231-234

Shafie, K.; Noorbaloochi, S.: Asymmetric unbiased estimation in location families. 95 StatDecn 13 307-314

Shafy, Mona Abdel; Osman, Magued I.: The compound Gompertz distribution. 95 EgpStJ 39 89-105

Shah, A. K.: Krishnamoorthy, K.; — —: Correction to "Testing equality of several normal means to a specified standard" (V27 p132-138). 95 JQualTek 27 402-402

Shah, Arvind K.: Krishnamoorthy, K.; — —: Testing equality of several normal means to a specified standard. 95 JQualTek 27 132-138
Kulkarni, Pandurang M.; — —: Testing the equality of several binomial proportions to a prespecified standard. 95 StPrLet 25 213-219
Selukar, Rajesh S.; — —; Mishra, Satya N.: A simple approach to simultaneous comparison of normal means. 95 JStCmpSm 52 174-177

Shah, Babubbai V.; LaVange, Lisa M.: Mixed models for survey data. 94 ASAProStEd 111-116

Shah, K. R.; Sinha, Bikas K.: Optimal designs with component-wise orthogonal row-column structures. 93 JCISS 18 68-78
Park, D. K.; — —: On connectedness of row-column designs. 95 CommStA 24 87-96

Shah, Kirti R.: Dey, Aloke; — —; Das, Ashish: Optimal block designs with minimal and nearly minimal number of units. 95 StSinica 5 547-558
Jacroux, Mike; Majumdar, Dibyen; — —: Efficient block designs in the presence of trends. 95 StSinica 5 605-615

Shah, S.; Green, J. R.: The distribution of Q: A powerful sibship test of association. 94 AnHuGene 58 163-173
— —; Green, J. R.: Disease susceptibility genes and the sib-pair method: A review of recent methodology. 94 AnHuGene 58 381-395
— —; Green, J. R.: Testing for haplotype concordance with incomplete parental data - Modified Q and correction for modified N. 93 AnHuGene 57 239-247
Green, J. R.; — —: Power comparison of various sibship tests of association. 93 AnHuGene 57 151-158
Green, J. R.; — —: The power of the N-test of haplotype concordance. 92 AnHuGene 56 331-338

Shah, S. M.; Ghosh, D. K.; Patel, A. H.: Construction of α-resolvable BIBDs using 1-resolvable and α-affine resolvable BIBDs. 93 UtilMath 43 219-224

Shahane, Aditi; George, Varghese; Johnson, William D.: Effect of bivariate regression toward the mean in uncontrolled clinical trials. 95 CommStA 24 2165-2181

Shahidullah, Mohammed: Smith, Stanley K.; — —: An evaluation of population projection errors for census tracts. 95 JASA 90 64-71

Shahzad, Naseer: Beg, Ismat; — —: Random fixed points for random multivalued operators defined on unbounded sets in Banach spaces. 95 StocAnAp 13 269-277

Shainin, Richard D.: A common sense approach to quality management. 95 ASQCTekCon 1163-1169

Shaked, Moshe; Shanthikumar, J. George: Hazard rate ordering of k-out-of-n systems. 95 StPrLet 23 1-8
— —; Shanthikumar, J. George; Tong, Y. L.: Parametric Schur convexity and arrangement monotonicity properties of partial sums. 95 JMultiAn 53 293-310
— —; Szekli, Ryszard: Comparison of replacement policies via point processes. 95 AdvAppPr 27 1079-1103
Li, Haijun; — —: On the first passage times for Markov processes with monotone convex transition kernels. 95 StocProc 58 205-216
Rocha-Martinez, José M.; — —: A discrete-time model of failure and repair. 95 AplStMDA 11 167-180
— —; Shanthikumar, J. George (Auth); Tong, Y. L. (Rev): Review of "Stochastic orders and their applications". 95 SIAMRvw 37 477-479

Shakeri, M.; Pattipati, K. R.; Kleinman, D. L.: Optimal measurement scheduling for state estimation. 95 IEEEAES 31 716-729

Sham, P. C.; Walters, E. E.; Neale, M. C.; Heath, A. C.; MacLean, C. J.; Kendler, K. S.: Logistic regression analysis of twin data: Estimation of parameters of the multifactorial liability-threshold model. 94 BehavGen 24 229-238
Curtis, D.; — —: Using risk calculation to implement an extended relative pair analysis. 94 AnHuGene 58 151-162

Shamlaye, C.: Yakir, Benjamin; Cox, Christopher; Tanner, Martin A.; Clarkson, Thomas W.; — —: Seasonal changes in methylmercury levels in human hair. 94 ASAProBays 35-39

Shamsunder, Sanyogita: Sadler, Brian M.; Giannakis, Georgios B.; — —: Noise subspace techniques in non-Gaussian noise using cumulants. 95 IEEEAES 31 1009-1018

Shanahan, P. E.: Hohls, T.; Clarke, G. P. Y.; — —; Gevers, H. O.: Additive and non-additive genetic correlations. II. Application to data obtained from a 12×12 diallel cross. 95 BiomtrcJ 37 937-943
Hohls, T.; Clarke, G. P. Y.; — —; Gevers, H. O.: An unbiased estimator for identifying lines useful for the improvement of elite single crosses, based on a combining ability model. 95 Heredity 75 155-163

Shanbhag, D. N.: Rao, C. R.; — —: Choquet-Deny type functional equations with applications to stochastic models. 94 Wiley:NY:UK 290p
Rao, C. Radhakrishna; — —: Characterizations based on regression properties: Improved versions of recent results. 95 SankhyaA 57 167-178
Rao, C. R.; — — (Auth); Lindvall, T. (Rev): Review of "Choquet-Deny type functional equations with applications to stochastic models". 95 ShrtBkRv 15/3 48-49
Rao, C. Radhakrishna; — — (Auth); Heinrich, L. (Rev): Review of "Choquet-Deny type functional equations with applications to stochastic models". 95 BiomtrcJ 37 726-726

Shanmugam, R.: Chattamvelli, R.; — —: Efficient computation of the noncentral χ^2 distribution. 95 CommStB 24 675-689

Shanmugam, Ramalingam: Goodness-of-fit test for length-biased data with discussion on prevalence, sensitivity and specificity of the length bias. 95 JStPlInf 48 277-290
Bethea, Robert M.; Rhinehart, R. Russell (Auth); — — (Rev): Review of "Applied engineering statistics". 95 JQualTek 27 173-173
Krzanowski, W. J.; Marriott, F. H. C. (Auth); — — (Rev): Review of "Multivariate analysis Part 1: Distributions, ordination and inference". 95 Technmcs 37 351-351
Sengupta, J. K. (Auth); — — (Rev): Review of "Econometrics of information and efficiency". 95 Ststcian 44 290-291

Shannon, William D.; Bryant, John; Logan, Theodore F.; Day, Roger: An application of decision theory to patient screening for an autologous tumour vaccine trial. 95 StatMed 14 2099-2110

Shanthikumar, J. George: Shaked, Moshe; — —: Hazard rate ordering of k-out-of-n systems. 95 StPrLet 23 1-8
Shaked, Moshe; — —; Tong, Y. L.: Parametric Schur convexity and arrangement monotonicity properties of partial sums. 95 JMultiAn 53 293-310
Shaked, Moshe; — — (Auth); Tong, Y. L. (Rev): Review of "Stochastic orders and their applications". 95 SIAMRvw 37 477-479

Shanubhogue, A.: Deshpande, J. V.; Gore, A. P.; — —: Statistical analysis of nonnormal data. 95 Wiley:NY:UK 240p

Shao, J.; Tu, D.: The jackknife and bootstrap. 95 Springer:Brln:NY 516p
— —; Tu, D. (Auth); Fedorov, V. V. (Rev): Review of "The jackknife and bootstrap". 96 ShrtBkRv 16/1 7-7

Shao, Peter Yi-Shi; Strawderman, William E.: Improving on the positive part of the UMVUE of a noncentrality parameter of a noncentral chi-square distribution. 95 JMultiAn 53 52-66

Shao, Q.-M.; Wang, D.: Small ball probabilities of Gaussian fields. 95 ProbTher 102 511-517

Shao, Qi-Man: Maximal inequalities for partial sums of ρ-mixing sequences. 95 AnlsProb 23 948-965
— —: Strong approximation theorems for independent random variables and their applications. 95 JMultiAn 52 107-130
— —: A Chung type law of the iterated logarithm for subsequences of a Wiener process. 95 StocProc 59 125-142
Csörgő, Miklós; Lin, Zheng-Yan; — —: On moduli of continuity for local times of Gaussian processes. 95 StocProc 58 1-21
Horváth, Lajos; — —: Limit theorems for the union-intersection test. 95 JStPlInf 44 133-148

Shao, Qi-man: Kuelbs, J.; Li, W. V.; — —: Small ball probabilities for Gaussian processes with stationary increments under Hölder norms. 95 JTherPrb 8 361-386

Shao, Qi-Man: Peligrad, Magda; — —: Estimation of the variance of partial sums for ρ-mixing random variables. 95 JMultiAn 52 140-157
Peligrad, Magda; — —: A note on the almost sure central limit theorem for weakly dependent random variables. 95 StPrLet 22 131-136

Shao, Yongzhao; Hahn, Marjorie G.: Limit theorems for the logarithm of sample spacings. 95 StPrLet 24 121-132

Shapiro, Alexander: Rubinstein, Reuven Y.; — — (Auth); Goodfriend, Jason H. (Rev): Review of "Discrete event systems: Sensitivity analysis and stochastic optimization by the score function method". 95 Technmcs 37 124-125

Shapiro, Gary: Comment on "Estimation issues in school surveys". 94 ASASrvyRsr 1128-1129

Shapiro, Larry S.; Brady, Michael: Rejecting outliers and estimating errors in an orthogonal outliers and estimating errors in an orthogonal-regression framework. 95 PTLnAMa 350 407-439

Shapiro, Samuel S.: Chen, Ling; — —: An alternative test for normality based on normalized spacings. 95 JStCmpSm 53 269-287

Sharma, Bhu Dev: Koschnick, Klaus-Uwe; — —: A generalization of the Sharma-Kaushik class of metrics. 93 JCISS 18 401-407

Sharma, Divakar: Kumar, Somesh; — —: Estimating the common location. 95 MaOpfStS 26 231-240

Sharma, M. K.: A note on non-orthogonal main effect plans. 95 SankhyaB 57 166-168

Sharp, Dan S.: Grove, John S.; — —; Wilkins, Lynne R.: Regression with random error in X: Increasing power of the test by non-linear adjustment for measurement error. 94 ASAProEpid 143-146

Shaw, Douglas V.: Olkin, Ingram; — —: Meta-analysis and its applications in horticultural science. 95 HortSci 30 1343-1348

Shaw, J.: Aitken, C. G. G.; — —; Talbot, M.: Contaminant detection in the visual inspection of seed samples. 95 ApplStat 44 431-440

Sheaffer, Susan E.; Jarvis, Robert L.: Bias in Canada goose population size estimates from sighting data. 95 JWildlMg 59 464-473

Shearer, Peter: — — (Auth); Raeside, Robert (Rev): Review of "Business forecasting and planning". 95 IntJFor 11 192-193

Sheather, S. J.: McKean, J. W.; — —; Hettmansperger, T. P.: The interpretability of LMS and LTS residual plots. 94 ASAProStEd 17-26
Ruppert, D.; — —; Wand, M. P.: An effective bandwidth selector for local least squares regression. 95 JASA 90 1257-1270

Sheather, Simon J.: George, Karen J.; McKean, Joseph W.; Schucany, William R.; — —: A comparison of confidence intervals from R-estimators in regression. 95 JStCmpSm 53 13-22

Shedler, G. S.: — — (Auth); Ripley, B. D. (Rev): Review of "Regenerative stochastic simulation". 94 BLndnMaS 26 410-411

Sheehan, Nuala: Jennison, Christopher; — —: Theoretical and empirical properties of the genetic algorithm as a numerical optimizer. 95 JCmpGrSt 4 296-318

Sheena, Yo: Unbiased estimator of risk for an orthogonally invariant estimator of a covariance matrix. 95 JJapnStS 25 35-48

Sheer, Andrew: Ellickson, B. (Auth); — — (Rev): Review of "Competitive equilibrium: Theory and applications". 95 JRSS-A 158 347-348

Sheesley, John H.; Thomas, Harold W.; Valenzuela, Carlos A.: Quantal response analysis of relief valve test data. 95 ASQCTekCon 741-748

Sheffield, R. M.: Zarnoch, S. J.; Anderson, R. L.; — —: Using the β-binomial distribution to characterize forest health. 95 CanJForR 25 462-469

Sheil, J.: Bissell, D. (Auth); — — (Rev): Review of "Statistical methods for SPC and TQM". 95 Ststcian 44 279-281

Sheiner, Lewis B.: Fattinger, Karin E.; — —; Verotta, Davide: A new method to explore the distribution of interindividual random effects in non-linear mixed effects models. 95 Biomtrcs 51 1236-1251

Sheldon, Andrew L.; Meffe, Gary K.: Path analysis of collective properties and habitat relationships of fish assemblages in coastal plain streams. 95 JFishRBC 52 23-33

Shelley, Mack C., II: By-elections and government stability in the United Kingdom and the United States: A transfer function approach. 94 ASAProSoSt 71-75

Shema, Sarah J.: Roberson, Paula K.; — —; Mundfrom, Daniel J.; Holmes, Talmage M.: Evaluating change and preference questions using paired Likert scale data. 94 ASAProSoSt 189-191

Shen, Larry Z.: On optimal B-robust influence functions in semiparametric models. 95 AnlsStat 23 968-989

Shen, Larry Zaiqian: Robust estimation for semiparametric exponential mixture models. 95 StSinica 5 333-349

Shen, Samuel S. P.: North, Gerald R.; Kim, Kwang-Y.; — —; Hardin, James W.: Detection of forced climate signals. Part I: Filter theory. 95 JClimate 8 401-408

Shen, Wei-Hsiung: Use of ranked set sampling for test of a normal mean. 94 CalcutSt 44 183-193

Shen, Xiaojun: Hu, Qing; Yang, Jingyu; Zhang, Qian; Liu, Ke; — —: An automatic seal imprint verification approach. 95 PattrnRc 28 1251-1266

Shen, Xiaotong: Wong, Wing Hung; — —: Probability inequalities for likelihood ratios and convergence rates of sieve MLEs. 95 AnlsStat 23 339-362

Shenfild, M.: Kogan, Y.; Liptser, R.; — —: State-dependent Beneš buffer model with fast loading and output rates. 95 AnlsApPr 5 97-120

Shenton, L. R.: Bowman, K. O.; — —: Estimation problems associated with the three parameter Gamma distribution. 95 CommStA 24 1355-1376

Shephard, Neil: De Jong, Piet; — —: The simulation smoother for time series models. 95 Biomtrka 82 339-350

Shepherd, Ursula L.: Mehlman, David W.; — —; Kelt, Douglas A.: Comment on "Stopping rules in principal components analysis: A comparison of heuristical and statistical approaches" (V74 p2204-2214). 95 Ecology 76 640-643

Shepp, L. A.; Shiryaev, A. N.: A new look at pricing of the "Russian option". 94 ThProbAp 39 103-119

Shepp, Larry: Assaf, David; — —; Yao, Yi-Ching: Optimal tracking of Brownian motion. 95 Stochatx 54 233-245

Sheppard, Lianne; Prentice, Ross L.: On the reliability and precision of within- and between-population estimates of relative rate parameters. 95 Biomtrcs 51 853-863

Prentice, Ross L.; — —: Aggregate data studies of disease risk factors. 95 Biomtrka 82 113-125

Shera, David: Lavori, Philip W.; Dawson, Ree; — —: A multiple imputation strategy for clinical trials with truncation of patient data. 95 StatMed 14 1913-1925

Sherali, Hanif D.; Alameddine, Amine; Glickman, Theodore S.: Biconvex models and algorithms for risk management problems. 94 AJMMgSci 14 197-228

Sheridan, John R.: Schnell, Daniel J.; Magee, Elvin; — —: A regression method for analysing ordinal data from intervention trials. 95 StatMed 14 1177-1189

Sherman, Malcolm J.: Dawes, Robyn M. (Auth); — — (Rev): Review of "House of cards: Psychology and psychotherapy built on myth". 95 JASA 90 1492-1492

Sherman, Peter J.: Wikle, Christopher K.; — —; Chen, Tsing-Chang: Identifying periodic components in atmospheric data using a family of minimum variance spectral estimators. 95 JClimate 8 2352-2363

Sherman, Simon A.: Sclove, Stanley L.; — —: Cluster analysis dihedral angles of amino acid residues in proteins. 94 ASAProBiop 399-404

Sherman, Stephanie L.: Feingold, Eleanor; Lamb, Neil E.; — —: Methods for genetic linkage analysis using trisomies. 95 AJHuGene 56 475-483

Sherwin, D. J.; Bossche, A. (Auth); Phillips, Michael J. (Rev): Review of "The reliability, availability and productiveness of systems". 95 Ststcian 44 412-413

Sherwin, David: Høyland, A.; Rausand, M. (Auth); — — (Rev): Review of "System reliability theory - Models and statistical methods". 95 Ststcian 44 543-544

Shettle, Carolyn: Mitchell, Susan; Mooney, Geraldine; — —: The impact of using priority mail in a mixed mode survey. 94 ASAProGovt 88-92

Shettle, Carolyn F.; Guenther, Patricia M.; Kaspryzk, Daniel; Gonzalez, Maria Elena: Investigating nonresponse in federal surveys. 94 ASASrvyRsr 972-976

Sheu, Yuan-Chung: On positive solutions of some nonlinear differential equations - A probabilistic approach. 95 StocProc 59 43-53

Shevchenko, Ivan P.; Lynch, John T.; Mattie, Angela S.; Reed-Fourquet, Lori L.: Verification of information in a large medical database using linkages with external databases. 95 StatMed 14 511-530 (Disc: p687-689)

Sheynin, O.: Helmert's work on the theory of errors. 95 AHistESc 49 73-104

— —: Density curves in the theory of errors. 95 AHistESc 49 163-196

— —: Bertrand's work on probability. 94 AHistESc 48 155-199

— —: Chebyshev's contribution to the theory of probability. 93 AHistESc 46 321-340

Shi, Daoji: Fisher information for a multivariate extreme value distribution. 95 Biomtrka 82 644-649

Shi, Dylan: Miescke, K. J.; — —: Optimum Markov replacement policies for using the most reliable components. 95 JStPlInf 45 331-346

Shi, J.: Rukhin, A. L.; — —: Recursive procedures for multiple decisions: Finite time memory and stepwise maximum likelihood procedure. 95 StHefte 36 155-162

Tyurin, Yu. N.; Yakovlev, Andrej Yu.; — —; Bass, L.: Testing a model of aging in animal experiments. 95 Biomtrcs 51 363-372

Shi, Jian: Multiparametric M-estimation and the random weighting method. 95 ActScNat 30 509-518

— —: Random weighting approximation for the distribution of standardized sample mean-multi dimensional case (Chinese). 95 ChJApPSt 10 233-242

— —; Zheng, Zhongguo: Edgeworth expansion and random weighted approximation in correlated model (Chinese). 95 KToChina 39 1925-1927

Shi, Jianqing: Wei, Bocheng; — —: Local influence analysis for regression transformation models (Chinese). 95 AMaAPknC 17 132-143

Shi, Ning-Zhong: The minimal L_1 isotonic regression. 95 CommStA 24 175-189

Shi, Peide: Asymptotic behaviour of nonparametric conditional quantile estimates for time series. 95 CandJSt 23 161-169

— —: On the rates of convergence of L_1-norm estimates for nonparametric regression (Chinese). 95 JSysSMaS 14 327-332

— —: Some new results of M-type regression spline estimators in a partial linear model. 95 KToChina 39 189-192

— —: Test of nonparametric components in partial linear models (Chinese). 95 KToChina 39 1358-1360

— —; Chen, Xiru: A note on the optimal global rates of convergence of M-estimator for nonparametric regression (Chinese). 95 ChinaAMA 15 171-176

— —; He, Xuming: The optimal rates of convergence on nonparametric quantile regression estimate on dependent data (Chinese). 95 KToChina 39 1537-1539

— —; Li, Guoying: Asymptotic normality of L_1-norm estimators for parametric components in a partly linear model (Chinese). 95 ChinaAMA 15 478-484

— —; Li, Guoying: Asymptotics of the minimum L_1-norm estimates in a partly linear model. 95 JSysSMSE 7 67-77

— —; Li, Guoying: Asymptotic normality of the M-estimator of parametric components in partial linear models (Chinese). 95 KToChina 39 1-2

— —; Li, Guoying: A note on the convergent rates of M-estimates for a partly linear model. 95 MaOpfStS 26 27-47

— —; Lin, Guoying: Global convergence rates of B-spline M-estimators in nonparametric regression. 95 StSinica 5 303-318

Shi, Z.: On the uniform Bahadur-Kiefer representation. 95 MaMethSt 4 39-55

— —: On transient Bessel processes and planar Brownian motion reflected at their future infima. 95 StocProc 60 87-102

— —: Lower limits of iterated Wiener processes. 95 StPrLet 23 259-270

Gruet, J.-C.; — —: Some asymptotic results for exponential functionals of Brownian motion. 95 JAppProb 32 930-940

Hu, Y.; — —: Sojourns and future infima of planar Brownian motion. 95 ProbTher 103 329-348

Hu, Y.; — —: The Csörgő-Révész modulus of non-differentiability of iterated Brownian motion. 95 StocProc 58 267-279

Hu, Y.; Pierre-Loti-Viaud, D.; — —: Laws of the iterated logarithm for iterated Wiener processes. 95 JTherPrb 8 303-319

Rogers, L. C. G.; — —: The value of an Asian option. 95 JAppProb 32 1077-1088

Shi, Zhan; Werner, Wendelin: Asymptotics for occupation times of half-lines by stable processes and perturbed reflecting Brownian motion. 95 Stochatx 55 71-85

Shibata, Takanori: Ishigami, Hideyuki; Fukuda, Toshio; — —; Arai, Fumihito: Structure optimization of fuzzy neural network by genetic algorithm. 95 FuzzySet 71 257-264

Shibata, Yoshisada: Cologne, John B.; — —: Optimal case-control matching in practice. 95 Epidmlgy 6 271-275

Shibayama, Tadashi; Shigemasu, Kazuo: An estimation procedure of parameters in a general item response model (Japanese). 94 JpJBeh 21/1 57-65

Shieh, Narn-Rueih: Some self-similar processes related to local times. 95 StPrLet 24 213-218

Shields, D. C.: Morton, N. E.; — —; Collins, A.: Genetic epidemiology of complex phenotypes. 91 AnHuGene 55 301-314

Shields, Denis C.; Marlow, Angela J.; Houlston, Richard S.; Eccles, Diana M.; Morton, Newton E.: Prediction of genetic risks from segregation analyses of morbid risks. 94 HumHrdty 44 52-55

Shiffler, R.; Adams, A. (Auth); Ziegel, Eric R. (Rev): Review of "Introductory business statistics with computer applications". 95 Technmcs 37 239-240

Shiga, Tokuzo: Cox, J. T.; Greven, Andreas; — —: Finite and infinite systems of interacting diffusions. 95 ProbTher 103 165-197

Shigemasu, K.: Press, S. James; — —: Posterior distribution for the number of factors. 94 ASAProBays 75-77

Shigemasu, Kazuo: Shibayama, Tadashi; — —: An estimation procedure of parameters in a general item response model (Japanese). 94 JpJBeh 21/1 57-65

Ueno, Maomi; Yoshimura, Osamu; — —: A latent trait model for reading process of a word (Japanese). 94 JpJBeh 21/1 48-56

Shih, Frank Y.: Moh, Jenlong; — —: A general purpose model for image operations based on multilayer perceptrons. 95 PattrnRc 28 1083-1090

Shih, Joanna H.: Sample size calculation for complex clinical trials with survival endpoints. 95 ConClnTr 16 395-407

— —; Louis, Thomas A.: Inferences on the association parameter in copula models for bivariate survival data. 95 Biomtrcs 51 1384-1399

— —; Louis, Thomas A.: Assessing gamma frailty models for clustered failure time data. 95 LiDaAnal 1 205-220

Shih, Neng Hui; Song, Wheyming Tina: Estimating simulation metamodels using correlated control variates. 95 CommStB 24 79-90

Shih, Weichung Joseph; Gould, A. Lawrence: Re-evaluating design specifications of longitudinal clinical trials without unblinding when the key response is rate of change. 95 StatMed 14 2239-2248

Shihadeh, Edward S.: Clogg, Clifford C.; — — (Auth); Andrich, David (Rev): Review of "Statistical models for ordinal variables". 95 JASA 90 1131-1131

Clogg, Clifford C.; — — (Auth); Richards, John D. (Rev): Review of "Statistical models for ordinal variables". 95 JQualTek 27 394-394

Clogg, Clifford C.; — — (Auth); Von Tress, Mark (Rev): Review of "Statistical methods for ordinal variables". 95 Technmcs 37 118-119

Shim, Hee-Jeong: Choi, Yong-Seok; — —: Applications of the supplementary principal component analysis for the 1982-1992 Korean pro baseball data (Korean). 95 KorApS 8/1 51-60

Shimberg, Michael Z.: Caldwell, Carol V.; Monsell, Brian C.; Piesto, Nancy A.; — —: Using seasonal adjustment models to correct for panel imbalance in the Monthly Wholesale Trade Survey. 94 ASAProBuEc 83-88

Shimizu, R.: Expansion of the scale mixture of the multivariate normal distribution with error bound evaluated in the L_1-norm. 95 JMultiAn 53 126-138

Shin, Bong Sup: Nam, Ho Soo; Song, Moon Sup; — —: A robust test for the parallelism of two regression lines (Korean). 95 KorApS 8/2 77-86

Shin, Dong Wan; Sarkar, Sahadeb: Estimation of the multivariate autoregressive moving average having parameter restrictions and an application to rotational sampling. 95 JTimSrAn 16 431-444

— —; Sarkar, Sahadeb: Estimation for stationary AR(1) models with nonconsecutively observed samples. 95 SankhyaA 57 287-298

Shin, DongWan: Lee, JongHyup; — —: Testing for a unit root in an ARIMA$(p, 1, q)$ signal observed with measurement error. 95 JKoreaSt 24 481-493

Shin, Hee-Choon: An approximate Rao-Scott modification factor in two-way tables with only known marginal DEFFS. 94 ASASrvyRsr 600-601

Shin, Min-Woong: Park, Taesung; — —: A practical extension of the generalized estimating equation approach for longitudinal data. 95 CommStA 24 2561-2579

Shin, Yang-Kyu: Double bootstrap confidence cones for spherical data based on prepivoting. 95 JKoreaSt 24 183-195

Shindo, Hisakazu: Quarshie, Leonard B.; — —: A control chart for attributes and its applications. 95 ASQCTekCon 378-388

Shinozaki, Nobuo: Some modifications of improved estimators of a normal variance. 95 AnInStMa 47 273-286

Shipley, Margaret F.; de Korvin, André: Rough set theory fuzzy belief functions related to statistical confidence: Application and evaluation for golf course closing. 95 StocAnAp 13 487-502

Shiraishi, T.; Konno, Y.: On construction of improved estimators in multiple-design multivariate linear models under general restriction. 95 AnInStMa 47 665-674

Shirakura, Teruhiro; Tazawa, Shinsei: Search designs for balanced 2^m factorial designs of resolution $2l+1$ when $l+s+1$-factor and higher order interactions are negligible. 93 JCISS 18 233-245

Shirland, Larry E.: — — (Auth); Mullins, Eamonn (Rev): Review of "Statistical quality control with microcomputer applications". 95 Chemolab 28 203-203

Shirley, Eryl A. C.: Correction to "Application of ranking methods to factorial and stratified experiments" (V36 p205-213). 95 ApplStat 44 135-137

Shiryaev, A. N.: Probability (Second edition). 95 Springer:Brln:NY 621p

— —: On some basic concepts and some basic stochastic models used in finance. 94 ThProbAp 39 1-13

— —; Kabanov, Yu. M.; Kramkov, D. O.; Mel'nikov, A. V.: Toward the theory of pricing of options of both European and American types. II. Continuous time.. 94 ThProbAp 39 61-102

— —; Kabanov, Yu. M.; Kramkov, O. D.; Mel'nikov, A. V.: Toward the theory of pricing of options of both European and American types. I. Discrete time.. 94 ThProbAp 39 14-60

Kramkov, D. O.; — —: On the rational pricing of the "Russian option" for the symmetrical binomial model of a (B,S)-market. 94 ThProbAp 39 153-162

Shepp, L. A.; — —: A new look at pricing of the "Russian option". 94 ThProbAp 39 103-119

Shiryayev, Albert N.: Föllmer, Hans; Protter, Philip; — —: Quadratic covariation and an extension of Itô's formula. 95 Bernoull 1 149-169

Shitan, Mahendran; Brockwell, Peter J.: An asymptotic test for separability of a spatial autoregressive model. 95 CommStA 24 2027-2040

Shiu, Elias S. W.: Gerber, Hans U.; — —: Option pricing by Esscher transforms. 95 TrSActur 46 99-140 (Disc: p141-191)

Gerber, Hans U.; — —: From perpetual strangles to Russian options. 94 InsrneME 15 121-126

Shiue, P. J.-S.: Niederreiter, H.; — — (Ed): Monte Carlo and quasi-Monte Carlo methods in scientific computing. 95 Springer:Brln:NY 372p

Shiver, A. W.: Helton, J. C.; Johnson, J. D.; McKay, M. D.; — —; Sprung, J. L.: Robustness of an uncertainty and sensitivity analysis of early exposure results with the MACCS reactor accident consequence model. 95 RelEngSS 48 129-148

Shnidman, David A.: Efficient computation of the circular error probability (CEP) integral. 95 IEEEAuCn 40 1472-1475

Shoemaker, Lewis H.: Tests for differences in dispersion based on quantiles. 95 AmerStat 49 179-182

— —: Two-sample tests for variability based on interquantile spreads. 94 ASAProQlPr 134-139

Shoesmith, Gary L.: Multiple cointegrating vectors, error correction, and forecasting with Litterman's model. 95 IntJFor 11 557-567

Pinder, Jonathan P.; — —: Modelling multivariate cointegrated systems: Insights from non-linear dynamics. 95 JForcast 14 311-324

Shope, P. A.: Hamilton, M. A.; Johnson, K. R.; Camper, A. K.; Stoodley, P.; Harkin, G. J.; Gillis, R. J.; — —: Analysis of bacterial spatial patterns at the initial stage of biofilm formation. 95 BiomtrcJ 37 393-408

Shore, Haim: Fitting a distribution by the first two moments (partial and complete). 95 CmpStDA 19 563-577

— —: Identifying a two-parameter distribution by the first two sample moments (partial and complete). 95 JStCmpSm 52 17-32

Short, Thomas H.; Moriarty, Helene; Cooley, Mary E.: Readability of educational materials for patients with cancer. 95 JStEd 3/2 15kb (jse/v3n2/datasets.short)

— —; Rossman, Allan J.; Hesterberg, Tim C.: Bayesian analyses of the Bradley-Terry model. 94 ASAProBays 117-122 (Disc: p134-135)

Rossman, Allan J.; — —: Conditional probability and education reform: Are they compatible?. 95 JStEd 3/2 23kb (jse/v3n2/rossman)

Shortland, C. F.: Roberts, G. O.; — —: The hazard rate tangent approximation for boundary hitting times. 95 AnlsApPr 5 446-460

Shotter, M. V.: Falk, R. (Auth); — — (Rev): Review of "Understanding probability and statistics: A book of problems". 95 Biomtrcs 51 386-386

Shoukri, M. M.; Martin, S. W.; Mian, I. U. H.: Maximum likelihood estimation of the kappa coefficient from models of matched binary responses. 95 StatMed 14 83-99

Shoukry, E. M.; Gharraph, M. K.; Hassan, N. A.: Multi-channel bi-level queueing system with heterogeneous servers. 94 EgpStJ 38 208-221

Shrestha, Laura B.; Preston, Samuel H.: Consistency of census and vital registration data on older Americans: 1970-1990. 95 SurvyM 21 167-177

Shreve, S. E.: Soner, H. M.; — —; Cvitanić, J.: There is no nontrivial hedging portfolio for option pricing with transaction costs. 95 AnlsApPr 5 327-355

Shrikhande, M. S.: The Delsarte polynomial of 5-(v,k,λ) designs with three intersection numbers. 93 JCISS 18 53-60

Shrikhande, Mohan S.: Ionin, Yury J.; — —: Five-designs with three intersection numbers. 95 JCombThA 69 36-50

Shu, Vincent: Hosmane, Balakrishna; — —; Morris, David: Finite sample properties of van Elteren test. 94 ASAProBiop 430-434

Shubov, Victor: Martin, Clyde; — —: Natural exponential families of probability distributions and exponential-polynomial approximation. 93 AplMaCmp 59 275-297

Shukla, G. K.: On the use of spatial information in the analysis of field experiments. 94 JIAgSt 46 337-348

Shukla, N. D.: Saxena, Sanjay Kumar; Nigam, A. K.; — —: Variance estimation for combined ratio estimator. 95 SankhyaB 57 85-92

Shukla, Praveen: Srivastava, Anil K.; Tracy, Derrick S.; — —: Concentration of estimators of coefficients in mixed regression model. 95 CommStA 24 797-811

Shumway, Robert H.: Jones, R. H. (Auth); — — (Rev): Review of "Longitudinal data with serial correlation: A state-space approach". 95 JStPlInf 47 395-397

Shun, Zhenming; McCullagh, Peter: Laplace approximation of high dimensional integrals. 95 JRSS-B 57 749-760

Shvets, Ye. M.: Velikoivanenko, O. I.; — —: On one discrete queueing model. 95 ThPrMaSt 50 55-60

Shwartz, Adam; Weiss, Alan: Large deviations for performance analysis. Queues, communication, and computing. 95 ChapmnHall:Lnd 576p

Feinberg, Eugene A.; — —: Constrained Markov decision models with weighted discounted rewards. 95 MaOpRsr 20 302-320

Shyur, Huan-Jyh: Luxhoj, James T.; — —: Reliability curve fitting for aging helicopter components. 95 RelEngSS 48 229-234

Siatkowski, I.: Orthogonality in a two-way elimination of heterogeneity design. 93 UtilMath 43 225-233

Siatkowski, Idzi: Some characterizations of a class of balanced two-way elimination of heterogeneity designs. 94 ListyBio 31 27-33

Sibisi, Sibusiso: Skilling, John; — —: Beyond maximum entropy. 94 ASAProStCp 71-79

Sibuya, Masaaki; Yamato, Hajime: Ordered and unordered random partitions of an integer and the GEM distribution. 95 StPrLet 25 177-183

Iiyama, Yuriko; Nishimura, Kazuo; — —: Power of record-breaking test (Japanese). 95 JpJAppSt 24 13-26

Ogawa, Naoko; Minami, Mihoko; — —: Characteristics of the linear smoothing spline (Japanese). 95 JpJAppSt 24 27-41

Siciliano, Roberta; van der Heijden, Peter G. M.: Simultaneous latent budget analysis of a set of two-way tables with constant-row-sum data. 94 Metron 52/1 155-180

Sid-Ahmed, M. A.: Abou-Nasr, M. A.; — —: Fast learning and efficient memory utilization with a prototype based neural classifier. 95 PattrnRc 28 581-593

Sidi, Moshe: Birman, Alexander; Gail, H. Richard; Hantler, Sidney L.; Rosberg, Zvi; — —: An optimal service policy for buffer systems. 95 JACM 42 641-657

Sidoravícus, Vladas; Vares, Maria Euália: Ergodicity of Spitzer's renewal model. 95 StocProc 55 119-130

Sidow, Arend; Nguyen, Trang; Speed, Terence P.: Estimating the fraction of invariable codons with a capture-recapture method. 92 JMolEvol 35 253-260

Siegal, Harvey A.: Wang, Jichuan; Fisher, James H.; — —; Falck, Russel S.; Carlson, Robert G.: Influence of measurement errors on HIV risk behavior analysis: A case study examining condom use among drug users. 95 StrEqMod 2 319-334

Siegel, Andrew F.; Morgan, Charles J.: Statistics and data analysis: An introduction (Second edition). 95 Wiley:NY:UK 656p

Siegmund, D.; Venkatraman, E. S.: Using the generalized likelihood ratio statistic for sequential detection of a change-point. 95 AnlsStat 23 255-271

Siegmund, David: Dupuis, Josée; Brown, Patrick O.; — —: Statistical methods for linkage analysis of complex traits from high-resolution maps of identity by descent. 95 Genetcs 140 843-856

Siegmund, David O.; Worsley, Keith J.: Testing for a signal with unknown location and scale in a stationary Gaussian random field. 95 AnlsStat 23 608-639

Sieniawska, R.; Śniady, P.: Design sensitivity of structures under non-Gaussian excitation. 95 ZAngMaMk 75 573-574S

Sievers, Gerald L.: Rao, Sudhakar H.; — —: A robust partial correlation measure. 95 JNnparSt 5 1-20

Sifi, Mohamed: Central limit theorem and infinitely divisible probabilities associated with partial differential operators. 95 JTherPrb 8 475-499

Sigman, Karl: Stationary marked point processes: An intuitive approach. 95 ChapmnHall:Lnd 200p

Bardhan, Indrajit; — —: Stationary regimes for inventory processes. 95 StocProc 56 77-86

Sigman, Richard: Zayatz, Laura; — —: Feasibility study of the use of Chromy's algorithm in Poisson-sample selection for the annual survey of manufactures. 94 ASASrvyRsr 641-646

Sijtsma, Klaas: Post, W. J. (Auth); — — (Rev): Review of "Nonparametric unfolding models. A latent structure approach". 95 JClass 12 153-156

Sikkel, D.; Hoogendoorn, A. W.: Models for monthly penetrations with incomplete panel data. 95 StNeerla 49 378-391

Siklos, Pierre L.: Granger, C. W. J.; — —: Systematic sampling, temporal aggregation, seasonal adjustment, and cointegration. Theory and evidence. 95 JEconmtx 66 357-369

Sil'vestrov, D. S.: Velikii, Yu. A.; Motsa, A. I.; — —: Dual processes and ergodic type theorems for Markov chains in the scheme of series. 94 ThProbAp 39 642-653

Silber, Jeffrey H.; Rosenbaum, Paul R.; Ross, Richard N.: Comparing the contributions of groups of predictors: Which outcomes vary with hospital rather than patient characteristics?. 95 JASA 90 7-18

Silberg, J.: Pickles, A.; Neale, M.; Simonoff, E.; Rutter, M.; Hewitt, J.; Meyer, J.; Crouchley, R.; — —; Eaves, L.: A simple method for censored age-of-onset data subject to recall bias: Mothers' reports of age of puberty in male twins. 94 BehavGen 24 457-468

Silberg, J. L.: Eaves, L. J.; — —; Hewitt, K.; Rutter, M.; Meyer, J. M.; Neale, M. C.; Pickles, A.: Analyzing twin resemblance in multisymptom data: Genetic applications of a latent class model for symptoms of conduct disorder in juvenile boys. 93 BehavGen 23 5-20

Silberman, Geoffrey L.: Maryak, John L.; Spall, James C.; — —: Uncertainties for recursive estimators in nonlinear state-space models, with applications to epidemiology. 95 Autmatca 31 1889-1892

Silman, Alan J.: Epidemiological studies. A practical guide. 95 CambridgeUniv 189p

Siltanen, Janet: Blackburn, Robert M.; — —; Jarman, Jennifer: The measurement of occupational gender segregation: Current problems and a new approach. 95 JRSS-A 158 319-331

Silva Mato, A.; Martín Andrés, A.: Optimal unconditional tables for comparing two independent proportions. 95 BiomtrcJ 37 821-836

Silvapulle, Mervyn J.: A Hotelling's T^2-type statistic for testing against one-sided hypotheses. 95 JMultiAn 55 312-319

— —; Silvapulle, Paramsothy: A score test against one-sided alternatives. 95 JASA 90 342-349

Silvapulle, Paramsothy: Silvapulle, Mervyn J.; — —: A score test against one-sided alternatives. 95 JASA 90 342-349

Silver, M.: The role of footnotes in a statistical metainformation system. 93 StatJUN 10 153-170

Silver, Mick: Scales of measurement and cluster analysis: An application concerning market segments in the babyfood market. 95 Ststcian 44 101-112

— —; Ioannidis, Christos: European inflation and its variability. 95 Ststcian 44 309-322

Silver, N. Clayton; Wadiak, Diane L.; Massey, Catherine J.: A Microsoft FORTRAN 77 program for testing the differences among independent first-order partial correlations. 95 EdPsMeas 55 245-248

Silver, R. N.; Martz, H. F.: Applications of quantum entropy to statistics. 94 ASAProStCp 61-70

Silverberg, Arthur: Wang, Chamont; — —: Comment on "Statistical values, quality, and certification" (V48 p65-87). 95 AmerStat 49 116-117

Silverman, B. W.: Incorporating parametric effects into functional principal components analysis. 95 JRSS-B 57 673-689

Stander, J.; — —: Minimax estimation of linear functionals, particularly in nonparametric regression and positron emission tomography. 95 CmpStQ 10 259-283

Scott, D. W. (Auth); — — (Rev): Review of "Multivariate density estimation". 95 Metrika 42 66-68

Green, P. J.; — — (Auth); Lange, Nicholas (Rev): Review of "Nonparametric regression and generalized linear models: A roughness penalty approach". 95 JASA 90 1489-1489

Green, P. J.; — — (Auth); Qui, Chunfu (Rev): Review of "Nonparametric regression and generalized linear models: A roughness penalty approach". 95 Technmcs 37 342-343

Silverstein, Jack W.: Strong convergence of the empirical distribution of eigenvalues of large dimensional random matrices. 95 JMultiAn 55 331-339

— —; Bai, Z. D.: On the empirical distribution of eigenvalues of a class of large dimensional random matrices. 95 JMultiAn 54 175-192

— —; Choi, Sang-Il: Analysis of the limiting spectral distribution of large dimensional random matrices. 95 JMultiAn 54 295-309

Sim, Ah Boon: Fisher, Gordon; — —: Some finite sample theory for bootstrap regression estimates. 95 JStPlInf 43 289-300

Sim, Songyong: Johnson, Richard A.; — —: Nonparametric tests for ordered c-sample contaminated distributions. 95 CommStA 24 861-879

Simakani, J. E.: Ikeda, N.; Elworthy, K. D. (Auth); — — (Rev): Review of "Asymptotic problems in probability theory: Wiener functionals and fractionals". 95 Ststcian 44 289-289

Simar, L.: Korostelev, A. P.; — —; Tsybakov, A. B.: Efficient estimation of monotone boundaries. 95 AnlsStat 23 476-489

Simeone, Bruno: Maravalle, Maurizio; — —: A spanning tree heuristic for regional clustering. 95 CommStA 24 625-639

Maravalle, Maurizio; — —: Correction to "A spanning tree heuristic for regional clustering" (V24 p625-639). 95 CommStA 24 2437-2437

Simioni, M.: Delecroix, M.; — —; Thomas-Agnan, C.: A shape constrained smoother: Simulation study. 95 CmpStQ 10 155-175

Simkins, Scott: Forecasting with vector autoregressive (VAR) models subject to business cycle restrictions. 95 IntJFor 11 569-583

Simó, Amelia: Ayala, Guillermo; — —: Bivariate random closed sets and nerve fibre degeneration. 95 AdvAppPr 27 293-305

Simon, Burton: Fleming, Philip J.; Schaeffer, Dennis; — —: Efficient Monte-Carlo simulation of a product-form model for a cellular system with dynamic resource sharing. 95 ACMMdSim 5 3-21

Simon, Richard: Faraggi, David; — —: Maximum likelihood neural network prediction models. 95 BiomtrcJ 37 713-725

Faraggi, David; — —: The maximum likelihood neural network as a statistical classification model. 95 JStPlInf 46 93-104

Faraggi, David; — —: A neural network model for survival data. 95 StatMed 14 73-82

Strauss, Nurith; — —: Investigating a sequence of randomized phase II trials to discover promising treatments. 95 StatMed 14 1479-1489

Simon, Richard M.: Thall, Peter F.; — —; Estey, Elihu H.: Bayesian sequential monitoring designs for single-arm clinical trials with multiple outcomes. 95 StatMed 14 357-379

Simon, Ronald: Wartenberg, Daniel; — —: Integrating epidemiologic data into risk assessment. 95 AJPubHlh 85 491-493

Simonelli, Italo: A limit theorem for a class of interacting particle systems. 95 AnlsProb 23 141-156

Simonian, Alain: Guillemin, Fabrice; — —: Transient characteristics of an M/M/∞ system. 95 AdvAppPr 27 862-888

Simonoff, E.: Pickles, A.; Neale, M.; — —; Rutter, M.; Hewitt, J.; Meyer, J.; Crouchley, R.; Silberg, J.; Eaves, L.: A simple method for censored age-of-onset data subject to recall bias: Mothers' reports of age of puberty in male twins. 94 BehavGen 24 457-468

Simonoff, Jeffrey S.: The anchor position of histograms and frequency polygons: Quantitative and qualitative smoothing. 95 CommStB 24 691-710

— —: Smoothing categorical data. 95 JStPlInf 47 41-69

— —: A simple, automatic and adaptive bivariate density estimator based on conditional densities. 95 StatComp 5 245-252

Chatterjee, Samprit; Handcock, Mark S.; — —: A casebook for a first course in statistics and data analysis. 95 Wiley:NY:UK 314p

Dong, Jianping; — —: A geometric combination estimator for d-dimensional ordinal sparse contingency tables. 95 AnlsStat 23 1143-1159

Hadi, Ali S.; — —: Reply to comment on "Procedures for the identification of multiple outliers in linear models". 95 JASA 90 811-812

Simonot, F.: Approximation of the stationary distribution of a stochastically monotone Markov chain (French). 95 StocProc 56 133-149

Simons, Gordon: Csörgő, Sándor; — —: Precision calculation of distributions for trimmed sums. 95 AnlsApPr 5 854-873

Simpson, James R.; Keats, J. Bert: Trade-off analysis versus constrained optimization with an economic control chart model. 95 IIETrn 27 765-772

— —; Keats, J. Bert: A sensitivity study and comparison of the CUSUM versus Shewhart control charts using an economic model. 94 ASAProQlPr 98-103

Simpson, Judy M.: Teaching statistics to non-specialists. 95 StatMed 14 199-208

— —; Klar, Neil; Donner, Allan: Accounting for cluster randomization: A review of primary prevention trials, 1990 through 1993. 95 AJPubHlh 85 1378-1383

Sinai, Y.: — — (Auth); Haughton, D. (Transl): Probability theory: An introductory course. 95 Springer:Brln:NY 138p

— — (Auth); Haughton, D. (Transl); Sedgwick, Philip (Rev): Review of "Probability theory: An introductory course". 95 Ststcian 44 142-142

Sinai, Yakov G.: — — (Auth); Jacobs, K. (Rev): Review of "Probability theory". 95 Metrika 42 77-78

Sinclair, Michael D.; Gastwirth, Joseph L.: On designs for measuring prevalence rates and error rates from diagnostic testing procedures. 94 ASAProBiop 148-153

Dillman, Don A.; Clark, Jon R.; — —: How prenotice letters, stamped return envelopes and reminder postcards affect mailback response rates for census questionnaires. 95 SurvyM 21 159-165

Singer, Eleanor: Discussion of "What respondents learn from questionnaires: The survey interview and the logic of conversation". 95 IntStRvw 63 169-172

Dillman, Don A.; — —; Clark, Jon R.; Treat, James B.: Effects of benefits appeals, mandatory appeals and variations in statements of confidentiality on completion rates for census questionnaires. 94 ASASrvyRsr 1333-1338

Singer, Hermann: Analytical score function for irregularly sampled continuous time stochastic processes with control variables and missing values. 95 EconTher 11 721-735

Singer, J. M.: Sen, P. K.; — — (Auth); MacGillivray, H. (Rev): Review of "Large sample methods in statistics". 95 Biomtrcs 51 381-382

Sen, P. K.; — — (Auth); Schmitz, N. (Rev): Review of "Large sample methods in statistics". 95 Metrika 42 274-274

Singer, Judith D.: Willett, John B.; — —: It's déjà vu all over again: Using multiple-spell discrete-time survival analysis. 95 JEdStat 20 41-67

Singh, A. C.; Rao, J. N. K.: On the adjustment of gross flow estimates for classification error with application to data from the Canadian Labour Force Survey. 95 JASA 90 478-488

Singh, Bahadur; Schell, Michael J.; Wright, F. T.: Approximations to the powers of the likelihood ratio tests: The loop ordering and slippage alternatives. 95 CommStB 24 91-109

Singh, Gopal K.: Hoyert, Donna L.; — —; Rosenberg, Harry M.: Sources of data on socio-economic differential mortality in the United States. 95 JOffStat 11 233-260

Singh, H. P.; Biradar, R. S.: A class of unbiased ratio estimators in two-phase sampling. 94 Statstca 54 349-359

Singh, Harshinder: Vijayasree, G.; Misra, Neeraj; — —: Componentwise estimation of ordered parameters of k (≥ 2) exponential populations. 95 AnInStMa 47 287-307

Singh, Housila P.: Katyar, N. P.; — —; Gangwar, D. K.: A class of almost unbiased regression-type estimators for finite population mean applying jackknife technique. 94 JIAgSt 46 183-189

Singh, Rajesh; — —: A Hartley-Ross type estimator for finite population mean when the variables are negatively correlated. 95 Metron 53 205-216

Singh, Jagbir; Kathuria, O. P.: Sampling on two occasions with two correlated characters. 94 JIAgSt 46 166-174

Singh, Karan: Rodgers, Jack; — —: Using microsimulation modeling techniques to estimate the effects of health reform on individual businesses. 94 ASAProGovt 39-44 (Disc: p51-52)

Singh, Karen P.: Fraser, Martin D.; Bartolucci, Alfred A.; Smith, William A.; — —: A Bayesian decision procedure for selecting prognostic variables associated with survival for data in which censoring is prevalent. 95 BiomtrcJ 37 463-479

Singh, Kesar: Liu, Regina Y.; — —: Using i.i.d. bootstrap inference for general non-i.i.d. models. 95 JStPlInf 43 67-75

Singh, M.; Ceccarelli, S.; Hamblin, J.: Estimation of heritability from varietal trials data. 93 ThApGene 86 437-441

— —; Hinkelmann, K.: Partial diallel crosses in incomplete blocks. 95 Biomtrcs 51 1302-1314

— —; Saxena, N. P.; Saxena, M. C.: Estimation of pinnule area in chickpea leaves: An errors-in-variables approach. 95 JAppStat 22 37-43

Singh, M. P.; Tambay, J. L.; Krawchuk, S.: The National Population Health Survey: Design and issues. 94 ASASrvyRsr 803-808

Singh, N.: Stochastic modeling of aggregates and products of variable failure rates. 95 IEEEReli 44 279-284

Singh, R.; Singh, S.; Mangat, N. S.; Tracy, D. S.: An improved two stage randomized response strategy. 95 StHefte 36 265-271

Mahajan, P. K.; Gupta, J. P.; — —: Determination of optimum strata boundaries for scrambled randomized response. 94 Statstca 54 375-381

Mangat, N. S.; — —; Singh, S.; Bellhouse, D. R.; Kashani, H. B.: On efficiency of using distinct respondents in a randomized response survey. 95 SurvyM 21 21-23

Singh, S.; Mangat, N. S.; — —: On estimation of mean/total of stigmatized quantitative variable. 94 Statstca 54 383-386

Singh, R. K.; Pandey, S. K.; Srivastava, V. K.: Correction to "A generalized class of shrinkage estimators in linear regression when disturbances are not normal" (V23 p2029-2046). 95 CommStA 24 859-859

Singh, R. Karan; Agnihotri, Monica; Verma, Sanjay: A generalized class of estimators utilizing Lindley-like mean correction in linear regression model. 94 Metron 52/3 147-160

— —; Chaturvedi, Ajit; Kumar, Somesh: Two stage and sequential procedures for estimating a linear function of k multinormal means. 94 JIndnStA 32 1-8

— —; Misra, Sheela; Pandey, S. K.: A generalized class of estimators in linear regression models with multivariate-t distributed error. 95 StPrLet 23 171-178

Singh, R. S.: Empirical Bayes linear loss hypothesis testing in a non-regular exponential family. 95 JStPlInf 43 107-120

Singh, Rajendra P.; Petroni, Rita J.; Allen, Tiwanda M.: Oversampling in panel surveys. 94 ASASrvyRsr 674-679 (Disc: p680-682)

Mulry, Mary H.; — —: New applications of sampling estimation in the 1995 Census Test. 94 ASASrvyRsr 742-747 (Disc: p754-756)

Singh, Rajesh; Singh, Housila P.: A Hartley-Ross type estimator for finite population mean when the variables are negatively correlated. 95 Metron 53 205-216

— —; Upadhyay, S. K.; Singh, Umesh: Bayes estimators for two parameter exponential distribution. 95 CommStA 24 227-240

Singh, Randhir: Goyal, R. C.; — —; Chhikara, R. S.: Estimation of crop yield using post-stratification based on satellite data. 94 JIAgSt 46 210-222

Singh, Ravindra; Singh, Sarjinder; Mangat, Naurang Singh: Mail survey design for sensitive quantitative variable. 95 Metron 53 43-54

Singh, S.; Mangat, N. S.; Singh, R.: On estimation of mean/total of stigmatized quantitative variable. 94 Statstca 54 383-386

Mangat, N. S.; Singh, R.; — —; Bellhouse, D. R.; Kashani, H. B.: On efficiency of using distinct respondents in a randomized response survey. 95 SurvyM 21 21-23

Singh, R.; — —; Mangat, N. S.; Tracy, D. S.: An improved two stage randomized response strategy. 95 StHefte 36 265-271

Singh, Sarjinder: Unrelated question randomized response sampling using continuous distributions. 94 JIAgSt 46 349-361

Singh, Ravindra; — —; Mangat, Naurang Singh: Mail survey design for sensitive quantitative variable. 95 Metron 53 43-54

Singh, Suresh: Ravishankar, Krishnamurthi; — —: Central limit theorem for time to broadcast in radio networks. 95 PrEngInf 9 201-209

Singh, Umesh: Singh, Rajesh; Upadhyay, S. K.; — —: Bayes estimators for two parameter exponential distribution. 95 CommStA 24 227-240

Singh, V.: Dube, M.; — —: Predictive efficiency of improved estimators in restricted regression models. 95 Test 4 323-331

Singhi, N. M.: Arasu, K. T.; Ray-Chaudhuri, D. K.; — —: Simple designs. 93 JCISS 18 130-135

Singpurwalla, N. D.: Gatsonis, C.; Hodges, J. S.; Kass, R. E.; — — (Ed): Case studies in Bayesian statistics. Volume II. 95 Springer:Brln:NY 364p

Singpurwalla, Nozer D.: Survival in dynamic environments. 95 StatSci 10 86-103

— —; Wilson, Simon P.: The exponentiation formula of reliability and survival: Does it always hold?. 95 LiDaAnal 1 187-194

— —; Wilson, Simon P.: Models for assessing the reliability of computer software. 92 ProIrStCl 177-194

Campodónico, Sylvia; — —: Inference and predictions from Poisson point processes incorporating expert knowledge. 95 JASA 90 220-226

Sinha, A. K.: Kaur, A.; Patil, G. P.; — —; Taillie, C.: Ranked set sampling: An annotated bibliography. 95 EnvEcoSt 2 25-54

Patil, G. P.; — —; Taillie, C.: Finite population corrections for ranked set sampling. 95 AnInStMa 47 621-636

Sinha, Atanu R.; Buchanan, Bruce S.: Assessing the stability of principal components using regression. 95 Psymtrka 60 355-369

Sinha, B. K.: Hedayat, A. S.; — — (Auth); Rust, Keith (Rev): Review of "Design and inference in finite population sampling". 94 JOffStat 10 226-228

Sinha, Bikas K.; Sinha, Himal K.: An application of bivariate exponential models and related inference. 95 JStPlInf 44 181-191

Pal, Nabendu; — —; Chaudhuri, Gopal; Chang, Ching-Hui: Estimation of a multivariate normal mean vector and local improvements. 95 MaOpfStS 26 1-17

Shah, K. R.; — —: Optimal designs with component-wise orthogonal row-column structures. 93 JCISS 18 68-78

Sinha, Bikas Kumar: Heiligers, Berthold; — —: Optimality aspects of Agrawal's designs: Part II. 95 StSinica 5 599-604

Pukelsheim, Friedrich; — —: Optimal block designs revisited: An approximate theory detour. 95 StPrLet 24 375-380

Sinha, Bimal K.: Fei, Heliang; — —; Wu, Zhong: Estimation of parameters in two-parameter Weibull and extreme-value distributions using ranked set sample. 94 JStatRsr 28 149-161

Li, Teng; — —: Tests of ordered hypotheses for gamma scale parameters. 95 JStPlInf 43 311-321

Li, Teng; — —: Tests of ordered hypotheses for gamma scale parameters. 95 JStPlInf 45 387-397

Sinha, Himal K.: Sinha, Bikas K.; — —: An application of bivariate exponential models and related inference. 95 JStPlInf 44 181-191

Sinha, R. K.: Madan, V. K.; — —: Digital smoothing of census data employing Fourier transforms. 95 CmpStDA 20 285-294

Sinha, S. K.; Guttman, Irwin: Acceptance sampling in presence of misclassification - A Bayesian approach. 94 JIndnStA 32 83-93

Sirken, Monroe G.: Contributions of statistical statespersons. 94 ASAProGovt 3-6

Sison, Christina P.; Glaz, Joseph: Correction to "Simultaneous confidence intervals and sample size determination for multinomial proportions" (V90 p366-369). 95 JASA 90 1494-1494

Sison, Cristina P.; Glaz, Joseph: Simultaneous confidence intervals and sample size determination for multinomial proportions. 95 JASA 90 366-369

Sitnikova, Tatyana; Rzhetsky, Andrey; Nei, Masatoshi: Interior-branch and bootstrap tests of phylogenetic trees. 95 MolBioEv 12 319-333

Sitter, R. R.: Rao, J. N. K.; — —: Variance estimation under two-phase sampling with application to imputation for missing data. 95 Biomtrka 82 453-460

Sitter, Randy R.; Torsney, Ben: Optimal designs for binary response experiments with two design variables. 95 StSinica 5 405-419

Sivazlian, B. D.: Gakis, K. G.; Rhee, H. K.; — —: Distributions and first moments of the busy and idle periods in controllable M/G/1 queueing models with simple and dyadic policy. 95 StocAnAp 13 47-81

Sjöberg, J.; Ljung, L.: Overtraining, regularization and searching for a minimum, with application to neural networks. 95 IJContrl 62 1391-1407

Sjöberg, Jonas; Zhang, Qinghua; Ljung, Lennart; Benveniste, Albert; Delyon, Bernard; Glorennec, Pierre-Yves; Hjalmarsson, Håkan; Juditsky, Anatoli: Nonlinear black-box modeling in system identification: A unified overview. 95 Autmatca 31 1691-1724

Juditsky, Anatoli; Hjalmarsson, Håkan; Benveniste, Albert; Delyon, Bernard; Ljung, Lennart; — —; Zhang, Qinghua: Nonlinear black-box models in system identification: Mathematical foundations. 95 Autmatca 31 1725-1750

Sjöström, Michael; Rännar, Stefan; Wieslander, Åke: Polypeptide sequence property relationships in Escherichia coli based on auto cross covariances. 95 Chemolab 29 295-305

Lindgren, Fredrik; Geladi, Paul; Berglund, Anders; — —; Wold, Svante: Interactive variable selection (IVS) for PLS. Part II: Chemical applications. 95 JChemtrc 9 331-342

Skalski, John R.: Hoffmann, Annette; — —: Inferential properties of an individual-based survival model using release-recapture data: Sample size, validity and power. 95 JAppStat 22 579-595

Skeels, Christopher L.: Some exact results for estimators of the coefficients on the exogenous variables in a single equation. 95 EconTher 11 484-497

: Instrumental variables estimation in misspecified single equations. 95 EconTher 11 498-529

— —; Taylor, Larry W.: On a simultaneous equations pre-test estimator. 95 JEconmtx 68 269-286

Skiadas, C. H.: Janssen, J.; — —: Dynamic modelling of life table data. 95 AplStMDA 11 35-49

Katsamaki, A.; — —: Analytic solution and estimation of parameters on a stochastic exponential model for a technological diffusion process. 95 AplStMDA 11 59-75

Skiadas, Christos H.: Janssen, Jacques; — —; Zopounidis, Constantin (Ed): Advances in stochastic modelling and data analysis. 95 Kluwer:MA:Hllnd 440p

Skilling, John; Sibisi, Sibusiso: Beyond maximum entropy. 94 ASAProStCp 71-79

Skinner, C. J.: Sample models and weights. 94 ASASrvyRsr 133-142

Nascimento Silva, P. L. D.; — —: Estimating distribution functions with auxiliary information using poststratification. 95 JOffStat 11 277-294

Skinner, Chris; Marsh, Catherine; Openshaw, Stan; Wymer, Colin: Disclosure control for census microdata. 94 JOffStat 10 31-51

Skinner, Jonathan B.: Crossover trials with asymmetric covariance within subject. 94 ASAProBiop 411-413

Skinner, Wilbert R.: Wu, Dong L.; Hays, Paul B.; — —: A least squares method for spectral analysis of space-time series. 95 JAtmsphr 52 3501-3511

Sklar, Michael G.; Armstrong, Ronald D.: A piecewise linear approximation procedure for L_p norm curve fitting. 95 JStCmpSm 52 323-335

Skolnick, Allison M.: Geske, Mark X.; Godbole, Anant P.; Schaffner, Andrew A.; — —; Wallstrom, Garrick L.: Compound Poisson approximations for word patterns under Markovian hypotheses. 95 JAppProb 32 877-892

Skorokhod, A. V.; Yadrenko, M. I.: I. I. Gikhman: A brief review of scientific and pedagogic activity. 95 ThPrMaSt 50 1-5

Hoppensteadt, F.; Salehi, H.; — —: Randomly perturbed Volterra integral equations and some applications. 95 Stochatx 54 89-125

Krylov, N. V.; Kuznetsov, S. E.; — —; Yushkevich, A. A.: E. B. Dynkin's 70th birthday. 94 ThProbAp 39 654-656

Skrzypek, J.; Karplus, W. (Ed); Sethi, Ishwar K. (Rev): Review of "Neural networks in vision and pattern recognition". 95 JClass 12 140-142

Slade, G.: Brydges, D. C.; — —: The diffusive phase of a model of self-interacting walks. 95 ProbTher 103 285-315

Slanta, John; Krenzke, Thomas: Applying the Lavallée and Hidiroglou method to obtain stratification boundaries for the Census Bureau's Annual Capital Expenditures Survey. 94 ASASrvyRsr 693-698

Slatkin, Montgomery: An exact test for neutrality based on the Ewens sampling distribution. 94 GeneRsr 64 71-74

Kirkpatrick, Mark; — —: Searching for evolutionary patterns in the shape of a phylogenetic tree. 93 Evolutn 47 1171-1181

Sleeper, Lynn: Lipsitz, Stuart R.; Fitzmaurice, Garret M.; — —; Zhao, L. P.: Estimation methods for the joint distribution of repeated binary observations. 95 Biomtrcs 51 562-570

Sloan, Jeff A.; Schwarz, Carl J.; Neden, Linda R.: Dataset on silver-maple-tree growth regulators. 95 CandJSt 23 325-329 (Disc: p329-332)

Slud, Eric: Bose, Sudip; — —: Maximin efficiency-robust tests and some extensions. 95 JStPlInf 46 105-121

Koutsoukos, Antonis; — —: Inaccuracy rates and Hodges-Lehmann large deviation rates for parametric inferences with nuisance parameters. 95 JStPlInf 48 47-68

Small, Christopher G.; McLeish, D. L. (Auth); Marshall, Guillermo; Allende, Renato (Rev): Review of "Hilbert space methods in probability and statistical inference". 95 JASA 90 1491-1492

Small, G. W.: Shaffer, R. E.; — —; Combs, R. J.; Knapp, R. B.; Kroutil, R. T.: Experimental design protocol for the pattern recognition analysis of bandpass filtered Fourier transform infrared interferograms. 95 Chemolab 29 89-108

Smallman-Raynor, M. R.: Cliff, A. D.; Haggett, P.; — —; Stroup, D. F.; Williamson, G. D.: The application of multidimensional scaling methods to epidemiological data. 95 StMedRes 4 102-123

Smeeton, Nigel: Selvin, Steve (Auth); — — (Rev): Review of "Practical biostatistical methods". 95 AJEpidm 141 229-230

Smeyers-Verbeke, J.: Vankeerberghen, P.; — —; Leardi, R.; Karr, C. L.; Massart, D. L.: Robust regression and outlier detection for non-linear models using genetic algorithms. 95 Chemolab 28 73-87

Smilde, Age: Tauler, Romá; — —; Kowalski, Bruce: Selectivity, local rank, three-way data analysis and ambiguity in multivariate curve resolution. 95 JChemtrc 9 31-58

Smilde, Age K.: Kiers, Henk A. L.; — —: Some theoretical results on second-order calibration methods for data with and without rank overlap. 95 JChemtrc 9 179-195

Smith, A. F. M.: Green, E. J.; — —; Strawderman, W. E.: Bayesian construction of thematic maps from satellite imagery. 94 ASASrvyRsr 188-193

Gutiérrez-Peña, E.; — —: Conjugate parameterizations for natural exponential families. 95 JASA 90 1347-1356

Stephens, D. A.; — —: Bayesian inference in multipoint gene mapping. 93 AnHuGene 57 65-82

Vounatsou, P.; — —: Bayesian analysis of ring-recovery data via Markov chain Monte Carlo simulation. 95 Biomtrcs 51 687-708

Bernardo, J. M.; — — (Auth); Lindley, D. V. (Rev): Review of "Bayesian theory". 95 JRSS-A 158 340-341

Freeman, P. R.; — — (Ed); Queen, Catriona (Rev): Review of "Aspects of uncertainty: A tribute to Dennis Lindley". 95 JAppStat 22 541-541

Smith, Adrian F. M.: Damien, Paul; Laud, Purushottam W.; — —: Approximate random variate generation from infinitely divisible distributions with applications to Bayesian inference. 95 JRSS-B 57 547-563

Bernardo, José M.; — — (Auth); Piccinato, Ludovico (Rev): Review of "Bayesian theory". 94 JItalStS 3 155-160

Smith, C.: Zhang, W.; — —: Simulation of marker-assisted selection utilizing linkage disequilibrium: The effects of several additional factors. 93 ThApGene 86 492-496

Zhang, W.; — —: Computer simulation of marker-assisted selection utilizing linkage disequilibrium. 92 ThApGene 83 813-820

Smith, C. A. B.: Edwards, A. W. F. (Auth); — — (Rev): Review of "Likelihood (Expanded edition)". 93 AnHuGene 57 315-321

Smith, D. M.: Iterative refinement of parameter estimates and residuals from a regression model fitted using QR decomposition methods. 95 CmpStDA 19 327-349

— —; Ridout, M. S.: Comment on "Algorithm AS 189: Maximum likelihood estimation of the parameters of the beta binomial distribution" (V32 p196-204). 95 ApplStat 44 545-547

Smith, David R.; Conroy, Michael J.; Brakhage, David H.: Efficiency of adaptive cluster sampling for estimating density of wintering waterfowl. 95 Biomtrcs 51 777-788

— —; Reinecke, Kenneth J.; Conroy, Michael J.; Brown, Michael W.; Nassar, James R.: Factors affecting visibility rate of waterfowl surveys in the Mississippi Alluvial Valley. 95 JWildlMg 59 515-527

Smith, Dennis E.; Burns, Kevin C.: Estimating a percentile of a contaminant concentration distribution when all observations are less than a detection limit. 94 ASAProEnvr 139-141

Smith, Eric P.; Rose, Kenneth A.: Model goodness-of-fit analysis using regression and related techniques. 95 EcolgMdl 77 49-64

See, K.; — —: Multiplicative effects in three-way analysis of variance. 94 ASAProStCp 225-230

Smith, Fraser: Mixed model analysis of incomplete longitudinal data from a high-dose trial of tacrine (Cognex) in Alzheimer's patients. 94 ASAProBiop 301-306

Smith, Fraser B.; Helms, Ronald W.: EM mixed model analysis of data from informatively censored normal distributions. 95 Biomtrcs 51 425-436

Smith, Fred: Bartholomew, David; Moore, Peter; — —; Allin, Paul: The measurement of unemployment in the UK. 95 JRSS-A 158 363-405 (Disc: p405-417)

Smith, J. Barry: Breslaw, Jon A.; — —: A simple and efficient method for estimating the magnitude and precision of welfare changes. 95 JApEcmtx 10 313-327

Breslaw, Jon A.; — —: Measuring welfare changes when quantity is constrained. 95 JBES 13 95-103

Smith, J. T.: Venables, W. N.; Ripley, B. D. (Auth); — — (Rev): Review of "Modern applied statistics with S-Plus". 95 ShrtBkRv 15/2 24-24

Smith, Jeremy; McAleer, Michael: Alternative procedures for converting qualitative response data to quantitative expectations: An application to Australian manufacturing. 95 JApEcmtx 10 165-185

— —; McAleer, Michael: The performance of alternative estimators in models with generated regressors when the expectations equation has reduced explanatory power. 95 MaCmpSim 39 343-346

Smith, Ken W.: Non-abelian Hadamard difference sets. 95 JCombThA 70 144-156

Smith, Lesbia F.: Public health role, aluminium and Alzheimer's disease. 95 EnvrMtrc 6 277-286 (Disc: p286-286)

Smith, Mary E.; Feild, Thomas C.: Veterans and VA system users classified by federal poverty guidelines and thresholds and the VA means test. 94 ASAProGovt 28-33

Smith, Murray D.: Belsley, D. A. (Ed); — — (Rev): Review of "Computational techniques for econometrics and economic analysis". 95 Ststcian 44 540-540

Smith, O. S.: Tivang, J. G.; Nienhuis, J.; — —: Estimation of sampling variance of molecular marker data using the bootstrap procedure. 94 ThApGene 89 259-264

Smith, P. W. F.: McDonald, J. W.; — —: Exact conditional tests of quasi-independence for triangular contingency tables: Estimating attained significance levels. 95 ApplStat 44 143-151

Smith, Perry F.: Muse, Alison G.; Mikl, Jaromir; — —: Evaluating the quality of anonymous record linkage using deterministic procedures with the New York State AIDS registry and a hospital discharge file. 95 StatMed 14 499-509 (Disc: p687-689)

Smith, Peter J.: A recursive formulation of the old problem of obtaining moments from cumulants and vice versa. 95 AmerStat 49 217-218

— —: On plotting renovated samples. 95 Biomtrcs 51 1147-1151

— —; Zhang, J.: Renovated scatterplots for censored data. 95 Biomtrka 82 447-452

Smith, Peter W. F.; McDonald, John W.: Exact conditional tests for incomplete contingency tables: Estimating attained significance levels. 95 StatComp 5 253-256

Ekholm, Anders; — —; McDonald, John W.: Marginal regression analysis of a multivariate binary response. 95 Biomtrka 82 847-854

Gajjar, Yogesh; Mehta, Cyrus; Patel, Nitin (Auth); — — (Rev): Review of "LOGXACT-TURBO (Version 1.1)". 95 ApplStat 44 139-141

Smith, Philip M.: Statistics Canada's environmental statistics program: Information for environmental analysis and assessment. 94 ASAProEnvr 26-35 (Disc: p36-37)

Smith, R. I.: Williams, E. R.; Matheson, A. C. (Auth); — — (Rev): Review of "Experimental design and analysis for use in tree improvement". 95 Biomtrcs 51 1190-1190

Smith, R. L.: Winkler, Gerhard (Auth); — — (Rev): Review of "Image analysis, random fields and dynamic Monte Carlo methods: A mathematical introduction". 95 ShrtBkRv 15/3 49-49

Smith, Richard: Lund, Robert; Hurd, Harry; Bloomfield, Peter; — —: Climatological time series with periodic correlation. 95 JClimate 8 2787-2809

Smith, Richard J.: Chesher, Andrew; — —: Bartlett corrections to likelihood ratio tests. 95 Biomtrka 82 433-436

Davidson, R.; Mackinnon, J. G. (Auth); — — (Rev): Review of "Estimation and inference in econometrics". 95 JApEcmtx 10 339-341

Smith, Robert T., Jr; Evans, B. Timothy; Fowler, John L.: The Commodity Flow Survey. 94 ASAProBuEc 89-94

Smith, Ron: Pesaran, M. Hashem; — —: The role of theory in econometrics. 95 JEconmtx 67 61-79

Pesaran, M. Hashem; — —: Estimating long-run relationships from dynamic heterogeneous panels. 95 JEconmtx 68 79-113

Smith, S. Jay; Caudill, Samuel P.; Steinberg, Karen K.; Thacker, Stephen B.: On combining dose-response data from epidemiological studies by meta-analysis. 95 StatMed 14 531-544 (Disc: p679-685)

Smith, S. P.: Differentiation of the Cholesky algorithm. 95 JCmpGrSt 4 134-147

Smith, Stanley K.; Shahidullah, Mohammed: An evaluation of population projection errors for census tracts. 95 JASA 90 64-71

Smith, Stephen M.: Distribution-free and robust statistical methods: Viable alternatives to parametric statistics?. 95 Ecology 76 1997-1998

Smith, T. M. F.: The statistical profession and the chartered statistician (CStat). 95 JOffStat 11 33-40

Gao, Sujuan; — —: On the nonexistence of a global nonnegative minimum bias invariant quadratic estimator of variance components. 95 StPrLet 25 117-120

Smith, Teresa C.; Spiegelhalter, David J.; Thomas, Andrew: Bayesian approaches to random-effects meta-analysis: A comparative study. 95 StatMed 14 2685-2699

Smith, Thomas J.: Stone, Roslyn A.; Marsh, Gary M.; Owens, Ada D.; Quinn, Margaret M.; — —: Retrospective estimation of airborne fiber exposure: Statistical aspects. 94 ASAProEpid 127-132

Smith, Thomas M.: Comment on "Education research using surveys". 94 ASAProSoSt 213-215

Smith, W. K.: Solow, A. R.; — —; Recchia, C.: A conditional test of independence of Markov chains. 95 BiomtrcJ 37 973-977

Smith, William A.: Fraser, Martin D.; Bartolucci, Alfred A.; — —; Singh, Karen P.: A Bayesian decision procedure for selecting prognostic variables associated with survival for data in which censoring is prevalent. 95 BiomtrcJ 37 463-479

Smith, William Boyce: Kshirsagar, Anant M.; — —: Growth curves. 95 Dekker:NY 359p

Smith, Wray: Ghosh, Dhiren; Kaufman, Steven F.; — —; Chang, Michael: Optimal periodicity of a survey: Sampling error, data deterioration, and cost. 94 ASASrvyRsr 1122-1127

Holt, Albert; Kaufman, Steven; Scheuren, Fritz; — —: Intersurvey consistency in school surveys. 94 ASASrvyRsr 1105-1110 (Disc: p1128-1129)

Smogor, Roy A.: Angermeier, Paul L.; — —: Estimating number of species and relative abundances in stream-fish communities: Effects of sampling effort and discontinuous spatial distributions. 95 JFishRBC 52 936-949

Smoleński, Włodzmierz: Millet, Annie; — —: Small perturbations of Gaussian regressors. 95 StPrLet 24 21-31

Smouse, Peter E.: Excoffier, Laurent; — —: Using allele frequencies and geographic subdivision to reconstruct gene trees within a species: Molecular variance parsimony. 94 Genetcs 136 343-359

Smyth, David J.: Dua, Pami; — —: Forecasting US home sales using BVAR models and survey data on households' buying attitudes for homes. 95 JForcast 14 217-227

Smyth, G. K.: Osborne, M. R.; — —: A modified prony algorithm for exponential function fitting. 95 SIAMSSC 16 119-138

Smythe, Robert T.: Krewski, Daniel; — —; Dewanji, Anup; Szyszkowicz, Mieczyslaw: Bootstrapping an empirical Bayes estimator of the distribution of historical controls in carcinogen bioassay. 95 JStPlInf 43 215-228

Snapinn, Steve: Sarkar, Sanat K.; — —; Wang, Wenjin: On improving the min test for the analysis of combination drug trials. 95 JStCmpSm 51 197-213

Snapinn, Steven M.: Chang, Myron N.; Gould, A. Lawrence; — —: P-values for group sequential testing. 95 Biomtrka 82 650-654

Snavely, Duane; Dobbins, Thomas: An exploratory data analysis of ambulatory blood pressure monitoring as screening criteria. 94 ASAProBiop 51-55

Snee, Ronald D.: Comment on "Mixture designs for four components in orthogonal blocks" (V35 p268-276). 95 Technmcs 37 131-132

Hare, Lynne; Hoerl, Roger W.; — —: Statistical thinking for business. 95 ASQCTekCon 787-788

Hare, Lynne B.; Hoerl, Roger W.; Hromi, John D.; — —: The role of statistical thinking in management. 95 QulPr 28/2 53-60

Snell, A. J.: Pierse, R. G.; — —: Temporal aggregation and the power of tests for a unit root. 95 JEconmtx 65 333-345

Snell, J. Laurie: The Journal of Statistics Education information service and other Internet resources for statistics teachers. 95 AmerStat 49 372-375

— — (Ed): Topics in contemporary probability. 95 CRCPress:FL 400p

Snell, Joyce: Jorgensen, B. (Auth); — — (Rev): Review of "The theory of linear models". 95 JRSS-A 158 195-195

Montgomery, D. C.; Runger, G. C. (Auth); — — (Rev): Review of "Applied statistics and probability for engineers". 95 JRSS-A 158 355-355

Sprent, P. (Auth); — — (Rev): Review of "Applied nonparametric statistical methods (Second edition)". 95 JRSS-A 158 355-356

Śniady, P.: Sieniawska, R.; — —: Design sensitivity of structures under non-Gaussian excitation. 95 ZAngMaMk 75 573-574S

Snijders, Tom: Frank, Ove; — —: Estimating the size of hidden populations using snowball sampling. 94 JOffStat 10 53-67

Snow, R. W.: Forster, D.; — —: An assessment of the use of hand-held computers during demographic surveys in developing countries. 95 SurvyM 21 179-184

Snyder, Richard: Seaver, William L.; — —: Process improvement in telecommunications with multivariate methods. 94 ASAProQlPr 235-240

Snyman, J. L. J.: Venter, J. H.; — —: A note on the generalised cross-validation criterion in linear model selection. 95 Biomtrka 82 215-219

So, Beong-Soo: On the optimal adaptive estimation in the semiparametric non-linear autoregressive time series model. 95 JKoreaSt 24 149-160

— —: Finite-sample, small-dispersion asymptotic optimality of the non-linear least squares estimator. 95 JKoreaSt 24 303-312

Sobel, E. L.: Tavaré, C. J.; — —; Gilles, F. H.: Misclassification of a prognostic dichotomous variable: Sample size and parameter estimate adjustment. 95 StatMed 14 1307-1314

Sobel, M. E.: Arminger, G.; Clogg, C. C.; — — (Ed); Kemp, G. C. R. (Rev): Review of "Handbook of statistical modeling for the social and behavioral sciences". 95 Biomtrcs 51 1586-1587

Sobel, Marc: Comparisons of estimators of a slippage probability. 95 CommStA 24 1039-1055

Sobel, Matthew J.: Lot sizes in serial manufacturing with random yields. 95 PrEngInf 9 151-157

Sobel, Michael E.: Arminger, Gerhard; Clogg, Clifford C.; — —: Handbook of statistical modeling for the social and behavioral sciences. 95 PlenumPubl:NY 618p

Arminger, Gerhard; Clogg, Clifford C.; — — (Ed); Dudgeon, Paul (Rev): Review of "Handbook of statistical modeling for the social and behavioural sciences". 95 Chance 8/4 32-36

Sobel, Milton: Ebneshahrashoob, M.; — —: Dirichlet analysis for inverse multinomial with both quota and quota-free cells. 95 CmpStDA 19 293-307

Sobol, Ilya M.: Saltelli, Andrea; — —: About the use of rank transformation in sensitivity analysis of model output. 95 RelEngSS 50 225-239

Sochorová, A.: Hájek, P.; — —; Zvárová, J.: GUHA for personal computers. 95 CmpStDA 19 149-153

Sohn, Joong Kweon; Kim, Heon Joo: Bayesian inference for the two-parameter exponential models: Type-II censored case. 95 JKoreaSt 24 313-335

Sokal, R.-S.; Rohlf, F.-J.: Biometry: The principles and practice of statistics in biological research. 95 Freeman:NY 918p

Solanky, Tumulesh K. S.: Mukhopadhyay, Nitis; — — (Auth); Chen, Pinyuen (Rev): Review of "Multistage selection and ranking procedures: Second-order asymptotics". 95 JASA 90 1130-1131

Solberg, Helge Erik: RefVal: A program implementing the recommendations of the International Federation of Clinical Chemistry on the statistical treatment of reference values. 95 CmpBiomd 48 247-256

Soliman, Naeem A.: Some restricted estimation and testing hypotheses problem with singularity assumption. 94 EgpStJ 38 231-248

Solka, Jeffrey L.; Poston, Wendy L.; Wegman, Edward J.: A visualization technique for studying the iterative estimation of mixture densities. 95 JCmpGrSt 4 180-198

Soller, M.: Darvasi, A.; — —: Optimum spacing of genetic markers for determining linkage between marker loci and quantitative trait loci. 94 ThApGene 89 351-357

Motro, U.; — —: Sequential sampling in determining linkage between marker loci and quantitative trait loci. 93 ThApGene 85 658-664

Solna, Knut; Switzer, Paul: An analysis of regional climate data. 92 ProIrStC1 231-241

Solo, Victor: Errors in variables and cointegration. 95 EconTher 11 60-80

Solomon, A. D.: Morris, M. D.; — —: Design and analysis for an inverse problem arising from an advection-dispersion process. 95 Technmcs 37 293-302

Solomon, P. J.: Ripley, B. D.; — —: Comment on "Statistical models for prevalent cohort data" (V49 p1-11). 95 Biomtrcs 51 373-374

Soloveitchik, M.: Boldrighini, C.; — —: Drift and diffusion for a mechanical system. 95 ProbTher 103 349-379

Solow, A. R.; Smith, W. K.; Recchia, C.: A conditional test of independence of Markov chains. 95 BiomtrcJ 37 973-977

Solow, Andrew R.: An exploratory analysis of a record of El Niño events, 1800-1987. 95 JASA 90 72-77

— —: Testing for change in the frequency of El Niño events. 95 JClimate 8 2563-2566

— —; Gaines, Arthur G.: An empirical Bayes approach to monitoring water quality. 95 EnvrMtrc 6 1-5

Thiébaux, H. Jean (Auth); — — (Rev): Review of "Statistical data analysis for ocean and atmospheric sciences". 95 AmerStat 49 399-399

Solozabal, J.: Iznaga, N.; Núñez, G.; — —; Morales, A.; Artaza, E.; Rubio, R.; Cardenas, E.: A personal computer-based system for parallel line analysis of bioassay data. 95 CmpBiomd 47 167-175

Soltani, A. R.: On spectral representation of multivariate stable processes. 94 ThProbAp 39 465-475

Soltani, A. Reza: More on conditional moments of stable random variables. 92 ProIrStC1 195-200

Som, Ranjan K.: Practical sampling techniques. 95 Dekker:NY 664p

Somers, Ian F.: Wang, You-Gan; Thomas, Mervyn R.; — —: A maximum likelihood approach for estimating growth from tag-recapture data. 95 JFishRBC 52 252-259

Sommer, Steve S.: Schaid, Daniel J.; — —: Comparison of statistics for candidate-gene association studies using cases and parents. 94 AJHuGene 55 402-409

Sommer, Suzanne; Staudte, Robert G.: Robust variable selection in regression in the presence of outliers and leverage points. 95 AstrlJSt 37 323-336

Sommers, John Paul: Analysis of the effects of imputation on variance estimates for the National Medical Expenditure Survey. 94 ASASrvyRsr 400-405

Son, M. S.: Hamdy, H. I.; — —: Accelerated sequential procedure to estimate the mean of unknown distribution. 94 ASASrvyRsr 1173-1178

Sone, Akio: Griffith, Daniel A.; — —: Trade-offs associated with normalizing constant computational simplifications for estimating spatial statistical models. 95 JStCmpSm 51 165-183

Soner, H. M.; Shreve, S. E.; Cvitanić, J.: There is no nontrivial hedging portfolio for option pricing with transaction costs. 95 AnlsApPr 5 327-355

Song, Frank: Aggarwal, Raj; Mohanty, Sunil; — —: Are survey forecasts of macroeconomic variables rational?. 95 JBusUChi 68 99-119

Song, Jae-Kee; Cho, Gyo-Young: A note on percentile residual life. 95 SankhyaA 57 333-335

Song, K. S.; Müller, H. G.; Clifford, A. J.; Furr, H. C.; Olson, J. A.: Estimating derivatives of pharmacokinetic response curves with varying bandwidths. 95 Biomtrcs 51 12-20

Song, Kai-Sheng: Rubin, Herman; — —: Exact computation of the asymptotic efficiency of maximum likelihood estimators of a discontinuous signal in a Gaussian white noise. 95 AnlsStat 23 732-739

Song, Mark: Morrice, Douglas J.; — —: An approximation method for a single server queue with cyclically indexed arrival and service rates. 95 AplMaCmp 71 247-269

Song, Moon Sup: Kim, Jinheum; — —: A goodness-of-fit test for the additive risk model with a binary covariate. 95 JKoreaSt 24 537-549

Kim, Soon-ock; — —: Distribution-free tests for symmetry based on the linear signed rank statistics. 95 JNnparSt 5 311-321

Nam, Ho Soo; — —; Shin, Bong Sup: A robust test for the parallelism of two regression lines (Korean). 95 KorApS 8/2 77-86

Song, Mun Sup: Lee, Eui Yong; — —; Park, Byung-Gu: A diffusion model for a system subject to random shocks. 95 JKoreaSt 24 141-147

Song, Qiang; Leland, Robert P.; Chissom, Brad S.: A new fuzzy time-series model of fuzzy number observations. 95 FuzzySet 73 341-348

Song, Renming: Feynman-Kac semigroup with discontinuous additive functionals. 95 JTherPrb 8 727-762

Song, S.: Hirsch, F.; — —: Symmetric Skorohod topology on n-variable functions and hierarchical Markov properties of n-parameter processes. 95 ProbTher 103 25-43

Hirsch, F.; — —: Markov properties of multiparameter processes and capacities. 95 ProbTher 103 45-71

Song, Seuck Heun: Consistency and bounds on the bias of S^2 in the linear regression model with moving average disturbances. 95 JKoreaSt 24 507-518

Song, Sung Y.: Class 3 association schemes whose symmetrizations have two classes. 95 JCombThA 70 1-29

Song, Wheyming Tina: Shih, Neng Hui; — —: Estimating simulation metamodels using correlated control variates. 95 CommStB 24 79-90

Songini, M.: Bernardinelli, L.; Clayton, D.; Pascutto, C.; Montomoli, C.; Ghislandi, M.; — —: Bayesian analysis of space-time variation in disease risk. 95 StatMed 14 2433-2443

Sønju, Torleif: Aalen, Odd O.; Bjertness, Espen; — —: Analysis of dependent survival data applied to lifetimes of amalgam fillings. 95 StatMed 14 1819-1829

Sonksen, P. M.: Wade, A. M.; Ades, A. E.; Salt, A. T.; Jayatunga, R.; — —: Age-related standards for ordinal data: Modelling the changes in visual acuity from 2 to 9 years of age. 95 StatMed 14 257-266

Sonksen, Peter: Young, Philip; Morgan, Byron; — —; Till, Sebastian; Williams, Charles: Using a mixture model to predict the occurrence of diabetic retinopathy. 95 StatMed 14 2599-2608

Soofi, Ehsan: Mazzuchi, Thomas A.; Soyer, Refik; — —: Lifetime modeling via Bayesian information distinguishability. 94 ASAProBays 238-245

Soofi, Ehsan S.: On the information theory and statistics. 92 ProIrStC1 203-229

— —; Ebrahimi, Nader; Habibullah, Mohamed: Information distinguishability with application to analysis of failure data. 95 JASA 90 657-668

Sørensen, D.: Andersen, H. H.; Højbjerre, M.; — —; Eriksen, P. S.: Linear and graphical models for the multivariate complex normal distribution. 95 Springer:Brln:NY 183p

Sorensen, D. A.: Wang, C. S.; Gianola, D.; — —; Jensen, J.; Christensen, A.; Rutledge, J. J.: Response to selection for litter size in Danish Landrace pigs: A Bayesian analysis. 94 ThApGene 88 220-230

Sørensen, Michael: Bibby, Bo Martin; — —: Martingale estimation functions for discretely observed diffusion processes. 95 Bernoull 1 17-39

Sorooshian, Soroosh: Hsu, Kuo-lin; Gupta, Hoshin Vijai; — —: Artificial neural network modeling of the rainfall-runoff process. 95 WaterReR 31 2517-2530

Souchon, Yves: Lamouroux, Nicolas; — —; Herouin, Eric: Predicting velocity frequency distributions in stream reaches. 95 WaterReR 31 2367–2375

Soundararajan, V.: Arumainayagam, S. D.; — —: Construction and evaluation of matched quick switching systems. 95 JAppStat 22 245–251

Arumainayagam, S. Devaraj; — —: Quick switching double sampling system indexed by the crossover point. 95 CommStB 24 765–773

Arumainayagam, S. Devaraj; — —: Construction and selection of quick switching double-sampling system: Sample size tightening. 95 JAppStat 22 105–119

Southwell, Colin: Anderson, David R.; — —: Estimates of macropod density from line transect surveys relative to analyst expertise. 95 JWildlMg 59 852–857

Souza, R. C.: Faria, A. E.; — —: A re-evaluation of the quasi-Bayes approach to the linear combination of forecasts. 95 JForcast 14 533–542

Sowers, R. B.: Multidimensional reaction-diffusion equations with white noise boundary perturbations. 94 AnlsProb 22 2071–2121

Sowey, Eric R.: Teaching statistics: Making it memorable. 95 JStEd 3/2 28kb (jse/v3n2/sowey)

Soyer, Refik: Mazzuchi, Thomas A.; — —; Soofi, Ehsan: Lifetime modeling via Bayesian information distinguishability. 94 ASAProBays 238–245

Neufelder, Ann M. (Auth); — — (Rev): Review of "Ensuring software reliability". 95 Technmcs 37 122–123

Spagon, Patrick D.: Reece, Jack E.; — —: Linked training and consulting in the semiconductor industry: The SEMATECH approach. 94 ASAProStEd 228–233

Spall, James C.: The Kantorovich inequality for error analysis of the Kalman filter with unknown noise distributions. 95 Autmatca 31 1513–1517

— —: Uncertainty bounds for parameter identification with small sample sizes. 95 PIEEEDecCt 3504–3515

— —: Uncertainty bounds for multivariate M-estimates in small samples. 94 ASAProPEng 126–131

— —; Cristion, John A.: Model-free control of nonlinear stochastic systems in discrete time. 95 PIEEEDecCt 2199–2204

Maryak, John L.; — —; Silberman, Geoffrey L.: Uncertainties for recursive estimators in nonlinear state-space models, with applications to epidemiology. 95 Autmatca 31 1889–1892

Bouleau, Nicolas; Lepingle, Dominique (Auth); — —; Nelson, J. Barry (Rev): Review of "Numerical methods for stochastic processes". 95 Technmcs 37 236–237

Spanos, Aris: On normality and the linear regression model. 95 EcomtrRv 14 195–203

— —: On theory testing in econometrics Modeling with nonexperimental data. 95 JEconmtx 67 189–226

Spar, Edward J.: Statistical statesmanship. 94 ASAProGovt 9–10

Sparks, Ross: Stacked boxes: A new goodness-of-fit graphical display. 95 NZndSt 30 48–61

— —; Ballantyne, Roy: Quality control methods in large-scale assessment procedures using 'double-marking' or 'partial double-marking'. 95 AstrlJSt 37 283–298

Sparrow, David: D'Agostino, Ralph B., Jr; — —; Weiss, Scott; Rosner, Bernard: Longitudinal models for analysis of respiratory function. 95 StatMed 14 2205–2216

Späth, Helmuth: — — (Auth); Antón, Vicente Núñez (Rev): Review of "Mathematical algorithms for linear regression". 95 JAppStat 22 318–319

Spatz, Chris: — — (Auth); Morse, David T. (Rev): Review of "Basic statistics: Tales of distributions (Fifth edition)". 95 EdPsMeas 55 140–146

Spaulding, JoAnn K.; Kimble, David L.: A decision theory model for stocking switch packs. 94 ASAProBuEc 205–210

Spector, Phil: — — (Auth); Ko, Daijin (Rev): Review of "An introduction to S and S-Plus". 95 Technmcs 37 232–232

Speed, T. P.: Jin, Kun; — —; Thomson, Glenys: Tests of random mating for a highly polymorphic locus: Application to HLA data. 95 Biomtrcs 51 1064–1076

Zhou, S.; Maller, R. A.; — —: On a shared allele test of random mating. 95 AstrlJSt 37 61–72

Speed, Terence P.: Sidow, Arend; Nguyen, Trang; — —: Estimating the fraction of invariable codons with a capture-recapture method. 92 JMolEvol 35 253–260

Zhao, Hongyu; — —; McPeek, Mary Sara: Statistical analysis of crossover interference using the chi-square model. 95 Genetcs 139 1045–1056

Zhao, Hongyu; McPeek, Mary Sara; — —: Statistical analysis of chromatid interference. 95 Genetcs 139 1057–1065

Speight, Joan D.: Jarjoura, David; — —: Statistical consulting as alchemy. 94 ASAProStEd 165–169

Spence, M. Anne; Bishop, D. Timothy; Boehnke, Michael; Elston, Robert C.; Falk, Catherine; Hodge, Susan E.: Methodological issues in linkage analyses for psychiatric disorders: Secular trends, assortative mating, bilineal pedigrees. 93 HumHrdty 43 166–172

Spencer, William J.; Tobias, Paul A.: Statistics in the semiconductor industry: A competitive necessity. 95 AmerStat 49 245–249

Spetsakis, Minas: Models of statistical visual motion estimation. 94 CVGIPIm 60 300–312

Spicer, D. V.: Pike, M. C.; — —: Hormones, breast cell proliferation and breast cancer risk. 94 ASAProEpid 24–33

Spiegelhalter, D. J.: Michie, D.; — —; Taylor, C. C. (Ed); Fulkerson, Bill (Rev): Review of "Machine learning, neural and statistical classification". 95 Technmcs 37 459–459

Spiegelhalter, David: Hand, D. J. (Ed); — — (Rev): Review of "Artificial intelligence frontiers in statistics: AI and statistics III". 95 JAppStat 22 186–186

Spiegelhalter, David J.: Smith, Teresa C.; — —; Thomas, Andrew: Bayesian approaches to random-effects meta-analysis: A comparative study. 95 StatMed 14 2685–2699

Spiegelman, Cliff: — — (Rev): Review of "SoftWindows, Mathematica and Data Desk for the Macintosh PowerPC". 95 Chemolab 28 208–209

The MathWorks Inc (Auth); — — (Rev): Review of "MATLAB for the Power Macintosh". 95 Chemolab 29 323–324

Spielman, Richard S.: Ewens, Warren J.; — —: The transmission/disequilibrium test: History, subdivision, and admixture. 95 AJHuGene 57 455–464

Spirer, Herbert; Spirer, Louise (Auth); Saphire, Diane G. (Rev): Review of "Data analysis for monitoring human rights". 95 AmerStat 49 322–322

Spirer, Louise: Spirer, Herbert; — — (Auth); Saphire, Diane G. (Rev): Review of "Data analysis for monitoring human rights". 95 AmerStat 49 322–322

Spisak, A. W.: Statistical process control of sampling frames. 95 SurvyM 21 185–190

Spizzichino, Fabio: Barlow, Richard E.; Clarotti, Carlo A.; — — (Ed); Gruber, Marvin H. J. (Rev): Review of "Reliability and decision making". 95 Technmcs 37 346–347

Spokoiny, V. G.: Lepskii, O. V.; — —: Local adaptation to inhomogeneous smoothness: Resolution level. 95 MaMethSt 4 239–258

Neumann, M. H.; — —: On the efficiency of wavelet estimators under arbitrary error distributions. 95 MaMethSt 4 137–166

Sprent, P.: — — (Auth); Randles, Ronald H. (Rev): Review of "Applied nonparametric statistical methods (Second edition)". 95 JAppStat 22 189–189

— — (Auth); Snell, Joyce (Rev): Review of "Applied nonparametric statistical methods (Second edition)". 95 JRSS-A 158 355–356

Springer, Janet A.; Liu, Peng T.; Albert, Richard H.; Chow, Lien P.: Construction of a spectrum lifetable. 94 ASAProSoSt 82–87

Liu, Peng T.; Albert, Richard H.; — —: Is one sample enough?. 94 ASAProBays 18–22

Sproule, Robert: Fu, James; — —: A generalization of the binomial distribution. 95 CommStA 24 2645–2658

Spruill, M. C.: Determinacy of means of order statistics. 95 SankhyaA 57 118–125

Sprung, J. L.: Helton, J. C.; Johnson, J. D.; McKay, M. D.; Shiver, A. W.; — —: Robustness of an uncertainty and sensitivity analysis of early exposure results with the MACCS reactor accident consequence model. 95 RelEngSS 48 129–148

Spurrier, John D.: More bounds for moments of some rank statistics. 95 CommStA 24 2679–2682

Lee, Robert Edwin; — —: Distribution-free multiple comparisons between successive treatments. 95 JNnparSt 5 261–273

Lee, Robert Edwin; — —: Successive comparisons between ordered treatments. 95 JStPlInf 43 323–330

— —; Edwards, Don; Thombs, Lori A. (Auth); Hartlaub, Bradley A. (Rev): Review of "Elementary statistics laboratory manual (MS-DOS Version)". 95 AmerStat 49 397–398

Sreehari, M.: Maximum stability and a generalization. 95 StPrLet 23 339–342

Sridhara, R.: Kanji, Gopal K. (Auth); — — (Rev): Review of "100 statistical tests". 95 StMedRes 4 269–269

Srinivasan, M. R.; Ponnuswamy, K. N.: Estimation of variance components based on partial triallel mating design. 95 BiomtrcJ 37 161–171

— —; Ponnuswamy, K. N.: Estimation of variance components based on triallel mating design. 93 ThApGene 85 593–597

Srinivasan, R.: Schultz, Delray J.; Parnes, Milton; — —: Further applications of D-complete designs to group testing. 93 JCISS 18 31–41

SriRam, M. G.: Dudewicz, Edward J.; van der Meulen, Edward C.; — —; Teoh, Nick K. W.: Entropy-based random number evaluation. 95 AJMMgSci 15 115–153

Sriram, T. N.: Datta, Somnath; — —: A modified bootstrap for branching processes with immigration. 95 StocProc 56 275–294

Srivastava, A. B. L.: A review of educational planning and statistics in India with focus on contributions of Professor P. K. Bose. 94 CalcutSt 44 251–264

Srivastava, A. K.: Improved predictors in survey sampling under a super population model. 95 StPrLet 25 249–252

Srivastava, Anil K.; Srivastava, Virendra K.; Ullah, Aman: The coefficient of determination and its adjusted version in linear regression models. 95 EcomtrRv 14 229–240

— —; Tracy, Derrick S.; Shukla, Praveen: Concentration of estimators of coefficients in mixed regression model. 95 CommStA 24 797–811

Srivastava, Virendra K.; — —: Improved estimation of regression coefficients and hypotheses testing. 94 Metron 52/3 161–170

Srivastava, Deo Kumar: Mudholkar, Govind S.; — —; Freimer, Marshall: The exponentiated Weibull family: A reanalysis of the bus-motor-failure data. 95 Technmcs 37 436–445

Mudholkar, Govind S.; — —; Lin, C. Thomas: Some p-variate adaptations of the Shapiro-Wilk test of normality. 95 CommStA 24 953–985

Srivastava, Jagdish N.: Nonadditivity in row-column designs. 93 JCISS 18 85-96

Srivastava, M.: Sen, A.; — —: Regression analysis. Theory, methods, and applications. 90 Springer:Brln:NY 347p

Srivastava, M. S.: Comparison of the inverse and classical estimators in multi-univariate linear calibration. 95 CommStA 24 2753-2767

— —: An economical control procedure for detecting shift in quality level. 94 ASAProQlPr 1-6

— —: Estimation of the intraclass correlation coefficient. 93 AnHuGene 57 159-165

— —; Wu, Y.: An improved version of Taguchi's on-line control procedure. 95 JStPlInf 43 133-145

Ghosh, M.; Carlin, B. P.; — —: Probability matching priors for linear calibration. 95 Test 4 333-357

Purkayastha, Sumitra; — —: Asymptotic distributions of some test criteria for the covariance matrix in elliptical distributions under local alternatives. 95 JMultiAn 55 165-186

Srivastava, Tanuja: Dhariyal, I. D.; Rathore, R. K. S.; — —; Munshi, P.; Rastogi, Richa: Design of optimal statistical filter for the discrete convolution backprojection method. 94 AJMMgSci 14 229-265

Srivastava, V. K.; Maekawa, Koichi: Efficiency properties of feasible generalized least squares estimators in SURE models under non-normal disturbances. 95 JEconmtx 66 99-121

Singh, R. K.; Pandey, S. K.; — —: Correction to "A generalized class of shrinkage estimators in linear regression when disturbances are not normal" (V23 p2029-2046). 95 CommStA 24 859-859

Vinod, H. D.; — —: Large sample asymptotic properties of the double k-class estimators in linear regression models. 95 EcomtrRv 14 75-100

Srivastava, Viren K.; Ohtani, Kazuhiro: A comparison of interval constrained least squares and mixed regression estimators. 95 CommStA 24 395-413

Srivastava, Virendra K.; Srivastava, Anil K.: Improved estimation of regression coefficients and hypotheses testing. 94 Metron 52/3 161-170

— —; Ullah, Aman: Stein-rule estimation in models with a lagged-dependent variable. 95 CommStA 24 1343-1353

Srivastava, Anil K.; — —; Ullah, Aman: The coefficient of determination and its adjusted version in linear regression models. 95 EcomtrRv 14 229-240

Ullah, Aman; — —; Roy, Nilanjana: Moments of the function of non-normal random vector with applications to econometric estimators and test statistics. 95 EcomtrRv 14 459-471

St-Georges, Mario: Nadeau, Simon; Décarie, Robert; Lambert, Daniel; — —: Nonlinear modeling of muskrat use of habitat. 95 JWildlMg 59 110-117

St. Clair, Joseph F.: Safety measurement: Towards continuous improvement. 95 ASQCTekCon 150-158

St. George, Ross: Hutchinson, P. T. (Auth); — — (Rev): Review of "Controversies in item response theory". 95 NZndSt 30 37-37

Hutchinson, P. T. (Auth); — — (Rev): Review of "Ability, partial information, guessing: Statistical modelling applied to multiple-choice tests". 95 NZndSt 30 38-38

St. Louis, Robert D.: Choudhury, Askar H.; Power, Simon; — —: A computationally attractive method for estimating the linear regression model with autoregressive-moving average disturbances. 94 Metron 52/1 41-51

Stack, Catharine: Holford, Theodore R.; — —: Study design for epidemiologic studies with measurement error. 95 StMedRes 4 339-358

Stack, Jim; Unwin, Antony R.: Selecting explanatory variables from survey data: A simple approach that explores for interactions. 95 Ststcian 44 17-29

Stadje, W.: Two asymptotic inequalities for the stochastic traveling salesman problem. 95 SankhyaA 57 33-40

— —: The busy periods of some queueing systems. 95 StocProc 55 159-167

— —: A note on the maximum of a random walk. 95 StPrLet 23 227-231

— —: A stochastic process arising in sequential experimentation. 95 StPrLet 23 359-365

Stadtmüller, U.; Tsybakov, A. B.: Nonparametric recursive variance estimation. 95 MaOpfStS 27 55-63

Stafford, James E.: Exact cumulant calculations for Pearson χ^2 and Zelterman statistics for r-way contingency tables. 95 JCmpGrSt 4 199-212

Stahel, W. A.: Morgenthaler, S.; Ronchetti, E.; — — (Ed); Lenz, H.-J. (Rev): Review of "New directions in statistical data analysis and robustness". 95 CmpStDA 19 711-712

Stahl, Gerhard: Huschens, Stefan; — —: Estimation in semiparametric models using an auxiliary model. 95 StHefte 36 313-326

Stahlecker, Peter: Schmidt, Karsten; — —: Reducing the maximum risk of regression estimators by polyhedral projection. 95 JStCmpSm 52 1-15

Stake, Robert E.: The art of case study research. 95 SagePubl:CA:Lnd 176p

Stallard, Nigel: Omar, Rumana Z.; — —; Whitehead, John: A parametric multistate model for the analysis of carcinogenicity experiments. 95 LiDaAnal 1 327-346

Stallings, R. L.: Bruno, W. J.; Knill, E.; Balding, D. J.; Bruce, D. C.; Doggett, N. A.; Sawhill, W. W.; — —; Whittaker, C. C.; Torney, D. C.: Efficient pooling designs for library screening. 95 Genomics 26 21-30

Stam, Piet: Jansen, Ritsert C.; — —: High resolution of quantitative traits into multiple loci via interval mapping. 94 Genetcs 136 1447-1455

Stamas, George: Scott, Stuart; — —; Sullivan, Thomas J.; Chester, Paul: Seasonal adjustment of hybrid economic time series. 94 ASAProBuEc 163-168

Stamate, Daniel: Luchian, Henri; — —: Answer-perturbation techniques for the protection of statistical databases. 95 StatComp 5 203-213

Stamp, Mark: Martin, Clyde F.; Allen, Linda J. S.; — —: Urn model simulations of a sexually transmitted disease epidemic. 95 AplMaCmp 71 179-199

Stander, J.; Silverman, B. W.: Minimax estimation of linear functionals, particularly in nonparametric regression and positron emission tomography. 95 CmpStQ 10 259-283

Stanek, Edward J., III; Buonaccorsi, John P.: Comment on "The mixed-model ANOVA: The truth, the computer packages, the books. Part I: Balanced data" (V47 p48-59). 95 AmerStat 49 323-324

Stanford, John L.; Vardeman, Stephen B. (Ed); Copeland, Karen A. F. (Rev): Review of "Statistical methods for physical science". 95 JQualTek 27 180-181

Stangl, Dalene K.: Prediction and decision making using Bayesian hierarchical models. 95 StatMed 14 2173-2190

Stapleton, James H.: Linear statistical models. 95 Wiley:NY:UK 449p

— — (Auth); White, L. V. (Rev): Review of "Linear statistical models". 96 ShrtBkRv 16/1 7-7

Starer, Amy: Brick, J. Michael; Waksberg, Joseph; Kulp, Dale; — —: Bias in list-assisted telephone samples. 95 PubOpinQ 59 218-235

Brick, J. Michael; Waksberg, Joseph; Kulp, Dale; — —: Bias in list-assisted telephone samples. 94 ASASrvyRsr 1293-1298

Stărică, Cătălin: Feigin, Paul D.; Resnick, Sidney I.; — —: Testing for independence in heavy tailed and positive innovation time series. 95 CommStSM 11 587-612

Stărică, Cătălin: Resnick, Sidney; — —: Consistency of Hill's estimator for dependent data. 95 JAppProb 32 139-167

Stark, D. B.: Reed, J. F.; — —: Robust analysis of variance: A simulation study. 95 JAppStat 22 87-104

Reed, J. F., III; — —: Monte Carlo generation of S_B and S_U distributions. 95 CmpBiomd 47 157-165

Stark, Dudley: Arratia, Richard; — —; Tavaré, Simon: Total variation asymptotics for Poisson process approximations of logarithmic combinatorial assemblies. 95 AnlsProb 23 1347-1388

Stark, P. B.; Parker, R. L.: Bounded-variable least-squares: An algorithm and applications. 95 CmpStQ 10 129-141

Stark, Philip B.: Hengartner, Nicolas W.; — —: Finite-sample confidence envelopes for shape-restricted densities. 95 AnlsStat 23 525-550

Stasny, Elizabeth A.; Reagan, Patricia B.: Adjusting pension fund membership for "don't know" responses. 94 ASAProBuEc 298-303

Staudte, Robert G.: Blyth, Colin R.; — —: Estimating statistical hypotheses. 95 StPrLet 23 45-52

Sommer, Suzanne; — —: Robust variable selection in regression in the presence of outliers and leverage points. 95 AstrlJSt 37 323-336

Stavropoulou, E.: Zopounidis, C.; Despotis, D. K.; — —: Multiattribute evaluation of Greek banking performance. 95 AplStMDA 11 97-107

Steckel-Berger, G.: Seeber, G. U. H.; Francis, B. J.; Hatzinger, R.; — —: Statistical modelling. 95 Springer:Brln:NY 327p

Stedinger, Jery R.: Fill, Heinz D.; — —: L moment and probability plot correlation coefficient goodness-of-fit tests for the Gumbel distribution and impact of autocorrelation. 95 WaterReR 31 225-229

Steel, M. F. J.: Koop, G.; — —; Osiewalski, J.: Posterior analysis of stochastic frontier models using Gibbs sampling. 95 CmpStQ 10 353-373

Steel, Mark F. J.: Fernández, Carmen; Osiewalski, Jacek; — —: Modeling and inference with v-spherical distributions. 95 JASA 90 1331-1340

Koop, Gary; Osiewalski, Jacek; — —: Bayesian long-run prediction in time series models. 95 JEconmtx 69 61-80

Steel, Philip M.; Konschnik, Carl A.: Administrative record matching for the 1992 economic censuses. 94 ASASrvyRsr 473-478

Steele, Kent P.; Zettler, Mark W.: A methodology for the simultaneous development of linear and nonlinear chemical reaction models from a linearly designed experimental program. 95 Chemolab 29 125-131

Steenland, Kyle: — — (Ed), Rushton, Lesley (Rev): Review of "Case studies in occupational epidemiology". 95 StatMed 14 1148-1148

Stefano, Benati: Cellular automata applications to geographical evolution simulation (Italian). 94 QStMaA 16 105-115

Stefanou, Despina: Papamichael, Emmanuel M.; Evmiridis, Nickolaos P.; Thanasoulias, Nickolaos; — —: Design of experiments for a precise estimation of the calibration curve of periodate in its determination by flow-injection analysis and chemiluminescence detection.. 95 Chemolab 30 227-237

Stefanov, Valeri T.: Mean passage times for tridiagonal transition matrices. 95 JAppProb 32 846-849

— —: On efficient sequential estimation for the simple single server queue. 95 RBrzlPrS 9 55-65

Stefanov, Valerie T.: Explicit limit results for minimal sufficient statistics and maximum likelihood estimators in some Markov processes: Exponential families approach. 95 AnlsStat 23 1073-1101

Stefanski, L. A.; Buzas, J. S.: Instrumental variable estimation in binary regression measurement error variables. 95 JASA 90 541-550

— —; Cook, J. R.: Simulation-extrapolation: The measurement error jackknife. 95 JASA 90 1247-1256

Stefański, Leonard A.: Carroll, Raymond J.; Ruppert, David; — —: Measurement error in nonlinear models. 95 ChapmnHall:Lnd 336p

Carroll, Raymond J.; Ruppert, David; — — (Auth); Draper, N. R. (Rev): Review of "Measurement error in nonlinear models". 96 ShrtBkRv 16/1 5-5

Stefey, Duane L.; Bradburn, Norman M. (Ed); Wolter, Kirk M. (Rev): Review of "Counting people in the information age". 95 Chance 8/3 44-49

Stehman, S. V.; Meredith, M. P.: Practical analysis of factorial experiments in forestry. 95 CanJForR 25 446-461

Overton, W. Scott; — —: Design implications of anticipated data uses for comprehensive environmental monitoring programmes. 95 EnvEcoSt 2 287-303

Stehman, Stephen V.: Overton, W. Scott; — —: The Horvitz-Thompson as a unifying perspective for probability sampling: With examples from natural resource sampling. 95 AmerStat 49 261-268

Steif, Jeffrey E.: Burton, Robert M.; — —: Quite weak Bernoulli with exponential rate and percolation for random fields. 95 StocProc 58 35-55

Steiger, W.: Härdle, W.; — —: [Algorithm AS 296] Optimal median smoothing. 95 ApplStat 44 258-264

Steigerwald, Douglas G.: Reply to comment on "Adaptive estimation in time series regression models". 95 JEconmtx 66 131-132

Steil, Garry M.: Bradley, David C.; — —; Bergman, Richard N.: OOPSEG: A data smoothing program for quantitation and isolation of random measurement error. 95 CmpBiomd 46 67-77

Stein, Alfred: Cui, Haiyan; — —; Myers, Don E.: Extension of spatial information, Bayesian kriging and updating of prior variogram parameters. 95 EnvrMtrc 6 373-384

Stein, D. L.: Newman, C. M.; — —: Random walk in a strongly inhomogeneous environment and invasion percolation. 95 AIHenrPB 31 249-261

Stein, Michael L.: Predicting integrals of stochastic processes. 95 AnlsApPr 5 158-170

— —: Fixed-domain asymptotics for spatial periodgrams. 95 JASA 90 1277-1288

— —: An approach to asymptotic inference for spatial point processes. 95 StSinica 5 221-234

Stein, Zena: Paik, Myunghee Cho; Begg, Melissa Dowd; El-Sadr, Wafaa; Gorman, Jack; — —: Difference in clinical implications of CD4 counts among HIV-infected homosexual men and injection drug using men and women. 95 StatMed 14 1889-1900

Steinberg, Karen K.: Smith, S. Jay; Caudill, Samuel P.; — —; Thacker, Stephen B.: On combining dose-response data from epidemiological studies by meta-analysis. 95 StatMed 14 531-544 (Disc: p679-685)

Steinberg, Lynne: Wainer, Howard; Dorans, Neil J.; Flaugher, Ronald; Green, Bert F.; Mislevy, Robert; — —; Thissen, David (Auth); Andrich, David (Rev): Review of "Computerized adaptive testing: A primer". 95 Psymtrka 60 615-620

Steinebach, J.: Alex, M.; — —: Invariance principles for renewal processes and some applications. 95 ThPrMaSt 50 23-54

Steinebach, Josef: Variance estimation based on invariance principles. 95 MaOpfStS 27 15-25

— —; Eastwood, Vera R.: On extreme value asymptotics for increments of renewal processes. 95 JStPlInf 45 301-312

Christ, Ralf; — —: Estimating the adjustment coefficient in an ARMA(p,q) risk model. 95 InsrneME 17 149-161

Steiner, R. Prasaad: Grady, Christopher S.; Looney, Stephen W.; — —: A study of biostatistics requirements in medical schools in the United States. 94 ASAProStEd 263-266

Steiner, S. H.; Wesolowsky, G. O.: Estimating the correlation between destructively measured variables using proof-loading. 95 Technmcs 37 94-101

— —; Wesolowsky, G. O.: Bivariate normal parameter estimation under destructive testing. 94 ASAProPEng 326-331

Steinhorst, R. Kirk; Keeler, Carolyn M.: Developing material for introductory statistics courses from a conceptual, active learning viewpoint. 95 JStEd 3/3 33kb (jse/v3n3/steinhorst)

Keeler, Carolyn M.; — —: Using small groups to promote active learning in the introductory statistics course: A report from the field. 95 JStEd 3/2 28kb (jse/v3n2/keeler)

Stenger, Horst: Gabler, Siegfried; — —: Improving the RHC-strategy. 95 StHefte 36 327-336

Stengle, Gilbert: Baryshnikov, Yuliy; Eisenberg, Bennett; — —: A necessary and sufficient condition for the existence of the limiting probability of a tie for first place. 95 StPrLet 23 203-209

Stengos, Thanasis: Li, Qi; — —: Adaptive estimation in the panel data error component model with heteroskedasticity of unknown form. 94 IntEconR 35 981-1000

Stephens, D. A.; Smith, A. F. M.: Bayesian inference in multipoint gene mapping. 93 AnHuGene 57 65-82

Stephens, David A.: Dellaportas, Petros; — —: Bayesian analysis of errors-in-variables regression models. 95 Biomtrcs 51 1085-1095

Stephens, Graeme L.: Evans, K. Franklin; Turk, Joseph; Wong, Takmeng; — —: A Bayesian approach to microwave precipitation profile retrieval. 95 JApMetrl 34 260-279

Stephens, Kenneth S.: How to perform skip-lot and chain sampling (Second edition). 95 ASQualCntrl:WI 64p

— —: How to perform continuous sampling (Second edition). 95 ASQualCntrl:-WI 73p

— — (Auth); Doganaksoy, Necip (Rev): Review of "Volume 4 - How to perform skip-lot and chain sampling (Second edition)". 95 JQualTek 27 394-395

Stępniak, C.; Niezgoda, M.: Inverting covariance matrices in unbalanced hierarchical models. 95 JStCmpSm 51 215-221

Sterling, T. D.; Rosenbaum, W. L.; Weinkam, J. J.: Publication decisions revisited: The effect of the outcome of statistical tests on the decision to publish and vice versa. 95 AmerStat 49 108-112

Sterling, Theodor D.: Rosenbaum, Wilfred L.; — —; Weinkam, James J.: Correcting standardized rate ratios for imprecise classification of a polychotomous exposure variable with limited data. 95 AJEpidm 141 442-445

Rosenbaum, Wilfred L.; — —; Weinkam, James J.: Correcting standardized rate ratios for imprecise classification of a polychotomous exposure variable with limited data. 95 AJEpidm 142 442-445

Stern, H.: Gelman, A.; Carlin, J.; — —; Rubin, D.: Bayesian data analysis. 95 ChapmnHall:Lnd 552p

Stern, H. S.: Gelman, A.; Carlin, J. B.; — —; Rubin, D. B. (Auth); Brown, P. J. (Rev): Review of "Bayesian data analysis". 96 ShrtBkRv 16/1 5-5

Stern, Hal S.: Who's number 1 in college football? ... And how might we decide?. 95 Chance 8/3 7-14

— —: Problems with logistic autoregression. 94 ASAProBiop 106-110

Glickman, Mark E.; — —: Inference from an autoregressive state-space model via iterative simulation. 94 ASAProSpts 72-76

Rubin, Donald B.; — —; Vehovar, Vasja: Handling "Don't know" survey responses: The case of the Slovenian plebiscite. 95 JASA 90 822-828

Stern, R.: Elliott, P.; Cuzick, J.; English, D.; — — (Ed); Dunn, Graham (Rev): Review of "Geographical and environmental epidemiology: Methods for small-area studies". 95 StMedRes 4 184-184

Sternad, Mikael: Öhrn, Kenth; Ahlén, Anders; — —: A probabilistic approach to multivariable robust filtering and open-loop control. 95 IEEEAuCn 40 405-418

Sterzik, Karl: Abt, Markus; Strehler, Erwin; Rosenbusch, Bernd; — —: Predicting the success of in vitro fertilization: Conventional semen analysis compared to the hamster ova penetration test. 95 StatMed 14 1807-1818

Stettner, Haro: Chui, Charles K. (Auth); — — (Rev): Review of "Multivariate splines". 95 CmpStQ 10 104-106

Stettner, Ł.: Di Masi, G. B.; — —: Bayesian ergodic adaptive control of discrete time Markov processes. 95 Stochatx 54 301-316

Steutel, F. W.: van Harn, K.; — —: Infinite divisibility and the waiting-time paradox. 95 CommStSM 11 527-540

Stevens, Cladd E.: Kiuchi, Amy S.; Hartigan, J. A.; Holford, Theodore R.; Rubinstein, Pablo; — —: Change points in the series of T4 counts prior to AIDS. 95 Biomtrcs 51 236-248

Stevens, Robert H.: Baker, Rose D.; — —: A random-effects model for analysis of infectious disease final-state data. 95 Biomtrcs 51 956-968

Stevenson, C. E.: Statistical models for cancer screening. 95 StMedRes 4 18-32

Stewart, Chris: Sarantis, Nicholas; — —: Structural, VAR and BVAR models of exchange rate determination: A comparison of their forecasting performance. 95 JForcast 14 201-215

Stewart, G. W.: Gauss, statistics and Gaussian elimination. 95 JCmpGrSt 4 1-11

Gauss, Karl Friedrich (Auth); — — (Transl): Theory of the combination of observations least subject to errors. Part one, part two, supplement. 95 SocApMathSim:PA 220p

Gauss, C. F. (Auth); — — (Transl); Stigler, S. M. (Rev): Review of "Theory of the combination of observations least subject to error". 96 ShrtBkRv 16/1 41-41

Stewart, John: Genetics and biology: A comment on the significance of the Elston-Stewart algorithm. 92 HumHrdty 42 9-15

Stewart, Lesley A.; Clarke, Michael J.: Practical methodology of meta-analyses (overviews) using updated individual patient data. 95 StatMed 14 2057-2079

Stewart, Sally E. A.: Yeung, David W. K.; — —: Stationary probability distributions of some Lotka-Volterra types of stochastic predation systems. 95 StocAnAp 13 503-516

Stewart, Walter F.: Correa-Villaseñor, Adolfo; — —; Franco-Marina, Francisco; Seacat, Hui: Bias from nondifferential misclassification in case-control studies with three exposure levels. 95 Epidmlgy 6 276-281

Stewart-Oaten, Allan: Rules and judgments in statistics: Three examples. 95 Ecology 76 2001-2009

Steyne, Dave: Drew, David; — —: The world's 100 largest cities: A case study of inequality. 95 TeachgSt 17/1 17-21

Sthapit, A. B.: Agrawal, M. C.; — —: Unbiased ratio-type variance estimation. 95 StPrLet 25 361-364

Stich, Andreas: The axiomatic derivation of a class of dynamic inequality measures. 95 Metron 53 31-41

Stigler, S. M.: Gauss, C. F. (Auth); Stewart, G. W. (Transl); — — (Rev): Review of "Theory of the combination of observations least subject to error". 96 ShrtBkRv 16/1 41-41

Stigler, Stephen M.: Mirowski, Philip (Ed); — — (Rev): Review of "Edgeworth on chance, economic hazard, and statistics". 95 JASA 90 803-804

Stine, Robert: Fox, John; — —: Introduction to the special issue on computing environments. 95 SocMethR 23 267-281

Stine, Robert A.: Graphical interpretation of variance inflation factors. 95 AmerStat 49 53-56

— —: Data analysis using Mathematica. 95 SocMethR 23 352-372

Stirzaker, D.: — — (Auth); Rees, Mike (Rev): Review of "Elementary probability". 95 JRSS-A 158 356-356

Stock, James H.: Cavanagh, Christopher L.; Elliott, Graham; — —: Inference in models with nearly integrated regressors. 95 EconTher 11 1131-1147

Stockford, Donald: Minority veterans' usage of VA medical care. 94 ASAProGovt 22-24

Stoica, Petre; Händel, Peter; Nehoral, Arye: Improved sequential MUSIC. 95 IEEEAES 31 1230-1239

Stoimenova, Eugenia: On the LFC of the natural classification rule under a new loss function. 95 StatDecn 13 39-51

Stokes, Lynne: Wright, M. Keith; — —; Dyer, James S.: Reliability and coherence of causal, diagnostic, and joint subjective probabilities. 95 DecisnSc 26 691-709

Wright, M. Keith; — —; Dyer, James S.: Reliability and coherence of causal, diagnostic, and joint subjective probabilities. 94 DecisnSc 25 691-709

Stokes, S. Lynne: Greenland, Betsy S.; — —: Repetitive testing in the presence of inspection errors. 95 Technmcs 37 102-111

Stokesr, Lynne: Parametric ranked set sampling. 95 AnInStMa 47 465-482

Stoll, Peter: Schmid, Bernhard; Polasek, Wolfgang; Weiner, Jacob; Krause, Andreas; — —: Modeling of discontinuous relationships in biology with censored regression. 94 ANatrls 143 494-507

Stolyar, A. L.: Coffman, E. G., Jr; Gilbert, E. N.; Greenberg, A. G.; Leighton, F. T.; Robert, Philippe; — —: Queues served by a rotating ring. 95 CommStSM 11 371-394

Stone, Charles J.: Kooperberg, Charles; — —; Truong, Young K.: Hazard regression. 95 JASA 90 78-94

Kooperberg, Charles; — —; Truong, Young K.: Logspline estimation of a possibly mixed spectral distribution. 95 JTimSrAn 16 359-388

Kooperberg, Charles; — —; Truong, Young K.: Rate of convergence for logspline spectral density estimation. 95 JTimSrAn 16 389-401

Stone, David: Gersch, Will; — —: Multivariate autoregressive time series modeling: One scalar autoregressive model at-a-time. 95 CommStA 24 2715-2733

Stone, Deborah: Chun, Young I.; — —: Using an establishment survey response model, and focus groups in the redesign of an establishment survey questionnaire. 94 ASASrvyRsr 1363-1367

Stone, Richard; Taylor, Mark: Time series models in statistical process control: Considerations of applicability. 95 Ststcian 44 227-234

Stone, Roslyn A.; Marsh, Gary M.; Owens, Ada D.; Quinn, Margaret M.; Smith, Thomas J.: Retrospective estimation of airborne fiber exposure: Statistical aspects. 94 ASAProEpid 127-132

Stoodley, P.: Hamilton, M. A.; Johnson, K. R.; Camper, A. K.; — —; Harkin, G. J.; Gillis, R. J.; Shope, P. A.: Analysis of bacterial spatial patterns at the initial stage of biofilm formation. 95 BiomtrcJ 37 393-408

Stoolmiller, Mike: Duncan, Terry E.; Duncan, Susan C.; Hops, Hyman; — —: An analysis of the relationship between parent and adolescent marijuana use via generalized estimating equation methodology. 95 MultBehR 30 317-339

Storer, Robert H.: Ku, Wenfu; — —; Georgakis, Christos: Disturbance detection and isolation by dynamic principal component analysis. 95 Chemolab 30 179-196

Storm, Andrew: Stochastic differential equations with a convex constraint. 95 Stochatx 53 241-274

Storms, Gert: On the robustness of maximum likelihood scaling for violations of the error model. 95 Psymtrka 60 247-258

Stow, C. A.; Carpenter, S. R.; Cottingham, K. L.: Resource vs. ratio-dependent consumer-resource models: A Bayesian perspective. 95 Ecology 76 1986-1990

Stoyan, D.; Stoyan, H. (Auth); Nagel, W. (Rev): Review of "Fractals, random shapes and point fields. Methods of geometrical statistics". 95 BiomtrcJ 37 978-978

Stoyan, Dietrich: Gavrikov, Vladimir; — —: The use of marked point processes in ecological and environmental forest studies. 95 EnvEcoSt 2 331-344

Stoyan, H.: Stoyan, D.; — — (Auth); Nagel, W. (Rev): Review of "Fractals, random shapes and point fields. Methods of geometrical statistics". 95 BiomtrcJ 37 978-978

Stoyanov, Jordan: Dependency measure for sets of random events or random variables. 95 StPrLet 23 13-20

Rosalsky, Andrew; — —; Presnell, Brett: An ergodic-type theorem á la Feller for nonintegrable strictly stationary continuous time processes. 95 StocAnAp 13 555-572

Straatman, H.: Schouten, L. J.; — —; Kiemeney, L. A.; Gimbrere, C. H.; Verbeek, A. L.: The capture-recapture method for estimation of cancer registry completeness: A useful tool?. 94 IJEpidem 23 1111-1116

Straf, Miron L.: What is statistical statesmanship?. 94 ASAProGovt 11-12

Fienberg, Stephen E.; Martin, Margaret E.; — —: The Committee on National Statistics: Fostering interactions between statisticians in academia and government. 95 IntStRvw 63 257-269

Strait, R. Scott: Decision analysis approach to competitive situations with a pure infinite regress. 94 DecisnSc 25 853-862

Stram, D. O.; Lee, J. W.: Correction to "Variance component testing in the longitudinal mixed effects model" (V50 p1171-1177). 95 Biomtrcs 51 1196-1196

Stramer, O.: Brockwell, P. J.; — —: On the approximation of continuous time threshold ARMA processes. 95 AnInStMa 47 1-20

Strasser, Sandra E.; Ozgur, Ceyhun: Undergraduate business statistics: A survey of topics and teaching methods. 95 Intrfc 25/3 95-103

Strauss, David J.: Lesch, Scott M.; — —; Rhoades, James D.: Spatial prediction of soil salinity using electromagnetic induction techniques 1. Statistical prediction models: A comparison of multiple linear regression and cokriging. 95 WaterReR 31 373-386

Lesch, Scott M.; — —; Rhoades, James D.: Spatial prediction of soil salinity using electromagnetic induction techniques. 2. An efficient spatial sampling algorithm suitable for multiple linear regression model identification and estimation. 95 WaterReR 31 387-398

Strauss, Nurith; Simon, Richard: Investigating a sequence of randomized phase II trials to discover promising treatments. 95 StatMed 14 1479-1489

Strawderman, W. E.: Green, E. J.; Smith, A. F. M.; — —: Bayesian construction of thematic maps from satellite imagery. 94 ASASrvyRsr 188-193

Strawderman, William E.: Green, Edwin J.; — —: A growth and yield model for slash pine. 94 ASAProBays 202-205

Natarajan, Jayalakshmi; — —: Sequential estimation of the variance of a normal distribution. 95 SqtlAnly 14 361-374

Shao, Peter Yi-Shi; — —: Improving on the positive part of the UMVUE of a noncentrality parameter of a noncentral chi-square distribution. 95 JMultiAn 53 52-66

Street, D. J.: Construction for orthogonal main effect plans. 94 UtilMath 45 115-123

Burgess, L.; — —: Algorithm for constructing orthogonal main effect plans. 94 UtilMath 46 33-48

Street, Deborah J.: Hand, D. J.; Daly, F.; Lunn, A. D.; McConway, K. J.; Ostrowski, E. (Ed); — — (Rev): Review of "A handbook of small data sets". 95 AstrlJSt 37 251-251

Strehler, Erwin: Abt, Markus; — —; Rosenbusch, Bernd; Sterzik, Karl: Predicting the success of in vitro fertilization: Conventional semen analysis compared to the hamster ova penetration test. 95 StatMed 14 1807-1818

Streit, Franz: Testing whether interpenetration is possible for the grains of a random set. 95 MaOpfStS 26 355-364

Streit, Ludwig: Hida, Takeyuki; Kuo, Hui-Hsuing; Potthoff, Jürgen; — — (Auth); Øksendal, Bernt (Rev): Review of "White noise. An infinite dimensional calculus". 95 BLndnMaS 27 189-191

Stricker, C.: Fernando, R. L.; — —; Elston, R. C.: An efficient algorithm to compute the posterior genotypic distribution for every member of a pedigree without loops. 93 ThApGene 87 89-93

Stricker, Christophe: Monat, Pascale; — —: Föllmer-Schweizer decomposition and mean-variance hedging for general claims. 95 AnlsProb 23 605-628

Strobeck, C.: Xie, C. Y.; Yeh, F. C.; Dancik, B. P.; — —: Joint estimation of immigration and mating system parameters in gymnosperms using the EM algorithm. 91 ThApGene 83 137-140

Strobl, Franz: On the reversed sub-martingale property of empirical discrepancies in arbitrary sample spaces. 95 JTherPrb 8 825-831

Stromberg, Arnold J.: Consistency of the least median of squares estimator in nonlinear regression. 95 CommStA 24 1971-1984

— —: Software for computing a robust efficient linear or nonlinear regression estimator. 94 ASAProStEd 6-11

Strömberg, Ulf: Examining an increasing trend with non-parametric methods. 94 JAppStat 21 575-587

Strong, Ann: Amy, Carolyn K.; — —: Teaching the concepts and tools of variation using body temperature. 95 QulPr 28/3 168-168

Strong, D. R.: Chen, R.; Hawaleshka, O.; — —: Analytical approach for determining the cost optimum raw material quality level. 95 JAppStat 22 375-388

Stroock, D. W.: — — (Auth); Coleman, Rodney (Rev): Review of "Probability theory, and analytic view". 95 JRSS-A 158 356-357

— — (Auth); Cranston, Michael (Rev): Review of "Probability theory: An analytic view". 95 JASA 90 808-808

Stroup, D. F.: Cliff, A. D.; Haggett, P.; Smallman-Raynor, M. R.; — —; Williamson, G. D.: The application of multidimensional scaling methods to epidemiological data. 95 StMedRes 4 102-123

Stroup, Donna: Klemm, Rebecca; — —: Meta-analysis for public health professionals. 94 ASAProEpid 107-111 (Disc: p125-126)

Stroup, Donna F.: Thacker, Stephen B.; — —; Rothenberg, Richard B.; Brownson, Ross C.: Public health surveillance for chronic conditions: A scientific basis for decisions. 95 StatMed 14 629-641 (Disc: p695-696)

Stroup, Nancy E.: Comment on "Differences in morbidity measures and risk factor identification using multiple data sources: The case of coronary heart disease". 95 StatMed 14 691-692

Stroup, Walter W.; Baenziger, P. Stephen; Mulitze, Dieter K.: Removing spatial variation from wheat yield trials: A comparison of methods. 94 CropSci 34 62-66

Stuart, A.; Ord, J. K. (Auth); Kemp, C. D. (Rev): Review of "Kendall's advanced theory of statistics, Volume I: Distribution theory (Sixth edition)". 95 Biomtrcs 51 1583-1584

Stuart, M. A.: Mayer, D. G.; Atzeni, M. G.; Swain, A. J.; — —: Models for the spatial dispersal of insect pests. 95 EnvrMtrc 6 497-503

Stuart, Michael: Changing the teaching of statistics. 95 Ststcian 44 45-54

Studden, W. J.: Dette, H.; Heiligers, B.; — —: Minimax designs in linear regression models. 95 AnlsStat 23 30-40

Studden, William J.: Dette, Holger; — —: Optimal designs for polynomial regression when the degree is not known. 95 StSinica 5 459-473

Stuetzle, Werner: Donnell, Deborah J.; Buja, Andreas; — —: Analysis of additive dependencies and concurvities using smallest additive principal components. 94 AnlsStat 22 1635-1673

Stufken, J.: Hedayat, A. S.; — —; Zhang, W. G.: Virtually balanced incomplete block designs for $v = 22$, $k = 8$, and $\lambda = 4$. 95 JComDsgn 3 195-201

Hedayat, A. S.; — —; Zhang, W. G.: Contingently and virtually balanced incomplete block designs and their efficiencies under various optimality criteria. 95 StSinica 5 575-591

Stufken, John: Combinatorial and statistical aspects of sampling plans to avoid the selection of adjacent units. 93 JCISS 18 149-160

Stukel, Janice: Anderson, Dan; Carlson, Richard R.; — —: The clinical indicators: Measurements of healthcare delivery. 94 ASAProQlPr 55-59

Sturm, K.-T.: Sharp estimates for capacities and applications to symmetric diffusions. 95 ProbTher 103 73-89

Sturm, Roland: Proportional hazard regression models for point processes: An analysis of nuclear power plant operations in Europe. 94 JAppStat 21 533-540

Stute, Winfried: The central limit theorem under random censorship. 95 AnlsStat 23 422-439

— —: U-statistic processes: A martingale approach. 94 AnlsProb 22 1725-1744

Stutzer, Michael: A Bayesian approach to diagnosis of asset pricing models. 95 JEconmtx 68 367-397

Su, John Q.: Chen, Hanfeng; — —: Simultaneous confidence bands for comparing diagnostic markers. 95 CommStA 24 375-393

Jung, Sin-Ho; — —: Non-parametric estimation for the difference or ratio of median failure times for paired observations. 95 StatMed 14 275-281

Suber, Marsha: Martin, Margaret; — —: Statistical education for quality managers. 94 ASAProBays 148-149

Subrahmonia, Jayashree; Keren, Daniel; Cooper, David B.: Recognition of objects in 2D images or 3D range data based on high degree implicit polynomial curves or surfaces, algebraic invariants, and asymptotic Bayesian methods. 94 ASAProPEng 44-49

Subramanian, R.; Anantharaman, V.: Reliability analysis of a complex standby redundant system. 95 RelEngSS 48 57-70

Succop, Paul A.: Jiang, Qi; — —: A study of the log-log and log-additive models for the relationship between blood lead and environmental lead. 94 ASAProEnvr 88-91

Wang, Ying; — —: Multiply-imputed estimation for multivariate missing data. 94 ASAProEnvr 181-186

Suchindran, C. M.: Pandey, Arvind; — —: Some analytical models to estimate maternal age at birth using age-specific fertility rates. 95 SankhyaB 57 142-150

Suchower, Stephan J.: Mullen, Gary L.; — —: Frequency hyperrectangles, error correcting codes and fractional replication plans. 93 JCISS 18 5-18

Sudman, Seymour; Phillips, Joan M.: Cognitive aspects of organizational reporting. 94 ASASrvyRsr 1351-1356

Sueyoshi, Glenn T.: A class of binary response models for grouped duration data. 95 JApEcmtx 10 411-431

Sueyoshi, Toshiyuke: Empirical regression quantile and leverage treatment method. 95 JOpRsJpn 38 34-54

Suganthan, P. N.; Teoh, E. K.; Mital, D. P.: Pattern recognition by graph matching using the Potts MFT neural networks. 95 PattrnRc 28 997-1009

Sugihara, K.: Okabe, A.; Boots, B.; — — (Auth); Lorz, U. (Rev): Review of "Spatial tessellations. Concepts and applications of Voronoi diagrams". 95 MaOpfStS 26 183-183

Sugimoto, Tamio: Yamamura, Kohji; — —: Estimation of the pest prevention ability of the import plant quarantine in Japan. 95 Biomtrcs 51 482-490

Sugiura, Nariaki: Approximating Bayes critical region for testing simple and tree ordered normal means. 94 JStatRsr 28 1-20

Sugrue, Timothy F.: McCrohan, Kevin F.; — —: Segmentation in the informal economy. 94 ASAProBuEc 259-264

Suh, Euy Hoon: Percentiles for the distributions belonging to the natural exponential families having power variance functions (Korean). 95 KorApS 8/1 133-149

Suissa, Samy; Blais, Lucie: Binary regression with continuous outcomes. 95 StatMed 14 247-255

Sukhatme, Shashikala; Lin, C. H.: Exact powers of the Wilcoxon test and the logrank test under random censoring. 95 JStPlInf 48 291-301

Sullivan, P. J.; Ye, H.: A prognosis for the sudden release of contaminant in an environmental flow. 95 EnvrMtrc 6 627-636

Chatwin, P. C.; — —: Some targets for turbulent diffusion research. 95 EnvrMtrc 6 685-687

Heagy, W. K.; — —: Fixed-point values of contaminant cloud dilution. 95 EnvrMtrc 6 637-641

Labropulu, F.; — —: Mean-square values of concentration in a contaminant cloud. 95 EnvrMtrc 6 619-625

Labropulu, F.; — —; Ye, H.: A contaminant-dilution heuristic model. 95 EnvrMtrc 6 677-684

Sullivan, Paul J.: The dilution of a contaminant substance released in an environmental flow. 95 SankhyaB 57 277-284

Chatwin, P. C.; Lewis, D. M.; — —: Turbulent dispersion and the beta distribution. 95 EnvrMtrc 6 395-402

Sullivan, Thomas J.: Scott, Stuart; Stamas, George; — —; Chester, Paul: Seasonal adjustment of hybrid economic time series. 94 ASAProBuEc 163-168

Sultan, K. S.: Mohie EL-Din, M. M.; — —: Recurrence relations for expectations of functions of order statistics for doubly truncated distributions and their applications. 95 CommStA 24 997-1010

Sultan, Khalaf Salman: El-Din, Moustafa Mohamed Mohie; Mahmoud, Mohamed Abdel Wahab; — —: On moment generating function of order statistics for doubly truncated exponential distribution. 95 Metron 53 171-183

Sumita, Ushio: Igaki, Nobuko; — —; Kowada, Masashi: Analysis of Markov renewal shock models. 95 JAppProb 32 821-831

Sumpf, D.: Rasch, D.; Tiku, M. L.; — — (Ed); Ahrens, H. (Rev. German): Review of "Elsevier's dictionary of biometry". 95 BiomtrcJ 37 896-896

Rasch, D.; Tiku, M. L.; — — (Ed); Kemp, A. W. (Rev): Review of "Elsevier's dictionary of biometry". 95 Biomtrcs 51 1186-1187

Sun, Don X.; Deng, Li; Wu, C. F. J.: Topics on hidden Markov models and their applications in speech recognition. 94 ASAProStCp 90-99

Sun, Dongchu: Integrable expansions for posterior distributions for a two-parameter exponential family. 94 AnlsStat 22 1808-1830

— —; Ye, Keying: Reference prior Bayesian analysis for normal mean products. 95 JASA 90 589-597

Sun, Feng-Bin: Kececioglu, Dimitri; — —: A general discrete-time dynamic programming model for the opportunistic replacement policy and its application to ball-bearing systems. 95 RelEngSS 47 175-185

Sun, He: Ette, Ene I.; — —; Ludden, Thomas M.: Design of population pharmacokinetic studies. 94 ASAProBiop 487-492

Sun, J.; Kalbfleisch, J. D.: Estimation of the mean function of point processes based on panel count data. 95 StSinica 5 279-289

Sun, Jianguo: Empirical estimation of a distribution function with truncated and doubly interval-censored data and its application to AIDS studies. 95 Biomtrcs 51 1096-1104

— —: A correlation principal component regression analysis of NIR data. 95 JChemtrc 9 21-29

Sun, Jiayang: Iterative estimates for a smoothing parameter. 95 StPrLet 24 347-356

Faraway, Julian J.; — —: Simultaneous confidence bands for linear regression with heteroscedastic errors. 95 JASA 90 1094-1098

Sun, Kai; Basu, Asit P.: A characterization of a bivariate geometric distribution. 95 StPrLet 23 307-311

Sun, Liuquan: Admissibility of linear estimation of regression coefficients in common growth curve models (Chinese). 95 AMaAPknC 17 628-630

Sun, Ping: Further study on the statistical test to detect spatial pattern. 95 BiomtrcJ 37 199-203

Sun, Shan: Perturbed empirical distribution functions and quantiles under dependence. 95 JTherPrb 8 763-777

Brunner, Edgar; Puri, Madan L.; — —: Nonparametric methods for stratified two-sample designs with application to multiclinic trials. 95 JASA 90 1004-1014

Sun, Yanqing: McKeague, Ian W.; Nikabadze, A. M.; — —: On omnibus test for independence of a survival time from a covariate. 95 AnlsStat 23 450-475

Sunder, Shyam: Friedman, Daniel; — — (Auth); Lamoureux, Christopher G. (Rev): Review of "Experimental methods: A primer for economists". 95 JFinance 50 1341-1345

Sundgren, B.: Statistical metainformation systems-pragmatics, semantics, syntactics. 93 StatJUN 10 121-142

— —: A proposal for a survey-oriented metainformation system. 92 StatJUN 9 139-174

Sundt, Bjørn: On some properties of De Pril transforms of counting distributions. 95 AstinB 25 19-31

— —; Teugels, Jozef L.: Ruin estimates under interest force. 95 InsrneME 16 7-22

Sung, Joo-Ho: Haberman, Steven; — —: Dynamic approaches to pension funding. 94 InsrneME 15 151-162

Sung, Nae Kyung: On a robust analysis of variance based on Winsorization (Korean). 95 KorApS 8/1 119-131

Sunter, Deborah; Kinack, Mark: Redesigning the Canadian Labour Force Survey questionnaire: Development and testing. 94 ASASrvyRsr 908-912

Superville, Claude R.: Monitoring measurable process data: Why not use the EWMA chart?. 95 QulPr 28/7 144-144

— —; Adams, Benjamin M.: Autocorrelated data: Control charts versus tracking signals. 94 ASAProQlPr 194-199

Suquet, Charles: Jacob, Pierre; — —: Estimating the edge of a Poisson process by orthogonal series. 95 JStPlInf 46 215-234

Suresh, K. K.; Balamurali, S.: Construction and selection of tightened-normal-tightened plans indexed by maximum allowable percentage defective. 94 JAppStat 21 589-595

Suresh, Ram: Peligard, Magda; — —: Estimation of variance of partial sums of an associated sequence of random variables. 95 StocProc 56 307-319

Surgailis, D.: Funaki, T.; — —; Woyczynski, W. A.: Gibbs-Cox random fields and Burgers turbulence. 95 AnlsApPr 5 461-492

Suter, Glenn: — — (Ed); Ziegel, Eric R. (Rev): Review of "Ecological risk assessment". 95 Technmcs 37 240-240

Suter, Natalie: Johnson, Robert; Malloy, Kay; Gerstein, Dean; — —; Nisenbaum, Susan: Increasing response rates in follow-up surveys of drug treatment patients. 94 ASASrvyRsr 1263-1268

Sutradhar, B. C.; MacNeil, I. B.; Dagum, E. B.: A simple test for stable seasonality. 95 JStPlInf 43 157-167

Sutradhar, Brajendra C.: A global test for the goodness of fit of Cox's proportional hazards model. 94 JStatRsr 28 123-148

— —; Das, Kalyan: Analysing exponential family based messy data. 95 CommStA 24 2683-2699

— —; Das, Kalyan: Tests for overdispersion in longitudinal data in the presence of outliers. 95 SankhyaA 57 299-315

Chu, David P. T.; — —: On Cochran's and Hartley's tests for homogeneity of variances when observations are autocorrelated. 95 CommStB 24 327-347

Das, Kalyan; — —: Score tests for fixed effects and overdispersion components in nonlinear models for repeated measures. 95 JMultiAn 54 126-146

Das, Kalyan; — —: Correction to "Score tests for fixed effects and overdispersion components in nonlinear models for repeated measures" (V54 p126-146). 95 JMultiAn 55 349-349

Suzukawa, Akio; Sato, Yoshiharu: A generalization of testing independence of sets of variates. 95 JJapnStS 25 71-80

Suzuki, Hisatoshi: Kusaka, Yasuo; — —: An analysis of new product diffusion process. 94 JOpRsJpn 38 188-211

Suzuki, Kazuyuki: Estimation of usage lifetime distribution from calendar lifetime data. 93 RStAJp 40/1 10-22

Svensson, Åke: On the asymptotic size and duration of a class of epidemic models. 95 JAppProb 32 11-24

— —: On the simultaneous distribution of size and costs of an epidemic in a closed multigroup population. 95 MaBisci 127 167-180

Svensson, E.: — — (Auth); Lesaffre, E. (Rev): Review of "Analysis of systematic and random differences between paired ordinal categorical data". 95 Biomtrcs 51 794-794

Svensson, Elisabeth: — — (Auth); Lindsey, J. K. (Rev): Review of "Analysis of systematic and random differences between paired ordinal categorical data". 95 StatMed 14 1726-1727

Svishchuk, A. V.: Korolyuk, V. S. (Auth); — — (Rev): Review of "Stochastic models for systems" (Russian). 95 ThPrMaSt 50 147-147

Swaim, Paul: Work experience, local labor markets, and dropping out of high school. 94 ASAProSoSt 207-212 (Disc: p213-215)

Swain, A. J.: Mayer, D. G.; Atzeni, M. G.; — —; Stuart, M. A.: Models for the spatial dispersal of insect pests. 95 EnvrMtrc 6 497-503

Swain, James J.: Graybill, Franklin A.; Iyer, Hariharan K. (Auth); — — (Rev): Review of "Regression analysis: Concepts and applications". 95 JQualTek 27 179-180

Swain, Larry: Comment on "The use of administrative records in government statistics". 94 ASAProGovt 212-214

Swallow, W. H.: Chen, C. L.; — —: Sensitivity analysis of variable-size group testing and its related continuous models. 95 BiomtrcJ 37 173-181

Swamy, Mamatha: Analysis of self and proxy responses in the assessment of health status. 94 ASASrvyRsr 719-724

Swank, W. T.: Chen, T. H.; Hornberger, G. M.; Jakeman, A. J.; — —: The performance of different loss models in the simulation of streamflow. 95 EnvrMtrc 6 479-484

Swanson, Norman R.; White, Halbert: A model-selection approach to assessing the information in the term structure using linear models and artificial neural network. 95 JBES 13 265-275

Swartz, T. B.; Villegas, C.; Martinez, C. J.: On semiparametric pivotal Bayesian inference for quantiles. 95 CommStA 24 2499-2515

Sweeney, E.: Betteley, G.; Mettrick, N.; — —; Wilson, D. (Auth); Shade, John (Rev): Review of "Using statistics in industry". 95 Ststcian 44 541-542

Sweet, Elizabeth M.; Alberti, Nicholas S.: Coverage implications of using the term 'usual residence'. 94 ASASrvyRsr 319-324

Sweeting, Trevor: Statistical models for nondestructive evaluation. 95 IntStRvw 63 199-214

Sweeting, Trevor J.: A framework for Bayesian and likelihood approximations in statistics. 95 Biomtrka 82 1-23

— —; Phillips, Michael J.: An application of nonstandard asymptotics to the analysis of repairable-systems data. 95 Technmcs 37 428-435

Swensson, B.: Särndel, C.-E.; — —; Wretman, J. (Auth); Penny, Richard N. (Rev): Review of "Model assisted survey sampling". 95 Ststcian 44 140-142

Swerdlow, Anthony; dos Santos Silva, Isabell (Auth); Yuen, Peter; Machin, David (Rev): Review of "Atlas of cancer incidence in England and Wales: 1968-85". 95 StatMed 14 1725-1726

Swift, A. Louise: Modelling and forecasting time series with a general non-normal distribution. 95 JForcast 14 45-66

Percival, D. A.; Walden, A. T. (Auth); — — (Rev): Review of "Spectral analysis for physical applications: Multitaper and conventional techniques". 95 JRSS-A 158 198-199

Swift, L.: Janacek, G.; — — (Auth); Davies, Neville (Rev): Review of "Time series - Forecasting, simulation, applications". 95 JRSS-A 158 193-195

Swift, Louise: Janacek, Gareth; — — (Auth); Findley, David F. (Rev): Review of "Time series: Forecasting, simulation, applications". 95 SIAMRvw 37 253-257

Swift, Michael B.: Simple confidence intervals for standardized rates based on the approximate bootstrap method. 95 StatMed 14 1875-1888

Swift, Randall J.: A class of harmonizable isotropic random fields. 95 JCISS 20 111-127

Swires-Hennessy, E.: Duncan, J. W.; Gross, A. C. (Auth); — — (Rev): Review of "Statistics for the 21st century: Proposals for improving statistics for better decision making". 95 Ststcian 44 138-139

Swishchuk, Anatoly V.: Korolyuk, Vladimir S.; — —: Evolution of systems in random media. 95 CRCPress:FL 368p

Switzer, Paul: Høst, Gudmund; Omre, Henning; — —: Spatial interpolation errors for monitoring data. 95 JASA 90 853-861

Solna, Knut; — —: An analysis of regional climate data. 92 ProIrStCl 231-241

Sydorenko, O. O.: On the distribution of supremum of a certain class of pre-Gaussian stochastic processes on a compact set in R^d. 95 ThPrMaSt 50 123-128

Symanowski, James T.: Koehler, Kenneth J.; — —: Constructing multivariate distributions with specific marginal distributions. 95 JMultiAn 55 261-282

Symeonides, Spyridon D.: Magdalinos, Michael A.; — —: Alternative size corrections for some GLS test statistics. The case of the AR(1) model. 95 JEconmtx 66 35-59

Szafarz, Ariane: Broze, Laurence; Gouriéroux, Christian; — —: Solutions of multivariate rational expectations models. 95 EconTher 11 229-257

Szalal, John Paul; Katic, Marko: Composite probability modelling of increasing resolution where diagnostic covariates are unmeasurable for some subjects. 95 PsychRpt 77 67-78

Szekely, Gabor: Rohatgi, Vijay K.; — —; Zsigri, G.: On aging life distribution classes: Further probability properties and tests. 94 JStatRsr 28 111-122

Székely, Gábor J.: Peña, Edsel A.; Rohatgi, Vijay K.; — —: Inference for a general type II censorship model. 95 MaOpfStS 26 241-252

Szekli, R.: Stochastic ordering and dependence in applied probability. 95 Springer:Brln:NY 194p

Daduna, H.; — —: Dependencies in Markovian networks. 95 AdvAppPr 27 226-254

— — (Auth); Teugels, J. L. (Rev): Review of "Stochastic ordering and dependence in applied probability". 95 ShrtBkRv 15/3 48-48

Szekli, Ryszard: Shaked, Moshe; — —: Comparison of replacement policies via point processes. 95 AdvAppPr 27 1079-1103

Szép, Tibor: Survival rates of Hungarian sand martins and their relationship with Sahel rainfall. 95 JAppStat 22 891-904

Szidarovszky, Ferenc: Hassett, Thomas F.; Dietrich, Duane L.; — —: Time-varying failure rates in the availability and reliability analysis of repairable systems. 95 IEEEReli 44 155-160

Sznitman, Alain-Sol: Crossing velocities and random lattice animals. 95 AnlsProb 23 1006-1023

Szpankowski, Wojciech: Sadowsky, John S.; — —: The probability of large queue lengths and waiting times in a heterogeneous multiserver queue I: Tight limits. 95 AdvAppPr 27 532-566

Szroeter, Jerzy: The exact power function of an exact test of a regression model against multiple separate alternatives. 95 CommStA 24 2329-2339

Szulga, Jerzy: de la Peña, V. H.; Montgomery-Smith, S. J.; — —: Contraction and decoupling inequalities for multilinear forms and U-statistics. 94 AnlsProb 22 1745-1765

Szwaczkowski, T.: Hyánek, J.; Moliński, K.; — —: Estimation of parameters in linear model with the presence of selection. 95 ListyBio 32 51-60

Szynal, D.: Grudzień, Z.; — —: Characterizations of distributions by moments of order statistics when the sample size is random. 95 ApplMath 23 305-318

Krajka, A.; — —: On monotone dependence functions of the quantile type. 95 ApplMath 23 51-72

Szyszkowicz, Mieczyslaw: Krewski, Daniel; Smythe, Robert T.; Dewanji, Anup; — —: Bootstrapping an empirical Bayes estimator of the distribution of historical controls in carcinogen bioassay. 95 JStPlInf 43 215-228

Taam, Winson: Some examples of applied statistics in automobile industry. 94 ASAProQlPr 258-263

Taavitsainen, Veli-Matti; Jalava, Juho-Pertti: Soft and harder multivariate modelling in developing the properties of titanium dioxide pigments. 95 Chemolab 29 307-319

Taba, S.: Crossa, J.; Hernandez, C. M.; Bretting, P.; Eberhart, S. A.; — —: Statistical genetic considerations for maintaining germ plasm collections. 93 ThApGene 86 673-678

Tabar, Laszlo: Duffy, Stephen W.; Chen, Hsiu-Hsi; — —; Day, Nicholas E.: Estimation of mean sojourn time in breast cancer screening using a Markov chain model of both entry to and exit from the preclinical detectable phase. 95 StatMed 14 1531-1543

Taniguchi, Masanobu; Puri, Madan L.: Higher order asymptotic theory for normalizing transformations of maximum likelihood estimators. 95 AnInStMa 47 581-600

Zhang, Guoqiang; — —: Nonparametric approach for discriminant analysis in time series. 95 JNnparSt 5 91-101

Taniguchi, Takeshi: Asymptotic stability theorems of semilinear stochastic evolution equations in Hilbert spaces. 95 Stochatx 53 41-52

Tanis, Elliot A.: International Statistical Institute (Auth); — — (Rev): Review of "Proceedings of the Fourth International Conference on Teaching Statistics". 95 AmerStat 49 219-220

Tanizaki, Hisashi: Asymptotically exact confidence intervals of CUSUM and CUSUMSQ tests: A numerical derivation using simulation technique. 95 CommStB 24 1019-1036

Mariano, Roberto S.; — —: Prediction of final data with use of preliminary and/or revised data. 95 JForcast 14 351-380

Tanner, J. Ernest: Leitch, Gordon; — —: Professional economic forecasts: Are they worth their costs?. 95 JForcast 14 143-157

Tanner, Martin A.: Reply to comment on "Applications of multiple imputation to the analysis of censored regression data". 95 Biomtrcs 51 362-362

Yakir, Benjamin; Cox, Christopher; — —; Clarkson, Thomas W.; Shamlaye, C.: Seasonal changes in methylmercury levels in human hair. 94 ASAProBays 35-39

Tanny, David: Madras, Neal; — —: Oscillating random walk with a moving boundary. 94 IsrelJMa 88 333-365

Tansel, Abdullah Uz: Query languages for statistical databases. 95 StatComp 5 59-72

Tanur, Judith M.: Conceptualizations of job search: Further evidence from verbatim responses. 94 ASASrvyRsr 512-516

Horvitz, Dan; Koshland, Daniel; Rubin, Donald; Gollin, Albert; Sawyer, Tom; — —: Pseudo-opinion polls: SLOP or useful data?. 95 Chance 8/2 16-25

Tapiero, Charles S.: Acceptance sampling in a producer-supplier conflicting environment: Risk neutral case. 95 AplStMDA 11 3-12

Taqqu, Murad S.: Cioczek-Georges, Renata; — —: Necessary conditions for the existence of conditional moments of stable random variables. 95 StocProc 56 233-246

Cioczek-Georges, Renata; — —: Form of the conditional variance for stable random variables. 95 StSinica 5 351-361

Cioczek-Georges, Renata; Mandelbrot, Benoit B.; Samorodnitsky, Gennady; — —: Stable fractal sums of pulses: The cylindrical case. 95 Bernoull 1 201-216

Kokoszka, Piotr S.; — —: Fractional ARIMA with stable innovations. 95 StocProc 60 19-47

Samorodnitsky, Gennady; — —: Lévy measures of infinitely divisible random vectors and Slepian inequalities. 94 AnlsProb 22 1930-1956

Willinger, Walter; — —; Leland, Will E.; Wilson, Daniel V.: Self-similarity in high-speed packet traffic: Analysis and modeling of ethernet traffic measurements. 95 StatSci 10 67-85

Samorodnitsky, Gennady; — — (Auth); Mikosch, Thomas (Rev): Review of "Stable non-Gaussian random processes: Stochastic models with infinite variance". 95 JASA 90 805-806

Tarboton, David G.: Hydrologic scenarios for severe sustained drought in the Southwestern United States. 95 WaterReB 31 803-813

Tardif, S.: Bellavance, F.; — —: A nonparametric approach to the analysis of three-treatment three-period crossover designs. 95 Biomtrka 82 865-875

Tardif, Serge: Larocque, Denis; — —: Development of an optimal test based on weighted rankings. 95 CommStB 24 367-383

Tardiff, Mark F.: Wolf, Dennis A.; — —; Ketelle, Richard H.: Characterizing groundwater at Oak Ridge National Laboratory hazardous waste sites. 94 ASAProEnvr 116-121

Taris, Toon: Dijkstra, Wil; — —: Measuring the agreement between sequences. 95 SocMethR 24 214-231 (Disc: p232-243)

Tarling, R.: — — (Auth); Walker, Monica (Rev): Review of "Analysing offending: Data, models and interpretations". 95 JRSS-A 158 357-358

Tarpey, Thaddeus: Principal points and self-consistent points of symmetric multivariate distributions. 95 JMultiAn 53 39-51

— —; Li, Luning; Flury, Bernard D.: Principal points and self-consistent points of elliptical distributions. 95 AnlsStat 23 103-112

Tartakovsky, A.: Asymptotic properties of CUSUM and Shiryaev's procedures for detecting a change in a nonhomogeneous Gaussian process. 95 MaMethSt 4 389-404

Tartar, M. E.; Lock, M. D. (Auth); Nason, Guy (Rev): Review of "Model-free curve estimation". 95 SIAMRvw 37 270-271

Tarter, Michael E.; Lock, Michael D. (Auth); Wharton, Robert M. (Rev): Review of "Model-free curve estimation". 95 Technmcs 37 460-460

Tauchen, George: Bansal, Ravi; Gallant, A. Ronald; Hussey, Robert; — —: Nonparametric estimation of structural models for high-frequency currency market data. 95 JEconmtx 66 251-287

Taudes, Alfred; Reepmeyer, Jan-Armin; Bensberg, Frank (Auth); Weinberger, Volker (Rev): Review of "UNIX: Introduction for DOS-users" (German). 95 CmpStDA 19 360-361

Taufik, A. M. Z.: Increasing the efficiency of the Monte-Carlo method in estimating error probabilities of likelihood-ratio tests (Russian). 94 VLenUM /2 105-108

Tauler, R.; Izquierdo-Ridorsa, A.; Gargallo, R.; Casassas, E.: Application of a new multivariate curve resolution procedure to the simultaneous analysis of several spectroscopic titrations of the copper(II)-polyinosinic acid system. 95 Chemolab 27 163-174

Tauler, Roma: Multivariate curve resolution applied to second order data. 95 Chemolab 30 133-146

Tauler, Romá; Smilde, Age; Kowalski, Bruce: Selectivity, local rank, three-way data analysis and ambiguity in multivariate curve resolution. 95 JChemtrc 9 31-58

Tavaré, C. J.; Sobel, E. L.; Gilles, F. H.: Misclassification of a prognostic dichotomous variable: Sample size and parameter estimate adjustment. 95 StatMed 14 1307-1314

Tavaré, Simon: Arratia, Richard; Stark, Dudley; — —: Total variation asymptotics for Poisson process approximations of logarithmic combinatorial assemblies. 95 AnlsProb 23 1347-1388

Griffiths, R. C.; — —: Unrooted genealogical tree probabilities in the infinitely-many-sites model. 95 MaBisci 127 77-98

Taverner, Fred: Lipshutz, Robert J.; — —; Hennessy, Kevin; Hartzell, George; Davis, Ron: DNA sequence confidence estimation. 94 Genomics 19 417-424

Tawn, Jonathan A.: Dixon, Mark J.; — —: A semi-parametric model for multivariate extreme values. 95 StatComp 5 215-225

Robinson, Michael E.; — —: Statistics for exceptional athletics records. 95 ApplStat 44 499-511

Taylor, Brian H.: Gutierrez, Roberto G.; Carroll, Raymond J.; Wang, Naisyin; Lee, Gyung-Hee; — —: Analysis of tomato root initiation using a normal mixture distribution. 95 Biomtrcs 51 1461-1468

Taylor, Bridget: Lynn, Peter; — —: On the bias and variance of samples of individuals: Comparison of the electoral registers and postcode address file as sampling frames. 95 Ststcian 44 173-194

Taylor, C. Barr: Fortmann, Stephen P.; Flora, June A.; Winkleby, Marilyn A.; Schooler, Caroline; — —; Farquhar, John W.: Community intervention trials: Reflections on the Stanford Five-City Project experience. 95 AJEpidm 142 576-586

Taylor, C. C.: McLachlan, G. J. (Auth); — — (Rev): Review of "Discriminant analysis and statistical pattern recognition". 95 JAppStat 22 185-185

Michie, D.; Spiegelhalter, D. J.; — — (Ed); Fulkerson, Bill (Rev): Review of "Machine learning, neural and statistical classification". 95 Technmcs 37 459-459

Taylor, Charles: Cressie, N. A. C. (Auth); — — (Rev): Review of "Statistics for spatial data". 95 StMedRes 4 91-91

Taylor, Douglas J.; Muller, Keith E.: Computing confidence bounds for power and sample size of the general linear univariate model. 95 AmerStat 49 43-47

— —; Muller, Keith E.: Computing defensible linear model power and sample size estimates for study planning. 94 ASAProBiop 383-388

Taylor, H. M.; Karlin, S. (Auth); Crowder, Martin (Rev): Review of "An introduction to stochastic modelling". 95 Ststcian 44 291-292

Taylor, J. A.: Mulquiney, J. E.; Norton, J. P.; Jakeman, A. J.; — —: Random walks in the Kalman filter: Implications for greenhouse gas flux deductions. 95 EnvrMtrc 6 473-478

Taylor, Jeremy M. G.: Semi-parametric estimation in failure time mixture models. 95 Biomtrcs 51 899-907

Kim, Dong K.; — —: The restricted EM algorithm for maximum likelihood estimation under linear restrictions on the parameters. 95 JASA 90 708-716

Wang, Yongxiao; — —: Flexible methods for analysing longitudinal data using piecewise cubic polynomials. 95 JStCmpSm 52 133-150

Wang, Yongxiao; — —: Inference for smooth curves in longitudinal data with application to an AIDS clinical trial. 95 StatMed 14 1205-1218

Taylor, Jim: A statistical analysis of the 1992 research assessment exercise. 95 JRSS-A 158 241-261

Taylor, Larry W.: Mudambi, Ram; — —: Some non-parametric tests for duration dependence: An application to UK business cycle data. 95 JAppStat 22 163-177

Skeels, Christopher L.; — —: On a simultaneous equations pre-test estimator. 95 JEconmtx 68 269-286

Taylor, Lowell J.: Kim, Taeil; — —: The employment effect in retail trade of California's 1988 minimum wage increase. 95 JBES 13/2 175-182

Taylor, Mark: Stone, Richard; — —: Time series models in statistical process control: Considerations of applicability. 95 Ststcian 44 227-234

Taylor, Mark P.: Cuthbertson, Keith; Hall, Stephen G.; — — (Auth); Ericsson, Neil R.; Irons, John S. (Rev): Review of "Applied econometric techniques". 95 EcomtrRv 14 121-132

Taylor, N. W.: Jones, K. F.; — —; Fosam, E. B.: Statistical approaches for the reliability of structural designs. 95 JAppStat 22 443-458

Taylor, P. G.: Bean, N. G.; — —: Maximal profit dimensioning and tariffing of loss networks. 95 PrEngInf 9 323-340

Bright, L.; — —: Calculating the equilibrium distribution in level dependent quasi-birth-and-death processes. 95 CommStSM 11 497-525

Coyle, A. J.; — —: Tight bounds on the sensitivity of generalised semi-Markov processes with a single generally distributed lifetime. 95 JAppProb 32 63-73

Henderson, W.; Pearce, C. E. M.; — —; van Dijk, N. M.: Insensitivity in discrete-time generalized semi-Markov processes allowing multiple events and probabilistic service scheduling. 95 AnlsApPr 5 78-96

Taylor, Robert L.: Inoue, Hiroshi; — —: A SLLN for arrays of rowwise exchangeable fuzzy random sets. 95 StocAnAp 13 461–470

Taylor, S. J.: Le Gall, J.-F.; Perkins, E. A.; — —: The packing measure of the support of super-Brownian motion. 95 StocProc 59 1–20

Olsen, L. (Auth); — — (Rev): Review of "Random geometrically graph directed self-similar multifractals". 95 Metrika 42 277–278

Taylor, Wayne A.: Reducing variation during design. 95 ASQCTekCon 217–226

— —: The secret behind variation transmission analysis: Why predicting variation may be better than observing it?. 94 ASAProQlPr 49–54

Tazawa, Shinsei: Shirakura, Teruhiro; — —: Search designs for balanced 2^m factorial designs of resolution $2l+1$ when $l+s+1$-factor and higher order interactions are negligible. 93 JCISS 18 233–245

Tehonica, Jason: Updating occupational compensation survey program data using the employment cost index. 94 ASASrvyRsr 1048–1052

Teich, Malvin C.: Turcott, Robert G.; Barker, Peter D. R.; — —: Long-duration correlation in the sequence of action potentials in an insect visual interneuron. 95 JStCmpSm 52 253–271

Teicher, Henry: Moments of randomly stopped sums-revisited. 95 JTherPrb 8 779–793

Tejada, J.: Manuel, C. M.; — —: Reciprocal cooperation in the prisoner's dilemma repeated with random horizon. 95 Top 3 97–116

Montero, J.; Pearman, A. D.; — —: Fuzzy multicriteria decision support for budget allocation in the transport sector. 95 Top 3 47–68

Tejedor, I. Herranz: Martín Andrés, A.; — —: Is Fisher's exact test very conservative?. 95 CmpStDA 19 579–591

Martín Andrés, A.; — —; Mato, A. Silva: The Wilcoxon, Spearman, Fisher, χ^2, Student and Pearson tests and 2×2 tables. 95 Ststcian 44 441–450

Tejerizo, María Luisa: Ferrándiz, Juan; López, Antonio; Llopis, Agustín; Morales, María; — —: Spatial interaction between neighbouring counties: Cancer mortality data in Valencia (Spain). 95 Biomtrcs 51 665–678

Telcs, András: Spectra of graphs and fractal dimensions II. 95 JTherPrb 8 77–96

Telfar, Grant: Haslett, Stephen; — —: Sample design for auditing error rates in claims data: A case study. 95 NZndSt 30 23–36

Ten Have, Thomas R.; Becker, Mark P.: Multivariate contingency tables and the analysis of exchangeability. 95 Biomtrcs 51 1001–1016

Diggle, Peter J.; Liang, Yung-Yee; Zeger, Scott L. (Auth); — — (Rev): Review of "Analysis of longitudinal data". 95 JASA 90 1123–1124

Tenenhaus, M.; Gauchi, J.-P.; Ménardo, C.: PLS regression and applications (French). 95 RStApp 43/1 7–63

Teoh, Choo B.: Chang, Tiao J.; — —: Use of the kriging method for studying characteristics of ground water droughts. 95 WaterReB 31 1001–1007

Teoh, E. K.: Suganthan, P. N.; — —; Mital, D. P.: Pattern recognition by graph matching using the Potts MFT neural networks. 95 PattrnRc 28 997–1009

Teoh, Nick K. W.: Dudewicz, Edward J.; van der Meulen, Edward C.; SriRam, M. G.; — —: Entropy-based random number evaluation. 95 AJMMgSci 15 115–153

Teplyaev, A. V.: Continual analogues of random polynomials which are orthogonal on a circle. 94 ThProbAp 39 476–489

Teräsvirta, T.: Granger, C. W. J.; — — (Auth); McKenzie, Sandra (Rev): Review of "Modelling nonlinear economic relationships". 95 JASA 90 1491–1491

Teräsvirta, Timo: Professor Clive W. J. Granger: An interview for the International Journal of Forecasting. 95 IntJFor 11 585–590

Terpessi, Cristina: Espa, Giuseppe; — —: Short and medium term rain forecasting in the Oristano basin (Italian). 94 QStMaA 16 145–160

Terraza, Michel: Cromwell, Jeff B.; Labys, Walter C.; — — (Auth); Downing, Darryl J. (Rev): Review of "Univariate tests for time series models". 95 Technmcs 37 346–346

Terrell, Dek: Flexibility and regularity properties of the asymptotically ideal production model. 95 EcomtrRv 14 1–17

Terrie, E. Walter: Counting public libraries in the United States. 94 ASAProGovt 58–63

Terrin, Norma: Chan, Ngai Hang; — —: Inference for unstable long-memory processes with applications to fractional unit root autoregressions. 95 AnlsStat 23 1662–1683

Terwilliger, Joseph D.: A powerful likelihood method for the analysis of linkage disequilibrium between trait loci and one or more polymorphic marker loci. 95 AJHuGene 56 777–787

Tessier, David: Dufour, Jean-Marie; Nsiri, Saïd; — —: Parsimonious autoregressive conditions for noncausality in multivariate ARMA models. 94 ASAProBuEc 129–134

Tetzlaff, Klaus: Schulz, Henrik; — —; Vogel, Lydia: Simple encoding of infrared spectra for pattern recognition. 1: Statistical examination of the effectiveness and information content by factor analysis. 95 JChemtrc 9 143–168

Teugels, J. L.: Rahimov, I. (Auth); — — (Rev): Review of "Random sums and branching stochastic processes". 95 ShrtBkRv 15/3 48–48

Szekli, R. (Auth); — — (Rev): Review of "Stochastic ordering and dependence in applied probability". 95 ShrtBkRv 15/3 48–48

Teugels, Jozef L.: Beirlant, Jan; Broniatowski, Michel; — —; Vynckier, Petra: The mean residual life function at great age: Applications to tail estimation. 95 JStPlInf 45 21–48

Sundt, Bjørn; — —: Ruin estimates under interest force. 95 InsrneME 16 7–22

Teuscher, F.; Guiard, V.: Sharp inequalities between skewness and kurtosis for unimodal distributions. 95 StPrLet 22 257–260

Tew, Jeffrey D.: Yunker, James M.; — —: Simulation optimization by genetic search. 94 MaCmpSim 37 17–28

Thacker, Stephen B.; Stroup, Donna F.; Rothenberg, Richard B.; Brownson, Ross C.: Public health surveillance for chronic conditions: A scientific basis for decisions. 95 StatMed 14 629–641 (Disc: p695–696)

Smith, S. Jay; Caudill, Samuel P.; Steinberg, Karen K.; — —: On combining dose-response data from epidemiological studies by meta-analysis. 95 StatMed 14 531–544 (Disc: p679–685)

Thall, Peter F.; Simon, Richard M.; Estey, Elihu H.: Bayesian sequential monitoring designs for single-arm clinical trials with multiple outcomes. 95 StatMed 14 357–379

Thanasoulias, Nickolaos: Papamichael, Emmanuel M.; Evmiridis, Nickolaos P.; — —; Stefanou, Despina: Design of experiments for a precise estimation of the calibration curve of periodate in its determination by flow-injection analysis and chemiluminescence detection.. 95 Chemolab 30 227–237

Thang, Dang Hung: The adjoint and the composition of random operators on a Hilbert space. 95 Stochatx 54 53–73

The MathWorks Inc: — — (Auth); Miller, Michael B. (Rev): Review of "MATLAB Version 4.2". 95 StrEqMod 2 175–185

— — (Auth); Spiegelman, Cliff (Rev): Review of "MATLAB for the Power Macintosh". 95 Chemolab 29 323–324

The Soft Warehouse Inc: — — (Auth); Nash, John C. (Rev): Review of "DERIVE (Version 2.55, Extended memory Version 2.55XM, and Version 2.60)". 95 AmerStat 49 93–99

Theiller, Gédéon: Feinberg, Max; Bugner, Emmanuel; — —; Houba, Victor; Kadijk, Fetze: Expression of the reference value for proficiency tests. 95 JChemtrc 9 197–209

Thelin, Bernt; Lundstedt, Torbjörn; Lundgren, Rolf; Olsson, Arne; Batra, Satish: Classification of Estradurin batches: Correlation between 31P NMR and a biological duration test for batch approval. 95 Chemolab 27 135–145

Lundstedt, Torbjörn; — —: A multivariate strategy for optimizing a two-step process. 95 Chemolab 29 255–261

Theobald, C. M.; Kirby, S. P. J.: Discrimination using covariance selection models in the automated allocation of human chromosomes. 95 Ststcian 44 235–240

Theodorescu, Radu: Bertin, Emile; — —: On a two-parameter family of discrete unimodal random variables. 95 SankhyaA 57 41–55

Bertin, Emile; — —: Preserving unimodality by mixing. 95 StPrLet 25 281–288

Therasse, P.: Rondel, R. K.; Varley, S. A.; Webb, C. F. (Ed); — — (Rev): Review of "Clinical data management". 95 StatMed 14 877–877

Therneau, Terry M.: Grambsch, Patricia M.; — —: Correction to "Proportional hazards tests and diagnostics based on weighted residuals" (V81 p515-526). 95 Biomtrka 82 668–668

Grambsch, Patricia M.; — —; Fleming, Thomas R.: Diagnostic plots to reveal functional form for covariates in multiplicative intensity models. 95 Biomtrcs 51 1469–1482

Theunissen, I.: Ritter, C.; Bouckaert, A.; van Lierde, M.; — —: Modelling mortality and morbidity of newborns. 95 StatMed 14 2191–2204

Theunissen, M.: Tan, E. S.; Imbos, Tj.; Does, R. J. M. M.; — —: An optimal, unbiased classification rule for mastery testing based on longitudinal data. 95 EdPsMeas 55 595–612

Theus, M.: Bernt, K.; — — (Rev): Review of "Mercator for Windows 4.0". 95 CmpStQ 10 307–308

Theus, Martin: Trellis displays vs. interactive graphics. 95 CmpStQ 10 113–127

Thiagarajah, K. Ranee: Homogeneity tests for scale parameters of 2-parameter exponential populations under time censoring. 95 IEEEReli 44 297–301

Thiagarajah, Kulathava Ranee: Comparisons of approximate confidence intervals for the parameters of extreme value distribution. 94 Metron 52/1 85–100

Thibaudeau, Yves; Kostanich, Donna; Parmer, Randall: The influence of implicit methodologies in the analysis of the CCO and the CPS. 94 ASASrvyRsr 730–735

Thiébaux, H. Jean: — — (Auth); Solow, Andrew R. (Rev): Review of "Statistical data analysis for ocean and atmospheric sciences". 95 AmerStat 49 399–399

Thiesson, Bo: Højsgaard, Søren; — —: BIFROST - Block recursive models induced from relevant knowledge, observations, and statistical techniques. 95 CmpStDA 19 155–175

Thioulouse, Jean; Chessel, Daniel; Champely, Stéphane: Multivariate analysis of spatial patterns: A unified approach to local and global structures. 95 EnvEcoSt 2 1–14

Thissen, David; Pommerich, Mary; Billeaud, Kathleen; Williams, Valerie S. L.: Item response theory for scores on tests including polytomous items with ordered responses. 95 ApPsychM 19 39–49

Wainer, Howard; Dorans, Neil J.; Flaugher, Ronald; Green, Bert F.; Mislevy, Robert; Steinberg, Lynne; — — (Auth); Andrich, David (Rev): Review of "Computerized adaptive testing: A primer". 95 Psymtrka 60 615–620

Thomas, Andrew: Smith, Teresa C.; Spiegelhalter, David J.; — —: Bayesian approaches to random-effects meta-analysis: A comparative study. 95 StatMed 14 2685–2699

Thomas, Bradley W.: Updating the Statistics Canada Farm Register with the Canadian Census of Agriculture. 94 ASAProGovt 189–194 (Disc: p212-214)

Thomas, D. L.: Manly, B. F. J.; McDonald, L. L.; — — (Auth); North, Philip M. (Rev): Review of "Resource selection by animals: Statistical design and analysis for field studies". 95 JAppStat 22 192-193

Thomas, Dana: Flint, Paul L.; Pollock, Kenneth H.; — —; Sedinger, James S.: Estimating prefledging survival : Allowing for brood mixing and dependence among brood mates. 95 JWildlMg 59 448-455

Thomas, David W.: Chen, Yan Qiu; Nixon, Mark S.; — —: Statistical geometrical features for texture classification. 95 PattrnRc 28 537-552

Thomas, Duncan C.: Comment on "When will nondifferential misclassification of an exposure reverse the direction of a trend?" (V140 p565-571). 95 AJEpidm 142 782-783

— —; Cortessis, Victoria: A Gibbs sampling approach to linkage analysis. 92 HumHrdty 42 63-76

Witte, John S.; — —; Langholz, Bryan: Comment on "Statistical significance testing in the American journal of epidemiology, 1970-1990" (V139 p1047-1052). 95 AJEpidm 142 101-101

Thomas, E.: Watt, T. A. (Auth); — — (Rev): Review of "Introductory statistics for biology students". 95 BiomtrcJ 37 96-96

Thomas, Edward V.: Incorporating auxiliary predictor variation in principal component regression models. 95 JChemtrc 9 471-481

Woodall, William H.; — —: Statistical process control with several components of common cause variability. 95 IIETrn 27 757-764

Thomas, G. E.: A note on chi-squared approximations for tests of rank interaction with ordered categorical data. 95 CommStB 24 439-459

Thomas, Harold W.: Sheesley, John H.; — —; Valenzuela, Carlos A.: Quantal response analysis of relief valve test data. 95 ASQCTekCon 741-748

Thomas, M.: Haßmann, S.; Reiss, R.-D.; — — (Auth); Anderson, C. W. (Rev): Review of "XTREMES". 95 CmpStQ 10 91-98

Thomas, M. R.: Krzanowski, W. J.; Jonathan, P.; McCarthy, W. V.; — —: Discriminant analysis with singular covariance matrices: Methods and applications to spectroscopic data. 95 ApplStat 44 101-115

Thomas, Mariamma: Biswas, Suddhendu; — —: On a continuous branching process and birth-death and immigration models for testing the efficacy of some control programmes for AIDS. 95 BiomtrcJ 37 235-249

Thomas, Mervyn R.: Wang, You-Gan; — —: Accounting for individual variability in the von Bertalanffy growth model. 95 JFishRBC 52 1368-1375

Wang, You-Gan; — —; Somers, Ian F.: A maximum likelihood approach for estimating growth from tag-recapture data. 95 JFishRBC 52 252-259

Thomas, Michael L.: Statistical process control in software development. 94 ASAProQlPr 140-145

Thomas, Roger: Campanelli, Pamela; Channell, Joanna; McAulay, Liz; Renouf, Antoinette; — —: A cognitive and linguistic exploration of the meaning of training: Implications for survey design. 94 ASASrvyRsr 1281-1286

Thomas, S. M.: Jones, M. E.; — —; Rogers, A.: Luria-Delbrück fluctuation experiments: Design and analysis. 94 Genetcs 136 1209-1216

Thomas-Agnan, C.: Delecroix, M.; Simioni, M.; — —: A shape constrained smoother: Simulation study. 95 CmpStQ 10 155-175

Thombs, Lori A.: Romano, Joseph P.; — —: Inference for autocorrelations by resampling. 94 ASAProStCp 1-10

Spurrier, John D.; Edwards, Don; — — (Auth); Hartlaub, Bradley A. (Rev): Review of "Elementary statistics laboratory manual (MS-DOS Version)". 95 AmerStat 49 397-398

Thome, Helmut: A Box-Jenkins approach to modeling outliers in time series analysis. 95 SocMethR 23 442-478

Thompson, Bruce: Exploring the replicability of a study's results: Bootstrap statistics for the multivariate case. 95 EdPsMeas 55 84-94

Huberty, Carl J. (Auth); — — (Rev): Review of "Applied discriminant analysis". 95 EdPsMeas 55 340-350

Thompson, E.: Lin, S.; — —; Wijsman, E.: An algorithm for Monte Carlo estimation of genotype probabilities on complex pedigrees. 94 AnHuGene 58 343-357

Thompson, Elizabeth: Lin, Shili; — —; Wijsman, Ellen: Finding noncommunicating sets for Markov chain Monte Carlo estimations on pedigrees. 94 AJHuGene 54 695-704

Thompson, Elizabeth A.: Geyer, Charles J.; — —: Annealing Markov chain Monte Carlo with applications to ancestral inference. 95 JASA 90 909-920

Thompson, Eric J.: The 1991 census of population in England and Wales. 95 JRSS-A 158 203-229 (Disc: p230-240)

Thompson, Jenny; Fisher, Robin: Two-sample McNemar tests for complex survey data. 94 ASASrvyRsr 608-613

— —; Parmer, Randall: A comparison of nonparametric methods with parametric methods for the CPS CATI/CAPI mode effects analysis. 94 ASASrvyRsr 614-619

Thompson, Mary E.: Valz, Paul D.; McLeod, A. Ian; — —: Cumulant generating function and tail probability approximations for Kendall's score with tied rankings. 95 AnlsStat 23 144-160

Thompson, Michael: Mertens, Bart; Fearn, Tom; — —: The efficient cross-validation of principal components applied to principal component regression. 95 StatComp 5 227-235

Thompson, Paul A.: Correspondence analysis in statistical package programs. 95 AmerStat 49 310-316

Thompson, Robin: Gilmour, Arthur R.; — —; Cullis, Brian R.: Average information REML: An efficient algorithm for variance parameter estimation in linear mixed models. 95 Biomtrcs 51 1440-1450

Thompson, Russell G.: Otto, Gordon H.; — —: The search for best practices in the U.S. petroleum industry. 94 ASAProQlPr 70-75

Thompson, Simon G.: Hughes, Michael D.; — —; Pocock, Stuart J.: Optimal sequential screening guidelines for quantitative risk factors measured with error. 95 JASA 90 19-26

Royston, Patrick; — —: Comparing non-nested regression models. 95 Biomtrcs 51 114-127

Thomson, Bruce: Burrows, William R.; Benjamin, Mario; Beauchamp, Stephen; Lord, Edward R.; McCollor, Douglas; — —: CART decision-tree statistical analysis and prediction of summer season maximum surface ozone for the Vancouver, Montreal, and Atlantic regions of Canada. 95 JApMetrl 34 1848-1862

Thomson, Glenys: Analysis of complex human genetic traits: An ordered-notation method and new tests for mode of inheritance. 95 AJHuGene 57 474-486

— —: Mapping disease genes: Family-based association studies. 95 AJHuGene 57 487-498

Jin, Kun; Speed, T. P.; — —: Tests of random mating for a highly polymorphic locus: Application to HLA data. 95 Biomtrcs 51 1064-1076

Thomson, P. C.: A hybrid paired and unpaired analysis for the comparison of proportions. 95 StatMed 14 1463-1470

Thomson, Peter: Gray, Alistair; — —: Design of moving-average trend filters using fidelity and smoothness criteria. 94 ASAProBuEc 157-162

Thoni, Janice: Herring, Susan; — —: Statistics before college. 94 ASAProStEd 218-221

Thorisson, Hermann: On time-and cycle-stationarity. 95 StocProc 55 183-209

Thorne, Jeffrey L.; Churchill, Gary A.: Estimation and reliability of molecular sequence alignments. 95 Biomtrcs 51 100-113

Thorogood, Jane: van Houwelingen, Hans C.; — —: Construction, validation and updating of a prognostic model for kidney graft survival. 95 StatMed 14 1999-2008

Thosar, Satish: Jaggia, Sanjiv; — —: Contested tender offers: An estimate of the hazard function. 95 JBES 13 113-119

Thran, Sara L.; Downes-Le Guin, Theodore; Berry, Sandra: Improving productivity in an annual survey of physicians. 94 ASASrvyRsr 577-582

Thursby, Jerry G.: Dezhbakhsh, Hashem; — —: A Monte Carlo comparison of tests based on the Durbin-Watson statistic with other autocorrelation tests in dynamic models. 95 EcomtrRv 14 347-365

Thurston, Sally: Zaslavsky, Alan; — —: Error analysis of food stamp microsimulation models. 94 ASASrvyRsr 535-540

Thurston, Sally W.; Zaslavsky, Alan M.: Using ethnographic data to evaluate dual system estimates. 94 ASASrvyRsr 254-259

Tian, Yubin: Cui, Hengjian; — —: Estimation of projection absolute median deviation and its application (Chinese). 95 JSysSMaS 14 63-72

Tiao, George C.: Niu, Xufeng; — —: Modeling satellite ozone data. 95 JASA 90 969-983

Tibshirani, R. J.: Efron, B.; — — (Auth); Brown, Jason (Rev): Review of "An introduction to the bootstrap". 95 JQualTek 27 90-91

Efron, B.; — — (Auth); Härdle, W. (Rev): Review of "An introduction to the bootstrap". 95 Metrika 42 267-268

Efron, B.; — — (Auth); Marriott, Paul (Rev): Review of "An introduction to the bootstrap". 95 JRSS-A 158 347-347

Tibshirani, Robert: Hastie, Trevor; — —: Generalized additive models for medical research. 95 StMedRes 4 187-196

Hastie, Trevor; Buja, Andreas; — —: Penalized discriminant analysis. 95 AnlsStat 23 73-102

Tibshirani, Robert J.: Efron, B.; — — (Auth); Brown, Morton B. (Rev): Review of "An introduction to the bootstrap". 95 StMedRes 4 91-92

Efron, Bradley; — — (Auth); Grimshaw, Scott D. (Rev): Review of "An introduction to the bootstrap". 95 Technmcs 37 341-341

Tichy, M.: — — (Auth); Csenki, A. (Rev): Review of "Applied methods of structural reliability". 95 Ststcian 44 414-415

Tidemann, Kevin: Esposito, Richard; Fox, James K.; Lin, Dongyow; — —: ARIES: Visual techniques for statistical data investigation at the Bureau of Labor Statistics. 94 ASAProStGr 7-10

Tiede, Manfred: The mean (German). 95 AllgeStA 79 289-318

Tiedge, J.: Schäbe, H.; — —: Kernel estimation for characteristics of pure jump processes. 95 StHefte 36 131-144

Tier, B.: Hoeschele, I.; — —; Graser, H.-U.: Multiple-trait genetic evaluation for one polychotomous trait and several continuous traits with missing data and unequal models. 95 JAnmlSci 73 1609-1627

Tier, Bruce: van Arendonk, Johan A. M.; — —; Kinghorn, Brian P.: Use of multiple genetic markers in prediction of breeding values. 94 Genetcs 137 319-329

Tier, Charles: Knessl, Charles; — —: Asymptotic properties of first passage times for tandem Jackson networks I: Buildup of large queue lengths. 95 CommStSM 11 133-162

Tierney, Camlin; McKnight, Barbara: Power of affected sibling method tests for linkage. 93 HumHrdty 43 276-287

Tricot, J.-M.; Latour, R.: An approach towards the comparison of profiles using a generalized concentration coefficient (French). 95 RStApp 43/3 35-53

Tricot, Jean-Marie; Lepage, Yves: Weighted concordance analysis. 95 BiomtrcJ 37 727-743

Triggs, Christopher M.: Andersson, Steen A.; Madigan, David; Perlman, Michael D.; — —: On the relation between conditional independence models determined by finite distributive lattices and by directed acyclic graphs. 95 JStPlInf 48 25-46

Trindade, David C.: Development of some new graphical analysis tools for designed experiments. 94 ASAProQlPr 115-117

Tripathi, Ram C.: Gupta, Pushpa L.; Gupta, Ramesh C.; — —: Inflated modified power series distributions with applications. 95 CommStA 24 2355-2374

Trippe, Carole; Cohen, Bob: Moving math from the mainframe to the PC: A better policy tool. 94 ASAProGovt 34-38 (Disc: p51-52)

Trippi, Robert R.: Refenes, Apostolos-Paul (Ed); — — (Rev): Review of "Neural networks in the capital markets". 95 IntJFor 11 603-604

— —; Turban, Efraim (Ed); Mangasarian, Olvi L. (Rev): Review of "Neural networks in finance and investing". 95 Intrfc 25/1 141-142

Tritchler, David: Interpreting the standardized difference. 95 Biomtrcs 51 351-353

Trivedi, Yagnesh C.; Kurz, Ludwik: Image restoration using recursive estimators. 95 IEEESyst 25 1470-1482

Trívez, F. Javier: Level shifts, temporary changes and forecasting. 95 JForcast 14 543-550

Troendle, J. F.; Yu, K. F.: A note on testing the Hardy-Weinberg law across strata. 94 AnHuGene 58 397-402

Troendle, James F.: A stepwise resampling method of multiple hypothesis testing. 95 JASA 90 370-378

Trosset, Michael W.; Sands, Barbara N.: On the choice of a discrepancy functional for model selection. 95 CommStA 24 2841-2863

Trost, Donald C.: Accomando, William P.; — —; Finman, Jeffrey S.: Moving from multinational clinical trials to supra-trials. 94 ASAProBiop 232-237

Trost, Robert P.: Joutz, Frederick L.; Maddala, G. S.; — —: An integrated Bayesian vector autoregression and error correction model for forecasting electricity consumption and prices. 95 JForcast 14 287-310

Trouvé, Alain: Asymptotical behaviour of several interacting annealing processes. 95 ProbTher 102 123-143

Trower, Jonathan: Moore, Kris; — —; Borowick, Kent: How birth order influences individual characteristics. 95 Chance 8/4 20-24

Moore, Kris; — —; Borowick, Kent: An empirical study of birth order. 94 ASAProBuEc 347-350

Tršinar, Irena: The central register of population of the Republic of Slovenia. 95 JOffStat 11 221-223

Truman, Benedict I.: Comment on "Study design and public health planning". 95 StatMed 14 695-696

Trumbo, Bruce E.: Some demonstration programs for use in teaching elementary probability and statistics: Parts 3 and 4. 95 JStEd 3/2 60kb (jse/v3n2/trumbo)

Gbur, Edward E., Jr; — —: Key words and phrases - The key to scholarly visibility and efficiency in an information explosion. 95 AmerStat 49 29-33

Truong, Young K.: Kooperberg, Charles; Stone, Charles J.; — —: Hazard regression. 95 JASA 90 78-94

Kooperberg, Charles; Stone, Charles J.; — —: Logspline estimation of a possibly mixed spectral distribution. 95 JTimSrAn 16 359-388

Kooperberg, Charles; Stone, Charles J.; — —: Rate of convergence for logspline spectral density estimation. 95 JTimSrAn 16 389-401

Truong-Van, B.: Invariance principles for semi-stationary sequence of linear processes and applications to ARMA process. 95 StocProc 58 155-172

Tsai, Chih-Ling: Azari, A. S.; — —: Non-parametric regression approach for model checking on the two-phase regression problem. 94 JAppStat 21 597-606

Hurvich, Clifford M.; — —: Model selection for extended quasi-likelihood models in small samples. 95 Biomtrcs 51 1077-1084

Hurvich, Clifford M.; — —: Relative rates of convergence for efficient model selection criteria in linear regression. 95 Biomtrka 82 418-425

Lyon, John; — —: Improved tests for the first-order autoregressive model with heteroscedasticity. 95 JStCmpSm 52 71-83

Tsai, Hsien-Tang; Moskowitz, Herbert; Tang, Jen: A one-sided single screening procedure based on individual unit misclassification error. 95 IIETrn 27 695-706

Tsai, Kao-Tai: Nonparametric analysis of the 2×2 cross-over design with multiple endpoints. 94 ASAProBiop 238-243

Tsai, Ming-Tien: A generalization of the Wishart distribution. 95 StPrLet 24 67-70

— —; Sen, Pranab K.: A test of quasi-independence in ordinal triangular contingency tables. 95 StSinica 5 767-780

Tsai, Peisung: Barlow, Richard E.; — —: Foundational issues concerning the analysis of censored data. 95 LiDaAnal 1 27-34

Tsai, Wei-Yann: Tang, Ming-Xin; — —; Marder, Karen; Mayeux, Richard: Linear rank tests for doubly censored data. 95 StatMed 14 2555-2563

Tsao, Min: Routledge, Rick; — —: Uniform validity of saddlepoint expansion on compact sets. 95 CandJSt 23 425-431

Tsao, S. L.: Gnanadesikan, R.; Kettenring, J. R.; — —: Weighting and selection of variables for cluster analysis. 95 JClass 12 113-136

Tsay, Ruey S.: Chen, Rong; Liu, Jun S.; — —: Additivity tests for nonlinear autoregression. 95 Biomtrka 82 369-383

Tschang, F. Ted; Dowlatabadi, Hadi: A Bayesian technique for refining the uncertainty in global energy model forecasts. 95 IntJFor 11 43-61

Tse, Siu-Keung; Chow, Shein-Chung: On model selection for standard curve in assay development. 95 JBiophSt 5 285-296

Tse, Y. K.: Lead-lag relationship between spot index and futures price of the Nikkei Stock Average. 95 JForcast 14 553-563

Tseng, Sheng-Tsaing; Hamada, Michael; Chiao, Chih-Hua: Using degradation data to improve fluorescent lamp reliability. 95 JQualTek 27 363-369

Tsiatis, A. A.; DeGruttola, Victor; Wulfsohn, M. S.: Modeling the relationship of survival to longitudinal data measured with error. Applications to survival and CD4 counts in patients with AIDS. 95 JASA 90 27-37

Tsiatis, Anastasios: Rabinowitz, Daniel; — —; Aragon, Jorge: Regression with interval-censored data. 95 Biomtrka 82 501-513

Tsiatis, Anastasios A.; Boucher, Hélène; Kim, Kyungmann: Sequential methods for parametric survival models. 95 Biomtrka 82 165-173

Kim, KyungMann; Boucher, Hélène; — —: Design and analysis of group sequential logrank tests in maximum duration versus information trials. 95 Biomtrcs 51 988-1000

Tsimikas, J.: Lenth, R. V.; Markatou, M.; — —: Robust tests based on the sample characteristic function. 95 AstrlJSt 37 45-60

Tsimikas, John; Ledolter, Johannes: New smoothing results for state space models. 94 ASAProBuEc 172-177

Tsodikov, Alexander D.; Asselain, Bernard; Fourque, Alain; Hoang, Thu; Yakovlev, Andrej Yu.: Discrete strategies of cancer post-treatment surveillance. Estimation and optimization problems. 95 Biomtrcs 51 437-447

Tsodikov, Alexandr D.: Yakovlev, Andrej Yu.; — —; Anisimov, Vladimir N.: A new model of aging: Specific versions and their application. 95 BiomtrcJ 37 435-448

Tsokos, Chris P.: Qiao, Hongzhu; — —: Estimation of the three parameter Weibull probability distribution. 95 MaCmpSim 39 173-185

Qiao, Hongzhu; — —: Parameter estimation of the Weibull probability distribution. 94 MaCmpSim 37 47-55

Tsong, Yi: Comparing reporting rates of adverse events between drugs with adjustment for year of marketing and secular trends in total reporting. 95 JBiophSt 5 95-114

— —; Hammerstrom, Thomas: Statistical issues in drug quality control based on dissolution testing. 94 ASAProBiop 295-300

— —; Hammerstrom, Thomas; Lin, Karl; Ong, Ting Eng: Dissolution test acceptance sampling plans. 95 JBiophSt 5 171-183

Chen, James J.; Hwang, Jing-Shiang; — —: Estimation of the shelf-life of drugs with mixed effects models. 95 JBiophSt 5 131-140

Tsou, Tai-Houn: Lii, Keh-Shin; — —: Bispectral analysis of randomly sampled data. 95 JTimSrAn 16 43-66

Tsou, Tsung-Shan; Royall, Richard M.: Robust likelihoods. 95 JASA 90 316-320

Tsoucas, Pantelis: Bhattacharya, Partha P.; Georgiadis, Leonidas; — —: Problems of adaptive optimization in multiclass M/GI/1 queues with Bernoulli feedback. 95 MaOpRsr 20 355-380

Tsubaki, Michiko: New weighted sum of χ^2 test for testing a linear hypothesis against a restricted alternative hypothesis. 95 CommStB 24 987-1004

Tsuchiya, Takahiro: A homogeneity analysis for exploring structural differences among several data sets: Its application to data from the Surveys of the Japanese National Character (Japanese). 95 PrInStMa 43 121-140

Komazawa, Tsutomu; — —: Exploratory data analysis for Hayashi's quantification method III by graphics: Its application of ordinal structure analysis to the Ninth Nationwide Survey of the Japanese National Character (Japanese). 95 PrInStMa 43 161-176

Tsui, Kwok-Leung: Comment on "Assessing geometric integrity through spherical regression techniques". 95 StSinica 5 210-215

Tsutakawa, Robert K.: Lee, Hyungseok; — —: Bayesian method for estimating multidimensional ability in item response analysis. 94 ASAProBays 7-12

Tsybakov, A. B.: Härdle, W.; — —: Additive nonparametric regression on principal components. 95 JNnparSt 5 157-184

Härdle, W.; Park, B. U.; — —: Estimation of non-sharp support boundaries. 95 JMultiAn 55 205-218

Korostelev, A. P.; Simar, L.; — —: Efficient estimation of monotone boundaries. 95 AnlsStat 23 476-489

Mammen, E.; — —: Asymptotical minimax recovery of sets with smooth boundaries. 95 AnlsStat 23 502-524

Maron, J. S.; — —: Visual error criteria for qualitative smoothing. 95 JASA 90 499-507

Stadtmüller, U.; — —: Nonparametric recursive variance estimation. 95 MaOpfStS 27 55-63

Tsykunova, S. Yu.: Ostrovskii, E. I.; — —: Asymptotic properties of distribution of maximum of Gaussian nonstationary process occurring in covariance statistic. 94 ThProbAp 39 527-534

Tu, D.: Shao, J.; — —: The jackknife and bootstrap. 95 Springer:Brln:NY 516p

Shao, J.; — — (Auth); Fedorov, V. V. (Rev): Review of "The jackknife and bootstrap". 96 ShrtBkRv 16/1 7-7

Tu, Dongsheng; Gross, Alan J.: Accurate confidence intervals for the ratio of specific occurrence/exposure rates in risk and survival analysis. 95 BiomtrcJ 37 611-626

Tu, Pierre D. V.: — — (Auth); Ray, W. D. (Rev): Review of "Dynamical systems: An introduction with applications in economics and biology (Second edition)". 95 IntJFor 11 491-492

Tu, Renjin: Mukerjee, Hari; — —: Order-restricted inferences in linear regression. 95 JASA 90 717-728

Tu, Wanzhu: Lin, Dennis K. J.; — —: Dual response surface optimization. 95 JQualTek 27 34-39

Tu, Xin Ming: Nonparametric estimation of survival distributions with censored initiating time, and censored and truncated terminating time: Application to transfusion data for acquired immune deficiency syndrome. 95 ApplStat 44 3-16

— —; Litvak, Eugene; Pagano, Marcello: On the informativeness and accuracy of pooled testing in estimating prevalence of a rare disease: Application to HIV screening. 95 Biomtrka 82 287-297

Tubert-Bitter, Pascale; Bloch, Daniel A.; Raynauld, Jean-Pierre: Comparing the bivariate effects of toxicity and efficacy of treatments. 95 StatMed 14 1129-1141 (Disc: p1143-1143)

Tuckel, Peter S.: Schlichting, Kurt C.; — —: A spatial analysis of contextual effects on voter participation in the 1992 presidential election. 94 ASASrvyRsr 1339-1344

Tucker, Clyde; Kojetin, Brian; Blass, Richard; Tucker, Ronald: Measuring the effects of the transition on field performance. 94 ASASrvyRsr 348-353 (Disc: p360-361)

Kojetin, Brian; — —; Cashman, Erin: Response to a government survey as political participation: The relation of economic and political conditions to refusal rates in the Current Population Survey. 94 ASASrvyRsr 1236-1241

Tucker, H. G.: Finkelstein, M.; Scheiberg, S.; — —: Integral transforms with infinitely divisible kernels. 94 ThProbAp 39 670-676

Tucker, Howard G.: Finkelstein, Mark; — —: The Wilcoxon rank-sum test in dichotomous regression analysis. 92 ProIrStCl 243-259

Tucker, John R.: Draper, David; Gaver, Donald, P., Jr; Goel, Prem K.; Greenhouse, Joel B.; Hedges, Larry V.; Morris, Carl N.; — —; Waternaux, Christine M. (Auth); Berlin, Jesse A. (Rev): Review of "Combining information: Statistical issues and opportunities for research". 95 StatMed 14 1611-1612

Tucker, Ronald: Tucker, Clyde; Kojetin, Brian; Blass, Richard; — —: Measuring the effects of the transition on field performance. 94 ASASrvyRsr 348-353 (Disc: p360-361)

Tuckwell, H. C.: Elementary applications of probability theory: With an introduction to stochastic differential equations (Second edition). 95 ChapmnHall:Lnd 320p

Tudor, Constantin: Prato, G. Da; — —: Periodic and almost periodic solutions for semilinear stochastic equations. 95 StocAnAp 13 13-33

Tudor, David; Zipfel, Andreas: Opportunities for multicenter clinical trials in joint US-European drug development. 94 ASAProBiop 229-231

Tudor, Gail E.; Morgan, Timothy M.: Analyzing survival data in conjunction with time-dependent surrogate endpoints in clinical trials. 95 CommStA 24 2293-2313

Tuene, Stig: Nortvedt, Ragnar; — —: Multivariate evaluation of feed for Atlantic halibut. 95 Chemolab 29 271-282

Tufte, E. R.: — — (Auth); Crescimanni, Annamaria (Rev): Review of "Envisioning information". 94 Metron 52/1 218-218

Tuljapurkar, Shripad; Lee, Ronald: Comment on "Inference from a deterministic population dynamics model for Bowhead whales". 95 JASA 90 424-425

Tulleken, R.: Kambezidis, H. D.; — —; Amanatidis, G. T.; Paliatsos, A. G.; Asimakopoulos, D. N.: Statistical evaluation of selected air pollutants in Athens, Greece. 95 EnvrMtrc 6 349-361

Tunstall-Pedoe, Hugh: Woodward, Mark; Laurent, Kirsty; — —: An analysis of risk factors for prevalent coronary heart disease by using the proportional odds model. 95 Ststcian 44 69-80

Turban, Efraim: Trippi, Robert R.; — — (Ed); Mangasarian, Olvi L. (Rev): Review of "Neural networks in finance and investing". 95 Intrfc 25/1 141-142

Turcott, Robert G.; Barker, Peter D. R.; Teich, Malvin C.: Long-duration correlation in the sequence of action potentials in an insect visual interneuron. 95 JStCmpSm 52 253-271

Turk, Joseph: Evans, K. Franklin; — —; Wong, Takmeng; Stephens, Graeme L.: A Bayesian approach to microwave precipitation profile retrieval. 95 JApMetrl 34 260-279

Turkheimer, Eric: Rosenfeld, Barry; — —: Multidimensional representation of decision-making in chronic schizophrenics. 95 MultBehR 30 199-211

Turkkan, N.: Pham-Gia, T.; — —: Comment on "Testing reliability in a stress-strength model when X and Y are normally distributed" (V34 p83-91). 95 Technmcs 37 130-131

Turkman, K. F.: Anderson, C. W.; — —: Sums and maxima of stationary sequences with heavy tailed distributions. 95 SankhyaA 57 1-10

Barnett, V.; — — (Ed): Statistics for the environment 2: Water related issues. 94 Wiley:NY:UK 406p

Barnett, V.; — — (Ed); Brizzi, M. (Rev. Italian): Review of "Statistics for the environment". 94 Statstca 54 397-398

Barnett, V.; — — (Ed); Kemp, A. W. (Rev): Review of "Statistics for the environment 2: Water related issues". 95 Biomtrcs 51 795-796

Turlach, B. A.: Härdle, W.; Klinke, S.; — — (Ed): XploRe: An interactive statistical computing environment. 95 Springer:Brln:NY 387p

Härdle, W.; Klinke, S.; — — (Ed); Murdoch, D. J. (Rev): Review of "XploRe: An interactive statistical computing environment". 96 ShrtBkRv 16/1 3-3

Turnbull, Bruce W.: Hsieh, Fushing; — —: A note on the local asymptotically minimax rate for estimating a crossing point in a diagnostic marker problem. 95 StPrLet 24 181-185

Waller, Lance A.; — —; Gustafsson, Göran; Hjalmars, Ulf; Andersson, Bertil: Detection and assessment of clusters of disease: An application to nuclear power plant facilities and childhood leukaemia in Sweden. 95 StatMed 14 3-16

Waller, Lance A.; — —; Hardin, J. Michael: Obtaining distribution functions by numerical inversion of characteristic functions with applications. 95 AmerStat 49 346-350

Turner, Charles F.: O'Reilly, James M.; Hubbard, Michael L.; Lessler, Judith T.; Biemer, Paul P.; — —: Audio and video computer assisted self-interviewing: Preliminary tests of new technologies for data collection. 94 JOffStat 10 197-214

Turner, Danny W.; Young, Dean M.; Seaman, John W., Jr: A Kolmogorov inequality for the sum of independent Bernoulli random variables with unequal means. 95 StPrLet 23 243-245

Turner, David L.; Wilson, Jeffrey R.: Monitoring cross country skiers. 94 ASAProEnvr 167-172

Turner, John R. G.: Lennon, Jack J.; — —: Predicting the spatial distribution of climate: Temperature in Great Britain. 95 JAnmlEco 64 370-392

Turner, Robert B.; Zipursky, Lorne S.: Quality tools help improve employee commitment. 95 ASQCTekCon 770-776

Turner, Ronald E.: What is the probability that your employees will succeed?. 95 QulPr 28/6 31-33

Turrero, Agustin: Relative efficiency of a censored experiment in terms of Fisher information matrix. 95 CommStA 24 1169-1191

Tusell, Fernando: Wang, Chamont (Auth); — — (Rev): Review of "Sense and nonsense of statistical inference". 95 JAppStat 22 320-320

Tutz, G.: Fahrmeir, L.; — — (Auth); Glosup, Jeffrey (Rev): Review of "Multivariate statistical modelling based on generalized linear models". 95 Technmcs 37 350-351

Tvrdík, Josef: Křivý, Ivan; — —: The controlled random search algorithm in optimizing regression models. 95 CmpStDA 20 229-234

Twardowska, Krystyna: An approximation theorem of Wong-Zakai type for nonlinear stochastic partial differential equations. 95 StocAnAp 13 601-626

Tweedie, R. L.; Mengersen, K. L.: Meta-analytic approaches to dose-response relationships, with application in studies of lung cancer and exposure to environmental tobacco smoke. 95 StatMed 14 545-569 (Disc: p679-685)

Mengersen, K. L.; — —; Biggerstaff, B.: The impact of method choice on meta-analysis. 95 AstrlJSt 37 19-44

Tyrrell, G.: Arndt, S.; — —; Wollson, R. F.; Flaum, M.; Andreasen, N. C.: Effects of errors in a multicenter medical study: Preventing misinterpreted data. 94 JPsyctrc 28 447-459

Tyurin, Yu. N.; Yakovlev, Andrej Yu.; Shi, J.; Bass, L.: Testing a model of aging in animal experiments. 95 Biomtrcs 51 363-372

Tzonou, Anastasia J.: Marinos, Aristides Th.; — —; Karantzas, Menelaos E.: Experimental quantiles of epidemiological indices in case-control studies with non-differential misclassification. 95 StatMed 14 1291-1306

Tzourio, C.; Maccario, J.; Alpérovitch, A.: Estimation of the proportion of genetic cases in late-onset diseases: An application to Alzheimer's disease. 95 StatMed 14 1353-1363

Ubøe, Jan: Complex valued multiparameter stochastic integrals. 95 JTherPrb 8 601-624

— —; Zhang, Tusheng: Gaussian tail estimates for Dirichlet processes. 95 Stochatx 52 295-302

— —; Zhang, Tusheng: A stability property of the stochastic heat equation. 95 StocProc 60 247-260

Uchida, Masayuki; Aki, Sigeo: Sooner and later waiting time problems in a two-state Markov chain. 95 AnInStMa 47 415-433

Uddin, Md. Borhan: Ferdous, J.; — —; Pandey, M.: Reliability estimation with Weibull inter failure times. 95 RelEngSS 50 285-296

Uddin, Nizam; Hossain, Syed A.: E-optimal nested row-column designs with unequal replications of treatments. 95 SankhyaB 57 112-121

— —; Whittinghill, Dexter C.: The robustness of binary and non-binary balanced nested row-column designs under the unavailability of blocks: A comparison. 95 AstrlJSt 37 111-124

Morgan, John P.; — —: Optimal, non-binary, variance balanced designs. 95 StSinica 5 535-546

Udevitz, Mark S.; Pollock, Kenneth H.: Using effort information with change-in-ratio data for population estimation. 95 Biomtrcs 51 471-481

Uebe, Götz: Belsley, D. A. (Ed); — — (Rev): Review of "Computational techniques for econometrics and economic analysis". 95 CmpStDA 19 359-360

Martin-Löf, Anders (Ed); — — (Rev): Review of "H. Cramér: Collected works. Volumes 1 and 2". 95 AllgeStA 79 332-332

Ueno, Maomi; Yoshimura, Osamu; Shigemasu, Kazuo: A latent trait model for reading process of a word (Japanese). 94 JpJBeh 21/1 48-56

Ukai, Y.: Hayashi, T.; — —: Change in genetic variance under selection in a self-fertilizing population. 94 Genetcs 136 693-704
Hayashi, T.; — —: Detection of additive and dominance effects of QTLs in interval mapping of F_2 RFLP data. 94 ThApGene 87 1021-1027

Ullah, Aman; Srivastava, Virendra K.; Roy, Nilanjana: Moments of the function of non-normal random vector with applications to econometric estimators and test statistics. 95 EcomtrRv 14 459-471
Srivastava, Anil K.; Srivastava, Virendra K.; — —: The coefficient of determination and its adjusted version in linear regression models. 95 EcomtrRv 14 229-240
Srivastava, Virendra K.; — —: Stein-rule estimation in models with a lagged-dependent variable. 95 CommStA 24 1343-1353
Takada, Hirokazu; — —; Chen, Yu-Min: Estimation of the seemingly unrelated regression model when the error covariance matrix is singular. 95 JAppStat 22 517-530

Umbach, David M.: Weinberg, Clarice R.; — —; Greenland, Sander: Reply to comments on "When will nondifferential misclassification of an exposure reverse the direction of a trend?". 95 AJEpidm 142 784-784

Umesh, U. N.: Predicting nominal variable relationships with multiple response. 95 JForcast 14 585-596

Umholtz, Robert L.: Are the Toronto Blue Jays really the world champions?. 94 ASAProSpts 42-45

Unnikrishnan, N. K.; Kunte, S.: Optimality of an analogue of Basu's estimator under a double sampling design. 95 SankhyaB 57 103-111

Unnithan, V. K. G.; Nair, N. Unnikrishnan: Minimum variance stratification. 95 CommStB 24 275-284

Uno, Chikara: A note on sequential estimation of the mean. 95 StPrLet 23 141-149
Isogai, Eiichi; — —: On the sequential point estimation of the mean of a gamma distribution. 95 StPrLet 22 287-293

Unwin, Antony R.: Stack, Jim; — —: Selecting explanatory variables from survey data: A simple approach that explores for interactions. 95 Ststcian 44 17-29

Upadhyay, S. K.: Singh, Rajesh; — —; Singh, Umesh: Bayes estimators for two parameter exponential distribution. 95 CommStA 24 227-240

Upadhyay, Satyanshu Kumar; Agrawal, Rekha: Bayesian prediction limits for the mean of future observations from the inverse Gaussian distribution. 94 Metron 52/1 101-111

Upsdell, Martin: Using a Bayesian smoother to determine root growth patterns. 94 ASAProBays 227-231

Upton, G. J. G.: Lawal, H. Bayo; — —: An algorithm for fitting models to $N \times N$ contingency tables having ordered categories. 95 CommStB 24 793-805

Upton, Graham J. G.; Guillen, Montserrat: Perfect cells, direct models and contingency table outliers. 95 CommStA 24 1843-1862

Urago, Sabine; Zerubia, Josiane; Berthod, Marc: A Markovian model for contour grouping. 95 PattrnRc 28 683-693

Urahama, K.; Furukawa, Y.: Gradient descent learning of nearest neighbor classifiers with outlier rejection. 95 PattrnRc 28 761-768

Urbain, Jean-Pierre: Partial versus full system modelling of cointegrated systems - An empirical illustration. 95 JEconmtx 69 177-210

Urbanek, Margrit: Long, Jeffrey C.; Williams, Robert C.; — —: An E-M algorithm and testing strategy for multiple-locus haplotypes. 95 AJHuGene 56 799-810

Urfer, W.; Schwarzenbach, F. H.; Kötting, J.; Müller, P.: Multistate models for monitoring individual trees in permanent observation plots. 94 EnvEcoSt 1 171-192 (Disc: p192-199)
— —; Scwarzenbach, F. H.; Kötting, J.; Müller, P.: Reply to comments on "Multistate models for monitoring individual trees in permanent observation plots". 94 EnvEcoSt 1 197-199

Urfer, Wolfgang; Becka, Michael: Statistical data analysis for identifying kinetic processes. 94 ASAProEnvr 14-18

Uri, T.; Wörgötter, A. (Ed): Econometrics of short and unreliable time series. 95 Physica:Ger 238p

Urquhart, N. S.: Larsen, D. P.; — —; Kugler, D. L.: Regional scale trend monitoring of indicators of trophic condition of lakes. 95 WaterReB 31 117-140

Uryasev, S. P.: Vesely, W. E.; — —; Samanta, P. K.: Failure of emergency diesel generators: A population analysis using empirical Bayes methods. 94 RelEngSS 46 221-229

Usami, Yoshihiro; Huzii, Mituaki: Estimation of coefficients of time series regression with a nonstationary error process. 95 JTimSrAn 16 105-118

Ushakov, I. A.: Gnedenko, B. V.; — —: Probabilistic reliability engineering. 95 Wiley:NY:UK 518p
Gnedenko, B. V.; — — (Auth); Goodwin, G. C. (Rev): Review of "Probabilistic reliability engineering". 96 ShrtBkRv 16/1 8-8

Ushakov, Igor A.; Harrison, Robert A. (Ed); Doganaksoy, Necip (Rev): Review of "Handbook on reliability engineering". 95 JASA 90 804-804

Ushakov, N. G.; Ushakova, A. P.: Estimations of the decomposition stability into identical components. 95 StPrLet 25 221-229

Ushakova, A. P.: Ushakov, N. G.; — —: Estimations of the decomposition stability into identical components. 95 StPrLet 25 221-229

Usher, John S.: Guttman, Irwin; Lin, Dennis K. J.; Reiser, B.; — —: Dependent masking and system life data analysis: Bayesian inference for two-component systems. 95 LiDaAnal 1 87-100
Reiser, B.; Guttman, I.; Lin, Dennis K. J.; Guess, Frank M.; — —: Bayesian inference for masked system lifetime data. 95 ApplStat 44 79-90

Üstünel, A. S.; Zakai, M.: Random rotations of the Wiener path. 95 ProbTher 103 409-429
Decreusefond, L.; — —: The Beneš equation and stochastic calculus of variations. 95 StocProc 57 273-284

Utev, Sergey: Lefèvre, Claude; — —: Poisson approximation for the final state of a generalized epidemic process. 95 AnlsProb 23 1139-1162
Lefèvre, Claude; — —: On a conjecture for the non-existence of the expectation of randomly stopped sums. 95 JAppProb 32 1138-1141

Utzet, Frederic: Sanchez, Alex; Ocaña, Jordi; — —: Sampling theory, estimation, and significance testing for Prevosti's estimate of genetic distance. 95 Biomtrcs 51 1216-1235

Uyenoyama, March K.: A generalized least-squares estimate for the origin of sporophytic self-incompatibility. 95 Genetcs 139 975-992

Vacca, E. Ann: Lewis, Paul J.; Wolfgang, Glenn S.; — —: 1992 Census of Agriculture coverage evaluation estimation. 94 ASASrvyRsr 1019-1024
Wolfgang, Glenn S.; Lewis, Paul J.; — —: Cognitive research for the 1997 Census of Agriculture report form. 94 ASASrvyRsr 503-508

Vach, W.: Classification trees. 95 CmpStQ 10 9-14
— — (Auth); Kianifard, Farid (Rev): Review of "Logistic regression with missing values in the covariates". 95 Technmcs 37 460-461

Vach, Werner; Blettner, Maria: Logistic regression with incompletely observed categorical covariates - Investigating the sensitivity against violation of the missing at random assumption. 95 StatMed 14 1315-1329

Vaillancourt, Jean: Asymptotics for multisample statistics. 95 CandJSt 23 171-184

Vajda, I.: Menéndez, M.; Morales, D.; Pardo, L.; — —: Divergence-based estimation and testing of statistical models of classification. 95 JMultiAn 54 329-354
Morales, D.; Pardo, L.; — —: Asymptotic divergence of estimates of discrete distributions. 95 JStPlInf 48 347-369

Vajda, Igor: Conditions equivalent to consistency of approximate MLE's for stochastic processes. 95 StocProc 56 35-56
Rukhin, Andrew L.; — —: Adaptive decision making for stochastic processes. 95 JStPlInf 45 313-329

Vajda, S.: Daykin, C. D.; Pentikainen, T.; Pesonen, M. (Auth); — — (Rev): Review of "Practical risk theory for actuaries". 95 Ststcian 44 282-283

Vakil, Farok; Rajan, Murli: A test of the three moment international capital asset pricing model. 94 ASAProBuEc 184-189

Valade, J. L.: Broderick, G.; Paris, J.; — —: A composite representation of pulp quality. 95 Chemolab 29 19-28

Valderrama, Mariano J.; Jimenez, Francisco; Gutierrez, Ramon; Martinez-Almecija, Alfredo: Estimation and filtering on a doubly stochastic Poisson process. 95 AplStMDA 11 13-24

Valenciano, F.: Calvo, E.; Tijs, S. H.; — —; Zarzuelo, J. M.: On the axiomatization of the τ-value. 95 Top 3 35-46

Valentine, Harry T.: Gregoire, Timothy G.; — —; Furnival, George M.: Sampling methods to estimate foliage and other characteristics of individual trees. 95 Ecology 76 1181-1194

Valenzuela, Carlos A.: Sheesley, John H.; Thomas, Harold W.; — —: Quantal response analysis of relief valve test data. 95 ASQCTekCon 741-748

Valiveti, R. S.; Oommen, B. J.; Zgierski, J. R.: Adaptive linear list reorganization under a generalized query system. 95 JAppProb 32 793-804

Valliant, Richard; Gentle, James E.: An application of mathematical programming to sample allocation. 94 ASASrvyRsr 683-688

Vallois, P.: Decomposing the Brownian path via the range process. 95 StocProc 55 211-226

Vallois, Pierre: Russo, Francesco; — —: The generalized covariation process and Ito formula. 95 StocProc 59 81-104

Valsecchi, M. G.: Marubini, E.; — —: Analysing survival data from clinical trials and observational studies. 94 Wiley:NY:UK 414p
Marubini, E.; — — (Auth); Juritz, J. M. (Rev): Review of "Analysing survival data from clinical trials and observational studies". 95 ShrtBkRv 15/2 23-23
Marubini, E.; — — (Auth); McLachlan, G. J. (Rev): Review of "Analysing survival data from clinical trials and observational studies". 95 Biomtrcs 51 1191-1191

Valsecchi, Maria Grazia: Marubini, Ettore; — —: Analyzing survival data from clinical trials and observational studies. 95 Wiley:NY:UK 424p

Valtonen, Esko: Alho, Juha M.; — —: Interval estimation of inverse dose-response. 95 Biomtrcs 51 491-501

Valz, Paul D.; McLeod, A. Ian; Thompson, Mary E.: Cumulant generating function and tail probability approximations for Kendall's score with tied rankings. 95 AnlsStat 23 144-160

van Arendonk, Johan A. M.; Tier, Bruce; Kinghorn, Brian P.: Use of multiple genetic markers in prediction of breeding values. 94 Genetcs 137 319-329
van der Beek, Sijne; — —: Criteria to optimize designs for detection and estimation of linkage between marker loci from segregating populations containing several families. 93 ThApGene 86 269-280

Van Bastelaer, Aloïs: Differences in the measurement of employment in the labour force surveys in the European Community. 94 JOffStat 10 277-305

van Belle, Gerald: Fisher, Lloyd D.; — — (Auth); D'Agostino, Ralph B. (Rev): Review of "Biostatistics: A methodology for the health sciences". 95 JASA 90 799-800

Fisher, Lloyd D.; — — (Auth); Machin, David (Rev): Review of "Biostatistics: A methodology for the health sciences". 95 StatMed 14 878-879

van Berkum, E. E. M.; Linssen, H. N.; Overdijk, D. A.: The reduction and ancillary classification of observations. 95 JStPlInf 45 347-372

Van Breukelen, Gerard J. P.: Psychometric and information processing properties of selected response time models. 95 Psymtrka 60 95-113

Van de Geer, John P.: — — (Auth); Fox, Richard (Rev): Review of "Multivariate analysis of categorical data: Theory and applications. Volumes 1 and 2". 95 JMarRsr 32 111-112

— — (Auth); Rose, Elizabeth L. (Rev): Review of "Multivariate analysis of categorical data: Theory and applications. Volumes 1 and 2". 95 StrEqMod 2 274-276

Van de Geer, S.: The method of sieves and minimum contrast estimators. 95 MaMethSt 4 20-38

van de Geer, Sara: Exponential inequalities for martingales, with application to maximum likelihood estimation for counting processes. 95 AnlsStat 23 1779-1801

van de Pol, F. J. R.: van Opstal, R.; — —: The transition from school to work in 1979-1987 in the Netherlands. 95 StNeerla 49 294-309

Van de Ven, Andrew H.: Huber, George P.; — — (Ed): Longitudinal field research methods. Studying processes of organizational change. 95 SagePubl:CA:Lnd 400p

van de Ven, R.; Weber, N. C.: Log-linear models for mean and dispersion in mixed Poisson regression models. 95 AstrlJSt 37 205-216

van Delden, A.: Weijers, E. P.; — —; Vugts, H. F.; Meesters, A. G. C. A.: Characteristics of convective turbulence in the surface layer investigated by principal component analysis. 95 JApMetrl 34 528-541

van den Berg, G. J.; Lindeboom, M.: Durations in panel data subject to attrition: A note on estimation in the case of a stock sample. 95 StNeerla 49 282-293

Van den Berg, Gerard J.: Koning, Pierre; Ridder, Geert; — —: Structural and frictional unemployment in an equilibrium search model with heterogeneous agents. 95 JApEcmtx 10 133-151S

van den Berg, J.; Gandolfi, A.: A triangle inequality for covariances of binary FKG random variables. 95 AnlsApPr 5 322-326

van den Berg, J. L.: — — (Auth); Newton, J. (Rev): Review of "Sojourn times in feedback and processor sharing queues". 95 JRSS-A 158 186-187

— — (Auth); Schassberger, R. (Rev): Review of "Sojourn times in feedback and processor sharing queues". 95 Metrika 42 146-147

van den Broek, Jan: A score test for zero inflation in a Poisson distribution. 95 Biomtrcs 51 738-743

van den Doel, I. T.; Kiviet, J. F.: Neglected dynamics in panel data models; consequences and detection in finite samples. 95 StNeerla 49 343-361

van den Hout, Wilbert; Blanc, Hans: Development and justification of the power-series algorithm for BMAP-systems. 95 CommStSM 11 471-496

van der Beek, Sijne; van Arendonk, Johan A. M.: Criteria to optimize designs for detection and estimation of linkage between marker loci from segregating populations containing several families. 93 ThApGene 86 269-280

van der Heijden, Peter G. M.; Mooijaart, Ab: Some new log-bilinear models for the analysis of asymmetry in a square contingency table. 95 SocMethR 24 7-29

Siciliano, Roberta; — —: Simultaneous latent budget analysis of a set of two-way tables with constant-row-sum data. 94 Metron 52/1 155-180

van der Klaauw, Wilbert: Imbens, Guido; — —: Evaluating the cost of conscription in The Netherlands. 95 JBES 13/2 207-215

van der Laan, Mark: Jewell, Nicholas P.; — —: Generalizations of current status data with applications. 95 LiDaAnal 1 101-109

van der Laan, Mark J.: An identity for the nonparametric maximum likelihood estimator in missing data and biased sampling models. 95 Bernoull 1 335-341

Gill, Richard D.; — —; Wellner, Jon A.: Inefficient estimators of the bivariate survival function for three models. 95 AIHenrPB 31 545-597

van der Laan, Paul: Subset selection for exponential populations: Determination of the selection constant. 95 BiomtrcJ 37 515-521

van der Linde, A.: Splines from a Bayesian point of view. 95 Test 4 63-81

— —; Witzko, K.-H.; Jöckel, K.-H.: Spatial-temporal analysis of mortality using splines. 95 Biomtrcs 51 1352-1360

van der Linde, Angelika: Computing the error for smoothing splines. 95 CmpStQ 10 143-154

van der Linden, W. J.; Zwarts, M. A.: Robustness of judgments in evaluation research. 95 TOnderwR 20 13-27

van der Linden, Wim J.: Berger, Martijn P. F.; — —: The design of tests with different optimality criteria (Dutch). 95 TOnderwR 20 79-92

van der Maas, J. H.: Luinge, H. J.; — —; Visser, T.: Partial least squares regression as a multivariate tool for the interpretation of infrared spectra. 95 Chemolab 28 129-138

van der Mei, Rob D.: Weststrate, Jan A.; — —: Waiting times in a two-queue model with exhaustive and Bernoulli service. 94 ZeitOpRs 40 289-303

van der Merwe, A. J.: Plessis, J. L. du; — —: A Bayesian approach to multivariate and conditional calibration. 95 CmpStDA 19 539-552

Plessis, J. L. du; — —: Bayesian calibration in meat science research-estimation of the pH in postmortem muscle. 95 JStCmpSm 53 1-11

Plessis, J. L. du; — —; Groenewald, P. C. N.: Reference priors for the multivariate calibration problem. 95 SoAfrStJ 29 155-168

van der Meulen, E. A.: Kroese, A. H.; — —; Poortema, K.; Schaafsma, W.: Distributional inference. 95 StNeerla 49 63-82

van der Meulen, E. C.: Carbonez, A.; Györfi, L.; — —: Partitioning-estimates of a regression function under random censoring. 95 StatDecn 13 21-37

van der Meulen, Edward C.: Dudewicz, Edward J.; — —; SriRam, M. G.; Teoh, Nick K. W.: Entropy-based random number evaluation. 95 AJMMgSci 15 115-153

van der Vaart, A. W.: Efficiency of infinite dimensional M-estimators. 95 StNeerla 49 9-30

van der Vaart, Aad: Maximum likelihood estimation with partially censored data. 94 AnlsStat 22 1896-1916

van der Voet, Hilko: Correction to "Comparing the predictive accuracy of models using a simple randomization test" (V25 p313-323). 95 Chemolab 28 315-315

van der Wal, A. J.: Application of fuzzy logic control in industry. 94 FuzzySet 74 33-41

van der Weide, J. A. M.: — — (Auth); Matthes, K. (Rev): Review of "Stochastic processes and point processes of excursions". 95 BiomtrcJ 37 764-764

van der Werf, J. H. J.: Meuwissen, T. H. E.; Engel, B.; — —: Maximizing selection efficiency for categorical traits. 95 JAnmlSci 73 1933-1939

van Deusen, Paul C.: Difference sampling as an alternative to importance sampling. 95 CanJForR 25 487-490

van Dijk, Herman K.; Monfort, Alain; Brown, Bryan W. (Ed): Econometric inference using simulation techniques. 95 Wiley:NY:UK 256p

Hoek, Henk; Lucas, André; — —: Classical and Bayesian aspects of robust unit root inference. 95 JEconmtx 69 27-59

van Dijk, N. M.: Henderson, W.; Pearce, C. E. M.; Taylor, P. G.; — —: Insensitivity in discrete-time generalized semi-Markov processes allowing multiple events and probabilistic service scheduling. 95 AnlsApPr 5 78-96

van Dongen, S.: How should we bootstrap allozyme data?. 95 Heredity 74 445-447

— —; Backeljau, T.: One- and two-sample tests for single-locus inbreeding coefficients using the bootstrap. 95 Heredity 74 129-135

Van Duijn, Cornelia M.: Commenges, Daniel; Jacqmin, Hélène; Letenneur, Luc; — —: Score test for familial aggregation in probands studies: Application to Alzheimer's disease. 95 Biomtrcs 51 542-551

van Duijn, Marijtje A. J.; Böckenholt, Ulf: Mixture models for the analysis of repeated count data. 95 ApplStat 44 473-485

— —; Jansen, Margo G. H.: Modeling repeated count data: Some extensions of the Rasch Poisson counts model. 95 JEdStat 20 241-258

van Dyk, David A.; Meng, Xiao-Li: Permuting CM steps within the ECM algorithm. 94 ASAProStCp 130-135

— —; Meng, Xiao-Li; Rubin, Donald B.: Maximum likelihood estimation via the ECM algorithm: Computing the asymptotic variance. 95 StSinica 5 55-75

van Eeden, Constance: Minimax estimation of a lower-bounded scale parameter of a gamma distribution for scale-invariant squared-error loss. 95 CandJSt 23 245-256

van Eeuwijk, F. A.: Comment on "The statistics of linear models: Back to basics" (V4 p221-234). 95 StatComp 5 93-95

— —; Keizer, L. C. P.: On the use of diagnostic biplots in model screening for genotype by environment tables. 95 StatComp 5 141-153

van Eeuwijk, Fred A.: Multiplicative interaction in generalized linear models. 95 Biomtrcs 51 1017-1032

Van Groenendaal, Willem J. H.: Kleijnen, Jack P. C.; — —: Two-stage versus sequential sample-size determination in regression analysis of simulation experiments. 95 AJMMgSci 15 83-114

van Harn, K.; Steutel, F. W.: Infinite divisibility and the waiting-time paradox. 95 CommStSM 11 527-540

van Houwelingen, Hans: Collett, D. (Auth); — — (Rev): Review of "Modelling survival data in medical research". 95 StatMed 14 1147-1148

van Houwelingen, Hans C.; Thorogood, Jane: Construction, validation and updating of a prognostic model for kidney graft survival. 95 StatMed 14 1999-2008

Houwing-Duistermaat, Jeanine J.; Derkx, Bert H. F.; Rosendaal, Frits R.; — —: Testing familial aggregation. 95 Biomtrcs 51 1292-1301

le Cessie, Saskia; — —: Testing the fit of a regression model via score tests in random effects models. 95 Biomtrcs 51 600-614

Schouten, Evert G.; Dekker, Jacqueline M.; Kok, Frans J.; LeCessie, Saskia; — —; Pool, Jan: Reply to comment on "Risk ratio and rate ratio estimation in case-cohort designs: Hypertension and cardiovascular mortality". 95 StatMed 14 1610-1610

Verweij, Pierre J. M.; — —: Time-dependent effects of fixed covariates in Cox regression. 95 Biomtrcs 51 1550-1556

van Houwelingen, J. C.: Zwinderman, A. H.; — —; Schweitzer, D.: Prediction of the next hypercalcemia free period: Application of random effect models with selection on first event. 95 StNeerla 49 310-323

van Lierde, M.: Ritter, C.; Bouckaert, A.; — —; Theunissen, I.: Modelling mortality and morbidity of newborns. 95 StatMed 14 2191-2204

van Lieshout, M. N. M.: Baddeley, A. J.; — —: Area-interaction point processes. 95 AnInStMa 47 601-619

Van Mechelen, Iven; De Boeck, Paul; Rosenberg, Seymour: The conjunctive model of hierarchical classes. 95 Psymtrka 60 505-521

van Mieghem, Jan A.: Dynamic scheduling with convex delay costs: The generalized $c\mu$ rule. 95 AnlsApPr 5 809-833

Van Mieghem, Jan A.: Avi-Itzhak, Hadar I.; — —; Rub, Leonardo: Multiple subclass pattern recognition: A maximin correlation approach. 95 IEEEPAMI 17 418-431

Van Ness, John W.: Yang, Jim J.; — —: Breakdown points for redescending M-estimates of location. 95 CommStA 24 1769-1787

van Noordwijk, A. J.: Bauchau, V.; — —: Comparison of survival estimates obtained from three different methods of recapture in the same population of the great tit. 95 JAppStat 22 1031-1037

van Noordwijk, Arie J.: On bias due to observer distribution in the analysis of data on natal dispersal in birds. 95 JAppStat 22 683-694

Van Nostrand, R. Craig: Modeling using ratios of polynomials. 94 ASAProPEng 97-102

van Oortmarssen, G. J.; Boer, Rob; Habbema, J. D. F.: Modelling issues in cancer screening. 95 StMedRes 4 33-54

van Opstal, R.; van de Pol, F. J. R.: The transition from school to work in 1979-1987 in the Netherlands. 95 StNeerla 49 294-309

Van Orden, Drew: Oehlert, Gary W.; Lee, Richard J.; — —. Statistical analysis of asbestos fibre counts. 95 EnvrMtrc 6 115-126

van Pul, M. C.: Barendregt, L. G.; — —: On the estimation of the parameters for the Littlewood model in software reliability. 95 StNeerla 49 165-184

van Pul, M. C. J.: — — (Auth); Reese, R. Allan (Rev): Review of "Statistical analysis of software reliability models". 95 JRSS-A 158 199-199

van Rooij, A. C. M.: Kortram, R. A.; — —; Lenstra, A. J.; Ridder, G.: Constructive identification of the mixed proportional hazards model. 95 StNeerla 49 269-281

Van Sanford, D. A.; Pfeiffer, T. W.; Cornelius, P. L.: Selection index based on genetic correlations among environments. 93 CropSci 33 1244-1248

Cornelius, P. L.; — —; Seyedsadr, M. S.: Clustering cultivars into groups without rank-change interactions. 93 CropSci 33 1193-1200

van Soest, A. H. O.: Charlier, E.; Melenberg, B.; — —: A smoothed maximum score estimator for the binary choice panel data model with an application to labour force participation. 95 StNeerla 49 324-342

van Staveren, Wija: Kaaks, Rudolf; Riboli, Elio; — —: Calibration of dietary intake measurements in prospective cohort studies. 95 AJEpidm 141 548-556

Kaaks, Rudolf; Riboli, Elio; — —: Sample size requirements for calibration studies of dietary intake measurements in prospective cohort investigations. 95 AJEpidm 141 557-565

Kaaks, Rudolf; Riboli, Elio; — —: Sample size requirements for calibration studies of dietary intake measurements in prospective cohort investigations. 95 AJEpidm 142 557-565

van Tienderen, Peter H.: Life cycle trade-offs in matrix population models. 95 Ecology 76 2482-2489

van Tongeren, Onno F. R.: Data analysis or simulation model: A critical evaluation of some methods. 95 EcolgMdl 78 51-60

van Veldhuizen, M.; Appelo, C. A. J.; Griffioen, J.: The (least squares) quotient algorithm as a rapid tool for obtaining sorption isotherms from column elution curves. 95 WaterReR 31 849-857

Van Vleck, L. D.: Variance of prediction error with mixed model equations when relationships are ignored. 93 ThApGene 85 545-549

VanBrackle, Anita S.: VanBrackle, Lewis; — —; Rothenberg, Lori F.: Students' internalized representation of simple linear regression. 94 ASAProStEd 100-101

VanBrackle, Lewis; VanBrackle, Anita S.; Rothenberg, Lori F.: Students' internalized representation of simple linear regression. 94 ASAProStEd 100-101

Vandebroek, Martina: Dhaene, Jan; — —: Recursions for the individual model. 95 InsrneME 16 31-38

Vandenbroucke, Jan P.: Comment on "A critical look at some popular meta-analytic methods" (V140 p290-296). 95 AJEpidm 142 1007-1008

Vangel, Mark G.: ANOVA estimates of variance components for a class of mixed models. 94 ASAProPEng 91-96

— —: One-sided β-content tolerance intervals for mixed models. 94 ASAProPEng 200-206

Vankeerberghen, P.; Smeyers-Verbeke, J.; Leardi, R.; Karr, C. L.; Massart, D. L.: Robust regression and outlier detection for non-linear models using genetic algorithms. 95 Chemolab 28 73-87

Griep, M. I.; Wakeling, I. N.; — —; Massart, D. L.: Comparison of semirobust and robust partial least squares procedures. 95 Chemolab 29 37-50

Vännman, Kerstin: On the distribution of the estimated mean from nonstandard mixtures of distributions. 95 CommStA 24 1569-1584

— —: A unified approach to capability indices. 95 StSinica 5 805-820

VanRaden, P. M.: Hoeschele, I.; — —: Bayesian analysis of linkage between genetic markers and quantitative trait loci. II. Combining prior knowledge with experimental evidence. 93 ThApGene 85 946-952

Hoeschele, I.; — —: Bayesian analysis of linkage between genetic markers and quantitative trait loci. I. Prior knowledge. 93 ThApGene 85 953-960

Vapnik, V. N.: The nature of statistical learning theory. 95 Springer:Brln:NY 188p

— — (Auth); Fedorov, V. V. (Rev): Review of "The nature of statistical learning theory". 96 ShrtBkRv 16/1 3-3

Varadhan, S. R. S.: Self diffusion of a tagged particle in equilibrium for asymmetric mean zero random walk with simple exclusion. 95 AIHenrPB 31 273-285

Vardeman, Stephen B.: Lei, Ding-Hwa; — —: LRT method of constructing a two-sided acceptance region for variables acceptance sampling and its comparison with other methods. 94 ASAProPEng 115-119

— — (Auth); Bisgaard, Soren (Rev): Review of "Statistics for engineering problem solving". 95 AmerStat 49 320-321

— — (Auth); Brugger, Richard M. (Rev): Review of "Statistics for engineering problem solving". 95 Technmcs 37 348-348

Stanford, John L.; — — (Ed); Copeland, Karen A. F. (Rev): Review of "Statistical methods for physical science". 95 JQualTek 27 180-181

Vares, Maria Euália: Sidoravićus, Vladas; — —: Ergodicity of Spitzer's renewal model. 95 StocProc 55 119-130

Vares, Maria Eulália: Martínez, Servet; — —: A Markov chain associated with the minimal quasi-stationary distribution of birth-death chains. 95 JAppProb 32 25-38

Varga, Nicholas: Morrell, Christopher H.; Peyrot, Mark F.; — —: An analysis of voting coalitions in the New York colonial legislature. 94 ASAProSoSt 66-70

Varga, T.: Gupta, A. K.; — —: Some inference problems for matrix variate elliptically contoured distributions. 95 MaOpfStS 26 219-229

Gupta, A. K.; — —: Normal mixture representations of matrix variate elliptically contoured distributions. 95 SankhyaA 57 68-78

Gupta, A. K.; — —: A new class of matrix variate elliptically contoured distributions. 94 JItalStS 3 255-270

Gupta, A. K.; — —: Moments and other expected values for matrix variate elliptically contoured distribution. 94 Statstca 54 361-373

Varga, Tamàs: Gupta, Arjun K.; — —: An empirical estimation procedure. 94 Metron 52/1 67-70

Varian, H. R.: Economic and financial modeling with Mathematica. 93 Springer:-Brln:NY 458p

Varis, Olli: Belief networks for modelling and assessment of environmental change. 95 EnvrMtrc 6 439-444

Varley, S. A.: Rondel, R. K.; — —; Webb, C. F. (Ed); Therasse, P. (Rev): Review of "Clinical data management". 95 StatMed 14 877-877

Varopoulos, N. Th.: Hardy-Littlewood theory on unimodular groups. 95 AIHenrPB 31 669-688

Vasantha, P.: Kumar, V. S. Sampath; — —: Continuous inspection of Markov processes with a clearance interval. 95 JAppStat 22 427-432

Vasin, Sergey A.: Kingkade, W. Ward; — —: Welfare consequences of mortality dynamics in the former Soviet Republics: Length of working life. 94 ASAProSoSt 94-99

Vasudeva, R.: Janardan, K. G.; — —; Mohan, N. R.: On some special birth processes in biology. 95 CommStA 24 1033-1038

Vatutin, V. A.: Branching processes with final types of particles and random trees. 94 ThProbAp 39 628-641

Yor, M. (Auth); — — (Rev): Review of "Some aspects of Brownian motion. Part I: Some special functionals". 94 ThProbAp 39 723-725

Vaughan, Timothy S.: The effect of sampling variability on statistical order point computation. 95 PrInvMgt 36 84-87

Vaupel, James W.: Yashin, Anatoli I.; — —; Iachine, Ivan A.: Correlated individual frailty: An advantageous approach to survival analysis of bivariate data. 95 MaPopStd 5 145-159

Vaurio, J. K.: Optimization of test and maintenance intervals based on risk and cost. 95 RelEngSS 49 23-36

Vaurio, Jussi K.: The probabilistic modeling of external common cause failure shocks in redundant systems. 95 RelEngSS 50 97-107

Vázquez-Barquero, José Luis: Pickles, Andrew; Dunn, Graham; — —: Screening for stratification in two-phase ('two-stage') epidemiological surveys. 95 StMedRes 4 73-89

Pickles, Andrew; Dunn, Graham; — —: Reply to comment on "Screening for stratification in two-phase ('two-stage') epidemiological surveys". 95 StMedRes 4 263-263

Vázquez-Polo, F. J.; Hernández Bastida, A.: Behaviour of the posterior error rate with partial prior information in auditing. 95 JAppStat 22 469-476

Veall, Michael R.; Zimmermann, Klaus F.: Comment on "Goodness-of-fit measures in binary choice models". 95 EcomtrRv 14 117-120

George, Peter J.; Oksanen, Ernest H.; — —: Analytic and bootstrap approaches to testing a market saturation hypothesis. 95 MaCmpSim 39 311-315

Vecchia, Dominic F.: Ahlbrandt, Robert; Iyer, Hari K.; — —: Optimum design of serial measurement trees. 95 JStPlInf 48 379-390

Veeh, J. A.: Hudson, W. N.; Mason, J. D.; — —: Operator stable probability measures: An overview. 94 ThProbAp 39 264-276

Veeh, Jerry Alan: Meerschaert, Mark M.; — —: Symmetry groups in R^d. 95 StPrLet 22 1-6

Jurek, Zbigniew J.; Mason, J. David (Auth); — — (Rev): Review of "Operator-limit distributions in probability theory". 95 BAmMaSoc 32 278-280

Veevers, Alan; Robinson, Geoff: Obituary: Richard A. Stone, 1961-1994. 95 JRSS-A 158 627-628

Vega, Juan: Sahai, Hardeo; — —: Training teachers to teach statistics in the secondary schools of Puerto Rico. 94 ASAProStEd 253-258

Vegas, Esteban: Ocaña, Jordi; — —: Variance reduction for Bernoulli response variables in simulation. 95 CmpStDA 19 631-640

Vehovar, Vasja: Field substitutions - A neglected option?. 94 ASASrvyRsr 589-594

Rubin, Donald B.; Stern, Hal S.; — —: Handling "Don't know" survey responses: The case of the Slovenian plebiscite. 95 JASA 90 822-828

Veinott, Arthur F., Jr: Bowker, Albert H.; Olkin, Ingram; — —: Gerald J. Lieberman. 95 PrEngInf 9 3-26

Huang, Ying; — —: Markov branching decision chains with interest-rate-dependent rewards. 95 PrEngInf 9 99-121

Velicer, Wayne F.: Ding, Lin; — —; Harlow, Lisa L.: Effects of estimation methods, number of indicators per factor, and improper solutions on structural equation modeling fit indices. 95 StrEqMod 2 119-144

Velikii, Yu. A.; Motsa, A. I.; Sil'vestrov, D. S.: Dual processes and ergodic type theorems for Markov chains in the scheme of series. 94 ThProbAp 39 642-653

Velikoivanenko, G. I.: On the existence of exponential moments of norms of random elements from the spaces $\text{sub}_{\phi}(\Omega)$. 95 ThPrMaSt 50 61-65

Velikoivanenko, O. I.; Shvets, Ye. M.: On one discrete queueing model. 95 ThPrMaSt 50 55-60

Velilla, Santiago: Diagnostics and robust estimation in multivariate data transformations. 95 JASA 90 945-951

Vella, V.: Bennett, S.; Radalowicz, A.; — —; Tomkins, A.: A computer simulation of household sampling schemes for health surveys in developing countries. 94 IJEpidem 23 1282-1291

Velleman, Paul F.: Hoaglin, David C.; — —: A critical look at some analyses of major league baseball salaries. 95 AmerStat 49 277-285

Venables, W. N.; Ripley, B. D.: Modern applied statistics with S-Plus. 94 Springer:Brln:NY 462p

Everitt, Brian S. (Auth); — — (Rev): Review of "A handbook of statistical analyses using S-Plus". 95 AstrlJSt 37 369-372

— —; Ripley, B. D. (Auth); Smith, J. T. (Rev): Review of "Modern applied statistics with S-Plus". 95 ShrtBkRv 15/2 24-24

Venetsanopoulos, Anastasios N.: Karayiannis, Nicolaos B.; — —: On the training and performance of high-order neural networks. 95 MaBisci 129 143-168

Venkatraman, E. S.: Siegmund, D.; — —: Using the generalized likelihood ratio statistic for sequential detection of a change-point. 95 AnlsStat 23 255-271

Venot, Alain: Coste, Joël; Fermanian, Jacques; — —: Methodological and statistical problems in the construction of composite measurement scales: A survey of six medical and epidemiological journals. 95 StatMed 14 331-345

Coste, Joël; Walter, Eric; — —: A new approach to selection and weighting of items in evaluative composite measurement scales. 95 StatMed 14 2565-2580

Venter, J. H.; Snyman, J. L. J.: A note on the generalised cross-validation criterion in linear model selection. 95 Biomtrka 82 215-219

Veraverbeke, N.: Aerts, M.; — —: Bootstrapping a nonparametric polytomous regression model. 95 MaMethSt 4 189-200

Janssen, P.; Marron, J. S.; — —; Sarle, W.: Scale measures for bandwidth selection. 95 JNnparSt 5 359-380

Breitung, K. W. (Auth); — — (Rev): Review of "Asymptotic approximations for probability integrals". 95 ShrtBkRv 15/2 26-26

Veraverbeke, Noël: Antoch, Jaromir; Hušková, Marie; — —: Change-point problem and bootstrap. 95 JNnparSt 5 123-144

Verbeek, A. L.: Schouten, L. J.; Straatman, H.; Kiemeney, L. A.; Gimbrere, C. H.; — —: The capture-recapture method for estimation of cancer registry completeness: A useful tool?. 94 IJEpidem 23 1111-1116

Verbeek, Marno: Alternative transformations to eliminate fixed effects. 95 EcomtrRv 14 205-211

Verboon, Peter; Gabriel, K. Ruben: Generalized Procrustes analysis with iterative weighting to achieve resistance. 95 BJMaStPs 48 57-73

Verbyla, A. P.: Diggle, Peter J.; Liang, Kung-Yee; Zeger, Scott L. (Auth); — — (Rev): Review of "Analysis of longitudinal data". 95 AstrlJSt 37 367-368

Verbyla, Arūnus P.: Gogel, Beverley J.; Cullis, Brian R.; — —: REML estimation of multiplicative effects in multienvironment variety trials. 95 Biomtrcs 51 744-749

Verdecchia, Arduino; Mariotto, Angela B.: A back-calculation method to estimate the age and period HIV infection intensity, considering the susceptible population. 95 StatMed 14 1513-1530

Verdegay, J. L.: Cadenas, J. M.; — —: PROBO: An interactive system in fuzzy linear programming. 95 FuzzySet 76 319-332

Verdinelli, Isabella; Wasserman, Larry: Computing Bayes factors using a generalization of the Savage-Dickey density ratio. 95 JASA 90 614-618

Verducci, Joseph S.: Huh, Myung-Hoe; Critchlow, Douglas E.; — —; Klecolt-Glaser, Janice; Glaser, Ronald; Malarkey, William B.: A symmetric analysis of paired rankings with application to temporal patterns of hormonal concentration. 95 Biomtrcs 51 1361-1371

Vere, D. T.; Griffith, G. R.: Modifying quantitative forecasts of livestock production using expert judgements. 95 JForcast 14 453-464

Vere-Jones, D.: Lomnitz, C. (Auth); — — (Rev): Review of "Fundamentals of earthquake prediction". 95 ShrtBkRv 15/2 22-22

Vere-Jones, David: Forecasting earthquakes and earthquake risk. 95 IntJFor 11 503-538

— —: The coming of age of statistical education. 95 IntStRvw 63 3-23

Verma, Sanjay: Singh, R. Karan; Agnihotri, Monica; — —: A generalized class of estimators utilizing Lindley-like mean correction in linear regression model. 94 Metron 52/3 147-160

Verma, Vijay: Kish, Leslie; Frankel, Martin R.; — —; Kaćiroti, Niko: Design effects for correlated $(P_i - P_j)$. 95 SurvyM 21 117-124

Veronese, Piero: Consonni, Guido; — —: A Bayesian method for combining results from several binomial experiments. 95 JASA 90 935-944

Verotta, Davide: Fattinger, Karin E.; Sheiner, Lewis B.; — —: A new method to explore the distribution of interindividual random effects in non-linear mixed effects models. 95 Biomtrcs 51 1236-1251

Verrazzani, L.: Corsini, G.; Gini, F.; Greco, M. V.; — —: Cramér-Rao bounds and estimation of the parameters of the Gumbel distribution. 95 IEEEAES 31 1202-1204

Verrier, E.; Colleau, J. J.; Foulley, J. L.: Long-term effects of selection based on the animal model BLUP in a finite population. 93 ThApGene 87 446-454

Vervaat, Wim: O'Brien, George L.; — —: Compactness in the theory of large deviations. 95 StocProc 57 1-10

Verweij, Pierre J. M.; van Houwelingen, Hans C.: Time-dependent effects of fixed covariates in Cox regression. 95 Biomtrcs 51 1550-1556

Verzani, John: Slow points in the support of historical Brownian motion. 95 AnlsProb 23 56-70

Vesely, W. E.; Uryasev, S. P.; Samanta, P. K.: Failure of emergency diesel generators: A population analysis using empirical Bayes methods. 94 RelEngSS 46 221-229

Vevea, Jack L.; Hedges, Larry V.: A general linear model for estimating effect size in the presence of publication bias. 95 Psymtrka 60 419-435

Viallefont, A.: Pradel, R.; Reboulet, A.-M.; — —: Testing hypotheses and estimating survival from capture histories with CR. 95 JAppStat 22 775-784

Viallefont, Anne; Cooch, Evan G.; Cooke, Fred: Estimation of trade-offs with capture-recapture models: A case study on the lesser snow goose. 95 JAppStat 22 847-861

Viana, Marlos: Olkin, Ingram; — —: Correlation analysis of extreme observations from a multivariate normal distribution. 95 JASA 90 1373-1379

Viana, Marlos A. G.: Combined Bayesian estimates for the equicorrelation coefficient. 95 CandJSt 23 43-53

Viano, M. C.; Deniau, Cl.; Oppenheim, G.: Long-range dependence and mixing for discrete time fractional processes. 95 JTimSrAn 16 323-338

Viano, Marie-Claude: Massé, Bruno; — —: Explicit and exponential bounds for a test on the coefficient of an AR (1) model. 95 StPrLet 25 365-371

Vicari, Donatella: Generalized partial correlation coefficients (Italian). 94 Statstca 54 491-499

Vichi, Maurizio: Rizzi, Alfredo; — —: Representation, synthesis, variability and data preprocessing of a three-way data set. 95 CmpStDA 19 203-222

Victorbabu, B. Re.; Narasimham, V. L.: A new type of slope rotatable central composite design. 94 JIAgSt 46 315-317

Vidakovic, Brani; DasGupta, Anirban: Lower bounds on Bayes risks for estimating a normal variance: With applications. 95 CandJSt 23 269-282

Insua, David Ríos; Ruggeri, Fabrizio; — —: Some results on posterior regret Γ-minimax estimation. 95 StatDecn 13 315-331

Mukhopadhyay, Saurabh; — —: Efficiency of linear Bayes rules for a normal mean: Skewed priors class. 95 Ststcian 44 389-397

Vidoni, Paolo: A simple predictive density based on the p^*-formula. 95 Biomtrka 82 855-863

Vidyashankar, A. N.: Athreya, K. B.; — —: Large deviation rates for branching processes. II. The multitype case. 95 AnlsApPr 5 566-576

Viel, Jean-François; Pobel, Dominique; Carré, André: Incidence of leukaemia in young people around the La Hague nuclear waste reprocessing plant: A sensitivity analysis. 95 StatMed 14 2459-2472

Lawson, Andrew B.; — —: Tests for directional space-time interaction in epidemiological data. 95 StatMed 14 2383-2391

Vieland, Veronica J.; Greenberg, David A.; Hodge, Susan E.: Adequacy of single-locus approximations for linkage analyses of oligogenic traits: Extension to multigenerational pedigree structures. 93 HumHrdty 43 329-336

— —; Hodge, Susan E.: Inherent intractability of the ascertainment problem for pedigree data: A general likelihood framework. 95 AJHuGene 56 33-43

Viet, Tran Quoc: Some adaptive estimators in simple regression models. 95 StatDecn 13 117-130

Vieu, P.: Boularan, J.; Ferre, L.; — —: A nonparametric model for unbalanced longitudinal data with application to geophysical data. 95 CmpStQ 10 285-298

Moncaup, S.; Sarda, P.; — —: Nonparametric estimators of an intensity function applied to meteorological data (French). 95 RStApp 43/4 77-88

Vieu, Philippe: Order choice in nonlinear autoregressive models. 95 MaOpfStS 26 307-328

Boularan, Joël; Ferré, Louis; — —: Location of particular points in non-parametric regression analysis. 95 AstrlJSt 37 161-168

Vigezzi, Michael J.: Lepeniotis, Stefanos S.; — —: Lowering manufacturing cost of material by formulating it through statistical modeling and design. 95 Chemolab 29 133-139

Vigneau, E.; Devaux, M. F.; Safar, M.: Application of Procrustean methods to mid- and near-infrared spectral data. 95 JChemtrc 9 125-135

Vijayan, K.; Mukhopadhyay, P.; Bhattacharyya, S.: On non-negative unbiased estimation of quadratic forms in finite population sampling. 95 AstrlJSt 37 169-178

Vijayasree, G.; Misra, Neeraj; Singh, Harshinder: Componentwise estimation of ordered parameters of k (≥ 2) exponential populations. 95 AnInStMa 47 287-307

Vilar Fernandez, J. M.: Gonzalez Manteiga, W.; — —: Testing linear regression models using non-parametric regression estimators when errors are non-independent. 95 CmpStDA 20 521-541

Vilar, J. A.: Kernel estimation of the regression function with random sampling times. 95 Test 4 137-178

Vilcassim, Naufel J.: Chintagunta, Pradeep K.; — —: A two-period repeated game advertising investment model for oligopolistic markets with an application to the beer industry. 95 DecisnSc 26 531-559

Villanueva, B.; Kennedy, B. W.: Index versus tandem selection after repeated generations of selection. 93 ThApGene 85 706-712

Villarroya, Angel; Ríos, Martín; Oller, Josep M.: Discriminant analysis algorithm based on a distance function and on a Bayesian decision. 95 Biomtrcs 51 908-919

Villegas, C.: Swartz, T. B.; — —; Martinez, C. J.: On semiparametric pivotal Bayesian inference for quantiles. 95 CommStA 24 2499-2515

Vinci, E.: Mann, Prem S. (Auth); — — (Rev): Review of "Introductory statistics" (Italian). 94 Metron 52/3 211-212

Vining, G. Geoff: Cornell, John A.; Randles, Ron H.; — —: Challenges in teaching short courses by an academic department: The University of Florida experience. 95 AmerStat 49 195-200

Vinod, H. D.: Double bootstrap for shrinkage estimators. 95 JEconmtx 68 287-302

— —; Basu, Parantap: Forecasting consumption, income and real interest rates from alternative state space models. 95 IntJFor 11 217-231

— —; McCullough, B. D.: Estimating cointegration parameters: An application of the double bootstrap. 95 JStPlInf 43 147-156

— —; Srivastava, V. K.: Large sample asymptotic properties of the double k-class estimators in linear regression models. 95 EcomtrRv 14 75-100

Maddala, G. S.; Rao, C. R.; — — (Ed); Gill, Len (Rev): Review of "Handbook of statistics (Volume 11) Econometrics". 95 IntJFor 11 189-191

Maddala, G. S.; Rao, C. R.; — — (Ed); Krämer, Walter (Rev. German): Review of "Econometrics (Handbook of statistics) (Volume II)". 95 AllgeStA 79 101-102

Vinogradov, O. P.: On certain asymptotic properties of waiting time in a multiserver queueing system with identical service times. 94 ThProbAp 39 714-718

Vinokurov, V. I.: Arbekov, I. M.; — —; Ryazanov, B. V.: On the problem of detecting random trajectories. 94 ThProbAp 39 547-557

Viollaz, Aido J.; Sfer, Ana M.; Salvatierra, Stella M.: An approximation of the unconditional error rates of the sample linear discriminant function. 95 CommStA 24 1941-1969

Viollaz, Aldo J.: Convergence in weighted supremum norms of the Skorokhod representation of the estimated empirical process. 95 CommStA 24 2829-2839

Rodríguez, Juan C.; — —: A Cramér-von Mises type goodness-of-fit test with asymmetric weight function. 95 CommStA 24 1095-1120

Viscusi, W. Kip; Born, Patricia: The general-liability reform experiments and the distribution of insurance-market outcomes. 95 JBES 13/2 183-188

Vishnyakov, A. N.: Bondarenko, M. V.; — —: On the Russian stock exchange. 94 ThProbAp 39 188-194

Visser, H.; Molenaar, J.: Trend estimation and regression analysis in climatological time series: An application of structural time series models and the Kalman filter. 95 JClimate 8 969-979

Visser, Michael: Magnac, Thierry; Robin, Jean-Marc; — —: Analysing incomplete individual employment histories using indirect inference. 95 JApEcmtx 10 153-169S

Visser, T.: Luinge, H. J.; van der Maas, J. H.; — —: Partial least squares regression as a multivariate tool for the interpretation of infrared spectra. 95 Chemolab 28 129-138

Viswanathan, S.: Foster, F. Douglas; — —: Can speculative trading explain the volume-volatility relation?. 95 JBES 13 379-396

Vitale, Richard A.: On the volume of parallel bodies: A probabilistic derivation of the Steiner formula. 95 AdvAppPr 27 97-101

Vitali, Ornello: Teaching statistics (Italian). 94 Statstca 54 387-396

Viveros, Román; Balasubramanian, K.; Balakrishnan, N.: Reply to comment on "Binomial and negative binomial analogues under correlated Bernoulli trials". 95 AmerStat 49 325-326

Vives, Josep: Imkeller, Peter; Perez-Abreu, Victor; — —: Chaos expansions of double intersection local time of Brownian motion in R^d and renormalization. 95 StocProc 56 1-34

Vogel, Frederic A.: The evolution and development of agricultural statistics at the United States Department of Agriculture. 95 JOffStat 11 161-180

Vogel, Lydia: Schulz, Henrik; Tetzlaff, Klaus; — —: Simple encoding of infrared spectra for pattern recognition. 1: Statistical examination of the effectiveness and information content by factor analysis. 95 JChemtrc 9 143-168

Vogel, Richard M.; Bolognese, Ralph A.: Storage-reliability-resilience-yield relations for over-year water supply systems. 95 WaterReR 31 645-654

Vogler, George P.; Wette, Reimut; McGue, Matthew K.; Rao, D. C.: Properties of alternative estimators of familial correlations under variable sibship size. 95 Biomtrcs 51 276-283

Vogt, Arthur: Simpson's paradox revisited. 95 Student 1 99-108

Vogt, W. P.: — — (Auth); Berg, Sven (Rev): Review of "Dictionary of statistics and methodology. A nontechnical guide for the social sciences". 94 JOffStat 10 225-226

Voiculescu, D.: Bercovici, H.; — —: Superconvergence to the central limit and failure of the Cramér theorem for free random variables. 95 ProbTher 103 215-222

Voinov, V. G.; Nikulin, M. S.: A review of the results on the Stein approach for estimators improvement. 95 Questo 19 65-91

Nikulin, M. S.; — —: On the problem of the means of weighted normal populations. 95 Questo 19 93-106

Voit, Eberhard O.: Yu, Shuiyang; — —: A simple, flexible failure model. 95 BiomtrcJ 37 595-609

Voit, M.: Wang, Z.; Yang, X. (Auth); — — (Rev): Review of "Birth and death processes and Markov chains". 95 Metrika 42 270-270

Vojnovic, D.; Campisi, B.; Mattei, A., Favretto, L.: Experimental mixture design to ameliorate the sensory quality evaluation of extra virgin olive oils. 95 Chemolab 27 205-210

Volkov, V. Ya.; Gladkov, Yu. M.: Reconfigurable combined forecasts in a non-stationary inflationary environment. 95 JForcast 14 395-403

Vollset, Stein E.: Tang, Man-Lai; Hirji, Karim F.; — —: Exact power computation for dose-response studies. 95 StatMed 14 2261-2272

von Collani, Elart: Dimitrov, Boyan; — —: Contorted uniform and Pareto distributions. 95 StPrLet 23 157-164

von Haeseler, Arndt; Churchill, Gary A.: Network models for sequence evolution. 93 JMolEvol 37 77-85

— —; Schöniger, Michael: Modelling evolution of autocorrelated sequences. 95 CmpStDA 20 457-458

von Rosen, Dietrich: Residuals in the growth curve model. 95 AnInStMa 47 129-136

Kollo, Tõnu; — —: Approximating by the Wishart distribution. 95 AnInStMa 47 767-783

Kollo, Tõnu; — —: Minimal moments and cumulants of symmetric matrices: An application to the Wishart distribution. 95 JMultiAn 55 149-164

von Storch, Hans: Zwiers, Francis W.; — —: Taking serial correlation into account in tests of the mean. 95 JClimate 8 336-351

Von Thurn, Dawn R.; Moore, Jeffrey C.: Results from a cognitive exploration of the 1993 American Housing Survey. 94 ASASrvyRsr 1210-1214

Von Tress, Mark: Repeated $2 \times k$ tables and the sandwich estimator. 94 ASAProBiop 256-261

Clogg, Clifford C.; Shihadeh, Edward S. (Auth); — — (Rev): Review of "Statistical methods for ordinal variables". 95 Technmcs 37 118-119

Vondracek, Bruce; Degan, Donald J.: Among-and within-transect variability in estimates of shad abundance made with hydroacoustics. 95 NAJFishM 15 933-939

Vondraček, Zoran: Some Markov processes with Brownian exit distributions. 95 ProbTher 101 393-407

Vos, Paul W.: Quasi-likelihood or extended quasi-likelihood? An information-geometric approach. 95 AnInStMa 47 49-64

Voss, Paul: Kale, Balkrishna; Besl, John; Palit, Charles; — —: Updating age-sex-specific net migration rates with limited data. 94 ASAProSoSt 116-121 (Disc: p122-123)

Voss, Theresa: Habibullah, Mohamed; Rogers, James; — —: Issues related to the bioequivalence of drug formulations. 94 ASAProStCp 222-224

Vostrikova, L.: Kutoyants, Yu.; — —: On non-consistency of estimators. 95 Stochatx 53 53-80

Vounatsou, P.; Smith, A. F. M.: Bayesian analysis of ring-recovery data via Markov chain Monte Carlo simulation. 95 Biomtrcs 51 687-708

Vowden, B. J.: Preece, D. A.; — —; Jones, R. Hughes; Roger, C. A.; Vowden, C. J.: Choreographing designs. 95 MaScitst 20 15-32

Vowden, C. J.: Preece, D. A.; Vowden, B. J.; Jones, R. Hughes; Roger, C. A.; — —: Choreographing designs. 95 MaScitst 20 15-32

Vregelaar, Jan M. Ten: On computing objective function and gradient in the context of least squares fitting a dynamic errors-in-variables model. 95 SIAMSSC 16 738-753

Vrscay, E. R.: Forte, B.; — —: Solving the inverse problem for measures using iterated function systems: A new approach. 95 AdvAppPr 27 800-820

Vugts, H. F.: Weijers, E. P.; van Delden, A.; — —; Meesters, A. G. C. A.: Characteristics of convective turbulence in the surface layer investigated by principal component analysis. 95 JApMetrl 34 528-541

Vuong, Quang: Gourieroux, Christian; Monfort, Alain (Auth); — — (Transl): Statistics and econometric models. Volumes 1 and 2. 95 CambridgeUniv 1062p

Vynckier, Petra: Beirlant, Jan; Broniatowski, Michel; Teugels, Jozef L.; — —: The mean residual life function at great age: Applications to tail estimation. 95 JStPlInf 45 21-48

Wabuyele, Busolo Wa; Harrington, Peter de B.: Quantitative comparison of bidirectional and optimal associative memories for background prediction of spectra. 95 Chemolab 29 51-61

Wacholder, Sholom: Design issues in case-control studies. 95 StMedRes 4 293-309

Wada, Cicilia Yuko; Sen, Pranab Kumar: Restricted alternative test in a parametric model with competing risk data. 95 JStPlInf 44 193-203

Waddell, Peter J.; Penny, David; Hendy, Michael D.; Arnold, Greg: The sampling distributions and covariance matrix of phylogenetic spectra. 94 MolBioEv 11 630-642

Wade, A. M.; Ades, A. E.; Salt, A. T.; Jayatunga, R.; Sonksen, P. M.: Age-related standards for ordinal data: Modelling the changes in visual acuity from 2 to 9 years of age. 95 StatMed 14 257-266

Wade, Mark R.: Woodall, William H.; Crowder, Stephen V.; — —: Comment on "Q charts". 95 JQualTek 27 328-332

Wade, Martcia: Kamlet, Mark S.; Paul, Nancy; Greenhouse, Joel; Kupfer, David; Frank, Ellen; — —: Cost utility analysis of maintenance treatment for recurrent depression. 95 ConClnTr 16 17-40

Wadiak, Diane L.: Silver, N. Clayton; — —; Massey, Catherine J.: A Microsoft FORTRAN 77 program for testing the differences among independent first-order partial correlations. 95 EdPsMeas 55 245-248

Wagner, Edward: Diehr, Paula; Martin, Donald; Koepsell, Thomas; Cheadle, Allen; — —; Psaty, Bruce: Optimal survey design for community-intervention evaluations: Cohort, cross-sectional, or both?. 94 ASAProEpid 34-43

Wagner, Gert: Rendtel, Ulrich; — —; Frick, Joachim: A method for checking longitudinal weights in panel surveys - The case of the German Socio-Economic Panel (German). 95 AllgeStA 79 252-277

Wagner, Udo; Geyer, Alois L. J.: A maximum entropy method for inverting Laplace transforms of probability density functions. 95 Biomtrka 82 887-892

Wagner, Wiesław; Brzeskwiniewicz, Henryk: Normality testing in a mixed model of a two-way nested classification. 95 BiomtrcJ 37 889-895

Wahba, Grace; Johnson, Donald R.; Gao, Feng; Gong, Jianjian: Adaptive tuning of numerical weather prediction models: Randomized GCV in three- and four-dimensional data assimilation. 95 MWeathr 11 3358-3369

— —; Wang, Yuedong: Behavior near zero of the distribution of GCV smoothing parameter estimates. 95 StPrLet 25 105-111

Wang, Yuedong; — —: Bootstrap confidence intervals for smoothing splines and their comparison to Bayesian confidence intervals. 95 JStCmpSm 51 263-279

Wang, Yuedong; — —; Chappell, Rick; Gu, Chong: Simulation studies of smoothing parameter estimates and Bayesian confidence intervals in Bernoulli SS ANOVA models. 95 CommStB 24 1037-1059

Wainer, H.: Holland, P. W.; — — (Ed); Mellenbergh, Gideon J. (Rev): Review of "Differential item functioning". 95 Psymtrka 60 621-623

Wainer, Howard: Visual revelations: A rose by another name. 95 Chance 8/2 46-51

— —: Graphical mysteries. 95 Chance 8/3 52-56

— —: Picturing an L.A. bus schedule. 95 Chance 8/4 44-48

— —; Dorans, Neil J.; Flaugher, Ronald; Green, Bert F.; Mislevy, Robert; Steinberg, Lynne; Thissen, David (Auth); Andrich, David (Rev): Review of "Computerized adaptive testing: A primer". 95 Psymtrka 60 615-620

Wakefield, Jon; Racine-Poon, Amy: An application of Bayesian population pharmacokinetic/pharmacodynamic models to dose recommendation. 95 StatMed 14 971-986

Wakeley, John: Substitution rate variation among sites in hypervariable region 1 of human mitochondrial DNA. 93 JMolEvol 37 613-623

Wakeling, I. N.: Griep, M. I.; — —; Vankeerberghen, P.; Massart, D. L.: Comparison of semirobust and robust partial least squares procedures. 95 Chemolab 29 37-50

Wakolbinger, A.: Elworthy, K. D.; Ikeda, N. (Ed); — — (Rev): Review of "Asymptotic problems in probability theory: Stochastic models and diffusions on fractals". 95 Metrika 42 69-70

Wakolbinger, Anton: Limits of spatial branching populations. 95 Bernoull 1 171-185 (Disc: p185-189)

Gorostiza, Luis G.; Hochberg, Kenneth J.; — —: Persistence of a critical super-2 process. 95 JAppProb 32 534-540

Waksberg, Joseph: Brick, J. Michael; — —; Kulp, Dale; Starer, Amy: Bias in list-assisted telephone samples. 95 PubOpinQ 59 218-235

Brick, J. Michael; — —; Kulp, Dale; Starer, Amy: Bias in list-assisted telephone samples. 94 ASASrvyRsr 1293-1298

Brick, Michael; — —; Keeter, Scott: Evaluating the use of data on interruptions in telephone service for nontelephone households. 94 ASASrvyRsr 19-28

Burke, John; Mohadjer, Leyla; Green, Jim; — —; Kirsch, Irwin S.; Kolstad, Andrew: Composite estimation in national and state surveys. 94 ASASrvyRsr 873-878

Khare, Meena; Ezzati-Rice, Trena M.; Mohadjer, Leyla K.; — —: An evaluation of nonresponse bias in NHANES III (1988-91). 94 ASASrvyRsr 949-954

Wakuta, Kazuyoshi: Vector-valued Markov decision processes and the systems of linear inequalities. 95 StocProc 56 159-169

Walckiers, Denise: Sartor, Francis; — —: Estimate of disease prevalence using drug consumption data. 95 AJEpidm 141 782-787

Walczak, B.: Outlier detection in multivariate calibration. 95 Chemolab 28 259-272

— —: Outlier detection in bilinear calibration. 95 Chemolab 29 63-73

— —; Massart, D. L.: Robust principal components regression as a detection tool for outliers. 95 Chemolab 27 41-54

Walden, A. T.; McCoy, E. J.; Percival, D. B.: The effective bandwidth of a multitaper spectral estimator. 95 Biomtrka 82 201-214

Percival, D. A.; — — (Auth); Swift, A. Louise (Rev): Review of "Spectral analysis for physical applications: Multitaper and conventional techniques". 95 JRSS-A 158 198-199

Wales, T. J.: Diewert, W. E.; — —: Flexible functional forms and tests of homogeneous separability. 95 JEconmtx 67 259-302

Walford, Nigel: Geographical data analysis. 95 Wiley:NY:UK 446p

Walker, A. M.: On results of Porat concerning asymptotic efficiency of sample covariances of Gaussian ARMA processes. 95 JTimSrAn 16 237-248

Anderson, T. W. (Auth); — — (Rev): Review of "The statistical analysis of time series". 95 JTimSrAn 16 357-358

Walker, Brett L.: Desante, David F.; Burton, Kenneth M.; Saracco, James F.; — —: Productivity indices and survival rate estimates from MAPS, a continent-wide programme of constant-effort mist-netting in North America. 95 JAppStat 22 935-947

Walker, Chip: Mendelsohn, Susan L.; — —; Bernstein, Hank: Conducting global consumer research: Considerations and pitfalls. 94 ASASrvyRsr 1196-1198

Walker, Esteban: Kotz, Samuel; Johnson, Norman L. (Auth); — — (Rev): Review of "Process capability indices". 95 JASA 90 1127-1127

Walker, John H.: Partially-ordered linear models. 94 ASAProStCp 152-157

Walker, Mary Beth: Brown, Bryan W.; — —: Stochastic specification in random production models of cost-minimizing firms. 95 JEconmtx 66 175-205

Walker, Monica: Tarling, R. (Auth); — — (Rev): Review of "Analysing offending: Data, models and interpretations". 95 JRSS-A 158 357-358

Walker, Stephen: Generating random variates from D-distributions via substitution sampling. 95 StatComp 5 311-315

Ratkowsky, D. A.; Evans, M. A.; Alldredge, J. R. (Auth); — — (Rev): Review of "Cross-over experiments: Design, analysis, and application". 95 JRSS-A 158 200-200

Walker, Vern R.: Direct inference, probability, and a conceptual gulf in risk communication. 95 RiskAnal 15 603-609

Wallace, C. S.: Allison, L.; — —: The posterior probability distribution of alignments and its application to parameter estimation of evolutionary trees and to optimization of multiple alignments. 94 JMolEvol 39 418-430

Wallace, Dennis D.: Examination of statistical procedures for evaluating air quality models. 94 ASAProEnvr 82-87

Wallace, Jean E.; Foster, Robert D.; Kendall, John E.: Improving control of polymer viscosity using time-series experimentation and statistical process control. 94 ASAProPEng 320-325

Wallace, William A.: Ji, Xiaoyan; Raghavachari, M.; — —: Approximate reasoning in belief networks: Statistical validity, accuracy, and determination of sample size. 94 ASAProBays 128-133 (Disc: p134-135)

Wallach, D.: Regional optimization of fertilization using a hierarchical linear model. 95 Biomtrcs 51 338-346

— —: Empirical Bayes optimal fertilizer decisions. 95 JAppStat 22 507-516

Wallenstein, Sylvan; Naus, Joseph; Glaz, Joseph: Power of the scan statistics. 94 ASAProEpid 70-75

Waller, Lance A.: Does the characteristic function numerically distinguish distributions?. 95 AmerStat 49 150-152

— —; Jacquez, Geoffrey M.: Disease models implicit in statistical tests of disease clustering. 95 Epidmlgy 6 584-590

— —; Lawson, Andrew B.: The power of focused tests to detect disease clustering. 95 StatMed 14 2291-2308

— —; Turnbull, Bruce W.; Gustafsson, Göran; Hjalmars, Ulf; Andersson, Bertil: Detection and assessment of clusters of disease: An application to nuclear power plant facilities and childhood leukaemia in Sweden. 95 StatMed 14 3-16

— —; Turnbull, Bruce W.; Hardin, J. Michael: Obtaining distribution functions by numerical inversion of characteristic functions with applications. 95 AmerStat 49 346-350

Conerly, Michael D.; Hardin, J. Michael; — —: When is Satterthwaite's approximation adequate?. 94 ASAProStCp 209-213

Waller, Niels G.: A DeFries and Fulker regression model for genetic nonadditivity. 94 BehavGen 24 149-154

— —; Muthén, Bengt O.: Genetic Tobit factor analysis: Quantitative genetic modeling with censored data. 92 BehavGen 22 265-292

— —; Reise, Steven P.: Genetic and environmental influences on item response pattern scalability. 92 BehavGen 22 135-152

Walliker, David: Hill, William G.; Babiker, Hamza A.; Ranford-Cartwright, Lisa C.; — —: Estimation of inbreeding coefficients from genotypic data on multiple alleles, and application to estimation of clonality in malaria parasites. 95 GeneRsr 65 53-61

Wallis, J. R.: Hosking, J. R. M.; — —: A comparison of unbiased and plotting-position estimators of L moments. 95 WaterReR 31 2019-2025

Wallis, Kenneth F.: — — (Ed); Holden, Ken (Rev): Review of "Macroeconomic modelling Vols I and II". 95 IntJFor 11 334-335

Wallis, W. D.: Lin, C.; — —; Zhu, L.: Generalized 4-profiles of Hadamard matrices. 93 JCISS 18 397-400

Lin, Cantian; — —: Barker sequences and circulant Hadamard matrices. 93 JCISS 18 19-25

Wallman, Katherine; Groves, Robert M.; Parsons, Joseph; Davis, Mary C.; Lapham, Susan J.: Increasing statistical awareness through graduate training: The joint program in survey methodology. 94 ASAProSoSt 39-44

Wallsten, Thomas S.: Moxey, Linda M.; Sanford, Anthony J. (Auth); Budescu, David V.; — — (Rev): Review of "Communicating quantities: A psychological perspective". 95 Chance 8/2 38-40

Wallstrom, Garrick L.: Geske, Mark X.; Godbole, Anant P.; Schaffner, Andrew A.; Skolnick, Allison M.; — —: Compound Poisson approximations for word patterns under Markovian hypotheses. 95 JAppProb 32 877-892

Walsh, Dennis P.: Equating Poisson and normal probability functions to derive Stirling's formula. 95 AmerStat 49 270-271

Walstad, William B.: Economic knowledge and public opinion on economic issues. 94 ASASrvyRsr 1380-1385

Walter, E.; Pronzato, L.: Identifying parametric models for experimental data (French). 94 Masson:NY:Paris 384p

Walter, Eric: Coste, Joël; — —; Venot, Alain: A new approach to selection and weighting of items in evaluative composite measurement scales. 95 StatMed 14 2565-2580

Walter, S. D.: Reply to comment on "A simple test for spatial pattern in regional health data". 95 StatMed 14 1393-1394

Walter, Stephen D.: Methods of reporting statistical results from medical research studies. 95 AJEpidm 141 896-906

Walters, E. E.: Sham, P. C.; — —; Neale, M. C.; Heath, A. C.; MacLean, C. J.; Kendler, K. S.: Logistic regression analysis of twin data: Estimation of parameters of the multifactorial liability-threshold model. 94 BehavGen 24 229-238

Wan, Alan T. K.: The optimal critical value of a pre-test for an inequality restriction in a mis-specified regression model. 95 AstrlJSt 37 73-81

— —: Estimation with multiple inequality restrictions in a model with multivariate t error terms. 94 ASAProBuEc 223-228

— —; Yang, Minxian: On the use of the F ratio in a misspecified model with an interval restriction. 95 JStCmpSm 52 151-161

Wan, Choi K.: Jaccard, James; — —: Measurement error in the analysis of interaction effects between continuous predictors using multiple regression: Multiple indicator and structural equation approaches. 95 PsychBu 117 348-357

Wan, Jim Y.; Wang, Whedy; Bromberg, Judith: A SAS macro for residual deviance of ordinal regression analysis. 94 CmpBiomd 45 307-310

Wan, Qihua: Some large sample properties of estimate of failure rates based on truncated sample (Chinese). 95 KToChina 39 2120-2122

Wand, M. P.: Fan, Jianqing; Heckman, Nancy E.; — —: Local polynomial kernel regression for generalized linear models and quasi-likelihood functions. 95 JASA 90 141-150

Herrmann, Eva; — —; Engel, Joachim; Gasser, Theo: A bandwidth selector for bivariate kernel regression. 95 JRSS-B 57 171-180

Ruppert, D.; Sheather, S. J.; — —: An effective bandwidth selector for local least squares regression. 95 JASA 90 1257-1270

— —; Jones, M. C. (Auth); Mackisack, Margaret S. (Rev): Review of "Kernel smoothing". 95 AstrlJSt 37 368-369

— —; Jones, M. C. (Auth); Ramsay, J. O. (Rev): Review of "Kernel smoothing". 95 ShrtBkRv 15/2 27-27

Wand, M.-P.; Jones, M.-C.: Kernel smoothing. 95 ChapmnHall:Lnd 224p

Wang, Ann-Lee: Chinese dominoes games. 95 TeachgSt 17/2 70-73

Wang, Antonia C.: Hwang, Ching C.; — —; Fisher, Alan C.: The application of Cox regression models to the analysis of a double-blind comparative trial and subsequent open label extension in HIV-infected subjects. 94 ASAProBiop 194-199

Wang, C. S.; Gianola, D.; Sorensen, D. A.; Jensen, J.; Christensen, A.; Rutledge, J. J.: Response to selection for litter size in Danish Landrace pigs: A Bayesian analysis. 94 ThApGene 88 220-230

Wang, C. Y.; Carroll, R. J.: On robust logistic case-control studies with response-dependent weights. 95 JStPlInf 43 331-340

— —; Wang, Suojin: On information matrices in case-control studies. 95 StPrLet 22 269-274

Carroll, R. J.; Knickerbocker, R. K.; — —: Dimension reduction in a semi-parametric regression model with errors in covariates. 95 AnlsStat 23 161-181

Carroll, R. J.; Wang, Suojin; — —: Prospective analysis of logistic case-control studies. 95 JASA 90 157-169

Wang, Chaiho C.: Criminal antitrust prosecution - 1960 to 1990. 94 ASAProBuEc 281-286

Wang, Chamant: — — (Auth); Healy, M. J. R. (Rev): Review of "Sense and nonsense of statistical inference". 95 StatMed 14 103-104

Wang, Chamont; Corley, Thomas: The longest chute is the worse chute: Comments and additions to "Mr. Markov plays chutes and ladders". 94 ASAProStEd 104-106

— —; Silverberg, Arthur: Comment on "Statistical values, quality, and certification" (V48 p65-87). 95 AmerStat 49 116-117

— — (Auth); Tusell, Fernando (Rev): Review of "Sense and nonsense of statistical inference". 95 JAppStat 22 320-320

Wang, Chengming; Qin, Yongsong: Linear empirical Bayes estimation applied in multiple linear regression (Chinese). 95 ChJApPSt 10 371-378

Wang, D.: Shao, Q.-M.; — —: Small ball probabilities of Gaussian fields. 95 ProbTher 102 511-517

Wang, Dao Yi: Pitt, Loren D.; Robeva, Raina; — —: An error analysis for the numerical calculation of certain random integrals: Part 1. 95 AnlsApPr 5 171-197

Wang, Feng-Yu: Estimates of the first Dirichlet eigenvalue by using diffusion processes. 95 ProbTher 101 363-369

Wang, Ferdinand T.: Automatic smoothing parameter selection for robust nonparametric regression. 94 ASAProStCp 106-111

Wang, Gang: Differential subordination and strong differential subordination for continuous-time martingales and related sharp inequalities. 95 AnlsProb 23 522-551

Wang, I-Jeng; Chong, Edwin K. P.; Kulkarni, Sanjeev R.: On equivalence of some noise conditions for stochastic approximation algorithms. 95 PIEEEDecCt 3849-3854

Wang, J.; Yang, J. B.; Sen, P.: Safety analysis and synthesis using fuzzy sets and evidential reasoning. 95 RelEngSS 47 103-118

Wang, J. C.; Wu, C. F. Jeff: A hidden projection property of Plackett-Burman and related designs. 95 StSinica 5 235-250

Wang, Ji-Hong; Xie, Yu-Long; Yu, Ru-Qin: Maximum sum of binary-coded residuals (MASBR) regression as a robust procedure for treatment of spectral data. 95 JChemtrc 9 373-387

Wang, Jia-gang: The asymptotic behavior of locally square integrable martingales. 95 AnlsProb 23 552-585

Wang, Jiang: Lo, Andrew W.; — —: Implementing option pricing models when assest returns are predictable. 95 JFinance 50 87-129

Wang, Jichuan; Fisher, James H.; Siegal, Harvey A.; Falck, Russel S.; Carlson, Robert G.: Influence of measurement errors on HIV risk behavior analysis: A case study examining condom use among drug users. 95 StrEqMod 2 319-334

Wang, Jinde: Asymptotic normality of L_1-estimators in nonlinear regression. 95 JMultiAn 54 227-238

— —: Correction to "Asymptotic normality of L_1-estimators in nonlinear regression" (V54 p227-238). 95 JMultiAn 55 350-350

Wang, Julia Qizhi: Lai, Tze Leung; — —: Asymptotic expansions for the distributions of stopped random walks and first passage times. 94 AnlsProb 22 1957-1992

Wang, Julian: — — (Transl); Kuk, A. Y. C. (Rev): Review of "Essentials of statistical methods". 95 AstrlJSt 37 127-128

Wang, Julian Z.: Selection of the k largest order statistics for the domain of attraction of the Gumbel distribution. 95 JASA 90 1055-1061

Wang, Jung-Chao: Construction of some asymmetrical resolution V designs. 94 ASAProQlPr 129-133

Wang, Liming: Nonparametric test for change-point problem in location model (Chinese). 95 ChJApPSt 10 379-386

Wang, Lingling; Chen, Yuan: Statistical analysis of life test and accelerated life test under type 1 censoring of a lognormal distribution (Chinese). 95 AMaAPknC 17 109-117

Wang, Liqun: Identifiability and estimation of censored errors-in-variables models. 94 ASAProBuEc 351-354

Wang, M.-C.; Brookmeyer, R.; Jewell, N. P.: Reply to comment on "Statistical models for prevalent cohort data". 95 Biomtrcs 51 374-375

Wang, Mao-Jiun J.: Chen, Toly; — —: Applying fuzzy set approach to signal detection theory. 95 FuzzySet 72 39-49

Wang, Mei-Cheng: Huang, Ying; — —: Estimating the occurrence rate for prevalent survival data in competing risks models. 95 JASA 90 1406-1415

Wang, Morgan C.: An object oriented computing environment for self-validating computing. 94 ASAProStCp 29-32

— —; Kennedy, William J.: A self-validating numerical method for computation of central and non-central F probabilities and percentiles. 95 StatComp 5 155-163

Bushman, Brad J.; — —: A procedure for combining sample correlation coefficients and vote counts to obtain an estimate and a confidence interval for the population correlation coefficient. 95 PsychBu 117 530-546

Wang, Naisyin; Ruppert, David: Nonparametric estimation of the transformation in the transform-both-sides regression model. 95 JASA 90 522-534

Gutierrez, Roberto G.; Carroll, Raymond J.; — —; Lee, Gyung-Hee; Taylor, Brian H.: Analysis of tomato root initiation using a normal mixture distribution. 95 Biomtrcs 51 1461-1468

Wang, Nin: Fan, Jincheng; — —; Mei, Changlin: The random weighting approximation for the distribution of the partial sum of a stationary linear series (Chinese). 95 AMaAPknC 17 65-75

Wang, Ouhon: Applications of numerical interval analysis to obtain self-validating results in a massively parallel computing environment. 94 ASAProStCp 21-28

Wang, P. C.; Chen, M. H.; Wei, S. C.: To analyze binary data from mixed-level orthogonal arrays. 95 CmpStDA 20 689-695

— —; Jan, H. W.: Designing two-level factorial experiments using orthogonal arrays when the run order is important. 95 Ststcian 44 379-388

Wang, Qiang: Correlation dimension estimates of global and local temperature data. 95 JApMetrl 34 2556-2564

Wang, Qihua: Some asymptotic behavior of the kernel estimator of a density function based on censored data (Chinese). 95 ChJApPSt 10 164-174

Wang, Qiying: Bootstrap approximation for L-statistics with unbounded score function (Chinese). 95 ChinaAMA 15 281-286

Wang, S. J.; Lee, Sik-Yum: A geometric approach of the generalized least-squares estimation in analysis of covariance structures. 95 StPrLet 24 39-47

Wang, S. T.; Klein, John P.; Moeschberger, M. L.: Semi-parametric estimation of covariate effects using the positive stable frailty model. 95 AplStMDA 11 121-133

Wang, Shaun: On two-sided compound binomial distributions. 95 InsrneME 17 35-41

Panjer, Harry H.; — —: Computational aspects of Sundt's generalized class. 95 AstinB 25 5-18

Wang, Shuan: Insurance pricing and increased limits ratemaking by proportional hazards transforms. 95 InsrneME 17 43-54

Wang, Song-Gui: Rao, J. N. K.; — —: On the power of F tests under regression models with nested error structure. 95 JMultiAn 53 237-246

— —; Chow, Shein-Chung (Auth); Dean, Angela (Rev): Review of "Advanced linear models: Theory and applications". 95 Technmcs 37 119-120

— —; Chow, Shein-Chung (Auth); Martin, Frank B. (Rev): Review of "Advanced linear models". 95 JASA 90 802-803

Wang, Songgui; Yang, Hu: Linear estimate under Pitman criterion (Chinese). 95 KToChina 39 1444-1447

Yan, Liqing; — —: Finite-sample properties of an adaptive ridge estimator (Chinese). 95 ChJApPSt 10 194-201

Wang, Suojin: Optimizing the smoothed bootstrap. 95 AnInStMa 47 65-80

— —: One-step saddlepoint approximations for quantiles. 95 CmpStDA 20 65-74

Carroll, R. J.; — —; Wang, C. Y.: Prospective analysis of logistic case-control studies. 95 JASA 90 157-169

Casady, Robert J.; Dorfman, Alan; — —: Confidence intervals for sub-domain means and totals. 94 ASASrvyRsr 1036-1041

Schucany, William R.; — —: Bootstrap methods for nonstandard mixtures. 94 ASAProPEng 59-64

Wang, C. Y.; — —: On information matrices in case-control studies. 95 StPrLet 22 269-274

Wang, Tianlin: Yang, Ziheng; — —: Mixed model analysis of DNA sequence evolution. 95 Biomtrcs 51 552-561

Wang, Tonghui; Wong, Chi Song: Cochran theorems for a multivariate elliptically contoured model. 95 JStPlInf 43 257-270

Wang, Weijing; Wells, Martin T.: Covariance structure for elliptical distributions. 94 ASAProSoSt 60-65

Wang, Wen-chung: Wilson, Mark; — —: Complex composites: Issues that arise in combining different modes of assessment. 95 ApPsychM 19 51-71

Wang, Wenjin: Sarkar, Sanat K.; Snapinn, Steve; — —: On improving the min test for the analysis of combination drug trials. 95 JStCmpSm 51 197-213

Wang, Wenping; Chow, Shein-Chung; Wei, William W. S.: On likelihood distance for outliers detection. 95 JBiophSt 5 307-322

— —; Hsuan, Francis; Chow, Shein-Chung: Patient compliance and the fluctuation of the serum drug concentration. 94 ASAProBiop 94-99

Ki, Fanny Y. C.; Liu, Jen-Pei; — —; Chow, Shein-Chung: The impact of outlying subjects on decision of bioequivalence. 95 JBiophSt 5 71-94

Wang, Wenyu; Begleiter, Henri; Porjesz, Bernice: Localization of multiple dipole sources of human cerebral potentials based on SCD and a model selection criterion. 94 ASAProBiop 217-222

Wang, Whedy: Wan, Jim Y.; — —; Bromberg, Judith: A SAS macro for residual deviance of ordinal regression analysis. 94 CmpBiomd 45 307-310

Wang, X.: Ramsay, J. O.; — —; Flanagan, R.: A functional data analysis of the pinch force of human fingers. 95 ApplStat 44 17-30

Wang, Xaingchen; Rao, M. Bhaskara; Li, Deli: Some results on strong limit theorems for (LB)-space-valued random variables. 95 StPrLet 23 247-251

Wang, Xiang Chen: Li, Deli; Rao, M. Bhaskara; — —: On the strong law of large numbers and the law of the logarithm for weighted sums of independent random variables with multidimensional indices. 95 JMultiAn 52 181-198

Wang, Xiangchen: Jiang, Tiefeng; Rao, M. Bhaskara; — —: Large deviations for moving average processes. 95 StocProc 59 309-320

Li, Deli; Rao, M. Bhaskara; Jiang, Tiefeng; — —: Complete convergence and almost sure convergence of weighted sums of random variables. 95 JTherPrb 8 49-76

Wang, Xiao-Qun: Ermakov, S. M.; — —: Random invariant cubature formulas (Russian). 94 VLenUM /4 109-112

Wang, Xuewen; Zhan, Jinlong; Chen, Jianbao: All admissible linear estimators of regression coefficients and parameters in variance component models. 95 AMaPeknE 37 653-662

Wang, Y.; Boukai, B.; Berger, J. O.: A Bayesian testing procedure with valid conditional frequentist interpretation. 94 ASAProBays 187-195

Krewski, D.; — —; Bartlett, S.; Krishnan, K.: Uncertainty, variability, and sensitivity analysis in physiological pharmacokinetic models. 95 JBiophSt 5 245-271

Fang, K.-T.; — — (Auth); Ripley, B. D. (Rev): Review of "Number-theoretic methods in statistics". 95 JRSS-A 158 189-190

Fang, Kai-Tai; — — (Auth); Ginebra, Josep (Rev): Review of "Number-theoretic methods in statistics". 95 JASA 90 1134-1134

Wang, Y. H.; Ji, Shuixin: Limit theorems for the number of occurrences of consecutive k successes in n Markovian trials. 95 JAppProb 32 727-735

— —; Yang, Zejiang: On a Markov multinomial distribution. 95 MaScitst 20 40-49

Wang, Yazhen: Jump and sharp cusp detection by wavelets. 95 Biomtrka 82 385-397

— —: The L_1 theory of estimation of monotone and unimodal densities. 95 JNnparSt 4 249-261

Wang, Ying; Succop, Paul A.: Multiply-imputed estimation for multivariate missing data. 94 ASAProEnvr 181-186

Wang, Yining: Gupta, Arjun K.; — —: A condition for the nonexistence of ancillary statistics. 95 StPrLet 23 367-369

Wang, Yong; Krewski, Daniel; Bartlett, Sheryl; Krishnan, Kannan: Analysis of uncertainty and sensitivity in physiologically based pharmacokinetic models. 94 ASAProBiop 100-105

Wang, Yongxiao; Taylor, Jeremy M. G.: Flexible methods for analysing longitudinal data using piecewise cubic polynomials. 95 JStCmpSm 52 133-150

— —; Taylor, Jeremy M. G.: Inference for smooth curves in longitudinal data with application to an AIDS clinical trial. 95 StatMed 14 1205-1218

Wang, You-Gan; Thomas, Mervyn R.: Accounting for individual variability in the von Bertalanffy growth model. 95 JFishRBC 52 1368-1375

— —; Thomas, Mervyn R.; Somers, Ian F.: A maximum likelihood approach for estimating growth from tag-recapture data. 95 JFishRBC 52 252-259

Wang, Yuchung; Lin, Dennis K. J.; Fang, Kai-Tai: Designing outer array points. 95 JQualTek 27 226-241

Wang, Yuedong; Wahba, Grace: Bootstrap confidence intervals for smoothing splines and their comparison to Bayesian confidence intervals. 95 JStCmpSm 51 263-279

— —; Wahba, Grace; Chappell, Rick; Gu, Chong: Simulation studies of smoothing parameter estimates and Bayesian confidence intervals in Bernoulli SS ANOVA models. 95 CommStB 24 1037-1059

Wahba, Grace; — —: Behavior near zero of the distribution of GCV smoothing parameter estimates. 95 StPrLet 25 105-111

Wang, Z.; Yang, X. (Auth); Voit, M. (Rev): Review of "Birth and death processes and Markov chains". 95 Metrika 42 270-270

Wang, Zhongzhi: Liu, Wen; — —: An extension of a theorem on gambling systems. 95 JMultiAn 55 125-132

Wang-Clow, Fong; Lange, Mary; Laird, Nan M.; Ware, James H.: A simulation study of estimators for rates of change in longitudinal studies with attrition. 95 StatMed 14 283-297

Ward, David H.: Schmutz, Joel A.; — —; Sedinger, James S.; Rexstad, Eric A.: Survival estimation and the effects of dependency among animals. 95 JAppStat 22 673-681

Ward, Joe H., Jr; Niland, Laura J.: Empowering high school students to exploit statistical models and software for research projects. 94 ASAProStEd 44-48

Ward, John: DiGaetano, Ralph; — —; Cadell, Diane; Ryaboy, Lana: Sample design of the National Pregnancy and Health Survey. 94 ASASrvyRsr 1100-1104

Ward, W. C.: Bennett, R. E.; — — (Ed); Adams, Robert (Rev): Review of "Construction versus choice in cognitive measurement". 95 Ststcian 44 137-137

Warde, William D.: Han, Geun-Shik; — —: Correlations in randomized response surveys. 94 ASASrvyRsr 491-496

Mehdi, J. (Auth); — — (Rev): Review of "Statistical methods: An introductory text". 95 AmerStat 49 113-114

Wardell, Don G.; Moskowitz, Herbert; Plante, Robert D.: Reply to comment on "Run-length distribution of special-cause control charts for correlated processes". 95 Technmcs 37 243-245

Wardrop, Robert L.: Simpson's paradox and the hot hand in basketball. 95 AmerStat 49 24-28

— — (Auth); Rossman, Allan J. (Rev): Review of "Statistics: Learning in the presence of variation". 95 AmerStat 49 237-238

Ware, James H.: Comment on "Multistate models for monitoring individual trees in permanent observation plots". 94 EnvEcoSt 1 196-197

Donnelly, Christl A.; Laird, Nan M.; — —: Prediction and creation of smooth curves for temporally correlated longitudinal data. 95 JASA 90 984-989

Wang-Clow, Fong; Lange, Mary; Laird, Nan M.; — —: A simulation study of estimators for rates of change in longitudinal studies with attrition. 95 StatMed 14 283-297

Wargenau, Manfred: Dose finding strategies involving interim analyses and unbalanced treatment allocation. 95 StatMed 14 901-907 (Disc: p909-909)

Warner, John: Bayesian bioassay designs: A simulation based approach. 94 ASAProBays 13-17

Warrack, Giles: The great streak. 95 Chance 8/3 41-43

Warsi, Ghulam: MacKinnon, David P.; — —; Dwyer, James H.: A simulation study of mediated effect measures. 95 MultBehR 30 41-62

Wartenberg, Daniel; Simon, Ronald: Integrating epidemiologic data into risk assessment. 95 AJPubHlh 85 491-493

Wascher, William: Neumark, David; — —: Minimum wage effects on employment and school enrollment. 95 JBES 13/2 199-206

Wasik, John L.: A comparative study of the use of the SAS and MINITAB statistical programs in teaching statistics. 94 ASAProStEd 234-237

Wasilkowski, Grzegorz W.: Ritter, Klaus; — —; Woźniakowski, Henryk: Multivariate integration and approximation for random fields satisfying Sacks-Ylvisaker conditions. 95 AnlsApPr 5 518-540

Wassell, J. T.; Kulczycki, G. W.; Moyer, E. S.: Frailty models of manufacturing effects. 95 LiDaAnal 1 161-170

Wassell, James T.; Kulczycki, Gregory W.; Moyer, Ernest S.: Modelling frailty in the manufacture of respiratory protective devices. 94 ASAProEpid 133-138

Wasserman, Gary S.: Hjorth, J. S. Urban (Auth); — —; Hohsen, Hassan (Rev): Review of "Computer intensive statistical methods: Validation, model selection and bootstrap". 95 JQualTek 27 268-270

Wasserman, L.; Clarke, B.: Information tradeoff. 95 Test 4 19-38

Wasserman, Larry: Greenhouse, Joel B.; — —: Robust Bayesian methods for monitoring clinical trials. 95 StatMed 14 1379-1391

Kass, Robert E.; — —: A reference Bayesian test for nested hypotheses and its relationship to the Schwarz criterion. 95 JASA 90 928-934

Ruggeri, Fabrizio; — —: Density based classes of priors: Infinitesimal properties and approximations. 95 JStPlInf 46 311-324

Verdinelli, Isabella; — —: Computing Bayes factors using a generalization of the Savage-Dickey density ratio. 95 JASA 90 614-618

Wasserman, Stanley: Anderson, Carolyn J.; — —: Log-multiplicative models for valued social relations. 95 SocMethR 24 96-127

Laumann, E. O.; Gagnon, J. H.; Michael, R. T.; Michaels, S. (Auth); — — (Rev): Review of "The social organization of sexuality". 95 Chance 8/2 40-42

Wasserstein, Sherry: Practical skills for a statistician. 95 Stats /12 14-15

Wassertheil-Smoller, S.: Biostatistics and epidemiology. A primer for health professionals. 95 Springer:Brln:NY 185p

Watanabe, Hiroshi: Li, Jian-Hua; — —: A Bayesian study of the correlation between standard scores of twins. 94 Behvmtrk 21 121-130

Watanabe, Masahide: Katayama, Ryu; Kuwata, Kaihei; Kajitani, Yuji; — —: Embedding dimension estimation of chaotic time series using self-generating radial basis function network. 95 FuzzySet 71 311-327

Watelet, Luc: Haber, Michael; Halloran, M. Elizabeth; Longini, Ira M., Jr; — —: Estimation of vaccine efficacy in non-randomly mixing populations. 95 BiomtrcJ 37 25-38

Waterloo Maple Software Inc: — — (Auth); Baglivo, Jenny (Rev): Review of "Maple V for the Macintosh (Version 5.0.2)". 95 AmerStat 49 86-92

— — (Auth); Hutton, John; Hutton, James (Rev): Review of "The Maple computer algebra system". 95 JApEcmtx 10 329-337

Waterman, Michael S.: Introduction to computational biology: Maps, sequences, and genomes. 95 ChapmnHall:Lnd 450p

Goldstein, Larry; — —: Poisson, compound Poisson and process approximations for testing statistical significance in sequence comparisons. 92 BMathBio 54 785-812

Waternaux, Christine M.: Draper, David; Gaver, Donald, P., Jr; Goel, Prem K.; Greenhouse, Joel B.; Hedges, Larry V.; Morris, Carl N.; Tucker, John R.; — — (Auth); Berlin, Jesse A. (Rev): Review of "Combining information: Statistical issues and opportunities for research". 95 StatMed 14 1611-1612

Waters, Howard R.: Dickson, David C. M.; Egídio dos Reis, Alfredo D.; — —: Some stable algorithms in ruin theory and their applications. 95 AstinB 25 153-175

Watier, Laurence; Richardson, Sylvia: Modelling of an epidemiological time series by a threshold autoregressive model. 95 Ststcian 44 353-364

Watkins, Ann E.: The law of averages. 95 Chance 8/2 28-32

Pereira-Mendoza, Lionel (Ed); — — (Rev): Review of "Introducing data analysis in the schools: Who should teach it and how?". 95 AmerStat 49 62-63

Watkins, James F.: Collett, D. (Auth); — — (Rev): Review of "Modelling survival data in medical research". 95 Ststcian 44 281-282

Watkins, Joseph: Heubach, Silvia; — —: A stochastic model for the movement of a white blood cell. 95 AdvAppPr 27 443-475

Watkins, Michael: Davis, Robert; Rosegrant, Susan; — —: Managing the link between measurement and compensation. 95 QulPr 28/2 101-106

Watson, G. S.: Detecting a change in the intercept in multiple regression. 95 StPrLet 23 69-72

Watson, Geoffrey S.: U_n^2 test for uniformity of discrete distributions. 95 JAppStat 22 273-276

Chikuse, Yasuko; — —: Large sample asymptotic theory of tests for uniformity on the Grassmann manifold. 95 JMultiAn 54 18-31

Watson, Jane M.: Conditional probability: Its place in the mathematics curriculum. 95 MathTchr 88 12-17

Watson, Mark W.: Horvath, Michael T. K.; — —: Testing for cointegration when some of the cointegrating vectors are prespecified. 95 EconTher 11 984-1014

Watson, Peter C.: Rudolfer, Stephan M.; — —; Lesaffre, Emmanuel: Are ordinal models useful for classification? A revised analysis. 95 JStCmpSm 52 105-132

Watson, Ricky M.: Alsup, Fred; — — (Auth); McLaughlin, Gregory C. (Rev): Review of "Practical statistical process control: A tool for quality manufacturing". 95 Technmcs 37 461-462

Watson, S. M.; Redfern, E.; Clark, S.; Tight, M.; Davies, N.: An influence method for outlier detection applied to time series traffic data. 95 JAppStat 22 135-149

Watson, Stephen R.: Reply to comment on "The meaning of probability in probabilistic risk analysis". 95 RelEngSS 49 207-209

Watt, T. A.: — — (Auth); Thomas, E. (Rev): Review of "Introductory statistics for biology students". 95 BiomtrcJ 37 96-96

Watts, Donald G.: Understanding canonical analysis. 95 JQualTek 27 40-44

Wayne, Andrew: Bayesianism and diverse evidence. 95 PhilSci 62 111-121

Weaver, Susan L.: Have, Thomas R. Ten; Landis, J. Richard; — —: Association models for periodontal disease progression: A comparison of methods for clustered binary data. 95 StatMed 14 413-429

Webb, Bridget S.: Cycle structures of automorphisms of 2-$(v,k,1)$ designs. 95 JComDsgn 3 341-348

Webb, C. F.: Rondel, R. K.; Varley, S. A.; — — (Ed); Therasse, P. (Rev): Review of "Clinical data management". 95 StatMed 14 877-877

Webb, James M.: Liang, Kun-Hsia; Krus, David J.; — —: K-fold crossvalidation in canonical analysis. 95 MultBehR 30 539-545

Webb, Roy H.: Forecasts of inflation from VAR models. 95 JForcast 14 267-285

Webber, Larry S.: Zucker, David M.; Lakatos, Edward; — —; Murray, David M.; McKinlay, Sonja M.; Feldman, Henry A.; Kelder, Steve H.; Nader, Philip R.: Statistical design of the child and adolescent trial for cardiovascular health (CATCH): Implications of cluster randomization. 95 ConClnTr 16 96-118

Weber, Christian E.: Cyclical output, cyclical unemployment, and Okun's coefficient: A new approach. 95 JApEcmtx 10 433-445

Weber, David P.: Uncertainty calculi for fault trees. 95 ASQCTekCon 1046-1053

Weber, G.: Pelizzon, L.; — —: The Italian term structure of interest rates and the currency devaluation of September 1992: A factor-ARCH analysis. 94 Statstca 54 313-327

Weber, N. C.: Ivanoff, B. G.; — —: Functional limit theorems for row and column exchangeable arrays. 95 JMultiAn 55 133-148

van de Ven, R.; — —: Log-linear models for mean and dispersion in mixed Poisson regression models. 95 AstrlJSt 37 205-216

Weber, Richard: Courcoubetis, Costas; — —: Effective bandwidths for stationary sources. 95 PrEngInf 9 285-296

Weber, Rudolf O.; Kaufmann, Pirmin: Automated classification scheme for wind fields. 95 JApMetrl 34 1133-1141

— —; Madden, Roland A.: Optimal averaging for the determination of global mean temperature: Experiments with model data. 95 JClimate 8 418-430

Wedel, Michel; DeSarbo, Wayne S.: A mixture likelihood approach for generalized linear models. 95 JClass 12 21-55

— —; Kamakura, Wagner A.; Desarbo, Wayne S.; Hofstede, Frenkel Ter: Implications for asymmetry, nonproportionality, and heterogeneity in brand switching from piece-wise exponential mixture hazard models. 95 JMarRsr 32 457-462

Weeks, D. E.: Li, C. C.; — —; Chakravarti, A.: Similarity of DNA fingerprints due to chance and relatedness. 93 HumHrdty 43 45-52

Weeks, Daniel E.: Invalidity of the Rao map function for three loci. 94 HumHrdty 44 178-180

— —; Lathrop, G. Mark; Ott, Jurg: Multipoint mapping under genetic interference. 93 HumHrdty 43 86-97

— —; Ott, Jurg; Lathrop, G. Mark: Detection of genetic interference: Simulation studies and mouse data. 94 Genetcs 136 1217-1226

Weeks, Michael: Duffer, Allen; Lessler, Judith; — —; Mosher, William: Effects of incentive payments on response rates and field costs in a pretest of a national CAPI survey. 94 ASASrvyRsr 1386-1392

Weeks, Michael F.: Lessler, Judith T.; — —; O'Reilly, James M.: Results from the National Survey of Family Growth cycle V pretest. 94 ASASrvyRsr 64-70

Weerahandi, S.: Exact statistical methods in data analysis. 95 Springer:Brln:NY 328p

Weerahandi, Samaradasa: ANOVA under unequal error variances. 95 Biomtrcs 51 589-599

— —; Moitra, Soumyo: Using survey data to predict adoption and switching for services. 95 JMarRsr 32 85-96

Weerakkody, Govinda J.; Givaruangsawat, Sumalee: Estimating the correlation coefficient in the presence of correlated observations from a bivariate normal population. 95 CommStA 24 1705-1719

Weerasinghe, A. P. N.: Athreya, K. B.; — —: Ergodic theorems for transient one-dimensional diffusions. 95 StocProc 58 173-185

Wefelmeyer, W.: Greenwood, P. E.; — —: Efficiency of empirical estimators for Markov chains. 95 AnlsStat 23 132-143

Wegman, Edward J.: Huge data sets and the frontiers of computational feasibility. 95 JCmpGrSt 4 281-295

Solka, Jeffrey L.; Poston, Wendy L.; — —: A visualization technique for studying the iterative estimation of mixture densities. 95 JCmpGrSt 4 180-198

Wehrle, Robert A.: Democracy, capitalism and the role of TQM and the statistician in public education reform. 94 ASAProQlPr 189-193

Wehrly, Thomas E.: Fisher, N. I. (Auth); — — (Rev): Review of "Statistical analysis of circular data". 95 Technmcs 37 229-230

Wei, Bocheng: Theory of geometry and asymptotic inference in exponential nonlinear models. 95 SciSPkng 24 1270-1278

— —; Feng, Yu: On calculating the moments of M-estimator in nonlinear regression (Chinese). 95 AMaAPknC 17 437-447

— —; Shi, Jianqing: Local influence analysis for regression transformation models (Chinese). 95 AMaAPknC 17 132-143

Wei, L.; Trenkler, G.: Mean square error matrix superiority of empirical Bayes estimators under misspecification. 95 Test 4 187-205

Wei, L. J.: Cheng, S. C.; — —; Ying, Z.: Analysis of transformation models with censored data. 95 Biomtrka 82 835-845

Ying, Z.; Jung, S. H.; — —: Survival analysis with median regression models. 95 JASA 90 178-184

Wei, Laisheng; Zhang, Shunpu: The convergence rates of empirical Bayes estimation in a multiple linear regression model. 95 AnInStMa 47 81-97

Wei, Paul; Zeidler, James; Ku, Walter: Analysis of multiframe target detection using pixel statistics. 95 IEEEAES 31 238-247

Wei, S. C.: Wang, P. C.; Chen, M. H.; — —: To analyze binary data from mixed-level orthogonal arrays. 95 CmpStDA 20 689-695

Wei, William W. S.: Wang, Wenping; Chow, Shein-Chung; — —: On likelihood distance for outliers detection. 95 JBiophSt 5 307-322

Weichselberger, A.: — — (Auth); Büning, H. (Rev): Review of "A new nonparametric goodness-of-fit test for location alternatives" (German). 95 Metrika 42 145-146

Weideman, William E.; Manry, Michael T.; Yau, Hung-Chun; Gong, Wei: Comparisons of a neural network and a nearest-neighbor classifier via the numeric handprint recognition problem. 95 IEEENrlN 6 1524-1530

Weidmer, Beverly: Issues and guidelines for translation in cross-cultural research. 94 ASASrvyRsr 1226-1231

Weightman, Rosina: Fielding, Katherine L.; Brettle, Raymond P.; Gore, Sheila M.; O'Brien, Fiona; Wyld, Rona; Robertson, J. Roy; — —: Heterosexual transmission of HIV analysed by generalized estimating equations. 95 StatMed 14 1365-1378

Weijers, E. P.; van Delden, A.; Vugts, H. F.; Meesters, A. G. C. A.: Characteristics of convective turbulence in the surface layer investigated by principal component analysis. 95 JApMetrl 34 528-541

Weil, Wolfgang: The estimation of mean shape and mean particle number in overlapping particle systems in the plane. 95 AdvAppPr 27 102-119

Weinberg, Clarice R.; Umbach, David M.; Greenland, Sander: Reply to comments on "When will nondifferential misclassification of an exposure reverse the direction of a trend?". 95 AJEpidm 142 784-784

— —; Wilcox, Allen J.: A model for estimating the potency and survival of human gametes in vivo. 95 Biomtrcs 51 405-412

Weinberger, Volker: Taudes, Alfred; Reepmeyer, Jan-Armin; Bensberg, Frank (Auth); — — (Rev): Review of "UNIX: Introduction for DOS-users" (German). 95 CmpStDA 19 360-361

Weiner, Jacob: Schmid, Bernhard; Polasek, Wolfgang; — —; Krause, Andreas; Stoll, Peter: Modeling of discontinuous relationships in biology with censored regression. 94 ANatrls 143 494-507

Weinkam, J. J.: Sterling, T. D.; Rosenbaum, W. L.; — —: Publication decisions revisited: The effect of the outcome of statistical tests on the decision to publish and vice versa. 95 AmerStat 49 108-112

Weinkam, James J.: Rosenbaum, Wilfred L.; Sterling, Theodor D.; — —: Correcting standardized rate ratios for imprecise classification of a polychotomous exposure variable with limited data. 95 AJEpidm 141 442-445

Rosenbaum, Wilfred L.; Sterling, Theodor D.; — —: Correcting standardized rate ratios for imprecise classification of a polychotomous exposure variable with limited data. 95 AJEpidm 142 442-445

Weinstein, Neil D.; Sandman, Peter M.; Hallman, William K.: Testing a visual display to explain small probabilities. 94 RiskAnal 14 895-896

Weir, B. S.: Cockerham, C. Clark; — —: Estimation of gene flow from F-statistics. 93 Evolutn 47 855-863

Gaut, Brandon S.; — —: Detecting substitution-rate heterogeneity among regions of a nucleotide sequence. 94 MolBioEv 11 620-629

Hill, W. G.; — —: Maximum-likelihood estimation of gene location by linkage disequilibrium. 94 AJHuGene 54 705-714

Kaplan, N. L.; Hill, W. G.; — —: Likelihood methods for locating disease genes in nonequilibrium populations. 95 AJHuGene 56 18-32

Zhu, J.; — —: Analysis of cytoplasmic and maternal effects. II. Genetic models for triploid endosperms. 94 ThApGene 89 160-166

Weir, Paula; Saavedra, Pedro J.: On-highway Diesel Fuel Price Survey. 94 ASAProGovt 144-148

O'Colmain, Benita J.; Saavedra, Pedro J.; — —: Variance estimation for EIA-877 propane prices. 94 ASASrvyRsr 232-235

Saavedra, Pedro J.; Hassett, Nancy C.; — —: Comparison of national and regional gasoline prices from two surveys. 94 ASASrvyRsr 1179-1183

Weisberg, S.; Welsh, A. H.: Adapting for the missing link. 94 AnlsStat 22 1674-1700

Cook, R. D.; — — (Auth); Craig, Peter (Rev): Review of "An introduction to regression graphics". 95 JAppStat 22 541-542

Cook, R. D.; — — (Auth); Pincus, R. (Rev): Review of "An introduction to regression graphics". 95 BiomtrcJ 37 298-298

Weisberg, Sandford: Cook, R. Dennis; — — (Auth); Cook, Dianne (Rev): Review of "An introduction to regression graphics". 95 JASA 90 1126-1127

Weisberg, Sanford: Cook, R. Dennis; — — (Auth); Goldman, Aaron (Rev): Review of "An introduction to regression graphics". 95 Technmcs 37 343-344

Weishaupt, Jürgen: Rieder, Ulrich; — —: Customer scheduling with incomplete information. 95 PrEngInf 9 269-284

Weiss, Alan: Shwartz, Adam; — —: Large deviations for performance analysis. Queues, communication, and computing. 95 ChapmnHall:Lnd 576p

Weiss, Benjamin: Furstenberg, Hillel; Peres, Yuval; — —: Perfect filtering and double disjointness. 95 AIHenrPB 31 453-465

Weiss, David G.: Williford, William O.; Krol, William F.; Bingham, Stephen F.; Collins, Joseph F.; — —: The multicenter clinical trials coordinating center statistician: 'More than a consultant'. 95 AmerStat 49 221-225

Weiss, David J.: Polychotomous or polytomous?. 95 ApPsychM 19 4-4

Weiss, G. H.: — — (Auth); Lawler, Gregory F. (Rev): Review of "Aspects and applications of the random walk". 95 SIAMRvw 37 470-471

Weiss, Gideon: On almost optimal priority rules for preemptive scheduling of stochastic jobs on parallel machines. 95 AdvAppPr 27 821-839

Weiss, Robert E.: The influence of variable selection: A Bayesian diagnostic perspective. 95 JASA 90 619-625

Liu, Honghu; — —; Wenger, Neil S.: Model selection in repeated measures random effects models. 94 ASAProBiop 371-376

Zhang, Fan; — —; Zeltzer, Lonnie: Analyzing long irregularly spaced repeated multivariate measures - A study of infant cardiac catherization stress responses. 94 ASAProBiop 365-370

Cleveland, William S. (Auth); — — (Rev): Review of "The elements of graphing data (Revised edition)". 95 JASA 90 1488-1489

Weiss, Scott: D'Agostino, Ralph B., Jr; Sparrow, David; — —; Rosner, Bernard: Longitudinal models for analysis of respiratory function. 95 StatMed 14 2205-2216

Weissfeld, Joel L.: Biomarkers, etiologic research, and primary disease prevention. 94 ASAProEpid 14-23

Weissfeld, Lisa A.: Dignam, James J.; — —; Anderson, Stewart J.: Methods for bounding the marginal survival distribution. 95 StatMed 14 1985-1998

Weissman, Ishay: The indices of the largest among n independent observations. 95 CommStSM 11 613-629

— —: Records from a power model of independent observations. 95 JAppProb 32 982-990

— —; Cohen, Uri: The extremal index and clustering of high values for derived stationary sequences. 95 JAppProb 32 972-981

Weitzman, Eben A.; Miles, Matthew B.: Computer programs for qualitative data analysis. A software sourcebook. 95 SagePubl:CA:Lnd 372p

Welch, Finis: Peracchi, Franco; — —: How representative are matched cross-sections? Evidence from the Current Population Survey. 95 JEconmtx 68 153-179

Welch, Michael: Tiller, Richard; — —: Predicting the national unemployment rate that the "old" CPS would have produced. 94 ASASrvyRsr 354-359 (Disc: p360-361)

Welch, William J.: Gao, Feng; Sacks, Jerome; — —: Predicting ozone levels and trends. 94 ASAProEnvr 77-81

Well, Arnold D.: Myers, Jerome L.; — —: Research design and statistical analysis. 95 ErlbaumA:NJ 728p

Wellens, Tracy: The cognitive evaluation of the nativity questions for the Current Population Survey. 94 ASASrvyRsr 1204-1209

Weller, Barry R.: Business Forecast Systems Inc. (Auth); — — (Rev): Review of "Forecast pro for windows extended edition (XE) (Version 2)". 95 IntJFor 11 175-187

Weller, J. I.; Wyler, A.: Power of different sampling strategies to detect quantitative trait loci variance effects. 92 ThApGene 83 582-588

de Koning, G. J.; — —: Efficiency of direct selection on quantitative trait loci for a two-trait breeding objective. 94 ThApGene 88 669-677

Weller, Joel I.: Hackett, Christine A.; — —: Genetic mapping of quantitative trait loci for traits with ordinal distributions. 95 Biomtrcs 51 1252-1263

Wellner, J. A.: Huang, J.; — —: Asymptotic normality of the NPMLE of linear functionals for interval censored data, case 1. 95 StNeerla 49 153-163

Groeneboom, P.; — — (Auth); Plachky, D. (Rev): Review of "Information bounds and nonparametric maximum likelihood estimation". 95 StatDecn 13 94-95

Wellner, Jon A.: Gill, Richard D.; van der Laan, Mark J.; — —: Inefficient estimators of the bivariate survival function for three models. 95 AIHenrPB 31 545-597

Wells, Martin T.: Casella, George; DiCiccio, Thomas J.; — —: Comment on "Inference". 95 StatSci 10 179-185

Fourdrinier, Dominique; — —: Estimation of a loss function for spherically symmetric distributions in the general linear model. 95 AnlsStat 23 571-592

Fourdrinier, Dominique; — —: Loss estimation for spherically symmetric distributions. 95 JMultiAn 53 311-331

Wang, Weijing; — —: Covariance structure for elliptical distributions. 94 ASAProSoSt 60-65

Welsh, A. H.: Weisberg, S.; — —: Adapting for the missing link. 94 AnlsStat 22 1674-1700

Welsh, Alan H.: Richardson, Alice M.; — —: Robust restricted maximum likelihood in mixed linear models. 95 Biomtrcs 51 1429-1439

Weng, C. S. Wayne; Liu, J. P.: Some pitfalls in sample size estimation for an anti-infective study. 94 ASAProBiop 56-60

Weng, Chung-Sing: Liu, Jen-Pei; — —: Bias of two one-sided tests procedures in assessment of bioequivalence. 95 StatMed 14 853-861

Wenger, Neil S.: Liu, Honghu; Weiss, Robert E.; — —: Model selection in repeated measures random effects models. 94 ASAProBiop 371-376

Wenli, Yang: MINQE(U, I) and UMVIQUE of the covariance matrix in the growth curve model. 95 MaOpfStS 26 49-59

Werking, George S.: Establishment surveys: Designing the survey operations of the future. 94 ASASrvyRsr 163-169

Wermuth, Nanny; Cox, D. R.: Derived variables calculated from similar joint responses: Some characteristics and examples. 95 CmpStDA 19 223-234

Wernecke, K. D.: Nagel, M.; — —; Fleischer, W. (Auth); Bartkowiak, Anna (Rev): Review of "Computer-aided data analysis" (German). 95 ListyBio 32 47-49

Wernecke, K.-D.: Discriminant and cluster analysis. 95 CmpStQ 10 15-20

Werner, Wendelin: On Brownian disconnection exponents. 95 Bernoull 1 371-380

— —: Green's formula, planar Brownian loop and Lévy's area (French). 95 StocProc 57 225-245

Meyre, Thierry; — —: On the occupation times of cones by Brownian motion. 95 ProbTher 101 409-419

Shi, Zhan; — —: Asymptotics for occupation times of half-lines by stable processes and perturbed reflecting Brownian motion. 95 Stochatx 55 71-85

Weron, A.: Computer-aided modeling and simulation of electrical circuits with α-stable noise. 95 ApplMath 23 83-93

Kalashnikov, V. V.; Zolotarev, V. M. (Ed); — — (Rev): Review of "Stability problems for stochastic models". 95 Metrika 42 272-273

Janicki, A.; — — (Auth); Lee, P. M. (Rev): Review of "Simulation and chaotic behaviour of α-stable stochastic processes". 95 JRSS-A 158 351-352

Weron, Aleksander: Janicki, Aleksander; — —: Simulation and chaotic behavior of α-stable stochastic processes. 95 Dekker:NY 376p

Janicki, Aleksander; — —: Computer simulation of attractors in stochastic models with α-stable noise. 95 MaCmpSim 39 9-19

Wertz, Wolfgang: Hurt, Jan; Novák, Břetislav; — —: Invariantly optimal look on reliability estimation in the exponential case. 95 StatDecn 13 193-197

Wesolowski, Jacek: Bivariate distributions via a Pareto conditional distribution and a regression function. 95 AnInStMa 47 177-183

— —: Bivariate discrete measures via a power series conditional distribution and a regression function. 95 JMultiAn 55 219-229

— —; Ahsanullah, Mohammad: Conditional specifications of multivariate Pareto and Student distributions. 95 CommStA 24 1023-1031

Wesolowsky, G. O.: Drezner, Zvi; — —: Multivariate screening procedures for quality cost minimization. 95 IIETrn 27 300-304

Steiner, S. H.; — —: Estimating the correlation between destructively measured variables using proof-loading. 95 Technmcs 37 94-101

Steiner, S. H.; — —: Bivariate normal parameter estimation under destructive testing. 94 ASAProPEng 326-331

West, Carol Taylor: Fullerton, Thomas M., Jr; — —: Assessing the historical accuracy of regional employment forecasts. 94 ASAProBuEc 333-338

West, Jerry; Hausken, Elvie Germino: Different approaches to counting early childhood program participation. 94 ASAProSoSt 141-146

West, Kenneth D.; Cho, Dongchul: The predictive ability of several models of exchange rate volatility. 95 JEconmtx 69 367-391

West, M.: Pole, A.; — —; Harrison, J. (Auth); Davies, Neville (Rev): Review of "Applied Bayesian forecasting and time series analysis". 95 JRSS-A 158 635-636

West, Mike: Bayesian inference in cyclical component dynamic linear models. 95 JASA 90 1301-1312

Escobar, Michael D.; — —: Bayesian density estimation and inference using mixtures. 95 JASA 90 577-588

West, Sandra A.; Kratzke, Diem-Tran; Robertson, Kenneth W.: Variance estimators for variables that have both observed and imputed values. 94 ASASrvyRsr 394-399

Westcott, Nancy E.: Changnon, Stanley A.; Gabriel, K. Ruben; — —; Czys, Robert R.: Exploratory analysis of seeding effects on rainfall: Illinois 1989. 95 JApMetrl 34 1215-1224

Westerlund, Tapio: Brink, Anders; — —: The joint problem of model structure determination and parameter estimation in quantitative IR spectroscopy. 95 Chemolab 29 29-36

Western, Bruce: Concepts and suggestions for robust regression analysis. 95 AJPolSci 39 786-817

Westfall, P. H.; Young, S. S. (Auth); Carpenter, James (Rev): Review of "Resampling-based multiple testing". 95 JAppStat 22 434-435

— —; Young, S. S. (Auth); Schlittgen, R. (Rev. German): Review of "Resampling - Base multiple testing". 95 AllgeStA 79 340-340

Westfall, Peter H.; Young, S. Stanley (Auth); Edwards, Don (Rev): Review of "Resampling-based multiple testing". 95 JQualTek 27 272-273

— —; Young, S. Stanley (Auth); Köhne, Stephan; Pigeot, Iris (Rev): Review of "Resampling-based multiple testing. Examples and methods for p-value adjustment". 95 CmpStDA 20 235-236

Weststrate, Jan A.; van der Mei, Rob D.: Waiting times in a two-queue model with exhaustive and Bernoulli service. 94 ZeitOpRs 40 289-303

Wette, Reimut: Vogler, George P.; — —; McGue, Matthew K.; Rao, D. C.: Properties of alternative estimators of familial correlations under variable sibship size. 95 Biomtrcs 51 276-283

Wetzel, Angela-Jo: Jenkins, Cleo R.; — —: The 1991-92 teacher follow-up survey reinterview and extensive reconciliation. 94 ASASrvyRsr 821-826

Wetzel, N.; Cook, R. Dennis: Exploring regression structure with graphics. 94 ASAProStGr 42-47

Whaley, Fredrick S.; Schoenfelder, John R.; Pinsky, Carl: Comparing a standard diagnostic procedure to a second procedure used in combination with the standard procedure. 94 ASAProBiop 272-275

Wharton, Robert; Czachor, James: An upper bound for the proportion of process output not meeting specifications. 94 ASAProQlPr 264-266

Wharton, Robert M.: Tarter, Michael E.; Lock, Michael D. (Auth); — — (Rev): Review of "Model-free curve estimation". 95 Technmcs 37 460-460

Wheeler, Donald J.: Modern extensions of interlaboratory testing. 94 ASAProPEng 35-43

Whelan, S. L.: Parkin, D. M.; Muir, C. S.; — —; Gao, Y. T.; Ferlay, J.; Powell, J. (Ed); Hill, Catherine (Rev): Review of "Cancer incidence in five continents. Volume VI". 95 StatMed 14 104-104

Wheldon, Tom E.: Mao, Jian-Hua; — —: A stochastic model for multistage tumorigenesis in developing and adult mice. 95 MaBisci 129 95-110

Whitaker, D.: A nested simulated annealing algorithm. 95 JStCmpSm 53 233-241

White, A. A.; Pickle, L. W.; Herrmann, D. J.; Croner, C. M.; Wilson, B. F.: Map design preferences associated with professional discipline. 94 ASAProStGr 54-59

White, A. P.: Liu, W. Z.; — —: A comparison of nearest neighbour and tree-based methods of non-parametric discriminant analysis. 95 JStCmpSm 53 41-50

White, Andrew A.: Comment on "Mapping and geographic display of data". 95 StatMed 14 697-699

Pickle, Linda W.; — —: Effects of the choice of age-adjustment method on maps of death rates. 95 StatMed 14 615-627 (Disc: p697-699)

White, D. J.: — — (Auth); Brooks, Stephen (Rev): Review of "Markov decision processes". 95 Ststcian 44 292-292

White, Donald B.; Foley, Marianne: Statistical modeling of long term cumulative effects. 94 ASAProBiop 328-333

White, G. C.: Burnham, K. P.; Anderson, D. R.; — —: Selection among open population capture-recapture models when capture probabilities are heterogeneous. 95 JAppStat 22 611-624

White, Gary C.: Anderson, David R.; — —; Burnham, Kenneth P.: Some specialized risk assessment methodologies for vertebrate populations. 95 EnvEcoSt 2 91-115

Burnham, Kenneth P.; — —; Anderson, David R.: Model selection strategy in the analysis of capture-recapture data. 95 Biomtrcs 51 888-898

White, Glenn D., Jr: A quality measures plan within the IRS: A case study. 94 ASAProQlPr 76-81

White, H.: — — (Auth); Davidson, James (Rev): Review of "Estimation, inference and specification analysis". 95 JApEcmtx 10 477-479

White, Halbert: Granger, Clive W. J.; King, Maxwell L.; — —: Comments on testing economic theories and the use of model selection criteria. 95 JEconmtx 67 173-187

Sakata, Shinichi; — —: An alternative definition of finite-sample breakdown point with applications to regression model estimators. 95 JASA 90 1099-1106

Swanson, Norman R.; — —: A model-selection approach to assessing the information in the term structure using linear models and artificial neural network. 95 JBES 13 265-275

White, L. V.: Stapleton, James H. (Auth); — — (Rev): Review of "Linear statistical models". 96 ShrtBkRv 16/1 7-7

White, Lynda: Hinkelmann, K.; Kempthorne, O. (Auth); — — (Rev): Review of "Design and analysis of experiments. Volume I, Introduction to experimental design". 95 JRSS-A 158 351-351

Whitehead, John; Brunier, Hazel: Bayesian decision procedures for dose determining experiments. 95 StatMed 14 885-893 (Disc: p897-899)

Omar, Rumana Z.; Stallard, Nigel; — —: A parametric multistate model for the analysis of carcinogenicity experiments. 95 LiDaAnal 1 327-346

Whitehead, P. G.: Sefton, C. E. M.; — —; Eatherall, A.; Littlewood, I. G.; Jakeman, A. J.: Dynamic response characteristics of the Plynlimon catchments and preliminary analysis of relationships to physical descriptors. 95 EnvrMtrc 6 465-472

Whiteman, Charles H.: Curry, David J.; Divakar, Suresh; Mathur, Sharat K.; — —: BVAR as a category management tool: An illustration and comparison with alternative techniques. 95 JForcast 14 181-199

Whitlock, Paula A.: Percus, Ora E.; — —: Theory and application of Marsaglia's monkey test for pseudorandom number generators. 95 ACMMdSim 5 87-100

Whitmore, G. A.: Estimating degradation by a Wiener diffusion process subject to measurement error. 95 LiDaAnal 1 307-319

Jewell, Nicholas P.; Kimber, Alan C.; Lee, Mei-Ling Ting; — — (Ed): Lifetime data: Models in reliability and survival analysis. 95 Kluwer:MA:Hllnd 424p

Whitmore, G. Alex: Cole, Bernard F.; Lee, Mei-Ling T.; — —; Zaslavsky, Alan M.: An empirical Bayes model for Markov-dependent binary sequences with randomly missing observations. 95 JASA 90 1364-1372

Whitmore, James B.: Exact moment formulae for the linear permutation statistic of order k. 94 ASAProBiop 262-266

Whitney, Paul: Woodward, Wayne A.; — —; Eslinger, Paul W.: Minimum Hellinger distance estimation of mixture proportions. 95 JStPlInf 48 303-319

Whitridge, Patricia J.; Poirier, Claude: Evaluation of the Canadian Potato Area Estimation Program. 94 ASASrvyRsr 1150-1155

Whitt, Ward: Berger, Arthur W.; — —: Maximum values in queueing processes. 95 PrEngInf 9 375-409

Choudhury, Gagan L.; Leung, Kin K.; — —: An algorithm to compute blocking probabilities in multi-rate multi-class multi-resource loss models. 95 AdvAppPr 27 1104-1143

Choudhury, Gagan L.; Leung, Kin K.; — —: Calculating normalization constants of closed queuing networks by numerically inverting their generating functions. 95 JACM 42 935-970

Whittaker, C. C.: Bruno, W. J.; Knill, E.; Balding, D. J.; Bruce, D. C.; Doggett, N. A.; Sawhill, W. W.; Stallings, R. L.; — —; Torney, D. C.: Efficient pooling designs for library screening. 95 Genomics 26 21-30

Whittaker, Joe: de Falguerolles, Antoine; Jmel, Said; — —: Correspondence analysis and association models constrained by a conditional independence graph. 95 Psymtrka 60 161-180

Milan, Luis; — —: Application of the parametric bootstrap to models that incorporate a singular value decomposition. 95 ApplStat 44 31-49

Whittemore, Alice S.: Logistic regression of family data from case-control studies. 95 Biomtrka 82 57-67

Whittinghill, D. C.: MacEachern, S. N.; Notz, W. I.; — —; Zhu, Y.: Robustness to the unavailability of data in the linear model, with applications. 95 JStPlInf 48 207-213

Whittinghill, Dexter C.: Uddin, Nizam; — —: The robustness of binary and non-binary balanced nested row-column designs under the unavailability of blocks: A comparison. 95 AstrlJSt 37 111-124

Whittle, P.: Large-deviation expressions for the distribution of first-passage coordinates. 95 AdvAppPr 27 692-710

Puterman, M. L. (Auth); — — (Rev): Review of "Markov decision processes". 95 JRSS-A 158 636-637

Wiatrowski, William: Ponikowski, Chester H.; Scheible, Paul; — —: The survey of 1992 health expenditures: A quick response survey. 94 ASASrvyRsr 282-287

Wichmann, K.: Caspary, W.; — — (Auth); Trenkler, G. (Rev): Review of "Linear models. Algebraic foundation and statistical applications" (German). 95 StHefte 36 190-190

Wickens, Thomas D.: — — (Auth); Hershberger, Scott L. (Rev): Review of "The geometry of multivariate statistics". 95 StrEqMod 2 379-382

Wickramaratne, Priya J.: Sample size determination in epidemiologic studies. 95 StMedRes 4 311-337

Widaman, Keith F.: Comment on "Testing for equivalence of measurement scales: Simple structure and metric invariance reconsidered". 95 MultBehR 30 101-106

Raykov, Tenko; — —: Issues in applied structural equation modeling research. 95 StrEqMod 2 289-318

Wieand, Sam: Jung, Sin-Ho; — —; Cha, Stephen S.: A statistic for comparing two correlated markers which are prognostic for time to an event. 95 StatMed 14 2217-2225

Wiens, Brian; Matthews, Holly: Parameters for analysis of wound healing in early clinical studies in humans. 94 ASAProBiop 377-382

Wiens, John A.: Crist, Thomas O.; — —: Individual movements and estimation of population size in darkling beetles (Coleoptera: Tenebrionidae). 95 JAnmlEco 64 733-746

Wier, B. S.: Zhu, J.; — —: Analysis of cytoplasmic and maternal effects. 1. A genetic model for diploid plant seeds and animals. 94 ThApGene 89 153-159

Wierich, Werner: Schwabe, Rainer; — —: D-optimal designs of experiments with non-interacting factors. 95 JStPlInf 44 371-384

Wieslander, Åke: Sjöström, Michael; Rännar, Stefan; — —: Polypeptide sequence property relationships in Escherichia coli based on auto cross covariances. 95 Chemolab 29 295-305

Wijesinha, Manel Cooray; Piantadosi, Steven: Dose-response models with covariates. 95 Biomtrcs 51 977-987

Wijsman, E.: Lin, S.; Thompson, E.; — —: An algorithm for Monte Carlo estimation of genotype probabilities on complex pedigrees. 94 AnHuGene 58 343-357

Wijsman, Ellen: Lin, Shili; Thompson, Elizabeth; — —: Finding noncommunicating sets for Markov chain Monte Carlo estimations on pedigrees. 94 AJHuGene 54 695-704

Wijsman, Ellen M.: Olson, Jane M.; — —: Design and sample-size considerations in the detection of linkage disequilibrium with a disease locus. 94 AJHuGene 55 574-580

Wijsman, R. A.: Hall, W. J.; — —; Ghosh, J. K.: Correction to "The relationship between sufficiency and invariance with applications in sequential analysis" (V36 p575-614). 95 AnlsStat 23 705-705

Wikle, Christopher K.; Sherman, Peter J.; Chen, Tsing-Chang: Identifying periodic components in atmospheric data using a family of minimum variance spectral estimators. 95 JClimate 8 2352-2363

Wilcox, Allen J.: Weinberg, Clarice R.; — —: A model for estimating the potency and survival of human gametes in vivo. 95 Biomtrcs 51 405-412

Wilcox, Rand R.: Statistics for the social sciences. 95 Academic:NY:Lnd 444p

— —: Three multiple comparison procedures for trimmed means. 95 BiomtrcJ 37 643-656

— —: ANOVA: The practical importance of heteroscedastic methods, using trimmed means versus means, and designing simulation studies. 95 BJMaStPs 48 99-114

— —: A regression smoother for resistant measures of location and scale. 95 BJMaStPs 48 189-204

— —: Some small-sample results on a bounded influence rank regression method. 95 CommStA 24 881-888

— —: Testing the hypothesis of independence between two sets of variates. 95 MultBehR 30 213-225

— —: Comparing two independent groups via multiple quantiles. 95 Ststcian 44 91-99

— —: Simulation results on solutions to the multivariate Behrens-Fisher problem via trimmed means. 95 Ststcian 44 213-225

Wild, C. J.: Continuous improvement of teaching: A case study in a large statistics course. 95 IntStRvw 63 49-68

Wildfire, Judith: Iannacchione, Vincent; — —; Lux, Linda; Hawes, Catherine: Covering the population of unlicensed board and care homes: Problems and solutions. 94 ASASrvyRsr 270-275

Wilf, Herbert S.: Scheinerman, Edward R.; — —: The rectilinear crossing number of a complete graph and Sylvester's "Four point problem" of geometric probability. 94 AmMaMnth 101 939-943

Wilhelm, A.: Pukelsheim, F.; — —: Comment on "Constrained optimization of experimental design". 95 MaOpfStS 26 168-172

Wilkens, Henry T.: Hsu, Margaretha M.; — —: Advertising-related aspirin consumption among under 45 and over 45 consumers. 94 ASAProBiop 276-278

Wilkins, Lynne R.: Grove, John S.; Sharp, Dan S.; — —: Regression with random error in X: Increasing power of the test by non-linear adjustment for measurement error. 94 ASAProEpid 143-146

Wilks, Daniel S.; Hamill, Thomas M.: Potential economic value of ensemble-based surface weather forecasts. 95 MWeathr 11 3565-3575

Willan, Andrew R.; O'Brien, Bernie J.: Cost-effectiveness ratios in clinical trials: From deterministic to stochastic models. 94 ASAProBiop 19-28

Willemain, Thomas R.: Runger, George C.; — —: Model-based and model-free control of autocorrelated processes. 95 JQualTek 27 283-292

Runger, George C.; — —; Prabhu, Sharad: Average run lengths for CUSUM control charts applied to residuals. 95 CommStA 24 273-282

Willenborg, L. C. R. J.: Testing the logical structure of questionnaires. 95 StNeerla 49 95-109

— —: Data editing and imputation from a computational point of view. 95 StNeerla 49 139-151

Willett, John B.; Singer, Judith D.: It's déjà vu all over again: Using multiple-spell discrete-time survival analysis. 95 JEdStat 20 41-67

Williams, Bill: Performance indices for on-ice hockey statistics. 94 ASAProSpts 50-54

Williams, Byron K.: Conroy, Michael J.; — —; Hines, James E.: Use of solicited bands and separation of hunting and natural mortality: A comment. 95 JWildlMg 59 619-621

Williams, Calvin: Everitt, Brian S. (Auth); — — (Rev): Review of "A handbook of statistical analyses using S-Plus". 95 JQualTek 27 89-90

Williams, Calvin L.: Alam, Khursheed; — —: A multivariate goodness-of-fit test for stochastically ordered distributions. 95 BiomtrcJ 37 945-956

Gibbons, J. D.; Chakraborti, S. (Auth); — — (Rev): Review of "Nonparametric statistical inference (Third edition)". 95 JAppStat 22 193-194

Williams, Charles: Young, Philip; Morgan, Byron; Sonksen, Peter; Till, Sebastian; — —: Using a mixture model to predict the occurrence of diabetic retinopathy. 95 StatMed 14 2599-2608

Williams, David: Rogers, L. C. G.; — —: Diffusions, Markov processes, and martingales. Volume 1: Foundations (Second edition). 94 Wiley:NY:UK 386p

Rogers, L. C. G.; — — (Auth); Embrechts, P. A. L. (Rev): Review of "Diffusions, Markov processes, and martingales. Volume 1: Foundations (Second edition)". 95 ShrtBkRv 15/2 28-28

Williams, David W.; Liebhold, Andrew M.: Detection of delayed density dependence: Effects of autocorrelation in an exogenous factor. 95 Ecology 76 1005-1008

Williams, E. R.: John, J. A.; — —: Cyclic and computer generated designs (Second edition). 95 ChapmnHall:Lnd 272p

— —; Matheson, A. C. (Auth); Nester, Marks R. (Rev): Review of "Experimental design and analysis for use in tree improvement". 95 AstrlJSt 37 247-248

— —; Matheson, A. C. (Auth); Smith, R. I. (Rev): Review of "Experimental design and analysis for use in tree improvement". 95 Biomtrcs 51 1190-1190

Williams, H. P.: — — (Auth); Appa, Gautam (Rev): Review of "Model solving in mathematical programming". 95 JRSS-A 158 202-202

Williams, Matthew R.: An extreme-value function model of the species incidence and species-area relations. 95 Ecology 76 2607-2616

Williams, R. J.: Chen, Z. Q.; — —; Zhao, Z.: On the existence of positive solutions for semilinear elliptic equations with Neumann boundary conditions. 95 ProbTher 101 251-276

Pitman, J. (Auth); — — (Rev): Review of "Probability". 95 Metrika 42 143-144

Williams, Rick L.: Product-limit survival functions with correlated survival times. 95 LiDaAnal 1 171-186

— —: Product-limit survival functions with correlated survival times. 94 ASAProBiop 211-216

Bieler, Gayle S.; — —: Cluster sampling techniques in quantal response teratology and developmental toxicity studies. 95 Biomtrcs 51 764-776

Williams, Robert C.: Long, Jeffrey C.; — —; Urbanek, Margrit: An E-M algorithm and testing strategy for multiple-locus haplotypes. 95 AJHuGene 56 799-810

Williams, Stanley C.: Hestir, Kevin; — —: Supports of doubly stochastic measures. 95 Bernoull 1 217-243

Williams, Valerie S. L.: Jones, Lyle V.; — —: Multiple comparisons for education data. 94 ASAProSoSt 6-8

Thissen, David; Pommerich, Mary; Billeaud, Kathleen; — —: Item response theory for scores on tests including polytomous items with ordered responses. 95 ApPsychM 19 39-49

Williams, W. H.: Clarkson, S. P.; — —: How much does poor reading lower math scores?. 94 ASAProStEd 203-208

Williamson, G. D.: Cliff, A. D.; Haggett, P.; Smallman-Raynor, M. R.; Stroup, D. F.; — —: The application of multidimensional scaling methods to epidemiological data. 95 StMedRes 4 102-123

Williamson, G. David: Rutter, Carolyn; — —: Small area analysis for public health data. 94 ASAProEpid 118-122 (Disc: p125-126)

Williamson, John M.; Kim, KyungMann; Lipsitz, Stuart R.: Analyzing bivariate ordinal data using a global odds ratio. 95 JASA 90 1432-1437

Williford, William O.; Krol, William F.; Bingham, Stephen F.; Collins, Joseph F.; Weiss, David G.: The multicenter clinical trials coordinating center statistician: 'More than a consultant'. 95 AmerStat 49 221-225

Willinger, Walter; Taqqu, Murad S.; Leland, Will E.; Wilson, Daniel V.: Self-similarity in high-speed packet traffic: Analysis and modeling of ethernet traffic measurements. 95 StatSci 10 67-85

Cutland, Nigel J.; Kopp, P. Ekkehard; — —: From discrete to continuous stochastic calculus. 95 Stochatx 52 173-192

Willis, Gordon B.: Rasinski, Kenneth A.; Baldwin, Alison K.; — —; Jobe, Jared B.: Risk and loss perceptions associated with survey reporting of sensitive behaviors. 94 ASASrvyRsr 497-502

Willms, J. Douglas: Raudenbush, Stephen W.; — —: The estimation of school effects. 95 JEdStat 20 307-335

Wilsnack, Richard W.: Harris, T. Robert; Klassen, Albert D.; — —; Wilsnack, Sharon C.: A model for the interaction of partnership dissolution and relationship quality as predicators of remission of problem drinking in women. 94 ASAProSoSt 232-234

Wilsnack, Sharon C.: Harris, T. Robert; Klassen, Albert D.; Wilsnack, Richard W.; — —: A model for the interaction of partnership dissolution and relationship quality as predicators of remission of problem drinking in women. 94 ASAProSoSt 232-234

Wilson, Alyson G.; Johnson, Valen E.: Using features to model prior structural information. 94 ASAProPEng 255-264

Wilson, B. F.: White, A. A.; Pickle, L. W.; Herrmann, D. J.; Croner, C. M.; — —: Map design preferences associated with professional discipline. 94 ASAProStGr 54-59

Wilson, D.: Betteley, G.; Mettrick, N.; Sweeney, E.; — — (Auth); Shade, John (Rev): Review of "Using statistics in industry". 95 Ststcian 44 541-542

Wilson, Daniel V.: Willinger, Walter; Taqqu, Murad S.; Leland, Will E.; — —: Self-similarity in high-speed packet traffic: Analysis and modeling of ethernet traffic measurements. 95 StatSci 10 67-85

Wilson, Hoyt G.: Gargiullo, Paul M.; Rothenberg, Richard B.; — —: Confidence intervals, hypothesis tests, and sample sizes for the prevented fraction in cross-sectional studies. 95 StatMed 14 51-72

Wilson, Jacqueline B.: Guadagno, Mary Ann; Abma, Joyce C.; Chandra, Anjani; London, Kathy A.; Peterson, Linda S.; Piccinino, Linda J.; — —: Women's reports of sensitive health conditions, attitudes, and behaviors. 94 ASAProGovt 82-87

Wilson, Janette: Longton, Barbara; Sailer, Peter; — —: How sales of capital assets were affected by the Tax Reform Act of 1986. 94 ASAProGovt 201-205 (Disc: p212-214)

Wilson, Jeffrey R.: Turner, David L.; — —: Monitoring cross country skiers. 94 ASAProEnvr 167-172

Yanez, N. David, III; — —: Comparison of quasi-likelihood models for overdispersion. 95 AstrlJSt 37 217-231

Wilson, Kenneth R.; Anderson, David R.: Continuous-time capture-recapture population estimation when capture probabilities vary over time. 95 EnvEcoSt 2 55-69

Wilson, Marcia H.: Dawson, Deanna K.; Sauer, John R.; Wood, Paul A.; Berlanga, Mauro; — —; Robbins, Chandler S.: Estimating bird species richness from capture and count data. 95 JAppStat 22 1063-1068

Wilson, Mark; Adams, Raymond J.: Rasch models for item bundles. 95 Psymtrka 60 181-198

— —; Wang, Wen-chung: Complex composites: Issues that arise in combining different modes of assessment. 95 ApPsychM 19 51-71

Wilson, Noel: Discussion on the symposium on the teaching of statistics in higher education in Ireland. 95 Ststcian 44 58-58

Wilson, P. David: Longitudinal data analysis for linear Gaussian models with random disturbed-highest-derivative-polynomial subject effects. 95 StatMed 14 1219-1233

Wilson, R. A.: Bosworth, D. L.; — —; Young, A. (Auth); Lievesley, Denise (Rev): Review of "Reviews of United Kingdom statistical sources, Volume 27, Research and development". 95 JRSS-A 158 631-632

Wilson, Rick L.; Hardgrave, Bill C.: Predicting graduate student success in an MBA program: Regression versus classification. 95 EdPsMeas 55 186-195

Wilson, S. R.: Comment on "Summarizing a set of radiocarbon determinations: A robust approach" (V43 p489-503). 95 ApplStat 44 407-407

Wilson, Simon P.: Singpurwalla, Nozer D.; — —: The exponentiation formula of reliability and survival: Does it always hold?. 95 LiDaAnal 1 187-194

Singpurwalla, Nozer D.; — —: Models for assessing the reliability of computer software. 92 ProIrStCl 177-194

Wilson, William J.: Supplementing a college preparatory course in statistics with emphasis on applied probability, quality control, and graphics. 94 ASAProStEd 42-43

Winch, Janice Kim; Avi-Itzhak, Benjamin: Ordering of tandem constant-service stations to minimize in-process stock cost. 95 PrEngInf 9 457-473

Windham, M. P.: Opitz, O.; Lausen, B.; Klar, R. (Ed); — — (Rev): Review of "Information and classification: Concepts, methods and applications". 95 JClass 12 157-158

Windham, Michael P.: Robustifying model fitting. 95 JRSS-B 57 599-609

Windig, W.; Margevich, D. E.; McKenna, W. P.: A novel tool for two-dimensional correlation spectroscopy. 95 Chemolab 28 109-128

Windmeijer, Frank A. G.: Goodness-of-fit measures in binary choice models. 95 EcomtrRv 14 101-116 (Disc: 117-120)

Winkelmann, Rainer: Duration dependence and dispersion in count-data models. 95 JBES 13 467-474

Winker, Peter: Identification of multivariate AR-models by threshold accepting. 95 CmpStDA 20 295-307

Winkleby, Marilyn A.: Fortmann, Stephen P.; Flora, June A.; — —; Schooler, Caroline; Taylor, C. Barr; Farquhar, John W.: Community intervention trials: Reflections on the Stanford Five-City Project experience. 95 AJEpidm 142 576-586

Winkler, Gerhard: Image analysis, random fields and dynamic Monte Carlo methods: A mathematical introduction. 95 Springer:Brln:NY 324p

— — (Auth); Smith, R. L. (Rev): Review of "Image analysis, random fields and dynamic Monte Carlo methods: A mathematical introduction". 95 ShrtBkRv 15/3 49-49

Winkler, Peter: Brightwell, Graham; Ott, Teunis J.; — —: Target shooting with programmed random variables. 95 AnlsApPr 5 834-853

Coffman, E. G., Jr; Poonen, Bjorn; — —: Packing random intervals. 95 ProbTher 102 105-121

Winkler, Robert L.: Clemen, Robert T.; Murphy, Allan H.; — —: Screening probability forecasts: Contrasts between choosing and combining. 95 IntJFor 11 133-145

Winkler, William E.: Advanced methods for record linkage. 94 ASASrvyRsr 467-472

Winne, Philip H.: Kosonen, Peter; — —: Effects of teaching statistical laws on reasoning about everyday problems. 95 JEdPsych 87 33-46

Winner, Larry: Graybill, Franklin A.; Iyer, Hariharan K. (Auth); — — (Rev): Review of "Regression analysis - Concepts and applications". 95 AmerStat 49 114-114

Winsberg, Suzanne: Carroll, J. Douglas; — —: Fitting an extended INDSCAL model to three-way proximity data. 95 JClass 12 57-71

Winter, Francine D.: Harris, Kenneth W.; Chamblee, Ronald F.; — —: Evaluation of an automated industry and occupation coding system. 94 ASAProSoSt 174-177

Winterstein, Scott R.: Pollock, Kenneth H.; Bunck, Christine M.; — —; Chen, Chiu-Lan: A capture-recapture survival analysis model for radio-tagged animals. 95 JAppStat 22 661-672

Wiper, Michael P.; French, Simon: Combining experts' opinions using a normal-Wishart model. 95 JForcast 14 25-34

Wirth, Heike: Müller, Walter; Blien, Uwe; — —: Identification risks of microdata: Evidence from experimental studies. 95 SocMethR 24 131-157

Wise, Barry M.; Haesloop, Dan: A non-linear dynamic model identification challenge problem. 95 Chemolab 30 91-96

— —; Holt, Bradley R.; Gallagher, Neal B.; Lee, Samuel: A comparison of neural networks, non-linear biased regression and a genetic algorithm for dynamic model identification. 95 Chemolab 30 81-89

Wise, Morgan D.: Fellingham, Gilbert W.; — —: EM for longitudinal data from multiple studies. 95 StSinica 5 97-107

Witt, Christine A.: Witt, Stephen F.; — —: Forecasting tourism demand: A review of empirical research. 95 IntJFor 11 447-475

Witt, L. Alan: Nye, Lendell G.; — —: Interpreting moderator effects: Substitute for the signed coefficient rule. 95 EdPsMeas 55 27-31

Witt, Lee D.; Naranjo, Joshua D.; McKean, Joseph W.: Influence functions for rank-based procedures in the linear model. 95 JNnparSt 5 339-358

Witt, Michael B.: Folsom, Ralph E.; — —: Testing a new attrition nonresponse adjustment method for SIPP. 94 ASASrvyRsr 428-433 (Disc: p440-442)

Witt, Stephen F.; Witt, Christine A.: Forecasting tourism demand: A review of empirical research. 95 IntJFor 11 447-475

Witte, John S.; Thomas, Duncan C.; Langholz, Bryan: Comment on "Statistical significance testing in the American journal of epidemiology, 1970-1990" (V139 p1047-1052). 95 AJEpidm 142 101-101

Wittke, P. H.: Campbell, L. L.; McKellips, A. L.; — —: Distributions and expectations of singular random variables. 95 JAppProb 32 1032-1040

Wittmann, Rainer: Prophet inequalities for dependent random variables. 95 Stochatx 52 283-293

Witzko, K.-H.: van der Linde, A.; — —; Jöckel, K.-H.: Spatial-temporal analysis of mortality using splines. 95 Biomtrcs 51 1352-1360

Wobus, Peter: de la Puente, Manuel; — —: An item nonresponse and log-linear analysis of the Spanish Language Forms Availability Test. 94 ASASrvyRsr 583-588

Woess, Wolfgang: Picardello, Massimo A.; — —: The full Martin boundary of the bi-tree. 94 AnlsProb 22 2203-2222

Wohar, Mark E.: Choi, Seungmook; — —: The expectations theory of interest rates: Cointegration and factor decomposition. 95 IntJFor 11 253-262

Wold, Svante: Chemometrics; what do we mean with it, and what do we want from it?. 95 Chemolab 30 109-115

Lindgren, Fredrik; Geladi, Paul; Berglund, Anders; Sjöström, Michael; — —: Interactive variable selection (IVS) for PLS. Part II: Chemical applications. 95 JChemtrc 9 331-342

Rännar, Stefan; Geladi, Paul; Lindgren, Fredrik; — —: A PLS kernel algorithm for data sets with many variables and few objects. Part II: Cross-validation, missing data and examples. 95 JChemtrc 9 459-470

Woldie, Mammo; Folks, J. Leroy: Calibration for inverse Gaussian regression. 95 CommStA 24 2609-2620

Wolf, Dennis A.; Tardiff, Mark F.; Ketelle, Richard H.: Characterizing groundwater at Oak Ridge National Laboratory hazardous waste sites. 94 ASAProEnvr 116-121

Wolf, Philip A.: D'Agostino, Ralph B.; Belanger, Albert J.; Markson, Elizabeth W.; Kelly-Hayes, Maggie; — —: Development of health risk appraisal functions in the presence of multiple indicators: The Framingham study nursing home institutionalization model. 95 StatMed 14 1757-1770

Wolfe, Douglas A.: Kim, Dongjae; — —: Distribution-free partially sequential tests for treatments versus control setting. 95 BiomtrcJ 37 259-271

Pan, Guohua; — —: Distribution-free test based on ranks for comparing groups with umbrella orderings. 95 JNnparSt 5 409-416

Wolfe, Robert A.: Lee, Seungyeoun; — —: A class of linear nonparametric rank tests for dependent censoring. 95 CommStA 24 687-712

Mauger, Elizabeth A.; — —; Port, Friedrich K.: Transient effects in the Cox proportional hazards regression model. 95 StatMed 14 1553-1565

Wolff, Hans: Amin, Raid W.; — —: The behaviour of EWMA-type quality control schemes in the case of mixture alternatives. 95 SqtlAnly 14 157-177

Wolff, Rodney C. L.: Hall, Peter; — —: Properties of invariant distributions and Lyapunov exponents for chaotic logistic maps. 95 JRSS-B 57 439-452

Hall, Peter; — —: Estimators of integrals of powers of density derivatives. 95 StPrLet 24 105-110

Wolfgang, Glenn S.; Lewis, Paul J.; Vacca, E. Ann: Cognitive research for the 1997 Census of Agriculture report form. 94 ASASrvyRsr 503-508

Lewis, Paul J.; — —; Vacca, E. Ann: 1992 Census of Agriculture coverage evaluation estimation. 94 ASASrvyRsr 1019-1024

Wolfinger, Russell: Rodriguez, Robert; Tobias, Randall; — —: Comment on "The statistics of linear models: Back to basics" (V4 p221-234). 95 StatComp 5 97-101

Wolfinger, Russell D.; Kass, Robert E.: Bayesian inference for variance component models. 94 ASAProBays 136-141

— —; Tobias, Randall D.: Joint estimation of location, dispersion, and random effects. 94 ASAProPEng 225-234

Wolford, Monica L.: A revised theory of longitudinal nonresponse. 94 ASASrvyRsr 960-965

Wolfram Research Inc: — — (Auth); Baglivo, Jenny (Rev): Review of "Mathematica for the Macintosh (Version 2.0.4)". 95 AmerStat 49 86-92

Wolfson, David B.: Joseph, Lawrence; — —; Berger, Roxane du: Sample size calculations for binomial proportions via highest posterior density intervals. 95 Ststcian 44 143-154

Joseph, Lawrence; — —; Berger, Roxane du: Some comments on Bayesian sample size determination. 95 Ststcian 44 167-171

Wolins, Leroy: A Monte Carlo study of constrained factor analysis using maximum likelihood and unweighted least squares. 95 EdPsMeas 55 545-557

Wollson, R. F.: Arndt, S.; Tyrrell, G.; — —; Flaum, M.; Andreasen, N. C.: Effects of errors in a multicenter medical study: Preventing misinterpreted data. 94 JPsyctrc 28 447-459

Wolpert, Robert L.: Comment on "Inference from a deterministic population dynamics model for Bowhead whales". 95 JASA 90 426-427

Berger, James O.; Brown, Lawrence D.; — —: A unified conditional frequentist and Bayesian test for fixed and sequential simple hypothesis testing. 94 AnlsStat 22 1787-1807

Wolstenholme, Linda C.: A nonparametric test of the weakest-link principle. 95 Technmcs 37 169-175

Wolter, Kirk M.: Edmonston, Barry; Schultze, Charles (Ed); — — (Rev): Review of "Modernizing the U.S. Census". 95 Chance 8/3 44-49

Stefey, Duane L.; Bradburn, Norman M. (Ed); — — (Rev): Review of "Counting people in the information age". 95 Chance 8/3 44-49

Wolters, J.: On the term structure of interest rates - Empirical results for Germany. 95 StHefte 36 193-214

Wolters, Jürgen: Hassler, Uwe; — —: Long memory in inflation rates: International evidence. 95 JBES 13 37-45

Wolthuis, Henk: Actuarial equivalence. 94 InsrneME 15 163-179

Wong, A.: On approximate inference for the two-parameter gamma model. 95 StHefte 36 49-59

— —: An averaging approach to prediction. 95 StHefte 36 253-264

Wong, A. C. M.: A simple and accurate inference procedure for the difference of failure intensities from two exponential distributions. 95 CommStA 24 2865-2879

Wong, Chi Song; Liu, Dongsheng: Moments of generalized Wishart distributions. 95 JMultiAn 52 280-294

Wang, Tonghui; — —: Cochran theorems for a multivariate elliptically contoured model. 95 JStPlInf 43 257-270

Wong, H.; Li, W. K.: Portmanteau test for conditional heteroscedasticity, using ranks of squared residuals. 95 JAppStat 22 121-134

Wong, Hoi Lam: Zhu, Li-Xing; — —; Fang, Kai-Tai: A test for multivariate normality based on sample entropy and projection pursuit. 95 JStPlInf 45 373-385

Wong, J. Y.: Kao, Chiang; — —: Several extensively tested random number generators. 95 CmpOpRsh 22 1035-1039

Wong, Kam L.: A new framework for part failure-rate prediction models. 95 IEEEReli 44 139-146

Wong, M. Y.: Robustness of testing the equality of two slopes in simple linear structural relationships. 95 JAppStat 22 3-8

— —: The statistical analysis of slope ratio assays using SAS software. 94 CmpBiomd 45 233-238

Wong, Paul: Pfeffermann, Danny; Morry, Marietta; — —: Estimation of the variances of X-11 ARIMA seasonally adjusted estimators for a multiplicative decomposition and heteroscedastic variances. 95 IntJFor 11 271-283

Wong, Raymond Sin-Kwok: Extensions in the use of log-multiplicative scaled association models in multiway contingency tables. 95 SocMethR 23 507-538

Wong, Takmeng: Evans, K. Franklin; Turk, Joseph; — —; Stephens, Graeme L.: A Bayesian approach to microwave precipitation profile retrieval. 95 JApMetrl 34 260-279

Wong, Weng Kee: A graphical approach for the construction of constrained D- and L-optimal designs using efficiency plots. 95 JStCmpSm 53 143-152

— —: Comment on "Constrained optimization of experimental design". 95 MaOpfStS 26 174-174

— —: On the equivalence of D- and G-optimal designs in heteroscedastic models. 95 StPrLet 25 317-321

Attanasio, Massimo; — —: Transformations for stabilizing spread in a time series model: A comparative study. 95 JStCmpSm 52 55-70

Huang, Mong-Na Lo; Chang, Fu-Chuen; — —: D-optimal designs for polynomial regression without an intercept. 95 StSinica 5 441-458

Wong, William: Ho, Chih-Chin; — —: Alternative imputation techniques for proportions of income variables for IRS compliance modeling. 94 ASASrvyRsr 388-393

Wong, Wing H.: Liu, Jun S.; — —; Kong, Augustine: Covariance structure and convergence rate of the Gibbs sampler with various scans. 95 JRSS-B 57 157-169

Wong, Wing Hung: Comment on "Bayesian computation and stochastic systems". 95 StatSci 10 52-53

— —; Shen, Xiaotong: Probability inequalities for likelihood ratios and convergence rates of sieve MLEs. 95 AnlsStat 23 339-362

Wood, Andrew T. A.: When is a truncated covariance function on the line a covariance function on the circle?. 95 StPrLet 24 157-164

Booth, James G.; — —: An example in which the Lugannani-Rice saddlepoint formula fails. 95 StPrLet 23 53-61

Booth, James G.; Butler, Ronald W.; Huzurbazar, S.; — —: Saddlepoint approximations for P-values of some tests of covariance matrices. 95 JStCmpSm 53 165-180

Wood, G. R.: Bridging the gap between town and gown. 94 ASAProQlPr 250-251

Wood, Paul A.: Dawson, Deanna K.; Sauer, John R.; — —; Berlanga, Mauro; Wilson, Marcia H.; Robbins, Chandler S.: Estimating bird species richness from capture and count data. 95 JAppStat 22 1063-1068

Wood, Phillip K.: Lawrence Erlbaum Associates Inc (Auth); — — (Rev): Review of "TETRAD II: Tools for causal modeling". 95 StrEqMod 2 277-287

Wu, Jiang-Lun: Albeverio, Sergio; — —: Stochastic anticipative calculus for functionals of a Gaussian generalized random field. 95 Stochatx 54 317-337

Wu, Ka Ho: Hamilton, David; — —: Confidence regions for parameters in the AR(1) model. 95 JTimSrAn 16 249-265

Wu, Lilian S. Y.: Ravishanker, Nalini; Dey, Dipak K.; — —: Forecasting, planning and contemporaneous outlier analysis for IBM regional revenue based on shrinkage estimation. 94 ASAProBuEc 211-216

Wu, Lilian S.-Y.: Ravishanker, Nalini; — —; Dey, Dipak K.: Shrinkage estimation in time series using a bootstrapped covariance estimate. 95 JStCmpSm 53 259-267

Wu, Liming: Moderate deviations of dependent random variables related to CLT. 95 AnlsProb 23 420-445

Wu, Margaret: Follmann, Dean; — —: An approximate generalized linear model with random effects for informative missing data. 95 Biomtrcs 51 151-168

Wu, Margaret C.: Zucker, David M.; Zerbe, Gary O.; — —: Inference for the association between coefficients in a multivariate growth curve model. 95 Biomtrcs 51 413-424

Wu, Qiguang: Necessary and sufficient condition for the existence of UMVU estimators in growth curve model. 95 KToChina 39 89-92

— —; He, Min: Several notes on admissible estimators of parameters and losses under truncated distribution families (Chinese). 95 JSysSMaS 14 168-176

Wu, Qiying: Consistency and asymptotic normality of maximum likelihood estimators for independent not identically distributed censored samples (Chinese). 95 ChJApPSt 10 405-419

Wu, Tiee-Jian: Adaptive root n estimates of integrated squared density derivatives. 95 AnlsStat 23 1474-1495

Wu, W. R.; Li, W. M.: A new approach for mapping quantitative trait loci using complete genetic marker linkage maps. 94 ThApGene 89 535-539

Wu, Y.; McGilchrist, C. A.: Recursive computation for repeated measures models with patterned variance matrices. 95 JStCmpSm 51 333-341

Srivastava, M. S.; — —: An improved version of Taguchi's on-line control procedure. 95 JStPlInf 43 133-145

Wu, Yaohua: Asymptotic behavior of M-estimator in linear models (Chinese). 95 AMaAPknC 17 401-410

Wu, Yow-Wu B.: Hierarchical linear models: A multilevel data analysis technique. 95 NurseRes 44 123-126

Wu, Yuan; McGilchrist, C. A.: Heterogeneous variances in repeated measures. 95 BiomtrcJ 37 523-530

Wu, Zhong: Fei, Heliang; Sinha, Bimal K.; — —: Estimation of parameters in two-parameter Weibull and extreme-value distributions using ranked set sample. 94 JStatRsr 28 149-161

Wulfert, Stefanie: Bischoff, Wolfgang; Fieger, Werner; — —: Minimax- and Γ-minimax estimation of a bounded normal mean under LINEX loss. 95 StatDecn 13 287-298

Wulff, Fredrik: Toompuu, Aleksander; — —: Spatial large-scale correlations for optimal interpolation of temperature, salinity and nutrient concentrations in the Gulf of Finland. 95 EnvrMtrc 6 55-72

Wulfsohn, M. S.: Tsiatis, A. A.; DeGruttola, Victor; — —: Modeling the relationship of survival to longitudinal data measured with error. Applications to survival and CD4 counts in patients with AIDS. 95 JASA 90 27-37

Wyld, Rona: Fielding, Katherine L.; Brettle, Raymond P.; Gore, Sheila M.; O'Brien, Fiona; — —; Robertson, J. Roy; Weightman, Rosina: Heterosexual transmission of HIV analysed by generalized estimating equations. 95 StatMed 14 1365-1378

Wyler, A.: Weller, J. I.; — —: Power of different sampling strategies to detect quantitative trait loci variance effects. 92 ThApGene 83 582-588

Wymer, Colin: Skinner, Chris; Marsh, Catherine; Openshaw, Stan; — —: Disclosure control for census microdata. 94 JOffStat 10 31-51

Wynn, H. P.: Giovagnoli, Alessandra; — —: Multivariate dispersion orderings. 95 StPrLet 22 325-332

Xia, Aihua: Weak convergence of Markov processes with extended generators. 94 AnlsProb 22 2183-2202

Brown, Timothy C.; — —: On metrics in point process approximation. 95 Stochatx 52 247 263

Brown, Timothy C.; — —: On Stein-Chen factors for Poisson approximation. 95 StPrLet 23 327-332

Xia, Yincun: On the density function of T_0^2 from elliptical distribution (Chinese). 95 ChJApPSt 10 119-124

Xiang, Qing; Chen, Yu Qing: On the size of the multiplier groups of cyclic difference sets. 95 JCombThA 69 168-169

Xiang, Xiaojing: Estimation of a quantile in some nonstandard cases. 95 AnInStMa 47 105-117

— —: A Berry-Esseen theorem for the kernel quantile estimator with application to studying the deficiency of quantile estimators. 95 AnInStMa 47 237-251

— —: Deficiency of the sample quantile estimator with respect to kernel quantile estimators for censored data. 95 AnlsStat 23 836-854

— —: A strong law of large number for L-statistics in the non-I.D. case. 95 CommStA 24 1813-1819

— —: Bahadur representation of the kernel quantile estimator under random censorship. 95 JMultiAn 54 193-209

— —: Estimation of conditional quantile density function. 95 JNnparSt 4 309-316

— —: On Bahadur-Kiefer representation of a kernel conditional quantile estimator. 95 JNnparSt 5 275-287

— —: A note on the bias of L-estimators and a bias reduction procedure. 95 StPrLet 23 123-127

Xiao, Baichun: Multiple and unacceptable solutions in nonlinear programming discriminant analysis. 94 AJMMgSci 14 267-287

Xiao, Y. M.: Falconer, K. J.; — —: Average densities of the image and zero set of stable processes. 95 StocProc 55 271-283

Xiao, Yimin: Dimension results for Gaussian vector fields and index-α stable fields. 95 AnlsProb 23 273-291

Xie, C. Y.; Yeh, F. C.; Dancik, B. P.; Strobeck, C.: Joint estimation of immigration and mating system parameters in gymnosperms using the EM algorithm. 91 ThApGene 83 137-140

Xie, Minyu: Admissibility of quadratic form estimate of variance matrix under some optimal criterion (Chinese). 95 AMaAPknC 17 149-153

Xie, Xuemin; Johnson, William D.: The rank tests for a completely randomized design with multivariate longitudinal responses and a randomized complete block design with multivariate responses. 94 ASAProBiop 250-255

Xie, Yingchao: Limit theorems of Hilbert valued semimartingales and Hilbert valued martingale measures. 95 StocProc 59 277-293

Xie, Yu-Long; Baeza-Baeza, Juan José; Ramis-Ramos, Guillermo: Kinetic spectrophotometric resolution of binary mixtures using three-way partial least squares. 95 Chemolab 27 211-220

— —; Liang, Yi-Zeng; Chen, Zhang-Guang; Huang, Zi-Hong; Yu, Ru-Qin: A nonlinearity tracking analysis algorithm for treatment of non-linearity in multivariate calibration. 95 Chemolab 27 21-32

Wang, Ji-Hong; — —; Yu, Ru-Qin: Maximum sum of binary-coded residuals (MASBR) regression as a robust procedure for treatment of spectral data. 95 JChemtrc 9 373-387

Xie, Yumin: Admissible estimates in the important class of estimates of the covariance matrix. 95 ChinaAMB 15 435-442

Xiong, J.: Kallianpur, G.; — —: Diffusion approximation of nuclear space-valued stochastic differential equations driven by Poisson random measures. 95 AnlsApPr 5 493-517

Xiong, Wei; Graffigne, Christine: Hierarchical Markov models: An application to motion estimation and segmentation. 94 QStMaA 16 45-67

Xiong, Xiaoping: A class of sequential conditional probability ratio tests. 95 JASA 90 1463-1473

Tan, Ming; — —: Sequential conditional probability ratio test approach to the design of clinical trials. 94 ASAProBiop 359-364

Xu, Jian-Lun; Prorok, Philip C.: Non-parametric estimation of the post-lead-time survival distribution of screen-detected cancer cases. 95 StatMed 14 2715-2725

— —; Yang, Grace L.: A note on a characterization of the exponential distribution based on a type II censored sample. 95 AnlsStat 23 769-773

Xu, Lei; Yuille, Alan L.: Robust principal component analysis by self-organizing rules based on statistical physics approach. 95 IEEENrlN 6 131-143

Jordan, Michael I.; — —: Convergence results for the EM approach to mixtures of experts architectures. 95 NuralNet 8 1409-1431

Xu, Luo: Large sample properties of the empirical Euclidean likelihood estimation for semiparametric model (Chinese). 95 ChJApPSt 10 344-352

Xu, Quanhua: Some results related to interpolation on Hardy spaces of regular martingales. 95 IsrelJMa 91 173-187

Xu, S.: Muir, W.; Nyquist, W. E.; — —: Alternative partitioning of the genotype-by-environment interaction. 92 ThApGene 84 193-200

Xu, Shizhong; Atchley, William R.; Fitch, Walter M.: Phylogenetic inference under the pure drift model. 94 MolBioEv 11 949-960

— —; Muir, W. M.: Selection index updating. 92 ThApGene 83 451-458

Xu, Susan H.: Hayya, Jack C.; — —; Ramasesh, Ranga V.; He, Xin X.: Order crossover in inventory systems. 95 CommStSM 11 279-309

Xu, Xingzhong: The choice of loss function (Chinese). 95 ChJApPSt 10 183-193

Xu, Yexiao J.: McDonald, James B.; — —: A generalization of the beta distribution with applications. 95 JEconmtx 66 133-152

McDonald, James B.; — —: Correction to "A generalization of the beta distribution with applications" (V66 p133-152). 95 JEconmtx 69 427-428

McDonald, James B.; — —: Some comparisons and applications of semi-parametric and flexible parametric estimation procedures in econometric models. 94 ASAProBuEc 229-234

Xu, Yuan: Galambos, Janos; — —: Bivariate extension of the method of polynomials for Bonferroni-type inequalities. 95 JMultiAn 52 131-139

Xuc, Liugcn: Asymptotically optimality and convergence rates of empirical Bayes estimator for the parameter of continuous one parameter exponential families (Chinese). 95 ChJApPSt 10 35-45

Xue, Zhiming; Gulas, Charles S.: Continuous improvement in college teaching: An application of statistical tools. 95 ASQCTekCon 267-275

Yadrenko, M. I.: Quasianalytic random fields. 95 ThPrMaSt 50 143-146

Skorokhod, A. V.; — —: I. I. Gikhman: A brief review of scientific and pedagogic activity. 95 ThPrMaSt 50 1-5

Yakir, Benjamin: A note on the run length to false alarm of a change-point detection policy. 95 AnlsStat 23 272-281

— —; Cox, Christopher; Tanner, Martin A.; Clarkson, Thomas W.; Shamlaye, C.: Seasonal changes in methylmercury levels in human hair. 94 ASAProBays 35-39

Yashchin, Emmanuel: Likelihood ratio methods for monitoring parameters of a nested random effect model. 95 JASA 90 729-738

— —: Comment on "Run-length distribution of special-cause control charts for correlated processes" (V36 p3-27). 95 Technmcs 37 243-243

— —: Estimating the current mean of a process subject to abrupt changes. 95 Technmcs 37 311-323

— —: Estimating the current mean of multivariate processes subject to abrupt changes. 94 ASAProPEng 81-86

Yashin, Anatoli I.; Vaupel, James W.; Iachine, Ivan A.: Correlated individual frailty: An advantageous approach to survival analysis of bivariate data. 95 MaPopStd 5 145-159

Yasui, Yutaka: Brookmeyer, Ron; — —: Statistical analysis of passive surveillance disease registry data. 95 Biomtrcs 51 831-842

Yatawara, Nihal; Lin, Hua: Multiple outlier detection in time series. 94 ASAProBuEc 310-315

Yates, J. R. W.: Chiano, M. N.; — —: Bootstrapping in human genetic linkage. 94 AnHuGene 58 129-143

Yates, P. K.: Kao, L.-J.; — —; Lewis, S. M.; Dean, A. M.: Approximate information matrices for estimating a given set of contrasts. 95 StSinica 5 593-598

Yatracos, Yannis G.: On the rare species of a population. 95 JStPlInf 48 321-329

Yau, Hung-Chun: Weideman, William E.; Manry, Michael T.; — —; Gong, Wei: Comparisons of a neural network and a nearest-neighbor classifier via the numeric handprint recognition problem. 95 IEEENrlN 6 1524-1530

Yau, K. K. W.: McGilchrist, C. A.; — —: The derivation of BLUP, ML, REML estimation methods for generalised linear mixed models. 95 CommStA 24 2963-2980

McGilchrist, C. A.; — —: Estimation of survival with time dependent frailty. 94 ASAProEpid 88-90

Yau, Sui K.: Regression and AMMI analyses of genotype × environment interactions: An empirical comparison. 95 AgronmyJ 87 121-126

Ycart, Bernard: Arcangeli, Rémi; — —: Almost sure convergence of smoothing D^m-splines for noisy data. 93 NumeriMa 66 281-294

Ye, H.: Labropulu, F.; Sullivan, P. J.; — —: A contaminant-dilution heuristic model. 95 EnvrMtrc 6 677-684

Sullivan, P. J.; — —: A prognosis for the sudden release of contaminant in an environmental flow. 95 EnvrMtrc 6 627-636

Ye, J. J.: Dempster, M. A. H.; — —: Impulse control of piecewise deterministic Markov processes. 95 AnlsApPr 5 399-423

Ye, K.: Selection of the reference priors for a balanced random effects model. 95 Test 4 1-17

Ye, Keying: EWMA control charts for AR(1) model using Bayes estimators. 94 ASAProBays 64-68

Sun, Dongchu; — —: Reference prior Bayesian analysis for normal mean products. 95 JASA 90 589-597

Ye, Yinyu: Bosch, Ronald J.; — —; Woodworth, George G.: A convergent algorithm for quantile regression with smoothing splines. 95 CmpStDA 19 613-630

Yeap, T. H.: Ahmed, N. U.; — —: Performance analysis of slotted ALOHA protocol for optical fiber local area networks with star topology. 95 StocAnAp 13 389-410

Yeh, F. C.: Xie, C. Y.; — —; Dancik, B. P.; Strobeck, C.: Joint estimation of immigration and mating system parameters in gymnosperms using the EM algorithm. 91 ThApGene 83 137-140

Yeh, Lam: An optimal inspection-repair-replacement policy for standby systems. 95 JAppProb 32 212-223

— —; Choy, S. T. B.: Bayesian variable sampling plans for the exponential distribution with uniformly distributed random censoring. 95 JStPlInf 47 277-293

Yellman, Ted W.; Murray, Thomas M.: Comment on "The meaning of probability in probabilistic safety analysis" (V45 p261-269). 95 RelEngSS 49 201-205

Yeung, David W. K.; Stewart, Sally E. A.: Stationary probability distributions of some Lotka-Volterra types of stochastic predation systems. 95 StocAnAp 13 503-516

Yeung, David Wing-Kay: Employment adjustments noise and the expected rate of inflation in a simple inflation-unemployment model. 95 StocAnAp 13 125-135

Yeung, Iris; Lam, James: Comparison of analog and digital simulation on some continuous time systems. 94 ASAProStEd 186-191

Yeung, Raymond W.: Takine, Tetsuya; Sengupta, Bhaskar; — —: A generalization of the matrix M/G/1 paradigm for Markov chains with a tree structure. 95 CommStSM 11 411-421

Yi, Hong-Wei: Zhao, Yi-Chun; — —: Random fixed point theorems for mappings satisfying weakly inwardness conditions. 95 StocAnAp 13 249-256

Yi, Qilong; Glickman, Lawrence T.: Computer simulation analysis of Sartwell's incubation period model for diseases with uncertain etiology - The effect of competing risk. 95 AJEpidm 142 363-368

Yi, W.: Bier, V. M.; — —: The performance of precursor-based estimators for rare event frequencies. 95 RelEngSS 50 241-251

Yi, Woojune: Bier, Vicki M.; — —: A Bayesian method for analyzing dependencies in precursor data. 95 IntJFor 11 25-41

Yin, G.; Zhu, Y. M.; Gao, A. J.: On characterization of a recursively defined process: Necessary and sufficient conditions, sensitivity. 95 Stochatx 52 265-282

Zhao, L.; — —: On variational inequality approach to Nash dynamic game of oil stockpiling. 95 StocAnAp 13 627-649

Yin, J.: Bennett, F. E.; — —: Some results on $(v, \{5, w^*\})$-PBDs. 95 JComDsgn 3 455-468

Yin, Jianxing: Furino, Steven; Greig, Malcolm; Miao, Ying; — —: Resolvable BIBDs with block size 8 and index 7. 94 JComDsgn 2 101-116

Ying, Jiangang: Multiplicative functionals of Lévy processes. 95 StocProc 58 319-327

Ying, Lisa H.; Pukelsheim, Friedrich; Draper, Norman R.: Slope rotatability over all directions designs. 95 JAppStat 22 331-341

— —; Pukelsheim, Friedrich; Draper, Norman R.: Slope rotatability over all directions designs for $k \geq 4$. 95 JAppStat 22 343-354

Ying, Z.; Jung, S. H.; Wei, L. J.: Survival analysis with median regression models. 95 JASA 90 178-184

Cheng, S. C.; Wei, L. J.; — —: Analysis of transformation models with censored data. 95 Biomtrka 82 835-845

Ying, Zhiliang: Gu, Minggao; — —: Group sequential methods for survival data using partial likelihood score processes with covariate adjustment. 95 StSinica 5 793-804

Lai, Tze Leung; — —; Zheng, Zukang: Asymptotic normality of a class of adaptive statistics with applications to synthetic data methods for censored regression. 95 JMultiAn 52 259-279

Lin, D. Y.; — —: Semiparametric analysis of general additive-multiplicative hazard models for counting processes. 95 AnlsStat 23 1712-1734

Lin, D. Y.; — —: Semiparametric inference for the accelerated life model with time-dependent covariates. 95 JStPlInf 44 47-63

Yip, Paul; Huggins, Richard: Some aspects of inference for chain binomial models. 95 StocAnAp 13 355-367

Yip, Paul S. F.: Fong, Daniel Y. T.; — —: A note on information loss in analyzing a mixture model of count data. 95 CommStA 24 3197-3209

Ylvisaker, D.: John, P. W. M.; Johnson, M. E.; Moore, L. M.; — —: Minimax distance designs in two-level factorial experiments. 95 JStPlInf 44 249-263

Ylvisaker, Donald: Mitchell, Toby J.; Morris, Max D.; — —: Two-level fractional factorials and Bayesian prediction. 95 StSinica 5 559-573

Yohai, Victor: Ferretti, Nélida; Kelmansky, Diana; — —: A goodness-of-fit test based on ranks for ARMA models. 95 CommStA 24 295-318

Yohai, Victor J.: Maronna, Ricardo A.; — —: The behavior of the Stahel-Donoho robust multivariate estimator. 95 JASA 90 330-341

Peña, Daniel; — —: The detection of influential subsets in linear regression by using an influence matrix. 95 JRSS-B 57 145-156

Peña, Daniel; — —: Correction to "The detection of influential subsets in linear regression by using an influence matrix" (V57 p145-156). 95 JRSS-B 57 611-611

Yokoyama, Ryozo: On the central limit theorem and law of the iterated logarithm for stationary processes with applications to linear processes. 95 StocProc 59 343-351

Yokoyama, Takahisa: LR test for random-effects covariance structure in a parallel profile model. 95 AnInStMa 47 309-320

Yokum, J. Thomas; Armstrong, J. Scott: Beyond accuracy: Comparison of criteria used to select forecasting methods. 95 IntJFor 11 591-597

Yong, Jiongmin: Duffie, Darrell; Ma, Jin; — —: Black's consol rate conjecture. 95 AnlsApPr 5 356-382

Yoo, Jaesin S.: Iyengar, Raghavan J.; — —; Zafra, Pablo M. R.: An application of some patterned matrices to the principal component analysis of linked variables. 95 PakJStA 11 23-33

Yoo, Seongmo; David, Herbert T.: Three remarks on Pitman domination. 95 JKoreaSt 24 361-373

Yor, M.: Embrechts, P.; Rogers, L. C. G.; — —: A proof of Dassios' representation of the α-quantile of Brownian motion with drift. 95 AnlsApPr 5 757-767

Rajeev, B.; — —: Local times and almost sure convergence of semi-martingales. 95 AIHenrPB 31 653-667

— — (Auth); Vatutin, V. A. (Rev): Review of "Some aspects of Brownian motion. Part I: Some special functionals". 94 ThProbAp 39 723-725

Yor, Marc: The distribution of Brownian quantiles. 95 JAppProb 32 405-416

York, Jeremy; Madigan, David; Heuch, Ivar; Lie, Rolv Terje: Birth defects registered by double sampling: A Bayesian approach incorporating covariates and model uncertainty. 95 ApplStat 44 227-242

Madigan, David; — —: Bayesian graphical models for discrete data. 95 IntStRvw 63 215-232

Yoshida, Michihiro: Morikawa, Toshihiko; — —: A useful testing strategy in phase III trials: Combined test of superiority and test of equivalence. 95 JBiophSt 5 297-306

Yoshihara, Ken-ichi: The Bahadur representation of sample quantiles for sequences of strongly mixing random variables. 95 StPrLet 24 299-304

Yoshikawa, S.: Tanaka, H.; Ishibuchi, H.; — —: Exponential possibility regression analysis. 95 FuzzySet 69 305-318

Yoshimura, Osamu: Ueno, Maomi; — —; Shigemasu, Kazuo: A latent trait model for reading process of a word (Japanese). 94 JpJBeh 21/1 48-56

Zhu, Wenwu; Galatsanos, Nikolas P.; Katsaggelos, Aggelos K.: Regularized multichannel restoration using cross-validation. 95 CVGIPGr 57 38-54

Zhu, Y.: MacEachern, S. N.; Notz, W. I.; Whittinghill, D. C.; — —: Robustness to the unavailability of data in the linear model, with applications. 95 JStPlInf 48 207-213

Zhu, Y. M.: Yin, G.; — —; Gao, A. J.: On characterization of a recursively defined process: Necessary and sufficient conditions, sensitivity. 95 Stochatx 52 265-282

Zhu, Yixin: Li, Huan; — —: On M/G/1 queues with exhaustive service and generalized vacations. 95 AdvAppPr 27 510-531

Zhu, Zhiwei: Koul, Hira L.; — —: Bahadur-Kiefer representations for GM-estimators in autoregression models. 95 StocProc 57 167-189

Zidek, J. V.: Duddek, C.; Le, N. D.; — —; Burnett, R. T.: Multivariate imputation in cross-sectional analysis of health effects associated with air pollution. 95 EnvEcoSt 2 191-212

Zidek, James V.: Hu, Feifang; — —: A bootstrap based on the estimating equations of the linear model. 95 Biomtrka 82 263-275

Ziegel, Eric R.: Box, G.; Jenkins, G.; Reinsel, G. (Auth); — — (Rev): Review of "Time series analysis: Forecasting and control". 95 Technmcs 37 238-239

Braun, H. (Ed); — — (Rev): Review of "The collected works of John W. Tukey, Volume VIII: Multiple comparisons". 95 Technmcs 37 242-242

Canavos, G.; Miller, D. (Auth); — — (Rev): Review of "Introduction to modern business statistics". 95 Technmcs 37 239-240

Cervetz, C. (Auth); — — (Rev): Review of "Developing new products with TQM". 95 Technmcs 37 241-241

Clayton, D.; Hills, M. (Auth); — — (Rev): Review of "Statistical models in epidemiology". 95 Technmcs 37 240-240

Collett, D. (Auth); — — (Rev): Review of "Modelling for survival data in medical research". 95 Technmcs 37 128-128

Cooper, H.; Hedges, L. (Ed); — — (Rev): Review of "The handbook of research synthesis". 95 Technmcs 37 242-242

Cothern, C. R.; Ross, N. P. (Ed); — — (Rev): Review of "Environmental statistics, assessment, and forecasting". 95 Technmcs 37 128-128

Cryer, J.; Miller, R. (Auth); — — (Rev): Review of "Statistics for business: Data analysis and modelling". 95 Technmcs 37 239-240

Deutsch, C.; Journel, A. (Auth); — — (Rev): Review of "GSLIB: Geostatistical software library and user's guide". 95 Technmcs 37 126-126

Dodson, Bryan (Auth); — — (Rev): Review of "Weibull analysis". 95 Technmcs 37 347-348

Gibbons, Robert (Auth); — — (Rev): Review of "Statistical methods for ground-water monitoring". 95 Technmcs 37 231-232

Hand, D.; Daly, F.; Lunn, A.; McConway, K.; Ostrowski, E. (Ed); — — (Rev): Review of "A handbook of small data sets". 95 Technmcs 37 242-242

Hand, D. J. (Ed); — — (Rev): Review of "Artificial intelligence frontiers in statistics". 95 Technmcs 37 127-127

Hand, D. J. (Ed); — — (Rev): Review of "AI and computer power: The impact of statistics". 95 Technmcs 37 241-242

Hayes, Bob (Auth); — — (Rev): Review of "Measuring customer satisfaction". 95 Technmcs 37 128-129

Hildrum, K.; Isaksson, T.; Naes, T.; Tandberg, A. (Ed); — — (Rev): Review of "Near infra-red spectroscopy: Bridging the gap between data analysis and NIR applications". 95 Technmcs 37 126-127

Joiner, Brian L. (Auth); — — (Rev): Review of "Fourth generation management: The new business consciousness". 95 Technmcs 37 113-113

Kowaka, M. (Ed); — — (Rev): Review of "Introduction to life prediction of industrial plant materials". 95 Technmcs 37 457-457

Mann, P. (Auth); — — (Rev): Review of "Statistics for business and economics". 95 Technmcs 37 239-240

McReynolds, Peter (Auth); — — (Rev): Review of "SPC made easy". 95 Technmcs 37 462-463

Morgan, M.; Henrion, M. (Auth); — — (Rev): Review of "Uncertainty". 95 Technmcs 37 237-238

Pitard, Francis (Auth); — — (Rev): Review of "Pierre Gy's sampling theory and sampling practice (Second edition)". 95 Technmcs 37 125-126

Rao, T. S. (Ed); — — (Rev): Review of "Developments in time series analysis". 95 Technmcs 37 127-128

Rauwendaal, Chris (Auth); — — (Rev): Review of "SPC in extrusion". 95 Technmcs 37 241-241

Rice, John (Auth); — — (Rev): Review of "Mathematical statistics and data analysis (Second edition)". 95 Technmcs 37 127-127

Scheaffer, R.; McClave, J. (Auth); — — (Rev): Review of "Probability and statistics for engineers". 95 Technmcs 37 239-239

Scheiner, S.; Gurevitch, J. (Ed); — — (Rev): Review of "Design and analysis of ecological experiments". 95 Technmcs 37 128-128

Selvin, Steve (Auth); — — (Rev): Review of "Statistical analysis of epidemiologic data". 95 Technmcs 37 240-241

Selvin, Steve (Auth); — — (Rev): Review of "Practical biostatistical methods". 95 Technmcs 37 241-241

Shiffler, R.; Adams, A. (Auth); — — (Rev): Review of "Introductory business statistics with computer applications". 95 Technmcs 37 239-240

Suter, Glenn (Ed); — — (Rev): Review of "Ecological risk assessment". 95 Technmcs 37 240-240

Ziegler, Andreas; Bachleitner, Reinhard; Arminger, Gerhard: Pseudo-maximum-likelihood-estimation and regression diagnostics for count data: Suicide by tourism? (German). 95 AllgeStA 79 170-195

Zieliński, R.: Estimating median and other quantiles in nonparametric models. 95 ApplMath 23 363-370

Zieliński, Ryszard: One-sample one-sided student t-test under contaminants. 95 MaOpfStS 27 143-150

Zimmerman, David J.: Levine, Phillip B.; — —: The benefit of additional high-school math and science classes for young men and women. 95 JBES 13/2 137-149

Zimmerman, Tamara Sue; Evans, Thomas D.; Tiller, Richard B.: State unemployment rate time series models. 94 ASASrvyRsr 1077-1082

Zimmerman, William B.; Chatwin, P. C.: Statistical fluctuations due to microscale mixing in a diffusion layer. 95 EnvrMtrc 6 665-675

Zimmermann, Helmuth: Rahlfs, Volker W.; — —: Nonparametric methods in standard statistics packages: The case of the Friedman analysis. 95 CmpStDA 20 101-110

Zimmermann, Klaus F.: Veall, Michael R.; — —: Comment on "Goodness-of-fit measures in binary choice models". 95 EcomtrRv 14 117-120

Zinde-Walsh, Victoria: Galbraith, John W.; — —: Transforming the error-components model for estimation with general ARMA disturbances. 95 JEconmtx 66 349-355

Davidson, Russell; MacKinnon, James G. (Auth); — — (Rev): Review of "Estimation and inference in econometrics". 95 EconTher 11 631-635

Zinn, Joel: Cuzick, Jack; Giné, Evarist; — —: Laws of large numbers for quadratic forms, maxima of products and truncated sums of i.i.d. random variables. 95 AnlsProb 23 292-333

Zio, E.: Biasing the transition probabilities in direct Monte Carlo. 95 RelEngSS 47 59-63

Zioutas, G.: Becker, G.; Camarinopoulos, L.; — —: An inhomogeneous state graph model and application for a phased mission and tolerable downtime problem. 95 RelEngSS 49 51-57

Zipfel, Andreas: Tudor, David; — —: Opportunities for multicenter clinical trials in joint US-European drug development. 94 ASAProBiop 229-231

Zipursky, Lorne S.: Turner, Robert B.; — —: Quality tools help improve employee commitment. 95 ASQCTekCon 770-776

Zitikis, Ričardas: Bentkus, Vidmantas; Götze, Friedrich; — —: Lower estimates of the convergence rate for U-statistics. 94 AnlsProb 22 1707-1714

Zmyślony, R.: Gnot, S.; Trenkler, G.; — —: Nonnegative minimum biased quadratic estimation in the linear regression models. 95 JMultiAn 54 113-125

Zocchi, S. S.: Atkinson, A. C.; Demetrio, C. G. B.; — —: Optimum dose levels when males and females differ in response. 95 ApplStat 44 213-226

Zoefel, P.: — — (Auth); Richter, W.-D. (Rev): Review of "Univariate analyses of variance" (German). 95 MaOpfStS 26 179-180

Zoglat, Abdelhak: Dabrowski, André Robert; — —: Strong invariance principles for triangular arrays of weakly dependent random variables. 95 CandJSt 23 299-310

Zoia, M. Grazia: Faliva, Mario; — —: Detecting and testing causality in linear econometric models. 94 JItalStS 3 61-76

Zolotarev, V. M.: On representation of densities of stable laws by special function. 94 ThProbAp 39 354-362

Kalashnikov, V. V.; — — (Ed); Weron, A. (Rev): Review of "Stability problems for stochastic models". 95 Metrika 42 272-273

Zopounidis, C.; Despotis, D. K.; Stavropoulou, E.: Multiattribute evaluation of Greek banking performance. 95 AplStMDA 11 97-107

Zopounidis, Constantin: Janssen, Jacques; Skiadas, Christos H.; — — (Ed): Advances in stochastic modelling and data analysis. 95 Kluwer:MA:Hllnd 440p

Zoppé, Alice: Principal points of univariate continuous distributions. 95 StatComp 5 127-132

Zörnig, Peter; Altmann, Gabriel: Unified representation of Zipf distributions. 95 CmpStDA 19 461-473

Zou, Guohua; Zhou, Guangya: Estimated loss for Stein's dominating estimator. 95 JMathChn 14 157-164

Zsigri, G.: Rohatgi, Vijay K.; Szekely, Gabor; — —: On aging life distribution classes: Further probability properties and tests. 94 JStatRsr 28 111-122

Zubkov, A. M.: Barbour, A. D.; Holst, Lars; Janson, Svante (Auth); — — (Rev): Review of "Poisson approximation". 94 ThProbAp 39 720-721

Zubovic, Yvonne: Wu, Cheng-Sheng Peter; — —: A large-scale Monte Carlo study of the Buckley-James estimator with censored data. 95 JStCmpSm 51 97-119

Zucchini, W.: MacDonald, I. L.; — —: Hidden Markov and other models for discrete-valued time series. 95 ChapmnHall:Lnd 180p

Zucker, David M.; Lakatos, Edward; Webber, Larry S.; Murray, David M.; McKinlay, Sonja M.; Feldman, Henry A.; Kelder, Steve H.; Nader, Philip R.: Statistical design of the child and adolescent trial for cardiovascular health (CATCH): Implications of cluster randomization. 95 ConClnTr 16 96-118

— —; Zerbe, Gary O.; Wu, Margaret C.: Inference for the association between coefficients in a multivariate growth curve model. 95 Biomtrcs 51 413-424

Zukang, Zheng: Two methods of estimating the mean survival time from censored samples. 95 SankhyaA 57 126-136

Zünd, Richard E.: Meier, Peter C.; — — (Auth); Hillyer, Martin (Rev): Review of "Statistical methods in analytic chemistry". 95 Technmcs 37 113–114

Zunica, L.: Romero, R.; Ferrer, A.; Capilla, C.; — —; Balasch, S.; Serra, V.; Alcover, R.: Teaching statistics to engineers: An innovative pedagogical experience. 95 JStEd 3/1 33kb (jse/v3n1/romero)

Zupan, J.: Lozano, J.; Novič, M.; Rius, F. X.; — —: Modelling metabolic energy by neural networks. 95 Chemolab 28 61–72

Zupan, Jure; Novič, Marjana; Gasteiger, Johann: Neural networks with counter-propagation learning strategy used for modelling. 95 Chemolab 27 175–187

Zuyev, Sergei: Baccelli, François; Klein, Maurice; — —: Perturbation analysis of functionals of random measures. 95 AdvAppPr 27 306–325

Zvárová, J.: Hájek, P.; Sochorová, A.; — —: GUHA for personal computers. 95 CmpStDA 19 149–153

Zwarts, M. A.: van der Linden, W. J.; — —: Robustness of judgments in evaluation research. 95 TOnderwR 20 13–27

Zwick, Rebecca: Test and item bias: An overview. 94 ASAProSoSt 9–16 (Disc: p27-29)

Zwiers, Francis W.; von Storch, Hans: Taking serial correlation into account in tests of the mean. 95 JClimate 8 336–351

Zwinderman, A. H.; van Houwelingen, J. C.; Schweitzer, D.: Prediction of the next hypercalcemia free period: Application of random effect models with selection on first event. 95 StNeerla 49 310–323

Zwirner, Walter W.: Mathematical programming in statistical analysis. 94 CalcutSt 44 229–237

SUBJECT INDEX

3. SUBJECT INDEX

Empirical Bayes estimators for the Burr type XII	accelerated life testing model based on type-2 censored data • Ali Mousa, M. A. M.	95 JStCmpSm 52 95-103
Optimum plans for step-stress	accelerated life testing (Chinese) • Cheng, Yiming	95 ChJApPSt 10 52-61
Weibull	accelerated life testing with unreported early failures • Glaser, R. E.	95 IEEEReli 44 31-36
Mukhopadhyay, Nitis A note on	accelerated sequential estimation of the mean of NEF-PVF distributions • Bose, Arup;	95 AnInStMa 47 99-104
Hamdy, H. I.; Son, M. S.	Accelerated sequential procedure to estimate the mean of unknown distribution •	94 ASASrvyRsr 1173-1178
mation via replicated piecewise stopping times and	accelerated stopping times in a class of two-parameter ... • Bose et al. ... interval esti-	95 SqtlAnly 14 287-305
Two-sample rank tests for	acceleration in cure models • Lee, Jae Won	95 StatMed 14 2111-2118
A quasi-Newton	acceleration of the EM algorithm • Lange, Kenneth	95 StSinica 5 1-18
Samejima, Fumiko	Acceleration model in the heterogeneous case of the general graded response model •	95 Psymtrka 60 549-572
	Acceptability profile	
	Estimating statistical hypotheses • Blyth, Colin R.; Staudte, Robert G.	95 StPrLet 23 45-52
chain sampling plans (ChSP-4 (c_1, c_2)) indexed by	acceptable quality level and limiting quality level • Raju, C. et al. Minimum risks	95 JAppStat 22 389-426
	Acceptance sampling	
S. D.; Soundararajan, V.	Construction and evaluation of matched quick switching systems • Arumainayagam,	95 JAppStat 22 245-251
mum allowable percentage defective • Suresh et al.	Construction and selection of tightened-normal-tightened plans indexed by maxi-	94 JAppStat 21 589-595
Balasooriya, Uditha	Failure-censored reliability sampling plans for the exponential distribution •	95 JStCmpSm 52 337-349
and LQL • Kuralmani, V.; Govindaraju, K.	Modified tables for the selection of double sampling attribute plan indexed by AQL	95 CommStA 24 1897-1921
Haberle, Katharine Rowse	Probability distributions for the Deming cost models • Kaminsky, Frank C.;	95 JQualTek 27 355-362
ChSP-4 (c_1, c_r) - Part 4 • Raju, C.; Murthy, M. N. N.	Procedures and tables for the construction and selection of chain sampling plans	95 JAppStat 22 261-271
Arumainayagam, S. Devaraj; Soundararajan, V.	Quick switching double sampling system indexed by the crossover point •	95 CommStB 24 765-773
Lee, Eun-joo	Three-level skip-lot sampling plan split by two stages (Korean) • Choi, Byoung-Chul;	95 KorApS 8/2 55-64
	Two-stage sampling for bounding a mean • Atwood, Corwin L.; Bryan, Mark F.	95 EnvrMtrc 6 73-83
a two-sided acceptance region for variables	acceptance sampling and its comparison with other methods • Lei et al. ... of constructing	94 ASAProPEng 115-119
Dissolution test	acceptance sampling plans • Tsong, Yi; Hammerstrom, Thomas; Lin, Karl; Ong, Ting Eng	95 JBiophSt 5 171-183
Guttman, Irwin	Acceptance sampling in presence of misclassification - A Bayesian approach • Sinha, S. K.;	94 JIndnStA 32 83-93
case • Tapiero, Charles S.	Acceptance sampling in a producer-supplier conflicting environment: Risk neutral	95 AplStMDA 11 3-12
of methods for univariate and multivariate	acceptance sampling by variables • Hamilton, David C.; Lesperance, Mary L. A comparison	95 Technmcs 37 329-339
Validating	access to care • Berk, Marc L.; Edwards, W. Sherman; Cantor, Joel C.	94 ASAProSoSt 240-244
Enhancing researcher	access to economic microdata at the Bureau of Labor Statistics • de Wolf, Virginia A.	94 ASAProGovt 164-169
Conflicts between the needs for	access to statistical information and demands for confidentiality • Fienberg, Stephen E.	94 JOffStat 10 115-132
of early exposure results with the MACCS reactor	accident consequence model • Helton et al. ... of an uncertainty and sensitivity analysis	95 RelEngSS 48 129-148
	Accident data	
Yi, Woojune	A Bayesian method for analyzing dependencies in precursor data • Bier, Vicki M.;	95 IntJFor 11 25-41
On the application of logistic regression on ski	accident data • Traweger, Christian M.	94 ASAProSpts 12-15
Models for automobile insurance	accidents and losses • Cooil, Bruce; Devinney, Timothy; McCormick, Douglas	94 ASAProBuEc 217-222
Böhning, Danmar Traffic	accident mapping in Bangkok metropolis: A case study • Ayuthya, Rampai Suksawasdi Na;	95 StatMed 14 2445-2458
"I always wanted to be a statistician" or The	accidental statistician • Legler, Julie	95 Stats /12 10-13
Prediction and	accommodation in econometric modelling • Dharmapala, Dhammika; McAleer, Michael	95 EnvrMtrc 6 551-556
Statistics for	accountants • Letchford, S.	95 ChapmnHall:Lnd 288p
tion of warming effects due to industrialization: An	accumulated intervention model with an application in Pohang, Korea • Ryoo et al. Detec-	95 JApMetrl 34 1466-1471
A note on the	accuracy of an approximate interval for the binomial parameter • Huwang, Longcheen	95 StPrLet 24 177-180
The asymptotic	accuracy of the bootstrap of U-quantiles • Arcones, Miguel A.	95 AnlsStat 23 1802-1822
Armstrong, J. Scott Beyond	accuracy: Comparison of criteria used to select forecasting methods • Yokum, J. Thomas;	95 IntJFor 11 591-597
reasoning in belief networks: Statistical validity,	accuracy, and determination of sample size (Disc: p134-135) • Ji, X. et al. Approximate	94 ASAProBays 128-133
The relative	accuracy of direct and indirect estimates of state poverty rates • Schirm, Allen L.	94 ASASrvyRsr 1071-1076
Self-employed nonfilers: The	accuracy of "enforcement-secured" returns (Disc: p212-214) • Rosage, Laura R.	94 ASAProGovt 206-211
	Accuracy of indices of industrial production in developing countries • Kmietowicz, Z. W.	95 Ststcian 44 295-307
Does reliable information improve the	accuracy of judgemental forecasts? • Remus, William; O'Connor, Marcus; Griggs, Kenneth	95 IntJFor 11 285-293
the reliability of indicators of learning curve model	accuracy • Moses, O. Douglas On	95 JCostAnl /1 63-101
Hilko Correction to "Comparing the predictive	accuracy of models using a simple randomization test" (V25 p313-323) • van der Voet,	95 Chemolab 28 315-315
• Obuchowski et al. Hypothesis testing of diagnostic	accuracy for multiple readers and multiple tests: An ANOVA approach with dependent ...	95 CommStB 24 285-308
al. Asymptotic behaviors of some measures of	accuracy in nonparametric curve estimation with dependent observations • Kim, T. Y. et	95 JMultiAn 53 67-93
Rudolph et al. Optimal experimental design and	accuracy of parameter estimation for nonlinear regression models used in long-term ... •	95 BiomtrcJ 37 183-190
screening • Tu et al. On the informativeness and	accuracy of pooled testing in estimating prevalence of a rare disease: Application to HIV	95 Biomtrka 82 287-297
in proportional hazards regression analysis II.	Accuracy and precision of regression estimates • Peduzzi et al. ... per independent variable	95 JClEpid 48 1503-1510
Bootstrap based analysis of neural network's	accuracy and prediction of clinical outcome • Katz, Alex S.; Katz, Simon; Lowe, Norma	94 ASAProBiop 177-181
Comparing predictive	accuracy • Diebold, Francis X.; Mariano, Roberto S.	95 JBES 13 253-263
in some binary regression models with prescribed	accuracy • Chang, Yuan-chin Ivan Estimation	95 JStPlInf 44 313-325
in regression with prescribed proportional	accuracy • Martinsek, Adam T. Estimating a slope parameter	95 StatDecn 13 363-377
data • Mocan, H. Naci; Azad, Sam	Accuracy and rationality of state General Fund Revenue forecasts: Evidence from panel	95 IntJFor 11 417-427
Assessing the historical	accuracy of regional employment forecasts • Fullerton, Thomas M., Jr; West, Carol Taylor	94 ASAProBuEc 333-338
likelihood tree • Hasegawa, M.; Kishino, H.	Accuracies of the simple methods for estimating the bootstrap probability of a maximum-	94 MolBioEv 11 142-145
On the	accuracy of the statistical distributions in GAUSS • Knüsel, Leo	95 CmpStDA 20 699-702
Spyros The impact of empirical	accuracy studies on time series analysis and forecasting • Fildes, Robert; Makridakis,	95 IntStRvw 63 289-308
A note on the	accuracy of two saddlepoint tail probability approximations • Booth, James	94 ASAProPEng 56-58
Estimating diagnostic test	accuracy using a "fuzzy gold standard" • Phelps, Charles E.; Hutson, Alan	95 MedDscnM 15 44-57
	Acid	
	• see also amino -, deoxyribonucleic -, DNA, ribonucleic -, RNA	
troscopic titrations of the copper(II)-polyinosinic	acid system • Tauler et al. ... procedure to the simultaneous analysis of several spec-	95 Chemolab 27 163-174
terminating time: Application to transfusion data for	acquired immune deficiency syndrome • Tu ... time, and censored and truncated	95 ApplStat 44 3-16
Sensitivity analysis, information, decision • Day et al.	Acquired immune deficiency syndrome predictions for England and Wales (1992-97):	95 JRSS-A 158 505-524
• Lin, Shang P.; Tang, Dei-In A hierarchical	active constraints search algorithm for optimal scaling of ordered categorical responses	95 JStCmpSm 53 197-209
Conventional null hypothesis testing in	active control equivalence studies • Ng, Tie-Hua	95 ConClnTr 16 356-358
of nine numerical techniques for detecting	active factors in unreplicated experimental designs • Benski, Claudio Applicability	94 ASAProStCp 214-217
keeler) • Keeler et al. Using small groups to promote	active learning in the introductory statistics course: A report from the field (jse/v3n2/	95 JStEd 3/2 28kb
introductory statistics courses from a conceptual,	active learning viewpoint (jse/v3n3/steinhorst) • Steinhorst et al. Developing material for	95 JStEd 3/3 33kb
	Actuarial equivalence • Wolthuis, Henk	94 InsrneME 15 163-179
of the principle underlying the conventional	actuarial estimator of q_x • Puzey, Anthony S. A reappraisal	95 InsrneME 17 125-132
Pitacco, Ermanno	Actuarial models for pricing disability benefits: Towards a unifying approach •	95 InsrneME 16 39-62
Allocation of solvency cost in group annuities:	Actuarial principles and cooperative game theory • Alegre, Antoni; Claramunt, M. Mercé	95 InsrneME 17 19-34
	Actuarial theory	
	On the preservation of some orderings of risks under convolution • Pellerey, Franco	95 InsrneME 16 23-30
	Recursions for the individual model • Dhaene, Jan; Vandebroek, Martina	95 InsrneME 16 31-38
Review of "Practical risk theory for	actuaries" • Daykin, C. D.; Pentikäinen, T.; Pesonen, E. (Auth); Hardy, M. R. (Rev)	95 JAppStat 22 547-548

to detect statistically significant departures from	additivity in chemical mixtures • Gennings et al. ... data from individual components	95 Biomtrcs 51 1264-1277
ation injury • Komornik, J.; Mikulecky, M.; Beno, M.	Additivity test for composition of binomial effects in chromosome aberrations after radi-	95 StatMed 14 1941-1947
	Additivity tests for nonlinear autoregression • Chen, Rong; Liu, Jun S.; Tsay, Ruey S.	95 Biomtrka 82 369-383
A note on using an unbiased weight matrix in the	ADF test statistic • Chan, Wai; Yung, Yiu-Fai; Bentler, Peter M.	95 MultBehR 30 453-459
The	adjoint and the composition of random operators on a Hilbert space • Thang, Dang Hung	95 Stochatx 54 53-73
normal probability density within the class of self-	adjoint (self-reciprocal) densities" (V22 p651-654) • Rossberg ... of the standard	95 MaOpfStS 26 387-387
	Adjusted individual control charts for short runs • Haugh, Larry D.; Pond, Andrew D.	95 ASQCTekCon 1117-1126
M. Curvature	adjusted optimal nonlinear designs for correlated responses • Hughes-Oliver, Jacqueline	94 ASAProBuEc 111-116
Barndorff-Nielsen, O. E. Stable and invariant	adjusted profile likelihood and directed likelihood for curved exponential models •	95 Biomtrka 82 489-499
Gagnon, D. R.; Ramaswamy, R. et al. Age-	adjusted survival curves with application in the Framingham study • Cupples, L. A.;	95 StatMed 14 1731-1744
cancer • Cole, B. F.; Gelber, R. D. et al. A quality-	adjusted survival meta-analysis of adjuvant chemotherapy for premenopausal breast	95 StatMed 14 1771-1784
A. Comparing treatments using quality-	adjusted survival: The Q-TWiST method • Gelber, R. D.; Cole, B. F.; Gelber, S.; Goldhirsch,	95 AmerStat 49 161-169
Aron Cost-effectiveness analysis using quality-	adjusted survival • Cole, Bernard F.; Gelber, Richard D.; Anderson, Keaven M.; Goldhirsch,	94 ASAProBiop 9-18
Ullah, Aman The coefficient of determination and its	adjusted version in linear regression models • Srivastava, Anil K.; Srivastava, Virendra K.;	95 EcomtrRv 14 229-240
	Adjusting for attrition in event-history analysis (Disc: p440-442) • Hill, Daniel H.	94 ASASrvyRsr 417-421
	Adjusting Moran's I for population density • Oden, Neal	95 StatMed 14 17-26
p440-442) • Rizzo, Lou; Kalton, Graham; Brick, Mike	Adjusting for panel nonresponse in the Survey of Income and Program Participation (Disc:	94 ASASrvyRsr 422-427
Reagan, Patricia B.	Adjusting pension fund membership for "don't know" responses • Stasny, Elizabeth A.;	94 ASAProBuEc 298-303
The effects of seasonally	adjusting a periodic autoregressive process • Franses, Philip Hans	95 CmpStDA 19 683-704
Reiser, B. Dynamic treatment allocation	adjusting for prognostic factors for more than two treatments • Dror, S.; Faraggi, D.;	95 Biomtrcs 51 1338-1343
obtained from a food frequency ... • Landin et al.	Adjusting for time trends when estimating the relationship between dietary intake	95 Biomtrcs 51 169-181
obtained from a food ... • Landin et al. Correction to	"Adjusting for time trends when estimating the relationship between dietary intake	95 Biomtrcs 51 1196-1196
	Adjustment	
	• see also Bartlett -, covariate -, dose -, nonresponse -, seasonal -	
spanning trees for tree metrics: Abridgements and	adjustments • Leclerc, Bruno Minimum	95 JClass 12 207-241
Estimating the	adjustment coefficient in an ARMA(p,q) risk model • Christ, Ralf; Steinebach, Josef	95 InsrneME 17 149-161
Ruin probabilities via local	adjustment coefficients • Asmussen, Søren; Nielsen, Hanne Mandrup	95 JAppProb 32 736-755
evidence • Aguirre-Torres, V. et al. The effect and	adjustment of complex surveys on chi-squared goodness-of-fit tests: Some Monte Carlo	94 ASASrvyRsr 602-607
proportional-integral control with constrained	adjustment • Box, George E. P.; Luceño, Alberto Discrete	95 Ststcian 44 479-495
Comparison of median survival times with	adjustment for covariates • Karrison, Theodore	95 StatMed 14 2537-2553
Comparison of Bartlett-type	adjustments for the efficient score statistic • Rao, C. Radhakrishna; Mukerjee, Rahul	95 JStPlInf 46 137-146
variable: Sample size and parameter estimate	adjustment • Tavaré, C. J.; Sobel, E. L. et al. Misclassification of a prognostic dichotomous	95 StatMed 14 1307-1314
Premium	adjustment by generalized adaptive exponential smoothing • Herkenrath, U.	94 InsrneME 15 203-217
the Canadian Labour Force ... • Singh et al. On the	adjustment of gross flow estimates for classification error with application to data from	95 JASA 90 478-488
The anchor and	adjustment heuristic in time-series forecasting • Lawrence, Michael; O'Connor, Marcus	95 JForcast 14 443-451
Statistical	adjustment of maize grain yield for sub-optimal plot stands • Kamidi, R. E.	95 ExpAgric 31 299-306
in X: Increasing power of the test by non-linear	adjustment for measurement error • Grove, J. S. et al. Regression with random error	94 ASAProEpid 143-146
model • Yeung, David Wing-Kay Employment	adjustments noise and the expected rate of inflation in a simple inflation-unemployment	95 StocAnAp 13 125-135
Alternative	adjustments where there are several levels of auxiliary information • Dupont, F.	95 SurvyM 21 125-135
P-value	adjustments for subgroup analyses of binary variables • Bristol, David R.	94 ASAProBiop 120-125
Score	adjustment for subjectively rated responses (Disc: p134-135) • Longford, N. T.	94 ASAProBays 123-127
indexes of the cost of shelter using hedonic quality	adjustment techniques • Moulton, Brent R. Interarea	95 JEconmtx 68 181-204
multiple testing. Examples and methods for p-value	adjustment" • Westfall, P. H.; Young, S. S. (Auth) et al. Review of "Resampling-based	95 CmpStDA 20 235-236
porting rates of adverse events between drugs with	adjustment for year of marketing and secular trends in total ... • Tsong Comparing re-	95 JBiophSt 5 95-114
A quality-adjusted survival meta-analysis of	adjuvant chemotherapy for premenopausal breast cancer • Cole, B. F.; Gelber, R. D. et al.	95 StatMed 14 1771-1784
Role of	administrative bodies in environmental studies • Ray, S. N.; Pandey, R. N.	95 SankhyaB 57 237-251
liner service (Disc: p212-214) • Himelein The use of	administrative data in the development of a longitudinal price index for inbound ocean	94 ASAProGovt 195-200
M.; Huang, Hertz Use of	administrative data in SIPP longitudinal estimation (Disc: p680-682) • Dorinski, Suzanne	94 ASASrvyRsr 656-661
The role of VA	administrative files in the National Survey of Veterans • Ahuja, Elizabeth M.	94 ASAProGovt 25-27
Comment on "The use of	administrative records in government statistics" • Swain, Larry	94 ASAProGovt 212-214
Konschnik, Carl A.	Administrative record matching for the 1992 economic censuses • Steel, Philip M.;	94 ASASrvyRsr 473-478
	Admissibility	
central chi-square distribution • Shao, P. Y.-S. et al.	Improving on the positive part of the UMVUE of a noncentrality parameter of a non-	95 JMultiAn 53 52-66
distribution • Chou, Jine-phone	Admissibility of conjugate Bayes estimators for the mean of a negative binomial	95 StatDecn 13 301-306
objective fixed-cost sample designs using an	admissibility criterion • Malec, Donald Selecting multiple-	95 JStPlInf 48 229-240
incomplete ellipsoidal restrictions • Lu, Chang-yu	Admissibility of inhomogeneous linear estimators in linear models with respect to	95 CommStA 24 1737-1742
the alternative • Andrews, D. W. K.; Ploberger, W.	Admissibility of the likelihood ratio test when a nuisance parameter is present only under	95 AnlsStat 23 1609-1629
(Chinese) • Qing, Hong General	admissibility of linear estimates of regression coefficients in growth curve model	95 ChJApPSt 10 265-271
models (Chinese) • Sun, Liuquan	Admissibility of linear estimation of regression coefficients in common growth curve	95 AMaAPknC 17 628-630
ellipsoidal restrictions • Lu, Changyu; Li, Weixin	Admissibility of linear estimator in linear models with respect to an incomplete	95 AMaPeknE 37 289-295
Govindarajulu, Z.	Admissibility and minimaxity of the UMVU estimator of $P\{X < Y\}$ • Yu, Qiqing;	95 AnlsStat 23 598-607
Nematollahi, N. On the	admissibility of ordered Poisson parameter under the entropy loss function • Parsian, A.;	95 CommStA 24 2451-2467
A note on	admissibility when precision is unbounded • Anderson, Charles; Pal, Nabendu	95 AnlsStat 23 593-597
	Admissibility of prediction intervals • Takada, Yoshikazu	95 AnInStMa 47 119-128
(Chinese) • Xie, Minyu	Admissibility of quadratic form estimate of variance matrix under some optimal criterion	95 AMaAPknC 17 149-153
	Admissibility: Survey of a concept in progress • Rukhin, Andrew L.	95 IntStRvw 63 95-115
Xie, Yumin	Admissible estimates in the important class of estimates of the covariance matrix •	95 ChinaAMB 15 435-442
De, Moloy On a class of	admissible estimators of the common mean of two univariate normal populations •	95 CalcutSt 45 103-109
(Chinese) • Wu, Qiguang; He, Min Several notes on	admissible estimators of parameters and losses under truncated distribution families	95 JSysSMaS 14 168-176
trary quadratic risk function • Baksalary, J. K. et al.	Admissible linear estimation in the general Gauss-Markov model with respect to an arbi-	95 JStPlInf 44 341-347
(Chinese) • Zhang, Shuanglin	Admissible linear estimation in linear models with respect to a restricted parameter set	95 AMaAPknC 17 470-472
component models • Wang, X.; Zhan, J.; Chen, J. All	admissible linear estimators of regression coefficients and parameters in variance	95 AMaPeknE 37 653-662
arbitrary parameter subset • Hoffmann, Kurt All	admissible linear estimators of the regression parameter vector in the case of an	95 JStPlInf 48 371-377
An analysis of the relationship between parent and	adolescent marijuana use via generalized estimating equation ... • Duncan et al.	95 MultBehR 30 317-339
• Zucker, D. M. et al. Statistical design of the child and	adolescent trial for cardiovascular health (CATCH): Implications of cluster randomization	95 ConClnTr 16 96-118
modeling of plasticity in fluid intelligence of aged	adults • Raykov, Tenko Multivariate structural	95 MultBehR 30 255-287
utilization measures for Medicaid beneficiaries:	Adult and child reports • Eisenhower et al. ... comparison of eligibility and health care	94 ASASrvyRsr 713-718
• Lindberg, M. S. et al. Estimating nest site fidelity of	adult female black brant with multi-state modeling and geographic information systems	95 JAppStat 22 725-735
for multistage tumorigenesis in developing and	adult mice • Mao, Jian-Hua; Wheldon, Tom E. A stochastic model	95 MaBisci 129 95-110
(Rev) Review of	"Adventures in stochastic processes" (German) • Resnick, Sidney I. (Auth); Matthes, K.	95 BiomtrcJ 37 272-272
Determination of the no-observed-	adverse-effect level by the AIC • Yanagawa, Takashi; Kikuchi, Yasuki	95 SankhyaB 57 285-297
in total ... • Tsong Comparing reporting rates of	adverse events between drugs with adjustment for year of marketing and secular trends	95 JBiophSt 5 95-114
evaluation of laboratory data to identify potential	adverse events • Chakravarty, Alaka G.; Harkins, Ralph D. Some approaches to the	94 ASAProBiop 73-76
• Chintagunta et al. A two-period repeated game	advertising investment model for oligopolistic markets with an application to the beer ...	95 DecisnSc 26 531-559

tial impact of marginal distributions on measures of	agreement • Agresti, Alan; Ghosh, Atalanta; Bini, Matilde Raking kappa: Describing poten-	95 BiomtrcJ 37 811-820
and marginal agreement: Two families of	agreement measures • Cook, R. J.; Farewell, V. T. Conditional inference for subject-specific	95 CandJSt 23 333-344
	Agreement models and probabilities for multiraters • Yu, Xujie	94 ASAProBiop 279-284
agreement studies: The case of nominal scale	agreement • Graham, Patrick Modelling covariate effects in observer	95 StatMed 14 299-310
agreement studies: The case of nominal scale	agreement" (V14 p299-310) • Guggenmoos-Holzmann ... covariate effects in observer	95 StatMed 14 2285-2286
server agreement studies: The case of nominal scale	agreement'" • Graham ... comment on "Comment on 'Modelling covariate effects in ob-	95 StatMed 14 2286-2288
Measuring the	agreement between sequences (Disc: p232-243) • Dijkstra, Wil; Taris, Toon	95 SocMethR 24 214-231
Modelling covariate effects in observer	agreement studies: The case of nominal scale agreement • Graham, Patrick	95 StatMed 14 299-310
on "Modelling covariate effects in observer	agreement studies: The case of nominal scale ... • Guggenmoos-Holzmann Comment	95 StatMed 14 2285-2286
on 'Modelling covariate effects in observer	agreement studies: The case of nominal scale agreement'" • Graham ... on "Comment	95 StatMed 14 2286-2288
An algorithm to find	agreement subtrees • Kubicka, Ewa; Kubicki, Grzegorz; McMorris, F. R.	95 JClass 12 91-99
ditional inference for subject-specific and marginal	agreement: Two families of agreement measures • Cook, Richard J.; Farewell, Vern T. Con-	95 CandJSt 23 333-344
Use of geostatistical techniques in the design of an	agricultural field experiment • Dulaney, Wayne P.; Lengnick, Laura L.; Hart, Galen F.	94 ASASrvyRsr 183-187
design and analysis. A practical approach for	agricultural and natural sciences • Hoshmand, A. Reza Experimental research	94 CRCPress:FL 416p
	Agricultural statistics on the environment • Clark, Cynthia Z. F.	94 ASAProEnvr 127-132
The evolution and development of	agricultural statistics at the United States Department of Agriculture • Vogel, Frederic A.	95 JOffStat 11 161-180
Measuring and reporting nonresponse rates in	agricultural surveys • Roy, Laurent	94 ASASrvyRsr 1156-1161
	Agriculture	
	Regional optimization of fertilization using a hierarchical linear model • Wallach, D.	95 Biomtrcs 51 338-346
capture methodology during the 1992 Census of	Agriculture • Ramos et al. Results from a feasibility test of an alternative automated data	94 ASASrvyRsr 769-774
Canada Farm Register with the Canadian Census of	Agriculture (Disc: p212-214) • Thomas, Bradley W. Updating the Statistics	94 ASAProGovt 189-194
Vacca, E. Ann 1992 Census of	Agriculture coverage evaluation estimation • Lewis, Paul J.; Wolfgang, Glenn S.;	94 ASASrvyRsr 1019-1024
statistics at the United States Department of	Agriculture • Vogel, Frederic A. The evolution and development of agricultural	95 JOffStat 11 161-180
Cognitive research for the 1997 Census of	Agriculture report form • Wolfgang, Glenn S.; Lewis, Paul J.; Vacca, E. Ann	94 ASASrvyRsr 503-508
Modelling: Keeping plant science and	agriculture students interested • Johnson, I. R.	95 EnvrMtrc 6 517-522
Review of	"AI and computer power: The impact of statistics" • Hand, D. J. (Ed); Ziegel, Eric R. (Rev)	95 Technmcs 37 241-242
of "Artificial intelligence frontiers in statistics:	AI and statistics III" • Hand, D. J. (Ed); Spiegelhalter, David (Rev) Review	95 JAppStat 22 186-186
of the no-observed-adverse-effect level by the	AIC • Yanagawa, Takashi; Kikuchi, Yasuki Determination	95 SankhyaB 57 285-297
Glosup, Jeffrey G.; Axelrod, Michael C. Use of the	AIC with the EM algorithm: A demonstration of a probability model selection technique •	94 ASAProPEng 120-125
	AIDS	
rates • Liao, Jiangang; Brookmeyer, Ron	An empirical Bayes approach to smoothing in backcalculation of HIV infection	95 Biomtrcs 51 579-588
tions to epidemic models • Gani, J.; Yakowitz, Sid	Error bounds for deterministic approximations to Markov processes, with applica-	95 JAppProb 32 1063-1076
	Extending the elements of tree-structured regression • Segal, Mark R.	95 StMedRes 4 219-236
data • Shih, Joanna H.; Louis, Thomas A.	Inferences on the association parameter in copula models for bivariate survival	95 Biomtrcs 51 1384-1399
doubly-censored and truncated data • Heisey et al.	Modelling age-specific survival in nesting studies, using a general approach for	95 Biomtrcs 51 51-60
Li, Gang	Nonparametric likelihood ratio estimation of probabilities for truncated data •	95 JASA 90 997-1003
for quantiles • Satten, Glen A.	Upper and lower bound distributions that give simultaneous confidence intervals	95 JASA 90 747-752
	• see also acquired immune deficiency syndrome	
Comment on "Surrogate markers in	AIDS and cancer trials" (V13 p1423-1435) • Day, N. E.; Duffy, S. W.	95 StatMed 14 2619-2620
A stochastic model for drug resistance in	AIDS chemotherapy and the HIV incubation distribution • Tan, Wai Y.	95 StPrLet 25 289-299
curves in longitudinal data with application to an	AIDS clinical trial • Wang, Yongxiao; Taylor, Jeremy M. G. Inference for smooth	95 StatMed 14 1205-1218
Schoenfeld, David A. (Ed)	AIDS clinical trials: Guidelines for design and analysis • Finkelstein, Dianne M.;	95 Wiley:NY:UK 472p
interval-censored observations with application to	AIDS data • Frydman ... estimation in a three-state duration-dependent Markov model from	95 Biomtrcs 51 502-511
Computing back-calculation estimates of	AIDS epidemic • Ding, Ye	95 StatMed 14 1505-1512
Altmann, Michael (Rev) Review of "Modeling the	AIDS epidemic: Planning, policy, and prediction" • Kaplan, E. H.; Brandeau, M. L. (Ed);	95 Chance 8/4 36-39
Evaluating causality in the	AIDS epidemic: The role of mathematical models • Kaldor, J. M.	95 EnvrMtrc 6 535-540
Ghani, Azra (Rev) Review of	"AIDS epidemiology: A quantitative approach" • Brookmeyer, R.; Gail, M. H. (Auth);	95 JRSS-A 158 343-343
DeGruttola, Victor (Rev) Review of	"AIDS epidemiology: A quantitative approach" • Brookmeyer, Ron; Gail, Mitchell H. (Auth);	95 JASA 90 797-798
Farrington, C. P. (Rev) Review of	"AIDS epidemiology: A quantitative approach" • Brookmeyer, Ron; Gail, Mitchell H. (Auth);	95 AJEpidm 141 180-181
Commenges, D. A direct approach for correcting	AIDS incidence: Variance formula and comparison with other methods • Alioum, A.;	95 StatMed 14 27-38
series with large state spaces, with application to	AIDS modeling • Yakowitz, S. Computational methods for Markov	95 MaBisci 127 99-121
to estimating changes in the incubation period of	AIDS • Bacchetti, P.; Segal, M. R. Survival trees with time-dependent covariates: Application	95 LiDaAnal 1 35-47
Change points in the series of T4 counts prior to	AIDS • Kiuchi, A. S.; Hartigan, J. A.; Holford, T. R.; Rubinstein, P.; Stevens, C. E.	95 Biomtrcs 51 236-248
testing the efficacy of some control programmes for	AIDS • Biswas et al. ... branching process and birth-death and immigration models for	95 BiomtrcJ 37 235-249
deterministic procedures with the New York State	AIDS registry and a hospital discharge file (Disc: p687-689) • Muse et al. ... linkage using	95 StatMed 14 499-509
Antonino Non-	AIDS related mortality in the human immunodeficiency virus infection • Salvaggio,	95 ListyBio 32 23-33
doubly interval-censored data and its application to	AIDS studies • Sun, J. Empirical estimation of a distribution function with truncated and	95 Biomtrcs 51 1096-1104
to survival and CD4 counts in patients with	AIDS • Tsiatis et al. ... of survival to longitudinal data measured with error. Applications	95 JASA 90 27-37
Total quality education at the U.S.	Air Force Academy • Tong, Terrence G. L.	94 ASAProQlPr 295-299
Paliatsos, A. G. et al. Statistical evaluation of selected	air pollutants in Athens, Greece • Kambezidis, H. D.; Tulleken, R.; Amanatidis, G. T.;	95 EnvrMtrc 6 349-361
	Air pollution	
Switzer, Paul	Spatial interpolation errors for monitoring data • Høst, Gudmund; Omre, Henning;	95 JASA 90 853-861
sectional analysis of health effects associated with	air pollution • Duddek, C.; Le, N. D.; Zidek, J. V. et al. Multivariate imputation in cross-	95 EnvEcoSt 2 191-212
Examination of statistical procedures for evaluating	air quality models • Wallace, Dennis D.	94 ASAProEnvr 82-87
regional ozone data for operational evaluation of an	air quality model • Sampson, P. D.; Guttorp, P.; Meiring, W. Spatio-temporal analysis of	94 ASAProEnvr 46-55
bias of surface shelter temperatures using upper-	air and satellite data. Part I: Development of models ... • Epperson et al. ... the urban	95 JApMetrl 34 340-357
Dynamic three-dimensional display of U.S.	air traffic • Eddy, William F.; Que, Shingo	95 JCmpGrSt 4 261-280
Quinn, M. M.; Smith, T. J. Retrospective estimation of	airborne fiber exposure: Statistical aspects • Stone, R. A.; Marsh, G. M.; Owens, A. D.;	94 ASAProEpid 127-132
Spatial modelling of snow water equivalent using	airborne and ground-based snow data • Carroll, S. S.; Day, G. N.; Cressie, N.; Carroll, T. R.	95 EnvrMtrc 6 127-139
An application of locally weighted regression to	airborne mercury deposition around an incinerator site • Opsomer, J.-D.; Agras, J. et al.	95 EnvrMtrc 6 205-219
conditional probability fields for source locations of	airborne pollutants • Hopke, P. K.; Li, C. L. et al. The use of bootstrapping to estimate	95 Chemolab 30 69-79
Schulz, Holger Fractal exponents for the upper	airways of mammalian lungs • Krause, Ekkehard; Bandt, Christoph; Schulz, Anne;	95 CmpStDA 20 583-590
A unified approach to asymptotic equivalence of	Aitken and feasible Aitken instrumental variables estimators • Mandy, D. M. et al.	94 IntEconR 35 957-979
A conversation with Hirotugu	Akaike • Findley, David F.; Parzen, Emanuel	95 StatSci 10 104-117
	Akaike information criterion	
Huang, Chuen-Chuen Joyce; Dayton, C. Mitchell	Detecting patterns of bivariate mean vectors using model-selection criteria •	95 BJMaStPs 48 129-147
Takashi; Kikuchi, Yasuki	Determination of the no-observed-adverse-effect level by the AIC • Yanagawa,	95 SankhyaB 57 285-297
nomial models • Macchiavelli, Raúl; Arnold, Steven F.	Difficulties with the use of penalized likelihood criteria in antedependence and poly-	95 CommStA 24 501-522
quantitative IR spectroscopy • Brink, A. et al.	The joint problem of model structure determination and parameter estimation in	95 Chemolab 29 29-36
Davis, T. M. E.; Martin, R. B.	Models of parasite clearance following treatment of falciparum malaria in humans •	95 EnvrMtrc 6 529-534
Hurvich, Clifford M.; Tsai, Chih-Ling	Model selection for extended quasi-likelihood models in small samples •	95 Biomtrcs 51 1077-1084
regression • Hurvich, Clifford M.; Tsai, Chih-Ling	Relative rates of convergence for efficient model selection criteria in linear	95 Biomtrka 82 418-425
ities are heterogeneous • Burnham, K. P. et al.	Selection among open population capture-recapture models when capture probabil-	95 JAppStat 22 611-624
Adachi, Kohei	Statistical classification procedures for polygraph tests of guilty knowledge •	95 Behvmtrk 22 49-66

Simulation and chaotic behavior of	α-stable stochastic processes • Janicki, Aleksander; Weron, Aleksander	95 Dekker:NY 376p
Review of "Simulation and chaotic behaviour of	α-stable stochastic processes" • Janicki, A.; Weron, A. (Auth); Lee, P. M. (Rev)	95 JRSS-A 158 351-352
Fessler, Jeffrey A. Convergence in norm for	alternating expectation-maximization (EM) type algorithms • Hero, Alfred O.;	95 StSinica 5 41-54
variance matrix in elliptical distributions under local	alternatives • Purkayastha et al. Asymptotic distributions of some test criteria for the co-	95 JMultiAn 55 165-186
procedure for estimating the parameters of	Altham's multiplicative generalization of the binomial distribution • From ... least-squares	95 StPrLet 25 193-199
Public health role,	aluminium and Alzheimer's disease (Disc: p286-286) • Smith, Lesbia F.	95 EnvrMtrc 6 277-286
Letterman, Raymond D. Factors regulating residual	aluminium concentrations in treated waters (Disc: p305-309) • Driscoll, Charles T.;	95 EnvrMtrc 6 287-305
Modelling market fundamentals: A model of the	aluminium market • Gilbert, Christopher L.	95 JApEcmtx 10 385-410
	Aluminium and the risk for Alzheimer's disease • McLachlan, D. R. C.	95 EnvrMtrc 6 233-275
	Alzheimer's disease	
and politics • Bonham, G. H.	Health and environment: Public health decisions based on evidence, psychology	95 EnvrMtrc 6 311-318
Public health role, aluminium and	Alzheimer's disease (Disc: p286-286) • Smith, Lesbia F.	95 EnvrMtrc 6 277-286
cases in late-onset diseases: An application to	Alzheimer's disease • Tzourio, C.; Maccario, J. et al. Estimation of the proportion of genetic	95 StatMed 14 1353-1363
ial aggregation in probands studies: Application to	Alzheimer's disease • Commenges, D.; Jacqmin, H.; Letenneur, L. et al. Score test for famil-	95 Biomtrcs 51 542-551
Estimating the probability for major gene	Alzheimer disease • Farrer, Lindsay A.; Cupples, L. Adrienne	94 AJHuGene 54 374-383
of repeated measurements with dropouts among	Alzheimer's disease patients using summary measures and meta ... • Talwalker Analysis	94 ASAProBiop 515-520
Aluminium and the risk for	Alzheimer's disease • McLachlan, D. R. C.	95 EnvrMtrc 6 233-275
data from a high-dose trial of tacrine (Cognex) in	Alzheimer's patients • Smith, Fraser Mixed model analysis of incomplete longitudinal	94 ASAProBiop 301-306
On	amarts with continuous time • Dzhvarsheishvili, I. A.	94 ThProbAp 39 512-519
of census and vital registration data on older	Americans: 1970-1990 • Shrestha, Laura B.; Preston, Samuel H. Consistency	95 SurvyM 21 167-177
Mercado; Ramnath, Bala Interactive polling and	Americans' comfort level with technology • Frankovic, Kathleen A.; Arnedt, Cheryl	94 ASASrvyRsr 1316-1321
Synthetic replication of	American contingent claims when portfolios are constrained • Bardhan, Indrajit	95 StocProc 57 149-165
Results from a cognitive exploration of the 1993	American Housing Survey • Von Thurn, Dawn R.; Moore, Jeffrey C.	94 ASASrvyRsr 1210-1214
of "The bell curve: Intelligence and class structure in	American life" • Herrnstein, R. J.; Murray, C. (Auth); Devlin, B.; Fienberg, S. E. et al. Review	95 JASA 90 1483-1488
theory of pricing of options of both European and	American types. I. Discrete time. • Shiryaev, A. N.; Kabanov, Yu. M. et al. Toward the	94 ThProbAp 39 14-60
theory of pricing of options of both European and	American types. II. Continuous time. • Shiryaev, A. N.; Kabanov, Yu. M. et al. Toward the	94 ThProbAp 39 61-102
Cluster analysis dihedral angles of	amino acid residues in proteins • Sclove, Stanley L.; Sherman, Simon A.	94 ASAProBiop 399-404
Heterogeneity in the substitution process of	amino acid sites of proteins coded for by mitochondrial DNA • Reeves, Jaxk H.	92 JMolEvol 35 17-31
Yau, Sui K. Regression and	AMMI analyses of genotype $\times$ environment interactions: An empirical comparison •	95 AgronmyJ 87 121-126
additive main effects and multiplicative interaction (	AMMI) analysis • Piepho ... prediction (BLUP) for regional yield trials: A comparison to	94 ThApGene 89 647-654
Statistical Software Inc et al. Review of	"AMOS, EQS, and LISREL for Windows: A comparative review" • Arbuckle, J.; BMDP	95 StrEqMod 2 79-91
approximation of the time-varying frequency and	amplitude • Katkovnik, Vladimir Nonparametric local polynomial	95 CommStA 24 3001-3025
ability design: A proposal for evaluating epidural	analgesia during labor" (V13 p2269-2278) • Baker et al. Correction to "The paired avail-	95 StatMed 14 1841-1841
Pharmacodynamic analysis of	analgesic clinical trials using empirical methods • Liu, Chui Yu; Sambol, Nancy C.	94 ASAProBiop 171-176
	Analysis	
	• see also curve fitting, data –	
	Analysis of covariance	
parallel • Larholt, Kay M.; Sampson, Allan R.	Effects of heteroscedasticity upon certain analyses when regression lines are not	95 Biomtrcs 51 731-737
	Impact of stratification imbalance on probability of type I error • Kazempour, Kazem	95 AmerStat 49 170-174
Fleming, Thomas R.	Model misspecification in proportional hazards regression • Anderson, Garnet L.;	95 Biomtrka 82 527-541
	Statistical adjustment of maize grain yield for sub-optimal plot stands • Kamidi, R. E.	95 ExpAgric 31 299-306
B.	When is it worth measuring a covariate in a randomized clinical trial? • Allison, David	95 JCCPsych 63 339-343
Non-parametric	analysis of covariance • Young, Stuart G.; Bowman, Adrian W.	95 Biomtrcs 51 920-931
In defence of	analysis of covariance: A reply to Chambless and Roeback • Senn, Stephen	95 StatMed 14 2283-2285
of the generalized least-squares estimation in	analysis of covariance structures • Wang, S. J.; Lee, Sik-Yum A geometric approach	95 StPrLet 24 39-47
	Analysis of variance	
	Additivity tests for nonlinear autoregression • Chen, Rong; Liu, Jun S.; Tsay, Ruey S.	95 Biomtrka 82 369-383
model • Mosier, Michael C.; Graybill, Franklin A.	Approximate confidence intervals for contrasts in the balanced three factor mixed	95 BiomtrcJ 37 273-297
Aitkin, Murray	Comment on "The statistics of linear models: Back to basics" (V4 p221-234) •	95 StatComp 5 85-86
Rodriguez, R.; Tobias, R.; Wolfinger, R.	Comment on "The statistics of linear models: Back to basics" (V4 p221-234) •	95 StatComp 5 97-101
Searle, Shayle R.	Comment on "The statistics of linear models: Back to basics" (V4 p221-234) •	95 StatComp 5 103-107
	A comparative study of some rank tests for interaction • Mansouri, H.; Chang, G.-H.	95 CmpStDA 19 85-96
	Detecting and handling heteroscedasticity in yield trial data • Piepho, Hans-Peter	95 CommStB 24 243-274
	Dose response studies. II. Analysis and interpretation • Ruberg, Stephen J.	95 JBiophSt 5 15-42
aclass correlation for paired data • Lyles, R. H. et al.	Effects of model misspecification in the estimation of variance components and intr-	95 StatMed 14 1693-1706
• Minkkinen, Pentti	Estimation of variance components from the results of interlaboratory comparisons	95 Chemolab 29 263-270
Mathieu, J.-R.	Generalized linear models with a small scale parameter • Dorkenoo, K. M.;	95 MaOpfStS 27 107-113
Novak, Thomas P.	MANOVAMAP: Graphical representation of MANOVA in marketing research •	95 JMarRsr 32 357-374
	A measure to evaluate the closeness of Satterthwaite's approximation • Khuri, A. I.	95 BiomtrcJ 37 547-563
	An overview of variance component estimation • Searle, Shayle R.	95 Metrika 42 215-230
	Partitioning error spaces in linear models (error contrasts) • Rady, El-Houssainy A.	95 EgpStJ 39 51-64
	Practical analysis of factorial experiments in forestry • Stehman, S. V.; Meredith, M. P.	95 CanJForR 25 446-461
	Randomization tests to compare means with unequal variation • Manly, Bryan F. J.	95 SankhyaB 57 200-222
Bachmaier, M.; Precht, M.	Robust confidence intervals for contrasts based upon a likelihood ratio test •	95 StHefte 36 215-236
Tsimikas, J.	Robust tests based on the sample characteristic function • Lenth, R. V.; Markatou, M.;	95 AstrlJSt 37 45-60
	Rules and judgments in statistics: Three examples • Stewart-Oaten, Allan	95 Ecology 76 2001-2009
Roes, Kit C. B.; Does, Ronald J. M. M.	Shewhart-type charts in nonstandard situations (Disc: p24-35; Reply: P36-40) •	95 Technmcs 37 15-24
incomplete block design (Polish) • Czajka, Stanisław	Statistical methods for the analysis of a series of multiple experiments carried out in	95 ListyBio 32 101-129
K.; Shah, Arvind K.	Testing equality of several normal means to a specified standard • Krishnamoorthy,	95 JQualTek 27 132-138
ities for useful partially ordered alternatives in the	analysis of variance • Hoshino, Naoto; Miyazaki, Haruo; Seki, Yoichi On the level probabil-	95 CommStA 24 2059-2071
Kuwada, Masahide; Hyodo, Yoshifumi	Analysis of variance of balanced fractional $r^m \times s^n$ factorial designs of resolution V •	95 JJapnStS 25 129-141
On a robust	analysis of variance based on Winsorization (Korean) • Sung, Nae Kyung	95 KorApS 8/1 119-131
R. A. et al. Are neighbour methods preferable to	analysis of variance for completely systematic designs? 'Silly designs are silly!' • Bailey,	95 Biomtrka 82 655-659
consequences for recurrent selection ... • Gallais	Analysis of variance components of testcross progenies in an autotetraploid species and	92 ThApGene 83 353-359
Rank tests for main and interaction effects in	analysis of variance • Marden, John I.; Muyot, Maria Emerita T.	95 JASA 90 1388-1398
Review of	"Analysis of variance in experimental design" • Lindman, H. R. (Auth); Fedorov, V. V. (Rev)	95 JQualTek 27 268-268
of maximum likelihood estimation in the mixed	analysis of variance model • Güven, B. Asymptotic properties	95 StHefte 36 175-182
The polynomial model in the multivariate	analysis of variance • Baumgartner, Cornelia	95 BiomtrcJ 37 299-318
resolution IV • Kuwada, Masahide	Analysis of variance of partially balanced fractional $2^{m_1+m_2}$ factorial designs of	93 JCISS 18 343-367
Checking the assumptions in mixed-model	analysis of variance: A residual analysis approach • Oman, Samuel D.	95 CmpStDA 20 309-330
Robust	analysis of variance: A simulation study • Reed, J. F.; Stark, D. B.	95 JAppStat 22 87-104
Review of "Univariate	analyses of variance" (German) • Zoefel, P. (Auth); Richter, W.-D. (Rev)	95 MaOpfStS 26 179-180
of marker-quantitative trait locus linkages using	analysis of variance • Knott, S. A. Prediction of the power of detection	94 ThApGene 89 318-322
(Rev) Review of "Quality control in	analytic chemistry (2nd ed)" • Kateman, G.; Buydens, L. (Auth); Bandeen-Roche, Karen J.	95 Technmcs 37 115-116

Review of "Statistical methods in	analytic chemistry" • Meier, Peter C.; Zünd, Richard E. (Auth); Hillyer, Martin (Rev)	95 Technmcs 37 113-114
A two-component model for measurement error in	analytical chemistry • Rocke, David M.; Lorenzato, Stefan	95 Technmcs 37 176-184
Review of "Information theory in	analytical chemistry" • Eckschlager, K.; Danzer, K. (Auth); Clegg, D. (Rev)	95 Chemolab 28 207-208
Forina, M. (Auth) et al. Review of "Chemometrics for	analytical chemistry Volume 1: PC-aided statistical data analysis" • Meloun, M.; Militky, J.;	95 Technmcs 37 114-115
Militky, J. et al. Review of "Chemometrics for	analytical chemistry, Volume 2: PC-aided regression and related methods" • Meloun, M.;	95 Chemolab 29 326-327
processes and the age of our most recent common	ancestor • O'Connell, Neil The genealogy of branching	95 AdvAppPr 27 418-442
Markov chain Monte Carlo with applications to	ancestral inference • Geyer, Charles J.; Thompson, Elizabeth A. Annealing	95 JASA 90 909-920
O'Connor, Marcus The	anchor and adjustment heuristic in time-series forecasting • Lawrence, Michael;	95 JForcast 14 443-451
smoothing • Simonoff, Jeffrey S. The	anchor position of histograms and frequency polygons: Quantitative and qualitative	95 CommStB 24 691-710
Gaskell, George; Wright, Daniel B. Weighing	anchors: Verbal and numeric labels for response scales • O'Muircheartaigh, Colm;	95 JOffStat 11 295-307
	Ancillarity	
	Information and conditional inference • Severini, Thomas A.	95 JASA 90 1341-1346
	The roles of conditioning in inference (Disc: p173-199) • Reid, N.	95 StatSci 10 138-157
	Simple counterexamples against the conditionality principle • Helland, Inge S.	95 AmerStat 49 351-356
exponential models • Barndorff-Nielsen, O. E.	Stable and invariant adjusted profile likelihood and directed likelihood for curved	95 Biomtrka 82 489-499
Overdijk, D. A. The reduction and	ancillary classification of observations • van Berkum, E. E. M.; Linssen, H. N.;	95 JStPlInf 45 347-372
A condition for the nonexistence of	ancillary statistics • Gupta, Arjun K.; Wang, Yining	95 StPrLet 23 367-369
Identifying regions of significance in	ANCOVA problems having non-homogeneous regressions • Hunka, Steve	95 BJMaStPs 48 161-188
Obituary: Oliver Duncan	Anderson, 1940-1995 • De Gooijer, Jan G.	95 IntJFor 11 195-196
Stationary parabolic	Anderson model and intermittency • Carmona, R. A.; Molchanov, S. A.	95 ProbTher 102 433-453
Statistical analysis of drug interactions in	anesthesia • Dexter, Franklin	95 JThBio 172 305-314
Fisher, Nicholas I. Estimating the	angle between the mean directions of two spherical distributions • Lewis, Toby;	95 AstrIJSt 37 179-191
Properties of catch rates used in analysis of	angler surveys • Jones, C. M.; Robson, D. S.; Lakkis, H. D.; Kressel, J.	95 TrAFish 124 911-928
Jones, C. M.; Brown, T. L. (Auth) et al. Review of	"Angler survey methods and their applications in fisheries management" • Pollock, K. H.;	95 Biomtrcs 51 385-385
D. K.; Overton, W. S.; Anthony, R. G. Estimation of	animal abundance when capture probabilities are low and heterogeneous • Rosenberg,	95 JWildlMg 59 252-261
J., II Bayesian estimation of	animal abundance for line transect sampling • Karunamuni, Rohana J.; Quinn, Terrance	95 Biomtrcs 51 1325-1337
Hidden Markov models and	animal behaviour • MacDonald, Iain L.	95 BiomtrcJ 37 701-712
Mehta Analysis of	animal carcinogenicity experiments with multiple tumor types • Lu, Ying; Malani, Hina	95 Biomtrcs 51 73-86
tumor types on experimentwise false positive rate in	animal carcinogenicity studies • Rahman Effects of observed correlation structure among	95 JBiophSt 5 273-283
estimation and the effects of dependency among	animals • Schmutz, Joel A.; Ward, David H.; Sedinger, James S.; Rexstad, Eric A. Survival	95 JAppStat 22 673-681
Testing a model of aging in	animal experiments • Tyurin, Yu. N.; Yakovlev, Andrej Yu.; Shi, J.; Bass, L.	95 Biomtrcs 51 363-372
Crossing velocities and random lattice	animals • Sznitman, Alain-Sol	95 AnlsProb 23 1006-1023
Long-term effects of selection based on the	animal model BLUP in a finite population • Verrier, E.; Colleau, J. J.; Foulley, J. L.	93 ThApGene 87 446-454
Pradel, Roger; Cooch, Evan; Cooke, Fred Transient	animals in a resident population of snow geese: Local emigration or heterogeneity? •	95 JAppStat 22 695-710
1. A genetic model for diploid plant seeds and	animals • Zhu, J.; Wier, B. S. Analysis of cytoplasmic and maternal effects.	94 ThApGene 89 153-159
L. et al. Review of "Resource selection by	animals: Statistical design and analysis for field studies" • Manly, B. F. J.; McDonald, L.	95 JAppStat 22 192-193
survival analysis model for radio-tagged	animals • Pollock, K. H.; Bunck, C. M.; Winterstein, S. R.; Chen, C.-L. A capture-recapture	95 JAppStat 22 661-672
Inequalities for the	anisotropic Poisson polytope • Mecke, J.	95 AdvAppPr 27 56-62
On second-order formulas in	anisotropic stereology • Beneš, Viktor	95 AdvAppPr 27 326-343
	Annealing	
Palma, Daniela	Optimal sampling designs for dependent spatial units • Benedetti, Roberto;	95 EnvrMtrc 6 101-114
	• see also simulated -	
Geyer, Charles J.; Thompson, Elizabeth A.	Annealing Markov chain Monte Carlo with applications to ancestral inference •	95 JASA 90 909-920
Asymptotical behaviour of several interacting	annealing processes • Trouvé, Alain	95 ProbTher 102 123-143
	Annihilation	
	A limit theorem for a class of interacting particle systems • Simonelli, Italo	95 AnlsProb 23 141-156
profiles by evolving projections and internal rank	annihilation • Grung et al. ... of embedded minor analytes in three-way multicomponent	95 Chemolab 29 213-221
M. Mercé Allocation of solvency cost in group	annuities: Actuarial principles and cooperative game theory • Alegre, Antoni; Claramunt,	95 InsrneME 17 19-34
notion of variation applied to Markov chains and	Anosov maps • Loviscach, J. A generalized	95 ProbTher 103 553-570
accuracy for multiple readers and multiple tests: An	ANOVA approach with dependent observations • Obuchowski et al. ... of diagnostic	95 CommStB 24 285-308
	ANOVA estimates of variance components for a class of mixed models • Vangel, Mark G.	94 ASAProPEng 91-96
	ANOVA estimator	
	Estimation of variance components in mixed linear models • Kubokawa, Tatsuya	95 JMultiAn 53 210-236
Molińska, Anna; Moliński, Krzysztof Improvement of	ANOVA estimators of variance components in block designs with random treatments •	95 StPrLet 22 281-285
Review of "Random factors in	ANOVA" • Jackson, Sally; Brashers, Dale E. (Auth); Iacobucci, Dawn (Rev)	95 JMarRsr 32 238-239
test and rank transformed F test in 2×2 factorial	ANOVA (Korean) • Choi, Young Hun Simulation analysis of type I error and power for F	95 KorApS 8/2 87-97
components • Shaban, Abdel Magid M.	ANOVA, MINQUE, PSD-MINQMBE, CANOVA and CMINQUE in estimating variance	94 Statstca 54 481-489
interpretation of the interaction terms in Mandel's	ANOVA model from Hirotsu's interaction elements • Miyakawa, Masami An	93 RStAJp 40/1 1-9
Kawasaki, Hidemi; Miyakawa, Masami An	ANOVA model and interaction element for three-way layout without replication •	95 JJapnStS 25 143-150
and Bayesian confidence intervals in Bernoulli SS	ANOVA models • Wang, Y. et al. Simulation studies of smoothing parameter estimates	95 CommStB 24 1037-1059
Robust Bayesian estimators in a one-way	ANOVA model • Bian, G.	95 Test 4 115-135
Bayes estimators of the error variance in one-way	ANOVA models • Datta, Gauri Sankar; Ghosh, Malay Hierarchical	95 JStPlInf 45 399-411
versus means, and designing simulation ... • Wilcox	ANOVA: The practical importance of heteroscedastic methods, using trimmed means	95 BJMaStPs 48 99-114
Brownie, Cavell	ANOVA and rank tests when the number of treatments is large • Boos, Dennis D.;	95 StPrLet 23 183-191
Paranjpe, S. A. Nested	ANOVA for repeated surveys with autocorrelated errors • Prayag, V. R.; Chiplonkar, S. A.;	95 BiomtrcJ 37 565-570
F ratios sharing a common denominator in	ANOVA table • Kang, Phillee Dependent	95 JKoreaSt 24 111-125
• Stanek, E. J. et al. Comment on "The mixed-model	ANOVA: The truth, the computer packages, the books. Part I: Balanced data" (V47 p48-59)	95 AmerStat 49 323-324
to comment on "Comment on 'The mixed-model	ANOVA: The truth, the computer packages, the books. Part I: Balanced ... • Schwarz Reply	95 AmerStat 49 324-324
	ANOVA under unequal error variances • Weerahandi, Samaradasa	95 Biomtrcs 51 589-599
Computer graphics approach to two-way	ANOVA (Korean) • Huh, Moon Yul	95 KorApS 8/1 75-87
An	Anscombe type robust regression statistic • Bradu, Dan; Hawkins, Douglas M.	95 CmpStDA 20 355-386
with the use of penalized likelihood criteria in	antedependence and polynomial models • Macchiavelli, Raúl; Arnold, Steven F. Difficulties	95 CommStA 24 501-522
Some pitfalls in sample size estimation for an	anti-infective study • Weng, C. S. Wayne; Liu, J. P.	94 ASAProBiop 56-60
Neal A. Percentile regression analysis of correlated	antibody responses • Moulton, Lawrence H.; Holt, Elizabeth A.; Job, Jayakaran S.; Halsey,	95 StatMed 14 777-787
with detection limits for regression analyses of	antibody response to vaccine • Moulton, Lawrence H.; Halsey, Neal A. A mixture model	95 Biomtrcs 51 1570-1578
Green formulas in	anticipating stochastic calculus • Delgado, Rosario; Sanz-Solé, Marta	95 StocProc 57 113-148
The Euler scheme for	anticipating stochastic differential equations • Ahn, Hyungsok; Kohatsu-Higa, Arturo	95 Stochatx 54 247-269
Smoothness of distributions for solutions of	anticipating stochastic differential equations • Caballero, M. E.; Fernández, B.; Nualart, D.	95 Stochatx 52 303-322
Khasminskii formulas for Lyapunov exponents via	anticipative calculus • Arnold, Ludwig; Imkeller, Peter Furstenberg-	95 Stochatx 54 127-168
Albeverio, Sergio; Wu, Jiang-Lun Stochastic	anticipative calculus for functionals of a Gaussian generalized random field •	95 Stochatx 54 317-337
Limit theorems for	anticipative stochastic differential equations (French) • Erraoui, M.; Ouknine, Y.	95 Stochatx 53 1-11
A class of micropulses and	antipersistent fractional Brownian motion • Ciozek-Georges, R.; Mandelbrot, B. B.	95 StocProc 60 1-18
	Antisymmetric functionals of reversible Markov processes • Goldstein, Sheldon	95 AIHenrPB 31 177-190

	Antithetic variable	
Song, Wheyming Tina	Estimating simulation metamodels using correlated control variates • Shih, Neng Hui;	95 CommStB 24 79-90
Criminal	antitrust prosecution - 1960 to 1990 • Wang, Chaiho C.	94 ASAProBuEc 281-286
Policy micro analysis and simulation using	APL • Burpee, Joseph E.	94 ASAProGovt 178-182
Data analysis using	APL2 and APL2STAT • Fox, John; Friendly, Michael	95 SocMethR 23 282-302
Data analysis using APL2 and	APL2STAT • Fox, John; Friendly, Michael	95 SocMethR 23 282-302
	A.P.O. rules in hierarchical Bayes regression models • Ghosh, Malay; Hoekstra, Robert M.	95 SqtlAnly 14 99-115
	Apportionment to whole numbers • Nolan, Frank	95 NZndSt 30 40-47
information • Mouchart, Michel; Scheihing, Eliana	Approximate Bayesian inferences: Numerical experiences when measuring the loss of	94 ASAProBays 221-226
intervals for standardized rates based on the	approximate bootstrap method • Swift, Michael B. Simple confidence	95 StatMed 14 1875-1888
model • Mosier, Michael C.; Graybill, Franklin A.	Approximate confidence intervals for contrasts in the balanced three factor mixed	95 BiomtrcJ 37 273-297
	Approximate confidence interval estimation of $P(Y < X)$ • Chan, Yiu-Man	94 ASAProPEng 146-150
Thiagarajah, Kulathava Ranee Comparisons of	approximate confidence intervals for the parameters of extreme value distribution •	94 Metron 52/1 85-100
Meeker, William Q.; Escobar, Luis A. Teaching about	approximate confidence regions based on maximum likelihood estimation •	95 AmerStat 49 48-53
Exact and	approximate D-optimal designs in polynomial regression • Happacher, Maximilian	95 Metrika 42 19-27
equality • Lix, Lisa M.; Keselman, H. J.	Approximate degrees of freedom tests: A unified perspective on testing for mean	95 PsychBu 117 547-560
Sebastian, G.	Approximate estimation of non-identifiable parameters in a convolution • Kale, B. K.;	95 StPrLet 25 373-378
designs • Alvo, Mayer; Cabilio, Paul	Approximate and exact distributions of rank tests for balanced incomplete block	95 CommStA 24 3073-3121
and highest posterior density regions with	approximate frequentist validity • Ghosh, J. K.; Mukerjee, Rahul On perturbed ellipsoidal	95 JRSS-B 57 761-769
data • Follmann, Dean; Wu, Margaret An	approximate generalized linear model with random effects for informative missing	95 Biomtrcs 51 151-168
A general purpose	approximate goodness-of-fit test • Chen, Gemai; Balakrishnan, N.	95 JQualTek 27 154-161
On	approximate inference for the two-parameter gamma model • Wong, A.	95 StHefte 36 49-59
Yates, P. K.; Lewis, S. M.; Dean, A. M.	Approximate information matrices for estimating a given set of contrasts • Kao, L.-J.;	95 StSinica 5 593-598
A note on the accuracy of an	approximate interval for the binomial parameter • Huwang, Longcheen	95 StPrLet 24 177-180
cubes • Gaffke, N.; Heiligers, B. Computing optimal	approximate invariant designs for cubic regression on multidimensional balls and	95 JStPlInf 47 347-376
Methods to	approximate joint uncertainty and variability in risk • Bogen, Kenneth T.	95 RiskAnal 15 411-419
On an	approximate likelihood for quantiles • Lavine, M.	95 Biomtrka 82 220-222
Liang, Kung-yee	Approximate likelihood ratios for general estimating functions • Hanfelt, John J.;	95 Biomtrka 82 461-477
observations • Mishra, M. N.; Bishwal, J. P. N.	Approximate maximum likelihood estimation for diffusion processes from discrete	95 Stochatx 52 1-13
• Johaadien, Hassan; Goutis, Constantinos	Approximate maximum likelihood estimation in the presence of extra-Poisson variation	95 CommStA 24 1625-1636
Consistency and asymptotic normality of an	approximate maximum likelihood estimator for discretely observed ... • Pedersen	95 Bernoull 1 257-279
Frigessi, Arnoldo; de Sá, Paula Gonzaga Fast	approximate maximum a posteriori restoration of multicolour images • Ferrari, Pablo A.;	95 JRSS-B 57 485-500
from DNA sequences with variable rates over sites:	Approximate methods • Yang, Ziheng Maximum likelihood phylogenetic estimation	94 JMolEvol 39 306-314
Conditions equivalent to consistency of	approximate MLE's for stochastic processes • Vajda, Igor	95 StocProc 56 35-56
applications to Bayesian inference • Damien, P. et al.	Approximate random variate generation from infinitely divisible distributions with	95 JRSS-B 57 547-563
DEFFS • Shin, Hee-Choon An	approximate Rao-Scott modification factor in two-way tables with only known marginal	94 ASASrvyRsr 600-601
tion of sample size (Disc: p134-135) • Ji, X. et al.	Approximate reasoning in belief networks: Statistical validity, accuracy, and determina-	94 ASAProBays 128-133
Noor, Islam; Ahmad, Naseer Exact and	approximate relative efficiency of generalized ridge-type shrinkage estimator •	95 PakJStA 11 17-22
Queueing networks with blocking: Exact and	approximate solutions • Perros, H. G.	94 OxfordUniv 288p
of "Queueing networks with blocking: Exact and	approximate solutions" • Perros, H. G. (Auth); Pitts, S. M. (Rev) Review	95 ShrtBkRv 15/3 46-46
Optimal block designs revisited: An	approximate theory detour • Pukelsheim, Friedrich; Sinha, Bikas Kumar	95 StPrLet 24 375-380
Goffinet, Bruno; Mangin, Brigitte	Approximate thresholds of interval mapping tests for QTL detection • Rebaï, Ahmed;	94 Genetcs 138 235-240
Power robustification of	approximately linear tests • Ehm, W.; Mammen, E.; Müller, D. W.	95 JASA 90 1025-1033
Sugiura, Nariaki	Approximating Bayes critical region for testing simple and tree ordered normal means •	94 JStatRsr 28 1-20
	Approximating the characteristic function of the Student's t distribution • Saleh, Amina A.	94 EgpStJ 38 177-195
Using power transformations when	approximating quantiles • Fearn, Dean H.; Nebenzahl, Elliott	95 CommStA 24 1073-1093
time-series properties versus asymptotic theory in	approximating sample serial correlation structure: Part 2 • Anderson Finite lengthed	95 JStCmpSm 52 167-171
Leonard, K. A.; An, A. B.; Nusser, S. M.; Breidt, F. J.	Approximating the variance of the survey regression estimator using poststratification •	94 ASASrvyRsr 222-227
	Approximating by the Wishart distribution • Kollo, Tõnu; von Rosen, Dietrich	95 AnInStMa 47 767-783
	Approximation	
	• see also names of specific types of approximations such as Poisson –, saddlepoint –, or stochastic –	
When is Satterthwaite's	approximation adequate? • Conerly, Michael D.; Hardin, J. Michael; Waller, Lance A.	94 ASAProStCp 209-213
Diffusion	approximation for an age-structured population • Bose, A.; Kaj, I.	95 AnlsApPr 5 140-157
Chen, Ping Randomly weighted	approximation and its application to empirical process (Chinese) • Zhang, Dixin;	95 KToChina 39 2232-2234
Halousková, Alena; Nedoma, Petr Recursive	approximation by ARX model: A tool for grey box modelling • Kárný, Miroslav;	95 IJAdCSgP 9 525-546
Moving-average representation of autoregressive	approximations • Bühlmann, Peter	95 StocProc 60 331-342
A test to detect inadequacy of Satterthwaite's	approximation in balanced mixed models • Khuri, André I.	95 MaOpfStS 27 45-54
Generalized Laplacian	approximations in Bayesian inference • Hsu, John S. J.	95 CandJSt 23 399-410
high-dimensional distribution based on bootstrap	approximation • Zhu, L.-X.; Fang, K.-T.; Bhatti, M. I.; Bentler, P. M. Testing sphericity of a	95 PakJStA 11 49-65
Neyman goodness-of-fit test and its bootstrap	approximation (Chinese) • Li, Guoying; Liu, Wanrong; Zha, Wenxing Projection pursuit	95 KToChina 39 678-681
Stramer, O. On the	approximation of continuous time threshold ARMA processes • Brockwell, P. J.;	95 AnInStMa 47 1-20
Edgeworth expansion and random weighted	approximation in correlated model (Chinese) • Shi, Jian; Zheng, Zhongguo	95 KToChina 39 1925-1927
Kouritzin, Michael A. Strong	approximation for cross-covariances of linear variables with long-range dependence •	95 StocProc 60 343-353
Rananand, Nol Markov	approximations to D-BMAPs: Information theoretic bounds on queueing performance •	95 CommStSM 11 713-734
Barabesi, Lucio; Fattorini, Lorenzo	Approximation of density functions by orthogonal series with grouped data •	94 JItalStS 3 181-200
Nagata, Yasushi On normal	approximation to the distribution of the estimated process capability index $\hat{C}_{pk}$ •	95 JJapnStS 25 159-171
Tail-specific linear	approximations for efficient bootstrap simulations • Hesterberg, Tim	95 JCmpGrSt 4 113-133
Strong	approximations for epidemic models • Ball, Frank; Donnelly, Peter	95 StocProc 55 1-21
Utev, Sergey Poisson	approximation for the final state of a generalized epidemic process • Lefèvre, Claude;	95 AnlsProb 23 1139-1162
Bootstrap	approximation for generalized U-processes (Chinese) • Zhang, Dixin; Li, Guoying	95 KToChina 39 1249-1252
Laplace	approximation of high dimensional integrals • Shun, Zhenming; McCullagh, Peter	95 JRSS-B 57 749-760
Khalikulov, S. I. On	approximation of homogeneous random fields by fields with bounded spectrum •	95 ThPrMaSt 50 129-132
Uniform local probability	approximations: Improvements on Berry-Esseen • Hahn, Marjorie G. W.; Klass, Michael J.	95 AnlsProb 23 446-463
On	approximations involving the beta distribution • Jóhannesson, Benedikt; Giri, Narayan	95 CommStB 24 489-503
Bootstrap	approximation for L-statistics with unbounded score function (Chinese) • Wang, Qiying	95 ChinaAMA 15 281-286
nondegenerate case • Bolthausen, E. et al. Laplace	approximations for large deviations of nonreversible Markov processes. The	95 AnlsProb 23 236-267
	Approximations of large population epidemic models • Garcia, Nancy Lopes	95 StocProc 60 147-160
Pinheiro, José C.; Bates, Douglas M.	Approximations to the log-likelihood function in the nonlinear mixed-effects model •	95 JCmpGrSt 4 12-35
Christopher A.	Approximations for marginal densities of M-estimators • Fan, Rocky Y. K.; Field,	95 CandJSt 23 185-197
service rates • Morrice, Douglas J.; Song, Mark An	approximation method for a single server queue with cyclically indexed arrival and	95 AplMaCmp 71 247-269
Yosida	approximations for multivalued stochastic differential equations • Pettersson, Roger	95 Stochatx 52 107-120
expectations • Joe, Harry	Approximations to multivariate normal rectangle probabilities based on conditional	95 JASA 90 957-964
son random measures • Kallianpur, G. et al. Diffusion	approximation of nuclear space-valued stochastic differential equations driven by Pois-	95 AnlsApPr 5 493-517

in the term structure using linear models and	artificial neural network • Swanson et al. ... approach to assessing the information	95 JBES 13 265-275
Gupta, Hoshin Vijai; Sorooshian, Soroosh	Artificial neural network modeling of the rainfall-runoff process • Hsu, Kuo-lin;	95 WaterReR 31 2517-2530
Bayesian multiperiod forecasts for	ARX models • Liu, Shu-Ing	95 AnInStMa 47 211-224
Nedoma, Petr Recursive approximation by	ARX model: A tool for grey box modelling • Kárný, Miroslav; Halousková, Alena;	95 IJAdCSgP 9 525-546
Leonard M.; González, María E. Is there a place in	ASA for statisticians who work for state/provincial or local governments? • Gaines,	94 ASAProGovt 149-152
Statistical analysis of	asbestos fibre counts • Oehlert, Gary W.; Lee, Richard J.; Van Orden, Drew	95 EnvrMtrc 6 115-126
mesothelioma incidence, time since exposure to	asbestos and level of exposure • Berry, G. Environmental	95 EnvrMtrc 6 221-228
C. A. A.; Coenegracht, P. M. J. Multicriteria steepest	ascent in a design space consisting of both mixture and process variables • Duineveld,	95 Chemolab 30 23-36
scaling input variables on the direction of steepest	ascent • Carroll, Steven S.; Monsees, Geoffrey A. The effect of	94 ASAProQlPr 267-269
Hodge, S. E. Inherent intractability of the	ascertainment problem for pedigree data: A general likelihood framework • Vieland, V. J.;	95 AJHuGene 56 33-43
The value of an	Asian option • Rogers, L. C. G.; Shi, Z.	95 JAppProb 32 1077-1088
How to	ask survey questions • Fink, Arlene	95 SagePubl:CA:Lnd 120p
	Asking sensitive questions in surveys • Hutchinson, Paul	95 TeachgSt 17/2 43-43
Wilkens, Henry T. Advertising-related	aspirin consumption among under 45 and over 45 consumers • Hsu, Margaretha M.;	94 ASAProBiop 276-278
On model selection for standard curve in	assay development • Tse, Siu-Keung; Chow, Shein-Chung	95 JBiophSt 5 285-296
methods for bias reduction in limiting dilution	assays • Mehrabi, Yadollah; Matthews, J. N. S. Likelihood-based	95 Biomtrcs 51 1543-1549
and detection limits for linear homoscedastic	assays • Dunne, Adrian Decision	95 StatMed 14 1949-1959
A multivariate tolerance interpretation of	assays of joint action • Frey, Michael	94 ASAProBiop 39-44
Analysis of	assays of the joint action of three drugs • Frey, Carolin; Frey, Michael	94 ASAProBiop 33-38
Pharmaceutical validation –	Assay and process • Chow, Shein-Chung	94 ASAProBiop 285-290
The statistical analysis of slope ratio	assays using SAS software • Wong, M. Y.	94 CmpBiomd 45 233-238
Implementing option pricing models when	assest returns are predictable • Lo, Andrew W.; Wang, Jiang	95 JFinance 50 87-129
Sailer, Peter; Wilson, Janette How sales of capital	assets were affected by the Tax Reform Act of 1986 (Disc: p212-214) • Longton, Barbara;	94 ASAProGovt 201-205
Efficient portfolios in the	asset liability context • Keel, Alex; Müller, Heinz H.	95 AstinB 25 33-48
A test of the three moment international capital	asset pricing model • Vakil, Farok; Rajan, Murli	94 ASAProBuEc 184-189
eralized method of moments in tests of conditional	asset pricing models • Ferson, W. E.; Foerster, S. R. Finite sample properties of the gen-	94 JFinEcon 36 29-55
A Bayesian approach to diagnosis of	asset pricing models • Stutzer, Michael	95 JEconmtx 68 367-397
An empirical model of	asset replacement in dairy production • Miranda, Mario J.; Schnitkey, Gary D.	95 JApEcmtx 10 41-55S
nonlinear model for the conditional expectations of	asset returns • Paruolo, P.; Pillati, M. A	94 Statstca 54 329-347
the instantaneous volatility and covariance of risky	assets • Chesney, Marc; Elliott, Robert J. Estimating	95 AplStMDA 11 51-58
testing and accessing complex statistical tools in an	ASSIST environment • Sanchez, Marcus John Advantages of developing,	94 ASAProStCp 141-146
	Association	
Greenhouse, Samuel W.	Biostatistical concepts and methods in the legal setting • Gastwirth, Joseph L.;	95 StatMed 14 1641-1653
	Comparing resemblance measures • Batagelj, Vladimir; Bren, Matevz	95 JClass 12 73-90
Koehler, Kenneth J.; Symanowski, James T.	Constructing multivariate distributions with specific marginal distributions •	95 JMultiAn 55 261-282
	Who discovered Spearman's rank correlation? • Lovie, A. D.	95 BJMaStPs 48 255-269
trends in multiway tables • Payne, C.; Payne, J. et al.	Testing the tightening bond hypothesis: An application of new models for analysing	95 JStPlInf 47 181-201
Zerbe, Gary O.; Wu, Margaret C. Inference for the	association between coefficients in a multivariate growth curve model • Zucker, David M.;	95 Biomtrcs 51 413-424
Correction to "Local divergence and	association" (V81 p579-584) • Blyth, S.	95 Biomtrka 82 667-667
Integrated probabilistic data	association-finite resolution • Mšicki, D.; Evans, R.	95 Autmatca 31 559-570
of a random field distribution function under	association • Roussas, George G. Asymptotic normality of a smooth estimate	95 StPrLet 24 77-90
Large sample results for tests of	association II. Multinomial and stratified sampling • Kulinskaya, Elena	95 CommStA 24 1121-1150
	Association of infinitely divisible random vectors • Samorodnitsky, Gennady	95 StocProc 55 45-55
Review of "Nonparametric measures of	association" • Gibbons, Jean Dickinson (Auth); Nelson, Lloyd S. (Rev)	95 JQualTek 27 93-93
The behaviour of nominal and ordinal partial	association measures • Olszak, Michael; Ritschard, Gilbert	95 Ststcian 44 195-212
Daniels's	association measures under right censoring • O'Quigley, John	95 AplStMDA 11 109-119
A.; Jmel, S. et al. Correspondence analysis and	association models constrained by a conditional independence graph • de Falguerolles,	95 Psymtrka 60 161-180
proach and correspondence analysis • Gokhale et al.	Association models for contingency tables via minimum discrimination information ap-	95 RBrzlPrS 9 83-98
Computation of maximum likelihood estimates in	association models • Haberman, Shelby J.	95 JASA 90 1438-1446
Extensions in the use of log-multiplicative scaled	association models in multiway contingency tables • Wong, Raymond Sin-Kwok	95 SocMethR 23 507-538
clustered binary data • Have, T. R. T. et al.	Association models for periodontal disease progression: A comparison of methods for	95 StatMed 14 413-429
Least squares estimation for	association models with linear constraints on the parameters (French) • Jmel, S.	95 RStApp 43/3 93-104
The EM algorithm for graphical	association models with missing data • Lauritzen, Steffen L.	95 CmpStDA 19 191-201
Cone order	association • Cohen, Arthur; Sackrowitz, H. B.; Samuel-Cahn, E.	95 JMultiAn 55 320-330
Louis, Thomas A. Inferences on the	association parameter in copula models for bivariate survival data • Shih, Joanna H.;	95 Biomtrcs 51 1384-1399
Class 3	association schemes whose symmetrizations have two classes • Song, Sung Y.	95 JCombThA 70 1-29
genes in a disease: Extension of the marker-	association-segregation χ^2 method • Dizier, M.-H. et al. Interactive effect of two candidate	94 AJHuGene 55 1042-1049
Fifty years of Calcutta Statistical	Association • Banerjee, P. K.	95 CalcutSt 45 1-23
of the First Scientific Meeting of the International	Association for Statistical Education" • Brunelli, L. et al. Review of "Proceedings	95 AmerStat 49 276-276
Mapping disease genes: Family-based	association studies • Thomson, Glenys	95 AJHuGene 57 487-498
et al. Inference of the strength of genotype-disease	association from studies comparing offspring with and without parental history ... • Tiret	93 AnHuGene 57 141-149
Comparison of statistics for candidate-gene	association studies using cases and parents • Schaid, Daniel J.; Sommer, Steve S.	94 AJHuGene 55 402-409
Schwede, Laurie; Ellis, Yukiko Exploring	associations between subjective and objective assessments of household membership •	94 ASASrvyRsr 325-330
The distribution of Q: A powerful sibship test of	association • Shah, S.; Green, J. R.	94 AnHuGene 58 163-173
Power comparison of various sibship tests of	association • Green, J. R.; Shah, S.	93 AnHuGene 57 151-158
Improved exact inference about conditional	association in three-way contingency tables • Kim, Donguk; Agresti, Alan	95 JASA 90 632-639
A test of the	association of a time-dependent state variable to survival • Cantor, Alan B.	95 CmpBiomd 46 101-105
Modeling manifest and latent dimensions of	association in two-way cross-classifications • Eliason, Scott R.	95 SocMethR 24 30-67
tion based on ranks and a nonparametric test of no	association between vectors • Cléroux, Robert; Lazraq, Aziz; Lepage, Yves Vector correla-	95 CommStA 24 713-733
pension to cytosine arabinoside: Statistical analysis,	associations with clinical outcome and experimental design ... • Minkin et al. ... in sus-	95 StatMed 14 1081-1095
titative comparison of bidirectional and optimal	associative memories for background prediction of spectra • Wabuyele, B. W. et al. Quan-	95 Chemolab 29 51-61
analyses for psychiatric disorders: Secular trends,	assortative mating, bilineal pedigrees • Spence et al. Methodological issues in linkage	93 HumHrdty 43 166-172
	Assumptions	
	• see also names of specific types of assumptions such as normality –	
approach to image restoration: Applications to	astronomical images • Molina, R. On the hierarchical Bayesian	94 IEEEPAMI 16 1122-1128
mator with doubly censored data and applications to	astronomy • Akritas, Michael G.; Murphy, Susan A.; LaValley, Michael P. The Theil-Sen esti-	95 JASA 90 170-177
cursion sets of random fields, with an application to	astrophysics • Worsley Boundary corrections for the expected Euler characteristic of ex-	95 AdvAppPr 27 943-959
Crossover trials with	asymmetric covariance within subject • Skinner, Jonathan B.	94 ASAProBiop 411-413
On e-optimal fractions of symmetric and	asymmetric factorials • Mukerjee, Rahul	95 StSinica 5 515-533
basis matrix multiplication. Part III. The use of	asymmetric and generalized eigenvalue equations • Alsberg ... method of postponed	95 Chemolab 30 223-225
Léonard, C. Large deviations of an	asymmetric hydrodynamical system of independent particles (French) • Kipnis, C.;	95 AIHenrPB 31 223-248
procedures in some nonexponential families using	asymmetric Linex loss function • Huang, Su-Yun Empirical Bayes testing	95 JStPlInf 46 293-309
Asymptotic efficiency of estimators under	asymmetric loss functions • Takagi, Yoshiji	95 JJapnStS 25 49-56
Real estate price prediction under	asymmetric loss • Cain, Michael; Janssen, Christian	95 AnInStMa 47 401-414

A nonparametric test for the regression function:	Asymptotic theory • Diebolt, Jean	95 JStPlInf 44 1-17
systems • Saikkonen, Pentti Problems with the	asymptotic theory of maximum likelihood estimation in integrated and cointegrated	95 EconTher 11 888-911
Taniguchi, Masanobu; Puri, Madan L. Higher order	asymptotic theory for normalizing transformations of maximum likelihood estimators •	95 AnInStMa 47 581-600
	Asymptotic theory for partly linear models • Gao, Jiti	95 CommStA 24 1985-2009
Liu, Hung-kung	Asymptotic theory for prediction through empirical linear models • Hwang, J. T. Gene;	94 ASAProPEng 207-216
Watson, Geoffrey S. Large sample	asymptotic theory of tests for uniformity on the Grassmann manifold • Chikuse, Yasuko;	95 JMultiAn 54 18-31
Some aspects of	asymptotic theory with applications to time series models • Jeganathan, P.	95 EconTher 11 818-887
Small-sample comparison of the exact and	asymptotic upper tail probabilities of chi-squared goodness-of-fit statistics: ... • Formann	95 JStCmpSm 51 369-384
estimation via the ECM algorithm: Computing the	asymptotic variance • van Dyk, D. A.; Meng, X.-L.; Rubin, D. B. Maximum likelihood	95 StSinica 5 55-75
Akahira, M. The Bhattacharyya type bound for the	asymptotic variance and the sequential discretized likelihood estimation procedure •	95 SqtlAnly 14 193-204
Kac's representation from an	asymptotic viewpoint • Horváth, Lajos	95 JStPlInf 46 27-45
	Asymptotical behaviour of several interacting annealing processes • Trouvé, Alain	95 ProbTher 102 123-143
Tsybakov, A. B.	Asymptotical minimax recovery of sets with smooth boundaries • Mammen, E.;	95 AnlsStat 23 502-524
	Asymptotically distribution-free procedure in a two-way layout • Choi, Young Hun	95 JKoreaSt 24 375-387
alternatives (Korean) • Kim, D. H.; Lim, D. H. An	asymptotically distribution-free test for parallelism of regression lines against umbrella	95 KorApS 8/1 105-117
Chen, Hung	Asymptotically efficient estimation in semiparametric generalized linear models •	95 AnlsStat 23 1102-1129
On the smallest possible	asymptotically efficient variance in semiparametric models • Liang, Hua	95 JSysSMSE 7 29-33
	Asymptotically ergodic Markov functionals of an ergodic process • Alimov, D.	94 ThProbAp 39 537-546
derivation using simulation technique • Tanizaki, H.	Asymptotically exact confidence intervals of CUSUM and CUSUMSQ tests: A numerical	95 CommStB 24 1019-1036
Flexibility and regularity properties of the	asymptotically ideal production model • Terrell, Dek	95 EcomtrRv 14 1-17
	Asymptotically Markov stochastic 1 • Sataev, E. A.	94 ThProbAp 39 448-464
lem • Hsieh, F.; Turnbull, B. W. A note on the local	asymptotically minimax rate for estimating a crossing point in a diagnostic marker prob-	95 StPrLet 24 181-185
On a modified bootstrap for certain	asymptotically nonnormal statistics • Datta, Somnath	95 StPrLet 24 91-98
Consistent and	asymptotically normal parameter estimates for hidden Markov models • Rydén, Tobias	94 AnlsStat 22 1884-1895
populations • Tahir, Mohamed; Rekab, Kamel	Asymptotically optimal empirical Bayes procedures for selecting good uniform $U(0,\theta_i)$	95 StocAnAp 13 233-247
Hickernell, Fred J. A kind of	asymptotically optimal sequential test under complex supposition • Chen, Jiading;	95 SciSPkng 24 673-683
parameter of continuous one parameter ... • Xuc	Asymptotically optimality and convergence rates of empirical Bayes estimator for the	95 ChJApPSt 10 35-45
Kuk, Anthony Y. C.	Asymptotically unbiased estimation in generalized linear models with random effects •	95 JRSS-B 57 395-407
Phillips, Michael J. An application of nonstandard	asymptotics to the analysis of repairable-systems data • Sweeting, Trevor J.;	95 Technmcs 37 428-435
location-scale families • Nagaev, A. V.	Asymptotics of the Bayes risk in discrimination between normal and uniform	95 MaMethSt 4 312-333
On the	asymptotics of constrained M-estimation • Geyer, Charles J.	94 AnlsStat 22 1993-2010
Absolute continuity, singular measures and	asymptotics for estimators • Dion, Jean-Pierre; Ferland, René	95 JStPlInf 43 235-242
	Asymptotics for the Euclidean TSP with power weighted edges • Yukich, J. E.	95 ProbTher 102 203-220
networks • Iscoe, I.; McDonald, D.	Asymptotics of exit times for Markov jump processes. II: Applications to Jackson	94 AnlsProb 22 2168-2182
	Asymptotics of forced-in variables from screening processes • Lian, Ie-Bin	95 CommStA 24 131-151
On extreme value	asymptotics for increments of renewal processes • Steinebach, Josef; Eastwood, Vera R.	95 JStPlInf 45 301-312
Review of "Inference and	asymptotics" • Barndorff-Nielsen, O. E.; Cox, D. R. (Auth); Corcoran, Stephen (Rev)	95 Ststcian 44 279-279
Review of "Inference and	asymptotics" • Barndorff-Nielsen, O. E.; Cox, D. R. (Auth); DiCiccio, Thomas J. (Rev)	95 JASA 90 1129-1130
Review of "Inference and	asymptotics" • Barndorff-Nielsen, O. E.; Cox, D. R. (Auth); Hingley, Peter J. (Rev)	95 Ststcian 44 407-408
Review of "Inference and	asymptotics" • Barndorff-Nielsen, O. E.; Cox, D. R. (Auth); Pfanzagl, J. (Rev)	95 Metrika 42 275-275
Babu, Gutti Jogesh	Asymptotics of k-mean clustering under non-i.i.d. sampling • Serinko, Regis J.;	95 StPrLet 24 57-66
Sharp	asymptotics of large deviations in R^d • Iltis, Michael	95 JTherPrb 8 501-522
Donald A.; Peters, Dawn Correction to "Higher-order	asymptotics and the likelihood principle: One-parameter models" (V81 p1-10) • Pierce,	95 Biomtrka 82 235-235
Li, Guoying	Asymptotics of the minimum L_1-norm estimates in a partly linear model • Shi, Peide;	95 JSysSMSE 7 67-77
	Asymptotics for multisample statistics • Vaillancourt, Jean	95 CandJSt 23 171-184
	Asymptotics of multivariate randomness statistics • Ghoudi, Kilani	95 JMultiAn 55 340-348
Hjort, Nils Lid; Fenstad, Grete Second-order	asymptotics for the number of times an estimator is more than ε from its target value •	95 JStPlInf 48 261-275
ing Brownian motion • Shi, Zhan; Werner, Wendelin	Asymptotics for occupation times of half-lines by stable processes and perturbed reflect-	95 Stochatx 55 71-85
selection and ranking procedures: Second-order	asymptotics" • Mukhopadhyay, N.; Solanky, T. K. S. (Auth) et al. Review of "Multistage	95 JASA 90 1130-1131
• Arratia, R.; Stark, D.; Tavaré, S. Total variation	asymptotics for Poisson process approximations of logarithmic combinatorial assemblies	95 AnlsProb 23 1347-1388
	Asymptotics of the run lengths of two control charts • Zhang, Jian	95 CommStA 24 2375-2403
Ferrari, Silvia L. P. Second order	asymptotics for score tests in generalised linear models • Cribari-Neto, Francisco;	95 Biomtrka 82 426-432
	Asymptotics of sliced inverse regression • Zhu, Li-Xing; Ng, Kai W.	95 StSinica 5 727-736
St. P. Exact	asymptotics for some probability distributions on compact manifolds • Richards, Donald	95 AnlsStat 23 1582-1586
Fixed-domain	asymptotics for spatial periodgrams • Stein, Michael L.	95 JASA 90 1277-1288
Malý, Marek The	asymptotics for studentized k-step M-estimators of location • Jurečková, Jana;	95 SqtlAnly 14 229-245
David	Asymptotics for the transformation kernel density estimator • Hössjer, Ola; Ruppert,	95 AnlsStat 23 1198-1222
	Asymptotics for transportation cost in high dimensions • Dobrić, V.; Yukich, J. E.	95 JTherPrb 8 97-118
Spectra, exit times and long time	asymptotics in the zero-white-noise limit • Mathieu, Pierre	95 Stochatx 55 1-20
Statistical evaluation of selected air pollutants in	Athens, Greece • Kambezidis, H. D.; Tulleken, R.; Amanatidis, G. T.; Paliatsos, A. G. et al.	95 EnvrMtrc 6 349-361
the NCAA's new standards affect minority student-	athletes? • Klein, Stephen P.; Bell, Robert M. How will	95 Chance 8/3 18-21
Gender and performance in	athletics • Chatterjee, Sangit; Laudato, Matthew	94 ASAProSpts 36-41
Statistics for exceptional	athletics records • Robinson, Michael E.; Tawn, Jonathan A.	95 ApplStat 44 499-511
Population study and management of	Atlantic Flyway Canada geese • Hestbeck, Jay B.	95 JAppStat 22 877-890
Multivariate evaluation of feed for	Atlantic halibut • Nortvedt, Ragnar; Tuene, Stig	95 Chemolab 29 271-282
surface ozone for the Vancouver, Montreal, and	Atlantic regions of Canada • Burrows et al. ... and prediction of summer season maximum	95 JApMetrl 34 1848-1862
Silva, I. (Auth); Yuen, P.; Machin, D. (Rev) Review of	"Atlas of cancer incidence in England and Wales: 1968-85" • Swerdlow, A.; dos Santos	95 StatMed 14 1725-1726
Estimating distribution maps from	atlas data using methods of statistical image analysis • Högmander, Harri; Møller, Jesper	95 Biomtrcs 51 393-404
The development of the series of U.S. cancer	atlases: Implications for future epidemiologic research (Disc: p697-699) • Mason, T. J.	95 StatMed 14 473-479
traffic shaper for statistical multiplexing in	ATM networks • Kim, K. S.; Lee, B. G. GA-based optimal dimensioning of three-level	95 AnlsTele 50 624-631
new model PDF for contaminants dispersing in the	atmosphere • Lewis, D. M.; Chatwin, P. C. A	95 EnvrMtrc 6 583-593
Wind energy resource estimation of the upper	atmosphere over Southern Africa • Bryukhan, Fedor F.; Diab, Roseanne D.	95 JApMetrl 34 2565-2571
Statistical properties of predictability from	atmospheric analogs and the existence of multiple flow regimes • Trevisan, Anna	95 JAtmsphr 52 3577-3592
estimation of error covariance parameters for	atmospheric data assimilation • Dee, Dick P. On-line	95 MWeathr 123 1128-1145
et al. Identifying periodic components in	atmospheric data using a family of minimum variance spectral estimators • Wikle, C. K.	95 JClimate 8 2352-2363
model for high concentrations in short-range	atmospheric dispersion • Mole, N.; Anderson, C. W. et al. A generalized Pareto distribution	95 EnvrMtrc 6 595-606
Dulal K. Upper and lower tolerance limits of	atmospheric ozone level and extreme value distribution • Dasgupta, Ratan; Bhaumik,	95 SankhyaB 57 182-199
Review of "Statistical data analysis for ocean and	atmospheric sciences" • Thiébaux, H. Jean (Auth); Solow, Andrew R. (Rev)	95 AmerStat 49 399-399
sion in the vicinity of an ... • Higson et al. Effect of	atmospheric stability on concentration fluctuations and wake retention times for disper-	95 EnvrMtrc 6 571-581
A note on the	attenuation of correlation • Bedrick, Edward J.	95 BJMaStPs 48 271-280
time-varying AR spectral estimation for ultrasound	attenuation measurement in biological tissues • Idier, J.; Giovannelli, J.-F. et al. Bayesian	94 ASAProBays 256-261
Phyllis S.	Attitudes and achievement in freshman math classes • Odell, Patricia M.; Schumacher,	94 ASAProStEd 197-202
region of South Africa • Glencross, M. J.; Cherian, V. I.	Attitudes toward applied statistics of postgraduate students in education in the Lebowa	95 PsychRpt 77 315-322
Women's reports of sensitive health conditions,	attitudes, and behaviors • Guadagno, M. A.; Abma, J. C.; Chandra, A.; London, K. A. et al.	94 ASAProGovt 82-87

Median estimation using	auxiliary information • Meeden, Glen	95 SurvyM 21 71-77
Estimating distribution functions with	auxiliary information using poststratification • Nascimento Silva, P. L. D.; Skinner, C. J.	95 JOffStat 11 277-294
Estimation in semiparametric models using an	auxiliary model • Huschens, Stefan; Stahl, Gerhard	95 StHefte 36 313-326
Edward V. Incorporating	auxiliary predictor variation in principal component regression models • Thomas,	95 JChemtrc 9 471-481
Unequal probability sampling using a transformed	auxiliary variable • Sahoo, Jagatananda; Sahoo, Lokanath; Mohanty, Suprasana	94 Metron 52/1 71-83
test estimators in double sampling with two	auxiliary variables • Das, G.; Bez, K. Preliminary	95 CommStA 24 1211-1226
Reliability, maintainability, and	availability assessment (Second edition) • Locks, Mitchell O.	95 ASQualCntrl:WI 348p
p2269-2278) • Baker et al. Correction to "The paired	availability design: A proposal for evaluating epidural analgesia during labor" (V13	95 StatMed 14 1841-1841
influence graph to reduce the size of Markovian	availability problems • Devooght, J.; Tombuyses, B. The use of the component	94 RelEngSS 46 237-244
Phillips, Michael J. (Rev) Review of "The reliability,	availability and productiveness of systems" • Sherwin, D. J.; Bossche, A. (Auth);	95 Ststcian 44 412-413
Szidarovszky, F. Time-varying failure rates in the	availability and reliability analysis of repairable systems • Hassett, T. F.; Dietrich, D. L.;	95 IEEEReli 44 155-160
Solving Markovian systems of O.D.E. for	availability and reliability calculations • Tombuyses, B.; Devooght, J.	95 RelEngSS 48 47-55
implementation • Csenki, Attila Mission	availability for repairable semi-Markov systems: Analytical results and computational	95 MaOpfStS 26 75-87
Limiting behavior of some measures of system	availability • Mi, Jie	95 JAppProb 32 482-493
Fowles, Richard Road closure to mitigate	avalanche danger: A case study for Little Cottonwood Canyon • Blattenberger, Gail;	95 IntJFor 11 159-174
	Average	
	• see also moving –, weighted –	
On the logarithmic	average of additive functionals • Csáki, E.; Földes, A.	95 StPrLet 22 261-268
policies • Gordienko, E.; Hernández-Lerma, O.	Average cost Markov control processes with weighted norms: Existence of canonical	95 ZastApMa 23 199-218
Gordienko, E.; Hernández-Lerma, O.	Average cost Markov control processes with weighted norms: Value iteration •	95 ZastApMa 23 219-237
	Average derivative estimation with errors-in-variables • Fan, Yanqin	95 JNnparSt 4 395-407
Consistent estimation of density-weighted	average derivative by orthogonal series method • Rao, B. L. S. Prakasa	95 StPrLet 22 205-212
	Average error in the central limit theorem for the cumulative processes • Roginsky, Allen	95 StPrLet 24 199-204
Weighted	average importance sampling and defensive mixture distributions • Hesterberg, Tim	95 Technmcs 37 185-194
linear mixed models • Gilmour, A. R. et al.	Average information REML: An efficient algorithm for variance parameter estimation in	95 Biomtrcs 51 1440-1450
The law of	averages • Watkins, Ann E.	95 Chance 8/2 28-32
Marcus, Michael B.; Rosen, Jay Logarithmic	averages for the local times of recurrent random walks and Levy processes •	95 StocProc 59 175-184
	Average run length	
Woodall, William H.; Drake, Laurel G.	Alarm rates for quality control charts • Margavio, Thomas M.; Conerly, Michael D.;	95 StPrLet 24 219-224
alternatives • Amin, Raid W.; Wolff, Hans	The behaviour of EWMA-type quality control schemes in the case of mixture	95 SqtlAnly 14 157-177
	Detecting changes in probabilities of a multi-component process • Khan, Rasul A.	95 SqtlAnly 14 375-388
moving average control charts • Gan, F. F.	Joint monitoring of process mean and variance using exponentially weighted	95 Technmcs 37 446-453
• Chandrasekaran, S.; English, J. R.; Disney, R. L.	Modeling and analysis of EWMA control schemes with variance-adjusted control limits	95 IIETrn 27 282-290
Runger, George C.	Statistical process control scheme design • Keats, J. Bert; Miskulin, John D.;	95 JQualTek 27 214-225
change-point • Siegmund, D.; Venkatraman, E. S.	Using the generalized likelihood ratio statistic for sequential detection of a	95 AnlsStat 23 255-271
Willemain, Thomas R.; Prabhu, Sharad	Average run lengths for CUSUM control charts applied to residuals • Runger, George C.;	95 CommStA 24 273-282
process (French) • Niang, N.; Saporta, G.	Average run length of a moving average control chart to detect shifts in the mean of a	95 RStApp 43/3 5-20
• Rigdon, S. E. An integral equation for the in-control	average run length of a multivariate exponentially weighted moving average control chart	95 JStCmpSm 52 351-365
• Rigdon, Steven E. A double-integral equation for the	average run length of a multivariate exponentially weighted moving average control chart	95 StPrLet 24 365-373
	Average sample number	
tightening • Arumainayagam, S. D. et al.	Construction and selection of quick switching double-sampling system: Sample size	95 JAppStat 22 105-119
An	averaging approach to prediction • Wong, A.	95 StHefte 36 253-264
data • Weber, Rudolf O.; Madden, Roland A. Optimal	averaging for the determination of global mean temperature: Experiments with model	95 JClimate 8 418-430
Yang, Jichuan Stochastic approximation with	averaging and feedback: Rapidly convergent "on-line" algorithms • Kushner, Harold J.;	95 IEEEAuCn 40 24-34
neighbours, partial least squares and weighted	averaging partial least squares) and classical approaches • Braak ... of inverse (k-nearest	95 Chemolab 28 165-180
Stochastic	averaging principle for systems with pathwise uniqueness • Golec, Janusz	95 StocAnAp 13 307-322
systems with zero switchover times: A heavy-traffic	averaging principle • Coffman, E. G., Jr; Puhalskii, A. A.; Reiman, M. I. Polling	95 AnlsApPr 5 681-719
Some results about	averaging in stochastic approximation • Le Breton, Alain; Novikov, Alexander	95 Metrika 42 153-171
An alternative	axiomatic approach to expected utility theory • Di Bacco, Mario; Regoli, Giuliana	95 Metron 53 21-29
The	axiomatic derivation of a class of dynamic inequality measures • Stich, Andreas	95 Metron 53 31-41
	Axiomatic of descriptive statistics (French) • Rousson, Valentin	95 Student 1 17-34
	Axiomatic price index theory: A survey • Balk, Bert M.	95 IntStRvw 63 69-93
Extending Blackwell's strengthening of	Azuma's martingale inequality • Ross, Sheldon M.	95 PrEngInf 9 493-496
On optimal	B-robust influence functions in semiparametric models • Shen, Larry Z.	95 AnlsStat 23 968-989
	***B*-spline**	
series • Shi, Peide	Asymptotic behaviour of nonparametric conditional quantile estimates for time	95 CandJSt 23 161-169
	Exact multivariate Bayesian bootstrap distributions of moments • Gasparini, Mauro	95 AnlsStat 23 762-768
inertia of two random variables in Hilbert space and	B-spline approximations with a forestry application (French) • Champely, S. et al. Co-	95 RStApp 43/1 91-107
Global convergence rates of	B-spline M-estimators in nonparametric regression • Shi, Peide; Lin, Guoying	95 StSinica 5 303-318
Hazard function estimation using	B-splines • Rosenberg, Philip S.	95 Biomtrcs 51 874-887
formal derivation of the expected behavior of prefix	b-trees • Mahmoud, Hosam M.; Orlandić, Ratko Toward a	95 PrEngInf 9 183-192
	Backcalculation	
Sensitivity analysis, information, decision • Day et al.	Acquired immune deficiency syndrome predictions for England and Wales (1992-97):	95 JRSS-A 158 505-524
HIV epidemic • Tan, W.-Y.; Lee, S. R.; Tang, S. C.	Characterization of HIV infection and seroconversion by a stochastic model of the	95 MaBisci 126 81-123
Salvaggio, Antonino	Non-AIDS related mortality in the human immunodeficiency virus infection •	95 ListyBio 32 23-33
An empirical Bayes approach to smoothing in	backcalculation of HIV infection rates • Liao, Jiangang; Brookmeyer, Ron	95 Biomtrcs 51 579-588
of marker-associated QTL effects in Monte Carlo	backcross generations using multiple regression • Moreno-Gonzalez, J. Estimates	92 ThApGene 85 423-434
	Backfitting	
marginal integration • Linton, O.; Nielsen, J. P.	A kernel method of estimating structured nonparametric regression based on	95 Biomtrka 82 93-100
	Backprojection	
positron emission tomography • O'Sullivan, Finbarr	A study of least squares and maximum likelihood for image reconstruction in	95 AnlsStat 23 1267-1300
optimal statistical filter for the discrete convolution	backprojection method • Dhariyal, I. D.; Rathore, R. K. S.; Srivastava, T. et al. Design of	94 AJMMgSci 14 229-265
	Backpropagation in time-series forecasting • Lachtermacher, Gerson; Fuller, J. David	95 JForcast 14 381-393
Hamadéne, S.; Lepeltier, J. P.	Backward equations, stochastic control and zero-sum stochastic differential games •	95 Stochatx 54 221-231
gamma-martingales with prescribed limit, using	backwards SDE • Darling, R. W. R. Constructing	95 AnlsProb 23 1234-1261
Solution of forward-	backward stochastic differential equations • Hu, Y.; Peng, S.	95 ProbTher 103 273-283
Adapted solutions of	backward stochastic differential equations with non-Lipschitz coefficients • Mao, Xuerong	95 StocProc 58 281-292
Product forms based on	backward traffic equations • Boucherie, Richard J.	95 JAppProb 32 508-518
Johnson, K. R.; Camper, A. K. et al. Analysis of	bacterial spatial patterns at the initial stage of biofilm formation • Hamilton, M. A.;	95 BiomtrcJ 37 393-408
semiparametric model • Liang, Hua On	Bahadur asymptotic efficiency of the maximum likelihood estimator for a generalized	95 StSinica 5 363-371
	Bahadur asymptotic efficiency in a semi-parametric regression model • Liang, Hua	95 ChinaAMB 15 129-140
	Bahadur exact slopes for a class of combinations of dependent tests • Pallini, Andrea	94 Metron 52/1 53-65
Bartoszewicz, Jarosław	Bahadur and Hodges-Lehmann approximate efficiencies of tests based on spacings •	95 StPrLet 23 211-220
Koul, Hira L.; Zhu, Zhiwei	Bahadur-Kiefer representations for GM-estimators in autoregression models •	95 StocProc 57 167-189

On	Bahadur-Kiefer representation of a kernel conditional quantile estimator • Xiang, Xiaojing	95 JNnparSt 5 275-287
On the uniform	Bahadur-Kiefer representation • Shi, Z.	95 MaMethSt 4 39-55
Zhou, Yong	Bahadur-Kiefer theorems for kernel smooth product-limit quantile estimator •	95 CommStA 24 2815-2828
A	Bahadur-Kiefer theorem beyond the largest observation • Einmahl, John H. J.	95 JMultiAn 55 29-38
Xiang, Xiaojing	Bahadur representation of the kernel quantile estimator under random censorship •	95 JMultiAn 54 193-209
variables • Yoshihara, Ken-ichi The	Bahadur representation of sample quantiles for sequences of strongly mixing random	95 StPrLet 24 299-304
of mixture-process designs via the Franklin-	Bailey algorithm • DuMouchel, William H.; Lane, Thomas P. Construction	94 ASAProQlPr 7-16
Obituary: Albert	Baines, 1927-1994 • Paddle, G. M.	95 JRSS-A 158 179-180
on variance components in regression models with	balanced $(Q-1)$-fold nested error structure • Yu, Q.-L.; Burdick, R. K. Confidence intervals	95 CommStA 24 1151-1167
interactions ... • Shirakura et al. Search designs for	balanced 2^m factorial designs of resolution $2l+1$ when $l+s+1$-factor and higher order	93 JCISS 18 233-245
On the number of constraints for	balanced arrays of strength four • Chopra, D. V.; Dios, R.	93 JCISS 18 203-208
	Balanced arrays of strength two and nested (r,λ)-designs • Kuriki, Shinji; Fuji-Hara, Ryoh	94 JComDsgn 2 407-414
A note on generalized binary proper efficiency-	balanced block designs • Kherwa, G. Ram; Prasad, Jagdish	95 CommStA 24 2797-2801
On a	balanced classification rule • Kim, Hea-Jung	95 JKoreaSt 24 453-470
	Balanced credibility estimation • Neuhaus, Walther	95 AplStMDA 11 217-230
The truth, the computer packages, the books. Part I:	Balanced data" (V47 p48-59) • Stanek, E. J. et al. Comment on "The mixed-model ANOVA:	95 AmerStat 49 323-324
The truth, the computer packages, the books. Part I:	Balanced data'" • Schwarz Reply to comment on "Comment on 'The mixed-model ANOVA:	95 AmerStat 49 324-324
Some methods for constructing efficiency	balanced designs • Ghosh, D. K.; Karmokar, P. K.	94 JIAgSt 46 305-309
Construction of variance	balanced designs through triangular PBIB designs • Ghosh, D. K.; Joshi, K. S.	95 CalcutSt 45 111-118
Optimal, non-binary, variance	balanced designs • Morgan, John P.; Uddin, Nizam	95 StSinica 5 535-546
Kageyama, S. Variance	balanced designs with unequal block sizes and unequal replications • Gupta, S.;	92 UtilMath 42 15-24
Analysis of variance of partially	balanced fractional $2^{m_1+m_2}$ factorial designs of resolution IV • Kuwada, Masahide	93 JCISS 18 343-367
Hyodo, Yoshifumi Analysis of variance of	balanced fractional $r^m \times s^n$ factorial designs of resolution V • Kuwada, Masahide;	95 JJapnStS 25 129-141
Damaraju; Heiberger, Richard M. Variance	balanced generalized nested row-column designs with empty diagonal cells • Raghavarao,	93 JCISS 18 42-52
	Balanced incomplete block design	
Mullin, Ronald C.	Block designs with large holes and α-resolvable BIBDs • Furino, Steven C.;	93 JComDsgn 1 101-112
	Construction and optimality of nearly linear trend-free designs • Chai, Feng-Shun	95 JStPlInf 48 113-129
	Estimation of variance components in mixed linear models • Kubokawa, Tatsuya	95 JMultiAn 53 210-236
	Existence results for near resolvable designs • Furino, Steven	95 JComDsgn 3 101-113
design • Hedayat, A. S.; Majumdar, Dibyen	Generating desirable sampling plans by the technique of trade-off in experimental	95 JStPlInf 44 237-247
Friedrich; Sinha, Bikas Kumar	Optimal block designs revisited: An approximate theory detour • Pukelsheim,	95 StPrLet 24 375-380
	Optimal regression designs under random block-effects models • Cheng, Ching-Shui	95 StSinica 5 485-497
Kumar	Optimality aspects of Agrawal's designs: Part II • Heiligers, Berthold; Sinha, Bikas	95 StSinica 5 599-604
Bailey, R. A.; Preece, D. A.; Rowley, C. A.	Randomization for a balanced superimposition of one Youden square on another •	95 JRSS-B 57 459-469
• Hedayat, A. S. et al. Contingently and virtually	balanced incomplete block designs and their efficiencies under various optimality criteria	95 StSinica 5 575-591
Stufken, J.; Zhang, W. G. Virtually	balanced incomplete block designs for $v=22$, $k=8$, and $\lambda=4$ • Hedayat, A. S.;	95 JComDsgn 3 195-201
of J. X. Lu's "An existence theory for resolvable	balanced incomplete block designs" • Lee, Therese C. Y.; Furino, Steven C. A translation	95 JComDsgn 3 321-340
and exact distributions of rank tests for	balanced incomplete block designs • Alvo, Mayer; Cabilio, Paul Approximate	95 CommStA 24 3073-3121
Stanley Warner's contributions to statistically	balanced information technology • Fienberg, Stephen E.; Jazairi, Nuri	95 SurvyM 21 7-12
inadequacy of Satterthwaite's approximation in	balanced mixed models • Khuri, André I. A test to detect	95 MaOpfStS 27 45-54
Uddin et al. The robustness of binary and non-binary	balanced nested row-column designs under the unavailability of blocks: A comparison •	95 AstrlJSt 37 111-124
Neighbor-	balanced orthogonal arrays • Cheng, Ching-Shui	93 JCISS 18 61-67
On a	balanced quadratic classification rule • Kim, Hea-Jung	95 CommStA 24 607-623
Selection of the reference priors for a	balanced random effects model • Ye, K.	95 Test 4 1-17
Eick, Stephen G. Adaptive assignment versus	balanced randomization in clinical trials: A decision analysis • Berry, Donald A.;	95 StatMed 14 231-246
Model specification tests for	balanced representation state space models • Dorfman, Jeffrey H.; Havenner, Arthur	95 CommStA 24 97-119
Developing a	balanced set of measures in health care • Lochner, Robert H.	95 ASQCTekCon 293-300
Regression diagnostic techniques to detect	balanced space-to-time ratios in STARMA models • Otranto, Edoardo; Gallo, Giampiero M.	94 Metron 52/3 129-145
Rowley, C. A. Randomization for a	balanced superimposition of one Youden square on another • Bailey, R. A.; Preece, D. A.;	95 JRSS-B 57 459-469
confidence intervals for contrasts in the	balanced three factor mixed model • Mosier, Michael C.; Graybill, Franklin A. Approximate	95 BiomtrcJ 37 273-297
Some characterizations of a class of	balanced two-way elimination of heterogeneity designs • Siatkowski, Idzi	94 ListyBio 31 27-33
Reinventing the population census:	Balancing privacy and efficiency. In memory of Roger A. Herriot • Alexander, Lois	94 ASAProGovt 1-2
Small	ball probabilities of Gaussian fields • Shao, Q.-M.; Wang, D.	95 ProbTher 102 511-517
norms • Kuelbs, J.; Li, W. V.; Shao, Qi-man Small	ball probabilities for Gaussian processes with stationary increments under Hölder	95 JTherPrb 8 361-386
	Ballot theorems revisited • Konstantopoulos, Takis	95 StPrLet 24 331-338
The Strassen law of the iterated logarithm in	Banach function spaces • Norvaiša, R.	95 StPrLet 25 1-8
equations of stochastic optimal control in	Banach space • Zhu, Qingxin; Ahmed, N. U. Dynamic programming	95 StocAnAp 13 369-387
bounds for the distributions of martingales in	Banach spaces • Pinelis, Iosif Optimum	94 AnlsProb 22 1679-1706
multivalued operators defined on unbounded sets in	Banach spaces • Beg, Ismat; Shahzad, Naseer Random fixed points for random	95 StocAnAp 13 269-277
orem, weak law of large numbers for martingales in	Banach spaces, and weak invariance principle - A quantitative ... • Anastassiou ... limit the-	95 JMultiAn 52 158-180
On refinement of	Banach-valued limit theorems for stable laws • Chuprunov, A. N.	94 ThProbAp 39 657-662
	Band	
	• see also confidence –, tolerance –	
Ligeralde, Antonio V.; Brown, Bryan W.	Band covariance matrix estimation using restricted residuals: A Monte Carlo analysis •	94 IntEconR 35 751-767
Ranges of posterior probability over a distribution	band • Basu, Sanjib	95 JStPlInf 44 149-166
Robust Bayesian analysis with distribution	bands • Basu, Sanjib; DasGupta, Anirban	95 StatDecn 13 333-349
Reconstruction of	band limited processes from irregular samples • Houdré, Christian	95 AnlsProb 23 674-696
space domain implementation of the turning	bands method • Dietrich, C. R. A simple and efficient	95 WaterReR 31 147-156
Williams, Byron K.; Hines, James E. Use of solicited	bands and separation of hunting and natural mortality: A comment • Conroy, Michael J.;	95 JWildlMg 59 619-621
Lévy	bandits: Multi-armed bandits driven by Lévy processes • Kaspi, Haya; Mandelbaum, Avi	95 AnlsApPr 5 541-565
On transforming an index for generalised	bandit problems • Glazebrook, K. D.; Greatrix, S.	95 JAppProb 32 168-182
policies with $O(\log n)$ regret for the multi-armed	bandit problem • Agrawal, Rajeev Sample mean based index	95 AdvAppPr 27 1054-1078
An exponential two-armed	bandit problem with one arm known under batch sampling • Hamada, Toshio	95 JJapnStS 25 205-216
Response surface	bandits • Ginebra, Josep; Clayton, Murray K.	95 JRSS-B 57 771-784
protocol for the pattern recognition analysis of	bandpass filtered Fourier transform infrared interferograms • Shaffer et al. ... design	95 Chemolab 29 89-108
	Bandwidth	
	Adaptive root n estimates of integrated squared density derivatives • Wu, Tiee-Jian	95 AnlsStat 23 1474-1495
Ruppert, David	Asymptotics for the transformation kernel density estimator • Hössjer, Ola;	95 AnlsStat 23 1198-1222
estimators • Hall, Peter; Patil, Prakash	Formulae for mean integrated squared error of nonlinear wavelet-based density	95 AnlsStat 23 905-928
Fraiman, Ricardo	Minimum distance density-based estimation • Cao, Ricardo; Cuevas, Antonio;	95 CmpStDA 20 611-631
Glad, Ingrid K.	Nonparametric density estimation with a parametric start • Hjort, Nils Lid;	95 AnlsStat 23 882-904
power of covariates in regression • Doksum, K. et al.	Nonparametric estimation of global functionals and a measure of the explanatory	95 AnlsStat 23 1443-1473
Titterington, D. M.	On partial local smoothing rules for curve estimation • Hall, Peter; Marron, J. Stephen;	95 Biomtrka 82 575-587
regression • Hurvich, Clifford M.; Tsai, Chih-Ling	Relative rates of convergence for efficient model selection criteria in linear	95 Biomtrka 82 418-425

Adaptive	bandwidth choice for kernel regression • Schucany, William R.	95 JASA 90 535-540
Smooth linear regression using sample-determined	bandwidth • González Manteiga, Wenceslao	95 SankhyaA 57 79-87
Local and variable	bandwidths and local linear regression • Jones, M. C.	95 MaOpfStS 27 65-71
The effective	bandwidth of a multitaper spectral estimator • Walden, A. T.; McCoy, E. J.; Percival, D. B.	95 Biomtrka 82 201-214
Adaptive order polynomial fitting:	Bandwidth robustification and bias reduction • Fan, Jianqing; Gijbels, Irène	95 JCmpGrSt 4 213-227
Fattorini, L. A note on	bandwidth selections for kernel density estimators at endpoints • Barabesi, L.;	94 Metron 52/3 43-56
Léger, Christian	Bandwidth selection for kernel distribution function estimation • Altman, Naomi;	95 JStPlInf 46 195-214
adaptation • Fan, Jianqing; Gijbels, Irène Data-driven	bandwidth selection in local polynomial fitting: Variable bandwidth and spatial	95 JRSS-B 57 371-394
Scale measures for	bandwidth selection • Janssen, P.; Marron, J. S.; Veraverbeke, N.; Sarle, W.	95 JNnparSt 5 359-380
	Bandwidth selection in nonparametric regression with general errors • Chu, C. K.	95 JStPlInf 44 265-275
Engel, Joachim; Gasser, Theo A	bandwidth selector for bivariate kernel regression • Herrmann, Eva; Wand, M. P.;	95 JRSS-B 57 171-180
Wand, M. P. An effective	bandwidth selector for local least squares regression • Ruppert, D.; Sheather, S. J.;	95 JASA 90 1257-1270
of the smoothness of the regression function on a	bandwidth selector in nonparametric regression • Chu, C. K. The effect	95 CommStB 24 61-77
selection in local polynomial fitting: Variable	bandwidth and spatial adaptation • Fan, Jianqing; Gijbels, Irène Data-driven bandwidth	95 JRSS-B 57 371-394
An invariance property of optimal spectral	bandwidths • Robinson, P. M.	95 StPrLet 24 345-346
Effective	bandwidths for stationary sources • Courcoubetis, Costas; Weber, Richard	95 PrEngInf 9 285-296
of pharmacokinetic response curves with varying	bandwidths • Song, K. S.; Müller, H. G.; Clifford, A. J. et al. Estimating derivatives	95 Biomtrcs 51 12-20
Traffic accident mapping in	Bangkok metropolis: A case study • Ayuthya, Rampai Suksawasdi Na; Böhning, Danmar	95 StatMed 14 2445-2458
Bridging the biostatistics-epidemiology gap: The	Bangladesh task • Sen, Pranab Kumar	94 JStatRsr 28 21-39
Barth, James R. Deposit insurance and	bank failures: A preliminary analysis • Cebula, Richard J.; Belton, Willie J., Jr; Chou, Ray Y.;	94 ASAProBuEc 293-297
Multiattribute evaluation of Greek	banking performance • Zopounidis, C.; Despotis, D. K.; Stavropoulou, E.	95 AplStMDA 11 97-107
The nonparametric translog with application to	banking scale and scope economies • McManus, Douglas A.	94 ASAProBuEc 117-122
Autoregressive-type models for the analysis of	bark beetle tracks • Preisler, Haiganoush K.; Akers, R. Patrick	95 Biomtrcs 51 259-267
	Barker sequences and circulant Hadamard matrices • Lin, Cantian; Wallis, W. D.	93 JCISS 18 19-25
Modelling and simulation of	barley kernel growth • Koesmarno, H. K.	95 EnvrMtrc 6 511-516
motions (French) • Mathieu, P.	Barlow-Yor inequalities for local times of the intersection of two planar Brownian	95 ProbTher 102 83-103
the movement among staging areas of the	barnacle goose (Branta leucopsis) • Schwarz, Carl James; Ganter, Barbara Estimating	95 JAppStat 22 711-724
of a Wiener process with drift between two elastic	barriers • Dominé, Marco Moments of the first-passage time	95 JAppProb 32 1007-1013
C. Successfully applying SPC to improve health care:	Barriers to improving quality and reducing liability • Benneyan, James C.; Kaminsky, Frank	95 ASQCTekCon 570-578
	Bartlett adjustment	
Ferrari, Silvia L. P.	An improved Lagrange multiplier test for heteroskedasticity • Cribari-Neto, Francisco;	95 CommStB 24 31-44
	Two-sample empirical likelihood method • Jing, Bing-Yi	95 StPrLet 24 315-319
A	Bartlett adjustment to the likelihood ratio test for a system of equations • Attfield, C. L. F.	95 JEconmtx 66 207-223
	Bartlett corrections to likelihood ratio tests • Chesher, Andrew; Smith, Richard J.	95 Biomtrka 82 433-436
Cribari-Neto, F.; Aubin, E. C. Q.; Ferrari, S. L. P.	Bartlett corrections for one-parameter exponential family models • Cordeiro, G. M.;	95 JStCmpSm 53 211-231
Mukerjee, Rahul Comparison of	Bartlett-type adjustments for the efficient score statistic • Rao, C. Radhakrishna;	95 JStPlInf 46 137-146
Performance of a	Bartlett-type modification for the deviance • Cordeiro, Gauss M.	95 JStCmpSm 51 385-403
component analysis for the 1982-1992 Korean pro	baseball data (Korean) • Choi, Y.-S.; Shim, H.-J. Applications of the supplementary principal	95 KorApS 8/1 51-60
Forecasting	baseball games • Kaigh, W. D.	95 Chance 8/2 33-37
Comments on modelling	baseball • Lindsey, George	94 ASAProSpts 21-24
A critical look at some analyses of major league	baseball salaries • Hoaglin, David C.; Velleman, Paul F.	95 AmerStat 49 277-285
• Herndon et al. The restricted cubic spline as	baseline hazard in the proportional hazards model with step function time-dependent ...	95 StatMed 14 2119-2129
I/II trials when subject selection is determined by	baseline marker level • Lin et al. ... marker data for assessing treatment effects in phase	95 Biomtrcs 51 1053-1063
Cochran Q test with Turbo	BASIC • Seuc, Armando H.	95 CmpBiomd 46 29-34
Simpson's paradox and the hot hand in	basketball • Wardrop, Robert L.	95 AmerStat 49 24-28
Optimality of an analogue of	Basu's estimator under a double sampling design • Unnikrishnan, N. K.; Kunte, S.	95 SankhyaB 57 103-111
between 31P NMR and a biological duration test for	batch approval • Thelin, B. et al. Classification of Estradurin batches: Correlation	95 Chemolab 27 135-145
	Batch arrival	
van den Hout, Wilbert; Blanc, Hans	Development and justification of the power-series algorithm for BMAP-systems •	95 CommStSM 11 471-496
performance • Rananand, Nol	Markov approximations to D-BMAPs: Information theoretic bounds on queueing	95 CommStSM 11 713-734
• Thelin, B. et al. Classification of Estradurin	batches: Correlation between 31P NMR and a biological duration test for batch approval	95 Chemolab 27 135-145
Statistical process control in chemical	batch processes (things are seldom what they seem) • Janis, Stuart J.	94 ASAProPEng 308-313
Multivariate SPC charts for monitoring	batch processes • Nomikos, Paul; MacGregor, John F.	95 Technmcs 37 41-59
Multi-way partial least squares in monitoring	batch processes • Nomikos, Paul; MacGregor, John F.	95 Chemolab 30 97-108
bandit problem with one arm known under	batch sampling • Hamada, Toshio An exponential two-armed	95 JJapnStS 25 205-216
	Bathtub failure rate and upside-down bathtub mean residual life • Mi, Jie	95 IEEEReli 44 388-391
Bathtub failure rate and upside-down	bathtub mean residual life • Mi, Jie	95 IEEEReli 44 388-391
	Bayes	
	• see also empirical –, hierarchical –	
	Bayes approach	
	An averaging approach to prediction • Wong, A.	95 StHefte 36 253-264
A re-evaluation of the quasi-	Bayes approach to the linear combination of forecasts • Faria, A. E.; Souza, R. C.	95 JForcast 14 533-542
(Auth); Neuberg, Leland G. (Rev) Review of	"Bayes or bust? – A critical examination of Bayesian confirmation theory" • Earman, J.	95 Ststcian 44 283-284
(Auth); Worrall, John (Rev) Review of	"Bayes or bust? A critical examination of Bayesian confirmation theory" • Earman, John	94 PhilSci 61 672-673
Approximating	Bayes critical region for testing simple and tree ordered normal means • Sugiura, Nariaki	94 JStatRsr 28 1-20
Teaching inference about proportions using	Bayes and discrete models (jse/v3n3/albert) • Albert, Jim	95 JStEd 3/3 58kb
A nonparametric bound for the	Bayes error • Ruiz, Alberto	95 PattrnRc 28 921-930
Using the Gibbs sampler to simulate from the	Bayes estimate of a decreasing density • Brunner, Lawrence J.	95 CommStA 24 215-226
Subhashis; Samanta, Tapas Asymptotic behaviour of	Bayes estimates and posterior distributions in multiparameter nonregular cases • Ghosal,	95 MaMethSt 4 361-388
Olcay	Bayes estimation in a mixture inverse Gaussian model • Gupta, Ramesh C.; Akman, H.	95 AnInStMa 47 493-503
Optimal robust	Bayes estimation • Li, Yuanzhang; Saxena, K. M. Lal	95 JStPlInf 46 365-380
	Bayes estimator	
	Balanced credibility estimation • Neuhaus, Walther	95 AplStMDA 11 217-230
	Comparisons of estimators of a slippage probability • Sobel, Marc	95 CommStA 24 1039-1055
Biggeri, A.; Braga, M.; Lagazio, C.; Marchi, M.	Exploratory tools for disease mapping in geographical epidemiology • Cislaghi, C.;	95 StatMed 14 2363-2381
	Mean estimation for a partitioned multinormal distribution • Dahel, S.	95 CommStA 24 601-606
scale-invariant squared-error loss • van Eeden, C.	Minimax estimation of a lower-bounded scale parameter of a gamma distribution for	95 CandJSt 23 245-256
	Minimax estimation of proportions under random sample size • Amrhein, Peter	95 JASA 90 1107-1111
	On non-consistency of estimators • Kutoyants, Yu.; Vostrikova, L.	95 Stochatx 53 53-80
Lawrence D.; Casella, George; Hwang, J. T. Gene	Optimal confidence sets, bioequivalence, and the Limaçon of Pascal • Brown,	95 JASA 90 880-889
	Selection of the reference priors for a balanced random effects model • Ye, K.	95 Test 4 1-17
comparison of the efficiency of Bayes and empirical	Bayes estimators • Gruber, Marvin H. J. The	94 ASAProPEng 132-135
Admissibility of conjugate	Bayes estimators for the mean of a negative binomial distribution • Chou, Jine-phone	95 StatDecn 13 301-306
A robust generalized	Bayes estimator of a multivariate normal mean • Karunamuni, R. J.; Schmuland, B.	95 MaMethSt 4 472-482

p27-46) • Dette, Holger Correction to	"Bayesian D-optimal and model robust designs in linear regression models" (V25	95 MaOpfStS 26 285-286
	Bayesian data analysis • Gelman, A.; Carlin, J.; Stern, H.; Rubin, D.	95 ChapmnHall:Lnd 552p
Brown, P. J. (Rev) Review of	"Bayesian data analysis" • Gelman, A.; Carlin, J. B.; Stern, H. S.; Rubin, D. B. (Auth);	96 ShrtBkRv 16/1 5-5
algorithm based on a distance function and on a	Bayesian decision • Villarroya, Angel; Ríos, Martín; Oller, Josep M. Discriminant analysis	95 Biomtrcs 51 908-919
Whitehead, John; Brunier, Hazel	Bayesian decision procedures for dose determining experiments (Disc: p897-899) •	95 StatMed 14 885-893
for data in which censoring is ... • Fraser et al. A	Bayesian decision procedure for selecting prognostic variables associated with survival	95 BiomtrcJ 37 463-479
West, Mike	Bayesian density estimation and inference using mixtures • Escobar, Michael D.;	95 JASA 90 577-588
	Bayesian design	
	D-optimum designs for heteroscedastic linear models • Atkinson, A. C.; Cook, R. D.	95 JASA 90 204-212
Parmigiani, Giovanni	Optimal design via curve fitting of Monte Carlo experiments • Müller, Peter;	95 JASA 90 1322-1330
ies • Goodman, S. N.; Zahurak, M. L.; Piantadosi, S.	Some practical improvements in the continual reassessment method for phase I stud-	95 StatMed 14 1149-1161
Optimal	Bayesian design of experiments for the one-way random effects model • Lohr, Sharon L.	95 Biomtrka 82 175-186
The influence of variable selection: A	Bayesian diagnostic perspective • Weiss, Robert E.	95 JASA 90 619-625
Hale, Michael Results and efficiency of	Bayesian dose adjustment in a clinical trial with binary endpoint • Reeve, Russell;	94 ASAProBiop 324-327
Stettner, Ł.	Bayesian ergodic adaptive control of discrete time Markov processes • Di Masi, G. B.;	95 Stochatx 54 301-316
Clayton, D.; Montomoli, C.	Bayesian estimates of disease maps: How important are priors? • Bernardinelli, L.;	95 StatMed 14 2411-2431
Combined	Bayesian estimates for the equicorrelation coefficient • Viana, Marlos A. G.	95 CandJSt 23 43-53
Regal, Ronald R.	Bayesian estimates of a population size • Zelterman, Daniel; Hook, Ernest B.;	94 ASAProBays 29-34
On the concept of perfect aggregation in	Bayesian estimation (Disc: p287-289) • Bier, Vicki M.	94 RelEngSS 46 271-281
Rohana J.; Quinn, Terrance J., II	Bayesian estimation of animal abundance for line transect sampling • Karunamuni,	95 Biomtrcs 51 1325-1337
absence of a gold standard • Joseph, L. et al.	Bayesian estimation of disease prevalence and the parameters of diagnostic tests in the	95 AJEpidm 141 263-272
	Bayesian estimation of ground-water levels • de Waal, D. J.; Groenewald, P. C. N.	95 SoAfrStJ 29 65-78
data • Dupuis, Jerome A.	Bayesian estimation of movement and survival probabilities from capture-recapture	95 Biomtrka 82 761-772
Bolfarine, Heleno	Bayesian estimation of the number of equiprobable classes in a population • Leite, José G.;	94 RBrzlPrS 8 127-133
Howlader, Hatem A.; Hossain, Anwar On	Bayesian estimation and prediction from Rayleigh based on type II censored data •	95 CommStA 24 2249-2259
Rodrigues, J.; Baba, M. Y.	Bayesian estimation of a simple regression model with measurement errors •	94 RBrzlPrS 8 107-118
Pruska, Krystyna	Bayesian estimation of a structural changes model with dependent Bernoulli variable •	95 ListyBio 32 15-21
expansions for the deterministic computation of	Bayesian estimators based on Markov random fields • Wu, C.-h.; Doerschuk, P. C. Cluster	95 IEEEPAMI 17 275-293
Robust	Bayesian estimators in a one-way ANOVA model • Bian, G.	95 Test 4 115-135
Rasson, J. P.	Bayesian filtering and supervised classification in image remote sensing • Granville, V.;	95 CmpStDA 20 203-225
	Bayesian forecasting	
alternative techniques • Curry, D. J.; Divakar, S. et al.	BVAR as a category management tool: An illustration and comparison with	95 JForcast 14 181-199
	A BVAR model for the Connecticut economy • Dua, Pami; Ray, Subhash C.	95 JForcast 14 167-180
Davies, Neville (Rev) Review of "Applied	Bayesian forecasting and time series analysis" • Pole, A.; West, M.; Harrison, J. (Auth);	95 JRSS-A 158 635-636
Testing for unit roots in a	Bayesian framework • Lubrano, Michel	95 JEconmtx 69 81-109
	Bayesian graphical models for discrete data • Madigan, David; York, Jeremy	95 IntStRvw 63 215-232
formation to enhance the predictive performance of	Bayesian graphical models • Madigan, D.; Gavrin, J.; Raftery, A. E. Eliciting prior in-	95 CommStA 24 2271-2292
Lewis, Roger J.; Berry, Donald A.	Bayesian group sequential clinical trial designs based on a quadratic loss function. •	94 ASAProBays 23-28
Rosner, Gary L.; Berry, Donald A. A	Bayesian group sequential design for a multiple arm randomized clinical trial •	95 StatMed 14 381-394
Prediction and decision making using	Bayesian hierarchical models • Stangl, Dalene K.	95 StatMed 14 2173-2190
Pierce, Rebecca L.	Bayesian HPM intervals for the number of nonconforming items in a lot • Dyer, Danny;	95 CmpStDA 20 159-172
Adaptive mixture estimation and unsupervised local	Bayesian image segmentation • Peng, Anrong; Pieczynski, Wojciech	95 CVGIPGr 57 389-399
New loss functions in	Bayesian imaging • Rue, Havard	95 JASA 90 900-908
	Bayesian inference	
under marginal probabilities • Kato, Takeshi	Asymptotic sufficiency of posterior densities with respect to Kullback information	95 StatDecn 13 253-271
	Complete classes of tests for regularly varying distributions • Kowalski, Jacek P.	95 AnInStMa 47 321-350
	Complete classes of tests for some nonparametric hypotheses • Kowalski, Jacek P.	95 StatDecn 13 167-179
Smith, A. F. M.	Conjugate parameterizations for natural exponential families • Gutiérrez-Peña, E.;	95 JASA 90 1347-1356
	Evaluation of operational reliability from ground test data • Ke, Huei-Yaw	95 RelEngSS 49 103-107
Bischoff, Wolfgang; Fieger, Werner; Wulfert, Stefanie	Minimax- and Γ-minimax estimation of a bounded normal mean under LINEX loss •	95 StatDecn 13 287-298
	Missing data imputation using the multivariate t distribution • Liu, Chuanhai	95 JMultiAn 53 139-158
Osiewalski, Jacek; Steel, Mark F. J.	Modeling and inference with v-spherical distributions • Fernández, Carmen;	95 JASA 90 1331-1340
	New loss functions in Bayesian imaging • Rue, Havard	95 JASA 90 900-908
alternative hypothesis • Tsubaki, Michiko	New weighted sum of χ^2 test for testing a linear hypothesis against a restricted	95 CommStB 24 987-1004
Müller, D. W.	Power robustification of approximately linear tests • Ehm, W.; Mammen, E.;	95 JASA 90 1025-1033
Srivastava, M. S.	Probability matching priors for linear calibration • Ghosh, M.; Carlin, B. P.;	95 Test 4 333-357
Turrero, Agustin	Relative efficiency of a censored experiment in terms of Fisher information matrix •	95 CommStA 24 1169-1191
	Sample-audit tax assessment for businesses: What's fair? • Press, S. James	95 JBES 13 357-359
	Tracking the output of a poorly known Markov process • Rishel, Raymond	95 Stochatx 52 147-156
Mukhopadhyay, Saurabh; Ghosh, Malay	On the uniform approximation of Laplace's prior by t-priors in location problems •	95 JMultiAn 54 284-294
Nonparametric	Bayesian inference for accelerated failure time model • Kuo, Lynn; Mallick, Bani	94 ASAProBays 90-95
infinitely divisible distributions with applications to	Bayesian inference • Damien, P. et al. Approximate random variate generation from	95 JRSS-B 57 547-563
Generalized Laplacian approximations in	Bayesian inference • Hsu, John S. J.	95 CandJSt 23 399-410
	Bayesian inference for conditionally specified models • Arnold, Barry C.	94 ASAProBays 165-168
	Bayesian inference in cyclical component dynamic linear models • West, Mike	95 JASA 90 1301-1312
Effect on	Bayesian inference of the degree of precision of recorded data • Tricker, A. R.	95 JAppStat 22 235-244
Kvam, Paul H.; Martz, Harry F.	Bayesian inference in a discrete shock model using confounded common cause data •	95 RelEngSS 48 19-25
Mukhopadhyay, Saurabh On nonparametric	Bayesian inference for the distribution of a random sample • Gelfand, Alan E.;	95 CandJSt 23 411-420
Brimacombe, Michael	Bayesian inference and likelihood selection: Empirically conditioned likelihood •	94 ASAProBays 40-45
Lin, Dennis K. J.; Guess, Frank M.; Usher, John S.	Bayesian inference for masked system lifetime data • Reiser, B.; Guttman, I.;	95 ApplStat 44 79-90
	Bayesian inference in multipoint gene mapping • Stephens, D. A.; Smith, A. F. M.	93 AnHuGene 57 65-82
Mouchart, Michel; Scheihing, Eliana Approximate	Bayesian inferences: Numerical experiences when measuring the loss of information •	94 ASAProBays 221-226
On semiparametric pivotal	Bayesian inference for quantiles • Swartz, T. B.; Villegas, C.; Martinez, C. J.	95 CommStA 24 2499-2515
	Bayesian inference for stable distributions • Buckle, D. J.	95 JASA 90 605-613
	Bayesian inference of threshold autoregressive models • Chen, Cathy W. S.; Lee, Jack C.	95 JTimSrAn 16 483-492
Dependent masking and system life data analysis:	Bayesian inference for two-component systems • Guttman, I.; Lin, D. K. J.; Reiser, B. et al.	95 LiDaAnal 1 87-100
Sohn, Joong Kweon; Kim, Heon Joo	Bayesian inference for the two-parameter exponential models: Type-II censored case •	95 JKoreaSt 24 313-335
On priors providing frequentist validity for	Bayesian inference • Datta, Gauri Sankar; Ghosh, Jayanta Kumar	95 Biomtrka 82 37-45
	Bayesian inference for variance component models • Wolfinger, Russell D.; Kass, Robert E.	94 ASAProBays 136-141
	Bayesian information criterion	
goodness-of-fit tests • Kallenberg, W. C. M. et al.	Consistency and Monte Carlo simulation of a data driven version of smooth	95 AnlsStat 23 1594-1608
Soofi, Ehsan Lifetime modeling via	Bayesian information distinguishability • Mazzuchi, Thomas A.; Soyer, Refik;	94 ASAProBays 238-245
Cheng; Mountain, Dean C.; Illman, Kathleen Ho A	Bayesian integration of end-use metering and conditional-demand analysis • Hsiao,	95 JBES 13 315-326
Larsen, Michael D. Data augmentation with	Bayesian iterative proportional fitting applied to a Census Bureau latent-class problem •	94 ASAProGovt 116-121

ments and estimation of population size in darkling	beetles (Coleoptera: Tenebrionidae) • Crist, Thomas O.; Wiens, John A. Individual move-	95 JAnmlEco 64 733-746
Autoregressive-type models for the analysis of bark	beetle tracks • Preisler, Haiganoush K.; Akers, R. Patrick	95 Biomtrcs 51 259-267
B. et al. Interplay between design and analysis for	behavioral intervention trials with community as the unit of randomization • Green, S.	95 AJEpidm 142 587-593
Experimental design: Procedures for the	behavioral sciences (Third edition) • Kirk, Roger E.	95 BrooksCole:CA 921p
Handbook of statistical modeling for the social and	behavioral sciences • Arminger, Gerhard; Clogg, Clifford C.; Sobel, Michael E.	95 PlenumPubl:NY 618p
“Handbook of statistical modeling for the social and	behavioral sciences” • Arminger, G.; Clogg, C. C.; Sobel, M. E. (Ed) et al. Review of	95 Biomtrcs 51 1586-1587
Sample size choice in	behavioral and social research: Classical versus Bayesian methods • Pollard, William E.	94 ASAProBays 206-210
Are people Bayesian? Uncovering	behavioral strategies • El-Gamal, Mahmoud A.; Grether, David M.	95 JASA 90 1137-1145
“Handbook of statistical modeling for the social and	behavioural sciences” • Arminger, G.; Clogg, C. C.; Sobel, M. E. (Ed) et al. Review of	95 Chance 8/4 32-36
Experiment design and statistical methods for	behavioural and social research • Boniface, D.	95 ChapmnHall:Lnd 272p
properties of central composite and Box-	Behnken designs • Borkowski, John J. Spherical prediction-variance	95 Technmcs 37 399-410
prediction variances for central composite and Box-	Behnken designs • Borkowski, John J. Minimum, maximum, and average spherical	95 CommStA 24 2581-2600
Some properties of Box-	Behnken designs • Jo, Jinnam; Hinkelmann, Klaus	93 JCISS 18 273-287
	Behrens-Fisher problem	
	ANOVA under unequal error variances • Weerahandi, Samaradasa	95 Biomtrcs 51 589-599
Woodroofe, Michael B.	Central limit theorems for doubly adaptive biased coin designs • Eisele, Jeffrey R.;	95 AnlsStat 23 234-254
	Coefficients of Lee-Gurland two-sample test on normal means • Lee, Austin F. S.	95 CommStA 24 1743-1768
Rachael L.; Muller, Keith E.	A comparison of power approximations for Satterthwaite's test • DiSantostefano,	95 CommStB 24 583-593
populations • Jordan, Scott M.; Krishnamoorthy, K.	Confidence regions for the common mean vector of several multivariate normal	95 CandJSt 23 283-297
	A measure to evaluate the closeness of Satterthwaite's approximation • Khuri, A. I.	95 BiomtrcJ 37 547-563
Voinov, V. G.	On the problem of the means of weighted normal populations • Nikulin, M. S.;	95 Questo 19 93-106
	Randomization tests to compare means with unequal variation • Manly, Bryan F. J.	95 SankhyaB 57 200-222
Simulation results on solutions to the multivariate	Behrens-Fisher problem via trimmed means • Wilcox, Rand R.	95 Ststcian 44 213-225
	Belief function	
II: Fuzzy logic • Mahler, Ronald P. S.	Combining ambiguous evidence with respect to ambiguous a priori knowledge. Part	95 FuzzySet 75 319-354
course closing • Shipley et al. Rough set theory fuzzy	belief functions related to statistical confidence: Application and evaluation for golf	95 StocAnAp 13 487-502
Graphic	belief modeling • Almond, Russell G.	95 ChapmnHall:Lnd 350p
Empirical model particularities and	belief in the natural rate hypothesis • Kim, Jinbang; De Marchi, Neil; Morgan, Mary S.	95 JEconmtx 67 81-102
	Belief network	
Steffen L.	The EM algorithm for graphical association models with missing data • Lauritzen,	95 CmpStDA 19 191-201
	Belief networks for modelling and assessment of environmental change • Varis, Olli	95 EnvrMtrc 6 439-444
for pairwise comparisons, binary sequences, and	belief networks” • Laskey, Kathryn B. Comment on “Bayesian analysis	94 ASAProBays 134-135
p134-135) • Ji, X. et al. Approximate reasoning in	belief networks: Statistical validity, accuracy, and determination of sample size (Disc:	94 ASAProBays 128-133
B.; Fienberg, S. E.; Resnick, D. P. et al. Wringing the	bell curve: A cautionary tale about the relationships among race, genes, and IQ • Devlin,	95 Chance 8/3 27-36
(Auth); Devlin, B.; Fienberg, S. E. et al. Review of “The	bell curve: Intelligence and class structure in American life” • Herrnstein, R. J.; Murray, C.	95 JASA 90 1483-1488
	Bell curves and monkey languages • Casti, John L.	95 Complxty 1 12-15
On strong uniform consistency of the Lynden-	Bell estimator for truncated data • Chen, Kani; Chao, Min-Te; Lo, Shaw-Hwa	95 AnlsStat 23 440-449
model: Some elements of an informed use of the	Belsley, Kuh and Welsch indicators (French) • Erkel-Rousse ... in an ordinary linear	95 RStApp 43/4 19-42
Calculation of	benchmark doses from continuous data • Crump, Kenny S.	95 RiskAnal 15 79-89
Shenfild, M. State-dependent	Beneš buffer model with fast loading and output rates • Kogan, Y.; Liptser, R.;	95 AnlsApPr 5 97-120
The	Beneš equation and stochastic calculus of variations • Decreusefond, L.; Üstünel, A. S.	95 StocProc 57 273-284
A direct computation of the	Beneš filter conditional density • Crisan, Dan	95 Stochatx 55 47-54
Spatial analysis of	benthic data from the Chesapeake Bay • Neerchal, Nagaraj K.; Shafer, Ronald W.	94 ASAProEnvr 204-208
of line transect methods to estimate abundance of	benthic stream fishes • Ensign, William E.; Angermeier, Paul L.; Dolloff, C. Andrew Use	95 JFishRBC 52 213-222
optimization in multiclass M/GI/1 queues with	Bernoulli feedback • Bhattacharya, P. P.; Georgiadis, L.; Tsoucas, P. Problems of adaptive	95 MaOpRsr 20 355-380
Limit distributions for the	Bernoulli meander • Takács, Lajos	95 JAppProb 32 375-395
On the limit distributions of sums of mixing	Bernoulli random variables • Dziubdziela, Wiesław	95 StPrLet 23 179-182
Kolmogorov inequality for the sum of independent	Bernoulli random variables with unequal means • Turner, D. W.; Young, D. M. et al. A	95 StPrLet 23 243-245
Variance reduction for	Bernoulli response variables in simulation • Ocaña, Jordi; Vegas, Esteban	95 CmpStDA 19 631-640
times in a two-queue model with exhaustive and	Bernoulli service • Weststrate, Jan A.; van der Mei, Rob D. Waiting	94 ZeitOpRs 40 289-303
	Bernoulli shift	
	Isomorphism of random walks • Aaronson, Jon; Keane, Michael	94 IsrelJMa 87 37-63
The foundation of the	Bernoulli Society • Kendall, David G.	95 Bernoull 1 1-2
estimates and Bayesian confidence intervals in	Bernoulli SS ANOVA models • Wang, Y. et al. Simulation studies of smoothing parameter	95 CommStB 24 1037-1059
	Bernoulli trial	
	Runs on a circle • Koutras, M. V.; Papadopoulos, G. K.; Papastavridis, S. G.	95 JAppProb 32 396-404
and negative binomial analogues under correlated	Bernoulli trials” (V48 p243-247) • Katz, Richard W. Comment on “Binomial	95 AmerStat 49 325-325
and negative binomial analogues under correlated	Bernoulli trials” • Viveros, R.; Balasubramanian, K. et al. Reply to comment on “Binomial	95 AmerStat 49 325-326
largest square of successes covering the origin for	Bernoulli trials on the lattice • Kopociński, B. The	95 StNeerla 49 31-39
	Bernoulli trials and number theory • Rawlings, Don	94 AmMaMnth 101 948-952
of a structural changes model with dependent	Bernoulli variable • Pruska, Krystyna Bayesian estimation	95 ListyBio 32 15-21
Steif, Jeffrey E. Quite weak	Bernoulli with exponential rate and percolation for random fields • Burton, Robert M.;	95 StocProc 58 35-55
The	Bernstein polynomial estimator of a smooth quantile function • Cheng, Cheng	95 StPrLet 24 321-330
A	Bernstein-type inequality for U-statistics and U-processes • Arcones, Miguel A.	95 StPrLet 22 239-247
	Berry-Esseen bound	
Euclidean space • Einmahl, Uwe	On the cluster set problem for the generalized law of the iterated logarithm in	95 AnlsProb 23 817-851
Götze, Friedrich; Zitikis, Ričardas	Lower estimates of the convergence rate for U-statistics • Bentkus, Vidmantas;	94 AnlsProb 22 1707-1714
	Strong approximation theorems for integrated kernel quantiles • Ralescu, S. S.	95 MaMethSt 4 201-215
local probability approximations: Improvements on	Berry-Esseen • Hahn, Marjorie G. W.; Klass, Michael J. Uniform	95 AnlsProb 23 446-463
coverage bounds for confidence intervals and	Berry-Esseen theorems for Edgeworth expansion • Hall, Peter; Jing, Bing-Yi Uniform	95 AnlsStat 23 363-375
deficiency of quantile estimators • Xiang, Xiaojing A	Berry-Esseen theorem for the kernel quantile estimator with application to studying the	95 AnInStMa 47 237-251
Accounting for individual variability in the von	Bertalanffy growth model • Wang, You-Gan; Thomas, Mervyn R.	95 JFishRBC 52 1368-1375
	Bertrand's work on probability • Sheynin, O.	94 AHistESc 48 155-199
	Bessel process	
	Long-term returns in stochastic interest rate models • Deelstra, G.; Delbaen, F.	95 InsrneME 17 163-169
On transient	Bessel processes and planar Brownian motion reflected at their future infima • Shi, Z.	95 StocProc 60 87-102
Arbitrage possibilities in	Bessel processes and their relations to local martingales • Delbaen, F.; Schachermayer, W.	95 ProbTher 102 357-366
A	best-choice problem with three stops and random termination • Majumdar, A. A. K.	95 SankhyaB 57 76-84
The	best-choice secretary problem with random freeze on jobs • Samuel-Cahn, Ester	95 StocProc 55 315-327
Rates of convergence for	best entropy estimates • Noll, Dominikus	95 StatDecn 13 141-165
models • Baksalary, J. K.; Gnot, S.; Kageyama, S.	Best estimation of variance components with arbitrary kurtosis in two-way layouts mixed	95 JStPlInf 44 65-75
Kumar, Narinder Selecting the	best exponential population when the scale parameters are bounded • Misra, Neeraj;	95 MaOpfStS 27 151-158
In search of the	‘best’ growth inhibitor (Disc: p329-332) • Lee, Hyun Suk; Philips, Bob	95 CandJSt 23 322-325

Bias correction

	On higher order kernels • Jones, M. C.	95 JNnparSt 5 215-221
sample survey inference via bootstrapping and	bias correction: The case of the ratio estimator • Chambers, R. L.; Dorfman, A. H. Robust	94 ASAProStCp 51-60
dispersion • Breslow, Norman E.; Lin, Xihong	Bias correction in generalised linear mixed models with a single component of	95 Biomtrka 82 81-91
Paula, Gilberto A.; Cordeiro, Gauss M.	Bias correction and improved residuals for non-exponential family nonlinear models •	95 CommStA 24 1193-1210
proportioned sample between risk and ... • Lee et al.	Bias correction in risk assessment when logistic regression is used with an unevenly	94 ASAProBuEc 304-309
A comparative study of some	bias correction techniques for kernel-based density estimators • Foster, Peter	95 JStCmpSm 51 137-152
Zasadzinski, J. A. Inherent	bias in correlation averaged images • Woodward, J. T.; Kono, C.; Madsen, L. L.;	95 JMicro 178 86-92
Heterogeneity	biases, distributional effects, and aggregate consumption • Fortin, Nicole M.	95 JApEcmtx 10 287-311
Comment on "Divergent	biases in ecologic and individual level studies" (V11 p1209-1223) • Cohen, Bernard L.	95 StatMed 14 327-328
Reply to comment on "Divergent	biases in ecologic and individual level studies" • Greenland, Sander	95 StatMed 14 328-330
budget projections: A nonparametric assessment of	bias and efficiency • Campbell, Bryan; Ghysels, Eric Federal	95 RvEconSt 77 17-31
Tiwari, Gayatri; Saxena, K. P. On	bias and efficiency of a preliminary test estimator of variance in a normal distribution •	95 Metron 53 163-169
On the	biases of error estimators in prediction problems • Hall, Peter	95 StPrLet 24 257-262
fish • Hayes, D. B. et al. Efficiency and	bias of estimators and sampling designs for determining length-weight relationships of	95 JFishRBC 52 84-92
Kiviet, Jan F. On	bias, inconsistency, and efficiency of various estimators in dynamic panel data models •	95 JEconmtx 68 53-78
reasonable between village ... • Fitch et al. Are the	biases and inefficiencies of the EPI method of sampling relatively unimportant with	94 ASASrvyRsr 288-292
the nonexistence of a global nonnegative minimum	bias invariant quadratic estimator of variance components • Gao, S.; Smith, T. M. F. On	95 StPrLet 25 117-120
"Defining, detecting, and correcting test and item	bias" • Holland, Paul W. Comment on	94 ASAProSoSt 27-29
A note on the	bias of L-estimators and a bias reduction procedure • Xiang, Xiaojing	95 StPrLet 23 123-127
Return rates in studies of life history evolution: Are	biases large? • Martin, Thomas E.; Clobert, Jean; Anderson, David R.	95 JAppStat 22 863-875
prevalence, sensitivity and specificity of the length	bias • Shanmugam, R. Goodness-of-fit test for length-biased data with discussion on	95 JStPlInf 48 277-290
Starer, Amy	Bias in list-assisted telephone samples • Brick, J. Michael; Waksberg, Joseph; Kulp, Dale;	94 ASASrvyRsr 1293-1298
Starer, Amy	Bias in list-assisted telephone samples • Brick, J. Michael; Waksberg, Joseph; Kulp, Dale;	95 PubOpinQ 59 218-235
for censored age-of-onset data subject to recall	bias: Mothers' reports of age of puberty in male twins • Pickles, A. et al. A simple method	94 BehavGen 24 457-468
Liu estimator and unbiased estimation of the	bias and MSE • Akdeniz, Fikri; Kaçranlar, Selahattin On the almost unbiased generalized	95 CommStA 24 1789-1797
of a design for quadratic model fits fearing a cubic	bias in multiple response case • Bae, Whasoo A study on a basis for the selection	95 JKoreaSt 24 31-44
LeClere, F. B. et al. Social correlates of response	bias in the National Health Interview Survey: Lessons for the redesign (Disc: p379-381) •	94 ASASrvyRsr 373-378
Waksberg, Joseph An evaluation of nonresponse	bias in NHANES III (1988-91) • Khare, Meena; Ezzati-Rice, Trena M.; Mohadjer, Leyla K.;	94 ASASrvyRsr 949-954
levels • Correa-Villaseñor, A.; Stewart, W. F. et al.	Bias from nondifferential misclassification in case-control studies with three exposure	95 Epidmlgy 6 276-281
van Noordwijk, Arie J. On	bias due to observer distribution in the analysis of data on natal dispersal in birds •	95 JAppStat 22 683-694
models • Kiviet, J. F.; Phillips, G. D. A.; Schipp, B. The	bias of OLS, GLS, and ZEF estimators in dynamic seemingly unrelated regression	95 JEconmtx 69 241-266
A note on the	bias due to omitted confounders • Drake, Christiana; McQuarrie, Allan	95 Biomtrka 82 633-638
Test and item	bias: An overview (Disc: p27-29) • Zwick, Rebecca	94 ASAProSoSt 9-16
estimating effect size in the presence of publication	bias • Vevea, Jack L.; Hedges, Larry V. A general linear model for	95 Psymtrka 60 419-435

Bias reduction

Joseph P.	Bias-corrected nonparametric spectral estimation • Politis, Dimitris N.; Romano,	95 JTimSrAn 16 67-103
	Local and variable bandwidths and local linear regression • Jones, M. C.	95 MaOpfStS 27 65-71
Cook, J. R.	Simulation-extrapolation: The measurement error jackknife • Stefanski, L. A.;	95 JASA 90 1247-1256
Likelihood-based methods for	bias reduction in limiting dilution assays • Mehrabi, Yadollah; Matthews, J. N. S.	95 Biomtrcs 51 1543-1549
Correction to	"Bias reduction of maximum likelihood estimates" (V80 p27-38) • Firth, David	95 Biomtrka 82 667-667
A simple	bias reduction method for density estimation • Jones, M. C.; Linton, O.; Nielsen, J. P.	95 Biomtrka 82 327-338
A note on the bias of L-estimators and a	bias reduction procedure • Xiang, Xiaojing	95 StPrLet 23 123-127
polynomial fitting: Bandwidth robustification and	bias reduction • Fan, Jianqing; Gijbels, Irène Adaptive order	95 JCmpGrSt 4 213-227
Lee, Lung-fei Asymptotic	bias in simulated maximum likelihood estimation of discrete choice models •	95 EconTher 11 437-483
in aging populations due to self-selection statistical	bias (selective survival bias) • Blauwet, M. B. et al. Changing risk factor relationships	94 ASAProEpid 76-81
ment of ... • Epperson et al. Estimating the urban	bias of surface shelter temperatures using upper-air and satellite data. Part I: Develop-	95 JApMetrl 34 340-357
to self-selection statistical bias (selective survival	bias) • Blauwet, M. B. et al. Changing risk factor relationships in aging populations due	94 ASAProEpid 76-81
Song, Seuck Heun Consistency and bounds on the	bias of S^2 in the linear regression model with moving average disturbances •	95 JKoreaSt 24 507-518
capture-recapture data • Hestbeck, Jay B.	Bias in transition-specific survival and movement probabilities estimated using	95 JAppStat 22 737-750
Weng, Chung-Sing	Bias of two one-sided tests procedures in assessment of bioequivalence • Liu, Jen-Pei;	95 StatMed 14 853-861
... • Cleophas Reply to comment on "Interaction	biases in two-period crossover studies: A modified analysis to test with more sensitivity"	95 BiomtrcJ 37 507-510
postcode address file as ... • Lynn et al. On the	bias and variance of samples of individuals: Comparison of the electoral registers and	95 Ststcian 44 173-194
	Bias-variance tradeoffs in functional estimation problems • Low, Mark G.	95 AnlsStat 23 824-835
the double-count technique to correct visibility	bias • Rivest, L.-P.; Potvin, F.; Crépeau, H. et al. Statistical methods for aerial surveys using	95 Biomtrcs 51 461-470
and controlling for the effects of volunteer	bias with pairs of relatives • Neale, Michael C.; Eaves, Lindon J. Estimating	93 BehavGen 23 271-278
Estimating telephone noncoverage	bias with a telephone survey • Keeter, Scott	95 PubOpinQ 59 196-217
Central limit theorems for doubly adaptive	biased coin designs • Eisele, Jeffrey R.; Woodroofe, Michael B.	95 AnlsStat 23 234-254
Minimax kernel density estimators with length	biased data • Wu, C. O.	95 MaMethSt 4 56-80
Shanmugam, R. Goodness-of-fit test for length-	biased data with discussion on prevalence, sensitivity and specificity of the length bias •	95 JStPlInf 48 277-290
Are OLS estimates of the return to schooling	biased downward? another look • Blackburn, McKinley L.; Neumark, David	95 RvEconSt 77 217-229
Zmyślony, R. Nonnegative minimum	biased quadratic estimation in the linear regression models • Gnot, S.; Trenkler, G.;	95 JMultiAn 54 113-125
et al. A comparison of neural networks, non-linear	biased regression and a genetic algorithm for dynamic model identification • Wise, B. M.	95 Chemolab 30 81-89
maximum likelihood estimator in missing data and	biased sampling models • van der Laan, Mark J. An identity for the nonparametric	95 Bernoull 1 335-341
J. A. Adaptive	biased urn randomization in small strata when blinding is impossible • Schouten, Hubert	95 Biomtrcs 51 1529-1535
A CUSUM test for symmetry against positive	biasedness • Aly, Emad-Eldin A. A.; Kochar, Subhash C.	94 CalcutSt 44 157-163
	Biasing the transition probabilities in direct Monte Carlo • Zio, E.	95 RelEngSS 47 59-63
Kaplan, David The impact of	BIB spiraling-induced missing data patterns on goodness-of-fit tests in factor analysis •	95 JEdStat 20 69-82
The existence of doubly near resolvable $(v,3,2)$-	BIBDs • Lamken, E. R.	94 JComDsgn 2 427-440
The existence of doubly resolvable $(v,3,2)$-	BIBDs • Lamken, E. R.	95 JCombThA 72 50-76
BIBDs using 1-resolvable and α-affine resolvable	BIBDs • Shah, S. M.; Ghosh, D. K.; Patel, A. H. Construction of α-resolvable	93 UtilMath 43 219-224
Block designs with large holes and α-resolvable	BIBDs • Furino, Steven C.; Mullin, Ronald C.	93 JComDsgn 1 101-112
Patel, A. H. Construction of α-resolvable	BIBDs using 1-resolvable and α-affine resolvable BIBDs • Shah, S. M.; Ghosh, D. K.;	93 UtilMath 43 219-224
Yin, Jianxing Resolvable	BIBDs with block size 8 and index 7 • Furino, Steven; Greig, Malcolm; Miao, Ying;	94 JComDsgn 2 101-116
Some new	BIBDs with $k = 6$ and $\lambda = 1$ • Abel, R. J. R.; Mills, W. H.	95 JComDsgn 3 381-391

Bibliography

information explosion • Gbur, E. E., Jr; Trumbo, B. E.	Key words and phrases – The key to scholarly visibility and efficiency in an	95 AmerStat 49 29-33
Ranked set sampling: An annotated	bibliography • Kaur, A.; Patil, G. P.; Sinha, A. K.; Taillie, C.	95 EnvEcoSt 2 25-54
Review of "The chronological annotated	bibliography of order statistics (Volumes IV-VII)" • Harter, H. L. (Auth); Hušková, M. (Rev)	95 MaOpfStS 26 182-182
A	bibliography on telephone survey methodology • Khurshid, Anwer; Sahai, Haredeo	95 JOffStat 11 325-367
Alameddine, Amine; Glickman, Theodore S.	Biconvex models and algorithms for risk management problems • Sherali, Hanif D.;	94 AJMMgSci 14 197-228
Moreno, Elías; Cano, Juan A. Classes of	bidimensional priors specified on a collection of sets: Bayesian robustness •	95 JStPlInf 46 325-334
Wabuyele, B. W. et al. Quantitative comparison of	bidirectional and optimal associative memories for background prediction of spectra •	95 Chemolab 29 51-61
statistical techniques • Højsgaard, S.; Thiesson, B.	BIFROST – Block recursive models induced from relevant knowledge, observations, and	95 CmpStDA 19 155-175

	Binomial distribution	
	A bivariate discrete Charlier series distribution • Papageorgiou, H.; Loukas, S.	95 BiomtrcJ 37 105-117
	Criterion of discrete unimodal mixtures (Korean) • Choi, Daiwoo	95 KorApS 8/1 159-167
	Hierarchical clustering method using probabilistic similarity measure • Miyano, Hisao	95 Behvmtrk 22 145-153
Silva Mato, A.; Martín Andrés, A.	Optimal unconditional tables for comparing two independent proportions •	95 BiomtrcJ 37 821-836
sampled patients • Lui, Kung-Jong	Sample size calculation for intraclass correlation in the presence of random loss of	95 BiomtrcJ 37 97-103
Lemon, G. H.; Norton, J. M.	Sample size requirements for rare or abundant attribute sampling • Arnold, Jesse C.;	95 CommStB 24 131-148
likelihood estimation of the parameters of the beta	binomial distribution" (V32 p196-204) • Smith et al. ... on "Algorithm AS 189: Maximum	95 ApplStat 44 545-547
Sheffield, R. M. Using the β-	binomial distribution to characterize forest health • Zarnoch, S. J.; Anderson, R. L.;	95 CanJForR 25 462-469
On two-sided compound	binomial distributions • Wang, Shaun	95 InsrneME 17 35-41
A generalization of the	binomial distribution • Fu, James; Sproule, Robert	95 CommStA 24 2645-2658
of Altham's multiplicative generalization of the	binomial distribution • From ... least-squares procedure for estimating the parameters	95 StPrLet 25 193-199
The structure of compounded trivariate	binomial distributions • Papageorgiou, H.; David, Katerina M.	95 BiomtrcJ 37 81-95
Mikulecky, M. et al. Additivity test for composition of	binomial effects in chromosome aberrations after radiation injury • Komornik, J.;	95 StatMed 14 1941-1947
Bayesian method for combining results from several	binomial experiments • Consonni, Guido; Veronese, Piero A	95 JASA 90 935-944
	Binomial logit analysis	
data • Garner, Thesia I.; Blanciforti, Laura A.	Household income reporting: An analysis of U.S. Consumer Expenditure Survey	94 JOffStat 10 69-91
pricing of the "Russian option" for the symmetrical	binomial model of a (B,S)-market • Kramkov, D. O.; Shiryaev, A. N. On the rational	94 ThProbAp 39 153-162
Some aspects of inference for chain	binomial models • Yip, Paul; Huggins, Richard	95 StocAnAp 13 355-367
p243-247) • Katz, Richard W. Comment on	"Binomial and negative binomial analogues under correlated Bernoulli trials" (V48	95 AmerStat 49 325-325
Balasubramanian, K. et al. Reply to comment on	"Binomial and negative binomial analogues under correlated Bernoulli trials" • Viveros, R.;	95 AmerStat 49 325-326
on the accuracy of an approximate interval for the	binomial parameter • Huwang, Longcheen A note	95 StPrLet 24 177-180
distance between the mean and the median of the	binomial and Poisson distributions • Hamza, K. The smallest uniform upper bound on the	95 StPrLet 23 21-25
on "Recursive schemes for calculating cumulative	binomial and Poisson probabilities" (V48 p136-138) • Nemenyi, Peter Comment	95 AmerStat 49 402-403
therapeutic equivalence based on the difference of	binomial probabilities • Roebruck et al. ... of tests and sample size formulae for proving	95 StatMed 14 1583-1594
Confidence limits for the ratio of two	binomial proportions based on likelihood scores: Non-iterative method • Nam, Jun-mo	95 BiomtrcJ 37 375-379
Berger, R. d. Sample size calculations for	binomial proportions via highest posterior density intervals • Joseph, L.; Wolfson, D. B.;	95 Ststcian 44 143-154
Testing the equality of several	binomial proportions to a prespecified standard • Kulkarni, Pandurang M.; Shah, Arvind K.	95 StPrLet 25 213-219
On properties of	binomial Q charts for attributes • Quesenberry, Charles P.	95 JQualTek 27 204-213
Generalized	binomial regression model • Famoye, Felix	95 BiomtrcJ 37 581-594
posterior distribution for a class of mixed models for	binomial responses • Natarajan, R.; McCulloch, C. E. A note on the existence of the	95 Biomtrka 82 639-643
	Binomial sampling	
per by Joseph, Wolfson and du Berger • Adcock, C. J.	The Bayesian approach to determination of sample sizes - Some comments on the pa-	95 Ststcian 44 155-161
Unbiased estimation in sequential	binomial sampling • Akahira, Masafumi; Takeuchi, Kei; Koike, Ken-ichi	92 RStAJp 39/4 1-13
of the likelihood ratio test that a mixture of two	binomials is a single binomial • Chernoff, Herman; Lander, Eric Asymptotic distribution	95 JStPlInf 43 19-40
Comment on "A	binomial test of model fit" (V1 p268-273) • Drezner, Zvi; Drezner, Tammy	95 StrEqMod 2 144-145
Comment on "A	binomial test of model fit" (V1 p268-273) • Hsu, Louis M.	95 StrEqMod 2 152-154
Comment on "A	binomial test of model fit" (V1 p268-273) • Raykov, Tenko; Penev, Spiridon	95 StrEqMod 2 146-151
Reply to comments on "A	binomial test of model fit" • Newman, Isadore; Fraas, John W.; Norfolk, Timothy	95 StrEqMod 2 155-162
stochastic models leading to the convolution of two	binomial variables • Ong, S. H. Some	95 StPrLet 22 161-166
Generation of over-dispersed and under-dispersed	binomial variates • Ahn, Hongshik; Chen, James J.	95 JCmpGrSt 4 55-64
Federico Describing extra-	binomial variation with partially correlated models • Luceño, Alberto; de Ceballos,	95 CommStA 24 1637-1653
	Bioassay	
assessment • Sen, Pranab Kumar; Margolin, Barry H.	Inhalation toxicology: Awareness, identifiability, statistical perspectives and risk	95 SankhyaB 57 252-276
	Methods to approximate joint uncertainty and variability in risk • Bogen, Kenneth T.	95 RiskAnal 15 411-419
to vaccine • Moulton, Lawrence H.; Halsey, Neal A.	A mixture model with detection limits for regression analyses of antibody response	95 Biomtrcs 51 1570-1578
designs with multivariate responses • Hanusz, Zofia	Relative potency of two preparations in two-way elimination of heterogeneity	95 Biomtrcs 51 1133-1139
	Statistical analysis of drug interactions in anesthesia • Dexter, Franklin	95 JThBio 172 305-314
	The statistical analysis of slope ratio assays using SAS software • Wong, M. Y.	94 CmpBiomd 45 233-238
data • El-Saidi, Mohammed A.	A symmetric extended logistic model with applications to experimental toxicity	95 BiomtrcJ 37 205-216
the distribution of historical controls in carcinogen	bioassay • Krewski, D.; Smythe, R. T. et al. Bootstrapping an empirical Bayes estimator of	95 JStPlInf 43 215-228
computer-based system for parallel line analysis of	bioassay data • Iznaga, N.; Núñez, G.; Solozabal, J.; Morales, A.; Artaza, E. et al. A personal	95 CmpBiomd 47 167-175
A-efficient designs for	bioassays • Mukerjee, Rahul; Gupta, Sudhir	95 JStPlInf 48 247-259
Bayesian	bioassay design: A simulation based approach • Warner, John	94 ASAProBays 13-17
approach to the optimal design of an LD50	bioassay • Markus, Richard A.; Frank, Jess; Groshen, Susan; Azen, Stanley P. An alternative	95 StatMed 14 841-852
estimating standard errors for estimators in quantal	bioassay • Cobb, E. Benton; Church, J. D. A discrete delta method for	95 BiomtrcJ 37 691-699
	Bioavailability	
data • Abdullah, M. E.; El-Sayed, Yousry M.	Design of crossover microcomputer program and application on drug bioequivalence	95 CmpBiomd 48 237-239
and politics • Bonham, G. H.	Health and environment: Public health decisions based on evidence, psychology	95 EnvrMtrc 6 311-318
a conference on analytical methods validation in	bioavailability, bioequivalence and pharmacokinetic studies • Kringle et al. ... from	94 ASAProBiop 510-514
population bioequivalence using the probability that	bioavailabilities are similar • Schall, Robert Assessment of individual and	95 Biomtrcs 51 615-626
Review of	"Biodiversity - Measurement and estimation" • Hawksworth, D. L. (Ed); Feu, Chris du (Rev)	95 TeachgSt 17/3 111-111
	Bioequivalence	
	Equivalence with respect to a control: Stepwise tests • Bofinger, Eve; Bofinger, Mark	95 JRSS-B 57 721-733
by an application • Schuster, Ernst	Preference of equivalence tests with standardized mean difference demonstrated	95 CmpStDA 19 353-356
Use of the repeated cross-over designs in assessing	bioequivalence (Disc: p1079-1080) • Liu, Jen-Pei	95 StatMed 14 1067-1078
of two one-sided tests procedures in assessment of	bioequivalence • Liu, Jen-Pei; Weng, Chung-Sing Bias	95 StatMed 14 853-861
microcomputer program and application on drug	bioequivalence data • Abdullah, M. E.; El-Sayed, Yousry M. Design of crossover	95 CmpBiomd 48 237-239
The impact of outlying subjects on decision of	bioequivalence • Ki, Fanny Y. C.; Liu, Jen-Pei; Wang, Wenping; Chow, Shein-Chung	95 JBiophSt 5 71-94
Voss, Theresa Issues related to the	bioequivalence of drug formulations • Habibullah, Mohamed; Rogers, James;	94 ASAProStCp 222-224
Hwang, J. T. Gene Optimal confidence sets,	bioequivalence, and the Limaçon of Pascal • Brown, Lawrence D.; Casella, George;	95 JASA 90 880-889
on analytical methods validation in bioavailability,	bioequivalence and pharmacokinetic studies • Kringle et al. ... from a conference	94 ASAProBiop 510-514
The assessment of individual	bioequivalence via tolerance intervals • Chinchilli, Vernon M.; Esinhart, James D.	94 ASAProBiop 29-32
Assessment of individual and population	bioequivalence using the probability that bioavailabilities are similar • Schall, Robert	95 Biomtrcs 51 615-626
Ten; Chinchilli, Vernon M. The analysis of	bioequivalence with respect to TMAX under a 2×2 crossover design • Have, Thomas R.	95 JBiophSt 5 185-199
sis of bacterial spatial patterns at the initial stage of	biofilm formation • Hamilton, M. A.; Johnson, K. R.; Camper, A. K.; Stoodley, P. et al. Analy-	95 BiomtrcJ 37 393-408
	Biography	
	Frank Yates, 1902-1994 - The work of a statistician • Healy, M. J. R.	95 IntStRvw 63 271-288
batches: Correlation between 31P NMR and a	biological duration test for batch approval • Thelin, B. et al. Classification of Estradurin	95 Chemolab 27 135-145
Mathematical models, statistical methods, and	biological experiments • Dennis, B. et al. Nonlinear demographic dynamics:	95 EcolgMon 65 261-281
Quantitative methods in	biological and medical sciences. A historical essay • Lancaster, H. O.	94 Springer:Brln:NY 297p
Oliver (Rev) Review of "Quantitative methods in	biological and medical science - A historical essay" • Lancaster, H. O. (Auth); Mayo,	95 AstrlJSt 37 372-374
V. T. (Rev) Review of "Quantitative methods in	biological and medical sciences. A historical essay" • Lancaster, H. O. (Auth); Farewell,	95 ShrtBkRv 15/2 22-22
Review of "Modelling	biological populations in space and time" • Renshaw, Eric (Auth); Zeh, Judy (Rev)	95 JASA 90 800-801

On the estimation of probabilities for	birth and death processes • Zeifman, A. I.	95 JAppProb 32 623-634
Garcia, Nancy Lopes	Birth and death processes as projections of higher-dimensional Poisson processes •	95 AdvAppPr 27 911-930
equilibrium distribution in level dependent quasi-	birth-and-death processes • Bright, L.; Taylor, P. G. Calculating the	95 CommStSM 11 497-525
Expected passage times in homogeneous quasi-	birth-and-death processes • Latouche, Guy; Ramaswami, V.	95 CommStSM 11 103-122
covariates and model uncertainty • York, J. et al.	Birth defects registered by double sampling: A Bayesian approach incorporating	95 ApplStat 44 227-242
the business area sample as a means of covering	birth employers and nonemployers in the Current Business ... • Konschnik ... of dropping	94 ASASrvyRsr 995-1000
Ponikowski, Chester H. Introducing	births into the employment cost index survey • Gessley, Mary K.; Kelly, E. Thomas;	94 ASASrvyRsr 797-802
Borowick, Kent How	birth order influences individual characteristics • Moore, Kris; Trower, Jonathan;	95 Chance 8/4 20-24
An empirical study of	birth order • Moore, Kris; Trower, Jonathan; Borowick, Kent	94 ASAProBuEc 347-350
	Birth process	
Pérez-Ocón, Rafael; Gámiz-Pérez, M. Luz	Conditions on the arrival process to obtain HNBUE survival using a shock model •	95 CommStA 24 931-944
absorbing state • Pakes, Anthony G.	Quasi-stationary laws for Markov processes: Examples of an always proximate	95 AdvAppPr 27 120-145
	Random record processes and state dependent thinning • Browne, Sid; Bunge, John	95 StocProc 55 131-142
On some special	birth processes in biology • Janardan, K. G.; Vasudeva, R.; Mohan, N. R.	95 CommStA 24 1033-1038
A note on variation in	birth processes • Ball, Frank	95 MaScitst 20 50-55
Some analytical models to estimate maternal age at	birth using age-specific fertility rates • Pandey, Arvind; Suchindran, C. M.	95 SankhyaB 57 142-150
using stochastic compartmental models with	births • Matis, J. H.; Zheng, Q.; Kiffe, T. R. Describing the spread of biological populations	95 MaBisci 126 215-247
	Bisexual branching diffusions • Mytnik, Leonid; Adler, Robert J.	95 AdvAppPr 27 980-1018
	Bispectral analysis of randomly sampled data • Lii, Keh-Shin; Tsou, Tai-Houn	95 JTimSrAn 16 43-66
	Bivariate autoregressive processes with finite and infinite variance errors • Knight, Keith	94 ASAProBuEc 28-34
structure on recovery in cluster analysis I. The	bivariate case • Donoghue, John R. The effects of within-group covariance	95 MultBehR 30 227-254
Optimal neighborhoods for local regression in the	bivariate case • Lenarduzzi, L.	95 JNnparSt 4 233-248
On a class of	bivariate compounded Poisson distributions • Papageorgiou, H.; David, Katerina M.	95 StPrLet 23 93-104
Mason, Robert L. A	bivariate control chart for paired measurements • Tracy, Nola D.; Young, John C.;	95 JQualTek 27 370-376
	Bivariate control charts for testing $\bar{X}$ data patterns • Chan, Lai K.; Li, Guo-Ying	95 CandJSt 23 85-99
Mason, R. L. A	bivariate control chart with meaningful principal components • Tracy, N. D.; Young, J. C.;	94 ASAProBays 157-158
	Bivariate copulas with quadratic sections • Quesada-Molina, J. J.; Rodríguez-Lallena, J. A.	95 JNnparSt 5 323-337
Kim, Kyungmann A	bivariate cumulative probit regression model for ordered categorical data •	95 StatMed 14 1341-1352
An advantageous approach to survival analysis of	bivariate data • Yashin, A. I.; Vaupel, J. W.; Iachine, I. A. Correlated individual frailty:	95 MaPopStd 5 145-159
A note on	bivariate Dawson-Sankoff-type bounds • Chen, Tuhao; Seneta, E.	95 StPrLet 24 99-104
A family of	bivariate densities constructed from marginals • Long, Dou; Krzysztofowicz, Roman	95 JASA 90 739-746
A simple, automatic and adaptive	bivariate density estimator based on conditional densities • Simonoff, Jeffrey S.	95 StatComp 5 245-252
A	bivariate discrete Charlier series distribution • Papageorgiou, H.; Loukas, S.	95 BiomtrcJ 37 105-117
Laird, Nan M. Regression models for a	bivariate discrete and continuous outcome with clustering • Fitzmaurice, Garrett M.;	95 JASA 90 845-852
function • Wesolowski, Jacek	Bivariate discrete measures via a power series conditional distribution and a regression	95 JMultiAn 55 219-229
	Bivariate distribution	
Conti, Pier Luigi	Asymptotic inference on a general measure of monotone dependence •	94 JItalStS 3 213-241
regression function • Wesolowski, Jacek	Bivariate discrete measures via a power series conditional distribution and a	95 JMultiAn 55 219-229
why is the rank correlation sensitive to choice of	bivariate distribution? • Hutchinson In the multifactorial model of disease transmission,	92 AnHuGene 56 159-161
Construction of	bivariate distributions by a generalised trivariate reduction technique • Lai, C. D.	95 StPrLet 25 265-270
Wesolowski, Jacek	Bivariate distributions via a Pareto conditional distribution and a regression function •	95 AnInStMa 47 177-183
Testing independence when the form of the	bivariate distribution is unspecified • Geisser, Seymour; Johnson, Wesley	95 StatMed 14 1621-1639
Bitter, P.; Bloch, D. A.; Raynauld, J.-P. Comparing the	bivariate effects of toxicity and efficacy of treatments (Disc: p1143-1143) • Tubert-	95 StatMed 14 1129-1141
Inferences for accelerated life tests considering a	bivariate exponential distribution • Achcar, Jorge Alberto	95 MaOpfStS 26 269-283
An application of	bivariate exponential models and related inference • Sinha, Bikas K.; Sinha, Himal K.	95 JStPlInf 44 181-191
of reliability under stress-strength model in a	bivariate exponential set-up • Jana, Prasanta Kumar; Roy, Dilip Estimation	94 CalcutSt 44 175-181
Galambos, Janos; Xu, Yuan	Bivariate extension of the method of polynomials for Bonferroni-type inequalities •	95 JMultiAn 52 131-139
Computation of	bivariate gamma and inverted beta distribution functions • Ong, S. H.	95 JStCmpSm 51 153-163
Consul, P. C.	Bivariate generalized Poisson distribution with some applications • Famoye, Felix;	95 Metrika 42 127-138
A characterization of a	bivariate geometric distribution • Sun, Kai; Basu, Asit P.	95 StPrLet 23 307-311
Characteristic properties of a	bivariate geometric law • Nair, N. Unnikrishnan; Asha, G.	94 JIndnStA 32 111-116
Estimation of parameters of the incomplete	bivariate hyper-Poisson distributions • Ahmad, Munir	92 ProIrStC1 157-162
A bandwidth selector for	bivariate kernel regression • Herrmann, Eva; Wand, M. P.; Engel, Joachim; Gasser, Theo	95 JRSS-B 57 171-180
Jensen, U. Parameter estimation for a	bivariate lifetime distribution in reliability with multivariate extensions • Heinrich, G.;	95 Metrika 42 49-65
Yakovlev, Andrej Yu. A	bivariate limiting distribution of tumor latency time • Rachev, Svetlozar T.; Wu, Chufang;	95 MaBisci 127 127-147
informants • Fitzmaurice, G. M.; Laird, N. M. et al.	Bivariate logistic regression analysis of childhood psychopathology ratings using multiple	95 AJEpidm 142 1194-1203
Dayton, C. Mitchell Detecting patterns of	bivariate mean vectors using model-selection criteria • Huang, Chuen-Chuen Joyce;	95 BJMaStPs 48 129-147
On the influence functions of certain	bivariate medians • Niinimaa, A.; Oja, H.	95 JRSS-B 57 565-574
	Bivariate normal distribution	
mality • Hössjer, Ola; Rousseeuw, Peter J.; Ruts, Ida	The repeated median intercept estimator: Influence function and asymptotic nor-	95 JMultiAn 52 45-72
A simple approximation for the	bivariate normal integral • Lin, Jinn-Tyan	95 PrEngInf 9 317-321
Wesolowsky, G. O.	Bivariate normal parameter estimation under destructive testing • Steiner, S. H.;	94 ASAProPEng 326-331
in the presence of correlated observations from a	bivariate normal population • Weerakkody, G. J. et al. Estimating the correlation coefficient	95 CommStA 24 1705-1719
moment-generating function of the minimum of	bivariate normal random variables" (V48 p124-125) • Sapra, Sunil K. Comment on "The	95 AmerStat 49 240-240
moment-generating function of the minimum of	bivariate normal random variables" • Cain, Michael Reply to comment on "The	95 AmerStat 49 240-240
Lipsitz, Stuart R. Analyzing	bivariate ordinal data using a global odds ratio • Williamson, John M.; Kim, KyungMann;	95 JASA 90 1432-1437
Characterization of a	bivariate Pareto distribution • Padamadan, Veenus; Nair, K. R. Muraleedharan	94 JIndnStA 32 15-20
	Bivariate Poisson distribution	
	Compound model for two dependent kinds of claim • Partrat, Christian	94 InsrneME 15 219-231
technique • Lai, C. D.	Construction of bivariate distributions by a generalised trivariate reduction	95 StPrLet 25 265-270
	Some remarks on the D compound Poisson distribution • Papageorgiou, H.	95 StHefte 36 371-375
	Statistical analysis on Markov-modulated Poisson processes • Ramesh, N. I.	95 EnvrMtrc 6 165-179
Smooth tests for the	bivariate Poisson distribution • Rayner, J. C. W.; Best, D. J.	95 AstrlJSt 37 233-245
"Estimation and testing for the parameters of the	bivariate Poisson populations based on a 2 by 2 sample" (V61 ... • Douglas Comment on	95 IntStRvw 63 117-118
Testing for independence between hitting times of	bivariate PQD processes • Ramalingam, T.; Ebrahimi, Nader	95 SankhyaA 57 202-220
Importance sampling to evaluate	bivariate probabilities with an application to sequential phase II trials • Conaway, Mark R.	95 JStCmpSm 51 249-262
Simó, Amelia	Bivariate random closed sets and nerve fibre degeneration • Ayala, Guillermo;	95 AdvAppPr 27 293-305
George, Varghese; Johnson, William D. Effect of	bivariate regression toward the mean in uncontrolled clinical trials • Shahane, Aditi;	95 CommStA 24 2165-2181
	Bivariate sequential designs for phase II trials • Conaway, Mark R.; Petroni, Gina R.	95 Biomtrcs 51 656-664
on the association parameter in copula models for	bivariate survival data • Shih, Joanna H.; Louis, Thomas A. Inferences	95 Biomtrcs 51 1384-1399
A Bayesian analysis of	bivariate survival data from a multicentre cancer clinical trial • Gustafson, Paul	95 StatMed 14 2523-2535
Wellner, Jon A. Inefficient estimators of the	bivariate survival function for three models • Gill, Richard D.; van der Laan, Mark J.;	95 AIHenrPB 31 545-597
Barmi, H. E.; Kochar, S. C. Likelihood ratio tests for	bivariate symmetry against ordered alternatives in a square contingency table •	95 StPrLet 22 167-173
A vector of quarters representation for	bivariate time series • Franses, Philip Hans	95 EcomtrRv 14 55-63

A class of symmetric	bivariate uniform distributions • Ferguson, T. S.	95 StHefte 36 31-40
Subrahmaniam Discrete	bivariate weighted distributions under multiplicative weight function • Kocherlakota,	95 CommStA 24 533-551
K.-H. A study on the efficiency comparison of the	black-box method with the randomized response technique (Korean) • Lee, H.-Y.; Hong,	95 KorApS 8/2 27-41
Hjalmarsson, H.; Benveniste, A. et al. Nonlinear	black-box models in system identification: Mathematical foundations • Juditsky, A.;	95 Automatca 31 1725-1750
Ljung, L.; Benveniste, A.; Delyon, B. et al. Nonlinear	black-box modeling in system identification: A unified overview • Sjöberg, J.; Zhang, Q.;	95 Automatca 31 1691-1724
interior-point Markov chains: A family of	black-box samplers • Chen, Ming-Hui; Schmeiser, Bruce W. Random-direction	94 ASAProBays 1-6
S. et al. Estimating nest site fidelity of adult female	black brant with multi-state modeling and geographic information systems • Lindberg, M.	95 JAppStat 22 725-735
	Black's consol rate conjecture • Duffie, Darrell; Ma, Jin; Yong, Jiongmin	95 AnlsApPr 5 356-382
Fienberg, Stephen E.	Black, white, and shades of gray (and brown and yellow) • Anderson, Margo;	95 Chance 8/1 15-18
	Blackwell	
	• see also Rao- –	
Yushkevich, A. A.	Blackwell optimal policies in a Markov decision process with a Borel state space •	94 ZeitOpRs 40 253-288
Extending	Blackwell's strengthening of Azuma's martingale inequality • Ross, Sheldon M.	95 PrEngInf 9 493-496
Cox regression models to the analysis of a double-	blind comparative trial and subsequent open label extension ... • Hwang et al. ... of	94 ASAProBiop 194-199
	Blind deconvolution of linear systems with multilevel nonstationary inputs • Li, Ta-Hsin	95 AnlsStat 23 690-704
	Blind deconvolution via sequential imputations • Liu, Jun S.; Cheng, Rong	95 JASA 90 567-576
biased urn randomization in small strata when	blinding is impossible • Schouten, Hubert J. A. Adaptive	95 Biomtrcs 51 1529-1535
Estimation subject to	block censoring • Baxter, Laurence A.	95 IEEEReli 44 489-495
row-column designs under the unavailability of	blocks: A comparison • Uddin et al. ... of binary and non-binary balanced nested	95 AstrlJSt 37 111-124
	Block design	
Yates, P. K.; Lewis, S. M.; Dean, A. M.	Approximate information matrices for estimating a given set of contrasts • Kao, L.-J.;	95 StSinica 5 593-598
	Construction and optimality of nearly linear trend-free designs • Chai, Feng-Shun	95 JStPlInf 48 113-129
	On modelling of experiments in natural sciences • Mejza, Stanisław	94 ListyBio 31 79-100
designs? 'Silly designs are silly!' • Bailey, R. A. et al.	Are neighbour methods preferable to analysis of variance for completely systematic	95 Biomtrka 82 655-659
	Optimal, non-binary, variance balanced designs • Morgan, John P.; Uddin, Nizam	95 StSinica 5 535-546
	On the PBB designs for multiresponse experiments • Brzeskwiniewicz, Henryk	95 BiomtrcJ 37 191-198
	On perpendicularity in designed experiments • Kala, Radosław	95 ListyBio 32 93-99
	• see also BIBD, balanced incomplete –, incomplete –, randomized –	
distributions of rank tests for ordered alternatives in	block designs • Black, Stephen; Mansouri, H. On exact	95 CmpStDA 20 265-274
on generalized binary proper efficiency-balanced	block designs • Kherwa, G. Ram; Prasad, Jagdish A note	95 CommStA 24 2797-2801
Kageyama, Sanpei Single replicate orthogonal	block designs for circulant partial diallel crosses • Gupta, Sudhir; Das, Ashish;	95 CommStA 24 2601-2607
Majorization and D-optimality for complete	block designs under correlations • Bhaumik, Dulal K.	95 JRSS-B 57 139-143
What really is experimental error in	block designs? • Gates, Charles E.	95 AmerStat 49 362-363
Analysis and planning of experimental	block designs (French) • Bergonzini, J.-Cl.	95 Masson:NY:Paris 372p
the randomization theory of experiments in nested	block designs • Caliński, Tadeusz On	94 ListyBio 31 45-77
Efficient	block designs in the presence of trends • Jacroux, Mike; Majumdar, Dibyen; Shah, Kirti R.	95 StSinica 5 605-615
Sinha, Bikas Kumar Optimal	block designs revisited: An approximate theory detour • Pukelsheim, Friedrich;	95 StPrLet 24 375-380
C.	Block designs with large holes and α-resolvable BIBDs • Furino, Steven C.; Mullin, Ronald	93 JComDsgn 1 101-112
Shah, Kirti R.; Das, Ashish Optimal	block designs with minimal and nearly minimal number of units • Dey, Aloke;	95 StSinica 5 547-558
longitudinal responses and a randomized complete	block design with multivariate responses • Xie et al. ... design with multivariate	94 ASAProBiop 250-255
of ANOVA estimators of variance components in	block designs with random treatments • Molińska, A.; Moliński, K. Improvement	95 StPrLet 22 281-285
Reinforced	block designs with two groups of treatments • Ceranka, B.; Krzyszkowska, J.	94 ListyBio 31 17-25
Optimal regression designs under random	block-effects models • Cheng, Ching-Shui	95 StSinica 5 485-497
Federal	block grants: What do statisticians have to contribute? • Adams, John; Burnam, Audrey	95 Chance 8/4 40-42
Covering with	blocks in the non-symmetric case • Móri, Tamás F.	95 JTherPrb 8 139-164
"Mixture designs for four components in orthogonal	blocks" (V35 p268-276) • Snee, Ronald D. Comment on	95 Technmcs 37 131-132
estimation for establishment surveys: Stumbling	blocks and progress • Hidiroglou, M. A. Sampling and	94 ASASrvyRsr 153-162
for generating and labelling all regular fractions or	blocks for q^{n-m}-designs • Laycock, P. J.; Rowley, P. J. A method	95 JRSS-B 57 191-204
techniques • Højsgaard, S.; Thiesson, B. BIFROST –	Block recursive models induced from relevant knowledge, observations, and statistical	95 CmpStDA 19 155-175
Resolvable BIBDs with	block size 8 and index 7 • Furino, Steven; Greig, Malcolm; Miao, Ying; Yin, Jianxing	94 JComDsgn 2 101-116
Variance balanced designs with unequal	block sizes and unequal replications • Gupta, S.; Kageyama, S.	92 UtilMath 42 15-24
Type S designs in unequal	blocks • Gupta, Sudhir; Kageyama, Sanpei	93 JCISS 18 97-112
problems in crop experimentation. I. The use of	blocks • Pearce, S. C. Some design	95 ExpAgric 31 191-203
Queueing networks with	blocking: Exact and approximate solutions • Perros, H. G.	94 OxfordUniv 288p
Review of "Queueing networks with	blocking: Exact and approximate solutions" • Perros, H. G. (Auth); Pitts, S. M. (Rev)	95 ShrtBkRv 15/3 46-46
G. L.; Leung, K. K.; Whitt, W. An algorithm to compute	blocking probabilities in multi-rate multi-class multi-resource loss models • Choudhury,	95 AdvAppPr 27 1104-1143
W. A physical device for demonstrating confounding,	blocking, and the role of randomization in uncovering a causal relationship • Oldford, R.	95 AmerStat 49 210-216
Jing, Bing-Yi On	blocking rules for the bootstrap with dependent data • Hall, Peter; Horowitz, Joel L.;	95 Biomtrka 82 561-574
problems in crop experimentation. II. Multiple	blocking systems • Pearce, S. C. Some design	95 ExpAgric 31 279-290
Bühlmann, Peter The	blockwise bootstrap for general empirical processes of stationary sequences •	95 StocProc 58 247-265
On the strong law of large numbers for	blockwise independent and blockwise orthogonal random variables • Gaposhkin, V. F.	94 ThProbAp 39 677-684
of large numbers for blockwise independent and	blockwise orthogonal random variables • Gaposhkin, V. F. On the strong law	94 ThProbAp 39 677-684
A stochastic model for the movement of a white	blood cell • Heubach, Silvia; Watkins, Joseph	95 AdvAppPr 27 443-475
log-additive models for the relationship between	blood lead and environmental lead • Jiang, Qi; Succop, Paul A. A study of the log-log and	94 ASAProEnvr 88-91
An exploratory data analysis of ambulatory	blood pressure monitoring as screening criteria • Snavely, Duane; Dobbins, Thomas	94 ASAProBiop 51-55
of exercise and calorie restriction on ambulatory	blood pressure in overweight, sedentary males: A time series analysis • Cox et al. Effects	95 EnvrMtrc 6 523-528
Are the Toronto	Blue Jays really the world champions? • Umholtz, Robert L.	94 ASAProSpts 42-45
effects of selection based on the animal model	BLUP in a finite population • Verrier, E.; Colleau, J. J.; Foulley, J. L. Long-term	93 ThApGene 87 446-454
C. A.; Yau, K. K. W. The derivation of	BLUP, ML, REML estimation methods for generalised linear mixed models • McGilchrist,	95 CommStA 24 2963-2980
• Piepho Best linear unbiased prediction (	BLUP) for regional yield trials: A comparison to additive main effects and multiplicative ...	94 ThApGene 89 647-654
Markov approximations to D-	BMAPs: Information theoretic bounds on queueing performance • Rananand, Nol	95 CommStSM 11 713-734
and justification of the power-series algorithm for	BMAP-systems • van den Hout, Wilbert; Blanc, Hans Development	95 CommStSM 11 471-496
Getson, Pamela (Rev) Review of	"BMDP New System for Windows (Version 1.11)" • BMDP Statistical Software Inc (Auth);	95 Chance 8/4 53-58
method of moments/instrumental variable (	BMOM/IV) analysis of mean and regression models • Zellner, Arnold Bayesian	94 ASAProBays 58-63
et al. Structural changes in the relationships of	body dimensions of four species of pulmonate slugs (Mollusca : Gastropoda : ... • Bhuyan	95 BiomtrcJ 37 1005-1015
Teaching the concepts and tools of variation using	body temperature • Amy, Carolyn K.; Strong, Ann	95 QulPr 28/3 168-168
Dalgleish, Lenard I. (Rev) Review of	"BOJA: A program for bootstrap and jackknife analysis" • Boomsma, A. (Auth);	95 StatComp 5 165-174
systems associated with spatially homogeneous	Boltzmann type equations • Léonard, C. On large deviations for particle	95 ProbTher 101 1-44
Payne, C.; Payne, J. et al. Testing the tightening	bond hypothesis: An application of new models for analysing trends in multiway tables •	95 JStPlInf 47 181-201
Considerations in the analysis of a single arm	bone marrow transplant clinical trial • Gonin, R.	94 ASAProBiop 493-497
The exact sampling distribution of the	Bonferroni concentration index • Giorgi, Giovanni Maria; Mondani, Riccardo	94 Metron 52/3 5-41
inbred strains • Belknap, J. K. Empirical estimates of	Bonferroni corrections for use in chromosome mapping studies with the BXD recombinant	92 BehavGen 22 677-684

The	bootstrap and Edgeworth expansion • Hall, P.	95 Springer:Brln:NY 352p
without ... • Zharkikh et al. Statistical properties of	bootstrap estimation of phylogenetic variability from nucleotide sequences: II. Four taxa	92 JMolEvol 35 356-366
Limit theory and	bootstrap for explosive and partially explosive autoregression • Datta, Somnath	95 StocProc 57 285-304
The blockwise	bootstrap for general empirical processes of stationary sequences • Bühlmann, Peter	95 StocProc 58 247-265
Datta, Somnath; McCormick, William P.	Bootstrap inference for a first-order autoregression with positive innovations •	95 JASA 90 1289-1300
Using i.i.d.	bootstrap inference for general non-i.i.d. models • Liu, Regina Y.; Singh, Kesar	95 JStPlInf 43 67-75
On the performance of iterated wild	bootstrap interval estimation of the mean response • Kim, Woochul; Ko, Duk-Hyun	95 JKoreaSt 24 551-562
Review of "An introduction to the	bootstrap" • Efron, B.; Tibshirani, R. J. (Auth); Brown, Jason (Rev)	95 JQualTek 27 90-91
Review of "An introduction to the	bootstrap" • Efron, B.; Tibshirani, R. J. (Auth); Härdle, W. (Rev)	95 Metrika 42 267-268
Review of "An introduction to the	bootstrap" • Efron, B.; Tibshirani, R. J. (Auth); Marriott, Paul (Rev)	95 JRSS-A 158 347-347
Review of "An introduction to the	bootstrap" • Efron, B.; Tibshirani, Robert J. (Auth); Brown, Morton B. (Rev)	95 StMedRes 4 91-92
Review of "An introduction to the	bootstrap" • Efron, Bradley; Tibshirani, Robert J. (Auth); Grimshaw, Scott D. (Rev)	95 Technmcs 37 341-341
The jackknife and	bootstrap • Shao, J.; Tu, D.	95 Springer:Brln:NY 516p
Review of "BOJA: A program for	bootstrap and jackknife analysis" • Boomsma, A. (Auth); Dalgleish, Lenard I. (Rev)	95 StatComp 5 165-174
Review of "The jackknife and	bootstrap" • Shao, J.; Tu, D. (Auth); Fedorov, V. V. (Rev)	96 ShrtBkRv 16/1 7-7
Accurate and efficient construction of	bootstrap likelihoods • Davison, A. C.; Hinkley, D. V.; Worton, B. J.	95 StatComp 5 257-264
for standardized rates based on the approximate	bootstrap method • Swift, Michael B. Simple confidence intervals	95 StatMed 14 1875-1888
Conditional	bootstrap methods for censored survival data • Kim, Ji-Hyun	95 JKoreaSt 24 197-218
Application of the	bootstrap methods in factor analysis • Ichikawa, Masanori; Konishi, Sadanori	95 Psymtrka 60 77-93
Goodness-of-fit tests for the Cox model via	bootstrap method • Burke, Murray D.; Yuen, Kam C.	95 JStPlInf 47 237-256
	Bootstrap methods for nonstandard mixtures • Schucany, William R.; Wang, Suojin	94 ASAProPEng 59-64
for generalized logit model of nominal type with	bootstrap method • Cho, JoongJae; Han, JeongHye Statistical estimation	95 JKoreaSt 24 1-18
Whittaker, Joe Application of the parametric	bootstrap to models that incorporate a singular value decomposition • Milan, Luis;	95 ApplStat 44 31-49
	Bootstrap for nonstandard cases • Babu, Gutti Jogesh	95 JStPlInf 43 197-203
On the asymptotic behaviour of the moving block	bootstrap for normalized sums of heavy-tail random variables • Lahiri, S. N.	95 AnlsStat 23 1331-1349
R. H. Equivalence of exponential decay rates for	bootstrap percolation like cellular automata • Andjel, E. D.; Mountford, T. S.; Schonmann,	95 AIHenrPB 31 13-25
Critical length for semi-oriented	bootstrap percolation • Mountford, T. S.	95 StocProc 56 185-205
Dunsmuir, William T. M. Improved	bootstrap prediction intervals for autoregressions • Breidt, F. Jay; Davis, Richard A.;	95 JTimSrAn 16 177-200
Accuracies of the simple methods for estimating the	bootstrap probability of a maximum-likelihood tree • Hasegawa, M.; Kishino, H.	94 MolBioEv 11 142-145
Change-point problem and	bootstrap • Antoch, Jaromir; Hušková, Marie; Veraverbeke, Noël	95 JNnparSt 5 123-144
models • DiCiccio, Thomas J.; Romano, Joseph P. On	bootstrap procedures for second-order accurate confidence limits in parametric	95 StSinica 5 141-160
variance of molecular marker data using the	bootstrap procedure • Tivang, J. G.; Nienhuis, J.; Smith, O. S. Estimation of sampling	94 ThApGene 89 259-264
Some finite sample theory for	bootstrap regression estimates • Fisher, Gordon; Sim, Ah Boon	95 JStPlInf 43 289-300
M. A residual	bootstrap for regression parameters in proportional hazards models • Loughin, Thomas	95 JStCmpSm 52 367-384
Léger, Christian Cross-validation, the	bootstrap, and related methods for tuning parameter selection • Altman, Naomi;	94 ASAProStCp 41-50
Some statistical mean value theorems related to the	bootstrap • Chao, M. T.; Lo, S. H.	95 StSinica 5 129-139
statistical methods: Validation, model selection and	bootstrap" • Hjorth, J. S. U. (Auth); Saldanha, R. A. (Rev) Review of "Computer intensive	95 Ststcian 44 288-288
statistical methods: Validation, model selection and	bootstrap" • Hjorth, J. S. U. (Auth); Wasserman, G. S. et al. Review of "Computer intensive	95 JQualTek 27 268-270
tistical methods: Validation, model selection, and	bootstrap" • Hjorth, J. S. U. (Auth); Lenth, R. V. (Rev) Review of "Computer intensive sta-	95 Technmcs 37 458-459
Estimation of transition probabilities and	bootstrap in a semiparametric Markov renewal model • Dabrowska, Dorota M.	95 JNnparSt 5 237-259
Stein confidence sets and the	bootstrap • Beran, Rudolf	95 StSinica 5 109-127
Double	bootstrap for shrinkage estimators • Vinod, H. D.	95 JEconmtx 68 287-302
Tail-specific linear approximations for efficient	bootstrap simulations • Hesterberg, Tim	95 JCmpGrSt 4 113-133
Optimizing the smoothed	bootstrap • Wang, Suojin	95 AnInStMa 47 65-80
Exploring the replicability of a study's results:	Bootstrap statistics for the multivariate case • Thompson, Bruce	95 EdPsMeas 55 84-94
Statistical process control via the subgroup	bootstrap • Seppala, Tomi; Moskowitz, Herbert; Plante, Robert; Tang, Jen	95 JQualTek 27 139-153
al. Predicting using Box-Jenkins, nonparametric, and	bootstrap techniques • García-Jurado, I.; González-Manteiga, W.; Prada-Sánchez, J. M. et	95 Technmcs 37 303-310
Nei, Masatoshi Interior-branch and	bootstrap tests of phylogenetic trees • Sitnikova, Tatyana; Rzhetsky, Andrey;	95 MolBioEv 12 319-333
The asymptotic accuracy of the	bootstrap of U-quantiles • Arcones, Miguel A.	95 AnlsStat 23 1802-1822
for single-locus inbreeding coefficients using the	bootstrap • van Dongen, S.; Backeljau, T. One- and two-sample tests	95 Heredity 74 129-135
estimation of drug abuse episodes using the	bootstrap • Hughes, Arthur L.; Brodsky, Marc D. Variance	94 ASASrvyRsr 212-217
Variance estimates in logistic regression using the	bootstrap • Friedl, Herwig; Tilg, Norbert	95 CommStA 24 473-486
Properties of the Schools and Staffing Survey's	bootstrap variance estimator (Disc: p1128-1129) • Kaufman, Steven	94 ASASrvyRsr 1116-1121
The weighted	bootstrap • Barbe, P.; Bertail, P.	95 Springer:Brln:NY 230p
Review of "The weighted	bootstrap" • Barbe, P.; Bertail, P. (Auth); Young, G. A. (Rev)	95 ShrtBkRv 15/3 45-45
On blocking rules for the	bootstrap with dependent data • Hall, Peter; Horowitz, Joel L.; Jing, Bing-Yi	95 Biomtrka 82 561-574
Shrinkage estimation in time series using a	bootstrapped covariance estimate • Ravishanker, Nalini; Wu, Lilian S.-Y.; Dey, Dipak K.	95 JStCmpSm 53 259-267
Dorfman, A. H. Robust sample survey inference via	bootstrapping and bias correction: The case of the ratio estimator • Chambers, R. L.;	94 ASAProStCp 51-60
	Bootstrapping binomial confidence intervals • Zheng, Xiaodong; Loh, Wei-Yin	95 JStPlInf 43 355-380
	Bootstrapping a consistent nonparametric goodness-of-fit test • Fan, Yanqin	95 EcomtrRv 14 367-382
Toward a defensible	bootstrapping • Mitchell, Sam	95 PhilSci 62 241-260
al. Confidence intervals in ridge regression by	bootstrapping the dependent variable: A simulation study • Crivelli, A.; Firinguetti, L. et	95 CommStB 24 631-652
carcinogen bioassay • Krewski, D. et al.	Bootstrapping an empirical Bayes estimator of the distribution of historical controls in	95 JStPlInf 43 215-228
of statistics based on quadratic entropy and	bootstrapping • Liu, Z. J.; Rao, C. Radhakrishna Asymptotic distribution	95 JStPlInf 43 1-18
pollutants • Hopke, P. K.; Li, C. L. et al. The use of	bootstrapping to estimate conditional probability fields for source locations of airborne	95 Chemolab 30 69-79
	Bootstrapping in human genetic linkage • Chiano, M. N.; Yates, J. R. W.	94 AnHuGene 58 129-143
Spline smoothed	bootstrapping for kernel estimator (Chinese) • Qian, Weimin	95 AMaAPknC 17 265-277
	Bootstrapping a method of phylogenetic inference • Navidi, W. C.	95 JStPlInf 43 169-184
	Bootstrapping a nonparametric polytomous regression model • Aerts, M.; Veraverbeke, N.	95 MaMethSt 4 189-200
	Bootstrapping quantile regression estimators • Hahn, Jinyong	95 EconTher 11 105-121
Weighted	bootstrapping for U-quantiles • Aerts, Marc; Janssen, Paul	95 StPrLet 22 317-323
Braun, W. John An illustration of	bootstrapping using video lottery terminal data (jse/v3n2/datasets.braun) •	95 JStEd 3/2 18kb
Lalonde, Robert	Bordeaux wine vintage quality and the weather • Ashenfelter, Orley; Ashmore, David;	95 Chance 8/4 7-14
Obituary: Sir John	Boreham, 1925-1994 • Erritt, John	95 JRSS-A 158 182-184
optimal policies in a Markov decision process with a	Borel state space • Yushkevich, A. A. Blackwell	94 ZeitOpRs 40 253-288
Improving two theorems of	Bose on difference families • Buratti, Marco	95 JComDsgn 3 15-24
India with focus on contributions of Professor P. K.	Bose • Srivastava, A. B. L. A review of educational planning and statistics in	94 CalcutSt 44 251-264
	Bound	
	• see also branch and –, confidence –, Cramér-Rao –, inequality, lower –, upper –	
procedure • Akahira, M. The Bhattacharyya type	bound for the asymptotic variance and the sequential discretized likelihood estimation	95 SqtlAnly 14 193-204
A nonparametric	bound for the Bayes error • Ruiz, Alberto	95 PattrnRc 28 921-930
disturbances • Song, Seuck Heun Consistency and	bounds on the bias of S^2 in the linear regression model with moving average	95 JKoreaSt 24 507-518
	Bounds for the breakdown point of the simplicial median • Chen, Zhiqiang	95 JMultiAn 55 1-13
Bayesian prediction	bounds for the Burr type XII failure model • Al-Hussaini, Essam K.; Jaheen, Zeinhum F.	95 CommStA 24 1829-1842

Reliability	bounds for coherent structures with independent components • Fu, J. C.; Koutras, M. V.	95 StPrLet 22 137-148
Hall, Peter; Jing, Bing-Yi Uniform coverage	bounds for confidence intervals and Berry-Esseen theorems for Edgeworth expansion •	95 AnlsStat 23 363-375
Optimum	bounds for the distributions of martingales in Banach spaces • Pinelis, Iosif	94 AnlsProb 22 1679-1706
A random linear functional approach to efficiency	bounds • Holly, Alberto	95 JEconmtx 65 235-261
Representation and	bounds for the general fractional factorial • Anderson, D. A.; Federer, W. T.	95 CommStA 24 363-373
and nonbinary linear codes meeting the Griesmer	bound • Hamada, N. A survey of recent work on characterization of minihypers in PG(t,q)	93 JCISS 18 161-191
Reliability	bounds for the L-class and Laplace order • Sengupta, Debasis	95 JAppProb 32 832-835
Kulasekera, K. B. A	bound on the L_1-error of a nonparametric density estimator with censored data •	95 StPrLet 23 233-238
Scott, Donald R.	Bounds for molecular weights of organic compounds from low resolution mass spectra •	95 Chemolab 27 129-133
application to simple ... • de la Peña Correction to "A	bound on the moment generating function of a sum of dependent variables with an	95 AIHenrPB 31 703-704
More	bounds for moments of some rank statistics • Spurrier, John D.	95 CommStA 24 2679-2682
Uncertainty	bounds for multivariate M-estimates in small samples • Spall, James C.	94 ASAProPEng 126-131
distributions • Rychlik, Tomasz	Bounds for order statistics based on dependent variables with given nonidentical	95 StPrLet 23 351-358
	Bounds on orthogonal arrays and resilient functions • Bierbrauer, Jürgen	95 JComDsgn 3 179-183
Uncertainty	bounds for parameter identification with small sample sizes • Spall, James C.	95 PIEEEDecCt 3504-3515
The moment bound is tighter than Chernoff's	bound for positive tail probabilities • Philips, Thomas K.; Nelson, Randolph	95 AmerStat 49 175-178
Local	bounds on power of T^2 in scale families • Jensen, D. R.	95 JStCmpSm 52 187-190
approximations to D-BMAPs: Information theoretic	bounds on queueing performance • Rananand, Nol Markov	95 CommStSM 11 713-734
A	bound on the rate of convergence for the discrete Gibbs sampler • Dinwoodie, I. H.	95 PrEngInf 9 211-215
distributed lifetime • Coyle, A. J.; Taylor, P. G. Tight	bounds on the sensitivity of generalised semi-Markov processes with a single generally	95 JAppProb 32 63-73
Claude Explicit and exponential	bounds for a test on the coefficient of an AR (1) model • Massé, Bruno; Viano, Marie-	95 StPrLet 25 365-371
A note on bivariate Dawson-Sankoff-type	bounds • Chen, Tuhao; Seneta, E.	95 StPrLet 24 99-104
Exact large deviation	bounds up to T_c for the Ising model in two dimensions • Ioffe, Dmitry	95 ProbTher 102 313-330
The full Martin	boundary of the bi-tree • Picardello, Massimo A.; Woess, Wolfgang	94 AnlsProb 22 2203-2222
method of line intercept sampling to eliminate	boundary bias • Gregoire, Timothy G.; Monkevich, Nicholas S. The reflection	94 EnvEcoSt 1 219-226
Chen, Bee-Lian; Gallo, Paul P.	Boundary calculation issues in clinical trials with two-sided stopping criteria •	94 ASAProBiop 115-119
Piotr W. Moments of the limiting distribution for the	boundary case in the first order autoregressive process • Monsour, Michael J.; Mikulski,	94 AJMMgSci 14 327-347
and Hidiroglou method to obtain stratification	boundaries for the Census Bureau's Annual Capital ... • Slanta et al. ... the Lavallée	94 ASASrvyRsr 693-698
On a stochastic delay difference equation with	boundary conditions and its Markov property • Baccin, Maria C.; Ferrante, Marco	95 StocProc 60 131-146
tions for semilinear elliptic equations with Neumann	boundary conditions • Chen, Z. Q.; Williams, R. J. et al. On the existence of positive solu-	95 ProbTher 101 251-276
	Boundary condition sets limits on Brass's model life table parameters • Mitra, S.	94 ASAProSoSt 55-59
Application to stochastic differential equations with	boundary conditions • Garnier, J. Stochastic invariant imbedding:	95 ProbTher 103 249-271
property for stochastic differential equations with	boundary conditions • Ferrante, M.; Nualart, D. Markov field	95 Stochatx 55 55-69
differential equations with reflecting convex	boundaries • Pettersson, Roger Approximations for stochastic	95 StocProc 59 295-308
fields, with an application to astrophysics • Worsley	Boundary corrections for the expected Euler characteristic of excursion sets of random	95 AdvAppPr 27 943-959
The Poisson	boundary of covering Markov operators • Kaimanovich, Vadim	95 IsrelJMa 89 77-134
	Boundary crossing	
passage times • Lai, Tze Leung; Wang, Julia Qizhi	Asymptotic expansions for the distributions of stopped random walks and first	94 AnlsProb 22 1957-1992
	A note on extreme values of locally stationary Gaussian processes • Hüsler, J.	95 JStPlInf 45 203-213
Martin	boundaries for discrete Markov chains • Aggoun, Lakhdar	95 StocAnAp 13 257-268
problem: Tests for the overlap of geographic	boundaries • Jacquez, Geoffrey M. The map comparison	95 StatMed 14 2343-2361
The hazard rate tangent approximation for	boundary hitting times • Roberts, G. O.; Shortland, C. F.	95 AnlsApPr 5 446-460
On the spectrum and Martin	boundary of homogeneous spaces • Northshield, S.	95 StPrLet 22 275-279
of a continuous Gaussian process over a U-shaped	boundary of increasing curvature • Balabushkin Approximation to the exit probability	95 JAppProb 32 429-442
to turbulence statistical studies in planetary	boundary layers" (V51 p1682-1690) • Mann, J. et al. Comment on "A definitive approach	95 JAtmsphr 52 3194-3196
to turbulence statistical studies in planetary	boundary layers" • Gluhovsky, A.; Agee, E. Reply to comment on "A definitive approach	95 JAtmsphr 52 3197-3198
Efficient estimation of monotone	boundaries • Korostelev, A. P.; Simar, L.; Tsybakov, A. B.	95 AnlsStat 23 476-489
Oscillating random walk with a moving	boundary • Madras, Neal; Tanny, David	94 IsrelJMa 88 333-365
reaction-diffusion equations with white noise	boundary perturbations • Sowers, R. B. Multidimensional	94 AnlsProb 22 2071-2121
Space-time	boundaries for random walks obtained from diffuse measures • Handelman, David	94 IsrelJMa 86 107-156
of 2-dimensional Markov chains in quadrants with	boundary reflection • Aspandiiarov, Sanjar On the convergence	95 Stochatx 53 275-303
Determination of optimum strata	boundaries for scrambled randomized response • Mahajan, P. K.; Gupta, J. P.; Singh, R.	94 Statstca 54 375-381
Asymptotical minimax recovery of sets with smooth	boundaries • Mammen, E.; Tsybakov, A. B.	95 AnlsStat 23 502-524
Estimation of non-sharp support	boundaries • Härdle, W.; Park, B. U.; Tsybakov, A. B.	95 JMultiAn 55 205-218
people out of our business": Pushing household	boundaries in survey data collection • Hainer, Peter C. "Strivin', gettin' over, and keepin'	94 ASASrvyRsr 337-341
conservative dynamical semigroups and a quantum	boundary theory • Bhat, B. V. Rajarama; Parthasarathy, K. R. Markov dilations of non-	95 AIHenrPB 31 601-651
Explicit analytic ruin probabilities for	bounded claims • De Vylder, F.; Marceau, E.	95 InsrneME 16 79-105
identification in non-linear conceptual models from	bounded-error data • Keesman, K. J. Model parameter and modelling error	95 EnvrMtrc 6 445-449
Random walks with stochastically	bounded increments: Renewal theory • Alsmeyer, Gerold	95 MatNach 172 13-31
A note on	bounded influence in Bayesian analysis • Pericchi, Luis R.; Sansó, Bruno	95 Biomtrka 82 223-225
Some small-sample results on a	bounded influence rank regression method • Wilcox, Rand R.	95 CommStA 24 881-888
theorem for (sub) Markov chains on $\{1,2,\dots\}$ with	bounded jumps • Kesten, Harry A ratio limit	95 AdvAppPr 27 652-691
properties in finite sample l_1 identification with	bounded noise • Kacewicz, B.; Milanese, M. Optimality	95 IJAdCSgP 9 87-96
Stefanie Minimax- and Γ-minimax estimation of a	bounded normal mean under LINEX loss • Bischoff, Wolfgang; Fieger, Werner; Wulfert,	95 StatDecn 13 287-298
population when the scale parameters are	bounded • Misra, Neeraj; Kumar, Narinder Selecting the best exponential	95 MaOpfStS 27 151-158
Liu, Ying-Hsing; Yang, Ming-Chung	Bounded risk conditions in simultaneous estimation of independent Poisson means •	95 JStPlInf 47 319-331
Mukhopadhyay, Nitis; Datta, Sujay On fine-tuned	bounded risk sequential point estimation of the mean of an exponential distribution •	95 SoAfrStJ 29 9-27
• van Eeden, C. Minimax estimation of a lower-	bounded scale parameter of a gamma distribution for scale-invariant squared-error loss	95 CandJSt 23 245-256
of homogeneous random fields by fields with	bounded spectrum • Khalikulov, S. I. On approximation	95 ThPrMaSt 50 129-132
of a stochastic integral in which the integrator is a	bounded submartingale • Hammack, William Sharp inequalities for the distribution	95 AnlsProb 23 223-235
Parker, R. L.	Bounded-variable least-squares: An algorithm and applications • Stark, P. B.;	95 CmpStQ 10 129-141
a deterministic population dynamics model for	Bowhead whales (Disc: p417-430) • Raftery, A. E.; Givens, G. H.; Zeh, J. E. Inference from	95 JASA 90 402-416
a deterministic population dynamics model for	Bowhead whales" • Schweder, Tore Comment On "Inference from	95 JASA 90 420-423
	Box	
	• see also black -, grey -	
iid samples does it take to see all the balls in a	box? • Sellke, Thomas M. How many	95 AnlsApPr 5 294-309
variance properties of central composite and	Box-Behnken designs • Borkowski, John J. Spherical prediction-	95 Technmcs 37 399-410
prediction variances for central composite and	Box-Behnken designs • Borkowski, John J. Minimum, maximum, and average spherical	95 CommStA 24 2581-2600
Some properties of	Box-Behnken designs • Jo, Jinnam; Hinkelmann, Klaus	93 JCISS 18 273-287
to "A large deviation-type approximation for the	'Box class' of likelihood ratio criteria" (V86 p437-440) • Jensen, Jens Ledet Correction	95 JASA 90 812-812
Concavity of	Box-Cox log-likelihood function • Kouider, Elies; Chen, Hanfeng	95 StPrLet 25 171-175

	Box-Cox transformation	
Rodgers, J. A., Jr; Linda, S. B.; Nesbitt, S. A.	Comparing aerial estimates with ground counts of nests in wood stork colonies •	95 JWildlMg 59 656-666
Velilla, Santiago	Diagnostics and robust estimation in multivariate data transformations •	95 JASA 90 945-951
Greene, W. H.; Greene, L. K.; Seaks, T. G.	Estimating the functional form of the independent variables in probit models •	95 ApplEcon 27 193-196
Katz, Richard W.; Parlange, Marc B.	Generalizations of chain-dependent processes: Application to hourly precipitation •	95 WaterReR 31 1331-1341
Traylor, L.	Non-regular maximum likelihood problems (Disc: p24-44) • Cheng, R. C. H.;	95 JRSS-B 57 3-24
	Tests following transformations • Chen, Hanfeng	95 AnlsStat 23 1587-1593
Moshe Quantile regression,	Box-Cox transformation model, and the U.S. wage structure, 1963-1987 • Buchinsky,	95 JEconmtx 65 109-154
Shrinkage estimation in nonlinear regression. The	Box-Cox transformation • Kim, Minbo; Hill, R. Carter	95 JEconmtx 66 1-33
Errors-in-variables and the	Box-Cox transformation • Edwards, Lloyd J.; Hamilton, Scott A.	95 CmpStDA 20 131-140
	Box-Jenkins	
Lin, Dennis K. J.	Robust recursive estimation for correlated observations • Guttman, Irwin;	95 StPrLet 23 79-92
A	Box-Jenkins approach to modeling outliers in time series analysis • Thome, Helmut	95 SocMethR 23 442-478
Manteiga, W. et al. Predicting using	Box-Jenkins, nonparametric, and bootstrap techniques • García-Jurado, I.; González-	95 Technmcs 37 303-310
function forecasting using neural networks and	Box-Jenkins techniques (Spanish) • de la Fuente et al. ... analysis of univariate and transfer-	95 Questo 19 187-215
Stacked	boxes: A new goodness-of-fit graphical display • Sparks, Ross	95 NZndSt 30 48-61
	Boxplot	
Rouncefield, Mary	The statistics of poverty and inequality (jse/v3n2/datasets.rouncefield) •	95 JStEd 3/2 10kb
Applicability of matrix and	bracket approach to stability study design • Lin, Tsae-Yun Daphne	94 ASAProBiop 142-147
Sir Austin	Bradford Hill: A personal view of his contribution to epidemiology • Doll, Sir Richard	95 JRSS-A 158 155-163
Before and after	Bradford Hill: Some trends in medical statistics • Armitage, Peter	95 JRSS-A 158 143-153
Hesterberg, Tim C. Bayesian analyses of the	Bradley-Terry model (Disc: p134-135) • Short, Thomas H.; Rossman, Allan J.;	94 ASAProBays 117-122
An application of	Bradley-Terry-type models to the measurement of pain • Matthews, J. N. S.; Morris, K. P.	95 ApplStat 44 243-255
estimation of covariate effects in multichannel	brain evoked potential data • Raz, J.; Cardenas, V.; Fletcher, D. Frequency domain	95 Biomtrcs 51 448-460
Lapointe, F.-J.; Casgrain, P. Modeling	brain evolution from behavior: A permutational regression approach • Legendre, P.;	94 Evolutn 48 1487-1499
A	brain-friendly first day of class • Zahn, Douglas A.	94 ASAProStEd 53-62
Nei, Masatoshi Interior-	branch and bootstrap tests of phylogenetic trees • Sitnikova, Tatyana; Rzhetsky, Andrey;	95 MolBioEv 12 319-333
Estimating errors and confidence intervals for	branch lengths in phylogenetic trees by a bootstrap approach • Dopazo, Joaquín	94 JMolEvol 38 300-304
hypotheses about evolutionary change on single	branches of a phylogeny using evolutionary contrasts • McPeek, Mark A. Testing	95 ANatrls 145 686-703
Veinott, Arthur F., Jr Markov	branching decision chains with interest-rate-dependent rewards • Huang, Ying;	95 PrEngInf 9 99-121
Bisexual	branching diffusions • Mytnik, Leonid; Adler, Robert J.	95 AdvAppPr 27 980-1018
Limits of spatial	branching populations (Disc: p185-189) • Wakolbinger, Anton	95 Bernoull 1 171-185
	Branching process	
M.	Conditioned superprocesses and their weighted occupation times • Krone, Stephen	95 StPrLet 22 59-69
	Controlled trees • Ceci, C.; Mazliak, L.	94 RBrzlPrS 8 93-105
Borovkov, K. A.	On crossing times for multidimensional walks with skip-free components •	95 JAppProb 32 991-1006
	Diffusion approximation for an age-structured population • Bose, A.; Kaj, I.	95 AnlsApPr 5 140-157
Peres, Yuval	Galton-Watson trees with the same mean have the same polar sets • Pemantle, Robin;	95 AnlsProb 23 1102-1124
its applications • Nishiyama, Yoichi	Local asymptotic normality of a sequential model for marked point processes and	95 AnInStMa 47 195-209
	Markov snakes and superprocesses • Dynkin, E. B.; Kuznetsov, S. E.	95 ProbTher 103 433-473
Klaus; Le Gall, Jean-François	A new approach to the single point catalytic super-Brownian motion • Fleischmann,	95 ProbTher 102 63-82
Scott D.	New results for self-similar trees with applications to river networks • Peckham,	95 WaterReR 31 1023-1029
absorbing state • Pakes, Anthony G.	Quasi-stationary laws for Markov processes: Examples of an always proximate	95 AdvAppPr 27 120-145
Fleischmann, Klaus; Li, Yi; Mueller, Carl	Singularity of super-Brownian local time at a point catalyst • Dawson, Donald A.;	95 AnlsProb 23 37-55
	Slow points in the support of historical Brownian motion • Verzani, John	95 AnlsProb 23 56-70
	A stochastic model for surname evolution • Islam, M. N.	95 BiomtrcJ 37 119-126
	Strong approximations for epidemic models • Ball, Frank; Donnelly, Peter	95 StocProc 55 1-21
The genealogy of	branching processes and the age of our most recent common ancestor • O'Connell, Neil	95 AdvAppPr 27 418-442
A multitype decomposable age-dependent	branching process and its applications • Lee, Chinsan; Yang, Grace L.	95 JAppProb 32 591-608
proofs of $L \log L$ criteria for mean behavior of	branching processes • Lyons, Russell; Pemantle, Robin; Peres, Yuval Conceptual	95 AnlsProb 23 1125-1138
some control ... • Biswas et al. On a continuous	branching process and birth-death and immigration models for testing the efficacy of	95 BiomtrcJ 37 235-249
A key limit theorem for critical	branching processes • Sagitov, Serik	95 StocProc 56 87-100
Analysis of transient effects for	branching processes • Borovkov, K. A.	94 ThProbAp 39 379-395
distributions in cell populations • Taib, Ziad	Branching processes and functional differential equations determining steady-size	95 JAppProb 32 1-10
Three papers on the history of	branching processes • Guttorp, Peter (Transl)	95 IntStRvw 63 233-245
Large deviation rates for	branching processes. II. The multitype case • Athreya, K. B.; Vidyashankar, A. N.	95 AnlsApPr 5 566-576
Review of "Statistical inference for	branching processes" • Guttorp, P. (Auth); Panaretos, J. (Rev)	95 Biomtrcs 51 383-384
Review of "Statistical inference for	branching processes" • Guttorp, Peter (Auth); Basawa, Ishwar V. (Rev)	95 JStPlInf 47 393-394
of discrete laws via mixed sums and Markov	branching processes • Pakes, Anthony G. Characterization	95 StocProc 55 285-300
	Branching processes as population dynamics • Jagers, Peter	95 Bernoull 1 191-200
	Branching processes. Proceedings of the First World Congress • Heyde, C. C. (Ed)	95 Springer:Brln:NY 180p
Renshaw, Eric Markov	branching processes regulated by emigration and large immigration • Chen, Anyue;	95 StocProc 57 339-359
for the limiting random variable in a supercritical	branching process • Hambly, B. M. On constant tail behaviour	95 JAppProb 32 267-273
find the extinction probability of a Galton-Watson	branching process • Komarov, S. V.; Krasulina, T. P. Using stochastic approximation to	94 VLenUM /3 34-39
	Branching processes with final types of particles and random trees • Vatutin, V. A.	94 ThProbAp 39 628-641
A modified bootstrap for	branching processes with immigration • Datta, Somnath; Sriram, I. N.	95 StocProc 56 275-294
Central limit theorem for martingales in BGWR	branching processes with some statistical applications • Dion, J. P.; Yanev, N. M.	95 MaMethSt 4 344-358
Minimal positions in a	branching random walk • McDiarmid, Colin	95 AnlsApPr 5 128-139
	Branching random walk with a critical branching part • Kesten, Harry	95 JTherPrb 8 921-962
Random sums and	branching stochastic processes • Rahimov, I.	95 Springer:Brln:NY 195p
Review of "Random sums and	branching stochastic processes" • Rahimov, I. (Auth); Teugels, J. L. (Rev)	95 ShrtBkRv 15/3 48-48
Allenby, Greg M.; Lenk, Peter J. Reassessing	brand loyalty, price sensitivity, and merchandising effects on consumer brand choice •	95 JBES 13 281-289
choice alternatives in memory: Probability models of	brand name recall • Hutchinson, J. Wesley; Raman, Kalyan; Mantrala, Murali K. Finding	94 JMarRsr 31 441-461
metry, nonproportionality, and heterogeneity in	brand switching from piece-wise exponential mixture hazard ... • Wedel et al. ... for asym-	95 JMarRsr 32 457-462
among staging areas of the barnacle goose (	Branta leucopsis) • Schwarz, Carl James; Ganter, Barbara Estimating the movement	95 JAppStat 22 711-724
Boundary condition sets limits on	Brass's model life table parameters • Mitra, S.	94 ASAProSoSt 55-59
between modernization and fertility: The case of	Brazil (Italian) • Drovandi A structural model with latent variables used to analyze the links	94 Statstca 54 521-538
The structure of technology in	Brazilian sugarcane production, 1975-87 • Rask, Kevin	95 JApEcmtx 10 221-232
al. Time-scales for Gaussian approximation and its	breakdown under a hierarchy of periodic spatial heterogeneities • Bhattacharya, R. N. et	95 Bernoull 1 81-123
	Breakdown point	
	An Anscombe type robust regression statistic • Bradu, Dan; Hawkins, Douglas M.	95 CmpStDA 20 355-386
A.; Yohai, Victor J.	The behavior of the Stahel-Donoho robust multivariate estimator • Maronna, Ricardo	95 JASA 90 330-341
• Hössjer, Ola	Exact computation of the least trimmed squares estimate in simple linear regression	95 CmpStDA 19 265-282
ate location parameter • Hössjer, O.; Croux, C.	Generalizing univariate signed rank statistics for testing and estimating a multivari-	95 JNnparSt 4 293-308

Bryan; Ghysels, Eric Federal	budget projections: A nonparametric assessment of bias and efficiency • Campbell,	95 RvEconSt 77 17-31
State-dependent Beneš	buffer model with fast loading and output rates • Kogan, Y.; Liptser, R.; Shenfild, M.	95 AnlsApPr 5 97-120
Analysis of the sensitivity of the GI/GI/m/B queue to	buffer size: The score function method • Kriman, Vladimir	95 CommStSM 11 171-194
Sidi, Moshe An optimal service policy for	buffer systems • Birman, Alexander; Gail, H. Richard; Hantler, Sidney L.; Rosberg, Zvi;	95 JACM 42 641-657
The Pentium	bug and statistical software • Gould, William	95 Chance 8/2 9-13
Factorisation of semi-martingales on principal fibre	bundles and the Faddeev-Popov procedure in gauge theories • Arnaudon, M.; Paycha, S.	95 Stochatx 53 81-107
Rasch models for item	bundles • Wilson, Mark; Adams, Raymond J.	95 Psymtrka 60 181-198
New CASIC technologies at the U.S.	Bureau of the Census • Nicholls, William L., II; Appel, Martin V.	94 ASASrvyRsr 757-762
of moving averages and statistical solutions of the	Burgers' equation • Hu, Yiming; Woyczynski, W. A. Limit behavior of quadratic forms	95 JMultiAn 52 15-44
Stochastic	Burgers equation with correlated noise • Prato, Giuseppe Da; Gatarek, Dariusz	95 Stochatx 52 29-41
Gibbs-Cox random fields and	Burgers turbulence • Funaki, T.; Surgailis, D.; Woyczynski, W. A.	95 AnlsApPr 5 461-492
	Burkholder inequality	
Zangeneh, Bijan Z.	Semilinear stochastic evolution equations with monotone nonlinearities •	95 Stochatx 53 129-174
Modeling and maximizing	burn-in effectiveness • Chien, Wei-Ting Kary; Kuo, Way	95 IEEEReli 44 19-25
of some goodness-of-fit test statistics using	Burr distribution and censored samples • Ashour, S. K. et al. The sampling distribution	94 EgpStJ 38 147-163
M. A. M. Empirical Bayes estimators for the	Burr type XII accelerated life testing model based on type-2 censored data • Ali Mousa,	95 JStCmpSm 52 95-103
Bayesian prediction bounds for the	Burr type XII failure model • Al-Hussaini, Essam K.; Jaheen, Zeinhum F.	95 CommStA 24 1829-1842
bers: Application of the analysis of a generalised	Burt sub-table (French) • Nodjiram, S. Analysis of a contingency table with negative num-	95 CAnDnnes 20 111-120
exponentiated Weibull family: A reanalysis of the	bus-motor-failure data • Mudholkar, G. S.; Srivastava, D. K.; Freimer, M. The	95 Technmcs 37 436-445
Picturing an L.A.	bus schedule • Wainer, Howard	95 Chance 8/4 44-48
Current ... • Konschnik The impact of dropping the	business area sample as a means of covering birth employers and nonemployers in the	94 ASASrvyRsr 995-1000
of "Fourth generation management: The new	business consciousness" • Joiner, Brian L. (Auth); Ziegel, Eric R. (Rev) Review	95 Technmcs 37 113-113
for Wales • Ioannidis, C.; Laws, J.; Matthews, K. et al.	Business cycle analysis and forecasting with a structural vector autoregression model	95 JForcast 14 251-265
for duration dependence: An application to UK	business cycle data • Mudambi, Ram; Taylor, Larry W. Some non-parametric tests	95 JAppStat 22 163-177
Berk et al. International interdependence of	business cycles in the manufacturing industry: The use of leading indicators for ... •	95 JForcast 14 1-23
Quarterly forecasts of the Italian	business cycle by means of monthly economic indicators • Parigi, G.; Schlitzer, G.	95 JForcast 14 117-141
with vector autoregressive (VAR) models subject to	business cycle restrictions • Simkins, Scott Forecasting	95 IntJFor 11 569-583
Frisch on testing of	business cycle theories • Boumans, Marcel	95 JEconmtx 67 129-147
Review of "Statistics for	business: Data analysis and modelling" • Cryer, J.; Miller, R. (Auth); Ziegel, Eric R. (Rev)	95 Technmcs 37 239-240
Review of "Statistics for	business and economics" • Mann, P. (Auth); Ziegel, Eric R. (Rev)	95 Technmcs 37 239-240
(Rev) Review of "Statistics for	business and economics" • Oltman, D. Olson; Lackritz, J. R. (Auth); Garcia, Eva Ferreira	95 JAppStat 22 437-437
Sample-audit tax assessment for	businesses: What's fair? • Press, S. James	95 JBES 13 357-359
Review of	"Business forecasting and planning" • Shearer, Peter (Auth); Raeside, Robert (Rev)	95 IntJFor 11 192-193
to estimate the effects of health reform on individual	businesses (Disc: p51-52) • Rodgers, J. et al. Using microsimulation modeling techniques	94 ASAProGovt 39-44
and variance estimation for the characteristics of	business owners survey • Kott, Phillip S. Reweighting	94 JOffStat 10 407-418
"Strivin', gettin' over, and keepin' people out of our	business": Pushing household boundaries in survey data collection • Hainer, Peter C.	94 ASASrvyRsr 337-341
Reflections on making statistics more effective in	business schools (MSMESB) (Disc: p319-320) • Roberts, Harry V.	94 ASAProBuEc 316-318
et al. Review of "Statistical tables for economic,	business and social studies (Second edition)" • Kmietowicz, Z. W.; Yannoulis, Y. (Auth)	95 JAppStat 22 439-440
Integrating cases into the	business statistics course • Nolan, James R.	94 ASAProStEd 123-125
Levine, D. M.; Ramsey, P. P. Reengineering the	business statistics course: Meeting the customer needs of the twenty-first century •	94 ASAProStEd 121-122
TQM in undergraduate	business statistics: What are the customers' needs? • Drapeau, Richard A.	94 ASAProQlPr 270-272
Liberal education in	business statistics • Bowerman, Bruce L.; O'Connell, Richard T.	94 ASAProStEd 259-262
Review of "Introduction to modern	business statistics" • Canavos, G.; Miller, D. (Auth); Ziegel, Eric R. (Rev)	95 Technmcs 37 239-240
Review of "Modern	business statistics" • Canavos, George C.; Miller, Don M. (Auth); Delaney, Nancy Jo (Rev)	95 AmerStat 49 397-397
Improving the presentation of	business statistics • Melton, Kim I.	94 ASAProQlPr 273-275
Ozgur, Ceyhun Undergraduate	business statistics: A survey of topics and teaching methods • Strasser, Sandra E.;	95 Intrfc 25/3 95-103
Ziegel, Eric R. (Rev) Review of "Introductory	business statistics with computer applications" • Shiffler, R.; Adams, A. (Auth);	95 Technmcs 37 239-240
birth employers and nonemployers in the Current	Business Surveys • Konschnik ... the business area sample as a means of covering	94 ASASrvyRsr 995-1000
Colledge, M. J.; Kott, P. S. (Ed)	Business survey methods • Cox, B. G.; Binder, D. A.; Chinnappa, B. N.; Christianson, A.;	95 Wiley:NY:UK 732p
between manufacturing production and different	business survey series in Sweden 1968-1992 • Bergström, Reinhold The relationship	95 IntJFor 11 379-393
Scheduling initial calls in a telephone survey of	businesses • Edwards, W. Sherman; Cunningham, Patricia	94 ASASrvyRsr 1368-1373
Statistical thinking for	business • Hare, Lynne; Hoerl, Roger W.; Snee, Ronald D.	95 ASQCTekCon 787-788
Gakis et al. Distributions and first moments of the	busy and idle periods in controllable M/G/1 queueing models with simple and dyadic ... •	95 StocAnAp 13 47-81
The	busy periods of some queueing systems • Stadje, W.	95 StocProc 55 159-167
using BVAR models and survey data on households'	buying attitudes for homes • Dua, Pami; Smyth, David J. Forecasting US home sales	95 JForcast 14 217-227
techniques • Curry, D. J.; Divakar, S. et al.	BVAR as a category management tool: An illustration and comparison with alternative	95 JForcast 14 181-199
A	BVAR model for the Connecticut economy • Dua, Pami; Ray, Subhash C.	95 JForcast 14 167-180
mance • Sarantis, N.; Stewart, C. Structural, VAR and	BVAR models of exchange rate determination: A comparison of their forecasting perfor-	95 JForcast 14 201-215
Smyth, David J. Forecasting US home sales using	BVAR models and survey data on households' buying attitudes for homes • Dua, Pami;	95 JForcast 14 217-227
for use in chromosome mapping studies with the	BXD recombinant inbred strains • Belknap Empirical estimates of Bonferroni corrections	92 BehavGen 22 677-684
David	Byar's approach to the analysis of survival data with covariates • Armitage, Peter	95 ConClnTr 16 259-269
Dave	Byar's contribution to epidemiology • Gail, Mitchell H.	95 ConClnTr 16 230-248
Some of David	Byar's work: In appreciation • Moses, Lincoln E.	95 ConClnTr 16 216-229
Nonparametric tests for ordered	c-sample contaminated distributions • Johnson, Richard A.; Sim, Songyong	95 CommStA 24 861-879
Hibbert, D. B.; Haddad, P. R. et al. Application of the	C4.5 classifier to building an expert system for ion chromatography • Mulholland, M.;	95 Chemolab 27 95-104
Uncertainty	calculi for fault trees • Weber, David P.	95 ASQCTekCon 1046-1053
formulas for Lyapunov exponents via anticipative	calculus • Arnold, Ludwig; Imkeller, Peter Furstenberg-Khasminskii	95 Stochatx 54 127-168
Review of "White noise. An infinite dimensional	calculus" • Hida, T.; Kuo, H.-H.; Potthoff, J.; Streit, L. (Auth); Øksendal, B. (Rev)	95 BLndnMaS 27 189-191
Wu, Jiang-Lun Stochastic anticipative	calculus for functionals of a Gaussian generalized random field • Albeverio, Sergio;	95 Stochatx 54 317-337
Differential	calculus relative to some point processes • Dermoune, A.	95 StPrLet 24 233-242
Green formulas in anticipating stochastic	calculus • Delgado, Rosario; Sanz-Solé, Marta	95 StocProc 57 113-148
measurement processes via classical stochastic	calculus • Barchielli, A.; Holevo, A. S. Constructing quantum	95 StocProc 58 293-317
From discrete to continuous stochastic	calculus • Cutland, Nigel J.; Kopp, P. Ekkehard; Willinger, Walter	95 Stochatx 52 173-192
Introduction to stochastic	calculus • Lawler, G. F.	95 ChapmnHall:Lnd 176p
Review of "Introduction to stochastic	calculus" • Lawler, G. F. (Auth); Riddler-Rowe, C. J. (Rev)	96 ShrtBkRv 16/1 8-8
Elements of queueing theory: Palm-martingale	calculus and stochastic recurrences • Baccelli, F.; Bremaud, P.	94 Springer:Brln:NY 256p
(Rev) Review of "Constrained optimization in the	calculus of variations and optimal control theory" • Gregory, J.; Lin, C. (Auth); Prvan, T.	95 AstrlJSt 37 250-250
The Beneš equation and stochastic	calculus of variations • Decreusefond, L.; Üstünel, A. S.	95 StocProc 57 273-284
Fifty years of	Calcutta Statistical Association • Banerjee, P. K.	95 CalcutSt 45 1-23
Estimation of usage lifetime distribution from	calendar lifetime data • Suzuki, Kazuyuki	93 RStAJp 40/1 10-22
A method for	calibrating false match rates in record linkage • Belin, Thomas R.; Rubin, Donald B.	95 JASA 90 694-707

	Calibration	
Dupont, F.	Alternative adjustments where there are several levels of auxiliary information •	95 SurvyM 21 125-135
dioimmunoassay (Disc: p953-953) • Rosen, O. et al.	Constructing a bootstrap confidence interval for the unknown concentration in ra-	95 StatMed 14 935-952
	A correlation principal component regression analysis of NIR data • Sun, Jianguo	95 JChemtrc 9 21-29
	Decision and detection limits for linear homoscedastic assays • Dunne, Adrian	95 StatMed 14 1949-1959
Nørgaard, Lars	Direct standardisation in multi wavelength fluorescence spectroscopy •	95 Chemolab 29 283-293
Davis, Richard A.; Dunsmuir, William T. M.	Improved bootstrap prediction intervals for autoregressions • Breidt, F. Jay;	95 JTimSrAn 16 177-200
	An improved support system for practical metrology • Huntley, Les	95 ASQCTekCon 142-149
models • Thomas, Edward V.	Incorporating auxiliary predictor variation in principal component regression	95 JChemtrc 9 471-481
Ferré, Louis; Vieu, Philippe	Location of particular points in non-parametric regression analysis • Boularan, Joël;	95 AstrlJSt 37 161-168
Cross-validation, missing data and ... • Rännar et al.	A PLS kernel algorithm for data sets with many variables and few objects. Part II:	95 JChemtrc 9 459-470
in the Unscrambler • De Vries, S.; Braak, Cajo J. F. Ter	Prediction error in partial least squares regression: A critique on the deviation used	95 Chemolab 30 239-245
analysis of Fourier transform ... • Lindblom et al.	Quantitative determination of stabiliser in a single base propellant by chemometric	95 Chemolab 29 243-254
algorithms • Vankeerberghen, P. et al.	Robust regression and outlier detection for non-linear models using genetic	95 Chemolab 28 73-87
pollutants (Disc: p145-167) • Gibbons, Robert D.	Some statistical and conceptual issues in the detection of low-level environmental	95 EnvEcoSt 2 125-145
	Univariate regression models with errors in both axes • Riu, J.; Rius, F. X.	95 JChemtrc 9 343-362
operation • Miller, Charles E.	The use of chemometric techniques in process analytical method development and	95 Chemolab 30 11-22
	Using guardbands to justify TURs less than 4:1 • Deaver, David	95 ASQCTekCon 136-141
Virtual manufacturing: Design,	calibration, and analysis • Heavlin, William D.; Krivokapić, Zoran	94 ASAProPEng 274-279
Outlier detection in bilinear	calibration • Walczak, B.	95 Chemolab 29 63-73
Bayesian approach to multivariate and conditional	calibration • Plessis, J. L. du; van der Merwe, A. J. A	95 CmpStDA 19 539-552
of experiments for a precise estimation of the	calibration curve of periodate in its determination by ... • Papamichael et al. Design	95 Chemolab 30 227-237
Riboli, Elio; van Staveren, Wija	Calibration of dietary intake measurements in prospective cohort studies • Kaaks, Rudolf;	95 AJEpidm 141 548-556
Ransom, K. P.; Oddy, V. H.	Calibration of empirical models: Simulation of pasture growth and decay • McPhee, M. J.;	95 EnvrMtrc 6 505-510
	Calibration error and the measurement of turbulence statistics • Derksen, R. W.	95 EnvrMtrc 6 651-658
Boggia, R.; Lanteri, S.; Leardi, R. et al. Transfer of	calibration function in near-infrared spectroscopy • Forina, M.; Drava, G.; Armanino, C.;	95 Chemolab 27 189-203
Cost effective	calibration intervals using Weibull analysis • Barringer, H. Paul	95 ASQCTekCon 1026-1038
	Calibration for inverse Gaussian regression • Woldie, Mammo; Folks, J. Leroy	95 CommStA 24 2609-2620
order standard addition method with application to	calibration of a kinetics-spectroscopic sensor for quantitation ... • Booksh et al. A second-	95 JChemtrc 9 263-282
and classical estimators in multi-univariate linear	calibration • Srivastava, M. S. Comparison of the inverse	95 CommStA 24 2753-2767
Probability matching priors for linear	calibration • Ghosh, M.; Carlin, B. P.; Srivastava, M. S.	95 Test 4 333-357
Robust linear	calibration • Kitsos, Christos P.; Müller, Christine H.	95 MaOpfStS 27 93-106
Plessis, J. L. du; van der Merwe, A. J. Bayesian	calibration in meat science research-estimation of the pH in postmortem muscle •	95 JStCmpSm 53 1-11
Age K. Some theoretical results on second-order	calibration methods for data with and without rank overlap • Kiers, Henk A. L.; Smilde,	95 JChemtrc 9 179-195
Outlier detection in multivariate	calibration • Walczak, B.	95 Chemolab 28 259-272
rithm for treatment of non-linearity in multivariate	calibration • Xie, Y.-L.; Liang, Y.-Z.; Chen, Z.-G. et al. A nonlinearity tracking analysis algo-	95 Chemolab 27 21-32
Design considerations for nonlinear	calibration • Bates, Douglas M.; Martin, Sandra J.	94 ASAProPEng 12-17
Reference priors for the multivariate	calibration problem • Plessis, J. L. du; van der Merwe, A. J.; Groenewald, P. C. N.	95 SoAfrStJ 29 155-168
Maximizing the use of auxiliary information for	calibration and regression estimation • Estevao, V.; Patak, Z.; Särndal, C.	94 ASASrvyRsr 42-48
Review of "Measurement, regression, and	calibration" • Brown, P. J. (Auth); Atkinson, Anthony (Rev)	95 JRSS-A 158 187-187
Review of "Measurement, regression, and	calibration" • Brown, P. J.; Fearn, Tom (Rev)	95 StMedRes 4 265-266
of a split-plot type • Lee, Kwan R.; Gagnon, Robert C.	Calibration and stability analysis with a simple mixed linear model when the experiment is	94 ASAProBiop 435-440
Kaaks, R. et al. Sample size requirements for	calibration studies of dietary intake measurements in prospective cohort investigations •	95 AJEpidm 141 557-565
Kaaks, R. et al. Sample size requirements for	calibration studies of dietary intake measurements in prospective cohort investigations •	95 AJEpidm 142 557-565
Non-linear methods for multivariate statistical	calibration and their use in palaeoecology: A comparison of inverse (*k*-nearest ... • Braak	95 Chemolab 28 165-180
Manuel Comment on "Functional comparative	calibration using an EM algorithm" (V48 p1263-1271) • Bolfarine, Heleno; Galea-Rojas,	95 Biomtrcs 51 1579-1580
Structural comparative	calibration using the EM algorithm • Bolfarine, Heleno; Galea Rojas, Manuel	95 JAppStat 22 277-292
A user-friendly spreadsheet program for	calibration using weighted regression • Gort, Steven M.; Hoogerbrugge, Ronald	95 Chemolab 28 193-199
and selection for linearity and interaction in	calibration with pure and multicomponent spectra • Brown, Philip J. Diagnostics	95 Chemolab 28 139-148
The employment effect in retail trade of	California's 1988 minimum wage increase • Kim, Taeil; Taylor, Lowell J.	95 JBES 13/2 175-182
Pradel, Roger Mark-resighting analysis of a	California gull population • Pugesek, Bruce H.; Nations, Chris; Diem, Kenneth L.;	95 JAppStat 22 625-639
frequency distribution of the numbers bet on the	California lottery • Finkelstein, Mark Estimating the	95 AplMaCmp 69 195-207
series analysis • Cox, K. et al. Effects of exercise and	calorie restriction on ambulatory blood pressure in overweight, sedentary males: A time	95 EnvrMtrc 6 523-528
The	Cambridge dictionary of statistics in the medical sciences • Everitt, B. S.	95 CambridgeUniv 274p
New	Cambridge statistical tables (Second edition) • Lindley, D. V.; Scott, W. F.	95 CambridgeUniv 96p
Employment and childbearing in	Canada • Rao, K. V.	94 ASAProSoSt 124-128
Gentleman, Jane F. Life expectancy in	Canada: An empirical comparison of different estimation methods • Ng, Edward;	94 ASAProSoSt 88-93
study and management of Atlantic Flyway	Canada geese • Hestbeck, Jay B. Population	95 JAppStat 22 877-890
Jarvis, Robert L. Bias in	Canada goose population size estimates from sighting data • Sheaffer, Susan E.;	95 JWildlMg 59 464-473
Trends in mortality from melanoma in	Canada and prediction of future rates • MacNeill, I. B.; Elwood, J. M.; Miller, D.; Mao, Y.	95 StatMed 14 821-839
the Vancouver, Montreal, and Atlantic regions of	Canada • Burrows et al. ... and prediction of summer season maximum surface ozone for	95 JApMetrl 34 1848-1862
C. Time series analysis of improved quality of life in	Canada: Social change, collective consciousness, and the TM-Sidhi program • Dillbeck, M.	95 PsychRpt 76 1171-1193
and volatility: The case of the United States and	Canada • Karolyi ... GARCH model of international transmissions of stock returns	95 JBES 13 11-25
Marketing of statistics in	Canada • Fellegi, I. P.	91 StatJUN 8 295-306
Modelling net undercoverage in the 1991	Canadian Census • Dick, Peter	95 SurvyM 21 45-54
A data collection methodology for the 2001	Canadian Census • Choudhry, G. H.	94 ASASrvyRsr 1134-1139
the Statistics Canada Farm Register with the	Canadian Census of Agriculture (Disc: p212-214) • Thomas, Bradley W. Updating	94 ASAProGovt 189-194
between profitability, size and concentration in	Canadian corp. • Lothian, Jack The relationship	94 ASAProBuEc 287-292
Sample allocation for the	Canadian Labour Force Survey • Mian, Ijaz U. H.; Laniel, Normand	94 ASASrvyRsr 913-918
of alternative sample designs for rural areas in the	Canadian Labour Force Survey • Mantel, H.; Laniel, N.; Duval, M.-C. et al. Cost modelling	94 ASASrvyRsr 925-930
classification error with application to data from the	Canadian Labour Force Survey • Singh et al. On the adjustment of gross flow estimates for	95 JASA 90 478-488
plementing computer-assisted interviewing for the	Canadian Labour Force Survey • Bench, J.; Clark, C.; Dufour, J.; Kaushal, R. Im-	94 ASAProGovt 93-98
Deborah; Kinack, Mark Redesigning the	Canadian Labour Force Survey questionnaire: Development and testing • Sunter,	94 ASASrvyRsr 908-912
Control charts for non-response rates in the	Canadian Labour Force Survey • Hapuarachchi, K. P.; Wronski, A.	94 ASASrvyRsr 937-942
estimation issues for income in the redesign of the	Canadian Labour Force Survey • Chen, E. J.; Gambino, J.; Laniel, N.; Lindeyer, J. Design and	94 ASASrvyRsr 919-924
Analysis of urban cluster size in the	Canadian Labour Force Survey • Laniel, Normand; Mohl, Chris	94 ASASrvyRsr 931-936
Raking ratio estimation: An application to the	Canadian Monthly Retail Trade Survey • Patak, Z.; Hidiroglou, M. A.	94 ASASrvyRsr 299-304
Evaluation of the	Canadian Potato Area Estimation Program • Whitridge, Patricia J.; Poirier, Claude	94 ASASrvyRsr 1150-1155
the generalized edit and imputation system for the	Canadian Survey of Family Expenditures: A feasibility study • Bérard, H. Use of	94 ASAProSoSt 178-183
puter assisted interviewing: The experience of the	Canadian survey of labour and income dynamics (Disc: ... • Grondin et al. ... using com-	94 ASASrvyRsr 839-844
Current and future approaches to editing	Canadian trade import data • Pursey, Stuart	94 ASAProGovt 105-109
New approach for constructing	Canadian working life tables, 1889-1987 • Belanger, A.; Larrivee, D.	92 StatJUN 9 27-50
Thomas J. The development of the series of U.S.	cancer atlases: Implications for future epidemiologic research (Disc: p697-699) • Mason,	95 StatMed 14 473-479
risk factors and post-menopausal breast	cancer • Lund, Eiliv Comparison of additive and multiplicative models for reproductive	95 StatMed 14 267-274

How to deal with the process capability	"catch-22" • Bailey, Steven P.	95 ASQCTekCon 1105-1116
Statistical analysis of	catch-at-age data with correlated errors • Myers, Ransom A.; Cadigan, Noel G.	95 JFishRBC 52 1265-1273
influence of error on population estimates from	catch-age models • Schnute, Jon T.; Richards, Laura J. The	95 JFishRBC 52 2063-2077
child and adolescent trial for cardiovascular health (	CATCH): Implications of cluster randomization • Zucker et al. Statistical design of the	95 ConClnTr 16 96-118
Kressel, J. Properties of	catch rates used in analysis of angler surveys • Jones, C. M.; Robson, D. S.; Lakkis, H. D.;	95 TrAFish 124 911-928
cination schedules: Availability for vaccination and	catch-up strategies • Rouderfer, Vladimir; Becker, Niels G. Assessment of two-dose vac-	95 MaBisci 129 41-66
On the	catch-up time method for analyzing cancer screening trials • Etzioni, Ruth; Self, Steven G.	95 Biomtrcs 51 31-43
Self, S. G. Correction to "On the	catch-up time method for analyzing cancer screening trials" (V51 p31-43) • Etzioni, R.;	95 Biomtrcs 51 1589-1589
Quantifying the effects of unequal	catchabilities on Jolly-Seber estimators via sample coverage • Hwang, W.-D.; Chao, Anne	95 Biomtrcs 51 128-141
and trend analysis in epidemiology: Alternatives to	categorical analysis • Greenland, Sander Dose-response	95 Epidmlgy 6 356-365
Logistic regression with incompletely observed	categorical covariates - Investigating the sensitivity against violation of the ... • Vach et al.	95 StatMed 14 1315-1329
	Categorical data	
Ritschard, Gilbert	The behaviour of nominal and ordinal partial association measures • Olszak, Michael;	95 Ststcian 44 195-212
exposure levels • Correa-Villaseñor, A. et al.	Bias from nondifferential misclassification in case-control studies with three	95 Epidmlgy 6 276-281
agreement measures • Cook, R. J.; Farewell, V. T.	Conditional inference for subject-specific and marginal agreement: Two families of	95 CandJSt 23 333-344
Takashi; Kikuchi, Yasuki	Determination of the no-observed-adverse-effect level by the AIC • Yanagawa,	95 SankhyaB 57 285-297
Menéndez, M.; Morales, D.; Pardo, L.; Vajda, I.	Divergence-based estimation and testing of statistical models of classification •	95 JMultiAn 54 329-354
	A note on the attenuation of correlation • Bedrick, Edward J.	95 BJMaStPs 48 271-280
for interactions • Stack, Jim; Unwin, Antony R.	Selecting explanatory variables from survey data: A simple approach that explores	95 Ststcian 44 17-29
Eskridge, K. M.	Statistical analysis of disease reaction data using nonparametric methods •	95 HortSci 30 478-481
	The 'unusual episode' data revisited • Dawson, Robert J. MacG.	95 JStEd 3/3 23kb
Variance estimation techniques in	categorical data analysis for survey data • Rai, Anil	94 JIAgSt 46 147-155
Lang, Joseph B. Marginal modelling of	categorical data from crossover experiments • Balagtas, Cecile C.; Becker, Mark P.;	95 ApplStat 44 63-77
Pereira, C. A. D. B. Bayesian methods for	categorical data under informative general censoring • Mimoso Paulino, C. D.;	95 Biomtrka 82 439-446
Graphical representation of quantified	categorical data: Its inherent problems • Nishisato, Shizuhiko	95 JStPlInf 43 121-132
Conceptual and visual models for	categorical data • Friendly, Michael	95 AmerStat 49 153-160
A random-effects model for ordered	categorical data • Crouchley, Robert	95 JASA 90 489-498
cumulative probit regression model for ordered	categorical data • Kim, Kyungmann A bivariate	95 StatMed 14 1341-1352
calculations for the Wilcoxon statistic for ordered	categorical data • Kolassa, John E. A comparison of size and power	95 StatMed 14 1577-1581
for tests of rank interaction with ordered	categorical data • Thomas, G. E. A note on chi-squared approximations	95 CommStB 24 439-459
and random differences between paired ordinal	categorical data" • Svensson, E. (Auth); Lesaffre, E. (Rev) Review of "Analysis of systematic	95 Biomtrcs 51 794-794
and random differences between paired ordinal	categorical data" • Svensson, E. (Auth); Lindsey, J. K. (Rev) Review of "Analysis of systematic	95 StatMed 14 1726-1727
mutational spectra: Some recommendations for	categorical data • Piegorsch, Walter W.; Bailer, A. John Statistical approaches for analyzing	94 Genetcs 136 403-416
Smoothing	categorical data • Simonoff, Jeffrey S.	95 JStPlInf 47 41-69
Fox, Richard (Rev) Review of "Multivariate analysis of	categorical data: Theory and applications. Volumes 1 and 2" • Van de Geer, John P. (Auth);	95 JMarRsr 32 111-112
Rose, E. L. (Rev) Review of "Multivariate analysis of	categorical data: Theory and applications. Volumes 1 and 2" • Van de Geer, J. P. (Auth);	95 StrEqMod 2 274-276
A one-step Gauss-Newton estimator for modelling	categorical data with extraneous variation • Morel, Jorge G.; Koehler, Kenneth J.	95 ApplStat 44 187-200
for estimating a nondecreasing function with	categorical data • Hsu, Y.-S. An optimal grouping	95 SochwJMa 21 253-261
Jacques A. (Auth); Eliason, Scott R. (Rev) Review of	"Categorical longitudinal data: Log-linear panel, trend, and cohort analysis" • Hagenaars,	95 JEdStat 20 103-104
Prediction functions for	categorical panel data • Gilula, Zvi; Haberman, Shelby J.	95 AnlsStat 23 1130-1142
Standardized estimates from	categorical regression models • Joffe, Marshall M.; Greenland, Sander	95 StatMed 14 2131-2141
Darroch, J. N. An analysis of	categorical repeated measurements in the presence of an external factor • McCloud, P. I.;	95 Biomtrka 82 47-55
ning and analysis of experiments with continuous or	categorical response • Toutenburg, H. Experimental design and model choice. The plan-	95 Physica:Ger 463p
ning and analysis of experiments with continuous or	categorical response • Toutenburg, H. Experimental design and model choice. The plan-	95 Springer:Brln:NY 463p
Analysing	categorical responses obtained from large clusters • Miller, Michael E.	95 ApplStat 44 173-186
search algorithm for optimal scaling of ordered	categorical responses • Lin, Shang P.; Tang, Dei-In A hierarchical active constraints	95 JStCmpSm 53 197-209
Maximizing selection efficiency for	categorical traits • Meuwissen, T. H. E.; Engel, B.; van der Werf, J. H. J.	95 JAnmlSci 73 1933-1939
	Categorical variables	
Christopher, Susan R.; Kupper, Lawrence L.	On the effects of predictor misclassification in multiple linear regression analysis •	95 CommStA 24 13-37
Gilula, Zvi; Haberman, Shelby J. Dispersion of	categorical variables and penalty functions: Derivation, estimation, and comparability •	95 JASA 90 1447-1452
Generating multivariate	categorical variates using the iterative proportional fitting algorithm • Gange, Stephen J.	95 AmerStat 49 134-138
Brown, Norman R. Strategies for estimating	category frequency: Effects of abstractness and distinctiveness • Conrad, Frederick G.;	94 ASASrvyRsr 1345-1350
quantitative variable to deal with multiple affection	categories in genetic linkage analysis • Curtis, D.; Gurling, H. M. D. Using a dummy	91 AnHuGene 55 321-327
Curry, D. J.; Divakar, S.; Mathur, S. K. et al. BVAR as a	category management tool: An illustration and comparison with alternative techniques •	95 JForcast 14 181-199
Effects of rank versus	category in measuring subjective social inequality • Hoffmeyer-Zlotnik, Jürgen H. P.	94 ASASrvyRsr 1322-1326
homogeneity for contingency tables with nominal	categories • Tomizawa, Sadao Measures of departure from marginal	95 Ststcian 44 425-439
Optimal scaling for ordered	categories • Pölz, W.	95 CmpStQ 10 37-41
models to $N \times N$ contingency tables having ordered	categories • Lawal, H. Bayo; Upton, G. J. G. An algorithm for fitting	95 CommStB 24 793-805
model for square contingency tables with ordered	categories • Tomizawa, Sadao A generalization of the marginal homogeneity	95 JEdStat 20 349-360
for square contingency tables with ordered	categories • Tomizawa, Sadao Measures of departure from global symmetry	95 Behvmtrk 22 91-98
Analysis of square tables with ordered	categories • Poon, Wai-Yin; Hung, Hin-Yan	94 ASAProStCp 292-298
ric methods with parametric methods for the CPS	CATI/CAPI mode effects analysis • Thompson, J.; Parmer, R. A comparison of nonparamet-	94 ASASrvyRsr 614-619
force surveys • Bergman, L. R. et al. Decentralised	CATI versus paper and pencil interviewing: Effects on the results in the Swedish labour	94 JOffStat 10 181-195
	Cauchy distribution	
	Complete classes of tests for regularly varying distributions • Kowalski, Jacek P.	95 AnInStMa 47 321-350
Wave front propagation for a	Cauchy problem with a fast component • Carmona, Sara C.	95 Stochatx 52 81-105
Liu, Shuangzhe; Neudecker, Heinz Matrix-trace	Cauchy-Schwarz inequalities and applications in canonical correlation analysis •	95 StHefte 36 287-298
of multimodal circular distributions of wrapped	Cauchy type and of von Mises type • Holmquist, Björn Simulation	95 CommStB 24 607-615
Review of "Statistical models for	causal analysis" • Retherford, R. D.; Choe, M. K. (Auth); Fearn, Tom (Rev)	95 JRSS-A 158 200-201
Hillmer, M. (Auth) et al. Review of "A	causal analysis with latent variables of macroeconomic interdependencies" (German) •	95 AllgeStA 79 224-227
Dyer, James S. Reliability and coherence of	causal, diagnostic, and joint subjective probabilities • Wright, M. Keith; Stokes, Lynne;	94 DecisnSc 25 691-709
Dyer, James S. Reliability and coherence of	causal, diagnostic, and joint subjective probabilities • Wright, M. Keith; Stokes, Lynne;	95 DecisnSc 26 691-709
	Causal diagrams for empirical research (Disc: p688-710) • Pearl, Judea	95 Biomtrka 82 669-688
Two-stage least squares estimation of average	causal effects in models with variable treatment intensity • Angrist, J. D.; Imbens, G. W.	95 JASA 90 431-442
	Causal inference	
	Probabilities and experiments • Cartwright, Nancy	95 JEconmtx 67 47-59
cation of covariance structure analysis ... • Maeda A	causal model of some aspects and determinants of satisfaction of the Japanese - An appli-	95 PrInStMa 43 141-160
Review of "TETRAD II: Tools for	causal modeling" • Lawrence Erlbaum Associates Inc (Auth); Wood, Phillip K. (Rev)	95 StrEqMod 2 277-287
	Causal network	
and statistical techniques • Højsgaard, S. et al.	BIFROST - Block recursive models induced from relevant knowledge, observations,	95 CmpStDA 19 155-175
and the role of randomization in uncovering a	causal relationship • Oldford A physical device for demonstrating confounding, blocking,	95 AmerStat 49 210-216

Linear rank tests for doubly	censored data • Tang, Ming-Xin; Tsai, Wei-Yann; Marder, Karen; Mayeux, Richard	95 StatMed 14 2555-2563
very small but related samples with application to	censored data estimation of PCB contamination • Johnson et al. Goodness-of-fit using	95 EnvrMtrc 6 341-348
with respect to kernel quantile estimators for	censored data • Xiang, Xiaojing Deficiency of the sample quantile estimator	95 AnlsStat 23 836-854
Wu, C. F. Analysis of	censored data from fractionated experiments: A Bayesian approach • Hamada, M.;	95 JASA 90 467-477
I. R.; Hogan, S. J.; Mallal, S. A. Analysis of doubly	censored data and the HIV incubation period distribution in Western Australia • James,	95 AstrlJSt 37 299-312
and prediction from Rayleigh based on type II	censored data • Howlader, Hatem A.; Hossain, Anwar On Bayesian estimation	95 CommStA 24 2249-2259
Regression with interval-	censored data • Rabinowitz, Daniel; Tsiatis, Anastasios; Aragon, Jorge	95 Biomtrka 82 501-513
Sequential testing with interval	censored data • Quan, Hui; Yu, Kai F.	95 JStCmpSm 51 239-247
Residual plots for the	censored data linear regression model • Hillis, Stephen L.	95 StatMed 14 2023-2036
definitions of nearest neighbour distances for	censored data on nearest neighbour kernel estimators of the ... • Dette et al. ... of different	95 JNnparSt 4 271-282
Maximum likelihood estimation with partially	censored data • van der Vaart, Aad	94 AnlsStat 22 1896-1916
ratio estimation of survival probabilities for	censored data • Li, Gang On nonparametric likelihood	95 StPrLet 25 95-104
Estimating a survival function with doubly	censored data in a proportional hazard model • Esteban, M. D.; Morales, D.	95 AplStMDA 11 145-157
Two-sample chi-square test for randomly	censored data (Korean) • Kim, Joo-Han; Kim, Jeong-Ran	95 KorApS 8/2 109-119
estimation of the center of symmetry with randomly	censored data • Yang, Song; Koul, Hira L. Minimum distance	95 Metrika 42 79-97
effect for the two-sample problem with right	censored data • Park, Hyo-Il; Park, Sang-Gue Quantile estimation of treatment	95 StPrLet 24 139-145
Renovated scatterplots for	censored data • Smith, Peter J.; Zhang, J.	95 Biomtrka 82 447-452
XII accelerated life testing model based on type-2	censored data • Ali Mousa, M. A. M. Empirical Bayes estimators for the Burr type	95 JStCmpSm 52 95-103
estimation of regression parameters from	censored data with a discrete covariate • Rahbar, M. H.; Gardiner, J. C. Nonparametric	95 StPrLet 24 13-20
confidence intervals for regression parameters from	censored data with a discrete covariate • Rahbar, Mohammad H. Sequential fixed width	95 SqtlAnly 14 143-156
L_1-error of a nonparametric density estimator with	censored data • Kulasekera, K. B. A bound on the	95 StPrLet 23 233-238
Carlo study of the Buckley-James estimator with	censored data • Wu, Cheng-Sheng Peter; Zubovic, Yvonne A large-scale Monte	95 JStCmpSm 51 97-119
confidence intervals for the renewal function with	censored data • Baxter, Laurence A.; Li, Linxiong Nonparametric	95 JNnparSt 4 317-326
A stochastic EM algorithm for mixtures with	censored data • Chauveau, Didier	95 JStPlInf 46 1-25
Analysis of transformation models with	censored data • Cheng, S. C.; Wei, L. J.; Ying, Z.	95 Biomtrka 82 835-845
factor analysis: Quantitative genetic modeling with	censored data • Waller, Niels G.; Muthén, Bengt O. Genetic Tobit	92 BehavGen 22 265-292
Identifiability and estimation of	censored errors-in-variables models • Wang, Liqun	94 ASAProBuEc 351-354
Relative efficiency of a	censored experiment in terms of Fisher information matrix • Turrero, Agustin	95 CommStA 24 1169-1191
parametric estimation of survival distributions with	censored initiating time, and censored and truncated terminating time: ... • Tu Non-	95 ApplStat 44 3-16
Type-I	censored life test plans in the exponential case • Jeong, Hyean-Seok; Yum, Bong-Jin	95 CommStB 24 187-205
mixed model analysis of data from informatively	censored normal distributions • Smith, Fraser B.; Helms, Ronald W. EM	95 Biomtrcs 51 425-436
duration-dependent Markov model from interval-	censored observations with application to AIDS data • Frydman ... in a three-state	95 Biomtrcs 51 502-511
of a Markov 'illness-death' process from interval-	censored observations, with application to diabetes survival data • Frydman ... estimation	95 Biomtrka 82 773-789
Hazard rate estimation for strong-mixing and	censored processes • Lecoutre, Jean Pierre; Ould-Saïd, Elias	95 JNnparSt 5 83-89
On the Fisher information in type-I	censored and quantal response data • Gertsbakh, I.	95 StPrLet 23 297-306
Estimating location parameters using	censored rank statistics • Öhman, Marie-Louise	95 BiomtrcJ 37 767-782
of multiple imputation to the analysis of	censored regression data" (V47 p1297-1309) • James, Ian R. Comment on "Applications	95 Biomtrcs 51 358-362
of multiple imputation to the analysis of	censored regression data" • Tanner, Martin A. Reply to comment on "Applications	95 Biomtrcs 51 362-362
parameter approach • Ioannatos, Petros E.	Censored regression estimation under unobserved heterogeneity: A stochastic	95 JBES 13 327-335
with applications to synthetic data methods for	censored regression • Lai, T. L. et al. Asymptotic normality of a class of adaptive statistics	95 JMultiAn 52 259-279
Recent contributions to	censored regression models • Rao, C. Radhakrishna; Zhao, L. C.	95 Metrika 42 203-213
Koul-DeWet-type estimation in	censored regression using the conditional empirical process • Yang, Song	95 JStPlInf 46 351-363
of discontinuous relationships in biology with	censored regression • Schmid, B.; Polasek, W.; Weiner, J.; Krause, A.; Stoll, P. Modeling	94 ANatrls 143 494-507
techniques for the comparative analysis of	censored reliability data: A review • Bohoris, George A. Parametric statistical	95 RelEngSS 48 149-155
Failure-	censored reliability sampling plans for the exponential distribution • Balasooriya, Uditha	95 JStCmpSm 52 337-349
where all observations are either left- or right-	censored • Andersen, P. K.; Rønn, B. B. A nonparametric test for comparing two samples	95 Biomtrcs 51 323-329
timators for independent not identically distributed	censored samples (Chinese) • Wu ... and asymptotic normality of maximum likelihood es-	95 ChJApPSt 10 405-419
test statistics using Burr distribution and	censored samples • Ashour, S. K. et al. The sampling distribution of some goodness-of-fit	94 EgpStJ 38 147-163
of the exponential distribution based on a type II	censored sample • Xu, Jian-Lun; Yang, Grace L. A note on a characterization	95 AnlsStat 23 769-773
exponential distribution based on doubly type II	censored samples • Raqab, Mohammad Z. On the maximum likelihood prediction of the	95 PakJStA 11 1-10
of the logistic distribution based on multiply type-II	censored samples • Balakrishnan, N. et al. Estimation of the mean and standard deviation	95 MaOpfStS 27 127-142
extreme value distribution based on multiply type-II	censored samples • Balakrishnan et al. ... of the location and scale parameters of the	95 CommStA 24 2105-2125
algorithm for generating progressive Type-II	censored samples • Balakrishnan, N.; Sandhu, R. A. A simple simulational	95 AmerStat 49 229-230
ratio test for the presence of immunes in a	censored sample • Zhou, S.; Maller, R. A. The likelihood	95 MaOpfStS 27 181-201
the Rayleigh scale parameter based on fairly large	censored samples • Adatia, A. Best linear unbiased estimator of	95 IEEEReli 44 302-309
Estimating the proportion of immunes in	censored samples: A simulation study • Ghitany, M. E.; Maller, R.; Zhou, S.	95 StatMed 14 39-49
Saunders distribution based on symmetrically	censored samples • Rieck, James R. Parameter estimation for the Birnbaum-	95 CommStA 24 1721-1736
Asis Kumar (Rev) Review of "Truncated and	censored samples. Theory and applications" • Cohen, A. Clifford (Auth); Chattopadhyay,	95 CalcutSt 45 133-133
methods of estimating the mean survival time from	censored samples • Zukang, Zheng Two	95 SankhyaA 57 126-136
statistics for discrete parents with applications to	censored sampling • Gan, Gaoxiong; Bain, Lee J. Distribution of order	95 JStPlInf 44 37-46
An implicitly defined parametric model for	censored survival data and covariates • Piantadosi, Steven; Crowley, John	95 Biomtrcs 51 249-258
for the presence of immune or cured individuals in	censored survival data • Maller, R. A.; Zhou, Sean Testing	95 Biomtrcs 51 1197-1205
Conditional bootstrap methods for	censored survival data • Kim, Ji-Hyun	95 JKoreaSt 24 197-218
pearance of Plasmodium ... • Gottschau et al. Interval	censored survival data and multistate compartmental models in the analysis of first ap-	95 StatMed 14 2727-2736
ing studies, using a general approach for doubly-	censored and truncated data • Heisey, D. M. et al. Modelling age-specific survival in nest-	95 Biomtrcs 51 51-60
Estimation subject to block	censoring • Baxter, Laurence A.	95 IEEEReli 44 489-495
at random in general sample spaces and random	censoring in continuous time • Jacobsen, Martin; Keiding, Niels Coarsening	95 AnlsStat 23 774-786
regression estimation in the presence of dependent	censoring • Rotnitzky, Andrea; Robins, James M. Semiparametric	95 Biomtrka 82 805-820
of linear nonparametric rank tests for dependent	censoring • Lee, Seungyeoun; Wolfe, Robert A. A class	95 CommStA 24 687-712
for categorical data under informative general	censoring • Mimoso Paulino, C. D.; Pereira, C. A. D. B. Bayesian methods	95 Biomtrka 82 439-446
the loss of information through the use of hybrid	censoring • Ismail, Moshira A. On	95 EgpStJ 39 65-88
extreme-value distribution under multiply type-II	censoring • Fei, H.; Kong, F. et al. Estimation for two-parameter Weibull distribution and	95 CommStA 24 2087-2104
Bandyopadhyay, U.; Chattopadhyay, G. Progressive	censoring under inverse sampling for nonparametric two-sample problems •	95 SqtlAnly 14 1-28
Local likelihood based on kernel	censoring • Copas, J. B.	95 JRSS-B 57 221-235
of life test and accelerated life test under type 1	censoring of a lognormal distribution (Chinese) • Wang, L.; Chen, Y. Statistical analysis	95 AMaAPknC 17 109-117
method for randomized trials with informative	censoring: Part II • Robins, James M. An analytic	95 LiDaAnal 1 417-434
method for randomized trials with informative	censoring: Part I • Robins, James M. An analytic	95 LiDaAnal 1 241-254
variables associated with survival for data in which	censoring is prevalent • Fraser et al. ... decision procedure for selecting prognostic	95 BiomtrcJ 37 463-479
distribution with uniformly distributed random	censoring • Yeh, Lam; Choy, S. T. B. Bayesian variable sampling plans for the exponential	95 JStPlInf 47 277-293
estimates of a regression function under random	censoring • Carbonez, A.; Györfi, L.; van der Meulen, E. C. Partitioning-	95 StatDecn 13 21-37
intervals for the length of the short*t* under random	censoring • Beirlant, J.; Einmahl, J. H. J. Asymptotic confidence	95 StNeerla 49 1-8
Wilcoxon test and the logrank test under random	censoring • Sukhatme, Shashikala; Lin, C. H. Exact powers of the	95 JStPlInf 48 291-301
Daniels's association measures under right	censoring • O'Quigley, John	95 AplStMDA 11 109-119
in the location scale model under randomly right-	censoring (Chinese) • Lu, Wuewen Estimation of the parameters	95 ChJApPSt 10 272-277

of 2-parameter exponential populations under time	censoring • Thiagarajah, K. Ranee	Homogeneity tests for scale parameters	95 IEEEReli 44 297-301
of exponential distribution under type I progressive	censoring with changing failure rates • Patel et al.	... likelihood estimators of parameters	95 CommStA 24 2421-2435
Tail processes under heavy random	censorship applications • Einmahl, John H. J.; Ruymgaart, Frits H.		95 JNnparSt 5 223-235
Inference for a general type II	censorship model • Peña, Edsel A.; Rohatgi, Vijay K.; Székely, Gábor J.		95 MaOpfStS 26 241-252
of the kernel quantile estimator under random	censorship • Xiang, Xiaojing	Bahadur representation	95 JMultiAn 54 193-209
in the proportional hazards model of random	censorship • Ghorai, J. K.	Large sample confidence interval for the quantiles	95 JStCmpSm 53 91-102
The central limit theorem under random	censorship • Stute, Winfried		95 AnlsStat 23 422-439
for cumulative hazard function in a mixed	censorship-truncation model with application to hazard estimators • Gu	... of increments	95 StPrLet 23 135-139
model • Bugaighis, Mohamed M. Exchange of	censorship types and its impact on the estimation of parameters of a Weibull regression		95 IEEEReli 44 496-499
	Census		
	Analysis of sample based capture-recapture experiments • Alho, Juha M.		94 JOffStat 10 245-256
in academia and government • Fienberg, S. E. et al.	The Committee on National Statistics: Fostering interactions between statisticians		95 IntStRvw 63 257-269
Fillion, Jean-Marc; Luc, Manchi; Nadeau, Christian	Imputing numeric and qualitative variables simultaneously • Bankier, Michael;		94 ASASrvyRsr 242-247
Zanutto, Elaine; Zaslavsky, Alan M.	Models for imputing nonsample households with sampled nonresponse follow-up •		94 ASASrvyRsr 236-241
Zaslavsky, Alan M.	Using ethnographic data to evaluate dual system estimates • Thurston, Sally W.;		94 ASASrvyRsr 254-259
Comment on "Alternative methods for the 2000	census" • Fay, Robert E.		94 ASASrvyRsr 90-92
Comment on "Alternative methods for the 2000	census" • Fellegi, Ivan P.		94 ASASrvyRsr 93-95
A private sector view of	census 2000 • Kintner, Hallie J.		94 ASAProSoSt 3-5
tomated data capture methodology during the 1992	Census of Agriculture • Ramos, M. et al.	Results from a feasibility test of an alternative au-	94 ASASrvyRsr 769-774
Statistics Canada Farm Register with the Canadian	Census of Agriculture (Disc: p212-214) • Thomas, Bradley W.	Updating the	94 ASAProGovt 189-194
Vacca, E. Ann 1992	Census of Agriculture coverage evaluation estimation • Lewis, Paul J.; Wolfgang, Glenn S.;		94 ASASrvyRsr 1019-1024
Cognitive research for the 1997	Census of Agriculture report form • Wolfgang, Glenn S.; Lewis, Paul J.; Vacca, E. Ann		94 ASASrvyRsr 503-508
Reinventing the population	census: Balancing privacy and efficiency. In memory of Roger A. Herriot • Alexander, Lois		94 ASAProGovt 1-2
method to obtain stratification boundaries for the	Census Bureau's Annual Capital Expenditures Survey • Slanta et al.	... and Hidiroglou	94 ASASrvyRsr 693-698
Developing metadata database at the	Census Bureau • Gillman, Daniel W.; Appel, Martin V.		94 ASASrvyRsr 781-784
The	Census Bureau, its funders, and its customers • Becker, Patricia C.		94 ASAProSoSt 1-2
Bayesian iterative proportional fitting applied to a	Census Bureau latent-class problem • Larsen, Michael D.	Data augmentation with	94 ASAProGovt 116-121
New CASIC technologies at the U.S. Bureau of the	Census • Nicholls, William L., II; Appel, Martin V.		94 ASASrvyRsr 757-762
Modelling net undercoverage in the 1991 Canadian	Census • Dick, Peter		95 SurvyM 21 45-54
data collection methodology for the 2001 Canadian	Census • Choudhry, G. H.	A	94 ASASrvyRsr 1134-1139
The role of questionnaire design in reducing	census coverage error (Disc: p754-756) • Martin, Elizabeth A.; Griffin, Deborah H.		94 ASASrvyRsr 736-741
Automated coding of	census data • Kalpić, Damir		94 JOffStat 10 449-463
Continuous measurement alternatives to	census data collection (Disc: p90-95) • Rust, Keith F.		94 ASASrvyRsr 80-89
Digital smoothing of	census data employing Fourier transforms • Madan, V. K.; Sinha, R. K.		95 CmpStDA 20 285-294
responding households from the 1990 Decennial	Census • Cecco, Kevin	Characteristics of responding and non-	94 ASAProGovt 76-81
Sampling and statistical estimation in the decennial	census (Disc: p90-95) • Bell, Robert M.		94 ASASrvyRsr 71-79
record matching for the 1992 economic	censuses • Steel, Philip M.; Konschnik, Carl A.	Administrative	94 ASASrvyRsr 473-478
Development of edit parameters for 1992 economic	census enterprise reports • Oh, Sungsoo; Paletz, Dave; Kim, Jay; Salyers, Eddie		94 ASASrvyRsr 1144-1149
and organization and their relationship to the	census of governments • Owens, Stephen D.	A study in library structure	94 ASAProGovt 64-69
Spanish and English: The 1990 Spanish Language	Census long form • Gabbard, S. M. et al.	Evaluating translations of survey instruments in	94 ASASrvyRsr 1232-1235
Disclosure control for	census microdata • Skinner, Chris; Marsh, Catherine; Openshaw, Stan; Wymer, Colin		94 JOffStat 10 31-51
	Census plus: An alternative coverage methodology • Schindler, Eric; Navarro, Alfredo		94 ASASrvyRsr 248-253
The 1991	census of population in England and Wales (Disc: p230-240) • Thompson, Eric J.		95 JRSS-A 158 203-229
Spoken language recognition of the Year 2000	Census Questionnaire Feasibility Test • Appel, Martin V.; Cole, Ronald		94 ASASrvyRsr 1310-1315
statements of confidentiality on completion rates for	census questionnaires • Dillman et al.	... appeals, mandatory appeals and variations in	94 ASASrvyRsr 1333-1338
postcards affect mailback response rates for	census questionnaires • Dillman et al.	... letters, stamped return envelopes and reminder	95 SurvyM 21 159-165
living in group quarters: What data from the 1990	Census show • Schwartz, Steven H.; Klein, Robert E.; Adashek, Daniel	Veterans	94 ASAProGovt 13-18
Comment on "The 1995	Census Test" • Wright, Tommy		94 ASASrvyRsr 754-756
applications of sampling estimation in the 1995	Census Test (Disc: p754-756) • Mulry, Mary H.; Singh, Rajendra P.	New	94 ASASrvyRsr 742-747
1994 National	Census Test: Quantitative comparisons of coverage • Pausche, Joan M.		94 ASASrvyRsr 313-318
An evaluation of population projection errors for	census tracts • Smith, Stanley K.; Shahidullah, Mohammed		95 JASA 90 64-71
rates and surrogate variables in the 1990 U.S.	Census • Kim, J. J.-I.; Zaslavsky, A.; Blodgett, R.	Between-state heterogeneity of undercount	95 SurvyM 21 55-62
Review of "Modernizing the U.S.	Census" • Edmonston, Barry; Schultze, Charles (Ed); Wolter, Kirk M. (Rev)		95 Chance 8/3 44-49
Review of "The 1991	Census User's Guide" • Dale, A.; Marsh, C. (Ed); Dorling, Daniel (Rev)		95 JRSS-A 158 188-189
Preston, Samuel H. Consistency of	census and vital registration data on older Americans: 1970-1990 • Shrestha, Laura B.;		95 SurvyM 21 167-177
The	centered normal conditionals distribution • Sarabia, José María		95 CommStA 24 2889-2900
The effect of different forms of	centering in hierarchical linear models • Kreft, Ita G. G.; de Leeuw, Jan; Aiken, Leona S.		95 MultBehR 30 1-21
Comment on "Algorithm AS 226: Computing non	central beta probabilities" (V36 p241-244) • Lam, Ming Long		95 ApplStat 44 551-552
Spherical prediction-variance properties of	central composite and Box-Behnken designs • Borkowski, John J.		95 Technmcs 37 399-410
and average spherical prediction variances for	central composite and Box-Behnken designs • Borkowski, John J.	Minimum, maximum,	95 CommStA 24 2581-2600
	Central composite design		
MacEachern, S. N.; Notz, W. I. et al.	Robustness to the unavailability of data in the linear model, with applications •		95 JStPlInf 48 207-213
Efficiency comparisons of	central composite designs • Jensen, D. R.		95 JStCmpSm 52 177-183
Finding maximum G-criterion values for	central composite designs on the hypercube • Borkowski, John J.		95 CommStA 24 2041-2058
A new type of slope rotatable	central composite design • Victorbabu, B. Re.; Narasimham, V. L.		94 JIAgSt 46 315-317
area-average rainfall during the CaPE experiment in	Central Florida • Duchon, C. E.; Renkevens, T. M., Crosson, W. L.	Estimation of daily	95 JApMetrl 34 2704-2714
Voiculescu, D. Superconvergence to the	central limit and failure of the Cramér theorem for free random variables • Bercovici, H.;		95 ProbTher 103 215-222
	Central limit theorem		
	Antisymmetric functionals of reversible Markov processes • Goldstein, Sheldon		95 AIHenrPB 31 177-190
	Asymptotic confidence bands for normal distributions • Lin, Yachen		95 CommStA 24 641-652
	Asymptotic distributions of records in nonstationary schemes • Nevzorov, Valery		95 JStPlInf 45 261-273
Györfi, László	Asymptotic normality of L_1-error in density estimation • Berlinet, Alain; Devroye, Luc;		95 MaOpfStS 26 329-343
tial counting process modelling ... • Capasso et al.	Asymptotic properties of the maximum likelihood estimators of parameters of a spa-		95 StocAnAp 13 279-294
	On the asymptotics of constrained M-estimation • Geyer, Charles J.		94 AnlsStat 22 1993-2010
	Asymptotics for multisample statistics • Vaillancourt, Jean		95 CandJSt 23 171-184
	Bell curves and monkey languages • Casti, John L.		95 Complxty 1 12-15
Bühlmann, Peter	The blockwise bootstrap for general empirical processes of stationary sequences •		95 StocProc 58 247-265
	Density-dependent limits for a nonlinear reaction-diffusion model • Blount, Douglas		94 AnlsProb 22 2040-2070
Galambos, J.	The development of the mathematical theory of extremes in the past half century •		94 ThProbAp 39 234-248
Walsh, Dennis P.	Equating Poisson and normal probability functions to derive Stirling's formula •		95 AmerStat 49 270-271
	Estimation of a quantile in some nonstandard cases • Xiang, Xiaojing		95 AnInStMa 47 105-117
Peligrad, Magda; Shao, Qi-Man	Estimation of the variance of partial sums for ρ-mixing random variables •		95 JMultiAn 52 140-157
Klüppelberg, Claudia; Mikosch, Thomas	Explosive Poisson shot noise processes with applications to risk reserves •		95 Bernoull 1 125-147
Prum, B.; Rodolphe, F.; de Turckheim, E.	Finding words with unexpected frequencies in deoxyribonucleic acid sequences •		95 JRSS-B 57 205-220

	Central limit theorem (cont.)	
Loviscach, J.	A generalized notion of variation applied to Markov chains and Anosov maps •	95 ProbTher 103 553-570
	On high level exceedance modeling and tail inference • Leadbetter, M. R.	95 JStPlInf 45 247-260
Massart, P.; Rio, E.	Invariance principles for absolutely regular empirical processes • Doukhan, P.;	95 AIHenrPB 31 393-427
variables • Hazod, W.	On the limit behavior of products of a random number of group valued random	94 ThProbAp 39 249-263
	Limit theorems for multitype epidemics • Andersson, Håkan; Djehiche, Boualem	95 StocProc 56 57-75
	On the logarithmic average of additive functionals • Csáki, E.; Földes, A.	95 StPrLet 22 261-268
mixing stationary random variables • Hsing, Tailen	A note on the asymptotic independence of the sum and maximum of strongly	95 AnlsProb 23 938-947
	A note on some negative dependence notions • Kim, Tae-Sung; Seo, Hye-Young	95 CommStA 24 845-858
Sun, Shan	Perturbed empirical distribution functions and quantiles under dependence •	95 JTherPrb 8 763-777
	The probabilities of large deviations for semimartingales • Grama, Ion G.	95 Stochatx 54 1-19
	On the rate of convergence in the martingale CLT • Račkauskas, Alfredas	95 StPrLet 23 221-226
(French) • Miclo, Laurent	Remarks about the ergodicity of simulated annealing algorithms on a graph	95 StocProc 58 329-360
walk with simple exclusion • Varadhan, S. R. S.	Self diffusion of a tagged particle in equilibrium for asymmetric mean zero random	95 AIHenrPB 31 273-285
	Stationary exponential families • Dinwoodie, I. H.	95 AnlsStat 23 327-337
distributions • Molchanov, Ilya S.	Statistics of the Boolean Model: From the estimation of means to the estimation of	95 AdvAppPr 27 63-86
Ruymgaart, Frits H.	Tail processes under heavy random censorship applications • Einmahl, John H. J.;	95 JNnparSt 5 223-235
	A uniform CLT for continuous martingales • Bae, Jongsig; Levental, Shlomo	95 JKoreaSt 24 225-231
Bae, Jongsig; Levental, Shlomo	Uniform CLT for Markov chains and its invariance principle: A martingale approach •	95 JTherPrb 8 549-570
conditions • Neuts, M. F.; Pearce, C. E. M. A	central limit theorem for a class of Markovian packet streams under heavy traffic	95 CommStSM 11 210-218
Average error in the	central limit theorem for the cumulative processes • Roginsky, Allen	95 StPrLet 24 199-204
On the almost sure	central limit theorem and domains of attraction • Berkes, István	95 ProbTher 102 1-17
Woodroofe, Michael B.	Central limit theorems for doubly adaptive biased coin designs • Eisele, Jeffrey R.;	95 AnlsStat 23 234-254
differential operators • Sifi, Mohamed	Central limit theorem and infinitely divisible probabilities associated with partial	95 JTherPrb 8 475-499
A	central limit theorem for the isotropic random sphere • Brown, Jason J.	95 AdvAppPr 27 642-651
Jackson networks • Anantharam, V. et al. A functional	central limit theorem for the jump counts of Markov processes with an application to	95 AdvAppPr 27 476-509
plications to linear processes • Yokoyama, R. On the	central limit theorem and law of the iterated logarithm for stationary processes with ap-	95 StocProc 59 343-351
Weight functions and pathwise local	central limit theorems • Horváth, Lajos; Khoshnevisan, Davar	95 StocProc 59 105-123
An elementary proof of the local	central limit theorem • Davis, Burgess; McDonald, David	95 JTherPrb 8 693-701
applications • Dion, J. P.; Yanev, N. M.	Central limit theorem for martingales in BGWR branching processes with some statistical	95 MaMethSt 4 344-358
	Central limit theorem in negative curvature • Ledrappier, François	95 AnlsProb 23 1219-1233
Sánchez de Naranjo, M. V. A	central limit theorem for non-linear functionals of stationary Gaussian vector processes •	95 StPrLet 22 223-230
Sample sizes and the	central limit theorem: The Poisson distribution as an illustration • Jolliffe, Ian T.	95 AmerStat 49 269-269
The	central limit theorem under random censorship • Stute, Winfried	95 AnlsStat 23 422-439
Dependent versions of a	central limit theorem for the squared length of a sample mean • Saikkonen, Pentti	95 StPrLet 22 185-194
A	central limit theorem for stationary random fields • Janž, M.; Lachout, P.	95 MaMethSt 4 463-471
Krishnamurthi; Singh, Suresh	Central limit theorem for time to broadcast in radio networks • Ravishankar,	95 PrEngInf 9 201-209
groups • Neuenschwander, Daniel A Gaussian	central limit theorem for trimmed products on simply connected step 2-nilpotent lie	95 JTherPrb 8 165-174
Alan Turing and the	central limit theorem • Zabell, S. L.	95 AmMaMnth 102 483-494
On the	central limit theorem for U-statistics under absolute regularity • Arcones, Miguel A.	95 StPrLet 24 245-249
weak invariance principle - A ... • Anastassiou	Central limit theorem, weak law of large numbers for martingales in Banach spaces, and	95 JMultiAn 52 158-180
Shao, Qi-Man A note on the almost sure	central limit theorem for weakly dependent random variables • Peligrad, Magda;	95 StPrLet 22 131-136
Hong, Dug Hun A	central limit theorem with random indices for martingale difference sequences •	95 JKoreaSt 24 219-224
self-validating numerical method for computation of	central and non-central F probabilities and percentiles • Wang, M. C.; Kennedy, W. J. A	95 StatComp 5 155-163
On some	central and non-central multivariate chi-square distributions • Royen, Thomas	95 StSinica 5 373-397
The	central register of population of the Republic of Slovenia • Tršinar, Irena	95 JOffStat 11 221-223
The British experience since reorganization of the	Central Statistical Office in July 1989 • Wright Presenting statistics to press and public:	91 StatJUN 8 313-320
Challenges facing the United Kingdom	Central Statistical Office • McLennan, William	95 JOffStat 11 21-31
On the new approximation to non-	central t-distributions • Akahira, Masafumi; Sato, Michikazu; Torigoe, Norio	95 JJapnStS 25 1-18
approximation to a percentage point of the non-	central t-distribution • Akahira, Masafumi A higher order	95 CommStB 24 595-605
model • Caudill, S. B.; Ford, J. M.; Kaserman, D. L.	Certificate-of-need regulation and the diffusion of innovations: A random coefficient	95 JApEcmtx 10 73-78
	Certification	
	The statistical profession and the chartered statistician (CStat) • Smith, T. M. F.	95 JOffStat 11 33-40
Comment on "Statistical values, quality, and	certification" (V48 p65-70) • Helm, Werner E.	95 AmerStat 49 240-241
Comment on "Statistical values, quality, and	certification" (V48 p65-87) • Campbell, J. M.	95 AmerStat 49 115-116
Comment on "Statistical values, quality, and	certification" (V48 p65-87) • Dallal, Gerard E.	95 AmerStat 49 116-116
Comment on "Statistical values, quality, and	certification" (V48 p65-87) • Higgins, James J.	95 AmerStat 49 116-116
Comment on "Statistical values, quality, and	certification" (V48 p65-87) • Wang, Chamont; Silverberg, Arthur	95 AmerStat 49 116-117
to comment on "Statistical values, quality, and	certification" • Imrey, Peter B. Reply	95 AmerStat 49 241-241
to comments on "Statistical values, quality, and	certification" • Imrey, Peter B. Reply	95 AmerStat 49 117-118
analysis of linearly combined order statistic	CFAR detectors • Nagle, Daniel T.; Saniie, Jafar Performance	95 IEEEAES 31 522-533
	Chain	
	• see also Markov -	
Some aspects of inference for	chain binomial models • Yip, Paul; Huggins, Richard	95 StocAnAp 13 355-367
Parlange, Marc B. Generalizations of	chain-dependent processes: Application to hourly precipitation • Katz, Richard W.;	95 WaterReR 31 1331-1341
Which stochastic model is underlying the	chain ladder method? • Mack, Thomas	94 InsrneME 15 133-138
effects in ecological interaction networks 1. The	chain rule approach • Higashi, Masahiko; Nakajima, Hisao Indirect	95 MaBisci 130 99-128
How to perform skip-lot and	chain sampling (Second edition) • Stephens, Kenneth S.	95 ASQualCntrl:WI 64p
Review of "Volume 4 - How to perform skip-lot and	chain sampling (Second edition)" • Stephens, Kenneth S. (Auth); Doganaksoy, Necip (Rev)	95 JQualTek 27 394-395
quality level • Raju, C. et al. Minimum risks	chain sampling plans (ChSP-4 (c_1, c_2)) indexed by acceptable quality level and limiting	95 JAppStat 22 389-426
and tables for the construction and selection of	chain sampling plans ChSP-4 (c_1, c_r) - Part 4 • Raju, C.; Murthy, M. N. N. Procedures	95 JAppStat 22 261-271
Markov branching decision	chains with interest-rate-dependent rewards • Huang, Ying; Veinott, Arthur F., Jr	95 PrEngInf 9 99-121
In defence of analysis of covariance: A reply to	Chambless and Roeback • Senn, Stephen	95 StatMed 14 2283-2285
Are the Toronto Blue Jays really the world	champions? • Umholtz, Robert L.	94 ASAProSpts 42-45
Secretary selection problem with a	chance constraint • Mukherjee, S. P.; Mandal, A.	94 JIndnStA 32 29-34
Review of "Edgeworth on	chance, economic hazard, and statistics" • Mirowski, Philip (Ed); Stigler, Stephen M. (Rev)	95 JASA 90 803-804
Similarity of DNA fingerprints due to	chance and relatedness • Li, C. C.; Weeks, D. E.; Chakravarti, A.	93 HumHrdty 43 45-52
	Change	
	• see also rate of -, structural -	
Group sequential comparison of	changes: Ad-hoc versus more exact method • Lee, Jae Won; DeMets, David L.	95 Biomtrcs 51 21-30
Testing for	change in the frequency of El Niño events • Solow, Andrew R.	95 JClimate 8 2563-2566
dependent covariates: Application to estimating	changes in the incubation period of AIDS • Bacchetti, P. et al. Survival trees with time-	95 LiDaAnal 1 35-47
Detecting a	change in the intercept in multiple regression • Watson, G. S.	95 StPrLet 23 69-72
A test for the detection of a	change in mean in a Gaussian model (French) • Ghorbanzadeh, Dariush	95 RStApp 43/2 67-76

	Characteristic properties of a bivariate geometric law • Nair, N. Unnikrishnan; Asha, G.	94 JIndnStA 32 111-116
The normal distribution -	Characterizations and applications • Bryc, W.	95 Springer:Brln:NY 139p
Review of "The normal distribution -	Characterizations and applications" • Bryc, W. (Auth); Leslie, J. R. (Rev)	96 ShrtBkRv 16/1 4-4
Rao, C. Radhakrishna; Shanbhag, D. N.	Characterizations based on regression properties: Improved versions of recent results •	95 SankhyaA 57 167-178
A	characterization of a bivariate geometric distribution • Sun, Kai; Basu, Asit P.	95 StPrLet 23 307-311
Muraleedharan	Characterization of a bivariate Pareto distribution • Padamadan, Veenus; Nair, K. R.	94 JIndnStA 32 15-20
Siatkowski, Idzi Some	characterizations of a class of balanced two-way elimination of heterogeneity designs •	94 ListyBio 31 27-33
Some measure theoretical	characterizations of compactness of metric spaces • Plachky, D.	95 SankhyaA 57 154-160
Ruiz, J. M. On	characterization of continuous distributions with adjacent order statistics • Franco, M.;	95 MaOpfStS 26 375-385
Navarro, J.	Characterization of discrete distributions using expected values • Ruiz, J. M.;	95 StHefte 36 237-252
Pakes, Anthony G.	Characterization of discrete laws via mixed sums and Markov branching processes •	95 StocProc 55 285-300
is random • Grudzień, Z.; Szynal, D.	Characterizations of distributions by moments of order statistics when the sample size	95 ZastApMa 23 305-318
Lin, G. D.	Characterization of distributions using equilibrium transformations • Huang, J. S.;	95 SankhyaA 57 179-185
Some	characterizations for the exponential class of distributions • Consul, P. C.	95 IEEEReli 44 403-407
Xu, Jian-Lun; Yang, Grace L. A note on a	characterization of the exponential distribution based on a type II censored sample •	95 AnlsStat 23 769-773
Khalil, Z.	Characterization of a general class of life-testing models • Ghitany, M. E.; El-Saidi, M. A.;	95 JAppProb 32 548-553
demic • Tan, Wai-Yuan; Lee, Sho Rong; Tang, Si Chin	Characterization of HIV infection and seroconversion by a stochastic model of the HIV epi-	95 MaBisci 126 81-123
mer bound • Hamada, N. A survey of recent work on	characterization of minihypers in PG(t,q) and nonbinary linear codes meeting the Gries-	93 JCISS 18 161-191
Conditional distributions and	characterizations of multivariate stable distribution • Nguyen, Truc T.	95 JMultiAn 53 181-193
Zeng, Wei-Bin On	characterization of multivariate stable distributions via random linear statistics •	95 JTherPrb 8 1-15
	Characterization of the negative binomial and gamma distributions • Ong, S. H.	95 BiomtrcJ 37 251-254
polynomials • Papathanasiou, V. A	characterization of the Pearson system of distributions and the associated orthogonal	95 AnInStMa 47 171-176
p294-298) • Meng, Xiao-Li Comment on "A	characterization of the Poisson distribution and the probability of winning a game" (V48	95 AmerStat 49 401-401
p294-298) • Sen, Ananda Comment on "A	characterization of the Poisson distribution and the probability of winning a game" (V48	95 AmerStat 49 401-401
Sapatinas, Theofanis	Characterizations of probability distributions based on discrete p-monotonicity •	95 StPrLet 24 339-344
sensitivity • Yin, G.; Zhu, Y. M.; Gao, A. J. On	characterization of a recursively defined process: Necessary and sufficient conditions,	95 Stochatx 52 265-282
mixtures of power-series distributions and related	characterizations • Sapatinas, Theofanis Identifiability of	95 AnInStMa 47 447-459
A generalization of covariance identity and related	characterizations • Cacoullos, T.; Papathanasiou, V.	95 MaMethSt 4 106-113
Rajiv Range surface	characterization and segmentation using neural networks • Ghosal, Sugata; Mehrotra,	95 PattrnRc 28 711-727
Texture analysis of imagery data for waste site	characterization • Irvine, John M.; Barbour, Garth	94 ASAProEnvr 122-126
	Characterization of some concepts of aging • Rojo, Javier	95 IEEEReli 44 285-290
adjoint (self-reciprocal) ... • Rossberg Correction to	"Characterization of the standard normal probability density within the class of self-	95 MaOpfStS 26 387-387
exponential • Dimitrov, B. et al. Definitions,	characterizations and structural properties of probability distributions similar to the	95 JStPlInf 43 271-287
systems • Kim, Hag-Min; Fotopoulos, S. B.	Characterization study of a stochastic process associated with times in queueing	94 ASAProPEng 163-168
On some	characterizations of the t-distribution • Arellano-Valle, Reinaldo B.; Bolfarine, Heleno	95 StPrLet 25 79-85
On a	characterization of variance functions of natural exponential families • Bshouty, D.	95 MaMethSt 4 92-98
The normal distribution:	Characterizations with applications • Bryc, W.	95 Springer:Brln:NY 140p
Using the β-binomial distribution to	characterize forest health • Zarnoch, S. J.; Anderson, R. L.; Sheffield, R. M.	95 CanJForR 25 462-469
Lepkowski, James M. Use of last physician visit to	characterize health care utilization (Disc: p379-381) • Jay, Gina M.; Belli, Robert F.;	94 ASASrvyRsr 368-372
Robert H. A comparison of methods that	characterize pulses in a time series • Mauger, David T.; Brown, Morton B.; Kushler,	95 StatMed 14 311-325
Wolf, Dennis A.; Tardiff, Mark F.; Ketelle, Richard H.	Characterizing groundwater at Oak Ridge National Laboratory hazardous waste sites •	94 ASAProEnvr 116-121
	Characterizing the idle-period distribution of GI/G/1 queues • Li, Jingwen; Ou, Jihong	95 JAppProb 32 247-255
	Characterizing Markov exchangeable sequences • Zabell, S. L.	95 JTherPrb 8 175-178
Lin, Dennis K. J.	Characterizing projected designs: Repeat and mirror-image runs • Draper, Norman R.;	95 CommStA 24 775-795
A bivariate discrete	Charlier series distribution • Papageorgiou, H.; Loukas, S.	95 BiomtrcJ 37 105-117
The statistical profession and the	chartered statistician (CStat) • Smith, T. M. F.	95 JOffStat 11 33-40
economical than the original HM (which is based on	Chatterjee's sampling) • Aoshima ... II (based on Healy's two-stage sampling) is more	94 AJMMgSci 14 167-196
	Chebychev inequality	
Allen, P. Geoffrey; Morzuch, Bernard J.	Comparing probability forecasts derived from theoretical distributions •	95 IntJFor 11 147-157
	Chebyshev's contribution to the theory of probability • Sheynin, O.	93 AHistESc 46 321-340
Review of "Intelligent software for	chemical analysis" • Buydens, Lutgarde M. C.; Schoenmakers, Peter J. (Ed); Pretsch, E. (Rev)	95 Chemolab 27 148-149
Interactive variable selection (IVS) for PLS. Part II:	Chemical applications • Lindgren, F.; Geladi, P.; Berglund, A.; Sjöström, M.; Wold, S.	95 JChemtrc 9 331-342
Statistical process control in	chemical batch processes (things are seldom what they seem) • Janis, Stuart J.	94 ASAProPEng 308-313
sessment of the risk from the exposure to fetotoxic	chemical compounds (Disc: p78-79) • Talwalker, S. et al. Qualitative and quantitative as-	95 EnvEcoSt 2 71-78
production: The statistical design and analysis of	chemical experiments" • Davies et al. Review of "Efficiency in research, development and	95 Technmcs 37 457-458
statistically significant departures from additivity in	chemical mixtures • Gennings et al. ... data from individual components to detect	95 Biomtrcs 51 1264-1277
simultaneous development of linear and nonlinear	chemical reaction models from a linearly designed ... • Steele et al. ... for the	95 Chemolab 29 125-131
D. (Rev) Review of "Large data analysis for the	chemical sciences: A guide to statistical techniques" • Graham, R. C. (Auth); Haaland, P.	95 Chemolab 27 245-246
Review of "Three-dimensional	chemical similarity searching" • Pepperrell, Catherine (Auth); Lewi, Paul J. (Rev)	95 Chemolab 28 205-207
	Chemistry	
analysis of several spectroscopic ... • Tauler et al.	Application of a new multivariate curve resolution procedure to the simultaneous	95 Chemolab 27 163-174
Zupan, Jure; Novič, Marjana; Gasteiger, Johann	Neural networks with counter-propagation learning strategy used for modelling •	95 Chemolab 27 175-187
Grieshaber, Mark V.; Schnur, Dora M.	Use experimental design in organic synthesis • Brannigan, Lawrence H.;	95 ChemTech 25 29-35
Review of "Quality control in analytic	chemistry (2nd ed)" • Kateman, G.; Buydens, L. (Auth); Bandeen-Roche, Karen J. (Rev)	95 Technmcs 37 115-116
Review of "Statistical methods in analytic	chemistry" • Meier, Peter C.; Zünd, Richard E. (Auth); Hillyer, Martin (Rev)	95 Technmcs 37 113-114
model for measurement error in analytical	chemistry • Rocke, David M.; Lorenzato, Stefan A two-component	95 Technmcs 37 176-184
Review of "Information theory in analytical	chemistry" • Eckschlager, K.; Danzer, K. (Auth); Clegg, D. (Rev)	95 Chemolab 28 207-208
its applications to stochastic problems in physics,	chemistry and electrical engineering • Coffey, W. T. et al. The Langevin equation and	95 WrldSciPub:Sing 500p
the analysis of experimental designs in formulation	chemistry (French) • Gauchi, J.-P. Use of PLS regression for	95 RStApp 43/1 65-89
dations of the International Federation of Clinical	Chemistry on the statistical treatment of reference values • Solberg ... the recommen-	95 CmpBiomd 48 247-256
(Auth) et al. Review of "Chemometrics for analytical	chemistry Volume 1: PC-aided statistical data analysis" • Meloun, M.; Militky, J.; Forina, M.	95 Technmcs 37 114-115
J. et al. Review of "Chemometrics for analytical	chemistry, Volume 2: PC-aided regression and related methods" • Meloun, M.; Militky,	95 Chemolab 29 326-327
mination of stabiliser in a single base propellant by	chemometric analysis of Fourier transform infrared spectra • Lindblom et al. ... deter-	95 Chemolab 29 243-254
Dunlop, P. J.; Bignell, C. M.; Jackson, J. F. et al.	Chemometric analysis of gas chromatographic data of oils from Eucalyptus species •	95 Chemolab 30 59-67
Review of "SCAN: Software for	chemometric analysis (Release 1 for Windows)" • Minitab Inc (Auth); Lavine, Barry K. (Rev)	95 Chemolab 30 253-255
Pattern recognition, reflections from a	chemometric point of view • Lewi, Paul J.	95 Chemolab 28 23-33
Miller, Charles E. The use of	chemometric techniques in process analytical method development and operation •	95 Chemolab 30 11-22
Meloun, M.; Militky, J. et al. Review of	"Chemometrics for analytical chemistry Volume 1: PC-aided statistical data analysis" •	95 Technmcs 37 114-115
methods" • Meloun, M.; Militky, J. et al. Review of	"Chemometrics for analytical chemistry, Volume 2: PC-aided regression and related	95 Chemolab 29 326-327
Review of "Practical guide to	chemometrics" • Haswell, Stephen John (Ed); Bayne, Charles K. (Rev)	95 Technmcs 37 230-231
Review of "Multivariate pattern recognition in	chemometrics, illustrated by case studies" • Brereton, R. G. (Ed); Nemeth, M. A. (Rev)	95 Technmcs 37 116-116
	Chemometrics; what do we mean with it, and what do we want from it? • Wold, Svante	95 Chemolab 30 109-115
chemometrics • Brown, Steven D. Has the	chemometrics revolution ended? Some views on the past, present and future of	95 Chemolab 30 49-58
Nonlinear models for assessing combinations of	chemotherapeutic drugs with resistance modifiers • Lam, Y. M.; Pym, J.; Campling, B. G.	94 ASAProBiop 394-398

A stochastic model for drug resistance in AIDS	chemotherapy and the HIV incubation distribution • Tan, Wai Y.	95 StPrLet 25 289-299
quality-adjusted survival meta-analysis of adjuvant	chemotherapy for premenopausal breast cancer • Cole, B. F.; Gelber, R. D. et al. A	95 StatMed 14 1771-1784
The moment bound is tighter than	Chernoff's bound for positive tail probabilities • Philips, Thomas K.; Nelson, Randolph	95 AmerStat 49 175-178
Spatial analysis of benthic data from the	Chesapeake Bay • Neerchal, Nagaraj K.; Shafer, Ronald W.	94 ASAProEnvr 204-208
	Chi-bar squared distribution	
Kochar, Subhash; Robertson, Tim	Likelihood ratio tests for symmetry against one-sided alternatives • Dykstra, Richard;	95 AnInStMa 47 719-730
parameters • Bhattacharya, Bhaskar	Restricted tests for and against the increasing failure rate ordering on multinomial	95 StPrLet 25 309-316
On some central and non-central multivariate	chi-square distributions • Royen, Thomas	95 StSinica 5 373-397
UMVUE of a noncentrality parameter of a noncentral	chi-square distribution • Shao, P. Y.-S. et al. Improving on the positive part of the	95 JMultiAn 53 52-66
analysis of crossover interference using the	chi-square model • Zhao, Hongyu; Speed, Terence P.; McPeek, Mary Sara Statistical	95 Genetcs 139 1045-1056
Glenn I.; Moser, E. Barry Aggregation and the Pearson	chi-square statistic for homogeneous proportions and distributions in ecology • Garson,	95 Ecology 76 2258-2269
between Fisher's exact test and Pearson's	chi-square test: A Bayesian perspective • Camilli, Gregory The relationship	95 Psymtrka 60 305-312
Two-sample	chi-square test for randomly censored data (Korean) • Kim, Joo-Han; Kim, Jeong-Ran	95 KorApS 8/2 109-119
Data driven versions of Pearson's	chi-square test for uniformity • Bogdan, Malgorzata	95 JStCmpSm 52 217-237
	Chi-squared approximation	
MANOVA models • Seo, T.; Kanda, T.; Fujikoshi, Y.	The effects of nonnormality on tests for dimensionality in canonical correlation and	95 JMultiAn 52 325-337
Thomas, G. E. A note on	chi-squared approximations for tests of rank interaction with ordered categorical data •	95 CommStB 24 439-459
	Chi-squared distribution	
	Performance of a Bartlett-type modification for the deviance • Cordeiro, Gauss M.	95 JStCmpSm 51 385-403
Confidence intervals for the noncentral	chi-squared distribution • Kent, John T.; Hainsworth, Timothy J.	95 JStPlInf 46 147-159
the exact and asymptotic upper tail probabilities of	chi-squared goodness-of-fit statistics: Pearson's χ^2, likelihood ... • Formann ... of	95 JStCmpSm 51 369-384
The effect and adjustment of complex surveys on	chi-squared goodness-of-fit tests: Some Monte Carlo evidence • Aguirre-Torres, V. et al.	94 ASASrvyRsr 602-607
	Chi-squared test	
Kh.; Axenovich, T. I.; Babenko, V. N.	Comparison of asymptotic tests of the Hardy-Weinberg distribution • Ginsburg, E.	91 AnHuGene 55 329-338
Lorenzen, Gunter	A new family of goodness-of-fit statistics for discrete multivariate data •	95 StPrLet 25 301-307
	Statistical foundation for hypothesis testing of image data • Kanatani, Kenichi	94 CVGIPIm 60 382-391
Some remarks on the	chi-squared test with both margins fixed • Park, Cheolyong	95 CommStA 24 653-661
randomization • Zucker et al. Statistical design of the	child and adolescent trial for cardiovascular health (CATCH): Implications of cluster	95 ConClnTr 16 96-118
Improving validity of parental reports of	child immunization status in a telephone survey • Haggard, Lois M.; Gray, Donald Z.	94 ASASrvyRsr 1258-1262
H.; Burgess, B. B. et al. Prospects for measuring the	child radio audience in the U.S. using the seven-day personal diary method • Patchen, R.	94 ASASrvyRsr 1220-1225
measures for Medicaid beneficiaries: Adult and	child reports • Eisenhower et al. ... comparison of eligibility and health care utilization	94 ASASrvyRsr 713-718
Employment and	childbearing in Canada • Rao, K. V.	94 ASAProSoSt 124-128
al. Random-digit dialing for control selection in	childhood cancer studies: The geographic proximity and demographics ... • Sakkinen et	95 AJPubHlh 85 555-557
et al. Spatial variation of natural radiation and	childhood leukaemia incidence in Great Britain • Richardson, S.; Monfort, C.; Green, M.	95 StatMed 14 2487-2501
An application to nuclear power plant facilities and	childhood leukaemia in Sweden • Waller et al. ... and assessment of clusters of disease:	95 StatMed 14 3-16
Different approaches to counting early	childhood program participation • West, Jerry; Hausken, Elvie Germino	94 ASAProSoSt 141-146
N. M. et al. Bivariate logistic regression analysis of	childhood psychopathology ratings using multiple informants • Fitzmaurice, G. M.; Laird,	95 AJEpidm 142 1194-1203
	Children's concepts of average and representativeness • Mokros, Jan; Russell, Susan Jo	95 JRsrMaEd 26 20-39
effect models: A comparison using clinical data in	children • Roe, D. J.; Cheng, M. et al. Estimating parameters in non-linear mixed	94 ASAProBiop 313-318
and cultural factors affecting diarrhea diseases in	children under two years in Egypt • Ashour et al. Discriminant analysis for social-economic	95 EgpStJ 39 1-11
	Chinese dominoes games • Wang, Ann-Lee	95 TeachgSt 17/2 70-73
Efficient computation of the noncentral	χ^2 distribution • Chattamvelli, R.; Shanmugam, R.	95 CommStB 24 675-689
PM algorithms for calculating minimum	χ^2 estimators with partially observed sub-tables • Ritov, Ya'acov	95 CmpStDA 20 19-33
of chi-squared goodness-of-fit statistics: Pearson's	χ^2, likelihood ratio, and power-divergence statistic ($\lambda = 2/3$) • Formann ... probabilities	95 JStCmpSm 51 369-384
Extension of the marker-association-segregation	χ^2 method • Dizier, M.-H. et al. Interactive effect of two candidate genes in a disease:	94 AJHuGene 55 1042-1049
computation of the exact distribution function of the	χ^2-and related test statistics in the equiprobable case • Kulmann, Hermann Notes on the	95 CmpStDA 19 707-710
Mato, A. Silva The Wilcoxon, Spearman, Fisher,	χ^2, Student and Pearson tests and 2×2 tables • Martín Andrés, A.; Tejedor, I. Herranz;	95 Ststcian 44 441-450
Tsubaki, Michiko New weighted sum of	χ^2 test for testing a linear hypothesis against a restricted alternative hypothesis •	95 CommStB 24 987-1004
Exact cumulant calculations for Pearson	χ^2 and Zelterman statistics for r-way contingency tables • Stafford, James E.	95 JCmpGrSt 4 199-212
Marsh, Herbert W. Δ2 and	χ^2I2 fit indices for structural equation models: A brief note of clarification •	95 StrEqMod 2 246-254
nucleotide substitution rates, with application to the	chloroplast genome • Muse et al. ... for comparing synonymous and nonsynonymous	94 MolBioEv 11 715-724
	CHODR - Using statistics to predict college hockey • Danehy, Timothy J.; Lock, Robin H.	95 Stats /13 10-14
and merchandising effects on consumer brand	choice • Allenby, Greg M.; Lenk, Peter J. Reassessing brand loyalty, price sensitivity,	95 JBES 13 281-289
	Choice model	
Schnitkey, Gary D.	An empirical model of asset replacement in dairy production • Miranda, Mario J.;	95 JApEcmtx 10 41-55S
	Polychotomous choices and self-selection (Italian) • Mealli, Fabrizia	94 Statstca 54 501-519
Goodness-of-fit measures in binary	choice models (Disc: 117-120) • Windmeijer, Frank A. G.	95 EcomtrRv 14 101-116
maximum likelihood estimation of discrete	choice models • Lee, Lung-fei Asymptotic bias in simulated	95 EconTher 11 437-483
Dichotomous	choice models and limited independent variables • Caudill, Steven B.	95 Ststcian 44 405-406
estimation of polychotomous and sequential	choice models • Lee, Lung-fei Semiparametric maximum likelihood	95 JEconmtx 65 381-428
localized misspecification error in retail store	choice models • Rust, Roland T.; Donthu, Naveen Capturing geographically	95 JMarRsr 32 103-110
of conditional choice probabilities in a binary	choice model under uncertainty • Ahn, Hyungtaik Nonparametric two-stage estimation	95 JEconmtx 67 337-378
computation of opportunity costs in polychotomous	choice models with selectivity • Lee, Lung-fei The	95 RvEconSt 77 423-435
smoothed maximum score estimator for the binary	choice panel data model with an application to labour force ... • Charlier et al. A	95 StNeerla 49 324-342
Differentiation of the	Cholesky algorithm • Smith, S. P.	95 JCmpGrSt 4 134-147
	Cholesky decomposition	
Barrlund, A.	Efficient implementation of some contextual classification methods • Flygare, A.-M.;	95 CmpStQ 10 327-338
Rao, C. R.; Shanbhag, D. N.	Choquet-Deny type functional equations with applications to stochastic models •	94 Wiley:NY:UK 290p
Rao, C. R.; Shanbhag, D. N. (Auth) et al. Review of	"Choquet-Deny type functional equations with applications to stochastic models" •	95 ShrtBkRv 15/3 48-49
Rao, C. R.; Shanbhag, D. N. (Auth) et al. Review of	"Choquet-Deny type functional equations with applications to stochastic models" •	95 BiomtrcJ 37 726-726
Vowden, C. J.	Choreographing designs • Preece, D. A.; Vowden, B. J.; Jones, R. Hughes; Roger, C. A.;	95 MaScitst 20 15-32
The ET interview: Gregory C.	Chow • Pagan, Adrian	95 EconTher 11 597-624
multidimensional indices • Hong, D. H.; Hwang, S. Y.	Chow-Robbins "fair" games problem for pairwise independent random variables with	95 BIMaASnc 23 7-11
Statistical analysis of	chromatid interference • Zhao, Hongyu; McPeek, Mary Sara; Speed, Terence P.	95 Genetcs 139 1057-1065
study of peak overlapping in multicomponent	chromatograms: Importance of the retention pattern • Pietrogrande, M. C. et al. Statistical	95 Chemolab 28 239-258
son, J. F.; Hibbert, D. B. Chemometric analysis of gas	chromatographic data of oils from Eucalyptus species • Dunlop, P. J.; Bignell, C. M.; Jack-	95 Chemolab 30 59-67
in window factor analysis of spectroscopic and	chromatographic data • Brereton et al. ... for determining the number of components	95 Chemolab 27 73-87
canonical correlation analysis for the evaluation of	chromatographic retention data • Cserháti, Tibor; Forgács, Esther Use of	95 Chemolab 28 305-313
rank analysis applied to multidetectional	chromatographic structures • Toft, Jostein Evolutionary	95 Chemolab 29 189-212
Lifetime resolved fluorescence detection in	chromatography. Global analysis of emission decay surfaces • Maeder, M.; Molloy, K. J.	95 Chemolab 28 99-107
On the resolution problem in hyphenated	chromatography • Manne, Rolf	95 Chemolab 27 89-94
Multivariate optimization strategy for liquid	chromatography. II. Exploring and finding optimal conditions in the search ... • Kaufmann	95 Chemolab 27 105-114
C4.5 classifier to building an expert system for ion	chromatography • Mulholland, M.; Hibbert, D. B.; Haddad, P. R. et al. Application of the	95 Chemolab 27 95-104
factor analysis as applied to high-performance liquid	chromatography • Elbergali et al. ... on the performance of methods for evolutionary	95 Chemolab 27 55-71

Clinical trial (cont.)

sequential design for a multiple arm randomized	clinical trial • Rosner, Gary L.; Berry, Donald A. A Bayesian group	95 StatMed 14 381-394
is it worth measuring a covariate in a randomized	clinical trial? • Allison, David B. When	95 JCCPsych 63 339-343
sequential methods for comparison of cure rates in	clinical trials • Lee, Jae Won; Sather, Harland N. Group	95 Biomtrcs 51 756-763
Assessing the robustness of	clinical trial results • Choudhury, Japobrata; Balch, Alfred H.	94 ASAProBiop 484-486
of the operating characteristics of group sequential	clinical trials • Elson, Paul J. Use of the fast Fourier transform algorithm in the calculation	95 CmpStDA 20 491-498
response when patients enter staggeredly in a	clinical trial • Biswas, Suddhendu; Chitkara, Sunita On the estimation of true treatment	95 BiomtrcJ 37 571-580
A simple measure in defining optimal strata in	clinical trials • Choi, Sung C.; Germanson, Teresa P.; Barnes, Thomas Y.	95 ConClnTr 16 164-171
Moving from multinational	clinical trials to supra-trials • Accomando, William P.; Trost, Donald C.; Finman, Jeffrey S.	94 ASAProBiop 232-237
analysis of a single arm bone marrow transplant	clinical trial • Gonin, R. Considerations in the	94 ASAProBiop 493-497
regression toward the mean in uncontrolled	clinical trials • Shahane, Aditi; George, Varghese; Johnson, William D. Effect of bivariate	95 CommStA 24 2165-2181
Pharmacodynamic analysis of analgesic	clinical trials using empirical methods • Liu, Chui Yu; Sambol, Nancy C.	94 ASAProBiop 171-176
and efficiency of Bayesian dose adjustment in a	clinical trial with binary endpoint • Reeve, Russell; Hale, Michael Results	94 ASAProBiop 324-327
sequential monitoring designs for single-arm	clinical trials with multiple outcomes • Thall, P. F.; Simon, R. M.; Estey, E. H. Bayesian	95 StatMed 14 357-379
Flexible monitoring of multi-arm	clinical trials with repeated measurements • Reboussin, David M.	94 ASAProBiop 111-114
Sample size calculation for complex	clinical trials with survival endpoints • Shih, Joanna H.	95 ConClnTr 16 395-407
Design and analysis of multiarm	clinical trials with survival endpoints • Liu, Ping-Yu; Dahlberg, Steve	95 ConClnTr 16 119-130
Shera, David A multiple imputation strategy for	clinical trials with truncation of patient data • Lavori, Philip W.; Dawson, Ree;	95 StatMed 14 1913-1925
Boundary calculation issues in	clinical trials with two-sided stopping criteria • Chen, Bee-Lian; Gallo, Paul P.	94 ASAProBiop 115-119
Re-evaluating design specifications of longitudinal	clinical trials without unblinding when the key response is rate of change • Shih, W. J. et al.	95 StatMed 14 2239-2248
on multiple alleles, and application to estimation of	clonality in malaria parasites • Hill et al. ... inbreeding coefficients from genotypic data	95 GeneRsr 65 53-61
linkage studies: Implications for the positional	cloning of human disease genes • Boehnke, Michael Limits of resolution of genetic	94 AJHuGene 54 379-390
On a class of multivariate distributions	closed under concomitance of order statistics • Balasubramanian, K.; Balakrishnan, N.	95 StPrLet 23 239-242
Haddad, John N. On the	closed form of the likelihood function of the first order moving average model •	95 Biomtrka 82 232-234
distribution of size and costs of an epidemic in a	closed multigroup population • Svensson, Åke On the simultaneous	95 MaBisci 127 167-180
Holland, Burt A new	closed multiple testing procedure for hierarchical families of hypotheses • Rom, Dror M.;	95 JStPlInf 46 265-275
Estimation of the size of a	closed population (Disc: p122-123) • Ahn, Chul W.; Collins, Mark	94 ASAProSoSt 112-115
et al. Calculating normalization constants of	closed queuing networks by numerically inverting their generating functions • Choudhury	95 JACM 42 935-970
On statistical inference for random	closed sets in compact subsets of R^d • Ferrándiz, J.; Montes, F.	95 AdvAppPr 27 344-349
Bivariate random	closed sets and nerve fibre degeneration • Ayala, Guillermo; Simó, Amelia	95 AdvAppPr 27 293-305
with unknown location under Pitman's measure of	closeness • Kourouklis ... powers of the scale parameter of an exponential distribution	95 JStPlInf 48 185-195
an exponential quantile under Pitman's measure of	closeness • Kourouklis, S. Estimation of	95 CandJSt 23 257-268
A measure to evaluate the	closeness of Satterthwaite's approximation • Khuri, A. I.	95 BiomtrcJ 37 547-563
Blattenberger, Gail; Fowles, Richard Road	closure to mitigate avalanche danger: A case study for Little Cottonwood Canyon •	95 IntJFor 11 159-174
values of concentration in a contaminant	cloud • Labropulu, F.; Sullivan, P. J. Mean-square	95 EnvrMtrc 6 619-625
Fixed-point values of contaminant	cloud dilution • Heagy, W. K.; Sullivan, P. J.	95 EnvrMtrc 6 637-641
A uniform	CLT for continuous martingales • Bae, Jongsig; Levental, Shlomo	95 JKoreaSt 24 225-231
Levental, Shlomo Uniform	CLT for Markov chains and its invariance principle: A martingale approach • Bae, Jongsig;	95 JTherPrb 8 549-570
On the rate of convergence in the martingale	CLT • Račkauskas, Alfredas	95 StPrLet 23 221-226
A	CLT for the periodograms of a ρ^*-mixing random field • Miller, Curtis	95 StocProc 60 313-330
deviations of dependent random variables related to	CLT • Wu, Liming Moderate	95 AnlsProb 23 420-445
	Cluster analysis	
Pirmin	Automated classification scheme for wind fields • Weber, Rudolf O.; Kaufmann,	95 JApMetrl 34 1133-1141
Sauer, John R.	Delineating bird populations using ring recoveries • Pendleton, Grey W.;	95 JAppStat 22 1049-1055
	On the enumeration of rectangular $(0,1)$-matrices • Hendrickson, John A., Jr	95 JStCmpSm 51 291-313
Milligan, Glenn W.	Hierarchical clustering algorithms with influence detection • Cheng, Richard;	95 EdPsMeas 55 237-244
Thaddeus; Li, Luning; Flury, Bernard D.	Principal points and self-consistent points of elliptical distributions • Tarpey,	95 AnlsStat 23 103-112
	Principal points of univariate continuous distributions • Zoppé, Alice	95 StatComp 5 127-132
	Rainfall discrimination and spatial variation using breakpoint data • Sansom, John	95 JClimate 8 624-636
Iyengar, R. N.; Basak, P.	Regionalization of Indian monsoon rainfall and long-term variability signals •	94 JClimlgy 14 1095-1114
Martuzzi, Marco; Shaddick, Gavin	Spatial statistical methods in environmental epidemiology: A critique • Elliott, Paul;	95 StMedRes 4 137-159
in Southern Italy • Macchiato, M. F. et al.	Time modelling and spatial clustering of daily ambient temperature: An application	95 EnvrMtrc 6 31-53
Silver, Mick Scales of measurement and	cluster analysis: An application concerning market segments in the babyfood market •	95 Ststcian 44 101-112
Sherman, Simon A.	Cluster analysis dihedral angles of amino acid residues in proteins • Sclove, Stanley L.;	94 ASAProBiop 399-404
Discriminant and	cluster analysis • Wernecke, K.-D.	95 CmpStQ 10 15-20
• Gong, X.; Richman, M. B. On the application of	cluster analysis to growing season precipitation data in North America east of the Rockies	95 JClimate 8 897-931
Hierarchical	cluster analysis • Fischer, J.	95 CmpStQ 10 21-27
Identifying influential observations in hierarchical	cluster analysis • Jolliffe, I. T.; Jones, B.; Morgan, B. J. T.	95 JAppStat 22 61-80
of within-group covariance structure on recovery in	cluster analysis I. The bivariate case • Donoghue, John R. The effects	95 MultBehR 30 227-254
Non-uniqueness and inversions in	cluster analysis • Morgan, Byron J. T.; Ray, Andrew P. G.	95 ApplStat 44 117-134
Univariate screening measures for	cluster analysis • Donoghue, John R.	95 MultBehR 30 385-427
Weighting and selection of variables for	cluster analysis • Gnanadesikan, R.; Kettenring, J. R.; Tsao, S. L.	95 JClass 12 113-136
Spatial disease	clusters: Detection and inference • Kulldorff, Martin; Nagarwalla, Neville	95 StatMed 14 799-810
Waller et al. Detection and assessment of	clusters of disease: An application to nuclear power plant facilities and childhood ... •	95 StatMed 14 3-16
Case-control designs for the detection of spatial	clusters of diseases • Biggeri, Annibale; Marchi, Marco	95 EnvrMtrc 6 385-393
concentrations and estimating density contour	clusters – An excess mass approach • Polonik, Wolfgang Measuring mass	95 AnlsStat 23 855-881
Markov random fields • Wu, C. H.; Doerschuk, P. C.	Cluster expansions for the deterministic computation of Bayesian estimators based on	95 IEEEPAMI 17 275-293
graph-theoretic method for organizing overlapping	clusters into trees, multiple trees, or extended trees • Carroll, J. D.; Corter, J. E. A	95 JClass 12 283-313
categorical responses obtained from large	clusters • Miller, Michael E. Analysing	95 ApplStat 44 173-186
Random-	cluster measures and uniform spanning trees • Häggström, Olle	95 StocProc 59 267-275
Testing procedures for detecting multiple	clusters • Glaz, Joseph	94 ASAProEpid 59-63
A simulation study of the weighted k-means	cluster procedure • Rost, D.	95 JStCmpSm 53 51-63
for cardiovascular health (CATCH): Implications of	cluster randomization • Zucker et al. Statistical design of the child and adolescent trial	95 ConClnTr 16 96-118
Simpson, J. M.; Klar, N.; Donner, A. Accounting for	cluster randomization: A review of primary prevention trials, 1990 through 1993 •	95 AJPubHlh 85 1378-1383
A new method of hierarchical	cluster of sample-difference sequence method (Chinese) • Chen, Qiansheng	95 KToChina 39 97-101
for $\Pr(X < Y) - \Pr(X > Y)$ estimated from simple	cluster samples • Edwardes, Michael D. deB. A confidence interval	95 Biomtrcs 51 571-578
of multivariate polytomous data from a two-stage	cluster sample • Marques et al. A model for distributions and correlation structure	94 ASAProBiop 346-351
	Cluster sampling	
a bivariate normal population • Weerakkody et al.	Estimating the correlation coefficient in the presence of correlated observations from	95 CommStA 24 1705-1719
	Random sampling from databases: A survey • Olken, Frank; Rotem, Doron	95 StatComp 5 25-42
Kim, Kyu Seong; Jeon, Jong Woo; Park, Hong Nai	A study on the sample design of the fishery household economy survey (Korean) •	95 KorApS 8/2 43-54
Estimation of a common odds ratio under binary	cluster sampling • Ahn, Chul; Odom-Maryon, Tamara	95 StatMed 14 1567-1576
Michael J.; Brakhage, David H. Efficiency of adaptive	cluster sampling for estimating density of wintering waterfowl • Smith, David R.; Conroy,	95 Biomtrcs 51 777-788
Optimal and robust strategies for	cluster sampling • Tam, S. M.	95 JASA 90 379-382
studies • Bieler, Gayle S.; Williams, Rick L.	Cluster sampling techniques in quantal response teratology and developmental toxicity	95 Biomtrcs 51 764-776

Einmahl, Uwe On the	cluster set problem for the generalized law of the iterated logarithm in Euclidean space •	95 AnlsProb 23 817-851
Analysis of urban	cluster size in the Canadian Labour Force Survey • Laniel, Normand; Mohl, Chris	94 ASASrvyRsr 931-936
Disease	cluster tests based on the maximum cell frequency • Grimson, Roger C.	94 ASAProEpid 64-69
Disease	cluster tests for uncertain health events • Jacquez, Geoffrey M.	94 ASAProEnvr 110-115
disease progression: A comparison of methods for	clustered binary data • Have, T. R. T.; Landis, J. R. et al. Association models for periodontal	95 StatMed 14 413-429
Assessing gamma frailty models for	clustered failure time data • Shih, Joanna H.; Louis, Thomas A.	95 LiDaAnal 1 205-220
Latent variable models for	clustered ordinal data • Qu, Yinsheng; Piedmonte, Marion R.; Medendorp, Sharon V.	95 Biomtrcs 51 268-275
	Clustering	
	Adjusting Moran's I for population density • Oden, Neal	95 StatMed 14 17-26
Nyquist, W. E.; Xu, S.	Alternative partitioning of the genotype-by-environment interaction • Muir, W.;	92 ThApGene 84 193-200
Oehler, Karen L.; Gray, Robert M.	Combining image compression and classification using vector quantization •	95 IEEEPAMI 17 461-473
Program • Murray, David M.	Design and analysis of community trials: Lessons from the Minnesota Heart Health	95 AJEpidm 142 569-575
	Estimation of within-group covariance • Townsend, Jeffrey L.	94 ASAProStCp 257-262
E. J.; Bellegarda, J. R.; Nahamoo, D.; Nathan, K. S.	A fast statistical mixture algorithm for on-line handwriting recognition • Bellegarda,	94 IEEEPAMI 16 1227-1233
	On hyperplanes of closest fit • Farebrother, Richard William	95 CmpStDA 19 53-58
cessing plant: A sensitivity analysis • Viel, J.-F. et al.	Incidence of leukaemia in young people around the La Hague nuclear waste repro-	95 StatMed 14 2459-2472
Hernández-Pajares, M.	The Kohonen self-organizing map method: An assessment • Murtagh, F.;	95 JClass 12 165-190
epidemiology • Lawson, Andrew B.	MCMC methods for putative pollution source problems in environmental	95 StatMed 14 2473-2485
period data recorded on the Catalan ... • Lana et al.	Minimum sample size for synoptic weather type classification. Application to winter	94 JClimlgy 14 1051-1060
Hadar I.; Van Mieghem, Jan A.; Rub, Leonardo	Multiple subclass pattern recognition: A maximin correlation approach • Avi-Itzhak,	95 IEEEPAMI 17 418-431
recognition applied to RIFS sensor ... • Kraus et al.	Optical reflectometric gas sensing: Classification of hydrocarbon vapours by pattern	95 Chemolab 30 211-221
proximity data • Young, Martin R.; DeSarbo, Wayne S.	A parametric procedure for ultrametric tree estimation from conditional rank order	95 Psymtrka 60 47-75
	A random-effects model for ordered categorical data • Crouchley, Robert	95 JASA 90 489-498
Bruno	A spanning tree heuristic for regional clustering • Maravalle, Maurizio; Simeone,	95 CommStA 24 625-639
Commenges, Daniel	Tests of homogeneity for generalized linear models • Jacqmin-Gadda, Hélène;	95 JASA 90 1237-1246
	Three-dimensional projection pursuit • Nason, Guy	95 ApplStat 44 411-430
	Three-way distances • Joly, S.; Le Calvé, G.	95 JClass 12 191-205
Hierarchical	clustering algorithms with influence detection • Cheng, Richard; Milligan, Glenn W.	95 EdPsMeas 55 237-244
	Clustering behavior of finite variance partial sum processes • Einmahl, U.; Goodman, V.	95 ProbTher 102 547-565
Single linkage	clustering and continuum percolation • Penrose, Mathew D.	95 JMultiAn 53 94-109
Van Sanford, D. A.; Seyedsadr, M. S.	Clustering cultivars into groups without rank-change interactions • Cornelius, P. L.;	93 CropSci 33 1193-1200
F.; La Rotonda, L. et al. Time modelling and spatial	clustering of daily ambient temperature: An application in Southern Italy • Macchiato, M.	95 EnvrMtrc 6 31-53
A new metric for data	clustering • Millar, David B.	94 ASAProStCp 251-256
The power of focused tests to detect disease	clustering • Waller, Lance A.; Lawson, Andrew B.	95 StatMed 14 2291-2308
of Stone's likelihood ratio tests for disease	clustering • Lumley, Thomas Efficient execution	95 CmpStDA 20 499-510
models implicit in statistical tests of disease	clustering • Waller, Lance A.; Jacquez, Geoffrey M. Disease	95 Epidmlgy 6 584-590
Hubert, Lawrence J. Additive two-mode	clustering: The error-variance approach revisited • Mirkin, Boris; Arabie, Phipps;	95 JClass 12 243-263
Mapping influence regions in hierarchical	clustering • Cheng, Richard; Milligan, Glenn W.	95 MultBehR 30 547-576
Cohen, Uri The extremal index and	clustering of high values for derived stationary sequences • Weissman, Ishay;	95 JAppProb 32 972-981
Rotondi, Renata; Drappo, Silvia A	clustering method for global optimization based on the kth nearest neighbour •	95 StatComp 5 317-326
Park, Nae-Hyun; Kim, Yung-Hwan A	clustering method using the Coulomb energy network (Korean) • Lee, Sukhoon;	95 KorApS 8/1 39-50
Hierarchical	clustering method using probabilistic similarity measure • Miyano, Hisao	95 Behvmtrk 22 145-153
A general fuzzy	clustering model based on aggregation operators • Sato, Mika; Sato, Yoshiharu	95 Behvmtrk 22 115-128
Gaussian parsimonious	clustering models • Celeux, Gilles; Govaert, Gérard	95 PattrnRc 28 781-793
Asymptotics of k-mean	clustering under non-i.i.d. sampling • Serinko, Regis J.; Babu, Gutti Jogesh	95 StPrLet 24 57-66
Hüsler, J. Extremes and	clustering on nonstationary max-AR(1) sequences • Alpuim, M. T.; Catkan, N. A.;	95 StocProc 56 171-184
A comparison of two statistics for detecting	clustering in one dimension • Anderson, N. H.; Titterington, D. M.	95 JStCmpSm 53 103-125
Modelling recall –	Clustering and order effects • Robertson, Chris	95 BJMaStPs 48 29-50
test of the goodness-of-fit for some pyramidal	clustering procedures • Capdevila, Carles; Arcas, Antoni Empirical significance	95 Questo 19 123-130
class of tests for detecting 'general' and 'focused'	clustering of rare diseases • Tango, Toshiro A	95 StatMed 14 2323-2334
A spanning tree heuristic for regional	clustering • Maravalle, Maurizio; Simeone, Bruno	95 CommStA 24 625-639
Correction to "A spanning tree heuristic for regional	clustering" (V24 p625-639) • Maravalle, Maurizio; Simeone, Bruno	95 CommStA 24 2437-2437
Gowda, K. Chidananda; Ravi, T. V. Divisive	clustering of symbolic objects using the concepts of both similarity and dissimilarity •	95 PattrnRc 28 1277-1282
Second-order analysis of space-time	clustering • Diggle, P. J.; Chetwynd, A. G.; Häggkvist, R.; Morris, S. E.	95 StMedRes 4 124-136
a bivariate discrete and continuous outcome with	clustering • Fitzmaurice, Garrett M.; Laird, Nan M. Regression models for	95 JASA 90 845-852
Permuting	CM steps within the ECM algorithm • van Dyk, David A.; Meng, Xiao-Li	94 ASAProStCp 130-135
ANOVA, MINQUE, PSD-MINQMBE, CANOVA and	CMINQUE in estimating variance components • Shaban, Abdel Magid M.	94 Statstca 54 481-489
Using	CMM data for process evaluation and improvement • Blazek, L. W.	94 ASAProPEng 26-34
time • Jacobsen, Martin; Keiding, Niels	Coarsening at random in general sample spaces and random censoring in continuous	95 AnlsStat 23 774-786
restees • Hyatt, R. R., Jr et al. The price and purity of	cocaine: The relationship to emergency room visits and death, and to drug use among ar-	95 StatMed 14 655-668
drug abuse treatment utilization among frequent	cocaine users • Colliver, James D.; Gfroerer, Joseph C. Predictors of	94 ASASrvyRsr 553-558
autocorrelated • Chu, D. P. T.; Sutradhar, B. C. On	Cochran's and Hartley's tests for homogeneity of variances when observations are	95 CommStB 24 327-347
A further look at the	Cochran-Mantel-Haenszel risk difference • Sato, Tosiya	95 ConClnTr 16 359-361
	Cochran Q test with Turbo BASIC • Seuc, Armando H.	95 CmpBiomd 46 29-34
Wong, Chi Song	Cochran theorems for a multivariate elliptically contoured model • Wang, Tonghui;	95 JStPlInf 43 257-270
Perfect	cocycles through stochastic differential equations • Arnold, Ludwig; Scheutzow, Michael	95 ProbTher 101 65-88
Frequency hyperrectangles, error correcting	codes and fractional replication plans • Mullen, Gary L.; Suchower, Stephan J.	93 JCISS 18 5-18
Designs through recoding of varietal and level	codes • Das, M. N.; Giri, N.; Ahmed, M.	95 StPrLet 23 371-380
to lag-sequential analysis when consecutive	codes may and cannot repeat • Bakeman, Roger; Quera, Vicenç Log-linear approaches	95 PsychBu 118 272-284
tion of minihypers in PG(t,q) and nonbinary linear	codes meeting the Griesmer bound • Hamada, N. A survey of recent work on characteriza-	93 JCISS 18 161-191
autoregressive models with cointegration and	codependence • Ahn, Sung K. Reduced rank estimation of vector	94 ASAProBuEc 151-156
Automated	coding of census data • Kalpić, Damir	94 JOffStat 10 449-463
comparison of two methods of automated industry	coding • Rowland, John H.; Kinack, Mark D. A	94 ASASrvyRsr 1130-1133
Evaluation of an automated industry and occupation	coding system • Harris, Kenneth W.; Chamblee, Ronald F.; Winter, Francine D.	94 ASAProSoSt 174-177
	Coding theory	
Sharma, Bhu Dev	A generalization of the Sharma-Kaushik class of metrics • Koschnick, Klaus-Uwe;	93 JCISS 18 401-407
linear codes meeting the Griesmer bound • Hamada	A survey of recent work on characterization of minihypers in PG(t,q) and nonbinary	93 JCISS 18 161-191
P. Estimating the fraction of invariable	codons with a capture-recapture method • Sidow, Arend; Nguyen, Trang; Speed, Terence	92 JMolEvol 35 253-260
	Coefficient	
	• see names of specific types of coefficients such as correlation –	
structural equation modeling • Miller, Michael B.	Coefficient alpha: A basic introduction from the perspectives of classical test theory and	95 StrEqMod 2 255-273
Srivastava, A. K.; Srivastava, V. K.; Ullah, A. The	coefficient of determination and its adjusted version in linear regression models •	95 EcomtrRv 14 229-240
approximation to the sampling distribution of the	coefficient of variation • Hürlimann, Werner A uniform	95 StPrLet 24 263-268
Improved estimation of the	coefficient of variation • Ahmed, S. E.	94 JAppStat 21 565-573

Yang, Minxian Testing for	cointegration: The effects of mis-specifying the lag length • Bewley, Ronald;	95 MaCmpSim 39 251-255
with applications to simultaneous equations and	cointegration • Johansen, Søren Identifying restrictions of linear equations	95 JEconmtx 69 111-132
exchange rates • Joseph, Nathan Lael	Cointegration, error-correction models, and forecasting using realigned foreign	95 JForcast 14 499-522
The expectations theory of interest rates:	Cointegration and factor decomposition • Choi, Seungmook; Wohar, Mark E.	95 IntJFor 11 253-262
behavior in Taiwan: Tests of functional forms and	cointegration • Hsing, Yu; Chang, Hui S. Household saving	94 ASAProBuEc 123-128
Testing for	cointegration at any frequency using spectral methods • Cubadda, Gianluca	94 JItalStS 3 37-50
Pi, Lynn K.	Cointegration of international stock market indices • Chou, Ray Y.; Ng, Victor K.;	94 ASAProBuEc 196-201
A periodic	cointegration model of quarterly consumption • Franses, Philip Hans; Kloek, Teun	95 AplStMDA 11 159-166
Mantani, Akihisa Estimating the rank of	co-integration when the order of a vector autoregression is unknown • Morimune, Kimio;	95 MaCmpSim 39 265-271
McCullough, B. D. Estimating	cointegration parameters: An application of the double bootstrap • Vinod, H. D.;	95 JStPlInf 43 147-156
Boswijk, H. Peter Efficient inference on	cointegration parameters in structural error correction models (Disc: p159-175) •	95 JEconmtx 69 133-158
Periodic	cointegration: Representation and inference • Boswijk, H. Peter; Franses, Philip Hans	95 RvEconSt 77 436-454
Nonstationary time series and	cointegration • Phillips, Peter C. B.	95 JApEcmtx 10 87-94
Michael T. K.; Watson, Mark W. Testing for	cointegration when some of the cointegrating vectors are prespecified • Horvath,	95 EconTher 11 984-1014
Testing for	cointegration in a system of equations • Choi, In; Ahn, Byung Chul	95 EconTher 11 952-983
Timmermann, Allan	Cointegration tests of present value models with a time-varying discount factor •	95 JApEcmtx 10 17-31
Symmetric tests for	cointegration • Deo, Rohit; Fuller, Wayne A.	94 ASAProBuEc 105-110
temporal aggregation, seasonal adjustment, and	cointegration. Theory and evidence • Granger, C. W. J.; Siklos, P. L. Systematic sampling,	95 JEconmtx 66 357-369
A statistical analysis of	cointegration for I(2) variables • Johansen, Søren	95 EconTher 11 25-59
A note on testing for seasonal	cointegration using principal components in the frequency domain • Cubadda, Gianluca	95 JTimSrAn 16 499-508
Errors in variables and	cointegration • Solo, Victor	95 EconTher 11 60-80
A comparison of multiple linear regression and	cokriging • Lesch et al. ... induction techniques 1. Statistical prediction models:	95 WaterReR 31 373-386
The	cold equations: Spearman and Wilson on factor indeterminacy • Lovie, P.; Lovie, A. D.	95 BJMaStPs 48 237-253
estimation of population size in darkling beetles (	Coleoptera: Tenebrionidae) • Crist, Thomas O.; Wiens, John A. Individual movements and	95 JAnmlEco 64 733-746
	Collapsibility	
pendence graph • de Falguerolles, A.; Jmel, S. et al.	Correspondence analysis and association models constrained by a conditional inde-	95 Psymtrka 60 161-180
	Collapsibility of logistic regression coefficients • Guo, Jianhua; Geng, Zhi	95 JRSS-B 57 263-267
Braun, H. I. (Ed); Young, Karen (Rev) Review of "The	collected works of John W. Tukey, Volume 8, Multiple comparisons: 1948-1983" •	95 JRSS-A 158 632-632
Braun, H. I. (Ed); Martin, M. A. (Rev) Review of "The	collected works of John W. Tukey: Volume VIII Multiple comparisons 1948-1983" •	95 Biomtrcs 51 380-381
Ziegel, Eric R. (Rev) Review of "The	collected works of John W. Tukey, Volume VIII: Multiple comparisons" • Braun, H. (Ed);	95 Technmcs 37 242-242
Review of "H. Cramér:	Collected works. Volumes 1 and 2" • Martin-Löf, Anders (Ed); Uebe, Götz (Rev)	95 AllgeStA 79 332-332
Harald Cramer	collected works. Volumes 1 and 2 • Martin-Löf, A. (Ed)	94 Springer:Brln:NY 2165p
The	collected works of Wassily Hoeffding • Fisher, N. I.; Sen, P. K.	94 Springer:Brln:NY 658p
of improved quality of life in Canada: Social change,	collective consciousness, and the TM-Sidhi program • Dillbeck, M. C. Time series analysis	95 PsychRpt 76 1171-1193
streams • Sheldon, A. L.; Meffe, G. K. Path analysis of	collective properties and habitat relationships of fish assemblages in coastal plain	95 JFishRBC 52 23-33
of collecting the occupation of persons with	college degrees • Hardy, Linda P.; Eisenhower, Donna L. Developing methods	94 ASAProGovt 110-115
ratization of diversion? The effect of community	colleges on educational attainment • Rouse, Cecilia Elena Democ-	95 JBES 13/2 217-224
Who's number 1 in	college football? ... And how might we decide? • Stern, Hal S.	95 Chance 8/3 7-14
CHODR – Using statistics to predict	college hockey • Danehy, Timothy J.; Lock, Robin H.	95 Stats /13 10-14
trol, and graphics • Wilson, W. J. Supplementing a	college preparatory course in statistics with emphasis on applied probability, quality con-	94 ASAProStEd 42-43
Statistics before	college • Herring, Susan; Thoni, Janice	94 ASAProStEd 218-221
Continuous improvement in	college teaching: An application of statistical tools • Xue, Zhiming; Gulas, Charles S.	95 ASQCTekCon 267-275
	Collinearity	
A simulation study • Crivelli, A.; Firinguetti, L. et al.	Confidence intervals in ridge regression by bootstrapping the dependent variable:	95 CommStB 24 631-652
	Graphical interpretation of variance inflation factors • Stine, Robert A.	95 AmerStat 49 53-56
models • Otranto, Edoardo; Gallo, Giampiero M.	Regression diagnostic techniques to detect balanced space-to-time ratios in STARMA	94 Metron 52/3 129-145
alternatives to classical regression in the case of	collinearity (French) • Palm, R.; Iemma, A. F. Some	95 RStApp 43/2 5-33
On the Hausdorff measure of multiple points and	collision points of super-Brownian motion • Serlet, Laurent	95 Stochatx 54 169-198
regression model for 20 year survival of	colon cancer in Norway • Zahl, Per-Henrik A proportional	95 StatMed 14 1249-1261
An analysis of voting coalitions in the New York	colonial legislature • Morrell, Christopher H.; Peyrot, Mark F.; Varga, Nicholas	94 ASAProSoSt 66-70
mark-sight population size estimation applied to	Colorado moose • Bowden, David C.; Kufeld, Roland C. Generalized	95 JWildlMg 59 840-851
work methods in the modelling and prediction of the	colour of young red wines from the Denomination of Origin ... • Fernández et al. ... net-	95 Chemolab 28 273-285
Campling, B. G. Nonlinear models for assessing	combinations of chemotherapeutic drugs with resistance modifiers • Lam, Y. M.; Pym, J.;	94 ASAProBiop 394-398
Gennings, C.; Chinchilli, V. M. Threshold models for	combination data from reproductive and developmental experiments • Schwartz, P. F.;	95 JASA 90 862-870
Bahadur exact slopes for a class of	combinations of dependent tests • Pallini, Andrea	94 Metron 52/1 53-65
On improving the min test for the analysis of	combination drug trials • Sarkar, Sanat K.; Snapinn, Steve; Wang, Wenjin	95 JStCmpSm 51 197-213
for estimating population characteristics of drug	combination effect • Gao, Min; Minkin, Salomon Nonlinear mixed effect model	94 ASAProBiop 160-165
Dong, Jianping; Simonoff, Jeffrey S. A geometric	combination estimator for d-dimensional ordinal sparse contingency tables •	95 AnlsStat 23 1143-1159
Mortera, J. Coherent	combination of experts' opinions (Disc: p293-313) • Dawid, A. P.; DeGroot, M. H.;	95 Test 4 263-293
of the quasi-Bayes approach to the linear	combination of forecasts • Faria, A. E.; Souza, R. C. A re-evaluation	95 JForcast 14 533-542
Confidence intervals on ratios of linear	combinations for non-disjoint sets of expected mean squares • Gui, R.; Graybill, F. A. et al.	95 JStPlInf 48 215-227
Gauss, K. F. (Auth) et al. Theory of the	combination of observations least subject to errors. Part one, part two, supplement •	95 SocApMathSim:PA 220p
(Transl); Stigler, S. M. (Rev) Review of "Theory of the	combination of observations least subject to error" • Gauss, C. F. (Auth); Stewart, G. W.	96 ShrtBkRv 16/1 41-41
On estimating a linear	combination of strata means with random sample sizes • He, Kun	95 JMultiAn 55 39-60
information from randomized clinical studies for	combination therapy • Hung, H. M. James; Dern, Philip On combining	94 ASAProBiop 77-81
Teaching the standard deviations of linear	combinations using graphics • Mansfield, Edward R.; Adams, Benjamin M.	94 ASAProStEd 80-83
diagnostic procedure to a second procedure used in	combination with the standard procedure • Whaley, F. S. et al. Comparing a standard	94 ASAProBiop 272-275
for Poisson process approximations of logarithmic	combinatorial assemblies • Arratia, R.; Stark, D.; Tavaré, S. Total variation asymptotics	95 AnlsProb 23 1347-1388
A simple	combinatorial proof of a result by Robertson and Pillers • Mukerjee, Hari	95 StPrLet 24 173-175
units • Stufken, John	Combinatorial and statistical aspects of sampling plans to avoid the selection of adjacent	93 JCISS 18 149-160
	Combinatorics	
	Barker sequences and circulant Hadamard matrices • Lin, Cantian; Wallis, W. D.	93 JCISS 18 19-25
Vowden, C. J.	Choreographing designs • Preece, D. A.; Vowden, B. J.; Jones, R. Hughes; Roger, C. A.;	95 MaScitst 20 15-32
	Class 3 association schemes whose symmetrizations have two classes • Song, Sung Y.	95 JCombThA 70 1-29
	Five-designs with three intersection numbers • Ionin, Yury J.; Shrikhande, Mohan S.	95 JCombThA 69 36-50
	Non-abelian Hadamard difference sets • Smith, Ken W.	95 JCombThA 70 144-156
Chen, Yu Qing	On the size of the multiplier groups of cyclic difference sets • Xiang, Qing;	95 JCombThA 69 168-169
models containing nuisance ... • Raktoe et al.	Some aspects on the design and analysis of fractional factorial designs with linear	93 JCISS 18 369-395
	Some new difference sets • Arasu, K. T.; Sehgal, Surinder K.	95 JCombThA 69 170-172
	Combined Bayesian estimates for the equicorrelation coefficient • Viana, Marlos A. G.	95 CandJSt 23 43-53
Robust minimum distance inference based on	combined distances • Park, Chanseok; Basu, Ayanendranath; Basu, Srabashi	95 CommStB 24 653-673
dependent action • Drescher et al. Assessment of the	combined effects of substances: The relationship between concentration addition and in-	95 Biomtrcs 51 716-730
Forecast-based monitoring schemes using	combined EWMA-Shewhart control charts • Adams, Benjamin M.; Lin, Winnie	94 ASAProQlPr 104-108
Gladkov, Yu. M. Reconfigurable	combined forecasts in a non-stationary inflationary environment • Volkov, V. Ya.;	95 JForcast 14 395-403

lation technique • Tanizaki, H. Asymptotically exact	confidence intervals of CUSUM and CUSUMSQ tests: A numerical derivation using simu-	95 CommStB 24 1019-1036
for calculating numerically standard deviations and	confidence intervals • Kragten, J. A standard scheme	95 Chemolab 28 89-97
Wang, Suojin	Confidence intervals for sub-domain means and totals • Casady, Robert J.; Dorfman, Alan;	94 ASASrvyRsr 1036-1041
Estimation and	confidence intervals for empirical mixing distributions • Link, William A.; Sauer, John R.	95 Biomtrcs 51 810-821
sion models • Hosmer, David W.; Lemeshow, Stanley	Confidence interval estimates of an index of quality performance based on logistic regres-	95 StatMed 14 2161-2172
Approximate	confidence interval estimation of $P(Y < X)$ • Chan, Yiu-Man	94 ASAProPEng 146-150
cross-sectional studies • Gargiullo, P. M. et al.	Confidence intervals, hypothesis tests, and sample sizes for the prevented fraction in	95 StatMed 14 51-72
tiple observations per subject • Buderer, N. F. et al.	Confidence intervals and hypothesis testing for screening parameters in the case of mul-	94 ASAProStEd 175-179
Einmahl, J. H. J. Asymptotic	confidence intervals for the length of the short$t$ under random censoring • Beirlant, J.;	95 StNeerla 49 1-8
Sanjib; DasGupta, Anirban Robustness of standard	confidence intervals for location parameters under departure from normality • Basu,	95 AnlsStat 23 1433-1442
ata, M. O.; Hariri, A. M. A. Point estimation and	confidence intervals of the M/M/2/N queue with balking and heterogeneity • Abou-El-	95 AJMMgSci 15 35-55
The profile Z plot and	confidence intervals in maximum likelihood analysis • Chen, Jian-Shen; Jennrich, Robert I.	94 ASAProStCp 124-129
Nagata, Yasushi Fixed-width sequential	confidence interval for the mean of a gamma distribution • Takada, Yoshikazu;	95 JStPlInf 44 277-289
Hainsworth, Timothy J.	Confidence intervals for the noncentral chi-squared distribution • Kent, John T.;	95 JStPlInf 46 147-159
Diagnostics for linearization	confidence intervals in nonlinear regression • Chen, Jian-Shen; Jennrich, Robert I.	95 JASA 90 1068-1074
Transformations for improving linearization	confidence intervals in nonlinear regression • Chen, Jian-Shen; Jennrich, Robert I.	95 JASA 90 1271-1276
Kulathava Ranee Comparisons of approximate	confidence intervals for the parameters of extreme value distribution • Thiagarajah,	94 Metron 52/1 85-100
cients and vote counts to obtain an estimate and a	confidence interval for the population correlation coefficient • Bushman et al. ... coeffi-	95 PsychBu 117 530-546
of saddle-point approximations to construct	confidence intervals for population means • Li, J.; Chen, M.-C. et al. Forestry applications	95 Biomtrcs 51 61-72
Hamdy, M. I. Distribution-free	confidence intervals for $\Pr(X < Y)$ based on independent samples of X and Y •	95 CommStB 24 1005-1017
Edwardes, Michael D. deB. A	confidence interval for $\Pr(X < Y) - \Pr(X > Y)$ estimated from simple cluster samples •	95 Biomtrcs 51 571-578
Constructing	confidence intervals for QTL location • Mangin, B.; Goffinet, B.; Rebaï, A.	94 Genetcs 138 1301-1308
censorship • Ghorai, J. K. Large sample	confidence interval for the quantiles in the proportional hazards model of random	95 JStCmpSm 53 91-102
lower bound distributions that give simultaneous	confidence intervals for quantiles • Satten, Glen A. Upper and	95 JASA 90 747-752
Schucany, W. R.; Sheather, S. J. A comparison of	confidence intervals from R-estimators in regression • George, K. J.; McKean, J. W.;	95 JStCmpSm 53 13-22
mean squares • Gui, R.; Graybill, F. A. et al.	Confidence intervals on ratios of linear combinations for non-disjoint sets of expected	95 JStPlInf 48 215-227
survival analysis • Tu, D.; Gross, A. J. Accurate	confidence intervals for the ratio of specific occurrence/exposure rates in risk and	95 BiomtrcJ 37 611-626
Accurate	confidence intervals from recapture experiments • Lloyd, Christopher J.	95 CommStA 24 1235-1253
	Confidence intervals versus regions • Draper, Norman R.; Guttman, Irwin	95 Ststcian 44 399-403
Better bootstrap	confidence intervals for regression curve estimation • Härdle, W.; Huet, S.; Jolivet, E.	95 MaOpfStS 26 287-306
ate • Rahbar, Mohammad H. Sequential fixed width	confidence intervals for regression parameters from censored data with a discrete covari-	95 SqtlAnly 14 143-156
Li, Linxiong Nonparametric	confidence intervals for the renewal function with censored data • Baxter, Laurence A.;	95 JNnparSt 4 317-326
simulation study • Crivelli, A.; Firinguetti, L. et al.	Confidence intervals in ridge regression by bootstrapping the dependent variable: A	95 CommStB 24 631-652
Lui, Kung-Jong	Confidence intervals for the risk ratio in cohort studies under inverse sampling •	95 BiomtrcJ 37 965-971
Sison, Cristina P.; Glaz, Joseph Simultaneous	confidence intervals and sample size determination for multinomial proportions •	95 JASA 90 366-369
p366-369) • Sison et al. Correction to "Simultaneous	confidence intervals and sample size determination for multinomial proportions" (V90	95 JASA 90 1494-1494
Kidd, Martin; Laubscher, Nico F. Robust	confidence intervals for scale and its application to the Rayleigh distribution •	95 SoAfrStJ 29 199-217
distance, with applications to simultaneous	confidence intervals • Brown, Mark The distribution of total variation	95 CmpOpRsh 22 373-381
intervals • Wang, Yuedong; Wahba, Grace Bootstrap	confidence intervals for smoothing splines and their comparison to Bayesian confidence	95 JStCmpSm 51 263-279
method • Swift, Michael B. Simple	confidence intervals for standardized rates based on the approximate bootstrap	95 StatMed 14 1875-1888
Likelihood ratio-based	confidence intervals in survival analysis • Murphy, S. A.	95 JASA 90 1399-1405
953) • Rosen, O.; Cohen, A. Constructing a bootstrap	confidence interval for the unknown concentration in radioimmunoassay (Disc: p953-	95 StatMed 14 935-952
$(Q-1)$-fold nested error structure • Yu, Q.-L. et al.	Confidence intervals on variance components in regression models with balanced	95 CommStA 24 1151-1167
Lower	confidence limit in censored case • Zheng, Zhongguo; Jin, Hua	95 JSysSMSE 7 40-44
Hall, C. B. Fast and accurate computation of the exact	confidence limits for the common odds ratio in several 2×2 tables • Liao, J. G.;	95 JCmpGrSt 4 173-179
Determining sample size for lower	confidence limits for C_p and C_{pk} • Franklin, LeRoy A.	94 ASAProQlPr 23-27
Notes on conditional	confidence limits under inverse sampling • Lui, Kung-Jong	95 StatMed 14 2051-2056
On bootstrap procedures for second-order accurate	confidence limits in parametric models • DiCiccio, Thomas J.; Romano, Joseph P.	95 StSinica 5 141-160
distribution • Lui, Kung-Jong	Confidence limits for the population prevalence rate based on the negative binomial	95 StatMed 14 1471-1477
Bootstrap	confidence limits on process capability indices • Collins, Alan J.	95 Ststcian 44 373-378
Non-iterative method • Nam, Jun-mo	Confidence limits for the ratio of two binomial proportions based on likelihood scores:	95 BiomtrcJ 37 375-379
Nelson, Wayne	Confidence limits for recurrence data – Applied to cost or number of product repairs •	95 Technmcs 37 147-157
The	confidence limits for the reliability in regression model • Lin, Huanbin; Chen, Jiading	95 ActScNat 30 23-29
Using nested designs: II.	Confidence limits for standard deviations • Nelson, Lloyd S.	95 JQualTek 27 265-267
Correction to "Using nested designs: II.	Confidence limits for standard deviations" (V27 p265-265) • Nelson, L. S.	95 JQualTek 27 403-403
disaggregate sector forecasting using consumer	confidence measures • Kumar, V.; Leone, Robert P.; Gaskins, John N. Aggregate and	95 IntJFor 11 361-377
	Confidence region	
	On the estimation of the influence curve • Cuevas, Antonio; Romo, Juan	95 CandJSt 23 1-9
mal means • Singh, R. K.; Chaturvedi, A.; Kumar, S.	Two stage and sequential procedures for estimating a linear function of k multinor-	94 JIndnStA 32 1-8
Belforte, G.	Uncertainty evaluation for estimates from recursive projection algorithms •	95 IJAdCSgP 9 97-106
Escobar, Luis A. Teaching about approximate	confidence regions based on maximum likelihood estimation • Meeker, William Q.;	95 AmerStat 49 48-53
N.; Romano, Joseph P. Large sample	confidence regions based on subsamples under minimal assumptions • Politis, Dimitris	94 AnlsStat 22 2031-2050
populations • Jordan, Scott M.; Krishnamoorthy, K.	Confidence regions for the common mean vector of several multivariate normal	95 CandJSt 23 283-297
On	confidence regions induced by the Wilcoxon rank sum test • Horstmann, Vibeke	95 StPrLet 23 333-338
tive confidence region from projections of an exact	confidence region in multiple linear regression" (V48 p120-124) • Nickerson ... conserva-	95 AmerStat 49 324-324
	Confidence regions for parameters in the AR(1) model • Hamilton, David; Wu, Ka Ho	95 JTimSrAn 16 249-265
The effect of model selection on	confidence regions and prediction regions (Disc: p550-559) • Kabaila, Paul	95 EconTher 11 537-549
Comment on "Construction of a conservative	confidence region from projections of an exact confidence region in ... • Nickerson	95 AmerStat 49 324-324
	Confidence set	
Ghosh, Jayanta Kumar	On priors providing frequentist validity for Bayesian inference • Datta, Gauri Sankar;	95 Biomtrka 82 37-45
Zen, M. M. A new general method for constructing	confidence sets in arbitrary dimensions: With applications • DasGupta, A.; Ghosh, J. K.;	95 AnlsStat 23 1408-1432
Casella, George; Hwang, J. T. Gene Optimal	confidence sets, bioequivalence, and the Limaçon of Pascal • Brown, Lawrence D.;	95 JASA 90 880-889
Stein	confidence sets and the bootstrap • Beran, Rudolf	95 StSinica 5 109-127
Complete classes for	confidence set estimation • Brown, L. D.; Cohen, Arthur	95 StSinica 5 291-301
	Confidence sets of fixed size with predetermined confidence level • Holm, Sture	95 CommStA 24 1521-1536
How reliable are contour curves?	Confidence sets for level contours • Lindgren, Georg; Rychlik, Igor	95 Bernoull 1 301-319
You can count on us – With	confidence • McLennan, Bill	95 JRSS-A 158 467-489
	Confidentiality	
Luchian, Henri; Stamate, Daniel	Answer-perturbation techniques for the protection of statistical databases •	95 StatComp 5 203-213
Walter; Blien, Uwe; Wirth, Heike	Identification risks of microdata: Evidence from experimental studies • Müller,	95 SocMethR 24 131-157
mandatory appeals and variations in statements of	confidentiality on completion rates for census questionnaires • Dillman et al. ... appeals,	94 ASASrvyRsr 1333-1338
access to statistical information and demands for	confidentiality • Fienberg, Stephen E. Conflicts between the needs for	94 JOffStat 10 115-132
	Confidentiality issues in videotaped data collection • Cederlund, Ann E.	94 ASASrvyRsr 1248-1251
"Bayes or bust? – A critical examination of Bayesian	confirmation theory" • Earman, J. (Auth); Neuberg, Leland G. (Rev) Review of	95 Ststcian 44 283-284

	Consistent variable selection in linear models • Zheng, Xiaodong; Loh, Wei-Yin	95 JASA 90 151-156
a concept of population in survey sampling	consistent with that in experimental statistics • Proctor, C. H. Developing	94 ASASrvyRsr 653-655
Black's	consol rate conjecture • Duffie, Darrell; Ma, Jin; Yong, Jiongmin	95 AnlsApPr 5 356-382
Discrete proportional-integral control with	constrained adjustment • Box, George E. P.; Luceño, Alberto	95 Ststcian 44 479-495
J. Correspondence analysis and association models	constrained by a conditional independence graph • de Falguerolles, A.; Jmel, S.; Whittaker,	95 Psymtrka 60 161-180
A graphical approach for the construction of	constrained D- and L-optimal designs using efficiency plots • Wong, Weng Kee	95 JStCmpSm 53 143-152
C.-C.; Cochran, J. K.; Lawrence, F. P. Statistically	constrained economic design of the EWMA control chart • Montgomery, D. C.; Torng, J.	95 JQualTek 27 250-256
D. C.; Lawrence, F. P. Implementing statistically	constrained economic EWMA control charts • Torng, J. C.-C.; Cochran, J. K.; Montgomery,	95 JQualTek 27 257-264
J. K. et al. Correction to "Implementing statistically	constrained economic EWMA control charts" (V27 p257-264) • Torng, J. C.-C.; Cochran,	95 JQualTek 27 402-402
Wolins, Leroy A Monte Carlo study of	constrained factor analysis using maximum likelihood and unweighted least squares •	95 EdPsMeas 55 545-557
Ohtani, Kazuhiro A comparison of interval	constrained least squares and mixed regression estimators • Srivastava, Viren K.;	95 CommStA 24 395-413
On the asymptotics of	constrained M-estimation • Geyer, Charles J.	94 AnlsStat 22 1993-2010
Eugene A.; Shwartz, Adam	Constrained Markov decision models with weighted discounted rewards • Feinberg,	95 MaOpRsr 20 302-320
of two-mode proximity matrices by sums of order-	constrained matrices • Hubert, Lawrence; Arabie, Phipps The approximation	95 Psymtrka 60 573-605
Marginally restricted	constrained optimal designs • Huang, Mong-Na; Chang, Hsiu-Fen	95 SankhyaB 57 128-141
	Constrained optimization	
	On a balanced classification rule • Kim, Hea-Jung	95 JKoreaSt 24 453-470
Gregory, J.; Lin, C. (Auth); Prvan, T. (Rev) Review of	"Constrained optimization in the calculus of variations and optimal control theory" •	95 AstrlJSt 37 250-250
Fedorov, Valery	Constrained optimization of experimental design (Disc: p149-178) • Cook, Dennis;	95 MaOpfStS 26 129-148
Keats, J. Bert Trade-off analysis versus	constrained optimization with an economic control chart model • Simpson, James R.;	95 IIETrn 27 765-772
Review of "The analysis of three-way arrays by	constrained PARAFAC methods" • Krijnen, W. P. (Auth); Klauer, K. C. (Rev)	95 JClass 12 147-149
of American contingent claims when portfolios are	constrained • Bardhan, Indrajit Synthetic replication	95 StocProc 57 149-165
Measuring welfare changes when quantity is	constrained • Breslaw, Jon A.; Smith, J. Barry	95 JBES 13 95-103
A shape	constrained smoother: Simulation study • Delecroix, M.; Simioni, M.; Thomas-Agnan, C.	95 CmpStQ 10 155-175
On the number of	constraints for balanced arrays of strength four • Chopra, D. V.; Dios, R.	93 JCISS 18 203-208
Secretary selection problem with a chance	constraint • Mukherjee, S. P.; Mandal, A.	94 JIndnStA 32 29-34
arrays of strength 2 and index 2 having maximal	constraints: Computational study • Yamamoto, S.; Fujii, Y. et al. Three-symbol orthogonal	93 JCISS 18 329-342
Stochastic differential equations with a convex	constraint • Storm, Andrew	95 Stochatx 53 241-274
Component analysis with different sets of	constraints on different dimensions • Takane, Yoshio; Kiers, Henk A. L.; de Leeuw, Jan	95 Psymtrka 60 259-280
linear least squares problem with linear inequality	constraints • Sagara, Nobuko; Fukushima, Masao A hybrid method for solving the non-	94 JOpRsJpn 38 55-69
Melenberg, Bertrand Nonnegativity	constraints and intratemporal uncertainty in a multi-good life-cycle model • Adang, Pim;	95 JApEcmtx 10 1-15
SWEEP operator for least-squares subject to linear	constraints • Neytchev, Plamen Nikolov	95 CmpStDA 20 599-609
Equality in	constraints and metric-setting measurement models • O'Brien, Robert M.; Reilly, Terence	95 StrEqMod 2 1-12
Testing for and against a set of linear inequality	constraints in a multinomial setting • Barmi, Hammou El; Dykstra, Richard	95 CandJSt 23 131-143
of principal components: Choice of normalization	constraints • Jolliffe, Ian T. Rotation	95 JAppStat 22 29-35
Nonparametric quantile estimation under order	constraints • Rojo, Javier	95 JNnparSt 5 185-200
Estimating equations, empirical likelihood and	constraints on parameters • Qin, Jing; Lawless, Jerry	95 CandJSt 23 145-159
estimation for association models with linear	constraints on the parameters (French) • Jmel, S. Least squares	95 RStApp 43/3 93-104
Lin, Shang P.; Tang, Dei-In A hierarchical active	constraints search algorithm for optimal scaling of ordered categorical responses •	95 JStCmpSm 53 197-209
Evolutionary analyses of DNA sequences subject to	constraints on secondary structure • Muse, Spencer V.	95 Genetcs 139 1429-1439
estimation procedure when modelling in terms of	constraints • Matthews, G. B.; Crowther, N. A. S. A maximum likelihood	95 SoAfrStJ 29 29-50
for Weibull distribution under expected test time	constraint • Bai, D. S. et al. Failure-censored accelerated life test sampling plans	95 RelEngSS 50 61-68
Di Crescenzo, A. G. et al. On a symmetry-based	constructive approach to probability densities for two-dimensional diffusion processes •	95 JAppProb 32 316-336
Menshikov, M. V. Topics in the	constructive theory of countable Markov chains • Fayolle, G.; Malyshev, V. A.;	95 CambridgeUniv 169p
Menshikov, M. V. (Auth) et al. Review of "Topics in the	constructive theory of countable Markov chains" • Fayolle, G.; Malyshev, V. A.;	96 ShrtBkRv 16/1 9-10
trials coordinating center statistician: 'More than a	consultant' • Williford, W. O.; Krol, W. F.; Bingham, S. F. et al. The multicenter clinical	95 AmerStat 49 221-225
The statistical	consultant in an organizational setting • Crain, Ray	94 ASAProStEd 222-227
In-house	consultation in an orthopaedic teaching hospital - STAT! • Peterson, Margaret G. E.	94 ASAProStEd 170-174
	Consulting	
multiple observations per subject • Buderer et al.	Confidence intervals and hypothesis testing for screening parameters in the case of	94 ASAProStEd 175-179
sample data • Froeschl, Karl Anton	A formal model evaluation approach to the analysis of treatment effects in paired	95 CmpStDA 19 493-517
	So many problems, so little time • Curren, Tom	95 Stats /12 7-9
	Practical skills for a statistician • Wasserstein, Sherry	95 Stats /12 14-15
	A simulation-oriented approach to disparate impact • Gardenier, Turkan K.	94 ASAProStEd 180-185
	Statistical ethics • Quinn, Linda M.	95 Stats /12 3-6
Tobias, Paul A.	Statistics in the semiconductor industry: A competitive necessity • Spencer, William J.;	95 AmerStat 49 245-249
Statistical	consulting as alchemy • Jarjoura, David; Speight, Joan D.	94 ASAProStEd 165-169
Improving the quality of journals while improving	consulting • Nick, Todd G.	94 ASAProQlPr 232-234
Statistical	consulting practicum: Opportunity for authentic learning • Derr, Janice A.	94 ASAProStEd 140-143
Spagon, Patrick D. Linked training and	consulting in the semiconductor industry: The SEMATECH approach • Reece, Jack E.;	94 ASAProStEd 228-233
Hasse diagrams in statistical	consulting and teaching • Lohr, Sharon L.	95 AmerStat 49 376-381
Tilley, B. C. A survey of biostatistical	consulting units throughout North America • Niland, J. C.; Odom-Maryon, T. L.; Lee, J.;	95 AmerStat 49 183-189
aspirin consumption among under 45 and over 45	consumers • Hsu, Margaretha M.; Wilkens, Henry T. Advertising-related	94 ASAProBiop 276-278
price sensitivity, and merchandising effects on	consumer brand choice • Allenby, Greg M.; Lenk, Peter J. Reassessing brand loyalty,	95 JBES 13 281-289
and disaggregate sector forecasting using	consumer confidence measures • Kumar, V.; Leone, Robert P.; Gaskins, John N. Aggregate	95 IntJFor 11 361-377
testing the homogeneity condition in the theory of	consumer demand • Keuzenkamp et al. Rejection without falsification: On the history of	95 JEconmtx 67 103-127
Forecasting the adoptions of new	consumer durable goods • Min, Chung-ki	94 ASAProBays 262-265
dents to the Current Population Survey (CPS) and	Consumer Expenditure Interview Survey (CEIS) • Kojetin Characteristics of nonrespon-	94 ASASrvyRsr 966-971
Household income reporting: An analysis of U.S.	Consumer Expenditure Survey data • Garner, Thesia I.; Blanciforti, Laura A.	94 JOffStat 10 69-91
between expenditures and low incomes in the U.S.	Consumer Expenditure Survey • Eltinge, J. L. et al. Assessment of reported differences	94 ASASrvyRsr 96-105
Evaluation of alternative rent estimators for the	consumer price index • Jacobson, Shawn	94 ASAProGovt 134-139
Stratification designs for the housing sample of the	consumer price index • Brown, Eugene F.; Johnson, William H.	94 ASAProGovt 140-143
"Computing elementary aggregates in the Swedish	Consumer Price Index" (V8 p129-147) • Reinsdorf, Marshall Comment on	94 JOffStat 10 103-107
"Computing elementary aggregates in the Swedish	Consumer Price Index" • Dalén, Jörgen Reply to comment on	94 JOffStat 10 107-108
Quantifying errors in the Swedish	consumer price index • Dalén, Jörgen	95 JOffStat 11 261-275
Variance estimation in the Swedish	consumer price index • Dalén, Jörgen; Ohlsson, Esbjörn	95 JBES 13 347-356
vacant units in the residential rent index of the U.S.	Consumer Price Index • Baskin, Robert M. Imputation of	94 ASASrvyRsr 382-387
Bernstein, Hank Conducting global	consumer research: Considerations and pitfalls • Mendelsohn, Susan L.; Walker, Chip;	94 ASASrvyRsr 1196-1198
Cottingham, K. L. Resource vs. ratio-dependent	consumer-resource models: A Bayesian perspective • Stow, C. A.; Carpenter, S. R.;	95 Ecology 76 1986-1990
Imputing income for an N-person	consumer unit • Chand, Nanak; Alexander, Charles H.	94 ASASrvyRsr 412-416
Wilkens, Henry T. Advertising-related aspirin	consumption among under 45 and over 45 consumers • Hsu, Margaretha M.;	94 ASAProBiop 276-278
biases, distributional effects, and aggregate	consumption • Fortin, Nicole M. Heterogeneity	95 JApEcmtx 10 287-311
Estimate of disease prevalence using drug	consumption data • Sartor, Francis; Walckiers, Denise	95 AJEpidm 141 782-787
Vinod, H. D.; Basu, Parantap Forecasting	consumption, income and real interest rates from alternative state space models •	95 IntJFor 11 217-231

Burt sub-table (French) • Nodjiram, S. Analysis of a	contingency table with negative numbers: Application of the analysis of a generalised	95 CAnDnnes 20 111-120
of departure from marginal homogeneity for	contingency tables with nominal categories • Tomizawa, Sadao Measures	95 Ststcian 44 425-439
of the marginal homogeneity model for square	contingency tables with ordered categories • Tomizawa, Sadao A generalization	95 JEdStat 20 349-360
of departure from global symmetry for square	contingency tables with ordered categories • Tomizawa, Sadao Measures	95 Behvmtrk 22 91-98
Sadao Double independence models for a 2×2	contingency table with the same classifications and their application to data • Tomizawa,	95 AllgeStA 79 140-145
Synthetic replication of American	contingent claims when portfolios are constrained • Bardhan, Indrajit	95 StocProc 57 149-165
various optimality criteria • Hedayat, A. S. et al.	Contingently and virtually balanced incomplete block designs and their efficiencies under	95 StSinica 5 575-591
Piantadosi, S. Some practical improvements in the	continual reassessment method for phase I studies • Goodman, S. N.; Zahurak, M. L.;	95 StatMed 14 1149-1161
finding study in cancer ... • Møller An extension of the	continual reassessment methods using a preliminary up-and-down design in a dose	95 StatMed 14 911-922
of some random variables connected with the	continued fraction expansion • Barbolosi, D.; Faivre, C. Metrical properties	95 IndgMatN 6 257-265
Bernadac, Evelyne Random	continued fractions and inverse Gaussian distribution on a symmetric cone •	95 JTherPrb 8 221-258
Seshadri, Vanamamalai A random	continued fraction in R^{d+1} with an inverse Gaussian distribution • Letac, Gérard;	95 Bernoull 1 381-393
of Fisher's exact test and Yates' correction for	continuity • Sahai et al. ... data involving a 2×2 contingency table: An overview	95 JBiophSt 5 43-70
Absolute	continuity of the distributions of additive arithmetic functions • Babu, Gutti Jogesh	95 SankhyaA 57 29-32
for Gaussian $S'(R^d)$-processes: Existence, path	continuity and examples • Bojdecki, Tomasz; Gorostiza, Luis G. Self-intersection local time	95 StocProc 60 191-226
Shao, Qi-Man On moduli of	continuity for local times of Gaussian processes • Csörgő, Miklós; Lin, Zheng-Yan;	95 StocProc 58 1-21
Absolute	continuity of one-sided random translations • Sato, Hiroshi; Tamashiro, Masakazu	95 StocProc 58 187-204
Ferland, René Absolute	continuity, singular measures and asymptotics for estimators • Dion, Jean-Pierre;	95 JStPlInf 43 235-242
independent increments • Yamazato, M. Absolute	continuity of transition probabilities of multidimensional processes with stationary	94 ThProbAp 39 347-354
efficacy of some control ... • Biswas et al. On a	continuous branching process and birth-death and immigration models for testing the	95 BiomtrcJ 37 235-249
The planning and analysis of experiments with	continuous or categorical response • Toutenburg Experimental design and model choice.	95 Physica:Ger 463p
The planning and analysis of experiments with	continuous or categorical response • Toutenburg Experimental design and model choice:	95 Springer:Brln:NY 463p
for stochastic reaction-diffusion equations with	continuous coefficients • Manthey, Ralf A strong existence theorem	95 Stochatx 52 15-27
Calculation of benchmark doses from	continuous data • Crump, Kenny S.	95 RiskAnal 15 79-89
estimator for randomized response sampling with	continuous distributions from a dichotomous population • Chua, T. C. et al. An alternative	95 CommStA 24 2919-2928
Principal points of univariate	continuous distributions • Zoppé, Alice	95 StatComp 5 127-132
question randomized response sampling using	continuous distributions • Singh, Sarjinder Unrelated	94 JIAgSt 46 349-361
On characterization of	continuous distributions with adjacent order statistics • Franco, M.; Ruiz, J. M.	95 MaOpfStS 26 375-385
Datta, Somnath; McCormick, William P. Some	continuous Edgeworth expansions for Markov chains with applications to bootstrap •	95 JMultiAn 52 83-106
case-control studies using dichotomous analysis of a	continuous exposure variable • Delpizzo et al. ... risk estimate and statistical power in	95 IJEpidem 24 851-862
Approximation to the exit probability of a	continuous Gaussian process over a U-shaped boundary of increasing ... • Balabushkin	95 JAppProb 32 429-442
Severtson, Forrestt The Elston-Stewart algorithm for	continuous genotypes and environmental factors • Elston, Robert C.; George, Varghese T.;	92 HumHrdty 42 16-27
decades of family change: Secondary analysis of	continuous government surveys • Harrop, Anne; Plewis, Ian Two	95 JRSS-A 158 91-106
Xue, Zhiming; Gulas, Charles S.	Continuous improvement in college teaching: An application of statistical tools •	95 ASQCTekCon 267-275
Safety measurement: Towards	continuous improvement • St. Clair, Joseph F.	95 ASQCTekCon 150-158
Wild, C. J.	Continuous improvement of teaching: A case study in a large statistics course •	95 IntStRvw 63 49-68
C. W. (Rev) Review of "Basic statistics tools for	continuous improvement (Third Edition)" • Kiemele, M. J.; Schmidt, S. R. (Auth); Champ,	95 JQualTek 27 174-175
Sampath; Vasantha, P.	Continuous inspection of Markov processes with a clearance interval • Kumar, V. S.	95 JAppStat 22 427-432
A time-	continuous Markov chain interest model with applications to insurance • Norberg, Ragnar	95 AplStMDA 11 245-256
A uniform CLT for	continuous martingales • Bae, Jongsig; Levental, Shlomo	95 JKoreaSt 24 225-231
Rust, Keith F.	Continuous measurement alternatives to census data collection (Disc: p90-95) •	94 ASASrvyRsr 80-89
A	continuous metric scaling solution for a random variable • Cuadras, C. M.; Fortiana, J.	95 JMultiAn 52 1-14
of variable-size group testing and its related	continuous models • Chen, C. L.; Swallow, W. H. Sensitivity analysis	95 BiomtrcJ 37 173-181
A minimax optimal estimator for	continuous monotone densities • Datta, Somnath	95 JStPlInf 46 183-193
of empirical Bayes estimator for the parameter of	continuous one parameter exponential families (Chinese) • Xuc ... and convergence rates	95 ChJApPSt 10 35-45
rates of monotone empirical Bayes tests for the	continuous one-parameter exponential family • Karunamuni, R. J. et al. On convergence	95 StatDecn 13 181-192
Regression models for a bivariate discrete and	continuous outcome with clustering • Fitzmaurice, Garrett M.; Laird, Nan M.	95 JASA 90 845-852
Binary regression with	continuous outcomes • Suissa, Samy; Blais, Lucie	95 StatMed 14 247-255
stage estimation of structural equation models with	continuous and polytomous variables • Lee, Sik-Yum; Poon, Wai-Yin; Bentler, P. M. A two-	95 BJMaStPs 48 339-358
A kind of rank test about intersection point of two	continuous populations (Chinese) • Li, Ningning; Chen, Guijing	95 KToChina 39 1543-1546
error in the analysis of interaction effects between	continuous predictors using multiple regression: Multiple ... • Jaccard et al. Measurement	95 PsychBu 117 348-357
al. Multivariate statistical methods for monitoring	continuous processes: Assessment of discrimination power of disturbance ... • Raich et	95 Chemolab 30 37-48
for the design of 2^p factorial experiments on	continuous processes • Saunders, I. W.; Eccleston, J. A.; Martin, R. J. An algorithm	95 AstrlJSt 37 353-365
	Continuous quality improvement in higher education • Hogg, Robert V.; Hogg, Mary C.	95 IntStRvw 63 35-48
applications • Biemer, Paul; Caspar, Rachel	Continuous quality improvement for survey operations: Some general principles and	94 JOffStat 10 307-326
parameters and quantiles of distributions of	continuous random variables • Castillo, Enrique; Hadi, Ali S. A method for estimating	95 CmpStDA 20 421-439
How to perform	continuous sampling (Second edition) • Stephens, Kenneth S.	95 ASQualCntrl:WI 73p
Local times of functions of	continuous semimartingales • Ouknine, Youssef; Rutkowski, Marek	95 StocAnAp 13 211-231
	Continuous speech recognition by a network of hidden Markov models • Pavešić, Nikola	95 JCmpInT 3 193-205
of the maximum pseudo-likelihood estimator of	continuous state space Gibbsian processes • Mase, Shigeru Consistency	95 AnlsApPr 5 603-612
From discrete to	continuous stochastic calculus • Cutland, Nigel J.; Kopp, P. Ekkehard; Willinger, Walter	95 Stochatx 52 173-192
sampling point for the quality assessment of	continuous streams • Barnett, Neil S.; Gomm, Ian S.; Armour, Len Location of the optimal	95 AstrlJSt 37 145-152
An error bound for a	continuous time approximation of a time-sharing queue • Falin, G.	95 CommStSM 11 163-169
vary over time • Wilson, K. R.; Anderson, D. R.	Continuous-time capture-recapture population estimation when capture probabilities	95 EnvEcoSt 2 55-69
in general sample spaces and random censoring in	continuous time • Jacobsen, Martin; Keiding, Niels Coarsening at random	95 AnlsStat 23 774-786
vexification over time • Eaves et al. A discounted-cost	continuous-time flexible manufacturing and operator scheduling model solved by decon-	95 PrEngInf 9 65-98
options of both European and American types. II.	Continuous time. • Shiryaev, A. N.; Kabanov, Yu. M. et al. Toward the theory of pricing of	94 ThProbAp 39 61-102
Gouweleeuw, José M. A	continuous-time job search model: General renewal processes • Boshuizen, Frans A.;	95 CommStSM 11 349-369
Coupling and harmonic functions in the case of	continuous time Markov processes • Cranston, Michael; Greven, Andreas	95 StocProc 60 261-286
ordination and strong differential subordination for	continuous-time martingales and related sharp inequalities • Wang, Gang Differential sub-	95 AnlsProb 23 522-551
sampling schemes for the spectral estimation of	continuous time processes • Lii, Keh-Shin; Masry, Elias On the selection of random	95 JTimSrAn 16 291-311
orem á la Feller for nonintegrable strictly stationary	continuous time processes • Rosalsky, A.; Stoyanov, J.; Presnell, B. An ergodic-type the-	95 StocAnAp 13 555-572
Keunkwan Analysis of a	continuous-time proportional hazard model using discrete duration data • Ryu,	95 EcomtrRv 14 299-313
H. Analytical score function for irregularly sampled	continuous time stochastic processes with control variables and missing values • Singer,	95 EconTher 11 721-735
of analog and digital simulation on some	continuous time systems • Yeung, Iris; Lam, James Comparison	94 ASAProStEd 186-191
On the approximation of	continuous time threshold ARMA processes • Brockwell, P. J.; Stramer, O.	95 AnInStMa 47 1-20
On amarts with	continuous time • Dzhvarsheishvili, I. A.	94 ThProbAp 39 512-519
evaluation for one polychotomous trait and several	continuous traits with missing data and unequal models • Hoeschele et al. ... genetic	95 JAnmlSci 73 1609-1627
Balakrishnan, N. (Auth); Chatfield, C. (Rev) Review of	"Continuous univariate distributions, Volume 1 (Second edition)" • Johnson, N. L.; Kotz, S.;	95 Ststcian 44 544-545
Balakrishnan, N. (Auth); Omori, Y. (Rev) Review of	"Continuous univariate distributions, Volume 1 (Second edition)" • Johnson, N. L.; Kotz, S.;	95 JASA 90 1490-1491
Kotz, S.; Balakrishnan, N. (Auth) et al. Review of	"Continuous univariate distributions, Volumes 1 and 2 (Second edition)" • Johnson, N. L.;	95 Biomtrcs 51 1582-1583
Balakrishnan, N. (Auth); Famoye, F. (Rev) Review of	"Continuous univariate distributions, Volume I (Second edition)" • Johnson, N. L.; Kotz, S.;	95 Technmcs 37 466-466
Balakrishnan, N. (Auth); Tricker, A. R. (Rev) Review of	"Continuous univariate distributions: Volume I (Second edition)" • Johnson, N. L.; Kotz, S.;	95 JAppStat 22 323-323
Kotz, Samuel; Balakrishnan, N.	Continuous univariate distributions. Volume 2 (Second edition) • Johnson, Norman L.;	95 Wiley:NY:UK 719p

Power of detecting the effect of a	continuous variable in a Cox model • Priore, Roger L.; Brady, Mark F.; Lawrence, David D.	94 ASAProBiop 200-205
Parameters and quantiles estimation for	continuously distributed random variables • Castillo, Enrique; Hadi, Ali S.	94 ASAProStCp 284-289
Single linkage clustering and	continuum percolation • Penrose, Mathew D.	95 JMultiAn 53 94-109
Equality of critical densities in	continuum percolation • Meester, Ronald	95 JAppProb 32 90-104
Brownian motion on the	continuum tree • Krebs, W. B.	95 ProbTher 101 421-433
	Contorted uniform and Pareto distributions • Dimitrov, Boyan; von Collani, Elart	95 StPrLet 23 157-164
mass concentrations and estimating density	contour clusters - An excess mass approach • Polonik, Wolfgang Measuring	95 AnlsStat 23 855-881
How reliable are	contour curves? Confidence sets for level contours • Lindgren, Georg; Rychlik, Igor	95 Bernoull 1 301-319
A Markovian model for	contour grouping • Urago, Sabine; Zerubia, Josiane; Berthod, Marc	95 PattrnRc 28 683-693
are contour curves? Confidence sets for level	contours • Lindgren, Georg; Rychlik, Igor How reliable	95 Bernoull 1 301-319
Flavio F.; de Macedo, Marcia M. A. Feasibility of	contour mapping epidemiological data with missing values (Disc: p697-699) • Nobre,	95 StatMed 14 605-613
	Contour plotting	
regressions • Hunka, Steve	Identifying regions of significance in ANCOVA problems having non-homogeneous	95 BJMaStPs 48 161-188
and solutions with an application to the demand for	contraceptive use in Tunisia • Bollen et al. ... endogenous explanatory variables: Tests	95 Dmogrphy 32 111-131
Peña, V. H.; Montgomery-Smith, S. J.; Szulga, Jerzy	Contraction and decoupling inequalities for multilinear forms and U-statistics • de la	94 AnlsProb 22 1745-1765
Franklin A. Approximate confidence intervals for	contrasts in the balanced three factor mixed model • Mosier, Michael C.; Graybill,	95 BiomtrcJ 37 273-297
Robust confidence intervals for	contrasts based upon a likelihood ratio test • Bachmaier, M.; Precht, M.	95 StHefte 36 215-236
Winkler, Robert L. Screening probability forecasts:	Contrasts between choosing and combining • Clemen, Robert T.; Murphy, Allan H.;	95 IntJFor 11 133-145
Partitioning error spaces in linear models (error	contrasts) • Rady, El-Houssainy A.	95 EgpStJ 39 51-64
The method of sieves and minimum	contrast estimators • Van de Geer, S.	95 MaMethSt 4 20-38
single branches of a phylogeny using evolutionary	contrasts • McPeek, Mark A. Testing hypotheses about evolutionary change on	95 ANatrls 145 686-703
information matrices for estimating a given set of	contrasts • Kao, L.-J.; Yates, P. K.; Lewis, S. M.; Dean, A. M. Approximate	95 StSinica 5 593-598
unreplicated factorials using the n smallest	contrasts • Kasperski, William J.; Schneider, Helmut; Johnson, Keith L. Analyzing	95 JQualTek 27 377-386
	Control	
	• see also adaptive –, case –, optimal –, process –, quality –, stochastic –	
On the optimal maintenance of systems and	control of arrivals in queues • Katehakis, Michael N.; Melolidakis, Costis	95 StocAnAp 13 137-164
Model-based and model-free	control of autocorrelated processes • Runger, George C.; Willemain, Thomas R.	95 JQualTek 27 283-292
trol chart • Rigdon An integral equation for the in-	control average run length of a multivariate exponentially weighted moving average con-	95 JStCmpSm 52 351-365
procedures for comparing several treatments with a	control based on binary data • Chuang-Stein, C.; Tong, D. M. Multiple comparisons	95 StatMed 14 2509-2522
Hu, Albert; Sachs, Emanuel Run by run	control: A Bayesian approach to the control of semiconductor fabrication processes •	94 ASAProPEng 245-254
ical Bayes estimator of the distribution of historical	controls in carcinogen bioassay • Krewski, D.; Smythe, R. T. et al. Bootstrapping an empir-	95 JStPlInf 43 215-228
Wymer, Colin Disclosure	control for census microdata • Skinner, Chris; Marsh, Catherine; Openshaw, Stan;	94 JOffStat 10 31-51
	Control chart	
Sheng	A cumulative score control scheme for detecting process shifts • Cheng, Chuen-	95 CommStA 24 755-774
LeRoy A.	Determining sample size for lower confidence limits for C_p and C_{pk} • Franklin,	94 ASAProQlPr 23-27
P.; Luceño, Alberto	Discrete proportional-integral control with constrained adjustment • Box, George E.	95 Ststcian 44 479-495
Erwin M.; Davis, Darwin J.; McWilliams, Thomas P.	Economic, statistical, and economic-statistical design of attribute charts • Saniga,	95 JQualTek 27 56-73
Emmanuel	Estimating the current mean of a process subject to abrupt changes • Yashchin,	95 Technmcs 37 311-323
Charles P.	Geometric Q charts for high quality processes (Disc: p316-343) • Quesenberry,	95 JQualTek 27 304-315
model • Yashchin, Emmanuel	Likelihood ratio methods for monitoring parameters of a nested random effect	95 JASA 90 729-738
control • Bailey, M.; Bowden, J.; Callahan, A. J.	Managing ship performance of naval gunfire support using statistical process	94 MilOpRes 1 5-11
C.; Willemain, Thomas R.	Model-based and model-free control of autocorrelated processes • Runger, George	95 JQualTek 27 283-292
MacGregor, John F.	Multivariate SPC charts for monitoring batch processes • Nomikos, Paul;	95 Technmcs 37 41-59
	New approach to process control • Bluvband, Zigmund; Grabov, Pavel; Ingman, Dov	95 ASQCTekCon 369-377
Yakir, Benjamin	A note on the run length to false alarm of a change-point detection policy •	95 AnlsStat 23 272-281
	Optimal estimation of sigma for X charts • Baran, Gary S.	94 ASAProQlPr 321-326
Roberts, Harry V.	The problem of misplaced control limits (Disc: p289-306) • Alwan, Layth C.;	95 ApplStat 44 269-278
	Proper and effective training of a pattern recognizer for cyclic data • Hwarng, H. Brian	95 IIETrn 27 746-756
	On properties of binomial Q charts for attributes • Quesenberry, Charles P.	95 JQualTek 27 204-213
Charles P.	On properties of Poisson Q charts for attributes (Disc: p316-343) • Quesenberry,	95 JQualTek 27 293-303
	On properties of Q charts for variables • Quesenberry, Charles P.	95 JQualTek 27 184-203
p645-658) • Lai, Tze Leung	Sequential changepoint detection in quality control and dynamical systems (Disc:	95 JRSS-B 57 613-644
	SPC: A new look at what to teach • Greer, Daniel A.; Halteman, Edward J.	94 ASAProBays 153-156
	Statistical education for quality managers • Martin, Margaret; Suber, Marsha	94 ASAProBays 148-149
Runger, George C.	Statistical process control scheme design • Keats, J. Bert; Miskulin, John D.;	95 JQualTek 27 214-225
Herbert; Plante, Robert; Tang, Jen	Statistical process control via the subgroup bootstrap • Seppala, Tomi; Moskowitz,	95 JQualTek 27 139-153
	Three dimensional modeling of quality concepts • Casey, William E.	95 ASQCTekCon 1134-1141
	• see also names of specific types of charts such as Shewhart –	
EWMA	control charts for AR(1) model using Bayes estimators • Ye, Keying	94 ASAProBays 64-68
A	control chart for attributes and its applications • Quarshie, Leonard B.; Shindo, Hisakazu	95 ASQCTekCon 378-388
estimation of the weighting factor for EWMA	control charts from autocorrelated data • Hembree, G. Barry Recursive	94 ASAProQlPr 39-43
multivariate exponentially weighted moving average	control chart • Rigdon, S. E. An integral equation for the in-control average run length of a	95 JStCmpSm 52 351-365
multivariate exponentially weighted moving average	control chart • Rigdon, Steven E. A double-integral equation for the average run length of a	95 StPrLet 24 365-373
using exponentially weighted moving average	control charts • Gan, F. F. Joint monitoring of process mean and variance	95 Technmcs 37 446-453
Bakir, Saad Nonparametric quality	control charts based on the sign statistic • Amin, Raid W.; Reynolds, Marion R., Jr;	95 CommStA 24 1597-1623
on "Run-length distribution of special-cause	control charts for correlated processes" (V36 p3-27) • Yashchin, Emmanuel Comment	95 Technmcs 37 243-243
ment on "Run-length distribution of special-cause	control charts for correlated processes" • Wardell, D. G.; Moskowitz, H. et al. Reply to com-	95 Technmcs 37 243-245
Using	control charts to corroborate bribery in jai alai • Charnes, John M.; Gitlow, Howard S.	95 AmerStat 49 386-389
	Control charts for customer satisfaction survey data • Kushler, Robert	94 ASAProQlPr 334-340
Average run length of a moving average	control chart to detect shifts in the mean of a process (French) • Niang, N.; Saporta, G.	95 RStApp 43/3 5-20
tistically constrained economic design of the EWMA	control chart • Montgomery, D. C.; Torng, J. C.-C.; Cochran, J. K.; Lawrence, F. P. Sta-	95 JQualTek 27 250-256
menting statistically constrained economic EWMA	control charts • Torng, J. C.-C.; Cochran, J. K.; Montgomery, D. C.; Lawrence, F. P. Imple-	95 JQualTek 27 257-264
menting statistically constrained economic EWMA	control charts" (V27 p257-264) • Torng, J. C.-C.; Cochran, J. K. et al. Correction to "Imple-	95 JQualTek 27 402-402
Effective use of	control charts in food processing • Perry, Robert L.	95 ASQCTekCon 588-597
Decomposition of T^2 for multivariate	control chart interpretation • Mason, Robert L.; Tracy, Nola D.; Young, John C.	95 JQualTek 27 99-108
An application of variable sampling interval	control charts • Baxley, Robert V., Jr	95 JQualTek 27 275-282
Evaluating properties of variable sampling interval	control charts • Reynolds, Marion R., Jr	95 SqtlAnly 14 59-97
Cumulative sum	control charts for the inverse Gaussian distribution • Nabar, S. P.; Bilgi, Shobha	94 JIndnStA 32 9-14
Small sample uni- and multi-variate	control charts for means • Scholz, F. W.; Tosch, T. J.	94 ASAProQlPr 17-22
versus constrained optimization with an economic	control chart model • Simpson, James R.; Keats, J. Bert Trade-off analysis	95 IIETrn 27 765-772
A cumulative sum	control chart for monitoring process variance • Chang, T. C.; Gan, F. F.	95 JQualTek 27 109-119
Correction to "A cumulative sum	control chart for monitoring process variance" (V27 p109-119) • Chang, T. C.; Gan, F. F.	95 JQualTek 27 402-402
Woodall, William H. The performance of	control charts for monitoring process variation • Lowry, Cynthia A.; Champ, Charles W.;	95 CommStB 24 409-437
Multi-process	control charts • Lipscomb, John R.	94 ASAProPEng 314-319

On the use of an SPSA-based model-free	controller in quality improvement • Rezayat, Fahimeh	95 Autmatca 31 913-915
Eaves, Lindon J. Estimating and	controlling for the effects of volunteer bias with pairs of relatives • Neale, Michael C.;	93 BehavGen 23 271-278
testing • Benjamini, Yoav; Hochberg, Yosef	Controlling the false discovery rate: A practical and powerful approach to multiple	95 JRSS-B 57 289-300
C.	Controlling the type I and type II errors in mapping quantitative trait loci • Jansen, Ritsert	94 Genetcs 138 871-881
Weijers, E. P.; van Delden, A. et al. Characteristics of	convective turbulence in the surface layer investigated by principal component analysis •	95 JApMetrl 34 528-541
	Convergence	
	• see also rate of –, speed of –	
S. Exponential	convergence for attractive reversible subcritical nearest particle systems • Mountford, T.	95 StocProc 59 235-249
Arsian, Olcay; Constable, Patrick D. L.; Kent, John T.	Convergence behavior of the EM algorithm for the multivariate t-distribution •	95 CommStA 24 2981-3000
Lecoutre, Jean-Pierre; Ould-Saïd, Elias	Convergence of the conditional Kaplan-Meier estimate under strong mixing •	95 JStPlInf 44 359-369
Poisson laws • Alekperov, Sh. O.; Kruglov, V. M.	Convergence of convolutions of concentration functions to degenerate, normal, and	94 ThProbAp 39 197-216
Gibbs sampler	convergence criteria • Zellner, Arnold; Min, Chung-ki	95 JASA 90 921-927
Di Masi, G. B. et al. A first order approximation for the	convergence of distributions of the Cox processes with fast Markov switchings •	95 Stochatx 54 211-219
in the Gibbs sampler: Issues of efficiency and	convergence • Dellaportas, Petros Random variate transformations	95 StatComp 5 133-140
Zabczyk, J.	Convergence to equilibrium for classical and quantum spin systems • Prato, G. Da;	95 ProbTher 103 529-552
Hawkins, Douglas M.	Convergence of the feasible solution algorithm for least median of squares regression •	95 CmpStDA 19 519-538
truncation model with application to hazard ... • Gu	Convergence of increments for cumulative hazard function in a mixed censorship-	95 StPrLet 23 135-139
	Convergence of independent particle systems • Hoffman, John R.; Rosenthal, Jeffrey S.	95 StocProc 56 295-305
	Convergence in international output • Bernard, Andrew B.; Durlauf, Steven N.	95 JApEcmtx 10 97-108
	Convergence of the iterative proportional fitting procedure • Rüschendorf, Ludger	95 AnlsStat 23 1160-1174
On the l^p	convergence of martingales associated with coverage (French) • Hua, Fan Ai	95 JAppProb 32 668-678
Hero, Alfred O.; Fessler, Jeffrey A.	Convergence in norm for alternating expectation-maximization (EM) type algorithms •	95 StSinica 5 41-54
Assessing and accelerating Gibbs sampler	convergence for the ordinal probit • Cowles, Mary Kathryn	94 ASAProBays 52-57
The present value of a series of cashflows:	Convergence in a random environment • Cairns, Andrew J. G.	95 AstinB 25 81-94
	Convergence of random sequences with independent random indices I. • Korolev, V. Yu.	94 ThProbAp 39 282-297
Lin, Guoying Global	convergence rates of B-spline M-estimators in nonparametric regression • Shi, Peide;	95 StSinica 5 303-318
Wei, Laisheng; Zhang, Shunpu The	convergence rates of empirical Bayes estimation in a multiple linear regression model •	95 AnInStMa 47 81-97
parameter ... • Xuc Asymptotically optimality and	convergence rates of empirical Bayes estimator for the parameter of continuous one	95 ChJApPSt 10 35-45
of prediction based selection criteria and the	convergence rates of estimators • Altman, Naomi; Léger, Christian On the optimality	94 ASAProStCp 112-117
Kong, Augustine Covariance structure and	convergence rate of the Gibbs sampler with various scans • Liu, Jun S.; Wong, Wing H.;	95 JRSS-B 57 157-169
	Convergence rates for Markov chains • Rosenthal, Jeffrey S.	95 SIAMRvw 37 387-405
Minorization conditions and	convergence rates for Markov chain Monte Carlo • Rosenthal, Jeffrey S.	95 JASA 90 558-566
Correction to "Minorization condition and	convergence rates for Markov chain Monte Carlo" (V90 p558-566) • Rosenthal, Jeffrey	95 JASA 90 1136-1136
exponential family • Karunamuni, R. J.; Yang, H. On	convergence rates of monotone empirical Bayes tests for the continuous one-parameter	95 StatDecn 13 181-192
(Chinese) • Hong, Shengyan; Cheng, Ping	Convergence rates for parameter estimation in semi-parametric regression models	95 ChJApPSt 10 62-71
• Miao, Baiqi; Zhang, Shuguang The exponential	convergence rates of the projection residue of U-statistics and their application (Chinese)	95 ChJApPSt 10 253-264
Probability inequalities for likelihood ratios and	convergence rates of sieve MLEs • Wong, Wing Hung; Shen, Xiaotong	95 AnlsStat 23 339-362
Lower estimates of the	convergence rate for U-statistics • Bentkus, Vidmantas; Götze, Friedrich; Zitikis, Ričardas	94 AnlsProb 22 1707-1714
Limit theorems involving restricted	convergence • Rossberg, H.-J.	94 ThProbAp 39 298-314
Another proof of a slow	convergence result of Birgé • Devroye, Luc	95 StPrLet 23 63-67
Jordan, Michael I.; Xu, Lei	Convergence results for the EM approach to mixtures of experts architectures •	95 NuralNet 8 1409-1431
Line transects, covariance functions and set	convergence • Cabo, A. J.; Baddeley, A. J.	95 AdvAppPr 27 585-605
	Convergence to stationary state for a Markov move-to-front scheme • Rodrigues, Eliane R.	95 JAppProb 32 768-776
Rao, C. R.; Zhao, L. C.	Convergence theorems for empirical cumulative quantile regression functions •	95 MaMethSt 4 81-91
p81-91) • Rao, C. R.; Zhao, L. C. Correction to	"Convergence theorems for empirical cumulative quantile regression functions" (V4	95 MaMethSt 4 359-359
regression estimation (Chinese) • He, Z.; He, L. The	convergence theorem and limit of the iterative algorithm for choosing k value in ridge	95 AMaAPknC 17 59-64
An L_2	convergence theorem for random affine mappings • Burton, Robert M.; Rösler, Uwe	95 JAppProb 32 183-192
Aspandiiarov, Sanjar On the	convergence of 2-dimensional Markov chains in quadrants with boundary reflection •	95 Stochatx 53 275-303
Skorohod's theorem and	convergence of types • Fazli, K.; Behboodian, J.	95 StPrLet 24 243-244
Hsing, T. Point process and partial sum	convergence for weakly dependent random variables with infinite variance • Davis, R. A.;	95 AnlsProb 23 879-917
Kratina, Ioana; Daffer, Peter	Convergence of weighted sums and laws of large numbers in $D([0,1];E)$ • Şchiopu-	95 JMultiAn 53 279-292
estimated empirical process • Viollaz, Aldo J.	Convergence in weighted supremum norms of the Skorokhod representation of the	95 CommStA 24 2829-2839
Ye, Yinyu; Woodworth, George G. A	convergent algorithm for quantile regression with smoothing splines • Bosch, Ronald J.;	95 CmpStDA 19 613-630
Charles S.; Coleman, S. C. The criteria for	convergent and discriminant validity in a multitrait-multimethod matrix • Reichardt,	95 MultBehR 30 513-538
proximation with averaging and feedback: Rapidly	convergent "on-line" algorithms • Kushner, Harold J.; Yang, Jichuan Stochastic ap-	95 IEEEAuCn 40 24-34
A note on the	convergent rates of M-estimates for a partly linear model • Shi, Peide; Li, Guoying	95 MaOpfStS 26 27-47
behavior of diffusion processes and their weakly	convergent sequences • Portenko ... idea concerning the methods for investigating local	95 ThPrMaSt 50 7-22
tionnaires: The survey interview and the logic of	conversation (Disc: p169-177) • Schwarz, Norbert What respondents learn from ques-	95 IntStRvw 63 153-168
tionnaires: The survey interview and the logic of	conversation" • Cantor, David Discussion of "What respondents learn from ques-	95 IntStRvw 63 173-177
tionnaires: The survey interview and the logic of	conversation" • Singer, Eleanor Discussion of "What respondents learn from ques-	95 IntStRvw 63 169-172
A	conversation with Frank Proschan • Hollander, Myles; Marshall, Albert W.	95 StatSci 10 118-133
A	conversation with Hirotugu Akaike • Findley, David F.; Parzen, Emanuel	95 StatSci 10 104-117
A	conversation with Joe Gani • Heyde, Chris	95 StatSci 10 214-230
problems for geometric probabilities involving	convex bodies • Bauer, Christina; Schneider, Rolf Extremal	95 AdvAppPr 27 20-34
for stochastic differential equations with reflecting	convex boundaries • Pettersson, Roger Approximations	95 StocProc 59 295-308
Minimax tests for	convex cones • Dümbgen, Lutz	95 AnInStMa 47 155-165
Stochastic differential equations with a	convex constraint • Storm, Andrew	95 Stochatx 53 241-274
Dynamic scheduling with	convex delay costs: The generalized $c\mu$ rule • van Mieghem, Jan A.	95 AnlsApPr 5 809-833
	Convex design theory	
Fedorov, Valery	Constrained optimization of experimental design (Disc: p149-178) • Cook, Dennis;	95 MaOpfStS 26 129-148
Itô's formula for multidimensional	convex functions • Qingji, Yang	95 Stochatx 52 157-171
A	convex hull-based estimator of home-range size • Worton, Bruce J.	95 Biomtrcs 51 1206-1215
convex domain • Nagaev, A. V. Some properties of	convex hulls generated by homogeneous Poisson point processes in an unbounded	95 AnInStMa 47 21-29
tent and the interior point proportion of a random	convex hull • Massé, B. Invariance principle for the deviation between the probability con-	95 JAppProb 32 1041-1047
in multivariate stratified design: An alternative to	convex programming • Rahim, M. A.; Jocelyn, Wisner Sample allocation	94 ASASrvyRsr 689-692
Estimates of some probabilities in multidimensional	convex records • Kałuszka, M.	95 ZastApMa 23 1-11
elementary renewal theorem for random compact	convex sets • Molchanov, Ilya S.; Omey, Edward; Kozarovitzky, Eugene An	95 AdvAppPr 27 931-942
passage times for Markov processes with monotone	convex transition kernels • Li, Haijun; Shaked, Moshe On the first	95 StocProc 58 205-216
Asymptotic design-cum-model approach for	convex weighting of direct and indirect components of small domain ... • Chaudhuri et al.	95 CommStA 24 2803-2813
Shanthikumar, J. G.; Tong, Y. L. Parametric Schur	convexity and arrangement monotonicity properties of partial sums • Shaked, M.;	95 JMultiAn 53 293-310
Pinedo, Michael A	convexity property of a Markov-modulated queueing loss system • Du, Charles;	95 PrEngInf 9 193-199

A family of partially	correlated Poisson models for overdispersion • Luceño, Alberto	95 CmpStDA 20 511-520
distribution of special-cause control charts for	correlated processes" (V36 p3-27) • Yashchin, Emmanuel Comment on "Run-length	95 Technmcs 37 243-243
distribution of special-cause control charts for	correlated processes" • Wardell, D. G. et al. Reply to comment on "Run-length	95 Technmcs 37 243-245
A method for the generation of	correlated random processes • Andronico, Maurizio; Casale, Salvatore; La Corte, Aurelio	95 JCmpInT 3 9-18
Curvature adjusted optimal nonlinear designs for	correlated responses • Hughes-Oliver, Jacqueline M.	94 ASAProBuEc 111-116
the parameters of the multilinear model with equi-	correlated responses • Khan, Shahjahan Inference about	94 JStatRsr 28 163-176
archical analysis of SUR models with extensions to	correlated serial errors and time-varying parameter models • Chib, S.; Greenberg, E. Hier-	95 JEconmtx 68 339-360
On permissible correlations for locally	correlated stationary processes • Donahue, Rafe M. J.; Brockwell, Peter J.; Davis, Richard	95 StPrLet 22 49-53
Product-limit survival functions with	correlated survival times • Williams, Rick L.	94 ASAProBiop 211-216
Product-limit survival functions with	correlated survival times • Williams, Rick L.	95 LiDaAnal 1 171-186
Interval mapping of quantitative trait loci employing	correlated trait complexes • Korol, Abraham B.; Ronin, Yefim I.; Kirzhner, Valery M.	95 Genetcs 140 1137-1147
medical method comparison ... • Rifkin Effects of	correlated and uncorrelated measurement error on linear regression and correlation in	95 StatMed 14 789-798
Removing cancer when it is	correlated with other causes of death • Carriére, Jacques F.	95 BiomtrcJ 37 339-350
redesign (Disc: p379-381) • LeClere, F. B. et al. Social	correlates of response bias in the National Health Interview Survey: Lessons for the	94 ASASrvyRsr 373-378

Correlation

	Bayesianism and diverse evidence • Wayne, Andrew	95 PhilSci 62 111-121
neurobehavioral phenotypes • Crusio, Wim E.	Bi- and multi-variate analyses of diallel crosses: A tool for the genetic dissection of	93 BehavGen 23 59-68
	Charles Spearman's contributions to test theory • Levy, Philip	95 BJMaStPs 48 221-235
	Correlation of interval-valued intuitionistic fuzzy sets • Bustince, H.; Burillo, P.	95 FuzzySet 74 237-244
matrix • Reichardt, Charles S.; Coleman, S. C.	The criteria for convergent and discriminant validity in a multitrait-multimethod	95 MultBehR 30 513-538
Massé, Bruno; Viano, Marie-Claude	Explicit and exponential bounds for a test on the coefficient of an AR (1) model •	95 StPrLet 25 365-371
study of complex characters • Blizard, David A.	Recombinant-inbred strains: General methodological considerations relevant to the	92 BehavGen 22 621-634
	On record indices and record times • Borovkov, Konstantin; Pfeifer, Dietmar	95 JStPlInf 45 65-79
	On some inequalities related to the range of some statistical indices • Benedetti, Carlo	94 Metron 52/1 189-215
	The use of Fisher's Z in Schmidt-Hunter-type meta-analyses • Law, Kenneth S.	95 JEdStat 20 287-306
	• see also names of specific types of correlation such as partial –	
Size and power of test statistics for gene	correlation in 2×2 contingency tables • Mannan, M.; Nassar, R.	95 BiomtrcJ 37 409-433
Thelin, B. et al. Classification of Estradurin batches:	Correlation between 31P NMR and a biological duration test for batch approval •	95 Chemolab 27 135-145
Olkin, Ingram; Viana, Marlos	Correlation analysis of extreme observations from a multivariate normal distribution •	95 JASA 90 1373-1379
Multiple subclass pattern recognition: A maximin	correlation approach • Avi-Itzhak, Hadar I.; Van Mieghem, Jan A.; Rub, Leonardo	95 IEEEPAMI 17 418-431
A note on the attenuation of	correlation • Bedrick, Edward J.	95 BJMaStPs 48 271-280
Inherent bias in	correlation averaged images • Woodward, J. T.; Kono, C.; Madsen, L. L.; Zasadzinski, J. A.	95 JMicro 178 86-92
Cléroux, Robert; Lazraq, Aziz; Lepage, Yves Vector	correlation based on ranks and a nonparametric test of no association between vectors •	95 CommStA 24 713-733
of ... • Fill et al. L moment and probability plot	correlation coefficient goodness-of-fit tests for the Gumbel distribution and impact	95 WaterReR 31 225-229
Hammond, Sean M.; Lienert, Gustav A. Modified phi	correlation coefficients for the multivariate analysis of ordinally scaled variables •	95 EdPsMeas 55 225-236
estimate and a confidence interval for the population	correlation coefficient • Bushman et al. ... coefficients and vote counts to obtain an	95 PsychBu 117 530-546
population • Weerakkody, G. J. et al. Estimating the	correlation coefficient in the presence of correlated observations from a bivariate normal	95 CommStA 24 1705-1719
and D-optimality for complete block designs under	correlations • Bhaumik, Dulal K. Majorization	95 JRSS-B 57 139-143
Wesolowsky, G. O. Estimating the	correlation between destructively measured variables using proof-loading • Steiner, S. H.;	95 Technmcs 37 94-101
	Correlation dimension estimates of global and local temperature data • Wang, Qiang	95 JApMetrl 34 2556-2564
Selection index based on genetic	correlations among environments • Van Sanford, D. A.; Pfeiffer, T. W.; Cornelius, P. L.	93 CropSci 33 1244-1248
McShane, Phil Comment on "Spurious	correlation and the fallacy of the ratio standard revisited" (V156 p379-392) •	95 JRSS-A 158 621-623
M.; Holder, Roger L. Comment on "Spurious	correlation and the fallacy of the ratio standard revisited" (V156 p379-392) • Nevill, Alan	95 JRSS-A 158 619-621
Reply to comments on "Spurious	correlation and the fallacy of the ratio standard revisited" • Kronmal, Richard A.	95 JRSS-A 158 623-625
Marked point process: Using	correlation functions to explore a spatial data set • Goulard, M.; Pagés, L.; Cabanettes, A.	95 BiomtrcJ 37 837-853
properties of the estimation of the inverse	correlation function (Chinese) • Chen, Min; Chang, Xuejiang The asymptotic	95 AMaAPknC 17 23-45
Simulation studies of Poisson	correlation functions • Guerrero-Cusumano, José-Luis	94 ASAProPEng 157-162
Additive and non-additive genetic	correlations. I. Properties of sampling distributions • Hohls, T.; Clarke, G. P. Y.	95 BiomtrcJ 37 919-935
Additive and non-additive genetic	correlations. II. Application to data obtained from a 12×12 diallel cross • Hohls, T. et al.	95 BiomtrcJ 37 937-943
Thomas W.; Woolson, Robert F. Using Kendall's τ_b	correlations to improve variable selection methods in case-control studies • O'Gorman,	95 Biomtrcs 51 1451-1460
with instrumental variables estimation when the	correlation between the instruments and the endogenous ... • Bound et al. Problems	95 JASA 90 443-450
Loadings and	correlations in the interpretation of principal components • Cadima, Jorge; Jolliffe, Ian T.	95 JAppStat 22 203-214
	Correlation of interval-valued intuitionistic fuzzy sets • Bustince, H.; Burillo, P.	95 FuzzySet 74 237-244
Hwang, Seok Yoon	Correlation of intuitionistic fuzzy sets in probability spaces • Hong, Dug Hun;	95 FuzzySet 75 77-81
Brockwell, Peter J.; Davis, Richard On permissible	correlations for locally correlated stationary processes • Donahue, Rafe M. J.;	95 StPrLet 22 49-53
analysis: Difference between using covariance and	correlation matrices • Fung, W. K.; Kwan, C. W. Sensitivity analysis in factor	95 Psymtrka 60 607-614
Displaying a	correlation matrix on the unit sphere in R^3 (French) • Falissard, B.	95 RStApp 43/2 35-48
Crowder, Martin On the use of a working	correlation matrix in using generalised linear models for repeated measures •	95 Biomtrka 82 407-410
related measurement error on linear regression and	correlation in medical method comparison studies • Rifkin Effects of correlated and uncor-	95 StatMed 14 789-798
and grouping on the estimated product-moment	correlation • Chai, John J.; Chen, Chung Joint effects of errors of measurement	95 CommStA 24 283-294
Multirelation – A	correlation among more than two variables • Drezner, Zvi	95 CmpStDA 19 283-292
to the pairwise estimator of a parent-offspring	correlation • Eliasziw, Michael; Donner, Allan Evaluation of a proposed modification	95 SankhyaB 57 151-157
in the Gulf ... • Toompuu et al. Spatial large-scale	correlations for optimal interpolation of temperature, salinity and nutrient concentrations	95 EnvrMtrc 6 55-72
Assessing cross-sectional	correlation in panel data • Frees, Edward W.	95 JEconmtx 69 393-414
Climatological time series with periodic	correlation • Lund, Robert; Hurd, Harry; Bloomfield, Peter; Smith, Richard	95 JClimate 8 2787-2809
A	correlation principal component regression analysis of NIR data • Sun, Jianguo	95 JChemtrc 9 21-29
Estimation of	correlation in randomized response • Bellhouse, D. R.	95 SurvyM 21 13-19
	Correlations in randomized response surveys • Han, Geun-Shik; Warde, William D.	94 ASASrvyRsr 491-496
	Correlations redux • Olkin, Ingram; Finn, Jeremy D.	95 PsychBu 118 155-164
R. G.; Barker, P. D. R.; Teich, M. C. Long-duration	correlation in the sequence of action potentials in an insect visual interneuron • Turcott,	95 JStCmpSm 52 253-271
Estimating the	correlation among siblings • Keen, K. J.	93 AnHuGene 57 297-305
Piepho, H.-P. Implication of	correlations among some common stability statistics – A Monte Carlo simulations •	95 ThApGene 90 457-461
A novel tool for two-dimensional	correlation spectroscopy • Windig, W.; Margevich, D. E.; McKenna, W. P.	95 Chemolab 28 109-128
A Bayesian study of the	correlation between standard scores of twins • Li, Jian-Hua; Watanabe, Hiroshi	94 Behvmtrk 21 121-130

Correlation structure

distribution • Kwong, Koon-Shing	Evaluation of one-sided percentage points of the singular multivariate normal	95 JStCmpSm 51 121-135
Marques et al. A model for distributions and	correlation structure of multivariate polytomous data from a two-stage cluster sample •	94 ASAProBiop 346-351
carcinogenicity studies • Rahman Effects of observed	correlation structure among tumor types on experimentwise false positive rate in animal	95 JBiophSt 5 273-283
Properties of alternative estimators of familial	correlations under variable sibship size • Vogler, G. P.; Wette, R.; McGue, M. K.; Rao, D. C.	95 Biomtrcs 51 276-283
Robust estimation of	correlation with applications to computer vision • Brunelli, R.; Messelodi, S.	95 PattrnRc 28 833-841
Rocio Estimating	correlations with missing data, a Bayesian approach • Gross, Alan L.; Torres-Quevedo,	95 Psymtrka 60 341-354
A maximum likelihood approach to	correlational outlier identification • Bacon, Donald R.	95 MultBehR 30 125-148
Success in renal transplantation: A	correlational study • Chadha, Janice; Katz, Barry	94 ASAProSoSt 251-256

Correspondence analysis

Arabie, Phipps; Hubert, Lawrence J.	Additive two-mode clustering: The error-variance approach revisited • Mirkin, Boris;	95 JClass 12 243-263
components • Donnell, D. J.; Buja, A.; Stuetzle, W.	Analysis of additive dependencies and concurvities using smallest additive principal	94 AnlsStat 22 1635-1673
coefficient (French) • Tricot, J.-M.; Latour, R.	An approach towards the comparison of profiles using a generalized concentration	95 RStApp 43/3 35-53
	The Eurovision song contest • Everitt, Brian	95 Stats /13 19-22
sets: Its application to data from the ... • Tsuchiya	A homogeneity analysis for exploring structural differences among several data	95 PrInStMa 43 121-140
structures • Thioulouse, J.; Chessel, D.; Champely, S.	Multivariate analysis of spatial patterns: A unified approach to local and global	95 EnvEcoSt 2 1-14
minimum discrimination information approach and	correspondence analysis • Gokhale et al. Association models for contingency tables via	95 RBrzlPrS 9 83-98
	Correspondence analysis of an artificial binary cylinder data • Endo, Hideki	95 StPrLet 25 231-240
Spherical trait in	correspondence analysis of artificial data • Okamoto, Masashi; Endo, Hideki	95 JJapnStS 25 183-191
	Correspondence analysis of an artificial torus data • Okamoto, Masashi	94 Behvmtrk 21 149-161
dence graph • de Falguerolles, A.; Jmel, S. et al.	Correspondence analysis and association models constrained by a conditional indepen-	95 Psymtrka 60 161-180
	Correspondence analysis in statistical package programs • Thompson, Paul A.	95 AmerStat 49 310-316
Influence and stability in non symmetrical	correspondence analysis • Balbi, Simona	94 Metron 52/3 111-128
of canonical coordinates and an alternative to	correspondence analysis using Hellinger distance • Rao, C. Radhakrishna A review	95 Questo 19 23-63
	Cosine products, Fourier transforms, and random sums • Morrison, Kent E.	95 AmMaMnth 102 716-724
Simplification of Milstar microwave and digital	cost analysis model • Gobreial, Hany S.	95 JParmc 15/1 92-117
	Cost effective calibration intervals using Weibull analysis • Barringer, H. Paul	95 ASQCTekCon 1026-1038
Designing a	cost-effective clinical trial • Hornberger, John C.; Brown, Byron Wm., Jr; Halpern, Jerry	95 StatMed 14 2249-2259
Richard D.; Anderson, Keaven M.; Goldhirsch, Aron	Cost-effectiveness analysis using quality-adjusted survival • Cole, Bernard F.; Gelber,	94 ASAProBiop 9-18
	Cost-effectiveness in public education • Bennett, Jay	95 Chance 8/1 38-41
Willan, Andrew R.; O'Brien, Bernie J.	Cost-effectiveness ratios in clinical trials: From deterministic to stochastic models •	94 ASAProBiop 19-28
A comparison of several outcome measures in	cost-effectiveness studies • Kogut, Sarah J. H.	94 ASAProBiop 267-271
Survey • Kalsbeek, W. D.; Botman, S. L. et al.	Cost-efficiency and the number of allowable call attempts in the National Health Interview	94 JOffStat 10 133-152
estimation of exposure to ... • Chen A minimum	cost estimator for the mean of positively skewed distributions with applications to	95 EnvrMtrc 6 181-193
A minimum	cost exponential CUSUM test • Boyd, Amy Hancock	95 CommStA 24 2929-2958

Cost function

	The structure of technology in Brazilian sugarcane production, 1975-87 • Rask, Kevin	95 JApEcmtx 10 221-232
in exponential-type processes for a polynomial	cost function • Magiera, Ryszard Bayes sequential estimation procedures	95 MaOpfStS 26 61-73
survey program data using the employment	cost index • Tehonica, Jason Updating occupational compensation	94 ASASrvyRsr 1048-1052
Introducing births into the employment	cost index survey • Gessley, Mary K.; Kelly, E. Thomas; Ponikowski, Chester H.	94 ASASrvyRsr 797-802
M.; Miller, S. The improvement of the Employment	Cost Index Survey: A total quality approach within the Bureau of Labor Statistics • Kelly,	94 ASAProQlPr 210-214
Coverage error and	cost issues in small area telephone surveys • Schejbal, Judith A.; Lavrakas, Paul J.	94 ASASrvyRsr 1287-1292
sequence of optimal tests to control manufacturing	costs (French) • Bergeret, F.; Chandon, Y. Search for a	95 RStApp 43/3 21-33
S. S.; Vigezzi, M. J. Lowering manufacturing	cost of material by formulating it through statistical modeling and design • Lepeniotis,	95 Chemolab 29 133-139
Multivariate screening procedures for quality	cost minimization • Drezner, Zvi; Wesolowsky, G. O.	95 IIETrn 27 300-304
specification in random production models of	cost-minimizing firms • Brown, Bryan W.; Walker, Mary Beth Stochastic	95 JEconmtx 66 175-205
Probability distributions for the Deming	cost models • Kaminsky, Frank C.; Haberle, Katharine Rowse	95 JQualTek 27 355-362
Survey • Mantel, H.; Laniel, N.; Duval, M.-C. et al.	Cost modelling of alternative sample designs for rural areas in the Canadian Labour Force	94 ASASrvyRsr 925-930
Analytical approach for determining the	cost optimum raw material quality level • Chen, R.; Hawaleshka, O.; Strong, D. R.	95 JAppStat 22 375-388
the 1992 FCRS • Perry et al. Redrawing the 1993 Farm	Cost and Returns Survey list sample to reduce overlap with three other 1993 surveys and	94 ASASrvyRsr 632-637
Paul, N.; Greenhouse, J.; Kupfer, D.; Frank, E. et al.	Cost utility analysis of maintenance treatment for recurrent depression • Kamlet, M. S.;	95 ConClnTr 16 17-40
Limit theorem on option replication	cost with transaction costs • Kusuoka, Shigeo	95 AnlsApPr 5 198-221
A stochastic target	costing model (German) • Abel, P.	95 ZAngMaMk 75 557-558S
Biometrical genetic analysis of the	cotwin control design • Duffy, David L.	94 BehavGen 24 341-344
A clustering method using the	Coulomb energy network (Korean) • Lee, Sukhoon; Park, Nae-Hyun; Kim, Yung-Hwan	95 KorApS 8/1 39-50

Count data

visit frequency • Moffatt, Peter G.	Grouped Poisson regression models: Theory and an application to public house	95 CommStA 24 2779-2796
mixture models • Dean, C. B.	Modified pseudo-likelihood estimator of the overdispersion parameter in Poisson	94 JAppStat 21 523-532
of empirical covariance matrices in the analysis of	count data • Dean, C. B.; Eaves, D. M.; Martinez, C. J. A comment on the use	95 JStPlInf 48 197-205
Estimating bird species richness from capture and	count data • Dawson, D. K.; Sauer, J. R.; Wood, P. A.; Berlanga, M.; Wilson, M. H. et al.	95 JAppStat 22 1063-1068
Modelling frequency and	count data • Lindsey, J. K.	95 ClrndnPr:UK 291p
Review of "Modelling frequency and	count data" • Lindsey, J. K. (Auth); Hand, D. J. (Rev)	96 ShrtBkRv 16/1 2-2
Duration dependence and dispersion in	count-data models • Winkelmann, Rainer	95 JBES 13 467-474
information loss in analyzing a mixture model of	count data • Fong, Daniel Y. T.; Yip, Paul S. F. A note on	95 CommStA 24 3197-3209
Prediction and control for a time-series	count data model • Brännäs, Kurt	95 IntJFor 11 263-270
mean function of point processes based on panel	count data • Sun, J.; Kalbfleisch, J. D. Estimation of the	95 StSinica 5 279-289
Tests for serial correlation and overdispersion in a	count data regression model • Johansson, Per	95 JStCmpSm 53 153-164
Mixture models for the analysis of repeated	count data • van Duijn, Marijtje A. J.; Böckenholt, Ulf	95 ApplStat 44 473-485
Overdispersion in	count data – A review (Korean) • Kim, Byung Soo; Oh, Kyong Joo; Park, Cheolyong	95 KorApS 8/2 147-161
Jansen, Margo G. H. Modeling repeated	count data: Some extensions of the Rasch Poisson counts model • van Duijn, Marijtje A. J.;	95 JEdStat 20 241-258
likelihood-estimation and regression diagnostics for	count data: Suicide by tourism? (German) • Ziegler, A. et al. Pseudo-maximum-	95 AllgeStA 79 170-195
Analysis of	count data using power variance function • Nakashima, Eiji	95 JJapnStS 25 193-204
Comment on "Analysis of data on duration and	counts" • Wooldridge, Jeffrey M.	94 ASAProBuEc 17-19
Uncertainty in Poisson event	counts and exposure time in rate estimation • Martz, Harry F.; Picard, Richard R.	95 RelEngSS 48 181-190
Statistical analysis of asbestos fibre	counts • Oehlert, Gary W.; Lee, Richard J.; Van Orden, Drew	95 EnvrMtrc 6 115-126
Paik et al. Difference in clinical implications of CD4	counts among HIV-infected homosexual men and injection drug using men and women •	95 StatMed 14 1889-1900
EM estimation for time series models involving	counts • Chan, K. S.; Ledolter, Johannes Monte Carlo	95 JASA 90 242-252
et al. A functional central limit theorem for the jump	counts of Markov processes with an application to Jackson networks • Anantharam, V.	95 AdvAppPr 27 476-509
Stephen A. Comparing aerial estimates with ground	counts of nests in wood stork colonies • Rodgers, James A., Jr; Linda, Stephen B.; Nesbitt,	95 JWildlMg 59 656-666
combining sample correlation coefficients and vote	counts to obtain an estimate and a confidence interval for ... • Bushman et al. ... for	95 PsychBu 117 530-546
with error. Applications to survival and CD4	counts in patients with AIDS • Tsiatis et al. ... of survival to longitudinal data measured	95 JASA 90 27-37
Stevens, C. E. Change points in the series of T4	counts prior to AIDS • Kiuchi, A. S.; Hartigan, J. A.; Holford, T. R.; Rubinstein, P.;	95 Biomtrcs 51 236-248
Regression models for repeated medical random	counts • Mackenzie, Gilbert	95 Ststcian 44 365-372
tistical methods for aerial surveys using the double-	count technique to correct visibility bias • Rivest, L.-P.; Potvin, F.; Crépeau, H. et al. Sta-	95 Biomtrcs 51 461-470
Quade, Dana Testing whether a	count variable belongs to the Katz family of distributions • Anstrom, Kevin J.;	94 ASAProStCp 194-199
You can	count on us – With confidence • McLennan, Bill	95 JRSS-A 158 467-489
cumulative sums for text analysis using word	counts • Bissell, A. F. Weighted	95 JRSS-A 158 525-545
Topics in the constructive theory of	countable Markov chains • Fayolle, G.; Malyshev, V. A.; Menshikov, M. V.	95 CambridgeUniv 169p
Review of "Topics in the constructive theory of	countable Markov chains" • Fayolle, G.; Malyshev, V. A.; Menshikov, M. V. (Auth) et al.	96 ShrtBkRv 16/1 9-10
A class of strong laws for functionals of	countable nonhomogeneous Markov chains • Liu, Wen; Liu, Guoxin	95 StPrLet 22 87-96
Weak convergence of conditioned processes on a	countable state space • Jacka, S. D.; Roberts, G. O.	95 JAppProb 32 902-916
Borgan, Ørnulf	Counter-matching: A stratified nested case-control sampling method • Langholz, Bryan;	95 Biomtrka 82 69-79
Gasteiger, Johann Neural networks with	counter-propagation learning strategy used for modelling • Zupan, Jure; Novič, Marjana;	95 Chemolab 27 175-187

Standard error of linear equating for the	counterbalanced design • Zeng, Lingjia; Cope, Ronald T.	95 JEdStat 20 337-348
Nobel, Andrew A	counterexample concerning uniform ergodic theorems for a class of functions •	95 StPrLet 24 165-168
Simple	counterexamples against the conditionality principle • Helland, Inge S.	95 AmerStat 49 351-356
The Rao-Blackwell theorem in stereology and some	counterexamples • Baddeley, A. J.; Cruz-Orive, L. M.	95 AdvAppPr 27 2-19
Inductive	countersupport • Dorn, G. J. W.	95 ZeitWiss 26 187-189
On some properties of De Pril transforms of	counting distributions • Sundt, Bjørn	95 AstinB 25 19-31
Different approaches to	counting early childhood program participation • West, Jerry; Hausken, Elvie Germino	94 ASAProSoSt 141-146
M.; Bowden, David C.; Gill, R. Bruce Aerial	counting experiments to estimate pronghorn density and herd structure • Pojar, Thomas	95 JWildlMg 59 117-128
	Counting gays and lesbians: A case for "new" survey methods • Edwards, Brad	94 ASASrvyRsr 1305-1309
Wolter, Kirk M. (Rev) Review of	"Counting people in the information age" • Stefey, Duane L.; Bradburn, Norman M. (Ed);	95 Chance 8/3 44-49
	Counting process	
fitting • Li, Gang; Doss, Hani	An approach to nonparametric regression for life history data using local linear	95 AnlsStat 23 787-823
	On the asymptotic size and duration of a class of epidemic models • Svensson, Åke	95 JAppProb 32 11-24
Seungyeoun; Wolfe, Robert A.	A class of linear nonparametric rank tests for dependent censoring • Lee,	95 CommStA 24 687-712
Pérez-Ocón, Rafael; Gámiz-Pérez, M. Luz	Conditions on the arrival process to obtain HNBUE survival using a shock model •	95 CommStA 24 931-944
Phayre, Sean; Renshaw, Eric	The evolution and measurement of a population of pairs • Jakeman, Eric;	95 JAppProb 32 1048-1062
application to Jackson networks • Anantharam et al.	A functional central limit theorem for the jump counts of Markov processes with an	95 AdvAppPr 27 476-509
	Martingales without tears • Lan, K. K. Gordon; Lachin, John M.	95 LiDaAnal 1 361-375
model • Borgan, Ø.; Goldstein, L.; Langholz, B.	Methods for the analysis of sampled cohort data in the Cox proportional hazards	95 AnlsStat 23 1749-1778
Lee, Chinsan; Yang, Grace L.	A multitype decomposable age-dependent branching process and its applications •	95 JAppProb 32 591-608
semi-Markov process • Csenki, Attila	The number of visits to a subset of the state space by a discrete-parameter	95 StPrLet 22 71-77
McKeague, Ian W.; Nikabadze, A. M.; Sun, Yanqing	On omnibus test for independence of a survival time from a covariate •	95 AnlsStat 23 450-475
Andersen, P. K.	Score test of homogeneity for survival data (Disc: p157-159) • Commenges, D.;	95 LiDaAnal 1 145-156
Jayant V.; Sengupta, Debasis	Testing the hypothesis of proportional hazards in two populations • Deshpande,	95 Biomtrka 82 251-261
A	counting process approach to stochastic interest • Møller, Christian Max	95 InsrneME 17 181-192
Statistical models based on	counting processes • Andersen, P. K.; Borgan, Ø.; Gill, R. D.; Keiding, N.	95 Springer:Brln:NY 767p
Review of "Statistical models based on	counting processes" • Andersen, P. K.; Borgan, O.; Gill, R. D.; Keiding, N. (Auth) et al.	95 AllgeStA 79 221-221
Nils Lid (Rev) Review of "Statistical models based on	counting processes" • Andersen, P. K.; Borgan, Ø.; Gill, R. D.; Keiding, N. (Auth); Hjort,	95 StNeerla 49 247-249
Review of "Statistical models based on	counting processes" • Andersen, P. K.; Borgan, Ø.; Gill, R. D.; Keiding, N. (Auth) et al.	95 SIAMRvw 37 475-476
application to maximum likelihood estimation for	counting processes • van de Geer, Sara Exponential inequalities for martingales, with	95 AnlsStat 23 1779-1801
general additive-multiplicative hazard models for	counting processes • Lin, D. Y.; Ying, Zhiliang Semiparametric analysis of	95 AnlsStat 23 1712-1734
likelihood estimators of parameters of a spatial	counting process modelling crystallization of polymers • Capasso et al. ... of the maximum	95 StocAnAp 13 279-294
Fitting parametric	counting processes by using log-linear models • Lindsey, J. K.	95 ApplStat 44 201-212
	Counting public libraries in the United States • Terrie, E. Walter	94 ASAProGovt 58-63
	Coupling	
	Stationary regimes for inventory processes • Bardhan, Indrajit; Sigman, Karl	95 StocProc 56 77-86
	Strong approximations for epidemic models • Ball, Frank; Donnelly, Peter	95 StocProc 55 1-21
Cranston, Michael; Greven, Andreas	Coupling and harmonic functions in the case of continuous time Markov processes •	95 StocProc 60 261-286
Review of "Lectures on the	coupling method" • Lindvall, T. (Auth); Borovkov, K. A. (Rev)	94 ThProbAp 39 721-723
Review of "Lectures on the	coupling method" • Lindvall, Torgny (Auth); O'Connell, Neil (Rev)	95 AnlsProb 23 1456-1460
	Coupling methods in reliability (French) • Cocozza-Thivent, C.; Roussignol, M.	95 AIHenrPB 31 119-141
Cao, Xi-Ren Event	coupling and performance sensitivity analysis of generalized semi-Markov processes •	95 AdvAppPr 27 741-769
Randles, R. H. et al. Challenges in teaching short	courses by an academic department: The University of Florida experience • Cornell, J. A.;	95 AmerStat 49 195-200
Probability theory. An advanced	course • Borkar, V. S.	95 Springer:Brln:NY 145p
Teaching short	courses in biostatistics and epidemiology • Kleinbaum, David G.	94 ASAProStEd 29-32
Teaching short	courses in biostatistics and epidemiology • Kleinbaum, David G.	95 AmerStat 49 200-203
Teaching graduate service	courses through case studies • Ramsey, Fred L.; Schafer, Daniel W.	94 ASAProStEd 27-28
Teaching critical review of the medical literature: A	course in clinical trials • Freeman, Katherine	94 ASAProStEd 156-161
confidence: Application and evaluation for golf	course closing • Shipley et al. Rough set theory fuzzy belief functions related to statistical	95 StocAnAp 13 487-502
et al. Developing material for introductory statistics	courses from a conceptual, active learning viewpoint (jse/v3n3/steinhorst) • Steinhorst	95 JStEd 3/3 33kb
Ten suggestions for effectively teaching short	courses to heterogeneous groups • Birch, Jeffrey B.	94 ASAProStEd 33-38
Ten suggestions for effectively teaching short	courses to heterogeneous groups • Birch, Jeffrey B.	95 AmerStat 49 190-195
Probability theory: An introductory	course • Sinai, Y. (Auth); Haughton, D. (Transl)	95 Springer:Brln:NY 138p
Review of "Probability theory: An introductory	course" • Sinai, Y. (Auth); Haughton, D. (Transl); Sedgwick, Philip (Rev)	95 Ststcian 44 142-142
sey, Patricia P. Reengineering the business statistics	course: Meeting the customer needs of the twenty-first century • Levine, David M.; Ram-	94 ASAProStEd 121-122
Review of "A	course on point processes" • Reiss, R.-D. (Auth); Matthes, Klaus (Rev)	95 BiomtrcJ 37 182-182
Review of "Video	course: Principal components" • Massart, D. L.; Lewi, P. J. (Auth); Kvalheim, O. M. (Rev)	95 Chemolab 27 147-147
Reliability always depends on probability of	course • Good, I. J.	95 JStCmpSm 52 192-193
An intermediate	course in probability • Gut, A.	95 Springer:Brln:NY 278p
Review of "An intermediate	course in probability" • Gut, A. (Auth); Embrechts, P. A. L. (Rev)	96 ShrtBkRv 16/1 4-4
King, Bruce (Rev) Review of "A first	course in probability models and statistical inference" • Creighton, J. H. C. (Auth);	95 AmerStat 49 113-113
promote active learning in the introductory statistics	course: A report from the field (jse/v3n2/keeler) • Keeler et al. Using small groups to	95 JStEd 3/2 28kb
Integrating cases into the business statistics	course • Nolan, James R.	94 ASAProStEd 123-125
Simonoff, Jeffrey S. A casebook for a first	course in statistics and data analysis • Chatterjee, Samprit; Handcock, Mark S.;	95 Wiley:NY:UK 314p
A data analysis based elementary statistics	course • Egge, Eric; Foley, Sean; Haskins, Loren; Johnson, Roger	94 ASAProStEd 144-149
to visualize concepts in an introductory statistics	course • Guffey, James M Using data and the computer to teach and	94 ASAProStEd 84-87
Projects in introductory statistics	courses • Ledolter, Johannes	95 AmerStat 49 364-367
of teaching: A case study in a large statistics	course • Wild, C. J. Continuous improvement	95 IntStRvw 63 49-68
Wilson, W. J. Supplementing a college preparatory	course in statistics with emphasis on applied probability, quality control, and graphics •	94 ASAProStEd 42-43
Lawson, J. S.; Hawks, V. D. Q^2: A TQM	course for undergraduates • Badrigal, J. L.; Richards, D. O.; Free, J. C.; Lee, T. N.;	94 ASAProQlPr 172-180
hazards model with step function time-dependent	covariables • Herndon et al. ... cubic spline as baseline hazard in the proportional	95 StatMed 14 2119-2129
program for multi-state Markov models with	covariables • Marshall, Guillermo; Guo, Wensheng; Jones, Richard H. MARKOV: A computer	95 CmpBiomd 47 147-156
Non-parametric analysis of	covariance • Young, Stuart G.; Bowman, Adrian W.	95 Biomtrcs 51 920-931
A triangle inequality for	covariances of binary FKG random variables • van den Berg, J.; Gandolfi, A.	95 AnlsApPr 5 322-326
Large-scale estimation of variance and	covariance components • Fraley, Chris; Burns, Patrick J.	95 SIAMSSC 16 192-209
in factor analysis: Difference between using	covariance and correlation matrices • Fung, W. K.; Kwan, C. W. Sensitivity analysis	95 Psymtrka 60 607-614
relationships in Escherichia coli based on auto cross	covariances • Sjöström, M.; Rännar, S.; Wieslander, Å. Polypeptide sequence property	95 Chemolab 29 295-305
estimation in time series using a bootstrapped	covariance estimate • Ravishanker, Nalini; Wu, Lilian S.-Y.; Dey, Dipak K. Shrinkage	95 JStCmpSm 53 259-267
On the nonparametric estimation of	covariance functions • Hall, Peter; Fisher, Nicholas I.; Hoffmann, Branka	94 AnlsStat 22 2115-2134
When is a truncated	covariance function on the line a covariance function on the circle? • Wood, Andrew T. A.	95 StPrLet 24 157-164
Line transects,	covariance functions and set convergence • Cabo, A. J.; Baddeley, A. J.	95 AdvAppPr 27 585-605
of Porat concerning asymptotic efficiency of sample	covariances of Gaussian ARMA processes • Walker, A. M. On results	95 JTimSrAn 16 237-248
Estimation of within-group	covariance • Townsend, Jeffrey L.	94 ASAProStCp 257-262
Houdré, Christian Some applications of	covariance identities and inequalities to functions of multivariate normal variables •	95 JASA 90 965-968

Houdré, Christian; Pérez-Abreu, Víctor	Covariance identities and inequalities for functionals on Wiener and Poisson spaces •	95 AnlsProb 23 400-419
A generalization of	covariance identity and related characterizations • Cacoullos, T.; Papathanasiou, V.	95 MaMethSt 4 106-113
Liu, Jinshan On optimality of the	covariance improved estimate: A simulation study (Chinese) • Chen, Maoxue;	95 ChJApPSt 10 1-4
Chan, Wai (Rev) Review of "Modelling	covariances and latent variables using EQS" • Dunn, G.; Everitt, B.; Pickles, A. (Auth);	95 SIAMRvw 37 462-464
Schimek, Michael G. (Rev) Review of "Modelling	covariances and latent variables using EQS" • Dunn, G.; Everitt, B.; Pickles, A. (Auth);	95 CmpStQ 10 196-196
Strong approximation for cross-	covariances of linear variables with long-range dependence • Kouritzin, Michael A.	95 StocProc 60 343-353
Martinez, C. J. A comment on the use of empirical	covariance matrices in the analysis of count data • Dean, C. B.; Eaves, D. M.;	95 JStPlInf 48 197-205
The influence in comparing	covariance matrices • Kim, Myung Geun	95 CommStA 24 1431-1441
Asymptotic distributions of some test criteria for the	covariance matrix in elliptical distributions under local alternatives • Purkayastha et al.	95 JMultiAn 55 165-186
estimation in linear models with nonscalar error	covariance matrix • Chaturvedi, Anoop A note on the Stein rule	95 SankhyaB 57 158-165
estimates in the important class of estimates of the	covariance matrix • Xie, Yumin Admissible	95 ChinaAMB 15 435-442
Ligeralde, Antonio V.; Brown, Bryan W. Band	covariance matrix estimation using restricted residuals: A Monte Carlo analysis •	94 IntEconR 35 751-767
of risk for an orthogonally invariant estimator of a	covariance matrix • Sheena, Yo Unbiased estimator	95 JJapnStS 25 35-48
estimation of regression coefficient matrix and	covariance matrix in growth curve models • Yang, Wenli Simultaneous	95 SciSPkng 24 474-484
MINQE(U,I) and UMVIQUE of the	covariance matrix in the growth curve model • Wenli, Yang	95 MaOpfStS 26 49-59
The effects of an outlier while modeling the	covariance matrix for incomplete longitudinal data • Grady, James J.; Boyarsky, Gregory	94 ASAProBiop 441-445
Model selection techniques for the	covariance matrix for incomplete longitudinal data • Grady, James J.; Helms, Ronald W.	95 StatMed 14 1397-1416
Jonathan, P. et al. Discriminant analysis with singular	covariance matrices: Methods and applications to spectroscopic data • Krzanowski, W. J.;	95 ApplStat 44 101-115
Arnold, G. The sampling distributions and	covariance matrix of phylogenetic spectra • Waddell, P. J.; Penny, D.; Hendy, M. D.;	94 MolBioEv 11 630-642
Moshe Estimating the asymptotic	covariance matrix for quantile regression models. A Monte Carlo study • Buchinsky,	95 JEconmtx 68 303-338
On the singularity of the sample	covariance matrix • Trenkler, Götz	95 JStCmpSm 52 172-173
Carlo study on two methods of calculating the MLE's	covariance matrix in a seemingly unrelated nonlinear regression • Jensen, Mark J. A Monte	95 EcomtrRv 14 315-330
ingly unrelated regression model when the error	covariance matrix is singular • Takada, H.; Ullah, A.; Chen, Y.-M. Estimation of the seem-	95 JAppStat 22 517-530
approximations for P-values of some tests of	covariance matrices • Booth, J. G.; Butler, R. W.; Huzurbazar, S.; Wood, A. T. A. Saddlepoint	95 JStCmpSm 53 165-180
Inverting	covariance matrices in unbalanced hierarchical models • Stępniak, C.; Niezgoda, M.	95 JStCmpSm 51 215-221
detection in growth curve model with unstructured	covariance matrix • Pan, Jian-Xin; Fang, Kai-Tai Multiple outlier	95 AnInStMa 47 137-153
Diagonalization of variance-	covariance matrices using F-G algorithm • Ali, A. K. A.; Lin, C. Y.	95 EgpStJ 39 12-25
Resistant h-plot for a sample variance-	covariance matrix • Choi, Yong-Seok	95 JKoreaSt 24 407-417
Estimate of	covariance matrix with additional information (Chinese) • Cui, Hengjian; Li, Guoying	95 KToChina 39 1354-1357
Estimation of a normal	covariance matrix with incomplete data under Stein's loss • Konno, Yoshihiko	95 JMultiAn 52 308-324
L. REML estimation of	covariance matrices with restricted parameter spaces • Calvin, James A.; Dykstra, Richard	95 JASA 90 321-329
site • Ota, Tatsuya; Nei, Masatoshi Variance and	covariances of the numbers of synonymous and nonsynonymous substitutions per	94 MolBioEv 11 613-619
On-line estimation of error	covariance parameters for atmospheric data assimilation • Dee, Dick P.	95 MWeathr 123 1128-1145
Bayes factors for zero partial	covariances • Giudici, Paolo	95 JStPlInf 46 161-174
for multivariate repeated measures with patterned	covariance • Crawford, Ann Marie K.; Johnson, William D. Statistical considerations	94 ASAProBiop 61-66
Improved minimax estimators of normal	covariance and precision matrices • Gupta, A. K.; Ofori-Nyarko, Samuel	95 MaOpfStS 26 19-25
Fernando, R. L.; Grossman, M.	Covariance between relatives in multibreed populations: Additive model • Lo, L. L.;	93 ThApGene 87 423-430
Estimating the instantaneous volatility and	covariance of risky assets • Chesney, Marc; Elliott, Robert J.	95 AplStMDA 11 51-58
Theobald, C. M.; Kirby, S. P. J. Discrimination using	covariance selection models in the automated allocation of human chromosomes •	95 Ststcian 44 235-240
of Gaussian nonstationary process occurring in	covariance statistic • Ostrovskii et al. Asymptotic properties of distribution of maximum	94 ThProbAp 39 527-534

Covariance structure

	Combined Bayesian estimates for the equicorrelation coefficient • Viana, Marlos A. G.	95 CandJSt 23 43-53
study • Pischke, Jörn-Steffen	Measurement error and earnings dynamics: Some estimates from the PSID validation	95 JBES 13 305-314
Puri, Madan L.	A multivariate Wald-Wolfowitz rank test against serial dependence • Hallin, Marc;	95 CandJSt 23 55-65
of satisfaction of the Japanese - An application of	covariance structure analysis to the Survey of Japanese ... • Maeda ... determinants	95 PrInStMa 43 141-160
A linear model with a special class of	covariance structure • Haq, M. Safiul; Ng, V. Ming	94 JStatRsr 28 189-198
Liu, Jun S.; Wong, Wing H.; Kong, Augustine	Covariance structure and convergence rate of the Gibbs sampler with various scans •	95 JRSS-B 57 157-169
Cross-validation criteria for	covariance structures • De Gooijer, Jan G.	95 CommStB 24 1-16
	Covariance structure for elliptical distributions • Wang, Weijing; Wells, Martin T.	94 ASAProSoSt 60-65
Selection of a	covariance structure for growth curves • Keramidas, Elaine M.; Lee, Jack C.	95 BiomtrcJ 37 783-797
likelihood estimation for unbalanced multilevel	covariance structure models via the EM algorithm • Raudenbush, Stephen W. Maximum	95 BJMaStPs 48 359-370
LR test for random-effects	covariance structure in a parallel profile model • Yokoyama, Takahisa	95 AnInStMa 47 309-320
al. Simultaneous genetic analysis of means and	covariance structure: Pearson-Lawley selection rules • Dolan, C. V.; Molenaar, P. C. M. et	94 BehavGen 24 17-24
John R. The effects of within-group	covariance structure on recovery in cluster analysis I. The bivariate case • Donoghue,	95 MultBehR 30 227-254
Multivariate analysis. Part 2: Classification,	covariance structures and repeated measurements • Krzanowski, W. J.; Marriott, F. H. C.	95 Arnold:Lnd:MD 280p
cation process via spatial modeling of the mean and	covariance surfaces • Hughes-Oliver et al. Robust improvement of a semi-conductor fabri-	94 ASAProPEng 235-244
Fernando, R. L. et al. Theory for modelling means and	covariances in a two-breed population with dominance inheritance • Lo, L. L.;	95 ThApGene 90 49-62
multivariate one-sided hypothesis with unknown	covariances • Frick, H. Comparing trials with multiple outcomes: The	95 BiomtrcJ 37 909-917
Crossover trials with asymmetric	covariance within subject • Skinner, Jonathan B.	94 ASAProBiop 411-413

Covariate

	Examining an increasing trend with non-parametric methods • Strömberg, Ulf	94 JAppStat 21 575-587
data using partial likelihood score processes with	covariate adjustment • Gu, Minggao; Ying, Zhiliang Group sequential methods for survival	95 StSinica 5 793-804
of median survival times with adjustment for	covariates • Karrison, Theodore Comparison	95 StatMed 14 2537-2553
P.; Segal, M. R. Survival trees with time-dependent	covariates: Application to estimating changes in the incubation period of AIDS • Bacchetti,	95 LiDaAnal 1 35-47
test for the additive risk model with a binary	covariate • Kim, Jinheum; Song, Moon Sup A goodness-of-fit	95 JKoreaSt 24 537-549
Nonparametric regression with censored	covariates • Dabrowska, Dorota M.	95 JMultiAn 54 253-283
repeated measures designs with a changing	covariate • Rashid, M. Mushfiqur; Nandram, Balgobin A rank-based analysis of one-way	95 JStPlInf 46 249-263
Time-dependent effects of fixed	covariates in Cox regression • Verweij, Pierre J. M.; van Houwelingen, Hans C.	95 Biomtrcs 51 1550-1556
Auxiliary	covariate data in failure time regression • Zhou, Haibo; Pepe, Margaret Sullivan	95 Biomtrka 82 139-149
A mean score method for missing and auxiliary	covariate data in regression models • Reilly, Marie; Pepe, Margaret Sullivan	95 Biomtrka 82 299-314
parametric model for censored survival data and	covariates • Piantadosi, Steven; Crowley, John An implicitly defined	95 Biomtrcs 51 249-258
for the accelerated life model with time-dependent	covariates • Lin, D. Y.; Ying, Zhiliang Semiparametric inference	95 JStPlInf 44 47-63
parameters from censored data with a discrete	covariate • Rahbar, M. H.; Gardiner, J. C. Nonparametric estimation of regression	95 StPrLet 24 13-20
sion parameters from censored data with a discrete	covariate • Rahbar, Mohammad H. Sequential fixed width confidence intervals for regres-	95 SqtlAnly 14 143-156
Fletcher, D. Frequency domain estimation of	covariate effects in multichannel brain evoked potential data • Raz, J.; Cardenas, V.;	95 Biomtrcs 51 448-460
Graham, Patrick Modelling	covariate effects in observer agreement studies: The case of nominal scale agreement •	95 StatMed 14 299-310
• Guggenmoos-Holzmann Comment on "Modelling	covariate effects in observer agreement studies: The case of nominal scale agreement" ...	95 StatMed 14 2285-2286
Reply to comment on "Comment on 'Modelling	covariate effects in observer agreement studies: The case of nominal scale ... • Graham	95 StatMed 14 2286-2288
Step-function	covariate effects in the proportional-hazards model • LeBlanc, Michael; Crowley, John	95 CandJSt 23 109-129
Moeschberger, M. L. Semi-parametric estimation of	covariate effects using the positive stable frailty model • Wang, S. T.; Klein, John P.;	95 AplStMDA 11 121-133
A critical look at methods for handling missing	covariates in epidemiologic regression analyses • Greenland, Sander; Finkle, William D.	95 AJEpidm 141 1255-1264
A critical look at methods for handling missing	covariates in epidemiologic regression analyses • Greenland, Sander; Finkle, William D.	95 AJEpidm 142 1255-1264
a semiparametric regression model with errors in	covariates • Carroll, R. J.; Knickerbocker, R. K.; Wang, C. Y. Dimension reduction in	95 AnlsStat 23 161-181
the univariate approach to unit root testing - Using	covariates to increase power • Hansen, Bruce E. Rethinking	95 EconTher 11 1148-1171

nonrespondents to the Current Population Survey (	CPS) and Consumer Expenditure Interview Survey (CEIS) • Kojetin, B. A. Characteristics of	94 ASASrvyRsr 966-971
Logistic regression analysis of	CPS overlap survey split panel data • Fisher, Robin	94 ASASrvyRsr 620-625
the national unemployment rate that the "old"	CPS would have produced (Disc: p360-361) • Tiller, Richard; Welch, Michael Predicting	94 ASASrvyRsr 354-359
Massam, H.	Craig-Sakamoto's theorem for the Wishart distributions on symmetric cones • Letac, G.;	95 AnInStMa 47 785-799
An accessible proof of	Craig's theorem in the general case • Driscoll, Michael F.; Krasnicka, Barbara	95 AmerStat 49 59-62
Review of "H.	Cramér: Collected works. Volumes 1 and 2" • Martin-Löf, Anders (Ed); Uebe, Götz (Rev)	95 AllgeStA 79 332-332
Harald	Cramer collected works. Volumes 1 and 2 • Martin-Löf, A. (Ed)	94 Springer:Brln:NY 2165p
Vector harmonizable processes: Wold and	Cramér decompositions • Kakihara, Yûichirô	95 StocAnAp 13 531-541
Harald	Cramér and the distribution of prime numbers • Granville, Andrew	95 ScanActJ 12-28
A survey of the life and works of Harald	Cramér • Grenander, Ulf	95 ScanActJ 2-5
Harald	Cramér and insurance mathematics • Martin-Löf, Anders	95 AplStMDA 11 271-276
diffusion • Schmidli, H.	Cramér-Lundberg approximations for ruin probabilities of risk processes perturbed by	95 InsrneME 16 135-149
Applications of the van Trees inequality: A Bayesian	Cramér-Rao bound • Gill, Richard D.; Levit, Boris Y.	95 Bernoull 1 59-79
	Cramer-Rao bounds for discrete-time nonlinear filtering problems • Doerschuk, Peter C.	95 IEEEAuCn 40 1465-1469
Corsini, G.; Gini, F.; Greco, M. V.; Verrazzani, L.	Cramér-Rao bounds and estimation of the parameters of the Gumbel distribution •	95 IEEEAES 31 1202-1204
	Cramér-Rao inequality	
	Bias-variance tradeoffs in functional estimation problems • Low, Mark G.	95 AnlsStat 23 824-835
	Intrinsic analysis of statistical estimation • Oller, J. M.; Corcuera, J. M.	95 AnlsStat 23 1562-1581
Nonexponential applications of a global	Cramér-Rao inequality • Gajek, Lesław; Kałuszka, Marek	95 MaOpfStS 26 111-122
convergence to the central limit and failure of the	Cramér theorem for free random variables • Bercovici, H.; Voiculescu, D. Super-	95 ProbTher 103 215-222
	Cramér-von Mises statistic	
	Detecting a change in the intercept in multiple regression • Watson, G. S.	95 StPrLet 23 69-72
Schmid, Friedrich; Trede, Mark	A distribution free test for the two sample problem for general alternatives •	95 CmpStDA 20 409-419
function (two dimensional case) • Chergui, Jamal D.	Estimation and tests of the discrete probability law based on the empirical generating	94 RACEMdrd 88 245-286
data • Yang, Song; Koul, Hira L.	Minimum distance estimation of the center of symmetry with randomly censored	95 Metrika 42 79-97
bution and censored samples • Ashour, S. K. et al.	The sampling distribution of some goodness-of-fit test statistics using Burr distri-	94 EgpStJ 38 147-163
	U_n^2 test for uniformity of discrete distributions • Watson, Geoffrey S.	95 JAppStat 22 273-276
Generalized	Cramér-von Mises tests of goodness of fit for doubly censored data • Ren, Jian-Jian	95 AnInStMa 47 525-549
Rodríguez, Juan C.; Viollaz, Aldo J. A	Cramér-von Mises type goodness-of-fit test with asymmetric weight function •	95 CommStA 24 1095-1120
The aftermath of	Cramér's work on stochastic processes • Masani, P. R.	95 ScanActJ 66-96
A view of Harald	Cramér: Working with the master in his later years • Leadbetter, M. R.	95 ScanActJ 6-11
impacts of "big bang" and the 1987 stock-market	crash • Masulis et al. ... stock-return dynamics on the London stock exchange: The	95 JBES 13 365-378
Spatial mark-recapture method in the estimation of	crayfish population size • Partanen, Päivi; Penttinen, Antti	95 BiomtrcJ 37 979-994
	Credibility	
	Modelling of discretized claim numbers in loss reserving • Hesselager, Ole	95 AstinB 25 119-135
Balanced	credibility estimation • Neuhaus, Walther	95 AplStMDA 11 217-230
A teacher's remark on exact	credibility • Gerber, Hans U.	95 AstinB 25 189-192
On the estimation of the	credibility factor: A Bayesian approach • Schnieper, René	95 AstinB 25 137-151
	Credibility interval	
van der Merwe, A. J.; Groenewald, P. C. N.	Reference priors for the multivariate calibration problem • Plessis, J. L. du;	95 SoAfrStJ 29 155-168
	Credible region	
parameters • Ghosh, Jayanta K.; Mukerjee, Rahul	Frequentist validity of highest posterior density regions in the presence of nuisance	95 StatDecn 13 131-139
Ghosh, Jayanta Kumar	On priors providing frequentist validity for Bayesian inference • Datta, Gauri Sankar;	95 Biomtrka 82 37-45
On posterior	credible sets based on the score statistic • Rao, C. Radhakrishna; Mukerjee, Rahul	95 StSinica 5 781-791
What is the worth of free casino	credit? • Orkin, Michael; Kakigi, Richard	95 AmMaMnth 102 3-8
	Cressie-Read statistic	
Lorenzen, Gunter	A new family of goodness-of-fit statistics for discrete multivariate data •	95 StPrLet 25 301-307
Elizabeth Respondent strategies for recall of	crime victimization incidents • Czaja, Ronald; Blair, Johnny; Bickart, Barbara; Eastman,	94 JOffStat 10 257-276
Branching random walk with a	critical branching part • Kesten, Harry	95 JTherPrb 8 921-962
A key limit theorem for	critical branching processes • Sagitov, Serik	95 StocProc 56 87-100
Equality of	critical densities in continuum percolation • Meester, Ronald	95 JAppProb 32 90-104
Time dependent	critical fluctuations of a one dimensional local mean field model • Fritz, J.; Rüdiger, B.	95 ProbTher 103 381-407
	Critical length for semi-oriented bootstrap percolation • Mountford, T. S.	95 StocProc 56 185-205
Zeitouni, Ofer	Critical phenomena for sequence matching with scoring • Dembo, Amir; Karlin, Samuel;	94 AnlsProb 22 1993-2021
	Critical random walk in random environment on trees • Pemantle, Robin; Peres, Yuval	95 AnlsProb 23 105-140
Approximating Bayes	critical region for testing simple and tree ordered normal means • Sugiura, Nariaki	94 JStatRsr 28 1-20
Teaching	critical review of the medical literature: A course in clinical trials • Freeman, Katherine	94 ASAProStEd 156-161
Persistence of a	critical super-2 process • Gorostiza, Luis G.; Hochberg, Kenneth J.; Wakolbinger, Anton	95 JAppProb 32 534-540
Lag order and	critical values of the augmented Dickey-Fuller test • Cheung, Yin-Wong; Lai, Kon S.	95 JBES 13 277-280
normality • Dunn, Charles L.	Critical values and powers for tests of uniformity of directions under multivariate	95 CommStA 24 2541-2560
model • Wan, Alan T. K. The optimal	critical value of a pre-test for an inequality restriction in a mis-specified regression	95 AstrlJSt 37 73-81
Improved upper bounds for the contact process	critical value • Liggett, Thomas M.	95 AnlsProb 23 697-723
Ramsey, P. P. Correction to "Updated version of the	critical values of the standardized fourth moment" (V44 p231-241) • Ramsey, P. H.;	95 JStCmpSm 53 127-127
Li, Yiqun Tables of	critical values for two group interim analyses • Hewett, John E.; Petroski, Gregory F.;	95 CommStB 24 825-852
Yin-Wong; Lai, Kon S. Estimating finite sample	critical values for unit root tests using pure random walk processes: A note • Cheung,	95 JTimSrAn 16 493-498
Rotational versions of the	Crofton formula • Jensen, Eva B. Vedel	95 AdvAppPr 27 87-96
Forecasting	crop acreages and yields in the face of and in spite of floods • Allen, Rich	94 ASAProEnvr 187-191
selfed half-sib or full-sib families in autogamous	crops • Burton, Joe W.; Carver, Brett F. Selection among S_1 families vs.	93 CropSci 33 21-28
multiplicative model to search for "separability" in	crop cultivar trials • Cornelius, P. L.; Seyedsadr, M.; Crossa, J. Using the shifted	92 ThApGene 84 161-172
Some design problems in	crop experimentation. I. The use of blocks • Pearce, S. C.	95 ExpAgric 31 191-203
Some design problems in	crop experimentation. II. Multiple blocking systems • Pearce, S. C.	95 ExpAgric 31 279-290
Some design problems in	crop experimentation. III. Non-orthogonality • Pearce, S. C.	95 ExpAgric 31 409-422
Optimal use of data in	crop model development and validation • Jones, P. N.; Carberry, P. S.	95 EnvrMtrc 6 491-496
Establishing a rejection procedure for	crop performance data • Bowman, Daryl T.; Rawlings, John O.	95 AgronmyJ 87 147-151
Large domain satellite based estimators of	crop planted area • Bellow, Michael E.	94 ASASrvyRsr 194-199
for intercropping experiments. Volume I: Two	crops" • Federer, W. T. (Auth); Rudolph, P. E. (Rev) Review of "Statistical design and analysis	95 BiomtrcJ 37 204-204
description of temporal yield variability of 10	crops in the United States • Eghball, Bahman; Power, James F. Fractal	95 AgronmyJ 87 152-156
Chhikara, R. S. Estimation of	crop yield using post-stratification based on satellite data • Goyal, R. C.; Singh, Randhir;	94 JIAgSt 46 210-222
epistatic variation for metrical traits. V. Triple test	cross analysis • Kumar, I. et al. A general method of detecting additive, dominance and	94 Heredity 72 563-569
lines useful for the improvement of elite single	crosses, based on a combining ability model • Hohls et al. ... estimator for identifying	95 Heredity 75 155-163
and latent dimensions of association in two-way	cross-classifications • Eliason, Scott R. Modeling manifest	95 SocMethR 24 30-67
Reich, R. M.; Czaplewski, R. L.; Bechtold, W. A. Spatial	cross-correlation of undisturbed, natural shortleaf pine stands in northern Georgia •	94 EnvEcoSt 1 201-217
Multiple regimes and	cross-country growth behaviour • Durlauf, Steven N.; Johnson, Paul A.	95 JApEcmtx 10 365-384
Monitoring	cross country skiers • Turner, David L.; Wilson, Jeffrey R.	94 ASAProEnvr 167-172

erty relationships in Escherichia coli based on auto	cross covariances • Sjöström, M.; Rännar, S.; Wieslander, Å. Polypeptide sequence prop-	95 Chemolab 29 295-305
Strong approximation for	cross-covariances of linear variables with long-range dependence • Kouritzin, Michael A.	95 StocProc 60 343-353
Issues and guidelines for translation in	cross-cultural research • Weidmer, Beverly	94 ASASrvyRsr 1226-1231
Use of the repeated	cross-over designs in assessing bioequivalence (Disc: p1079-1080) • Liu, Jen-Pei	95 StatMed 14 1067-1078
Nonparametric analysis of the 2 × 2	cross-over design with multiple endpoints • Tsai, Kao-Tai	94 ASAProBiop 238-243
Evans, M. A.; Alldredge, J. R. (Auth) et al. Review of	"Cross-over experiments. Design, analysis and application" • Ratkowsky, D. A.;	95 CmpStQ 10 99-100
A.; Alldredge, J. R. (Auth); Walker, S. (Rev) Review of	"Cross-over experiments: Design, analysis, and application" • Ratkowsky, D. A.; Evans, M.	95 JRSS-A 158 200-200
(Scenario evaluation and analysis through repeated	cross impact handling): A new method for scenario analysis with an ... • Sapio SEARCH	95 IntJFor 11 113-131
Meng, Xiao-Li Exploring	cross-match estimators with multiply-imputed data sets (Disc: p906-907) • Barnard, John;	94 ASASrvyRsr 894-899
Elsen, Jean-Michel Mapping quantitative trait loci in	crosses between outbred lines using least squares • Haley, Chris S.; Knott, Sara A.;	94 Genetcs 136 1195-1207
Genetic models for predicting maize single-	cross performance in unbalanced yield trial data • Bernardo, Rex	95 CropSci 35 141-147
Prediction of maize single-	cross performance using RFLPs and information from related hybrids • Bernardo, Rex	94 CropSci 34 20-25
Family structure and income: A first analysis on	cross-section data (Italian) • Angeli, A.; Cocchi, D.; Pasquini, L.; Samoggia, A.	94 Statstca 54 435-452
Welch, Finis How representative are matched	cross-sections? Evidence from the Current Population Survey • Peracchi, Franco;	95 JEconmtx 68 153-179
The errors in the variables problem in the	cross-section of expected stock returns • Kim, Dongcheol	95 JFinance 50 1605-1634
Le, N. D.; Zidek, J. V. et al. Multivariate imputation in	cross-sectional analysis of health effects associated with air pollution • Duddek, C.;	95 EnvEcoSt 2 191-212
for community-intervention evaluations: Cohort,	cross-sectional, or both? • Diehr, P.; Martin, D.; Koepsell, T. et al. Optimal survey design	94 ASAProEpid 34-43
Assessing	cross-sectional correlation in panel data • Frees, Edward W.	95 JEconmtx 69 393-414
hypotheses about direction of causation using	cross-sectional family data • Heath, A. C.; Kessler, R. C.; Neale, M. C. et al. Testing	93 BehavGen 23 29-50
Martin, A. R. et al. Estimating foetal growth rate from	cross-sectional samples of field data with application to pilot-whales • Rothery, P.;	95 ApplStat 44 163-172
estimation of population attributable risk under	cross-sectional sampling • Basu, Srabashi; Landis, J. Richard Model-based	95 AJEpidm 141 1338-1343
estimation of population attributable risk under	cross-sectional sampling • Basu, Srabashi; Landis, J. Richard Model-based	95 AJEpidm 142 1338-1343
tests, and sample sizes for the prevented fraction in	cross-sectional studies • Gargiullo, P. M. et al. Confidence intervals, hypothesis	95 StatMed 14 51-72
the weight share method • Lavallée, Pierre	Cross-sectional weighting of longitudinal surveys of individuals and households using	95 SurvyM 21 25-32
Review of	"Cross-over trials in clinical research" • Senn, S. (Auth); Jones, Byron (Rev)	95 JAppStat 22 317-318
The case for	cross-over trials in phase III (Disc: p1053-1055) • Jones, B.; Lewis, J. A.	95 StatMed 14 1025-1038
	Cross-validation	
	Bandwidth selection in nonparametric regression with general errors • Chu, C. K.	95 JStPlInf 44 265-275
	A Bayesian analysis for a class of penalised likelihood estimates • Barry, Daniel	95 CommStA 24 1057-1071
Wahba, Grace; Wang, Yuedong	Behavior near zero of the distribution of GCV smoothing parameter estimates •	95 StPrLet 25 105-111
additive main effects and multiplicative ... • Piepho	Best linear unbiased prediction (BLUP) for regional yield trials: A comparison to	94 ThApGene 89 647-654
surface ozone for the Vancouver, ... • Burrows et al.	CART decision-tree statistical analysis and prediction of summer season maximum	95 JApMetrl 34 1848-1862
regression • Kim, Choongrak; Jeon, Jong-Woo	Detecting influential observations on the smoothing parameter in nonparametric	95 JKoreaSt 24 495-506
to spectroscopic data • Krzanowski, W. J. et al.	Discriminant analysis with singular covariance matrices: Methods and applications	95 ApplStat 44 101-115
nonparametric regression • Chu, C. K.	The effect of the smoothness of the regression function on a bandwidth selector in	95 CommStB 24 61-77
• Wang, Yongxiao; Taylor, Jeremy M. G.	Flexible methods for analysing longitudinal data using piecewise cubic polynomials	95 JStCmpSm 52 133-150
obtained from quantile spline ... • O'Connell et al.	A generalized linear classification model with a smooth link function and predictors	95 JStPlInf 47 153-164
	Imputation of missing values in spatio-temporal solar radiation data • Glasbey, C. A.	95 EnvrMtrc 6 363-371
	Kernel estimation of relative risk • Kelsall, Julia E.; Diggle, Peter J.	95 Bernoull 1 3-16
	A kernel estimator for discrete distributions • Rajagopalan, Balaji; Lall, Upmanu	95 JNnparSt 4 409-426
	On linear discriminant analysis with adaptive ridge classification rules • Loh, Wei-Liem	95 JMultiAn 53 264-278
estimation • Granville, Vincent; Rasson, Jean Paul	Multivariate discriminant analysis and maximum penalized likelihood density	95 JRSS-B 57 501-517
(French) • Moncaup, S.; Sarda, P.; Vieu, P.	Nonparametric estimators of an intensity function applied to meteorological data	95 RStApp 43/4 77-88
Nychka, D.; Gray, G.; Haaland, P.; Martin, D. et al.	A nonparametric regression approach to syringe grading for quality improvement •	95 JASA 90 1171-1178
Carberry, P. S.	Optimal use of data in crop model development and validation • Jones, P. N.;	95 EnvrMtrc 6 491-496
	Practical estimation of multivariate densities using wavelet methods • Tribouley, K.	95 StNeerla 49 41-62
	Projection pursuit discriminant analysis • Polzehl, Jörg	95 CmpStDA 20 141-157
smoothing • Cheng, Bing; Kay, Jim	A regularization method for one-dimensional edge detection and edge-preserving	95 CmpStQ 10 53-69
and multiplicative interaction model for ... • Piepho	Robustness of statistical tests for multiplicative terms in the additive main effects	95 ThApGene 90 438-443
Switzer, Paul	Spatial interpolation errors for monitoring data • Høst, Gudmund; Omre, Henning;	95 JASA 90 853-861
Altman, Naomi; Léger, Christian	Cross-validation, the bootstrap, and related methods for tuning parameter selection •	94 ASAProStCp 41-50
nent analysis • Krzanowski, Wojtek J.; Kline, Paul	Cross-validation for choosing the number of important components in principal compo-	95 MultBehR 30 149-165
	Cross-validation criteria for covariance structures • De Gooijer, Jan G.	95 CommStB 24 1-16
A note on the generalised	cross-validation criterion in linear model selection • Venter, J. H.; Snyman, J. L. J.	95 Biomtrka 82 215-219
recovery problems ... • Girard The fast Monte-Carlo	cross-validation and C_L procedures: Comments, new results and application to image	95 CmpStQ 10 205-231
Kramer, Mark A. Hazard function modeling using	cross validation: From data collection to model selection • Tan, Jonathan S.;	95 RelEngSS 49 155-169
sets with many variables and few objects. Part II:	Cross-validation, missing data and examples • Rännar et al. ... kernel algorithm for data	95 JChemtrc 9 459-470
Mertens, B.; Fearn, T.; Thompson, M. The efficient	cross-validation of principal components applied to principal component regression •	95 StatComp 5 227-235
approach in nonparametric smoothing and	cross-validation • Chaudhuri, Probal; Dewanji, Anup On a likelihood-based	95 StPrLet 22 7-15
Regularized multichannel restoration using	cross-validation • Zhu, Wenwu; Galatsanos, Nikolas P.; Katsaggelos, Aggelos K.	95 CVGIPGr 57 38-54
	Crossing	
	• see also boundary –, level –	
The expected number of level	crossings for certain symmetric α-stable processes • Michna, Zbigniew; Rychlik, Igor	95 CommStSM 11 1-19
probability • Scheinerman, E. R. et al. The rectilinear	crossing number of a complete graph and Sylvester's "Four point problem" of geometric	94 AmMaMnth 101 939-943
Review of "Time series analysis by higher order	crossings" • Kedem, Benjamin (Auth); Frey, Michael (Rev)	95 Technmcs 37 235-235
local asymptotically minimax rate for estimating a	crossing point in a diagnostic marker problem • Hsieh, F.; Turnbull, B. W. A note on the	95 StPrLet 24 181-185
	Crossing time	
	Crossing velocities and random lattice animals • Sznitman, Alain-Sol	95 AnlsProb 23 1006-1023
On	crossing times for multidimensional walks with skip-free components • Borovkov, K. A.	95 JAppProb 32 991-1006
On the typical level	crossing time and path • Nyrhinen, Harri	95 StocProc 58 121-137
	Crossing velocities and random lattice animals • Sznitman, Alain-Sol	95 AnlsProb 23 1006-1023
Extending a Bayesian analysis of the two-period	crossover to accommodate missing data • Grieve, Andrew P.	95 Biomtrka 82 277-286
	Crossover design	
Weng, Chung-Sing	Bias of two one-sided tests procedures in assessment of bioequivalence • Liu, Jen-Pei;	95 StatMed 14 853-861
Rosenfeld, Daniel	Estimation of the effect of operational seeding on rain amounts in Israel • Nirel, Ronit;	95 JApMetrl 34 2220-2229
Liu, Jen-Pei; Wang, Wenping; Chow, Shein-Chung	The impact of outlying subjects on decision of bioequivalence • Ki, Fanny Y. C.;	95 JBiophSt 5 71-94
Norrie, John; Ahmadi, Susan	Model inconsistency, illustrated by the Cox proportional hazards model • Ford, Ian;	95 StatMed 14 735-746
	Repeated measurements designs with unequal period sizes • Afsarinejad, Kasra	94 JItalStS 3 161-168
bioequivalence with respect to TMAX under a 2 × 2	crossover design • Have, Thomas R. Ten; Chinchilli, Vernon M. The analysis of	95 JBiophSt 5 185-199
to the analysis of three-treatment three-period	crossover designs • Bellavance, F.; Tardif, S. A nonparametric approach	95 Biomtrka 82 865-875
Analyzing multivariate data in	crossover designs using permutation tests • Johnson, William D.; Mercante, Donald E.	94 ASAProBiop 244-249
Marginal modelling of categorical data from	crossover experiments • Balagtas, Cecile C.; Becker, Mark P.; Lang, Joseph B.	95 ApplStat 44 63-77
Evans, M. A.; Alldredge, J. R. (Auth) et al. Review of	"Crossover experiments: Design, analysis and application" • Ratkowsky, D. A.;	95 JAppStat 22 188-188
McPeek, Mary Sara Statistical analysis of	crossover interference using the chi-square model • Zhao, Hongyu; Speed, Terence P.;	95 Genetcs 139 1045-1056

He, Xin X. Order	crossover in inventory systems • Hayya, Jack C.; Xu, Susan H.; Ramasesh, Ranga V.;	95 CommStSM 11 279-309
Abdullah, M. E.; El-Sayed, Yousry M. Design of	crossover microcomputer program and application on drug bioequivalence data •	95 CmpBiomd 48 237-239
switching double sampling system indexed by the	crossover point • Arumainayagam, S. Devaraj; Soundararajan, V. Quick	95 CommStB 24 765-773
A delta procedure for the analysis of	crossover and related designs in drug trials • Hwang, Dar-Shong; Hsuan, Francis	94 ASAProBiop 307-312
Essential efficacy data analysis for a 2×2	crossover study • Huster, William; Brunelle, Rocco; Enas, Greg	94 ASAProBiop 503-509
Robins, J. M. Control sampling strategies for case-	crossover studies: An assessment of relative efficiency • Mittleman, M. A.; Maclure, M.;	95 AJEpidm 142 91-98
regression methods for the analysis of a multiperiod	crossover study to compare two treatments • Davis et al. Survey and conditional logistic	94 ASAProBiop 446-451
to comment on "Interaction biases in two-period	crossover studies: A modified analysis to test with more sensitivity" ... • Cleophas Reply	95 BiomtrcJ 37 507-510
Interim analyses in 2×2	crossover trials • Cook, Richard J.	95 Biomtrcs 51 932-945
	Crossover trials with asymmetric covariance within subject • Skinner, Jonathan B.	94 ASAProBiop 411-413
K-fold	crossvalidation in canonical analysis • Liang, Kun-Hsia; Krus, David J.; Webb, James M.	95 MultBehR 30 539-545
parameters of a spatial counting process modelling	crystallization of polymers • Capasso et al. ... of the maximum likelihood estimators of	95 StocAnAp 13 279-294
Shi, Z. The	Csörgő-Révész modulus of non-differentiability of iterated Brownian motion • Hu, Y.;	95 StocProc 58 267-279
statistical profession and the chartered statistician (	CStat) • Smith, T. M. F. The	95 JOffStat 11 33-40
A statistical comparison of	Cuban socioeconomic development • Romeu, Jorge Luis	94 ASAProSoSt 159-163
Random invariant	cubature formulas (Russian) • Ermakov, S. M.; Wang, Xiao-Qun	94 VLenUM /4 109-112
for cubic regression on multidimensional balls and	cubes • Gaffke, N.; Heiligers, B. Computing optimal approximate invariant designs	95 JStPlInf 47 347-376
of a design for quadratic model fits fearing a	cubic bias in multiple response case • Bae, Whasoo A study on a basis for the selection	95 JKoreaSt 24 31-44
	Cubic lattices • Pearce, S. C.	95 JAppStat 22 355-373
Optimal and robust invariant designs for	cubic multiple regression • Gaffke, Norbert; Heiligers, Berthold	95 Metrika 42 29-48
for analysing longitudinal data using piecewise	cubic polynomials • Wang, Yongxiao; Taylor, Jeremy M. G. Flexible methods	95 JStCmpSm 52 133-150
puting optimal approximate invariant designs for	cubic regression on multidimensional balls and cubes • Gaffke, N.; Heiligers, B. Com-	95 JStPlInf 47 347-376
Djojosugito, Rianto A. A	cubic smoothing spline based lack of fit test for nonlinear regression models •	95 CommStA 24 2183-2197
time-dependent ... • Herndon et al. The restricted	cubic spline as baseline hazard in the proportional hazards model with step function	95 StatMed 14 2119-2129
Probabilities and mechanics of a	cuboctahedral die (Korean) • Chae, K. C.; Lee, C. S.	95 KorApS 8/2 179-185
Bayesian	cuboid prediction intervals: An application to tensile-strength prediction • Nandram, B.	95 JStPlInf 44 167-180
Beginning statistics with	cuisenaire rods • Hollingsworth, Caroline	95 TeachgSt 17/2 38-39
Assessing	cultivar adaptability by multiple comparison with the best • Piepho, Hans-Peter	95 AgronmyJ 87 1225-1227
A.; Seyedsadr, M. S. Clustering	cultivars into groups without rank-change interactions • Cornelius, P. L.; Van Sanford, D.	93 CropSci 33 1193-1200
based ... • Hühn et al. Nonparametric analysis of	cultivar performance trials: Experimental results and comparison of different procedures	95 AgronmyJ 87 627-632
fusion method for grouping environments without	cultivar rank change • Crossa, J.; Cornelius, P. L. et al. A shifted multiplicative model	95 CropSci 35 54-62
model to search for "separability" in crop	cultivar trials • Cornelius, P. L.; Seyedsadr, M.; Crossa, J. Using the shifted multiplicative	92 ThApGene 84 161-172
effects and multiplicative interaction model for	cultivar trials • Piepho ... of statistical tests for multiplicative terms in the additive main	95 ThApGene 90 438-443
effects and multiplicative interaction model for	cultivar trials • Cornelius, P. L. Statistical tests and retention of terms in the additive main	93 CropSci 33 1186-1193
et al. Discriminant analysis for social-economic and	cultural factors affecting diarrhea diseases in children under two years in Egypt • Ashour	95 EgpStJ 39 1-11
Issues and guidelines for translation in cross-	cultural research • Weidmer, Beverly	94 ASASrvyRsr 1226-1231
tables • Stafford, James E. Exact	cumulant calculations for Pearson χ^2 and Zelterman statistics for r-way contingency	95 JCmpGrSt 4 199-212
with tied rankings • Valz, P. D.; McLeod, A. I. et al.	Cumulant generating function and tail probability approximations for Kendall's score	95 AnlsStat 23 144-160
formulae for the determination of moments and	cumulants • Rahman, M. S.; Nahar, S. General	95 PakJStA 11 11-15
Tõnu; von Rosen, Dietrich Minimal moments and	cumulants of symmetric matrices: An application to the Wishart distribution • Kollo,	95 JMultiAn 55 149-164
subspace techniques in non-Gaussian noise using	cumulants • Sadler, Brian M.; Giannakis, Georgios B.; Shamsunder, Sanyogita Noise	95 IEEEAES 31 1009-1018
of the old problem of obtaining moments from	cumulants and vice versa • Smith, Peter J. A recursive formulation	95 AmerStat 49 217-218
Comment on "Recursive schemes for calculating	cumulative binomial and Poisson probabilities" (V48 p136-138) • Nemenyi, Peter	95 AmerStat 49 402-403
Statistical modeling of long term	cumulative effects • White, Donald B.; Foley, Marianne	94 ASAProBiop 328-333
hazard ... • Gu Convergence of increments for	cumulative hazard function in a mixed censorship-truncation model with application to	95 StPrLet 23 135-139
case study • Henderson, W. G.; Moritz, T. et al. Use of	cumulative meta-analysis in the design, monitoring, and final analysis of a clinical trial: A	95 ConClnTr 16 331-341
Cox, Christopher Location-scale	cumulative odds models for ordinal data: A generalized non-linear model approach •	95 StatMed 14 1191-1203
Mills, Terence C.; Roberts, Jennifer Testing	cumulative prediction errors in event study methodology • Coutts, J. Andrew;	95 JForcast 14 107-115
A bivariate	cumulative probit regression model for ordered categorical data • Kim, Kyungmann	95 StatMed 14 1341-1352
Average error in the central limit theorem for the	cumulative processes • Roginsky, Allen	95 StPrLet 24 199-204
Convergence theorems for empirical	cumulative quantile regression functions • Rao, C. R.; Zhao, L. C.	95 MaMethSt 4 81-91
Correction to "Convergence theorems for empirical	cumulative quantile regression functions" (V4 p81-91) • Rao, C. R.; Zhao, L. C.	95 MaMethSt 4 359-359
A	cumulative score control scheme for detecting process shifts • Cheng, Chuen-Sheng	95 CommStA 24 755-774
On the HNBUE property in a class of correlated	cumulative shock models • Pérez-Ocón, Rafael; Gámiz-Pérez, M. Luz	95 AdvAppPr 27 1186-1188
Bilgi, Shobha	Cumulative sum control charts for the inverse Gaussian distribution • Nabar, S. P.;	94 JIndnStA 32 9-14
A	cumulative sum control chart for monitoring process variance • Chang, T. C.; Gan, F. F.	95 JQualTek 27 109-119
Chang, T. C.; Gan, F. F. Correction to "A	cumulative sum control chart for monitoring process variance" (V27 p109-119) •	95 JQualTek 27 402-402
Weighted	cumulative sums for text analysis using word counts • Bissell, A. F.	95 JRSS-A 158 525-545
Two-sample rank tests for acceleration in	cure models • Lee, Jae Won	95 StatMed 14 2111-2118
Group sequential methods for comparison of	cure rates in clinical trials • Lee, Jae Won; Sather, Harland N.	95 Biomtrcs 51 756-763
covering birth employers and nonemployers in the	Current Business Surveys • Konschnik ... dropping the business area sample as a means of	94 ASASrvyRsr 995-1000
Estimating the	current mean of multivariate processes subject to abrupt changes • Yashchin, Emmanuel	94 ASAProPEng 81-86
Estimating the	current mean of a process subject to abrupt changes • Yashchin, Emmanuel	95 Technmcs 37 311-323
Effects of methodological change in the	Current Population Survey • Dippo, Cathryn S.; Kostanich, Donna L.; Polivka, Anne E.	94 ASASrvyRsr 260-263
Comment on	"Current Population Survey" • Copeland, Kennon R.	94 ASASrvyRsr 360-361
Kojetin Characteristics of nonrespondents to the	Current Population Survey (CPS) and Consumer Expenditure Interview Survey (CEIS) •	94 ASASrvyRsr 966-971
are matched cross-sections? Evidence from the	Current Population Survey • Peracchi, Franco; Welch, Finis How representative	95 JEconmtx 68 153-179
cognitive evaluation of the nativity questions for the	Current Population Survey • Wellens, Tracy The	94 ASASrvyRsr 1204-1209
nomic and political conditions to refusal rates in the	Current Population Survey • Kojetin et al. ... political participation: The relation of eco-	94 ASASrvyRsr 1236-1241
Composite weights for the	Current Population Survey • Lent, Janice; Miller, Stephen; Cantwell, Patrick	94 ASASrvyRsr 867-872
Rank regression for	current status data • Aragón, Jorge; Quiróz, Adolfo J.	95 StPrLet 24 251-256
Generalizations of	current status data with applications • Jewell, Nicholas P.; van der Laan, Mark	95 LiDaAnal 1 101-109
Conditional probability: Its place in the mathematics	curriculum • Watson, Jane M.	95 MathTchr 88 12-17
Jacqueline M.	Curvature adjusted optimal nonlinear designs for correlated responses • Hughes-Oliver,	94 ASAProBuEc 111-116
process over a U-shaped boundary of increasing	curvature • Balabushkin Approximation to the exit probability of a continuous Gaussian	95 JAppProb 32 429-442
	Curvature measures for multiresponse regression models • Guay, M.	95 Biomtrka 82 411-417
Central limit theorem in negative	curvature • Ledrappier, François	95 AnlsProb 23 1219-1233
	Curve	
	• see also concentration –, growth –, learning –, survival –	
Fienberg, S. E.; Resnick, D. P. et al. Wringing the bell	curve: A cautionary tale about the relationships among race, genes, and IQ • Devlin, B.;	95 Chance 8/3 27-36
II: A more complicated PCA example comparing	curves • Gunter, Bert Graphical methods and principles for data analysis	95 QulPr 28/12 114-115
principles for data analysis II: Graphically comparing	curves • Gunter, Bert Graphical methods and	95 QulPr 28/8 141-143
Review of "Model-free	curve estimation" • Tartar, M. E.; Lock, M. D. (Auth); Nason, Guy (Rev)	95 SIAMRvw 37 270-271
Review of "Model-free	curve estimation" • Tarter, Michael E.; Lock, Michael D. (Auth); Wharton, Robert M. (Rev)	95 Technmcs 37 460-460

Sampling theory for	cytonuclear disequilibria • Asmussen, Marjorie A.; Basten, Christopher J.	94 Genetcs 138 1351-1363
Weir, B. S. Analysis of	cytoplasmic and maternal effects. II. Genetic models for triploid endosperms • Zhu, J.;	94 ThApGene 89 160-166
animals • Zhu, J.; Wier, B. S. Analysis of	cytoplasmic and maternal effects. 1. A genetic model for diploid plant seeds and	94 ThApGene 89 153-159
Minkin et al. Assessing sensitivity in suspension to	cytosine arabinoside: Statistical analysis, associations with clinical outcome and ... •	95 StatMed 14 1081-1095
Obituary:	Czesław Platt, 1925-1995 (Polish) • Mejza, Stanisław	95 ListyBio 32 140-140
classification through adaptive Delaunay-based	C^0 spline approximation • Cubanski, David; Cyganski, David Multivariate	95 IEEEPAMI 17 403-417
Girard The fast Monte-Carlo cross-validation and	C_L procedures: Comments, new results and application to image recovery problems ... •	95 CmpStQ 10 205-231
More comments on	C_p • Mallows, Colin L.	95 Technmcs 37 362-372
	C_p-criterion	
the minimax regret approach • Droge, B.; Georg, T.	On selecting the smoothing parameter of least squares regression estimates using	95 StatDecn 13 1-20
sample size for lower confidence limits for	C_p and C_{pk} • Franklin, LeRoy A. Determining	94 ASAProQlPr 23-27
sample size for lower confidence limits for C_p and	C_{pk} • Franklin, LeRoy A. Determining	94 ASAProQlPr 23-27
of the estimated process capability index	$\hat{C}_{pk}$ • Nagata, Yasushi On normal approximation to the distribution	95 JJapnStS 25 159-171
distribution of the process capability index	C_{pmk} • Chen, Sy-Mien; Hsu, Nai-Feng The asymptotic	95 CommStA 24 1279-1291
Further applications of	D-complete designs to group testing • Schultz, Delray J.; Parnes, Milton; Srinivasan, R.	93 JCISS 18 31-41
Some remarks on the	D compound Poisson distribution • Papageorgiou, H.	95 StHefte 36 371-375
for the efficiency of designs with respect to the	D-criterion when the observations are correlated • Bischoff, Wolfgang Lower bounds	95 MaOpfStS 27 27-44
A geometric combination estimator for	d-dimensional ordinal sparse contingency tables • Dong, Jianping; Simonoff, Jeffrey S.	95 AnlsStat 23 1143-1159
Generating random variates from	D-distributions via substitution sampling • Walker, Stephen	95 StatComp 5 311-315
On the equivalence of	D- and G-optimal designs in heteroscedastic models • Wong, Weng Kee	95 StPrLet 25 317-321
approach for the construction of constrained	D- and L-optimal designs using efficiency plots • Wong, Weng Kee A graphical	95 JStCmpSm 53 143-152
Phan-Than-Luu, R.	D-optimal designs • de Aguiar, P. F.; Bourguignon, B.; Khots, M. S.; Massart, D. L.;	95 Chemolab 30 199-210
Wierich, Werner	D-optimal designs of experiments with non-interacting factors • Schwabe, Rainer;	95 JStPlInf 44 371-384
Haines, Linda M. Bayesian	D-optimal designs for the exponential growth model • Mukhopadhyay, Saurabh;	95 JStPlInf 44 385-397
Exact and approximate	D-optimal designs in polynomial regression • Happacher, Maximilian	95 Metrika 42 19-27
Chang, Fu-Chuen; Wong, Weng Kee	D-optimal designs for polynomial regression without an intercept • Huang, Mong-Na Lo;	95 StSinica 5 441-458
Dette, Holger Correction to "Bayesian	D-optimal and model robust designs in linear regression models" (V25 p27-46) •	95 MaOpfStS 26 285-286
Mukerjee, Rahul	D-optimal saturated main effect plans for $2 \times s_2 \times s_3$ factorials • Chatterjee, Kashinath;	93 JCISS 18 116-122
	D-optimality	
in its determination by ... • Papamichael et al.	Design of experiments for a precise estimation of the calibration curve of periodate	95 Chemolab 30 227-237
de Wet, T.; Fresen, J. L.	An experimental design for evaluating sonar detection performance • de Jongh, P. J.;	95 SoAfrStJ 29 79-103
in dimethylsulfoxide - An ... • Feuer et al.	Free radical polymerization of nonlinear optical side chain methacrylate copolymers	95 Chemolab 29 113-124
Lohr, Sharon L.	Optimal Bayesian design of experiments for the one-way random effects model •	95 Biomtrka 82 175-186
	Optimal designs for mixture amount experiments • Hilgers, Ralf-Dieter; Bauer, Peter	95 JStPlInf 48 241-246
Mukerjee, Rahul	On optimal designs with restricted circular string property • Huda, S.; Khan, I. H.;	95 CmpStDA 19 75-83
	Optimal regression designs under random block-effects models • Cheng, Ching-Shui	95 StSinica 5 485-497
Federer, W. T.	Representation and bounds for the general fractional factorial • Anderson, D. A.;	95 CommStA 24 363-373
Majorization and	D-optimality for complete block designs under correlations • Bhaumik, Dulal K.	95 JRSS-B 57 139-143
	D-optimum designs for heteroscedastic linear models • Atkinson, A. C.; Cook, R. D.	95 JASA 90 204-212
Spurious regressions between I(	d) processes • Marmol, Francesc	95 JTimSrAn 16 313-321
of the running maximum and its location of	D-valued Markov processes • Ferger, Dietmar The joint distribution	95 JAppProb 32 842-845
An empirical model of asset replacement in	dairy production • Miranda, Mario J.; Schnitkey, Gary D.	95 JApEcmtx 10 41-55S
Optimal control of	dams • Abdel-Hameed, Mohamed; Nakhi, Yasmin	95 AplStMDA 11 25-34
some income distributions based on multiplicative	damage models • Fosam, E. B.; Sapatinas, T. Characterisations of	95 AstrlJSt 37 89-93
D. G.; Lausen, B.; Sauerbrei, W.; Schumacher, M.	Dangers of using "optimal" cutpoints in the evaluation of prognostic factors • Altman,	94 JNCncrI 86 829-835
	Daniels's association measures under right censoring • O'Quigley, John	95 AplStMDA 11 109-119
Jensen, J. et al. Response to selection for litter size in	Danish Landrace pigs: A Bayesian analysis • Wang, C. S.; Gianola, D.; Sorensen, D. A.;	94 ThApGene 88 220-230
movements and estimation of population size in	darkling beetles (Coleoptera: Tenebrionidae) • Crist, Thomas O.; Wiens, John A. Individual	95 JAnmlEco 64 733-746
	Darwin's log: A toy model of speciation and extinction • Aldous, David	95 JAppProb 32 279-295
Rogers, L. C. G.; Yor, M. A proof of	Dassios' representation of the α-quantile of Brownian motion with drift • Embrechts, P.;	95 AnlsApPr 5 757-767
	Data	
	Birds of Britain • Feu, Chris du	95 TeachgSt 17/3 104-106
Cadish, Marc; Clerc, Jean-Thomas; Pretsch, Ernö	Hypermedia - A new concept for information management • Macher, Christian L.;	95 Chemolab 28 213-228
	• see names of specific types of data such as binary -, categorical -, cohort -, missing -, survival -	
	Data analysis	
Lalonde, Robert	Bordeaux wine vintage quality and the weather • Ashenfelter, Orley; Ashmore, David;	95 Chance 8/4 7-14
	Data features • Davies, P. L.	95 StNeerla 49 185-245
Jing, Yihua	The effect of thyroxine and prolactin on the molting of newts: A factorial study •	95 Stats /12 24-27
	The statistician and the computer • Jeffers, J. N. R.	95 JAppStat 22 227-234
	• see also descriptive statistics, categorical -, exploratory -	
Essential efficacy	data analysis for a 2×2 crossover study • Huster, William; Brunelle, Rocco; Enas, Greg	94 ASAProBiop 503-509
Anna (Rev) Review of "Computer-aided	data analysis" (German) • Nagel, M.; Wernecke, K. D.; Fleischer, W. (Auth); Bartkowiak,	95 ListyBio 32 47-49
Kowalski, Bruce Selectivity, local rank, three-way	data analysis and ambiguity in multivariate curve resolution • Tauler, Romá; Smilde, Age;	95 JChemtrc 9 31-58
Johnson, Roger A	data analysis based elementary statistics course • Egge, Eric; Foley, Sean; Haskins, Loren;	94 ASAProStEd 144-149
Bayesian	data analysis • Gelman, A.; Carlin, J.; Stern, H.; Rubin, D.	95 ChapmnHall:Lnd 552p
Reiser, B. et al. Dependent masking and system life	data analysis: Bayesian inference for two-component systems • Guttman, I.; Lin, D. K. J.;	95 LiDaAnal 1 87-100
Review of "Bayesian	data analysis" • Gelman, A.; Carlin, J. B.; Stern, H. S.; Rubin, D. B. (Auth); Brown, P. J. (Rev)	96 ShrtBkRv 16/1 5-5
C. (Auth); Haaland, P. D. (Rev) Review of "Large	data analysis for the chemical sciences: A guide to statistical techniques" • Graham, R.	95 Chemolab 27 245-246
index distribution: An application of functional	data analysis • Grambsch, P. M.; Randall, B. L.; Bostick, R. M. et al. Modeling the labeling	95 JASA 90 813-821
Geographical	data analysis • Walford, Nigel	95 Wiley:NY:UK 446p
and SPSS for Windows - A beginner's guide to	data analysis (Second edition)" • Foster, J. J. (Auth); Hawkins, A. (Rev) Review of "SPSS/PC+	95 TeachgSt 17/2 79-79
Review of "The	data analysis handbook" • Frank, I. E.; Todeschini, R. (Auth); Rius, F. Xavier (Rev)	95 Chemolab 29 327-327
Statistical	data analysis for identifying kinetic processes • Urfer, Wolfgang; Becka, Michael	94 ASAProEnvr 14-18
Graphical methods and principles for	data analysis II: Austria first? Zambia last? • Gunter, Bert	95 QulPr 28/4 103-105
Graphical methods and principles for	data analysis II: Dot plots of multiway data • Gunter, Bert	95 QulPr 28/6 88-95
Graphical methods and principles for	data analysis II: Graphically comparing curves • Gunter, Bert	95 QulPr 28/8 141-143
Graphical methods and principles for	data analysis II: Graphically comparing many curves using PCA • Gunters, Bert	95 QulPr 28/10 129-137
Graphical methods and principles for	data analysis II: A more complicated PCA example comparing curves • Gunter, Bert	95 QulPr 28/12 114-115
Review of	"Data analysis: An introduction" • Nolan, B. (Auth); Fielding, Antony (Rev)	95 Ststcian 44 546-547
Statistics and	data analysis: An introduction (Second edition) • Siegel, Andrew F.; Morgan, Charles J.	95 Wiley:NY:UK 656p
Graphical	data analysis in linear regression • Jamshidian, Mortaza	92 ProIrStC1 51-122
undergraduate students modern methods of	data analysis • Deppa, Brant D. Teaching	94 ASAProStEd 209-214

Exact statistical methods in data analysis • Weerahandi, S. 95 Springer:Brln:NY 328p
Review of "Statistics for business: Data analysis and modelling" • Cryer, J.; Miller, R. (Auth); Ziegel, Eric R. (Rev) 95 Technmcs 37 239-240
Advances in stochastic modelling and data analysis • Janssen, Jacques; Skiadas, Christos H.; Zopounidis, Constantin (Ed) 95 Kluwer:MA:Hllnd 440p
Saphire, Diane G. (Rev) Review of "Data analysis for monitoring human rights" • Spirer, Herbert; Spirer, Louise (Auth); 95 AmerStat 49 322-322
Applied multivariate data analysis • Everitt, Brian S.; Dunn, Graham 95 Wiley:NY:UK 304p
to tensor products with applications to multiway data analysis • Burdick, Donald S. An introduction 95 Chemolab 28 229-237
infra-red spectroscopy: Bridging the gap between data analysis and NIR applications" • Hildrum, K.; Isaksson, T. et al. Review of "Near 95 Technmcs 37 126-127
Solow, Andrew R. (Rev) Review of "Statistical data analysis for ocean and atmospheric sciences" • Thiébaux, H. Jean (Auth); 95 AmerStat 49 399-399
MODA 4 - Advances in model-oriented data analysis • Kitsos, Christos P.; Müller, Werner G. (Ed) 95 Physica:Ger 297p
A functional data analysis of the pinch force of human fingers • Ramsay, J. O.; Wang, X.; Flanagan, R. 95 ApplStat 44 17-30
H.-J. (Rev) Review of "New directions in statistical data analysis and robustness" • Morgenthaler, S.; Ronchetti, E.; Stahel, W. A. (Ed); Lenz, 95 CmpStDA 19 711-712
Watkins, Ann E. (Rev) Review of "Introducing data analysis in the schools: Who should teach it and how?" • Pereira-Mendoza, Lionel (Ed); 95 AmerStat 49 62-63
van Tongeren, Onno F. R. Data analysis or simulation model: A critical evaluation of some methods • 95 EcolgMdl 78 51-60
use of conditional probability functions in range data analysis and simulation • Lambert, David K.; Harris, Thomas R. The 93 JRngeMgt 46 157-160
analytical chemistry Volume 1: PC-aided statistical data analysis" • Meloun, M.; Militky, J.; Forina, M. (Auth) et al. Review of "Chemometrics for 95 Technmcs 37 114-115
A casebook for a first course in statistics and data analysis • Chatterjee, Samprit; Handcock, Mark S.; Simonoff, Jeffrey S. 95 Wiley:NY:UK 314p
Environmental statistics and data analysis • Ott, Wayne R. 95 CRCPress:FL 336p
Review of "Environmental statistics and data analysis" • Ott, Wayne R. (Auth); Lesser, Virginia M. (Rev) 95 AmerStat 49 398-399
Review of "Mathematical statistics and data analysis (Second edition)" • Rice, John (Auth); Ziegel, Eric R. (Rev) 95 Technmcs 37 127-127
Variance estimation techniques in categorical data analysis for survey data • Rai, Anil 94 JIAgSt 46 147-155
Problems and prediction in survival-data analysis • Henderson, Robin 95 StatMed 14 161-184
Hierarchical linear models: A multilevel data analysis technique • Wu, Yow-Wu B. 95 NurseRes 44 123-126
Data analysis using APL2 and APL2STAT • Fox, John; Friendly, Michael 95 SocMethR 23 282-302
Data analysis using GAUSS and Markov • Long, J. Scott; Noss, Brian 95 SocMethR 23 303-328
Data analysis using Lisp-Stat • Tierney, Luke 95 SocMethR 23 329-351
Data analysis using Mathematica • Stine, Robert A. 95 SocMethR 23 352-372
Data analysis using S-Plus • Schulman, Daniel A.; Campbell, Alec D.; Kostello, Eric C. 95 SocMethR 23 392-413
Data analysis using SAS • Hallahan, Charlie 95 SocMethR 23 373-391
of error covariance parameters for atmospheric data assimilation • Dee, Dick P. On-line estimation 95 MWeathr 123 1128-1145

Data augmentation

Missing data imputation using the multivariate t distribution • Liu, Chuanhai 95 JMultiAn 53 139-158
Bureau latent-class problem • Larsen, Michael D. Data augmentation with Bayesian iterative proportional fitting applied to a Census 94 ASAProGovt 116-121
Keep track of your measures by using a measures data base • Dickinson, Teresa 95 QulPr 28/9 168-168
Review of "A data-based approach to statistics" • Iman, Ronald L. (Auth); Edgeman, Rick L. (Rev) 95 JQualTek 27 88-89
Review of "A data-based approach to statistics" • Iman, Ronald L. (Auth); Halvorsen, Katherine (Rev) 95 AmerStat 49 321-321
Recognition of objects in 2D images or 3D range data based on high degree implicit polynomial curves or surfaces, ... • Subrahmonia et al. 94 ASAProPEng 44-49
Double bootstrap confidence cones for spherical data based on prepivoting • Shin, Yang-Kyu 95 JKoreaSt 24 183-195

Data collection

Design issues in case-control studies • Wacholder, Sholom 95 StMedRes 4 293-309
A data collection methodology for the 2001 Canadian Census • Choudhry, G. H. 94 ASASrvyRsr 1134-1139
function modeling using cross validation: From data collection to model selection • Tan, Jonathan S.; Kramer, Mark A. Hazard 95 RelEngSS 49 155-169
of library data • Kindel, Carrol Benner Electronic data collection at the National Center for Education Statistics: Successes in the collection 94 JOffStat 10 93-102
empirical validity of dependent and independent data collection in a panel survey • Hill, Daniel H. The relative 94 JOffStat 10 359-380
Nony, P. et al. Dose-ranging trials: Guidelines for data collection and standardized descriptions • Boissel, J.-P.; Durieu, I.; Girard, P.; 95 ConClnTr 16 319-330
Preliminary tests of new technologies for data collection • O'Reilly, J. M. et al. Audio and video computer assisted self-interviewing: 94 JOffStat 10 197-214
Confidentiality issues in videotaped data collection • Cederlund, Ann E. 94 ASASrvyRsr 1248-1251
A storage model for data communication systems • Prabhu, N. U.; Pacheco, A. 94 ThProbAp 39 604-627
course • Guffey, James M. Using data and the computer to teach and to visualize concepts in an introductory statistics 94 ASAProStEd 84-87
When the metadata exceed the data: Data management with uncertain data • Klensin, John C. 95 StatComp 5 73-84
Perron, Pierre Unit root tests in ARMA models with data-dependent methods for the selection of the truncation lag • Ng, Serena; 95 JASA 90 268-281
Review of "SoftWindows, Mathematica and Data Desk for the Macintosh PowerPC" • Spiegelman, Cliff (Rev) 95 Chemolab 28 208-209
Optimal periodicity of a survey: Sampling error, data deterioration, and cost • Ghosh, D.; Kaufman, S. F.; Smith, W.; Chang, M. 94 ASASrvyRsr 1122-1127

Data disclosure

confidentiality • Fienberg, Stephen E. Conflicts between the needs for access to statistical information and demands for 94 JOffStat 10 115-132
spatial adaptation • Fan, Jianqing; Gijbels, Irène Data-driven bandwidth selection in local polynomial fitting: Variable bandwidth and 95 JRSS-B 57 371-394
Boente, G.; Fraiman, R. Asymptotic distribution of data-driven smoothers in density and regression estimation under dependence • 95 CandJSt 23 383-397
Data driven versions of Pearson's chi-square test for uniformity • Bogdan, Malgorzata 95 JStCmpSm 52 217-237
Consistency and Monte Carlo simulation of a data driven version of smooth goodness-of-fit tests • Kallenberg, W. C. M.; Ledwina, T. 95 AnlsStat 23 1594-1608
Data editing and imputation from a computational point of view • Willenborg, L. C. R. J. 95 StNeerla 49 139-151
Data editing in a mixed DBMS environment • Cox, N. W. P.; Croot, D. A. 91 StatJUN 8 117-136
Evaluation of data editing procedures: Results of a simulation approach • Garcia Rubio, E.; Peirats, V. 91 StatJUN 8 175-190
SAS use in data editing • Ferguson, D. P. 91 StatJUN 8 167-174
McKay, R. B.; Robison, E. L.; Malik, A. B. Touch-tone data entry for household surveys: Research findings and possible applications • 94 ASASrvyRsr 509-511
Converting mail reporters to touchtone data entry • Rosen, Richard J.; Clayton, Richard L.; Pivetz, Lynda L. 94 ASASrvyRsr 763-768

Data envelopment analysis

efficiency • Lovell, C. A. K. Linear programming approaches to the measurement and analysis of productive 94 Top 2 175-248
Validation and generalization of DEA and its uses • Banker, R. D.; Cooper, W. W. 94 Top 2 249-314
Identifying parametric models for experimental data (French) • Walter, E.; Pronzato, L. 94 Masson:NY:Paris 384p
Data features • Davies, P. L. 95 StNeerla 49 185-245
Probabilistic linkage of large public health data files (Disc: p687-689) • Jaro, Matthew A. 95 StatMed 14 491-498
Review of "Clinical data management" • Rondel, R. K.; Varley, S. A.; Webb, C. F. (Ed); Therasse, P. (Rev) 95 StatMed 14 877-877
When the metadata exceed the data: Data management with uncertain data • Klensin, John C. 95 StatComp 5 73-84
Data, metadata and information • Hand, D. J. 93 StatJUN 10 143-152

Data quality

Shrestha, Laura B.; Preston, Samuel H. Consistency of census and vital registration data on older Americans: 1970-1990 • 95 SurvyM 21 167-177
Design issues in case-control studies • Wacholder, Sholom 95 StMedRes 4 293-309
Arndt, S.; Tyrrell, G.; Wollson, R. F.; Flaum, M. et al. Effects of errors in a multicenter medical study: Preventing misinterpreted data • 94 JPsyctrc 28 447-459
Data quality in a CAPI survey: Keying errors • Dielman, Lynn; Couper, Mick P. 95 JOffStat 11 141-146
questions, questioning answers: Evaluating data quality in an establishment survey • Goldenberg, Karen L. Answering 94 ASASrvyRsr 1357-1362
Canadian survey of labour and ... • Grondin et al. Data quality of income data using computer assisted interviewing: The experience of the 94 ASASrvyRsr 839-844
of final data with use of preliminary and/or revised data • Mariano, Roberto S.; Tanizaki, Hisashi Prediction 95 JForcast 14 351-380

Minimax testing of hypotheses on the distribution	density for ellipsoids in l_p • Ingster, Yu. I.	94 ThProbAp 39 417-436
approach to nonparametric deconvolution of a	density estimate • Barry, J.; Diggle, P. Choosing the smoothing parameter in a Fourier	95 JNnparSt 4 223-232
Kernel type	density estimates for positive valued random variables • Bagai, Isha; Rao, B. L. S. Prakasa	95 SankhyaA 57 56-67
	Density estimation	
Hinkley, D. V.; Worton, B. J.	Accurate and efficient construction of bootstrap likelihoods • Davison, A. C.;	95 StatComp 5 257-264
qualitative smoothing • Simonoff, Jeffrey S.	The anchor position of histograms and frequency polygons: Quantitative and	95 CommStB 24 691-710
	Another proof of a slow convergence result of Birgé • Devroye, Luc	95 StPrLet 23 63-67
Barabesi, Lucio; Fattorini, Lorenzo	Approximation of density functions by orthogonal series with grouped data •	94 JItalStS 3 181-200
tomography and applications • Chang, I-S. et al.	Asymptotic consistency of the maximum likelihood estimate in positron emission	94 AnlsStat 22 1871-1883
Mason, D. M.	On the asymptotic normality of L_p-norms of empirical functionals • Beirlant, J.;	95 MaMethSt 4 1-19
Rohana J.; Quinn, Terrance J., II	Bayesian estimation of animal abundance for line transect sampling • Karunamuni,	95 Biomtrcs 51 1325-1337
Joseph P.	Bias-corrected nonparametric spectral estimation • Politis, Dimitris N.; Romano,	95 JTimSrAn 16 67-103
	Bootstrapping a consistent nonparametric goodness-of-fit test • Fan, Yanqin	95 EcomtrRv 14 367-382
Anderson, N. H.; Titterington, D. M.	A comparison of two statistics for detecting clustering in one dimension •	95 JStCmpSm 53 103-125
Belkacem	Computationally efficient classes of higher-order kernel functions • Abdous,	95 CandJSt 23 21-27
Weron, A.	Computer-aided modeling and simulation of electrical circuits with α-stable noise •	95 ApplMath 23 83-93
noise • Janicki, A.	Computer simulation of a nonlinear model for electrical circuits with α-stable	95 ApplMath 23 95-105
Shaw, J.; Talbot, M.	Contaminant detection in the visual inspection of seed samples • Aitken, C. G. G.;	95 ApplStat 44 431-440
Christian H.	Deconvolving a density from contaminated dependent observations • Hesse,	95 AnInStMa 47 645-663
H.	Deconvolving a density from partially contaminated observations • Hesse, Christian	95 JMultiAn 55 246-260
	The detection of change in spatial processes • Martin, E. B.; Scott, E. M.	95 EnvrMtrc 6 141-154
function • Pham, Dinh Tuan	On the discretisation error in the computation of the empirical characteristic	95 JStCmpSm 53 129-141
Niemiro, W.	Estimation of nuisance parameters for inference based on least absolute deviations •	95 ApplMath 22 515-529
a noisy image • Godtliebsen, Fred; Chu, Chih-Kang	Estimation of the number of true gray levels, their values, and relative frequencies in	95 JASA 90 890-899
Nicolas W.; Stark, Philip B.	Finite-sample confidence envelopes for shape-restricted densities • Hengartner,	95 AnlsStat 23 525-550
	On higher order kernels • Jones, M. C.	95 JNnparSt 5 215-221
	Improved Monte Carlo inference for models with additive error • Hazelton, Martin	95 StatComp 5 343-350
Arcones, Miguel A.; Giné, Evarist	On the law of the iterated logarithm for canonical U-statistics and processes •	95 StocProc 58 217-245
	Local likelihood based on kernel censoring • Copas, J. B.	95 JRSS-B 57 221-235
pricing • Kim, B.-D.; Blattberg, R. C.; Rossi, P. E.	Modeling the distribution of price sensitivity and implications for optimal retail	95 JBES 13 291-303
Diggle, Peter J.	Non-parametric estimation of spatial variation in relative risk • Kelsall, Julia E.;	95 StatMed 14 2335-2342
	Nonparametric kernel estimation for semiparametric models • Andrews, Donald W. K.	95 EconTher 11 560-596
agnostic marker problem • Hsieh, F.; Turnbull, B. W.	A note on the local asymptotically minimax rate for estimating a crossing point in a di-	95 StPrLet 24 181-185
	Practical estimation of multivariate densities using wavelet methods • Tribouley, K.	95 StNeerla 49 41-62
	Projection pursuit discriminant analysis • Polzehl, Jörg	95 CmpStDA 20 141-157
siglio, Loredana; Patil, Ganapati P.; Taillie, Charles	Propagation functions for monitoring distributional changes • Kaur, Amarjot; Di Con-	95 EnvEcoSt 2 239-269
maps • Hall, Peter; Wolff, Rodney C. L.	Properties of invariant distributions and Lyapunov exponents for chaotic logistic	95 JRSS-B 57 439-452
	Rates of convergence for best entropy estimates • Noll, Dominikus	95 StatDecn 13 141-165
	Robustness of minimum disparity estimators in linear regression models • Pak, Rojin	95 JKoreaSt 24 349-360
Sarle, W.	Scale measures for bandwidth selection • Janssen, P.; Marron, J. S.; Veraverbeke, N.;	95 JNnparSt 5 359-380
	Searching for structure in curve samples • Gasser, Theo; Kneip, Alois	95 JASA 90 1179-1188
Ramsay, J. O.	A similarity-based smoothing approach to nondimensional item analysis •	95 Psymtrka 60 323-339
Pfeifer, C.	Some remarks on optimal kernel functions • Granovsky, B. L.; Müller, H. G.;	95 StatDecn 13 101-116
	Strong convergence of sums of φ-mixing random variables • Liebscher, E.	95 MaMethSt 4 216-229
Zhu, Li-Xing; Wong, Hoi Lam; Fang, Kai-Tai	A test for multivariate normality based on sample entropy and projection pursuit •	95 JStPlInf 45 373-385
Solka, J. L.; Poston, W. L.; Wegman, E. J.	A visualization technique for studying the iterative estimation of mixture densities •	95 JCmpGrSt 4 180-198
	On wavelet methods for estimating smooth functions • Hall, Peter; Patil, Prakash	95 Bernoull 1 41-58
Iain M.; Kerkyacharian, Gérard; Picard, Dominique	Wavelet shrinkage: Asymptopia? (Disc: p337-369) • Donoho, David L.; Johnstone,	95 JRSS-B 57 301-337
	• see also nonparametric –, kernel –	
Smoothing spline	density estimation: Conditional distribution • Gu, Chong	95 StSinica 5 709-726
Asymptotic normality of L_1-error in	density estimation • Berlinet, Alain; Devroye, Luc; Györfi, László	95 MaOpfStS 26 329-343
Bayesian	density estimation and inference using mixtures • Escobar, Michael D.; West, Mike	95 JASA 90 577-588
Maximum smoothed likelihood	density estimation for inverse problems • Eggermont, P. P. B.; LaRiccia, V. N.	95 AnlsStat 23 199-220
analysis and maximum penalized likelihood	density estimation • Granville, Vincent; Rasson, Jean Paul Multivariate discriminant	95 JRSS-B 57 501-517
Maximum smoothed likelihood	density estimation • Eggermont, P. P. B.; Lariccia, V. N.	95 JNnparSt 4 211-222
	Density estimation under long-range dependence • Csörgő, Sándor; Mielniczuk, Jan	95 AnlsStat 23 990-999
A simple bias reduction method for	density estimation • Jones, M. C.; Linton, O.; Nielsen, J. P.	95 Biomtrka 82 327-338
Wolfgang	Density estimation under qualitative assumptions in higher dimensions • Polonik,	95 JMultiAn 55 61-81
On the strong uniform consistency of	density estimation for strongly dependent sequences • Ho, Hwai-Chung	95 StPrLet 22 149-156
A two-stage rank test using	density estimation • Albers, Willem	95 AnInStMa 47 675-691
A simple, automatic and adaptive bivariate	density estimator based on conditional densities • Simonoff, Jeffrey S.	95 StatComp 5 245-252
some bias correction techniques for kernel-based	density estimators • Foster, Peter A comparative study of	95 JStCmpSm 51 137-152
integrated squared error of nonlinear wavelet-based	density estimators • Hall, Peter; Patil, Prakash Formulae for mean	95 AnlsStat 23 905-928
logarithm for the multivariate nearest neighbor	density estimators • Ralescu, Stefan S. The law of the iterated	95 JMultiAn 53 159-179
A minimax result for a class of nonparametric	density estimators • Engel, Joachim; Gasser, Theo	95 JNnparSt 4 327-334
Uniform convergence of	density estimators on spheres • Bertrand-Retali, M.; Ait-Hennani, L.	95 ZastApMa 22 427-446
On L_1-consistency of kernel-type	density estimator for stationary Markov processes • Mishra, M. N.; Rao, B. L. S. Prakasa	95 CommStA 24 581-591
Optimal asymptotic quadratic error of	density estimators for strong mixing or chaotic data • Bosq, Denis	95 StPrLet 22 339-347
A bound on the L_1-error of a nonparametric	density estimator with censored data • Kulasekera, K. B.	95 StPrLet 23 233-238
Prakasa Kernel-type	density and failure rate estimation for associated sequences • Bagai, Isha; Rao, B. L. S.	95 AnInStMa 47 253-266
Hyndman, Rob J. Highest-	density forecast regions for non-linear and non-normal time series models •	95 JForcast 14 431-441
asymptotic behavior of the kernel estimator of a	density function based on censored data (Chinese) • Wang, Qihua Some	95 ChJApPSt 10 164-174
Estimation of quantile	density function based on regression quantiles • Dodge, Yadolah; Jurečková, Jana	95 StPrLet 23 73-78
Dual	density functions • Good, I. J.	95 JStCmpSm 52 193-194
Lorenzo Approximation of	density functions by orthogonal series with grouped data • Barabesi, Lucio; Fattorini,	94 JItalStS 3 181-200
Estimation of conditional quantile	density function • Xiang, Xiaojing	95 JNnparSt 4 309-316
On the	density function of T_0^2 from elliptical distribution (Chinese) • Xia, Yincun	95 ChJApPSt 10 119-124
Estimation of integral functionals of a	density • Birgé, Lucien; Massart, Pascal	95 AnlsStat 23 11-29
Aerial counting experiments to estimate pronghorn	density and herd structure • Pojar, Thomas M.; Bowden, David C.; Gill, R. Bruce	95 JWildlMg 59 117-128
Average	densities of the image and zero set of stable processes • Falconer, K. J.; Xiao, Y. M.	95 StocProc 55 271-283
Southwell, Colin Estimates of macropod	density from line transect surveys relative to analyst expertise • Anderson, David R.;	95 JWildlMg 59 852-857
Approximations for marginal	densities of M-estimators • Fan, Rocky Y. K.; Field, Christopher A.	95 CandJSt 23 185-197
Estimating coyote	density from mark-resight surveys • Hein, Eric W.; Andelt, William F.	95 JWildlMg 59 164-169
paradox and estimating the distances between the	densities of measures with the same marginals • Khalfin et al. ... computer tomography	94 AnlsProb 22 2235-2241

for studying the iterative estimation of mixture	densities • Solka, Jeffrey L.; Poston, Wendy L.; Wegman, Edward J. A visualization technique	95 JCmpGrSt 4 180-198
optimal estimator for continuous monotone	densities • Datta, Somnath A minimax	95 JStPlInf 46 183-193
A limit theorem for the entropy	density of nonhomogeneous Markov information source • Liu, Wen; Yang, Weiguo	95 StPrLet 22 295-301
Deconvolving a	density from partially contaminated observations • Hesse, Christian H.	95 JMultiAn 55 246-260
Adjusting Moran's I for population	density • Oden, Neal	95 StatMed 14 17-26
	Density probability plots • Jones, M. C.; Daly, F.	95 CommStB 24 911-927
factors using a generalization of the Savage-Dickey	density ratio • Verdinelli, Isabella; Wasserman, Larry Computing Bayes	95 JASA 90 614-618
within the class of self-adjoint (self-reciprocal)	densities" (V22 p651-654) • Rossberg ... of the standard normal probability density	95 MaOpfStS 26 387-387
Asymptotic distribution of data-driven smoothers in	density and regression estimation under dependence • Boente, G.; Fraiman, R.	95 CandJSt 23 383-397
confidence envelopes for shape-restricted	densities • Hengartner, Nicolas W.; Stark, Philip B. Finite-sample	95 AnlsStat 23 525-550
On representation of	densities of stable laws by special function • Zolotarev, V. M.	94 ThProbAp 39 354-362
The joint	density of two functionals of Brownian motion • Abadir, K. M.	95 MaMethSt 4 449-462
L_1 theory of estimation of monotone and unimodal	densities • Wang, Yazhen The	95 JNnparSt 4 249-261
with error terms that have symmetric unimodal	densities • Brunner, Lawrence J. Bayesian linear regression	95 JNnparSt 4 335-348
Consistent estimation of	density-weighted average derivative by orthogonal series method • Rao, B. L. S. Prakasa	95 StPrLet 22 205-212
ficiency of adaptive cluster sampling for estimating	density of wintering waterfowl • Smith, David R.; Conroy, Michael J.; Brakhage, David H. Ef-	95 Biomtrcs 51 777-788
Semiparametric efficient estimation of a conditional	density with missing or mismeasured covariates • Robins, J. M.; Hsieh, F.; Newey, W.	95 JRSS-B 57 409-424
Existence and uniqueness of the smoothest	density with prescribed moments • Hong, Changkon; Kim, Choongrak	95 JKoreaSt 24 233-242
"Characterization of the standard normal probability	density within the class of self-adjoint (self-reciprocal) ... • Rossberg Correction to	95 MaOpfStS 26 387-387
in inhibiting the development of supragingival	dental plaque • Hadgu, Alula Analysis of three mouth rinses	94 ASAProEpid 101-106
Statistical strategies for event rate comparisons in	dental studies • Davies, Glenn M.; Koch, Gary G.; Beck, James	94 ASAProBiop 420-424
On weak lumpability of	denumerable Markov chains • Ledoux, James	95 StPrLet 25 329-339
Finding words with unexpected frequencies in	deoxyribonucleic acid sequences • Prum, B.; Rodolphe, F.; de Turckheim, E.	95 JRSS-B 57 205-220
components to detect statistically significant	departures from additivity in chemical mixtures • Gennings et al. ... data from individual	95 Biomtrcs 51 1264-1277
Tomizawa, Sadao Measures of	departure from global symmetry for square contingency tables with ordered categories •	95 Behvmtrk 22 91-98
On the weak convergence of	departures from an infinite series of ·/M/1 queues • Mountford, T.; Prabhakar, B.	95 AnlsApPr 5 121-127
Tomizawa, Sadao Measures of	departure from marginal homogeneity for contingency tables with nominal categories •	95 Ststcian 44 425-439
confidence intervals for location parameters under	departure from normality • Basu, Sanjib; DasGupta, Anirban Robustness of standard	95 AnlsStat 23 1433-1442
M. N.; McNickle, D. C.; Salles, M. C. B.	Departure processes from a loss system with two stations and one server • Magalhães,	94 RBrzlPrS 8 135-146
	Dependence	
	• see also names of specific types of dependence such as positive - or weak -	
Some non-parametric tests for duration	dependence: An application to UK business cycle data • Mudambi, Ram; Taylor, Larry W.	95 JAppStat 22 163-177
Stochastic ordering and	dependence in applied probability • Szekli, R.	95 Springer:Brln:NY 194p
Review of "Stochastic ordering and	dependence in applied probability" • Szekli, R. (Auth); Teugels, J. L. (Rev)	95 ShrtBkRv 15/3 48-48
prefledging survival : Allowing for brood mixing and	dependence among brood mates • Flint, P. L.; Pollock, K. H.; Thomas, D. et al. Estimating	95 JWildlMg 59 448-455
Duration	dependence and dispersion in count-data models • Winkelmann, Rainer	95 JBES 13 467-474
Identifying state	dependence in duration models (Disc: p17-19) • McCall, Brian P.	94 ASAProBuEc 14-16
Liebhold, Andrew M. Detection of delayed density	dependence: Effects of autocorrelation in an exogenous factor • Williams, David W.;	95 Ecology 76 1005-1008
in density and regression estimation under	dependence • Boente, G.; Fraiman, R. Asymptotic distribution of data-driven smoothers	95 CandJSt 23 383-397
On monotone	dependence functions of the quantile type • Krajka, A.; Szynal, D.	95 ZastApMa 23 51-72
based on local means and local medians under	dependence • Boente, Graciela; Fraiman, Ricardo Asymptotic distribution of smoothers	95 JMultiAn 54 77-90
inference on a general measure of monotone	dependence • Conti, Pier Luigi Asymptotic	94 JItalStS 3 213-241
Grzegorzewski, P. The robustness against	dependence of nonparametric tests for the two-sample location problem •	95 ZastApMa 22 469-476
A note on some negative	dependence notions • Kim, Tae-Sung; Seo, Hye-Young	95 CommStA 24 845-858
process • Krug, T. et al. Information loss on the	dependence parameter due to incomplete data from a one-parameter spatial Gaussian	95 CommStA 24 3055-3072
et al. A semiparametric estimation procedure of	dependence parameters in multivariate families of distributions • Genest, C.; Ghoudi, K.	95 Biomtrka 82 543-552
empirical distribution functions and quantiles under	dependence • Sun, Shan Perturbed	95 JTherPrb 8 763-777
abilities in $2 \times k$ contingency tables under negative	dependence restrictions with various sampling schemes • Oh ... estimation of cell prob-	95 CommStA 24 2127-2143
Simultaneous spike-trains and stochastic	dependence • Das, Shubhabrata; Sen, Pranab Kumar	95 SankhyaB 57 32-47
Fung, Wing K. A statistical analysis of	dependence of voter shares under a two-vote system: The Hong Kong experience •	94 ASAProSoSt 76-81
tad, Eric A. Survival estimation and the effects of	dependency among animals • Schmutz, Joel A.; Ward, David H.; Sedinger, James S.; Rexs-	95 JAppStat 22 673-681
D. J.; Buja, A.; Stuetzle, W. Analysis of additive	dependencies and concurvities using smallest additive principal components • Donnell,	94 AnlsStat 22 1635-1673
	Dependencies in Markovian networks • Daduna, H.; Szekli, R.	95 AdvAppPr 27 226-254
	Dependency measure for sets of random events or random variables • Stoyanov, Jordan	95 StPrLet 23 13-20
A Bayesian method for analyzing	dependencies in precursor data • Bier, Vicki M.; Yi, Woojune	95 IntJFor 11 25-41
on randomization when observations are spatially	dependent and the analysis uses generalized least-squares • Martin, R. J. Some results	95 CommStA 24 2199-2210
estimation of a structural changes model with	dependent Bernoulli variable • Pruska, Krystyna Bayesian	95 ListyBio 32 15-21
L. T. et al. An empirical Bayes model for Markov-	dependent binary sequences with randomly missing observations • Cole, B. F.; Lee, M.-	95 JASA 90 1364-1372
Recursive estimation for hidden Markov models: A	dependent case • Elliott, Robert J.	95 StocAnAp 13 437-460
regression estimation in the presence of	dependent censoring • Rotnitzky, Andrea; Robins, James M. Semiparametric	95 Biomtrka 82 805-820
A class of linear nonparametric rank tests for	dependent censoring • Lee, Seungyeoun; Wolfe, Robert A.	95 CommStA 24 687-712
Estimates of marginal survival for	dependent competing risks based on an assumed copula • Zheng, Ming; Klein, John P.	95 Biomtrka 82 127-138
Jacobson, T. Modeling unemployment duration in a	dependent competing risks framework: Identification and estimation • Carling, K.;	95 LiDaAnal 1 111-122
Statistical methods for	dependent competing risks • Moeschberger, M. L.; Klein, John P.	95 LiDaAnal 1 195-204
bounds for the joint survival function in the	dependent competing risks setup • Deshpande, J. V.; Karia, S. R. Confidence	95 CommStA 24 2315-2327
Reliability for a system with	dependent components • Derman, Cyrus; Ross, Sheldon	95 PrEngInf 9 59-63
S. R.; Cottingham, K. L. Resource vs. ratio-	dependent consumer-resource models: A Bayesian perspective • Stow, C. A.; Carpenter,	95 Ecology 76 1986-1990
on nonparametric quantile regression estimate on	dependent data (Chinese) • Shi, Peide; He, Xuming The optimal rates of convergence	95 KToChina 39 1537-1539
Consistency of Hill's estimator for	dependent data • Resnick, Sidney; Stărică, Cătălin	95 JAppProb 32 139-167
On blocking rules for the bootstrap with	dependent data • Hall, Peter; Horowitz, Joel L.; Jing, Bing-Yi	95 Biomtrka 82 561-574
	Dependent decrement theory (Disc: p67-74) • Carriere, Jacques F.	95 TrSActur 46 45-66
butions in the first order autoregressive model with	dependent errors • Nabeya, Seiji; Perron, Pierre Approximations to some exact distri-	95 EcomtrRv 14 421-457
	Dependent F ratios sharing a common denominator in ANOVA table • Kang, Phillee	95 JKoreaSt 24 111-125
logarithm for non-instantaneous filters of strongly	dependent Gaussian sequences • Ho, Hwai-Chung The law of the iterated	95 JTherPrb 8 347-360
Kernel estimation in a nonparametric marker	dependent hazard model • Nielsen, Jens P.; Linton, Oliver B.	95 AnlsStat 23 1735-1748
Laws of large numbers for	dependent heterogeneous processes • de Jong, R. M.	95 EconTher 11 347-358
The relative empirical validity of	dependent and independent data collection in a panel survey • Hill, Daniel H.	94 JOffStat 10 359-380
Compound model for two	dependent kinds of claim • Partrat, Christian	94 InsrneME 15 219-231
A maximal inequality and	dependent Marcinkiewicz-Zygmund strong laws • Rio, Emmanuel	95 AnlsProb 23 918-937
Semiparametric estimation in a three-state duration-	dependent Markov model from interval-censored observations with ... • Frydman	95 Biomtrcs 51 502-511
systems • Guttman, I.; Lin, D. K. J.; Reiser, B. et al.	Dependent masking and system life data analysis: Bayesian inference for two-component	95 LiDaAnal 1 87-100
Nonparametric regression under long-range	dependent normal errors • Csörgő, Sándor; Mielniczuk, Jan	95 AnlsStat 23 1000-1014
normal means with a simply ordered alternative and	dependent observations • Lei, Xingye; Peng, Y. B.; Wright, F. T. Testing homogeneity of	95 CmpStDA 20 173-183
Deconvolving a density from contaminated	dependent observations • Hesse, Christian H.	95 AnInStMa 47 645-663

An optimum	design for estimating the first derivative • Erickson, Roy V.; Fabian, Václav; Mařík, Jan	95 AnlsStat 23 1234-1247
Maximin efficient	designs for estimating nonlinear aspects in linear models • Müller, Christine H.	95 JStPlInf 44 117-132
Lawrence, F. P. Statistically constrained economic	design of the EWMA control chart • Montgomery, D. C.; Torng, J. C.-C.; Cochran, J. K.;	95 JQualTek 27 250-256
A statistical	design of experiments approach for texture description • Ganesan, L.; Bhattacharyya, P.	95 PattrnRc 28 99-105
Holle, Rolf (Rev) Review of	"Design of experiments. A no-name approach" • Lorenzen, T. J.; Anderson, V. L. (Auth);	95 CmpStDA 19 101-102
Conniffe, D. (Rev) Review of	"Design of experiments: A no-name approach" • Lorenzen, T. J.; Anderson, V. L. (Auth);	95 JRSS-A 158 195-197
L. (Auth); O'Brien, C. J. (Rev) Review of	"Design of experiments: A no-name approach" • Lorenzen, Thomas J.; Anderson, Virgil	95 AstrlJSt 37 125-127
its determination by ... • Papamichael et al.	Design of experiments for a precise estimation of the calibration curve of periodate in	95 Chemolab 30 227-237
A train the trainer program for	design of experiments • Eckes, George P.; Orr, Carolyn	95 ASQCTekCon 348-355
Review of "Reliability improvement with	design of experiments" • Condra, Lloyd W. (Auth); Lenz, H.-J. (Rev)	95 CmpStDA 19 605-605
	Design issues in case-control studies • Wacholder, Sholom	95 StMedRes 4 293-309
	Design issues in cohort studies • Prentice, Ross L.	95 StMedRes 4 273-292
The National Population Health Survey:	Design and issues • Singh, M. P.; Tambay, J. L.; Krawchuk, S.	94 ASASrvyRsr 803-808
A new method for generating the	design matrix of a linear regression model • Ilallawa, Abdel M.; Azzam, Abdul-Mordy H.	95 EgpStJ 39 106-119
	Design optimality	
Interview Survey • Kalsbeek, W. D.; Botman, S. L. et al.	Cost-efficiency and the number of allowable call attempts in the National Health	94 JOffStat 10 133-152
Johnson, M. E.; Moore, L. M.; Ylvisaker, D.	Minimax distance designs in two-level factorial experiments • John, P. W. M.;	95 JStPlInf 44 249-263
Some	design problems in crop experimentation. I. The use of blocks • Pearce, S. C.	95 ExpAgric 31 191-203
Some	design problems in crop experimentation. II. Multiple blocking systems • Pearce, S. C.	95 ExpAgric 31 279-290
Some	design problems in crop experimentation. III. Non-orthogonality • Pearce, S. C.	95 ExpAgric 31 409-422
Comment on "Study	design and public health planning" • Truman, Benedict I.	95 StatMed 14 695-696
to analysis of variance for completely systematic	designs? 'Silly designs are silly!' • Bailey, R. A. et al. Are neighbour methods preferable	95 Biomtrka 82 655-659
Research	design and statistical analysis • Myers, Jerome L.; Well, Arnold D.	95 ErlbaumA:NJ 728p
of matrix and bracket approach to stability study	design • Lin, Tsae-Yun Daphne Applicability	94 ASAProBiop 142-147
How to	design surveys • Fink, Arlene	95 SagePubl:CA:Lnd 96p
van der Linden, Wim J. The	design of tests with different optimality criteria (Dutch) • Berger, Martijn P. F.;	95 TOnderwR 20 79-92
method of parameter design and tolerance	design • Li, Wenlian; Wu, C. F. J. An integrated	94 ASAProQlPr 44-48
Bai, Do Sun; Kwon, Hyuck Moo Economic	design of a two-stage screening procedure with a prescribed outgoing quality •	95 Metrika 42 1-18
Type S	designs in unequal blocks • Gupta, Sudhir; Kageyama, Sanpei	93 JCISS 18 97-112
Some new results of uniform	design (Chinese) • Fang, Kaitai; Li, Jiukun	95 KToChina 39 1921-1924
Efficiency of check-plot	designs in unreplicated field trials • Chandra, S.	94 ThApGene 88 618-620
designs for binary response experiments with two	design variables • Sitter, Randy R.; Torsney, Ben Optimal	95 StSinica 5 405-419
On symmetric	designs with parameters (176, 50, 14) • Janko, Zvonimir	95 JCombThA 72 310-314
Kramer, E. Construction of new symmetric	designs with parameters (66, 26, 10) • Matulić-Bedenić, I.; Horvatić-Baldasar, K.;	95 JComDsgn 3 405-410
and variance of all possible lines and hybrids from	designs with partially inbred progenies • Naciri, Y.; Gallais, A. Predicting mean	94 ThApGene 89 693-697
Economic	design of $\bar{X}$ control chart when Taguchi's loss function is considered • Yang, Su-Fen	95 ASQCTekCon 871-878
programming • Rajgopal, Jayant; Mazumdar, Mainak	Designing component test plans for series system reliability via mathematical	95 Technmcs 37 195-212
Halpern, Jerry	Designing a cost-effective clinical trial • Hornberger, John C.; Brown, Byron Wm., Jr;	95 StatMed 14 2249-2259
Mikkelson, Gordon; Yule, James	Designing a customer satisfaction survey • Chun, Young I.; Devens, Richard;	94 ASAProQlPr 204-209
	Designing for minimally dependent observations • Torsney, Ben; Alahmadi, Abdulhadi M.	95 StSinica 5 499-514
On efficient	designing of nonlinear experiments • Chaudhuri, Probal; Mykland, Per A.	95 StSinica 5 421-440
Determinant formulas with applications to	designing when the observations are correlated • Bischoff, Wolfgang	95 AnInStMa 47 385-399
	Designing outer array points • Wang, Yuchung; Lin, Dennis K. J.; Fang, Kai-Tai	95 JQualTek 27 226-241
Mathiason, David J.	Designing and running undergraduate statistical laboratories • Halavin, James J.;	94 ASAProStEd 132-136
methods, using trimmed means versus means, and	designing simulation studies • Wilcox ... The practical importance of heteroscedastic	95 BJMaStPs 48 99-114
Establishment surveys:	Designing the survey operations of the future • Werking, George S.	94 ASASrvyRsr 163-169
is important • Wang, P. C.; Jan, H. W.	Designing two-level factorial experiments using orthogonal arrays when the run order	95 Ststcian 44 379-388
	Desktop documentation: Individual income tax microdata, SOL • David, Martin	94 ASASrvyRsr 775-780
Bivariate normal parameter estimation under	destructive testing • Steiner, S. H.; Wesolowsky, G. O.	94 ASAProPEng 326-331
Estimating the correlation between	destructively measured variables using proof-loading • Steiner, S. H.; Wesolowsky, G. O.	95 Technmcs 37 94-101
	Detection limit	
Rocke, David M.; Lorenzato, Stefan	A two-component model for measurement error in analytical chemistry •	95 Technmcs 37 176-184
distribution when all observations are less than a	detection limit • Smith, D. E. et al. Estimating a percentile of a contaminant concentration	94 ASAProEnvr 139-141
Decision and	detection limits for linear homoscedastic assays • Dunne, Adrian	95 StatMed 14 1949-1959
Lawrence H.; Halsey, Neal A. A mixture model with	detection limits for regression analyses of antibody response to vaccine • Moulton,	95 Biomtrcs 51 1570-1578
analysis of linearly combined order statistic CFAR	detectors • Nagle, Daniel T.; Saniie, Jafar Performance	95 IEEEAES 31 522-533
A random shock model for a linearly	deteriorating system • Lee, Jiyeon; Lee, Eui Yong	95 JKoreaSt 24 471-479
periodicity of a survey: Sampling error, data	deterioration, and cost • Ghosh, D.; Kaufman, S. F.; Smith, W.; Chang, M. Optimal	94 ASASrvyRsr 1122-1127
and maintenance policies of devices subject to	deterioration" (V19 p917-931) • Abdel-Hameed, M. Correction to "Inspection	95 AdvAppPr 27 584-584
	Determinacy of means of order statistics • Spruill, M. C.	95 SankhyaA 57 118-125
correlated • Bischoff, Wolfgang	Determinant formulas with applications to designing when the observations are	95 AnInStMa 47 385-399
Computation of	determinantal subset influence in regression • Barrett, Bruce E.; Gray, J. Brian	94 ASAProStCp 164-169
models • Gani, J.; Yakowitz, Sid Error bounds for	deterministic approximations to Markov processes, with applications to epidemic	95 JAppProb 32 1063-1076
C.-h.; Doerschuk, P. C. Cluster expansions for the	deterministic computation of Bayesian estimators based on Markov random fields • Wu,	95 IEEEPAMI 17 275-293
tification of seismic records based on established	deterministic formulations • Dargahi-Noubary, G. R. Stochastic modeling and iden-	95 JTimSrAn 16 201-219
A semi-	deterministic generalization of a (2×2) contingency table • Seneta, E.	95 ThPrMaSt 50 117-121
Impulse control of piecewise	deterministic Markov processes • Dempster, M. A. H.; Ye, J. J.	95 AnlsApPr 5 399-423
A. E.; Givens, G. H.; Zeh, J. E. Inference from a	deterministic population dynamics model for Bowhead whales (Disc: p417-430) • Raftery,	95 JASA 90 402-416
Comment On "Inference from a	deterministic population dynamics model for Bowhead whales" • Schweder, Tore	95 JASA 90 420-423
the quality of anonymous record linkage using	deterministic procedures with the New York State AIDS registry ... • Muse et al. Evaluating	95 StatMed 14 499-509
State space identification of linear	deterministic rainfall-runoff models • Ramos, José; Mallants, Dirk; Feyen, Jan	95 WaterReR 31 1519-1531
Cost-effectiveness ratios in clinical trials: From	deterministic to stochastic models • Willan, Andrew R.; O'Brien, Bernie J.	94 ASAProBiop 19-28
Testing for	deterministic trends in correlated data • Gan, Nianci; Pantula, Sastry G.	94 ASAProBuEc 141-146
Unit root tests for seasonal models with	deterministic trends • Cho, Sinsup; Park, Young J.; Ahn, Sung K.	95 StPrLet 25 27-35
ian term structure of interest rates and the currency	devaluation of September 1992: A factor-ARCH analysis • Pelizzon, L.; Weber, G. The Ital-	94 Statstca 54 313-327
Accuracy of indices of industrial production in	developing countries • Kmietowicz, Z. W.	95 Ststcian 44 295-307
computers during demographic surveys in	developing countries • Forster, D.; Snow, R. W. An assessment of the use of hand-held	95 SurvyM 21 179-184
household sampling schemes for health surveys in	developing countries • Bennett, S.; Radalowicz, A.; Vella, V. et al. A computer simulation of	94 IJEpidem 23 1282-1291
Statistics: A key technology for	development - How well have we used the key? • Parikh, Kirit S.	95 CalcutSt 45 25-59
abilities • Cardon, L. R.; Fulker, D. W. A model of	developmental change in hierarchical phenotypes with application to specific cognitive	94 BehavGen 24 1-16
models for combination data from reproductive and	developmental experiments • Schwartz, P. F.; Gennings, C.; Chinchilli, V. M. Threshold	95 JASA 90 862-870
Chen, James J. Modeling	developmental toxic effects through recursive partitioning • Ahn, Hongshik;	94 ASAProBiop 130-135
Senchaudhuri, Pralay	Developmental toxicity modeling for risk assessment • Catalano, Paul J.; Ryan, Louise M.;	94 ASAProEnvr 38-45
changeable binary data: Applications to clinical and	developmental toxicity studies • Bowman, D. et al. A saturated model for analyzing ex-	95 JASA 90 871-879

Lag order and critical values of the augmented	Dickey-Fuller test • Cheung, Yin-Wong; Lai, Kon S.	95 JBES 13 277-280
Alternative strategies for 'augmenting' the	Dickey-Fuller test: Size-robustness in the face of pre-testing • Dods, J. L.; Giles, D. E. A.	95 JStCmpSm 53 243-258
Review of "Elsevier's	dictionary of biometry" • Rasch, D.; Tiku, M. L.; Sumpf, D. (Ed); Ahrens, H. (Rev. German)	95 BiomtrcJ 37 896-896
Review of "Elsevier's	dictionary of biometry" • Rasch, D.; Tiku, M. L.; Sumpf, D. (Ed); Kemp, A. W. (Rev)	95 Biomtrcs 51 1186-1187
A	dictionary of epidemiology (Third edition) • Last, John M. (Ed)	95 OxfordUniv 208p
The Cambridge	dictionary of statistics in the medical sciences • Everitt, B. S.	95 CambridgeUniv 274p
Vogt, W. P. (Auth); Berg, Sven (Rev) Review of	"Dictionary of statistics and methodology. A nontechnical guide for the social sciences" •	94 JOffStat 10 225-226
Uryasev, S. P.; Samanta, P. K. Failure of emergency	diesel generators: A population analysis using empirical Bayes methods • Vesely, W. E.;	94 RelEngSS 46 221-229
usual food intake distributions from 24-hour	dietary intake data • Nusser, Sarah M. Estimating	94 ASAProEpid 49-58
Sample size requirements for calibration studies of	dietary intake measurements in prospective cohort investigations • Kaaks, R. et al.	95 AJEpidm 141 557-565
Sample size requirements for calibration studies of	dietary intake measurements in prospective cohort investigations • Kaaks, R. et al.	95 AJEpidm 142 557-565
van Staveren, Wija Calibration of	dietary intake measurements in prospective cohort studies • Kaaks, Rudolf; Riboli, Elio;	95 AJEpidm 141 548-556
trends when estimating the relationship between	dietary intake obtained from a food frequency questionnaire ... • Landin et al. ... for time	95 Biomtrcs 51 169-181
trends when estimating the relationship between	dietary intake obtained from a food frequency questionnaire ... • Landin et al. ... for time	95 Biomtrcs 51 1196-1196
Problems in estimating usual	dietary intakes • Guenther, Patricia M.; Kott, Phillip S.	94 ASAProEpid 44-48
evaluation • Potenza, Maria T.; Dorans, Neil J.	DIF assessment for polytomously scored items: A framework for classification and	95 ApPsychM 19 23-37
Lord_DIF: A computer program to compute Lord's	DIF statistics • Li, Yuan Hwang	95 ApPsychM 19 72-72
for proving therapeutic equivalence based on the	difference of binomial probabilities • Roebruck et al. ... of tests and sample size formulae	95 StatMed 14 1583-1594
and injection drug using men and women • Paik et al.	Difference in clinical implications of CD4 counts among HIV-infected homosexual men	95 StatMed 14 1889-1900
Tests for	differences in dispersion based on quantiles • Shoemaker, Lewis H.	95 AmerStat 49 179-182
Ferrante, Marco On a stochastic delay	difference equation with boundary conditions and its Markov property • Baccin, Maria C.;	95 StocProc 60 131-146
Stochastic vector	difference equations with stationary coefficients • Glasserman, Paul; Yao, David D.	95 JAppProb 32 851-866
Survey • Eltinge, J. L. et al. Assessment of reported	differences between expenditures and low incomes in the U.S. Consumer Expenditure	94 ASASrvyRsr 96-105
A simple and accurate inference procedure for the	difference of failure intensities from two exponential distributions • Wong, A. C. M.	95 CommStA 24 2865-2879
Improving two theorems of Bose on	difference families • Buratti, Marco	95 JComDsgn 3 15-24
for comparative ratings in studies of within-family	differences • Eaves, Lindon J.; Neale, Michael C.; Meyer, Joanne M. A model	91 BehavGen 21 531-536
A Microsoft FORTRAN 77 program for testing the	differences among independent first-order partial correlations • Silver, N. C. et al.	95 EdPsMeas 55 245-248
The probability of divorce, based on the age	difference of the married couple • Gentleman, Jane F.; Park, Evelyn	94 ASAProSoSt 216-221
European Community • Van Bastelaer, Aloïs	Differences in the measurement of employment in the labour force surveys in the	94 JOffStat 10 277-305
Time dependent problems and	difference methods • Gustafsson, Bertil; Kreiss, Heinz-Otto; Oliger, Joseph	95 Wiley:NY:UK 642p
A statistical analysis of structural	differences in minority household electricity demand • Poyer, David A.; Earl, Elizabeth	94 ASAProGovt 221-226
sources: The case of coronary ... • Madans et al.	Differences in morbidity measures and risk factor identification using multiple data	95 StatMed 14 643-653
Saddlepoint approximations for the	difference of order statistics • Harvill, Jane L.; Newton, H. Joseph	95 Biomtrka 82 226-231
to evaluate saddlepoint approximations for the	difference of order statistics • Harvill, Jane L.; Newton, H. Joseph Using symbolic math	95 CommStB 24 781-791
(Rev) Review of "Analysis of systematic and random	differences between paired ordinal categorical data" • Svensson, E. (Auth); Lesaffre, E.	95 Biomtrcs 51 794-794
(Rev) Review of "Analysis of systematic and random	differences between paired ordinal categorical data" • Svensson, E. (Auth); Lindsey, J. K.	95 StatMed 14 1726-1727
Su, John Q. Non-parametric estimation for the	difference or ratio of median failure times for paired observations • Jung, Sin-Ho;	95 StatMed 14 275-281
A further look at the Cochran-Mantel-Haenszel risk	difference • Sato, Tosiya	95 ConClnTr 16 359-361
	Difference sampling as an alternative to importance sampling • van Deusen, Paul C.	95 CanJForR 25 487-490
limit theorem with random indices for martingale	difference sequences • Hong, Dug Hun A central	95 JKoreaSt 24 219-224
A new method of hierarchical cluster of sample-	difference sequence method (Chinese) • Chen, Qiansheng	95 KToChina 39 97-101
On the size of the multiplier groups of cyclic	difference sets • Xiang, Qing; Chen, Yu Qing	95 JCombThA 69 168-169
Non-abelian Hadamard	difference sets • Smith, Ken W.	95 JCombThA 70 144-156
Some new	difference sets • Arasu, K. T.; Sehgal, Surinder K.	95 JCombThA 69 170-172
A homogeneity analysis for exploring structural	differences among several data sets: Its application to data from the Surveys ... • Tsuchiya	95 PrInStMa 43 121-140
Interpreting the standardized	difference • Tritchler, David	95 Biomtrcs 51 351-353
Testing a time series for	difference stationarity • McCabe, B. P. M.; Tremayne, A. R.	95 AnlsStat 23 1015-1028
Testing for trend stationarity versus	difference stationarity • Arellano, Consuelo; Pantula, Sastry G.	95 JTimSrAn 16 147-164
weighted contribution of transition and transversion	differences • Pollock et al. ... methods for constructing evolutionary distances from a	95 MolBioEv 12 713-717
Kwan, C. W. Sensitivity analysis in factor analysis:	Difference between using covariance and correlation matrices • Fung, W. K.;	95 Psymtrka 60 607-614
Models in	differences with measurement errors (Italian) • Lucchetti, R.	93 GiornEco 52 403-415
Bayesian analysis of fractionally	differenced ARIMA processes using Gibbs sampling • Pai, Jeffrey S.; Ravishanker, Nalini	94 ASAProBays 69-74
I. V. Maximum likelihood estimation for a fractionally	differenced autoregressive model on a two-dimensional lattice • Sethuraman, S.; Basawa,	95 JStPlInf 44 219-235
Hashem A non-nested test of level-	differenced versus log-differenced stationary models • Pesaran, Bahram; Pesaran, M.	95 EcomtrRv 14 213-227
On fractionally	differenced periodic processes • Hui, Y. V.; Li, W. K.	95 SankhyaB 57 19-31
A non-nested test of level-differenced versus log-	differenced stationary models • Pesaran, Bahram; Pesaran, M. Hashem	95 EcomtrRv 14 213-227
Identification of fractional	differencing autoregressive models • Li, Guibin	95 CommStA 24 2635-2643
Comparison of various estimates of	differencing order in an ARIMA model (Chinese) • Li, Guibin	95 ChJApPSt 10 353-362
• Beran Maximum likelihood estimation of the	differencing parameter for invertible short and long memory autoregressive integrated ...	95 JRSS-B 57 659-672
Estimation of the fractionally	differencing parameter with the R/S method • Hauser, Michael A.; Reschenhofer, Erhard	95 CmpStDA 20 569-579
A	differencing test • Franses, Philip Hans	95 EcomtrRv 14 183-193
Hadamard	differentiability on $D[0,1]^p$ • Ren, Jian-Jian; Sen, Pranab Kumar	95 JMultiAn 55 14-28
The Csörgő-Révész modulus of non-	differentiability of iterated Brownian motion • Hu, Y.; Shi, Z.	95 StocProc 58 267-279
	Differentiability of the matrices R and G in the matrix-analytic method • He, Qi-Ming	95 CommStSM 11 123-132
	Differential calculus relative to some point processes • Dermoune, A.	95 StPrLet 24 233-242
	Differential equation	
Kaj, I.; Salminen, P.	On the ultimate value of local time of one-dimensional super-Brownian motion •	95 StocProc 59 21-42
	• see also partial –, stochastic –	
Taib, Ziad Branching processes and functional	differential equations determining steady-size distributions in cell populations •	95 JAppProb 32 1-10
approximation of nuclear space-valued stochastic	differential equations driven by Poisson random measures • Kallianpur, G. et al. Diffusion	95 AnlsApPr 5 493-517
W. N.; Lee, E. B. (Ed); Zacharias, K. (Rev) Review of	"Differential equations, dynamical systems, and control science" • Elworthy, K. D.; Everitt,	95 BiomtrcJ 37 127-128
Stability of a class of stochastic integro-	differential equations • Drozdov, Aleksey	95 StocAnAp 13 517-530
A class of quasilinear stochastic partial	differential equations of McKean-Vlasov type with mass conservation • Kotelenez, Peter	95 ProbTher 102 159-188
	Differential equations for moments of present values in life insurance • Norberg, Ragnar	95 InsrneME 17 171-180
Stochastic partial	differential equations • Etheridge, Alison (Ed)	95 CambridgeUniv 347p
measures for white noise driven stochastic partial	differential equations • Prato, G. Da Invariant	95 StocAnAp 13 295-305
of Wong-Zakai type for nonlinear stochastic partial	differential equations • Twardowska, Krystyna An approximation theorem	95 StocAnAp 13 601-626
orem in Hölder norm for parabolic stochastic partial	differential equations • Bally, V.; Millet, A.; Sanz-Solé, M. Approximation and support the-	95 AnlsProb 23 178-222
Stability properties of stochastic partial	differential equations • Lindstrøm, Tom; Øksendal, Bernt	95 StocAnAp 13 177-204
of Wong-Zakai type for stochastic partial	differential equations • Brzeźniak, Zdzisław; Flandoli, Franco Almost sure approximation	95 StocProc 55 329-358
Implicit scheme for quasi-linear parabolic partial	differential equations perturbed by space-time white noise • Gyöngy, I.; Nualart, D.	95 StocProc 58 57-72
On positive solutions of some nonlinear	differential equations – A probabilistic approach • Sheu, Yuan-Chung	95 StocProc 59 43-53
instability phenomenon of the nonlinear stochastic	differential equation (French) • Roynette, B. An	95 Stochatx 53 175-194
Stochastic	differential games: The linear quadratic zero sum case • Ardanuy, R.	95 SankhyaA 57 161-165

stochastic control and zero-sum stochastic	differential games • Hamadéne, S.; Lepeltier, J. P. Backward equations,	95 Stochatx 54 221-231
Marriott, Paul (Rev) Review of	“Differential geometry and statistics” • Murray, Michael K.; Rice, John W. (Auth);	95 BLndnMaS 27 619-620
The behavior of solutions of stochastic	differential inequalities • Assing, S.; Manthey, R.	95 ProbTher 103 493-514
	Differential item functioning	
evaluation • Potenza, Maria T.; Dorans, Neil J.	DIF assessment for polytomously scored items: A framework for classification and	95 ApPsychM 19 23-37
• Allen, Nancy L.; Donoghue, John R. Detecting	differential item functioning: Current methods and continuing problems (Disc: p27-29)	94 ASAProSoSt 17-26
(Rev) Review of	“Differential item functioning” • Holland, P. W.; Wainer, H. (Ed); Mellenbergh, Gideon J.	95 Psymtrka 60 621-623
ological indices in case-control studies with non-	differential misclassification • Marinos, A. Th. et al. Experimental quantiles of epidemi-	95 StatMed 14 1291-1306
Harry M. Sources of data on socio-economic	differential mortality in the United States • Hoyert, Donna L.; Singh, Gopal K.; Rosenberg,	95 JOffStat 11 233-260
divisible probabilities associated with partial	differential operators • Sifi, Mohamed Central limit theorem and infinitely	95 JTherPrb 8 475-499
of “Completeness of root functions of regular	differential operators” • Yakubov, S. (Auth); Rehberg, J. (Rev) Review	95 BiomtrcJ 37 118-118
• Wang, Gang Differential subordination and strong	differential subordination for continuous-time martingales and related sharp inequalities	95 AnlsProb 23 522-551
tingales and related sharp inequalities • Wang, Gang	Differential subordination and strong differential subordination for continuous-time mar-	95 AnlsProb 23 522-551
lie, C. Dose-response models for the probabilities of	differential toxic effects in developmental toxicity studies • Talwalker, S.; Patil, G. P.; Tail-	94 JStatRsr 28 221-232
Ma, C. Y.; Bayne, C. K.	Differentiation of aroclors in environmental samples using linear discrimination •	94 ASAProEnvr 142-143
	Differentiation of the Cholesky algorithm • Smith, S. P.	95 JCmpGrSt 4 134-147
boundaries for random walks obtained from	diffuse measures • Handelman, David Space-time	94 IsrelJMa 86 107-156
A	diffuse prior limit in semiparametric binary regression • Newton, Michael A.	94 ASAProBays 181-186
	Diffusion	
	Further applications of a general rate conservation law • Bardhan, Indrajit	95 StocProc 60 113-130
Medhin, N. G.	Derivation of optimality conditions for a stochastic control problem • Ladde, G. S.;	95 StocAnAp 13 165-176
Sullivan, Paul J.	The dilution of a contaminant substance released in an environmental flow •	95 SankhyaB 57 277-284
Marc; Elliott, Robert J.	Estimating the instantaneous volatility and covariance of risky assets • Chesney,	95 AplStMDA 11 51-58
	Hardy-Littlewood theory on unimodular groups • Varopoulos, N. Th.	95 AIHenrPB 31 669-688
	Large deviations for Langevin spin glass dynamics • Arous, G. B.; Guionnet, A.	95 ProbTher 102 455-509
	Markov population replacement processes • Gerontidis, Ioannis I.	95 AdvAppPr 27 711-740
Kabanov, Yu. M.; Runggaldier, W. J.	Mean-variance hedging of options on stocks with Markov volatilities • Di Masi, G. B.;	94 ThProbAp 39 172-182
	A new look at pricing of the “Russian option” • Shepp, L. A.; Shiryaev, A. N.	94 ThProbAp 39 103-119
	Optimal choice of production capacity in a random market • Arntzen, Halvard	95 Stochatx 55 87-120
	The radial part of a γ-martingale and a non-implosion theorem • Kendall, Wilfrid S.	95 AnlsProb 23 479-500
Newman, C. M.; Stein, D. L.	Random walk in a strongly inhomogeneous environment and invasion percolation •	95 AIHenrPB 31 249-261
	The scaling limit for a stochastic PDE and the separation of phases • Funaki, T.	95 ProbTher 102 221-288
Burdzy, Krzysztof; Toby, Ellen	A Skorohod-type lemma and a decomposition of reflected Brownian motion •	95 AnlsProb 23 586-604
	Some distributed algorithms revisited • Louchard, Guy	95 CommStSM 11 563-586
	Spectra of graphs and fractal dimensions II • Telcs, András	95 JTherPrb 8 77-96
Watkins, Joseph	A stochastic model for the movement of a white blood cell • Heubach, Silvia;	95 AdvAppPr 27 443-475
	Diffusion approximation	
Kaj, I.; Salminen, P.	On the ultimate value of local time of one-dimensional super-Brownian motion •	95 StocProc 59 21-42
	Diffusion approximation for an age-structured population • Bose, A.; Kaj, I.	95 AnlsApPr 5 140-157
by Poisson random measures • Kallianpur, G. et al.	Diffusion approximation of nuclear space-valued stochastic differential equations driven	95 AnlsApPr 5 493-517
The lifetime of conditioned	diffusions associated with some degenerate elliptic operators • Gao, Ping	95 StocProc 57 319-337
Bisexual branching	diffusions • Mytnik, Leonid; Adler, Robert J.	95 AdvAppPr 27 980-1018
problem for a reaction-diffusion equation with a fast	diffusion component • Carmona, Sara C. An asymptotic	95 Stochatx 52 43-80
Jaime Asymptotic laws for one-dimensional	diffusions conditioned to nonabsorption • Collet, Pierre; Martínez, Servet; San Martín,	95 AnlsProb 23 1300-1314
Ergodic theorems for transient one-dimensional	diffusions • Athreya, K. B.; Weerasinghe, A. P. N.	95 StocProc 58 173-185
A strong existence theorem for stochastic reaction-	diffusion equations with continuous coefficients • Manthey, Ralf	95 Stochatx 52 15-27
An asymptotic problem for a reaction-	diffusion equation with a fast diffusion component • Carmona, Sara C.	95 Stochatx 52 43-80
Multidimensional reaction-	diffusion equations with white noise boundary perturbations • Sowers, R. B.	94 AnlsProb 22 2071-2121
lems in probability theory: Stochastic models and	diffusions on fractals” • Elworthy, K. D.; Ikeda, N. (Ed) et al. Review of “Asymptotic prob-	95 Ststcian 44 139-140
lems in probability theory: Stochastic models and	diffusions on fractals” • Elworthy, K. D.; Ikeda, N. (Ed) et al. Review of “Asymptotic prob-	95 Metrika 42 69-70
Strong Feller property and irreducibility for	diffusions on Hilbert spaces • Peszat, Szymon; Zabczyk, Jerzy	95 AnlsProb 23 157-172
Measurement and filtering of quantum	diffusion II • Belavkin, V. P.	94 ThProbAp 39 363-378
man, D. L. Certificate-of-need regulation and the	diffusion of innovations: A random coefficient model • Caudill, S. B.; Ford, J. M.; Kaser-	95 JApEcmtx 10 73-78
Finite and infinite systems of interacting	diffusions • Cox, J. T.; Greven, Andreas; Shiga, Tokuzo	95 ProbTher 103 165-197
Statistical fluctuations due to microscale mixing in a	diffusion layer • Zimmerman, William B.; Chatwin, P. C.	95 EnvrMtrc 6 665-675
Subdiffusive fluctuations for internal	diffusion limited aggregation • Lawler, Gregory F.	95 AnlsProb 23 71-86
edition) • Rogers, L. C. G.; Williams, David	Diffusions, Markov processes, and martingales. Volume 1: Foundations (Second	94 Wiley:NY:UK 386p
edition)” • Rogers, L. C. G. et al. Review of	“Diffusions, Markov processes, and martingales. Volume 1: Foundations (Second	95 ShrtBkRv 15/2 28-28
Drift and	diffusion for a mechanical system • Boldrighini, C.; Soloveitchik, M.	95 ProbTher 103 349-379
Density-dependent limits for a nonlinear reaction-	diffusion model • Blount, Douglas	94 AnlsProb 22 2040-2070
two-sample U-statistics and a two-species reaction-	diffusion model • Penrose, Mathew D. Generalised	95 StocProc 55 57-64
Park, Byung-Gu A	diffusion model for a system subject to random shocks • Lee, Eui Yong; Song, Mun Sup;	95 JKoreaSt 24 141-147
A Liapunov-type theorem for Nelson	diffusions and nonattainability of nodes • Posilicano, Andrea	95 StocAnAp 13 111-124
for ruin probabilities of risk processes perturbed by	diffusion • Schmidli, H. Cramér-Lundberg approximations	95 InsrneME 16 135-149
Growth and	diffusion phenomena: Mathematical frameworks and applications • Banks, R. B.	94 Springer:Brln:NY 455p
	Diffusion process	
Stramer, O.	On the approximation of continuous time threshold ARMA processes • Brockwell, P. J.;	95 AnInStMa 47 1-20
Jacobs, Patricia A.	Asymptotic properties of stochastic greedy bin-packing • Gaver, Donald P.;	95 CommStSM 11 333-348
	Brownian motion on simple fractal spaces • Nyberg, Svein Olav	95 Stochatx 55 21-45
	Conditional propagation of chaos and a class of quasilinear PDE's • Zheng, Weian	95 AnlsProb 23 1389-1413
laws • Borodin, A. N.; Freidlin, M. I.	Fast oscillating random perturbations of dynamical systems with conservation	95 AIHenrPB 31 485-525
Shortland, C. F.	The hazard rate tangent approximation for boundary hitting times • Roberts, G. O.;	95 AnlsApPr 5 446-460
Lo, Andrew W.; Wang, Jiang	Implementing option pricing models when assest returns are predictable •	95 JFinance 50 87-129
Flandoli, Franco; Gatarek, Dariusz	Martingale and stationary solutions for stochastic Navier-Stokes equations •	95 ProbTher 102 367-391
	Modelling radio-tracking data • Worton, Bruce J.	95 EnvEcoSt 2 15-23
• Gutiérrez Jáimez, R.; Román, P. R.; Ruiz, F. T.	A note on the Volterra integral equation for the first-passage-time probability density	95 JAppProb 32 635-648
	Optimal control of dams • Abdel-Hameed, Mohamed; Nakhi, Yasmin	95 AplStMDA 11 25-34
	Sample path behaviour in connection with generalized arcsine laws • Bertoin, Jean	95 ProbTher 103 317-327
Liptser, R.; Shenfild, M.	State-dependent Beneš buffer model with fast loading and output rates • Kogan, Y.;	95 AnlsApPr 5 97-120
Taukir	Statistical inference for switching Brownian motions • Frydman, Halina; Hussain,	95 SankhyaA 57 11-28
	Stochastic differential equations with a convex constraint • Storm, Andrew	95 Stochatx 53 241-274
	A theory of risk, return and solvency • Powers, Michael R.	95 InsrneME 17 101-118
periodic spatial heterogeneities • Bhattacharya et al.	Time-scales for Gaussian approximation and its breakdown under a hierarchy of	95 Bernoull 1 81-123
	Weak convergence of Markov processes with extended generators • Xia, Aihua	94 AnlsProb 22 2183-2202

	Dirichlet-multinomial model	
outcomes • Thall, P. F.; Simon, R. M.; Estey, E. H.	Bayesian sequential monitoring designs for single-arm clinical trials with multiple	95 StatMed 14 357-379
	Dirichlet polynomial	
approximation • Martin, Clyde; Shubov, Victor	Natural exponential families of probability distributions and exponential-polynomial	93 AplMaCmp 59 275-297
	Dirichlet prior	
sampling • Doss, Hani	Bayesian nonparametric estimation for incomplete data via successive substitution	94 AnlsStat 22 1763-1786
	Exact multivariate Bayesian bootstrap distributions of moments • Gasparini, Mauro	95 AnlsStat 23 762-768
	Dirichlet process	
	A nonparametric Bayes method for isotonic regression • Lavine, M.; Mockus, A.	95 JStPlInf 46 235-248
Gelfand, Alan E.; Mukhopadhyay, Saurabh	On nonparametric Bayesian inference for the distribution of a random sample •	95 CandJSt 23 411-420
	Nonparametric Bayesian methods in hierarchical models • Escobar, Michael D.	95 JStPlInf 43 97-106
A large deviation result for a class of	Dirichlet processes • Zheng, Weian	95 ProbTher 101 237-249
Gaussian tail estimates for	Dirichlet processes • Ubøe, Jan; Zhang, Tusheng	95 Stochatx 52 295-302
Actuarial models for pricing	disability benefits: Towards a unifying approach • Pitacco, Ermanno	95 InsrneME 16 39-62
for sampling elderly minorities and persons with	disabilities (Disc: p695-696) • Ezzati, T. M.; Hoffman, K. et al. A dual frame design	95 StatMed 14 571-583
Leone, Robert P.; Gaskins, John N. Aggregate and	disaggregate sector forecasting using consumer confidence measures • Kumar, V.;	95 IntJFor 11 361-377
Martínez, J. A recursive ARIMA-based procedure for	disaggregating a time series variable using concurrent data • Guerrero, V. M.;	95 Test 4 359-376
systems: Some of Tomás Havránek's interests and	disambiguating word senses • Gale, William A. Cognition, computation, and formal	95 CmpStDA 19 135-148
Assessing public health consequences of a flood	disaster • Eckoff, Ronald D.	94 ASAProEnvr 192-193
Stan; Wymer, Colin	Disclosure control for census microdata • Skinner, Chris; Marsh, Catherine; Openshaw,	94 JOffStat 10 31-51
An agenda for research in statistical	disclosure limitation • Cox, Lawrence H.; Zayatz, Laura V.	95 JOffStat 11 205-220
On Brownian	disconnection exponents • Werner, Wendelin	95 Bernoull 1 371-380
A new method for detecting undocumented	discontinuities in climatological time series • Easterling, David R.; Peterson, Thomas C.	95 JClimlgy 15 369-377
Feynman-Kac semigroup with	discontinuous additive functionals • Song, Renming	95 JTherPrb 8 727-762
Weiner, J.; Krause, A.; Stoll, P. Modeling of	discontinuous relationships in biology with censored regression • Schmid, B.; Polasek, W.;	94 ANatrls 143 494-507
efficiency of maximum likelihood estimators of a	discontinuous signal in a Gaussian white noise • Rubin et al. ... of the asymptotic	95 AnlsStat 23 732-739
fish communities: Effects of sampling effort and	discontinuous spatial distributions • Angermeier et al. ... relative abundances in stream-	95 JFishRBC 52 936-949
tests of present value models with a time-varying	discount factor • Timmermann, Allan Cointegration	95 JApEcmtx 10 17-31
solved by deconvexification over time • Eaves et al. A	discounted-cost continuous-time flexible manufacturing and operator scheduling model	95 PrEngInf 9 65-98
Constrained Markov decision models with weighted	discounted rewards • Feinberg, Eugene A.; Shwartz, Adam	95 MaOpRsr 20 302-320
the reversed sub-martingale property of empirical	discrepancies in arbitrary sample spaces • Strobl, Franz On	95 JTherPrb 8 825-831
On the choice of a	discrepancy functional for model selection • Trosset, Michael W.; Sands, Barbara N.	95 CommStA 24 2841-2863
Kocherlakota, Subrahmaniam	Discrete bivariate weighted distributions under multiplicative weight function •	95 CommStA 24 533-551
A bivariate	discrete Charlier series distribution • Papageorgiou, H.; Loukas, S.	95 BiomtrcJ 37 105-117
in simulated maximum likelihood estimation of	discrete choice models • Lee, Lung-fei Asymptotic bias	95 EconTher 11 437-483
Regression models for a bivariate	discrete and continuous outcome with clustering • Fitzmaurice, Garrett M.; Laird, Nan M.	95 JASA 90 845-852
Willinger, Walter From	discrete to continuous stochastic calculus • Cutland, Nigel J.; Kopp, P. Ekkehard;	95 Stochatx 52 173-192
T. et al. Design of optimal statistical filter for the	discrete convolution backprojection method • Dhariyal, I. D.; Rathore, R. K. S.; Srivastava,	94 AJMMgSci 14 229-265
of regression parameters from censored data with a	discrete covariate • Rahbar, Mohammad H.; Gardiner, Joseph C. Nonparametric estimation	95 StPrLet 24 13-20
regression parameters from censored data with a	discrete covariate • Rahbar, Mohammad H. Sequential fixed width confidence intervals for	95 SqtlAnly 14 143-156
Bayesian graphical models for	discrete data • Madigan, David; York, Jeremy	95 IntStRvw 63 215-232
Logistic regression and other	discrete data models for serially correlated observations • Azzalini, A.	94 JItalStS 3 169-179
Cobb, E. Benton; Church, J. D. A	discrete delta method for estimating standard errors for estimators in quantal bioassay •	95 BiomtrcJ 37 691-699
	Discrete distribution	
Kocherlakota, Subrahmaniam	Discrete bivariate weighted distributions under multiplicative weight function •	95 CommStA 24 533-551
Asymptotic divergence of estimates of	discrete distributions • Morales, D.; Pardo, L.; Vajda, I.	95 JStPlInf 48 347-369
A kernel estimator for	discrete distributions • Rajagopalan, Balaji; Lall, Upmanu	95 JNnparSt 4 409-426
A family of	discrete distributions • Ambagaspitiya, R. S.	95 InsrneME 16 107-127
Goodness of fit testing for ordered	discrete distributions by resampling techniques • Pesarin, Fortunato	94 Metron 52/3 57-71
procedure for estimating the parameters of some	discrete distributions • From, Steven G. A two-stage	95 JStCmpSm 53 23-39
Subarau A class of	discrete distributions suited to fitting very long-tailed data • Ong, S. H.; Muthaloo,	95 CommStB 24 929-945
U_n^2 test for uniformity of	discrete distributions • Watson, Geoffrey S.	95 JAppStat 22 273-276
Review of "Univariate	discrete distributions" • Johnson, N. L.; Kotz, S.; Kemp, A. W. (Auth); Consul, P. C. (Rev)	95 Metrika 42 72-73
Characterization of	discrete distributions using expected values • Ruiz, J. M.; Navarro, J.	95 StHefte 36 237-252
continuous-time proportional hazard model using	discrete duration data • Ryu, Keunkwan Analysis of a	95 EcomtrRv 14 299-313
function method" • Rubinstein, R. Y. et al. Review of	"Discrete event systems: Sensitivity analysis and stochastic optimization by the score	95 Technmcs 37 124-125
Review of "Monotone structure in	discrete-event systems" • Glasserman, Paul; Yao, David D. (Auth); Haas, Peter J. (Rev)	95 JASA 90 808-809
Madabhushi (Rev) Review of "Monotone structure in	discrete-event systems" • Glasserman, Paul; Yao, David D. (Auth); Raghavachari,	95 Technmcs 37 467-468
On estimating mixing densities in	discrete exponential family models • Zhang, Cun-Hui	95 AnlsStat 23 929-945
A bound on the rate of convergence for the	discrete Gibbs sampler • Dinwoodie, I. H.	95 PrEngInf 9 211-215
vector quantization for speech recognition with	discrete hidden Markov Model • Huo, Qiang; Chan, Chorkin Contextual	95 PattrnRc 28 513-517
A Wold-like decomposition of two-dimensional	discrete homogeneous random fields • Francos, Joseph M.; Meiri, A. Zvi; Porat, Boaz	95 AnlsApPr 5 248-260
Review of	"Discrete latent variable models" • Heinen, Ton (Auth); Andersen, E. B. (Rev)	95 Psymtrka 60 449-450
Characterization of	discrete laws via mixed sums and Markov branching processes • Pakes, Anthony G.	95 StocProc 55 285-300
Minification processes with	discrete marginals • Kalamkar, V. A.	95 JAppProb 32 692-706
Martin boundaries for	discrete Markov chains • Aggoun, Lakhdar	95 StocAnAp 13 257-268
Wesolowski, Jacek Bivariate	discrete measures via a power series conditional distribution and a regression function •	95 JMultiAn 55 219-229
	Discrete Mittag-Leffler distributions • Pillai, R. N.; Jayakumar, K.	95 StPrLet 23 271-274
inference about proportions using Bayes and	discrete models (jse/v3n3/albert) • Albert, Jim Teaching	95 JStEd 3/3 58kb
A new family of goodness-of-fit statistics for	discrete multivariate data • Lorenzen, Gunter	95 StPrLet 25 301-307
likelihood estimation for diffusion processes from	discrete observations • Mishra, M. N.; Bishwal, J. P. N. Approximate maximum	95 Stochatx 52 1-13
acterizations of probability distributions based on	discrete p-monotonicity • Sapatinas, Theofanis Char-	95 StPrLet 24 339-344
number of visits to a subset of the state space by a	discrete-parameter semi-Markov process • Csenki, Attila The	95 StPrLet 22 71-77
Distribution of order statistics for	discrete parents with applications to censored sampling • Gan, Gaoxiong; Bain, Lee J.	95 JStPlInf 44 37-46
Conceptual notes on models for	discrete polytomous item responses • Mellenbergh, Gideon J.	95 ApPsychM 19 91-100
case) • Chergui, Jamal D. Estimation and tests of the	discrete probability law based on the empirical generating function (two dimensional	94 RACEMdrd 88 245-286
Luceño, Alberto	Discrete proportional-integral control with constrained adjustment • Box, George E. P.;	95 Ststcian 44 479-495
Statistical artifacts in the ratio of	discrete quantities • Johnston, Roger G.; Schroder, Shayla D.; Mallawaaratchy, A. Rajika	95 AmerStat 49 285-291
On one	discrete queueing model • Velikoivanenko, O. I.; Shvets, Ye. M.	95 ThPrMaSt 50 55-60
on a single occupancy Fermi-Dirac model for the	discrete range distribution • Efird et al. A non-parametric two-sample survival test based	94 ASAProBiop 166-170
Martz, Harry F. Bayesian inference in a	discrete shock model using confounded common cause data • Kvam, Paul H.;	95 RelEngSS 48 19-25
on autocovariance estimation in the presence of	discrete spectra • Houdré, Christian; Kedem, Benjamin A note	95 StPrLet 24 1-8
State aggregation and	discrete-state Markov chains embedded in a class of point processes • Cao, Xi-Ren	95 JAppProb 32 39-51
Review of "Markov decision processes:	Discrete stochastic dynamic programming" • Puterman, M. L. (Auth); Collins, Sean (Rev)	95 JAppStat 22 545-546

Random walks in a quarter plane with zero	drifts: Transience and recurrence • Asymont, I. M.; Fayolle, G.; Menshikov, M. V.	95 JAppProb 32 941-955
of the first-passage time of a Wiener process with	drift between two elastic barriers • Dominé, Marco Moments	95 JAppProb 32 1007-1013
The role of the	drift in I(2) systems • Paruolo, Paolo	94 JItalStS 3 93-123
of one-dimensional Brownian motion with a	drift • Povel, Tobias On weak convergence of conditional survival measure	95 AnlsApPr 5 222-238
of the α-quantile of Brownian motion with	drift • Embrechts, P.; Rogers, L. C. G.; Yor, M. A proof of Dassios' representation	95 AnlsApPr 5 757-767
And not a drop to drink! Perceived threats to clean	drinking water • Tomazic, Terry; Katz, Barry	94 ASAProEnvr 173-178
quality as predicators of remission of problem	drinking in women • Harris et al. ... interaction of partnership dissolution and relationship	94 ASAProSoSt 232-234
Darryl Characteristics of persons who reported	driving under the influence of alcohol or other drugs • Greenblatt, Janet C.; Bertolucci,	94 ASASrvyRsr 547-552
And not a	drop to drink! Perceived threats to clean drinking water • Tomazic, Terry; Katz, Barry	94 ASAProEnvr 173-178
Modeling the	drop-out mechanism in repeated-measures studies • Little, Roderick J. A.	95 JASA 90 1112-1121
for longitudinal binary responses with informative	drop-outs • Fitzmaurice, G. M.; Molenberghs, G.; Lipsitz, S. R. Regression models	95 JRSS-B 57 691-704
Talwalker Analysis of repeated measurements with	dropouts among Alzheimer's disease patients using summary measures and meta ... •	94 ASAProBiop 515-520
Software for handling	dropouts in clinical trials • Entsuah, Richard	94 ASAProBiop 188-193
Work experience, local labor markets, and	dropping out of high school (Disc: p213-215) • Swaim, Paul	94 ASAProSoSt 207-212
quadratic entropy indices for diversity studies of	drosophilid assemblages • Izsák, János; Papp, László Application of the	95 EnvEcoSt 2 213-224
Hydrologic scenarios for severe sustained	drought in the Southwestern United States • Tarboton, David G.	95 WaterReB 31 803-813
Bradley, A. Allen Errors in streamflow	drought statistics reconstructed from tree ring data • Brockway, Charles G.;	95 WaterReR 31 2279-2293
method for studying characteristics of ground water	droughts • Chang, Tiao J.; Teoh, Choo B. Use of the kriging	95 WaterReB 31 1001-1007
Modeling trends in emergency room	drug abuse episodes • Fairchild, Charles K.	94 ASASrvyRsr 305-309
Variance estimation of	drug abuse episodes using the bootstrap • Hughes, Arthur L.; Brodsky, Marc D.	94 ASASrvyRsr 212-217
estimation for the National Household Survey of	Drug Abuse • Folsom, Ralph E.; Liu, Jun Small area	94 ASASrvyRsr 565-570
Gfroerer, Joseph C. Predictors of	drug abuse treatment utilization among frequent cocaine users • Colliver, James D.;	94 ASASrvyRsr 553-558
A new modeling approach for joint action of	drugs • Jia, Xinwei; Rao, M. Bhaskara	94 ASAProBiop 154-159
microcomputer program and application on	drug bioequivalence data • Abdullah, M. E.; El-Sayed, Yousry M. Design of crossover	95 CmpBiomd 48 237-239
model for estimating population characteristics of	drug combination effect • Gao, Min; Minkin, Salomon Nonlinear mixed effect	94 ASAProBiop 160-165
Patient compliance and the fluctuation of the serum	drug concentration • Wang, Wenping; Hsuan, Francis; Chow, Shein-Chung	94 ASAProBiop 94-99
Estimate of disease prevalence using	drug consumption data • Sartor, Francis; Walckiers, Denise	95 AJEpidm 141 782-787
Moeller, John F. Prescription	drug coverage and the demand for outpatient prescription drugs by the elderly •	94 ASAProSoSt 147-152
Statistical designs for	drug developments • Hedayat, Samad	92 ProIrStC1 39-50
for multicenter clinical trials in joint US-European	drug development • Tudor, David; Zipfel, Andreas Opportunities	94 ASAProBiop 229-231
and evaluation of drug safety from clinical trials in	drug development: Issues of international harmonization • O'Neill ... in the planning	95 StatMed 14 1117-1127
and the demand for outpatient prescription	drugs by the elderly • Moeller, John F. Prescription drug coverage	94 ASAProSoSt 147-152
Marc L.; Mueller, Curt D. Developing a prescription	drug expenditure model: The effect of alternative assumptions • Schur, Claudia L.; Berk,	94 ASAProSoSt 245-250
Issues related to the bioequivalence of	drug formulations • Habibullah, Mohamed; Rogers, James; Voss, Theresa	94 ASAProStCp 222-224
Statistical analysis of	drug interactions in anesthesia • Dexter, Franklin	95 JThBio 172 305-314
driving under the influence of alcohol or other	drugs • Greenblatt, Janet C.; Bertolucci, Darryl Characteristics of persons who reported	94 ASASrvyRsr 547-552
Statistical issues in	drug quality control based on dissolution testing • Tsong, Yi; Hammerstrom, Thomas	94 ASAProBiop 295-300
Favorite data sets from early phases of	drug research - Part 3 • Bradstreet, Thomas E.	94 ASAProStEd 247-252
A stochastic model for	drug resistance in AIDS chemotherapy and the HIV incubation distribution • Tan, Wai Y.	95 StPrLet 25 289-299
Statistical concepts in the planning and evaluation of	drug safety from clinical trials in drug development: Issues of international ... • O'Neill	95 StatMed 14 1117-1127
On stability designs in	drug shelf-life estimation • Ju, Huey L.; Chow, Shein-Chung	95 JBiophSt 5 201-214
Analysis of assays of the joint action of three	drugs • Frey, Carolin; Frey, Michael	94 ASAProBiop 33-38
Increasing response rates in follow-up surveys of	drug treatment patients • Johnson, R.; Malloy, K.; Gerstein, D.; Suter, N.; Nisenbaum, S.	94 ASASrvyRsr 1263-1268
the min test for the analysis of combination	drug trials • Sarkar, Sanat K.; Snapinn, Steve; Wang, Wenjin On improving	95 JStCmpSm 51 197-213
for the analysis of crossover and related designs in	drug trials • Hwang, Dar-Shong; Hsuan, Francis A delta procedure	94 ASAProBiop 307-312
An alternative model for joint action of two	drugs • Jia, Xinwei; Rao, M. Bhaskara	94 ASAProQlPr 305-309
tionship to emergency room visits and death, and to	drug use among arrestees • Hyatt, R. R., Jr et al. The price and purity of cocaine: The rela-	95 StatMed 14 655-668
Ratio estimation of hard-core	drug use • Wright, Douglas; Gfroerer, Joseph; Epstein, Joan	94 ASASrvyRsr 541-546
The validity of self-reported	drug use among the incarcerated • Cantor, David; Gaertner, Gregory; Keil, Linda	94 ASASrvyRsr 1088-1093
A case study examining condom use among	drug users • Wang, J. et al. Influence of measurement errors on HIV risk behavior analysis:	95 StrEqMod 2 319-334
among HIV-infected homosexual men and injection	drug using men and women • Paik et al. Difference in clinical implications of CD4 counts	95 StatMed 14 1889-1900
paring reporting rates of adverse events between	drugs with adjustment for year of marketing and secular trends in total ... • Tsong Com-	95 JBiophSt 5 95-114
Estimation of the shelf-life of	drugs with mixed effects models • Chen, James J.; Hwang, Jing-Shiang; Tsong, Yi	95 JBiophSt 5 131-140
for assessing combinations of chemotherapeutic	drugs with resistance modifiers • Lam, Y. M.; Pym, J.; Campling, B. G. Nonlinear models	94 ASAProBiop 394-398
An evaluation of an estimation procedure from a	DSRDD telephone sample design or how well do sample weights work • Palit, Charles D.	94 ASASrvyRsr 647-652
estimation of missing data in a time series using the	dual autocorrelation function (Spanish) • Nieto, Fabio H. Note on the	94 Estd 46 85-103
	Dual density functions • Good, I. J.	95 JStCmpSm 52 193-194
p695-696) • Ezzati, T. M.; Hoffman, K. et al. A	dual frame design for sampling elderly minorities and persons with disabilities (Disc:	95 StatMed 14 571-583
digit dialing sampling schemes for single and	dual frame surveys • Biemer, Paul; Akin, Don The efficiency of list-assisted random	94 ASASrvyRsr 1-10
	Dual likelihood • Mykland, Per Aslak	95 AnlsStat 23 396-421
Velikii, Yu. A.; Motsa, A. I.; Sil'vestrov, D. S.	Dual processes and ergodic type theorems for Markov chains in the scheme of series •	94 ThProbAp 39 642-653
	Dual response surface optimization • Lin, Dennis K. J.; Tu, Wanzhu	95 JQualTek 27 34-39
	Dual scaling	
Nishisato, Shizuhiko	Graphical representation of quantified categorical data: Its inherent problems •	95 JStPlInf 43 121-132
A sequential	dual of Stein's two-sample procedure • Arghami, N. R.; Billard, L.	92 ProIrStC1 11-18
Using ethnographic data to evaluate	dual system estimates • Thurston, Sally W.; Zaslavsky, Alan M.	94 ASASrvyRsr 254-259
A general	duality approach to I-projections • Bhattacharya, Bhaskar; Dykstra, Richard	95 JStPlInf 47 203-216
Bitmead, Robert R.	Duality between the discrete-time Kalman filter and LQ control law • Blackmore, Perry A.;	95 IEEEAuCn 40 1442-1444
invariance, predictive invariance, and the	duality paradox • Millsap, Roger E. Measurement	95 MultBehR 30 577-605
A general	duality theorem for marginal problems • Ramachandran, D.; Rüschendorf, L.	95 ProbTher 101 311-319
analysis • Curtis, D.; Gurling, H. M. D. Using a	dummy quantitative variable to deal with multiple affection categories in genetic linkage	91 AnHuGene 55 321-327
properties of the estimators of coefficients in mixed	dummy-variable error components model • Ghazal, G. A.; Mira, S. I. The exact finite sample	95 EgpStJ 39 26-50
in time series: Some Monte Carlo evidence on	dummy variable plot approaches • Nebebe, F.; Kwan, A. C. C. Detecting 'potential' outliers	95 JStPlInf 43 87-95
	Dunnett procedure	
ments: Why do we feel differently? • Proschan et al.	Multiple comparisons with control in a single experiment versus separate experi-	95 AmerStat 49 144-149
A new formula for tail probabilities of	Dunnett's T with unequal sample sizes • Cheng, Peter H.; Meng, Cliff Y. K.	95 CommStA 24 523-532
in biometry - A festschrift in honor of Charles W.	Dunnett" • Hoppe, F. M. et al. Review of "Multiple comparisons, selection, and applications	95 AstrlJSt 37 251-252
in biometry: A festschrift in honor of Charles W.	Dunnett" • Hoppe, F. M. et al. Review of "Multiple comparisons, selection, and applications	95 Biomtrcs 51 380-380
A statistical model for	duplicate tournaments in games of luck and skill • Yu, Philip L. H.; Cowan, Richard	95 AstrlJSt 37 1-17
Forecasting the adoptions of new consumer	durable goods • Min, Chung-ki	94 ASAProBays 262-265
On the asymptotic size and	duration of a class of epidemic models • Svensson, Åke	95 JAppProb 32 11-24
Turcott, R. G.; Barker, P. D. R.; Teich, M. C. Long-	duration correlation in the sequence of action potentials in an insect visual interneuron •	95 JStCmpSm 52 253-271
Comment on "Analysis of data on	duration and counts" • Wooldridge, Jeffrey M.	94 ASAProBuEc 17-19
proportional hazard model using discrete	duration data • Ryu, Keunkwan Analysis of a continuous-time	95 EcomtrRv 14 299-313

A class of binary response models for grouped	duration data • Sueyoshi, Glenn T.	95 JApEcmtx 10 411-431
Determining the	duration of a demonstration life-test before all units fail • Doganaksoy, Necip	95 IEEEReli 44 26-30
Taylor, Larry W. Some non-parametric tests for	duration dependence: An application to UK business cycle data • Mudambi, Ram;	95 JAppStat 22 163-177
	Duration dependence and dispersion in count-data models • Winkelmann, Rainer	95 JBES 13 467-474
Carling, K.; Jacobson, T. Modeling unemployment	duration in a dependent competing risks framework: Identification and estimation •	95 LiDaAnal 1 111-122
Frydman Semiparametric estimation in a three-state	duration-dependent Markov model from interval-censored observations with ... •	95 Biomtrcs 51 502-511
of group sequential logrank tests in maximum	duration versus information trials • Kim, K.; Boucher, H.; Tsiatis, A. A. Design and analysis	95 Biomtrcs 51 988-1000
	Duration model	
Thosar, Satish	Contested tender offers: An estimate of the hazard function • Jaggia, Sanjiv;	95 JBES 13 113-119
Identifying state dependence in	duration models (Disc: p17-19) • McCall, Brian P.	94 ASAProBuEc 14-16
Fitting Weibull	duration models with random effects • Morris, Carl; Christiansen, Cindy	95 LiDaAnal 1 347-359
sample • van den Berg, G. J.; Lindeboom, M.	Durations in panel data subject to attrition: A note on estimation in the case of a stock	95 StNeerla 49 282-293
Correlation between 31P NMR and a biological	duration test for batch approval • Thelin, B. et al. Classification of Estradurin batches:	95 Chemolab 27 135-145
	Durbin-Watson revisited • Axelrod, Michael C.; Glosup, Jeffrey G.	94 ASAProStCp 176-181
et al. A Monte Carlo comparison of tests based on the	Durbin-Watson statistic with other autocorrelation tests in dynamic models • Dezhbakhsh	95 EcomtrRv 14 347-365
	Durbin-Watson test	
	Detecting a change in the intercept in multiple regression • Watson, G. S.	95 StPrLet 23 69-72
Ng, V. Ming	A linear model with a special class of covariance structure • Haq, M. Safiul;	94 JStatRsr 28 189-198
rors • King, M. L.; Harris, D. C. The application of the	Durbin-Watson test to the dynamic regression model under normal and non-normal er-	95 EcomtrRv 14 487-510
M/G/1 queueing models with simple and	dyadic policy • Gakis et al. ... and first moments of the busy and idle periods in controllable	95 StocAnAp 13 47-81
Jones, Clifton T. A	dynamic analysis of interfuel substitution in U.S. industrial energy demand •	95 JBES 13 459-465
Marked point processes on the real line: The	dynamic approach • Last, Günter; Brandt, Andreas	95 Springer:Brln:NY 490p
of "Marked point processes on the real line: The	dynamic approach" • Last, Günter; Brandt, Andreas (Auth); Embrechts, P. A. L. (Rev) Review	96 ShrtBkRv 16/1 10-10
	Dynamic approaches to pension funding • Haberman, Steven; Sung, Joo-Ho	94 InsrneME 15 151-162
Méléard, Sylvie	Dynamic asymptotic results for a generalized star-shaped loss network • Graham, Carl;	95 AnlsApPr 5 666-680
	Dynamic component detection in a multifactor model for stock returns • Costa, Michele	94 JItalStS 3 25-36
Introduction: The microeconometrics of	dynamic decision making • Kapteyn, Arie; Kiefer, Nicholas; Rust, John	95 JApEcmtx 10 1-7S
Testing, encompassing, and simulating	dynamic econometric models • Gouriéroux, Christian; Monfort, Alain	95 EconTher 11 195-228
	Dynamic econometrics • Hendry, David	95 OxfordUniv 904p
Survival in	dynamic environments • Singpurwalla, Nozer D.	95 StatSci 10 86-103
Optimal	dynamic equilibrium in insurance policies • Page, Dominique; Tremolieres, Raymond	95 AplStMDA 11 181-198
and gradient in the context of least squares fitting a	dynamic errors-in-variables model • Vregelaar, J. M. T. On computing objective function	95 SIAMSSC 16 738-753
daughter/father and ... • Hershberger et al.	Dynamic factor analysis: An application to emotional response patterns underlying	95 StrEqMod 2 31-52
control and related minimax design problems - A	dynamic game approach (Second edition) • Basar, T.; Bernhard, P. H^∞-optimal	95 Birkhs:Basl:MA 411p
control and related minimax design problems - A	dynamic game approach (Second edition)" • Basar, T. et al. Review of "H^∞-optimal	96 ShrtBkRv 16/1 11-11
On variational inequality approach to Nash	dynamic game of oil stockpiling • Zhao, L.; Yin, G.	95 StocAnAp 13 627-649
Lambert, P.	Dynamic generalized linear models and repeated measurements • Lindsey, J. K.;	95 JStPlInf 47 129-139
	Dynamic graphics	
Hurley, Catherine	Grand tour and projection pursuit • Cook, Dianne; Buja, Andreas; Cabrera, Javier;	95 JCmpGrSt 4 155-172
Interactive	dynamic graphics for exploratory survival analysis • Atkinson, E. Neely	95 AmerStat 49 77-84
of misconceptions in learning statistical power with	dynamic graphics as a remedial tool • Yu, Chong Ho; Behrens, John T. Identification	94 ASAProStEd 242-246
Estimating long-run relationships from	dynamic heterogeneous panels • Pesaran, M. Hashem; Smith, Ron	95 JEconmtx 68 79-113
The axiomatic derivation of a class of	dynamic inequality measures • Stich, Andreas	95 Metron 53 31-41
Bayesian inference in cyclical component	dynamic linear models • West, Mike	95 JASA 90 1301-1312
Spline smoother as a	dynamic linear model • Moyeed, R. A.	95 AstrlJSt 37 193-204
teams competing in pairwise comparisons using the	dynamic linear model (Disc: p134-135) • Peck, R. R.; Madrigal, J. L. et al. Ranking	94 ASAProBays 112-116
aggregation algorithm for the solution of stochastic	dynamic minimax problems • Breton, Michèle; Hachem, Saeb El A scenario	95 Stochatx 53 305-322
	Dynamic model	
Arellano, Manuel; Bover, Olympia	Another look at the instrumental variable estimation of error-components models •	95 JEconmtx 68 29-51
with dependent errors • Nabeya, Seiji; Perron, Pierre	Approximations to some exact distributions in the first order autoregressive model	95 EcomtrRv 14 421-457
van Tongeren, Onno F. R.	Data analysis or simulation model: A critical evaluation of some methods •	95 EcolgMdl 78 51-60
systems • Jacobson, Tor	On the determination of lag order in vector autoregressions of cointegrated	95 CmpStQ 10 177-192
equations and cointegration • Johansen, Søren	Identifying restrictions of linear equations with applications to simultaneous	95 JEconmtx 69 111-132
structural econometric analysis • Chirinko, Robert S.	Nonconvexities, labor hoarding, technology shocks, and procyclical productivity. A	95 JEconmtx 66 61-98
with star topology • Ahmed, N. U.; Yeap, T. H.	Performance analysis of slotted ALOHA protocol for optical fiber local area networks	95 StocAnAp 13 389-410
Virendra K.; Ullah, Aman	Stein-rule estimation in models with a lagged-dependent variable • Srivastava,	95 CommStA 24 1343-1353
linear biased regression and a genetic algorithm for	dynamic model identification • Wise, B. M. et al. A comparison of neural networks, non-	95 Chemolab 30 81-89
A non-linear	dynamic model identification challenge problem • Wise, Barry M.; Haesloop, Dan	95 Chemolab 30 91-96
Watson statistic with other autocorrelation tests in	dynamic models • Dezhbakhsh et al. ... Carlo comparison of tests based on the Durbin-	95 EcomtrRv 14 347-365
	Dynamic modelling of life table data • Janssen, J.; Skiadas, C. H.	95 AplStMDA 11 35-49
Image analysis, random fields and	dynamic Monte Carlo methods: A mathematical introduction • Winkler, Gerhard	95 Springer:Brln:NY 324p
Review of "Image analysis, random fields and	dynamic Monte Carlo methods: A mathematical introduction" • Winkler, G. (Auth) et al.	95 ShrtBkRv 15/3 49-49
Majetić, Dubravko	Dynamic neural network for prediction and identification of nonlinear dynamic systems •	95 JCmpInT 3 99-106
	Dynamic noncooperative game theory (Second edition) • Basar, Tamer; Olsder, Geert Jan	95 Academic:NY:Lnd 519p
Efficient estimation of models for	dynamic panel data • Ahn, Seung C.; Schmidt, Peter	95 JEconmtx 68 5-27
consistency, and efficiency of various estimators in	dynamic panel data models • Kiviet, Jan F. On bias, in-	95 JEconmtx 68 53-78
model • Kim, S. S.; Park, S. H.	Dynamic plots for displaying the roles of variables and observations in regression	95 CmpStDA 19 401-418
The multiprocess	dynamic Poisson model • Bolstad, William M.	95 JASA 90 227-232
Disturbance detection and isolation by	dynamic principal component analysis • Ku, Wenfu; Storer, Robert H.; Georgakis, Christos	95 Chemolab 30 179-196
Nonlinear time series models for multivariable	dynamic processes • Çinar, Ali	95 Chemolab 30 147-158
	Dynamic programming	
	A best-choice problem with three stops and random termination • Majumdar, A. A. K.	95 SankhyaB 57 76-84
	A class of Bayesian models for optimal exploration • Glazebrook, K. D.; Boys, R. J.	95 JRSS-B 57 705-720
	Customer scheduling with incomplete information • Rieder, Ulrich; Weishaupt, Jürgen	95 PrEngInf 9 269-284
model solved by deconvexification ... • Eaves et al.	A discounted-cost continuous-time flexible manufacturing and operator scheduling	95 PrEngInf 9 65-98
	Optimal control of two competing diseases or species • Kyriakidis, E. G.	95 MaScitst 20 56-66
Bar-Lev, S. K.; Parlar, M.; Perry, D.	Optimal sequential decisions for incomplete identification of group-testable items •	95 SqtlAnly 14 41-57
David, Israel	A sequential assignment match process with general renewal arrival times •	95 PrEngInf 9 475-492
Wakuta, Kazuyoshi	Vector-valued Markov decision processes and the systems of linear inequalities •	95 StocProc 56 159-169
Zhu, Qingxin; Ahmed, N. U.	Dynamic programming equations of stochastic optimal control in Banach space •	95 StocAnAp 13 369-387
to ... • Kececioglu et al. A general discrete-time	dynamic programming model for the opportunistic replacement policy and its application	95 RelEngSS 47 175-185
of "Markov decision processes: Discrete stochastic	dynamic programming" • Puterman, M. L. (Auth); Collins, Sean (Rev) Review	95 JAppStat 22 545-546
of "Markov decision processes: Discrete stochastic	dynamic programming" • Puterman, M. L. (Auth); Iancu, C. (Rev) Review	95 JAppStat 22 547-547
of "Markov decision processes: Discrete stochastic	dynamic programming" • Puterman, M. L. (Auth); Plachky, D. (Rev) Review	95 StatDecn 13 93-93

Review of "Design and analysis of	ecological experiments" • Scheiner, S.; Gurevitch, J. (Ed); Ziegel, Eric R. (Rev)	95 Technmcs 37 128-128
Evaluation of methods for	ecological inference • Cleave, N.; Brown, P. J.; Payne, C. D.	95 JRSS-A 158 55-72
Nakajima, Hisao Indirect effects in	ecological interaction networks 1. The chain rule approach • Higashi, Masahiko;	95 MaBisci 130 99-128
On	ecological regression and ridge estimation • Füle, Erika	95 CommStB 24 385-398
Review of	"Ecological risk assessment" • Suter, Glenn (Ed); Ziegel, Eric R. (Rev)	95 Technmcs 37 240-240
Statistical ecology and related	ecological statistics - 25 years • Patil, G. P.	94 ASAProEnvr 1-7
Statistical ecology and related	ecological statistics - 25 years • Patil, G. P.	95 EnvEcoSt 2 81-89
	Ecology	
	The definition of the niche by fuzzy set theory • Cao, Guangxia	95 EcolgMdl 77 65-71
	Measuring resilience in stochastic systems • Ives, Anthony R.	95 EcolgMon 65 217-233
Smith, Eric P.; Rose, Kenneth A.	Model goodness-of-fit analysis using regression and related techniques •	95 EcolgMdl 77 49-64
(Disc: p192-199) • Urfer, W. et al.	Multistate models for monitoring individual trees in permanent observation plots	94 EnvEcoSt 1 171-192
	Spatial analysis by distance indices • Perry, Joe N.	95 JAnmlEco 64 303-314
Capture-recapture and evolutionary	ecology: A difficult wedding? • Clobert, J.	95 JAppStat 22 989-1008
for homogeneous proportions and distributions in	ecology • Garson, G. I.; Moser, E. B. Aggregation and the Pearson chi-square statistic	95 Ecology 76 2258-2269
Trends in statistical	ecology and environmental statistics • MacNeill, Ian B.	94 ASAProEnvr 8-9
models to address questions in evolutionary	ecology • Nichols, James D.; Kendall, William L. The use of multi-state capture-recapture	95 JAppStat 22 835-846
Applications of statistical	ecology in forestry - Past accomplishments and future outlook • Burkhardt, Harold E.	94 ASAProEnvr 10-13
(Ed)	Ecology of infectious diseases in natural populations • Grenfell, Bryan; Dobson, Andrew	95 CambridgeUniv 532p
model for the variance to mean power law in	ecology • Kendal, Wayne S. A probabilistic	95 EcolgMdl 80 293-297
effects growth models for repeated measures in	ecology • Schabenberger, Oliver Nonlinear mixed	94 ASAProEnvr 156-161
Statistical	ecology and related ecological statistics - 25 years • Patil, G. P.	94 ASAProEnvr 1-7
Statistical	ecology and related ecological statistics - 25 years • Patil, G. P.	95 EnvEcoSt 2 81-89
time pricing of electricity for residential customers:	Econometric analysis of an experiment • Aubin, C.; Fougére, D.; Husson, E.; Ivaldi, M. Real-	95 JApEcmtx 10 171-191S
shocks, and procyclical productivity. A structural	econometric analysis • Chirinko, Robert S. Nonconvexities, labor hoarding, technology	95 JEconmtx 66 61-98
Integrating	econometric and environmetric modelling • Powell, Alan A.	95 EnvrMtrc 6 429-437
Berndt, Ernst R.; Griliches, Zvi; Rappaport, Neal J.	Econometric estimates of price indexes for personal computers in the 1990's •	95 JEconmtx 68 243-268
of non-normal random vector with applications to	econometric estimators and test statistics • Ullah, A. et al. Moments of the function	95 EcomtrRv 14 459-471
of early twentieth-century U.S. homicide rates: An	econometric forecasting approach • Eckberg, Douglas Lee Estimates	95 Dmogrphy 32 1-16
(Rev) Review of "The history of	econometric ideas" (Disc: p384-397) • Morgan, Mary S. (Auth); Neuberg, Leland Gerson	95 EconTher 11 371-383
Brown, Bryan W. (Ed)	Econometric inference using simulation techniques • van Dijk, Herman K.; Monfort, Alain;	95 Wiley:NY:UK 256p
	Econometric model	
Vere, D. T.; Griffith, G. R.	Modifying quantitative forecasts of livestock production using expert judgements •	95 JForcast 14 453-464
Testing, encompassing, and simulating dynamic	econometric models • Gouriéroux, Christian; Monfort, Alain	95 EconTher 11 195-228
Detecting and testing causality in linear	econometric models • Faliva, Mario; Zoia, M. Grazia	94 JItalStS 3 61-76
and flexible parametric estimation procedures in	econometric models • McDonald et al. ... and applications of semi-parametric	94 ASAProBuEc 229-234
Vuong, Quang (Transl) Statistics and	econometric models. Volumes 1 and 2 • Gourieroux, Christian; Monfort, Alain (Auth);	95 CambridgeUniv 1062p
Prediction and accommodation in	econometric modelling • Dharmapala, Dhammika; McAleer, Michael	95 EnvrMtrc 6 551-556
(Rev) Review of "The	econometric modelling of financial time series" • Mills, Terence C. (Auth); Lee, Bong-Soo	95 JFinance 50 387-390
icsson, Neil R.; Irons, John S. (Rev) Review of "Applied	econometric techniques" • Cuthbertson, Keith; Hall, Stephen G.; Taylor, Mark P. (Auth); Er-	95 EcomtrRv 14 121-132
	Econometrics	
models • Kiviet, J. F.; Phillips, G. D. A.; Schipp, B.	The bias of OLS, GLS, and ZEF estimators in dynamic seemingly unrelated regression	95 JEconmtx 69 241-266
(Disc: p159-175) • Boswijk, H. Peter	Efficient inference on cointegration parameters in structural error correction models	95 JEconmtx 69 133-158
Kim, Jinbang; De Marchi, Neil; Morgan, Mary S.	Empirical model particularities and belief in the natural rate hypothesis •	95 JEconmtx 67 81-102
	Errors in variables and cointegration • Solo, Victor	95 EconTher 11 60-80
	Estimation of cointegrated systems with I(2) processes • Kitamura, Yuichi	95 EconTher 11 1-24
	The ET interview: Gregory C. Chow • Pagan, Adrian	95 EconTher 11 597-624
Terrell, Dek	Flexibility and regularity properties of the asymptotically ideal production model •	95 EcomtrRv 14 1-17
	Frisch on testing of business cycle theories • Boumans, Marcel	95 JEconmtx 67 129-147
ity, and other irregular features • Dorsey, R. E. et al.	Genetic algorithms for estimation problems with multiple optima, nondifferentiabil-	95 JBES 13 53-66
	Identification of multivariate AR-models by threshold accepting • Winker, Peter	95 CmpStDA 20 295-307
equations and cointegration • Johansen, Søren	Identifying restrictions of linear equations with applications to simultaneous	95 JEconmtx 69 111-132
	Interactive statistics tutorials in Stata (jse/v3n3/ferrall) • Ferrall, Christopher	95 JStEd 3/3 28kb
data • Davies, Anthony; Lahiri, Kajal	A new framework for analyzing survey forecasts using three-dimensional panel	95 JEconmtx 68 205-227
illustration • Urbain, Jean-Pierre	Partial versus full system modelling of cointegrated systems - An empirical	95 JEconmtx 69 177-210
An example of likelihood inference in ... • Juselius	Do purchasing power parity and uncovered interest rate parity hold in the long run?	95 JEconmtx 69 211-240
the theory of consumer demand • Keuzenkamp et al.	Rejection without falsification: On the history of testing the homogeneity condition in	95 JEconmtx 67 103-127
	The significance of testing empirical non-nested models • McAleer, Michael	95 JEconmtx 67 149-171
Gouriéroux, Christian; Szafarz, Ariane	Solutions of multivariate rational expectations models • Broze, Laurence;	95 EconTher 11 229-257
	A statistical analysis of cointegration for I(2) variables • Johansen, Søren	95 EconTher 11 25-59
likelihood using principal components • Onishi, H.	A user-knowledge-based variable selection method for limited information maximum	95 CmpStDA 19 379-399
Review of "Handbook of statistics (Volume 11)	Econometrics" • Maddala, G. S.; Rao, C. R.; Vinod, H. D. (Ed); Gill, Len (Rev)	95 IntJFor 11 189-191
Review of "Applied	econometrics" • Kalirajan, K. P. (Auth); Senroy, Sugata (Rev)	95 CalcutSt 45 135-136
Review of "Nonlinear dynamics chaos and	econometrics" • Pesaran, M. Hashem; Potter, Simon M. (Ed); Adcock, Chris (Rev)	95 IntJFor 11 599-601
Dynamic	econometrics • Hendry, David	95 OxfordUniv 904p
Review of "Computational techniques for	econometrics and economic analysis" • Belsley, D. A. (Ed); Smith, Murray D. (Rev)	95 Ststcian 44 540-540
Review of "Computational techniques for	econometrics and economic analysis" • Belsley, D. A. (Ed); Uebe, Götz (Rev)	95 CmpStDA 19 359-360
Review of "The formation of	econometrics" • Duo, Q. (Auth); Costa, M. (Rev. Italian)	94 Statstca 54 540-541
H. D. (Ed); Krämer, Walter (Rev. German) Review of	"Econometrics (Handbook of statistics) (Volume II)" • Maddala, G. S.; Rao, C. R.; Vinod,	95 AllgeStA 79 101-102
Review of "Estimation and inference in	econometrics" • Davidson, R.; Mackinnon, J. G. (Auth); Smith, Richard J. (Rev)	95 JApEcmtx 10 339-341
(Rev) Review of "Estimation and inference in	econometrics" • Davidson, Russell; MacKinnon, James G. (Auth); Zinde-Walsh, Victoria	95 EconTher 11 631-635
Ramalingam (Rev) Review of	"Econometrics of information and efficiency" • Sengupta, J. K. (Auth); Shanmugam,	95 Ststcian 44 290-291
The	econometrics of learning in financial markets • Bossaerts, Peter	95 EconTher 11 151-189
On theory testing in	econometrics Modeling with nonexperimental data • Spanos, Aris	95 JEconmtx 67 189-226
Review of "Learning and practicing	econometrics" • Griffiths, W. E.; Hill, R. C.; Judge, G. G. (Auth); Phillips, Robert F. (Rev)	95 IntJFor 11 331-333
H. D. (Ed); Krämer, Walter (Rev. German) Review of	"Econometrics (Handbook of statistics) (Volume II)" • Maddala, G. S.; Rao, C. R.; Vinod,	95 AllgeStA 79 101-102
	Econometrics of short and unreliable time series • Uri, T.; Wörgötter, A. (Ed)	95 Physica:Ger 238p
On tests and significance in	econometrics • Keuzenkamp, Hugo A.; Magnus, Jan R.	95 JEconmtx 67 5-24
Randomization tests in	econometrics • Kennedy, Peter E.	95 JBES 13 85-94
Editors' introduction - The significance of testing in	econometrics • Keuzenkamp, Hugo A.; Magnus, Jan R.	95 JEconmtx 67 1-3
Three ways to think about testing in	econometrics • Mirowski, Philip	95 JEconmtx 67 25-46
The role of theory in	econometrics • Pesaran, M. Hashem; Smith, Ron	95 JEconmtx 67 61-79
Topics in advanced	econometrics. Volume 2: Linear and nonlinear simultaneous equations • Dhrymes, P. J.	94 Springer:Brln:NY 401p
analysis of stock prices, inflation and real	economic activity • Raj, Baldev; Amoako-Adu, Ben Cointegration	94 ASAProBuEc 190-195

of "Computational techniques for econometrics and	economic analysis" • Belsley, D. A. (Ed); Smith, Murray D. (Rev) Review	95 Ststcian 44 540-540
of "Computational techniques for econometrics and	economic analysis" • Belsley, D. A. (Ed); Uebe, Götz (Rev) Review	95 CmpStDA 19 359-360
Y. (Auth) et al. Review of "Statistical tables for	economic, business and social studies (Second edition)" • Kmietowicz, Z. W.; Yannoulis,	95 JAppStat 22 439-440
Administrative record matching for the 1992	economic censuses • Steel, Philip M.; Konschnik, Carl A.	94 ASASrvyRsr 473-478
Development of edit parameters for 1992	economic census enterprise reports • Oh, Sungsoo; Paletz, Dave; Kim, Jay; Salyers, Eddie	94 ASASrvyRsr 1144-1149
analysis versus constrained optimization with an	economic control chart model • Simpson, James R.; Keats, J. Bert Trade-off	95 IIETrn 27 765-772
Comparing extinction risk and	economic cost in wildlife conservation planning • Haight, Robert G.	95 EcolgApp 5 767-775
Ashour et al. Discriminant analysis for social-	economic and cultural factors affecting diarrhea diseases in children under two years ... •	95 EgpStJ 39 1-11
Review of "Forecasting financial and	economic cycles" • Niemira, Michael P.; Klein, Philip A. (Auth); Mullineux, Andy W. (Rev)	95 IntJFor 11 191-192
J. K.; Lawrence, F. P. Statistically constrained	economic design of the EWMA control chart • Montgomery, D. C.; Torng, J. C.-C.; Cochran,	95 JQualTek 27 250-256
quality • Bai, Do Sun; Kwon, Hyuck Moo	Economic design of a two-stage screening procedure with a prescribed outgoing	95 Metrika 42 1-18
Yang, Su-Fen	Economic design of $\bar{X}$ control chart when Taguchi's loss function is considered •	95 ASQCTekCon 871-878
of "Population growth, income distribution and	economic development" • Heerink, Nico (Auth); Ioannides, Chris (Rev) Review	95 IntJFor 11 333-334
Rosenberg, Harry M. Sources of data on socio-	economic differential mortality in the United States • Hoyert, Donna L.; Singh, Gopal K.;	95 JOffStat 11 233-260
Implementing statistically constrained	economic EWMA control charts • Torng, J. C.-C.; Cochran, J. K.; Montgomery, D. C. et al.	95 JQualTek 27 257-264
rection to "Implementing statistically constrained	economic EWMA control charts" (V27 p257-264) • Torng, J. C.-C.; Cochran, J. K. et al. Cor-	95 JQualTek 27 402-402
	Economic and financial modeling with Mathematica • Varian, H. R.	93 Springer:Brln:NY 458p
Professional	economic forecasts: Are they worth their costs? • Leitch, Gordon; Tanner, J. Ernest	95 JForcast 14 143-157
Review of "Edgeworth on chance,	economic hazard, and statistics" • Mirowski, Philip (Ed); Stigler, Stephen M. (Rev)	95 JASA 90 803-804
of the Italian business cycle by means of monthly	economic indicators • Parigi, G.; Schlitzer, G. Quarterly forecasts	95 JForcast 14 117-141
	Economic knowledge and public opinion on economic issues • Walstad, William B.	94 ASASrvyRsr 1380-1385
Enhancing researcher access to	economic microdata at the Bureau of Labor Statistics • de Wolf, Virginia A.	94 ASAProGovt 164-169
	Economic model	
able state variables • Deaton, Angus; Laroque, Guy	Estimating a nonlinear rational expectations commodity price model with unobserv-	95 JApEcmtx 10 9-40S
	Modelling the supply of fishing effort • Oxley, Les; Roberts, Colin J.	95 EnvrMtrc 6 541-545
experiment • Aubin, C.; Fougére, D.; Husson, E. et al.	Real-time pricing of electricity for residential customers: Econometric analysis of an	95 JApEcmtx 10 171-191S
Horvath, Michael T. K.; Watson, Mark W.	Testing for cointegration when some of the cointegrating vectors are prespecified •	95 EconTher 11 984-1014
use of pooled coefficients in a structural regional	economic model • Rickman, Dan S. A Bayesian analysis of the	95 IntJFor 11 477-490
the CUSUM versus Shewhart control charts using an	economic model • Simpson, James R.; Keats, J. Bert A sensitivity study and comparison of	94 ASAProQlPr 98-103
in panel surveys - The case of the German Socio-	Economic Panel (German) • Rendtel, U. et al. A method for checking longitudinal weights	95 AllgeStA 79 252-277
Berndt, E. R. et al. High-tech capital formation and	economic performance in U.S. manufacturing industries: An exploratory analysis •	95 JEconmtx 65 9-43
survey as political participation: The relation of	economic and political conditions to refusal rates in the ... • Kojetin et al. ... government	94 ASASrvyRsr 1236-1241
Review of "Modelling nonlinear	economic relationships" • Granger, C. W. J.; Teräsvirta, T. (Auth); McKenzie, Sandra (Rev)	95 JASA 90 1491-1491
Robert H. Establishment microdata for	economic research and policy analysis: Looking beyond the aggregates • McGuckin,	95 JBES 13 121-126
Uncertainty about the persistence of	economic shocks • Miller, John P.; Newbold, Paul	95 JBES 13 435-440
M.; Davis, Darwin J.; McWilliams, Thomas P.	Economic, statistical, and economic-statistical design of attribute charts • Saniga, Erwin	95 JQualTek 27 56-73
Maxwell L.; White, Halbert Comments on testing	economic theories and the use of model selection criteria • Granger, Clive W. J.; King,	95 JEconmtx 67 173-187
Temporal aggregation and	economic time series • Rossana, Robert J.; Seater, John J.	95 JBES 13 441-451
Seasonal adjustment of hybrid	economic time series • Scott, Stuart; Stamas, George; Sullivan, Thomas J.; Chester, Paul	94 ASAProBuEc 163-168
Aggregation and the long run properties of	economic time series • Kőrösi, Gábor; Lovrics, László; Mátyás, László	95 MaCmpSim 39 279-286
Carvalho, José L. Analysis of	economic time series. A synthesis (Revised edition) • Nerlove, Marc; Grether, David M.;	95 Academic:NY:Lnd 468p
Hamill, Thomas M. Potential	economic value of ensemble-based surface weather forecasts • Wilks, Daniel S.;	95 MWeathr 11 3565-3575
Timmermann, Allan An assessment of the	economic value of non-linear foreign exchange rate forecasts • Satchell, Steve;	95 JForcast 14 477-497
An	economical control procedure for detecting shift in quality level • Srivastava, M. S.	94 ASAProQlPr 1-6
II (based on Healy's two-stage sampling) is more	economical than the original HM (which is based on Chatterjee's ... • Aoshima ... method	94 AJMMgSci 14 167-196
	Economics	
Z. W.	Accuracy of indices of industrial production in developing countries • Kmietowicz,	95 Ststcian 44 295-307
George, P. J.; Oksanen, E. H.; Veall, M. R.	Analytic and bootstrap approaches to testing a market saturation hypothesis •	95 MaCmpSim 39 311-315
	Axiomatic price index theory: A survey • Balk, Bert M.	95 IntStRvw 63 69-93
Timmermann, Allan	Cointegration tests of present value models with a time-varying discount factor •	95 JApEcmtx 10 17-31
selectivity • Lee, Lung-fei	The computation of opportunity costs in polychotomous choice models with	95 RvEconSt 77 423-435
Peracchi, Franco	The conditional distribution of excess returns: An empirical analysis • Foresi, Silverio;	95 JASA 90 451-466
inflation-unemployment model • Yeung, D. W.-K.	Employment adjustments noise and the expected rate of inflation in a simple	95 StocAnAp 13 125-135
Vinod, H. D.; McCullough, B. D.	Estimating cointegration parameters: An application of the double bootstrap •	95 JStPlInf 43 147-156
Kiefer, Nicholas M.; Neumann, George R.	Estimation of equilibrium wage distributions with heterogeneity • Bowlus, Audra J.;	95 JApEcmtx 10 119-131S
Nawata, Kazumitsu	Estimation of sample-selection models by the maximum likelihood method •	95 MaCmpSim 39 299-303
	European inflation and its variability • Silver, Mick; Ioannidis, Christos	95 Ststcian 44 309-322
	Forecasts of inflation from VAR models • Webb, Roy H.	95 JForcast 14 267-285
space models • Vinod, H. D.; Basu, Parantap	Forecasting consumption, income and real interest rates from alternative state	95 IntJFor 11 217-231
Fortin, Nicole M.	Heterogeneity biases, distributional effects, and aggregate consumption •	95 JApEcmtx 10 287-311
	A household-based, nonparametric test of demand theory • Famulari, Melissa	95 RvEconSt 77 372-382
States • Provencher, Bill	An investigation of the harvest decision of timber firms in the south-east United	95 JApEcmtx 10 57-74S
Hjalmarsson, Lennart	Labour-use efficiency in Swedish social insurance offices • Kumbhakar, Subal C.;	95 JApEcmtx 10 33-47
Knott, Michael	The measurement of price elasticities - The BT experience • Cracknell, David;	95 IntJFor 11 321-329
using linear models and artificial ... • Swanson et al.	A model-selection approach to assessing the information in the term structure	95 JBES 13 265-275
distributions • Mills, Terence C.	Modelling skewness and kurtosis in the London Stock Exchange FT-SE index return	95 Ststcian 44 323-332
Marseille fish market • Härdle, W.; Kirman, A.	Nonclassical demand - A model-free examination of price-quantity relations in the	95 JEconmtx 67 227-257
model • Adang, Pim; Melenberg, Bertrand	Nonnegativity constraints and intratemporal uncertainty in a multi-good life-cycle	95 JApEcmtx 10 1-15
	A nonparametric conditional moment test for structural stability • Hidalgo, Javier	95 EconTher 11 671-698
	Nonparametric estimation of welfare changes • Breslaw, Jon A.	95 EcomtrRv 14 163-182
Blackburn, McKinley L.; Neumark, David	Are OLS estimates of the return to schooling biased downward? another look •	95 RvEconSt 77 217-229
capacity • Hashimoto, Hideo	On the optimal extraction rate of a non-renewable resource with an extraction	95 EnvrMtrc 6 547-550
industry • Rust, John; Rothwell, Geoffrey	Optimal response to a shift in regulatory regime: The case of the US nuclear power	95 JApEcmtx 10 75-118S
	From perpetual strangles to Russian options • Gerber, Hans U.; Shiu, Elias S. W.	94 InsrneME 15 121-126
Steel, M. F. J.; Osiewalski, J.	Posterior analysis of stochastic frontier models using Gibbs sampling • Koop, G.;	95 CmpStQ 10 353-373
Abramson, Bruce; Finizza, Anthony	Probabilistic forecasts from probabilistic models: A case study in the oil market •	95 IntJFor 11 63-72
	The role of theory in econometrics • Pesaran, M. Hashem; Smith, Ron	95 JEconmtx 67 61-79
	Separability test for the electricity supply industry • Lee, Byung-Joo	95 JApEcmtx 10 49-60
Brown, Bryan W.; Walker, Mary Beth	Stochastic specification in random production models of cost-minimizing firms •	95 JEconmtx 66 175-205
heterogeneous agents • Koning, P.; Ridder, G. et al.	Structural and frictional unemployment in an equilibrium search model with	95 JApEcmtx 10 133-151S
	On tests and significance in econometrics • Keuzenkamp, Hugo A.; Magnus, Jan R.	95 JEconmtx 67 5-24
	Three ways to think about testing in econometrics • Mirowski, Philip	95 JEconmtx 67 25-46
systems: An introduction with applications in	economics and biology (Second edition)" • Tu, P. D. V. (Auth) et al. Review of "Dynamical	95 IntJFor 11 491-492
Review of "Statistics for business and	economics" • Mann, P. (Auth); Ziegel, Eric R. (Rev)	95 Technmcs 37 239-240

symposium on the teaching of statistics in higher	education in Ireland • Cran, G. W. Discussion of the	95 Ststcian 44 57-58
symposium on the teaching of statistics in higher	education in Ireland • McClean, Sally Discussion on the	95 Ststcian 44 56-57
symposium on the teaching of statistics in higher	education in Ireland • Wilson, Noel Discussion on the	95 Ststcian 44 58-58
toward applied statistics of postgraduate students in	education in the Lebowa region of South Africa • Glencross, M. J.; Cherian, V. I. Attitudes	95 PsychRpt 77 315-322
Cost-effectiveness in public	education • Bennett, Jay	95 Chance 8/1 38-41
Statistical	education for quality managers • Martin, Margaret; Suber, Marsha	94 ASAProBays 148-149
Short, Thomas H. Conditional probability and	education reform: Are they compatible? (jse/v3n2/rossman) • Rossman, Allan J.;	95 JStEd 3/2 23kb
and the role of TQM and the statistician in public	education reform • Wehrle, Robert A. Democracy, capitalism	94 ASAProQlPr 189-193
Comment on	"Education research using surveys" • Smith, Thomas M.	94 ASAProSoSt 213-215
The coming of age of statistical	education • Vere-Jones, David	95 IntStRvw 63 3-23
of the International Association for Statistical	Education" • Brunelli, L. et al. Review of "Proceedings of the First Scientific Meeting	95 AmerStat 49 276-276
Comment on "Statistics	education" • Garfield, Joan B.	95 AmerStat 49 18-20
Comment on "Statistics	education" • Morris, Carl N.	95 AmerStat 49 21-23
Electronic data collection at the National Center for	Education Statistics: Successes in the collection of library data • Kindel, Carrol Benner	94 JOffStat 10 93-102
Total quality	education at the U.S. Air Force Academy • Tong, Terrence G. L.	94 ASAProQlPr 295-299
of diversion? The effect of community colleges on	educational attainment • Rouse, Cecilia Elena Democratization	95 JBES 13/2 217-224
Hierarchical models for	educational data: An overview • Morris, Carl N.	95 JEdStat 20 190-200
H.; Moriarty, Helene; Cooley, Mary E. Readability of	educational materials for patients with cancer (jse/v3n2/datasets.short) • Short, Thomas	95 JStEd 3/2 15kb
K. Bose • Srivastava, A. B. L. A review of	educational planning and statistics in India with focus on contributions of Professor P.	94 CalcutSt 44 251-264
Adding value to the value-added	educational production function specification (Disc: p213-215) • Akerhielm, Karen	94 ASAProSoSt 192-195
Methods for uncertainty in	educational testing • Longford, N. T.	95 Springer:Brln:NY 285p
Review of "Methods for uncertainty in	educational testing" • Longford, N. T. (Auth); Ramsay, J. O. (Rev)	96 ShrtBkRv 16/1 5-6
Models for uncertainty in	educational testing • Longford, N.	95 Springer:Brln:NY 300p
	Effective dose	
Malani, Hina Mehta	Analysis of animal carcinogenicity experiments with multiple tumor types • Lu, Ying;	95 Biomtrcs 51 73-86
	Effective population size	
overlapping generations • Jorde, P. E.; Ryman, N.	Temporal allele frequency change and estimation of effective size in populations with	95 Genetcs 139 1077-1090
Temporal allele frequency change and estimation of	effective size in populations with overlapping generations • Jorde, Per Erik; Ryman, Nils	95 Genetcs 139 1077-1090
Enas, Greg Essential	efficacy data analysis for a 2×2 crossover study • Huster, William; Brunelle, Rocco;	94 ASAProBiop 503-509
Chi, George Y. H.; O'Neill, Robert T.	Efficacy evaluation for monotherapies in two-by-two factorial trials • Hung, H. M. James;	95 Biomtrcs 51 1483-1493
model methods for evaluating treatment	efficacy in multicenter clinical trials • Davis, Charles S.; Chung, Younghae Randomization	95 Biomtrcs 51 1163-1174
Longini, Ira M., Jr; Watelet, Luc Estimation of vaccine	efficacy in non-randomly mixing populations • Haber, Michael; Halloran, M. Elizabeth;	95 BiomtrcJ 37 25-38
Rubenstein, Linda M. Analysis of the	efficacy of rhIL-1ra for the treatment of sepsis • Lemke, Jon H.; LaBrecque, John F.;	94 ASAProBiop 340-345
birth-death and immigration models for testing the	efficacy of some control programmes for AIDS • Biswas et al. ... branching process and	95 BiomtrcJ 37 235-249
Comparing the bivariate effects of toxicity and	efficacy of treatments (Disc: p1143-1143) • Tubert-Bitter, P.; Bloch, D. A.; Raynauld, J.-P.	95 StatMed 14 1129-1141
From here to there (dosing to	efficacy) with neural nets • Minor, James M.; Namini, Hamid; Cruickshank, Scott	94 ASAProBiop 334-339
	Efficiency	
• Vos, Paul W.	Quasi-likelihood or extended quasi-likelihood? An information-geometric approach	95 AnInStMa 47 49-64
	• see also asymptotic –, relative –	
Smith, D. R.; Conroy, M. J.; Brakhage, D. H.	Efficiency of adaptive cluster sampling for estimating density of wintering waterfowl •	95 Biomtrcs 51 777-788
	Efficiency-balance	
designs • Siatkowski, Idzi	Some characterizations of a class of balanced two-way elimination of heterogeneity	94 ListyBio 31 27-33
A note on generalized binary proper	efficiency-balanced block designs • Kherwa, G. Ram; Prasad, Jagdish	95 CommStA 24 2797-2801
Some methods for constructing	efficiency balanced designs • Ghosh, D. K.; Karmokar, P. K.	94 JIAgSt 46 305-309
The comparison of the	efficiency of Bayes and empirical Bayes estimators • Gruber, Marvin H. J.	94 ASAProPEng 132-135
Reeve, Russell; Hale, Michael Results and	efficiency of Bayesian dose adjustment in a clinical trial with binary endpoint •	94 ASAProBiop 324-327
A nonparametric assessment of bias and	efficiency • Campbell, Bryan; Ghysels, Eric Federal budget projections:	95 RvEconSt 77 17-31
relationships of fish • Hayes, D. B. et al.	Efficiency and bias of estimators and sampling designs for determining length-weight	95 JFishRBC 52 84-92
A random linear functional approach to	efficiency bounds • Holly, Alberto	95 JEconmtx 65 235-261
Maximizing selection	efficiency for categorical traits • Meuwissen, T. H. E.; Engel, B.; van der Werf, J. H. J.	95 JAnmlSci 73 1933-1939
	Efficiency of check-plot designs in unreplicated field trials • Chandra, S.	94 ThApGene 88 618-620
(Korean) • Lee, H.-Y.; Hong, K.-H. A study on the	efficiency comparison of the black-box method with the randomized response technique	95 KorApS 8/2 27-41
	Efficiency comparisons of central composite designs • Jensen, D. R.	95 JStCmpSm 52 177-183
transformations in the Gibbs sampler: Issues of	efficiency and convergence • Dellaportas, Petros Random variate	95 StatComp 5 133-140
rakhmedov, Sh. A. Large deviations and intermediate	efficiency of decomposable statistics in a multinomial scheme • Ivchenko, G. I.; Mi-	95 MaMethSt 4 294-311
correlated • Bischoff, Wolfgang Lower bounds for the	efficiency of designs with respect to the D-criterion when the observations are	95 MaOpfStS 27 27-44
de Koning, G. J.; Weller, J. I.	Efficiency of direct selection on quantitative trait loci for a two-trait breeding objective •	94 ThApGene 88 669-677
	Efficiency of empirical estimators for Markov chains • Greenwood, P. E.; Wefelmeyer, W.	95 AnlsStat 23 132-143
Nonparametric tests of regularity, Farrell	efficiency, and goodness-of-fit • Färe, Rolf; Grosskopf, Shawna	95 JEconmtx 69 415-425
Predictive	efficiency of improved estimators in restricted regression models • Dube, M.; Singh, V.	95 Test 4 323-331
	Efficiency indices for second-order designs • Jensen, D. R.	95 BiomtrcJ 37 65-80
	Efficiency of infinite dimensional M-estimators • van der Vaart, A. W.	95 StNeerla 49 9-30
Review of "Econometrics of information and	efficiency" • Sengupta, J. K. (Auth); Shanmugam, Ramalingam (Rev)	95 Ststcian 44 290-291
and phrases – The key to scholarly visibility and	efficiency in an information explosion • Gbur, Edward E., Jr; Trumbo, Bruce E. Key words	95 AmerStat 49 29-33
What is sample design	efficiency? Let's let the customer decide • Chromy, James R.	94 ASAProQlPr 226-231
Saurabh; Vidakovic, Brani	Efficiency of linear Bayes rules for a normal mean: Skewed priors class • Mukhopadhyay,	95 Ststcian 44 389-397
frame surveys • Biemer, Paul; Akin, Don The	efficiency of list-assisted random digit dialing sampling schemes for single and dual	94 ASASrvyRsr 1-10
the population census: Balancing privacy and	efficiency. In memory of Roger A. Herriot • Alexander, Lois Reinventing	94 ASAProGovt 1-2
tests (Russian) • Taufik, A. M. Z. Increasing the	efficiency of the Monte-Carlo method in estimating error probabilities of likelihood-ratio	94 VLenUM /2 105-108
linear discriminant analysis, ... • Sánchez et al.	Efficiency of multi-layered feed-forward neural networks on classification in relation to	95 Chemolab 28 287-303
Rotnitzky, Andrea Semiparametric	efficiency in multivariate regression models with missing data • Robins, James M.;	95 JASA 90 122-129
Survey • Kalsbeek, W. D.; Botman, S. L. et al. Cost-	efficiency and the number of allowable call attempts in the National Health Interview	94 JOffStat 10 133-152
of constrained D- and L-optimal designs using	efficiency plots • Wong, Weng Kee A graphical approach for the construction	95 JStCmpSm 53 143-152
Lefkopoulou, Myrto; Ryan, Louise M.	Efficiency and power of tests for multiple binary outcomes • Legler, Julie M.;	95 JASA 90 680-693
Tiwari, Gayatri; Saxena, K. P. On bias and	efficiency of a preliminary test estimator of variance in a normal distribution •	95 Metron 53 163-169
to the measurement and analysis of productive	efficiency • Lovell, C. A. K. Linear programming approaches	94 Top 2 175-248
der non-normal disturbances • Srivastava, V. K. et al.	Efficiency properties of feasible generalized least squares estimators in SURE models un-	95 JEconmtx 66 99-121
UK universities • Glass, J. C.; McKillop, D. G. et al.	Efficiency in the provision of university teaching and research: An empirical analysis of	95 JApEcmtx 10 61-72
of chemical experiments" • Davies, L. et al. Review of	"Efficiency in research, development and production: The statistical design and analysis	95 Technmcs 37 457-458
Maximin	efficiency-robust tests and some extensions • Bose, Sudip; Slud, Eric	95 JStPlInf 46 105-121
Dalabehera, M.; Sahoo, L. N.	Efficiencies of six almost unbiased ratio estimators under a particular model •	95 StHefte 36 61-67
Loss of	efficiency in spoilt experiments • Hunt, D. N.; Li, E.	95 JAppStat 22 309-315
Lennart Labour-use	efficiency in Swedish social insurance offices • Kumbhakar, Subal C.; Hjalmarsson,	95 JApEcmtx 10 33-47
Bahadur and Hodges-Lehmann approximate	efficiencies of tests based on spacings • Bartoszewicz, Jarosław	95 StPrLet 23 211-220

A note on the	efficiencies of two almost unbiased predictive ratio estimators • Sahoo, J.; Sahoo, L. N.	95 StHefte 36 273-278
Singh, R.; Singh, S.; Bellhouse, D. R. et al. On	efficiency of using distinct respondents in a randomized response survey • Mangat, N. S.;	95 SurvyM 21 21-23
On bias, inconsistency, and	efficiency of various estimators in dynamic panel data models • Kiviet, Jan F.	95 JEconmtx 68 53-78
tually balanced incomplete block designs and their	efficiencies under various optimality criteria • Hedayat, A. S. et al. Contingently and vir-	95 StSinica 5 575-591
Spokoiny, V. G. On the	efficiency of wavelet estimators under arbitrary error distributions • Neumann, M. H.;	95 MaMethSt 4 137-166
Second order asymptotic	efficiency for $Y = \theta_1 + g(T) + \epsilon$ (Chinese) • Liang, Hua; Yuan, Yahua	95 ChJApPSt 10 133-141
Chien, Steve; Gratch, Jonathan; Burl, Michael On the	efficient allocation of resources for hypothesis evaluation: A statistical approach •	95 IEEEPAMI 17 652-665
D. W. et al. Equivalence of several methods for	efficient best subsets selection in generalized linear models • Jovanovic, B. D.; Hosmer,	95 CmpStDA 20 59-64
Tail-specific linear approximations for	efficient bootstrap simulations • Hesterberg, Tim	95 JCmpGrSt 4 113-133
Methods of constructing A-	efficient BTIUB designs • Angelis, L.; Moyssiadis, C.; Kageyama, S.	93 UtilMath 44 5-15
Computationally	efficient classes of higher-order kernel functions • Abdous, Belkacem	95 CandJSt 23 21-27
	Efficient computation of the circular error probability (CEP) integral • Shnidman, David A.	95 IEEEAuCn 40 1472-1475
	Efficient computation of the noncentral χ^2 distribution • Chattamvelli, R.; Shanmugam, R.	95 CommStB 24 675-689
Correction to	"Efficient confidence bounds for ROC curves" (V13 p1551-1561) • Schäfer, H.	95 StatMed 14 875-875
Worton, B. J. Accurate and	efficient construction of bootstrap likelihoods • Davison, A. C.; Hinkley, D. V.;	95 StatComp 5 257-264
gression • Mertens, B.; Fearn, T.; Thompson, M. The	efficient cross-validation of principal components applied to principal component re-	95 StatComp 5 227-235
On	efficient designing of nonlinear experiments • Chaudhuri, Probal; Mykland, Per A.	95 StSinica 5 421-440
Sitharama; Deng, Weian An	efficient edge detection algorithm using relaxation labeling technique • Iyengar, S.	95 PattrnRc 28 519-536
Robins, J. M.; Hsieh, F.; Newey, W. Semiparametric	efficient estimation of a conditional density with missing or mismeasured covariates •	95 JRSS-B 57 409-424
	Efficient estimation of models for dynamic panel data • Ahn, Seung C.; Schmidt, Peter	95 JEconmtx 68 5-27
	Efficient estimation of monotone boundaries • Korostelev, A. P.; Simar, L.; Tsybakov, A. B.	95 AnlsStat 23 476-489
Correction to "On	efficient estimation in regression models" (V21 p1486-1521) • Schick, Anton	95 AnlsStat 23 1862-1863
Thomas	Efficient execution of Stone's likelihood ratio tests for disease clustering • Lumley,	95 CmpStDA 20 499-510
Selecting	efficient experiment designs for mixed 2-level and 3-level factors • Rand, Daniel R.	94 ASAProPEng 332-338
Tobias, Randall D.; Garratt, Mark	Efficient experimental design with marketing research applications • Kuhfeld, Warren F.;	94 JMarRsr 31 545-557
choose? • Hennig, Christian	Efficient high-breakdown-point estimators in robust regression: Which function to	95 StatDecn 13 221-241
Barrlund, A.	Efficient implementation of some contextual classification methods • Flygare, A.-M.;	95 CmpStQ 10 327-338
(Disc: p159-175) • Boswijk, H. Peter	Efficient inference on cointegration parameters in structural error correction models	95 JEconmtx 69 133-158
	Efficient IV estimation in nonstationary regression • Kitamura, Yuichi; Phillips, Peter C. B.	95 EconTher 11 1095-1130
Software for computing a robust	efficient linear or nonlinear regression estimator • Stromberg, Arnold J.	94 ASAProStEd 6-11
models • Dahlhaus, Rainer	Efficient location and regression estimation for long range dependent regression	95 AnlsStat 23 1029-1047
Sid-Ahmed, M. A. Fast learning and	efficient memory utilization with a prototype based neural classifier • Abou-Nasr, M. A.;	95 PattrnRc 28 581-593
Breslaw, Jon A.; Smith, J. Barry A simple and	efficient method for estimating the magnitude and precision of welfare changes •	95 JApEcmtx 10 313-327
Highly robust regression methods which are also	efficient • Croux, Christophe; Rousseeuw, Peter J.	94 ASAProStEd 1-5
Relative rates of convergence for	efficient model selection criteria in linear regression • Hurvich, Clifford M.; Tsai, Chih-Ling	95 Biomtrka 82 418-425
dynamic resource sharing • Fleming, P. J. et al.	Efficient Monte-Carlo simulation of a product-form model for a cellular system with	95 ACMMdSim 5 3-21
Sahu, Sujit K.; Carlin, Bradley P.	Efficient parametrisations for normal linear mixed models • Gelfand, Alan E.;	95 Biomtrka 82 479-488
Bruce, D. C.; Doggett, N. A.; Sawhill, W. W. et al.	Efficient pooling designs for library screening • Bruno, W. J.; Knill, E.; Balding, D. J.;	95 Genomics 26 21-30
	Efficient portfolios in the asset liability context • Keel, Alex; Müller, Heinz H.	95 AstinB 25 33-48
Comparison of Bartlett-type adjustments for the	efficient score statistic • Rao, C. Radhakrishna; Mukerjee, Rahul	95 JStPlInf 46 137-146
On	efficient sequential estimation for the simple single server queue • Stefanov, Valeri T.	95 RBrzlPrS 9 55-65
A simple and	efficient space domain implementation of the turning bands method • Dietrich, C. R.	95 WaterReR 31 147-156
using electromagnetic induction techniques. 2. An	efficient spatial sampling algorithm suitable for multiple ... • Lesch et al. ... salinity	95 WaterReR 31 387-398
hazards • Sasieni, Peter D.	Efficiently weighted estimating equations with application to proportional excess	95 LiDaAnal 1 49-57
in stream-fish communities: Effects of sampling	effort and discontinuous spatial distributions • Angermeier et al. ... relative abundances	95 JFishRBC 52 936-949
The 'Golden	Egg' • Kimber, Hilary	95 TeachgSt 17/2 34-37
ing diarrhea diseases in children under two years in	Egypt • Ashour et al. Discriminant analysis for social-economic and cultural factors affect-	95 EgpStJ 39 1-11
Variance estimation for	EIA-877 propane prices • O'Colmain, Benita J.; Saavedra, Pedro J.; Weir, Paula	94 ASASrvyRsr 232-235
	Eigenvalue	
	Correspondence analysis of an artificial torus data • Okamoto, Masashi	94 Behvmtrk 21 149-161
Bai, Z. D. On the empirical distribution of	eigenvalues of a class of large dimensional random matrices • Silverstein, Jack W.;	95 JMultiAn 54 175-192
	Eigenvalue decomposition	
	Gaussian parsimonious clustering models • Celeux, Gilles; Govaert, Gérard	95 PattrnRc 28 781-793
spectroscopic and ... • Brereton et al. Use of	eigenvalues for determining the number of components in window factor analysis of	95 Chemolab 27 73-87
tion. Part III. The use of asymmetric and generalized	eigenvalue equations • Alsberg ... by the method of postponed basis matrix multiplica-	95 Chemolab 30 223-225
	Eigenvalues and expansion of regular graphs • Kahale, Nabil	95 JACM 42 1091-1106
A simple proof and refinement of Wielandt's	eigenvalue inequality • Dümbgen, Lutz	95 StPrLet 25 113-115
Strong convergence of the empirical distribution of	eigenvalues of large dimensional random matrices • Silverstein, Jack W.	95 JMultiAn 55 331-339
Estimates of the first Dirichlet	eigenvalue by using diffusion processes • Wang, Feng-Yu	95 ProbTher 101 363-369
Obituary: Churchill	Eisenhart, 1913-1994 • Cameron, Joseph M.; Rosenblatt, Joan R.	95 AmerStat 49 243-244
Obituary: Churchill	Eisenhart, 1913-1994 • Plackett, Robin	95 JRSS-A 158 338-338
An exploratory analysis of a record of	El Niño events, 1800-1987 • Solow, Andrew R.	95 JASA 90 72-77
Testing for change in the frequency of	El Niño events • Solow, Andrew R.	95 JClimate 8 2563-2566
time of a Wiener process with drift between two	elastic barriers • Dominé, Marco Moments of the first-passage	95 JAppProb 32 1007-1013
The measurement of price	elasticities - The BT experience • Cracknell, David; Knott, Michael	95 IntJFor 11 321-329
demand for outpatient prescription drugs by the	elderly • Moeller, John F. Prescription drug coverage and the	94 ASAProSoSt 147-152
Assessment of cardiovascular risk factors in the	elderly: The Framingham heart study • Larson, Martin G.	95 StatMed 14 1745-1756
Hoffman, K. et al. A dual frame design for sampling	elderly minorities and persons with disabilities (Disc: p695-696) • Ezzati, T. M.;	95 StatMed 14 571-583
fer function approach • Shelley, Mack C., II By-	elections and government stability in the United Kingdom and the United States: A trans-	94 ASAProSoSt 71-75
on voter participation in the 1992 presidential	election • Schlichting, Kurt C.; Tuckel, Peter S. A spatial analysis of contextual effects	94 ASASrvyRsr 1339-1344
ance of samples of individuals: Comparison of the	electoral registers and postcode address file as sampling frames • Lynn et al. ... and vari-	95 Ststcian 44 173-194
Judgmental and statistical methods of peak	electric load management • Anderson, Elizabeth A.	95 IntJFor 11 295-305
Computer simulation of a nonlinear model for	electrical circuits with α-stable noise • Janicki, A.	95 ZastApMa 23 95-105
Computer-aided modeling and simulation of	electrical circuits with α-stable noise • Weron, A.	95 ZastApMa 23 83-93
to stochastic problems in physics, chemistry and	electrical engineering • Coffey, W. T. et al. The Langevin equation and its applications	95 WrldSciPub:Sing 500p
Regression-based short term prediction of	electrical load for a power system • Polzehl, Jörg	94 ASAProPEng 285-290
gression and error correction model for forecasting	electricity consumption and prices • Joutz, F. L. et al. An integrated Bayesian vector autore-	95 JForcast 14 287-310
Gaines, Leonard M. Measuring relative	electricity cost sensitivity in manufacturing establishments in New York State •	94 ASAProGovt 183-188
of structural differences in minority household	electricity demand • Poyer, David A.; Earl, Elizabeth A statistical analysis	94 ASAProGovt 221-226
models to forecast nuclear net generation of	electricity • Kundra, Inderjit Time-series	94 ASAProBuEc 327-332
Fougére, D.; Husson, E. et al. Real-time pricing of	electricity for residential customers: Econometric analysis of an experiment • Aubin, C.;	95 JApEcmtx 10 171-191S
Separability test for the	electricity supply industry • Lee, Byung-Joo	95 JApEcmtx 10 49-60
	Electrifying statistics for your students • Rycraft, Michael	95 TeachgSt 17/2 67-69
• Lesch et al. Spatial prediction of soil salinity using	electromagnetic induction techniques 1. Statistical prediction models: A comparison ...	95 WaterReR 31 373-386

Lesch et al. Spatial prediction of soil salinity using	electromagnetic induction techniques. 2. An efficient spatial sampling algorithm ... •	95 WaterReR 31 387-398
the collection of library data • Kindel, Carrol Benner	Electronic data collection at the National Center for Education Statistics: Successes in	94 JOffStat 10 93-102
	Electronic publication	
information explosion • Gbur, E. E., Jr; Trumbo, B. E.	Key words and phrases - The key to scholarly visibility and efficiency in an	95 AmerStat 49 29-33
	Electronic services in statistics • Krause, Andreas	95 CmpStDA 19 595-604
IntroStat: A hypertext-based design for an	electronic textbook to introduce biomedical statistics • Okada, Mihoko; O'Brien, Myles	95 CmpBiomd 46 265-276
Omey, Edward; Kozarovitzky, Eugene An	elementary renewal theorem for random compact convex sets • Molchanov, Ilya S.;	95 AdvAppPr 27 931-942
Robust	elementwise estimation of a dispersion matrix • Mehrotra, Devan V.	95 Biomtrcs 51 1344-1351
Mitigating	elevation-induced errors in satellite telemetry locations • Keating, Kim A.	95 JWildlMg 59 801-808
Exact	Elfving-minimax designs for quadratic regression • Krafft, Olaf; Schaefer, Martin	95 StSinica 5 475-484
Girón, Javier Prior	elicitation and Bayesian robustness in a noisy Poisson process • Moreno, Eias;	94 ASAProBays 217-220
Herren, L. T.; Clemen, R. T. The use of probability	elicitation in the high-level nuclear waste regulation program • DeWispelare, A. R.;	95 IntJFor 11 5-24
models • Madigan, D.; Gavrin, J.; Raftery, A. E.	Eliciting prior information to enhance the predictive performance of Bayesian graphical	95 CommStA 24 2271-2292
Eisenhower et al. A records-survey comparison of	eligibility and health care utilization measures for Medicaid beneficiaries: Adult and ... •	94 ASASrvyRsr 713-718
Gauss, statistics and Gaussian	elimination • Stewart, G. W.	95 JCmpGrSt 4 1-11
characterizations of a class of balanced two-way	elimination of heterogeneity designs • Siatkowski, Idzi Some	94 ListyBio 31 27-33
Orthogonality in a two-way	elimination of heterogeneity design • Siatkowski, I.	93 UtilMath 43 225-233
On the E-optimality of some two-way	elimination of heterogeneity designs • Brzeskwiniewicz, Henryk	95 StPrLet 25 9-13
Relative potency of two preparations in two-way	elimination of heterogeneity designs with multivariate responses • Hanusz, Zofia	95 Biomtrcs 51 1133-1139
Heuristic algorithms for finding inexpensive	elimination schemes • Harbron, Chris	95 StatComp 5 275-287
to evaluate populations for alleles to improve an	elite hybrid • Fabrizius, M. A.; Openshaw, S. J. Methods	94 ThApGene 88 653-661
for identifying lines useful for the improvement of	elite single crosses, based on a combining ability model • Hohls et al. ... estimator	95 Heredity 75 155-163
On minimax filtering over	ellipsoids • Belitser, E. N.; Levit, B. Y.	95 MaMethSt 4 259-273
testing of hypotheses on the distribution density for	ellipsoids in l_p • Ingster, Yu. I. Minimax	94 ThProbAp 39 417-436
Fixed diameter confidence	ellipsoid in time-sequential models • Aras, Girish; Gardiner, Joseph	95 SqtlAnly 14 29-39
Ghosh, J. K.; Mukerjee, Rahul On perturbed	ellipsoidal and highest posterior density regions with approximate frequentist validity •	95 JRSS-B 57 761-769
in linear models with respect to an incomplete	ellipsoidal restrictions • Lu, Changyu; Li, Weixin Admissibility of linear estimator	95 AMaPeknE 37 289-295
in linear models with respect to incomplete	ellipsoidal restrictions • Lu, Chang-yu Admissibility of inhomogeneous linear estimators	95 CommStA 24 1737-1742
On the existence of positive solutions for semilinear	elliptic equations with Neumann boundary conditions • Chen, Z. Q.; Williams, R. J. et al.	95 ProbTher 101 251-276
Quasilinear stochastic	elliptic equations with reflection • Nualart, David; Tindel, Samy	95 StocProc 57 73-82
diffusions associated with some degenerate	elliptic operators • Gao, Ping The lifetime of conditioned	95 StocProc 57 319-337
of number of lattice points in random narrow	elliptic strip • Bleher, Pavel M.; Lebowitz, Joel L. Variance	95 AIHenrPB 31 27-58
	Elliptical distribution	
Tarpey, Thaddeus	Principal points and self-consistent points of symmetric multivariate distributions •	95 JMultiAn 53 39-51
Bolfarine, Heleno	On some characterizations of the t-distribution • Arellano-Valle, Reinaldo B.;	95 StPrLet 25 79-85
Correction to "Characteristic functions of a class of	elliptical distributions" (V49 p164-178) • Kotz, Samuel; Ostrovskii, Iosif	95 JMultiAn 52 352-354
of some test criteria for the covariance matrix in	elliptical distributions under local alternatives • Purkayastha et al. ... distributions	95 JMultiAn 55 165-186
Principal points and self-consistent points of	elliptical distributions • Tarpey, Thaddeus; Li, Luning; Flury, Bernard D.	95 AnlsStat 23 103-112
Covariance structure for	elliptical distributions • Wang, Weijing; Wells, Martin T.	94 ASAProSoSt 60-65
On the density function of T_0^2 from	elliptical distribution (Chinese) • Xia, Yincun	95 ChJApPSt 10 119-124
A new class of matrix variate	elliptically contoured distributions • Gupta, A. K.; Varga, T.	94 JItalStS 3 255-270
Some inference problems for matrix variate	elliptically contoured distributions • Gupta, A. K.; Varga, T.	95 MaOpfStS 26 219-229
Normal mixture representations of matrix variate	elliptically contoured distributions • Gupta, A. K.; Varga, T.	95 SankhyaA 57 68-78
and other expected values for matrix variate	elliptically contoured distribution • Gupta, A. K.; Varga, T. Moments	94 Statstca 54 361-373
Estimation of scale matrix of	elliptically contoured matrix distributions • Li, Run-Ze; Fang, Kai-Tai	95 StPrLet 24 289-297
Cochran theorems for a multivariate	elliptically contoured model • Wang, Tonghui; Wong, Chi Song	95 JStPlInf 43 257-270
karyotyping of human chromosomes based on	elliptically symmetric statistical distributions • Ritter, G. et al. Automatic context-sensitive	95 PattrnRc 28 823-831
The number of loci needed for	ELOD calculations • Lewis, C. M.; Cannings, C.	92 AnHuGene 56 59-69
(Rev. German) Review of	"Elsevier's dictionary of biometry" • Rasch, D.; Tiku, M. L.; Sumpf, D. (Ed); Ahrens, H.	95 BiomtrcJ 37 896-896
(Rev) Review of	"Elsevier's dictionary of biometry" • Rasch, D.; Tiku, M. L.; Sumpf, D. (Ed); Kemp, A. W.	95 Biomtrcs 51 1186-1187
Elston, R. C.; George, V. T.; Severtson, F. The	Elston-Stewart algorithm for continuous genotypes and environmental factors •	92 HumHrdty 42 16-27
and biology: A comment on the significance of the	Elston-Stewart algorithm • Stewart, John Genetics	92 HumHrdty 42 9-15
tool for obtaining sorption isotherms from column	elution curves • van Veldhuizen, M. et al. The (least squares) quotient algorithm as a rapid	95 WaterReR 31 849-857
applied to ... • Elbergali et al. Influence of predicted	elution regions on the performance of methods for evolutionary factor analysis as	95 Chemolab 27 55-71
	EM algorithm	
mixture models • Belin, Thomas R.; Rubin, Donald B.	The analysis of repeated-measures data on schizophrenic reaction times using	95 StatMed 14 747-768
Desarbo, W. S.; Ramaswamy, V.; Chatterjee, R.	Analyzing constant-sum multiple criterion data: A segment-level approach •	95 JMarRsr 32 222-232
Möttönen, J.; Oja, H.; Krause, U.; Rantakallio, P.	Application of random coefficient regression model to myopia data: A case study •	95 BiomtrcJ 37 657-672
variation • Johaadien, Hassan; Goutis, Constantinos	Approximate maximum likelihood estimation in the presence of extra-Poisson	95 CommStA 24 1625-1636
Holford, T. R.; Rubinstein, P.; Stevens, C. E.	Change points in the series of T4 counts prior to AIDS • Kiuchi, A. S.; Hartigan, J. A.;	95 Biomtrcs 51 236-248
rates • Liao, Jiangang; Brookmeyer, Ron	An empirical Bayes approach to smoothing in backcalculation of HIV infection	95 Biomtrcs 51 579-588
Cheng, Philip E.	Equipercentile equating via data-imputation techniques • Liou, Michelle;	95 Psymtrka 60 119-136
Churchill, Gary A.	Estimation and reliability of molecular sequence alignments • Thorne, Jeffrey L.;	95 Biomtrcs 51 100-113
Elliott, Robert J.	Finite-dimensional models for hidden Markov chains • Aggoun, Lakhdar;	95 AdvAppPr 27 146-160
Carlson, James E.	Full-information factor analysis for polytomous item responses • Muraki, Eiji;	95 ApPsychM 19 73-90
Rubin, Donald B.; Stern, Hal S.; Vehovar, Vasja	Handling "Don't know" survey responses: The case of the Slovenian plebiscite •	95 JASA 90 822-828
Jansen, Ritsert C.; Stam, Piet	High resolution of quantitative traits into multiple loci via interval mapping •	94 Genetcs 136 1447-1455
tion • Bello, A. L.	Imputation techniques in regression analysis: Looking closely at their implementa-	95 CmpStDA 20 45-57
Gouri K.	Inference procedures for the linear failure rate model • Sen, Ananda; Bhattacharyya,	95 JStPlInf 46 59-76
Burns, Patrick J.	Large-scale estimation of variance and covariance components • Fraley, Chris;	95 SIAMSSC 16 192-209
ships • Bolfarine, Heleno; Galea Rojas, Manuel	Maximum likelihood estimation of simultaneous pairwise linear structural relation-	95 BiomtrcJ 37 673-689
Renée Marie	Maximum likelihood with multiparameter models of substitution • Tillier, Elisabeth	94 JMolEvol 39 409-417
models • Reilly, Marie; Pepe, Margaret Sullivan	A mean score method for missing and auxiliary covariate data in regression	95 Biomtrka 82 299-314
Rubin, Donald B.	A method for calibrating false match rates in record linkage • Belin, Thomas R.;	95 JASA 90 694-707
Weinberg, Clarice R.; Wilcox, Allen J.	A model for estimating the potency and survival of human gametes in vivo •	95 Biomtrcs 51 405-412
doubly-censored and truncated data • Heisey et al.	Modelling age-specific survival in nesting studies, using a general approach for	95 Biomtrcs 51 51-60
McClean, Sally; Devine, Colum	A nonparametric maximum likelihood estimator for incomplete renewal data •	95 Biomtrka 82 791-803
sub-tables • Ritov, Ya'acov	PM algorithms for calculating minimum χ^2 estimators with partially observed	95 CmpStDA 20 19-33
Nummi, Tapio	Prediction and inverse estimation in repeated-measures models • Liski, Erkki P.;	95 JStPlInf 47 141-151
Calvin, James A.; Dykstra, Richard L.	REML estimation of covariance matrices with restricted parameter spaces •	95 JASA 90 321-329
models • Böhning, Dankmar	A review of reliable maximum likelihood algorithms for semiparametric mixture	95 JStPlInf 47 5-28
Pickles, A.; Dunn, G.; Vázquez-Barquero, J. L.	Screening for stratification in two-phase ('two-stage') epidemiological surveys •	95 StMedRes 4 73-89
data • Siciliano, Roberta; van der Heijden, Peter G. M.	Simultaneous latent budget analysis of a set of two-way tables with constant-row-sum	94 Metron 52/1 155-180
Haruhiko	Structural model of ability distribution in the item response theory • Ogasawara,	95 Behvmtrk 22 37-48

	EM algorithm (cont.)	
Solka, J. L.; Poston, W. L.; Wegman, E. J.	A visualization technique for studying the iterative estimation of mixture densities •	95 JCmpGrSt 4 180-198
A quasi-Newton acceleration of the	EM algorithm • Lange, Kenneth	95 StSinica 5 1-18
A. Applications of the	EM algorithm to the analysis of life length data • Albert, José Ramon G.; Baxter, Laurence	95 ApplStat 44 323-341
Jeffrey G.; Axelrod, Michael C. Use of the AIC with the	EM algorithm: A demonstration of a probability model selection technique • Glosup,	94 ASAProPEng 120-125
A gradient algorithm locally equivalent to the	EM algorithm • Lange, Kenneth	95 JRSS-B 57 425-437
The	EM algorithm for graphical association models with missing data • Lauritzen, Steffen L.	95 CmpStDA 19 191-201
estimation from incomplete panel data using the	EM algorithm and the Kalman smoother • Jansen et al. Longitudinal LISREL model	95 StNeerla 49 362-377
parameters • Kim, D. K.; Taylor, J. M. G. The restricted	EM algorithm for maximum likelihood estimation under linear restrictions on the	95 JASA 90 708-716
A stochastic	EM algorithm for mixtures with censored data • Chauveau, Didier	95 JStPlInf 46 1-25
multilevel covariance structure models via the	EM algorithm • Raudenbush, Stephen W. Maximum likelihood estimation for unbalanced	95 BJMaStPs 48 359-370
Kent, John T. Convergence behavior of the	EM algorithm for the multivariate t-distribution • Arsian, Olcay; Constable, Patrick D. L.;	95 CommStA 24 2981-3000
Information geometry of the EM and	em algorithms for neural networks • Amari, Shun-ichi	95 NuralNet 8 1379-1408
The	EM algorithm for the quasi-likelihood regression model • Paik, Myunghee Cho	95 CommStA 24 1403-1430
struction algorithm (ISRA) as an alternative to the	EM algorithm for solving positive linear inverse problems • Archer et al. ... space recon-	95 StSinica 5 77-96
on "Functional comparative calibration using an	EM algorithm" (V48 p1263-1271) • Bolfarine, Heleno; Galea-Rojas, Manuel Comment	95 Biomtrcs 51 1579-1580
Structural comparative calibration using the	EM algorithm • Bolfarine, Heleno; Galea Rojas, Manuel	95 JAppStat 22 277-292
of age-specific HIV incidence estimates using the	EM algorithm • Marschner, Ian C. Computation	95 JStCmpSm 53 299-312
system parameters in gymnosperms using the	EM algorithm • Xie, C. Y.; Yeh, F. C. et al. Joint estimation of immigration and mating	91 ThApGene 83 137-140
Convergence results for the	EM approach to mixtures of experts architectures • Jordan, Michael I.; Xu, Lei	95 NuralNet 8 1409-1431
Information geometry of the	EM and em algorithms for neural networks • Amari, Shun-ichi	95 NuralNet 8 1379-1408
Monte Carlo	EM estimation for time series models involving counts • Chan, K. S.; Ledolter, Johannes	95 JASA 90 242-252
ML estimation of the t distribution using	EM and its extensions, ECM and ECME • Liu, Chuanhai; Rubin, Donald B.	95 StSinica 5 19-39
	EM for longitudinal data from multiple studies • Fellingham, Gilbert W.; Wise, Morgan D.	95 StSinica 5 97-107
Smith, Fraser B.; Helms, Ronald W.	EM mixed model analysis of data from informatively censored normal distributions •	95 Biomtrcs 51 425-436
in norm for alternating expectation-maximization (	EM) type algorithms • Hero, Alfred O.; Fessler, Jeffrey A. Convergence	95 StSinica 5 41-54
State aggregation and discrete-state Markov chains	embedded in a class of point processes • Cao, Xi-Ren	95 JAppProb 32 39-51
and ... • Grung et al. Detection and quantification of	embedded minor analytes in three-way multicomponent profiles by evolving projections	95 Chemolab 29 213-221
	Embedding and asymptotic expansions for martingales • Mykland, P. A.	95 ProbTher 103 475-492
function network • Katayama, R.; Kuwata, K. et al.	Embedding dimension estimation of chaotic time series using self-generating radial basis	95 FuzzySet 71 311-327
A note on the	embedding of discrete-time ARMA processes • Brockwell, Peter J.	95 JTimSrAn 16 451-460
Variety of hypercube	embeddings of the equidistant metric and designs • Deza, Michel; Laurent, Monique	93 JCISS 18 289-315
Vesely, W. E.; Uryasev, S. P.; Samanta, P. K. Failure of	emergency diesel generators: A population analysis using empirical Bayes methods •	94 RelEngSS 46 221-229
Modeling trends in	emergency room drug abuse episodes • Fairchild, Charles K.	94 ASASrvyRsr 305-309
The price and purity of cocaine: The relationship to	emergency room visits and death, and to drug use among arrestees • Hyatt, R. R., Jr et al.	95 StatMed 14 655-668
capture-recapture samples to estimate temporary	emigration and breeding proportions • Kendall, W. L. et al. On the use of secondary	95 JAppStat 22 751-762
in a resident population of snow geese: Local	emigration or heterogeneity? • Pradel, Roger; Cooch, Evan; Cooke, Fred Transient animals	95 JAppStat 22 695-710
Markov branching processes regulated by	emigration and large immigration • Chen, Anyue; Renshaw, Eric	95 StocProc 57 339-359
detection in chromatography. Global analysis of	emission decay surfaces • Maeder, M.; Molloy, K. J. Lifetime resolved fluorescence	95 Chemolab 28 99-107
Prediction of automotive	emissions from gasoline composition • Rocke, David M.; Pallas, Michelle	94 ASAProPEng 269-273
Dynamic factor analysis: An application to	emotional response patterns underlying daughter/father and ... • Hershberger et al.	95 StrEqMod 2 31-52
Makridakis, Spyros The impact of	empirical accuracy studies on time series analysis and forecasting • Fildes, Robert;	95 IntStRvw 63 289-308
The conditional distribution of excess returns: An	empirical analysis • Foresi, Silverio; Peracchi, Franco	95 JASA 90 451-466
measure ζ and the principle of transfers: An	empirical analysis (Italian) • Polisicchio, M. The global concentration	93 GiornEco 52 291-308
provision of university teaching and research: An	empirical analysis of UK universities • Glass, J. C.; McKillop, D. G. et al. Efficiency in the	95 JApEcmtx 10 61-72
Bayesian model selection and prediction with	empirical applications (Disc: p333-365) • Phillips, Peter C. B.	95 JEconmtx 69 289-331
	Empirical Bayes	
Aalen, Odd O.; Bjertness, Espen; Sønju, Torleif	Analysis of dependent survival data applied to lifetimes of amalgam fillings •	95 StatMed 14 1819-1829
Holford, T. R.; Rubinstein, P.; Stevens, C. E.	Change points in the series of T4 counts prior to AIDS • Kiuchi, A. S.; Hartigan, J. A.;	95 Biomtrcs 51 236-248
applications • Majumdar, Suman	Consistency of posterior mixtures in the Gaussian family on a Hilbert space and its	95 JMultiAn 55 187-204
• van Houwelingen, Hans C.; Thorogood, Jane	Construction, validation and updating of a prognostic model for kidney graft survival	95 StatMed 14 1999-2008
Link, William A.; Sauer, John R.	Estimation and confidence intervals for empirical mixing distributions •	95 Biomtrcs 51 810-821
	Estimation of a stationary Markov chain • Billard, L.; Meshkani, M. R.	95 JASA 90 307-315
	General estimators for the reliability of qualitative data • Cooil, Bruce; Rust, Roland T.	95 Psymtrka 60 199-220
	Mixed model analysis of DNA sequence evolution • Yang, Ziheng; Wang, Tianlin	95 Biomtrcs 51 552-561
	Modelling net undercoverage in the 1991 Canadian Census • Dick, Peter	95 SurvyM 21 45-54
	A pooling procedure for the exponential distribution • Chiou, Paul	95 CommStA 24 445-458
G. An	empirical Bayes approach to monitoring water quality • Solow, Andrew R.; Gaines, Arthur	95 EnvrMtrc 6 1-5
Liao, Jiangang; Brookmeyer, Ron An	empirical Bayes approach to smoothing in backcalculation of HIV infection rates •	95 Biomtrcs 51 579-588
Chengming; Qin, Yongsong Linear	empirical Bayes estimation applied in multiple linear regression (Chinese) • Wang,	95 ChJApPSt 10 371-378
Zhang, Shunpu The convergence rates of	empirical Bayes estimation in a multiple linear regression model • Wei, Laisheng;	95 AnInStMa 47 81-97
Dominique; Robert, Christian P. Intrinsic losses for	empirical Bayes estimation: A note on normal and Poisson cases • Fourdrinier,	95 StPrLet 23 35-44
The comparison of the efficiency of Bayes and	empirical Bayes estimators • Gruber, Marvin H. J.	94 ASAProPEng 132-135
type-2 censored data • Ali Mousa, M. A. M.	Empirical Bayes estimators for the Burr type XII accelerated life testing model based on	95 JStCmpSm 52 95-103
• Krewski, D.; Smythe, R. T. et al. Bootstrapping an	empirical Bayes estimator of the distribution of historical controls in carcinogen bioassay	95 JStPlInf 43 215-228
Mean square error matrix superiority of	empirical Bayes estimators under misspecification • Wei, L.; Trenkler, G.	95 Test 4 187-205
Asymptotically optimality and convergence rates of	empirical Bayes estimator for the parameter of continuous one parameter ... • Xuc	95 ChJApPSt 10 35-45
Nonparametric	empirical Bayes growth curve analysis • Altman, N. S.; Casella, G.	95 JASA 90 508-515
Singh, R. S.	Empirical Bayes linear loss hypothesis testing in a non-regular exponential family •	95 JStPlInf 43 107-120
diesel generators: A population analysis using	empirical Bayes methods • Vesely, W. E.; Uryasev, S. P.; Samanta, P. K. Failure of emergency	94 RelEngSS 46 221-229
observations • Cole, B. F.; Lee, M.-L. T. et al. An	empirical Bayes model for Markov-dependent binary sequences with randomly missing	95 JASA 90 1364-1372
	Empirical Bayes optimal fertilizer decisions • Wallach, D.	95 JAppStat 22 507-516
Mohamed; Rekab, Kamel Asymptotically optimal	empirical Bayes procedures for selecting good uniform $U(0,\theta_i)$ populations • Tahir,	95 StocAnAp 13 233-247
Woo-Chul; Han, Kyung Soo On the performance of	empirical Bayes simultaneous interval estimates for all pairwise comparisons • Kim,	95 JKoreaSt 24 161-181
R. J.; Yang, H. On convergence rates of monotone	empirical Bayes tests for the continuous one-parameter exponential family • Karunamuni,	95 StatDecn 13 181-192
Choosing between two unknown intervals: An	empirical Bayes testing approach to a classification problem • Brouaye, F.	95 CommStA 24 3163-3181
Linex loss function • Huang, Su-Yun	Empirical Bayes testing procedures in some nonexponential families using asymmetric	95 JStPlInf 46 293-309
Meta-analysis for the ROC studies using	empirical Bayesian method • Zhou, X. H.	94 ASAProBays 102-105
Robust scale estimation based on the	empirical characteristic function • Markatou, Marianthi; Horowitz, Joel L.; Lenth, Russell V.	95 StPrLet 25 185-192
the discretisation error in the computation of the	empirical characteristic function • Pham, Dinh Tuan On	95 JStCmpSm 53 129-141
estimation in the error-components model using the	empirical characteristic function • Markatou, Marianthi; Horowitz, Joel L. Robust scale	95 CandJSt 23 369-381
specification • Geoffard, P.-Y.; Philipson, T. The	empirical content of canonical models of infectious diseases: The proportional hazard	95 Biomtrka 82 101-111
Martinez, C. J. A comment on the use of	empirical covariance matrices in the analysis of count data • Dean, C. B.; Eaves, D. M.;	95 JStPlInf 48 197-205
Convergence theorems for	empirical cumulative quantile regression functions • Rao, C. R.; Zhao, L. C.	95 MaMethSt 4 81-91

Causal diagrams for	empirical research (Disc: p688-710) • Pearl, Judea	95 Biomtrka 82 669-688
Forecasting tourism demand: A review of	empirical research • Witt, Stephen F.; Witt, Christine A.	95 IntJFor 11 447-475
procedures • Capdevila, Carles; Arcas, Antoni	Empirical significance test of the goodness-of-fit for some pyramidal clustering	95 Questo 19 123-130
Computer simulations and	empirical testing of sociological theory • Jacobsen, Chanoch; Bronson, Richard	95 SocMethR 23 479-506
	Empirical threshold values for quantitative trait mapping • Churchill, G. A.; Doerge, R. W.	94 Genetcs 138 963-971
Hill, Daniel H. The relative	empirical validity of dependent and independent data collection in a panel survey •	94 JOffStat 10 359-380
Quality tools help improve	employee commitment • Turner, Robert B.; Zipursky, Lorne S.	95 ASQCTekCon 770-776
What is the probability that your	employees will succeed? • Turner, Ronald E.	95 QulPr 28/6 31-33
Overview of the 1994 National	Employer Health Insurance Survey (NEHIS) • Hing, Esther	94 ASAProGovt 122-127
Moriarity, Chris National	Employer Health Insurance Survey sample design • Marker, David A.; Bryant, Edward C.;	94 ASASrvyRsr 264-269
business area sample as a means of covering birth	employers and nonemployers in the Current Business Surveys • Konschnik ... the	94 ASASrvyRsr 995-1000
unemployment model • Yeung, David Wing-Kay	Employment adjustments noise and the expected rate of inflation in a simple inflation-	95 StocAnAp 13 125-135
	Employment and childbearing in Canada • Rao, K. V.	94 ASAProSoSt 124-128
H. Introducing births into the	employment cost index survey • Gessley, Mary K.; Kelly, E. Thomas; Ponikowski, Chester	94 ASASrvyRsr 797-802
tics • Kelly, M.; Miller, S. The improvement of the	Employment Cost Index Survey: A total quality approach within the Bureau of Labor Statis-	94 ASAProQlPr 210-214
compensation survey program data using the	employment cost index • Tehonica, Jason Updating occupational	94 ASASrvyRsr 1048-1052
Commercial payroll software and the quality of	employment data • Goldenberg, Karen L.; Moore, Susan E.; Rosen, Richard J.	94 ASASrvyRsr 1025-1030
Kim, Taeil; Taylor, Lowell J. The	employment effect in retail trade of California's 1988 minimum wage increase •	95 JBES 13/2 175-182
Zaccarin, S. Woman roles between family and	employment: Empirical evidence from individual data (Italian) • Egidi, V.; Sabbadini, L. L.;	94 Statstca 54 411-433
Assessing the historical accuracy of regional	employment forecasts • Fullerton, Thomas M., Jr; West, Carol Taylor	94 ASAProBuEc 333-338
Visser, Michael Analysing incomplete individual	employment histories using indirect inference • Magnac, Thierry; Robin, Jean-Marc;	95 JApEcmtx 10 153-169S
Differences in the measurement of	employment in the labour force surveys in the European Community • Van Bastelaer, Aloïs	94 JOffStat 10 277-305
Statistical issues arising in equal	employment litigation (Disc: p54-54) • Gastwirth, Joseph L.	94 ASAProBuEc 44-53
Minimum wage effects on	employment and school enrollment • Neumark, David; Wascher, William	95 JBES 13/2 199-206
mation and variance estimation in the Occupational	Employment Statistics Survey • Tou, A.; Robertson, K. A study of imputation effects on esti-	94 ASAProGovt 99-104
Enchautegui, María E. Impacts of immigrants on	employment and wage growth: Evidence of positive effects of U.S. Immigration •	94 ASAProSoSt 30-38
generalized nested row-column designs with	empty diagonal cells • Raghavarao, Damaraju; Heiberger, Richard M. Variance balanced	93 JCISS 18 42-52
fectiveness and information ... • Schulz et al. Simple	encoding of infrared spectra for pattern recognition. 1: Statistical examination of the ef-	95 JChemtrc 9 143-168
He, Weizhong On	encoding values for data recorded as $X > C$ • Lachenbruch, Peter A.; Clements, Philip J.;	95 BiomtrcJ 37 855-867
Monfort, Alain Testing,	encompassing, and simulating dynamic econometric models • Gouriéroux, Christian;	95 EconTher 11 195-228
	Encounter probability	
Udevitz, Mark S.; Pollock, Kenneth H.	Using effort information with change-in-ratio data for population estimation •	95 Biomtrcs 51 471-481
A correction for	'end effects' in subjective rating experiments • Irvine, John M.	94 ASAProSoSt 184-188
Illman, Kathleen Ho A Bayesian integration of	end-use metering and conditional-demand analysis • Hsiao, Cheng; Mountain, Dean C.;	95 JBES 13 315-326
Optimal design in	end-use metering experiments • Bartels, Robert; Fiebig, Denzil G.	95 MaCmpSim 39 305-309
Establishment surveys from the	end user perspectives • McGuckin, Robert H.	94 ASASrvyRsr 146-152
Measurement error or	endogeneity: Sorting out simultaneity • Romeo, Charles J.	94 ASAProBays 196-201
demand for ... • Bollen et al. Binary outcomes and	endogenous explanatory variables: Tests and solutions with an application to the	95 Dmogrphy 32 111-131
the correlation between the instruments and the	endogenous explanatory variable is weak • Bound et al. ... variables estimation when	95 JASA 90 443-450
	Endogenous sampling	
	Optimal stock/flow panels • Lancaster, Tony; Imbens, Guido	95 JEconmtx 66 325-348
Analysis of ordinal data in a study of	endometrial cancer under a matched pairs case-control design • Ganguly, S. S. et al.	95 StatMed 14 1545-1552
Representation of Wiener space	endomorphisms preserving martingales (French) • Attal, Stéphane	95 AIHenrPB 31 467-484
and maternal effects. II. Genetic models for triploid	endosperms • Zhu, J.; Weir, B. S. Analysis of cytoplasmic	94 ThApGene 89 160-166
dose adjustment in a clinical trial with binary	endpoint • Reeve, Russell; Hale, Michael Results and efficiency of Bayesian	94 ASAProBiop 324-327
Multivariate tests for multiple	endpoints in clinical trials • Follmann, Dean	95 StatMed 14 1163-1175
data in conjunction with time-dependent surrogate	endpoints in clinical trials • Tudor, Gail E.; Morgan, Timothy M. Analyzing survival	95 CommStA 24 2293-2313
selections for kernel density estimators at	endpoints • Barabesi, L.; Fattorini, L. A note on bandwidth	94 Metron 52/3 43-56
parameter estimates for distributions with finite	endpoint • Falk, Michael Some best	95 MaOpfStS 27 115-125
On improving some methods for multiple	endpoints • Tang, Dei-In; Lin, Shang P.	94 ASAProBiop 466-474
analysis of the 2×2 cross-over design with multiple	endpoints • Tsai, Kao-Tai Nonparametric	94 ASAProBiop 238-243
on "Procedure for comparing samples with multiple	endpoints" (V40 p1079-1087) • Sankoh, A. J.; Huque, M. F. Comment	95 Biomtrcs 51 1580-1581
the hypothesis that matters for multiple primary	endpoints • Capizzi, Thomas P.; Zhang, Ji Testing	94 ASAProBiop 460-465
procedure for survival studies with surrogate	endpoints • Flandre, Philippe; O'Quigley, John A two-stage	95 Biomtrcs 51 969-976
calculation for complex clinical trials with survival	endpoints • Shih, Joanna H. Sample size	95 ConClnTr 16 395-407
and analysis of multiarm clinical trials with survival	endpoints • Liu, Ping-Yu; Dahlberg, Steve Design	95 ConClnTr 16 119-130
analysis of interfuel substitution in U.S. industrial	energy demand • Jones, Clifton T. A dynamic	95 JBES 13 459-465
	Energy information administration guidelines for statistical graphs • Rutchik, Robert H.	94 ASAProStGr 48-53
technique for refining the uncertainty in global	energy model forecasts • Tschang, F. Ted; Dowlatabadi, Hadi A Bayesian	95 IntJFor 11 43-61
A clustering method using the Coulomb	energy network (Korean) • Lee, Sukhoon; Park, Nae-Hyun; Kim, Yung-Hwan	95 KorApS 8/1 39-50
Modelling metabolic	energy by neural networks • Lozano, J.; Novič, M.; Rius, F. X.; Zupan, J.	95 Chemolab 28 61-72
Bryukhan, Fedor F.; Diab, Roseanne D. Wind	energy resource estimation of the upper atmosphere over Southern Africa •	95 JApMetrl 34 2565-2571
Self-employed nonfilers: The accuracy of	"enforcement-secured" returns (Disc: p212-214) • Rosage, Laura R.	94 ASAProGovt 206-211
Nonlinear errors in variables: Estimation of some	Engel curves • Hausman, J. A.; Newey, W. K.; Powell, J. L.	95 JEconmtx 65 205-233
General purpose technologies:	'Engines of growth'? • Bresnahan, Timothy F.; Trajtenberg, M.	95 JEconmtx 65 83-108
Statistics for environmental	engineers • Berthouex, Paul Mac; Brown, Linfield C.	94 CRCPress:FL 352p
A.; Capilla, C.; Zunica, L. et al. Teaching statistics to	engineers: An innovative pedagogical experience (jse/v3n1/romero) • Romero, R.; Ferrer,	95 JStEd 3/1 33kb
Reliability analysis for	engineers. An introduction • Leitch, Roger D.	95 OxfordUniv 248p
Review of "Applied statistics and probability for	engineers" • Montgomery, D. C.; Runger, G. C. (Auth); Snell, Joyce (Rev)	95 JRSS-A 158 355-355
Review of "Applied statistics and probability for	engineers" • Montgomery, Douglas C.; Runger, George; Alexander, Melvin T. (Rev)	95 Technmcs 37 455-457
Spreadsheet analysis for	engineers and scientists • Bloch, Sylvan C.	95 Wiley:NY:UK 336p
Statistical methods for	engineers and scientists • Bethea, Robert M.; Duran, Benjamin S.; Boullion, Thomas L.	95 Dekker:NY 672p
Nelson, L. S. (Rev) Review of "Statistical methods for	engineers and scientists (Third edition)" • Bethea, R. M.; Duran, B. S.; Boullion, T. L. (Auth);	95 JQualTek 27 392-393
D. (Rev) Review of "Probability and statistics for	engineers (Fourth edition)" • Scheaffer, Richard; McClave, James T. (Auth); Herrin, Gary	95 AmerStat 49 238-239
Review of "Probability and statistics for	engineers" • Scheaffer, R.; McClave, J. (Auth); Ziegel, Eric R. (Rev)	95 Technmcs 37 239-239
Review of "Probability and statistics for	engineers" • Scheaffer, Richard L.; McClave, James T. (Auth); Rigdon, Steven E. (Rev)	95 JQualTek 27 178-179
problems in physics, chemistry and electrical	engineering • Coffey, W. T. et al. The Langevin equation and its applications to stochastic	95 WrldSciPub:Sing 500p
Review of "Statistics in	engineering: A practical approach" • Metcalfe, Andrew V. (Auth); Larsen, Greg (Rev)	95 Technmcs 37 455-455
Review of "Statistics in	engineering: A practical approach" • Metcalfe, Andrew V. (Auth); Prvan, Tania (Rev)	95 AstrlJSt 37 128-128
Review of "Statistics for	engineering problem solving" • Vardeman, Stephen B. (Auth); Bisgaard, Soren (Rev)	95 AmerStat 49 320-321
Review of "Statistics for	engineering problem solving" • Vardeman, Stephen B. (Auth); Brugger, Richard M. (Rev)	95 Technmcs 37 348-348
Choosing the EWMA parameter in	engineering process control • Luceño, Alberto	95 JQualTek 27 162-168
Introduction to random processes in	engineering • Balakrishnan, A. V.	95 Wiley:NY:UK 402p
Metrics and models in software quality	engineering • Kan, Stephen H.	95 ASQualCntrl:WI 344p

Epidemiology

Choi, Seungmook; Wohar, Mark E. The	expectations theory of interest rates: Cointegration and factor decomposition •	95 IntJFor 11 253-262
Toward a formal derivation of the	expected behavior of prefix b-trees • Mahmoud, Hosam M.; Orlandić, Ratko	95 PrEngInf 9 183-192
Chinya	Expected deadlock time in a multiprocessing system • Compton, Kevin J.; Ravishankar,	95 JACM 42 562-583
astrophysics • Worsley Boundary corrections for the	expected Euler characteristic of excursion sets of random fields, with an application to	95 AdvAppPr 27 943-959
goodness-of-fit test for the latent class model when	expected frequencies are small • Reiser, Mark; Lin, Ginnie A	94 ASAProSoSt 49-54
of linear combinations for non-disjoint sets of	expected mean squares • Gui, R.; Graybill, F. A. et al. Confidence intervals on ratios	95 JStPlInf 48 215-227
Michna, Zbigniew; Rychlik, Igor The	expected number of level crossings for certain symmetric α-stable processes •	95 CommStSM 11 1-19
Latouche, Guy; Ramaswami, V.	Expected passage times in homogeneous quasi-birth-and-death processes •	95 CommStSM 11 103-122
Monica An investigation of the standard errors of	expected a posteriori ability estimates • De Ayala, R. J.; Schafer, William D.; Sava-Bolesta,	95 BJMaStPs 48 385-405
Gillett, Raphael The	expected power of F and t tests conditional on information from an earlier experiment •	95 BJMaStPs 48 371-384
Wing-Kay Employment adjustments noise and the	expected rate of inflation in a simple inflation-unemployment model • Yeung, David	95 StocAnAp 13 125-135
The time variation of	expected returns and volatility in foreign-exchange markets • Bekaert, Geert	95 JBES 13 397-408
Clark, Gordon M. Estimating	expected sojourn times for a Markov renewal process • Rossetti, Manuel D.;	95 CommStA 24 553-579
in the variables problem in the cross-section of	expected stock returns • Kim, Dongcheol The errors	95 JFinance 50 1605-1634
test sampling plans for Weibull distribution under	expected test time constraint • Bai, D. S.; Chun, Y. R. et al. Failure-censored accelerated life	95 RelEngSS 50 61-68
An alternative axiomatic approach to	expected utility theory • Di Bacco, Mario; Regoli, Giuliana	95 Metron 53 21-29
Varga, T. Moments and other	expected values for matrix variate elliptically contoured distribution • Gupta, A. K.;	94 Statstca 54 361-373
Bertino, Salvatore The minimum of the	expected value of the product of three random variables in the Fréchet class •	94 JItalStS 3 201-211
Characterization of discrete distributions using	expected values • Ruiz, J. M.; Navarro, J.	95 StHefte 36 237-252
Relating quantiles and	expectiles under weighted-symmetry • Abdous, Belkacem; Remillard, Bruno	95 AnInStMa 47 371-384
ison of household and medical provider reported	expenditures in the 1987 NMES • Cohen, Steven B.; Carlson, Barbara Lepidus A compar-	94 JOffStat 10 3-29
Paulin, Geoffrey D.; Ferraro, David L. Do	expenditures explain income?: A study of variables for income imputation •	94 ASASrvyRsr 106-115
imputation system for the Canadian Survey of Family	Expenditures: A feasibility study • Bérard, H. Use of the generalized edit and	94 ASAProSoSt 178-183
strategies for imputing premiums and predicting	expenditures under health care reform • Doyle, Pat; Farley, Dean Alternative	94 ASASrvyRsr 517-522
the Current Population Survey (CPS) and Consumer	Expenditure Interview Survey (CEIS) • Kojetin, B. A. Characteristics of nonrespondents to	94 ASASrvyRsr 966-971
et al. Assessment of reported differences between	expenditures and low incomes in the U.S. Consumer Expenditure Survey • Eltinge, J. L.	94 ASASrvyRsr 96-105
L.; Mueller, Curt D. Developing a prescription drug	expenditure model: The effect of alternative assumptions • Schur, Claudia L.; Berk, Marc	94 ASAProSoSt 245-250
Wiatrowski, William The survey of 1992 health	expenditures: A quick response survey • Ponikowski, Chester H.; Scheible, Paul;	94 ASASrvyRsr 282-287
boundaries for the Census Bureau's Annual Capital	Expenditures Survey • Slanta et al. ... and Hidiroglou method to obtain stratification	94 ASASrvyRsr 693-698
income reporting: An analysis of U.S. Consumer	Expenditure Survey data • Garner, Thesia I.; Blanciforti, Laura A. Household	94 JOffStat 10 69-91
on variance estimates for the National Medical	Expenditure Survey • Sommers, John Paul Analysis of the effects of imputation	94 ASASrvyRsr 400-405

Experiment

- see also design of experiments

Testing a model of aging in animal	experiments • Tyurin, Yu. N.; Yakovlev, Andrej Yu.; Shi, J.; Bass, L.	95 Biomtrcs 51 363-372
A statistical design of	experiments approach for texture description • Ganesan, L.; Bhattacharyya, P.	95 PattrnRc 28 99-105
Exploratory designs for computational	experiments • Morris, Max D.; Mitchell, Toby J.	95 JStPlInf 43 381-402
Variance components and computer	experiments • Heavlin, William D.	94 ASAProPEng 103-108
of "Numerical solution of SDE through computer	experiments" • Kloeden, P. E.; Platen, E.; Schurz, H. (Auth); Hall, T. (Rev) Review	95 Technmcs 37 125-125
Melgaard, H.	Experiment design for grey box identification • Sadegh, P.; Holst, J.; Madsen, H.;	95 IJAdCSgP 9 491-507
Selecting efficient	experiment designs for mixed 2-level and 3-level factors • Rand, Daniel R.	94 ASAProPEng 332-338
Review of "Optimal design of	experiments" • Pukelsheim, F. (Auth); Titterington, D. M. (Rev)	95 Metrika 42 74-75
Review of "Optimal design of	experiments" • Pukelsheim, Friedrich (Auth); Mejza, Stanisław (Rev. Polish)	94 ListyBio 31 121-125
Review of "Optimal design of	experiments" • Pukelsheim, Friedrich (Auth); Tobias, Randall D. (Rev)	95 JQualTek 27 93-94
A train the trainer program for design of	experiments • Eckes, George P.; Orr, Carolyn	95 ASQCTekCon 348-355
Boniface, D.	Experiment design and statistical methods for behavioural and social research •	95 ChapmnHall:Lnd 272p
Review of "Reliability improvement with design of	experiments" • Condra, Lloyd W. (Auth); Lenz, H.-J. (Rev)	95 CmpStDA 19 605-605
Hessian direction for analysis of data from designed	experiments • Cheng, Ching-Shui; Li, Ker-Chau A study of the method of principal	95 StSinica 5 617-639
Graphical analysis for a large designed	experiment • Freeny, Anne E.; Landwehr, James M.	95 Technmcs 37 1-14
On perpendicularity in designed	experiments • Kala, Radosław	95 ListyBio 32 93-99
Breakdown points for designed	experiments • Müller, Christine H.	95 JStPlInf 45 413-427
of some new graphical analysis tools for designed	experiments • Trindade, David C. Development	94 ASAProQlPr 115-117
Process and product optimization using designed	experiments • Myers, R. H.; Montgomery, D. C. Response surface methodology:	95 Wiley:NY:UK 720p
techniques in the design of an agricultural field	experiment • Dulaney, Wayne P.; Lengnick, Laura L.; Hart, Galen F. Use of geostatistical	94 ASASrvyRsr 183-187
use of spatial information in the analysis of field	experiments • Shukla, G. K. On the	94 JIAgSt 46 337-348
W. Edwards Deming's red bead	experiment fosters profound knowledge • Boardman, Thomas J.	95 Stats /13 3-9
On the randomization theory of	experiments in nested block designs • Caliński, Tadeusz	94 ListyBio 31 45-77
Holle, Rolf (Rev) Review of "Design of	experiments. A no-name approach" • Lorenzen, T. J.; Anderson, V. L. (Auth);	95 CmpStDA 19 101-102
Conniffe, D. (Rev) Review of "Design of	experiments: A no-name approach" • Lorenzen, T. J.; Anderson, V. L. (Auth);	95 JRSS-A 158 195-197
O'Brien, C. J. (Rev) Review of "Design of	experiments: A no-name approach" • Lorenzen, Thomas J.; Anderson, Virgil L. (Auth);	95 AstrlJSt 37 125-127
Optimal Bayesian design of	experiments for the one-way random effects model • Lohr, Sharon L.	95 Biomtrka 82 175-186
determination by ... • Papamichael et al. Design of	experiments for a precise estimation of the calibration curve of periodate in its	95 Chemolab 30 227-237
Probabilities and	experiments • Cartwright, Nancy	95 JEconmtx 67 47-59
R. E.; Santner, T. J. et al. Design and analysis of	experiments for statistical selection, screening, and multiple comparisons • Bechhofer,	95 Wiley:NY:UK 325p
Review of "Design and analysis of	experiments, Volume I: Introduction to experimental design" • Hinkelmann, K. et al.	95 Biomtrcs 51 378-379
Review of "Design and analysis of	experiments. Volume I, Introduction to experimental design" • Hinkelmann, K. et al.	95 JRSS-A 158 351-351
Review of "Design and analysis of	experiments, Volume 1: Introduction to experimental design" • Hinkelmann, K. et al.	95 JQualTek 27 175-176
Review of "Design and analysis of	experiments, Volume 1: Introduction to experimental design" • Hinkelmann, K. et al.	95 Technmcs 37 348-349
Review of "Design and analysis of	experiments. Volume 1. Introduction to experimental design" • Kempthorne, O. et al.	95 Chemolab 28 203-205
sign and model choice. The planning and analysis of	experiments with continuous or categorical response • Toutenburg, H. Experimental de-	95 Physica:Ger 463p
sign and model choice: The planning and analysis of	experiments with continuous or categorical response • Toutenburg, H. Experimental de-	95 Springer:Brln:NY 463p
Optimal designs for	experiments with a large number of factors • Jeong, Yousceek	95 JStPlInf 46 381-393
D-optimal designs of	experiments with non-interacting factors • Schwabe, Rainer; Wierich, Werner	95 JStPlInf 44 371-384
Lee, Hyunshik; Rancourt, Eric; Särndal, Carl E.	Experiments with variance estimation from survey data with imputed values •	94 JOffStat 10 231-243
Analysis and planning of	experimental block designs (French) • Bergonzini, J.-Cl.	95 Masson:NY:Paris 372p
Response surface methodology for non-equispaced	experimental configurations applied to a discrete and stochastic simulator ... • Giribone	94 JAppStat 21 549-564
Identifying parametric models for	experimental data (French) • Walter, E.; Pronzato, L.	94 Masson:NY:Paris 384p

Experimental design

Chadjiconstantinidis, S.; Moyassiadis, C.	The A-optimal saturated weighing design for $N = 17$ observations •	92 UtilMath 42 3-13
Gaffke, Norbert; Heiligers, Berthold	Algorithms for optimal design with application to multiple polynomial regression •	95 Metrika 42 173-190
	Barker sequences and circulant Hadamard matrices • Lin, Cantian; Wallis, W. D.	93 JCISS 18 19-25
tions for medicinal products • CPMP Working Party	Biostatistical methodology in clinical trials in applications for marketing authoriza-	95 StatMed 14 1659-1682
design • Gholston, Sampson; Batson, Robert G.	Case study: Precision improvement of a digital ball comparator via parameter	94 ASAProPEng 302-307
symbol orthogonal arrays of ... • Yamamoto et al.	Classification of orthogonal 24-run 2^5 factorial designs derivable from saturated two-	95 TRUMTk 31/1 39-53
DuMouchel, William H.; Lane, Thomas P.	Construction of mixture-process designs via the Franklin-Bailey algorithm •	94 ASAProQlPr 7-16

Experimental design (cont.)

Kachlicka, Danuta; Mejza, Stanisław	Cox's design in repeated row-column design with split plots and a control •	95 ListyBio 32 81-92
	Criteria for supersaturated designs • Deng, Lih-Yuan; Lin, Dennis K. J.	94 ASAProQlPr 124-128
	Cubic lattices • Pearce, S. C.	95 JAppStat 22 355-373
Phan-Than-Luu, R.	*D*-optimal designs • de Aguiar, P. F.; Bourguignon, B.; Khots, M. S.; Massart, D. L.;	95 Chemolab 30 199-210
process • Morris, M. D.; Solomon, A. D.	Design and analysis for an inverse problem arising from an advection-dispersion	95 Technmcs 37 293-302
	Design considerations for nonlinear calibration • Bates, Douglas M.; Martin, Sandra J.	94 ASAProPEng 12-17
Ahmed, M.	Designs through recoding of varietal and level codes • Das, M. N.; Giri, N.;	95 StPrLet 23 371-380
order is important • Wang, P. C.; Jan, H. W.	Designing two-level factorial experiments using orthogonal arrays when the run	95 Ststcian 44 379-388
	Dose response studies. I. Some design considerations • Ruberg, Stephen J.	95 JBiophSt 5 1-14
Brzeskwiniewicz, Henryk	On the *E*-optimality of some two-way elimination of heterogeneity designs •	95 StPrLet 25 9-13
	Efficiency of check-plot designs in unreplicated field trials • Chandra, S.	94 ThApGene 88 618-620
data • Grieve, Andrew P.	Extending a Bayesian analysis of the two-period crossover to accommodate missing	95 Biomtrka 82 277-286
	Factor screening by sequential bifurcation • Bettonvil, Bert	95 CommStB 24 165-185
	Five-designs with three intersection numbers • Ionin, Yury J.; Shrikhande, Mohan S.	95 JCombThA 69 36-50
	Frank Yates, 1902-1994 - The work of a statistician • Healy, M. J. R.	95 IntStRvw 63 271-288
	Hasse diagrams in statistical consulting and teaching • Lohr, Sharon L.	95 AmerStat 49 376-381
factorial design • Byar, D. P.; Freedman, L. S. et al.	Identifying which sets of parameters are simultaneously estimable in an incomplete	95 Ststcian 44 451-456
Algina, James; Oshima, T. C.	An improved general approximation test for the main effect in a split-plot design •	95 BJMaStPs 48 149-160
Michael A.	Inference for effect-saturated fractional factorials • Haaland, Perry D.; O'Connell,	95 Technmcs 37 82-93
Wu, C. F. J.	An integrated method of parameter design and tolerance design • Li, Wenlian;	94 ASAProQlPr 44-48
Tobias, Randall D.	Joint estimation of location, dispersion, and random effects • Wolfinger, Russell D.;	94 ASAProPEng 225-234
Chowdhury, S. Roy	Linear graphs for 2^{k-p} fractional factorial under split plot design • Aggarwal, M. L.;	94 JStatRsr 28 73-87
Anderson, Christine M.; Wu, C. F. Jeff	Measuring location effects from factorial experiments with a directional response •	95 IntStRvw 63 345-363
Manly, Bryan F. J.	Measuring selectivity from multiple choice feeding-preference experiments •	95 Biomtrcs 51 709-715
variables • Duineveld, C. A. A.; Coenegracht, P. M. J.	Multicriteria steepest ascent in a design space consisting of both mixture and process	95 Chemolab 30 23-36
Thelin, Bernt	A multivariate strategy for optimizing a two-step process • Lundstedt, Torbjörn;	95 Chemolab 29 255-261
	Natural and quasi-experiments in economics • Meyer, Bruce D.	95 JBES 13/2 151-161
	Neighbor-balanced orthogonal arrays • Cheng, Ching-Shui	93 JCISS 18 61-67
designs? 'Silly designs are silly!' • Bailey, R. A. et al.	Are neighbour methods preferable to analysis of variance for completely systematic	95 Biomtrka 82 655-659
	On even and odd latin squares • Janssen, Jeannette C. M.	95 JCombThA 69 173-181
	Optimal choice of a new observation in a linear model • Sengupta, Debasis	95 SankhyaA 57 137-153
	Optimal designs on the refinement of the *n*-way classification model • Ting, C.-P.	93 UtilMath 43 141-154
Shah, K. R.; Sinha, Bikas K.	Optimal designs with component-wise orthogonal row-column structures •	93 JCISS 18 68-78
	Optimal weighing designs and some new weighing matrices • Koukouvinos, Christos	95 StPrLet 25 37-44
and quantitative trait loci • Darvasi, A.; Soller, M.	Optimum spacing of genetic markers for determining linkage between marker loci	94 ThApGene 89 351-357
	Partial diallel crosses in incomplete blocks • Singh, M.; Hinkelmann, K.	95 Biomtrcs 51 1302-1314
Brekke, Kris	Partition of variation: A new method for σ reduction • Little, Thomas A.;	95 ASQCTekCon 208-216
	Patterns in residuals for 2^n designs • John, Peter W. M.	95 JAppStat 22 253-259
ization in uncovering a causal relationship • Oldford	A physical device for demonstrating confounding, blocking, and the role of random-	95 AmerStat 49 210-216
A.; Meeker, William Q., Jr	Planning accelerated life tests with two or more experimental factors • Escobar, Luis	95 Technmcs 37 411-427
Bailey, R. A.; Preece, D. A.; Rowley, C. A.	Randomization for a balanced superimposition of one Youden square on another •	95 JRSS-B 57 459-469
	What really is experimental error in block designs? • Gates, Charles E.	95 AmerStat 49 362-363
applied to a discrete and stochastic ... • Giribone	Response surface methodology for non-equispaced experimental configurations	94 JAppStat 21 549-564
der, Shyam (Auth); Lamoureux, Christopher G. (Rev)	Review of "Experimental methods: A primer for economists" • Friedman, Daniel; Sun-	95 JFinance 50 1341-1345
	A review of robust process design approaches • Grize, Yves L.	95 JChemtrc 9 239-262
under the unavailability of blocks: A ... • Uddin et al.	The robustness of binary and non-binary balanced nested row-column designs	95 AstrlJSt 37 111-124
nonlinear regression - A case study • Rasch, Dieter	The robustness against parameter variation of exact locally optimum designs in	95 CmpStDA 20 441-453
	Screening by conference designs • Elster, Clemens; Neumaier, Arnold	95 Biomtrka 82 589-602
design method • Saltelli, A.; Andres, T. H. et al.	Sensitivity analysis of model output. Performance of the iterated fractional factorial	95 CmpStDA 20 387-407
Charles E., Jr	Sets of mutually orthogonal Latin squares with "like subsquares," II • Roberts,	95 JCombThA 70 159-164
Chen, Yu Qing	On the size of the multiplier groups of cyclic difference sets • Xiang, Qing;	95 JCombThA 69 168-169
Park, Sung Hyun; Lee, Min Soo	Slope-rotatability over all directions in third order response surface models •	95 JKoreaSt 24 519-536
	Some design problems in crop experimentation. I. The use of blocks • Pearce, S. C.	95 ExpAgric 31 191-203
	Some new difference sets • Arasu, K. T.; Sehgal, Surinder K.	95 JCombThA 69 170-172
analysis uses generalized least-squares • Martin, R. J.	Some results on randomization when observations are spatially dependent and the	95 CommStA 24 2199-2210
	Statistical designs for drug developments • Hedayat, Samad	92 ProIrStC1 39-50
Yu, Philip L. H.; Cowan, Richard	A statistical model for duplicate tournaments in games of luck and skill •	95 AstrlJSt 37 1-17
cubic bias in multiple response case • Bae, Whasoo	A study on a basis for the selection of a design for quadratic model fits fearing a	95 JKoreaSt 24 31-44
	Two-stage parameter designs for quality improvement • Jeong, Yousceek	94 ASAProQlPr 215-219
	Type S designs in unequal blocks • Gupta, Sudhir; Kageyama, Sanpei	93 JCISS 18 97-112
Grieshaber, Mark V.; Schnur, Dora M.	Use experimental design in organic synthesis • Brannigan, Lawrence H.;	95 ChemTech 25 29-35
cells • Raghavarao, Damaraju; Heiberger, Richard M.	Variance balanced generalized nested row-column designs with empty diagonal	93 JCISS 18 42-52
Cornell, John A.	Wagon wheel designs with residual effects from neighboring treatments •	93 JCISS 18 255-272
ceramic fibre composites, using a fractional 2^n	experimental design • Jones et al. ... and improvement of a process for vacuum-formed	95 JAppStat 22 459-467
	Experimental design for additive linear models • Schwabe, Rainer	94 ASAProPEng 280-284
Principles of	experimental design and analysis • Garcia-Diaz, A.; Phillips, D. T.	94 ChapmnHall:Lnd 384p
eson, A. C. (Auth); Nester, Marks R. (Rev) Review of	"Experimental design and analysis for use in tree improvement" • Williams, E. R.; Math-	95 AstrlJSt 37 247-248
Matheson, A. C. (Auth); Smith, R. I. (Rev) Review of	"Experimental design and analysis for use in tree improvement" • Williams, E. R.;	95 Biomtrcs 51 1190-1190
Germain-Tourbier, S.	Experimental design: Construction and analysis (French) • Benoist, D.; Tourbier, Y.;	95 Lavoisier:Paris ?p
methacrylate copolymers in dimethylsulfoxide - An	experimental design • Feuer et al. ... polymerization of nonlinear optical side chain	95 Chemolab 29 113-124
de Wet, T.; Fresen, J. L. An	experimental design for evaluating sonar detection performance • de Jongh, P. J.;	95 SoAfrStJ 29 79-103
Improving fabric finishing through	experimental design • Artiles-Léon, Noel; Novoa-Ramírez, Clara M.; Domenech, Carolina	95 ASQCTekCon 952-962
Use of PLS regression for the analysis of	experimental designs in formulation chemistry (French) • Gauchi, J.-P.	95 RStApp 43/1 65-89
analysis of experiments, Volume 1: Introduction to	experimental design" • Hinkelmann, K.; Kempthorne, O. (Auth) et al. Review of "Design and	95 JQualTek 27 175-176
analysis of experiments, Volume 1: Introduction to	experimental design" • Hinkelmann, K.; Kempthorne, O. (Auth) et al. Review of "Design and	95 Technmcs 37 348-349
analysis of experiments. Volume 1. Introduction to	experimental design" • Kempthorne, O.; Hinkelmann, K. (Auth) et al. Review of "Design and	95 Chemolab 28 203-205
analysis of experiments, Volume I: Introduction to	experimental design" • Hinkelmann, K.; Kempthorne, O. (Auth) et al. Review of "Design and	95 Biomtrcs 51 378-379
analysis of experiments. Volume I, Introduction to	experimental design" • Hinkelmann, K.; Kempthorne, O. (Auth) et al. Review of "Design and	95 JRSS-A 158 351-351
Transformation for	experimental design in off-line quality control (German) • Schweitzer, Walter	95 AllgeStA 79 117-139
continuous or categorical response • Toutenburg, H.	Experimental design and model choice. The planning and analysis of experiments with	95 Physica:Ger 463p
continuous or categorical response • Toutenburg, H.	Experimental design and model choice: The planning and analysis of experiments with	95 Springer:Brln:NY 463p
exchange algorithm for constructing exact optimal	experimental designs • Meyer, Ruth K.; Nachtsheim, Christopher J. The coordinate-	95 Technmcs 37 60-69
Constrained optimization of	experimental design (Disc: p149-178) • Cook, Dennis; Fedorov, Valery	95 MaOpfStS 26 129-148

surfaces: Recent developments in optimum	experimental design • Atkinson, A. C. Beyond response	95 Chemolab 28 35-47
Review of "Optimum	experimental designs" • Atkinson, A. C.; Donev, A. N. (Auth); Laycock, P. J. (Rev)	95 Biomtrcs 51 379-379
Schnur, Dora M. Use	experimental design in organic synthesis • Brannigan, Lawrence H.; Grieshaber, Mark V.;	95 ChemTech 25 29-35
analysis, associations with clinical outcome and	experimental design (Disc: p1097-1097) • Minkin et al. ... cytosine arabinoside: Statistical	95 StatMed 14 1081-1095
Kirk, Roger E.	Experimental design: Procedures for the behavioral sciences (Third edition) •	95 BrooksCole:CA 921p
Fourier transform infrared ... • Shaffer et al.	Experimental design protocol for the pattern recognition analysis of bandpass filtered	95 Chemolab 29 89-108
sampling plans by the technique of trade-off in	experimental design • Hedayat, A. S.; Majumdar, Dibyen Generating desirable	95 JStPlInf 44 237-247
for detecting active factors in unreplicated	experimental designs • Benski, Claudio Applicability of nine numerical techniques	94 ASAProStCp 214-217
Review of "Analysis of variance in	experimental design" • Lindman, H. R. (Auth); Fedorov, V. V. (Rev)	95 JQualTek 27 268-268
Tobias, Randall D.; Garratt, Mark Efficient	experimental design with marketing research applications • Kuhfeld, Warren F.;	94 JMarRsr 31 545-557
What really is	experimental error in block designs? • Gates, Charles E.	95 AmerStat 49 362-363
tative trait loci. II. Combining prior knowledge with	experimental evidence • Hoeschele et al. ... linkage between genetic markers and quanti-	93 ThApGene 85 946-952
Planning accelerated life tests with two or more	experimental factors • Escobar, Luis A.; Meeker, William Q., Jr	95 Technmcs 37 411-427
The treatment of related	experimental factors by sliding levels • Hamada, M.; Wu, C. F. J.	95 JQualTek 27 45-55
(Auth); Lamoureux, Christopher G. (Rev) Review of	"Experimental methods: A primer for economists" • Friedman, Daniel; Sunder, Shyam	95 JFinance 50 1341-1345
olive oils • Vojnovic, D.; Campisi, B.; Mattei, A. et al.	Experimental mixture design to ameliorate the sensory quality evaluation of extra virgin	95 Chemolab 27 205-210
(Rev) Review of "Modeling	experimental and observational data" • Lunneborg, Clifford E. (Auth); Handcock, Mark S.	95 AmerStat 49 238-238
chemical reaction models from a linearly designed	experimental program • Steele et al. ... simultaneous development of linear and nonlinear	95 Chemolab 29 125-131
differential misclassification • Marinos, A. Th. et al.	Experimental quantiles of epidemiological indices in case-control studies with non-	95 StatMed 14 1291-1306
natural sciences • Hoshmand, A. Reza	Experimental research design and analysis. A practical approach for agricultural and	94 CRCPress:FL 416p
parametric analysis of cultivar performance trials:	Experimental results and comparison of different procedures based ... • Hühn et al. Non-	95 AgronmyJ 87 627-632
in survey sampling consistent with that in	experimental statistics • Proctor, C. H. Developing a concept of population	94 ASASrvyRsr 653-655
Identification risks of microdata: Evidence from	experimental studies • Müller, Walter; Blien, Uwe; Wirth, Heike	95 SocMethR 24 131-157
extended logistic model with applications to	experimental toxicity data • El-Saidi, Mohammed A. A symmetric	95 BiomtrcJ 37 205-216
in control/noise factors can't be controlled during	experimentation • Coons, Lori; LaLonde, Steven Robust design when variability	94 ASAProPEng 65-70
Some design problems in crop	experimentation. I. The use of blocks • Pearce, S. C.	95 ExpAgric 31 191-203
Some design problems in crop	experimentation. II. Multiple blocking systems • Pearce, S. C.	95 ExpAgric 31 279-290
Some design problems in crop	experimentation. III. Non-orthogonality • Pearce, S. C.	95 ExpAgric 31 409-422
A stochastic process arising in sequential	experimentation • Stadje, W.	95 StPrLet 23 359-365
and cause and effect inference in social	experimentation • Neuberg, Leland G. Randomization	94 ASAProSoSt 153-158
ing control of polymer viscosity using time-series	experimentation and statistical process control • Wallace, J. E.; Foster, R. D. et al. Improv-	94 ASAProPEng 320-325
Review of "Multi-vari chart and analysis - A pre-	experimentation technique" • Perez Wilson, Mario (Auth); Papp, Zoltan (Rev)	95 Technmcs 37 121-121
Creating	experimenters by design • Schneider, Michael T.	95 ASQCTekCon 365-366
served correlation structure among tumor types on	experimentwise false positive rate in animal carcinogenicity ... • Rahman Effects of ob-	95 JBiophSt 5 273-283
results for the EM approach to mixtures of	experts architectures • Jordan, Michael I.; Xu, Lei Convergence	95 NuralNet 8 1409-1431
quantitative forecasts of livestock production using	expert judgements • Vere, D. T.; Griffith, G. R. Modifying	95 JForcast 14 453-464
tions from Poisson point processes incorporating	expert knowledge • Campodónico, Sylvia; Singpurwalla, Nozer D. Inference and predic-	95 JASA 90 220-226
	Expert opinion	
DeWispelare, A. R.; Herren, L. T.; Clemen, R. T.	The use of probability elicitation in the high-level nuclear waste regulation program •	95 IntJFor 11 5-24
Mallick, Bani K.; Dey, Dipak K. Modeling	expert opinion arising as a partial probabilistic specification • Gelfand, Alan E.;	95 JASA 90 598-604
Gåsemyr, Jørund; Natvig, Bent Using	expert opinions in Bayesian prediction of component lifetimes in a shock model •	95 MaOpRsr 20 227-242
Coherent combination of	experts' opinions (Disc: p293-313) • Dawid, A. P.; DeGroot, M. H.; Mortera, J.	95 Test 4 263-293
Combining	experts' opinions using a normal-Wishart model • Wiper, Michael P.; French, Simon	95 JForcast 14 25-34
	Expert system	
organising neural networks • Mulholland, M. et al.	A comparison of classification in artificial intelligence, induction versus a self-	95 Chemolab 30 117-128
Steffen L.	The EM algorithm for graphical association models with missing data • Lauritzen,	95 CmpStDA 19 191-201
Application of the C4.5 classifier to building an	expert system for ion chromatography • Mulholland, M.; Hibbert, D. B.; Haddad, P. R. et al.	95 Chemolab 27 95-104
density from line transect surveys relative to analyst	expertise • Anderson, David R.; Southwell, Colin Estimates of macropod	95 JWildlMg 59 852-857
An investigation on measures of	explained variation in survival analysis • Graf, Erika; Schumacher, Martin	95 Ststcian 44 497-507
timation of global functionals and a measure of the	explanatory power of covariates in regression • Doksum, K. et al. Nonparametric es-	95 AnlsStat 23 1443-1473
sors when the expectations equation has reduced	explanatory power • Smith et al. ... alternative estimators in models with generated regres-	95 MaCmpSim 39 343-346
Nonparametric test on dimensionality of	explanatory variables (Korean) • Seo, H. S.	95 KorApS 8/2 65-75
an overspecified regression model with interrelated	explanatory variables • Elston, D. A.; Proe, M. F. Smoothing regression coefficients in	95 ApplStat 44 395-406
• Stack, Jim; Unwin, Antony R. Selecting	explanatory variables from survey data: A simple approach that explores for interactions	95 Ststcian 44 17-29
Bollen et al. Binary outcomes and endogenous	explanatory variables: Tests and solutions with an application to the demand for ... •	95 Dmogrphy 32 111-131
between the instruments and the endogenous	explanatory variable is weak • Bound et al. ... variables estimation when the correlation	95 JASA 90 443-450
nonparametric approach in survival models with	explanatory variables • Escolano, Sylvie; Golmard, Jean-Louis; Mallet, Alain A fully	95 CommStA 24 3027-3054
performance in U.S. manufacturing industries: An	exploratory analysis • Berndt, E. R. et al. High-tech capital formation and economic	95 JEconmtx 65 9-43
An	exploratory analysis of a record of El Niño events, 1800-1987 • Solow, Andrew R.	95 JASA 90 72-77
Gabriel, K. R.; Westcott, N. E.; Czys, R. R.	Exploratory analysis of seeding effects on rainfall: Illinois 1989 • Changnon, S. A.;	95 JApMetrl 34 1215-1224
	Exploratory data analysis	
	Classification trees • Vach, W.	95 CmpStQ 10 9-14
Velleman, Paul F.	A critical look at some analyses of major league baseball salaries • Hoaglin, David C.;	95 AmerStat 49 277-285
	Exploring Cook's statistic graphically • Hines, R. J. O'Hara; Hines, W. G. S.	95 AmerStat 49 389-394
considerations • Pedersen, F.; Bengtsson, E. et al.	An extended strategy for exploratory multivariate image analysis including noise	95 JChemtrc 9 389-409
	Fast and simple scatterplot smoothing • Härdle, W.; Marron, J. S.	95 CmpStDA 20 1-17
Hurley, Catherine	Grand tour and projection pursuit • Cook, Dianne; Buja, Andreas; Cabrera, Javier;	95 JCmpGrSt 4 155-172
James M.	Graphical analysis for a large designed experiment • Freeny, Anne E.; Landwehr,	95 Technmcs 37 1-14
	GUHA for personal computers • Hájek, P.; Sochorová, A.; Zvárová, J.	95 CmpStDA 19 149-153
and complete) • Shore, Haim	Identifying a two-parameter distribution by the first two sample moments (partial	95 JStCmpSm 52 17-32
	Investigating the geometry of a p-dimensional data set • Dawkins, Brian P.	95 JASA 90 350-359
	Mathematical programming in statistical analysis • Zwirner, Walter W.	94 CalcutSt 44 229-237
Grambsch, P. M.; Randall, B. L.; Bostick, R. M. et al.	Modeling the labeling index distribution: An application of functional data analysis •	95 JASA 90 813-821
	A regression smoother for resistant measures of location and scale • Wilcox, Rand R.	95 BJMaStPs 48 189-204
criteria • Snavely, Duane; Dobbins, Thomas An	exploratory data analysis of ambulatory blood pressure monitoring as screening	94 ASAProBiop 51-55
application of ordinal structure ... • Komazawa et al.	Exploratory data analysis for Hayashi's quantification method III by graphics: Its	95 PrInStMa 43 161-176
(German) • Polasek, W. EDA.	Exploratory data analysis. Introduction to descriptive statistics (Second edition)	94 Springer:Brln:NY 345p
(German)" • Polasek, W. (Auth) et al. Review of "EDA.	Exploratory data analysis. Introduction to descriptive statistics (Second edition)	95 ShrtBkRv 15/2 23-23
Projection pursuit	exploratory data analysis • Posse, Christian	95 CmpStDA 20 669-687
	Exploratory designs for computational experiments • Morris, Max D.; Mitchell, Toby J.	95 JStPlInf 43 381-402
Bengtsson, E.; Nordin, B. An extended strategy for	exploratory multivariate image analysis including noise considerations • Pedersen, F.;	95 JChemtrc 9 389-409
Tools for two-dimensional	exploratory projection pursuit • Posse, Christian	95 JCmpGrSt 4 83-100
Interactive dynamic graphics for	exploratory survival analysis • Atkinson, E. Neely	95 AmerStat 49 77-84
Biggeri, A.; Braga, M.; Lagazio, C.; Marchi, M.	Exploratory tools for disease mapping in geographical epidemiology • Cislaghi, C.;	95 StatMed 14 2363-2381

sign to ameliorate the sensory quality evaluation of	extra virgin olive oils • Vojnovic, D.; Campisi, B.; Mattei, A. et al. Experimental mixture de-	95 Chemolab 27 205-210
A method for fitting Poisson regression model with	extra zeros • Jovanovic, Borko D.; Hosmer, David W.; Murray, James	94 ASAProEpid 97-100
Hideo On the optimal	extraction rate of a non-renewable resource with an extraction capacity • Hashimoto,	95 EnvrMtrc 6 547-550
estimator for modelling categorical data with	extraneous variation • Morel, Jorge G.; Koehler, Kenneth J. A one-step Gauss-Newton	95 ApplStat 44 187-200
Environmental risk assessment model for low dose	extrapolation • Janardan, K. G.	95 CommStA 24 2621-2633
Simulation-	extrapolation: The measurement error jackknife • Stefanski, L. A.; Cook, J. R.	95 JASA 90 1247-1256
Little, Jani Simplicity and complexity in	extrapolative population forecasting models • McNown, Robert; Rogers, Andrei;	95 MaPopStd 5 235-257
the tail parameter in the domain of attraction of an	extremal distribution • Alves, M. I. Fraga Estimation of	95 JStPlInf 45 143-173
Weissman, Ishay; Cohen, Uri The	extremal index and clustering of high values for derived stationary sequences •	95 JAppProb 32 972-981
The sample mid-range and symmetrized	extremal laws • Bingham, N. H.	95 StPrLet 23 281-288
Schneider, Rolf	Extremal problems for geometric probabilities involving convex bodies • Bauer, Christina;	95 AdvAppPr 27 20-34
On the influence of the	extremal term in a (simple) U-statistic • Mijnheer, Joop	95 ThPrMaSt 50 103-106
Weak limit results for the	extremes of a class of shot noise processes • Homble, Patrick; McCormick, William P.	95 JAppProb 32 707-726
Catkan, N. A.; Hüsler, J.	Extremes and clustering on nonstationary max-AR(1) sequences • Alpuim, M. T.;	95 StocProc 56 171-184
A note on the asymptotic behavior of conditional	extremes • Gangopadhyay, Ashis K.	95 StPrLet 25 163-170
A cautionary note on the analysis of	extreme data with Cox regression • Hsieh, F. Y.	95 AmerStat 49 226-228
The development of the mathematical theory of	extremes in the past half century • Galambos, J.	94 ThProbAp 39 234-248
	Extremes of interarrival times of a Poisson process under conditioning • Abay, A.	95 ZastApMa 23 73-82
	Extremes of Markov sequences • Haiman, George; Kiki, Maxime; Puri, Madan L.	95 JStPlInf 45 185-201
Viana, Marlos Correlation analysis of	extreme observations from a multivariate normal distribution • Olkin, Ingram;	95 JASA 90 1373-1379
Sums of independent triangular arrays and	extreme order statistics • Janssen, Arnold	94 AnlsProb 22 1766-1793
LAN of	extreme order statistics • Falk, Michael	95 AnInStMa 47 693-717
Sensitivity analysis of	extreme precipitation events • Katz, Richard W.; Acero, Juan Garrido	94 JClimlgy 14 985-999
Empirical models for evaluating errors in fitting	extremes of a probability distribution • Bai, Jun; Jakeman, Anthony J.; McAleer, Michael	95 MaCmpSim 39 1-7
	Extremes of a random number of variables from periodic sequences • Ferreira, Helena	95 JStPlInf 45 133-141
Thomas (Rev) Review of "Law of small numbers:	Extremes and rare events" • Falk, Michael; Hüsler, Jurg; Reiss, Rolf-Dieter (Auth); Mikosch,	95 JASA 90 1133-1134
Schmid, W.	Extreme sums of strictly stationary sequences of M-dependent variables • Flak, T.;	95 SankhyaA 57 186-201

Extreme value

a single binomial • Chernoff, Herman; Lander, Eric	Asymptotic distribution of the likelihood ratio test that a mixture of two binomials is	95 JStPlInf 43 19-40
stationary random variables • Hsing, Tailen	On the asymptotic independence of the sum and rare values of weakly dependent	95 StocProc 60 49-63
	A Bahadur-Kiefer theorem beyond the largest observation • Einmahl, John H. J.	95 JMultiAn 55 29-38
Galambos, J.	The development of the mathematical theory of extremes in the past half century •	94 ThProbAp 39 234-248
and Vermont • Dingman, S. L.; Lawlor, S. C.	Estimating low-flow quantiles from drainage-basin characteristics in New Hampshire	95 WaterReB 31 243-256
est inventory (French) • Pierrat, J. C.; Houllier, F. et al.	Estimating the mean of the r largest values in a finite population: Application to for-	95 Biomtrcs 51 679-686
Palutikof, J. P.	Estimation of extreme wind speeds with very long return periods • Dukes, M. D. G.;	95 JApMetrl 34 1950-1961
spheric dispersion • Mole, N.; Anderson, C. W. et al.	A generalized Pareto distribution model for high concentrations in short-range atmo-	95 EnvrMtrc 6 595-606
	On high level exceedance modeling and tail inference • Leadbetter, M. R.	95 JStPlInf 45 247-260
Gore, S. D.; Patil, G. P.	Identifying extremely large values using composite sample data (Disc: p240-245) •	94 EnvEcoSt 1 227-240
	The indices of the largest among n independent observations • Weissman, Ishay	95 CommStSM 11 613-629
point proportion of a random convex hull • Massé, B.	Invariance principle for the deviation between the probability content and the interior	95 JAppProb 32 1041-1047
Joshi, S. N.; Nagaraja, H. N.	Joint distribution of maxima of concomitants of subsets of order statistics •	95 Bernoull 1 245-255
	Large quantile estimation in a multivariate setting • de Haan, Laurens; Huang, Xin	95 JMultiAn 53 247-263
Kou, Shing-Gang	Limits of first passage times to rare sets in regenerative processes • Glasserman, Paul;	95 AnlsApPr 5 424-445
	Maximum values in queueing processes • Berger, Arthur W.; Whitt, Ward	95 PrEnglnf 9 375-409
J.; Broniatowski, M.; Teugels, J. L.; Vynckier, P.	The mean residual life function at great age: Applications to tail estimation • Beirlant,	95 JStPlInf 45 21-48
with periodic intensity • Ballerini, Rocco	Nonconstant levels for t-year return periods for two-dimensional Poisson processes	95 StPrLet 22 43-47
mixing stationary random variables • Hsing, Tailen	A note on the asymptotic independence of the sum and maximum of strongly	95 AnlsProb 23 938-947
distribution • Wang, Julian Z.	Selection of the k largest order statistics for the domain of attraction of the Gumbel	95 JASA 90 1055-1061
Garrido	Sensitivity analysis of extreme precipitation events • Katz, Richard W.; Acero, Juan	94 JClimlgy 14 985-999
Nebenzahl, Elliott	Using power transformations when approximating quantiles • Fearn, Dean H.;	95 CommStA 24 1073-1093
Patrick; McCormick, William P.	Weak limit results for the extremes of a class of shot noise processes • Homble,	95 JAppProb 32 707-726
Eastwood, Vera R. On	extreme value asymptotics for increments of renewal processes • Steinebach, Josef;	95 JStPlInf 45 301-312

Extreme value distribution

	Asymptotic distributions of records in nonstationary schemes • Nevzorov, Valery	95 JStPlInf 45 261-273
Allen, P. Geoffrey; Morzuch, Bernard J.	Comparing probability forecasts derived from theoretical distributions •	95 IntJFor 11 147-157
Hosking, J. R. M.; Wallis, J. R.	A comparison of unbiased and plotting-position estimators of L moments •	95 WaterReR 31 2019-2025
	Detecting changes in linear regressions • Horváth, Lajos	95 MaOpfStS 26 189-208
	Distance in variation and a Fisher-type information • Papadatos, N.; Papathanasiou, V.	95 MaMethSt 4 230-237
Gertsbakh, I.	On the Fisher information in type-I censored and quantal response data •	95 StPrLet 23 297-306
vectors • Kobus, Maria	Generalized Poisson distributions as limits of sums for arrays of dependent random	95 JMultiAn 52 199-244
Gumbel distribution and impact of ... • Fill et al.	L moment and probability plot correlation coefficient goodness-of-fit tests for the	95 WaterReR 31 225-229
	Limit theorems for the time of completion of Johnson-Mehl tessellations • Chiu, S. N.	95 AdvAppPr 27 889-910
	Limit theorems for the union-intersection test • Horváth, Lajos; Shao, Qi-Man	95 JStPlInf 44 133-148
	Maximum stability and a generalization • Sreehari, M.	95 StPrLet 23 339-342
P. C. N.; Kemp, C. J.	Perturbation of the normal model in linear regression • de Waal, D. J.; Groenewald,	95 SoAfrStJ 29 105-130
cated distributions and their ... • Mohie EL-Din et al.	Recurrence relations for expectations of functions of order statistics for doubly trun-	95 CommStA 24 997-1010
	The sample mid-range and symmetrized extremal laws • Bingham, N. H.	95 StPrLet 23 281-288
Tawn, Jonathan A.	A semi-parametric model for multivariate extreme values • Dixon, Mark J.;	95 StatComp 5 215-225
	Statistics for exceptional athletics records • Robinson, Michael E.; Tawn, Jonathan A.	95 ApplStat 44 499-511
Anderson, C. W.; Turkman, K. F.	Sums and maxima of stationary sequences with heavy tailed distributions •	95 SankhyaA 57 1-10
A unified criterion for the domain of attraction of	extreme-value distributions • de Haan, L.	94 ThProbAp 39 323-329
timation of the location and scale parameters of the	extreme value distribution based on multiply type-II censored ... • Balakrishnan et al. Es-	95 CommStA 24 2105-2125
tolerance limits of atmospheric ozone level and	extreme value distribution • Dasgupta, Ratan; Bhaumik, Dulal K. Upper and lower	95 SankhyaB 57 182-199
timation for two-parameter Weibull distribution and	extreme-value distribution under multiply type-II censoring • Fei, H.; Kong, F.; Tang, Y. Es-	95 CommStA 24 2087-2104
Fisher information for a multivariate	extreme value distribution • Shi, Daoji	95 Biomtrka 82 644-649
confidence intervals for the parameters of	extreme value distribution • Thiagarajah, Kulathava Ranee Comparisons of approximate	94 Metron 52/1 85-100
mation of parameters in two-parameter Weibull and	extreme-value distributions using ranked set sample • Fei, H.; Sinha, B. K.; Wu, Z. Esti-	94 JStatRsr 28 149-161
and tests of hypothesis in simple linear model with	extreme-value errors • Saleh, A. K. Md. E. et al. Optimum spacings, ABLUE	94 JStatRsr 28 53-63

Extreme value estimator

Datta, Somnath; McCormick, William P.	Bootstrap inference for a first-order autoregression with positive innovations •	95 JASA 90 1289-1300
Williams, Matthew R. An	extreme-value function model of the species incidence and species-area relations •	95 Ecology 76 2607-2616
K-record values and the	extreme-value index • Berred, Mohamed	95 JStPlInf 45 49-63
A note on	extreme values of locally stationary Gaussian processes • Hüsler, J.	95 JStPlInf 45 203-213
Kim, Keeyoung Flows and some	extreme values in multiple server open Jackson network • Park, YouSung; Lee, Haeyong;	95 JKoreaSt 24 389-405

A semi-parametric model for multivariate	extreme values • Dixon, Mark J.; Tawn, Jonathan A.	95 StatComp 5 215-225
Strong theorems on the	extreme values of stationary Poisson processes • Révész, P.	95 JStPlInf 45 291-300
of a Khinchin-type decomposition theorem for	extreme values • Pancheva, E. On a problem	94 ThProbAp 39 329-336
	Extreme value theory	
distribution • Alves, M. I. Fraga	Estimation of the tail parameter in the domain of attraction of an extremal	95 JStPlInf 45 143-173
Schmid, W.	Extreme sums of strictly stationary sequences of M-dependent variables • Flak, T.;	95 SankhyaA 57 186-201
	K-record values and the extreme-value index • Berred, Mohamed	95 JStPlInf 45 49-63
	On a subclass of regularly varying functions • Omey, E.	95 JStPlInf 45 275-290
Kaufmann, Edgar	Von Mises conditions, δ-neighborhoods and rates of convergence for maxima •	95 StPrLet 25 63-70
	Extreme value theory for stochastic processes • Rootzén, Holger	95 ScanActJ 54-65
Estimation of	extreme wind speeds with very long return periods • Dukes, M. D. G.; Palutikof, J. P.	95 JApMetrl 34 1950-1961
Patil, G. P. Identifying	extremely large values using composite sample data (Disc: p240-245) • Gore, S. D.;	94 EnvEcoSt 1 227-240
Review of "SPC in	extrusion" • Rauwendaal, Chris (Auth); Ziegel, Eric R. (Rev)	95 Technmcs 37 241-241
	***F* distribution**	
	Bootstrap for nonstandard cases • Babu, Gutti Jogesh	95 JStPlInf 43 197-203
On the doubly noncentral	F distribution • Chattamvelli, R.	95 CmpStDA 20 481-489
agonalization of variance-covariance matrices using	F-G algorithm • Ali, A. K. A.; Lin, C. Y. Di-	95 EgpStJ 39 12-25
method for computation of central and non-central	F probabilities and percentiles • Wang, M. C.; Kennedy, W. J. A self-validating numerical	95 StatComp 5 155-163
Yang, Minxian On the use of the	F ratio in a misspecified model with an interval restriction • Wan, Alan T. K.;	95 JStCmpSm 52 151-161
Dependent	F ratios sharing a common denominator in ANOVA table • Kang, Phillee	95 JKoreaSt 24 111-125
Conditional p-values for the	F-statistic in a forward selection procedure • Grechanovsky, Eugene; Pinsker, Ilia	95 CmpStDA 20 239-263
Estimation of gene flow from	F-statistics • Cockerham, C. Clark; Weir, B. S.	93 Evolutn 47 855-863
formulas for linear and circular consecutive-k-of-n:	F systems • Peköz, Erol A.; Ross, Sheldon M. A simple derivation of exact reliability	95 JAppProb 32 554-557
The expected power of	F and t tests conditional on information from an earlier experiment • Gillett, Raphael	95 BJMaStPs 48 371-384
	***F*-test**	
	A simple test for stable seasonality • Sutradhar, B. C.; MacNeil, I. B.; Dagum, E. B.	95 JStPlInf 43 157-167
Simulation analysis of type I error and power for	F test and rank transformed F test in 2×2 factorial ANOVA (Korean) • Choi, Young Hun	95 KorApS 8/2 87-97
On the power of	F tests under regression models with nested error structure • Rao, J. N. K.; Wang, Song-Gui	95 JMultiAn 53 237-246
Generalized	F-tests for unbalanced nested designs under heteroscedasticity • Ananda, Malwane M. A.	95 AnInStMa 47 731-742
Local transformation of	facial features • Gordon, A. D.	95 JAppStat 22 179-184
	Factor	
	• see also names of specific types of factors such as Bayes –	
	Factor analysis	
ponent analysis • Krzanowski, Wojtek J.; Kline, Paul	Cross-validation for choosing the number of important components in principal com-	95 MultBehR 30 149-165
	Cross-validation criteria for covariance structures • De Gooijer, Jan G.	95 CommStB 24 1-16
Costa, Michele	Dynamic component detection in a multifactor model for stock returns •	94 JItalStS 3 25-36
Inagaki, Atsushi	Entromin and entromax: Criteria for factor rotation based on entropy (Japanese) •	94 JpJBeh 21/2 10-20
Waller, Niels G.; Reise, Steven P.	Genetic and environmental influences on item response pattern scalability •	92 BehavGen 22 135-152
M.-C.	Goodness-of-fit in item response models • McDonald, Roderick P.; Mok, Magdalena	95 MultBehR 30 23-40
number of factors to retain • Glorfeld, Louis W.	An improvement on Horn's parallel analysis methodology for selecting the correct	95 EdPsMeas 55 377-393
Terence J.; Tanaka, J. S.	Incremental fit index changes for nested structural equation models • La Du,	95 MultBehR 30 289-316
1992: A factor-ARCH analysis • Pelizzon, L. et al.	The Italian term structure of interest rates and the currency devaluation of September	94 Statstca 54 313-327
direct methods • Grung, Bjørn; Kvalheim, Olav M.	Resolution of multicomponent profiles with partial selectivity. A comparison of	95 Chemolab 29 75-87
resolution • Tauler, R.; Smilde, A.; Kowalski, B.	Selectivity, local rank, three-way data analysis and ambiguity in multivariate curve	95 JChemtrc 9 31-58
invariance problem • Bove, Giuseppe	Some methods for the simultaneous analysis of data matrices and the factorial	94 Metron 52/3 73-87
Galea Rojas, Manuel	Structural comparative calibration using the EM algorithm • Bolfarine, Heleno;	95 JAppStat 22 277-292
of factorial invariance • Kaplan, David; George, Rani	A study of the power associated with testing factor mean differences under violations	95 StrEqMod 2 101-118
varimax rotations • Berge, Jos M. F. ten	Suppressing permutations or rigid planar rotations: A remedy against nonoptimal	95 Psymtrka 60 437-446
on the performance of methods for evolutionary	factor analysis as applied to high-performance liquid ... • Elbergali et al. ... regions	95 Chemolab 27 55-71
Joseph, Stephen; Lewis, Christopher Alan	Factor analysis of the Center for Epidemiological Studies-Depression Scale •	95 PsychRpt 76 40-42
of the effectiveness and information content by	factor analysis • Schulz et al. ... spectra for pattern recognition. 1: Statistical examination	95 JChemtrc 9 143-168
Spearman and the origin and development of	factor analysis • Bartholomew, D. J.	95 BJMaStPs 48 211-220
Fung, W. K.; Kwan, C. W. Sensitivity analysis in	factor analysis: Difference between using covariance and correlation matrices •	95 Psymtrka 60 607-614
Dion, K. L.; Kilik, L. Using and misusing	factor analysis to explore group cohesion • Cota, A. A.; Evans, C. R.; Longman, R. S.;	95 JClPsych 51 308-316
Application of the bootstrap methods in	factor analysis • Ichikawa, Masanori; Konishi, Sadanori	95 Psymtrka 60 77-93
Identifiability of full, marginal, and conditional	factor analysis models • Ihara, Masamori; Kano, Yutaka	95 StPrLet 23 343-350
of factor scores in a two-level confirmatory	factor analysis model • Lee, Sik-Yum; Poon, Wai-Yin Estimation	95 CmpStDA 20 275-284
condition for identification of confirmatory	factor analysis models of factor complexity one • Reilly, Terence A necessary and sufficient	95 SocMethR 23 421-441
Full-information	factor analysis for polytomous item responses • Muraki, Eiji; Carlson, James E.	95 ApPsychM 19 73-90
	Factor analysis and principal components • Schneeweiss, Hans; Mathes, Harald	95 JMultiAn 55 105-124
Muthén, Bengt O. Genetic Tobit	factor analysis: Quantitative genetic modeling with censored data • Waller, Niels G.;	92 BehavGen 22 265-292
(Auth); Jackson, J. Edward (Rev) Review of "Statistical	factor analysis and related methods: Theory and applications" • Basilevsky, Alexander	95 Technmcs 37 227-228
determining the number of components in window	factor analysis of spectroscopic and chromatographic data • Brereton et al. ... for	95 Chemolab 27 73-87
of Cu(II) • Esteves da Silva, J. C. G. et al. Evolving	factor analysis of synchronous fluorescence spectra of humic substances in the presence	95 Chemolab 27 115-128
missing data patterns on goodness of fit tests in	factor analysis • Kaplan, David The impact of BIB spiraling-induced	95 JEdStat 20 69-82
rand, Mihkel; Koel, Mihkel Application of evolving	factor analysis in thermochromatography • Elomaa, Matti; Lochmüller, Charles H.; Kalju-	95 Chemolab 30 129-132
A Monte Carlo study of constrained	factor analysis using maximum likelihood and unweighted least squares • Wolins, Leroy	95 EdPsMeas 55 545-557
matography of oil shale • Koel, M. et al.	Factor analytical resolution of simultaneous, dynamic thermal process in thermochro-	95 Chemolab 30 173-178
and the currency devaluation of September 1992: A	factor-ARCH analysis • Pelizzon, L.; Weber, G. The Italian term structure of interest rates	94 Statstca 54 313-327
Rashid, M. Mushfiqur Analysis of two-	factor experiments with repeated measures on both factors based on R-estimators •	94 ASAProQlPr 220-225
factorial designs of resolution $2l+1$ when $l+s+1$-	factor and higher order interactions are negligible • Shirakura et al. ... for balanced 2^m	93 JCISS 18 233-245
experiment designs for mixed 2-level and 3-level	factors • Rand, Daniel R. Selecting efficient	94 ASAProPEng 332-338
intervals for contrasts in the balanced three	factor mixed model • Mosier, Michael C.; Graybill, Franklin A. Approximate confidence	95 BiomtrcJ 37 273-297
Posterior distribution for the number of	factors • Press, S. James; Shigemasu, K.	94 ASAProBays 75-77
designs for experiments with a large number of	factors • Jeong, Yousceek Optimal	95 JStPlInf 46 381-393
methodology for selecting the correct number of	factors to retain • Glorfeld, Louis W. An improvement on Horn's parallel analysis	95 EdPsMeas 55 377-393
saturated main effect plans for $2 \times s_2 \times s_3$	factorials • Chatterjee, Kashinath; Mukerjee, Rahul D-optimal	93 JCISS 18 116-122
effect plans with higher revealing power for 2^7	factorials • Ghosh, Subir Sequential construction of new main	93 JCISS 18 217-231
construction of main effect plus two plans for 2^m	factorials • Mukerjee, R.; Chatterjee, K. An application of Hadamard matrices for the	94 UtilMath 45 213-218
Search designs of resolution 3.1(2) for 2^m	factorials • Gupta, B. C.; Carvajal, S. S. R.	93 JCISS 18 247-254
for F test and rank transformed F test in 2×2	factorial ANOVA (Korean) • Choi, Young Hun Simulation analysis of type I error and power	95 KorApS 8/2 87-97
e-optimal fractions of symmetric and asymmetric	factorials • Mukerjee, Rahul On	95 StSinica 5 515-533

	Factorial design	
Ahmed, M.	Designs through recoding of varietal and level codes • Das, M. N.; Giri, N.;	95 StPrLet 23 371-380
factors • Ghosh, Subir; Derderian, Edna	Determination of robust design against noise factors and in presence of signal	95 CommStB 24 309-326
q^{n-m}-designs • Laycock, P. J.; Rowley, P. J.	A method for generating and labelling all regular fractions or blocks for	95 JRSS-B 57 191-204
	On modelling of experiments in natural sciences • Mejza, Stanisław	94 ListyBio 31 79-100
ing optimal conditions in the search ... • Kaufmann	Multivariate optimization strategy for liquid chromatography. II. Exploring and find-	95 Chemolab 27 105-114
categorical data • Thomas, G. E.	A note on chi-squared approximations for tests of rank interaction with ordered	95 CommStB 24 439-459
	The polynomial model in the multivariate analysis of variance • Baumgartner, Cornelia	95 BiomtrcJ 37 299-318
applied to a discrete and stochastic ... • Giribone	Response surface methodology for non-equispaced experimental configurations	94 JAppStat 21 549-564
MacEachern, S. N.; Notz, W. I. et al.	Robustness to the unavailability of data in the linear model, with applications •	95 JStPlInf 48 207-213
	Some projection properties of orthogonal arrays • Cheng, Ching-Shui	95 AnlsStat 23 1223-1233
	Some split-plot designs in series of experiments • Mejza, Iwona	94 ListyBio 31 3-16
designed experiments • Cheng, C.-S.; Li, K.-C.	A study of the method of principal Hessian direction for analysis of data from	95 StSinica 5 617-639
Classification of orthogonal 24-run 2^5	factorial designs derivable from saturated two-symbol orthogonal ... • Yamamoto et al.	95 TRUMTk 31/1 39-53
are simultaneously estimable in an incomplete	factorial design • Byar, D. P.; Freedman, L. S. et al. Identifying which sets of parameters	95 Ststcian 44 451-456
• Shirakura et al. Search designs for balanced 2^m	factorial designs of resolution $2l+1$ when $l+s+1$-factor and higher order interactions ...	93 JCISS 18 233-245
of variance of partially balanced fractional $2^{m_1+m_2}$	factorial designs of resolution IV • Kuwada, Masahide Analysis	93 JCISS 18 343-367
Analysis of variance of balanced fractional $r^m \times s^n$	factorial designs of resolution V • Kuwada, Masahide; Hyodo, Yoshifumi	95 JJapnStS 25 129-141
Stopping rules for sequences of	factorial designs • Gilmour, S. G.; Mead, R.	95 ApplStat 44 343-355
A permutation test for effects in an unreplicated	factorial design • Loughin, Thomas M.; Noble, William	94 ASAProPEng 181-186
On confounding in	factorial experiments • Kane, S. P.	94 JIAgSt 46 330-333
Martin, R. J. An algorithm for the design of 2^p	factorial experiments on continuous processes • Saunders, I. W.; Eccleston, J. A.;	95 AstrlJSt 37 353-365
Practical analysis of	factorial experiments in forestry • Stehman, S. V.; Meredith, M. P.	95 CanJForR 25 446-461
Minimax distance designs in two-level	factorial experiments • John, P. W. M.; Johnson, M. E.; Moore, L. M.; Ylvisaker, D.	95 JStPlInf 44 249-263
The computer analysis of	factorial experiments: In memoriam - Frank Yates • Nelder, John A.; Lane, Peter W.	95 AmerStat 49 382-385
• Radson, Darrell; Herrin, Gary D. Augmenting a	factorial experiment when one factor is an uncontrollable random variable: A case study	95 Technmcs 37 70-81
Wang, P. C.; Jan, H. W. Designing two-level	factorial experiments using orthogonal arrays when the run order is important •	95 Ststcian 44 379-388
Measuring location effects from	factorial experiments with a directional response • Anderson, Christine M.; Wu, C. F. Jeff	95 IntStRvw 63 345-363
several three-factor interactions in general	factorials • Chatterjee, K.; Mukerjee, R. Search designs for searching	92 UtilMath 42 181-191
the simultaneous analysis of data matrices and the	factorial invariance problem • Bove, Giuseppe Some methods for	94 Metron 52/3 73-87
testing factor mean differences under violations of	factorial invariance • Kaplan, David; George, Rani A study of the power associated with	95 StrEqMod 2 101-118
	Factorial moment expansion for stochastic systems • Błaszczyszyn, B.	95 StocProc 56 321-335
Correction to "Application of ranking methods to	factorial and stratified experiments" (V36 p205-213) • Shirley, Eryl A. C.	95 ApplStat 44 135-137
thyroxine and prolactin on the molting of newts: A	factorial study • Jing, Yihua The effect of	95 Stats /12 24-27
Efficacy evaluation for monotherapies in two-by-two	factorial trials • Hung, H. M. James; Chi, George Y. H.; O'Neill, Robert T.	95 Biomtrcs 51 1483-1493
Johnson, Keith L. Analyzing unreplicated	factorials using the n smallest contrasts • Kasperski, William J.; Schneider, Helmut;	95 JQualTek 27 377-386
cedure in gauge theories • Arnaudon, M.; Paycha, S.	Factorisation of semi-martingales on principal fibre bundles and the Faddeev-Popov pro-	95 Stochatx 53 81-107
	Factorization theory for probability distributions • Bondesson, Lennart	95 ScanActJ 44-53
semi-martingales on principal fibre bundles and the	Faddeev-Popov procedure in gauge theories • Arnaudon, M.; Paycha, S. Factorisation of	95 Stochatx 53 81-107
Dharmadhikari, A. D. Prediction of	failure - A case study • Joshi, B. H.; Naik-Nimbalkar, U. V.; Deshmukh, S. R.;	95 JAppStat 22 485-493
analysis for the systems with common cause	failures • Pan, Zhijie; Nonaka, Yasuo Importance	95 RelEngSS 50 297-300
expected test time constraint • Bai, D. S. et al.	Failure-censored accelerated life test sampling plans for Weibull distribution under	95 RelEngSS 50 61-68
Balasooriya, Uditha	Failure-censored reliability sampling plans for the exponential distribution •	95 JStCmpSm 52 337-349
Superconvergence to the central limit and	failure of the Cramér theorem for free random variables • Bercovici, H.; Voiculescu, D.	95 ProbTher 103 215-222
distinguishability with application to analysis of	failure data • Soofi, Ehsan S.; Ebrahimi, Nader; Habibullah, Mohamed Information	95 JASA 90 657-668
How pooling	failure data may reverse increasing failure rates • Gurland, John; Sethuraman, Jayaram	95 JASA 90 1416-1423
Murthy, D. N. P. Modeling	failure-data by mixture of 2 Weibull distributions: A graphical approach • Jiang, R.;	95 IEEEReli 44 477-488
Weibull family: A reanalysis of the bus-motor-	failure data • Mudholkar, G. S.; Srivastava, D. K.; Freimer, M. The exponentiated	95 Technmcs 37 436-445
A queueing system with random	failures and delayed repairs • Madan, K. C.	94 JIndnStA 32 39-48
Cao, Jin Methods for the estimation of	failure distributions and rates from automobile warranty data • Lawless, Jerry; Hu, Joan;	95 LiDaAnal 1 227-240
accelerated life testing with unreported early	failures • Glaser, R. E. Weibull	95 IEEEReli 44 31-36
methods • Vesely, W. E.; Uryasev, S. P.; Samanta, P. K.	Failure of emergency diesel generators: A population analysis using empirical Bayes	94 RelEngSS 46 221-229
accurate inference procedure for the difference of	failure intensities from two exponential distributions • Wong, A. C. M. A simple and	95 CommStA 24 2865-2879
data analyses for products with multiple-modes of	failure (Korean) • Bai, Do Sun; Choi, In Su; Hwang, Yong Keun Field	95 KorApS 8/1 89-104
	Failure model	
	Survival in dynamic environments • Singpurwalla, Nozer D.	95 StatSci 10 86-103
A simple, flexible	failure model • Yu, Shuiyang; Voit, Eberhard O.	95 BiomtrcJ 37 595-609
issues in the Bayesian analysis of a Weibull	failure model • Papadopoulos, Alex S.; Hoekstra, Robert M. Some robustness	95 JStCmpSm 52 85-94
Bayesian prediction bounds for the Burr type XII	failure model • Al-Hussaini, Essam K.; Jaheen, Zeinhum F.	95 CommStA 24 1829-1842
Correction to "Tests on early	failures - The precedence life test" (V25 p140-143) • Nelson, L. S.	95 JQualTek 27 402-402
Barth, James R. Deposit insurance and bank	failures: A preliminary analysis • Cebula, Richard J.; Belton, Willie J., Jr; Chou, Ray Y.;	94 ASAProBuEc 293-297
	Failure rate	
Rohatgi, Vijay K.; Szekely, Gabor; Zsigri, G.	On aging life distribution classes: Further probability properties and tests •	94 JStatRsr 28 111-122
	A characterization of a bivariate geometric distribution • Sun, Kai; Basu, Asit P.	95 StPrLet 23 307-311
pital rather than patient characteristics? • Silber et al.	Comparing the contributions of groups of predictors: Which outcomes vary with hos-	95 JASA 90 7-18
A.; Gouweleeuw, José M.	A continuous-time job search model: General renewal processes • Boshuizen, Frans	95 CommStSM 11 349-369
	Ergodicity of Spitzer's renewal model • Sidoravićus, Vladas; Vares, Maria Eulália	95 StocProc 55 119-130
Singpurwalla, Nozer D.; Wilson, Simon P.	The exponentiation formula of reliability and survival: Does it always hold? •	95 LiDaAnal 1 187-194
Gupta, Ramesh C.; Akman, H. Olcay	Mean residual life function for certain types of non-monotonic ageing •	95 CommStSM 11 219-225
	Reliability reports - Product, customer, and process • Conte, John A.	95 ASQCTekCon 733-740
Akman, H. Olcay	On the reliability studies of a weighted inverse Gaussian model • Gupta, Ramesh C.;	95 JStPlInf 48 69-83
Yang, J. B.; Sen, P.	Safety analysis and synthesis using fuzzy sets and evidential reasoning • Wang, J.;	95 RelEngSS 47 103-118
Marlow, Norman A.; Tortorella, Michael	Some general characteristics of two-state reliability models for maintained systems •	95 JAppProb 32 805-820
Kochar, Subhash C.; Kirmani, S. N. U. A.	Some results on normalized spacings from restricted families of distributions •	95 JStPlInf 46 47-57
nonparametrics • Sen, Pranab Kumar	Statistical analysis of some reliability models: Parametrics, semi-parametrics and	95 JStPlInf 43 41-66
Jarosław	Stochastic order relations and the total time on test transform • Bartoszewicz,	95 StPrLet 22 103-110
Jayant V.; Sengupta, Debasis	Testing the hypothesis of proportional hazards in two populations • Deshpande,	95 Biomtrka 82 251-261
Dietrich, D. L.; Szidarovszky, F. Time-varying	failure rates in the availability and reliability analysis of repairable systems • Hassett, T. F.;	95 IEEEReli 44 155-160
Some large sample properties of estimate of	failure rates based on truncated sample (Chinese) • Wan, Qihua	95 KToChina 39 2120-2122
under type I progressive censoring with changing	failure rates • Patel et al. ... likelihood estimators of parameters of exponential distribution	95 CommStA 24 2421-2435
A Bayesian model for estimating the	failure rate for different groups • de Waal, D. J.; Kemp, C. J.	95 SoAfrStJ 29 131-154
Kernel-type density and	failure rate estimation for associated sequences • Bagai, Isha; Rao, B. L. S. Prakasa	95 AnInStMa 47 253-266
Inference procedures for the linear	failure rate model • Sen, Ananda; Bhattacharyya, Gouri K.	95 JStPlInf 46 59-76
Restricted tests for and against the increasing	failure rate ordering on multinomial parameters • Bhattacharya, Bhaskar	95 StPrLet 25 309-316

Optimal winsorizing cutoffs for a stratified	finite population estimator • Kokic, P. N.; Bell, P. A.	94 JOffStat 10 419-435
of almost unbiased regression-type estimators for	finite population mean applying jackknife technique • Katyar, N. P. et al. A class	94 JIAgSt 46 183-189
Rank-based predictive inference for the	finite population mean of a small area • Rashid, M. Mushfiqur; Nandram, Balgobin	94 ASASrvyRsr 1059-1064
Singh, Housila P. A Hartley-Ross type estimator for	finite population mean when the variables are negatively correlated • Singh, Rajesh;	95 Metron 53 205-216
minimax estimators for two-stage sampling from a	finite population under measurement error models • Mukhopadhyay, P. Bayes and	95 CommStA 24 663-674
Maximum likelihood prediction in	finite populations • Rodrigues, Josemar; Elian, Silvia Nagib	95 RBrzlPrS 9 43-54
Small-sample properties of	finite population ratios • Sankoh, Abdul J.	95 CommStB 24 897-909
A (Bayesian) note on non-random samples from	finite populations • Bayarri, M. J.; Font, B.	94 ASAProBays 246-251
negative unbiased estimation of quadratic forms in	finite population sampling • Vijayan, K.; Mukhopadhyay, P.; Bhattacharyya, S. On non-	95 AstrlJSt 37 169-178
Review of "Design and inference in	finite population sampling" • Hedayat, A. S.; Sinha, B. K. (Auth); Rust, Keith (Rev)	94 JOffStat 10 226-228
Using a diagnostic test in a	finite population • Englund, Jan-Eric	94 MaScitst 19 108-118
Optimal estimation of a	finite population total under randomized response surveys • Arnab, Raghunath	95 MaOpfStS 27 175-180
Integrated probabilistic data association-	finite resolution • Mšicki, D.; Evans, R.	95 Autmatca 31 559-570
	Finite-sample bias	
Angrist, Joshua D.; Krueger, Alan B.	Split-sample instrumental variables estimates of the return to schooling •	95 JBES 13/2 225-235
Shinichi; White, Halbert An alternative definition of	finite-sample breakdown point with applications to regression model estimators • Sakata,	95 JASA 90 1099-1106
W.; Stark, Philip B.	Finite-sample confidence envelopes for shape-restricted densities • Hengartner, Nicolas	95 AnlsStat 23 525-550
note • Cheung, Yin-Wong; Lai, Kon S. Estimating	finite sample critical values for unit root tests using pure random walk processes: A	95 JTimSrAn 16 493-498
panel data models; consequences and detection in	finite samples • van den Doel, I. T.; Kiviet, J. F. Neglected dynamics in	95 StNeerla 49 343-361
Reinsel, Gregory C.	Finite sample forecast results for vector autoregressive moving average models •	95 JForcast 14 405-412
Optimality properties in	finite sample l_1 identification with bounded noise • Kacewicz, B.; Milanese, M.	95 IJAdCSgP 9 87-96
autoregressions • Toda, Hiro Y.	Finite sample performance of likelihood ratio tests for cointegrating ranks in vector	95 EconTher 11 1015-1032
	Finite sample properties of adaptive regression estimators • Ng, Pin T.	95 EcomtrRv 14 267-297
Wang, Songgui	Finite-sample properties of an adaptive ridge estimator (Chinese) • Yan, Liqing;	95 ChJApPSt 10 194-201
components model • Ghazal, G. A. et al. The exact	finite sample properties of the estimators of coefficients in mixed dummy-variable error	95 EgpStJ 39 26-50
asset pricing models • Ferson, W. E.; Foerster, S. R.	Finite sample properties of the generalized method of moments in tests of conditional	94 JFinEcon 36 29-55
IGARCH(1, 1) models: A Monte Carlo ... • Lumsdaine	Finite-sample properties of the maximum likelihood estimator in GARCH(1, 1) and	95 JBES 13 1-10
Morris, David	Finite sample properties of van Elteren test • Hosmane, Balakrishna; Shu, Vincent;	94 ASAProBiop 430-434
estimator • So, Beong-Soo	Finite-sample, small-dispersion asymptotic optimality of the non-linear least squares	95 JKoreaSt 24 303-312
Some	finite sample theory for bootstrap regression estimates • Fisher, Gordon; Sim, Ah Boon	95 JStPlInf 43 289-300
Multidimensional random subshifts of	finite type and their large deviations • Kifer, Y.	95 ProbTher 103 223-248
Clustering behavior of	finite variance partial sum processes • Einmahl, U.; Goodman, V.	95 ProbTher 102 547-565
Kadane, Joseph B.; O'Hagan, Anthony Using	finitely additive probability: Uniform distributions on the natural numbers •	95 JASA 90 626-631
salinity and nutrient concentrations in the Gulf of	Finland • Toompuu et al. ... correlations for optimal interpolation of temperature,	95 EnvrMtrc 6 55-72
	Finlay-Wilkinson model	
Gogel, Beverley J.; Cullis, Brian R.; Verbyla, Arūnus P.	REML estimation of multiplicative effects in multienvironment variety trials •	95 Biomtrcs 51 744-749
Marketing of statistical service: Some	Finnish experiences • Laihonen, A.	91 StatJUN 8 307-312
methods for non-response and their application to	Finnish income data • Laaksonen, S. Adjustment	92 StatJUN 9 125-138
of Israeli industry: Evidence from individual	firm data • Bregman, Arie; Fuss, Melvyn; Regev, Haim The production and cost structure	95 JEconmtx 65 45-81
R and D and productivity in French manufacturing	firms • Hall, Bronwyn H.; Mariesse, Jacques Exploring the relationship between	95 JEconmtx 65 263-293
in random production models of cost-minimizing	firms • Brown, Bryan W.; Walker, Mary Beth Stochastic specification	95 JEconmtx 66 175-205
	Firm productivity in Israeli industry 1979-1988 • Griliches, Zvi; Regev, Haim	95 JEconmtx 65 175-203
An investigation of the harvest decision of timber	firms in the south-east United States • Provencher, Bill	95 JApEcmtx 10 57-74S
Ford, J. M.; Gropper, D. M. Frontier estimation and	firm-specific inefficiency measures in the presence of heteroscedasticity • Caudill, S. B.;	95 JBES 13 105-111
multistate compartmental models in the analysis of	first appearance of Plasmodium falciparum parasites in infants • Gottschau et al. ... and	95 StatMed 14 2727-2736
	1st - Class! • Feu, Chris du	95 TeachgSt 17/1 13-15
	First crossing time	
two-dimensional diffusion ... • Di Crescenzo et al.	On a symmetry-based constructive approach to probability densities for	95 JAppProb 32 316-336
A brain-friendly	first day of class • Zahn, Douglas A.	94 ASAProStEd 53-62
Tricks of the trade: Getting going the	first day • Lock, Robin H.	94 ASAProStEd 49-52
What should you do on the	first day? Motivating and setting goals • Blanchard, Fletcher A.	94 ASAProStEd 63-65
	First (?) occurrence of common terms in mathematical statistics • David, H. A.	95 AmerStat 49 121-133
with fast Markov switchings • Di Masi, G. B. et al. A	first order approximation for the convergence of distributions of the Cox processes	95 Stochatx 54 211-219
William P. Bootstrap inference for a	first-order autoregression with positive innovations • Datta, Somnath; McCormick,	95 JASA 90 1289-1300
Ergodicity of nonlinear	first order autoregressive models • Bhattacharya, Rabi N.; Lee, Chanho	95 JTherPrb 8 207-219
Approximations to some exact distributions in the	first order autoregressive model with dependent errors • Nabeya, Seiji; Perron, Pierre	95 EcomtrRv 14 421-457
Improved tests for the	first-order autoregressive model with heteroscedasticity • Lyon, John; Tsai, Chih-Ling	95 JStCmpSm 52 71-83
the limiting distribution for the boundary case in the	first order autoregressive process • Monsour, Michael J.; Mikulski, Piotr W. Moments of	94 AJMMgSci 14 327-347
On the closed form of the likelihood function of the	first order moving average model • Haddad, John N.	95 Biomtrka 82 232-234
for testing the differences among independent	first-order partial correlations • Silver, N. C. et al. A Microsoft FORTRAN 77 program	95 EdPsMeas 55 245-248
Numerical solution of likelihood equations for a	first-order response surface model for mixture paired comparisons • Hassan, Tawfik	94 EgpStJ 38 196-207
Correction to "Another look at	first-order saturated designs: The p-efficient designs" (V35 p384-292) • Lin, Dennis K. J.	95 Technmcs 37 473-473
of alternative estimators for the probit model with	first-order serial correlation • Dagenais, Marcel G. The performance	95 JStCmpSm 51 343-351
Large-deviation expressions for the distribution of	first-passage coordinates • Whittle, P.	95 AdvAppPr 27 692-710
Supercritical behaviors in	first-passage percolation • Zhang, Yu	95 StocProc 59 251-266
	First passage time	
	A decomposition theorem for infinite stochastic matrices • Heyman, Daniel P.	95 JAppProb 32 893-901
Song, Mun Sup; Park, Byung-Gu	A diffusion model for a system subject to random shocks • Lee, Eui Yong;	95 JKoreaSt 24 141-147
	Dynamic modelling of life table data • Janssen, J.; Skiadas, C. H.	95 AplStMDA 11 35-49
Yuan, Xue-Ming	First passage times and their algorithms for Markov processes • Hsu, Guang-Hui;	95 CommStSM 11 195-209
Li, Haijun; Shaked, Moshe On the	first passage times for Markov processes with monotone convex transition kernels •	95 StocProc 58 205-216
A note on the Volterra integral equation for the	first-passage-time probability density • Gutiérrez Jáimez, R.; Román, P. R.; Ruiz, F. T.	95 JAppProb 32 635-648
Lefebvre, Mario Solutions to	first passage time problems by the method of separation of variables (French) •	95 JAppProb 32 337-348
Kou, Shing-Gang Limits of	first passage times to rare sets in regenerative processes • Glasserman, Paul;	95 AnlsApPr 5 424-445
of convergence for common subsequences and	first passage time • Rhee, WanSoo T. On rates	95 AnlsApPr 5 44-48
Knessl, C.; Tier, C. Asymptotic properties of	first passage times for tandem Jackson networks I: Buildup of large queue lengths •	95 CommStSM 11 133-162
for the distributions of stopped random walks and	first passage times • Lai, Tze Leung; Wang, Julia Qizhi Asymptotic expansions	94 AnlsProb 22 1957-1992
Dominé, Marco Moments of the	first-passage time of a Wiener process with drift between two elastic barriers •	95 JAppProb 32 1007-1013
the existence of the limiting probability of a tie for	first place • Baryshnikov, Y.; Eisenberg, B. et al. A necessary and sufficient condition for	95 StPrLet 23 203-209
Branching processes. Proceedings of the	First World Congress • Heyde, C. C. (Ed)	95 Springer:Brln:NY 180p
of collective properties and habitat relationships of	fish assemblages in coastal plain streams • Sheldon, A. L.; Meffe, G. K. Path analysis	95 JFishRBC 52 23-33
of species and relative abundances in stream-	fish communities: Effects of sampling effort and ... • Angermeier et al. ... number	95 JFishRBC 52 936-949
ination of price-quantity relations in the Marseille	fish market • Härdle, Wolfgang; Kirman, Alan Nonclassical demand - A model-free exam-	95 JEconmtx 67 227-257
for determining length-weight relationships of	fish • Hayes, D. B. et al. Efficiency and bias of estimators and sampling designs	95 JFishRBC 52 84-92

on "On the limitations of comparing mean square	forecast errors" • Clements, Michael P.; Hendry, David F. Reply to comment	95 JForcast 14 73-75
Evaluation of probability	forecasts of events • Ogata, Yosihiko	95 IntJFor 11 539-541
and rationality of state General Fund Revenue	forecasts: Evidence from panel data • Mocan, H. Naci; Azad, Sam Accuracy	95 IntJFor 11 417-427
use of earnings forecasts in predicting stock returns:	Forecast horizon effects • Bandyopadhyay, S. P.; Brown, L. D.; Richardson, G. D. Analysts'	95 IntJFor 11 429-445
	Forecasts of inflation from VAR models • Webb, Roy H.	95 JForcast 14 267-285
Parigi, G.; Schlitzer, G. Quarterly	forecasts of the Italian business cycle by means of monthly economic indicators •	95 JForcast 14 117-141
information improve the accuracy of judgemental	forecasts? • Remus, William; O'Connor, Marcus; Griggs, Kenneth Does reliable	95 IntJFor 11 285-293
Modifying quantitative	forecasts of livestock production using expert judgements • Vere, D. T.; Griffith, G. R.	95 JForcast 14 453-464
Song, Frank Are survey	forecasts of macroeconomic variables rational? • Aggarwal, Raj; Mohanty, Sunil;	95 JBusUChi 68 99-119
Restricted	forecasts of missing observations in univariate time series • Guerrero, Víctor M.	94 Estd 46 1-23
for refining the uncertainty in global energy model	forecasts • Tschang, F. Ted; Dowlatabadi, Hadi A Bayesian technique	95 IntJFor 11 43-61
Comment on "Testing the rationality of price	forecasts: New evidence from panel data" (V80 p714-735) • Bonham, Carl; Cohen, Richard	95 AmEcRv 85 284-289
to comment on "Testing the rationality of price	forecasts: New evidence from panel data" • Keane, Michael P.; Runkle, David E. Reply	95 AmEcRv 85 290-290
Reconfigurable combined	forecasts in a non-stationary inflationary environment • Volkov, V. Ya.; Gladkov, Yu. M.	95 JForcast 14 395-403
Time-series models to	forecast nuclear net generation of electricity • Kundra, Inderjit	94 ASAProBuEc 327-332
Brown, L. D. et al. Analysts' use of earnings	forecasts in predicting stock returns: Forecast horizon effects • Bandyopadhyay, S. P.;	95 IntJFor 11 429-445
Inc. (Auth); Weller, Barry R. (Rev) Review of	"Forecast pro for windows extended edition (XE) (Version 2)" • Business Forecast Systems	95 IntJFor 11 175-187
Finizza, Anthony Probabilistic	forecasts from probabilistic models: A case study in the oil market • Abramson, Bruce;	95 IntJFor 11 63-72
economic value of non-linear foreign exchange rate	forecasts • Satchell, Steve; Timmermann, Allan An assessment of the	95 JForcast 14 477-497
Highest-density	forecast regions for non-linear and non-normal time series models • Hyndman, Rob J.	95 JForcast 14 431-441
Finite sample	forecast results for vector autoregressive moving average models • Reinsel, Gregory C.	95 JForcast 14 405-412
monitoring of state tax collections and revenue	forecasts • Nelson, Ray D.; Cornia, Gary C. Intrayear	94 ASAProBuEc 339-344
The decomposition of	forecast in seasonal ARIMA models • Espasa, Antoni; Peña, Daniel	95 JForcast 14 565-583
Effects of feedback on probabilistic	forecasts of stock prices • Önkal, Dilek; Muradoğlu, Gulnur	95 IntJFor 11 307-319
	Forecast of unemployment in Switzerland • Frey, Béatrice	95 Student 1 1-15
A new framework for analyzing survey	forecasts using three-dimensional panel data • Davies, Anthony; Lahiri, Kajal	95 JEconmtx 68 205-227
economic value of ensemble-based surface weather	forecasts • Wilks, Daniel S.; Hamill, Thomas M. Potential	95 MWeathr 11 3565-3575
with misspecified ARCH models II: Making the right	forecast with the wrong model • Nelson, Daniel B.; Foster, Dean P. Filtering and forecasting	95 JEconmtx 67 303-335
Professional economic	forecasts: Are they worth their costs? • Leitch, Gordon; Tanner, J. Ernest	95 JForcast 14 143-157
	Forecasting	
	Analysing geographically-related disease data • Cliff, A. D.	95 StMedRes 4 93-101
	Assessment and propagation of model uncertainty (Disc: p71-97) • Draper, David	95 JRSS-B 57 45-70
economic model • Rickman, Dan S.	A Bayesian analysis of the use of pooled coefficients in a structural regional	95 IntJFor 11 477-490
	Causality in the long run • Granger, Clive W. J.; Lin, Jin-Lung	95 EconTher 11 530-536
predictions • Gilbert, Paul D.	Combining VAR estimation and state space model reduction for simple good	95 JForcast 14 229-250
	Comparing predictive accuracy • Diebold, Francis X.; Mariano, Roberto S.	95 JBES 13 253-263
and a Bayesian approach • Liu, Shu-Ing	Comparison of forecasts for ARMA models between a random coefficient approach	95 CommStA 24 319-333
model with an application in Pohang, ... • Ryoo et al.	Detection of warming effects due to industrialization: An accumulated intervention	95 JApMetrl 34 1466-1471
Shahidullah, Mohammed	An evaluation of population projection errors for census tracts • Smith, Stanley K.;	95 JASA 90 64-71
Department of Agriculture • Vogel, Frederic A.	The evolution and development of agricultural statistics at the United States	95 JOffStat 11 161-180
Campbell, Bryan; Ghysels, Eric	Federal budget projections: A nonparametric assessment of bias and efficiency •	95 RvEconSt 77 17-31
above, a given value • Bolger, Fergus; Harvey, Nigel	Judging the probability that the next point in an observed time-series will be below, or	95 JForcast 14 597-607
Anderson, Elizabeth A.	Judgmental and statistical methods of peak electric load management •	95 IntJFor 11 295-305
• Dee, Dick P.	On-line estimation of error covariance parameters for atmospheric data assimilation	95 MWeathr 123 1128-1145
using linear models and artificial ... • Swanson et al.	A model-selection approach to assessing the information in the term structure	95 JBES 13 265-275
Chatfield, Chris	Model uncertainty, data mining and statistical inference (Disc: p444-466) •	95 JRSS-A 158 419-444
	The multiprocess dynamic Poisson model • Bolstad, William M.	95 JASA 90 227-232
Timmermann, Allan	On the optimality of adaptive expectations: Muth revisited • Satchell, Steve;	95 IntJFor 11 407-416
	Predicting nominal variable relationships with multiple response • Umesh, U. N.	95 JForcast 14 585-596
Cho, Dongchul	The predictive ability of several models of exchange rate volatility • West, Kenneth D.;	95 JEconmtx 69 367-391
Canyon • Blattenberger, Gail; Fowles, Richard	Road closure to mitigate avalanche danger: A case study for Little Cottonwood	95 IntJFor 11 159-174
	Forecasting the adoptions of new consumer durable goods • Min, Chung-ki	94 ASAProBays 262-265
facturing industry: The use of leading indicators for	forecasting and analysis • Berk et al. ... interdependence of business cycles in the manu-	95 JForcast 14 1-23
accuracy studies on time series analysis and	forecasting • Fildes, Robert; Makridakis, Spyros The impact of empirical	95 IntStRvw 63 289-308
century U.S. homicide rates: An econometric	forecasting approach • Eckberg, Douglas Lee Estimates of early twentieth-	95 Dmogrphy 32 1-16
Environmental statistics, assessment, and	forecasting • Cothern, C. Richard; Ross, N. Phillip (Ed)	94 CRCPress:FL 432p
of "Environmental statistics, assessment, and	forecasting" • Cothern, C. R.; Ross, N. P. (Ed); Ziegel, Eric R. (Rev) Review	95 Technmcs 37 128-128
	Forecasting baseball games • Kaigh, W. D.	95 Chance 8/2 33-37
Level shifts, temporary changes and	forecasting • Trívez, F. Javier	95 JForcast 14 543-550
Chaos and	forecasting • Tong, H. (Ed)	95 WrldSciPub:Sing 356p
	Forecasting in cointegrated systems • Clements, Michael P.; Hendry, David F.	95 JApEcmtx 10 127-146
Interval	forecasting in cointegrated systems • Reimers, Hans-Eggert	95 StHefte 36 349-369
models • Vinod, H. D.; Basu, Parantap	Forecasting consumption, income and real interest rates from alternative state space	95 IntJFor 11 217-231
Review of "Time series analysis:	Forecasting and control" • Box, G.; Jenkins, G.; Reinsel, G. (Auth); Ziegel, Eric R. (Rev)	95 Technmcs 37 238-239
	Forecasting crop acreages and yields in the face of and in spite of floods • Allen, Rich	94 ASAProEnvr 187-191
J. (Rev) Review of "Technological	forecasting for decision making (Third edition)" • Martino, Joseph P. (Auth); Fields, Paul	95 Intrfc 25/1 137-139
	Forecasting earthquakes and earthquake risk • Vere-Jones, David	95 IntJFor 11 503-538
vector autoregression and error correction model for	forecasting electricity consumption and prices • Joutz, F. L. et al. An integrated Bayesian	95 JForcast 14 287-310
Kuan, Chung-Ming; Liu, Tung	Forecasting exchange rates using feedforward and recurrent neural networks •	95 JApEcmtx 10 347-364
approach • Patterson, K. D.	Forecasting the final vintage of real personal disposable income: A state space	95 IntJFor 11 395-405
production • Patterson A state space approach to	forecasting the final vintage of revised data with an application to the index of industrial	95 JForcast 14 337-350
Mullineux, Andy W. (Rev) Review of	"Forecasting financial and economic cycles" • Niemira, Michael P.; Klein, Philip A. (Auth);	95 IntJFor 11 191-192
Review of "Neural network time series	forecasting of financial markets" • Azoff, E. Michael (Auth); Meade, Nigel (Rev)	95 IntJFor 11 601-602
	Forecasting growth with time series models • Peña, Daniel	95 JForcast 14 97-105
	Forecasting hurricane tracks using the Kalman filter • Bril, Gordon	95 EnvrMtrc 6 7-16
	Forecasting industrial production using non-linear methods • Byers, J. D.; Peel, D. A.	95 JForcast 14 325-336
An interview for the International Journal of	Forecasting • Teräsvirta, Timo Professor Clive W. J. Granger:	95 IntJFor 11 585-590
accuracy: Comparison of criteria used to select	forecasting methods • Yokum, J. Thomas; Armstrong, J. Scott Beyond	95 IntJFor 11 591-597
and complexity in extrapolative population	forecasting models • McNown, Robert; Rogers, Andrei; Little, Jani Simplicity	95 MaPopStd 5 235-257
An evaluation study of	forecasting models • Yu, Ching C.; Absalom, Sandra T.; Plunkert, Patricia A.	94 ASAProGovt 173-177
Vector autoregression modelling and	forecasting • Holden, Ken	95 JForcast 14 159-166
Model-free	forecasting for nonlinear time series (with application to exchange rates) • Wu, Berlin	95 CmpStDA 19 433-459
Short and medium term rain	forecasting in the Oristano basin (Italian) • Espa, Giuseppe; Terpessi, Cristina	94 QStMaA 16 145-160
exchange rate determination: A comparison of their	forecasting performance • Sarantis, N.; Stewart, C. Structural, VAR and BVAR models of	95 JForcast 14 201-215

On some predictors of	future order statistics • Raqab, Mohammad Z.; Nagaraja, Haikady N.	95 Metron 53 185-204
ecology in forestry - Past accomplishments and	future outlook • Burkhardt, Harold E. Applications of statistical	94 ASAProEnvr 10-13
perspective • Legreton, Jean-Dominique The	future of population dynamic studies using marked individuals: A statistician's	95 JAppStat 22 1009-1030
Lead-lag relationship between spot index and	futures price of the Nikkei Stock Average • Tse, Y. K.	95 JForcast 14 553-563
from melanoma in Canada and prediction of	future rates • MacNeill, I. B.; Elwood, J. M.; Miller, D.; Mao, Y. Trends in mortality	95 StatMed 14 821-839
Basic concepts of the	fuzzied statistical decision and its superiority (Chinese) • Li, Xihe; Liu, Rongbing	95 AMaAPknC 17 512-521
course closing • Shipley, M. F. et al. Rough set theory	fuzzy belief functions related to statistical confidence: Application and evaluation for golf	95 StocAnAp 13 487-502
A general	fuzzy clustering model based on aggregation operators • Sato, Mika; Sato, Yoshiharu	95 Behvmtrk 22 115-128
	Fuzzy data distortion • Panda, S. K.; Nagabhushanam, A.	95 CmpStDA 19 553-562
Statistical tests for	fuzzy data • Römer, Christoph; Kandel, Abraham	95 FuzzySet 72 1-26
A qualitative simulation approach for	fuzzy dynamical models • Bonarini, Andrea; Bontempi, Gianluca	94 ACMMdSim 4 285-313
Estimating diagnostic test accuracy using a	"fuzzy gold standard" • Phelps, Charles E.; Hutson, Alan	95 MedDscnM 15 44-57
PROBO: An interactive system in	fuzzy linear programming • Cadenas, J. M.; Verdegay, J. L.	95 FuzzySet 76 319-332
	Fuzzy logic	
	Uncertainty calculi for fault trees • Weber, David P.	95 ASQCTekCon 1046-1053
Application of	fuzzy logic control in industry • van der Wal, A. J.	94 FuzzySet 74 33-41
	Fuzzy logic: Great hope or grating hype? • Laviolette, Michael	95 Chance 8/4 15-19
respect to ambiguous a priori knowledge. Part II:	Fuzzy logic • Mahler, Ronald P. S. Combining ambiguous evidence with	95 FuzzySet 75 319-354
Image processing and object classification with	fuzzy logic (Italian) • Bellacicco, Antonio	94 QStMaA 16 19-26
Woodall, W. H. A probabilistic and statistical view of	fuzzy methods (Disc: p262-292) • Laviolette, M.; Seaman, J. W., Jr; Barrett, J. D.;	95 Technmcs 37 249-261
L. I. Improving classification performance using	fuzzy MLP and two-level selective partitioning of the feature space • Mitra, S.; Kuncheva,	95 FuzzySet 70 1-13
Montero, J.; Pearman, A. D.; Tejada, J.	Fuzzy multicriteria decision support for budget allocation in the transport sector •	95 Top 3 47-68
Takanori; Arai, Fumihito Structure optimization of	fuzzy neural network by genetic algorithm • Ishigami, Hideyuki; Fukuda, Toshio; Shibata,	95 FuzzySet 71 257-264
	Fuzzy number	
data • Savchuk, V. P.	Estimation of structures reliability for non-precise limit state models and vague	95 RelEngSS 47 47-58
A SLLN for arrays of rowwise exchangeable	fuzzy random sets • Inoue, Hiroshi; Taylor, Robert L.	95 StocAnAp 13 461-470
Randomly weighted sums for exchangeable	fuzzy random variables • Inoue, Hiroshi	95 FuzzySet 69 347-354
Statistical tests optimally meeting certain	fuzzy requirements on the power function and on the sample size • Arnold, Bernhard F.	95 FuzzySet 75 365-372
Applying	fuzzy set approach to signal detection theory • Chen, Toly; Wang, Mao-Jiun J.	95 FuzzySet 72 39-49
Safety analysis and synthesis using	fuzzy sets and evidential reasoning • Wang, J.; Yang, J. B.; Sen, P.	95 RelEngSS 47 103-118
Random	fuzzy sets and fuzzy martingales • Li, Lushu	95 FuzzySet 69 181-192
Correlation of interval-valued intuitionistic	fuzzy sets • Bustince, H.; Burillo, P.	95 FuzzySet 74 237-244
Probabilities, possibilities, and	fuzzy sets • Drakopoulos, John A.	95 FuzzySet 75 1-15
Correlation of intuitionistic	fuzzy sets in probability spaces • Hong, Dug Hun; Hwang, Seok Yoon	95 FuzzySet 75 77-81
The definition of the niche by	fuzzy set theory • Cao, Guangxia	95 EcolgMdl 77 65-71
Review of "Statistical applications using	fuzzy sets" • Manton, K. G.; Woodbury, M. A.; Tolley, H. D. (Auth); Gordon, A. D. (Rev)	95 Biomtrcs 51 793-794
Review of "Statistical applications using	fuzzy sets" • Manton, K. G.; Woodbury, M. A.; Tolley, H. D. (Auth); Haberman, S. J. (Rev)	95 JASA 90 1131-1133
Review of "Statistical applications using	fuzzy sets" • Manton, K. G.; Woodbury, M. A.; Tolley, H. D. (Auth); Kandel, A. (Rev)	95 Technmcs 37 237-237
	Fuzzy surface reconstruction • Pereira, Ricardo Alberto Marques; Fedrizzi, Michele	94 QStMaA 16 71-77
Neural nets for	fuzzy systems • Buckley, James J.; Hayashi, Yoichi	95 FuzzySet 71 265-276
Chissom, Brad S. A new	fuzzy time-series model of fuzzy number observations • Song, Qiang; Leland, Robert P.;	95 FuzzySet 73 341-348
molecular-marker, quantitative trait data from large	F_2 populations • Wright, A. J.; Mowers, R. P. Multiple regression for	94 ThApGene 89 305-312
genetic markers showing segregation distortion. 2.	F_2 populations • Lorieux, M.; Perrier, X. et al. Maximum-likelihood models for mapping	95 ThApGene 90 81-89
dominance effects of QTLs in interval mapping of	F_2 RFLP data • Hayashi, T.; Ukai, Y. Detection of additive and	94 ThApGene 87 1021-1027
Finding maximum	G-criterion values for central composite designs on the hypercube • Borkowski, John J.	95 CommStA 24 2041-2058
Application of Hoaglin's	G and H curves to process specifications • Finnegan, G. Paul; Heavlin, William D.	94 ASAProQlPr 146-150
Differentiability of the matrices R and	G in the matrix-analytic method • He, Qi-Ming	95 CommStSM 11 123-132
Response time distributions in tandem	g-networks • Harrison, Peter G.; Pitel, Edwige	95 JAppProb 32 224-246
On the equivalence of D- and	G-optimal designs in heteroscedastic models • Wong, Weng Kee	95 StPrLet 25 317-321
Excursions of the workload process in	G/GI/1 queues • Guillemin, Fabrice M.; Mazumdar, Ravi R.	95 StocProc 58 139-154
A	G/SM/1 queue with vacations depending on service times • Machihara, Fumiaki	95 CommStSM 11 671-690
in ATM networks • Kim, Kyeong Soo; Lee, Byeong Gi	GA-based optimal dimensioning of three-level traffic shaper for statistical multiplexing	95 AnlsTele 50 624-631
ing statistics with data of historical significance:	Galileo's gravity and motion experiments (jse/v3n1/datasets.dickey) • Dickey et al. Teach-	95 JStEd 3/1 20kb
approximation to find the extinction probability of a	Galton-Watson branching process • Komarov, S. V.; Krasulina, T. P. Using stochastic	94 VLenUM /3 34-39
	Galton-Watson process	
	Analysis of transient effects for branching processes • Borovkov, K. A.	94 ThProbAp 39 379-395
Lyons, Russell; Pemantle, Robin; Peres, Yuval	Conceptual proofs of $L\log L$ criteria for mean behavior of branching processes •	95 AnlsProb 23 1125-1138
	Three papers on the history of branching processes • Guttorp, Peter (Transl)	95 IntStRvw 63 233-245
Peres, Yuval	Galton-Watson trees with the same mean have the same polar sets • Pemantle, Robin;	95 AnlsProb 23 1102-1124
A rule of thumb (not only) for	gamblers • Kozek, Andrzej S.	95 StocProc 55 169-181
	Gambler's ruin	
	A game with four players • Chang, Derek K.	95 StPrLet 23 111-115
	Gambling	
Finkelstein, Mark	Estimating the frequency distribution of the numbers bet on the California lottery •	95 AplMaCmp 69 195-207
	Longest runs in coin tossing • Binswanger, K.; Embrechts, P.	94 InsrneME 15 139-149
	A national lottery in the UK • Moore, Peter G.	94 JAppStat 21 607-622
	What is the worth of free casino credit? • Orkin, Michael; Kakigi, Richard	95 AmMaMnth 102 3-8
Correction to "Universal schemes for prediction,	gambling and portfolio selection" (V20 p901-941) • Algoet, Paul H.	95 AnlsProb 23 474-478
The theory of	gambling and statistical logic (Revised edition) • Epstein, Richard A.	95 Academic:NY:Lnd 450p
Donegan, Kevin Comment on "Wanna bet? On	gambling strategies that may or may not work in a casino" (V47 p108-111) •	95 AmerStat 49 402-402
An extension of a theorem on	gambling systems • Liu, Wen; Wang, Zhongzhi	95 JMultiAn 55 125-132
beer ... • Chintagunta et al. A two-period repeated	game advertising investment model for oligopolistic markets with an application to the	95 DecisnSc 26 531-559
and related minimax design problems - A dynamic	game approach (Second edition) • Basar, T.; Bernhard, P. H^∞-optimal control	95 Birkhs:Basl:MA 411p
and related minimax design problems - A dynamic	game approach (Second edition)" • Basar, T. et al. Review of "H^∞-optimal control	96 ShrtBkRv 16/1 11-11
Forecasting baseball	games • Kaigh, W. D.	95 Chance 8/2 33-37
control and zero-sum stochastic differential	games • Hamadéne, S.; Lepeltier, J. P. Backward equations, stochastic	95 Stochatx 54 221-231
Chinese dominoes	games • Wang, Ann-Lee	95 TeachgSt 17/2 70-73
Stochastic differential	games: The linear quadratic zero sum case • Ardanuy, R.	95 SankhyaA 57 161-165
A statistical model for duplicate tournaments in	games of luck and skill • Yu, Philip L. H.; Cowan, Richard	95 AstrlJSt 37 1-17
On variational inequality approach to Nash dynamic	game of oil stockpiling • Zhao, L.; Yin, G.	95 StocAnAp 13 627-649
A stochastic	game of optimal stopping and order selection • Gnedin, Alexander V.; Krengel, Ulrich	95 AnlsApPr 5 310-321
dices • Hong, D. H.; Hwang, S. Y. Chow-Robbins "fair"	games problem for pairwise independent random variables with multidimensional in-	95 BlMaASnc 23 7-11
Life and hard times of a statistician: From	games to statistics • Hogg, Robert V.	95 Stats /13 24-25

	Game theory	
case • Tapiero, Charles S.	Acceptance sampling in a producer-supplier conflicting environment: Risk neutral	95 AplStMDA 11 3-12
Zarzuelo, J. M.	On the axiomatization of the τ-value • Calvo, E.; Tijs, S. H.; Valenciano, F.;	95 Top 3 35-46
multidimensional indices • Hong, D. H.; Hwang, S. Y.	Chow-Robbins "fair" games problem for pairwise independent random variables with	95 BlMaASnc 23 7-11
	Cooperation indices and coalitional value • Amer, R.; Carreras, F.	95 Top 3 117-135
Strait, R. Scott	Decision analysis approach to competitive situations with a pure infinite regress •	94 DecisnSc 25 853-862
Yannis A.; Lazar, Aurel A.	On the existence of equilibria in noncooperative optimal flow control • Korilis,	95 JACM 42 584-613
	Optimal stopping in 'The Showcase Showdown' • Coe, Paul R.; Butterworth, William	95 AmerStat 49 271-275
	Stochastic differential games: The linear quadratic zero sum case • Ardanuy, R.	95 SankhyaA 57 161-165
Krengel, Ulrich	A stochastic game of optimal stopping and order selection • Gnedin, Alexander V.;	95 AnlsApPr 5 310-321
	Game theory (Third edition) • Owen, Guillermo	95 Academic:NY:Lnd 472p
annuities: Actuarial principles and cooperative	game theory • Alegre, Antoni; Claramunt, M. Mercé Allocation of solvency cost in group	95 InsrneME 17 19-34
Essays in	game theory: In honor of Michael Maschler • Megiddo, N. (Ed)	94 Springer:Brln:NY 203p
Introduction to	game theory • Morris, P.	94 Springer:Brln:NY 230p
Dynamic noncooperative	game theory (Second edition) • Basar, Tamer; Olsder, Geert Jan	95 Academic:NY:Lnd 519p
Putting	game theory to the test • Pool, R.	95 Science 267 1591-1593
software release policy based on a two-person	game of timing • Zeephongsekul, P.; Chiera, C. Optimal	95 JAppProb 32 470-481
distribution and the probability of winning a	game" (V48 p294-298) • Meng, Xiao-Li Comment on "A characterization of the Poisson	95 AmerStat 49 401-401
distribution and the probability of winning a	game" (V48 p294-298) • Sen, Ananda Comment on "A characterization of the Poisson	95 AmerStat 49 401-401
A	game with four players • Chang, Derek K.	95 StPrLet 23 111-115
for estimating the potency and survival of human	gametes in vivo • Weinberg, Clarice R.; Wilcox, Allen J. A model	95 Biomtrcs 51 405-412
	Gamma distribution	
rejection of a preliminary test • Chiou, P.; Han, C.-P.	Conditional interval estimation of the exponential scale parameter following	94 JStatRsr 28 65-72
distributions • Vännman, Kerstin	On the distribution of the estimated mean from nonstandard mixtures of	95 CommStA 24 1569-1584
• Bai, Jun; Jakeman, Anthony J.; McAleer, Michael	Empirical models for evaluating errors in fitting extremes of a probability distribution	95 MaCmpSim 39 1-7
rates differ over nucleotide sites • Yang, Ziheng	Evaluation of several methods for estimating phylogenetic trees when substitution	95 JMolEvol 40 689-697
	Generalized unimodality • Alamatsaz, M. H.	92 ProIrStCl 1-10
	On the HIV incubation distribution under AZT treatment • Tan, W. Y.	95 BiomtrcJ 37 319-338
cleotide sites • Gu, Xun; Fu, Yun-Xin; Li, Wen-Hsiung	Maximum likelihood estimation of the heterogeneity of substitution rate among nu-	95 MolBioEv 12 546-557
rates over sites: Approximate methods • Yang, Z.	Maximum likelihood phylogenetic estimation from DNA sequences with variable	94 JMolEvol 39 306-314
streamflows • Chebaane, M.; Salas, J. D.; Boes, D. C.	Product periodic autoregressive processes for modeling intermittent monthly	95 WaterReR 31 1513-1518
Prescott, P.	On Rohlf's method for the detection of outliers in multivariate data • Caroni, C.;	95 JMultiAn 52 295-307
	Some characterizations for the exponential class of distributions • Consul, P. C.	95 IEEEReli 44 403-407
mitochondrial DNA • Wakeley, John	Substitution rate variation among sites in hypervariable region 1 of human	93 JMolEvol 37 613-623
Lin, Gwo Dong	On weak convergence within a Lorenz ordering family of distributions •	95 JAppProb 32 1028-1031
Characterization of the negative binomial and	gamma distributions • Ong, S. H.	95 BiomtrcJ 37 251-254
The generalized	Gamma distribution: Its hazard rate and stress-strength model • Pham, T.; Almhana, J.	95 IEEEReli 44 392-397
On the sequential point estimation of the mean of a	gamma distribution • Isogai, Eiichi; Uno, Chikara	95 StPrLet 22 287-293
sequential confidence interval for the mean of a	gamma distribution • Takada, Yoshikazu; Nagata, Yasushi Fixed-width	95 JStPlInf 44 277-289
parameter estimation in the three-parameter	gamma distribution • Hirose, Hideo Maximum likelihood	95 CmpStDA 20 343-354
problems associated with the three parameter	Gamma distribution • Bowman, K. O.; Shenton, L. R. Estimation	95 CommStA 24 1355-1376
estimation of a lower-bounded scale parameter of a	gamma distribution for scale-invariant squared-error loss • van Eeden, C. Minimax	95 CandJSt 23 245-256
Assessing	gamma frailty models for clustered failure time data • Shih, Joanna H.; Louis, Thomas A.	95 LiDaAnal 1 205-220
Haruhiko The	gamma-gamma regression for the distribution model of event times • Ogasawara,	95 JpnPsyRs 37 70-79
Computation of bivariate	gamma and inverted beta distribution functions • Ong, S. H.	95 JStCmpSm 51 153-163
Exit probability estimates for	Γ-martingales on manifolds • Le, Huiling	95 Stochatx 54 199-209
The radial part of a	γ-martingale and a non-implosion theorem • Kendall, Wilfrid S.	95 AnlsProb 23 479-500
Constructing	gamma-martingales with prescribed limit, using backwards SDE • Darling, R. W. R.	95 AnlsProb 23 1234-1261
Fieger, Werner; Wulfert, Stefanie Minimax- and	Γ-minimax estimation of a bounded normal mean under LINEX loss • Bischoff, Wolfgang;	95 StatDecn 13 287-298
Some results on posterior regret	Γ-minimax estimation • Insua, David Ríos; Ruggeri, Fabrizio; Vidakovic, Brani	95 StatDecn 13 315-331
Hierarchical Bayesian models and	Γ-minimaxity • George, Robert; Goel, Prem K.	94 ASAProBays 96-101
On approximate inference for the two-parameter	gamma model • Wong, A.	95 StHefte 36 49-59
The gamma-	gamma regression for the distribution model of event times • Ogasawara, Haruhiko	95 JpnPsyRs 37 70-79
Tests of ordered hypotheses for	gamma scale parameters • Li, Teng; Sinha, Bimal K.	95 JStPlInf 43 311-321
Tests of ordered hypotheses for	gamma scale parameters • Li, Teng; Sinha, Bimal K.	95 JStPlInf 45 387-397
Comparison of	Gamma survival curves (Spanish) • Beamonte, Eduardo; Bermúdez, José D.	95 Questo 19 171-186
A conversation with Joe	Gani • Heyde, Chris	95 StatSci 10 214-230
	Gap test	
Prescott, P.	On Rohlf's method for the detection of outliers in multivariate data • Caroni, C.;	95 JMultiAn 52 295-307
	GARCH model	
impacts of "big bang" and the 1987 ... • Masulis et al.	Overnight and daytime stock-return dynamics on the London stock exchange: The	95 JBES 13 365-378
Cho, Dongchul	The predictive ability of several models of exchange rate volatility • West, Kenneth D.;	95 JEconmtx 69 367-391
Bekaert, Geert	The time variation of expected returns and volatility in foreign-exchange markets •	95 JBES 13 397-408
Mele, A.	A two-factor model of the term structure with optimal filtering behaviour • Fornari, F.;	94 Statstca 54 293-312
the United States and Canada • Karolyi A multivariate	GARCH model of international transmissions of stock returns and volatility: The case of	95 JBES 13 11-25
Detecting parameter shift in	GARCH models • Chu, Chia-Shang James	95 EcomtrRv 14 241-266
Maximum likelihood estimation of a	GARCH-stable model • Liu, Shi-Miin; Brorsen, B. Wade	95 JApEcmtx 10 273-285
properties of the maximum likelihood estimator in	GARCH(1,1) and IGARCH(1,1) models: A Monte Carlo ... • Lumsdaine Finite-sample	95 JBES 13 1-10
Better subset regression using the nonnegative	garrote • Breiman, Leo	95 Technmcs 37 373-384
Jackson, J. F.; Hibbert, D. B. Chemometric analysis of	gas chromatographic data of oils from Eucalyptus species • Dunlop, P. J.; Bignell, C. M.;	95 Chemolab 30 59-67
in the Kalman filter: Implications for greenhouse	gas flux deductions • Mulquiney, J. E.; Norton, J. P.; Jakeman, A. J. et al. Random walks	95 EnvrMtrc 6 473-478
sensor signals • Kraus, G. et al. Optical reflectometric	gas sensing: Classification of hydrocarbon vapours by pattern recognition applied to RIFS	95 Chemolab 30 211-221
Prediction of automotive emissions from	gasoline composition • Rocke, David M.; Pallas, Michelle	94 ASAProPEng 269-273
Comparison of national and regional	gasoline prices from two surveys • Saavedra, Pedro J.; Hassett, Nancy C.; Weir, Paula	94 ASASrvyRsr 1179-1183
sions of four species of pulmonate slugs (Mollusca :	Gastropoda : Pulmonata) • Bhuyan et al. ... changes in the relationships of body dimen-	95 BiomtrcJ 37 1005-1015
A globally	gated polling system with a dormant server • Borst, S. C.	95 PrEngInf 9 239-254
Goyret, Jorge Principal component analysis of rain	gauge data: Detecting outliers (Spanish) • López, Carlos; González, Elizabeth;	94 Estd 46 25-54
Curbelo, R. Principal component analysis of rain	gauge data: Treating missing observations (Spanish) • López, C.; González, J. F.;	94 Estd 46 55-83
fibre bundles and the Faddeev-Popov procedure in	gauge theories • Arnaudon, M.; Paycha, S. Factorisation of semi-martingales on principal	95 Stochatx 53 81-107
On the accuracy of the statistical distributions in	GAUSS • Knüsel, Leo	95 CmpStDA 20 699-702
K. et al. Admissible linear estimation in the general	Gauss-Markov model with respect to an arbitrary quadratic risk function • Baksalary, J.	95 JStPlInf 44 341-347
Data analysis using	GAUSS and Markov • Long, J. Scott; Noss, Brian	95 SocMethR 23 303-328
Morel, Jorge G.; Koehler, Kenneth J. A one-step	Gauss-Newton estimator for modelling categorical data with extraneous variation •	95 ApplStat 44 187-200
	Gauss, statistics and Gaussian elimination • Stewart, G. W.	95 JCmpGrSt 4 1-11

	Gaussian	
	• see also inverse- –, non- –, normal distribution	
geneities • Bhattacharya, R. N. et al. Time-scales for	Gaussian approximation and its breakdown under a hierarchy of periodic spatial hetero-	95 Bernoull 1 81-123
of model selection criteria: The nonzero mean	Gaussian AR(∞) case • Karagrigoriou, Alexandros Asymptotic efficiency	95 CommStA 24 911-930
asymptotic efficiency of sample covariances of	Gaussian ARMA processes • Walker, A. M. On results of Porat concerning	95 JTimSrAn 16 237-248
Prediction and non-	Gaussian autoregressive stationary sequences • Rosenblatt, Murray	95 AnlsApPr 5 239-247
2-nilpotent lie groups • Neuenschwander, Daniel A	Gaussian central limit theorem for trimmed products on simply connected step	95 JTherPrb 8 165-174
	Sub-Gaussian distribution	
	Consistent variable selection in linear models • Zheng, Xiaodong; Loh, Wei-Yin	95 JASA 90 151-156
Mean motion and	Gaussian dynamical system (French) • de la Rue, Thierry	95 ProbTher 102 45-56
Gauss, statistics and	Gaussian elimination • Stewart, G. W.	95 JCmpGrSt 4 1-11
	Gaussian estimation	
moving average model • Yap, S. F.; Reinsel, G. C.	Estimation and testing for unit roots in a partially nonstationary vector autoregressive	95 JASA 90 253-267
Design sensitivity of structures under non-	Gaussian excitation • Sieniawska, R.; Śniady, P.	95 ZAngMaMk 75 573-574S
Consistency of posterior mixtures in the	Gaussian family on a Hilbert space and its applications • Majumdar, Suman	95 JMultiAn 55 187-204
	Gaussian field	
Molchanov, S. A.	Stationary parabolic Anderson model and intermittency • Carmona, R. A.;	95 ProbTher 102 433-453
Small ball probabilities of	Gaussian fields • Shao, Q.-M.; Wang, D.	95 ProbTher 102 511-517
V. D.; Piterbarg, V. I. High level excursions of	Gaussian fields and the weakly optimal choice of the smoothing parameter. I • Konakov,	95 MaMethSt 4 421-434
Stochastic anticipative calculus for functionals of a	Gaussian generalized random field • Albeverio, Sergio; Wu, Jiang-Lun	95 Stochatx 54 317-337
Power of Hotelling's T^2 in	Gaussian mixtures • Jensen, D. R.	95 JStCmpSm 52 184-187
biomarker data • Doksum, K. A.; Normand, S.-L. T.	Gaussian models for degradation processes - Part I: Methods for the analysis of	95 LiDaAnal 1 131-144
A test for the detection of a change in mean in a	Gaussian model (French) • Ghorbanzadeh, Dariush	95 RStApp 43/2 67-76
structure of mean vector of compound symmetric	Gaussian model • Gupta, A. K.; Nagar, D. K. On testing	95 CommStA 24 3123-3136
Wilson, P. David Longitudinal data analysis for linear	Gaussian models with random disturbed-highest-derivative-polynomial subject effects •	95 StatMed 14 1219-1233
Sanyogita Noise subspace techniques in non-	Gaussian noise using cumulants • Sadler, Brian M.; Giannakis, Georgios B.; Shamsunder,	95 IEEEAES 31 1009-1018
totic properties of distribution of maximum of	Gaussian nonstationary process occurring in covariance statistic • Ostrovskii et al. Asymp-	94 ThProbAp 39 527-534
	Gaussian parsimonious clustering models • Celeux, Gilles; Govaert, Gérard	95 PattrnRc 28 781-793
	Gaussian process	
	A Bayesian model for growth curve analysis • Barry, Daniel	95 Biomtrcs 51 639-655
States • Tarboton, David G.	Hydrologic scenarios for severe sustained drought in the Southwestern United	95 WaterReB 31 803-813
Gill, Richard D.; van der Laan, Mark J.; Wellner, Jon A.	Inefficient estimators of the bivariate survival function for three models •	95 AIHenrPB 31 545-597
	Nonparametric regression with censored covariates • Dabrowska, Dorota M.	95 JMultiAn 54 253-283
Poskitt, D. S.	Periodogram-based estimators of fractal properties • Chan, Grace; Hall, Peter;	95 AnlsStat 23 1684-1711
normal sequences • Hüsler, J.; Kratz, M.	Rate of Poisson approximation of the number of exceedances of nonstationary	95 StocProc 55 301-313
Rudiuk, Edmund M.	The sampling distribution of the serial correlation coefficient • Provost, Serge B.;	95 AJMMgSci 15 57-81
Su, John Q.	Simultaneous confidence bands for comparing diagnostic markers • Chen, Hanfeng;	95 CommStA 24 375-393
	Small perturbations of Gaussian regressors • Millet, Annie; Smoleński, Włodzmierz	95 StPrLet 24 21-31
Ramalingam, T.; Ebrahimi, Nader	Testing for independence between hitting times of bivariate PQD processes •	95 SankhyaA 57 202-220
	Testing multivariate symmetry • Heathcote, C. R.; Rachev, S. T.; Cheng, B.	95 JMultiAn 54 91-112
circle? • Wood, Andrew T. A.	When is a truncated covariance function on the line a covariance function on the	95 StPrLet 24 157-164
Rootzén, Holger Small values of	Gaussian processes and functional laws of the iterated logarithm • Monrad, Ditlev;	95 ProbTher 101 173-192
On the law of the iterated logarithm for	Gaussian processes • Arcones, Miguel A.	95 JTherPrb 8 877-903
dures for detecting a change in a nonhomogeneous	Gaussian process • Tartakovsky, A. Asymptotic properties of CUSUM and Shiryaev's proce-	95 MaMethSt 4 389-404
due to incomplete data from a one-parameter spatial	Gaussian process • Krug, T.; Martin, R. J. Information loss on the dependence parameter	95 CommStA 24 3055-3072
A note on extreme values of locally stationary	Gaussian processes • Hüsler, J.	95 JStPlInf 45 203-213
deviations of empirical measures for stationary	Gaussian processes • Bryc, Włodzimierz; Dembo, Amir On large	95 StocProc 58 23-34
On moduli of continuity for local times of	Gaussian processes • Csörgő, Miklós; Lin, Zheng-Yan; Shao, Qi-Man	95 StocProc 58 1-21
proximation to the exit probability of a continuous	Gaussian process over a U-shaped boundary of increasing curvature • Balabushkin Ap-	95 JAppProb 32 429-442
Shao, Qi-man Small ball probabilities for	Gaussian processes with stationary increments under Hölder norms • Kuelbs, J.; Li, W. V.;	95 JTherPrb 8 361-386
	Gaussian quadrature	
model • Pinheiro, José C.; Bates, Douglas M.	Approximations to the log-likelihood function in the nonlinear mixed-effects	95 JCmpGrSt 4 12-35
with unknown location and scale in a stationary	Gaussian random field • Siegmund, David O.; Worsley, Keith J. Testing for a signal	95 AnlsStat 23 608-639
	Gaussian random functions • Lifshits, M. A.	95 Kluwer:MA:Hllnd 352p
G.; Taqqu, M. S. (Auth) et al. Review of "Stable non-	Gaussian random processes: Stochastic models with infinite variance" • Samorodnitsky,	95 JASA 90 805-806
Orlicz norm of a linear form of independent pre-	Gaussian random variables • Kurchenko, O. O. Some inequalities for the	95 ThPrMaSt 50 99-102
M. Pre-	Gaussian random vectors and their applications • Kozachenko, Yu. V.; Mokliatchouk, O.	95 ThPrMaSt 50 89-98
Small perturbations of	Gaussian regressors • Millet, Annie; Smoleński, Włodzmierz	95 StPrLet 24 21-31
Gorostiza, L. G. Self-intersection local time for	Gaussian $S'(R^d)$-processes: Existence, path continuity and examples • Bojdecki, T.;	95 StocProc 60 191-226
Talagrand, Michel Lim inf results for	Gaussian samples and Chung's functional LIL • Kuelbs, James; Li, Wenbo V.;	94 AnlsProb 22 1879-1903
Exact behavior of	Gaussian seminorms • Dembo, Amir; Mayer-Wolf, Eddy; Zeitouni, Ofer	95 StPrLet 23 275-280
	Gaussian semiparametric estimation of long range dependence • Robinson, P. M.	95 AnlsStat 23 1630-1661
for non-instantaneous filters of strongly dependent	Gaussian sequences • Ho, Hwai-Chung The law of the iterated logarithm	95 JTherPrb 8 347-360
Neglecting observations in	Gaussian sequences of statistical experiments • Marohn, F.	95 StatDecn 13 83-92
theorems for nonlinear functionals of a stationary	Gaussian sequence of vectors • Arcones, Miguel A. Limit	94 AnlsProb 22 2242-2274
A simulation technique of a non-	Gaussian spatial process • De Cesare, L.; Posa, D.	95 CmpStDA 20 543-555
model discrimination and Bayes factors for linear	Gaussian state space models • Frühwirth-Schnatter, Sylvia Bayesian	95 JRSS-B 57 237-246
distribution of supremum of a certain class of pre-	Gaussian stochastic processes on a compact set in R^d • Sydorenko, O. O. On the	95 ThPrMaSt 50 123-128
	Gaussian tail estimates for Dirichlet processes • Ubøe, Jan; Zhang, Tusheng	95 Stochatx 52 295-302
Variance inequalities for functions of	Gaussian variables • Houdré, Christian; Kagan, Abram	95 JTherPrb 8 23-30
Dimension results for	Gaussian vector fields and index-α stable fields • Xiao, Yimin	95 AnlsProb 23 273-291
theorem for non-linear functionals of stationary	Gaussian vector processes • Sánchez de Naranjo, M. V. A central limit	95 StPrLet 22 223-230
a discrete-time nonlinear system with generalized	Gaussian white noise • Ferrante, Marco; Giummolé, Federica Finite dimensional filters for	95 Stochatx 53 195-211
likelihood estimators of a discontinuous signal in a	Gaussian white noise • Rubin et al. ... of the asymptotic efficiency of maximum	95 AnlsStat 23 732-739
Counting	gays and lesbians: A case for "new" survey methods • Edwards, Brad	94 ASASrvyRsr 1305-1309
The feasibility of latent variables applied to	GC-MS data • Brakstad, Frode	95 Chemolab 29 157-176
Behavior near zero of the distribution of	GCV smoothing parameter estimates • Wahba, Grace; Wang, Yuedong	95 StPrLet 25 105-111
numerical weather prediction models: Randomized	GCV in three- and four-dimensional data assimilation • Wahba, G. et al. Adaptive tuning of	95 MWeathr 11 3358-3369
Chi, George Y. H. Using the	GEE method to construct linear models for ordinal responses • Nuri, Walid A.;	94 ASAProBiop 82-87
study and management of Atlantic Flyway Canada	geese • Hestbeck, Jay B. Population	95 JAppStat 22 877-890
Transient animals in a resident population of snow	geese: Local emigration or heterogeneity? • Pradel, Roger; Cooch, Evan; Cooke, Fred	95 JAppStat 22 695-710
unordered random partitions of an integer and the	GEM distribution • Sibuya, Masaaki; Yamato, Hajime Ordered and	95 StPrLet 25 177-183
	Gender and performance in athletics • Chatterjee, Sangit; Laudato, Matthew	94 ASAProSpts 36-41
Jarman, J. The measurement of occupational	gender segregation: Current problems and a new approach • Blackburn, R. M.; Siltanen, J.;	95 JRSS-A 158 319-331

Estimating the probability for major	gene Alzheimer disease • Farrer, Lindsay A.; Cupples, L. Adrienne	94 AJHuGene 54 374-383
Comparison of statistics for candidate-	gene association studies using cases and parents • Schaid, Daniel J.; Sommer, Steve S.	94 AJHuGene 55 402-409
estimating the linkage of marker genes to viability	genes controlling inbreeding depression • Fu, Y. B.; Ritland, K. On	94 ThApGene 88 925-932
Size and power of test statistics for	gene correlation in 2×2 contingency tables • Mannan, M.; Nassar, R.	95 BiomtrcJ 37 409-433
Simple test statistics for major	gene detection: A numerical comparison • Le Roy, Pascale; Elsen, J. M.	92 ThApGene 83 635-644
Dizier, M.-H. et al. Interactive effect of two candidate	genes in a disease: Extension of the marker-association-segregation χ^2 method •	94 AJHuGene 55 1042-1049
for the positional cloning of human disease	genes • Boehnke, Michael Limits of resolution of genetic linkage studies: Implications	94 AJHuGene 54 379-390
Estimation, variance and optimal sampling of	gene diversity. I. Haploid locus • Pons, O.; Petit, R. J.	95 ThApGene 90 462-470
Mapping disease	genes: Family-based association studies • Thomson, Glenys	95 AJHuGene 57 487-498
Estimation of	gene flow from F-statistics • Cockerham, C. Clark; Weir, B. S.	93 Evolutn 47 855-863
	Gene flow technique for stage-structured populations • Muralidharan, K.; Jain, J. P.	95 BiomtrcJ 37 229-234
cautionary tale about the relationships among race,	genes, and IQ • Devlin, B.; Fienberg, S. E.; Resnick, D. P. et al. Wringing the bell curve: A	95 Chance 8/3 27-36
Maximum-likelihood estimation of	gene location by linkage disequilibrium • Hill, W. G.; Weir, B. S.	94 AJHuGene 54 705-714
Detecting marker inconsistencies in human	gene mapping • Ott, Jurg	93 HumHrdty 43 25-30
Bayesian inference in multipoint	gene mapping • Stephens, D. A.; Smith, A. F. M.	93 AnHuGene 57 65-82
methods and optimal experimental design for	gene mapping by radiation hybrids • Lange, K.; Boehnke, M. Bayesian	92 AnHuGene 56 119-144
Likelihood methods for locating disease	genes in nonequilibrium populations • Kaplan, N. L.; Hill, W. G.; Weir, B. S.	95 AJHuGene 56 18-32
use of mixture models to detect effects of major	genes on quantitative characters in a plant breeding experiment • Jiang, C. et al. The	94 Genetcs 136 383-394
Disease susceptibility	genes and the sib-pair method: A review of recent methodology • Shah, S.; Green, J. R.	94 AnHuGene 58 381-395
Proportion of	genes survived in offspring conditional on inheritance of flanking markers • Guo, Sun-Wei	94 Genetcs 138 953-962
quencies and geographic subdivision to reconstruct	gene trees within a species: Molecular variance parsimony • Excoffier et al. Using allele fre-	94 Genetcs 136 343-359
On estimating the linkage of marker	genes to viability genes controlling inbreeding depression • Fu, Y. B.; Ritland, K.	94 ThApGene 88 925-932
Tavaré, Simon Unrooted	genealogical tree probabilities in the infinitely-many-sites model • Griffiths, R. C.;	95 MaBisci 127 77-98
O'Connell, Neil The	genealogy of branching processes and the age of our most recent common ancestor •	95 AdvAppPr 27 418-442
On transforming an index for	generalised bandit problems • Glazebrook, K. D.; Greatrix, S.	95 JAppProb 32 168-182
negative numbers: Application of the analysis of a	generalised Burt sub-table (French) • Nodjiram, S. Analysis of a contingency table with	95 CAnDnnes 20 111-120
	Generalised components of variance models • McGilchrist, C. A.; Ragunathan, T. E.	95 CommStA 24 349-361
Snyman, J. L. J. A note on the	generalised cross-validation criterion in linear model selection • Venter, J. H.;	95 Biomtrka 82 215-219
	Generalised geometric random variables • Kinney, John J.; Kepner, James L.	95 TeachgSt 17/1 7-9
of BLUP, ML, REML estimation methods for	generalised linear mixed models • McGilchrist, C. A.; Yau, K. K. W. The derivation	95 CommStA 24 2963-2980
Norman E.; Lin, Xihong Bias correction in	generalised linear mixed models with a single component of dispersion • Breslow,	95 Biomtrka 82 81-91
On the use of a working correlation matrix in using	generalised linear models for repeated measures • Crowder, Martin	95 Biomtrka 82 407-410
Second order asymptotics for score tests in	generalised linear models • Cribari-Neto, Francisco; Ferrari, Silvia L. P.	95 Biomtrka 82 426-432
A. J.; Taylor, P. G. Tight bounds on the sensitivity of	generalised semi-Markov processes with a single generally distributed lifetime • Coyle,	95 JAppProb 32 63-73
Construction of bivariate distributions by a	generalised trivariate reduction technique • Lai, C. D.	95 StPrLet 25 265-270
Penrose, Mathew D.	Generalised two-sample U-statistics and a two-species reaction-diffusion model •	95 StocProc 55 57-64
Premium adjustment by	generalized adaptive exponential smoothing • Herkenrath, U.	94 InsrneME 15 203-217
	Generalized additive models for medical research • Hastie, Trevor; Tibshirani, Robert	95 StMedRes 4 187-196
Logistic regression versus a	generalized additive model: A practical example (French) • Cans, C.; Lavergne, C.	95 RStApp 43/2 77-90
type Sturm-Liouville problems • Gregory, J. et al. A	generalized approximation theory for quadratic forms: Application to randomized spline	95 JTherPrb 8 703-715
spline type focal/conjugate point ... • Gregory et al. A	generalized approximation theory for quadratic forms II: Application to randomized	95 JTherPrb 8 963-971
Multivariate simultaneous	generalized ARCH • Engle, Robert F.; Kroner, Kenneth F.	95 EconTher 11 122-150
Sample path behaviour in connection with	generalized arcsine laws • Bertoin, Jean	95 ProbTher 103 317-327
Schmuland, B. A robust	generalized Bayes estimator of a multivariate normal mean • Karunamuni, R. J.;	95 MaMethSt 4 472-482
Prasad, Jagdish A note on	generalized binary proper efficiency-balanced block designs • Kherwa, G. Ram;	95 CommStA 24 2797-2801
	Generalized binomial regression model • Famoye, Felix	95 BiomtrcJ 37 581-594
error • Singh, R. Karan; Misra, Sheela; Pandey, S. K. A	generalized class of estimators in linear regression models with multivariate-t distributed	95 StPrLet 23 171-178
model • Singh, R. K.; Agnihotri, M.; Verma, S. A	generalized class of estimators utilizing Lindley-like mean correction in linear regression	94 Metron 52/3 147-160
not normal" (V23 ... • Singh et al. Correction to "A	generalized class of shrinkage estimators in linear regression when disturbances are	95 CommStA 24 859-859
Computational aspects of Sundt's	generalized class • Panjer, Harry H.; Wang, Shaun	95 AstinB 25 5-18
Dynamic scheduling with convex delay costs: The	generalized $c\mu$ rule • van Mieghem, Jan A.	95 AnlsApPr 5 809-833
towards the comparison of profiles using a	generalized concentration coefficient (French) • Tricot, J.-M.; Latour, R. An approach	95 RStApp 43/3 35-53
A cautionary note on the use of	generalized Cook-type measures • Fung, Wing K.	95 CmpStDA 19 321-326
The	generalized covariation process and Ito formula • Russo, Francesco; Vallois, Pierre	95 StocProc 59 81-104
Ren, Jian-Jian	Generalized Cramér-von Mises tests of goodness of fit for doubly censored data •	95 AnInStMa 47 525-549
Step-down test for	generalized dispersion • Gupta, S. Das	94 CalcutSt 44 223-227
A feasibility study • Bérard, H. Use of the	generalized edit and imputation system for the Canadian Survey of Family Expenditures:	94 ASAProSoSt 178-183
multiplication. Part III. The use of asymmetric and	generalized eigenvalue equations • Alsberg ... by the method of postponed basis matrix	95 Chemolab 30 223-225
Poisson approximation for the final state of a	generalized epidemic process • Lefèvre, Claude; Utev, Sergey	95 AnlsProb 23 1139-1162
	Generalized estimating equation	
design • Have, Thomas R. Ten; Chinchilli, Vernon M.	The analysis of bioequivalence with respect to TMAX under a 2×2 crossover	95 JBiophSt 5 185-199
of missing data • Robins, J. M.; Rotnitzky, A. et al.	Analysis of semiparametric regression models for repeated outcomes in the presence	95 JASA 90 106-121
Ying, Z.	Analysis of transformation models with censored data • Cheng, S. C.; Wei, L. J.;	95 Biomtrka 82 835-845
Kim, KyungMann; Lipsitz, Stuart R.	Analyzing bivariate ordinal data using a global odds ratio • Williamson, John M.;	95 JASA 90 1432-1437
tions • Gange, S. J.; Linton, K. L. P.; Scott, A. J. et al.	A comparison of methods for correlated ordinal measures with ophthalmic applica-	95 StatMed 14 1961-1974
Lefkopoulou, Myrto; Ryan, Louise M.	Efficiency and power of tests for multiple binary outcomes • Legler, Julie M.;	95 JASA 90 680-693
Chen, James J.; George, E. Olusegun	Estimating variance functions in developmental toxicity studies • Bowman, Dale;	95 Biomtrcs 51 1523-1528
Olusegun; Bowman, Dale	A full likelihood procedure for analysing exchangeable binary data • George, E.	95 Biomtrcs 51 512-523
algorithm • Gange, Stephen J.	Generating multivariate categorical variates using the iterative proportional fitting	95 AmerStat 49 134-138
R.; Medendorp, Sharon V.	Latent variable models for clustered ordinal data • Qu, Yinsheng; Piedmonte, Marion	95 Biomtrcs 51 268-275
Li, Bing	Projected partial likelihood and its application to longitudinal data • Murphy, Susan;	95 Biomtrka 82 399-406
Fitzmaurice, Garrett M.; Laird, Nan M.	Regression models for a bivariate discrete and continuous outcome with clustering •	95 JASA 90 845-852
Fitzmaurice, G. M.; Molenberghs, G.; Lipsitz, S. R.	Regression models for longitudinal binary responses with informative drop-outs •	95 JRSS-B 57 691-704
and developmental toxicity studies • Bowman et al.	A saturated model for analyzing exchangeable binary data: Applications to clinical	95 JASA 90 871-879
Robins, James M.; Rotnitzky, Andrea	Semiparametric efficiency in multivariate regression models with missing data •	95 JASA 90 122-129
observations • May, Warren L.; Johnson, William D.	Some applications of the analysis of multivariate normal data with missing	95 JBiophSt 5 215-228
et al. Heterosexual transmission of HIV analysed by	generalized estimating equations • Fielding, K. L.; Brettle, R. P.; Gore, S. M.; O'Brien, F.	95 StatMed 14 1365-1378
Shin, Min-Woong A practical extension of the	generalized estimating equation approach for longitudinal data • Park, Taesung;	95 CommStA 24 2561-2579
Applications to the analysis of ... • Albert et al. A	generalized estimating equations approach for spatially correlated binary data:	95 Biomtrcs 51 627-638
Small sample characteristics of	generalized estimating equations • Gunsolley, J. C.; Getchell, C.; Chinchilli, V. M.	95 CommStB 24 869-878
between parent and adolescent marijuana use via	generalized estimating equation methodology • Duncan et al. ... of the relationship	95 MultBehR 30 317-339
Särndal, C. E. Methodological principles for a	generalized estimation system at Statistics Canada • Estevao, V.; Hidiroglou, M. A.;	95 JOffStat 11 181-204
Ananda, Malwane M. A.	Generalized F-tests for unbalanced nested designs under heteroscedasticity •	95 AnInStMa 47 731-742
The	generalized fluctuation test: A unifying view • Kuan, Chung-Ming; Hornik, Kurt	95 EcomtrRv 14 135-161

	Generalized 4-profiles of Hadamard matrices • Lin, C.; Wallis, W. D.; Zhu, L.	93 JCISS 18 397-400
Almhana, J. The	generalized Gamma distribution: Its hazard rate and stress-strength model • Pham, T.;	95 IEEEReli 44 392-397
filters for a discrete-time nonlinear system with	generalized Gaussian white noise • Ferrante, M.; Giummolé, F. Finite dimensional	95 Stochatx 53 195-211
gle and product moments of order statistics from a	generalized half logistic distribution with applications to ... • Balakrishnan et al. ... for sin-	95 JStCmpSm 52 385-398
Maria L. R.	Generalized Hellinger transforms as information measures • Goel, Prem K.; Padilla,	94 ASAProBays 78-83
	Generalized inverse	
	On group inverses of matrices with order and rank perturbations • Rao, P. S. S. N. V. P.	94 CalcutSt 44 209-222
	Generalized Laplacian approximations in Bayesian inference • Hsu, John S. J.	95 CandJSt 23 399-410
On the cluster set problem for the	generalized law of the iterated logarithm in Euclidean space • Einmahl, Uwe	95 AnlsProb 23 817-851
	Generalized least squares	
with unknown covariances • Frick, H.	Comparing trials with multiple outcomes: The multivariate one-sided hypothesis	95 BiomtrcJ 37 909-917
under general restriction • Shiraishi, T.; Konno, Y.	On construction of improved estimators in multiple-design multivariate linear models	95 AnInStMa 47 665-674
	Curvature measures for multiresponse regression models • Guay, M.	95 Biomtrka 82 411-417
tions on structural equation modeling ... • Ding et al.	Effects of estimation methods, number of indicators per factor, and improper solu-	95 StrEqMod 2 119-144
heterogeneous treatment effect • Hsieh, Fushing	The empirical process approach for semiparametric two-sample models with	95 JRSS-B 57 735-748
Lee, Sik-Yum; Poon, Wai-Yin	Estimation of factor scores in a two-level confirmatory factor analysis model •	95 CmpStDA 20 275-284
Pantula, Sastry G.	Estimation of nonlinear random coefficient models • Ramos, Rogelio Q.;	95 StPrLet 24 49-56
matrix is singular • Takada, H.; Ullah, A.; Chen, Y.-M.	Estimation of the seemingly unrelated regression model when the error covariance	95 JAppStat 22 517-530
	Local influence in structural equation models • Cadigan, N. G.	95 StrEqMod 2 13-30
the notation of the reticular action mode • Toyoda	A new identification rule and an estimator for the simultaneous equation model using	94 Behvmtrk 21 163-175
matrix • Chaturvedi, Anoop	A note on the Stein rule estimation in linear models with nonscalar error covariance	95 SankhyaB 57 158-165
Palma, Daniela	Optimal sampling designs for dependent spatial units • Benedetti, Roberto;	95 EnvrMtrc 6 101-114
	Polynomial measurement error modeling • Moon, Myung-Sang; Gunst, Richard F.	95 CmpStDA 19 1-21
Rao, J. N. K.; Wang, Song-Gui	On the power of F tests under regression models with nested error structure •	95 JMultiAn 53 237-246
	A regression approach for estimating the center of symmetry • Hsieh, Fushing	95 StPrLet 22 157-160
Ofir, Chezy; Khuri, André I.	Replicated paired comparisons at the individual level • Bechtel, Gordon G.;	95 BJMaStPs 48 115-127
system • Lee, Byung-Joo	Seemingly unrelated regression on the autoregressive AR(p) singular equation	95 EcomtrRv 14 65-74
Robins, James M.; Rotnitzky, Andrea	Semiparametric efficiency in multivariate regression models with missing data •	95 JASA 90 122-129
Lin, Guan-Chyun	Stochastic regression model with heteroscedastic disturbance • Chang, Der-Shin;	95 AnInStMa 47 351-369
relations with a comprehensive ... • Muthén et al.	Technical aspects of Muthén's LISCOMP approach to estimation of latent variable	95 Psymtrka 60 489-503
mous variables • Lee, S.-Y.; Poon, W.-Y.; Bentler, P. M.	A two-stage estimation of structural equation models with continuous and polyto-	95 BJMaStPs 48 339-358
Uyenoyama, March K. A	generalized least-squares estimate for the origin of sporophytic self-incompatibility •	95 Genetcs 139 975-992
Lee, Sik-Yum A geometric approach of the	generalized least-squares estimation in analysis of covariance structures • Wang, S. J.;	95 StPrLet 24 39-47
observations, unequal ... • Isogawa Slope-range and	generalized least squares: Estimation of the structural relationship with replicated	95 JJapnStS 25 57-70
Srivastava, V. K. et al. Efficiency properties of feasible	generalized least squares estimators in SURE models under non-normal disturbances •	95 JEconmtx 66 99-121
Statistical properties of the ordinary least-squares,	generalized least-squares, and minimum-evolution methods of ... • Rzhetsky et al.	92 JMolEvol 35 367-375
On the numerical implementation of the	generalized least squares procedure for ARMA estimation • Salau, M. O.	95 CommStB 24 111-130
are spatially dependent and the analysis uses	generalized least-squares • Martin, R. J. Some results on randomization when observations	95 CommStA 24 2199-2210
Siegmund, D.; Venkatraman, E. S. Using the	generalized likelihood ratio statistic for sequential detection of a change-point •	95 AnlsStat 23 255-271
tained from quantile spline fits ... • O'Connell et al. A	generalized linear classification model with a smooth link function and predictors ob-	95 JStPlInf 47 153-164
	Generalized linear model	
	Adapting for the missing link • Weisberg, S.; Welsh, A. H.	94 AnlsStat 22 1674-1700
Niggemeyer, B.; Radtke, M.; Reich, A.	Applications of risk theory and multivariate analysis in insurance practice •	95 AplStMDA 11 231-244
Dey, Dipak K.	Bayesian analysis of outlier problems using divergence measures • Peng, Fengchun;	95 CandJSt 23 199-213
pital rather than patient characteristics? • Silber et al.	Comparing the contributions of groups of predictors: Which outcomes vary with hos-	95 JASA 90 7-18
markers • Jansen, R. C.	A general mixture model for mapping quantitative trait loci by using molecular	92 ThApGene 85 252-260
Yang, Ching-Ching	Generalized regression trees • Chaudhuri, Probal; Lo, Wen-Da; Loh, Wei-Yin;	95 StSinica 5 641-666
insurance • Renshaw, A. E.; Haberman, S.	On the graduations associated with a multiple state model for permanent health	95 InsrneME 17 1-17
Stefanski, L. A.; Buzas, J. S.	Instrumental variable estimation in binary regression measurement error variables •	95 JASA 90 541-550
Chaudhuri, Probal; Dewanji, Anup	On a likelihood-based approach in nonparametric smoothing and cross-validation •	95 StPrLet 22 7-15
(French) • Cans, C.; Lavergne, C.	Logistic regression versus a generalized additive model: A practical example	95 RStApp 43/2 77-90
van der Werf, J. H. J.	Maximizing selection efficiency for categorical traits • Meuwissen, T. H. E.; Engel, B.;	95 JAnmlSci 73 1933-1939
• Matthews, G. B.; Crowther, N. A. S.	A maximum likelihood estimation procedure when modelling in terms of constraints	95 SoAfrStJ 29 29-50
Kanter, Marek	How to measure the error of an M-estimate and report it conservatively •	95 JMultiAn 54 32-76
coprotheres • Piper, S. E.	A model of the ring-recovery reporting process for the Cape Griffon Gyps	95 JAppStat 22 641-659
	Performance of a Bartlett-type modification for the deviance • Cordeiro, Gauss M.	95 JStCmpSm 51 385-403
	Poisson regression analysis in clinical research • Kianifard, Farid; Gallo, Paul P.	95 JBiophSt 5 115-129
els with selection on first event • Zwinderman et al.	Prediction of the next hypercalcemia free period: Application of random effect mod-	95 StNeerla 49 310-323
• Vos, Paul W.	Quasi-likelihood or extended quasi-likelihood? An information-geometric approach	95 AnInStMa 47 49-64
	Semiparametric regression functionals • LeBlanc, Michael; Crowley, John	95 JASA 90 95-105
Müller, Hans-Georg; Zhao, Peng-Liang	On a semiparametric variance function model and a test for heteroscedasticity •	95 AnlsStat 23 946-967
expressed as proportions or ... • Dyke et al.	Sensitivity (stability) analysis of multiple variety trials, with special reference to data	95 ExpAgric 31 75-87
Sutradhar, Brajendra C.; Das, Kalyan	Tests for overdispersion in longitudinal data in the presence of outliers •	95 SankhyaA 57 299-315
le Cessie, Saskia; van Houwelingen, Hans C.	Testing the fit of a regression model via score tests in random effects models •	95 Biomtrcs 51 600-614
A mixture likelihood approach for	generalized linear models • Wedel, Michel; DeSarbo, Wayne S.	95 JClass 12 21-55
of "Multivariate statistical modelling based on	generalized linear models" • Fahrmeir, L.; Tutz, G. (Auth); Glosup, Jeffrey (Rev) Review	95 Technmcs 37 350-351
Correction for covariate measurement error in	generalized linear models – A bootstrap approach • Haukka, Jari K.	95 Biomtrcs 51 1127-1132
Overdispersion diagnostics for	generalized linear models • Lambert, Diane; Roeder, Kathryn	95 JASA 90 1225-1236
Unspecified link functions in	generalized linear models (French) • Bonneu, M.	95 RStApp 43/2 49-66
Tests of homogeneity for	generalized linear models • Jacqmin-Gadda, Hélène; Commenges, Daniel	95 JASA 90 1237-1246
Multiplicative interaction in	generalized linear models • van Eeuwijk, Fred A.	95 Biomtrcs 51 1017-1032
Second-order asymptotic efficiency of PMLE in	generalized linear models • Liang, Hua	95 StPrLet 24 273-279
E.; Wand, M. P. Local polynomial kernel regression for	generalized linear models and quasi-likelihood functions • Fan, Jianqing; Heckman, Nancy	95 JASA 90 141-150
Dynamic	generalized linear models and repeated measurements • Lindsey, J. K.; Lambert, P.	95 JStPlInf 47 129-139
Influence and residuals in restricted	generalized linear models • Paula, Gilberto A.	95 JStCmpSm 51 315-331
Review of "Nonparametric regression and	generalized linear models: A roughness penalty approach" • Green, P. J. et al.	95 JASA 90 1489-1489
Review of "Nonparametric regression and	generalized linear models: A roughness penalty approach" • Green, P. J. et al.	95 Technmcs 37 342-343
methods for efficient best subsets selection in	generalized linear models • Jovanovic, B. D.; Hosmer, D. W. et al. Equivalence of several	95 CmpStDA 20 59-64
totically efficient estimation in semiparametric	generalized linear models • Chen, Hung Asymp-	95 AnlsStat 23 1102-1129
One-sided tests in	generalized linear models with parallel regression lines • Paula, Gilberto A.; Sen, Pranab K.	95 Biomtrcs 51 1494-1501
Asymptotically unbiased estimation in	generalized linear models with random effects • Kuk, Anthony Y. C.	95 JRSS-B 57 395-407
Follmann, Dean; Wu, Margaret An approximate	generalized linear model with random effects for informative missing data •	95 Biomtrcs 51 151-168
	Generalized linear models with a small scale parameter • Dorkenoo, K. M.; Mathieu, J.-R.	95 MaOpfStS 27 107-113
Kaçranlar, Selahattin On the almost unbiased	generalized Liu estimator and unbiased estimation of the bias and MSE • Akdeniz, Fikri;	95 CommStA 24 1789-1797

	Generalized log-normal distributions with reliability application • Chen, Gemai	95 CmpStDA 19 309-319
Computation of probabilities of a	generalized log-series and related distributions • Ong, S. H.	95 CommStA 24 253-271
Han, JeongHye Statistical estimation for	generalized logit model of nominal type with bootstrap method • Cho, JoongJae;	95 JKoreaSt 24 1-18
Bowden, David C.; Kufeld, Roland C.	Generalized mark-sight population size estimation applied to Colorado moose •	95 JWildlMg 59 840-851
study of tests for non-nested models estimated by	generalized method of moments • Arkonac, Seyhan Z.; Higgins, Matthew L. A Monte Carlo	95 CommStB 24 745-763
W. E.; Foerster, S. R. Finite sample properties of the	generalized method of moments in tests of conditional asset pricing models • Ferson,	94 JFinEcon 36 29-55
On the	generalized negative binomial distribution • Consul, P. C.; Famoye, Felix	95 CommStA 24 459-472
Damaraju; Heiberger, Richard M. Variance balanced	generalized nested row-column designs with empty diagonal cells • Raghavarao,	93 JCISS 18 42-52
and stochastic Navier-Stokes equations with a	generalized noise • Flandoli, F. et al. Time discretization of Ornstein-Uhlenbeck equations	95 Stochatx 55 141-165
cumulative odds models for ordinal data: A	generalized non-linear model approach • Cox, Christopher Location-scale	95 StatMed 14 1191-1203
Asymptotic confidence bands for	generalized nonlinear regression models • Cox, Christopher; Ma, Guangqin	95 Biomtrcs 51 142-150
A	generalized notion of variation applied to Markov chains and Anosov maps • Loviscach, J.	95 ProbTher 103 553-570
A concept of	generalized order statistics • Kamps, Udo	95 JStPlInf 48 1-23
Review of "A concept of	generalized order statistics" • Kamps, U. (Auth); Plachky, D. (Rev)	95 StatDecn 13 402-403
dispersion • Mole, N.; Anderson, C. W. et al. A	generalized Pareto distribution model for high concentrations in short-range atmospheric	95 EnvrMtrc 6 595-606
	Generalized partial correlation coefficients (Italian) • Vicari, Donatella	94 Statstca 54 491-499
vectors • Kobus, Maria	Generalized Poisson distributions as limits of sums for arrays of dependent random	95 JMultiAn 52 199-244
Bivariate	generalized Poisson distribution with some applications • Famoye, Felix; Consul, P. C.	95 Metrika 42 127-138
Bootstrap based goodness of fit tests for the	generalized Poisson model • Henze, Norbert; Klar, Bernhard	95 CommStA 24 1875-1896
Verboon, Peter; Gabriel, K. Ruben	Generalized Procrustes analysis with iterative weighting to achieve resistance •	95 BJMaStPs 48 57-73
	Generalized quasilinearization of stochastic initial value problem • McRae, F. A.	95 StocAnAp 13 205-210
Adaptive linear list reorganization under a	generalized query system • Valiveti, R. S.; Oommen, B. J.; Zgierski, J. R.	95 JAppProb 32 793-804
Dubreuil, Guylaine; Tremblay, Johanne The use of	generalized raking procedures to improve the quality of small domain estimation •	94 ASASrvyRsr 293-298
anticipative calculus for functionals of a Gaussian	generalized random field • Albeverio, Sergio; Wu, Jiang-Lun Stochastic	95 Stochatx 54 317-337
Yang, Ching-Ching	Generalized regression trees • Chaudhuri, Probal; Lo, Wen-Da; Loh, Wei-Yin;	95 StSinica 5 641-666
	Generalized ridge regression estimators under the LINEX loss function • Ohtani, K.	95 StHefte 36 99-110
	Generalized ridge regression estimators under the LINEX loss function • Ohtani, K.	95 StHefte 36 99-110
Exact and approximate relative efficiency of	generalized ridge-type shrinkage estimator • Noor, Islam; Ahmad, Naseer	95 PakJStA 11 17-22
• Henderson et al. Insensitivity in discrete-time	generalized semi-Markov processes allowing multiple events and probabilistic service ...	95 AnlsApPr 5 78-96
coupling and performance sensitivity analysis of	generalized semi-Markov processes • Cao, Xi-Ren Event	95 AdvAppPr 27 741-769
Miyazawa, M. et al. On the structure of an insensitive	generalized semi-Markov process with reallocation and point-process input •	95 AdvAppPr 27 203-225
efficiency of the maximum likelihood estimator for a	generalized semiparametric model • Liang, Hua On Bahadur asymptotic	95 StSinica 5 363-371
Wright, Peter (Rev) Review of "A handbook of	generalized special functions for statistical and physical sciences" • Mathai, A. M. (Auth);	95 AstrlJSt 37 249-250
Dynamic asymptotic results for a	generalized star-shaped loss network • Graham, Carl; Méléard, Sylvie	95 AnlsApPr 5 666-680
Bootstrap approximation for	generalized U-processes (Chinese) • Zhang, Dixin; Li, Guoying	95 KToChina 39 1249-1252
	Generalized unimodality • Alamatsaz, M. H.	92 ProIrStC1 1-10
On M/G/1 queues with exhaustive service and	generalized vacations • Li, Huan; Zhu, Yixin	95 AdvAppPr 27 510-531
Improved estimation of	generalized variance and precision • Gupta, A. K.; Ofori-Nyarko, S.	95 MaOpfStS 26 99-109
Newbold, Paul A	generalized variance ratio test of ARIMA$(p,1,q)$ model specification • Miller, John P.;	95 JTimSrAn 16 403-413
Stochastic integration of processes with finite	generalized variations. I • Towghi, Nasser	95 AnlsProb 23 629-667
Moments of	generalized Wishart distributions • Wong, Chi Song; Liu, Dongsheng	95 JMultiAn 52 280-294
	Generating function	
•	• see also cumulant -, moment -, probability -	
the discrete probability law based on the empirical	generating function (two dimensional case) • Chergui, Jamal D. Estimation and tests of	94 RACEMdrd 88 245-286
queuing networks by numerically inverting their	generating functions • Choudhury et al. Calculating normalization constants of closed	95 JACM 42 935-970
125) • Sapra, Sunil K. Comment on "The moment-	generating function of the minimum of bivariate normal random variables" (V48 p124-	95 AmerStat 49 240-240
Cain, Michael Reply to comment on "The moment-	generating function of the minimum of bivariate normal random variables" •	95 AmerStat 49 240-240
rankings • Valz, P. D.; McLeod, A. I. et al. Cumulant	generating function and tail probability approximations for Kendall's score with tied	95 AnlsStat 23 144-160
La Corte, Aurelio A method for the	generation of correlated random processes • Andronico, Maurizio; Casale, Salvatore;	95 JCmpInT 3 9-18
Chen, James J.	Generation of over-dispersed and under-dispersed binomial variates • Ahn, Hongshik;	95 JCmpGrSt 4 55-64
Pseudorandom number	generation by nonlinear methods • Eichenauer-Herrmann, Jürgen	95 IntStRvw 63 247-255
[Algorithm AS 299]	Generation of simplex lattice points • Chasalow, Scott D.; Brand, Richard J.	95 ApplStat 44 534-545
Using neural networks to learn intractable	generative models • Hinton, Geoffrey E.; Dayan, Peter; Neal, Radford M.; Zemel, Richard S.	94 ASAProStCp 11-20
A.; Tommasini, L. et al. Comparison of random	generators for Monte Carlo estimates of Weibull parameters • Montanari, G. C.; Cavallini,	95 Metron 53 55-77
Marsaglia's monkey test for pseudorandom number	generators • Percus, Ora E.; Whitlock, Paula A. Theory and application of	95 ACMMdSim 5 87-100
S. P.; Samanta, P. K. Failure of emergency diesel	generators: A population analysis using empirical Bayes methods • Vesely, W. E.; Uryasev,	94 RelEngSS 46 221-229
	Genetic algorithm	
	Application of fuzzy logic control in industry • van der Wal, A. J.	94 FuzzySet 74 33-41
	Messy simulated annealing • Kvasnička, V.; Pospíchal, J.	95 JChemtrc 9 309-322
	Simulation optimization by genetic search • Yunker, James M.; Tew, Jeffrey D.	94 MaCmpSim 37 17-28
yarns • de Weijer, A. P.; Buydens, L. et al.	Spectral curve fitting of infrared spectra obtained from semi-crystalline polyester	95 Chemolab 28 149-164
neural networks, non-linear biased regression and a	genetic algorithm for dynamic model identification • Wise, B. M. et al. A comparison of	95 Chemolab 30 81-89
and other irregular features • Dorsey, R. E. et al.	Genetic algorithms for estimation problems with multiple optima, nondifferentiability,	95 JBES 13 53-66
Structure optimization of fuzzy neural network by	genetic algorithm • Ishigami, Hideyuki; Fukuda, Toshio; Shibata, Takanori; Arai, Fumihito	95 FuzzySet 71 257-264
Theoretical and empirical properties of the	genetic algorithm as a numerical optimizer • Jennison, Christopher; Sheehan, Nuala	95 JCmpGrSt 4 296-318
	Genetic algorithms: An overview • Mitchell, Melanie	95 Complxty 1 31-39
and outlier detection for non-linear models using	genetic algorithms • Vankeerberghen, P.; Smeyers-Verbeke, J. et al. Robust regression	95 Chemolab 28 73-87
tifactorial transmission • Li, Z.; Bonney, G. E. et al.	Genetic analysis combining path analysis with regressive models: The TAU model of mul-	94 HumHrdty 44 305-311
Biometrical	genetic analysis of the cotwin control design • Duffy, David L.	94 BehavGen 24 341-344
Dolan, C. V.; Molenaar, P. C. M. et al. Simultaneous	genetic analysis of means and covariance structure: Pearson-Lawley selection rules •	94 BehavGen 24 17-24
Analyzing twin resemblance in multisymptom data:	Genetic applications of a latent class model for symptoms of conduct ... • Eaves et al.	93 BehavGen 23 5-20
of quantitative trait loci • Eaves, Lindon J. Effect of	genetic architecture on the power of human linkage studies to resolve the contribution	94 Heredity 72 175-192
systems • Arunachalam, V. A	genetic basis behind combining ability and breeding values in mono- and di-genic	95 BiomtrcJ 37 217-228
Maccario, J. et al. Estimation of the proportion of	genetic cases in late-onset diseases: An application to Alzheimer's disease • Tzourio, C.;	95 StatMed 14 1353-1363
C. M.; Bretting, P.; Eberhart, S. A.; Taba, S. Statistical	genetic considerations for maintaining germ plasm collections • Crossa, J.; Hernandez,	93 ThApGene 86 673-678
Cornelius, P. L. Selection index based on	genetic correlations among environments • Van Sanford, D. A.; Pfeiffer, T. W.;	93 CropSci 33 1244-1248
Additive and non-additive	genetic correlations. I. Properties of sampling distributions • Hohls, T.; Clarke, G. P. Y.	95 BiomtrcJ 37 919-935
T.; Clarke, G. P. Y. et al. Additive and non-additive	genetic correlations. II. Application to data obtained from a 12×12 diallel cross • Hohls,	95 BiomtrcJ 37 937-943
analyses of diallel crosses: A tool for the	genetic dissection of neurobehavioral phenotypes • Crusio, Wim E. Bi- and multi-variate	93 BehavGen 23 59-68
and significance testing for Prevosti's estimate of	genetic distance • Sanchez, A.; Ocaña, J.; Utzet, F. Sampling theory, estimation,	95 Biomtrcs 51 1216-1235
familial disease data with variable age at onset and	genetic and environmental effects • Gauderman, W. James A method for simulating	95 StatComp 5 237-243
Waller, Niels G.; Reise, Steven P.	Genetic and environmental influences on item response pattern scalability •	92 BehavGen 22 135-152

Genotype-environment interaction (cont.)

	Multiplicative interaction in generalized linear models • van Eeuwijk, Fred A.	95 Biomtrcs 51 1017-1032
Lu, Hsiu Ying	PC-SAS program for estimating Hühn's nonparametric stability statistics •	95 AgronmyJ 87 888-891
and multiplicative interaction model for ... • Piepho	Robustness of statistical tests for multiplicative terms in the additive main effects	95 ThApGene 90 438-443
cultivar rank change • Crossa, J.; Cornelius, P. L. et al.	A shifted multiplicative model fusion method for grouping environments without	95 CropSci 35 54-62
incomplete block design (Polish) • Czajka, Stanisław	Statistical methods for the analysis of a series of multiple experiments carried out in	95 ListyBio 32 101-129
trials • Cornelius, P. L.; Seyedsadr, M.; Crossa, J.	Using the shifted multiplicative model to search for "separability" in crop cultivar	92 ThApGene 84 161-172
Alternative partitioning of the	genotype-by-environment interaction • Muir, W.; Nyquist, W. E.; Xu, S.	92 ThApGene 84 193-200
the use of diagnostic biplots in model screening for	genotype by environment tables • van Eeuwijk, F. A.; Keizer, L. C. P. On	95 StatComp 5 141-153
Forrestt The Elston-Stewart algorithm for continuous	genotypes and environmental factors • Elston, Robert C.; George, Varghese T.; Severtson,	92 HumHrdty 42 16-27
An algorithm for Monte Carlo estimation of	genotype probabilities on complex pedigrees • Lin, S.; Thompson, E.; Wijsman, E.	94 AnHuGene 58 343-357
Regression and AMMI analyses of	genotype × environment interactions: An empirical comparison • Yau, Sui K.	95 AgronmyJ 87 121-126
different ... • Hühn et al. Relationships between	genotype × environment interactions and rank orders for a set of genotypes tested in	93 ThApGene 86 943-950
required to obtain sufficient progeny with a desired	genotype at two quantitative trait loci • Schmitz et al. ... the minimum sample size	93 ThApGene 87 136-144
Hazard and probabilities of unknown	genotypes • Jeanpierre, M.	92 AnHuGene 56 325-330
of an exact test of Hardy-Weinberg proportions of	genotypic data at a multiallelic locus • Chakraborty, Ranajit; Zhong, Yixi Statistical power	94 HumHrdty 44 1-9
Hill et al. Estimation of inbreeding coefficients from	genotypic data on multiple alleles, and application to estimation of clonality in ... •	95 GeneRsr 65 53-61
al. An efficient algorithm to compute the posterior	genotypic distribution for every member of a pedigree without loops • Fernando, R. L. et	93 ThApGene 87 89-93
loci: Guidelines for the selection of sibling pairs for	genotyping • Eaves, Lindon; Meyer, Joanne Locating human quantitative trait	94 BehavGen 24 443-456
Japan's financial deregulation and linkage of the	Gensaki and Euroyen deposit markets • Lin, Wen-Ling	95 JApEcmtx 10 447-467
Free energy for Brownian and	geodesic homology • Le Jan, Y.	95 ProbTher 102 57-61
in a five-dimensional shape space: Looking at	geodesics • Kendall, David G. How to look at objects	95 AdvAppPr 27 35-43
map comparison problem: Tests for the overlap of	geographic boundaries • Jacquez, Geoffrey M. The	95 StatMed 14 2343-2361
Comment on "Mapping and	geographic display of data" • White, Andrew A.	95 StatMed 14 697-699

Geographic information system

trophes • Myers, W. L.; Johnson, G. D.; Patil, G. P.	Rapid mobilization of spatial/temporal information in the context of natural catas-	94 ASAProStGr 25-31
female black brant with multi-state modeling and	geographic information systems • Lindberg, M. S. et al. Estimating nest site fidelity of adult	95 JAppStat 22 725-735
Andes et al. Linking public health data using	geographic information system techniques: Alaskan community characteristics and ... •	95 StatMed 14 481-490
control selection in childhood cancer studies: The	geographic proximity and demographics within matched sets • Sakkinen et al. ... for	95 AJPubHlh 85 555-557
	Geographic sampling with a distance threshold • Gallego, F. Javier; Fuentes, Montserrat	94 QStMaA 16 93-101
• Excoffier et al. Using allele frequencies and	geographic subdivision to reconstruct gene trees within a species: Molecular variance ...	94 Genetcs 136 343-359
	Geographical data analysis • Walford, Nigel	95 Wiley:NY:UK 446p
Elliott, P.; Cuzick, J.; English, D. et al. Review of	"Geographical and environmental epidemiology: Methods for small-area studies" •	95 StMedRes 4 184-184
Exploratory tools for disease mapping in	geographical epidemiology • Cislaghi, C.; Biggeri, A.; Braga, M.; Lagazio, C.; Marchi, M.	95 StatMed 14 2363-2381
Cellular automata applications to	geographical evolution simulation (Italian) • Stefano, Benati	94 QStMaA 16 105-115
and smart spatial pattern exploration tools for	geographical information systems applications • Openshaw, Stan Developing automated	95 Ststcian 44 3-16
Rust, Roland T.; Donthu, Naveen Capturing	geographically localized misspecification error in retail store choice models •	95 JMarRsr 32 103-110
Analysing	geographically-related disease data • Cliff, A. D.	95 StMedRes 4 93-101
structures • Wang, S. J.; Lee, Sik-Yum A	geometric approach of the generalized least-squares estimation in analysis of covariance	95 StPrLet 24 39-47
or extended quasi-likelihood? An information-	geometric approach • Vos, Paul W. Quasi-likelihood	95 AnInStMa 47 49-64
Haines, Linda M. A	geometric approach to optimal design for one-parameter non-linear models •	95 JRSS-B 57 575-598
A	geometric approach to spatial time series analysis (Italian) • Benedetti, Roberto	94 QStMaA 16 27-43
Dong, Jianping; Simonoff, Jeffrey S. A	geometric combination estimator for d-dimensional ordinal sparse contingency tables •	95 AnlsStat 23 1143-1159

Geometric distribution

Ong, S. H.	Computation of probabilities of a generalized log-series and related distributions •	95 CommStA 24 253-271
k successes • Aki, Sigeo; Hirano, Katuomi	Joint distributions of numbers of success-runs and failures until the first consecutive	95 AnInStMa 47 225-235
tie for first place • Baryshnikov, Y.; Eisenberg, B. et al.	A necessary and sufficient condition for the existence of the limiting probability of a	95 StPrLet 23 203-209
A characterization of a bivariate	geometric distribution • Sun, Kai; Basu, Asit P.	95 StPrLet 23 307-311
Multivariate exponential and	geometric distributions with limited memory • Marshall, Albert W.; Olkin, Ingram	95 JMultiAn 53 110-125
Lee, Chanho On	geometric ergodicity of nonlinear autoregressive models • Bhattacharya, Rabi;	95 StPrLet 22 311-315
Chapman, G. R.; Chen, G.; Kim, P. T. Assessing	geometric integrity through spherical regression techniques (Disc: p191-220) •	95 StSinica 5 173-191
Characteristic properties of a bivariate	geometric law • Nair, N. Unnikrishnan; Asha, G.	94 JIndnStA 32 111-116
	Geometric Markov chains • Rising, William	95 JAppProb 32 349-374
Bonham, Charles D. Least squares regression vs.	geometric mean regression for ecotoxicology studies • Angleton, George M.;	95 AplMaCmp 72 21-32
Mean size-and-shapes and mean shapes: A	geometric point of view • Lee, Huiling	95 AdvAppPr 27 44-55

Geometric probability

	Estimates of some probabilities in multidimensional convex records • Kałuszka, M.	95 ApplMath 23 1-11
of excursion sets, with applications to ... • Worsley	Estimating the number of peaks in a random field using the Hadwiger characteristic	95 AnlsStat 23 640-669
Extremal problems for	geometric probabilities involving convex bodies • Bauer, Christina; Schneider, Rolf	95 AdvAppPr 27 20-34
plete graph and Sylvester's "Four point problem" of	geometric probability • Scheinerman, E. R. et al. The rectilinear crossing number of a com-	94 AmMaMnth 101 939-943
	Geometric Q charts for high quality processes (Disc: p316-343) • Quesenberry, Charles P.	95 JQualTek 27 304-315
	Geometric quality inspection • Chen, Gemai	94 ASAProPEng 50-55
Generalised	geometric random variables • Kinney, John J.; Kepner, James L.	95 TeachgSt 17/1 7-9
one • Pistone, G.; Sempi, C. An infinite-dimensional	geometric structure on the space of all the probability measures equivalent to a given	95 AnlsStat 23 1543-1561
Thomas, David W. Statistical	geometrical features for texture classification • Chen, Yan Qiu; Nixon, Mark S.;	95 PattrnRc 28 537-552
random shapes and point fields. Methods of	geometrical statistics" • Stoyan, D.; Stoyan, H. (Auth); Nagel, W. (Rev) Review of "Fractals,	95 BiomtrcJ 37 978-978
Nicol, David On the minimum of independent	geometrically distributed random variables • Ciardo, Gianfranco; Leemis, Lawrence M.;	95 StPrLet 23 313-326
Review of "Random	geometrically graph directed self-similar multifractals" • Olsen, L. (Auth); Taylor, S. J. (Rev)	95 Metrika 42 277-278
Mixture models: Theory,	geometry and applications • Lindsay, Bruce G.	95 IMathStat:CA 163p
Theory of	geometry and asymptotic inference in exponential nonlinear models • Wei, Bocheng	95 SciSPkng 24 1270-1278
Information	geometry of the EM and em algorithms for neural networks • Amari, Shun-ichi	95 NuralNet 8 1379-1408
(Rev) Review of "The	geometry of multivariate statistics" • Wickens, Thomas D. (Auth); Hershberger, Scott L.	95 StrEqMod 2 379-382
The role of	geometry in pairwise and mutual independence • Schwertman, Neil C.; Kiser, Terry L.	95 AmerStat 49 56-59
Investigating the	geometry of a p-dimensional data set • Dawkins, Brian P.	95 JASA 90 350-359
Infinite divisibility of sub-stable processes. Part I.	Geometry of sub-spaces of L_α-space • Misiewicz, Jolanta K.	95 StocProc 56 100-116
Review of "Differential	geometry and statistics" • Murray, Michael K.; Rice, John W. (Auth); Marriott, Paul (Rev)	95 BLndnMaS 27 619-620
unbalanced longitudinal data with application to	geophysical data • Boularan, J.; Ferre, L.; Vieu, P. A nonparametric model for	95 CmpStQ 10 285-298
turbed, natural shortleaf pine stands in northern	Georgia • Reich, R. M.; Czaplewski, R. L.; Bechtold, W. A. Spatial cross-correlation of undis-	94 EnvEcoSt 1 201-217
Ziegel, Eric R. (Rev) Review of "GSLIB:	Geostatistical software library and user's guide" • Deutsch, C.; Journel, A. (Auth);	95 Technmcs 37 126-126
Wayne P.; Lengnick, Laura L.; Hart, Galen F. Use of	geostatistical techniques in the design of an agricultural field experiment • Dulaney,	94 ASASrvyRsr 183-187
Quasi-linear	geostatistical theory for inversing • Kitanidis, Peter K.	95 WaterReR 31 2411-2419

	Gittins index	
Greatrix, S.	On transforming an index for generalised bandit problems • Glazebrook, K. D.;	95 JAppProb 32 168-182
Large deviations for Langevin spin	glass dynamics • Arous, G. B.; Guionnet, A.	95 ProbTher 102 455-509
surface methodology within the framework of	GLM • Khuri, André I. Response	93 JCISS 18 193-202
resolved fluorescence detection in chromatography.	Global analysis of emission decay surfaces • Maeder, Marcel; Molloy, Kirsten Jane Lifetime	95 Chemolab 28 99-107
On local and	global asymptotic normality • Pfanzagl, J.	95 MaMethSt 4 115-136
(Italian) • Polisicchio, M. The	global concentration measure ζ and the principle of transfers: An empirical analysis	93 GiornEco 52 291-308
Walker, Chip; Bernstein, Hank Conducting	global consumer research: Considerations and pitfalls • Mendelsohn, Susan L.;	94 ASASrvyRsr 1196-1198
Shi, Peide; Lin, Guoying	Global convergence rates of B-spline M-estimators in nonparametric regression •	95 StSinica 5 303-318
Nonexponential applications of a	global Cramér-Rao inequality • Gajek, Lesław; Kałuszka, Marek	95 MaOpfStS 26 111-122
A Bayesian technique for refining the uncertainty in	global energy model forecasts • Tschang, F. Ted; Dowlatabadi, Hadi	95 IntJFor 11 43-61
Doksum, K. et al. Nonparametric estimation of	global functionals and a measure of the explanatory power of covariates in regression •	95 AnlsStat 23 1443-1473
Correlation dimension estimates of	global and local temperature data • Wang, Qiang	95 JApMetrl 34 2556-2564
Optimal averaging for the determination of	global mean temperature: Experiments with model data • Weber, R. O.; Madden, R. A.	95 JClimate 8 418-430
Locating	global minima in optimization problems by a random-cost approach • Berg, Bernd A.	93 Nature 361 708-710
• Gao, S.; Smith, T. M. F. On the nonexistence of a	global nonnegative minimum bias invariant quadratic estimator of variance components	95 StPrLet 25 117-120
Analyzing bivariate ordinal data using a	global odds ratio • Williamson, John M.; Kim, KyungMann; Lipsitz, Stuart R.	95 JASA 90 1432-1437
A clustering method for	global optimization based on the kth nearest neighbour • Rotondi, Renata; Drappo, Silvia	95 StatComp 5 317-326
Shi, Peide; Chen, Xiru A note on the optimal	global rates of convergence of M-estimator for nonparametric regression (Chinese) •	95 ChinaAMA 15 171-176
The	global statistics for global environment problems • Takeuchi, Kei	95 SankhyaB 57 173-181
Csörgő, Miklós; Földes, Antónia; Révész, Pál	Global Strassen-type theorems for iterated Brownian motions • Csáki, Endre;	95 StocProc 59 321-341
of spatial patterns: A unified approach to local and	global structures • Thioulouse, J.; Chessel, D.; Champely, S. Multivariate analysis	95 EnvEcoSt 2 1-14
Sadao Measures of departure from	global symmetry for square contingency tables with ordered categories • Tomizawa,	95 Behvmtrk 22 91-98
Brajendra C. A	global test for the goodness of fit of Cox's proportional hazards model • Sutradhar,	94 JStatRsr 28 123-148
Heij, Christiaan	Global total least squares modeling of multivariable time series • Roorda, Berend;	95 IEEEAuCn 40 50-63
A	globally gated polling system with a dormant server • Borst, S. C.	95 PrEngInf 9 239-254
et al. The exact distribution of the OLS and	GLS estimators in regression with an integrated regressor and correlated ... • Hisamatsu	95 MaCmpSim 39 273-277
ity result for GMM estimation, with applications to	GLS prediction and conditional moment tests • Ahn, Seung C.; Schmidt, Peter A separabil-	95 EcomtrRv 14 19-34
Spyridon D. Alternative size corrections for some	GLS test statistics. The case of the AR(1) model • Magdalinos, Michael A.; Symeonides,	95 JEconmtx 66 35-59
Phillips, G. D. A.; Schipp, B. The bias of OLS,	GLS, and ZEF estimators in dynamic seemingly unrelated regression models • Kiviet, J. F.;	95 JEconmtx 69 241-266
Bahadur-Kiefer representations for	GM-estimators in autoregression models • Koul, Hira L.; Zhu, Zhiwei	95 StocProc 57 167-189
One-step	GM-estimators for orthogonal regression • Abdullah, Mokhtar B.	95 JStCmpSm 52 301-309
Seung C.; Schmidt, Peter A separability result for	GMM estimation, with applications to GLS prediction and conditional moment tests • Ahn,	95 EcomtrRv 14 19-34
• Sevcik, Karel; Brooker, Peter I. Application of the	gnostical theory of uncertain data to spatial estimation, identification and quantification	95 EnvrMtrc 6 457-463
A nonlinear approach to US	GNP • Potter, Simon M.	95 JApEcmtx 10 109-125
Perceptions of	God's role in the origin of life on earth • Núñez, Megan; Kearns, Brian; Ronen, Amit	95 Stats /13 28-29
parameters of diagnostic tests in the absence of a	gold standard • Joseph, L. et al. Bayesian estimation of disease prevalence and the	95 AJEpidm 141 263-272
Estimating diagnostic test accuracy using a "fuzzy	gold standard" • Phelps, Charles E.; Hutson, Alan	95 MedDscnM 15 44-57
The	'Golden Egg' • Kimber, Hilary	95 TeachgSt 17/2 34-37
tistical confidence: Application and evaluation for	golf course closing • Shipley et al. Rough set theory fuzzy belief functions related to sta-	95 StocAnAp 13 487-502
The GRIP: A performance statistic for tournament	golf • Agard, David B.	94 ASAProSpts 1-6
The compound	Gompertz distribution • Shafy, Mona Abdel; Osman, Magued I.	95 EgpStJ 39 89-105
K. Estimating the	Goodman, Keyfitz, Pullum kinship equations: An alternative procedure • Burch, Thomas	95 MaPopStd 5 161-170
	Goodness-of-fit	
mial populations: A unified study • Menéndez et al.	Asymptotic behaviour and statistical applications of divergence measures in multino-	95 StHefte 36 1-29
Bartoszewicz, Jarosław	Bahadur and Hodges-Lehmann approximate efficiencies of tests based on spacings •	95 StPrLet 23 211-220
when parameters are estimated • Kulinskaya, Elena	Coefficients of the asymptotic distribution of the Kolmogorov-Smirnov statistic	95 JNnparSt 5 43-60
estimated empirical process • Viollaz, Aldo J.	Convergence in weighted supremum norms of the Skorokhod representation of the	95 CommStA 24 2829-2839
	Data driven versions of Pearson's chi-square test for uniformity • Bogdan, Malgorzata	95 JStCmpSm 52 217-237
Mudholkar, G. S.; Srivastava, D. K.; Freimer, M.	The exponentiated Weibull family: A reanalysis of the bus-motor-failure data •	95 Technmcs 37 436-445
	Generalized log-normal distributions with reliability application • Chen, Gemai	95 CmpStDA 19 309-319
Kochar, Subhash; Robertson, Tim	Likelihood ratio tests for symmetry against one-sided alternatives • Dykstra, Richard;	95 AnInStMa 47 719-730
Böckenholt, Ulf	Mixture models for the analysis of repeated count data • van Duijn, Marijtje A. J.;	95 ApplStat 44 473-485
Hennemuth, R. C.	Modelling and analysis of recruitment distributions • Taillie, C.; Patil, G. P.;	95 EnvEcoSt 2 315-329
Li, Gang	Nonparametric likelihood ratio estimation of probabilities for truncated data •	95 JASA 90 997-1003
data • Li, Gang	On nonparametric likelihood ratio estimation of survival probabilities for censored	95 StPrLet 25 95-104
regression problem • Azari, A. S.; Tsai, Chih-Ling	Non-parametric regression approach for model checking on the two-phase	94 JAppStat 21 597-606
	A nonparametric test of the weakest-link principle • Wolstenholme, Linda C.	95 Technmcs 37 169-175
McKeague, Ian W.; Nikabadze, A. M.; Sun, Yanqing	On omnibus test for independence of a survival time from a covariate •	95 AnlsStat 23 450-475
Taneichi, Nobuhiro; Sekiya, Yuri	Power approximations of the test of homogeneity for multinomial populations •	95 JJapnStS 25 97-109
Saito, Takayuki	Relationships between redundancy analysis and canonical regression analysis •	95 JJapnStS 25 111-119
	On a shared allele test of random mating • Zhou, S.; Maller, R. A.; Speed, T. P.	95 AstrlJSt 37 61-72
	Smooth tests for the bivariate Poisson distribution • Rayner, J. C. W.; Best, D. J.	95 AstrlJSt 37 233-245
Pardo, L.; Morales, D.; Salicrú, M.	Some statistical applications of (r,s)-directed divergences • Menéndez, M. L.;	92 UtilMath 42 115-127
	Statistical inference for Voronoi images • Archambault, S.; Moore, M.	95 CandJSt 23 11-20
	A test of fit for lattice distributions • Epps, T. W.	95 CommStA 24 1455-1479
le Cessie, Saskia; van Houwelingen, Hans C.	Testing the fit of a regression model via score tests in random effects models •	95 Biomtrcs 51 600-614
Maller, R. A.; Zhou, Sean	Testing for the presence of immune or cured individuals in censored survival data •	95 Biomtrcs 51 1197-1205
mation • Zhu, L.-X.; Fang, K.-T.; Bhatti, M. I. et al.	Testing sphericity of a high-dimensional distribution based on bootstrap approxi-	95 PakJStA 11 49-65
distribution • Fleischer, Karlheinz	Two tests of pseudo random number generators for independence and uniform	95 JStCmpSm 52 311-322
H. B.; Sackrowitz, H. B.	Unbiased testing in exponential family regression • Cohen, Arthur; Kemperman, J.	94 AnlsStat 22 1931-1946
	Weighted concordance analysis • Tricot, Jean-Marie; Lepage, Yves	95 BiomtrcJ 37 727-743
Rose, Kenneth A. Model	goodness-of-fit analysis using regression and related techniques • Smith, Eric P.;	95 EcolgMdl 77 49-64
A global test for the	goodness of fit of Cox's proportional hazards model • Sutradhar, Brajendra C.	94 JStatRsr 28 123-148
Generalized Cramér-von Mises tests of	goodness of fit for doubly censored data • Ren, Jian-Jian	95 AnInStMa 47 525-549
tests of regularity, Farrell efficiency, and	goodness-of-fit • Färe, Rolf; Grosskopf, Shawna Nonparametric	95 JEconmtx 69 415-425
Stacked boxes: A new	goodness-of-fit graphical display • Sparks, Ross	95 NZndSt 30 48-61
	Goodness-of-fit in item response models • McDonald, Roderick P.; Mok, Magdalena M.-C.	95 MultBehR 30 23-40
A note on smooth tests of	goodness of fit for location-scale families • Boulerice, Bernard; Ducharme, Gilles R.	95 Biomtrka 82 437-438
A. G.	Goodness-of-fit measures in binary choice models (Disc: 117-120) • Windmeijer, Frank	95 EcomtrRv 14 101-116
	Goodness-of-fit for multivariate linear regression models • Haroun, Ahmed H.	94 EgpStJ 38 138-146
Cohen, Ayala; Barnett, Ofra Assessing	goodness-of-fit of parametric regression models for lifetime data – Graphical methods •	95 StatMed 14 1785-1795
Arcas, Antoni Empirical significance test of the	goodness-of-fit for some pyramidal clustering procedures • Capdevila, Carles;	95 Questo 19 123-130
A new family of	goodness-of-fit statistics for discrete multivariate data • Lorenzen, Gunter	95 StPrLet 25 301-307

asymptotic upper tail probabilities of chi-squared	goodness-of-fit statistics: Pearson's χ^2, likelihood ratio, and ... • Formann ... exact and	95 JStCmpSm 51 369-384
Song, Moon Sup A	goodness-of-fit test for the additive risk model with a binary covariate • Kim, Jinheum;	95 JKoreaSt 24 537-549
A general purpose approximate	goodness-of-fit test • Chen, Gemai; Balakrishnan, N.	95 JQualTek 27 154-161
Yohai, Victor A	goodness-of-fit test based on ranks for ARMA models • Ferretti, Nélida; Kelmansky, Diana;	95 CommStA 24 295-318
Wanrong; Zha, Wenxing Projection pursuit Neyman	goodness-of-fit test and its bootstrap approximation (Chinese) • Li, Guoying; Liu,	95 KToChina 39 678-681
Yuen, Kam C.	Goodness-of-fit tests for the Cox model via bootstrap method • Burke, Murray D.;	95 JStPlInf 47 237-256
of BIB spiraling-induced missing data patterns on	goodness-of-fit tests in factor analysis • Kaplan, David The impact	95 JEdStat 20 69-82
Bootstrap based	goodness of fit tests for the generalized Poisson model • Henze, Norbert; Klar, Bernhard	95 CommStA 24 1875-1896
moment and probability plot correlation coefficient	goodness-of-fit tests for the Gumbel distribution and impact of ... • Fill et al. L	95 WaterReR 31 225-229
probability • Basu, S.; Basu, A. Comparison of several	goodness-of-fit tests for the kappa statistic based on exact power and coverage	95 StatMed 14 347-356
Reiser, Mark; Lin, Ginnie A	goodness-of-fit test for the latent class model when expected frequencies are small •	94 ASAProSoSt 49-54
and specificity of the length bias • Shanmugam, R.	Goodness-of-fit test for length-biased data with discussion on prevalence, sensitivity	95 JStPlInf 48 277-290
Büning, H. (Rev) Review of "A new nonparametric	goodness-of-fit test for location alternatives" (German) • Weichselberger, A. (Auth);	95 Metrika 42 145-146
Bootstrapping a consistent nonparametric	goodness-of-fit test • Fan, Yanqin	95 EcomtrRv 14 367-382
Anderson, T. W.	Goodness-of-fit tests for probability distributions and spectral distributions •	95 PrEngInf 9 27-37
Carlo simulation of a data driven version of smooth	goodness-of-fit tests • Kallenberg, Wilbert C. M.; Ledwina, Teresa Consistency and Monte	95 AnlsStat 23 1594-1608
and adjustment of complex surveys on chi-squared	goodness-of-fit tests: Some Monte Carlo evidence • Aguirre-Torres, V. et al. The effect	94 ASASrvyRsr 602-607
Ashour, S. K. et al. The sampling distribution of some	goodness-of-fit test statistics using Burr distribution and censored samples •	94 EgpStJ 38 147-163
Williams, Calvin L. A multivariate	goodness-of-fit test for stochastically ordered distributions • Alam, Khursheed;	95 BiomtrcJ 37 945-956
A Cramér-von Mises type	goodness-of-fit test with asymmetric weight function • Rodríguez, Juan C.; Viollaz, Aldo J.	95 CommStA 24 1095-1120
Pesarin, Fortunato	Goodness of fit testing for ordered discrete distributions by resampling techniques •	94 Metron 52/3 57-71
timation of PCB contamination • Johnson, R. A. et al.	Goodness-of-fit using very small but related samples with application to censored data es-	95 EnvrMtrc 6 341-348
Forecasting the adoptions of new consumer durable	goods • Min, Chung-ki	94 ASAProBays 262-265
an algorithm in the theory of markets in indivisible	goods • Frieze, Alan; Pittel, Boris G. Probabilistic analysis of	95 AnlsApPr 5 768-808
the movement among staging areas of the barnacle	goose (Branta leucopsis) • Schwarz, Carl James; Ganter, Barbara Estimating	95 JAppStat 22 711-724
Bias in Canada	goose population size estimates from sighting data • Sheaffer, Susan E.; Jarvis, Robert L.	95 JWildlMg 59 464-473
recapture models: A case study on the lesser snow	goose • Viallefont, A.; Cooch, E. G.; Cooke, F. Estimation of trade-offs with capture-	95 JAppStat 22 847-861
interactions between statisticians in academia and	government • Fienberg, S. E. et al. The Committee on National Statistics: Fostering	95 IntStRvw 63 257-269
organization and their relationship to the census of	governments • Owens, Stephen D. A study in library structure and	94 ASAProGovt 64-69
statisticians who work for state/provincial or local	governments? • Gaines, Leonard M.; González, María E. Is there a place in ASA for	94 ASAProGovt 149-152
What the	government needs (Disc: p18-23) • Ross, N. Phillip	95 AmerStat 49 7-9
approach • Shelley, Mack C., II By-elections and	government stability in the United Kingdom and the United States: A transfer function	94 ASAProSoSt 71-75
Training	government statisticians in Zimbabwe - An update • Woodward, Mark	94 JOffStat 10 215-220
Comment on "The use of administrative records in	government statistics" • Swain, Larry	94 ASAProGovt 212-214
of family change: Secondary analysis of continuous	government surveys • Harrop, Anne; Plewis, Ian Two decades	95 JRSS-A 158 91-106
conditions to refusal ... • Kojetin et al. Response to a	government survey as political participation: The relation of economic and political	94 ASASrvyRsr 1236-1241
residual volume (RV) by a regression built from	Gower distances • Bartkowiak, Anna; Liebhart, Jerzy Estimation of the spirometric	95 BiomtrcJ 37 131-149
Neerchal, Nagaraj K. Classification of	grab samples using composite sampling: An improved strategy • Sengupta, Soma;	94 CalcutSt 44 195-208
model in the heterogeneous case of the general	graded response model • Samejima, Fumiko Acceleration	95 Psymtrka 60 549-572
properties of two common classes of IRT models for	graded responses • Andrich, David Distinctive and incompatible	95 ApPsychM 19 101-119
precision, and the nondichotomization of	graded responses (Disc: p27-46) • Andrich, David Models for measurement,	95 Psymtrka 60 7-26
A	gradient algorithm locally equivalent to the EM algorithm • Lange, Kenneth	95 JRSS-B 57 425-437
Vregelaar On computing objective function and	gradient in the context of least squares fitting a dynamic errors-in-variables model •	95 SIAMSSC 16 738-753
Urahama, K.; Furukawa, Y.	Gradient descent learning of nearest neighbor classifiers with outlier rejection •	95 PattrnRc 28 761-768
Likelihood ratio	gradient estimation for stochastic recursions • Glynn, Peter W.; L'Ecuyer, Pierre	95 AdvAppPr 27 1019-1053
a stationary non-equilibrium distribution for a non-	gradient interacting particle system • Kipnis, C. et al. Macroscopic properties of	95 AIHenrPB 31 191-221
The conjugate	gradient method for queueing networks • Moustafa, Magdi S.	95 AplMaCmp 71 289-294
Bruce H.; Tomer, Adrian Determination of selection	gradients using multiple regression versus structural equation models (SEM) • Pugesek,	95 BiomtrcJ 37 449-462
M. A nonparametric regression approach to syringe	grading for quality improvement • Nychka, D.; Gray, G.; Haaland, P.; Martin, D.; O'Connell,	95 JASA 90 1171-1178
Teaching	graduate service courses through case studies • Ramsey, Fred L.; Schafer, Daniel W.	94 ASAProStEd 27-28
Wilson, Rick L.; Hardgrave, Bill C. Predicting	graduate student success in an MBA program: Regression versus classification •	95 EdPsMeas 55 186-195
M. et al. Increasing statistical awareness through	graduate training: The joint program in survey methodology • Wallman, K.; Groves, R.	94 ASAProSoSt 39-44
Renshaw, A. E.; Haberman, S. On the	graduations associated with a multiple state model for permanent health insurance •	95 InsrneME 17 1-17
and updating of a prognostic model for kidney	graft survival • van Houwelingen, Hans C.; Thorogood, Jane Construction, validation	95 StatMed 14 1999-2008
Testing whether interpenetration is possible for the	grains of a random set • Streit, Franz	95 MaOpfStS 26 355-364
Distribution of the	grammatical forms in the Italian language (Italian) • Rizzi, Alfredo	94 Statstca 54 275-291
Hurley, Catherine	Grand tour and projection pursuit • Cook, Dianne; Buja, Andreas; Cabrera, Javier;	95 JCmpGrSt 4 155-172
Professor Clive W. J.	Granger: An interview for the International Journal of Forecasting • Teräsvirta, Timo	95 IntJFor 11 585-590
get when you combine a Pentium and a research	grant? • Jones, Albyn C. What do you	95 Chance 8/2 7-8
Federal block	grants: What do statisticians have to contribute? • Adams, John; Burnam, Audrey	95 Chance 8/4 40-42
	Graph	
	Percolation of arbitrary words in $\{0,1\}^N$ • Benjamini, Itai; Kesten, Harry	95 AnlsProb 23 1024-1060
Bruno	A spanning tree heuristic for regional clustering • Maravalle, Maurizio; Simeone,	95 CommStA 24 625-639
	• see also directed –, random –	
Chowdhury, S. Roy Linear	graphs for 2^{k-p} fractional factorial under split plot design • Aggarwal, M. L.;	94 JStatRsr 28 73-87
by finite distributive lattices and by directed acyclic	graphs • Andersson et al. ... between conditional independence models determined	95 JStPlInf 48 25-46
ergodicity of simulated annealing algorithms on a	graph (French) • Miclo, Laurent Remarks about the	95 StocProc 58 329-360
Review of "Random geometrically	graph directed self-similar multifractals" • Olsen, L. (Auth); Taylor, S. J. (Rev)	95 Metrika 42 277-278
Perfect domination in	graphs • Cockayne, E. J.; Hartnell, B. L.; Hedetniemi, S. T.; Laskar, R.	93 JCISS 18 136-148
of sums of random variables indexed by edges of	graphs • Karpelevich, F. I. On estimation of maxima	94 ThProbAp 39 696-702
Spectra of	graphs and fractal dimensions II • Telcs, András	95 JTherPrb 8 77-96
models constrained by a conditional independence	graph • de Falguerolles, A.; Jmel, S.; Whittaker, J. Correspondence analysis and association	95 Psymtrka 60 161-180
Percolation and minimal spanning forests in infinite	graphs • Alexander, Kenneth S.	95 AnlsProb 23 87-104
Mital, D. P. Pattern recognition by	graph matching using the Potts MFT neural networks • Suganthan, P. N.; Teoh, E. K.;	95 PattrnRc 28 997-1009
Becker, G. et al. An inhomogeneous state	graph model and application for a phased mission and tolerable downtime problem •	95 RelEngSS 49 51-57
straight-line representations of stationary random	graphs • Burton, Robert M.; Choi, Kyeong; Meester, Ronald Stationary	95 IndgMatN 6 267-277
The use of the component influence	graph to reduce the size of Markovian availability problems • Devooght, J.; Tombuyses, B.	94 RelEngSS 46 237-244
Eigenvalues and expansion of regular	graphs • Kahale, Nabil	95 JACM 42 1091-1106
Some simple	graphs for simultaneous comparisons • Marr, Ray L.	94 ASAProPEng 109-114
information administration guidelines for statistical	graphs • Rutchik, Robert H. Energy	94 ASAProStGr 48-53
et al. The rectilinear crossing number of a complete	graph and Sylvester's "Four point problem" of geometric probability • Scheinerman, E. R.	94 AmMaMnth 101 939-943
or extended trees • Carroll, J. D.; Corter, J. E. A	graph-theoretic method for organizing overlapping clusters into trees, multiple trees,	95 JClass 12 283-313

privacy and efficiency. In memory of Roger A.	Herriot • Alexander, Lois Reinventing the population census: Balancing	94 ASAProGovt 1-2
Obituary: Amato	Herzel, 1923-1994 • Leti, Giuseppe	94 Metron 52/1 5-15
Li, Ker-Chau A study of the method of principal	Hessian direction for analysis of data from designed experiments • Cheng, Ching-Shui;	95 StSinica 5 617-639
	Heterogeneity	
Halloran, M. E.; Longini, I. M., Jr; Watelet, L.	Estimation of vaccine efficacy in non-randomly mixing populations • Haber, M.;	95 BiomtrcJ 37 25-38
Lix, Lisa M.	Robust and powerful nonorthogonal analyses • Keselman, H. J.; Carriere, K. C.;	95 Psymtrka 60 395-418
Louis, Thomas A.	Survival analysis using a scale change random effects model • Anderson, Jon E.;	95 JASA 90 669-679
intervals of the M/M/2/N queue with balking and	heterogeneity • Abou-El-ata, M. O.; Hariri, A. M. A. Point estimation and confidence	95 AJMMgSci 15 35-55
A capture-recapture model with	heterogeneity and behavioural response • Norris, James L., III; Pollock, Kenneth H.	95 EnvEcoSt 2 305-313
M.	Heterogeneity biases, distributional effects, and aggregate consumption • Fortin, Nicole	95 JApEcmtx 10 287-311
Implications for asymmetry, nonproportionality, and	heterogeneity in brand switching from piece-wise exponential mixture ... • Wedel et al.	95 JMarRsr 32 457-462
of a class of balanced two-way elimination of	heterogeneity designs • Siatkowski, Idzi Some characterizations	94 ListyBio 31 27-33
Orthogonality in a two-way elimination of	heterogeneity design • Siatkowski, I.	93 UtilMath 43 225-233
On the E-optimality of some two-way elimination of	heterogeneity designs • Brzeskwiniewicz, Henryk	95 StPrLet 25 9-13
of two preparations in two-way elimination of	heterogeneity designs with multivariate responses • Hanusz, Zofia Relative potency	95 Biomtrcs 51 1133-1139
Avena sterilis: The effects of spatial and temporal	heterogeneity and of dispersal • González-Andujar et al. ... control of the seed bank of	95 JAppEcol 32 578-587
population of snow geese: Local emigration or	heterogeneity? • Pradel, Roger; Cooch, Evan; Cooke, Fred Transient animals in a resident	95 JAppStat 22 695-710
A combined test of linkage	heterogeneity • Goldstein, Darlene R.	94 AJHuGene 55 841-848
Benefits of	heterogeneity in meta-analysis of data from epidemiologic studies • Berlin, Jesse A.	95 AJEpidm 142 383-387
Colditz, G. A.; Burdick, E.; Mosteller, F.	Heterogeneity in meta-analysis of data from epidemiologic studies: A commentary •	95 AJEpidm 142 371-382
Kleczkowski, A.; Gilligan, C. A.; Bolker, B. M. Spatial	heterogeneity, nonlinear dynamics and chaos in infectious diseases • Grenfell, B. T.;	95 StMedRes 4 160-183
A method to detect and estimate population	heterogeneity • Fountain, Robert L.	94 ASAProEpid 82-87
Detecting substitution-rate	heterogeneity among regions of a nucleotide sequence • Gaut, Brandon S.; Weir, B. S.	94 MolBioEv 11 620-629
its breakdown under a hierarchy of periodic spatial	heterogeneities • Bhattacharya, R. N. et al. Time-scales for Gaussian approximation and	95 Bernoull 1 81-123
Censored regression estimation under unobserved	heterogeneity: A stochastic parameter approach • Ioannatos, Petros E.	95 JBES 13 327-335
mitochondrial DNA • Reeves, Jaxk H.	Heterogeneity in the substitution process of amino acid sites of proteins coded for by	92 JMolEvol 35 17-31
Wen-Hsiung Maximum likelihood estimation of the	heterogeneity of substitution rate among nucleotide sites • Gu, Xun; Fu, Yun-Xin; Li,	95 MolBioEv 12 546-557
Gaines, Steven D. Extending nondirectional	heterogeneity tests to evaluate simply ordered alternative hypotheses • Rice, William R.;	94 PNatAcSc 91 225-226
Kim, J. J.-I.; Zaslavsky, A.; Blodgett, R. Between-state	heterogeneity of undercount rates and surrogate variables in the 1990 U.S. Census •	95 SurvyM 21 55-62
Power of a test that is robust against variance	heterogeneity • Overall, John E.; Atlas, Robert S.	95 PsychRpt 77 155-159
Estimation of equilibrium wage distributions with	heterogeneity • Bowlus, Audra J.; Kiefer, Nicholas M.; Neumann, George R.	95 JApEcmtx 10 119-131S
unemployment in an equilibrium search model with	heterogeneous agents • Koning, P.; Ridder, G.; Van den Berg, G. J. Structural and frictional	95 JApEcmtx 10 133-151S
equations: Modeling transport and mass transfer in	heterogeneous aquifers • Harvey, C. F.; Gorelick, S. M. Temporal moment-generating	95 WaterReR 31 1895-1911
Acceleration model in the	heterogeneous case of the general graded response model • Samejima, Fumiko	95 Psymtrka 60 549-572
suggestions for effectively teaching short courses to	heterogeneous groups • Birch, Jeffrey B. Ten	94 ASAProStEd 33-38
suggestions for effectively teaching short courses to	heterogeneous groups • Birch, Jeffrey B. Ten	95 AmerStat 49 190-195
abundance when capture probabilities are low and	heterogeneous • Rosenberg, D. K.; Overton, W. S.; Anthony, R. G. Estimation of animal	95 JWildlMg 59 252-261
of large queue lengths and waiting times in a	heterogeneous multiserver queue II: Positive recurrence and ... • Sadowsky The probability	95 AdvAppPr 27 567-583
of large queue lengths and waiting times in a	heterogeneous multiserver queue I: Tight limits • Sadowsky, J. S. et al. The probability	95 AdvAppPr 27 532-566
Estimating long-run relationships from dynamic	heterogeneous panels • Pesaran, M. Hashem; Smith, Ron	95 JEconmtx 68 79-113
recapture models when capture probabilities are	heterogeneous • Burnham, K. P. et al. Selection among open population capture-	95 JAppStat 22 611-624
Laws of large numbers for dependent	heterogeneous processes • de Jong, R. M.	95 EconTher 11 347-358
Multi-channel bi-level queueing system with	heterogeneous servers • Shoukry, E. M.; Gharraph, M. K.; Hassan, N. A.	94 EgpStJ 38 208-221
for semiparametric two-sample models with	heterogeneous treatment effect • Hsieh, Fushing The empirical process approach	95 JRSS-B 57 735-748
	Heterogeneous variances in repeated measures • Wu, Yuan; McGilchrist, C. A.	95 BiomtrcJ 37 523-530
Stochastic regression model with	heteroscedastic disturbance • Chang, Der-Shin; Lin, Guan-Chyun	95 AnInStMa 47 351-369
confidence bands for linear regression with	heteroscedastic errors • Faraway, Julian J.; Sun, Jiayang Simultaneous	95 JASA 90 1094-1098
D-optimum designs for	heteroscedastic linear models • Atkinson, A. C.; Cook, R. D.	95 JASA 90 204-212
studies • Wilcox ANOVA: The practical importance of	heteroscedastic methods, using trimmed means versus means, and designing simulation	95 BJMaStPs 48 99-114
returns by a threshold autoregressive conditional	heteroscedastic model • Li, W. K.; Lam, K. Modelling asymmetry in stock	95 Ststcian 44 333-341
On the equivalence of D- and G-optimal designs in	heteroscedastic models • Wong, Weng Kee	95 StPrLet 25 317-321
László; Endrényi, László	Heteroscedastic nonlinear regression when there are possible outliers • Tóthfalusi,	94 ASAProStCp 239-244
Healy's two-stage sampling) is more ... • Aoshima In	heteroscedastic simultaneous inference, the heteroscedastic method II (based on	94 AJMMgSci 14 167-196
estimators for a multiplicative decomposition and	heteroscedastic variances • Pfeffermann et al. ... of X-11 ARIMA seasonally adjusted	95 IntJFor 11 271-283
	Heteroscedasticity	
equality • Lix, Lisa M.; Keselman, H. J.	Approximate degrees of freedom tests: A unified perspective on testing for mean	95 PsychBu 117 547-560
	Asymptotically distribution-free procedure in a two-way layout • Choi, Young Hun	95 JKoreaSt 24 375-387
analysis • Ligeralde, Antonio V.; Brown, Bryan W.	Band covariance matrix estimation using restricted residuals: A Monte Carlo	94 IntEconR 35 751-767
Kim, Dongcheol	The errors in the variables problem in the cross-section of expected stock returns •	95 JFinance 50 1605-1634
IGARCH(1,1) models: A Monte Carlo ... • Lumsdaine	Finite-sample properties of the maximum likelihood estimator in GARCH(1,1) and	95 JBES 13 1-10
responses • Khan, Shahjahan	Inference about the parameters of the multilinear model with equi-correlated	94 JStatRsr 28 163-176
volatility: The case of the United States ... • Karolyi	A multivariate GARCH model of international transmissions of stock returns and	95 JBES 13 11-25
Godfrey, L. G.; Orme, C. D.	The sensitivity of some general checks to omitted variables in the linear model •	94 IntEconR 35 489-506
Brown, Bryan W.; Walker, Mary Beth	Stochastic specification in random production models of cost-minimizing firms •	95 JEconmtx 66 175-205
simulation experiments • Kleijnen, J. P. C. et al.	Two-stage versus sequential sample-size determination in regression analysis of	95 AJMMgSci 15 83-114
Larholt, Kay M.; Sampson, Allan R. Effects of	heteroscedasticity upon certain analyses when regression lines are not parallel •	95 Biomtrcs 51 731-737
F-tests for unbalanced nested designs under	heteroscedasticity • Ananda, Malwane M. A. Generalized	95 AnInStMa 47 731-742
specific inefficiency measures in the presence of	heteroscedasticity • Caudill, S. B.; Ford, J. M.; Gropper, D. M. Frontier estimation and firm-	95 JBES 13 105-111
parametric variance function model and a test for	heteroscedasticity • Müller, Hans-Georg; Zhao, Peng-Liang On a semi-	95 AnlsStat 23 946-967
A likelihood ratio type test for	heteroscedasticity • Chaubey, Yogendra P.; Talukder, Enayet	94 ASAProStCp 170-175
Portmanteau test for conditional	heteroscedasticity, using ranks of squared residuals • Wong, H.; Li, W. K.	95 JAppStat 22 121-134
tests for the first-order autoregressive model with	heteroscedasticity • Lyon, John; Tsai, Chih-Ling Improved	95 JStCmpSm 52 71-83
Detecting and handling	heteroscedasticity in yield trial data • Piepho, Hans-Peter	95 CommStB 24 243-274
Fielding, K. L.; Brettle, R. P.; Gore, S. M. et al.	Heterosexual transmission of HIV analysed by generalized estimating equations •	95 StatMed 14 1365-1378
Heinz; Polasek, Wolfgang; Liu, Shuangzhe The	heteroskedastic linear regression model and the Hadamard product. A note • Neudecker,	95 JEconmtx 68 361-366
Two-step estimation of	heteroskedastic sample selection models • Donald, Stephen G.	95 JEconmtx 65 347-380
An improved Lagrange multiplier test for	heteroskedasticity • Cribari-Neto, Francisco; Ferrari, Silvia L. P.	95 CommStB 24 31-44
in the panel data error component model with	heteroskedasticity of unknown form • Li, Qi; Stengos, Thanasis Adaptive estimation	94 IntEconR 35 981-1000
A test for the presence of conditional	heteroskedasticity within ARCH-M framework • Bera, Anil K.; Ra, Sungsup	95 EcomtrRv 14 473-485
	Heuristic algorithms for finding inexpensive elimination schemes • Harbron, Chris	95 StatComp 5 275-287
A	heuristic for conceptual change • Arntzenius, Frank	95 PhilSci 62 357-369
in principal components analysis: A comparison of	heuristical and statistical approaches" (V74 p2204-2214) • Mehlman et al. ... rules	95 Ecology 76 640-643
in principal components analysis: A comparison of	heuristical and statistical approaches" • Jackson Reply to comment on "Stopping rules	95 Ecology 76 644-645
	Hidden assumptions: The use of vignettes in cognitive interviewing • Gerber, Eleanor R.	94 ASASrvyRsr 1269-1274

Finite-dimensional models for	hidden Markov chains • Aggoun, Lakhdar; Elliott, Robert J.	95 AdvAppPr 27 146-160
	Hidden Markov models and animal behaviour • MacDonald, Iain L.	95 BiomtrcJ 37 701-712
Deng, Li; Wu, C. F. J. Topics on	hidden Markov models and their applications in speech recognition • Sun, Don X.;	94 ASAProStCp 90-99
Recursive estimation for	hidden Markov models: A dependent case • Elliott, Robert J.	95 StocAnAp 13 437-460
quantization for speech recognition with discrete	hidden Markov Model • Huo, Qiang; Chan, Chorkin Contextual vector	95 PattrnRc 28 513-517
and asymptotically normal parameter estimates for	hidden Markov models • Rydén, Tobias Consistent	94 AnlsStat 22 1884-1895
	Hidden Markov models and their estimation • Nádas, Arthur; Mercer, Robert L.	94 ASAProStCp 80-89
Moore, John B.	Hidden Markov models: Estimation and control • Elliott, Robert J.; Aggoun, Lakhdar;	95 Springer:Brln:NY 361p
Moore, John B. (Auth); Goodwin, C. G. (Rev) Review of	"Hidden Markov models: Estimation and control" • Elliott, Robert J.; Aggoun, Lakhdar;	95 ShrtBkRv 15/2 29-29
Continuous speech recognition by a network of	hidden Markov models • Pavešić, Nikola	95 JCmplnT 3 193-205
Estimating the order of	hidden Markov models • Rydén, Tobias	95 MaOpfStS 26 345-354
Zucchini, W.	Hidden Markov and other models for discrete-valued time series • MacDonald, I. L.;	95 ChapmnHall:Lnd 180p
	Hidden Markov random fields • Künsch, Hans; Geman, Stuart; Kehagias, Athanasios	95 AnlsApPr 5 577-602
Use of phase for the detection of	hidden periodicities • Reschenhofer, Erhard	95 BiomtrcJ 37 957-964
Estimating the size of	hidden populations using snowball sampling • Frank, Ove; Snijders, Tom	94 JOffStat 10 53-67
Wu, C. F. Jeff A	hidden projection property of Plackett-Burman and related designs • Wang, J. C.;	95 StSinica 5 235-250
responses • Lin, Shang P.; Tang, Dei-In A	hierarchical active constraints search algorithm for optimal scaling of ordered categorical	95 JStCmpSm 53 197-209
varying parameter models • Chib, S.; Greenberg, E.	Hierarchical analysis of SUR models with extensions to correlated serial errors and time-	95 JEconmtx 68 339-360
	Hierarchical Bayes	
Annibale; Marchi, Marco	Case-control designs for the detection of spatial clusters of diseases • Biggeri,	95 EnvrMtrc 6 385-393
Holford, T. R.; Rubinstein, P.; Stevens, C. E.	Change points in the series of T4 counts prior to AIDS • Kiuchi, A. S.; Hartigan, J. A.;	95 Biomtrcs 51 236-248
wise comparisons • Kim, Woo-Chul; Han, Kyung Soo	On the performance of empirical Bayes simultaneous interval estimates for all pair-	95 JKoreaSt 24 161-181
Britain • Richardson, S.; Monfort, C.; Green, M. et al.	Spatial variation of natural radiation and childhood leukaemia incidence in Great	95 StatMed 14 2487-2501
and outliers • Datta, G. S.; Lahiri, P. Robust	hierarchical Bayes estimation of small area characteristics in the presence of covariates	95 JMultiAn 54 310-328
Datta, Gauri Sankar; Ghosh, Malay	Hierarchical Bayes estimators of the error variance in one-way ANOVA models •	95 JStPlInf 45 399-411
A.P.O. rules in	hierarchical Bayes regression models • Ghosh, Malay; Hoekstra, Robert M.	95 SqtlAnly 14 99-115
images • Molina, R. On the	hierarchical Bayesian approach to image restoration: Applications to astronomical	94 IEEEPAMI 16 1122-1128
the losses caused by an industrial intervention: A	hierarchical Bayesian approach • Arjas, Elja; Liu, Liping Assessing	95 ApplStat 44 357-368
	Hierarchical Bayesian models and Γ-minimaxity • George, Robert; Goel, Prem K.	94 ASAProBays 96-101
The conjunctive model of	hierarchical classes • Van Mechelen, Iven; De Boeck, Paul; Rosenberg, Seymour	95 Psymtrka 60 505-521
	Hierarchical cluster analysis • Fischer, J.	95 CmpStQ 10 21-27
Identifying influential observations in	hierarchical cluster analysis • Jolliffe, I. T.; Jones, B.; Morgan, B. J. T.	95 JAppStat 22 61-80
A new method of	hierarchical cluster of sample-difference sequence method (Chinese) • Chen, Qiansheng	95 KToChina 39 97-101
	Hierarchical clustering	
Carroll, J. Douglas	Minimax length links of a dissimilarity matrix and minimum spanning trees •	95 Psymtrka 60 371-374
Milligan, Glenn W.	Hierarchical clustering algorithms with influence detection • Cheng, Richard;	95 EdPsMeas 55 237-244
	Hierarchical clustering method using probabilistic similarity measure • Miyano, Hisao	95 Behvmtrk 22 145-153
Mapping influence regions in	hierarchical clustering • Cheng, Richard; Milligan, Glenn W.	95 MultBehR 30 547-576
Zhang, Q.	Hierarchical decision making in stochastic manufacturing systems • Sethi, S. P.;	94 Birkhs:Basl:MA 19p
Zhang, Q. (Auth); Goodwin, G. C. (Rev) Review of	"Hierarchical decision making in stochastic manufacturing systems" • Sethi, S. P.;	95 ShrtBkRv 15/2 29-29
A new closed multiple testing procedure for	hierarchical families of hypotheses • Rom, Dror M.; Holland, Burt	95 JStPlInf 46 265-275
ing precision through modelling: An illustration with	hierarchical kappa • Basu, Srabashi; Raychaudhuri, Aparna; Basu, Ayanendranath Improv-	95 CommStB 24 399-408
The effect of different forms of centering in	hierarchical linear models • Kreft, Ita G. G.; de Leeuw, Jan; Aiken, Leona S.	95 MultBehR 30 1-21
	Hierarchical linear models: A multilevel data analysis technique • Wu, Yow-Wu B.	95 NurseRes 44 123-126
Regional optimization of fertilization using a	hierarchical linear model • Wallach, D.	95 Biomtrcs 51 338-346
Xiong, Wei; Graffigne, Christine	Hierarchical Markov models: An application to motion estimation and segmentation •	94 QStMaA 16 45-67
Skorohod topology on n-variable functions and	hierarchical Markov properties of n-parameter processes • Hirsch, F.; Song, S. Symmetric	95 ProbTher 103 25-43
	Hierarchical model	
Steffen L.	The EM algorithm for graphical association models with missing data • Lauritzen,	95 CmpStDA 19 191-201
	The estimation of school effects • Raudenbush, Stephen W.; Willms, J. Douglas	95 JEdStat 20 307-335
• Moulton, Brent R.	Interarea indexes of the cost of shelter using hedonic quality adjustment techniques	95 JEconmtx 68 181-204
p201-227, 234-238) • Rogosa, David; Saner, Hilary	Longitudinal data analysis examples with random coefficient models (Disc:	95 JEdStat 20 149-170
Kreft, Ita G. G.	Questioning multilevel models (Disc: p201-227, 239-240) • de Leeuw, Jan;	95 JEdStat 20 171-189
Donegan, H. A.; McMaster, T. B. M.	Reassessment of consistency criteria in judgment matrices • Dodd, F. J.;	95 Ststcian 44 31-41
Prediction and decision making using Bayesian	hierarchical models • Stangl, Dalene K.	95 StatMed 14 2173-2190
	Hierarchical models for educational data: An overview • Morris, Carl N.	95 JEdStat 20 190-200
Conditional Bayesian significance levels:	Hierarchical models • Brimacombe, Michael	94 ASAProBays 169-174
Nonparametric Bayesian methods in	hierarchical models • Escobar, Michael D.	95 JStPlInf 43 97-106
Inverting covariance matrices in unbalanced	hierarchical models • Stępniak, C.; Niezgoda, M.	95 JStCmpSm 51 215-221
Inference and	hierarchical modeling in the social sciences (Disc: p201-233) • Draper, David	95 JEdStat 20 115-147
Fulker, D. W. A model of developmental change in	hierarchical phenotypes with application to specific cognitive abilities • Cardon, L. R.;	94 BehavGen 24 1-16
Hennig, Christian Efficient	high-breakdown-point estimators in robust regression: Which function to choose? •	95 StatDecn 13 221-241
for the computation of multivariate estimators with	high breakdown point • Juan, Jesus; Prieto, Francisco J. A subsampling method	95 JCmpGrSt 4 319-334
Asymptotics for transportation cost in	high dimensions • Dobrić, V.; Yukich, J. E.	95 JTherPrb 8 97-118
	High-dimension data	
mass approach • Polonik, Wolfgang	Measuring mass concentrations and estimating density contour clusters – An excess	95 AnlsStat 23 855-881
	High-dimensional data	
Roosen, Charles B.	An introduction to multivariate adaptive regression splines • Friedman, Jerome H.;	95 StMedRes 4 197-217
	Projection pursuit exploratory data analysis • Posse, Christian	95 CmpStDA 20 669-687
and predictors obtained from quantile spline fits to	high-dimensional data • O'Connell et al. ... model with a smooth link function	95 JStPlInf 47 153-164
Bhatti, M. I.; Bentler, P. M. Testing sphericity of a	high-dimensional distribution based on bootstrap approximation • Zhu, L.-X.; Fang, K.-T.;	95 PakJStA 11 49-65
Laplace approximation of	high dimensional integrals • Shun, Zhenming; McCullagh, Peter	95 JRSS-B 57 749-760
Random	high-dimensional orders • Bollobás, Béla; Brightwell, Graham	95 AdvAppPr 27 161-184
On	high level exceedance modeling and tail inference • Leadbetter, M. R.	95 JStPlInf 45 247-260
parameter. I • Konakov, V. D.; Piterbarg, V. I.	High level excursions of Gaussian fields and the weakly optimal choice of the smoothing	95 MaMethSt 4 421-434
Clemen, R. T. The use of probability elicitation in the	high-level nuclear waste regulation program • DeWispelare, A. R.; Herren, L. T.;	95 IntJFor 11 5-24
ods for evolutionary factor analysis as applied to	high-performance liquid chromatography • Elbergali et al. ... on the performance of meth-	95 Chemolab 27 55-71
	High-resolution genetic mapping of complex traits • Kruglyak, Leonid; Lander, Eric S.	95 AJHuGene 56 1212-1223
methods for linkage analysis of complex traits from	high-resolution maps of identity by descent • Dupuis, J.; Brown, P. O. et al. Statistical	95 Genetcs 140 843-856
Jansen, Ritsert C.; Stam, Piet	High resolution of quantitative traits into multiple loci via interval mapping •	94 Genetcs 136 1447-1455
local labor markets, and dropping out of	high school (Disc: p213-215) • Swaim, Paul Work experience,	94 ASAProSoSt 207-212
Zimmerman, David J. The benefit of additional	high-school math and science classes for young men and women • Levine, Phillip B.;	95 JBES 13/2 137-149
Ward, Joe H., Jr; Niland, Laura J. Empowering	high school students to exploit statistical models and software for research projects •	94 ASAProStEd 44-48
An exploratory analysis • Berndt, E. R.; Morrison, C. J.	High-tech capital formation and economic performance in U.S. manufacturing industries:	95 JEconmtx 65 9-43

The extremal index and clustering of	high values for derived stationary sequences • Weissman, Ishay; Cohen, Uri		95 JAppProb 32 972-981
Moving to	higher dimensions 3 ≠ 2 • Balch, Alfred H.		94 ASAProBiop 45-50
Density estimation under qualitative assumptions in	higher dimensions • Polonik, Wolfgang		95 JMultiAn 55 61-81
Fleischmann, Klaus Super-Brownian motions in	higher dimensions with absolutely continuous measure states • Dawson, Donald A.;		95 JTherPrb 8 179-206
Birth and death processes as projections of	higher-dimensional Poisson processes • Garcia, Nancy Lopes		95 AdvAppPr 27 911-930
Exact tests of significance in	higher dimensional tables • Zelterman, Daniel; Chan, Ivan Siu-Fung; Mielke, Paul W., Jr		95 AmerStat 49 357-361
Continuous quality improvement in	higher education • Hogg, Robert V.; Hogg, Mary C.		95 IntStRvw 63 35-48
of the symposium on the teaching of statistics in	higher education in Ireland • Boland, Philip J.	Discussion	95 Ststcian 44 55-56
of the symposium on the teaching of statistics in	higher education in Ireland • Cran, G. W.	Discussion	95 Ststcian 44 57-58
on the symposium on the teaching of statistics in	higher education in Ireland • McClean, Sally	Discussion	95 Ststcian 44 56-57
on the symposium on the teaching of statistics in	higher education in Ireland • Wilson, Noel	Discussion	95 Ststcian 44 58-58
Akahira, Masafumi A	higher order approximation to a percentage point of the non-central t-distribution •		95 CommStB 24 595-605
estimators • Taniguchi, Masanobu; Puri, Madan L.	Higher order asymptotic theory for normalizing transformations of maximum likelihood		95 AnInStMa 47 581-600
10) • Pierce, Donald A.; Peters, Dawn Correction to	"Higher-order asymptotics and the likelihood principle: One-parameter models" (V81 p1-		95 Biomtrka 82 235-235
Review of "Time series analysis by	higher order crossings" • Kedem, Benjamin (Auth); Frey, Michael (Rev)		95 Technmcs 37 235-235
designs of resolution $2l+1$ when $l+s+1$-factor and	higher order interactions are negligible • Shirakura et al.	... for balanced 2^m factorial	93 JCISS 18 233-245
On	higher order kernels • Jones, M. C.		95 JNnparSt 5 215-221
Computationally efficient classes of	higher-order kernel functions • Abdous, Belkacem		95 CandJSt 23 21-27
construction of new main effect plans with	higher revealing power for 2^7 factorials • Ghosh, Subir	Sequential	93 JCISS 18 217-231
Hyndman, Rob J.	Highest-density forecast regions for non-linear and non-normal time series models •		95 JForcast 14 431-441
for linear Gaussian models with random disturbed-	highest-derivative-polynomial subject effects • Wilson, P. David	Longitudinal data analysis	95 StatMed 14 1219-1233
size calculations for binomial proportions via	highest posterior density intervals • Joseph, L.; Wolfson, D. B.; Berger, R. d.	Sample	95 Ststcian 44 143-154
Jayanta K.; Mukerjee, Rahul Frequentist validity of	highest posterior density regions in the presence of nuisance parameters • Ghosh,		95 StatDecn 13 131-139
Mukerjee, Rahul On perturbed ellipsoidal and	highest posterior density regions with approximate frequentist validity • Ghosh, J. K.;		95 JRSS-B 57 761-769
On-	highway Diesel Fuel Price Survey • Weir, Paula; Saavedra, Pedro J.		94 ASAProGovt 144-148
of posterior mixtures in the Gaussian family on a	Hilbert space and its applications • Majumdar, Suman	Consistency	95 JMultiAn 55 187-204
et al. Co-inertia of two random variables in	Hilbert space and B-spline approximations with a forestry application (French) • Champely		95 RStApp 43/1 91-107
Feller property and irreducibility for diffusions on	Hilbert spaces • Peszat, Szymon; Zabczyk, Jerzy	Strong	95 AnlsProb 23 157-172
of semilinear stochastic evolution equations in	Hilbert spaces • Taniguchi, Takeshi	Asymptotic stability theorems	95 Stochatx 53 41-52
McKean-Vlasov stochastic evolution equation in	Hilbert space • Ahmed, N. U.; Ding, X.	A semilinear	95 StocProc 60 65-85
measures for stochastic semilinear equations on	Hilbert spaces • Chojnowska-Michalik, A.; Goldys, B.	Existence, uniqueness and invariant	95 ProbTher 102 331-356
Stochastic analysis and random maps in	Hilbert space • Dorogovtsev, A. A.		94 VSP:Neth 110p
D. L. (Auth); Marshall, G.; Allende, R. (Rev) Review of	"Hilbert space methods in probability and statistical inference" • Small, C. G.; McLeish,		95 JASA 90 1491-1492
and the composition of random operators on a	Hilbert space • Thang, Dang Hung	The adjoint	95 Stochatx 54 53-73
Limit theorems of	Hilbert valued semimartingales and Hilbert valued martingale measures • Xie, Yingchao		95 StocProc 59 277-293
Consistency of	Hill's estimator for dependent data • Resnick, Sidney; Stărică, Cătălin		95 JAppProb 32 139-167
Sir Austin Bradford	Hill: A personal view of his contribution to epidemiology • Doll, Sir Richard		95 JRSS-A 158 155-163
Before and after Bradford	Hill: Some trends in medical statistics • Armitage, Peter		95 JRSS-A 158 143-153
interaction terms in Mandel's ANOVA model from	Hirotsu's interaction elements • Miyakawa, Masami	An interpretation of the	93 RStAJp 40/1 1-9
	Histogram		
	Folded empirical distribution function curves - Mountain plots • Monti, Katherine L.		95 AmerStat 49 342-345
	Low-storage quantile estimation • Hurley, C.; Modarres, R.		95 CmpStQ 10 311-325
	A note on variance reduction • Berlinet, Alain		95 StPrLet 25 357-360
datasets.short) • Short, T. H.; Moriarty, H. et al.	Readability of educational materials for patients with cancer (jse/v3n2/		95 JStEd 3/2 15kb
Schroder, Shayla D.; Mallawaaratchy, A. Rajika	Statistical artifacts in the ratio of discrete quantities • Johnston, Roger G.;		95 AmerStat 49 285-291
Ait-Hennani, L.	Uniform convergence of density estimators on spheres • Bertrand-Retali, M.;		95 ApplMath 22 427-446
	Histogram estimate		
Györfi, László	Asymptotic normality of L_1-error in density estimation • Berlinet, Alain; Devroye, Luc;		95 MaOpfStS 26 329-343
A general approach to nonparametric	histogram estimation • Jacob, Pierre; Oliveira, Paulo Eduardo		95 MaOpfStS 27 73-92
Simonoff, Jeffrey S. The anchor position of	histograms and frequency polygons: Quantitative and qualitative smoothing •		95 CommStB 24 691-710
West, Carol Taylor Assessing the	historical accuracy of regional employment forecasts • Fullerton, Thomas M., Jr;		94 ASAProBuEc 333-338
Slow points in the support of	historical Brownian motion • Verzani, John		95 AnlsProb 23 56-70
an empirical Bayes estimator of the distribution of	historical controls in carcinogen bioassay • Krewski, D.; Smythe, R. T. et al.	Bootstrapping	95 JStPlInf 43 215-228
	Historical controls and modern survival analysis • Keiding, Niels		95 LiDaAnal 1 19-25
methods in biological and medical sciences. A	historical essay • Lancaster, H. O.	Quantitative	94 Springer:Brln:NY 297p
methods in biological and medical science - A	historical essay" • Lancaster, H. O. (Auth); Mayo, Oliver (Rev)	Review of "Quantitative	95 AstrlJSt 37 372-374
methods in biological and medical sciences. A	historical essay" • Lancaster, H. O. (Auth); Farewell, V. T. (Rev)	Review of "Quantitative	95 ShrtBkRv 15/2 22-22
lection is determined by baseline ... • Lin et al. Use of	historical marker data for assessing treatment effects in phase I/II trials when subject se-		95 Biomtrcs 51 1053-1063
Official labour statistics: A	historical perspective • Davidson, Roger		95 JRSS-A 158 165-173
Sampling with unequal probabilities: A	historical review • Hanif, Muhammad; Beg, M. A.; Khawaja, Iffat		92 ProIrStC1 19-37
Dickey et al. Teaching statistics with data of	historical significance: Galileo's gravity and motion experiments (jse/v3n1/ ... •		95 JStEd 3/1 20kb
	History		
	"I always wanted to be a statistician" or The accidental statistician • Legler, Julie		95 Stats /12 10-13
	Bertrand's work on probability • Sheynin, O.		94 AHistESc 48 155-199
	Before and after Bradford Hill: Some trends in medical statistics • Armitage, Peter		95 JRSS-A 158 143-153
	Charles Spearman's contributions to test theory • Levy, Philip		95 BJMaStPs 48 221-235
	Chebyshev's contribution to the theory of probability • Sheynin, O.		93 AHistESc 46 321-340
Lovie, A. D.	The cold equations: Spearman and Wilson on factor indeterminacy • Lovie, P.;		95 BJMaStPs 48 237-253
	The coming of age of statistical education • Vere-Jones, David		95 IntStRvw 63 3-23
	Who discovered Spearman's rank correlation? • Lovie, A. D.		95 BJMaStPs 48 255-269
W. F.	Fiducial inference and the fundamental theorem of natural selection • Edwards, A.		95 Biomtrcs 51 799-809
	Fifty years of Calcutta Statistical Association • Banerjee, P. K.		95 CalcutSt 45 1-23
	First (?) occurrence of common terms in mathematical statistics • David, H. A.		95 AmerStat 49 121-133
	Harald Cramér and insurance mathematics • Martin-Löf, Anders		95 AplStMDA 11 271-276
	Helmert's work on the theory of errors • Sheynin, O.		95 AHistESc 49 73-104
	Pivotal models and the fiducial argument • Barnard, George A.		95 IntStRvw 63 309-323
	The problem of points • Takács, Lajos		94 MaScitst 19 119-139
	The roads travelled • Billard, Lynne		95 Biomtrcs 51 1-11
	Spearman and the origin and development of factor analysis • Bartholomew, D. J.		95 BJMaStPs 48 211-220
Three papers on the	history of branching processes • Guttorp, Peter (Transl)		95 IntStRvw 63 233-245
On queueing networks with signals and	history-dependent routing • Chao, Xiuli; Pinedo, Michael		95 PrEngInf 9 341-354
ies comparing offspring with and without parental	history of disease • Tiret et al.	... the strength of genotype-disease association from stud-	93 AnHuGene 57 141-149
Leland Gerson (Rev) Review of "The	history of econometric ideas" (Disc: p384-397) • Morgan, Mary S. (Auth); Neuberg,		95 EconTher 11 371-383
Return rates in studies of life	history evolution: Are biases large? • Martin, Thomas E.; Clobert, Jean; Anderson, David R.		95 JAppStat 22 863-875

eling physicochemical properties of polyaromatic hydrocarbons • Todeschini et al. ... Part 2. Theory development and applications on mod- 95 Chemolab 27 221-229
Optical reflectometric gas sensing: Classification of hydrocarbon vapours by pattern recognition applied to RIFS sensor ... • Kraus et al. 95 Chemolab 30 211-221
Large deviations from a hydrodynamic scaling limit for a nongradient system • Quastel, Jeremy 95 AnlsProb 23 724-742
Kipnis, C.; Landim, C. Large deviations from the hydrodynamical limit of mean zero asymmetric zero range processes • Benois, O.; 95 StocProc 55 65-89
Large deviations of an asymmetric hydrodynamical system of independent particles (French) • Kipnis, C.; Léonard, C. 95 AlHenrPB 31 223-248
Tarboton, David G. Hydrologic scenarios for severe sustained drought in the Southwestern United States • 95 WaterReB 31 803-813

Hydrology

Donald A. Tree-ring reconstruction of Upper Gila River discharge • Meko, David; Graybill, 95 WaterReB 31 605-616
planning under uncertainty with an application to hydropower generation • Alvarez, M.; Cuevas, C. M.; Escudero, L. F. et al. Network 94 Top 2 25-58
of the proportion of diploid males in populations of Hymenoptera • Owen, Robin E.; Packer, Laurence Estimation 94 Heredity 72 219-227
of parameters of the incomplete bivariate hyper-Poisson distributions • Ahmad, Munir Estimation 92 ProIrStC1 157-162
Hyper-weakly harmonizable processes and operator families • Leitner, M. 95 StocAnAp 13 471-485
Hyperbolic distributions in finance • Eberlein, Ernst; Keller, Ulrich 95 Bernoull 1 281-299
Hypercontractivity for spin systems with infinite support (French) • Laroche, Etienne 95 ProbTher 101 89-132
values for central composite designs on the hypercube • Borkowski, John J. Finding maximum G-criterion 95 CommStA 24 2041-2058
Laurent, Monique Variety of hypercube embeddings of the equidistant metric and designs • Deza, Michel; 93 JCISS 18 289-315
Lee, Joo-Ho Two-stage Latin hypercube sampling and its application (Korean) • Im, Mi-Jeong; Kwon, Woo-Joo; 95 KorApS 8/2 99-108

Hypergeometric distribution

ratio in several 2 × 2 tables • Liao, J. G.; Hall, C. B. Fast and accurate computation of the exact confidence limits for the common odds 95 JCmpGrSt 4 173-179
Minimax estimation of proportions under random sample size • Amrhein, Peter 95 JASA 90 1107-1111
tics: Parts 3 and 4 (jse/v3n2/trumbo) • Trumbo, B. E. Some demonstration programs for use in teaching elementary probability and statis- 95 JStEd 3/2 60kb
A regression characterisation of the Meixner hypergeometric distribution • Pusz, J. 95 AstrlJSt 37 83-87
Brainerd, Barron A generalization of the hypergeometric distribution suggested by a model for the structure of a play • 95 MaScitst 20 33-39

Hypergeometric function

On the doubly noncentral F distribution • Chattamvelli, R. 95 CmpStDA 20 481-489
The limiting distribution of the t ratio under a unit root • Abadir, Karim M. 95 EconTher 11 775-793

Hypergeometric sampling

Pierce, Rebecca L. Bayesian HPM intervals for the number of nonconforming items in a lot • Dyer, Danny; 95 CmpStDA 20 159-172
Hypergeometric series $\Phi(a,b,z)$: Further developments (Italian) • Passaquindici, Maria 94 Metron 52/3 197-210
Harmonic analysis of probability measures on hypergroups • Bloom, W. R.; Heyer, H. 94 DGruyter:NY:Grm 601p
of "Harmonic analysis of probability measures on hypergroups" • Bloom, W. R.; Heyer, H. (Auth); Barnett, C. (Rev) Review 95 ShrtBkRv 15/3 49-49
Cadish, Marc; Clerc, Jean-Thomas; Pretsch, Ernö Hypermedia – A new concept for information management • Macher, Christian L.; 95 Chemolab 28 213-228
Majorization methods on hyperplanes and their applications • Kakiuchi, Itsuro; Kimura, Miyoshi 95 JStPlInf 47 217-235
On hyperplanes of closest fit • Farebrother, Richard William 95 CmpStDA 19 53-58
Suchower, Stephan J. Frequency hyperrectangles, error correcting codes and fractional replication plans • Mullen, Gary L.; 93 JCISS 18 5-18
and rate ratio estimation in case-cohort designs: Hypertension and cardiovascular mortality" (V12 p1733-1745) • Edwardes ... on "Risk ratio 95 StatMed 14 1609-1609
tio and rate ratio estimation in case-cohort designs: Hypertension and cardiovascular mortality" • Schouten et al. Reply to comment on "Risk ra- 95 StatMed 14 1610-1610
Okada, Mihoko; O'Brien, Myles IntroStat: A hypertext-based design for an electronic textbook to introduce biomedical statistics • 95 CmpBiomd 46 265-276
Substitution rate variation among sites in hypervariable region 1 of human mitochondrial DNA • Wakeley, John 93 JMolEvol 37 613-623
Kessler, R. C.; Neale, M. C.; Hewitt, J. K. et al. Testing hypotheses about direction of causation using cross-sectional family data • Heath, A. C.; 93 BehavGen 23 29-50
Minimax testing of hypotheses on the distribution density for ellipsoids in l_p • Ingster, Yu. I. 94 ThProbAp 39 417-436
Minimax nonparametric testing of hypotheses on the distribution density • Ermakov, M. S. 94 ThProbAp 39 396-416
Reboulet, A.-M.; Viallefont, A. Testing hypotheses and estimating survival from capture histories with CR • Pradel, R.; 95 JAppStat 22 775-784
evolutionary contrasts • McPeek, Mark A. Testing hypotheses about evolutionary change on single branches of a phylogeny using 95 ANatrls 145 686-703
testing procedure for hierarchical families of hypotheses • Rom, Dror M.; Holland, Burt A new closed multiple 95 JStPlInf 46 265-275
Tests of ordered hypotheses for gamma scale parameters • Li, Teng; Sinha, Bimal K. 95 JStPlInf 43 311-321
Tests of ordered hypotheses for gamma scale parameters • Li, Teng; Sinha, Bimal K. 95 JStPlInf 45 387-397
approximations for word patterns under Markovian hypotheses • Geske, M. X.; Godbole, A. P.; Schaffner, A. A. et al. Compound Poisson 95 JAppProb 32 877-892
Testing the hypothesis that matters for multiple primary endpoints • Capizzi, Thomas P.; Zhang, Ji 94 ASAProBiop 460-465
Complete classes of tests for some nonparametric hypotheses • Kowalski, Jacek P. 95 StatDecn 13 167-179
Some restricted estimation and testing hypotheses problem with singularity assumption • Soliman, Naeem A. 94 EgpStJ 38 231-248
T^2-type statistic for testing against one-sided hypotheses • Silvapulle, Mervyn J. A Hotelling's 95 JMultiAn 55 312-319
Optimum spacings, ABLUE and tests of hypothesis in simple linear model with extreme-value errors • Saleh, A. K. Md. E. et al. 94 JStatRsr 28 53-63
Estimating statistical hypotheses • Blyth, Colin R.; Staudte, Robert G. 95 StPrLet 23 45-52
Hill, Robert H. Multiple hypothesis tests in multiple investigations (Disc: p701-702) • Caudill, Samuel P.; 95 StatMed 14 585-589
Gargiullo, P. M. et al. Confidence intervals, hypothesis tests, and sample sizes for the prevented fraction in cross-sectional studies • 95 StatMed 14 51-72
Consistent nonparametric hypothesis tests with an application to Slutsky symmetry • Lewbel, Arthur 95 JEconmtx 67 379-401

Hypothesis testing

Bayes factors • Kass, Robert E.; Raftery, Adrian E. 95 JASA 90 773-795
Cone order association • Cohen, Arthur; Sackrowitz, H. B.; Samuel-Cahn, E. 95 JMultiAn 55 320-330
Schaafsma, W. Distributional inference • Kroese, A. H.; van der Meulen, E. A.; Poortema, K.; 95 StNeerla 49 63-82
Kim, Jinbang; De Marchi, Neil; Morgan, Mary S. Empirical model particularities and belief in the natural rate hypothesis • 95 JEconmtx 67 81-102
Frequentist post-data inference • Goutis, Constantinos; Casella, George 95 IntStRvw 63 325-344
Zeitouni, Ofer A general classification rule for probability measures • Kulkarni, Sanjeev R.; 95 AnlsStat 23 1393-1407
Walter, Stephen D. Methods of reporting statistical results from medical research studies • 95 AJEpidm 141 896-906
Lawrence D.; Casella, George; Hwang, J. T. Gene Optimal confidence sets, bioequivalence, and the Limaçon of Pascal • Brown, 95 JASA 90 880-889
Order-restricted inferences in linear regression • Mukerjee, Hari; Tu, Renjin 95 JASA 90 717-728
Randomization tests in econometrics • Kennedy, Peter E. 95 JBES 13 85-94
trimmed means • Wilcox, Rand R. Simulation results on solutions to the multivariate Behrens-Fisher problem via 95 Ststcian 44 213-225
Statistical tests for fuzzy data • Römer, Christoph; Kandel, Abraham 95 FuzzySet 72 1-26
Conventional null hypothesis testing in active control equivalence studies • Ng, Tie-Hua 95 ConClnTr 16 356-358
squares neural network learning algorithms for hypothesis testing • Pados, Dimitris A.; Papantoni-Kazakos, P. New nonleast- 95 IEEENrlN 6 596-609
Improved estimation of regression coefficients and hypotheses testing • Srivastava, Virendra K.; Srivastava, Anil K. 94 Metron 52/3 161-170
ANOVA approach with ... • Obuchowski et al. Hypothesis testing of diagnostic accuracy for multiple readers and multiple tests: An 95 CommStB 24 285-308
Multiple Bayes factors for testing hypotheses • Bertolino, Francesco; Piccinato, Ludovico; Racugno, Walter 95 JASA 90 213-219
Statistical foundation for hypothesis testing of image data • Kanatani, Kenichi 94 CVGIPIm 60 382-391
L. A bootstrap approach to hypothesis testing in least absolute value regression • Dielman, Terry E.; Rose, Elizabeth 95 CmpStDA 20 119-130
practical guide to resampling methods for testing hypotheses" • Good, Ph. (Auth); Rudolph, P. E. (Rev) Review of "Permutation tests. A 95 BiomtrcJ 37 150-150
practical guide to resampling methods for testing hypotheses" • Good, Philip (Auth); Niknian, Minoo (Rev) Review of "Permutation tests: A 95 Technmcs 37 341-342
A stepwise resampling method of multiple hypothesis testing • Troendle, James F. 95 JASA 90 370-378
Empirical Bayes linear loss hypothesis testing in a non-regular exponential family • Singh, R. S. 95 JStPlInf 43 107-120
missing information • Pedlow, S. et al. Exploring hypothesis-testing procedures with multiply-imputed data under unequal fractions of 94 ASASrvyRsr 595-599
subject • Buderer, N. F. et al. Confidence intervals and hypothesis testing for screening parameters in the case of multiple observations per 94 ASAProStEd 175-179
tist and Bayesian test for fixed and sequential simple hypothesis testing • Berger, J. O.; Brown, L. D.; Wolpert, R. L. A unified conditional frequen- 94 AnlsStat 22 1787-1807

On near	ignorance classes • Sansó, B.; Pericchi, L. R.	94 RBrzlPrS 8 119-126
Using	i.i.d. bootstrap inference for general non-i.i.d. models • Liu, Regina Y.; Singh, Kesar	95 JStPlInf 43 67-75
Tightness of products of	I.I.D. random matrices. II • Mukherjea, Arunava	94 AnlsProb 22 2223-2234
forms, maxima of products and truncated sums of	i.i.d. random variables • Cuzick, J.; Giné, E.; Zinn, J. Laws of large numbers for quadratic	95 AnlsProb 23 292-333
A strong law for weighted sums of	i.i.d. random variables • Cuzick, Jack	95 JTherPrb 8 625-641
Limiting curves for	I.I.D. records • Deuschel, Jean-Dominique; Zeitouni, Ofer	95 AnlsProb 23 852-878
Asymptotics of k-mean clustering under non-	i.i.d. sampling • Serinko, Regis J.; Babu, Gutti Jogesh	95 StPrLet 24 57-66
... • Frydman Nonparametric estimation of a Markov	'illness-death' process from interval-censored observations, with application to diabetes	95 Biomtrka 82 773-789
Review of "Understanding log-linear analysis with	ILOG: An interactive approach" • Bakeman, R.; Robinson, B. F. (Auth); Rice, J. C. (Rev)	95 EdPsMeas 55 350-355
• Subrahmonia et al. Recognition of objects in 2D	images or 3D range data based on high degree implicit polynomial curves or surfaces, ...	94 ASAProPEng 44-49
	Image analysis	
Tsybakov, A. B.	Asymptotical minimax recovery of sets with smooth boundaries • Mammen, E.;	95 AnlsStat 23 502-524
Green, Peter; Higdon, David; Mengersen, Kerrie	Bayesian computation and stochastic systems (Disc: p41-66) • Besag, Julian;	95 StatSci 10 3-41
random fields, with an application to ... • Worsley	Boundary corrections for the expected Euler characteristic of excursion sets of	95 AdvAppPr 27 943-959
of excursion sets, with applications to ... • Worsley	Estimating the number of peaks in a random field using the Hadwiger characteristic	95 AnlsStat 23 640-669
	Fuzzy surface reconstruction • Pereira, Ricardo Alberto Marques; Fedrizzi, Michele	94 QStMaA 16 71-77
• Xiong, Wei; Graffigne, Christine	Hierarchical Markov models: An application to motion estimation and segmentation	94 QStMaA 16 45-67
Madsen, L. L.; Zasadzinski, J. A.	Inherent bias in correlation averaged images • Woodward, J. T.; Kono, C.;	95 JMicro 178 86-92
EM algorithm for solving positive ... • Archer et al.	The iterative image space reconstruction algorithm (ISRA) as an alternative to the	95 StSinica 5 77-96
	Partition models in the analysis of autoradiographic images • Aykroyd, R. G.	95 ApplStat 44 441-454
struction • Hurn, Merrilee; Jennison, Christopher	A study of simulated annealing and a revised cascade algorithm for image recon-	95 StatComp 5 175-190
random field • Siegmund, David O.; Worsley, Keith J.	Testing for a signal with unknown location and scale in a stationary Gaussian	95 AnlsStat 23 608-639
	Image analysis for the biological sciences • Glasbey, C. A.; Horgan, G. W.	95 Wiley:NY:UK 218p
An extended strategy for exploratory multivariate	image analysis including noise considerations • Pedersen, F.; Bengtsson, E.; Nordin, B.	95 JChemtrc 9 389-409
introduction • Winkler, Gerhard	Image analysis, random fields and dynamic Monte Carlo methods: A mathematical	95 Springer:Brln:NY 324p
introduction" • Winkler, G. (Auth) et al. Review of	"Image analysis, random fields and dynamic Monte Carlo methods: A mathematical	95 ShrtBkRv 15/3 49-49
maps from atlas data using methods of statistical	image analysis • Högmander, Harri; Møller, Jesper Estimating distribution	95 Biomtrcs 51 393-404
Partition models in the analysis of autoradiographic	images • Aykroyd, R. G.	95 ApplStat 44 441-454
Inherent bias in correlation averaged	images • Woodward, J. T.; Kono, C.; Madsen, L. L.; Zasadzinski, J. A.	95 JMicro 178 86-92
Bernard Estimation of edge parameters and	image blur using polynomial transforms • Kayargadde, Vishwakumara; Martens, Jean-	94 CVGIPGr 56 442-461
Gray, Robert M. Combining	image compression and classification using vector quantization • Oehler, Karen L.;	95 IEEEPAMI 17 461-473
Statistical foundation for hypothesis testing of	image data • Kanatani, Kenichi	94 CVGIPIm 60 382-391
least squares based on multiple homomorphic	images • Keller-McNulty, Sallie Error-free	94 ASAProStCp 33-40
An interactive	image index • Eddy, William F.; Mockus, Audris	94 ASAProStGr 1-6
Landsat	image interpretation • Zaki, M.; Tohamy, F. El; Hassan, A.	95 CmpStDA 20 75-97
	Image of "Made in Switzerland" (French) • Brossard, Hubert	95 Student 1 51-61
istic of excursion sets, with applications to medical	images • Worsley ... the number of peaks in a random field using the Hadwiger character-	95 AnlsStat 23 640-669
Bayesian segmentation methodology for parametric	image models • LaValle, Steven M.; Hutchinson, Seth A. A	95 IEEEPAMI 17 211-217
	Image-modeling Gibbs priors • Levitan, Emanuel; Chan, Michael; Herman, Gabor T.	95 CVGIPGr 57 117-130
maximum a posteriori restoration of multicolour	images • Ferrari, Pablo A.; Frigessi, Arnoldo; de Sá, Paula Gonzaga Fast approximate	95 JRSS-B 57 485-500
their values, and relative frequencies in a noisy	image • Godtliebsen, Fred; Chu, Chih-Kang Estimation of the number of true gray levels,	95 JASA 90 890-899
A general purpose model for	image operations based on multilayer perceptrons • Moh, Jenlong; Shih, Frank Y.	95 PattrnRc 28 1083-1090
	Image processing	
Peng, Anrong; Pieczynski, Wojciech	Adaptive mixture estimation and unsupervised local Bayesian image segmentation •	95 CVGIPGr 57 389-399
Kayargadde, Vishwakumara; Martens, Jean-Bernard	Estimation of edge parameters and image blur using polynomial transforms •	94 CVGIPGr 56 442-461
mentations • LaValle, Steven M.; Hutchinson, Seth A.	A framework for constructing probability distributions on the space of image seg-	95 CVGIPIm 61 203-230
Moh, Jenlong; Shih, Frank Y.	A general purpose model for image operations based on multilayer perceptrons •	95 PattrnRc 28 1083-1090
	Image-modeling Gibbs priors • Levitan, Emanuel; Chan, Michael; Herman, Gabor T.	95 CVGIPGr 57 117-130
Rama	Image understanding and statistics: What each can do for the other? • Chellappa,	94 ASAProStGr 32-41
morphological filters • Loce, R. P.; Dougherty, E. R.	Mean-absolute-error representation and optimization of computational-	95 CVGIPGr 57 27-37
Maragos, Petros	Min-max classifiers: Learnability, design and application • Yang, Ping-Fai;	95 PattrnRc 28 879-899
	Models of statistical visual motion estimation • Spetsakis, Minas	94 CVGIPIm 60 300-312
Ghosal, Sugata; Mehrotra, Rajiv	Range surface characterization and segmentation using neural networks •	95 PattrnRc 28 711-727
Galatsanos, Nikolas P.; Katsaggelos, Aggelos K.	Regularized multichannel restoration using cross-validation • Zhu, Wenwu;	95 CVGIPGr 57 38-54
Hanson, Allen R.	Robust methods for estimating pose and a sensitivity analysis • Kumar, Rakesh;	94 CVGIPIm 60 313-342
Bhattacharyya, P.	A statistical design of experiments approach for texture description • Ganesan, L.;	95 PattrnRc 28 99-105
	Statistical foundation for hypothesis testing of image data • Kanatani, Kenichi	94 CVGIPIm 60 382-391
	Image processing and object classification with fuzzy logic (Italian) • Bellacicco, Antonio	94 QStMaA 16 19-26
annealing and a revised cascade algorithm for	image reconstruction • Hurn, Merrilee; Jennison, Christopher A study of simulated	95 StatComp 5 175-190
Comparison of	image reconstruction algorithms • Ferryman, Thomas A.; Press, S. James	94 ASAProBays 159-164
study of least squares and maximum likelihood for	image reconstruction in positron emission tomography • O'Sullivan, Finbarr A	95 AnlsStat 23 1267-1300
Comments, new results and application to	image recovery problems (Disc: p233-258) • Girard ... cross-validation and C_L procedures:	95 CmpStQ 10 205-231
Depth and	image recovery using a MRF model • Kapoor, S.; Mundkur, P. Y.; Desai, U. B.	94 IEEEPAMI 16 1117-1122
Bayesian filtering and supervised classification in	image remote sensing • Granville, V.; Rasson, J. P.	95 CmpStDA 20 203-225
Comparison of simulated annealing algorithms for	image restoration • Lamotte, J. L.; Alt, R.	94 MaCmpSim 37 1-15
On the hierarchical Bayesian approach to	image restoration: Applications to astronomical images • Molina, R.	94 IEEEPAMI 16 1122-1128
	Image restoration using recursive estimators • Trivedi, Yagnesh C.; Kurz, Ludwik	95 IEEESyst 25 1470-1482
acterizing projected designs: Repeat and mirror-	image runs • Draper, Norman R.; Lin, Dennis K. J. Char-	95 CommStA 24 775-795
mixture estimation and unsupervised local Bayesian	image segmentation • Peng, Anrong; Pieczynski, Wojciech Adaptive	95 CVGIPGr 57 389-399
structing probability distributions on the space of	image segmentations • LaValle, Steven M.; Hutchinson, Seth A. A framework for con-	95 CVGIPIm 61 203-230
solving positive linear ... • Archer et al. The iterative	image space reconstruction algorithm (ISRA) as an alternative to the EM algorithm for	95 StSinica 5 77-96
Morii, Fujiki An	image thresholding method using a minimum weighted squared-distortion criterion •	95 PattrnRc 28 1063-1071
	Image understanding and statistics: What each can do for the other? • Chellappa, Rama	94 ASAProStGr 32-41
Statistical inference for Voronoi	images • Archambault, S.; Moore, M.	95 CandJSt 23 11-20
Average densities of the	image and zero set of stable processes • Falconer, K. J.; Xiao, Y. M.	95 StocProc 55 271-283
Texture analysis of	imagery data for waste site characterization • Irvine, John M.; Barbour, Garth	94 ASAProEnvr 122-126
construction of thematic maps from satellite	imagery • Green, E. J.; Smith, A. F. M.; Strawderman, W. E. Bayesian	94 ASASrvyRsr 188-193
New loss functions in Bayesian	imaging • Rue, Havard	95 JASA 90 900-908
Bayesian approach to synthetic magnetic resonance	imaging • Glad, Ingrid K.; Sebastiani, Giovanni A	95 Biomtrka 82 237-250
ing seasonal adjustment models to correct for panel	imbalance in the Monthly Wholesale Trade Survey • Caldwell, C. V.; Monsell, B. C. et al. Us-	94 ASAProBuEc 83-88
Impact of stratification	imbalance on probability of type I error • Kazempour, Kazem	95 AmerStat 49 170-174
Immigration • Enchautegui, María E. Impacts of	immigrants on employment and wage growth: Evidence of positive effects of U.S.	94 ASAProSoSt 30-38
Poverty migration for U.S. States:	Immigration impacts • Frey, William H.	94 ASAProSoSt 135-140
processes regulated by emigration and large	immigration • Chen, Anyue; Renshaw, Eric Markov branching	95 StocProc 57 339-359

Incomplete data

Sun, J.; Kalbfleisch, J. D.	Estimation of the mean function of point processes based on panel count data •	95 StSinica 5 279-289
	Psychometric latent response models • Maris, Eric	95 Psymtrka 60 523-547
tions on the parameters • Kim, D. K.; Taylor, J. M. G.	The restricted EM algorithm for maximum likelihood estimation under linear restric-	95 JASA 90 708-716
attrition • Wang-Clow, F.; Lange, M.; Laird, N. M. et al.	A simulation study of estimators for rates of change in longitudinal studies with	95 StatMed 14 283-297
	A stochastic EM algorithm for mixtures with censored data • Chauveau, Didier	95 JStPlInf 46 1-25
formation loss on the dependence parameter due to	incomplete data from a one-parameter spatial Gaussian process • Krug, T.; Martin, R. J. In-	95 CommStA 24 3055-3072
Testing ordered alternatives in the presence of	incomplete data • Alvo, Mayer; Cabilio, Paul	95 JASA 90 1015-1024
Estimation of a normal covariance matrix with	incomplete data under Stein's loss • Konno, Yoshihiko	95 JMultiAn 52 308-324
Bayesian nonparametric estimation for	incomplete data via successive substitution sampling • Doss, Hani	94 AnlsStat 22 1763-1786
The analysis of	incomplete data using stochastic covariates • Rai, S. N.; Matthews, D. E.	95 CandJSt 23 29-42
Forecasting with	incomplete data • Lee, C. P.	95 CommStA 24 1255-1269
parametric estimation of a survival function from	incomplete and doubly censored data • Bravo, J. I.; De Fuentes, Í.; Fernández, A. J. A semi-	95 CommStA 24 2735-2752
linear estimator in linear models with respect to an	incomplete ellipsoidal restrictions • Lu, Changyu; Li, Weixin Admissibility of	95 AMaPeknE 37 289-295
linear estimators in linear models with respect to	incomplete ellipsoidal restrictions • Lu, Chang-yu Admissibility of inhomogeneous	95 CommStA 24 1737-1742
of parameters are simultaneously estimable in an	incomplete factorial design • Byar, D. P.; Freedman, L. S. et al. Identifying which sets	95 Ststcian 44 451-456
Optimal sequential decisions for	incomplete identification of group-testable items • Bar-Lev, S. K.; Parlar, M.; Perry, D.	95 SqtlAnly 14 41-57
Robin, Jean-Marc; Visser, Michael Analysing	incomplete individual employment histories using indirect inference • Magnac, Thierry;	95 JApEcmtx 10 153-169S
Customer scheduling with	incomplete information • Rieder, Ulrich; Weishaupt, Jürgen	95 PrEngInf 9 269-284
patients • Smith, Fraser Mixed model analysis of	incomplete longitudinal data from a high-dose trial of tacrine (Cognex) in Alzheimer's	94 ASAProBiop 301-306
an outlier while modeling the covariance matrix for	incomplete longitudinal data • Grady, James J.; Boyarsky, Gregory The effects of	94 ASAProBiop 441-445
selection techniques for the covariance matrix for	incomplete longitudinal data • Grady, James J.; Helms, Ronald W. Model	95 StatMed 14 1397-1416
Analysis of	incomplete multivariate data by simulation • Schafer, Joseph L.	95 ChapmnHall:Lnd 300p
the extended-maximum likelihood method, and	incomplete multivariate normal data • Haider The expectation-maximization method,	94 ASAProQlPr 281-285
G. et al. Longitudinal LISREL model estimation from	incomplete panel data using the EM algorithm and the Kalman smoother • Jansen, R. A. R.	95 StNeerla 49 362-377
Models for monthly penetrations with	incomplete panel data • Sikkel, D.; Hoogendoorn, A. W.	95 StNeerla 49 378-391
Green, J. R. Testing for haplotype concordance with	incomplete parental data – Modified Q and correction for modified N • Shah, S.;	93 AnHuGene 57 239-247
A nonparametric maximum likelihood estimator for	incomplete renewal data • McClean, Sally; Devine, Colum	95 Biomtrka 82 791-803
lation of the ... • Vach et al. Logistic regression with	incompletely observed categorical covariates – Investigating the sensitivity against vio-	95 StatMed 14 1315-1329
Kiviet, Jan F. On bias,	inconsistency, and efficiency of various estimators in dynamic panel data models •	95 JEconmtx 68 53-78
Detecting marker	inconsistencies in human gene mapping • Ott, Jurg	93 HumHrdty 43 25-30
Ahmadi, Susan Model	inconsistency, illustrated by the Cox proportional hazards model • Ford, Ian; Norrie, John;	95 StatMed 14 735-746
	Inconsistent maximum likelihood estimators for the Rasch model • Ghosh, Malay	95 StPrLet 23 165-170
Statistical analysis of observations of	increasing dimension • Girko, Vyacheslav L.	95 Kluwer:MA:Hllnd 286p
Restricted tests for and against the	increasing failure rate ordering on multinomial parameters • Bhattacharya, Bhaskar	95 StPrLet 25 309-316
Hawkes, Alan G.; Jalali, Assad The	increasing failure rate property of consecutive-k-out-of-n systems • Cui, Lirong;	95 PrEngInf 9 217-225
How pooling failure data may reverse	increasing failure rates • Gurland, John; Sethuraman, Jayaram	95 JASA 90 1416-1423
Dolsen, Dana E.; Machlis, Gary E.	Increasing response to personally-delivered mail-back questionnaires • Dillman, Don A.;	95 JOffStat 11 129-139
Malloy, K.; Gerstein, D.; Suter, N.; Nisenbaum, S.	Increasing response rates in follow-up surveys of drug treatment patients • Johnson, R.;	94 ASASrvyRsr 1263-1268
A lim inf result for the	increments of the Wiener process under the L_2-norm • Grill, K.	95 ProbTher 102 567-573
On the HIV	incubation distribution under AZT treatment • Tan, W. Y.	95 BiomtrcJ 37 319-338
drug resistance in AIDS chemotherapy and the HIV	incubation distribution • Tan, Wai Y. A stochastic model for	95 StPrLet 25 289-299
covariates: Application to estimating changes in the	incubation period of AIDS • Bacchetti, P.; Segal, M. R. Survival trees with time-dependent	95 LiDaAnal 1 35-47
Analysis of doubly censored data and the HIV	incubation period distribution in Western Australia • James, I. R.; Hogan, S. J.; Mallal, S. A.	95 AstrlJSt 37 299-312
Computer simulation analysis of Sartwell's	incubation period model for diseases with uncertain etiology – The effect of ... • Yi et al.	95 AJEpidm 142 363-368

Independence

Jordan	Dependency measure for sets of random events or random variables • Stoyanov,	95 StPrLet 23 13-20
	A semi-deterministic generalization of a (2×2) contingency table • Seneta, E.	95 ThPrMaSt 50 117-121
	• see also names of specific types of independence such as conditional – or quasi –	
for panel data models under conditional mean	independence assumptions • Wooldridge, Jeffrey M. Selection corrections	95 JEconmtx 68 115-132
et al. Nondetectability of restriction fragments and	independence of DNA fragment sizes within and between loci in RFLP ... • Chakraborty	94 AJHuGene 54 391-401
M. A caveat concerning	independence estimating equations with multivariate binary data • Fitzmaurice, Garrett	95 Biomtrcs 51 309-317
Carmen Sensitivity to prior	independence via Farlie-Gumbel-Morgenstern model • De la Horra, Julian; Fernandez,	95 CommStA 24 987-996
Seymour; Johnson, Wesley Testing	independence when the form of the bivariate distribution is unspecified • Geisser,	95 StatMed 14 1621-1639
Resnick, Sidney I.; Stărică, Cătălin Testing for	independence in heavy tailed and positive innovation time series • Feigin, Paul D.;	95 CommStSM 11 587-612
Ebrahimi, Nader Testing for	independence between hitting times of bivariate PQD processes • Ramalingam, T.;	95 SankhyaA 57 202-220
A conditional test of	independence of Markov chains • Solow, A. R.; Smith, W. K.; Recchia, C.	95 BiomtrcJ 37 973-977
their application to data • Tomizawa, Sadao Double	independence models for a 2×2 contingency table with the same classifications and	95 AllgeStA 79 140-145
The role of geometry in pairwise and mutual	independence • Schwertman, Neil C.; Kiser, Terry L.	95 AmerStat 49 56-59
A generalization of testing	independence of sets of variates • Suzukawa, Akio; Sato, Yoshiharu	95 JJapnStS 25 71-80
Sun, Yanqing On omnibus test for	independence of a survival time from a covariate • McKeague, Ian W.; Nikabadze, A. M.;	95 AnlsStat 23 450-475
Testing the hypothesis of	independence between two sets of variates • Wilcox, Rand R.	95 MultBehR 30 213-225
Two tests of pseudo random number generators for	independence and uniform distribution • Fleischer, Karlheinz	95 JStCmpSm 52 311-322
relationship between concentration addition and	independent action • Drescher et al. ... of the combined effects of substances: The	95 Biomtrcs 51 716-730
M. et al. A Kolmogorov inequality for the sum of	independent Bernoulli random variables with unequal means • Turner, D. W.; Young, D.	95 StPrLet 23 243-245
On the strong law of large numbers for blockwise	independent and blockwise orthogonal random variables • Gaposhkin, V. F.	94 ThProbAp 39 677-684
tests for the two sample problem with two	independent competing risks • Bagai, Isha; Deshpande, J. V. Nonparametric	95 JNnparSt 5 21-31
Reliability bounds for coherent structures with	independent components • Fu, J. C.; Koutras, M. V.	95 StPrLet 22 137-148
The relative empirical validity of dependent and	independent data collection in a panel survey • Hill, Daniel H.	94 JOffStat 10 359-380
gression in case-control studies: The effect of using	independent as dependent variables • Nagelkerke, N. J. D.; Moses, S. et al. Logistic re-	95 StatMed 14 769-775
On the existence of sequences of s by s	independent exchangeable or stationary random variables (French) • Bretagnolle, J. et al.	95 AlHenrPB 31 325-350
77 program for testing the differences among	independent first-order partial correlations • Silver, N. C. et al. A Microsoft FORTRAN	95 EdPsMeas 55 245-248
Lawrence M.; Nicol, David On the minimum of	independent geometrically distributed random variables • Ciardo, Gianfranco; Leemis,	95 StPrLet 23 313-326
Montgomery, Alan L.; Muller, Mervin E. A system-	independent graphical user interface for statistical software • Liu, Lon-Mu; Chan, Ki-Kan;	95 CmpStDA 19 23-44
Comparing two	independent groups via multiple quantiles • Wilcox, Rand R.	95 Ststcian 44 91-99
The law of large numbers for multiple sums of	independent identically distributed random variables • Klesov, O. I.	95 ThPrMaSt 50 77-87
of multidimensional processes with stationary	independent increments • Yamazato, M. Absolute continuity of transition probabilities	94 ThProbAp 39 347-354
Random variables with	independent integer and fractional parts • Marsaglia, G.	95 StNeerla 49 133-137
On the single-server queue with	independent MAP/G and M/GI input streams • Takine, Tetsuya; Hasegawa, Toshiharu	95 CommStSM 11 227-234
regression estimators when errors are non-	independent • Gonzalez Manteiga et al. ... linear regression models using non-parametric	95 CmpStDA 20 521-541
the number of change points in a sequence of	independent normal random variables • Lee, Chung-Bow Estimating	95 StPrLet 25 241-248
normality of maximum likelihood estimators for	independent not identically distributed censored samples ... • Wu ... and asymptotic	95 ChJApPSt 10 405-419
Records from a power model of	independent observations • Weissman, Ishay	95 JAppProb 32 982-990
The indices of the largest among n	independent observations • Weissman, Ishay	95 CommStSM 11 613-629

Convergence of	independent particle systems • Hoffman, John R.; Rosenthal, Jeffrey S.		95 StocProc 56 295-305
of an asymmetric hydrodynamical system of	independent particles (French) • Kipnis, C.; Léonard, C.	Large deviations	95 AIHenrPB 31 223-248
risk conditions in simultaneous estimation of	independent Poisson means • Liu, Ying-Hsing; Yang, Ming-Chung	Bounded	95 JStPlInf 47 319-331
inequalities for the Orlicz norm of a linear form of	independent pre-Gaussian random variables • Kurchenko, O. O.	Some	95 ThPrMaSt 50 99-102
Optimal unconditional tables for comparing two	independent proportions • Silva Mato, A.; Martín Andrés, A.		95 BiomtrcJ 37 821-836
Convergence of random sequences with	independent random indices I. • Korolev, V. Yu.		94 ThProbAp 39 282-297
Strong approximation theorems for	independent random variables and their applications • Shao, Qi-Man		95 JMultiAn 52 107-130
strong law of large numbers for sums of pairwise	independent random variables • Martikainen, Alexander	On the	95 StPrLet 25 21-26
Limit theorems of probability theory: Sequences of	independent random variables • Petrov, V. V.		95 OxfordUniv 292p
"Limit theorems of probability theory: Sequences of	independent random variables" • Petrov, V. V. (Auth); Pyke, R. (Rev)	Review of	95 ShrtBkRv 15/3 46-46
the almost certain rate of convergence of series of	independent random variables • Nam, Eunwoo; Rosalsky, Andrew	On	95 JKoreaSt 24 91-109
On symmetric sum of sub-	independent random variables • Hamedani, G. G.		95 CalcutSt 45 119-124
Chow-Robbins "fair" games problem for pairwise	independent random variables with multidimensional indices • Hong, D. H.; Hwang, S. Y.		95 BIMaASnc 23 7-11
and the law of the logarithm for weighted sums of	independent random variables with multidimensional indices • Li et al.	... large numbers	95 JMultiAn 52 181-198
Large deviations for	independent random walks on the line • Lee, Tzong-Yow		95 AnlsProb 23 1315-1331
confidence intervals for Pr($X < Y$) based on	independent samples of X and Y • Hamdy, M. I.	Distribution-free	95 CommStB 24 1005-1017
Stationarity of	independent sequences • Lacaze, B.		95 StPrLet 23 9-11
individual sequential experiments available from	independent studies • Mukhopadhyay et al.	... exponential clinical trials by combining	94 Metron 52/1 17-40
Meta-analysis: Reconciling the results of	independent studies • Olkin, Ingram		95 StatMed 14 457-472
Poisson approximation for unbounded functions, I:	Independent summands • Barbour, A. D.; Chen, Louis H. Y.; Choi, K. P.		95 StSinica 5 749-766
Coffman, Edward G., Jr; Downey, Peter Scheduling	independent tasks to minimize the makespan on identical machines • Bruno, John;		95 PrEngInf 9 447-456
A note on combining	independent tests for a common mean • Zhou, Leping; Mathew, Thomas		93 JCISS 18 79-84
On combining	independent tests in linear models • Jordan, Scott M.; Krishnamoorthy, K.		95 StPrLet 23 117-122
Sums of	independent triangular arrays and extreme order statistics • Janssen, Arnold		94 AnlsProb 22 1766-1793
Dichotomous choice models and limited	independent variables • Caudill, Steven B.		95 Ststcian 44 405-406
Seaks, Terry G. Estimating the functional form of the	independent variables in probit models • Greene, William H.; Greene, Laura Knapp;		95 ApplEcon 27 193-196
strategy • Concato, J. et al. Importance of events per	independent variable in proportional hazards analysis I. Background, goals, and general		95 JClEpid 48 1495-1501
cision of ... • Peduzzi et al. Importance of events per	independent variable in proportional hazards regression analysis II. Accuracy and pre-		95 JClEpid 48 1503-1510
	Indeterminate Markov systems • Hartfiel, D. J.		95 AplMaCmp 72 51-59
	Index		
	• see also names of specific types of indices such as Gini - or process capability -		
Cohen, Uri The extremal	index and clustering of high values for derived stationary sequences • Weissman, Ishay;		95 JAppProb 32 972-981
Trajstman, A. C.	Indices of disease severity in diseases that cause abscesses in target organs •		95 BiomtrcJ 37 995-1004
vintage of revised data with an application to the	index of industrial production • Patterson A state space approach to forecasting the final		95 JForcast 14 337-350
(Rev. German) Review of "Essays in	index number theory (Volume I)" • Diewert, W. E.; Nakamura, A. (Ed); Krämer, Walter		95 AllgeStA 79 97-97
Kennedy, B. W.	Index versus tandem selection after repeated generations of selection • Villanueva, B.;		93 ThApGene 85 706-712
Arcas, Antoni	Indexed pyramids and pyramidal dissimilarities (Spanish) • Capdevila, Carles;		95 Questo 19 131-151
A review of educational planning and statistics in	India with focus on contributions of Professor P. K. Bose • Srivastava, A. B. L.		94 CalcutSt 44 251-264
Regionalization of	Indian monsoon rainfall and long-term variability signals • Iyengar, R. N.; Basak, P.		94 JClimlgy 14 1095-1114
	Indifference zone approach		
	An optimal sequential selection procedure • Dragalin, V.		95 CommStA 24 429-443
Poston, C. A. et al. Solving the ranking and selection	indifference-zone formulation for normal distributions using computer software •		94 ASAProStCp 290-291
approach for convex weighting of direct and	indirect components of small domain predictors • Chaudhuri et al.	... design-cum-model	95 CommStA 24 2803-2813
Higashi, Masahiko; Nakajima, Hisao	Indirect effects in ecological interaction networks 1. The chain rule approach •		95 MaBisci 130 99-128
The relative accuracy of direct and	indirect estimates of state poverty rates • Schirm, Allen L.		94 ASASrvyRsr 1071-1076
	Indirect estimators in U.S. federal programs • Schaible, W. L.		95 Springer:Brln:NY 195p
incomplete individual employment histories using	indirect inference • Magnac, Thierry; Robin, Jean-Marc; Visser, Michael	Analysing	95 JApEcmtx 10 153-169S
D. The assessment of	individual bioequivalence via tolerance intervals • Chinchilli, Vernon M.; Esinhart, James		94 ASAProBiop 29-32
analysis of an algorithm in the theory of markets in	indivisible goods • Frieze, Alan; Pittel, Boris G.	Probabilistic	95 AnlsApPr 5 768-808
Fitting an extended	INDSCAL model to three-way proximity data • Carroll, J. Douglas; Winsberg, Suzanne		95 JClass 12 57-71
obtain robust products: Application to circuits with	inductance and resistance (Spanish) • Prat et al. Determination of design factor values to		95 Questo 19 153-169
comparison of classification in artificial intelligence,	induction versus a self-organising neural networks • Mulholland, M. et al.	A	95 Chemolab 30 117-128
tial prediction of soil salinity using electromagnetic	induction techniques 1. Statistical prediction models: A comparison ... • Lesch et al.	Spa-	95 WaterReR 31 373-386
prediction of soil salinity using electromagnetic	induction techniques. 2. An efficient spatial sampling algorithm ... • Lesch et al.	Spatial	95 WaterReR 31 387-398
	Inductive countersupport • Dorn, G. J. W.		95 ZeitWiss 26 187-189
A dynamic analysis of interfuel substitution in U.S.	industrial energy demand • Jones, Clifton T.		95 JBES 13 459-465
Assessing the losses caused by an	industrial intervention: A hierarchical Bayesian approach • Arjas, Elja; Liu, Liping		95 ApplStat 44 357-368
empirical linear prediction to quality assurance in	industrial manufacturing • Hwang, J. T. Gene; Liu, Hung-kung	Application of	94 ASAProPEng 175-180
Review of "Introduction to life prediction of	industrial plant materials" • Kowaka, M. (Ed); Ziegel, Eric R. (Rev)		95 Technmcs 37 457-457
Accuracy of indices of	industrial production in developing countries • Kmietowicz, Z. W.		95 Ststcian 44 295-307
of revised data with an application to the index of	industrial production • Patterson A state space approach to forecasting the final vintage		95 JForcast 14 337-350
Forecasting	industrial production using non-linear methods • Byers, J. D.; Peel, D. A.		95 JForcast 14 325-336
Deming's impact on	industrial statistics: Some reflections • Hahn, Gerald J.		95 AmerStat 49 336-341
Firm productivity in Israeli	industry 1979-1988 • Griliches, Zvi; Regev, Haim		95 JEconmtx 65 175-203
Some examples of applied statistics in automobile	industry • Taam, Winson		94 ASAProQlPr 258-263
oligopolistic markets with an application to the beer	industry • Chintagunta et al.	... repeated game advertising investment model for	95 DecisnSc 26 531-559
Statistics in the semiconductor	industry: Challenges in volume manufacturing (Disc: p9-11) • Gordon, Robert D.		94 ASAProPEng 6-8
A comparison of two methods of automated	industry coding • Rowland, John H.; Kinack, Mark D.		94 ASASrvyRsr 1130-1133
Statistics in the semiconductor	industry: A competitive necessity • Spencer, William J.; Tobias, Paul A.		95 AmerStat 49 245-249
Application of fuzzy logic control in	industry • van der Wal, A. J.		94 FuzzySet 74 33-41
The production and cost structure of Israeli	industry: Evidence from individual firm data • Bregman, Arie; Fuss, Melvyn; Regev, Haim		95 JEconmtx 65 45-81
Providing effective training	industry - An experience with robust design • Callanan, Terrance P.; Altland, Hank W.		94 ASAProStEd 126-131
and economic performance in U.S. manufacturing	industries: An exploratory analysis • Berndt, E. R. et al.	High-tech capital formation	95 JEconmtx 65 9-43
Statistics in the semiconductor	industry: Implementation strategy (Disc: p9-11) • Czitrom, A. Veronica		94 ASAProPEng 1-5
of equity capital in the property-liability insurance	industry • Cummins, J. David; Lamm-Tennant, Joan	Capital structure and the cost	94 InsrneME 15 187-201
Winter, Francine D. Evaluation of an automated	industry and occupation coding system • Harris, Kenneth W.; Chamblee, Ronald F.;		94 ASAProSoSt 174-177
The search for best practices in the U.S. petroleum	industry • Otto, Gordon H.; Thompson, Russell G.		94 ASAProQlPr 70-75
regulatory regime: The case of the US nuclear power	industry • Rust, John; Rothwell, Geoffrey	Optimal response to a shift in	95 JApEcmtx 10 75-118S
training and consulting in the semiconductor	industry: The SEMATECH approach • Reece, Jack E.; Spagon, Patrick D.	Linked	94 ASAProStEd 228-233
difficulties in applying TQM to a particular service	industry • Fricker, Ronald D., Jr	The inherent	94 ASAProQlPr 181-188
Review of "Using statistics in	industry" • Betteley, G.; Mettrick, N.; Sweeney, E.; Wilson, D. (Auth); Shade, John (Rev)		95 Ststcian 44 541-542
Separability test for the electricity supply	industry • Lee, Byung-Joo		95 JApEcmtx 10 49-60
pendence of business cycles in the manufacturing	industry: The use of leading indicators for forecasting and ... • Berk et al.	... interde-	95 JForcast 14 1-23
between village ... • Fitch et al. Are the biases and	inefficiencies of the EPI method of sampling relatively unimportant with reasonable		94 ASASrvyRsr 288-292

Gropper, D. M. Frontier estimation and firm-specific	inefficiency measures in the presence of heteroscedasticity • Caudill, S. B.; Ford, J. M.;	95 JBES 13 105-111
van der Laan, Mark J.; Wellner, Jon A.	Inefficient estimators of the bivariate survival function for three models • Gill, Richard D.;	95 AIHenrPB 31 545-597
	Inequality	
	• see also names of specific inequalities such as Bonferroni –, Cauchy-Schwarz –, Cramér-Rao –, or triangle –	
	Inequalities for the anisotropic Poisson polytope • Mecke, J.	95 AdvAppPr 27 56-62
On variational	inequality approach to Nash dynamic game of oil stockpiling • Zhao, L.; Yin, G.	95 StocAnAp 13 627-649
Applications of the van Trees	inequality: A Bayesian Cramér-Rao bound • Gill, Richard D.; Levit, Boris Y.	95 Bernoull 1 59-79
the nonlinear least squares problem with linear	inequality constraints • Sagara, Nobuko; Fukushima, Masao A hybrid method for solving	94 JOpRsJpn 38 55-69
Testing for and against a set of linear	inequality constraints in a multinomial setting • Barmi, Hammou El; Dykstra, Richard	95 CandJSt 23 131-143
for societal decisions involving distributional	inequalities • Keller, L. Robin; Sarin, Rakesh K. Fair processes	95 RiskAnal 15 49-59
proof and refinement of Wielandt's eigenvalue	inequality • Dümbgen, Lutz A simple	95 StPrLet 25 113-115
Variance	inequalities for functions of Gaussian variables • Houdré, Christian; Kagan, Abram	95 JTherPrb 8 23-30
Some applications of covariance identities and	inequalities to functions of multivariate normal variables • Houdré, Christian	95 JASA 90 965-968
Pérez-Abreu, Víctor Covariance identities and	inequalities for functionals on Wiener and Poisson spaces • Houdré, Christian;	95 AnlsProb 23 400-419
The missing factor in Hoeffding's	inequalities • Talagrand, Michel	95 AIHenrPB 31 689-702
Measuring interdistributional income	inequality – Inferential procedures (German) • Schmid, Friedrich	95 AllgeStA 79 278-288
one-sided tests for hypotheses about linear	inequalities • Liu, Huimei; Berger, Roger L. Uniformly more powerful,	95 AnlsStat 23 55-72
decision processes and the systems of linear	inequalities • Wakuta, Kazuyoshi Vector-valued Markov	95 StocProc 56 159-169
On some	inequalities of local time • Chou, Ching-Sung	95 JTherPrb 8 17-22
Blackwell's strengthening of Azuma's martingale	inequality • Ross, Sheldon M. Extending	95 PrEngInf 9 493-496
counting processes • van de Geer, Sara Exponential	inequalities for martingales, with application to maximum likelihood estimation for	95 AnlsStat 23 1779-1801
The axiomatic derivation of a class of dynamic	inequality measures • Stich, Andreas	95 Metron 53 31-41
S. J.; Szulga, Jerzy Contraction and decoupling	inequalities for multilinear forms and U-statistics • de la Peña, V. H.; Montgomery-Smith,	94 AnlsProb 22 1745-1765
The statistics of poverty and	inequality (jse/v3n2/datasets.rouncefield) • Rouncefield, Mary	95 JStEd 3/2 10kb
On some	inequalities related to the range of some statistical indices • Benedetti, Carlo	94 Metron 52/1 189-215
The optimal critical value of a pre-test for an	inequality restriction in a mis-specified regression model • Wan, Alan T. K.	95 AstrlJSt 37 73-81
Estimation with multiple	inequality restrictions in a model with multivariate t error terms • Wan, Alan T. K.	94 ASAProBuEc 223-228
Guiard, V. Sharp	inequalities between skewness and kurtosis for unimodal distributions • Teuscher, F.;	95 StPrLet 22 257-260
of infinitely divisible random vectors and Slepian	inequalities • Samorodnitsky, Gennady; Taqqu, Murad S. Lévy measures	94 AnlsProb 22 1930-1956
rank versus category in measuring subjective social	inequality • Hoffmeyer-Zlotnik, Jürgen H. P. Effects of	94 ASASrvyRsr 1322-1326
The world's 100 largest cities: A case study of	inequality • Drew, David; Steyne, Dave	95 TeachgSt 17/1 17-21
Binder et al. Estimating some measures of income	inequality from survey data: An application of the estimating equations approach •	95 SurvyM 21 137-145
An exponential	inequality for U-statistics with applications to testing • Bickel, Peter J.; Ritov, Ya'acov	95 PrEngInf 9 39-52
	Inequalities for variances of some functions of random variables • Nollau, V.	95 StHefte 36 163-174
Stochastic	inertial manifold • Bensoussan, Alain; Flandoli, Franco	95 Stochatx 53 13-39
spaced repeated multivariate measures – A study of	infant cardiac catherization stress responses • Zhang, F. et al. Analyzing long irregularly	94 ASAProBiop 365-370
techniques: Alaskan community characteristics and	infant mortality (Disc: p687-689) • Andes et al. ... using geographic information system	95 StatMed 14 481-490
appearance of Plasmodium falciparum parasites in	infants • Gottschau et al. ... and multistate compartmental models in the analysis of first	95 StatMed 14 2727-2736
in clinical implications of CD4 counts among HIV-	infected homosexual men and injection drug using men and women • Paik et al. Difference	95 StatMed 14 1889-1900
trial and subsequent open label extension in HIV-	infected subjects • Hwang et al. ... models to the analysis of a double-blind comparative	94 ASAProBiop 194-199
... method to estimate the age and period HIV	infection intensity, considering the susceptible population • Verdecchia et al.	95 StatMed 14 1513-1530
approach to smoothing in backcalculation of HIV	infection rates • Liao, Jiangang; Brookmeyer, Ron An empirical Bayes	95 Biomtrcs 51 579-588
Lee, Sho Rong; Tang, Si Chin Characterization of HIV	infection and seroconversion by a stochastic model of the HIV epidemic • Tan, Wai-Yuan;	95 MaBisci 126 81-123
mortality in the human immunodeficiency virus	infection • Salvaggio, Antonino Non-AIDS related	95 ListyBio 32 23-33
heterogeneity, nonlinear dynamics and chaos in	infectious diseases • Grenfell, B. T.; Kleczkowski, A.; Gilligan, C. A.; Bolker, B. M. Spatial	95 StMedRes 4 160-183
A random-effects model for analysis of	infectious disease final-state data • Baker, Rose D.; Stevens, Robert H.	95 Biomtrcs 51 956-968
distribution on transmission and control of highly	infectious diseases • Becker, Niels G.; Dietz, Klaus The effect of household	95 MaBisci 127 207-219
Ecology of	infectious diseases in natural populations • Grenfell, Bryan; Dobson, Andrew (Ed)	95 CambridgeUniv 532p
The empirical content of canonical models of	infectious diseases: The proportional hazard specification • Geoffard, P.-Y.; Philipson, T.	95 Biomtrka 82 101-111
rate parameter and its relationship with latent and	infectious periods in plant disease epidemics • Dayananda, P. W. A. et al. Estimation of	95 Biomtrcs 51 284-292
an epidemic model with several different types of	infective in a large population • Ball, Frank; Clancy, Damian The final outcome of	95 JAppProb 32 579-590
	Inference	
Linssen, H. N.; Overdijk, D. A.	The reduction and ancillary classification of observations • van Berkum, E. E. M.;	95 JStPlInf 45 347-372
	• see also names of specific types of inference such as nonparametric –	
Ying, Zhiliang Semiparametric	inference for the accelerated life model with time-dependent covariates • Lin, D. Y.;	95 JStPlInf 44 47-63
Achcar, Jorge Alberto	Inferences for accelerated life tests considering a bivariate exponential distribution •	95 MaOpfStS 26 269-283
chain Monte Carlo with applications to ancestral	inference • Geyer, Charles J.; Thompson, Elizabeth A. Annealing Markov	95 JASA 90 909-920
alized half logistic distribution with applications to	inference • Balakrishnan et al. ... and product moments of order statistics from a gener-	95 JStCmpSm 52 385-398
Zucker, David M.; Zerbe, Gary O.; Wu, Margaret C.	Inference for the association between coefficients in a multivariate growth curve model •	95 Biomtrcs 51 413-424
Shih, Joanna H.; Louis, Thomas A.	Inferences on the association parameter in copula models for bivariate survival data •	95 Biomtrcs 51 1384-1399
Stephen (Rev) Review of	"Inference and asymptotics" • Barndorff-Nielsen, O. E.; Cox, D. R. (Auth); Corcoran,	95 Ststcian 44 279-279
Thomas J. (Rev) Review of	"Inference and asymptotics" • Barndorff-Nielsen, O. E.; Cox, D. R. (Auth); DiCiccio,	95 JASA 90 1129-1130
J. (Rev) Review of	"Inference and asymptotics" • Barndorff-Nielsen, O. E.; Cox, D. R. (Auth); Hingley, Peter	95 Ststcian 44 407-408
Review of	"Inference and asymptotics" • Barndorff-Nielsen, O. E.; Cox, D. R. (Auth); Pfanzagl, J. (Rev)	95 Metrika 42 275-275
	Inference for autocorrelations by resampling • Romano, Joseph P.; Thombs, Lori A.	94 ASAProStCp 1-10
Glickman, Mark E.; Stern, Hal S.	Inference from an autoregressive state-space model via iterative simulation •	94 ASAProSpts 72-76
Basu, Srabashi Robust minimum distance	inference based on combined distances • Park, Chanseok; Basu, Ayanendranath;	95 CommStB 24 653-673
p173-199) • Liang, Kyung-Yee; Zeger, Scott L.	Inference based on estimating functions in the presence of nuisance parameters (Disc:	95 StatSci 10 158-173
Estimation of nuisance parameters for	inference based on least absolute deviations • Niemiro, W.	95 ZastApMa 22 515-529
Chambers, R. L. et al. Robust sample survey	inference via bootstrapping and bias correction: The case of the ratio estimator •	94 ASAProStCp 51-60
Review of "Statistical	inference for branching processes" • Guttorp, P. (Auth); Panaretos, J. (Rev)	95 Biomtrcs 51 383-384
Review of "Statistical	inference for branching processes" • Guttorp, Peter (Auth); Basawa, Ishwar V. (Rev)	95 JStPlInf 47 393-394
Some aspects of	inference for chain binomial models • Yip, Paul; Huggins, Richard	95 StocAnAp 13 355-367
p159-175) • Boswijk, H. Peter Efficient	inference on cointegration parameters in structural error correction models (Disc:	95 JEconmtx 69 133-158
The roles of conditioning in	inference (Disc: p173-199) • Reid, N.	95 StatSci 10 138-157
Frequentist post-data	inference • Goutis, Constantinos; Casella, George	95 IntStRvw 63 325-344
Spatial disease clusters: Detection and	inference • Kulldorff, Martin; Nagarwalla, Neville	95 StatMed 14 799-810
p417-430) • Raftery, A. E.; Givens, G. H.; Zeh, J. E.	Inference from a deterministic population dynamics model for Bowhead whales (Disc:	95 JASA 90 402-416
Schweder, Tore Comment On	"Inference from a deterministic population dynamics model for Bowhead whales" •	95 JASA 90 420-423
Distributional	inference • Kroese, A. H.; van der Meulen, E. A.; Poortema, K.; Schaafsma, W.	95 StNeerla 49 63-82
Evaluation of methods for ecological	inference • Cleave, N.; Brown, P. J.; Payne, C. D.	95 JRSS-A 158 55-72
Review of "Estimation and	inference in econometrics" • Davidson, R.; Mackinnon, J. G. (Auth); Smith, Richard J. (Rev)	95 JApEcmtx 10 339-341
Zinde-Walsh, Victoria (Rev) Review of "Estimation and	inference in econometrics" • Davidson, Russell; MacKinnon, James G. (Auth);	95 EconTher 11 631-635

	Finite and	infinite systems of interacting diffusions • Cox, J. T.; Greven, Andreas; Shiga, Tokuzo	95 ProbTher 103 165-197
	Bivariate autoregressive processes with finite and	infinite variance errors • Knight, Keith	94 ASAProBuEc 28-34
	Parameter estimation for ARMA models with	infinite variance innovations • Mikosch, T.; Gadrich, T.; Klüppelberg, C.; Adler, R. J.	95 AnlsStat 23 305-326
	Model selection for	infinite variance time series • Glendinning, R. H.	95 CommStA 24 889-910
	Gaussian random processes: Stochastic models with	infinite variance" • Samorodnitsky, G.; Taqqu, M. S. (Auth) et al. Review of "Stable non-	95 JASA 90 805-806
	gence for weakly dependent random variables with	infinite variance • Davis, Richard A.; Hsing, Tailen Point process and partial sum conver-	95 AnlsProb 23 879-917

Infinitely divisible distribution

Pakes, Anthony G.		Characterization of discrete laws via mixed sums and Markov branching processes •	95 StocProc 55 285-300
		Factorization theory for probability distributions • Bondesson, Lennart	95 ScanActJ 44-53
Penrose, Mathew D.		Generalised two-sample U-statistics and a two-species reaction-diffusion model •	95 StocProc 55 57-64
$L\alpha$-space • Misiewicz, Jolanta K.		Infinite divisibility of sub-stable processes. Part I. Geometry of sub-spaces of	95 StocProc 56 100-116
modelling • Rachev, S. T.; Samorodnitsky, G.		Limit laws for a stochastic process and random recursion arising in probabilistic	95 AdvAppPr 27 185-202
Veeh, J. A.		Operator stable probability measures: An overview • Hudson, W. N.; Mason, J. D.;	94 ThProbAp 39 264-276
Samorodnitsky, Gennady		Sample quantiles of heavy tailed stochastic processes • Embrechts, Paul;	95 StocProc 59 217-233
		Sums of independent triangular arrays and extreme order statistics • Janssen, Arnold	94 AnlsProb 22 1766-1793
Sara C.		Wave front propagation for a Cauchy problem with a fast component • Carmona,	95 Stochatx 52 81-105
	Multimodal convolutions of unimodal	infinitely divisible distributions • Sato, K.	94 ThProbAp 39 336-347
	Approximate random variate generation from	infinitely divisible distributions with applications to Bayesian inference • Damien, P. et al.	95 JRSS-B 57 547-563
	Integral transforms with	infinitely divisible kernels • Finkelstein, M.; Scheiberg, S.; Tucker, H. G.	94 ThProbAp 39 670-676
Sifi, Mohamed	Central limit theorem and	infinitely divisible probabilities associated with partial differential operators •	95 JTherPrb 8 475-499
	Association of	infinitely divisible random vectors • Samorodnitsky, Gennady	95 StocProc 55 45-55
Taqqu, Murad S.	Lévy measures of	infinitely divisible random vectors and Slepian inequalities • Samorodnitsky, Gennady;	94 AnlsProb 22 1930-1956
Samorodnitsky, Gennady	Functionals of	infinitely divisible stochastic processes with exponential tails • Braverman, Michael;	95 StocProc 56 207-231
	Random walks of	infinitely many particles • Révész, P.	94 WrldSciPub:Sing 191p
	Review of "Random walks of	infinitely many particles" • Révész, P. (Auth); Heyde, C. C. (Rev)	95 ShrtBkRv 15/3 47-47
	Unrooted genealogical tree probabilities in the	infinitely-many-sites model • Griffiths, R. C.; Tavaré, Simon	95 MaBisci 127 77-98
	Density based classes of priors:	Infinitesimal properties and approximations • Ruggeri, Fabrizio; Wasserman, Larry	95 JStPlInf 46 311-324
	The risk	inflation criterion for multiple regression • Foster, Dean P.; George, Edward I.	94 AnlsStat 22 1947-1975
	Graphical interpretation of variance	inflation factors • Stine, Robert A.	95 AmerStat 49 53-56
	A score test for zero	inflation in a Poisson distribution • van den Broek, Jan	95 Biomtrcs 51 738-743
	Long memory in	inflation rates: International evidence • Hassler, Uwe; Wolters, Jürgen	95 JBES 13 37-45
	Cointegration analysis of stock prices,	inflation and real economic activity • Raj, Baldev; Amoako-Adu, Ben	94 ASAProBuEc 190-195
	adjustments noise and the expected rate of	inflation in a simple inflation-unemployment model • Yeung, David Wing-Kay Employment	95 StocAnAp 13 125-135
	Forecasts of	inflation from VAR models • Webb, Roy H.	95 JForcast 14 267-285
	European	inflation and its variability • Silver, Mick; Ioannidis, Christos	95 Ststcian 44 309-322
	figurable combined forecasts in a non-stationary	inflationary environment • Volkov, V. Ya.; Gladkov, Yu. M. Recon-	95 JForcast 14 395-403

Inflection point

		On qualitative smoothness of kernel density estimates • Mammen, E.	95 MaOpfStS 26 253-267

Influence

Glenn W.		Mapping influence regions in hierarchical clustering • Cheng, Richard; Milligan,	95 MultBehR 30 547-576
Shi, Jianqing	Local	influence analysis for regression transformation models (Chinese) • Wei, Bocheng;	95 AMaAPknC 17 132-143
	A note on bounded	influence in Bayesian analysis • Pericchi, Luis R.; Sansó, Bruno	95 Biomtrka 82 223-225
	On the estimation of the	influence curve • Cuevas, Antonio; Romo, Juan	95 CandJSt 23 1-9
	Hierarchical clustering algorithms with	influence detection • Cheng, Richard; Milligan, Glenn W.	95 EdPsMeas 55 237-244

Influence diagnostics

Anderson, Jon E.; Dayton, J. David		Instructional regression modules using XLISP-STAT (jse/v3n1/anderson) •	95 JStEd 3/1 37kb

Influence diagram

Abramson, Bruce; Finizza, Anthony		Probabilistic forecasts from probabilistic models: A case study in the oil market •	95 IntJFor 11 63-72

Influence function

Van Ness, John W.		Breakdown points for redescending M-estimates of location • Yang, Jim J.;	95 CommStA 24 1769-1787
Frangos, Chris C.; Schucany, William R.		Improved bootstrap confidence intervals in certain toxicological experiments •	95 CommStA 24 829-844
Zaiqian		Robust estimation for semiparametric exponential mixture models • Shen, Larry	95 StSinica 5 333-349
Marianthi; Horowitz, Joel L.; Lenth, Russell V.		Robust scale estimation based on the empirical characteristic function • Markatou,	95 StPrLet 25 185-192
characteristic function • Markatou, M.; Horowitz, J. L.		Robust scale estimation in the error-components model using the empirical	95 CandJSt 23 369-381
Pacifico, M. Perone; Salinetti, G.		Robust statistics: From classical to Bayesian analysis • Conigliani, C.; Dall'aglio, M.;	94 Metron 52/3 89-109
Tsimikas, J.		Robust tests based on the sample characteristic function • Lenth, R. V.; Markatou, M.;	95 AstrlJSt 37 45-60
Rousseeuw, Peter J.; Croux, Christophe; Hössjer, Ola		Sensitivity functions and numerical analysis of the repeated median slope •	95 CmpStQ 10 71-90
Ruts, Ida	The repeated median intercept estimator:	Influence function and asymptotic normality • Hössjer, Ola; Rousseeuw, Peter J.;	95 JMultiAn 52 45-72
	On the	influence functions of certain bivariate medians • Niinimaa, A.; Oja, H.	95 JRSS-B 57 565-574
	Empirical	influence functions for the Cox regression model • Dubas, Françoise	95 Student 1 87-98
Naranjo, Joshua D.; McKean, Joseph W.		Influence functions for rank-based procedures in the linear model • Witt, Lee D.;	95 JNnparSt 5 339-358
	On optimal B-robust	influence functions in semiparametric models • Shen, Larry Z.	95 AnlsStat 23 968-989
Tombuyses, B.	The use of the component	influence graph to reduce the size of Markovian availability problems • Devooght, J.;	94 RelEngSS 46 237-244
	Deletion	influence and masking in regression • Lawrance, A. J.	95 JRSS-B 57 181-189
	influential subsets in linear regression by using an	influence matrix • Peña, Daniel; Yohai, Victor J. The detection of	95 JRSS-B 57 145-156
	influential subsets in linear regression by using an	influence matrix" (V57 p145-156) • Peña, D.; Yohai, V. J. Correction to "The detection of	95 JRSS-B 57 611-611
	A unifying representation of some case-deletion	influence measures in univariate and multivariate linear regression • Hadi, A. S. et al.	95 JStPlInf 46 123-135
Redfern, E.; Clark, S.; Tight, M.; Davies, N.	An	influence method for outlier detection applied to time series traffic data • Watson, S. M.;	95 JAppStat 22 135-149
	Local	influence in multivariate regression • Kim, Myung Geun	95 CommStA 24 1271-1278
	Assessment of	influence in non-linear measurement error models • Zhao, Yuejen; Lee, Andy H.	95 JAppStat 22 215-225
	Some small-sample results on a bounded	influence rank regression method • Wilcox, Rand R.	95 CommStA 24 881-888
	Mapping	influence regions in hierarchical clustering • Cheng, Richard; Milligan, Glenn W.	95 MultBehR 30 547-576
	Computation of determinantal subset	influence in regression • Barrett, Bruce E.; Gray, J. Brian	94 ASAProStCp 164-169
		Influence and residuals in restricted generalized linear models • Paula, Gilberto A.	95 JStCmpSm 51 315-331
		Influence and stability in non symmetrical correspondence analysis • Balbi, Simona	94 Metron 52/3 111-128
	Local	influence in structural equation models • Cadigan, N. G.	95 StrEqMod 2 13-30

Influential observation

		A cautionary note on the use of generalized Cook-type measures • Fung, Wing K.	95 CmpStDA 19 321-326
		Diagnostics in linear discriminant analysis • Fung, Wing K.	95 JASA 90 952-956
		Exploring Cook's statistic graphically • Hines, R. J. O'Hara; Hines, W. G. S.	95 AmerStat 49 389-394
correlation matrices • Fung, W. K.; Kwan, C. W.		Sensitivity analysis in factor analysis: Difference between using covariance and	95 Psymtrka 60 607-614
Fung, Wing K.	Detecting	influential observations for estimated probabilities in multiple discriminant analysis •	95 CmpStDA 20 557-568
	Detection of	influential observations in functional errors-in-variables models • Abdullah, Mokhtar B.	95 CommStA 24 1585-1595
Morgan, B. J. T.	Identifying	influential observations in hierarchical cluster analysis • Jolliffe, I. T.; Jones, B.;	95 JAppStat 22 61-80

for the identification of multiple outliers and	influential observations • Fung, Wing K. A stepwise method	95 SoAfrStJ 29 51-64
Kim, Choongrak; Jeon, Jong-Woo Detecting	influential observations on the smoothing parameter in nonparametric regression •	95 JKoreaSt 24 495-506
Yohai, Victor J. The detection of	influential subsets in linear regression by using an influence matrix • Peña, Daniel;	95 JRSS-B 57 145-156
Peña, D.; Yohai, V. J. Correction to "The detection of	influential subsets in linear regression by using an influence matrix" (V57 p145-156) •	95 JRSS-B 57 611-611

Information

	• see also names of specific types of information such as Fisher – or Akaike –	
Review of "Counting people in the	information age" • Stefey, Duane L.; Bradburn, Norman M. (Ed); Wolter, Kirk M. (Rev)	95 Chance 8/3 44-49
where there are several levels of auxiliary	information • Dupont, F. Alternative adjustments	95 SurvyM 21 125-135
and quantile estimation in the presence of auxiliary	information • Zhang, Biao M-estimation	95 JStPlInf 44 77-94
Median estimation using auxiliary	information • Meeden, Glen	95 SurvyM 21 71-77
P.; Wellner, J. A. (Auth); Plachky, D. (Rev) Review of	"Information bounds and nonparametric maximum likelihood estimation" • Groeneboom,	95 StatDecn 13 94-95
Särndal, C. Maximizing the use of auxiliary	information for calibration and regression estimation • Estevao, V.; Patak, Z.;	94 ASASrvyRsr 42-48
B.; Klar, R. (Ed); Windham, M. P. (Rev) Review of	"Information and classification: Concepts, methods and applications" • Opitz, O.; Lausen,	95 JClass 12 157-158
	Information and conditional inference • Severini, Thomas A.	95 JASA 90 1341-1346
1: Statistical examination of the effectiveness and	information content by factor analysis • Schulz et al. ... spectra for pattern recognition.	95 JChemtrc 9 143-168

Information criterion

Kanjilal, P. P.	On a method of identification of best subset model from full AR-model • Sarkar, A.;	95 CommStA 24 1551-1567
From data to	information: How to define the context? • Goldschmidt, H. M. J.; Lent, R. W.	95 Chemolab 28 181-192
Lifetime modeling via Bayesian	information distinguishability • Mazzuchi, Thomas A.; Soyer, Refik; Soofi, Ehsan	94 ASAProBays 238-245
Ebrahimi, Nader; Habibullah, Mohamed	Information distinguishability with application to analysis of failure data • Soofi, Ehsan S.;	95 JASA 90 657-668
Review of "Econometrics of	information and efficiency" • Sengupta, J. K. (Auth); Shanmugam, Ramalingam (Rev)	95 Ststcian 44 290-291
Review of "Envisioning	information" • Tufte, E. R. (Auth); Crescimanni, Annamaria (Rev)	94 Metron 52/1 218-218
totals and distribution functions using auxiliary	information at the estimation stage • Rao, J. N. K. Estimating	94 JOffStat 10 153-165
– The key to scholarly visibility and efficiency in an	information explosion • Gbur, Edward E., Jr; Trumbo, Bruce E. Key words and phrases	95 AmerStat 49 29-33
Quasi-likelihood or extended quasi-likelihood? An	information-geometric approach • Vos, Paul W.	95 AnInStMa 47 49-64
	Information geometry of the EM and em algorithms for neural networks • Amari, Shun-ichi	95 NuralNet 8 1379-1408
Customer scheduling with incomplete	information • Rieder, Ulrich; Weishaupt, Jürgen	95 PrEngInf 9 269-284
	Information inequalities for the minimax risk • Sato, Michikazu; Akahira, Masafumi	95 JJapnStS 25 151-158
Yip, Paul S. F. A note on	information loss in analyzing a mixture model of count data • Fong, Daniel Y. T.;	95 CommStA 24 3197-3209
parameter spatial Gaussian process • Krug, T. et al.	Information loss on the dependence parameter due to incomplete data from a one-	95 CommStA 24 3055-3072
Numerical experiences when measuring the loss of	information • Mouchart, Michel; Scheihing, Eliana Approximate Bayesian inferences:	94 ASAProBays 221-226

Information management

	Changes in statistical technology • Keller, Wouter J.	95 JOffStat 11 115-127
events • Chute, C. G.; Yang, Y.	An overview of statistical methods for the classification and retrieval of patient	95 MInfoMed 34 104-110
On	information matrices in case-control studies • Wang, C. Y.; Wang, Suojin	95 StPrLet 22 269-274
Lewis, S. M.; Dean, A. M. Approximate	information matrices for estimating a given set of contrasts • Kao, L.-J.; Yates, P. K.;	95 StSinica 5 593-598
of a censored experiment in terms of Fisher	information matrix • Turrero, Agustin Relative efficiency	95 CommStA 24 1169-1191
multiplier and likelihood ratio tests when the	information matrix is singular • El-Helbawy, A. T.; Hassan, T. On the Wald, Lagrangian	94 JItalStS 3 51-60
Anderson, Donald A. The spectrum of the	information matrix for s^n parallel flats fractions when s is prime • BuHamra, Sana;	95 JStPlInf 47 333-345
Data, metadata and	information • Hand, D. J.	93 StatJUN 10 143-152
Phelan, Michael J. Asymptotic	information for parametric estimation from an equilibrium particle process •	95 StPrLet 24 193-198
Gerard J. P. Psychometric and	information processing properties of selected response time models • Van Breukelen,	95 Psymtrka 60 95-113
	Information-processing slope scores can be reliable! • Damos, Diane L.; Carter, Robert C.	95 HumanFac 37 424-429
mixed models • Gilmour, A. R. et al. Average	information REML: An efficient algorithm for variance parameter estimation in linear	95 Biomtrcs 51 1440-1450

Information retrieval

information explosion • Gbur, E. E., Jr; Trumbo, B. E.	Key words and phrases – The key to scholarly visibility and efficiency in an	95 AmerStat 49 29-33
Maximizing	information in several orthonormal directions • Constantine, Gregory M.	93 JCISS 18 123-129
the entropy density of nonhomogeneous Markov	information source • Liu, Wen; Yang, Weiguo A limit theorem for	95 StPrLet 22 295-301
Some limit properties of nonhomogeneous Markov	information sources • Liu, Wen; Yang, Weiguo	95 JCISS 20 279-292
	Information from subjects by variables tables (French) • Antoni, A.; Dhorne, T.	95 RStApp 43/4 43-61
spatial pattern exploration tools for geographical	information systems applications • Openshaw, Stan Developing automated and smart	95 Ststcian 44 3-16

Information technology

	Using IT to investigate world statistics • Hudson, Brian	95 TeachgSt 17/2 57-60
Warner's contributions to statistically balanced	information technology • Fienberg, Stephen E.; Jazairi, Nuri Stanley	95 SurvyM 21 7-12
et al. A model-selection approach to assessing the	information in the term structure using linear models and artificial neural ... • Swanson	95 JBES 13 265-275
Markov approximations to D-BMAPs:	Information theoretic bounds on queueing performance • Rananand, Nol	95 CommStSM 11 713-734

Information theory

Kenneth P.; White, Gary C.; Anderson, David R.	Model selection strategy in the analysis of capture-recapture data • Burnham,	95 Biomtrcs 51 888-898
Clegg, D. (Rev) Review of	"Information theory in analytical chemistry" • Eckschlager, K.; Danzer, K. (Auth);	95 Chemolab 28 207-208
On the	information theory and statistics • Soofi, Ehsan S.	92 ProIrStC1 203-229
	Information tradeoff • Wasserman, L.; Clarke, B.	95 Test 4 19-38
quential logrank tests in maximum duration versus	information trials • Kim, K.; Boucher, H.; Tsiatis, A. A. Design and analysis of group se-	95 Biomtrcs 51 988-1000
Distance in variation and a Fisher-type	information • Papadatos, N.; Papathanasiou, V.	95 MaMethSt 4 230-237
Estimating distribution functions with auxiliary	information using poststratification • Nascimento Silva, P. L. D.; Skinner, C. J.	95 JOffStat 11 277-294
An analytic method for randomized trials with	informative censoring: Part II • Robins, James M.	95 LiDaAnal 1 417-434
An analytic method for randomized trials with	informative censoring: Part I • Robins, James M.	95 LiDaAnal 1 241-254
models for longitudinal binary responses with	informative drop-outs • Fitzmaurice, G. M.; Molenberghs, G.; Lipsitz, S. R. Regression	95 JRSS-B 57 691-704
Bayesian methods for categorical data under	informative general censoring • Mimoso Paulino, C. D.; Pereira, C. A. D. B.	95 Biomtrka 82 439-446
generalized linear model with random effects for	informative missing data • Follmann, Dean; Wu, Margaret An approximate	95 Biomtrcs 51 151-168
EM mixed model analysis of data from	informatively censored normal distributions • Smith, Fraser B.; Helms, Ronald W.	95 Biomtrcs 51 425-436
... in an ordinary linear model: Some elements of an	informed use of the Belsley, Kuh and Welsch indicators (French) • Erkel-Rousse	95 RStApp 43/4 19-42
Hildrum, K.; Isaksson, T. et al. Review of "Near	infra-red spectroscopy: Bridging the gap between data analysis and NIR applications" •	95 Technmcs 37 126-127
analysis of bandpass filtered Fourier transform	infrared interferograms • Shaffer et al. ... design protocol for the pattern recognition	95 Chemolab 29 89-108
sion as a multivariate tool for the interpretation of	infrared spectra • Luinge, H. J.; van der Maas, J. H.; Visser, T. Partial least squares regres-	95 Chemolab 28 129-138
L.; Kateman, G. et al. Spectral curve fitting of	infrared spectra obtained from semi-crystalline polyester yarns • de Weijer, A. P.; Buydens,	95 Chemolab 28 149-164
and information ... • Schulz et al. Simple encoding of	infrared spectra for pattern recognition. 1: Statistical examination of the effectiveness	95 JChemtrc 9 143-168
by chemometric analysis of Fourier transform	infrared spectra • Lindblom et al. ... determination of stabiliser in a single base propellant	95 Chemolab 29 243-254
assessment • Sen, Pranab Kumar; Margolin, Barry H.	Inhalation toxicology: Awareness, identifiability, statistical perspectives and risk	95 SankhyaB 57 252-276
in a two-breed population with dominance	inheritance • Lo, L. L.; Fernando, R. L. et al. Theory for modelling means and covariances	95 ThApGene 90 49-62
of genes survived in offspring conditional on	inheritance of flanking markers • Guo, Sun-Wei Proportion	94 Genetcs 138 953-962
ordered-notation method and new tests for mode of	inheritance • Thomson, Glenys Analysis of complex human genetic traits: An	95 AJHuGene 57 474-486
Random walk in a strongly	inhomogeneous environment and invasion percolation • Newman, C. M.; Stein, D. L.	95 AIHenrPB 31 249-261
restrictions • Lu, Chang-yu Admissibility of	inhomogeneous linear estimators in linear models with respect to incomplete ellipsoidal	95 CommStA 24 1737-1742
Local adaptation to	inhomogeneous smoothness: Resolution level • Lepskii, O. V.; Spokoiny, V. G.	95 MaMethSt 4 239-258

Diaconis, P. Hammersley's	interacting particle process and longest increasing subsequences • Aldous, D.;	95 ProbTher 103 199-213
A limit theorem for a class of	interacting particle systems • Simonelli, Italo	95 AnlsProb 23 141-156
and structure of interfaces in a Kawasaki + Glauber	interacting particle system • Giacomin, G. Onset	95 ProbTher 103 1-24
non-equilibrium distribution for a non-gradient	interacting particle system • Kipnis, C. et al. Macroscopic properties of a stationary	95 AlHenrPB 31 191-221
Large deviations for long range	interacting particle systems with jumps • Léonard, Christian	95 AlHenrPB 31 289-323
Large deviations for empirical process of mean-field	interacting particle system with unbounded jumps • Feng, Shui	94 AnlsProb 22 2122-2151
The diffusive phase of a model of self-	interacting walks • Brydges, D. C.; Slade, G.	95 ProbTher 103 285-315
	Interaction	
development • International Working Group ...	Capture-recapture and multiple-record systems estimation 1: History and theoretical	95 AJEpidm 141 1047-1058
	Detecting and handling heteroscedasticity in yield trial data • Piepho, Hans-Peter	95 CommStB 24 243-274
Wasserman, Stanley	Log-multiplicative models for valued social relations • Anderson, Carolyn J.;	95 SocMethR 24 96-127
	Practical analysis of factorial experiments in forestry • Stehman, S. V.; Meredith, M. P.	95 CanJForR 25 446-461
Häggkvist, R.; Morris, S. E.	Second-order analysis of space-time clustering • Diggle, P. J.; Chetwynd, A. G.;	95 StMedRes 4 124-136
Saltelli, Andrea; Sobol, Ilya M.	About the use of rank transformation in sensitivity analysis of model output •	95 RelEngSS 50 225-239
parison to additive main effects and multiplicative	interaction (AMMI) analysis • Piepho ... prediction (BLUP) for regional yield trials: A com-	94 ThApGene 89 647-654
more sensitivity" ... • Cleophas Reply to comment on	"Interaction biases in two-period crossover studies: A modified analysis to test with	95 BiomtrcJ 37 507-510
Diagnostics and selection for linearity and	interaction in calibration with pure and multicomponent spectra • Brown, Philip J.	95 Chemolab 28 139-148
cultivars into groups without rank-change	interactions • Cornelius, P. L.; Van Sanford, D. A.; Seyedsadr, M. S. Clustering	93 CropSci 33 1193-1200
Rank tests for main and	interaction effects in analysis of variance • Marden, John I.; Muyot, Maria Emerita T.	95 JASA 90 1388-1398
• Jaccard et al. Measurement error in the analysis of	interaction effects between continuous predictors using multiple regression: Multiple ...	95 PsychBu 117 348-357
Bayesian logit cohort models with age-by-period	interaction effects and an over-dispersion parameter and their application ... • Nakamura	95 PrInStMa 43 99-119
Miyakawa, Masami An ANOVA model and	interaction element for three-way layout without replication • Kawasaki, Hidemi;	95 JJapnStS 25 143-150
and AMMI analyses of genotype × environment	interactions: An empirical comparison • Yau, Sui K. Regression	95 AgronmyJ 87 121-126
partitioning of the genotype-by-environment	interaction • Muir, W.; Nyquist, W. E.; Xu, S. Alternative	92 ThApGene 84 193-200
Tests for directional space-time	interaction in epidemiological data • Lawson, Andrew B.; Viel, Jean-François	95 StatMed 14 2383-2391
structure models based on latent variates involving	interactions among the exogenous constructs • Baumgartner et al. ... testing of moment	95 SocMethR 24 187-213
survey data: A simple approach that explores for	interactions • Stack, Jim; Unwin, Antony R. Selecting explanatory variables from	95 Ststcian 44 17-29
Nigel; Saab, Elias; Montgomery-Smith, Stephen (Ed)	Interaction between functional analysis, harmonic analysis, and probability • Kalton,	95 Dekker:NY 494p
Search designs for searching several three-factor	interactions in general factorials • Chatterjee, K.; Mukerjee, R.	92 UtilMath 42 181-191
Multiplicative	interaction in generalized linear models • van Eeuwijk, Fred A.	95 Biomtrcs 51 1017-1032
	Interaction levels in an autoregressive graphical model • Lagona, Francesco	94 Metron 52/1 113-141
Stochastic models of host-macroparasite	interaction • Isham, Valerie	95 AnlsApPr 5 720-740
terms in the additive main effects and multiplicative	interaction model for cultivar trials • Piepho ... of statistical tests for multiplicative	95 ThApGene 90 438-443
terms in the additive main effects and multiplicative	interaction model for cultivar trials • Cornelius, P. L. Statistical tests and retention of	93 CropSci 33 1186-1193
lution $2l+1$ when $l+s+1$-factor and higher order	interactions are negligible • Shirakura et al. ... for balanced 2^m factorial designs of reso-	93 JCISS 18 233-245
Ferrándiz, J.; López, A.; Llopis, A. et al. Spatial	interaction between neighbouring counties: Cancer mortality data in Valencia (Spain) •	95 Biomtrcs 51 665-678
Indirect effects in ecological	interaction networks 1. The chain rule approach • Higashi, Masahiko; Nakajima, Hisao	95 MaBisci 130 99-128
of problem ... • Harris et al. A model for the	interaction of partnership dissolution and relationship quality as predicators of remission	94 ASAProSoSt 232-234
Area-	interaction point processes • Baddeley, A. J.; van Lieshout, M. N. M.	95 AnInStMa 47 601-619
Stephen R. Mean-variance	interactions in process improvement and capacity design • Buss, Arnold H.; Lawrence,	95 IIETrn 27 339-347
A parsimonious estimating technique for	interaction and quadratic latent variables • Ping, Robert A., Jr	95 JMarRsr 32 336-347
Relationships between genotype × environment	interactions and rank orders for a set of genotypes tested in different ... • Hühn et al.	93 ThApGene 86 943-950
Analyzing	interaction: Sequential analysis with SDIS and GSEQ • Bakeman, Roger; Quera, Vicenç	95 CambridgeUniv 166p
The Committee on National Statistics: Fostering	interactions between statisticians in academia and government • Fienberg, S. E. et al.	95 IntStRvw 63 257-269
Miyakawa, Masami An interpretation of the	interaction terms in Mandel's ANOVA model from Hirotsu's interaction elements •	93 RStAJp 40/1 1-9
A comparative study of some rank tests for	interaction • Mansouri, H.; Chang, G.-H.	95 CmpStDA 19 85-96
On the	interactions of unit roots and exogeneity • Hendry, David F.	95 EcomtrRv 14 383-419
on chi-squared approximations for tests of rank	interaction with ordered categorical data • Thomas, G. E. A note	95 CommStB 24 439-459
of "Understanding log-linear analysis with ILOG: An	interactive approach" • Bakeman, R.; Robinson, B. F. (Auth); Rice, J. C. (Rev) Review	95 EdPsMeas 55 350-355
	Interactive dynamic graphics for exploratory survival analysis • Atkinson, E. Neely	95 AmerStat 49 77-84
association-segregation χ^2 method • Dizier et al.	Interactive effect of two candidate genes in a disease: Extension of the marker-	94 AJHuGene 55 1042-1049
Trellis displays vs.	interactive graphics • Theus, Martin	95 CmpStQ 10 113-127
Look first, then model: Exploring the data using the	interactive graphics of PC-ISP • Huber, Thomas M.; Nagel, Matthias	95 CmpStQ 10 29-36
An	interactive image index • Eddy, William F.; Mockus, Audris	94 ASAProStGr 1-6
Arnedt, Cheryl Mercado; Ramnath, Bala	Interactive polling and Americans' comfort level with technology • Frankovic, Kathleen A.;	94 ASASrvyRsr 1316-1321
analytic approach to Fleming-Viot processes with	interactive selection • Overbeck, Ludger; Röckner, Michael; Schmuland, Byron An	95 AnlsProb 23 1-36
	Interactive spatial data analysis • Bailey, Trevor C.; Gatrell, Anthony C.	95 Wiley:NY:UK 432p
XploRe: An	interactive statistical computing environment • Härdle, W.; Klinke, S.; Turlach, B. A. (Ed)	95 Springer:Brln:NY 387p
Murdoch, D. J. (Rev) Review of "XploRe: An	interactive statistical computing environment" • Härdle, W.; Klinke, S.; Turlach, B. A. (Ed);	96 ShrtBkRv 16/1 3-3
	Interactive statistics tutorials in Stata (jse/v3n3/ferrall) • Ferrall, Christopher	95 JStEd 3/3 28kb
PROBO: An	interactive system in fuzzy linear programming • Cadenas, J. M.; Verdegay, J. L.	95 FuzzySet 76 319-332
Ostroff, D. J. (Auth) et al. Review of "Statistics. An	interactive text for the health and life sciences" • Krishnamurty, G. B.; Kasovia-Schmitt, P.;	95 StatMed 14 2621-2622
Geladi, P.; Berglund, A.; Sjöström, M.; Wold, S.	Interactive variable selection (IVS) for PLS. Part II: Chemical applications • Lindgren, F.;	95 JChemtrc 9 331-342
Extremes of	interarrival times of a Poisson process under conditioning • Abay, A.	95 ZastApMa 23 73-82
Rousseeuw, Peter J.; Ruts, Ida The repeated median	intercept estimator: Influence function and asymptotic normality • Hössjer, Ola;	95 JMultiAn 52 45-72
Detecting a change in the	intercept in multiple regression • Watson, G. S.	95 StPrLet 23 69-72
Nicholas S. The reflection method of line	intercept sampling to eliminate boundary bias • Gregoire, Timothy G.; Monkevich,	94 EnvEcoSt 1 219-226
designs for polynomial regression without an	intercept • Huang, Mong-Na Lo; Chang, Fu-Chuen; Wong, Weng Kee D-optimal	95 StSinica 5 441-458
	Interclass correlation	
correlation • Eliasziw, Michael; Donner, Allan	Evaluation of a proposed modification to the pairwise estimator of a parent-offspring	95 SankhyaB 57 151-157
(Rev) Review of "Statistical design and analysis for	intercropping experiments. Volume I: Two crops" • Federer, W. T. (Auth); Rudolph, P. E.	95 BiomtrcJ 37 204-204
robustness of recombination frequency estimates in	intercrosses with dominant markers • Säll, Torbjörn; Nilsson, Nils-Otto The	94 Genetcs 137 589-596
indicators for forecasting ... • Berk et al. International	interdependence of business cycles in the manufacturing industry: The use of leading	95 JForcast 14 1-23
likelihood with RNA sequences: Addressing the	interdependence of sites • Tillier, E. R. M.; Collins, R. A. Neighbor joining and maximum	95 MolBioEv 12 7-15
stochastic differential equation for the force of	interest • Parker, Gary A second order	95 InsrneME 16 211-224
Ruin estimates under	interest force • Sundt, Bjørn; Teugels, Jozef L.	95 InsrneME 16 7-22
A time-continuous Markov chain	interest model with applications to insurance • Norberg, Ragnar	95 AplStMDA 11 245-256
Forecasting consumption, income and real	interest rates from alternative state space models • Vinod, H. D.; Basu, Parantap	95 IntJFor 11 217-231
Nonstationarity of regressors and tests on real-	interest-rate behavior • Mishkin, Frederic S.	95 JBES 13 47-51
E. The expectations theory of	interest rates: Cointegration and factor decomposition • Choi, Seungmook; Wohar, Mark	95 IntJFor 11 253-262
Pelizzon, L.; Weber, G. The Italian term structure of	interest rates and the currency devaluation of September 1992: A factor-ARCH analysis •	94 Statstca 54 313-327
Markov branching decision chains with	interest-rate-dependent rewards • Huang, Ying; Veinott, Arthur F., Jr	95 PrEngInf 9 99-121
On the term structure of	interest rates – Empirical results for Germany • Wolters, J.	95 StHefte 36 193-214
Long-term returns in stochastic	interest rate models • Deelstra, G.; Delbaen, F.	95 InsrneME 17 163-169

Juselius Do purchasing power parity and uncovered	interest rate parity hold in the long run? An example of likelihood inference in a ... •	95 JEconmtx 69 211-240
Equity-linked life insurance: A model with stochastic	interest rates • Nielsen, J. Aase; Sandmann, Klaus	95 InsrneME 16 225-253
A counting process approach to stochastic	interest • Møller, Christian Max	95 InsrneME 17 181-192
Presutti, E. Brownian fluctuations of the	interface in the $D = 1$ Ginzburg-Landau equation with noise • Brassesco, S.; De Masi, A.;	95 AlHenrPB 31 81-118
Onset and structure of	interfaces in a Kawasaki + Glauber interacting particle system • Giacomin, G.	95 ProbTher 103 1-24
Hybrid zones and voter model	interfaces • Cox, J. T.; Durrett, R.	95 Bernoull 1 343-370
Large time behavior of	interface solutions to the heat equation with Fisher-Wright white noise • Tribe, Roger	95 ProbTher 102 289-311
Mervin E. A system-independent graphical user	interface for statistical software • Liu, Lon-Mu; Chan, Ki-Kan; Montgomery, Alan L.; Muller,	95 CmpStDA 19 23-44
Statistical analysis of chromatid	interference • Zhao, Hongyu; McPeek, Mary Sara; Speed, Terence P.	95 Genetcs 139 1057-1065
Multipoint mapping under genetic	interference • Weeks, Daniel E.; Lathrop, G. Mark; Ott, Jurg	93 HumHrdty 43 86-97
Lathrop, G. Mark Detection of genetic	interference: Simulation studies and mouse data • Weeks, Daniel E.; Ott, Jurg;	94 Genetcs 136 1217-1226
McPeek, Mary Sara Statistical analysis of crossover	interference using the chi-square model • Zhao, Hongyu; Speed, Terence P.;	95 Genetcs 139 1045-1056
of bandpass filtered Fourier transform infrared	interferograms • Shaffer et al. ... design protocol for the pattern recognition analysis	95 Chemolab 29 89-108
	Interim analysis	
Lecoutre, Bruno; Derzko, Gérard; Grouin, Jean-Marie	Bayesian predictive approach for inference about proportions (Disc: p1065-1065) •	95 StatMed 14 1057-1063
when the key response is rate of change • Shih et al.	Re-evaluating design specifications of longitudinal clinical trials without unblinding	95 StatMed 14 2239-2248
	Interim analyses in 2 × 2 crossover trials • Cook, Richard J.	95 Biomtrcs 51 932-945
a clinical trial very early based on unplanned	interim analyses: A group sequential approach • Emerson, Scott S. Stopping	95 Biomtrcs 51 1152-1162
Tables of critical values for two group	interim analyses • Hewett, John E.; Petroski, Gregory F.; Li, Yiqun	95 CommStB 24 825-852
Manfred Dose finding strategies involving	interim analyses and unbalanced treatment allocation (Disc: p909-909) • Wargenau,	95 StatMed 14 901-907
A new method to explore the distribution of	interindividual random effects in non-linear mixed effects models • Fattinger, K. E. et al.	95 Biomtrcs 51 1236-1251
Rzhetsky, Andrey; Nei, Masatoshi	Interior-branch and bootstrap tests of phylogenetic trees • Sitnikova, Tatyana;	95 MolBioEv 12 319-333
Schmeiser, Bruce W. Random-direction	interior-point Markov chains: A family of black-box samplers • Chen, Ming-Hui;	94 ASAProBays 1-6
deviation between the probability content and the	interior point proportion of a random convex hull • Massé, B. Invariance principle for the	95 JAppProb 32 1041-1047
of variance components from the results of	interlaboratory comparisons • Minkkinen, Pentti Estimation	95 Chemolab 29 263-270
Modern extensions of	interlaboratory testing • Wheeler, Donald J.	94 ASAProPEng 35-43
G. I.; Mirakhmedov, Sh. A. Large deviations and	intermediate efficiency of decomposable statistics in a multinomial scheme • Ivchenko,	95 MaMethSt 4 294-311
A parametric model for survival and	intermediate event times • Epstein, Leonardo D.; Muñoz, Alvaro	94 ASAProEpid 91-96
On stability of	intermediate order statistics • Gather, Ursula; Tomkins, Jim	95 JStPlInf 45 175-183
Papadatos, N.	Intermediate order statistics with applications to nonparametric estimation •	95 StPrLet 22 231-238
	Intermediate phase for the contact process on a tree • Durrett, Rick; Schinazi, Rinaldo	95 AnlsProb 23 668-673
Mason, David M.	Intermediate sums and stochastic compactness of maxima • Csörgő, Sándor;	95 JStPlInf 45 81-90
Stationary parabolic Anderson model and	intermittency • Carmona, R. A.; Molchanov, S. A.	95 ProbTher 102 433-453
periodic autoregressive processes for modeling	intermittent monthly streamflows • Chebaane, M.; Salas, J. D.; Boes, D. C. Product	95 WaterReR 31 1513-1518
Subdiffusive fluctuations for	internal diffusion limited aggregation • Lawler, Gregory F.	95 AnlsProb 23 71-86
ticomponent profiles by evolving projections and	internal rank annihilation • Grung et al. ... of embedded minor analytes in three-way mul-	95 Chemolab 29 213-221
Anita S.; Rothenberg, Lori F. Students'	internalized representation of simple linear regression • VanBrackle, Lewis; VanBrackle,	94 ASAProStEd 100-101
"Proceedings of the First Scientific Meeting of the	International Association for Statistical Education" • Brunelli, L. et al. Review of	95 AmerStat 49 276-276
A test of the three moment	international capital asset pricing model • Vakil, Farok; Rajan, Murli	94 ASAProBuEc 184-189
Long memory in inflation rates:	International evidence • Hassler, Uwe; Wolters, Jürgen	95 JBES 13 37-45
program implementing the recommendations of the	International Federation of Clinical Chemistry on the statistical ... • Solberg RefVal: A	95 CmpBiomd 48 247-256
from clinical trials in drug development: Issues of	international harmonization • O'Neill ... in the planning and evaluation of drug safety	95 StatMed 14 1117-1127
Scheerens, Jaap	International indicators on teachers within the framework of the OECD-INES •	94 ASAProGovt 159-163
of leading indicators for forecasting ... • Berk et al.	International interdependence of business cycles in the manufacturing industry: The use	95 JForcast 14 1-23
Professor Clive W. J. Granger: An interview for the	International Journal of Forecasting • Teräsvirta, Timo	95 IntJFor 11 585-590
Convergence in	international output • Bernard, Andrew B.; Durlauf, Steven N.	95 JApEcmtx 10 97-108
A response model for the	international price program • Ferguson, Gwyn R.; Ciapponi, Tamara M.	94 ASASrvyRsr 1007-1012
Cointegration of	international stock market indices • Chou, Ray Y.; Ng, Victor K.; Pi, Lynn K.	94 ASAProBuEc 196-201
	Internet	
	Electronic services in statistics • Krause, Andreas	95 CmpStDA 19 595-604
Statistics Education information service and other	Internet resources for statistics teachers • Snell, J. Laurie The Journal of	95 AmerStat 49 372-375
How to start surfing the	Internet • Hopke, Philip K.	95 Chemolab 30 1-9
the sequence of action potentials in an insect visual	interneuron • Turcott, R. G.; Barker, P. D. R.; Teich, M. C. Long-duration correlation in	95 JStCmpSm 52 253-271
	Interoutput times in an M/G/1/PS queue with instant feedback • Grishechkin, S. A.	94 ThProbAp 39 692-696
Testing whether	interpenetration is possible for the grains of a random set • Streit, Franz	95 MaOpfStS 26 355-364
Bickmore, W. A. Models of DNA replication timing in	interphase nuclei: An exercise in inferring process from state • Carothers, A. D.;	95 Biomtrcs 51 750-755
Spatial	interpolation errors for monitoring data • Høst, Gudmund; Omre, Henning; Switzer, Paul	95 JASA 90 853-861
Some results related to	interpolation on Hardy spaces of regular martingales • Xu, Quanhua	95 IsrelJMa 91 173-187
Moklyachuk, M. P. Minimax	interpolation of random fields that are time homogeneous and isotropic on a sphere •	95 ThPrMaSt 50 107-115
et al. Spatial large-scale correlations for optimal	interpolation of temperature, salinity and nutrient concentrations in the ... • Toompuu	95 EnvrMtrc 6 55-72
Two-sample tests for variability based on	interquantile spreads • Shoemaker, Lewis H.	94 ASAProQlPr 134-139
	Interquartile range	
order statistics • Harvill, Jane L.; Newton, H. Joseph	Using symbolic math to evaluate saddlepoint approximations for the difference of	95 CommStB 24 781-791
Keeter, S. Evaluating the use of data on	interruptions in telephone service for nontelephone households • Brick, M.; Waksberg, J.;	94 ASASrvyRsr 19-28
Abreu, V.; Vives, J. Chaos expansions of double	intersection local time of Brownian motion in R^d and renormalization • Imkeller, P.; Perez-	95 StocProc 56 1-34
examples • Bojdecki, T.; Gorostiza, L. G. Self-	intersection local time for Gaussian $S'(R^d)$-processes: Existence, path continuity and	95 StocProc 60 191-226
Delsarte polynomial of 5-(v, k, λ) designs with three	intersection numbers • Shrikhande, M. S. The	93 JCISS 18 53-60
Five-designs with three	intersection numbers • Ionin, Yury J.; Shrikhande, Mohan S.	95 JCombThA 69 36-50
A kind of rank test about	intersection point of two continuous populations (Chinese) • Li, Ningning; Chen, Guijing	95 KToChina 39 1543-1546
Simultaneous inference using finite	intersection tests: A better mousetrap • Timm, Neil H.	95 MultBehR 30 461-512
mott, Michael P.; Mudholkar, Anil Robust finite-	intersection tests for homogeneity of ordered variances • Mudholkar, Govind S.; McDer-	95 JStPlInf 43 185-195
Limit theorems for the union-	intersection test • Horváth, Lajos; Shao, Qi-Man	95 JStPlInf 44 133-148
Union-	intersection testing for outliers in multivariate normal data • Caroni, C.; Prescott, P.	95 JStCmpSm 51 185-196
Barlow-Yor inequalities for local times of the	intersection of two planar Brownian motions (French) • Mathieu, P.	95 ProbTher 102 83-103
	Intersection-union test	
Liu, Huimei; Berger, Roger L.	Uniformly more powerful, one-sided tests for hypotheses about linear inequalities •	95 AnlsStat 23 55-72
Public library statistics: An	interstate service analysis • Liu, Ben-Chieh	94 ASAProGovt 215-220
Kaufman, Steven; Scheuren, Fritz; Smith, Wray	Intersurvey consistency in school surveys (Disc: p1128-1129) • Holt, Albert;	94 ASASrvyRsr 1105-1110
	Interval	
	• see also Bayesian –, confidence –, prediction –, sampling –, tolerance –	
A fixed-width	interval for $1/\beta$ in simple linear regression • Coleman, Daniel A.	95 JStPlInf 44 291-312
ronment • Wang, Ouhon Applications of numerical	interval analysis to obtain self-validating results in a massively parallel computing envi-	94 ASAProStCp 21-28
of a distribution function with truncated and doubly	interval-censored data and its application to AIDS studies • Sun, J. Empirical estimation	95 Biomtrcs 51 1096-1104
normality of the NPMLE of linear functionals for	interval censored data, case 1 • Huang, J.; Wellner, J. A. Asymptotic	95 StNeerla 49 153-163

Regression with	interval-censored data • Rabinowitz, Daniel; Tsiatis, Anastasios; Aragon, Jorge	95 Biomtrka 82 501-513
Sequential testing with	interval censored data • Quan, Hui; Yu, Kai F.	95 JStCmpSm 51 239-247
three-state duration-dependent Markov model from	interval-censored observations with application to AIDS data • Frydman ... in a	95 Biomtrcs 51 502-511
estimation of a Markov 'illness-death' process from	interval-censored observations, with application to diabetes ... • Frydman Nonparametric	95 Biomtrka 82 773-789
first appearance of Plasmodium ... • Gottschau et al.	Interval censored survival data and multistate compartmental models in the analysis of	95 StatMed 14 2727-2736
Ohtani, Kazuhiro A comparison of	interval constrained least squares and mixed regression estimators • Srivastava, Viren K.;	95 CommStA 24 395-413
the performance of empirical Bayes simultaneous	interval estimates for all pairwise comparisons • Kim, Woo-Chul; Han, Kyung Soo On	95 JKoreaSt 24 161-181
samples • Fellingham et al. Sensitivity of point and	interval estimates to distributional assumptions in longitudinal data analysis of small	95 CommStB 24 617-630
different estimation methods ... • Scazzero et al.	Interval estimates for the optimum point of a quadratic logistic curve – A comparison of	95 CommStB 24 207-226
	Interval estimation	
experiments available from ... • Mukhopadhyay et al.	Comparing exponential clinical trials by combining individual sequential	94 Metron 52/1 17-40
	Estimating location parameters using censored rank statistics • Öhman, Marie-Louise	95 BiomtrcJ 37 767-782
	Estimating products in forensic identification using DNA profiles • Balding, David J.	95 JASA 90 839-844
Li, Gang	Nonparametric likelihood ratio estimation of probabilities for truncated data •	95 JASA 90 997-1003
	Optimal robust Bayes estimation • Li, Yuanzhang; Saxena, K. M. Lal	95 JStPlInf 46 365-380
	Pivotal models and the fiducial argument • Barnard, George A.	95 IntStRvw 63 309-323
Chaturvedi, Ajit; Gupta, Rahul	Second-order approximations to two classes of sequential estimation procedures •	95 MaOpfStS 27 159-173
experiments • Schwartz, P. F.; Gennings, C. et al.	Threshold models for combination data from reproductive and developmental	95 JASA 90 862-870
the point estimation and the improvement on the	interval estimation for the disturbance variance in a linear ... • Nagata ... on	95 CommStA 24 1687-1704
model • Chiou, Paul; Han, Chien-Pai	Interval estimation of error variance following a preliminary test in one-way random	95 CommStB 24 817-824
pre-test • Chiou, Paul; Han, Chien-Pai Conditional	interval estimation of the exponential location parameter following rejection of a	95 CommStA 24 1481-1492
test • Chiou, Paul; Han, Chien-Pai Conditional	interval estimation of the exponential scale parameter following rejection of a preliminary	94 JStatRsr 28 65-72
	Interval estimation of inverse dose-response • Alho, Juha M.; Valtonen, Esko	95 Biomtrcs 51 491-501
procedure • Gastaldi, Tommaso	Interval estimation from a life-test through a sequential Kolmogorov-Smirnov type	95 IEEEReli 44 470-476
On the performance of iterated wild bootstrap	interval estimation of the mean response • Kim, Woochul; Ko, Duk-Hyun	95 JKoreaSt 24 551-562
Correction to "Sample size determination for	interval estimation of multinomial probabilities" (V47 p203-206) • Bromaghin, Jeffrey F.	95 AmerStat 49 403-403
times in a class of ... • Bose et al. Sequential	interval estimation via replicated piecewise stopping times and accelerated stopping	95 SqtlAnly 14 287-305
Nam, Jun-mo	Interval estimation and significance testing for cyclic trends in seasonality studies •	95 Biomtrcs 51 1411-1417
Huwang, Longcheen	Interval estimation in structural errors-in-variables model with partial replication •	95 JMultiAn 55 230-245
	Interval forecasting in cointegrated systems • Reimers, Hans-Eggert	95 StHefte 36 349-369
Multipoint linkage analysis using sib pairs: An	interval mapping approach for dichotomous outcomes • Olson, Jane M.	95 AJHuGene 56 788-798
of additive and dominance effects of QTLs in	interval mapping of F_2 RFLP data • Hayashi, T.; Ukai, Y. Detection	94 ThApGene 87 1021-1027
resolution of quantitative traits into multiple loci via	interval mapping • Jansen, Ritsert C.; Stam, Piet High	94 Genetcs 136 1447-1455
A sib-pair approach to	interval mapping of quantitative trait loci • Fulker, David W.; Cardon, Lon R.	94 AJHuGene 54 1092-1103
Abraham B.; Ronin, Yefim I.; Kirzhner, Valery M.	Interval mapping of quantitative trait loci employing correlated trait complexes • Korol,	95 Genetcs 140 1137-1147
Fulker, David W. The power of	interval mapping of quantitative trait loci, using selected sib pairs • Cardon, Lon R.;	94 AJHuGene 55 825-833
Cardon, L. R. Multipoint	interval mapping of quantitative trait loci, using sib pairs • Fulker, D. W.; Cherny, S. S.;	95 AJHuGene 56 1224-1233
Approximate thresholds of	interval mapping tests for QTL detection • Rebaï, Ahmed; Goffinet, Bruno; Mangin, Brigitte	94 Genetcs 138 235-240
Packing random	intervals • Coffman, E. G., Jr; Poonen, Bjorn; Winkler, Peter	95 ProbTher 102 105-121
An integral equation approach to the	interval reliability of systems modelled by finite semi-Markov processes • Csenki, A.	95 RelEngSS 47 37-45
use of the F ratio in a misspecified model with an	interval restriction • Wan, Alan T. K.; Yang, Minxian On the	95 JStCmpSm 52 151-161
arising from iterated random subdivisions of an	interval • Johnson, N. L.; Kotz, S. Use of moments in studies of limit distributions	95 StPrLet 24 111-119
A likelihood approach to calculating risk support	intervals • Leal, Suzanne M.; Ott, Jurg	94 AJHuGene 54 913-917
Exponential waiting time for filling a large	interval in the symmetric simple exclusion process • Ferrari, P. A.; Galves, A.; Liggett, T. M.	95 AIHenrPB 31 155-175
method: Adaptive random search with translating	intervals • Edissonov, Ivan The new ARSTI optimization	94 AJMMgSci 14 143-166
Cost effective calibration	intervals using Weibull analysis • Barringer, H. Paul	95 ASQCTekCon 1026-1038
Correlation of	interval-valued intuitionistic fuzzy sets • Bustince, H.; Burillo, P.	95 FuzzySet 74 237-244
distributions and tail behavior in stationary	interval-valued probability models • Sadrolhefazi, A.; Fine, T. L. Finite-dimensional	94 AnlsStat 22 1840-1870
Bayesian analysis of an	intervened Poisson distribution • Scollnik, David P. M.	95 CommStA 24 735-754
Optimal survey design for community-	intervention evaluations: Cohort, cross-sectional, or both? • Diehr, P.; Martin, D. et al.	94 ASAProEpid 34-43
Assessing the losses caused by an industrial	intervention: A hierarchical Bayesian approach • Arjas, Elja; Liu, Liping	95 ApplStat 44 357-368
effects due to industrialization: An accumulated	intervention model with an application in Pohang, Korea • Ryoo et al. Detection of warming	95 JApMetrl 34 1466-1471
the matches in a paired t-test for community	interventions when the number of pairs is small • Diehr, P.; Martin, D. C. et al. Breaking	95 StatMed 14 1491-1504
Symposium on community	intervention trials • Koepsell, Thomas D.; Diehr, Paula H.; Cheadle, Allen; Kristal, Alan	95 AJEpidm 142 594-599
regression method for analysing ordinal data from	intervention trials • Schnell, Daniel J.; Magee, Elvin; Sheridan, John R. A	95 StatMed 14 1177-1189
S. P.; Flora, J. A.; Winkleby, M. A. et al. Community	intervention trials: Reflections on the Stanford Five-City Project experience • Fortmann,	95 AJEpidm 142 576-586
Interplay between design and analysis for behavioral	intervention trials with community as the unit of randomization • Green, S. B. et al.	95 AJEpidm 142 587-593
The ET	interview: Gregory C. Chow • Pagan, Adrian	95 EconTher 11 597-624
Professor Clive W. J. Granger: An	interview for the International Journal of Forecasting • Teräsvirta, Timo	95 IntJFor 11 585-590
respondents learn from questionnaires: The survey	interview and the logic of conversation (Disc: p169-177) • Schwarz, Norbert What	95 IntStRvw 63 153-168
respondents learn from questionnaires: The survey	interview and the logic of conversation" • Cantor, David Discussion of "What	95 IntStRvw 63 173-177
respondents learn from questionnaires: The survey	interview and the logic of conversation" • Singer, Eleanor Discussion of "What	95 IntStRvw 63 169-172
Jack, Susan S. Combining National Health	Interview Survey datasets: Issues and approaches (Disc: p679-693) • Botman, Steven L.;	95 StatMed 14 669-677
Population Survey (CPS) and Consumer Expenditure	Interview Survey (CEIS) • Kojetin, Brian A. Characteristics of nonrespondents to the Current	94 ASASrvyRsr 966-971
of allowable call attempts in the National Health	Interview Survey • Kalsbeek, W. D.; Botman, S. L. et al. Cost-efficiency and the number	94 JOffStat 10 133-152
correlates of response bias in the National Health	Interview Survey: Lessons for the redesign (Disc: p379-381) • LeClere, F. B. et al. Social	94 ASASrvyRsr 373-378
Comment on "Redesign of the National Health	Interview Survey questionnaire" • Fowler, Floyd Jackson, Jr	94 ASASrvyRsr 379-381
State estimates for the National Health	Interview Survey using a random digit dialing supplement • Malec, Donald; Parsons, Van L.	94 ASASrvyRsr 1190-1195
Norris, Douglas A.; Hatcher, Jeff The impact of	interviewer characteristics on response in a national survey of violence against women •	94 ASASrvyRsr 449-454
	Interviewer error	
	Data quality in a CAPI survey: Keying errors • Dielman, Lynn; Couper, Mick P.	95 JOffStat 11 141-146
Measuring	interviewer performance using CAPI • Couper, Mick P.; Sadosky, Sally A.; Hansen, Sue Ellen	94 ASASrvyRsr 845-850
The effect of	interviewer variance on domain comparisons • Davis, Peter; Scott, Alastair	95 SurvyM 21 99-106
hane; Kaushal, Ritu Implementing computer-assisted	interviewing for the Canadian Labour Force Survey • Bench, Jill; Clark, Colleen; Dufour, Jo-	94 ASAProGovt 93-98
assumptions: The use of vignettes in cognitive	interviewing • Gerber, Eleanor R. Hidden	94 ASASrvyRsr 1269-1274
Understanding the standardized/non-standardized	interviewing controversy • Beatty, Paul	95 JOffStat 11 147-160
et al. Decentralised CATI versus paper and pencil	interviewing: Effects on the results in the Swedish labour force surveys • Bergman, L. R.	94 JOffStat 10 181-195
quality of income data using computer assisted	interviewing: The experience of the Canadian survey of labour and ... • Grondin et al. Data	94 ASASrvyRsr 839-844
Comment on "Computer assisted personal	interviewing" • Connett, Bill	94 ASASrvyRsr 861-862
al. Audio and video computer assisted self-	interviewing: Preliminary tests of new technologies for data collection • O'Reilly, J. M. et	94 JOffStat 10 197-214
	Intra-block analysis	
Tadeusz	On the randomization theory of experiments in nested block designs • Caliński,	94 ListyBio 31 45-77

	Intraclass correlation	
binary responses • Shoukri, M. M.; Martin, S. W. et al.	Maximum likelihood estimation of the kappa coefficient from models of matched	95 StatMed 14 83-99
• Vogler, G. P.; Wette, R.; McGue, M. K.; Rao, D. C.	Properties of alternative estimators of familial correlations under variable sibship size	95 Biomtrcs 51 276-283
Estimation of the	intraclass correlation coefficient • Srivastava, M. S.	93 AnHuGene 57 159-165
in the estimation of variance components and	intraclass correlation for paired data • Lyles, R. H. et al. Effects of model misspecification	95 StatMed 14 1693-1706
Sample size calculation for	intraclass correlation in the presence of random loss of sampled patients • Lui, Kung-Jong	95 BiomtrcJ 37 97-103
work • Vieland, Veronica J.; Hodge, Susan E. Inherent	intractability of the ascertainment problem for pedigree data: A general likelihood frame-	95 AJHuGene 56 33-43
Zemel, Richard S. Using neural networks to learn	intractable generative models • Hinton, Geoffrey E.; Dayan, Peter; Neal, Radford M.;	94 ASAProStCp 11-20
Bertrand Nonnegativity constraints and	intratemporal uncertainty in a multi-good life-cycle model • Adang, Pim; Melenberg,	95 JApEcmtx 10 1-15
	Intrinsic analysis of statistical estimation • Oller, J. M.; Corcuera, J. M.	95 AnlsStat 23 1562-1581
On conditional and	intrinsic autoregressions • Besag, Julian; Kooperberg, Charles	95 Biomtrka 82 733-746
statistics • Okada, Mihoko; O'Brien, Myles	IntroStat: A hypertext-based design for an electronic textbook to introduce biomedical	95 CmpBiomd 46 265-276
Probability,	intuition and a spreadsheet • Graham, Alan	95 TeachgSt 17/3 84-87
Correlation of	intuitionistic fuzzy sets in probability spaces • Hong, Dug Hun; Hwang, Seok Yoon	95 FuzzySet 75 77-81
Correlation of interval-valued	intuitionistic fuzzy sets • Bustince, H.; Burillo, P.	95 FuzzySet 74 237-244
	Intuitive biostatistics • Motulsky, Harvey	95 OxfordUniv 384p
	Invariance	
	A note on stationary linearly positive quadrant dependent sequences • Kim, Tae-sung	95 JKoreaSt 24 249-256
Linssen, H. N.; Overdijk, D. A.	The reduction and ancillary classification of observations • van Berkum, E. E. M.;	95 JStPlInf 45 347-372
Gratiane	Strong approximation of the number of renewal paced record times • Ennadifi,	95 JStPlInf 45 113-132
regression • Chen, Jian-Shen; Jennrich, Robert I.	Transformations for improving linearization confidence intervals in nonlinear	95 JASA 90 1271-1276
factor mean differences under violations of factorial	invariance • Kaplan, David; George, Rani A study of the power associated with testing	95 StrEqMod 2 101-118
Measurement	invariance, predictive invariance, and the duality paradox • Millsap, Roger E.	95 MultBehR 30 577-605
Massart, P.; Rio, E.	Invariance principles for absolutely regular empirical processes • Doukhan, P.;	95 AIHenrPB 31 393-427
Variance estimation based on	invariance principles • Steinebach, Josef	95 MaOpfStS 27 15-25
point proportion of a random convex hull • Massé, B.	Invariance principle for the deviation between the probability content and the interior	95 JAppProb 32 1041-1047
Uniform CLT for Markov chains and its	invariance principle: A martingale approach • Bae, Jongsig; Levental, Shlomo	95 JTherPrb 8 549-570
random domains • Osada, H.; Saitoh, T. An	invariance principle for non-symmetric Markov processes and reflecting diffusions in	95 ProbTher 101 45-63
for martingales in Banach spaces, and weak	invariance principle - A quantitative study • Anastassiou ... weak law of large numbers	95 JMultiAn 52 158-180
Steinebach, J.	Invariance principles for renewal processes and some applications • Alex, M.;	95 ThPrMaSt 50 23-54
to ARMA process • Truong-Van, B.	Invariance principles for semi-stationary sequence of linear processes and applications	95 StocProc 58 155-172
Seo, Hye-Young An	invariance principle for stationary strong mixing random fields • Kim, Tae-Sung;	95 JKoreaSt 24 281-292
Dabrowski, André Robert; Zoglat, Abdelhak Strong	invariance principles for triangular arrays of weakly dependent random variables •	95 CandJSt 23 299-310
analysis of data matrices and the factorial	invariance problem • Bove, Giuseppe Some methods for the simultaneous	94 Metron 52/3 73-87
An	invariance property of optimal spectral bandwidths • Robinson, P. M.	95 StPrLet 24 345-346
measurement scales: Simple structure and metric	invariance reconsidered (Disc: p77-123) • Labouvie, E. et al. Testing for equivalence of	95 MultBehR 30 63-76
	Invariance of the reference prior under reparametrization • Yang, R.	95 Test 4 83-94
estimation in nonlinear regression under shape	invariance • Kneip, Alois; Engel, Joachim Model	95 AnlsStat 23 551-570
Group	invariance in statistical inference • Giri, Narayan C.	95 WrldSciPub:Sing 250p
rection to "The relationship between sufficiency and	invariance with applications in sequential analysis" (V36 p575-614) • Hall, W. J. et al. Cor-	95 AnlsStat 23 705-705
models • Barndorff-Nielsen, O. E. Stable and	invariant adjusted profile likelihood and directed likelihood for curved exponential	95 Biomtrka 82 489-499
implicit polynomial curves or surfaces, algebraic	invariants, and asymptotic Bayesian methods • Subrahmonia et al. ... based on high degree	94 ASAProPEng 44-49
Random	invariant cubature formulas (Russian) • Ermakov, S. M.; Wang, Xiao-Qun	94 VLenUM /4 109-112
Optimal and robust	invariant designs for cubic multiple regression • Gaffke, Norbert; Heiligers, Berthold	95 Metrika 42 29-48
Heiligers, B. Computing optimal approximate	invariant designs for cubic regression on multidimensional balls and cubes • Gaffke, N.;	95 JStPlInf 47 347-376
Wolff, Rodney C. L. Properties of	invariant distributions and Lyapunov exponents for chaotic logistic maps • Hall, Peter;	95 JRSS-B 57 439-452
On the	invariant estimation of a normal variance ratio • Madi, Mohamed T.	95 JStPlInf 44 349-357
Unbiased estimator of risk for an orthogonally	invariant estimator of a covariance matrix • Sheena, Yo	95 JJapnStS 25 35-48
conditions • Garnier, J. Stochastic	invariant imbedding: Application to stochastic differential equations with boundary	95 ProbTher 103 249-271
Mohammad-Djafari, Ali Scale	invariant Markov models for linear inverse problems • Brette, Stéphane; Idier, Jérôme;	94 ASAProBays 266-270
estimation of the chaotic function and the	invariant measure of a dynamical system • Bosq, D.; Guégan, D. Nonparametric	95 StPrLet 25 201-212
Michalik, A.; Goldys, B. Existence, uniqueness and	invariant measures for stochastic semilinear equations on Hilbert spaces • Chojnowska-	95 ProbTher 102 331-356
Prato, G. Da	Invariant measures for white noise driven stochastic partial differential equations •	95 StocAnAp 13 295-305
modeling ... • Todeschini et al. Weighted holistic	invariant molecular descriptors. Part 2. Theory development and applications on	95 Chemolab 27 221-229
Noninformative priors for maximal	invariant parameter in group models • Datta, G. S.; Ghosh, J. K.	95 Test 4 95-114
Estimating probabilities from	invariant permutation distributions • Pallini, Andrea	94 JItalStS 3 77-91
nonexistence of a global nonnegative minimum bias	invariant quadratic estimator of variance components • Gao, Sujuan; Smith, T. M. F. On the	95 StPrLet 25 117-120
M. Strategies in testing for an	invariant second-order factor structure: A comparison of EQS and LISREL • Byrne, Barbara	95 StrEqMod 2 53-72
Improved	invariant set estimation for general-scale families • Goutis, Constantinos; Casella, George	95 JStPlInf 44 327-340
the moment matrices of designs over permutation-	invariant sets • Cheng, Ching-Shui Complete class results for	95 AnlsStat 23 41-54
On the asymptotic and	invariant σ-algebras of random walks on locally compact groups • Jaworski, Wojciech	95 ProbTher 101 147-171
scale parameter of a gamma distribution for scale-	invariant squared-error loss • van Eeden, C. Minimax estimation of a lower-bounded	95 CandJSt 23 245-256
Novák, Břetislav; Wertz, Wolfgang	Invariantly optimal look on reliability estimation in the exponential case • Hurt, Jan;	95 StatDecn 13 193-197
Dalenius, Tore Controlling	invasion of privacy in surveys of change over time - A non-technical review •	95 JOffStat 11 107-113
values in a finite population: Application to forest	inventory (French) • Pierrat, J. C.; Houllier, F. et al. Estimating the mean of the r largest	95 Biomtrcs 51 679-686
Stationary regimes for	inventory processes • Bardhan, Indrajit; Sigman, Karl	95 StocProc 56 77-86
Order crossover in	inventory systems • Hayya, Jack C.; Xu, Susan H.; Ramasesh, Ranga V.; He, Xin X.	95 CommStSM 11 279-309
Comparison of the	inverse and classical estimators in multi-univariate linear calibration • Srivastava, M. S.	95 CommStA 24 2753-2767
Revealed preference of the Federal Reserve: Using	inverse-control theory to interpret the policy equation of a vector autoregression • Salemi	95 JBES 13 419-433
The asymptotic properties of the estimation of the	inverse correlation function (Chinese) • Chen, Min; Chang, Xuejiang	95 AMaAPknC 17 23-45
Interval estimation of	inverse dose-response • Alho, Juha M.; Valtonen, Esko	95 Biomtrcs 51 491-501
Prediction and	inverse estimation in repeated-measures models • Liski, Erkki P.; Nummi, Tapio	95 JStPlInf 47 141-151
	Inverse Gaussian distribution	
scale mixtures • Carpenter, Mark; Hebert, Jaimie L.	Estimating the minimum and maximum location parameters for two IG-exponential	95 CommStA 24 1227-1233
Wood, Andrew T. A.	An example in which the Lugannani-Rice saddlepoint formula fails • Booth, James G.;	95 StPrLet 23 53-61
Cumulative sum control charts for the	inverse Gaussian distribution • Nabar, S. P.; Bilgi, Shobha	94 JIndnStA 32 9-14
estimators for a measure of dispersion of an	inverse Gaussian distribution • Pandey, B. N.; Malik, H. J.; Dubey, P. K. Bayesian shrinkage	95 CommStA 24 2261-2270
limits for the mean of future observations from the	inverse Gaussian distribution • Upadhyay, S. K.; Agrawal, R. Bayesian prediction	94 Metron 52/1 101-111
sampling plan for the scale parameter of the	inverse Gaussian distribution • Nabar, Smita A sequential	94 JIndnStA 32 117-121
Random continued fractions and	inverse Gaussian distribution on a symmetric cone • Bernadac, Evelyne	95 JTherPrb 8 221-258
A random continued fraction in R^{d+1} with an	inverse Gaussian distribution • Letac, Gérard; Seshadri, Vanamamalai	95 Bernoull 1 381-393
Bayes estimation in a mixture	inverse Gaussian model • Gupta, Ramesh C.; Akman, H. Olcay	95 AnInStMa 47 493-503
On the reliability studies of a weighted	inverse Gaussian model • Gupta, Ramesh C.; Akman, H. Olcay	95 JStPlInf 48 69-83
Calibration for	inverse Gaussian regression • Woldie, Mammo; Folks, J. Leroy	95 CommStA 24 2609-2620

for (sub) Markov chains on $\{1, 2, \ldots\}$ with bounded	jumps • Kesten, Harry	A ratio limit theorem	95 AdvAppPr 27 652-691
V. et al. A functional central limit theorem for the	jump counts of Markov processes with an application to Jackson networks • Anantharam,		95 AdvAppPr 27 476-509
Green, Peter J. Reversible	jump Markov chain Monte Carlo computation and Bayesian model determination •		95 Biomtrka 82 711-732
Stability and control for linear systems with	jump Markov perturbations • Morozan, T.		95 StocAnAp 13 91-110
rate of growth of Lévy processes with no positive	jumps • Bertoin, Jean	On the local	95 StocProc 55 91-100
	Jump process		
	Large deviations for empirical process of mean-field interacting particle system with		94 AnlsProb 22 2122-2151
unbounded jumps • Feng, Shui			
	jump processes. II: Applications to Jackson networks • Iscoe, I.; McDonald, D.		94 AnlsProb 22 2168-2182
Asymptotics of exit times for Markov	jump processes • Kersting, G.; Klebaner, F. C.	Sharp conditions	95 AnlsProb 23 268-272
for nonexplosions and explosions in Markov	jump processes • Schäbe, H.; Tiedge, J.		95 StHefte 36 131-144
Kernel estimation for characteristics of pure	jump processes • Djehiche, Boualem; Kaj, Ingemar		95 AnlsProb 23 1414-1438
The rate function for some measure-valued	Jump and sharp cusp detection by wavelets • Wang, Yazhen		95 Biomtrka 82 385-397
interacting particle system with unbounded	jumps • Feng, Shui	Large deviations for empirical process of mean-field	94 AnlsProb 22 2122-2151
for long range interacting particle systems with	jumps • Léonard, Christian	Large deviations	95 AIHenrPB 31 289-323
Review of	"Junior Pinpoint" • Longman Logotron (Auth); Graham, Alan (Rev)		95 TeachgSt 17/2 75-75
ponentwise estimation of ordered parameters of	k (≥ 2) exponential populations • Vijayasree, G.; Misra, Neeraj; Singh, Harshinder	Com-	95 AnInStMa 47 287-307
Projection-method Mantel-Haenszel estimator for	$K \times J$ tables • Yanagawa, Takashi; Fujii, Yoshinori		95 JASA 90 649-656
Large sample asymptotic properties of the double	k-class estimators in linear regression models • Vinod, H. D.; Srivastava, V. K.		95 EcomtrRv 14 75-100
Webb, James M.	K-fold crossvalidation in canonical analysis • Liang, Kun-Hsia; Krus, David J.;		95 MultBehR 30 539-545
Temporal analogues to spatial	K functions • Cannon, Ann; Cressie, Noel		95 BiomtrcJ 37 351-373
Wang, Julian Z. Selection of the	k largest order statistics for the domain of attraction of the Gumbel distribution •		95 JASA 90 1055-1061
A simulation study of the weighted	k-means cluster procedure • Rost, D.		95 JStCmpSm 53 51-63
Asymptotics of	k-mean clustering under non-i.i.d. sampling • Serinko, Regis J.; Babu, Gutti Jogesh		95 StPrLet 24 57-66
procedures for estimating a linear function of	k multinormal means • Singh, R. K.; Chaturvedi, A.; Kumar, S.	Two stage and sequential	94 JIndnStA 32 1-8
formulas for linear and circular consecutive-	k-of-n: F systems • Peköz, Erol A.; Ross, Sheldon M.	A simple derivation of exact reliability	95 JAppProb 32 554-557
The increasing failure rate property of consecutive-	k-out-of-n systems • Cui, Lirong; Hawkes, Alan G.; Jalali, Assad		95 PrEngInf 9 217-225
Hazard rate ordering of	k-out-of-n systems • Shaked, Moshe; Shanthikumar, J. George		95 StPrLet 23 1-8
use in palaeoecology: A comparison of inverse (	k-nearest neighbours, partial least squares and weighted ... • Braak	... and their	95 Chemolab 28 165-180
on the finite sum of the binary sequence of order	k • Dhar, Sunil K.; Jiang, Xulun	Probability bounds	95 JAppProb 32 1014-1027
for the linear permutation statistic of order	k • Whitmore, James B.	Exact moment formulae	94 ASAProBiop 262-266
between any two ranked location parameters of	k populations • Liu	On the construction of exact lower confidence bounds on the distance	95 CommStA 24 2073-2085
	K-record values and the extreme-value index • Berred, Mohamed		95 JStPlInf 45 49-63
A comparison of tests for the	k-sample, non-decreasing alternative • Mahrer, Joanne M.; Magel, Rhonda C.		95 StatMed 14 863-871
Statistical analysis appropriate for	K samples of two variables • Nagarsenker, Brahamanand; Nagarsenker, Panna		94 ASAProStCp 218-221
The asymptotics for studentized	k-step M-estimators of location • Jurečková, Jana; Malý, Marek		95 SqtlAnly 14 229-245
success-runs and failures until the first consecutive	k successes • Aki, Sigeo; Hirano, Katuomi	Joint distributions of numbers of	95 AnInStMa 47 225-235
for the number of occurrences of consecutive	k successes in n Markovian trials • Wang, Y. H.; Ji, Shuixin	Limit theorems	95 JAppProb 32 727-735
procedure for all-pairwise comparisons of	k treatments based on the range statistic • Liu, Wei	A group sequential	95 Biomtrcs 51 946-955
and limit of the iterative algorithm for choosing	k value in ridge regression estimation (Chinese) • He, Z.; He, L.	The convergence theorem	95 AMaAPknC 17 59-64
deviation principles for the Hopfield model and the	Kac-Hopfield model • Bovier, Anton; Gayrard, Véronique; Picco, Pierre	Large	95 ProbTher 101 511-546
	Kac's representation from an asymptotic viewpoint • Horváth, Lajos		95 JStPlInf 46 27-45
Feynman-	Kac semigroup with discontinuous additive functionals • Song, Renming		95 JTherPrb 8 727-762
	Kalman filter		
and four-dimensional data ... • Wahba et al.	Adaptive tuning of numerical weather prediction models: Randomized GCV in three-		95 MWeathr 11 3358-3369
	Forecasting with incomplete data • Lee, C. P.		95 CommStA 24 1255-1269
	Integrated probabilistic data association-finite resolution • Mšicki, D.; Evans, R.		95 Autmatca 31 559-570
• Dee, Dick P.	On-line estimation of error covariance parameters for atmospheric data assimilation		95 MWeathr 123 1128-1145
Naniwa, Sadao; Ishiguro, Makio	A multivariate stochastic model with non-stationary trend component • Kato, Hiroko;		95 AplStMDA 11 77-95
Lin, Dennis K. J.	Robust recursive estimation for correlated observations • Guttman, Irwin;		95 StPrLet 23 79-92
	The simulation smoother for time series models • De Jong, Piet; Shephard, Neil		95 Biomtrka 82 339-350
	Spline smoother as a dynamic linear model • Moyeed, R. A.		95 AstrlJSt 37 193-204
	Square-root RTS smoothing algorithms • Park, PooGyeon; Kailath, Thomas		95 IJContrl 62 1049-1060
tion to the index of industrial production • Patterson	A state space approach to forecasting the final vintage of revised data with an applica-		95 JForcast 14 337-350
tions to epidemiology • Maryak, J. L.; Spall, J. C. et al.	Uncertainties for recursive estimators in nonlinear state-space models, with applica-		95 Autmatca 31 1889-1892
J. P.; Jakeman, A. J.; Taylor, J. A. Random walks in the	Kalman filter: Implications for greenhouse gas flux deductions • Mulquiney, J. E.; Norton,		95 EnvrMtrc 6 473-478
Duality between the discrete-time	Kalman filter and LQ control law • Blackmore, Perry A.; Bitmead, Robert R.		95 IEEEAuCn 40 1442-1444
application of structural time series models and the	Kalman filter • Visser et al.	... and regression analysis in climatological time series: An	95 JClimate 8 969-979
the observational error are correlated • Nieto et al.	Kalman filter for singular and conditional state-space models when the system state and		95 StPrLet 22 303-310
Forecasting hurricane tracks using the	Kalman filter • Bril, Gordon		95 EnvrMtrc 6 7-16
The Kantorovich inequality for error analysis of the	Kalman filter with unknown noise distributions • Spall, James C.		95 Autmatca 31 1513-1517
New square-root algorithms for	Kalman filtering • Park, PooGyeon; Kailath, Thomas		95 IEEEAuCn 40 895-899
On practical implementation of robust	Kalman filtering • Romera, Rosario; Cipra, Tomas		95 CommStB 24 461-488
complete panel data using the EM algorithm and the	Kalman smoother • Jansen, R. A. R. G. et al.	Longitudinal LISREL model estimation from in-	95 StNeerla 49 362-377
On two recent papers of Y.	Kanazawa • Jones, M. C.		95 StPrLet 24 269-271
distributions • Spall, James C. The	Kantorovich inequality for error analysis of the Kalman filter with unknown noise		95 Autmatca 31 1513-1517
Convergence of the conditional	Kaplan-Meier estimate under strong mixing • Lecoutre, Jean-Pierre; Ould-Saïd, Elias		95 JStPlInf 44 359-369
	Kaplan-Meier estimator		
Xiang, Xiaojing	Bahadur representation of the kernel quantile estimator under random censorship •		95 JMultiAn 54 193-209
	The central limit theorem under random censorship • Stute, Winfried		95 AnlsStat 23 422-439
	Conditional bootstrap methods for censored survival data • Kim, Ji-Hyun		95 JKoreaSt 24 197-218
copula • Zheng, Ming; Klein, John P.	Estimates of marginal survival for dependent competing risks based on an assumed		95 Biomtrka 82 127-138
brood mates • Flint, P. L.; Pollock, K. H. et al.	Estimating prefledging survival : Allowing for brood mixing and dependence among		95 JWildlMg 59 448-455
Malani, Hina Mehta	A modification of the redistribution to the right algorithm using disease markers •		95 Biomtrka 82 515-526
data • Li, Gang	On nonparametric likelihood ratio estimation of survival probabilities for censored		95 StPrLet 25 95-104
	Product-limit survival functions with correlated survival times • Williams, Rick L.		95 LiDaAnal 1 171-186
censored data • Park, Hyo-Il; Park, Sang-Gue	Quantile estimation of treatment effect for the two-sample problem with right		95 StPrLet 24 139-145
	Semi-parametric estimation in failure time mixture models • Taylor, Jeremy M. G.		95 Biomtrcs 51 899-907
data with a discrete covariate • Rahbar, M. H.	Sequential fixed width confidence intervals for regression parameters from censored		95 SqtlAnly 14 143-156
functions • Rojo, Javier	On the weak convergence of certain estimators of stochastically ordered survival		95 JNnparSt 4 349-363
Extensions of the	Kaplan-Meier estimator • Chien, Wei-Ting Kary; Kuo, Way		95 CommStB 24 953-964
W. et al. Maximum likelihood estimation of the	kappa coefficient from models of matched binary responses • Shoukri, M. M.; Martin, S.		95 StatMed 14 83-99
• Agresti, Alan; Ghosh, Atalanta; Bini, Matilde Raking	kappa: Describing potential impact of marginal distributions on measures of agreement		95 BiomtrcJ 37 811-820
through modelling: An illustration with hierarchical	kappa • Basu, Srabashi; Raychaudhuri, Aparna; Basu, Ayanendranath	Improving precision	95 CommStB 24 399-408

	Kappa statistic	
agreement coefficient • Lehmann, M. et al.	Comparison between exact and parametric distributions of multiple inter-raters	95 CmpBiomd 47 113-121
	When quality is a matter of taste, use reliability indexes • Futrell, David	95 QulPr 28/5 81-86
	Weighted concordance analysis • Tricot, Jean-Marie; Lepage, Yves	95 BiomtrcJ 37 727-743
Comparison of several goodness-of-fit tests for the	kappa statistic based on exact power and coverage probability • Basu, S.; Basu, A.	95 StatMed 14 347-356
tions • Ritter, G. et al. Automatic context-sensitive	karyotyping of human chromosomes based on elliptically symmetric statistical distribu-	95 PattrnRc 28 823-831
Testing whether a count variable belongs to the	Katz family of distributions • Anstrom, Kevin J.; Quade, Dana	94 ASAProStCp 194-199
Onset and structure of interfaces in a	Kawasaki + Glauber interacting particle system • Giacomin, G.	95 ProbTher 103 1-24
Stuart, A.; Ord, J. K. (Auth) et al. Review of	"Kendall's advanced theory of statistics, Volume I: Distribution theory (Sixth edition)" •	95 Biomtrcs 51 1583-1584
proof of a long-standing conjecture of D. G.	Kendall concerning the shapes of certain large random polygons • Miles, R. E. A heuristic	95 AdvAppPr 27 397-417
function and tail probability approximations for	Kendall's score with tied rankings • Valz, P. D.; McLeod, A. I. et al. Cumulant generating	95 AnlsStat 23 144-160
	Kendall's tau	
Sibuya, Masaaki	Power of record-breaking test (Japanese) • Iiyama, Yuriko; Nishimura, Kazuo;	95 JpJAppSt 24 13-26
	A robust partial correlation measure • Rao, Sudhakar H.; Sievers, Gerald L.	95 JNnparSt 5 1-20
O'Gorman, Thomas W.; Woolson, Robert F. Using	Kendall's τ_b correlations to improve variable selection methods in case-control studies •	95 Biomtrcs 51 1451-1460
validation, missing data and ... • Rännar et al. A PLS	kernel algorithm for data sets with many variables and few objects. Part II: Cross-	95 JChemtrc 9 459-470
study of some bias correction techniques for	kernel-based density estimators • Foster, Peter A comparative	95 JStCmpSm 51 137-152
M.-C.; Jéquier, J.-C. Reference values obtained by	kernel-based estimation of quantile regressions • Ducharme, G. R.; Gannoun, A.; Guertin,	95 Biomtrcs 51 1105-1116
Using Monte Carlo simulation to evaluate	kernel-based home range estimators • Worton, Bruce J.	95 JWildlMg 59 794-800
Local likelihood based on	kernel censoring • Copas, J. B.	95 JRSS-B 57 221-235
On Bahadur-Kiefer representation of a	kernel conditional quantile estimator • Xiang, Xiaojing	95 JNnparSt 5 275-287
On qualitative smoothness of	kernel density estimates • Mammen, E.	95 MaOpfStS 26 253-267
A note on bandwidth selections for	kernel density estimators at endpoints • Barabesi, L.; Fattorini, L.	94 Metron 52/3 43-56
Consistency of the local	kernel density estimator • Givens, Geof H.	95 StPrLet 25 55-61
Asymptotics for the transformation	kernel density estimator • Hössjer, Ola; Ruppert, David	95 AnlsStat 23 1198-1222
Minimax	kernel density estimators with length biased data • Wu, C. O.	95 MaMethSt 4 56-80
Bandwidth selection for	kernel distribution function estimation • Altman, Naomi; Léger, Christian	95 JStPlInf 46 195-214
Integral transforms with infinitely divisible	kernels • Finkelstein, M.; Scheiberg, S.; Tucker, H. G.	94 ThProbAp 39 670-676
A class of unbiased	kernel estimates of a probability density function • Rychlik, T.	95 ZastApMa 22 485-497
Improved variable window	kernel estimates of probability densities • Hall, Peter; Hu, Tien Chung; Marron, J. S.	95 AnlsStat 23 1-10
	Kernel estimation	
Tsybakov, A. B.	Additive nonparametric regression on principal components • Härdle, W.;	95 JNnparSt 5 157-184
models • Gao, Jiti	Asymptotic properties of some estimators for partly linear stationary autoregressive	95 CommStA 24 2011-2026
	Asymptotic theory for partly linear models • Gao, Jiti	95 CommStA 24 1985-2009
Miklós; Horváth, Lajos	On the distance between smoothed empirical and quantile processes • Csörgő,	95 AnlsStat 23 113-131
tables • Dong, Jianping; Simonoff, Jeffrey S.	A geometric combination estimator for d-dimensional ordinal sparse contingency	95 AnlsStat 23 1143-1159
ing parameter. I • Konakov, V. D.; Piterbarg, V. I.	High level excursions of Gaussian fields and the weakly optimal choice of the smooth-	95 MaMethSt 4 421-434
Nielsen, Jens P.; Linton, Oliver B.	Kernel estimation in a nonparametric marker dependent hazard model •	95 AnlsStat 23 1735-1748
estimation • Cox, Dennis D.; Kim, Tae Yoon	Moment bounds for mixing random variables useful in nonparametric function	95 StocProc 56 151-158
Nicholas I.; Hoffmann, Branka	On the nonparametric estimation of covariance functions • Hall, Peter; Fisher,	94 AnlsStat 22 2115-2134
siglio, Loredana; Patil, Ganapati P.; Taillie, Charles	Propagation functions for monitoring distributional changes • Kaur, Amarjot; Di Con-	95 EnvEcoSt 2 239-269
	Smoothing categorical data • Simonoff, Jeffrey S.	95 JStPlInf 47 41-69
	Kernel estimation for characteristics of pure jump processes • Schäbe, H.; Tiedge, J.	95 StHefte 36 131-144
Linton, Oliver B.	Kernel estimation in a nonparametric marker dependent hazard model • Nielsen, Jens P.;	95 AnlsStat 23 1735-1748
	Kernel estimation of the regression function with random sampling times • Vilar, J. A.	95 Test 4 137-178
	Kernel estimation of relative risk • Kelsall, Julia E.; Diggle, Peter J.	95 Bernoull 1 3-16
On multivariate	kernel estimation for samples from weighted distributions • Ahmad, Ibrahim A.	95 StPrLet 22 121-129
Nonparametric	kernel estimation for semiparametric models • Andrews, Donald W. K.	95 EconTher 11 560-596
	Kernel estimator	
	Adaptive root n estimates of integrated squared density derivatives • Wu, Tiee-Jian	95 AnlsStat 23 1474-1495
M.	The approximate distribution of nonparametric regression estimates • Robinson, P.	95 StPrLet 23 193-201
tion with dependent observations • Kim, T. Y. et al.	Asymptotic behaviors of some measures of accuracy in nonparametric curve estima-	95 JMultiAn 53 67-93
tion under dependence • Boente, G.; Fraiman, R.	Asymptotic distribution of data-driven smoothers in density and regression estima-	95 CandJSt 23 383-397
Mason, D. M.	On the asymptotic normality of L_p-norms of empirical functionals • Beirlant, J.;	95 MaMethSt 4 1-19
	Bandwidth selection in nonparametric regression with general errors • Chu, C. K.	95 JStPlInf 44 265-275
West, Mike	Bayesian density estimation and inference using mixtures • Escobar, Michael D.;	95 JASA 90 577-588
Huet, S.; Jolivet, E.	Better bootstrap confidence intervals for regression curve estimation • Härdle, W.;	95 MaOpfStS 26 287-306
Veraverbeke, N.	Bootstrapping a nonparametric polytomous regression model • Aerts, M.;	95 MaMethSt 4 189-200
	Comparison of regression curves using quasi-residuals • Kulasekera, K. B.	95 JASA 90 1085-1093
Fan, Yanqin; Gencay, Ramazan	A consistent nonparametric test of symmetry in linear regression models •	95 JASA 90 551-557
	Density estimation under long-range dependence • Csörgő, Sándor; Mielniczuk, Jan	95 AnlsStat 23 990-999
nonparametric regression • Chu, C. K.	The effect of the smoothness of the regression function on a bandwidth selector in	95 CommStB 24 61-77
Sheather, S. J.; Wand, M. P.	An effective bandwidth selector for local least squares regression • Ruppert, D.;	95 JASA 90 1257-1270
Monte Carlo study • Buchinsky, Moshe	Estimating the asymptotic covariance matrix for quantile regression models. A	95 JEconmtx 68 303-338
• Song, K. S.; Müller, H. G.; Clifford, A. J. et al.	Estimating derivatives of pharmacokinetic response curves with varying bandwidths	95 Biomtrcs 51 12-20
Kim, W. C.; Park, B. U.	On estimating integrated squared spectral density derivatives • Lee, Y. H.; Cho, S.;	95 JStPlInf 48 165-184
Hui	On estimating mixing densities in discrete exponential family models • Zhang, Cun-	95 AnlsStat 23 929-945
Kokic, P. N.; Moon, L.; Gooday, J.; Chambers, R. L.	Estimating temporal farm income distributions using spatial smoothing techniques •	95 AstrlJSt 37 129-143
	Estimation of integral functionals of a density • Birgé, Lucien; Massart, Pascal	95 AnlsStat 23 11-29
Dodge, Yadolah; Jurečková, Jana	Estimation of quantile density function based on regression quantiles •	95 StPrLet 23 73-78
	Estimation of a quantile in some nonstandard cases • Xiang, Xiaojing	95 AnInStMa 47 105-117
C. L.	Estimators of integrals of powers of density derivatives • Hall, Peter; Wolff, Rodney	95 StPrLet 24 105-110
Pierre; Ould-Saïd, Elias	Hazard rate estimation for strong-mixing and censored processes • Lecoutre, Jean	95 JNnparSt 5 83-89
Rao, B. L. S. Prakasa	Kernel-type density and failure rate estimation for associated sequences • Bagai, Isha;	95 AnInStMa 47 253-266
estimators • Ralescu, Stefan S.	The law of the iterated logarithm for the multivariate nearest neighbor density	95 JMultiAn 53 159-179
Hsing, Tailen	Limit theorems for stable processes with application to spectral density estimation •	95 StocProc 57 39-71
Gasser, Theo	A minimax result for a class of nonparametric density estimators • Engel, Joachim;	95 JNnparSt 4 327-334
Fraiman, Ricardo	Minimum distance density-based estimation • Cao, Ricardo; Cuevas, Antonio;	95 CmpStDA 20 611-631
Engel, Joachim	Model estimation in nonlinear regression under shape invariance • Kneip, Alois;	95 AnlsStat 23 551-570
• Holiday, David B.	Near optimal weights in nonparametric regression under some common restrictions	95 StPrLet 22 33-42
Glad, Ingrid K.	Nonparametric density estimation with a parametric start • Hjort, Nils Lid;	95 AnlsStat 23 882-904
	Nonparametric drift estimation from ergodic samples • Arfi, Mounir	95 JNnparSt 5 381-389
regression model • Wang, Naisyin; Ruppert, David	Nonparametric estimation of the transformation in the transform-both-sides	95 JASA 90 522-534
Gulati, Sneh; Padgett, W. J.	Nonparametric function estimation from inversely sampled record-breaking data •	95 CandJSt 23 359-368

issues for income in the redesign of the Canadian	Labour Force Survey • Chen, E. J.; Gambino, J.; Laniel, N. et al. Design and estimation	94 ASASrvyRsr 919-924
Analysis of urban cluster size in the Canadian	Labour Force Survey • Laniel, Normand; Mohl, Chris	94 ASASrvyRsr 931-936
in the measurement of employment in the	labour force surveys in the European Community • Van Bastelaer, Aloïs Differences	94 JOffStat 10 277-305
Kinack, Mark Redesigning the Canadian	Labour Force Survey questionnaire: Development and testing • Sunter, Deborah;	94 ASASrvyRsr 908-912
interviewing: Effects on the results in the Swedish	labour force surveys • Bergman, L. R. et al. Decentralised CATI versus paper and pencil	94 JOffStat 10 181-195
viewing: The experience of the Canadian survey of	labour and income dynamics (Disc: p861-862) • Grondin et al. ... computer assisted inter-	94 ASASrvyRsr 839-844
Official	labour statistics: A historical perspective • Davidson, Roger	95 JRSS-A 158 165-173
limited dependent variables: A case study of female	labour supply • Lee, M.-J. Semi-parametric estimation of simultaneous equations with	95 JApEcmtx 10 187-200
Hjalmarsson, Lennart	Labour-use efficiency in Swedish social insurance offices • Kumbhakar, Subal C.;	95 JApEcmtx 10 33-47
	Lack-of-fit	
O'Brien, Ralph G.; Heft, Marc W.	Discrimination indexes and models for studying sensory functioning in aging •	95 JAppStat 22 9-27
tensen, M. Modelling the growth of human placental	lactogen in maternal serum during late pregnancy • Lundbye-Christensen, S.; Chris-	95 RBrzlPrS 9 1-24
and additions to "Mr. Markov plays chutes and	ladders" • Wang, C.; Corley, T. The longest chute is the worse chute: Comments	94 ASAProStEd 104-106
	Ladder height distributions with marks • Asmussen, Søren; Schmidt, Volker	95 StocProc 58 105-119
Which stochastic model is underlying the chain	ladder method? • Mack, Thomas	94 InsrneME 15 133-138
Spitzer's condition and	ladder variables in random walks • Doney, R. A.	95 ProbTher 101 577-580
for cointegration: The effects of mis-specifying the	lag length • Bewley, Ronald; Yang, Minxian Testing	95 MaCmpSim 39 251-255
Lai, Kon S.	Lag order and critical values of the augmented Dickey-Fuller test • Cheung, Yin-Wong;	95 JBES 13 277-280
On the determination of	lag order in vector autoregressions of cointegrated systems • Jacobson, Tor	95 CmpStQ 10 177-192
Tse, Y. K. Lead-	lag relationship between spot index and futures price of the Nikkei Stock Average •	95 JForcast 14 553-563
Quera, Vicenç Log-linear approaches to	lag-sequential analysis when consecutive codes may and cannot repeat • Bakeman, Roger;	95 PsychBu 118 272-284
methods for the selection of the truncation	lag • Ng, Serena; Perron, Pierre Unit root tests in ARMA models with data-dependent	95 JASA 90 268-281
	Lag window	
Joseph P.	Bias-corrected nonparametric spectral estimation • Politis, Dimitris N.; Romano,	95 JTimSrAn 16 67-103
McCoy, E. J.; Percival, D. B.	The effective bandwidth of a multitaper spectral estimator • Walden, A. T.;	95 Biomtrka 82 201-214
	Lagged dependent variable	
Arellano, Manuel; Bover, Olympia	Another look at the instrumental variable estimation of error-components models •	95 JEconmtx 68 29-51
models • Kiviet, Jan F.	On bias, inconsistency, and efficiency of various estimators in dynamic panel data	95 JEconmtx 68 53-78
	Efficient estimation of models for dynamic panel data • Ahn, Seung C.; Schmidt, Peter	95 JEconmtx 68 5-27
Pesaran, M. Hashem; Smith, Ron	Estimating long-run relationships from dynamic heterogeneous panels •	95 JEconmtx 68 79-113
Stein-rule estimation in models with a	lagged-dependent variable • Srivastava, Virendra K.; Ullah, Aman	95 CommStA 24 1343-1353
	Lagrange multiplier test	
	Detecting parameter shift in GARCH models • Chu, Chia-Shang James	95 EcomtrRv 14 241-266
P. An improved	Lagrange multiplier test for heteroskedasticity • Cribari-Neto, Francisco; Ferrari, Silvia L.	95 CommStB 24 31-44
	LaGrange multiplier test for multiple unit roots • Park, Young J.; Cho, Sinsup	94 ASAProBuEc 147-150
Cho, Sinsup Relations between the	Lagrange multiplier tests and t-statistics for seasonal unit roots • Park, Young Jin;	95 JKoreaSt 24 77-90
El-Helbawy, Abdalla T.; Hassan, Tawfik On the Wald,	Lagrangian multiplier and likelihood ratio tests when the information matrix is singular •	94 JItalStS 3 51-60
Turbulent diffusion as a stochastic	Lagrangian process • Rodean, Howard C.	95 EnvrMtrc 6 659-663
The Great Salt	Lake: A barometer of low-frequency climatic variability • Lall, Upmanu; Mann, Michael	95 WaterReR 31 2503-2515
monitoring of indicators of trophic condition of	lakes • Larsen, D. P.; Urquhart, N. S.; Kugler, D. L. Regional scale trend	95 WaterReB 31 117-140
The uniform, logistic, and	lambda distributions • Appleton, David R.	95 CommStB 24 947-951
	LAN of extreme order statistics • Falk, Michael	95 AnInStMa 47 693-717
for future directions for studies of marked migratory	landbirds from the perspective of a practitioner in population ... • Desante Suggestions	95 JAppStat 22 949-965
Fish shape analysis using	landmarks • Glasbey, C. A.; Horgan, G. W.; Gibson, G. J.; Hitchcock, D.	95 BiomtrcJ 37 481-495
et al. Response to selection for litter size in Danish	Landrace pigs: A Bayesian analysis • Wang, C. S.; Gianola, D.; Sorensen, D. A.; Jensen, J.	94 ThApGene 88 220-230
	Landsat image interpretation • Zaki, M.; Tohamy, F. El; Hassan, A.	95 CmpStDA 20 75-97
electrical engineering • Coffey, W. T. et al. The	Langevin equation and its applications to stochastic problems in physics, chemistry and	95 WrldSciPub:Sing 500p
Okabe et al. Application of the theory of KM_2O-	Langevin equations to the non-linear prediction problem for the one-dimensional ... •	95 JMaJapan 47 349-367
Large deviations for	Langevin spin glass dynamics • Arous, G. B.; Guionnet, A.	95 ProbTher 102 455-509
Transforms for numerical estimation in	Langmuir • Betne, Prabha; Katti, S. K.	94 ASAProQlPr 286-290
Optimal design for	Langmuir with two inputs • Betne, Prabha; Katti, S. K.	94 ASAProBiop 425-429
ments in Spanish and English: The 1990 Spanish	Language Census long form • Gabbard, S. M. et al. Evaluating translations of survey instru-	94 ASASrvyRsr 1232-1235
nonresponse and log-linear analysis of the Spanish	Language Forms Availability Test • de la Puente, Manuel; Wobus, Peter An item	94 ASASrvyRsr 583-588
Distribution of the grammatical forms in the Italian	language (Italian) • Rizzi, Alfredo	94 Statstca 54 275-291
Bell curves and monkey	languages • Casti, John L.	95 Complxty 1 12-15
Appel, Martin V.; Cole, Ronald Spoken	language recognition of the Year 2000 Census Questionnaire Feasibility Test •	94 ASASrvyRsr 1310-1315
Query	languages for statistical databases • Tansel, Abdullah Uz	95 StatComp 5 59-72
of semantic classification trees to natural	language understanding • Kuhn, Roland; De Mori, Renato The application	95 IEEEPAMI 17 449-460
	Laplace approximation of high dimensional integrals • Shun, Zhenming; McCullagh, Peter	95 JRSS-B 57 749-760
nondegenerate case • Bolthausen, E. et al.	Laplace approximations for large deviations of nonreversible Markov processes. The	95 AnlsProb 23 236-267
	Laplace expansion	
dispersion • Breslow, Norman E.; Lin, Xihong	Bias correction in generalised linear mixed models with a single component of	95 Biomtrka 82 81-91
Reliability bounds for the L-class and	Laplace order • Sengupta, Debasis	95 JAppProb 32 832-835
Pierre-Simon	Laplace, philosophical essay on probabilities • Dale, Andrew I. (Transl)	95 Springer:Brln:NY 270p
On the uniform approximation of	Laplace's prior by t-priors in location problems • Mukhopadhyay, Saurabh; Ghosh, Malay	95 JMultiAn 54 284-294
A maximum entropy method for inverting	Laplace transforms of probability density functions • Wagner, Udo; Geyer, Alois L. J.	95 Biomtrka 82 887-892
Generalized	Laplacian approximations in Bayesian inference • Hsu, John S. J.	95 CandJSt 23 399-410
Graham, R. C. (Auth); Haaland, P. D. (Rev) Review of	"Large data analysis for the chemical sciences: A guide to statistical techniques" •	95 Chemolab 27 245-246
	Large deviation	
	Adaptive decision making for stochastic processes • Rukhin, Andrew L.; Vajda, Igor	95 JStPlInf 45 313-329
component • Carmona, Sara C.	An asymptotic problem for a reaction-diffusion equation with a fast diffusion	95 Stochatx 52 43-80
location-scale families • Nagaev, A. V.	Asymptotics of the Bayes risk in discrimination between normal and uniform	95 MaMethSt 4 312-333
cation to large deviations • Chen, L. H. Y.; Roos, M.	Compound Poisson approximation for unbounded functions on a group, with appli-	95 ProbTher 103 515-528
process • Ferrari, P. A.; Galves, A.; Liggett, T. M.	Exponential waiting time for filling a large interval in the symmetric simple exclusion	95 AIHenrPB 31 155-175
	Finding the conjugate of Markov fluid processes • Mandjes, Michel; Ridder, Ad	95 PrEngInf 9 297-315
	Limiting curves for I.I.D. records • Deuschel, Jean-Dominique; Zeitouni, Ofer	95 AnlsProb 23 852-878
	Many multivariate records • Goldie, Charles M.; Resnick, Sidney I.	95 StocProc 59 185-216
	Occupation measures for Markov chains • Dinwoodie, I. H.; Ney, Peter	95 JTherPrb 8 679-691
with infinite variance • Davis, R. A.; Hsing, T.	Point process and partial sum convergence for weakly dependent random variables	95 AnlsProb 23 879-917
Barbour, A. D.; Chen, Louis H. Y.; Choi, K. P.	Poisson approximation for unbounded functions, I: Independent summands •	95 StSinica 5 749-766
Mokliatchouk, O. M.	Pre-Gaussian random vectors and their applications • Kozachenko, Yu. V.;	95 ThPrMaSt 50 89-98
Kaj, Ingemar	The rate function for some measure-valued jump processes • Djehiche, Boualem;	95 AnlsProb 23 1414-1438
Hanne Mandrup	Ruin probabilities via local adjustment coefficients • Asmussen, Søren; Nielsen,	95 JAppProb 32 736-755
problem • Agrawal, Rajeev	Sample mean based index policies with $O(\log n)$ regret for the multi-armed bandit	95 AdvAppPr 27 1054-1078

	Latent variable	
Toft, Jostein	Evolutionary rank analysis applied to multidetectional chromatographic structures •	95 Chemolab 29 189-212
conditional asset pricing models • Ferson, W. E. et al.	Finite sample properties of the generalized method of moments in tests of	94 JFinEcon 36 29-55
the notation of the reticular action mode • Toyoda	A new identification rule and an estimator for the simultaneous equation model using	94 Behvmtrk 21 163-175
Viswanathan, S.	Can speculative trading explain the volume-volatility relation? • Foster, F. Douglas;	95 JBES 13 379-396
The feasibility of	latent variables applied to GC-MS data • Brakstad, Frode	95 Chemolab 29 157-176
Modeling the processes of development via	latent variable growth curve methodology • Duncan, Terry E.; Duncan, Susan C.	95 StrEqMod 2 187-213
Review of "A causal analysis with	latent variables of macroeconomic interdependencies" (German) • Hillmer, M. (Auth) et al.	95 AllgeStA 79 224-227
Medendorp, Sharon V.	Latent variable models for clustered ordinal data • Qu, Yinsheng; Piedmonte, Marion R.;	95 Biomtrcs 51 268-275
Review of "Discrete	latent variable models" • Heinen, Ton (Auth); Andersen, E. B. (Rev)	95 Psymtrka 60 449-450
predictive validity in the context of selection and	latent variable modelling: A Monte Carlo study • Hsu Sampling behaviour in estimating	95 BJMaStPs 48 75-97
estimating technique for interaction and quadratic	latent variables • Ping, Robert A., Jr A parsimonious	95 JMarRsr 32 336-347
of Muthén's LISCOMP approach to estimation of	latent variable relations with a comprehensive measurement ... • Muthén et al. ... aspects	95 Psymtrka 60 489-503
Brazil (Italian) • Drovandi, S. A structural model with	latent variables used to analyze the links between modernization and fertility: The case of	94 Statstca 54 521-538
Review of "Modelling covariances and	latent variables using EQS" • Dunn, G.; Everitt, B.; Pickles, A. (Auth); Chan, Wai (Rev)	95 SIAMRvw 37 462-464
(Rev) Review of "Modelling covariances and	latent variables using EQS" • Dunn, G.; Everitt, B.; Pickles, A. (Auth); Schimek, Michael G.	95 CmpStQ 10 196-196
Review of "Loglinear models with	latent variables" • Hagenaars, Jacques A. (Auth); Hershberger, Scott L. (Rev)	95 StrEqMod 2 172-174
testing of moment structure models based on	latent variates involving interactions among the ... • Baumgartner et al. ... and	95 SocMethR 24 187-213
	Latin hypercube sampling	
	Exploratory designs for computational experiments • Morris, Max D.; Mitchell, Toby J.	95 JStPlInf 43 381-402
the MACCS reactor accident ... • Helton et al.	Robustness of an uncertainty and sensitivity analysis of early exposure results with	95 RelEngSS 48 129-148
Lee, Joo-Ho Two-stage	Latin hypercube sampling and its application (Korean) • Im, Mi-Jeong; Kwon, Woo-Joo;	95 KorApS 8/2 99-108
	Latin square	
Laywine, C.	Complete sets of orthogonal frequency squares and affine resolvable designs •	93 UtilMath 43 161-170
	Construction for orthogonal main effect plans • Street, D. J.	94 UtilMath 45 115-123
	Nonadditivity in row-column designs • Srivastava, Jagdish N.	93 JCISS 18 85-96
designs • Bellavance, F.; Tardif, S.	A nonparametric approach to the analysis of three-treatment three-period crossover	95 Biomtrka 82 865-875
On even and odd	latin squares • Janssen, Jeannette C. M.	95 JCombThA 69 173-181
Hedayat, A. S.; Parker, E. T. A pair of orthogonal	Latin squares of order 10 with four shared parallel transversals • Brown, J. W.;	93 JCISS 18 113-115
Sets of mutually orthogonal	Latin squares with "like subsquares," II • Roberts, Charles E., Jr	95 JCombThA 70 159-164
Crossing velocities and random	lattice animals • Sznitman, Alain-Sol	95 AnlsProb 23 1006-1023
Unbiasedness of the likelihood ratio test for	lattice conditional independence models • Andersson, Steen A.; Perlman, Michael D.	95 JMultiAn 53 1-17
Testing	lattice conditional independence models • Andersson, Steen Arne; Perlman, Michael D.	95 JMultiAn 53 18-38
Cubic	lattices • Pearce, S. C.	95 JAppStat 22 355-373
	Lattice design	
	Some design problems in crop experimentation. III. Non-orthogonality • Pearce, S. C.	95 ExpAgric 31 409-422
autoregressive model on a two-dimensional	lattice • Sethuraman, S. et al. Maximum likelihood estimation for a fractionally differenced	95 JStPlInf 44 219-235
pendence models determined by finite distributive	lattices and by directed acyclic graphs • Andersson et al. ... between conditional inde-	95 JStPlInf 48 25-46
A test of fit for	lattice distributions • Epps, T. W.	95 CommStA 24 1455-1479
Variance of number of	lattice points in random narrow elliptic strip • Bleher, Pavel M.; Lebowitz, Joel L.	95 AIHenrPB 31 27-58
[Algorithm AS 299] Generation of simplex	lattice points • Chasalow, Scott D.; Brand, Richard J.	95 ApplStat 44 534-545
Kratina, Ioana	Lattices of random sets and progressivity • Ivanoff, Gail B.; Merzbach, Ely; Schiopu-	95 StPrLet 22 97-102
Heavy traffic analysis of a random walk on a	lattice semi-strip • Falin, Gennadi	95 CommStSM 11 395-409
covering the origin for Bernoulli trials on the	lattice • Kopociński, B. The largest square of successes	95 StNeerla 49 31-39
theorem for nonhomogeneous random walk on a	lattice • Zhizhina, E. A.; Minlos, A. A local limit	94 ThProbAp 39 490-503
Bureau's Annual Capital ... • Slanta et al. Applying the	Lavallée and Hidiroglou method to obtain stratification boundaries for the Census	94 ASASrvyRsr 693-698
	Law	
	• see also probability distribution, legal statistics, limit -, zero-one -	
	• see also names of specific probability distributions such as stable -	
of dynamical systems with conservation	laws • Borodin, A. N.; Freidlin, M. I. Fast oscillating random perturbations	95 AIHenrPB 31 485-525
	Law of the iterated logarithm	
	The asymptotic behavior of locally square integrable martingales • Wang, Jia-gang	95 AnlsProb 23 552-585
	Asymptotic confidence bands for normal distributions • Lin, Yachen	95 CommStA 24 641-652
	Asymptotics for multisample statistics • Vaillancourt, Jean	95 CandJSt 23 171-184
Xiang, Xiaojing	Bahadur representation of the kernel quantile estimator under random censorship •	95 JMultiAn 54 193-209
Goodman, V.	Clustering behavior of finite variance partial sum processes • Einmahl, U.;	95 ProbTher 102 547-565
	A conditional least squares approach to bilinear time series estimation • Grahn, T.	95 JTimSrAn 16 509-529
	Estimation of conditional quantile density function • Xiang, Xiaojing	95 JNnparSt 4 309-316
	Estimation of a quantile in some nonstandard cases • Xiang, Xiaojing	95 AnInStMa 47 105-117
	On the fractal nature of empirical increments • Deheuvels, Paul; Mason, David M.	95 AnlsProb 23 355-387
Csörgő, Miklós; Földes, Antónia; Révész, Pál	Global Strassen-type theorems for iterated Brownian motions • Csáki, Endre;	95 StocProc 59 321-341
Li, Wenbo V.; Talagrand, Michel	Lim inf results for Gaussian samples and Chung's functional LIL • Kuelbs, James;	94 AnlsProb 22 1879-1903
Kou, Shing-Gang	Limits of first passage times to rare sets in regenerative processes • Glasserman, Paul;	95 AnlsApPr 5 424-445
	On the local rate of growth of Lévy processes with no positive jumps • Bertoin, Jean	95 StocProc 55 91-100
performance • Rananand, Nol	Markov approximations to D-BMAPs: Information theoretic bounds on queueing	95 CommStSM 11 713-734
Mason, David M.	Nonstandard local empirical processes indexed by sets • Deheuvels, Paul;	95 JStPlInf 45 91-112
	Rates of convergence for tail series • Tomkins, R. J.	95 JStPlInf 43 77-85
(French) • Miclo, Laurent	Remarks about the ergodicity of simulated annealing algorithms on a graph	95 StocProc 58 329-360
Joo, Sang Yeol; Kim, Yun Kyong	A sharp result of random upper functions for Levy processes • Kim, Gwan Young;	95 JKoreaSt 24 65-76
Hölder norms • Kuelbs, J.; Li, W. V.; Shao, Qi-man	Small ball probabilities for Gaussian processes with stationary increments under	95 JTherPrb 8 361-386
Shi, Z.	Some asymptotic results for exponential functionals of Brownian motion • Gruet, J.-C.;	95 JAppProb 32 930-940
dependence • Kouritzin, Michael A.	Strong approximation for cross-covariances of linear variables with long-range	95 StocProc 60 343-353
applications • Shao, Qi-Man	Strong approximation theorems for independent random variables and their	95 JMultiAn 52 107-130
ables • Dabrowski, André Robert; Zoglat, Abdelhak	Strong invariance principles for triangular arrays of weakly dependent random vari-	95 CandJSt 23 299-310
of independent random variables with ... • Li et al.	On the strong law of large numbers and the law of the logarithm for weighted sums	95 JMultiAn 52 181-198
The Strassen	law of the iterated logarithm in Banach function spaces • Norvaiša, R.	95 StPrLet 25 1-8
Some asymptotic results related to the	law of iterated logarithm for Brownian motion • Chan, Terence	95 JTherPrb 8 643-667
Giné, Evarist On the	law of the iterated logarithm for canonical U-statistics and processes • Arcones, Miguel A.;	95 StocProc 58 217-245
Tobbal, Khelifa A functional	law of the iterated logarithm for the Dekkers-Einmahl-de Haan tail index estimator •	95 JNnparSt 5 145-156
Arcones, Miguel A. The	law of the iterated logarithm for empirical processes under absolute regularity •	95 JTherPrb 8 433-451
On the cluster set problem for the generalized	law of the iterated logarithm in Euclidean space • Einmahl, Uwe	95 AnlsProb 23 817-851
Small values of Gaussian processes and functional	laws of the iterated logarithm • Monrad, Ditlev; Rootzén, Holger	95 ProbTher 101 173-192
On the	law of the iterated logarithm for Gaussian processes • Arcones, Miguel A.	95 JTherPrb 8 877-903
Shi, Z.	Laws of the iterated logarithm for iterated Wiener processes • Hu, Y.; Pierre-Loti-Viaud, D.;	95 JTherPrb 8 303-319

statistics: Techniques from research on classroom	learning and assessment • Busk, Patricia L. Improving the quality of teaching	94 ASAProQlPr 164-171
consulting practicum: Opportunity for authentic	learning • Derr, Janice A. Statistical	94 ASAProStEd 140-143
through hands-on experiments and cooperative	learning • Bilotti-Aliaga, Martha Exploring probability and statistics	94 ASAProStEd 150-155
On the reliability of indicators of	learning curve model accuracy • Moses, O. Douglas	95 JCostAnl /1 63-101
Abou-Nasr, M. A.; Sid-Ahmed, M. A. Fast	learning and efficient memory utilization with a prototype based neural classifier •	95 PattrnRc 28 581-593
The econometrics of	learning in financial markets • Bossaerts, Peter	95 EconTher 11 151-189
Keeler et al. Using small groups to promote active	learning in the introductory statistics course: A report from the field (jse/v3n2/keeler) •	95 JStEd 3/2 28kb
Gradient descent	learning of nearest neighbor classifiers with outlier rejection • Urahama, K.; Furukawa, Y.	95 PattrnRc 28 761-768
C. (Ed); Fulkerson, Bill (Rev) Review of "Machine	learning, neural and statistical classification" • Michie, D.; Spiegelhalter, D. J.; Taylor, C.	95 Technmcs 37 459-459
Phillips, Robert F. (Rev) Review of	"Learning and practicing econometrics" • Griffiths, W. E.; Hill, R. C.; Judge, G. G. (Auth);	95 IntJFor 11 331-333
Review of "Statistics:	Learning in the presence of variation" • Wardrop, Robert L. (Auth); Rossman, Allan J. (Rev)	95 AmerStat 49 237-238
Lower bounds in pattern recognition and	learning • Devroye, Luc; Lugosi, Gábor	95 PattrnRc 28 1011-1018
implementation • Maeda, Y.; Hirano, H.; Kanata, Y. A	learning rule of neural networks via simultaneous perturbation and its hardware	95 Autmatca 31 251-259
Successful customers of statistics at a distant	learning site • Bolland, Thomas W.	94 ASAProQlPr 300-304
component extraction using recursive least squares	learning • Bannour, Sami; Azimi-Sadjadi, Mahmood R. Principal	95 IEEENrlN 6 457-469
Behrens, John T. Identification of misconceptions in	learning statistical power with dynamic graphics as a remedial tool • Yu, Chong Ho;	94 ASAProStEd 242-246
STEPS for	learning statistics • Bibby, Katherine; Davies, Neville	95 TeachgSt 17/3 107-110
Neural networks with counter-propagation	learning strategy used for modelling • Zupan, Jure; Novič, Marjana; Gasteiger, Johann	95 Chemolab 27 175-187
Survival analysis. A self-	learning text • Kleinbaum, D. G.	95 Springer:Brln:NY 250p
Review of "Logistic regression: A self	learning text" • Kleinbaum, D. G. (Auth); Howel, Denise (Rev)	95 Ststcian 44 410-411
Review of "Logistic regression: A self-	learning text" • Kleinbaum, D. G. (Auth); Kianifard, Farid (Rev)	95 Technmcs 37 116-117
The nature of statistical	learning theory • Vapnik, V. N.	95 Springer:Brln:NY 188p
Review of "The nature of statistical	learning theory" • Vapnik, V. N. (Auth); Fedorov, V. V. (Rev)	96 ShrtBkRv 16/1 3-3
ductory statistics courses from a conceptual, active	learning viewpoint (jse/v3n3/steinhorst) • Steinhorst et al. Developing material for intro-	95 JStEd 3/3 33kb
	Least absolute deviation	
Nolan, D.	A general Akaike-type criterion for model selection in robust regression • Burman, P.;	95 Biomtrka 82 877-886
Zhao, L. C.	Recent contributions to censored regression models • Rao, C. Radhakrishna;	95 Metrika 42 203-213
	Robust nonstationary regression • Phillips, Peter C. B.	95 EconTher 11 912-951
	Survival analysis with median regression models • Ying, Z.; Jung, S. H.; Wei, L. J.	95 JASA 90 178-184
response • Ducharme, Gilles R.	Uniqueness of the least-distances estimator in regression models with multivariate	95 CandJSt 23 421-424
of nuisance parameters for inference based on	least absolute deviations • Niemiro, W. Estimation	95 ZastApMa 22 515-529
	Least absolute deviation estimation of a shift • Bai, Jushan	95 EconTher 11 403-436
Asymptotic properties of nonlinear	least absolute deviation estimators • Kim, Hae Kyung; Choi, Seung Hoe	95 JKoreaSt 24 127-139
A bootstrap approach to hypothesis testing in	least absolute value regression • Dielman, Terry E.; Rose, Elizabeth L.	95 CmpStDA 20 119-130
Rose, Elizabeth L. Pre-test estimators in	least absolute value regression with autocorrelated errors • Dielman, Terry E.;	94 ASAProBuEc 95-100
Ducharme, Gilles R. Uniqueness of the	least-distances estimator in regression models with multivariate response •	95 CandJSt 23 421-424
	Least median of squares	
	An Anscombe type robust regression statistic • Bradu, Dan; Hawkins, Douglas M.	95 CmpStDA 20 355-386
Carrizosa, Emilio; Plastria, Frank	The determination of a "least quantile of squares regression line" for all quantiles •	95 CmpStDA 20 467-479
	On hyperplanes of closest fit • Farebrother, Richard William	95 CmpStDA 19 53-58
	Linear programming approach to LMS-estimation • Boček, P.; Lachout, P.	95 CmpStDA 19 129-134
random variables • Castillo, Enrique; Hadi, Ali S.	A method for estimating parameters and quantiles of distributions of continuous	95 CmpStDA 20 421-439
	Robust estimation of mean and variance in fisheries • Chen, Y.; Jackson, D. A.	95 TrAFish 124 401-412
Massart, D. L.	Robust principal components regression as a detection tool for outliers • Walczak, B.;	95 Chemolab 27 41-54
algorithms • Vankeerberghen, P. et al.	Robust regression and outlier detection for non-linear models using genetic	95 Chemolab 28 73-87
Consistency of the	least median of squares estimator in nonlinear regression • Stromberg, Arnold J.	95 CommStA 24 1971-1984
Fernández, M. C. et al. Regression by	least median squares, a methodological contribution to titration analysis • Ortiz-	95 Chemolab 27 231-243
Convergence of the feasible solution algorithm for	least median of squares regression • Hawkins, Douglas M.	95 CmpStDA 19 519-538
	Least norm squares estimation	
space • Kundu, Debasis	Consistency of the undamped exponential signal model on a restricted parameter	95 CommStA 24 241-251
Plastria, Frank The determination of a	"least quantile of squares regression line" for all quantiles • Carrizosa, Emilio;	95 CmpStDA 20 467-479
	Least squares	
variables • Houdré, Christian	On the almost sure convergence of series of stationary and related nonstationary	95 AnlsProb 23 1204-1218
models • Kiviet, Jan F.	On bias, inconsistency, and efficiency of various estimators in dynamic panel data	95 JEconmtx 68 53-78
and a Bayesian approach • Liu, Shu-Ing	Comparison of forecasts for ARMA models between a random coefficient approach	95 CommStA 24 319-333
weighted contribution of transition ... • Pollock et al.	A comparison of two methods for constructing evolutionary distances from a	95 MolBioEv 12 713-717
moving average disturbances • Song, Seuck Heun	Consistency and bounds on the bias of S^2 in the linear regression model with	95 JKoreaSt 24 507-518
	On a downdating formula for regression • Chan, N. N.	95 JStPlInf 46 347-350
	An empirical estimation procedure • Gupta, Arjun K.; Varga, Tamàs	94 Metron 52/1 67-70
D.; Clark, Gordon M.	Estimating expected sojourn times for a Markov renewal process • Rossetti, Manuel	95 CommStA 24 553-579
process • Usami, Yoshihiro; Huzii, Mituaki	Estimation of coefficients of time series regression with a nonstationary error	95 JTimSrAn 16 105-118
ear model • Fourdrinier, Dominique; Wells, Martin T.	Estimation of a loss function for spherically symmetric distributions in the general lin-	95 AnlsStat 23 571-592
rates differ over nucleotide sites • Yang, Ziheng	Evaluation of several methods for estimating phylogenetic trees when substitution	95 JMolEvol 40 689-697
	Fast and simple scatterplot smoothing • Härdle, W.; Marron, J. S.	95 CmpStDA 20 1-17
	Gauss, statistics and Gaussian elimination • Stewart, G. W.	95 JCmpGrSt 4 1-11
	A general duality approach to I-projections • Bhattacharya, Bhaskar; Dykstra, Richard	95 JStPlInf 47 203-216
regression model • Singh, R. K.; Agnihotri, M. et al.	A generalized class of estimators utilizing Lindley-like mean correction in linear	94 Metron 52/3 147-160
	Generalized ridge regression estimators under the LINEX loss function • Ohtani, K.	95 StHefte 36 99-110
Srivastava, Virendra K.; Srivastava, Anil K.	Improved estimation of regression coefficients and hypotheses testing •	94 Metron 52/3 161-170
Gouri K.	Inference procedures for the linear failure rate model • Sen, Ananda; Bhattacharyya,	95 JStPlInf 46 59-76
Gao, Jiti	The laws of the iterated logarithm of some estimates in partly linear models •	95 StPrLet 25 153-162
	Least absolute deviation estimation of a shift • Bai, Jushan	95 EconTher 11 403-436
Robinson, P. M.	Log-periodogram regression of time series with long range dependence •	95 AnlsStat 23 1048-1072
observations are correlated • Bischoff, Wolfgang	Lower bounds for the efficiency of designs with respect to the D-criterion when the	95 MaOpfStS 27 27-44
Bhaumik, Dulal K.	Majorization and D-optimality for complete block designs under correlations •	95 JRSS-B 57 139-143
misspecification • Wei, L.; Trenkler, G.	Mean square error matrix superiority of empirical Bayes estimators under	95 Test 4 187-205
	The method of sieves and minimum contrast estimators • Van de Geer, S.	95 MaMethSt 4 20-38
Engel, Joachim	Model estimation in nonlinear regression under shape invariance • Kneip, Alois;	95 AnlsStat 23 551-570
	Models of statistical visual motion estimation • Spetsakis, Minas	94 CVGIPIm 60 300-312
	A note on the Kuks-Olman estimator • Hoffmann, Kurt	95 MaOpfStS 26 185-187
	Who's number 1 in college football? ... And how might we decide? • Stern, Hal S.	95 Chance 8/3 7-14
	Optimal choice of a new observation in a linear model • Sengupta, Debasis	95 SankhyaA 57 137-153
when the expectations equation has ... • Smith et al.	The performance of alternative estimators in models with generated regressors	95 MaCmpSim 39 343-346
Fitch, Walter M.	Phylogenetic inference under the pure drift model • Xu, Shizhong; Atchley, William R.;	94 MolBioEv 11 949-960

	Least squares (cont.)	
Dube, M.; Singh, V.	Predictive efficiency of improved estimators in restricted regression models •	95 Test 4 323-331
	Quadratic distance estimators for the zeta family • Doray, Louis G.; Luong, Andrew	95 InsrneME 16 255-260
Ramanathan, S.	Recursive inference in a general linear model • Bhimasankaram, P.; Sengupta, D.;	95 SankhyaA 57 227-255
	On the reliability of indicators of learning curve model accuracy • Moses, O. Douglas	95 JCostAnl /1 63-101
Müller, Hans-Georg; Zhao, Peng-Liang	On a semiparametric variance function model and a test for heteroscedasticity •	95 AnlsStat 23 946-967
	Simplification of Milstar microwave and digital cost analysis model • Gobreial, Hany S.	95 JParmc 15/1 92-117
Gupta, A. K.; Varga, T.	Some inference problems for matrix variate elliptically contoured distributions •	95 MaOpfStS 26 219-229
Fitzpatrick, Ben G.	Statistical tests of fit in estimation problems for structured population modeling •	95 QApplMa 53 105-128
Virendra K.; Ullah, Aman	Stein-rule estimation in models with a lagged-dependent variable • Srivastava,	95 CommStA 24 1343-1353
	Unit root tests based on M estimators • Lucas, André	95 EconTher 11 331-346
	• see also names of specific types of least squares such as generalized - or partial -	
Bounded-variable	least-squares: An algorithm and applications • Stark, P. B.; Parker, R. L.	95 CmpStQ 10 129-141
Linear models:	Least squares and alternatives • Rao, C. Radhakrishna; Toutenburg, Helge	95 Springer:Brln:NY 352p
Draper, N. R. (Rev) Review of "Linear models:	Least squares and alternatives" • Rao, C. Radhakrishna; Toutenburg, Helge (Auth);	96 ShrtBkRv 16/1 7-8
A conditional	least squares approach to bilinear time series estimation • Grahn, T.	95 JTimSrAn 16 509-529
Error-free	least squares based on multiple homomorphic images • Keller-McNulty, Sallie	94 ASAProStCp 33-40
March K. A generalized	least-squares estimate for the origin of sporophytic self-incompatibility • Uyenoyama,	95 Genetcs 139 975-992
A geometric approach of the generalized	least-squares estimation in analysis of covariance structures • Wang, S. J.; Lee, Sik-Yum	95 StPrLet 24 39-47
parameters (French) • Jmel, S.	Least squares estimation for association models with linear constraints on the	95 RStApp 43/3 93-104
intensity • Angrist, J. D.; Imbens, G. W. Two-stage	least squares estimation of average causal effects in models with variable treatment	95 JASA 90 431-442
An upper bound for the recursive	least squares estimation error • Niederliński, A.	95 IEEEAuCn 40 1655-1661
	Least squares estimator	
Schmidt, Karsten; Stahlecker, Peter	Reducing the maximum risk of regression estimators by polyhedral projection •	95 JStCmpSm 52 1-15
dispersion asymptotic optimality of the non-linear	least squares estimator • So, Beong-Soo Finite-sample, small-	95 JKoreaSt 24 303-312
Perturbation analysis for an orthogonal	least-squares estimator • Chapman, G. R.	95 CandJSt 23 101-108
of measurement errors under survey conditions on	least squares estimator of regression coefficients • Agarwal, R. C. et al. The effect	94 JIAgSt 46 290-297
ing objective function and gradient in the context of	least squares fitting a dynamic errors-in-variables model • Vregelaar, J. M. T. On comput-	95 SIAMSSC 16 738-753
Arabie, Phipps Iterative projection strategies for the	least squares fitting of tree structures to proximity data • Hubert, Lawrence;	95 BJMaStPs 48 281-317
Rzhetsky et al. Statistical properties of the ordinary	least-squares, generalized least-squares, and minimum-evolution methods of ... •	92 JMolEvol 35 367-375
tially dependent and the analysis uses generalized	least-squares • Martin, R. J. Some results on randomization when observations are spa-	95 CommStA 24 2199-2210
Principal component extraction using recursive	least squares learning • Bannour, Sami; Azimi-Sadjadi, Mahmood R.	95 IEEENrlN 6 457-469
Isidro Fitting a linear regression model by combining	least squares and least absolute value estimation • Allende, Sira; Bouza, Carlos; Romero,	95 Questo 19 107-121
tomography • O'Sullivan, Finbarr A study of	least squares and maximum likelihood for image reconstruction in positron emission	95 AnlsStat 23 1267-1300
input-output tables (Arabic) • Kerim, A. et al. RAS and	least squares methods versus other mathematical techniques in developing and updating	95 EgpStJ 39 120-143
Hays, Paul B.; Skinner, Wilbert R. A	least squares method for spectral analysis of space-time series • Wu, Dong L.;	95 JAtmsphr 52 3501-3511
A comparison of interval constrained	least squares and mixed regression estimators • Srivastava, Viren K.; Ohtani, Kazuhiro	95 CommStA 24 395-413
Global total	least squares modeling of multivariable time series • Roorda, Berend; Heij, Christiaan	95 IEEEAuCn 40 50-63
Takane, Yoshio (Rev) Review of	"Least squares optimization in multivariate analysis" • Berge, Jos M. F. ten (Auth);	95 Psymtrka 60 447-448
column elution curves • van Veldhuizen et al. The (	least squares) quotient algorithm as a rapid tool for obtaining sorption isotherms from	95 WaterReR 31 849-857
Georg, T. On selecting the smoothing parameter of	least squares regression estimates using the minimax regret approach • Droge, B.;	95 StatDecn 13 1-20
Angleton, George M.; Bonham, Charles D.	Least squares regression vs. geometric mean regression for ecotoxicology studies •	95 AplMaCmp 72 21-32
An effective bandwidth selector for local	least squares regression • Ruppert, D.; Sheather, S. J.; Wand, M. P.	95 JASA 90 1257-1270
SWEEP operator for	least-squares subject to linear constraints • Neytchev, Plamen Nikolov	95 CmpStDA 20 599-609
analysis using maximum likelihood and unweighted	least squares • Wolins, Leroy A Monte Carlo study of constrained factor	95 EdPsMeas 55 545-557
trait loci in crosses between outbred lines using	least squares • Haley, Chris S.; Knott, Sara A.; Elsen, Jean-Michel Mapping quantitative	94 Genetcs 136 1195-1207
view of "Theory of the combination of observations	least subject to error" • Gauss, C. F. (Auth); Stewart, G. W. (Transl); Stigler, S. M. (Rev) Re-	96 ShrtBkRv 16/1 41-41
(Transl) Theory of the combination of observations	least subject to errors. Part one, part two, supplement • Gauss, K. F. (Auth); Stewart, G. W.	95 SocApMathSim:PA 220p
	Least trimmed squares	
Song, Moon Sup; Shin, Bong Sup	A robust test for the parallelism of two regression lines (Korean) • Nam, Ho Soo;	95 KorApS 8/2 77-86
Exact computation of the	least trimmed squares estimate in simple linear regression • Hössjer, Ola	95 CmpStDA 19 265-282
of postgraduate students in education in the	Lebowa region of South Africa • Glencross, M. J. et al. Attitudes toward applied statistics	95 PsychRpt 77 315-322
Coefficients of	Lee-Gurland two-sample test on normal means • Lee, Austin F. S.	95 CommStA 24 1743-1768
two samples where all observations are either	left- or right-censored • Andersen, P. K.; Rønn, B. B. A nonparametric test for comparing	95 Biomtrcs 51 323-329
Marek	Left and right linear innovations for a multivariate S α S random variable • Rutkowski,	95 StPrLet 22 175-184
Optimal lengths of	left-turn lanes • Applebaugh, W.; Chadha, S. S.	94 ASAProStEd 102-103
as a value enhancing technique when faced with	legal liability • MacMinn, Richard D.; Brockett, Patrick L. Corporate spin-offs	95 InsrneME 16 63-68
Biostatistical concepts and methods in the	legal setting • Gastwirth, Joseph L.; Greenhouse, Samuel W.	95 StatMed 14 1641-1653
of voting coalitions in the New York colonial	legislature • Morrell, Christopher H.; Peyrot, Mark F.; Varga, Nicholas An analysis	94 ASAProSoSt 66-70
length bias • Shanmugam, R. Goodness-of-fit test for	length-biased data with discussion on prevalence, sensitivity and specificity of the	95 JStPlInf 48 277-290
Minimax kernel density estimators with	length biased data • Wu, C. O.	95 MaMethSt 4 56-80
Optimal	lengths of left-turn lanes • Applebaugh, W.; Chadha, S. S.	94 ASAProStEd 102-103
Correction to "The rate of convergence of the mean	length of the longest common subsequence" (V4 p1074-1082) • Alexander, K. S.	95 AnlsApPr 5 327-327
errors and confidence intervals for branch	lengths in phylogenetic trees by a bootstrap approach • Dopazo, Joaquín Estimating	94 JMolEvol 38 300-304
versions of a central limit theorem for the squared	length of a sample mean • Saikkonen, Pentti Dependent	95 StPrLet 22 185-194
Critical	length for semi-oriented bootstrap percolation • Mountford, T. S.	95 StocProc 56 185-205
Factors affecting hospital admissions and	length of stay: Some preliminary results • Savoca, Elizabeth	94 ASAProSoSt 222-226
Counting gays and	lesbians: A case for "new" survey methods • Edwards, Brad	94 ASASrvyRsr 1305-1309
among staging areas of the barnacle goose (Branta	leucopsis) • Schwarz, Carl James; Ganter, Barbara Estimating the movement	95 JAppStat 22 711-724
Spatial variation of natural radiation and childhood	leukaemia incidence in Great Britain • Richardson, S.; Monfort, C.; Green, M. et al.	95 StatMed 14 2487-2501
tion to nuclear power plant facilities and childhood	leukaemia in Sweden • Waller et al. ... and assessment of clusters of disease: An applica-	95 StatMed 14 3-16
sensitivity analysis • Viel, J.-F. et al. Incidence of	leukaemia in young people around the La Hague nuclear waste reprocessing plant: A	95 StatMed 14 2459-2472
reliable are contour curves? Confidence sets for	level contours • Lindgren, Georg; Rychlik, Igor How	95 Bernoull 1 301-319
	Level crossing	
Georg; Rychlik, Igor	How reliable are contour curves? Confidence sets for level contours • Lindgren,	95 Bernoull 1 301-319
Rychlik, Igor The expected number of	level crossings for certain symmetric α-stable processes • Michna, Zbigniew;	95 CommStSM 11 1-19
On the typical	level crossing time and path • Nyrhinen, Harri	95 StocProc 58 121-137
	Level shifts, temporary changes and forecasting • Trívez, F. Javier	95 JForcast 14 543-550
	Levenberg algorithm	
Smyth, G. K.	A modified prony algorithm for exponential function fitting • Osborne, M. R.;	95 SIAMSSC 16 119-138
	Leverage outlier	
	Deletion influence and masking in regression • Lawrance, A. J.	95 JRSS-B 57 181-189
regression • Kim, Choongrak; Jeon, Jong-Woo	Detecting influential observations on the smoothing parameter in nonparametric	95 JKoreaSt 24 495-506
in regression in the presence of outliers and	leverage points • Sommer, Suzanne; Staudte, Robert G. Robust variable selection	95 AstrlJSt 37 323-336

Tang, Y. et al. Maximum likelihood estimation in the	life testing for the two-parameter exponential distribution with replacement (Chinese) •	95 ChJApPSt 10 399-404
Construction of a spectrum	lifetable • Springer, Janet A.; Liu, Peng T.; Albert, Richard H.; Chow, Lien P.	94 ASAProSoSt 82-87
	Lifetime	
	Infinite divisibility and the waiting-time paradox • van Harn, K.; Steutel, F. W.	95 CommStSM 11 527-540
	Sequential testing with interval censored data • Quan, Hui; Yu, Kai F.	95 JStCmpSm 51 239-247
Analysis of dependent survival data applied to	lifetimes of amalgam fillings • Aalen, Odd O.; Bjertness, Espen; Sønju, Torleif	95 StatMed 14 1819-1829
Gao, Ping The	lifetime of conditioned diffusions associated with some degenerate elliptic operators •	95 StocProc 57 319-337
goodness-of-fit of parametric regression models for	lifetime data - Graphical methods • Cohen, Ayala; Barnett, Ofra Assessing	95 StatMed 14 1785-1795
Alan C.; Lee, Mei-Ling Ting; Whitmore, G. A. (Ed)	Lifetime data: Models in reliability and survival analysis • Jewell, Nicholas P.; Kimber,	95 Kluwer:MA:Hllnd 424p
Bayesian inference for masked system	lifetime data • Reiser, B.; Guttman, I.; Lin, Dennis K. J.; Guess, Frank M.; Usher, John S.	95 ApplStat 44 79-90
Modeling	lifetime data with application to fatigue models • Castillo, Enrique; Hadi, Ali S.	95 JASA 90 1041-1054
Markov processes with a single generally distributed	lifetime • Coyle, A. J.; Taylor, P. G. Tight bounds on the sensitivity of generalised semi-	95 JAppProb 32 63-73
	Lifetime distribution	
Choi, In Su; Hwang, Yong Keun	Field data analyses for products with multiple-modes of failure (Korean) • Bai, Do Sun;	95 KorApS 8/1 89-104
S. R.; Dharmadhikari, A. D.	Prediction of failure - A case study • Joshi, B. H.; Naik-Nimbalkar, U. V.; Deshmukh,	95 JAppStat 22 485-493
censored data • Park, Hyo-Il; Park, Sang-Gue	Quantile estimation of treatment effect for the two-sample problem with right	95 StPrLet 24 139-145
Ruymgaart, Frits H.	Tail processes under heavy random censorship applications • Einmahl, John H. J.;	95 JNnparSt 5 223-235
Estimation of usage	lifetime distribution from calendar lifetime data • Suzuki, Kazuyuki	93 RStAJp 40/1 10-22
Parameter estimation for a bivariate	lifetime distribution in reliability with multivariate extensions • Heinrich, G.; Jensen, U.	95 Metrika 42 49-65
Soyer, Refik; Soofi, Ehsan	Lifetime modeling via Bayesian information distinguishability • Mazzuchi, Thomas A.;	94 ASAProBays 238-245
decay surfaces • Maeder, Marcel; Molloy, Kirsten Jane	Lifetime resolved fluorescence detection in chromatography. Global analysis of emission	95 Chemolab 28 99-107
System state monitoring and	lifetime scales - II • Kordonsky, Kh. B.; Gertsbakh, I.	95 RelEngSS 49 145-154
System state monitoring and	lifetime scales - I • Kordonsky, Kh. B.; Gertsbakh, I. B.	95 RelEngSS 47 1-14
opinions in Bayesian prediction of component	lifetimes in a shock model • Gåsemyr, Jørund; Natvig, Bent Using expert	95 MaOpRsr 20 227-242
Rolski, T. Correction to	"Light traffic approximations in many-server queues" (V24 p202-218) • Daley, D. J.;	95 AdvAppPr 27 584-584
The mean waiting time of a GI/G/1 queue in	light traffic via random thinning • Nananukul, Soracha; Gong, Wei-Bo	95 JAppProb 32 256-266
	Likelihood	
parents • Schaid, Daniel J.; Sommer, Steve S.	Comparison of statistics for candidate-gene association studies using cases and	94 AJHuGene 55 402-409
models • Rydén, Tobias	Consistent and asymptotically normal parameter estimates for hidden Markov	94 AnlsStat 22 1884-1895
member of a pedigree without loops • Fernando et al.	An efficient algorithm to compute the posterior genotypic distribution for every	93 ThApGene 87 89-93
Neale, Michael C.; Eaves, Lindon J.	Estimating and controlling for the effects of volunteer bias with pairs of relatives •	93 BehavGen 23 271-278
Owen, Robin E.; Packer, Laurence	Estimation of the proportion of diploid males in populations of Hymenoptera •	94 Heredity 72 219-227
times • Luschgy, Harald	Local asymptotic quadraticity of stochastic process models based on stopping	95 StocProc 57 305-317
	Martingale expansions and second order inference • Mykland, Per Aslak	95 AnlsStat 23 707-731
	Simple counterexamples against the conditionality principle • Helland, Inge S.	95 AmerStat 49 351-356
	• see also names of specific kinds of likelihood such as partial - or quasi -	
A framework for Bayesian and	likelihood approximations in statistics • Sweeting, Trevor J.	95 Biomtrka 82 1-23
under the robust design • Kendall, W. L. et al. A	likelihood-based approach to capture-recapture estimation of demographic parameters	95 Biomtrcs 51 293-308
Chaudhuri, Probal; Dewanji, Anup On a	likelihood-based approach in nonparametric smoothing and cross-validation •	95 StPrLet 22 7-15
Dykstra, Richard; Kochar, Subhash; Robertson, Tim	Likelihood based inference for cause specific hazard rates under order restrictions •	95 JMultiAn 54 163-174
Local	likelihood based on kernel censoring • Copas, J. B.	95 JRSS-B 57 221-235
Yadollah; Matthews, J. N. S.	Likelihood-based methods for bias reduction in limiting dilution assays • Mehrabi,	95 Biomtrcs 51 1543-1549
Accurate and efficient construction of bootstrap	likelihoods • Davison, A. C.; Hinkley, D. V.; Worton, B. J.	95 StatComp 5 257-264
Maximum smoothed	likelihood density estimation for inverse problems • Eggermont, P. P. B.; LaRiccia, V. N.	95 AnlsStat 23 199-220
Maximum smoothed	likelihood density estimation • Eggermont, P. P. B.; Lariccia, V. N.	95 JNnparSt 4 211-222
Wei, William W. S. On	likelihood distance for outliers detection • Wang, Wenping; Chow, Shein-Chung;	95 JBiophSt 5 307-322
Dual	likelihood • Mykland, Per Aslak	95 AnlsStat 23 396-421
comparisons • Hassan, Tawfik Numerical solution of	likelihood equations for a first-order response surface model for mixture paired	94 EgpStJ 38 196-207
Quasi profile and directed	likelihoods from estimating functions • Barndorff-Nielsen, O. E.	95 AnInStMa 47 461-464
ascertainment problem for pedigree data: A general	likelihood framework • Vieland, Veronica J.; Hodge, Susan E. Inherent intractability of the	95 AJHuGene 56 33-43
On the closed form of the	likelihood function of the first order moving average model • Haddad, John N.	95 Biomtrka 82 232-234
Concavity of Box-Cox log-	likelihood function • Kouider, Elies; Chen, Hanfeng	95 StPrLet 25 171-175
Bates, Douglas M. Approximations to the log-	likelihood function in the nonlinear mixed-effects model • Pinheiro, José C.;	95 JCmpGrSt 4 12-35
Nadkarni, Prakash M. Fast computation of genetic	likelihoods on human pedigree data • Goradia, Tushar M.; Lange, Kenneth; Miller, Perry L.;	92 HumHrdty 42 42-62
	Likelihood as an index of fit • Di Bacco, Mario; Regoli, Giuliana	94 JItalStS 3 243-253
est rate parity hold in the long run? An example of	likelihood inference in a multivariate time-series model • Juselius ... and uncovered inter-	95 JEconmtx 69 211-240
Kaplan, N. L.; Hill, W. G.; Weir, B. S.	Likelihood methods for locating disease genes in nonequilibrium populations •	95 AJHuGene 56 18-32
	Likelihood principle	
	Estimation of cointegrated systems with I(2) processes • Kitamura, Yuichi	95 EconTher 11 1-24
hypothesis testing • Berger, J. O.; Brown, L. D. et al.	A unified conditional frequentist and Bayesian test for fixed and sequential simple	94 AnlsStat 22 1787-1807
Correction to "Higher-order asymptotics and the	likelihood principle: One-parameter models" (V81 p1-10) • Pierce, D. A.; Peters, D.	95 Biomtrka 82 235-235
Bowman, Dale A full	likelihood procedure for analysing exchangeable binary data • George, E. Olusegun;	95 Biomtrcs 51 512-523
On an approximate	likelihood for quantiles • Lavine, M.	95 Biomtrka 82 220-222
	Likelihood ratio	
	Adaptive decision making for stochastic processes • Rukhin, Andrew L.; Vajda, Igor	95 JStPlInf 45 313-329
method • Kriman, Vladimir	Analysis of the sensitivity of the GI/GI/m/B queue to buffer size: The score function	95 CommStSM 11 171-194
eter nonregular cases • Ghosal, S.; Samanta, T.	Asymptotic behaviour of Bayes estimates and posterior distributions in multiparam-	95 MaMethSt 4 361-388
Cribari-Neto, F.; Aubin, E. C. Q.; Ferrari, S. L. P.	Bartlett corrections for one-parameter exponential family models • Cordeiro, G. M.;	95 JStCmpSm 53 211-231
	Detecting change points and monitoring biomedical data • Zhang, Heping	95 CommStA 24 1307-1324
Gaut, Brandon S.; Weir, B. S.	Detecting substitution-rate heterogeneity among regions of a nucleotide sequence •	94 MolBioEv 11 620-629
	An extension of a theorem on gambling systems • Liu, Wen; Wang, Zhongzhi	95 JMultiAn 55 125-132
IGARCH(1,1) models: A Monte Carlo ... • Lumsdaine	Finite-sample properties of the maximum likelihood estimator in GARCH(1,1) and	95 JBES 13 1-10
substitution rates, with application ... • Muse et al.	A likelihood approach for comparing synonymous and nonsynonymous nucleotide	94 MolBioEv 11 715-724
L'Ecuyer, Pierre	Likelihood ratio gradient estimation for stochastic recursions • Glynn, Peter W.;	95 AdvAppPr 27 1019-1053
Taniguchi, Masanobu	Nonparametric approach for discriminant analysis in time series • Zhang, Guoqiang;	95 JNnparSt 5 91-101
• Gupta, A. K.; Varga, T.	Normal mixture representations of matrix variate elliptically contoured distributions	95 SankhyaA 57 68-78
	Performance of a Bartlett-type modification for the deviance • Cordeiro, Gauss M.	95 JStCmpSm 51 385-403
Ryazanov, B. V.	On the problem of detecting random trajectories • Arbekov, I. M.; Vinokurov, V. I.;	94 ThProbAp 39 547-557
assumption • Soliman, Naeem A.	Some restricted estimation and testing hypotheses problem with singularity	94 EgpStJ 38 231-248
setting • Barmi, Hammou El; Dykstra, Richard	Testing for and against a set of linear inequality constraints in a multinomial	95 CandJSt 23 131-143
in a plant breeding experiment • Jiang, C. et al.	The use of mixture models to detect effects of major genes on quantitative characters	94 Genetcs 136 383-394
	Likelihood ratio-based confidence intervals in survival analysis • Murphy, S. A.	95 JASA 90 1399-1405
variables • Liu, Wen; Liu, Zikuan	Likelihood ratio and a class of strong laws for the sequence of integer-valued random	95 StPrLet 22 249-256
Probability inequalities for	likelihood ratios and convergence rates of sieve MLEs • Wong, Wing Hung; Shen, Xiaotong	95 AnlsStat 23 339-362

	Limit distribution	
chain (French) • Simonot, F.	Approximation of the stationary distribution of a stochastically monotone Markov	95 StocProc 56 133-149
Penrose, Mathew D.	Generalised two-sample U-statistics and a two-species reaction-diffusion model •	95 StocProc 55 57-64
man L.; Kotz, Samuel Use of moments in studies of	limit distributions arising from iterated random subdivisions of an interval • Johnson, Nor-	95 StPrLet 24 111-119
	Limit distributions for the Bernoulli meander • Takács, Lajos	95 JAppProb 32 375-395
	Limit distributions of conditional U-statistics • Rao, B. L. S. Prakasa; Sen, Arusharka	95 JTherPrb 8 261-301
Karlin, Samuel; Zeitouni, Ofer	Limit distribution of maximal non-aligned two-sequence segmental score • Dembo, Amir;	94 AnlsProb 22 2022-2039
Review of "Operator-	limit distributions in probability theory" • Jurek, Z. H.; Mason, D. J. (Auth); Hazod, W. (Rev)	95 Metrika 42 265-267
Review of "Operator-	limit distributions in probability theory" • Jurek, Z. J.; Mason, J. D. (Auth); Blower, G. (Rev)	95 JRSS-A 158 352-353
Veeh, Jerry Alan (Rev) Review of "Operator-	limit distributions in probability theory" • Jurek, Zbigniew J.; Mason, J. David (Auth);	95 BAmMaSoc 32 278-280
On the	limit distributions of sums of mixing Bernoulli random variables • Dziubdziela, Wiesław	95 StPrLet 23 179-182
Kou, Shing-Gang	Limits of first passage times to rare sets in regenerative processes • Glasserman, Paul;	95 AnlsApPr 5 424-445
modelling • Rachev, S. T.; Samorodnitsky, G.	Limit laws for a stochastic process and random recursion arising in probabilistic	95 AdvAppPr 27 185-202
queueing networks: A unified approach via fluid	limit models • Dai, J. G. On positive Harris recurrence of multiclass	95 AnlsApPr 5 49-77
Yang, Weiguo Some	limit properties of nonhomogeneous Markov information sources • Liu, Wen;	95 JCISS 20 279-292
human disease genes • Boehnke, Michael	Limits of resolution of genetic linkage studies: Implications for the positional cloning of	94 AJHuGene 54 379-390
McCormick, William P. Weak	limit results for the extremes of a class of shot noise processes • Homble, Patrick;	95 JAppProb 32 707-726
Markov processes: Exponential ... • Stefanov Explicit	limit results for minimal sufficient statistics and maximum likelihood estimators in some	95 AnlsStat 23 1073-1101
	Limits of spatial branching populations (Disc: p185-189) • Wakolbinger, Anton	95 Bernoull 1 171-185
Estimation of structures reliability for non-precise	limit state models and vague data • Savchuk, V. P.	95 RelEngSS 47 47-58
Generalized Poisson distributions as	limits of sums for arrays of dependent random vectors • Kobus, Maria	95 JMultiAn 52 199-244
	Limit theorem	
	Asymptotic behaviour of record times for renewal paced records • Ennadifi, G.	95 StatDecn 13 379-397
	On the asymptotic size and duration of a class of epidemic models • Svensson, Åke	95 JAppProb 32 11-24
	Asymptotics for transportation cost in high dimensions • Dobrić, V.; Yukich, J. E.	95 JTherPrb 8 97-118
modelling • Rachev, S. T.; Samorodnitsky, G.	Limit laws for a stochastic process and random recursion arising in probabilistic	95 AdvAppPr 27 185-202
death chains • Martínez, Servet; Vares, Maria Eulália	A Markov chain associated with the minimal quasi-stationary distribution of birth-	95 JAppProb 32 25-38
Horsthemke, Benedikt; Rüttermann, Markus	On the maximum entropy principle for a class of stochastic processes •	95 StocProc 56 117-132
expansion • Barbolosi, D.; Faivre, C.	Metrical properties of some random variables connected with the continued fraction	95 IndgMatN 6 257-265
Sriram, T. N.	A modified bootstrap for branching processes with immigration • Datta, Somnath;	95 StocProc 56 275-294
	Nonparametric truncated sequential change-point detection • Gombay, E.	95 StatDecn 13 71-82
• Novak, S. Yu.	Poisson approximation for the number of long match patterns in random sequences	94 ThProbAp 39 593-603
Koutras, M. V.	Reliability bounds for coherent structures with independent components • Fu, J. C.;	95 StPrLet 22 137-148
	On a subclass of regularly varying functions • Omey, E.	95 JStPlInf 45 275-290
combinatorial assemblies • Arratia, R.; Stark, D. et al.	Total variation asymptotics for Poisson process approximations of logarithmic	95 AnlsProb 23 1347-1388
Hahn, Marjorie G. W.; Klass, Michael J.	Uniform local probability approximations: Improvements on Berry-Esseen •	95 AnlsProb 23 446-463
	• see also central –, functional –, strong –	
Ouknine, Y.	Limit theorems for anticipative stochastic differential equations (French) • Erraoui, M.;	95 Stochatx 53 1-11
A	limit theorem for a class of interacting particle systems • Simonelli, Italo	95 AnlsProb 23 141-156
A key	limit theorem for critical branching processes • Sagitov, Serik	95 StocProc 56 87-100
Two	limit theorems for diffusion type stochastic equations • Pisanets, S. I.	94 ThProbAp 39 708-713
	Limit theorems for diffusions with a random potential • Mathieu, Pierre	95 StocProc 60 103-111
Liu, Wen; Yang, Weiguo A	limit theorem for the entropy density of nonhomogeneous Markov information source •	95 StPrLet 22 295-301
systems I" (V46 p79-106) • Mikami, T. Correction to	"Limit theorems on the exit problems for small random perturbations of dynamical	95 Stochatx 52 323-323
measures • Xie, Yingchao	Limit theorems of Hilbert valued semimartingales and Hilbert valued martingale	95 StocProc 59 277-293
	Limit theorems involving restricted convergence • Rossberg, H.-J.	94 ThProbAp 39 298-314
Bhaskara; Li, Deli Some results on strong	limit theorems for (LB)-space-valued random variables • Wang, Xiangchen; Rao, M.	95 StPrLet 23 247-251
	Limit theorems for the logarithm of sample spacings • Shao, Yongzhao; Hahn, Marjorie G.	95 StPrLet 24 121-132
Chernov, N. I.	Limit theorems and Markov approximations for chaotic dynamical systems •	95 ProbTher 101 321-362
A ratio	limit theorem for (sub) Markov chains on $\{1, 2, \ldots\}$ with bounded jumps • Kesten, Harry	95 AdvAppPr 27 652-691
	Limit theorems for multitype epidemics • Andersson, Håkan; Djehiche, Boualem	95 StocProc 56 57-75
Minlos, A. A local	limit theorem for nonhomogeneous random walk on a lattice • Zhizhina, E. A.;	94 ThProbAp 39 490-503
Arcones, Miguel A.	Limit theorems for nonlinear functionals of a stationary Gaussian sequence of vectors •	94 AnlsProb 22 2242-2274
On the local	limit theorem for non-uniformly ergodic Markov chains • Séva, Marc	95 JAppProb 32 52-62
trials • Wang, Y. H.; Ji, Shuixin	Limit theorems for the number of occurrences of consecutive k successes in n Markovian	95 JAppProb 32 727-735
	Limit theorem on option replication cost with transaction costs • Kusuoka, Shigeo	95 AnlsApPr 5 198-221
	Limit theorems of probability theory • Petrov, Valentin V.	95 OxfordUniv 304p
Petrov, V. V.	Limit theorems of probability theory: Sequences of independent random variables •	95 OxfordUniv 292p
Petrov, V. V. (Auth); Pyke, R. (Rev) Review of	"Limit theorems of probability theory: Sequences of independent random variables" •	95 ShrtBkRv 15/3 46-46
Holst, Lars A	limit theorem for the replication time of a DNA molecule • Cowan, Richard; Chiu, S. N.;	95 JAppProb 32 296-303
Olley, Steven A	limit theorem for a smooth class of semiparametric estimators • Pakes, Ariel;	95 JEconmtx 65 295-332
On refinement of Banach-valued	limit theorems for stable laws • Chuprunov, A. N.	94 ThProbAp 39 657-662
Hsing, Tailen	Limit theorems for stable processes with application to spectral density estimation •	95 StocProc 57 39-71
Rudnicki, Ryszard	Limit theorems for stochastically perturbed dynamical systems • Łoskot, Krzysztof;	95 JAppProb 32 459-469
	Limit theorems for the time of completion of Johnson-Mehl tessellations • Chiu, S. N.	95 AdvAppPr 27 889-910
	Limit theorems for the union intersection test • Horváth, Lajos; Shao, Qi-Man	95 JStPlInf 44 133-148
Pflug, G. Non-standard	limit theorems for urn models and stochastic approximation procedures • Kaniovski, Yu.;	95 CommStSM 11 79-102
Datta, Somnath	Limit theory and bootstrap for explosive and partially explosive autoregression •	95 StocProc 57 285-304
Establishing release	limits with a random slopes stability model • Grimes, James A.; Foust, Laura B.	94 ASAProBiop 498-502
estimation of simultaneous equations with	limited dependent variables: A case study of female labour supply • Lee Semi-parametric	95 JApEcmtx 10 187-200
user-knowledge-based variable selection method for	limited information maximum likelihood using principal components • Onishi, Haruo A	95 CmpStDA 19 379-399
	Limiting behavior of some measures of system availability • Mi, Jie	95 JAppProb 32 482-493
Weak	limiting behaviour of a simple tail Pareto-index estimator • Bacro, J. N.; Brito, M.	95 JStPlInf 45 7-19
	Limiting curves for I.I.D. records • Deuschel, Jean-Dominique; Zeitouni, Ofer	95 AnlsProb 23 852-878
Likelihood-based methods for bias reduction in	limiting dilution assays • Mehrabi, Yadollah; Matthews, J. N. S.	95 Biomtrcs 51 1543-1549
Monsour, M. J.; Mikulski, P. W. Moments of the	limiting distribution for the boundary case in the first order autoregressive process •	94 AJMMgSci 14 327-347
Fu, James C. Exact and	limiting distributions of the number of successions in a random permutation •	95 AnInStMa 47 435-446
Some properties of a	limiting distribution in Quicksort • Tan, Kok Hooi; Hadjicostas, Petros	95 StPrLet 25 87-94
The	limiting distribution of the t ratio under a unit root • Abadir, Karim M.	95 EconTher 11 775-793
Yakovlev, Andrej Yu. A bivariate	limiting distribution of tumor latency time • Rachev, Svetlozar T.; Wu, Chufang;	95 MaBisci 127 127-147
(c_1, c_2)) indexed by acceptable quality level and	limiting quality level • Raju, C. et al. Minimum risks chain sampling plans (ChSP-4	95 JAppStat 22 389-426
of statistical methods in mathematical modelling in	limnology • Krambeck, H.-J. Application and abuse	95 EcolgMdl 78 7-15
of "Aspects of uncertainty: A tribute to Dennis	Lindley" • Freeman, P. R.; Smith, A. F. M. (Ed); Queen, Catriona (Rev) Review	95 JAppStat 22 541-541
Verma, S. A generalized class of estimators utilizing	Lindley-like mean correction in linear regression model • Singh, R. K.; Agnihotri, M.;	94 Metron 52/3 147-160

Line transect

Pojar, Thomas M.; Bowden, David C.; Gill, R. Bruce	Aerial counting experiments to estimate pronghorn density and herd structure •	95 JWildlMg 59 117-128
Lesser, Virginia M.	Conditional and unconditional estimators of population size • Ramsey, Fred L.;	95 EnvEcoSt 2 181-190
Gregoire, Timothy G.; Monkevich, Nicholas S.	The reflection method of line intercept sampling to eliminate boundary bias •	94 EnvEcoSt 1 219-226
Review of	"Linear algebra and linear models" • Bapat, R. B. (Auth); Drygas, H. (Rev)	95 Metrika 42 78-78
Tail-specific	linear approximations for efficient bootstrap simulations • Hesterberg, Tim	95 JCmpGrSt 4 113-133
normality of pseudo-LS estimator for partly	linear autoregression models • Gao, Jiti; Liang, Hua Asymptotic	95 StPrLet 23 27-34
Vidakovic, Brani Efficiency of	linear Bayes rules for a normal mean: Skewed priors class • Mukhopadhyay, Saurabh;	95 Ststcian 44 389-397
A	linear birth-and-death predator-prey process • Coffey, John	95 JAppProb 32 274-277
inverse and classical estimators in multi-univariate	linear calibration • Srivastava, M. S. Comparison of the	95 CommStA 24 2753-2767
Probability matching priors for	linear calibration • Ghosh, M.; Carlin, B. P.; Srivastava, M. S.	95 Test 4 333-357
Robust	linear calibration • Kitsos, Christos P.; Müller, Christine H.	95 MaOpfSt 27 93-106
A simple derivation of exact reliability formulas for	linear and circular consecutive-k-of-n: F systems • Peköz, Erol A.; Ross, Sheldon M.	95 JAppProb 32 554-557
acterization of minihypers in PG(t, q) and nonbinary	linear codes meeting the Griesmer bound • Hamada, N. A survey of recent work on char-	93 JCISS 18 161-191
A re-evaluation of the quasi-Bayes approach to the	linear combination of forecasts • Faria, A. E.; Souza, R. C.	95 JForcast 14 533-542
Graybill, F. A. et al. Confidence intervals on ratios of	linear combinations for non-disjoint sets of expected mean squares • Gui, R.;	95 JStPlInf 48 215-227
On estimating a	linear combination of strata means with random sample sizes • He, Kun	95 JMultiAn 55 39-60
Teaching the standard deviations of	linear combinations using graphics • Mansfield, Edward R.; Adams, Benjamin M.	94 ASAProStEd 80-83
Wooff, David A. Bayes	linear computation: Concepts, implementation and programs • Goldstein, Michael;	95 StatComp 5 327-341
squares estimation for association models with	linear constraints on the parameters (French) • Jmel, S. Least	95 RStApp 43/3 93-104
SWEEP operator for least-squares subject to	linear constraints • Neytchev, Plamen Nikolov	95 CmpStDA 20 599-609
On the M/G/1 retrial queueing system with	linear control policy • Martín, M.; Gómez-Corral, A.	95 Top 3 285-305
State space identification of	linear deterministic rainfall-runoff models • Ramos, José; Mallants, Dirk; Feyen, Jan	95 WaterReR 31 1519-1531
Diagnostics in	linear discriminant analysis • Fung, Wing K.	95 JASA 90 952-956
... neural networks on classification in relation to	linear discriminant analysis, quadratic discriminant analysis ... • Sánchez et al.	95 Chemolab 28 287-303
On	linear discriminant analysis with adaptive ridge classification rules • Loh, Wei-Liem	95 JMultiAn 53 264-278
McLachlan et al. Asymptotic relative efficiency of the	linear discriminant function under partial nonrandom classification of the training data •	95 JStCmpSm 52 415-426
of the unconditional error rates of the sample	linear discriminant function • Viollaz, A. J.; Sfer, A. M.; Salvatierra, S. M. An approximation	95 CommStA 24 1941-1969
Correction to "The location	linear discriminant with covariates" (V23 p1027-1046) • Leung, C. Y.	95 CommStB 24 1061-1062
of aroclors in environmental samples using	linear discrimination • Ma, C. Y.; Bayne, C. K. Differentiation	94 ASAProEnvr 142-143
Detecting and testing causality in	linear econometric models • Faliva, Mario; Zoia, M. Grazia	94 JItalStS 3 61-76
Wang, Chengming; Qin, Yongsong	Linear empirical Bayes estimation applied in multiple linear regression (Chinese) •	95 ChJApPSt 10 371-378
Standard error of	linear equating for the counterbalanced design • Zeng, Lingjia; Cope, Ronald T.	95 JEdStat 20 337-348
	Linear estimate under Pitman criterion (Chinese) • Wang, Songgui; Yang, Hu	95 KToChina 39 1444-1447
General admissibility of	linear estimates of regression coefficients in growth curve model (Chinese) • Qing, Hong	95 ChJApPSt 10 265-271
quadratic risk function • Baksalary et al. Admissible	linear estimation in the general Gauss-Markov model with respect to an arbitrary	95 JStPlInf 44 341-347
Zhang, Shuanglin Admissible	linear estimation in linear models with respect to a restricted parameter set (Chinese) •	95 AMaAPknC 17 470-472
Sun, Liuquan Admissibility of	linear estimation of regression coefficients in common growth curve models (Chinese) •	95 AMaAPknC 17 628-630
Asymptotic design unbiasedness and robustness of	linear estimation under superpopulation models: A selection procedure ... • Casas et al.	94 Questo 18 173-188
Lu, Changyu; Li, Weixin Admissibility of	linear estimator in linear models with respect to an incomplete ellipsoidal restrictions •	95 AMaPeknE 37 289-295
Lu, Chang-yu Admissibility of inhomogeneous	linear estimators in linear models with respect to incomplete ellipsoidal restrictions •	95 CommStA 24 1737-1742
models • Wang, X.; Zhan, J.; Chen, J. All admissible	linear estimators of regression coefficients and parameters in variance component	95 AMaPeknE 37 653-662
subset • Hoffmann, Kurt All admissible	linear estimators of the regression parameter vector in the case of an arbitrary parameter	95 JStPlInf 48 371-377
Inference procedures for the	linear failure rate model • Sen, Ananda; Bhattacharyya, Gouri K.	95 JStPlInf 46 59-76

Linear filter

Li, Ta-Hsin	Blind deconvolution of linear systems with multilevel nonstationary inputs •	95 AnlsStat 23 690-704
regression for life history data using local	linear fitting • Li, Gang; Doss, Hani An approach to nonparametric	95 AnlsStat 23 787-823
Some inequalities for the Orlicz norm of a	linear form of independent pre-Gaussian random variables • Kurchenko, O. O.	95 ThPrMaSt 50 99-102
stage and sequential procedures for estimating a	linear function of k multinormal means • Singh, R. K.; Chaturvedi, A.; Kumar, S. Two	94 JIndnStA 32 1-8
A random	linear functional approach to efficiency bounds • Holly, Alberto	95 JEconmtx 65 235-261
Asymptotic normality of the NPMLE of	linear functionals for interval censored data, case 1 • Huang, J.; Wellner, J. A.	95 StNeerla 49 153-163
mography • Stander, J. et al. Minimax estimation of	linear functionals, particularly in nonparametric regression and positron emission to-	95 CmpStQ 10 259-283
fects • Wilson, P. David Longitudinal data analysis for	linear Gaussian models with random disturbed-highest-derivative-polynomial subject ef-	95 StatMed 14 1219-1233
Bayesian model discrimination and Bayes factors for	linear Gaussian state space models • Frühwirth-Schnatter, Sylvia	95 JRSS-B 57 237-246
Chowdhury, S. Roy	Linear graphs for 2^{k-p} fractional factorial under split plot design • Aggarwal, M. L.;	94 JStatRsr 28 73-87
H. H.; Højberre, M.; Sørensen, D.; Eriksen, P. S.	Linear and graphical models for the multivariate complex normal distribution • Andersen,	95 Springer:Brln:NY 183p
Decision and detection limits for	linear homoscedastic assays • Dunne, Adrian	95 StatMed 14 1949-1959

Linear hypothesis

• see also analysis of variance, general –, linear model, regression

New weighted sum of χ^2 test for testing a	linear hypothesis against a restricted alternative hypothesis • Tsubaki, Michiko	95 CommStB 24 987-1004
Pordzik, P. R. On the robustness of the	linear hypothesis test procedure in the singular linear model with implied restrictions •	95 StHefte 36 69-75
Richard Testing for and against a set of	linear inequality constraints in a multinomial setting • Barmi, Hammou El; Dykstra,	95 CandJSt 23 131-143
solving the nonlinear least squares problem with	linear inequality constraints • Sagara, Nobuko; Fukushima, Masao A hybrid method for	94 JOpRsJpn 38 55-69
powerful, one-sided tests for hypotheses about	linear inequalities • Liu, Huimei; Berger, Roger L. Uniformly more	95 AnlsStat 23 55-72
Markov decision processes and the systems of	linear inequalities • Wakuta, Kazuyoshi Vector-valued	95 StocProc 56 159-169
Left and right	linear innovations for a multivariate S α S random variable • Rutkowski, Marek	95 StPrLet 22 175-184
Las Vegas algorithms for	linear and integer programming when the dimension is small • Clarkson, Kenneth L.	95 JACM 42 488-499
Scale invariant Markov models for	linear inverse problems • Brette, Stéphane; Idier, Jérôme; Mohammad-Djafari, Ali	94 ASAProBays 266-270
alternative to the EM algorithm for solving positive	linear inverse problems • Archer et al. ... space reconstruction algorithm (ISRA) as an	95 StSinica 5 77-96
Zgierski, J. R. Adaptive	linear list reorganization under a generalized query system • Valiveti, R. S.; Oommen, B. J.;	95 JAppProb 32 793-804
Empirical Bayes	linear loss hypothesis testing in a non-regular exponential family • Singh, R. S.	95 JStPlInf 43 107-120
algorithm for variance parameter estimation in	linear mixed models • Gilmour, A. R. et al. Average information REML: An efficient	95 Biomtrcs 51 1440-1450
BLUP, ML, REML estimation methods for generalised	linear mixed models • McGilchrist, C. A.; Yau, K. K. W. The derivation of	95 CommStA 24 2963-2980
Efficient parametrisations for normal	linear mixed models • Gelfand, Alan E.; Sahu, Sujit K.; Carlin, Bradley P.	95 Biomtrka 82 479-488
Lin, Xihong Bias correction in generalised	linear mixed models with a single component of dispersion • Breslow, Norman E.;	95 Biomtrka 82 81-91

Linear model

	An analysis of new product diffusion process • Kusaka, Yasuo; Suzuki, Hisatoshi	94 JOpRsJpn 38 188-211
Kawasaki, Hidemi; Miyakawa, Masami	An ANOVA model and interaction element for three-way layout without replication •	95 JJapnStS 25 143-150
model • Mosier, Michael C.; Graybill, Franklin A.	Approximate confidence intervals for contrasts in the balanced three factor mixed	95 BiomtrcJ 37 273-297
	On the biases of error estimators in prediction problems • Hall, Peter	95 StPrLet 24 257-262
	A comparative study of some rank tests for interaction • Mansouri, H.; Chang, G.-H.	95 CmpStDA 19 85-96
Nelder, John A.; Lane, Peter W.	The computer analysis of factorial experiments: In memoriam – Frank Yates •	95 AmerStat 49 382-385
correlated • Bischoff, Wolfgang	Determinant formulas with applications to designing when the observations are	95 AnInStMa 47 385-399
Wong, Weng Kee	On the equivalence of D- and G-optimal designs in heteroscedastic models •	95 StPrLet 25 317-321

	Linear model (cont.)	
tations of various classes in a sample of DNA ... • Fu	Estimating effective population size or mutation rate using the frequencies of mu-	94 Genetcs 138 1375-1386
Philips, Robert	Estimating the number of aberrant laboratories • Olkin, Ingram; Guttman, Irwin;	95 PrEngInf 9 133-150
Ananda, Malwane M. A.	Generalized F-tests for unbalanced nested designs under heteroscedasticity •	95 AnInStMa 47 731-742
	Interaction levels in an autoregressive graphical model • Lagona, Francesco	94 Metron 52/1 113-141
	Limit theorems for the union-intersection test • Horváth, Lajos; Shao, Qi-Man	95 JStPlInf 44 133-148
Wenli, Yang	MINQE(U,I) and UMVIQUE of the covariance matrix in the growth curve model •	95 MaOpfStS 26 49-59
	On near ignorance classes • Sansó, B.; Pericchi, L. R.	94 RBrzlPrS 8 119-126
Pesaran, Bahram; Pesaran, M. Hashem	A non-nested test of level-differenced versus log-differenced stationary models •	95 EcomtrRv 14 213-227
	Outlier detection in multivariate calibration • Walczak, B.	95 Chemolab 28 259-272
	Predictive model selection • Laud, Purushottam W.; Ibrahim, Joseph G.	95 JRSS-B 57 247-262
	Real estate price prediction under asymmetric loss • Cain, Michael; Janssen, Christian	95 AnInStMa 47 401-414
mental variables estimators • Mandy, D. M. et al.	A unified approach to asymptotic equivalence of Aitken and feasible Aitken instru-	94 IntEconR 35 957-979
	• see also dynamic –, general –, generalized –, log- –, mixed –	
Experimental design for additive	linear models • Schwabe, Rainer	94 ASAProPEng 280-284
Review of "Advanced	linear models" • Wang, Song-Gui; Chow, Shein-Chung (Auth); Martin, Frank B. (Rev)	95 JASA 90 802-803
Wichmann, K. (Auth); Trenkler, G. (Rev) Review of	"Linear models. Algebraic foundation and statistical applications" (German) • Caspary, W.;	95 StHefte 36 190-190
of LP-mixingale sequence and its application in	linear models (Chinese) • Hu, Shuhe Sums of double arrays	95 AMaAPknC 17 248-254
the information in the term structure using	linear models and artificial neural network • Swanson et al. ... approach to assessing	95 JBES 13 265-275
efficient designs for estimating nonlinear aspects in	linear models • Müller, Christine H. Maximin	95 JStPlInf 44 117-132
Comment on "The statistics of	linear models: Back to basics" (V4 p221-234) • Aitkin, Murray	95 StatComp 5 85-86
Comment on "The statistics of	linear models: Back to basics" (V4 p221-234) • Gower, John	95 StatComp 5 91-92
Comment on "The statistics of	linear models: Back to basics" (V4 p221-234) • Lindsey, J. K.	95 StatComp 5 87-89
Wolfinger, Russell Comment on "The statistics of	linear models: Back to basics" (V4 p221-234) • Rodriguez, Robert; Tobias, Randall;	95 StatComp 5 97-101
Comment on "The statistics of	linear models: Back to basics" (V4 p221-234) • Searle, Shayle R.	95 StatComp 5 103-107
Comment on "The statistics of	linear models: Back to basics" (V4 p221-234) • van Eeuwijk, F. A.	95 StatComp 5 93-95
Reply to comments on "The statistics of	linear models: Back to basics" • Nelder, J. A.	95 StatComp 5 109-111
sign and analysis of fractional factorial designs with	linear models containing nuisance parameters • Raktoe, B. L. et al. Some aspects on the de-	93 JCISS 18 369-395
Bayesian inference in cyclical component dynamic	linear models • West, Mike	95 JASA 90 1301-1312
Spline smoother as a dynamic	linear model • Moyeed, R. A.	95 AstrlJSt 37 193-204
in pairwise comparisons using the dynamic	linear model (Disc: p134-135) • Peck, R. R.; Madrigal, J. L. et al. Ranking teams competing	94 ASAProBays 112-116
Asymptotic theory for prediction through empirical	linear models • Hwang, J. T. Gene; Liu, Hung-kung	94 ASAProPEng 207-216
bootstrap based on the estimating equations of the	linear model • Hu, Feifang; Zidek, James V. A	95 Biomtrka 82 263-275
"Using results from replicated studies to estimate	linear models" (V17 p341-362) • Becker, Betsy Jane Corrections to	95 JEdStat 20 100-102
Hedges, Larry V. A general	linear model for estimating effect size in the presence of publication bias • Vevea, Jack L.;	95 Psymtrka 60 419-435
A note on the smoothness of L^1-estimators for the	linear model • Ellis, Steven P.	95 SankhyaA 57 221-226
Asymptotic behavior of M-estimator in	linear models (Chinese) • Wu, Yaohua	95 AMaAPknC 17 401-410
ibration and stability analysis with a simple mixed	linear model when the experiment is of a split-plot type • Lee, K. R.; Gagnon, R. C. Cal-	94 ASAProBiop 435-440
spherically symmetric distributions in the general	linear model • Fourdrinier, Dominique; Wells, Martin T. Estimation of a loss function for	95 AnlsStat 23 571-592
Recursive inference in a general	linear model • Bhimasankaram, P.; Sengupta, D.; Ramanathan, S.	95 SankhyaA 57 227-255
improved estimators in multiple-design multivariate	linear models under general restriction • Shiraishi, T.; Konno, Y. On construction of	95 AnInStMa 47 665-674
estimators under spherical symmetry for the general	linear model • Cellier, Dominique; Fourdrinier, Dominique Shrinkage	95 JMultiAn 52 338-351
D-optimum designs for heteroscedastic	linear models • Atkinson, A. C.; Cook, R. D.	95 JASA 90 204-212
effect of different forms of centering in hierarchical	linear models • Kreft, Ita G. G.; de Leeuw, Jan; Aiken, Leona S. The	95 MultBehR 30 1-21
optimization of fertilization using a hierarchical	linear model • Wallach, D. Regional	95 Biomtrcs 51 338-346
and its applications to equivalent hypotheses in	linear models • Chan, Nai N.; Li, Kim-Hung A matrix identity	95 CommStA 24 2769-2777
difficulties of estimation of dispersion parameters in	linear models – An illustration • Infante, A. M. The	95 StHefte 36 183-189
	Linear models: Least squares and alternatives • Rao, C. Radhakrishna; Toutenburg, Helge	95 Springer:Brln:NY 352p
Helge (Auth); Draper, N. R. (Rev) Review of	"Linear models: Least squares and alternatives" • Rao, C. Radhakrishna; Toutenburg,	96 ShrtBkRv 16/1 7-8
Estimation of variance components in mixed	linear models • Kubokawa, Tatsuya	95 JMultiAn 53 210-236
Robust restricted maximum likelihood in mixed	linear models • Richardson, Alice M.; Welsh, Alan H.	95 Biomtrcs 51 1429-1439
A Bayesian analysis of the multivariate mixed	linear model • Jelenkowska, T. H.	95 CommStA 24 3183-3196
estimators of the mean vector in normal mixed	linear models • Bilodeau, Martin Minimax	95 JMultiAn 52 73-82
Hierarchical	linear models: A multilevel data analysis technique • Wu, Yow-Wu B.	95 NurseRes 44 123-126
Optimal choice of a new observation in a	linear model • Sengupta, Debasis	95 SankhyaA 57 137-153
Partially-ordered	linear models • Walker, John H.	94 ASAProStCp 152-157
Using the GEE method to construct	linear models for ordinal responses • Nuri, Walid A.; Chi, George Y. H.	94 ASAProBiop 82-87
for the identification of multiple outliers in	linear models"(V88 p1264-72) • Ryan, Thomas P. Comment on "Procedures	95 JASA 90 811-811
for the identification of multiple outliers in	linear models" • Hadi, Ali S.; Simonoff, Jeffrey S. Reply to comment on "Procedures	95 JASA 90 811-812
estimators for parametric components in a partly	linear model (Chinese) • Shi, Peide; Li, Guoying Asymptotic normality of L_1-norm	95 ChinaAMA 15 478-484
on the convergent rates of M-estimates for a partly	linear model • Shi, Peide; Li, Guoying A note	95 MaOpfStS 26 27-47
of the minimum L_1-norm estimates in a partly	linear model • Shi, Peide; Li, Guoying Asymptotics	95 JSysSMSE 7 67-77
the iterated logarithm of some estimates in partly	linear models • Gao, Jiti The laws of	95 StPrLet 25 153-162
Asymptotic theory for partly	linear models • Gao, Jiti	95 CommStA 24 1985-2009
Muller, Keith E. Computing defensible	linear model power and sample size estimates for study planning • Taylor, Douglas J.;	94 ASAProBiop 383-388
functions for rank-based procedures in the	linear model • Witt, Lee D.; Naranjo, Joshua D.; McKean, Joseph W Influence	95 JNnparSt 5 339-358
Detection of outliers in the	linear model when the regression parameters are stochastic • Bhandary, Madhusudan	95 CommStA 24 153-173
note on the generalised cross-validation criterion in	linear model selection • Venter, J. H.; Snyman, J. L. J. A	95 Biomtrka 82 215-219
Consistent variable selection in	linear models • Zheng, Xiaodong; Loh, Wei-Yin	95 JASA 90 151-156
Effect of measurement error in testing separate	linear models • Edwards, Lloyd J.	95 JStCmpSm 52 205-216
Detection of multicollinearity in an ordinary	linear model: Some elements of an informed use of the Belsley, Kuh and ... • Erkel-Rousse	95 RStApp 43/4 19-42
Partitioning error spaces in	linear models (error contrasts) • Rady, El-Houssainy A.	95 EgpStJ 39 51-64
On combining independent tests in	linear models • Jordan, Scott M.; Krishnamoorthy, K.	95 StPrLet 23 117-122
Dean, Angela (Rev) Review of "Advanced	linear models: Theory and applications" • Wang, Song-Gui; Chow, Shein-Chung (Auth);	95 Technmcs 37 119-120
Review of "The theory of	linear models" • Jorgensen, B. (Auth); Snell, Joyce (Rev)	95 JRSS-A 158 195-195
Review of "The theory of	linear models" • Jorgensen, Bent (Auth); Borowiak, Dale (Rev)	95 Technmcs 37 228-229
Review of "The theory of	linear models" • Jorgensen, Bent (Auth); Prvan, Tania (Rev)	95 AstrlJSt 37 248-248
Review of "The theory of	linear models" • Jørgensen, B. (Auth); Böhning, D. (Rev)	95 Biomtrcs 51 791-792
Review of "The theory of	linear models" • Jørgensen, B. (Auth); Drum, Melinda L. (Rev)	95 StMedRes 4 266-267
Review of "The theory of	linear models" • Jørgensen, B. (Auth); Martin, R. J. (Rev)	95 JAppStat 22 543-544
Review of "The theory of	linear models" • Jørgenson, B. (Auth); Neuhaus, G. (Rev)	95 Metrika 42 276-277
On an identity derived from unbiasedness in	linear models • McCulloch, C. E.; Searle, S. R.	95 AmerStat 49 39-42
of some general checks to omitted variables in the	linear model • Godfrey, L. G.; Orme, C. D. The sensitivity	94 IntEconR 35 489-506
Zhu, Y. Robustness to the unavailability of data in the	linear model, with applications • MacEachern, S. N.; Notz, W. I.; Whittinghill, D. C.;	95 JStPlInf 48 207-213

Nonstandard	local empirical processes indexed by sets • Deheuvels, Paul; Mason, David M.	95 JStPlInf 45 91-112
On	local and global asymptotic normality • Pfanzagl, J.	95 MaMethSt 4 115-136
analysis of spatial patterns: A unified approach to	local and global structures • Thioulouse, J.; Chessel, D.; Champely, S. Multivariate	95 EnvEcoSt 2 1-14
for statisticians who work for state/provincial or	local governments? • Gaines, Leonard M.; González, María E. Is there a place in ASA	94 ASAProGovt 149-152
Estimation of a multivariate normal mean vector and	local improvements • Pal, N.; Sinha, B. K.; Chaudhuri, G.; Chang, C.-H.	95 MaOpfStS 26 1-17
Shi, Jianqing	Local influence analysis for regression transformation models (Chinese) • Wei, Bocheng;	95 AMaAPknC 17 132-143
	Local influence in multivariate regression • Kim, Myung Geun	95 CommStA 24 1271-1278
	Local influence in structural equation models • Cadigan, N. G.	95 StrEqMod 2 13-30
Consistency of the	local kernel density estimator • Givens, Geof H.	95 StPrLet 25 55-61
Work experience,	local labor markets, and dropping out of high school (Disc: p213-215) • Swaim, Paul	94 ASAProSoSt 207-212
An effective bandwidth selector for	local least squares regression • Ruppert, D.; Sheather, S. J.; Wand, M. P.	95 JASA 90 1257-1270
	Local likelihood based on kernel censoring • Copas, J. B.	95 JRSS-B 57 221-235
Minlos, A. A	local limit theorem for nonhomogeneous random walk on a lattice • Zhizhina, E. A.;	94 ThProbAp 39 490-503
On the	local limit theorem for non-uniformly ergodic Markov chains • Séva, Marc	95 JAppProb 32 52-62
nonparametric regression for life history data using	local linear fitting • Li, Gang; Doss, Hani An approach to	95 AnlsStat 23 787-823
Stopping and set-indexed	local martingales • Ivanoff, B. Gail; Merzbach, Ely	95 StocProc 57 83-98
in Bessel processes and their relations to	local martingales • Delbaen, F.; Schachermayer, W. Arbitrage possibilities	95 ProbTher 102 357-366
dependent critical fluctuations of a one dimensional	local mean field model • Fritz, J.; Rüdiger, B. Time	95 ProbTher 103 381-407
Asymptotic distribution of smoothers based on	local means and local medians under dependence • Boente, Graciela; Fraiman, Ricardo	95 JMultiAn 54 77-90
Katkovnik, Vladimir Nonparametric	local polynomial approximation of the time-varying frequency and amplitude •	95 CommStA 24 3001-3025
Gijbels, Irène Data-driven bandwidth selection in	local polynomial fitting: Variable bandwidth and spatial adaptation • Fan, Jianqing;	95 JRSS-B 57 371-394
functions • Fan, J.; Heckman, N. E.; Wand, M. P.	Local polynomial kernel regression for generalized linear models and quasi-likelihood	95 JASA 90 141-150
Vladimir	Local polynomial periodogram for time-varying frequency estimation • Katkovnik,	95 SoAfrStJ 29 169-198
deposition • Haas, Timothy C.	Local prediction of a spatio-temporal process with an application to wet sulfate	95 JASA 90 1189-1199
Klass, Michael J. Uniform	local probability approximations: Improvements on Berry-Esseen • Hahn, Marjorie G. W.;	95 AnlsProb 23 446-463
Romá; Smilde, Age; Kowalski, Bruce Selectivity,	local rank, three-way data analysis and ambiguity in multivariate curve resolution • Tauler,	95 JChemtrc 9 31-58
On the	local rate of growth of Lévy processes with no positive jumps • Bertoin, Jean	95 StocProc 55 91-100
	Local regression	
	A new estimator of entropy • Correa, Juan C.	95 CommStA 24 2439-2449
Optimal neighborhoods for	local regression in the bivariate case • Lenarduzzi, L.	95 JNnparSt 4 233-248
	Local sensitivity, functional derivatives and nonlinear posterior quantities • Basu, Sanjib	94 ASAProBays 175-180
Lawrence, Charles E. Bayesian models for multiple	local sequence alignment and Gibbs sampling strategies • Liu, Jun S.; Neuwald, Andrew F.;	95 JASA 90 1156-1170
Splines as	local smoothers • Nychka, Douglas	95 AnlsStat 23 1175-1197
D. M. On partial	local smoothing rules for curve estimation • Hall, Peter; Marron, J. Stephen; Titterington,	95 Biomtrka 82 575-587
Correlation dimension estimates of global and	local temperature data • Wang, Qiang	95 JApMetrl 34 2556-2564
	Local time	
reflecting Brownian motion • Shi, Z.; Werner, W.	Asymptotics for occupation times of half-lines by stable processes and perturbed	95 Stochatx 55 71-85
Krone, Stephen M.	Comparing Fleming-Viot and Dawson-Watanabe processes • Ethier, S. N.;	95 StocProc 60 171-190
Weerasinghe, A. P. N.	Ergodic theorems for transient one-dimensional diffusions • Athreya, K. B.;	95 StocProc 58 173-185
	Limit distributions for the Bernoulli meander • Takács, Lajos	95 JAppProb 32 375-395
Protter, Philip; Shiryayev, Albert N.	Quadratic covariation and an extension of Itô's formula • Föllmer, Hans;	95 Bernoull 1 149-169
(French) • Leuridan, Christophe	Ray-Knight theorems and the Itô measure for Brownian motion over the torus R/Z	95 Stochatx 53 109-128
Fourati, S.	Representation of measures for additive functionals between two times (French) •	95 AIHenrPB 31 527-544
Khoshnevisan, Davar	Weight functions and pathwise local central limit theorems • Horváth, Lajos;	95 StocProc 59 105-123
	Local times and almost sure convergence of semi-martingales • Rajeev, B.; Yor, M.	95 AIHenrPB 31 653-667
On the	local time of the Brownian motion • Takács, Lajos	95 AnlsApPr 5 741-756
Vives, J. Chaos expansions of double intersection	local time of Brownian motion in R^d and renormalization • Imkeller, P.; Perez-Abreu, V.;	95 StocProc 56 1-34
Laws of the iterated logarithm for	local times of the empirical process • Bass, Richard F.; Khoshnevisan, Davar	95 AnlsProb 23 388-399
Rutkowski, Marek	Local times of functions of continuous semimartingales • Ouknine, Youssef;	95 StocAnAp 13 211-231
On moduli of continuity for	local times of Gaussian processes • Csörgő, Miklós; Lin, Zheng-Yan; Shao, Qi-Man	95 StocProc 58 1-21
Bojdecki, T.; Gorostiza, L. G. Self-intersection	local time for Gaussian $S'(R^d)$-processes: Existence, path continuity and examples •	95 StocProc 60 191-226
On some inequalities of	local time • Chou, Ching-Sung	95 JTherPrb 8 17-22
Barlow-Yor inequalities for	local times of the intersection of two planar Brownian motions (French) • Mathieu, P.	95 ProbTher 102 83-103
of the H^p norm of a martingale for maxima of	local times (French) • Leuridan, Christophe Control	95 AnlsProb 23 1289-1299
On the ultimate value of	local time of one-dimensional super-Brownian motion • Kaj, I.; Salminen, P.	95 StocProc 59 21-42
Mueller, Carl Singularity of super-Brownian	local time at a point catalyst • Dawson, Donald A.; Fleischmann, Klaus; Li, Yi;	95 AnlsProb 23 37-55
Rosen, Jay Logarithmic averages for the	local times of recurrent random walks and Levy processes • Marcus, Michael B.;	95 StocProc 59 175-184
Some self-similar processes related to	local times • Shieh, Narn-Rueih	95 StPrLet 24 213-218
	Local transformation of facial features • Gordon, A. D.	95 JAppStat 22 179-184
	Local and variable bandwidths and local linear regression • Jones, M. C.	95 MaOpfStS 27 65-71
Law of the iterated logarithm and	local variations at zero of the sticky Brownian motion • Amir, Madjid; Knight, Frank B.	95 StPrLet 23 289-295
a model selection criterion • Wang, W. et al.	Localization of multiple dipole sources of human cerebral potentials based on SCD and	94 ASAProBiop 217-222
Donthu, Naveen Capturing geographically	localized misspecification error in retail store choice models • Rust, Roland T.;	95 JMarRsr 32 103-110
and invariant σ-algebras of random walks on	locally compact groups • Jaworski, Wojciech On the asymptotic	95 ProbTher 101 147-171
Davis, Richard On permissible correlations for	locally correlated stationary processes • Donahue, Rafe M. J.; Brockwell, Peter J.;	95 StPrLet 22 49-53
A gradient algorithm	locally equivalent to the EM algorithm • Lange, Kenneth	95 JRSS-B 57 425-437
partial information: An algorithm for computing a	locally optimal periodic policy • Hordijk et al. Undiscounted Markov decision chains with	94 ZeitÖpRs 40 163-181
robustness against parameter variation of exact	locally optimum designs in nonlinear regression - A case study • Rasch, Dieter The	95 CmpStDA 20 441-453
The asymptotic behavior of	locally square integrable martingales • Wang, Jia-gang	95 AnlsProb 23 552-585
A note on extreme values of	locally stationary Gaussian processes • Hüsler, J.	95 JStPlInf 45 203-213
Opsomer, J.-D.; Agras, J. et al. An application of	locally weighted regression to airborne mercury deposition around an incinerator site •	95 EnvrMtrc 6 205-219
Asymptotic properties of	locally weighted regression • Hwang, Ruey-Ching	95 JNnparSt 5 303-310
to estimate conditional probability fields for source	locations of airborne pollutants • Hopke, P. K.; Li, C. L. et al. The use of bootstrapping	95 Chemolab 30 69-79
of "A new nonparametric goodness-of-fit test for	location alternatives" (German) • Weichselberger, A. (Auth); Büning, H. (Rev) Review	95 Metrika 42 145-146
A trimmed mean of	location of an AR(∞) stationary process • Lee, Sangyeol	95 JStPlInf 48 131-140
Estimating the common	location • Kumar, Somesh; Sharma, Divakar	95 MaOpfStS 26 231-240
joint distribution of the running maximum and its	location of D-valued Markov processes • Ferger, Dietmar The	95 JAppProb 32 842-845
Joint estimation of	location, dispersion, and random effects • Wolfinger, Russell D.; Tobias, Randall D.	94 ASAProPEng 225-234
Christine M.; Wu, C. F. Jeff Measuring	location effects from factorial experiments with a directional response • Anderson,	95 IntStRvw 63 345-363
Breakdown points for redescending M-estimates of	location • Yang, Jim J.; Van Ness, John W.	95 CommStA 24 1769-1787
asymptotics for studentized k-step M-estimators of	location • Jurečková, Jana; Malý, Marek The	95 SqtlAnly 14 229-245
Asymmetric unbiased estimation in	location families • Shafie, K.; Noorbaloochi, S.	95 StatDecn 13 307-314
Correction to "The	location linear discriminant with covariates" (V23 p1027-1046) • Leung, C. Y.	95 CommStB 24 1061-1062
Maximum-likelihood estimation of gene	location by linkage disequilibrium • Hill, W. G.; Weir, B. S.	94 AJHuGene 54 705-714

	Log-density ratio	
Trevor J.	A framework for Bayesian and likelihood approximations in statistics • Sweeting,	95 Biomtrka 82 1-23
A non-nested test of level-differenced versus	log-differenced stationary models • Pesaran, Bahram; Pesaran, M. Hashem	95 EcomtrRv 14 213-227
	Log-gamma distribution	
models • Paula, Gilberto A.; Cordeiro, Gauss M.	Bias correction and improved residuals for non-exponential family nonlinear	95 CommStA 24 1193-1210
Concavity of Box-Cox	log-likelihood function • Kouider, Elies; Chen, Hanfeng	95 StPrLet 25 171-175
Bates, Douglas M. Approximations to the	log-likelihood function in the nonlinear mixed-effects model • Pinheiro, José C.;	95 JCmpGrSt 4 12-35
Wobus, Peter An item nonresponse and	log-linear analysis of the Spanish Language Forms Availability Test • de la Puente, Manuel;	94 ASASrvyRsr 583-588
(Auth); Rice, J. C. (Rev) Review of "Understanding	log-linear analysis with ILOG: An interactive approach" • Bakeman, R.; Robinson, B. F.	95 EdPsMeas 55 350-355
cannot repeat • Bakeman, Roger; Quera, Vicenç	Log-linear approaches to lag-sequential analysis when consecutive codes may and	95 PsychBu 118 272-284
	Log-linear model	
factor • McCloud, P. I.; Darroch, J. N.	An analysis of categorical repeated measurements in the presence of an external	95 Biomtrka 82 47-55
	Analysis of contingency tables by marginal scores • Gokhale, D. V.; Klein, Ruben	95 RBrzlPrS 9 25-42
variation • Johaadien, Hassan; Goutis, Constantinos	Approximate maximum likelihood estimation in the presence of extra-Poisson	95 CommStA 24 1625-1636
E.	Asymptotic standard error of equipercentile equating • Liou, Michelle; Cheng, Philip	95 JEdStat 20 259-286
	Bayesian graphical models for discrete data • Madigan, David; York, Jeremy	95 IntStRvw 63 215-232
development • International Working Group ...	Capture-recapture and multiple-record systems estimation 1: History and theoretical	95 AJEpidm 141 1047-1058
pendence graph • de Falguerolles, A.; Jmel, S. et al.	Correspondence analysis and association models constrained by a conditional inde-	95 Psymtrka 60 161-180
Greene, W. H.; Greene, L. K.; Seaks, T. G.	Estimating the functional form of the independent variables in probit models •	95 ApplEcon 27 193-196
	A family of partially correlated Poisson models for overdispersion • Luceño, Alberto	95 CmpStDA 20 511-520
Gertsbakh, I.	On the Fisher information in type-I censored and quantal response data •	95 StPrLet 23 297-306
algorithm • Gange, Stephen J.	Generating multivariate categorical variates using the iterative proportional fitting	95 AmerStat 49 134-138
Basu, S.; Raychaudhuri, A.; Basu, A.	Improving precision through modelling: An illustration with hierarchical kappa •	95 CommStB 24 399-408
	Logistic regression of family data from case-control studies • Whittemore, Alice S.	95 Biomtrka 82 57-67
the sensitivity against violation of ... • Vach et al.	Logistic regression with incompletely observed categorical covariates - Investigating	95 StatMed 14 1315-1329
Evans, Geoffrey; Payne, Clive	Modelling the class-party relationship in Britain, 1964-92 • Heath, Anthony;	95 JRSS-A 158 563-574
scale agreement • Graham, Patrick	Modelling covariate effects in observer agreement studies: The case of nominal	95 StatMed 14 299-310
Thomas R.; Becker, Mark P.	Multivariate contingency tables and the analysis of exchangeability • Ten Have,	95 Biomtrcs 51 1001-1016
Pesaran, Bahram; Pesaran, M. Hashem	A non-nested test of level-differenced versus log-differenced stationary models •	95 EcomtrRv 14 213-227
	Performance of a Bartlett-type modification for the deviance • Cordeiro, Gauss M.	95 JStCmpSm 51 385-403
	Prediction functions for categorical panel data • Gilula, Zvi; Haberman, Shelby J.	95 AnlsStat 23 1130-1142
trends in multiway tables • Payne, C.; Payne, J. et al.	Testing the tightening bond hypothesis: An application of new models for analysing	95 JStPlInf 47 181-201
surveys • Harrop, Anne; Plewis, Ian	Two decades of family change: Secondary analysis of continuous government	95 JRSS-A 158 91-106
Logit models and related quasi-symmetric	log-linear models for comparing responses to similar items in a survey • Agresti, Alan	95 SocMethR 24 68-95
van de Ven, R.; Weber, N. C.	Log-linear models for mean and dispersion in mixed Poisson regression models •	95 AstrlJSt 37 205-216
Fitting parametric counting processes by using	log-linear models • Lindsey, J. K.	95 ApplStat 44 201-212
samples from exponential families using Poisson	log-linear modelling, and model comparison using Bayes and ... • Aitkin ... choice in single	95 StatComp 5 113-120
R. (Rev) Review of "Categorical longitudinal data:	Log-linear panel, trend, and cohort analysis" • Hagenaars, Jacques A. (Auth); Eliason, Scott	95 JEdStat 20 103-104
	Log-linear regression	
	On the reliability of indicators of learning curve model accuracy • Moses, O. Douglas	95 JCostAnl /1 63-101
	Weibull accelerated life testing with unreported early failures • Glaser, R. E.	95 IEEEReli 44 31-36
mental lead • Jiang, Qi; Succop, Paul A. A study of the	log-log and log-additive models for the relationship between blood lead and environ-	94 ASAProEnvr 88-91
	Log-logistic distribution	
Richard A.	Measuring skewness with respect to the mode • Arnold, Barry C.; Groeneveld,	95 AmerStat 49 34-38
Brüederl, Josef; Diekmann, Andreas The	log-logistic rate model: Two generalizations with an application to demographic data •	95 SocMethR 24 158-186
Stanley	Log-multiplicative models for valued social relations • Anderson, Carolyn J.; Wasserman,	95 SocMethR 24 96-127
Wong, Raymond Sin-Kwok Extensions in the use of	log-multiplicative scaled association models in multiway contingency tables •	95 SocMethR 23 507-538
Generalized	log-normal distributions with reliability application • Chen, Gemai	95 CmpStDA 19 309-319
	Log-periodogram regression of time series with long range dependence • Robinson, P. M.	95 AnlsStat 23 1048-1072
Computation of probabilities of a generalized	log-series and related distributions • Ong, S. H.	95 CommStA 24 253-271
Useful relationship between the	log-survival function and truncated moments, with application • El-Arishy, S.	95 StHefte 36 145-154
Darwin's	log: A toy model of speciation and extinction • Aldous, David	95 JAppProb 32 279-295
The	log transformation is special • Keene, Oliver N.	95 StatMed 14 811-819
Limit theorems for the	logarithm of sample spacings • Shao, Yongzhao; Hahn, Marjorie G.	95 StPrLet 24 121-132
the strong law of large numbers and the law of the	logarithm for weighted sums of independent random variables with ... • Li et al. On	95 JMultiAn 52 181-198
On the	logarithmic average of additive functionals • Csáki, E.; Földes, A.	95 StPrLet 22 261-268
Marcus, Michael B.; Rosen, Jay	Logarithmic averages for the local times of recurrent random walks and Levy processes •	95 StocProc 59 175-184
asymptotics for Poisson process approximations of	logarithmic combinatorial assemblies • Arratia, R.; Stark, D.; Tavaré, S. Total variation	95 AnlsProb 23 1347-1388
neous multiserver queue II: Positive recurrence and	logarithmic limits • Sadowsky ... of large queue lengths and waiting times in a heteroge-	95 AdvAppPr 27 567-583
	Logic	
	• see also fuzzy -	
from questionnaires: The survey interview and the	logic of conversation (Disc: p169-177) • Schwarz, Norbert What respondents learn	95 IntStRvw 63 153-168
from questionnaires: The survey interview and the	logic of conversation" • Cantor, David Discussion of "What respondents learn	95 IntStRvw 63 173-177
from questionnaires: The survey interview and the	logic of conversation" • Singer, Eleanor Discussion of "What respondents learn	95 IntStRvw 63 169-172
The theory of gambling and statistical	logic (Revised edition) • Epstein, Richard A.	95 Academic:NY:Lnd 450p
Testing the	logical structure of questionnaires • Willenborg, L. C. R. J.	95 StNeerla 49 95-109
Problems with	logistic autoregression • Stern, Hal S.	94 ASAProBiop 106-110
Prospective analysis of	logistic case-control studies • Carroll, R. J.; Wang, Suojin; Wang, C. Y.	95 JASA 90 157-169
Carroll, R. J. On robust	logistic case-control studies with response-dependent weights • Wang, C. Y.;	95 JStPlInf 43 331-340
val estimates for the optimum point of a quadratic	logistic curve - A comparison of different estimation methods via ... • Scazzero et al. Inter-	95 CommStB 24 207-226
Estimation of the mean and standard deviation of the	logistic distribution based on multiply type-II censored samples • Balakrishnan, N. et al.	95 MaOpfStS 27 127-142
Review of "Handbook of the	logistic distribution" • Balakrishnan, N. (Ed); Howlader, Hatema A. (Rev)	95 JAppStat 22 189-191
parameter by absolute values of order statistics in	logistic distribution • Kantam, R. R. L.; Rao, A. V. et al. Linear unbiased estimation of scale	94 JStatRsr 28 199-219
moments of order statistics from a generalized half	logistic distribution with applications to inference • Balakrishnan et al. ... and product	95 JStCmpSm 52 385-398
The uniform,	logistic, and lambda distributions • Appleton, David R.	95 CommStB 24 947-951
distributions and Lyapunov exponents for chaotic	logistic maps • Hall, Peter; Wolff, Rodney C. L. Properties of invariant	95 JRSS-B 57 439-452
	Logistic model	
pital rather than patient characteristics? • Silber et al.	Comparing the contributions of groups of predictors: Which outcomes vary with hos-	95 JASA 90 7-18
plant disease epidemics • Dayananda, P. W. A. et al.	Estimation of rate parameter and its relationship with latent and infectious periods in	95 Biomtrcs 51 284-292
	Generalized additive models for medical research • Hastie, Trevor; Tibshirani, Robert	95 StMedRes 4 187-196
Fielding, K. L.; Brettle, R. P.; Gore, S. M. et al.	Heterosexual transmission of HIV analysed by generalized estimating equations •	95 StatMed 14 1365-1378
Greenland, Sander	Standardized estimates from categorical regression models • Joffe, Marshall M.;	95 StatMed 14 2131-2141
Haruhiko	Structural model of ability distribution in the item response theory • Ogasawara,	95 Behvmtrk 22 37-48
Multivariate	logistic models • Glonek, G. F. V.; McCullagh, P.	95 JRSS-B 57 533-546

Transient characteristics of an	$M/M/\infty$ system • Guillemin, Fabrice; Simonian, Alain	95 AdvAppPr 27 862-888
Matrix analytical methods for	M/PH/1 retrial queues • Diamond, Jeffrey E.; Alfa, Attahiru Sule	95 CommStSM 11 447-470
Testing AR(1) against	MA(1) disturbances in an error component model • Baltagi, Badi H.; Li, Qi	95 JEconmtx 68 133-151
sensitivity analysis of early exposure results with the	MACCS reactor accident consequence model • Helton et al. ... of an uncertainty and	95 RelEngSS 48 129-148
	Machine learning	
Kuhn, Roland; De Mori, Renato	The application of semantic classification trees to natural language understanding •	95 IEEEPAMI 17 449-460
and statistical techniques • Højsgaard, S. et al.	BIFROST – Block recursive models induced from relevant knowledge, observations,	95 CmpStDA 19 155-175
approach • Chien, S.; Gratch, J.; Burl, M.	On the efficient allocation of resources for hypothesis evaluation: A statistical	95 IEEEPAMI 17 652-665
Taylor, C. C. (Ed); Fulkerson, Bill (Rev) Review of	"Machine learning, neural and statistical classification" • Michie, D.; Spiegelhalter, D. J.;	95 Technmcs 37 459-459
Onsager-	Machlup functional for some smooth norms on Wiener space • Capitaine, Mireille	95 ProbTher 102 189-201
Review of "Mathematica for the	Macintosh (Version 2.0.4)" • Wolfram Research Inc (Auth); Baglivo, Jenny (Rev)	95 AmerStat 49 86-92
Review of "MATLAB for the Power	Macintosh" • The MathWorks Inc (Auth); Spiegelman, Cliff (Rev)	95 Chemolab 29 323-324
"SoftWindows, Mathematica and Data Desk for the	Macintosh PowerPC" • Spiegelman, Cliff (Rev) Review of	95 Chemolab 28 208-209
Review of "Maple V for the	Macintosh (Version 5.0.2)" • Waterloo Maple Software Inc (Auth); Baglivo, Jenny (Rev)	95 AmerStat 49 86-92
A	macro-editing application developed for PC-SAS • Lindstrom, K.	91 StatJUN 8 155-166
Granquist, L.	Macro-editing – A review of some methods for rationalizing the editing of survey data •	91 StatJUN 8 137-154
Bromberg, Judith A SAS	macro for residual deviance of ordinal regression analysis • Wan, Jim Y.; Wang, Whedy;	94 CmpBiomd 45 307-310
Review of "A causal analysis with latent variables of	macroeconomic interdependencies" (German) • Hillmer, M. (Auth) et al.	95 AllgeStA 79 224-227
quences for environmental statistics • Alfsen Use of	macroeconomic models in analysis of environmental problems in Norway, and conse-	92 StatJUN 9 51-72
• Review of	"Macroeconomic modelling Vols I and II" • Wallis, Kenneth F. (Ed); Holden, Ken (Rev)	95 IntJFor 11 334-335
Review of	"Macroeconomic modelling of the long run" • Hargreaves, Colin (Ed); Barker, Terry (Rev)	95 JApEcmtx 10 211-212
Are survey forecasts of	macroeconomic variables rational? • Aggarwal, Raj; Mohanty, Sunil; Song, Frank	95 JBusUChi 68 99-119
David R.; Southwell, Colin Estimates of	macropod density from line transect surveys relative to analyst expertise • Anderson,	95 JWildlMg 59 852-857
interacting particle system • Kipnis, C. et al.	Macroscopic properties of a stationary non-equilibrium distribution for a non-gradient	95 AIHenrPB 31 191-221
	Magnetic resonance image	
a noisy image • Godtliebsen, Fred; Chu, Chih-Kang	Estimation of the number of true gray levels, their values, and relative frequencies in	95 JASA 90 890-899
A Bayesian approach to synthetic	magnetic resonance imaging • Glad, Ingrid K.; Sebastiani, Giovanni	95 Biomtrka 82 237-250
	Mahalanobis distance	
problem in discriminant analysis • Markham, Ina S.	Combining neural networks and statistical predictions to solve the classification	95 DecisnSc 26 229-242
Lanteri, S.; Sarabia, L.	Distance and class space in the UNEQ class-modeling technique • Forina, M.;	95 JChemtrc 9 69-89
Donald R.	A maximum likelihood approach to correlational outlier identification • Bacon,	95 MultBehR 30 125-148
Generalization of the	Mahalanobis distance in the mixed case • Bar-Hen, Avner; Daudin, J.-J.	95 JMultiAn 53 332-342
mosome classification using monotonic functions of	Mahalanobis distance and the transportation method • Kleinschmidt et al. Improved chro-	94 ZeitOpRs 40 305-323
Increasing response to personally-delivered	mail-back questionnaires • Dillman, Don A.; Dolsen, Dana E.; Machlis, Gary E.	95 JOffStat 11 129-139
The impact of using priority	mail in a mixed mode survey • Mitchell, Susan; Mooney, Geraldine; Shettle, Carolyn	94 ASAProGovt 88-92
Pivetz, Lynda L. Converting	mail reporters to touchtone data entry • Rosen, Richard J.; Clayton, Richard L.;	94 ASASrvyRsr 763-768
Mangat, Naurang Singh	Mail survey design for sensitive quantitative variable • Singh, Ravindra; Singh, Sarjinder;	95 Metron 53 43-54
The influence of motivational appeals on	mailback response • Treat, James B.	94 ASASrvyRsr 571-576
return envelopes and reminder postcards affect	mailback response rates for census questionnaires • Dillman et al. ... letters, stamped	95 SurvyM 21 159-165
	Main effect design	
R.; Lin, Dennis K. J.	Characterizing projected designs: Repeat and mirror-image runs • Draper, Norman	95 CommStA 24 775-795
Federer, W. T.	Representation and bounds for the general fractional factorial • Anderson, D. A.;	95 CommStA 24 363-373
scale changes on meta-analysis of multiplicative and	main effects models • Kanetkar, V.; Evans, M. G.; Everell, S. A.; Irvine, D. et al. The effect of	95 EdPsMeas 55 206-224
for regional yield trials: A comparison to additive	main effects and multiplicative interaction (AMMI) analysis • Piepho ... prediction (BLUP)	94 ThApGene 89 647-654
tests for multiplicative terms in the additive	main effects and multiplicative interaction model for cultivar trials • Piepho ... of statistical	95 ThApGene 90 438-443
Statistical tests and retention of terms in the additive	main effects and multiplicative interaction model for cultivar trials • Cornelius, P. L.	93 CropSci 33 1186-1193
D-optimal saturated	main effect plans for $2 \times s_2 \times s_3$ factorials • Chatterjee, Kashinath; Mukerjee, Rahul	93 JCISS 18 116-122
A note on non-orthogonal	main-effect plans • Sharma, M. K.	95 SankhyaB 57 166-168
Algorithm for constructing orthogonal	main effect plans • Burgess, L.; Street, D. J.	94 UtilMath 46 33-48
Construction for orthogonal	main effect plans • Street, D. J.	94 UtilMath 45 115-123
Sequential construction of new	main effect plans with higher revealing power for 2^7 factorials • Ghosh, Subir	93 JCISS 18 217-231
of Hadamard matrices for the construction of	main effect plus two plans for 2^m factorials • Mukerjee, R.; Chatterjee, K. An application	94 UtilMath 45 213-218
An improved general approximation test for the	main effect in a split-plot design • Algina, James; Oshima, T. C.	95 BJMaStPs 48 149-160
Emerita T. Rank tests for	main and interaction effects in analysis of variance • Marden, John I.; Muyot, Maria	95 JASA 90 1388-1398
Moving math from the	mainframe to the PC: A better policy tool (Disc: p51-52) • Trippe, Carole; Cohen, Bob	94 ASAProGovt 34-38
Reliability,	maintainability, and availability assessment (Second edition) • Locks, Mitchell O.	95 ASQualCntrl:WI 348p
Lynch, James (Rev) Review of "Reliability,	maintainability and supportability: A probabilistic approach" • Knezevic, J. (Auth);	95 Technmcs 37 122-122
Lester R. The	maintenance of ethical principles in marketing research • Dutka, Solomon; Frankel,	94 ASASrvyRsr 143-145
Optimization of test and	maintenance intervals based on risk and cost • Vaurio, J. K.	95 RelEngSS 49 23-36
Hameed, M. Correction to "Inspection and	maintenance policies of devices subject to deterioration" (V19 p917-931) • Abdel-	95 AdvAppPr 27 584-584
Melolidakis, Costis On the optimal	maintenance of systems and control of arrivals in queues • Katehakis, Michael N.;	95 StocAnAp 13 137-164
Kupfer, D.; Frank, E.; Wade, M. Cost utility analysis of	maintenance treatment for recurrent depression • Kamlet, M. S.; Paul, N.; Greenhouse, J.;	95 ConClnTr 16 17-40
Statistical adjustment of	maize grain yield for sub-optimal plot stands • Kamidi, R. E.	95 ExpAgric 31 299-306
Genetic models for predicting	maize single-cross performance in unbalanced yield trial data • Bernardo, Rex	95 CropSci 35 141-147
Bernardo, Rex Prediction of	maize single-cross performance using RFLPs and information from related hybrids •	94 CropSci 34 20-25
Patrick J. F.; Mathar, Rudolf; Heiser, Willem J. The	majorization approach to multidimensional scaling for Minkowski distances • Groenen,	95 JClass 12 3-19
Bhaumik, Dulal K.	Majorization and D-optimality for complete block designs under correlations •	95 JRSS-B 57 139-143
Kimura, Miyoshi	Majorization methods on hyperplanes and their applications • Kakiuchi, Itsuro;	95 JStPlInf 47 217-235
A	majorization theorem for the number of distinct outcomes in an urn model • Hwang, F. K.	93 JCISS 18 26-30
Scheduling independent tasks to minimize the	makespan on identical machines • Bruno, John; Coffman, Edward G., Jr; Downey, Peter	95 PrEngInf 9 447-456
clearance following treatment of falciparum	malaria in humans • Davis, T. M. E.; Martin, R. B. Models of parasite	95 EnvrMtrc 6 529-534
alleles, and application to estimation of clonality in	malaria parasites • Hill et al. ... inbreeding coefficients from genotypic data on multiple	95 GeneRsr 65 53-61
Optimum dose levels when	males and females differ in response • Atkinson, A. C.; Demetrio, C. G. B.; Zocchi, S. S.	95 ApplStat 44 213-226
Estimation of the proportion of diploid	males in populations of Hymenoptera • Owen, Robin E.; Packer, Laurence	94 Heredity 72 219-227
ambulatory blood pressure in overweight, sedentary	males: A time series analysis • Cox, K. et al. Effects of exercise and calorie restriction on	95 EnvrMtrc 6 523-528
to recall bias: Mothers' reports of age of puberty in	male twins • Pickles, A. et al. A simple method for censored age-of-onset data subject	94 BehavGen 24 457-468
The	Malliavin calculus and related topics • Nualart, D.	95 Springer:Brln:NY 266p
Review of "The	Malliavin calculus and related topics" • Nualart, D. (Auth); Paulauskas, V. (Rev)	96 ShrtBkRv 16/1 8-8
Fractal exponents for the upper airways of	mammalian lungs • Krause, Ekkehard; Bandt, Christoph; Schulz, Anne; Schulz, Holger	95 CmpStDA 20 583-590
May the best	man win? • Appleton, David R.	95 Ststcian 44 529-538
	Management	
	• see also data –, database –, information –, risk –	
from the perspective of a practitioner in population	management and conservation • Desante ... for studies of marked migratory landbirds	95 JAppStat 22 949-965
survey methods and their applications in fisheries	management" • Pollock, K. H.; Jones, C. M.; Brown, T. L. (Auth) et al. Review of "Angler	95 Biomtrcs 51 385-385

Organizational needs and	management issues for establishment survey programs • Clark, Cynthia Z. F.	94 ASASrvyRsr 178-182
(Rev) Review of "Fourth generation	management: The new business consciousness" • Joiner, Brian L. (Auth); Ziegel, Eric R.	95 Technmcs 37 113-113
Review of "Handbook in operations research and	management science: Marketing" • Eliashberg, J.; Lilien, G. (Ed); Arabie, Phipps (Rev)	95 JClass 12 143-144
Achieving measurable improvement in service	management • Keller, Paul A.; Runger, George C.	95 ASQCTekCon 330-337
The role of statistical thinking in	management • Hare, Lynne B.; Hoerl, Roger W.; Hromi, John D.; Snee, Ronald D.	95 QulPr 28/2 53-60
and manufacturing ... • Battles et al. Demand-side	management: Utility programs offered and participants in the residential, commercial,	94 ASAProBuEc 20-27
Statistical education for quality	managers • Martin, Margaret; Suber, Marsha	94 ASAProBays 148-149
Review of "Statistical thinking for	managers (Third edition)" • Hildebrand, D. K.; Ott, L. (Auth); Garin, Araceli (Rev)	95 JAppStat 22 438-439
	Managing risks • Nash, John C.; Nash, Mary M.	95 Chance 8/4 25-31
Bailey, Michael; Bowden, John; Callahan, Alexander J.	Managing ship performance of naval gunfire support using statistical process control •	94 MilOpRes 1 5-11
for ... • Dillman et al. Effects of benefits appeals,	mandatory appeals and variations in statements of confidentiality on completion rates	94 ASASrvyRsr 1333-1338
An interpretation of the interaction terms in	Mandel's ANOVA model from Hirotsu's interaction elements • Miyakawa, Masami	93 RStAJp 40/1 1-9
Eliason, Scott R. Modeling	manifest and latent dimensions of association in two-way cross-classifications •	95 SocMethR 24 30-67
for some probability distributions on compact	manifolds • Richards, Donald St. P. Exact asymptotics	95 AnlsStat 23 1582-1586
Brownian motion on a Lipschitz Riemannian	manifold composed of different materials • Zheng, Weian	95 Stochatx 54 75-87
harmonic functions on groups and on covering	manifolds • Bougerol, Philippe; Elie, Laure Existence of positive	95 AIHenrPB 31 59-80
theory of tests for uniformity on the Grassmann	manifold • Chikuse, Yasuko; Watson, Geoffrey S. Large sample asymptotic	95 JMultiAn 54 18-31
Stochastic inertial	manifold • Bensoussan, Alain; Flandoli, Franco	95 Stochatx 53 13-39
Exit probability estimates for Γ-martingales on	manifolds • Le, Huiling	95 Stochatx 54 199-209
Radial part of Brownian motion on a Riemannian	manifold • Liao, M.; Zheng, W. A.	95 AnlsProb 23 173-177
terms for the radial parts of semimartingales on	manifolds • Le, Huiling; Barden, Dennis Itô correction	95 ProbTher 101 133-146
	Mann-Whitney statistic	
Govindarajulu, Z.	Admissibility and minimaxity of the UMVU estimator of $P\{X < Y\}$ • Yu, Qiqing;	95 AnlsStat 23 598-607
samples • Edwardes, Michael D. deB.	A confidence interval for $\Pr(X < Y) - \Pr(X > Y)$ estimated from simple cluster	95 Biomtrcs 51 571-578
	MANOVA	
Konno, Yoshihiko	Estimation of a normal covariance matrix with incomplete data under Stein's loss •	95 JMultiAn 52 308-324
A decision rule for dimension in the context of	MANOVA • De, Anindya K.; Ghosh, Jayanta K.	95 SankhyaA 57 256-266
MANOVAMAP: Graphical representation of	MANOVA in marketing research • Novak, Thomas P.	95 JMarRsr 32 357-374
tests for dimensionality in canonical correlation and	MANOVA models • Seo, T.; Kanda, T.; Fujikoshi, Y. The effects of nonnormality on	95 JMultiAn 52 325-337
Novak, Thomas P.	MANOVAMAP: Graphical representation of MANOVA in marketing research •	95 JMarRsr 32 357-374
	Mantel-Haenszel	
year of marketing and secular trends in ... • Tsong	Comparing reporting rates of adverse events between drugs with adjustment for	95 JBiophSt 5 95-114
	Dave Byar's contribution to epidemiology • Gail, Mitchell H.	95 ConClnTr 16 230-248
Odom-Maryon, Tamara	Estimation of a common odds ratio under binary cluster sampling • Ahn, Chul;	95 StatMed 14 1567-1576
estimators of the common odds ratio • Yu, Kai Fun	A necessary and sufficient condition for the strong consistency of a family of	95 CandJSt 23 215-225
clinical trials • Davis, Charles S.; Chung, Younghae	Randomization model methods for evaluating treatment efficacy in multicenter	95 Biomtrcs 51 1163-1174
Germanson, Teresa P.; Barnes, Thomas Y.	A simple measure in defining optimal strata in clinical trials • Choi, Sung C.;	95 ConClnTr 16 164-171
	Some neglected problems in IRT • Fischer, Gerhard H.	95 Psymtrka 60 459-487
May, Warren L.; Johnson, William D.	Testing quasi-independence in stratified triangular contingency tables •	95 BiomtrcJ 37 531-546
Projection-method	Mantel-Haenszel estimator for $K\ 2 \times J$ tables • Yanagawa, Takashi; Fujii, Yoshinori	95 JASA 90 649-656
A further look at the Cochran-	Mantel-Haenszel risk difference • Sato, Tosiya	95 ConClnTr 16 359-361
Moyer, Ernest S. Modelling frailty in the	manufacture of respiratory protective devices • Wassell, James T.; Kulczycki, Gregory W.;	94 ASAProEpid 133-138
in Poisson-sample selection for the annual survey of	manufactures • Zayatz, L.; Sigman, R. Feasibility study of the use of Chromy's algorithm	94 ASASrvyRsr 641-646
expectations: An application to Australian	manufacturing • Smith et al. ... for converting qualitative response data to quantitative	95 JApEcmtx 10 165-185
applied to a discrete and stochastic simulator of a	manufacturing company • Giribone ... for non-equispaced experimental configurations	94 JAppStat 21 549-564
Search for a sequence of optimal tests to control	manufacturing costs (French) • Bergeret, F.; Chandon, Y.	95 RStApp 43/3 21-33
• Lepeniotis, Stefanos S.; Vigezzi, Michael J. Lowering	manufacturing cost of material by formulating it through statistical modeling and design	95 Chemolab 29 133-139
Virtual	manufacturing: Design, calibration, and analysis • Heavlin, William D.; Krivokapić, Zoran	94 ASAProPEng 274-279
Frailty models of	manufacturing effects • Wassell, J. T.; Kulczycki, G. W.; Moyer, E. S.	95 LiDaAnal 1 161-170
Measuring relative electricity cost sensitivity in	manufacturing establishments in New York State • Gaines, Leonard M.	94 ASAProGovt 183-188
between R and D and productivity in French	manufacturing firms • Hall, Bronwyn H.; Mariesse, Jacques Exploring the relationship	95 JEconmtx 65 263-293
linear prediction to quality assurance in industrial	manufacturing • Hwang, J. T. Gene; Liu, Hung-kung Application of empirical	94 ASAProPEng 175-180
capital formation and economic performance in U.S.	manufacturing industries: An exploratory analysis • Berndt, E. R.; Morrison, C. J. High-tech	95 JEconmtx 65 9-43
national interdependence of business cycles in the	manufacturing industry: The use of leading indicators for forecasting ... • Berk et al. Inter-	95 JForcast 14 1-23
al. A discounted-cost continuous-time flexible	manufacturing and operator scheduling model solved by deconvexification ... • Eaves et	95 PrEngInf 9 65-98
Bergström, Reinhold The relationship between	manufacturing production and different business survey series in Sweden 1968-1992 •	95 IntJFor 11 379-393
tical statistical process control: A tool for quality	manufacturing" • Alsup, F.; Watson, R. M. (Auth); McLaughlin, G. C. (Rev) Review of "Prac-	95 Technmcs 37 461-462
and participants in the residential, commercial, and	manufacturing sectors of the U.S. economy • Battles et al. ... Utility programs offered	94 ASAProBuEc 20-27
TQM implementation strategies in European	manufacturing and service organizations • Kartha, C. P.	94 ASAProQlPr 66-69
Hierarchical decision making in stochastic	manufacturing systems • Sethi, S. P.; Zhang, Q.	94 Birkhs:Basl:MA 19p
of "Hierarchical decision making in stochastic	manufacturing systems" • Sethi, S. P.; Zhang, Q. (Auth); Goodwin, G. C. (Rev) Review	95 ShrtBkRv 15/2 29-29
the semiconductor industry: Challenges in volume	manufacturing (Disc: p9-11) • Gordon, Robert D. Statistics in	94 ASAProPEng 6-8
Lot sizes in serial	manufacturing with random yields • Sobel, Matthew J.	95 PrEngInf 9 151-157
	Many-one comparisons	
Cheng, Peter H.; Meng, Cliff Y. K.	A new formula for tail probabilities of Dunnett's T with unequal sample sizes •	95 CommStA 24 523-532
of variation applied to Markov chains and Anosov	maps • Loviscach, J. A generalized notion	95 ProbTher 103 553-570
Møller, Jesper Estimating distribution	maps from atlas data using methods of statistical image analysis • Högmander, Harri;	95 Biomtrcs 51 393-404
Jacquez, Geoffrey M. The	map comparison problem: Tests for the overlap of geographic boundaries •	95 StatMed 14 2343-2361
Effects of the choice of age-adjustment method on	maps of death rates (Disc: p697-699) • Pickle, Linda W.; White, Andrew A.	95 StatMed 14 615-627
W.; Herrmann, D. J.; Croner, C. M.; Wilson, B. F.	Map design preferences associated with professional discipline • White, A. A.; Pickle, L.	94 ASAProStGr 54-59
Invalidity of the Rao	map function for three loci • Weeks, Daniel E.	94 HumHrdty 44 178-180
On the single-server queue with independent	MAP/G and M/GI input streams • Takine, Tetsuya; Hasegawa, Toshiharu	95 CommStSM 11 227-234
Stochastic analysis and random	maps in Hilbert space • Dorogovtsev, A. A.	94 VSP:Neth 110p
analysis of complex traits from high-resolution	maps of identity by descent • Dupuis, J.; Brown, P. O. et al. Statistical methods for linkage	95 Genetcs 140 843-856
Bayesian estimates of disease	maps: How important are priors? • Bernardinelli, L.; Clayton, D.; Montomoli, C.	95 StatMed 14 2411-2431
trait loci using complete genetic marker linkage	maps • Wu, W. R.; Li, W. M. A new approach for mapping quantitative	94 ThApGene 89 535-539
and Lyapunov exponents for chaotic logistic	maps • Hall, Peter; Wolff, Rodney C. L. Properties of invariant distributions	95 JRSS-B 57 439-452
The Kohonen self-organizing	map method: An assessment • Murtagh, F.; Hernández-Pajares, M.	95 JClass 12 165-190
two-point lod scores into a summary multipoint	map • Curtis, David; Gurling, Hugh A procedure for combining	93 HumHrdty 43 173-185
Bayesian Poisson rate estimation using parameter	maps • Buck, James R.; Pignatiello, Joseph J., Jr; Raz, Tzvi	95 IIETrn 27 405-412
Bayesian construction of thematic	maps from satellite imagery • Green, E. J.; Smith, A. F. M.; Strawderman, W. E.	94 ASASrvyRsr 188-193
Introduction to computational biology:	Maps, sequences, and genomes • Waterman, Michael S.	95 ChapmnHall:Lnd 450p
On spatial analysis and	map transformations • Palma, Daniela	94 QStMaA 16 127-143
Flow time in the	MAP/G/1 queue with customer batching and setup times • He, Qi-Ming; Jewkes, E.	95 CommStSM 11 691-711

	Marketing of statistics in Canada • Fellegi, I. P.	91 StatJUN 8 295-306
in large-scale assessment procedures using 'double-	marking' or 'partial double-marking' • Sparks, R.; Ballantyne, R. Quality control methods	95 AstrlJSt 37 283-298
	Markov	
	• see also Gauss- –, hidden –, semi- –, Markoff	
Large exceedances for uniformly recurrent	Markov-additive processes and strong-mixing stationary processes • Zajic, Tim	95 JAppProb 32 679-691
Limit theorems and	Markov approximations for chaotic dynamical systems • Chernov, N. I.	95 ProbTher 101 321-362
performance • Rananand, Nol	Markov approximations to D-BMAPs: Information theoretic bounds on queueing	95 CommStSM 11 713-734
Veinott, Arthur F., Jr	Markov branching decision chains with interest-rate-dependent rewards • Huang, Ying;	95 PrEngInf 9 99-121
Chen, Anyue; Renshaw, Eric	Markov branching processes regulated by emigration and large immigration •	95 StocProc 57 339-359
acterization of discrete laws via mixed sums and	Markov branching processes • Pakes, Anthony G. Char-	95 StocProc 55 285-300
	Markov chain	
Dion, Jean-Pierre; Ferland, René	Absolute continuity, singular measures and asymptotics for estimators •	95 JStPlInf 43 235-242
	Asymptotically Markov stochastic 1 • Sataev, E. A.	94 ThProbAp 39 448-464
	Bayesian inference in cyclical component dynamic linear models • West, Mike	95 JASA 90 1301-1312
models • Frühwirth-Schnatter, Sylvia	Bayesian model discrimination and Bayes factors for linear Gaussian state space	95 JRSS-B 57 237-246
• Geske, M. X.; Godbole, A. P.; Schaffner, A. A. et al.	Compound Poisson approximations for word patterns under Markovian hypotheses	95 JAppProb 32 877-892
Kjærulff, Uffe	dHugin: A computational system for dynamic time-sliced Bayesian networks •	95 IntJFor 11 89-111
Norberg, Ragnar	Differential equations for moments of present values in life insurance •	95 InsrneME 17 171-180
Palutikof, J. P.	Estimation of extreme wind speeds with very long return periods • Dukes, M. D. G.;	95 JApMetrl 34 1950-1961
R., Jr	Evaluating properties of variable sampling interval control charts • Reynolds, Marion	95 SqtlAnly 14 59-97
permutation • Fu, James C.	Exact and limiting distributions of the number of successions in a random	95 AnInStMa 47 435-446
	An exploratory analysis of a record of El Niño events, 1800-1987 • Solow, Andrew R.	95 JASA 90 72-77
Hutchinson, J. W.; Raman, K.; Mantrala, M. K.	Finding choice alternatives in memory: Probability models of brand name recall •	94 JMarRsr 31 441-461
Prum, B.; Rodolphe, F.; de Turckheim, E.	Finding words with unexpected frequencies in deoxyribonucleic acid sequences •	95 JRSS-B 57 205-220
Katz, Richard W.; Parlange, Marc B.	Generalizations of chain-dependent processes: Application to hourly precipitation •	95 WaterReR 31 1331-1341
	On group inverses of matrices with order and rank perturbations • Rao, P. S. S. N. V. P.	94 CalcutSt 44 209-222
	Heavy traffic analysis of a random walk on a lattice semi-strip • Falin, Gennadi	95 CommStSM 11 395-409
time-varying parameter models • Chib, S. et al.	Hierarchical analysis of SUR models with extensions to correlated serial errors and	95 JEconmtx 68 339-360
k successes • Aki, Sigeo; Hirano, Katuomi	Joint distributions of numbers of success-runs and failures until the first consecutive	95 AnInStMa 47 225-235
Dembo, Amir; Zajic, Tim	Large deviations: From empirical mean and measure to partial sums process •	95 StocProc 57 191-224
L'Ecuyer, Pierre	Likelihood ratio gradient estimation for stochastic recursions • Glynn, Peter W.;	95 AdvAppPr 27 1019-1053
Gómez-Corral, A.	On the M/G/1 retrial queueing system with linear control policy • Martín, M.;	95 Top 3 285-305
	How many iid samples does it take to see all the balls in a box? • Sellke, Thomas M.	95 AnlsApPr 5 294-309
	On a Markov multinomial distribution • Wang, Y. H.; Yang, Zejiang	95 MaScitst 20 40-49
	Mixture models for time series • Jalali, Assad; Pemberton, John	95 JAppProb 32 123-138
convergence • D'Aristotile, Anthony J.	The nearest neighbor random walk on subspaces of a vector space and rate of	95 JTherPrb 8 321-346
	Optimal search for a moving target • MacPhee, I. M.; Jordan, B. P.	95 PrEngInf 9 159-182
	Oscillating random walk with a moving boundary • Madras, Neal; Tanny, David	94 IsrelJMa 88 333-365
Saloff-Coste, L.	Precise estimates on the rate at which certain diffusions tend to equilibrium •	94 MaZeit 217 641-677
Frieze, Alan; Pittel, Boris G.	Probabilistic analysis of an algorithm in the theory of markets in indivisible goods •	95 AnlsApPr 5 768-808
tiserver queue II: Positive recurrence ... • Sadowsky	The probability of large queue lengths and waiting times in a heterogeneous mul-	95 AdvAppPr 27 567-583
tiserver queue I: Tight limits • Sadowsky, J. S. et al.	The probability of large queue lengths and waiting times in a heterogeneous mul-	95 AdvAppPr 27 532-566
	On record indices and record times • Borovkov, Konstantin; Pfeifer, Dietmar	95 JStPlInf 45 65-79
	Return probabilities for random walk on a half-line • Lalley, Steven P.	95 JTherPrb 8 571-599
	Runs on a circle • Koutras, M. V.; Papadopoulos, G. K.; Papastavridis, S. G.	95 JAppProb 32 396-404
Kersting, G.; Klebaner, F. C.	Sharp conditions for nonexplosions and explosions in Markov jump processes •	95 AnlsProb 23 268-272
Stephen; McClean, Sally	Simulation of wind speed for the modelling of pine saw-fly mortality • Robinson,	95 Ststcian 44 469-477
Ong, S. H.	Some stochastic models leading to the convolution of two binomial variables •	95 StPrLet 22 161-166
	A storage model for data communication systems • Prabhu, N. U.; Pacheco, A.	94 ThProbAp 39 604-627
Vogel, Richard M.; Bolognese, Ralph A.	Storage-reliability-resilience-yield relations for over-year water supply systems •	95 WaterReR 31 645-654
tems • Hassett, T. F.; Dietrich, D. L.; Szidarovszky, F.	Time-varying failure rates in the availability and reliability analysis of repairable sys-	95 IEEEReli 44 155-160
Roberts, G. O.	Weak convergence of conditioned processes on a countable state space • Jacka, S. D.;	95 JAppProb 32 902-916
Tail events of simulated annealing	Markov chains • Niemiro, Wojciech	95 JAppProb 32 867-876
A generalized notion of variation applied to	Markov chains and Anosov maps • Loviscach, J.	95 ProbTher 103 553-570
Runs, scans and urn model distributions: A unified	Markov chain approach • Koutras, M. V.; Alexandrou, V. A.	95 AnInStMa 47 743-766
chains • Martínez, Servet; Vares, Maria Eulália A	Markov chain associated with the minimal quasi-stationary distribution of birth-death	95 JAppProb 32 25-38
Rates of convergence for the move-to-root	Markov chain for binary search trees • Dobrow, Robert P.; Fill, James Allen	95 AnlsApPr 5 20-36
Martin capacity for	Markov chains • Benjamini, Itai; Pemantle, Robin; Peres, Yuval	95 AnlsProb 23 1332-1346
Topics in the constructive theory of countable	Markov chains • Fayolle, G.; Malyshev, V. A.; Menshikov, M. V.	95 CambridgeUniv 169p
of "Topics in the constructive theory of countable	Markov chains" • Fayolle, G.; Malyshev, V. A.; Menshikov, M. V. (Auth) et al. Review	96 ShrtBkRv 16/1 9-10
On weak lumpability of denumerable	Markov chains • Ledoux, James	95 StPrLet 25 329-339
	Markov chain designs for one-per-stratum sampling • Breidt, F. Jay	95 SurvyM 21 63-70
Martin boundaries for discrete	Markov chains • Aggoun, Lakhdar	95 StocAnAp 13 257-268
State aggregation and discrete-state	Markov chains embedded in a class of point processes • Cao, Xi-Ren	95 JAppProb 32 39-51
the local limit theorem for non-uniformly ergodic	Markov chains • Séva, Marc On	95 JAppProb 32 52-62
Efficiency of empirical estimators for	Markov chains • Greenwood, P. E.; Wefelmeyer, W.	95 AnlsStat 23 132-143
	Markov chains for exploring posterior distributions (Disc: p1728-1762) • Tierney, Luke	94 AnlsStat 22 1701-1728
Random-direction interior-point	Markov chains: A family of black-box samplers • Chen, Ming-Hui; Schmeiser, Bruce W.	94 ASAProBays 1-6
On additive functionals of	Markov chains • Csáki, Endre; Csörgő, Miklós	95 JTherPrb 8 905-919
Ratio ergodicity for non-homogeneous	Markov chains in general state spaces • Fleischer, I.; Joffe, A.	95 JTherPrb 8 31-37
Geometric	Markov chains • Rising, William	95 JAppProb 32 349-374
A conditional test of independence of	Markov chains • Solow, A. R.; Smith, W. K.; Recchia, C.	95 BiomtrcJ 37 973-977
Conditions for integrability of	Markov chains • Hamza, K.; Klebaner, F. C.	95 JAppProb 32 541-547
A time-continuous	Markov chain interest model with applications to insurance • Norberg, Ragnar	95 AplStMDA 11 245-256
Levental, Shlomo Uniform CLT for	Markov chains and its invariance principle: A martingale approach • Bae, Jongsig;	95 JTherPrb 8 549-570
Occupation measures for	Markov chains • Dinwoodie, I. H.; Ney, Peter	95 JTherPrb 8 679-691
sojourn time in breast cancer screening using a	Markov chain model of both entry to and exit from the ... • Duffy et al. ... of mean	95 StatMed 14 1531-1543
stationary distribution of a stochastically monotone	Markov chain (French) • Simonot, F. Approximation of the	95 StocProc 56 133-149
	Markov chain Monte Carlo	
pedigrees • Lin, S.; Thompson, E.; Wijsman, E.	An algorithm for Monte Carlo estimation of genotype probabilities on complex	94 AnHuGene 58 343-357
	Bayes factors • Kass, Robert E.; Raftery, Adrian E.	95 JASA 90 773-795
trial • Gustafson, Paul	A Bayesian analysis of bivariate survival data from a multicentre cancer clinical	95 StatMed 14 2523-2535
Stephens, David A.	Bayesian analysis of errors-in-variables regression models • Dellaportas, Petros;	95 Biomtrcs 51 1085-1095
P. M.	Bayesian analysis of two overdispersed Poisson regression models • Scollnik, David	95 CommStA 24 2901-2918

of tightened-normal-tightened plans indexed by	maximum allowable percentage defective • Suresh, K. K. et al. Construction and selection	94 JAppStat 21 589-595
Box-Behnken designs • Borkowski, John J. Minimum,	maximum, and average spherical prediction variances for central composite and	95 CommStA 24 2581-2600
Disease cluster tests based on the	maximum cell frequency • Grimson, Roger C.	94 ASAProEpid 64-69
and analysis of group sequential logrank tests in	maximum duration versus information trials • Kim, K.; Boucher, H.; Tsiatis, A. A. Design	95 Biomtrcs 51 988-1000
	Maximum entropy	
	Calibration error and the measurement of turbulence statistics • Derksen, R. W.	95 EnvrMtrc 6 651-658
Payne, C. D.	Evaluation of methods for ecological inference • Cleave, N.; Brown, P. J.;	95 JRSS-A 158 55-72
heterogeneous aquifers • Harvey, C. F. et al.	Temporal moment-generating equations: Modeling transport and mass transfer in	95 WaterReR 31 1895-1911
Beyond	maximum entropy • Skilling, John; Sibisi, Sibusiso	94 ASAProStCp 71-79
functions • Wagner, Udo; Geyer, Alois L. J. A	maximum entropy method for inverting Laplace transforms of probability density	95 Biomtrka 82 887-892
Rüttermann, Markus On the	maximum entropy principle for a class of stochastic processes • Horsthemke, Benedikt;	95 StocProc 56 117-132
	Maximum entropy principles for disordered spins • Seppäläinen, Timo	95 ProbTher 101 547-576
Borkowski, John J. Finding	maximum G-criterion values for central composite designs on the hypercube •	95 CommStA 24 2041-2058
et al. Asymptotic properties of distribution of	maximum of Gaussian nonstationary process occurring in covariance statistic • Ostrovskii	94 ThProbAp 39 527-534
and SEQOPR: Computer programs for design of	maximum information trials based on group sequential logrank tests • Kim, K. SEQPWR	95 CmpBiomd 46 143-153
	Maximum likelihood	
LuValle, Michael J.	An accelerated life test model based on reliability kinetics • Meeker, William Q.;	95 Technmcs 37 133-146
	Adaptive control of partially observed linear stochastic systems • Bertrand, Pierre	95 Stochatx 54 21-51
Peiris, S.	Analysis of short time series with an over-dispersion model • Rai, S. N.; Abraham, B.;	95 CommStA 24 335-348
Konishi, Sadanori	Application of the bootstrap methods in factor analysis • Ichikawa, Masanori;	95 Psymtrka 60 77-93
model • Pinheiro, José C.; Bates, Douglas M.	Approximations to the log-likelihood function in the nonlinear mixed-effects	95 JCmpGrSt 4 12-35
under marginal probabilities • Kato, Takeshi	Asymptotic sufficiency of posterior densities with respect to Kullback information	95 StatDecn 13 253-271
Chen, Hung	Asymptotically efficient estimation in semiparametric generalized linear models •	95 AnlsStat 23 1102-1129
	On the asymptotics of constrained M-estimation • Geyer, Charles J.	94 AnlsStat 22 1993-2010
Haiganoush K.; Akers, R. Patrick	Autoregressive-type models for the analysis of bark beetle tracks • Preisler,	95 Biomtrcs 51 259-267
	On the behavior of a Capon-type spectral density estimator • Ioannidis, Evangelos E.	94 AnlsStat 22 2089-2114
Consul, P. C.	Bivariate generalized Poisson distribution with some applications • Famoye, Felix;	95 Metrika 42 127-138
parameter approach • Ioannatos, Petros E.	Censored regression estimation under unobserved heterogeneity: A stochastic	95 JBES 13 327-335
	Common canonical variates • Neuenschwander, Beat E.; Flury, Bernard D.	95 Biomtrka 82 553-560
	A comparative study of multidimensional sampling plans • Dutta, K.; Goswami, V.	94 JIAgSt 46 254-266
	Comparing non-nested regression models • Royston, Patrick; Thompson, Simon G.	95 Biomtrcs 51 114-127
value distribution • Thiagarajah, Kulathava Ranee	Comparisons of approximate confidence intervals for the parameters of extreme	94 Metron 52/1 85-100
	Comparisons of estimators of a slippage probability • Sobel, Marc	95 CommStA 24 1039-1055
Kruglyak, Leonid; Lander, Eric S.	Complete multipoint sib-pair analysis of qualitative and quantitative traits •	95 AJHuGene 57 439-454
• Vajda, Igor	Conditions equivalent to consistency of approximate MLE's for stochastic processes	95 StocProc 56 35-56
Wu, Ka Ho	Confidence regions for parameters in the AR(1) model • Hamilton, David;	95 JTimSrAn 16 249-265
	Consistency of the MLE for a general sequential design problem • Datta, Somnath	95 SankhyaA 57 88-99
• Hero, Alfred O.; Fessler, Jeffrey A.	Convergence in norm for alternating expectation-maximization (EM) type algorithms	95 StSinica 5 41-54
Corsini, G.; Gini, F.; Greco, M. V.; Verrazzani, L.	Cramér-Rao bounds and estimation of the parameters of the Gumbel distribution •	95 IEEEAES 31 1202-1204
	Detecting changes in linear regressions • Horváth, Lajos	95 MaOpfStS 26 189-208
selection • Keightley, Peter D.; Bulfield, Grahame	Detection of quantitative trait loci from frequency changes of marker alleles under	93 GeneRsr 62 195-203
Doganaksoy, Necip	Determining the duration of a demonstration life-test before all units fail •	95 IEEEReli 44 26-30
tion problems • Tsodikov, A. D.; Asselain, B. et al.	Discrete strategies of cancer post-treatment surveillance. Estimation and optimiza-	95 Biomtrcs 51 437-447
sampling • Gan, Gaoxiong; Bain, Lee J.	Distribution of order statistics for discrete parents with applications to censored	95 JStPlInf 44 37-46
tions on structural equation modeling ... • Ding et al.	Effects of estimation methods, number of indicators per factor, and improper solu-	95 StrEqMod 2 119-144
Jiroušek, Radim; Přeučil, Stanislav	On the effective implementation of the iterative proportional fitting procedure •	95 CmpStDA 19 177-189
	Efficiency of infinite dimensional M-estimators • van der Vaart, A. W.	95 StNeerla 49 9-30
	An empirical estimation procedure • Gupta, Arjun K.; Varga, Tamàs	94 Metron 52/1 67-70
tion to pilot-whales • Rothery, P.; Martin, A. R. et al.	Estimating foetal growth rate from cross-sectional samples of field data with applica-	95 ApplStat 44 163-172
Sidow, Arend; Nguyen, Trang; Speed, Terence P.	Estimating the fraction of invariable codons with a capture-recapture method •	92 JMolEvol 35 253-260
multiply type-II censored ... • Balakrishnan et al.	Estimation of the mean and standard deviation of the logistic distribution based on	95 MaOpfStS 27 127-142
	Estimation of non-occurrence probability in periodic model • Chen, Ling; Mi, Jie	95 CommStA 24 1493-1505
Yamamura, Kohji; Sugimoto, Tamio	Estimation of the pest prevention ability of the import plant quarantine in Japan •	95 Biomtrcs 51 482-490
	Estimation in the polytomous logistic regression model • Rom, Meir; Cohen, Ayala	95 JStPlInf 43 341-353
• Shibayama, Tadashi; Shigemasu, Kazuo	An estimation procedure of parameters in a general item response model (Japanese)	94 JpJBeh 21/1 57-65
	Estimation of sample size with grouped data • Blumenthal, Saul; Dahiya, Ram C.	95 JStPlInf 44 95-115
common unknown scale parameter • Carpenter et al.	Estimation of the smaller and larger of two exponential location parameters with a	95 CalcutSt 45 61-71
death data • Ahn, Hongshik; Kodell, Ralph L.	Estimation and testing of tumor incidence rates in experiments lacking cause-of-	95 BiomtrcJ 37 745-763
channel kinetics • Pulford, G. W.; Gallant, J. C. et al.	Evaluation and estimation of various Markov models with applications to membrane	95 BiomtrcJ 37 39-63
	Fisher information for a multivariate extreme value distribution • Shi, Daoji	95 Biomtrka 82 644-649
potential data • Raz, J.; Cardenas, V.; Fletcher, D.	Frequency domain estimation of covariate effects in multichannel brain evoked	95 Biomtrcs 51 448-460
	Generalised components of variance models • McGilchrist, C. A.; Ragunathan, T. E.	95 CommStA 24 349-361
	Generalized log-normal distributions with reliability application • Chen, Gemai	95 CmpStDA 19 309-319
	On the generalized negative binomial distribution • Consul, P. C.; Famoye, Felix	95 CommStA 24 459-472
	A gradient algorithm locally equivalent to the EM algorithm • Lange, Kenneth	95 JRSS-B 57 425-437
by mitochondrial DNA • Reeves, Jaxk H.	Heterogeneity in the substitution process of amino acid sites of proteins coded for	92 JMolEvol 35 17-31
Piantadosi, Steven; Crowley, John	An implicitly defined parametric model for censored survival data and covariates •	95 Biomtrcs 51 249-258
model • Zucker, D. M.; Zerbe, G. O.; Wu, M. C.	Inference for the association between coefficients in a multivariate growth curve	95 Biomtrcs 51 413-424
Gouri K.	Inference procedures for the linear failure rate model • Sen, Ananda; Bhattacharyya,	95 JStPlInf 46 59-76
	On information matrices in case-control studies • Wang, C. Y.; Wang, Suojin	95 StPrLet 22 269-274
rare disease: Application to HIV screening • Tu et al.	On the informativeness and accuracy of pooled testing in estimating prevalence of a	95 Biomtrka 82 287-297
La Tona, Letizia	A jackknife estimator of the scale parameter of L_p-normal distribution (Italian) •	94 Metron 52/1 181-188
	Local influence in structural equation models • Cadigan, N. G.	95 StrEqMod 2 13-30
	Local likelihood based on kernel censoring • Copas, J. B.	95 JRSS-B 57 221-235
Rukhin, Andrew L.	A lower bound for error probability in change-point estimation • Hu, Inchi;	95 StSinica 5 319-331
	Mean estimation for a partitioned multinormal distribution • Dahel, S.	95 CommStA 24 601-606
Kanter, Marek	How to measure the error of an M-estimate and report it conservatively •	95 JMultiAn 54 32-76
random variables • Castillo, Enrique; Hadi, Ali S.	A method for estimating parameters and quantiles of distributions of continuous	95 CmpStDA 20 421-439
	The method of sieves and minimum contrast estimators • Van de Geer, S.	95 MaMethSt 4 20-38
Gasser, Theo	A minimax result for a class of nonparametric density estimators • Engel, Joachim;	95 JNnparSt 4 327-334
	Mixed model analysis of DNA sequence evolution • Yang, Ziheng; Wang, Tianlin	95 Biomtrcs 51 552-561
Liu, Chuanhai; Rubin, Donald B.	ML estimation of the t distribution using EM and its extensions, ECM and ECME •	95 StSinica 5 19-39
cess from state • Carothers, A. D.; Bickmore, W. A.	Models of DNA replication timing in interphase nuclei: An exercise in inferring pro-	95 Biomtrcs 51 750-755
	Models of statistical visual motion estimation • Spetsakis, Minas	94 CVGIPIm 60 300-312

Maximum likelihood (cont.)

traits with missing data and ... • Hoeschele et al.	Multiple-trait genetic evaluation for one polychotomous trait and several continuous	95 JAnmlSci 73 1609-1627
	Multivariate simultaneous generalized ARCH • Engle, Robert F.; Kroner, Kenneth F.	95 EconTher 11 122-150
	Network models for sequence evolution • von Haeseler, Arndt; Churchill, Gary A.	93 JMolEvol 37 77-85
	Nonlinear smoothing for random fields • Aihara, Shin Ichi; Bagchi, Arunabha	95 StocProc 55 143-158
	Optimal rate of convergence for finite mixture models • Chen, Jiahua	95 AnlsStat 23 221-233
	Optimization using simulated annealing • Brooks, S. P.; Morgan, B. J. T.	95 Ststcian 44 241-257
	An overview of variance component estimation • Searle, Shayle R.	95 Metrika 42 215-230
multivariate extensions • Heinrich, G.; Jensen, U.	Parameter estimation for a bivariate lifetime distribution in reliability with	95 Metrika 42 49-65
	Performance of a Bartlett-type modification for the deviance • Cordeiro, Gauss M.	95 JStCmpSm 51 385-403
	Polynomial measurement error modeling • Moon, Myung-Sang; Gunst, Richard F.	95 CmpStDA 19 1-21
• Vogler, G. P.; Wette, R.; McGue, M. K.; Rao, D. C.	Properties of alternative estimators of familial correlations under variable sibship size	95 Biomtrcs 51 276-283
	A quasi-Newton acceleration of the EM algorithm • Lange, Kenneth	95 StSinica 5 1-18
Stone, Charles J.; Truong, Young K.	Rate of convergence for logspline spectral density estimation • Kooperberg, Charles;	95 JTimSrAn 16 389-401
matrices • Wu, Y.; McGilchrist, C. A.	Recursive computation for repeated measures models with patterned variance	95 JStCmpSm 51 333-341
	Regression models for repeated medical random counts • Mackenzie, Gilbert	95 Ststcian 44 365-372
	A review of semiparametric mixture models • Lindsay, Bruce G.; Lesperance, Mary L.	95 JStPlInf 47 29-39
Storms, Gert	On the robustness of maximum likelihood scaling for violations of the error model •	95 Psymtrka 60 247-258
markers • Säll, Torbjörn; Nilsson, Nils-Otto	The robustness of recombination frequency estimates in intercrosses with dominant	94 Genetcs 137 589-596
and developmental toxicity studies • Bowman et al.	A saturated model for analyzing exchangeable binary data: Applications to clinical	95 JASA 90 871-879
Pickles, A.; Dunn, G.; Vázquez-Barquero, J. L.	Screening for stratification in two-phase ('two-stage') epidemiological surveys •	95 StMedRes 4 73-89
	Second-order asymptotic efficiency of PMLE in generalized linear models • Liang, Hua	95 StPrLet 24 273-279
Boucher, Hélène; Kim, Kyungmann	Sequential methods for parametric survival models • Tsiatis, Anastasios A.;	95 Biomtrka 82 165-173
Schuenemeyer, John H.	Shift detection in two-way arrays • Altekar, Maneesha; Lucas, James M.;	95 CommStB 24 711-731
Kim, Minbo; Hill, R. Carter	Shrinkage estimation in nonlinear regression. The Box-Cox transformation •	95 JEconmtx 66 1-33
Gupta, A. K.; Varga, T.	Some inference problems for matrix variate elliptically contoured distributions •	95 MaOpfStS 26 219-229
Lo, S. H.	Some statistical mean value theorems related to the bootstrap • Chao, M. T.;	95 StSinica 5 129-139
method • Cho, JoongJae; Han, JeongHye	Statistical estimation for generalized logit model of nominal type with bootstrap	95 JKoreaSt 24 1-18
Ferrándiz, J.; Montes, F.	On statistical inference for random closed sets in compact subsets of R^d •	95 AdvAppPr 27 344-349
	Statistical tests of models of DNA substitution • Goldman, Nick	93 JMolEvol 36 182-198
	A stochastic EM algorithm for mixtures with censored data • Chauveau, Didier	95 JStPlInf 46 1-25
Galea Rojas, Manuel	Structural comparative calibration using the EM algorithm • Bolfarine, Heleno;	95 JAppStat 22 277-292
data • El-Saidi, Mohammed A.	A symmetric extended logistic model with applications to experimental toxicity	95 BiomtrcJ 37 205-216
Yang, Minxian	Tests for cointegration based on canonical correlation analysis • Bewley, Ronald;	95 JASA 90 990-996
estimating spatial statistical models • Griffith et al.	Trade-offs associated with normalizing constant computational simplifications for	95 JStCmpSm 51 165-183
Håkan	Twenty-one ML estimators for model selection • Gustafsson, Fredrik; Hjalmarsson,	95 Autmatca 31 1377-1392
Rocke, David M.; Lorenzato, Stefan	A two-component model for measurement error in analytical chemistry •	95 Technmcs 37 176-184
mous variables • Lee, S.-Y.; Poon, W.-Y.; Bentler, P. M.	A two-stage estimation of structural equation models with continuous and polyto-	95 BJMaStPs 48 339-358
	On the use of artificial regressions in certain microeconometric models • Orme, Chris	95 EconTher 11 290-305
Tilg, Norbert	Variance estimates in logistic regression using the bootstrap • Friedl, Herwig;	95 CommStA 24 473-486
functions • Rojo, Javier	On the weak convergence of certain estimators of stochastically ordered survival	95 JNnparSt 4 349-363
	• see also names of specific types of maximum likelihood such as restricted –	
A review of reliable	maximum likelihood algorithms for semiparametric mixture models • Böhning, Dankmar	95 JStPlInf 47 5-28
The profile Z plot and confidence intervals in	maximum likelihood analysis • Chen, Jian-Shen; Jennrich, Robert I.	94 ASAProStCp 124-129
A	maximum likelihood approach to correlational outlier identification • Bacon, Donald R.	95 MultBehR 30 125-148
Wang, You-Gan; Thomas, Mervyn R.; Somers, Ian F. A	maximum likelihood approach for estimating growth from tag-recapture data •	95 JFishRBC 52 252-259
Computation of	maximum likelihood estimates in association models • Haberman, Shelby J.	95 JASA 90 1438-1446
Asymptotic properties of	maximum likelihood estimates in a class of space-time regression models • Niu, Xufeng	95 JMultiAn 55 82-104
Kutoyants, Yu. Expansion of distribution of	maximum likelihood estimate for misspecified diffusion type observation • Dermoune, A.;	95 Stochatx 52 121-145
Chang, I-S. et al. Asymptotic consistency of the	maximum likelihood estimate in positron emission tomography and applications •	94 AnlsStat 22 1871-1883
Correction to "Bias reduction of	maximum likelihood estimates" (V80 p27-38) • Firth, David	95 Biomtrka 82 667-667
Golding, Brian	Maximum-likelihood estimates of selection coefficients from DNA sequence data •	93 Evolutn 47 1420-1431
The law of iterated logarithms of	maximum likelihood estimate for Weibull processes (Chinese) • Fang, Xiangzhong	95 ChJApPSt 10 314-319
about approximate confidence regions based on	maximum likelihood estimation • Meeker, William Q.; Escobar, Luis A. Teaching	95 AmerStat 49 48-53
Kale, Mandar; Deshmukh, S. R.	Maximum likelihood estimation in birth-death process allowing catastrophes •	95 CommStA 24 1863-1873
negative dependence restrictions with ... • Oh On	maximum likelihood estimation of cell probabilities in $2 \times k$ contingency tables under	95 CommStA 24 2127-2143
inequalities for martingales, with application to	maximum likelihood estimation for counting processes • van de Geer, Sara Exponential	95 AnlsStat 23 1779-1801
long memory autoregressive integrated ... • Beran	Maximum likelihood estimation of the differencing parameter for invertible short and	95 JRSS-B 57 659-672
Mishra, M. N.; Bishwal, J. P. N. Approximate	maximum likelihood estimation for diffusion processes from discrete observations •	95 Stochatx 52 1-13
Asymptotic bias in simulated	maximum likelihood estimation of discrete choice models • Lee, Lung-fei	95 EconTher 11 437-483
variance • van Dyk, D. A.; Meng, X.-L.; Rubin, D. B.	Maximum likelihood estimation via the ECM algorithm: Computing the asymptotic	95 StSinica 5 55-75
Agrò, G.	Maximum likelihood estimation for the exponential power function parameters •	95 CommStB 24 523-536
two-dimensional lattice • Sethuraman, S. et al.	Maximum likelihood estimation for a fractionally differenced autoregressive model on a	95 JStPlInf 44 219-235
	Maximum likelihood estimation of a GARCH-stable model • Liu, Shi-Miin; Brorsen, B. Wade	95 JApEcmtx 10 273-285
Weir, B. S.	Maximum-likelihood estimation of gene location by linkage disequilibrium • Hill, W. G.;	94 AJHuGene 54 705-714
cleotide sites • Gu, Xun; Fu, Yun-Xin; Li, Wen-Hsiung	Maximum likelihood estimation of the heterogeneity of substitution rate among nu-	95 MolBioEv 12 546-557
Pentti Problems with the asymptotic theory of	maximum likelihood estimation in integrated and cointegrated systems • Saikkonen,	95 EconTher 11 888-911
responses • Shoukri, M. M.; Martin, S. W. et al.	Maximum likelihood estimation of the kappa coefficient from models of matched binary	95 StatMed 14 83-99
tribution with replacement (Chinese) • Tang, Y. et al.	Maximum likelihood estimation in the life testing for the two-parameter exponential dis-	95 ChJApPSt 10 399-404
Taylor, J. M. G. The restricted EM algorithm for	maximum likelihood estimation under linear restrictions on the parameters • Kim, D. K.;	95 JASA 90 708-716
	Maximum likelihood estimation and Mathematica • Currie, Iain D.	95 ApplStat 44 379-394
Asymptotic properties of	maximum likelihood estimation in the mixed analysis of variance model • Güven, B.	95 StHefte 36 175-182
Review of "Information bounds and nonparametric	maximum likelihood estimation" • Groeneboom, P.; Wellner, J. A. (Auth); Plachky, D. (Rev)	95 StatDecn 13 94-95
(V32 ... • Smith et al. Comment on "Algorithm AS 189:	Maximum likelihood estimation of the parameters of the beta binomial distribution"	95 ApplStat 44 545-547
Lee, Lung-fei Semiparametric	maximum likelihood estimation of polychotomous and sequential choice models •	95 JEconmtx 65 381-428
Hassan; Goutis, Constantinos Approximate	maximum likelihood estimation in the presence of extra-Poisson variation • Johaadien,	95 CommStA 24 1625-1636
Matthews, G. B.; Crowther, N. A. S. A	maximum likelihood estimation procedure when modelling in terms of constraints •	95 SoAfrStJ 29 29-50
tourism? (German) • Ziegler, A. et al. Pseudo-	maximum-likelihood-estimation and regression diagnostics for count data: Suicide by	95 AllgeStA 79 170-195
Güven, Bilgehan	Maximum likelihood estimation in simple linear regression with one fold nested error •	95 CommStA 24 121-130
• Bolfarine, Heleno; Galea Rojas, Manuel	Maximum likelihood estimation of simultaneous pairwise linear structural relationships	95 BiomtrcJ 37 673-689
Alberto Exact	maximum likelihood estimation of stationary vector ARMA models • Mauricio, José	95 JASA 90 282-291
process • Garcia, Nancy Lopes	Maximum likelihood estimation and suboptimal stopping time for a Poisson point	95 RBrzlPrS 9 67-82
via the EM algorithm • Raudenbush, Stephen W.	Maximum likelihood estimation for unbalanced multilevel covariance structure models	95 BJMaStPs 48 359-370
	Maximum likelihood estimation with partially censored data • van der Vaart, Aad	94 AnlsStat 22 1896-1916

An episode in the history of	measurement • Hand, David	95 TeachgSt 17/2 54-56
E.	Measurement invariance, predictive invariance, and the duality paradox • Millsap, Roger	95 MultBehR 30 577-605
of latent variable relations with a comprehensive	measurement model • Muthén et al. ... of Muthén's LISCOMP approach to estimation	95 Psymtrka 60 489-503
A performance	measurement model • Rose, Kenneth H.	95 QulPr 28/2 63-66
Equality in constraints and metric-setting	measurement models • O'Brien, Robert M.; Reilly, Terence	95 StrEqMod 2 1-12
proach • Blackburn, R. M.; Siltanen, J.; Jarman, J. The	measurement of occupational gender segregation: Current problems and a new ap-	95 JRSS-A 158 319-331
Functional errors-in-variables models in	measurement optimization experiments • Liggett, Walter	94 ASAProPEng 193-199
An application of Bradley-Terry-type models to the	measurement of pain • Matthews, J. N. S.; Morris, K. P.	95 ApplStat 44 243-255
within latent trait dimensions: Application of mixed-	measurement to personality assessment • Reise, S. P. et al. Modeling qualitative variation	95 MultBehR 30 341-358
p27-46) • Andrich, David Models for	measurement, precision, and the nondichotomization of graded responses (Disc:	95 Psymtrka 60 7-26
The	measurement of price elasticities - The BT experience • Cracknell, David; Knott, Michael	95 IntJFor 11 321-329
Constructing quantum	measurement processes via classical stochastic calculus • Barchielli, A.; Holevo, A. S.	95 StocProc 58 293-317
Review of	"Measurement, regression, and calibration" • Brown, P. J. (Auth); Atkinson, Anthony (Rev)	95 JRSS-A 158 187-187
Review of	"Measurement, regression, and calibration" • Brown, P. J.; Fearn, Tom (Rev)	95 StMedRes 4 265-266
	Measurement scale	
segments in the babyfood market • Silver, Mick	Scales of measurement and cluster analysis: An application concerning market	95 Ststcian 44 101-112
and weighting of items in evaluative composite	measurement scales • Coste, Joël; Walter, Eric; Venot, Alain A new approach to selection	95 StatMed 14 2565-2580
123) • Labouvie, E. et al. Testing for equivalence of	measurement scales: Simple structure and metric invariance reconsidered (Disc: p77-	95 MultBehR 30 63-76
problems in the construction of composite	measurement scales: A survey of six medical and ... • Coste et al. ... and statistical	95 StatMed 14 331-345
Optimum design of serial	measurement trees • Ahlbrandt, Robert; Iyer, Hari K.; Vecchia, Dominic F.	95 JStPlInf 48 379-390
van der Merwe, A. J. Bayesian calibration in	meat science research-estimation of the pH in postmortem muscle • Plessis, J. L. du;	95 JStCmpSm 53 1-11
Drift and diffusion for a	mechanical system • Boldrighini, C.; Soloveitchik, M.	95 ProbTher 103 349-379
Note on generalizations of	Mecke's formula and extensions of $H = \lambda G$ • Miyazawa, Masakiyo	95 JAppProb 32 105-122
Depuydt, C. E.; Comhaire, F. H.	MedCalc: A new computer program for medical statistics • Schoonjans, F.; Zalata, A.;	95 CmpBiomd 48 257-262
How statisticians can help the news	media do a better job • Bailar, John C., III	95 Chance 8/1 24-24
How the news	media can help statisticians do a better job • Bailar, John C., III	95 Chance 8/1 25-25+
Evolution of systems in random	media • Korolyuk, Vladimir S.; Swishchuk, Anatoly V.	95 CRCPress:FL 368p
of misuses of statistical graphics in Korean mass	media and a study for statistical and institutional solutions to misuses ... • Jang A survey	95 KorApS 8/2 1-26
	Median absolute difference	
lite data. Part I: Development of ... • Epperson et al.	Estimating the urban bias of surface shelter temperatures using upper-air and satel-	95 JApMetrl 34 340-357
bound on the distance between the mean and the	median of the binomial and Poisson distributions • Hamza, K. The smallest uniform upper	95 StPrLet 23 21-25
On the influence functions of certain bivariate	medians • Niinimaa, A.; Oja, H.	95 JRSS-B 57 565-574
On the asymptotic efficiency of the sample circular	median • Purkayastha, Sumitra	95 StatDecn 13 243-252
An almost sure representation of sample circular	median • Purkayastha, Sumitra	95 JStPlInf 46 77-91
of smoothers based on local means and local	medians under dependence • Boente, Graciela; Fraiman, Ricardo Asymptotic distribution	95 JMultiAn 54 77-90
Estimation of projection absolute	median deviation and its application (Chinese) • Cui, Hengiian; Tian, Yubin	95 JSysSMaS 14 63-72
	Median estimation using auxiliary information • Meeden, Glen	95 SurvyM 21 71-77
The storage complexity of some estimators of the	median • Finkelstein, Mark; Scheinberg, Stephen	94 AplMaCmp 60 15-24
parametric estimation for the difference or ratio of	median failure times for paired observations • Jung, Sin-Ho; Su, John Q. Non-	95 StatMed 14 275-281
Rousseeuw, Peter J.; Ruts, Ida The repeated	median intercept estimator: Influence function and asymptotic normality • Hössjer, Ola;	95 JMultiAn 52 45-72
Estimating the asymptotic dispersion of the L_1	median • Bose, Arup	95 AnInStMa 47 267-271
Mean unbiased	medians, median unbiased means, and symmetric random walks • Mattner, L.	95 MaMethSt 4 99-105
Estimating	median and other quantiles in nonparametric models • Zieliński, R.	95 ZastApMa 23 363-370
demonstration of an optimality property of the	median • Lee, Yoong-Sin Graphical	95 AmerStat 49 369-372
	Median and quartiles from grouped data • Evans, Gwyn	95 TeachgSt 17/3 99-100
Survival analysis with	median regression models • Ying, Z.; Jung, S. H.; Wei, L. J.	95 JASA 90 178-184
Bounds for the breakdown point of the simplicial	median • Chen, Zhiqiang	95 JMultiAn 55 1-13
functions and numerical analysis of the repeated	median slope • Rousseeuw, Peter J.; Croux, Christophe; Hössjer, Ola Sensitivity	95 CmpStQ 10 71-90
[Algorithm AS 296] Optimal	median smoothing • Härdle, W.; Steiger, W.	95 ApplStat 44 258-264
Robustness of the half-space	median • Chen, Zhiqiang	95 JStPlInf 46 175-181
Consistency of the least	median of squares estimator in nonlinear regression • Stromberg, Arnold J.	95 CommStA 24 1971-1984
M. C.; Herrero-Gutiérrez, A. Regression by least	median squares, a methodological contribution to titration analysis • Ortiz-Fernández,	95 Chemolab 27 231-243
of the feasible solution algorithm for least	median of squares regression • Hawkins, Douglas M. Convergence	95 CmpStDA 19 519-538
Comparison of	median survival times with adjustment for covariates • Karrison, Theodore	95 StatMed 14 2537-2553
A simulation study of	mediated effect measures • MacKinnon, David P.; Warsi, Ghulam; Dwyer, James H.	95 MultBehR 30 41-62
eligibility and health care utilization measures for	Medicaid beneficiaries: Adult and child reports • Eisenhower et al. ... comparison of	94 ASASrvyRsr 713-718
Minority veterans' usage of VA	medical care • Stockford, Donald	94 ASAProGovt 22-24
Ryaboy, Svetlana Imputation of	medical cost and payment data • England, Amy M.; Hubbell, Katie A.; Judkins, David R.;	94 ASASrvyRsr 406-411
I. P. et al. Verification of information in a large	medical database using linkages with external databases (Disc: p687-689) • Shevchenko,	95 StatMed 14 511-530
Statistical considerations for survival analysis from	medical device clinical studies • Lao, Chang S.	95 JBiophSt 5 159-170
Use of field data in reliability studies of	medical devices • Kim, John S.; Mattson, LeRoy T.	94 ASAProQlPr 200-203
of composite measurement scales: A survey of six	medical and epidemiological journals • Coste et al. ... problems in the construction	95 StatMed 14 331-345
imputation on variance estimates for the National	Medical Expenditure Survey • Sommers, John Paul Analysis of the effects of	94 ASASrvyRsr 400-405
characteristic of excursion sets, with applications to	medical images • Worsley ... the number of peaks in a random field using the Hadwiger	95 AnlsStat 23 640-669
Teaching critical review of the	medical literature: A course in clinical trials • Freeman, Katherine	94 ASAProStEd 156-161
Statistics in the	medical literature: 1 • Altman, Douglas	95 StatMed 14 873-874
ment error on linear regression and correlation in	medical method comparison studies • Rifkin ... of correlated and uncorrelated measure-	95 StatMed 14 789 798
Barbara Lepidus A comparison of household and	medical provider reported expenditures in the 1987 NMES • Cohen, Steven B.; Carlson,	94 JOffStat 10 3-29
Regression models for repeated	medical random counts • Mackenzie, Gilbert	95 Ststcian 44 365-372
Quality of last doctor visit reports: A comparison of	medical record and survey data (Disc: p379-381) • Jay, G. M.; Belli, R. F. et al.	94 ASASrvyRsr 362-367
The biostatistician in	medical research: Allocating time and effort • Lesser, Martin L.; Parker, Robert A.	95 StatMed 14 1683-1692
Review of "Modelling for survival data in	medical research" • Collett, D. (Auth); Ziegel, Eric R. (Rev)	95 Technmcs 37 128-128
Review of "Modelling survival data in	medical research" • Collett, D. (Auth); Abrams, Keith (Rev)	95 JAppStat 22 434-434
Review of "Modelling survival data in	medical research" • Collett, D. (Auth); D'Agostino, Ralph B. (Rev)	95 JASA 90 1125-1126
Review of "Modelling survival data in	medical research" • Collett, D. (Auth); Manatunga, Amita (Rev)	95 StMedRes 4 268-268
Review of "Modelling survival data in	medical research" • Collett, D. (Auth); Ray, W. D. (Rev)	95 JRSS-A 158 188-188
Review of "Modelling survival data in	medical research" • Collett, D. (Auth); Scott, E. M. (Rev)	95 Ststcian 44 137-138
Review of "Modelling survival data in	medical research" • Collett, D. (Auth); Watkins, James F. (Rev)	95 Ststcian 44 281-282
Review of "Modelling survival data in	medical research" • Collett, D. (Auth); van Houwelingen, Hans (Rev)	95 StatMed 14 1147-1148
Statistics in	medical research. Developments in clinical trials • Gehan, E. A.; Lemak, N. A.	94 PlenumPubl:NY 214p
Statistics in	medical research: Developments in clinical trials • Gehan, Edmund A.; Lemak, Noreen A.	94 PlenumPubl:NY 224p
De Stavola, B. L. (Rev) Review of "Statistics in	medical research. Developments in clinical trials" • Gehan, E. A.; Lemak, N. A. (Auth);	95 ShrtBkRv 15/2 22-22
Generalized additive models for	medical research • Hastie, Trevor; Tibshirani, Robert	95 StMedRes 4 187-196
Methods of reporting statistical results from	medical research studies • Walter, Stephen D.	95 AJEpidm 141 896-906

Teaching undergraduates and	medical residents how to critique journal articles • Norton, H. James	94 ASAProStEd 162-164
R. Prasaad A study of biostatistics requirements in	medical schools in the United States • Grady, Christopher S.; Looney, Stephen W.; Steiner,	94 ASAProStEd 263-266
Quantitative methods in biological and	medical sciences. A historical essay • Lancaster, H. O.	94 Springer:Brln:NY 297p
Review of "Quantitative methods in biological and	medical science – A historical essay" • Lancaster, H. O. (Auth); Mayo, Oliver (Rev)	95 AstrlJSt 37 372-374
Review of "Quantitative methods in biological and	medical sciences. A historical essay" • Lancaster, H. O. (Auth); Farewell, V. T. (Rev)	95 ShrtBkRv 15/2 22-22
The Cambridge dictionary of statistics in the	medical sciences • Everitt, B. S.	95 CambridgeUniv 274p
Life as an academic	medical statistician and how to survive it • Pocock, Stuart J.	95 StatMed 14 209-222
MedCalc: A new computer program for	medical statistics • Schoonjans, F.; Zalata, A.; Depuydt, C. E.; Comhaire, F. H.	95 CmpBiomd 48 257-262
Before and after Bradford Hill: Some trends in	medical statistics • Armitage, Peter	95 JRSS-A 158 143-153
Flaum, M. et al. Effects of errors in a multicenter	medical study: Preventing misinterpreted data • Arndt, S.; Tyrrell, G.; Wollson, R. F.;	94 JPsyctrc 28 447-459
in applications for marketing authorizations for	medicinal products • CPMP Working Party Biostatistical methodology in clinical trials	95 StatMed 14 1659-1682
biostatistics. An introduction to evidence-based	medicine • Dunn, Graham; Everitt, Brian Clinical	95 Wiley:NY:UK 154p
biostatistics: An introduction to evidence-based	medicine • Dunn, G.; Everitt, B. Clinical	95 Arnold:Lnd:MD 154p
biostatistics: An introduction to evidence-based	medicine" • Dunn, G.; Everitt, B. (Auth); Lagakos, S. W. (Rev) Review of "Clinical	95 ShrtBkRv 15/3 41-42
Application of artificial neural networks to clinical	medicine • Baxt, William G.	95 Lancet 346 1135-1138
Statistical bases of reference values in laboratory	medicine • Harris, Eugene K.; Boyd, James C.	95 Dekker:NY 384p
Statistical issues in the regulation of	medicines • Lewis, John A.	95 StatMed 14 127-136
tures: An analysis of survival rates • Senar, J. C. et al.	Mediterranean house sparrows (Passer domesticus) are not used to freezing tempera-	95 JAppStat 22 1069-1074
Terpessi, Cristina Short and	medium term rain forecasting in the Oristano basin (Italian) • Espa, Giuseppe;	94 QStMaA 16 145-160
Convergence of the conditional Kaplan-	Meier estimate under strong mixing • Lecoutre, Jean-Pierre; Ould-Saïd, Elias	95 JStPlInf 44 359-369
Extensions of the Kaplan-	Meier estimator • Chien, Wei-Ting Kary; Kuo, Way	95 CommStB 24 953-964
Two statistical tests for	meiotic breakpoint analysis • Plaetke, Rosemarie; Schachtel, Gabriel A.	95 AJHuGene 56 508-518
A regression characterisation of the	Meixner hypergeometric distribution • Pusz, J.	95 AstrlJSt 37 83-87
Miller, D.; Mao, Y. Trends in mortality from	melanoma in Canada and prediction of future rates • MacNeill, I. B.; Elwood, J. M.;	95 StatMed 14 821-839
comparison of bidirectional and optimal associative	memories for background prediction of spectra • Wabuyele, B. W. et al. Quantitative	95 Chemolab 29 51-61
and geometric distributions with limited	memory • Marshall, Albert W.; Olkin, Ingram Multivariate exponential	95 JMultiAn 53 110-125
Hurvich, Clifford M.; Ray, Bonnie K. Estimation of the	memory parameter for nonstationary or noninvertible fractionally integrated processes •	95 JTimSrAn 16 17-41
procedures for multiple decisions: Finite time	memory and stepwise maximum likelihood procedure • Rukhin, A. L.; Shi, J. Recursive	95 StHefte 36 155-162
Sid-Ahmed, M. A. Fast learning and efficient	memory utilization with a prototype based neural classifier • Abou-Nasr, M. A.;	95 PattrnRc 28 581-593
of CD4 counts among HIV-infected homosexual	men and injection drug using men and women • Paik et al. ... in clinical implications	95 StatMed 14 1889-1900
	Men's jobs and women's jobs: A pilot study • Marcis, John G.; Lawrimore, Kay W.	94 ASAProBuEc 247-252
Replaying the 1994 Wimbledon	men's singles final • Croucher, John S.	95 NZndSt 30 2-8
transitions among older African American and white	men • Burr, J. A.; Massagli, M. P.; Mutchler, J. E.; Pienta, A. Labor force status and	94 ASAProSoSt 129-134
using a piecewise nonlinear mixed-effects model in	men with prostate cancer • Morrell, C. H. et al. Estimating unknown transition times	95 JASA 90 45-53
high-school math and science classes for young	men and women • Levine, Phillip B.; Zimmerman, David J. The benefit of additional	95 JBES 13/2 137-149
O'Hanesian, Mary Ann An analysis for	menstrual data with time-varying covariates • Murphy, Susan A.; Bentley, Gillian R.;	95 StatMed 14 1843-1857
Minitab	menus – Do they help or hinder? • Reed, Margaret; Garven, Frances	95 TeachgSt 17/1 22-25
Review of	"Mercator for Windows 4.0" • Bernt, K.; Theus, M. (Rev)	95 CmpStQ 10 307-308
Reassessing brand loyalty, price sensitivity, and	merchandising effects on consumer brand choice • Allenby, Greg M.; Lenk, Peter J.	95 JBES 13 281-289
plication of locally weighted regression to airborne	mercury deposition around an incinerator site • Opsomer, J.-D.; Agras, J. et al. An ap-	95 EnvrMtrc 6 205-219
Berry, G. Environmental	mesothelioma incidence, time since exposure to asbestos and level of exposure •	95 EnvrMtrc 6 221-228
Analysing exponential family based	messy data • Sutradhar, Brajendra C.; Das, Kalyan	95 CommStA 24 2683-2699
	Messy simulated annealing • Kvasnička, V.; Pospíchal, J.	95 JChemtrc 9 309-322

Meta-analysis

	Bahadur exact slopes for a class of combinations of dependent tests • Pallini, Andrea	94 Metron 52/1 53-65
p1113-1116) • Enas, Gregory G.; Goldstein, David J.	Defining, monitoring and combining safety information in clinical trials (Disc:	95 StatMed 14 1099-1111
bias • Vevea, Jack L.; Hedges, Larry V.	A general linear model for estimating effect size in the presence of publication	95 Psymtrka 60 419-435
Samuel P.; Hill, Robert H.	Multiple hypothesis tests in multiple investigations (Disc: p701-702) • Caudill,	95 StatMed 14 585-589
estimation of a mixture component • Li, Zhaohai	A multiplicative random effects model for meta-analysis with application to	95 Biomtrcs 51 864-873
an estimate and a confidence ... • Bushman et al.	A procedure for combining sample correlation coefficients and vote counts to obtain	95 PsychBu 117 530-546
tobacco smoke (Disc: p413-424) • Gross, Alan J.	Uncertainties in lung cancer risk estimates reported for exposure to environmental	95 EnvrMtrc 6 403-412
Gelber, R. D. et al. A quality-adjusted survival	meta-analysis of adjuvant chemotherapy for premenopausal breast cancer • Cole, B. F.;	95 StatMed 14 1771-1784
V.	Meta-analysis and its applications in horticultural science • Olkin, Ingram; Shaw, Douglas	95 HortSci 30 1343-1348
The impact of method choice on	meta-analysis • Mengersen, K. L.; Tweedie, R. L.; Biggerstaff, B.	95 AstrlJSt 37 19-44
Andrew Bayesian approaches to random-effects	meta-analysis: A comparative study • Smith, Teresa C.; Spiegelhalter, David J.; Thomas,	95 StatMed 14 2685-2699
Burdick, E.; Mosteller, F. Heterogeneity in	meta-analysis of data from epidemiologic studies: A commentary • Colditz, G. A.;	95 AJEpidm 142 371-382
Benefits of heterogeneity in	meta-analysis of data from epidemiologic studies • Berlin, Jesse A.	95 AJEpidm 142 383-387
• Henderson, W. G.; Moritz, T. et al. Use of cumulative	meta-analysis in the design, monitoring, and final analysis of a clinical trial: A case study	95 ConClnTr 16 331-341
Comment on	"Meta-analysis for dose-response models" • DuMouchel, William	95 StatMed 14 679-685
disease patients using summary measures and	meta analysis • Talwalker ... of repeated measurements with dropouts among Alzheimer's	94 ASAProBiop 515-520
A random-effects regression model for	meta-analysis • Berkey, C. S.; Hoaglin, D. C.; Mosteller, F.; Colditz, G. A.	95 StatMed 14 395-411
Everell, S. A. et al. The effect of scale changes on	meta-analysis of multiplicative and main effects models • Kanetkar, V.; Evans, M. G.;	95 EdPsMeas 55 206-224
Clarke, Michael J. Practical methodology of	meta-analyses (overviews) using updated individual patient data • Stewart, Lesley A.;	95 StatMed 14 2057-2079
Stroup, Donna	Meta-analysis for public health professionals (Disc: p125-126) • Klemm, Rebecca;	94 ASAProEpid 107-111
	Meta-analysis: Reconciling the results of independent studies • Olkin, Ingram	95 StatMed 14 457-472
	Meta-analysis for the ROC studies using empirical Bayesian method • Zhou, X. H.	94 ASAProBays 102-105
	Meta-analysis of screening and diagnostic tests • Hasselblad, Vic; Hedges, Larry V.	95 PsychBu 117 167-178
	Meta-analysis software: A comparative review • Normand, Sharon-Lise T.	95 AmerStat 49 298-309
dose-response data from epidemiological studies by	meta-analysis (Disc: p679-685) • Smith, S. J.; Caudill, S. P. et al. On combining	95 StatMed 14 531-544
The use of Fisher's Z in Schmidt-Hunter-type	meta-analyses • Law, Kenneth S.	95 JEdStat 20 287-306
	Meta-analysis: Weighing the evidence • Jones, David R.	95 StatMed 14 137-149
A multiplicative random effects model for	meta-analysis with application to estimation of a mixture component • Li, Zhaohai	95 Biomtrcs 51 864-873
lung cancer and exposure to ... • Tweedie et al.	Meta-analytic approaches to dose-response relationships, with application in studies of	95 StatMed 14 545-569
Comment on "A critical look at some popular	meta-analytic methods" (V140 p290-296) • Vandenbroucke, Jan P.	95 AJEpidm 142 1007-1008
to comment on "A critical look at some popular	meta-analytic methods" • Greenland, Sander Reply	95 AJEpidm 142 1008-1009
Writing	meta-analytic reviews • Rosenthal, Robert	95 PsychBu 118 183-192
Statistical	meta-knowledge and -data • Nordbotten, S.	93 StatJUN 10 101-112
Modelling	metabolic energy by neural networks • Lozano, J.; Novič, M.; Rius, F. X.; Zupan, J.	95 Chemolab 28 61-72
Picado, Ana; Mendonça, Elsa	Metabolic responses in the assessment of pollution effects • Pires, Ana; Branco, Joao;	95 EnvrMtrc 6 155-163
Developing	metadata database at the Census Bureau • Gillman, Daniel W.; Appel, Martin V.	94 ASASrvyRsr 781-784
When the	metadata exceed the data: Data management with uncertain data • Klensin, John C.	95 StatComp 5 73-84
Data,	metadata and information • Hand, D. J.	93 StatJUN 10 143-152
Modelling	metadata • Darius, P.; Boucneau, M.; De Greef, P.; De Feber, E.; Froeschl, K.	93 StatJUN 10 171-180
	Metadata and official statistics • Franchet, Y.	93 StatJUN 10 97-100

The unified file system and its	metadata part • Saijets, M.	93 StatJUN 10 201-208
Towards an operative view of semantic	metadata • Froeschl, K. A.	93 StatJUN 10 181-194
User requirements for statistical	metainformation • Podehl, M.	93 StatJUN 10 113-120
A proposal for a survey-oriented	metainformation system • Sundgren, B.	92 StatJUN 9 139-174
Statistical	metainformation systems-pragmatics, semantics, syntactics • Sundgren, B.	93 StatJUN 10 121-142
The role of footnotes in a statistical	metainformation system • Silver, M.	93 StatJUN 10 153-170
Implementing a statistical	metainformation system • Malmborg, E.; Lisagor, L.	93 StatJUN 10 195-200
Estimating simulation	metamodels using correlated control variates • Shih, Neng Hui; Song, Wheyming Tina	95 CommStB 24 79-90
estimators of an intensity function applied to	meteorological data (French) • Moncaup, S.; Sarda, P.; Vieu, P. Nonparametric	95 RStApp 43/4 77-88
spatial correlations from spatial-temporal	meteorological data • Gunst, Richard F. Estimating	95 JClimate 8 2454-2470
effect of station drift on the estimation of shifts in	meteorological records due to site changes • Rhoades, D. A.; Neill, W. A. The	95 JClimlgy 15 207-220
applications of a metric on subspaces to satellite	meteorology • Crone, L. J.; Crosby, D. S. Statistical	95 Technmcs 37 324-328
Kathleen Ho A Bayesian integration of end-use	metering and conditional-demand analysis • Hsiao, Cheng; Mountain, Dean C.; Illman,	95 JBES 13 315-326
Optimal design in end-use	metering experiments • Bartels, Robert; Fiebig, Denzil G.	95 MaCmpSim 39 305-309
ical polymerization of nonlinear optical side chain	methacrylate copolymers in dimethylsulfoxide – An experimental ... • Feuer et al. Free rad-	95 Chemolab 29 113-124
	Method of moments	
C. B.	Efficient IV estimation in nonstationary regression • Kitamura, Yuichi; Phillips, Peter	95 EconTher 11 1095-1130
data • Davies, Anthony; Lahiri, Kajal	A new framework for analyzing survey forecasts using three-dimensional panel	95 JEconmtx 68 205-227
cycle data • Mudambi, Ram; Taylor, Larry W.	Some non-parametric tests for duration dependence: An application to UK business	95 JAppStat 22 163-177
Viswanathan, S.	Can speculative trading explain the volume-volatility relation? • Foster, F. Douglas;	95 JBES 13 379-396
The	method of moments in computer tomography • Klebanov, L. B.; Rachev, S. T.	95 MaScitst 20 1-14
for non-nested models estimated by generalized	method of moments • Arkonac, S. Z.; Higgins, M. L. A Monte Carlo study of tests	95 CommStB 24 745-763
models • Zellner, Arnold Bayesian	method of moments/instrumental variable (BMOM/IV) analysis of mean and regression	94 ASAProBays 58-63
S. R. Finite sample properties of the generalized	method of moments in tests of conditional asset pricing models • Ferson, W. E.; Foerster,	94 JFinEcon 36 29-55
	Method of moments using Monte Carlo simulation • Gelman, Andrew	95 JCmpGrSt 4 36-54
Shamlaye, C. Seasonal changes in	methylmercury levels in human hair • Yakir, B.; Cox, C.; Tanner, M. A.; Clarkson, T. W.;	94 ASAProBays 35-39
Minimum spanning trees for tree	metrics: Abridgements and adjustments • Leclerc, Bruno	95 JClass 12 207-241
A generalization of the Sharma-Kaushik class of	metrics • Koschnick, Klaus-Uwe; Sharma, Bhu Dev	93 JCISS 18 401-407
A new	metric for data clustering • Millar, David B.	94 ASAProStCp 251-256
Variety of hypercube embeddings of the equidistant	metric and designs • Deza, Michel; Laurent, Monique	93 JCISS 18 289-315
	Metrics and models in software quality engineering • Kan, Stephen H.	95 ASQualCntrl:WI 344p
On	metrics in point process approximation • Brown, Timothy C.; Xia, Aihua	95 Stochatx 52 247-263
to Poisson approximation with respect to the point	metric • Roos, Bero A semigroup approach	95 StPrLet 24 305-314
Probability	metrics and recursive algorithms • Rachev, S. T.; Rüschendorf, L.	95 AdvAppPr 27 770-799
A continuous	metric scaling solution for a random variable • Cuadras, C. M.; Fortiana, J.	95 JMultiAn 52 1-14
Equality in constraints and	metric-setting measurement models • O'Brien, Robert M.; Reilly, Terence	95 StrEqMod 2 1-12
theoretical characterizations of compactness of	metric spaces • Plachky, D. Some measure	95 SankhyaA 57 154-160
Constantin, G. (Rev) Review of "Probability	metrics and the stability of stochastic models" • Rachev, Svetlozar T. (Auth);	95 BLndnMaS 27 87-89
Statistical applications of a	metric on subspaces to satellite meteorology • Crone, L. J.; Crosby, D. S.	95 Technmcs 37 324-328
expansion • Barbolosi, D.; Faivre, C.	Metrical properties of some random variables connected with the continued fraction	95 IndgMatN 6 257-265
additive, dominance and epistatic variation for	metrical traits. V. Triple test cross analysis • Kumar, I. et al. A general method of detecting	94 Heredity 72 563-569
An improved support system for practical	metrology • Huntley, Les	95 ASQCTekCon 142-149
	Metropolis algorithm	
Geyer, Charles J.; Thompson, Elizabeth A.	Annealing Markov chain Monte Carlo with applications to ancestral inference •	95 JASA 90 909-920
Vounatsou, P.; Smith, A. F. M.	Bayesian analysis of ring-recovery data via Markov chain Monte Carlo simulation •	95 Biomtrcs 51 687-708
Green, Peter; Higdon, David; Mengersen, Kerrie	Bayesian computation and stochastic systems (Disc: p41-66) • Besag, Julian;	95 StatSci 10 3-41
Lee, Jack C.	Bayesian inference of threshold autoregressive models • Chen, Cathy W. S.;	95 JTimSrAn 16 483-492
Tierney, Luke	Markov chains for exploring posterior distributions (Disc: p1728-1762) •	94 AnlsStat 22 1701-1728
Rosenthal, Jeffrey S.	Minorization conditions and convergence rates for Markov chain Monte Carlo •	95 JASA 90 558-566
Dinwoodie, I. H.	A probability inequality for the occupation measure of a reversible Markov chain •	95 AnlsApPr 5 37-43
	Regeneration in Markov chain samplers • Mykland, Per; Tierney, Luke; Yu, Bin	95 JASA 90 233-241
Traffic accident mapping in Bangkok	metropolis: A case study • Ayuthya, Rampai Suksawasdi Na; Böhning, Danmar	95 StatMed 14 2445-2458
Understanding the	Metropolis-Hastings algorithm • Chib, Siddhartha; Greenberg, Edward	95 AmerStat 49 327-335
Adaptive rejection	Metropolis sampling within Gibbs sampling • Gilks, W. R.; Best, N. G.; Tan, K. K. C.	95 ApplStat 44 455-472
Exequiel; de Ita, Celia	Mexico's policy framework for environmental reporting (Disc: p36-37) • Ezcurra,	94 ASAProEnvr 19-25
Policy	micro analysis and simulation using APL • Burpee, Joseph E.	94 ASAProGovt 178-182
Review of "EViews –	Micro TSP for Windows" • Quantitative Micro Software (Auth); Noble, Nicholas R. (Rev)	95 IntJFor 11 497-499
Review of "Statistical quality control with	microcomputer applications" • Shirland, Larry E. (Auth); Mullins, Eamonn (Rev)	95 Chemolab 28 203-203
El-Sayed, Yousry M. Design of crossover	microcomputer program and application on drug bioequivalence data • Abdullah, M. E.;	95 CmpBiomd 48 237-239
	Microdata access	
confidentiality • Fienberg, Stephen E.	Conflicts between the needs for access to statistical information and demands for	94 JOffStat 10 115-132
Enhancing researcher access to economic	microdata at the Bureau of Labor Statistics • de Wolf, Virginia A.	94 ASAProGovt 164-169
Disclosure control for census	microdata • Skinner, Chris; Marsh, Catherine; Openshaw, Stan; Wymer, Colin	94 JOffStat 10 31-51
McGuckin, Robert H. Establishment	microdata for economic research and policy analysis: Looking beyond the aggregates •	95 JBES 13 121-126
Identification risks of	microdata: Evidence from experimental studies • Müller, Walter; Blien, Uwe; Wirth, Heike	95 SocMethR 24 131-157
Desktop documentation: Individual income tax	microdata, SOL • David, Martin	94 ASASrvyRsr 775-780
On the use of artificial regressions in certain	microeconometric models • Orme, Chris	95 EconTher 11 290-305
Rust, John Introduction: The	microeconometrics of dynamic decision making • Kapteyn, Arie; Kiefer, Nicholas;	95 JApEcmtx 10 1-7S
Mandelbrot, B. B. A class of	micropulses and antipersistent fractional Brownian motion • Ciozek-Georges, R.;	95 StocProc 60 1-18
Statistical fluctuations due to	microscale mixing in a diffusion layer • Zimmerman, William B.; Chatwin, P. C.	95 EnvrMtrc 6 665-675
Comment on	"Microsimulating federal program choices" • Cohen, Michael L.	94 ASAProGovt 51-52
A multiple imputation approach to	microsimulation • Atrostic, B. K.	94 ASASrvyRsr 529-534
Error analysis of food stamp	microsimulation models • Zaslavsky, Alan; Thurston, Sally	94 ASASrvyRsr 535-540
ual businesses (Disc: p51-52) • Rodgers et al. Using	microsimulation modeling techniques to estimate the effects of health reform on individ-	94 ASAProGovt 39-44
Simplification of Milstar	microwave and digital cost analysis model • Gobreial, Hany S.	95 JParmc 15/1 92-117
meng; Stephens, Graeme L. A Bayesian approach to	microwave precipitation profile retrieval • Evans, K. Franklin; Turk, Joseph; Wong, Tak-	95 JApMetrl 34 260-279
Application of Procrustean methods to	mid- and near-infrared spectral data • Vigneau, E.; Devaux, M. F.; Safar, M.	95 JChemtrc 9 125-135
	Mid-*P* confidence intervals: A brief review • Berry, G.; Armitage, P.	95 Ststcian 44 417-423
The sample	mid-range and symmetrized extremal laws • Bingham, N. H.	95 StPrLet 23 281-288
An alternative estimator in	Midzuno scheme for multiple characteristics • Bedi, Pradeep Kumar	95 CommStB 24 17-30
Charles; Voss, Paul Updating age-sex-specific net	migration rates with limited data (Disc: p122-123) • Kale, Balkrishna; Besl, John; Palit,	94 ASAProSoSt 116-121
Poverty	migration for U.S. States: Immigration impacts • Frey, William H.	94 ASAProSoSt 135-140
gestions for future directions for studies of marked	migratory landbirds from the perspective of a practitioner in population ... • Desante Sug-	95 JAppStat 22 949-965
spatial preferences in settlement site selection in	migratory pastoralism • Schwartz, H. J.; Mosler, C.; Hary, I.; Pielert, V. Factors affecting	95 EnvrMtrc 6 485-490

Simplification of	Milstar microwave and digital cost analysis model • Gobreial, Hany S.	95 JParmc 15/1 92-117
Petros	Min-max classifiers: Learnability, design and application • Yang, Ping-Fai; Maragos,	95 PattrnRc 28 879-899
Wang, Wenjin On improving the	min test for the analysis of combination drug trials • Sarkar, Sanat K.; Snapinn, Steve;	95 JStCmpSm 51 197-213
	Minification processes with discrete marginals • Kalamkar, V. A.	95 JAppProb 32 692-706
N. A survey of recent work on characterization of	minihypers in PG(t,q) and nonbinary linear codes meeting the Griesmer bound • Hamada,	93 JCISS 18 161-191
Locating global	minima in optimization problems by a random-cost approach • Berg, Bernd A.	93 Nature 361 708-710
confidence regions based on subsamples under	minimal assumptions • Politis, Dimitris N.; Romano, Joseph P. Large sample	94 AnlsStat 22 2031-2050
On the almost sure	minimal growth rate of partial sum maxima • Klass, Michael J.; Zhang, Cun-Hui	94 AnlsProb 22 1857-1878
The	minimal L_1 isotonic regression • Shi, Ning-Zhong	95 CommStA 24 175-189
Martin On the	minimal martingale measure and the Föllmer-Schweizer decomposition • Schweizer,	95 StocAnAp 13 573-599
distribution • Kollo, Tõnu; von Rosen, Dietrich	Minimal moments and cumulants of symmetric matrices: An application to the Wishart	95 JMultiAn 55 149-164
Optimal block designs with	minimal and nearly minimal number of units • Dey, Aloke; Shah, Kirti R.; Das, Ashish	95 StSinica 5 547-558
	Minimal positions in a branching random walk • McDiarmid, Colin	95 AnlsApPr 5 128-139
Maria Eulália A Markov chain associated with the	minimal quasi-stationary distribution of birth-death chains • Martínez, Servet; Vares,	95 JAppProb 32 25-38
Some results on randomly stopped	minimal repair processes • Kirmani, S. N. U. A.; Gupta, Ramesh C.	95 CommStSM 11 631-644
Percolation and	minimal spanning forests in infinite graphs • Alexander, Kenneth S.	95 AnlsProb 23 87-104
processes: ... • Stefanov Explicit limit results for	minimal sufficient statistics and maximum likelihood estimators in some Markov	95 AnlsStat 23 1073-1101
Designing for	minimally dependent observations • Torsney, Ben; Alahmadi, Abdulhadi M.	95 StSinica 5 499-514
	Minimax	
Tsybakov, A. B.	Efficient estimation of monotone boundaries • Korostelev, A. P.; Simar, L.;	95 AnlsStat 23 476-489
Konno, Yoshihiko	Estimation of a normal covariance matrix with incomplete data under Stein's loss •	95 JMultiAn 52 308-324
Datta, Gauri Sankar; Ghosh, Malay	Hierarchical Bayes estimators of the error variance in one-way ANOVA models •	95 JStPlInf 45 399-411
	Minimax vs. Bayes prediction • Blackwell, D.	95 PrEngInf 9 53-58
	Minimax designs in linear regression models • Dette, H.; Heiligers, B.; Studden, W. J.	95 AnlsStat 23 30-40
Bernhard, P. H^∞-optimal control and related	minimax design problems - A dynamic game approach (Second edition) • Basar, T.;	95 Birkhs:Basl:MA 411p
Review of "H^∞-optimal control and related	minimax design problems - A dynamic game approach (Second edition)" • Basar, T. et al.	96 ShrtBkRv 16/1 11-11
Exact Elfving-	minimax designs for quadratic regression • Krafft, Olaf; Schaefer, Martin	95 StSinica 5 475-484
Johnson, M. E.; Moore, L. M.; Ylvisaker, D.	Minimax distance designs in two-level factorial experiments • John, P. W. M.;	95 JStPlInf 44 249-263
Some results on posterior regret Γ-	minimax estimation • Insua, David Ríos; Ruggeri, Fabrizio; Vidakovic, Brani	95 StatDecn 13 315-331
positron emission tomography • Stander, J. et al.	Minimax estimation of linear functionals, particularly in nonparametric regression and	95 CmpStQ 10 259-283
scale-invariant squared-error loss • van Eeden, C.	Minimax estimation of a lower-bounded scale parameter of a gamma distribution for	95 CandJSt 23 245-256
scale model (with an application to the second-order	minimax estimation) • Gajek, L. et al. Lower bounds for the asymptotic Bayes risk in the	94 AnlsStat 22 1831-1839
the entropy loss function • Pal, N.; Ling, C. Improved	minimax estimation of powers of the variance of a multivariate normal distribution under	95 StPrLet 24 205-211
	Minimax estimation of proportions under random sample size • Amrhein, Peter	95 JASA 90 1107-1111
	Minimax estimator	
means • Liu, Ying-Hsing; Yang, Ming-Chung	Bounded risk conditions in simultaneous estimation of independent Poisson	95 JStPlInf 47 319-331
Vidakovic, Brani; DasGupta, Anirban	Lower bounds on Bayes risks for estimating a normal variance: With applications •	95 CandJSt 23 269-282
	On minimax filtering over ellipsoids • Belitser, E. N.; Levit, B. Y.	95 MaMethSt 4 259-273
Schmidt, Karsten; Stahlecker, Peter	Reducing the maximum risk of regression estimators by polyhedral projection •	95 JStCmpSm 52 1-15
	Shrinkage estimation of a mean vector in a two-sample problem • Jin, Chun	95 CommStA 24 2881-2888
Iain M.; Kerkyacharian, Gérard; Picard, Dominique	Wavelet shrinkage: Asymptopia? (Disc: p337-369) • Donoho, David L.; Johnstone,	95 JRSS-B 57 301-337
	Minimax estimators of the mean vector in normal mixed linear models • Bilodeau, Martin	95 JMultiAn 52 73-82
Ofori-Nyarko, Samuel Improved	minimax estimators of normal covariance and precision matrices • Gupta, A. K.;	95 MaOpfStS 26 19-25
error models • Mukhopadhyay, P. Bayes and	minimax estimators for two-stage sampling from a finite population under measurement	95 CommStA 24 663-674
On	minimax filtering over ellipsoids • Belitser, E. N.; Levit, B. Y.	95 MaMethSt 4 259-273
Bischoff, Wolfgang; Fieger, Werner; Wulfert, Stefanie	Minimax- and Γ-minimax estimation of a bounded normal mean under LINEX loss •	95 StatDecn 13 287-298
sphere • Moklyachuk, M. P.	Minimax interpolation of random fields that are time homogeneous and isotropic on a	95 ThPrMaSt 50 107-115
	Minimax kernel density estimators with length biased data • Wu, C. O.	95 MaMethSt 4 56-80
Carroll, J. Douglas	Minimax length links of a dissimilarity matrix and minimum spanning trees •	95 Psymtrka 60 371-374
	Minimax nonparametric testing of hypotheses on the distribution density • Ermakov, M. S.	94 ThProbAp 39 396-416
A	minimax optimal estimator for continuous monotone densities • Datta, Somnath	95 JStPlInf 46 183-193
algorithm for the solution of stochastic dynamic	minimax problems • Breton, Michèle; Hachem, Saeb El A scenario aggregation	95 Stochatx 53 305-322
Turnbull, B. W. A note on the local asymptotically	minimax rate for estimating a crossing point in a diagnostic marker problem • Hsieh, F.;	95 StPrLet 24 181-185
Asymptotical	minimax recovery of sets with smooth boundaries • Mammen, E.; Tsybakov, A. B.	95 AnlsStat 23 502-524
of least squares regression estimates using the	minimax regret approach • Droge, B.; Georg, T. On selecting the smoothing parameter	95 StatDecn 13 1-20
Gasser, Theo A	minimax result for a class of nonparametric density estimators • Engel, Joachim;	95 JNnparSt 4 327-334
Information inequalities for the	minimax risk • Sato, Michikazu; Akahira, Masafumi	95 JJapnStS 25 151-158
	Minimax tests for convex cones • Dümbgen, Lutz	95 AnInStMa 47 155-165
Ingster, Yu. I.	Minimax testing of hypotheses on the distribution density for ellipsoids in l_p •	94 ThProbAp 39 417-436
	Minimaxity	
Wells, Martin T.	Loss estimation for spherically symmetric distributions • Fourdrinier, Dominique;	95 JMultiAn 53 311-331
Krengel, Ulrich	A stochastic game of optimal stopping and order selection • Gnedin, Alexander V.;	95 AnlsApPr 5 310-321
Hierarchical Bayesian models and Γ-	minimaxity • George, Robert; Goel, Prem K.	94 ASAProBays 96-101
criterion • Bhaumik, Dulal K.; Mathew, Thomas	Minimaxity of randomized optimal designs with respect to a general optimality	95 SankhyaB 57 122-127
Admissibility and	minimaxity of the UMVU estimator of $P\{X<Y\}$ • Yu, Qiqing; Govindarajulu, Z.	95 AnlsStat 23 598-607
Multivariate screening procedures for quality cost	minimization • Drezner, Zvi; Wesolowsky, G. O.	95 IIETrn 27 300-304
Cattiaux, P.; Léonard, C. Correction to	"Minimization of the Kullback information of diffusion processes" (V30 p83-132) •	95 AIHenrPB 31 705-707
Downey, Peter Scheduling independent tasks to	minimize the makespan on identical machines • Bruno, John; Coffman, Edward G., Jr;	95 PrEngInf 9 447-456
Ordering of tandem constant-service stations to	minimize in-process stock cost • Winch, Janice Kim; Avi-Itzhak, Benjamin	95 PrEngInf 9 457-473
specification in random production models of cost-	minimizing firms • Brown, Bryan W.; Walker, Mary Beth Stochastic	95 JEconmtx 66 175-205
Single machine stochastic scheduling:	Minimizing the number of tardy jobs • Lin, Chen-Sin	94 AJMMgSci 14 289-300
T. M. F. On the nonexistence of a global nonnegative	minimum bias invariant quadratic estimator of variance components • Gao, Sujuan; Smith,	95 StPrLet 25 117-120
Trenkler, G.; Zmyślony, R. Nonnegative	minimum biased quadratic estimation in the linear regression models • Gnot, S.;	95 JMultiAn 54 113-125
ment on "The moment-generating function of the	minimum of bivariate normal random variables" (V48 p124-125) • Sapra, Sunil K. Com-	95 AmerStat 49 240-240
on "The moment-generating function of the	minimum of bivariate normal random variables" • Cain, Michael Reply to comment	95 AmerStat 49 240-240
PM algorithms for calculating	minimum χ^2 estimators with partially observed sub-tables • Ritov, Ya'acov	95 CmpStDA 20 19-33
The method of sieves and	minimum contrast estimators • Van de Geer, S.	95 MaMethSt 4 20-38
applications to estimation of exposure to ... • Chen A	minimum cost estimator for the mean of positively skewed distributions with	95 EnvrMtrc 6 181-193
A	minimum cost exponential CUSUM test • Boyd, Amy Hancock	95 CommStA 24 2929-2958
et al. Association models for contingency tables via	minimum discrimination information approach and correspondence analysis • Gokhale	95 RBrzlPrS 9 83-98
Robustness of	minimum disparity estimators in linear regression models • Pak, Rojin	95 JKoreaSt 24 349-360
Fraiman, Ricardo	Minimum distance density-based estimation • Cao, Ricardo; Cuevas, Antonio;	95 CmpStDA 20 611-631
Yang, Song; Koul, Hira L.	Minimum distance estimation of the center of symmetry with randomly censored data •	95 Metrika 42 79-97
Millar, P. W.	Minimum distance estimation in random coefficient regression models • Beran, R.;	94 AnlsStat 22 1976-1992

	Missing data (cont.)	
longitudinal data • Park, Taesung; Shin, Min-Woong	A practical extension of the generalized estimating equation approach for	95 CommStA 24 2561-2579
Fitzmaurice, G. M.; Molenberghs, G.; Lipsitz, S. R.	Regression models for longitudinal binary responses with informative drop-outs •	95 JRSS-B 57 691-704
sured covariates • Robins, J. M.; Hsieh, F.; Newey, W.	Semiparametric efficient estimation of a conditional density with missing or mismea-	95 JRSS-B 57 409-424
Rotnitzky, Andrea; Robins, James M.	Semiparametric regression estimation in the presence of dependent censoring •	95 Biomtrka 82 805-820
Cabilio, Paul	Testing ordered alternatives in the presence of incomplete data • Alvo, Mayer;	95 JASA 90 1015-1024
of the two-period crossover to accommodate	missing data • Grieve, Andrew P. Extending a Bayesian analysis	95 Biomtrka 82 277-286
Estimating correlations with	missing data, a Bayesian approach • Gross, Alan L.; Torres-Quevedo, Rocio	95 Psymtrka 60 341-354
nonparametric maximum likelihood estimator in	missing data and biased sampling models • van der Laan, Mark J. An identity for the	95 Bernoull 1 335-341
Hayman analysis of diallel tables with complete or	missing data • Rodriguez et al. ... likelihood model-fitting approach to conducting a	93 BehavGen 23 69-76
variables and few objects. Part II: Cross-validation,	missing data and examples • Rännar et al. A PLS kernel algorithm for data sets with many	95 JChemtrc 9 459-470
sampling with application to imputation for	missing data • Rao, J. N. K.; Sitter, R. R. Variance estimation under two-phase	95 Biomtrka 82 453-460
	Missing data imputation using the multivariate t distribution • Liu, Chuanhai	95 JMultiAn 53 139-158
linear model with random effects for informative	missing data • Follmann, Dean; Wu, Margaret An approximate generalized	95 Biomtrcs 51 151-168
Effect of non-random	missing data mechanisms in clinical trials • Choi, S. C.; Lu, I. L.	95 StatMed 14 2675-2684
Rank correlation methods for	missing data • Alvo, Mayer; Cabilio, Paul	95 CandJSt 23 345-358
Multiply-imputed estimation for multivariate	missing data • Wang, Ying; Succop, Paul A.	94 ASAProEnvr 181-186
The impact of BIB spiraling-induced	missing data patterns on goodness-of-fit tests in factor analysis • Kaplan, David	95 JEdStat 20 69-82
models for repeated outcomes in the presence of	missing data • Robins, J. M.; Rotnitzky, A. et al. Analysis of semiparametric regression	95 JASA 90 106-121
Nieto, Fabio H. Note on the estimation of	missing data in a time series using the dual autocorrelation function (Spanish) •	94 Estd 46 85-103
chotomous trait and several continuous traits with	missing data and unequal models • Hoeschele et al. ... genetic evaluation for one poly-	95 JAnmlSci 73 1609-1627
Nonparametric regression estimation with	missing data • Chu, C. K.; Cheng, P. E.	95 JStPlInf 48 85-99
EM algorithm for graphical association models with	missing data • Lauritzen, Steffen L. The	95 CmpStDA 19 191-201
efficiency in multivariate regression models with	missing data • Robins, James M.; Rotnitzky, Andrea Semiparametric	95 JASA 90 122-129
The	missing factor in Hoeffding's inequalities • Talagrand, Michel	95 AIHenrPB 31 689-702
in Cox's regression model for competing risks with	missing failure types • Hemmi, Isao Parameter estimation	95 JJapnStS 25 81-96
multiply-imputed data under unequal fractions of	missing information • Pedlow, S. et al. Exploring hypothesis-testing procedures with	94 ASASrvyRsr 595-599
Adapting for the	missing link • Weisberg, S.; Welsh, A. H.	94 AnlsStat 22 1674-1700
The	missing linkage in ISO 9001: What's being done about it? • Marquardt, Donald W.	95 ASQCTekCon 1056-1063
Martínez, Octavio; Curnow, Robert N.	Missing markers when estimating quantitative trait loci using regression mapping •	94 Heredity 73 198-206
efficient estimation of a conditional density with	missing or mismeasured covariates • Robins, J. M.; Hsieh, F.; Newey, W. Semiparametric	95 JRSS-B 57 409-424
Markov-dependent binary sequences with randomly	missing observations • Cole, B. F.; Lee, M.-L. T. et al. An empirical Bayes model for	95 JASA 90 1364-1372
exponential signals when some observations are	missing • Kundu, Debasis; Kundu, Rameswar Consistent estimates of super imposed	95 JStPlInf 44 205-218
pal component analysis of rain gauge data: Treating	missing observations (Spanish) • López, Carlos; González, Juan F.; Curbelo, Rosario Princi-	94 Estd 46 55-83
Restricted forecasts of	missing observations in univariate time series • Guerrero, Víctor M.	94 Estd 46 1-23
of the analysis of multivariate normal data with	missing observations • May, Warren L.; Johnson, William D. Some applications	95 JBiophSt 5 215-228
Investigating the sensitivity against violation of the	missing at random assumption • Vach et al. ... observed categorical covariates -	95 StatMed 14 1315-1329
risks survival data when some failure types are	missing • Goetghebeur, Els; Ryan, Louise Analysis of competing	95 Biomtrka 82 821-833
A permutation paired test allowing for	missing values • Maritz, J. S.	95 AstrlJSt 37 153-159
Review of "Logistic regression with	missing values in the covariates" • Vach, W. (Auth); Kianifard, Farid (Rev)	95 Technmcs 37 460-461
	Missing values, imputation and error rate estimators in discriminant analysis • Bello, A. L.	95 AstrlJSt 37 95-104
Imputation of	missing values in spatio-temporal solar radiation data • Glasbey, C. A.	95 EnvrMtrc 6 363-371
time stochastic processes with control variables and	missing values • Singer, H. Analytical score function for irregularly sampled continuous	95 EconTher 11 721-735
of contour mapping epidemiological data with	missing values (Disc: p697-699) • Nobre, Flavio F.; de Macedo, Marcia M. A. Feasibility	95 StatMed 14 605-613
computational implementation • Csenki, Attila	Mission availability for repairable semi-Markov systems: Analytical results and	95 MaOpfStS 26 75-87
affecting visibility rate of waterfowl surveys in the	Mississippi Alluvial Valley • Smith, D. R.; Reinecke, K. J.; Conroy, M. J. et al. Factors	95 JWildlMg 59 515-527
	Misspecification	
	A differencing test • Franses, Philip Hans	95 EcomtrRv 14 183-193
Capturing geographically localized	misspecification error in retail store choice models • Rust, Roland T.; Donthu, Naveen	95 JMarRsr 32 103-110
paired data • Lyles, R. H. et al. Effects of model	misspecification in the estimation of variance components and intraclass correlation for	95 StatMed 14 1693-1706
superiority of empirical Bayes estimators under	misspecification • Wei, L.; Trenkler, G. Mean square error matrix	95 Test 4 187-205
Model	misspecification and multipoint linkage analysis • Risch, Neil; Giuffra, Luis	92 HumHrdty 42 77-92
Thomas R. Model	misspecification in proportional hazards regression • Anderson, Garnet L.; Fleming,	95 Biomtrka 82 527-541
D. B.; Foster, D. P. Filtering and forecasting with	misspecified ARCH models II: Making the right forecast with the wrong model • Nelson,	95 JEconmtx 67 303-335
of distribution of maximum likelihood estimate for	misspecified diffusion type observation • Dermoune, A.; Kutoyants, Yu. Expansion	95 Stochatx 52 121-145
On the use of the F ratio in a	misspecified model with an interval restriction • Wan, Alan T. K.; Yang, Minxian	95 JStCmpSm 52 151-161
Instrumental variables estimation in	misspecified single equations • Skeels, Christopher L.	95 EconTher 11 498-529
a continent-wide programme of constant-effort	mist-netting in North America • Desante et al. ... and survival rate estimates from MAPS,	95 JAppStat 22 935-947
Barkan, Philip The role of variation,	mistakes, and complexity in producing nonconformities • Hinckley, C. Martin;	95 JQualTek 27 242-249
stitutional solutions to misuses ... • Jang A survey of	misuses of statistical graphics in Korean mass media and a study for statistical and in-	95 KorApS 8/2 1-26
Longman, R. S.; Dion, K. L.; Kilik, L. Using and	misusing factor analysis to explore group cohesion • Cota, A. A.; Evans, C. R.;	95 JClPsych 51 308-316
process of amino acid sites of proteins coded for by	mitochondrial DNA • Reeves, Jaxk H. Heterogeneity in the substitution	92 JMolEvol 35 17-31
among sites in hypervariable region 1 of human	mitochondrial DNA • Wakeley, John Substitution rate variation	93 JMolEvol 37 613-623
Discrete	Mittag-Leffler distributions • Pillai, R. N.; Jayakumar, K.	95 StPrLet 23 271-274
properties of maximum likelihood estimation in the	mixed analysis of variance model • Güven, B. Asymptotic	95 StHefte 36 175-182
Generalization of the Mahalanobis distance in the	mixed case • Bar-Hen, Avner; Daudin, J.-J.	95 JMultiAn 53 332-342
of increments for cumulative hazard function in a	mixed censorship-truncation model with application to hazard ... • Gu Convergence	95 StPrLet 23 135-139
Data editing in a	mixed DBMS environment • Cox, N. W. P.; Croot, D. A.	91 StatJUN 8 117-136
sample properties of the estimators of coefficients in	mixed dummy-variable error components model • Ghazal, G. A.; Mira, S. I. The exact finite	95 EgpStJ 39 26-50
Nonlinear	mixed effects growth models for repeated measures in ecology • Schabenberger, Oliver	94 ASAProEnvr 156-161
Cheng, M. et al. Estimating parameters in non-linear	mixed effect models: A comparison using clinical data in children • Roe, D. J.;	94 ASAProBiop 313-318
effect • Gao, Min; Minkin, Salomon Nonlinear	mixed effect model for estimating population characteristics of drug combination	94 ASAProBiop 160-165
to "Variance component testing in the longitudinal	mixed effects model" (V50 p1171-1177) • Stram, D. O.; Lee, J. W. Correction	95 Biomtrcs 51 1196-1196
transition times using a piecewise nonlinear	mixed-effects model in men with prostate cancer • Morrell, C. H. et al. Estimating unknown	95 JASA 90 45-53
study on different approaches for nonlinear	mixed effects models • Zhang, Donghui Simulation	94 ASAProBiop 67-72
Model building in nonlinear	mixed effects models • Pinheiro, José C.; Bates, Douglas M.; Lindstrom, Mary J.	94 ASAProBiop 1-8
of interindividual random effects in non-linear	mixed effects models • Fattinger, K. E. et al. A new method to explore the distribution	95 Biomtrcs 51 1236-1251
to the log-likelihood function in the nonlinear	mixed-effects model • Pinheiro, José C.; Bates, Douglas M. Approximations	95 JCmpGrSt 4 12-35
Estimation of the shelf-life of drugs with	mixed effects models • Chen, James J.; Hwang, Jing-Shiang; Tsong, Yi	95 JBiophSt 5 131-140
iterative algorithm for estimation in nonlinear	mixed-effect models with an evaluation in population ... • Mentré et al. A two-step	95 JBiophSt 5 141-158
	Mixed factorial design	
V • Kuwada, Masahide; Hyodo, Yoshifumi	Analysis of variance of balanced fractional $r^m \times s^n$ factorial designs of resolution	95 JJapnStS 25 129-141
To analyze binary data from	mixed-level orthogonal arrays • Wang, P. C.; Chen, M. H.; Wei, S. C.	95 CmpStDA 20 689-695
Estimation of variance components in	mixed linear models • Kubokawa, Tatsuya	95 JMultiAn 53 210-236

The method of	moments in computer tomography • Klebanov, L. B.; Rachev, S. T.	95 MaScitst 20 1-14
Modelling the evolution of	moments of contaminant concentration in turbulent flows • Clarke, Liz; Mole, Nils	95 EnvrMtrc 6 607-617
and grouping on the estimated product-	moment correlation • Chai, John J.; Chen, Chung Joint effects of errors of measurement	95 CommStA 24 283-294
General formulae for the determination of	moments and cumulants • Rahman, M. S.; Nahar, S.	95 PakJStA 11 11-15
tion • Kollo, Tõnu; von Rosen, Dietrich Minimal	moments and cumulants of symmetric matrices: An application to the Wishart distribu-	95 JMultiAn 55 149-164
formulation of the old problem of obtaining	moments from cumulants and vice versa • Smith, Peter J. A recursive	95 AmerStat 49 217-218
multivariate Bayesian bootstrap distributions of	moments • Gasparini, Mauro Exact	95 AnlsStat 23 762-768
	Moment estimator	
statistical applications • Dion, J. P.; Yanev, N. M.	Central limit theorem for martingales in BGWR branching processes with some	95 MaMethSt 4 344-358
ing time interval omission • Ball, F. G.; Davies, S. S.	Statistical inference for a two-state Markov model of a single ion channel, incorporat-	95 JRSS-B 57 269-287
Factorial	moment expansion for stochastic systems • Błaszczyszyn, B.	95 StocProc 56 321-335
barriers • Dominé, Marco	Moments of the first-passage time of a Wiener process with drift between two elastic	95 JAppProb 32 1007-1013
Exact	moment formulae for the linear permutation statistic of order k • Whitmore, James B.	94 ASAProBiop 262-266
of the critical values of the standardized fourth	moment" (V44 p231-241) • Ramsey, P. H.; Ramsey, P. P. Correction to "Updated version	95 JStCmpSm 53 127-127
estimators and test statistics • Ullah, A. et al.	Moments of the function of non-normal random vector with applications to econometric	95 EcomtrRv 14 459-471
	Moments of generalized Wishart distributions • Wong, Chi Song; Liu, Dongsheng	95 JMultiAn 52 280-294
aquifers • Harvey, C. F.; Gorelick, S. M. Temporal	moment-generating equations: Modeling transport and mass transfer in heterogeneous	95 WaterReR 31 1895-1911
	Moment generating function	
expectations • Joe, Harry	Approximations to multivariate normal rectangle probabilities based on conditional	95 JASA 90 957-964
	A note on the maximum of a random walk • Stadje, W.	95 StPrLet 23 227-231
(V48 p124-125) • Sapra, Sunil K. Comment on "The	moment-generating function of the minimum of bivariate normal random variables"	95 AmerStat 49 240-240
Cain, Michael Reply to comment on "The	moment-generating function of the minimum of bivariate normal random variables" •	95 AmerStat 49 240-240
bution • El-Din, M. M. M.; Mahmoud, M. A. W. et al. On	moment generating function of order statistics for doubly truncated exponential distri-	95 Metron 53 171-183
simple ... • de la Peña Correction to "A bound on the	moment generating function of a sum of dependent variables with an application to	95 AIHenrPB 31 703-704
Zellner, Arnold Bayesian method of	moments/instrumental variable (BMOM/IV) analysis of mean and regression models •	94 ASAProBays 58-63
A test of the three	moment international capital asset pricing model • Vakil, Farok; Rajan, Murli	94 ASAProBuEc 184-189
density, distribution functions and inverse	moments • Chattamvelli, R.; Jones, M. C. Recurrence relations for noncentral	95 JStCmpSm 52 289-299
of unbiased and plotting-position estimators of L	moments • Hosking, J. R. M.; Wallis, J. R. A comparison	95 WaterReR 31 2019-2025
sive process • Monsour, Michael J.; Mikulski, Piotr W.	Moments of the limiting distribution for the boundary case in the first order autoregres-	94 AJMMgSci 14 327-347
On calculating the	moments of M-estimator in nonlinear regression (Chinese) • Wei, Bocheng; Feng, Yu	95 AMaAPknC 17 437-447
Complete class results for the	moment matrices of designs over permutation-invariant sets • Cheng, Ching-Shui	95 AnlsStat 23 41-54
	Moments of measures attracted to operator semi-stable laws • Scheffler, Hans-Peter	95 StPrLet 24 187-192
nested models estimated by generalized method of	moments • Arkonac, Seyhan Z.; Higgins, Matthew L. A Monte Carlo study of tests for non-	95 CommStB 24 745-763
the maximum likelihood method and the method of	moments (Japanese) • Yanagimoto, T. Estimation based on estimating equations - From	95 JpJAppSt 24 1-12
On the existence of exponential	moments of norms of random elements from the spaces sub$_\phi(\Omega)$ • Velikoivanenko, G. I.	95 ThPrMaSt 50 61-65
Mitchell, Ann F. S. Correction to "A note on posterior	moments for a normal mean with double-exponential prior" (V56 p605-610) •	95 JRSS-B 57 471-471
The power terms of the asymptotic series for	moments of the number of zeros of the Wong process on a large time ... • Miroshin	95 VLenUM /2 36-42
Characterizations of distributions by	moments of order statistics when the sample size is random • Grudzień, Z.; Szynal, D.	95 ZastApMa 23 305-318
• Gupta, A. K.; Varga, T.	Moments and other expected values for matrix variate elliptically contoured distribution	94 Statstca 54 361-373
of the smoothest density with prescribed	moments • Hong, Changkon; Kim, Choongrak Existence and uniqueness	95 JKoreaSt 24 233-242
Differential equations for	moments of present values in life insurance • Norberg, Ragnar	95 InsrneME 17 171-180
distribution and impact of ... • Fill et al. L	moment and probability plot correlation coefficient goodness-of-fit tests for the Gumbel	95 WaterReR 31 225-229
	Moments of randomly stopped sums-revisited • Teicher, Henry	95 JTherPrb 8 779-793
On the existence of	moments of ratios of quadratic forms • Roberts, Leigh A.	95 EconTher 11 750-774
Recurrence relations for	moments of record values • Kamps, Udo	95 JStPlInf 45 225-234
The precision of approximations to	moments of sample moments • Anderson, Oliver D.	95 JStCmpSm 52 163-165
two-parameter distribution by the first two sample	moments (partial and complete) • Shore, Haim Identifying a	95 JStCmpSm 52 17-32
Two-	moment scaling formulae for aggregating examination marks • Daley, D. J.	95 AstrlJSt 37 253-272
More bounds for	moments of some rank statistics • Spurrier, John D.	95 CommStA 24 2679-2682
essary conditions for the existence of conditional	moments of stable random variables • Cioczek-Georges, Renata; Taqqu, Murad S. Nec-	95 StocProc 56 233-246
More on conditional	moments of stable random variables • Soltani, A. Reza	92 ProIrStC1 195-200
Specification, estimation, and testing of	moment structure models based on latent variates involving ... • Baumgartner et al.	95 SocMethR 24 187-213
interval • Johnson, Norman L.; Kotz, Samuel Use of	moments in studies of limit distributions arising from iterated random subdivisions of an	95 StPrLet 24 111-119
sample properties of the generalized method of	moments in tests of conditional asset pricing models • Ferson, W. E.; Foerster, S. R. Finite	94 JFinEcon 36 29-55
with applications to GLS prediction and conditional	moment tests • Ahn, Seung C.; Schmidt, Peter A separability result for GMM estimation,	95 EcomtrRv 14 19-34
A nonparametric conditional	moment test for structural stability • Hidalgo, Javier	95 EconTher 11 671-698
The α-β model for concentration	moments in turbulent flows • Mole, Nils	95 EnvrMtrc 6 559-569
Fitting a distribution by the first two	moments (partial and complete) • Shore, Haim	95 CmpStDA 19 563-577
Method of	moments using Monte Carlo simulation • Gelman, Andrew	95 JCmpGrSt 4 36-54
Monte ... • Kamgar-Parsi et al. Distribution and	moments of the weighted sum of uniform random variables, with applications in reducing	95 JStCmpSm 52 399-414
between the log-survival function and truncated	moments, with application • El-Arishy, S. Useful relationship	95 StHefte 36 145-154
Money, output, and prices: Evidence from a new	monetary aggregate • Rotemberg, Julio J.; Driscoll, John C.; Poterba, James M.	95 JBES 13 67-83
	Monetary policy	
Julio J.; Driscoll, John C.; Poterba, James M.	Money, output, and prices: Evidence from a new monetary aggregate • Rotemberg,	95 JBES 13 67-83
Estimation of	monetary value of uninterrupted telephone service • Kimble, David L.	94 ASAProBuEc 202-204
J.; Driscoll, John C.; Poterba, James M.	Money, output, and prices: Evidence from a new monetary aggregate • Rotemberg, Julio	95 JBES 13 67-83
hypothesis of the long-run income velocity of	money and parameter stability of the equilibrium relationship • Raj, Baldev Institutional	95 JApEcmtx 10 233-253
and the chaotic structure of the velocity of	money • Serletis, Apostolos Random walks, breaking trend functions,	95 JBES 13 453-458
Bell curves and	monkey languages • Casti, John L.	95 Complxty 1 12-15
Theory and application of Marsaglia's	monkey test for pseudorandom number generators • Percus, Ora E.; Whitlock, Paula A.	95 ACMMdSim 5 87-100
behind combining ability and breeding values in	mono- and di-genic systems • Arunachalam, V. A genetic basis	95 BiomtrcJ 37 217-228
O'Neill, Robert T. Efficacy evaluation for	monotherapies in two-by-two factorial trials • Hung, H. M. James; Chi, George Y. H.;	95 Biomtrcs 51 1483-1493
Matrán, C. Consistency of L_p-best	monotone approximations • Cuesta-Albertos, J. A.; Domínguez-Menchero, J. S.;	95 JStPlInf 47 295-318
Efficient estimation of	monotone boundaries • Korostelev, A. P.; Simar, L.; Tsybakov, A. B.	95 AnlsStat 23 476-489
the first passage times for Markov processes with	monotone convex transition kernels • Li, Haijun; Shaked, Moshe On	95 StocProc 58 205-216
estimators for multivariate normal distribution with	monotone data • Fujisawa, Hironori A note on the maximum likelihood	95 CommStA 24 1377-1382
A minimax optimal estimator for continuous	monotone densities • Datta, Somnath	95 JStPlInf 46 183-193
On	monotone dependence functions of the quantile type • Krajka, A.; Szynal, D.	95 ZastApMa 23 51-72
Asymptotic inference on a general measure of	monotone dependence • Conti, Pier Luigi	94 JItalStS 3 213-241
Karunamuni, R. J.; Yang, H. On convergence rates of	monotone empirical Bayes tests for the continuous one-parameter exponential family •	95 StatDecn 13 181-192
analysis of proportional hazards models built from	monotone functions • Gelfand, Alan E.; Mallick, Bani K. Bayesian	95 Biomtrcs 51 843-852
of the stationary distribution of a stochastically	monotone Markov chain (French) • Simonot, F. Approximation	95 StocProc 56 133-149
Semilinear stochastic evolution equations with	monotone nonlinearities • Zangeneh, Bijan Z.	95 Stochatx 53 129-174
Haas, Peter J. (Rev) Review of	"Monotone structure in discrete-event systems" • Glasserman, Paul; Yao, David D. (Auth);	95 JASA 90 808-809

Frequency domain estimation of covariate effects in	multichannel brain evoked potential data • Raz, J.; Cardenas, V.; Fletcher, D.	95 Biomtrcs 51 448-460
Katsaggelos, Aggelos K. Regularized	multichannel restoration using cross-validation • Zhu, Wenwu; Galatsanos, Nikolas P.;	95 CVGIPGr 57 38-54
Tsoucas, P. Problems of adaptive optimization in	multiclass M/GI/1 queues with Bernoulli feedback • Bhattacharya, P. P.; Georgiadis, L.;	95 MaOpRsr 20 355-380
On positive Harris recurrence of	multiclass queueing networks: A unified approach via fluid limit models • Dai, J. G.	95 AnlsApPr 5 49-77
Fluid approximations and stability of	multiclass queueing networks: Work-conserving disciplines • Chen, Hong	95 AnlsApPr 5 637-665
stratified two-sample designs with application to	multiclinic trials • Brunner, Edgar; Puri, Madan L.; Sun, Shan Nonparametric methods for	95 JASA 90 1004-1014
	Multicollinearity	
Cheng; Mountain, Dean C.; Illman, Kathleen Ho	A Bayesian integration of end-use metering and conditional-demand analysis • Hsiao,	95 JBES 13 315-326
	Diagnostics statistics in QSAR • Mager, Peter P.	95 JChemtrc 9 211-221
estimator • Noor, Islam; Ahmad, Naseer	Exact and approximate relative efficiency of generalized ridge-type shrinkage	95 PakJStA 11 17-22
Jin, Chun; Hanumara, R. C.	Unbiased ridge estimation with prior information and ridge trace • Crouse, Robert H.;	95 CommStA 24 2341-2354
Belsley, Kuh and ... • Erkel-Rousse Detection of	multicollinearity in an ordinary linear model: Some elements of an informed use of the	95 RStApp 43/4 19-42
M. C. et al. Statistical study of peak overlapping in	multicomponent chromatograms: Importance of the retention pattern • Pietrogrande,	95 Chemolab 28 239-258
singular value decomposition based algorithm for	multicomponent exponential fitting of NMR relaxation signals • Lupu, M.; Todor, D. A	95 Chemolab 29 11-17
Detecting changes in probabilities of a	multi-component process • Khan, Rasul A.	95 SqtlAnly 14 375-388
tification of embedded minor analytes in three-way	multicomponent profiles by evolving projections and internal ... • Grung et al. ... and quan-	95 Chemolab 29 213-221
Rank mapping of three-way	multicomponent profiles • Grung, Bjørn; Kvalheim, Olav M.	95 Chemolab 29 223-232
Grung, Bjørn; Kvalheim, Olav M. Resolution of	multicomponent profiles with partial selectivity. A comparison of direct methods •	95 Chemolab 29 75-87
linearity and interaction in calibration with pure and	multicomponent spectra • Brown, Philip J. Diagnostics and selection for	95 Chemolab 28 139-148
Pearman, A. D.; Tejada, J. Fuzzy	multicriteria decision support for budget allocation in the transport sector • Montero, J.;	95 Top 3 47-68
variables • Duineveld, C. A. A.; Coenegracht, P. M. J.	Multicriteria steepest ascent in a design space consisting of both mixture and process	95 Chemolab 30 23-36
K. Bayesian method for estimating	multidimensional ability in item response analysis • Lee, Hyungseok; Tsutakawa, Robert	94 ASAProBays 7-12
proximate invariant designs for cubic regression on	multidimensional balls and cubes • Gaffke, N.; Heiligers, B. Computing optimal ap-	95 JStPlInf 47 347-376
Itô's formula for	multidimensional convex functions • Qingji, Yang	95 Stochatx 52 157-171
Estimates of some probabilities in	multidimensional convex records • Kałuszka, M.	95 ZastApMa 23 1-11
Exploring	multidimensional data with the flipped empirical distribution function • Huh, Moon Yul	95 JCmpGrSt 4 335-343
Improvement of some	multidimensional estimates by reduction of dimensionality • Ferré, Louis	95 JMultiAn 54 147-162
for pairwise independent random variables with	multidimensional indices • Hong, D. H.; Hwang, S. Y. Chow-Robbins "fair" games problem	95 BIMaASnc 23 7-11
sums of independent random variables with	multidimensional indices • Li et al. ... numbers and the law of the logarithm for weighted	95 JMultiAn 52 181-198
Optimal estimation of a finite	multidimensional population total • Mukerjee, Rahul; Sengupta, S.	95 JStPlInf 48 339-346
Absolute continuity of transition probabilities of	multidimensional processes with stationary independent increments • Yamazato, M.	94 ThProbAp 39 347-354
	Multidimensional random subshifts of finite type and their large deviations • Kifer, Y.	95 ProbTher 103 223-248
• Sowers, R. B.	Multidimensional reaction-diffusion equations with white noise boundary perturbations	94 AnlsProb 22 2071-2121
Rosenfeld, Barry; Turkheimer, Eric	Multidimensional representation of decision-making in chronic schizophrenics •	95 MultBehR 30 199-211
A comparative study of	multidimensional sampling plans • Dutta, K.; Goswami, V.	94 JIAgSt 46 254-266
	Multidimensional scaling	
Takane, Yoshio; Kiers, Henk A. L.; de Leeuw, Jan	Component analysis with different sets of constraints on different dimensions •	95 Psymtrka 60 259-280
Winsberg, Suzanne	Fitting an extended INDSCAL model to three-way proximity data • Carroll, J. Douglas;	95 JClass 12 57-71
Rosenfeld, Barry; Turkheimer, Eric	Multidimensional representation of decision-making in chronic schizophrenics •	95 MultBehR 30 199-211
	A new method for multidimensional unfolding • Hojo, Hiroshi	94 Behvmtrk 21 131-147
Storms, Gert	On the robustness of maximum likelihood scaling for violations of the error model •	95 Psymtrka 60 247-258
	Multidimensional scaling • Cox, Trevor F.; Cox, Michael A.	95 ChapmnHall:Lnd 232p
Smallman-Raynor, M. R. et al. The application of	multidimensional scaling methods to epidemiological data • Cliff, A. D.; Haggett, P.;	95 StMedRes 4 102-123
Heiser, Willem J. The majorization approach to	multidimensional scaling for Minkowski distances • Groenen, Patrick J. F.; Mathar, Rudolf;	95 JClass 12 3-19
preference data" (Spanish) • Arce, C. et al. Review of	"Multidimensional scaling. A multivariate technique for the analysis of proximity and	95 JClass 12 137-139
	Multidimensional scaling of a three-way table (French) • Dosse, M. Bennani	95 RStApp 43/4 63-75
finding optimal conditions in the search area of the	multidimensional solvent space • Kaufmann ... liquid chromatography. II. Exploring and	95 Chemolab 27 105-114
Calculation of	multidimensional stable densities • Nolan, John P.; Rajput, Balram	95 CommStB 24 551-566
A new method for	multidimensional unfolding • Hojo, Hiroshi	94 Behvmtrk 21 131-147
On crossing times for	multidimensional walks with skip-free components • Borovkov, K. A.	95 JAppProb 32 991-1006
Dynamic component detection in a	multifactor model for stock returns • Costa, Michele	94 JItalStS 3 25-36
"Random geometrically graph directed self-similar	multifractals" • Olsen, L. (Auth); Taylor, S. J. (Rev) Review of	95 Metrika 42 277-278
Review of "Random graph directed self-similar	multifractals" • Olsen, Lars (Auth); Patzschke, Norbert (Rev)	95 BiomtrcJ 37 374-374
constraints and intratemporal uncertainty in a	multi-good life-cycle model • Adang, Pim; Melenberg, Bertrand Nonnegativity	95 JApEcmtx 10 1-15
of size and costs of an epidemic in a closed	multigroup population • Svensson, Åke On the simultaneous distribution	95 MaBisci 127 167-180
purpose model for image operations based on	multilayer perceptrons • Moh, Jenlong; Shih, Frank Y. A general	95 PattrnRc 28 1083-1090
discriminant analysis, ... • Sánchez et al. Efficiency of	multi-layered feed-forward neural networks on classification in relation to linear	95 Chemolab 28 287-303
Robustness analysis of radial base function and	multi-layered feed-forward neural network models • Derks, E. P. P. A.; Pastor, M. S. S. et al.	95 Chemolab 28 49-60
Maximum likelihood estimation for unbalanced	multilevel covariance structure models via the EM algorithm • Raudenbush, Stephen W.	95 BJMaStPs 48 359-370
Hierarchical linear models: A	multilevel data analysis technique • Wu, Yow-Wu B.	95 NurseRes 44 123-126
Questioning	multilevel models (Disc: p201-227, 239-240) • de Leeuw, Jan; Kreft, Ita G. G.	95 JEdStat 20 171-189
An assessment of estimation procedures for	multilevel models with binary responses • Rodríguez, Germán; Goldman, Noreen	95 JRSS-A 158 73-89
Blind deconvolution of linear systems with	multilevel nonstationary inputs • Li, Ta-Hsin	95 AnlsStat 23 690-704
	Multilevel statistical models (Second edition) • Goldstein, Harvey	95 Arnold:Lnd:MD 192p
Review of	"Multilevel statistical models" • Goldstein, Harvey (Auth); Hand, D. J. (Rev)	95 ShrtBkRv 15/3 43-44
Jerzy Contraction and decoupling inequalities for	multilinear forms and U-statistics • de la Peña, V. H.; Montgomery-Smith, S. J.; Szulga,	94 AnlsProb 22 1745-1765
Inference about the parameters of the	multilinear model with equi-correlated responses • Khan, Shahjahan	94 JStatRsr 28 163-176
Drava, Giuliana Correct and incorrect use of	multilinear regression • Sergent, Michelle; Mathieu, Didier; Phan-Tan-Luu, Roger;	95 Chemolab 27 153-162
convergent and discriminant validity in a multitrait-	multimethod matrix • Reichardt, Charles S.; Coleman, S. C. The criteria for	95 MultBehR 30 513-538
Holmquist, Björn Simulation of	multimodal circular distributions of wrapped Cauchy type and of von Mises type •	95 CommStB 24 607-615
	Multimodal convolutions of unimodal infinitely divisible distributions • Sato, K.	94 ThProbAp 39 336-347
Finman, Jeffrey S. Moving from	multinational clinical trials to supra-trials • Accomando, William P.; Trost, Donald C.;	94 ASAProBiop 232-237
	Multinomial distribution	
Handley, J.	Comparison of three estimators of the number of species • Bunge, J.; Fitzpatrick, M.;	95 JAppStat 22 45-59
Lipsitz, S. R.; Fitzmaurice, G. M.; Sleeper, L. et al.	Estimation methods for the joint distribution of repeated binary observations •	95 Biomtrcs 51 562-570
Payne, C. D.	Evaluation of methods for ecological inference • Cleave, N.; Brown, P. J.;	95 JRSS-A 158 55-72
Hwang, F. K.	A majorization theorem for the number of distinct outcomes in an urn model •	93 JCISS 18 26-30
Faraggi, David; Simon, Richard	The maximum likelihood neural network as a statistical classification model •	95 JStPlInf 46 93-104
model • van Duijn, Marijtje A. J.; Jansen, Margo G. H.	Modeling repeated count data: Some extensions of the Rasch Poisson counts	95 JEdStat 20 241-258
chi-squared goodness-of-fit statistics: ... • Formann	Small-sample comparison of the exact and asymptotic upper tail probabilities of	95 JStCmpSm 51 369-384
	Some remarks on the chi-squared test with both margins fixed • Park, Cheolyong	95 CommStA 24 653-661
On a Markov	multinomial distribution • Wang, Y. H.; Yang, Zejiang	95 MaScitst 20 40-49
Extension of a negative	multinomial model • Dhar, Sunil K.	95 CommStA 24 39-57
Recursively simulating	multinomial multiperiod probit probabilities • Geweke, John; Keane, Michael	94 ASAProBuEc 1-6

dard deviation of the logistic distribution based on	multiply type-II censored samples • Balakrishnan, N. et al. Estimation of the mean and stan-	95 MaOpfStS 27 127-142
eters of the extreme value distribution based on	multiply type-II censored samples • Balakrishnan et al. ... of the location and scale param-	95 CommStA 24 2105-2125
distribution and extreme-value distribution under	multiply type-II censoring • Fei, H.; Kong, F.; Tang, Y. Estimation for two-parameter Weibull	95 CommStA 24 2087-2104
Bayesian inference in	multipoint gene mapping • Stephens, D. A.; Smith, A. F. M.	93 AnHuGene 57 65-82
Cherny, S. S.; Cardon, L. R.	Multipoint interval mapping of quantitative trait loci, using sib pairs • Fulker, D. W.;	95 AJHuGene 56 1224-1233
Model misspecification and	multipoint linkage analysis • Risch, Neil; Giuffra, Luis	92 HumHrdty 42 77-92
mapping • Kruglyak, L.; Daly, M. J.; Lander, E. S. Rapid	multipoint linkage analysis of recessive traits in nuclear families, including homozygosity	95 AJHuGene 56 519-527
dichotomous outcomes • Olson, Jane M.	Multipoint linkage analysis using sib pairs: An interval mapping approach for	95 AJHuGene 56 788-798
for combining two-point lod scores into a summary	multipoint map • Curtis, David; Gurling, Hugh A procedure	93 HumHrdty 43 173-185
Ott, Jurg	Multipoint mapping under genetic interference • Weeks, Daniel E.; Lathrop, G. Mark;	93 HumHrdty 43 86-97
Lander, Eric S. Complete	multipoint sib-pair analysis of qualitative and quantitative traits • Kruglyak, Leonid;	95 AJHuGene 57 439-454
	Multi-process control charts • Lipscomb, John R.	94 ASAProPEng 314-319
The	multiprocess dynamic Poisson model • Bolstad, William M.	95 JASA 90 227-232
Expected deadlock time in a	multiprocessing system • Compton, Kevin J.; Ravishankar, Chinya	95 JACM 42 562-583
An algorithm to compute blocking probabilities in	multi-rate multi-class multi-resource loss models • Choudhury, G. L.; Leung, K. K. et al.	95 AdvAppPr 27 1104-1143
Agreement models and probabilities for	multiraters • Yu, Xujie	94 ASAProBiop 279-284
	Multirelation - A correlation among more than two variables • Drezner, Zvi	95 CmpStDA 19 283-292
On the PBB designs for	multiresponse experiments • Brzeskwiniewicz, Henryk	95 BiomtrcJ 37 191-198
Curvature measures for	multiresponse regression models • Guay, M.	95 Biomtrka 82 411-417
	Multiscaling and skew separation in regional floods • Dawdy, David R.; Gupta, Vijay K.	95 WaterReR 31 2761-2767
	Multiserver queue	
method • Kriman, Vladimir	Analysis of the sensitivity of the GI/GI/m/B queue to buffer size: The score function	95 CommStSM 11 171-194
queue lengths and waiting times in a heterogeneous	multiserver queue II: Positive recurrence and logarithmic limits • Sadowsky ... of large	95 AdvAppPr 27 567-583
Light-traffic approximations for Markov-modulated	multi-server queues • Błaszczyszyn, Bartłomiej; Frey, Andreas; Schmidt, Volker	95 CommStSM 11 423-445
queue lengths and waiting times in a heterogeneous	multiserver queue I: Tight limits • Sadowsky, J. S.; Szpankowski, W. The probability of large	95 AdvAppPr 27 532-566
certain asymptotic properties of waiting time in a	multiserver queueing system with identical service times • Vinogradov, O. P. On	94 ThProbAp 39 714-718
of Moolgavkar, Venzon, and Knudson, and of the	multistage model of Armitage and Doll • Little ... two-mutation model of carcinogenesis	95 Biomtrcs 51 1278-1291
	Multistage process	
	Lot sizes in serial manufacturing with random yields • Sobel, Matthew J.	95 PrEngInf 9 151-157
Mukhopadhyay, N. et al. Review of	"Multistage selection and ranking procedures: Second-order asymptotics" •	95 JASA 90 1130-1131
A stochastic model for	multistage tumorigenesis in developing and adult mice • Mao, Jian-Hua; Wheldon, Tom E.	95 MaBisci 129 95-110
Nichols, James D.; Kendall, William L. The use of	multi-state capture-recapture models to address questions in evolutionary ecology •	95 JAppStat 22 835-846
Gottschau et al. Interval censored survival data and	multistate compartmental models in the analysis of first appearance of Plasmodium ... •	95 StatMed 14 2727-2736
Jones, Richard H. MARKOV: A computer program for	multi-state Markov models with covariates • Marshall, Guillermo; Guo, Wensheng;	95 CmpBiomd 47 147-156
program fitting proportional odds model for	multi-state Markov process • Guo, Wensheng; Marshall, Guillermo ORDMKV: A computer	95 CmpBiomd 46 257-263
Stallard, Nigel; Whitehead, John A parametric	multistate model for the analysis of carcinogenicity experiments • Omar, Rumana Z.;	95 LiDaAnal 1 327-346
	Multi-state models and diabetic retinopathy • Marshall, Guillermo; Jones, Richard H.	95 StatMed 14 1975-1983
p192-199) • Urfer, W.; Schwarzenbach, F. H. et al.	Multistate models for monitoring individual trees in permanent observation plots (Disc:	94 EnvEcoSt 1 171-192
ing nest site fidelity of adult female black brant with	multi-state modeling and geographic information systems • Lindberg, M. S. et al. Estimat-	95 JAppStat 22 725-735
criteria for convergent and discriminant validity in a	multitrait-multimethod matrix • Reichardt, Charles S.; Coleman, S. C. The	95 MultBehR 30 513-538
deviation rates for branching processes. II. The	multitype case • Athreya, K. B.; Vidyashankar, A. N. Large	95 AnlsApPr 5 566-576
Lee, Chinsan; Yang, Grace L. A	multitype decomposable age-dependent branching process and its applications •	95 JAppProb 32 591-608
Limit theorems for	multitype epidemics • Andersson, Håkan; Djehiche, Boualem	95 StocProc 56 57-75
Nonlinear master equation of	multitype particle systems • Feng, Shui	95 StocProc 57 247-271
of the inverse and classical estimators in	multi-univariate linear calibration • Srivastava, M. S. Comparison	95 CommStA 24 2753-2767
Shahzad, Naseer Random fixed points for random	multivalued operators defined on unbounded sets in Banach spaces • Beg, Ismat;	95 StocAnAp 13 269-277
Yosida approximations for	multivalued stochastic differential equations • Pettersson, Roger	95 Stochatx 52 107-120
(Auth); Papp, Zoltan (Rev) Review of	"Multi-vari chart and analysis - A pre-experimentation technique" • Perez Wilson, Mario	95 Technmcs 37 121-121
The	multi-vari chart: An underutilized quality tool • Zaciewski, Robert D.; Németh, Lou	95 QulPr 28/10 81-83
Nonlinear time series models for	multivariable dynamic processes • Çinar, Ali	95 Chemolab 30 147-158
Sternad, Mikael A probabilistic approach to	multivariable robust filtering and open-loop control • Öhrn, Kenth; Ahlén, Anders;	95 IEEEAuCn 40 405-418
Global total least squares modeling of	multivariable time series • Roorda, Berend; Heij, Christiaan	95 IEEEAuCn 40 50-63
A comparison of methods for univariate and	multivariate acceptance sampling by variables • Hamilton, David C.; Lesperance, Mary L.	95 Technmcs 37 329-339
An introduction to	multivariate adaptive regression splines • Friedman, Jerome H.; Roosen, Charles B.	95 StMedRes 4 197-217
III. The use of ... • Alsberg Speed improvement of	multivariate algorithms by the method of postponed basis matrix multiplication. Part	95 Chemolab 30 223-225
	Multivariate analysis	
components • Donnell, D. J.; Buja, A.; Stuetzle, W.	Analysis of additive dependencies and concurvities using smallest additive principal	94 AnlsStat 22 1635-1673
likelihood estimators • Taniguchi, M.; Puri, M. L.	Higher order asymptotic theory for normalizing transformations of maximum	95 AnInStMa 47 581-600
Jones, B.; Morgan, B. J. T.	Identifying influential observations in hierarchical cluster analysis • Jolliffe, I. T.;	95 JAppStat 22 61-80
Cadima, Jorge; Jolliffe, Ian T.	Loadings and correlations in the interpretation of principal components •	95 JAppStat 22 203-214
Kiers, Henk A. L.	Maximization of sums of quotients of quadratic forms and some generalizations •	95 Psymtrka 60 221-245
Despotis, D. K.; Stavropoulou, E.	Multiattribute evaluation of Greek banking performance • Zopounidis, C.;	95 AplStMDA 11 97-107
discriminant analysis • Krzanowski, W. J.	Selection of variables, and assessment of their performance, in mixed-variable	95 CmpStDA 19 419-431
invariance problem • Bove, Giuseppe	Some methods for the simultaneous analysis of data matrices and the factorial	94 Metron 52/3 73-87
Govind S.; Srivastava, Deo Kumar; Lin, C. Thomas	Some p-variate adaptations of the Shapiro-Wilk test of normality • Mudholkar,	95 CommStA 24 953-985
data (Italian) • Egidi, V.; Sabbadini, L. L.; Zaccarin, S.	Woman roles between family and employment: Empirical evidence from individual	94 Statstca 54 411-433
	Multivariate analysis and its applications • Anderson, T. W.; Fang, K. T.; Olkin, I. (Ed)	95 Arnold:Lnd:MD 471p
Van de Geer, J. P. (Auth); Fox, R. (Rev) Review of	"Multivariate analysis of categorical data: Theory and applications. Volumes 1 and 2" •	95 JMarRsr 32 111-112
Van de Geer, J. P. (Auth); Rose, E. L. (Rev) Review of	"Multivariate analysis of categorical data: Theory and applications. Volumes 1 and 2" •	95 StrEqMod 2 274-276
Recent advances in descriptive	multivariate analysis • Krzanowski, W. J. (Ed)	95 OxfordUniv 376p
Review of "Recent advances in descriptive	multivariate analysis" • Krzanowski, W. (Ed); Hand, D. J. (Rev)	95 ShrtBkRv 15/3 44-44
neurobehavioral phenotypes • Crusio, Wim E. Bi- and	multi-variate analyses of diallel crosses: A tool for the genetic dissection of	93 BehavGen 23 59-68
Applications of risk theory and	multivariate analysis in insurance practice • Niggemeyer, B.; Radtke, M.; Reich, A.	95 AplStMDA 11 231-244
Methods of	multivariate analysis • Rencher, Alvin C.	95 Wiley:NY:UK 627p
Review of "Least squares optimization in	multivariate analysis" • Berge, Jos M. F. ten (Auth); Takane, Yoshio (Rev)	95 Psymtrka 60 447-448
Modified phi correlation coefficients for the	multivariate analysis of ordinally scaled variables • Hammond, Sean M.; Lienert, Gustav A.	95 EdPsMeas 55 225-236
Marriott, F. H. C. (Auth) et al. Review of	"Multivariate analysis Part 1: Distributions, ordination and inference" • Krzanowski, W. J.;	95 Technmcs 37 351-351
J.; Marriott, F. H. C. (Auth) et al. Review of	"Multivariate analysis, Part 1: Distributions, ordination, and inference" • Krzanowski, W.	95 JASA 90 1490-1490
measurements • Krzanowski, W. J.; Marriott, F. H. C.	Multivariate analysis. Part 2: Classification, covariance structures and repeated	95 Arnold:Lnd:MD 280p
structures • Thioulouse, J.; Chessel, D.; Champely, S.	Multivariate analysis of spatial patterns: A unified approach to local and global	95 EnvEcoSt 2 1-14
	Multivariate analysis of variance	
	The analysis of repeated measures: A practical review with examples • Everitt, B. S.	95 Ststcian 44 113-135
Wong, Chi Song	Cochran theorems for a multivariate elliptically contoured model • Wang, Tonghui;	95 JStPlInf 43 257-270
	Residuals in the growth curve model • von Rosen, Dietrich	95 AnInStMa 47 129-136

The polynomial model in the	multivariate analysis of variance • Baumgartner, Cornelia	95 BiomtrcJ 37 299-318
Identification of	multivariate AR-models by threshold accepting • Winker, Peter	95 CmpStDA 20 295-307
autoregressive conditions for noncausality in	multivariate ARMA models • Dufour, Jean-Marie; Nsiri, Saïd; Tessier, David Parsimonious	94 ASAProBuEc 129-134
Local asymptotic normality of	multivariate ARMA processes with a linear trend • Garel, Bernard; Hallin, Marc	95 AnInStMa 47 551-579
The order determination methods for a	multivariate autoregressive model (Polish) • Koziara, Jan	95 ListyBio 32 1-14
cation to rotational ... • Shin et al. Estimation of the	multivariate autoregressive moving average having parameter restrictions and an appli-	95 JTimSrAn 16 431-444
Almahmeed, M. A. Bayesian classification with	multivariate autoregressive sources that might have different orders • Shaarawy, S. M.;	95 CommStB 24 567-581
at-a-time • Gersch, Will; Stone, David	Multivariate autoregressive time series modeling: One scalar autoregressive model	95 CommStA 24 2715-2733
Exact	multivariate Bayesian bootstrap distributions of moments • Gasparini, Mauro	95 AnlsStat 23 762-768
Simulation results on solutions to the	multivariate Behrens-Fisher problem via trimmed means • Wilcox, Rand R.	95 Ststcian 44 213-225
Comment on "An extension of Yule's Q to	multivariate binary data" (V50 p847-852) • de Falguerolles, Antoine	95 Biomtrcs 51 1184-1185
to "Comment on 'An extension of Yule's Q to	multivariate binary data'" (V51 p1184-1185) • de Falguerolles, A. Correction	95 Biomtrcs 51 1589-1589
Reply to comment on "An extension of Yule's Q to	multivariate binary data" • Lipsitz, Stuart R.; Fitzmaurice, Garrett	95 Biomtrcs 51 1185-1185
concerning independence estimating equations with	multivariate binary data • Fitzmaurice, Garrett M. A caveat	95 Biomtrcs 51 309-317
Marginal regression analysis of a	multivariate binary response • Ekholm, Anders; Smith, Peter W. F.; McDonald, John W.	95 Biomtrka 82 847-854
Outlier detection in	multivariate calibration • Walczak, B.	95 Chemolab 28 259-272
analysis algorithm for treatment of non-linearity in	multivariate calibration • Xie, Y.-L.; Liang, Y.-Z.; Chen, Z.-G. et al. A nonlinearity tracking	95 Chemolab 27 21-32
C. N. Reference priors for the	multivariate calibration problem • Plessis, J. L. du; van der Merwe, A. J.; Groenewald, P.	95 SoAfrStJ 29 155-168
of a study's results: Bootstrap statistics for the	multivariate case • Thompson, Bruce Exploring the replicability	95 EdPsMeas 55 84-94
Gange, Stephen J. Generating	multivariate categorical variates using the iterative proportional fitting algorithm •	95 AmerStat 49 134-138
On some central and non-central	multivariate chi-square distributions • Royen, Thomas	95 StSinica 5 373-397
Cubanski, David; Cyganski, David	Multivariate classification through adaptive Delaunay-based C^0 spline approximation •	95 IEEEPAMI 17 403-417
P.; Shoesmith, Gary L. Modelling	multivariate cointegrated systems: Insights from non-linear dynamics • Pinder, Jonathan	95 JForcast 14 311-324
Eriksen, P. S. Linear and graphical models for the	multivariate complex normal distribution • Andersen, H. H.; Højberre, M.; Sørensen, D.;	95 Springer:Brln:NY 183p
A Bayesian approach to	multivariate and conditional calibration • Plessis, J. L. du; van der Merwe, A. J.	95 CmpStDA 19 539-552
R.; Becker, Mark P.	Multivariate contingency tables and the analysis of exchangeability • Ten Have, Thomas	95 Biomtrcs 51 1001-1016
Decomposition of T^2 for	multivariate control chart interpretation • Mason, Robert L.; Tracy, Nola D.; Young, John C.	95 JQualTek 27 99-108
Small sample uni- and	multi-variate control charts for means • Scholz, F. W.; Tosch, T. J.	94 ASAProQlPr 17-22
for determining sample sizes of the operations of	multivariate control charts • Jolayemi, Joel K. A power function model	95 CmpStDA 20 633-641
A review of	multivariate control charts • Lowry, Cynthia A.; Montgomery, Douglas C.	95 IIETrn 27 800-810
local rank, three-way data analysis and ambiguity in	multivariate curve resolution • Tauler, Romá; Smilde, Age; Kowalski, Bruce Selectivity,	95 JChemtrc 9 31-58
	Multivariate curve resolution applied to second order data • Tauler, Roma	95 Chemolab 30 133-146
scopic titrations ... • Tauler et al. Application of a new	multivariate curve resolution procedure to the simultaneous analysis of several spectro-	95 Chemolab 27 163-174
Applied	multivariate data analysis • Everitt, Brian S.; Dunn, Graham	95 Wiley:NY:UK 304p
Mercante, Donald E. Analyzing	multivariate data in crossover designs using permutation tests • Johnson, William D.;	94 ASAProBiop 244-249
new family of goodness-of-fit statistics for discrete	multivariate data • Lorenzen, Gunter A	95 StPrLet 25 301-307
in the presence of group structure, for general	multivariate data" (VII p195-207) • Krzanowski, W. J. Correction to "Ordination	95 JClass 12 2-2
On Rohlf's method for the detection of outliers in	multivariate data • Caroni, C.; Prescott, P.	95 JMultiAn 52 295-307
Analysis of incomplete	multivariate data by simulation • Schafer, Joseph L.	95 ChapmnHall:Lnd 300p
Diagnostics and robust estimation in	multivariate data transformations • Velilla, Santiago	95 JASA 90 945-951
Review of	"Multivariate density estimation" • Scott, D. W. (Auth); Silverman, B. W. (Rev)	95 Metrika 42 66-68
Practical estimation of	multivariate densities using wavelet methods • Tribouley, K.	95 StNeerla 49 41-62
• Granville, Vincent; Rasson, Jean Paul	Multivariate discriminant analysis and maximum penalized likelihood density estimation	95 JRSS-B 57 501-517
	Multivariate dispersion orderings • Giovagnoli, Alessandra; Wynn, H. P.	95 StPrLet 22 325-332
Balasubramanian, K.; Balakrishnan, N. On a class of	multivariate distributions closed under concomitance of order statistics •	95 StPrLet 23 239-242
A strategy for constructing	multivariate distributions • Chakak, Abderrahmane; Koehler, Kenneth J.	95 CommStB 24 537-550
points and self-consistent points of symmetric	multivariate distributions • Tarpey, Thaddeus Principal	95 JMultiAn 53 39-51
Symanowski, James T. Constructing	multivariate distributions with specific marginal distributions • Koehler, Kenneth J.;	95 JMultiAn 55 261-282
Cochran theorems for a	multivariate elliptically contoured model • Wang, Tonghui; Wong, Chi Song	95 JStPlInf 43 257-270
Volume 6" • Patil, G. P.; Rao, C. R. (Ed) et al. Review of	"Multivariate environmental statistics. North-Holland Series in Statistics and Probability	95 BiomtrcJ 37 255-256
Review of	"Multivariate environmental statistics" • Patil, G. P.; Rao, C. R. (Ed); Seber, G. A. F. (Rev)	95 Biomtrcs 51 1189-1190
The behavior of the Stahel-Donoho robust	multivariate estimator • Maronna, Ricardo A.; Yohai, Victor J.	95 JASA 90 330-341
A subsampling method for the computation of	multivariate estimators with high breakdown point • Juan, Jesus; Prieto, Francisco J.	95 JCmpGrSt 4 319-334
	Multivariate evaluation of feed for Atlantic halibut • Nortvedt, Ragnar; Tuene, Stig	95 Chemolab 29 271-282
Approximation rates for	multivariate exceedances • Kaufmann, E.; Reiss, R.-D.	95 JStPlInf 45 235-245
Marshall, Albert W.; Olkin, Ingram	Multivariate exponential and geometric distributions with limited memory •	95 JMultiAn 53 110-125
equation for the in-control average run length of a	multivariate exponentially weighted moving average control chart • Rigdon An integral	95 JStCmpSm 52 351-365
integral equation for the average run length of a	multivariate exponentially weighted moving average control chart • Rigdon, S. E. A double-	95 StPrLet 24 365-373
for a bivariate lifetime distribution in reliability with	multivariate extensions • Heinrich, G.; Jensen, U. Parameter estimation	95 Metrika 42 49-65
Fisher information for a	multivariate extreme value distribution • Shi, Daoji	95 Biomtrka 82 644-649
A semi-parametric model for	multivariate extreme values • Dixon, Mark J.; Tawn, Jonathan A.	95 StatComp 5 215-225
recent developments for regression analysis of	multivariate failure time data • Liang, K.-Y.; Self, S. G.; Bandeen-Roche, K. J. et al. Some	95 LiDaAnal 1 403-415
estimation procedure of dependence parameters in	multivariate families of distributions • Genest, C.; Ghoudi, K. et al. A semiparametric	95 Biomtrka 82 543-552
The case of the United States and Canada • Karolyi A	multivariate GARCH model of international transmissions of stock returns and volatility:	95 JBES 13 11-25
Khursheed; Williams, Calvin L. A	multivariate goodness-of-fit test for stochastically ordered distributions • Alam,	95 BiomtrcJ 37 945-956
ference for the association between coefficients in a	multivariate growth curve model • Zucker, David M.; Zerbe, Gary O.; Wu, Margaret C. In-	95 Biomtrcs 51 413-424
Nordin, B. An extended strategy for exploratory	multivariate image analysis including noise considerations • Pedersen, F.; Bengtsson, E.;	95 JChemtrc 9 389-409
pollution • Duddek, C.; Le, N. D.; Zidek, J. V. et al.	Multivariate imputation in cross-sectional analysis of health effects associated with air	95 EnvEcoSt 2 191-212
Refined approximations to permutation tests for	multivariate inference • Kazi-Aoual, F.; Hitier, S.; Sabatier, R.; Lebreton, J.-D.	95 CmpStDA 20 643-656
conditions • Ritter, K.; Wasilkowski, G. W. et al.	Multivariate integration and approximation for random fields satisfying Sacks-Ylvisaker	95 AnlsApPr 5 518-540
Ibrahim A. On	multivariate kernel estimation for samples from weighted distributions • Ahmad,	95 StPrLet 22 121-129

Multivariate linear model

Garel, Bernard; Hallin, Marc	Local asymptotic normality of multivariate ARMA processes with a linear trend •	95 AnInStMa 47 551-579
struction of improved estimators in multiple-design	multivariate linear models under general restriction • Shiraishi, T.; Konno, Y. On con-	95 AnInStMa 47 665-674
Goodness-of-fit for	multivariate linear regression models • Haroun, Ahmed H.	94 EgpStJ 38 138-146
case-deletion influence measures in univariate and	multivariate linear regression • Hadi, A. S. et al. A unifying representation of some	95 JStPlInf 46 123-135
A note on a unified test for	multivariate linear relationships • Isogawa, Yoshiko	95 StPrLet 24 147-151
	Multivariate Liouville distributions, IV • Gupta, Rameshwar D.; Richards, Donald St. P.	95 JMultiAn 54 1-17
signed rank statistics for testing and estimating a	multivariate location parameter • Hössjer, Ola; Croux, Christophe Generalizing univariate	95 JNnparSt 4 293-308
	Multivariate logistic models • Glonek, G. F. V.; McCullagh, P.	95 JRSS-B 57 533-546
rank tests for a completely randomized design with	multivariate longitudinal responses and a randomized complete block ... • Xie et al. The	94 ASAProBiop 250-255
	Multivariate l_p-norm symmetric distributions • Yue, Xinnian; Ma, Chunsheng	95 StPrLet 24 281-288
Uncertainty bounds for	multivariate M-estimates in small samples • Spall, James C.	94 ASAProPEng 126-131
F. et al. Analyzing long irregularly spaced repeated	multivariate measures – A study of infant cardiac catherization stress responses • Zhang,	94 ASAProBiop 365-370

	Multivariate tests for multiple endpoints in clinical trials • Follmann, Dean	95 StatMed 14 1163-1175
of repeated measurements: Univariate tests,	multivariate tests, or both? • Keselman, H. J.; Keselman, Joanne C.; Lix, Lisa M. The analysis	95 BJMaStPs 48 319-338
Review of "Elements of	multivariate time series analysis" • Reinsel, C. (Auth); Lütkepohl, H. (Rev)	95 Metrika 42 147-148
Review of "Elements of	multivariate time series analysis" • Reinsel, Gregory C. (Auth); Timm, Neil H. (Rev)	95 Technmcs 37 118-118
long run? An example of likelihood inference in a	multivariate time-series model • Juselius ... and uncovered interest rate parity hold in the	95 JEconmtx 69 211-240
A	multivariate tolerance interpretation of assays of joint action • Frey, Michael	94 ASAProBiop 39-44
J. H.; Visser, T. Partial least squares regression as a	multivariate tool for the interpretation of infrared spectra • Luinge, H. J.; van der Maas,	95 Chemolab 28 129-138
Asymptotic normality of	multivariate trimmed means • Arcones, Miguel A.	95 StPrLet 25 45-53
Decoupling inequalities for the tail probabilities of	multivariate U-statistics • de la Peña, Victor H.; Montgomery-Smith, S. J.	95 AnlsProb 23 806-816
Puri, Madan L. A	multivariate Wald-Wolfowitz rank test against serial dependence • Hallin, Marc;	95 CandJSt 23 55-65
On	multivariate weighted distributions • Jain, Kanchan; Nanda, Asok K.	95 CommStA 24 2517-2539
The trait-state-error model for	multiwave data • Kenny, David A.; Zautra, Alex	95 JCCPsych 63 52-59
of log-multiplicative scaled association models in	multiway contingency tables • Wong, Raymond Sin-Kwok Extensions in the use	95 SocMethR 23 507-538
introduction to tensor products with applications to	multiway data analysis • Burdick, Donald S. An	95 Chemolab 28 229-237
and principles for data analysis II: Dot plots of	multiway data • Gunter, Bert Graphical methods	95 QulPr 28/6 88-95
MacGregor, John F.	Multi-way partial least squares in monitoring batch processes • Nomikos, Paul;	95 Chemolab 30 97-108
application of new models for analysing trends in	multiway tables • Payne, C.; Payne, J. et al. Testing the tightening bond hypothesis: An	95 JStPlInf 47 181-201
Propensity to abuse - Propensity to	murder? • Merz, Jon F.; Caulkins, Jonathan P.	95 Chance 8/2 14-14
research-estimation of the pH in postmortem	muscle • Plessis, J. L. du; van der Merwe, A. J. Bayesian calibration in meat science	95 JStCmpSm 53 1-11
Improved sequential	MUSIC • Stoica, Petre; Händel, Peter; Nehoral, Arye	95 IEEEAES 31 1230-1239
Mario Nonlinear modeling of	muskrat use of habitat • Nadeau, Simon; Décarie, Robert; Lambert, Daniel; St-Georges,	95 JWildlMg 59 110-117
Hines, W. G. S. Comment on	"Mutations, perturbations and evolutionarily stable strategies" (V19 p204-209) •	95 JAppProb 32 836-839
estimator of effective population size or	mutation rate • Fu, Yun-Xin A phylogenetic	94 Genetcs 136 685-692
quences • Fu Estimating effective population size or	mutation rate using the frequencies of mutations of various classes in a sample of DNA se-	94 Genetcs 138 1375-1386
of carcinogenesis of Moolgavkar, ... • Little Are two	mutations sufficient to cause cancer? Some generalizations of the two-mutation model	95 Biomtrcs 51 1278-1291
Bailer, A. John Statistical approaches for analyzing	mutational spectra: Some recommendations for categorical data • Piegorsch, Walter W.;	94 Genetcs 136 403-416
On the optimality of adaptive expectations:	Muth revisited • Satchell, Steve; Timmermann, Allan	95 IntJFor 11 407-416
Muthén et al. Technical aspects of	Muthén's LISCOMP approach to estimation of latent variable relations with a ... •	95 Psymtrka 60 489-503
The role of geometry in pairwise and	mutual independence • Schwertman, Neil C.; Kiser, Terry L.	95 AmerStat 49 56-59
Sets of	mutually orthogonal Latin squares with "like subsquares," II • Roberts, Charles E., Jr	95 JCombThA 70 159-164
	***MV*-optimality**	
Mukerjee, Rahul	On optimal designs with restricted circular string property • Huda, S.; Khan, I. H.;	95 CmpStDA 19 75-83
	MVP, LVP, and PGP: A statistical analysis of Toronto in the World Series • Bennett, Jay M.	94 ASAProSpts 66-71
of random coefficient regression model to	myopia data: A case study • Möttönen, J.; Oja, H.; Krause, U.; Rantakallio, P. Application	95 BiomtrcJ 37 657-672
Graphical	mysteries • Wainer, Howard	95 Chance 8/3 52-56
of cards: Psychology and psychotherapy built on	myth" • Dawes, Robyn M. (Auth); Sherman, Malcolm J. (Rev) Review of "House	95 JASA 90 1492-1492
Iterative solution of the	$M^{[x]}/C_2/S/N$ queues • Moustafa, Magdi S.; El-Sayed, Khaled M. F.	94 AplMaCmp 66 313-323
An	N-class structured priority queue with vacations • Katsaros, A.; Langaris, C.	95 CommStSM 11 235-248
Adaptive root	n estimates of integrated squared density derivatives • Wu, Tiee-Jian	95 AnlsStat 23 1474-1495
for analyzing the probability of M-out-of-	N events • Haskin, F. Eric; Radke, George E., Jr; Evanoff, Javon Algorithms	95 RelEngSS 47 153-162
formulas for linear and circular consecutive-k-of-	n: F systems • Peköz, Erol A.; Ross, Sheldon M. A simple derivation of exact reliability	95 JAppProb 32 554-557
The indices of the largest among	n independent observations • Weissman, Ishay	95 CommStSM 11 613-629
of occurrences of consecutive k successes in	n Markovian trials • Wang, Y. H.; Ji, Shuixin Limit theorems for the number	95 JAppProb 32 727-735
data - Modified Q and correction for modified	N • Shah, S.; Green, J. R. Testing for haplotype concordance with incomplete parental	93 AnHuGene 57 239-247
Imputing income for an	N-person consumer unit • Chand, Nanak; Alexander, Charles H.	94 ASASrvyRsr 412-416
Analyzing unreplicated factorials using the	n smallest contrasts • Kasperski, William J.; Schneider, Helmut; Johnson, Keith L.	95 JQualTek 27 377-386
The power of the	N-test of haplotype concordance • Green, J. R.; Shah, S.	92 AnHuGene 56 331-338
An algorithm for fitting models to	$N \times N$ contingency tables having ordered categories • Lawal, H. Bayo; Upton, G. J. G.	95 CommStB 24 793-805
Hirsch, F.; Song, S. Symmetric Skorohod topology on	n-variable functions and hierarchical Markov properties of n-parameter processes •	95 ProbTher 103 25-43
Optimal designs on the refinement of the	n-way classification model • Ting, C.-P.	93 UtilMath 43 141-154
Comment on "Environmental assessments in the	NAFTA era" • Bartelmus, Peter	94 ASAProEnvr 36-37
Stochastic	Nagumo's viability theorem • Aubin, J. P.; Prato, G. Da	95 StocAnAp 13 1-11
On variational inequality approach to	Nash dynamic game of oil stockpiling • Zhao, L.; Yin, G.	95 StocAnAp 13 627-649
	Nash equilibrium	
Zeephongsekul, P.; Chiera, C.	Optimal software release policy based on a two-person game of timing •	95 JAppProb 32 470-481
Yin, G.	On variational inequality approach to Nash dynamic game of oil stockpiling • Zhao, L.;	95 StocAnAp 13 627-649
to observer distribution in the analysis of data on	natal dispersal in birds • van Noordwijk, Arie J. On bias due	95 JAppStat 22 683-694
on response rates and field costs in a pretest of a	national CAPI survey • Duffer, A.; Lessler, J.; Weeks, M. et al. Effects of incentive payments	94 ASASrvyRsr 1386-1392
1994	National Census Test: Quantitative comparisons of coverage • Pausche, Joan M.	94 ASASrvyRsr 313-318
Kindel, Carrol Benner Electronic data collection at the	National Center for Education Statistics: Successes in the collection of library data •	94 JOffStat 10 93-102
Sample surveys at	National Center for Health Statistics • Choi, Jai Won; Curtin, Lester R.	94 ASASrvyRsr 1162-1167
Yosiyuki A statistical research of the Japanese	national character: General trend of opinion over past forty years (Japanese) • Sakamoto,	95 PrInStMa 43 5-26
plication to the data from the study of the Japanese	national character (Japanese) • Nakamura ... an over-dispersion parameter and their ap-	95 PrInStMa 43 99-119
structure analysis to the Survey of Japanese	National Character (Japanese) • Maeda ... of the Japanese - An application of covariance	95 PrInStMa 43 141-160
application to data from the Surveys of the Japanese	National Character (Japanese) • Tsuchiya ... differences among several data sets: Its	95 PrInStMa 43 121-140
to the Ninth Nationwide Survey of the Japanese	National Character (Japanese) • Komazawa et al. ... application of ordinal structure analysis	95 PrInStMa 43 161-176
Comparative study of	national character (Japanese) • Hayashi, Chikio; Hayashi, Fumi	95 PrInStMa 43 27-80
Overview of the 1994	National Employer Health Insurance Survey (NEHIS) • Hing, Esther	94 ASAProGovt 122-127
Bryant, Edward C.; Moriarity, Chris	National Employer Health Insurance Survey sample design • Marker, David A.;	94 ASASrvyRsr 264-269
The labor-market effects of introducing	national health insurance: Evidence from Canada • Gruber, Jonathan; Hanratty, Maria	95 JBES 13/2 163-173
and the number of allowable call attempts in the	National Health Interview Survey • Kalsbeek, W. D.; Botman, S. L. et al. Cost-efficiency	94 JOffStat 10 133-152
Botman, Steven L.; Jack, Susan S. Combining	National Health Interview Survey datasets: Issues and approaches (Disc: p679-693) •	95 StatMed 14 669-677
F. B. et al. Social correlates of response bias in the	National Health Interview Survey: Lessons for the redesign (Disc: p379-381) • LeClere,	94 ASASrvyRsr 373-378
Comment on "Redesign of the	National Health Interview Survey questionnaire" • Fowler, Floyd Jackson, Jr	94 ASASrvyRsr 379-381
Donald; Parsons, Van L. State estimates for the	National Health Interview Survey using a random digit dialing supplement • Malec,	94 ASASrvyRsr 1190-1195
Modeling of response propensity in the Third	National Health and Nutrition Examination Survey • Ezzati-Rice, Trena M.; Khare, Meena	94 ASASrvyRsr 955-959
Schutz, Robert W. Overtime in the	National Hockey League: Is it a valid tie-breaking procedure? • Liu, Yuanlong;	94 ASAProSpts 55-60
Small area estimation for the	National Household Survey of Drug Abuse • Folsom, Ralph E.; Liu, Jun	94 ASASrvyRsr 565-570
Richard H. Characterizing groundwater at Oak Ridge	National Laboratory hazardous waste sites • Wolf, Dennis A.; Tardiff, Mark F.; Ketelle,	94 ASAProEnvr 116-121
A	national lottery in the UK • Moore, Peter G.	94 JAppStat 21 607-622
effects of imputation on variance estimates for the	National Medical Expenditure Survey • Sommers, John Paul Analysis of the	94 ASASrvyRsr 400-405
Edmonston, B. et al. U.S. Population projections for	national origin groups: Taking into account ethnicity and exogamy (Disc: p122-123) •	94 ASAProSoSt 100-105
Krawchuk, S. The	National Population Health Survey: Design and issues • Singh, M. P.; Tambay, J. L.;	94 ASASrvyRsr 803-808
Ryaboy, Lana Sample design of the	National Pregnancy and Health Survey • DiGaetano, Ralph; Ward, John; Cadell, Diane;	94 ASASrvyRsr 1100-1104

Hassett, Nancy C.; Weir, Paula Comparison of	national and regional gasoline prices from two surveys • Saavedra, Pedro J.;	94 ASASrvyRsr 1179-1183
	National sample surveys for development planning - An overview • Seal, K. C.	94 CalcutSt 44 239-250
Kolstad, A. Composite estimation in	national and state surveys • Burke, J.; Mohadjer, L.; Green, J.; Waksberg, J.; Kirsch, I. S.;	94 ASASrvyRsr 873-878
government • Fienberg, S. E. et al. The Committee on	National Statistics: Fostering interactions between statisticians in academia and	95 IntStRvw 63 257-269
O'Reilly, James M. Results from the	National Survey of Family Growth cycle V pretest • Lessler, Judith T.; Weeks, Michael F.;	94 ASASrvyRsr 64-70
resulting from a sample size reduction in a large	national survey • Botman, Steven L. Estimation and analysis effects	94 ASASrvyRsr 1168-1172
Improving coverage in a	national survey of teachers • Royce, Daniel; Schreiner, Irwin	94 ASASrvyRsr 827-832
The role of VA administrative files in the	National Survey of Veterans • Ahuja, Elizabeth M.	94 ASAProGovt 25-27
An overview of the	National Survey of Veterans • Krumhaus, Susan G.; Klemm, Rebecca J.	94 ASAProGovt 19-21
of interviewer characteristics on response in a	national survey of violence against women • Norris, Douglas A.; Hatcher, Jeff The impact	94 ASASrvyRsr 449-454
Tiller, Richard; Welch, Michael Predicting the	national unemployment rate that the "old" CPS would have produced (Disc: p360-361) •	94 ASASrvyRsr 354-359
Privacy issues and the	National Voter Registration Act of 1993 • Montjoy, Robert S.	94 ASAProSoSt 170-173
Non-response to sensitive questions:	Nationalism in Germany • Krebs, Dagmar	94 ASASrvyRsr 1199-1203
application of ordinal structure analysis to the Ninth	Nationwide Survey of the Japanese National Character ... • Komazawa et al. ... Its	95 PrInStMa 43 161-176
The cognitive evaluation of the	nativity questions for the Current Population Survey • Wellens, Tracy	94 ASASrvyRsr 1204-1209
	Natural experiment	
van der Klaauw, Wilbert	Evaluating the cost of conscription in The Netherlands • Imbens, Guido;	95 JBES 13/2 207-215
On a characterization of variance functions of	natural exponential families • Bshouty, D.	95 MaMethSt 4 92-98
Conjugate parameterizations for	natural exponential families • Gutiérrez-Peña, E.; Smith, A. F. M.	95 JASA 90 1347-1356
Percentiles for the distributions belonging to the	natural exponential families having power variance functions (Korean) • Suh, Euy Hoon	95 KorApS 8/1 133-149
approximation • Martin, Clyde; Shubov, Victor	Natural exponential families of probability distributions and exponential-polynomial	93 AplMaCmp 59 275-297
Bansal, Naveen K. On selecting the best	natural exponential families with quadratic variance function • Abughalous, Mansour M.;	95 StPrLet 25 341-349
The application of semantic classification trees to	natural language understanding • Kuhn, Roland; De Mori, Renato	95 IEEEPAMI 17 449-460
additive probability: Uniform distributions on the	natural numbers • Kadane, Joseph B.; O'Hagan, Anthony Using finitely	95 JASA 90 626-631
	Natural and quasi-experiments in economics • Meyer, Bruce D.	95 JBES 13/2 151-161
Empirical model particularities and belief in the	natural rate hypothesis • Kim, Jinbang; De Marchi, Neil; Morgan, Mary S.	95 JEconmtx 67 81-102
tive for probability sampling: With examples from	natural resource sampling • Overton et al. The Horvitz-Thompson as a unifying perspec-	95 AmerStat 49 261-268
analysis. A practical approach for agricultural and	natural sciences • Hoshmand, A. Reza Experimental research design and	94 CRCPress:FL 416p
On modelling of experiments in	natural sciences • Mejza, Stanisław	94 ListyBio 31 79-100
Fiducial inference and the fundamental theorem of	natural selection • Edwards, A. W. F.	95 Biomtrcs 51 799-809
Callahan, A. J. Managing ship performance of	naval gunfire support using statistical process control • Bailey, M.; Bowden, J.;	94 MilOpRes 1 5-11
Martingale and stationary solutions for stochastic	Navier-Stokes equations • Flandoli, Franco; Gatarek, Dariusz	95 ProbTher 102 367-391
Foias and Hope statistical solutions for stochastic	Navier-Stokes equations • Capiński, Marek; Cutland, Nigel J.	95 Stochatx 52 193-205
of Ornstein-Uhlenbeck equations and stochastic	Navier-Stokes equations with a generalized noise • Flandoli, F. et al. Time discretization	95 Stochatx 55 141-165
A better beta model of	navigation • McLaren, I. P. L.	95 BJMaStPs 48 51-55
Shock models with	NBRU properties • Abouammoh, A. M.; Hendi, M. I.; Khalique, A.	95 PakJStA 11 35-47
M. How will the	NCAA's new standards affect minority student-athletes? • Klein, Stephen P.; Bell, Robert	95 Chance 8/3 18-21
Review of	"NCSS 6.0 for Windows" • NCSS Statistical Software (Auth); Lovie, Pat (Rev)	95 BJMaStPs 48 407-408
On	near ignorance classes • Sansó, B.; Pericchi, L. R.	94 RBrzlPrS 8 119-126
tions" • Hildrum, K.; Isaksson, T. et al. Review of	"Near infra-red spectroscopy: Bridging the gap between data analysis and NIR applica-	95 Technmcs 37 126-127
Pattern recognition analysis of	near-infrared spectra by robust distance method • Cho, Junghwan; Gemperline, Paul J.	95 JChemtrc 9 169-178
Application of Procrustean methods to mid- and	near-infrared spectral data • Vigneau, E.; Devaux, M. F.; Safar, M.	95 JChemtrc 9 125-135
Leardi, R. et al. Transfer of calibration function in	near-infrared spectroscopy • Forina, M.; Drava, G.; Armanino, C.; Boggia, R.; Lanteri, S.;	95 Chemolab 27 189-203
asymptotic distributions of some test statistics in	near-integrated AR processes • Larsson, Rolf The	95 EconTher 11 306-330
Holiday, David B.	Near optimal weights in nonparametric regression under some common restrictions •	95 StPrLet 22 33-42
The existence of doubly	near resolvable $(v, 3, 2)$-BIBDs • Lamken, E. R.	94 JComDsgn 2 427-440
Existence results for	near resolvable designs • Furino, Steven	95 JComDsgn 3 101-113
	Nearest neighbor	
fitting • Li, Gang; Doss, Hani	An approach to nonparametric regression for life history data using local linear	95 AnlsStat 23 787-823
	Area-interaction point processes • Baddeley, A. J.; van Lieshout, M. N. M.	95 AnInStMa 47 601-619
dependence • Boente, Graciela; Fraiman, Ricardo	Asymptotic distribution of smoothers based on local means and local medians under	95 JMultiAn 54 77-90
Gasser, Theo	A minimax result for a class of nonparametric density estimators • Engel, Joachim;	95 JNnparSt 4 327-334
Hadar I.; Van Mieghem, Jan A.; Rub, Leonardo	Multiple subclass pattern recognition: A maximin correlation approach • Avi-Itzhak,	95 IEEEPAMI 17 418-431
	A note on strong convergence rates in nonparametric regression • Cheng, Philip E.	95 StPrLet 24 357-364
Bai, Zhidong	Optimal strong convergence rates in nonparametric regression • Cheng, Ph. E.;	95 MaMethSt 4 405-420
Stroup, W. W.; Baenziger, P. S.; Mulitze, D. K.	Removing spatial variation from wheat yield trials: A comparison of methods •	94 CropSci 34 62-66
W. E. et al. Comparisons of a neural network and a	nearest-neighbor classifier via the numeric handprint recognition problem • Weideman,	95 IEEENrlN 6 1524-1530
Gradient descent learning of	nearest neighbor classifiers with outlier rejection • Urahama, K.; Furukawa, Y.	95 PattrnRc 28 761-768
law of the iterated logarithm for the multivariate	nearest neighbor density estimators • Ralescu, Stefan S. The	95 JMultiAn 53 159-179
D'Aristotile, Anthony J. The	nearest neighbor random walk on subspaces of a vector space and rate of convergence •	95 JTherPrb 8 321-346
approach • Bagui, S. C.; Mehra, K. L.; Rao, M. S. A	nearest neighbour classification rule for multiple observations based on a sub-sample	95 SankhyaA 57 316-332
• Dette et al. The impact of different definitions of	nearest neighbour distances for censored data on nearest neighbour kernel estimators ...	95 JNnparSt 4 271-282
	Nearest-neighbour estimation of semiparametric regression models • Robinson, P. M.	95 JNnparSt 5 33-41
shik Estimation of the variance in the presence of	nearest neighbour imputation (Disc: p906-907) • Rancourt, Eric; Särndal, Carl; Lee, Hyun-	94 ASASrvyRsr 888-893
method for global optimization based on the kth	nearest neighbour • Rotondi, Renata; Drappo, Silvia A clustering	95 StatComp 5 317-326
use in palaeoecology: A comparison of inverse (k-	nearest neighbours, partial least squares and weighted averaging ... • Braak ... and their	95 Chemolab 28 165-180
Liu, W. Z.; White, A. P. A comparison of	nearest neighbour and tree-based methods of non-parametric discriminant analysis •	95 JStCmpSm 53 41-50
convergence for attractive reversible subcritical	nearest particle systems • Mountford, T. S. Exponential	95 StocProc 59 235-249
ables • Cioczek-Georges, Renata; Taqqu, Murad S.	Necessary conditions for the existence of conditional moments of stable random vari-	95 StocProc 56 233-246
Tang, Shanjian General	necessary conditions for partially observed optimal stochastic controls • Li, Xunjing;	95 JAppProb 32 1118-1137
for first place • Baryshnikov, Y.; Eisenberg, B. et al. A	necessary and sufficient condition for the existence of the limiting probability of a tie	95 StPrLet 23 203-209
model • Wu, Qiguang	Necessary and sufficient condition for the existence of UMVU estimators in growth curve	95 KToChina 39 89-92
models of factor complexity one • Reilly, Terence A	necessary and sufficient condition for identification of confirmatory factor analysis	95 SocMethR 23 421-441
characterization of a recursively defined process:	Necessary and sufficient conditions, sensitivity • Yin, G.; Zhu, Y. M.; Gao, A. J. On	95 Stochatx 52 265-282
nonuniform topologies • Deheuvels, P. et al.	Necessary and sufficient conditions for the Strassen law of the iterated logarithm in	94 AnlsProb 22 1838-1856
of the common odds ratio • Yu, Kai Fun A	necessary and sufficient condition for the strong consistency of a family of estimators	95 CandJSt 23 215-225
accelerated sequential estimation of the mean of	NEF-PVF distributions • Bose, Arup; Mukhopadhyay, Nitis A note on	95 AnInStMa 47 99-104
corrected sequential estimation for the mean of	NEF-PVF distributions • Bose, Arup; Boukai, Benzion Bias	95 SqtlAnly 14 307-319
Katz, Richard W. Comment on "Binomial and	negative binomial analogues under correlated Bernoulli trials" (V48 p243-247) •	95 AmerStat 49 325-325
manian, K. et al. Reply to comment on "Binomial and	negative binomial analogues under correlated Bernoulli trials" • Viveros, R.; Balasubra-	95 AmerStat 49 325-326
	Negative binomial distribution	
Ong, S. H.	Computation of bivariate gamma and inverted beta distribution functions •	95 JStCmpSm 51 153-163
	A family of discrete distributions • Ambagaspitiya, R. S.	95 InsrneME 16 107-127
	Minification processes with discrete marginals • Kalamkar, V. A.	95 JAppProb 32 692-706

On some central and	non-central multivariate chi-square distributions • Royen, Thomas	95 StSinica 5 373-397
On the new approximation to	non-central t-distributions • Akahira, Masafumi; Sato, Michikazu; Torigoe, Norio	95 JJapnStS 25 1-18
order approximation to a percentage point of the	non-central t-distribution • Akahira, Masafumi A higher	95 CommStB 24 595-605
	Noncentrality parameter	
Nelder, John A.; Lane, Peter W.	The computer analysis of factorial experiments: In memoriam - Frank Yates •	95 AmerStat 49 382-385
Hainsworth, Timothy J.	Confidence intervals for the noncentral chi-squared distribution • Kent, John T.;	95 JStPlInf 46 147-159
Rao, J. N. K.; Wang, Song-Gui	On the power of F tests under regression models with nested error structure •	95 JMultiAn 53 237-246
Improving on the positive part of the UMVUE of a	noncentrality parameter of a noncentral chi-square distribution • Shao, P. Y.-S. et al.	95 JMultiAn 53 52-66
seille fish market • Härdle, Wolfgang; Kirman, Alan	Nonclassical demand - A model-free examination of price-quantity relations in the Mar-	95 JEconmtx 67 227-257
Shili; Thompson, Elizabeth; Wijsman, Ellen Finding	noncommunicating sets for Markov chain Monte Carlo estimations on pedigrees • Lin,	94 AJHuGene 54 695-704
On	non-consistency of estimators • Kutoyants, Yu.; Vostrikova, L.	95 Stochatx 53 53-80
structural econometric analysis • Chirinko, Robert S.	Nonconvexities, labor hoarding, technology shocks, and procyclical productivity. A	95 JEconmtx 66 61-98
Dynamic	noncooperative game theory (Second edition) • Basar, Tamer; Olsder, Geert Jan	95 Academic:NY:Lnd 519p
Estimating telephone	noncoverage bias with a telephone survey • Keeter, Scott	95 PubOpinQ 59 196-217
A comparison of tests for the k-sample,	non-decreasing alternative • Mahrer, Joanne M.; Magel, Rhonda C.	95 StatMed 14 863-871
Statistical models for	nondestructive evaluation • Sweeting, Trevor	95 IntStRvw 63 199-214
Estimating	"non-detects" for inclusion in linear regression • Bolognese, James A.; Nguyen, Ha H.	94 ASAProBiop 414-419
The Csörgő-Révész modulus of	non-differentiability of iterated Brownian motion • Hu, Y.; Shi, Z.	95 StocProc 58 267-279
for estimation problems with multiple optima,	nondifferentiability, and other irregular features • Dorsey, R. E. et al. Genetic algorithms	95 JBES 13 53-66
Correa-Villaseñor, A.; Stewart, W. F. et al. Bias from	nondifferential misclassification in case-control studies with three exposure levels •	95 Epidmlgy 6 276-281
p565-571) • Birkett, N. J. Comment on "When will	nondifferential misclassification of an exposure reverse the direction of a trend?" (V140	95 AJEpidm 142 783-784
p565-571) • Thomas, D. C. Comment on "When will	nondifferential misclassification of an exposure reverse the direction of a trend?" (V140	95 AJEpidm 142 782-783
Weinberg et al. Reply to comments on "When will	nondifferential misclassification of an exposure reverse the direction of a trend?" •	95 AJEpidm 142 784-784
epidemiological indices in case-control studies with	non-differential misclassification • Marinos, A. Th. et al. Experimental quantiles of	95 StatMed 14 1291-1306
intervals on ratios of linear combinations for	non-disjoint sets of expected mean squares • Gui, R.; Graybill, F. A. et al. Confidence	95 JStPlInf 48 215-227
et al. Macroscopic properties of a stationary	non-equilibrium distribution for a non-gradient interacting particle system • Kipnis, C.	95 AIHenrPB 31 191-221
Likelihood methods for locating disease genes in	nonequilibrium populations • Kaplan, N. L.; Hill, W. G.; Weir, B. S.	95 AJHuGene 56 18-32
Giribone Response surface methodology for	non-equispaced experimental configurations applied to a discrete and stochastic ... •	94 JAppStat 21 549-564
A condition for the	nonexistence of ancillary statistics • Gupta, Arjun K.; Wang, Yining	95 StPrLet 23 367-369
Utev, Sergey On a conjecture for the	non-existence of the expectation of randomly stopped sums • Lefèvre, Claude;	95 JAppProb 32 1138-1141
variance components • Gao, S.; Smith, T. M. F. On the	nonexistence of a global nonnegative minimum bias invariant quadratic estimator of	95 StPrLet 25 117-120
On theory testing in econometrics Modeling with	nonexperimental data • Spanos, Aris	95 JEconmtx 67 189-226
Bias correction and improved residuals for	non-exponential family nonlinear models • Paula, Gilberto A.; Cordeiro, Gauss M.	95 CommStA 24 1193-1210
Empirical Bayes testing procedures in some	nonexponential families using asymmetric Linex loss function • Huang, Su-Yun	95 JStPlInf 46 293-309
Rosage, Laura R. Self-employed	nonfilers: The accuracy of "enforcement-secured" returns (Disc: p212-214) •	94 ASAProGovt 206-211
Prediction and	non-Gaussian autoregressive stationary sequences • Rosenblatt, Murray	95 AnlsApPr 5 239-247
Design sensitivity of structures under	non-Gaussian excitation • Sieniawska, R.; Śniady, P.	95 ZAngMaMk 75 573-574S
sunder, Sanyogita Noise subspace techniques in	non-Gaussian noise using cumulants • Sadler, Brian M.; Giannakis, Georgios B.; Sham-	95 IEEEAES 31 1009-1018
	Non-Gaussian process	
Li, Ta-Hsin	Blind deconvolution of linear systems with multilevel nonstationary inputs •	95 AnlsStat 23 690-704
noise distributions • Spall, James C.	The Kantorovich inequality for error analysis of the Kalman filter with unknown	95 Autmatca 31 1513-1517
Samorodnitsky, G. et al. Review of "Stable	non-Gaussian random processes: Stochastic models with infinite variance" •	95 JASA 90 805-806
A simulation technique of a	non-Gaussian spatial process • De Cesare, L.; Posa, D.	95 CmpStDA 20 543-555
and lower bounds on the rate of convergence for	nonhomogeneous birth and death processes • Zeifman, A. I. Upper	95 StocProc 59 157-173
Shiryaev's procedures for detecting a change in a	nonhomogeneous Gaussian process • Tartakovsky Asymptotic properties of CUSUM and	95 MaMethSt 4 389-404
A class of strong laws for functionals of countable	nonhomogeneous Markov chains • Liu, Wen; Liu, Guoxin	95 StPrLet 22 87-96
Ratio ergodicity for	non-homogeneous Markov chains in general state spaces • Fleischer, I.; Joffe, A.	95 JTherPrb 8 31-37
Tail events of some	nonhomogeneous Markov chains • Niemiro, Wojciech; Pokarowski, Piotr	95 AnlsApPr 5 261-293
A limit theorem for the entropy density of	nonhomogeneous Markov information source • Liu, Wen; Yang, Weiguo	95 StPrLet 22 295-301
Some limit properties of	nonhomogeneous Markov information sources • Liu, Wen; Yang, Weiguo	95 JCISS 20 279-292
A unified approach to the	nonhomogeneous Poisson process in reliability modeling • Kuo, Lynn; Yang, Tae	94 ASAProPEng 169-174
A local limit theorem for	nonhomogeneous random walk on a lattice • Zhizhina, E. A.; Minlos, A.	94 ThProbAp 39 490-503
regions of significance in ANCOVA problems having	non-homogeneous regressions • Hunka, Steve Identifying	95 BJMaStPs 48 161-188
A strong law of large number for L-statistics in the	non-I.D. case • Xiang, Xiaojing	95 CommStA 24 1813-1819
Order statistics from	non-identical power function random variables • Balakrishnan, N.; Balasubramanian, K.	95 CommStA 24 1443-1454
Approximate estimation of	non-identifiable parameters in a convolution • Kale, B. K.; Sebastian, G.	95 StPrLet 25 373-378
for repeated binary data with outcome subject to	non-ignorable non-response • Baker, Stuart G. Marginal regression	95 Biomtrcs 51 1042-1052
Using i.i.d. bootstrap inference for general	non-i.i.d. models • Liu, Regina Y.; Singh, Kesar	95 JStPlInf 43 67-75
Asymptotics of k-mean clustering under	non-i.i.d. sampling • Serinko, Regis J.; Babu, Gutti Jogesh	95 StPrLet 24 57-66
The radial part of a γ-martingale and a	non-implosion theorem • Kendall, Wilfrid S.	95 AnlsProb 23 479-500
	Noninformative prior	
	Information tradeoff • Wasserman, L.; Clarke, B.	95 Test 4 19-38
Ghosh, Jayanta Kumar	On priors providing frequentist validity for Bayesian inference • Datta, Gauri Sankar;	95 Biomtrka 82 37-45
Ghosh, J. K.	Noninformative priors for maximal invariant parameter in group models • Datta, G. S.;	95 Test 4 95-114
Some remarks on	noninformative priors • Datta, Gauri Sankar; Ghosh, Malay	95 JASA 90 1357-1363
for random coefficient regression models using	noninformative priors • Yang, Ruoyong; Chen, Ming-Hui Bayesian analysis	95 JMultiAn 55 283-311
model • Belzile, Éric; Angers, Jean-François Choice of	noninformative priors for the variance components of an unbalanced one way random	95 CommStA 24 1325-1341
The law of the iterated logarithm for	non-instantaneous filters of strongly dependent Gaussian sequences • Ho, Hwai-Chung	95 JTherPrb 8 347-360
D-optimal designs of experiments with	non-interacting factors • Schwabe, Rainer; Wierich, Werner	95 JStPlInf 44 371-384
of the memory parameter for nonstationary or	noninvertible fractionally integrated processes • Hurvich, C. M.; Ray, B. K. Estimation	95 JTimSrAn 16 17-41
binomial proportions based on likelihood scores:	Non-iterative method • Nam, Jun-mo Confidence limits for the ratio of two	95 BiomtrcJ 37 375-379
random error in X: Increasing power of the test by	non-linear adjustment for measurement error • Grove, J. S. et al. Regression with	94 ASAProEpid 143-146
A	nonlinear approach to US GNP • Potter, Simon M.	95 JApEcmtx 10 109-125
Nonparametric estimation and identification of	nonlinear ARCH time series • Masry, Elias; Tjøstheim, Dag	95 EconTher 11 258-289
Maximin efficient designs for estimating	nonlinear aspects in linear models • Müller, Christine H.	95 JStPlInf 44 117-132
Additivity tests for	nonlinear autoregression • Chen, Rong; Liu, Jun S.; Tsay, Ruey S.	95 Biomtrka 82 369-383
Order choice in	nonlinear autoregressive models • Vieu, Philippe	95 MaOpfStS 26 307-328
On geometric ergodicity of	nonlinear autoregressive models • Bhattacharya, Rabi; Lee, Chanho	95 StPrLet 22 311-315
optimal adaptive estimation in the semiparametric	non-linear autoregressive time series model • So, Beong-Soo On the	95 JKoreaSt 24 149-160
Wise, B. M. et al. A comparison of neural networks,	non-linear biased regression and a genetic algorithm for dynamic model identification •	95 Chemolab 30 81-89
Juditsky, A.; Hjalmarsson, H.; Benveniste, A. et al.	Nonlinear black-box models in system identification: Mathematical foundations •	95 Autmatca 31 1725-1750
Zhang, Q.; Ljung, L.; Benveniste, A.; Delyon, B. et al.	Nonlinear black-box modeling in system identification: A unified overview • Sjöberg, J.;	95 Autmatca 31 1691-1724
Design considerations for	nonlinear calibration • Bates, Douglas M.; Martin, Sandra J.	94 ASAProPEng 12-17
ogy for the simultaneous development of linear and	nonlinear chemical reaction models from a linearly designed ... • Steele et al. A methodol-	95 Chemolab 29 125-131

parameter and modelling error identification in	non-linear conceptual models from bounded-error data • Keesman, K. J. Model	95 EnvrMtrc 6 445-449
ical experiments • Dennis, B.; Desharnais, R. A. et al.	Nonlinear demographic dynamics: Mathematical models, statistical methods, and biolog-	95 EcolgMon 65 261-281
Curvature adjusted optimal	nonlinear designs for correlated responses • Hughes-Oliver, Jacqueline M.	94 ASAProBuEc 111-116
On positive solutions of some	nonlinear differential equations - A probabilistic approach • Sheu, Yuan-Chung	95 StocProc 59 43-53
Homogenization and propagation of chaos to a	nonlinear diffusion with sticky reflection • Graham, Carl	95 ProbTher 101 291-302
Haesloop, Dan A	non-linear dynamic model identification challenge problem • Wise, Barry M.;	95 Chemolab 30 91-96
neural network for prediction and identification of	nonlinear dynamic systems • Majetić, Dubravko Dynamic	95 JCmpInT 3 99-106
(Ed); Adcock, Chris (Rev) Review of	"Nonlinear dynamics chaos and econometrics" • Pesaran, M. Hashem; Potter, Simon M.	95 IntJFor 11 599-601
Gilligan, C. A.; Bolker, B. M. Spatial heterogeneity,	nonlinear dynamics and chaos in infectious diseases • Grenfell, B. T.; Kleczkowski, A.;	95 StMedRes 4 160-183
multivariate cointegrated systems: Insights from	non-linear dynamics • Pinder, Jonathan P.; Shoesmith, Gary L. Modelling	95 JForcast 14 311-324
Sandra (Rev) Review of "Modelling	nonlinear economic relationships" • Granger, C. W. J.; Teräsvirta, T. (Auth); McKenzie,	95 JASA 90 1491-1491
Newey, W. K.; Powell, J. L.	Nonlinear errors in variables: Estimation of some Engel curves • Hausman, J. A.;	95 JEconmtx 65 205-233
Jungsywan H. Estimation of linear and	nonlinear errors-in-variables models using validation data • Lee, Lung-fei; Sepanski,	95 JASA 90 130-140
Lee, Lung-Fei Semiparametric estimation of	nonlinear errors-in-variables models with validation study • Sepanski, Jungsywan H.;	95 JNnparSt 4 365-394
Review of	"Nonlinear estimation" • Ross, G. J. S. (Auth); Droge, B. (Rev)	95 MaOpfStS 26 181-182
On efficient designing of	nonlinear experiments • Chaudhuri, Probal; Mykland, Per A.	95 StSinica 5 421-440
Cramer-Rao bounds for discrete-time	nonlinear filtering problems • Doerschuk, Peter C.	95 IEEEAuCn 40 1465-1469
Ergodicity of	nonlinear first order autoregressive models • Bhattacharya, Rabi N.; Lee, Chanho	95 JTherPrb 8 207-219
An assessment of the economic value of	non-linear foreign exchange rate forecasts • Satchell, Steve; Timmermann, Allan	95 JForcast 14 477-497
Andersson et al. A method for variance estimation of	non-linear functions of totals in surveys - Theory and software implementation •	94 JOffStat 10 395-405
Limit theorems for	nonlinear functionals of a stationary Gaussian sequence of vectors • Arcones, Miguel A.	94 AnlsProb 22 2242-2274
V. A central limit theorem for	non-linear functionals of stationary Gaussian vector processes • Sánchez de Naranjo, M.	95 StPrLet 22 223-230
Asymptotic estimation of a	non-linear infinite filter. Application to the estimation of Volterra series • Bosq, Denis	95 JNnparSt 5 61-73
Asymptotic properties of	nonlinear least absolute deviation estimators • Kim, Hae Kyung; Choi, Seung Hoe	95 JKoreaSt 24 127-139
	Nonlinear least squares	
volution of a density estimate • Barry, J.; Diggle, P.	Choosing the smoothing parameter in a Fourier approach to nonparametric decon-	95 JNnparSt 4 223-232
relationships of fish • Hayes, D. B. et al.	Efficiency and bias of estimators and sampling designs for determining length-weight	95 JFishRBC 52 84-92
Lee, Lung-fei; Sepanski, Jungsywan H.	Estimation of linear and nonlinear errors-in-variables models using validation data •	95 JASA 90 130-140
Kuan, Chung-Ming; Liu, Tung	Forecasting exchange rates using feedforward and recurrent neural networks •	95 JApEcmtx 10 347-364
isotherms from column ... • van Veldhuizen et al.	The (least squares) quotient algorithm as a rapid tool for obtaining sorption	95 WaterReR 31 849-857
deposition • Haas, Timothy C.	Local prediction of a spatio-temporal process with an application to wet sulfate	95 JASA 90 1189-1199
Smyth, G. K.	A modified prony algorithm for exponential function fitting • Osborne, M. R.;	95 SIAMSSC 16 119-138
Asymptotic properties of	nonlinear least squares estimates in stochastic regression models • Lai, Tze Leung	94 AnlsStat 22 1917-1930
small-dispersion asymptotic optimality of the	non-linear least squares estimator • So, Beong-Soo Finite-sample,	95 JKoreaSt 24 303-312
Fukushima, Masao A hybrid method for solving the	nonlinear least squares problem with linear inequality constraints • Sagara, Nobuko;	94 JOpRsJpn 38 55-69
An algorithm for	nonlinear L_1 curve-fitting based on the smooth approximation • Yang, Ziqiang	95 CmpStDA 19 45-52
	Nonlinear master equation of multitype particle systems • Feng, Shui	95 StocProc 57 247-271
Assessment of influence in	non-linear measurement error models • Zhao, Yuejen; Lee, Andy H.	95 JAppStat 22 215-225
Pseudorandom number generation by	nonlinear methods • Eichenauer-Herrmann, Jürgen	95 IntStRvw 63 247-255
A comparison of inverse (k-nearest ... • Braak	Non-linear methods for multivariate statistical calibration and their use in palaeoecology:	95 Chemolab 28 165-180
Forecasting industrial production using	non-linear methods • Byers, J. D.; Peel, D. A.	95 JForcast 14 325-336
Schabenberger, Oliver	Nonlinear mixed effects growth models for repeated measures in ecology •	94 ASAProEnvr 156-161
Simulation study on different approaches for	nonlinear mixed effects models • Zhang, Donghui	94 ASAProBiop 67-72
Model building in	nonlinear mixed effects models • Pinheiro, José C.; Bates, Douglas M.; Lindstrom, Mary J.	94 ASAProBiop 1-8
Cheng, M. et al. Estimating parameters in	non-linear mixed effect models: A comparison using clinical data in children • Roe, D. J.;	94 ASAProBiop 313-318
the distribution of interindividual random effects in	non-linear mixed effects models • Fattinger, K. E. et al. A new method to explore	95 Biomtrcs 51 1236-1251
combination effect • Gao, Min; Minkin, Salomon	Nonlinear mixed effect model for estimating population characteristics of drug	94 ASAProBiop 160-165
Approximations to the log-likelihood function in the	nonlinear mixed-effects model • Pinheiro, José C.; Bates, Douglas M.	95 JCmpGrSt 4 12-35
mating unknown transition times using a piecewise	nonlinear mixed-effects model in men with prostate cancer • Morrell, C. H. et al. Esti-	95 JASA 90 45-53
A two-step iterative algorithm for estimation in	nonlinear mixed-effect models with an evaluation in population ... • Mentré et al.	95 JBiophSt 5 141-158
	Non-linear mixed regression models • Burnett, Richard T.; Ross, W. H.; Krewski, Daniel	95 EnvrMtrc 6 85-99
	Nonlinear model	
Satchell, Steve; Timmermann, Allan	An assessment of the economic value of non-linear foreign exchange rate forecasts •	95 JForcast 14 477-497
Pontier, J.; Borms, J.; Duquet, W.	The benefit of modelling growth curves for longitudinal data (French) • Abidi, H.;	95 RStApp 43/3 55-72
reaction models from a linearly ... • Steele et al.	A methodology for the simultaneous development of linear and nonlinear chemical	95 Chemolab 29 125-131
yield response • Burt, Oscar R.	Pooling experiments over time to quantify the dynamics of dryland winter wheat	95 AgronmyJ 87 156-164
odds models for ordinal data: A generalized	non-linear model approach • Cox, Christopher Location-scale cumulative	95 StatMed 14 1191-1203
modifiers • Lam, Y. M.; Pym, J.; Campling, B. G.	Nonlinear models for assessing combinations of chemotherapeutic drugs with resistance	94 ASAProBiop 394-398
Pillati, M. A	nonlinear model for the conditional expectations of asset returns • Paruolo, P.;	94 Statstca 54 329-347
Computer simulation of a	nonlinear model for electrical circuits with α-stable noise • Janicki, A.	95 ZastApMa 23 95-105
Measurement error in	nonlinear models • Carroll, Raymond J.; Ruppert, David; Stefanski, Leonard A.	95 ChapmnHall:Lnd 336p
Draper, N. R. (Rev) Review of "Measurement error in	nonlinear models" • Carroll, Raymond J.; Ruppert, David; Stefanski, Leonard A. (Auth);	96 ShrtBkRv 16/1 5-5
Large sample properties of M-estimators in	nonlinear models (Chinese) • Jun, Shao	95 ChJApPSt 10 125-132
geometry and asymptotic inference in exponential	nonlinear models • Wei, Bocheng Theory of	95 SciSPkng 24 1270-1278
and improved residuals for non-exponential family	nonlinear models • Paula, Gilberto A.; Cordeiro, Gauss M. Bias correction	95 CommStA 24 1193-1210
A	non-linear model of an open flow adsorption process • Argyelan, J.; Nassar, R.	95 EnvrMtrc 6 195-203
approach to optimal design for one-parameter	non-linear models • Haines, Linda M. A geometric	95 JRSS-B 57 575-598
for fixed effects and overdispersion components in	nonlinear models for repeated measures • Das, Kalyan; Sutradhar, Brajendra C. Score tests	95 JMultiAn 54 126-146
for fixed effects and overdispersion components in	nonlinear models for repeated measures" (V54 p126-146) • Das et al. ... to "Score tests	95 JMultiAn 55 349-349
	Nonlinear models for repeated measurement data • Davidian, Marie; Giltinan, David M.	95 ChapmnHall:Lnd 384p
	Nonlinear models for repeated measurement data. • Davidian, M.; Giltinan, D. M.	95 ChapmnHall:Lnd 359p
Hand, D. J. (Rev) Review of	"Nonlinear models for repeated measurement data." • Davidian, M.; Giltinan, D. M. (Auth);	96 ShrtBkRv 16/1 6-7
Ferreira, Irene	Nonlinear model selection for finding receptor sites • Amaratunga, Dhammika;	94 ASAProBiop 126-129
et al. Robust regression and outlier detection for	non-linear models using genetic algorithms • Vankeerberghen, P.; Smeyers-Verbeke, J.	95 Chemolab 28 73-87
Lambert, Daniel; St-Georges, Mario	Nonlinear modeling of muskrat use of habitat • Nadeau, Simon; Décarie, Robert;	95 JWildlMg 59 110-117
Highest-density forecast regions for	non-linear and non-normal time series models • Hyndman, Rob J.	95 JForcast 14 431-441
Feuer et al. Free radical polymerization of	nonlinear optical side chain methacrylate copolymers in dimethylsulfoxide - An ... •	95 Chemolab 29 113-124
A	nonlinear parking problem • Keener, Robert; Lerche, Hans Rudolf; Woodroofe, Michael	95 SqtlAnly 14 247-272
Local sensitivity, functional derivatives and	nonlinear posterior quantities • Basu, Sanjib	94 ASAProBays 175-180
of the theory of KM_2O-Langevin equations to the	non-linear prediction problem for the one-dimensional strictly ... • Okabe et al. Application	95 JMaJapan 47 349-367
	Nonlinear programming	
quantitative IR spectroscopy • Brink, A. et al.	The joint problem of model structure determination and parameter estimation in	95 Chemolab 29 29-36
	An optimal inspection-repair-replacement policy for standby systems • Yeh, Lam	95 JAppProb 32 212-223
Multiple and unacceptable solutions in	nonlinear programming discriminant analysis • Xiao, Baichun	94 AJMMgSci 14 267-287

Estimation of	nonlinear random coefficient models • Ramos, Rogelio Q.; Pantula, Sastry G.	95 StPrLet 24 49-56
• Deaton, Angus; Laroque, Guy Estimating a	nonlinear rational expectations commodity price model with unobservable state variables	95 JApEcmtx 10 9-40S
Density-dependent limits for a	nonlinear reaction-diffusion model • Blount, Douglas	94 AnlsProb 22 2040-2070
	Nonlinear regression	
regression model estimators • Sakata, S.; White, H.	An alternative definition of finite-sample breakdown point with applications to	95 JASA 90 1099-1106
Kim, Hae Kyung; Choi, Seung Hoe	Asymptotic properties of nonlinear least absolute deviation estimators •	95 JKoreaSt 24 127-139
Saurabh; Haines, Linda M.	Bayesian D-optimal designs for the exponential growth model • Mukhopadhyay,	95 JStPlInf 44 385-397
Mayekawa, Shin-ichi	An efficient algorithm for feed-forward neural network regression analysis •	95 Behvmtrk 22 67-90
distances • Bartkowiak, Anna; Liebhart, Jerzy	Estimation of the spirometric residual volume (RV) by a regression built from Gower	95 BiomtrcJ 37 131-149
Krzysztofowicz, Roman	A family of bivariate densities constructed from marginals • Long, Dou;	95 JASA 90 739-746
Simon, Richard	Maximum likelihood neural network prediction models • Faraggi, David;	95 BiomtrcJ 37 713-725
linkage maps • Wu, W. R.; Li, W. M.	A new approach for mapping quantitative trait loci using complete genetic marker	94 ThApGene 89 535-539
	Non-parametric analysis of covariance • Young, Stuart G.; Bowman, Adrian W.	95 Biomtrcs 51 920-931
	A nonparametric test for the regression function: Asymptotic theory • Diebolt, Jean	95 JStPlInf 44 1-17
neural networks • Sjöberg, J.; Ljung, L.	Overtraining, regularization and searching for a minimum, with application to	95 IJContrl 62 1391-1407
	Polynomial measurement error modeling • Moon, Myung-Sang; Gunst, Richard F.	95 CmpStDA 19 1-21
for repeated measures • Das, K.; Sutradhar, B. C.	Score tests for fixed effects and overdispersion components in nonlinear models	95 JMultiAn 54 126-146
experiments (jse/v3n1/ ... • Dickey et al.	Teaching statistics with data of historical significance: Galileo's gravity and motion	95 JStEd 3/1 20kb
Robust	non-linear regression analysis • Roy, Tapon	95 JChemtrc 9 451-457
Shrinkage estimation in	nonlinear regression. The Box-Cox transformation • Kim, Minbo; Hill, R. Carter	95 JEconmtx 66 1-33
variation of exact locally optimum designs in	nonlinear regression - A case study • Rasch, Dieter The robustness against parameter	95 CmpStDA 20 441-453
tion to the asymptotic distribution • Hougaard, Philip	Nonlinear regression and curved exponential families. Improvement of the approxima-	95 Metrika 42 191-202
Software for computing a robust efficient linear or	nonlinear regression estimator • Stromberg, Arnold J.	94 ASAProStEd 6-11
Asymptotic normality of L_1-estimators in	nonlinear regression • Wang, Jinde	95 JMultiAn 54 227-238
to "Asymptotic normality of L_1-estimators in	nonlinear regression" (V54 p227-238) • Wang, Jinde Correction	95 JMultiAn 55 350-350
On calculating the moments of M-estimator in	nonlinear regression (Chinese) • Wei, Bocheng; Feng, Yu	95 AMaAPknC 17 437-447
of the least median of squares estimator in	nonlinear regression • Stromberg, Arnold J. Consistency	95 CommStA 24 1971-1984
Diagnostics for linearization confidence intervals in	nonlinear regression • Chen, Jian-Shen; Jennrich, Robert I.	95 JASA 90 1068-1074
for improving linearization confidence intervals in	nonlinear regression • Chen, Jian-Shen; Jennrich, Robert I. Transformations	95 JASA 90 1271-1276
Modern	nonlinear regression methods • Frank, Ildiko E.	95 Chemolab 27 1-19
Asymptotic confidence bands for generalized	nonlinear regression models • Cox, Christopher; Ma, Guangqin	95 Biomtrcs 51 142-150
A cubic smoothing spline based lack of fit test for	nonlinear regression models • Djojosugito, Rianto A.	95 CommStA 24 2183-2197
design and accuracy of parameter estimation for	nonlinear regression models used in long-term selection • Rudolph et al. ... experimental	95 BiomtrcJ 37 183-190
Testing outliers in	nonlinear regression • Kahng, Myung-Wook	95 JKoreaSt 24 419-437
László Heteroscedastic	nonlinear regression when there are possible outliers • Tóthfalusi, László; Endrényi,	94 ASAProStCp 239-244
Model estimation in	nonlinear regression under shape invariance • Kneip, Alois; Engel, Joachim	95 AnlsStat 23 551-570
MLE's covariance matrix in a seemingly unrelated	nonlinear regression • Jensen, M. J. A Monte Carlo study on two methods of calculating the	95 EcomtrRv 14 315-330
in advanced econometrics. Volume 2: Linear and	nonlinear simultaneous equations • Dhrymes, P. J. Topics	94 Springer:Brln:NY 401p
	Nonlinear smoothing for random fields • Aihara, Shin Ichi; Bagchi, Arunabha	95 StocProc 55 143-158
A	nonlinear state-space model of Chinook salmon population dynamics • Rein, Steven R.	94 ASAProEnvr 150-155
C. et al. Uncertainties for recursive estimators in	nonlinear state-space models, with applications to epidemiology • Maryak, J. L.; Spall, J.	95 Automatca 31 1889-1892
An instability phenomenon of the	nonlinear stochastic differential equation (French) • Roynette, B.	95 Stochatx 53 175-194
de Socio, L. M. (Auth); Elworthy, K. D. (Rev) Review of	"Nonlinear stochastic evolution problems in applied sciences" • Bellomo, N.; Brzezniak, Z.;	95 Metrika 42 140-141
de Socio, L. M. (Auth); Renshaw, Eric (Rev) Review of	"Nonlinear stochastic evolution problems in applied sciences" • Bellomo, N.; Brzezniak, Z.;	94 BLndnMaS 26 409-410
Issues in	nonlinear stochastic grey box identification • Bohlin, Torsten; Graebe, Stefan F.	95 IJAdCSgP 9 465-490
Phase transitions of some	non-linear stochastic models • Feng, Shui	95 JAppProb 32 193-201
An approximation theorem of Wong-Zakai type for	nonlinear stochastic partial differential equations • Twardowska, Krystyna	95 StocAnAp 13 601-626
Model-free control of	nonlinear stochastic systems in discrete time • Spall, James C.; Cristion, John A.	95 PIEEEDecCt 2199-2204
Modeling	nonlinear systems, bridging theory and technology • Rey, Timothy D.; Duever, Thomas A.	94 ASAProQlPr 310-315
Federica Finite dimensional filters for a discrete-time	nonlinear system with generalized Gaussian white noise • Ferrante, Marco; Giummolé,	95 Stochatx 53 195-211
	Nonlinear time series	
Rajarshi, M. B.	Filtering and smoothing via estimating functions • Naik-Nimbalkar, U. V.;	95 JASA 90 301-306
Newton, H. Joseph	Saddlepoint approximations for the difference of order statistics • Harvill, Jane L.;	95 Biomtrka 82 226-231
Computational aspects of linear and	non-linear time series analysis • Moeanaddin, Rahim	92 ProIrStC1 123-155
A study on a regression model with	nonlinear time series errors (Korean) • Hwang, Sun Young	95 KorApS 8/2 187-200
Model-free forecasting for	nonlinear time series (with application to exchange rates) • Wu, Berlin	95 CmpStDA 19 433-459
On the estimation of	nonlinear time series models • Hili, Quagnina	95 Stochatx 52 207-226
Invertibility of	non-linear time series models • De Gooijer, Jan G.; Brännäs, Kurt	95 CommStA 24 2701-2714
	Nonlinear time series models for multivariable dynamic processes • Çinar, Ali	95 Chemolab 30 147-158
	Non-linear time series models for ordinal data • Hoffmann, Raymond G.	94 ASAProStCp 245-250
D. An interesting	non-linear transformation of the theoretical autocorrelation function • Anderson, Oliver	95 JStCmpSm 52 165-167
Formulae for mean integrated squared error of	nonlinear wavelet-based density estimators • Hall, Peter; Patil, Prakash	95 AnlsStat 23 905-928
	Nonlinearity index	
Chen, Jian-Shen; Jennrich, Robert I.	Diagnostics for linearization confidence intervals in nonlinear regression •	95 JASA 90 1068-1074
stochastic evolution equations with monotone	nonlinearities • Zangeneh, Bijan Z. Semilinear	95 Stochatx 53 129-174
earity tracking analysis algorithm for treatment of	non-linearity in multivariate calibration • Xie, Y.-L.; Liang, Y.-Z.; Chen, Z.-G. et al. A nonlin-	95 Chemolab 27 21-32
calibration • Xie, Y.-L.; Liang, Y.-Z.; Chen, Z.-G. et al. A	nonlinearity tracking analysis algorithm for treatment of non-linearity in multivariate	95 Chemolab 27 21-32
of backward stochastic differential equations with	non-Lipschitz coefficients • Mao, Xuerong Adapted solutions	95 StocProc 58 281-292
Mean residual life function for certain types of	non-monotonic ageing • Gupta, Ramesh C.; Akman, H. Olcay	95 CommStSM 11 219-225
to "Mean residual life function for certain types of	non-monotonic ageing" (V11 p219-225) • Gupta, Ramesh C.; Akman, H. Olcay Correction	95 CommStSM 11 561-562
Change point estimation in	non-monotonic aging models • Mitra, Murari; Basu, Sujit K.	95 AnInStMa 47 483-491
Better subset regression using the	nonnegative garrote • Breiman, Leo	95 Technmcs 37 373-384
A mixed random walk on	nonnegative matrices: A law of large numbers • Högnäs, Göran; Mukherjea, Arunava	95 JTherPrb 8 973-990
S.; Smith, T. M. F. On the nonexistence of a global	nonnegative minimum bias invariant quadratic estimator of variance components • Gao,	95 StPrLet 25 117-120
Gnot, S.; Trenkler, G.; Zmyślony, R.	Nonnegative minimum biased quadratic estimation in the linear regression models •	95 JMultiAn 54 113-125
On conditioning a random walk to stay	nonnegative • Bertoin, J.; Doney, R. A.	94 AnlsProb 22 2152-2167
Vijayan, K.; Mukhopadhyay, P.; Bhattacharyya, S. On	non-negative unbiased estimation of quadratic forms in finite population sampling •	95 AstrlJSt 37 169-178
model • Adang, Pim; Melenberg, Bertrand	Nonnegativity constraints and intratemporal uncertainty in a multi-good life-cycle	95 JApEcmtx 10 1-15
	Nonnested hypotheses	
parameter approach • Ioannatos, Petros E.	Censored regression estimation under unobserved heterogeneity: A stochastic	95 JBES 13 327-335
Granger, Clive W. J.; King, Maxwell L.; White, Halbert	Comments on testing economic theories and the use of model selection criteria •	95 JEconmtx 67 173-187
Christian; Monfort, Alain	Testing, encompassing, and simulating dynamic econometric models • Gouriéroux,	95 EconTher 11 195-228
The significance of testing empirical	non-nested models • McAleer, Michael	95 JEconmtx 67 149-171
Higgins, Matthew L. A Monte Carlo study of tests for	non-nested models estimated by generalized method of moments • Arkonac, Seyhan Z.;	95 CommStB 24 745-763

A	nonparametric test for the regression function: Asymptotic theory • Diebolt, Jean	95 JStPlInf 44 1-17
Grosskopf, Shawna	Nonparametric tests of regularity, Farrell efficiency, and goodness-of-fit • Färe, Rolf;	95 JEconmtx 69 415-425
Rosenberger, W. F.; Lachin, J. M.; Bain, R. P.	Nonparametric test of stochastic ordering for multiple longitudinal measures •	95 JBiophSt 5 235-243
Ramazan A consistent	nonparametric test of symmetry in linear regression models • Fan, Yanqin; Gencay,	95 JASA 90 551-557
The robustness against dependence of	nonparametric tests for the two-sample location problem • Grzegorzewski, P.	95 ZastApMa 22 469-476
risks • Bagai, Isha; Deshpande, J. V.	Nonparametric tests for the two sample problem with two independent competing	95 JNnparSt 5 21-31
A	nonparametric test of the weakest-link principle • Wolstenholme, Linda C.	95 Technmcs 37 169-175
Minimax	nonparametric testing of hypotheses on the distribution density • Ermakov, M. S.	94 ThProbAp 39 396-416
Ruppert, David	Nonparametric transformations for both sides of a regression model • Nychka, Douglas;	95 JRSS-B 57 519-532
McManus, Douglas A. The	nonparametric translog with application to banking scale and scope economies •	94 ASAProBuEc 117-122
	Nonparametric truncated sequential change-point detection • Gombay, E.	95 StatDecn 13 71-82
Progressive censoring under inverse sampling for	nonparametric two-sample problems • Bandyopadhyay, Uttam; Chattopadhyay, Gopaldeb	95 SqtlAnly 14 1-28
for the discrete range distribution • Efird, J. T. et al. A	non-parametric two-sample survival test based on a single occupancy Fermi-Dirac model	94 ASAProBiop 166-170
choice model under uncertainty • Ahn, Hyungtaik	Nonparametric two-stage estimation of conditional choice probabilities in a binary	95 JEconmtx 67 337-378
Sijtsma, Klaas (Rev) Review of	"Nonparametric unfolding models. A latent structure approach" • Post, W. J. (Auth);	95 JClass 12 153-156
Statistical sirens: The allure of	nonparametrics • Johnson, Douglas H.	95 Ecology 76 1998-2000
models: Parametrics, semi-parametrics and	nonparametrics • Sen, Pranab Kumar Statistical analysis of some reliability	95 JStPlInf 43 41-66
Using survey participants to estimate the impact of	nonparticipation • Lin, I-fen; Schaeffer, Nora Cate	95 PubOpinQ 59 236-258
Generation of the typical cell of a	non-Poissonian Johnson-Mehl tessellation • Heinrich, Lothar; Schüle, Eik	95 CommStSM 11 541-560
Estimation of structures reliability for	non-precise limit state models and vague data • Savchuk, V. P.	95 RelEngSS 47 47-58
orthogonal arrays having factors with a large	non-prime power number of levels • Mandeli, John P. Construction of asymmetrical	95 JStPlInf 47 377-391
of the linear discriminant function under partial	nonrandom classification of the training data • McLachlan et al. ... relative efficiency	95 JStCmpSm 52 415-426
Effect of	non-random missing data mechanisms in clinical trials • Choi, S. C.; Lu, I. L.	95 StatMed 14 2675-2684
A (Bayesian) note on	non-random samples from finite populations • Bayarri, M. J.; Font, B.	94 ASAProBays 246-251
Quantiles in	nonrandom samples and observational studies • Rosenbaum, Paul R.	95 JASA 90 1424-1431
Sample size calculations for	non-randomized studies • Lipsitz, Stuart R.; Parzen, Michael	95 Ststcian 44 81-90
M., Jr; Watelet, Luc Estimation of vaccine efficacy in	non-randomly mixing populations • Haber, Michael; Halloran, M. Elizabeth; Longini, Ira	95 BiomtrcJ 37 25-38
and posterior distributions in multiparameter	nonregular cases • Ghosal, S.; Samanta, T. Asymptotic behaviour of Bayes estimates	95 MaMethSt 4 361-388
Empirical Bayes linear loss hypothesis testing in a	non-regular exponential family • Singh, R. S.	95 JStPlInf 43 107-120
	Non-regular maximum likelihood problems (Disc: p24-44) • Cheng, R. C. H.; Traylor, L.	95 JRSS-B 57 3-24
	Non-regular statistical estimation • Akahira, M.; Takeuchi, K.	95 Springer:Brln:NY 183p
On the optimal extraction rate of a	non-renewable resource with an extraction capacity • Hashimoto, Hideo	95 EnvrMtrc 6 547-550
view Survey (CEIS) • Kojetin, B. A. Characteristics of	nonrespondents to the Current Population Survey (CPS) and Consumer Expenditure Inter-	94 ASASrvyRsr 966-971
Characteristics of responding and	nonresponding households from the 1990 Decennial Census • Cecco, Kevin	94 ASAProGovt 76-81
	Nonresponse	
Interview Survey • Kalsbeek, W. D.; Botman, S. L. et al.	Cost-efficiency and the number of allowable call attempts in the National Health	94 JOffStat 10 133-152
	Field substitutions - A neglected option? • Vehovar, Vasja	94 ASASrvyRsr 589-594
Downes-Le Guin, Theodore; Berry, Sandra	Improving productivity in an annual survey of physicians • Thran, Sara L.;	94 ASASrvyRsr 577-582
	The influence of motivational appeals on mailback response • Treat, James B.	94 ASASrvyRsr 571-576
A.	Modeling the drop-out mechanism in repeated-measures studies • Little, Roderick J.	95 JASA 90 1112-1121
survey • Kott, Phillip S.	Reweighting and variance estimation for the characteristics of business owners	94 JOffStat 10 407-418
	A split questionnaire survey design • Raghunathan, Trivellore E.; Grizzle, James E.	95 JASA 90 55-63
Deville, Jean-Claude; Särndal, Carl-Erik	Variance estimation for the regression imputed Horvitz-Thompson estimator •	94 JOffStat 10 381-394
Theoretical motivation for post-survey	nonresponse adjustment in household surveys • Groves, Robert M.; Couper, Mick P.	95 JOffStat 11 93-106
Witt, Michael B. Testing a new attrition	nonresponse adjustment method for SIPP (Disc: p440-442) • Folsom, Ralph E.;	94 ASASrvyRsr 428-433
(Disc: p680-682) • Allen, T. M.; Petroni, R. J. Mover	nonresponse adjustment research for the Survey of Income and Program Participation	94 ASASrvyRsr 662-667
Model-based reweighting for	nonresponse adjustment • Binder, David A.; Michaud, Sylvie; Poirier, Claude	94 ASASrvyRsr 989-994
Adjustment methods for	non-response and their application to Finnish income data • Laaksonen, S.	92 StatJUN 9 125-138
hadjer, Leyla K.; Waksberg, Joseph An evaluation of	nonresponse bias in NHANES III (1988-91) • Khare, Meena; Ezzati-Rice, Trena M.; Mo-	94 ASASrvyRsr 949-954
Kalsbeek, William D.; Durham, Todd A.	Nonresponse and its effects in a followup telephone survey of low income women •	94 ASASrvyRsr 943-948
Botman, Steven L.; Basiotis, Peter	Nonresponse in federal demographic surveys: 1981-1991 • Johnson, Ayah E.;	94 ASASrvyRsr 983-988
Daniel; Gonzalez, Maria Elena Investigating	nonresponse in federal surveys • Shettle, Carolyn F.; Guenther, Patricia M.; Kaspryzk,	94 ASASrvyRsr 972-976
for imputing nonsample households with sampled	nonresponse follow-up • Zanutto, Elaine; Zaslavsky, Alan M. Models	94 ASASrvyRsr 236-241
de la Puente, Manuel; Wobus, Peter An item	nonresponse and log-linear analysis of the Spanish Language Forms Availability Test •	94 ASASrvyRsr 583-588
A revised theory of longitudinal	nonresponse • Wolford, Monica L.	94 ASASrvyRsr 960-965
binary data with outcome subject to non-ignorable	non-response • Baker, Stuart G. Marginal regression for repeated	95 Biomtrcs 51 1042-1052
Measuring and reporting	nonresponse rates in agricultural surveys • Roy, Laurent	94 ASASrvyRsr 1156-1161
Wronski, A. Control charts for	non-response rates in the Canadian Labour Force Survey • Hapuarachchi, K. P.;	94 ASASrvyRsr 937-942
	Non-response to sensitive questions: Nationalism in Germany • Krebs, Dagmar	94 ASASrvyRsr 1199-1203
Lou; Kalton, Graham; Brick, Mike Adjusting for panel	nonresponse in the Survey of Income and Program Participation (Disc: p440-442) • Rizzo,	94 ASASrvyRsr 422-427
Laplace approximations for large deviations of	nonreversible Markov processes. The nondegenerate case • Bolthausen, E. et al.	95 AnlsProb 23 236-267
an unevenly proportioned sample between risk and	non-risk groups • Lee et al. ... in risk assessment when logistic regression is used with	94 ASAProBuEc 304-309
	Nonsampling error	
	Quantifying errors in the Swedish consumer price index • Dalén, Jörgen	95 JOffStat 11 261-275
conversation (Disc: p169-177) • Schwarz, Norbert	What respondents learn from questionnaires: The survey interview and the logic of	95 IntStRvw 63 153-168
Surveys • Dietz, Elizabeth Reducing	nonsampling error in the Bureau of Labor Statistics Occupational Compensation	94 ASASrvyRsr 1042-1047
Review of	"Non-sampling errors in surveys" • Kessler, J. T., Kelsbeck, W. D. (Auth); Fielding, A. (Rev)	95 Ststcian 44 289-290
Estimation of	non-sharp support boundaries • Härdle, W.; Park, B. U.; Tsybakov, A. B.	95 JMultiAn 55 205-218
Teaching statistics to	non-specialists • Simpson, Judy M.	95 StatMed 14 199-208
Phillips, Michael J. An application of	nonstandard asymptotics to the analysis of repairable-systems data • Sweeting, Trevor J.;	95 Technmcs 37 428-435
Bootstrap for	nonstandard cases • Babu, Gutti Jogesh	95 JStPlInf 43 197-203
Estimation of a quantile in some	nonstandard cases • Xiang, Xiaojing	95 AnInStMa 47 105-117
Nonparametric tests for	nonstandard change-point problems • Ferger, D.	95 AnlsStat 23 1848-1861
Kaniovski, Yu.; Pflug, G.	Non-standard limit theorems for urn models and stochastic approximation procedures •	95 CommStSM 11 79-102
M.	Nonstandard local empirical processes indexed by sets • Deheuvels, Paul; Mason, David	95 JStPlInf 45 91-112
Survey of linear programming for standard and	nonstandard Markovian control problems. Part II: Applications • Kallenberg, L. C. M.	94 ZeitOpRs 40 127-143
On the distribution of the estimated mean from	nonstandard mixtures of distributions • Vännman, Kerstin	95 CommStA 24 1569-1584
Bootstrap methods for	nonstandard mixtures • Schucany, William R.; Wang, Suojin	94 ASAProPEng 59-64
Review of	"Non-standard rank tests" • Janssen, A.; Mason, D. M. (Auth); Jurečková, Jana (Rev)	95 SIAMRvw 37 261-262
M. M. Shewhart-type charts in	nonstandard situations (Disc: p24-35; Reply: P36-40) • Roes, Kit C. B.; Does, Ronald J.	95 Technmcs 37 15-24
Understanding the standardized/	non-standardized interviewing controversy • Beatty, Paul	95 JOffStat 11 147-160
S.	Nonstationarity of regressors and tests on real-interest-rate behavior • Mishkin, Frederic	95 JBES 13 47-51
Y. H. et al. Testing the arbitrage pricing theory and	non-stationarity of the risk premium: An error components non-linear approach • Chen,	94 ASAProBuEc 178-183
A new test for	nonstationarity against the stable alternative • Abadir, Karim M.	95 EconTher 11 81-104

on estimation and testing for a unit root in the	nonstationary autoregressive moving-average model • Yap, S. F.; Reinsel, G. C. Results	95 JTimSrAn 16 339-353
Some exact inference procedures for stationary and	nonstationary autoregressive processes • Dufour, Jean-Marie; Torrés, Olivier	94 ASAProBuEc 135-140
of coefficients of time series regression with a	nonstationary error process • Usami, Yoshihiro; Huzii, Mituaki Estimation	95 JTimSrAn 16 105-118
Reconfigurable combined forecasts in a	non-stationary inflationary environment • Volkov, V. Ya.; Gladkov, Yu. M.	95 JForcast 14 395-403
deconvolution of linear systems with multilevel	nonstationary inputs • Li, Ta-Hsin Blind	95 AnlsStat 23 690-704
Extremes and clustering on	nonstationary max-AR(1) sequences • Alpuim, M. T.; Catkan, N. A.; Hüsler, J.	95 StocProc 56 171-184
	Nonstationary multivariate spatial prediction • Haas, Timothy C.	94 ASAProEnvr 194-203
Ray, B. K. Estimation of the memory parameter for	nonstationary or noninvertible fractionally integrated processes • Hurvich, C. M.;	95 JTimSrAn 16 17-41
approximation of the number of exceedances of	nonstationary normal sequences • Hüsler, J.; Kratz, M. Rate of Poisson	95 StocProc 55 301-313
	Nonstationary process	
	The aftermath of Cramér's work on stochastic processes • Masani, P. R.	95 ScanActJ 66-96
similar to the exponential • Dimitrov, B. et al.	Definitions, characterizations and structural properties of probability distributions	95 JStPlInf 43 271-287
job search theory • Belzil, Christian	Estimating models of labor market transitions based on alternative specifications of	95 JStPlInf 43 243-255
	A numerical Bayesian test for cointegration of AR processes • Dorfman, Jeffrey H.	95 JEconmtx 66 289-324
	On a theorem of K. Schmidt • Bradley, Richard C.	95 StPrLet 24 9-12
	A vector of quarters representation for bivariate time series • Franses, Philip Hans	95 EcomtrRv 14 55-63
properties of distribution of maximum of Gaussian	nonstationary process occurring in covariance statistic • Ostrovskii, E. I. et al. Asymptotic	94 ThProbAp 39 527-534
estimators of cyclic functional parameters for some	nonstationary processes • Dehay, Dominique Asymptotic behavior of	95 StatDecn 13 273-286
Efficient IV estimation in	nonstationary regression • Kitamura, Yuichi; Phillips, Peter C. B.	95 EconTher 11 1095-1130
Robust	nonstationary regression • Phillips, Peter C. B.	95 EconTher 11 912-951
in demand systems when the regressors are	nonstationary • Ng, Serena Testing for homogeneity	95 JApEcmtx 10 147-163
Asymptotic distributions of records in	nonstationary schemes • Nevzorov, Valery	95 JStPlInf 45 261-273
	Nonstationary time series and cointegration • Phillips, Peter C. B.	95 JApEcmtx 10 87-94
A multivariate stochastic model with	non-stationary trend component • Kato, Hiroko; Naniwa, Sadao; Ishiguro, Makio	95 AplStMDA 11 77-95
sure convergence of series of stationary and related	nonstationary variables • Houdré, Christian On the almost	95 AnlsProb 23 1204-1218
Estimation and testing for unit roots in a partially	nonstationary vector autoregressive moving average model • Yap, S. F.; Reinsel, G. C.	95 JASA 90 253-267
Covering with blocks in the	non-symmetric case • Móri, Tamás F.	95 JTherPrb 8 139-164
Review of "Introduction to the	non-symmetric Dirichlet forms" • Ma, Z. M.; Röckner, M. (Auth); Fukushima, M. (Rev)	95 Metrika 42 70-72
Osada, H.; Saitoh, T. An invariance principle for	non-symmetric Markov processes and reflecting diffusions in random domains •	95 ProbTher 101 45-63
Zooplankton proportion estimates from	non-uniform sample volumes • Billheimer, Dean; Guttorp, Peter	95 EnvEcoSt 2 117-124
for the Strassen law of the iterated logarithm in	nonuniform topologies • Deheuvels, P.; Lifshits, M. A. Necessary and sufficient conditions	94 AnlsProb 22 1838-1856
On the local limit theorem for	non-uniformly ergodic Markov chains • Séva, Marc	95 JAppProb 32 52-62
	Non-uniqueness and inversions in cluster analysis • Morgan, Byron J. T.; Ray, Andrew P. G.	95 ApplStat 44 117-134
optimal stopping problem for random walks with	non-zero drift • Roters, Markus An	95 JAppProb 32 956-959
The basis of	norms for vague quantifiers • O'Muircheartaigh, Colm; Gaskell, George	94 ASASrvyRsr 1327-1332
	Normal	
	• see also bivariate –, inverse –, log- –, multivariate –	
	Normal approximation	
	Asymptotic expansions at work • Jensen, Jens Ledet	95 ScanActJ 143-152
	• see also Gaussian approximation	
$\hat{C}_{pk}$ • Nagata, Yasushi On	normal approximation to the distribution of the estimated process capability index	95 JJapnStS 25 159-171
On the	normal approximation for the sample p-quantile • Koti, Kallappa M.	94 ASAProStCp 188-193
On the	normal approximations of studentized U-statistic • Maesono, Yoshihiko	95 JJapnStS 25 19-33
The centered	normal conditionals distribution • Sarabia, José María	95 CommStA 24 2889-2900
Estimation of a	normal covariance matrix with incomplete data under Stein's loss • Konno, Yoshihiko	95 JMultiAn 52 308-324
Improved minimax estimators of	normal covariance and precision matrices • Gupta, A. K.; Ofori-Nyarko, Samuel	95 MaOpfStS 26 19-25
Asymptotic confidence bands for	normal distributions • Lin, Yachen	95 CommStA 24 641-652
model analysis of data from informatively censored	normal distributions • Smith, Fraser B.; Helms, Ronald W. EM mixed	95 Biomtrcs 51 425-436
The	normal distribution – Characterizations and applications • Bryc, W.	95 Springer:Brln:NY 139p
(Rev) Review of "The	normal distribution – Characterizations and applications" • Bryc, W. (Auth); Leslie, J. R.	96 ShrtBkRv 16/1 4-4
The	normal distribution: Characterizations with applications • Bryc, W.	95 Springer:Brln:NY 140p
and graphical models for the multivariate complex	normal distribution • Andersen, H. H.; Højbjerre, M.; Sørensen, D.; Eriksen, P. S. Linear	95 Springer:Brln:NY 183p
A jackknife estimator of the scale parameter of L_p-	normal distribution (Italian) • La Tona, Letizia	94 Metron 52/1 181-188
sequential procedure for estimating the mean of a	normal distribution • Tahir, Mohamed On the asymptotic regret of a	95 JStPlInf 48 331-338
program for visualizing comparisons between two	normal distributions (jse/v3n1/mathieson) • Mathieson, K.; Doane, D. P.; Tracy, R. L. A	95 JStEd 3/1 23kb
and selection indifference-zone formulation for	normal distributions using computer software • Poston, C. A. et al. Solving the ranking	94 ASAProStCp 290-291
Sequential estimation of the variance of a	normal distribution • Natarajan, Jayalakshmi; Strawderman, William E.	95 SqtlAnly 14 361-374
of a preliminary test estimator of variance in a	normal distribution • Tiwari, Gayatri; Saxena, K. P. On bias and efficiency	95 Metron 53 163-169
sion matrix of a weighted sample from a p-variate	normal distribution • Coelho, C. Brief note on the approximate distribution of the disper-	94 JIndnStA 32 77-81
A robust estimate of the location parameter for a	normal distribution with asymmetrical littering (Russian) • Dubinko, T. Yu.	95 VLenUM /2 94-96
tor for the case of two SUR equations with compound	normal disturbances • Hasegawa, Hikaru On small sample properties of Zellner's estima-	95 CommStB 24 45-59
regression under long-range dependent	normal errors • Csörgő, Sándor; Mielniczuk, Jan Nonparametric	95 AnlsStat 23 1000-1014
Efficient parametrisations for	normal linear mixed models • Gelfand, Alan E.; Sahu, Sujit K.; Carlin, Bradley P.	95 Biomtrka 82 479-488
A simple approach to simultaneous comparison of	normal means • Selukar, Rajesh S.; Shah, Arvind K.; Mishra, Satya N.	95 JStCmpSm 52 174-177
A note on the symmetry of	normal mean hypotheses and its implications • Hu, Xiaomi	95 StPrLet 25 15-20
Minimax- and Γ-minimax estimation of a bounded	normal mean under LINEX loss • Bischoff, Wolfgang; Fieger, Werner; Wulfert, Stefanie	95 StatDecn 13 287-298
critical region for testing simple and tree ordered	normal means • Sugiura, Nariaki Approximating Bayes	94 JStatRsr 28 1-20
Reference prior Bayesian analysis for	normal mean products • Sun, Dongchu; Ye, Keying	95 JASA 90 589-597
Efficiency of linear Bayes rules for a	normal mean: Skewed priors class • Mukhopadhyay, Saurabh; Vidakovic, Brani	95 Ststcian 44 389-397
Testing equality of several	normal means to a specified standard • Krishnamoorthy, K.; Shah, Arvind K.	95 JQualTek 27 132-138
Shah, A. K. Correction to "Testing equality of several	normal means to a specified standard" (V27 p132-138) • Krishnamoorthy, K.;	95 JQualTek 27 402-402
Coefficients of Lee-Gurland two-sample test on	normal means • Lee, Austin F. S.	95 CommStA 24 1743-1768
Use of ranked set sampling for test of a	normal mean • Shen, Wei-Hsiung	94 CalcutSt 44 183-193
Correction to "A note on posterior moments for a	normal mean with double-exponential prior" (V56 p605-610) • Mitchell, Ann F. S.	95 JRSS-B 57 471-471
Peng, Y. B.; Wright, F. T. Testing homogeneity of	normal means with a simply ordered alternative and dependent observations • Lei, X.;	95 CmpStDA 20 173-183
Minimax estimators of the mean vector in	normal mixed linear models • Bilodeau, Martin	95 JMultiAn 52 73-82
	Normal mixture	
Hamel, Denis; Massé, Jean-Claude	A Monte Carlo study of the robustness of prediction in growth curve models •	95 JStCmpSm 52 33-53
B. H. Analysis of tomato root initiation using a	normal mixture distribution • Gutierrez, R. G.; Carroll, R. J.; Wang, N.; Lee, G.-H.; Taylor,	95 Biomtrcs 51 1461-1468
lihood ratio test of one versus two components in a	normal mixture model fitted to grouped and truncated data • McLachlan et al. ... the like-	95 CommStB 24 965-985
Gupta, A. K.; Varga, T.	Normal mixture representations of matrix variate elliptically contoured distributions •	95 SankhyaA 57 68-78
Perturbation of the	normal model in linear regression • de Waal, D. J.; Groenewald, P. C. N.; Kemp, C. J.	95 SoAfrStJ 29 105-130
Watson test to the dynamic regression model under	normal and non-normal errors • King, M. L.; Harris, D. C. The application of the Durbin-	95 EcomtrRv 14 487-510
in linear regression when disturbances are not	normal" (V23 p2029-2046) • Singh et al. ... to "A generalized class of shrinkage estimators	95 CommStA 24 859-859

Ester Constructing tests for	normal order-restricted inference • Cohen, Arthur; Sackrowitz, Harold B.; Samuel-Cahn,	95 Bernoull 1 321-333
Consistent and asymptotically	normal parameter estimates for hidden Markov models • Rydén, Tobias	94 AnlsStat 22 1884-1895
	Normal plot	
	Patterns in residuals for 2^n designs • John, Peter W. M.	95 JAppStat 22 253-259
losses for empirical Bayes estimation: A note on	normal and Poisson cases • Fourdrinier, Dominique; Robert, Christian P. Intrinsic	95 StPrLet 23 35-44
of concentration functions to degenerate,	normal, and Poisson laws • Alekperov, Sh. O.; Kruglov, V. M. Convergence of convolutions	94 ThProbAp 39 197-216
Single test that two samples came from the same	normal population • Nelson, Lloyd S.	95 JQualTek 27 84-87
estimators of the common mean of two univariate	normal populations • De, Moloy On a class of admissible	95 CalcutSt 45 103-109
On the problem of the means of weighted	normal populations • Nikulin, M. S.; Voinov, V. G.	95 Questo 19 93-106
On fitting a modified limiting	normal to the posterior distribution • Ladalla, Joseph N.	95 JStPlInf 43 229-234
Calculating	normal probabilities • Bagby, Richard J.	95 AmMaMnth 102 46-49
Correction to "Characterization of the standard	normal probability density within the class of self-adjoint (self-reciprocal) ... • Rossberg	95 MaOpfStS 26 387-387
Equating Poisson and	normal probability functions to derive Stirling's formula • Walsh, Dennis P.	95 AmerStat 49 270-271
of change points in a sequence of independent	normal random variables • Lee, Chung-Bow Estimating the number	95 StPrLet 25 241-248
of the number of exceedances of nonstationary	normal sequences • Hüsler, J.; Kratz, M. Rate of Poisson approximation	95 StocProc 55 301-313
K. et al. Construction and selection of tightened-	normal-tightened plans indexed by maximum allowable percentage defective • Suresh, K.	94 JAppStat 21 589-595
of the Bayes risk in discrimination between	normal and uniform location-scale families • Nagaev, A. V. Asymptotics	95 MaMethSt 4 312-333
of sums and the sum of maxima for correlated	normal variables • Bain, Lee J.; Gan, Gaoxiong Expectations of the maximum	95 CommStB 24 775-780
Simulation of truncated	normal variables • Robert, Christian P.	95 StatComp 5 121-125
Some modifications of improved estimators of a	normal variance • Shinozaki, Nobuo	95 AnInStMa 47 273-286
On the invariant estimation of a	normal variance ratio • Madi, Mohamed T.	95 JStPlInf 44 349-357
Lower bounds on Bayes risks for estimating a	normal variance: With applications • Vidakovic, Brani; DasGupta, Anirban	95 CandJSt 23 269-282
Combining experts' opinions using a	normal-Wishart model • Wiper, Michael P.; French, Simon	95 JForcast 14 25-34
	Normality	
	• see also asymptotic –, goodness-of-fit, normal distribution, nonnormality	
How to use tests for univariate	normality to assess multivariate normality • Looney, Stephen W.	95 AmerStat 49 64-70
An alternative test for	normality based on normalized spacings • Chen, Ling; Shapiro, Samuel S.	95 JStCmpSm 53 269-287
for location parameters under departure from	normality • Basu, Sanjib; DasGupta, Anirban Robustness of standard confidence intervals	95 AnlsStat 23 1433-1442
On	normality and the linear regression model • Spanos, Aris	95 EcomtrRv 14 195-203
Runway landing distribution and tests for	normality • Andersen, John M.	94 ASAProBays 150-152
Comment on "Algorithm AS 181: The W-test for	normality" (V31 p176-180) • Royston, Patrick	95 ApplStat 44 547-551
p-variate adaptations of the Shapiro-Wilk test of	normality • Mudholkar, Govind S.; Srivastava, Deo Kumar; Lin, C. Thomas Some	95 CommStA 24 953-985
Brzeskwiniewicz, Henryk	Normality testing in a mixed model of a two-way nested classification • Wagner, Wiesław;	95 BiomtrcJ 37 889-895
Tests for	normality using estimated score function • Bera, Anil K.; Ng, Pin T.	95 JStCmpSm 52 273-287
erating functions • Choudhury, G. L. et al. Calculating	normalization constants of closed queuing networks by numerically inverting their gen-	95 JACM 42 935-970
Rotation of principal components: Choice of	normalization constraints • Jolliffe, Ian T.	95 JAppStat 22 29-35
An alternative test for normality based on	normalized spacings • Chen, Ling; Shapiro, Samuel S.	95 JStCmpSm 53 269-287
Kirmani, S. N. U. A. Some results on	normalized spacings from restricted families of distributions • Kochar, Subhash C.;	95 JStPlInf 46 47-57
behaviour of the moving block bootstrap for	normalized sums of heavy-tail random variables • Lahiri, S. N. On the asymptotic	95 AnlsStat 23 1331-1349
els • Griffith, D. A. et al. Trade-offs associated with	normalizing constant computational simplifications for estimating spatial statistical mod-	95 JStCmpSm 51 165-183
Puri, Madan L. Higher order asymptotic theory for	normalizing transformations of maximum likelihood estimators • Taniguchi, Masanobu;	95 AnInStMa 47 581-600
ability in a stress-strength model when X and Y are	normally distributed" (V34 p83-91) • Pham-Gia, T.; Turkkan, N. Comment on "Testing reli-	95 Technmcs 37 130-131
programme of constant-effort mist-netting in	North America • Desante et al. ... and survival rate estimates from MAPS, a continent-wide	95 JAppStat 22 935-947
How useful are recoveries of	North American passerines for survival analyses? • Francis, Charles M.	95 JAppStat 22 1075-1081
Asymptotic normality of the	NPMLE of linear functionals for interval censored data, case 1 • Huang, J.; Wellner, J. A.	95 StNeerla 49 153-163
The existence of doubly near resolvable	$(v,3,2)$-BIBDs • Lamken, E. R.	94 JComDsgn 2 427-440
Cycle structures of automorphisms of 2-	$(v,k,1)$ designs • Webb, Bridget S.	95 JComDsgn 3 341-348
The Delsarte polynomial of 5-	(v,k,λ) designs with three intersection numbers • Shrikhande, M. S.	93 JCISS 18 53-60
Some results on	$(v,\{5,w^*\})$-PBDs • Bennett, F. E.; Yin, J.	95 JComDsgn 3 455-468
trol signals Bayesian updating of the probability of	nuclear attack • Paté-Cornell, M. E. et al. Probabilistic interpretation of command and con-	95 RelEngSS 47 27-36
signals: Bayesian updating of the probability of	nuclear attack" (V47 p27-36) • Clarotti ... interpretation of command and control	95 RelEngSS 48 157-158
signals: Bayesian updating of the probability of	nuclear attack" • Paté-Cornell et al. ... "Probabilistic interpretation of command and control	95 RelEngSS 48 159-160
multipoint linkage analysis of recessive traits in	nuclear families, including homozygosity mapping • Kruglyak, L.; Daly, M. J. et al. Rapid	95 AJHuGene 56 519-527
Seuchter, S. A.; Baur, M. P. Linkage analysis in	nuclear families. 1: Optimality criteria for affected sib-pair tests • Knapp, M.;	94 HumHrdty 44 37-43
Knapp, M.; Seuchter, S. A. et al. Linkage analysis in	nuclear families. 2: Relationship between affected sib-pair tests and lod score analysis •	94 HumHrdty 44 44-51
Time-series models to forecast	nuclear net generation of electricity • Kundra, Inderjit	94 ASAProBuEc 327-332
to a shift in regulatory regime: The case of the US	nuclear power industry • Rust, John; Rothwell, Geoffrey Optimal response	95 JApEcmtx 10 75-118S
assessment of clusters of disease: An application to	nuclear power plant facilities and childhood leukaemia in Sweden • Waller et al. ... and	95 StatMed 14 3-16
models for point processes: An analysis of	nuclear power plant operations in Europe • Sturm, Roland Proportional hazard regression	94 JAppStat 21 533-540
sures • Kallianpur et al. Diffusion approximation of	nuclear space-valued stochastic differential equations driven by Poisson random mea-	95 AnlsApPr 5 493-517
The use of probability elicitation in the high-level	nuclear waste regulation program • DeWispelare, A. R.; Herren, L. T.; Clemen, R. T.	95 IntJFor 11 5-24
of leukaemia in young people around the La Hague	nuclear waste reprocessing plant: A sensitivity analysis • Viel, J.-F. et al. Incidence	95 StatMed 14 2459-2472
and comparative analysis of models for random	nucleation • Møller, Jesper Generation of Johnson-Mehl crystals	95 AdvAppPr 27 367-383
Models of DNA replication timing in interphase	nuclei: An exercise in inferring process from state • Carothers, A. D.; Bickmore, W. A.	95 Biomtrcs 51 750-755
Estimation of evolutionary distances between	nucleotide sequences • Zharkikh, Andrey	94 JMolEvol 39 315-329
strap estimation of phylogenetic variability from	nucleotide sequences: II. Four taxa without a molecular clock • Zharkikh et al. ... of boot-	92 JMolEvol 35 356-366
substitution-rate heterogeneity among regions of a	nucleotide sequence • Gaut, Brandon S.; Weir, B. S. Detecting	94 MolBioEv 11 620-629
trees when substitution rates differ over	nucleotide sites • Yang, Ziheng Evaluation of several methods for estimating phylogenetic	95 JMolEvol 40 689-697
of the heterogeneity of substitution rate among	nucleotide sites • Gu, Xun; Fu, Yun-Xin; Li, Wen-Hsiung Maximum likelihood estimation	95 MolBioEv 12 546-557
Estimating the pattern of	nucleotide substitution • Yang, Ziheng	94 JMolEvol 39 105-111
for comparing synonymous and nonsynonymous	nucleotide substitution rates, with application to the ... • Muse et al. ... approach	94 MolBioEv 11 715-724
A.; Nei, M. Unbiased estimates of the number of	nucleotide substitutions when substitution rate varies among different sites • Rzhetsky,	94 JMolEvol 38 295-299
of words with unusual occurrence frequency in	nucleotidic sequences • Colosimo, A.; Morante, S. et al. An improved method for detection	93 JThBio 165 659-672
factorial designs with linear models containing	nuisance parameters • Raktoe et al. Some aspects on the design and analysis of fractional	93 JCISS 18 369-395
Estimation of	nuisance parameters for inference based on least absolute deviations • Niemiro, W.	95 ZastApMa 22 515-529
based on estimating functions in the presence of	nuisance parameters (Disc: p173-199) • Liang, Kyung-Yee; Zeger, Scott L. Inference	95 StatSci 10 158-173
highest posterior density regions in the presence of	nuisance parameters • Ghosh, Jayanta K.; Mukerjee, Rahul Frequentist validity of	95 StatDecn 13 131-139
W. Admissibility of the likelihood ratio test when a	nuisance parameter is present only under the alternative • Andrews, D. W. K.; Ploberger,	95 AnlsStat 23 1609-1629
large deviation rates for parametric inferences with	nuisance parameters • Koutsoukos, A.; Slud, E. Inaccuracy rates and Hodges-Lehmann	95 JStPlInf 48 47-68
a two-way layout • Kao, N. L. et al. Small sample	null distribution of two nonparametric tests for comparing treatments with a control in	95 CommStB 24 149-164
Conventional	null hypothesis testing in active control equivalence studies • Ng, Tie-Hua	95 ConClnTr 16 356-358
S. Analysis of a contingency table with negative	numbers: Application of the analysis of a generalised Burt sub-table (French) • Nodjiram,	95 CAnDnnes 20 111-120
Who's	number 1 in college football? ... And how might we decide? • Stern, Hal S.	95 Chance 8/3 7-14
Review of	"Number-theoretic methods in statistics" • Fang, K.-T.; Wang, Y. (Auth); Ripley, B. D. (Rev)	95 JRSS-A 158 189-190

	Odds ratio (cont.)	
Fujii, Yoshinori	Projection-method Mantel-Haenszel estimator for $K\ 2 \times J$ tables • Yanagawa, Takashi;	95 JASA 90 649-656
Estimation of a common	odds ratio under binary cluster sampling • Ahn, Chul; Odom-Maryon, Tamara	95 StatMed 14 1567-1576
consistency of a family of estimators of the common	odds ratio • Yu, Kai Fun A necessary and sufficient condition for the strong	95 CandJSt 23 215-225
exposure data for improving the estimate of an	odds ratio • Dahm, P. F.; Gail, M. H. et al. Determining the value of additional surrogate	95 StatMed 14 2581-2598
Analyzing bivariate ordinal data using a global	odds ratio • Williamson, John M.; Kim, KyungMann; Lipsitz, Stuart R.	95 JASA 90 1432-1437
A model for binary time series data with serial	odds ratio patterns • Fitzmaurice, Garrett M.; Lipsitz, Stuart R.	95 ApplStat 44 51-61
surement error • Forbes, A. B. et al. Estimators of	odds ratio regression parameters in matched case-control studies with covariate mea-	95 JASA 90 1075-1084
tion of the exact confidence limits for the common	odds ratio in several 2×2 tables • Liao, J. G.; Hall, Charles B. Fast and accurate computa-	95 JCmpGrSt 4 173-179
Solving Markovian systems of	O.D.E. for availability and reliability calculations • Tombuyses, B.; Devooght, J.	95 RelEngSS 48 47-55
indicators on teachers within the framework of the	OECD-INES • Scheerens, Jaap International	94 ASAProGovt 159-163
a system of international education indicators: The	OECD INES project • Bottani, Norberto Creating	94 ASAProGovt 153-158
Review of "Analysing	offending: Data, models and interpretations" • Tarling, R. (Auth); Walker, Monica (Rev)	95 JRSS-A 158 357-358
	Official labour statistics: A historical perspective • Davidson, Roger	95 JRSS-A 158 165-173
	Official statistics	
	You can count on us – With confidence • McLennan, Bill	95 JRSS-A 158 467-489
	Official statistics and democracy • Bodin, Jean-Louis	94 ASAProGovt 170-172
Ten years of the Journal of	Official Statistics • Scheuren, Fritz	95 JOffStat 11 5-9
Metadata and	official statistics • Franchet, Y.	93 StatJUN 10 97-100
Statistical statesmanship in	official statistics • Fellegi, Ivan P.	94 ASAProGovt 7-8
A classification of types of use of	official statistics • Goedegebuure, R. V.; Overgaag, P. H. A.	91 StatJUN 8 217-232
Proportion of genes survived in	offspring conditional on inheritance of flanking markers • Guo, Sun-Wei	94 Genetcs 138 953-962
modification to the pairwise estimator of a parent-	offspring correlation • Eliasziw, Michael; Donner, Allan Evaluation of a proposed	95 SankhyaB 57 151-157
Sample-size determination for parent-	offspring regression of a composite trait • Orzack, Steven Hecht; Chappell, Rick	95 Biomtrcs 51 822-830
disease association from studies comparing	offspring with and without parental history of disease • Tiret et al. ... strength of genotype-	93 AnHuGene 57 141-149
analysis of gas chromatographic data of	oils from Eucalyptus species • Dunlop, P. J.; Bignell, C. M.; Jackson, J. F. et al. Chemometric	95 Chemolab 30 59-67
from probabilistic models: A case study in the	oil market • Abramson, Bruce; Finizza, Anthony Probabilistic forecasts	95 IntJFor 11 63-72
the sensory quality evaluation of extra virgin olive	oils • Vojnovic, D.; Campisi, B.; Mattei, A. et al. Experimental mixture design to ameliorate	95 Chemolab 27 205-210
namic thermal process in thermochromatography of	oil shale • Koel, M.; Kaljurand, M. et al. Factor analytical resolution of simultaneous, dy-	95 Chemolab 30 173-178
inequality approach to Nash dynamic game of	oil stockpiling • Zhao, L.; Yin, G. On variational	95 StocAnAp 13 627-649
Review of	"Ökonometrie (Fourth edition)" • Schneeweiß, Hans (Auth); Trenkler, Dietrich (Rev)	95 CmpStDA 19 360-360
Cyclical output, cyclical unemployment, and	Okun's coefficient: A new approach • Weber, Christian E.	95 JApEcmtx 10 433-445
single-locus approximations for linkage analyses of	oligogenic traits: Extension to multigenerational ... • Vieland et al. Adequacy of	93 HumHrdty 43 329-336
... repeated game advertising investment model for	oligopolistic markets with an application to the beer industry • Chintagunta et al.	95 DecisnSc 26 531-559
liorate the sensory quality evaluation of extra virgin	olive oils • Vojnovic, D.; Campisi, B.; Mattei, A. et al. Experimental mixture design to ame-	95 Chemolab 27 205-210
McKinley L.; Neumark, David Are	OLS estimates of the return to schooling biased downward? another look • Blackburn,	95 RvEconSt 77 217-229
– ... • Hisamatsu et al. The exact distribution of the	OLS and GLS estimators in regression with an integrated regressor and correlated errors	95 MaCmpSim 39 273-277
Kiviet, J. F.; Phillips, G. D. A.; Schipp, B. The bias of	OLS, GLS, and ZEF estimators in dynamic seemingly unrelated regression models •	95 JEconmtx 69 241-266
A note on the bias due to	omitted confounders • Drake, Christiana; McQuarrie, Allan	95 Biomtrka 82 633-638
The sensitivity of some general checks to	omitted variables in the linear model • Godfrey, L. G.; Orme, C. D.	94 IntEconR 35 489-506
Nikabadze, A. M.; Sun, Yanqing On	omnibus test for independence of a survival time from a covariate • McKeague, Ian W.;	95 AnlsStat 23 450-475
of two statistics for detecting clustering in	one dimension • Anderson, N. H.; Titterington, D. M. A comparison	95 JStCmpSm 53 103-125
convergence of conditional survival measure of	one-dimensional Brownian motion with a drift • Povel, Tobias On weak	95 AnlsApPr 5 222-238
Servet; San Martín, Jaime Asymptotic laws for	one-dimensional diffusions conditioned to nonabsorption • Collet, Pierre; Martínez,	95 AnlsProb 23 1300-1314
Ergodic theorems for transient	one-dimensional diffusions • Athreya, K. B.; Weerasinghe, A. P. N.	95 StocProc 58 173-185
Kay, Jim A regularization method for	one-dimensional edge detection and edge-preserving smoothing • Cheng, Bing;	95 CmpStQ 10 53-69
Martingale methods for random walks in a	one-dimensional random environment • Butov, A. A.	94 ThProbAp 39 558-572
the comparison of probability laws of a subset of	one-dimensional real variates (Italian) • Freddi, Augusto On	95 Metron 53 147-161
tions to the non-linear prediction problem for the	one-dimensional strictly stationary time series • Okabe et al. ... of KM_2O-Langevin equa-	95 JMaJapan 47 349-367
On the ultimate value of local time of	one-dimensional super-Brownian motion • Kaj, I.; Salminen, P.	95 StocProc 59 21-42
Bayes estimator for the parameter of continuous	one parameter exponential families (Chinese) • Xuc ... and convergence rates of empirical	95 ChJApPSt 10 35-45
monotone empirical Bayes tests for the continuous	one-parameter exponential family • Karunamuni, R. J.; Yang, H. On convergence rates of	95 StatDecn 13 181-192
Aubin, E. C. Q.; Ferrari, S. L. P. Bartlett corrections for	one-parameter exponential family models • Cordeiro, G. M.; Cribari-Neto, F.;	95 JStCmpSm 53 211-231
Comment on "A large sample test for	one-parameter families of copulas" (V23 p1311-1317) • Carrière, J. F.	95 CommStA 24 2959-2962
A look-up table for the corresponding estimated	one-parameter logistic model scale score and the unbiased domain ... • Li LTDOMAIN:	95 ApPsychM 19 50-50
asymptotics and the likelihood principle:	One-parameter models" (V81 p1-10) • Pierce, D. A.; Peters, D. Correction to "Higher-order	95 Biomtrka 82 235-235
A geometric approach to optimal design for	one-parameter non-linear models • Haines, Linda M.	95 JRSS-B 57 575-598
pendence parameter due to incomplete data from a	one-parameter spatial Gaussian process • Krug, T.; Martin, R. J. Information loss on the de-	95 CommStA 24 3055-3072
Is	one sample enough? • Liu, Peng T.; Albert, Richard H.; Springer, Janet A.	94 ASAProBays 18-22
A formula for deficiency:	One sample L- and R-tests. I • Bening, V. E.	95 MaMethSt 4 167-188
A formula for deficiency:	One-sample L- and R-tests. II • Bening, V. E.	95 MaMethSt 4 274-293
	One-sample one-sided student t-test under contaminants • Zieliński, Ryszard	95 MaOpfStS 27 143-150
uniform scale distributions against two-sided and	one-sided alternatives • Cohen, Arthur; Fygenson, Mendel Testing homogeneity of	95 JASA 90 1062-1067
Likelihood ratio tests for symmetry against	one-sided alternatives • Dykstra, Richard; Kochar, Subhash; Robertson, Tim	95 AnInStMa 47 719-730
A score test against	one-sided alternatives • Silvapulle, Mervyn J.; Silvapulle, Paramsothy	95 JASA 90 342-349
	One-sided β-content tolerance intervals for mixed models • Vangel, Mark G.	94 ASAProPEng 200-206
Bean, Judy A. On the sample size for	one-sided equivalence of sensitivities based upon McNemar's test • Lu, Ying;	95 StatMed 14 1831-1839
On some asymptotic properties of U statistics and	one-sided estimates • Boxe, Arup; Dasgupta, Ratan	94 AnlsProb 22 1715-1724
A Hotelling's T^2-type statistic for testing against	one-sided hypotheses • Silvapulle, Mervyn J.	95 JMultiAn 55 312-319
trials with multiple outcomes: The multivariate	one-sided hypothesis with unknown covariances • Frick, H. Comparing	95 BiomtrcJ 37 909-917
A universal	one-sided law of the iterated logarithm • Mason, David M.	94 AnlsProb 22 1826-1837
Kwong, Koon-Shing Evaluation of	one-sided percentage points of the singular multivariate normal distribution •	95 JStCmpSm 51 121-135
Absolute continuity of	one-sided random translations • Sato, Hiroshi; Tamashiro, Masakazu	95 StocProc 58 187-204
Tsai, Hsien-Tang; Moskowitz, Herbert; Tang, Jen A	one-sided single screening procedure based on individual unit misclassification error •	95 IIETrn 27 695-706
Paula, Gilberto A.; Sen, Pranab K.	One-sided tests in generalized linear models with parallel regression lines •	95 Biomtrcs 51 1494-1501
Uniformly more powerful,	one-sided tests for hypotheses about linear inequalities • Liu, Huimei; Berger, Roger L.	95 AnlsStat 23 55-72
Sing Bias of two	one-sided tests procedures in assessment of bioequivalence • Liu, Jen-Pei; Weng, Chung-	95 StatMed 14 853-861
On some	one-sided tests in the two-sample problem • Bajorski, Piotr	95 CommStA 24 75-86
Chen, Ling A	one-sided trimmed mean estimator for the mean of positively skewed distributions •	95 AstrlJSt 37 313-322
variation • Morel, Jorge G.; Koehler, Kenneth J. A	one-step Gauss-Newton estimator for modelling categorical data with extraneous	95 ApplStat 44 187-200
	One-step GM-estimators for orthogonal regression • Abdullah, Mokhtar B.	95 JStCmpSm 52 301-309
	One-step saddlepoint approximations for quantiles • Wang, Suojin	95 CmpStDA 20 65-74
Robust Bayesian estimators in a	one-way ANOVA model • Bian, G.	95 Test 4 115-135
Hierarchical Bayes estimators of the error variance in	one-way ANOVA models • Datta, Gauri Sankar; Ghosh, Malay	95 JStPlInf 45 399-411

One-way classification

Yokoyama, Takahisa	LR test for random-effects covariance structure in a parallel profile model •	95 AnInStMa 47 309-320
Optimal Bayesian design of experiments for the	one-way random effects model • Lohr, Sharon L.	95 Biomtrka 82 175-186
MINQUE and the PSD-MINQMBE in the unbalanced	one-way random effects model • Shaban, Abdel Magid M. Comparison between the	94 EgpStJ 38 222-230
of error variance following a preliminary test in	one-way random model • Chiou, Paul; Han, Chien-Pai Interval estimation	95 CommStB 24 817-824
priors for the variance components of an unbalanced	one-way random model • Belzile, Éric; Angers, Jean-François Choice of noninformative	95 CommStA 24 1325-1341
Nandram, Balgobin A rank-based analysis of	one-way repeated measures designs with a changing covariate • Rashid, M. Mushfiqur;	95 JStPlInf 46 249-263
	Onsager-Machlup functional for some smooth norms on Wiener space • Capitaine, Mireille	95 ProbTher 102 189-201
measurement error • Bradley, D. C.; Steil, G. M. et al.	OOPSEG: A data smoothing program for quantitation and isolation of random	95 CmpBiomd 46 67-77
A non-linear model of an	open flow adsorption process • Argyelan, J.; Nassar, R.	95 EnvrMtrc 6 195-203
Flows and some extreme values in multiple server	open Jackson network • Park, YouSung; Lee, Haeyong; Kim, Keeyoung	95 JKoreaSt 24 389-405
of a double-blind comparative trial and subsequent	open label extension in HIV-infected subjects • Hwang et al. ... models to the analysis	94 ASAProBiop 194-199
approach to multivariable robust filtering and	open-loop control • Öhrn, Kenth; Ahlén, Anders; Sternad, Mikael A probabilistic	95 IEEEAuCn 40 405-418
Threshold variable selection in	open-loop threshold autoregressive models • Chen, Rong	95 JTimSrAn 16 461-481
• Burnham, K. P. et al. Selection among	open population capture-recapture models when capture probabilities are heterogeneous	95 JAppStat 22 611-624
system for the analysis of mark-recapture data from	open populations • Arnason, A. Neil; Schwarz, Carl J. POPAN-4: Enhancements to a	95 JAppStat 22 785-800
Fourier transform algorithm in the calculation of the	operating characteristics of group sequential clinical trials • Elson, Paul J. Use of the fast	95 CmpStDA 20 491-498
Spatio-temporal analysis of regional ozone data for	operational evaluation of an air quality model • Sampson, P. D.; Guttorp, P.; Meiring, W.	94 ASAProEnvr 46-55
Evaluation of	operational reliability from ground test data • Ke, Huei-Yaw	95 RelEngSS 49 103-107
Estimation of the effect of	operational seeding on rain amounts in Israel • Nirel, Ronit; Rosenfeld, Daniel	95 JApMetrl 34 2220-2229
Reid, D. H.	Operational time and a fundamental problem of insurance in a data-rich environment •	95 AplStMDA 11 257-269
Towards an	operative view of semantic metadata • Froeschl, K. A.	93 StatJUN 10 181-194
Random fixed points for random multivalued	operators defined on unbounded sets in Banach spaces • Beg, Ismat; Shahzad, Naseer	95 StocAnAp 13 269-277
Hyper-weakly harmonizable processes and	operator families • Leitner, M.	95 StocAnAp 13 471-485
Hazod, W. (Rev) Review of	“Operator-limit distributions in probability theory” • Jurek, Z. H.; Mason, D. J. (Auth);	95 Metrika 42 265-267
Blower, G. (Rev) Review of	“Operator-limit distributions in probability theory” • Jurek, Z. J.; Mason, J. D. (Auth);	95 JRSS-A 158 352-353
(Auth); Veeh, Jerry Alan (Rev) Review of	“Operator-limit distributions in probability theory” • Jurek, Zbigniew J.; Mason, J. David	95 BAmMaSoc 32 278-280
continuous-time flexible manufacturing and	operator scheduling model solved by deconvexification ... • Eaves et al. A discounted-cost	95 PrEngInf 9 65-98
Moments of measures attracted to	operator semi-stable laws • Scheffler, Hans-Peter	95 StPrLet 24 187-192
Veeh, J. A.	Operator stable probability measures: An overview • Hudson, W. N.; Mason, J. D.;	94 ThProbAp 39 264-276
of methods for correlated ordinal measures with	ophthalmic applications • Gange, S. J.; Linton, K. L. P.; Scott, A. J. et al. A comparison	95 StatMed 14 1961-1974

Opinion poll

	Exact distribution of the sample lead • Kmietowicz, Z. W.	95 JAppStat 22 151-161
Correction to “The significance of the lead in	opinion polls” (V17 p9-30) • Kmietowicz, Z. W.	95 JAppStat 22 551-551
Sawyer, T.; Tanur, J. M. Pseudo-	opinion polls: SLOP or useful data? • Horvitz, D.; Koshland, D.; Rubin, D.; Gollin, A.;	95 Chance 8/2 16-25
Review of “The Phantom respondents:	Opinion surveys and political representation” • Brehm, John (Auth); Parsley, Teresa L. (Rev)	94 JOffStat 10 351-355
Performance analysis of slotted ALOHA protocol for	optical fiber local area networks with star topology • Ahmed, N. U.; Yeap, T. H.	95 StocAnAp 13 389-410
recognition applied to RIFS sensor ... • Kraus et al.	Optical reflectometric gas sensing: Classification of hydrocarbon vapours by pattern	95 Chemolab 30 211-221
Feuer et al. Free radical polymerization of nonlinear	optical side chain methacrylate copolymers in dimethylsulfoxide – An experimental ... •	95 Chemolab 29 113-124
Lohr, Sharon L.	Optimal Bayesian design of experiments for the one-way random effects model •	95 Biomtrka 82 175-186
Sinha, Bikas Kumar	Optimal block designs revisited: An approximate theory detour • Pukelsheim, Friedrich;	95 StPrLet 24 375-380
Shah, Kirti R.; Das, Ashish	Optimal block designs with minimal and nearly minimal number of units • Dey, Aloke;	95 StSinica 5 547-558
for liquid chromatography. II. Exploring and finding	optimal conditions in the search area of the multidimensional ... • Kaufmann ... strategy	95 Chemolab 27 105-114
Dynamic programming equations of stochastic	optimal control in Banach space • Zhu, Qingxin; Ahmed, N. U.	95 StocAnAp 13 369-387
	Optimal control of dams • Abdel-Hameed, Mohamed; Nakhi, Yasmin	95 AplStMDA 11 25-34
(Second edition) • Basar, T.; Bernhard, P. H^∞-	optimal control and related minimax design problems – A dynamic game approach	95 Birkhs:Basl:MA 411p
(Second edition)” • Basar, T. et al. Review of “H^∞-	optimal control and related minimax design problems – A dynamic game approach	96 ShrtBkRv 16/1 11-11
Nearly	optimal controls for singularly perturbed wideband noise systems • Ramachandran, K. M.	95 StocAnAp 13 335-353
optimization in the calculus of variations and	optimal control theory” • Gregory, J.; Lin, C. (Auth); Prvan, T. (Rev) Review of “Constrained	95 AstrlJSt 37 250-250
	Optimal control of two competing diseases or species • Kyriakidis, E. G.	95 MaScitst 20 56-66
form estimate of variance matrix under some	optimal criterion (Chinese) • Xie, Minyu Admissibility of quadratic	95 AMaAPknC 17 149-153

Optimal design

	A-efficient designs for bioassays • Mukerjee, Rahul; Gupta, Sudhir	95 JStPlInf 48 247-259
Brand, Richard J.	[Algorithm AS 299] Generation of simplex lattice points • Chasalow, Scott D.;	95 ApplStat 44 534-545
models • Lai, Tze Leung	Asymptotic properties of nonlinear least squares estimates in stochastic regression	94 AnlsStat 22 1917-1930
able: A case study • Radson, Darrell; Herrin, Gary D.	Augmenting a factorial experiment when one factor is an uncontrollable random vari-	95 Technmcs 37 70-81
	Breakdown points for designed experiments • Müller, Christine H.	95 JStPlInf 45 413-427
invariant sets • Cheng, Ching-Shui	Complete class results for the moment matrices of designs over permutation-	95 AnlsStat 23 41-54
designs • Meyer, Ruth K.; Nachtsheim, Christopher J.	The coordinate-exchange algorithm for constructing exact optimal experimental	95 Technmcs 37 60-69
Abdulhadi M.	Designing for minimally dependent observations • Torsney, Ben; Alahmadi,	95 StSinica 5 499-514
Lee, Robert Edwin; Spurrier, John D.	Distribution-free multiple comparisons between successive treatments •	95 JNnparSt 5 261-273
Shah, Kirti R.	Efficient block designs in the presence of trends • Jacroux, Mike; Majumdar, Dibyen;	95 StSinica 5 605-615
	Minimax designs in linear regression models • Dette, H.; Heiligers, B.; Studden, W. J.	95 AnlsStat 23 30-40
Hackl, P.; Müller, W. G.	Optimal and practicable designs for measuring plaque pH-profiles • Fedorov, V. V.;	95 StatMed 14 2609-2617
Demetrio, C. G. B.; Zocchi, S. S.	Optimum dose levels when males and females differ in response • Atkinson, A. C.;	95 ApplStat 44 213-226
	Robust linear calibration • Kitsos, Christos P.; Müller, Christine H.	95 MaOpfStS 27 93-106
D. C.; Torng, J. C.-C.; Cochran, J. K.; Lawrence, F. P.	Statistically constrained economic design of the EWMA control chart • Montgomery,	95 JQualTek 27 250-256
	Three-stage designs for seed testing experiments • Ridout, M. S.	95 ApplStat 44 153-162
Sitter, Randy R.; Torsney, Ben	Optimal designs for binary response experiments with two design variables •	95 StSinica 5 405-419
Marginally restricted constrained	optimal designs • Huang, Mong-Na; Chang, Hsiu-Fen	95 SankhyaB 57 128-141
Parmigiani, Giovanni	Optimal design via curve fitting of Monte Carlo experiments • Müller, Peter;	95 JASA 90 1322-1330
	Optimal design in end-use metering experiments • Bartels, Robert; Fiebig, Denzil G.	95 MaCmpSim 39 305-309
Review of	“Optimal design of experiments” • Pukelsheim, F. (Auth); Titterington, D. M. (Rev)	95 Metrika 42 74-75
Polish) Review of	“Optimal design of experiments” • Pukelsheim, Friedrich (Auth); Mejza, Stanisław (Rev.	94 ListyBio 31 121-125
Review of	“Optimal design of experiments” • Pukelsheim, Friedrich (Auth); Tobias, Randall D. (Rev)	95 JQualTek 27 93-94
	Optimal designs for experiments with a large number of factors • Jeong, Yousceek	95 JStPlInf 46 381-393
	Optimal designs for identifying the degree of a polynomial regression • Dette, Holger	95 AnlsStat 23 1248-1266
	Optimal design for Langmuir with two inputs • Betne, Prabha; Katti, S. K.	94 ASAProBiop 425-429
Azen, Stanley P. An alternative approach to the	optimal design of an LD50 bioassay • Markus, Richard A.; Frank, Jess; Groshen, Susan;	95 StatMed 14 841-852
	Optimal designs for mixture amount experiments • Hilgers, Ralf-Dieter; Bauer, Peter	95 JStPlInf 48 241-246
A geometric approach to	optimal design for one-parameter non-linear models • Haines, Linda M.	95 JRSS-B 57 575-598
(Chinese) • Guan, Yingnan; Zhang, Chongqi I_λ-	optimal designs of parameter estimation for the additive mixture model of 2-degree	95 AMaAPknC 17 161-172
Determination of	optimal design parameters of S control chart • Yang, Su-Fen	94 ASAProQlPr 327-333
Park, Hee-chang; Lee, Suk-hoon	Optimal design of partially step-stress life testing for the series systems (Korean) •	95 KorApS 8/2 121-132

Studden, William J.	Optimal designs for polynomial regression when the degree is not known • Dette, Holger;	95 StSinica 5 459-473
	Optimal designs on the refinement of the n-way classification model • Ting, C.-P.	93 UtilMath 43 141-154
A V-	optimal design for Scheffé's polynomial model • Liu, Shuangzhe; Neudecker, Heinz	95 StPrLet 23 253-258
for the construction of constrained D- and L-	optimal designs using efficiency plots • Wong, Weng Kee A graphical approach	95 JStCmpSm 53 143-152
Heiligers, Berthold Algorithms for	optimal design with application to multiple polynomial regression • Gaffke, Norbert;	95 Metrika 42 173-190
Sinha, Bikas K.	Optimal designs with component-wise orthogonal row-column structures • Shah, K. R.;	93 JCISS 18 68-78
Mathew, Thomas Minimaxity of randomized	optimal designs with respect to a general optimality criterion • Bhaumik, Dulal K.;	95 SankhyaB 57 122-127
Mukerjee, Rahul On	optimal designs with restricted circular string property • Huda, S.; Khan, I. H.;	95 CmpStDA 19 75-83
Tahir, Mohamed; Rekab, Kamel Asymptotically	optimal empirical Bayes procedures for selecting good uniform $U(0,\theta_i)$ populations •	95 StocAnAp 13 233-247
Sengupta, S.	Optimal estimation of a finite multidimensional population total • Mukerjee, Rahul;	95 JStPlInf 48 339-346
Arnab, Raghunath	Optimal estimation of a finite population total under randomized response surveys •	95 MaOpfStS 27 175-180
	Optimal estimation of sigma for X charts • Baran, Gary S.	94 ASAProQlPr 321-326
sion models used in long-term ... • Rudolph et al.	Optimal experimental design and accuracy of parameter estimation for nonlinear regres-	95 BiomtrcJ 37 183-190
exchange algorithm for constructing exact	optimal experimental designs • Meyer, R. K.; Nachtsheim, C. J. The coordinate-	95 Technmcs 37 60-69
Boehnke, M. Bayesian methods and	optimal experimental design for gene mapping by radiation hybrids • Lange, K.;	92 AnHuGene 56 119-144
A two-factor model of the term structure with	optimal filtering behaviour • Fornari, F.; Mele, A.	94 Statstca 54 293-312
An	optimal inspection-repair-replacement policy for standby systems • Yeh, Lam	95 JAppProb 32 212-223
	Optimal interval	
Casella, George	Improved invariant set estimation for general-scale families • Goutis, Constantinos;	95 JStPlInf 44 327-340
information: An algorithm for computing a locally	optimal periodic policy • Hordijk et al. Undiscounted Markov decision chains with partial	94 ZeitOpRs 40 163-181
Blackwell	optimal policies in a Markov decision process with a Borel state space • Yushkevich, A. A.	94 ZeitOpRs 40 253-288
Wei; Adachi, Kohichi; Kowada, Masasi Dynamically	optimal replacement policy for a shock model in a Markov random environment • Feng,	94 JOpRsJpn 37 255-270
	Optimal robust Bayes estimation • Li, Yuanzhang; Saxena, K. M. Lal	95 JStPlInf 46 365-380
approximation and a neural network model for	optimal sample size in vendor selection • Houshmand, A. A. et al. An analytical	95 JStCmpSm 53 65-78
Daniela	Optimal sampling designs for dependent spatial units • Benedetti, Roberto; Palma,	95 EnvrMtrc 6 101-114
Estimation, variance and	optimal sampling of gene diversity. I. Haploid locus • Pons, O.; Petit, R. J.	95 ThApGene 90 462-470
S.; Gomm, Ian S.; Armour, Len Location of the	optimal sampling point for the quality assessment of continuous streams • Barnett, Neil	95 AstrlJSt 37 145-152
Bar-Lev, S. K.; Parlar, M.; Perry, D.	Optimal sequential decisions for incomplete identification of group-testable items •	95 SqtlAnly 14 41-57
error • Hughes, M. D.; Thompson, S. G.; Pocock, S. J.	Optimal sequential screening guidelines for quantitative risk factors measured with	95 JASA 90 19-26
An	optimal sequential selection procedure • Dragalin, V.	95 CommStA 24 429-443
A kind of asymptotically	optimal sequential test under complex supposition • Chen, Jiading; Hickernell, Fred J.	95 SciSPkng 24 673-683
P.; Chiera, C.	Optimal software release policy based on a two-person game of timing • Zeephongsekul,	95 JAppProb 32 470-481
	Optimal stopping	
	The best-choice secretary problem with random freeze on jobs • Samuel-Cahn, Ester	95 StocProc 55 315-327
sampling • Hamada, Toshio	An exponential two-armed bandit problem with one arm known under batch	95 JJapnStS 25 205-216
	From perpetual strangles to Russian options • Gerber, Hans U.; Shiu, Elias S. W.	94 InsrneME 15 121-126
	Prophet inequalities for dependent random variables • Wittmann, Rainer	95 Stochatx 52 283-293
A stochastic game of	optimal stopping and order selection • Gnedin, Alexander V.; Krengel, Ulrich	95 AnlsApPr 5 310-321
An	optimal stopping problem for random walks with non-zero drift • Roters, Markus	95 JAppProb 32 956-959
	Optimal stopping in 'The Showcase Showdown' • Coe, Paul R.; Butterworth, William	95 AmerStat 49 271-275
Development of an	optimal test based on weighted rankings • Larocque, Denis; Tardif, Serge	95 CommStB 24 367-383
Search for a sequence of	optimal tests to control manufacturing costs (French) • Bergeret, F.; Chandon, Y.	95 RStApp 43/3 21-33
The robustness of point	optimal testing for Rosenberg random regression coefficients • Brooks, Robert	95 EcomtrRv 14 35-53
	Optimal two-sided group sequential tests • Eales, John D.; Jennison, Christopher	95 SqtlAnly 14 273-286
	Optimality	
correlated • Bischoff, Wolfgang	Determinant formulas with applications to designing when the observations are	95 AnInStMa 47 385-399
Derivation of	optimality conditions for a stochastic control problem • Ladde, G. S.; Medhin, N. G.	95 StocAnAp 13 165-176
Linkage analysis in nuclear families. 1:	Optimality criteria for affected sib-pair tests • Knapp, M.; Seuchter, S. A.; Baur, M. P.	94 HumHrdty 44 37-43
Discrete-time Markov control processes: Basic	optimality criteria • Hernández-Lerma, O.; Lasserre, J.-B.	95 Springer:Brln:NY 216p
The design of tests with different	optimality criteria (Dutch) • Berger, Martijn P. F.; van der Linden, Wim J.	95 TOnderwR 20 79-92
block designs and their efficiencies under various	optimality criteria • Hedayat, A. S. et al. Contingently and virtually balanced incomplete	95 StSinica 5 575-591
optimal designs with respect to a general	optimality criterion • Bhaumik, Dulal K.; Mathew, Thomas Minimaxity of randomized	95 SankhyaB 57 122-127
On asymptotic	optimality of estimating functions • Chen, K.; Heyde, C. C.	95 JStPlInf 48 101-112
Finite-sample, small-dispersion asymptotic	optimality of the non-linear least squares estimator • So, Beong-Soo	95 JKoreaSt 24 303-312
Estimation of parameters in survey sampling:	Optimality • Godambe, V. P.	95 CandJSt 23 227-243
Asymptotic	optimality of the Wald sequential test • Koell, Christophe	95 SqtlAnly 14 341-360
	Optimization	
Yang, Ziqiang	An algorithm for nonlinear L_1 curve-fitting based on the smooth approximation •	95 CmpStDA 19 45-52
Alameddine, Amine; Glickman, Theodore S.	Biconvex models and algorithms for risk management problems • Sherali, Hanif D.;	94 AJMMgSci 14 197-228
Parker, R. L.	Bounded-variable least-squares: An algorithm and applications • Stark, P. B.;	95 CmpStQ 10 129-141
a dynamic errors-in-variables model • Vregelaar	On computing objective function and gradient in the context of least squares fitting	95 SIAMSSC 16 738-753
ity, and other irregular features • Dorsey, R. E. et al.	Genetic algorithms for estimation problems with multiple optima, nondifferentiabil-	95 JBES 13 53-66
Li, Susan	The role of cardinal utilities in multiple objective programming • Huang, Zhimin;	94 AJMMgSci 14 301-325
	A stochastic target costing model (German) • Abel, P.	95 ZAngMaMk 75 557-558S
	Tools for two-dimensional exploratory projection pursuit • Posse, Christian	95 JCmpGrSt 4 83-100
	• see also global –, process –, stochastic –	
Hörchner, U. et al. Simulated-annealing-based	optimization algorithms: Fundamentals and wavelength selection applications •	95 JChemtrc 9 283-308
[Algorithm AS 298] A hybrid	optimization algorithm • Brooks, Stephen P.	95 ApplStat 44 530-533
(Auth) et al. Review of "Network models in	optimization and their applications in practice" • Glover, F.; Kingman, D.; Phillips, N. V.	95 JClass 12 150-152
Lin, C. (Auth); Prvan, T. (Rev) Review of "Constrained	optimization in the calculus of variations and optimal control theory" • Gregory, J.;	95 AstrlJSt 37 250-250
Edward R. Mean-absolute-error representation and	optimization of computational-morphological filters • Loce, Robert P.; Dougherty,	95 CVGIPGr 57 27-37
Constrained	optimization of experimental design (Disc: p149-178) • Cook, Dennis; Fedorov, Valery	95 MaOpfStS 26 129-148
Ivan The new ARSTI	optimization method: Adaptive random search with translating intervals • Edissonov,	94 AJMMgSci 14 143-166
Review of "Markov models and	optimization" • Davis, M. H. A. (Auth); Collins, Sean (Rev)	95 JAppStat 22 186-187
Review of "Markov models and	optimization" • Davis, M. H. A. (Auth); Dassios, Angelos (Rev)	95 JASA 90 1134-1135
Review of "Markov models and	optimization" • Davis, M. H. A. (Auth); Hoffman, Lorrie L. (Rev)	95 Technmcs 37 236-236
Review of "Markov models and	optimization" • Davis, M. H. A. (Auth); Lenhart, Suzanne (Rev)	95 SIAMRvw 37 262-263
Georgiadis, L.; Tsoucas, P. Problems of adaptive	optimization in multiclass M/GI/1 queues with Bernoulli feedback • Bhattacharya, P. P.;	95 MaOpRsr 20 355-380
Review of "Least squares	optimization in multivariate analysis" • Berge, Jos M. F. ten (Auth); Takane, Yoshio (Rev)	95 Psymtrka 60 447-448
cancer post-treatment surveillance. Estimation and	optimization problems • Tsodikov, A. D.; Asselain, B. et al. Discrete strategies of	95 Biomtrcs 51 437-447
Locating global minima in	optimization problems by a random-cost approach • Berg, Bernd A.	93 Nature 361 708-710
Dual response surface	optimization • Lin, Dennis K. J.; Tu, Wanzhu	95 JQualTck 27 34-39
B. (Rev) Review of	"Optimization techniques in statistics" • Rustagi, Jagdish S. (Auth); Fotopoulos, Stergios	95 Technmcs 37 466-467
Review of "Probability, statistics and	optimization: A tribute to Peter Whittle" • Kelly, F. P. (Ed); Bingham, N. H. (Rev)	95 JRSS-A 158 353-354

A useful property of	order statistics and its application to UMVU estimation • Boukai, B.	95 StHefte 36 279-283
number of runs and some other two-sample rank	order statistics for arbitrary sample sizes • Saran, Jagdish Joint distributions based on the	95 JStPlInf 47 257-275
Rychlik, Tomasz Bounds for	order statistics based on dependent variables with given nonidentical distributions •	95 StPrLet 23 351-358
of "The chronological annotated bibliography of	order statistics (Volumes IV-VII)" • Harter, H. L. (Auth); Hušková, M. (Rev) Review	95 MaOpfStS 26 182-182
distributions closed under concomitance of	order statistics • Balasubramanian, K.; Balakrishnan, N. On a class of multivariate	95 StPrLet 23 239-242
Saddlepoint approximations for the difference of	order statistics • Harvill, Jane L.; Newton, H. Joseph	95 Biomtrka 82 226-231
saddlepoint approximations for the difference of	order statistics • Harvill, Jane L.; Newton, H. Joseph Using symbolic math to evaluate	95 CommStB 24 781-791
Gan, Gaoxiong; Bain, Lee J. Distribution of	order statistics for discrete parents with applications to censored sampling •	95 JStPlInf 44 37-46
Selection of the k largest	order statistics for the domain of attraction of the Gumbel distribution • Wang, Julian Z.	95 JASA 90 1055-1061
Recurrence relations for expectations of functions of	order statistics for doubly truncated distributions and their ... • Mohie EL-Din et al.	95 CommStA 24 997-1010
On moment generating function of	order statistics for doubly truncated exponential distribution • El-Din, M. M. M. et al.	95 Metron 53 171-183
recurrence relations between product moments of	order statistics of a doubly truncated Pareto distribution • Khurana, A. P.; Jha, V. D. Some	95 SankhyaB 57 1-9
Sums of independent triangular arrays and extreme	order statistics • Janssen, Arnold	94 AnlsProb 22 1766-1793
LAN of extreme	order statistics • Falk, Michael	95 AnInStMa 47 693-717
On some predictors of future	order statistics • Raqab, Mohammad Z.; Nagaraja, Haikady N.	95 Metron 53 185-204
A concept of generalized	order statistics • Kamps, Udo	95 JStPlInf 48 1-23
Review of "A concept of generalized	order statistics" • Kamps, U. (Auth); Plachky, D. (Rev)	95 StatDecn 13 402-403
rence relations for single and product moments of	order statistics from a generalized half logistic distribution ... • Balakrishnan et al. Recur-	95 JStCmpSm 52 385-398
On stability of intermediate	order statistics • Gather, Ursula; Tomkins, Jim	95 JStPlInf 45 175-183
estimation of scale parameter by absolute values of	order statistics in logistic distribution • Kantam, R. R. L.; Rao, A. V. et al. Linear unbiased	94 JStatRsr 28 199-219
Determinacy of means of	order statistics • Spruill, M. C.	95 SankhyaA 57 118-125
Balasubramanian, K.	Order statistics from non-identical power function random variables • Balakrishnan, N.;	95 CommStA 24 1443-1454
On recurrence relations for	order statistics • David, H. A.	95 StPrLet 24 133-138
Characterizations of distributions by moments of	order statistics when the sample size is random • Grudzień, Z.; Szynal, D.	95 ZastApMa 23 305-318
of maxima of concomitants of subsets of	order statistics • Joshi, S. N.; Nagaraja, H. N. Joint distribution	95 Bernoull 1 245-255
The asymptotic distributions for sums of	order statistics (I) • Cheng, Shidong	95 ActScNat 30 323-338
Maximum variance of	order statistics • Papadatos, Nickos	95 AnInStMa 47 185-193
The Fourier transform of	order statistics with applications to Lorentz spaces • Dilworth, S. J.; Koldobsky, A. L.	95 IsrelJMa 92 411-425
Intermediate	order statistics with applications to nonparametric estimation • Papadatos, N.	95 StPrLet 22 231-238
Optimal	order for two servers in tandem • Yamazaki, Genji; Ito, Hiroshi	95 AnInStMa 47 31-48
On exact distributions of rank tests for	ordered alternatives in block designs • Black, Stephen; Mansouri, H.	95 CmpStDA 20 265-274
Testing homogeneity of normal means with a simply	ordered alternative and dependent observations • Lei, Xingye; Peng, Y. B.; Wright, F. T.	95 CmpStDA 20 173-183
rectional heterogeneity tests to evaluate simply	ordered alternative hypotheses • Rice, William R.; Gaines, Steven D. Extending nondi-	94 PNatAcSc 91 225-226
Testing	ordered alternatives in the presence of incomplete data • Alvo, Mayer; Cabilio, Paul	95 JASA 90 1015-1024
Likelihood ratio tests for bivariate symmetry against	ordered alternatives in a square contingency table • Barmi, H. E.; Kochar, S. C.	95 StPrLet 22 167-173
A robust surveillance scheme for stochastically	ordered alternatives • Gordon, Louis; Pollak, Moshe	95 AnlsStat 23 1350-1375
Wright, F. T. Testing homogeneity with an	ordered alternative in a two-factor layout by combining p-values • Moonesinghe, Ramal;	95 AnInStMa 47 505-523
Nonparametric tests for	ordered c-sample contaminated distributions • Johnson, Richard A.; Sim, Songyong	95 CommStA 24 861-879
A random-effects model for	ordered categorical data • Crouchley, Robert	95 JASA 90 489-498
A bivariate cumulative probit regression model for	ordered categorical data • Kim, Kyungmann	95 StatMed 14 1341-1352
power calculations for the Wilcoxon statistic for	ordered categorical data • Kolassa, John E. A comparison of size and	95 StatMed 14 1577-1581
approximations for tests of rank interaction with	ordered categorical data • Thomas, G. E. A note on chi-squared	95 CommStB 24 439-459
constraints search algorithm for optimal scaling of	ordered categorical responses • Lin, Shang P.; Tang, Dei-In A hierarchical active	95 JStCmpSm 53 197-209
Optimal scaling for	ordered categories • Pölz, W.	95 CmpStQ 10 37-41
fitting models to $N \times N$ contingency tables having	ordered categories • Lawal, H. Bayo; Upton, G. J. G. An algorithm for	95 CommStB 24 793-805
model for square contingency tables with	ordered categories • Tomizawa, Sadao A generalization of the marginal homogeneity	95 JEdStat 20 349-360
global symmetry for square contingency tables with	ordered categories • Tomizawa, Sadao Measures of departure from	95 Behvmtrk 22 91-98
Analysis of square tables with	ordered categories • Poon, Wai-Yin; Hung, Hin-Yan	94 ASAProStCp 292-298
Goodness of fit testing for	ordered discrete distributions by resampling techniques • Pesarin, Fortunato	94 Metron 52/3 57-71
multivariate goodness-of-fit test for stochastically	ordered distributions • Alam, Khursheed; Williams, Calvin L. A	95 BiomtrcJ 37 945-956
Tests of	ordered hypotheses for gamma scale parameters • Li, Teng; Sinha, Bimal K.	95 JStPlInf 43 311-321
Tests of	ordered hypotheses for gamma scale parameters • Li, Teng; Sinha, Bimal K.	95 JStPlInf 45 387-397
Partially-	ordered linear models • Walker, John H.	94 ASAProStCp 152-157
Bayes critical region for testing simple and tree	ordered normal means • Sugiura, Nariaki Approximating	94 JStatRsr 28 1-20
Analysis of complex human genetic traits: An	ordered-notation method and new tests for mode of inheritance • Thomson, Glenys	95 AJHuGene 57 474-486
Singh, Harshinder Componentwise estimation of	ordered parameters of k (≥ 2) exponential populations • Vijayasree, G.; Misra, Neeraj;	95 AnInStMa 47 287-307
On the admissibility of	ordered Poisson parameter under the entropy loss function • Parsian, A.; Nematollahi, N.	95 CommStA 24 2451-2467
for scores on tests including polytomous items with	ordered responses • Thissen, D.; Pommerich, M.; Billeaud, K. et al. Item response theory	95 ApPsychM 19 39-49
H-Akman, Olcay Estimation of reliability under	ordered restriction on the parameters • Gupta, Ramesh C.; Gupta, Rameshwar D.;	95 CommStA 24 1799-1812
convergence of certain estimators of stochastically	ordered survival functions • Rojo, Javier On the weak	95 JNnparSt 4 349-363
Successive comparisons between	ordered treatments • Lee, Robert Edwin; Spurrier, John D.	95 JStPlInf 43 323-330
Some inadmissibility results for estimating	ordered uniform scale parameters • Misra, Neeraj; Dhariyal, Ishwari D.	95 CommStA 24 675-685
Sibuya, Masaaki; Yamato, Hajime	Ordered and unordered random partitions of an integer and the GEM distribution •	95 StPrLet 25 177-183
Robust finite-intersection tests for homogeneity of	ordered variances • Mudholkar, Govind S.; McDermott, Michael P.; Mudholkar, Anil	95 JStPlInf 43 185-195
Hesselager, O.	Ordering claim size distributions and mixed Poisson probabilities • Kaas, R.;	95 InsrneME 17 193-201
Multivariate dispersion	orderings • Giovagnoli, Alessandra; Wynn, H. P.	95 StPrLet 22 325-332
On weak convergence within a Lorenz	ordering family of distributions • Lin, Gwo Dong	95 JAppProb 32 1028-1031
Hazard rate	ordering of k-out-of-n systems • Shaked, Moshe; Shanthikumar, J. George	95 StPrLet 23 1-8
tests for and against the increasing failure rate	ordering on multinomial parameters • Bhattacharya, Bhaskar Restricted	95 StPrLet 25 309-316
On the preservation of some	orderings of risks under convolution • Pellerey, Franco	95 InsrneME 16 23-30
to the powers of the likelihood ratio tests: The loop	ordering and slippage alternatives • Singh, B.; Schell, M. J.; Wright, F. T. Approximations	95 CommStB 24 91-109
Winch, Janice Kim; Avi-Itzhak, Benjamin	Ordering of tandem constant-service stations to minimize in-process stock cost •	95 PrEngInf 9 457-473
Inference for likelihood ratio	ordering in the two-sample problem • Dykstra, Richard; Kochar, Subhash; Robertson, Tim	95 JASA 90 1034-1040
based on ranks for comparing groups with umbrella	orderings • Pan, Guohua; Wolfe, Douglas A. Distribution-free test	95 JNnparSt 5 409-416
systematic and random differences between paired	ordinal categorical data" • Svensson, E. (Auth); Lesaffre, E. (Rev) Review of "Analysis of	95 Biomtrcs 51 794-794
systematic and random differences between paired	ordinal categorical data" • Svensson, E. (Auth); Lindsey, J. K. (Rev) Review of "Analysis of	95 StatMed 14 1726-1727
	Ordinal data	
contingency table • Barmi, H. E.; Kochar, S. C.	Likelihood ratio tests for bivariate symmetry against ordered alternatives in a square	95 StPrLet 22 167-173
Latent variable models for clustered	ordinal data • Qu, Yinsheng; Piedmonte, Marion R.; Medendorp, Sharon V.	95 Biomtrcs 51 268-275
Location-scale cumulative odds models for	ordinal data: A generalized non-linear model approach • Cox, Christopher	95 StatMed 14 1191-1203
A regression method for analysing	ordinal data from intervention trials • Schnell, Daniel J.; Magee, Elvin; Sheridan, John R.	95 StatMed 14 1177-1189
Non-linear time series models for	ordinal data • Hoffmann, Raymond G.	94 ASAProStCp 245-250
Ades, A. E. et al. Age-related standards for	ordinal data: Modelling the changes in visual acuity from 2 to 9 years of age • Wade, A. M.;	95 StatMed 14 257-266
• Ganguly, S. S.; Naik-Nimbalkar, U. V. Analysis of	ordinal data in a study of endometrial cancer under a matched pairs case-control design	95 StatMed 14 1545-1552

Lipsitz, Stuart R. Analyzing bivariate	ordinal data using a global odds ratio • Williamson, John M.; Kim, KyungMann;	95 JASA 90 1432-1437
mapping of quantitative trait loci for traits with	ordinal distributions • Hackett, Christine A.; Weller, Joel I. Genetic	95 Biomtrcs 51 1252-1263
et al. A comparison of methods for correlated	ordinal measures with ophthalmic applications • Gange, S. J.; Linton, K. L. P.; Scott, A. J.	95 StatMed 14 1961-1974
Predictive comparisons in	ordinal models • de Moraes, Aipore R.; Dunsmore, Ian R.	95 CommStA 24 2145-2164
Watson, Peter C.; Lesaffre, Emmanuel Are	ordinal models useful for classification? A revised analysis • Rudolfer, Stephan M.;	95 JStCmpSm 52 105-132
The behaviour of nominal and	ordinal partial association measures • Olszak, Michael; Ritschard, Gilbert	95 Ststcian 44 195-212
Isotonic	ordinal probabilistic models (ISOP) • Scheiblechner, Hartmann	95 Psymtrka 60 281-304
accelerating Gibbs sampler convergence for the	ordinal probit • Cowles, Mary Kathryn Assessing and	94 ASAProBays 52-57
Modelling	ordinal recurrent events • Berridge, Damon M.	95 JStPlInf 47 71-78
A SAS macro for residual deviance of	ordinal regression analysis • Wan, Jim Y.; Wang, Whedy; Bromberg, Judith	94 CmpBiomd 45 307-310
Group truncated	ordinal regression • O'Neill, Terence J.; Barry, Simon C.	95 StPrLet 22 195-203
Analysis of neurologic outcomes using	ordinal regression • Hwang, Lie-Ju; Alves, Wayne	94 ASAProBiop 352-354
the GEE method to construct linear models for	ordinal responses • Nuri, Walid A.; Chi, George Y. H. Using	94 ASAProBiop 82-87
power loss associated with categorization and	ordinal scores in dose-response and trend analysis • Greenland, Sander Avoiding	95 Epidmlgy 6 450-454
geometric combination estimator for d-dimensional	ordinal sparse contingency tables • Dong, Jianping; Simonoff, Jeffrey S. A	95 AnlsStat 23 1143-1159
tification method III by graphics: Its application of	ordinal structure analysis to the Ninth Nationwide Survey ... • Komazawa et al. ... quan-	95 PrInStMa 43 161-176
A test of quasi-independence in	ordinal triangular contingency tables • Tsai, Ming-Tien; Sen, Pranab K.	95 StSinica 5 767-780
Review of "Statistical methods for	ordinal variables" • Clogg, Clifford C.; Shihadeh, Edward S. (Auth); Von Tress, Mark (Rev)	95 Technmcs 37 118-119
Review of "Statistical models for	ordinal variables" • Clogg, Clifford C.; Shihadeh, Edward S. (Auth); Andrich, David (Rev)	95 JASA 90 1131-1131
Review of "Statistical models for	ordinal variables" • Clogg, Clifford C.; Shihadeh, Edward S. (Auth); Richards, John D. (Rev)	95 JQualTek 27 394-394
coefficients for the multivariate analysis of	ordinally scaled variables • Hammond, Sean M.; Lienert, Gustav A. Modified phi correlation	95 EdPsMeas 55 225-236
of ... • Rzhetsky et al. Statistical properties of the	ordinary least-squares, generalized least-squares, and minimum-evolution methods	92 JMolEvol 35 367-375
of "Multivariate analysis Part 1: Distributions,	ordination and inference" • Krzanowski, W. J.; Marriott, F. H. C. (Auth) et al. Review	95 Technmcs 37 351-351
of "Multivariate analysis, Part 1: Distributions,	ordination, and inference" • Krzanowski, W. J.; Marriott, F. H. C. (Auth) et al. Review	95 JASA 90 1490-1490
p195-207) • Krzanowski, W. J. Correction to	"Ordination in the presence of group structure, for general multivariate data" (VII	95 JClass 12 2-2
process • Guo, Wensheng; Marshall, Guillermo	ORDMKV: A computer program fitting proportional odds model for multi-state Markov	95 CmpBiomd 46 257-263
Bounds for molecular weights of	organic compounds from low resolution mass spectra • Scott, Donald R.	95 Chemolab 27 129-133
methods for the estimation of solution properties of	organic pollutants • Dunn, W. J., III Computational	95 Chemolab 29 1-10
Use experimental design in	organic synthesis • Brannigan, Lawrence H.; Grieshaber, Mark V.; Schnur, Dora M.	95 ChemTech 25 29-35
Short and medium term rain forecasting in the	Oristano basin (Italian) • Espa, Giuseppe; Terpessi, Cristina	94 QStMaA 16 145-160
Kurchenko, O. O. Some inequalities for the	Orlicz norm of a linear form of independent pre-Gaussian random variables •	95 ThPrMaSt 50 99-102
Stable processes with sample paths in	Orlicz spaces • Norvaiša, Rimas; Samorodnitsky, Gennady	94 AnlsProb 22 1904-1929
noise • Flandoli, F. et al. Time discretization of	Ornstein-Uhlenbeck equations and stochastic Navier-Stokes equations with a generalized	95 Stochatx 55 141-165
	Ornstein-Uhlenbeck process	
Jacobs, Patricia A.	Asymptotic properties of stochastic greedy bin-packing • Gaver, Donald P.;	95 CommStSM 11 333-348
	Estimating the rate of phenotypic evolution from comparative data • Martins, Emí P.	94 ANatrls 144 193-209
Jimenez, F.; Gutierrez, R.; Martinez-Almecija, A.	Estimation and filtering on a doubly stochastic Poisson process • Valderrama, M. J.;	95 AplStMDA 11 13-24
Hilbert spaces • Chojnowska-Michalik, A.; Goldys, B.	Existence, uniqueness and invariant measures for stochastic semilinear equations on	95 ProbTher 102 331-356
Samorodnitsky, Gennady; Taqqu, Murad S.	Lévy measures of infinitely divisible random vectors and Slepian inequalities •	94 AnlsProb 22 1930-1956
	Limit theorems for multitype epidemics • Andersson, Håkan; Djehiche, Boualem	95 StocProc 56 57-75
	Loss networks under diverse routing: The symmetric star network • Hunt, P. J.	95 AdvAppPr 27 255-272
Bibby, Bo Martin; Sørensen, Michael	Martingale estimation functions for discretely observed diffusion processes •	95 Bernoull 1 17-39
	Modelling radio-tracking data • Worton, Bruce J.	95 EnvEcoSt 2 15-23
Watkins, Joseph	A stochastic model for the movement of a white blood cell • Heubach, Silvia;	95 AdvAppPr 27 443-475
Khoshnevisan, Davar	Weight functions and pathwise local central limit theorems • Horváth, Lajos;	95 StocProc 59 105-123
On large increments of infinite series of	Ornstein-Uhlenbeck processes • Lin, Z. Y.	95 StocProc 60 161-169
arrays of ... • Yamamoto et al. Classification of	orthogonal 24-run 2^5 factorial designs derivable from saturated two-symbol orthogonal	95 TRUMTk 31/1 39-53
	Orthogonal array	
shafts • Davis, T. P.	Analysis of an experiment aimed at improving the reliability of transmission centre	95 LiDaAnal 1 275-306
Morgan, J. P.	Construction and optimality of affine-resolvable designs • Bailey, R. A.; Monod, H.;	95 Biomtrka 82 187-200
	Designing outer array points • Wang, Yuchung; Lin, Dennis K. J.; Fang, Kai-Tai	95 JQualTek 27 226-241
	On e-optimal fractions of symmetric and asymmetric factorials • Mukerjee, Rahul	95 StSinica 5 515-533
Michael A.	Inference for effect-saturated fractional factorials • Haaland, Perry D.; O'Connell,	95 Technmcs 37 82-93
Neighbor-balanced	orthogonal arrays • Cheng, Ching-Shui	93 JCISS 18 61-67
Mandeli, John P. Construction of asymmetrical	orthogonal arrays having factors with a large non-prime power number of levels •	95 JStPlInf 47 377-391
To analyze binary data from mixed-level	orthogonal arrays • Wang, P. C.; Chen, M. H.; Wei, S. C.	95 CmpStDA 20 689-695
Some projection properties of	orthogonal arrays • Cheng, Ching-Shui	95 AnlsStat 23 1223-1233
Bounds on	orthogonal arrays and resilient functions • Bierbrauer, Jürgen	95 JComDsgn 3 179-183
Designing two-level factorial experiments using	orthogonal arrays when the run order is important • Wang, P. C.; Jan, H. W.	95 Ststcian 44 379-388
study • Yamamoto, S.; Fujii, Y. et al. Three-symbol	orthogonal arrays of strength 2 and index 2 having maximal constraints: Computational	93 JCISS 18 329-342
Some	orthogonal arrays with variable symbols • Dey, A.	93 JCISS 18 209-215
on "Mixture designs for four components in	orthogonal blocks" (V35 p268-276) • Snee, Ronald D. Comment	95 Technmcs 37 131-132
Kageyama, Sanpei Single replicate	orthogonal block designs for circulant partial diallel crosses • Gupta, Sudhir; Das, Ashish;	95 CommStA 24 2601-2607
	Orthogonal canonical variates for discrimination and classification • Krzanowski, W. J.	95 JChemtrc 9 509-520
analogues of random polynomials which are	orthogonal on a circle • Teplyaev, A. V. Continual	94 ThProbAp 39 476-489
On the structure of PLS in	orthogonal designs • Langsrud, Øyvind; Næs, Tormod	95 JChemtrc 9 483-487
Complete sets of	orthogonal frequency squares and affine resolvable designs • Laywine, C.	93 UtilMath 43 161-170
Hedayat, A. S.; Parker, E. T. A pair of	orthogonal Latin squares of order 10 with four shared parallel transversals • Brown, J. W.;	93 JCISS 18 113-115
Sets of mutually	orthogonal Latin squares with "like subsquares," II • Roberts, Charles E., Jr	95 JCombThA 70 159-164
Perturbation analysis for an	orthogonal least-squares estimator • Chapman, G. R.	95 CandJSt 23 101-108
Algorithm for constructing	orthogonal main effect plans • Burgess, L.; Street, D. J.	94 UtilMath 46 33-48
Construction for	orthogonal main effect plans • Street, D. J.	94 UtilMath 45 115-123
On a class of strongly threefold	orthogonal matrices of index eight • Pesotan, H.	93 JCISS 18 317-328
Rejecting outliers and estimating errors in an	orthogonal outliers and estimating errors in an orthogonal-regression ... • Shapiro et al.	95 PTLnAMa 350 407-439
	Orthogonal polynomial	
Bhattacharyya, P.	A statistical design of experiments approach for texture description • Ganesan, L.;	95 PattrnRc 28 99-105
Pearson system of distributions and the associated	orthogonal polynomials • Papathanasiou, V. A characterization of the	95 AnInStMa 47 171-176
Müller, H. G. [Algorithm AS 297]	Orthogonal polynomial and hybrid estimators for nonparametric regression • Azari, A. S.;	95 ApplStat 44 264-268
numbers for blockwise independent and blockwise	orthogonal random variables • Gaposhkin, V. F. On the strong law of large	94 ThProbAp 39 677-684
One-step GM-estimators for	orthogonal regression • Abdullah, Mokhtar B.	95 JStCmpSm 52 301-309
Optimal designs with component-wise	orthogonal row-column structures • Shah, K. R.; Sinha, Bikas K.	93 JCISS 18 68-78
Weighted	orthogonal series estimates of the regression curve (Chinese) • Yang, Ying	95 JSysSMaS 14 337-349
of density-weighted average derivative by	orthogonal series method • Rao, B. L. S. Prakasa Consistent estimation	95 StPrLet 22 205-212
Estimating the edge of a Poisson process by	orthogonal series • Jacob, Pierre; Suquet, Charles	95 JStPlInf 46 215-234

	Overdispersion	
Alberto; de Ceballos, Federico	Describing extra-binomial variation with partially correlated models • Luceño,	95 CommStA 24 1637-1653
	Duration dependence and dispersion in count-data models • Winkelmann, Rainer	95 JBES 13 467-474
	Generalized binomial regression model • Famoye, Felix	95 BiomtrcJ 37 581-594
Frangos, Chris C.; Schucany, William R.	Improved bootstrap confidence intervals in certain toxicological experiments •	95 CommStA 24 829-844
van de Ven, R.; Weber, N. C.	Log-linear models for mean and dispersion in mixed Poisson regression models •	95 AstrlJSt 37 205-216
Böckenholt, Ulf	Mixture models for the analysis of repeated count data • van Duijn, Marijtje A. J.;	95 ApplStat 44 473-485
Daniel	Non-linear mixed regression models • Burnett, Richard T.; Ross, W. H.; Krewski,	95 EnvrMtrc 6 85-99
	A note on variation in birth processes • Ball, Frank	95 MaScitst 20 50-55
variation • Morel, Jorge G.; Koehler, Kenneth J.	A one-step Gauss-Newton estimator for modelling categorical data with extraneous	95 ApplStat 44 187-200
	Poisson regression analysis in clinical research • Kianifard, Farid; Gallo, Paul P.	95 JBiophSt 5 115-129
longitudinal data • Park, Taesung; Shin, Min-Woong	A practical extension of the generalized estimating equation approach for	95 CommStA 24 2561-2579
chemical compounds (Disc: ... • Talwalker et al.	Qualitative and quantitative assessment of the risk from the exposure to fetotoxic	95 EnvEcoSt 2 71-78
Andersen, P. K.	Score test of homogeneity for survival data (Disc: p157-159) • Commenges, D.;	95 LiDaAnal 1 145-156
experiments • Schwartz, P. F.; Gennings, C. et al.	Threshold models for combination data from reproductive and developmental	95 JASA 90 862-870
Tilg, Norbert	Variance estimates in logistic regression using the bootstrap • Friedl, Herwig;	95 CommStA 24 473-486
tradhar, Brajendra C. Score tests for fixed effects and	overdispersion components in nonlinear models for repeated measures • Das, Kalyan; Su-	95 JMultiAn 54 126-146
et al. Correction to "Score tests for fixed effects and	overdispersion components in nonlinear models for repeated measures" (V54 ... • Das	95 JMultiAn 55 349-349
Tests for serial correlation and	overdispersion in a count data regression model • Johansson, Per	95 JStCmpSm 53 153-164
Park, Cheolyong	Overdispersion in count data - A review (Korean) • Kim, Byung Soo; Oh, Kyong Joo;	95 KorApS 8/2 147-161
Kathryn	Overdispersion diagnostics for generalized linear models • Lambert, Diane; Roeder,	95 JASA 90 1225-1236
Das, Kalyan Tests for	overdispersion in longitudinal data in the presence of outliers • Sutradhar, Brajendra C.;	95 SankhyaA 57 299-315
A family of partially correlated Poisson models for	overdispersion • Luceño, Alberto	95 CmpStDA 20 511-520
Comparison of quasi-likelihood models for	overdispersion • Yanez, N. David, III; Wilson, Jeffrey R.	95 AstrlJSt 37 217-231
Modified pseudo-likelihood estimator of the	overdispersion parameter in Poisson mixture models • Dean, C. B.	94 JAppStat 21 523-532
Tests for detecting	overdispersion in Poisson models • Lee, Sunho; Park, Cheolyong; Kim, Byung Soo	95 CommStA 24 2405-2420
p1134-1139) • Guibert, J. Correction to	"Overflow probability upper bound in fluid queues with general on-off sources" (V31	95 JAppProb 32 278-278
Testing whether	overhauls renew the system • Purohit, S. G.	94 JIndnStA 32 57-68
The map comparison problem: Tests for the	overlap of geographic boundaries • Jacquez, Geoffrey M.	95 StatMed 14 2343-2361
calibration methods for data with and without rank	overlap • Kiers, Henk A. L.; Smilde, Age K. Some theoretical results on second-order	95 JChemtrc 9 179-195
transportation algorithm for maximizing the	overlap between surveys • Ernst, Lawrence R.; Ikeda, Michael M. A reduced-size	95 SurvyM 21 147-157
Logistic regression analysis of CPS	overlap survey split panel data • Fisher, Robin	94 ASASrvyRsr 620-625
Farm Cost and Returns Survey list sample to reduce	overlap with three other 1993 surveys and the 1992 FCRS • Perry et al. Redrawing the 1993	94 ASASrvyRsr 632-637
Corter, J. E. A graph-theoretic method for organizing	overlapping clusters into trees, multiple trees, or extended trees • Carroll, J. D.;	95 JClass 12 283-313
and estimation of effective size in populations with	overlapping generations • Jorde, Per Erik; Ryman, Nils Temporal allele frequency change	95 Genetcs 139 1077-1090
Pietrogrande, M. C. et al. Statistical study of peak	overlapping in multicomponent chromatograms: Importance of the retention pattern •	95 Chemolab 28 239-258
of mean shape and mean particle number in	overlapping particle systems in the plane • Weil, Wolfgang The estimation	95 AdvAppPr 27 102-119
	Oversampling	
	Problems of screening for poverty status • Moeller, John F.; Mathiowetz, Nancy A.	94 JOffStat 10 327-337
Allen, Tiwanda M.	Oversampling in panel surveys (Disc: p680-682) • Singh, Rajendra P.; Petroni, Rita J.;	94 ASASrvyRsr 674-679
Proe, M. F. Smoothing regression coefficients in an	overspecified regression model with interrelated explanatory variables • Elston, D. A.;	95 ApplStat 44 395-406
Liu, Yuanlong; Schutz, Robert W.	Overtime in the National Hockey League: Is it a valid tie-breaking procedure? •	94 ASAProSpts 55-60
	Overtime or shootout: Deciding ties in hockey • Hurley, William	95 Chance 8/1 19-22
networks • Sjöberg, J.; Ljung, L.	Overtraining, regularization and searching for a minimum, with application to neural	95 IJContrl 62 1391-1407
Meiring, W. Spatio-temporal analysis of regional	ozone data for operational evaluation of an air quality model • Sampson, P. D.; Guttorp, P.;	94 ASAProEnvr 46-55
Modeling satellite	ozone data • Niu, Xufeng; Tiao, George C.	95 JASA 90 969-983
Upper and lower tolerance limits of atmospheric	ozone level and extreme value distribution • Dasgupta, Ratan; Bhaumik, Dulal K.	95 SankhyaB 57 182-199
Predicting	ozone levels and trends • Gao, Feng; Sacks, Jerome; Welch, William J.	94 ASAProEnvr 77-81
and prediction of summer season maximum surface	ozone for the Vancouver, Montreal, and Atlantic regions of ... • Burrows et al. ... analysis	95 JApMetrl 34 1848-1862
Mid-	P confidence intervals: A brief review • Berry, G.; Armitage, P.	95 Ststcian 44 417-423
"Another look at first-order saturated designs: The	p-efficient designs" (V35 p384-292) • Lin, Dennis K. J. Correction to	95 Technmcs 37 473-473
of probability distributions based on discrete	p-monotonicity • Sapatinas, Theofanis Characterizations	95 StPrLet 24 339-344
On the normal approximation for the sample	p-quantile • Koti, Kallappa M.	94 ASAProStCp 188-193
	***P* value**	
	A stepwise resampling method of multiple hypothesis testing • Troendle, James F.	95 JASA 90 370-378
based multiple testing. Examples and methods for	p-value adjustment" • Westfall, P. H.; Young, S. S. (Auth) et al. Review of "Resampling-	95 CmpStDA 20 235-236
	P-value adjustments for subgroup analyses of binary variables • Bristol, David R.	94 ASAProBiop 120-125
alternative in a two-factor layout by combining	p-values • Moonesinghe, Ramal; Wright, F. T. Testing homogeneity with an ordered	95 AnInStMa 47 505-523
Pinsker, Ilia Conditional	p-values for the F-statistic in a forward selection procedure • Grechanovsky, Eugene;	95 CmpStDA 20 239-263
Snapinn, Steven M.	P-values for group sequential testing • Chang, Myron N.; Gould, A. Lawrence;	95 Biomtrka 82 650-654
Metha, Cyrus R.; Patel, Nitin R. Estimating exact	p values by the method of control variates or Monte Carlo rescue • Senchaudhuri, Pralay;	95 JASA 90 640-648
(German) • Böker, Fred; Dannenberg, Olaf What can	P-values say about the quality of test statistics for comparative simulation studies?	95 AllgeStA 79 233-251
Wood, A. T. A. Saddlepoint approximations for	P-values of some tests of covariance matrices • Booth, J. G.; Butler, R. W.; Huzurbazar, S.;	95 JStCmpSm 53 165-180
Srivastava, Deo Kumar; Lin, C. Thomas Some	p-variate adaptations of the Shapiro-Wilk test of normality • Mudholkar, Govind S.;	95 CommStA 24 953-985
of the dispersion matrix of a weighted sample from a	p-variate normal distribution • Coelho, Carlos Brief note on the approximate distribution	94 JIndnStA 32 77-81
Nonparametric methods in standard statistics	packages: The case of the Friedman analysis • Rahlfs, Volker W.; Zimmermann, Helmuth	95 CmpStDA 20 101-110
Correspondence analysis in statistical	package programs • Thompson, Paul A.	95 AmerStat 49 310-316
A review of five statistical	packages for Windows • Lurie, Philip M.	95 AmerStat 49 99-107
A central limit theorem for a class of Markovian	packet streams under heavy traffic conditions • Neuts, M. F.; Pearce, C. E. M.	95 CommStSM 11 210-218
Taqqu, M. S. et al. Self-similarity in high-speed	packet traffic: Analysis and modeling of ethernet traffic measurements • Willinger, W.;	95 StatSci 10 67-85
Asymptotic properties of stochastic greedy bin-	packing • Gaver, Donald P.; Jacobs, Patricia A.	95 CommStSM 11 333-348
	Packing dimension	
	Dimension results for Gaussian vector fields and index-α stable fields • Xiao, Yimin	95 AnlsProb 23 273-291
Taylor, S. J. The	packing measure of the support of super-Brownian motion • Le Gall, J.-F.; Perkins, E. A.;	95 StocProc 59 1-20
	Packing random intervals • Coffman, E. G., Jr; Poonen, Bjorn; Winkler, Peter	95 ProbTher 102 105-121
silver ... • Camacho et al. Effects of the regulators	paclobutrazol and flurprimidol on the growth of terminal sprouts formed on trimmed	95 CandJSt 23 312-322
Bradley-Terry-type models to the measurement of	pain • Matthews, J. N. S.; Morris, K. P. An application of	95 ApplStat 44 243-255
management system for an H.U.D. lead-based	paint abatement study • Pan, U. W.; Clark, S.; Buncher, R. et al. Implementing a database	94 ASAProEnvr 162-166
The evolution and measurement of a population of	pairs • Jakeman, Eric; Phayre, Sean; Renshaw, Eric	95 JAppProb 32 1048-1062
p2269-2278) • Baker, S. G. et al. Correction to "The	paired availability design: A proposal for evaluating epidural analgesia during labor" (V13	95 StatMed 14 1841-1841
	Paired comparison	
Matthews, J. N. S.; Morris, K. P.	An application of Bradley-Terry-type models to the measurement of pain •	95 ApplStat 44 243-255
	May the best man win? • Appleton, David R.	95 Ststcian 44 529-538
	Temporal analogues to spatial K functions • Cannon, Ann; Cressie, Noel	95 BiomtrcJ 37 351-373

Masahide	Analysis of variance of	partially balanced fractional $2^{m_1+m_2}$ factorial designs of resolution IV • Kuwada,	93 JCISS 18 343-367
		Partially balanced incomplete block design	
		Constructions of group divisible designs • Duan, Xiaoping; Kageyama, Sanpei	95 JJapnStS 25 121-128
Ghosh, D. K.; Joshi, K. S.		Construction of variance balanced designs through triangular PBIB designs •	95 CalcutSt 45 111-118
		A note on non-orthogonal main-effect plans • Sharma, M. K.	95 SankhyaB 57 166-168
	Methods of constructing nested	partially balanced incomplete block designs • Banerjee, S.; Kageyama, S.	93 UtilMath 43 3-6
	Maximum likelihood estimation with	partially censored data • van der Vaart, Aad	94 AnlsStat 22 1896-1916
	Deconvolving a density from	partially contaminated observations • Hesse, Christian H.	95 JMultiAn 55 246-260
	Describing extra-binomial variation with	partially correlated models • Luceño, Alberto; de Ceballos, Federico	95 CommStA 24 1637-1653
	A family of	partially correlated Poisson models for overdispersion • Luceño, Alberto	95 CmpStDA 20 511-520
	Exchangeable and	partially exchangeable random partitions • Pitman, Jim	95 ProbTher 102 145-158
	Limit theory and bootstrap for explosive and	partially explosive autoregression • Datta, Somnath	95 StocProc 57 285-304
	of all possible lines and hybrids from designs with	partially inbred progenies • Naciri, Y.; Gallais, A. Predicting mean and variance	94 ThApGene 89 693-697
G. C.	Estimation and testing for unit roots in a	partially nonstationary vector autoregressive moving average model • Yap, S. F.; Reinsel,	95 JASA 90 253-267
	Adaptive control of	partially observed linear stochastic systems • Bertrand, Pierre	95 Stochatx 54 21-51
	General necessary conditions for	partially observed optimal stochastic controls • Li, Xunjing; Tang, Shanjian	95 JAppProb 32 1118-1137
	for calculating minimum χ^2 estimators with	partially observed sub-tables • Ritov, Ya'acov PM algorithms	95 CmpStDA 20 19-33
Seki, Y.	On the level probabilities for useful	partially ordered alternatives in the analysis of variance • Hoshino, N.; Miyazaki, H.;	95 CommStA 24 2059-2071
		Partially-ordered linear models • Walker, John H.	94 ASAProStCp 152-157
Wolfe, Douglas A.	Distribution-free	partially sequential tests for treatments versus control setting • Kim, Dongjae;	95 BiomtrcJ 37 259-271
Lee, Suk-hoon	Optimal design of	partially step-stress life testing for the series systems (Korean) • Park, Hee-chang;	95 KorApS 8/2 121-132
Varadhan, S. R. S.	Self diffusion of a tagged	particle in equilibrium for asymmetric mean zero random walk with simple exclusion •	95 AlHenrPB 31 273-285
	Symmetric two-	particle exclusion-eating processes • Liu, Xijian	95 AnlsProb 23 1439-1455
	asymmetric hydrodynamical system of independent	particles (French) • Kipnis, C.; Léonard, C. Large deviations of an	95 AlHenrPB 31 223-248
	Random walks of infinitely many	particles • Révész, P.	94 WrldSciPub:Sing 191p
	Review of "Random walks of infinitely many	particles" • Révész, P. (Auth); Heyde, C. C. (Rev)	95 ShrtBkRv 15/3 47-47
	The estimation of mean shape and mean	particle number in overlapping particle systems in the plane • Weil, Wolfgang	95 AdvAppPr 27 102-119
	for parametric estimation from an equilibrium	particle process • Phelan, Michael J. Asymptotic information	95 StPrLet 24 193-198
	Branching processes with final types of	particles and random trees • Vatutin, V. A.	94 ThProbAp 39 628-641
	Stereological estimation of	particle size distributions • Mase, Shigeru	95 AdvAppPr 27 350-366
Léonard, C.	On large deviations for	particle systems associated with spatially homogeneous Boltzmann type equations •	95 ProbTher 101 1-44
	Convergence of independent	particle systems • Hoffman, John R.; Rosenthal, Jeffrey S.	95 StocProc 56 295-305
	Nonlinear master equation of multitype	particle systems • Feng, Shui	95 StocProc 57 247-271
	for attractive reversible subcritical nearest	particle systems • Mountford, T. S. Exponential convergence	95 StocProc 59 235-249
	Ordered and unordered random	partitions of an integer and the GEM distribution • Sibuya, Masaaki; Yamato, Hajime	95 StPrLet 25 177-183
		Partition models in the analysis of autoradiographic images • Aykroyd, R. G.	95 ApplStat 44 441-454
	Exchangeable and partially exchangeable random	partitions • Pitman, Jim	95 ProbTher 102 145-158
		Partition of variation: A new method for σ reduction • Little, Thomas A.; Brekke, Kris	95 ASQCTekCon 208-216
	Mean estimation for a	partitioned multinormal distribution • Dahel, S.	95 CommStA 24 601-606
		Partitioning error spaces in linear models (error contrasts) • Rady, El-Houssainy A.	95 EgpStJ 39 51-64
Györfi, L.; van der Meulen, E. C.		Partitioning-estimates of a regression function under random censoring • Carbonez, A.;	95 StatDecn 13 21-37
	performance using fuzzy MLP and two-level selective	partitioning of the feature space • Mitra, S.; Kuncheva, L. I. Improving classification	95 FuzzySet 70 1-13
Xu, S.	Alternative	partitioning of the genotype-by-environment interaction • Muir, W.; Nyquist, W. E.;	92 ThApGene 84 193-200
	developmental toxic effects through recursive	partitioning • Ahn, Hongshik; Chen, James J. Modeling	94 ASAProBiop 130-135
	Asymptotic normality of pseudo-LS estimator for	partly linear autoregression models • Gao, Jiti; Liang, Hua	95 StPrLet 23 27-34
	L_1-norm estimators for parametric components in a	partly linear model (Chinese) • Shi, Peide; Li, Guoying Asymptotic normality of	95 ChinaAMA 15 478-484
	note on the convergent rates of M-estimates for a	partly linear model • Shi, Peide; Li, Guoying A	95 MaOpfStS 26 27-47
	Asymptotics of the minimum L_1-norm estimates in a	partly linear model • Shi, Peide; Li, Guoying	95 JSysSMSE 7 67-77
	laws of the iterated logarithm of some estimates in	partly linear models • Gao, Jiti The	95 StPrLet 25 153-162
	Asymptotic theory for	partly linear models • Gao, Jiti	95 CommStA 24 1985-2009
	Asymptotic properties of some estimators for	partly linear stationary autoregressive models • Gao, Jiti	95 CommStA 24 2011-2026
• Harris et al.	A model for the interaction of	partnership dissolution and relationship quality as predicators of remission of problem ...	94 ASAProSoSt 232-234
	The statisticians role in quality improvement	partnerships • Hoff, Jennifer; Johnson, Thomas; Lu, J. C.	94 ASAProQlPr 252-257
	Statistics of income	partnership studies: Evaluation of preliminary estimates • McMahon, Paul B.	94 ASASrvyRsr 1001-1006
	Modelling the class-	party relationship in Britain, 1964-92 • Heath, Anthony; Evans, Geoffrey; Payne, Clive	95 JRSS-A 158 563-574
	confidence sets, bioequivalence, and the Limaçon of	Pascal • Brown, Lawrence D.; Casella, George; Hwang, J. T. Gene Optimal	95 JASA 90 880-889
		Passage time	
Michna, Zbigniew; Rychlik, Igor		The expected number of level crossings for certain symmetric α-stable processes •	95 CommStSM 11 1-19
Ramaswami, V.	Expected	passage times in homogeneous quasi-birth-and-death processes • Latouche, Guy;	95 CommStSM 11 103-122
	Mean	passage times for tridiagonal transition matrices • Stefanov, Valeri T.	95 JAppProb 32 846-849
	movements and capture parameters of resting	passerines in standardized capture-recapture studies • Kaiser, A. Estimating turnover,	95 JAppStat 22 1039-1047
	How useful are recoveries of North American	passerines for survival analyses? • Francis, Charles M.	95 JAppStat 22 1075-1081
	phasis on the epidemiology studies ... • Bayard et al.	Passive smoking and lung cancer: The U.S. EPS's weight-of-evidence analysis, with em-	94 ASAProEnvr 56-64
	Statistical analysis of	passive surveillance disease registry data • Brookmeyer, Ron; Yasui, Yutaka	95 Biomtrcs 51 831-842
	Calibration of empirical models: Simulation of	pasture growth and decay • McPhee, M. J.; Ransom, K. P.; Oddy, V. H.	95 EnvrMtrc 6 505-510
		Path analysis	
		Biometrical genetic analysis of the cotwin control design • Duffy, David L.	94 BehavGen 24 341-344
	equation models (SEM) • Pugesek, B. H.; Tomer, A.	Determination of selection gradients using multiple regression versus structural	95 BiomtrcJ 37 449-462
	cific cognitive abilities • Cardon, L. R.; Fulker, D. W.	A model of developmental change in hierarchical phenotypes with application to spe-	94 BehavGen 24 1-16
	coastal plain streams • Sheldon, A. L.; Meffe, G. K.	Path analysis of collective properties and habitat relationships of fish assemblages in	95 JFishRBC 52 23-33
Li, Z.; Bonney, G. E. et al.	Genetic analysis combining	path analysis with regressive models: The TAU model of multifactorial transmission •	94 HumHrdty 44 305-311
	time for Gaussian $S'(R^d)$-processes: Existence,	path continuity and examples • Bojdecki, Tomasz; Gorostiza, Luis G. Self-intersection local	95 StocProc 60 191-226
	Brownian motion with drift and the pricing of related	path-dependent options • Dassios, Angelos The distribution of the quantile of a	95 AnlsApPr 5 389-398
	Some formulae for a new type of	path-dependent option • Akahori, Jirô	95 AnlsApPr 5 383-388
	Decomposing the Brownian	path via the range process • Vallois, P.	95 StocProc 55 211-226
	On the typical level crossing time and	path • Nyrhinen, Harri	95 StocProc 58 121-137
	Random rotations of the Wiener	path • Üstünel, A. S.; Zakai, M.	95 ProbTher 103 409-429
	Host dispersal and the spatial spread of insect	pathogens • Dwyer, Greg; Elkinton, Joseph S.	95 Ecology 76 1262-1275
	Weight functions and	pathwise local central limit theorems • Horváth, Lajos; Khoshnevisan, Davar	95 StocProc 59 105-123
	On	pathwise stochastic integration • Karandikar, Rajeeva L.	95 StocProc 57 11-18
	Stochastic averaging principle for systems with	pathwise uniqueness • Golec, Janusz	95 StocAnAp 13 307-322

intractability of the ascertainment problem for	pedigree data: A general likelihood framework • Vieland, V. J.; Hodge, S. E. Inherent	95 AJHuGene 56 33-43
Fast computation of genetic likelihoods on human	pedigree data • Goradia, Tushar M.; Lange, Kenneth; Miller, Perry L.; Nadkarni, Prakash M.	92 HumHrdty 42 42-62
sets for Markov chain Monte Carlo estimations on	pedigrees • Lin, Shili; Thompson, Elizabeth; Wijsman, Ellen Finding noncommunicating	94 AJHuGene 54 695-704
approach for assessing genetic linkage in	pedigrees • Amos, Christopher I. Robust variance-components	94 AJHuGene 54 535-543
of oligogenic traits: Extension to multigenerational	pedigree structures • Vieland et al. ... single-locus approximations for linkage analyses	93 HumHrdty 43 329-336
The relative efficiency and power of small-	pedigree studies of the heritability of a quantitative trait • Schork, N. J.; Schork, M. A.	93 HumHrdty 43 1-11
terior genotypic distribution for every member of a	pedigree without loops • Fernando, R. L. et al. An efficient algorithm to compute the pos-	93 ThApGene 87 89-93
	Peeling	
	Heuristic algorithms for finding inexpensive elimination schemes • Harbron, Chris	95 StatComp 5 275-287
A Bayesian analysis for a class of	penalised likelihood estimates • Barry, Daniel	95 CommStA 24 1057-1071
	Penalized discriminant analysis • Hastie, Trevor; Buja, Andreas; Tibshirani, Robert	95 AnlsStat 23 73-102
	Penalized likelihood	
Bayesian confidence intervals • Wang, Y.; Wahba, G.	Bootstrap confidence intervals for smoothing splines and their comparison to	95 JStCmpSm 51 263-279
rates • Liao, Jiangang; Brookmeyer, Ron	An empirical Bayes approach to smoothing in backcalculation of HIV infection	95 Biomtrcs 51 579-588
random variables • Lee, Chung-Bow	Estimating the number of change points in a sequence of independent normal	95 StPrLet 25 241-248
	Estimating the order of hidden Markov models • Rydén, Tobias	95 MaOpfStS 26 345-354
• Wang, Yongxiao; Taylor, Jeremy M. G.	Flexible methods for analysing longitudinal data using piecewise cubic polynomials	95 JStCmpSm 52 133-150
trial • Wang, Yongxiao; Taylor, Jeremy M. G.	Inference for smooth curves in longitudinal data with application to an AIDS clinical	95 StatMed 14 1205-1218
Douglas; Ruppert, David	Nonparametric transformations for both sides of a regression model • Nychka,	95 JRSS-B 57 519-532
	Smoothing categorical data • Simonoff, Jeffrey S.	95 JStPlInf 47 41-69
lated explanatory variables • Elston, D. A.; Proe, M. F.	Smoothing regression coefficients in an overspecified regression model with interre-	95 ApplStat 44 395-406
	Smoothing spline density estimation: Conditional distribution • Gu, Chong	95 StSinica 5 709-726
van Houwelingen, Hans C.	Time-dependent effects of fixed covariates in Cox regression • Verweij, Pierre J. M.;	95 Biomtrcs 51 1550-1556
Raúl; Arnold, Steven F. Difficulties with the use of	penalized likelihood criteria in antedependence and polynomial models • Macchiavelli,	95 CommStA 24 501-522
Multivariate discriminant analysis and maximum	penalized likelihood density estimation • Granville, Vincent; Rasson, Jean Paul	95 JRSS-B 57 501-517
Variational solution of	penalized likelihood problems and smooth curve estimation • Mächler, Martin B.	95 AnlsStat 23 1496-1517
Influence on smoothness in	penalized likelihood regression for binary data • Jernigan, Robert W.; O'Connell, Julie A.	94 ASAProStCp 118-123
	Penalized quasi-likelihood	
dispersion • Breslow, Norman E.; Lin, Xihong	Bias correction in generalised linear mixed models with a single component of	95 Biomtrka 82 81-91
	Penalty	
	• see also roughness –	
	Penalty function	
	Consistent variable selection in linear models • Zheng, Xiaodong; Loh, Wei-Yin	95 JASA 90 151-156
Models for monthly	penetrations with incomplete panel data • Sikkel, D.; Hoogendoorn, A. W.	95 StNeerla 49 378-391
Obituary: Geoffrey	Penrice, 1923-1994 • Boreham, John	95 JRSS-A 158 180-181
Reagan, Patricia B. Adjusting	pension fund membership for "don't know" responses • Stasny, Elizabeth A.;	94 ASAProBuEc 298-303
Dynamic approaches to	pension funding • Haberman, Steven; Sung, Joo-Ho	94 InsrneME 15 151-162
	Pension funding with time delays and the optimal spread period • Haberman, Steven	95 AstinB 25 177-187
The	Pentium bug and statistical software • Gould, William	95 Chance 8/2 9-13
What do you get when you combine a	Pentium and a research grant? • Jones, Albyn C.	95 Chance 8/2 7-8
And not a drop to drink!	Perceived threats to clean drinking water • Tomazic, Terry; Katz, Barry	94 ASAProEnvr 173-178
A higher order approximation to a	percentage point of the non-central t-distribution • Akahira, Masafumi	95 CommStB 24 595-605
Evaluation of one-sided	percentage points of the singular multivariate normal distribution • Kwong, Koon-Shing	95 JStCmpSm 51 121-135
	Percentile	
Edward C.; SriRam, M. G.; Teoh, Nick K. W.	Entropy-based random number evaluation • Dudewicz, Edward J.; van der Meulen,	95 AJMMgSci 15 115-153
a detection limit • Smith, D. E. et al. Estimating a	percentile of a contaminant concentration distribution when all observations are less than	94 ASAProEnvr 139-141
power variance functions (Korean) • Suh, Euy Hoon	Percentiles for the distributions belonging to the natural exponential families having	95 KorApS 8/1 133-149
dose of a compound via regression modelling and	percentile estimation (Disc: p933-933) • Filloon ... the minimum therapeutically effective	95 StatMed 14 925-932
Filar, Jerzy A.; Krass, Dmitry; Ross, Keith W.	Percentile performance criteria for limiting average Markov decision processes •	95 IEEEAuCn 40 2-10
tation of central and non-central F probabilities and	percentiles • Wang, M. C.; Kennedy, W. J. A self-validating numerical method for compu-	95 StatComp 5 155-163
Holt, Elizabeth A.; Job, Jayakaran S.; Halsey, Neal A.	Percentile regression analysis of correlated antibody responses • Moulton, Lawrence H.;	95 StatMed 14 777-787
A note on	percentile residual life • Song, Jae-Kee; Cho, Gyo-Young	95 SankhyaA 57 333-335
model for image operations based on multilayer	perceptrons • Moh, Jenlong; Shih, Frank Y. A general purpose	95 PattrnRc 28 1083-1090
	Percolation	
Peres, Yuval	Critical random walk in random environment on trees • Pemantle, Robin;	95 AnlsProb 23 105-140
Piza, Marcelo S. T.	Divergence of shape fluctuations in two dimensions • Newman, Charles M.;	95 AnlsProb 23 977-1005
Peres, Yuval	Galton-Watson trees with the same mean have the same polar sets • Pemantle, Robin;	95 AnlsProb 23 1102-1124
	Martin capacity for Markov chains • Benjamini, Itai; Pemantle, Robin; Peres, Yuval	95 AnlsProb 23 1332-1346
	Percolation of arbitrary words in $\{0,1\}^N$ • Benjamini, Itai; Kesten, Harry	95 AnlsProb 23 1024-1060
Critical length for semi-oriented bootstrap	percolation • Mountford, T. S.	95 StocProc 56 185-205
Single linkage clustering and continuum	percolation • Penrose, Mathew D.	95 JMultiAn 53 94-109
Equality of critical densities in continuum	percolation • Meester, Ronald	95 JAppProb 32 90-104
strongly inhomogeneous environment and invasion	percolation • Newman, C. M.; Stein, D. L. Random walk in a	95 AIHenrPB 31 249-261
Equivalence of exponential decay rates for bootstrap	percolation like cellular automata • Andjel, E. D.; Mountford, T. S.; Schonmann, R. H.	95 AIHenrPB 31 13-25
	Percolation and minimal spanning forests in infinite graphs • Alexander, Kenneth S.	95 AnlsProb 23 87-104
time growth models, with applications to oriented	percolation • Liggett, Thomas M. Survival of discrete	95 AnlsApPr 5 613-636
Supercritical behaviors in first-passage	percolation • Zhang, Yu	95 StocProc 59 251-266
Quite weak Bernoulli with exponential rate and	percolation for random fields • Burton, Robert M.; Steif, Jeffrey E.	95 StocProc 58 35-55
On the concept of	perfect aggregation in Bayesian estimation (Disc: p287-289) • Bier, Vicki M.	94 RelEngSS 46 271-281
Guillen, Montserrat	Perfect cells, direct models and contingency table outliers • Upton, Graham J. G.;	95 CommStA 24 1843-1862
Michael	Perfect cocycles through stochastic differential equations • Arnold, Ludwig; Scheutzow,	95 ProbTher 101 65-88
Laskar, R.	Perfect domination in graphs • Cockayne, E. J.; Hartnell, B. L.; Hedetniemi, S. T.;	93 JCISS 18 136-148
Benjamin	Perfect filtering and double disjointness • Furstenberg, Hillel; Peres, Yuval; Weiss,	95 AIHenrPB 31 453-465
	Performance analysis of general tracking algorithms • Guo, Lei; Ljung, Lennart	95 IEEEAuCn 40 1388-1402
T.; Saniie, Jafar	Performance analysis of linearly combined order statistic CFAR detectors • Nagle, Daniel	95 IEEEAES 31 522-533
ster Iteratively reweighted partial least squares: A	performance analysis by Monte Carlo simulation • Cummins, David J.; Andrews, C. Web-	95 JChemtrc 9 489-507
Weiss, Alan Large deviations for	performance analysis. Queues, communication, and computing • Shwartz, Adam;	95 ChapmnHall:Lnd 576p
with star topology • Ahmed, N. U.; Yeap, T. H.	Performance analysis of slotted ALOHA protocol for optical fiber local area networks	95 StocAnAp 13 389-410
	Performance measurement	
	Safety measurement: Towards continuous improvement • St. Clair, Joseph F.	95 ASQCTekCon 150-158
a precise estimation of the calibration curve of	periodate in its determination by flow-injection analysis ... • Papamichael et al. ... for	95 Chemolab 30 227-237
Fixed-domain asymptotics for spatial	periodograms • Stein, Michael L.	95 JASA 90 1277-1288
Tudor, Constantin	Periodic and almost periodic solutions for semilinear stochastic equations • Prato, G. Da;	95 StocAnAp 13 13-33
Parameter estimation for	periodic ARMA models • Adams, G. J.; Goodwin, G. C.	95 JTimSrAn 16 127-145

Hans On	periodic autoregressions and structural breaks in seasonal time series • Franses, Philip	95 EnvrMtrc 6 451-455
The effects of seasonally adjusting a	periodic autoregressive process • Franses, Philip Hans	95 CmpStDA 19 683-704
Chebaane, M.; Salas, J. D.; Boes, D. C. Product	periodic autoregressive processes for modeling intermittent monthly streamflows •	95 WaterReR 31 1513-1518
Kloek, Teun A	periodic cointegration model of quarterly consumption • Franses, Philip Hans;	95 AplStMDA 11 159-166
Hans	Periodic cointegration: Representation and inference • Boswijk, H. Peter; Franses, Philip	95 RvEconSt 77 436-454
timators • Wikle, C. K.; Sherman, P. J. et al. Identifying	periodic components in atmospheric data using a family of minimum variance spectral es-	95 JClimate 8 2352-2363
Climatological time series with	periodic correlation • Lund, Robert; Hurd, Harry; Bloomfield, Peter; Smith, Richard	95 JClimate 8 2787-2809
Moving average filters and	periodic integration • Franses, Philip Hans; Paap, Richard	95 MaCmpSim 39 245-249
periods for two-dimensional Poisson processes with	periodic intensity • Ballerini, Rocco Nonconstant levels for t-year return	95 StPrLet 22 43-47
Estimation of non-occurrence probability in	periodic model • Chen, Ling; Mi, Jie	95 CommStA 24 1493-1505
tion: An algorithm for computing a locally optimal	periodic policy • Hordijk et al. Undiscounted Markov decision chains with partial informa-	94 ZeitOpRs 40 163-181
On fractionally differenced	periodic processes • Hui, Y. V.; Li, W. K.	95 SankhyaB 57 19-31
Extremes of a random number of variables from	periodic sequences • Ferreira, Helena	95 JStPlInf 45 133-141
proximation and its breakdown under a hierarchy of	periodic spatial heterogeneities • Bhattacharya, R. N. et al. Time-scales for Gaussian ap-	95 Bernoull 1 81-123
Use of phase for the detection of hidden	periodicities • Reschenhofer, Erhard	95 BiomtrcJ 37 957-964
Kaufman, S. F.; Smith, W.; Chang, M. Optimal	periodicity of a survey: Sampling error, data deterioration, and cost • Ghosh, D.;	94 ASASrvyRsr 1122-1127
	Periodogram	
integrated processes • Hurvich, C. M.; Ray, B. K.	Estimation of the memory parameter for nonstationary or noninvertible fractionally	95 JTimSrAn 16 17-41
Jürgen	Long memory in inflation rates: International evidence • Hassler, Uwe; Wolters,	95 JBES 13 37-45
Mikosch, T.; Gadrich, T.; Klüppelberg, C.; Adler, R. J.	Parameter estimation for ARMA models with infinite variance innovations •	95 AnlsStat 23 305-326
	Spectral density estimation for stationary stable random fields • Sabre, R.	95 ApplMath 23 107-133
Ouliaris, Sam	Spectral tests of the martingale hypothesis for exchange rates • Fong, Wai Mun;	95 JApEcmtx 10 255-271
	Testing equality of variances for paired time series • Beran, Jan; Gasser, Theo	95 JTimSrAn 16 165-176
	Use of phase for the detection of hidden periodicities • Reschenhofer, Erhard	95 BiomtrcJ 37 957-964
Poskitt, D. S.	Periodogram-based estimators of fractal properties • Chan, Grace; Hall, Peter;	95 AnlsStat 23 1684-1711
The	periodogram regression: Correction and comments • Hassler, Uwe	95 CommStA 24 3137-3146
Log-	periodogram regression of time series with long range dependence • Robinson, P. M.	95 AnlsStat 23 1048-1072
A CLT for the	periodograms of a ρ^*-mixing random field • Miller, Curtis	95 StocProc 60 313-330
Local polynomial	periodogram for time-varying frequency estimation • Katkovnik, Vladimir	95 SoAfrStJ 29 169-198
Have, T. R. T. et al. Association models for	periodontal disease progression: A comparison of methods for clustered binary data •	95 StatMed 14 413-429
	Permutation	
Aldous, D.; Diaconis, P.	Hammersley's interacting particle process and longest increasing subsequences •	95 ProbTher 103 199-213
Estimating probabilities from invariant	permutation distributions • Pallini, Andrea	94 JItalStS 3 77-91
results for the moment matrices of designs over	permutation-invariant sets • Cheng, Ching-Shui Complete class	95 AnlsStat 23 41-54
A	permutation paired test allowing for missing values • Maritz, J. S.	95 AstrlJSt 37 153-159
of the number of successions in a random	permutation • Fu, James C. Exact and limiting distributions	95 AnInStMa 47 435-446
Berge, Jos M. F. ten Suppressing	permutations or rigid planar rotations: A remedy against nonoptimal varimax rotations •	95 Psymtrka 60 437-446
Exact moment formulae for the linear	permutation statistic of order k • Whitmore, James B.	94 ASAProBiop 262-266
	Permutation test	
	Additivity tests for nonlinear autoregression • Chen, Rong; Liu, Jun S.; Tsay, Ruey S.	95 Biomtrka 82 369-383
Joseph	Comparison tests for dendrograms: A comparative evaluation • Lapointe, François-	95 JClass 12 265-282
Doerge, R. W.	Empirical threshold values for quantitative trait mapping • Churchill, G. A.;	94 Genetcs 138 963-971
	Estimating probabilities from invariant permutation distributions • Pallini, Andrea	94 JItalStS 3 77-91
Legendre, P.; Lapointe, F.-J.; Casgrain, P.	Modeling brain evolution from behavior: A permutational regression approach •	94 Evolutn 48 1487-1499
	Nonparametric test on dimensionality of explanatory variables (Korean) • Seo, H. S.	95 KorApS 8/2 65-75
at a multiallelic locus • Chakraborty, R.; Zhong, Y.	Statistical power of an exact test of Hardy-Weinberg proportions of genotypic data	94 HumHrdty 44 1-9
Philippe; O'Quigley, John	A two-stage procedure for survival studies with surrogate endpoints • Flandre,	95 Biomtrcs 51 969-976
Thomas S.; Pazdan, James L. A	permutation test analogue to Tarone's test for trend in survival analysis • Graves,	95 JStCmpSm 53 79-89
Noble, William A	permutation test for effects in an unreplicated factorial design • Loughin, Thomas M.;	94 ASAProPEng 181-186
Two	permutation tests of equality of variances • Baker, Rose D.	95 StatComp 5 289-296
Lebreton, J.-D. Refined approximations to	permutation tests for multivariate inference • Kazi-Aoual, F.; Hitier, S.; Sabatier, R.;	95 CmpStDA 20 643-656
Good, Ph. (Auth); Rudolph, P. E. (Rev) Review of	"Permutation tests. A practical guide to resampling methods for testing hypotheses" •	95 BiomtrcJ 37 150-150
Good, Philip (Auth); Niknian, Minoo (Rev) Review of	"Permutation tests: A practical guide to resampling methods for testing hypotheses" •	95 Technmcs 37 341-342
multivariate data in crossover designs using	permutation tests • Johnson, William D.; Mercante, Donald E. Analyzing	94 ASAProBiop 244-249
Modeling brain evolution from behavior: A	permutational regression approach • Legendre, P.; Lapointe, F.-J.; Casgrain, P.	94 Evolutn 48 1487-1499
On	perpendicularity in designed experiments • Kala, Radosław	95 ListyBio 32 93-99
Wakolbinger, Anton	Persistence of a critical super-2 process • Gorostiza, Luis G.; Hochberg, Kenneth J.;	95 JAppProb 32 534-540
Uncertainty about the	persistence of economic shocks • Miller, John P.; Newbold, Paul	95 JBES 13 435-440
O'Sullivan, Finbarr	Persistent oscillation patterns in climatological fields • Kooperberg, Charles;	94 ASASrvyRsr 200-205
Econometric estimates of price indexes for	personal computers in the 1990's • Berndt, Ernst R.; Griliches, Zvi; Rappaport, Neal J.	95 JEconmtx 68 243-268
Núñez, G.; Solozabal, J.; Morales, A. et al. A	personal computer-based system for parallel line analysis of bioassay data • Iznaga, N.;	95 CmpBiomd 47 167-175
GUHA for	personal computers • Hájek, P.; Sochorová, A.; Zvárová, J.	95 CmpStDA 19 149-153
Forecasting the final vintage of real	personal disposable income: A state space approach • Patterson, K. D.	95 IntJFor 11 395-405
Can we get respondents to use their	personal income records • Bogen, Karen; Moore, Jeffrey C.; Marquis, Kent H.	94 ASASrvyRsr 1252-1257
The distribution of	personal income: Revisited • McDonald, James B.; Mantrala, Anand	95 JApEcmtx 10 201-204
	Personal probability	
DeGroot, M. H.; Mortera, J.	Coherent combination of experts' opinions (Disc: p293-313) • Dawid, A. P.;	95 Test 4 263-293
dimensions: Application of mixed-measurement to	personality assessment • Reise, S. P. et al. Modeling qualitative variation within latent trait	95 MultBehR 30 341-358
	Perturbation	
MANOVA models • Seo, T.; Kanda, T.; Fujikoshi, Y.	The effects of nonnormality on tests for dimensionality in canonical correlation and	95 JMultiAn 52 325-337
processes • Cao, Xi-Ren	Event coupling and performance sensitivity analysis of generalized semi-Markov	95 AdvAppPr 27 741-769
associated fluid queueing ... • Kontovasilis et al.	Markov-modulated traffic with nearly complete decomposability characteristics and	95 AdvAppPr 27 1144-1185
Klein, Maurice; Zuyev, Sergei	Perturbation analysis of functionals of random measures • Baccelli, François;	95 AdvAppPr 27 306-325
	Perturbation analysis for an orthogonal least-squares estimator • Chapman, G. R.	95 CandJSt 23 101-108
On unbounded hazard rates for smoothed	perturbation analysis • Fu, Michael C.; Hu, Jian-Qiang	95 JAppProb 32 659-667
diffusion equations with white noise boundary	perturbations • Sowers, R. B. Multidimensional reaction-	94 AnlsProb 22 2071-2121
theorems on the exit problems for small random	perturbations of dynamical systems I" (V46 p79-106) • Mikami, T. Correction to "Limit	95 Stochatx 52 323-323
Freidlin, M. I. Fast oscillating random	perturbations of dynamical systems with conservation laws • Borodin, A. N.;	95 AIHenrPB 31 485-525
Comment on "Mutations,	perturbations and evolutionarily stable strategies" (V19 p204-209) • Hines, W. G. S.	95 JAppProb 32 836-839
Small	perturbations of Gaussian regressors • Millet, Annie; Smoleński, Włodzmierz	95 StPrLet 24 21-31
learning rule of neural networks via simultaneous	perturbation and its hardware implementation • Maeda, Y.; Hirano, H.; Kanata, Y. A	95 Automatca 31 251-259
and control for linear systems with jump Markov	perturbations • Morozan, T. Stability	95 StocAnAp 13 91-110
Kemp, C. J.	Perturbation of the normal model in linear regression • de Waal, D. J.; Groenewald, P. C. N.;	95 SoAfrStJ 29 105-130
On group inverses of matrices with order and rank	perturbations • Rao, P. S. S. N. V. P.	94 CalcutSt 44 209-222

Equating	Poisson and normal probability functions to derive Stirling's formula • Walsh, Dennis P.	95 AmerStat 49 270-271
On the admissibility of ordered	Poisson parameter under the entropy loss function • Parsian, A.; Nematollahi, N.	95 CommStA 24 2451-2467
walla, Nozer D. Inference and predictions from	Poisson point processes incorporating expert knowledge • Campodónico, Sylvia; Singpur-	95 JASA 90 220-226
estimation and suboptimal stopping time for a	Poisson point process • Garcia, Nancy Lopes Maximum likelihood	95 RBrzlPrS 9 67-82
of convex hulls generated by homogeneous	Poisson point processes in an unbounded convex domain • Nagaev, A. V. Some properties	95 AnInStMa 47 21-29
Inequalities for the anisotropic	Poisson polytope • Mecke, J.	95 AdvAppPr 27 56-62
tion and testing for the parameters of the bivariate	Poisson populations based on a 2 by 2 sample" (V61 p361-367) • Douglas ... on "Estima-	95 IntStRvw 63 117-118
schemes for calculating cumulative binomial and	Poisson probabilities" (V48 p136-138) • Nemenyi, Peter Comment on "Recursive	95 AmerStat 49 402-403
Ordering claim size distributions and mixed	Poisson probabilities • Kaas, R.; Hesselager, O.	95 InsrneME 17 193-201
	Poisson process	
process • Phelan, Michael J.	Asymptotic information for parametric estimation from an equilibrium particle	95 StPrLet 24 193-198
hybrids • Lange, K.; Boehnke, M.	Bayesian methods and optimal experimental design for gene mapping by radiation	92 AnHuGene 56 119-144
Pinedo, Michael	A convexity property of a Markov-modulated queueing loss system • Du, Charles;	95 PrEngInf 9 193-199
	Covering with blocks in the non-symmetric case • Móri, Tamás F.	95 JTherPrb 8 139-164
Song, Mun Sup; Park, Byung-Gu	A diffusion model for a system subject to random shocks • Lee, Eui Yong;	95 JKoreaSt 24 141-147
	Duration dependence and dispersion in count-data models • Winkelmann, Rainer	95 JBES 13 467-474
	Equality of critical densities in continuum percolation • Meester, Ronald	95 JAppProb 32 90-104
systems in the plane • Weil, Wolfgang	The estimation of mean shape and mean particle number in overlapping particle	95 AdvAppPr 27 102-119
with fast Markov switchings • Di Masi, G. B. et al.	A first order approximation for the convergence of distributions of the Cox processes	95 Stochatx 54 211-219
nucleation • Møller, Jesper	Generation of Johnson-Mehl crystals and comparative analysis of models for random	95 AdvAppPr 27 367-383
of certain large random polygons • Miles, R. E.	A heuristic proof of a long-standing conjecture of D. G. Kendall concerning the shapes	95 AdvAppPr 27 397-417
lattice • Kopociński, B.	The largest square of successes covering the origin for Bernoulli trials on the	95 StNeerla 49 31-39
	On metrics in point process approximation • Brown, Timothy C.; Xia, Aihua	95 Stochatx 52 247-263
	Modelling of discretized claim numbers in loss reserving • Hesselager, Ole	95 AstinB 25 119-135
Diggle, Peter J.	Non-parametric estimation of spatial variation in relative risk • Kelsall, Julia E.;	95 StatMed 14 2335-2342
(French) • Moncaup, S.; Sarda, P.; Vieu, P.	Nonparametric estimators of an intensity function applied to meteorological data	95 RStApp 43/4 77-88
with star topology • Ahmed, N. U.; Yeap, T. H.	Performance analysis of slotted ALOHA protocol for optical fiber local area networks	95 StocAnAp 13 389-410
Goudie, I. B. J.	A plant-capture approach for achieving complete coverage of a population •	95 CommStA 24 1293-1305
	Random record processes and state dependent thinning • Browne, Sid; Bunge, John	95 StocProc 55 131-142
	A random shock model for a linearly deteriorating system • Lee, Jiyeon; Lee, Eui Yong	95 JKoreaSt 24 471-479
	The rank of the present excursion • Scheffer, Carel L.	95 StocProc 55 101-118
	Shock models with NBRU properties • Abouammoh, A. M.; Hendi, M. I.; Khalique, A.	95 PakJStA 11 35-47
	Single linkage clustering and continuum percolation • Penrose, Mathew D.	95 JMultiAn 53 94-109
Gupta, Ramesh C.	Some results on randomly stopped minimal repair processes • Kirmani, S. N. U. A.;	95 CommStSM 11 631-644
Nadeau, C.	Some simple robust methods for the analysis of recurrent events • Lawless, J. F.;	95 Technmcs 37 158-168
	Temporal analogues to spatial K functions • Cannon, Ann; Cressie, Noel	95 BiomtrcJ 37 351-373
Feigin, Paul D.; Resnick, Sidney I.; Stărică, Cătălin	Testing for independence in heavy tailed and positive innovation time series •	95 CommStSM 11 587-612
Mountford, T.; Prabhakar, B.	On the weak convergence of departures from an infinite series of ·/M/1 queues •	95 AnlsApPr 5 121-127
Patrick; McCormick, William P.	Weak limit results for the extremes of a class of shot noise processes • Homble,	95 JAppProb 32 707-726
Stark, D.; Tavaré, S. Total variation asymptotics for	Poisson process approximations of logarithmic combinatorial assemblies • Arratia, R.;	95 AnlsProb 23 1347-1388
Extremes of interarrival times of a	Poisson process under conditioning • Abay, A.	95 ZastApMa 23 73-82
processes as projections of higher-dimensional	Poisson processes • Garcia, Nancy Lopes Birth and death	95 AdvAppPr 27 911-930
Statistical analysis on Markov-modulated	Poisson processes • Ramesh, N. I.	95 EnvrMtrc 6 165-179
single-server queue subject to a Markov-modulated	Poisson process • Du, Qing A monotonicity result for a	95 JAppProb 32 1103-1111
Prior elicitation and Bayesian robustness in a noisy	Poisson process • Moreno, Elías; Girón, Javier	94 ASAProBays 217-220
Estimating the edge of a	Poisson process by orthogonal series • Jacob, Pierre; Suquet, Charles	95 JStPlInf 46 215-234
The favorite point of a	Poisson process • Khoshnevisan, Davar; Lewis, Thomas M.	95 StocProc 57 19-38
Review of	"Poisson processes" • Kingman, J. F. C. (Auth); Booker, Jane (Rev)	95 Technmcs 37 124-124
Review of	"Poisson processes" • Kingman, J. F. C. (Auth); Kemp, A. W. (Rev)	94 BLndnMaS 26 612-614
theorems on the extreme values of stationary	Poisson processes • Révész, P. Strong	95 JStPlInf 45 291-300
Estimation and filtering on a doubly stochastic	Poisson process • Valderrama, M. J.; Jimenez, F.; Gutierrez, R.; Martinez-Almecija, A.	95 AplStMDA 11 13-24
levels for t-year return periods for two-dimensional	Poisson processes with periodic intensity • Ballerini, Rocco Nonconstant	95 StPrLet 22 43-47
On properties of	Poisson Q charts for attributes (Disc: p316-343) • Quesenberry, Charles P.	95 JQualTek 27 293-303
valued stochastic differential equations driven by	Poisson random measures • Kallianpur, G. et al. Diffusion approximation of nuclear space-	95 AnlsApPr 5 493-517
A stopping rule for a compound	Poisson random variable • Randolph, Paul; Şahinoğlu, Mehmet	95 AplStMDA 11 135-143
Raz, Tzvi Bayesian	Poisson rate estimation using parameter maps • Buck, James R.; Pignatiello, Joseph J., Jr;	95 IIETrn 27 405-412
	Poisson regression	
	The design and analysis of randomized trials with recurrent events • Cook, Richard J.	95 StatMed 14 2081-2098
	Fitting parametric counting processes by using log-linear models • Lindsey, J. K.	95 ApplStat 44 201-212
Britain • Richardson, S.; Monfort, C.; Green, M. et al.	Spatial variation of natural radiation and childhood leukaemia incidence in Great	95 StatMed 14 2487-2501
	Poisson regression analysis in clinical research • Kianifard, Farid; Gallo, Paul P.	95 JBiophSt 5 115-129
Instrumental variable estimation of	Poisson regression models (Disc: p17-19) • Mullahy, John	94 ASAProBuEc 7-13
Log-linear models for mean and dispersion in mixed	Poisson regression models • van de Ven, R.; Weber, N. C.	95 AstrlJSt 37 205-216
Bayesian analysis of two overdispersed	Poisson regression models • Scollnik, David P. M.	95 CommStA 24 2901-2918
Moffatt, Peter G. Grouped	Poisson regression models: Theory and an application to public house visit frequency •	95 CommStA 24 2779-2796
Murray, James A method for fitting	Poisson regression model with extra zeros • Jovanovic, Borko D.; Hosmer, David W.;	94 ASAProEpid 97-100
Feasibility study of the use of Chromy's algorithm in	Poisson-sample selection for the annual survey of manufactures • Zayatz, L.; Sigman, R.	94 ASASrvyRsr 641-646
	Poisson search models and the evaluation of stopping rules • Bather, John	95 SqtlAnly 14 205-216
Mikosch, Thomas Explosive	Poisson shot noise processes with applications to risk reserves • Klüppelberg, Claudia;	95 Bernoull 1 125-147
and inequalities for functionals on Wiener and	Poisson spaces • Houdré, Christian; Pérez-Abreu, Víctor Covariance identities	95 AnlsProb 23 400-419
likelihood estimation in the presence of extra-	Poisson variation • Johaadien, Hassan; Goutis, Constantinos Approximate maximum	95 CommStA 24 1625-1636
Generation of the typical cell of a non-	Poissonian Johnson-Mehl tessellation • Heinrich, Lothar; Schüle, Eik	95 CommStSM 11 541-560
trees with the same mean have the same	polar sets • Pemantle, Robin; Peres, Yuval Galton-Watson	95 AnlsProb 23 1102-1124
Kojetin et al. Response to a government survey as	political participation: The relation of economic and political conditions to refusal ... •	94 ASASrvyRsr 1236-1241
of "The Phantom respondents: Opinion surveys and	political representation" • Brehm, John (Auth); Parsley, Teresa L. (Rev) Review	94 JOffStat 10 351-355
decisions based on evidence, psychology and	politics • Bonham, G. H. Health and environment: Public health	95 EnvrMtrc 6 311-318
Correction to "The significance of the lead in opinion	polls" (V17 p9-30) • Kmietowicz, Z. W.	95 JAppStat 22 551-551
Sawyer, Tom; Tanur, Judith M. Pseudo-opinion	polls: SLOP or useful data? • Horvitz, Dan; Koshland, Daniel; Rubin, Donald; Gollin, Albert;	95 Chance 8/2 16-25
	Polling	
Kmietowicz, Z. W.	Significance of the sample lead: Some theoretical and empirical results •	95 CommStA 24 1507-1519
Arnedt, Cheryl Mercado; Ramnath, Bala Interactive	polling and Americans' comfort level with technology • Frankovic, Kathleen A.;	94 ASASrvyRsr 1316-1321
A globally gated	polling system with a dormant server • Borst, S. C.	95 PrEngInf 9 239-254
Coffman, E. G., Jr; Puhalskii, A. A.; Reiman, M. I.	Polling systems with zero switchover times: A heavy-traffic averaging principle •	95 AnlsApPr 5 681-719
probability fields for source locations of airborne	pollutants • Hopke, P. K.; Li, C. L. et al. The use of bootstrapping to estimate conditional	95 Chemolab 30 69-79

sos, A. G. et al. Statistical evaluation of selected air	pollutants in Athens, Greece • Kambezidis, H. D.; Tulleken, R.; Amanatidis, G. T.; Paliat-	95 EnvrMtrc 6 349-361
issues in the detection of low-level environmental	pollutants (Disc: p145-167) • Gibbons, Robert D. Some statistical and conceptual	95 EnvEcoSt 2 125-145
for the estimation of solution properties of organic	pollutants • Dunn, W. J., III Computational methods	95 Chemolab 29 1-10
analysis of health effects associated with air	pollution • Duddek, C.; Le, N. D. et al. Multivariate imputation in cross-sectional	95 EnvEcoSt 2 191-212
Metabolic responses in the assessment of	pollution effects • Pires, Ana; Branco, Joao; Picado, Ana; Mendonça, Elsa	95 EnvrMtrc 6 155-163
MCMC methods for putative	pollution source problems in environmental epidemiology • Lawson, Andrew B.	95 StatMed 14 2473-2485
	Polya distribution	
Pierce, Rebecca L.	Bayesian HPM intervals for the number of nonconforming items in a lot • Dyer, Danny;	95 CmpStDA 20 159-172
cations on modeling physicochemical properties of	polyaromatic hydrocarbons • Todeschini et al. ... Part 2. Theory development and appli-	95 Chemolab 27 221-229
The computation of opportunity costs in	polychotomous choice models with selectivity • Lee, Lung-fei	95 RvEconSt 77 423-435
	Polychotomous choices and self-selection (Italian) • Mealli, Fabrizia	94 Statstca 54 501-519
dardized rate ratios for imprecise classification of a	polychotomous exposure variable with limited data • Rosenbaum et al. Correcting stan-	95 AJEpidm 141 442-445
dardized rate ratios for imprecise classification of a	polychotomous exposure variable with limited data • Rosenbaum et al. Correcting stan-	95 AJEpidm 142 442-445
	Polychotomous or polytomous? • Weiss, David J.	95 ApPsychM 19 4-4
Semiparametric maximum likelihood estimation of	polychotomous and sequential choice models • Lee, Lung-fei	95 JEconmtx 65 381-428
et al. Multiple-trait genetic evaluation for one	polychotomous trait and several continuous traits with missing data and ... • Hoeschele	95 JAnmlSci 73 1609-1627
On the use of circular tetrads in preference	polygon (Japanese) • Makino, Toji; Konno, Toshiharu	94 JpJBeh 21/2 1-9
The anchor position of histograms and frequency	polygons: Quantitative and qualitative smoothing • Simonoff, Jeffrey S.	95 CommStB 24 691-710
concerning the shapes of certain large random	polygons • Miles, R. E. A heuristic proof of a long-standing conjecture of D. G. Kendall	95 AdvAppPr 27 397-417
Statistical classification procedures for	polygraph tests of guilty knowledge • Adachi, Kohei	95 Behvmtrk 22 49-66
the maximum risk of regression estimators by	polyhedral projection • Schmidt, Karsten; Stahlecker, Peter Reducing	95 JStCmpSm 52 1-15
spatial counting process modelling crystallization of	polymers • Capasso et al. ... of the maximum likelihood estimators of parameters of a	95 StocAnAp 13 279-294
Use of statistical design to develop flame retardant	polymer formulations • Babcock, Laura M.; Altekar, Maneesha	95 Chemolab 29 141-146
Wallace, J. E.; Foster, R. D. et al. Improving control of	polymer viscosity using time-series experimentation and statistical process control •	94 ASAProPEng 320-325
foxide - An experimental ... • Feuer et al. Free radical	polymerization of nonlinear optical side chain methacrylate copolymers in dimethylsul-	95 Chemolab 29 113-124
Tests of random mating for a highly	polymorphic locus: Application to HLA data • Jin, Kun; Speed, T. P.; Thomson, Glenys	95 Biomtrcs 51 1064-1076
disequilibrium between trait loci and one or more	polymorphic marker loci • Terwilliger ... likelihood method for the analysis of linkage	95 AJHuGene 56 777-787
	Polynomial	
	• see also names of specific types of polynomials such as orthogonal –	
of probability distributions and exponential-	polynomial approximation • Martin, Clyde; Shubov, Victor Natural exponential families	93 AplMaCmp 59 275-297
Vladimir Nonparametric local	polynomial approximation of the time-varying frequency and amplitude • Katkovnik,	95 CommStA 24 3001-3025
Bivariate extension of the method of	polynomials for Bonferroni-type inequalities • Galambos, Janos; Xu, Yuan	95 JMultiAn 52 131-139
analysing longitudinal data using piecewise cubic	polynomials • Wang, Yongxiao; Taylor, Jeremy M. G. Flexible methods for	95 JStCmpSm 52 133-150
or 3D range data based on high degree implicit	polynomial curves or surfaces, algebraic invariants, and ... • Subrahmonia et al. ... images	94 ASAProPEng 44-49
Gijbels, Irène Adaptive order	polynomial fitting: Bandwidth robustification and bias reduction • Fan, Jianqing;	95 JCmpGrSt 4 213-227
Irène Data-driven bandwidth selection in local	polynomial fitting: Variable bandwidth and spatial adaptation • Fan, Jianqing; Gijbels,	95 JRSS-B 57 371-394
functions • Fan, J.; Heckman, N. E.; Wand, M. P. Local	polynomial kernel regression for generalized linear models and quasi-likelihood	95 JASA 90 141-150
Sengupta, Arindam Time-spaced	polynomial martingales generated by a discrete-time martingale • Goswami, A.;	95 JTherPrb 8 417-432
	Polynomial measurement error modeling • Moon, Myung-Sang; Gunst, Richard F.	95 CmpStDA 19 1-21
penalized likelihood criteria in antedependence and	polynomial models • Macchiavelli, Raúl; Arnold, Steven F. Difficulties with the use of	95 CommStA 24 501-522
The	polynomial model in the multivariate analysis of variance • Baumgartner, Cornelia	95 BiomtrcJ 37 299-318
A V-optimal design for Scheffé's	polynomial model • Liu, Shuangzhe; Neudecker, Heinz	95 StPrLet 23 253-258
Continual analogues of random	polynomials which are orthogonal on a circle • Teplyaev, A. V.	94 ThProbAp 39 476-489
Local	polynomial periodogram for time-varying frequency estimation • Katkovnik, Vladimir	95 SoAfrStJ 29 169-198
Modeling using ratios of	polynomials • Van Nostrand, R. Craig	94 ASAProPEng 97-102
How many zeros of a random	polynomial are real? • Edelman, Alan; Kostlan, Eric	95 BAmMaSoc 32 1-37
Optimal designs for identifying the degree of a	polynomial regression • Dette, Holger	95 AnlsStat 23 1248-1266
Optimal designs for	polynomial regression when the degree is not known • Dette, Holger; Studden, William J.	95 StSinica 5 459-473
Exact and approximate D-optimal designs in	polynomial regression • Happacher, Maximilian	95 Metrika 42 19-27
for optimal design with application to multiple	polynomial regression • Gaffke, Norbert; Heiligers, Berthold Algorithms	95 Metrika 42 173-190
Testing for	polynomial regression using nonparametric regression techniques • Jayasuriya, Bodhini	94 ASAProBuEc 241-246
Wong, Weng Kee D-optimal designs for	polynomial regression without an intercept • Huang, Mong-Na Lo; Chang, Fu-Chuen;	95 StSinica 5 441-458
models with random disturbed-highest-derivative-	polynomial subject effects • Wilson, P. David Longitudinal data analysis for linear Gaussian	95 StatMed 14 1219-1233
Estimation of edge parameters and image blur using	polynomial transforms • Kayargadde, Vishwakumara; Martens, Jean-Bernard	94 CVGIPGr 56 442-461
Bilinear forms and zonal	polynomials • Mathai, A. M.; Provost, S. B.; Hayakawa, T.	95 Springer:Brln:NY 376p
variances • Sjöström, M.; Rännar, S.; Wieslander, Å.	Polypeptide sequence property relationships in Escherichia coli based on auto cross co-	95 Chemolab 29 295-305
tributions and correlation structure of multivariate	polytomous data from a two-stage cluster sample • Marques et al. A model for dis-	94 ASAProBiop 346-351
Full-information factor analysis for	polytomous item responses • Muraki, Eiji; Carlson, James E.	95 ApPsychM 19 73-90
Conceptual notes on models for discrete	polytomous item responses • Mellenbergh, Gideon J.	95 ApPsychM 19 91-100
Item response theory for scores on tests including	polytomous items with ordered responses • Thissen, D.; Pommerich, M.; Billeaud, K. et al.	95 ApPsychM 19 39-49
Estimation in the	polytomous logistic regression model • Rom, Meir; Cohen, Ayala	95 JStPlInf 43 341-353
Polychotomous or	polytomous? • Weiss, David J.	95 ApPsychM 19 4-4
Residual analysis in the	polytomous Rasch model • Andersen, Erling B.	95 Psymtrka 60 375-393
Bootstrapping a nonparametric	polytomous regression model • Aerts, M.; Veraverbeke, N.	95 MaMethSt 4 189-200
of structural equation models with continuous and	polytomous variables • Lee, Sik-Yum; Poon, Wai-Yin; Bentler, P. M. A two-stage estimation	95 BJMaStPs 48 339-358
Potenza, Maria T.; Dorans, Neil J. DIF assessment for	polytomously scored items: A framework for classification and evaluation •	95 ApPsychM 19 23-37
Inequalities for the anisotropic Poisson	polytope • Mecke, J.	95 AdvAppPr 27 56-62
A Bayesian analysis of the use of	pooled coefficients in a structural regional economic model • Rickman, Dan S.	95 IntJFor 11 477-490
Poststratification of	pooled survey data • Fahimi, Mansour	94 ASASrvyRsr 879-881
Tu et al. On the informativeness and accuracy of	pooled testing in estimating prevalence of a rare disease: Application to HIV screening •	95 Biomtrka 82 287-297
Doggett, N. A.; Sawhill, W. W. et al. Efficient	pooling designs for library screening • Bruno, W. J.; Knill, E.; Balding, D. J.; Bruce, D. C.;	95 Genomics 26 21-30
response • Burt, Oscar R.	Pooling experiments over time to quantify the dynamics of dryland winter wheat yield	95 AgronmyJ 87 156-164
Jayaram How	pooling failure data may reverse increasing failure rates • Gurland, John; Sethuraman,	95 JASA 90 1416-1423
A	pooling methodology for coefficient of variation • Ahmed, S. E.	95 SankhyaB 57 57-75
A	pooling procedure for the exponential distribution • Chiou, Paul	95 CommStA 24 445-458
populations • Arnason, A. Neil; Schwarz, Carl J.	POPAN-4: Enhancements to a system for the analysis of mark-recapture data from open	95 JAppStat 22 785-800
	Population	
	• see also closed –, demography, finite –, super- –	
	• see also names of specific probability distributions such as normal –	
Landis, J. Richard Model-based estimation of	population attributable risk under cross-sectional sampling • Basu, Srabashi;	95 AJEpidm 141 1338-1343
Landis, J. Richard Model-based estimation of	population attributable risk under cross-sectional sampling • Basu, Srabashi;	95 AJEpidm 142 1338-1343
of variability for estimates of absolute risk from a	population-based case-control study on breast cancer • Benichou, J. A complete analysis	95 BiomtrcJ 37 3-24
inference for estimates of absolute risk derived from	population-based case-control studies • Benichou, Jacques; Gail, Mitchell H. Methods of	95 Biomtrcs 51 182-194
Alexander, Lois Reinventing the	population census: Balancing privacy and efficiency. In memory of Roger A. Herriot •	94 ASAProGovt 1-2

to unit root testing - Using covariates to increase	power • Hansen, Bruce E. Rethinking the univariate approach	95 EconTher 11 1148-1171
in regulatory regime: The case of the US nuclear	power industry • Rust, John; Rothwell, Geoffrey Optimal response to a shift	95 JApEcmtx 10 75-118S
Cardon, Lon R.; Fulker, David W. The	power of interval mapping of quantitative trait loci, using selected sib pairs •	94 AJHuGene 55 825-833
A probabilistic model for the variance to mean	power law in ecology • Kendal, Wayne S.	95 EcolgMdl 80 293-297
B.; Schell, M. J.; Wright, F. T. Approximations to the	powers of the likelihood ratio tests: The loop ordering and slippage alternatives • Singh,	95 CommStB 24 91-109
trend analysis • Greenland, Sander Avoiding	power loss associated with categorization and ordinal scores in dose-response and	95 Epidmlgy 6 450-454
Review of "MATLAB for the	Power Macintosh" • The MathWorks Inc (Auth); Spiegelman, Cliff (Rev)	95 Chemolab 29 323-324
Records from a	power model of independent observations • Weissman, Ishay	95 JAppProb 32 982-990
The	power of modified instrumental variables tests for unit roots • Li, Hongyi	95 Biomtrka 82 660-666
The	power of the N-test of haplotype concordance • Green, J. R.; Shah, S.	92 AnHuGene 56 331-338
arrays having factors with a large non-prime	power number of levels • Mandeli, John P. Construction of asymmetrical orthogonal	95 JStPlInf 47 377-391
Can purchasing	power parity help forecast the dollar? • Cochran, Steven J.; Defina, Robert H.	95 JForcast 14 523-532
likelihood inference in a ... • Juselius Do purchasing	power parity and uncovered interest rate parity hold in the long run? An example of	95 JEconmtx 69 211-240
of clusters of disease: An application to nuclear	power plant facilities and childhood leukaemia in Sweden • Waller et al. ... and assessment	95 StatMed 14 3-16
models for point processes: An analysis of nuclear	power plant operations in Europe • Sturm, Roland Proportional hazard regression	94 JAppStat 21 533-540
A maximum test for scale: Type I error rates and	power • Algina, James; Blair, R. Clifford; Coombs, William T.	95 JEdStat 20 27-39
Sibuya, Masaaki	Power of record-breaking test (Japanese) • Iiyama, Yuriko; Nishimura, Kazuo;	95 JpJAppSt 24 13-26
	Power robustification of approximately linear tests • Ehm, W.; Mammen, E.; Müller, D. W.	95 JASA 90 1025-1033
Computing defensible linear model	power and sample size estimates for study planning • Taylor, Douglas J.; Muller, Keith E.	94 ASAProBiop 383-388
	Power and sample size in experiments involving repeated measures • Pollard, William E.	94 ASAProSoSt 45-48
Muller, Keith E. Computing confidence bounds for	power and sample size of the general linear univariate model • Taylor, Douglas J.;	95 AmerStat 49 43-47
under Pitman's measure of ... • Kourouklis Estimating	powers of the scale parameter of an exponential distribution with unknown location	95 JStPlInf 48 185-195
	Power of the scan statistics • Wallenstein, Sylvan; Naus, Joseph; Glaz, Joseph	94 ASAProEpid 70-75
Development and justification of the	power-series algorithm for BMAP-systems • van den Hout, Wilbert; Blanc, Hans	95 CommStSM 11 471-496
Bivariate discrete measures via a	power series conditional distribution and a regression function • Wesolowski, Jacek	95 JMultiAn 55 219-229
Identifiability of mixtures of	power-series distributions and related characterizations • Sapatinas, Theofanis	95 AnInStMa 47 447-459
Tripathi, Ram C. Inflated modified	power series distributions with applications • Gupta, Pushpa L.; Gupta, Ramesh C.;	95 CommStA 24 2355-2374
The effect of unequal variances on the	power of several two-sample tests • O'Gorman, Thomas W.	95 CommStB 24 853-867
J.; Schork, M. Anthony The relative efficiency and	power of small-pedigree studies of the heritability of a quantitative trait • Schork, Nicholas	93 HumHrdty 43 1-11
Modified stationarity tests with improved	power in small samples • Breitung, J.	95 StHefte 36 77-95
short term prediction of electrical load for a	power system • Polzehl, Jörg Regression-based	94 ASAProPEng 285-290
process on a large time interval ... • Miroshin The	power terms of the asymptotic series for moments of the number of zeros of the Wong	95 VLenUM /2 36-42
	Power of a test	
through 1993 • Simpson, J. M.; Klar, N.; Donner, A.	Accounting for cluster randomization: A review of primary prevention trials, 1990	95 AJPubHlh 85 1378-1383
statistics? • Smith, Stephen M.	Distribution-free and robust statistical methods: Viable alternatives to parametric	95 Ecology 76 1997-1998
studies? (German) • Böker, Fred; Dannenberg, Olaf	What can P-values say about the quality of test statistics for comparative simulation	95 AllgeStA 79 233-251
Chappell, David	Quick power estimates for comparing proportions • Hopkins, Kenneth D.;	94 EdPsMeas 54 903-912
treatments • Nam, Jun-Mo	Sample size determination in stratified trials to establish the equivalence of two	95 StatMed 14 2037-2049
Type II errors • Mapstone, Bruce D.	Scalable decision rules for environmental impact studies: Effect size, Type I, and	95 EcolgApp 5 401-410
	Testing the mean of skewed distributions • Chen, Ling	95 JASA 90 767-772
	On the use of a pilot sample for sample size determination • Browne, Richard H.	95 StatMed 14 1933-1940
B.	When is it worth measuring a covariate in a randomized clinical trial? • Allison, David	95 JCCPsych 63 339-343
An upper bound for the	power of a test • Bismarck, M.; Gajek, L.	95 CommStA 24 813-828
Ryan, Louise M. Efficiency and	power of tests for multiple binary outcomes • Legler, Julie M.; Lefkopoulou, Myrto;	95 JASA 90 680-693
Regression with random error in X: Increasing	power of the test by non-linear adjustment for measurement error • Grove, J. S. et al.	94 ASAProEpid 143-146
Atlas, Robert S.	Power of a test that is robust against variance heterogeneity • Overall, John E.;	95 PsychRpt 77 155-159
Nassar, R. Size and	power of test statistics for gene correlation in 2×2 contingency tables • Mannan, M.;	95 BiomtrcJ 37 409-433
L. Critical values and	powers for tests of uniformity of directions under multivariate normality • Dunn, Charles	95 CommStA 24 2541-2560
Temporal aggregation and the	power of tests for a unit root • Pierse, R. G.; Snell, A. J.	95 JEconmtx 65 333-345
	Power transformation	
	Concavity of Box-Cox log-likelihood function • Kouider, Elies; Chen, Hanfeng	95 StPrLet 25 171-175
Using	power transformations when approximating quantiles • Fearn, Dean H.; Nebenzahl, Elliott	95 CommStA 24 1073-1093
G.; Danell, Ö. Needs and benefits of empirical	power transformations for production and quality traits in forest tree breeding • Jansson,	93 ThApGene 87 487-497
Visualizing	power and type I and II error • Daone, David P.; Tracy, Ronald L.; Mathieson, Kieran D.	94 ASAProStEd 66-69
Local bounds on	power of T^2 in scale families • Jensen, D. R.	95 JStCmpSm 52 187-190
comparison procedures • Klockars, A. J. et al.	Power of unweighted and weighted versions of simultaneous and sequential multiple-	95 PsychBu 118 300-307
ing release-recapture data: Sample size, validity and	power • Hoffmann, A. et al. Inferential properties of an individual-based survival model us-	95 JAppStat 22 579-595
	Power variance function	
Bose, Arup; Mukhopadhyay, Nitis	A note on accelerated sequential estimation of the mean of NEF-PVF distributions •	95 AnInStMa 47 99-104
belonging to the natural exponential families having	power variance functions (Korean) • Suh, Euy Hoon Percentiles for the distributions	95 KorApS 8/1 133-149
Analysis of count data using	power variance function • Nakashima, Eiji	95 JJapnStS 25 193-204
tion • Pal, N. et al. Improved minimax estimation of	powers of the variance of a multivariate normal distribution under the entropy loss func-	95 StPrLet 24 205-211
Asymptotics for the Euclidean TSP with	power weighted edges • Yukich, J. E.	95 ProbTher 102 203-220
Sukhatme, Shashikala; Lin, C. H. Exact	powers of the Wilcoxon test and the logrank test under random censoring •	95 JStPlInf 48 291-301
tification of misconceptions in learning statistical	power with dynamic graphics as a remedial tool • Yu, Chong Ho; Behrens, John T. Iden-	94 ASAProStEd 242-246
Mathematica and Data Desk for the Macintosh	PowerPC" • Spiegelman, Cliff (Rev) Review of "SoftWindows,	95 Chemolab 28 208-209
for independence between hitting times of bivariate	PQD processes • Ramalingam, T.; Ebrahimi, Nader Testing	95 SankhyaA 57 202-220
Statistical consulting	practicum: Opportunity for authentic learning • Derr, Janice A.	94 ASAProStEd 140-143
Synthetic estimates for small areas in	PRAMS • Johnson, Christopher H.	94 ASASrvyRsr 1065-1070
The Shewhart, the Hodges-Lehman chart and the	pre-control: Some comparisons • Talwar, Prem P.	94 ASAProQlPr 151-154
Review of "Multi-vari chart and analysis - A	pre-experimentation technique" • Perez Wilson, Mario (Auth); Papp, Zoltan (Rev)	95 Technmcs 37 121-121
for the Orlicz norm of a linear form of independent	pre-Gaussian random variables • Kurchenko, O. O. Some inequalities	95 ThPrMaSt 50 99-102
O. M.	Pre-Gaussian random vectors and their applications • Kozachenko, Yu. V.; Mokliatchouk,	95 ThPrMaSt 50 89-98
the distribution of supremum of a certain class of	pre-Gaussian stochastic processes on a compact set in R^d • Sydorenko, O. O. On	95 ThPrMaSt 50 123-128
Krzanowski, W. J. Application of wavelets to the	pre-processing of underwater sounds • Powell, K. J.; Sapatinas, T.; Bailey, T. C.;	95 StatComp 5 265-273
	Pre-test estimator	
loss function and nonnormal error terms • Hasegawa	On risk comparisons of some estimators in a linear regression model under Inagaki's	95 CommStA 24 1655-1685
On a simultaneous equations	pre-test estimator • Skeels, Christopher L.; Taylor, Larry W.	95 JEconmtx 68 269-286
Dielman, Terry E.; Rose, Elizabeth L.	Pre-test estimators in least absolute value regression with autocorrelated errors •	94 ASAProBuEc 95-100
The optimal critical value of a	pre-test for an inequality restriction in a mis-specified regression model • Wan, Alan T. K.	95 AstrlJSt 37 73-81
location parameter following rejection of a	pre-test • Chiou, Paul; Han, Chien-Pai Conditional interval estimation of the exponential	95 CommStA 24 1481-1492
the Dickey-Fuller test: Size-robustness in the face of	pre-testing • Dods, Johannah L.; Giles, David E. A. Alternative strategies for 'augmenting'	95 JStCmpSm 53 243-258
Correction to "Tests on early failures - The	precedence life test" (V25 p140-143) • Nelson, L. S.	95 JQualTek 27 402-402
application of cluster analysis to growing season	precipitation data in North America east of the Rockies • Gong, X.; Richman, M. B. On the	95 JClimate 8 897-931

Sensitivity analysis of extreme	precipitation events • Katz, Richard W.; Acero, Juan Garrido	94 JClimlgy 14 985-999
chain-dependent processes: Application to hourly	precipitation • Katz, Richard W.; Parlange, Marc B. Generalizations of	95 WaterReR 31 1331-1341
A Bayesian approach to microwave	precipitation profile retrieval • Evans, K. F.; Turk, J.; Wong, T.; Stephens, G. L.	95 JApMetrl 34 260-279
Saloff-Coste, L.	Precise estimates on the rate at which certain diffusions tend to equilibrium •	94 MaZeit 217 641-677
Papamichael et al. Design of experiments for a	precise estimation of the calibration curve of periodate in its determination by ... •	95 Chemolab 30 227-237
Estimation of structures reliability for non-	precise limit state models and vague data • Savchuk, V. P.	95 RelEngSS 47 47-58
The	precision of approximations to moments of sample moments • Anderson, Oliver D.	95 JStCmpSm 52 163-165
	Precision calculation of distributions for trimmed sums • Csörgő, Sándor; Simons, Gordon	95 AnlsApPr 5 854-873
Sampson; Batson, Robert G. Case study:	Precision improvement of a digital ball comparator via parameter design • Gholston,	94 ASAProPEng 302-307
	Precision mapping of quantitative trait loci • Zeng, Zhao-Bang	94 Genetcs 136 1457-1468
minimax estimators of normal covariance and	precision matrices • Gupta, A. K.; Ofori-Nyarko, Samuel Improved	95 MaOpfStS 26 19-25
chaudhuri, Aparna; Basu, Ayanendranath Improving	precision through modelling: An illustration with hierarchical kappa • Basu, Srabashi; Ray-	95 CommStB 24 399-408
Andrich, David Models for measurement,	precision, and the nondichotomization of graded responses (Disc: p27-46) •	95 Psymtrka 60 7-26
Effect on Bayesian inference of the degree of	precision of recorded data • Tricker, A. R.	95 JAppStat 22 235-244
tional hazards regression analysis II. Accuracy and	precision of regression estimates • Peduzzi et al. ... per independent variable in propor-	95 JClEpid 48 1503-1510
Estimators with prescribed	precision in stochastic regression models • Konev, Victor; Lai, Tze Leung	95 SqtlAnly 14 179-192
A note on admissibility when	precision is unbounded • Anderson, Charles; Pal, Nabendu	95 AnlsStat 23 593-597
Improved estimation of generalized variance and	precision • Gupta, A. K.; Ofori-Nyarko, S.	95 MaOpfStS 26 99-109
efficient method for estimating the magnitude and	precision of welfare changes • Breslaw, Jon A.; Smith, J. Barry A simple and	95 JApEcmtx 10 313-327
Sheppard, L.; Prentice, R. L. On the reliability and	precision of within- and between-population estimates of relative rate parameters •	95 Biomtrcs 51 853-863
chain model of both entry to and exit from the	preclinical detectable phase • Duffy et al. ... time in breast cancer screening using a Markov	95 StatMed 14 1531-1543
The performance of	precursor-based estimators for rare event frequencies • Bier, V. M.; Yi, W.	95 RelEngSS 50 241-251
A Bayesian method for analyzing dependencies in	precursor data • Bier, Vicki M.; Yi, Woojune	95 IntJFor 11 25-41
butions of some Lotka-Volterra types of stochastic	predation systems • Yeung, David W. K.; Stewart, Sally E. A. Stationary probability distri-	95 StocAnAp 13 503-516

Predator-prey model

Stow, C. A.; Carpenter, S. R.; Cottingham, K. L.	Resource vs. ratio-dependent consumer-resource models: A Bayesian perspective •	95 Ecology 76 1986-1990
A linear birth-and-death	predator-prey process • Coffey, John	95 JAppProb 32 274-277
A. Detecting small-scale spatial patterns in lotic	predator-prey relationships: Statistical methods and a case study • Muotka, T.; Penttinen,	94 JFishRBC 51 2210-2218
partnership dissolution and relationship quality as	predicators of remission of problem drinking in women • Harris et al. ... interaction of	94 ASAProSoSt 232-234
Using survey data to	predict adoption and switching for services • Weerahandi, Samaradasa; Moitra, Soumyo	95 JMarRsr 32 85-96
CHODR – Using statistics to	predict college hockey • Danehy, Timothy J.; Lock, Robin H.	95 Stats /13 10-14
Till, S.; Williams, C. Using a mixture model to	predict the occurrence of diabetic retinopathy • Young, P.; Morgan, B.; Sonksen, P.;	95 StatMed 14 2599-2608
Trevisan, Anna Statistical properties of	predictability from atmospheric analogs and the existence of multiple flow regimes •	95 JAtmsphr 52 3577-3592
option pricing models when assest returns are	predictable • Lo, Andrew W.; Wang, Jiang Implementing	95 JFinance 50 87-129
error: General discussion and an example	predicting defoliation • Buonaccorsi, John P. Prediction in the presence of measurement	95 Biomtrcs 51 1562-1569
	Predicting dew formation • Boulaevskaia, Vera	95 Stats /12 21-22
Alternative strategies for imputing premiums and	predicting expenditures under health care reform • Doyle, Pat; Farley, Dean	94 ASASrvyRsr 517-522
A practical approach for	predicting fatigue reliability under random cyclic loading • Tang, Jun; Zhao, Jie	95 RelEngSS 50 7-15
• Wilson, Rick L.; Hardgrave, Bill C.	Predicting graduate student success in an MBA program: Regression versus classification	95 EdPsMeas 55 186-195
	Predicting integrals of stochastic processes • Stein, Michael L.	95 AnlsApPr 5 158-170
Genetic models for	predicting maize single-cross performance in unbalanced yield trial data • Bernardo, Rex	95 CropSci 35 141-147
inbred progenies • Naciri, Y.; Gallais, A.	Predicting mean and variance of all possible lines and hybrids from designs with partially	94 ThApGene 89 693-697
(Disc: p360-361) • Tiller, Richard; Welch, Michael	Predicting the national unemployment rate that the "old" CPS would have produced	94 ASASrvyRsr 354-359
	Predicting nominal variable relationships with multiple response • Umesh, U. N.	95 JForcast 14 585-596
	Predicting ozone levels and trends • Gao, Feng; Sacks, Jerome; Welch, William J.	94 ASAProEnvr 77-81
Multiplicative risk model for	predicting rates (Disc: p125-126) • Davis, Alan B.; Larimore, Walter L.	94 ASAProEpid 123-124
	Predicting the response of populations to environmental change • Ives, Anthony R.	95 Ecology 76 926-941
LaBree, Laurie Use of neural networks in	predicting the risk of coronary artery disease • Lapuerta, Pablo; Azen, Stanley P.;	95 CmpBimdR 28 38-52
Balgobin A Gibbs sampling approach for	predicting shelf-life of a military ration with sensory data • Chen, Ming-Hui; Nandram,	94 ASAProBays 84-89
Lennon, Jack J.; Turner, John R. G.	Predicting the spatial distribution of climate: Temperature in Great Britain •	95 JAnmlEco 64 370-392
et al. Analysts' use of earnings forecasts in	predicting stock returns: Forecast horizon effects • Bandyopadhyay, S. P.; Brown, L. D.	95 IntJFor 11 429-445
to the hamster ova penetration test • Abt, M. et al.	Predicting the success of in vitro fertilization: Conventional semen analysis compared	95 StatMed 14 1807-1818
and satellite data. Part I: Development of models	predicting surface shelter temperatures • Epperson et al. ... temperatures using upper-air	95 JApMetrl 34 340-357
I.; González-Manteiga, W.; Prada-Sánchez, J. M. et al.	Predicting using Box-Jenkins, nonparametric, and bootstrap techniques • García-Jurado,	95 Technmcs 37 303-310
secret behind variation transmission analysis: Why	predicting variation may be better than observing it? • Taylor, Wayne A. The	94 ASAProQlPr 49-54
Souchon, Yves; Herouin, Eric	Predicting velocity frequency distributions in stream reaches • Lamouroux, Nicolas;	95 WaterReR 31 2367-2375

Prediction

endosperms • Zhu, J.; Weir, B. S.	Analysis of cytoplasmic and maternal effects. II. Genetic models for triploid	94 ThApGene 89 160-166
seeds and animals • Zhu, J.; Wier, B. S.	Analysis of cytoplasmic and maternal effects. 1. A genetic model for diploid plant	94 ThApGene 89 153-159
	Assessment and propagation of model uncertainty (Disc: p71-97) • Draper, David	95 JRSS-B 57 45-70
Cheng; Mountain, Dean C.; Illman, Kathleen Ho	A Bayesian integration of end-use metering and conditional-demand analysis • Hsiao,	95 JBES 13 315-326
	Better subset regression using the nonnegative garrote • Breiman, Leo	95 Technmcs 37 373-384
Willett, John B.; Singer, Judith D.	It's déjà vu all over again: Using multiple-spell discrete-time survival analysis •	95 JEdStat 20 41-67
Sullivan, Paul J.	The dilution of a contaminant substance released in an environmental flow •	95 SankhyaB 57 277-284
basis function network • Katayama, R. et al.	Embedding dimension estimation of chaotic time series using self-generating radial	95 FuzzySet 71 311-327
distances • Bartkowiak, Anna; Liebhart, Jerzy	Estimation of the spirometric residual volume (RV) by a regression built from Gower	95 BiomtrcJ 37 131-149
Elliott, Graham; Stock, James H.	Inference in models with nearly integrated regressors • Cavanagh, Christopher L.;	95 EconTher 11 1131-1147
from bounded-error data • Keesman, K. J.	Model parameter and modelling error identification in non-linear conceptual models	95 EnvrMtrc 6 445-449
	Modelling and simulation of barley kernel growth • Koesmarno, H. K.	95 EnvrMtrc 6 511-516
Öhrn, Kenth; Ahlén, Anders; Sternad, Mikael	A probabilistic approach to multivariable robust filtering and open-loop control •	95 IEEEAuCn 40 405-418
D. A.; Pfeiffer, T. W.; Cornelius, P. L.	Selection index based on genetic correlations among environments • Van Sanford,	93 CropSci 33 1244-1248
pollutants (Disc: p145-167) • Gibbons, Robert D.	Some statistical and conceptual issues in the detection of low-level environmental	95 EnvEcoSt 2 125-145
albert) • Albert, Jim	Teaching inference about proportions using Bayes and discrete models (jse/v3n3/	95 JStEd 3/3 58kb
McKenzie, C. R.	Testing for multivariate normality in simultaneous equations models • Jarque, C. M.;	95 MaCmpSim 39 323-328
McAleer, Michael	Prediction and accommodation in econometric modelling • Dharmapala, Dhammika;	95 EnvrMtrc 6 551-556
An averaging approach to	prediction • Wong, A.	95 StHefte 36 253-264
Real estate price	prediction under asymmetric loss • Cain, Michael; Janssen, Christian	95 AnInStMa 47 401-414
Pallas, Michelle	Prediction of automotive emissions from gasoline composition • Rocke, David M.;	94 ASAProPEng 269-273
Altman, Naomi; Léger, Christian On the optimality of	prediction based selection criteria and the convergence rates of estimators •	94 ASAProStCp 112-117
Minimax vs. Bayes	prediction • Blackwell, D.	95 PrEngInf 9 53-58
Two-level fractional factorials and Bayesian	prediction • Mitchell, Toby J.; Morris, Max D.; Ylvisaker, Donald	95 StSinica 5 559-573
tiplicative interaction ... • Piepho Best linear unbiased	prediction (BLUP) for regional yield trials: A comparison to additive main effects and mul-	94 ThApGene 89 647-654
Jaheen, Zeinhum F. Bayesian	prediction bounds for the Burr type XII failure model • Al-Hussaini, Essam K.;	95 CommStA 24 1829-1842
Use of multiple genetic markers in	prediction of breeding values • van Arendonk, Johan A. M.; Tier, Bruce; Kinghorn, Brian P.	94 Genetcs 137 319-329

Review of	"Predictive inference: An introduction" • Geisser, Seymour (Auth); Lawera, Martin (Rev)	95 Technmcs 37 121-121
hamane (Rev) Review of	"Predictive inference: An introduction" • Geisser, Seymour (Auth); Mouhab, Abderra-	95 JASA 90 1489-1490
Bayesian	predictive inference for multivariate sample surveys • Nandram, Balgobin	94 JOffStat 10 167-179
Measurement invariance,	predictive invariance, and the duality paradox • Millsap, Roger E.	95 MultBehR 30 577-605
	Predictive model selection • Laud, Purushottam W.; Ibrahim, Joseph G.	95 JRSS-B 57 247-262
A. E. Eliciting prior information to enhance the	predictive performance of Bayesian graphical models • Madigan, D.; Gavrin, J.; Raftery,	95 CommStA 24 2271-2292
A note on the efficiencies of two almost unbiased	predictive ratio estimators • Sahoo, J.; Sahoo, L. N.	95 StHefte 36 273-278
	Predictive stop-loss premiums and Student's *t*-distribution • Hürlimann, Werner	95 InsrneME 16 151-159
Carlo study • Hsu Sampling behaviour in estimating	predictive validity in the context of selection and latent variable modelling: A Monte	95 BJMaStPs 48 75-97
of direct and indirect components of small domain	predictors • Chaudhuri et al. ... design-cum-model approach for convex weighting	95 CommStA 24 2803-2813
Colliver, James D.; Gfroerer, Joseph C.	Predictors of drug abuse treatment utilization among frequent cocaine users •	94 ASASrvyRsr 553-558
On some	predictors of future order statistics • Raqab, Mohammad Z.; Nagaraja, Haikady N.	95 Metron 53 185-204
Kupper, Lawrence L. On the effects of	predictor misclassification in multiple linear regression analysis • Christopher, Susan R.;	95 CommStA 24 13-37
classification model with a smooth link function and	predictors obtained from quantile spline fits to high- ... • O'Connell et al. ... linear	95 JStPlInf 47 153-164
et al. Comparing the contributions of groups of	predictors: Which outcomes vary with hospital rather than patient characteristics? • Silber	95 JASA 90 7-18
Graphics for studying net effects of regression	predictors • Cook, R. Dennis	95 StSinica 5 689-708
Improved	predictors in survey sampling under a super population model • Srivastava, A. K.	95 StPrLet 25 249-252
analysis of interaction effects between continuous	predictors using multiple regression: Multiple indicator and ... • Jaccard et al. ... in the	95 PsychBu 117 348-357
Incorporating auxiliary	predictor variation in principal component regression models • Thomas, Edward V.	95 JChemtrc 9 471-481
On almost optimal priority rules for	preemptive scheduling of stochastic jobs on parallel machines • Weiss, Gideon	95 AdvAppPr 27 821-839
D. J.; Croner, C. M.; Wilson, B. F. Map design	preferences associated with professional discipline • White, A. A.; Pickle, L. W.; Herrmann,	94 ASAProStGr 54-59
variate technique for the analysis of proximity and	preference data" (Spanish) • Arce, C. et al. Review of "Multidimensional scaling. A multi-	95 JClass 12 137-139
application • Schuster, Ernst	Preference of equivalence tests with standardized mean difference demonstrated by an	95 CmpStDA 19 353-356
Measuring selectivity from multiple choice feeding-	preference experiments • Manly, Bryan F. J.	95 Biomtrcs 51 709-715
equation of a vector ... • Salemi Revealed	preference of the Federal Reserve: Using inverse-control theory to interpret the policy	95 JBES 13 419-433
On the use of circular tetrads in	preference polygon (Japanese) • Makino, Toji; Konno, Toshiharu	94 JpJBeh 21/2 1-9
from, D. J.; Holmes, T. M. Evaluating change and	preference questions using paired Likert scale data • Roberson, P. K.; Shema, S. J.; Mund-	94 ASAProSoSt 189-191
C.; Hary, I.; Pielert, V. Factors affecting spatial	preferences in settlement site selection in migratory pastoralism • Schwartz, H. J.; Mosler,	95 EnvrMtrc 6 485-490
a formal derivation of the expected behavior of	prefix *b*-trees • Mahmoud, Hosam M.; Orlandić, Ratko Toward	95 PrEngInf 9 183-192
Ryaboy, Lana Sample design of the National	Pregnancy and Health Survey • DiGaetano, Ralph; Ward, John; Cadell, Diane;	94 ASASrvyRsr 1100-1104
placental lactogen in maternal serum during late	pregnancy • Lundbye-Christensen, S.; Christensen, M. Modelling the growth of human	95 RBrzlPrS 9 1-24
Prediction of final data with use of	preliminary and/or revised data • Mariano, Roberto S.; Tanizaki, Hisashi	95 JForcast 14 351-380
	Preliminary test estimator	
Tomkins, R. J.	Estimating lognormal means using uncertain prior information • Ahmad, S. E.;	95 PakJStA 11 67-92
	Improved shrinkage estimation of relative potency • Ahmed, S. E.	95 BiomtrcJ 37 627-638
	A pooling methodology for coefficient of variation • Ahmed, S. E.	95 SankhyaB 57 57-75
	A pooling procedure for the exponential distribution • Chiou, Paul	95 CommStA 24 445-458
Bez, K.	Preliminary test estimators in double sampling with two auxiliary variables • Das, G.;	95 CommStA 24 1211-1226
Saxena, K. P. On bias and efficiency of a	preliminary test estimator of variance in a normal distribution • Tiwari, Gayatri;	95 Metron 53 163-169
and video computer assisted self-interviewing:	Preliminary tests of new technologies for data collection • O'Reilly, J. M. et al. Audio	94 JOffStat 10 197-214
Interval estimation of error variance following a	preliminary test in one-way random model • Chiou, Paul; Han, Chien-Pai	95 CommStB 24 817-824
exponential scale parameter following rejection of a	preliminary test • Chiou, Paul; Han, Chien-Pai Conditional interval estimation of the	94 JStatRsr 28 65-72
survival meta-analysis of adjuvant chemotherapy for	premenopausal breast cancer • Cole, B. F.; Gelber, R. D.; Goldhirsch, A. A quality-adjusted	95 StatMed 14 1771-1784
	Premium adjustment by generalized adaptive exponential smoothing • Herkenrath, U.	94 InsrneME 15 203-217
trage pricing theory and non-stationarity of the risk	premium: An error components non-linear approach • Chen, Y. H. et al. Testing the arbi-	94 ASAProBuEc 178-183
Farley, Dean Alternative strategies for imputing	premiums and predicting expenditures under health care reform • Doyle, Pat;	94 ASASrvyRsr 517-522
	Premium principle	
transforms • Wang, Shuan	Insurance pricing and increased limits ratemaking by proportional hazards	95 InsrneME 17 43-54
Predictive stop-loss	premiums and Student's *t*-distribution • Hürlimann, Werner	95 InsrneME 16 151-159
Stochastic compartmental models with	Prendville growth mechanisms • Parthasarathy, P. R.; Kumar, B. Krishna	95 MaBisci 125 51-60
response rates for census ... • Dillman et al. How	prenotice letters, stamped return envelopes and reminder postcards affect mailback	95 SurvyM 21 159-165
confidence cones for spherical data based on	prepivoting • Shin, Yang-Kyu Double bootstrap	95 JKoreaSt 24 183-195
of the multiplicative scatter correction method for	preprocessing spectroscopic data • Helland, I. S.; Næs, T.; Isaksson, T. Related versions	95 Chemolab 29 233-241
Representation, synthesis, variability and data	preprocessing of a three-way data set • Rizzi, Alfredo; Vichi, Maurizio	95 CmpStDA 19 203-222
elderly • Moeller, John F.	Prescription drug coverage and the demand for outpatient prescription drugs by the	94 ASAProSoSt 147-152
dia L.; Berk, Marc L.; Mueller, Curt D. Developing a	prescription drug expenditure model: The effect of alternative assumptions • Schur, Clau-	94 ASAProSoSt 245-250
The software taxi: An easy-to-use tool for statistical	presentations • Nash, John C.	94 ASAProStCp 269-271
contextual effects on voter participation in the 1992	presidential election • Schlichting, Kurt C.; Tuckel, Peter S. A spatial analysis of	94 ASASrvyRsr 1339-1344
fice in July 1989 • Wright, J. B. Presenting statistics to	press and public: The British experience since reorganization of the Central Statistical Of-	91 StatJUN 8 313-320
	PRESS statistic	
efficient spatial sampling algorithm ... • Lesch et al.	Spatial prediction of soil salinity using electromagnetic induction techniques. 2. An	95 WaterReR 31 387-398
event histories, and incentives in the SNFG cycle 5	pretest • Mosher, William D.; Pratt, William F. CAPI,	94 ASASrvyRsr 59-63
payments on response rates and field costs in a	pretest of a national CAPI survey • Duffer, A.; Lessler, J.; Weeks, M. et al. Effects of incentive	94 ASASrvyRsr 1386-1392
from the National Survey of Family Growth cycle V	pretest • Lessler, Judith T.; Weeks, Michael F.; O'Reilly, James M. Results	94 ASASrvyRsr 64-70
Laboratory • Bergman, Lars R.	Pretesting procedures at Statistics Sweden's Measurement, Evaluation and Development	95 JOffStat 11 309-323
	Prevalence	
	Using a diagnostic test in a finite population • Englund, Jan-Eric	94 MaScitst 19 108-118
Alternative approaches for estimating	prevalence in epidemiologic surveys with two waves of respondents • Brenner, Hermann	95 AJEpidm 142 1236-1245
Joseph, L. et al. Bayesian estimation of disease	prevalence and the parameters of diagnostic tests in the absence of a gold standard •	95 AJEpidm 141 263-272
relative risk • Mittleman Estimation of exposure	prevalence in a population at risk using data from cases and an external estimate of the	95 Epidmlgy 6 551-553
ness and accuracy of pooled testing in estimating	prevalence of a rare disease: Application to HIV screening • Tu et al. On the informative-	95 Biomtrka 82 287-297
Confidence limits for the population	prevalence rate based on the negative binomial distribution • Lui, Kung-Jong	95 StatMed 14 1471-1477
Gastwirth, Joseph L. On designs for measuring	prevalence rates and error rates from diagnostic testing procedures • Sinclair, Michael D.;	94 ASAProBiop 148-153
test for length-biased data with discussion on	prevalence, sensitivity and specificity of the length bias • Shanmugam, R. Goodness-of-fit	95 JStPlInf 48 277-290
Dennis, Michael L.; Bray, Robert M. Estimating the	prevalence of substance abuse among people who are homeless • Iachan, Ronaldo;	94 ASASrvyRsr 1094-1099
Estimate of disease	prevalence using drug consumption data • Sartor, Francis; Walckiers, Denise	95 AJEpidm 141 782-787
with survival for data in which censoring is	prevalent • Fraser et al. ... decision procedure for selecting prognostic variables associated	95 BiomtrcJ 37 463-479
Comment on "Statistical models for	prevalent cohort data" (V49 p1-11) • Ripley, B. D.; Solomon, P. J.	95 Biomtrcs 51 373-374
Reply to comment on "Statistical models for	prevalent cohort data" • Wang, M.-C.; Brookmeyer, R.; Jewell, N. P.	95 Biomtrcs 51 374-375
Laurent, K. et al. An analysis of risk factors for	prevalent coronary heart disease by using the proportional odds model • Woodward, M.;	95 Ststcian 44 69-80
Estimating the occurrence rate for	prevalent survival data in competing risks models • Huang, Ying; Wang, Mei-Cheng	95 JASA 90 1406-1415
intervals, hypothesis tests, and sample sizes for the	prevented fraction in cross-sectional studies • Gargiullo, P. M. et al. Confidence	95 StatMed 14 51-72
Sugimoto, Tamio Estimation of the pest	prevention ability of the import plant quarantine in Japan • Yamamura, Kohji;	95 Biomtrcs 51 482-490
Biomarkers, etiologic research, and primary disease	prevention • Weissfeld, Joel L.	94 ASAProEpid 14-23

for cluster randomization: A review of primary	prevention trials, 1990 through 1993 • Simpson, J. M.; Klar, N.; Donner, A. Accounting	95 AJPubHlh 85 1378-1383
On the role, design, and analysis of disease	prevention trials • Prentice, Ross L.	95 ConClnTr 16 249-258
pling theory, estimation, and significance testing for	Prevosti's estimate of genetic distance • Sanchez, Alex; Ocaña, Jordi; Utzet, Frederic Sam-	95 Biomtrcs 51 1216-1235
model for forecasting electricity consumption and	prices • Joutz, F. L. et al. An integrated Bayesian vector autoregression and error correction	95 JForcast 14 287-310
The measurement of	price elasticities – The BT experience • Cracknell, David; Knott, Michael	95 IntJFor 11 321-329
Poterba, James M. Money, output, and	prices: Evidence from a new monetary aggregate • Rotemberg, Julio J.; Driscoll, John C.;	95 JBES 13 67-83
Richard Comment on "Testing the rationality of	price forecasts: New evidence from panel data" (V80 p714-735) • Bonham, Carl; Cohen,	95 AmEcRv 85 284-289
Reply to comment on "Testing the rationality of	price forecasts: New evidence from panel data" • Keane, Michael P.; Runkle, David E.	95 AmEcRv 85 290-290
istrative data in the development of a longitudinal	price index for inbound ocean liner service (Disc: p212-214) • Himelein The use of admin-	94 ASAProGovt 195-200
Rappaport, Neal J. Econometric estimates of	price indexes for personal computers in the 1990's • Berndt, Ernst R.; Griliches, Zvi;	95 JEconmtx 68 243-268
Axiomatic	price index theory: A survey • Balk, Bert M.	95 IntStRvw 63 69-93
a nonlinear rational expectations commodity	price model with unobservable state variables • Deaton, Angus; Laroque, Guy Estimating	95 JApEcmtx 10 9-40S
relationship between spot index and futures	price of the Nikkei Stock Average • Tse, Y. K. Lead-lag	95 JForcast 14 553-563
A multiple linear regression model for the selling	price of oceanfront lots • Judge, John J.	94 ASAProBuEc 345-346
Models for option	prices • Rachev, S. T.; Ruschendorf, L.	94 ThProbAp 39 120-152
Real estate	price prediction under asymmetric loss • Cain, Michael; Janssen, Christian	95 AnInStMa 47 401-414
A response model for the international	price program • Ferguson, Gwyn R.; Ciapponi, Tamara M.	94 ASASrvyRsr 1007-1012
Variance estimation for EIA-877 propane	prices • O'Colmain, Benita J.; Saavedra, Pedro J.; Weir, Paula	94 ASASrvyRsr 232-235
drug use among arrestees • Hyatt, R. R., Jr et al. The	price and purity of cocaine: The relationship to emergency room visits and death, and to	95 StatMed 14 655-668
The production-theoretic measurement of input	price and quantity indices • Fisher, Franklin M.	95 JEconmtx 65 155-174
Nonclassical demand – A model-free examination of	price-quantity relations in the Marseille fish market • Härdle, Wolfgang; Kirman, Alan	95 JEconmtx 67 227-257
Rossi, P. E. Modeling the distribution of	price sensitivity and implications for optimal retail pricing • Kim, B.-D.; Blattberg, R. C.;	95 JBES 13 291-303
Lenk, Peter J. Reassessing brand loyalty,	price sensitivity, and merchandising effects on consumer brand choice • Allenby, Greg M.;	95 JBES 13 281-289
On-highway Diesel Fuel	Price Survey • Weir, Paula; Saavedra, Pedro J.	94 ASAProGovt 144-148
Comparison of national and regional gasoline	prices from two surveys • Saavedra, Pedro J.; Hassett, Nancy C.; Weir, Paula	94 ASASrvyRsr 1179-1183
General framework for	pricing derivative securities • Musiela, Marek	95 StocProc 55 227-251
Actuarial models for	pricing disability benefits: Towards a unifying approach • Pitacco, Ermanno	95 InsrneME 16 39-62
Aubin, C.; Fougére, D.; Husson, E. et al. Real-time	pricing of electricity for residential customers: Econometric analysis of an experiment •	95 JApEcmtx 10 171-191S
Option	pricing by Esscher transforms (Disc: p141-191) • Gerber, Hans U.; Shiu, Elias S. W.	95 TrSActur 46 99-140
Wang, Shuan Insurance	pricing and increased limits ratemaking by proportional hazards transforms •	95 InsrneME 17 43-54
Kabanov, Yu. M. et al. Toward the theory of	pricing of options of both European and American types. I. Discrete time. • Shiryaev, A. N.;	94 ThProbAp 39 14-60
A. N.; Kabanov, Yu. M. et al. Toward the theory of	pricing of options of both European and American types. II. Continuous time. • Shiryaev,	94 ThProbAp 39 61-102
changes of probability measure and option	pricing • Geman, Hélyette; Karoui, Nicole El; Rochet, Jean-Charles Changes of numéraire,	95 JAppProb 32 443-458
analysis for Monte Carlo simulation of option	pricing • Fu, Michael C.; Hu, Jian-Qiang Sensitivity	95 PrEngInf 9 417-446
the quantile of a Brownian motion with drift and the	pricing of related path-dependent options • Dassios, Angelos The distribution of	95 AnlsApPr 5 389-398
price sensitivity and implications for optimal retail	pricing • Kim, Byung-Do; Blattberg, Robert C.; Rossi, Peter E. Modeling the distribution of	95 JBES 13 291-303
A new look at	pricing of the "Russian option" • Shepp, L. A.; Shiryaev, A. N.	94 ThProbAp 39 103-119
Kramkov, D. O.; Shiryaev, A. N. On the rational	pricing of the "Russian option" for the symmetrical binomial model of a (B,S)-market •	94 ThProbAp 39 153-162
There is no nontrivial hedging portfolio for option	pricing with transaction costs • Soner, H. M.; Shreve, S. E.; Cvitanić, J.	95 AnlsApPr 5 327-355
	Primary data • Bramwell, Andrew	95 TeachgSt 17/3 88-89
Biomarkers, etiologic research, and	primary disease prevention • Weissfeld, Joel L.	94 ASAProEpid 14-23
Testing the hypothesis that matters for multiple	primary endpoints • Capizzi, Thomas P.; Zhang, Ji	94 ASAProBiop 460-465
Accounting for cluster randomization: A review of	primary prevention trials, 1990 through 1993 • Simpson, Judy M.; Klar, Neil; Donner, Allan	95 AJPubHlh 85 1378-1383
Graphing in the	primary school: Algorithm versus comprehension • Pereira-Mendoza, Lionel	95 TeachgSt 17/1 2-6
Harald Cramér and the distribution of	prime numbers • Granville, Andrew	95 ScanActJ 12-28
orthogonal arrays having factors with a large non-	prime power number of levels • Mandeli, John P. Construction of asymmetrical	95 JStPlInf 47 377-391
matrix for s^n parallel flats fractions when s is	prime • BuHamra, Sana; Anderson, Donald A. The spectrum of the information	95 JStPlInf 47 333-345
	Prime time for Bayes • Kadane, Joseph B.	95 ConClnTr 16 313-318

Principal components

Mason, Robert L.	A bivariate control chart for paired measurements • Tracy, Nola D.; Young, John C.;	95 JQualTek 27 370-376
Dunlop, P. J.; Bignell, C. M.; Jackson, J. F. et al.	Chemometric analysis of gas chromatographic data of oils from Eucalyptus species •	95 Chemolab 30 59-67
a forestry application (French) • Champely, S. et al.	Co-inertia of two random variables in Hilbert space and B-spline approximations with	95 RStApp 43/1 91-107
	A combined theory for PCA and PLS • Höskuldsson, Agnar	95 JChemtrc 9 91-123
Fortiana, J.	A continuous metric scaling solution for a random variable • Cuadras, C. M.;	95 JMultiAn 52 1-14
	Displaying a correlation matrix on the unit sphere in R^3 (French) • Falissard, B.	95 RStApp 43/2 35-48
	An efficient algorithm for weighted PCA • Krijnen, W. P.; Kiers, H. A. L.	95 CmpStQ 10 299-306
considerations • Pedersen, F.; Bengtsson, E. et al.	An extended strategy for exploratory multivariate image analysis including noise	95 JChemtrc 9 389-409
Dickson, R. A.	Familial spinal shape • Mardia, K. V.; Dryden, I. L.; Hurn, M. A.; Li, Q.; Millner, P. A.;	94 JAppStat 21 623-641
curves using PCA • Gunters, Bert	Graphical methods and principles for data analysis II: Graphically comparing many	95 QulPr 28/10 129-137
	Introduction to multivariate methodology, an alternative way? • Christie, Olav H. J.	95 Chemolab 29 177-188
Hernández-Pajares, M.	The Kohonen self-organizing map method: An assessment • Murtagh, F.;	95 JClass 12 165-190
period data recorded on the Catalan ... • Lana et al.	Minimum sample size for synoptic weather type classification. Application to winter	94 JClimlgy 14 1051-1060
Grambsch, P. M.; Randall, B. L.; Bostick, R. M. et al.	Modeling the labeling index distribution: An application of functional data analysis •	95 JASA 90 813-821
of discrimination power of ... • Raich et al.	Multivariate statistical methods for monitoring continuous processes: Assessment	95 Chemolab 30 37-48
recognition applied to RIFS sensor ... • Kraus et al.	Optical reflectometric gas sensing: Classification of hydrocarbon vapours by pattern	95 Chemolab 30 211-221
	Orthogonal canonical variates for discrimination and classification • Krzanowski, W. J.	95 JChemtrc 9 509-520
in coastal plain streams • Sheldon, A. L.; Meffe, G. K.	Path analysis of collective properties and habitat relationships of fish assemblages	95 JFishRBC 52 23-33
	PLS regression and applications (French) • Tenenhaus, M.; Gauchi, J.-P.; Ménardo, C.	95 RStApp 43/1 7-63
Thaddeus; Li, Luning; Flury, Bernard D.	Principal points and self-consistent points of elliptical distributions • Tarpey,	95 AnlsStat 23 103-112
Tarpey, Thaddeus	Principal points and self-consistent points of symmetric multivariate distributions •	95 JMultiAn 53 39-51
Iyengar, R. N.; Basak, P.	Regionalization of Indian monsoon rainfall and long-term variability signals •	94 JClimlgy 14 1095-1114
empirical comparison • Yau, Sui K.	Regression and AMMI analyses of genotype × environment interactions: An	95 AgronmyJ 87 121-126
using Hellinger distance • Rao, C. Radhakrishna	A review of canonical coordinates and an alternative to correspondence analysis	95 Questo 19 23-63
matrix multiplication. Part III. The use of ... • Alsberg	Speed improvement of multivariate algorithms by the method of postponed basis	95 Chemolab 30 223-225
	Two-moment scaling formulae for aggregating examination marks • Daley, D. J.	95 AstrlJSt 37 253-272
dencies and concurvities using smallest additive	principal components • Donnell, D. J.; Buja, A.; Stuetzle, W. Analysis of additive depen-	94 AnlsStat 22 1635-1673

Principal component analysis

duration test for batch approval • Thelin, B. et al.	Classification of Estradurin batches: Correlation between 31P NMR and a biological	95 Chemolab 27 135-145
	A composite representation of pulp quality • Broderick, G.; Paris, J.; Valade, J. L.	95 Chemolab 29 19-28
the presence of Cu(II) • Esteves da Silva, J. C. G. et al.	Evolving factor analysis of synchronous fluorescence spectra of humic substances in	95 Chemolab 27 115-128
	Image of "Made in Switzerland" (French) • Brossard, Hubert	95 Student 1 51-61
structures • Thioulouse, J.; Chessel, D.; Champely, S.	Multivariate analysis of spatial patterns: A unified approach to local and global	95 EnvEcoSt 2 1-14
MacGregor, John F.	Multivariate SPC charts for monitoring batch processes • Nomikos, Paul;	95 Technmcs 37 41-59
data • Cserháti, Tibor; Forgács, Esther	Use of canonical correlation analysis for the evaluation of chromatographic retention	95 Chemolab 28 305-313

Y.-S.; Shim, H.-J. Applications of the supplementary	principal component analysis for the 1982-1992 Korean pro baseball data (Korean) • Choi,	95 KorApS 8/1 51-60
• Mehlman et al. Comment on "Stopping rules in	principal components analysis: A comparison of heuristical and statistical approaches" ...	95 Ecology 76 640-643
Jackson Reply to comment on "Stopping rules in	principal components analysis: A comparison of heuristical and statistical approaches" •	95 Ecology 76 644-645
Selection of components in	principal component analysis: A comparison of methods • Ferré, Louis	95 CmpStDA 19 669-682
choosing the number of important components in	principal component analysis • Krzanowski, Wojtek J.; Kline, Paul Cross-validation for	95 MultBehR 30 149-165
Disturbance detection and isolation by dynamic	principal component analysis • Ku, Wenfu; Storer, Robert H.; Georgakis, Christos	95 Chemolab 30 179-196
Factor analysis and	principal components • Schneeweiss, Hans; Mathes, Harald	95 JMultiAn 55 105-124
Incorporating parametric effects into functional	principal components analysis • Silverman, B. W.	95 JRSS-B 57 673-689
turbulence in the surface layer investigated by	principal component analysis • Weijers, E. P. et al. Characteristics of convective	95 JApMetrl 34 528-541
An application of some patterned matrices to the	principal component analysis of linked variables • Iyengar, R. J.; Yoo, J. S.; Zafra, P. M. R.	95 PakJStA 11 23-33
López, Carlos; González, Elizabeth; Goyret, Jorge	Principal component analysis of rain gauge data: Detecting outliers (Spanish) •	94 Estd 46 25-54
• López, Carlos; González, Juan F.; Curbelo, Rosario	Principal component analysis of rain gauge data: Treating missing observations (Spanish)	94 Estd 46 55-83
approach • Xu, Lei; Yuille, Alan L. Robust	principal component analysis by self-organizing rules based on statistical physics	95 IEEENrlN 6 131-143
Standardization and transformation in	principal component analysis, with applications to archaeometry • Baxter, M. J.	95 ApplStat 44 513-527
Thompson, M. The efficient cross-validation of	principal components applied to principal component regression • Mertens, B.; Fearn, T.;	95 StatComp 5 227-235
Želimir Structure of	principal component based neural network models of dynamic systems • Kurtanjek,	95 JCmplnT 3 1-8
Rotation of	principal components: Choice of normalization constraints • Jolliffe, Ian T.	95 JAppStat 22 29-35
Review of "Video course:	Principal components" • Massart, D. L.; Lewi, P. J. (Auth); Kvalheim, O. M. (Rev)	95 Chemolab 27 147-147
	Principal component decomposition of non-parametric tests • Janssen, Arnold	95 ProbTher 101 193-209
Azimi-Sadjadi, Mahmood R.	Principal component extraction using recursive least squares learning • Bannour, Sami;	95 IEEENrlN 6 457-469
A note on testing for seasonal cointegration using	principal components in the frequency domain • Cubadda, Gianluca	95 JTimSrAn 16 499-508
Review of "A user's guide to	principal components" • Jackson, J. Edward (Auth); Millsap, Roger E. (Rev)	95 JEdStat 20 105-107
Loadings and correlations in the interpretation of	principal components • Cadima, Jorge; Jolliffe, Ian T.	95 JAppStat 22 203-214
A bivariate control chart with meaningful	principal components • Tracy, N. D.; Young, J. C.; Mason, R. L.	94 ASAProBays 157-158
	Principal component regression	
	Diagnostics statistics in QSAR • Mager, Peter P.	95 JChemtrc 9 211-221
Brockway, Charles G.; Bradley, A. Allen	Errors in streamflow drought statistics reconstructed from tree ring data •	95 WaterReR 31 2279-2293
Crosby, D. S.	Statistical applications of a metric on subspaces to satellite meteorology • Crone, L. J.;	95 Technmcs 37 324-328
A correlation	principal component regression analysis of NIR data • Sun, Jianguo	95 JChemtrc 9 21-29
Massart, D. L. Robust	principal components regression as a detection tool for outliers • Walczak, B.;	95 Chemolab 27 41-54
Incorporating auxiliary predictor variation in	principal component regression models • Thomas, Edward V.	95 JChemtrc 9 471-481
Additive nonparametric regression on	principal components • Härdle, W.; Tsybakov, A. B.	95 JNnparSt 5 157-184
Likelihood ratio tests for	principal components • Dümbgen, Lutz	95 JMultiAn 52 245-258
for limited information maximum likelihood using	principal components • Onishi, Haruo A user-knowledge-based variable selection method	95 CmpStDA 19 379-399
Assessing the stability of	principal components using regression • Sinha, Atanu R.; Buchanan, Bruce S.	95 Psymtrka 60 355-369
	Principal coordinates	
Fortiana, J.	A continuous metric scaling solution for a random variable • Cuadras, C. M.;	95 JMultiAn 52 1-14
M.; Paycha, S. Factorisation of semi-martingales on	principal fibre bundles and the Faddeev-Popov procedure in gauge theories • Arnaudon,	95 Stochatx 53 81-107
Ching-Shui; Li, Ker-Chau A study of the method of	principal Hessian direction for analysis of data from designed experiments • Cheng,	95 StSinica 5 617-639
Li, Luning; Flury, Bernard D.	Principal points and self-consistent points of elliptical distributions • Tarpey, Thaddeus;	95 AnlsStat 23 103-112
Tarpey, Thaddeus	Principal points and self-consistent points of symmetric multivariate distributions •	95 JMultiAn 53 39-51
	Principal points of univariate continuous distributions • Zoppé, Alice	95 StatComp 5 127-132
Uniqueness of	principal points for univariate distributions • Li, Luning; Flury, Bernard	95 StPrLet 25 323-327
Statistics as	principled argument • Abelson, Robert P.	95 ErlbaumA:NJ 232p
Change points in the series of T4 counts	prior to AIDS • Kiuchi, A. S.; Hartigan, J. A.; Holford, T. R.; Rubinstein, P.; Stevens, C. E.	95 Biomtrcs 51 236-248
of linear Bayes rules for a normal mean: Skewed	priors class • Mukhopadhyay, Saurabh; Vidakovic, Brani Efficiency	95 Ststcian 44 389-397
	Prior distribution	
	On an approximate likelihood for quantiles • Lavine, M.	95 Biomtrka 82 220-222
	On a balanced quadratic classification rule • Kim, Hea-Jung	95 CommStA 24 607-623
Upadhyay, S. K.; Singh, Umesh	Bayes estimators for two parameter exponential distribution • Singh, Rajesh;	95 CommStA 24 227-240
parameters • Ghosh, Jayanta K.; Mukerjee, Rahul	Frequentist validity of highest posterior density regions in the presence of nuisance	95 StatDecn 13 131-139
linear regression model with and Edgeworth series	prior distribution • Chaturvedi, A.; Hasegawa, H. et al. Bayesian predictive analysis of the	95 CommStA 24 2469-2484
Girón, Javier	Prior elicitation and Bayesian robustness in a noisy Poisson process • Moreno, Elías;	94 ASAProBays 217-220
for a normal mean with double-exponential	prior" (V56 p605-610) • Mitchell, Ann F. S. Correction to "A note on posterior moments	95 JRSS-B 57 471-471
Image-modeling Gibbs	priors • Levitan, Emanuel; Chan, Michael; Herman, Gabor T.	95 CVGIPGr 57 117-130
estimates of disease maps: How important are	priors? • Bernardinelli, L.; Clayton, D.; Montomoli, C. Bayesian	95 StatMed 14 2411-2431
Fernandez, Carmen Sensitivity to	prior independence via Farlie-Gumbel-Morgenstern model • De la Horra, Julian;	95 CommStA 24 987-996
Larry Density based classes of	priors: Infinitesimal properties and approximations • Ruggeri, Fabrizio; Wasserman,	95 JStPlInf 46 311-324
Behaviour of the posterior error rate with partial	prior information in auditing • Vázquez-Polo, F. J.; Hernández Bastida, A.	95 JAppStat 22 469-476
Madigan, D.; Gavrin, J.; Raftery, A. E. Eliciting	prior information to enhance the predictive performance of Bayesian graphical models •	95 CommStA 24 2271-2292
Unbiased ridge estimation with	prior information and ridge trace • Crouse, Robert H.; Jin, Chun; Hanumara, R. C.	95 CommStA 24 2341-2354
Estimating lognormal means using uncertain	prior information • Ahmad, S. E.; Tomkins, R. J.	95 PakJStA 11 67-92
James-Stein estimators of the mean based on	prior knowledge • Chang, Yuan-Tsung Two-stage	95 CommStA 24 2211-2227
genetic markers and quantitative trait loci. I.	Prior knowledge • Hoeschele, I.; VanRaden, P. M. Bayesian analysis of linkage between	93 ThApGene 85 953-960
Ginter, James L. Incorporating	prior knowledge into the analysis of conjoint studies • Allenby, Greg M.; Arora, Neeraj;	95 JMarRsr 32 152-162
markers and quantitative trait loci. II. Combining	prior knowledge with experimental evidence • Hoeschele et al. ... linkage between genetic	93 ThApGene 85 946-952
A diffuse	prior limit in semiparametric binary regression • Newton, Michael A.	94 ASAProBays 181-186
Probability matching	priors for linear calibration • Ghosh, M.; Carlin, B. P.; Srivastava, M. S.	95 Test 4 333-357
Noninformative	priors for maximal invariant parameter in group models • Datta, G. S.; Ghosh, J. K.	95 Test 4 95-114
Some remarks on noninformative	priors • Datta, Gauri Sankar; Ghosh, Malay	95 JASA 90 1357-1363
coefficient regression models using noninformative	priors • Yang, Ruoyong; Chen, Ming-Hui Bayesian analysis for random	95 JMultiAn 55 283-311
Ghosh, Jayanta Kumar On	priors providing frequentist validity for Bayesian inference • Datta, Gauri Sankar;	95 Biomtrka 82 37-45
Cano, Juan A. Classes of bidimensional	priors specified on a collection of sets: Bayesian robustness • Moreno, Elías;	95 JStPlInf 46 325-334
Using features to model	prior structural information • Wilson, Alyson G.; Johnson, Valen E.	94 ASAProPEng 255-264
On the uniform approximation of Laplace's	prior by t-priors in location problems • Mukhopadhyay, Saurabh; Ghosh, Malay	95 JMultiAn 54 284-294
Single observation unbiased	priors • Meng, Xiao-Li; Zaslavsky, Alan M.	94 ASAProBays 46-51
Éric; Angers, Jean-François Choice of noninformative	priors for the variance components of an unbalanced one-way random model • Belzile,	95 CommStA 24 1325-1341
information, Bayesian kriging and updating of	prior variogram parameters • Cui, Haiyan; Stein, Alfred; Myers, Don E. Extension of spatial	95 EnvrMtrc 6 373-384
Bayesian multivariate normal analysis with a Wishart	prior • Bekker, A.; Roux, J. J. J.	95 CommStA 24 2485-2497
A practical method to	prioritize problem investigations • Lu, Ming-Wei; Rudy, Richard J.	95 ASQCTekCon 57-65
Carolyn The impact of using	priority mail in a mixed mode survey • Mitchell, Susan; Mooney, Geraldine; Shettle,	94 ASAProGovt 88-92
An N-class structured	priority queue with vacations • Katsaros, A.; Langaris, C.	95 CommStSM 11 235-248
Weiss, Gideon On almost optimal	priority rules for preemptive scheduling of stochastic jobs on parallel machines •	95 AdvAppPr 27 821-839
Reciprocal cooperation in the	prisoner's dilemma repeated with random horizon • Manuel, C. M.; Tejada, J.	95 Top 3 97-116

Reinventing the population census: Balancing	privacy and efficiency. In memory of Roger A. Herriot • Alexander, Lois	94 ASAProGovt 1-2
	Privacy issues and the National Voter Registration Act of 1993 • Montjoy, Robert S.	94 ASAProSoSt 170-173
	Privacy issues and the use of voter contact lists • Traugott, Michael W.	94 ASAProSoSt 164-169
Controlling invasion of	privacy in surveys of change over time - A non-technical review • Dalenius, Tore	95 JOffStat 11 107-113
E.; Schwanz, Dennis J. Improving the coverage of	private elementary-secondary schools • Jackson, Betty J.; Frazier, Richard L.; King, Karen	94 ASASrvyRsr 833-838
Teacher quality in public and	private schools (Disc: p213-215) • Ballou, Dale; Podgursky, Michael	94 ASAProSoSt 196-200
	Probabilistic techniques in analysis • Bass, R. F.	95 Springer:Brln:NY 392p
Review of	"Probabilistic techniques in analysis" • Bass, R. F. (Auth); Embrechts, P. A. L. (Rev)	95 ShrtBkRv 15/3 43-43
	Probability	
Iyengar, S. Sitharama; Deng, Weian	An efficient edge detection algorithm using relaxation labeling technique •	95 PattrnRc 28 519-536
	A note on bivariate Dawson-Sankoff-type bounds • Chen, Tuhao; Seneta, E.	95 StPrLet 24 99-104
	The problem of points • Takács, Lajos	94 MaScitst 19 119-139
	• see also conditional -, geometric -, marginal -, tail -, transition -	
	Probability (Second edition) • Shiryaev, A. N.	95 Springer:Brln:NY 621p
between functional analysis, harmonic analysis, and	probability • Kalton, Nigel; Saab, Elias; Montgomery-Smith, Stephen (Ed) Interaction	95 Dekker:NY 494p
Review of "Theoretical	probability for applications" • Port, Sidney C. (Auth); Goldie, Charles M. (Rev)	95 JASA 90 806-808
Stochastic ordering and dependence in applied	probability • Szekli, R.	95 Springer:Brln:NY 194p
of "Stochastic ordering and dependence in applied	probability" • Szekli, R. (Auth); Teugels, J. L. (Rev) Review	95 ShrtBkRv 15/3 48-48
Klass, Michael J. Uniform local	probability approximations: Improvements on Berry-Esseen • Hahn, Marjorie G. W.;	95 AnlsProb 23 446-463
Sensitivity analysis for	probability assessments in Bayesian networks • Laskey, Kathryn Blackmond	95 IEEESyst 25 901-909
Central limit theorem and infinitely divisible	probabilities associated with partial differential operators • Sifi, Mohamed	95 JTherPrb 8 475-499
Approximations to multivariate normal rectangle	probabilities based on conditional expectations • Joe, Harry	95 JASA 90 957-964
on "Algorithm AS 226: Computing non central beta	probabilities" (V36 p241-244) • Lam, Ming Long Comment	95 ApplStat 44 551-552
two-stage estimation of conditional choice	probabilities in a binary choice model under uncertainty • Ahn, Hyungtaik Nonparametric	95 JEconmtx 67 337-378
Jiang, Xulun	Probability bounds on the finite sum of the binary sequence of order k • Dhar, Sunil K.;	95 JAppProb 32 1014-1027
Green, brown, and	probability • Chung, Kai Lai	95 WrldSciPub:Sing 120p
Algorithmic	probability. A collection of problems • Neuts, Marcel F.	95 ChapmnHall:Lnd 448p
Review of "Algorithmic	probability. A collection of problems" • Neuts, Marcel F. (Auth); Coleman, R. (Rev)	96 ShrtBkRv 16/1 4-4
Secondary students' concepts of	probability • Madsen, Richard W.	95 TeachgSt 17/3 90-93
Direct inference,	probability, and a conceptual gulf in risk communication • Walker, Vern R.	95 RiskAnal 15 603-609
Topics in contemporary	probability • Snell, J. Laurie (Ed)	95 CRCPress:FL 400p
Reliability always depends on	probability of course • Good, I. J.	95 JStCmpSm 52 192-193
An intermediate course in	probability • Gut, A.	95 Springer:Brln:NY 278p
Review of "An intermediate course in	probability" • Gut, A. (Auth); Embrechts, P. A. L. (Rev)	96 ShrtBkRv 16/1 4-4
kappa statistic based on exact power and coverage	probability • Basu, S.; Basu, A. Comparison of several goodness-of-fit tests for the	95 StatMed 14 347-356
Improved variable window kernel estimates of	probability densities • Hall, Peter; Hu, Tien Chung; Marron, J. S.	95 AnlsStat 23 1-10
A class of unbiased kernel estimates of a	probability density function • Rychlik, T.	95 ZastApMa 22 485-497
entropy method for inverting Laplace transforms of	probability density functions • Wagner, Udo; Geyer, Alois L. J. A maximum	95 Biomtrka 82 887-892
Volterra integral equation for the first-passage-time	probability density • Gutiérrez Jáimez, R.; Román, P. Román; Ruiz, F. Torres A note on the	95 JAppProb 32 635-648
On a symmetry-based constructive approach to	probability densities for two-dimensional diffusion processes • Di Crescenzo, A. G. et al.	95 JAppProb 32 316-336
to "Characterization of the standard normal	probability density within the class of self-adjoint (self-reciprocal) ... • Rossberg Correction	95 MaOpfStS 26 387-387
Characterizations of	probability distributions based on discrete p-monotonicity • Sapatinas, Theofanis	95 StPrLet 24 339-344
Exact asymptotics for some	probability distributions on compact manifolds • Richards, Donald St. P.	95 AnlsStat 23 1582-1586
Katharine Rowse	Probability distributions for the Deming cost models • Kaminsky, Frank C.; Haberle,	95 JQualTek 27 355-362
Shubov, Victor Natural exponential families of	probability distributions and exponential-polynomial approximation • Martin, Clyde;	93 AplMaCmp 59 275-297
models for evaluating errors in fitting extremes of a	probability distribution • Bai, Jun; Jakeman, Anthony J.; McAleer, Michael Empirical	95 MaCmpSim 39 1-7
tions, characterizations and structural properties of	probability distributions similar to the exponential • Dimitrov, B.; Chukova, S. et al. Defini-	95 JStPlInf 43 271-287
Hutchinson, Seth A. A framework for constructing	probability distributions on the space of image segmentations • LaValle, Steven M.;	95 CVGIPIm 61 203-230
Goodness-of-fit tests for	probability distributions and spectral distributions • Anderson, T. W.	95 PrEngInf 9 27-37
Factorization theory for	probability distributions • Bondesson, Lennart	95 ScanActJ 44-53
Parameter estimation of the Weibull	probability distribution • Qiao, Hongzhu; Tsokos, Chris P.	94 MaCmpSim 37 47-55
Estimation of the three parameter Weibull	probability distribution • Qiao, Hongzhu; Tsokos, Chris P.	95 MaCmpSim 39 173-185
Jane F.; Park, Evelyn The	probability of divorce, based on the age difference of the married couple • Gentleman,	94 ASAProSoSt 216-221
Rossman, Allan J.; Short, Thomas H. Conditional	probability and education reform: Are they compatible? (jse/v3n2/rossman) •	95 JStEd 3/2 23kb
Review of "Elementary	probability" • Stirzaker, D. (Auth); Rees, Mike (Rev)	95 JRSS-A 158 356-356
A. R.; Herren, L. T.; Clemen, R. T. The use of	probability elicitation in the high-level nuclear waste regulation program • DeWispelare,	95 IntJFor 11 5-24
	Probabilities and experiments • Cartwright, Nancy	95 JEconmtx 67 47-59
The use of bootstrapping to estimate conditional	probability fields for source locations of airborne pollutants • Hopke, P. K.; Li, C. L. et al.	95 Chemolab 30 69-79
Murphy, Allan H.; Winkler, Robert L. Screening	probability forecasts: Contrasts between choosing and combining • Clemen, Robert T.;	95 IntJFor 11 133-145
Morzuch, Bernard J. Comparing	probability forecasts derived from theoretical distributions • Allen, P. Geoffrey;	95 IntJFor 11 147-157
Evaluation of	probability forecasts of events • Ogata, Yosihiko	95 IntJFor 11 539-541
	Probability forecasting	
A new method for scenario analysis with ... • Sapio	SEARCH (Scenario evaluation and analysis through repeated cross impact handling):	95 IntJFor 11 113-131
	Probability forecasting • Abramson, Bruce; Clemen, Robert	95 IntJFor 11 1-4
Equating Poisson and normal	probability functions to derive Stirling's formula • Walsh, Dennis P.	95 AmerStat 49 270-271
Harris, Thomas R. The use of conditional	probability functions in range data analysis and simulation • Lambert, David K.;	93 JRngeMgt 46 157-160
and Sylvester's "Four point problem" of geometric	probability • Scheinerman, E. R. et al. The rectilinear crossing number of a complete graph	94 AmMaMnth 101 939-943
	Probabilities of hands in 1 person seven-card stud (Korean) • Kim, T. S.; Chae, K. C.	95 KorApS 8/2 163-177
Wong, Wing Hung; Shen, Xiaotong	Probability inequalities for likelihood ratios and convergence rates of sieve MLEs •	95 AnlsStat 23 339-362
Dinwoodie, I. H. A	probability inequality for the occupation measure of a reversible Markov chain •	95 AnlsApPr 5 37-43
Maximal	probability inequalities for stochastic processes • Móricz, F.	95 MatNach 172 239-248
Asymptotic approximations for	probability integrals • Breitung, K. W.	94 Springer:Brln:NY 146p
Review of "Asymptotic approximations for	probability integrals" • Breitung, K. W. (Auth); Plachky, D. (Rev)	95 StatDecn 13 399-399
Review of "Asymptotic approximations for	probability integrals" • Breitung, K. W. (Auth); Veraverbeke, N. (Rev)	95 ShrtBkRv 15/2 26-26
	Probability integral transform	
Krzysztofowicz, Roman	A family of bivariate densities constructed from marginals • Long, Dou;	95 JASA 90 739-746
Integration and	probability • Malliavin, Paul; Airault, Helene; Kay, Leslie; Letac, Gerard	95 Springer:Brln:NY 322p
Barnett, C. (Rev) Review of "Integration and	probability" • Malliavin, Paul; Airault, Helene; Kay, Leslie; Letac, Gerard (Auth);	95 ShrtBkRv 15/3 46-46
	Probability, intuition and a spreadsheet • Graham, Alan	95 TeachgSt 17/3 84-87
Chergui, Jamal D. Estimation and tests of the discrete	probability law based on the empirical generating function (two dimensional case) •	94 RACEMdrd 88 245-286
On the comparison of	probability laws of a subset of one-dimensional real variates (Italian) • Freddi, Augusto	95 Metron 53 147-161
random methods for Cox processes with unknown	probability law • Crétois, E. Estimation of reduced Palm distributions by	95 ZastApMa 23 247-259

	Probability-matching prior	
Ghosh, Jayanta Kumar	On priors providing frequentist validity for Bayesian inference • Datta, Gauri Sankar;	95 Biomtrka 82 37-45
	Probability matching priors for linear calibration • Ghosh, M.; Carlin, B. P.; Srivastava, M. S.	95 Test 4 333-357
	Probability measure	
	Symmetry groups in R^d • Meerschaert, Mark M.; Veeh, Jerry Alan	95 StPrLet 22 1-6
	Probability and measure (Third edition) • Billingsley, Patrick	95 Wiley:NY:UK 593p
Phase retrieval for	probability measures on abelian groups • Carnal, H.; Fel'dman, G. M.	95 JTherPrb 8 717-725
geometric structure on the space of all the	probability measures equivalent to a given one • Pistone, G. et al. An infinite-dimensional	95 AnlsStat 23 1543-1561
	Probability measures on groups and related structures XI • Heyer, H. (Ed)	95 WrldSciPub:Sing 460p
Harmonic analysis of	probability measures on hypergroups • Bloom, W. R.; Heyer, H.	94 DGruyter:NY:Grm 601p
Review of "Harmonic analysis of	probability measures on hypergroups" • Bloom, W. R.; Heyer, H. (Auth); Barnett, C. (Rev)	95 ShrtBkRv 15/3 49-49
Weak convergence of	probability measures in L_p spaces (Russian) • Baushev, A. N.	94 VLenUM /4 15-20
Jean-Charles Changes of numéraire, changes of	probability measure and option pricing • Geman, Hélyette; Karoui, Nicole El; Rochet,	95 JAppProb 32 443-458
Operator stable	probability measures: An overview • Hudson, W. N.; Mason, J. D.; Veeh, J. A.	94 ThProbAp 39 264-276
Review of	"Probability and measure (Third edition)" • Billingsley, Patrick (Auth); Krämer, Walter (Rev)	95 CmpStDA 20 703-703
A general classification rule for	probability measures • Kulkarni, Sanjeev R.; Zeitouni, Ofer	95 AnlsStat 23 1393-1407
matrices • Högnäs, Göran; Mukherjee, Arunava	Probability measures on semigroups: Convolution products, random walks, and random	95 PlenumPubl:NY 400p
	Probability metric	
L.	A general duality theorem for marginal problems • Ramachandran, D.; Rüschendorf,	95 ProbTher 101 311-319
	Probability metrics and recursive algorithms • Rachev, S. T.; Rüschendorf, L.	95 AdvAppPr 27 770-799
Constantin, G. (Rev) Review of	"Probability metrics and the stability of stochastic models" • Rachev, Svetlozar T. (Auth);	95 BLndnMaS 27 87-89
Murali K. Finding choice alternatives in memory:	Probability models of brand name recall • Hutchinson, J. Wesley; Raman, Kalyan; Mantrala,	94 JMarRsr 31 441-461
linear modelling, and model comparison ... • Aitkin	Probability model choice in single samples from exponential families using Poisson log-	95 StatComp 5 113-120
A family of joint	probability models for cost and schedule uncertainties • Garvey, Paul R.	95 JCostAnl /1 156-200
	Probability models on horse race outcomes • Ali, Mukhtar M.	94 ASAProSpts 7-11
Review of "Introduction to	probability models (Fifth edition)" • Ross, S. M. (Auth); Jolliffe, Flavia (Rev)	95 JRSS-A 158 201-202
Ridge regression estimation of the linear	probability model • Gana, Rajaram	95 JAppStat 22 537-539
AIC with the EM algorithm: A demonstration of a	probability model selection technique • Glosup, Jeffrey G.; Axelrod, Michael C. Use of the	94 ASAProPEng 120-125
Review of "A first course in	probability models and statistical inference" • Creighton, J. H. C. (Auth); King, Bruce (Rev)	95 AmerStat 49 113-113
and tail behavior in stationary interval-valued	probability models • Sadrolhefazi, Amir; Fine, Terrence L. Finite-dimensional distributions	94 AnlsStat 22 1840-1870
Calculating normal	probabilities • Bagby, Richard J.	95 AmMaMnth 102 46-49
Rodney (Rev) Review of "Studies in applied	probability: Papers in honour of Lajos Takács" • Galambos, J.; Gani, J. (Ed); Coleman,	95 JRSS-A 158 634-635
Review of "Studies in applied	probability: Papers in honour of Lajos Takács" • Galambos, J.; Gani, J. (Ed); Haigh, J. (Rev)	95 Biomtrcs 51 1187-1187
Conditional	probability: Its place in the mathematics curriculum • Watson, Jane M.	95 MathTchr 88 12-17
The pleasures of	probability • Isaac, R. E.	95 Springer:Brln:NY 241p
Review of "The pleasures of	probability" • Isaac, R. (Auth); Coleman, R. (Rev)	95 ShrtBkRv 15/3 42-42
and impact of ... • Fill et al. L moment and	probability plot correlation coefficient goodness-of-fit tests for the Gumbel distribution	95 WaterReR 31 225-229
Density	probability plots • Jones, M. C.; Daly, F.	95 CommStB 24 911-927
	Probabilities, possibilities, and fuzzy sets • Drakopoulos, John A.	95 FuzzySet 75 1-15
Reply to comment on "The meaning of	probability in probabilistic risk analysis" • Watson, Stephen R.	95 RelEngSS 49 207-209
Murray, Thomas M. Comment on "The meaning of	probability in probabilistic safety analysis" (V45 p261-269) • Yellman, Ted W.;	95 RelEngSS 49 201-205
tory course in statistics with emphasis on applied	probability, quality control, and graphics • Wilson, W. J. Supplementing a college prepara-	94 ASAProStEd 42-43
Return	probabilities for random walk on a half-line • Lalley, Steven P.	95 JTherPrb 8 571-599
Sequential conditional	probability ratio test approach to the design of clinical trials • Tan, Ming; Xiong, Xiaoping	94 ASAProBiop 359-364
A class of sequential conditional	probability ratio tests • Xiong, Xiaoping	95 JASA 90 1463-1473
Review of "Quantum	probability and related topics (Volume 7)" • Accardi, L. (Ed); Lindblad, G. (Rev)	95 Metrika 42 268-269
Review of	"Probability" • Kari, A. F. (Auth); Dette, H. (Rev)	95 Metrika 42 275-276
Review of	"Probability" • Karr, A. F. (Auth); Schmidt, M. (Rev. German)	95 AllgeStA 79 333-334
Review of	"Probability" • Pitman, J. (Auth); Williams, R. J. (Rev)	95 Metrika 42 143-144
Saddlepoint approximations for the	probability of ruin in finite time • Barndorff-Nielsen, Ole E.; Schmidli, Hanspeter	95 ScanActJ 169-186
	Probability sampling	
itoring programmes • Overton, W. S.; Stehman, S. V.	Design implications of anticipated data uses for comprehensive environmental mon-	95 EnvEcoSt 2 287-303
estimation stage • Rao, J. N. K.	Estimating totals and distribution functions using auxiliary information at the	94 JOffStat 10 153-165
Sahoo, Lokanath; Mohanty, Suprasana Unequal	probability sampling using a transformed auxiliary variable • Sahoo, Jagatananda;	94 Metron 52/1 71-83
The Horvitz-Thompson as a unifying perspective for	probability sampling: With examples from natural resource sampling • Overton et al.	95 AmerStat 49 261-268
Correlation of intuitionistic fuzzy sets in	probability spaces • Hong, Dug Hun; Hwang, Seok Yoon	95 FuzzySet 75 77-81
Allende, R. (Rev) Review of "Hilbert space methods in	probability and statistical inference" • Small, C. G.; McLeish, D. L. (Auth); Marshall, G.;	95 JASA 90 1491-1492
(Rev) Review of "Understanding	probability and statistics – A book of problems" • Peters, A. K. (Auth); Bullard, Peter R.	95 Ststcian 44 412-412
Review of "Understanding	probability and statistics: A book of problems" • Falk, R. (Auth); Shotter, M. V. (Rev)	95 Biomtrcs 51 386-386
Review of "Understanding	probability and statistics: A book of problems" • Falk, R. (Auth); Woodward, P. W. (Rev)	95 Ststcian 44 284-285
James T. (Auth); Herrin, Gary D. (Rev) Review of	"Probability and statistics for engineers (Fourth edition)" • Scheaffer, Richard; McClave,	95 AmerStat 49 238-239
R. (Rev) Review of	"Probability and statistics for engineers" • Scheaffer, R.; McClave, J. (Auth); Ziegel, Eric	95 Technmcs 37 239-239
Rigdon, Steven E. (Rev) Review of	"Probability and statistics for engineers" • Scheaffer, Richard L.; McClave, James T. (Auth);	95 JQualTek 27 178-179
Mukherjee, S. P. et al. Review of "Essays on	probability and statistics: Festschrift in honour of Professor Anil Kumar Bhattacharyya" •	95 CalcutSt 45 131-132
Mukherjee, S. P. et al. Review of "Essays on	probability and statistics: Festschrift in honour of Professor Anil Kumar Bhattacharyya" •	95 SankhyaA 57 166-166
Bilotti-Aliaga, Martha Exploring	probability and statistics through hands-on experiments and cooperative learning •	94 ASAProStEd 150-155
Halvorsen, K. T. (Rev) Review of "Introduction to	probability and statistics (Ninth edition)" • Mendenhall, W.; Beaver, R. J. (Auth);	95 AmerStat 49 398-398
Review of "Introduction to	probability and statistics (Second edition)" • Giri, N. C. (Auth); Bandopadhyay, T. (Rev)	94 CalcutSt 44 267-268
Bingham, N. H. (Rev) Review of	"Probability, statistics and optimization: A tribute to Peter Whittle" • Kelly, F. P. (Ed);	95 JRSS-A 158 353-354
stration programs for use in teaching elementary	probability and statistics: Parts 3 and 4 (jse/v3n2/trumbo) • Trumbo, B. E. Some demon-	95 JStEd 3/2 60kb
Issues in assessing conceptual understanding in	probability and statistics (jse/v3n1/konold) • Konold, Clifford	95 JStEd 3/1 27kb
Aparna V. (Rev) Review of "Concepts in	probability and stochastic modeling" • Higgins, J.; Keller-McNulty, S. (Auth); Huzurbazar,	95 Technmcs 37 467-467
Nickerson, David M. (Rev) Review of "Concepts in	probability and stochastic modeling" • Higgins, James J.; Keller-McNulty, Sallie (Auth);	95 AmerStat 49 321-321
performance modelling • Nelson, R.	Probability, stochastic processes, and queueing theory: The mathematics of computer	95 Springer:Brln:NY 592p
	Probability theory • Rotar, Vladimir	95 WrldSciPub:Sing 350p
Advanced	probability theory (Second edition) • Galambos, Janos	95 Dekker:NY 461p
	Probability theory. An advanced course • Borkar, V. S.	95 Springer:Brln:NY 145p
Review of	"Probability theory, and analytic view" • Stroock, D. W. (Auth); Coleman, Rodney (Rev)	95 JRSS-A 158 356-357
Review of	"Probability theory: An analytic view" • Stroock, D. W. (Auth); Cranston, Michael (Rev)	95 JASA 90 808-808
Chebyshev's contribution to the theory of	probability • Sheynin, O.	93 AHistESc 46 321-340
Review of "Operator-limit distributions in	probability theory" • Jurek, Z. H.; Mason, D. J. (Auth); Hazod, W. (Rev)	95 Metrika 42 265-267
Review of "Operator-limit distributions in	probability theory" • Jurek, Z. J.; Mason, J. D. (Auth); Blower, G. (Rev)	95 JRSS-A 158 352-353
Review of "Operator-limit distributions in	probability theory" • Jurek, Zbigniew J.; Mason, J. David (Auth); Veeh, Jerry Alan (Rev)	95 BAmMaSoc 32 278-280
	Probability theory: An introductory course • Sinai, Y. (Auth); Haughton, D. (Transl)	95 Springer:Brln:NY 138p

Statistical	process control of multiple stream processes • Mortell, Robert R.; Runger, George C.	95 JQualTek 27 1-12
Statistical	process control of sampling frames • Spisak, A. W.	95 SurvyM 21 185-190
Statistical	process control scheme design • Keats, J. Bert; Miskulin, John D.; Runger, George C.	95 JQualTek 27 214-225
Statistical	process control in software development • Thomas, Michael L.	94 ASAProQlPr 140-145
and analysis in pharmaceutical science: Validation,	process controls, and stability • Chow, Shein-Chung; Liu, Jen-Pei Statistical design	95 Dekker:NY 584p
and analysis in pharmaceutical science: Validation,	process controls, and stability" • Chow, S.-C. et al. Review of "Statistical design	95 Biomtrcs 51 1190-1191
using time-series experimentation and statistical	process control • Wallace, J. E.; Foster, R. D. et al. Improving control of polymer viscosity	94 ASAProPEng 320-325
formance of naval gunfire support using statistical	process control • Bailey, Michael; Bowden, John; Callahan, Alexander J. Managing ship per-	94 MilOpRes 1 5-11
Reynolds, Marion R., Jr; Arnold, Jesse C.	Process control strategies using variable sampling intervals • Sethuraman, Shanthi;	94 ASAProQlPr 109-114
Plante, Robert; Tang, Jen Statistical	process control via the subgroup bootstrap • Seppala, Tomi; Moskowitz, Herbert;	95 JQualTek 27 139-153
McLaughlin, G. C. (Rev) Review of "Practical statistical	process control: A tool for quality manufacturing" • Alsup, F.; Watson, R. M. (Auth);	95 Technmcs 37 461-462
Review of "Bilateral plans for inspection and	process control by variables" • Panizzon, G. (Auth); Iacobini, Alberto (Rev)	94 Metron 52/1 216-217
Review of "Bilateral plans for inspection and	process control by variables" • Panizzon, G. (Auth); Magagnoli, U. (Rev. Italian)	94 Statstca 54 539-540
Review of "Bilateral plans for inspection and	process control by variables" • Panizzon, Gaetano (Auth); Copeland, Karen A. F. (Rev)	95 JQualTek 27 88-88
Review of "Bilateral plans for inspection and	process control by variables" • Panizzon, Gaetano (Auth); Godsey, E. B. (Rev)	95 Technmcs 37 463-464
H.; Thomas, Edward V. Statistical	process control with several components of common cause variability • Woodall, William	95 IIETrn 27 757-764
Using CMM data for	process evaluation and improvement • Blazek, L. W.	94 ASAProPEng 26-34
Mean-variance interactions in	process improvement and capacity design • Buss, Arnold H.; Lawrence, Stephen R.	95 IIETrn 27 339-347
L.; Snyder, Richard	Process improvement in telecommunications with multivariate methods • Seaver, William	94 ASAProQlPr 235-240
Abetti, Frank; Savage, Phillip	Process improvement utilizing computer simulation: Case study • Anderson, DeAnn;	95 ASQCTekCon 713-724
charts • Gan, F. F. Joint monitoring of	process mean and variance using exponentially weighted moving average control	95 Technmcs 37 446-453
An integrated scheme for	process monitoring and diagnosis • Chang, Shing I.	95 ASQCTekCon 725-732
An upper bound for the proportion of	process output not meeting specifications • Wharton, Robert; Czachor, James	94 ASAProQlPr 264-266
Montgomery, D. C. Response surface methodology:	Process and product optimization using designed experiments • Myers, R. H.;	95 Wiley:NY:UK 720p
A cumulative score control scheme for detecting	process shifts • Cheng, Chuen-Sheng	95 CommStA 24 755-774
Application of Hoaglin's G and H curves to	process specifications • Finnegan, G. Paul; Heavlin, William D.	94 ASAProQlPr 146-150
	Process variable	
Cornell, John A.	Fitting models to data from mixture experiments containing other factors •	95 JQualTek 27 13-33
in a design space consisting of both mixture and	process variables • Duineveld, C. A. A.; Coenegracht, P. M. J. Multicriteria steepest ascent	95 Chemolab 30 23-36
A cumulative sum control chart for monitoring	process variance • Chang, T. C.; Gan, F. F.	95 JQualTek 27 109-119
to "A cumulative sum control chart for monitoring	process variance" (V27 p109-119) • Chang, T. C.; Gan, F. F. Correction	95 JQualTek 27 402-402
The performance of control charts for monitoring	process variation • Lowry, Cynthia A.; Champ, Charles W.; Woodall, William H.	95 CommStB 24 409-437
Review of "Sojourn times in feedback and	processor sharing queues" • van den Berg, J. L. (Auth); Newton, J. (Rev)	95 JRSS-A 158 186-187
Review of "Sojourn times in feedback and	processor sharing queues" • van den Berg, J. L. (Auth); Schassberger, R. (Rev)	95 Metrika 42 146-147
	Procrustean method	
	Mean size-and-shapes and mean shapes: A geometric point of view • Lee, Huiling	95 AdvAppPr 27 44-55
Safar, M. Application of	Procrustean methods to mid- and near-infrared spectral data • Vigneau, E.; Devaux, M. F.;	95 JChemtrc 9 125-135
	Procrustes analysis	
Hitchcock, D.	Fish shape analysis using landmarks • Glasbey, C. A.; Horgan, G. W.; Gibson, G. J.;	95 BiomtrcJ 37 481-495
Ko, Daijin	M-estimates of rigid body motion on the sphere and in Euclidean space • Chang, Ted;	95 AnlsStat 23 1823-1847
Gabriel, K. Ruben Generalized	Procrustes analysis with iterative weighting to achieve resistance • Verboon, Peter;	95 BJMaStPs 48 57-73
	Procrustes fitting	
Silverman, B. W.	Incorporating parametric effects into functional principal components analysis •	95 JRSS-B 57 673-689
	Procrustes rotation	
Dickson, R. A.	Familial spinal shape • Mardia, K. V.; Dryden, I. L.; Hurn, M. A.; Li, Q.; Millner, P. A.;	94 JAppStat 21 623-641
vexities, labor hoarding, technology shocks, and	procyclical productivity. A structural econometric analysis • Chirinko, Robert S. Noncon-	95 JEconmtx 66 61-98
Acceptance sampling in a	producer-supplier conflicting environment: Risk neutral case • Tapiero, Charles S.	95 AplStMDA 11 3-12
A system for measuring short-term	producibility • Keenan, Thomas M.	95 ASQCTekCon 50-56
Reliability reports -	Product, customer, and process • Conte, John A.	95 ASQCTekCon 733-740
	Product design	
Wierich, Werner	D-optimal designs of experiments with non-interacting factors • Schwabe, Rainer;	95 JStPlInf 44 371-384
Hinckley, C. Martin; Barkan, Philip	The role of variation, mistakes, and complexity in producing nonconformities •	95 JQualTek 27 242-249
An analysis of new	product diffusion process • Kusaka, Yasuo; Suzuki, Hisatoshi	94 JOpRsJpn 38 188-211
	Product estimator	
Arcones, Miguel A.; Giné, Evarist	On the law of the iterated logarithm for canonical U-statistics and processes •	95 StocProc 58 217-245
q_x • Puzey, Anthony S.	A reappraisal of the principle underlying the conventional actuarial estimator of	95 InsrneME 17 125-132
	Product forms based on backward traffic equations • Boucherie, Richard J.	95 JAppProb 32 508-518
et al. Efficient Monte-Carlo simulation of a	product-form model for a cellular system with dynamic resource sharing • Fleming, P. J.	95 ACMMdSim 5 3-21
Bahadur-Kiefer theorems for kernel smooth	product-limit quantile estimator • Zhou, Yong	95 CommStA 24 2815-2828
	Product-limit survival functions with correlated survival times • Williams, Rick L.	94 ASAProBiop 211-216
	Product-limit survival functions with correlated survival times • Williams, Rick L.	95 LiDaAnal 1 171-186
Reference prior Bayesian analysis for normal mean	products • Sun, Dongchu; Ye, Keying	95 JASA 90 589-597
On symmetrization of	product measures • Mashayekhi, M.	95 MaMethSt 4 334-343
of measurement and grouping on the estimated	product-moment correlation • Chai, John J.; Chen, Chung Joint effects of errors	95 CommStA 24 283-294
A. P.; Jha, V. D. Some recurrence relations between	product moments of order statistics of a doubly truncated Pareto distribution • Khurana,	95 SankhyaB 57 1-9
al. Recurrence relations for single and	product moments of order statistics from a generalized half logistic ... • Balakrishnan et	95 JStCmpSm 52 385-398
	Product multinomial model	
	Weighted concordance analysis • Tricot, Jean-Marie; Lepage, Yves	95 BiomtrcJ 37 727-743
streamflows • Chebaane, M.; Salas, J. D.; Boes, D. C.	Product periodic autoregressive processes for modeling intermittent monthly	95 WaterReR 31 1513-1518
Myers, M. N.; Schafer, E. T. An improved approach to	product quality through testing • Hillis, M. A.; Minkus, G. W.; DeCaluwe, C. L.;	95 ATTTJ 74/3 40-48
for recurrence data - Applied to cost or number of	product repairs • Nelson, Wayne Confidence limits	95 Technmcs 37 147-157
Field data analyses for repairable	products (Korean) • Bai, Do Sun; Yun, Hyung Je; Choi, In Su	95 KorApS 8/2 133-145
dul L. (Rev) Review of "Tables of integrals, series, and	products (Fifth edition)" • Gradshteyn, I. S.; Ryzhik, I. M. (Auth); Jeffrey, A. (Ed); Bello, Ab-	95 Ststcian 44 287-288
A Gaussian central limit theorem for trimmed	products on simply connected step 2-nilpotent lie groups • Neuenschwander, Daniel	95 JTherPrb 8 165-174
The minimum of the expected value of the	product of three random variables in the Fréchet class • Bertino, Salvatore	94 JItalStS 3 201-211
of large numbers for quadratic forms, maxima of	products and truncated sums of i.i.d. random variables • Cuzick, J.; Giné, E.; Zinn, J. Laws	95 AnlsProb 23 292-333
Two new direct	product-type constructions for resolvable group-divisible designs • Rees, Rolf S.	93 JComDsgn 1 15-26
Review of "Developing new	products with TQM" • Cervetz, C. (Auth); Ziegel, Eric R. (Rev)	95 Technmcs 37 241-241
Optimal choice of	production capacity in a random market • Arntzen, Halvard	95 Stochatx 55 87-120
Bregman, Arie; Fuss, Melvyn; Regev, Haim The	production and cost structure of Israeli industry: Evidence from individual firm data •	95 JEconmtx 65 45-81
Adding value to the value-added educational	production function specification (Disc: p213-215) • Akerhielm, Karen	94 ASAProSoSt 192-195
Estimation of	production function using quadratic programming • Mukherjee, S. P.; Dasgupta, U.	94 JIAgSt 46 223-230
Deductible insurance and	production • Machnes, Yaffa	95 InsrneME 17 119-123
Stochastic specification in random	production models of cost-minimizing firms • Brown, Bryan W.; Walker, Mary Beth	95 JEconmtx 66 175-205

Dynamic reliability models with conditional	proportional hazards • Hollander, Myles; Peña, Edsel A.	95 LiDaAnal 1 377-401
Computer program for the	proportional hazards measurement error model • Nakamura, Tsuyoshi; Akazawa, Kouhei	94 CmpBiomd 45 203-212
Mallick, Bani K. Bayesian analysis of	proportional hazards models built from monotone functions • Gelfand, Alan E.;	95 Biomtrcs 51 843-852
for the analysis of sampled cohort data in the Cox	proportional hazards model • Borgan, Ø.; Goldstein, L.; Langholz, B. Methods	95 AnlsStat 23 1749-1778
A global test for the goodness of fit of Cox's	proportional hazards model • Sutradhar, Brajendra C.	94 JStatRsr 28 123-148
Model inconsistency, illustrated by the Cox	proportional hazards model • Ford, Ian; Norrie, John; Ahmadi, Susan	95 StatMed 14 735-746
a survival function with doubly censored data in a	proportional hazard model • Esteban, M. D.; Morales, D. Estimating	95 AplStMDA 11 145-157
Step-function covariate effects in the	proportional-hazards model • LeBlanc, Michael; Crowley, John	95 CandJSt 23 109-129
Ridder, G. Constructive identification of the mixed	proportional hazards model • Kortram, R. A.; van Rooij, A. C. M.; Lenstra, A. J.;	95 StNeerla 49 269-281
A residual bootstrap for regression parameters in	proportional hazards models • Loughin, Thomas M.	95 JStCmpSm 52 367-384
sample confidence interval for the quantiles in the	proportional hazards model of random censorship • Ghorai, J. K. Large	95 JStCmpSm 53 91-102
Analysis of a continuous-time	proportional hazard model using discrete duration data • Ryu, Keunkwan	95 EcomtrRv 14 299-313
restricted cubic spline as baseline hazard in the	proportional hazards model with step function time-dependent ... • Herndon et al. The	95 StatMed 14 2119-2129
Importance of events per independent variable in	proportional hazards regression analysis II. Accuracy and precision of ... • Peduzzi et al.	95 JClEpid 48 1503-1510
Model misspecification in	proportional hazards regression • Anderson, Garnet L.; Fleming, Thomas R.	95 Biomtrka 82 527-541
Port, Friedrich K. Transient effects in the Cox	proportional hazards regression model • Mauger, Elizabeth A.; Wolfe, Robert A.;	95 StatMed 14 1553-1565
power plant operations in Europe • Sturm, Roland	Proportional hazard regression models for point processes: An analysis of nuclear	94 JAppStat 21 533-540
of canonical models of infectious diseases: The	proportional hazard specification • Geoffard, P.-Y.; Philipson, T. The empirical content	95 Biomtrka 82 101-111
• Grambsch, P. M.; Therneau, T. M. Correction to	"Proportional hazards tests and diagnostics based on weighted residuals" (V81 p515-526)	95 Biomtrka 82 668-668
pricing and increased limits ratemaking by	proportional hazards transforms • Wang, Shuan Insurance	95 InsrneME 17 43-54
Testing the hypothesis of	proportional hazards in two populations • Deshpande, Jayant V.; Sengupta, Debasis	95 Biomtrka 82 251-261
Luceño, Alberto Discrete	proportional-integral control with constrained adjustment • Box, George E. P.;	95 Ststcian 44 479-495
	Proportional mixing	
Halloran, M. E.; Longini, I. M., Jr; Watelet, L.	Estimation of vaccine efficacy in non-randomly mixing populations • Haber, M.;	95 BiomtrcJ 37 25-38
	Proportional odds	
(Disc: p192-199) • Urfer, W. et al.	Multistate models for monitoring individual trees in permanent observation plots	94 EnvEcoSt 1 171-192
Guillermo ORDMKV: A computer program fitting	proportional odds model for multi-state Markov process • Guo, Wensheng; Marshall,	95 CmpBiomd 46 257-263
Estimating the real parameter in a two-sample	proportional odds model • Wu, Colin O.	95 AnlsStat 23 376-395
for prevalent coronary heart disease by using the	proportional odds model • Woodward, M.; Laurent, K. et al. An analysis of risk factors	95 Ststcian 44 69-80
Zahl, Per-Henrik A	proportional regression model for 20 year survival of colon cancer in Norway •	95 StatMed 14 1249-1261
A conversation with Frank	Proschan • Hollander, Myles; Marshall, Albert W.	95 StatSci 10 118-133
Wang, C. Y.	Prospective analysis of logistic case-control studies • Carroll, R. J.; Wang, Suojin;	95 JASA 90 157-169
ibration studies of dietary intake measurements in	prospective cohort investigations • Kaaks, R. et al. Sample size requirements for cal-	95 AJEpidm 141 557-565
ibration studies of dietary intake measurements in	prospective cohort investigations • Kaaks, R. et al. Sample size requirements for cal-	95 AJEpidm 142 557-565
Calibration of dietary intake measurements in	prospective cohort studies • Kaaks, Rudolf; Riboli, Elio; van Staveren, Wija	95 AJEpidm 141 548-556
wise nonlinear mixed-effects model in men with	prostate cancer • Morrell, C. H. et al. Estimating unknown transition times using a piece-	95 JASA 90 45-53
Stephen F. A	protein alignment scoring system sensitive at all evolutionary distances • Altschul,	93 JMolEvol 36 290-300
in the substitution process of amino acid sites of	proteins coded for by mitochondrial DNA • Reeves, Jaxk H. Heterogeneity	92 JMolEvol 35 17-31
analysis dihedral angles of amino acid residues in	proteins • Sclove, Stanley L.; Sherman, Simon A. Cluster	94 ASAProBiop 399-404
infrared ... • Shaffer et al. Experimental design	protocol for the pattern recognition analysis of bandpass filtered Fourier transform	95 Chemolab 29 89-108
a place in ASA for statisticians who work for state/	provincial or local governments? • Gaines, Leonard M.; González, María E. Is there	94 ASAProGovt 149-152
laws for Markov processes: Examples of an always	proximate absorbing state • Pakes, Anthony G. Quasi-stationary	95 AdvAppPr 27 120-145
metric tree estimation from conditional rank order	proximity data • Young, Martin R.; DeSarbo, Wayne S. A parametric procedure for ultra-	95 Psymtrka 60 47-75
for the least squares fitting of tree structures to	proximity data • Hubert, Lawrence; Arabie, Phipps Iterative projection strategies	95 BJMaStPs 48 281-317
Fitting an extended INDSCAL model to three-way	proximity data • Carroll, J. Douglas; Winsberg, Suzanne	95 JClass 12 57-71
in childhood cancer studies: The geographic	proximity and demographics within matched sets • Sakkinen et al. ... for control selection	95 AJPubHlh 85 555-557
Arabie, Phipps The approximation of two-mode	proximity matrices by sums of order-constrained matrices • Hubert, Lawrence;	95 Psymtrka 60 573-605
scaling. A multivariate technique for the analysis of	proximity and preference data" (Spanish) • Arce, C. et al. Review of "Multidimensional	95 JClass 12 137-139
Analysis of self and	proxy responses in the assessment of health status • Swamy, Mamatha	94 ASASrvyRsr 719-724
Shaban, Abdel Magid M. ANOVA, MINQUE,	PSD-MINQMBE, CANOVA and CMINQUE in estimating variance components •	94 Statstca 54 481-489
Comparison between the MINQUE and the	PSD-MINQMBE in the unbalanced one way random effects model • Shaban, Abdel Magid M.	94 EgpStJ 38 222-230
Fleischer, Karlheinz Two tests of	pseudo random number generators for independence and uniform distribution •	95 JStCmpSm 52 311-322
	Pseudo-likelihood	
	Analysis of count data using power variance function • Nakashima, Eiji	95 JJapnStS 25 193-204
	The centered normal conditionals distribution • Sarabia, José María	95 CommStA 24 2889-2900
Choi, In Su; Hwang, Yong Keun	Field data analyses for products with multiple-modes of failure (Korean) • Bai, Do Sun;	95 KorApS 8/1 89-104
	Interaction levels in an autoregressive graphical model • Lagona, Francesco	94 Metron 52/1 113-141
failure types • Hemmi, Isao	Parameter estimation in Cox's regression model for competing risks with missing	95 JJapnStS 25 81-96
families of distributions • Genest, C. et al.	A semiparametric estimation procedure of dependence parameters in multivariate	95 Biomtrka 82 543-552
	Unspecified link functions in generalized linear models (French) • Bonneu, M.	95 RStApp 43/2 49-66
observations • Särkkä, Aila	Pseudo-likelihood approach for Gibbs point processes in connection with field	95 MaOpfStS 26 89-97
Mase, Shigeru Consistency of the maximum	pseudo-likelihood estimator of continuous state space Gibbsian processes •	95 AnlsApPr 5 603-612
models • Dean, C. B. Modified	pseudo-likelihood estimator of the overdispersion parameter in Poisson mixture	94 JAppStat 21 523-532
Asymptotic normality of	pseudo-LS estimator for partly linear autoregression models • Gao, Jiti; Liang, Hua	95 StPrLet 23 27-34
Suicide by tourism? (German) • Ziegler, A. et al.	Pseudo-maximum-likelihood-estimation and regression diagnostics for count data:	95 AllgeStA 79 170-195
Donald; Gollin, Albert; Sawyer, Tom; Tanur, Judith M.	Pseudo-opinion polls: SLOP or useful data? • Horvitz, Dan; Koshland, Daniel; Rubin,	95 Chance 8/2 16-25
Göller, Gernot Analysis of	pseudo-profiles in organ pharmacokinetics and toxicokinetics • Mager, Harry;	95 StatMed 14 1009-1024
	Pseudorandom number generation by nonlinear methods • Eichenauer-Herrmann, Jürgen	95 IntStRvw 63 247-255
and application of Marsaglia's monkey test for	pseudorandom number generators • Percus, Ora E.; Whitlock, Paula A. Theory	95 ACMMdSim 5 87-100
Digital inversive	pseudorandom numbers • Eichenauer-Herrmann, Jürgen; Niederreiter, Harald	94 ACMMdSim 4 339-349
and earnings dynamics: Some estimates from the	PSID validation study • Pischke, Jörn-Steffen Measurement error	95 JBES 13 305-314
A. et al. Methodological issues in linkage analyses for	psychiatric disorders: Secular trends, assortative mating, bilineal pedigrees • Spence, M.	93 HumHrdty 43 166-172
T. S. (Rev) Review of "Communicating quantities: A	psychological perspective" • Moxey, L. M.; Sanford, A. J. (Auth); Budescu, D. V.; Wallsten,	95 Chance 8/2 38-40
Perspectives for investigating data of a	psychological survey • Schweizer, Karl	95 CmpStQ 10 1-8
Review of "Statistical methods for	psychology (Third edition)" • Howell, D. C. (Auth); Grimm, Laurence G. (Rev)	95 JAppStat 22 194-195
Public health decisions based on evidence,	psychology and politics • Bonham, G. H. Health and environment:	95 EnvrMtrc 6 311-318
Malcolm J. (Rev) Review of "House of cards:	Psychology and psychotherapy built on myth" • Dawes, Robyn M. (Auth); Sherman,	95 JASA 90 1492-1492
Tennis in Lilliput: A fable on sports and	psychology • Jackson, David	95 Chance 8/3 37-40
Van Breukelen, Gerard J. P.	Psychometric and information processing properties of selected response time models •	95 Psymtrka 60 95-113
	Psychometric latent response models • Maris, Eric	95 Psymtrka 60 523-547
Douglas; Green, Paul E.	Psychometric methods in marketing research: Part 1, Conjoint analysis • Carroll, J.	95 JMarRsr 32 385-391
Review of	"Psychometric theory" • Nunnally, J. C.; Bernstein, I. H. (Auth); Berge, Jos M. F. ten (Rev)	95 Psymtrka 60 313-315

Quality control (cont.)

empirical power transformations for production and	quality traits in forest tree breeding • Jansson, G.; Danell, Ö. Needs and benefits of	93 ThApGene 87 487-497
An empirical Bayes approach to monitoring water	quality • Solow, Andrew R.; Gaines, Arthur G.	95 EnvrMtrc 6 1-5
Bordeaux wine vintage	quality and the weather • Ashenfelter, Orley; Ashmore, David; Lalonde, Robert	95 Chance 8/4 7-14
	Quantal assay	
Scallan, A. J.	Asymmetric tolerance distributions: An application of two-piece distributions •	95 JAppStat 22 81-85
for estimating standard errors for estimators in	quantal bioassay • Cobb, E. Benton; Church, J. D. A discrete delta method	95 BiomtrcJ 37 691-699
	Quantal response	
	Calculation of benchmark doses from continuous data • Crump, Kenny S.	95 RiskAnal 15 79-89
Valenzuela, Carlos A.	Quantal response analysis of relief valve test data • Sheesley, John H.; Thomas, Harold W.;	95 ASQCTekCon 741-748
On the Fisher information in type-I censored and	quantal response data • Gertsbakh, I.	95 StPrLet 23 297-306
Williams, Rick L. Cluster sampling techniques in	quantal response teratology and developmental toxicity studies • Bieler, Gayle S.;	95 Biomtrcs 51 764-776
evolving projections ... • Grung et al. Detection and	quantification of embedded minor analytes in three-way multicomponent profiles by	95 Chemolab 29 213-221
certain data to spatial estimation, identification and	quantification • Sevcik, Karel; Brooker, Peter I. Application of the gnostical theory of un-	95 EnvrMtrc 6 457-463
et al. Exploratory data analysis for Hayashi's	quantification method III by graphics: Its application of ordinal structure ... • Komazawa	95 PrInStMa 43 161-176
The basis of norms for vague	quantifiers • O'Muircheartaigh, Colm; Gaskell, George	94 ASASrvyRsr 1327-1332
Pooling experiments over time to	quantify the dynamics of dryland winter wheat yield response • Burt, Oscar R.	95 AgronmyJ 87 156-164
coverage • Hwang, W.-D.; Chao, Anne	Quantifying the effects of unequal catchabilities on Jolly-Seber estimators via sample	95 Biomtrcs 51 128-141
	Quantifying errors in the Swedish consumer price index • Dalén, Jörgen	95 JOffStat 11 261-275
	Quantile	
French, Simon	Combining experts' opinions using a normal-Wishart model • Wiper, Michael P.;	95 JForcast 14 25-34
	Examining an increasing trend with non-parametric methods • Strömberg, Ulf	94 JAppStat 21 575-587
Papadatos, N.	Intermediate order statistics with applications to nonparametric estimation •	95 StPrLet 22 231-238
	Methods to approximate joint uncertainty and variability in risk • Bogen, Kenneth T.	95 RiskAnal 15 411-419
Chen, Huifen; Schmeiser, Bruce W.	Monte Carlo estimation for guaranteed-coverage non-normal tolerance intervals •	95 JStCmpSm 51 223-238
	New square-root algorithms for Kalman filtering • Park, PooGyeon; Kailath, Thomas	95 IEEEAuCn 40 895-899
	Sampling distributions of relative poverty statistics • Preston, Ian	95 ApplStat 44 91-99
Using power transformations when approximating	quantiles • Fearn, Dean H.; Nebenzahl, Elliott	95 CommStA 24 1073-1093
One-step saddlepoint approximations for	quantiles • Wang, Suojin	95 CmpStDA 20 65-74
Tests for differences in dispersion based on	quantiles • Shoemaker, Lewis H.	95 AmerStat 49 179-182
The distribution of Brownian	quantiles • Yor, Marc	95 JAppProb 32 405-416
A proof of Dassios' representation of the α-	quantile of Brownian motion with drift • Embrechts, P.; Rogers, L. C. G.; Yor, M.	95 AnlsApPr 5 757-767
options • Dassios, Angelos The distribution of the	quantile of a Brownian motion with drift and the pricing of related path-dependent	95 AnlsApPr 5 389-398
Jana Estimation of	quantile density function based on regression quantiles • Dodge, Yadolah; Jurečková,	95 StPrLet 23 73-78
Estimation of conditional	quantile density function • Xiang, Xiaojing	95 JNnparSt 4 309-316
Perturbed empirical distribution functions and	quantiles under dependence • Sun, Shan	95 JTherPrb 8 763-777
Hadi, Ali S. A method for estimating parameters and	quantiles of distributions of continuous random variables • Castillo, Enrique;	95 CmpStDA 20 421-439
Hesterberg, Tim Correction to "Saddlepoint	quantiles and distribution curves, with bootstrap applications" (V9 p207-211) •	95 CmpStQ 10 193-193
S. Lawrence; Lawlor, Stephen C. Estimating low-flow	quantiles from drainage-basin characteristics in New Hampshire and Vermont • Dingman,	95 WaterReB 31 243-256
misclassification • Marinos, A. Th. et al. Experimental	quantiles of epidemiological indices in case-control studies with non-differential	95 StatMed 14 1291-1306
Asymptotic behaviour of nonparametric conditional	quantile estimates for time series • Shi, Peide	95 CandJSt 23 161-169
	Quantile estimation	
Hosking, J. R. M.; Wallis, J. R.	A comparison of unbiased and plotting-position estimators of L moments •	95 WaterReR 31 2019-2025
Hadi, Ali S.	Modeling lifetime data with application to fatigue models • Castillo, Enrique;	95 JASA 90 1041-1054
	Multiscaling and skew separation in regional floods • Dawdy, David R.; Gupta, Vijay K.	95 WaterReR 31 2761-2767
Hadi, Ali S. Parameters and	quantiles estimation for continuously distributed random variables • Castillo, Enrique;	94 ASAProStCp 284-289
Large	quantile estimation in a multivariate setting • de Haan, Laurens; Huang, Xin	95 JMultiAn 53 247-263
Nonparametric	quantile estimation under order constraints • Rojo, Javier	95 JNnparSt 5 185-200
M-estimation and	quantile estimation in the presence of auxiliary information • Zhang, Biao	95 JStPlInf 44 77-94
Low-storage	quantile estimation • Hurley, C.; Modarres, R.	95 CmpStQ 10 311-325
data • Park, Hyo-Il; Park, Sang-Gue	Quantile estimation of treatment effect for the two-sample problem with right censored	95 StPrLet 24 139-145
Kiefer representation of a kernel conditional	quantile estimator • Xiang, Xiaojing On Bahadur-	95 JNnparSt 5 275-287
theorems for kernel smooth product-limit	quantile estimator • Zhou, Yong Bahadur-Kiefer	95 CommStA 24 2815-2828
Bahadur representation of the kernel	quantile estimator under random censorship • Xiang, Xiaojing	95 JMultiAn 54 193-209
Xiang, Xiaojing A Berry-Esseen theorem for the kernel	quantile estimator with application to studying the deficiency of quantile estimators •	95 AnInStMa 47 237-251
Xiang, Xiaojing Deficiency of the sample	quantile estimator with respect to kernel quantile estimators for censored data •	95 AnlsStat 23 836-854
Relating	quantiles and expectiles under weighted-symmetry • Abdous, Belkacem; Remillard, Bruno	95 AnInStMa 47 371-384
	Quantile function	
Ramesh C.	On some transforms of distribution functions and their relationships • Gupta,	95 MaOpfStS 26 123-128
	A universal Chung-type law of the iterated logarithm • Einmahl, Uwe; Mason, David M.	94 AnlsProb 22 1803-1825
	A universal one-sided law of the iterated logarithm • Mason, David M.	94 AnlsProb 22 1826-1837
The Bernstein polynomial estimator of a smooth	quantile function • Cheng, Cheng	95 StPrLet 24 321-330
Bounding shares and	quantiles from grouped data when group means are known • Schmid, Friedrich	94 Metron 52/1 143-153
Sample	quantiles of heavy tailed stochastic processes • Embrechts, Paul; Samorodnitsky, Gennady	95 StocProc 59 217-233
On semiparametric pivotal Bayesian inference for	quantiles • Swartz, T. B.; Villegas, C.; Martinez, C. J.	95 CommStA 24 2499-2515
that give simultaneous confidence intervals for	quantiles • Satten, Glen A. Upper and lower bound distributions	95 JASA 90 747-752
approximation theorems for integrated kernel	quantiles • Ralescu, S. S. Strong	95 MaMethSt 4 201-215
when the populations have a common known	quantile • Kumar, N.; Gill, A. N.; Mehta, G. P. Selection procedures for scale parameters	94 Metron 52/3 171-196
Empirical regression	quantile and leverage treatment method • Sueyoshi, Toshiyuke	95 JOpRsJpn 38 34-54
On an approximate likelihood for	quantiles • Lavine, M.	95 Biomtrka 82 220-222
Comparing two independent groups via multiple	quantiles • Wilcox, Rand R.	95 Ststcian 44 91-99
Estimating median and other	quantiles in nonparametric models • Zieliński, R.	95 ZastApMa 23 363-370
	Quantiles in nonrandom samples and observational studies • Rosenbaum, Paul R.	95 JASA 90 1424-1431
On the normal approximation for the sample p-	quantile • Koti, Kallappa M.	94 ASAProStCp 188-193
Estimation of an exponential	quantile under Pitman's measure of closeness • Kourouklis, S.	95 CandJSt 23 257-268
	Quantile process	
	A regression approach for estimating the center of symmetry • Hsieh, Fushing	95 StPrLet 22 157-160
Zhao, L. C.	Strassen's law of the iterated logarithm for the lorenz curves • Rao, C. Radhakrishna;	95 JMultiAn 54 239-252
On the distance between smoothed empirical and	quantile processes • Csörgő, Miklós; Horváth, Lajos	95 AnlsStat 23 113-131
Large sample confidence interval for the	quantiles in the proportional hazards model of random censorship • Ghorai, J. K.	95 JStCmpSm 53 91-102
	Quantile regression	
Nolan, D.	A general Akaike-type criterion for model selection in robust regression • Burman, P.;	95 Biomtrka 82 877-886
outcomes • Viscusi, W. Kip; Born, Patricia	The general-liability reform experiments and the distribution of insurance-market	95 JBES 13/2 183-188
Rogers, William H.	Quantile regression as an alternative to robust regression • Gould, William W.;	94 ASAProStEd 12-16
1963-1987 • Buchinsky, Moshe	Quantile regression, Box-Cox transformation model, and the U.S. wage structure,	95 JEconmtx 65 109-154

The optimal rates of convergence on nonparametric	quantile regression estimate on dependent data (Chinese) • Shi, Peide; He, Xuming	95 KToChina 39 1537-1539
ence values obtained by kernel-based estimation of	quantile regressions • Ducharme, G. R.; Gannoun, A.; Guertin, M.-C.; Jéquier, J.-C. Refer-	95 Biomtrcs 51 1105-1116
Bootstrapping	quantile regression estimators • Hahn, Jinyong	95 EconTher 11 105-121
Convergence theorems for empirical cumulative	quantile regression functions • Rao, C. R.; Zhao, L. C.	95 MaMethSt 4 81-91
to "Convergence theorems for empirical cumulative	quantile regression functions" (V4 p81-91) • Rao, C. R.; Zhao, L. C. Correction	95 MaMethSt 4 359-359
Estimating the asymptotic covariance matrix for	quantile regression models. A Monte Carlo study • Buchinsky, Moshe	95 JEconmtx 68 303-338
George G. A convergent algorithm for	quantile regression with smoothing splines • Bosch, Ronald J.; Ye, Yinyu; Woodworth,	95 CmpStDA 19 613-630
Autoregression	quantiles and related rank-scores processes • Koul, Hira L.; Saleh, A. K. Md. E.	95 AnlsStat 23 670-689
The Bahadur representation of sample	quantiles for sequences of strongly mixing random variables • Yoshihara, Ken-ichi	95 StPrLet 24 299-304
Resampling inference for	quantile shares • Kovačević, Milorad S.; Yung, Wesley; Pandher, Gurupdesh S.	94 ASASrvyRsr 206-211
Estimation of a	quantile in some nonstandard cases • Xiang, Xiaojing	95 AnInStMa 47 105-117
smooth link function and predictors obtained from	quantile spline fits to high-dimensional data • O'Connell et al. ... model with a	95 JStPlInf 47 153-164
The determination of a "least	quantile of squares regression line" for all quantiles • Carrizosa, Emilio; Plastria, Frank	95 CmpStDA 20 467-479
On monotone dependence functions of the	quantile type • Krajka, A.; Szynal, D.	95 ZastApMa 23 51-72
The asymptotic accuracy of the bootstrap of U-	quantiles • Arcones, Miguel A.	95 AnlsStat 23 1802-1822
Weighted bootstrapping for U-	quantiles • Aerts, Marc; Janssen, Paul	95 StPrLet 22 317-323
Review of "AIDS epidemiology: A	quantitative approach" • Brookmeyer, R.; Gail, M. H. (Auth); Ghani, Azra (Rev)	95 JRSS-A 158 343-343
(Rev) Review of "AIDS epidemiology: A	quantitative approach" • Brookmeyer, Ron; Gail, Mitchell H. (Auth); DeGruttola, Victor	95 JASA 90 797-798
Review of "AIDS epidemiology: A	quantitative approach" • Brookmeyer, Ron; Gail, Mitchell H. (Auth); Farrington, C. P. (Rev)	95 AJEpidm 141 180-181
(Disc: p78-79) • Talwalker, S. et al. Qualitative and	quantitative assessment of the risk from the exposure to fetotoxic chemical compounds	95 EnvEcoSt 2 71-78
Unrelated question model with	quantitative attribute by stratified double sampling (Korean) • Lee, Gi-Sung; Hong, Ki-Hak	95 KorApS 8/1 27-38
mixture models to detect effects of major genes on	quantitative characters in a plant breeding experiment • Jiang, C.; Pan, X. et al. The use of	94 Genetcs 136 383-394
ground prediction of spectra • Wabuyele, B. W. et al.	Quantitative comparison of bidirectional and optimal associative memories for back-	95 Chemolab 29 51-61
1994 National Census Test:	Quantitative comparisons of coverage • Pausche, Joan M.	94 ASASrvyRsr 313-318
ysis of Fourier transform infrared ... • Lindblom et al.	Quantitative determination of stabiliser in a single base propellant by chemometric anal-	95 Chemolab 29 243-254
for converting qualitative response data to	quantitative expectations: An application to Australian ... • Smith et al. ... procedures	95 JApEcmtx 10 165-185
Griffith, G. R. Modifying	quantitative forecasts of livestock production using expert judgements • Vere, D. T.;	95 JForcast 14 453-464
Genetic Tobit factor analysis:	Quantitative genetic modeling with censored data • Waller, Niels G.; Muthén, Bengt O.	92 BehavGen 22 265-292
	Quantitative genetics	
Jaggi, Seema; Agarwal, K. N.	Augmented partial diallel design for estimating combining ability of parents •	95 BiomtrcJ 37 879-887
Barnes, R. D.	The breeding seedling orchard in the multiple population breeding strategy •	95 SilvGene 44 81-88
Van Sanford, D. A.; Seyedsadr, M. S.	Clustering cultivars into groups without rank-change interactions • Cornelius, P. L.;	93 CropSci 33 1193-1200
trial data • Bernardo, Rex	Genetic models for predicting maize single-cross performance in unbalanced yield	95 CropSci 35 141-147
related hybrids • Bernardo, Rex	Prediction of maize single-cross performance using RFLPs and information from	94 CropSci 34 20-25
Orzack, Steven Hecht; Chappell, Rick	Sample-size determination for parent-offspring regression of a composite trait •	95 Biomtrcs 51 822-830
D. A.; Pfeiffer, T. W.; Cornelius, P. L.	Selection index based on genetic correlations among environments • Van Sanford,	93 CropSci 33 1244-1248
crops • Burton, Joe W.; Carver, Brett F.	Selection among S_1 families vs. selfed half-sib or full-sib families in autogamous	93 CropSci 33 21-28
interaction model for cultivar trials • Cornelius, P. L.	Statistical tests and retention of terms in the additive main effects and multiplicative	93 CropSci 33 1186-1193
Choosing parents • Panter, D. M.; Allen, F. L.	Using best linear unbiased predictions to enhance breeding for yield in soybean: I.	95 CropSci 35 397-405
ture determination and parameter estimation in	quantitative IR spectroscopy • Brink, A.; Westerlund, T. The joint problem of model struc-	95 Chemolab 29 29-36
Lancaster, H. O.	Quantitative methods in biological and medical sciences. A historical essay •	94 Springer:Brln:NY 297p
Lancaster, H. O. (Auth); Mayo, Oliver (Rev) Review of	"Quantitative methods in biological and medical science - A historical essay" •	95 AstrlJSt 37 372-374
H. O. (Auth); Farewell, V. T. (Rev) Review of	"Quantitative methods in biological and medical sciences. A historical essay" • Lancaster,	95 ShrtBkRv 15/2 22-22
position of histograms and frequency polygons:	Quantitative and qualitative smoothing • Simonoff, Jeffrey S. The anchor	95 CommStB 24 691-710
Epidemiology and	quantitative risk assessment: A bridge from science to policy • Hertz-Picciotto, Irva	95 AJPubHlh 85 484-491
S. J. Optimal sequential screening guidelines for	quantitative risk factors measured with error • Hughes, M. D.; Thompson, S. G.; Pocock,	95 JASA 90 19-26
estimate maternally controlled genetic variation in	quantitative seed characters • Foolad, M. R.; Jones, R. A. Models to	92 ThApGene 83 360-366
Reduced rank regression: With applications to	quantitative structure-activity relationships • Schmidli, H.	95 Springer:Brln:NY 179p
in Banach spaces, and weak invariance principle - A	quantitative study • Anastassiou ... theorem, weak law of large numbers for martingales	95 JMultiAn 52 158-180
Multiple regression for molecular-marker,	quantitative trait data from large F_2 populations • Wright, A. J.; Mowers, R. P.	94 ThApGene 89 305-312
of small-pedigree studies of the heritability of a	quantitative trait • Schork, N. J.; Schork, M. A. The relative efficiency and power	93 HumHrdty 43 1-11
High resolution of	quantitative traits into multiple loci via interval mapping • Jansen, Ritsert C.; Stam, Piet	94 Genetcs 136 1447-1455
human linkage studies to resolve the contribution of	quantitative trait loci • Eaves, Lindon J. Effect of genetic architecture on the power of	94 Heredity 72 175-192
S.; Knott, Sara A.; Elsen, Jean-Michel Mapping	quantitative trait loci in crosses between outbred lines using least squares • Haley, Chris	94 Genetcs 136 1195-1207
Yefim I.; Kirzhner, Valery M. Interval mapping of	quantitative trait loci employing correlated trait complexes • Korol, Abraham B.; Ronin,	95 Genetcs 140 1137-1147
Keightley, Peter D.; Bulfield, Grahame Detection of	quantitative trait loci from frequency changes of marker alleles under selection •	93 GeneRsr 62 195-203
Eaves, Lindon; Meyer, Joanne Locating human	quantitative trait loci: Guidelines for the selection of sibling pairs for genotyping •	94 BehavGen 24 443-456
analysis of linkage between genetic markers and	quantitative trait loci. I. Prior knowledge • Hoeschele, I.; VanRaden, P. M. Bayesian	93 ThApGene 85 953-960
analysis of linkage between genetic markers and	quantitative trait loci. II. Combining prior knowledge with ... • Hoeschele et al. Bayesian	93 ThApGene 85 946-952
for determining linkage between marker loci and	quantitative trait loci • Darvasi, A.; Soller, M. Optimum spacing of genetic markers	94 ThApGene 89 351-357
in determining linkage between marker loci and	quantitative trait loci • Motro, U.; Soller, M. Sequential sampling	93 ThApGene 85 658-664
A nonparametric approach for mapping	quantitative trait loci • Kruglyak, Leonid; Lander, Eric S.	95 Genetcs 139 1421-1428
Controlling the type I and type II errors in mapping	quantitative trait loci • Jansen, Ritsert C.	94 Genetcs 138 871-881
Multiple trait analysis of genetic mapping for	quantitative trait loci • Jiang, Changjian; Zeng, Zhao-Bang	95 Genetcs 140 1111-1127
A sib-pair approach to interval mapping of	quantitative trait loci • Fulker, David W.; Cardon, Lon R.	94 AJHuGene 54 1092-1103
Precision mapping of	quantitative trait loci • Zeng, Zhao-Bang	94 Genetcs 136 1457-1468
Weller, Joel I. Genetic mapping of	quantitative trait loci for traits with ordinal distributions • Hackett, Christine A.;	95 Biomtrcs 51 1252-1263
sufficient progeny with a desired genotype at two	quantitative trait loci • Schmitz et al. ... the minimum sample size required to obtain	93 ThApGene 87 136-144
Efficiency of direct selection on	quantitative trait loci for a two-trait breeding objective • de Koning, G. J.; Weller, J. I.	94 ThApGene 88 669-677
A new approach for mapping	quantitative trait loci using complete genetic marker linkage maps • Wu, W. R.; Li, W. M.	94 ThApGene 89 535-539
the locations and the sizes of the effects of	quantitative trait loci using flanking markers • Martínez, O.; Curnow, R. N. Estimating	92 ThApGene 85 480-488
A general mixture model for mapping	quantitative trait loci by using molecular markers • Jansen, R. C.	92 ThApGene 85 252-260
Missing markers when estimating	quantitative trait loci using regression mapping • Martínez, Octavio; Curnow, Robert N.	94 Heredity 73 198-206
The power of interval mapping of	quantitative trait loci, using selected sib pairs • Cardon, Lon R.; Fulker, David W.	94 AJHuGene 55 825-833
Multipoint interval mapping of	quantitative trait loci, using sib pairs • Fulker, D. W.; Cherny, S. S.; Cardon, L. R.	95 AJHuGene 56 1224-1233
Power of different sampling strategies to detect	quantitative trait loci variance effects • Weller, J. I.; Wyler, A.	92 ThApGene 83 582-588
Prediction of the power of detection of marker-	quantitative trait locus linkages using analysis of variance • Knott, S. A.	94 ThApGene 89 318-322
Empirical threshold values for	quantitative trait mapping • Churchill, G. A.; Doerge, R. W.	94 Genetcs 138 963-971
multipoint sib-pair analysis of qualitative and	quantitative traits • Kruglyak, Leonid; Lander, Eric S. Complete	95 AJHuGene 57 439-454
• Curtis, D.; Gurling, H. M. D. Using a dummy	quantitative variable to deal with multiple affection categories in genetic linkage analysis	91 AnHuGene 55 321-327
Mail survey design for sensitive	quantitative variable • Singh, Ravindra; Singh, Sarjinder; Mangat, Naurang Singh	95 Metron 53 43-54
On estimation of mean/total of stigmatized	quantitative variable • Singh, S.; Mangat, N. S.; Singh, R.	94 Statstca 54 383-386
Chan, Chorkin Contextual vector	quantization for speech recognition with discrete hidden Markov Model • Huo, Qiang;	95 PattrnRc 28 513-517
image compression and classification using vector	quantization • Oehler, Karen L.; Gray, Robert M. Combining	95 IEEEPAMI 17 461-473

The effects of	quantized data • Jewell, William S.	95 AplStMDA 11 201-216
of nonconservative dynamical semigroups and a	quantum boundary theory • Bhat, B. V. Rajarama; Parthasarathy, K. R. Markov dilations	95 AIHenrPB 31 601-651
Measurement and filtering of	quantum diffusion II • Belavkin, V. P.	94 ThProbAp 39 363-378
Applications of	quantum entropy to statistics • Silver, R. N.; Martz, H. F.	94 ASAProStCp 61-70
Holevo, A. S. Constructing	quantum measurement processes via classical stochastic calculus • Barchielli, A.;	95 StocProc 58 293-317
Review of	"Quantum probability and related topics (Volume 7)" • Accardi, L. (Ed); Lindblad, G. (Rev)	95 Metrika 42 268-269
Convergence to equilibrium for classical and	quantum spin systems • Prato, G. Da; Zabczyk, J.	95 ProbTher 103 529-552
of the pest prevention ability of the import plant	quarantine in Japan • Yamamura, Kohji; Sugimoto, Tamio Estimation	95 Biomtrcs 51 482-490
The	quarter-of-circumference law • Girko, V. L.; Repin, K. Yu.	95 ThPrMaSt 50 67-70
A periodic cointegration model of	quarterly consumption • Franses, Philip Hans; Kloek, Teun	95 AplStMDA 11 159-166
indicators • Parigi, G.; Schlitzer, G.	Quarterly forecasts of the Italian business cycle by means of monthly economic	95 JForcast 14 117-141
Median and	quartiles from grouped data • Evans, Gwyn	95 TeachgSt 17/3 99-100
	Quasi profile and directed likelihoods from estimating functions • Barndorff-Nielsen, O. E.	95 AnInStMa 47 461-464
	Quasianalytic random fields • Yadrenko, M. I.	95 ThPrMaSt 50 143-146
A re-evaluation of the	quasi-Bayes approach to the linear combination of forecasts • Faria, A. E.; Souza, R. C.	95 JForcast 14 533-542
the equilibrium distribution in level dependent	quasi-birth-and-death processes • Bright, L.; Taylor, P. G. Calculating	95 CommStSM 11 497-525
Expected passage times in homogeneous	quasi-birth-and-death processes • Latouche, Guy; Ramaswami, V.	95 CommStSM 11 103-122
Natural and	quasi-experiments in economics • Meyer, Bruce D.	95 JBES 13/2 151-161
Sen, Pranab K. A test of	quasi-independence in ordinal triangular contingency tables • Tsai, Ming-Tien;	95 StSinica 5 767-780
Johnson, William D. Testing	quasi-independence in stratified triangular contingency tables • May, Warren L.;	95 BiomtrcJ 37 531-546
levels • McDonald et al. Exact conditional tests of	quasi-independence for triangular contingency tables: Estimating attained significance	95 ApplStat 44 143-151
	Quasi-likelihood	
data • Ryu, Keunkwan	Analysis of a continuous-time proportional hazard model using discrete duration	95 EcomtrRv 14 299-313
Liang, Kung-yee	Approximate likelihood ratios for general estimating functions • Hanfelt, John J.;	95 Biomtrka 82 461-477
	On asymptotic optimality of estimating functions • Chen, K.; Heyde, C. C.	95 JStPlInf 48 101-112
data • Dean, C. B.; Eaves, D. M.; Martinez, C. J.	A comment on the use of empirical covariance matrices in the analysis of count	95 JStPlInf 48 197-205
van de Ven, R.; Weber, N. C.	Log-linear models for mean and dispersion in mixed Poisson regression models •	95 AstrlJSt 37 205-216
Fitzmaurice, Garrett M.; Laird, Nan M.	Regression models for a bivariate discrete and continuous outcome with clustering •	95 JASA 90 845-852
Amos, Christopher I.	Robust variance-components approach for assessing genetic linkage in pedigrees •	94 AJHuGene 54 535-543
latent variable modelling: A Monte Carlo study • Hsu	Sampling behaviour in estimating predictive validity in the context of selection and	95 BJMaStPs 48 75-97
experiments • Schwartz, P. F.; Gennings, C. et al.	Threshold models for combination data from reproductive and developmental	95 JASA 90 862-870
Vos, Paul W.	Quasi-likelihood or extended quasi-likelihood? An information-geometric approach •	95 AnInStMa 47 49-64
kernel regression for generalized linear models and	quasi-likelihood functions • Fan, J.; Heckman, N. E.; Wand, M. P. Local polynomial	95 JASA 90 141-150
Comparison of	quasi-likelihood models for overdispersion • Yanez, N. David, III; Wilson, Jeffrey R.	95 AstrlJSt 37 217-231
Model selection for extended	quasi-likelihood models in small samples • Hurvich, Clifford M.; Tsai, Chih-Ling	95 Biomtrcs 51 1077-1084
The EM algorithm for the	quasi-likelihood regression model • Paik, Myunghee Cho	95 CommStA 24 1403-1430
	Quasi-linear geostatistical theory for inversing • Kitanidis, Peter K.	95 WaterReR 31 2411-2419
• Gyöngy, István; Nualart, David Implicit scheme for	quasi-linear parabolic partial differential equations perturbed by space-time white noise	95 StocProc 58 57-72
Conditional propagation of chaos and a class of	quasilinear PDE's • Zheng, Weian	95 AnlsProb 23 1389-1413
	Quasilinear stochastic elliptic equations with reflection • Nualart, David; Tindel, Samy	95 StocProc 57 73-82
conservation • Kotelenez, Peter A class of	quasilinear stochastic partial differential equations of McKean-Vlasov type with mass	95 ProbTher 102 159-188
Generalized	quasilinearization of stochastic initial value problem • McRae, F. A.	95 StocAnAp 13 205-210
Monte Carlo and	quasi-Monte Carlo methods in scientific computing • Niederreiter, H.; Shiue, P. J.-S. (Ed)	95 Springer:Brln:NY 372p
A	quasi-Newton acceleration of the EM algorithm • Lange, Kenneth	95 StSinica 5 1-18
Comparison of regression curves using	quasi-residuals • Kulasekera, K. B.	95 JASA 90 1085-1093
A Markov chain associated with the minimal	quasi-stationary distribution of birth-death chains • Martínez, Servet; Vares, Maria Eulália	95 JAppProb 32 25-38
Martinez, S.; Picco, P. Existence of	quasi-stationary distributions. A renewal dynamical approach • Ferrari, P. A.; Kesten, H.;	95 AnlsProb 23 501-521
state • Pakes, Anthony G.	Quasi-stationary laws for Markov processes: Examples of an always proximate absorbing	95 AdvAppPr 27 120-145
survey • Agresti, Alan Logit models and related	quasi-symmetric log-linear models for comparing responses to similar items in a	95 SocMethR 24 68-95
	Quasi-symmetry	
factor • McCloud, P. I.; Darroch, J. N.	An analysis of categorical repeated measurements in the presence of an external	95 Biomtrka 82 47-55
A decade of	questions • Schaeffer, Nora Cate	95 JOffStat 11 79-92
Lee, Gi-Sung; Hong, Ki-Hak Unrelated	question model with quantitative attribute by stratified double sampling (Korean) •	95 KorApS 8/1 27-38
Non-response to sensitive	questions: Nationalism in Germany • Krebs, Dagmar	94 ASASrvyRsr 1199-1203
Goldenberg, Karen L. Answering	questions, questioning answers: Evaluating data quality in an establishment survey •	94 ASASrvyRsr 1357-1362
Sarjinder Unrelated	question randomized response sampling using continuous distributions • Singh,	94 JIAgSt 46 349-361
How to ask survey	questions • Fink, Arlene	95 SagePubl:CA:Lnd 120p
Asking sensitive	questions in surveys • Hutchinson, Paul	95 TeachgSt 17/2 43-43
Holmes, T. M. Evaluating change and preference	questions using paired Likert scale data • Roberson, P. K.; Shema, S. J.; Mundfrom, D. J.;	94 ASAProSoSt 189-191
	Questioning multilevel models (Disc: p201-227, 239-240) • de Leeuw, Jan; Kreft, Ita G. G.	95 JEdStat 20 171-189
response to personally-delivered mail-back	questionnaires • Dillman, Don A.; Dolsen, Dana E.; Machlis, Gary E. Increasing	95 JOffStat 11 129-139
of confidentiality on completion rates for census	questionnaires • Dillman et al. ... appeals, mandatory appeals and variations in statements	94 ASASrvyRsr 1333-1338
postcards affect mailback response rates for census	questionnaires • Dillman et al. ... letters, stamped return envelopes and reminder	95 SurvyM 21 159-165
	Questionnaire design	
Glenn S.; Lewis, Paul J.; Vacca, E. Ann	Cognitive research for the 1997 Census of Agriculture report form • Wolfgang,	94 ASASrvyRsr 503-508
European Community • Van Bastelaer, Aloïs	Differences in the measurement of employment in the labour force surveys in the	94 JOffStat 10 277-305
Development Laboratory • Bergman, Lars R.	Pretesting procedures at Statistics Sweden's Measurement, Evaluation and	95 JOffStat 11 309-323
Blair, Johnny; Bickart, Barbara; Eastman, Elizabeth	Respondent strategies for recall of crime victimization incidents • Czaja, Ronald;	94 JOffStat 10 257-276
Elizabeth A.; Griffin, Deborah H. • The role of	questionnaire design in reducing census coverage error (Disc: p754-756) • Martin,	94 ASASrvyRsr 736-741
Some effects of	questionnaire design on reported rates of victimization • Paton, D.	94 ASASrvyRsr 443-448
Research on establishment-survey	questionnaire design • Phipps, Polly A.; Butani, Shail J.; Chun, Young I.	95 JBES 13 337-346
Redesigning the Canadian Labour Force Survey	questionnaire: Development and testing • Sunter, Deborah; Kinack, Mark	94 ASASrvyRsr 908-912
language recognition of the Year 2000 Census	Questionnaire Feasibility Test • Appel, Martin V.; Cole, Ronald Spoken	94 ASASrvyRsr 1310-1315
in the development and testing of a traffic safety	questionnaire • Schechter, S. et al. Cognitive issues and methodological implications	94 ASASrvyRsr 1215-1219
Testing the logical structure of	questionnaires • Willenborg, L. C. R. J.	95 StNeerla 49 95-109
A split	questionnaire survey design • Raghunathan, Trivellore E.; Grizzle, James E.	95 JASA 90 55-63
groups in the redesign of an establishment survey	questionnaire • Chun, Y. I. et al. Using an establishment survey response model, and focus	94 ASASrvyRsr 1363-1367
"Redesign of the National Health Interview Survey	questionnaire" • Fowler, Floyd Jackson, Jr Comment on	94 ASASrvyRsr 379-381
Schwarz, Norbert What respondents learn from	questionnaires: The survey interview and the logic of conversation (Disc: p169-177) •	95 IntStRvw 63 153-168
Discussion of "What respondents learn from	questionnaires: The survey interview and the logic of conversation" • Cantor, David	95 IntStRvw 63 173-177
Discussion of "What respondents learn from	questionnaires: The survey interview and the logic of conversation" • Singer, Eleanor	95 IntStRvw 63 169-172
tween dietary intake obtained from a food frequency	questionnaire and true average intake • Landin et al. ... estimating the relationship be-	95 Biomtrcs 51 169-181
tween dietary intake obtained from a food frequency	questionnaire and true average intake" (V51 p169-181) • Landin et al. ... relationship be-	95 Biomtrcs 51 1196-1196

Controlled	queueing systems • Kitaev, Mikhail Yu.; Rykov, Vladimir V.	95 CRCPress:FL 304p
The busy periods of some	queueing systems • Stadje, W.	95 StocProc 55 159-167
of a stochastic process associated with times in	queueing systems • Kim, Hag-Min; Fotopoulos, S. B. Characterization study	94 ASAProPEng 163-168
Hassan, N. A. Multi-channel bi-level	queueing system with heterogeneous servers • Shoukry, E. M.; Gharraph, M. K.;	94 EgpStJ 38 208-221
properties of waiting time in a multiserver	queueing system with identical service times • Vinogradov, O. P. On certain asymptotic	94 ThProbAp 39 714-718
On the M/G/1 retrial	queueing system with linear control policy • Martín, M.; Gómez-Corral, A.	95 Top 3 285-305
A	queueing system with random failures and delayed repairs • Madan, K. C.	94 JIndnStA 32 39-48
	Queueing theory	
	Ballot theorems revisited • Konstantopoulos, Takis	95 StPrLet 24 331-338
Borovkov, K. A.	On crossing times for multidimensional walks with skip-free components •	95 JAppProb 32 991-1006
Magalhães, M. N.; McNickle, D. C.; Salles, M. C. B.	Departure processes from a loss system with two stations and one server •	94 RBrzlPrS 8 135-146
	Differentiability of the matrices R and G in the matrix-analytic method • He, Qi-Ming	95 CommStSM 11 123-132
	Factorial moment expansion for stochastic systems • Błaszczyszyn, B.	95 StocProc 56 321-335
events and probabilistic service ... • Henderson et al.	Insensitivity in discrete-time generalized semi-Markov processes allowing multiple	95 AnlsApPr 5 78-96
Steinebach, J.	Invariance principles for renewal processes and some applications • Alex, M.;	95 ThPrMaSt 50 23-54
Dembo, Amir; Zajic, Tim	Large deviations: From empirical mean and measure to partial sums process •	95 StocProc 57 191-224
	Optimal order for two servers in tandem • Yamazaki, Genji; Ito, Hiroshi	95 AnInStMa 47 31-48
Ranga V.; He, Xin X.	Order crossover in inventory systems • Hayya, Jack C.; Xu, Susan H.; Ramasesh,	95 CommStSM 11 279-309
measurements • Willinger, W.; Taqqu, M. S. et al.	Self-similarity in high-speed packet traffic: Analysis and modeling of ethernet traffic	95 StatSci 10 67-85
death processes • Zeifman, A. I.	Upper and lower bounds on the rate of convergence for nonhomogeneous birth and	95 StocProc 59 157-173
Probability, stochastic processes, and	queueing theory: The mathematics of computer performance modelling • Nelson, R.	95 Springer:Brln:NY 592p
Bremaud, P. Elements of	queueing theory: Palm-martingale calculus and stochastic recurrences • Baccelli, F.;	94 Springer:Brln:NY 256p
Advances in	queuing. Models, methods, and problems • Dshalalow, Jewgeni H. (Ed)	95 CRCPress:FL 384p
et al. Calculating normalization constants of closed	queuing networks by numerically inverting their generating functions • Choudhury, G. L.	95 JACM 42 935-970
Review of "Mathematical methods in	queuing theory" • Kalashnikov, V. V. (Auth); Gerontidis, Ioannis I. (Rev)	95 Ststcian 44 409-410
Some properties of a limiting distribution in	Quicksort • Tan, Kok Hooi; Hadjicostas, Petros	95 StPrLet 25 87-94
Dirichlet analysis for inverse multinomial with both	quota and quota-free cells • Ebneshahrashoob, M.; Sobel, Milton	95 CmpStDA 19 293-307
	Quota-sampling	
Ebneshahrashoob, M.; Sobel, Milton	Dirichlet analysis for inverse multinomial with both quota and quota-free cells •	95 CmpStDA 19 293-307
curves • van Veldhuizen, M. et al. The (least squares)	quotient algorithm as a rapid tool for obtaining sorption isotherms from column elution	95 WaterReR 31 849-857
Maximization of sums of	quotients of quadratic forms and some generalizations • Kiers, Henk A. L.	95 Psymtrka 60 221-245
Lee, T. N.; Lawson, J. S.; Hawks, V. D.	Q^2: A TQM course for undergraduates • Badrigal, J. L.; Richards, D. O.; Free, J. C.;	94 ASAProQlPr 172-180
and labelling all regular fractions or blocks for	q^{n-m}-designs • Laycock, P. J.; Rowley, P. J. A method for generating	95 JRSS-B 57 191-204
underlying the conventional actuarial estimator of	q_x • Puzey, Anthony S. A reappraisal of the principle	95 InsrneME 17 125-132
$\bar{X}$ and	R control charts for skewed populations • Bai, D. S.; Choi, I. S.	95 JQualTek 27 120-131
Jacques Exploring the relationship between	R and D and productivity in French manufacturing firms • Hall, Bronwyn H.; Mariesse,	95 JEconmtx 65 263-293
	***R*-estimator**	
Rashid, M. Mushfiqur	Robust analysis of two-way models with repeated measures on both factors •	95 Test 4 39-62
with repeated measures on both factors based on	R-estimators • Rashid, M. Mushfiqur Analysis of two-factor experiments	94 ASAProQlPr 220-225
A comparison of confidence intervals from	R-estimators in regression • George, K. J.; McKean, J. W.; Schucany, W. R.; Sheather, S. J.	95 JStCmpSm 53 13-22
repeated measures incomplete block designs using	R-estimators • Rashid, M. Mushfiqur Analysis of	95 StPrLet 22 213-221
Differentiability of the matrices	R and G in the matrix-analytic method • He, Qi-Ming	95 CommStSM 11 123-132
J. C.; Houllier, F. et al. Estimating the mean of the	r largest values in a finite population: Application to forest inventory (French) • Pierrat,	95 Biomtrcs 51 679-686
of the fractionally differencing parameter with the	R/S method • Hauser, Michael A.; Reschenhofer, Erhard Estimation	95 CmpStDA 20 569-579
A formula for deficiency: One-sample L- and	R-tests. II • Bening, V. E.	95 MaMethSt 4 274-293
A formula for deficiency: One sample L- and	R-tests. I • Bening, V. E.	95 MaMethSt 4 167-188
for Pearson χ^2 and Zelterman statistics for	r-way contingency tables • Stafford, James E. Exact cumulant calculations	95 JCmpGrSt 4 199-212
Balanced arrays of strength two and nested	(r,λ)-designs • Kuriki, Shinji; Fuji-Hara, Ryoh	94 JComDsgn 2 407-414
Some statistical applications of	(r,s)-directed divergences • Menéndez, M. L.; Pardo, L.; Morales, D.; Salicrú, M.	92 UtilMath 42 115-127
the Itô measure for Brownian motion over the torus	R/Z (French) • Leuridan, Christophe Ray-Knight theorems and	95 Stochatx 53 109-128
A cautionary tale about the relationships among	race, genes, and IQ • Devlin, B.; Fienberg, S. E.; Resnick, D. P. et al. Wringing the bell curve:	95 Chance 8/3 27-36
Probability models on horse	race outcomes • Ali, Mukhtar M.	94 ASAProSpts 7-11
P. P. A.; Pastor, M. S. S. et al. Robustness analysis of	radial base function and multi-layered feed-forward neural network models • Derks, E.	95 Chemolab 28 49-60
mation of chaotic time series using self-generating	radial basis function network • Katayama, R.; Kuwata, K. et al. Embedding dimension esti-	95 FuzzySet 71 311-327
	Radial part of Brownian motion on a Riemannian manifold • Liao, M.; Zheng, W. A.	95 AnlsProb 23 173-177
The	radial part of a γ-martingale and a non-implosion theorem • Kendall, Wilfrid S.	95 AnlsProb 23 479-500
Itô correction terms for the	radial parts of semimartingales on manifolds • Le, Huiling; Barden, Dennis	95 ProbTher 101 133-146
Green, M.; Draper, G. et al. Spatial variation of natural	radiation and childhood leukaemia incidence in Great Britain • Richardson, S.; Monfort, C.;	95 StatMed 14 2487-2501
of missing values in spatio-temporal solar	radiation data • Glasbey, C. A. Imputation	95 EnvrMtrc 6 363-371
optimal experimental design for gene mapping by	radiation hybrids • Lange, K.; Boehnke, M. Bayesian methods and	92 AnHuGene 56 119-144
binomial effects in chromosome aberrations after	radiation injury • Komornik, J.; Mikulecky, M.; Beno, M. Additivity test for composition of	95 StatMed 14 1941-1947
Burgess, B. B. et al. Prospects for measuring the child	radio audience in the U.S. using the seven-day personal diary method • Patchen, R. H.;	94 ASASrvyRsr 1220-1225
Central limit theorem for time to broadcast in	radio networks • Ravishankar, Krishnamurthi; Singh, Suresh	95 PrEngInf 9 201-209
A capture-recapture survival analysis model for	radio-tagged animals • Pollock, K. H.; Bunck, C. M., Winterstein, S. R.; Chen, C.-L.	95 JAppStat 22 661-672
Modelling	radio-tracking data • Worton, Bruce J.	95 EnvEcoSt 2 15-23
Comment on "Summarizing a set of	radiocarbon determinations: A robust approach" (V43 p489-503) • Wilson, S. R.	95 ApplStat 44 407-407
Reply to comment on "Summarizing a set of	radiocarbon determinations: A robust approach" • Christen, J. A.	95 ApplStat 44 408-408
fidence interval for the unknown concentration in	radioimmunoassay (Disc: p953-953) • Rosen, O.; Cohen, A. Constructing a bootstrap con-	95 StatMed 14 935-952
Modeling dose and local control in	radiotherapy • Chappell, Rick; Nondahl, David M.; Fowler, John F.	95 JASA 90 829-838
Asymptotic properties of kernel estimators of the	Radon-Nikodym derivative with applications to discriminant analysis • Gijbels, I. et al.	95 StSinica 5 261-278
Estimation of the effect of operational seeding on	rain amounts in Israel • Nirel, Ronit; Rosenfeld, Daniel	95 JApMetrl 34 2220-2229
Short and medium term	rain forecasting in the Oristano basin (Italian) • Espa, Giuseppe; Terpessi, Cristina	94 QStMaA 16 145-160
Goyret, Jorge Principal component analysis of	rain gauge data: Detecting outliers (Spanish) • López, Carlos; González, Elizabeth;	94 Estd 46 25-54
F.; Curbelo, Rosario Principal component analysis of	rain gauge data: Treating missing observations (Spanish) • López, Carlos; González, Juan	94 Estd 46 55-83
Crosson, W. L. Estimation of daily area-average	rainfall during the CaPE experiment in Central Florida • Duchon, C. E.; Renkevens, T. M.;	95 JApMetrl 34 2704-2714
	Rainfall discrimination and spatial variation using breakpoint data • Sansom, John	95 JClimate 8 624-636
Robert R. Exploratory analysis of seeding effects on	rainfall: Illinois 1989 • Changnon, Stanley A.; Gabriel, K. Ruben; Westcott, Nancy E.; Czys,	95 JApMetrl 34 1215-1224
Regionalization of Indian monsoon	rainfall and long-term variability signals • Iyengar, R. N.; Basak, P.	94 JClimlgy 14 1095-1114
State space identification of linear deterministic	rainfall-runoff models • Ramos, José; Mallants, Dirk; Feyen, Jan	95 WaterReR 31 1519-1531
Artificial neural network modeling of the	rainfall-runoff process • Hsu, Kuo-lin; Gupta, Hoshin Vijai; Sorooshian, Soroosh	95 WaterReR 31 2517-2530
sand martins and their relationship with Sahel	rainfall • Szép, Tibor Survival rates of Hungarian	95 JAppStat 22 891-904
agreement • Agresti, A.; Ghosh, A.; Bini, M.	Raking kappa: Describing potential impact of marginal distributions on measures of	95 BiomtrcJ 37 811-820
Tremblay, Johanne The use of generalized	raking procedures to improve the quality of small domain estimation • Dubreuil, Guylaine;	94 ASASrvyRsr 293-298
Patak, Z.; Hidiroglou, M. A.	Raking ratio estimation: An application to the Canadian Monthly Retail Trade Survey •	94 ASASrvyRsr 299-304

An L_2 convergence theorem for random affine mappings • Burton, Robert M.; Rösler, Uwe 95 JAppProb 32 183-192
Random arrays and functionals with multivariate rotational symmetries • Kallenberg, O. 95 ProbTher 103 91-141
the sensitivity against violation of the missing at random assumption • Vach et al. ... observed categorical covariates – Investigating 95 StatMed 14 1315-1329

Random balance design

Generating Systematic supersaturated designs • Lin, Dennis K. J. 95 Technmcs 37 213-225
Optimal regression designs under random block-effects models • Cheng, Ching-Shui 95 StSinica 5 485-497
exponential distribution with uniformly distributed random censoring • Yeh, Lam; Choy, S. T. B. Bayesian variable sampling plans for the 95 JStPlInf 47 277-293
estimates of a regression function under random censoring • Carbonez, A.; Györfi, L.; van der Meulen, E. C. Partitioning- 95 StatDecn 13 21-37
intervals for the length of the shortt under random censoring • Beirlant, J.; Einmahl, J. H. J. Asymptotic confidence 95 StNeerla 49 1-8
of the Wilcoxon test and the logrank test under random censoring • Sukhatme, Shashikala; Lin, C. H. Exact powers 95 JStPlInf 48 291-301
Tail processes under heavy random censorship applications • Einmahl, John H. J.; Ruymgaart, Frits H. 95 JNnparSt 5 223-235
resentation of the kernel quantile estimator under random censorship • Xiang, Xiaojing Bahadur rep- 95 JMultiAn 54 193-209
the quantiles in the proportional hazards model of random censorship • Ghorai, J. K. Large sample confidence interval for 95 JStCmpSm 53 91-102
The central limit theorem under random censorship • Stute, Winfried 95 AnlsStat 23 422-439
On statistical inference for random closed sets in compact subsets of R^d • Ferrándiz, J.; Montes, F. 95 AdvAppPr 27 344-349
Bivariate random closed sets and nerve fibre degeneration • Ayala, Guillermo; Simó, Amelia 95 AdvAppPr 27 293-305
Random-cluster measures and uniform spanning trees • Häggström, Olle 95 StocProc 59 267-275
of forecasts for ARMA models between a random coefficient approach and a Bayesian approach • Liu, Shu-Ing Comparison 95 CommStA 24 319-333
A note on the strong mixing property for a random coefficient autoregressive process • Lee, Sangyeol 95 JKoreaSt 24 243-248
Review of "Random coefficient methods" • Longford, Nicholas T. (Auth); Gallo, Paul P. (Rev) 95 Technmcs 37 120-120
On testing for equicorrelation in random coefficient models • Bhatti, M. I.; Barry, A. M. 95 BiomtrcJ 37 383-391
regulation and the diffusion of innovations: A random coefficient model • Caudill, S. B.; Ford, J. M.; Kaserman, D. L. Certificate-of-need 95 JApEcmtx 10 73-78
Estimation of nonlinear random coefficient models • Ramos, Rogelio Q.; Pantula, Sastry G. 95 StPrLet 24 49-56
Review of "Random coefficient models" • Longford, N. T. (Auth); Rodriguez-Poo, Juan M. (Rev) 95 JAppStat 22 544-545
Review of "Random coefficient models" • Longford, Nicholas T. (Auth); Albert, Paul S. (Rev) 95 JASA 90 801-802
Longitudinal data analysis examples with random coefficient models (Disc: p201-227, 234-238) • Rogosa, David; Saner, Hilary 95 JEdStat 20 149-170

Random coefficient regression

Calculation of unconditional and conditional reference intervals for foetal size and growth from longitudinal measurements • Royston 95 StatMed 14 1417-1436
The effect of different forms of centering in hierarchical linear models • Kreft, Ita G. G.; de Leeuw, Jan; Aiken, Leona S. 95 MultBehR 30 1-21
Longitudinal data analysis for linear Gaussian models with random disturbed-highest-derivative-polynomial subject ... • Wilson 95 StatMed 14 1219-1233
Modeling the distribution of price sensitivity and implications for optimal retail pricing • Kim, B.-D.; Blattberg, R. C.; Rossi, P. E. 95 JBES 13 291-303
Overdispersion diagnostics for generalized linear models • Lambert, Diane; Roeder, Kathryn 95 JASA 90 1225-1236
Minimum distance estimation in random coefficient regression models • Beran, R.; Millar, P. W. 94 AnlsStat 22 1976-1992
Hannu; Krause, Ulf; Rantakallio, Paula Application of random coefficient regression model to myopia data: A case study • Möttönen, Jyrki; Oja, 95 BiomtrcJ 37 657-672
Chen, Ming-Hui Bayesian analysis for random coefficient regression models using noninformative priors • Yang, Ruoyong; 95 JMultiAn 55 283-311
Prediction in random coefficient regression • Beran, Rudolf 95 JStPlInf 43 205-213
An elementary renewal theorem for random compact convex sets • Molchanov, Ilya S.; Omey, Edward; Kozarovitzky, Eugene 95 AdvAppPr 27 931-942
Bernadac, Evelyne Random continued fractions and inverse Gaussian distribution on a symmetric cone • 95 JTherPrb 8 221-258
Seshadri, Vanamamalai A random continued fraction in R^{d+1} with an inverse Gaussian distribution • Letac, Gérard; 95 Bernoull 1 381-393
bility content and the interior point proportion of a random convex hull • Massé, B. Invariance principle for the deviation between the proba- 95 JAppProb 32 1041-1047
global minima in optimization problems by a random-cost approach • Berg, Bernd A. Locating 93 Nature 361 708-710
Regression models for repeated medical random counts • Mackenzie, Gilbert 95 Ststcian 44 365-372
approach for predicting fatigue reliability under random cyclic loading • Tang, Jun; Zhao, Jie A practical 95 RelEngSS 50 7-15
fre, E. (Rev) Review of "Analysis of systematic and random differences between paired ordinal categorical data" • Svensson, E. (Auth); Lesaf- 95 Biomtrcs 51 794-794
sey, J. K. (Rev) Review of "Analysis of systematic and random differences between paired ordinal categorical data" • Svensson, E. (Auth); Lind- 95 StatMed 14 1726-1727
proximity and demographics ... • Sakkinen et al. Random-digit dialing for control selection in childhood cancer studies: The geographic 95 AJPubHlh 85 555-557
Dynamics of 'list-assisted' random digit dialing (RDD) frame coverage • Kulp, Dale W. 94 ASASrvyRsr 11-18
Akin, Don The efficiency of list-assisted random digit dialing sampling schemes for single and dual frame surveys • Biemer, Paul; 94 ASASrvyRsr 1-10
for the National Health Interview Survey using a random digit dialing supplement • Malec, Donald; Parsons, Van L. State estimates 94 ASASrvyRsr 1190-1195
Chen, Ming-Hui; Schmeiser, Bruce W. Random-direction interior-point Markov chains: A family of black-box samplers • 94 ASAProBays 1-6
tudinal data analysis for linear Gaussian models with random disturbed-highest-derivative-polynomial subject effects • Wilson, P. David Longi- 95 StatMed 14 1219-1233
Markov processes and reflecting diffusions in random domains • Osada, H.; Saitoh, T. An invariance principle for non-symmetric 95 ProbTher 101 45-63
LR test for random-effects covariance structure in a parallel profile model • Yokoyama, Takahisa 95 AnInStMa 47 309-320
Joint estimation of location, dispersion, and random effects • Wolfinger, Russell D.; Tobias, Randall D. 94 ASAProPEng 225-234
An approximate generalized linear model with random effects for informative missing data • Follmann, Dean; Wu, Margaret 95 Biomtrcs 51 151-168
David J.; Thomas, Andrew Bayesian approaches to random-effects meta-analysis: A comparative study • Smith, Teresa C.; Spiegelhalter, 95 StatMed 14 2685-2699

Random effects model

Accuracy and rationality of state General Fund Revenue forecasts: Evidence from panel data • Mocan, H. Naci; Azad, Sam 95 IntJFor 11 417-427
Adaptive estimation in the panel data error component model with heteroskedasticity of unknown form • Li, Qi; Stengos, Thanasis 94 IntEconR 35 981-1000
Aggregate data studies of disease risk factors • Prentice, Ross L.; Sheppard, Lianne 95 Biomtrka 82 113-125
Analysis of censored correlated observations • Oskrochi, G. Reza 95 JStPlInf 47 165-180
ANOVA, MINQUE, PSD-MINQMBE, CANOVA and CMINQUE in estimating variance components • Shaban, Abdel Magid M. 94 Statstca 54 481-489
An application of Bradley-Terry-type models to the measurement of pain • Matthews, J. N. S.; Morris, K. P. 95 ApplStat 44 243-255
An assessment of estimation procedures for multilevel models with binary responses • Rodríguez, Germán; Goldman, Noreen 95 JRSS-A 158 73-89
Asymptotic theory for the frailty model • Murphy, S. A. 95 AnlsStat 23 182-198
Bayesian analysis of a protocol for sampling preserved tumor specimens • Heitjan, Daniel F. 95 JASA 90 38-44
Confidence intervals on ratios of linear combinations for non-disjoint sets of expected mean squares • Gui, R.; Graybill, F. A. et al. 95 JStPlInf 48 215-227
Efficient parametrisations for normal linear mixed models • Gelfand, Alan E.; Sahu, Sujit K.; Carlin, Bradley P. 95 Biomtrka 82 479-488
EM for longitudinal data from multiple studies • Fellingham, Gilbert W.; Wise, Morgan D. 95 StSinica 5 97-107
Hierarchical models for educational data: An overview • Morris, Carl N. 95 JEdStat 20 190-200
The impact of method choice on meta-analysis • Mengersen, K. L.; Tweedie, R. L.; Biggerstaff, B. 95 AstrlJSt 37 19-44
A jackknife method for estimation of variance components • Lavergne, Christian 95 MaOpfStS 27 1-13
Laplace approximation of high dimensional integrals • Shun, Zhenming; McCullagh, Peter 95 JRSS-B 57 749-760
Meta-analysis software: A comparative review • Normand, Sharon-Lise T. 95 AmerStat 49 298-309
Multivariate survival models for repeated and correlated events • Crouchley, Robert; Pickles, Andrew R. 95 JStPlInf 47 95-110
Non-linear mixed regression models • Burnett, Richard T.; Ross, W. H.; Krewski, Daniel 95 EnvrMtrc 6 85-99
Partial likelihood analysis of within-unit variances in repeated measurement experiments • Chinchilli, V. M.; Esinhart, J. D.; Miller, W. G. 95 Biomtrcs 51 205-216
Reassessing brand loyalty, price sensitivity, and merchandising effects on consumer brand choice • Allenby, Greg M.; Lenk, Peter J. 95 JBES 13 281-289
Robust estimation of mean squared error of small area estimators • Lahiri, P.; Rao, J. N. K. 95 JASA 90 758-766
Smoothing splines for longitudinal data • Anderson, Stewart J.; Jones, Richard H. 95 StatMed 14 1235-1248
Some remarks on noninformative priors • Datta, Gauri Sankar; Ghosh, Malay 95 JASA 90 1357-1363
Some specification tests for probit models estimated on panel data • Lechner, Michael 95 JBES 13 475-488
Tests of homogeneity for generalized linear models • Jacqmin-Gadda, Hélène; Commenges, Daniel 95 JASA 90 1237-1246
Utilizing concentration-response data from individual components to detect statistically significant departures from ... • Gennings et al. 95 Biomtrcs 51 1264-1277

Random graph

de la Peña, Victor H.; Montgomery-Smith, S. J.	Decoupling inequalities for the tail probabilities of multivariate U-statistics •	95 AnlsProb 23 806-816
	On the enumeration of rectangular $(0,1)$-matrices • Hendrickson, John A., Jr	95 JStCmpSm 51 291-313
Norbert (Rev) Review of	"Random graph directed self-similar multifractals" • Olsen, Lars (Auth); Patzschke,	95 BiomtrcJ 37 374-374
Stationary straight-line representations of stationary	random graphs • Burton, Robert M.; Choi, Kyeong; Meester, Ronald	95 IndgMatN 6 267-277
	Random high-dimensional orders • Bollobás, Béla; Brightwell, Graham	95 AdvAppPr 27 161-184
cooperation in the prisoner's dilemma repeated with	random horizon • Manuel, C. M.; Tejada, J. Reciprocal	95 Top 3 97-116
A central limit theorem with	random indices for martingale difference sequences • Hong, Dug Hun	95 JKoreaSt 24 219-224
analysis for the numerical calculation of certain	random integrals: Part 1 • Pitt, Loren D.; Robeva, Raina; Wang, Dao Yi An error	95 AnlsApPr 5 171-197
	Random integration rules for statistical computation • Genz, Alan; Monahan, John	94 ASAProStCp 182-187
Packing	random intervals • Coffman, E. G., Jr; Poonen, Bjorn; Winkler, Peter	95 ProbTher 102 105-121
	Random invariant cubature formulas (Russian) • Ermakov, S. M.; Wang, Xiao-Qun	94 VLenUM /4 109-112
Crossing velocities and	random lattice animals • Sznitman, Alain-Sol	95 AnlsProb 23 1006-1023
Allan R. Multiple population models for multivariate	random length data - With applications in clinical trials • Barnhart, Huiman X.; Sampson,	95 Biomtrcs 51 195-204
A	random linear functional approach to efficiency bounds • Holly, Alberto	95 JEconmtx 65 235-261
acterization of multivariate stable distributions via	random linear statistics • Zeng, Wei-Bin On char-	95 JTherPrb 8 1-15
for intraclass correlation in the presence of	random loss of sampled patients • Lui, Kung-Jong Sample size calculation	95 BiomtrcJ 37 97-103
Stochastic analysis and	random maps in Hilbert space • Dorogovtsev, A. A.	94 VSP:Neth 110p
Optimal choice of production capacity in a	random market • Arntzen, Halvard	95 Stochatx 55 87-120
Speed, T. P.; Thomson, Glenys Tests of	random mating for a highly polymorphic locus: Application to HLA data • Jin, Kun;	95 Biomtrcs 51 1064-1076
On a shared allele test of	random mating • Zhou, S.; Maller, R. A.; Speed, T. P.	95 AstrlJSt 37 61-72

Random matrix

Rösler, Uwe	An L_2 convergence theorem for random affine mappings • Burton, Robert M.;	95 JAppProb 32 183-192
	How many zeros of a random polynomial are real? • Edelman, Alan; Kostlan, Eric	95 BAmMaSoc 32 1-37
Yao, David D.	Stochastic vector difference equations with stationary coefficients • Glasserman, Paul;	95 JAppProb 32 851-866
of eigenvalues of a class of large dimensional	random matrices • Silverstein, Jack W.; Bai, Z. D. On the empirical distribution	95 JMultiAn 54 175-192
limiting spectral distribution of large dimensional	random matrices • Silverstein, Jack W.; Choi, Sang-Il Analysis of the	95 JMultiAn 54 295-309
distribution of eigenvalues of large dimensional	random matrices • Silverstein, Jack W. Strong convergence of the empirical	95 JMultiAn 55 331-339
Tightness of products of I.I.D.	random matrices. II • Mukherjea, Arunava	94 AnlsProb 22 2223-2234
Convolutions of	random measures on compact groups • Rubshtein, Ben-Zion	95 JTherPrb 8 523-538
Perturbation analysis of functionals of	random measures • Baccelli, François; Klein, Maurice; Zuyev, Sergei	95 AdvAppPr 27 306-325
stochastic differential equations driven by Poisson	random measures • Kallianpur, G. et al. Diffusion approximation of nuclear space-valued	95 AnlsApPr 5 493-517
smoothing program for quantitation and isolation of	random measurement error • Bradley, D. C.; Steil, G. M.; Bergman, R. N. OOPSEG: A data	95 CmpBiomd 46 67-77
Evolution of systems in	random media • Korolyuk, Vladimir S.; Swishchuk, Anatoly V.	95 CRCPress:FL 368p
Estimation of reduced Palm distributions by	random methods for Cox processes with unknown probability law • Crétois, E.	95 ZastApMa 23 247-259
Effect of non-	random missing data mechanisms in clinical trials • Choi, S. C.; Lu, I. L.	95 StatMed 14 2675-2684
variance following a preliminary test in one-way	random model • Chiou, Paul; Han, Chien-Pai Interval estimation of error	95 CommStB 24 817-824
the variance components of an unbalanced one-way	random model • Belzile, Éric; Angers, Jean-François Choice of noninformative priors for	95 CommStA 24 1325-1341
Variance of number of lattice points in	random narrow elliptic strip • Bleher, Pavel M.; Lebowitz, Joel L.	95 AIHenrPB 31 27-58
crystals and comparative analysis of models for	random nucleation • Møller, Jesper Generation of Johnson-Mehl	95 AdvAppPr 27 367-383
SriRam, M. G.; Teoh, Nick K. W. Entropy-based	random number evaluation • Dudewicz, Edward J.; van der Meulen, Edward C.;	95 AJMMgSci 15 115-153

Random number generator

Edward C.; SriRam, M. G.; Teoh, Nick K. W.	Entropy-based random number evaluation • Dudewicz, Edward J.; van der Meulen,	95 AJMMgSci 15 115-153
Karlheinz Two tests of pseudo	random number generators for independence and uniform distribution • Fleischer,	95 JStCmpSm 52 311-322
Several extensively tested	random number generators • Kao, Chiang; Wong, J. Y.	95 CmpOpRsh 22 1035-1039
On the limit behavior of products of a	random number of group valued random variables • Hazod, W.	94 ThProbAp 39 249-263
Extremes of a	random number of variables from periodic sequences • Ferreira, Helena	95 JStPlInf 45 133-141
The adjoint and the composition of	random operators on a Hilbert space • Thang, Dang Hung	95 Stochatx 54 53-73
Exchangeable and partially exchangeable	random partitions • Pitman, Jim	95 ProbTher 102 145-158
Yamato, Hajime Ordered and unordered	random partitions of an integer and the GEM distribution • Sibuya, Masaaki;	95 StPrLet 25 177-183
	Random permanents • Borovskikh, Yu. V.; Korolyuk, V. S.	94 VSP:Neth 194p
distributions of the number of successions in a	random permutation • Fu, James C. Exact and limiting	95 AnInStMa 47 435-446
to "Limit theorems on the exit problems for small	random perturbations of dynamical systems I" (V46 p79-106) • Mikami, T. Correction	95 Stochatx 52 323-323
Freidlin, M. I. Fast oscillating	random perturbations of dynamical systems with conservation laws • Borodin, A. N.;	95 AIHenrPB 31 485-525
model • Pierre Loti Viaud, Daniel	Random perturbations of recursive sequences with an application to an epidemic	95 JAppProb 32 559-578
G. Kendall concerning the shapes of certain large	random polygons • Miles, R. E. A heuristic proof of a long-standing conjecture of D.	95 AdvAppPr 27 397-417
Continual analogues of	random polynomials which are orthogonal on a circle • Teplyaev, A. V.	94 ThProbAp 39 476-489
How many zeros of a	random polynomial are real? • Edelman, Alan; Kostlan, Eric	95 BAmMaSoc 32 1-37
Limit theorems for diffusions with a	random potential • Mathieu, Pierre	95 StocProc 60 103-111
Almost sure behaviour of evolution with	random potential • Noble, John M.	95 StocAnAp 13 323-334
A method for the generation of correlated	random processes • Andronico, Maurizio; Casale, Salvatore; La Corte, Aurelio	95 JCmpInT 3 9-18
Introduction to	random processes in engineering • Balakrishnan, A. V.	95 Wiley:NY:UK 402p
M. S. (Auth) et al. Review of "Stable non-Gaussian	random processes: Stochastic models with infinite variance" • Samorodnitsky, G.; Taqqu,	95 JASA 90 805-806
Stochastic specification in	random production models of cost-minimizing firms • Brown, Bryan W.; Walker, Mary Beth	95 JEconmtx 66 175-205
Correction to "Functional limit theorems for	random quadratic forms" (V37 p81-98) • Mikosch, T.	95 StocProc 59 353-353
Record values,	random record models and concomitants • Ahsanullah, M.	94 JStatRsr 28 89-109
	Random record processes and state dependent thinning • Browne, Sid; Bunge, John	95 StocProc 55 131-142
sky, Gennady Limit laws for a stochastic process and	random recursion arising in probabilistic modelling • Rachev, Svetlozar T.; Samorodnit-	95 AdvAppPr 27 185-202

Random regression

	Bayesian multiperiod forecasts for ARX models • Liu, Shu-Ing	95 AnInStMa 47 211-224
robustness of point optimal testing for Rosenberg	random regression coefficients • Brooks, Robert The	95 EcomtrRv 14 35-53
Uniformly consistent estimation in	random regression model (Chinese) • An, Hongzhi; Hickernell, F. J.; Zhu, Lixing	95 KToChina 39 1829-1832
	Random rotations of the Wiener path • Üstünel, A. S.; Zakai, M.	95 ProbTher 103 409-429
Bayesian inference for the distribution of a	random sample • Gelfand, Alan E.; Mukhopadhyay, Saurabh On nonparametric	95 CandJSt 23 411-420
A (Bayesian) note on non-	random samples from finite populations • Bayarri, M. J.; Font, B.	94 ASAProBays 246-251
Minimax estimation of proportions under	random sample size • Amrhein, Peter	95 JASA 90 1107-1111
a linear combination of strata means with	random sample sizes • He, Kun On estimating	95 JMultiAn 55 39-60
	Random sampling from databases: A survey • Olken, Frank; Rotem, Doron	95 StatComp 5 25-42
	Random sampling versus representative samples • Schield, Milo	94 ASAProStEd 107-110
Lii, Keh-Shin; Masry, Elias On the selection of	random sampling schemes for the spectral estimation of continuous time processes •	95 JTimSrAn 16 291-311
Kernel estimation of the regression function with	random sampling times • Vilar, J. A.	95 Test 4 137-178
dependent variables with an application to simple	random sampling without replacement" (V30 p197-211) • de la Peña ... of a sum of	95 AIHenrPB 31 703-704
The controlled	random search algorithm in optimizing regression models • Křivý, Ivan; Tvrdík, Josef	95 CmpStDA 20 229-234
The new ARSTI optimization method: Adaptive	random search with translating intervals • Edissonov, Ivan	94 AJMMgSci 14 143-166

on strong limit theorems for (LB)-space-valued	random variables • Wang, Xaingchen; Rao, M. Bhaskara; Li, Deli Some results	95 StPrLet 23 247-251
	Random variables with independent integer and fractional parts • Marsaglia, G.	95 StNeerla 49 133-137
and partial sum convergence for weakly dependent	random variables with infinite variance • Davis, Richard A.; Hsing, Tailen Point process	95 AnlsProb 23 879-917
"fair" games problem for pairwise independent	random variables with multidimensional indices • Hong, D. H.; Hwang, S. Y. Chow-Robbins	95 BIMaASnc 23 7-11
of the logarithm for weighted sums of independent	random variables with multidimensional indices • Li et al. ... large numbers and the law	95 JMultiAn 52 181-198
inequality for the sum of independent Bernoulli	random variables with unequal means • Turner, D. W.; Young, D. M. et al. A Kolmogorov	95 StPrLet 23 243-245
Generating	random variates from D-distributions via substitution sampling • Walker, Stephen	95 StatComp 5 311-315
Bayesian inference • Damien, P. et al. Approximate	random variate generation from infinitely divisible distributions with applications to	95 JRSS-B 57 547-563
convergence • Dellaportas, Petros	Random variate transformations in the Gibbs sampler: Issues of efficiency and	95 StatComp 5 133-140
Pre-Gaussian	random vectors and their applications • Kozachenko, Yu. V.; Mokliatchouk, O. M.	95 ThPrMaSt 50 89-98
as limits of sums for arrays of dependent	random vectors • Kobus, Maria Generalized Poisson distributions	95 JMultiAn 52 199-244
On stochastic ordering of	random vectors • Cohen, Arthur; Sackrowitz, Harold B.	95 JAppProb 32 960-965
et al. Moments of the function of non-normal	random vector with applications to econometric estimators and test statistics • Ullah, A.	95 EcomtrRv 14 459-471
	Random walk	
Zhang, Cun-Hui	On the almost sure minimal growth rate of partial sum maxima • Klass, Michael J.;	94 AnlsProb 22 1857-1878
Park, Heon Jin; Fuller, Wayne A.	Alternative estimators and unit root tests for the autoregressive process •	95 JTimSrAn 16 415-429
chain (French) • Simonot, F.	Approximation of the stationary distribution of a stochastically monotone Markov	95 StocProc 56 133-149
	Convergence rates for Markov chains • Rosenthal, Jeffrey S.	95 SIAMRvw 37 387-405
	Covering with blocks in the non-symmetric case • Móri, Tamás F.	95 JTherPrb 8 139-164
Willinger, Walter	From discrete to continuous stochastic calculus • Cutland, Nigel J.; Kopp, P. Ekkehard;	95 Stochatx 52 173-192
	Eigenvalues and expansion of regular graphs • Kahale, Nabil	95 JACM 42 1091-1106
Bougerol, Philippe; Elie, Laure	Existence of positive harmonic functions on groups and on covering manifolds •	95 AIHenrPB 31 59-80
	The full Martin boundary of the bi-tree • Picardello, Massimo A.; Woess, Wolfgang	94 AnlsProb 22 2203-2222
	Hybrid zones and voter model interfaces • Cox, J. T.; Durrett, R.	95 Bernoull 1 343-370
order statistics for arbitrary sample sizes • Saran, J.	Joint distributions based on the number of runs and some other two-sample rank	95 JStPlInf 47 257-275
(French) • Kipnis, C.; Léonard, C.	Large deviations of an asymmetric hydrodynamical system of independent particles	95 AIHenrPB 31 223-248
	Limit distributions for the Bernoulli meander • Takács, Lajos	95 JAppProb 32 375-395
	On the local limit theorem for non-uniformly ergodic Markov chains • Séva, Marc	95 JAppProb 32 52-62
	The noisy voter model • Granovsky, Boris L.; Madras, Neal	95 StocProc 55 23-43
Michael	A nonlinear parking problem • Keener, Robert; Lerche, Hans Rudolf; Woodroofe,	95 SqtlAnly 14 247-272
Lefèvre, Claude; Utev, Sergey	Poisson approximation for the final state of a generalized epidemic process •	95 AnlsProb 23 1139-1162
Letac, Gérard; Seshadri, Vanamamalai	A random continued fraction in R^{d+1} with an inverse Gaussian distribution •	95 Bernoull 1 381-393
of a (B,S)-market • Kramkov, D. O.; Shiryaev, A. N.	On the rational pricing of the "Russian option" for the symmetrical binomial model	94 ThProbAp 39 153-162
	A rule of thumb (not only) for gamblers • Kozek, Andrzej S.	95 StocProc 55 169-181
	Some distributed algorithms revisited • Louchard, Guy	95 CommStSM 11 563-586
	Spectra of graphs and fractal dimensions II • Telcs, András	95 JTherPrb 8 77-96
	Symmetric two-particle exclusion-eating processes • Liu, Xijian	95 AnlsProb 23 1439-1455
	On two-sided compound binomial distributions • Wang, Shaun	95 InsrneME 17 35-41
Khoshnevisan, Davar	Weight functions and pathwise local central limit theorems • Horváth, Lajos;	95 StocProc 59 105-123
Review of "Aspects and applications of the	random walk" • Weiss, G. H. (Auth); Lawler, Gregory F. (Rev)	95 SIAMRvw 37 470-471
Minimal positions in a branching	random walk • McDiarmid, Colin	95 AnlsApPr 5 128-139
money • Serletis, Apostolos	Random walks, breaking trend functions, and the chaotic structure of the velocity of	95 JBES 13 453-458
Occupation measure for	random walk on the circle • Dinwoodie, I. H.	95 JTherPrb 8 669-677
L_2-lower bounds for a special class of	random walks • Porod, Ursula	95 ProbTher 101 277-289
expansions for the distributions of stopped	random walks and first passage times • Lai, Tze Leung; Wang, Julia Qizhi Asymptotic	94 AnlsProb 22 1957-1992
Return probabilities for	random walk on a half-line • Lalley, Steven P.	95 JTherPrb 8 571-599
	Random walks of infinitely many particles • Révész, P.	94 WrldSciPub:Sing 191p
Review of	"Random walks of infinitely many particles" • Révész, P. (Auth); Heyde, C. C. (Rev)	95 ShrtBkRv 15/3 47-47
Isomorphism of	random walks • Aaronson, Jon; Keane, Michael	94 IsrelJMa 87 37-63
of renewal sequences and isomorphism of	random walks • Aaronson, Jon; Liggett, Thomas; Picco, Pierre Equivalence	94 IsrelJMa 87 65-76
Mulquiney, J. E.; Norton, J. P.; Jakeman, A. J. et al.	Random walks in the Kalman filter: Implications for greenhouse gas flux deductions •	95 EnvrMtrc 6 473-478
A local limit theorem for nonhomogeneous	random walk on a lattice • Zhizhina, E. A.; Minlos, A.	94 ThProbAp 39 490-503
Heavy traffic analysis of a	random walk on a lattice semi-strip • Falin, Gennadi	95 CommStSM 11 395-409
Logarithmic averages for the local times of recurrent	random walks and Levy processes • Marcus, Michael B.; Rosen, Jay	95 StocProc 59 175-184
Large deviations for independent	random walks on the line • Lee, Tzong-Yow	95 AnlsProb 23 1315-1331
On the asymptotic and invariant σ-algebras of	random walks on locally compact groups • Jaworski, Wojciech	95 ProbTher 101 147-171
A note on the maximum of a	random walk • Stadje, W.	95 StPrLet 23 227-231
Mukherjea, Arunava A mixed	random walk on nonnegative matrices: A law of large numbers • Högnäs, Göran;	95 JTherPrb 8 973-990
Space-time boundaries for	random walks obtained from diffuse measures • Handelman, David	94 IsrelJMa 86 107-156
Martingale methods for	random walks in a one-dimensional random environment • Butov, A. A.	94 ThProbAp 39 558-572
sample critical values for unit root tests using pure	random walk processes: A note • Cheung, Yin-Wong; Lai, Kon S. Estimating finite	95 JTimSrAn 16 493-498
Asymont, I. M.; Fayolle, G.; Menshikov, M. V.	Random walks in a quarter plane with zero drifts: Transience and recurrence •	95 JAppProb 32 941-955
Critical	random walk in random environment on trees • Pemantle, Robin; Peres, Yuval	95 AnlsProb 23 105-140
	Random walks and random environments. Volume 1: Random walks • Hughes, Barry D.	95 OxfordUniv 654p
D. (Auth); Révész, P. (Rev) Review of	"Random walks and random environments. Volume 1: Random walks" • Hughes, Barry	95 ShrtBkRv 15/3 47-47
measures on semigroups: Convolution products,	random walks, and random matrices • Högnäs, Göran; Mukherjee, Arunava Probability	95 PlenumPubl:NY 400p
Enlargement of obstacles for the simple	random walk • Antal, Peter	95 AnlsProb 23 1061-1101
On conditioning a	random walk to stay nonnegative • Bertoin, J.; Doney, R. A.	94 AnlsProb 22 2152-2167
Newman, C. M.; Stein, D. L.	Random walk in a strongly inhomogeneous environment and invasion percolation •	95 AIHenrPB 31 249-261
Anthony J. The nearest neighbor	random walk on subspaces of a vector space and rate of convergence • D'Aristotile,	95 JTherPrb 8 321-346
medians, median unbiased means, and symmetric	random walks • Mattner, L. Mean unbiased	95 MaMethSt 4 99-105
Spitzer's condition and ladder variables in	random walks • Doney, R. A.	95 ProbTher 101 577-580
Branching	random walk with a critical branching part • Kesten, Harry	95 JTherPrb 8 921-962
Oscillating	random walk with a moving boundary • Madras, Neal; Tanny, David	94 IsrelJMa 88 333-365
An optimal stopping problem for	random walks with non-zero drift • Roters, Markus	95 JAppProb 32 956-959
particle in equilibrium for asymmetric mean zero	random walk with simple exclusion • Varadhan, S. R. S. Self diffusion of a tagged	95 AIHenrPB 31 273-285
Gerold	Random walks with stochastically bounded increments: Renewal theory • Alsmeyer,	95 MatNach 172 13-31
Kwang Sik; Park, Hee Joo On a stopping rule for the	random walks with time stationary random distribution function • Hong, Dug Hun; Oh,	95 JKoreaSt 24 293-301
On the range of reversible	random walks on Z^2 in a random environment • Zhou, Xian Yin	95 JTherPrb 8 453-473
Zhongguo Edgeworth expansion and	random weighted approximation in correlated model (Chinese) • Shi, Jian; Zheng,	95 KToChina 39 1925-1927
linear series (Chinese) • Fan, J.; Wang, N.; Mei, C. The	random weighting approximation for the distribution of the partial sum of a stationary	95 AMaAPknC 17 65-75
mean-multi dimensional case (Chinese) • Shi, Jian	Random weighting approximation for the distribution of standardized sample	95 ChJApPSt 10 233-242
Multiparametric M-estimation and the	random weighting method • Shi, Jian	95 ActScNat 30 509-518
Lot sizes in serial manufacturing with	random yields • Sobel, Matthew J.	95 PrEngInf 9 151-157

hydrodynamical limit of mean zero asymmetric zero	range processes • Benois, O.; Kipnis, C.; Landim, C. Large deviations from the	95 StocProc 55 65-89
comparisons of k treatments based on the	range statistic • Liu, Wei A group sequential procedure for all-pairwise	95 Biomtrcs 51 946-955
Sugata; Mehrotra, Rajiv	Range surface characterization and segmentation using neural networks • Ghosal,	95 PattrnRc 28 711-727
Evolutionary	rank analysis applied to multidetectional chromatographic structures • Toft, Jostein	95 Chemolab 29 189-212
ponent profiles by evolving projections and internal	rank annihilation • Grung et al. ... of embedded minor analytes in three-way multicom-	95 Chemolab 29 213-221
A goodness-of-fit test based on	ranks for ARMA models • Ferretti, Nélida; Kelmansky, Diana; Yohai, Victor	95 CommStA 24 295-318
Rashid, M. Mushfiqur; Nandram, Balgobin A	rank-based analysis of one-way repeated measures designs with a changing covariate •	95 JStPlInf 46 249-263
Rashid, M. Mushfiqur; Nandram, Balgobin	Rank-based predictive inference for the finite population mean of a small area •	94 ASASrvyRsr 1059-1064
and comparison of different procedures based on	ranks • Hühn et al. ... analysis of cultivar performance trials: Experimental results	95 AgronmyJ 87 627-632
McKean, Joseph W. Influence functions for	rank-based procedures in the linear model • Witt, Lee D.; Naranjo, Joshua D.;	95 JNnparSt 5 339-358
Jürgen H. P. Effects of	rank versus category in measuring subjective social inequality • Hoffmeyer-Zlotnik,	94 ASASrvyRsr 1322-1326
method for grouping environments without cultivar	rank change • Crossa, J.; Cornelius, P. L. et al. A shifted multiplicative model fusion	95 CropSci 35 54-62
Clustering cultivars into groups without	rank-change interactions • Cornelius, P. L.; Van Sanford, D. A.; Seyedsadr, M. S.	93 CropSci 33 1193-1200
Morimune, Kimio; Mantani, Akihisa Estimating the	rank of co-integration when the order of a vector autoregression is unknown •	95 MaCmpSim 39 265-271
Distribution-free test based on	ranks for comparing groups with umbrella orderings • Pan, Guohua; Wolfe, Douglas A.	95 JNnparSt 5 409-416
	Rank correlation	
	Hadamard differentiability on $D[0,1]^p$ • Ren, Jian-Jian; Sen, Pranab Kumar	95 JMultiAn 55 14-28
simulations • Piepho, H.-P.	Implication of correlations among some common stability statistics - A Monte Carlo	95 ThApGene 90 457-461
Cabilio, Paul	Testing ordered alternatives in the presence of incomplete data • Alvo, Mayer;	95 JASA 90 1015-1024
	Rank correlation methods for missing data • Alvo, Mayer; Cabilio, Paul	95 CandJSt 23 345-358
tifactorial model of disease transmission, why is the	rank correlation sensitive to choice of bivariate distribution? • Hutchinson, T. P. In the mul-	92 AnHuGene 56 159-161
Who discovered Spearman's	rank correlation? • Lovie, A. D.	95 BJMaStPs 48 255-269
Analyzing and modeling	rank data • Marden, John I.	95 ChapmnHall:Lnd 330p
Nonparametric	rank discrimination method • Ćwik, Jan; Mielniczuk, Jan	95 CmpStDA 19 59-74
Ahn, Sung K. Reduced	rank estimation of vector autoregressive models with cointegration and codependence •	94 ASAProBuEc 151-156
	Rank estimator	
Rashid, M. Mushfiqur	Analysis of repeated measures incomplete block designs using R-estimators •	95 StPrLet 22 213-221
	A robust partial correlation measure • Rao, Sudhakar H.; Sievers, Gerald L.	95 JNnparSt 5 1-20
A note on chi-squared approximations for tests of	rank interaction with ordered categorical data • Thomas, G. E.	95 CommStB 24 439-459
	Rank mapping of three-way multicomponent profiles • Grung, Bjørn; Kvalheim, Olav M.	95 Chemolab 29 223-232
Multivariate spatial sign and	rank methods • Möttönen, Jyrki; Oja, Hannu	95 JNnparSt 5 201-213
Lazraq, A.; Lepage, Y. Vector correlation based on	ranks and a nonparametric test of no association between vectors • Cléroux, R.;	95 CommStA 24 713-733
for ultrametric tree estimation from conditional	rank order proximity data • Young, Martin R.; DeSarbo, Wayne S. A parametric procedure	95 Psymtrka 60 47-75
on the number of runs and some other two-sample	rank order statistics for arbitrary sample sizes • Saran, Jagdish Joint distributions based	95 JStPlInf 47 257-275
between genotype × environment interactions and	rank orders for a set of genotypes tested in different ... • Hühn et al. Relationships	93 ThApGene 86 943-950
calibration methods for data with and without	rank overlap • Kiers, Henk A. L.; Smilde, Age K. Some theoretical results on second-order	95 JChemtrc 9 179-195
On group inverses of matrices with order and	rank perturbations • Rao, P. S. S. N. V. P.	94 CalcutSt 44 209-222
The	rank of the present excursion • Scheffer, Carel L.	95 StocProc 55 101-118
	Rank regression for current status data • Aragón, Jorge; Quiróz, Adolfo J.	95 StPrLet 24 251-256
Some small-sample results on a bounded influence	rank regression method • Wilcox, Rand R.	95 CommStA 24 881-888
Schmidli, H. Reduced	rank regression: With applications to quantitative structure-activity relationships •	95 Springer:Brln:NY 179p
Autoregression quantiles and related	rank-scores processes • Koul, Hira L.; Saleh, A. K. Md. E.	95 AnlsStat 23 670-689
estimation in a simple linear regression using	rank set sampling • Muttlak, Hassen A. Parameters	95 BiomtrcJ 37 799-810
test for conditional heteroscedasticity, using	ranks of squared residuals • Wong, H.; Li, W. K. Portmanteau	95 JAppStat 22 121-134
	Rank statistic	
Lee, Robert Edwin; Spurrier, John D.	Distribution-free multiple comparisons between successive treatments •	95 JNnparSt 5 261-273
Kim, Soon-ock; Song, Moon Sup	Distribution-free tests for symmetry based on the linear signed rank statistics •	95 JNnparSt 5 311-321
	A formula for deficiency: One sample L- and R-tests. I • Bening, V. E.	95 MaMethSt 4 167-188
	A formula for deficiency: One-sample L- and R-tests. II • Bening, V. E.	95 MaMethSt 4 274-293
families of distributions • Genest, C. et al.	A semiparametric estimation procedure of dependence parameters in multivariate	95 Biomtrka 82 543-552
	Weighted bootstrapping for U-quantiles • Aerts, Marc; Janssen, Paul	95 StPrLet 22 317-323
	• see also aligned -, log- -, linear -, signed -	
	• see also names of specific types of rank statistics such as Wilcoxon -	
Estimating location parameters using censored	rank statistics • Öhman, Marie-Louise	95 BiomtrcJ 37 767-782
More bounds for moments of some	rank statistics • Spurrier, John D.	95 CommStA 24 2679-2682
The Wilcoxon	rank-sum test in dichotomous regression analysis • Finkelstein, Mark; Tucker, Howard G.	92 ProIrStC1 243-259
Edgeworth approximations for	rank sum test statistics • Kolassa, John E.	95 StPrLet 24 169-171
On confidence regions induced by the Wilcoxon	rank sum test • Horstmann, Vibeke	95 StPrLet 23 333-338
	Rank test	
O'Gorman, Thomas W.	The effect of unequal variances on the power of several two-sample tests •	95 CommStB 24 853-867
Senchaudhuri, Pralay; Metha, Cyrus R.; Patel, Nitin R.	Estimating exact p values by the method of control variates or Monte Carlo rescue •	95 JASA 90 640-648
multiclinic trials • Brunner, E.; Puri, M. L.; Sun, S.	Nonparametric methods for stratified two-sample designs with application to	95 JASA 90 1004-1014
Johnson, Richard A.; Sim, Songyong	Nonparametric tests for ordered c-sample contaminated distributions •	95 CommStA 24 861-879
Müller, D. W.	Power robustification of approximately linear tests • Ehm, W.; Mammen, E.;	95 JASA 90 1025-1033
Rashid, M. Mushfiqur	Robust analysis of two-way models with repeated measures on both factors •	95 Test 4 39-62
	On some one-sided tests in the two-sample problem • Bajorski, Piotr	95 CommStA 24 75-86
Two-sample	rank tests for acceleration in cure models • Lee, Jae Won	95 StatMed 14 2111-2118
Approximate and exact distributions of	rank tests for balanced incomplete block designs • Alvo, Mayer; Cabilio, Paul	95 CommStA 24 3073-3121
and a randomized complete block ... • Xie et al. The	rank tests for a completely randomized design with multivariate longitudinal responses	94 ASAProBiop 250-255
A class of linear nonparametric	rank tests for dependent censoring • Lee, Seungyeoun; Wolfe, Robert A.	95 CommStA 24 687-712
Linear	rank tests derived from a superpopulation model • Bouza, Carlos N.	95 BiomtrcJ 37 497-506
Mayeux, Richard Linear	rank tests for doubly censored data • Tang, Ming-Xin; Tsai, Wei-Yann; Marder, Karen;	95 StatMed 14 2555-2563
A comparative study of some	rank tests for interaction • Mansouri, H.; Chang, G.-H.	95 CmpStDA 19 85-96
Chen, Guijing A kind of	rank test about intersection point of two continuous populations (Chinese) • Li, Ningning;	95 KToChina 39 1543-1546
Muyot, Maria Emerita T.	Rank tests for main and interaction effects in analysis of variance • Marden, John I.;	95 JASA 90 1388-1398
Review of "Non-standard	rank tests" • Janssen, A.; Mason, D. M. (Auth); Jurečková, Jana (Rev)	95 SIAMRvw 37 261-262
An example of a two-sided Wilcoxon signed	rank test which is not unbiased • Amrhein, Peter	95 AnInStMa 47 167-170
ANOVA and	rank tests when the number of treatments is large • Boos, Dennis D.; Brownie, Cavell	95 StPrLet 23 183-191
On exact distributions of	rank tests for ordered alternatives in block designs • Black, Stephen; Mansouri, H.	95 CmpStDA 20 265-274
Kim, Dong Hee; Lim, Dong Hoon	Rank tests for parallelism of regression lines against umbrella alternatives •	95 JNnparSt 5 289-302
A brief survey of LMP	rank tests for random effects in one and two-way models • Govindarajulu, Z.	95 CommStA 24 59-73
A multivariate Wald-Wolfowitz	rank test against serial dependence • Hallin, Marc; Puri, Madan L.	95 CandJSt 23 55-65
Koenker, R. W. Robust	rank tests of the unit root hypothesis based on regression rank scores • Hasan, M. N.;	94 ASAProBuEc 35-43
A two-stage	rank test using density estimation • Albers, Willem	95 AnInStMa 47 675-691

	Rank tests with weighted data • Blodgett, Robert J.	94 ASAProStCp 200-204
Romá; Smilde, Age; Kowalski, Bruce Selectivity, local	rank, three-way data analysis and ambiguity in multivariate curve resolution • Tauler,	95 JChemtrc 9 31-58
	Rank transformation	
applications to archaeometry • Baxter, M. J.	Standardization and transformation in principal component analysis, with	95 ApplStat 44 513-527
Sobol, Ilya M. About the use of	rank transformation in sensitivity analysis of model output • Saltelli, Andrea;	95 RelEngSS 50 225-239
analysis of type I error and power for F test and	rank transformed F test in 2×2 factorial ANOVA (Korean) • Choi, Young Hun Simulation	95 KorApS 8/2 87-97
of likelihood ratio tests for cointegrating	ranks in vector autoregressions • Toda, Hiro Y. Finite sample performance	95 EconTher 11 1015-1032
	Ranked data	
	A new method for multidimensional unfolding • Hojo, Hiroshi	94 Behvmtrk 21 131-147
Biplot of	ranked data • Han, Sang-Tae; Huh, Myung-Hoe	95 JKoreaSt 24 439-451
confidence bounds on the distance between any two	ranked location parameters of k populations • Liu, Wei On the construction of exact lower	95 CommStA 24 2073-2085
The	ranked-set sample sign test • Hettmansperger, Thomas P.	95 JNnparSt 4 263-270
Weibull and extreme-value distributions using	ranked set sample • Fei, H.; Sinha, B. K.; Wu, Z. Estimation of parameters in two-parameter	94 JStatRsr 28 149-161
Taillie, C.	Ranked set sampling: An annotated bibliography • Kaur, A.; Patil, G. P.; Sinha, A. K.;	95 EnvEcoSt 2 25-54
Finite population corrections for	ranked set sampling • Patil, G. P.; Sinha, A. K.; Taillie, C.	95 AnInStMa 47 621-636
Editorial:	Ranked set sampling • Patil, G. P.	95 EnvEcoSt 2 271-285
Parametric	ranked set sampling • Stokesr, Lynne	95 AnInStMa 47 465-482
Use of	ranked set sampling for test of a normal mean • Shen, Wei-Hsiung	94 CalcutSt 44 183-193
	Ranking	
Tracy, Nola D.; Young, John C.	Decomposition of T^2 for multivariate control chart interpretation • Mason, Robert L.;	95 JQualTek 27 99-108
Stoimenova, Eugenia	On the LFC of the natural classification rule under a new loss function •	95 StatDecn 13 39-51
Mulekar, Madhuri S.	Selecting best negative binomial population using two distance measures •	95 StatDecn 13 53-70
Some further results on	ranking the importance of system components • Meng, Fan C.	95 RelEngSS 47 97-101
A. C. Correction to "Application of	ranking methods to factorial and stratified experiments" (V36 p205-213) • Shirley, Eryl	95 ApplStat 44 135-137
An evaluation of three methods for	ranking paired contestants • Blake, Jonathan E.; Collings, Bruce Jay	94 ASAProSpts 31-35
(Auth) et al. Review of "Multistage selection and	ranking procedures: Second-order asymptotics" • Mukhopadhyay, N.; Solanky, T. K. S.	95 JASA 90 1130-1131
	Ranking and selection	
designs • Alvo, Mayer; Cabilio, Paul	Approximate and exact distributions of rank tests for balanced incomplete block	95 CommStA 24 3073-3121
populations (Korean) • Hwang, Hyung T.	The likelihood ratio test for detecting the best treatment among several exponential	95 KorApS 8/1 151-157
	A random-size subset approach to the selection problem • Liu, W.	95 JStPlInf 48 153-164
computer software • Poston, C. A. et al. Solving the	ranking and selection indifference-zone formulation for normal distributions using	94 ASAProStCp 290-291
(Disc: p134-135) • Peck, R. R.; Madrigal, J. L. et al.	Ranking teams competing in pairwise comparisons using the dynamic linear model	94 ASAProBays 112-116
approximations for Kendall's score with tied	rankings • Valz, P. D.; McLeod, A. I. et al. Cumulant generating function and tail probability	95 AnlsStat 23 144-160
Development of an optimal test based on weighted	rankings • Larocque, Denis; Tardif, Serge	95 CommStB 24 367-383
Critchlow, D. E. et al. A symmetric analysis of paired	rankings with application to temporal patterns of hormonal concentration • Huh, M.-H.;	95 Biomtrcs 51 1361-1371
	Rao-Blackwell theorem	
	Complete classes for confidence set estimation • Brown, L. D.; Cohen, Arthur	95 StSinica 5 291-301
	Conditioning in Markov chain Monte Carlo • Geyer, Charles J.	95 JCmpGrSt 4 148-154
	Intrinsic analysis of statistical estimation • Oller, J. M.; Corcuera, J. M.	95 AnlsStat 23 1562-1581
Cruz-Orive, L. M. The	Rao-Blackwell theorem in stereology and some counterexamples • Baddeley, A. J.;	95 AdvAppPr 27 2-19
	Rao-Hartley-Cochran estimator	
	Improving the RHC-strategy • Gabler, Siegfried; Stenger, Horst	95 StHefte 36 327-336
Invalidity of the	Rao map function for three loci • Weeks, Daniel E.	94 HumHrdty 44 178-180
Shin, Hee-Choon An approximate	Rao-Scott modification factor in two-way tables with only known marginal DEFFS •	94 ASASrvyRsr 600-601
An expanded table of probability values for	Rao's spacing test • Russell, Gerald S.; Levitin, Daniel J.	95 CommStB 24 879-888
on conclusions about effects of human activities on	raptors • Schueck, Linda S.; Marzluff, John M. Influence of weather	95 JWildlMg 59 674-682
Sample size requirements for	rare or abundant attribute sampling • Arnold, Jesse C.; Lemon, G. H.; Norton, J. M.	95 CommStB 24 131-148
of pooled testing in estimating prevalence of a	rare disease: Application to HIV screening • Tu et al. On the informativeness and accuracy	95 Biomtrka 82 287-297
for detecting 'general' and 'focused' clustering of	rare diseases • Tango, Toshiro A class of tests	95 StatMed 14 2323-2334
Review of "Law of small numbers: Extremes and	rare events" • Falk, Michael; Hüsler, Jurg; Reiss, Rolf-Dieter (Auth); Mikosch, Thomas (Rev)	95 JASA 90 1133-1134
The performance of precursor-based estimators for	rare event frequencies • Bier, V. M.; Yi, W.	95 RelEngSS 50 241-251
Fast simulation of	rare events in queueing and reliability models • Heidelberger, Philip	95 ACMMdSim 5 43-85
Limits of first passage times to	rare sets in regenerative processes • Glasserman, Paul; Kou, Shing-Gang	95 AnlsApPr 5 424-445
On the	rare species of a population • Yatracos, Yannis G.	95 JStPlInf 48 321-329
On the asymptotic independence of the sum and	rare values of weakly dependent stationary random variables • Hsing, Tailen	95 StocProc 60 49-63
Second class particles in the	rarefaction fan • Ferrari, P. A.; Kipnis, C.	95 AIHenrPB 31 143-154
updating input-output tables (Arabic) • Kerim et al.	RAS and least squares methods versus other mathematical techniques in developing and	95 EgpStJ 39 120-143
	Rasch model	
graded responses • Andrich, David	Distinctive and incompatible properties of two common classes of IRT models for	95 ApPsychM 19 101-119
	On the enumeration of rectangular $(0,1)$-matrices • Hendrickson, John A., Jr	95 JStCmpSm 51 291-313
responses to similar items in a survey • Agresti, Alan	Logit models and related quasi-symmetric log-linear models for comparing	95 SocMethR 24 68-95
responses (Disc: p27-46) • Andrich, David	Models for measurement, precision, and the nondichotomization of graded	95 Psymtrka 60 7-26
	Psychometric latent response models • Maris, Eric	95 Psymtrka 60 523-547
	Some neglected problems in IRT • Fischer, Gerhard H.	95 Psymtrka 60 459-487
Inconsistent maximum likelihood estimators for the	Rasch model • Ghosh, Malay	95 StPrLet 23 165-170
Molenaar, I. W.	Rasch models: Foundations, recent developments, and applications • Fischer, G. H.;	95 Springer:Brln:NY 436p
	Rasch models for item bundles • Wilson, Mark; Adams, Raymond J.	95 Psymtrka 60 181-198
Residual analysis in the polytomous	Rasch model • Andersen, Erling B.	95 Psymtrka 60 375-393
repeated count data: Some extensions of the	Rasch Poisson counts model • van Duijn, Marijtje A. J.; Jansen, Margo G. H. Modeling	95 JEdStat 20 241-258
	Rate	
	• see also names of specific types of rates such as decay –, failure –, hazard –, or interest –	
tumor types on experimentwise false positive	rate in animal carcinogenicity studies • Rahman ... correlation structure among	95 JBiophSt 5 273-283
M.; Ware, J. H. A simulation study of estimators for	rates of change in longitudinal studies with attrition • Wang-Clow, F.; Lange, M.; Laird, N.	95 StatMed 14 283-297
trials without unblinding when the key response is	rate of change • Shih, W. J. et al. Re-evaluating design specifications of longitudinal clinical	95 StatMed 14 2239-2248
Further applications of a general	rate conservation law • Bardhan, Indrajit	95 StocProc 60 113-130
	Rates of convergence for best entropy estimates • Noll, Dominikus	95 StatDecn 13 141-165
On	rates of convergence for common subsequences and first passage time • Rhee, WanSoo T.	95 AnlsApPr 5 44-48
A bound on the	rate of convergence for the discrete Gibbs sampler • Dinwoodie, I. H.	95 PrEngInf 9 211-215
Hurvich, Clifford M.; Tsai, Chih-Ling Relative	rates of convergence for efficient model selection criteria in linear regression •	95 Biomtrka 82 418-425
Jeffrey S.	Rates of convergence for everywhere-positive Markov chains • Baxter, J. R.; Rosenthal,	95 StPrLet 22 333-338
Optimal	rate of convergence for finite mixture models • Chen, Jiahua	95 AnlsStat 23 221-233
Jeffrey S.	Rates of convergence for Gibbs sampling for variance component models • Rosenthal,	95 AnlsStat 23 740-761
Stone, Charles J.; Truong, Young K.	Rate of convergence for logspline spectral density estimation • Kooperberg, Charles;	95 JTimSrAn 16 389-401

How much does poor	reading lower math scores? • Clarkson, S. P.; Williams, W. H.	94 ASAProStEd 203-208
Kazuo A latent trait model for	reading process of a word (Japanese) • Ueno, Maomi; Yoshimura, Osamu; Shigemasu,	94 JpJBeh 21/1 48-56
	Real estate price prediction under asymmetric loss • Cain, Michael; Janssen, Christian	95 AnInStMa 47 401-414
Estimating the	real parameter in a two-sample proportional odds model • Wu, Colin O.	95 AnlsStat 23 376-395
How many zeros of a random polynomial are	real? • Edelman, Alan; Kostlan, Eric	95 BAmMaSoc 32 1-37
experiment • Aubin, C.; Fougére, D.; Husson, E. et al.	Real-time pricing of electricity for residential customers: Econometric analysis of an	95 JApEcmtx 10 171-191S
of probability laws of a subset of one-dimensional	real variates (Italian) • Freddi, Augusto On the comparison	95 Metron 53 147-161
size (Disc: p134-135) • Ji, X. et al. Approximate	reasoning in belief networks: Statistical validity, accuracy, and determination of sample	94 ASAProBays 128-133
Effects of teaching statistical laws on	reasoning about everyday problems • Kosonen, Peter; Winne, Philip H.	95 JEdPsych 87 33-46
and synthesis using fuzzy sets and evidential	reasoning • Wang, J.; Yang, J. B.; Sen, P. Safety analysis	95 RelEngSS 47 103-118
Uncertain	reasoning and forecasting • Dagum, P.; Galper, A.; Horvitz, E.; Seiver, A.	95 IntJFor 11 73-87
Evidential foundations of probabilistic	reasoning • Schum, David A.	94 Wiley:NY:UK 545p
Some practical improvements in the continual	reassessment method for phase I studies • Goodman, S. N.; Zahurak, M. L.; Piantadosi, S.	95 StatMed 14 1149-1161
in cancer ... • Møller An extension of the continual	reassessment methods using a preliminary up-and-down design in a dose finding study	95 StatMed 14 911-922
method for censored age-of-onset data subject to	recall bias: Mothers' reports of age of puberty in male twins • Pickles, A. et al. A simple	94 BehavGen 24 457-468
Modelling	recall – Clustering and order effects • Robertson, Chris	95 BJMaStPs 48 29-50
Eastman, Elizabeth Respondent strategies for	recall of crime victimization incidents • Czaja, Ronald; Blair, Johnny; Bickart, Barbara;	94 JOffStat 10 257-276
tives in memory: Probability models of brand name	recall • Hutchinson, J. Wesley; Raman, Kalyan; Mantrala, Murali K. Finding choice alterna-	94 JMarRsr 31 441-461
Or estimating population size using the capture/	recapture method • Drummey, Kevin Ward How many dinosaurs are there?	94 ASAProStEd 215-217
Bayesian analysis of a multiple-	recapture model • Garthwaite, Paul H.; Yu, Keming; Hope, Peter B.	95 CommStA 24 2229-2247
estimates obtained from three different methods of	recapture in the same population of the great tit • Bauchau, V. et al. Comparison of survival	95 JAppStat 22 1031-1037
	Receiver operating characteristic	
	Applying fuzzy set approach to signal detection theory • Chen, Toly; Wang, Mao-Jiun J.	95 FuzzySet 72 39-49
ates are unmeasurable for some ... • Szalal et al.	Composite probability modelling of increasing resolution where diagnostic covari-	95 PsychRpt 77 67-78
heterogeneous treatment effect • Hsieh, Fushing	The empirical process approach for semiparametric two-sample models with	95 JRSS-B 57 735-748
Hutson, Alan	Estimating diagnostic test accuracy using a "fuzzy gold standard" • Phelps, Charles E.;	95 MedDscnM 15 44-57
	Evaluating multiple diagnostic tests with partial verification • Baker, Stuart G.	95 Biomtrcs 51 330-337
An ANOVA approach with ... • Obuchowski et al.	Hypothesis testing of diagnostic accuracy for multiple readers and multiple tests:	95 CommStB 24 285-308
model approach • Cox, Christopher	Location-scale cumulative odds models for ordinal data: A generalized non-linear	95 StatMed 14 1191-1203
van Lierde, M.; Theunissen, I.	Modelling mortality and morbidity of newborns • Ritter, C.; Bouckaert, A.;	95 StatMed 14 2191-2204
Su, John Q.	Simultaneous confidence bands for comparing diagnostic markers • Chen, Hanfeng;	95 CommStA 24 375-393
Lindgren, B. R. Construction and comparison of two	receiver operating characteristic curves derived from the same samples • Le, C. T.;	95 BiomtrcJ 37 869-877
Nonlinear model selection for finding	receptor sites • Amaratunga, Dhammika; Ferreira, Irene	94 ASAProBiop 126-129
Daly, M. J. et al. Rapid multipoint linkage analysis of	recessive traits in nuclear families, including homozygosity mapping • Kruglyak, L.;	95 AJHuGene 56 519-527
Manuel, C. M.; Tejada, J.	Reciprocal cooperation in the prisoner's dilemma repeated with random horizon •	95 Top 3 97-116
density within the class of self-adjoint (self-	reciprocal) densities" (V22 p651-654) • Rossberg ... of the standard normal probability	95 MaOpfStS 26 387-387
Designs through	recoding of varietal and level codes • Das, M. N.; Giri, N.; Ahmed, M.	95 StPrLet 23 371-380
Proper and effective training of a pattern	recognizer for cyclic data • Hwarng, H. Brian	95 IIETrn 27 746-756
use in chromosome mapping studies with the BXD	recombinant inbred strains • Belknap Empirical estimates of Bonferroni corrections for	92 BehavGen 22 677-684
study of complex characters • Blizard, David A.	Recombinant-inbred strains: General methodological considerations relevant to the	92 BehavGen 22 621-634
Säll, Torbjörn; Nilsson, Nils-Otto The robustness of	recombination frequency estimates in intercrosses with dominant markers •	94 Genetcs 137 589-596
Sarah P.; Feldman, Marcus W. On the evolution of	recombination in haploids and diploids II. Stochastic models • Bergman, Aviv; Otto,	95 Complxty 1 49-57
Volkov, V. Ya.; Gladkov, Yu. M.	Reconfigurable combined forecasts in a non-stationary inflationary environment •	95 JForcast 14 395-403
linear ... • Archer et al. The iterative image space	reconstruction algorithm (ISRA) as an alternative to the EM algorithm for solving positive	95 StSinica 5 77-96
	Reconstruction of band limited processes from irregular samples • Houdré, Christian	95 AnlsProb 23 674-696
On a	reconstruction of a Markov chain • Grosek, Otokar	95 JCISS 20 85-93
Fuzzy surface	reconstruction • Pereira, Ricardo Alberto Marques; Fedrizzi, Michele	94 QStMaA 16 71-77
Tree-ring	reconstruction of Upper Gila River discharge • Meko, David; Graybill, Donald A.	95 WaterReB 31 605-616
function estimation from inversely sampled	record-breaking data • Gulati, Sneh; Padgett, W. J. Nonparametric	95 CandJSt 23 359-368
function estimation from inversely sampled	record-breaking data • Gulati, Sneh; Padgett, W. J. Smooth nonparametric	94 ASAProPEng 140-145
Power of	record-breaking test (Japanese) • Iiyama, Yuriko; Nishimura, Kazuo; Sibuya, Masaaki	95 JpJAppSt 24 13-26
Limiting curves for I.I.D.	records • Deuschel, Jean-Dominique; Zeitouni, Ofer	95 AnlsProb 23 852-878
On	record indices and record times • Borovkov, Konstantin; Pfeifer, Dietmar	95 JStPlInf 45 65-79
	Record linkage	
	An application of stochastic optimization to combine two files • Griffin, Richard A.	94 ASASrvyRsr 485-490
database (Disc: p697-699) • Mackinnon, D. P. et al.	Development and applications of a city-level alcohol availability and alcohol problems	95 StatMed 14 591-604
	Linking individuals in a capital gains panel for tax policy analysis • Hostetter, Susan C.	94 ASASrvyRsr 461-466
Jaro, Matthew A.	Probabilistic linkage of large public health data files (Disc: p687-689) •	95 StatMed 14 491-498
Liu, Tzen-Ping	Statistical matching of survey datafiles: A simulation study • Kovačević, Milorad S.;	94 ASASrvyRsr 479-484
databases (Disc: p687-689) • Shevchenko, I. P. et al.	Verification of information in a large medical database using linkages with external	95 StatMed 14 511-530
Advanced methods for	record linkage • Winkler, William E.	94 ASASrvyRsr 467-472
A method for calibrating false match rates in	record linkage • Belin, Thomas R.; Rubin, Donald B.	95 JASA 90 694-707
Muse et al. Evaluating the quality of anonymous	record linkage using deterministic procedures with the New York State AIDS registry ... •	95 StatMed 14 499-509
Functional laws of the iterated logarithm for	records • Imlahi, A.	95 JStPlInf 45 215-324
Many multivariate	records • Goldie, Charles M.; Resnick, Sidney I.	95 StocProc 59 185-216
Asymptotic distributions of	records in nonstationary schemes • Nevzorov, Valery	95 JStPlInf 45 261-273
beneficiaries: Adult and ... • Eisenhower et al. A	records-survey comparison of eligibility and health care utilization measures for Medicaid	94 ASASrvyRsr 713-718
	Record time	
	Asymptotic distributions of records in nonstationary schemes • Nevzorov, Valery	95 JStPlInf 45 261-273
	Functional laws of the iterated logarithm for records • Imlahi, A.	95 JStPlInf 45 215-324
approximation of the number of renewal paced	record times • Ennadifi, Gratiane Strong	95 JStPlInf 45 113-132
Asymptotic behaviour of	record times for renewal paced records • Ennadifi, G.	95 StatDecn 13 379-397
K-	record values and the extreme-value index • Berred, Mohamed	95 JStPlInf 45 49-63
Recurrence relations for moments of	record values • Kamps, Udo	95 JStPlInf 45 225-234
	Record values, random record models and concomitants • Ahsanullah, M.	94 JStatRsr 28 89-109
Estimating survival rates from	recoveries of birds ringed as young: A case study • Francis, Charles M.	95 JAppStat 22 567-577
Bayesian analysis of ring-	recovery data via Markov chain Monte Carlo simulation • Vounatsou, P.; Smith, A. F. M.	95 Biomtrcs 51 687-708
	Recovery of interblock information	
sizes 2 and 3 • Federer, Walter T.	A simple procedure for constructing experiment designs with incomplete blocks of	95 BiomtrcJ 37 899-907
age variation in survival and reporting rates for	recovery models • Catchpole, E. A.; Freeman, S. N.; Morgan, B. J. T. Modelling	95 JAppStat 22 597-609
How useful are	recoveries of North American passerines for survival analyses? • Francis, Charles M.	95 JAppStat 22 1075-1081
Delineating bird populations using ring	recoveries • Pendleton, Grey W.; Sauer, John R.	95 JAppStat 22 1049-1055
of British pigeons and doves by analysis of ringing	recoveries • Aebischer, N. J. Investigating the effects of hunting on the survival	95 JAppStat 22 923-934
Modelling and analysis of	recruitment distributions • Taillie, C.; Patil, G. P.; Hennemuth, R. C.	95 EnvEcoSt 2 315-329

Regression (cont.)

de Wet, T.; Fresen, J. L.	An experimental design for evaluating sonar detection performance • de Jongh, P. J.;	95 SoAfrStJ 29 79-103
	Forecasting growth with time series models • Peña, Daniel	95 JForcast 14 97-105
Bandt, Christoph; Schulz, Anne; Schulz, Holger	Fractal exponents for the upper airways of mammalian lungs • Krause, Ekkehard;	95 CmpStDA 20 583-590
Xu, Yexiao J.	A generalization of the beta distribution with applications • McDonald, James B.;	95 JEconmtx 66 133-152
	The generalized fluctuation test: A unifying view • Kuan, Chung-Ming; Hornik, Kurt	95 EcomtrRv 14 135-161
	Implementing partial least squares • Denham, M. C.	95 StatComp 5 191-202
Nye, Lendell G.; Witt, L. Alan	Interpreting moderator effects: Substitute for the signed coefficient rule •	95 EdPsMeas 55 27-31
of different estimation methods ... • Scazzero et al.	Interval estimates for the optimum point of a quadratic logistic curve - A comparison	95 CommStB 24 207-226
Wu, Cheng-Sheng Peter; Zubovic, Yvonne	A large-scale Monte Carlo study of the Buckley-James estimator with censored data •	95 JStCmpSm 51 97-119
	Least absolute deviation estimation of a shift • Bai, Jushan	95 EconTher 11 403-436
	Linear programming approach to LMS-estimation • Boček, P.; Lachout, P.	95 CmpStDA 19 129-134
Müller, Christine H.	Maximin efficient designs for estimating nonlinear aspects in linear models •	95 JStPlInf 44 117-132
	More comments on C_p • Mallows, Colin L.	95 Technmcs 37 362-372
matrix • Pan, Jian-Xin; Fang, Kai-Tai	Multiple outlier detection in growth curve model with unstructured covariance	95 AnInStMa 47 137-153
	Multirelation - A correlation among more than two variables • Drezner, Zvi	95 CmpStDA 19 283-292
Ahmad, Ibrahim A.	On multivariate kernel estimation for samples from weighted distributions •	95 StPrLet 22 121-129
	Multivariate pHd analysis (Korean) • Lee, Yong Goo	95 KorApS 8/1 61-74
Newey, W. K.; Powell, J. L.	Nonlinear errors in variables: Estimation of some Engel curves • Hausman, J. A.;	95 JEconmtx 65 205-233
	Nonparametric kernel estimation for semiparametric models • Andrews, Donald W. K.	95 EconTher 11 560-596
	Outlier detection in bilinear calibration • Walczak, B.	95 Chemolab 29 63-73
	Partially-ordered linear models • Walker, John H.	94 ASAProStCp 152-157
	Penalized discriminant analysis • Hastie, Trevor; Buja, Andreas; Tibshirani, Robert	95 AnlsStat 23 73-102
	Performance analysis of general tracking algorithms • Guo, Lei; Ljung, Lennart	95 IEEEAuCn 40 1388-1402
Sklar, Michael G.; Armstrong, Ronald D.	A piecewise linear approximation procedure for L_p norm curve fitting •	95 JStCmpSm 52 323-335
	Precision mapping of quantitative trait loci • Zeng, Zhao-Bang	94 Genetcs 136 1457-1468
time • Cambanis, S.; Fakhre-Zakeri, I.	On prediction of heavy-tailed autoregressive sequences: Forward versus reversed	94 ThProbAp 39 217-233
	Pyrrho's lemma, or have it your way • Dijkstra, Theo K.	95 Metrika 42 119-125
	Randomization tests in econometrics • Kennedy, Peter E.	95 JBES 13 85-94
Remillard, Bruno	Relating quantiles and expectiles under weighted-symmetry • Abdous, Belkacem;	95 AnInStMa 47 371-384
	Searching for structure in curve samples • Gasser, Theo; Kneip, Alois	95 JASA 90 1179-1188
assumptions • Wooldridge, Jeffrey M.	Selection corrections for panel data models under conditional mean independence	95 JEconmtx 68 115-132
• Wang, S. T.; Klein, John P.; Moeschberger, M. L.	Semi-parametric estimation of covariate effects using the positive stable frailty model	95 AplStMDA 11 121-133
study • Sepanski, Jungsywan H.; Lee, Lung-Fei	Semiparametric estimation of nonlinear errors-in-variables models with validation	95 JNnparSt 4 365-394
	Semiparametric models: An evaluation • Editors	95 StNeerla 49 111-125
design method • Saltelli, A.; Andres, T. H. et al.	Sensitivity analysis of model output. Performance of the iterated fractional factorial	95 CmpStDA 20 387-407
unselected samples with multiple ... • Rodgers et al.	A simple algebraic demonstration of the validity of DeFries-Fulker analysis in	94 BehavGen 24 259-262
	A simple message for autocorrelation correctors: Don't • Mizon, Grayham E.	95 JEconmtx 69 267-288
Taylor, Larry W.	On a simultaneous equations pre-test estimator • Skeels, Christopher L.;	95 JEconmtx 68 269-286
	Smoothing spline density estimation: Conditional distribution • Gu, Chong	95 StSinica 5 709-726
Novikov, Alexander	Some results about averaging in stochastic approximation • Le Breton, Alain;	95 Metrika 42 153-171
observations • Fung, Wing K.	A stepwise method for the identification of multiple outliers and influential	95 SoAfrStJ 29 51-64
the incubation period of AIDS • Bacchetti, P. et al.	Survival trees with time-dependent covariates: Application to estimating changes in	95 LiDaAnal 1 35-47
Barry, A. M.	On testing for equicorrelation in random coefficient models • Bhatti, M. I.;	95 BiomtrcJ 37 383-391
	Transforms for numerical estimation in Langmuir • Betne, Prabha; Katti, S. K.	94 ASAProQlPr 286-290
Wan, Alan T. K.; Yang, Minxian	On the use of the F ratio in a misspecified model with an interval restriction •	95 JStCmpSm 52 151-161
	Validation studies in epidemiology • Altpeter, Ekkehardt	95 Student 1 35-49
	• see also linear -, multiple -, nonlinear -, stepwise -	
Agras, J. et al. An application of locally weighted	regression to airborne mercury deposition around an incinerator site • Opsomer, J.-D.;	95 EnvrMtrc 6 205-219
Quantile	regression as an alternative to robust regression • Gould, William W.; Rogers, William H.	94 ASAProStEd 12-16
comparison • Yau, Sui K.	Regression and AMMI analyses of genotype × environment interactions: An empirical	95 AgronmyJ 87 121-126
Neal A. A mixture model with detection limits for	regression analyses of antibody response to vaccine • Moulton, Lawrence H.; Halsey,	95 Biomtrcs 51 1570-1578
between redundancy analysis and canonical	regression analysis • Saito, Takayuki Relationships	95 JJapnStS 25 111-119
models and ... • Visser et al. Trend estimation and	regression analysis in climatological time series: An application of structural time series	95 JClimate 8 969-979
K. (Auth); Winner, Larry (Rev) Review of	"Regression analysis - Concepts and applications" • Graybill, Franklin A.; Iyer, Hariharan	95 AmerStat 49 114-114
K. (Auth); Barrett, Bruce (Rev) Review of	"Regression analysis: Concepts and applications" • Graybill, Franklin A.; Iyer, Hariharan	95 Technmcs 37 229-229
K. (Auth); Swain, James J. (Rev) Review of	"Regression analysis: Concepts and applications" • Graybill, Franklin A.; Iyer, Hariharan	95 JQualTek 27 179-180
beth A.; Job, Jayakaran S.; Halsey, Neal A. Percentile	regression analysis of correlated antibody responses • Moulton, Lawrence H.; Holt, Eliza-	95 StatMed 14 777-787
The Wilcoxon rank-sum test in dichotomous	regression analysis • Finkelstein, Mark; Tucker, Howard G.	92 ProIrStC1 243-259
Graphics for	regression analysis display and diagnostics • Greer, Daniel A.	94 ASAProStGr 66-70
for handling missing covariates in epidemiologic	regression analyses • Greenland, Sander; Finkle, William D. A critical look at methods	95 AJEpidm 141 1255-1264
for handling missing covariates in epidemiologic	regression analyses • Greenland, Sander; Finkle, William D. A critical look at methods	95 AJEpidm 142 1255-1264
Gauchi, J.-P. Use of PLS	regression for the analysis of experimental designs in formulation chemistry (French) •	95 RStApp 43/1 65-89
Imputation techniques in	regression analysis: Looking closely at their implementation • Bello, A. L.	95 CmpStDA 20 45-57
McDonald, John W. Marginal	regression analysis of a multivariate binary response • Ekholm, Anders; Smith, Peter W. F.;	95 Biomtrka 82 847-854
Roche, K. J. et al. Some recent developments for	regression analysis of multivariate failure time data • Liang, K.-Y.; Self, S. G.; Bandeen-	95 LiDaAnal 1 403-415
efficient algorithm for feed-forward neural network	regression analysis • Mayekawa, Shin-ichi An	95 Behvmtrk 22 67-90
A correlation principal component	regression analysis of NIR data • Sun, Jianguo	95 JChemtrc 9 21-29
A SAS macro for residual deviance of ordinal	regression analysis • Wan, Jim Y.; Wang, Whedy; Bromberg, Judith	94 CmpBiomd 45 307-310
Exponential possibility	regression analysis • Tanaka, H.; Ishibuchi, H.; Yoshikawa, S.	95 FuzzySet 69 305-318
	Regression analysis of repeated survey data (with available software) • Kott, Phillip S.	94 ASASrvyRsr 116-123
versus sequential sample-size determination in	regression analysis of simulation experiments • Kleijnen, J. P. C. et al. Two-stage	95 AJMMgSci 15 83-114
	Regression analysis. Theory, methods, and applications • Sen, A.; Srivastava, M.	90 Springer:Brln:NY 347p
PLS	regression and applications (French) • Tenenhaus, M.; Gauchi, J.-P.; Ménardo, C.	95 RStApp 43/1 7-63
QTL analysis: A simple 'marker-	regression' approach • Kearsey, M. J.; Hyne, V.	94 ThApGene 89 698-702
Seemingly unrelated	regression on the autoregressive AR(p) singular equation system • Lee, Byung-Joo	95 EcomtrRv 14 65-74
Polzehl, Jörg	Regression-based short term prediction of electrical load for a power system •	94 ASAProPEng 285-290
Influence on smoothness in penalized likelihood	regression for binary data • Jernigan, Robert W.; O'Connell, Julie A.	94 ASAProStCp 118-123
Buchinsky, Moshe Quantile	regression, Box-Cox transformation model, and the U.S. wage structure, 1963-1987 •	95 JEconmtx 65 109-154
of the spirometric residual volume (RV) by a	regression built from Gower distances • Bartkowiak, Anna; Liebhart, Jerzy Estimation	95 BiomtrcJ 37 131-149
Review of "Measurement,	regression, and calibration" • Brown, P. J. (Auth); Atkinson, Anthony (Rev)	95 JRSS-A 158 187-187
Review of "Measurement,	regression, and calibration" • Brown, P. J.; Fearn, Tom (Rev)	95 StMedRes 4 265-266
Some alternatives to classical	regression in the case of collinearity (French) • Palm, R.; Iemma, A. F.	95 RStApp 43/2 5-33
applications to synthetic data methods for censored	regression • Lai, T. L. et al. Asymptotic normality of a class of adaptive statistics with	95 JMultiAn 52 259-279

continuous relationships in biology with censored	regression • Schmid, B.; Polasek, W.; Weiner, J.; Krause, A.; Stoll, P. Modeling of dis-	94 ANatrls 143 494-507
On the use of artificial	regressions in certain microeconometric models • Orme, Chris	95 EconTher 11 290-305
A	regression characterisation of the Meixner hypergeometric distribution • Pusz, J.	95 AstrlJSt 37 83-87
graduate student success in an MBA program:	Regression versus classification • Wilson, Rick L.; Hardgrave, Bill C. Predicting	95 EdPsMeas 55 186-195
Seeking the evolutionary	regression coefficient: An analysis of what comparative methods measure • Pagel, Mark	93 JThBio 164 191-205
Admissibility of linear estimation of	regression coefficients in common growth curve models (Chinese) • Sun, Liuquan	95 AMaAPknC 17 628-630
survey conditions on least squares estimator of	regression coefficients • Agarwal, R. C. et al. The effect of measurement errors under	94 JIAgSt 46 290-297
General admissibility of linear estimates of	regression coefficients in growth curve model (Chinese) • Qing, Hong	95 ChJApPSt 10 265-271
K. Improved estimation of	regression coefficients and hypotheses testing • Srivastava, Virendra K.; Srivastava, Anil	94 Metron 52/3 161-170
Simultaneous estimation of	regression coefficient matrix and covariance matrix in growth curve models • Yang, Wenli	95 SciSPkng 24 474-484
tory variables • Elston, D. A.; Proe, M. F. Smoothing	regression coefficients in an overspecified regression model with interrelated explana-	95 ApplStat 44 395-406
Zhan, J.; Chen, J. All admissible linear estimators of	regression coefficients and parameters in variance component models • Wang, X.;	95 AMaPeknE 37 653-662
Prediction in random coefficient	regression • Beran, Rudolf	95 JStPlInf 43 205-213
of point optimal testing for Rosenberg random	regression coefficients • Brooks, Robert The robustness	95 EcomtrRv 14 35-53
cipal components applied to principal component	regression • Mertens, B.; Fearn, T.; Thompson, M. The efficient cross-validation of prin-	95 StatComp 5 227-235
Sample-size determination for parent-offspring	regression of a composite trait • Orzack, Steven Hecht; Chappell, Rick	95 Biomtrcs 51 822-830
The periodogram	regression: Correction and comments • Hassler, Uwe	95 CommStA 24 3137-3146
a measure of the explanatory power of covariates in	regression • Doksum, K.; Samarov, A. Nonparametric estimation of global functionals and	95 AnlsStat 23 1443-1473
Cajo J. F. Ter Prediction error in partial least squares	regression: A critique on the deviation used in the Unscrambler • De Vries, S.; Braak,	95 Chemolab 30 239-245
Rank	regression for current status data • Aragón, Jorge; Quiróz, Adolfo J.	95 StPrLet 24 251-256
Weighted orthogonal series estimates of the	regression curve (Chinese) • Yang, Ying	95 JSysSMaS 14 337-349
Better bootstrap confidence intervals for	regression curve estimation • Härdle, W.; Huet, S.; Jolivet, E.	95 MaOpfStS 26 287-306
Comparison of	regression curves using quasi-residuals • Kulasekera, K. B.	95 JASA 90 1085-1093
Spurious	regressions between I(*d*) processes • Marmol, Francesc	95 JTimSrAn 16 313-321
of multiple imputation to the analysis of censored	regression data" (V47 p1297-1309) • James, Ian R. Comment on "Applications	95 Biomtrcs 51 358-362
of multiple imputation to the analysis of censored	regression data" • Tanner, Martin A. Reply to comment on "Applications	95 Biomtrcs 51 362-362
Optimal	regression designs under random block-effects models • Cheng, Ching-Shui	95 StSinica 5 485-497
Robust principal components	regression as a detection tool for outliers • Walczak, B.; Massart, D. L.	95 Chemolab 27 41-54
models • Otranto, Edoardo; Gallo, Giampiero M.	Regression diagnostic techniques to detect balanced space-to-time ratios in STARMA	94 Metron 52/3 129-145
	Regression diagnostics	
Lee, Andy H.	Assessment of influence in non-linear measurement error models • Zhao, Yuejen;	95 JAppStat 22 215-225
	A cautionary note on the use of generalized Cook-type measures • Fung, Wing K.	95 CmpStDA 19 321-326
	Graphical interpretation of variance inflation factors • Stine, Robert A.	95 AmerStat 49 53-56
	Graphical summaries for influence of multiple observations • Fung, Wing K.	95 CommStA 24 415-427
Ramanathan, S.	Recursive inference in a general linear model • Bhimasankaram, P.; Sengupta, D.;	95 SankhyaA 57 227-255
errors in an orthogonal-regression ... • Shapiro et al.	Rejecting outliers and estimating errors in an orthogonal outliers and estimating	95 PTLnAMa 350 407-439
	Renovated scatterplots for censored data • Smith, Peter J.; Zhang, J.	95 Biomtrka 82 447-452
Pseudo-maximum-likelihood-estimation and	regression diagnostics for count data: Suicide by tourism? (German) • Ziegler, A. et al.	95 AllgeStA 79 170-195
The gamma-gamma	regression for the distribution model of event times • Ogasawara, Haruhiko	95 JpnPsyRs 37 70-79
for the numerical solution of seemingly unrelated	regression equations models • Kontoghiorghes, E. J. et al. An alternative approach	95 CmpStDA 19 369-377
coefficients in estimating two seemingly unrelated	regression equations (Chinese) • Liu, Jinshan A method of undetermined	95 ChJApPSt 10 78-83
Some finite sample theory for bootstrap	regression estimates • Fisher, Gordon; Sim, Ah Boon	95 JStPlInf 43 289-300
rates of convergence on nonparametric quantile	regression estimate on dependent data (Chinese) • Shi, Peide; He, Xuming The optimal	95 KToChina 39 1537-1539
selecting the smoothing parameter of least squares	regression estimates using the minimax regret approach • Droge, B.; Georg, T. On	95 StatDecn 13 1-20
the use of auxiliary information for calibration and	regression estimation • Estevao, V.; Patak, Z.; Särndal, C. Maximizing	94 ASASrvyRsr 42-48
distribution of data-driven smoothers in density and	regression estimation under dependence • Boente, Graciela; Fraiman, Ricardo Asymptotic	95 CandJSt 23 383-397
Efficient location and	regression estimation for long range dependent regression models • Dahlhaus, Rainer	95 AnlsStat 23 1029-1047
Distant long-range dependent sums and	regression estimation • Csörgő, Sándor; Mielniczuk, Jan	95 StocProc 59 143-155
approach • Ioannatos, Petros E. Censored	regression estimation under unobserved heterogeneity: A stochastic parameter	95 JBES 13 327-335
	Regression estimator	
Dupont, F.	Alternative adjustments where there are several levels of auxiliary information •	95 SurvyM 21 125-135
Sahoo, L. N.; Mohanty, S.	Empirical comparison of some regression and regression-type strategies • Sahoo, J.;	95 StHefte 36 337-347
• Nascimento Silva, P. L. D.; Skinner, C. J.	Estimating distribution functions with auxiliary information using poststratification	95 JOffStat 11 277-294
Estevao, V.; Hidiroglou, M. A.; Särndal, C. E.	Methodological principles for a generalized estimation system at Statistics Canada •	95 JOffStat 11 181-204
Hadi, Ali S.	Modeling lifetime data with application to fatigue models • Castillo, Enrique;	95 JASA 90 1041-1054
Bez, K.	Preliminary test estimators in double sampling with two auxiliary variables • Das, G.;	95 CommStA 24 1211-1226
Jagatananda; Sahoo, Lokanath; Mohanty, Suprasana	Unequal probability sampling using a transformed auxiliary variable • Sahoo,	94 Metron 52/1 71-83
Finite sample properties of adaptive	regression estimators • Ng, Pin T.	95 EcomtrRv 14 267-297
Generalized ridge	regression estimators under the LINEX loss function • Ohtani, K.	95 StHefte 36 99-110
of interval constrained least squares and mixed	regression estimators • Srivastava, Viren K.; Ohtani, Kazuhiro A comparison	95 CommStA 24 395-413
Reducing the maximum risk of	regression estimators by polyhedral projection • Schmidt, Karsten; Stahlecker, Peter	95 JStCmpSm 52 1-15
Bootstrapping quantile	regression estimators • Hahn, Jinyong	95 EconTher 11 105-121
of confidence intervals from R-estimators in	regression • George, K. J.; McKean, J. W.; Schucany, W. R.; Sheather, S. J. A comparison	95 JStCmpSm 53 13-22
Breidt, F. J. Approximating the variance of the survey	regression estimator using poststratification • Leonard, K. A.; An, A. B.; Nusser, S. M.;	94 ASASrvyRsr 222-227
Unbiased testing in exponential family	regression • Cohen, Arthur; Kemperman, J. H. B.; Sackrowitz, H. B.	94 AnlsStat 22 1931-1946
On a downdating formula for	regression • Chan, N. N.	95 JStPlInf 46 347-350
outliers and estimating errors in an orthogonal-	regression framework • Shapiro et al. ... outliers and estimating errors in an orthogonal	95 PTLnAMa 350 407-439
A nonparametric test for the	regression function: Asymptotic theory • Diebolt, Jean	95 JStPlInf 44 1-17
The effect of the smoothness of the	regression function on a bandwidth selector in nonparametric regression • Chu, C. K.	95 CommStB 24 61-77
via a Pareto conditional distribution and a	regression function • Wesolowski, Jacek Bivariate distributions	95 AnInStMa 47 177-183
via a power series conditional distribution and a	regression function • Wesolowski, Jacek Bivariate discrete measures	95 JMultiAn 55 219-229
theorems for empirical cumulative quantile	regression functions • Rao, C. R.; Zhao, L. C. Convergence	95 MaMethSt 4 81-91
theorems for empirical cumulative quantile	regression functions" (V4 p81-91) • Rao, C. R.; Zhao, L. C. Correction to "Convergence	95 MaMethSt 4 359-359
E. C. Partitioning-estimates of a	regression function under random censoring • Carbonez, A.; Györfi, L.; van der Meulen,	95 StatDecn 13 21-37
M-type estimators of	regression function with applications • Zhu, Li-Xing; Ng, Kai W.	95 StPrLet 25 133-144
Kernel estimation of the	regression function with random sampling times • Vilar, J. A.	95 Test 4 137-178
Calibration for inverse Gaussian	regression • Woldie, Mammo; Folks, J. Leroy	95 CommStA 24 2609-2620
Heckman, N. E.; Wand, M. P. Local polynomial kernel	regression for generalized linear models and quasi-likelihood functions • Fan, J.;	95 JASA 90 141-150
comparison of neural networks, non-linear biased	regression and a genetic algorithm for dynamic model identification • Wise, B. M. et al. A	95 Chemolab 30 81-89
M.; Bonham, Charles D. Least squares	regression vs. geometric mean regression for ecotoxicology studies • Angleton, George	95 AplMaCmp 72 21-32
Review of "An introduction to	regression graphics" • Cook, R. D.; Weisberg, S. (Auth); Craig, Peter (Rev)	95 JAppStat 22 541-542
Review of "An introduction to	regression graphics" • Cook, R. D.; Weisberg, S. (Auth); Pincus, R. (Rev)	95 BiomtrcJ 37 298-298
Review of "An introduction to	regression graphics" • Cook, R. Dennis; Weisberg, Sandford (Auth); Cook, Dianne (Rev)	95 JASA 90 1126-1127
Review of "An introduction to	regression graphics" • Cook, R. Dennis; Weisberg, Sanford (Auth); Goldman, Aaron (Rev)	95 Technmcs 37 343-344

On the power of F tests under	regression models with nested error structure • Rao, J. N. K.; Wang, Song-Gui	95 JMultiAn 53 237-246
A study on a	regression model with nonlinear time series errors (Korean) • Hwang, Sun Young	95 KorApS 8/2 187-200
Estimation in some binary	regression models with prescribed accuracy • Chang, Yuan-chin Ivan	95 JStPlInf 44 313-325
therapeutically effective dose of a compound via	regression modelling and percentile estimation (Disc: p933-933) • Filloon ... the minimum	95 StatMed 14 925-932
Dayton, J. David Instructional	regression modules using XLISP-STAT (jse/v3n1/anderson) • Anderson, Jon E.;	95 JStEd 3/1 37kb
optimal approximate invariant designs for cubic	regression on multidimensional balls and cubes • Gaffke, N.; Heiligers, B. Computing	95 JStPlInf 47 347-376
Correct and incorrect use of multilinear	regression • Sergent, Michelle; Mathieu, Didier; Phan-Tan-Luu, Roger; Drava, Giuliana	95 Chemolab 27 153-162
van der Maas, J. H.; Visser, T. Partial least squares	regression as a multivariate tool for the interpretation of infrared spectra • Luinge, H. J.;	95 Chemolab 28 129-138
in ANCOVA problems having non-homogeneous	regressions • Hunka, Steve Identifying regions of significance	95 BJMaStPs 48 161-188
Efficient IV estimation in nonstationary	regression • Kitamura, Yuichi; Phillips, Peter C. B.	95 EconTher 11 1095-1130
Robust nonstationary	regression • Phillips, Peter C. B.	95 EconTher 11 912-951
Group truncated ordinal	regression • O'Neill, Terence J.; Barry, Simon C.	95 StPrLet 22 195-203
Analysis of neurologic outcomes using ordinal	regression • Hwang, Lie-Ju; Alves, Wayne	94 ASAProBiop 352-354
One-step GM-estimators for orthogonal	regression • Abdullah, Mokhtar B.	95 JStCmpSm 52 301-309
H.; Gardiner, Joseph C. Nonparametric estimation of	regression parameters from censored data with a discrete covariate • Rahbar, Mohammad	95 StPrLet 24 13-20
H. Sequential fixed width confidence intervals for	regression parameters from censored data with a discrete covariate • Rahbar, Mohammad	95 SqtlAnly 14 143-156
• Forbes, A. B.; Santner, T. J. Estimators of odds ratio	regression parameters in matched case-control studies with covariate measurement error	95 JASA 90 1075-1084
A residual bootstrap for	regression parameters in proportional hazards models • Loughin, Thomas M.	95 JStCmpSm 52 367-384
Ding, Shuliang Some results on estimating the	regression parameters in seemingly unrelated regression system (SURS) (Chinese) •	95 ChJApPSt 10 243-252
Detection of outliers in the linear model when the	regression parameters are stochastic • Bhandary, Madhusudan	95 CommStA 24 153-173
All admissible linear estimators of the	regression parameter vector in the case of an arbitrary parameter subset • Hoffmann, Kurt	95 JStPlInf 48 371-377
A simple iteration algorithm for PLS	regression • Zhu, Eryi; Barnes, Ramon M.	95 JChemtrc 9 363-372
Graphics for studying net effects of	regression predictors • Cook, R. Dennis	95 StSinica 5 689-708
Staudte, Robert G. Robust variable selection in	regression in the presence of outliers and leverage points • Sommer, Suzanne;	95 AstrlJSt 37 323-336
Shanbhag, D. N. Characterizations based on	regression properties: Improved versions of recent results • Rao, C. Radhakrishna;	95 SankhyaA 57 167-178
Exact Elfving-minimax designs for quadratic	regression • Krafft, Olaf; Schaefer, Martin	95 StSinica 5 475-484
	Regression quantile	
extreme-value errors • Saleh, A. K. Md. E. et al.	Optimum spacings, ABLUE and tests of hypothesis in simple linear model with	94 JStatRsr 28 53-63
Estimation of quantile density function based on	regression quantiles • Dodge, Yadolah; Jurečková, Jana	95 StPrLet 23 73-78
obtained by kernel-based estimation of quantile	regressions • Ducharme, G. R.; Gannoun, A.; Guertin, M.-C.; Jéquier, J.-C. Reference values	95 Biomtrcs 51 1105-1116
Empirical	regression quantile and leverage treatment method • Sueyoshi, Toshiyuke	95 JOpRsJpn 38 34-54
rank tests of the unit root hypothesis based on	regression rank scores • Hasan, M. N.; Koenker, R. W. Robust	94 ASAProBuEc 35-43
Empirical comparison of some	regression and regression-type strategies • Sahoo, J.; Sahoo, L. N.; Mohanty, S.	95 StHefte 36 337-347
for analytical chemistry, Volume 2: PC-aided	regression and related methods" • Meloun, M.; Militky, J. et al. Review of "Chemometrics	95 Chemolab 29 326-327
Model goodness-of-fit analysis using	regression and related techniques • Smith, Eric P.; Rose, Kenneth A.	95 EcolgMdl 77 49-64
• Baker, Stuart G. Marginal	regression for repeated binary data with outcome subject to non-ignorable non-response	95 Biomtrcs 51 1042-1052
On ecological	regression and ridge estimation • Füle, Erika	95 CommStB 24 385-398
al. Maximum sum of binary-coded residuals (MASBR)	regression as a robust procedure for treatment of spectral data • Wang, J.-H.; Xie, Y.-L. et	95 JChemtrc 9 373-387
Review of "Subset selection in	regression" • Miller, A. J. (Auth); Droge, B. (Rev)	95 MaOpfStS 26 179-179
Adaptive signal	regression • Land, Stephanie R.; Friedman, Jerome H.	94 ASAProStCp 100-105
An introduction to multivariate adaptive	regression splines • Friedman, Jerome H.; Roosen, Charles B.	95 StMedRes 4 197-217
Some new results of M-type	regression spline estimators in a partial linear model • Shi, Peide	95 KToChina 39 189-192
effective bandwidth selector for local least squares	regression • Ruppert, D.; Sheather, S. J.; Wand, M. P. An	95 JASA 90 1257-1270
solution algorithm for least median of squares	regression • Hawkins, Douglas M. Convergence of the feasible	95 CmpStDA 19 519-538
Exploring	regression structure with graphics • Wetzel, N.; Cook, R. Dennis	94 ASAProStGr 42-47
Extending the elements of tree-structured	regression • Segal, Mark R.	95 StMedRes 4 219-236
Model selection in tree-structured	regression • Miller, Thomas W.	94 ASAProStCp 158-163
Assessing geometric integrity through spherical	regression techniques (Disc: p191-220) • Chapman, G. Robert; Chen, Gemai; Kim, Peter T.	95 StSinica 5 173-191
Auxiliary covariate data in failure time	regression • Zhou, Haibo; Pepe, Margaret Sullivan	95 Biomtrka 82 139-149
Log-periodogram	regression of time series with long range dependence • Robinson, P. M.	95 AnlsStat 23 1048-1072
Local influence analysis for	regression transformation models (Chinese) • Wei, Bocheng; Shi, Jianqing	95 AMaAPknC 17 132-143
	Regression tree	
James, Ian R.; French, Martyn A. H.; Mallal, Simon A.	Statistical issues in the evaluation of markers of HIV progression • Segal, Mark R.;	95 IntStRvw 63 179-197
Generalized	regression trees • Chaudhuri, Probal; Lo, Wen-Da; Loh, Wei-Yin; Yang, Ching-Ching	95 StSinica 5 641-666
Katyar, N. P. et al. A class of almost unbiased	regression-type estimators for finite population mean applying jackknife technique •	94 JIAgSt 46 183-189
the stability of principal components using	regression • Sinha, Atanu R.; Buchanan, Bruce S. Assessing	95 Psymtrka 60 355-369
Koul-DeWet-type estimation in censored	regression using the conditional empirical process • Yang, Song	95 JStPlInf 46 351-363
Better subset	regression using the nonnegative garrote • Breiman, Leo	95 Technmcs 37 373-384
to hypothesis testing in least absolute value	regression • Dielman, Terry E.; Rose, Elizabeth L. A bootstrap approach	95 CmpStDA 20 119-130
Asymptotic properties of locally weighted	regression • Hwang, Ruey-Ching	95 JNnparSt 5 303-310
spreadsheet program for calibration using weighted	regression • Gort, Steven M.; Hoogerbrugge, Ronald A user-friendly	95 Chemolab 28 193-199
Breidt, F. Jay; Fuller, Wayne A.	Regression weighting methods for SIPP data (Disc: p440-442) • An, Anthony B.;	94 ASASrvyRsr 434-439
Schmidli, H. Reduced rank	regression: With applications to quantitative structure-activity relationships •	95 Springer:Brln:NY 179p
Pre-test estimators in least absolute value	regression with autocorrelated errors • Dielman, Terry E.; Rose, Elizabeth L.	94 ASAProBuEc 95-100
exact distribution of the OLS and GLS estimators in	regression with an integrated regressor and correlated errors – ... • Hisamatsu et al. The	95 MaCmpSim 39 273-277
Aragon, Jorge	Regression with interval-censored data • Rabinowitz, Daniel; Tsiatis, Anastasios;	95 Biomtrka 82 501-513
Time series	regression with mixtures of integrated processes • Chang, Yoosoon; Phillips, Peter C. B.	95 EconTher 11 1033-1094
Estimation of coefficients of time series	regression with a nonstationary error process • Usami, Yoshihiro; Huzii, Mituaki	95 JTimSrAn 16 105-118
A convergent algorithm for quantile	regression with smoothing splines • Bosch, Ronald J.; Ye, Yinyu; Woodworth, George G.	95 CmpStDA 19 613-630
Compound	regressive models for family data • Bonney, George Ebow	92 HumHrdty 42 28-41
et al. Genetic analysis combining path analysis with	regressive models: The TAU model of multifactorial transmission • Li, Z.; Bonney, G. E.	94 HumHrdty 44 305-311
and GLS estimators in regression with an integrated	regressor and correlated errors – Comparison of numerical ... • Hisamatsu et al. ... OLS	95 MaCmpSim 39 273-277
Carroll, R. J.; Li, Ker-Chau Binary	regressors in dimension reduction models: A new look at treatment comparisons •	95 StSinica 5 667-688
alternative estimators in models with generated	regressors when the expectations equation has reduced ... • Smith et al. ... of	95 MaCmpSim 39 343-346
Small perturbations of Gaussian	regressors • Millet, Annie; Smoleński, Włodzmierz	95 StPrLet 24 21-31
Inference in models with nearly integrated	regressors • Cavanagh, Christopher L.; Elliott, Graham; Stock, James H.	95 EconTher 11 1131-1147
for homogeneity in demand systems when the	regressors are nonstationary • Ng, Serena Testing	95 JApEcmtx 10 147-163
Nonstationarity of	regressors and tests on real-interest-rate behavior • Mishkin, Frederic S.	95 JBES 13 47-51
	Regret	
parameter exponential family • Karunamuni et al.	On convergence rates of monotone empirical Bayes tests for the continuous one-	95 StatDecn 13 181-192
	A note on sequential estimation of the mean • Uno, Chikara	95 StPrLet 23 141-149
wise stopping number • Bose, A.; Mukhopadhyay, N.	Sequential estimation of the mean of an exponential distribution via replicated piece-	95 StatDecn 13 351-361
squares regression estimates using the minimax	regret approach • Droge, B.; Georg, T. On selecting the smoothing parameter of least	95 StatDecn 13 1-20
Some results on posterior	regret Γ-minimax estimation • Insua, David Ríos; Ruggeri, Fabrizio; Vidakovic, Brani	95 StatDecn 13 315-331

Sample mean based index policies with $O(\log n)$	regret for the multi-armed bandit problem • Agrawal, Rajeev	95 AdvAppPr 27 1054-1078
Tahir, Mohamed On the asymptotic	regret of a sequential procedure for estimating the mean of a normal distribution •	95 JStPlInf 48 331-338
Review of "Completeness of root functions of	regular differential operators" • Yakubov, S. (Auth); Rehberg, J. (Rev)	95 BiomtrcJ 37 118-118
Invariance principles for absolutely	regular empirical processes • Doukhan, P.; Massart, P.; Rio, E.	95 AlHenrPB 31 393-427
A method for generating and labelling all	regular fractions or blocks for q^{n-m}-designs • Laycock, P. J.; Rowley, P. J.	95 JRSS-B 57 191-204
Eigenvalues and expansion of	regular graphs • Kahale, Nabil	95 JACM 42 1091-1106
results related to interpolation on Hardy spaces of	regular martingales • Xu, Quanhua Some	95 IsrelJMa 91 173-187
"General pattern theory: A mathematical study of	regular structures" • Grenander, U. (Auth); Keenan, Daniel (Rev) Review of	95 SIAMRvw 37 258-261
"General pattern theory: A mathematical theory of	regular structures" • Grenander, U. (Auth); Ripley, B. D. (Rev) Review of	95 JRSS-A 158 635-635
logarithm for empirical processes under absolute	regularity • Arcones, Miguel A. The law of the iterated	95 JTherPrb 8 433-451
central limit theorem for U-statistics under absolute	regularity • Arcones, Miguel A. On the	95 StPrLet 24 245-249
Nonparametric tests of	regularity, Farrell efficiency, and goodness-of-fit • Färe, Rolf; Grosskopf, Shawna	95 JEconmtx 69 415-425
Flexibility and	regularity properties of the asymptotically ideal production model • Terrell, Dek	95 EcomtrRv 14 1-17
smoothing • Cheng, Bing; Kay, Jim A	regularization method for one-dimensional edge detection and edge-preserving	95 CmpStQ 10 53-69
Sjöberg, J.; Ljung, L. Overtraining,	regularization and searching for a minimum, with application to neural networks •	95 IJContrl 62 1391-1407
analysis, quadratic discriminant analysis and	regularized discriminant analysis • Sánchez et al. ... in relation to linear discriminant	95 Chemolab 28 287-303
Variably	regularized discriminant analysis • Higbee, Kenneth T.	94 ASAProStCp 263-268
Nikolas P.; Katsaggelos, Aggelos K.	Regularized multichannel restoration using cross-validation • Zhu, Wenwu; Galatsanos,	95 CVGIPGr 57 38-54
Complete classes of tests for	regularly varying distributions • Kowalski, Jacek P.	95 AnInStMa 47 321-350
On a subclass of	regularly varying functions • Omey, E.	95 JStPlInf 45 275-290
Ford, J. M.; Kaserman, D. L. Certificate-of-need	regulation and the diffusion of innovations: A random coefficient model • Caudill, S. B.;	95 JApEcmtx 10 73-78
Statistical issues in the	regulation of medicines • Lewis, John A.	95 StatMed 14 127-136
Geoffrey Optimal response to a shift in	regulatory regime: The case of the US nuclear power industry • Rust, John; Rothwell,	95 JApEcmtx 10 75-118S
The 1991-92 teacher follow-up survey	reinterview and extensive reconciliation • Jenkins, Cleo R.; Wetzel, Angela-Jo	94 ASASrvyRsr 821-826
Roger A. Herriot • Alexander, Lois	Reinventing the population census: Balancing privacy and efficiency. In memory of	94 ASAProGovt 1-2
an orthogonal-regression framework • Shapiro et al.	Rejecting outliers and estimating errors in an orthogonal outliers and estimating errors in	95 PTLnAMa 350 407-439
Establishing a	rejection procedure for crop performance data • Bowman, Daryl T.; Rawlings, John O.	95 AgronmyJ 87 147-151
	Rejection sampling	
convergence • Dellaportas, Petros	Random variate transformations in the Gibbs sampler: Issues of efficiency and	95 StatComp 5 133-140
A	rejection technique for sampling from T-concave distributions • Hörmann, Wolfgang	95 ACMMaSfw 21 182-193
the theory of consumer demand • Keuzenkamp et al.	Rejection without falsification: On the history of testing the homogeneity condition in	95 JEconmtx 67 103-127
for case-crossover studies: An assessment of	relative efficiency • Mittleman, M. A.; Maclure, M.; Robins, J. M. Control sampling strategies	95 AJEpidm 142 91-98
Turrero, Agustin	Relative efficiency of a censored experiment in terms of Fisher information matrix •	95 CommStA 24 1169-1191
Ahmad, Naseer Exact and approximate	relative efficiency of generalized ridge-type shrinkage estimator • Noor, Islam;	95 PakJStA 11 17-22
tion of the training data • McLachlan et al. Asymptotic	relative efficiency of the linear discriminant function under partial nonrandom classifica-	95 JStCmpSm 52 415-426
trait • Schork, Nicholas J.; Schork, M. Anthony The	relative efficiency and power of small-pedigree studies of the heritability of a quantitative	93 HumHrdty 43 1-11
	Relative efficiency of two two-stage sample designs • Chaudhary, Mohammad A.	94 ASASrvyRsr 228-231
survey • Hill, Daniel H. The	relative empirical validity of dependent and independent data collection in a panel	94 JOffStat 10 359-380
dependent random sequences in τ-topology and the	relative entropy • Hu A note on the relationship between the rate functions for stationary	95 StocProc 56 351-355
of the number of true gray levels, their values, and	relative frequencies in a noisy image • Godtliebsen, Fred; Chu, Chih-Kang Estimation	95 JASA 90 890-899
Farrington, C. P.	Relative incidence estimation from case series for vaccine safety evaluation •	95 Biomtrcs 51 228-235
Improved shrinkage estimation of	relative potency • Ahmed, S. E.	95 BiomtrcJ 37 627-638
with multivariate responses • Hanusz, Zofia	Relative potency of two preparations in two-way elimination of heterogeneity designs	95 Biomtrcs 51 1133-1139
Hurvich, Clifford M.; Tsai, Chih-Ling	Relative rates of convergence for efficient model selection criteria in linear regression •	95 Biomtrka 82 418-425
sion of within- and between-population estimates of	relative rate parameters • Sheppard, Lianne; Prentice, Ross L. On the reliability and preci-	95 Biomtrcs 51 853-863
	Relative risk	
Sullivan	Auxiliary covariate data in failure time regression • Zhou, Haibo; Pepe, Margaret	95 Biomtrka 82 139-149
scores: Non-iterative method • Nam, Jun-mo	Confidence limits for the ratio of two binomial proportions based on likelihood	95 BiomtrcJ 37 375-379
sion analyses • Greenland, Sander; Finkle, William D.	A critical look at methods for handling missing covariates in epidemiologic regres-	95 AJEpidm 141 1255-1264
Tests of loglinear and linear	relative risks for Cox's model • Baade, Ingrid A.; Pettitt, Anthony N.	95 Biomtrcs 51 1502-1513
data from cases and an external estimate of the	relative risk • Mittleman Estimation of exposure prevalence in a population at risk using	95 Epidmlgy 6 551-553
Kernel estimation of	relative risk • Kelsall, Julia E.; Diggle, Peter J.	95 Bernoull 1 3-16
levels for ... • Lao Statistical analysis for estimating	relative risk of first cancer progression by monitoring carcinoembryonic antigen (CEA)	94 ASAProBiop 206-210
Genetic	relative risk methods using cases and their parents • Schaid, Daniel J.	94 ASAProEpid 1-7
Non-parametric estimation of spatial variation in	relative risk • Kelsall, Julia E.; Diggle, Peter J.	95 StatMed 14 2335-2342
sampling • Knaub, James R., Jr	Relative standard error for a ratio of variables at an aggregate level under model	94 ASASrvyRsr 310-312
The	relative survival rate • Henson, Donald Earl; Ries, Lynn A.	95 Cancer 76 1687-1688
Optimal software	release policy based on a two-person game of timing • Zeephongsekul, P.; Chiera, C.	95 JAppProb 32 470-481
erties of an individual-based survival model using	release-recapture data: Sample size, validity and power • Hoffmann et al. Inferential prop-	95 JAppStat 22 579-595
	Reliability	
data • Sweeting, Trevor J.; Phillips, Michael J.	An application of nonstandard asymptotics to the analysis of repairable-systems	95 Technmcs 37 428-435
data • Kvam, Paul H.; Martz, Harry F.	Bayesian inference in a discrete shock model using confounded common cause	95 RelEngSS 48 19-25
Lin, Dennis K. J.; Guess, Frank M.; Usher, John S.	Bayesian inference for masked system lifetime data • Reiser, B.; Guttman, I.;	95 ApplStat 44 79-90
case • Sohn, Joong Kweon; Kim, Heon Joo	Bayesian inference for the two-parameter exponential models: Type-II censored	95 JKoreaSt 24 313-335
	Biasing the transition probabilities in direct Monte Carlo • Zio, E.	95 RelEngSS 47 59-63
Asha, G.	Characteristic properties of a bivariate geometric law • Nair, N. Unnikrishnan;	94 JIndnStA 32 111-116
	Characterization of some concepts of aging • Rojo, Javier	95 IEEEReli 44 285-290
	Cost effective calibration intervals using Weibull analysis • Barringer, H. Paul	95 ASQCTekCon 1026-1038
component systems • Guttman, I.; Lin, D. K. J. et al.	Dependent masking and system life data analysis: Bayesian inference for two-	95 LiDaAnal 1 87-100
on type-2 censored data • Ali Mousa, M. A. M.	Empirical Bayes estimators for the Burr type XII accelerated life testing model based	95 JStCmpSm 52 95-103
	Estimating products in forensic identification using DNA profiles • Balding, David J.	95 JASA 90 839-844
kernels • Li, Haijun; Shaked, Moshe	On the first passage times for Markov processes with monotone convex transition	95 StocProc 58 205-216
Gardiner, Joseph	Fixed diameter confidence ellipsoid in time-sequential models • Aras, Girish;	95 SqtlAnly 14 29-39
Tsai, Peisung	Foundational issues concerning the analysis of censored data • Barlow, Richard E.;	95 LiDaAnal 1 27-34
Pham, T.; Almhana, J.	The generalized Gamma distribution: Its hazard rate and stress-strength model •	95 IEEEReli 44 392-397
Nonaka, Yasuo	Importance analysis for the systems with common cause failures • Pan, Zhijie;	95 RelEngSS 50 297-300
Hawkes, Alan G.; Jalali, Assad	The increasing failure rate property of consecutive-k-out-of-n systems • Cui, Lirong;	95 PrEngInf 9 217-225
edge • Campodónico, Sylvia; Singpurwalla, Nozer D.	Inference and predictions from Poisson point processes incorporating expert knowl-	95 JASA 90 220-226
Ehsan S.; Ebrahimi, Nader; Habibullah, Mohamed	Information distinguishability with application to analysis of failure data • Soofi,	95 JASA 90 657-668
Maller, R. A.	The likelihood ratio test for the presence of immunes in a censored sample • Zhou, S.;	95 MaOpfStS 27 181-201
computational implementation • Csenki, Attila	Mission availability for repairable semi-Markov systems: Analytical results and	95 MaOpfStS 26 75-87
Smith, Eric P.; Rose, Kenneth A.	Model goodness-of-fit analysis using regression and related techniques •	95 EcolgMdl 77 49-64
Chen, Huifen; Schmeiser, Bruce W.	Monte Carlo estimation for guaranteed-coverage non-normal tolerance intervals •	95 JStCmpSm 51 223-238
	A new framework for part failure-rate prediction models • Wong, Kam L.	95 IEEEReli 44 139-146

	Reliability (cont.)	
Zeephongsekul, P.; Chiera, C.	Optimal software release policy based on a two-person game of timing •	95 JAppProb 32 470-481
sums • Shaked, M.; Shanthikumar, J. G.; Tong, Y. L.	Parametric Schur convexity and arrangement monotonicity properties of partial	95 JMultiAn 53 293-310
A.; Meeker, William Q., Jr	Planning accelerated life tests with two or more experimental factors • Escobar, Luis	95 Technmcs 37 411-427
systems • Vaurio, Jussi K.	The probabilistic modeling of external common cause failure shocks in redundant	95 RelEngSS 50 97-107
power plant operations in Europe • Sturm, Roland	Proportional hazard regression models for point processes: An analysis of nuclear	94 JAppStat 21 533-540
Harold W.; Valenzuela, Carlos A.	Quantal response analysis of relief valve test data • Sheesley, John H.; Thomas,	95 ASQCTekCon 741-748
cated distributions and their ... • Mohie EL-Din et al.	Recurrence relations for expectations of functions of order statistics for doubly trun-	95 CommStA 24 997-1010
Papadopoulos, Alex S.; Hoekstra, Robert M.	Some robustness issues in the Bayesian analysis of a Weibull failure model •	95 JStCmpSm 52 85-94
Nadeau, C.	Some simple robust methods for the analysis of recurrent events • Lawless, J. F.;	95 Technmcs 37 158-168
	Statistical models for nondestructive evaluation • Sweeting, Trevor	95 IntStRvw 63 199-214
Jarosław	Stochastic order relations and the total time on test transform • Bartoszewicz,	95 StPrLet 22 103-110
	Survival in dynamic environments • Singpurwalla, Nozer D.	95 StatSci 10 86-103
	Trend change in mean residual life • Lim, Jae-Hak; Park, Dong Ho	95 IEEEReli 44 291-296
	Reliability always depends on probability of course • Good, I. J.	95 JStCmpSm 52 192-193
Anantharaman, V.	Reliability analysis of a complex standby redundant system • Subramanian, R.;	95 RelEngSS 48 57-70
	Reliability analysis for engineers. An introduction • Leitch, Roger D.	95 OxfordUniv 248p
Time-varying failure rates in the availability and	reliability analysis of repairable systems • Hassett, T. F.; Dietrich, D. L.; Szidarovszky, F.	95 IEEEReli 44 155-160
Generalized log-normal distributions with	reliability application • Chen, Gemai	95 CmpStDA 19 309-319
Phillips, Michael J. (Rev) Review of "The	reliability, availability and productiveness of systems" • Sherwin, D. J.; Bossche, A. (Auth);	95 Ststcian 44 412-413
Koutras, M. V.	Reliability bounds for coherent structures with independent components • Fu, J. C.;	95 StPrLet 22 137-148
	Reliability bounds for the L-class and Laplace order • Sengupta, Debasis	95 JAppProb 32 832-835
Markovian systems of O.D.E. for availability and	reliability calculations • Tombuyses, B.; Devooght, J. Solving	95 RelEngSS 48 47-55
Wright, M. Keith; Stokes, Lynne; Dyer, James S.	Reliability and coherence of causal, diagnostic, and joint subjective probabilities •	94 DecisnSc 25 691-709
Wright, M. Keith; Stokes, Lynne; Dyer, James S.	Reliability and coherence of causal, diagnostic, and joint subjective probabilities •	95 DecisnSc 26 691-709
Models for assessing the	reliability of computer software • Singpurwalla, Nozer D.; Wilson, Simon P.	92 ProIrStC1 177-194
Jyh	Reliability curve fitting for aging helicopter components • Luxhoj, James T.; Shyur, Huan-	95 RelEngSS 48 229-234
Problem of estimation in the analysis of	reliability data • Jain, Sudha; Jain, R. K.	94 ASAProPEng 136-139
techniques for the comparative analysis of censored	reliability data: A review • Bohoris, George A. Parametric statistical	95 RelEngSS 48 149-155
Fabio (Ed); Gruber, Marvin H. J. (Rev) Review of	"Reliability and decision making" • Barlow, Richard E.; Clarotti, Carlo A.; Spizzichino,	95 Technmcs 37 346-347
(Rev) Review of "Handbook on	reliability engineering" • Ushakov, Igor A.; Harrison, Robert A. (Ed); Doganaksoy, Necip	95 JASA 90 804-804
Probabilistic	reliability engineering • Gnedenko, B. V.; Ushakov, I. A.	95 Wiley:NY:UK 518p
Review of "Probabilistic	reliability engineering" • Gnedenko, B. V.; Ushakov, I. A. (Auth); Goodwin, G. C. (Rev)	96 ShrtBkRv 16/1 8-8
Review of	"Reliability engineering" • Aggarwal, K. K. (Auth); Khamis, Imad H. (Rev)	95 Technmcs 37 234-235
On the uncertainty of a system	reliability estimate • Lindqvist, Bo	94 RelEngSS 46 263-269
Invariantly optimal look on	reliability estimation in the exponential case • Hurt, Jan; Novák, Břetislav; Wertz, Wolfgang	95 StatDecn 13 193-197
Bayesian Weibull	reliability estimation • de Souza, Daniel I., Jr; Lamberson, Leonard R.	95 IIETrn 27 311-320
Pandey, M.	Reliability estimation with Weibull inter failure times • Ferdous, J.; Uddin, Md. Borhan;	95 RelEngSS 50 285-296
B. S.; Carroll, R. J.; Koenig, K. L. et al. Estimating the	reliability of an exposure variable in the presence of confounders • Kim, M. Y.; Pasternack,	95 StatMed 14 1437-1446
Ross, Sheldon M. A simple derivation of exact	reliability formulas for linear and circular consecutive-k-of-n: F systems • Peköz, Erol A.;	95 JAppProb 32 554-557
On the calculation of the	reliability of general load sharing systems • Lee, Shiowjen; Durham, S.; Lynch, J.	95 JAppProb 32 777-792
Evaluation of operational	reliability from ground test data • Ke, Huei-Yaw	95 RelEngSS 49 103-107
Joint	reliability-importance of components • Armstrong, Michael J.	95 IEEEReli 44 408-412
Lenz, H.-J. (Rev) Review of	"Reliability improvement with design of experiments" • Condra, Lloyd W. (Auth);	95 CmpStDA 19 605-605
When quality is a matter of taste, use	reliability indexes • Futrell, David	95 QulPr 28/5 81-86
On the	reliability of indicators of learning curve model accuracy • Moses, O. Douglas	95 JCostAnl /1 63-101
An accelerated life test model based on	reliability kinetics • Meeker, William Q.; LuValle, Michael J.	95 Technmcs 37 133-146
degradation data to improve fluorescent lamp	reliability • Tseng, Sheng-Tsaing; Hamada, Michael; Chiao, Chih-Hua Using	95 JQualTek 27 363-369
The	reliability of the locality pay estimates • Coleman, Joan L.; James, Penny L.	94 ASASrvyRsr 1031-1035
Growth rate	reliability in longitudinal measurement • Woodruff, David; Houston, Mark	94 EdPsMeas 54 897-902
O.	Reliability, maintainability, and availability assessment (Second edition) • Locks, Mitchell	95 ASQualCntrl:WI 348p
(Auth); Lynch, James (Rev) Review of	"Reliability, maintainability and supportability: A probabilistic approach" • Knezevic, J.	95 Technmcs 37 122-122
Designing component test plans for series system	reliability via mathematical programming • Rajgopal, Jayant; Mazumdar, Mainak	95 Technmcs 37 195-212
Coupling methods in	reliability (French) • Cocozza-Thivent, C.; Roussignol, M.	95 AIHenrPB 31 119-141
Some general characteristics of two-state	reliability models for maintained systems • Marlow, Norman A.; Tortorella, Michael	95 JAppProb 32 805-820
Kumar Statistical analysis of some	reliability models: Parametrics, semi-parametrics and nonparametrics • Sen, Pranab	95 JStPlInf 43 41-66
Fast simulation of rare events in queueing and	reliability models • Heidelberger, Philip	95 ACMMdSim 5 43-85
Review of "Statistical analysis of software	reliability models" • van Pul, M. C. J. (Auth); Reese, R. Allan (Rev)	95 JRSS-A 158 199-199
Dynamic	reliability models with conditional proportional hazards • Hollander, Myles; Peña, Edsel A.	95 LiDaAnal 1 377-401
	Reliability modeling involving two Weibull distributions • Jiang, R.; Murthy, D. N. P.	95 RelEngSS 47 187-198
to the nonhomogeneous Poisson process in	reliability modeling • Kuo, Lynn; Yang, Tae A unified approach	94 ASAProPEng 169-174
Estimation and	reliability of molecular sequence alignments • Thorne, Jeffrey L.; Churchill, Gary A.	95 Biomtrcs 51 100-113
Estimation of structures	reliability for non-precise limit state models and vague data • Savchuk, V. P.	95 RelEngSS 47 47-58
Rameshwar D.; H-Akman, Olcay Estimation of	reliability under ordered restriction on the parameters • Gupta, Ramesh C.; Gupta,	95 CommStA 24 1799-1812
rameters • Sheppard, Lianne; Prentice, Ross L. On the	reliability and precision of within- and between-population estimates of relative rate pa-	95 Biomtrcs 51 853-863
General estimators for the	reliability of qualitative data • Cooil, Bruce; Rust, Roland T.	95 Psymtrka 60 199-220
A practical approach for predicting fatigue	reliability under random cyclic loading • Tang, Jun; Zhao, Jie	95 RelEngSS 50 7-15
The confidence limits for the	reliability in regression model • Lin, Huanbin; Chen, Jiading	95 ActScNat 30 23-29
	Reliability reports - Product, customer, and process • Conte, John A.	95 ASQCTekCon 733-740
Bolognese, Ralph A. Storage-	reliability-resilience-yield relations for over-year water supply systems • Vogel, Richard M.;	95 WaterReR 31 645-654
Failure-censored	reliability sampling plans for the exponential distribution • Balasooriya, Uditha	95 JStCmpSm 52 337-349
Bayesian computation of software	reliability • Kuo, Lynn; Yang, Tae	95 JCmpGrSt 4 65-82
Review of "Ensuring software	reliability" • Neufelder, Ann M. (Auth); Soyer, Refik (Rev)	95 Technmcs 37 122-123
parameters for the Littlewood model in software	reliability • Barendregt, L. G.; van Pul, M. C. On the estimation of the	95 StNeerla 49 165-184
Kumar; Roy, Dilip Estimation of	reliability under stress-strength model in a bivariate exponential set-up • Jana, Prasanta	94 CalcutSt 44 175-181
• Pham-Gia, T.; Turkkan, N. Comment on "Testing	reliability in a stress-strength model when X and Y are normally distributed" (V34 p83-91)	95 Technmcs 37 130-131
Statistical approaches for the	reliability of structural designs • Jones, K. F.; Taylor, N. W.; Fosam, E. B.	95 JAppStat 22 443-458
Young, Murray Composite	reliability in structural equations modeling • Bacon, Donald R.; Sauer, Paul L.;	95 EdPsMeas 55 394-406
Review of "Applied methods of structural	reliability" • Tichy, M. (Auth); Csenki, A. (Rev)	95 Ststcian 44 414-415
Use of field data in	reliability studies of medical devices • Kim, John S.; Mattson, LeRoy T.	94 ASAProQlPr 200-203
Akman, H. Olcay On the	reliability studies of a weighted inverse Gaussian model • Gupta, Ramesh C.;	95 JStPlInf 48 69-83
The exponentiation formula of	reliability and survival: Does it always hold? • Singpurwalla, Nozer D.; Wilson, Simon P.	95 LiDaAnal 1 187-194
Whitmore, G. A. (Ed) Lifetime data: Models in	reliability and survival analysis • Jewell, Nicholas P.; Kimber, Alan C.; Lee, Mei-Ling Ting;	95 Kluwer:MA:Hllnd 424p
An integral equation approach to the interval	reliability of systems modelled by finite semi-Markov processes • Csenki, A.	95 RelEngSS 47 37-45

Review of "Models for	repeated measurements" • Lindsey, J. K. (Auth); Corcoran, Stephen (Rev)	95 JAppStat 22 435-435
Review of "Models for	repeated measurements" • Lindsey, J. K. (Auth); Lipsitz, Stuart R. (Rev)	95 JASA 90 1124-1125
Darroch, J. N. An analysis of categorical	repeated measurements in the presence of an external factor • McCloud, P. I.;	95 Biomtrka 82 47-55
Part 2: Classification, covariance structures and	repeated measurements • Krzanowski, W. J.; Marriott, F. H. C. Multivariate analysis.	95 Arnold:Lnd:MD 280p
Keselman, Joanne C.; Lix, Lisa M. The analysis of	repeated measurements: Univariate tests, multivariate tests, or both? • Keselman, H. J.;	95 BJMaStPs 48 319-338
mary measures and meta ... • Talwalker Analysis of	repeated measurements with dropouts among Alzheimer's disease patients using sum-	94 ASAProBiop 515-520
Flexible monitoring of multi-arm clinical trials with	repeated measurements • Reboussin, David M.	94 ASAProBiop 111-114
Hössjer, Ola; Rousseeuw, Peter J.; Ruts, Ida The	repeated median intercept estimator: Influence function and asymptotic normality •	95 JMultiAn 52 45-72
Sensitivity functions and numerical analysis of the	repeated median slope • Rousseeuw, Peter J.; Croux, Christophe; Hössjer, Ola	95 CmpStQ 10 71-90
Regression models for	repeated medical random counts • Mackenzie, Gilbert	95 Ststcian 44 365-372
• Zhang, F. et al. Analyzing long irregularly spaced	repeated multivariate measures - A study of infant cardiac catherization stress responses	94 ASAProBiop 365-370
P. Analysis of semiparametric regression models for	repeated outcomes in the presence of missing data • Robins, J. M.; Rotnitzky, A.; Zhao, L.	95 JASA 90 106-121
Mejza, Stanisław Cox's design in	repeated row-column design with split plots and a control • Kachlicka, Danuta;	95 ListyBio 32 81-92
Regression analysis of	repeated survey data (with available software) • Kott, Phillip S.	94 ASASrvyRsr 116-123
Paranjpe, S. A. Nested ANOVA for	repeated surveys with autocorrelated errors • Prayag, V. R.; Chiplonkar, S. A.;	95 BiomtrcJ 37 565-570
Reciprocal cooperation in the prisoner's dilemma	repeated with random horizon • Manuel, C. M.; Tejada, J.	95 Top 3 97-116
Stokes, S. Lynne	Repetitive testing in the presence of inspection errors • Greenland, Betsy S.;	95 Technmcs 37 102-111
superpopulation parameters: Should one use with-	replacement estimators? • Korn, Edward L.; Graubard, Barry I. Variance estimation for	94 ASASrvyRsr 124-132
dynamic programming model for the opportunistic	replacement policy and its application to ball-bearing systems • Kececioglu et al. ...	95 RelEngSS 47 175-185
Comparison of	replacement policies via point processes • Shaked, Moshe; Szekli, Ryszard	95 AdvAppPr 27 1079-1103
Adachi, K.; Kowada, M. Dynamically optimal	replacement policy for a shock model in a Markov random environment • Feng, W.;	94 JOpRsJpn 37 255-270
An optimal inspection-repair-	replacement policy for standby systems • Yeh, Lam	95 JAppProb 32 212-223
Optimum Markov	replacement policies for using the most reliable components • Miescke, K. J.; Shi, Dylan	95 JStPlInf 45 331-346
for the two-parameter exponential distribution with	replacement (Chinese) • Tang, Y.; Fei, H. Maximum likelihood estimation in the life testing	95 ChJApPSt 10 399-404
an application to simple random sampling without	replacement" (V30 p197-211) • de la Peña ... of a sum of dependent variables with	95 AIHenrPB 31 703-704
Estimation of the structural relationship with	replicated observations, unequal means and variances • Isogawa ... least squares:	95 JJapnStS 25 57-70
Khuri, André I.	Replicated paired comparisons at the individual level • Bechtel, Gordon G.; Ofir, Chezy;	95 BJMaStPs 48 115-127
tion of the mean of an exponential distribution via	replicated piecewise stopping number • Bose, A.; Mukhopadhyay, N. Sequential estima-	95 StatDecn 13 351-361
Bose et al. Sequential interval estimation via	replicated piecewise stopping times and accelerated stopping times in a class of ... •	95 SqtlAnly 14 287-305
Corrections to "Using results from	replicated studies to estimate linear models" (V17 p341-362) • Becker, Betsy Jane	95 JEdStat 20 100-102
Bardhan, Indrajit Synthetic	replication of American contingent claims when portfolios are constrained •	95 StocProc 57 149-165
Limit theorem on option	replication cost with transaction costs • Kusuoka, Shigeo	95 AnlsApPr 5 198-221
in structural errors-in-variables model with partial	replication • Huwang, Longcheen Interval estimation	95 JMultiAn 55 230-245
perrectangles, error correcting codes and fractional	replication plans • Mullen, Gary L.; Suchower, Stephan J. Frequency hy-	93 JCISS 18 5-18
A limit theorem for the	replication time of a DNA molecule • Cowan, Richard; Chiu, S. N.; Holst, Lars	95 JAppProb 32 296-303
Carothers, A. D.; Bickmore, W. A. Models of DNA	replication timing in interphase nuclei: An exercise in inferring process from state •	95 Biomtrcs 51 750-755
E-optimal nested row-column designs with unequal	replications of treatments • Uddin, Nizam; Hossain, Syed A.	95 SankhyaB 57 112-121
Testing regression lack of fit without	replication: A tutorial around Minitab's XLOF procedures • Lawrance, A. J.	94 JAppStat 21 541-548
designs with unequal block sizes and unequal	replications • Gupta, S.; Kageyama, S. Variance balanced	92 UtilMath 42 15-24
interaction element for three-way layout without	replication • Kawasaki, Hidemi; Miyakawa, Masami An ANOVA model and	95 JJapnStS 25 143-150
Survey • Peracchi, Franco; Welch, Finis How	representative are matched cross-sections? Evidence from the Current Population	95 JEconmtx 68 153-179
Random sampling versus	representative samples • Schield, Milo	94 ASAProStEd 107-110
Children's concepts of average and	representativeness • Mokros, Jan; Russell, Susan Jo	95 JRsrMaEd 26 20-39
V. M. Threshold models for combination data from	reproductive and developmental experiments • Schwartz, P. F.; Gennings, C.; Chinchilli,	95 JASA 90 862-870
of additive and multiplicative models for	reproductive risk factors and post-menopausal breast cancer • Lund, Eiliv Comparison	95 StatMed 14 267-274
C. V. Issues and approaches for assessing risks from	reproductive toxicants: DBCP as a case study • Mazumdar, S.; Mattison, D. R.; Damaraju,	95 SankhyaB 57 223-236
Introduction to motor activity data (	reproductive toxicology) • Beyer, Bruce K.	94 ASAProBiop 482-483
User	requirements for statistical metainformation • Podehl, M.	93 StatJUN 10 113-120
	Resampling	
	Bayesian statistics using Mathematica • Cook, Peyton; Broemeling, Lyle D.	95 AmerStat 49 70-76
Zidek, James V.	A bootstrap based on the estimating equations of the linear model • Hu, Feifang;	95 Biomtrka 82 263-275
	Bootstrap confidence limits on process capability indices • Collins, Alan J.	95 Ststcian 44 373-378
Veraverbeke, N.	Bootstrapping a nonparametric polytomous regression model • Aerts, M.;	95 MaMethSt 4 189-200
Bayesian alternative • Cohen, A.; Makov, U. et al.	Computer intensive methods for inference on parameters of complex models - A	95 JStPlInf 47 79-94
Lefkopoulou, Myrto; Ryan, Louise M.	Efficiency and power of tests for multiple binary outcomes • Legler, Julie M.;	95 JASA 90 680-693
Davis, Richard A.; Dunsmuir, William T. M.	Improved bootstrap prediction intervals for autoregressions • Breidt, F. Jay;	95 JTimSrAn 16 177-200
lutionary rates • Adell, José Carlos; Dopazo, Joaquín	Monte Carlo simulation in phylogenies: An application to test the constancy of evo-	94 JMolEvol 38 305-309
Loughin, Thomas M.	A residual bootstrap for regression parameters in proportional hazards models •	95 JStCmpSm 52 367-384
	Two permutation tests of equality of variances • Baker, Rose D.	95 StatComp 5 289-296
harvest decisions using the sampling/importance	resampling algorithm • McAllister et al. A Bayesian approach to stock assessment and	94 JFishRBC 51 2673-2687
Mammen, Enno On general	resampling algorithms and their performance in distribution estimation • Hall, Peter;	94 AnlsStat 22 2011-2030
Inference for autocorrelations by	resampling • Romano, Joseph P.; Thombs, Lori A.	94 ASAProStCp 1-10
(Rev. German) Review of	"Resampling - Base multiple testing" • Westfall, P. H.; Young, S. S. (Auth); Schlittgen, R.	95 AllgeStA 79 340-340
Westfall, P. H.; Young, S. S. (Auth) et al. Review of	"Resampling-based multiple testing. Examples and methods for p-value adjustment" •	95 CmpStDA 20 235-236
James (Rev) Review of	"Resampling-based multiple testing" • Westfall, P. H.; Young, S. S. (Auth); Carpenter,	95 JAppStat 22 434-435
Edwards, Don (Rev) Review of	"Resampling-based multiple testing" • Westfall, Peter H.; Young, S. Stanley (Auth);	95 JQualTek 27 272-273
Correction to "Review of	'Resampling-based multiple testing'" (V27 p272-273) • Edwards, D.	95 JQualTek 27 403-403
Pandher, Gurupdesh S.	Resampling inference for quantile shares • Kovačević, Milorad S.; Yung, Wesley;	94 ASASrvyRsr 206-211
	Resampling methods for complex surveys • Rao, J. N. K.	94 ASASrvyRsr 35-41
A stepwise	resampling method of multiple hypothesis testing • Troendle, James F.	95 JASA 90 370-378
Some recent applications of	resampling methods to survey data • Rao, J. N. K.	92 ProIrStC1 163-175
Review of "Permutation tests. A practical guide to	resampling methods for testing hypotheses" • Good, Ph. (Auth); Rudolph, P. E. (Rev)	95 BiomtrcJ 37 150-150
Review of "Permutation tests: A practical guide to	resampling methods for testing hypotheses" • Good, Philip (Auth); Niknian, Minoo (Rev)	95 Technmcs 37 341-342
Some	resampling procedures under symmetry • Jing, Bing-Yi	95 AstrIJSt 37 337-344
Review of	"Resampling stats (Version 3.14)" • Bruce, Peter (Auth); Reeves, Jane (Rev)	95 TeachgSt 17/3 101-103
of fit testing for ordered discrete distributions by	resampling techniques • Pesarin, Fortunato Goodness	94 Metron 52/3 57-71
Fieller's problems and	resampling techniques • Hwang, J. T. Gene	95 StSinica 5 161-171
Robb, William H.	Resampling variance estimates for complex survey designs - A simulation study •	94 ASASrvyRsr 218-221
A statistical analysis of the 1992	research assessment exercise • Taylor, Jim	95 JRSS-A 158 241-261
shot noise processes with applications to risk	reserves • Klüppelberg, Claudia; Mikosch, Thomas Explosive Poisson	95 Bernoull 1 125-147
The effect of the retention limit on the risk	reserve • Centeno, Maria de Lourdes	95 AstinB 25 67-74
temporary mobility patterns and reporting of usual	residence • Bates, Nancy; Gerber, Eleanor A typology of	94 ASASrvyRsr 331-336
Coverage implications of using the term 'usual	residence' • Sweet, Elizabeth M.; Alberti, Nicholas S.	94 ASASrvyRsr 319-324
Teaching undergraduates and medical	residents how to critique journal articles • Norton, H. James	94 ASAProStEd 162-164

... Utility programs offered and participants in the	residential, commercial, and manufacturing sectors of the ... • Battles et al.	94 ASAProBuEc 20-27
Husson, E. et al. Real-time pricing of electricity for	residential customers: Econometric analysis of an experiment • Aubin, C.; Fougére, D.;	95 JApEcmtx 10 171-191S
Imputation of vacant units in the	residential rent index of the U.S. Consumer Price Index • Baskin, Robert M.	94 ASASrvyRsr 382-387
Patterns in	residuals for 2^n designs • John, Peter W. M.	95 JAppStat 22 253-259
T.; Letterman, Raymond D. Factors regulating	residual aluminium concentrations in treated waters (Disc: p305-309) • Driscoll, Charles	95 EnvrMtrc 6 287-305
assumptions in mixed-model analysis of variance: A	residual analysis approach • Oman, Samuel D. Checking the	95 CmpStDA 20 309-330
Bayesian	residual analysis for binary response regression models • Albert, Jim; Chib, Siddhartha	95 Biomtrka 82 747-759
	Residual analysis in the polytomous Rasch model • Andersen, Erling B.	95 Psymtrka 60 375-393
run lengths for CUSUM control charts applied to	residuals • Runger, George C.; Willemain, Thomas R.; Prabhu, Sharad Average	95 CommStA 24 273-282
of the sequential empirical processes of	residuals in ARMA models • Bai, Jushan Weak convergence	94 AnlsStat 22 2051-2061
	Residuals based estimators of the covariogram • Grondona, Martin O.; Cressie, Noel	95 MaOpfStS 26 209-218
Loughin, Thomas M. A	residual bootstrap for regression parameters in proportional hazards models •	95 JStCmpSm 52 367-384
Bromberg, Judith A SAS macro for	residual deviance of ordinal regression analysis • Wan, Jim Y.; Wang, Whedy;	94 CmpBiomd 45 307-310
Wagon wheel designs with	residual effects from neighboring treatments • Cornell, John A.	93 JCISS 18 255-272
	Residuals in the growth curve model • von Rosen, Dietrich	95 AnInStMa 47 129-136
Peña, Edsel A. Properties of hazard-based	residuals and implications in model diagnostics • Baltazar-Aban, Inmaculada;	95 JASA 90 185-197
Akman, H. Olcay Mean	residual life function for certain types of non-monotonic ageing • Gupta, Ramesh C.;	95 CommStSM 11 219-225
Ramesh C.; Akman, H. Olcay Correction to "Mean	residual life function for certain types of non-monotonic ageing" (V11 p219-225) • Gupta,	95 CommStSM 11 561-562
towski, M.; Teugels, J. L.; Vynckier, P. The mean	residual life function at great age: Applications to tail estimation • Beirlant, J.; Bronia-	95 JStPlInf 45 21-48
Bathtub failure rate and upside-down bathtub mean	residual life • Mi, Jie	95 IEEEReli 44 388-391
Trend change in mean	residual life • Lim, Jae-Hak; Park, Dong Ho	95 IEEEReli 44 291-296
A note on percentile	residual life • Song, Jae-Kee; Cho, Gyo-Young	95 SankhyaA 57 333-335
J.-H.; Xie, Y.-L. et al. Maximum sum of binary-coded	residuals (MASBR) regression as a robust procedure for treatment of spectral data • Wang,	95 JChemtrc 9 373-387
Band covariance matrix estimation using restricted	residuals: A Monte Carlo analysis • Ligeralde, Antonio V.; Brown, Bryan W.	94 IntEconR 35 751-767
Gauss M. Bias correction and improved	residuals for non-exponential family nonlinear models • Paula, Gilberto A.; Cordeiro,	95 CommStA 24 1193-1210
	Residual plot	
	Seeing a curve in multiple regression • Berk, Kenneth N.; Booth, David E.	95 Technmcs 37 385-398
	Residual plots for the censored data linear regression model • Hillis, Stephen L.	95 StatMed 14 2023-2036
The interpretability of LMS and LTS	residual plots • McKean, J. W.; Sheather, S. J.; Hettmansperger, T. P.	94 ASAProStEd 17-26
Comparison of regression curves using quasi-	residuals • Kulasekera, K. B.	95 JASA 90 1085-1093
Iterative refinement of parameter estimates and	residuals from a regression model fitted using QR decomposition methods • Smith, D. M.	95 CmpStDA 19 327-349
Influence and	residuals in restricted generalized linear models • Paula, Gilberto A.	95 JStCmpSm 51 315-331
heteroscedasticity, using ranks of squared	residuals • Wong, H.; Li, W. K. Portmanteau test for conditional	95 JAppStat 22 121-134
Liebhart, Jerzy Estimation of the spirometric	residual volume (RV) by a regression built from Gower distances • Bartkowiak, Anna;	95 BiomtrcJ 37 131-149
hazards tests and diagnostics based on weighted	residuals" (V81 p515-526) • Grambsch, P. M.; Therneau, T. M. Correction to "Proportional	95 Biomtrka 82 668-668
Cluster analysis dihedral angles of amino acid	residues in proteins • Sclove, Stanley L.; Sherman, Simon A.	94 ASAProBiop 399-404
exponential convergence rates of the projection	residue of U-statistics and their application (Chinese) • Miao, Baiqi; Zhang, Shuguang The	95 ChJApPSt 10 253-264
Estimating coyote density from mark-	resight surveys • Hein, Eric W.; Andelt, William F.	95 JWildlMg 59 164-169
Diem, Kenneth L.; Pradel, Roger Mark-	resighting analysis of a California gull population • Pugesek, Bruce H.; Nations, Chris;	95 JAppStat 22 625-639
Bounds on orthogonal arrays and	resilient functions • Bierbrauer, Jürgen	95 JComDsgn 3 179-183
analysis with iterative weighting to achieve	resistance • Verboon, Peter; Gabriel, K. Ruben Generalized Procrustes	95 BJMaStPs 48 57-73
Application to circuits with inductance and	resistance (Spanish) • Prat et al. ... of design factor values to obtain robust products:	95 Questo 19 153-169
	Resistant h-plot for a sample variance-covariance matrix • Choi, Yong-Seok	95 JKoreaSt 24 407-417
A regression smoother for	resistant measures of location and scale • Wilcox, Rand R.	95 BJMaStPs 48 189-204
balanced fractional $2^{m_1+m_2}$ factorial designs of	resolution IV • Kuwada, Masahide Analysis of variance of partially	93 JCISS 18 343-367
methods • Grung, Bjørn; Kvalheim, Olav M.	Resolution of multicomponent profiles with partial selectivity. A comparison of direct	95 Chemolab 29 75-87
On the	resolution problem in hyphenated chromatography • Manne, Rolf	95 Chemolab 27 89-94
of balanced fractional $r^m \times s^n$ factorial designs of	resolution V • Kuwada, Masahide; Hyodo, Yoshifumi Analysis of variance	95 JJapnStS 25 129-141
The existence of doubly near	resolvable $(v, 3, 2)$-BIBDs • Lamken, E. R.	94 JComDsgn 2 427-440
The existence of doubly	resolvable $(v, 3, 2)$-BIBDs • Lamken, E. R.	95 JCombThA 72 50-76
A translation of J. X. Lu's "An existence theory for	resolvable balanced incomplete block designs" • Lee, Therese C. Y.; Furino, Steven C.	95 JComDsgn 3 321-340
Block designs with large holes and α-	resolvable BIBDs • Furino, Steven C.; Mullin, Ronald C.	93 JComDsgn 1 101-112
Ghosh, D. K.; Patel, A. H. Construction of α-	resolvable BIBDs using 1-resolvable and α-affine resolvable BIBDs • Shah, S. M.;	93 UtilMath 43 219-224
Miao, Ying; Yin, Jianxing	Resolvable BIBDs with block size 8 and index 7 • Furino, Steven; Greig, Malcolm;	94 JComDsgn 2 101-116
Construction and optimality of affine-	resolvable designs • Bailey, R. A.; Monod, H.; Morgan, J. P.	95 Biomtrka 82 187-200
sets of orthogonal frequency squares and affine	resolvable designs • Laywine, C. Complete	93 UtilMath 43 161-170
A construction of	resolvable designs • Kageyama, S.; Miao, Y.	94 UtilMath 46 49-54
Existence results for near	resolvable designs • Furino, Steven	95 JComDsgn 3 101-113
Cyclically	resolvable designs and triple whist tournaments • Anderson, I.; Finizio, N. J.	93 JComDsgn 1 347-358
Two new direct product-type constructions for	resolvable group-divisible designs • Rees, Rolf S.	93 JComDsgn 1 15-26
Stow, C. A.; Carpenter, S. R.; Cottingham, K. L.	Resource vs. ratio-dependent consumer-resource models: A Bayesian perspective •	95 Ecology 76 1986-1990
Comparison of smoothing methods when applied to	respiratory data collected in mice exposed to an aerosolized ... • Boylstein et al.	94 ASAProBiop 136-141
Rosner, Bernard Longitudinal models for analysis of	respiratory function • D'Agostino, Ralph B., Jr; Sparrow, David; Weiss, Scott;	95 StatMed 14 2205-2216
Modelling frailty in the manufacture of	respiratory protective devices • Wassell, James T.; Kulczycki, Gregory W.; Moyer, Ernest S.	94 ASAProEpid 133-138
tion (Disc: p169-177) • Schwarz, Norbert What	respondents learn from questionnaires: The survey interview and the logic of conversa-	95 IntStRvw 63 153-168
conversation" • Cantor, David Discussion of "What	respondents learn from questionnaires: The survey interview and the logic of	95 IntStRvw 63 173-177
conversation" • Singer, Eleanor Discussion of "What	respondents learn from questionnaires: The survey interview and the logic of	95 IntStRvw 63 169-172
Parsley, Teresa L. (Rev) Review of "The Phantom	respondents: Opinion surveys and political representation" • Brehm, John (Auth);	94 JOffStat 10 351-355
Bellhouse, D. R. et al. On efficiency of using distinct	respondents in a randomized response survey • Mangat, N. S.; Singh, R.; Singh, S.;	95 SurvyM 21 21-23
Income statistics from survey data: Effects of	respondent rounding • Rowe, Geoff; Gribble, Steve	94 ASAProBuEc 77-82
Romuald, K. S. et al. The effect of varying the	respondent selection script on respondent self-selection in RDD telephone surveys •	94 ASASrvyRsr 1299-1304
Blair, Johnny; Bickart, Barbara; Eastman, Elizabeth	Respondent strategies for recall of crime victimization incidents • Czaja, Ronald;	94 JOffStat 10 257-276
Marquis, Kent H. Can we get	respondents to use their personal income records • Bogen, Karen; Moore, Jeffrey C.;	94 ASASrvyRsr 1252-1257
in epidemiologic surveys with two waves of	respondents • Brenner, Hermann Alternative approaches for estimating prevalence	95 AJEpidm 142 1236-1245
Cecco, Kevin Characteristics of	responding and nonresponding households from the 1990 Decennial Census •	94 ASAProGovt 76-81
	Response	
	• see names of specific types of responses such as item - or quantal -	
	Response bias	
	Estimation of correlation in randomized response • Bellhouse, D. R.	95 SurvyM 21 13-19
Lee, Kay-O; Lee, Gi-Sung	A practical plan of randomized response technique (Korean) • Ryu, Jea-Bok;	95 KorApS 8/1 9-26
p379-381) • LeClere, F. B. et al. Social correlates of	response bias in the National Health Interview Survey: Lessons for the redesign (Disc:	94 ASASrvyRsr 373-378
H. C. et al. Estimating derivatives of pharmacokinetic	response curves with varying bandwidths • Song, K. S.; Müller, H. G.; Clifford, A. J.; Furr,	95 Biomtrcs 51 12-20
On robust logistic case-control studies with	response-dependent weights • Wang, C. Y.; Carroll, R. J.	95 JStPlInf 43 331-340

Estimation with multiple inequality	restrictions in a model with multivariate t error terms • Wan, Alan T. K.	94 ASAProBuEc 223-228
for maximum likelihood estimation under linear	restrictions on the parameters • Kim, D. K.; Taylor, J. M. G. The restricted EM algorithm	95 JASA 90 708-716
Estimation of reliability under ordered	restriction on the parameters • Gupta, Ramesh C.; Gupta, Rameshwar D.; H-Akman, Olcay	95 CommStA 24 1799-1812
contingency tables under negative dependence	restrictions with various sampling schemes • Oh ... estimation of cell probabilities in $2 \times k$	95 CommStA 24 2127-2143
An approximation to the distribution of the	resultant from a Fisher distribution • Sarkar, I. C.; Mudholkar, G. S.; Raubertas, R. F.	95 CommStB 24 227-241
tion of price sensitivity and implications for optimal	retail pricing • Kim, Byung-Do; Blattberg, Robert C.; Rossi, Peter E. Modeling the distribu-	95 JBES 13 291-303
geographically localized misspecification error in	retail store choice models • Rust, Roland T.; Donthu, Naveen Capturing	95 JMarRsr 32 103-110
The employment effect in	retail trade of California's 1988 minimum wage increase • Kim, Taeil; Taylor, Lowell J.	95 JBES 13/2 175-182
J.; Mayda, J. E. Redesign of the Hungarian	Retail Trade and Catering Survey: Methodological issues • Choudhry, G. H.; Colledge, M.	94 ASAProBuEc 71-76
estimation: An application to the Canadian Monthly	Retail Trade Survey • Patak, Z.; Hidiroglou, M. A. Raking ratio	94 ASASrvyRsr 299-304
analysis for the evaluation of chromatographic	retention data • Cserháti, Tibor; Forgács, Esther Use of canonical correlation	95 Chemolab 28 305-313
The effect of the	retention limit on the risk reserve • Centeno, Maria de Lourdes	95 AstinB 25 67-74
multicomponent chromatograms: Importance of the	retention pattern • Pietrogrande, M. C. et al. Statistical study of peak overlapping in	95 Chemolab 28 239-258
cultivar trials • Cornelius, P. L. Statistical tests and	retention of terms in the additive main effects and multiplicative interaction model for	93 CropSci 33 1186-1193
stability on concentration fluctuations and wake	retention times for dispersion in the vicinity of an isolated ... • Higson et al. ... atmospheric	95 EnvrMtrc 6 571-581
power • Hansen, Bruce E.	Rethinking the univariate approach to unit root testing - Using covariates to increase	95 EconTher 11 1148-1171
taneous equation model using the notation of the	reticular action mode • Toyoda, H. A new identification rule and an estimator for the simul-	94 Behvmtrk 21 163-175
Stochastic decomposition for	retrial queues • Artalejo, J. R.; Falin, J. I.	94 Top 2 329-342
Matrix analytical methods for M/PH/1	retrial queues • Diamond, Jeffrey E.; Alfa, Attahiru Sule	95 CommStSM 11 447-470
On the M/G/1	retrial queueing system with linear control policy • Martín, M.; Gómez-Corral, A.	95 Top 3 285-305
of statistical methods for the classification and	retrieval of patient events • Chute, C. G.; Yang, Y. An overview	95 MInfoMed 34 104-110
Phase	retrieval for probability measures on abelian groups • Carnal, H.; Fel'dman, G. M.	95 JTherPrb 8 717-725
approach to microwave precipitation profile	retrieval • Evans, K. F.; Turk, J.; Wong, T.; Stephens, G. L. A Bayesian	95 JApMetrl 34 260-279
Marsh, G. M.; Owens, A. D.; Quinn, M. M.; Smith, T. J.	Retrospective estimation of airborne fiber exposure: Statistical aspects • Stone, R. A.;	94 ASAProEpid 127-132
kurtosis in the London Stock Exchange FT-SE index	return distributions • Mills, Terence C. Modelling skewness and	95 Ststcian 44 323-332
Dillman et al. How prenotice letters, stamped	return envelopes and reminder postcards affect mailback response rates for census ... •	95 SurvyM 21 159-165
	Return probabilities for random walk on a half-line • Lalley, Steven P.	95 JTherPrb 8 571-599
Clobert, Jean; Anderson, David R.	Return rates in studies of life history evolution: Are biases large? • Martin, Thomas E.;	95 JAppStat 22 863-875
A theory of risk,	return and solvency • Powers, Michael R.	95 InsrneME 17 101-118
Wiener-Wintner	return-times ergodic theorem • Assani, I.; Lesigne, E.; Rudolph, D.	95 IsrelJMa 92 375-395
The time variation of expected	returns and volatility in foreign-exchange markets • Bekaert, Geert	95 JBES 13 397-408
policy equation of a vector autoregression • Salemi	Revealed preference of the Federal Reserve: Using inverse-control theory to interpret the	95 JBES 13 419-433
contemporaneous outlier analysis for IBM regional	revenue based on shrinkage estimation • Ravishanker, N. et al. Forecasting, planning and	94 ASAProBuEc 211-216
Intrayear monitoring of state tax collections and	revenue forecasts • Nelson, Ray D.; Cornia, Gary C.	94 ASAProBuEc 339-344
Accuracy and rationality of state General Fund	Revenue forecasts: Evidence from panel data • Mocan, H. Naci; Azad, Sam	95 IntJFor 11 417-427
will nondifferential misclassification of an exposure	reverse the direction of a trend?" (V140 p565-571) • Birkett, N. J. Comment on "When	95 AJEpidm 142 783-784
will nondifferential misclassification of an exposure	reverse the direction of a trend?" (V140 p565-571) • Thomas, D. C. Comment on "When	95 AJEpidm 142 782-783
will nondifferential misclassification of an exposure	reverse the direction of a trend?" • Weinberg, C. R. et al. Reply to comments on "When	95 AJEpidm 142 784-784
How pooling failure data may	reverse increasing failure rates • Gurland, John; Sethuraman, Jayaram	95 JASA 90 1416-1423
Uniform convergence of	reversed martingales • Peškir, Goran	95 JTherPrb 8 387-415
Strobl, Franz On the	reversed sub-martingale property of empirical discrepancies in arbitrary sample spaces •	95 JTherPrb 8 825-831
autoregressive sequences: Forward versus	reversed time • Cambanis, S.; Fakhre-Zakeri, I. On prediction of heavy-tailed	94 ThProbAp 39 217-233
determination • Green, Peter J.	Reversible jump Markov chain Monte Carlo computation and Bayesian model	95 Biomtrka 82 711-732
inequality for the occupation measure of a	reversible Markov chain • Dinwoodie, I. H. A probability	95 AnlsApPr 5 37-43
of positive dependence with time for stationary	reversible Markov chains • Hu, Taizhong; Joe, Harry Monotonicity	95 PrEngInf 9 227-237
Antisymmetric functionals of	reversible Markov processes • Goldstein, Sheldon	95 AIHenrPB 31 177-190
On the range of	reversible random walks on Z^2 in a random environment • Zhou, Xian Yin	95 JTherPrb 8 453-473
Exponential convergence for attractive	reversible subcritical nearest particle systems • Mountford, T. S.	95 StocProc 59 235-249
Poirier, Claude Model-based	reweighting for nonresponse adjustment • Binder, David A.; Michaud, Sylvie;	94 ASASrvyRsr 989-994
Kott, Phillip S.	Reweighting and variance estimation for the characteristics of business owners survey •	94 JOffStat 10 407-418
dominance effects of QTLs in interval mapping of F_2	RFLP data • Hayashi, T.; Ukai, Y. Detection of additive and	94 ThApGene 87 1021-1027
Prediction of maize single-cross performance using	RFLPs and information from related hybrids • Bernardo, Rex	94 CropSci 34 20-25
of DNA fragment sizes within and between loci in	RFLP typing of DNA • Chakraborty et al. ... of restriction fragments and independence	94 AJHuGene 54 391-401
Improving the	RHC-strategy • Gabler, Siegfried; Stenger, Horst	95 StHefte 36 327-336
Investigating the geometry of a	ρ-dimensional data set • Dawkins, Brian P.	95 JASA 90 350-359
	ρ-mixing	
variables • Peligrad, Magda; Shao, Qi-Man	A note on the almost sure central limit theorem for weakly dependent random	95 StPrLet 22 131-136
Estimation of the variance of partial sums for	ρ-mixing random variables • Peligrad, Magda; Shao, Qi-Man	95 JMultiAn 52 140-157
Maximal inequalities for partial sums of	ρ-mixing sequences • Shao, Qi-Man	95 AnlsProb 23 948-965
A CLT for the periodograms of a	ρ^*-mixing random field • Miller, Curtis	95 StocProc 60 313-330
and a fundamental problem of insurance in a data-	rich environment • Reid, D. H. Operational time	95 AplStMDA 11 257-269
M.; Wilson, M. H. et al. Estimating bird species	richness from capture and count data • Dawson, D. K.; Sauer, J. R.; Wood, P. A.; Berlanga,	95 JAppStat 22 1063-1068
On linear discriminant analysis with adaptive	ridge classification rules • Loh, Wei-Liem	95 JMultiAn 53 264-278
On ecological regression and	ridge estimation • Füle, Erika	95 CommStB 24 385-398
Hanumara, R. C. Unbiased	ridge estimation with prior information and ridge trace • Crouse, Robert H.; Jin, Chun;	95 CommStA 24 2341-2354
Finite-sample properties of an adaptive	ridge estimator (Chinese) • Yan, Liqing; Wang, Songgui	95 ChJApPSt 10 194-201
	Ridge regression	
bias and MSE • Akdeniz, Fikri; Kaçranlar, Selahattin	On the almost unbiased generalized Liu estimator and unbiased estimation of the	95 CommStA 24 1789-1797
	A Bayesian analysis for a class of penalised likelihood estimates • Barry, Daniel	95 CommStA 24 1057-1071
	Double bootstrap for shrinkage estimators • Vinod, H. D.	95 JEconmtx 68 287-302
Ilallawa, Abdel M.; Azzam, Abdul-Mordy H.	A new method for generating the design matrix of a linear regression model •	95 EgpStJ 39 106-119
lated explanatory variables • Elston, D. A.; Proe, M. F.	Smoothing regression coefficients in an overspecified regression model with interre-	95 ApplStat 44 395-406
A.; Firinguetti, L. et al. Confidence intervals in	ridge regression by bootstrapping the dependent variable: A simulation study • Crivelli,	95 CommStB 24 631-652
	Ridge regression estimation of the linear probability model • Gana, Rajaram	95 JAppStat 22 537-539
of the iterative algorithm for choosing k value in	ridge regression estimation (Chinese) • He, Z.; He, L. The convergence theorem and limit	95 AMaAPknC 17 59-64
Generalized	ridge regression estimators under the LINEX loss function • Ohtani, K.	95 StHefte 36 99-110
and approximate relative efficiency of generalized	ridge-type shrinkage estimator • Noor, Islam; Ahmad, Naseer Exact	95 PakJStA 11 17-22
	Ridit analysis	
	Analysis of contingency tables by marginal scores • Gokhale, D. V.; Klein, Ruben	95 RBrzlPrS 9 25-42
Schnell, Daniel J.; Magee, Elvin; Sheridan, John R.	A regression method for analysing ordinal data from intervention trials •	95 StatMed 14 1177-1189
	Riemann-Stieltjes integration	
Towghi, Nasser	Stochastic integration of processes with finite generalized variations. I •	95 AnlsProb 23 629-667
Brownian motion on a Lipschitz	Riemannian manifold composed of different materials • Zheng, Weian	95 Stochatx 54 75-87
Radial part of Brownian motion on a	Riemannian manifold • Liao, M.; Zheng, W. A.	95 AnlsProb 23 173-177

treatment effect for the two-sample problem with	right censored data • Park, Hyo-Il; Park, Sang-Gue Quantile estimation of	95 StPrLet 24 139-145
samples where all observations are either left- or	right-censored • Andersen, P. K.; Rønn, B. B. A nonparametric test for comparing two	95 Biomtrcs 51 323-329
Daniels's association measures under	right censoring • O'Quigley, John	95 AplStMDA 11 109-119
in the location scale model under randomly	right-censoring (Chinese) • Lu, Wuewen Estimation of the parameters	95 ChJApPSt 10 272-277
Left and	right linear innovations for a multivariate S α S random variable • Rutkowski, Marek	95 StPrLet 22 175-184
Distributions slanted to the	right • Rösler, U.	95 StNeerla 49 83-93
The	rights and wrongs of control charts (Disc: p289-306) • Caulcutt, Roland	95 ApplStat 44 279-288
M-estimates of	rigid body motion on the sphere and in Euclidean space • Chang, Ted; Ko, Daijin	95 AnlsStat 23 1823-1847
ten Suppressing permutations or	rigid planar rotations: A remedy against nonoptimal varimax rotations • Berge, Jos M. F.	95 Psymtrka 60 437-446
Smith, A. F. M. Bayesian analysis of	ring-recovery data via Markov chain Monte Carlo simulation • Vounatsou, P.;	95 Biomtrcs 51 687-708
MATLAB: An environment for analyzing	ring-recovery and recapture data • Catchpole, E. A.	95 JAppStat 22 801-816
A model of the	ring-recovery reporting process for the Cape Griffon Gyps coprotheres • Piper, S. E.	95 JAppStat 22 641-659
Delineating bird populations using	ring recoveries • Pendleton, Grey W.; Sauer, John R.	95 JAppStat 22 1049-1055
S. Separation of hunting and natural mortality using	ring-return models: An overview • Pollock, Kenneth H.; Conroy, Michael J.; Hearn, William	95 JAppStat 22 557-566
Stolyar, A. L. Queues served by a rotating	ring • Coffman, E. G., Jr; Gilbert, E. N.; Greenberg, A. G.; Leighton, F. T.; Robert, Philippe;	95 CommStSM 11 371-394
Estimating survival rates from recoveries of birds	ringed as young: A case study • Francis, Charles M.	95 JAppStat 22 567-577
scheme perspective • Baillie, Stephen R. Uses of	ringing data for the conservation and management of bird populations: A ringing	95 JAppStat 22 967-987
survival of British pigeons and doves by analysis of	ringing recoveries • Aebischer, N. J. Investigating the effects of hunting on the	95 JAppStat 22 923-934
young red wines from the Denomination of Origin	'Rioja' • Fernández et al. ... methods in the modelling and prediction of the colour of	95 Chemolab 28 273-285
	Risk	
distribution • Chou, Jine-phone	Admissibility of conjugate Bayes estimators for the mean of a negative binomial	95 StatDecn 13 301-306
	Deductible insurance and production • Machnes, Yaffa	95 InsrneME 17 119-123
	Dynamic approaches to pension funding • Haberman, Steven; Sung, Joo-Ho	94 InsrneME 15 151-162
Lawson, Andrew B.	The power of focused tests to detect disease clustering • Waller, Lance A.;	95 StatMed 14 2291-2308
• Bardhan, Indrajit	Synthetic replication of American contingent claims when portfolios are constrained	95 StocProc 57 149-165
Discrete time. • Shiryaev, A. N.; Kabanov, Yu. M. et al.	Toward the theory of pricing of options of both European and American types. I.	94 ThProbAp 39 14-60
tinuous time. • Shiryaev, A. N.; Kabanov, Yu. M. et al.	Toward the theory of pricing of options of both European and American types. II. Con-	94 ThProbAp 39 61-102
Suksawasdi Na; Böhning, Danmar	Traffic accident mapping in Bangkok metropolis: A case study • Ayuthya, Rampai	95 StatMed 14 2445-2458
	• see also absolute -, Bayes -, competing -, relative -, decision theory, insurance	
Aluminium and the	risk for Alzheimer's disease • McLachlan, D. R. C.	95 EnvrMtrc 6 233-275
	Risk analysis	
Keller, L. Robin; Sarin, Rakesh K.	Fair processes for societal decisions involving distributional inequalities •	95 RiskAnal 15 49-59
• Reid, D. H.	Operational time and a fundamental problem of insurance in a data-rich environment	95 AplStMDA 11 257-269
prospective cohort investigations • Kaaks, R. et al.	Sample size requirements for calibration studies of dietary intake measurements in	95 AJEpidm 141 557-565
on "The meaning of probability in probabilistic	risk analysis" • Watson, Stephen R. Reply to comment	95 RelEngSS 49 207-209
nursing ... • D'Agostino et al. Development of health	risk appraisal functions in the presence of multiple indicators: The Framingham study	95 StatMed 14 1757-1770
	Risk assessment	
analysis • Greenland, Sander	Dose-response and trend analysis in epidemiology: Alternatives to categorical	95 Epidmlgy 6 356-365
Cupples, L. Adrienne	Estimating the probability for major gene Alzheimer disease • Farrer, Lindsay A.;	94 AJHuGene 54 374-383
tobacco smoke (Disc: p413-424) • Gross, Alan J.	Uncertainties in lung cancer risk estimates reported for exposure to environmental	95 EnvrMtrc 6 403-412
Martz, Harry F.; Picard, Richard R.	Uncertainty in Poisson event counts and exposure time in rate estimation •	95 RelEngSS 48 181-190
Epidemiology and quantitative	risk assessment: A bridge from science to policy • Hertz-Picciotto, Irva	95 AJPubHlh 85 484-491
Review of "Ecological	risk assessment" • Suter, Glenn (Ed); Ziegel, Eric R. (Rev)	95 Technmcs 37 240-240
Integrating epidemiologic data into	risk assessment • Wartenberg, Daniel; Simon, Ronald	95 AJPubHlh 85 491-493
between risk and ... • Lee et al. Bias correction in	risk assessment when logistic regression is used with an unevenly proportioned sample	94 ASAProBuEc 304-309
Flatman, George T. Comment on	"Risk assessment of lung cancer due to environmental tobacco smoke (ETS)" •	94 ASAProEnvr 74-76
Gary C.; Burnham, Kenneth P. Some specialized	risk assessment methodologies for vertebrate populations • Anderson, David R.; White,	95 EnvEcoSt 2 91-115
Environmental	risk assessment model for low dose extrapolation • Janardan, K. G.	95 CommStA 24 2621-2633
Developmental toxicity modeling for	risk assessment • Catalano, Paul J.; Ryan, Louise M.; Senchaudhuri, Pralay	94 ASAProEnvr 38-45
ness, identifiability, statistical perspectives and	risk assessment • Sen, Pranab Kumar; Margolin, Barry H. Inhalation toxicology: Aware-	95 SankhyaB 57 252-276
Sequential nonparametric estimation with assigned	risk • Efromovich, Sam	95 AnlsStat 23 1376-1392
J. et al. Influence of measurement errors on HIV	risk behavior analysis: A case study examining condom use among drug users • Wang,	95 StrEqMod 2 319-334
Using	risk calculation to implement an extended relative pair analysis • Curtis, D.; Sham, P. C.	94 AnHuGene 58 151-162
breast cell proliferation and breast cancer	risk • Pike, M. C.; Spicer, D. V. Hormones,	94 ASAProEpid 24-33
ing quality level • Raju, C.; Murthy, M. N. N. Minimum	risks chain sampling plans (ChSP-4 (c_1, c_2)) indexed by acceptable quality level and limit-	95 JAppStat 22 389-426
inference, probability, and a conceptual gulf in	risk communication • Walker, Vern R. Direct	95 RiskAnal 15 603-609
function and nonnormal error terms • Hasegawa On	risk comparisons of some estimators in a linear regression model under Inagaki's loss	95 CommStA 24 1655-1685
Liu, Ying-Hsing; Yang, Ming-Chung Bounded	risk conditions in simultaneous estimation of independent Poisson means •	95 JStPlInf 47 319-331
On the preservation of some orderings of	risks under convolution • Pellerey, Franco	95 InsrneME 16 23-30
Use of neural networks in predicting the	risk of coronary artery disease • Lapuerta, Pablo; Azen, Stanley P.; LaBree, Laurie	95 CmpBimdR 28 38-52
of test and maintenance intervals based on	risk and cost • Vaurio, J. K. Optimization	95 RelEngSS 49 23-36
Model-based estimation of population attributable	risk under cross-sectional sampling • Basu, Srabashi; Landis, J. Richard	95 AJEpidm 141 1338-1343
Model-based estimation of population attributable	risk under cross-sectional sampling • Basu, Srabashi; Landis, J. Richard	95 AJEpidm 142 1338-1343
Review of "Stochastic orders and decision under	risk" • Mosler, K.; Scarsini, M. (Ed); Heyer, H. (Rev)	95 Metrika 42 139-139
Handbook for environmental	risk decisionmaking • Cothern, C. Richard (Ed)	95 CRCPress:FL 416p
Methods of inference for estimates of absolute	risk derived from population-based case-control studies • Benichou, J.; Gail, M. H.	95 Biomtrcs 51 182-194
A further look at the Cochran-Mantel-Haenszel	risk difference • Sato, Tosiya	95 ConClnTr 16 359-361
Bayesian analysis of space-time variation in disease	risk • Bernardinelli, L.; Clayton, D.; Pascutto, C.; Montomoli, C.; Ghislandi, M.; Songini, M.	95 StatMed 14 2433-2443
Forecasting earthquakes and earthquake	risk • Vere-Jones, David	95 IntJFor 11 503-538
Comparing extinction	risk and economic cost in wildlife conservation planning • Haight, Robert G.	95 EcolgApp 5 767-775
• Gross, Alan J. Uncertainties in lung cancer	risk estimates reported for exposure to environmental tobacco smoke (Disc: p413-424)	95 EnvrMtrc 6 403-412
et al. Qualitative and quantitative assessment of the	risk from the exposure to fetotoxic chemical compounds (Disc: p78-79) • Talwalker, S.	95 EnvEcoSt 2 71-78
	Risk factor	
multiple informants • Fitzmaurice, G. M. et al.	Bivariate logistic regression analysis of childhood psychopathology ratings using	95 AJEpidm 142 1194-1203
Aggregate data studies of disease	risk factors • Prentice, Ross L.; Sheppard, Lianne	95 Biomtrka 82 113-125
Assessment of cardiovascular	risk factors in the elderly: The Framingham heart study • Larson, Martin G.	95 StatMed 14 1745-1756
Madans et al. Differences in morbidity measures and	risk factor identification using multiple data sources: The case of coronary heart ... •	95 StatMed 14 643-653
sequential screening guidelines for quantitative	risk factors measured with error • Hughes, M. D.; Thompson, S. G.; Pocock, S. J. Optimal	95 JASA 90 19-26
additive and multiplicative models for reproductive	risk factors and post-menopausal breast cancer • Lund, Eiliv Comparison of	95 StatMed 14 267-274
Woodward, M.; Laurent, K. et al. An analysis of	risk factors for prevalent coronary heart disease by using the proportional odds model •	95 Ststcian 44 69-80
tive survival bias) • Blauwet, M. B. et al. Changing	risk factor relationships in aging populations due to self-selection statistical bias (selec-	94 ASAProEpid 76-81
for living probabilistic safety assessment and	risk follow-up • Arjas, Elja; Holmberg, Jan Marked point process framework	95 RelEngSS 49 59-73

	Risk function	
Sinha, B. K.; Chaudhuri, G.; Chang, C.-H.	Estimation of a multivariate normal mean vector and local improvements • Pal, N.;	95 MaOpfStS 26 1-17
Gupta, A. K.; Ofori-Nyarko, Samuel	Improved minimax estimators of normal covariance and precision matrices •	95 MaOpfStS 26 19-25
Markov model with respect to an arbitrary quadratic	risk function • Baksalary, J. K. et al. Admissible linear estimation in the general Gauss-	95 JStPlInf 44 341-347
Does Markov-modulation increase the	risk? • Asmussen, Søren; Frey, Andreas; Rolski, Tomasz; Schmidt, Volker	95 AstinB 25 49-66
The	risk inflation criterion for multiple regression • Foster, Dean P.; George, Edward I.	94 AnlsStat 22 1947-1975
Public health methods - Attributable	risk as a link between causality and public health action • Northridge, Mary E.	95 AJPubHlh 85 1202-1204
Rasinski, K. A.; Baldwin, A. K.; Willis, G. B.; Jobe, J. B.	Risk and loss perceptions associated with survey reporting of sensitive behaviors •	94 ASASrvyRsr 497-502
S. Biconvex models and algorithms for	risk management problems • Sherali, Hanif D.; Alameddine, Amine; Glickman, Theodore	94 AJMMgSci 14 197-228
Analyzing uncertainty for	risk management • Castrup, Howard	95 ASQCTekCon 120-135
Managing	risks • Nash, John C.; Nash, Mary M.	95 Chance 8/4 25-31
Wirth, Heike Identification	risks of microdata: Evidence from experimental studies • Müller, Walter; Blien, Uwe;	95 SocMethR 24 131-157
Information inequalities for the minimax	risk • Sato, Michikazu; Akahira, Masafumi	95 JJapnStS 25 151-158
	Risk model	
Mikosch, T.	Delay in claim settlement and ruin probability approximations • Klüppelberg, C.;	95 ScanActJ 154-168
the adjustment coefficient in an ARMA(p,q)	risk model • Christ, Ralf; Steinebach, Josef Estimating	95 InsrneME 17 149-161
Multiplicative	risk model for predicting rates (Disc: p125-126) • Davis, Alan B.; Larimore, Walter L.	94 ASAProEpid 123-124
A goodness-of-fit test for the additive	risk model with a binary covariate • Kim, Jinheum; Song, Moon Sup	95 JKoreaSt 24 537-549
Baek, Hoh Yoo Estimators with nondecreasing	risk in a multivariate normal distribution • Kim, Byung Hwee; Koh, Tae Wook;	95 JKoreaSt 24 257-266
The choice of test for detecting raised disease	risk near a point source • Bithell, J. F.	95 StatMed 14 2309-2322
in a producer-supplier conflicting environment:	Risk neutral case • Tapiero, Charles S. Acceptance sampling	95 AplStMDA 11 3-12
Unbiased estimator of	risk for an orthogonally invariant estimator of a covariance matrix • Sheena, Yo	95 JJapnStS 25 35-48
analysis of variability for estimates of absolute	risk from a population-based case-control study on breast cancer • Benichou, J. A complete	95 BiomtrcJ 37 3-24
	Risk portfolio	
	Some alternatives for the individual model • Kaas, R.; Gerber, H. U.	94 InsrneME 15 127-132
arbitrage pricing theory and non-stationarity of the	risk premium: An error components non-linear approach • Chen, Y. H. et al. Testing the	94 ASAProBuEc 178-183
approximations for ruin probabilities of	risk processes perturbed by diffusion • Schmidli, H. Cramér-Lundberg	95 InsrneME 16 135-149
Confidence intervals for the	risk ratio in cohort studies under inverse sampling • Lui, Kung-Jong	95 BiomtrcJ 37 965-971
vascular mortality" (V12 ... • Edwardes Comment on	"Risk ratio and rate ratio estimation in case-cohort designs: Hypertension and cardio-	95 StatMed 14 1609-1609
Schouten et al. Reply to comment on	"Risk ratio and rate ratio estimation in case-cohort designs: Hypertension and ... •	95 StatMed 14 1610-1610
	Risk reduction	
	On the invariant estimation of a normal variance ratio • Madi, Mohamed T.	95 JStPlInf 44 349-357
Peter Reducing the maximum	risk of regression estimators by polyhedral projection • Schmidt, Karsten; Stahlecker,	95 JStCmpSm 52 1-15
Damaraju, C. V. Issues and approaches for assessing	risks from reproductive toxicants: DBCP as a case study • Mazumdar, S.; Mattison, D. R.;	95 SankhyaB 57 223-236
Poisson shot noise processes with applications to	risk reserves • Klüppelberg, Claudia; Mikosch, Thomas Explosive	95 Bernoull 1 125-147
The effect of the retention limit on the	risk reserve • Centeno, Maria de Lourdes	95 AstinB 25 67-74
Detection and estimation of J-shaped	risk-response relationships • Goetghebeur, Els J. T.; Pocock, Stuart J.	95 JRSS-A 158 107-121
A theory of	risk, return and solvency • Powers, Michael R.	95 InsrneME 17 101-118
Review of	"Risk" • Adams, John (Auth); Senn, Stephen (Rev)	95 StatMed 14 1727-1728
risk • Tomasson, Helgi	Risk scores from logistic regression: Unbiased estimates of relative and attributable	95 StatMed 14 1331-1339
R. S.; Eccles, D. M.; Morton, N. E. Prediction of genetic	risks from segregation analyses of morbid risks • Shields, D. C.; Marlow, A. J.; Houlston,	94 HumHrdty 44 52-55
Mukhopadhyay, N.; Datta, S. On fine-tuned bounded	risk sequential point estimation of the mean of an exponential distribution •	95 SoAfrStJ 29 9-27
A likelihood approach to calculating	risk support intervals • Leal, Suzanne M.; Ott, Jurg	94 AJHuGene 54 913-917
the ratio of specific occurrence/exposure rates in	risk and survival analysis • Tu, Dongsheng; Gross, Alan J. Accurate confidence intervals for	95 BiomtrcJ 37 611-626
	Risk theory	
	The effects of quantized data • Jewell, William S.	95 AplStMDA 11 201-216
	Explicit analytic ruin probabilities for bounded claims • De Vylder, F.; Marceau, E.	95 InsrneME 16 79-105
	Harald Cramér and insurance mathematics • Martin-Löf, Anders	95 AplStMDA 11 271-276
Steinebach, J.	Invariance principles for renewal processes and some applications • Alex, M.;	95 ThPrMaSt 50 23-54
• Paulsen, Jostein	Optimal per claim deductibility in insurance with the possibility of risky investments	95 InsrneME 17 133-147
	Order relations for some distributions • Hesselager, Ole	95 InsrneME 16 129-134
Nielsen, Ole E.; Schmidli, Hanspeter	Saddlepoint approximations for the probability of ruin in finite time • Barndorff-	95 ScanActJ 169-186
Hardy, M. R. (Rev) Review of "Practical	risk theory for actuaries" • Daykin, C. D.; Pentikäinen, T.; Pesonen, E. (Auth);	95 JAppStat 22 547-548
Coutts, Stewart M. (Rev) Review of "Practical	risk theory for actuaries" • Daykin, C. D.; Pentikäinen, T.; Pesonen, M. (Auth);	95 JRSS-A 158 633-633
Neuhaus, Walther (Rev) Review of "Practical	risk theory for actuaries" • Daykin, C. D.; Pentikäinen, T.; Pesonen, M. (Auth);	95 AstinB 25 75-76
Vajda, S. (Rev) Review of "Practical	risk theory for actuaries" • Daykin, C. D.; Pentikainen, T.; Pesonen, M. (Auth);	95 Ststcian 44 282-283
Reich, A. Applications of	risk theory and multivariate analysis in insurance practice • Niggemeyer, B.; Radtke, M.;	95 AplStMDA 11 231-244
	Risk theory of the second and third kind • Embrechts, Paul	95 ScanActJ 35-43
Estimation of exposure prevalence in a population at	risk using data from cases and an external estimate of the relative risk • Mittleman	95 Epidmlgy 6 551-553
Reduction of	risk using restricted estimators • Rueda, Cristina; Salvador, Bonifacio	95 CommStA 24 1011-1022
to approximate joint uncertainty and variability in	risk • Bogen, Kenneth T. Methods	95 RiskAnal 15 411-419
the instantaneous volatility and covariance of	risky assets • Chesney, Marc; Elliott, Robert J. Estimating	95 AplStMDA 11 51-58
deductibility in insurance with the possibility of	risky investments • Paulsen, Jostein Optimal per claim	95 InsrneME 17 133-147
Tree-ring reconstruction of Upper Gila	River discharge • Meko, David; Graybill, Donald A.	95 WaterReB 31 605-616
results for self-similar trees with applications to	river networks • Peckham, Scott D. New	95 WaterReR 31 1023-1029
Neighbor joining and maximum likelihood with	RNA sequences: Addressing the interdependence of sites • Tillier, E. R. M.; Collins, R. A.	95 MolBioEv 12 7-15
A simple combinatorial proof of a result by	Robertson and Pillers • Mukerjee, Hari	95 StPrLet 24 173-175
Review of	"Robust and adaptive tests" (German) • Büning, Herbert (Auth); Rieder, Helmut (Rev)	95 StHefte 36 272-272
	Robust analysis	
Mukhopadhyay, Saurabh; Vidakovic, Brani	Efficiency of linear Bayes rules for a normal mean: Skewed priors class •	95 Ststcian 44 389-397
Rashid, M. Mushfiqur	Robust analysis of two-way models with repeated measures on both factors •	95 Test 4 39-62
On a	robust analysis of variance based on Winsorization (Korean) • Sung, Nae Kyung	95 KorApS 8/1 119-131
	Robust analysis of variance: A simulation study • Reed, J. F.; Stark, D. B.	95 JAppStat 22 87-104
marizing a set of radiocarbon determinations: A	robust approach" (V43 p489-503) • Wilson, S. R. Comment on "Sum-	95 ApplStat 44 407-407
marizing a set of radiocarbon determinations: A	robust approach" • Christen, J. A. Reply to comment on "Sum-	95 ApplStat 44 408-408
Review of	"Robust asymptotic statistics" • Rieder, H. (Auth); Plachky, D. (Rev)	95 StatDecn 13 93-94
Review of	"Robust asymptotic statistics" • Rieder, Helmut (Auth); Beran, Rudolf (Rev)	95 JASA 90 1128-1129
Optimal	robust Bayes estimation • Li, Yuanzhang; Saxena, K. M. Lal	95 JStPlInf 46 365-380
	Robust Bayesian analysis	
Ruggeri, Fabrizio; Vidakovic, Brani	Some results on posterior regret Γ-minimax estimation • Insua, David Ríos;	95 StatDecn 13 315-331
	Robust Bayesian analysis with distribution bands • Basu, Sanjib; DasGupta, Anirban	95 StatDecn 13 333-349
	Robust Bayesian estimators in a one-way ANOVA model • Bian, G.	95 Test 4 115-135
Wasserman, Larry	Robust Bayesian methods for monitoring clinical trials • Greenhouse, Joel B.;	95 StatMed 14 1379-1391
Bachmaier, M.; Precht, M.	Robust confidence intervals for contrasts based upon a likelihood ratio test •	95 StHefte 36 215-236

Review of "Algorithms, routines, and S functions, for	robust statistics" • Marazzi, A. (Auth); Iglewicz, B.; Heiberger, R. M.; Moore, D. et al.	95 Technmcs 37 351-353
Optimal and	robust strategies for cluster sampling • Tam, S. M.	95 JASA 90 379-382
Pollak, Moshe A	robust surveillance scheme for stochastically ordered alternatives • Gordon, Louis;	95 AnlsStat 23 1350-1375
Tsimikas, J.	Robust tests based on the sample characteristic function • Lenth, R. V.; Markatou, M.;	95 AstrlJSt 37 45-60
A	robust test for homoscedasticity in a two-way layout • Piepho, Hans-Peter	95 BiomtrcJ 37 151-160
Song, Moon Sup; Shin, Bong Sup A	robust test for the parallelism of two regression lines (Korean) • Nam, Ho Soo;	95 KorApS 8/2 77-86
Maximin efficiency-	robust tests and some extensions • Bose, Sudip; Slud, Eric	95 JStPlInf 46 105-121
	Robust testing	
Kimura, Miyoshi	Majorization methods on hyperplanes and their applications • Kakiuchi, Itsuro;	95 JStPlInf 47 217-235
Classical and Bayesian aspects of	robust unit root inference • Hoek, Henk; Lucas, André; van Dijk, Herman K.	95 JEconmtx 69 27-59
Lucas, André An outlier	robust unit root test with an application to the extended Nelson-Plosser data •	95 JEconmtx 66 153-173
Sommer, Suzanne; Staudte, Robert G.	Robust variable selection in regression in the presence of outliers and leverage points •	95 AstrlJSt 37 323-336
Amos, Christopher I.	Robust variance-components approach for assessing genetic linkage in pedigrees •	94 AJHuGene 54 535-543
Power of a test that is	robust against variance heterogeneity • Overall, John E.; Atlas, Robert S.	95 PsychRpt 77 155-159
Power	robustification of approximately linear tests • Ehm, W.; Mammen, E.; Müller, D. W.	95 JASA 90 1025-1033
Adaptive order polynomial fitting: Bandwidth	robustification and bias reduction • Fan, Jianqing; Gijbels, Irène	95 JCmpGrSt 4 213-227
	Robustifying model fitting • Windham, Michael P.	95 JRSS-B 57 599-609
	Robustness	
Brownie, Cavell	ANOVA and rank tests when the number of treatments is large • Boos, Dennis D.;	95 StPrLet 23 183-191
	Assessment and propagation of model uncertainty (Disc: p71-97) • Draper, David	95 JRSS-B 57 45-70
	Asymptotic normality of multivariate trimmed means • Arcones, Miguel A.	95 StPrLet 25 45-53
Rodrigues, J.; Baba, M. Y.	Bayesian estimation of a simple regression model with measurement errors •	94 RBrzlPrS 8 107-118
A.; Yohai, Victor J.	The behavior of the Stahel-Donoho robust multivariate estimator • Maronna, Ricardo	95 JASA 90 330-341
	Bounds for the breakdown point of the simplicial median • Chen, Zhiqiang	95 JMultiAn 55 1-13
Sands, Barbara N.	On the choice of a discrepancy functional for model selection • Trosset, Michael W.;	95 CommStA 24 2841-2863
	A comparative study of some rank tests for interaction • Mansouri, H.; Chang, G.-H.	95 CmpStDA 19 85-96
Srivastava, Anil K.; Tracy, Derrick S.; Shukla, Praveen	Concentration of estimators of coefficients in mixed regression model •	95 CommStA 24 797-811
Arsian, Olcay; Constable, Patrick D. L.; Kent, John T.	Convergence behavior of the EM algorithm for the multivariate t-distribution •	95 CommStA 24 2981-3000
parallel • Larholt, Kay M.; Sampson, Allan R.	Effects of heteroscedasticity upon certain analyses when regression lines are not	95 Biomtrcs 51 731-737
	Empirical influence functions for the Cox regression model • Dubas, Françoise	95 Student 1 87-98
and the method of moments ... • Yanagimoto	Estimation based on estimating equations - From the maximum likelihood method	95 JpJAppSt 24 1-12
• Hössjer, Ola	Exact computation of the least trimmed squares estimate in simple linear regression	95 CmpStDA 19 265-282
estimators • Hall, Peter; Patil, Prakash	Formulae for mean integrated squared error of nonlinear wavelet-based density	95 AnlsStat 23 905-928
	Fractional Bayes factors for model comparison (Disc: p118-138) • O'Hagan, Anthony	95 JRSS-B 57 99-118
Kelmansky, Diana; Yohai, Victor	A goodness-of-fit test based on ranks for ARMA models • Ferretti, Nélida;	95 CommStA 24 295-318
Naranjo, Joshua D.; McKean, Joseph W.	Influence functions for rank-based procedures in the linear model • Witt, Lee D.;	95 JNnparSt 5 339-358
method of moments • Arkonac, S. Z.; Higgins, M. L.	A Monte Carlo study of tests for non-nested models estimated by generalized	95 CommStB 24 745-763
Ludovico; Racugno, Walter	Multiple Bayes factors for testing hypotheses • Bertolino, Francesco; Piccinato,	95 JASA 90 213-219
Alvarez, M.; Cuevas, C. M.; Escudero, L. F. et al.	Network planning under uncertainty with an application to hydropower generation •	94 Top 2 25-58
Dette, Holger; Studden, William J.	Optimal designs for polynomial regression when the degree is not known •	95 StSinica 5 459-473
Heiligers, Berthold	Optimal and robust invariant designs for cubic multiple regression • Gaffke, Norbert;	95 Metrika 42 29-48
Wang, C. Y.	Prospective analysis of logistic case-control studies • Carroll, R. J.; Wang, Suojin;	95 JASA 90 157-169
	Reducing variation during design • Taylor, Wayne A.	95 ASQCTekCon 217-226
Basu, Ayanendranath; Basu, Srabashi	Robust minimum distance inference based on combined distances • Park, Chanseok;	95 CommStB 24 653-673
control • Luo, N.; de la Sen, M.	Robust stabilization of internally delayed uncertain systems via sliding mode	95 Top 3 1-34
	A robust test for homoscedasticity in a two-way layout • Piepho, Hans-Peter	95 BiomtrcJ 37 151-160
models • Chen, C. L.; Swallow, W. H.	Sensitivity analysis of variable-size group testing and its related continuous	95 BiomtrcJ 37 173-181
factorials • Ghosh, Subir	Sequential construction of new main effect plans with higher revealing power for 2^7	93 JCISS 18 217-231
Sutradhar, Brajendra C.; Das, Kalyan	Tests for overdispersion in longitudinal data in the presence of outliers •	95 SankhyaA 57 299-315
R.	Testing the hypothesis of independence between two sets of variates • Wilcox, Rand	95 MultBehR 30 213-225
	Three multiple comparison procedures for trimmed means • Wilcox, Rand R.	95 BiomtrcJ 37 643-656
	Two permutation tests of equality of variances • Baker, Rose D.	95 StatComp 5 289-296
	An upper bound for the power of a test • Bismarck, M.; Gajek, L.	95 CommStA 24 813-828
tion and breeding proportions • Kendall, W. L. et al.	On the use of secondary capture-recapture samples to estimate temporary emigra-	95 JAppStat 22 751-762
	Validation studies in epidemiology • Altpeter, Ekkehardt	95 Student 1 35-49
of "New directions in statistical data analysis and	robustness" • Morgenthaler, S.; Ronchetti, E.; Stahel, W. A. (Ed); Lenz, H.-J. (Rev) Review	95 CmpStDA 19 711-712
work models • Derks, E. P. P. A.; Pastor, M. S. S. et al.	Robustness analysis of radial base function and multi-layered feed-forward neural net-	95 Chemolab 28 49-60
priors specified on a collection of sets: Bayesian	robustness • Moreno, Elías; Cano, Juan A. Classes of bidimensional	95 JStPlInf 46 325-334
availability of blocks: A comparison • Uddin et al. The	robustness of binary and non-binary balanced nested row-column designs under the un-	95 AstrlJSt 37 111-124
Assessing the	robustness of clinical trial results • Choudhury, Japobrata; Balch, Alfred H.	94 ASAProBiop 484-486
problem • Grzegorzewski, P. The	robustness against dependence of nonparametric tests for the two-sample location	95 ZastApMa 22 469-476
	Robustness of the Ewens sampling formula • Joyce, Paul	95 JAppProb 32 609-622
for 'augmenting' the Dickey-Fuller test: Size-	robustness in the face of pre-testing • Dods, J. L.; Giles, D. E. A. Alternative strategies	95 JStCmpSm 53 243-258
Loss	robustness via Fisher-weighted squared-error loss function • Makov, Udi E.	95 InsrneME 16 1-6
	Robustness of the half-space median • Chen, Zhiqiang	95 JStPlInf 46 175-181
Alex S.; Hoekstra, Robert M. Some	robustness issues in the Bayesian analysis of a Weibull failure model • Papadopoulos,	95 JStCmpSm 52 85-94
	Robustness of judgments in evaluation research • van der Linden, W. J.; Zwarts, M. A.	95 TOnderwR 20 13-27
• Casas et al. Asymptotic design unbiasedness and	robustness of linear estimation under superpopulation models: A selection procedure ...	94 Questo 18 173-188
implied restrictions • Pordzik, P. R. On the	robustness of the linear hypothesis test procedure in the singular linear model with	95 StHefte 36 69-75
On the	robustness of maximum likelihood scaling for violations of the error model • Storms, Gert	95 Psymtrka 60 247-258
	Robustness of minimum disparity estimators in linear regression models • Pak, Rojin	95 JKoreaSt 24 349-360
Prior elicitation and Bayesian	robustness in a noisy Poisson process • Moreno, Eías; Girón, Javier	94 ASAProBays 217-220
	Robustness: Where are we now? • Huber, Peter J.	95 Student 1 75-86
regression - A case study • Rasch, Dieter The	robustness against parameter variation of exact locally optimum designs in nonlinear	95 CmpStDA 20 441-453
Brooks, Robert The	robustness of point optimal testing for Rosenberg random regression coefficients •	95 EcomtrRv 14 35-53
A Monte Carlo study of the	robustness of prediction in growth curve models • Hamel, Denis; Massé, Jean-Claude	95 JStCmpSm 52 33-53
markers • Säll, Torbjörn; Nilsson, Nils-Otto The	robustness of recombination frequency estimates in intercrosses with dominant	94 Genetcs 137 589-596
from normality • Basu, Sanjib; DasGupta, Anirban	Robustness of standard confidence intervals for location parameters under departure	95 AnlsStat 23 1433-1442
multiplicative interaction model for ... • Piepho	Robustness of statistical tests for multiplicative terms in the additive main effects and	95 ThApGene 90 438-443
	Robustness of Stein's two stage procedure • Oh, Hyun Sook; DasGupta, Anirban	95 SqtlAnly 14 321-334
Savas; Amemiya, Yasuo Asymptotic	robustness for the structural equation analysis of several populations • Papadopoulos,	94 ASAProBuEc 65-70
individuals • Bunck, C. M.; Chen, C.-L.; Pollock, K. H.	Robustness of survival estimates from ratio-telemetry studies with uncertain relocation of	95 JWildlMg 59 790-794
• Wong, M. Y.	Robustness of testing the equality of two slopes in simple linear structural relationships	95 JAppStat 22 3-8
MacEachern, S. N.; Notz, W. I. et al.	Robustness to the unavailability of data in the linear model, with applications •	95 JStPlInf 48 207-213

MACCS reactor accident ... • Helton et al.	Robustness of an uncertainty and sensitivity analysis of early exposure results with the	95 RelEngSS 48 129-148
Correction to "Efficient confidence bounds for	ROC curves" (V13 p1551-1561) • Schäfer, H.	95 StatMed 14 875-875
	ROC curves with multiple marker measurements • Murtaugh, Paul A.	95 Biomtrcs 51 1514-1522
Meta-analysis for the	ROC studies using empirical Bayesian method • Zhou, X. H.	94 ASAProBays 102-105
Prescott, P. On	Rohlf's method for the detection of outliers in multivariate data • Caroni, C.;	95 JMultiAn 52 295-307
	Root	
	• see names of specific types of roots such as characteristic – or square –	
Review of "Completeness of	root functions of regular differential operators" • Yakubov, S. (Auth); Rehberg, J. (Rev)	95 BiomtrcJ 37 118-118
Visual revelations: A	rose by another name • Wainer, Howard	95 Chance 8/2 46-51
The robustness of point optimal testing for	Rosenberg random regression coefficients • Brooks, Robert	95 EcomtrRv 14 35-53
	Rotatability	
Park, Sung Hyun; Lee, Min Soo	Slope-rotatability over all directions in third order response surface models •	95 JKoreaSt 24 519-536
Draper, Norman R. Slope	rotatability over all directions designs for $k \geq 4$ • Ying, Lisa H.; Pukelsheim, Friedrich;	95 JAppStat 22 343-354
Draper, Norman R. Slope	rotatability over all directions designs • Ying, Lisa H.; Pukelsheim, Friedrich;	95 JAppStat 22 331-341
Lee, Min Soo Slope-	rotatability over all directions in third order response surface models • Park, Sung Hyun;	95 JKoreaSt 24 519-536
A new type of slope	rotatable central composite design • Victorbabu, B. Re.; Narasimham, V. L.	94 JIAgSt 46 315-317
	Rotatable design	
mensional balls and cubes • Gaffke, N.; Heiligers, B.	Computing optimal approximate invariant designs for cubic regression on multidi-	95 JStPlInf 47 347-376
designed experiments • Cheng, C.-S.; Li, K.-C.	A study of the method of principal Hessian direction for analysis of data from	95 StSinica 5 617-639
Robert, Philippe; Stolyar, A. L. Queues served by a	rotating ring • Coffman, E. G., Jr; Gilbert, E. N.; Greenberg, A. G.; Leighton, F. T.;	95 CommStSM 11 371-394
Entromin and entromax: Criteria for factor	rotation based on entropy (Japanese) • Inagaki, Atsushi	94 JpJBeh 21/2 10-20
	Rotation of principal components: Choice of normalization constraints • Jolliffe, Ian T.	95 JAppStat 22 29-35
Suppressing permutations or rigid planar	rotations: A remedy against nonoptimal varimax rotations • Berge, Jos M. F. ten	95 Psymtrka 60 437-446
Random	rotations of the Wiener path • Üstünel, A. S.; Zakai, M.	95 ProbTher 103 409-429
On the	rotational dimension of stochastic matrices • Kalpazidou, S.	95 AnlsProb 23 966-975
having parameter restrictions and an application to	rotational sampling • Shin et al. ... of the multivariate autoregressive moving average	95 JTimSrAn 16 431-444
Random arrays and functionals with multivariate	rotational symmetries • Kallenberg, O.	95 ProbTher 103 91-141
	Rotational versions of the Crofton formula • Jensen, Eva B. Vedel	95 AdvAppPr 27 87-96
evaluation for golf course closing • Shipley et al.	Rough set theory fuzzy belief functions related to statistical confidence: Application and	95 StocAnAp 13 487-502
regression and generalized linear models: A	roughness penalty approach" • Green, P. J. et al. Review of "Nonparametric	95 JASA 90 1489-1489
regression and generalized linear models: A	roughness penalty approach" • Green, P. J. et al. Review of "Nonparametric	95 Technmcs 37 342-343
statistics from survey data: Effects of respondent	rounding • Rowe, Geoff; Gribble, Steve Income	94 ASAProBuEc 77-82
networks with signals and history-dependent	routing • Chao, Xiuli; Pinedo, Michael On queueing	95 PrEngInf 9 341-354
Loss networks under diverse	routing: The symmetric star network • Hunt, P. J.	95 AdvAppPr 27 255-272
	Row-column design	
	Orthogonality in a two-way elimination of heterogeneity design • Siatkowski, I.	93 UtilMath 43 225-233
	On perpendicularity in designed experiments • Kala, Radosław	95 ListyBio 32 93-99
designs with multivariate responses • Hanusz, Zofia	Relative potency of two preparations in two-way elimination of heterogeneity	95 Biomtrcs 51 1133-1139
designs • Siatkowski, Idzi	Some characterizations of a class of balanced two-way elimination of heterogeneity	94 ListyBio 31 27-33
Pearce, S. C.	Some design problems in crop experimentation. II. Multiple blocking systems •	95 ExpAgric 31 279-290
On connectedness of	row-column designs • Park, D. K.; Shah, K. R.	95 CommStA 24 87-96
Correction to "Nested	row-column designs" (V38 p197-202) • Ipinyomi, R. A.; Adeleke, B. L.	95 Ststcian 44 547-547
Nonadditivity in	row-column designs • Srivastava, Jagdish N.	93 JCISS 18 85-96
bustness of binary and non-binary balanced nested	row-column designs under the unavailability of blocks: A comparison • Uddin et al. The ro-	95 AstrlJSt 37 111-124
Richard M. Variance balanced generalized nested	row-column designs with empty diagonal cells • Raghavarao, Damaraju; Heiberger,	93 JCISS 18 42-52
Cox's design in repeated	row-column design with split plots and a control • Kachlicka, Danuta; Mejza, Stanisław	95 ListyBio 32 81-92
Hossain, Syed A. *E*-optimal nested	row-column designs with unequal replications of treatments • Uddin, Nizam;	95 SankhyaB 57 112-121
Functional limit theorems for	row and column exchangeable arrays • Ivanoff, B. G.; Weber, N. C.	95 JMultiAn 55 133-148
Optimal designs with component-wise orthogonal	row-column structures • Shah, K. R.; Sinha, Bikas K.	93 JCISS 18 68-78
analysis of a set of two-way tables with constant-	row-sum data • Siciliano, Roberta; van der Heijden, Peter G. M. Simultaneous latent budget	94 Metron 52/1 155-180
A SLLN for arrays of	rowwise exchangeable fuzzy random sets • Inoue, Hiroshi; Taylor, Robert L.	95 StocAnAp 13 461-470
Square-root	RTS smoothing algorithms • Park, PooGyeon; Kailath, Thomas	95 IJContrl 62 1049-1060
Scoring patterns in	rugby league • Croucher, John	95 TeachgSt 17/2 47-49
	Ruin estimates under interest force • Sundt, Bjørn; Teugels, Jozef L.	95 InsrneME 16 7-22
Saddlepoint approximations for the probability of	ruin in finite time • Barndorff-Nielsen, Ole E.; Schmidli, Hanspeter	95 ScanActJ 169-186
	Ruin probability	
	The effect of the retention limit on the risk reserve • Centeno, Maria de Lourdes	95 AstinB 25 67-74
Steinebach, Josef	Estimating the adjustment coefficient in an ARMA(p,q) risk model • Christ, Ralf;	95 InsrneME 17 149-161
Rolski, Tomasz; Schmidt, Volker	Does Markov-modulation increase the risk? • Asmussen, Søren; Frey, Andreas;	95 AstinB 25 49-66
	A note on the solution of practical ruin problems • De Vylder, F.; Goovaerts, M. J.	94 InsrneME 15 181-186
	Risk theory of the second and third kind • Embrechts, Paul	95 ScanActJ 35-43
	Ruin estimates under interest force • Sundt, Bjørn; Teugels, Jozef L.	95 InsrneME 16 7-22
Egídio dos Reis, Alfredo D.; Waters, Howard R.	Some stable algorithms in ruin theory and their applications • Dickson, David C. M.;	95 AstinB 25 153-175
Delay in claim settlement and	ruin probability approximations • Klüppelberg, C.; Mikosch, T.	95 ScanActJ 154-168
Explicit analytic	ruin probabilities for bounded claims • De Vylder, F.; Marceau, E.	95 InsrneME 16 79-105
Stochastic differential equations for	ruin probabilities • Møller, Christian Max	95 JAppProb 32 74-89
Mandrup	Ruin probabilities via local adjustment coefficients • Asmussen, Søren; Nielsen, Hanne	95 JAppProb 32 736-755
Cramér-Lundberg approximations for	ruin probabilities of risk processes perturbed by diffusion • Schmidli, H.	95 InsrneME 16 135-149
A note on the solution of practical	ruin problems • De Vylder, F.; Goovaerts, M. J.	94 InsrneME 15 181-186
Waters, Howard R. Some stable algorithms in	ruin theory and their applications • Dickson, David C. M.; Egídio dos Reis, Alfredo D.;	95 AstinB 25 153-175
Yamamoto et al. Classification of orthogonal 24-	run 2^5 factorial designs derivable from saturated two-symbol orthogonal arrays of ... •	95 TRUMTk 31/1 39-53
	Runs on a circle • Koutras, M. V.; Papadopoulos, G. K.; Papastavridis, S. G.	95 JAppProb 32 396-404
Longest	runs in coin tossing • Binswanger, K.; Embrechts, P.	94 InsrneME 15 139-149
Joint distributions of numbers of success-	runs and failures until the first consecutive k successes • Aki, Sigeo; Hirano, Katuomi	95 AnInStMa 47 225-235
projected designs: Repeat and mirror-image	runs • Draper, Norman R.; Lin, Dennis K. J. Characterizing	95 CommStA 24 775-795
	Run length	
	• see also average –, median –	
p3-27) • Yashchin, Emmanuel Comment on	"Run-length distribution of special-cause control charts for correlated processes" (V36	95 Technmcs 37 243-243
Wardell, D. G. et al. Reply to comment on	"Run-length distribution of special-cause control charts for correlated processes" •	95 Technmcs 37 243-245
A note on the	run length to false alarm of a change-point detection policy • Yakir, Benjamin	95 AnlsStat 23 272-281
On the	run length of a Shewhart chart for correlated data • Schmid, W.	95 StHefte 36 111-130
Asymptotics of the	run lengths of two control charts • Zhang, Jian	95 CommStA 24 2375-2403
experiments using orthogonal arrays when the	run order is important • Wang, P. C.; Jan, H. W. Designing two-level factorial	95 Ststcian 44 379-388
processes • Hu, Albert; Sachs, Emanuel	Run by run control: A Bayesian approach to the control of semiconductor fabrication	94 ASAProPEng 245-254

Koutras, M. V.; Alexandrou, V. A.	Runs, scans and urn model distributions: A unified Markov chain approach •	95 AnInStMa 47 743-766
Jagdish Joint distributions based on the number of	runs and some other two-sample rank order statistics for arbitrary sample sizes • Saran,	95 JStPlInf 47 257-275
SRS	runs and spacings tests to assess randomness • Kaigh, W. D.; Schuster, E. F.	94 ASAProStCp 272-277
	Run test	
	Runs on a circle • Koutras, M. V.; Papadopoulos, G. K.; Papastavridis, S. G.	95 JAppProb 32 396-404
Substance abuse among	runaway and homeless youth • Iachan, Ronald; Ringwalt, Chris L.; Greene, Jody M.	94 ASASrvyRsr 1083-1087
The joint distribution of the	running maximum and its location of D-valued Markov processes • Ferger, Dietmar	95 JAppProb 32 842-845
space identification of linear deterministic rainfall-	runoff models • Ramos, José; Mallants, Dirk; Feyen, Jan State	95 WaterReR 31 1519-1531
Artificial neural network modeling of the rainfall-	runoff process • Hsu, Kuo-lin; Gupta, Hoshin Vijai; Sorooshian, Soroosh	95 WaterReR 31 2517-2530
	Runway landing distribution and tests for normality • Andersen, John M.	94 ASAProBays 150-152
Cost modelling of alternative sample designs for	rural areas in the Canadian Labour Force Survey • Mantel, H.; Laniel, N.; Duval, M.-C. et al.	94 ASASrvyRsr 925-930
A new look at pricing of the	"Russian option" • Shepp, L. A.; Shiryaev, A. N.	94 ThProbAp 39 103-119
From perpetual strangles to	Russian options • Gerber, Hans U.; Shiu, Elias S. W.	94 InsrneME 15 121-126
Shiryaev, A. N. On the rational pricing of the	"Russian option" for the symmetrical binomial model of a (B,S)-market • Kramkov, D. O.;	94 ThProbAp 39 153-162
On the	Russian stock exchange • Bondarenko, M. V.; Vishnyakov, A. N.	94 ThProbAp 39 188-194
	R^2	
Frank A. G.	Goodness-of-fit measures in binary choice models (Disc: 117-120) • Windmeijer,	95 EcomtrRv 14 101-116
Displaying a correlation matrix on the unit sphere in	R^3 (French) • Falissard, B.	95 RStApp 43/2 35-48
Sharp asymptotics of large deviations in	R^d • Iltis, Michael	95 JTherPrb 8 501-522
Symmetry groups in	R^d • Meerschaert, Mark M.; Veeh, Jerry Alan	95 StPrLet 22 1-6
double intersection local time of Brownian motion in	R^d and renormalization • Imkeller, P.; Perez-Abreu, V.; Vives, J. Chaos expansions of	95 StocProc 56 1-34
stochastic processes on a compact set in	R^d • Sydorenko, O. O. On the distribution of supremum of a certain class of pre-Gaussian	95 ThPrMaSt 50 123-128
for random closed sets in compact subsets of	R^d • Ferrándiz, J.; Montes, F. On statistical inference	95 AdvAppPr 27 344-349
A random continued fraction in	R^{d+1} with an inverse Gaussian distribution • Letac, Gérard; Seshadri, Vanamamalai	95 Bernoull 1 381-393
Analysis of variance of balanced fractional	$r^m \times s^n$ factorial designs of resolution V • Kuwada, Masahide; Hyodo, Yoshifumi	95 JJapnStS 25 129-141
Left and right linear innovations for a multivariate	S α S random variable • Rutkowski, Marek	95 StPrLet 22 175-184
Determination of optimal design parameters of	S control chart • Yang, Su-Fen	94 ASAProQlPr 327-333
Type	S designs in unequal blocks • Gupta, Sudhir; Kageyama, Sanpei	93 JCISS 18 97-112
	***S*-estimator**	
choose? • Hennig, Christian	Efficient high-breakdown-point estimators in robust regression: Which function to	95 StatDecn 13 221-241
Review of "Algorithms, routines and	S functions for robust statistics" • Marazzi, A. (Auth); Anderson, A. J. B. (Rev)	95 Biomtrcs 51 385-386
Moore, D. et al. Review of "Algorithms, routines, and	S functions, for robust statistics" • Marazzi, A. (Auth); Iglewicz, B.; Heiberger, R. M.;	95 Technmcs 37 351-353
Data analysis using	S-Plus • Schulman, Daniel A.; Campbell, Alec D.; Kostello, Eric C.	95 SocMethR 23 392-413
Review of "A handbook of statistical analyses using	S-Plus" • Everitt, B. S. (Auth); Kimber, Alan (Rev)	95 Ststcian 44 408-409
Review of "A handbook of statistical analyses using	S-Plus" • Everitt, B. S. (Auth); Ruxton, Graeme D. (Rev)	95 JAppStat 22 545-545
Review of "A handbook of statistical analyses using	S-Plus" • Everitt, Brian S. (Auth); Ko, Daijin (Rev)	95 Technmcs 37 232-233
Review of "A handbook of statistical analyses using	S-Plus" • Everitt, Brian S. (Auth); Venables, W. N. (Rev)	95 AstrlJSt 37 369-372
Review of "A handbook of statistical analyses using	S-Plus" • Everitt, Brian S. (Auth); Williams, Calvin (Rev)	95 JQualTek 27 89-90
Review of "A handbook of statistical analysis using	S-Plus" • Everitt, B. S. (Auth); Davies, Neville (Rev)	95 JRSS-A 158 633-634
Modern applied statistics with	S-Plus • Venables, W. N.; Ripley, B. D.	94 Springer:Brln:NY 462p
Review of "Modern applied statistics with	S-Plus" • Venables, W. N.; Ripley, B. D. (Auth); Smith, J. T. (Rev)	95 ShrtBkRv 15/2 24-24
matrix for s^n parallel flats fractions when	s is prime • BuHamra, Sana; Anderson, Donald A. The spectrum of the information	95 JStPlInf 47 333-345
Klopotowski, A. On the existence of sequences of	s by s independent exchangeable or stationary random variables (French) • Bretagnolle, J.;	95 AIHenrPB 31 325-350
Review of "An introduction to	S and S-Plus" • Spector, Phil (Auth); Ko, Daijin (Rev)	95 Technmcs 37 232-232
tiza, Luis G. Self-intersection local time for Gaussian	$S'(R^d)$-processes: Existence, path continuity and examples • Bojdecki, Tomasz; Goros-	95 StocProc 60 191-226
Analysis of adaptive step-size	SA algorithms for parameter tracking • Kushner, Harold J.; Yang, J.	95 IEEEAuCn 40 1403-1410
and approximation for random fields satisfying	Sacks-Ylvisaker conditions • Ritter, K.; Wasilkowski, G. W. et al. Multivariate integration	95 AnlsApPr 5 518-540
Li, J.; Chen, M.-C. et al. Forestry applications of	saddle-point approximations to construct confidence intervals for population means •	95 Biomtrcs 51 61-72
On the exit law from	saddle points • Day, Martin V.	95 StocProc 60 287-311
	Saddlepoint approximation	
	On approximate inference for the two-parameter gamma model • Wong, A.	95 StHefte 36 49-59
Field, Christopher A.	Approximations for marginal densities of M-estimators • Fan, Rocky Y. K.;	95 CandJSt 23 185-197
• Larsson, Rolf	The asymptotic distributions of some test statistics in near-integrated AR processes	95 EconTher 11 306-330
	Asymptotic expansions at work • Jensen, Jens Ledet	95 ScanActJ 143-152
	Bootstrapping binomial confidence intervals • Zheng, Xiaodong; Loh, Wei-Yin	95 JStPlInf 43 355-380
mation to the asymptotic distribution • Hougaard, P.	Nonlinear regression and curved exponential families. Improvement of the approxi-	95 Metrika 42 191-202
	Saddlepoint approximations • Jensen, Jens Ledet	95 ClrndnPr:UK 329p
Newton, H. Joseph	Saddlepoint approximations for the difference of order statistics • Harvill, Jane L.;	95 Biomtrka 82 226-231
Newton, H. Joseph Using symbolic math to evaluate	saddlepoint approximations for the difference of order statistics • Harvill, Jane L.;	95 CommStB 24 781-791
	Saddlepoint approximation in the linear structural relationship model • Penev, Spiridon	95 CommStB 24 349-366
J. G.; Butler, R. W.; Huzurbazar, S.; Wood, A. T. A.	Saddlepoint approximations for P-values of some tests of covariance matrices • Booth,	95 JStCmpSm 53 165-180
Ole E.; Schmidli, Hanspeter	Saddlepoint approximations for the probability of ruin in finite time • Barndorff-Nielsen,	95 ScanActJ 169-186
One-step	saddlepoint approximations for quantiles • Wang, Suojin	95 CmpStDA 20 65-74
Uniform validity of	saddlepoint expansion on compact sets • Routledge, Rick; Tsao, Min	95 CandJSt 23 425-431
An example in which the Lugannani-Rice	saddlepoint formula fails • Booth, James G.; Wood, Andrew T. A.	95 StPrLet 23 53-61
p207-211) • Hesterberg, Tim Correction to	"Saddlepoint quantiles and distribution curves, with bootstrap applications" (V9	95 CmpStQ 10 193-193
A note on the accuracy of two	saddlepoint tail probability approximations • Booth, James	94 ASAProPEng 56-58
on "The meaning of probability in probabilistic	safety analysis" (V45 p261-269) • Yellman, Ted W.; Murray, Thomas M. Comment	95 RelEngSS 49 201-205
Anomalies in Bayesian launch range	safety analysis • Philipson, Lloyd L.	95 RelEngSS 49 355-357
Yang, J. B.; Sen, P.	Safety analysis and synthesis using fuzzy sets and evidential reasoning • Wang, J.;	95 RelEngSS 47 103-118
tance and distributional sensitivity in probabilistic	safety assessment • Park, C. K. et al. A new approach for measuring uncertainty impor-	94 RelEngSS 46 253-261
point process framework for living probabilistic	safety assessment and risk follow-up • Arjas, Elja; Holmberg, Jan Marked	95 RelEngSS 49 59-73
incidence estimation from case series for vaccine	safety evaluation • Farrington, C. P. Relative	95 Biomtrcs 51 228-235
David J. Defining, monitoring and combining	safety information in clinical trials (Disc: p1113-1116) • Enas, Gregory G.; Goldstein,	95 StatMed 14 1099-1111
	Safety measurement: Towards continuous improvement • St. Clair, Joseph F.	95 ASQCTekCon 150-158
Reliable design of	safety monitoring systems accounting for the uncertainty zone • Tanaka, Kenji	95 RelEngSS 48 75-84
cations in the development and testing of a traffic	safety questionnaire • Schechter, S. et al. Cognitive issues and methodological impli-	94 ASASrvyRsr 1215-1219
Hungarian sand martins and their relationship with	Sahel rainfall • Szép, Tibor Survival rates of	95 JAppStat 22 891-904
Brownian martingales and	Sakai's conjecture (French) • Piau, Didier	95 AIHenrPB 31 429-452
look at some analyses of major league baseball	salaries • Hoaglin, David C.; Velleman, Paul F. A critical	95 AmerStat 49 277-285
Longton, Barbara; Sailer, Peter; Wilson, Janette How	sales of capital assets were affected by the Tax Reform Act of 1986 (Disc: p212-214) •	94 ASAProGovt 201-205
Dua, Pami; Smyth, David J. Forecasting US home	sales using BVAR models and survey data on households' buying attitudes for homes •	95 JForcast 14 217-227
for optimal interpolation of temperature,	salinity and nutrient concentrations in the Gulf of Finland • Toompuu et al. ... correlations	95 EnvrMtrc 6 55-72
comparison ... • Lesch et al. Spatial prediction of soil	salinity using electromagnetic induction techniques 1. Statistical prediction models: A	95 WaterReR 31 373-386

Statistical validity, accuracy, and determination of	sample size (Disc: p134-135) • Ji, X. et al. Approximate reasoning in belief networks:	94 ASAProBays 128-133
Some comments on Bayesian	sample size determination • Joseph, Lawrence; Wolfson, David B.; Berger, Roxane du	95 Ststcian 44 167-171
	Sample size determination in Bayesian statistics – A commentary • Phan-Gia, T.	95 Ststcian 44 163-166
Ahnn, Sang; Anderson, Stewart J.	Sample size determination for comparing more than two survival distributions •	95 StatMed 14 2273-2282
	Sample size determination in epidemiologic studies • Wickramaratne, Priya J.	95 StMedRes 4 311-337
p203-206) • Bromaghin, Jeffrey F. Correction to	"Sample size determination for interval estimation of multinomial probabilities" (V47	95 AmerStat 49 403-403
Simultaneous confidence intervals and	sample size determination for multinomial proportions • Sison, Cristina P.; Glaz, Joseph	95 JASA 90 366-369
rection to "Simultaneous confidence intervals and	sample size determination for multinomial proportions" (V90 p366-369) • Sison et al. Cor-	95 JASA 90 1494-1494
Federer, W. T.; Scully, B. T. The effect of linkage on	sample size determination for multiple trait selection • Schwager, S. J.; Mutschler, M. A.;	93 ThApGene 86 964-974
Orzack, Steven Hecht; Chappell, Rick	Sample-size determination for parent-offspring regression of a composite trait •	95 Biomtrcs 51 822-830
Kleijnen, J. P. C. et al. Two-stage versus sequential	sample-size determination in regression analysis of simulation experiments •	95 AJMMgSci 15 83-114
treatments • Nam, Jun-Mo	Sample size determination in stratified trials to establish the equivalence of two	95 StatMed 14 2037-2049
Fuller, Howard T.	Sample size estimates for 2^{k-p} designs with binary responses • Bisgaard, Søren;	95 JQualTek 27 344-354
Computing defensible linear model power and	sample size estimates for study planning • Taylor, Douglas J.; Muller, Keith E.	94 ASAProBiop 383-388
Some pitfalls in	sample size estimation for an anti-infective study • Weng, C. S. Wayne; Liu, J. P.	94 ASAProBiop 56-60
Rochon, James Comment on	"Sample size estimation for repeated measurement designs" (V15 p100-123, 165-172) •	95 ConClnTr 16 449-450
Doyle, Suzanne R. Reply to comment on	"Sample size estimation for repeated measurement designs" • Overall, John E.;	95 ConClnTr 16 451-452
Power and	sample size in experiments involving repeated measures • Pollard, William E.	94 ASAProSoSt 45-48
binomial ... • Roebruck et al. Comparison of tests and	sample size formulae for proving therapeutic equivalence based on the difference of	95 StatMed 14 1583-1594
requirements on the power function and on the	sample size • Arnold, Bernhard F. Statistical tests optimally meeting certain fuzzy	95 FuzzySet 75 365-372
Computing confidence bounds for power and	sample size of the general linear univariate model • Taylor, Douglas J.; Muller, Keith E.	95 AmerStat 49 43-47
Determining	sample size for lower confidence limits for C_p and C_{pk} • Franklin, LeRoy A.	94 ASAProQlPr 23-27
Lu, Ying; Bean, Judy A. On the	sample size for one-sided equivalence of sensitivities based upon McNemar's test •	95 StatMed 14 1831-1839
A power function model for determining	sample sizes of the operations of multivariate control charts • Jolayemi, Joel K.	95 CmpStDA 20 633-641
classification of a prognostic dichotomous variable:	Sample size and parameter estimate adjustment • Tavaré, C. J.; Sobel, E. L. et al. Mis-	95 StatMed 14 1307-1314
Confidence intervals, hypothesis tests, and	sample sizes for the prevented fraction in cross-sectional studies • Gargiullo, P. M. et al.	95 StatMed 14 51-72
Minimax estimation of proportions under random	sample size • Amrhein, Peter	95 JASA 90 1107-1111
by moments of order statistics when the	sample size is random • Grudzień, Z.; Szynal, D. Characterizations of distributions	95 ZastApMa 23 305-318
a linear combination of strata means with random	sample sizes • He, Kun On estimating	95 JMultiAn 55 39-60
Estimation and analysis effects resulting from a	sample size reduction in a large national survey • Botman, Steven L.	94 ASASrvyRsr 1168-1172
Schmitz et al. Determining the minimum	sample size required to obtain sufficient progeny with a desired genotype at two ... •	93 ThApGene 87 136-144
prospective cohort investigations • Kaaks, R. et al.	Sample size requirements for calibration studies of dietary intake measurements in	95 AJEpidm 141 557-565
prospective cohort investigations • Kaaks, R. et al.	Sample size requirements for calibration studies of dietary intake measurements in	95 AJEpidm 142 557-565
Lemon, G. H.; Norton, J. M.	Sample size requirements for rare or abundant attribute sampling • Arnold, Jesse C.;	95 CommStB 24 131-148
C. J. The Bayesian approach to determination of	sample sizes – Some comments on the paper by Joseph, Wolfson and du Berger • Adcock,	95 Ststcian 44 155-161
recorded on the Catalan ... • Lana et al. Minimum	sample size for synoptic weather type classification. Application to winter period data	94 JClimlgy 14 1051-1060
lection of quick switching double-sampling system:	Sample size tightening • Arumainayagam, S. D.; Soundararajan, V. Construction and se-	95 JAppStat 22 105-119
Planning and revising the	sample size for a trial (Disc: p1053-1055) • Gould, A. Lawrence	95 StatMed 14 1039-1051
for tail probabilities of Dunnett's T with unequal	sample sizes • Cheng, Peter H.; Meng, Cliff Y. K. A new formula	95 CommStA 24 523-532
based survival model using release-recapture data:	Sample size, validity and power • Hoffmann, A. et al. Inferential properties of an individual-	95 JAppStat 22 579-595
CUSUM charts with variable	sample size and variable interval size • Arnold, Jesse C.; Reynolds, Marion R.	94 ASAProQlPr 87-92
proximation and a neural network model for optimal	sample size in vendor selection • Houshmand, A. A.; Panganamamula, S. An analytical ap-	95 JStCmpSm 53 65-78
Estimation of	sample size with grouped data • Blumenthal, Saul; Dahiya, Ram C.	95 JStPlInf 44 95-115
property of empirical discrepancies in arbitrary	sample spaces • Strobl, Franz On the reversed sub-martingale	95 JTherPrb 8 825-831
Keiding, Niels Coarsening at random in general	sample spaces and random censoring in continuous time • Jacobsen, Martin;	95 AnlsStat 23 774-786
Limit theorems for the logarithm of	sample spacings • Shao, Yongzhao; Hahn, Marjorie G.	95 StPrLet 24 121-132
	Sample survey	
	On a class of strongly threefold orthogonal matrices of index eight • Pesotan, H.	93 JCISS 18 317-328
	A decade of questions • Schaeffer, Nora Cate	95 JOffStat 11 79-92
Kaćiroti, Niko	Design effects for correlated $(P_i - P_j)$ • Kish, Leslie; Frankel, Martin R.; Verma, Vijay;	95 SurvyM 21 117-124
(Disc: p695-696) • Ezzati, T. M.; Hoffman, K. et al.	A dual frame design for sampling elderly minorities and persons with disabilities	95 StatMed 14 571-583
	Dynamics of 'list-assisted' random digit dialing (RDD) frame coverage • Kulp, Dale W.	94 ASASrvyRsr 11-18
Scott, Alastair	The effect of interviewer variance on domain comparisons • Davis, Peter;	95 SurvyM 21 99-106
Sahoo, L. N.; Mohanty, S.	Empirical comparison of some regression and regression-type strategies • Sahoo, J.;	95 StHefte 36 337-347
households • Brick, M.; Waksberg, J.; Keeter, S.	Evaluating the use of data on interruptions in telephone service for nontelephone	94 ASASrvyRsr 19-28
Department of Agriculture • Vogel, Frederic A.	The evolution and development of agricultural statistics at the United States	95 JOffStat 11 161-180
	Frank Yates, 1902-1994 – The work of a statistician • Healy, M. J. R.	95 IntStRvw 63 271-288
estimation • Estevao, V.; Patak, Z.; Särndal, C.	Maximizing the use of auxiliary information for calibration and regression	94 ASASrvyRsr 42-48
	Median estimation using auxiliary information • Meeden, Glen	95 SurvyM 21 71-77
	Methods for design effects • Kish, Leslie	95 JOffStat 11 55-77
Estevao, V.; Hidiroglou, M. A.; Särndal, C. E.	Methodological principles for a generalized estimation system at Statistics Canada •	95 JOffStat 11 181-204
	Minimum variance stratification • Unnithan, V. K. G.; Nair, N. Unnikrishnan	95 CommStB 24 275-284
Bell, P. A.	Optimal winsorizing cutoffs for a stratified finite population estimator • Kokic, P. N.;	94 JOffStat 10 419-435
	Outlier robust Horvitz-Thompson estimators • Hulliger, Beat	95 SurvyM 21 79-87
Lee, Kay-O; Lee, Gi-Sung	A practical plan of randomized response technique (Korean) • Ryu, Jea-Bok;	95 KorApS 8/1 9-26
Donald; Gollin, Albert; Sawyer, Tom; Tanur, Judith M.	Pseudo-opinion polls: SLOP or useful data? • Horvitz, Dan; Koshland, Daniel; Rubin,	95 Chance 8/2 16-25
sampling • Knaub, James R., Jr	Relative standard error for a ratio of variables at an aggregate level under model	94 ASASrvyRsr 310-312
surveys • Groves, Robert M.; Couper, Mick P.	Theoretical motivation for post-survey nonresponse adjustment in household	95 JOffStat 11 93-106
(Korean) • Lee, Gi-Sung; Hong, Ki-Hak	Unrelated question model with quantitative attribute by stratified double sampling	95 KorApS 8/1 27-38
estimation • Dubreuil, Guylaine; Tremblay, Johanne	The use of generalized raking procedures to improve the quality of small domain	94 ASASrvyRsr 293-298
How to	sample in surveys • Fink, Arlene	95 SagePubl:CA:Lnd 96p
National	sample surveys for development planning – An overview • Seal, K. C.	94 CalcutSt 44 239-250
weighted and unweighted estimates from a	sample survey • Korn, Edward L.; Graubard, Barry I. Examples of differing	95 AmerStat 49 291-295
estimator • Chambers, R. L.; Dorfman, A. H. Robust	sample survey inference via bootstrapping and bias correction: The case of the ratio	94 ASAProStCp 51-60
Madow, W. G. (Auth) et al. Review of	"Sample survey methods and theory (Volume I and II)" • Hansen, M. H.; Hurwitz, W. N.;	95 AllgeStA 79 335-337
W. G. (Auth); Fielding, Antony (Rev) Review of	"Sample survey methods and theory (Volume I, 2)" • Hansen, M. H.; Hurwitz, W. N.; Madow,	95 JRSS-A 158 350-350
Bayesian predictive inference for multivariate	sample surveys • Nandram, Balgobin	94 JOffStat 10 167-179
	Sample surveys at National Center for Health Statistics • Choi, Jai Won; Curtin, Lester R.	94 ASASrvyRsr 1162-1167
Visitor	sample surveys • Iachan, Ronaldo; Kemp, Suzanne, S.	95 SurvyM 21 89-96
How many iid	samples does it take to see all the balls in a box? • Sellke, Thomas M.	95 AnlsApPr 5 294-309
Comment on "Three curious properties of the	sample variance and autocovariance for stationary processes with unknown ... • Anderson	95 AmerStat 49 118-118
Resistant h-plot for a	sample variance-covariance matrix • Choi, Yong-Seok	95 JKoreaSt 24 407-417
intervals for $\Pr(X < Y)$ based on independent	samples of X and Y • Hamdy, M. I. Distribution-free confidence	95 CommStB 24 1005-1017
Langholz, B. Methods for the analysis of	sampled cohort data in the Cox proportional hazards model • Borgan, Ø.; Goldstein, L.;	95 AnlsStat 23 1749-1778

interior-point Markov chains: A family of black-box	samplers • Chen, Ming-Hui; Schmeiser, Bruce W. Random-direction	94 ASAProBays 1-6
Regeneration in Markov chain	samplers • Mykland, Per; Tierney, Luke; Yu, Bin	95 JASA 90 233-241
	Sampling	
	Linear rank tests derived from a superpopulation model • Bouza, Carlos N.	95 BiomtrcJ 37 497-506
Snell, A. J.	Temporal aggregation and the power of tests for a unit root • Pierse, R. G.;	95 JEconmtx 65 333-345
	• see also names of types of sampling such as acceptance –, cluster –, or importance –	
Difference	sampling as an alternative to importance sampling • van Deusen, Paul C.	95 CanJForR 25 487-490
bandit problem with one arm known under batch	sampling • Hamada, Toshio An exponential two-armed	95 JJapnStS 25 205-216
variable modelling: A Monte Carlo study • Hsu	Sampling behaviour in estimating predictive validity in the context of selection and latent	95 BJMaStPs 48 75-97
for discrete parents with applications to censored	sampling • Gan, Gaoxiong; Bain, Lee J. Distribution of order statistics	95 JStPlInf 44 37-46
How to perform skip-lot and chain	sampling (Second edition) • Stephens, Kenneth S.	95 ASQualCntrl:WI 64p
of "Volume 4 - How to perform skip-lot and chain	sampling (Second edition)" • Stephens, Kenneth S. (Auth); Doganaksoy, Necip (Rev) Review	95 JQualTek 27 394-395
the original HM (which is based on Chatterjee's	sampling) • Aoshima ... II (based on Healy's two-stage sampling) is more economical than	94 AJMMgSci 14 167-196
How to perform continuous	sampling (Second edition) • Stephens, Kenneth S.	95 ASQualCntrl:WI 73p
	Sampling design	
Sampath; Vasantha, P.	Continuous inspection of Markov processes with a clearance interval • Kumar, V. S.	95 JAppStat 22 427-432
review • Dalenius, Tore	Controlling invasion of privacy in surveys of change over time - A non-technical	95 JOffStat 11 107-113
Yamamura, Kohji; Sugimoto, Tamio	Estimation of the pest prevention ability of the import plant quarantine in Japan •	95 Biomtrcs 51 482-490
	Improving the RHC-strategy • Gabler, Siegfried; Stenger, Horst	95 StHefte 36 327-336
Sengupta, S.	Optimal estimation of a finite multidimensional population total • Mukerjee, Rahul;	95 JStPlInf 48 339-346
	Phased sampling for soil remediation • Englund, Evan J.; Heravi, Naser	94 EnvEcoSt 1 247-263
Larsen, D. P.; Urquhart, N. S.; Kugler, D. L.	Regional scale trend monitoring of indicators of trophic condition of lakes •	95 WaterReB 31 117-140
Beg, M. A.; Khawaja, Iffat	Sampling with unequal probabilities: A historical review • Hanif, Muhammad;	92 ProIrStC1 19-37
	Visitor sample surveys • Iachan, Ronaldo; Kemp, Suzanne, S.	95 SurvyM 21 89-96
Analysis of large health surveys: Accounting for the	sampling design • Korn, Edward L.; Graubard, Barry I.	95 JRSS-A 158 263-295
Optimal	sampling designs for dependent spatial units • Benedetti, Roberto; Palma, Daniela	95 EnvrMtrc 6 101-114
Efficiency and bias of estimators and	sampling designs for determining length-weight relationships of fish • Hayes, D. B. et al.	95 JFishRBC 52 84-92
Oregon • Crone, Paul R.	Sampling design and statistical considerations for the commercial groundfish fishery of	95 JFishRBC 52 716-732
Mondani, Riccardo The exact	sampling distribution of the Bonferroni concentration index • Giorgi, Giovanni Maria;	94 Metron 52/3 5-41
Giorgi, G. M.; Mondani, R.	Sampling distribution of the Bonferroni inequality index from exponential population •	95 SankhyaB 57 10-18
A uniform approximation to the	sampling distribution of the coefficient of variation • Hürlimann, Werner	95 StPrLet 24 263-268
Penny, David; Hendy, Michael D.; Arnold, Greg The	sampling distributions and covariance matrix of phylogenetic spectra • Waddell, Peter J.;	94 MolBioEv 11 630-642
An exact test for neutrality based on the Ewens	sampling distribution • Slatkin, Montgomery	94 GeneRsr 64 71-74
Sensitivity of posterior quantities to the	sampling distribution: More than one observation • Basu, Sanjib	94 ASAProBays 106-111
non-additive genetic correlations. I. Properties of	sampling distributions • Hohls, T.; Clarke, G. P. Y. Additive and	95 BiomtrcJ 37 919-935
	Sampling distributions of relative poverty statistics • Preston, Ian	95 ApplStat 44 91-99
Rudiuk, Edmund M. The	sampling distribution of the serial correlation coefficient • Provost, Serge B.;	95 AJMMgSci 15 57-81
censored samples • Ashour, S. K. et al. The	sampling distribution of some goodness-of-fit test statistics using Burr distribution and	94 EgpStJ 38 147-163
On the	sampling distribution of swap-rate • Kikuchi, Kenichi; Mayekawa, Shin-ichi	95 Behvmtrk 22 185-204
abundances in stream-fish communities: Effects of	sampling effort and discontinuous spatial distributions • Angermeier et al. ... relative	95 JFishRBC 52 936-949
Wray; Chang, Michael Optimal periodicity of a survey:	Sampling error, data deterioration, and cost • Ghosh, Dhiren; Kaufman, Steven F.; Smith,	94 ASASrvyRsr 1122-1127
Review of "Non-	sampling errors in surveys" • Kessler, J. T.; Kelsbeck, W. D. (Auth); Fielding, A. (Rev)	95 Ststcian 44 289-290
Singh, Rajendra P. New applications of	sampling estimation in the 1995 Census Test (Disc: p754-756) • Mulry, Mary H.;	94 ASASrvyRsr 742-747
Hidiroglou, M. A.	Sampling and estimation for establishment surveys: Stumbling blocks and progress •	94 ASASrvyRsr 153-162
Statistical process control of	sampling frames • Spisak, A. W.	95 SurvyM 21 185-190
the electoral registers and postcode address file as	sampling frames • Lynn et al. ... and variance of samples of individuals: Comparison of	95 Ststcian 44 173-194
Estimation, variance and optimal	sampling of gene diversity. I. Haploid locus • Pons, O.; Petit, R. J.	95 ThApGene 90 462-470
to stock assessment and harvest decisions using the	sampling/importance resampling algorithm • McAllister, M. K. et al. A Bayesian approach	94 JFishRBC 51 2673-2687
An application of variable	sampling interval control charts • Baxley, Robert V., Jr	95 JQualTek 27 275-282
Evaluating properties of variable	sampling interval control charts • Reynolds, Marion R., Jr	95 SqtlAnly 14 59-97
Process control strategies using variable	sampling intervals • Sethuraman, Shanthi; Reynolds, Marion R., Jr; Arnold, Jesse C.	94 ASAProQlPr 109-114
Counter-matching: A stratified nested case-control	sampling method • Langholz, Bryan; Borgan, Ørnulf	95 Biomtrka 82 69-79
Gregoire, T. G.; Valentine, H. T.; Furnival, G. M.	Sampling methods to estimate foliage and other characteristics of individual trees •	95 Ecology 76 1181-1194
A comparison of algorithms for	sampling models • Clyde, Merlise A.; DeSimone, Heather; Parmigiani, Giovanni	94 ASAProBays 211-216
likelihood estimator in missing data and biased	sampling models • van der Laan, Mark J. An identity for the nonparametric maximum	95 Bernoull 1 335-341
Combinatorial and statistical aspects of	sampling plans to avoid the selection of adjacent units • Stufken, John	93 JCISS 18 149-160
level • Raju, C.; Murthy, M. N. N. Minimum risks chain	sampling plans (ChSP-4 (c_1, c_2)) indexed by acceptable quality level and limiting quality	95 JAppStat 22 389-426
Using known populations of pronghorn to evaluate	sampling plans and estimators • Kraft, K. M.; Johnson, D. H.; Samuelson, J. M.; Allen, S. H.	95 JWildlMg 59 129-137
Failure-censored reliability	sampling plans for the exponential distribution • Balasooriya, Uditha	95 JStCmpSm 52 337-349
censoring • Yeh, Lam; Choy, S. T. B. Bayesian variable	sampling plans for the exponential distribution with uniformly distributed random	95 JStPlInf 47 277-293
A comparative study of multidimensional	sampling plans • Dutta, K.; Goswami, V.	94 JIAgSt 46 254-266
for quality control by attribute. Part I: Simple	sampling plans (French) • Parent, E.; Chaouche, A. et al. In support of Bayesian statistics	95 RStApp 43/4 5-18
Majumdar, Dibyen Generating desirable	sampling plans by the technique of trade-off in experimental design • Hedayat, A. S.;	95 JStPlInf 44 237-247
Bai, D. S. et al. Failure-censored accelerated life test	sampling plans for Weibull distribution under expected test time constraint •	95 RelEngSS 50 61-68
Gomm, Ian S.; Armour, Len Location of the optimal	sampling point for the quality assessment of continuous streams • Barnett, Neil S.;	95 AstrlJSt 37 145-152
estimation of quadratic forms in finite population	sampling • Vijayan, K.; Mukhopadhyay, P.; Bhattacharyya, S. On non-negative unbiased	95 AstrlJSt 37 169-178
Review of "Design and inference in finite population	sampling" • Hedayat, A. S.; Sinha, B. K. (Auth); Rust, Keith (Rev)	94 JOffStat 10 226-228
Bayesian analysis of a protocol for	sampling preserved tumor specimens • Heitjan, Daniel F.	95 JASA 90 38-44
the biases and inefficiencies of the EPI method of	sampling relatively unimportant with reasonable between village ... • Fitch et al. Are	94 ASASrvyRsr 288-292
James H. (Rev) Review of "Telephone survey methods:	Sampling, selection and supervision (Second edition)" • Lavrakas, Paul J. (Auth); Frey,	94 JOffStat 10 349-351
in a simple linear regression using rank set	sampling • Muttlak, Hassen A. Parameters estimation	95 BiomtrcJ 37 799-810
	Sampling from spatial databases • Olken, Frank; Rotem, Doron	95 StatComp 5 43-57
Selection techniques in	sampling spatial units • Arbia, Giuseppe	94 QStMaA 16 81-91
Bell, Robert M.	Sampling and statistical estimation in the decennial census (Disc: p90-95) •	94 ASASrvyRsr 71-79
Mittleman, M. A.; Maclure, M.; Robins, J. M. Control	sampling strategies for case-crossover studies: An assessment of relative efficiency •	95 AJEpidm 142 91-98
Wyler, A. Power of different	sampling strategies to detect quantitative trait loci variance effects • Weller, J. I.;	92 ThApGene 83 582-588
Practical	sampling techniques • Som, Ranjan K.	95 Dekker:NY 664p
One way to build an estimator with applications to	sampling theory • Woodruff, Steve	94 ASASrvyRsr 1013-1018
Christopher J.	Sampling theory for cytonuclear disequilibria • Asmussen, Marjorie A.; Basten,	94 Genetcs 138 1351-1363
distance • Sanchez, A.; Ocaña, J.; Utzet, F.	Sampling theory, estimation, and significance testing for Prevosti's estimate of genetic	95 Biomtrcs 51 1216-1235
Ziegel, Eric R. (Rev) Review of "Pierre Gy's	sampling theory and sampling practice (Second edition)" • Pitard, Francis (Auth);	95 Technmcs 37 125-126
	Sampling on two occasions with two correlated characters • Singh, Jagbir; Kathuria, O. P.	94 JIAgSt 46 166-174
Unrelated question randomized response	sampling using continuous distributions • Singh, Sarjinder	94 JIAgSt 46 349-361

Tests of ordered hypotheses for gamma	scale parameters • Li, Teng; Sinha, Bimal K.	95 JStPlInf 45 387-397
A sequential sampling plan for the	scale parameter of the inverse Gaussian distribution • Nabar, Smita	94 JIndnStA 32 117-121
A jackknife estimator of the	scale parameter of L_p-normal distribution (Italian) • La Tona, Letizia	94 Metron 52/1 181-188
Estimation of location and	scale parameters under order restrictions • Kubokawa, Tatsuya; Saleh, A. K. Md. E.	94 JStatRsr 28 41-51
Gill, A. N.; Mehta, G. P. Selection procedures for	scale parameters when the populations have a common known quantile • Kumar, N.;	94 Metron 52/3 171-196
Generalized linear models with a small	scale parameter • Dorkenoo, K. M.; Mathieu, J.-R.	95 MaOpfStS 27 107-113
comments on maximum likelihood estimators of	scale parameters or its square • Gupta, Rameshwar D.; Khatree, Ravindra Some	94 JIndnStA 32 123-131
Thiagarajah, K. Ranee Homogeneity tests for	scale parameters of 2-parameter exponential populations under time censoring •	95 IEEEReli 44 297-301
missibility results for estimating ordered uniform	scale parameters • Misra, Neeraj; Dhariyal, Ishwari D. Some inad-	95 CommStA 24 675-685
tial location parameters with a common unknown	scale parameter • Carpenter, M. et al. Estimation of the smaller and larger of two exponen-	95 CalcutSt 45 61-71
administration mode on responses to numeric rating	scales • Sangster, Roberta L.; Rockwood, Todd H.; Dillman, Don A. The influence of	94 ASASrvyRsr 708-712
anchors: Verbal and numeric labels for response	scales • O'Muircheartaigh, Colm; Gaskell, George; Wright, Daniel B. Weighing	95 JOffStat 11 295-307
responding estimated one-parameter logistic model	scale score and the unbiased domain score for each number-correct ... • Li ... for the cor-	95 ApPsychM 19 50-50
Testing for a signal with unknown location and	scale in a stationary Gaussian random field • Siegmund, David O.; Worsley, Keith J.	95 AnlsStat 23 608-639
Urquhart, N. S.; Kugler, D. L. Regional	scale trend monitoring of indicators of trophic condition of lakes • Larsen, D. P.;	95 WaterReB 31 117-140
A maximum test for	scale: Type I error rates and power • Algina, James; Blair, R. Clifford; Coombs, William T.	95 JEdStat 20 27-39
	Scaling	
	• see also affine –, dual –, metric –, multidimensional –	
A likelihood approach to	scaling examination marks • Biggins, J. D.; Yue, K. K.	95 AstrlJSt 37 273-282
Two-moment	scaling formulae for aggregating examination marks • Daley, D. J.	95 AstrlJSt 37 253-272
Monsees, Geoffrey A. The effect of	scaling input variables on the direction of steepest ascent • Carroll, Steven S.;	94 ASAProQlPr 267-269
Large deviations from a hydrodynamic	scaling limit for a nongradient system • Quastel, Jeremy	95 AnlsProb 23 724-742
The	scaling limit for a stochastic PDE and the separation of phases • Funaki, T.	95 ProbTher 102 221-288
active constraints search algorithm for optimal	scaling of ordered categorical responses • Lin, Shang P.; Tang, Dei-In A hierarchical	95 JStCmpSm 53 197-209
Optimal	scaling for ordered categories • Pölz, W.	95 CmpStQ 10 37-41
A continuous metric	scaling solution for a random variable • Cuadras, C. M.; Fortiana, J.	95 JMultiAn 52 1-14
On the robustness of maximum likelihood	scaling for violations of the error model • Storms, Gert	95 Psymtrka 60 247-258
Lavine, Barry K. (Rev) Review of	"SCAN: Software for chemometric analysis (Release 1 for Windows)" • Minitab Inc (Auth);	95 Chemolab 30 253-255
	Scan statistic	
Anderson, N. H.; Titterington, D. M.	A comparison of two statistics for detecting clustering in one dimension •	95 JStCmpSm 53 103-125
Power of the	scan statistics • Wallenstein, Sylvan; Naus, Joseph; Glaz, Joseph	94 ASAProEpid 70-75
Alexandrou, V. A. Runs,	scans and urn model distributions: A unified Markov chain approach • Koutras, M. V.;	95 AnInStMa 47 743-766
convergence rate of the Gibbs sampler with various	scans • Liu, Jun S.; Wong, Wing H.; Kong, Augustine Covariance structure and	95 JRSS-B 57 157-169
Isaksson, T. Related versions of the multiplicative	scatter correction method for preprocessing spectroscopic data • Helland, I. S.; Næs, T.;	95 Chemolab 29 233-241
Renovated	scatterplots for censored data • Smith, Peter J.; Zhang, J.	95 Biomtrka 82 447-452
Fast and simple	scatterplot smoothing • Härdle, W.; Marron, J. S.	95 CmpStDA 20 1-17
The visual separability of plotting symbols in	scatterplots • Tremmel, Lothar	95 JCmpGrSt 4 101-112
sources of human cerebral potentials based on	SCD and a model selection criterion • Wang, W. et al. Localization of multiple dipole	94 ASAProBiop 217-222
imputation of the 1983 and 1989 waves of the	SCF • Kennickell, Arthur B.; McManus, Douglas A. Multiple	94 ASASrvyRsr 523-528
A family of joint probability models for cost and	schedule uncertainties • Garvey, Paul R.	95 JCostAnl /1 156-200
Bruno, John; Coffman, Edward G., Jr; Downey, Peter	Scheduling independent tasks to minimize the makespan on identical machines •	95 PrEngInf 9 447-456
Cunningham, Patricia	Scheduling initial calls in a telephone survey of businesses • Edwards, W. Sherman;	94 ASASrvyRsr 1368-1373
Single machine stochastic	scheduling: Minimizing the number of tardy jobs • Lin, Chen-Sin	94 AJMMgSci 14 289-300
time flexible manufacturing and operator	scheduling model solved by deconvexification over time • Eaves et al. ... continuous-	95 PrEngInf 9 65-98
On the value of adaptive solutions to stochastic	scheduling problems • Glazebrook, K. D.; Owen, R. W.	95 MaOpRsr 20 65-89
allowing multiple events and probabilistic service	scheduling • Henderson et al. ... in discrete-time generalized semi-Markov processes	95 AnlsApPr 5 78-96
Optimal measurement	scheduling for state estimation • Shakeri, M.; Pattipati, K. R.; Kleinman, D. L.	95 IEEEAES 31 716-729
On almost optimal priority rules for preemptive	scheduling of stochastic jobs on parallel machines • Weiss, Gideon	95 AdvAppPr 27 821-839
Dynamic	scheduling with convex delay costs: The generalized $c\mu$ rule • van Mieghem, Jan A.	95 AnlsApPr 5 809-833
Customer	scheduling with incomplete information • Rieder, Ulrich; Weishaupt, Jürgen	95 PrEngInf 9 269-284
A V-optimal design for	Scheffé's polynomial model • Liu, Shuangzhe; Neudecker, Heinz	95 StPrLet 23 253-258
comparisons • Schwertman et al. A more practical	Scheffé-type multiple comparison procedure for commonly encountered numbers of	95 JStCmpSm 53 181-196
representation of decision-making in chronic	schizophrenics • Rosenfeld, Barry; Turkheimer, Eric Multidimensional	95 MultBehR 30 199-211
The analysis of repeated-measures data on	schizophrenic reaction times using mixture models • Belin, Thomas R.; Rubin, Donald B.	95 StatMed 14 747-768
The use of Fisher's Z in	Schmidt-Hunter-type meta-analyses • Law, Kenneth S.	95 JEdStat 20 287-306
On a theorem of K.	Schmidt • Bradley, Richard C.	95 StPrLet 24 9-12
Trumbo, B. E. Key words and phrases - The key to	scholarly visibility and efficiency in an information explosion • Gbur, E. E., Jr;	95 AmerStat 49 29-33
Graphing in the primary	school: Algorithm versus comprehension • Pereira-Mendoza, Lionel	95 TeachgSt 17/1 2-6
Fink, S. Some data issues in	school-based surveys • Kasprzyk, D.; Gruber, K.; Salvucci, S.; Saba, M.; Zhang, F.;	94 ASASrvyRsr 815-820
on making statistics more effective in business	schools (MSMESB) (Disc: p319-320) • Roberts, Harry V. Reflections	94 ASAProBuEc 316-318
The estimation of	school effects • Raudenbush, Stephen W.; Willms, J. Douglas	95 JEdStat 20 307-335
Minimum wage effects on employment and	school enrollment • Neumark, David; Wascher, William	95 JBES 13/2 199-206
local labor markets, and dropping out of high	school (Disc: p213-215) • Swaim, Paul Work experience,	94 ASAProSoSt 207-212
man, David J. The benefit of additional high-	school math and science classes for young men and women • Levine, Phillip B.; Zimmer-	95 JBES 13/2 137-149
Teacher quality in public and private	schools (Disc: p213-215) • Ballou, Dale; Podgursky, Michael	94 ASAProSoSt 196-200
teachers to teach statistics in the secondary	schools of Puerto Rico • Sahai, Hardeo; Vega, Juan Training	94 ASAProStEd 253-258
the coverage of private elementary-secondary	schools • Jackson, Betty J.; Frazier, Richard L.; King, Karen E.; Schwanz, Dennis J. Improving	94 ASASrvyRsr 833-838
(Rev) Review of "Introducing data analysis in the	schools: Who should teach it and how?" • Pereira-Mendoza, Lionel (Ed); Watkins, Ann E.	95 AmerStat 49 62-63
Kaufman, Steven Properties of the	Schools and Staffing Survey's bootstrap variance estimator (Disc: p1128-1129) •	94 ASASrvyRsr 1116-1121
Ward, Joe H., Jr; Niland, Laura J. Empowering high	school students to exploit statistical models and software for research projects •	94 ASAProStEd 44-48
Smith, Wray Intersurvey consistency in	school surveys (Disc: p1128-1129) • Holt, Albert; Kaufman, Steven; Scheuren, Fritz;	94 ASASrvyRsr 1105-1110
Comment on "Estimation issues in	school surveys" • Shapiro, Gary	94 ASASrvyRsr 1128-1129
A study of biostatistics requirements in medical	schools in the United States • Grady, C. S.; Looney, S. W.; Steiner, R. P.	94 ASAProStEd 263-266
The transition from	school to work in 1979-1987 in the Netherlands • van Opstal, R.; van de Pol, F. J. R.	95 StNeerla 49 294-309
Are OLS estimates of the return to	schooling biased downward? another look • Blackburn, McKinley L.; Neumark, David	95 RvEconSt 77 217-229
instrumental variables estimates of the return to	schooling • Angrist, Joshua D.; Krueger, Alan B. Split-sample	95 JBES 13/2 225-235
From Brownian motion to	Schrödinger equation • Chung, K. L.; Zhao, Z.	95 Springer:Brln:NY 287p
Review of "From Brownian motion to	Schrödinger equation" • Chung, K. L.; Zhao, Z. (Auth); Nagasawa, M. (Rev)	95 ShrtBkRv 15/3 49-49
An application of the	Schur-Cohn algorithm to time series analysis • Barnard, Roger W.; Chanda, Kamal C.	95 JTimSrAn 16 445-449
Shanthikumar, J. G.; Tong, Y. L. Parametric	Schur convexity and arrangement monotonicity properties of partial sums • Shaked, M.;	95 JMultiAn 53 293-310
for nested hypotheses and its relationship to the	Schwarz criterion • Kass, Robert E.; Wasserman, Larry A reference Bayesian test	95 JASA 90 928-934
The benefit of additional high-school math and	science classes for young men and women • Levine, Phillip B.; Zimmerman, David J.	95 JBES 13/2 137-149
and quantitative risk assessment: A bridge from	science to policy • Hertz-Picciotto, Irva Epidemiology	95 AJPubHlh 85 484-491
Public health surveillance for chronic conditions: A	scientific basis for decisions (Disc: p695-696) • Thacker, S. B.; Stroup, D. F. et al.	95 StatMed 14 629-641

bon vapours by pattern recognition applied to RIFS	sensor signals • Kraus et al. Optical reflectometric gas sensing: Classification of hydrocar-	95 Chemolab 30 211-221
for predicting shelf-life of a military ration with	sensory data • Chen, Ming-Hui; Nandram, Balgobin A Gibbs sampling approach	94 ASAProBays 84-89
Discrimination indexes and models for studying	sensory functioning in aging • O'Brien, Ralph G.; Heft, Marc W.	95 JAppStat 22 9-27
et al. Experimental mixture design to ameliorate the	sensory quality evaluation of extra virgin olive oils • Vojnovic, D.; Campisi, B.; Mattei, A.	95 Chemolab 27 205-210
Using the shifted multiplicative model to search for	"separability" in crop cultivar trials • Cornelius, P. L.; Seyedsadr, M.; Crossa, J.	92 ThApGene 84 161-172
Flexible functional forms and tests of homogeneous	separability • Diewert, W. E.; Wales, T. J.	95 JEconmtx 67 259-302
The visual	separability of plotting symbols in scatterplots • Tremmel, Lothar	95 JCmpGrSt 4 101-112
conditional moment tests • Ahn, S. C.; Schmidt, P. A	separability result for GMM estimation, with applications to GLS prediction and	95 EcomtrRv 14 19-34
An asymptotic test for	separability of a spatial autoregressive model • Shitan, Mahendran; Brockwell, Peter J.	95 CommStA 24 2027-2040
	Separability test for the electricity supply industry • Lee, Byung-Joo	95 JApEcmtx 10 49-60
of the efficacy of rhIL-1ra for the treatment of	sepsis • Lemke, Jon H.; LaBrecque, John F.; Rubenstein, Linda M. Analysis	94 ASAProBiop 340-345
sequential logrank tests • Kim, K. SEQPWR and	SEQOPR: Computer programs for design of maximum information trials based on group	95 CmpBiomd 46 143-153
based on group sequential logrank tests • Kim, K.	SEQPWR and SEQOPR: Computer programs for design of maximum information trials	95 CmpBiomd 46 143-153
	Sequence	
	• see also names of specific types of sequences such as random –	
unexpected frequencies in deoxyribonucleic acid	sequences • Prum, B.; Rodolphe, F.; de Turckheim, E. Finding words with	95 JRSS-B 57 205-220
Neighbor joining and maximum likelihood with RNA	sequences: Addressing the interdependence of sites • Tillier, E. R. M.; Collins, R. A.	95 MolBioEv 12 7-15
Measuring the agreement between	sequences (Disc: p232-243) • Dijkstra, Wil; Taris, Toon	95 SocMethR 24 214-231
Lawrence, C. E. Bayesian models for multiple local	sequence alignment and Gibbs sampling strategies • Liu, J. S.; Neuwald, A. F.;	95 JASA 90 1156-1170
Sums of double arrays of LP-mixingale	sequence and its application in linear models (Chinese) • Hu, Shuhe	95 AMaAPknC 17 248-254
and clustering on nonstationary max-AR(1)	sequences • Alpuim, M. T.; Catkan, N. A.; Hüsler, J. Extremes	95 StocProc 56 171-184
density and failure rate estimation for associated	sequences • Bagai, Isha; Rao, B. L. S. Prakasa Kernel-type	95 AnInStMa 47 253-266
Modelling evolution of autocorrelated	sequences • von Haeseler, Arndt; Schöniger, Michael	95 CmpStDA 20 457-458
"Bayesian analysis for pairwise comparisons, binary	sequences, and belief networks" • Laskey, Kathryn B. Comment on	94 ASAProBays 134-135
Recurring-with-carry	sequences • Koç, Cemal	95 JAppProb 32 966-971
Barker	sequences and circulant Hadamard matrices • Lin, Cantian; Wallis, W. D.	93 JCISS 18 19-25
approximations for testing statistical significance in	sequence comparisons • Goldstein, L. et al. Poisson, compound Poisson and process	92 BMathBio 54 785-812
of diffusion processes and their weakly convergent	sequences • Portenko ... idea concerning the methods for investigating local behavior	95 ThPrMaSt 50 7-22
on stationary linearly positive quadrant dependent	sequences • Kim, Tae-sung A note	95 JKoreaSt 24 249-256
of density estimation for strongly dependent	sequences • Ho, Hwai-Chung On the strong uniform consistency	95 StPrLet 22 149-156
with random indices for martingale difference	sequences • Hong, Dug Hun A central limit theorem	95 JKoreaSt 24 219-224
Network models for	sequence evolution • von Haeseler, Arndt; Churchill, Gary A.	93 JMolEvol 37 77-85
On a class of exchangeable	sequences • Gnedin, Alexander V.	95 StPrLet 25 351-355
Characterizing Markov exchangeable	sequences • Zabell, S. L.	95 JTherPrb 8 175-178
Stopping rules for	sequences of factorial designs • Gilmour, S. G.; Mead, R.	95 ApplStat 44 343-355
filters of strongly dependent Gaussian	sequences • Ho, Hwai-Chung The law of the iterated logarithm for non-instantaneous	95 JTherPrb 8 347-360
Introduction to computational biology: Maps,	sequences, and genomes • Waterman, Michael S.	95 ChapmnHall:Lnd 450p
timation of phylogenetic variability from nucleotide	sequences: II. Four taxa without a molecular clock • Zharkikh et al. ... of bootstrap es-	92 JMolEvol 35 356-366
Estimating the number of change points in a	sequence of independent normal random variables • Lee, Chung-Bow	95 StPrLet 25 241-248
Limit theorems of probability theory:	Sequences of independent random variables • Petrov, V. V.	95 OxfordUniv 292p
Review of "Limit theorems of probability theory:	Sequences of independent random variables" • Petrov, V. V. (Auth); Pyke, R. (Rev)	95 ShrtBkRv 15/3 46-46
Stationarity of independent	sequences • Lacaze, B.	95 StPrLet 23 9-11
Picco, Pierre Equivalence of renewal	sequences and isomorphism of random walks • Aaronson, Jon; Liggett, Thomas;	94 IsrelJMa 87 65-76
Critical phenomena for	sequence matching with scoring • Dembo, Amir; Karlin, Samuel; Zeitouni, Ofer	94 AnlsProb 22 1993-2021
iterated logarithm for stationary strongly mixing	sequences • Rio, Emmanuel The functional law of the	95 AnlsProb 23 1188-1203
number of exceedances of nonstationary normal	sequences • Hüsler, J.; Kratz, M. Rate of Poisson approximation of the	95 StocProc 55 301-313
of evolutionary distances between nucleotide	sequences • Zharkikh, Andrey Estimation	94 JMolEvol 39 315-329
heterogeneity among regions of a nucleotide	sequence • Gaut, Brandon S.; Weir, B. S. Detecting substitution-rate	94 MolBioEv 11 620-629
with unusual occurrence frequency in nucleotidic	sequences • Colosimo, A.; Morante, S. et al. An improved method for detection of words	93 JThBio 165 659-672
Chandon, Y. Search for a	sequence of optimal tests to control manufacturing costs (French) • Bergeret, F.;	95 RStApp 43/3 21-33
Probability bounds on the finite sum of the binary	sequence of order k • Dhar, Sunil K.; Jiang, Xulun	95 JAppProb 32 1014-1027
of a random number of variables from periodic	sequences • Ferreira, Helena Extremes	95 JStPlInf 45 133-141
Sjöström, M.; Rännar, S.; Wieslander, Å. Polypeptide	sequence property relationships in Escherichia coli based on auto cross covariances •	95 Chemolab 29 295-305
of variance of partial sums of an associated	sequence of random variables • Peligard, Magda; Suresh, Ram Estimation	95 StocProc 56 307-319
Bretagnolle, J.; Klopotowski, A. On the existence of	sequences of s by s independent exchangeable or stationary random variables (French) •	95 AlHenrPB 31 325-350
Neglecting observations in Gaussian	sequences of statistical experiments • Marohn, F.	95 StatDecn 13 83-92
The Bahadur representation of sample quantiles for	sequences of strongly mixing random variables • Yoshihara, Ken-ichi	95 StPrLet 24 299-304
the rate functions for stationary dependent random	sequences in τ-topology and the relative entropy • Hu A note on the relationship between	95 StocProc 56 351-355
for nonlinear functionals of a stationary Gaussian	sequence of vectors • Arcones, Miguel A. Limit theorems	94 AnlsProb 22 2242-2274
Random perturbations of recursive	sequences with an application to an epidemic model • Pierre Loti Viaud, Daniel	95 JAppProb 32 559-578
empirical Bayes model for Markov-dependent binary	sequences with randomly missing observations • Cole, B. F.; Lee, M.-L. T. et al. An	95 JASA 90 1364-1372
	Sequential	
	• see also group –, two-stage, multi-stage	
	Sequential allocation rules for multi-armed clinical trials • Coad, D. S.	95 JStCmpSm 52 239-251
	Sequential analysis	
Malý, Marek	The asymptotics for studentized k-step M-estimators of location • Jurečková, Jana;	95 SqtlAnly 14 229-245
	A class of Bayesian models for optimal exploration • Glazebrook, K. D.; Boys, R. J.	95 JRSS-B 57 705-720
experiments available from ... • Mukhopadhyay et al.	Comparing exponential clinical trials by combining individual sequential	94 Metron 52/1 17-40
class • Mita, Haruyoshi	Tail probability of the sequential maximum likelihood estimator for the exponential	95 JJapnStS 25 173-182
tween sufficiency and invariance with applications in	sequential analysis" (V36 p575-614) • Hall, W. J. et al. Correction to "The relationship be-	95 AnlsStat 23 705-705
Quera, Vicenç Log-linear approaches to lag-	sequential analysis when consecutive codes may and cannot repeat • Bakeman, Roger;	95 PsychBu 118 272-284
Analyzing interaction:	Sequential analysis with SDIS and GSEQ • Bakeman, Roger; Quera, Vicenç	95 CambridgeUniv 166p
A	sequential assignment match process with general renewal arrival times • David, Israel	95 PrEngInf 9 475-492
Factor screening by	sequential bifurcation • Bettonvil, Bert	95 CommStB 24 165-185
Unbiased estimation in	sequential binomial sampling • Akahira, Masafumi; Takeuchi, Kei; Koike, Ken-ichi	92 RStAJp 39/4 1-13
Nonparametric truncated	sequential change-point detection • Gombay, E.	95 StatDecn 13 71-82
p645-658) • Lai, Tze Leung	Sequential changepoint detection in quality control and dynamical systems (Disc:	95 JRSS-B 57 613-644
likelihood estimation of polychotomous and	sequential choice models • Lee, Lung-fei Semiparametric maximum	95 JEconmtx 65 381-428
Berry, Donald A. Bayesian group	sequential clinical trial designs based on a quadratic loss function. • Lewis, Roger J.;	94 ASAProBays 23-28
calculation of the operating characteristics of group	sequential clinical trials • Elson, Paul J. Use of the fast Fourier transform algorithm in the	95 CmpStDA 20 491-498
Tan, Ming; Xiong, Xiaoping	Sequential conditional probability ratio test approach to the design of clinical trials •	94 ASAProBiop 359-364
A class of	sequential conditional probability ratio tests • Xiong, Xiaoping	95 JASA 90 1463-1473
Nagata, Yasushi Fixed-width	sequential confidence interval for the mean of a gamma distribution • Takada, Yoshikazu;	95 JStPlInf 44 277-289
factorials • Ghosh, Subir	Sequential construction of new main effect plans with higher revealing power for 2^7	93 JCISS 18 217-231

Parlar, M.; Perry, D. Optimal	sequential decisions for incomplete identification of group-testable items • Bar-Lev, S. K.;	95 SqtlAnly 14 41-57
	Sequential design	
Woodroofe, Michael B.	Central limit theorems for doubly adaptive biased coin designs • Eisele, Jeffrey R.;	95 AnlsStat 23 234-254
	On efficient designing of nonlinear experiments • Chaudhuri, Probal; Mykland, Per A.	95 StSinica 5 421-440
	Response surface bandits • Ginebra, Josep; Clayton, Murray K.	95 JRSS-B 57 771-784
Bivariate	sequential designs for phase II trials • Conaway, Mark R.; Petroni, Gina R.	95 Biomtrcs 51 656-664
Consistency of the MLE for a general	sequential design problem • Datta, Somnath	95 SankhyaA 57 88-99
Using the generalized likelihood ratio statistic for	sequential detection of a change-point • Siegmund, D.; Venkatraman, E. S.	95 AnlsStat 23 255-271
type bound for the asymptotic variance and the	sequential discretized likelihood estimation procedure • Akahira, M. The Bhattacharyya	95 SqtlAnly 14 193-204
Rosenberger, William F. Toxicity in	sequential dose-response experiments • Flournoy, Nancy; Durham, Stephen D.;	95 SqtlAnly 14 217-227
A	sequential dual of Stein's two-sample procedure • Arghami, N. R.; Billard, L.	92 ProIrStC1 11-18
Weak convergence of the	sequential empirical processes of residuals in ARMA models • Bai, Jushan	94 AnlsStat 22 2051-2061
	Sequential estimation	
Kim, Sung Lai	Asymptotic properties of the stopping times in a certain sequential procedure •	95 JKoreaSt 24 337-347
Martinsek, Adam T.	Estimating a slope parameter in regression with prescribed proportional accuracy •	95 StatDecn 13 363-377
Chang, Yuan-chin Ivan	Estimation in some binary regression models with prescribed accuracy •	95 JStPlInf 44 313-325
rection to "Asymptotic expansions in the problem of	sequential estimation of an autoregressive parameter" (V37 p381-383) • Malinovskii Cor-	94 ThProbAp 39 719-719
stopping number • Bose, Arup; Mukhopadhyay, Nitis	Sequential estimation of the mean of an exponential distribution via replicated piecewise	95 StatDecn 13 351-361
Nitis A note on accelerated	sequential estimation of the mean of NEF-PVF distributions • Bose, Arup; Mukhopadhyay,	95 AnInStMa 47 99-104
Bias corrected	sequential estimation for the mean of NEF-PVF distributions • Bose, Arup; Boukai, Benzion	95 SqtlAnly 14 307-319
A note on	sequential estimation of the mean • Uno, Chikara	95 StPrLet 23 141-149
Second-order approximations to two classes of	sequential estimation procedures • Chaturvedi, Ajit; Gupta, Rahul	95 MaOpfStS 27 159-173
function • Magiera, Ryszard Bayes	sequential estimation procedures in exponential-type processes for a polynomial cost	95 MaOpfStS 26 61-73
On efficient	sequential estimation for the simple single server queue • Stefanov, Valeri T.	95 RBrzlPrS 9 55-65
Strawderman, William E.	Sequential estimation of the variance of a normal distribution • Natarajan, Jayalakshmi;	95 SqtlAnly 14 361-374
... exponential clinical trials by combining individual	sequential experiments available from independent studies • Mukhopadhyay et al.	94 Metron 52/1 17-40
A stochastic process arising in	sequential experimentation • Stadje, W.	95 StPrLet 23 359-365
with a discrete covariate • Rahbar, Mohammad H.	Sequential fixed width confidence intervals for regression parameters from censored data	95 SqtlAnly 14 143-156
Blind deconvolution via	sequential imputations • Liu, Jun S.; Cheng, Rong	95 JASA 90 567-576
	Sequential inference	
Pignatiello, Joseph J., Jr; Raz, Tzvi	Bayesian Poisson rate estimation using parameter maps • Buck, James R.;	95 IIETrn 27 405-412
	Some adaptive estimators in simple regression models • Viet, Tran Quoc	95 StatDecn 13 117-130
	A stepwise resampling method of multiple hypothesis testing • Troendle, James F.	95 JASA 90 370-378
stopping times in a class of ...• Bose et al.	Sequential interval estimation via replicated piecewise stopping times and accelerated	95 SqtlAnly 14 287-305
Interval estimation from a life-test through a	sequential Kolmogorov-Smirnov type procedure • Gastaldi, Tommaso	95 IEEEReli 44 470-476
Tail probability of the	sequential maximum likelihood estimator for the exponential class • Mita, Haruyoshi	95 JJapnStS 25 173-182
Hélène; Kim, Kyungmann	Sequential methods for parametric survival models • Tsiatis, Anastasios A.; Boucher,	95 Biomtrka 82 165-173
Local asymptotic normality of a	sequential model for marked point processes and its applications • Nishiyama, Yoichi	95 AnInStMa 47 195-209
Fixed diameter confidence ellipsoid in time-	sequential models • Aras, Girish; Gardiner, Joseph	95 SqtlAnly 14 29-39
Thall, P. F.; Simon, R. M.; Estey, E. H. Bayesian	sequential monitoring designs for single-arm clinical trials with multiple outcomes •	95 StatMed 14 357-379
Toward vital status sweeps: A case history in	sequential monitoring • Hallstrom, Al; McBride, Ruth; Moore, Rich	95 StatMed 14 1927-1931
Monika A simulation study used to design the	sequential monitoring plan for a clinical trial • Brooks, Maria Mori; Hallstrom, Al; Peckova,	95 StatMed 14 2227-2237
Rosenberger, William F.; Lachin, John M.	Sequential monitoring of survival data with the Wilcoxon statistic • Lan, K. K. Gordon;	95 Biomtrcs 51 1175-1183
Comment on "Unbiased estimation for	sequential multinomial sampling plans" (V12 p253-259) • Sato, Michikazu	95 SqtlAnly 14 335-340
and weighted versions of simultaneous and	sequential multiple-comparison procedures • Klockars, A. J. et al. Power of unweighted	95 PsychBu 118 300-307
Improved	sequential MUSIC • Stoica, Petre; Händel, Peter; Nehoral, Arye	95 IEEEAES 31 1230-1239
	Sequential nonparametric estimation with assigned risk • Efromovich, Sam	95 AnlsStat 23 1376-1392
bivariate probabilities with an application to	sequential phase II trials • Conaway, Mark R. Importance sampling to evaluate	95 JStCmpSm 51 249-262
Nitis; Datta, Sujay On fine-tuned bounded risk	sequential point estimation of the mean of an exponential distribution • Mukhopadhyay,	95 SoAfrStJ 29 9-27
Uno, Chikara On the	sequential point estimation of the mean of a gamma distribution • Isogai, Eiichi;	95 StPrLet 22 287-293
Second-order approximations in the time-	sequential point estimation methodologies for the mean of an ... • Mukhopadhyay	95 SqtlAnly 14 133-142
	Sequential probability ratio test	
Bilgi, Shobha	Cumulative sum control charts for the inverse Gaussian distribution • Nabar, S. P.;	94 JIndnStA 32 9-14
	On time-sequential test for a class of distributions • Liu, Jeng-Fu	95 StSinica 5 251-260
	Sequential procedure	
	Fixed-width simultaneous confidence intervals for all-pairwise comparisons • Liu, W.	95 CmpStDA 20 35-44
Datta, Sujay On fine-tuning a purely	sequential procedure and associated second-order properties • Mukhopadhyay, Nitis;	95 SankhyaA 57 100-117
properties of the stopping times in a certain	sequential procedure • Kim, Sung Lai Asymptotic	95 JKoreaSt 24 337-347
Son, M. S. Accelerated	sequential procedure to estimate the mean of unknown distribution • Hamdy, H. I.;	94 ASASrvyRsr 1173-1178
Singh, R. K.; Chaturvedi, A.; Kumar, S. Two stage and	sequential procedures for estimating a linear function of k multinormal means •	94 JIndnStA 32 1-8
On the asymptotic regret of a	sequential procedure for estimating the mean of a normal distribution • Tahir, Mohamed	95 JStPlInf 48 331-338
Kleijnen, J. P. C. et al. Two-stage versus	sequential sample-size determination in regression analysis of simulation experiments •	95 AJMMgSci 15 83-114
loci • Motro, U.; Soller, M.	Sequential sampling in determining linkage between marker loci and quantitative trait	93 ThApGene 85 658-664
Nabar, Smita A	sequential sampling plan for the scale parameter of the inverse Gaussian distribution •	94 JIndnStA 32 117-121
Hughes, M. D.; Thompson, S. G.; Pocock, S. J. Optimal	sequential screening guidelines for quantitative risk factors measured with error •	95 JASA 90 19-26
An optimal	sequential selection procedure • Dragalin, V.	95 CommStA 24 429-443
frequentist and Bayesian test for fixed and	sequential simple hypothesis testing • Berger, J. O.; Brown, L. D. et al. A unified conditional	94 AnlsStat 22 1787-1807
	Sequential stochastic optimization • Cairoli, R.; Dalang, Robert C.	95 Wiley:NY:UK 350p
	Sequential test	
Els J. T.; Pocock, Stuart J.	Detection and estimation of J-shaped risk-response relationships • Goetghebeur,	95 JRSS-A 158 107-121
On time-	sequential test for a class of distributions • Liu, Jeng-Fu	95 StSinica 5 251-260
A kind of asymptotically optimal	sequential test under complex supposition • Chen, Jiading; Hickernell, Fred J.	95 SciSPkng 24 673-683
A comparison of two	sequential tests for μ/σ • McWilliams, Thomas P.	95 SqtlAnly 14 117-131
Distribution-free partially	sequential tests for treatments versus control setting • Kim, Dongjae; Wolfe, Douglas A.	95 BiomtrcJ 37 259-271
Asymptotic optimality of the Wald	sequential test • Koell, Christophe	95 SqtlAnly 14 341-360
of exponential dispersion models: Application to	sequential testing • Massam, Hélène An exact decomposition for a class	95 CandJSt 23 67-84
	Sequential testing with interval censored data • Quan, Hui; Yu, Kai F.	95 JStCmpSm 51 239-247
Review of "Optimally	sequentially planned decision procedures" • Schmitz, N. (Auth); Cressie, N. (Rev)	95 Metrika 42 141-143
lations in fluctuating environments: The case of the	Serengeti wildebeest • Pascual, Miguel A.; Hilborn, Ray Conservation of harvested popu-	95 JAppEcol 32 468-480
	Serial correlation	
	A minimum cost exponential CUSUM test • Boyd, Amy Hancock	95 CommStA 24 2929-2958
	Prediction and control for a time-series count data model • Brännäs, Kurt	95 IntJFor 11 263-270
Baltagi, Badi H.; Li, Qi	Testing AR(1) against MA(1) disturbances in an error component model •	95 JEconmtx 68 133-151
errors are ... • Gonzalez Manteiga et al.	Testing linear regression models using non-parametric regression estimators when	95 CmpStDA 20 521-541

How to look at objects in a five-dimensional	shape space: Looking at geodesics • Kendall, David G.	95 AdvAppPr 27 35-43
Familial spinal	shape • Mardia, K. V.; Dryden, I. L.; Hurn, M. A.; Li, Q.; Millner, P. A.; Dickson, R. A.	94 JAppStat 21 623-641
	Shape theory	
Kendall, David G.	How to look at objects in a 5-dimensional shape space. I: Looking at distributions •	94 ThProbAp 39 277-281
Understanding variable	shapes in 3-D • Grenander, Ulf; Keenan, Daniel	95 CommStSM 11 51-78
	Shapiro-Wilk test	
distribution • Wang, Julian Z.	Selection of the k largest order statistics for the domain of attraction of the Gumbel	95 JASA 90 1055-1061
Looney, Stephen W.	How to use tests for univariate normality to assess multivariate normality •	95 AmerStat 49 64-70
Lin, C. Thomas Some p-variate adaptations of the	Shapiro-Wilk test of normality • Mudholkar, Govind S.; Srivastava, Deo Kumar;	95 CommStA 24 953-985
Bounding	shares and quantiles from grouped data when group means are known • Schmid, Friedrich	94 Metron 52/1 143-153
Resampling inference for quantile	shares • Kovačević, Milorad S.; Yung, Wesley; Pandher, Gurupdesh S.	94 ASASrvyRsr 206-211
On a	shared allele test of random mating • Zhou, S.; Maller, R. A.; Speed, T. P.	95 AstrlJSt 37 61-72
Guo, Sun-Wei Proportion of genome	shared identical by descent by relatives: Concept, computation, and applications •	95 AJHuGene 56 1468-1476
of orthogonal Latin squares of order 10 with four	shared parallel transversals • Brown, J. W.; Hedayat, A. S.; Parker, E. T. A pair	93 JCISS 18 113-115
Computation of identity-by-descent proportions	shared by two siblings • Guo, Sun-Wei	94 AJHuGene 54 1104-1109
A generalization of the	Sharma-Kaushik class of metrics • Koschnick, Klaus-Uwe; Sharma, Bhu Dev	93 JCISS 18 401-407
	Sharp asymptotics of large deviations in R^d • Iltis, Michael	95 JTherPrb 8 501-522
Kersting, G.; Klebaner, F. C.	Sharp conditions for nonexplosions and explosions in Markov jump processes •	95 AnlsProb 23 268-272
Jump and	sharp cusp detection by wavelets • Wang, Yazhen	95 Biomtrka 82 385-397
	Sharp estimates for capacities and applications to symmetric diffusions • Sturm, K.-T.	95 ProbTher 103 73-89
a bounded submartingale • Hammack, William	Sharp inequalities for the distribution of a stochastic integral in which the integrator is	95 AnlsProb 23 223-235
nation for continuous-time martingales and related	sharp inequalities • Wang, Gang Differential subordination and strong differential subordi-	95 AnlsProb 23 522-551
Teuscher, F.; Guiard, V.	Sharp inequalities between skewness and kurtosis for unimodal distributions •	95 StPrLet 22 257-260
Joo, Sang Yeol; Kim, Yun Kyong A	sharp result of random upper functions for Levy processes • Kim, Gwan Young;	95 JKoreaSt 24 65-76
Estimation of non-	sharp support boundaries • Härdle, W.; Park, B. U.; Tsybakov, A. B.	95 JMultiAn 55 205-218
Tree rings in the Northern	Shawangunks • Heiner, Karlyn S.; Heiner, Karl W.	95 Chance 8/1 30-37
On the distribution of bubbles of the Brownian	sheet • Khoshnevisan, Davar	95 AnlsProb 23 786-805
and application to the supremum of the Brownian	sheet • Florit, Carme; Nualart, David A local criterion for smoothness of densities	95 StPrLet 22 25-31
Tsong, Yi Estimation of the	shelf-life of drugs with mixed effects models • Chen, James J.; Hwang, Jing-Shiang;	95 JBiophSt 5 131-140
On stability designs in drug	shelf-life estimation • Ju, Huey L.; Chow, Shein-Chung	95 JBiophSt 5 201-214
A Gibbs sampling approach for predicting	shelf-life of a military ration with sensory data • Chen, Ming-Hui; Nandram, Balgobin	94 ASAProBays 84-89
Epperson et al. Estimating the urban bias of surface	shelter temperatures using upper-air and satellite data. Part I: Development of ... •	95 JApMetrl 34 340-357
Interarea indexes of the cost of	shelter using hedonic quality adjustment techniques • Moulton, Brent R.	95 JEconmtx 68 181-204
	Shewhart chart	
A.; Champ, Charles W.; Woodall, William H.	The performance of control charts for monitoring process variation • Lowry, Cynthia	95 CommStB 24 409-437
On the run length of a	Shewhart chart for correlated data • Schmid, W.	95 StHefte 36 111-130
monitoring schemes using combined EWMA-	Shewhart control charts • Adams, Benjamin M.; Lin, Winnie Forecast-based	94 ASAProQlPr 104-108
sitivity study and comparison of the CUSUM versus	Shewhart control charts using an economic model • Simpson, James R.; Keats, J. Bert A sen-	94 ASAProQlPr 98-103
Talwar, Prem P. The	Shewhart, the Hodges-Lehman chart and the pre-control: Some comparisons •	94 ASAProQlPr 151-154
Roes, Kit C. B.; Does, Ronald J. M. M.	Shewhart-type charts in nonstandard situations (Disc: p24-35; Reply: P36-40) •	95 Technmcs 37 15-24
John H.	Shift detection in two-way arrays • Altekar, Maneesha; Lucas, James M.; Schuenemeyer,	94 ASAProPEng 76-80
John H.	Shift detection in two-way arrays • Altekar, Maneesha; Lucas, James M.; Schuenemeyer,	95 CommStB 24 711-731
Least absolute deviation estimation of a	shift • Bai, Jushan	95 EconTher 11 403-436
Detecting parameter	shift in GARCH models • Chu, Chia-Shang James	95 EcomtrRv 14 241-266
length of a moving average control chart to detect	shifts in the mean of a process (French) • Niang, N.; Saporta, G. Average run	95 RStApp 43/3 5-20
Simultaneous detection of	shift in means and variances • Flury, Bernard D.; Nel, Daan G.; Pienaar, Inet	95 JASA 90 1474-1481
The effect of station drift on the estimation of	shifts in meteorological records due to site changes • Rhoades, D. A.; Neill, W. A.	95 JClimlgy 15 207-220
score control scheme for detecting process	shifts • Cheng, Chuen-Sheng A cumulative	95 CommStA 24 755-774
An economical control procedure for detecting	shift in quality level • Srivastava, M. S.	94 ASAProQlPr 1-6
Rothwell, Geoffrey Optimal response to a	shift in regulatory regime: The case of the US nuclear power industry • Rust, John;	95 JApEcmtx 10 75-118S
Self-modelling with random	shift and scale parameters and a free-knot spline shape function • Lindstrom, Mary J.	95 StatMed 14 2009-2021
Sustainability of the deficit process with structural	shifts • Quintos, Carmela E.	95 JBES 13 409-417
Level	shifts, temporary changes and forecasting • Trívez, F. Javier	95 JForcast 14 543-550
rank change • Crossa, J.; Cornelius, P. L. et al. A	shifted multiplicative model fusion method for grouping environments without cultivar	95 CropSci 35 54-62
Cornelius, P. L.; Seyedsadr, M.; Crossa, J. Using the	shifted multiplicative model to search for "separability" in crop cultivar trials •	92 ThApGene 84 161-172
Bowden, J.; Callahan, A. J. Managing	ship performance of naval gunfire support using statistical process control • Bailey, M.;	94 MilOpRes 1 5-11
Tartakovsky, A. Asymptotic properties of CUSUM and	Shiryaev's procedures for detecting a change in a nonhomogeneous Gaussian process •	95 MaMethSt 4 389-404
Uncertainty about the persistence of economic	shocks • Miller, John P.; Newbold, Paul	95 JBES 13 435-440
HNBUE property in a class of correlated cumulative	shock models • Pérez-Ocón, Rafael; Gámiz-Pérez, M. Luz On the	95 AdvAppPr 27 1186-1188
in Bayesian prediction of component lifetimes in a	shock model • Gåsemyr, Jørund; Natvig, Bent Using expert opinions	95 MaOpRsr 20 227-242
A random	shock model for a linearly deteriorating system • Lee, Jiyeon; Lee, Eui Yong	95 JKoreaSt 24 471-479
Dynamically optimal replacement policy for a	shock model in a Markov random environment • Feng, W.; Adachi, K.; Kowada, M.	94 JOpRsJpn 37 255-270
Analysis of Markov renewal	shock models • Igaki, Nobuko; Sumita, Ushio; Kowada, Masashi	95 JAppProb 32 821-831
Bayesian inference in a discrete	shock model using confounded common cause data • Kvam, Paul H.; Martz, Harry F.	95 RelEngSS 48 19-25
the arrival process to obtain HNBUE survival using a	shock model • Pérez-Ocón, Rafael; Gámiz-Pérez, M. Luz Conditions on	95 CommStA 24 931-944
	Shock models with NBRU properties • Abouammoh, A. M.; Hendi, M. I.; Khalique, A.	95 PakJStA 11 35-47
S. Nonconvexities, labor hoarding, technology	shocks, and procyclical productivity. A structural econometric analysis • Chirinko, Robert	95 JEconmtx 66 61-98
of the totally asymmetric simple exclusion process:	Shock profiles • Janowski, Steven A. Exact solution	94 RBrzlPrS 8 85-91
A diffusion model for a system subject to random	shocks • Lee, Eui Yong; Song, Mun Sup; Park, Byung-Gu	95 JKoreaSt 24 141-147
modeling of external common cause failure	shocks in redundant systems • Vaurio, Jussi K. The probabilistic	95 RelEngSS 50 97-107
Imposing views, imposing	shoes: A statistician as a sole model • Alper, Paul; Raymond, Robert L.	95 AmerStat 49 317-319
Overtime or	shootout: Deciding ties in hockey • Hurley, William	95 Chance 8/1 19-22
A	short method for constructing an abridged life table • Brookmeyer, Ron	95 AJEpidm 141 991-992
Pareto distribution model for high concentrations in	short-range atmospheric dispersion • Mole, N.; Anderson, C. W. et al. A generalized	95 EnvrMtrc 6 595-606
Adjusted individual control charts for	short runs • Haugh, Larry D.; Pond, Andrew D.	95 ASQCTekCon 1117-1126
Regression-based	short term prediction of electrical load for a power system • Polzehl, Jörg	94 ASAProPEng 285-290
A system for measuring	short-term producibility • Keenan, Thomas M.	95 ASQCTekCon 50-56
Analysis of	short time series: Correcting for autocorrelation • Bence, James R.	95 EcolgMon 65 628-639
Analysis of	short time series with an over-dispersion model • Rai, S. N.; Abraham, B.; Peiris, S.	95 CommStA 24 335-348
Econometrics of	short and unreliable time series • Uri, T.; Wörgötter, A. (Ed)	95 Physica:Ger 238p
Teacher	shortages and teacher quality (Disc: p213-215) • Ingersoll, Richard M.	94 ASAProSoSt 201-206
Anthony F.; Yang, Zhenlin	Shortest prediction intervals for the Birnbaum-Saunders distribution • Desmond,	95 CommStA 24 1383-1401
Spatial cross-correlation of undisturbed, natural	shortleaf pine stands in northern Georgia • Reich, R. M.; Czaplewski, R. L.; Bechtold, W. A.	94 EnvEcoSt 1 201-217
Asymptotic confidence intervals for the length of the	short$t$ under random censoring • Beirlant, J.; Einmahl, J. H. J.	95 StNeerla 49 1-8

Testing for a unit root in an ARIMA$(p, 1, q)$	signal observed with measurement error • Lee, JongHyup; Shin, DongWan	95 JKoreaSt 24 481-493
Hardin, James W. Detection of forced climate	signals. Part I: Filter theory • North, Gerald R.; Kim, Kwang-Y.; Shen, Samuel S. P.;	95 JClimate 8 401-408
	Signal processing	
Sapatinas, T.; Bailey, T. C.; Krzanowski, W. J.	Application of wavelets to the pre-processing of underwater sounds • Powell, K. J.;	95 StatComp 5 265-273
systems • Majetić, Dubravko	Dynamic neural network for prediction and identification of nonlinear dynamic	95 JCmpInT 3 99-106
channel kinetics • Pulford, G. W.; Gallant, J. C. et al.	Evaluation and estimation of various Markov models with applications to membrane	95 BiomtrcJ 37 39-63
filtered Fourier transform infrared ... • Shaffer et al.	Experimental design protocol for the pattern recognition analysis of bandpass	95 Chemolab 29 89-108
p645-658) • Lai, Tze Leung	Sequential changepoint detection in quality control and dynamical systems (Disc:	95 JRSS-B 57 613-644
Review of "Fundamentals of statistical	signal processing: Estimation theory" • Kay, Steven M. (Auth); Sengupta, Sailes K. (Rev)	95 Technmcs 37 465-466
A design tool to evaluate average time to	signal properties of adaptive $\bar{X}$ charts • Prabhu, S. S.; Montgomery, D. C.; Runger, G. C.	95 JQualTek 27 74-83
Adaptive	signal regression • Land, Stephanie R.; Friedman, Jerome H.	94 ASAProStCp 100-105
ticomponent exponential fitting of NMR relaxation	signals • Lupu, M.; Todor, D. A singular value decomposition based algorithm for mul-	95 Chemolab 29 11-17
by pattern recognition applied to RIFS sensor	signals • Kraus et al. ... reflectometric gas sensing: Classification of hydrocarbon vapours	95 Chemolab 30 211-221
Consistent estimates of super imposed exponential	signals when some observations are missing • Kundu, Debasis; Kundu, Rameswar	95 JStPlInf 44 205-218
Autocorrelated data: Control charts versus tracking	signals • Superville, Claude R.; Adams, Benjamin M.	94 ASAProQlPr 194-199
of Indian monsoon rainfall and long-term variability	signals • Iyengar, R. N.; Basak, P. Regionalization	94 JClimlgy 14 1095-1114
Siegmund, David O.; Worsley, Keith J. Testing for a	signal with unknown location and scale in a stationary Gaussian random field •	95 AnlsStat 23 608-639
Interpreting moderator effects: Substitute for the	signed coefficient rule • Nye, Lendell G.; Witt, L. Alan	95 EdPsMeas 55 27-31
	Signed rank statistic	
	Some adaptive estimators in simple regression models • Viet, Tran Quoc	95 StatDecn 13 117-130
free tests for symmetry based on the linear	signed rank statistics • Kim, Soon-ock; Song, Moon Sup Distribution-	95 JNnparSt 5 311-321
Hössjer, O.; Croux, C. Generalizing univariate	signed rank statistics for testing and estimating a multivariate location parameter •	95 JNnparSt 4 293-308
An example of a two-sided Wilcoxon	signed rank test which is not unbiased • Amrhein, Peter	95 AnInStMa 47 167-170
On tests and	significance in econometrics • Keuzenkamp, Hugo A.; Magnus, Jan R.	95 JEconmtx 67 5-24
David; McDavitt, Chris	Significance editing in the Australian Survey of Average Weekly Earnings • Lawrence,	94 JOffStat 10 437-447
Mielke, Paul W., Jr Exact tests of	significance in higher dimensional tables • Zelterman, Daniel; Chan, Ivan Siu-Fung;	95 AmerStat 49 357-361
incomplete contingency tables: Estimating attained	significance levels • Smith, Peter W. F.; McDonald, John W. Exact conditional tests for	95 StatComp 5 253-256
triangular contingency tables: Estimating attained	significance levels • McDonald et al. Exact conditional tests of quasi-independence for	95 ApplStat 44 143-151
Conditional Bayesian	significance levels: Hierarchical models • Brimacombe, Michael	94 ASAProBays 169-174
	Significance probability	
	Estimating statistical hypotheses • Blyth, Colin R.; Staudte, Robert G.	95 StPrLet 23 45-52
Capdevila, Carles; Arcas, Antoni Empirical	significance test of the goodness-of-fit for some pyramidal clustering procedures •	95 Questo 19 123-130
1052) • Witte, J. S. et al. Comment on "Statistical	significance testing in the American journal of epidemiology, 1970-1990" (V139 p1047-	95 AJEpidm 142 101-101
Steven N. Reply to comment on "Statistical	significance testing in the American journal of epidemiology, 1970-1990" • Goodman,	95 AJEpidm 142 103-103
Sander Reply to comment on "Statistical	significance testing in the American journal of epidemiology, 1970-1990" • Greenland,	95 AJEpidm 142 102-103
Savitz, D. A. Reply to comment on "Statistical	significance testing in the American journal of epidemiology, 1970-1990" • Poole, C.;	95 AJEpidm 142 102-102
Interval estimation and	significance testing for cyclic trends in seasonality studies • Nam, Jun-mo	95 Biomtrcs 51 1411-1417
Editors' introduction - The	significance of testing in econometrics • Keuzenkamp, Hugo A.; Magnus, Jan R.	95 JEconmtx 67 1-3
The	significance of testing empirical non-nested models • McAleer, Michael	95 JEconmtx 67 149-171
Utzet, F. Sampling theory, estimation, and	significance testing for Prevosti's estimate of genetic distance • Sanchez, A.; Ocaña, J.;	95 Biomtrcs 51 1216-1235
The	significant-digit phenomenon • Hill, Theodore P.	95 AmMaMnth 102 322-327
Reporting negative contamination levels on	silicon wafers - Is this nonsense? • Pankratz, Peter C.; Coria, José C.	94 ASAProPEng 71-75
of variance for completely systematic designs?	'Silly designs are silly!' • Bailey, R. A. et al. Are neighbour methods preferable to analysis	95 Biomtrka 82 655-659
Extensions of multiple testing procedures based on	Simes' test • Hochberg, Yosef; Rom, Dror	95 JStPlInf 48 141-152
A	similarity-based smoothing approach to nondimensional item analysis • Ramsay, J. O.	95 Psymtrka 60 323-339
Viewpoint: Trend assessment by	similarity - A demonstration • Ratliff, Raymond D.	93 JRngeMgt 46 139-141
of symbolic objects using the concepts of both	similarity and dissimilarity • Gowda, K. Chidananda; Ravi, T. V. Divisive clustering	95 PattrnRc 28 1277-1282
Chakravarti, A.	Similarity of DNA fingerprints due to chance and relatedness • Li, C. C.; Weeks, D. E.;	93 HumHrdty 43 45-52
Hierarchical clustering method using probabilistic	similarity measure • Miyano, Hisao	95 Behvmtrk 22 145-153
Review of "Three-dimensional chemical	similarity searching" • Pepperrell, Catherine (Auth); Lewi, Paul J. (Rev)	95 Chemolab 28 205-207
DNA profile	similarity in a subdivided population • Li, Ching Chun; Chakravarti, Aravinda	94 HumHrdty 44 100-109
Review of	"SIMPC 3.81" • DON Econmetrics (Auth); Black, R. J. (Rev)	95 JApEcmtx 10 469-476
	Simple designs • Arasu, K. T.; Ray-Chaudhuri, D. K.; Singhi, N. M.	93 JCISS 18 130-135
Exact solution of the totally asymmetric	simple exclusion process: Shock profiles • Janowski, Steven A.	94 RBrzlPrS 8 85-91
time for filling a large interval in the symmetric	simple exclusion process • Ferrari, P. A.; Galves, A.; Liggett, T. M. Exponential waiting	95 AIHenrPB 31 155-175
for asymmetric mean zero random walk with	simple exclusion • Varadhan, S. R. S. Self diffusion of a tagged particle in equilibrium	95 AIHenrPB 31 273-285
quentist and Bayesian test for fixed and sequential	simple hypothesis testing • Berger, J. O.; Brown, L. D. et al. A unified conditional fre-	94 AnlsStat 22 1787-1807
spacings, ABLUE and tests of hypothesis in	simple linear model with extreme-value errors • Saleh, A. K. Md. E. et al. Optimum	94 JStatRsr 28 53-63
A fixed-width interval for $1/\beta$ in	simple linear regression • Coleman, Daniel A.	95 JStPlInf 44 291-312
of the least trimmed squares estimate in	simple linear regression • Hössjer, Ola Exact computation	95 CmpStDA 19 265-282
Students' internalized representation of	simple linear regression • VanBrackle, Lewis; VanBrackle, Anita S.; Rothenberg, Lori F.	94 ASAProStEd 100-101
Updating	simple linear regression • Klotz, Jerome H.	95 StSinica 5 399-403
Parameters estimation in a	simple linear regression using rank set sampling • Muttlak, Hassen A.	95 BiomtrcJ 37 799-810
Maximum likelihood estimation in	simple linear regression with one fold nested error • Güven, Bilgehan	95 CommStA 24 121-130
Robustness of testing the equality of two slopes in	simple linear structural relationships • Wong, M. Y.	95 JAppStat 22 3-8
sum of dependent variables with an application to	simple random sampling without replacement" (V30 p197-211) • de la Peña ... of a	95 AIHenrPB 31 703-704
Enlargement of obstacles for the	simple random walk • Antal, Peter	95 AnlsProb 23 1061-1101
Some adaptive estimators in	simple regression models • Viet, Tran Quoc	95 StatDecn 13 117-130
Bayesian estimation of a	simple regression model with measurement errors • Rodrigues, J.; Baba, M. Y.	94 RBrzlPrS 8 107-118
	Simplex-centroid design	
Heinz	A V-optimal design for Scheffé's polynomial model • Liu, Shuangzhe; Neudecker,	95 StPrLet 23 253-258
[Algorithm AS 299] Generation of	simplex lattice points • Chasalow, Scott D.; Brand, Richard J.	95 ApplStat 44 534-545
Bounds for the breakdown point of the	simplicial median • Chen, Zhiqiang	95 JMultiAn 55 1-13
Robert; Rogers, Andrei; Little, Jani	Simplicity and complexity in extrapolative population forecasting models • McNown,	95 MaPopStd 5 235-257
cal and asymptotic behavior of Brownian motion on	simply connected nilpotent Lie groups • Neuenschwander, Daniel; Schott, René On the lo-	95 JTherPrb 8 795-806
central limit theorem for trimmed products on	simply connected step 2-nilpotent lie groups • Neuenschwander, Daniel A Gaussian	95 JTherPrb 8 165-174
F. T. Testing homogeneity of normal means with a	simply ordered alternative and dependent observations • Lei, Xingye; Peng, Y. B.; Wright,	95 CmpStDA 20 173-183
nondirectional heterogeneity tests to evaluate	simply ordered alternative hypotheses • Rice, William R.; Gaines, Steven D. Extending	94 PNatAcSc 91 225-226
	Simpson's paradox and the hot hand in basketball • Wardrop, Robert L.	95 AmerStat 49 24-28
	Simpson's paradox revisited • Vogt, Arthur	95 Student 1 99-108
	Simulated annealing	
	[Algorithm AS 298] A hybrid optimization algorithm • Brooks, Stephen P.	95 ApplStat 44 530-533
	Analog optimization with Wong's stochastic neural network • Kesidis, George	95 IEEENrlN 6 258-260
Desai, U. B.	Depth and image recovery using a MRF model • Kapoor, S.; Mundkur, P. Y.;	94 IEEEPAMI 16 1117-1122

Simulated annealing (cont.)

Pablo A.; Frigessi, Arnoldo; de Sá, Paula Gonzaga	Fast approximate maximum a posteriori restoration of multicolour images • Ferrari,	95 JRSS-B 57 485-500
Berg, Bernd A.	Locating global minima in optimization problems by a random-cost approach •	93 Nature 361 708-710
estimation of evolutionary trees ... • Allison et al.	The posterior probability distribution of alignments and its application to parameter	94 JMolEvol 39 418-430
Pokarowski, Piotr	Tail events of some nonhomogeneous Markov chains • Niemiro, Wojciech;	95 AnlsApPr 5 261-293
Remarks about the ergodicity of	simulated annealing algorithms on a graph (French) • Miclo, Laurent	95 StocProc 58 329-360
Comparison of	simulated annealing algorithms for image restoration • Lamotte, J. L.; Alt, R.	94 MaCmpSim 37 1-15
A nested	simulated annealing algorithm • Whitaker, D.	95 JStCmpSm 53 233-241
Stochastic ordering for	simulated annealing analysis (German) • Tran, M. T.	95 ZAngMaMk 75 579-580S
lection applications • Hörchner, Uwe; Kalivas, John H.	Simulated-annealing-based optimization algorithms: Fundamentals and wavelength se-	95 JChemtrc 9 283-308
Tail events of	simulated annealing Markov chains • Niemiro, Wojciech	95 JAppProb 32 867-876
Messy	simulated annealing • Kvasnička, V.; Pospíchal, J.	95 JChemtrc 9 309-322
Hurn, Merrilee; Jennison, Christopher A study of	simulated annealing and a revised cascade algorithm for image reconstruction •	95 StatComp 5 175-190
Optimization using	simulated annealing • Brooks, S. P.; Morgan, B. J. T.	95 Ststcian 44 241-257
A piecewise Markovian model for	simulated annealing with stochastic cooling schedules • Kolonko, M.	95 JAppProb 32 649-658
Testing, encompassing, and	simulating dynamic econometric models • Gouriéroux, Christian; Monfort, Alain	95 EconTher 11 195-228
effects • Gauderman, W. James A method for	simulating familial disease data with variable age at onset and genetic and environmental	95 StatComp 5 237-243
Recursively	simulating multinomial multiperiod probit probabilities • Geweke, John; Keane, Michael	94 ASAProBuEc 1-6

Simulation

Bayesian alternative • Cohen, A.; Makov, U. et al.	Computer intensive methods for inference on parameters of complex models - A	95 JStPlInf 47 79-94
	Grappling with complexity • Segel, Lee A.	95 Complxty 1 18-25
datasets.braun) • Braun, W. John	An illustration of bootstrapping using video lottery terminal data (jse/v3n2/	95 JStEd 3/2 18kb
Casale, Salvatore; La Corte, Aurelio	A method for the generation of correlated random processes • Andronico, Maurizio;	95 JCmpInT 3 9-18
	Replaying the 1994 Wimbledon men's singles final • Croucher, John S.	95 NZndSt 30 2-8
tics: Parts 3 and 4 (jse/v3n2/trumbo) • Trumbo, B. E.	Some demonstration programs for use in teaching elementary probability and statis-	95 JStEd 3/2 60kb
tance of the retention pattern • Pietrogrande et al.	Statistical study of peak overlapping in multicomponent chromatograms: Impor-	95 Chemolab 28 239-258
	Stereology for some classes of polyhedrons • Ohser, J.; Mücklich, F.	95 AdvAppPr 27 384-396
probability functions in range data analysis and	simulation • Lambert, David K.; Harris, Thomas R. The use of conditional	93 JRngeMgt 46 157-160
etiology - The effect of ... • Yi et al. Computer	simulation analysis of Sartwell's incubation period model for diseases with uncertain	95 AJEpidm 142 363-368
2×2 factorial ANOVA (Korean) • Choi, Young Hun	Simulation analysis of type I error and power for F test and rank transformed F test in	95 KorApS 8/2 87-97
Review of “Time series - Forecasting,	simulation, applications” • Janacek, G.; Swift, L. (Auth); Davies, Neville (Rev)	95 JRSS-A 158 193-195
Review of “Time series: Forecasting,	simulation, applications” • Janacek, Gareth; Swift, Louise (Auth); Findley, David F. (Rev)	95 SIAMRvw 37 253-257
A qualitative	simulation approach for fuzzy dynamical models • Bonarini, Andrea; Bontempi, Gianluca	94 ACMMdSim 4 285-313
Evaluation of data editing procedures: Results of a	simulation approach • Garcia Rubio, E.; Peirats, V.	91 StatJUN 8 175-190
Weron, Aleksander Computer	simulation of attractors in stochastic models with α-stable noise • Janicki, Aleksander;	95 MaCmpSim 39 9-19
Modelling and	simulation of barley kernel growth • Koesmarno, H. K.	95 EnvrMtrc 6 511-516
Bayesian bioassay designs: A	simulation based approach • Warner, John	94 ASAProBays 13-17
linear approximations for efficient bootstrap	simulations • Hesterberg, Tim Tail-specific	95 JCmpGrSt 4 113-133
squares: A performance analysis by Monte Carlo	simulation • Cummins, David J.; Andrews, C. Webster Iteratively reweighted partial least	95 JChemtrc 9 489-507
of ring-recovery data via Markov chain Monte Carlo	simulation • Vounatsou, P.; Smith, A. F. M. Bayesian analysis	95 Biomtrcs 51 687-708
variables, with applications in reducing Monte Carlo	simulations • Kamgar-Parsi et al. ... moments of the weighted sum of uniform random	95 JStCmpSm 52 399-414
some common stability statistics - A Monte Carlo	simulations • Piepho, H.-P. Implication of correlations among	95 ThApGene 90 457-461
Method of moments using Monte Carlo	simulation • Gelman, Andrew	95 JCmpGrSt 4 36-54
Process improvement utilizing computer	simulation: Case study • Anderson, DeAnn; Abetti, Frank; Savage, Phillip	95 ASQCTekCon 713-724
Weron, Aleksander	Simulation and chaotic behavior of α-stable stochastic processes • Janicki, Aleksander;	95 Dekker:NY 376p
Weron, A. (Auth); Lee, P. M. (Rev) Review of	“Simulation and chaotic behaviour of α-stable stochastic processes” • Janicki, A.;	95 JRSS-A 158 351-352
C. M.; Ledwina, Teresa Consistency and Monte Carlo	simulation of a data driven version of smooth goodness-of-fit tests • Kallenberg, Wilbert	95 AnlsStat 23 1594-1608
Analysis of incomplete multivariate data by	simulation • Schafer, Joseph L.	95 ChapmnHall:Lnd 300p
Computer-aided modeling and	simulation of electrical circuits with α-stable noise • Weron, A.	95 ZastApMa 23 83-93
Bronson, Richard Computer	simulations and empirical testing of sociological theory • Jacobsen, Chanoch;	95 SocMethR 23 479-506
Using Monte Carlo	simulation to evaluate kernel-based home range estimators • Worton, Bruce J.	95 JWildlMg 59 794-800
automata applications to geographical evolution	simulation (Italian) • Stefano, Benati Cellular	94 QStMaA 16 105-115
sample-size determination in regression analysis of	simulation experiments • Kleijnen, J. P. C. et al. Two-stage versus sequential	95 AJMMgSci 15 83-114
Cook, J. R.	Simulation-extrapolation: The measurement error jackknife • Stefanski, L. A.;	95 JASA 90 1247-1256
Bennett, S.; Radalowicz, A.; Vella, V. et al. A computer	simulation of household sampling schemes for health surveys in developing countries •	94 IJEpidem 23 1282-1291
an autoregressive state-space model via iterative	simulation • Glickman, Mark E.; Stern, Hal S. Inference from	94 ASAProSpts 72-76
Smith, C. Computer	simulation of marker-assisted selection utilizing linkage disequilibrium • Zhang, W.;	92 ThApGene 83 813-820
several additional factors • Zhang, W.; Smith, C.	Simulation of marker-assisted selection utilizing linkage disequilibrium: The effects of	93 ThApGene 86 492-496
Wheyming Tina Estimating	simulation metamodels using correlated control variates • Shih, Neng Hui; Song,	95 CommStB 24 79-90
A comparison of different estimation methods via	simulation • Scazzero et al. ... for the optimum point of a quadratic logistic curve -	95 CommStB 24 207-226
Data analysis or	simulation model: A critical evaluation of some methods • van Tongeren, Onno F. R.	95 EcolgMdl 78 51-60
ating system operation • Román, J. et al. Stochastic	simulation and Monte Carlo methods applied to the assessment of hydro-thermal gener-	94 Top 2 1-24
Mises type • Holmquist, Björn	Simulation of multimodal circular distributions of wrapped Cauchy type and of von	95 CommStB 24 607-615
Computer	simulation of a nonlinear model for electrical circuits with α-stable noise • Janicki, A.	95 ZastApMa 23 95-105
	Simulation optimization by genetic search • Yunker, James M.; Tew, Jeffrey D.	94 MaCmpSim 37 17-28
Sensitivity analysis for Monte Carlo	simulation of option pricing • Fu, Michael C.; Hu, Jian-Qiang	95 PrEngInf 9 417-446
A	simulation-oriented approach to disparate impact • Gardenier, Turkan K.	94 ASAProStEd 180-185
Calibration of empirical models:	Simulation of pasture growth and decay • McPhee, M. J.; Ransom, K. P.; Oddy, V. H.	95 EnvrMtrc 6 505-510
Adell, José Carlos; Dopazo, Joaquín Monte Carlo	simulation in phylogenies: An application to test the constancy of evolutionary rates •	94 JMolEvol 38 305-309
sharing • Fleming, P. J. et al. Efficient Monte-Carlo	simulation of a product-form model for a cellular system with dynamic resource	95 ACMMdSim 5 3-21
Fast	simulation of rare events in queueing and reliability models • Heidelberger, Philip	95 ACMMdSim 5 43-85
A martingale approach to regenerative	simulation • Glynn, Peter W.; Iglehart, Donald L.	95 PrEngInf 9 123-131
means • Wilcox, Rand R.	Simulation results on solutions to the multivariate Behrens-Fisher problem via trimmed	95 Ststcian 44 213-225
Stamp, Mark Urn model	simulations of a sexually transmitted disease epidemic • Martin, Clyde F.; Allen, Linda J. S.;	95 AplMaCmp 71 179-199
The	simulation smoother for time series models • De Jong, Piet; Shephard, Neil	95 Biomtrka 82 339-350
Comparison of analog and digital	simulation on some continuous time systems • Yeung, Iris; Lam, James	94 ASAProStEd 186-191
Review of “Regenerative stochastic	simulation” • Shedler, G. S. (Auth); Ripley, B. D. (Rev)	94 BLndnMaS 26 410-411
The performance of different loss models in the	simulation of streamflow • Chen, T. H.; Hornberger, G. M.; Jakeman, A. J.; Swank, W. T.	95 EnvrMtrc 6 479-484
about the quality of test statistics for comparative	simulation studies? (German) • Böker, Fred; Dannenberg, Olaf What can P-values say	95 AllgeStA 79 233-251
Statistical matching of survey datafiles: A	simulation study • Kovačević, Milorad S.; Liu, Tzen-Ping	94 ASASrvyRsr 479-484
variance estimates for complex survey designs - A	simulation study • Robb, William H. Resampling	94 ASASrvyRsr 218-221
using trimmed means versus means, and designing	simulation studies • Wilcox ANOVA: The practical importance of heteroscedastic methods,	95 BJMaStPs 48 99-114
Zhang, Donghui	Simulation study on different approaches for nonlinear mixed effects models •	94 ASAProBiop 67-72
optimality of the covariance improved estimate: A	simulation study (Chinese) • Chen, Maoxue; Liu, Jinshan On	95 ChJApPSt 10 1-4

Wang-Clow, F.; Lange, M.; Laird, N. M.; Ware, J. H. A	simulation study of estimators for rates of change in longitudinal studies with attrition •	95 StatMed 14 283-297
Dwyer, James H. A	simulation study of mediated effect measures • MacKinnon, David P.; Warsi, Ghulam;	95 MultBehR 30 41-62
Detection of genetic interference:	Simulation studies and mouse data • Weeks, Daniel E.; Ott, Jurg; Lathrop, G. Mark	94 Genetcs 136 1217-1226
	Simulation studies of Poisson correlation functions • Guerrero-Cusumano, José-Luis	94 ASAProPEng 157-162
the proportion of immunes in censored samples: A	simulation study • Ghitany, M. E.; Maller, R.; Zhou, S. Estimating	95 StatMed 14 39-49
A shape constrained smoother:	Simulation study • Delecroix, M.; Simioni, M.; Thomas-Agnan, C.	95 CmpStQ 10 155-175
in Bernoulli SS ANOVA models • Wang, Y. et al.	Simulation studies of smoothing parameter estimates and Bayesian confidence intervals	95 CommStB 24 1037-1059
Brooks, Maria Mori; Hallstrom, Al; Peckova, Monika A	simulation study used to design the sequential monitoring plan for a clinical trial •	95 StatMed 14 2227-2237
by bootstrapping the dependent variable: A	simulation study • Crivelli, A.; Firinguetti, L. et al. Confidence intervals in ridge regression	95 CommStB 24 631-652
Robust analysis of variance: A	simulation study • Reed, J. F.; Stark, D. B.	95 JAppStat 22 87-104
Bayesian classification of vasculitis: A	simulation study • Heller, I.; Isakov, A.; Blinder-Weiner, S.; Topilsky, M.	95 MInfoMed 34 259-265
A	simulation study of the weighted *k*-means cluster procedure • Rost, D.	95 JStCmpSm 53 51-63
A	simulation technique of a non-Gaussian spatial process • De Cesare, L.; Posa, D.	95 CmpStDA 20 543-555
and CUSUMSQ tests: A numerical derivation using	simulation technique • Tanizaki, H. Asymptotically exact confidence intervals of CUSUM	95 CommStB 24 1019-1036
Econometric inference using	simulation techniques • van Dijk, Herman K.; Monfort, Alain; Brown, Bryan W. (Ed)	95 Wiley:NY:UK 256p
	Simulation of truncated normal variables • Robert, Christian P.	95 StatComp 5 121-125
Policy micro analysis and	simulation using APL • Burpee, Joseph E.	94 ASAProGovt 178-182
reduction for Bernoulli response variables in	simulation • Ocaña, Jordi; Vegas, Esteban Variance	95 CmpStDA 19 631-640
McClean, Sally	Simulation of wind speed for the modelling of pine saw-fly mortality • Robinson, Stephen;	95 Ststcian 44 469-477
Balakrishnan, N.; Sandhu, R. A. A simple	simulational algorithm for generating progressive Type-II censored samples •	95 AmerStat 49 229-230
configurations applied to a discrete and stochastic	simulator of a manufacturing company • Giribone ... for non-equispaced experimental	94 JAppStat 21 549-564
Measurement error or endogeneity: Sorting out	simultaneity • Romeo, Charles J.	94 ASAProBays 196-201
Bove, Giuseppe Some methods for the	simultaneous analysis of data matrices and the factorial invariance problem •	94 Metron 52/3 73-87
new multivariate curve resolution procedure to the	simultaneous analysis of several spectroscopic titrations of ... • Tauler et al. ... of a	95 Chemolab 27 163-174
Some simple graphs for	simultaneous comparisons • Marr, Ray L.	94 ASAProPEng 109-114
Mishra, Satya N. A simple approach to	simultaneous comparison of normal means • Selukar, Rajesh S.; Shah, Arvind K.;	95 JStCmpSm 52 174-177
Su, John Q.	Simultaneous confidence bands for comparing diagnostic markers • Chen, Hanfeng;	95 CommStA 24 375-393
Faraway, Julian J.; Sun, Jiayang	Simultaneous confidence bands for linear regression with heteroscedastic errors •	95 JASA 90 1094-1098
Fixed-width	simultaneous confidence intervals for all-pairwise comparisons • Liu, W.	95 CmpStDA 20 35-44
of total variation distance, with applications to	simultaneous confidence intervals • Brown, Mark The distribution	95 CmpOpRsh 22 373-381
Upper and lower bound distributions that give	simultaneous confidence intervals for quantiles • Satten, Glen A.	95 JASA 90 747-752
proportions • Sison, Cristina P.; Glaz, Joseph	Simultaneous confidence intervals and sample size determination for multinomial	95 JASA 90 366-369
portions" (V90 p366-369) • Sison et al. Correction to	"Simultaneous confidence intervals and sample size determination for multinomial pro-	95 JASA 90 1494-1494
Pienaar, Inet	Simultaneous detection of shift in means and variances • Flury, Bernard D.; Nel, Daan G.;	95 JASA 90 1474-1481
early designed ... • Steele et al. A methodology for the	simultaneous development of linear and nonlinear chemical reaction models from a lin-	95 Chemolab 29 125-131
population • Svensson, Åke On the	simultaneous distribution of size and costs of an epidemic in a closed multigroup	95 MaBisci 127 167-180
Kaljurand, M. et al. Factor analytical resolution of	simultaneous, dynamic thermal process in thermochromatography of oil shale • Koel, M.;	95 Chemolab 30 173-178
restrictions of linear equations with applications to	simultaneous equations and cointegration • Johansen, Søren Identifying	95 JEconmtx 69 111-132
Estimation and inference in	simultaneous equations • Kadiyala, K. Rao; Oberhelman, Dennis	94 ASAProBuEc 235-240
Testing for multivariate normality in	simultaneous equations models • Jarque, C. M.; McKenzie, C. R.	95 MaCmpSim 39 323-328
A new identification rule and an estimator for the	simultaneous equation model using the notation of the reticular action mode • Toyoda, H.	94 Behvmtrk 21 163-175
econometrics. Volume 2: Linear and nonlinear	simultaneous equations • Dhrymes, P. J. Topics in advanced	94 Springer:Brln:NY 401p
On a	simultaneous equations pre-test estimator • Skeels, Christopher L.; Taylor, Larry W.	95 JEconmtx 68 269-286
supply • Lee, M.-J. Semi-parametric estimation of	simultaneous equations with limited dependent variables: A case study of female labour	95 JApEcmtx 10 187-200
Chung Bounded risk conditions in	simultaneous estimation of independent Poisson means • Liu, Ying-Hsing; Yang, Ming-	95 JStPlInf 47 319-331
Younshik; Kim, Chansoo; Dey, Dipak K.	Simultaneous estimation of Poisson means under weighted entropy loss • Chung,	94 CalcutSt 44 165-174
growth curve models • Yang, Wenli	Simultaneous estimation of regression coefficient matrix and covariance matrix in	95 SciSPkng 24 474-484
Multivariate	simultaneous generalized ARCH • Engle, Robert F.; Kroner, Kenneth F.	95 EconTher 11 122-150
selection rules • Dolan, C. V.; Molenaar, P. C. M. et al.	Simultaneous genetic analysis of means and covariance structure: Pearson-Lawley	94 BehavGen 24 17-24
	Simultaneous inference	
Rom, Dror M.; Holland, Burt	A new closed multiple testing procedure for hierarchical families of hypotheses •	95 JStPlInf 46 265-275
Dunnett, Charles W.; Tamhane, Ajit C.	Step-up multiple testing of parameters with unequally correlated estimates •	95 Biomtrcs 51 217-227
Spurrier, John D.	Successive comparisons between ordered treatments • Lee, Robert Edwin;	95 JStPlInf 43 323-330
sampling) is more ... • Aoshima In heteroscedastic	simultaneous inference, the heteroscedastic method II (based on Healy's two-stage	94 AJMMgSci 14 167-196
	Simultaneous inference using finite intersection tests: A better mousetrap • Timm, Neil H.	95 MultBehR 30 461-512
Kyung Soo On the performance of empirical Bayes	simultaneous interval estimates for all pairwise comparisons • Kim, Woo-Chul; Han,	95 JKoreaSt 24 161-181
data • Siciliano, Roberta; van der Heijden, Peter G. M.	Simultaneous latent budget analysis of a set of two-way tables with constant-row-sum	94 Metron 52/1 155-180
Manuel Maximum likelihood estimation of	simultaneous pairwise linear structural relationships • Bolfarine, Heleno; Galea Rojas,	95 BiomtrcJ 37 673-689
Kanata, Y. A learning rule of neural networks via	simultaneous perturbation and its hardware implementation • Maeda, Y.; Hirano, H.;	95 Autmatca 31 251-259
Power of unweighted and weighted versions of	simultaneous and sequential multiple-comparison procedures • Klockars, A. J. et al.	95 PsychBu 118 300-307
Kumar	Simultaneous spike-trains and stochastic dependence • Das, Shubhabrata; Sen, Pranab	95 SankhyaB 57 32-47
J.-D.; Morgan, B. J. T.; Pradel, R.; Freeman, S. N. A	simultaneous survival rate analysis of dead recovery and live recapture data • Lebreton,	95 Biomtrcs 51 1418-1428
et al. Identifying which sets of parameters are	simultaneously estimable in an incomplete factorial design • Byar, D. P.; Freedman, L. S.	95 Ststcian 44 451-456
Imputing numeric and qualitative variables	simultaneously • Bankier, Michael; Fillion, Jean-Marc; Luc, Manchi; Nadeau, Christian	94 ASASrvyRsr 242-247
Elihu H. Bayesian sequential monitoring designs for	single-arm clinical trials with multiple outcomes • Thall, Peter F.; Simon, Richard M.; Estey,	95 StatMed 14 357-379
identifying lines useful for the improvement of elite	single crosses, based on a combining ability model • Hohls et al. An unbiased estimator for	95 Heredity 75 155-163
Genetic models for predicting maize	single-cross performance in unbalanced yield trial data • Bernardo, Rex	95 CropSci 35 141-147
Bernardo, Rex Prediction of maize	single-cross performance using RFLPs and information from related hybrids •	94 CropSci 34 20-25
random digit dialing sampling schemes for	single and dual frame surveys • Biemer, Paul; Akin, Don The efficiency of list-assisted	94 ASASrvyRsr 1-10
inference for a two-state Markov model of a	single ion channel, incorporating time interval omission • Ball, F. G.; Davies, S. S. Statistical	95 JRSS-B 57 269-287
	Single linkage clustering and continuum percolation • Penrose, Mathew D.	95 JMultiAn 53 94-109
generational pedigree ... • Vieland et al. Adequacy of	single-locus approximations for linkage analyses of oligogenic traits: Extension to multi-	93 HumHrdty 43 329-336
One- and two-sample tests for	single-locus inbreeding coefficients using the bootstrap • van Dongen, S.; Backeljau, T.	95 Heredity 74 129-135
Lin, Chen-Sin	Single machine stochastic scheduling: Minimizing the number of tardy jobs •	94 AJMMgSci 14 289-300
non-parametric two-sample survival test based on a	single occupancy Fermi-Dirac model for the discrete range distribution • Efird et al. A	94 ASAProBiop 166-170
comparison ... • Aitkin Probability model choice in	single samples from exponential families using Poisson log-linear modelling, and model	95 StatComp 5 113-120
A monotonicity result for a	single-server loss system • Chao, Xiuli; Dai, Liyi	95 JAppProb 32 1112-1117
	Single server queue	
	The busy periods of some queueing systems • Stadje, W.	95 StocProc 55 159-167
On efficient sequential estimation for the simple	single server queue • Stefanov, Valeri T.	95 RBrzlPrS 9 55-65
A monotonicity result for a	single-server queue subject to a Markov-modulated Poisson process • Du, Qing	95 JAppProb 32 1103-1111
Song, Mark An approximation method for a	single server queue with cyclically indexed arrival and service rates • Morrice, Douglas J.;	95 AplMaCmp 71 247-269
Hasegawa, Toshiharu On the	single-server queue with independent MAP/G and M/GI input streams • Takine, Tetsuya;	95 CommStSM 11 227-234
error are correlated • Nieto et al. Kalman filter for	singular and conditional state-space models when the system state and the observational	95 StPrLet 22 303-310

Sawyer, Tom; Tanur, Judith M. Pseudo-opinion polls:	SLOP or useful data? • Horvitz, Dan; Koshland, Daniel; Rubin, Donald; Gollin, Albert;	95 Chance 8/2 16-25
Bahadur exact	slopes for a class of combinations of dependent tests • Pallini, Andrea	94 Metron 52/1 53-65
and numerical analysis of the repeated median	slope • Rousseeuw, Peter J.; Croux, Christophe; Hössjer, Ola Sensitivity functions	95 CmpStQ 10 71-90
T. Estimating a	slope parameter in regression with prescribed proportional accuracy • Martinsek, Adam	95 StatDecn 13 363-377
with replicated observations, unequal ... • Isogawa	Slope-range and generalized least squares: Estimation of the structural relationship	95 JJapnStS 25 57-70
The statistical analysis of	slope ratio assays using SAS software • Wong, M. Y.	94 CmpBiomd 45 233-238
Draper, Norman R.	Slope rotatability over all directions designs • Ying, Lisa H.; Pukelsheim, Friedrich;	95 JAppStat 22 331-341
Friedrich; Draper, Norman R.	Slope rotatability over all directions designs for $k \geq 4$ • Ying, Lisa H.; Pukelsheim,	95 JAppStat 22 343-354
Park, Sung Hyun; Lee, Min Soo	Slope-rotatability over all directions in third order response surface models •	95 JKoreaSt 24 519-536
A new type of	slope rotatable central composite design • Victorbabu, B. Re.; Narasimham, V. L.	94 JIAgSt 46 315-317
Information-processing	slope scores can be reliable! • Damos, Diane L.; Carter, Robert C.	95 HumanFac 37 424-429
Robustness of testing the equality of two	slopes in simple linear structural relationships • Wong, M. Y.	95 JAppStat 22 3-8
Establishing release limits with a random	slopes stability model • Grimes, James A.; Foust, Laura B.	94 ASAProBiop 498-502
Ahmed, N. U.; Yeap, T. H. Performance analysis of	slotted ALOHA protocol for optical fiber local area networks with star topology •	95 StocAnAp 13 389-410
The central register of population of the Republic of	Slovenia • Tršinar, Irena	95 JOffStat 11 221-223
"Don't know" survey responses: The case of the	Slovenian plebiscite • Rubin, Donald B.; Stern, Hal S.; Vehovar, Vasja Handling	95 JASA 90 822-828
	Slow points in the support of historical Brownian motion • Verzani, John	95 AnlsProb 23 56-70
of body dimensions of four species of pulmonate	slugs (Mollusca : Gastropoda : Pulmonata) • Bhuyan et al. ... changes in the relationships	95 BiomtrcJ 37 1005-1015
Tree	slugs • Feu, Chris du	95 TeachgSt 17/2 52-53
parametric hypothesis tests with an application to	Slutsky symmetry • Lewbel, Arthur Consistent non-	95 JEconmtx 67 379-401
Williamson, G. David	Small area analysis for public health data (Disc: p125-126) • Rutter, Carolyn;	94 ASAProEpid 118-122
Lahiri, P. Robust hierarchical Bayes estimation of	small area characteristics in the presence of covariates and outliers • Datta, G. S.;	95 JMultiAn 54 310-328
Liu, Jun	Small area estimation for the National Household Survey of Drug Abuse • Folsom, Ralph E.;	94 ASASrvyRsr 565-570
Robust estimation of mean squared error of	small area estimators • Lahiri, P.; Rao, J. N. K.	95 JASA 90 758-766
inference for the finite population mean of a	small area • Rashid, M. Mushfiqur; Nandram, Balgobin Rank-based predictive	94 ASASrvyRsr 1059-1064
Estimation of fertility rates for	small areas in postcensal years (Disc: p122-123) • Bousfield, Marie V.	94 ASAProSoSt 106-111
Synthetic estimates for	small areas in PRAMS • Johnson, Christopher H.	94 ASASrvyRsr 1065-1070
and environmental epidemiology: Methods for	small-area studies" • Elliott, P.; Cuzick, J.; English, D. et al. Review of "Geographical	95 StMedRes 4 184-184
Coverage error and cost issues in	small area telephone surveys • Schejbal, Judith A.; Lavrakas, Paul J.	94 ASASrvyRsr 1287-1292
	Small ball probabilities of Gaussian fields • Shao, Q.-M.; Wang, D.	95 ProbTher 102 511-517
norms • Kuelbs, J.; Li, W. V.; Shao, Qi-man	Small ball probabilities for Gaussian processes with stationary increments under Hölder	95 JTherPrb 8 361-386
raking procedures to improve the quality of	small domain estimation • Dubreuil, Guylaine; Tremblay, Johanne The use of generalized	94 ASASrvyRsr 293-298
weighting of direct and indirect components of	small domain predictors • Chaudhuri et al. ... design-cum-model approach for convex	95 CommStA 24 2803-2813
(Auth); Mikosch, Thomas (Rev) Review of "Law of	small numbers: Extremes and rare events" • Falk, Michael; Hüsler, Jurg; Reiss, Rolf-Dieter	95 JASA 90 1133-1134
tional assumptions in longitudinal data analysis of	small samples • Fellingham et al. Sensitivity of point and interval estimates to distribu-	95 CommStB 24 617-630
Getchell, C.; Chinchilli, V. M.	Small sample characteristics of generalized estimating equations • Gunsolley, J. C.;	95 CommStB 24 869-878
chi-squared goodness-of-fit statistics: ... • Formann	Small-sample comparison of the exact and asymptotic upper tail probabilities of	95 JStCmpSm 51 369-384
Uncertainty bounds for multivariate M-estimates in	small samples • Spall, James C.	94 ASAProPEng 126-131
selection for extended quasi-likelihood models in	small samples • Hurvich, Clifford M.; Tsai, Chih-Ling Model	95 Biomtrcs 51 1077-1084
with a control in a two-way layout • Kao, N. L. et al.	Small sample null distribution of two nonparametric tests for comparing treatments	95 CommStB 24 149-164
Modified stationarity tests with improved power in	small samples • Breitung, J.	95 StHefte 36 77-95
	Small-sample properties of finite population ratios • Sankoh, Abdul J.	95 CommStB 24 897-909
compound normal disturbances • Hasegawa, H. On	small sample properties of Zellner's estimator for the case of two SUR equations with	95 CommStB 24 45-59
Some	small-sample results on a bounded influence rank regression method • Wilcox, Rand R.	95 CommStA 24 881-888
bounds for parameter identification with	small sample sizes • Spall, James C. Uncertainty	95 PIEEEDecCt 3504-3515
	Small sample uni- and multi-variate control charts for means • Scholz, F. W.; Tosch, T. J.	94 ASAProQlPr 17-22
Multivariate nonparametric control charts using	small samples • Kapatou, Alexandra; Reynolds, Marion R., Jr	94 ASAProQlPr 241-246
parameter • Carpenter, M.; Pal, N. Estimation of the	smaller and larger of two exponential location parameters with a common unknown scale	95 CalcutSt 45 61-71
of additive dependencies and concurvities using	smallest additive principal components • Donnell, D. J.; Buja, A.; Stuetzle, W. Analysis	94 AnlsStat 22 1635-1673
Analyzing unreplicated factorials using the n	smallest contrasts • Kasperski, William J.; Schneider, Helmut; Johnson, Keith L.	95 JQualTek 27 377-386
On the	smallest possible asymptotically efficient variance in semiparametric models • Liang, Hua	95 JSysSMSE 7 29-33
binomial and Poisson distributions • Hamza, Kais The	smallest uniform upper bound on the distance between the mean and the median of the	95 StPrLet 23 21-25
of lung cancer due to environmental tobacco	smoke (ETS)" • Flatman, George T. Comment on "Risk assessment	94 ASAProEnvr 74-76
lung cancer and exposure to environmental tobacco	smoke (Disc: p679-685) • Tweedie et al. ... relationships, with application in studies of	95 StatMed 14 545-569
reported for exposure to environmental tobacco	smoke (Disc: p413-424) • Gross, Alan J. Uncertainties in lung cancer risk estimates	95 EnvrMtrc 6 403-412
the epidemiology studies ... • Bayard et al. Passive	smoking and lung cancer: The U.S. EPS's weight-of-evidence analysis, with emphasis on	94 ASAProEnvr 56-64
for nonlinear L_1 curve-fitting based on the	smooth approximation • Yang, Ziqiang An algorithm	95 CmpStDA 19 45-52
Asymptotical minimax recovery of sets with	smooth boundaries • Mammen, E.; Tsybakov, A. B.	95 AnlsStat 23 502-524
A limit theorem for a	smooth class of semiparametric estimators • Pakes, Ariel; Olley, Steven	95 JEconmtx 65 295-332
solution of penalized likelihood problems and	smooth curve estimation • Mächler, Martin B. Variational	95 AnlsStat 23 1496-1517
Wang, Yongxiao; Taylor, Jeremy M. G. Inference for	smooth curves in longitudinal data with application to an AIDS clinical trial •	95 StatMed 14 1205-1218
Nan M.; Ware, James H. Prediction and creation of	smooth curves for temporally correlated longitudinal data • Donnelly, Christl A.; Laird,	95 JASA 90 984-989
Roussas, George G. Asymptotic normality of a	smooth estimate of a random field distribution function under association •	95 StPrLet 24 77-90
On wavelet methods for estimating	smooth functions • Hall, Peter; Patil, Prakash	95 Bernoull 1 41-58
Monte Carlo simulation of a data driven version of	smooth goodness-of-fit tests • Kallenberg, Wilbert C. M.; Ledwina, Teresa Consistency and	95 AnlsStat 23 1594-1608
Wenceslao	Smooth linear regression using sample-determined bandwidth • González Manteiga,	95 SankhyaA 57 79-87
A generalized linear classification model with a	smooth link function and predictors obtained from quantile spline fits ... • O'Connell et al.	95 JStPlInf 47 153-164
data • Gulati, Sneh; Padgett, W. J.	Smooth nonparametric function estimation from inversely sampled record-breaking	94 ASAProPEng 140-145
Onsager-Machlup functional for some	smooth norms on Wiener space • Capitaine, Mireille	95 ProbTher 102 189-201
Bahadur-Kiefer theorems for kernel	smooth product-limit quantile estimator • Zhou, Yong	95 CommStA 24 2815-2828
The Bernstein polynomial estimator of a	smooth quantile function • Cheng, Cheng	95 StPrLet 24 321-330
	Smooth tests for the bivariate Poisson distribution • Rayner, J. C. W.; Best, D. J.	95 AstrlJSt 37 233-245
Ducharme, Gilles R. A note on	smooth tests of goodness of fit for location-scale families • Boulerice, Bernard;	95 Biomtrka 82 437-438
Optimizing the	smoothed bootstrap • Wang, Suojin	95 AnInStMa 47 65-80
Spline	smoothed bootstrapping for kernel estimator (Chinese) • Qian, Weimin	95 AMaAPknC 17 265-277
On the distance between	smoothed empirical and quantile processes • Csörgő, Miklós; Horváth, Lajos	95 AnlsStat 23 113-131
LaRiccia, V. N. Maximum	smoothed likelihood density estimation for inverse problems • Eggermont, P. P. B.;	95 AnlsStat 23 199-220
Maximum	smoothed likelihood density estimation • Eggermont, P. P. B.; Lariccia, V. N.	95 JNnparSt 4 211-222
cation to labour force participation • Charlier et al. A	smoothed maximum score estimator for the binary choice panel data model with an appli-	95 StNeerla 49 324-342
On unbounded hazard rates for	smoothed perturbation analysis • Fu, Michael C.; Hu, Jian-Qiang	95 JAppProb 32 659-667
Fraiman, Ricardo Asymptotic distribution of	smoothers based on local means and local medians under dependence • Boente, Graciela;	95 JMultiAn 54 77-90
Fraiman, R. Asymptotic distribution of data-driven	smoothers in density and regression estimation under dependence • Boente, G.;	95 CandJSt 23 383-397
Using a Bayesian	smoother to determine root growth patterns • Upsdell, Martin	94 ASAProBays 227-231
Spline	smoother as a dynamic linear model • Moyeed, R. A.	95 AstrlJSt 37 193-204

panel data using the EM algorithm and the Kalman	smoother • Jansen, R. A. R. G. et al. Longitudinal LISREL model estimation from incomplete	95 StNeerla 49 362-377
Splines as local	smoothers • Nychka, Douglas	95 AnlsStat 23 1175-1197
A regression	smoother for resistant measures of location and scale • Wilcox, Rand R.	95 BJMaStPs 48 189-204
A shape constrained	smoother: Simulation study • Delecroix, M.; Simioni, M.; Thomas-Agnan, C.	95 CmpStQ 10 155-175
The simulation	smoother for time series models • De Jong, Piet; Shephard, Neil	95 Biomtrka 82 339-350
	Smoothing	
Hinkley, D. V.; Worton, B. J.	Accurate and efficient construction of bootstrap likelihoods • Davison, A. C.;	95 StatComp 5 257-264
	Adaptive bandwidth choice for kernel regression • Schucany, William R.	95 JASA 90 535-540
	An adaptive expansion method for regression • LeBlanc, Michael	95 StSinica 5 737-748
estimators • Foster, Peter	A comparative study of some bias correction techniques for kernel-based density	95 JStCmpSm 51 137-152
Marschner, Ian C.	Computation of age-specific HIV incidence estimates using the EM algorithm •	95 JStCmpSm 53 299-312
Kim, Kwang-Y.; Shen, Samuel S. P.; Hardin, James W.	Detection of forced climate signals. Part I: Filter theory • North, Gerald R.;	95 JClimate 8 401-408
• Song, K. S.; Müller, H. G.; Clifford, A. J. et al.	Estimating derivatives of pharmacokinetic response curves with varying bandwidths	95 Biomtrcs 51 12-20
and application to image recovery ... • Girard	The fast Monte-Carlo cross-validation and C_L procedures: Comments, new results	95 CmpStQ 10 205-231
Elliott, Robert J.	Finite-dimensional models for hidden Markov chains • Aggoun, Lakhdar;	95 AdvAppPr 27 146-160
	Local and variable bandwidths and local linear regression • Jones, M. C.	95 MaOpfStS 27 65-71
	A minimax optimal estimator for continuous monotone densities • Datta, Somnath	95 JStPlInf 46 183-193
	Modern nonlinear regression methods • Frank, Ildiko E.	95 Chemolab 27 1-19
	Nonparametric rank discrimination method • Ćwik, Jan; Mielniczuk, Jan	95 CmpStDA 19 59-74
Öhrn, Kenth; Ahlén, Anders; Sternad, Mikael	A probabilistic approach to multivariable robust filtering and open-loop control •	95 IEEEAuCn 40 405-418
Víctor M.	Restricted forecasts of missing observations in univariate time series • Guerrero,	94 Estd 46 1-23
Sarle, W.	Scale measures for bandwidth selection • Janssen, P.; Marron, J. S.; Veraverbeke, N.;	95 JNnparSt 5 359-380
	Seeing a curve in multiple regression • Berk, Kenneth N.; Booth, David E.	95 Technmcs 37 385-398
Nielsen, J. P.	A simple bias reduction method for density estimation • Jones, M. C.; Linton, O.;	95 Biomtrka 82 327-338
	Splines from a Bayesian point of view • van der Linde, A.	95 Test 4 63-81
Øksendal, Bernt	Stability properties of stochastic partial differential equations • Lindstrøm, Tom;	95 StocAnAp 13 177-204
	Stereological estimation of particle size distributions • Mase, Shigeru	95 AdvAppPr 27 350-366
Mächler, Martin B.	Variational solution of penalized likelihood problems and smooth curve estimation •	95 AnlsStat 23 1496-1517
Iain M.; Kerkyacharian, Gérard; Picard, Dominique	Wavelet shrinkage: Asymptopia? (Disc: p337-369) • Donoho, David L.; Johnstone,	95 JRSS-B 57 301-337
Square-root RTS	smoothing algorithms • Park, PooGyeon; Kailath, Thomas	95 IJContrl 62 1049-1060
A similarity-based	smoothing approach to nondimensional item analysis • Ramsay, J. O.	95 Psymtrka 60 323-339
An empirical Bayes approach to	smoothing in backcalculation of HIV infection rates • Liao, Jiangang; Brookmeyer, Ron	95 Biomtrcs 51 579-588
	Smoothing categorical data • Simonoff, Jeffrey S.	95 JStPlInf 47 41-69
Digital	smoothing of census data employing Fourier transforms • Madan, V. K.; Sinha, R. K.*	95 CmpStDA 20 285-294
On a likelihood-based approach in nonparametric	smoothing and cross-validation • Chaudhuri, Probal; Dewanji, Anup	95 StPrLet 22 7-15
Almost sure convergence of	smoothing D^m-splines for noisy data • Arcangeli, Rémi; Ycart, Bernard	93 NumeriMa 66 281-294
Filtering and	smoothing via estimating functions • Naik-Nimbalkar, U. V.; Rajarshi, M. B.	95 JASA 90 301-306
adjustment by generalized adaptive exponential	smoothing • Herkenrath, U. Premium	94 InsrneME 15 203-217
Kernel	smoothing • Wand, M.-P.; Jones, M.-C.	95 ChapmnHall:Lnd 224p
Review of "Kernel	smoothing" • Wand, M. P.; Jones, M. C. (Auth); Mackisack, Margaret S. (Rev)	95 AstrlJSt 37 368-369
Review of "Kernel	smoothing" • Wand, M. P.; Jones, M. C. (Auth); Ramsay, J. O. (Rev)	95 ShrtBkRv 15/2 27-27
[Algorithm AS 296] Optimal median	smoothing • Härdle, W.; Steiger, W.	95 ApplStat 44 258-264
aerosolized ... • Boylstein et al. Comparison of	smoothing methods when applied to respiratory data collected in mice exposed to an	94 ASAProBiop 136-141
Gaussian fields and the weakly optimal choice of the	smoothing parameter. I • Konakov, V. D.; Piterbarg, V. I. High level excursions of	95 MaMethSt 4 421-434
models • Wang, Y. et al. Simulation studies of	smoothing parameter estimates and Bayesian confidence intervals in Bernoulli SS ANOVA	95 CommStB 24 1037-1059
Behavior near zero of the distribution of GCV	smoothing parameter estimates • Wahba, Grace; Wang, Yuedong	95 StPrLet 25 105-111
Iterative estimates for a	smoothing parameter • Sun, Jiayang	95 StPrLet 24 347-356
estimate • Barry, J.; Diggle, P. Choosing the	smoothing parameter in a Fourier approach to nonparametric deconvolution of a density	95 JNnparSt 4 223-232
approach • Droge, B.; Georg, T. On selecting the	smoothing parameter of least squares regression estimates using the minimax regret	95 StatDecn 13 1-20
Detecting influential observations on the	smoothing parameter in nonparametric regression • Kim, Choongrak; Jeon, Jong-Woo	95 JKoreaSt 24 495-506
Automatic	smoothing parameter selection for robust nonparametric regression • Wang, Ferdinand T.	94 ASAProStCp 106-111
dimensional edge detection and edge-preserving	smoothing • Cheng, Bing; Kay, Jim A regularization method for one-	95 CmpStQ 10 53-69
Bradley, D. C.; Steil, G. M. et al. OOPSEG: A data	smoothing program for quantitation and isolation of random measurement error •	95 CmpBiomd 46 67-77
Visual error criteria for qualitative	smoothing • Maron, J. S.; Tsybakov, A. B.	95 JASA 90 499-507
frequency polygons: Quantitative and qualitative	smoothing • Simonoff, Jeffrey S. The anchor position of histograms and	95 CommStB 24 691-710
Nonlinear	smoothing for random fields • Aihara, Shin Ichi; Bagchi, Arunabha	95 StocProc 55 143-158
explanatory variables • Elston, D. A.; Proe, M. F.	Smoothing regression coefficients in an overspecified regression model with interrelated	95 ApplStat 44 395-406
New	smoothing results for state space models • Tsimikas, John; Ledolter, Johannes	94 ASAProBuEc 172-177
On partial local	smoothing rules for curve estimation • Hall, Peter; Marron, J. Stephen; Titterington, D. M.	95 Biomtrka 82 575-587
Fast and simple scatterplot	smoothing • Härdle, W.; Marron, J. S.	95 CmpStDA 20 1-17
	Smoothing spline	
Wahba, Grace; Wang, Yuedong	Behavior near zero of the distribution of GCV smoothing parameter estimates •	95 StPrLet 25 105-111
Bosch, Ronald J.; Ye, Yinyu; Woodworth, George G.	A convergent algorithm for quantile regression with smoothing splines •	95 CmpStDA 19 613-630
Hong, Changkon; Kim, Choongrak	Existence and uniqueness of the smoothest density with prescribed moments •	95 JKoreaSt 24 233-242
highest-derivative-polynomial subject ... • Wilson	Longitudinal data analysis for linear Gaussian models with random disturbed-	95 StatMed 14 1219-1233
Douglas; Ruppert, David	Nonparametric transformations for both sides of a regression model • Nychka,	95 JRSS-B 57 519-532
Rianto A. A cubic	smoothing spline based lack of fit test for nonlinear regression models • Djojosugito,	95 CommStA 24 2183-2197
Wahba, G. Bootstrap confidence intervals for	smoothing splines and their comparison to Bayesian confidence intervals • Wang, Y.;	95 JStCmpSm 51 263-279
	Smoothing spline density estimation: Conditional distribution • Gu, Chong	95 StSinica 5 709-726
Computing the error for	smoothing splines • van der Linde, Angelika	95 CmpStQ 10 143-154
Characteristics of the linear	smoothing spline (Japanese) • Ogawa, Naoko; Minami, Mihoko; Sibuya, Masaaki	95 JpJAppSt 24 27-41
	Smoothing splines for longitudinal data • Anderson, Stewart J.; Jones, Richard H.	95 StatMed 14 1235-1248
convergent algorithm for quantile regression with	smoothing splines • Bosch, Ronald J.; Ye, Yinyu; Woodworth, George G. A	95 CmpStDA 19 613-630
temporal farm income distributions using spatial	smoothing techniques • Kokic, P. N.; Moon, L.; Gooday, J.; Chambers, R. L. Estimating	95 AstrlJSt 37 129-143
of moving-average trend filters using fidelity and	smoothness criteria • Gray, Alistair; Thomson, Peter Design	94 ASAProBuEc 157-162
Florit, Carme; Nualart, David A local criterion for	smoothness of densities and application to the supremum of the Brownian sheet •	95 StPrLet 22 25-31
• Caballero, M. E.; Fernández, B.; Nualart, D.	Smoothness of distributions for solutions of anticipating stochastic differential equations	95 Stochatx 52 303-322
On qualitative	smoothness of kernel density estimates • Mammen, E.	95 MaOpfStS 26 253-267
A note on the	smoothness of L^1-estimators for the linear model • Ellis, Steven P.	95 SankhyaA 57 221-226
O'Connell, Julie A. Influence on	smoothness in penalized likelihood regression for binary data • Jernigan, Robert W.;	94 ASAProStCp 118-123
regression • Chu, C. K. The effect of the	smoothness of the regression function on a bandwidth selector in nonparametric	95 CommStB 24 61-77
Local adaptation to inhomogeneous	smoothness: Resolution level • Lepskii, O. V.; Spokoiny, V. G.	95 MaMethSt 4 239-258
Adapting to unknown	smoothness via wavelet shrinkage • Donoho, David L.; Johnstone, Iain M.	95 JASA 90 1200-1224
Review of "Statistics in action	SMP 16-19 Mathematics" • School Mathematics Project (Auth); Endean, Sarah (Rev)	95 TeachgSt 17/2 79-79

The Brownian	snake and solutions of $\Delta u = u^2$ in a domain • Le Gall, Jean-François	95 ProbTher 102 393-432
Markov	snakes and superprocesses • Dynkin, E. B.; Kuznetsov, S. E.	95 ProbTher 103 433-473
CAPI, event histories, and incentives in the	SNFG cycle 5 pretest • Mosher, William D.; Pratt, William F.	94 ASASrvyRsr 59-63
Transient animals in a resident population of	snow geese: Local emigration or heterogeneity? • Pradel, Roger; Cooch, Evan; Cooke, Fred	95 JAppStat 22 695-710
recapture models: A case study on the lesser	snow goose • Viallefont, A.; Cooch, E. G.; Cooke, F. Estimation of trade-offs with capture-	95 JAppStat 22 847-861
G. N.; Cressie, N.; Carroll, T. R. Spatial modelling of	snow water equivalent using airborne and ground-based snow data • Carroll, S. S.; Day,	95 EnvrMtrc 6 127-139
Estimating the size of hidden populations using	snowball sampling • Frank, Ove; Snijders, Tom	94 JOffStat 10 53-67
ground advantage of individual clubs in English	soccer • Clarke, Stephen R.; Norman, John M. Home	95 Ststcian 44 509-521
Analysis of the 1986 and 1994 World Cup	soccer tournament • McGarry, Tim; Schutz, Robert W.	94 ASAProSpts 61-65
Handbook of statistical modeling for the	social and behavioral sciences • Arminger, Gerhard; Clogg, Clifford C.; Sobel, Michael E.	95 PlenumPubl:NY 618p
Review of "Handbook of statistical modeling for the	social and behavioral sciences" • Arminger, G.; Clogg, C. C.; Sobel, M. E. (Ed) et al.	95 Biomtrcs 51 1586-1587
Review of "Handbook of statistical modeling for the	social and behavioural sciences" • Arminger, G.; Clogg, C. C.; Sobel, M. E. (Ed) et al.	95 Chance 8/4 32-36
series analysis of improved quality of life in Canada:	Social change, collective consciousness, and the TM-Sidhi program • Dillbeck, M. C. Time	95 PsychRpt 76 1171-1193
the estimation and projection of demographic and	social characteristics" • Becker, Patricia C. Comment on "Statistical advances in	94 ASAProSoSt 122-123
the redesign (Disc: p379-381) • LeClere, F. B. et al.	Social correlates of response bias in the National Health Interview Survey: Lessons for	94 ASASrvyRsr 373-378
years ... • Ashour et al. Discriminant analysis for	social-economic and cultural factors affecting diarrhea diseases in children under two	95 EgpStJ 39 1-11
Review of "Spatial data analysis in the	social and environmental sciences" • Haining, Robert P. (Auth); Chang, Hyejung (Rev)	95 Chance 8/1 42-43
Randomization and cause and effect inference in	social experimentation • Neuberg, Leland G.	94 ASAProSoSt 153-158
	Social factors and health: The causation-selection issue revisited • Goldman, Noreen	94 PNatAcSc 91 1251-1255
of rank versus category in measuring subjective	social inequality • Hoffmeyer-Zlotnik, Jürgen H. P. Effects	94 ASASrvyRsr 1322-1326
Labour-use efficiency in Swedish	social insurance offices • Kumbhakar, Subal C.; Hjalmarsson, Lennart	95 JApEcmtx 10 33-47
Review of "Algebraic models for	social networks" • Pattison, P. (Auth); Prescott, P. (Rev)	95 JRSS-A 158 197-198
S. (Auth); Wasserman, Stanley (Rev) Review of "The	social organization of sexuality" • Laumann, E. O.; Gagnon, J. H.; Michael, R. T.; Michaels,	95 Chance 8/2 40-42
Log-multiplicative models for valued	social relations • Anderson, Carolyn J.; Wasserman, Stanley	95 SocMethR 24 96-127
design and statistical methods for behavioural and	social research • Boniface, D. Experiment	95 ChapmnHall:Lnd 272p
Sample size choice in behavioral and	social research: Classical versus Bayesian methods • Pollard, William E.	94 ASAProBays 206-210
and methodology. A nontechnical guide for the	social sciences" • Vogt, W. P. (Auth); Berg, Sven (Rev) Review of "Dictionary of statistics	94 JOffStat 10 225-226
Inference and hierarchical modeling in the	social sciences (Disc: p201-233) • Draper, David	95 JEdStat 20 115-147
Statistics for the	social sciences • Wilcox, Rand R.	95 Academic:NY:Lnd 444p
Review of "Understanding	social science statistics: A spreadsheet approach" • Bakeman, R. (Auth); Goda, David (Rev)	95 Ststcian 44 539-540
of "Statistical tables for economic, business and	social studies (Second edition)" • Kmietowicz, Z. W.; Yannoulis, Y. (Auth) et al. Review	95 JAppStat 22 439-440
	Social trends and social change • Nissel, Muriel	95 JRSS-A 158 491-504
	Social values do change? (Japanese) • Nisihira, Sigeki	95 PrInStMa 43 81-97
Gopal K.; Rosenberg, Harry M. Sources of data on	socio-economic differential mortality in the United States • Hoyert, Donna L.; Singh,	95 JOffStat 11 233-260
weights in panel surveys - The case of the German	Socio-Economic Panel (German) • Rendtel, U. et al. A method for checking longitudinal	95 AllgeStA 79 252-277
A statistical comparison of Cuban	socioeconomic development • Romeu, Jorge Luis	94 ASAProSoSt 159-163
	Socioeconomic status and suicide • Kung, H. C.; Liu, X. H.; Chan, J.	94 ASAProSoSt 235-239
Computer simulations and empirical testing of	sociological theory • Jacobsen, Chanoch; Bronson, Richard	95 SocMethR 23 479-506
Statistics for	sociologists: Teaching techniques that work • Potter, Alisa M.	95 TeachSoc 23 259-263
pigments • Taavitsainen, V.-M.; Jalava, J.-P.	Soft and harder multivariate modelling in developing the properties of titanium dioxide	95 Chemolab 29 307-319
	Software	
	1st - Class! • Feu, Chris du	95 TeachgSt 17/1 13-15
	On the accuracy of the statistical distributions in GAUSS • Knüsel, Leo	95 CmpStDA 20 699-702
responses • Rodríguez, Germán; Goldman, Noreen	An assessment of estimation procedures for multilevel models with binary	95 JRSS-A 158 73-89
	What do you get when you combine a Pentium and a research grant? • Jones, Albyn C.	95 Chance 8/2 7-8
Rodriguez, R.; Tobias, R.; Wolfinger, R.	Comment on "The statistics of linear models: Back to basics" (V4 p221-234) •	95 StatComp 5 97-101
	Comparison of programs MULT, ESTIMATE and BROWNIE • Conroy, Michael J.	95 JAppStat 22 763-773
Kruglyak, Leonid; Lander, Eric S.	Complete multipoint sib-pair analysis of qualitative and quantitative traits •	95 AJHuGene 57 439-454
Rahim	Computational aspects of linear and non-linear time series analysis • Moeanaddin,	92 ProIrStC1 123-155
	Computer graphics approach to two-way ANOVA (Korean) • Huh, Moon Yul	95 KorApS 8/1 75-87
	Correspondence analysis in statistical package programs • Thompson, Paul A.	95 AmerStat 49 310-316
	Data analysis using APL2 and APL2STAT • Fox, John; Friendly, Michael	95 SocMethR 23 282-302
	Data analysis using GAUSS and Markov • Long, J. Scott; Noss, Brian	95 SocMethR 23 303-328
	Data analysis using Lisp-Stat • Tierney, Luke	95 SocMethR 23 329-351
	Data analysis using Mathematica • Stine, Robert A.	95 SocMethR 23 352-372
	Data analysis using S-Plus • Schulman, Daniel A.; Campbell, Alec D.; Kostello, Eric C.	95 SocMethR 23 392-413
	Data analysis using SAS • Hallahan, Charlie	95 SocMethR 23 373-391
sample data • Froeschl, Karl Anton	A formal model evaluation approach to the analysis of treatment effects in paired	95 CmpStDA 19 493-517
Study • Klein, John P.; Keiding, Niels; Kreiner, Svend	Graphical models for panel studies, illustrated on data from the Framingham Heart	95 StatMed 14 1265-1290
	GUHA for personal computers • Hájek, P.; Sochorová, A.; Zvárová, J.	95 CmpStDA 19 149-153
Young, Forrest W.; Lubinsky, David J.	Guiding data analysts with visual statistical strategies (Disc: p251-260) •	95 JCmpGrSt 4 229-250
	Implementing partial least squares • Denham, M. C.	95 StatComp 5 191-202
	Interactive dynamic graphics for exploratory survival analysis • Atkinson, E. Neely	95 AmerStat 49 77-84
Stine, Robert	Introduction to the special issue on computing environments • Fox, John;	95 SocMethR 23 267-281
ical statistics • Okada, Mihoko; O'Brien, Myles	IntroStat: A hypertext-based design for an electronic textbook to introduce biomed-	95 CmpBiomd 46 265-276
	Lord_DIF: A computer program to compute Lord's DIF statistics • Li, Yuan Hwang	95 ApPsychM 19 72-72
model scale score and the unbiased domain ... • Li	LTDOMAIN: A look-up table for the corresponding estimated one-parameter logistic	95 ApPsychM 19 50-50
Catchpole, E. A.	MATLAB: An environment for analyzing ring-recovery and recapture data •	95 JAppStat 22 801-816
	Maximum likelihood estimation and Mathematica • Currie, Iain D.	95 ApplStat 44 379-394
	Minitab menus - Do they help or hinder? • Reed, Margaret; Garven, Frances	95 TeachgSt 17/1 22-25
analysis • Rahlfs, Volker W.; Zimmermann, Helmuth	Nonparametric methods in standard statistics packages: The case of the Friedman	95 CmpStDA 20 101-110
	A note on beta coefficients in SPSS • Kockläuner, Gerhard	95 AllgeStA 79 115-116
measurement error • Bradley, D. C.; Steil, G. M. et al.	OOPSEG: A data smoothing program for quantitation and isolation of random	95 CmpBiomd 46 67-77
	PcGive Professional 8 • Escribano, Alvaro	95 JApEcmtx 10 79-86
open populations • Arnason, A. Neil; Schwarz, Carl J.	POPAN-4: Enhancements to a system for the analysis of mark-recapture data from	95 JAppStat 22 785-800
programme of constant-effort ... • Desante et al.	Productivity indices and survival rate estimates from MAPS, a continent-wide	95 JAppStat 22 935-947
mozygosity mapping • Kruglyak, L.; Daly, M. J. et al.	Rapid multipoint linkage analysis of recessive traits in nuclear families, including ho-	95 AJHuGene 56 519-527
Pandey, M.	Reliability estimation with Weibull inter failure times • Ferdous, J.; Uddin, Md. Borhan;	95 RelEngSS 50 285-296
Inc (Auth); Getson, Pamela (Rev)	Review of "BMDP New System for Windows (Version 1.11)" • BMDP Statistical Software	95 Chance 8/4 53-58
	A review of five statistical packages for Windows • Lurie, Philip M.	95 AmerStat 49 99-107
	Review of "InStat" • GraphPad Software (Auth); Lachenbruch, Peter A. (Rev)	95 Chance 8/3 57-60
	Review of "Junior Pinpoint" • Longman Logotron (Auth); Graham, Alan (Rev)	95 TeachgSt 17/2 75-75
	Review of "NCSS 6.0 for Windows" • NCSS Statistical Software (Auth); Lovie, Pat (Rev)	95 BJMaStPs 48 407-408
	Review of "Pinpoint - Datafiles" • Longman Logotron (Auth); Feu, Chris du (Rev)	95 TeachgSt 17/2 75-76

	Software (cont.)	
Lachenbruch, Peter A. (Rev)	Review of "Survey Pro for Windows (Version 1.8)" • Apian Software (Auth);	95 Chance 8/4 58-59
nonlinear regression - A case study • Rasch, Dieter	The robustness against parameter variation of exact locally optimum designs in	95 CmpStDA 20 441-453
Shih, Joanna H.	Sample size calculation for complex clinical trials with survival endpoints •	95 ConClnTr 16 395-407
Wang, Whedy; Bromberg, Judith	A SAS macro for residual deviance of ordinal regression analysis • Wan, Jim Y.;	94 CmpBiomd 45 307-310
	How to start surfing the Internet • Hopke, Philip K.	95 Chemolab 30 1-9
Cantor, Alan B.	A test of the association of a time-dependent state variable to survival •	95 CmpBiomd 46 101-105
Pradel, R.; Reboulet, A.-M.; Viallefont, A.	Testing hypotheses and estimating survival from capture histories with CR •	95 JAppStat 22 775-784
procedures • Lawrance, A. J.	Testing regression lack of fit without replication: A tutorial around Minitab's XLOF	94 JAppStat 21 541-548
Gort, Steven M.; Hoogerbrugge, Ronald	A user-friendly spreadsheet program for calibration using weighted regression •	95 Chemolab 28 193-199
analysis of repeated survey data (with available	software) • Kott, Phillip S. Regression	94 ASASrvyRsr 116-123
Pretsch, E. (Rev) Review of "Intelligent	software for chemical analysis" • Buydens, Lutgarde M. C.; Schoenmakers, Peter J. (Ed);	95 Chemolab 27 148-149
Lavine, Barry K. (Rev) Review of "SCAN:	Software for chemometric analysis (Release 1 for Windows)" • Minitab Inc (Auth);	95 Chemolab 30 253-255
Meta-analysis	software: A comparative review • Normand, Sharon-Lise T.	95 AmerStat 49 298-309
Models for assessing the reliability of computer	software • Singpurwalla, Nozer D.; Wilson, Simon P.	92 ProIrStCl 177-194
formulation for normal distributions using computer	software • Poston, C. A. et al. Solving the ranking and selection indifference-zone	94 ASAProStCp 290-291
Stromberg, Arnold J.	Software for computing a robust efficient linear or nonlinear regression estimator •	94 ASAProStEd 6-11
ULM, a	software for conservation and evolutionary biologists • Legendre, S.; Clobert, J.	95 JAppStat 22 817-834
Statistical process control in	software development • Thomas, Michael L.	94 ASAProQlPr 140-145
The Statistical	Software Guide '94/95 • Koch, Armin; Haag, Uwe (Ed)	95 CmpStDA 19 237-262
	Software for handling dropouts in clinical trials • Entsuah, Richard	94 ASAProBiop 188-193
non-linear functions of totals in surveys - Theory and	software implementation • Andersson, C. et al. A method for variance estimation of	94 JOffStat 10 395-405
Review of "GSLIB: Geostatistical	software library and user's guide" • Deutsch, C.; Journel, A. (Auth); Ziegel, Eric R. (Rev)	95 Technmcs 37 126-126
Rosen, Richard J. Commercial payroll	software and the quality of employment data • Goldenberg, Karen L.; Moore, Susan E.;	94 ASASrvyRsr 1025-1030
Metrics and models in	software quality engineering • Kan, Stephen H.	95 ASQualCntrl:WI 344p
Chiera, C. Optimal	software release policy based on a two-person game of timing • Zeephongsekul, P.;	95 JAppProb 32 470-481
	Software reliability	
Peña, Edsel A.	Dynamic reliability models with conditional proportional hazards • Hollander, Myles;	95 LiDaAnal 1 377-401
Bayesian computation of	software reliability • Kuo, Lynn; Yang, Tae	95 JCmpGrSt 4 65-82
Review of "Ensuring	software reliability" • Neufelder, Ann M. (Auth); Soyer, Refik (Rev)	95 Technmcs 37 122-123
Review of "Statistical analysis of	software reliability models" • van Pul, M. C. J. (Auth); Reese, R. Allan (Rev)	95 JRSS-A 158 199-199
of the parameters for the Littlewood model in	software reliability • Barendregt, L. G.; van Pul, M. C. On the estimation	95 StNeerla 49 165-184
school students to exploit statistical models and	software for research projects • Ward, Joe H., Jr; Niland, Laura J. Empowering high	94 ASAProStEd 44-48
statistical analysis of slope ratio assays using SAS	software • Wong, M. Y. The	94 CmpBiomd 45 233-238
Computer programs for qualitative data analysis. A	software sourcebook • Weitzman, Eben A.; Miles, Matthew B.	95 SagePubl:CA:Lnd 372p
Statistical issues in spreadsheet	software • Nash, John C.; Quon, Tony K.; Gianini, Jacqueline	94 ASAProStEd 238-241
The Pentium bug and statistical	software • Gould, William	95 Chance 8/2 9-13
independent graphical user interface for statistical	software • Liu, Lon-Mu; Chan, Ki-Kan; Montgomery, Alan L.; Muller, Mervin E. A system-	95 CmpStDA 19 23-44
Student editions of forecasting	software: A survey • Rycroft, Robert S.	95 IntJFor 11 337-351
of quality control and improvement statistical	software system • Jarrett, Jeffrey Review	95 CmpStDA 20 333-339
The	software taxi: An easy-to-use tool for statistical presentations • Nash, John C.	94 ASAProStCp 269-271
Cliff (Rev) Review of	"SoftWindows, Mathematica and Data Desk for the Macintosh PowerPC" • Spiegelman,	95 Chemolab 28 208-209
cations to estimation of exposure to contaminated	soils • Chen ... cost estimator for the mean of positively skewed distributions with appli-	95 EnvrMtrc 6 181-193
Phased sampling for	soil remediation • Englund, Evan J.; Heravi, Naser	94 EnvEcoSt 1 247-263
A comparison ... • Lesch et al. Spatial prediction of	soil salinity using electromagnetic induction techniques 1. Statistical prediction models:	95 WaterReR 31 373-386
algorithm ... • Lesch et al. Spatial prediction of	soil salinity using electromagnetic induction techniques. 2. An efficient spatial sampling	95 WaterReR 31 387-398
	Sojourns and future infima of planar Brownian motion • Hu, Y.; Shi, Z.	95 ProbTher 103 329-348
	Sojourn time	
Hasegawa, Toshiharu	A conformance measure for queues • Takine, Tetsuya; Sengupta, Bhaskar;	95 CommStSM 11 645-670
	On one discrete queueing model • Velikoivanenko, O. I.; Shvets, Ye. M.	95 ThPrMaSt 50 55-60
	Response time distributions in tandem g-networks • Harrison, Peter G.; Pitel, Edwige	95 JAppProb 32 224-246
and point-process input • Miyazawa, M. et al.	On the structure of an insensitive generalized semi-Markov process with reallocation	95 AdvAppPr 27 203-225
and exit from the ... • Duffy et al. Estimation of mean	sojourn time in breast cancer screening using a Markov chain model of both entry to	95 StatMed 14 1531-1543
Newton, J. (Rev) Review of	"Sojourn times in feedback and processor sharing queues" • van den Berg, J. L. (Auth);	95 JRSS-A 158 186-187
Schassberger, R. (Rev) Review of	"Sojourn times in feedback and processor sharing queues" • van den Berg, J. L. (Auth);	95 Metrika 42 146-147
Estimating expected	sojourn times for a Markov renewal process • Rossetti, Manuel D.; Clark, Gordon M.	95 CommStA 24 553-579
documentation: Individual income tax microdata,	SOL • David, Martin Desktop	94 ASASrvyRsr 775-780
Imputation of missing values in spatio-temporal	solar radiation data • Glasbey, C. A.	95 EnvrMtrc 6 363-371
Alegre, Antoni; Claramunt, M. Mercé Allocation of	solvency cost in group annuities: Actuarial principles and cooperative game theory •	95 InsrneME 17 19-34
A theory of risk, return and	solvency • Powers, Michael R.	95 InsrneME 17 101-118
ditions in the search area of the multidimensional	solvent space • Kaufmann ... liquid chromatography. II. Exploring and finding optimal con-	95 Chemolab 27 105-114
An experimental design for evaluating	sonar detection performance • de Jongh, P. J.; de Wet, T.; Fresen, J. L.	95 SoAfrStJ 29 79-103
The Eurovision	song contest • Everitt, Brian	95 Stats /13 19-22
Aki, Sigeo	Sooner and later waiting time problems in a two-state Markov chain • Uchida, Masayuki;	95 AnInStMa 47 415-433
quotient algorithm as a rapid tool for obtaining	sorption isotherms from column elution curves • van Veldhuizen et al. The (least squares)	95 WaterReR 31 849-857
Measurement error or endogeneity:	Sorting out simultaneity • Romeo, Charles J.	94 ASAProBays 196-201
students in education in the Lebowa region of	South Africa • Glencross, M. J. et al. Attitudes toward applied statistics of postgraduate	95 PsychRpt 77 315-322
cer screening illustrated by a specific model using	South Australian cancer registry data • O'Neill et al. ... the technical features of breast can-	95 StMedRes 4 55-72
resource estimation of the upper atmosphere over	Southern Africa • Bryukhan, Fedor F.; Diab, Roseanne D. Wind energy	95 JApMetrl 34 2565-2571
ing of daily ambient temperature: An application in	Southern Italy • Macchiato, M. F.; La Rotonda, L. et al. Time modelling and spatial cluster-	95 EnvrMtrc 6 31-53
scenarios for severe sustained drought in the	Southwestern United States • Tarboton, David G. Hydrologic	95 WaterReB 31 803-813
consequences of mortality dynamics in the former	Soviet Republics: Length of working life • Kingkade, W. Ward; Vasin, Sergey A. Welfare	94 ASAProSoSt 94-99
predictions to enhance breeding for yield in	soybean: I. Choosing parents • Panter, D. M.; Allen, F. L. Using best linear unbiased	95 CropSci 35 397-405
	Space	
	• see names of specific types of spaces such as Hilbert - or state -	
A simple and efficient	space domain implementation of the turning bands method • Dietrich, C. R.	95 WaterReR 31 147-156
David	Space-time boundaries for random walks obtained from diffuse measures • Handelman,	94 IsrelJMa 86 107-156
Second-order analysis of	space-time clustering • Diggle, P. J.; Chetwynd, A. G.; Häggkvist, R.; Morris, S. E.	95 StMedRes 4 124-136
Species diversity in	space and time • Rosenzweig, Michael L.	95 CambridgeUniv 456p
Tests for directional	space-time interaction in epidemiological data • Lawson, Andrew B.; Viel, Jean-François	95 StatMed 14 2383-2391
	Space-time model	
models • Otranto, Edoardo; Gallo, Giampiero M.	Regression diagnostic techniques to detect balanced space-to-time ratios in STARMA	94 Metron 52/3 129-145
Review of "Modelling biological populations in	space and time" • Renshaw, Eric (Auth); Zeh, Judy (Rev)	95 JASA 90 800-801
A	space-time process model for the evolution of DNA sequences • Yang, Ziheng	95 Genetcs 139 993-1005

Regression diagnostic techniques to detect balanced	space-to-time ratios in STARMA models • Otranto, Edoardo; Gallo, Giampiero M.	94 Metron 52/3 129-145
of maximum likelihood estimates in a class of	space-time regression models • Niu, Xufeng Asymptotic properties	95 JMultiAn 55 82-104
A least squares method for spectral analysis of	space-time series • Wu, Dong L.; Hays, Paul B.; Skinner, Wilbert R.	95 JAtmsphr 52 3501-3511
moli, C.; Ghislandi, M. et al. Bayesian analysis of	space-time variation in disease risk • Bernardinelli, L.; Clayton, D.; Pascutto, C.; Monto-	95 StatMed 14 2433-2443
parabolic partial differential equations perturbed by	space-time white noise • Gyöngy, István; Nualart, David Implicit scheme for quasi-linear	95 StocProc 58 57-72
• Saleh, A. K. Md. E.; Hassanein, K. M. et al. Optimum	spacings, ABLUE and tests of hypothesis in simple linear model with extreme-value errors	94 JStatRsr 28 53-63
approximate efficiencies of tests based on	spacings • Bartoszewicz, Jarosław Bahadur and Hodges-Lehmann	95 StPrLet 23 211-220
trait loci • Darvasi, A.; Soller, M. Optimum	spacing of genetic markers for determining linkage between marker loci and quantitative	94 ThApGene 89 351-357
alternative test for normality based on normalized	spacings • Chen, Ling; Shapiro, Samuel S. An	95 JStCmpSm 53 269-287
A. Some results on normalized	spacings from restricted families of distributions • Kochar, Subhash C.; Kirmani, S. N. U.	95 JStPlInf 46 47-57
Limit theorems for the logarithm of sample	spacings • Shao, Yongzhao; Hahn, Marjorie G.	95 StPrLet 24 121-132
SRS runs and	spacings tests to assess randomness • Kaigh, W. D.; Schuster, E. F.	94 ASAProStCp 272-277
An expanded table of probability values for Rao's	spacing test • Russell, Gerald S.; Levitin, Daniel J.	95 CommStB 24 879-888
An analysis of the international tourism demand in	Spain • González, Pilar; Moral, Paz	95 IntJFor 11 233-251
data recorded on the Catalan coast (North-east	Spain) • Lana et al. ... for synoptic weather type classification. Application to winter period	94 JClimlgy 14 1051-1060
Evaluating translations of survey instruments in	Spanish and English: The 1990 Spanish Language Census long form • Gabbard, S. M. et al.	94 ASASrvyRsr 1232-1235
vey instruments in Spanish and English: The 1990	Spanish Language Census long form • Gabbard, S. M. et al. Evaluating translations of sur-	94 ASASrvyRsr 1232-1235
An item nonresponse and log-linear analysis of the	Spanish Language Forms Availability Test • de la Puente, Manuel; Wobus, Peter	94 ASASrvyRsr 583-588
Percolation and minimal	spanning forests in infinite graphs • Alexander, Kenneth S.	95 AnlsProb 23 87-104
	Spanning tree	
	Percolation and minimal spanning forests in infinite graphs • Alexander, Kenneth S.	95 AnlsProb 23 87-104
A	spanning tree heuristic for regional clustering • Maravalle, Maurizio; Simeone, Bruno	95 CommStA 24 625-639
Simeone, Bruno Correction to "A	spanning tree heuristic for regional clustering" (V24 p625-639) • Maravalle, Maurizio;	95 CommStA 24 2437-2437
length links of a dissimilarity matrix and minimum	spanning trees • Carroll, J. Douglas Minimax	95 Psymtrka 60 371-374
Minimum	spanning trees for tree metrics: Abridgements and adjustments • Leclerc, Bruno	95 JClass 12 207-241
Random-cluster measures and uniform	spanning trees • Häggström, Olle	95 StocProc 59 267-275
survival rates • Senar, J. C. et al. Mediterranean house	sparrows (Passer domesticus) are not used to freezing temperatures: An analysis of	95 JAppStat 22 1069-1074
combination estimator for d-dimensional ordinal	sparse contingency tables • Dong, Jianping; Simonoff, Jeffrey S. A geometric	95 AnlsStat 23 1143-1159
in local polynomial fitting: Variable bandwidth and	spatial adaptation • Fan, Jianqing; Gijbels, Irène Data-driven bandwidth selection	95 JRSS-B 57 371-394
	Spatial analysis	
D.; Pascutto, C.; Montomoli, C.; Ghislandi, M. et al.	Bayesian analysis of space-time variation in disease risk • Bernardinelli, L.; Clayton,	95 StatMed 14 2433-2443
Clayton, D.; Montomoli, C.	Bayesian estimates of disease maps: How important are priors? • Bernardinelli, L.;	95 StatMed 14 2411-2431
	On conditional and intrinsic autoregressions • Besag, Julian; Kooperberg, Charles	95 Biomtrka 82 733-746
Jacquez, Geoffrey M.	Disease models implicit in statistical tests of disease clustering • Waller, Lance A.;	95 Epidmlgy 6 584-590
• Duchon, C. E.; Renkevens, T. M.; Crosson, W. L.	Estimation of daily area-average rainfall during the CaPE experiment in Central Florida	95 JApMetrl 34 2704-2714
Biggeri, A.; Braga, M.; Lagazio, C.; Marchi, M.	Exploratory tools for disease mapping in geographical epidemiology • Cislaghi, C.;	95 StatMed 14 2363-2381
Jacquez, Geoffrey M.	The map comparison problem: Tests for the overlap of geographic boundaries •	95 StatMed 14 2343-2361
S.	A probabilistic model for the variance to mean power law in ecology • Kendal, Wayne	95 EcolgMdl 80 293-297
Cruz-Orive, L. M.	The Rao-Blackwell theorem in stereology and some counterexamples • Baddeley, A. J.;	95 AdvAppPr 27 2-19
Andrew B.; Viel, Jean-François	Tests for directional space-time interaction in epidemiological data • Lawson,	95 StatMed 14 2383-2391
Suksawasdi Na; Böhning, Danmar	Traffic accident mapping in Bangkok metropolis: A case study • Ayuthya, Rampai	95 StatMed 14 2445-2458
Gavrikov, Vladimir; Stoyan, Dietrich	The use of marked point processes in ecological and environmental forest studies •	95 EnvEcoSt 2 331-344
Shafer, Ronald W.	Spatial analysis of benthic data from the Chesapeake Bay • Neerchal, Nagaraj K.;	94 ASAProEnvr 204-208
election • Schlichting, Kurt C.; Tuckel, Peter S. A	spatial analysis of contextual effects on voter participation in the 1992 presidential	94 ASASrvyRsr 1339-1344
Christl A. The	spatial analysis of covariates in a study of environmental epidemiology • Donnelly,	95 StatMed 14 2393-2409
	Spatial analysis by distance indices • Perry, Joe N.	95 JAnmlEco 64 303-314
On	spatial analysis and map transformations • Palma, Daniela	94 QStMaA 16 127-143
	Spatial analysis for monitoring forest health • Roesch, Francis A.	94 ASAProEnvr 104-109
A portmanteau test for	spatial ARMA models • Etchison, Tonya; Brownie, Cavell; Pantula, Sastry G.	95 Biomtrcs 51 1536-1542
An asymptotic test for separability of a	spatial autoregressive model • Shitan, Mahendran; Brockwell, Peter J.	95 CommStA 24 2027-2040
Limits of	spatial branching populations (Disc: p185-189) • Wakolbinger, Anton	95 Bernoull 1 171-185
Case-control designs for the detection of	spatial clusters of diseases • Biggeri, Annibale; Marchi, Marco	95 EnvrMtrc 6 385-393
Macchiato, M. F. et al. Time modelling and	spatial clustering of daily ambient temperature: An application in Southern Italy •	95 EnvrMtrc 6 31-53
Estimating	spatial correlations from spatial-temporal data • Gunst, Richard F.	94 ASAProEnvr 98-103
Estimating	spatial correlations from spatial-temporal meteorological data • Gunst, Richard F.	95 JClimate 8 2454-2470
maximum likelihood estimators of parameters of a	spatial counting process modelling crystallization of polymers • Capasso et al. ... of the	95 StocAnAp 13 279-294
gia • Reich, R. M.; Czaplewski, R. L.; Bechtold, W. A.	Spatial cross-correlation of undisturbed, natural shortleaf pine stands in northern Geor-	94 EnvEcoSt 1 201-217
	Spatial data	
	The ability of wet deposition networks to detect temporal trends • Oehlert, Gary W.	95 EnvrMtrc 6 327-339
	Adjusting Moran's I for population density • Oden, Neal	95 StatMed 14 17-26
birds • van Noordwijk, Arie J.	On bias due to observer distribution in the analysis of data on natal dispersal in	95 JAppStat 22 683-694
p697-699) • Pickle, Linda W.; White, Andrew A.	Effects of the choice of age-adjustment method on maps of death rates (Disc:	95 StatMed 14 615-627
analysis • Högmander, Harri; Møller, Jesper	Estimating distribution maps from atlas data using methods of statistical image	95 Biomtrcs 51 393-404
699) • Nobre, Flavio F.; de Macedo, Marcia M. A.	Feasibility of contour mapping epidemiological data with missing values (Disc: p697-	95 StatMed 14 605-613
	Interaction levels in an autoregressive graphical model • Lagona, Francesco	94 Metron 52/1 113-141
community characteristics and ... • Andes et al.	Linking public health data using geographic information system techniques: Alaskan	95 StatMed 14 481-490
Neville	Spatial disease clusters: Detection and inference • Kulldorff, Martin; Nagarwalla,	95 StatMed 14 799-810
locations of airborne pollutants • Hopke, P. K. et al.	The use of bootstrapping to estimate conditional probability fields for source	95 Chemolab 30 69-79
Interactive	spatial data analysis • Bailey, Trevor C.; Gatrell, Anthony C.	95 Wiley:NY:UK 432p
Chang, Hyejung (Rev) Review of	"Spatial data analysis in the social and environmental sciences" • Haining, Robert P. (Auth);	95 Chance 8/1 42-43
A bird's eye view of	spatial data and applications • Arbia, Giuseppe	94 QStMaA 16 5-16
process: Using correlation functions to explore a	spatial data set • Goulard, M.; Pagés, L.; Cabanettes, A. Marked point	95 BiomtrcJ 37 837-853
Review of "Statistics for	spatial data" • Cressie, N. A. C. (Auth); Taylor, Charles (Rev)	95 StMedRes 4 91-91
Sampling from	spatial databases • Olken, Frank; Rotem, Doron	95 StatComp 5 43-57
	Spatial disease clusters: Detection and inference • Kulldorff, Martin; Nagarwalla, Neville	95 StatMed 14 799-810
Models for the	spatial dispersal of insect pests • Mayer, D. G.; Atzeni, M. G.; Swain, A. J.; Stuart, M. A.	95 EnvrMtrc 6 497-503
Turner, John R. G. Predicting the	spatial distribution of climate: Temperature in Great Britain • Lennon, Jack J.;	95 JAnmlEco 64 370-392
nities: Effects of sampling effort and discontinuous	spatial distributions • Angermeier et al. ... relative abundances in stream-fish commu-	95 JFishRBC 52 936-949
plication of the gnostical theory of uncertain data to	spatial estimation, identification and quantification • Sevcik, Karel; Brooker, Peter I. Ap-	95 EnvrMtrc 6 457-463
due to incomplete data from a one-parameter	spatial Gaussian process • Krug, T. et al. Information loss on the dependence parameter	95 CommStA 24 3055-3072
of individual feature surface area with the vertical	spatial grid • Cruz-Orive, L. M.; Howard, C. V. Estimation	95 JMicro 178 146-151
T.; Kleczkowski, A.; Gilligan, C. A.; Bolker, B. M.	Spatial heterogeneity, nonlinear dynamics and chaos in infectious diseases • Grenfell, B.	95 StMedRes 4 160-183
and its breakdown under a hierarchy of periodic	spatial heterogeneities • Bhattacharya, R. N. et al. Time-scales for Gaussian approximation	95 Bernoull 1 81-123
On the use of	spatial information in the analysis of field experiments • Shukla, G. K.	94 JIAgSt 46 337-348

Cui, Haiyan; Stein, Alfred; Myers, Don E. Extension of	spatial information, Bayesian kriging and updating of prior variogram parameters •	95 EnvrMtrc 6 373-384
(Spain) • Ferrándiz, J.; López, A.; Llopis, A. et al.	Spatial interaction between neighbouring counties: Cancer mortality data in Valencia	95 Biomtrcs 51 665-678
Switzer, Paul	Spatial interpolation errors for monitoring data • Høst, Gudmund; Omre, Henning;	95 JASA 90 853-861
Temporal analogues to	spatial K functions • Cannon, Ann; Cressie, Noel	95 BiomtrcJ 37 351-373
trient concentrations in the Gulf ... • Toompuu et al.	Spatial large-scale correlations for optimal interpolation of temperature, salinity and nu-	95 EnvrMtrc 6 55-72
Partanen, Päivi; Penttinen, Antti	Spatial mark-recapture method in the estimation of crayfish population size •	95 BiomtrcJ 37 979-994
The development of	spatial models for community evaluation • Millar, Alexander B.; Gruenewald, Paul J.	94 ASAProSoSt 227-231
ment of a semi-conductor fabrication process via	spatial modeling of the mean and covariance surfaces • Hughes-Oliver et al. ... improve-	94 ASAProPEng 235-244
• Carroll, S. S.; Day, G. N.; Cressie, N.; Carroll, T. R.	Spatial modelling of snow water equivalent using airborne and ground-based snow data	95 EnvrMtrc 6 127-139
Further study on the statistical test to detect	spatial pattern • Sun, Ping	95 BiomtrcJ 37 199-203
Openshaw, Stan Developing automated and smart	spatial pattern exploration tools for geographical information systems applications •	95 Ststcian 44 3-16
Camper, A. K.; Stoodley, P. et al. Analysis of bacterial	spatial patterns at the initial stage of biofilm formation • Hamilton, M. A.; Johnson, K. R.;	95 BiomtrcJ 37 393-408
• Muotka, T.; Penttinen, A. Detecting small-scale	spatial patterns in lotic predator-prey relationships: Statistical methods and a case study	94 JFishRBC 51 2210-2218
Comment on "A simple test for	spatial pattern in regional health data" (V13 p1037-1044) • Lai, Dejian	95 StatMed 14 1393-1393
Reply to comment on "A simple test for	spatial pattern in regional health data" • Walter, S. D.	95 StatMed 14 1393-1394
Chessel, D.; Champely, S. Multivariate analysis of	spatial patterns: A unified approach to local and global structures • Thioulouse, J.;	95 EnvEcoSt 2 1-14
Fixed-domain asymptotics for	spatial periodgrams • Stein, Michael L.	95 JASA 90 1277-1288
An approach to asymptotic inference for	spatial point processes • Stein, Michael L.	95 StSinica 5 221-234
Nonstationary multivariate	spatial prediction • Haas, Timothy C.	94 ASAProEnvr 194-203
prediction models: A comparison ... • Lesch et al.	Spatial prediction of soil salinity using electromagnetic induction techniques 1. Statistical	95 WaterReR 31 373-386
efficient spatial sampling algorithm ... • Lesch et al.	Spatial prediction of soil salinity using electromagnetic induction techniques. 2. An	95 WaterReR 31 387-398
Mosler, C.; Hary, I.; Pielert, V. Factors affecting	spatial preferences in settlement site selection in migratory pastoralism • Schwartz, H. J.;	95 EnvrMtrc 6 485-490
	Spatial process	
	A class of harmonizable isotropic random fields • Swift, Randall J.	95 JCISS 20 111-127
	Hybrid zones and voter model interfaces • Cox, J. T.; Durrett, R.	95 Bernoull 1 343-370
The detection of change in	spatial processes • Martin, E. B.; Scott, E. M.	95 EnvrMtrc 6 141-154
A simulation technique of a non-Gaussian	spatial process • De Cesare, L.; Posa, D.	95 CmpStDA 20 543-555
Multivariate	spatial sign and rank methods • Möttönen, Jyrki; Oja, Hannu	95 JNnparSt 5 201-213
timating temporal farm income distributions using	spatial smoothing techniques • Kokic, P. N.; Moon, L.; Gooday, J.; Chambers, R. L. Es-	95 AstrlJSt 37 129-143
Host dispersal and the	spatial spread of insect pathogens • Dwyer, Greg; Elkinton, Joseph S.	95 Ecology 76 1262-1275
Martuzzi, Marco; Shaddick, Gavin	Spatial statistical methods in environmental epidemiology: A critique • Elliott, Paul;	95 StMedRes 4 137-159
stant computational simplifications for estimating	spatial statistical models • Griffith, D. A. et al. Trade-offs associated with normalizing con-	95 JStCmpSm 51 165-183
	Spatial statistics	
	Area-interaction point processes • Baddeley, A. J.; van Lieshout, M. N. M.	95 AnInStMa 47 601-619
Green, Peter; Higdon, David; Mengersen, Kerrie	Bayesian computation and stochastic systems (Disc: p41-66) • Besag, Julian;	95 StatSci 10 3-41
Gibbsian processes • Mase, Shigeru	Consistency of the maximum pseudo-likelihood estimator of continuous state space	95 AnlsApPr 5 603-612
Pitt, Loren D.; Robeva, Raina; Wang, Dao Yi	An error analysis for the numerical calculation of certain random integrals: Part 1 •	95 AnlsApPr 5 171-197
	Quasi-linear geostatistical theory for inversing • Kitanidis, Peter K.	95 WaterReR 31 2411-2419
Practical handbook of	spatial statistics • Arlinghaus, Sandra L.; Griffith, Daniel A. (Ed)	95 CRCPress:FL 336p
Jöckel, K.-H.	Spatial-temporal analysis of mortality using splines • van der Linde, A.; Witzko, K.-H.;	95 Biomtrcs 51 1352-1360
of the seed bank of Avena sterilis: The effects of	spatial and temporal heterogeneity and of dispersal • González-Andujar et al. ... control	95 JAppEcol 32 578-587
Johnson, G. D.; Patil, G. P. Rapid mobilization of	spatial/temporal information in the context of natural catastrophes • Myers, W. L.;	94 ASAProStGr 25-31
B.; Sugihara, K. (Auth); Lorz, U. (Rev) Review of	"Spatial tessellations. Concepts and applications of Voronoi diagrams" • Okabe, A.; Boots,	95 MaOpfStS 26 183-183
A geometric approach to	spatial time series analysis (Italian) • Benedetti, Roberto	94 QStMaA 16 27-43
Optimal sampling designs for dependent	spatial units • Benedetti, Roberto; Palma, Daniela	95 EnvrMtrc 6 101-114
Selection techniques in sampling	spatial units • Arbia, Giuseppe	94 QStMaA 16 81-91
	Spatial variability	
	Statistics for environmental change • Barnett, Vic; Riley, Janet	95 ExpAgric 31 117-130
Britain • Richardson, S.; Monfort, C.; Green, M. et al.	Spatial variation of natural radiation and childhood leukaemia incidence in Great	95 StatMed 14 2487-2501
Non-parametric estimation of	spatial variation in relative risk • Kelsall, Julia E.; Diggle, Peter J.	95 StatMed 14 2335-2342
Rainfall discrimination and	spatial variation using breakpoint data • Sansom, John	95 JClimate 8 624-636
Baenziger, P. Stephen; Mulitze, Dieter K. Removing	spatial variation from wheat yield trials: A comparison of methods • Stroup, Walter W.;	94 CropSci 34 62-66
A generalized estimating equations approach for	spatially correlated binary data: Applications to the analysis of ... • Albert et al.	95 Biomtrcs 51 627-638
results on randomization when observations are	spatially dependent and the analysis uses generalized least-squares • Martin, R. J. Some	95 CommStA 24 2199-2210
deviations for particle systems associated with	spatially homogeneous Boltzmann type equations • Léonard, C. On large	95 ProbTher 101 1-44
ity model • Sampson, P. D.; Guttorp, P.; Meiring, W.	Spatio-temporal analysis of regional ozone data for operational evaluation of an air qual-	94 ASAProEnvr 46-55
Local prediction of a	spatio-temporal process with an application to wet sulfate deposition • Haas, Timothy C.	95 JASA 90 1189-1199
Imputation of missing values in	spatio-temporal solar radiation data • Glasbey, C. A.	95 EnvrMtrc 6 363-371
Spot check your	SPC applications • Glaccum, John F.	95 QulPr 28/5 184-184
Multivariate	SPC charts for monitoring batch processes • Nomikos, Paul; MacGregor, John F.	95 Technmcs 37 41-59
Review of	"SPC in extrusion" • Rauwendaal, Chris (Auth); Ziegel, Eric R. (Rev)	95 Technmcs 37 241-241
the pieces together: Integrated implementation of	SPC • Ayers, Christopher V. Putting	94 ASAProPEng 291-294
Benneyan, J. C.; Kaminsky, F. C. Successfully applying	SPC to improve health care: Barriers to improving quality and reducing liability •	95 ASQCTekCon 570-578
Review of	"SPC made easy" • McReynolds, Peter (Auth); Ziegel, Eric R. (Rev)	95 Technmcs 37 462-463
	SPC: A new look at what to teach • Greer, Daniel A.; Halteman, Edward J.	94 ASAProBays 153-156
Quesenberry, C. P. Correction to	"SPC Q charts for start-up processes and short or long runs" (V23 p213-224) •	95 JQualTek 27 402-402
Review of "Statistical methods for	SPC and TQM" • Bissell, D. (Auth); Davies, Neville (Rev)	95 JRSS-A 158 341-342
Review of "Statistical methods for	SPC and TQM" • Bissell, D. (Auth); Sheil, J. (Rev)	95 Ststcian 44 279-281
Using	SPC in two Mexican poultry processing plants • Elizundia, José M.; Hau, Guadalupe	95 ASQCTekCon 579-587
Charles	Spearman's contributions to test theory • Levy, Philip	95 BJMaStPs 48 221-235
Tejedor, I. Herranz; Mato, A. Silva The Wilcoxon,	Spearman, Fisher, χ^2, Student and Pearson tests and 2×2 tables • Martín Andrés, A.;	95 Ststcian 44 441-450
	Spearman and the origin and development of factor analysis • Bartholomew, D. J.	95 BJMaStPs 48 211-220
Who discovered	Spearman's rank correlation? • Lovie, A. D.	95 BJMaStPs 48 255-269
	Spearman's rho	
hypotheses • Rice, William R.; Gaines, Steven D.	Extending nondirectional heterogeneity tests to evaluate simply ordered alternative	94 PNatAcSc 91 225-226
The cold equations:	Spearman and Wilson on factor indeterminacy • Lovie, P.; Lovie, A. D.	95 BJMaStPs 48 237-253
Darwin's log: A toy model of	speciation and extinction • Aldous, David	95 JAppProb 32 279-295
ponents of testcross progenies in an autotetraploid	species and consequences for recurrent selection with a tester • Gallais ... variance com-	92 ThApGene 83 353-359
A note on the analysis of	species co-occurrences • Manly, Bryan F. J.	95 Ecology 76 1109-1115
Optimal control of two competing diseases or	species • Kyriakidis, E. G.	95 MaScitst 20 56-66
	Species diversity in space and time • Rosenzweig, Michael L.	95 CambridgeUniv 456p
gas chromatographic data of oils from Eucalyptus	species • Dunlop, P. J.; Bignell, C. M.; Jackson, J. F.; Hibbert, D. B. Chemometric analysis of	95 Chemolab 30 59-67
An extreme-value function model of the	species incidence and species-area relations • Williams, Matthew R.	95 Ecology 76 2607-2616
subdivision to reconstruct gene trees within a	species: Molecular variance parsimony • Excoffier et al. ... frequencies and geographic	94 Genetcs 136 343-359

general approximation test for the main effect in a	split-plot design • Algina, James; Oshima, T. C. An improved	95 BJMaStPs 48 149-160
Linear graphs for 2^{k-p} fractional factorial under	split plot design • Aggarwal, M. L.; Chowdhury, S. Roy	94 JStatRsr 28 73-87
Some	split-plot designs in series of experiments • Mejza, Iwona	94 ListyBio 31 3-16
comparison of inference procedures in unbalanced	split-plot designs • Remmenga, Marta D.; Johnson, Dallas E. A	95 JStCmpSm 51 353-367
mixed linear model when the experiment is of a	split-plot type • Lee, K. R.; Gagnon, R. C. Calibration and stability analysis with a simple	94 ASAProBiop 435-440
A	split questionnaire survey design • Raghunathan, Trivellore E.; Grizzle, James E.	95 JASA 90 55-63
Joshua D.; Krueger, Alan B.	Split-sample instrumental variables estimates of the return to schooling • Angrist,	95 JBES 13/2 225-235
Three-level skip-lot sampling plan	split by two stages (Korean) • Choi, Byoung-Chul; Lee, Eun-joo	95 KorApS 8/2 55-64
Loss of efficiency in	spoilt experiments • Hunt, D. N.; Li, E.	95 JAppStat 22 309-315
Appel, Martin V.; Cole, Ronald	Spoken language recognition of the Year 2000 Census Questionnaire Feasibility Test •	94 ASASrvyRsr 1310-1315
generalized least-squares estimate for the origin of	sporophytic self-incompatibility • Uyenoyama, March K. A	95 Genetcs 139 975-992
	Sports	
Schutz, Robert W.	Analysis of the 1986 and 1994 World Cup soccer tournament • McGarry, Tim;	94 ASAProSpts 61-65
	On the application of logistic regression on ski accident data • Traweger, Christian M.	94 ASAProSpts 12-15
Korean pro baseball data (Korean) • Choi, Y.-S. et al.	Applications of the supplementary principal component analysis for the 1982-1992	95 KorApS 8/1 51-60
Lock, Robin H.	CHODR – Using statistics to predict college hockey • Danehy, Timothy J.;	95 Stats /13 10-14
	Comments on modelling baseball • Lindsey, George	94 ASAProSpts 21-24
Velleman, Paul F.	A critical look at some analyses of major league baseball salaries • Hoaglin, David C.;	95 AmerStat 49 277-285
Dobson, S. M.; Goddard, J. A.	The demand for professional league football in England and Wales, 1925-92 •	95 Ststcian 44 259-277
Collings, Bruce Jay	An evaluation of three methods for ranking paired contestants • Blake, Jonathan E.;	94 ASAProSpts 31-35
	Figure skating: What is a 5.8 really? • Russell, Edmund L., III	94 ASAProSpts 25-30
	Forecasting baseball games • Kaigh, W. D.	95 Chance 8/2 33-37
	Gender and performance in athletics • Chatterjee, Sangit; Laudato, Matthew	94 ASAProSpts 36-41
	The great streak • Warrack, Giles	95 Chance 8/3 41-43
	The GRIP: A performance statistic for tournament golf • Agard, David B.	94 ASAProSpts 1-6
Norman, John M.	Home ground advantage of individual clubs in English soccer • Clarke, Stephen R.;	95 Ststcian 44 509-521
	May the best man win? • Appleton, David R.	95 Ststcian 44 529-538
of draw percentages among players • Kuk, A. Y. C.	Modelling paired comparison data with large numbers of draws and large variability	95 Ststcian 44 523-528
Bennett, Jay M.	MVP, LVP, and PGP: A statistical analysis of Toronto in the World Series •	94 ASAProSpts 66-71
	Who's number 1 in college football? ... And how might we decide? • Stern, Hal S.	95 Chance 8/3 7-14
Myers, Casey; Rasp, John	On the odds of an extended hitting streak • Coelho, Robby; Grunow, Bryan;	94 ASAProSpts 46-49
Liu, Yuanlong; Schutz, Robert W.	Overtime in the National Hockey League: Is it a valid tie-breaking procedure? •	94 ASAProSpts 55-60
	Overtime or shootout: Deciding ties in hockey • Hurley, William	95 Chance 8/1 19-22
	Performance indices for on-ice hockey statistics • Williams, Bill	94 ASAProSpts 50-54
	Probability models on horse race outcomes • Ali, Mukhtar M.	94 ASAProSpts 7-11
	Replaying the 1994 Wimbledon men's singles final • Croucher, John S.	95 NZndSt 30 2-8
and evaluation for golf course closing • Shipley et al.	Rough set theory fuzzy belief functions related to statistical confidence: Application	95 StocAnAp 13 487-502
	Scoring patterns in rugby league • Croucher, John	95 TeachgSt 17/2 47-49
	Simpson's paradox and the hot hand in basketball • Wardrop, Robert L.	95 AmerStat 49 24-28
	Are the Toronto Blue Jays really the world champions? • Umholtz, Robert L.	94 ASAProSpts 42-45
Gitlow, Howard S.	Using control charts to corroborate bribery in jai alai • Charnes, John M.;	95 AmerStat 49 386-389
P.; Bell, Robert M.	How will the NCAA's new standards affect minority student-athletes? • Klein, Stephen	95 Chance 8/3 18-21
The systematic study of "statistics in	sports": Do we need a framework? • Schutz, Robert W.	94 ASAProSpts 16-20
Tennis in Lilliput: A fable on	sports and psychology • Jackson, David	95 Chance 8/3 37-40
	Spot check your SPC applications • Glaccum, John F.	95 QulPr 28/5 184-184
Lead-lag relationship between	spot index and futures price of the Nikkei Stock Average • Tse, Y. K.	95 JForcast 14 553-563
tests for variability based on interquantile	spreads • Shoemaker, Lewis H. Two-sample	94 ASAProQlPr 134-139
Pension funding with time delays and the optimal	spread period • Haberman, Steven	95 AstinB 25 177-187
Kee Transformations for stabilizing	spread in a time series model: A comparative study • Attanasio, Massimo; Wong, Weng	95 JStCmpSm 52 55-70
surprises on stock prices, trading volumes, and	spreads • Maddala et al. ... component panel data model to study the effect of earnings	95 JEconmtx 68 229-242
	Spreadsheet analysis for engineers and scientists • Bloch, Sylvan C.	95 Wiley:NY:UK 336p
of "Understanding social science statistics: A	spreadsheet approach" • Bakeman, R. (Auth); Goda, David (Rev) Review	95 Ststcian 44 539-540
Probability, intuition and a	spreadsheet • Graham, Alan	95 TeachgSt 17/3 84-87
Hoogerbrugge, Ronald A user-friendly	spreadsheet program for calibration using weighted regression • Gort, Steven M.;	95 Chemolab 28 193-199
Statistical issues in	spreadsheet software • Nash, John C.; Quon, Tony K.; Gianini, Jacqueline	94 ASAProStEd 238-241
Forecasting	spring flow time series • Cascini, Egidio; Cascini, Leonardo	94 JItalStS 3 1-23
On the use of an	SPSA-based model-free controller in quality improvement • Rezayat, Fahimeh	95 Autmatca 31 913-915
A note on beta coefficients in	SPSS • Kockläuner, Gerhard	95 AllgeStA 79 115-116
Review of	"SPSS/PC+ made simple" • Kinnear, P. R.; Gray, C. D. (Auth); Carpenter, James (Rev)	95 JAppStat 22 436-437
Foster, J. J. (Auth); Hawkins, A. (Rev) Review of	"SPSS/PC+ and SPSS for Windows – A beginner's guide to data analysis (Second edition)" •	95 TeachgSt 17/2 79-79
Review of	"SPSS for Windows made simple" • Kinnear, P. R.; Gray, C. D. (Auth); Carpenter, James (Rev)	95 JAppStat 22 436-437
	Spurious break • Nunes, Luis C.; Kuan, Chung-Ming; Newbold, Paul	95 EconTher 11 736-749
McShane, Phil Comment on	"Spurious correlation and the fallacy of the ratio standard revisited" (V156 p379-392) •	95 JRSS-A 158 621-623
Nevill, Alan M.; Holder, Roger L. Comment on	"Spurious correlation and the fallacy of the ratio standard revisited" (V156 p379-392) •	95 JRSS-A 158 619-621
Reply to comments on	"Spurious correlation and the fallacy of the ratio standard revisited" • Kronmal, Richard A.	95 JRSS-A 158 623-625
	Spurious regressions between I(d) processes • Marmol, Francesc	95 JTimSrAn 16 313-321
The law of large numbers and	$\sqrt{2}$ • Liggett, Thomas M.; Petersen, Peter	95 AmMaMnth 102 31-35
	Square	
	• see chi –, Latin –, least –, mean –, Youden –	
bivariate symmetry against ordered alternatives in a	square contingency table • Barmi, H. E.; Kochar, S. C. Likelihood ratio tests for	95 StPrLet 22 167-173
bilinear models for the analysis of asymmetry in a	square contingency table • van der Heijden, Peter G. M.; Mooijaart, Ab Some new log-	95 SocMethR 24 7-29
of the marginal homogeneity model for	square contingency tables with ordered categories • Tomizawa, Sadao A generalization	95 JEdStat 20 349-360
Measures of departure from global symmetry for	square contingency tables with ordered categories • Tomizawa, Sadao	95 Behvmtrk 22 91-98
Consistency of the least median of	squares estimator in nonlinear regression • Stromberg, Arnold J.	95 CommStA 24 1971-1984
The asymptotic behavior of locally	square integrable martingales • Wang, Jia-gang	95 AnlsProb 23 552-585
Papantoni-Kazakos, P. New nonleast-	squares neural network learning algorithms for hypothesis testing • Pados, Dimitris A.;	95 IEEENrlN 6 596-609
The determination of a "least quantile of	squares regression line" for all quantiles • Carrizosa, Emilio; Plastria, Frank	95 CmpStDA 20 467-479
New	square-root algorithms for Kalman filtering • Park, PooGyeon; Kailath, Thomas	95 IEEEAuCn 40 895-899
	Square-root RTS smoothing algorithms • Park, PooGyeon; Kailath, Thomas	95 IJContrl 62 1049-1060
The largest	square of successes covering the origin for Bernoulli trials on the lattice • Kopociński, B.	95 StNeerla 49 31-39
Analysis of	square tables with ordered categories • Poon, Wai-Yin; Hung, Hin-Yan	94 ASAProStCp 292-298
Loss robustness via Fisher-weighted	squared-error loss function • Makov, Udi E.	95 InsrneME 16 1-6
rameter of a gamma distribution for scale-invariant	squared-error loss • van Eeden, C. Minimax estimation of a lower-bounded scale pa-	95 CandJSt 23 245-256
for conditional heteroscedasticity, using ranks of	squared residuals • Wong, H.; Li, W. K. Portmanteau test	95 JAppStat 22 121-134

	Stacked boxes: A new goodness-of-fit graphical display • Sparks, Ross	95 NZndSt 30 48-61
Properties of the Schools and	Staffing Survey's bootstrap variance estimator (Disc: p1128-1129) • Kaufman, Steven	94 ASASrvyRsr 1116-1121
	Stage	
	• see two -, three -, multi -	
itoring carcinoembryonic antigen (CEA) levels for	stage IV breast cancer patients • Lao ... relative risk of first cancer progression by mon-	94 ASAProBiop 206-210
Gene flow technique for	stage-structured populations • Muralidharan, K.; Jain, J. P.	95 BiomtrcJ 37 229-234
A	stage-theory model of cognitive development • Preece, Peter F. W.; Read, Kenneth L. Q.	95 BJMaStPs 48 1-7
Statistical tests for random effects in	staggered nested designs • Khattree, Ravindra; Naik, Dayanand N.	95 JAppStat 22 495-505
of true treatment response when patients enter	staggeredly in a clinical trial • Biswas, Suddhendu; Chitkara, Sunita On the estimation	95 BiomtrcJ 37 571-580
The behavior of the	Stahel-Donoho robust multivariate estimator • Maronna, Ricardo A.; Yohai, Victor J.	95 JASA 90 330-341
census ... • Dillman et al. How prenotice letters,	stamped return envelopes and reminder postcards affect mailback response rates for	95 SurvyM 21 159-165
On model selection for	standard curve in assay development • Tse, Siu-Keung; Chow, Shein-Chung	95 JBiophSt 5 285-296
Using nested designs: I. Estimation of	standard deviations • Nelson, Lloyd S.	95 JQualTek 27 169-171
Using nested designs: II. Confidence limits for	standard deviations • Nelson, Lloyd S.	95 JQualTek 27 265-267
to "Using nested designs: II. Confidence limits for	standard deviations" (V27 p265-265) • Nelson, L. S. Correction	95 JQualTek 27 403-403
Adams, Benjamin M. Teaching the	standard deviations of linear combinations using graphics • Mansfield, Edward R.;	94 ASAProStEd 80-83
• Balakrishnan, N. et al. Estimation of the mean and	standard deviation of the logistic distribution based on multiply type-II censored samples	95 MaOpfStS 27 127-142
Asymptotic	standard error of equipercentile equating • Liou, Michelle; Cheng, Philip E.	95 JEdStat 20 259-286
A discrete delta method for estimating	standard errors for estimators in quantal bioassay • Cobb, E. Benton; Church, J. D.	95 BiomtrcJ 37 691-699
D.; Sava-Bolesta, Monica An investigation of the	standard errors of expected a posteriori ability estimates • De Ayala, R. J.; Schafer, William	95 BJMaStPs 48 385-405
Cope, Ronald T.	Standard error of linear equating for the counterbalanced design • Zeng, Lingjia;	95 JEdStat 20 337-348
Knaub, James R., Jr Relative	standard error for a ratio of variables at an aggregate level under model sampling •	94 ASASrvyRsr 310-312
Statistical	standards: A global perspective • Schilling, Edward G.	94 JOffStat 10 339-346
A Bayesian study of the correlation between	standard scores of twins • Li, Jian-Hua; Watanabe, Hiroshi	94 Behvmtrk 21 121-130
Direct	standardisation in multi wavelength fluorescence spectroscopy • Nørgaard, Lars	95 Chemolab 29 283-293
Methods for maternal age-	standardization of the incidence of congenital abnormalities • Carothers, Andrew D.	95 StatMed 14 1797-1806
to archaeometry • Baxter, M. J.	Standardization and transformation in principal component analysis, with applications	95 ApplStat 44 513-527
and capture parameters of resting passerines in	standardized capture-recapture studies • Kaiser, A. Estimating turnover, movements	95 JAppStat 22 1039-1047
ranging trials: Guidelines for data collection and	standardized descriptions • Boissel, J.-P.; Durieu, I.; Girard, P.; Nony, P. et al. Dose-	95 ConClnTr 16 319-330
Interpreting the	standardized difference • Tritchler, David	95 Biomtrcs 51 351-353
Greenland, Sander	Standardized estimates from categorical regression models • Joffe, Marshall M.;	95 StatMed 14 2131-2141
to "Updated version of the critical values of the	standardized fourth moment" (V44 p231-241) • Ramsey, P. H.; Ramsey, P. P. Correction	95 JStCmpSm 53 127-127
Preference of equivalence tests with	standardized mean difference demonstrated by an application • Schuster, Ernst	95 CmpStDA 19 353-356
Comparing	standardized mortality ratios • Armstrong, Ben G.	95 AnEpidm 5 60-64
Understanding the	standardized/non-standardized interviewing controversy • Beatty, Paul	95 JOffStat 11 147-160
Simple confidence intervals for	standardized rates based on the approximate bootstrap method • Swift, Michael B.	95 StatMed 14 1875-1888
able with limited data • Rosenbaum et al. Correcting	standardized rate ratios for imprecise classification of a polychotomous exposure vari-	95 AJEpidm 141 442-445
able with limited data • Rosenbaum et al. Correcting	standardized rate ratios for imprecise classification of a polychotomous exposure vari-	95 AJEpidm 142 442-445
weighting approximation for the distribution of	standardized sample mean-multi dimensional case (Chinese) • Shi, Jian Random	95 ChJApPSt 10 233-242
Reliability analysis of a complex	standby redundant system • Subramanian, R.; Anantharaman, V.	95 RelEngSS 48 57-70
An optimal inspection-repair-replacement policy for	standby systems • Yeh, Lam	95 JAppProb 32 212-223
Community intervention trials: Reflections on the	Stanford Five-City Project experience • Fortmann, S. P.; Flora, J. A.; Winkleby, M. A. et al.	95 AJEpidm 142 576-586
networks under diverse routing: The symmetric	star network • Hunt, P. J. Loss	95 AdvAppPr 27 255-272
Dynamic asymptotic results for a generalized	star-shaped loss network • Graham, Carl; Méléard, Sylvie	95 AnlsApPr 5 666-680
protocol for optical fiber local area networks with	star topology • Ahmed, N. U.; Yeap, T. H. Performance analysis of slotted ALOHA	95 StocAnAp 13 389-410
niques to detect balanced space-to-time ratios in	STARMA models • Otranto, Edoardo; Gallo, Giampiero M. Regression diagnostic tech-	94 Metron 52/3 129-145
Correction to "SPC Q charts for	start-up processes and short or long runs" (V23 p213-224) • Quesenberry, C. P.	95 JQualTek 27 402-402
Carling, Kenneth	Starting values for the iterative maximum likelihood estimator in survival analysis •	95 JAppStat 22 531-535
Interactive statistics tutorials in	Stata (jse/v3n3/ferrall) • Ferrall, Christopher	95 JStEd 3/3 28kb
	State	
	• see finite -, steady -	
Identifying	state dependence in duration models (Disc: p17-19) • McCall, Brian P.	94 ASAProBuEc 14-16
Liptser, R.; Shenfild, M.	State-dependent Beneš buffer model with fast loading and output rates • Kogan, Y.;	95 AnlsApPr 5 97-120
Random record processes and	state dependent thinning • Browne, Sid; Bunge, John	95 StocProc 55 131-142
supplement • Malec, Donald; Parsons, Van L.	State estimates for the National Health Interview Survey using a random digit dialing	94 ASASrvyRsr 1190-1195
Optimal measurement scheduling for	state estimation • Shakeri, M.; Pattipati, K. R.; Kleinman, D. L.	95 IEEEAES 31 716-729
there a place in ASA for statisticians who work for	state/provincial or local governments? • Gaines, Leonard M.; González, María E. Is	94 ASAProGovt 149-152
	State space	
Rajarshi, M. B.	Filtering and smoothing via estimating functions • Naik-Nimbalkar, U. V.;	95 JASA 90 301-306
Schnute, Jon T.; Richards, Laura J.	The influence of error on population estimates from catch-age models •	95 JFishRBC 52 2063-2077
of "Longitudinal data with serial correlation: A	state-space approach" • Jones, R. H. (Auth); Glimm, Ekkehard (Rev) Review	95 BiomtrcJ 37 172-172
of "Longitudinal data with serial correlation: A	state-space approach" • Jones, R. H. (Auth); Laird, N. M. (Rev) Review	95 Metrika 42 68-69
of "Longitudinal data with serial correlation: A	state-space approach" • Jones, R. H. (Auth); Lange, N. (Rev) Review	95 Biomtrcs 51 1193-1194
of "Longitudinal data with serial correlation: A	state-space approach" • Jones, R. H. (Auth); Schlittgen, R. (Rev. German) Review	95 AllgeStA 79 332-333
of "Longitudinal data with serial correlation: A	state-space approach" • Jones, R. H. (Auth); Shumway, Robert H. (Rev) Review	95 JStPlInf 47 395-397
the index of industrial production • Patterson, K. D. A	state space approach to forecasting the final vintage of revised data with an application to	95 JForcast 14 337-350
final vintage of real personal disposable income: A	state space approach • Patterson, K. D. Forecasting the	95 IntJFor 11 395-405
policies in a Markov decision process with a Borel	state space • Yushkevich, A. A. Blackwell optimal	94 ZeitOpRs 40 253-288
of conditioned processes on a countable	state space • Jacka, S. D.; Roberts, G. O. Weak convergence	95 JAppProb 32 902-916
The number of visits to a subset of the	state space by a discrete-parameter semi-Markov process • Csenki, Attila	95 StPrLet 22 71-77
for non-homogeneous Markov chains in general	state spaces • Fleischer, I.; Joffe, A. Ratio ergodicity	95 JTherPrb 8 31-37
pseudo-likelihood estimator of continuous	state space Gibbsian processes • Mase, Shigeru Consistency of the maximum	95 AnlsApPr 5 603-612
Mallants, Dirk; Feyen, Jan	State space identification of linear deterministic rainfall-runoff models • Ramos, José;	95 WaterReR 31 1519-1531
income and real interest rates from alternative	state space models • Vinod, H. D.; Basu, Parantap Forecasting consumption,	95 IntJFor 11 217-231
A nonlinear	state-space model of Chinook salmon population dynamics • Rein, Steven R.	94 ASAProEnvr 150-155
discrimination and Bayes factors for linear Gaussian	state space models • Frühwirth-Schnatter, Sylvia Bayesian model	95 JRSS-B 57 237-246
Inference from an autoregressive	state-space model via iterative simulation • Glickman, Mark E.; Stern, Hal S.	94 ASAProSpts 72-76
Combining VAR estimation and	state space model reduction for simple good predictions • Gilbert, Paul D.	95 JForcast 14 229-250
specification tests for balanced representation	state space models • Dorfman, Jeffrey H.; Havenner, Arthur Model	95 CommStA 24 97-119
New smoothing results for	state space models • Tsimikas, John; Ledolter, Johannes	94 ASAProBuEc 172-177
Nieto et al. Kalman filter for singular and conditional	state-space models when the system state and the observational error are correlated •	95 StPrLet 22 303-310
Uncertainties for recursive estimators in nonlinear	state-space models, with applications to epidemiology • Maryak, J. L.; Spall, J. C. et al.	95 Automatca 31 1889-1892
Computational methods for Markov series with large	state spaces, with application to AIDS modeling • Yakowitz, S.	95 MaBisci 127 99-121
A test of the association of a time-dependent	state variable to survival • Cantor, Alan B.	95 CmpBiomd 46 101-105

tations commodity price model with unobservable	state variables • Deaton, Angus; Laroque, Guy Estimating a nonlinear rational expec-	95 JApEcmtx 10 9-40S
Statistical	statesmanship in official statistics • Fellegi, Ivan P.	94 ASAProGovt 7-8
Statistical	statesmanship • Spar, Edward J.	94 ASAProGovt 9-10
What is statistical	statesmanship? • Straf, Miron L.	94 ASAProGovt 11-12
Contributions of statistical	statespersons • Sirken, Monroe G.	94 ASAProGovt 3-6
	Stationarity	
under association • Roussas, George G.	Asymptotic normality of a smooth estimate of a random field distribution function	95 StPrLet 24 77-90
	On the estimation of nonlinear time series models • Hili, Quagnina	95 Stochatx 52 207-226
	Ladder height distributions with marks • Asmussen, Søren; Schmidt, Volker	95 StocProc 58 105-119
Houdré, Christian; Kedem, Benjamin	A note on autocovariance estimation in the presence of discrete spectra •	95 StPrLet 24 1-8
	Residuals based estimators of the covariogram • Grondona, Martin O.; Cressie, Noel	95 MaOpfStS 26 209-218
	Temporal analogues to spatial K functions • Cannon, Ann; Cressie, Noel	95 BiomtrcJ 37 351-373
On time-and cycle-	stationarity • Thorisson, Hermann	95 StocProc 55 183-209
Testing a time series for difference	stationarity • McCabe, B. P. M.; Tremayne, A. R.	95 AnlsStat 23 1015-1028
Testing for trend	stationarity versus difference stationarity • Arellano, Consuelo; Pantula, Sastry G.	95 JTimSrAn 16 147-164
	Stationarity of independent sequences • Lacaze, B.	95 StPrLet 23 9-11
Modified	stationarity tests with improved power in small samples • Breitung, J.	95 StHefte 36 77-95
	Stationary	
	• see also quasi- -, non- -, stochastic process, time series	
Sarkar, Sahadeb Estimation for	stationary AR(1) models with nonconsecutively observed samples • Shin, Dong Wan;	95 SankhyaA 57 287-298
properties of some estimators for partly linear	stationary autoregressive models • Gao, Jiti Asymptotic	95 CommStA 24 2011-2026
Stochastic vector difference equations with	stationary coefficients • Glasserman, Paul; Yao, David D.	95 JAppProb 32 851-866
type theorem á la Feller for nonintegrable strictly	stationary continuous time processes • Rosalsky, A.; Stoyanov, J.; Presnell, B. An ergodic-	95 StocAnAp 13 555-572
on the relationship between the rate functions for	stationary dependent random sequences in τ-topology and the relative ... • Hu A note	95 StocProc 56 351-355
between queue length and waiting time in a	stationary discrete-time queue • Takahashi, Yoshitaka; Miyazawa, Masakiyo Relationship	95 CommStSM 11 249-271
	Stationary distribution	
	Intermediate phase for the contact process on a tree • Durrett, Rick; Schinazi, Rinaldo	95 AnlsProb 23 668-673
Asmussen, Søren	Stationary distributions for fluid flow models with or without Brownian noise •	95 CommStSM 11 21-49
Approximation of the	stationary distribution of a stochastically monotone Markov chain (French) • Simonot, F.	95 StocProc 56 133-149
	Stationary exponential families • Dinwoodie, I. H.	95 AnlsStat 23 327-337
A note on extreme values of locally	stationary Gaussian processes • Hüsler, J.	95 JStPlInf 45 203-213
On large deviations of empirical measures for	stationary Gaussian processes • Bryc, Włodzimierz; Dembo, Amir	95 StocProc 58 23-34
for a signal with unknown location and scale in a	stationary Gaussian random field • Siegmund, David O.; Worsley, Keith J. Testing	95 AnlsStat 23 608-639
Limit theorems for nonlinear functionals of a	stationary Gaussian sequence of vectors • Arcones, Miguel A.	94 AnlsProb 22 2242-2274
central limit theorem for non-linear functionals of	stationary Gaussian vector processes • Sánchez de Naranjo, M. V. A	95 StPrLet 22 223-230
Small ball probabilities for Gaussian processes with	stationary increments under Hölder norms • Kuelbs, J.; Li, W. V.; Shao, Qi-man	95 JTherPrb 8 361-386
probabilities of multidimensional processes with	stationary independent increments • Yamazato, M. Absolute continuity of transition	94 ThProbAp 39 347-354
Finite-dimensional distributions and tail behavior in	stationary interval-valued probability models • Sadrolhefazi, Amir; Fine, Terrence L.	94 AnlsStat 22 1840-1870
state • Pakes, Anthony G. Quasi-	stationary laws for Markov processes: Examples of an always proximate absorbing	95 AdvAppPr 27 120-145
imation for the distribution of the partial sum of a	stationary linear series (Chinese) • Fan, J.; Wang, N.; Mei, C. The random weighting approx-	95 AMaAPknC 17 65-75
A note on	stationary linearly positive quadrant dependent sequences • Kim, Tae-sung	95 JKoreaSt 24 249-256
	Stationary marked point processes: An intuitive approach • Sigman, Karl	95 ChapmnHall:Lnd 200p
Estimation of a	stationary Markov chain • Billard, L.; Meshkani, M. R.	95 JASA 90 307-315
L_1-consistency of kernel-type density estimator for	stationary Markov processes • Mishra, M. N.; Rao, B. L. S. Prakasa On	95 CommStA 24 581-591
test of level-differenced versus log-differenced	stationary models • Pesaran, Bahram; Pesaran, M. Hashem A non-nested	95 EcomtrRv 14 213-227
Kipnis, C. et al. Macroscopic properties of a	stationary non-equilibrium distribution for a non-gradient interacting particle system •	95 AlHenrPB 31 191-221
Torrés, Olivier Some exact inference procedures for	stationary and nonstationary autoregressive processes • Dufour, Jean-Marie;	94 ASAProBuEc 135-140
	Stationary parabolic Anderson model and intermittency • Carmona, R. A.; Molchanov, S. A.	95 ProbTher 102 433-453
Strong theorems on the extreme values of	stationary Poisson processes • Révész, P.	95 JStPlInf 45 291-300
Statistical inference in	stationary populations • Nagelkerke, Nico J. D.	95 Biomtrcs 51 347-350
systems • Yeung, David W. K.; Stewart, Sally E. A.	Stationary probability distributions of some Lotka-Volterra types of stochastic predation	95 StocAnAp 13 503-516
	Stationary process	
	The aftermath of Cramér's work on stochastic processes • Masani, P. R.	95 ScanActJ 66-96
	Convolutions of random measures on compact groups • Rubshtein, Ben-Zion	95 JTherPrb 8 523-538
	Forecasting growth with time series models • Peña, Daniel	95 JForcast 14 97-105
	Geometric Markov chains • Rising, William	95 JAppProb 32 349-374
	Hidden Markov random fields • Künsch, Hans; Geman, Stuart; Kehagias, Athanasios	95 AnlsApPr 5 577-602
	Mixture models for time series • Jalali, Assad; Pemberton, John	95 JAppProb 32 123-138
Taniguchi, Masanobu	Nonparametric approach for discriminant analysis in time series • Zhang, Guoqiang;	95 JNnparSt 5 91-101
Miyazawa, Masakiyo	Note on generalizations of Mecke's formula and extensions of $H = \lambda G$ •	95 JAppProb 32 105-122
Weiss, Benjamin	Perfect filtering and double disjointness • Furstenberg, Hillel; Peres, Yuval;	95 AlHenrPB 31 453-465
	Phase transitions of some non-linear stochastic models • Feng, Shui	95 JAppProb 32 193-201
Rosenthal, Jeffrey S.	Rates of convergence for everywhere-positive Markov chains • Baxter, J. R.;	95 StPrLet 22 333-338
Stone, Charles J.; Truong, Young K.	Rate of convergence for logspline spectral density estimation • Kooperberg, Charles;	95 JTimSrAn 16 389-401
	Stationary exponential families • Dinwoodie, I. H.	95 AnlsStat 23 327-337
A trimmed mean of location of an AR(∞)	stationary process • Lee, Sangyeol	95 JStPlInf 48 131-140
On permissible correlations for locally correlated	stationary processes • Donahue, Rafe M. J.; Brockwell, Peter J.; Davis, Richard	95 StPrLet 22 49-53
Markov-additive processes and strong-mixing	stationary processes • Zajic, Tim Large exceedances for uniformly recurrent	95 JAppProb 32 679-691
limit theorem and law of the iterated logarithm for	stationary processes with applications to linear processes • Yokoyama, R. On the central	95 StocProc 59 343-351
ties of the sample variance and autocovariance for	stationary processes with unknown mean" (V47 p274-276) • Anderson ... curious proper-	95 AmerStat 49 118-118
On a stopping rule for the random walks with time	stationary random distribution function • Hong, Dug Hun; Oh, Kwang Sik; Park, Hee Joo	95 JKoreaSt 24 293-301
A central limit theorem for	stationary random fields • Janž, M.; Lachout, P.	95 MaMethSt 4 463-471
of the sum and rare values of weakly dependent	stationary random variables • Hsing, Tailen On the asymptotic independence	95 StocProc 60 49-63
sequences of s by s independent exchangeable or	stationary random variables (French) • Bretagnolle, J.; Klopotowski, A. On the existence of	95 AlHenrPB 31 325-350
of the sum and maximum of strongly mixing	stationary random variables • Hsing, Tailen A note on the asymptotic independence	95 AnlsProb 23 938-947
	Stationary regimes for inventory processes • Bardhan, Indrajit; Sigman, Karl	95 StocProc 56 77-86
On the almost sure convergence of series of	stationary and related nonstationary variables • Houdré, Christian	95 AnlsProb 23 1204-1218
Monotonicity of positive dependence with time for	stationary reversible Markov chains • Hu, Taizhong; Joe, Harry	95 PrEngInf 9 227-237
Prediction and non-Gaussian autoregressive	stationary sequences • Rosenblatt, Murray	95 AnlsApPr 5 239-247
index and clustering of high values for derived	stationary sequences • Weissman, Ishay; Cohen, Uri The extremal	95 JAppProb 32 972-981
Truong-Van, B. Invariance principles for semi-	stationary sequence of linear processes and applications to ARMA process •	95 StocProc 58 155-172
Extreme sums of strictly	stationary sequences of M-dependent variables • Flak, T.; Schmid, W.	95 SankhyaA 57 186-201
bootstrap for general empirical processes of	stationary sequences • Bühlmann, Peter The blockwise	95 StocProc 58 247-265
Sums and maxima of	stationary sequences with heavy tailed distributions • Anderson, C. W.; Turkman, K. F.	95 SankhyaA 57 1-10

	Stochastic approximation (cont.)	
Hu, Jian-Qiang	Sensitivity analysis for Monte Carlo simulation of option pricing • Fu, Michael C.;	95 PrEngInf 9 417-446
algorithms • Kushner, Harold J.; Yang, Jichuan	Stochastic approximation with averaging and feedback: Rapidly convergent "on-line"	95 IEEEAuCn 40 24-34
Rezayat, Fahimeh	On the use of an SPSA-based model-free controller in quality improvement •	95 Automatca 31 913-915
R. On equivalence of some noise conditions for	stochastic approximation algorithms • Wang, I-Jeng; Chong, Edwin K. P.; Kulkarni, Sanjeev	95 PIEEEDecCt 3849-3854
Some results about averaging in	stochastic approximation • Le Breton, Alain; Novikov, Alexander	95 Metrika 42 153-171
process • Komarov, S. V.; Krasulina, T. P. Using	stochastic approximation to find the extinction probability of a Galton-Watson branching	94 VLenUM /3 34-39
Non-standard limit theorems for urn models and	stochastic approximation procedures • Kaniovski, Yu.; Pflug, G.	95 CommStSM 11 79-102
algorithms • Kushner, Harold J.; Yang, Jichuan	Stochastic approximation with averaging and feedback: Rapidly convergent "on-line"	95 IEEEAuCn 40 24-34
	Stochastic calculus	
	Differential calculus relative to some point processes • Dermoune, A.	95 StPrLet 24 233-242
	A simplified exposition of the Health, Jarrow and Morton model • Carverhill, Andrew	95 Stochatx 53 227-240
Green formulas in anticipating	stochastic calculus • Delgado, Rosario; Sanz-Solé, Marta	95 StocProc 57 113-148
quantum measurement processes via classical	stochastic calculus • Barchielli, A.; Holevo, A. S. Constructing	95 StocProc 58 293-317
From discrete to continuous	stochastic calculus • Cutland, Nigel J.; Kopp, P. Ekkehard; Willinger, Walter	95 Stochatx 52 173-192
Introduction to	stochastic calculus • Lawler, G. F.	95 ChapmnHall:Lnd 176p
Review of "Introduction to	stochastic calculus" • Lawler, G. F. (Auth); Riddler-Rowe, C. J. (Rev)	96 ShrtBkRv 16/1 8-8
The Beneš equation and	stochastic calculus of variations • Decreusefond, L.; Üstünel, A. S.	95 StocProc 57 273-284
Intermediate sums and	stochastic compactness of maxima • Csörgő, Sándor; Mason, David M.	95 JStPlInf 45 81-90
	Stochastic control	
	Adaptive control of partially observed linear stochastic systems • Bertrand, Pierre	95 Stochatx 54 21-51
	Controlled trees • Ceci, C.; Mazliak, L.	94 RBrzlPrS 8 93-105
	Dynamic approaches to pension funding • Haberman, Steven; Sung, Joo-Ho	94 InsrneME 15 151-162
	Optimal choice of production capacity in a random market • Arntzen, Halvard	95 Stochatx 55 87-120
problem • Agrawal, Rajeev	Sample mean based index policies with $O(\log n)$ regret for the multi-armed bandit	95 AdvAppPr 27 1054-1078
necessary conditions for partially observed optimal	stochastic controls • Li, Xunjing; Tang, Shanjian General	95 JAppProb 32 1118-1137
Derivation of optimality conditions for a	stochastic control problem • Ladde, G. S.; Medhin, N. G.	95 StocAnAp 13 165-176
Lepeltier, J. P. Backward equations,	stochastic control and zero-sum stochastic differential games • Hamadéne, S.;	95 Stochatx 54 221-231
	Stochastic decomposition for retrial queues • Artalejo, J. R.; Falin, J. I.	94 Top 2 329-342
Baccin, Maria C.; Ferrante, Marco On a	stochastic delay difference equation with boundary conditions and its Markov property •	95 StocProc 60 131-146
Simultaneous spike-trains and	stochastic dependence • Das, Shubhabrata; Sen, Pranab Kumar	95 SankhyaB 57 32-47
	Stochastic differential equation	
observations • Mishra, M. N.; Bishwal, J. P. N.	Approximate maximum likelihood estimation for diffusion processes from discrete	95 Stochatx 52 1-13
Stramer, O.	On the approximation of continuous time threshold ARMA processes • Brockwell, P. J.;	95 AnInStMa 47 1-20
Nualart, David; Sanz-Sole, Marta	Approximation and support theorems in modulus spaces • Gyöngy, István;	95 ProbTher 101 495-509
Hamadéne, S.; Lepeltier, J. P.	Backward equations, stochastic control and zero-sum stochastic differential games •	95 Stochatx 54 221-231
	Black's consol rate conjecture • Duffie, Darrell; Ma, Jin; Yong, Jiongmin	95 AnlsApPr 5 356-382
Weron, A.	Computer-aided modeling and simulation of electrical circuits with α-stable noise •	95 ApplMath 23 83-93
Janicki, Aleksander; Weron, Aleksander	Computer simulation of attractors in stochastic models with α-stable noise •	95 MaCmpSim 39 9-19
noise • Janicki, A.	Computer simulation of a nonlinear model for electrical circuits with α-stable	95 ApplMath 23 95-105
mator for discretely observed diffusion ... • Pedersen	Consistency and asymptotic normality of an approximate maximum likelihood esti-	95 Bernoull 1 257-279
Darling, R. W. R.	Constructing gamma-martingales with prescribed limit, using backwards SDE •	95 AnlsProb 23 1234-1261
Barchielli, A.; Holevo, A. S.	Constructing quantum measurement processes via classical stochastic calculus •	95 StocProc 58 293-317
Teplyaev, A. V.	Continual analogues of random polynomials which are orthogonal on a circle •	94 ThProbAp 39 476-489
Norberg, Ragnar	Differential equations for moments of present values in life insurance •	95 InsrneME 17 171-180
calculus • Arnold, Ludwig; Imkeller, Peter	Furstenberg-Khasminskii formulas for Lyapunov exponents via anticipative	95 Stochatx 54 127-168
Li, Xunjing; Tang, Shanjian	General necessary conditions for partially observed optimal stochastic controls •	95 JAppProb 32 1118-1137
	Large deviations for Langevin spin glass dynamics • Arous, G. B.; Guionnet, A.	95 ProbTher 102 455-509
Takanobu, Satoshi	Multiple stochastic integrals appearing in the stochastic Taylor expansions •	95 JMaJapan 47 67-92
	Nonlinear smoothing for random fields • Aihara, Shin Ichi; Bagchi, Arunabha	95 StocProc 55 143-158
the plane • Deriyeva, O. M.	Some properties of measures corresponding to random fields of diffusion type on	95 ThPrMaSt 50 71-76
	Stochastic analysis of fractional Brownian motions • Lin, S. J.	95 Stochatx 55 121-140
Gatarek, Dariusz	Stochastic Burgers equation with correlated noise • Prato, Giuseppe Da;	95 Stochatx 52 29-41
	Stochastic Nagumo's viability theorem • Aubin, J. P.; Prato, G. Da	95 StocAnAp 13 1-11
Elworthy, K. D.; Zabczyk, J.	Strong Feller property for stochastic semilinear equations • Prato, G. Da;	95 StocAnAp 13 35-45
Fritz, J.; Rüdiger, B.	Time dependent critical fluctuations of a one dimensional local mean field model •	95 ProbTher 103 381-407
	Utility maximization with partial information • Lakner, Peter	95 StocProc 56 247-273
The Euler scheme for anticipating	stochastic differential equations • Ahn, Hyungsok; Kohatsu-Higa, Arturo	95 Stochatx 54 247-269
of distributions for solutions of anticipating	stochastic differential equations • Caballero, M. E.; Fernández, B.; Nualart, D. Smoothness	95 Stochatx 52 303-322
Limit theorems for anticipative	stochastic differential equations (French) • Erraoui, M.; Ouknine, Y.	95 Stochatx 53 1-11
Solution of forward-backward	stochastic differential equations • Hu, Y.; Peng, S.	95 ProbTher 103 273-283
Perfect cocycles through	stochastic differential equations • Arnold, Ludwig; Scheutzow, Michael	95 ProbTher 101 65-88
Pardoux, Étienne; Protter, Philip Stratonovich	stochastic differential equations driven by general semimartingales • Kurtz, Thomas G.;	95 AIHenrPB 31 351-377
Diffusion approximation of nuclear space-valued	stochastic differential equations driven by Poisson random measures • Kallianpur, G. et al.	95 AnlsApPr 5 493-517
A second order	stochastic differential equation for the force of interest • Parker, Gary	95 InsrneME 16 211-224
Numerical integration of	stochastic differential equations • Milstein, G. N.	95 Kluwer:MA:Hllnd 184p
Øksendal, B.	Stochastic differential equations: An introduction with applications (Fourth edition) •	95 Springer:Brln:NY 271p
tions of probability theory: With an introduction to	stochastic differential equations (Second edition) • Tuckwell, H. C. Elementary applica-	95 ChapmnHall:Lnd 320p
Yosida approximations for multivalued	stochastic differential equations • Pettersson, Roger	95 Stochatx 52 107-120
An instability phenomenon of the nonlinear	stochastic differential equation (French) • Roynette, B.	95 Stochatx 53 175-194
Markov field property of	stochastic differential equations • Alabert, Aureli; Ferrante, Marco; Nualart, David	95 AnlsProb 23 1262-1288
	Stochastic differential equations for ruin probabilities • Møller, Christian Max	95 JAppProb 32 74-89
Römisch, W. (Rev) Review of "Numerical solution of	stochastic differential equations" • Kloeden, P. E.; Platen, E. (Auth); Gelbrich, M.;	95 SIAMRvw 37 272-275
On the support of the solutions of	stochastic differential equations • Gyöngy, I.	94 ThProbAp 39 519-523
Stability radii of some	stochastic differential equations • Morozan, T.	95 Stochatx 54 281-291
Markov field property for	stochastic differential equations with boundary conditions • Ferrante, M.; Nualart, D.	95 Stochatx 55 55-69
	Stochastic differential equations with a convex constraint • Storm, Andrew	95 Stochatx 53 241-274
Adapted solutions of backward	stochastic differential equations with non-Lipschitz coefficients • Mao, Xuerong	95 StocProc 58 281-292
Approximations for	stochastic differential equations with reflecting convex boundaries • Pettersson, Roger	95 StocProc 59 295-308
	Stochastic differential games: The linear quadratic zero sum case • Ardanuy, R.	95 SankhyaA 57 161-165
The behavior of solutions of	stochastic differential inequalities • Assing, S.; Manthey, R.	95 ProbTher 103 493-514
scenario aggregation algorithm for the solution of	stochastic dynamic minimax problems • Breton, Michèle; Hachem, Saeb El A	95 Stochatx 53 305-322
Review of "Markov decision processes: Discrete	stochastic dynamic programming" • Puterman, M. L. (Auth); Collins, Sean (Rev)	95 JAppStat 22 545-546
Review of "Markov decision processes: Discrete	stochastic dynamic programming" • Puterman, M. L. (Auth); Iancu, C. (Rev)	95 JAppStat 22 547-547

Review of "Markov decision processes: Discrete	stochastic dynamic programming" • Puterman, M. L. (Auth); Plachky, D. (Rev)	95 StatDecn 13 93-93
Review of "Markov decision processes: Discrete	stochastic dynamic programming" • Puterman, Martin L. (Auth); Baxter, Laurence A. (Rev)	95 Technmcs 37 353-353
	Stochastic equation	
Zabczyk, J.	Convergence to equilibrium for classical and quantum spin systems • Prato, G. Da;	95 ProbTher 103 529-552
and almost periodic solutions for semilinear	stochastic equations • Prato, G. Da; Tudor, Constantin Periodic	95 StocAnAp 13 13-33
Two limit theorems for diffusion type	stochastic equations • Pisanets, S. I.	94 ThProbAp 39 708-713
	Stochastic evolution equation	
Peszat, Szymon; Zabczyk, Jerzy	Strong Feller property and irreducibility for diffusions on Hilbert spaces •	95 AnlsProb 23 157-172
Asymptotic stability theorems of semilinear	stochastic evolution equations in Hilbert spaces • Taniguchi, Takeshi	95 Stochatx 53 41-52
A semilinear McKean-Vlasov	stochastic evolution equation in Hilbert space • Ahmed, N. U.; Ding, X.	95 StocProc 60 65-85
Semilinear	stochastic evolution equations with monotone nonlinearities • Zangeneh, Bijan Z.	95 Stochatx 53 129-174
de Socio, L. M. (Auth) et al. Review of "Nonlinear	stochastic evolution problems in applied sciences" • Bellomo, N.; Brzezniak, Z.;	95 Metrika 42 140-141
L. M. (Auth); Renshaw, Eric (Rev) Review of "Nonlinear	stochastic evolution problems in applied sciences" • Bellomo, N.; Brzezniak, Z.; de Socio,	94 BLndnMaS 26 409-410
	Stochastic frontier	
heteroscedasticity • Caudill, S. B.; Ford, J. M. et al.	Frontier estimation and firm-specific inefficiency measures in the presence of	95 JBES 13 105-111
Posterior analysis of	stochastic frontier models using Gibbs sampling • Koop, G.; Steel, M. F. J.; Osiewalski, J.	95 CmpStQ 10 353-373
On the	stochastic Fubini theorem • Bichteler, K.; Lin, S. J.	95 Stochatx 54 271-279
Krengel, Ulrich A	stochastic game of optimal stopping and order selection • Gnedin, Alexander V.;	95 AnlsApPr 5 310-321
	Stochastic geometry	
	Limit theorems for the time of completion of Johnson-Mehl tessellations • Chiu, S. N.	95 AdvAppPr 27 889-910
Kendall, David G.	How to look at objects in a five-dimensional shape space: Looking at geodesics •	95 AdvAppPr 27 35-43
Asymptotic properties of	stochastic greedy bin-packing • Gaver, Donald P.; Jacobs, Patricia A.	95 CommStSM 11 333-348
Issues in nonlinear	stochastic grey box identification • Bohlin, Torsten; Graebe, Stefan F.	95 IJAdCSgP 9 465-490
The	stochastic group • Poole, David G.	95 AmMaMnth 102 798-801
A stability property of the	stochastic heat equation • Ubøe, Jan; Zhang, Tusheng	95 StocProc 60 247-260
	Stochastic integral	
	The asymptotic behavior of locally square integrable martingales • Wang, Jia-gang	95 AnlsProb 23 552-585
renormalization • Imkeller, P.; Perez-Abreu, V. et al.	Chaos expansions of double intersection local time of Brownian motion in R^d and	95 StocProc 56 1-34
Guo, Sun-Wei	Computation of identity-by-descent proportions shared by two siblings •	94 AJHuGene 54 1104-1109
Peña, V. H.; Montgomery-Smith, S. J.; Szulga, Jerzy	Contraction and decoupling inequalities for multilinear forms and U-statistics • de la	94 AnlsProb 22 1745-1765
de la Peña, Victor H.; Montgomery-Smith, S. J.	Decoupling inequalities for the tail probabilities of multivariate U-statistics •	95 AnlsProb 23 806-816
Willinger, Walter	From discrete to continuous stochastic calculus • Cutland, Nigel J.; Kopp, P. Ekkehard;	95 Stochatx 52 173-192
Monat, Pascale; Stricker, Christophe	Föllmer-Schweizer decomposition and mean-variance hedging for general claims •	95 AnlsProb 23 605-628
Vallois, Pierre	The generalized covariation process and Ito formula • Russo, Francesco;	95 StocProc 59 81-104
	On L^p stochastic representations • Lin, S. J.	95 StPrLet 24 225-227
markers • Guo, Sun-Wei	Proportion of genes survived in offspring conditional on inheritance of flanking	94 Genetcs 138 953-962
	Some EATA properties for marked point processes • Kofman, D.; Korezlioglu, H.	95 JAppProb 32 922-929
	Variance-optimal hedging in discrete time • Schweizer, Martin	95 MaOpRsr 20 1-64
Multiple	stochastic integrals appearing in the stochastic Taylor expansions • Takanobu, Satoshi	95 JMaJapan 47 67-92
Weak convergence of	stochastic integrals driven by martingale measure • Cho, Nhansook	95 StocProc 59 55-79
	Stochastic integral equation	
Hoppensteadt, F.; Salehi, H.; Skorokhod, A. V.	Randomly perturbed Volterra integral equations and some applications •	95 Stochatx 54 89-125
	Stochastic integral finance	
Schachermayer, Walter	The no-arbitrage property under a change of numéraire • Delbaen, Freddy;	95 Stochatx 53 213-226
William Sharp inequalities for the distribution of a	stochastic integral in which the integrator is a bounded submartingale • Hammack,	95 AnlsProb 23 223-235
Complex valued multiparameter	stochastic integrals • Ubøe, Jan	95 JTherPrb 8 601-624
On Chung's law of the iterated logarithm for some	stochastic integrals • Rémillard, Bruno	94 AnlsProb 22 1794-1802
Itô formula for an extended	stochastic integral with nonanticipating kernel • Norin, N. V.	94 ThProbAp 39 573-592
On pathwise	stochastic integration • Karandikar, Rajeeva L.	95 StocProc 57 11-18
	Stochastic integration of processes with finite generalized variations. I • Towghi, Nasser	95 AnlsProb 23 629-667
Stability of a class of	stochastic integro-differential equations • Drozdov, Aleksey	95 StocAnAp 13 517-530
A counting process approach to	stochastic interest • Møller, Christian Max	95 InsrneME 17 181-192
Long-term returns in	stochastic interest rate models • Deelstra, G.; Delbaen, F.	95 InsrneME 17 163-169
Equity-linked life insurance: A model with	stochastic interest rates • Nielsen, J. Aase; Sandmann, Klaus	95 InsrneME 16 225-253
A possibilistic linear program is equivalent to a	stochastic linear program in a special case • Inuiguchi, Masahiro; Sakawa, Masatoshi	95 FuzzySet 76 309-317
On the rotational dimension of	stochastic matrices • Kalpazidou, S.	95 AnlsProb 23 966-975
A decomposition theorem for infinite	stochastic matrices • Heyman, Daniel P.	95 JAppProb 32 893-901
	Stochastic model	
	Darwin's log: A toy model of speciation and extinction • Aldous, David	95 JAppProb 32 279-295
Review of	"Stochastic models for systems" (Russian) • Korolyuk, V. S. (Auth); Svishchuk, A. V. (Rev)	95 ThPrMaSt 50 147-147
Ishiguro, Makio A multivariate	stochastic model with non-stationary trend component • Kato, Hiroko; Naniwa, Sadao;	95 AplStMDA 11 77-95
	Stochastic modeling of aggregates and products of variable failure rates • Singh, N.	95 IEEEReli 44 279-284
Catlin, Sandra; Assunção, Renato; Abkowitz, Janis L.	Stochastic modeling of early hematopoiesis • Newton, Michael A.; Guttorp, Peter;	95 JASA 90 1146-1155
deterministic formulations • Dargahi-Noubary, G. R.	Stochastic modeling and identification of seismic records based on established	95 JTimSrAn 16 201-219
Review of "Concepts in probability and	stochastic modeling" • Higgins, J.; Keller-McNulty, S. (Auth); Huzurbazar, Aparna V. (Rev)	95 Technmcs 37 467-467
M. (Rev) Review of "Concepts in probability and	stochastic modeling" • Higgins, James J.; Keller-McNulty, Sallie (Auth); Nickerson, David	95 AmerStat 49 321-321
	Stochastic modeling of scientific data • Guttorp, Peter	95 ChapmnHall:Lnd 392p
Review of	"Stochastic modeling of scientific data" • Guttorp, Peter (Auth); Embrechts, P. A. L. (Rev)	96 ShrtBkRv 16/1 1-2
Zopounidis, Constantin (Ed) Advances in	stochastic modelling and data analysis • Janssen, Jacques; Skiadas, Christos H.;	95 Kluwer:MA:Hllnd 440p
Review of "An introduction to	stochastic modelling" • Taylor, H. M.; Karlin, S. (Auth); Crowder, Martin (Rev)	95 Ststcian 44 291-292
	Stochastic Nagumo's viability theorem • Aubin, J. P.; Prato, G. Da	95 StocAnAp 13 1-11
Martingale and stationary solutions for	stochastic Navier-Stokes equations • Flandoli, Franco; Gatarek, Dariusz	95 ProbTher 102 367-391
Foias and Hope statistical solutions for	stochastic Navier-Stokes equations • Capiński, Marek; Cutland, Nigel J.	95 Stochatx 52 193-205
discretization of Ornstein-Uhlenbeck equations and	stochastic Navier-Stokes equations with a generalized noise • Flandoli, F. et al. Time	95 Stochatx 55 141-165
Analog optimization with Wong's	stochastic neural network • Kesidis, George	95 IEEENrlN 6 258-260
Asymptotically Markov	stochastic 1 • Sataev, E. A.	94 ThProbAp 39 448-464
Dynamic programming equations of	stochastic optimal control in Banach space • Zhu, Qingxin; Ahmed, N. U.	95 StocAnAp 13 369-387
	Stochastic optimization	
Parmigiani, Giovanni	Optimal design via curve fitting of Monte Carlo experiments • Müller, Peter;	95 JASA 90 1322-1330
An application of	stochastic optimization to combine two files • Griffin, Richard A.	94 ASASrvyRsr 485-490
of "Discrete event systems: Sensitivity analysis and	stochastic optimization by the score function method" • Rubinstein, R. Y. et al. Review	95 Technmcs 37 124-125
Sequential	stochastic optimization • Cairoli, R.; Dalang, Robert C.	95 Wiley:NY:UK 350p
	Stochastic order relations and the total time on test transform • Bartoszewicz, Jarosław	95 StPrLet 22 103-110

Local asymptotic quadraticity of	stochastic process models based on stopping times • Luschgy, Harald	95 StocProc 57 305-317
Matthes, K. (Rev) Review of	"Stochastic processes and point processes of excursions" • van der Weide, J. A. M. (Auth);	95 BiomtrcJ 37 764-764
modelling • Nelson, R. Probability,	stochastic processes, and queueing theory: The mathematics of computer performance	95 Springer:Brln:NY 592p
lozar T.; Samorodnitsky, Gennady Limit laws for a	stochastic process and random recursion arising in probabilistic modelling • Rachev, Svet-	95 AdvAppPr 27 185-202
Review of	"Stochastic process (Second edition)" • Medhi, J. (Auth); Basu, A. K. (Rev)	94 CalcutSt 44 265-266
Simulation and chaotic behavior of α-stable	stochastic processes • Janicki, Aleksander; Weron, Aleksander	95 Dekker:NY 376p
of "Simulation and chaotic behaviour of α-stable	stochastic processes" • Janicki, A.; Weron, A. (Auth); Lee, P. M. (Rev) Review	95 JRSS-A 158 351-352
Sample quantiles of heavy tailed	stochastic processes • Embrechts, Paul; Samorodnitsky, Gennady	95 StocProc 59 217-233
Extreme value theory for	stochastic processes • Rootzén, Holger	95 ScanActJ 54-65
function for irregularly sampled continuous time	stochastic processes with control variables and missing values • Singer, H. Analytical score	95 EconTher 11 721-735
Gennady Functionals of infinitely divisible	stochastic processes with exponential tails • Braverman, Michael; Samorodnitsky,	95 StocProc 56 207-231
The aftermath of Cramér's work on	stochastic processes • Masani, P. R.	95 ScanActJ 66-96
	Stochastic programming	
problems • Breton, Michèle; Hachem, Saeb El	A scenario aggregation algorithm for the solution of stochastic dynamic minimax	95 Stochatx 53 305-322
A strong existence theorem for	stochastic reaction-diffusion equations with continuous coefficients • Manthey, Ralf	95 Stochatx 52 15-27
	Stochastic relaxation	
	Understanding variable shapes in 3-D • Grenander, Ulf; Keenan, Daniel	95 CommStSM 11 51-78
Single machine	stochastic scheduling: Minimizing the number of tardy jobs • Lin, Chen-Sin	94 AJMMgSci 14 289-300
On the value of adaptive solutions to	stochastic scheduling problems • Glazebrook, K. D.; Owen, R. W.	95 MaOpRsr 20 65-89
Modeling and analysis of	stochastic systems • Kulkarni, Vidyadhar G.	95 ChapmnHall:Lnd 624p
Review of "Modeling and analysis of	stochastic systems" • Kulkarni, Vidyadhar G. (Auth); Leslie, J. R. (Rev)	96 ShrtBkRv 16/1 4-5
Mengersen, Kerrie Bayesian computation and	stochastic systems (Disc: p41-66) • Besag, Julian; Green, Peter; Higdon, David;	95 StatSci 10 3-41
Model-free control of nonlinear	stochastic systems in discrete time • Spall, James C.; Cristion, John A.	95 PIEEEDecCt 2199-2204
Complex	stochastic systems and engineering • Titterington, D. M. (Ed)	95 OxfordUniv 312p
Factorial moment expansion for	stochastic systems • Błaszczyszyn, B.	95 StocProc 56 321-335
Adaptive control of partially observed linear	stochastic systems • Bertrand, Pierre	95 Stochatx 54 21-51
Measuring resilience in	stochastic systems • Ives, Anthony R.	95 EcolgMon 65 217-233
Random walks with	stochastically bounded increments: Renewal theory • Alsmeyer, Gerold	95 MatNach 172 13-31
Approximation of the stationary distribution of a	stochastically monotone Markov chain (French) • Simonot, F.	95 StocProc 56 133-149
A robust surveillance scheme for	stochastically ordered alternatives • Gordon, Louis; Pollak, Moshe	95 AnlsStat 23 1350-1375
A multivariate goodness-of-fit test for	stochastically ordered distributions • Alam, Khursheed; Williams, Calvin L.	95 BiomtrcJ 37 945-956
On the weak convergence of certain estimators of	stochastically ordered survival functions • Rojo, Javier	95 JNnparSt 4 349-363
Limit theorems for	stochastically perturbed dynamical systems • Łoskot, Krzysztof; Rudnicki, Ryszard	95 JAppProb 32 459-469
rithm • McAllister, M. K. et al. A Bayesian approach to	stock assessment and harvest decisions using the sampling/importance resampling algo-	94 JFishRBC 51 2673-2687
Optimal	stock/flow panels • Lancaster, Tony; Imbens, Guido	95 JEconmtx 66 325-348
Cointegration of international	stock market indices • Chou, Ray Y.; Ng, Victor K.; Pi, Lynn K.	94 ASAProBuEc 196-201
Effects of feedback on probabilistic forecasts of	stock prices • Önkal, Dilek; Muradoğlu, Gulnur	95 IntJFor 11 307-319
Cointegration analysis of	stock prices, inflation and real economic activity • Raj, Baldev; Amoako-Adu, Ben	94 ASAProBuEc 190-195
model to study the effect of earnings surprises on	stock prices, trading volumes, and spreads • Maddala et al. ... component panel data	95 JEconmtx 68 229-242
the 1987 ... • Masulis et al. Overnight and daytime	stock-return dynamics on the London stock exchange: The impacts of "big bang" and	95 JBES 13 365-378
variables problem in the cross-section of expected	stock returns • Kim, Dongcheol The errors in the	95 JFinance 50 1605-1634
Analysts' use of earnings forecasts in predicting	stock returns: Forecast horizon effects • Bandyopadhyay, S. P.; Brown, L. D. et al.	95 IntJFor 11 429-445
component detection in a multifactor model for	stock returns • Costa, Michele Dynamic	94 JItalStS 3 25-36
Lam, K. Modelling asymmetry in	stock returns by a threshold autoregressive conditional heteroscedastic model • Li, W. K.;	95 Ststcian 44 333-341
ate GARCH model of international transmissions of	stock returns and volatility: The case of the United States and Canada • Karolyi A multivari-	95 JBES 13 11-25
Mean-variance hedging of options on	stocks with Markov volatilities • Di Masi, G. B.; Kabanov, Yu. M.; Runggaldier, W. J.	94 ThProbAp 39 172-182
Obituary: Richard A.	Stone, 1961-1994 • Veevers, Alan; Robinson, Geoff	95 JRSS-A 158 627-628
Efficient execution of	Stone's likelihood ratio tests for disease clustering • Lumley, Thomas	95 CmpStDA 20 499-510
Predictive	stop-loss premiums and Student's t-distribution • Hürlimann, Werner	95 InsrneME 16 151-159
Asymptotic expansions for the distributions of	stopped random walks and first passage times • Lai, Tze Leung; Wang, Julia Qizhi	94 AnlsProb 22 1957-1992
the non-existence of the expectation of randomly	stopped sums • Lefèvre, Claude; Utev, Sergey On a conjecture for	95 JAppProb 32 1138-1141
Moments of randomly	stopped sums-revisited • Teicher, Henry	95 JTherPrb 8 779-793
	Stopping	
	• see also early –, optimal –	
sequential approach • Emerson, Scott S.	Stopping a clinical trial very early based on unplanned interim analyses: A group	95 Biomtrcs 51 1152-1162
p286-295) • Bring, Johan Comment on	"Stopping a clinical trial very early because of toxicity: Summarizing the evidence" (V14	95 ConClnTr 16 131-132
calculation issues in clinical trials with two-sided	stopping criteria • Chen, Bee-Lian; Gallo, Paul P. Boundary	94 ASAProBiop 115-119
an exponential distribution via replicated piecewise	stopping number • Bose, Arup; Mukhopadhyay, Nitis Sequential estimation of the mean of	95 StatDecn 13 351-361
	Stopping rule	
vendor selection • Houshmand, A. A. et al.	An analytical approximation and a neural network model for optimal sample size in	95 JStCmpSm 53 65-78
Robert M.	A.P.O. rules in hierarchical Bayes regression models • Ghosh, Malay; Hoekstra,	95 SqtlAnly 14 99-115
	A best-choice problem with three stops and random termination • Majumdar, A. A. K.	95 SankhyaB 57 76-84
cretized likelihood estimation procedure • Akahira	The Bhattacharyya type bound for the asymptotic variance and the sequential dis-	95 SqtlAnly 14 193-204
the evidence" (V14 p286-295) • Bring, Johan	Comment on "Stopping a clinical trial very early because of toxicity: Summarizing	95 ConClnTr 16 131-132
A.; Gouweleeuw, José M.	A continuous-time job search model: General renewal processes • Boshuizen, Frans	95 CommStSM 11 349-369
Valeri T.	On efficient sequential estimation for the simple single server queue • Stefanov,	95 RBrzlPrS 9 55-65
Lai, Tze Leung	Estimators with prescribed precision in stochastic regression models • Konev, Victor;	95 SqtlAnly 14 179-192
its applications • Nishiyama, Yoichi	Local asymptotic normality of a sequential model for marked point processes and	95 AnInStMa 47 195-209
	A new look at pricing of the "Russian option" • Shepp, L. A.; Shiryaev, A. N.	94 ThProbAp 39 103-119
Goudie, I. B. J.	A plant-capture approach for achieving complete coverage of a population •	95 CommStA 24 1293-1305
	Some adaptive estimators in simple regression models • Viet, Tran Quoc	95 StatDecn 13 117-130
Discrete time. • Shiryaev, A. N.; Kabanov, Yu. M. et al.	Toward the theory of pricing of options of both European and American types. I.	94 ThProbAp 39 14-60
hypothesis testing • Berger, J. O.; Brown, L. D. et al.	A unified conditional frequentist and Bayesian test for fixed and sequential simple	94 AnlsStat 22 1787-1807
Mehmet A	stopping rule for a compound Poisson random variable • Randolph, Paul; Şahinoğlu,	95 AplStMDA 11 135-143
Poisson search models and the evaluation of	stopping rules • Bather, John	95 SqtlAnly 14 205-216
tistical approaches" ... • Mehlman et al. Comment on	"Stopping rules in principal components analysis: A comparison of heuristical and sta-	95 Ecology 76 640-643
tistical approaches" • Jackson Reply to comment on	"Stopping rules in principal components analysis: A comparison of heuristical and sta-	95 Ecology 76 644-645
Hong, Dug Hun; Oh, Kwang Sik; Park, Hee Joo On a	stopping rule for the random walks with time stationary random distribution function •	95 JKoreaSt 24 293-301
	Stopping rules for sequences of factorial designs • Gilmour, S. G.; Mead, R.	95 ApplStat 44 343-355
	Stopping and set-indexed local martingales • Ivanoff, B. Gail; Merzbach, Ely	95 StocProc 57 83-98
	Stopping time	
	On amarts with continuous time • Dzhvarsheishvili, I. A.	94 ThProbAp 39 512-519
sums • Lefèvre, Claude; Utev, Sergey	On a conjecture for the non-existence of the expectation of randomly stopped	95 JAppProb 32 1138-1141
Cranston, Michael; Greven, Andreas	Coupling and harmonic functions in the case of continuous time Markov processes •	95 StocProc 60 261-286

	Studentized residual	
	Influence and residuals in restricted generalized linear models • Paula, Gilberto A.	95 JStCmpSm 51 315-331
On the normal approximations of	studentized U-statistic • Maesono, Yoshihiko	95 JJapnStS 25 19-33
forms: Application to randomized spline type	Sturm-Liouville problems • Gregory et al. ... approximation theory for quadratic	95 JTherPrb 8 703-715
Some properties of the	stuttering Poisson distribution • Moothathu, T. S.; Kumar, C. Satheesh	95 CalcutSt 45 125-130
Van Mieghem, Jan A.; Rub, Leonardo Multiple	subclass pattern recognition: A maximin correlation approach • Avi-Itzhak, Hadar I.;	95 IEEEPAMI 17 418-431
On a	subclass of regularly varying functions • Omey, E.	95 JStPlInf 45 275-290
	Subdiffusive fluctuations for internal diffusion limited aggregation • Lawler, Gregory F.	95 AnlsProb 23 71-86
P-value adjustments for	subgroup analyses of binary variables • Bristol, David R.	94 ASAProBiop 120-125
Statistical process control via the	subgroup bootstrap • Seppala, Tomi; Moskowitz, Herbert; Plante, Robert; Tang, Jen	95 JQualTek 27 139-153
a search method for identification of clinical	subgroups • Marshall, Roger J. A program to implement	95 StatMed 14 2645-2659
Ellis, Yukiko Exploring associations between	subjective and objective assessments of household membership • Schwede, Laurie;	94 ASASrvyRsr 325-330
	Subjective probability	
DeWispelare, A. R.; Herren, L. T.; Clemen, R. T.	The use of probability elicitation in the high-level nuclear waste regulation program •	95 IntJFor 11 5-24
	Wisdom updated • Falk, A.	95 PhilSci 62 389-403
and coherence of causal, diagnostic, and joint	subjective probabilities • Wright, M. Keith; Stokes, Lynne; Dyer, James S. Reliability	94 DecisnSc 25 691-709
and coherence of causal, diagnostic, and joint	subjective probabilities • Wright, M. Keith; Stokes, Lynne; Dyer, James S. Reliability	95 DecisnSc 26 691-709
Review of	"Subjective probability" • Wright, G.; Ayton, P. (Ed); Lee, Peter M. (Rev)	95 JRSS-A 158 637-638
A correction for 'end effects' in	subjective rating experiments • Irvine, John M.	94 ASAProSoSt 184-188
Effects of rank versus category in measuring	subjective social inequality • Hoffmeyer-Zlotnik, Jürgen H. P.	94 ASASrvyRsr 1322-1326
Score adjustment for	subjectively rated responses (Disc: p134-135) • Longford, N. T.	94 ASAProBays 123-127
	Submartingale	
martingales and related sharp inequalities • Wang	Differential subordination and strong differential subordination for continuous-time	95 AnlsProb 23 522-551
integral in which the integrator is a bounded	submartingale • Hammack, William Sharp inequalities for the distribution of a stochastic	95 AnlsProb 23 223-235
On the Doléans function of set-indexed	submartingales • De Giosa, Marcello; Mininni, Rosamaria	95 StPrLet 24 71-75
Maximum likelihood estimation and	suboptimal stopping time for a Poisson point process • Garcia, Nancy Lopes	95 RBrzlPrS 9 67-82
related sharp inequalities • Wang, Gang Differential	subordination and strong differential subordination for continuous-time martingales and	95 AnlsProb 23 522-551
On estimating domain totals over a	subpopulation • He, Kun	95 AnInStMa 47 637-644
surement program to develop estimates for small	subpopulations using double sampling • Bredfeldt et al. ... of IRS's tax compliance mea-	94 ASAProGovt 128-133
Large sample confidence regions based on	subsamples under minimal assumptions • Politis, Dimitris N.; Romano, Joseph P.	94 AnlsStat 22 2031-2050
breakdown point • Juan, Jesus; Prieto, Francisco J. A	subsampling method for the computation of multivariate estimators with high	95 JCmpGrSt 4 319-334
of the mean length of the longest common	subsequence" (V4 p1074-1082) • Alexander, K. S. Correction to "The rate of convergence	95 AnlsApPr 5 327-327
On rates of convergence for common	subsequences and first passage time • Rhee, WanSoo T.	95 AnlsApPr 5 44-48
interacting particle process and longest increasing	subsequences • Aldous, D.; Diaconis, P. Hammersley's	95 ProbTher 103 199-213
A Chung type law of the iterated logarithm for	subsequences of a Wiener process • Shao, Qi-Man	95 StocProc 59 125-142
Fellingham, G. W.; Beus, C. M. Using	subsequent studies to recover estimability (Disc: p125-126) • Davis, A. B.; Larimore, W. L.;	94 ASAProEpid 112-117
On a method of identification of best	subset model from full AR-model • Sarkar, A.; Kanjilal, P. P.	95 CommStA 24 1551-1567
Better	subset regression using the nonnegative garrote • Breiman, Leo	95 Technmcs 37 373-384
	Subset selection	
	Better subset regression using the nonnegative garrote • Breiman, Leo	95 Technmcs 37 373-384
	Dual response surface optimization • Lin, Dennis K. J.; Tu, Wanzhu	95 JQualTek 27 34-39
Chatfield, Chris	Model uncertainty, data mining and statistical inference (Disc: p444-466) •	95 JRSS-A 158 419-444
	More comments on C_p • Mallows, Colin L.	95 Technmcs 37 362-372
	Outlier detection in multivariate calibration • Walczak, B.	95 Chemolab 28 259-272
	Predictive model selection • Laud, Purushottam W.; Ibrahim, Joseph G.	95 JRSS-B 57 247-262
Huang, Deng-Yuan; Lee, Ren Fen	Selecting the largest capability index from several quality control processes •	95 JStPlInf 46 335-346
known quantile • Kumar, N.; Gill, A. N.; Mehta, G. P.	Selection procedures for scale parameters when the populations have a common	94 Metron 52/3 171-196
van der Laan, Paul	Subset selection for exponential populations: Determination of the selection constant •	95 BiomtrcJ 37 515-521
Equivalence of several methods for efficient best	subsets selection in generalized linear models • Jovanovic, B. D.; Hosmer, D. W. et al.	95 CmpStDA 20 59-64
Review of	"Subset selection in regression" • Miller, A. J. (Auth); Droge, B. (Rev)	95 MaOpfStS 26 179-179
Using dimension-reduction	subspaces to identify important inputs in models of physical systems • Cook, R. Dennis	94 ASAProPEng 18-25
Statistical applications of a metric on	subspaces to satellite meteorology • Crone, L. J.; Crosby, D. S.	95 Technmcs 37 324-328
Georgios B.; Shamsunder, Sanyogita Noise	subspace techniques in non-Gaussian noise using cumulants • Sadler, Brian M.; Giannakis,	95 IEEEAES 31 1009-1018
The nearest neighbor random walk on	subspaces of a vector space and rate of convergence • D'Aristotile, Anthony J.	95 JTherPrb 8 321-346
Sets of mutually orthogonal Latin squares with "like	subsquares," II • Roberts, Charles E., Jr	95 JCombThA 70 159-164
Bray, Robert M. Estimating the prevalence of	substance abuse among people who are homeless • Iachan, Ronaldo; Dennis, Michael L.;	94 ASASrvyRsr 1094-1099
Greene, Jody M.	Substance abuse among runaway and homeless youth • Iachan, Ronald; Ringwalt, Chris L.;	94 ASASrvyRsr 1083-1087
Estimating	substance abuse treatment need from the NHSDA • Epstein, Joan F.; Gfroerer, Joseph C.	94 ASASrvyRsr 559-564
Statistical tests of models of DNA	substitution • Goldman, Nick	93 JMolEvol 36 182-198
diagnostic statistical tests of models for DNA	substitution • Goldman, Nick Simple	93 JMolEvol 37 650-661
Maximum likelihood with multiparameter models of	substitution • Tillier, Elisabeth Renée Marie	94 JMolEvol 39 409-417
Field	substitutions – A neglected option? • Vehovar, Vasja	94 ASASrvyRsr 589-594
Estimating the pattern of nucleotide	substitution • Yang, Ziheng	94 JMolEvol 39 105-111
Reeves, Jaxk H. Heterogeneity in the	substitution process of amino acid sites of proteins coded for by mitochondrial DNA •	92 JMolEvol 35 17-31
methods for estimating phylogenetic trees when	substitution rates differ over nucleotide sites • Yang, Ziheng Evaluation of several	95 JMolEvol 40 689-697
S.; Weir, B. S. Detecting	substitution-rate heterogeneity among regions of a nucleotide sequence • Gaut, Brandon	94 MolBioEv 11 620-629
likelihood estimation of the heterogeneity of	substitution rate among nucleotide sites • Gu, Xun; Fu, Yun-Xin; Li, Wen-Hsiung Maximum	95 MolBioEv 12 546-557
DNA • Wakeley, John	Substitution rate variation among sites in hypervariable region 1 of human mitochondrial	93 JMolEvol 37 613-623
ing synonymous and nonsynonymous nucleotide	substitution rates, with application to the chloroplast genome • Muse et al. ... for compar-	94 MolBioEv 11 715-724
Generating random variates from D-distributions via	substitution sampling • Walker, Stephen	95 StatComp 5 311-315
estimation for incomplete data via successive	substitution sampling • Doss, Hani Bayesian nonparametric	94 AnlsStat 22 1763-1786
the numbers of synonymous and nonsynonymous	substitutions per site • Ota, Tatsuya; Nei, Masatoshi Variance and covariances of	94 MolBioEv 11 613-619
Unbiased estimates of the number of nucleotide	substitutions when substitution rate varies among different sites • Rzhetsky, A.; Nei, M.	94 JMolEvol 38 295-299
A dynamic analysis of interfuel	substitution in U.S. industrial energy demand • Jones, Clifton T.	95 JBES 13 459-465
An algorithm to find agreement	subtrees • Kubicka, Ewa; Kubicki, Grzegorz; McMorris, F. R.	95 JClass 12 91-99
The largest square of	successes covering the origin for Bernoulli trials on the lattice • Kopociński, B.	95 StNeerla 49 31-39
	Successful customers of statistics at a distant learning site • Bolland, Thomas W.	94 ASAProQlPr 300-304
Exact and limiting distributions of the number of	successions in a random permutation • Fu, James C.	95 AnInStMa 47 435-446
D.	Successive comparisons between ordered treatments • Lee, Robert Edwin; Spurrier, John	95 JStPlInf 43 323-330
A	successive filtering technique for identifying long-term trends • Assimakopoulos, V.	95 JForcast 14 35-43
nonparametric estimation for incomplete data via	successive substitution sampling • Doss, Hani Bayesian	94 AnlsStat 22 1763-1786
Distribution-free multiple comparisons between	successive treatments • Lee, Robert Edwin; Spurrier, John D.	95 JNnparSt 5 261-273

	Sufficiency	
Linssen, H. N.; Overdijk, D. A.	The reduction and ancillary classification of observations • van Berkum, E. E. M.;	95 JStPlInf 45 347-372
W. J. et al. Correction to "The relationship between	sufficiency and invariance with applications in sequential analysis" (V36 p575-614) • Hall,	95 AnlsStat 23 705-705
probabilities • Kato, Takeshi Asymptotic	sufficiency of posterior densities with respect to Kullback information under marginal	95 StatDecn 13 253-271
ByoungSeon A Sanov-type proof of the joint	sufficiency of the sample mean and the sample variance • Kim, Chul Eung; Choi,	95 JKoreaSt 24 563-568
Baryshnikov, Y.; Eisenberg, B. et al. A necessary and	sufficient condition for the existence of the limiting probability of a tie for first place •	95 StPrLet 23 203-209
Wu, Qiguang Necessary and	sufficient condition for the existence of UMVU estimators in growth curve model •	95 KToChina 39 89-92
complexity one • Reilly, Terence A necessary and	sufficient condition for identification of confirmatory factor analysis models of factor	95 SocMethR 23 421-441
of a recursively defined process: Necessary and	sufficient conditions, sensitivity • Yin, G.; Zhu, Y. M.; Gao, A. J. On characterization	95 Stochatx 52 265-282
gies • Deheuvels, P.; Lifshits, M. A. Necessary and	sufficient conditions for the Strassen law of the iterated logarithm in nonuniform topolo-	94 AnlsProb 22 1838-1856
odds ratio • Yu, Kai Fun A necessary and	sufficient condition for the strong consistency of a family of estimators of the common	95 CandJSt 23 215-225
E. A necessary and	sufficient identification rule for structural models estimated in practice • Rigdon, Edward	95 MultBehR 30 359-383
Stefanov Explicit limit results for minimal	sufficient statistics and maximum likelihood estimators in some Markov processes: ... •	95 AnlsStat 23 1073-1101
The structure of technology in Brazilian	sugarcane production, 1975-87 • Rask, Kevin	95 JApEcmtx 10 221-232
Socioeconomic status and	suicide • Kung, H. C.; Liu, X. H.; Chan, J.	94 ASAProSoSt 235-239
and regression diagnostics for count data:	Suicide by tourism? (German) • Ziegler, A. et al. Pseudo-maximum-likelihood-estimation	95 AllgeStA 79 170-195
spatio-temporal process with an application to wet	sulfate deposition • Haas, Timothy C. Local prediction of a	95 JASA 90 1189-1199
	Sum	
	• see also cumulative -, partial -, trimmed -, weighted -, zero- -	
Generalized Poisson distributions as limits of	sums for arrays of dependent random vectors • Kobus, Maria	95 JMultiAn 52 199-244
Probability bounds on the finite	sum of the binary sequence of order k • Dhar, Sunil K.; Jiang, Xulun	95 JAppProb 32 1014-1027
Random	sums and branching stochastic processes • Rahimov, I.	95 Springer:Brln:NY 195p
Review of "Random	sums and branching stochastic processes" • Rahimov, I. (Auth); Teugels, J. L. (Rev)	95 ShrtBkRv 15/3 48-48
Rakhimov, G. M. On approximation of	sums of conditionally independent random variables • Khusanbayev, Ya. M.;	95 ThPrMaSt 50 133-142
"A bound on the moment generating function of a	sum of dependent variables with an application to simple ... • de la Peña Correction to	95 AlHenrPB 31 703-704
The first divisible	sum • Barbour, A. D.; Grübel, R.	95 JTherPrb 8 39-47
(Chinese) • Hu, Shuhe	Sums of double arrays of LP-mixingale sequence and its application in linear models	95 AMaAPknC 17 248-254
Randomly weighted	sums for exchangeable fuzzy random variables • Inoue, Hiroshi	95 FuzzySet 69 347-354
of the moving block bootstrap for normalized	sums of heavy-tail random variables • Lahiri, S. N. On the asymptotic behaviour	95 AnlsStat 23 1331-1349
quadratic forms, maxima of products and truncated	sums of i.i.d. random variables • Cuzick, J.; Giné, E.; Zinn, J. Laws of large numbers for	95 AnlsProb 23 292-333
A strong law for weighted	sums of i.i.d. random variables • Cuzick, Jack	95 JTherPrb 8 625-641
Young, D. M. et al. A Kolmogorov inequality for the	sum of independent Bernoulli random variables with unequal means • Turner, D. W.;	95 StPrLet 23 243-245
The law of large numbers for multiple	sums of independent identically distributed random variables • Klesov, O. I.	95 ThPrMaSt 50 77-87
On symmetric	sum of sub-independent random variables • Hamedani, G. G.	95 CalcutSt 45 119-124
numbers and the law of the logarithm for weighted	sums of independent random variables with multidimensional ... • Li et al. ... law of large	95 JMultiAn 52 181-198
	Sums of independent triangular arrays and extreme order statistics • Janssen, Arnold	94 AnlsProb 22 1766-1793
Characterization of discrete laws via mixed	sums and Markov branching processes • Pakes, Anthony G.	95 StocProc 55 285-300
C. W.; Turkman, K. F.	Sums and maxima of stationary sequences with heavy tailed distributions • Anderson,	95 SankhyaA 57 1-10
A note on the asymptotic independence of the	sum and maximum of strongly mixing stationary random variables • Hsing, Tailen	95 AnlsProb 23 938-947
On the limit distributions of	sums of mixing Bernoulli random variables • Dziubdziela, Wiesław	95 StPrLet 23 179-182
approximation of two-mode proximity matrices by	sums of order-constrained matrices • Hubert, Lawrence; Arabie, Phipps The	95 Psymtrka 60 573-605
The asymptotic distributions for	sums of order statistics (I) • Cheng, Shidong	95 ActScNat 30 323-338
On the strong law of large numbers for	sums of pairwise independent random variables • Martikainen, Alexander	95 StPrLet 25 21-26
Strong convergence of	sums of φ-mixing random variables • Liebscher, E.	95 MaMethSt 4 216-229
Samorodnitsky, G.; Taqqu, M. S. Stable fractal	sums of pulses: The cylindrical case • Cioczek-Georges, R.; Mandelbrot, B. B.;	95 Bernoull 1 201-216
Maximization of	sums of quotients of quadratic forms and some generalizations • Kiers, Henk A. L.	95 Psymtrka 60 221-245
Cosine products, Fourier transforms, and random	sums • Morrison, Kent E.	95 AmMaMnth 102 716-724
On the almost sure behaviour of	sums of random variables • Petrov, Valentin V.	95 StPrLet 24 229-231
On estimation of maxima of	sums of random variables indexed by edges of graphs • Karpelevich, F. I.	94 ThProbAp 39 696-702
and almost sure convergence of weighted	sums of random variables • Li, D.; Rao, M. B.; Jiang, T.; Wang, X. Complete convergence	95 JTherPrb 8 49-76
On the asymptotic independence of the	sum and rare values of weakly dependent stationary random variables • Hsing, Tailen	95 StocProc 60 49-63
Distant long-range dependent	sums and regression estimation • Csörgő, Sándor; Mielniczuk, Jan	95 StocProc 59 143-155
Moments of randomly stopped	sums-revisited • Teicher, Henry	95 JTherPrb 8 779-793
Intermediate	sums and stochastic compactness of maxima • Csörgő, Sándor; Mason, David M.	95 JStPlInf 45 81-90
of the expectation of randomly stopped	sums • Lefèvre, Claude; Utev, Sergey On a conjecture for the non-existence	95 JAppProb 32 1138-1141
Extreme	sums of strictly stationary sequences of M-dependent variables • Flak, T.; Schmid, W.	95 SankhyaA 57 186-201
Expectations of the maximum of	sums and the sum of maxima for correlated normal variables • Bain, Lee J.; Gan, Gaoxiong	95 CommStB 24 775-780
for unbounded functions, I: Independent	summands • Barbour, A. D.; Chen, Louis H. Y.; Choi, K. P. Poisson approximation	95 StSinica 5 749-766
a clinical trial very early because of toxicity:	Summarizing the evidence" (V14 p286-295) • Bring, Johan Comment on "Stopping	95 ConClnTr 16 131-132
	Summary measures	
	The analysis of repeated measures: A practical review with examples • Everitt, B. S.	95 Ststcian 44 113-135
Computational aspects of	Sundt's generalized class • Panjer, Harry H.; Wang, Shaun	95 AstinB 25 5-18
Li, Yi; Mueller, Carl Singularity of	super-Brownian local time at a point catalyst • Dawson, Donald A.; Fleischmann, Klaus;	95 AnlsProb 23 37-55
A new approach to the single point catalytic	super-Brownian motion • Fleischmann, Klaus; Le Gall, Jean-François	95 ProbTher 102 63-82
the ultimate value of local time of one-dimensional	super-Brownian motion • Kaj, I.; Salminen, P. On	95 StocProc 59 21-42
On the connected components of the support of	super Brownian motion and of its exit measure • Abraham, Romain	95 StocProc 60 227-245
states • Dawson, Donald A.; Fleischmann, Klaus	Super-Brownian motions in higher dimensions with absolutely continuous measure	95 JTherPrb 8 179-206
measure of multiple points and collision points of	super-Brownian motion • Serlet, Laurent On the Hausdorff	95 Stochatx 54 169-198
Some dimension results for	super-Brownian motion • Serlet, Laurent	95 ProbTher 101 371-391
The packing measure of the support of	super-Brownian motion • Le Gall, J.-F.; Perkins, E. A.; Taylor, S. J.	95 StocProc 59 1-20
self-similar (super) processes • Adler, R. J. et al.	Super fractional Brownian motion, fractional super Brownian motion and related	95 AnlsProb 23 743-766
Debasis; Kundu, Rameswar Consistent estimates of	super imposed exponential signals when some observations are missing • Kundu,	95 JStPlInf 44 205-218
Improved predictors in survey sampling under a	super population model • Srivastava, A. K.	95 StPrLet 25 249-252
Persistence of a critical	super-2 process • Gorostiza, Luis G.; Hochberg, Kenneth J.; Wakolbinger, Anton	95 JAppProb 32 534-540
variables • Bercovici, H.; Voiculescu, D.	Superconvergence to the central limit and failure of the Cramér theorem for free random	95 ProbTher 103 215-222
	Supercritical behaviors in first-passage percolation • Zhang, Yu	95 StocProc 59 251-266
tail behaviour for the limiting random variable in a	supercritical branching process • Hambly, B. M. On constant	95 JAppProb 32 267-273
	Superdiffusion	
approach • Sheu, Yuan-Chung	On positive solutions of some nonlinear differential equations - A probabilistic	95 StocProc 59 43-53
al. Asymptotic properties of the EVLP estimation for	superimposed exponential signals in noise (Chinese) • Bai, Z.; Chen, X.; Krishnaiah, P. R. et	95 ChJApPSt 10 148-163
Rowley, C. A. Randomization for a balanced	superimposition of one Youden square on another • Bailey, R. A.; Preece, D. A.;	95 JRSS-B 57 459-469
Mean square error matrix	superiority of empirical Bayes estimators under misspecification • Wei, L.; Trenkler, G.	95 Test 4 187-205
testing strategy in phase III trials: Combined test of	superiority and test of equivalence • Morikawa, Toshihiko; Yoshida, Michihiro A useful	95 JBiophSt 5 297-306

	Survey (cont.)	
R. J.	Data editing and imputation from a computational point of view • Willenborg, L. C.	95 StNeerla 49 139-151
Pratt, William F.; Rasinski, Kenneth	Design and results of the Women's Health Study • Tourangeau, Roger; Jobe, Jared B.;	94 ASASrvyRsr 49-58
the collection of library data • Kindel, Carrol Benner	Electronic data collection at the National Center for Education Statistics: Successes in	94 JOffStat 10 93-102
Iachan, Ronaldo; Dennis, Michael L.; Bray, Robert M.	Estimating the prevalence of substance abuse among people who are homeless •	94 ASASrvyRsr 1094-1099
Gfroerer, Joseph C.	Estimating substance abuse treatment need from the NHSDA • Epstein, Joan F.;	94 ASASrvyRsr 559-564
• King, Karen Ellen; Kaufman, Steve	Estimation issues related to the student component of the SASS (Disc: p1128-1129)	94 ASASrvyRsr 1111-1115
Paulin, Geoffrey D.; Ferraro, David L.	Do expenditures explain income?: A study of variables for income imputation •	94 ASASrvyRsr 106-115
Marquis, Kent H.; Moore, Jeffrey C.; Bogen, Karen	An experiment to reduce measurement error in the SIPP: Preliminary results •	94 ASASrvyRsr 725-729
membership • Schwede, Laurie; Ellis, Yukiko	Exploring associations between subjective and objective assessments of household	94 ASASrvyRsr 325-330
fractions of missing information • Pedlow, S. et al.	Exploring hypothesis-testing procedures with multiply-imputed data under unequal	94 ASASrvyRsr 595-599
Angeli, A.; Cocchi, D.; Pasquini, L.; Samoggia, A.	Family structure and income: A first analysis on cross-section data (Italian) •	94 Statstca 54 435-452
Frazier, R. L.; King, K. E.; Schwanz, D. J.	Improving the coverage of private elementary-secondary schools • Jackson, B. J.;	94 ASASrvyRsr 833-838
Don A.; Dolsen, Dana E.; Machlis, Gary E.	Increasing response to personally-delivered mail-back questionnaires • Dillman,	95 JOffStat 11 129-139
Sangster, R. L.; Rockwood, T. H.; Dillman, D. A.	The influence of administration mode on responses to numeric rating scales •	94 ASASrvyRsr 708-712
Cahoon, Lawrence; Fay, Robert; Kostanich, Donna	Measurement of different design effects (Disc: p360-361) • Butani, Shail;	94 ASASrvyRsr 342-347
David; Moore, Peter; Smith, Fred; Allin, Paul	The measurement of unemployment in the UK (Disc: p405-417) • Bartholomew,	95 JRSS-A 158 363-405
Tucker, C.; Kojetin, B.; Blass, R.; Tucker, R.	Measuring the effects of the transition on field performance (Disc: p360-361) •	94 ASASrvyRsr 348-353
(Disc: p360-361) • Tiller, Richard; Welch, Michael	Predicting the national unemployment rate that the "old" CPS would have produced	94 ASASrvyRsr 354-359
Colliver, James D.; Gfroerer, Joseph C.	Predictors of drug abuse treatment utilization among frequent cocaine users •	94 ASASrvyRsr 553-558
mailback response rates for census ... • Dillman et al.	How prenotice letters, stamped return envelopes and reminder postcards affect	95 SurvyM 21 159-165
	Problems of screening for poverty status • Moeller, John F.; Mathiowetz, Nancy A.	94 JOffStat 10 327-337
Epstein, Joan	Ratio estimation of hard-core drug use • Wright, Douglas; Gfroerer, Joseph;	94 ASASrvyRsr 541-546
Medicaid beneficiaries: Adult ... • Eisenhower et al.	A records-survey comparison of eligibility and health care utilization measures for	94 ASASrvyRsr 713-718
	The reliability of the locality pay estimates • Coleman, Joan L.; James, Penny L.	94 ASASrvyRsr 1031-1035
Pandher, Gurupdesh S.	Resampling inference for quantile shares • Kovačević, Milorad S.; Yung, Wesley;	94 ASASrvyRsr 206-211
	Sample models and weights • Skinner, C. J.	94 ASASrvyRsr 133-142
	Social values do change? (Japanese) • Nisihira, Sigeki	95 PrInStMa 43 81-97
	Some effects of questionnaire design on reported rates of victimization • Paton, D.	94 ASASrvyRsr 443-448
Fienberg, Stephen E.; Jazairi, Nuri	Stanley Warner's contributions to statistically balanced information technology •	95 SurvyM 21 7-12
over past forty years (Japanese) • Sakamoto, Yosiyuki	A statistical research of the Japanese national character: General trend of opinion	95 PrInStMa 43 5-26
McMahon, Paul B.	Statistics of income partnership studies: Evaluation of preliminary estimates •	94 ASASrvyRsr 1001-1006
response technique (Korean) • Lee, H.-Y.; Hong, K.-H.	A study on the efficiency comparison of the black-box method with the randomized	95 KorApS 8/2 27-41
Ringwalt, Chris L.; Greene, Jody M.	Substance abuse among runaway and homeless youth • Iachan, Ronald;	94 ASASrvyRsr 1083-1087
Folsom, Ralph E.; Witt, Michael B.	Testing a new attrition nonresponse adjustment method for SIPP (Disc: p440-442) •	94 ASASrvyRsr 428-433
Bates, Nancy; Gerber, Eleanor	A typology of temporary mobility patterns and reporting of usual residence •	94 ASASrvyRsr 331-336
	Underreporting of harassment in public places • Northrup, David A.	94 ASASrvyRsr 455-460
Beatty, Paul	Understanding the standardized/non-standardized interviewing controversy •	95 JOffStat 11 147-160
Dorinski, Suzanne M.; Huang, Hertz	Use of administrative data in SIPP longitudinal estimation (Disc: p680-682) •	94 ASASrvyRsr 656-661
Gaertner, Gregory; Keil, Linda	The validity of self-reported drug use among the incarcerated • Cantor, David;	94 ASASrvyRsr 1088-1093
Brodsky, Marc D.	Variance estimation of drug abuse episodes using the bootstrap • Hughes, Arthur L.;	94 ASASrvyRsr 212-217
Pedro J.; Weir, Paula	Variance estimation for EIA-877 propane prices • O'Colmain, Benita J.; Saavedra,	94 ASASrvyRsr 232-235
replacement estimators? • Korn, E. L.; Graubard, B. I.	Variance estimation for superpopulation parameters: Should one use with-	94 ASASrvyRsr 124-132
	• see also opinion poll, panel data, sample -	
	• see also names of specific surveys such as Current Population -	
	• see also names of specific types of surveys such as establishment - or telephone -	
Nonresponse in federal demographic	surveys: 1981-1991 • Johnson, Ayah E.; Botman, Steven L.; Basiotis, Peter	94 ASASrvyRsr 983-988
Scheible, Paul; Wiatrowski, William The	survey of 1992 health expenditures: A quick response survey • Ponikowski, Chester H.;	94 ASASrvyRsr 282-287
Analysis of large health	surveys: Accounting for the sampling design • Korn, Edward L.; Graubard, Barry I.	95 JRSS-A 158 263-295
and reporting nonresponse rates in agricultural	surveys • Roy, Laurent Measuring	94 ASASrvyRsr 1156-1161
Fink, Steven Some data issues in school-based	surveys • Kasprzyk, Daniel; Gruber, Kerry; Salvucci, Sameena; Saba, Mehrdad; Zhang, Fan;	94 ASASrvyRsr 815-820
Odom-Maryon, T. L.; Lee, J.; Tilley, B. C. A	survey of biostatistical consulting units throughout North America • Niland, J. C.;	95 AmerStat 49 183-189
and nonemployers in the Current Business	Surveys • Konschnik ... the business area sample as a means of covering birth employers	94 ASASrvyRsr 995-1000
Scheduling initial calls in a telephone	survey of businesses • Edwards, W. Sherman; Cunningham, Patricia	94 ASASrvyRsr 1368-1373
Controlling invasion of privacy in	surveys of change over time - A non-technical review • Dalenius, Tore	95 JOffStat 11 107-113
Torres et al. The effect and adjustment of complex	surveys on chi-squared goodness-of-fit tests: Some Monte Carlo evidence • Aguirre-	94 ASASrvyRsr 602-607
ciaries: Adult and ... • Eisenhower et al. A records-	survey comparison of eligibility and health care utilization measures for Medicaid benefi-	94 ASASrvyRsr 713-718
Analysis of domain means in complex	surveys • Bellhouse, D. R.; Rao, J. N. K.	94 ASASrvyRsr 29-34
Resampling methods for complex	surveys • Rao, J. N. K.	94 ASASrvyRsr 35-41
crossover study to compare two ... • Davis et al.	Survey and conditional logistic regression methods for the analysis of a multiperiod	94 ASAProBiop 446-451
	Survey data	
• Nascimento Silva, P. L. D.; Skinner, C. J.	Estimating distribution functions with auxiliary information using poststratification	95 JOffStat 11 277-294
Donna L.; Singh, Gopal K.; Rosenberg, Harry M.	Sources of data on socio-economic differential mortality in the United States • Hoyert,	95 JOffStat 11 233-260
	Testing the logical structure of questionnaires • Willenborg, L. C. R. J.	95 StNeerla 49 95-109
techniques in categorical data analysis for	survey data • Rai, Anil Variance estimation	94 JIAgSt 46 147-155
How to analyze	survey data • Fink, Arlene	95 SagePubl:CA:Lnd 112p
timating some measures of income inequality from	survey data: An application of the estimating equations approach • Binder, D. A. et al. Es-	95 SurvyM 21 137-145
of our business": Pushing household boundaries in	survey data collection • Hainer, Peter C. "Strivin', gettin' over, and keepin' people out	94 ASASrvyRsr 337-341
Two-sample McNemar tests for complex	survey data • Thompson, Jenny; Fisher, Robin	94 ASASrvyRsr 608-613
of some methods for rationalizing the editing of	survey data • Granquist, L. Macro-editing - A review	91 StatJUN 8 137-154
Income statistics from	survey data: Effects of respondent rounding • Rowe, Geoff; Gribble, Steve	94 ASAProBuEc 77-82
Forecasting US home sales using BVAR models and	survey data on households' buying attitudes for homes • Dua, Pami; Smyth, David J.	95 JForcast 14 217-227
Some recent applications of resampling methods to	survey data • Rao, J. N. K.	92 ProIrStC1 163-175
Mixed models for	survey data • Shah, Babubbai V.; LaVange, Lisa M.	94 ASAProStEd 111-116
Poststratification of pooled	survey data • Fahimi, Mansour	94 ASASrvyRsr 879-881
Moitra, Soumyo Using	survey data to predict adoption and switching for services • Weerahandi, Samaradasa;	95 JMarRsr 32 85-96
visit reports: A comparison of medical record and	survey data (Disc: p379-381) • Jay, G. M.; Belli, R. F.; Lepkowski, J. M. Quality of last doctor	94 ASASrvyRsr 362-367
Regression analysis of repeated	survey data (with available software) • Kott, Phillip S.	94 ASASrvyRsr 116-123
Control charts for customer satisfaction	survey data • Kushler, Robert	94 ASAProQlPr 334-340
Analyzing imputed	survey data sets with model assisted estimators (Disc: p906-907) • Fay, Robert E.	94 ASASrvyRsr 900-905
Antony R. Selecting explanatory variables from	survey data: A simple approach that explores for interactions • Stack, Jim; Unwin,	95 Ststcian 44 17-29
Experiments with variance estimation from	survey data with imputed values • Lee, Hyunshik; Rancourt, Eric; Särndal, Carl E.	94 JOffStat 10 231-243
Statistical matching of	survey datafiles: A simulation study • Kovačević, Milorad S.; Liu, Tzen-Ping	94 ASASrvyRsr 479-484
How to design	surveys • Fink, Arlene	95 SagePubl:CA:Lnd 96p

• Diehr, P.; Martin, D.; Koepsell, T. et al. Optimal	survey design for community-intervention evaluations: Cohort, cross-sectional, or both?	94 ASAProEpid 34-43
of the meaning of training: Implications for	survey design • Campanelli, P.; Channell, J. et al. A cognitive and linguistic exploration	94 ASASrvyRsr 1281-1286
The National Population Health	Survey: Design and issues • Singh, M. P.; Tambay, J. L.; Krawchuk, S.	94 ASASrvyRsr 803-808
A split questionnaire	survey design • Raghunathan, Trivellore E.; Grizzle, James E.	95 JASA 90 55-63
Mangat, Naurang Singh Mail	survey design for sensitive quantitative variable • Singh, Ravindra; Singh, Sarjinder;	95 Metron 53 43-54
Resampling variance estimates for complex	survey designs - A simulation study • Robb, William H.	94 ASASrvyRsr 218-221
use of hand-held computers during demographic	surveys in developing countries • Forster, D.; Snow, R. W. An assessment of the	95 SurvyM 21 179-184
of household sampling schemes for health	surveys in developing countries • Bennett, S.; Radalowicz, A. et al. A computer simulation	94 IJEpidem 23 1282-1291
O'Reilly, James M. Results from the National	Survey of Family Growth cycle V pretest • Lessler, Judith T.; Weeks, Michael F.;	94 ASASrvyRsr 64-70
Elena Investigating nonresponse in federal	surveys • Shettle, Carolyn F.; Guenther, Patricia M.; Kaspryzk, Daniel; Gonzalez, Maria	94 ASASrvyRsr 972-976
Song, Frank Are	survey forecasts of macroeconomic variables rational? • Aggarwal, Raj; Mohanty, Sunil;	95 JBusUChi 68 99-119
A new framework for analyzing	survey forecasts using three-dimensional panel data • Davies, Anthony; Lahiri, Kajal	95 JEconmtx 68 205-227
Secondary analysis of continuous government	surveys • Harrop, Anne; Plewis, Ian Two decades of family change:	95 JRSS-A 158 91-106
The	survey handbook • Fink, Arlene	95 SagePubl:CA:Lnd 144p
Sample design of the National Pregnancy and Health	Survey • DiGaetano, Ralph; Ward, John; Cadell, Diane; Ryaboy, Lana	94 ASASrvyRsr 1100-1104
Cross-sectional weighting of longitudinal	surveys of individuals and households using the weight share method • Lavallée, Pierre	95 SurvyM 21 25-32
form • Gabbard, S. M. et al. Evaluating translations of	survey instruments in Spanish and English: The 1990 Spanish Language Census long	94 ASASrvyRsr 1232-1235
What respondents learn from questionnaires: The	survey interview and the logic of conversation (Disc: p169-177) • Schwarz, Norbert	95 IntStRvw 63 153-168
"What respondents learn from questionnaires: The	survey interview and the logic of conversation" • Cantor, David Discussion of	95 IntStRvw 63 173-177
"What respondents learn from questionnaires: The	survey interview and the logic of conversation" • Singer, Eleanor Discussion of	95 IntStRvw 63 169-172
sisted interviewing: The experience of the Canadian	survey of labour and income dynamics (Disc: p861-862) • Grondin et al. ... computer as-	94 ASASrvyRsr 839-844
Nonresponse and its effects in a followup telephone	survey of low income women • Kalsbeek, William D.; Durham, Todd A.	94 ASASrvyRsr 943-948
C. M.; Brown, T. L. (Auth) et al. Review of "Angler	survey methods and their applications in fisheries management" • Pollock, K. H.; Jones,	95 Biomtrcs 51 385-385
Colledge, M. J.; Kott, P. S. (Ed) Business	survey methods • Cox, B. G.; Binder, D. A.; Chinnappa, B. N.; Christianson, A.;	95 Wiley:NY:UK 732p
Counting gays and lesbians: A case for "new"	survey methods • Edwards, Brad	94 ASASrvyRsr 1305-1309
J. (Auth); Frey, James H. (Rev) Review of "Telephone	survey methods: Sampling, selection and supervision (Second edition)" • Lavrakas, Paul	94 JOffStat 10 349-351
through graduate training: The joint program in	survey methodology • Wallman, K.; Groves, R. M. et al. Increasing statistical awareness	94 ASAProSoSt 39-44
A bibliography on telephone	survey methodology • Khurshid, Anwer; Sahai, Haredeo	95 JOffStat 11 325-367
The impact of using priority mail in a mixed mode	survey • Mitchell, Susan; Mooney, Geraldine; Shettle, Carolyn	94 ASAProGovt 88-92
from a sample size reduction in a large national	survey • Botman, Steven L. Estimation and analysis effects resulting	94 ASASrvyRsr 1168-1172
P. Theoretical motivation for post-	survey nonresponse adjustment in household surveys • Groves, Robert M.; Couper, Mick	95 JOffStat 11 93-106
Caspar, Rachel Continuous quality improvement for	survey operations: Some general principles and applications • Biemer, Paul;	94 JOffStat 10 307-326
A proposal for a	survey-oriented metainformation system • Sundgren, B.	92 StatJUN 9 139-174
algorithm for maximizing the overlap between	surveys • Ernst, Lawrence R.; Ikeda, Michael M. A reduced-size transportation	95 SurvyM 21 147-157
Nora Cate Using	survey participants to estimate the impact of nonparticipation • Lin, I-fen; Schaeffer,	95 PubOpinQ 59 236-258
refusal ... • Kojetin et al. Response to a government	survey as political participation: The relation of economic and political conditions to	94 ASASrvyRsr 1236-1241
Review of "The Phantom respondents: Opinion	surveys and political representation" • Brehm, John (Auth); Parsley, Teresa L. (Rev)	94 JOffStat 10 351-355
for investigating data of a psychological	survey • Schweizer, Karl Perspectives	95 CmpStQ 10 1-8
How to ask	survey questions • Fink, Arlene	95 SagePubl:CA:Lnd 120p
Asking sensitive questions in	surveys • Hutchinson, Paul	95 TeachgSt 17/2 43-43
Research on establishment-	survey questionnaire design • Phipps, Polly A.; Butani, Shail J.; Chun, Young I.	95 JBES 13 337-346
S. M.; Breidt, F. J. Approximating the variance of the	survey regression estimator using poststratification • Leonard, K. A.; An, A. B.; Nusser,	94 ASASrvyRsr 222-227
How to measure	survey reliability and validity • Litwin, Mark S.	95 SagePubl:CA:Lnd 96p
How to report on	surveys • Fink, Arlene	95 SagePubl:CA:Lnd 112p
Jobe, J. B. Risk and loss perceptions associated with	survey reporting of sensitive behaviors • Rasinski, K. A.; Baldwin, A. K.; Willis, G. B.;	94 ASASrvyRsr 497-502
Vehovar, Vasja Handling "Don't know"	survey responses: The case of the Slovenian plebiscite • Rubin, Donald B.; Stern, Hal S.;	95 JASA 90 822-828
Correlations in randomized response	surveys • Han, Geun-Shik; Warde, William D.	94 ASASrvyRsr 491-496
distinct respondents in a randomized response	survey • Mangat, N. S.; Singh, R.; Singh, S.; Bellhouse, D. R. et al. On efficiency of using	95 SurvyM 21 21-23
finite population total under randomized response	surveys • Arnab, Raghunath Optimal estimation of a	95 MaOpfStS 27 175-180
How to sample in	surveys • Fink, Arlene	95 SagePubl:CA:Lnd 96p
National Employer Health Insurance	Survey sample design • Marker, David A.; Bryant, Edward C.; Moriarity, Chris	94 ASASrvyRsr 264-269
	Survey sampling	
components of small domain ... • Chaudhuri et al.	Asymptotic design-cum-model approach for convex weighting of direct and indirect	95 CommStA 24 2803-2813
registers and postcode address file as ... • Lynn et al.	On the bias and variance of samples of individuals: Comparison of the electoral	95 Ststcian 44 173-194
	Estimation in the presence of measurement error (Disc: p142-147) • Fuller, Wayne A.	95 IntStRvw 63 121-141
Mangat, N. S.; Tracy, D. S.	An improved two stage randomized response strategy • Singh, R.; Singh, S.;	95 StHefte 36 265-271
Butani, Shail J.; Chun, Young I.	Research on establishment-survey questionnaire design • Phipps, Polly A.;	95 JBES 13 337-346
Ohlsson, Esbjörn	Variance estimation in the Swedish consumer price index • Dalén, Jörgen;	95 JBES 13 347-356
	Survey sampling • Kish, Leslie	95 Wiley:NY:UK 661p
(Rev) Review of "Model assisted	survey sampling" • Särndel, C.-E.; Swensson, B.; Wretman, J. (Auth); Penny, Richard N.	95 Ststcian 44 140-142
Developing a concept of population in	survey sampling consistent with that in experimental statistics • Proctor, C. H.	94 ASASrvyRsr 653-655
Smith, W.; Chang, M. Optimal periodicity of a	survey: Sampling error, data deterioration, and cost • Ghosh, D.; Kaufman, S. F.;	94 ASASrvyRsr 1122-1127
Estimation of parameters in	survey sampling: Optimality • Godambe, V. P.	95 CandJSt 23 227-243
Improved predictors in	survey sampling under a super population model • Srivastava, A. K.	95 StPrLet 25 249-252
manufacturing production and different business	survey series in Sweden 1968-1992 • Bergström, Reinhold The relationship between	95 IntJFor 11 379-393
Improving coverage in a national	survey of teachers • Royce, Daniel; Schreiner, Irwin	94 ASASrvyRsr 827-832
error and cost issues in small area telephone	surveys • Schejbal, Judith A.; Lavrakas, Paul J. Coverage	94 ASASrvyRsr 1287-1292
anything" increase response rates in telephone	surveys? • Pinkleton, B.; Reagan, J.; Aaronson, D.; Ramo, E. Does "I'm not selling	94 ASASrvyRsr 1242-1247
script on respondent self-selection in RDD telephone	surveys • Romuald, K. S.; Haggard, L. M. The effect of varying the respondent selection	94 ASASrvyRsr 1299-1304
reports of child immunization status in a telephone	survey • Haggard, Lois M.; Gray, Donald Z. Improving validity of parental	94 ASASrvyRsr 1258-1262
telephone noncoverage bias with a telephone	survey • Keeter, Scott Estimating	95 PubOpinQ 59 196-217
ance estimation of non-linear functions of totals in	surveys - Theory and software implementation • Andersson, C. et al. A method for vari-	94 JOffStat 10 395-405
Comment on "Education research using	surveys" • Smith, Thomas M.	94 ASAProSoSt 213-215
The role of VA administrative files in the National	Survey of Veterans • Ahuja, Elizabeth M.	94 ASAProGovt 25-27
An overview of the National	Survey of Veterans • Krumhaus, Susan G.; Klemm, Rebecca J.	94 ASAProGovt 19-21
interviewer characteristics on response in a national	survey of violence against women • Norris, Douglas A.; Hatcher, Jeff The impact of	94 ASASrvyRsr 449-454
Nested ANOVA for repeated	surveys with autocorrelated errors • Prayag, V. R.; Chiplonkar, S. A.; Paranjpe, S. A.	95 BiomtrcJ 37 565-570
Cost-effectiveness analysis using quality-adjusted	survival • Cole, Bernard F.; Gelber, Richard D.; Anderson, Keaven M.; Goldhirsch, Aron	94 ASAProBiop 9-18
Pollock, K. H. et al. Estimating prefledging	survival : Allowing for brood mixing and dependence among brood mates • Flint, P. L.;	95 JWildlMg 59 448-455
The exponentiation formula of reliability and	survival: Does it always hold? • Singpurwalla, Nozer D.; Wilson, Simon P.	95 LiDaAnal 1 187-194
	Survival analysis	
	Analysis of censored correlated observations • Oskrochi, G. Reza	95 JStPlInf 47 165-180
	Analysis of fracture toughness data • Crowder, Martin	95 LiDaAnal 1 59-71
Robins, James M.	An analytic method for randomized trials with informative censoring: Part II •	95 LiDaAnal 1 417-434

Survival analysis (cont.)

	Applying Cox regression to competing risks • Lunn, Mary; McNeil, Don	95 Biomtrcs 51 524-532
fitting • Li, Gang; Doss, Hani	An approach to nonparametric regression for life history data using local linear	95 AnlsStat 23 787-823
Graphical methods • Cohen, Ayala; Barnett, Ofra	Assessing goodness-of-fit of parametric regression models for lifetime data –	95 StatMed 14 1785-1795
data methods for censored regression • Lai et al.	Asymptotic normality of a class of adaptive statistics with applications to synthetic	95 JMultiAn 52 259-279
	Asymptotic theory for the frailty model • Murphy, S. A.	95 AnlsStat 23 182-198
Gelfand, Alan E.; Mallick, Bani K.	Bayesian analysis of proportional hazards models built from monotone functions •	95 Biomtrcs 51 843-852
tributed random censoring • Yeh, Lam; Choy, S. T. B.	Bayesian variable sampling plans for the exponential distribution with uniformly dis-	95 JStPlInf 47 277-293
Seungyeoun; Wolfe, Robert A.	A class of linear nonparametric rank tests for dependent censoring • Lee,	95 CommStA 24 687-712
continuous time • Jacobsen, Martin; Keiding, Niels	Coarsening at random in general sample spaces and random censoring in	95 AnlsStat 23 774-786
Gelber, R. D.; Cole, B. F.; Gelber, S.; Goldhirsch, A.	Comparing treatments using quality-adjusted survival: The Q-TWiST method •	95 AmerStat 49 161-169
Koehler, Kenneth J.; Symanowski, James T.	Constructing multivariate distributions with specific marginal distributions •	95 JMultiAn 55 261-282
truncation model with application to hazard ... • Gu	Convergence of increments for cumulative hazard function in a mixed censorship-	95 StPrLet 23 135-139
Bryan; Borgan, Ørnulf	Counter-matching: A stratified nested case-control sampling method • Langholz,	95 Biomtrka 82 69-79
	Daniels's association measures under right censoring • O'Quigley, John	95 AplStMDA 11 109-119
Els J. T.; Pocock, Stuart J.	Detection and estimation of *J*-shaped risk-response relationships • Goetghebeur,	95 JRSS-A 158 107-121
	Dual likelihood • Mykland, Per Aslak	95 AnlsStat 23 396-421
Smith, Fraser B.; Helms, Ronald W.	EM mixed model analysis of data from informatively censored normal distributions •	95 Biomtrcs 51 425-436
hazard specification • Geoffard, P.-Y.; Philipson, T.	The empirical content of canonical models of infectious diseases: The proportional	95 Biomtrka 82 101-111
Ghitany, M. E.; Maller, R.; Zhou, S.	Estimating the proportion of immunes in censored samples: A simulation study •	95 StatMed 14 39-49
	Extending the elements of tree-structured regression • Segal, Mark R.	95 StMedRes 4 219-236
Cindy	Fitting Weibull duration models with random effects • Morris, Carl; Christiansen,	95 LiDaAnal 1 347-359
Moyer, E. S.	Frailty models of manufacturing effects • Wassell, J. T.; Kulczycki, G. W.;	95 LiDaAnal 1 161-170
van der Laan, Mark	Generalizations of current status data with applications • Jewell, Nicholas P.;	95 LiDaAnal 1 101-109
	Generalized additive models for medical research • Hastie, Trevor; Tibshirani, Robert	95 StMedRes 4 187-196
Yuen, Kam C.	Goodness-of-fit tests for the Cox model via bootstrap method • Burke, Murray D.;	95 JStPlInf 47 237-256
Study • Klein, John P.; Keiding, Niels; Kreiner, Svend	Graphical models for panel studies, illustrated on data from the Framingham Heart	95 StatMed 14 1265-1290
Székely, Gábor J.	Inference for a general type II censorship model • Peña, Edsel A.; Rohatgi, Vijay K.;	95 MaOpfStS 26 241-252
analysis of ringing recoveries • Aebischer, N. J.	Investigating the effects of hunting on the survival of British pigeons and doves by	95 JAppStat 22 923-934
Nielsen, Jens P.; Linton, Oliver B.	Kernel estimation in a nonparametric marker dependent hazard model •	95 AnlsStat 23 1735-1748
Nations, Chris; Diem, Kenneth L.; Pradel, Roger	Mark-resighting analysis of a California gull population • Pugesek, Bruce H.;	95 JAppStat 22 625-639
Marshall, G.; Guo, W.; Jones, R. H.	MARKOV: A computer program for multi-state Mařkov models with covariables •	95 CmpBiomd 47 147-156
	Maximin efficiency-robust tests and some extensions • Bose, Sudip; Slud, Eric	95 JStPlInf 46 105-121
model • Borgan, Ø.; Goldstein, L.; Langholz, B.	Methods for the analysis of sampled cohort data in the Cox proportional hazards	95 AnlsStat 23 1749-1778
Malani, Hina Mehta	A modification of the redistribution to the right algorithm using disease markers •	95 Biomtrka 82 515-526
	Multi-state models and diabetic retinopathy • Marshall, Guillermo; Jones, Richard H.	95 StatMed 14 1975-1983
Tsodikov, Alexandr D.; Anisimov, Vladimir N.	A new model of aging: Specific versions and their application • Yakovlev, Andrej Yu.;	95 BiomtrcJ 37 435-448
recurrence of breast cancer • dos Santos, D. M. et al.	Nonparametric hazard versus nonparametric frailty distribution in modelling	95 JStPlInf 47 111-127
Markov process • Guo, W.; Marshall, G.	ORDMKV: A computer program fitting proportional odds model for multi-state	95 CmpBiomd 46 257-263
failure types • Hemmi, Isao	Parameter estimation in Cox's regression model for competing risks with missing	95 JJapnStS 25 81-96
Omar, Rumana Z.; Stallard, Nigel; Whitehead, John	A parametric multistate model for the analysis of carcinogenicity experiments •	95 LiDaAnal 1 327-346
	Population study and management of Atlantic Flyway Canada geese • Hestbeck, Jay B.	95 JAppStat 22 877-890
Aragon, Jorge	Regression with interval-censored data • Rabinowitz, Daniel; Tsiatis, Anastasios;	95 Biomtrka 82 501-513
Clobert, Jean; Anderson, David R.	Return rates in studies of life history evolution: Are biases large? • Martin, Thomas E.;	95 JAppStat 22 863-875
based on group sequential logrank tests • Kim, K.	SEQPWR and SEQOPR: Computer programs for design of maximum information trials	95 CmpBiomd 46 143-153
	A simple, flexible failure model • Yu, Shuiyang; Voit, Eberhard O.	95 BiomtrcJ 37 595-609
Liang, K.-Y.; Self, S. G.; Bandeen-Roche, K. J. et al.	Some recent developments for regression analysis of multivariate failure time data •	95 LiDaAnal 1 403-415
James, Ian R.; French, Martyn A. H.; Mallal, Simon A.	Statistical issues in the evaluation of markers of HIV progression • Segal, Mark R.;	95 IntStRvw 63 179-197
Klein, John P.	Statistical methods for dependent competing risks • Moeschberger, M. L.;	95 LiDaAnal 1 195-204
Crowley, John	Step-function covariate effects in the proportional-hazards model • LeBlanc, Michael;	95 CandJSt 23 109-129
geneity? • Pradel, Roger; Cooch, Evan; Cooke, Fred	Transient animals in a resident population of snow geese: Local emigration or hetero-	95 JAppStat 22 695-710
van de Pol, F. J. R.	The transition from school to work in 1979-1987 in the Netherlands • van Opstal, R.;	95 StNeerla 49 294-309
individual frailty: An advantageous approach to	survival analysis of bivariate data • Yashin, A. I.; Vaupel, J. W.; Iachine, I. A. Correlated	95 MaPopStd 5 145-159
for the iterative maximum likelihood estimator in	survival analysis • Carling, Kenneth Starting values	95 JAppStat 22 531-535
Interactive dynamic graphics for exploratory	survival analysis • Atkinson, E. Neely	95 AmerStat 49 77-84
Likelihood ratio-based confidence intervals in	survival analysis • Murphy, S. A.	95 JASA 90 1399-1405
Statistical considerations for	survival analysis from medical device clinical studies • Lao, Chang S.	95 JBiophSt 5 159-170
Winterstein, S. R.; Chen, C.-L. A capture-recapture	survival analysis model for radio-tagged animals • Pollock, K. H.; Bunck, C. M.;	95 JAppStat 22 661-672
Historical controls and modern	survival analysis • Keiding, Niels	95 LiDaAnal 1 19-25
are recoveries of North American passerines for	survival analyses? • Francis, Charles M. How useful	95 JAppStat 22 1075-1081
	Survival analysis: A practical approach • Parmar, Mahesh K. B.; Machin, David	95 Wiley:NY:UK 255p
Juritz, J. M. (Rev) Review of	"Survival analysis: A practical approach" • Parmar, Mahesh K. B.; Machin, David (Auth);	96 ShrtBkRv 16/1 2-3
A. (Ed) Lifetime data: Models in reliability and	survival analysis • Jewell, Nicholas P.; Kimber, Alan C.; Lee, Mei-Ling Ting; Whitmore, G.	95 Kluwer:MA:Hllnd 424p
of specific occurrence/exposure rates in risk and	survival analysis • Tu, Dongsheng; Gross, Alan J. Accurate confidence intervals for the ratio	95 BiomtrcJ 37 611-626
Multiple time scales in	survival analysis • Oakes, David	95 LiDaAnal 1 7-18
	Survival analysis. A self-learning text • Kleinbaum, D. G.	95 Springer:Brln:NY 250p
Trees and splines in	survival analysis • Intrator, Orna; Kooperberg, Charles	95 StMedRes 4 237-261
Rainer (Rev) Review of	"Survival analysis: State of the art" • Klein, John P.; Goel, Prem K. (Auth); Schlittgen,	95 CmpStDA 20 592-593
vu all over again: Using multiple-spell discrete-time	survival analysis • Willett, John B.; Singer, Judith D. It's déjà	95 JEdStat 20 41-67
test analogue to Tarone's test for trend in	survival analysis • Graves, Thomas S.; Pazdan, James L. A permutation	95 JStCmpSm 53 79-89
Louis, Thomas A.	Survival analysis using a scale change random effects model • Anderson, Jon E.;	95 JASA 90 669-679
investigation on measures of explained variation in	survival analysis • Graf, Erika; Schumacher, Martin An	95 Ststcian 44 497-507
	Survival analysis with median regression models • Ying, Z.; Jung, S. H.; Wei, L. J.	95 JASA 90 178-184
due to self-selection statistical bias (selective	survival bias) • Blauwet, M. B. et al. Changing risk factor relationships in aging populations	94 ASAProEpid 76-81
Investigating the effects of hunting on the	survival of British pigeons and doves by analysis of ringing recoveries • Aebischer, N. J.	95 JAppStat 22 923-934
Testing hypotheses and estimating	survival from capture histories with CR • Pradel, R.; Reboulet, A.-M.; Viallefont, A.	95 JAppStat 22 775-784
A proportional regression model for 20 year	survival of colon cancer in Norway • Zahl, Per-Henrik	95 StatMed 14 1249-1261
Comparison of Gamma	survival curves (Spanish) • Beamonte, Eduardo; Bermúdez, José D.	95 Questo 19 171-186
	Survival curves subjected to occasional insults • Zelterman, Daniel; Curtsinger, James W.	95 Biomtrcs 51 1140-1146
Ramaswamy, R.; D'Agostino, R. B. Age-adjusted	survival curves with application in the Framingham study • Cupples, L. A.; Gagnon, D. R.;	95 StatMed 14 1731-1744
	Survival data	
censored data • Rao, B. L. S. Prakasa; Sen, Arusharka	An application of the 'martingale methods' to nonparametric regression based on	95 StatDecn 13 201-220
data • Cai, Jianwen; Prentice, Ross L.	Estimating equations for hazard ratio parameters based on correlated failure time	95 Biomtrka 82 151-164
	Fitting parametric counting processes by using log-linear models • Lindsey, J. K.	95 ApplStat 44 201-212

	Survival probability	
	Asymptotic behavior of basic contact process with rapid stirring • Konno, Norio	95 JTherPrb 8 833-876
Bayesian estimation of movement and	survival probabilities from capture-recapture data • Dupuis, Jerome A.	95 Biomtrka 82 761-772
On nonparametric likelihood ratio estimation of	survival probabilities for censored data • Li, Gang	95 StPrLet 25 95-104
Comparing treatments using quality-adjusted	survival: The Q-TWiST method • Gelber, R. D.; Cole, B. F.; Gelber, S.; Goldhirsch, A.	95 AmerStat 49 161-169
B. J. T.; Pradel, R.; Freeman, S. N. A simultaneous	survival rate analysis of dead recovery and live recapture data • Lebreton, J.-D.; Morgan,	95 Biomtrcs 51 1418-1428
not used to freezing temperatures: An analysis of	survival rates • Senar, J. C. et al. Mediterranean house sparrows (Passer domesticus) are	95 JAppStat 22 1069-1074
	Survival rates of breeding common gulls in Estonia • Rattiste, Kalev; Lilleleht, Vilju	95 JAppStat 22 1057-1062
netting ... • Desante et al. Productivity indices and	survival rate estimates from MAPS, a continent-wide programme of constant-effort mist-	95 JAppStat 22 935-947
Szép, Tibor	Survival rates of Hungarian sand martins and their relationship with Sahel rainfall •	95 JAppStat 22 891-904
Estimating	survival rates from recoveries of birds ringed as young: A case study • Francis, Charles M.	95 JAppStat 22 567-577
The relative	survival rate • Henson, Donald Earl; Ries, Lynn A.	95 Cancer 76 1687-1688
Morgan, B. J. T. Modelling age variation in	survival and reporting rates for recovery models • Catchpole, E. A.; Freeman, S. N.;	95 JAppStat 22 597-609
A two-stage procedure for	survival studies with surrogate endpoints • Flandre, Philippe; O'Quigley, John	95 Biomtrcs 51 969-976
tribution • Efird et al. A non-parametric two-sample	survival test based on a single occupancy Fermi-Dirac model for the discrete range dis-	94 ASAProBiop 166-170
Random thoughts on	survival (Disc: 127-128) • Cox, David R.	95 Student 1 125-126
Two methods of estimating the mean	survival time from censored samples • Zukang, Zheng	95 SankhyaA 57 126-136
On omnibus test for independence of a	survival time from a covariate • McKeague, Ian W.; Nikabadze, A. M.; Sun, Yanqing	95 AnlsStat 23 450-475
Comparison of median	survival times with adjustment for covariates • Karrison, Theodore	95 StatMed 14 2537-2553
incubation period of AIDS • Bacchetti, P.; Segal, M. R.	Survival trees with time-dependent covariates: Application to estimating changes in the	95 LiDaAnal 1 35-47
Conditions on the arrival process to obtain HNBUE	survival using a shock model • Pérez-Ocón, Rafael; Gámiz-Pérez, M. Luz	95 CommStA 24 931-944
association of a time-dependent state variable to	survival • Cantor, Alan B. A test of the	95 CmpBiomd 46 101-105
Estimation of	survival with time dependent frailty • McGilchrist, C. A.; Yau, K. K. W.	94 ASAProEpid 88-90
Green, J. R. Disease	susceptibility genes and the sib-pair method: A review of recent methodology • Shah, S.;	94 AnHuGene 58 381-395
and period HIV infection intensity, considering the	susceptible population • Verdecchia et al. A back-calculation method to estimate the age	95 StatMed 14 1513-1530
	Sustainability of the deficit process with structural shifts • Quintos, Carmela E.	95 JBES 13 409-417
On the sampling distribution of	swap-rate • Kikuchi, Kenichi; Mayekawa, Shin-ichi	95 Behvmtrk 22 185-204
production and different business survey series in	Sweden 1968-1992 • Bergström, Reinhold The relationship between manufacturing	95 IntJFor 11 379-393
power plant facilities and childhood leukaemia in	Sweden • Waller et al. ... and assessment of clusters of disease: An application to nuclear	95 StatMed 14 3-16
Pretesting procedures at Statistics	Sweden's Measurement, Evaluation and Development Laboratory • Bergman, Lars R.	95 JOffStat 11 309-323
on "Computing elementary aggregates in the	Swedish Consumer Price Index" (V8 p129-147) • Reinsdorf, Marshall Comment	94 JOffStat 10 103-107
on "Computing elementary aggregates in the	Swedish Consumer Price Index" • Dalén, Jörgen Reply to comment	94 JOffStat 10 107-108
Quantifying errors in the	Swedish consumer price index • Dalén, Jörgen	95 JOffStat 11 261-275
Variance estimation in the	Swedish consumer price index • Dalén, Jörgen; Ohlsson, Esbjörn	95 JBES 13 347-356
pencil interviewing: Effects on the results in the	Swedish labour force surveys • Bergman, L. R. et al. Decentralised CATI versus paper and	94 JOffStat 10 181-195
Labour-use efficiency in	Swedish social insurance offices • Kumbhakar, Subal C.; Hjalmarsson, Lennart	95 JApEcmtx 10 33-47
Moore, Rich Toward vital status	sweeps: A case history in sequential monitoring • Hallstrom, Al; McBride, Ruth;	95 StatMed 14 1927-1931
	SWEEP operator for least-squares subject to linear constraints • Neytchev, Plamen Nikolov	95 CmpStDA 20 599-609
Statistical inference for	switching Brownian motions • Frydman, Halina; Hussain, Taukir	95 SankhyaA 57 11-28
Devaraj; Soundararajan, V. Quick	switching double sampling system indexed by the crossover point • Arumainayagam, S.	95 CommStB 24 765-773
Construction and selection of quick	switching double-sampling system: Sample size tightening • Arumainayagam, S. D. et al.	95 JAppStat 22 105-119
distributions of the Cox processes with fast Markov	switchings • Di Masi, G. B. et al. A first order approximation for the convergence of	95 Stochatx 54 211-219
try, nonproportionality, and heterogeneity in brand	switching from piece-wise exponential mixture hazard models • Wedel et al. ... asymme-	95 JMarRsr 32 457-462
Using survey data to predict adoption and	switching for services • Weerahandi, Samaradasa; Moitra, Soumyo	95 JMarRsr 32 85-96
Construction and evaluation of matched quick	switching systems • Arumainayagam, S. D.; Soundararajan, V.	95 JAppStat 22 245-251
Image of "Made in	Switzerland" (French) • Brossard, Hubert	95 Student 1 51-61
Forecast of unemployment in	Switzerland • Frey, Béatrice	95 Student 1 1-15
rectilinear crossing number of a complete graph and	Sylvester's "Four point problem" of geometric probability • Scheinerman, E. R. et al. The	94 AmMaMnth 101 939-943
statistics • Harvill, Jane L.; Newton, H. Joseph Using	symbolic math to evaluate saddlepoint approximations for the difference of order	95 CommStB 24 781-791
Chidananda; Ravi, T. V. Divisive clustering of	symbolic objects using the concepts of both similarity and dissimilarity • Gowda, K.	95 PattrnRc 28 1277-1282
	Symmetric α-stable distributions	
Rutkowski, Marek	Left and right linear innovations for a multivariate S α S random variable •	95 StPrLet 22 175-184
The expected number of level crossings for certain	symmetric α-stable processes • Michna, Zbigniew; Rychlik, Igor	95 CommStSM 11 1-19
concentration • Huh, M.-H.; Critchlow, D. E. et al. A	symmetric analysis of paired rankings with application to temporal patterns of hormonal	95 Biomtrcs 51 1361-1371
On e-optimal fractions of	symmetric and asymmetric factorials • Mukerjee, Rahul	95 StSinica 5 515-533
A class of	symmetric bivariate uniform distributions • Ferguson, T. S.	95 StHefte 36 31-40
Covering with blocks in the non-	symmetric case • Móri, Tamás F.	95 JTherPrb 8 139-164
fractions and inverse Gaussian distribution on a	symmetric cone • Bernadac, Evelyne Random continued	95 JTherPrb 8 221-258
theorem for the Wishart distributions on	symmetric cones • Letac, G.; Massam, H. Craig-Sakamoto's	95 AnInStMa 47 785-799
On	symmetric designs with parameters (176, 50, 14) • Janko, Zvonimir	95 JCombThA 72 310-314
K.; Kramer, E. Construction of new	symmetric designs with parameters (66, 26, 10) • Matulić-Bedenić, I.; Horvatić-Baldasar,	95 JComDsgn 3 405-410
Sharp estimates for capacities and applications to	symmetric diffusions • Sturm, K.-T.	95 ProbTher 103 73-89
Review of "Introduction to the non-	symmetric Dirichlet forms" • Ma, Z. M.; Röckner, M. (Auth); Fukushima, M. (Rev)	95 Metrika 42 70-72
Martin T. Estimation of a loss function for spherically	symmetric distributions in the general linear model • Fourdrinier, Dominique; Wells,	95 AnlsStat 23 571-592
Multivariate l_p-norm	symmetric distributions • Yue, Xinnian; Ma, Chunsheng	95 StPrLet 24 281-288
Loss estimation for spherically	symmetric distributions • Fourdrinier, Dominique; Wells, Martin T.	95 JMultiAn 53 311-331
Shrinkage positive rule estimators for spherically	symmetric distributions • Cellier, Dominique; Fourdrinier, Dominique	95 JMultiAn 53 194-209
El-Saidi, Mohammed A. A	symmetric extended logistic model with applications to experimental toxicity data •	95 BiomtrcJ 37 205-216
On testing structure of mean vector of compound	symmetric Gaussian model • Gupta, A. K.; Nagar, D. K.	95 CommStA 24 3123-3136
Saitoh, T. An invariance principle for non-	symmetric Markov processes and reflecting diffusions in random domains • Osada, H.;	95 ProbTher 101 45-63
Dietrich Minimal moments and cumulants of	symmetric matrices: An application to the Wishart distribution • Kollo, Tõnu; von Rosen,	95 JMultiAn 55 149-164
Principal points and self-consistent points of	symmetric multivariate distributions • Tarpey, Thaddeus	95 JMultiAn 53 39-51
unbiased medians, median unbiased means, and	symmetric random walks • Mattner, L. Mean	95 MaMethSt 4 99-105
waiting time for filling a large interval in the	symmetric simple exclusion process • Ferrari, P. A.; Galves, A.; Liggett, T. M. Exponential	95 AIHenrPB 31 155-175
of n-parameter processes • Hirsch, F.; Song, S.	Symmetric Skorohod topology on n-variable functions and hierarchical Markov properties	95 ProbTher 103 25-43
Loss networks under diverse routing: The	symmetric star network • Hunt, P. J.	95 AdvAppPr 27 255-272
ing of human chromosomes based on elliptically	symmetric statistical distributions • Ritter, G. et al. Automatic context-sensitive karyotyp-	95 PattrnRc 28 823-831
On	symmetric sum of sub-independent random variables • Hamedani, G. G.	95 CalcutSt 45 119-124
	Symmetric tests for cointegration • Deo, Rohit; Fuller, Wayne A.	94 ASAProBuEc 105-110
	Symmetric two-particle exclusion-eating processes • Liu, Xijian	95 AnlsProb 23 1439-1455
linear regression with error terms that have	symmetric unimodal densities • Brunner, Lawrence J. Bayesian	95 JNnparSt 4 335-348
On	symmetrization of product measures • Mashayekhi, M.	95 MaMethSt 4 334-343
Class 3 association schemes whose	symmetrizations have two classes • Song, Sung Y.	95 JCombThA 70 1-29
The sample mid-range and	symmetrized extremal laws • Bingham, N. H.	95 StPrLet 23 281-288

The expected power of F and	t tests conditional on information from an earlier experiment • Gillett, Raphael	95 BJMaStPs 48 371-384
One-sample one-sided student	t-test under contaminants • Zieliński, Ryszard	95 MaOpfStS 27 143-150
A new formula for tail probabilities of Dunnett's	T with unequal sample sizes • Cheng, Peter H.; Meng, Cliff Y. K.	95 CommStA 24 523-532
Ballerini, Rocco Nonconstant levels for	t-year return periods for two-dimensional Poisson processes with periodic intensity •	95 StPrLet 22 43-47
Stevens, C. E. Change points in the series of	T4 counts prior to AIDS • Kiuchi, A. S.; Hartigan, J. A.; Holford, T. R.; Rubinstein, P.;	95 Biomtrcs 51 236-248
	Table	
	• see also names of specific types of tables such as contingency - or life -	
limits for the common odds ratio in several 2×2	tables • Liao, J. G.; Hall, Charles B. Fast and accurate computation of the exact confidence	95 JCmpGrSt 4 173-179
Fisher, χ^2, Student and Pearson tests and 2×2	tables • Martín Andrés, A.; Tejedor, I. Herranz; Mato, A. Silva The Wilcoxon, Spearman,	95 Ststcian 44 441-450
method Mantel-Haenszel estimator for K $2 \times J$	tables • Yanagawa, Takashi; Fujii, Yoshinori Projection-	95 JASA 90 649-656
F ratios sharing a common denominator in ANOVA	table • Kang, Phillee Dependent	95 JKoreaSt 24 111-125
Optimal unconditional	tables for comparing two independent proportions • Silva Mato, A.; Martín Andrés, A.	95 BiometrcJ 37 821-836
4 • Raju, C.; Murthy, M. N. Narasimha Procedures and	tables for the construction and selection of chain sampling plans ChSP-4 (c_1, c_r) - Part	95 JAppStat 22 261-271
Combining 2×2	tables that contain structural zeros • Johnson, William D.; May, Warren L.	95 StatMed 14 1901-1911
unbiased domain score ... • Li LTDOMAIN: A look-up	table for the corresponding estimated one-parameter logistic model scale score and the	95 ApPsychM 19 50-50
Exact tests of significance in higher dimensional	tables • Zelterman, Daniel; Chan, Ivan Siu-Fung; Mielke, Paul W., Jr	95 AmerStat 49 357-361
Yannoulis, Y. (Auth) et al. Review of "Statistical	tables for economic, business and social studies (Second edition)" • Kmietowicz, Z. W.;	95 JAppStat 22 439-440
in model screening for genotype by environment	tables • van Eeuwijk, F. A.; Keizer, L. C. P. On the use of diagnostic biplots	95 StatComp 5 141-153
(Auth); Jeffrey, A. (Ed); Bello, Abdul L. (Rev) Review of	"Tables of integrals, series, and products (Fifth edition)" • Gradshteyn, I. S.; Ryzhik, I. M.	95 Ststcian 44 287-288
The joint structure of two totally matched	tables (French) • Torre, F.; Chessel, D.	95 RStApp 43/1 109-121
of new models for analysing trends in multiway	tables • Payne, C.; Payne, J. et al. Testing the tightening bond hypothesis: An application	95 JStPlInf 47 181-201
minimum χ^2 estimators with partially observed sub-	tables • Ritov, Ya'acov PM algorithms for calculating	95 CmpStDA 20 19-33
techniques in developing and updating input-output	tables (Arabic) • Kerim, A. et al. RAS and least squares methods versus other mathematical	95 EgpStJ 39 120-143
Repeated $2 \times k$	tables and the sandwich estimator • Von Tress, Mark	94 ASAProBiop 256-261
Kuralmani, V.; Govindaraju, K. Modified	tables for the selection of double sampling attribute plan indexed by AQL and LQL •	95 CommStA 24 1897-1921
New Cambridge statistical	tables (Second edition) • Lindley, D. V.; Scott, W. F.	95 CambridgeUniv 96p
Information from subjects by variables	tables (French) • Antoni, A.; Dhorne, T.	95 RStApp 43/4 43-61
Multidimensional scaling of a three-way	table (French) • Dosse, M. Bennani	95 RStApp 43/4 63-75
approach to conducting a Hayman analysis of diallel	tables with complete or missing data • Rodriguez et al. ... likelihood model-fitting	93 BehavGen 23 69-76
Analysis of square	tables with ordered categories • Poon, Wai-Yin; Hung, Hin-Yan	94 ASAProStCp 292-298
complete longitudinal data from a high-dose trial of	tacrine (Cognex) in Alzheimer's patients • Smith, Fraser Mixed model analysis of in-	94 ASAProBiop 301-306
	Tactical aspects of search and rescue operations • Meloche, J.	95 StPrLet 25 271-279
Visions,	tactics, and TQM for academic programs in quality • Barker, Anne M.	94 ASAProQlPr 294-294
likelihood approach for estimating growth from	tag-recapture data • Wang, You-Gan; Thomas, Mervyn R.; Somers, Ian F. A maximum	95 JFishRBC 52 252-259
capture-recapture survival analysis model for radio-	tagged animals • Pollock, K. H.; Bunck, C. M.; Winterstein, S. R.; Chen, C.-L. A	95 JAppStat 22 661-672
exclusion • Varadhan, S. R. S. Self diffusion of a	tagged particle in equilibrium for asymmetric mean zero random walk with simple	95 AIHenrPB 31 273-285
An improved version of	Taguchi's on-line control procedure • Srivastava, M. S.; Wu, Y.	95 JStPlInf 43 133-145
Economic design of $\bar{X}$ control chart when	Taguchi's loss function is considered • Yang, Su-Fen	95 ASQCTekCon 871-878
	Taguchi method	
shafts • Davis, T. P.	Analysis of an experiment aimed at improving the reliability of transmission centre	95 LiDaAnal 1 275-306
	Creating experimenters by design • Schneider, Michael T.	95 ASQCTekCon 365-366
	D-optimum designs for heteroscedastic linear models • Atkinson, A. C.; Cook, R. D.	95 JASA 90 204-212
	Designing outer array points • Wang, Yuchung; Lin, Dennis K. J.; Fang, Kai-Tai	95 JQualTek 27 226-241
circuits with inductance and resistance ... • Prat et al.	Determination of design factor values to obtain robust products: Application to	95 Questo 19 153-169
	Reducing variation during design • Taylor, Wayne A.	95 ASQCTekCon 217-226
Mak, T. K.	A regression approach for discovering small variation around a target • Chan, L. K.;	95 ApplStat 44 369-377
	A review of robust process design approaches • Grize, Yves L.	95 JChemtrc 9 239-262
	Robust parameter design: A systematic approach • Grize, Yves-L.	94 ASAProPEng 217-224
Schweitzer, Walter	Transformation for experimental design in off-line quality control (German) •	95 AllgeStA 79 117-139
Wu, C. F. J.	The treatment of related experimental factors by sliding levels • Hamada, M.;	95 JQualTek 27 45-55
Terrence L. Finite-dimensional distributions and	tail behavior in stationary interval-valued probability models • Sadrolhefazi, Amir; Fine,	94 AnlsStat 22 1840-1870
Hambly, B. M. On constant	tail behaviour for the limiting random variable in a supercritical branching process •	95 JAppProb 32 267-273
Cheng, Ping The asymptotic	tail behaviours of projection pursuit-type Kolmogorov statistics • Zhu, Li-Xing;	95 StPrLet 25 253-263
Gaussian	tail estimates for Dirichlet processes • Ubøe, Jan; Zhang, Tusheng	95 Stochatx 52 295-302
residual life function at great age: Applications to	tail estimation • Beirlant, J.; Broniatowski, M.; Teugels, J. L.; Vynckier, P. The mean	95 JStPlInf 45 21-48
	Tail events of simulated annealing Markov chains • Niemiro, Wojciech	95 JAppProb 32 867-876
Piotr	Tail events of some nonhomogeneous Markov chains • Niemiro, Wojciech; Pokarowski,	95 AnlsApPr 5 261-293
divisible stochastic processes with exponential	tails • Braverman, Michael; Samorodnitsky, Gennady Functionals of infinitely	95 StocProc 56 207-231
iterated logarithm for the Dekkers-Einmahl-de Haan	tail index estimator • Tobbal, Khelifa A functional law of the	95 JNnparSt 5 145-156
On high level exceedance modeling and	tail inference • Leadbetter, M. R.	95 JStPlInf 45 247-260
Estimation of the	tail parameter in the domain of attraction of an extremal distribution • Alves, M. I. Fraga	95 JStPlInf 45 143-173
Weak limiting behaviour of a simple	tail Pareto-index estimator • Bacro, J. N.; Brito, M.	95 JStPlInf 45 7-19
	Tail probability	
Black, Stephen; Mansouri, H.	On exact distributions of rank tests for ordered alternatives in block designs •	95 CmpStDA 20 265-274
Wood, Andrew T. A.	An example in which the Lugannani-Rice saddlepoint formula fails • Booth, James G.;	95 StPrLet 23 53-61
Garrido	Sensitivity analysis of extreme precipitation events • Katz, Richard W.; Acero, Juan	94 JClimlgy 14 985-999
McLeod, A. I. et al. Cumulant generating function and	tail probability approximations for Kendall's score with tied rankings • Valz, P. D.;	95 AnlsStat 23 144-160
A note on the accuracy of two saddlepoint	tail probability approximations • Booth, James	94 ASAProPEng 56-58
comparison of the exact and asymptotic upper	tail probabilities of chi-squared goodness-of-fit statistics: ... • Formann Small-sample	95 JStCmpSm 51 369-384
Meng, Cliff Y. K. A new formula for	tail probabilities of Dunnett's T with unequal sample sizes • Cheng, Peter H.;	95 CommStA 24 523-532
S. J. Decoupling inequalities for the	tail probabilities of multivariate U-statistics • de la Peña, Victor H.; Montgomery-Smith,	95 AnlsProb 23 806-816
bound is tighter than Chernoff's bound for positive	tail probabilities • Philips, Thomas K.; Nelson, Randolph The moment	95 AmerStat 49 175-178
class • Mita, Haruyoshi	Tail probability of the sequential maximum likelihood estimator for the exponential	95 JJapnStS 25 173-182
Ruymgaart, Frits H.	Tail processes under heavy random censorship applications • Einmahl, John H. J.;	95 JNnparSt 5 223-235
block bootstrap for normalized sums of heavy-	tail random variables • Lahiri, S. N. On the asymptotic behaviour of the moving	95 AnlsStat 23 1331-1349
On the probability of reaching a given head to	tail ratio • Lengyel, Tamás	95 JCISS 20 273-278
Rates of convergence for	tail series • Tomkins, R. J.	95 JStPlInf 43 77-85
	Tail-specific linear approximations for efficient bootstrap simulations • Hesterberg, Tim	95 JCmpGrSt 4 113-133
of discrete distributions suited to fitting very long-	tailed data • Ong, S. H.; Muthaloo, Subarau A class	95 CommStB 24 929-945
and maxima of stationary sequences with heavy	tailed distributions • Anderson, C. W.; Turkman, K. F. Sums	95 SankhyaA 57 1-10
Stărică, Cătălin Testing for independence in heavy	tailed and positive innovation time series • Feigin, Paul D.; Resnick, Sidney I.;	95 CommStSM 11 587-612
Sample quantiles of heavy	tailed stochastic processes • Embrechts, Paul; Samorodnitsky, Gennady	95 StocProc 59 217-233
Household saving behavior in	Taiwan: Tests of functional forms and cointegration • Hsing, Yu; Chang, Hui S.	94 ASAProBuEc 123-128
in applied probability: Papers in honour of Lajos	Takács" • Galambos, J.; Gani, J. (Ed); Coleman, Rodney (Rev) Review of "Studies	95 JRSS-A 158 634-635

	Teaching (cont.)	
	Integrating cases into the business statistics course • Nolan, James R.	94 ASAProStEd 123-125
	Interactive statistics tutorials in Stata (jse/v3n3/ferrall) • Ferrall, Christopher	95 JStEd 3/3 28kb
(jse/v3n1/konold) • Konold, Clifford	Issues in assessing conceptual understanding in probability and statistics	95 JStEd 3/1 27kb
	On large-sample tests concerning proportions • Chan, Wai-sum	95 TeachgSt 17/1 16-16
plays chutes and ladders" • Wang, C.; Corley, T.	The longest chute is the worse chute: Comments and additions to "Mr. Markov	94 ASAProStEd 104-106
	A mathematical investigation of quality control • Edgerton, Richard T.	95 MathTchr 88 200-202
	Maximum likelihood estimation and Mathematica • Currie, Iain D.	95 ApplStat 44 379-394
	Median and quartiles from grouped data • Evans, Gwyn	95 TeachgSt 17/3 99-100
	Minitab menus - Do they help or hinder? • Reed, Margaret; Garven, Frances	95 TeachgSt 17/1 22-25
	Modelling: Keeping plant science and agriculture students interested • Johnson, I. R.	95 EnvrMtrc 6 517-522
	More equal than others? • Wright, Helen	95 TeachgSt 17/1 26-28
	A multiple regression project • Johnson, Roger	95 TeachgSt 17/2 64-66
	Optimal lengths of left-turn lanes • Applebaugh, W.; Chadha, S. S.	94 ASAProStEd 102-103
	Optimal stopping in 'The Showcase Showdown' • Coe, Paul R.; Butterworth, William	95 AmerStat 49 271-275
ization in uncovering a causal relationship • Oldford	A physical device for demonstrating confounding, blocking, and the role of random-	95 AmerStat 49 210-216
	Play your cards right • Fletcher, Mike	95 TeachgSt 17/2 74-74
	How much does poor reading lower math scores? • Clarkson, S. P.; Williams, W. H.	94 ASAProStEd 203-208
	Primary data • Bramwell, Andrew	95 TeachgSt 17/3 88-89
	Probability, intuition and a spreadsheet • Graham, Alan	95 TeachgSt 17/3 84-87
Sardy, Hyman	Are the problems the problem in introductory statistics? • Grosof, Miriam Schapiro;	94 ASAProStEd 192-196
	Projects in introductory statistics courses • Ledolter, Johannes	95 AmerStat 49 364-367
Callanan, Terrance P.; Altland, Hank W.	Providing effective training industry - An experience with robust design •	94 ASAProStEd 126-131
Lee, T. N.; Lawson, J. S.; Hawks, V. D.	Q^2: A TQM course for undergraduates • Badrigal, J. L.; Richards, D. O.; Free, J. C.;	94 ASAProQlPr 172-180
	Random sampling versus representative samples • Schield, Milo	94 ASAProStEd 107-110
datasets.short) • Short, T. H.; Moriarty, H. et al.	Readability of educational materials for patients with cancer (jse/v3n2/	95 JStEd 3/2 15kb
twenty-first century • Levine, D. M.; Ramsey, P. P.	Reengineering the business statistics course: Meeting the customer needs of the	94 ASAProStEd 121-122
	Reflections on the past 15 years • Garfield, Joan	95 TeachgSt 17/2 77-78
Pereira-Mendoza, Lionel (Ed); Watkins, Ann E. (Rev)	Review of "Introducing data analysis in the schools: Who should teach it and how?" •	95 AmerStat 49 62-63
illustration • Jolliffe, Ian T.	Sample sizes and the central limit theorem: The Poisson distribution as an	95 AmerStat 49 269-269
	Scoring patterns in rugby league • Croucher, John	95 TeachgSt 17/2 47-49
	Secondary students' concepts of probability • Madsen, Richard W.	95 TeachgSt 17/3 90-93
	Setting objectives in scientific studies • Riley, Janet	95 TeachgSt 17/2 61-63
Fletcher A.	What should you do on the first day? Motivating and setting goals • Blanchard,	94 ASAProStEd 63-65
	Statistical consulting practicum: Opportunity for authentic learning • Derr, Janice A.	94 ASAProStEd 140-143
Gianini, Jacqueline	Statistical issues in spreadsheet software • Nash, John C.; Quon, Tony K.;	94 ASAProStEd 238-241
	Statistical tools and statistical literacy: The case of the average • Gal, Iddo	95 TeachgSt 17/3 97-99
	Statistics before college • Herring, Susan; Thoni, Janice	94 ASAProStEd 218-221
Meeker, William Q.	Statistics education fin de Siécle • Moore, David S.; Cobb, George W.; Garfield, Joan;	95 AmerStat 49 250-260
	Statistics and the mathematical processes • Begg, Andy	95 TeachgSt 17/2 40-43
Rouncefield, Mary	The statistics of poverty and inequality (jse/v3n2/datasets.rouncefield) •	95 JStEd 3/2 10kb
	STEPS for learning statistics • Bibby, Katherine; Davies, Neville	95 TeachgSt 17/3 107-110
	Student editions of forecasting software: A survey • Rycroft, Robert S.	95 IntJFor 11 337-351
VanBrackle, Anita S.; Rothenberg, Lori F.	Students' internalized representation of simple linear regression • VanBrackle, Lewis;	94 ASAProStEd 100-101
	How students learn statistics • Garfield, Joan	95 IntStRvw 63 25-34
	Successful customers of statistics at a distant learning site • Bolland, Thomas W.	94 ASAProQlPr 300-304
probability, quality control, and graphics • Wilson	Supplementing a college preparatory course in statistics with emphasis on applied	94 ASAProStEd 42-43
	Don't get t out of proportion! • Goodall, Gerald	95 TeachgSt 17/2 50-51
Orr, Carolyn	A train the trainer program for design of experiments • Eckes, George P.;	95 ASQCTekCon 348-355
	Tree slugs • Feu, Chris du	95 TeachgSt 17/2 52-53
	Tricks of the trade: Getting going the first day • Lock, Robin H.	94 ASAProStEd 49-52
	The 'unusual episode' data revisited • Dawson, Robert J. MacG.	95 JStEd 3/3 23kb
	Using computers to teach theoretical statistics • Carter, Lucette; Mougeot, Mathilde	95 TeachgSt 17/2 44-46
statistics course • Guffey, James M.	Using data and the computer to teach and to visualize concepts in an introductory	94 ASAProStEd 84-87
report from the field (jse/v3n2/keeler) • Keeler et al.	Using small groups to promote active learning in the introductory statistics course: A	95 JStEd 3/2 28kb
Mathieson, Kieran; Doane, David P.	Visualizing and describing the shape of distributions • Tracy, Ronald L.;	94 ASAProStEd 70-74
Mathieson, Kieran D.	Visualizing power and type I and II error • Daone, David P.; Tracy, Ronald L.;	94 ASAProStEd 66-69
Minstrell, J.; Hunt, E.; Madigan, D.; Traynor, C.	VITAL: An intelligent tutoring system for statistics • Clarkson, D. B.; Donnell, D.;	94 ASAProStEd 88-93
Steyne, Dave	The world's 100 largest cities: A case study of inequality • Drew, David;	95 TeachgSt 17/1 17-21
Continuous improvement in college	teaching: An application of statistical tools • Xue, Zhiming; Gulas, Charles S.	95 ASQCTekCon 267-275
estimation • Meeker, William Q.; Escobar, Luis A.	Teaching about approximate confidence regions based on maximum likelihood	95 AmerStat 49 48-53
Continuous improvement of	teaching: A case study in a large statistics course • Wild, C. J.	95 IntStRvw 63 49-68
Strong, Ann	Teaching the concepts and tools of variation using body temperature • Amy, Carolyn K.;	95 QulPr 28/3 168-168
Hasse diagrams in statistical consulting and	teaching • Lohr, Sharon L.	95 AmerStat 49 376-381
Freeman, Katherine	Teaching critical review of the medical literature: A course in clinical trials •	94 ASAProStEd 156-161
Trumbo Some demonstration programs for use in	teaching elementary probability and statistics: Parts 3 and 4 (jse/v3n2/trumbo) •	95 JStEd 3/2 60kb
Daniel W.	Teaching graduate service courses through case studies • Ramsey, Fred L.; Schafer,	94 ASAProStEd 27-28
In-house consultation in an orthopaedic	teaching hospital - STAT! • Peterson, Margaret G. E.	94 ASAProStEd 170-174
• Albert, Jim	Teaching inference about proportions using Bayes and discrete models (jse/v3n3/albert)	95 JStEd 3/3 58kb
business statistics: A survey of topics and	teaching methods • Strasser, Sandra E.; Ozgur, Ceyhun Undergraduate	95 Intrfc 25/3 95-103
D. G. et al. Efficiency in the provision of university	teaching and research: An empirical analysis of UK universities • Glass, J. C.; McKillop,	95 JApEcmtx 10 61-72
• Cornell, J. A.; Randles, R. H. et al. Challenges in	teaching short courses by an academic department: The University of Florida experience	95 AmerStat 49 195-200
	Teaching short courses in biostatistics and epidemiology • Kleinbaum, David G.	94 ASAProStEd 29-32
	Teaching short courses in biostatistics and epidemiology • Kleinbaum, David G.	95 AmerStat 49 200-203
Ten suggestions for effectively	teaching short courses to heterogeneous groups • Birch, Jeffrey B.	94 ASAProStEd 33-38
Ten suggestions for effectively	teaching short courses to heterogeneous groups • Birch, Jeffrey B.	95 AmerStat 49 190-195
Edward R.; Adams, Benjamin M.	Teaching the standard deviations of linear combinations using graphics • Mansfield,	94 ASAProStEd 80-83
Winne, Philip H. Effects of	teaching statistical laws on reasoning about everyday problems • Kosonen, Peter;	95 JEdPsych 87 33-46
	Teaching statistics (Italian) • Vitali, Ornello	94 Statstca 54 387-396
Changing the	teaching of statistics • Stuart, Michael	95 Ststcian 44 45-54
of the Fourth International Conference on	Teaching Statistics" • International Statistical Institute (Auth) et al. Review of "Proceedings	95 AmerStat 49 219-220
romero) • Romero, R.; Ferrer, A.; Capilla, C. et al.	Teaching statistics to engineers: An innovative pedagogical experience (jse/v3n1/	95 JStEd 3/1 33kb
Discussion of the symposium on the	teaching of statistics in higher education in Ireland • Boland, Philip J.	95 Ststcian 44 55-56
Discussion of the symposium on the	teaching of statistics in higher education in Ireland • Cran, G. W.	95 Ststcian 44 57-58

Time series (cont.)

Toxicity

concentration addition and ... • Drescher et al.	Assessment of the combined effects of substances: The relationship between	95 Biomtrcs 51 716-730
Day, Roger Incorporating	toxicity considerations into the design of two-stage phase II clinical trials • Bryant, John;	95 Biomtrcs 51 1372-1383
logistic model with applications to experimental	toxicity data • El-Saidi, Mohammed A. A symmetric extended	95 BiomtrcJ 37 205-216
Raynauld, J.-P. Comparing the bivariate effects of	toxicity and efficacy of treatments (Disc: p1143-1143) • Tubert-Bitter, P.; Bloch, D. A.;	95 StatMed 14 1129-1141
Pralay Developmental	toxicity modeling for risk assessment • Catalano, Paul J.; Ryan, Louise M.; Senchaudhuri,	94 ASAProEnvr 38-45
D.; Rosenberger, William F.	Toxicity in sequential dose-response experiments • Flournoy, Nancy; Durham, Stephen	95 SqtlAnly 14 217-227
data: Applications to clinical and developmental	toxicity studies • Bowman, D. et al. A saturated model for analyzing exchangeable binary	95 JASA 90 871-879
bilities of differential toxic effects in developmental	toxicity studies • Talwalker, S.; Patil, G. P.; Taillie, C. Dose-response models for the proba-	94 JStatRsr 28 221-232
Estimating variance functions in developmental	toxicity studies • Bowman, Dale; Chen, James J.; George, E. Olusegun	95 Biomtrcs 51 1523-1528
in quantal response teratology and developmental	toxicity studies • Bieler, Gayle S.; Williams, Rick L. Cluster sampling techniques	95 Biomtrcs 51 764-776
on "Stopping a clinical trial very early because of	toxicity: Summarizing the evidence" (V14 p286-295) • Bring, Johan Comment	95 ConClnTr 16 131-132
of pseudo-profiles in organ pharmacokinetics and	toxicokinetics • Mager, Harry; Göller, Gernot Analysis	95 StatMed 14 1009-1024
Improved bootstrap confidence intervals in certain	toxicological experiments • Frangos, Chris C.; Schucany, William R.	95 CommStA 24 829-844

Toxicology

	Aluminium and the risk for Alzheimer's disease • McLachlan, D. R. C.	95 EnvrMtrc 6 233-275
Sen, Pranab Kumar; Margolin, Barry H. Inhalation	toxicology: Awareness, identifiability, statistical perspectives and risk assessment •	95 SankhyaB 57 252-276
Introduction to motor activity data (reproductive	toxicology) • Beyer, Bruce K.	94 ASAProBiop 482-483
Darwin's log: A	toy model of speciation and extinction • Aldous, David	95 JAppProb 32 279-295

TQM

	An analysis of the thoracic surgical literature • Ferraris, Victor A.	94 ASAProQlPr 158-163
	Bridging the gap between town and gown • Wood, G. R.	94 ASAProQlPr 250-251
	Estimating telephone demand: Validation of previous models • Lee, Robin	94 ASAProQlPr 82-86
	A quality measures plan within the IRS: A case study • White, Glenn D., Jr	94 ASAProQlPr 76-81
	What is sample design efficiency? Let's let the customer decide • Chromy, James R.	94 ASAProQlPr 226-231
Thompson, Russell G.	The search for best practices in the U.S. petroleum industry • Otto, Gordon H.;	94 ASAProQlPr 70-75
Johnson, Thomas; Lu, J. C.	The statisticians role in quality improvement partnerships • Hoff, Jennifer;	94 ASAProQlPr 252-257
LeRoy T.	Use of field data in reliability studies of medical devices • Kim, John S.; Mattson,	94 ASAProQlPr 200-203
Visions, tactics, and	TQM for academic programs in quality • Barker, Anne M.	94 ASAProQlPr 294-294
Lawson, J. S.; Hawks, V. D. Q^2: A	TQM course for undergraduates • Badrigal, J. L.; Richards, D. O.; Free, J. C.; Lee, T. N.;	94 ASAProQlPr 172-180
Kartha, C. P.	TQM implementation strategies in European manufacturing and service organizations •	94 ASAProQlPr 66-69
The inherent difficulties in applying	TQM to a particular service industry • Fricker, Ronald D., Jr	94 ASAProQlPr 181-188
Review of "Statistical methods for SPC and	TQM" • Bissell, D. (Auth); Davies, Neville (Rev)	95 JRSS-A 158 341-342
Review of "Statistical methods for SPC and	TQM" • Bissell, D. (Auth); Sheil, J. (Rev)	95 Ststcian 44 279-281
Democracy, capitalism and the role of	TQM and the statistician in public education reform • Wehrle, Robert A.	94 ASAProQlPr 189-193
Drapeau, Richard A.	TQM in undergraduate business statistics: What are the customers' needs? •	94 ASAProQlPr 270-272
Review of "Developing new products with	TQM" • Cervetz, C. (Auth); Ziegel, Eric R. (Rev)	95 Technmcs 37 241-241
Liu, Shuangzhe; Neudecker, Heinz Matrix-	trace Cauchy-Schwarz inequalities and applications in canonical correlation analysis •	95 StHefte 36 287-298
David	Trace operators, Feynman distributions, and multiparameter white noise • Betounes,	95 JTherPrb 8 119-138
ridge estimation with prior information and ridge	trace • Crouse, Robert H.; Jin, Chun; Hanumara, R. C. Unbiased	95 CommStA 24 2341-2354
Performance analysis of general	tracking algorithms • Guo, Lei; Ljung, Lennart	95 IEEEAuCn 40 1388-1402
Xie, Y.-L.; Liang, Y.-Z.; Chen, Z.-G. et al. A nonlinearity	tracking analysis algorithm for treatment of non-linearity in multivariate calibration •	95 Chemolab 27 21-32
Optimal	tracking of Brownian motion • Assaf, David; Shepp, Larry; Yao, Yi-Ching	95 Stochatx 54 233-245
Modelling radio-	tracking data • Worton, Bruce J.	95 EnvEcoSt 2 15-23
	Tracking the output of a poorly known Markov process • Rishel, Raymond	95 Stochatx 52 147-156
of adaptive step-size SA algorithms for parameter	tracking • Kushner, Harold J.; Yang, J. Analysis	95 IEEEAuCn 40 1403-1410
Autocorrelated data: Control charts versus	tracking signals • Superville, Claude R.; Adams, Benjamin M.	94 ASAProQlPr 194-199
model • Simpson, James R.; Keats, J. Bert	Trade-off analysis versus constrained optimization with an economic control chart	95 IIETrn 27 765-772
The employment effect in retail	trade of California's 1988 minimum wage increase • Kim, Taeil; Taylor, Lowell J.	95 JBES 13/2 175-182
Mayda, J. E. Redesign of the Hungarian Retail	Trade and Catering Survey: Methodological issues • Choudhry, G. H.; Colledge, M. J.;	94 ASAProBuEc 71-76
desirable sampling plans by the technique of	trade-off in experimental design • Hedayat, A. S.; Majumdar, Dibyen Generating	95 JStPlInf 44 237-247
Tricks of the	trade: Getting going the first day • Lock, Robin H.	94 ASAProStEd 49-52
Current and future approaches to editing Canadian	trade import data • Pursey, Stuart	94 ASAProGovt 105-109
mating spatial statistical models • Griffith, D. A. et al.	Trade-offs associated with normalizing constant computational simplifications for esti-	95 JStCmpSm 51 165-183
Life cycle	trade-offs in matrix population models • van Tienderen, Peter H.	95 Ecology 76 2482-2489
Problems,	trade-offs, and solutions for CAPI surveys (Disc: p861-862) • Gardenier, John S.	94 ASASrvyRsr 857-860
Viallefont, A.; Cooch, E. G.; Cooke, F. Estimation of	trade-offs with capture-recapture models: A case study on the lesser snow goose •	95 JAppStat 22 847-861
An application to the Canadian Monthly Retail	Trade Survey • Patak, Z.; Hidiroglou, M. A. Raking ratio estimation:	94 ASASrvyRsr 299-304
for panel imbalance in the Monthly Wholesale	Trade Survey • Caldwell, C. V. et al. Using seasonal adjustment models to correct	94 ASAProBuEc 83-88
Bias-variance	tradeoffs in functional estimation problems • Low, Mark G.	95 AnlsStat 23 824-835
Information	tradeoff • Wasserman, L.; Clarke, B.	95 Test 4 19-38
Can speculative	trading explain the volume-volatility relation? • Foster, F. Douglas; Viswanathan, S.	95 JBES 13 379-396
the effect of earnings surprises on stock prices,	trading volumes, and spreads • Maddala et al. ... component panel data model to study	95 JEconmtx 68 229-242

Traffic

• see also heavy –, light –, queueing theory, telecommunications

Suksawasdi Na; Böhning, Danmar	Traffic accident mapping in Bangkok metropolis: A case study • Ayuthya, Rampai	95 StatMed 14 2445-2458
Dynamic three-dimensional display of U.S. air	traffic • Eddy, William F.; Que, Shingo	95 JCmpGrSt 4 261-280
Taqqu, M. S. et al. Self-similarity in high-speed packet	traffic: Analysis and modeling of ethernet traffic measurements • Willinger, W.;	95 StatSci 10 67-85
Heavy	traffic analysis of a random walk on a lattice semi-strip • Falin, Gennadi	95 CommStSM 11 395-409
Bartłomiej; Frey, Andreas; Schmidt, Volker Light-	traffic approximations for Markov-modulated multi-server queues • Błaszczyszyn,	95 CommStSM 11 423-445
systems with zero switchover times: A heavy-	traffic averaging principle • Coffman, E. G., Jr; Puhalskii, A. A.; Reiman, M. I. Polling	95 AnlsApPr 5 681-719
a class of Markovian packet streams under heavy	traffic conditions • Neuts, M. F.; Pearce, C. E. M. A central limit theorem for	95 CommStSM 11 210-218
method for outlier detection applied to time series	traffic data • Watson, S. M.; Redfern, E.; Clark, S.; Tight, M.; Davies, N. An influence	95 JAppStat 22 135-149
Product forms based on backward	traffic equations • Boucherie, Richard J.	95 JAppProb 32 508-518
arrival driven queueing models: Admission control,	traffic policing, abandonments, and correlated arrivals • Milito, Rodolfo A. On	95 AplMaCmp 70 77-94
implications in the development and testing of a	traffic safety questionnaire • Schechter, S. et al. Cognitive issues and methodological	94 ASASrvyRsr 1215-1219
Gi GA-based optimal dimensioning of three-level	traffic shaper for statistical multiplexing in ATM networks • Kim, Kyeong Soo; Lee, Byeong	95 AnlsTele 50 624-631
analysis of single resource loss systems in heavy	traffic, with applications to integrated networks • Bean, N. G. et al. Asymptotic	95 AdvAppPr 27 273-292
ing models • Kontovasilis et al. Markov-modulated	traffic with nearly complete decomposability characteristics and associated fluid queue-	95 AdvAppPr 27 1144-1185
Reece, Jack E.; Spagon, Patrick D. Linked	training and consulting in the semiconductor industry: The SEMATECH approach •	94 ASAProStEd 228-233
of neural network classifiers and discriminative	training criteria • Ney, Hermann On the probabilistic interpretation	95 IEEEPAMI 17 107-119
under partial nonrandom classification of the	training data • McLachlan et al. ... relative efficiency of the linear discriminant function	95 JStCmpSm 52 415-426
tools to the masses: Common problems in	training design and execution • Finn, L. M.; Garhart, C. A.; Reynard, S. E. Getting statistical	95 ASQCTekCon 356-364
	Training government statisticians in Zimbabwe – An update • Woodward, Mark	94 JOffStat 10 215-220

	Trend-free design	
order is important • Wang, P. C.; Jan, H. W.	Designing two-level factorial experiments using orthogonal arrays when the run	95 Ststcian 44 379-388
Shah, Kirti R.	Efficient block designs in the presence of trends • Jacroux, Mike; Majumdar, Dibyen;	95 StSinica 5 605-615
Construction and optimality of nearly linear	trend-free designs • Chai, Feng-Shun	95 JStPlInf 48 113-129
Random walks, breaking	trend functions, and the chaotic structure of the velocity of money • Serletis, Apostolos	95 JBES 13 453-458
Predicting ozone levels and	trends • Gao, Feng; Sacks, Jerome; Welch, William J.	94 ASAProEnvr 77-81
of multivariate ARMA processes with a linear	trend • Garel, Bernard; Hallin, Marc Local asymptotic normality	95 AnInStMa 47 551-579
Before and after Bradford Hill: Some	trends in medical statistics • Armitage, Peter	95 JRSS-A 158 143-153
Kugler, D. L. Regional scale	trend monitoring of indicators of trophic condition of lakes • Larsen, D. P.; Urquhart, N. S.;	95 WaterReB 31 117-140
esis: An application of new models for analysing	trends in multiway tables • Payne, C.; Payne, J. et al. Testing the tightening bond hypoth-	95 JStPlInf 47 181-201
research of the Japanese national character: General	trend of opinion over past forty years (Japanese) • Sakamoto, Yosiyuki A statistical	95 PrInStMa 43 5-26
Efficient block designs in the presence of	trends • Jacroux, Mike; Majumdar, Dibyen; Shah, Kirti R.	95 StSinica 5 605-615
Selecting a model for detecting the presence of a	trend • Woodward, Wayne A.; Gray, H. L.	95 JClimate 8 1929-1937
estimation and significance testing for cyclic	trends in seasonality studies • Nam, Jun-mo Interval	95 Biomtrcs 51 1411-1417
Social	trends and social change • Nissel, Muriel	95 JRSS-A 158 491-504
Testing for	trend stationarity versus difference stationarity • Arellano, Consuelo; Pantula, Sastry G.	95 JTimSrAn 16 147-164
	Trends in statistical ecology and environmental statistics • MacNeill, Ian B.	94 ASAProEnvr 8-9
A permutation test analogue to Tarone's test for	trend in survival analysis • Graves, Thomas S.; Pazdan, James L.	95 JStCmpSm 53 79-89
of wet deposition networks to detect temporal	trends • Oehlert, Gary W. The ability	95 EnvrMtrc 6 327-339
filtering technique for identifying long-term	trends • Assimakopoulos, V. A successive	95 JForcast 14 35-43
Comment on "A Wilcoxon-type test for	trend" (V4 p87-90) • Lui, Kung-Jong	95 StatMed 14 445-446
with adjustment for year of marketing and secular	trends in total reporting • Tsong ... reporting rates of adverse events between drugs	95 JBiophSt 5 95-114
Examining an increasing	trend with non-parametric methods • Strömberg, Ulf	94 JAppStat 21 575-587
	Trending multiple time series • Phillips, Peter C. B.	95 EconTher 11 811-817
Estimation of variance components based on	triallel mating design • Srinivasan, M. R.; Ponnuswamy, K. N.	93 ThApGene 85 593-597
Estimation of variance components based on partial	triallel mating design • Srinivasan, M. R.; Ponnuswamy, K. N.	95 BiomtrcJ 37 161-171
Gandolfi, A. A	triangle inequality for covariances of binary FKG random variables • van den Berg, J.;	95 AnlsApPr 5 322-326
Sums of independent	triangular arrays and extreme order statistics • Janssen, Arnold	94 AnlsProb 22 1766-1793
integrability of vector-valued random series and	triangular arrays • Rosiński, Jan Remarks on strong exponential	95 AnlsProb 23 464-473
Zoglat, Abdelhak Strong invariance principles for	triangular arrays of weakly dependent random variables • Dabrowski, André Robert;	95 CandJSt 23 299-310
al. Exact conditional tests of quasi-independence for	triangular contingency tables: Estimating attained significance levels • McDonald, J. W. et	95 ApplStat 44 143-151
A test of quasi-independence in ordinal	triangular contingency tables • Tsai, Ming-Tien; Sen, Pranab K.	95 StSinica 5 767-780
Testing quasi-independence in stratified	triangular contingency tables • May, Warren L.; Johnson, William D.	95 BiomtrcJ 37 531-546
Construction of variance balanced designs through	triangular PBIB designs • Ghosh, D. K.; Joshi, K. S.	95 CalcutSt 45 111-118
Estimation in mixed models via layer	triangular transformation • Öfversten, Jukka	95 CmpStDA 20 657-667
	Tricks of the trade: Getting going the first day • Lock, Robin H.	94 ASAProStEd 49-52
Mean passage times for	tridiagonal transition matrices • Stefanov, Valeri T.	95 JAppProb 32 846-849
A one-sided	trimmed mean estimator for the mean of positively skewed distributions • Chen, Ling	95 AstrlJSt 37 313-322
A	trimmed mean of location of an AR(∞) stationary process • Lee, Sangyeol	95 JStPlInf 48 131-140
importance of heteroscedastic methods, using	trimmed means versus means, and designing simulation studies • Wilcox ... The practical	95 BJMaStPs 48 99-114
Asymptotic normality of multivariate	trimmed means • Arcones, Miguel A.	95 StPrLet 25 45-53
to the multivariate Behrens-Fisher problem via	trimmed means • Wilcox, Rand R. Simulation results on solutions	95 Ststcian 44 213-225
Three multiple comparison procedures for	trimmed means • Wilcox, Rand R.	95 BiomtrcJ 37 643-656
On the variance of the	trimmed mean • Capéraà, Philippe; Rivest, Louis-Paul	95 StPrLet 22 79-85
method • Hove, H.; Liang, Y.-Z.; Kvalheim, O. M.	Trimmed object projections: A nonparametric robust latent-structure decomposition	95 Chemolab 27 33-40
Daniel A Gaussian central limit theorem for	trimmed products on simply connected step 2-nilpotent lie groups • Neuenschwander,	95 JTherPrb 8 165-174
Exact computation of the least	trimmed squares estimate in simple linear regression • Hössjer, Ola	95 CmpStDA 19 265-282
Precision calculation of distributions for	trimmed sums • Csörgő, Sándor; Simons, Gordon	95 AnlsApPr 5 854-873
The law of the iterated logarithm for	trimmed sums (Russian) • Egorov, V. A.; Pozdnyakov, V. I.	94 VLenUM /4 34-38
The strong law of large numbers for	trimmed sums (Russian) • Pozdnyakov, V. I.	94 VLenUM /2 20-25
Kreher, Donald L. Partial Steiner	triple systems with equal-sized holes • Colbourn, Charles J.; Cusack, Charles A.;	95 JCombThA 70 56-65
inance and epistatic variation for metrical traits. V.	Triple test cross analysis • Kumar, I. et al. A general method of detecting additive, dom-	94 Heredity 72 563-569
Cyclically resolvable designs and	triple whist tournaments • Anderson, I.; Finizio, N. J.	93 JComDsgn 1 347-358
and maternal effects. II. Genetic models for	triploid endosperms • Zhu, J.; Weir, B. S. Analysis of cytoplasmic	94 ThApGene 89 160-166
Methods for genetic linkage analysis using	trisomies • Feingold, Eleanor; Lamb, Neil E.; Sherman, Stephanie L.	95 AJHuGene 56 475-483
The structure of compounded	trivariate binomial distributions • Papageorgiou, H.; David, Katerina M.	95 BiomtrcJ 37 81-95
of bivariate distributions by a generalised	trivariate reduction technique • Lai, C. D. Construction	95 StPrLet 25 265-270
Regional scale trend monitoring of indicators of	trophic condition of lakes • Larsen, D. P.; Urquhart, N. S.; Kugler, D. L.	95 WaterReB 31 117-140
Chattopadhyay, Asis Kumar (Rev) Review of	"Truncated and censored samples. Theory and applications" • Cohen, A. Clifford (Auth);	95 CalcutSt 45 133-133
Wood, Andrew T. A. When is a	truncated covariance function on the line a covariance function on the circle? •	95 StPrLet 24 157-164
using a general approach for doubly-censored and	truncated data • Heisey, D. M. et al. Modelling age-specific survival in nesting studies,	95 Biomtrcs 51 51-60
consistency of the Lynden-Bell estimator for	truncated data • Chen, Kani; Chao, Min-Te; Lo, Shaw-Hwa On strong uniform	95 AnlsStat 23 440-449
in a normal mixture model fitted to grouped and	truncated data • McLachlan et al. ... the likelihood ratio test of one versus two components	95 CommStB 24 965-985
likelihood ratio estimation of probabilities for	truncated data • Li, Gang Nonparametric	95 JASA 90 997-1003
tations of functions of order statistics for doubly	truncated distributions and their applications • Mohie EL-Din et al. ... relations for expec-	95 CommStA 24 997-1010
missible estimators of parameters and losses under	truncated distribution families (Chinese) • Wu, Qiguang; He, Min Several notes on ad-	95 JSysSMaS 14 168-176
Empirical estimation of a distribution function with	truncated and doubly interval-censored data and its application to AIDS studies • Sun, J.	95 Biomtrcs 51 1096-1104
generating function of order statistics for doubly	truncated exponential distribution • El-Din, M. M. M.; Mahmoud, M. A. W. et al. On moment	95 Metron 53 171-183
Parametric and semiparametric estimates in general	truncated families (Chinese) • Chen, Guijing; Zhao, Zhongbo; Ding, Yuanyao	95 JSysSMaS 14 228-241
	Truncated logistic regression • O'Neill, Terence J.; Barry, Simon C.	95 Biomtrcs 51 533-541
relationship between the log-survival function and	truncated moments, with application • El-Arishy, S. Useful	95 StHefte 36 145-154
Simulation of	truncated normal variables • Robert, Christian P.	95 StatComp 5 121-125
Group	truncated ordinal regression • O'Neill, Terence J.; Barry, Simon C.	95 StPrLet 22 195-203
product moments of order statistics of a doubly	truncated Pareto distribution • Khurana, A. P. et al. Some recurrence relations between	95 SankhyaB 57 1-9
properties of estimate of failure rates based on	truncated sample (Chinese) • Wan, Qihua Some large sample	95 KToChina 39 2120-2122
Nonparametric	truncated sequential change-point detection • Gombay, E.	95 StatDecn 13 71-82
bers for quadratic forms, maxima of products and	truncated sums of i.i.d. random variables • Cuzick, J.; Giné, E.; Zinn, J. Laws of large num-	95 AnlsProb 23 292-333
with censored initiating time, and censored and	truncated terminating time: Application to transfusion data for ... • Tu ... distributions	95 ApplStat 44 3-16
	Truncation	
Navarro, J.	Characterization of discrete distributions using expected values • Ruiz, J. M.;	95 StHefte 36 237-252
Shahane, A.; George, V.; Johnson, W. D.	Effect of bivariate regression toward the mean in uncontrolled clinical trials •	95 CommStA 24 2165-2181
	On the generalized negative binomial distribution • Consul, P. C.; Famoye, Felix	95 CommStA 24 459-472
Wang, Xiangchen	Large deviations for moving average processes • Jiang, Tiefeng; Rao, M. Bhaskara;	95 StocProc 59 309-320
Boukai, B.	A useful property of order statistics and its application to UMVU estimation •	95 StHefte 36 279-283

application to the beer industry • Chintagunta et al. A	two-period repeated game advertising investment model for oligopolistic markets with an	95 DecisnSc 26 531-559
Optimal software release policy based on a	two-person game of timing • Zeephongsekul, P.; Chiera, C.	95 JAppProb 32 470-481
	Two-phase estimation by imputation • McVey, Anita; Breidt, F. Jay; Fuller, Wayne A.	94 ASASrvyRsr 1053-1058
regression approach for model checking on the	two-phase regression problem • Azari, A. S.; Tsai, Chih-Ling Non-parametric	94 JAppStat 21 597-606
A class of unbiased ratio estimators in	two-phase sampling • Singh, H. P.; Biradar, R. S.	94 Statstca 54 349-359
Sitter, R. R. Variance estimation under	two-phase sampling with application to imputation for missing data • Rao, J. N. K.;	95 Biomtrka 82 453-460
Barquero, J. L. Screening for stratification in	two-phase ('two-stage') epidemiological surveys • Pickles, A.; Dunn, G.; Vázquez-	95 StMedRes 4 73-89
Comment on "Screening for stratification in	two-phase ('two-stage') epidemiological surveys" (V4 p73-89) • Hand, David J.	95 StMedRes 4 263-263
Reply to comment on "Screening for stratification in	two-phase ('two-stage') epidemiological surveys" • Pickles, A.; Dunn, G. et al.	95 StMedRes 4 263-263
Single test that	two samples came from the same normal population • Nelson, Lloyd S.	95 JQualTek 27 84-87
Kim, Jeong-Ran	Two-sample chi-square test for randomly censored data (Korean) • Kim, Joo-Han;	95 KorApS 8/2 109-119
Sun, Shan Nonparametric methods for stratified	two-sample designs with application to multiclinic trials • Brunner, Edgar; Puri, Madan L.;	95 JASA 90 1004-1014
	Two-sample empirical likelihood method • Jing, Bing-Yi	95 StPrLet 24 315-319
against dependence of nonparametric tests for the	two-sample location problem • Grzegorzewski, P. The robustness	95 ZastApMa 22 469-476
	Two-sample McNemar tests for complex survey data • Thompson, Jenny; Fisher, Robin	94 ASASrvyRsr 608-613
The empirical process approach for semiparametric	two-sample models with heterogeneous treatment effect • Hsieh, Fushing	95 JRSS-B 57 735-748
	Two-sample problem	
of X and Y • Hamdy, M. I.	Distribution-free confidence intervals for Pr($X < Y$) based on independent samples	95 CommStB 24 1005-1017
A distribution free test for the	two sample problem for general alternatives • Schmid, Friedrich; Trede, Mark	95 CmpStDA 20 409-419
censoring under inverse sampling for nonparametric	two-sample problems • Bandyopadhyay, Uttam; Chattopadhyay, Gopaldeb Progressive	95 SqtlAnly 14 1-28
Inference for likelihood ratio ordering in the	two-sample problem • Dykstra, Richard; Kochar, Subhash; Robertson, Tim	95 JASA 90 1034-1040
On some one-sided tests in the	two-sample problem • Bajorski, Piotr	95 CommStA 24 75-86
Shrinkage estimation of a mean vector in a	two-sample problem • Jin, Chun	95 CommStA 24 2881-2888
Quantile estimation of treatment effect for the	two-sample problem with right censored data • Park, Hyo-Il; Park, Sang-Gue	95 StPrLet 24 139-145
Nonparametric tests for the	two sample problem with two independent competing risks • Bagai, Isha; Deshpande, J. V.	95 JNnparSt 5 21-31
A sequential dual of Stein's	two-sample procedure • Arghami, N. R.; Billard, L.	92 ProIrStCl 11-18
Estimating the real parameter in a	two-sample proportional odds model • Wu, Colin O.	95 AnlsStat 23 376-395
tions based on the number of runs and some other	two-sample rank order statistics for arbitrary sample sizes • Saran, Jagdish Joint distribu-	95 JStPlInf 47 257-275
	Two-sample rank tests for acceleration in cure models • Lee, Jae Won	95 StatMed 14 2111-2118
range distribution • Efird, J. T. et al. A non-parametric	two-sample survival test based on a single occupancy Fermi-Dirac model for the discrete	94 ASAProBiop 166-170
Coefficients of Lee-Gurland	two-sample test on normal means • Lee, Austin F. S.	95 CommStA 24 1743-1768
effect of unequal variances on the power of several	two-sample tests • O'Gorman, Thomas W. The	95 CommStB 24 853-867
van Dongen, S.; Backeljau, T. One- and	two-sample tests for single-locus inbreeding coefficients using the bootstrap •	95 Heredity 74 129-135
	Two-sample tests for variability based on interquantile spreads • Shoemaker, Lewis H.	94 ASAProQlPr 134-139
Generalised	two-sample U-statistics and a two-species reaction-diffusion model • Penrose, Mathew D.	95 StocProc 55 57-64
other ... • Lei et al. LRT method of constructing a	two-sided acceptance region for variables acceptance sampling and its comparison with	94 ASAProPEng 115-119
On	two-sided compound binomial distributions • Wang, Shaun	95 InsrneME 17 35-41
Optimal	two-sided group sequential tests • Eales, John D.; Jennison, Christopher	95 SqtlAnly 14 273-286
homogeneity of uniform scale distributions against	two-sided and one-sided alternatives • Cohen, Arthur; Fygenson, Mendel Testing	95 JASA 90 1062-1067
Boundary calculation issues in clinical trials with	two-sided stopping criteria • Chen, Bee-Lian; Gallo, Paul P.	94 ASAProBiop 115-119
An example of a	two-sided Wilcoxon signed rank test which is not unbiased • Amrhein, Peter	95 AnInStMa 47 167-170
structure of multivariate polytomous data from a	two-stage cluster sample • Marques et al. A model for distributions and correlation	94 ASAProBiop 346-351
	Two-stage design	
	Interim analyses in 2 × 2 crossover trials • Cook, Richard J.	95 Biomtrcs 51 932-945
uncertainty • Ahn, Hyungtaik Nonparametric	two-stage estimation of conditional choice probabilities in a binary choice model under	95 JEconmtx 67 337-378
ables • Lee, Sik-Yum; Poon, Wai-Yin; Bentler, P. M. A	two-stage estimation of structural equation models with continuous and polytomous vari-	95 BJMaStPs 48 339-358
	Two-stage estimator	
	Analysis of short time series: Correcting for autocorrelation • Bence, James R.	95 EcolgMon 65 628-639
variable error components model • Ghazal et al.	The exact finite sample properties of the estimators of coefficients in mixed dummy-	95 EgpStJ 39 26-50
Jayalakshmi; Strawderman, William E.	Sequential estimation of the variance of a normal distribution • Natarajan,	95 SqtlAnly 14 361-374
Stochastic orders of magnitude associated with	two-stage estimators of fractional ARIMA systems • Wright, J. H.	95 JTimSrAn 16 119-125
Tsung	Two-stage James-Stein estimators of the mean based on prior knowledge • Chang, Yuan-	95 CommStA 24 2211-2227
Kwon, Woo-Joo; Lee, Joo-Ho	Two-stage Latin hypercube sampling and its application (Korean) • Im, Mi-Jeong;	95 KorApS 8/2 99-108
	Two-stage least squares	
Angrist, Joshua D.; Krueger, Alan B.	Split-sample instrumental variables estimates of the return to schooling •	95 JBES 13/2 225-235
treatment intensity • Angrist, J. D.; Imbens, G. W.	Two-stage least squares estimation of average causal effects in models with variable	95 JASA 90 431-442
	Two-stage parameter designs for quality improvement • Jeong, Yousceek	94 ASAProQlPr 215-219
toxicity considerations into the design of	two-stage phase II clinical trials • Bryant, John; Day, Roger Incorporating	95 Biomtrcs 51 1372-1383
	Two-stage procedure	
populations • De, Moloy	On a class of admissible estimators of the common mean of two univariate normal	95 CalcutSt 45 103-109
Hunsberger, Sally A.	Designed extension of studies based on conditional power • Proschan, Michael A.;	95 Biomtrcs 51 1315-1324
	Empirical Bayes optimal fertilizer decisions • Wallach, D.	95 JAppStat 22 507-516
Kuan, Chung-Ming; Liu, Tung	Forecasting exchange rates using feedforward and recurrent neural networks •	95 JApEcmtx 10 347-364
geophysical data • Boularan, J.; Ferre, L.; Vieu, P.	A nonparametric model for unbalanced longitudinal data with application to	95 CmpStQ 10 285-298
the probability of nuclear attack • Paté-Cornell et al.	Probabilistic interpretation of command and control signals Bayesian updating of	95 RelEngSS 47 27-36
Taylor, Larry W.	On a simultaneous equations pre-test estimator • Skeels, Christopher L.;	95 JEconmtx 68 269-286
L. Q.	A stage-theory model of cognitive development • Preece, Peter F. W.; Read, Kenneth	95 BJMaStPs 48 1-7
From, Steven G. A	two-stage procedure for estimating the parameters of some discrete distributions •	95 JStCmpSm 53 23-39
Comment on "Stein's	two-stage procedure and exact consistency" (82ScanActJ p110-122) • Schmitz, Norbert	95 ScanActJ 187-187
Robustness of Stein's	two stage procedure • Oh, Hyun Sook; DasGupta, Anirban	95 SqtlAnly 14 321-334
O'Quigley, John A	two-stage procedure for survival studies with surrogate endpoints • Flandre, Philippe;	95 Biomtrcs 51 969-976
Tracy, D. S. An improved	two stage randomized response strategy • Singh, R.; Singh, S.; Mangat, N. S.;	95 StHefte 36 265-271
A	two-stage rank test using density estimation • Albers, Willem	95 AnInStMa 47 675-691
	Two-stage regression	
series • Easterling, David R.; Peterson, Thomas C.	A new method for detecting undocumented discontinuities in climatological time	95 JClimlgy 15 369-377
	Two-stage sampling	
	Bayesian predictive inference for multivariate sample surveys • Nandram, Balgobin	94 JOffStat 10 167-179
Healy's two-stage sampling) is more ... • Aoshima	In heteroscedastic simultaneous inference, the heteroscedastic method II (based on	94 AJMMgSci 14 167-196
models • Reilly, Marie; Pepe, Margaret Sullivan	A mean score method for missing and auxiliary covariate data in regression	95 Biomtrka 82 299-314
fishery of Oregon • Crone, Paul R.	Sampling design and statistical considerations for the commercial groundfish	95 JFishRBC 52 716-732
	Two-stage sampling for bounding a mean • Atwood, Corwin L.; Bryan, Mark F.	95 EnvrMtrc 6 73-83
Mukhopadhyay, P. Bayes and minimax estimators for	two-stage sampling from a finite population under measurement error models •	95 CommStA 24 663-674
the heteroscedastic method II (based on Healy's	two-stage sampling) is more economical than the original HM ... • Aoshima ... inference,	94 AJMMgSci 14 167-196
Kwon, Hyuck Moo Economic design of a	two-stage screening procedure with a prescribed outgoing quality • Bai, Do Sun;	95 Metrika 42 1-18

Undercount

Urn model

structure of a play • Brainerd, Barron	A generalization of the hypergeometric distribution suggested by a model for the	95 MaScitst 20 33-39
V. A. Runs, scans and	urn model distributions: A unified Markov chain approach • Koutras, M. V.; Alexandrou,	95 AnInStMa 47 743-766
theorem for the number of distinct outcomes in an	urn model • Hwang, F. K. A majorization	93 JCISS 18 26-30
Allen, Linda J. S.; Stamp, Mark	Urn model simulations of a sexually transmitted disease epidemic • Martin, Clyde F.;	95 AplMaCmp 71 179-199
Non-standard limit theorems for	urn models and stochastic approximation procedures • Kaniovski, Yu.; Pflug, G.	95 CommStSM 11 79-102
Adaptive biased	urn randomization in small strata when blinding is impossible • Schouten, Hubert J. A.	95 Biomtrcs 51 1529-1535
Total quality education at the	U.S. Air Force Academy • Tong, Terrence G. L.	94 ASAProQlPr 295-299
Dynamic three-dimensional display of	U.S. air traffic • Eddy, William F.; Que, Shingo	95 JCmpGrSt 4 261-280
New CASIC technologies at the	U.S. Bureau of the Census • Nicholls, William L., II; Appel, Martin V.	94 ASASrvyRsr 757-762
Mason, Thomas J. The development of the series of	U.S. cancer atlases: Implications for future epidemiologic research (Disc: p697-699) •	95 StatMed 14 473-479
dercount rates and surrogate variables in the 1990	U.S. Census • Kim, J. J.-I.; Zaslavsky, A.; Blodgett, R. Between-state heterogeneity of un-	95 SurvyM 21 55-62
Review of "Modernizing the	U.S. Census" • Edmonston, Barry; Schultze, Charles (Ed); Wolter, Kirk M. (Rev)	95 Chance 8/3 44-49
Household income reporting: An analysis of	U.S. Consumer Expenditure Survey data • Garner, Thesia I.; Blanciforti, Laura A.	94 JOffStat 10 69-91
between expenditures and low incomes in the	U.S. Consumer Expenditure Survey • Eltinge, J. L. et al. Assessment of reported differences	94 ASASrvyRsr 96-105
of vacant units in the residential rent index of the	U.S. Consumer Price Index • Baskin, Robert M. Imputation	94 ASASrvyRsr 382-387
commercial, and manufacturing sectors of the	U.S. economy • Battles et al. ... Utility programs offered and participants in the residential,	94 ASAProBuEc 20-27
Bayard et al. Passive smoking and lung cancer: The	U.S. EPS's weight-of-evidence analysis, with emphasis on the epidemiology studies ... •	94 ASAProEnvr 56-64
Opportunities for multicenter clinical trials in joint	US-European drug development • Tudor, David; Zipfel, Andreas	94 ASAProBiop 229-231
Indirect estimators in	U.S. federal programs • Schaible, W. L.	95 Springer:Brln:NY 195p
A nonlinear approach to	US GNP • Potter, Simon M.	95 JApEcmtx 10 109-125
homes • Dua, Pami; Smyth, David J. Forecasting	US home sales using BVAR models and survey data on households' buying attitudes for	95 JForcast 14 217-227
Estimates of early twentieth-century	U.S. homicide rates: An econometric forecasting approach • Eckberg, Douglas Lee	95 Dmogrphy 32 1-16
and wage growth: Evidence of positive effects of	U.S. Immigration • Enchautegui, María E. Impacts of immigrants on employment	94 ASAProSoSt 30-38
A dynamic analysis of interfuel substitution in	U.S. industrial energy demand • Jones, Clifton T.	95 JBES 13 459-465
tech capital formation and economic performance in	U.S. manufacturing industries: An exploratory analysis • Berndt, E. R.; Morrison, C. J. High-	95 JEconmtx 65 9-43
to a shift in regulatory regime: The case of the	US nuclear power industry • Rust, John; Rothwell, Geoffrey Optimal response	95 JApEcmtx 10 75-118S
The search for best practices in the	U.S. petroleum industry • Otto, Gordon H.; Thompson, Russell G.	94 ASAProQlPr 70-75
exogamy (Disc: p122-123) • Edmonston, B. et al.	U.S. Population projections for national origin groups: Taking into account ethnicity and	94 ASAProSoSt 100-105
Poverty migration for	U.S. States: Immigration impacts • Frey, William H.	94 ASAProSoSt 135-140
for measuring the child radio audience in the	U.S. using the seven-day personal diary method • Patchen, R. H. et al. Prospects	94 ASASrvyRsr 1220-1225
regression, Box-Cox transformation model, and the	U.S. wage structure, 1963-1987 • Buchinsky, Moshe Quantile	95 JEconmtx 65 109-154
distribution models and their applications to the	U.S.A. • Dagum, Camilo Human capital, income and wealth	94 ASAProBuEc 253-258
Greenhouse, J.; Kupfer, D.; Frank, E.; Wade, M. Cost	utility analysis of maintenance treatment for recurrent depression • Kamlet, M. S.; Paul, N.;	95 ConClnTr 16 17-40
	Utility maximization with partial information • Lakner, Peter	95 StocProc 56 247-273
Battles et al. Demand-side management:	Utility programs offered and participants in the residential, commercial, and ... •	94 ASAProBuEc 20-27
An alternative axiomatic approach to expected	utility theory • Di Bacco, Mario; Regoli, Giuliana	95 Metron 53 21-29
Statistical assessment of the	utility of tumor markers • George, Stephen L.	94 ASAProEpid 8-13
	U_n^2 test for uniformity of discrete distributions • Watson, Geoffrey S.	95 JAppStat 22 273-276
A	V-optimal design for Scheffé's polynomial model • Liu, Shuangzhe; Neudecker, Heinz	95 StPrLet 23 253-258
Modeling and inference with	v-spherical distributions • Fernández, Carmen; Osiewalski, Jacek; Steel, Mark F. J.	95 JASA 90 1331-1340
The existence of doubly resolvable	$(v, 3, 2)$-BIBDs • Lamken, E. R.	95 JCombThA 72 50-76
The role of	VA administrative files in the National Survey of Veterans • Ahuja, Elizabeth M.	94 ASAProGovt 25-27
Minority veterans' usage of	VA medical care • Stockford, Donald	94 ASAProGovt 22-24
test • Smith, Mary E.; Feild, Thomas C. Veterans and	VA system users classified by federal poverty guidelines and thresholds and the VA means	94 ASAProGovt 28-33
A G/SM/1 queue with	vacations depending on service times • Machihara, Fumiaki	95 CommStSM 11 671-690
a cyclic-service tandem queue with multiple server	vacations and exhaustive services • Katayama, Tsuyoshi Analysis of	94 JOpRsJpn 38 249-253
queues with exhaustive service and generalized	vacations • Li, Huan; Zhu, Yixin On M/G/1	95 AdvAppPr 27 510-531
An M/G/1	vacation model with two service modes • Nishimura, Shoichi; Jiang, Yong	95 PrEngInf 9 355-374
An N-class structured priority queue with	vacations • Katsaros, A.; Langaris, C.	95 CommStSM 11 235-248
Vladimir; Becker, Niels G. Assessment of two-dose	vaccination schedules: Availability for vaccination and catch-up strategies • Rouderfer,	95 MaBisci 129 41-66
abeth; Longini, Ira M., Jr; Watelet, Luc Estimation of	vaccine efficacy in non-randomly mixing populations • Haber, Michael; Halloran, M. Eliz-	95 BiomtrcJ 37 25-38
for regression analyses of antibody response to	vaccine • Moulton, Lawrence H.; Halsey, Neal A. A mixture model with detection limits	95 Biomtrcs 51 1570-1578
Relative incidence estimation from case series for	vaccine safety evaluation • Farrington, C. P.	95 Biomtrcs 51 228-235
to patient screening for an autologous tumour	vaccine trial • Shannon, W. D.; Bryant, J.; Logan, T. F. et al. An application of decision theory	95 StatMed 14 2099-2110
reliability for non-precise limit state models and	vague data • Savchuk, V. P. Estimation of structures	95 RelEngSS 47 47-58
The basis of norms for	vague quantifiers • O'Muircheartaigh, Colm; Gaskell, George	94 ASASrvyRsr 1327-1332
neighbouring counties: Cancer mortality data in	Valencia (Spain) • Ferrándiz, J.; López, A.; Llopis, A. et al. Spatial interaction between	95 Biomtrcs 51 665-678
proximate reasoning in belief networks: Statistical	validity, accuracy, and determination of sample size (Disc: p134-135) • Ji, X. et al. Ap-	94 ASAProBays 128-133
On priors providing frequentist	validity for Bayesian inference • Datta, Gauri Sankar; Ghosh, Jayanta Kumar	95 Biomtrka 82 37-45
Hsu Sampling behaviour in estimating predictive	validity in the context of selection and latent variable modelling: A Monte Carlo study •	95 BJMaStPs 48 75-97
A simple algebraic demonstration of the	validity of DeFries-Fulker analysis in unselected samples with multiple ... • Rodgers et al.	94 BehavGen 24 259-262
The relative empirical	validity of dependent and independent data collection in a panel survey • Hill, Daniel H.	94 JOffStat 10 359-380
density regions with approximate frequentist	validity • Ghosh, J. K.; Mukerjee, Rahul On perturbed ellipsoidal and highest posterior	95 JRSS-B 57 761-769
Ghosh, Jayanta K.; Mukerjee, Rahul Frequentist	validity of highest posterior density regions in the presence of nuisance parameters •	95 StatDecn 13 131-139
Teacher evaluation forms: Their	validity or lack there of • Katti, S. K.; Kuruppu, Janaki	94 ASAProStEd 94-99
The criteria for convergent and discriminant	validity in a multitrait-multimethod matrix • Reichardt, Charles S.; Coleman, S. C.	95 MultBehR 30 513-538
Haggard, Lois M.; Gray, Donald Z. Improving	validity of parental reports of child immunization status in a telephone survey •	94 ASASrvyRsr 1258-1262
model using release-recapture data: Sample size,	validity and power • Hoffmann et al. Inferential properties of an individual-based survival	95 JAppStat 22 579-595
disease in the Saskatchewan health ... • Rawson et al.	Validity of the recording of ischaemic heart disease and chronic obstructive pulmonary	95 StatMed 14 2627-2643
How to measure survey reliability and	validity • Litwin, Mark S.	95 SagePubl:CA:Lnd 96p
Uniform	validity of saddlepoint expansion on compact sets • Routledge, Rick; Tsao, Min	95 CandJSt 23 425-431
Gregory; Keil, Linda The	validity of self-reported drug use among the incarcerated • Cantor, David; Gaertner,	94 ASASrvyRsr 1088-1093
Establishing factor	validity using variable reduction in confirmatory factor analysis • Hofmann, Rich	95 EdPsMeas 55 572-582
Akerhielm, Karen Adding value to the	value-added educational production function specification (Disc: p213-215) •	94 ASAProSoSt 192-195
Quantal response analysis of relief	valve test data • Sheesley, John H.; Thomas, Harold W.; Valenzuela, Carlos A.	95 ASQCTekCon 741-748
Finite sample properties of	van Elteren test • Hosmane, Balakrishna; Shu, Vincent; Morris, David	94 ASAProBiop 430-434
Applications of the	van Trees inequality: A Bayesian Cramér-Rao bound • Gill, Richard D.; Levit, Boris Y.	95 Bernoull 1 59-79
performance • Sarantis, N.; Stewart, C. Structural,	VAR and BVAR models of exchange rate determination: A comparison of their forecasting	95 JForcast 14 201-215
Gilbert, Paul D. Combining	VAR estimation and state space model reduction for simple good predictions •	95 JForcast 14 229-250
Forecasts of inflation from	VAR models • Webb, Roy H.	95 JForcast 14 267-285
Forecasting with vector autoregressive (	VAR) models subject to business cycle restrictions • Simkins, Scott	95 IntJFor 11 569-583
Two-sample tests for	variability based on interquantile spreads • Shoemaker, Lewis H.	94 ASAProQlPr 134-139
control with several components of common cause	variability • Woodall, William H.; Thomas, Edward V. Statistical process	95 IIETrn 27 757-764

information REML: An efficient algorithm for	variance parameter estimation in linear mixed models • Gilmour, A. R. et al. Average	95 Biomtrcs 51 1440-1450
to reconstruct gene trees within a species: Molecular	variance parsimony • Excoffier et al. Using allele frequencies and geographic subdivision	94 Genetcs 136 343-359
Suresh, Ram Estimation of	variance of partial sums of an associated sequence of random variables • Peligard, Magda;	95 StocProc 56 307-319
Clustering behavior of finite	variance partial sum processes • Einmahl, U.; Goodman, V.	95 ProbTher 102 547-565
Estimation of the	variance of partial sums for ρ-mixing random variables • Peligrad, Magda; Shao, Qi-Man	95 JMultiAn 52 140-157
The effect of unequal	variances on the power of several two-sample tests • O'Gorman, Thomas W.	95 CommStB 24 853-867
ignored • Van Vleck, L. D.	Variance of prediction error with mixed model equations when relationships are	93 ThApGene 85 545-549
Eric; Särndal, Carl; Lee, Hyunshik Estimation of the	variance in the presence of nearest neighbour imputation (Disc: p906-907) • Rancourt,	94 ASASrvyRsr 888-893
CUSUM charts-for-	variance procedure and its ARL approximation • Chow, William	94 ASAProQlPr 93-97
sum control chart for monitoring process	variance • Chang, T. C.; Gan, F. F. A cumulative	95 JQualTek 27 109-119
sum control chart for monitoring process	variance" (V27 p109-119) • Chang, T. C.; Gan, F. F. Correction to "A cumulative	95 JQualTek 27 402-402
Spherical prediction-	variance properties of central composite and Box-Behnken designs • Borkowski, John J.	95 Technmcs 37 399-410
On the invariant estimation of a normal	variance ratio • Madi, Mohamed T.	95 JStPlInf 44 349-357
	Variance reduction	
dynamic resource sharing • Fleming, P. J. et al.	Efficient Monte-Carlo simulation of a product-form model for a cellular system with	95 ACMMdSim 5 3-21
	Estimating probabilities from invariant permutation distributions • Pallini, Andrea	94 JItalStS 3 77-91
	Fast simulation of rare events in queueing and reliability models • Heidelberger, Philip	95 ACMMdSim 5 43-85
sequential phase II trials • Conaway, Mark R.	Importance sampling to evaluate bivariate probabilities with an application to	95 JStCmpSm 51 249-262
	Random invariant cubature formulas (Russian) • Ermakov, S. M.; Wang, Xiao-Qun	94 VLenUM /4 109-112
Tim	Tail-specific linear approximations for efficient bootstrap simulations • Hesterberg,	95 JCmpGrSt 4 113-133
Hesterberg, Tim	Weighted average importance sampling and defensive mixture distributions •	95 Technmcs 37 185-194
Vegas, Esteban	Variance reduction for Bernoulli response variables in simulation • Ocaña, Jordi;	95 CmpStDA 19 631-640
A note on	variance reduction • Berlinet, Alain	95 StPrLet 25 357-360
• Lynch, Thomas B.	Variance reduction in Ueno's method and cylinder sampling for forest volume estimation	95 CanJForR 25 1783-1794
Heterogeneous	variances in repeated measures • Wu, Yuan; McGilchrist, C. A.	95 BiomtrcJ 37 523-530
Miller, W. G. Partial likelihood analysis of within-unit	variances in repeated measurement experiments • Chinchilli, V. M.; Esinhart, J. D.;	95 Biomtrcs 51 205-216
address file as ... • Lynn et al. On the bias and	variance of samples of individuals: Comparison of the electoral registers and postcode	95 Ststcian 44 173-194
On the smallest possible asymptotically efficient	variance in semiparametric models • Liang, Hua	95 JSysSMSE 7 29-33
Inequalities for	variances of some functions of random variables • Nollau, V.	95 StHefte 36 163-174
	Variance stabilization	
regression model • Wang, Naisyin; Ruppert, David	Nonparametric estimation of the transformation in the transform-both-sides	95 JASA 90 522-534
Form of the conditional	variance for stable random variables • Cioczek-Georges, Renata; Taqqu, Murad S.	95 StSinica 5 351-361
A. B.; Nusser, S. M.; Breidt, F. J. Approximating the	variance of the survey regression estimator using poststratification • Leonard, K. A.; An,	94 ASASrvyRsr 222-227
Bias-	variance tradeoffs in functional estimation problems • Low, Mark G.	95 AnlsStat 23 824-835
On the	variance of the trimmed mean • Capéraà, Philippe; Rivest, Louis-Paul	95 StPrLet 22 79-85
multiple comparisons with the best for unequal	variances • Matejcik, Frank J.; Nelson, Barry L. One-stage	94 ASAProBuEc 55-60
Joint monitoring of process mean and	variance using exponentially weighted moving average control charts • Gan, F. F.	95 Technmcs 37 446-453
atively unimportant with reasonable between village	variances? • Fitch et al. Are the biases and inefficiencies of the EPI method of sampling rel-	94 ASASrvyRsr 288-292
On estimation of the wavelet	variance • Percival, Donald P.	95 Biomtrka 82 619-631
bounds on Bayes risks for estimating a normal	variance: With applications • Vidakovic, Brani; DasGupta, Anirban Lower	95 CandJSt 23 269-282
sition and ... • Pfeffermann et al. Estimation of the	variances of X-11 ARIMA seasonally adjusted estimators for a multiplicative decompo-	95 IntJFor 11 271-283
Yin, G. On	variational inequality approach to Nash dynamic game of oil stockpiling • Zhao, L.;	95 StocAnAp 13 627-649
Mächler, Martin B.	Variational solution of penalized likelihood problems and smooth curve estimation •	95 AnlsStat 23 1496-1517
Designs through recoding of	varietal and level codes • Das, M. N.; Giri, N.; Ahmed, M.	95 StPrLet 23 371-380
Estimation of heritability from	varietal trials data • Singh, M.; Ceccarelli, S.; Hamblin, J.	93 ThApGene 86 437-441
Laurent, Monique	Variety of hypercube embeddings of the equidistant metric and designs • Deza, Michel;	93 JCISS 18 289-315
	Variety trial	
Green, Peter; Higdon, David; Mengersen, Kerrie	Bayesian computation and stochastic systems (Disc: p41-66) • Besag, Julian;	95 StatSci 10 3-41
of multiplicative effects in multienvironment	variety trials • Gogel, Beverley J.; Cullis, Brian R.; Verbyla, Arūnus P. REML estimation	95 Biomtrcs 51 744-749
Dyke et al. Sensitivity (stability) analysis of multiple	variety trials, with special reference to data expressed as proportions or percentages •	95 ExpAgric 31 75-87
planar rotations: A remedy against nonoptimal	varimax rotations • Berge, Jos M. F. ten Suppressing permutations or rigid	95 Psymtrka 60 437-446
	Variogram	
Guns+, Richard F.	Estimating spatial correlations from spatial-temporal meteorological data •	95 JClimate 8 2454-2470
streams • Barnett, Neil S.; Gomm, Ian S.; Armour, Len	Location of the optimal sampling point for the quality assessment of continuous	95 AstrlJSt 37 145-152
	Quasi-linear geostatistical theory for inversing • Kitanidis, Peter K.	95 WaterReR 31 2411-2419
Donnelly, Christl A.	The spatial analysis of covariates in a study of environmental epidemiology •	95 StatMed 14 2393-2409
Statistical prediction models: A ... • Lesch et al.	Spatial prediction of soil salinity using electromagnetic induction techniques 1.	95 WaterReR 31 373-386
Chang, Tiao J.; Teoh, Choo B.	Use of the kriging method for studying characteristics of ground water droughts •	95 WaterReB 31 1001-1007
Gavrikov, Vladimir; Stoyan, Dietrich	The use of marked point processes in ecological and environmental forest studies •	95 EnvEcoSt 2 331-344
information, Bayesian kriging and updating of prior	variogram parameters • Cui, Haiyan; Stein, Alfred; Myers, Don E. Extension of spatial	95 EnvrMtrc 6 373-384
Bayesian classification of	vasculitis: A simulation study • Heller, I.; Isakov, A.; Blinder-Weiner, S.; Topilsky, M.	95 MInfoMed 34 259-265
	Vector	
	• see also names of specific types of vectors such mean - or unit -	
Exact maximum likelihood estimation of stationary	vector ARMA models • Mauricio, José Alberto	95 JASA 90 282-291
	Vector autoregression	
sures • Kumar, V.; Leone, Robert P.; Gaskins, John N.	Aggregate and disaggregate sector forecasting using consumer confidence mea-	95 IntJFor 11 361-377
regression models • Niu, Xufeng	Asymptotic properties of maximum likelihood estimates in a class of space-time	95 JMultiAn 55 82-104
time-varying parameter models • Chib, S. et al.	Hierarchical analysis of SUR models with extensions to correlated serial errors and	95 JEconmtx 68 339-360
	Interval forecasting in cointegrated systems • Reimers, Hans-Eggert	95 StHefte 36 349-369
Average • Tse, Y. K.	Lead-lag relationship between spot index and futures price of the Nikkei Stock	95 JForcast 14 553-563
Pinder, Jonathan P.; Shoesmith, Gary L.	Modelling multivariate cointegrated systems: Insights from non-linear dynamics •	95 JForcast 14 311-324
model • Shoesmith, Gary L.	Multiple cointegrating vectors, error correction, and forecasting with Litterman's	95 IntJFor 11 557-567
	The role of the drift in I(2) systems • Paruolo, Paolo	94 JItalStS 3 93-123
Bewley, Ronald; Yang, Minxian	Testing for cointegration: The effects of mis-specifying the lag length •	95 MaCmpSim 39 251-255
Horvath, Michael T. K.; Watson, Mark W.	Testing for cointegration when some of the cointegrating vectors are prespecified •	95 EconTher 11 984-1014
Bekaert, Geert	The time variation of expected returns and volatility in foreign-exchange markets •	95 JBES 13 397-408
	Uncertainty about the persistence of economic shocks • Miller, John P.; Newbold, Paul	95 JBES 13 435-440
On the determination of lag order in	vector autoregressions of cointegrated systems • Jacobson, Tor	95 CmpStQ 10 177-192
control theory to interpret the policy equation of a	vector autoregression • Salemi Revealed preference of the Federal Reserve: Using inverse-	95 JBES 13 419-433
and prices • Joutz, F. L. et al. An integrated Bayesian	vector autoregression and error correction model for forecasting electricity consumption	95 JForcast 14 287-310
cycle analysis and forecasting with a structural	vector autoregression model for Wales • Ioannidis, C.; Laws, J.; Matthews, K. et al. Business	95 JForcast 14 251-265
	Vector autoregression modelling and forecasting • Holden, Ken	95 JForcast 14 159-166
of likelihood ratio tests for cointegrating ranks in	vector autoregressions • Toda, Hiro Y. Finite sample performance	95 EconTher 11 1015-1032
the rank of co-integration when the order of a	vector autoregression is unknown • Morimune, Kimio; Mantani, Akihisa Estimating	95 MaCmpSim 39 265-271

Taku Statistical inference in	vector autoregressions with possibly integrated processes • Toda, Hiro Y.; Yamamoto,	95 JEconmtx 66 225-250
Reduced rank estimation of	vector autoregressive models with cointegration and codependence • Ahn, Sung K.	94 ASAProBuEc 151-156
testing for unit roots in a partially nonstationary	vector autoregressive moving average model • Yap, S. F.; Reinsel, G. C. Estimation and	95 JASA 90 253-267
Finite sample forecast results for	vector autoregressive moving average models • Reinsel, Gregory C.	95 JForcast 14 405-412
Forecasting with	vector autoregressive (VAR) models subject to business cycle restrictions • Simkins, Scott	95 IntJFor 11 569-583
vectors • Cléroux, Robert; Lazraq, Aziz; Lepage, Yves	Vector correlation based on ranks and a nonparametric test of no association between	95 CommStA 24 713-733
Stochastic	vector difference equations with stationary coefficients • Glasserman, Paul; Yao, David D.	95 JAppProb 32 851-866
Multiple cointegrating	vectors, error correction, and forecasting with Litterman's model • Shoesmith, Gary L.	95 IntJFor 11 557-567
Dimension results for Gaussian	vector fields and index-α stable fields • Xiao, Yimin	95 AnlsProb 23 273-291
	Vector harmonizable processes: Wold and Cramér decompositions • Kakihara, Yûichirô	95 StocAnAp 13 531-541
Some aspects of forecasting with	vector moving average processes • Peiris, M. Shelton	94 CalcutSt 44 151-156
for cointegration when some of the cointegrating	vectors are prespecified • Horvath, Michael T. K.; Watson, Mark W. Testing	95 EconTher 11 984-1014
for non-linear functionals of stationary Gaussian	vector processes • Sánchez de Naranjo, M. V. A central limit theorem	95 StPrLet 22 223-230
Huo, Qiang; Chan, Chorkin Contextual	vector quantization for speech recognition with discrete hidden Markov Model •	95 PattrnRc 28 513-517
image compression and classification using	vector quantization • Oehler, Karen L.; Gray, Robert M. Combining	95 IEEEPAMI 17 461-473
A	vector of quarters representation for bivariate time series • Franses, Philip Hans	95 EcomtrRv 14 55-63
nearest neighbor random walk on subspaces of a	vector space and rate of convergence • D'Aristotile, Anthony J. The	95 JTherPrb 8 321-346
Wakuta, Kazuyoshi	Vector-valued Markov decision processes and the systems of linear inequalities •	95 StocProc 56 159-169
Remarks on strong exponential integrability of	vector-valued random series and triangular arrays • Rosiński, Jan	95 AnlsProb 23 464-473
Herouin, Eric Predicting	velocity frequency distributions in stream reaches • Lamouroux, Nicolas; Souchon, Yves;	95 WaterReR 31 2367-2375
Institutional hypothesis of the long-run income	velocity of money and parameter stability of the equilibrium relationship • Raj, Baldev	95 JApEcmtx 10 233-253
trend functions, and the chaotic structure of the	velocity of money • Serletis, Apostolos Random walks, breaking	95 JBES 13 453-458
Crossing	velocities and random lattice animals • Sznitman, Alain-Sol	95 AnlsProb 23 1006-1023
Construction of simple and observable	Venn diagrams for any finite number of sets • Hu, Stephen N.	94 ASAProStGr 60-65
Wright, Daniel B. Weighing anchors:	Verbal and numeric labels for response scales • O'Muircheartaigh, Colm; Gaskell, George;	95 JOffStat 11 295-307
tualizations of job search: Further evidence from	verbatim responses • Tanur, Judith M. Concep-	94 ASASrvyRsr 512-516
databases (Disc: p687-689) • Shevchenko, I. P. et al.	Verification of information in a large medical database using linkages with external	95 StatMed 14 511-530
Evaluating multiple diagnostic tests with partial	verification • Baker, Stuart G.	95 Biomtrcs 51 330-337
The complexity of probabilistic	verification • Courcoubetis, Costas; Yannakakis, Mihalis	95 JACM 42 857-907
Optimization of reliability	verification test strategies • Dodson, Bryan	95 ASQCTekCon 1039-1045
basin characteristics in New Hampshire and	Vermont • Dingman, S. L.; Lawlor, S. C. Estimating low-flow quantiles from drainage-	95 WaterReB 31 243-256
specialized risk assessment methodologies for	vertebrate populations • Anderson, David R.; White, Gary C.; Burnham, Kenneth P. Some	95 EnvEcoSt 2 91-115
Unbiased estimation of capillary length from	vertical slices • Batra, S.; König, M. F.; Cruz-Orive, L. M.	95 JMicro 178 152-159
of individual feature surface area with the	vertical spatial grid • Cruz-Orive, L. M.; Howard, C. V. Estimation	95 JMicro 178 146-151
Steven H.; Klein, Robert E.; Adashek, Daniel	Veterans living in group quarters: What data from the 1990 Census show • Schwartz,	94 ASAProGovt 13-18
of VA administrative files in the National Survey of	Veterans • Ahuja, Elizabeth M. The role	94 ASAProGovt 25-27
An overview of the National Survey of	Veterans • Krumhaus, Susan G.; Klemm, Rebecca J.	94 ASAProGovt 19-21
Minority	veterans' usage of VA medical care • Stockford, Donald	94 ASAProGovt 22-24
the VA means test • Smith, Mary E.; Feild, Thomas C.	Veterans and VA system users classified by federal poverty guidelines and thresholds and	94 ASAProGovt 28-33
On estimating the linkage of marker genes to	viability genes controlling inbreeding depression • Fu, Y. B.; Ritland, K.	94 ThApGene 88 925-932
Stochastic Nagumo's	viability theorem • Aubin, J. P.; Prato, G. Da	95 StocAnAp 13 1-11
Randomly perturbed	vibrations • Elshamy, M.	95 JAppProb 32 417-428
effects of questionnaire design on reported rates of	victimization • Paton, D. Some	94 ASASrvyRsr 443-448
data collection • O'Reilly, J. M. et al. Audio and	video computer assisted self-interviewing: Preliminary tests of new technologies for	94 JOffStat 10 197-214
(Rev) Review of	"Video course: Principal components" • Massart, D. L.; Lewi, P. J. (Auth); Kvalheim, O. M.	95 Chemolab 27 147-147
An illustration of bootstrapping using	video lottery terminal data (jse/v3n2/datasets.braun) • Braun, W. John	95 JStEd 3/2 18kb
Confidentiality issues in	videotaped data collection • Cederlund, Ann E.	94 ASASrvyRsr 1248-1251
for scenario analysis with an application to the	Videotel service in Italy • Sapio ... through repeated cross impact handling): A new method	95 IntJFor 11 113-131
Hidden assumptions: The use of	vignettes in cognitive interviewing • Gerber, Eleanor R.	94 ASASrvyRsr 1269-1274
Bordeaux wine	vintage quality and the weather • Ashenfelter, Orley; Ashmore, David; Lalonde, Robert	95 Chance 8/4 7-14
Forecasting the final	vintage of real personal disposable income: A state space approach • Patterson, K. D.	95 IntJFor 11 395-405
A state space approach to forecasting the final	vintage of revised data with an application to the index of industrial ... • Patterson	95 JForcast 14 337-350
the robustness of maximum likelihood scaling for	violations of the error model • Storms, Gert On	95 Psymtrka 60 247-258
sociated with testing factor mean differences under	violations of factorial invariance • Kaplan, David; George, Rani A study of the power as-	95 StrEqMod 2 101-118
covariates - Investigating the sensitivity against	violation of the missing at random assumption • Vach et al. ... observed categorical	95 StatMed 14 1315-1329
Graphical methods for assessing	violations of the proportional hazards assumption in Cox regression • Hess, Kenneth R.	95 StatMed 14 1707-1723
characteristics on response in a national survey of	violence against women • Norris, Douglas A.; Hatcher, Jeff The impact of interviewer	94 ASASrvyRsr 449-454
Krivokapić, Zoran	Virtual manufacturing: Design, calibration, and analysis • Heavlin, William D.;	94 ASAProPEng 274-279
ity criteria • Hedayat, A. S. et al. Contingently and	virtually balanced incomplete block designs and their efficiencies under various optimal-	95 StSinica 5 575-591
Stufken, J.; Zhang, W. G.	Virtually balanced incomplete block designs for $v = 22$, $k = 8$, and $\lambda = 4$ • Hedayat, A. S.;	95 JComDsgn 3 195-201
Foster, R. D. et al. Improving control of polymer	viscosity using time-series experimentation and statistical process control • Wallace, J. E.;	94 ASAProPEng 320-325
surveys using the double-count technique to correct	visibility bias • Rivest, L.-P.; Potvin, F.; Crépeau, H.; Daigle, G. Statistical methods for aerial	95 Biomtrcs 51 461-470
Key words and phrases - The key to scholarly	visibility and efficiency in an information explosion • Gbur, Edward E., Jr; Trumbo, Bruce E.	95 AmerStat 49 29-33
Stochastic	visibility in random fields • Zacks, Shelemyahu	94 Springer:Brln:NY 175p
Review of "Stochastic	visibility in random fields" • Zacks, Shelemyahu (Auth); Heyde, C. C. (Rev)	95 ShrtBkRv 15/3 48-48
Reinecke, K. J.; Conroy, M. J. et al. Factors affecting	visibility rate of waterfowl surveys in the Mississippi Alluvial Valley • Smith, D. R.;	95 JWildlMg 59 515-527
of correlation with applications to computer	vision • Brunelli, R.; Messelodi, S. Robust estimation	95 PattrnRc 28 833-841
Review of "Neural networks in	vision and pattern recognition" • Skrzypek, J.; Karplus, W. (Ed); Sethi, Ishwar K. (Rev)	95 JClass 12 140-142
	Visitor sample surveys • Iachan, Ronaldo; Kemp, Suzanne, S.	95 SurvyM 21 89-96
standards for ordinal data: Modelling the changes in	visual acuity from 2 to 9 years of age • Wade, A. M.; Ades, A. E. et al. Age-related	95 StatMed 14 257-266
Hallman, William K. Testing a	visual display to explain small probabilities • Weinstein, Neil D.; Sandman, Peter M.;	94 RiskAnal 14 895-896
	Visual error criteria for qualitative smoothing • Maron, J. S.; Tsybakov, A. B.	95 JASA 90 499-507
Contaminant detection in the	visual inspection of seed samples • Aitken, C. G. G.; Shaw, J.; Talbot, M.	95 ApplStat 44 431-440
in the sequence of action potentials in an insect	visual interneuron • Turcott, R. G.; Barker, P. D. R.; Teich, M. C. Long-duration correlation	95 JStCmpSm 52 253-271
Conceptual and	visual models for categorical data • Friendly, Michael	95 AmerStat 49 153-160
Models of statistical	visual motion estimation • Spetsakis, Minas	94 CVGIPIm 60 300-312
	Visual revelations: A rose by another name • Wainer, Howard	95 Chance 8/2 46-51
The	visual separability of plotting symbols in scatterplots • Tremmel, Lothar	95 JCmpGrSt 4 101-112
Guiding data analysts with	visual statistical strategies (Disc: p251-260) • Young, Forrest W.; Lubinsky, David J.	95 JCmpGrSt 4 229-250
Esposito, R.; Fox, J. K.; Lin, D.; Tidemann, K. ARIES:	Visual techniques for statistical data investigation at the Bureau of Labor Statistics •	94 ASAProStGr 7-10
	Visualization	
	Huge data sets and the frontiers of computational feasibility • Wegman, Edward J.	95 JCmpGrSt 4 281-295
Jeffrey L.; Poston, Wendy L.; Wegman, Edward J. A	visualization technique for studying the iterative estimation of mixture densities • Solka,	95 JCmpGrSt 4 180-198
Using data and the computer to teach and to	visualize concepts in an introductory statistics course • Guffey, James M.	94 ASAProStEd 84-87

Mathieson, K.; Doane, D. P. et al. A program for	visualizing comparisons between two normal distributions (jse/v3n1/mathieson) •	95 JStEd 3/1 23kb
Review of	"Visualizing data" • Cleveland, W. S. (Auth); Bowman, Adrian (Rev)	95 JRSS-A 158 344-345
Review of	"Visualizing data" • Cleveland, William S. (Auth); Royston, Patrick (Rev)	95 StatMed 14 1395-1396
Kieran; Doane, David P.	Visualizing and describing the shape of distributions • Tracy, Ronald L.; Mathieson,	94 ASAProStEd 70-74
Mathieson, Kieran D.	Visualizing power and type I and II error • Daone, David P.; Tracy, Ronald L.;	94 ASAProStEd 66-69
The FU-PLOT: A graphical method for	visualizing the timing of follow-up in longitudinal studies • Lesser, M. L.; Kohn, N. E. et al.	95 AmerStat 49 139-143
Minstrell, J.; Hunt, E.; Madigan, D.; Traynor, C.	VITAL: An intelligent tutoring system for statistics • Clarkson, D. B.; Donnell, D.;	94 ASAProStEd 88-93
Samuel H. Consistency of census and	vital registration data on older Americans: 1970-1990 • Shrestha, Laura B.; Preston,	95 SurvyM 21 167-177
McBride, Ruth; Moore, Rich Toward	vital status sweeps: A case history in sequential monitoring • Hallstrom, Al;	95 StatMed 14 1927-1931
of international transmissions of stock returns and	volatility: The case of the United States and Canada • Karolyi A multivariate GARCH model	95 JBES 13 11-25
Forecasting	volatility in commodity markets • Kroner, Kenneth F.; Kneafsey, Kevin P.; Claessens, Stijn	95 JForcast 14 77-95
Estimating the instantaneous	volatility and covariance of risky assets • Chesney, Marc; Elliott, Robert J.	95 AplStMDA 11 51-58
The time variation of expected returns and	volatility in foreign-exchange markets • Bekaert, Geert	95 JBES 13 397-408
hedging of options on stocks with Markov	volatilities • Di Masi, G. B.; Kabanov, Yu. M.; Runggaldier, W. J. Mean-variance	94 ThProbAp 39 172-182
predictive ability of several models of exchange rate	volatility • West, Kenneth D.; Cho, Dongchul The	95 JEconmtx 69 367-391
Can speculative trading explain the volume-	volatility relation? • Foster, F. Douglas; Viswanathan, S.	95 JBES 13 379-396
	Volterra equation	
	Stability of a class of stochastic integro-differential equations • Drozdov, Aleksey	95 StocAnAp 13 517-530
Potthoff, Jürgen Stochastic	Volterra equations with singular kernels • Cochran, W. George; Lee, Jung-Soon;	95 StocProc 56 337-349
Jáimez, R.; Román, P. R.; Ruiz, F. T. A note on the	Volterra integral equation for the first-passage-time probability density • Gutiérrez	95 JAppProb 32 635-648
Skorokhod, A. V. Randomly perturbed	Volterra integral equations and some applications • Hoppensteadt, F.; Salehi, H.;	95 Stochatx 54 89-125
infinite filter. Application to the estimation of	Volterra series • Bosq, Denis Asymptotic estimation of a non-linear	95 JNnparSt 5 61-73
Stationary probability distributions of some Lotka-	Volterra types of stochastic predation systems • Yeung, David W. K.; Stewart, Sally E. A.	95 StocAnAp 13 503-516
Estimating and controlling for the effects of	volunteer bias with pairs of relatives • Neale, Michael C.; Eaves, Lindon J.	93 BehavGen 23 271-278
Accounting for individual variability in the	von Bertalanffy growth model • Wang, You-Gan; Thomas, Mervyn R.	95 JFishRBC 52 1368-1375
Kaufmann, Edgar	Von Mises conditions, δ-neighborhoods and rates of convergence for maxima •	95 StPrLet 25 63-70
	von Mises distribution	
data methods for censored regression • Lai et al.	Asymptotic normality of a class of adaptive statistics with applications to synthetic	95 JMultiAn 52 259-279
Haiganoush K.; Akers, R. Patrick	Autoregressive-type models for the analysis of bark beetle tracks • Preisler,	95 Biomtrcs 51 259-267
likelihood ratio test for the mean direction in the	von Mises distribution • Yamamoto, Eiji; Yanagimoto, Takemi A modified	95 CommStA 24 2659-2678
A large deviation principle for m-variate	von Mises-statistics and U-statistics • Eichelsbacher, Peter; Löwe, Matthias	95 JTherPrb 8 807-824
distributions of wrapped Cauchy type and of	von Mises type • Holmquist, Björn Simulation of multimodal circular	95 CommStB 24 607-615
"Spatial tessellations. Concepts and applications of	Voronoi diagrams" • Okabe, A.; Boots, B.; Sugihara, K. (Auth); Lorz, U. (Rev) Review of	95 MaOpfStS 26 183-183
Statistical inference for	Voronoi images • Archambault, S.; Moore, M.	95 CandJSt 23 11-20
for combining sample correlation coefficients and	vote counts to obtain an estimate and a confidence ... • Bushman et al. A procedure	95 PsychBu 117 530-546
analysis of dependence of voter shares under a two-	vote system: The Hong Kong experience • Fung, Wing K. A statistical	94 ASAProSoSt 76-81
Privacy issues and the use of	voter contact lists • Traugott, Michael W.	94 ASAProSoSt 164-169
Hybrid zones and	voter model interfaces • Cox, J. T.; Durrett, R.	95 Bernoull 1 343-370
The noisy	voter model • Granovsky, Boris L.; Madras, Neal	95 StocProc 55 23-43
A	voter model in random environment (French) • Saada, Ellen	95 AIHenrPB 31 263-271
A spatial analysis of contextual effects on	voter participation in the 1992 presidential election • Schlichting, Kurt C.; Tuckel, Peter S.	94 ASASrvyRsr 1339-1344
arising from the long range contact and long range	voter processes • Müller, C.; Tribe, R. Stochastic p.d.e.'s	95 ProbTher 102 519-545
Privacy issues and the National	Voter Registration Act of 1993 • Montjoy, Robert S.	94 ASAProSoSt 170-173
A statistical analysis of dependence of	voter shares under a two-vote system: The Hong Kong experience • Fung, Wing K.	94 ASAProSoSt 76-81
Peyrot, Mark F.; Varga, Nicholas An analysis of	voting coalitions in the New York colonial legislature • Morrell, Christopher H.;	94 ASAProSoSt 66-70
Review of "Probabilistic	voting theory" • Coughlin, P. J. (Auth); Tanguiane, A. S. (Rev)	95 JClass 12 145-146
Comment on "Algorithm AS 181: The	W-test for normality" (V31 p176-180) • Royston, Patrick	95 ApplStat 44 547-551
Reporting negative contamination levels on silicon	wafers - Is this nonsense? • Pankratz, Peter C.; Coria, José C.	94 ASAProPEng 71-75
Neumann, George R. Estimation of equilibrium	wage distributions with heterogeneity • Bowlus, Audra J.; Kiefer, Nicholas M.;	95 JApEcmtx 10 119-131S
Minimum	wage effects on employment and school enrollment • Neumark, David; Wascher, William	95 JBES 13/2 199-206
Impacts of immigrants on employment and	wage growth: Evidence of positive effects of U.S. Immigration • Enchautegui, María E.	94 ASAProSoSt 30-38
effect in retail trade of California's 1988 minimum	wage increase • Kim, Taeil; Taylor, Lowell J. The employment	95 JBES 13/2 175-182
Box-Cox transformation model, and the U.S.	wage structure, 1963-1987 • Buchinsky, Moshe Quantile regression,	95 JEconmtx 65 109-154
An empirical investigation of	Wagner's hypothesis by using a model occurrence framework • Koop, Gary; Poirier, Dale J.	95 JRSS-A 158 123-141
	Wagon wheel designs with residual effects from neighboring treatments • Cornell, John A.	93 JCISS 18 255-272
Optimizing the number of customers in a closed	waiting queue with several customer classes (German) • Schmidt, R.	95 ZAngMaMk 75 571-572S
	Waiting time	
permutation • Fu, James C.	Exact and limiting distributions of the number of successions in a random	95 AnInStMa 47 435-446
	A G/SM/1 queue with vacations depending on service times • Machihara, Fumiaki	95 CommStSM 11 671-690
Zhu, Yixin	On M/G/1 queues with exhaustive service and generalized vacations • Li, Huan;	95 AdvAppPr 27 510-531
Alfa, Attahiru Sule	Matrix analytical methods for M/PH/1 retrial queues • Diamond, Jeffrey E.;	95 CommStSM 11 447-470
	An N-class structured priority queue with vacations • Katsaros, A.; Langaris, C.	95 CommStSM 11 235-248
Coffman, E. G., Jr; Puhalskii, A. A.; Reiman, M. I.	Polling systems with zero switchover times: A heavy-traffic averaging principle •	95 AnlsApPr 5 681-719
Takine, Tetsuya; Hasegawa, Toshiharu	On the single-server queue with independent MAP/G and M/GI input streams •	95 CommStSM 11 227-234
Ferrari, P. A.; Galves, A.; Liggett, T. M. Exponential	waiting time for filling a large interval in the symmetric simple exclusion process •	95 AIHenrPB 31 155-175
Gong, Wei-Bo The mean	waiting time of a GI/G/1 queue in light traffic via random thinning • Nananukul, Soracha;	95 JAppProb 32 256-266
Sadowsky The probability of large queue lengths and	waiting times in a heterogeneous multiserver queue II: Positive recurrence and ... •	95 AdvAppPr 27 567-583
The probability of large queue lengths and	waiting times in a heterogeneous multiserver queue I: Tight limits • Sadowsky, J. S. et al.	95 AdvAppPr 27 532-566
Li, Jingwen; Chou, Fee Seng	Waiting time in M/G/1 queues with impolite arrival disciplines • Özekici, Süleyman;	95 PrEngInf 9 255-267
O. P. On certain asymptotic properties of	waiting time in a multiserver queueing system with identical service times • Vinogradov,	94 ThProbAp 39 714-718
Infinite divisibility and the	waiting-time paradox • van Harn, K.; Steutel, F. W.	95 CommStSM 11 527-540
Sooner and later	waiting time problems in a two-state Markov chain • Uchida, Masayuki; Aki, Sigeo	95 AnInStMa 47 415-433
Masakiyo Relationship between queue length and	waiting time in a stationary discrete-time queue • Takahashi, Yoshitaka; Miyazawa,	95 CommStSM 11 249-271
Jan A.; van der Mei, Rob D.	Waiting times in a two-queue model with exhaustive and Bernoulli service • Weststrate,	94 ZeitOpRs 40 289-303
spheric stability on concentration fluctuations and	wake retention times for dispersion in the vicinity of an ... • Higson et al. Effect of atmo-	95 EnvrMtrc 6 571-581
A best possible improvement of	Wald's equation for stochastic processes • Roters, Markus	95 Stochatx 54 293-300
gular • El-Helbawy, Abdalla T.; Hassan, Tawfik On the	Wald, Lagrangian multiplier and likelihood ratio tests when the information matrix is sin-	94 JItalStS 3 51-60
Asymptotic optimality of the	Wald sequential test • Koell, Christophe	95 SqtlAnly 14 341-360
	Wald statistic	
analysis • Ligeralde, Antonio V.; Brown, Bryan W.	Band covariance matrix estimation using restricted residuals: A Monte Carlo	94 IntEconR 35 751-767
IGARCH(1,1) models: A Monte Carlo ... • Lumsdaine	Finite-sample properties of the maximum likelihood estimator in GARCH(1,1) and	95 JBES 13 1-10
	Monte Carlo response surfaces: A comparative approach • Power, Simon	95 JStPlInf 43 301-310
Mukerjee, Rahul	On posterior credible sets based on the score statistic • Rao, C. Radhakrishna;	95 StSinica 5 781-791
Size corrections for the	Wald statistic in structural equation models • Oya, Kosuke	95 MaCmpSim 39 317-322

	Weighted least squares	
populations • Jordan, Scott M.; Krishnamoorthy, K.	Confidence regions for the common mean vector of several multivariate normal	95 CandJSt 23 283-297
spatial adaptation • Fan, Jianqing; Gijbels, Irène	Data-driven bandwidth selection in local polynomial fitting: Variable bandwidth and	95 JRSS-B 57 371-394
models • Dahlhaus, Rainer	Efficient location and regression estimation for long range dependent regression	95 AnlsStat 23 1029-1047
	Estimation in the polytomous logistic regression model • Rom, Meir; Cohen, Ayala	95 JStPlInf 43 341-353
Welsh, Alan H.	Robust restricted maximum likelihood in mixed linear models • Richardson, Alice M.;	95 Biomtrcs 51 1429-1439
attrition • Wang-Clow, F.; Lange, M.; Laird, N. M. et al.	A simulation study of estimators for rates of change in longitudinal studies with	95 StatMed 14 283-297
Lebreton, J.-D.; Morgan, B. J. T.; Pradel, R. et al.	A simultaneous survival rate analysis of dead recovery and live recapture data •	95 Biomtrcs 51 1418-1428
tive generalization of the binomial ... • From A	weighted least-squares procedure for estimating the parameters of Altham's multiplica-	95 StPrLet 25 193-199
	Weighted maximum likelihood	
Chaudhuri, Probal; Dewanji, Anup	On a likelihood-based approach in nonparametric smoothing and cross-validation •	95 StPrLet 22 7-15
Optimally	weighted means in stratified sampling • Qian, Jiahe	94 ASASrvyRsr 863-866
average run length of a multivariate exponentially	weighted moving average control chart • Rigdon An integral equation for the in-control	95 JStCmpSm 52 351-365
average run length of a multivariate exponentially	weighted moving average control chart • Rigdon, S. E. A double-integral equation for the	95 StPrLet 24 365-373
of process mean and variance using exponentially	weighted moving average control charts • Gan, F. F. Joint monitoring	95 Technmcs 37 446-453
Average cost Markov control processes with	weighted norms: Existence of canonical policies • Gordienko, E.; Hernández-Lerma, O.	95 ZastApMa 23 199-218
Average cost Markov control processes with	weighted norms: Value iteration • Gordienko, E.; Hernández-Lerma, O.	95 ZastApMa 23 219-237
On the problem of the means of	weighted normal populations • Nikulin, M. S.; Voinov, V. G.	95 Questo 19 93-106
Conditioned superprocesses and their	weighted occupation times • Krone, Stephen M.	95 StPrLet 22 59-69
	Weighted orthogonal series estimates of the regression curve (Chinese) • Yang, Ying	95 JSysSMaS 14 337-349
An efficient algorithm for	weighted PCA • Krijnen, W. P.; Kiers, H. A. L.	95 CmpStQ 10 299-306
Development of an optimal test based on	weighted rankings • Larocque, Denis; Tardif, Serge	95 CommStB 24 367-383
Opsomer, J.-D. et al. An application of locally	weighted regression to airborne mercury deposition around an incinerator site •	95 EnvrMtrc 6 205-219
Asymptotic properties of locally	weighted regression • Hwang, Ruey-Ching	95 JNnparSt 5 303-310
spreadsheet program for calibration using	weighted regression • Gort, Steven M.; Hoogerbrugge, Ronald A user-friendly	95 Chemolab 28 193-199
portional hazards tests and diagnostics based on	weighted residuals" (V81 p515-526) • Grambsch, P. M.; Therneau, T. M. Correction to "Pro-	95 Biomtrka 82 668-668
proximate distribution of the dispersion matrix of a	weighted sample from a p-variate normal distribution • Coelho, C. Brief note on the ap-	94 JIndnStA 32 77-81
An image thresholding method using a minimum	weighted squared-distortion criterion • Morii, Fujiki	95 PattrnRc 28 1063-1071
Loss robustness via Fisher-	weighted squared-error loss function • Makov, Udi E.	95 InsrneME 16 1-6
hypothesis • Tsubaki, Michiko New	weighted sum of χ^2 test for testing a linear hypothesis against a restricted alternative	95 CommStB 24 987-1004
Randomly	weighted sums for exchangeable fuzzy random variables • Inoue, Hiroshi	95 FuzzySet 69 347-354
A strong law for	weighted sums of i.i.d. random variables • Cuzick, Jack	95 JTherPrb 8 625-641
of large numbers and the law of the logarithm for	weighted sums of independent random variables with ... • Li et al. ... strong law	95 JMultiAn 52 181-198
Daffer, Peter Convergence of	weighted sums and laws of large numbers in $D([0,1];E)$ • Şchiopu-Kratina, Ioana;	95 JMultiAn 53 279-292
convergence and almost sure convergence of	weighted sums of random variables • Li, D.; Rao, M. B.; Jiang, T.; Wang, X. Complete	95 JTherPrb 8 49-76
Kamgar-Parsi et al. Distribution and moments of the	weighted sum of uniform random variables, with applications in reducing Monte ... •	95 JStCmpSm 52 399-414
process • Viollaz, Aldo J. Convergence in	weighted supremum norms of the Skorokhod representation of the estimated empirical	95 CommStA 24 2829-2839
Relating quantiles and expectiles under	weighted-symmetry • Abdous, Belkacem; Remillard, Bruno	95 AnInStMa 47 371-384
Graubard, Barry I. Examples of differing	weighted and unweighted estimates from a sample survey • Korn, Edward L.;	95 AmerStat 49 291-295
Klockars, A. J. et al. Power of unweighted and	weighted versions of simultaneous and sequential multiple-comparison procedures •	95 PsychBu 118 300-307
Generalized Procrustes analysis with iterative	weighting to achieve resistance • Verboon, Peter; Gabriel, K. Ruben	95 BJMaStPs 48 57-73
ries (Chinese) • Fan, J.; Wang, N.; Mei, C. The random	weighting approximation for the distribution of the partial sum of a stationary linear se-	95 AMaAPknC 17 65-75
dimensional case (Chinese) • Shi, Jian Random	weighting approximation for the distribution of standardized sample mean-multi	95 ChJApPSt 10 233-242
Asymptotic design-cum-model approach for convex	weighting of direct and indirect components of small domain ... • Chaudhuri et al.	95 CommStA 24 2803-2813
Recursive estimation of the	weighting factor for EWMA control charts from autocorrelated data • Hembree, G. Barry	94 ASAProQlPr 39-43
Venot, Alain A new approach to selection and	weighting of items in evaluative composite measurement scales • Coste, Joël; Walter, Eric;	95 StatMed 14 2565-2580
method • Lavallée, Pierre Cross-sectional	weighting of longitudinal surveys of individuals and households using the weight share	95 SurvyM 21 25-32
Multiparametric M-estimation and the random	weighting method • Shi, Jian	95 ActScNat 30 509-518
Fuller, Wayne A. Regression	weighting methods for SIPP data (Disc: p440-442) • An, Anthony B.; Breidt, F. Jay;	94 ASASrvyRsr 434-439
Comment on "Sample design and	weighting research in panel surveys" • Michaud, Sylvie; Lavallée, Pierre	94 ASASrvyRsr 680-682
Hall, John W.	Weighting sample data when multiple sample frames are used • Carlson, Barbara Lepidus;	94 ASASrvyRsr 882-887
	Weighting schemes for household panel surveys • Kalton, Graham; Brick, J. Michael	94 ASASrvyRsr 785-790
	Weighting schemes for household panel surveys • Kalton, Graham; Brick, J. Michael	95 SurvyM 21 33-44
Kettenring, J. R.; Tsao, S. L.	Weighting and selection of variables for cluster analysis • Gnanadesikan, R.;	95 JClass 12 113-136
Nonparametric estimation of	welfare changes • Breslaw, Jon A.	95 EcomtrRv 14 163-182
for estimating the magnitude and precision of	welfare changes • Breslaw, Jon A.; Smith, J. Barry A simple and efficient method	95 JApEcmtx 10 313-327
Measuring	welfare changes when quantity is constrained • Breslaw, Jon A.; Smith, J. Barry	95 JBES 13 95-103
working life • Kingkade, W. Ward; Vasin, Sergey A.	Welfare consequences of mortality dynamics in the former Soviet Republics: Length of	94 ASAProSoSt 94-99
pact of policy changes on women's labor supply and	welfare participation (Disc: p51-52) • Jacobson et al. Using panel data to simulate the im-	94 ASAProGovt 45-50
The ability of	wet deposition networks to detect temporal trends • Oehlert, Gary W.	95 EnvrMtrc 6 327-339
of a spatio-temporal process with an application to	wet sulfate deposition • Haas, Timothy C. Local prediction	95 JASA 90 1189-1199
time to quantify the dynamics of dryland winter	wheat yield response • Burt, Oscar R. Pooling experiments over	95 AgronmyJ 87 156-164
Mulitze, Dieter K. Removing spatial variation from	wheat yield trials: A comparison of methods • Stroup, Walter W.; Baenziger, P. Stephen;	94 CropSci 34 62-66
Cyclically resolvable designs and triple	whist tournaments • Anderson, I.; Finizio, N. J.	93 JComDsgn 1 347-358
A stochastic model for the movement of a	white blood cell • Heubach, Silvia; Watkins, Joseph	95 AdvAppPr 27 443-475
and transitions among older African American and	white men • Burr, J. A.; Massagli, M. P.; Mutchler, J. E.; Pienta, A. Labor force status	94 ASAProSoSt 129-134
Multidimensional reaction-diffusion equations with	white noise boundary perturbations • Sowers, R. B.	94 AnlsProb 22 2071-2121
Invariant measures for	white noise driven stochastic partial differential equations • Prato, G. Da	95 StocAnAp 13 295-305
time nonlinear system with generalized Gaussian	white noise • Ferrante, Marco; Giummolé, Federica Finite dimensional filters for a discrete-	95 Stochatx 53 195-211
estimators of a discontinuous signal in a Gaussian	white noise • Rubin et al. ... of the asymptotic efficiency of maximum likelihood	95 AnlsStat 23 732-739
Streit, L. (Auth); Øksendal, B. (Rev) Review of	"White noise. An infinite dimensional calculus" • Hida, T.; Kuo, H.-H.; Potthoff, J.;	95 BLndnMaS 27 189-191
exit times and long time asymptotics in the zero-	white-noise limit • Mathieu, Pierre Spectra,	95 Stochatx 55 1-20
Feynman distributions, and multiparameter	white noise • Betounes, David Trace operators,	95 JTherPrb 8 119-138
tial differential equations perturbed by space-time	white noise • Gyöngy, István; Nualart, David Implicit scheme for quasi-linear parabolic par-	95 StocProc 58 57-72
solutions to the heat equation with Fisher-Wright	white noise • Tribe, Roger Large time behavior of interface	95 ProbTher 102 289-311
E. Black,	white, and shades of gray (and brown and yellow) • Anderson, Margo; Fienberg, Stephen	95 Chance 8/1 15-18
statistics and optimization: A tribute to Peter	Whittle" • Kelly, F. P. (Ed); Bingham, N. H. (Rev) Review of "Probability,	95 JRSS-A 158 353-354
Isotonic estimation and rates of convergence in	Wicksell's problem • Groeneboom, Piet; Jongbloed, Geurt	95 AnlsStat 23 1518-1542
	Wide sense expectation	
Lee, Lung-fei; Sepanski, Jungsywan H.	Estimation of linear and nonlinear errors-in-variables models using validation data •	95 JASA 90 130-140
Nearly optimal controls for singularly perturbed	wideband noise systems • Ramachandran, K. M.	95 StocAnAp 13 335-353
On	Wieand's theorem • Kallenberg, Wilbert C. M.; Koning, Alex J.	95 StPrLet 25 121-132
A simple proof and refinement of	Wielandt's eigenvalue inequality • Dümbgen, Lutz	95 StPrLet 25 113-115
Estimating degradation by a	Wiener diffusion process subject to measurement error • Whitmore, G. A.	95 LiDaAnal 1 307-319

in the modelling and prediction of the colour of	young red wines from the Denomination of Origin 'Rioja' • Fernández et al. ... methods	95 Chemolab 28 273-285
Substance abuse among runaway and homeless	youth • Iachan, Ronald; Ringwalt, Chris L.; Greene, Jody M.	94 ASASrvyRsr 1083-1087
Comment on "An extension of	Yule's Q to multivariate binary data" (V50 p847-852) • de Falguerolles, Antoine	95 Biomtrcs 51 1184-1185
Correction to "Comment on 'An extension of	Yule's Q to multivariate binary data'" (V51 p1184-1185) • de Falguerolles, A.	95 Biomtrcs 51 1589-1589
Reply to comment on "An extension of	Yule's Q to multivariate binary data" • Lipsitz, Stuart R.; Fitzmaurice, Garrett	95 Biomtrcs 51 1185-1185
Jennrich, Robert I. The profile	Z plot and confidence intervals in maximum likelihood analysis • Chen, Jian-Shen;	94 ASAProStCp 124-129
The use of Fisher's	Z in Schmidt-Hunter-type meta-analyses • Law, Kenneth S.	95 JEdStat 20 287-306
Extension of a uniqueness theorem for	Zakai's equation (French) • Bel, Liliane; Robbiano, Luc	95 Stochatx 52 227-245
An approximation theorem of Wong-	Zakai type for nonlinear stochastic partial differential equations • Twardowska, Krystyna	95 StocAnAp 13 601-626
Franco Almost sure approximation of Wong-	Zakai type for stochastic partial differential equations • Brzeźniak, Zdzisław; Flandoli,	95 StocProc 55 329-358
and principles for data analysis II: Austria first?	Zambia last? • Gunter, Bert Graphical methods	95 QulPr 28/4 103-105
Phillips, G. D. A.; Schipp, B. The bias of OLS, GLS, and	ZEF estimators in dynamic seemingly unrelated regression models • Kiviet, J. F.;	95 JEconmtx 69 241-266
bances • Hasegawa On small sample properties of	Zellner's estimator for the case of two SUR equations with compound normal distur-	95 CommStB 24 45-59
Exact cumulant calculations for Pearson χ^2 and	Zelterman statistics for r-way contingency tables • Stafford, James E.	95 JCmpGrSt 4 199-212
The Lévy transform and	zeros in Brownian motion (French) • Malric, M.	95 ProbTher 101 227-236
Wang, Yuedong Behavior near	zero of the distribution of GCV smoothing parameter estimates • Wahba, Grace;	95 StPrLet 25 105-111
stopping problem for random walks with non-	zero drift • Roters, Markus An optimal	95 JAppProb 32 956-959
Random walks in a quarter plane with	zero drifts: Transience and recurrence • Asymont, I. M.; Fayolle, G.; Menshikov, M. V.	95 JAppProb 32 941-955
for fitting Poisson regression model with extra	zeros • Jovanovic, Borko D.; Hosmer, David W.; Murray, James A method	94 ASAProEpid 97-100
	Zero inflated Poisson	
Gupta, Ramesh C.; Tripathi, Ram C.	Inflated modified power series distributions with applications • Gupta, Pushpa L.;	95 CommStA 24 2355-2374
A score test for	zero inflation in a Poisson distribution • van den Broek, Jan	95 Biomtrcs 51 738-743
On the enumeration of rectangular	$(0,1)$-matrices • Hendrickson, John A., Jr	95 JStCmpSm 51 291-313
Percolation of arbitrary words in	$\{0,1\}^N$ • Benjamini, Itai; Kesten, Harry	95 AnlsProb 23 1024-1060
Bayes factors for	zero partial covariances • Giudici, Paolo	95 JStPlInf 46 161-174
How many	zeros of a random polynomial are real? • Edelman, Alan; Kostlan, Eric	95 BAmMaSoc 32 1-37
tagged particle in equilibrium for asymmetric mean	zero random walk with simple exclusion • Varadhan, S. R. S. Self diffusion of a	95 AIHenrPB 31 273-285
the hydrodynamical limit of mean zero asymmetric	zero range processes • Benois, O.; Kipnis, C.; Landim, C. Large deviations from	95 StocProc 55 65-89
Average densities of the image and	zero set of stable processes • Falconer, K. J.; Xiao, Y. M.	95 StocProc 55 271-283
Law of the iterated logarithm and local variations at	zero of the sticky Brownian motion • Amir, Madjid; Knight, Frank B.	95 StPrLet 23 289-295
Combining 2×2 tables that contain structural	zeros • Johnson, William D.; May, Warren L.	95 StatMed 14 1901-1911
Stochastic differential games: The linear quadratic	zero sum case • Ardanuy, R.	95 SankhyaA 57 161-165
Backward equations, stochastic control and	zero-sum stochastic differential games • Hamadéne, S.; Lepeltier, J. P.	95 Stochatx 54 221-231
Puhalskii, A. A.; Reiman, M. I. Polling systems with	zero switchover times: A heavy-traffic averaging principle • Coffman, E. G., Jr;	95 AnlsApPr 5 681-719
Spectra, exit times and long time asymptotics in the	zero-white-noise limit • Mathieu, Pierre	95 Stochatx 55 1-20
the asymptotic series for moments of the number of	zeros of the Wong process on a large time interval (Russian) • Miroshin The power terms of	95 VLenUM /2 36-42
Quadratic distance estimators for the	zeta family • Doray, Louis G.; Luong, Andrew	95 InsrneME 16 255-260
The global concentration measure	ζ and the principle of transfers: An empirical analysis (Italian) • Polisicchio, M.	93 GiornEco 52 291-308
Training government statisticians in	Zimbabwe - An update • Woodward, Mark	94 JOffStat 10 215-220
Unified representation of	Zipf distributions • Zörnig, Peter; Altmann, Gabriel	95 CmpStDA 19 461-473
Bilinear forms and	zonal polynomials • Mathai, A. M.; Provost, S. B.; Hayakawa, T.	95 Springer:Brln:NY 376p
Guttorp, Peter	Zooplankton proportion estimates from non-uniform sample volumes • Billheimer, Dean;	95 EnvEcoSt 2 117-124
On the range of reversible random walks on	Z^2 in a random environment • Zhou, Xian Yin	95 JTherPrb 8 453-473

JOURNAL AND EDITED BOOK ABBREVIATIONS

4. JOURNAL AND EDITED BOOK ABBREVIATIONS

ACMMaSfw	ACM Transactions on Mathematical Software * ACM:NY
ACMMdSim	ACM Transactions on Modeling and Computer Simulation * ACM:NY
ACMTPLS	ACM Transactions on Programming Languages and Systems * ACM:NY
ActScNat	Acta Scientiarum Naturaliam Universitatis Pekinensis (Chinese) * UnivBeijing
AdvAppPr	Advances in Applied Probability * AppProb:UK
AdvMatCh	Advances in Mathematics (Chinese) * UnivBeijing
AgronmyJ	Agronomy Journal * AmerSocAgrny:WI
AHistESc	Archive for History of Exact Sciences * Springer:Brln:NY
AIHenrPB	Annales de l'Institut Henri Poincare, Section B, Calcul des Probabilities et Statistique * Gauther-Villars, CDR:France
AJClnOnc	American Journal of Clinical Oncology * RavenP:NY
AJEpidm	American Journal of Epidemiology * JHopkinsPubHlth
AJHuGene	American Journal of Human Genetics * American Society of Human Genetics, UnivChicagoPr
AJMMgSci	American Journal of Mathematical and Management Sciences * AmerSciPr:OH
AJPolSci	American Journal of Political Science * UnivTexas
AJPubHlh	American Journal of Public Health * AmerPubHlhA:DC
AllgeStA	Allgemeines Statistisches Archiv * Deutsche Statistische Gesellschaft, Vandenhoeck:WG
AMaAPknC	Acta Mathematicae Applicatae Sinica (Chinese series) * Beijing
AMaAPknE	Acta Mathematicae Applicatae Sinica (English series) * Beijing
AmAntiqt	American Antiquity * AmerSArch:DC
AMaPeknE	Acta Mathematica Sinica (English Series) * PlenumPubl:NY
AmEcRv	American Economic Review * AmerEcon:Nash
AmEdRsrJ	American Educational Research Journal * AmerEdRsrA:DC
AmerStat	The American Statistician * AmerStatA:VA
AmMaMnth	The American Mathematical Monthly * MathAAmer
ANatrls	American Naturalist * UnivChicagoPr
AnEpidm	Annals of Epidemiology * ElsevierNHllnd
AnHuGene	Annals of Human Genetics * CambridgeUniv
AnInMed	Annals of Internal Medicine * AmerCPhyscans:PA
AnInStMa	Annals of the Institute of Statistical Mathematics * IStatMath:Japan
AnlsApPr	Annals of Applied Probability * IMathStat:CA
AnlsProb	The Annals of Probability * IMathStat:CA
AnlsSci	Annals of Science: An International Review of the History of Science and Technology from the Thirteenth Century * TaylrFrncs:Lnd
AnlsStat	The Annals of Statistics * IMathStat:CA
AnlsTele	Annales des Telecommunications * CNatlTelecom:Fr
AplMaCmp	Applied Mathematics and Computation * ElsevierNHllnd
AplStMDA	Applied Stochastic Models and Data Analysis * Wiley:NY:UK
ApplEcon	Applied Economics * ChapmnHall:Lnd
ApplMath	Applicationes Mathematicae [Formerly: Zastosowania Matematyki] * PolishSciPr:Plnd
ApplStat	Applied Statistics * RoyalStSoc:UK
ApPsychM	Applied Psychological Measurement * ApPsychMeas:MN
ARContrl	Automation and Remote Control * PlenumPubl:NY
ArMaSwdn	Arkiv for Matematik * MittagLeffler
ASAProBays	ASA Proceedings of the Section on Bayesian Statistical Science * AmerStatA:VA
ASAProBiop	ASA Proceedings of Biopharmaceutical Section * AmerStatA:VA
ASAProBuEc	ASA Proceedings of Business and Economic Statistics Section * AmerStatA:VA
ASAProEnvr	ASA Proceedings of the Section on Statistics and the Environment * AmerStatA:VA
ASAProEpid	ASA Proceedings of the Section on Epidemiology * AmerStatA:VA
ASAProGovt	ASA Proceedings of Government Statistics Section * AmerStatA:VA
ASAProPEng	ASA Proceedings of Section on Physical and Engineering Sciences * AmerStatA:VA
ASAProQlPr	ASA Proceedings of the Quality and Productivity Section * AmerStatA:VA
ASAProSoSt	ASA Proceedings of Social Statistics Section * AmerStatA:VA
ASAProSpts	ASA Proceedings of Section on Statistics in Sports * AmerStatA:VA
ASAProStCp	ASA Proceedings of Statistical Computing Section * AmerStatA:VA
ASAProStEd	ASA Proceedings of Statistical Education Section * AmerStatA:VA
ASAProStGr	ASA Proceedings of Statistical Graphics Section * AmerStatA:VA
ASASrvyRsr	ASA Proceedings of Survey Research Methods Section * AmerStatA:VA
AScintst	American Scientist * SciRsrSoc:CT
ASQCTekCon	ASQC Technical Conference Transactions * ASQualCntrl:WI
AstinB	Astin Bulletin. A Journal of the International Actuarial Association * IActurSoc:Blgm
ASTM	ASTM Standardization News * AmSocTstMat:PA
AstrlJSt	The Australian Journal of Statistics * StatSocA:Astrl
ATTTJ	AT&T Technical Journal * ATTBellLab:NJ

Auk	Auk. A Quarterly Journal of Ornithology * AmerOrnithU:OH
Autmatca	Automatica * PergamonP:NY:Oxf
BAmMaSoc	Bulletin of the American Mathematical Society * AmerMathS:RI
BehavGen	Behavior Genetics * PlenumPubl:NY
Behvmtrk	Behaviormetrika, Journal of the Behaviormetric Society of Japan * IStatMath:Japan
Bernoull	Bernoulli * ChapmnHall:Lnd
BIMaASnc	Bulletin of the Institute of Mathematics * InstMath:Taiwan
BiomtrcJ	Biometrical Journal. Journal of Mathematical Methods in Biosciences. [Continues: Biometrische Zeitschrift. Zeitschrift für mathematische Methoden in den Biowissenschaften] * Akademie:Brln
Biomtrcs	Biometrics * BiometricS:DC
Biomtrka	Biometrika * MathImprlC:Lnd
BJHistSc	The British Journal for the History of Science * Blackwell:Oxf
BJMaStPs	The British Journal of Mathematical and Statistical Psychology * BrPsychSoc
BJPhilSc	The British Journal for the Philosophy of Science * OxfordUniv
BLndnMaS	The Bulletin of the London Mathematical Society * LondonMaSoc:UK
BMathBio	Bulletin of Mathematical Biology * PergamonP:NY:Oxf
CalcutSt	Calcutta Statistical Association Bulletin * UnivCalcutStat
Cancer	Cancer * Lippincott:PA
CandJSt	The Canadian Journal of Statistics * StatSocC:Cnda
CAnDnnes	Cahiers de l'Analyse des Donnees * Dunod:Paris
CanJForR	Canadian Journal of Forest Research * NatlRsrC:Cnda
Centaur	Centaurus: International Magazine of the History of Mathematics, Science and Technology * Hunksgaard:Dnmk
Chance	Chance, New Directions for Statistics and Computers * AmerStatA:VA
Chemolab	Chemometrics and Intelligent Laboratory Systems * ElsevierNHllnd
ChemTech	Chemical Technology * AmerChemS:DC
ChinaAMA	Chinese Annals of Mathematics A (Chinese) * ShngSciPub:China
ChinaAMB	Chinese Annals of Mathematics, Series B * BaltzerS:Bsl
ChinaJMa	Chinese Journal of Mathematics * MathSoc:Taiwan
ChJApPSt	Chinese Journal of Applied Probability and Statistics * EChinaNUniv
Circltn	Circulation * AmerHeartA:TX
CmpBimdR	Computers and Biomedical Research * Academic:NY:Lnd
CmpBiomd	Computer Methods and Programs in Biomedicine * ElsevierNHllnd
CmpOpRsh	Computers and Operations Research * PergamonP:NY:Oxf
CmpStDA	Computational Statistics and Data Analysis * ElsevierNHllnd
CmpStQ	Computational Statistics [Formerly: Computational Statistics Quarterly] * Physica:Ger
CmpSurvy	Computing Surveys * ACM:NY
ColMaJ	The College Mathematics Journal * MathAAmer
CommPAMa	Communications on Pure and Applied Mathematics * WileyISci:NJ
CommStA	Communications in Statistics, Part A-Theory and Methods [Split from: Communications in Statistics] * Dekker:NY
CommStB	Communications in Statistics, Part B-Simulation and Computation [Split from: Communications in Statistics] * Dekker:NY
CommStSM	Communications in Statistics-Stochastic Models * Dekker:NY
Complxty	Complexity * Wiley:NY:UK
ConClnTr	Controlled Clinical Trials * ElsevierNHllnd
CropSci	Crop Science * AmerSocAgrny:WI
CVGIPGr	CVGIP: Graphical Models and Image Processing * Academic:NY:Lnd
CVGIPIm	CVGIP: Image Understanding * Academic:NY:Lnd
DecisnSc	Decision Sciences * StanfordUniv:CA
Dmogrphy	Demography * PopAssocAmer:VA
EcolgApp	Ecological Applications * EcolgSocAmer:AZ
EcolgMdl	Ecological Modelling * ElsevierNHllnd
EcolgMon	Ecological Monographs * EcolgSocAmer:AZ
Ecology	Ecology * EcolgSocAmer:AZ
EcomtrRv	Econometric Reviews * Dekker:NY
Econmica	Economica * LondonSchEcon:UK
EconTher	Econometric Theory * CambridgeUniv
EdPsMeas	Educational and Psychological Measurement * SagePubl:CA:Lnd
EdPsych	Educational Psychologist * AmerPsychAssoc
EgpStJ	The Egyptian Statistical Journal * CairoUniv:Egp
EnvEcoSt	Environmental and Ecological Statistics * ChapmnHall:Lnd
EnvrMtrc	EnvironMetrics * Wiley:NY:UK
EnvrnBeh	Environment and Behavior * SagePubl:CA:Lnd
Epidmlgy	Epidemiology * Blackwell:Oxf
Estd	Estadistica * IASI:Panama
EvalProg	Evaluation and Program Planning * PergamonP:NY:Oxf

Evolutn	Evolution * AllenPr:KS
ExpAgric	Experimental Agriculture * CambridgeUniv
FuzzySet	Fuzzy Sets and Systems * ElsevierNHllnd
GeneRsr	Genetical Research * CambridgeUniv
Genetcs	Genetics * GenetcSocAmer:TX
Genomics	Genomics * Academic:NY:Lnd
GiornEco	Giornale degli Economisti e Annali di Economia * UnivBocconi:Itly
Heredity	Heredity * Longman:UK
Herptol	Herpetologica * HerptLeag
HistMat	Historia Mathematica * Academic:NY:Lnd
HistSci	History of Science * SciHistPub:UK
HortSci	Horticultural Science * AmerSocHort:VA
HumanBio	Human Biology * WayneStUnivPr:MI
HumanFac	Human Factors * HumanFacSoc:CA
HumHrdty	Human Heredity * Karger:Bsl
HustnJMa	Houston Journal of Mathematics * UnivHouston:TX
IEEEAES	IEEE Transactions on Aerospace and Electronic Systems * IEEE:NY
IEEEAuCn	IEEE Transactions on Automatic Control * IEEE:NY
IEEENrlN	IEEE Transactions on Neural Networks * IEEE:NY
IEEEPAMI	IEEE Transactions on Pattern Analysis and Machine Intelligence * IEEE:NY
IEEEReli	IEEE Transactions on Reliability * IEEE:NY
IEEESyst	IEEE Transactions on Systems, Man, Cybernetics * IEEE:NY
IIETrn	IIE Transactions * AmerIndEng:GA
IJAdCSgP	International Journal of Adaptive Control and Signal Processing * Wiley:NY:UK
IJContrl	International Journal of Control * TaylrFrncs:Lnd
IJEpidem	International Journal of Epidemiology * OxfordUniv
IJOrthdn	International Journal of Orthodontics * FedOrthAssoc:WI
IllinJMa	Illinois Journal of Mathematics * UnivIllnoisPr:IL
IndgMatN	Indagationes Mathematicae (New Series) * ElsevierNHllnd
InsrneME	Insurance Mathematics and Economics * ElsevierNHllnd
IntEconR	International Economic Review * UnivPennPr, UnivOsaka:Jpn
IntJFor	International Journal of Forecasting * NHollndElsevier
Intrfc	Interfaces, Journal of The Institute of Management Sciences and the Operations Research Society of America * InstMgmtSc:RI
IntStRvw	International Statistical Review * IntlStatI:Vorbg
Isis	Isis: Official Journal of the History of Science Society * HistSciSoc:PA
IsrelJMa	Israel Journal of Mathematics * MagnesPr:Isrel
JAbnPsy	Journal of Abnormal Psychology * AmerPsychAssoc
JACM	Journal of the Association for Computing Machinery * ACM:NY
JAMA	Journal of the American Medical Association * AmerMedAssoc:IL
JAnmlEco	Journal of Animal Ecology * Blackwell:Oxf
JAnmlSci	Journal of Animal Science * AmerAnimalSci:IL
JApEcmtx	Journal of Applied Econometrics * Wiley:NY:UK
JApMetrl	Journal of Applied Meteorology [Split from: Journal of Climate and Applied Meteorology] * AmerMeteorS:MA
JAppEcol	Journal of Applied Ecology * Blackwell:Oxf
JAppProb	Journal of Applied Probability * AppProb:UK
JAppStat	Journal of Applied Statistics * CarfaxPub:FL
JASA	Journal of the American Statistical Association * AmerStatA:VA
JAtmsphr	Journal of the Atmospheric Sciences * AmerMeteorS:MA
JBES	Journal of Business and Economic Statistics * AmerStatA:VA
JBiophSt	Journal of Biopharmaceutical Statistics * Dekker:NY
JBusFcst	Journal of Business Forecasting, Methods and Systems * GracewayPub:NY
JBusUChi	Journal of Business of the University of Chicago * UnivChicagoPr
JCCPsych	Journal of Consulting and Clinical Psychology * AmerPsychAssoc
JChemtrc	Journal of Chemometrics * Wiley:NY:UK
JCISS	Journal of Combinatorics, Information and System Sciences * Forum for Interdisciplinary Mathematics, PrintsInd:India
JClass	Journal of Classification * ClassSocNA:Cnda
JClEpid	Journal of Clinical Epidemiology, [Formerly: Journal of Chronic Diseases] * PergamonP:NY:Oxf
JClimate	Journal of Climate [Split from: Journal of Climate and Applied Meteorology] * AmerMeteorS:MA
JClimlgy	International Journal of Climatology * Wiley:NY:UK
JClPsych	Journal of Clinical Psychology * ClincPsychPub:VT
JCmpGrSt	Journal of Computational and Graphical Statistics * AmerStatA:VA
JCmpInT	Journal of Computing and Information Technology * UnivComputrCnt
JCombThA	Journal of Combinatorial Theory, Series A * Academic:NY:Lnd
JCombThB	Journal of Combinatorial Theory, Series B * Academic:NY:Lnd
JComDsgn	Journal of Combinatorial Design * Wiley:NY:UK

JConsumR Journal of Consumer Research * UCLAMngmnt:CA
JCostAnl Journal of Cost Analysis * SocCostEA:VA
JCrmJust Journal of Criminal Justice * PergamonP:NY:Oxf
JEconLit Journal of Economic Literature * AmerEcon:Nash
JEconmtx Journal of Econometrics * ElsevierNHllnd
JEconPer Journal of Economic Perspectives * AmerEcon:Nash
JEdMeas Journal of Educational Measurement * NatCMsurmntEd:DC
JEdPsych Journal of Educational Psychology * AmerPsychAssoc
JEdStat Journal of Educational and Behavioral Statistics [Formerly: Journal of Educational Statistics] * AmerEdRsrA:DC
JFinance The Journal of Finance * NYUnivGBusnss:NY
JFinEcon Journal of Financial Economics * ElsevierNHllnd
JFishRBC Canadian Journal of Fisheries and Aquatic Science * CanadianGovtPubl
JForcast Journal of Forecasting * Wiley:NY:UK
JHeredty Journal of Heredity * AmerGeneticA:DC
JIAgSt Journal of the Indian Society of Agricultural Statistics * IndAgriStl:India
JIndnStA Journal of the Indian Statistical Association * UnivPonaSt:India
JItalStS Journal of the Italian Statistical Society * Giardini:Itly
JJapnStS Journal of the Japan Statistical Society * IStatMath:Japan
JKoreaSt Journal of the Korean Statistical Society * KoreaStSoc:Kor
JMaJapan Journal of the Mathematical Society of Japan * MathSocJapn:Japn
JMarkt Journal of Marketing * AmerMarktA:IL
JMarRsr Journal of Marketing Research * AmerMarktA:IL
JMathChn Journal of Mathematics (Chinese) * WahanUniv:China
JMicro Journal of Microscopy * Blackwell:Oxf
JMolEvol Journal of Molecular Evolution * Springer:Brln:NY
JMRsExp Journal of Mathematical Research and Exposition (Chinese) * DalianUniv:China
JMultiAn Journal of Multivariate Analysis * Academic:NY:Lnd
JNCncrI Journal of the National Cancer Institute * NCancerInst:MD
JNnparSt Journal of Nonparametric Statistics * GrdnBreach:Lnd
JNOknSt Jahrbücher für Nationalökonomie und Statistik * VEB-GFischer
JOffStat Journal of Official Statistics * StatSweden:Swdn
JOpRsJpn Journal of the Operations Research Society of Japan * OperRsrSoc:Japn
JParmc Journal of Parametrics * ISocParAn:MD
JpJAppSt Japanese Journal of Applied Statistics * Japanese Society of Applied Statistics, KeioUnivMath:Jpn
JpJBeh Japanese Journal of Behaviormetrics * The Behaviormetric Society of Japan, IStatMath:Japan
JpnPsyRs Japanese Psychological Research * JpnPsychA:Japn
JPsyctrc Journal of Psychiatric Research * PergamonP:NY:Oxf
JQualTek Journal of Quality Technology * ASQualCntrl:WI
JRngeMgt Journal of Range Management * SocRangeMngmt:CO
JRsrMaEd Journal for Research in Mathematics Education * NCTM:VA
JRSS-A Journal of the Royal Statistical Society, Series A, General * RoyalStSoc:UK
JRSS-B Journal of the Royal Statistical Society, Series B, Methodological * RoyalStSoc:UK
JStatRsr Journal of Statistical Research * UnivDacca:Bngl
JStCmpSm Journal of Statistical Computation and Simulation * GrdnBreach:Lnd
JStEd Journal of Statistics Education * NCarolinaStU:NC [e-mail: listserv@jse.stat.ncsu.edu]
JStPhycs Journal of Statistical Physics * PlenumPubl:NY
JStPlInf Journal of Statistical Planning and Inference * ElsevierNHllnd
JSysSMaS Journal of Systems Sciences and Mathematical Sciences * Xitong Kexue Yu Shuzue, Beijing, China
JSysSMSE Journal of Systems Science and Mathematical Science (English) * Beijing
JThBio Journal of Theoretical Biology * Academic:NY:Lnd
JTherPrb Journal of Theoretical Probability * PlenumPubl:NY
JTimSrAn Journal of Time Series Analysis * BaslBlkwell:UK
JWildlMg Journal of Wildlife Management * WildlifeSoc:MD
KorApS Korean Journal of Applied Statistics * KoreaStSoc:Kor
KToChina Kexue Tongbao (China) * Acad Sinica, Peking People's Republic of China, Transl: China Program, Consultants Bureau, PlenumPubl:NY
Lancet Lancet * LittleBrown:MA
LiDaAnal Lifetime Data Analysis * Kluwer:MA:Hllnd
LinAlAp Linear Algebra and its Applications * ElsevierNHllnd
ListyBio Listy Biometryczne/Biometrical Letters * PolBiomSoc:Plnd
MaBisci Mathematical Biosciences * ElsevierNHllnd
MaCmpSim Mathematics and Computers in Simulation * ElsevierNHllnd
MaMethSt Mathematical Methods of Statistics * AllertonPr:NY
MaOpfStS Statistics [Formerly: Mathematische Operationsforschung und Statistik, Series Statistics] * GrdnBreach:Lnd
MaOpRsr Mathematics of Operations Research * InstMgmtSc:RI
MaPopStd Mathematical Population Studies * GrdnBreach:Lnd

MaScand	Mathematica Scandinavica * UnivArhus:Dnmk
MaScitst	The Mathematical Scientist * AppProb:UK
MathTchr	Mathematics Teacher * NCTM:VA
MatNach	Mathematische Nachrichten * Akademie:Brln
MayoClnP	Mayo Clinic Proceedings * MayoClinic:MN
MaZeit	Mathematische Zeitschrift * Springer:Brln:NY
MedDscnM	Medical Decision Making * Birkhs:Basl:MA
Metrika	Metrika * Physica:Ger
Metron	Metron * UnivRomaSap:Itly
MilOpRes	Military Operations Research * MilOperRsrSoc:VA
MInfoMed	Methods of Information in Medicine * Schattauer:Ger
MolBioEv	Molecular Biology and Evolution * UnivChicagoPr
MultBehR	Multivariate Behavioral Research * ErlbaumA:NJ
MWeathr	Monthly Weather Review * AmerMeteorS:MA
NAJFishM	North American Journal of Fisheries Management * AmerFishS:MD
Nature	Nature * Macmill:NY:Lnd
NDFLogic	Notre Dame Journal of Formal Logic * UnivNotreDame:IN
NEngJMd	New England Journal of Medicine * MassMedSoc:MA
Networks	Networks * WileyISci:NJ
NoteEcon	Note Economiche * Monte dei Paschi di Siena Piazza Salimbeni 3, 53100 Siena, Italy
NumeriMa	Numerische Mathematik * Springer:Brln:NY
NuralNet	Neural Networks * ElsevierNHllnd
NurseRes	Nursing Research * AmerJNurseCo:NY
NZndSt	New Zealand Statistician * NZlndStatAssn:NZ
ORSpktrm	OR Spektrum * Springer:Brln:NY
PakJStA	Pakistan Journal of Statistics, Series A [Formerly: Pakistan Journal of Statistics] * IslamSOSS:Pak
PattrnRc	The Journal of the Pattern Recognition Society * PergamonP:NY:Oxf
PattrnRL	Pattern Recognition Letters * ElsevierNHllnd
PercpMtr	Perceptual and Motor Skills * Missoula, MT
PhilSci	Philosophy of Science * MichStUnivPhl:MI
Phytopgy	Phytopathology * AmerPhytoSoc:MN
PIEEEDecCt	Proceedings of IEEE Conference on Decision and Control * IEEE:NY
PNatAcSc	Proceedings of the National Academy of Sciences * NatlAcdmyP:DC
PrEngInf	Probability in the Engineering and Informational Sciences * CambridgeUniv
PrInStMa	Proceedings of the Institute of Statistical Mathematics * IStatMath:Japan
PrInvMgt	Production and Inventory Management * AmProdICSoc:VA
ProbTher	Probability Theory and Related Fields [Continuation of: Zeitschrift für Wahrscheinlichkeitstheorie und Verwandte Gebiete] * Springer:Brln:NY
ProIrStC1	Proceedings of the First Iranian Statistics Conference * IsfahanUniv:Iran
PsychBu	Psychological Bulletin * AmerPsychAssoc
PsychRpt	Psychological Reports * PsychReports:MT
Psymtrka	Psychometrika * WmMaryPsych:VA
PTLnAMa	Philosophical Transactions of the Royal Society of London, Series A, Math and Physical Science * RoyalSoc:UK
PublcIn	The Public Interest * NatnlAffairs:DC
PubOpinQ	Public Opinion Quarterly * UnivChicagoPr
QApplMa	Quarterly of Applied Mathematics * AmerMathS:RI
QStEco	Quaderni di Statistica e Econometria * CentSpRic:Itly
QStMaA	Quaderni di Statistica e Matematica Applicata alle Scienze Economico-Sociali * UnivTrento:Itly
QualQnty	Quality and Quantity * Kluwer:MA:Hllnd
Questo	Questiio * Centre de Calcul de la Universitat Politecnica de Catalunya, Avda. Dr. Gregorio Maranon, 08028 Barcelona, SPAIN
QulPr	Quality Progress * ASQualCntrl:WI
RACEMdrd	Revista de la Real Academia di Ciencias Exactas, Fisicas y Naturales * Valverde 22, 28013 Madrid, Spain
RBrzlPrS	Revista Brasileira de Probabilidade e Estatistica (The Brazilian Journal of Probability and Statistics) * BrazilStSoc:Brz
RelEngSS	Reliability Engineering & System Safety * ElsevierNHllnd
RiskAnal	Risk Analysis * PlenumPubl:NY
RndStrAl	Random Structures and Algorithms * Wiley:NY:UK
RStAJp	Reports of Statistical Application Research * UJapnSciEng:Japn
RStApp	Revue de Statistique Appliquee - CERESTA * CERESTA:Paris
RvEconSt	The Review of Economics and Statistics * ElsevierNHllnd
SankhyaA	Sankhyā, Series A, Indian Journal of Statistics * IndnStI:India
SankhyaB	Sankhyā, Series B, Indian Journal of Statistics * IndnStI:India
ScanActJ	Scandinavian Actuarial Journal * ScandUnivP:Nor
SchoolMa	School Science and Mathematics * BowlGrnSUniv:OH
Science	Science * AmerAAdvSci:DC

ScifcAm Scientific American * SciAmerInc:NY
SciSPkng Scientia Sinica, (Chung-kuo ko Hsueh) * Ko Hseuh Chu Pan She, Chung-kuo kuo Chi Shu Tien, P.X. Box 399, Peking, People's Republic of China
ShrtBkRv Short Book Reviews * IntlStatI:Vorbg
SIAMCmpu SIAM Journal on Computing * SIAM:PA
SIAMMaAn SIAM Journal on Mathematical Analysis * SIAM:PA
SIAMRvw SIAM Review * SIAM:PA
SIAMSSC SIAM Journal on Scientific and Statistical Computing [Formerly: SIAM Journal on Scientific Computing] * SIAM:PA
SilvGene Silvae Genetica * Sauerlaender:Ger
SkyTlscp Sky and Telescope * SkyPubl:MA
SoAfrStJ South African Statistical Journal * SoAfrStatAssn:SA
SochwJMa Soochow Journal of Mathematics * UnivSochw:Taiwan
SocMethR Sociological Methods & Research * SagePubl:CA:Lnd
SqtlAnly Sequential Analysis * Dekker:NY
StatComp Statistics and Computing * ChapmnHall:Lnd
StatDecn Statistics & Decisions * Oldenbourg:Grm
StatJUN Statistical Journal of the U.N. Economic Commission for Europe * ElsevierNHllnd
StatMed Statistics in Medicine * Wiley:NY:UK
Stats Stats. The Magazine for Students of Statistics * AmerStatA:VA
StatSci Statistical Science * IMathStat:CA
Statstca Statistica * CLUEB:Itly
StHefte Statistical Papers [Formerly: Statistischen Hefte] * Springer:Brln:NY
StMedRes Statistical Methods in Medical Research * Arnold:Lnd:MD
StNeerla Statistica Neerlandica * Blackwell:Oxf
StocAnAp Stochastic Analysis and Applications * Dekker:NY
Stochatx Stochastics * GrdnBreach:Lnd
StocProc Stochastic Processes and their Applications * ElsevierNHllnd
StPrLet Statistics & Probability Letters * ElsevierNHllnd
StrEqMod Structural Equation Modeling * ErlbaumA:NJ
StSinica Statistica Sinica * InStSci:Taiwan
Ststcian The Statistician * RoyalStSoc:UK
Student Student * PrAcadmNeu:Switz
StudHist Studies in History and Philosophy of Science * PergamonP:NY:Oxf
SurvyM Survey Methodology * StatCanada:Ont
Synthese Synthese, International Journal for Epistemology, Methodology and Philosophy of Science * DReidel:MA:Hllnd
TeachgSt Teaching Statistics * UnivSheffield
TeachSoc Teaching Sociology * SagePubl:CA:Lnd
Technmcs Technometrics * AmerStatA:VA
Test Test [Formerly: Trabajos de Estadistica] * SEIO:Spain
ThApGene Theoretical and Applied Genetics * Springer:Brln:NY
ThPrMaSt Theory of Probability and Mathematical Statistics (Transl of Teorija Verojatnostei i Matematiceskaja Statistika) * AmerMathS:RI
ThProbAp Theory of Probability and its Applications (Transl of Teorija Verojatnostei i ee Primenenija) * SIAM:PA
TmkngJMa Tamkang Journal of Mathematics * UnivTmkng:Taiwan
TOnderwR Tijdschrift Voor Onderwijs Research * InstOnderw:Neth
Top Top [Formerly: Trabajos de Investigacion Operativa] * SEIO:Spain
TrAFish Transactions of the American Fisheries Society * AmerFishS:MD
TrAmMaS Transactions of the American Mathematical Society * AmerMathS:RI
TrSActur Transactions of the Society of Actuaries * SocActuaries:IL
TRUMTk TRU Mathematics * SciUnivTkyo:Japn
UtilMath Utilitas Mathematica * UnivManitoba
VLenUM Vestnik St. Petersburg University [Formerly: Vestnik Leningrads kogo Universiteta, Mat, Mehanika, Astronomija] * Leningradskii University, St. Petersburg, Russia
WaterReB Water Resources Bulletin * AmerWaterResA:MN
WaterReR Water Resources Research * AmerGeophU:DC
WildlfSB Wildlife Society Bulletin * WildlifeSoc:MD
ZAngMaMk Zeitschrift für Angewandte Mathematik und Mechanik Ingenieurwissenschaftliche Forschunggarbeiten * Akademie:Brln
ZeitOpRs Zeitschrift für Operations Research * Physica:Ger
ZeitSozl Zeitschrift für Soziologie * EnkeVerlag:Ger
ZeitWiss Zeitschrift für Allgemeine Wissenschaftstheorie-Journal for General Philosophy of Science * SteinerVerlg:Ger
ZExpAngP Zeitschrift für Experimentelle und Angewandte Psychologie * PsychVerlag:Ger

PUBLISHER ABBREVIATIONS

5. PUBLISHER ABBREVIATIONS

Academic:NY:Lnd	Academic Press * New York; London
ACM:NY	Association for Computing Machinery * New York
Akademie:Brln	Akademie-Verlag * Berlin
AllenPr:KS	Allen Press * Lawrence, KS
AllertonPr:NY	Allerton Press * New York
AmerAAdvSci:DC	American Association for Advancement of Science * Washington
AmerAnimalSci:IL	American Society of Animal Science * Champaign, IL
AmerChemS:DC	American Chemical Society * Washington
AmerCPhyscans:PA	American College of Physicians * Philadelphia
AmerEcon:Nash	American Economic Association * Nashville
AmerEdRsrA:DC	American Educational Research Association * Washington
AmerFishS:MD	American Fisheries Society * Bethesda, MD
AmerGeneticA:DC	American Genetic Association * Washington
AmerGeophU:DC	American Geophysical Union * Washington
AmerHeartA:TX	American Heart Association, Inc. * Dallas, TX
AmerIndEng:GA	American Institute of Industrial Engineers Inc. * Norcross, GA
AmerJNurseCo:NY	American Journal of Nursing Co * New York
AmerMarktA:IL	American Marketing Association * Chicago
AmerMathS:RI	American Mathematical Society * Providence
AmerMedAssoc:IL	American Medical Association * Chicago
AmerMeteorS:MA	American Meteorological Society * Boston
AmerOrnithU:OH	American Ornithologists Union * Columbus, OH
AmerPhytoSoc:MN	The American Phytopathological Society * St. Paul, MN
AmerPsychAssoc	American Psychological Association * Washington
AmerPubHlhA:DC	American Public Health Association * Washington
AmerSArch:DC	Society for American Archaeology * Washington
AmerSciPr:OH	American Sciences Press * Syracuse
AmerSocAgrny:WI	American Society of Agronomy * Madison
AmerSocHort:VA	American Society for Horticultural Science * Alexandria, VA
AmerStatA:VA	American Statistical Association * Alexandria, VA
AmerWaterResA:MN	American Water Resources Association * Minneapolis, MN
AmProdICSoc:VA	American Production and Inventory Control Society, Inc. * Falls Church, VA
AmSocTstMat:PA	American Society for Testing and Materials * Philadelphia
AppProb:UK	Applied Probability Trust * Sheffield
ApPsychMeas:MN	Applied Psychological Measurement Inc. * University of Minnesota, Minneapolis
Arnold:Lnd:MD	Edward Arnold Publishers Ltd * London; Baltimore
ASQualCntrl:WI	American Society for Quality Control * Milwaukee
ATTBellLab:NJ	AT&T Bell Laboratories, Circulation Department * Short Hills, NJ
BaltzerS:Bsl	J. C. Baltzer AG Scientific Publishing Co. * Basel
BaslBlkwell:UK	Basil Blackwell, Ltd. * Oxford
BiometricS:DC	International Biometric Society * Alexandria, VA
Birkhs:Basl:MA	Birkhaeuser-Verlag * Basel; Cambridge, MA
Blackwell:Oxf	Blackwell Scientific Publications Ltd * Oxford
BowlGrnSUniv:OH	Bowling Green State University * Bowling Green, OH
BrazilStSoc:Brz	Brazilian Statistical Association * Sao Paulo
BrooksCole:CA	Brooks/Cole Publishing Co * Monterey, CA
BrPsychSoc	British Psychological Society * London
CairoUniv:Egp	Institute of Statistical Studies and Research [Cairo University] * Giza, Egypt
CambridgeUniv	Cambridge University Press * Cambridge, UK
CanadianGovtPubl	Canadian Government Publications Centre * Quebec
CarfaxPub:FL	Carfax Publishing Company * Dunnellon, FL
CDR:France	Centrale des Revues Dunod-Gauthier-Villars * Montrouge
CentSpRic:Itly	Centro Specializzazione e Ricerche * Naples
CERESTA:Paris	CERESTA [Inst. de Statistique des Université de Paris] * Paris
ChapmnHall:Lnd	Chapman & Hall Ltd * London; New York
ClassSocNA:Cnda	Classification Society of North America * St. Johns
ClincPsychPub:VT	Clinical Psychology Publishing Company * Brandon, VT
ClrndnPr:UK	Clarendon Press [Oxford University Press] * Oxford
CLUEB:Itly	Cooperativa Libraria Universitaria Editrice * Bologna, Italy
CNatlTelecom:Fr	Centre National d'Etudes des Telecommunications * Issy-les-Moulineaux, France
CRCPress:FL	CRC Press Inc. * Boca Raton, FL
DalianUniv:China	Institute of Applied Mathematics, Dalian University of Science Technology * China
Dekker:NY	Marcel Dekker Inc * New York
DGruyter:NY:Grm	Walter de Gruyter and Company * Hawthorne, NY; Berlin

DReidel:MA:Hllnd	D. Reidel Publishing Co * Dordrecht; Hingham, MA
Dunod:Paris	Dunod-Gauthier-Villars [Centrale des Revues Dunod-Gauthier-Villars] * Montrouge
EChinaNUniv	East China Normal University * Shanghai
EcolgSocAmer:AZ	Ecological Society of America [Center for Environmental Studies, Arizona State University] * Tempe
ElsevierNHllnd	[Elsevier Science Publishing Co., New York; North-Holland Publishing Co., Amsterdam] * New York; Amsterdam
EnkeVerlag:Ger	Ferdinand Enke Verlag * Stuttgart, Germany
ErlbaumA:NJ	Lawrence Erlbaum Associates * Hillsdale, NJ
FedOrthAssoc:WI	Federation of Orthodontic Associations * Milwaukee
Freeman:NY	W. H. Freeman & Co * New York
GenetcSocAmer:TX	Genetics Society of America * Austin, TX
Giardini:Itly	Giardini Editori * Agnano Pisano, Italy
GracewayPub:NY	Graceway Publishing Company * Flushing, NY
GrdnBreach:Lnd	Gordon and Breach Science Publishers Inc. * London; New York
Hachette:Fr	Hachette * Paris
HerptLeag	The Herpetologists' League Inc [Department of Biology, University of Southwestern Louisiana] * Lafayette, LA
HistSciSoc:PA	History of Science Society * Department of History and Sociology of Science, University of Pennsylvania
HumanFacSoc:CA	Human Factors Society Inc * Santa Monica, CA
Hunksgaard:Dnmk	Hunksgaard Ltd * Copenhagen
IActurSoc:Blgm	International Actuarial Association * Leuven, Belgium
IASI:Panama	Inter-American Statistical Institute * Panamá City
IEEE:NY	Institute of Electrical and Electronics Engineers Inc * New York
IMathStat:CA	Institute of Mathematical Statistics * Hayward
IndAgriStI:India	Indian Agricultural Statistical Research Institute * New Delhi
IndnStI:India	Indian Statistical Institute * Calcutta
InstMath:Taiwan	Institute of Mathematics [Academia Sinica, Taipei] * Taipei
InstMgmtSc:RI	Institute of Management Science and the Operations Research Society of America * Providence
InstOnderw:Neth	Institute Onderwijskunde * Nijmegen, Netherlands
InStSci:Taiwan	Institute of Statistical Science [Academia Sinica, Taipei] * Taipei
IntlStatI:Vorbg	International Statistical Institute * Voorburg, The Netherlands
IsfahanUniv:Iran	Isfahan University of Technology * Isfahan, Iran
IslamSOSS:Pak	Islamic Society of Statistical Sciences * Lahore
ISocParAn:MD	International Society of Parametric Analysts * Germantown, MD
IStatMath:Japan	Institute of Statistical Mathematics * Tokyo
JHopkinsPubHlth	Johns Hopkins University School of Hygiene and Public Health * Baltimore, MD
JpnPsychA:Japn	Japanese Psychological Association * Tokyo
Karger:Bsl	Karger, S. Karger AG, Medical & Scientific Publ. * Basel
KeioUnivMath:Jpn	Department of Mathematics, Keio University * Yokohama, Japan
Kluwer:MA:Hllnd	Kluwer Academic Publishers Group * Dordrecht; Norwell, MA
KoreaStSoc:Kor	The Korean Statistical Society * Seoul
Lavoisier:Paris	Lavoisier * Paris
Lippincott:PA	J. B. Lippincott Co. * Philadelphia
LittleBrown:MA	Little, Brown and Co * Boston
LondonMaSoc:UK	London Mathematical Society * London
LondonSchEcon:UK	London School of Economics & Political Science * London
Longman:UK	Longman Group Ltd * Harlow; Edinburgh
Macmill:NY:Lnd	Macmillan Publishing Co Inc * New York; London
MagnesPr:Isrel	The Magnes Press, Hebrew University * Jerusalem
MassMedSoc:MA	Massachusetts Medical Society * Waltham, MA
Masson:NY:Paris	Masson Publishing * Paris; New York
MathAAmer	Mathematical Association of America * Washington
MathImprlC:Lnd	Department of Mathematics, Imperial College * London
MathSoc:Taiwan	Math Research Center and the Math Society of the Republic of China, National Taiwan University * Taipei
MathSocJapn:Japn	Mathematical Society of Japan * University of Tokyo, Faculty of Science
MayoClinic:MN	Mayo Clinic * Rochester, MN
MichStUnivPhl:MI	Philosophy Department, Michigan State University * East Lansing, MI
MilOperRsrSoc:VA	Military Operations Research Society * Alexandria, VA
MittagLeffler	Institut Mittag-Leffler * Djursholm, Sweden
NatCMsurmntEd:DC	National Council on Measurement in Education * Washington, D.C.
NatlAcdmyP:DC	National Academy Press * Washington
NatlRsrC:Cnda	National Research Council of Canada * Ottawa
NatnlAffairs:DC	National Affairs Inc. * Washington
NCancerInst:MD	National Cancer Institute, Bloch Cancer Information Center * Bethesda, MD
NCarolinaStU:NC	Department of Statistics, North Carolina State University * Raleigh, NC

NCTM:VA	National Council of Teachers of Mathematics Inc * Reston, VA
NHollndElsevier	[Elsevier Science Publishing Co., New York; North-Holland Publishing Co., Amsterdam] * New York; Amsterdam
NYUnivGBusnss:NY	Graduate School of Business, New York University * New York
NZIndStatAssn:NZ	New Zealand Statistical Association * Wellington, New Zealand
Oldenbourg:Grm	R. Oldenbourg Verlag * Munich
OperRsrSoc:Japn	Operations Research Society of Japan * Tokyo
OxfordUniv	Oxford University Press * Oxford
PergamonP:NY:Oxf	Pergamon Journals Inc. * Elmsford, NY; Oxford
Physica:Ger	Physica-Verlag Ges.m.b.H * Heidelberg
PlenumPubl:NY	Plenum Publishing Corp * New York; London
PolBiomSoc:Plnd	Polish Biometric Society * Wrocław
PolishSciPr:Plnd	Panstwowe Wydawnictwo Naukowe [Polish Science Publ] * Warsaw
PopAssocAmer:VA	Population Association of America * Alexandria, VA
PrAcadmNeu:Switz	Presses Académiques Neuchâtael * Neuchâtel, Switzerland
PrintsInd:India	Prints India * New Delhi
PsychReports:MT	Psychological Reports * Missoula, MT
PsychVerlag:Ger	Verlag für Psychology — Dr. C. J. Hogrefe * Göttingen
PyrczakPubl:CA	Pyrczak Publishing * Los Angeles
RavenP:NY	Raven Press * New York
RoyalSoc:UK	The Royal Society * London
RoyalStSoc:UK	The Royal Statistical Society * London
SagePubl:CA:Lnd	Sage Publications Inc * Newbury Park, CA; London
Sauerlaender:Ger	J.D. Sauerlaender's Verlag * Frankfurt, Germany
ScandUnivP:Nor	Scandinavian University Press * Oslo
Schattauer:Ger	Friedrich Karl Schattauer Verlag GmbH * Stuttgart, Germany
SciAmerInc:NY	Scientific American Inc * New York
SciHistPub:UK	Science History Publications Ltd * St. Giles, UK
SciRsrSoc:CT	The Scientific Research Society Inc. * New Haven, CT
SciUnivTkyo:Japn	Science University of Tokyo * Tokyo
SEIO:Spain	Spanish Statistical and Operations Research Society * Madrid
ShngSciPub:China	Shanghai Scientific and Technological Literature Publishing House * Shanghai
SIAM:PA	SIAM [Society for Industrial and Applied Mathematics] * Philadelphia
SkyPubl:MA	Sky Publishing Corp * Cambridge, MA
SoAfrStatAssn:SA	South African Statistical Association * Pretoria, Republic of South Africa
SocActuaries:IL	Society of Actuaries * Chicago
SocApMathSim:PA	The Society of Applied Mathematics, Modeling and Simulation * Pittsburgh, PA
SocCostEA:VA	Society for Cost Estimating and Analysis * Alexandria, VA
SocRangeMngmt:CO	Society for Range Management * Denver, CO
Springer:Brln:NY	Springer-Verlag Inc * Berlin; New York
StanfordUniv:CA	Stanford University * Stanford, CA
StatCanada:Ont	Statistics Canada * Ottawa
StatSocA:Astrl	Statistical Society of Australia [Australian National University] * Canberra
StatSocC:Cnda	Statistical Society of Canada * McGill University, Montreal
StatSweden:Swdn	Statistics Sweden * Stockholm
SteinerVerlg:Ger	Franz Steiner Verlag Wieshaben GMbH * Stuttgart, Germany
TaylrFrncs:Lnd	Taylor & Francis Ltd * London
UCLAMngmnt:CA	Graduate School of Management, UCLA * Los Angeles
UJapnSciEng:Japn	Union of Japanese Scientists & Engineers * Tokyo
UnivArhus:Dnmk	Aarhus Universitet [Matematisk Institut, Department of Theoretical Statistics] * Aarhus, Denmark
UnivBeijing	Press of Beijing University * Beijing
UnivBocconi:Itly	Universita Bocconi * Milan
UnivCalcutStat	Department of Statistics, Calcutta University * Calcutta
UnivChicagoPr	University of Chicago Press * Chicago
UnivComputrCnt	University Computing Centre * Zagreb
UnivDacca:Bngl	Institute of Statistical Research and Training, University of Dacca * Dacca, Bangladesh
UnivHouston:TX	University of Houston, Department of Mathematics * Houston, TX
UnivIllnoisPr:IL	University of Illinois Press * Champaign, IL
UnivManitoba	University of Manitoba Press * Winnipeg, Manitoba
UnivNotreDame:IN	University of Notre Dame * South Bend, IN
UnivOsaka:Jpn	University of Osaka Prefecture * Osaka, Japan
UnivPennPr	University of Pennsylvania Press * Philadelphia
UnivPonaSt:India	Department of Statistics, University of Poona * Poona
UnivRomaSap:Itly	Departmento di Statistica, Probabilità e Statistiche Applicate, Universitá degli Studi di Roma la Sapienza * Rome
UnivSheffield	Department of Probability and Statistics, University of Sheffield * Sheffield, UK
UnivSochw:Taiwan	College of Science, Soochow University * Taipei

UnivTexas	University of Texas Press * Austin, TX
UnivTmkng:Taiwan	Research Institute of Mathematics, Tamkang University * Taipei
UnivTrento:Itly	University of Trento, Dipartimento di Metodi Quantitativi * Trento, Italy
Vandenhoeck:WG	Vandenhoeck and Ruprecht * Göttingen
VEB-GFischer	VEB Gustav Fischer Verlag * Stuttgart
VSP:Neth	VSP * Zeist, The Netherlands
WahanUniv:China	Institute of Mathematics, Wahan University * Wahan, China
WayneStUnivPr:MI	Wayne State University Press * Detroit, MI
WildlifeSoc:MD	Wildlife Society * Bethesda, MD
Wiley:NY:UK	John Wiley & Sons * New York; Chichester
WileyISci:NJ	Wiley-Interscience * Somerset, NJ
WmMaryPsych:VA	Department of Psychology, College of William and Mary * Williamsburg, VA
WrldSciPub:Sing	World Scientific Publications * Singapore; River Edge, NJ